2015 28th IEEE International Conference on Micro Electro Mechanical Systems

(MEMS 2015)

Estoril, Portugal
18-22 January 2015

Pages 1-568

IEEE Catalog Number:	CFP15MEM-POD
ISBN:	978-1-4799-7956-1

Copyright © 2015 by the Institute of Electrical and Electronic Engineers, Inc
All Rights Reserved

Copyright and Reprint Permissions: Abstracting is permitted with credit to the source. Libraries are permitted to photocopy beyond the limit of U.S. copyright law for private use of patrons those articles in this volume that carry a code at the bottom of the first page, provided the per-copy fee indicated in the code is paid through Copyright Clearance Center, 222 Rosewood Drive, Danvers, MA 01923.

For other copying, reprint or republication permission, write to IEEE Copyrights Manager, IEEE Service Center, 445 Hoes Lane, Piscataway, NJ 08854. All rights reserved.

***This publication is a representation of what appears in the IEEE Digital Libraries. Some format issues inherent in the e-media version may also appear in this print version.**

IEEE Catalog Number: CFP15MEM-POD
ISBN 13:   978-1-4799-7956-1

Additional Copies of This Publication Are Available From:

Curran Associates, Inc
57 Morehouse Lane
Red Hook, NY 12571 USA
Phone: (845) 758-0400
Fax: (845) 758-2633
E-mail: curran@proceedings.com
Web: www.proceedings.com

2015 28th IEEE International Conference on Micro Electro Mechanical Systems (MEMS 2015)

Estoril, Portugal
18-22 January 2015

IEEE Catalog Number: CFP15MEM-POD
ISBN: 978-1-47997-956-1

MEMS 2015 Program Schedule

Monday, 19 January

07:55   **Opening and Welcome Address**
Jürgen Brugger, *EPFL Lausanne, SWITZERLAND*
Wouter van der Wijngaart, *KTH Royal Institute of Technology, SWEDEN*

Invited Plenary Speaker I
Session Chairs:
J. Brugger, *EPFL Lausanne, SWITZERLAND*
H. Toshiyoshi, *University of Tokyo, JAPAN*

08:15   **THE LONG PATH FROM MEMS RESONATORS TO TIMING PRODUCTS** 1
E. Ng[1], Y. Yang[1], V.A. Hong[1], C.H. Ahn[1], D.B. Heinz[1], I. Flader[1], Y. Chen[1], C.L.M. Everhart[1], B. Kim[2],
R. Melamud[2], R.N. Candler[2], M.A. Hopcroft[2], J.C. Salvia[2], S. Yoneoka[2], A.B. Graham[2], M. Agarwal[2],
M.W. Messana[2], K.L. Chen[2], H.K. Lee[2], S. Wang[2], G. Bahl[2], V. Qu[2], C.F. Chiang[2], **Thomas W. Kenny, Ph.D.**[1],
A. Partridge[3], M. Lutz[3], G. Yama[4] and G.J. O'Brien[4]
[1]*Stanford University, USA*, [2]*PhD Alumni of Stanford University, USA*, [3]*SiTime Inc, USA, and* [4]*Robert Bosch RTC, USA*

Session I - Micro and Nanofluidics
Session Chairs:
B. Stoeber, *University of British Columbia, CANADA*
F.G. Tseng, *National Tsing Hua University, TAIWAN*

09:30   **GRAPHENE OXIDE MEMBRANES FOR PHASE-SELECTIVE MICROFLUIDIC FLOW CONTROL** 2
J. Gaughran, D. Boyle, J. Murphy, and J. Ducrée   * Award Nominee
Dublin City University, IRELAND

We investigate the unique properties of Graphene Oxide (GO) as a barrier selective to the solvent and state of aggregation of the fluid. To this end, we developed novel processes for the assembly of GO as membranes into polymeric microfluidic systems. We show that GO completely blocks pressurized air and organic solutions while it is permeable to water. These GO membranes are then employed as a flow control element in a microfluidic system.

09:45   **STRUCTURE-BASED SUPERHYDROPHOBICITY FOR SERUM DROPLETS** 6
T. Liu and C.-J. Kim
University of California, Los Angeles, USA

We report that superhydrophobic (SHPo) surfaces based purely on surface structuring shows a robust super-repellency under a prolonged contact with serum droplet as an example of protein-rich biological fluids. In contrast, normal SHPo surfaces, which are based on surface chemistry and surface structuring, lose repellency and eventually get wetted by the same tests. This is the first report of a SHPo surface not degraded by a biological fluid.

10:00   **MICROFABRICATED LIQUID CHAMBER UTILIZING SOLVENT-DRYING**
FOR IN-SITU TEM IMAGING OF NANOPARTICLE SELF-ASSEMBLY 8
W.C. Lee[1,2], J. Park[3,4,5], D.A. Weitz[5], S. Takeuchi[1,2], and A.P. Alivisatos[3,4]
[1]*University of Tokyo, JAPAN*, [2]*Japan Science and Technology Agency (JST), JAPAN*
[3]*University of California, Berkeley, USA*, [4]*Lawrence Berkeley National Laboratory, USA, and* [5]*Harvard University, USA*

We present a microfabricated liquid-sample chamber for real-time TEM (Transmission Electron Microscopy) of nanoscale processes driven by liquid evaporation. The present chamber (TEM liquid-cell) uses intended leakage/failure in its bonding process in order to generate solvent-drying during in-situ TEM imaging. The captured real-time nanometer-scale TEM movies visualize critical steps in the self-assembly process of 2 dimensional nanoparticle arrays.

10:15   **SYNTHETIC MICROFLUIDIC PAPER** 10
J. Hansson, H. Yasuga, T. Haraldsson, and W. van der Wijngaart
KTH Royal Institute of Technology, SWEDEN

We demonstrate a polymer synthetic microfluidic paper with the aim to combine the high surface area of paper or nitrocellulose with the repeatability, controlled structure, and transparency of polymer micropillars for lateral flow devices. It consists of a dense, high aspect ratio, stiff polymer micropillar array with thin slanted pillars that are interlocked, and is manufactured with multidirectional UV lithography in Off-Stochiometry-Thiol-Ene-Epoxy polymer.

10:30 **Break & Exhibit Inspection**

Session II - Actuators
Session Chairs:
M. Kohl, *Karlsruhe Institute of Technology, GERMANY*
F. Niklaus, *KTH - Royal Institute of Technology, SWEDEN*

11:00 **ASSEMBLED COMB-DRIVE XYZ-MICROSTAGE WITH LARGE DISPLACEMENTS FOR LOW TEMPERATURE MEASUREMENT SYSTEMS** 14
G. Xue, M. Toda, and T. Ono
Tohoku University, JAPAN

In this research, we report the novel design, fabrication and testing of an assembled comb-drive XYZ-microstage that produces highly decoupled motions into X-, Y-, and Z-directions for the 3D scanning stage of magnetic resonance force microscopy. It is demonstrated that the assembled XYZ-microstage can achieve large displacements of 20.4 micrometre in X direction, 25.2 micrometre in Y direction and 58.5 micrometre in Z direction.

11:15 **PNEUMATIC BALLOON ACTUATOR WITH TUNABLE BENDING POINTS** 18
L. Zheng[1], S. Yoshida[1], Y. Morimoto[1,2], H. Onoe[1,2], and S. Takeuchi[1,2]
[1]*University of Tokyo, JAPAN and* [2]*Japan Science and Technology Agency (JST), JAPAN*

We propose a pneumatic balloon actuator capable of controlling its bending points. Local stiffness of the balloon actuator can be controlled by injected low-melting-point-alloy. The bending point can be changed depending on the position of the rigid low-melting-point-alloy. We believe that the proposed actuation mechanism will be useful in designing highly flexible actuators for soft robotics.

11:30 **SELF-LIFTING ARTIFICIAL INSECT WINGS VIA ELECTROSTATIC FLAPPING ACTUATORS** 22
X. Yan[1,2], M. Qi[1], and L. Lin[3]
[1]*Beihang University, CHINA,* [2]*Collaborative Innovation Center of Advanced Aero-Engine, CHINA, and*
[3]*University of California, Berkeley, USA*

We present a self-lifting artificial insect wings using electrostatic actuation for the first time. Excited by a DC power source, biomimetic flapping motions have been generated to lift the artificial wings under an operation frequency of 50-70Hz. Three achievements have been accomplished: (1) first successful demonstration of electrostatic flying wings; (2) low power consumption as compared to other actuation schemes; and (3) self-adjustable rotating wing design to provide the lifting force.

11:45 **SELF-ASSEMBLED HYDROGEL MICROSPRING FOR SOFT ACTUATOR** 26
K. Yoshida and H. Onoe
Keio University, JAPAN

We found that the hydrogel microspring was formed by extruding sodium alginate pre-gel solution into calcium chloride using a bevel-tip microfluidic capillary. The formation of the microspring depends on the diameter of the capillary, flow rate and tip angle. As an example of soft actuator, the microspring including magnetic colloids was actuated by applying magnetic fields. We believe that the microspring will be used to various application including mechanical elements using soft materials.

12:00 **Lunch & Exhibit Inspection**

13:00 **Poster/Oral Session I**
Poster presentations are listed by topic category with their assigned number starting on page 14.

15:00 **Break & Exhibit Inspection**

Session III - Gyroscopes

Session Chairs:
F. Ayazi, *Georgia Institute of Technology, USA*
A. Seshia, *University of Cambridge, UK*

15:30   **PARAMETRIC DRIVE OF A TOROIDAL MEMS RATE INTEGRATING GYROSCOPE DEMONSTRATING < 20 PPM SCALE FACTOR STABILITY** **29**

D. Senkal[1], E.J. Ng[2], V. Hong[2], Y. Yang[2], C.H. Ahn[2], T.W. Kenny[2], and A.M. Shkel[1]
[1]University of California, Irvine, USA and [2]Stanford University, USA

In this paper, we report parametric drive of a MEMS rate integrating gyroscope for reduction of drifts induced by drive electronics, resulting in < 20 ppm scale factor stability. Due to the parametric pumping effect, energy added to each (x & y) mode is proportional to the existing amplitude of the respective mode. As a result, errors associated with finding the orientation of the standing wave and x-y drive gain drift can be bypassed, demonstrating > 20x improvement in scale factor stability.

15:45   **A 7PPM, 6°/HR FREQUENCY-OUTPUT MEMS GYROSCOPE** **33**

I.I. Izyumin[1], M.H. Kline[1], Y.C. Yeh[1], B. Eminoglu[1], C.H. Ahn[2],
V.A. Hong[2], Y. Yang[2], E.J. Ng[2], T.W. Kenny[2], and B.E. Boser[1]
[1]University of California, Berkeley, USA and [2]Stanford University, USA

We report the first frequency-output MEMS gyroscope to achieve <7 ppm scale factor accuracy and <6 deg/hr bias stability with a 3.24mm^2 transducer. By employing continuous-time mode reversal, the rate measurement is made insensitive to the resonant frequency of the transducer. The scale factor is almost entirely ratiometric; scale factor sensitivity to transducer and circuit parameters is significantly reduced compared to conventional open-loop and force-rebalance operating modes.

16:00   **LARGE FULL SCALE, LINEARITY AND CROSS-AXIS REJECTION IN LOW-POWER 3-AXIS GYROSCOPES BASED ON NANOSCALE PIEZORESISTORS** **37**

S. Dellea[1], F. Giacci[1], A. Longoni[1], P. Rey[2], A. Berthelot[2], and G. Langfelder[1]
[1]Politecnico di Milano, ITALY and [2]CEA-Leti, FRANCE

The work presents 3-axis rate gyroscopes based on nanoscale piezoresistive readout and eutectic bonding between a bottom wafer, where the sensor is formed, and a cap wafer, where routing and pads are fabricated. The design features a central levered sense frame, to maximize symmetry, compactness and vibration rejection. Operation on a ±3000 dps full-scale shows competitive performance, with linearity errors <0.2% and cross-axis rejections 50x better than state-of-the-art consumer gyroscopes.

16:15   **THE ELECTROMECHANICAL RESPONSE OF A SELF-EXCITED MEMS FRANKLIN OSCILLATOR** **41**

S. Shmulevich, I. Hotzen, and D. Elata
Technion - Israel Institute of Technology, ISRAEL

We present a self-excited MEMS Franklin oscillator, which responds in steady state vibrations when subjected to a dc voltage. The system is constructed from a floating rotor, which transfers charge between a source and drain electrodes. Current flows through the system only when the rotor is in transition. Surprisingly, at contact of the rotor with either the source or the drain electrodes, there is no current, and the charge transfer mechanism is essentially a recombination of opposite charges.

Session IV - Photonics

Session Chairs:
J. Conde, *Instituto Superior Tecnico, PORTUGAL*
C. Lee, *National University of Singapore, SINGAPORE*

16:30   **A NANOMACHINED TUNABLE OSCILLATOR CONTROLLED BY ELECTROSTATIC AND OPTICAL FORCE** **45**

J. G. Huang[1,2,3], B. Dong[2,3], H. Cai[2], Y. D. Gu[3], J. H. Wu[1], T. N. Chen[1],
Z. C. Yang[4], Y. F. Jin[4], Y. L. Hao[4], D. L. Kwong[3], and A. Q. Liu[3]
[1]Xi'an Jiaotong University, CHINA, [2]Agency for Science, Technology and Research (A*STAR), SINGAPORE, [3]Nanyang Technological University, SINGAPORE, and [4]Peking University, CHINA

We develop a miniaturized electrically tunable optomechanical oscillator, whose frequencies can be electrostatically tuned by as much as 10%. By taking advantage of the optical and the electrical spring, the oscillator achieves a high tuning sensitivity without resorting to mechanical tension. Particularly, the high-Q optical cavity greatly enhances the system sensitivity, making it extremely sensitive to the motional signal, which is often overwhelmed by spurious coupling or background.

16:45   **NANO-OPTOMECHANICAL STATIC RANDOM ACCESS MEMORY (SRAM)** **49**

B. Dong[1,2], H. Cai[2], Y. D. Gu[2], Z. C. Yang[3], Y. F. Jin[3], Y. L. Hao[3], D. L. Kwong[2], and A. Q. Liu[1]
[1]Nanyang Technological University, SINGAPORE, [2]Agency for Science, Technology and Research (A*STAR), SINGAPORE, and [3]Peking University, CHINA

We develop an on chip NEMS optomechanical SRAM, which is integrated with light modulation system on a single silicon chip. In particular, a doubly-clamped silicon beam shows bistability due to the non-linear optical gradient force generated from a ring resonator. The memory states are assigned with two stable deformation positions, which can be switched by modulating the control light's power with the integrated optical modulator.

17:00 **A LOW-POWER MEMS TUNABLE PHOTONIC RING RESONATOR FOR RECONFIGURABLE OPTICAL NETWORKS** 53

C. Errando-Herranz, F. Niklaus, G. Stemme, and K.B. Gylfason

KTH Royal Institute of Technology, SWEDEN

* Award Nominee

We experimentally demonstrate a low-power MEMS tunable photonic ring resonator with 10 selectable channels for wavelength selection in reconfigurable optical networks. The tuning is achieved by changing the geometry of a silicon slot-waveguide ring resonator by vertical electrostatic parallel-plate actuation. Our device provides static power dissipation below 0.1 µW, a wavelength tuning range of 1 nm, and a bandwidth of 0.1 nm, i.e. 10-8 watts per selectable channel, the lowest number reported.

17:15 **NEMS OPTICAL CROSS CONNECT (OXC) DRIVEN BY OPTICL FORCE** 57

H. Cai[1], J. X. Lin[2], J. H. Wu[2], B. Dong[1,3], Y. D. Gu[1], Z. C. Yang[4], Y. F. Jin[4], Y. L. Hao[4], D. L. Kwong[1], and A. Q. Liu[3]

[1]*Agency for Science, Technology and Research (A*STAR), SINGAPORE,*
[2]*Xi'an Jiaotong University, CHINA,* [3]*Nanyang Technological University, SINGAPORE, and* [4]*Peking University, CHINA*

We report a nano-silicon-photonic optical cross connect driven by optical gradient force, which demonstrates the all-optical OXC system on silicon. A switching time of 170 ns is experimentally demonstrated, which is much faster than that of conventional optical switches. In addition, the proposed switch system has the advantages of compact size (35 µm x 35 µm for switching element), high extinction ratio and low power consumption.

17:30 **Adjourn for the Day**

Tuesday 20, January

08:25   **Announcements**

Invited Plenary Speaker II
Session Chairs:
H. Toshiyoshi, *University of Tokyo, JAPAN*
J. Brugger, *EPFL Lausanne, SWITZERLAND*

08:30   **THE FUTURE OF MEMS SENSORS IN OUR CONNECTED WORLD** 61
Gerhard Lammel, Ph.D.
Bosch Sensortec GmbH, GERMANY

Session V - Micro-Optics
Session Chairs:
I.-J. Cho, *Korea Institute of Science and Technology, SOUTH KOREA*
B. Legrand, *LAAS-CNRS, FRANCE*

09:15   **FABRICATION AND CHARACTERIZATION OF A NEW VARIFOCAL LIQUID LENS
WITH EMBEDDED PZT ACTUATORS FOR HIGH OPTICAL PERFORMANCES** 65
S. Nicolas[1], M. Allain[1], C. Bridoux[1], S. Fanget[1], S. Lesecq[1], M. Zarudniev[1], S. Bolis[2],
A. Pouydebasque[2], and F. Jacquet[2]
[1]CEA, LETI, MINATEC Campus, FRANCE and [2]WAVELENS, FRANCE

This paper reports the fabrication and characterizations of a compact Varifocal microlens with an embedded MEMS actuator. Optical aperture is typically between 1.5 mm to 3 mm diameter with a total component thickness down to 400μm. High power efficiency (< 0.1 μW), high speed response time (down to 1 msec), high electro-optical performances (10 diopters optical power variation at 10V) and good optical quality (wavefront error lower than 50 nm) are reported through this paper.

09:30   **IMPLEMENTATION OF NANOPOROUS ANODIC ALUMINUM OXIDE
LAYER WITH DIFFERENT POROSITIES FOR INTERFEROMETRIC
RGB COLOR PIXELS AS HANDHELD DISPLAY APPLICATION** 69
P.-H. Lo[1], G.-L. Luo[2], and W. Fang[1]
[1]National Tsing Hua University, TAIWAN and [2]Asia Pacific Microsystems Inc., TAIWAN

This study employs the nanoporous anodic aluminum oxide (np-AAO) for interferometric modulation color pixels. Advantages of the proposed color-pixel are, (1) porosity of np-AAO layer can be adjusted to modulate therefractive index, (2) color-pixels of different porosities can be implemented on single np-AAO layer for different colors modulation, (3) the morphology of np-AAO with Al half-reflector could scatter reflect light to enhance view-angle, and (4) anti-stiction coating is not required.

09:45   **LOW-NOISE ALN-ON-SI RESONANT INFRARED DETECTORS USING A
COMMERCIAL FOUNDRY MEMS FABRICATION PROCESS** 73
V.J. Gokhale[1], C. Figueroa[1], J.M.L. Tsai[2], and M. Rais-Zadeh[1]
[1]University of Michigan, USA and [2]Invensense Inc,, USA

This work presents the first measured results for resonant AlN-based IR detectors fabricated in a proprietary AlN MEMS process. Resonators fabricated in the first fabrication run achieved high electromechanical performance (Q of ~1400 at 115 MHz), an infrared responsivity of 10.7%/W, and low noise, as evidenced by an NEDT of 51 mK and an NEP of 52.7 pW/Hz^0.5. The resonators are fabricated in a hybrid MEMS/CMOS wafer level packaged die, allowing for CMOS-based routing and readout.

10:00   **Break & Exhibit Inspection**

Session VI - Novel Fabrication
Session Chairs:
A. Dietzel, *Technische Universität Braunschweig, GERMANY*
H. Moon, *University of Texas, Arlington, USA*

10:30   **HARD MASK FREE DRIE OF CRYSTALLINE SI NANOBARREL WITH
6.7NM WALL THICKNESS AND 50:1 ASPECT RATIO** 77
P. Liu, F. Yang, W. Wang, W. Wang, K. Luo, Y. Wang, and D. Zhang   *** Award Nominee**
Peking University, CHINA

Crystal Si barrel with wall thickness of 6.7nm and aspect ratio of 50:1 were fabricated using IC/MEMS compatible process. PR mask was generated by e-beam lithography then etched into Si by fluorine based DRIE directly, without hard mask transfer. To achieve high anisotropic etching, a model of ions transportation around barrels in DRIE was established and studied. This tactic with high repeatability and manufacturability provides an arsenal for the next generation of 3D nano devices fabrication.

10:45   **SELF-HEALING METAL WIRE USING AN ELECTRIC FIELD TRAPPING OF GOLD NANOPARTICLES FOR FLEXIBLE DEVICES** 81

T. Koshi and E. Iwase
Waseda University, JAPAN

     * **Award Nominee**

We developed a self-healing metal wire for flexible devices. A cracked metal wire on a stretchable substrate can get its conductivity again by the self-healing ability using an electric field trapping of gold nanoparticles. First, we analyzed the electric field trapping. Next, we fabricated cracked wires on a glass substrate and verified the self-healing by experiments. Finally, we demonstrated the self-healing on the stretchable substrate to show a usefulness for flexible devices.

11:00   **CONTROLLED FABRICATION OF NANOSCALE GAPS USING STICTION** 85

F. Niroui, E.M. Sletten, P.B. Deotare, A.I. Wang, T.M. Swager, J.H. Lang, and V. Bulović
Massachusetts Institute of Technology, USA

Utilizing stiction, a common mode of failure in electromechanical systems, our work develops a method for the controlled fabrication of nanometer-thin gaps between electrodes. In this approach, through nanoscale force control, stiction promotes formation of nanogaps of controlled widths within the range as small as sub-15 nm. Our work demonstrates that through modifications of device design, the nanogaps can be optimized for applications in nanoelectromechanical and molecular devices.

11:15   **DIFFERENTIAL MICRO-PIRANI GAUGE FOR MONITORING MEMS WAFER-LEVEL PACKAGE** 89

Y.-C. Chen, W.-C. Lin, H.-S. Wang, C.-C. Fan, K.C.-H. Lin, B.C.S. Chou, and M.C.-M. Liu
Taiwan Semiconductor Manufacturing Company, TAIWAN

We proposed a multiple-sensor-solution, where two Pirani gauges were constructed under different pressures; one in sealed micro-cavity for measuring pressures and the other one in opened micro-cavity as a reference. The differential scheme compensates errors, allowing accurate pressure determinations, and it was successfully used in examining reliabilities and monitoring processes of wafer-level packages.

11:30   **MICROFABRICATED ELECTROSTATIC PLANAR LENS ARRAY AND EXTRACTORS FOR MULTI-FOCUSED ION BEAM SYSTEM USING IONIC LIQUID ION SOURCE EMITTER ARRAY** 93

R. Yoshida, M. Hara, H. Oguchi, and H. Kuwano
Tohoku University, JAPAN

To develop multi-focused ion beam system, we fabricated the electrostatic planar extractors and lenses, and integrated those with the field-emission ion source emitter array. Focusing ability of the integrated device was verified by confirming divergence angle reduction to ~70% of that without focusing effect. In addition, to find suitable ionic liquid for Si etching, we adopted various ionic liquid as an ion source and investigated etching characteristics by microscopy and mass spectrometry.

11:45   **NOVEL IONIC LIQUID - POLYMER COMPOSITE AND AN APPROACH FOR ITS PATTERNING BY CONVENTIONAL PHOTOLITHOGRAPHY** 97

N.A. Bakhtina[1], A. Voigt[2], N. MacKinnon[1], G. Ahrens[2], G. Gruetzner[2], and J.G. Korvink[1]
[1]University of Freiburg-IMTEK, GERMANY and [2]micro resist technology GmbH, GERMANY

     * **Award Nominee**

We report a novel composite material based on an ionic liquid and a photoresist. In addition, an approach for the patterning of the composite material by conventional photolithography is introduced. The unique properties of the material are utilized for direct manufacturing of highly transparent, electrically conductive microcomponents.

12:00   **IEEE 2015 Andrew S. Grove Award Recipient**
Dr. Masayoshi Esashi, Ph.D.
Tohoku University, JAPAN

12:15   **Lunch & Exhibit Inspection**

13:15   **Poster/Oral Session II**
Poster presentations are listed by topic category with their assigned number starting on page 14.

15:15   **Break & Exhibit Inspection**

Session VII - Power & Energy I
Session Chairs:
H. Kim, *University of Utah, USA*
P. Woias, *Albert-Ludwig-University Freiburg, GERMANY*

15:30 **TRIBOELECTRIFICATION BASED ACTIVE SENSOR FOR POLYMER DISTINGUISHING** 102
B. Meng, X.L. Cheng, M.D. Han, H.T. Chen, F.Y. Zhu, and H.X. Zhang * **Award Nominee**
Peking University, CHINA

We present a novel sensor to realize polymer distinguishing based on triboelectrification and electrostatic induction. Multiple cells of single friction layer and electrode are integrated on a flexible substrate. For different polymer groups, the friction layers can be selected according to the triboelectric serials. As an example, the distinguishing of PDMS, PE and PET has been well demonstrated by employing PI and PS friction layers, showing potential applications in robotics and industry use.

15:45 **SKIN BASED FLEXIBLE TRIBOELECTRIC NANOGENERATORS
WITH MOTION SENSING CAPABILITY** 106
L. Dhakar, F.E.H. Tay, and C. Lee * **Award Nominee**
National University of Singapore, SINGAPORE

This paper presents a novel triboelectric nanogenerator (TENG) using outermost layer of human skin i.e. epidermis as an active triboelectric layer for device operation. The human skin also has an advantage of high tendency to lose electrons relative to PDMS, which leads to improved performance of triboelectric mechanism. TENG is also demonstrated as a wearable self-powered sensor to track human motion/activity.

16:00 **FLEXIBLE TRIBOELECTRIC AND PIEZOELECTRIC COUPLING NANOGENERATOR
BASED ON ELECTROSPINNING P(VDF-TRFE) NANOWIRES** 110
X. Wang, B. Yang, J. Liu, Q. He, H. Guo, C.S. Yang, and X. Chen
Shanghai Jiaotong University, CHINA

We fabricate and characterize a triboelectric and piezoelectric coupling nanogenerator based on MEMS technology which is more flexible and thickness controllable than conventional process fabricated. The electrospinning PVDF-TrFE nanofibers and MWCNTs doped PDMS films are function and friction layers. To characterize the sandwich-shaped nanogenerator's performance, its open-circuit output voltage and energy power density under two kinds of energy generation mechanism are tested and discussed.

16:15 **HIGHLY RELIABLE MEMS RELAY WITH TWO-STEP SPRING SYSTEM
AND HEAT SINK INSULATOR FOR POWER APPLICATIONS** 114
Y.-H. Yoon, Y.-H. Song, S.-D. Ko, C.-H. Han, G.-S. Yun, M.-H. Seo, and J.-B. Yoon
Korea Advanced Institute of Science and Technology (KAIST), SOUTH KOREA

This paper reports remarkably reliable MEMS relays having a unique two-step spring system and a heat sink insulator. The two-step spring system is designed to reduce Joule-heating by lowering contact resistance. The heat sink insulator is proposed for efficiently removing heat generated in the contact area. These two features are adopted in the MEMS relay for minimizing thermal damage in high current level, thus enhancing reliability significantly.

Session VIII - Power & Energy II
Session Chairs:
H. Kim, *University of Utah, USA*
P. Woias, *Albert-Ludwig-University Freiburg, GERMANY*

16:30 **SOLIDIFIED IONIC LIQUID FOR HIGH-POWER VIBRATIONAL ENERGY HARVESTERS** 118
S. Yamada[1], H. Mitsuya[2], S. Ono[3], K. Miwa[3], and H. Fujita[1]
[1]*University of Tokyo, JAPAN,* [2]*SAGINOMIYA SEISAKUSHO, Inc., JAPAN, and*
[3]*Central Research Institute of Electric Power Industry, JAPAN*

We propose a high power-output vibrational energy harvester based on ionic liquid. Ionic liquid enables very large capacitance (1.0-10 µF cm-2) on the electrode at bias voltage less than 1.9 V due to its extremely thin (~ 1 nm) electrical double layer. By mechanical squeezing and drawing the ionic liquid solidified with a polymer addictive between a pair of electrodes at 15 Hz, we stably obtained the current output of 22 µAp-p cm-2 at 1.5 V.

16:45 **FERROFLUID LIQUID SPRING FOR VIBRATION ENERGY HARVESTING** 122
Y. Wang, Q. Zhang, L. Zhao, and E.S. Kim
University of Southern California, Los Angeles, USA

We developed a ferrofluid liquid spring to suspend a magnet array for harvesting vibration energy. For the first time, the concept of ferrofluidbased suspension is demonstrated for low resonant frequency and high reliability. A ferrofluid liquid spring has reduced the resonant frequency of a microfabricated electromagnetic energy harvester to around 340 Hz. 36 nW is delivered into a load of 2.3 Ω from 7 g acceleration.

17:00 **AN ELECTROSTATIC ENERGY HARVESTER EXPLOITING VARIABLE-AREA WATER ELECTRODE BY RESPIRATION** 126

M.-H. Seo, D.-H. Choi, C.-H. Han, J.-Y. Yoo, and J.-B. Yoon

Korea Advanced Institute of Science and Technology (KAIST), SOUTH KOREA

* **Award Nominee**

This paper reports an electrostatic energy-harvester exploiting water-layer formed by respiration as a variable-area electrode. We discover that electrically conductive water-layer (~45 Mohms/sq under 5 V) is instantly formed on a silicon-dioxide surface by exhaled-breath. We adopt this layer as variable capacitive electrodes for electrostatic energy-harvesting. The capacitance change was evaluated using a theoretical modeling and finite-element-method (FEM) simulation, and theoretical power-generation was estimated (~2 $\mu W/cm^2$ at 1 V). We then fabricated the prototype device and verified the capacitance change experimentally. Finally, the prototype showed charging and discharging characteristics by respiration successfully for being used as an energy-harvester driven by human-breath solely.

17:15 **SILICON ANODE SUPPORTED BY CARBON SCAFFOLD FOR HIGH PERFORMANCE LITHIUM ION MICRO-BATTERY** 130

X. Li[1,2], X. Wang[1,2], and S. Li[1,2]

[1]Tsinghua National Laboratory for Information Science and Technology, CHINA and [2]Tsinghua University, CHINA

This paper reports a novel Si/void/C anode for lithium ion batteries, with high specific capacity and superior cyclability. SiNPs at anode make a significant contribution to specific capacity increase. Nano void spaces provide enough space for expansion and contraction of SiNPs to ensure a long cycle life. The porous carbon scaffold is obtained from a photoresist (SU-8) with SiNPs templates. As such, it's possible to implement direct prototyping of three dimensional (3D) micro-battery on chip.

17:30 **Adjourn for the Day**

Wednesday 21, January

08:25   **Announcements**

Invited Plenary Speaker III
Session Chairs:
W. van der Wijngaart, *KTH - Royal Institute of Technology, SWEDEN*
X. Wang, *Tsinghua University, CHINA*

08:30   **STATUS AND FUTURE TRENDS OF THE MINIATURIZATION OF MASS SPECTROMETRY** 134
Richard R.A. Syms, Ph.D.
Imperial College London, UK

Session IX - Acoustic Sensors
Session Chairs:
Y.-K. Kim, *Seoul National University, SOUTH KOREA*
T. Seki, *OMRON Corporation, JAPAN*

09:15   **SHORT-RANGE AND HIGH-RESOLUTION ULTRASOUND IMAGING
USING AN 8 MHZ ALUMINUM NITRIDE PMUT ARRAY** 140
Y. Lu[1], H.-Y. Tang[2], S. Fung[1], B.E. Boser[2], and D.A. Horsley[1]   * **Award Nominee**
[1]University of California, Davis, USA and [2]University of California, Berkeley, USA

We demonstrate short-range (~mm) and high-resolution (<100µm) imaging based on piezoelectric micromachined ultrasonic transducers (PMUTs) and a 1.8V interface ASIC. The PMUTs use piezoelectric Aluminum Nitride (AlN), which has the advantages of low-temperature (<400°C) deposition and compatibility with CMOS fabrication but has a relatively low piezoelectric constant (e31=-0.5C/m2), making detection of ultrasound signals from tiny (50µm) PMUTs a challenging task.

09:30   **MEASUREMENT OF SURFACE ACOUSTIC WAVES PROPAGATION
USING A PIEZORESISTIVE CANTILEVER ARRAY** 144
N. Minh-Dung, P. Quang-Khang, N. Thanh-Vinh, K. Matsumoto, and I. Shimoyama
University of Tokyo, JAPAN

We presents an approach for measuring the propagation of surface acoustic waves (SAW), using a piezo-resistive cantilever array. SAW was measured by cantilevers designed at the liquid-substrate interaction area by utilizing the structure of liquid-cantilever-air, which is highly sensitive and is able to measure acoustic wave at high frequency. Experiment results demonstrate that the measurable range was from 0.1 MHz up to 100 MHz.

09:45   **BROADBAND PIEZOELECTRIC MICROMACHINED ULTRASONIC
TRANSDUCERS BASED ON DUAL RESONANCE MODES** 146
Y. Lu[1], O. Rozen[1], H.-Y. Tang[2], G.L. Smith[3], S. Fung[1], B.E. Boser[2], R.G. Polcawich[3], and D.A. Horsley[1]
[1]University of California, Davis, USA, [2]University of California, Berkeley, USA, and [3]US Army Research Lab, USA

We demonstrate broadband PZT pMUTs that achieve a 97% fractional bandwidth by utilizing a thinner structure excited at two adjacent mechanical vibration modes. A front side XeF2 release process reduces fabrication cost and allows higher fill factor relative to pMUTs requiring through wafer DRIE. To reduce the frequency variations that result from non-uniform or inaccurately-controlled release etching, a 4µm thick metal layer defines the effective boundary of each pMUT.

10:00   **WHEN CAPACITIVE TRANSDUCTION MEETS THE THERMOMECHANICAL LIMIT:
TOWARDS FEMTO-NEWTON FORCE SENSORS AT VERY HIGH FREQUENCY** 150
S. Houmadi[1,2], B. Legrand[3,4], J.P. Salvetat[2], B. Walter[5], E. Mairiaux[5], J.P. Aimé[1], D. Ducatteau[5],
P. Merzeau[2], L. Buisson[2], J. Elezgaray[1], D. Théron[5], and M. Faucher[5]
*[1]CNRS, CBMN, FRANCE, [2]CNRS, CRPP, FRANCE, [3]CNRS, LAAS, FRANCE,
[4]University de Toulouse, FRANCE, and [5]IEMN, CNRS, FRANCE*

We show that the capacitive transduction associated with a microwave detection scheme achieves the measurement of the thermomechanical noise spectrum of high-frequency (>10 MHz) high-stiffness (>105 N/m) microresonators, reaching the outstanding displacement resolution of 1fm/√Hz. This paves the way for vibrating sensors with exquisite force resolution in the fN/√Hz range, enabling large-bandwidth measurements of mechanical interactions at small scale and rheology of fluids at high frequency.

10:15   **Break & Exhibit Inspection**

Session X - Medical Microdevices

Session Chairs:
A. Hierlemann, *ETH Zurich, SWITZERLAND*
W. Li, *Michigan State University, USA*

10:45 MEMS OXYGEN TRANSPORTER TO TREAT RETINAL ISCHEMIA 154
D. Kang[1], K. Murali[2], N. Scianmarello[1], J. Park[1], J.H.-C. Chang[1], Y. Liu[1], * **Award Nominee**
K.-T. Chang[1], Y.-C. Tai[1], and M.S. Humayun[2]
[1]California Institute of Technology, USA and [2]University of Southern California, Los Angeles, USA

A paradigm shift of treating diabetic retinopathy is proposed in the sense of using MEMS devices to bring more oxygen to retina. A passive MEMS oxygen transporter was designed, built, and tested both in mechanical models and pig eyes to confirm its feasibility. The results predict that the proposed approach can even treat a complete retinal ischemia, although our current on-going live animal experiments must finish before going for human trials.

**11:00 A NEW MONOLITHICALLY INTEGRATED MULTI-FUNCTIONAL MEMS
NEURAL PROBE FOR OPTICAL STIMULATION AND DRUG DELIVERY** 158
Y. Son[1], H.J. Lee[1], J. Kim[1], C.J. Lee[1], E.-S. Yoon[1], T.G. Kim[2], and I.-J. Cho[1]
[1]Korea Institute of Science and Technology (KIST), SOUTH KOREA and [2]Korea University, SOUTH KOREA

We present a monolithically integrated multifunctional MEMS neural probe by integrating both an embedded microfluidic channel for drug delivery and a SU-8 waveguide for optical stimulation using controlled glass reflow process. Using our multifunctional neural probe, we conducted successful in vivo experiments and recorded neural spike signals from individual neurons with a good SNR. In this work, we present distinctive changes in neural signals induced by both optical and chemical stimulation.

**11:15 MINIATURIZED 3×3 OPTICAL FIBER ARRAY FOR OPTOGENETICS WITH INTEGRATED
460 NM LIGHT SOURCES AND FLEXIBLE ELECTRICAL INTERCONNECTION** 162
M. Schwaerzle, P. Elmlinger, O. Paul, and P. Ruther
University of Freiburg-IMTEK, GERMANY

We report on the design, fabrication, assembly, and optical and thermal characterization of a novel MEMS-based optical probe array for optogenetic research in neuroscience. Nine high-efficiency light emitting diodes (LED) are integrated as a 3×3 array in a micromachined silicon (Si) housing that ensures the mechanical stability and precise alignment of the nine 5-mm-long optical fibers. The overall housing volume of less than 2.2 mm^3 is compatible with chronic implantation.

**11:30 FLEXIBLE END-EFFECTOR INTEGRATED WITH SCANNING ACTUATOR AND OPTICAL
WAVEGUIDE FOR ENDOSCOPIC FLUORESCENCE IMAGING DIAGNOSIS** 166
Y. Muramatsu, T. Kobayashi, and S. Konishi
Ritsumeikan University, JAPAN

This paper presents a flexible end-effector integrated with a scanning actuator and optical waveguide for endoscopic fluorescence imaging diagnosis. Pneumatic balloon actuator (PBA) is used as the scanning actuator for the end-effector in consideration of its soft and safe features. SU-8 optical waveguides are integrated onto the PBA structure made of PDMS. Our developed device has successfully scanned, excited, and detected fluorescence beads distributed in a pseudo tissue.

**11:45 FULLY IMPLANTABLE AND RESORBABLE WIRELESS MEDICAL
DEVICES FOR POSTSURGICAL INFECTION ABATEMENT** 168
H. Tao[1], S.W. Hwang[2], B. Marelli[3], B. An[3], J.E. Moreau[3], M. Yang[3],
M.A. Brenckle[3], S. Kim[2], D. Kaplan[3], J.A. Rogers[2], and F.G. Omenetto[3]
*[1]Shanghai Institute of Microsystem and Information Technology, CAS, CHINA,
[2]University of Illinois, Urbana-Champaign, USA, and [3]Tufts University, USA*

We present a therapeutic application of a microfabricated implantable and resorbable medical device by demonstrating in vivo elimination of bacterial infection by wireless activation of the device after implantation. The device disappears upon its completion, requiring no retrieval.

12:00 MEMS 2016 Announcement

12:15 Lunch & Exhibit Inspection

13:15 Poster/Oral Session III
Poster presentations are listed by topic category with their assigned number starting on page 14.

15:15 Break & Exhibit Inspection

Session XI.a - Resonant Sensors

Session Chairs:
C. Nguyen, *University of California, Berkeley, USA*
L. Sorenson, *HRL Laboratories, LLC, USA*

15:30   TUNING THE FIRST INSTABILITY WINDOW OF A MEMS MEISSNER PARAMETRIC RESONATOR USING A LINEAR ELECTROSTATIC ANTI-SPRING 172
S. Shmulevich, I. Hotzen, and D. Elata
Technion - Israel Institute of Technology, ISRAEL

We demonstrate frequency tuning of the first instability window of a MEMS Meissner parametric resonator. We achieve parametric excitation by time-modulation of a negative electrostatic stiffness. In our device this negative electrostatic stiffness is not affected by motion. In contrast, all state of the art MEMS parametric resonators are either detrimentally affected by a nonlinear electrostatic stiffness, or are more sensitive to fabrication tolerances relative to our design.

15:45   A VISCOMETER BASED ON VIBRATION OF DROPLETS ON A PIEZORESISTIVE CANTILEVER ARRAY 176
N. Thanh-Vinh, K. Matsumoto, and I. Shimoyama
University of Tokyo, JAPAN

This paper reports a method to measure viscosity based on the resonant vibration of droplets on a piezoresistive cantilever array. We demonstrate that viscosity of small droplets (3 µL) can be estimated from the attenuation rate of the cantilever output during free-decay of the droplet vibration. Moreover, we show that the optimized location of a cantilever should be on the periphery of the contact area where the force change is largest.

16:00   HOLLOW MEMS MASS SENSORS FOR REAL-TIME PARTICLES WEIGHING AND SIZING FROM A FEW 10 NM TO THE µM SCALE 180
C. Hadji[1,2], C. Berthet[1,2], F. Baléras[1,2], M. Cochet[1,2], B. Icard[1,2], and V. Agache[1,2]
[1]*University Grenoble, FRANCE and* [2]*CEA, LETI, MINATEC, FRANCE*

We report hollow MEMS plate oscillators for mass sensing in liquid with an expected mass resolution of 3 femtograms. The performances reached by our sensors – 10,000-range Q factor and ppb-range frequency stability – make them amenable to individual particles metrology from a few 10 nanometers up to the micrometre diameter range. Our devices are operated in air inside a customized "plug and play" test platform and do not need to work in vacuum sealed package.

16:15   DEVELOPMENT OF MICROFLUIDIC RESONATORS VIA SILICON-ON-NOTHING TECHNIQUE 184
J. Kim and J. Lee
Sogang University, SOUTH KOREA

We report wafer-level batch fabrication of microfluidic resonators based on Silicon-on-Nothing (SON) structures resulting from high temperature argon annealing of silicon wafers with periodic cylindrical pits. Besides the process optimization of SON structures, elemental fabrication techniques such as planarization, device release, metallization, and packaging are developed. These techniques reduce fabrication cost and time significantly and enable switching of the resonator materials.

Session XI.b - Cell Handling

Session Chairs:
N. Tas, *University of Twente, THE NETHERLANDS*
W. Wang, *Peking University, CHINA*

15:30   A MICROFABRICATED, FLOW DRIVEN MILL FOR THE MECHANICAL LYSIS OF ALGAE 188
J. Millis, L. Connell, S.D. Collins, and R.L. Smith
University of Maine, USA

We report the micrfabrication, computer modeling and experimental testing of a novel, new means of mechanically lysing algae using a flow-driven grinding mill. The mill is demonstrated to have 96% efficiency in lysing the dinoflagellate genus Alexandrium, a neurotoxin producing algae, responsible for Red Tide and paralytic shellfish poisoning. Lysate DNA is demonstrated to be viable through successful PCR amplification using primers specific to Alexandrium.

15:45   CLUSTER SIZING OF CANCER CELLS BY RAIL-BASED SERIAL GAP FILTRATION IN STOPPED-FLOW, CONTINUOUS SEDIMENTATION MODE 192
M. Glynn, C. Nwankire, D. Kinahan, and J. Ducrée
Dublin City University, IRELAND

We have developed a centrifugal microfluidic strategy for the isolation and sizing analysis of multicellular clusters from a blood sample. The strategy is based on passing the sample over a size exclusion rail that gates entry of clusters to underlying bins, allowing estimation of clustering extent in the sample.

16:00   TUNABLE DIELECTROPHORESIS FOR SHEATHLESS 3D FOCUSING 196
Y.-C. Kung, D.L. Clemens, B.-Y. Lee, and P.-Y. Chiou
University of California, Los Angeles, USA

We report on a novel tunable insulator-based dielectrophoresis (TiDEP) for three-dimensional, sheathless, single-stream cell and bacteria focusing. For the first time, objects as small as sub-micron sized infectious bacteria are continuously focused in the center of a channel without sheath flows. Compared to prior DEP works, this new TiDEP can provide an extremely long DEP interaction distance to migrate small objects with weak DEP forces to a focused single stream at high speed flows.

16:15 **CELL MANIPULATION METHOD BASED ON VIBRATIN-INDUCED LOCAL FLOW CONTROL IN OPEN CHIP ENVIRONMENT** 200
T. Hayakawa, S. Sakuma, and F. Arai
Nagoya University, JAPAN

We present a novel cell manipulation method using vibration-induced local flow control in open-chip environment. By applying circular vibration to micropillars on a chip, whirling flow is induced around it. This phenomenon is theoretically analyzed considering an effect of convective flow. Based on it, we show two important applications; transport and 3D rotation of oocytes for 3D observation of oocyte. We succeeded in the transportation of oocytes with 25 μm/s and rotation with 184 degrees/s.

Session XII.a - Magnetic & Resonant Sensors
Session Chairs:
M. Rais-Zadeh, *University of Michigan, USA*
M. Sasaki, *Toyota Technological Institute, JAPAN*

16:30 **ULTRA SENSITIVE LORENTZ FORCE MEMS MAGNETOMETER WITH PICO-TESLA LIMIT OF DETECTION** 204
V. Kumar, M. Mahdavi, X. Guo, E. Mehdizadeh, and S. Pourkamali
University of Texas, Dallas, USA   * **Award Nominee**

This work presents ultra-high sensitivities for Lorentz Force resonant MEMS magnetometers enabled by internal thermal-piezoresistive vibration amplification. Up to 2400X sensitivity amplification has been demonstrated with the noise floor calculated to be as low as 18 pt/√Hz. This is by far the most sensitive MEMS Lorentz force magnetometer demonstrated to date.

16:45 **HIGH FREQUENCY MICROWAVE ON-CHIP INDUCTORS USING INCREASED FERROMAGNETIC RESONANCE FREQUENCY OF MAGNETIC FILMS** 208
K. Koh[1], D.S. Gardner[2], C. Yang[1], K.P. O'Brien[2], N. Tayebi[2], and L. Lin[1]
[1]*University of California, Berkeley, USA and* [2]*Intel Corporation, USA*

The fabrication and characterization of high frequency on-chip inductors using sputtered magnetic films with an improved frequency range is presented. Reducing the sputtering power in the deposition process was found to result in smoother film surfaces and stronger uniaxial magnetic anisotropy and increased the FMR of CoZrTaB from 1.48 GHz to 2.13 GHz. A magnetic-core, on-chip inductor was fabricated using the CoZrTaB films. Results have shown 150% higher inductance and a larger Q-factor up to 1.2 GHz as compared to an air-core inductor.

17:00 **FUSION OF CANTILEVER AND DIAPHRAGM PRESSURE SENSORS ACCORDING TO FREQUENCY CHARACTERISTICS** 212
R. Watanabe, N. Minh-Dung, H. Takahashi, T. Takahata, K. Matsumoto, and I. Shimoyama
University of Tokyo, JAPAN

This paper reports on an approach to measure sensitive barometric pressure by fusing a cantilever-based differential pressure sensor (DPS) and a commercial available diaphragm-based absolute pressure sensor (APS). At high frequency, the DPS can detect smaller absolute pressure change than the APS. At low frequency, the APS show absolute pressure of less drift than the DPS. By utilizing the DPS and APS at each advantageous frequency, we propose high sensitive measurement of barometric pressure.

17:15 **DYNAMICALLY-BALANCED FOLDED-BEAM SUSPENSIONS** 215
S. Shmulevich, I. Hotzen, and D. Elata
Technion - Israel Institute of Technology, ISRAEL

We present a design methodology and experimental evidence of a dynamically balanced folded-beam suspension. This suspension responds as a linear spring at the fundamental resonance, which is in sharp contrast to the response of standard folded-beam suspensions. The dynamic response of standard suspensions becomes strongly nonlinear for motions larger than the width of the flexure beams. The resonance response of the new dynamically-balanced suspension is linear over a wider range of motions.

Session XII.b - BioSensing
Session Chairs:
E. Iwase, *Waseda University, JAPAN*
R. Yokokawa, *Kyoto University, JAPAN*

16:30 **MEMBRANE-BASED CHEMOMECHANICAL TRANSDUCER FOR THE DETECTION OF APTAMER-PROTEIN BINDING** 219
J.-K Choi[1] and J. Lee[2]
[1]*Small Machines Incorporation, SOUTH KOREA and* [2]*Seoul National University, SOUTH KOREA*

We report a membrane-based chemomechanical transducer for the sensitive detection of surface molecular reaction through a highly reliable common mode rejection (CMR) technique. Chemomechanical transduction, originally based on the micro-cantilever, offers potential benefits: label-free assay, and real-time monitoring of molecular interaction via mechanical deformation. Here we show clear-cut detection of molecular binding using a membrane transducer fabricated with conventional MEMS technology.

16:45   **HIGHLY SENSITIVE SERS DIAGNOSIS FOR BACTERIA BY THREE DIMENSIONAL NANO-MUSHROOMS AND NANO-STARS-ARRAY SANDWICHED ON BACTERIAL AGGREGATION** 223
C.-W. Lee[1], J.-K. Wu[1], and F.-G. Tseng[1,2]
[1]National Tsing Hua University, TAIWAN and [2]Research Center for Applied Sciences, Academia Sinica, TAIWAN

This paper reports a highly sensitive SERS Diagnosis system by incorporating three dimensional Nano-Mushrooms and Nano-Stars-Array sandwiched on Bacterial Aggregation. Through the action of ACEOF and nano-mushroom/bacteria/nano-stars-array self-aggregation process, the signal can be much enhanced by 5 orders of magnitude in 5 minutes. Detection limit can approach 1 bacterium/ml from the analysis result.

17:00   **SIMULTANEOUS IMPEDANCE SPECTROSCOPY AND STIMULATION OF HUMAN IPS-DERIVED CARDIAC 3D SPHEROIDS IN HANGING-DROP NETWORKS** 226
S.C. Bürgel[1], Y. Schmid[1], I. Agarkova[2], D.A. Fluri[2], J.M. Kelm[2], A. Hierlemann[1], and O. Frey[1]
[1]ETH Zurich, SWITZERLAND and [2]InSphero AG, SWITZERLAND

We present a platform for in-situ electrical impedance spectroscopy (EIS) measurements and electrical stimulation of human iPS-derived cardiac 3D microtissue spheroids in hanging drop networks. Electrical stimulations and EIS measurements were performed in parallel through the same electrodes. Stimulation was performed with sine-wave signals. Our results reveal beating frequency modulation upon tuning the stimulation signal amplitude.

17:15   **DROPLET FLOW FOCUSING FOR MOLECULAR BINDING DETECTION** 230
S. Kim and J. Lee
Seoul National University, SOUTH KOREA

We report the detection of interfacial molecular binding via droplet generation in a flow focusing device. We introduce the detection of DNA hybridization based on the mode of droplet production caused by interfacial tension shift at the oil-water boundary. Our report includes a molecular protocol to functionalize the interface with single-strand (ss) DNA as receptor, and flow condition tuning for an unambiguous distinction of complementary binding.

17:30   **Adjourn for the Day**

19:00 -   **Conference Banquet**
22:00

Thursday 22, January

08:40 **Announcements**

Invited Plenary Speaker IV
Session Chairs:
X. Wang, *Tsinghua University, CHINA*
W. van der Wijngaart, *KTH - Royal Institute of Technology, SWEDEN*

08:45 **SEMICONDUCTOR IC PACKAGING, THE NEXT WAVE** 234
CP Hung, Ph.D.
Advanced Semiconductor Engineering Inc., TAIWAN

Session XIII - BioMEMS
Session Chairs:
T. Kawano, *Toyohashi University of Technology, JAPAN*
N. Roxhed, *KTH - Royal Institute of Technology, SWEDEN*

09:30 **ROTATIONAL CHAMBERS ON FLUIDIC CHANNELS FOR THE REPETITIVE
FORMATION OF OPTICALLY OBSERVABLE LIPID-BILAYER MEMBRANES** 235
F. Tomoike[1], T. Tonooka[1], T. Osaki[2], and S. Takeuchi[1]
[1]*University of Tokyo, JAPAN and* [2]*Kanagawa Academy of Science and Technology, JAPAN*

We develop a device adapted for repetitive formation of horizontal lipid bilayer membranes. This device enables simultaneous optical and electrophysiological measurements of the membranes. We integrated a rotational chamber on a fluidic channel via parylene micropores. The rotational motion enables us to form/reform the bilayer repeatedly. The formation of the bilayers was confirmed from the bilayer thickness and nanopore incorporation into the bilayer.

09:45 **MICROTUBULE SORTING WITHIN A GIVEN ELECTRIC
FIELD BY DESIGNING FLEXURAL RIGIDITY** 238
N. Isozaki[1], H. Shintaku[1], H. Kotera[1], E. Meyhöfer[2], and R. Yokokawa[1]
[1]*Kyoto University, JAPAN and* [2]*University of Michigan, USA*

We propose a method to control gliding directions of kinesin-propelled microtubules (MTs) corresponding to their flexural rigidity (EI). We prepared two kinds of MTs having different EI and their trajectories within an electric field were clearly separated. Therefore, this study demonstrated the EI-altered MTs can be workhorses to sort/concentrate various combinations of molecules with techniques of loading MTs with target molecules and capturing the sorted MTs.

10:00 **CARVING OF PROTEIN CRYSTAL BY HIGH-SPEED MICRO-BUBBLE
JET USING MICRO-FLUIDIC PLATFORM** 242
S. Takasawa[1], T. Syu[1], and Y. Yamanishi[1,2]
[1]*Shibaura Institute of Technology, JAPAN and* [2]*Japan Science and Technology Agency (JST), JAPAN*

This paper reports a novel processing method for protein crystal with electric-induced-bubble. This minimally invasive micro-processing method overcomes the difficulties of processing fragile material such as protein crystal under water. The combination of electrically-induced bubble knife and glass capillary provide effective carving of protein crystal by ablation crystal and suction of chips respectively. Also, we successfully made a new micro-fluidic platform for processing protein crystal.

10:15 **Break & Exhibit Inspection**

Session XIV - Tactile and Force Sensors
Session Chairs:
N. Miki, *Keio University, JAPAN*
E. Sarajlic, *SmartTip B.V., THE NETHERLANDS*

10:45 A NOVEL CONFIGURATION OF TACTILE SENSOR TO ACQUIRE THE CORRELATION BETWEEN SURFACE ROUGHNESS AND FRICTIONAL FORCE 245
R. Kozai, K. Terao, T. Suzuki, F. Shimokawa, and H. Takao
Kagawa University, JAPAN

We propose a novel configuration of MEMS tactile sensor that can interact with micro surface roughness at a high resolution are proposed and reported for quantification of the fingertip sense. Two-axis movements of the contactor-tip are independently detected by the two independent suspensions. In the evaluation experiments, correlation between the surface shape and the local frictional force were successfully obtained at the same time and at the same point for many samples for the first time.

11:00 TUNNELING PIEZORESISTIVE TACTILE SENSING ARRAY FABRICATED BY A NOVEL FABRICATION PROCESS WITH MEMBRANE FILTERS 249
C.-W. Ma, T.-H. Ling, and Y.-J. Yang
National Taiwan University, TAIWAN

In this work, a highly-sensitive tactile sensor array using the tunneling piezoresistive effect is presented. The sensing element, which is made of multi-wall carbon nanotubes and polydimethylsiloxane conductive polymer, was patterned with microdome structures by a novel fabrication process on a membrane filter substrate. The tunneling piezoresistive effects of the interlocked microdome structures with different MWCNT concentrations are demonstrated.

11:15 A TACTILE SENSOR WITH THE REFERENCE PLANE FOR DETECTION ABILITIES OF FRICTIONAL FORCE AND HUMAN BODY HARDNESS AIMED TO MEDICAL APPLICATIONS 253
Y. Maeda, K. Terao, T. Suzuki, F. Shimokawa, and H. Takao
Kagawa University, JAPAN

In this study, a highly sensitive tactile sensor with detection abilities of both human body hardness and frictional force is reported. Employing the new structure, hardness signal becomes less sensitive to the contact pressure, and hardness is stably measured even under an unstable contact force. Surface frictional force was successfully measured in real time. Also, shore A hardness was successfully measured in the range from 1HS to 54HS corresponding to the human organs.

11:30 6-AXIS FORCE/TORQUE SENSOR FOR SPIKE PINS OF SPORTS SHOES 257
H. Ishido, H. Takahashi, A. Nakai, T. Takahata, K. Matsumoto, and I. Shimoyama
University of Tokyo, JAPAN

This paper reports on the method of force measurement of spike pins for sports shoes. It is important to measure force acting on spike pins, because they are related to the increase of GRF (ground reaction force) in running. We fabricated a $2 \times 2 \times 0.3$ mm^3 size 6-axis force/torque sensor chip that can be embedded in spike pins. The sensor chip consists of 6 straight piezo-resistive beams. We calibrated the spike-pin-shaped sensor, and confirmed that 6-axis force/torque was able to be detected.

11:45 Award Ceremony

12:00 IEEE MEMS 2015 Conference Adjourns

Poster/Oral Presentations

M – Monday (13:00 - 15:00) **T** – Tuesday (13:15 - 15:15)
W – Wednesday (13:15 - 15:15)

Generic MEMS and Nanotechnologies
Generic MEMS and NEMS Manufacturing Techniques

M-001 3D PRINTED RF PASSIVE COMPONENTS BY LIQUID METAL FILLING 261
C. Yang[1], S.-Y. Wu[1,2], C. Glick[1], Y.S. Choi[3], W. Hsu[2], and L. Lin[1]
[1]*University of California, Berkeley, USA,* [2]*National Chiao Tung University, TAIWAN, and*
[3]*Samsung Electronics Co., Ltd., SOUTH KOREA*

We present a novel method to form three-dimensional (3D) micro-scale electrical devices by using 3D printing and a liquid-metal-filling technique. Various RF passive components including inductors, capacitors and resistors are fabricated and characterized as proof-of-concepts. This technique establishes an innovative way to form arbitrary 3D structures with highly efficient, labor-saving metallization process.

**T-002 A CONVENIENT METHOD TO FABRICATE MULTILAYER
INTERCONNECTIONS FOR MICRODEVICES** 265
J. Li, S. Chen, and C.-J. Kim
University of California, Los Angeles, USA

We report a new method to fabricate multilayer interconnection without wet or dry etching or deposition of insulating layers. Electrical connections are completed by merely depositing metal layers and anodizing them after lithographically defining a photoresist. Without the need to etch metal layers or deposit and pattern insulation layers, the overall process is simple, cheap, safe, and of low temperature.

**W-003 DEVELOPMENT OF AN ATMOSPHERIC PRESSURE AIR MICROPLASMA
JET FOR THE SELECTIVE ETCHING OF PARYLENE-C FILM** 268
H. Guo, J. Liu, Z. Wang, X.Z. Wang, X. Chen, B. Yang, and C. Yang
Shanghai Jiao Tong University, CHINA

We develop a novel and simple process device based on the atmospheric pressure air microplasma jet for the selective etching of parylene-C film. The main feature of this process device is that it can be easily integrated with roll-to-roll systems for large-scale manufacturing of flexible electronic devices due to its operating at ambient conditions. In order to realize the selective etching, a quartz glass microtube (100 μm I.D) is employed to fabricate the microplasma jet source.

**M-004 ELECTROPLATED STENCIL REINFORCED WITH ARCH STRUCTURES
FOR PRINTING FINE AND LONG CONDUCTIVE PASTE** 272
P.-H. Chen and C.-H. Lin
National Sun Yat-sen University, TAIWAN

This study presents an electroplated stencil reinforced with arch structures and a surrounding buffer reservoir for printing conductive paste of fine and long lines. The developed reinforced stencil successfully solves the problems came with the conventional stencil structure including limited printable line width and ease of facture. This work presents a novel process for fabricating a thin yet robust MEMS-based stencil by using two AZ4620 layers and one SU-8 layer as the electroplating molds.

**T-005 FABRICATION AND CHARACTERIZATION OF SINGLE CRYSTALLINE
4H-SIC MEMS DEVICES WITH N-P-N HOMOEPITAXIAL STRUCTURE** 276
F. Zhao and A.V. Lim
Washington State University, Vancouver, USA

We report single crystalline 4H-SiC MEMS with homoepitaxial n-p-n structure. Single crystalline fully exploits the superior material properties of SiC for operations in harsh environments. The n-p-n structure makes electrostatic actuation applicable which is essentially important for applications of resonators and actuators to sensor devices, and extends the capability of monolithic integration between SiC MEMS and electronic devices/circuits with n-p-n configurations such as BJTs and MOSFETs.

**W-006 INTEGRATION OF DISTRIBUTED GE ISLANDS ONTO SI WAFERS BY ADHESIVE
WAFER BONDING AND LOW-TEMPERATURE GE EXFOLIATION** 280
F. Forsberg[1], N. Roxhed[1], C. Colinge[2], G. Stemme[1], and F. Niklaus[1]
[1]*KTH Royal Institute of Technology, SWEDEN and* [2]*California State University, USA*

We present a novel and highly efficient wafer-level batch transfer process for populating silicon (Si) wafers with distributed islands of 1 micrometer-thick single-crystalline germanium (Ge). This is achieved by transferring Ge from a Si donor wafer containing thick Ge dies to a Si target wafer by adhesive wafer bonding and subsequent low-temperature Ge exfoliation.

M-007 MULTILAYER ETCH MASKS FOR 3-DIMENSIONAL FABRICATION OF ROBUST SILICON CARBIDE MICROSTRUCTURES 284
K.M. Dowling, A.J. Suria, A. Shankar, C.A. Chapin, and D.G. Senesky
Stanford University, USA

This paper demonstrates the fabrication of 3-D microstructures in 4H-silicon carbide (4H-SiC) substrates for the first time using a plasma etch and multilayer etch masks. This process was developed using a variety of thin film masks and demonstrated SiC etch rates as high as ~1 μm/min, a SiC:Ni etch selectivity as high as 60:1, and aspect ratio dependent etch characteristics. The microfabrication of complex SiC microstructures (mechanical gears, Lego®-like bricks, and poker chips) is presented.

T-008 NEMS BY MULTILAYER SIDEWALL TRANSFER LITHOGRAPHY 288
D. Liu, R.R.A. Syms, and M.M. Ahmad
Imperial College London, UK

We report an extension of a recently demonstrated technique to fabricate NEMS based on sidewall transfer lithography (STL). The process uses two STL steps to form intersecting nanoscale features such as suspension beams, breaking an important restriction of single-layer STL NEMS. The new process only requires optical lithography, making it suitable for low-cost mass parallel fabrication for complex NEMS on wafer scale. Current nanoscale features have a width of 100 nm and 50:1 aspect ratio.

W-009 NEW SCALABLE MICROFABRICATION METHOD FOR SURFACE ION TRAPS AND EXPERIMENTAL RESULTS WITH TRAPPED IONS 292
S. Hong[1], M. Lee[1], H. Cheon[1], J. Ahn[2], M. Kim[2], T. Kim[2], and D.D. Cho[1]
[1]Seoul National University, SOUTH KOREA and [2]SK Telecom, SOUTH KOREA

This paper presents a new microfabrication method for surface ion traps and experimental results with trapped ions. Using SiO_2 timed-etch method or copper sacrificial layer method, the ion trap chips with electrode overhang structures are fabricated. The ion trap chips are implemented in a 1×10^{-11} Torr vacuum environment for ion trapping experiments. Successful results in trapping strings of $^{171}Yb^+$ and $^{174}Yb^+$ ions as well as manipulating $^{171}Yb^+$ ions for qubit operation are demonstrated.

M-010 POST-RELEASE STRESS-ENGINEERING OF SURFACE-MICROMACHINED MEMS STRUCTURES USING EVAPORATED CHROMIUM AND IN-SITU FABRICATED RECONFIGURABLE SHADOW MASKS 296
R. Majumdar, V. Foroutan, and I. Paprotny
University of Illinois, Chicago, USA

We develop a novel post-release stress engineering process to provide an out-of-plane curvature to initially plane MEMS microstructures. The stressor layer(Chromium) is applied post-release and patterned appropriately with help of in-situ fabricated reconfigurable shadow masks. In addition to avoiding photolithography step, these shadow masks also provide variable coverage for underlying structures. A modified model is also shown which considers non-uniform Cr thickness on released structures.

T-011 SELF-ASSEMBLY OF MICROCOMPONENTS USING THE ENTROPIC EFFECT 300
U. Okabe, T. Okano, and H. Suzuki
Chuo University, JAPAN

We propose the use of the entropic effect (the depletion volume effect) for the self-assembly of microcomponents. When the solution contains macromolecule at relatively high concentration, microcomponents formed assembled structures. The bonding energy is not originated from the surface; it is generated by increasing the translational entropy of macromolecules in the solution. We expect that use of the depletion volume effect promotes the search of the global free-energy minima in the system by avoiding being trapped to the local minima in the self-assembly process.

W-012 THREE-DIMENSIONAL INTEGRATION OF SUSPENDED SINGLE-CRYSTALLINE SILICON MEMS ARRAYS WITH CMOS 304
Z. Song[1], Y. Du[1], M. Liu[1], S. Yang[1], D. Wu[1], and Z. Wang[1,2]
[1]Tsinghua University, CHINA and [2]Innovation Center for MicroNanoelectronics and Integrated System, CHINA

We present a generic three-dimensional (3-D) integration method to fabricate suspended single-crystalline silicon (SCS) MEMS arrays on CMOS. This method is applicable to a large variety of SCS MEMS including accelerometers, gyroscopes, micromirrors, RF MEMS switches, and resonators. Key challenges including fabrication process, mechanical reliability, and residue stress induced deflection have been addressed.

M-013 TRAJECTORY CONTROL OF MEMS FALLING OBJECT FABRICATED BY SU-8 MULTILAYER STRUCTURE 308
H. Yamane and S. Nagasawa
Shibaura Institute of Technology, JAPAN

We propose a trajectory control method for a MEMS falling object. The MEMS falling object is consisted of two units, an autorotation part and a non-rotation part. By using large falling objects, aerodynamics of the falling object was characterized. Then the MEMS falling object was designed considering with aerodynamics. The MEMS falling object was fabricated with a method of the SU-8 multi-layer structure. A MEMS autorotation part was fabricated and it rotated successfully in the wind-tunnel.

T-014 WAFER-SCALE INTEGRATION OF CARBON NANOTUBE TRANSISTORS AS PROCESS MONITORS FOR SENSING APPLICATIONS 312
K. Chikkadi, W. Liu, C. Roman, M. Haluska, and C. Hierold
ETH Zurich, SWITZERLAND

We report on the fabrication of carbon nanotube transistors designed as process control monitors for applications such as gas and pressure sensing. We demonstrate the concept for an integration process used for gas sensor fabrication on a 100 mm wafer. Our analysis on 4463 (including 2702 semiconducting) devices allows the extraction of distributions of threshold voltage, minimum device resistance, hysteresis width, process yield and wafer uniformity data on a 100 mm wafer.

Manufacturing for Bio- and Medical MEMS and Microfluidics

W-015 FABRICATION OF A MONOLITHIC CARBON MOLD FOR PRODUCING A MIXED-SCALE PDMS CHANNEL NETWORK USING A SINGLE MOLDING PROCESS 316
Y. Lee, Y. Lim, and H. Shin
Ulsan National Institute of Science & Technology (UNIST), SOUTH KOREA

We introduce a batch fabrication technique for a mixed-scale monolithic carbon mold producing a mixed-scale PDMS channel. A SU-8 structure fabricated by UV lithography was converted to a carbon structure trough the pyrolysis. The carbon mold dimension could be easily controlled from micrometer-scale to nanometer-scale when the pyrolysis accompanies enormous volume reduction. The mixed-scale PDMS channel network has a functionality for the nanochannel electroporation (NEP) without roof collapse.

M-016 FABRICATION OF PATTERNABLE NANOPILLARS FOR MICROFLUIDIC SERS DEVICES BASED ON GAP-INDUCED UNEVEN ETCHING 320
Y. Wang[1,2], L.C. Tang[1,3], H.Y. Mao[1,2], C. Lei[1,3], W. Ou[1,2], J.J. Xiong[3], Y. Ou[1,2], A.J. Ming[1,2], D. Li[4], and D.P. Chen[1,2]
[1]Chinese Academy of Sciences, CHINA, [2]Jiangsu R&D Center for Internet of Things, CHINA, [3]North University of China, CHINA, and [4]Stanford University, UNITED STATES

We report a novel, simple and time-saving lithography-free approach for fabricating patternable nanopillars. The key technique of the approach is to introduce a gap by covering it with a cap, which contains through holes and the material on its lower surface has a similar etching rate with the substrate. By adjusting sizes and profiles of the perforations, nanopillars with desirable patterns can be obtained. Thus a new way for fabricating microfluidic SERS devices is further developed.

T-017 A LOW-COST AND LABEL-FREE ALPHA-FETOPROTEIN SENSOR BASED ON SELF-ASSEMBLED GRAPHENE ON SHRINK POLYMER 324
S. Sando, B. Zhang, and T. Cui
University of Minnesota, USA

We develop a shrink-induced grapheme sensor for label-free biomolecule detection. While Enzyme-linked immunosorbent assay (ELISA) is the most popular method to detect specific proteins in the current medicine, label-free biosensors have attracted great attention due to simplicity and ease of use. The sensor described in this work demonstrates the ability to detect alpha-fetoproteins (AFP), one of the most important tumor markers associated with liver cancer and ovarian cancer.

W-018 HIGH-TOPOGRAPHY SURFACE FUNCTIONALIZATION BASED ON PARYLENE-C PEEL-OFF FOR PATTERNED CELL GROWTH 328
F. Larramendy[1], D. Serien[2], S. Yoshida[2], L. Jalabert[2], S. Takeuchi[2], and O. Paul[1]
[1]University of Freiburg - IMTEK, GERMANY and [2]University of Tokyo, JAPAN

We develop a new technique for patterning functionalization layers on substrates with high topography. The method is based on a parylene-C template shaped by a structured, sacrificial photoresist layer and attached to the substrate where functionalization is not intended. We successfully demonstrate the technique with the guided growth of PC12 cells on honeycomb-shaped protein patterns on micropillars and microwells.

M-019 LASER TREATED GLASS PLATFORM WITH RAPID WICKING-DRIVEN TRANSPORT AND PARTICLE SEPARATION CAPABILITIES 332
M. Ochoa, H. Jiang, R. Rahimi, and B. Ziaie
Purdue University, USA

Wicking and particle separation are two required capabilities for many microfluidics and lab-on-a-chip devices, but they often require multiple materials and structures (e.g., paper, polymer filters) which are difficult to integrate with established microfabrication techniques and materials. In this work, we combine both properties into a single glass platform with a straightforward and economical fabrication process. By laser machining soda lime glass with a specific power and laser speed, we create channels defined by an array of micro cracks (3–4 μm) which provide particle separation properties and simultaneously enable rapid liquid transport (up to 24.2 mm/s) as a result of capillary forces from the crevices and laser-induced surface hydrophilization.

T-020 LIPOSOME ARRANGEMENT CONNECTED WITH AVIDIN-BIOTIN COMPLEX FOR CONSTRUCTING FUNCTIONAL SYNTHETIC TISSUE 336
H. Hamano[1], T. Osaki[2], and S. Takeuchi[1]
[1]University of Tokyo, JAPAN and [2]Kanagawa Academy of Science and Technology (KAST), JAPAN

This paper describes a method to arrange liposomes into organized structures with biochemical binding of avidin biotin complex, inspired by the biological anchoring junction. This approach enhances the stability of the liposome structure, which would provide an improved model for a liposome-based tissue in synthetic biology.

W-021 MAGNESIUM-EMBEDDED LIVE CELL FILTER FOR CTC ISOLATION 340
Y. Liu[1], J. Park[1], T. Xu[2], Y. Xu[2], J.H.-C Chang[1], D. Kang[1], X. Zhang[1], A. Goldkorn[2], and Y.C. Tai[1]
[1]California Institute of Technology, USA and [2]University of Southern California, USA

We develop a novel Magnesium-embedded cell filter for Circulating Tumor Cell (CTC) capture, release and isolation. The new and novel feature is the use of thin-film Mg to release the captured CTCs based on the fact that any Cl- containing culture medium can readily etch Mg away. The releasing and the isolation of each individual CTC are demonstrated. The top PA-C filter pieces break apart from the bottom after Mg completely dissolves, enabling captured CTCs to detach from the filter.

**M-022 MICRO FLUIDIC CHAMBER WITH THIN SI WINDOWS FOR
OBSERVATION OF BIOLOGICAL SAMPLES IN VACUUM** 344
H. Hayashi, M. Toda, and T. Ono
Tohoku University, JAPAN

A micro fluidic chamber with 178 nm-thick thin Si windows on a micro channel has been developed. Using this windows, the aquatic structure inside of the channel has been monitored by scanning electron microscopy. Secondary electrons from a sample in the channel are able to be detected in vacuum with an acceleration voltage of 15 kV. The micro fluidic chamber is possibly applied to cell imaging via the single crystal Si thin window in vacuum using magnetic resonance force microscopy.

**T-023 RAPID, LOW COST FABRICATION OF CIRCULAR CROSS-SECTION
MICROCHANNELS BY THERMAL AIR MOLDING** 348
T.-Q. Nguyen and W.-T. Park
Seoul National University of Science and Technology, SOUTH KOREA

This paper demonstrates a simple fabrication process of polydimethylsiloxane(PDMS) circular cross section microfluidic channels by using a PDMS master mold and thermal air molding. Based on this technique, circular cross section microchannel can be easily produced in a wide range of dimensions from 10μm to 500μm with simple bench top equipment. This technique can create perfect circular channels without any plasma activated bonding and alignment process. We can also apply this technique to fabricate micro concaves for spheroid culture, micro nozzles for droplet generation, and micro patch clamps for cell immobilization.

W-024 SKIN-EQUIVALENT INTEGRATED WITH PERFUSABLE CHANNELS ON CURVED SURFACE 351
N. Mori[1], Y. Morimoto[1,2], and S. Takeuchi[1,2]
[1]University of Tokyo, JAPAN and [2]Japan Science and Technology Agency (JST), JAPAN

We constructed a skin-equivalent on a curved surface. The skin-equivalent consists of not only the epidermis/dermis but also perfusable channels. We embedded an anchoring structure in our device to prevent horizontal contraction of the tissue. Owing to perfusion, we can culture epidermis at the air-liquid interface for cornification. Our method enables the skin-equivalent construction on a curved surface that is necessary for the construction of 3D skin surface such as biohybrid robots' skin.

**M-025 TFT DISPLAY PANEL TECHNOLOGY AS A BASE FOR BIOLOGICAL CELLS
ELECTRICAL MANIPULATION - APPLICATION TO DIELECTROPHORESIS** 354
A. Tixier-Mita, B.-D. Segard, Y.-J. Kim, Y. Matsunaga, H. Fujita, and H. Toshiyoshi
University of Tokyo, JAPAN

This paper reports for the first time the use of TFT (Thin Film Transistor) technology of display panels for biological cells electrical manipulation. This technology allows to have high density distributed transparent micro-electrodes, independently controllable, covering centimeter-size glass substrates. This technology is much superior to usual micro-technology used for Multielectrode Arrays (MEAs).The chosen application, to demonstrate the capability of such technology, is dielectrophoresis.

Materials for MEMS and NEMS

**T-026 3D MORPHOLOGY RECONSTRUCTION OF HIGH ASPECT RATIO MEMS
STRUCTURE BY USING AUTOFLUORESCENCE OF PARYLENE C** 358
L. Zhang, Y. Liu, F. Yang, W. Wang, D. Zhang, and Z. Li
Peking University, CHINA

We reported a MEMS fabrication compatible, damage free method for in-process 3D morphology reconstruction of high aspect ratio microstructure. As a novel morphology tracer, Parylene C thin film was conformally deposited onto the structure and annealed at high temperature under N2. By scanning with a confocal microscopy, 3D morphology of the microstructure was reconstructed from the autofluorescence information of Parylene C.

**W-027 A NANOWIRE GAUGE FACTOR EXTRACTION METHOD FOR
MATERIAL COMPARISON AND IN-LINE MONITORING** 361
I. Ouerghi, J. Philippe, C. Ladner, P. Scheiblin, L. Duraffourg, S. Hentz, and T. Ernst
CEA-LETI, FRANCE

We propose a new extraction method of gauge factor of nanowires for in-line monitoring of this parameter and piezoresistive material properties comparisons. Unlike conventional techniques which are destructive and suffer from reproducibility issues, this method allows a direct measurement of the GF locally at the nanoscale and at the wafer level. GFs have been reliably measured on a wide range of silicon-based NEMS resonators with different designs,crystalline structure and doping level.

M-028 A THREE-STEP MODEL OF BLACK SILICON FORMATION IN DEEP REACTIVE ION ETCHING PROCESS 365

F. Zhu[1], C. Wang[2], X. Zhang[1], X. Zhao[2], and H. Zhang[1]

[1]*Peking University, CHINA and* [2]*Nankai University, CHINA*

A three-step model used for modeling and simulation of black silicon formation in DRIE (Deep Reactive Ion Etching) process is presented. It combines quantum mechanics, sheath dynamics and diffusion theory together based on plasma environment. The simulation results show very good coincidence with experimental SEM images, proving the applicability of this theory and it's very promising to make black silicon formation in DRIE process to be controllable.

T-029 ANOMALOUS RESISTANCE CHANGE OF ULTRASTRAINED INDIVIDUAL MWCNT USING MEMS-BASED STRAIN ENGINEERING 369

K. Yamauchi, T. Kuno, K. Sugano, and Y. Isono

Kobe University, JAPAN

This research clarified the anomalous electric resistance change of ultrastrained multi-walled carbon nanotube (MWCNT), as well as its mechanical properties, using the in-situ SEM nanomanipulation system with Electrostatically Actuated NAnotensile Testing device (EANAT). Although the resistance change ratio was almost constant during the interlayer sliding of MWCNT, it showed a sharp raise at the end of the sliding in spite of the MWCNT not breaking mechanically.

W-030 FABRICATION OF TETRAPOD-SHAPED AL/NI MICROPARTICLES WITH TUNABLE SELF-PROPAGATING EXOTHERMIC FUNCTION 373

K. Inoue[1], T. Fujito[1], K. Fujita[2], Y. Kuroda[2], K. Takane[2], and T. Namazu[1]

[1]*University of Hyogo, JAPAN and* [2]*Gauss Co., Ltd., JAPAN*

This paper reports tetrapod-shaped Al/Ni microparticles fabricated by injection-molding and electroless-plating. We realize, for the first time, fabricating the 3D microtetrapods with self-propagating exothermic function that is able to tunable by changing Al powder's diameter and simultaneously by keeping porous Al tetrapod's void fraction constant. The maximum surface temperature and high-temperature duration during the reaction differ from sputtered Al/Ni multilayer film's exothermic performances. The tetrapod's exothermic performances can be freely controlled in response to the applications.

M-031 HIGH-PRODUCTIVE FABRICATION METHOD OF FLEXIBLE PIEZOELECTRIC SUBSTRATE 377

H. Hida, S. Yagami, Y. Sakurai, and I. Kanno

Kobe University, JAPAN

This paper reports a simple and high-productive fabrication method of flexible piezoelectric substrate using a new transfer technique. We experimentally clarified that Pb(Zr,Ti)O3 (PZT) thin films deposited on metal substrates can be transfer to PDMS substrates by using metal wet etching process. By characterizing crystal structures and electric properties, we confirmed that transferred PZT thin films have piezoelectric properties.

T-032 MECHANICAL CHARACTERIZATION OF THIN FILMS USING A MEMS DEVICE INSIDE SEM 381

C. Cao, B. Chen, T. Filleter, and Y. Sun

University of Toronto, CANADA

A MEMS device was developed for mechanical characterization of 2D ultra-thin films. The device utilizes electrothermal actuators to apply uniaxial tension. The robust design makes the device capable of withstanding both dry and wet transfer of 2D ultra-thin film materials onto the suspended structures on the device. Fracture stress of thin graphene oxide (GO) films was measured.

W-033 ROOM TEMPERATURE SYNTHESIS OF SILICON DIOXIDE THIN FILMS FOR MEMS AND SILICON SURFACE TEXTURING 385

A. Ashok and P. Pal

Indian Institute of Technology Hyderabad, INDIA

In this paper, SiO2 thin films deposited at room temperature using anodic oxidation method are explored for the fabrication of MEMS components and the surface texturing for solar cells applications. The anodic oxide is used as structural layer for the formation of freestanding structures (e.g. cantilever) of nanometer thickness on Si{100} wafers using anisotropic etchants. Further, the oxide film is employed as mask in KOH to texturize the silicon surface without using lithography.

M-034 SIZE EFFECT ON BRITTLE-DUCTILE TRANSITION TEMPERATURE OF SILICON BY MEANS OF TENSILE TESTING 389

A. Uesugi, Y. Hirai, T. Tsuchiya, and O. Tabata

Kyoto University, JAPAN

We report the size effect on BDTT of single crystal silicon using different width specimens (120-μm long, 5-μm thick and 4 or 9-μm wide). The BDTT was characterized by tensile testing in vacuum up to 600 °C with IR light heating. The fractured specimens showed slips on (111) planes at 500 °C and above, which indicated that the BDTT of micrometer-sized silicon decreased from millimeter-sized specimens. We found that the length along slip-propagation direction might dominate the BDTT.

T-035 STICTION FORCES AND REDUCTION BY DYNAMIC CONTACT IN ULTRA-CLEAN ENCAPSULATED MEMS DEVICES 393

D.B. Heinz[1], V.A. Hong[1], T.S. Kimbrell[1], J. Stehle[2], C.H. Ahn[1], E.J. Ng[1], Y. Yang[1], G. Yama[2], G.J. O'Brien[2], and T.W. Kenny[1]

[1]*Stanford University, USA and* [2]*Robert Bosch RTC, USA*

We demonstrate the consistent and manageable surface adhesion and stiction forces in epitaxially encapsulate MEMS devices. Data from over 2000 test structures of 80 design variations from three different fabrication runs were gathered. The measured adhesion forces (18- 25uN) are small enough for inertial sensors and are independent of contact geometry. In addition, we demonstrate anti-stiction bump stops with springs for a sliding contact and reduce the probability of stiction by over 50%.

W-036 STUDY OF THE HYBRID PARYLENE/PDMS MATERIAL 397
D. Kang[1], S. Matsuki[2], and Y.-C. Tai[1]
[1]*California Institute of Technology, USA and* [2]*Northeastern University, USA*

This paper reports the mechanical behaviors and barrier properties of hybrid PDMS-Parylene materials and presents a novel approach of implementing in-situ heating deposition to facilitate the diffusion and penetration of parylene coatings into PDMS, demonstrating enhanced pore sealing capability, for which a mathematical model was proposed and the average PDMS pore size was determined.

M-037 SYNTHESIS OF CARBON NANOTUBES-NI COMPOSITE FOR
MICROMECHANICAL ELEMENTS APPLICATION 401
Z. An, M. Toda, G. Yamamoto, T. Hashida, and T. Ono
Tohoku University, JAPAN

We present the fabrication and characterization of a silicon micromirror with carbon nanotubes -nickel composite beams. A novel electroplating method is developed for synthesis of the CNTs-Ni composite. The maximum variation of the resonant frequency of the fabricated micromirror during a long term stability test is about 0.3%, and its scanning angle is about 20o. It shows the potential ability of the CNTs-Ni composite for micromechanical elements application.

Packaging and Assembly

T-038 3D STRUCTURAL FORAMATION UTILIZING GLASS TRANSITION OF A PARYLENE FILM 405
T. Kan, A. Isozaki, H. Takahashi, K. Matsumoto, and I. Shimoyama
University of Tokyo, JAPAN

We reports on a fabrication of 3D micro structures supported by a Parylene thin film. The 3D structure formation procedure starts with an out-of-plane actuation of the Si micro structure coated with a Parylene film. At the same time, the environmental temperature was elevated above the glass transition temperature of the Parylene, and then cooled down below Tg. The rearrangement of the Parylene film happens above Tg, and the consolidated Parylene after the cooling down maintains the structure.

W-039 A NOVEL FABRICATION AND WAFER LEVEL HERMETIC
SEALING METHOD FOR SOI-MEMS DEVICES USING SOI CAP WAFERS 409
M.M. Torunbalci, S.E. Alper, and T. Akin
Middle East Technical University (METU), TURKEY

This paper presents a novel and inherently simple all-silicon (glass-free) fabrication and hermetic packaging method developed for SOI-MEMS devices, enabling lead transfer using vertical feedthroughs formed on an SOI cap wafer. The processes of the SOI cap wafer and the SOI-MEMS wafer require a total of five inherently simple mask steps, providing a combined process and packaging yield as high as 95%.

M-040 BONDING MECHANISM IN THE VELCRO CONCEPT SI-SI
LOW TEMPERATURE DIRECT BONDING TECHNIQUE 413
S. Keshavarzi[1,2], U. Mescheder[1], and H. Reinecke[2]
[1]*Furtwangen University - IAF, GERMANY and* [2]*University of Freiburg - IMTEK, GERMANY*

In this paper, we present the bonding mechanism of two Velcro-like (needle-like) surfaces for low temperature Si-Si direct bonding at ambient environment based on capillary force approach. The model considers both deformation and interaction mechanisms of the needles during bonding which makes it superior to other presented models.

T-041 FET PROPERTIES OF SINGLE-WALLED CARBON NANOTUBES
INDIVIDUALLY ASSEMBLED UTILIZING SINGLE STRAND DNA 417
K. Hokazono, Y. Hirai, T. Tsuchiya, and O. Tabata
Kyoto University, JAPAN

A new assembly process for isolated single-walled carbon nanotubes (SWCNTs) on MEMS structures and the electrical properties of SWCNT field effect transistor (FET) are reported. Mono-dispersed SWCNT solution is prepared by biotin modified single strand DNA's wrapping and the tubes are assembled onto gold electrodes using biotin-avidin bindings. The isolated SWCNT bridges over electrode gaps are successfully demonstrated. The Id-Vg curves of SWCNT show both conductor and semiconductor properties.

W-042 IMPACT-INDUCED HARDENING PACKAGE FOR
TACTILE SENSORS USING DILATANT FLUID 421
T. Takahata, K. Matsumoto, and I. Shimoyama
University of Tokyo, JAPAN

To realize high-sensitive and shock-resistant tactile sensor, the sensing element was surrounded by dilatant fluid, which is soft to a static force and hard to an impact, force. The applied static force was concentrated to the sensor, whereas the impact force was dispersed to the substrate. We have experimentally shown that the shock-resistant nature of the sensor with dilatant fluid package was 4 to 16 times as large as that of without the fluid.

M-043 MICRO DEVICES INTEGRATION WITH STRETCHABLE SPRING EMBEDDED IN LONG PDMS-FIBER FOR FLEXIBLE ELECTRONICS 423

W.-L. Sung, C.-L. Cheng, and W. Fang
National Tsing Hua University, TAIWAN

This study presents a PDMS (Polydimethylsiloxane) fiber integrated with multi-devices scheme using stretchable electroplating copper spring. Each device was located on the node and embedded in PDMS-fiber. Thus, devices are mechanically connected by PDMS-fiber and electrically connected by inner stretchable spring. Advantages of this approach: (1) length magnification by stretchable spring; (2) thicker stretchable spring embedded in PDMS provides well mechanical/electrical characteristics; (3) node acts as a hub for devices implementation and integration; (4) partially stretched spring could reduce the resistance variation by external loads.

T-044 PREPARATION OF WAFER LEVEL GLASS-EMBEDDED HIGH-ASPECT-RATIO PASSIVES USING A GLASS REFLOW PROCESS 427

M. Ma, J. Shang, and B. Luo
Southeast University, CHINA

We develop an innovative, uncomplicated and inexpensive method based on a glass reflow process to fabricate void-free glass-embedded passives for 3D MEMS packaging. The embedded structures include cylindrical, annular cylindrical and coaxial cylindrical conductive through-holes, plate and coaxial torus trench capacitors, square spiral, circular spiral and folding type trench inductors, and filters. These void-free structures have vertical and smooth sidewall, large signal pathways.

Micro- and Nanofluidics
Lab-on-Chip Medical Diagnostic Devices

W-045 "CELL PINBALL": WHAT IS THE PHYSICS? 431

R. Murakami[1], M. Kaneko[1], S. Sakuma[2], and F. Arai[2]
[1]Osaka University, JAPAN and [2]Nagoya University, JAPAN

During the deformability test of Red Blood Cell (RBC) by utilizing a micro fluidic channel, we found an interesting phenomenon where some RBCs behave just like elastic pinball. This phenomenon is called "Cell Pinball". Through visualization, we found that the RBC being in cell pinball mode rotates around the perpendicular axis to the flow line and its direction is one-to-one relationship with the moving direction. We also found that the rotating axis exists slightly behind the center of gravity

M-046 A FULLY MONOLITHIC MICROFLUIDIC DEVICE FOR COUNTING BLOOD CELLS FROM RAW BLOOD 435

J. Nguyen[1], Y. Wei[1], Y. Zheng[1], C. Wang[2], and Y. Sun[1]
[1]University of Toronto, CANADA and [2]Mount Sinai Hospital, CANADA

We develop a monolithic microfluidic device capable of on-chip sample preparation for complete blood count from raw blood. For the first time, on-chip sample processing (e.g. dilution, lysis, and filtration) and downstream measurements were fully integrated to enable sample preparation and single cell analysis from raw blood on a single device. RBC and WBC concentration, WBC differential, mean corpuscular volume and cell distribution width are determined by electrical impedance measurements.

T-047 A MICROWELL DEVICE FOR MEASUREMENT OF MEMBRANE TRANSPORT OF ADHERENT CELLS 439

Y. Okada, M. Tsugane, and H. Suzuki
Chuo University, JAPAN

We developed the microwell device for measurement of membrane transport for single adherent cells. When cells were cultured on the microwells with ~10 μm opening, they spread to form the closed picoliter space. Thus, molecules exported from cells accumulate in such a space and be detected by imaging. We show that, by employing horizontal microwell design, materials exported from the cell membrane can be visualized. Efflux of the cancer drug transported by the multidrug resistance protein was tested.

W-048 ALGINATE HYDROGEL BASED 3-DIMENSIONAL CELL CULTURE AND CHEMICAL SCREENING PLATFORM USING DIGITAL MICROFLUIDICS 443

S.M. George and H. Moon
University of Texas, Arlington, USA

We develop a method for creating arrays of individually addressable cell seeded calcium alginate hydrogels for 3D cell culture using electrowetting on dielectric (EWOD) digital microfluidics (DMF). Combined with EWOD DMF's multiplexing abilities, we demonstrate how a single integrated DMF device is capable of forming cell seeded alginate hydrogels, generating different concentrations of chemicals and delivering these to different gels to observe the effect of chemical concentrations on 3D tissue

M-049 AN ELECTROCHEMICAL MICROFLUIDIC PAPER-BASED GLUCOSE SENSOR INTEGRATING ZINC OXIDE NANOWIRES 447

X. Li, C. Zhao, and X. Liu
McGill University, CANADA

We develop an electrochemical microfluidic paper-based analytical device, featuring a working electrode decorated with semiconductor zinc oxide nanowires (ZnO NWs), for glucose detection in human serum. The integration of ZnO NWs into the paper device is realized via facile, low-cost hydrothermal synthesis of ZnO NWs on paper, and leads to superior analytical performance (high sensitivity and low limit of detection) and enhanced device stability (by removing light-sensitive electron mediators).

T-050 DRY REAGENT STORAGE IN DISSOLVABLE FILMS AND LIQUID TRIGGERED RELEASE FOR PROGRAMMED MULTI-STEP LAB-ON-CHIP-DIAGNOSTICS 451

G. Lenk, G. Stemme, and N. Roxhed

KTH Royal Institute of Technology, SWEDEN

A capillary driven lab on a chip system using dissolvable films for on-chip reagent storage, volume metering and timing of a multi-step sequence is demonstrated. Activation of the chip with a single liquid causes rehydration of four different reagents stored in dissolvable polymer layers and sequential release of the reagents to a common reaction zone. This capillary driven, single-liquid triggered multi-reagent sequence can potentially be used for multi-step PoC immunoassays.

W-051 ON-CHIP DETECTION OF WILD 3R, 4R AND MUTANT 4R TAU THROUGH KINESIN-MICROTUBULE BINDING 455

S.P. Subramaniyan[1], M.C. Tarhan[2], S.L. Karsten[3], H. Fujita[2], H. Shintaku[1], H. Kotera[1], and R. Yokokawa[1]

[1]Kyoto University, JAPAN, [2]University of Tokyo, JAPAN, and [3]NeuroInDx Inc., USA

We report demonstration of on-chip tau detection based on difference of landing rate and binding density of microtubules on kinesin surface. Tau detection device comprises of a MT reservoir, channel and collector region with overhung structures. We assayed MTs decorated with three tau types in the kinesin coated device. Since the increase of fluorescent intensity at collector regions reflected the type of tau decorated on MTs, thus by measuring FI we distinguish wild 3R, 4R and P301L mutant tau.

M-052 THE CAPILLARY NUMBER EFFECT ON THE CAPTURE EFFICIENCY OF CANCER CELLS ON COMPOSITE MICROFLUIDIC FILTRATION CHIPS 459

C. Zhao[1], R. Xu[1], K. Song[1], D. Liu[2], S. Ma[1], C. Tang[1], C. Liang[1], Y. Zohar[3], and Y.-K. Lee[1]

[1]Hong Kong University of Science and Technology (HKUST), HONG KONG,
[2]Guangzhou First Municipal People's Hospital, CHINA, and [3]University of Arizona, USA

We present a systematic study of the Capillary number effect on the capture efficiency of cancer cells on a composite microfluidic filtration chip. A phase diagram for the capture efficiency of microfiltration chips as a function of normalized cell diameter and Capillary number has been obtained, which will be useful for the designing the next generation of microfiltration devices for isolating circulating tumor cells.

Materials for Bio- and Medical MEMS and Microfluidics

T-053 3D CULTURE OF MOUSE IPSCS IN HYDROGEL CORE-SHELL MICROFIBERS 463

K. Ikeda[1,2,3], T. Okitsu[1,2], H. Onoe[1,2], and S. Takeuchi[1,2]

[1]University of Tokyo, JAPAN, [2]Japan Science and Technology Agency (JST), JAPAN, and
[3]University of Tsukuba, JAPAN

This paper reports the culturing and expansion of mouse induced pluripotent stem cells (iPSCs) in hydrogel core-shell microfibers; the core consists of iPSCs with or without extracellular matrix (ECM) proteins, and the shell is composed of calcium alginate. We revealed mouse iPSCs cultured in the micro-scale space with ECM proteins sustain their pluriotency efficiently. This 3D culture system may be useful tool to expand iPSCs for clinical use.

W-054 ALIGNMENT OF COLLAGEN NANOFIBERS IN 2D SUBSTRATES USING CYCLIC STRETCH 465

E.R. Nam[1,2], W.C. Lee[1,2], and S. Takeuchi[1,2]

[1]University of Tokyo, JAPAN and [2]Japan Science and Technology Agency (JST), JAPAN

In this work, collagen nanofibers are self-aligned in fully 2-dimensional substrates by applying cyclic stretch during the gelation process of collagen solution, and the fabricated collagen sheet induce the alignment of cells without any mechanical force within the cultivation period. We believe that our new aligned collagen sheet contributes to the regenerative medicine, which needs a scaffold that has biological structures and microenvironment.

M-055 BONDING-FRIENDLY PCPDMS: DEPOSITING PARYLENE C INTO PDMS MATRIX AT AN ELEVATED TEMPERATURE 467

Y. Liu[1], L. Zhang[1], W. Wang[1,2,3], and W. Wu[1,2,3]

[1]Peking University, CHINA, [2]National Key Laboratory of Science and Technology on Micro/Nano Fabrication, CHINA,
and [3]Innovation Center for MicroNanoelectronics and Integrated System, CHINA

This paper reported a simple and effective process of bonding-friendly Parylene C-caulked PDMS (pcPDMS) for low-permeability required microfluidics. Parylene C was deposited into PDMS matrix at an elevated temperature (higher than 135°C) to caulk the permeable sites. The so-prepared pcPDMS can be directly bonded with oxygen plasma treatment just as pristine PDMS. SEM EDAX and Laser scanning confocal microscopy (LSCM) were introduced to characterize the Parylene C caulked status in the PDMS.

T-056 CHEMICALLY RESPONSIVE PROTEIN-PHOTORESIST HYBRID ACTUATOR 470

D. Serien[1] and S. Takeuchi[1,2]

[1]University of Tokyo, JAPAN and [2]Japan Science and Technology Agency (JST), JAPAN

We report the multiphoton fabrication of hybrid microstructures of photoresist and chemically responsive protein hydrogel for microactuation, such as a lever and a rotary stepper.

W-057 CORE-SHELL MICROPARTICLE SYNTHESIS IN DROPLET MICROFLUIDICS USING A SINGLE STEP POLYMERIZATION 472

X. Zhou, Y. Sun, A. Finne-Wistrand, W. van der Wijngaart, and T. Haraldsson

KTH Royal Institute of Technology, SWEDEN

We present, for the first time, a method for the synthesis of core-shell microparticles in a single polymerization step using two-phase droplet microfluidics. We verify the successful generation of core-shell microparticles using the novel synthesis approach.

M-058 VASCULAR NETWORK FORMATION FOR A LONG-TERM SPHEROID CULTURE BY CO-CULTURING ENDOTHELIAL CELLS AND FIBROBLASTS 476

T. Hayashi[1], H. Takigawa-Imamura[2], K. Nishiyama, H. Shintaku[1], H. Kotera[1], T. Miura[2], and R. Yokokawa[1]

[1]Kyoto University, JAPAN, [2]Kyushu University, JAPAN, and [3]Kumamoto University, JAPAN

We developed a PDMS microfluidic device to create a vascular network for a long-term spheroid culture, which consists of co-culture of human umbilical vein endothelial cells (HUVEC) and normal human lung fibroblasts (LF). Although network formation has been reported in several microfluidic devices, we successfully visualized that HUVEC networks formed by the co-culture with LFs reached a LF-based spheroid. Moreover, perfusability of the network was evaluated by injecting fluorescent microbeads.

T-059 HIGHLY CONTROLLABLE THREE-DIMENSIONAL SHEATH FLOW DEVICE FOR FABRICATION OF ARTIFICIAL CAPILLARY VESSELS 480

J. Ito, R. Sekine, D.H. Yoon, Y. Nakamura, H. Oku, H. Nansai, T. Chikasawa, T. Goto, T. Sekiguchi, N. Takeda, and S. Shoji

Waseda University, JAPAN

We developed a highly controllable three-dimensional (3D) sheath flow device for fabrication of artificial capillary vessels. In order to fabricate double-layer coaxial Core-Sheath microfibers applicable to long micro capillary vessels, three step sheath injection type 3D flow device which realizes wide core and sheath structure variations by simply flow rate control was applied, and vascular endothelial cells embedded to inside of the fabricated microfiber were cultured.

W-060 MODELING NON-WETTING PERFORMANCES OF SUPERLYOPHOBIC SURFACES BASED ON LOCAL CONTACT LINE 484

Z. Wang and T. Wu

Shenzhen Institutes of Advanced Technology, Chinese Academy of Sciences, CHINA

Superlyophobic surfaces (SLS) are promising as the novel universal platform for microfluidics due to the unique super-repellency and extremely low adhesion for almost all liquids. This paper proposed the pressure stability criteria and CAH estimation based on the local contact line analysis and were in excellent agreement with the experimental results. The achievements may shed new light for designing high-performance SLS for microfluidics and other MEMS fields.

Microfluidics and Nanofluidics

M-061 MICROFLUIDIC SELECTION OF APTAMERS USING COMBINED ELECTROKINETIC AND HYDRODYNAMIC MANIPULATION 488

T. Olsen, J. Zhu[1], J. Kim[1], R. Pei[2], M.N. Stojanovic[1], and Q. Lin[1]

[1]Columbia University, USA and [2]Chinese Academy of Sciences, CHINA

We present a microfluidic device that is capable of closed-loop, multi-round SELEX without manual intervention or use of off-chip instruments, as demonstrated by selection of DNA aptamers against the protein IgE with high affinity (KD = 12 nM) in a rapid manner (4 rounds in 10 hours).

T-062 A NOVEL DENSITY-BASED DIELECTROPHORETIC PARTICLE FOCUSING TECHNIQUE FOR DIGITAL MICROFLUIDICS 492

E. Samiei, H. Rezaei Nejad, and M. Hoorfar

University of British Columbia, CANADA

A new particle focusing technique based on negative dielectrophoretic (DEP) manipulation of non-buoyant particles is developed for digital microfluidic (DMF) platforms. This technique is compatible with conventional DMF electrode designs and does not require geometrical modification. Non buoyant particles can be concentrated on an electrode, followed by droplet splitting, resulting in two daughter droplets one with a high, and the other with a very low concentration of the particles.

W-063 ACTIVE CONTROL OF CHEERIOS EFFECT FOR DIELECTRIC FLUID 496

J. Yuan and S.K. Cho

University of Pittsburgh, USA

Using di-electrowetting we achieved on-demand control of Cheerios effect for dielectric(non-conductive) fluids. Additionally, our theory and experiment discovered that the tilting angle of wall is critical in this control. The present control would provide an efficient tool in many micro/nano particle manipulation/processes on phase interfaces.

M-064 BIOMECHANICAL CANAL SENSORS INSPIRED BY CANAL NEUROMASTS FOR ULTRA SENSITIVE FLOW SENSING 500

A.G.P. Kottapalli[1,2], M. Asadnia[1,2], J.M. Miao[1], and M.S. Triantafyllou[2,3]

[1]Nanyang Technological University, SINGAPORE,
[2]Singapore-MIT Alliance for Research and Technology, SINGAPORE, and
[3]Massachusetts Institute of Technology, USA

Fishes use their mechanosensory lateral-line system to detect minute disturbances underwater. The lateral-lines consist of superficial and canal neuromast (SN and CN) sensory sub-systems. In this paper, for the first time, we present the design, fabrication and experimental characterization of arrays of zero-powered and ultrasensitive MEMS piezoelectric haircell sensors encapsulated into biomimetic canals.

T-065 BAKING-POWDER DRIVEN CENTRIPETAL PUMPING CONTROLLED BY EVENT-TRIGGERING OF FUNCTIONAL LIQUIDS 504

D.J. Kinahan[1], R. Burger[2], A. Vembadi[1], N.A. Kilcawley[1], D. Lawlor[1], M.T. Glynn[1] and J. Ducrée[1]

[1]Dublin City University, IRELAND and [2]Technical University of Denmark (DTU), DENMARK

This paper reports radially inbound pumping by the event-triggered addition of water to on-board stored baking powder in combination with valving by an immiscible, high-specific weight liquid on a centrifugal microfluidic platform. This technology allows making efficient use of precious real estate near the center of rotation by enabling the placement of early sample preparation steps as well as reagent reservoirs at the spacious, high-field region on the perimeter.

W-066 CONTINUOUS-FLOW DIELECTROPHORETIC SORTING OF PARTICLES VIA 3D SILICON ELECTRODES FEATURING CASTELLATED SIDEWALLS 508

X. Xing and L. Yobas

Hong Kong University of Science and Technology (HKUST), HONG KONG

This paper for the first time describes continuous-flow dielectrophoretic (DEP) separation of particles using a simple microfluidic design incorporating 3D silicon electrodes featuring castellated sidewalls. The 3D electrodes generates non-uniform electric field along the channel depth which drive the particles into distinct layer. Meanwhile, continuous flow transport the separated particles into different outlets simultaneously.

M-067 ELECTRON BEAM SWITCHED TRAPPING AND RELEASE OF NANOPARTICLES ON NANOPORE ARRAY 512

T. Hoshino and K. Mabuchi

University of Tokyo, JAPAN

We demonstrated switching of trap and release of nanoparticles using an inverted-electron beam lithography (I-EBL). 240-nm nanobeads suspended in pure water were trapped and released on nanopore array. The nanopores in a silicon nitride membrane generated trapping flows for the beads by infinitesimal leakages toward directly connected high vacuum via the nanopores. The incident electron beam selectively induced release of the trapping beads into the solution by Coulomb force.

T-068 ENCODING AND MANIPULATING MICROCOMPONENT ON ELECTROMICROFLUIDIC PLATFORM 516

M.-Y. Chiang[1], S.-Y. Chen[1], and S.-K. Fan[2]   * Award Nominee

[1]National Chiao Tung University, TAIWAN and [2]National Taiwan University, TAIWAN

Microcomponent encoding and manipulation were performed on an electromicrofluidic platform using electrowetting and dielectrophoresis to drive particles and cells on micrometer scale for encoding and liquid or solid microcomponents on millimeter scale for larger structures assembly. 3D cell culture with reorganized fibroblasts in hydrogel microcomponents is demonstrated on the platform. The technology is applicable to heterogeneous structure formation and alternative 3D bioprinting.

W-069 IN VITRO DYNAMIC FERTILIZATION BY USING EWOD DEVICE 519

L.-Y. Chung[1], H.-H. Shen[1], Y.-H. Chung[1], C.-C. Chen[1], C.-H. Hsu[2], H.-Y. Huang[3], and D.-J. Yao[1]

[1]National Tsing Hua University, TAIWAN, [2]National Health Research Institutes, TAIWAN, and [3]Chang Gung Memorial Hospital, TAIWAN

The result of EWOD chip culturing 2-cell to Blastocyst (B.C.) is shown on Table 1. The probability of 2-cell to 4-cell is 76.19%, 2-cell to 8-cell is 42.85%, and 2-cell to B.C. is 33.33%. This result is proved that the chip biocompatibility of EWOD has kept as a stable ratio. Figure 3 shows oocyte and sperm droplet mixed inside EWOD chip with the process from (A) to (D) with smooth situation prospectively. Table 2 is the comparison result of EWOD (Dynamic) and traditional IVF (Static) development.

M-070 LIPOPHILIC-MEMBRANE BASED ROUTING FOR CENTRIFUGAL AUTOMATION OF HETEROGENEOUS IMMUNOASSAYS 523

R. Mishra[1], R. Alam[2], D.J. Kinahan[1], K. Anderson[2], and J. Ducrée[1]

[1]Dublin City University, IRELAND and [2]Arizona State University, USA

We have devised strategic routing of flow from the reaction chamber in heterogeneous bead based ELISA to a distinct optical measurement chamber using lipophilic film valves which remain intact in aqueous solutions and selectively dissolve only when exposed to an ancillary, oleophilic solvent. We have integrated this routing feature on a "Lab-on-a-Disc" platform for the multi-step detection of anti-p53 antibodies from whole blood using event-triggered rotational flow control.

T-071 LOW RESISTANCE LIQUID MOTION FOR ENERGY HARVESTING 527

A. Goswami, S. Gowda, A. Tripathy, D. Roy, V. Bharadwaja, and P. Sen

Indian Institute of Science, INDIA

We demonstrate low resistance motion of liquid bulge on a well-defined path for energy harvesting application. A liquid bulge which arises due to an instability in a pre-wetted strip moves with very small hysteresis due to absence of a contact line. The pre-wetted strip confines the bulge and defines it motion path. Resistance to initiate motion of a bulge was studied experimentally and compared to other cases. An electrostatic energy harvesting device based on bulge motion is also demonstrated.

W-072 MICROFLUIDIC SWITCHING DEVICES SHOWING CONTROLLABLE HYSTERESIS 531

S. Ortiz and J.A. Lell

Comisión Nacional de Energía Atómica, ARGENTINA

We present a microfluidic MEMS device that is capable of switching among different states, thus behaving as an effective flip-flop. The devices consist of linear microfabricated deLaval nozzles with three exit channels, and we analyze the response as a function of input and output pressures, feed gas, and dimensions. In all cases, we have seen the appearance of a vortex, whose direction of swirl changes sign according to the input pressure, and showing hysteresis.

M-073 MICROFLUIDIC HANGING-DROP PLATFORM FOR PARALLEL CLOSED-LOOP MULTI-TISSUE EXPERIMENTS 535

S. Rismani Yazdi[1,2], A. Shadmani[1], A. Hierlemann[1], and O. Frey[1]
[1]ETH Zurich, SWITZERLAND and [2]Politecnico de Milano, ITALY

We present a new on-chip pumping approach for microfluidic hanging-drop networks that are used for experiments with 3D microtissue spheroids. Several independent hanging drop networks can be operated in parallel with only one single pneumatic actuation line. The pump concept enables closed-loop medium circulation between different organ models for body-on-a-chip applications and allows for multiple simultaneous assays in parallel.

T-074 ONE CORE-FIVE SHEATHS COAXIAL FLOW FORMATION USING MULTILAYER STACKED FLOW FOCUSING STRUCTURE 539

D.H. Yoon, J. Ito, N. Takeda, T. Sekiguchi, and S. Shoji
Waseda University, JAPAN

We proposed multilayer coaxial sheath flow formation by stacking multilayers of a single flow focusing structure. One core and five sheaths are simply formed with low diffusion between different core and sheaths. Only one point alignment for the sheath area is relatively free from misalignment, and the number of samples is infinitely expandable by increase in the number of stacking layers. The coaxial sheath flow is useful for biological fiber formation such as artificial blood vessel.

W-075 RECONFIGURABLE MICROFLUIDIC DILUTION GENERATION FOR QUANTITATIVE ASSAY 543

J. Fan[1], B. Li[2], and T. Pan[1]
[1]University of California, Davis, USA and [2]University of Science and Technology of China, CHINA

We report the first reconfigurable microfluidic dilution generator, producing discrete logarithmic dilution concentrations from a fixed sample volume of 10uL, without assistance of continuous fluid pumps or vacuum source. This portable chip serves as a facile tool for automatic generation of standard curves, indicating that it could be potentially employed for running generic quantitative assays for daily monitoring tasks in fields and biochemical laboratories.

M-076 SELF-MIXING BY ON-CHIP PREPARATION OF AQUEOUS TWO PHASE SYSTEMS AND ITS INFLUENCE ON EXTRACTION KINETICS 547

P.A.L. Wijethunga and H. Moon
University of Texas, Arlington, USA

This paper introduces an advantageous self-mixing phenomenon created on a digital microfluidic (DMF) device, and its influence on enhancing on-chip extraction kinetics, for the first time, highlighting a significant mixing capability in the absence of forced mixing. Such self mixing could contribute to achieve portable micro fluidics where only basic operations should be implemented and powered.

T-077 SINGLE-LAYER MICROFLUIDIC CURRENT SOURCE *VIA* OPTOFLUIDIC LITHOGRAPHY 551

C.C. Glick, S. Peng, M. Chung, K. Korner, M. Veale, C. Liu, J. Moore, A. Chu, A. Buckley, K. Iwai, R.D. Sochol, and L. Lin
University of California, Berkeley, USA

We develop and test a microfluidic current source which auto-regulates fluidic flow rate. We construct the device using optofluidic lithography in a single-layer PDMS channel.

W-078 THIN-FILM EDGE ELECTRODE LITHOGRAPHY ENABLING LOW-COST COLLECTIVE TRANSFER OF NANOPATTERNS 555

Y. Li[1], A. Goryu[1], K. Chen[2], H. Toshiyoshi[2], and H. Fujita[2]
[1]Toshiba Corporation, JAPAN and [2]University of Tokyo, JAPAN

This paper reports a new lithography method using thin-film edge electrodes (TEEs) to collectively transfer nanopatterns by generating oxide on the substrate surface via an electrochemical reaction (ECR). Nanometric thick TEEs are formed on the sidewall of insulating structures. ECR-based oxide patterns have the same width and shape as the TEEs because ECR is induced only between the conductor and the substrate. Oxide nanopatterns of 300nm and 70nm wide were collectively transferred on Si substrate in millimeter-scale area

M-079 ULTRAFINE PARTICLE COUNTER USING A MEMS-BASED PARTICLE PROCESSING CHIP 559

H.-L. Kim, J.S. Han, S.-M. Lee, H.B. Kwon, J. Hwang, and Y.-J. Kim
Yonsei University, SOUTH KOREA

We develop a microfluidic chip based ultrafine particle counter which is more compact and cost-effective than commercially available particle detection instruments. Unlike a conventional liquid-based microfluidic chip, the proposed particle processing chip handles a mixture of particles and air. We also develop a signal processing circuit which can process output signal from the microfluidic chip.

T-080 VERTICAL MEMBRANE MICROVALVES IN PDMS 563

J. Hansson, M. Hillmering, T. Haraldsson, and W. van der Wijngaart
KTH Royal Institute of Technology, SWEDEN

We present the design, realization and evaluation of the first leak-tight vertical membrane pneumatic microvalve. In comparison to horizontal membrane valves, our novel design features a 3D, instead of 2D, microchannel design, in which a vertical membrane actuates a vertical flow channel section. The valve closes under similar pneumatic control pressures to those for horizontal membrane microvalves but allows for a flow throughput per footprint area that is increased two orders of magnitude.

Bio and Medical MEMS
Biochemical Sensors

W-081 3D HUMAN CARDIAC MUSCLE ON A CHIP: QUANTIFICATION OF CONTRACTILE FORCE OF HUMAN IPS-DERIVED CARDIOMYOCYTES 566
Y. Morimoto[1,2], S. Mori[1,2], and S. Takeuchi[1,2]
[1]University of Tokyo, JAPAN and [2]Japan Science and Technology Agency (JST), JAPAN

We propose a method for constructing fiber-type 3D tissue of human iPS-derived cardiomyocytes and quantifying its contractile force in response to the addition of drug. By culturing the cardiomyocytes in micropatterned hydrogel with anchors, we successfully obtained the fibers with aligned cardiomyocytes and fixed the fiber edges to the anchors. The contraction of aligned fibers in a single direction provides us to measure the contractile force reproducibly.

M-082 A MICROFLUIDIC APTASENSOR INTEGRATING SPECIFIC ENRICHMENT WITH A GRAPHENE NANOSENSOR FOR LABEL-FREE DETECTION OF SMALL BIOMOLECULES 569
J. Yang[1], C. Wang[1], Y. Zhu[1], G. Liu[2], and Q. Lin[1]
[1]Columbia University, USA and [2]Nankai University, CHINA

We develop a microfluidic biosensor that combines aptamer-based specific enrichment and graphene conductance-based nanosensing on a single microchip, allowing label-free, specific, and quantitative detection of small biomolecules at low concentrations via aptamer-based competitive binding assay.

T-083 A MICROSIZED MICROBIAL FUEL CELL BASED BIOSENSOR FOR FAST AND SENSITIVE DETECTION OF TOXIC SUBSTANCES IN WATER 573
H. Lee, W. Yang, X. Wei, A. Fraiwan, and S. Choi
State University of New York-Binghamton, USA

We demonstrate a microliter-sized (140 μL) microbial fuel cell (MFC)-based biosensor integrated with electrochemical sensing functionality and air-bubble trap, in which microorganisms act as the sensor for toxic substances in water. The small-scale MFC biosensor (i) reduces measurement time by increasing the probability of cell attachment and biofilm formation in the micro-sized chamber and (ii) enhances sensitivity by preventing air-bubbles on the sensing surface.

W-084 A PAPER-BASED 48-WELL MICROBIAL FUEL CELL ARRAY FOR RAPID AND HIGH-THROUGHPUT SCREENING OF ELECTROCHEMICALLY ACTIVE BACTERIA 577
G. Choi[1], A. Fraiwan[1], D.J. Hassett[2], and S. Choi[1]
[1]State University of New York-Binghamton, USA and [2]University of Cincinnati College of Medicine, USA

We demonstrate the use of paper-based sensing platform for rapid and high-throughput characterization of microbial electricity-generating capabilities. A 48-well microbial fuel cell (MFC) array was fabricated on paper substrates, providing 48 high-throughput measurements and highly comparable performance characteristics in a reliable and reproducible manner. Within just 15 minutes, we successfully determined the electricity generation capacity of ten bacterial species with two controls.

M-085 AN INTEGRATED MICROFLUIDIC SYSTEM WITH FIELD-EFFECT-TRANSISTOR-BASED BIOSENSORS FOR AUTOMATIC HIGHLY-SENSITIVE C-REACTIVE PROTEIN MEASUREMENT 581
C.-H. Chu, W.-H. Chang, W.-J. Kao, C.-L. Lin, K.-W. Chang, Y.-L. Wang, and G.-B. Lee
National Tsing Hua University, TAIWAN

In this study, a new microfluidic device with a new methodology for measuring field-effect-transistor (FET)-based biosensors is presented. Not only can the proposed system work in a solution with physiological salt concentration but it also detects C-reactive protein with ultra-high sensitivity in an automatic fashion. This is the first time that a FET-based biosensor can effectively and automatically detect proteins in a physiological salt concentration without decreasing the sensitivity.

T-086 BATCH-FABRICATED HYDROGEL/POLYMERIC-MAGNET BILAYER FOR WIRELESS CHEMICAL SENSING 585
J.H. Park, A. Kim, and B. Ziaie
Purdue University, USA

We introduce a fabrication and wireless chemical sensing scheme using a hydrogel/polymeric-magnet bilayer. Polymeric permanent magnets are batch fabricated/integrated on top of a hydrogel thin film. The swelling/shrinking of the hydrogel in response to chemical stimuli is remotely detected by a giant magneto resistance (GMR) sensor. The described device is the first integrated wireless hydrogel/polymeric-magnet transducer with potential applications in biomedical and environmental sensing areas

W-087 CELL-LADEN HINGED MICROPLATES FOR MEASURING THE CONTRACTILE FORCES OF CARDIOMYOCYTES 589
H. Matsumoto[1], S. Yoshida[1], Y. Morimoto[1,2], N. Mori[1], D. Serien[1], and S. Takeuchi[1,2]
[1]University of Tokyo, JAPAN and [2]Japan Science and Technology Agency (JST), JAPAN

We report a method to measure contractile forces of cardiomyocytes at cellular level using microplates; pairs of microplates are connected by a flexible hinge at the center. Cardiomyocytes repeatedly contract and expand, and thereby fold the microplates at the flexible hinge. By measuring the change of the angle between folded microplates, we estimate contractile forces of cardiomyocytes. We believe that this method is a useful tool to study the dynamics of cardiomyocytes.

M-088 DROP TO MEASURE: A NOVEL NANOFLUIDIC CRYSTAL SENSING SCHEME WITH IMPROVED CHIP-TO-CHIP DATA CONSISTENCY FOR PORTABLE BIOCHEMICAL DETECTION 593

B. Wang[1], W. Zhao[1], R. Zhang[1], and W. Wang[1,2,3]

[1]Peking University, CHINA, [2]National Key Laboratory of Science and Technology on Micro/Nano Fabrication, CHINA, and [3]Innovation Center for MicroNanoelectronics and Integrated System, CHINA

We proposed a novel "drop to measure" nanofluidic crystal sensing scheme with improved chip-to-chip data consistency. Nanoparticles were self-assembled in a confined space guided by a well-designed surface chemistry treatment. The electrical readouts from different chips (n=5) varied within 4.8%. Biotin (using streptavidin-modified nanoparticles) and Pb^{2+} (using DNAzyme probed nanoparticles) were successfully detected by the present nanofluidic crystal sensor with a limit of detection of 1 nM.

T-089 ELECTRICAL DETECTION OF PESTICIDE VAPORS BY BIOLOGICAL NANOPORES WITH DNA APTAMERS 596

A. Nobukawa[1,2], T. Osaki[1,2], T. Tonooka[1], Y. Morimoto[1], and S. Takeuchi[1,2]

[1]University of Tokyo, JAPAN and [2]Kanagawa Academy of Science and Technology (KAST), JAPAN

We developed a vapor sensor using two robust biological molecules: A biological nanopore formed in a lipid bilayer and a DNA aptamer. The aptamer selectively binds to the target molecule, while the target molecule-aptamer complex clogs at the nanopore and blocks ionic current under electrical detection. A feasibility test was performed using a vapor phase sample, omethoate, demonstrating long-and-deep current blockades.

W-090 HIGH-PERFORMANCE AND LOW-COST LUNG CANCER SENSOR ARRAY BASED ON SELF-ASSEMBLED GRAPHENE 600

B. Zhang and T. Cui

University of Minnesota, USA

We develop a lung cancer sensor array (LCSA) based on layer-by-layer (LbL) self assembled grapheme, showing features including high performance and low cost in lung cancer biomarker detection due to graphene material properties in nature, self assembly technique, and multiple antigens detection within a single chip.

M-091 NOBEL DETECTION PLATFORM FOR ALZHEIMER'S AMYLOID-BETA USING MAGNETIC BEADS IN ELECTROCHEMICAL IMPEDANCE SPECTROSCOPY 604

K.-S. Shin, M.J. Kim, S.H. Lee, and J.Y. Kang

Korea Institute of Science and Technology (KIST), SOUTH KOREA

In this work, we proposed noble detection platform to detect Alzheimer's amyloid-beta (A-beta) using pre-treated magnetic beads in Electrochemical Impedance Spectroscopy (EIS), for the first time. Without any immobilization on the electrodes of the EIS device, it shows ability to detect a few pg/ml of amyloid-beta oligomers and compared to the result of a conventional ELISA, which allows to simplify the measurement procedure, recycle the device by only changing magnet beads.

T-092 ULTRASENSITIVE SURFACE-ENHANCED RAMAN SPECTROSCOPY USING DIRECTIONALLY ARRAYED GOLD NANOPARTICLE DIMERS 608

K. Sugano[1], D. Matsui[2], T. Tsuchiya[2], and O. Tabata[2]

[1]Kobe University, JAPAN and [2]Kyoto University, JAPAN

This paper reports an ultrasensitive nanostructure for surface-enhanced Raman spectroscopy (SERS). The gold nanoparticle dimer, which has been reported as the highest Raman enhancing structure, was directionally arrayed on a substrate for the first time, in order to match all dimers direction to polarization direction of the incident light. The strong enhancement can be achieved at all dimers. Optimizing the dimer arrangement, 10 pM limit of detection and 0.5 s rapid detection were achieved.

W-093 WATER-PROOF 'μ-DIVING SUIT' DRESSED ON RESONANT BIOCHEMICAL SENSOR FOR ONLINE DETECTION IN SOLUTION 612

H. Yu, Y. Chen, P. Xu, F. Yu, and X. Li

Shanghai Institute of Microsystem and Information Technology, Chinese Academy of Sciences, CHINA

This paper reports a new method to ensure resonant micro-sensor long-time resonating in solution for real-timebiochemical sensing/analysis. By design of a water-proof 'diving-suit' for the cantilever resonator and an antileakagenarrow 'slit' to free the cantilever vibration, only the sensing-region of the cantilever contacts to analytesolution, while the other parts remained in air for free resonance. The sensor experimentally realizes liquid-phase detection to ppb-level pesticide residue.

Medical Microsystems (Probes, Implantables, Minimally Invasive, Etc.)

M-094 A POLYCRYSTALLINE DIAMOND-BASED, HYBRID NEURAL INTERFACING PROBE FOR OPTOGENETICS 616

B. Fan[1], K.-Y. Kwon[1], R. Rechenberg[2], A. Khomenko[1], M. Haq[1], M.F. Becker[2], A.J. Weber[1], and W. Li[1]

[1]Michigan State University, USA and [2]Fraunhofer USA-CCL, USA

This paper reports a hybrid optoelectronic neural interfacing probe, combining microscale light emitting diode (μLED) and microelectrodes on a polycrystalline diamond (PCD) substrate for optogenetic stimulation and electrical recording of neural activity. PCD has superior thermal conductivity, which allows rapid dissipation of localized LED heat to a larger area to improve heat exchange with surrounding perfused tissues, and thus significantly reduce the risk of thermal damage to nerve tissue.

T-095 AN IMPLANTABLE TIME OF FLIGHT FLOW SENSOR 620
L. Yu, B.J. Kim, and E. Meng
University of Southern California, Los Angeles, USA

A micro time of flight (TOF) electrochemical impedance (EI) flow sensor was developed for characterization of in vivo flow dynamics. The transducer utilizes EI measurement between electrode pairs to monitor the passage of an electrolytically generated gas bubble within flowing solution. Biocompatible construction, low power consumption, and low profile thin film format make it ideally suited for chronic *in vivo* monitoring with immediate application in monitoring of hydrocephalus.

W-096 APPLICATION OF PERIODIC LOADS ON CELLS FROM MAGNETIC MICROPILLAR ARRAYS IMPEDES CELLULAR MIGRATION 624
F. Khademolhosseini[1], C.-C. Liu[1,2], C.J. Lim[1,2], and M. Chiao[1]
[1]University of British Columbia, CANADA and [2]Child and Family Research Institute, CANADA

We conduct an experimental study on the application of active micropillar structures to control cell migration. In contrast to passive micropillar structures which cause no significant alterations in cell migration rates, active micropillar structures actuated at 1 Hz decrease cell migration rates by up to 80%. The magnetic micropillar structures presented can be actuated remotely, making them a viable candidate for the development of smart materials for tissue engineering applications in vivo.

M-097 CELL MOTILITY REGULATION ON STEPPED MICRO PILLAR ARRAY DEVICE (SMPAD) WITH DISCRETE STIFFNESS GRADIENT 628
S. Lee, B. Saha, and J. Lee
Seoul National University, SOUTH KOREA

We report a micro pillar array device that provides discrete rigidity gradient to a cell with constant focal adhesion area. This goal is achieved through the use of "stepped" micro pillar array device (SMPAD) whose top area in contact with a cell is kept constant while the diameter of pillar bodies vary for variable mechanical stiffness. We show manipulating cell behavior using this simple, artificial platform that produces a pure physical stimulus.

T-098 FIBERED REFLECTIVE MICRO OBJECTIVES FOR MINIATURIZED SCANNING CONFOCAL FLUORESCENCE MICROSCOPY 632
S. Xie[1], E. Shaffer[1], L. Jacot-Descombes[1], D. Joss[1], B. Rachet[1], D. Kosanic[2], and J. Brugger[1]
[1]École Polytechnique Fédérale de Lausanne (EPFL), SWITZERLAND and
[2]SamanTree Technologies AG, SWITZERLAND

We report on a design, fabrication and characterization of a novel micro-optical system for imaging based on a miniaturized reflective objective, which is fabricated by combing two additive micro-fabrication techniques, inkjet printing to create the spherical mirror shape and stencil lithography for local metal deposition. This novel fabrication process produces reflective micro-objectives of different optical properties tailored for targeted bio-imaging application.

W-099 FLEXIBLE SILICON-POLYMER NEURAL PROBE RIGIDIFIED BY DISSOLVABLE INSERTION VEHICLE FOR HIGH-RESOLUTION NEURAL RECORDING WITH IMPROVED DURATION 636
F. Barz[1], P. Ruther[1], S. Takeuchi[2], and O. Paul[1]
[1]University of Freiburg-IMTEK, GERMANY and [2]University of Tokyo, JAPAN

We present a novel concept for flexible, intracortical neural probes delivered into the neural tissue by bio-dissolvable insertion vehicles. A completely implantable, silicon-based electrode array constitutes the probe tip. It is interfaced by a flexible ribbon cable that reduces stiffness and volume of the probe system. This is expected to increase the longevity of high resolution neural recording. The probes are encased in the insertion vehicles by means of a centrifuge-based molding process.

M-100 FLOW SPEED MEASUREMENT WITH DOPPLER EFFECT USING ULTRASONIC RECEIVER FOR SMALL-SIZED SMART CATHETER 640
R. Matsui, Y. Takei, N. Minh-Dung, T. Takahata, K. Matsumoto, and I. Shimoyama
University of Tokyo, JAPAN

We propose a wide range frequency receiver which has "liquid / piezoresistive cantilever / air" multilayer structure. This structure can measure acoustic waves from Hz to MHz order frequency because the cantilever vibrates obeying the surface waves on liquid. Experimental results demonstrated that our device can measure flow speed in a cylindrical pipe ranging from 6 to 25 mm/s, which is equal to blood flow speed of an arteriole, with MHz order Doppler Effect within one percent error.

T-101 HUMAN ADIPOSE-DERIVED STEM CELL FIBER FOR BREAST RECONSTRUCTION 643
A.Y. Hsiao[1], T. Okitsu[1,2], and S. Takeuchi[1,2]
[1]University of Tokyo, JAPAN and [2]Japan Science and Technology Agency (JST), JAPAN

We describe the construction and differentiation of human adipose-derived stem cell (ADSC) fibers into the adipocyte lineage for breast reconstruction. Human ADSCs cultured as fiber-shaped constructs were induced for adipogenic differentiation. Accumulation of lipid droplets of significant size was observed in the cells, and viability assay showed that most of the cells were alive. These findings suggest the use of ADSC fibers as a promising approach for breast reconstruction.

W-102 MEASURING THE PROPAGATING TEETH VIBRATION OF HUMAN CHEWING 646
C. Suzuki, Y. Takei, T. Takahata, K. Matsumoto, and I. Shimoyama
University of Tokyo, JAPAN

We measured the propagation waves of teeth's vibrations when chewing food. Human senses texture of chewing food by teeth's vibrations. Therefore, measuring the teeth's vibrations will allow us to quantify the food texture which human actually senses. We made the acoustic sensor that is small enough to be attached to teeth. For sensor evaluation, we conducted the rice cracker chewing test with our sensor attached to the real scale 3D jaw model, and propagation waves of around 500Hz are observed.

M-103 MICRO-NEEDLE-BASED ELECTRO TACTILE DISPLAY TO PRESENT VARIOUS TACTILE SENSATION 649

N. Kitamura[1] and N. Miki[1,2]

[1]*Keio University, JAPAN and* [2]*Japan Science and Technology Agency (JST), JAPAN*

In prior work, micro-needle electrodes were developed, however they could only stimulate the tactile receptors that located very close to the surface and therefore, they could only present stinging sensation. We developed a newly electrotactile display which has a micro-needle electrode array and a counter flat electrode. The display can stimulate all the tactile receptors, which resulted in even lower voltage required to tactile stimulation and successful display of various tactile sensation.

T-104 MICROMACHINED ULTRASOUND TRANSDUCER ARRAY FOR CELL STIMULATION WITH HIGH SPATIAL RESOLUTION 651

K. Ko[1], J.-H. Lee[1], H.J. Lee[1], S.-J. Oh[1], Y.E. Chun[1], T.S. Kim[1], C.J. Lee[1], E.-S. Yoon[1], K.-S. Yun[2], and I.-J. Cho[1]

[1]*Korea Institute of Science and Technology (KIST), SOUTH KOREA and* [2]*Sogang University, SOUTH KOREA*

We present a piezoelectric micromachined ultrasonic transducer (pMUT) array for localized stimulating on cultured cells or brain slice with high spatial resolution for the first time. We observed an increase in the level of Ca2+ in more than 15 percent of TRPA1 expressing HEK293T cells under ultrasound irradiation, which confirms that TRPA1 channel in HEK293T cells is activated by ultrasound which produced mechanical stress on cells.

W-105 MINIMALLY INVASIVE NEEDLE-FREE BUBBLE INJECTOR FOR GENE THERAPY 655

K. Takahashi[1], S. Omi[1], and Y. Yamanishi[1,2]

[1]*Shibaura Institute of Technology, JAPAN and* [2]*Japan Science and Technology Agency (JST), JAPAN*

We have successfully developed minimally-invasive needle-free bubble injector designed for the usage in air. The novelty is that the minimally-invasiveness of injection whose resolution is less than 10 μm, and injection can be possible without any pain. The injector can be used for any kind of materials with various hardness, owing to the strong impact of cavitation phenomenon when the high-speed micro-bubbles are collapsed. The developed injector can be used for wide range of biomedical study.

M-106 OPTIMIZATION OF DRUG COCKTAIL ON AN INTEGRATED MICROFLUIDIC SYSTEM 658

W.-Y. Huang[1], K. Wang[2], and G.-B. Lee[1]

[1]*National Tsing Hua University, TAIWAN and* [2]*Academia Sinica, TAIWAN*

The present study demonstrates a new microfluidic platform capable of automatically dispensing a small amount of drugs to expedite screening of drug cocktails. It could significantly decrease manual bias and enhance the throughput of drug cocktail formulation on an automated and minituriazed microfluidic system.

T-107 MEMS ELECTROCHEMICAL PATENCY SENSOR FOR DETECTION OF HYDROCEPHALUS SHUNT OBSTRUCTION 662

B.J. Kim, W. Jin, L. Yu, and E. Meng

University of Southern California, USA

We present the first Parylene-based electrochemical (EC)-MEMS patency sensor module for direct and quantitative diagnosis of patency in hydrocephalus drainage shunts. The impact of electrode size, temperature, flow conditions, and H_2O_2 plasma sterilization on sensor functionality was evaluated and sensor operation in the presence of static and dynamic obstruction was demonstrated. This device will enable simple quantitative monitoring of shunt state and more importantly, a more accurate and timely diagnosis of shunt failure.

W-108 PDMS BALLOON PUMP WITH A MICROFLUIDIC REGULATOR FOR THE CONTINUOUS DRUG SUPPLY IN LOW FLOW RATE 666

Y. Mukouyama, Y. Morimoto, S. Habasaki, T. Okitsu, and S. Takeuchi

University of Tokyo, JAPAN

This paper describes small sized balloon pump for providing liquid in low flow rate without batteries. The balloon pump is composed of a balloon tank and a microfluidic regulator with a micro valve. The balloon tank can work as a driving source to pump liquid. By connecting the micro valve to the balloon tank, we achieved extremely low flow rate of the liquid. Therefore, our system will be applicable to implantable passive pumps for the continuous drug supply in low flow rate without batteries.

M-109 PULSE WAVE MEASUREMENT IN HUMAN USING PIEZORESISTIVE CANTILEVER ON LIQUID 670

T. Kaneko, N. Minh-Dung, P. Quang-Khang, Y. Takei, T. Takahata, K. Matsumoto, and I. Shimoyama

University of Tokyo, JAPAN

We propose a device that can measure pulse waves at various points on human body with high sensitivity. Pulse wave velocity was calculated from a synchronized measurement on two points. The device has a piezoresistive cantilever placed on silicone oil. Pressure waves from arteries can be well conveyed to the cantilever through human issues, for the human-skin-like acoustic impedance of the silicone oil. The SNR of the device was ~80 dB in 10–100 Hz, when excited ~1 μm of displacement.

T-110 THE MICRO SADDLE COIL WITH SWITCHABLE SENSITIVITY FOR MAGNETIC RESONANCE IMAGING 674

K. Murashige and T. Dohi

Chuo University, JAPAN

We fabricated a micro saddle coil with switchable sensitivity for MRI (magnetic resonance imaging). Since the coil is embedded in polydimethylsiloxane (PDMS) tube, the saddle-shaped coil deforms to planar shape by pushing. By placing the saddle-shaped coil in the luminal tissue, we can take large area MR images. By deforming the coil, the sensitive area is concentrated in one side and the sensitivity becomes higher. Therefore we can take both large area MR images and high sensitive MR images.

W-111 ULTRA-SENSITIVE AND STRETCHABLE STRAIN SENSOR BASED ON PIEZOELECTRIC POLYMERIC NANOFIBERS 678

M. Asadnia[1], A.G.P. Kottapalli[2], J.M. Miao[1], and M.S. Triantafyllou[3]
[1]*Nanyang Technological University, SINGAPORE,*
[2]*Singapore for MIT Alliance for Research and Technology (SMART), SINGAPORE, and*
[3]*Massachusetts Institute of Technology (MIT), USA*

There have been increasing demands for stretchable and high-sensitivity sensors for use in structure health monitoring, human motion capture, sport performance monitoring and rehabilitation. Here, we present a novel, highly stretchable, self-powered and ultra-sensitive strain sensor based on piezoelectric PVDF electrospun nanofiber. Complete studies on mechanical and piezoelectric characteristics of the single PVDF nanofiber are presented.

M-112 ULTRACOMPACT OPTOFLEX NEURAL PROBES FOR HIGH-RESOLUTION ELECTROPHYSIOLOGY AND OPTOGENETIC STIMULATION 682

M. Chamanzar[1], D.J. Denman[2], T.J. Blanche[2], and M.M. Maharbiz[1]
[1]*University of California, Berkeley, USA and* [2]*Allen Institute for Brain Science, USA*

Here we report on our recent development of high-density neural probes for high resolution, multiscale electrophysiology. Our 64-channel hybrid silicon-parylene probes provide at least three-fold better spatiotemporal resolution compared to the state of the art and minimize the tethering force on the brain tissue by two orders of magnitude. We demonstrate, for the first time, the design of ultracompact polymer optical waveguides that can be monolithically integrated with our neural probes.

T-113 VERTICALLY ALIGNED EXTRACELLULAR MICROPROBE ARRAYS/(111) INTEGRATED WITH (100)-SILICON MOSFET AMPLIFIERS 686

H. Makino, K. Asai, M. Tanaka, S. Yamagiwa, H. Sawahata, I. Akita, M. Ishida, and T. Kawano
Toyohashi University of Technology, JAPAN

We report a heterogeneous integration of vertically aligned extracellular microscale silicon (Si)-probe arrays/(111) with MOSFET amplifiers/(100), by IC processes and subsequent vapor–liquid–solid (VLS) growth of Si-probes. To improve the extracellular recording capability of the microprobe with a high impedance of > 1 Mohm at 1 kHz, here we integrated (100)-Si source follower buffer amplifiers by ~700 degree VLS growth compatible (100)-Si MOSFET technology.

W-114 WEARABLE FLEXIBLE MICRO ELECTRODE FOR ADULT ZEBRAFISH LONG TERM ECG MONITORING 690

X. Zhang[1], T. Beebe[2], Y. Liu[1], J. Park[1], T. Hsiai[2], and Y.-C. Tai[1]
[1]*California Institute of Technology, USA and* [2]*University of California, Los Angeles, USA*

All published adult zebrafish ECG recorded to this date have been done acutely with anesthetized fish. This work presents, for the first time a wearable flexible parylene (PA) micro-electrode that monitors the Adult Zebrafish ECG in longer term. We show here the design, fabrication and testing of the flexible electrode along with a micro-molded ultrasoft density adjusted silicone jacket, allowing ECG recording to be carried under water, in the fish's natural habitat with no need for anesthesia.

Nanobiotechnology

M-115 A METHOD FOR CONTROLLING MICROTUBULE VELOCITY USING LIGHT IRRADIANCE ON A PATTERNED GOLD SURFACE 694

T. Nakahara, H. Shintaku, H. Kotera, and R. Yokokawa
Kyoto University, JAPAN

We report a method to control the velocity of gliding microtubules by light irradiance and a gold pattern. The irradiance controlled a temperature in the assay condition by heat transfer from the gold pattern. The result showed that the velocity of microtubule increased approximately 1.8 folds from initial velocity at the irradiance of 13.5 W/cm2 in the gold pattern. This is first demonstration to perform the control and the switching velocity on the patterned gold surface.

T-116 CULTURING AND PROBING PHYSICAL BEHAVIOR OF INDIVIDUAL BREAST CANCER CELLS ON SiC MICRODISK RESONATORS 698

H. Jia, X. Wu, H. Tang, Z.-R. Lu, and P.X.-L. Feng   * Award Nominee
Case Western Reserve University, USA

This work describes the first exploration of directly culturing and measuring breast cancer cells, at single-cell level, by using silicon carbide (SiC) microdisk resonators. Enabled by the superior biocompatibility of SiC, individual breast cancer cells are observed to attach and spread on device surface within 3hrs of culturing. Multimode responses of SiC microdisk resonators (20-30µm in diameters) to single MDA-MB-231 cell loading are characterized by taking advantage of their robust high-frequency multimodality in water and biological solutions.

W-117 EARLY CHARACTERIZATION METHOD OF PLANT ROOT ADAPTABILITY TO SOIL ENVIRONMENTS 702

K. Ozoe[1], H. Hida[1], I. Kanno[1], T. Higashiyama[2,3], and M. Notaguchi[2,3]
[1]*Kobe University, JAPAN,* [2]*Japan Science and Technology Agency (JST), JAPAN, and* [3]*Nagoya University, JAPAN*

This paper reports a microfluidic platform for studying physical mechanisms of plant root at early growth stage. To measure driving force of root growth precisely and quantitatively, we developed a silicon microchannel device integrated with force displacement sensor which mimics a barrier in soil. By using developed microsystem, we successfully measured the driving forces of root growth in three kinds of plants including Arabidopsis thaliana known as model organism.

Physical Sensors
Fluidic Sensors (Flow, Pressure, Density, Viscosity, Etc.)

M-118 A CANTILEVER WITH COMB STRUCTURE MODELED BY A BRISTLED WING OF THRIPS FOR SLIGHT AIR LEAK 706
H. Takahashi, A. Isozaki, K. Matsumoto, and I. Shimoyama
University of Tokyo, JAPAN

This paper reports a cantilever with comb structure, mimicking a bristled wing of thrips which acts as a continuous membrane wing because of the effects of low Reynolds numbers. The comb structures are formed at the edges of the cantilever and its surrounding. When differential pressure is applied to the cantilever, both comb structures act as airflow suppression through the gap. The leakage of the fabricated comb cantilever was smaller than the normal cantilever.

T-119 AIRFLOW SHEAR STRESS SENSOR USING SIDE-WALL DOPED PIEZORESISTIVE PLATE 710
R. Kazama, H. Takahashi, T. Takahata, K. Matsumoto, and I. Shimoyama
University of Tokyo, JAPAN

We report an airflow wall shear stress sensor consisting of a plate and surrounding membrane with narrow gap. The plate is supported by side-wall doped beams, which can detect the horizontal deformation of the plate due to airflow wall shear stress. The sensor structure does not disturb target airflow circumstance because of flat surface, and measures shear stress directly. Wind tunnel test shows our sensor was able to measure laminar airflow shear stress with the resolution under 1.0 Pa.

W-120 MICRO TRIPLE-HOT-WIRE ANEMOMETER ON SMALL SIZED GLASS TUBE FABRICATED IN 5DOF UV LITHOGRAPHY SYSTEM 714
S. Liu[1], Z. Yang[2], Y. Zhang[3], F. Xue[1], S. Pan[1], J. Miao[1], and L.K. Norford[4]
[1]Nanyang Technological University, SINGAPORE, [2]Shanghai Jiao Tong University, CHINA,
[3]National Institute of Advanced Industrial Science and Technology, JAPAN, and
[4]Massachusetts Institute of Technology, USA

We develop novel designed and fabricated micro airflow sensors based on the hot-wire sensing principle, i.e. gas cooling of electrically heated resistance. With three micro Ti/Pt hot-wire components fabricated on a glass tube in five degrees of freedom (5DOF) UV lithography system with multi-layer alignment, the sensors on a cylindrical base have demonstrated high sensitivity, fast response time and ability to detect wind speed and direction.

M-121 MULTI ROOF TILE-SHAPED VIBRATION MODES IN MEMS CANTILEVER SENSORS FOR LIQUID MONITORING PURPOSES 718
G. Pfusterschmied[1], M. Kucera[1,2], V. Ruiz-Díez[3], A. Bittner[1], J.L. Sánchez-Rojas[3], and U. Schmid[1]
[1]Vienna University of Technology, AUSTRIA, [2]AC2T research GmbH, AUSTRIA, and
[3]Universidad de Castilla-La Mancha, SPAIN

We realized piezoelectrically self-actuated self-sensing cantilever sensors for liquid monitoring purposes excited in higher roof tile-shaped modes. This advanced class of vibration modes supports very high Q-factors in liquid media and very high volume strain values which result in combination with an optimized electrode design in very high strain related conductance peaks. Therefore, precise fluid property measurements even for highly viscous liquids like D500 (~ 430 cP) are feasible.

T-122 ON-CHIP PRESSURE SENSING BY VISUALIZING PDMS DEFORMATION USING MICROBEADS 722
C.-H.D. Tsai and M. Kaneko
Osaka University, JAPAN

A novel pressure sensing technique is proposed here for measuring local pressure inside a microfluidic device. By the proposed method, the local pressure can be directly "seen" without any wire foils but simply microbeads patterns. The experimental results show that microbeads pattern is stable and repeatable where the variation for the same given pressure is less than 1%. The correlation between the pressure obtained from the proposed method and a commercial pressure connected outside is 0.995.

Force and Displacement Sensors (Tactile, Force, Torque, Stress and Strain Sensor)

W-123 3-AXIS ALL ELASTOMER MEMS TACTILE SENSOR 726
A. Charalambides, J. Cheng, T. Li, and S. Bergbreiter
University of Maryland, College Park, USA

This paper reports the first 3-axis (normal and shear force) all-elastomer capacitive MEMS tactile sensor. Sensor area is 1.5 x 1.5 mm and uses vertical capacitive structures with 20 μm electrode gaps to achieve high shear force sensitivities of 8.8 fF/N, shear force resolutions of 50 mN, and shear range up to 700 mN; this aligns with the developed finite element prediction. Fabrication utilizes a simple elastomer molding process with reusable DRIE silicon molds for inexpensive manufacturing.

M-124 6-AXIS FORCE-TORQUE SENSOR CHIP COMPOSED OF 16 PIEZORESISTIVE BEAMS 730
A. Nakai[1], Y. Morishita[2], K. Matsumoto[1], and I. Shimoyama[1]
[1]University of Tokyo, JAPAN and [2]Touchence Inc., JAPAN

We propose the 6-axis force-torque sensor chip composed of 16 piezoresistive beams whose area, 2 mm square in size, is one-third of that of the state of the art, which will enhance the mounting density of sensor array and also reduce the cost in case of volume production. This paper will show the design, a part of the fabrication process in MNOIC, 8-inch MEMS foundry in Japan, especially ion doping method by oblique ion implantation, the calibration method and experimental results.

T-125 A 0.25mm³ ATOMIC FORCE MICROSCOPE ON-A-CHIP 732

N. Sarkar[1,2], D. Strathearn[1,2], G. Lee[1,2], M. Olfat[1,2], and R.R. Mansour[1,2]

[1]University of Waterloo, CANADA and [2]ICSPI Corp, CANADA

We report the highest resolution achieved with a single-chip atomic force microscope (sc-AFM). Images of a 20nm AFM calibration standard were obtained to show, for the first time, that a single-chip instrument may obtain a vertical resolution comparable to state-of-the-art instruments at a minuscule fraction of the size (1/1,000,000) and cost (1/1000). The reported performance represents a four-fold improvement in resolution when compared to previously reported sc-AFMs.

W-126 AN INSTRUMENTED TOOTH 736

F. Becker[1], C. Sander[1], F. Schmidt[2], B. Lapatki[2], and O. Paul[1]

[1]University of Freiburg - IMTEK, GERMANY and [2]University of Ulm, GERMANY

We developed a tool for orthodontic research and education, namely an instrumented tooth (IT) that allows measuring all six applied force and moment components. The core component is an 11.6-mm-high and 3.5-mm-diameter sensor module based on a CMOS stress sensor chip sandwiched between two metal pins. In the IT, the sensor module constitutes the root. The stiff and robust sensor module is capable of measuring orthodontically relevant forces up to 60 N and moments up to 10 Ncm in all directions.

**M-127 ASYMMETRIC FAN-SHAPE-ELECTRODE FOR
HIGH-ANGLE-DETECTION-ACCURACY TACTILE SENSOR** 740

S.-T. Chuang, T.-Y. Chen, Y.-C. Chung, R. Chen, and C.-Y. Lo

National Tsing Hua University, TAIWAN

This paper reports an up to 95.9% angle detection accuracy enhancement for capacitive tactile sensors, which entails asymmetric and intentionally shifted electrodes. The asymmetric electrodes containing one fan- and one square-shape in capacitors reduced unexpected and rotational-shift induced errors, by keeping the same overlap area of the electrodes. The minimal angle detection resolution was improved from 5.8 to 0.3-degree, making the tactile sensor practical and reliable in artificial skins.

**T-128 CARBON NANOTUBES-ECOFLEX NANOCOMPOSITE FOR
STRAIN SENSING WITH ULTRA-HIGH STRETCHABILITY** 744

M. Amjadi[1] and I. Park[2]

*[1]Electronics & Telecommunications Research Institute (ETRI), SOUTH KOREA and
[2]Korea Advanced Institute of Science & Technology (KAIST), SOUTH KOREA*

We developed ultra-stretchable, flexible and very soft conductors based on the carbon nanotubes (CNTs)-siliconrubber (Ecoflex®) nanocomposite thin films. Highly stretchable conductors were utilized as skin-mountable and wearable strain sensors. The resistance of the CNTs-Ecoflex nanocomposite thin film was fully recovered undercyclic loading/unloading for strains as large as 510%. Finally, motion detection of finger and wrist joints was conducted using CNTs-Ecoflex nanocomposite thin film.

**W-129 DEVELOPMENT OF A MINIATURIZED LASER DOPPLER VELOCIMETER
FOR USE AS A SLIP SENSOR FOR ROBOT HAND CONTROL** 748

N. Morita[1], H. Nogami[1], Y. Hayashida[1], E. Higurashi[2], T. Ito[3], and R. Sawada[1]

[1]Kyushu University, JAPAN, [2]University of Tokyo, JAPAN, and [3]Kyushu Institute of Technology, JAPAN

We have developed a miniaturized laser Doppler velocimeter (LDV), designed for use as a slip sensor in the control of a robot hand. This sensor is only 1/10,000th of the volume of commercial LDVs, which enables the sensor to be attached to a robot hand. Our LDV was able to detect scattering objects moving at velocities ranging from 10 μm/s to 20,000 μm/s. The output of this sensor is independent of the type of material measured, which included aluminum, cardboard, or rough-surface black plastic.

**M-130 MICROCANTILEVER SYSTEM INCORPORATING INTERNAL RESONANCE
FOR MULTI-HARMONIC ATOMIC FORCE MICROSCOPY** 752

C. Pettit[1], B. Jeong[2], H. Keum[1], J. Lee[1], J. Kim[1], S. Kim[2], D.M. McFarland[2], L.A. Bergman[2],
A.F. Vakakis[2], and H. Cho[1]

[1]Texas Tech University, USA and [2]University of Illinois, Urbana-Champaign, USA

We report a new design concept of a micromechanical cantilever system incorporating the internal resonance during dynamic mode AFM. The passive amplification of nth harmonic triggered through the mechanism of 1:n internal resonance enables AFM to utilize multiple harmonics. Detailed theoretical and experimental studies of the proposed design demonstrate that the multi-harmonic AFM is capable of simultaneous topography and compositional mapping with 10-fold enhanced sensitivity.

**T-131 PRINTABLE FLEXIBLE TACTILE PRESSURE AND TEMPERATURE
SENSORS WITH HIGH SELECTIVITY AGAINST BENDING** 756

K. Kanao, S. Harada, Y. Yamamoto, W. Honda, T. Arie, S. Akita, and K. Takei

Osaka Prefecture University, JAPAN

Conventional flexible tactile sensors detect the bending of a substrate in addition to a tactile pressure, and that is the one of bottlenecks to realize stable operation of flexible device such as an artificial electronic skin. To achieve high sensitivity, a cantilever type strain sensor and a temperature sensor in a flexible substrate are developed by using a fully printed method.

M-132 SOFT FLEXION SENSORS INTEGRATING STRETCHABLE METAL CONDUCTORS ON A SILICONE SUBSTRATE FOR SMART GLOVE APPLICATIONS 760
H.O. Michaud, J. Teixidor and S.P. Lacour
École Polytechnique Fédérale de Lausanne (EPFL), SWITZERLAND

We have designed and implemented a sensory skin that monitors in real time finger flexure (3 sensors per finger) of a user's hand. Compared to current technologies, the electronic skin is made entirely of stretchable materials integrating silicone rubber, low resistivity liquid metal interconnects and highly strain sensitive, microstructured thin metal films. We incorporated the skin on a textile glove and demonstrated its function as an interface for finger motion detection.

Gas and Chemical Sensors

W-133 A CIRCUMFERENTIALLY GROWN ZNO NANOWIRE FOREST ON A SUSPENDED CARBON NANOWIRE FOR A HIGHLY SENSITIVE GAS SENSOR 764
Y. Lim, Y. Lee, J. Lee, and H. Shin
Ulsan National Institute of Science and Technology (UNIST), SOUTH KOREA

We develop a suspended ZnO nanowire forest as a highly sensitive gas sensor. The nanowires were grown selectively on a suspended single glassy carbon nanowire using hydrothermal method so that the detrimental effects from the substrate inclusive of contamination, stagnant layer and limited mass transfer could be alleviated. The novel geometry of the radially grown ZnO nanowires resembling burs of a chestnut is expected to enhance the gas sensing capability because of enhanced mass transfer.

M-134 A CMOS CAPACITIVE VERTICAL-PARALLEL-PLATE-ARRAY HUMIDITY SENSOR WITH RF-AEROGEL FILL-IN FOR SENSITIVITY AND RESPONSE TIME IMPROVEMENT 767
V.P.J. Chung, C.-L. Cheng, M.-C. Yip, and W. Fang,
National Tsing Hua University, TAIWAN

This paper reports a high-sensitivity and high-speed capacitive humidity sensor with resorcinol-formaldehyde (RF) aerogel fill-in. A novel capacitive vertical parallel-plate (VPP) array topology was designed and implemented based on standard TSMC 0.18μm CMOS process and subsequent in-house post-processes.

T-135 AN INTEGRATED CHROMATOGRAPHY CHIP FOR RAPID GAS SEPARATION AND DETECTION 771
M. Akbar, H. Shakeel, and M. Agah
Virginia Tech, USA

This paper reports the first implementation of a highly sensitive micro helium discharge photoionization detector in a silicon-glass architecture and its monolithic integration with a separation column. The new detector requires a two-mask fabrication process, is universal, non-destructive, low power (<5mW), and insensitive to flow and temperature variations. It has yielded a minimum detection limit of ~10pg which is on par with the widely used destructive flame ionization detector.

W-136 FABRICATION OF SILVER NANOPARTICILES ON CYLINDRICAL SURFACE OF U-SHAPED FIBER ATR SENSOR BY MATERIAL REDUCTION 775
D. Li, C. Sun, S. Yu, H. Yu, and K. Xu
Tianjin University, CHINA

An implantable fiber ATR sensor enhanced by silver nanoparticles on circumferential surface was presented for continuous glucose monitoring. U-shaped structure was addressed to increase optical length for sensitivity enhancement. A novel method to fabricate silver nanoparticles on circumferential surface of fiber sensor based on chemical reduction of its silver halide material directly without any preliminary nanoparticles synthesis and the following covalent bond or self-assembly was proposed.

M-137 INTRINSIC ZnO NANOWIRES WITH NEW SENSING MECHANISM OF SULFURATION-DESULFURATION TWO-STEP REACTION FOR HIGH PERFORMANCE SENSING TO ppb-LEVEL H_2S GAS 779
P. Xu[1], H. Huang[2], D. Zheng[2], and X. Li[1]
[1]Chinese Academy of Sciences, CHINA and [2]Shanghai Institute of Technology, CHINA

The paper reports a novel H2S sensing-effect for intrinsic ZnO nanowires (NWs) chemiresistive sensor. Herein 50nm-diameter ZnO-NWs are found and verified to be sulfurized by H2S to form ZnS that can be latterly desulfurized back to ZnO by ambient oxygen, which is different from conventional ZnO sensing-mechanism where resistance of semiconductor ZnO is changed via electron depletion-layer variation by surface adsorbed ambient oxygen. The ZnO-NWs have realized detection to 50ppb H2S.

T-138 IONIC-GEL-COATED FABRIC AS FLEXIBLE HUMIDITY SENSOR 783
Y. Takei, K. Matsumoto, and I. Shimoyama
University of Tokyo, JAPAN

We fabricated flexible humidity sensor which responds 10 times faster than commercial CMOS humidity sensor. Our sensor is based on Ionic-Gel-coated Fabric (IG-Fabric). IG-Fabric has wide surface area and high gas permeability so that gases can be easily absorbed and detached. As a demonstration, we fabricate flu-mask-type IG-fabric humidity sensor and measured the relative humidity change caused by human breath.

W-139 DOG-BONE RESONATOR WITH HIGH-Q IN LIQUID FOR LOW-COST QUICK 'TEST-PAPER' DETECTION OF ANALYTE DROPLET 785
F. Yu, P. Xu, J. Wang, and X. Li
Shanghai Institute of Microsystem and Information Technology, Chinese Academy of Sciences, CHINA

Imitating on-site liquid-droplet analysis/assay with test-papers, a novel tri-beam structure dog-bone resonator is proposed for low-cost quick detection of trace-amount biochemical liquid sample. Effective depression to signal feed-through effect is realized by independent piezoresistive readout with the specifically designed central beam, thereby, the new tri-beam resonator exhibits high-Q resonance of extension-mode in liquid and liquid-droplet detection of ppb-level mercury ion in water.

Inertial Sensors (Gyros, Accelerometers, Resonators, Etc.)

M-140 A DYNAMICALLY MODE-MATCHED PIEZOELECTRICALLY TRANSDUCED HIGH-FREQUENCY FLEXURAL DISK GYROSCOPE 789
M. Hodjat-Shamami, A. Norouzpour-Shirazi, R. Tabrizian, and F. Ayazi
Georgia Institute of Technology, USA

This paper presents, for the first time, the design and implementation of a dynamically mode-matched high-frequency piezoelectric silicon disk gyroscope utilizing a unique pair of in-plane flexural gyroscopic modes. A linear bidirectional frequency tuning scheme compatible with piezo-only transduction is introduced to achieve dynamic mode-matching via electromechanical feedback. A fabricated AlN-on-Silicon solid disk gyroscope was frequency tuned by 500 ppm to yield a sensitivity of 410 pA/°/s.

T-141 A TEMPERATURE-STABLE MEMS OSCILLATOR ON AN OVENIZED MICRO-PLATFORM USING A PLL-BASED HEATER CONTROL SYSTEM 793
Z. Wu and M. Rais-Zadeh
University of Michigan, USA

In this work, an oxide-refill process is used to realize passive TCF compensation for silicon MEMS resonators as well as integrated thermal isolation structures. The technology enables fabrication of a low-power ovenized micro-platform on which multiple MEMS devices can be integrated. Intrinsic frequency drifts of two MEMS resonators are utilized for temperature sensing, and closed-loop oven control is realized by phase-locking two MEMS oscillators at a specific temperature.

W-142 ALN PIEZOELECTRIC ON SILICON MEMS RESONATOR WITH BOOSTED Q USING PLANAR PATTERNED PHONONIC CRYSTALS ON ANCHORS 797
H. Zhu and J.E.-Y. Lee
City University of Hong Kong, HONG KONG

We report an approach to suppress anchor loss in thin-film piezoelectric-on-silicon MEMS resonators by patterning 2D phononic crystals (PnCs) externally on the anchors. According to our measurements, adding the PnCs helps to double the unloaded quality factor (Q), while reducing the motional resistance by half. The results suggest significant reduction of acoustic leakage to the substrate by the PnCs.

M-143 BATCH-FABRICATED HIGH Q-FACTOR MICROCRYSTALLINE DIAMOND CYLINDRICAL RESONATOR 801
D. Saito[1], C. Yang[1], A. Heidari[2], H. Najar[2], L. Lin[1], and D.A. Horsley[2]
[1]University of California, Berkeley, USA and [2]University of California, Davis, USA

We report, for the first time, a 1.5mm batch-fabricated polycrystalline diamond Cylindrical Resonator (CR) for gyroscope applications. A quality factor (Q) of 313,100 is measured at the 23kHz 2theta elliptical wineglass modes, producing a ring-down time of 4.32seconds. Annealing CRs at 700°C in a nitrogen atmosphere improved Q from 75,000 to over 300,000 with an excellent frequency mismatch of 3Hz (130ppm) between the 2theta degenerate wineglass modes without applying any tuning voltage.

T-144 DESIGN, FABRICATION, AND CHARACTERIZATION OF A MICROMACHINED GLASS-BLOWN SPHERICAL RESONATOR WITH IN-SITU INTEGRATED SILICON ELECTRODES AND ALD TUNGSTEN INTERIOR COATING 805
J. Giner[1], J.M. Gray[2], J. Gertsch[2], V.M. Bright[2], and A.M. Shkel[1]
[1]University of California, Irvine, USA and [2]University of Colorado, Boulder, USA

The paper reports on design, fabrication, and characterization of micromachined spherical resonators with integrated high-aspect ratio silicon electrodes. The electrical connection of the spherical, non-conductive micromechanical resonator is possible thanks to Atomic Layer Deposition (ALD) of Tungsten inside the shell. Operating frequencies in the range of MHz have been measured providing one of the highest frequencies in spherical resonator shells up to date.

W-145 IN-SITU OVENIZATION OF LAMÉ-MODE SILICON RESONATORS FOR TEMPERATURE COMPENSATION 809
Y. Chen, E.J. Ng, Y. Yang, C.H. Ahn, I. Flader, and T.W. Kenny
Stanford University, USA

We demonstrates an inside-encapsulation ovenization method for the temperature compensation of Lamé-mode epi-sealed silicon resonators. With this method, the square Lamé-mode resonator itself acts both as a thermometer and a heater, which allows for simultaneous in situ sensing and control of the operating temperature. In this device, only the resonating element is heated, minimizing the time constant and the heating power.

M-146 ON-CHIP CHARACTERIZATION OF STRESS EFFECTS ON GYROSCOPE ZERO RATE OUTPUT AND SCALE FACTOR 813
E. Tatar, T. Mukherjee, and G.K. Fedder
Carnegie Mellon University, USA

This paper presents stress effects on a vacuum packaged MEMS gyroscope zero rate output (ZRO), scale factor (SF), and resonance frequencies by using on-chip environmental sensors measuring the temperature and stress separately, for the first time. Environmental sensors comprise released SOI-silicon resistors. Experimental results show that a system model can be established to compensate the gyroscope ZRO using the environmental sensor outputs.

Manufacturing Techniques for Physical Sensors

T-147 A FACILE FABRICATION TECHNIQUE FOR STRETCHABLE INTERCONNECTS AND TRANSDUCERS VIA LASER CARBONIZATION 817
R. Rahimi, M. Ochoa, W. Yu, and B. Ziaie
Purdue University, USA

We present a facile, low-cost approach to fabricate highly porous conductive carbon patterns on elastomeric substrates using laser carbonization of polyimide and subsequent transfer of locally pyrolyzed features onto a PDMS sheet. Using this technique, we fabricated stretchable interconnects and an array of piezoresistive tactile sensors. Characterizations of the stretchable patterns showed linear sensitivities of $8.912k\Omega/\varepsilon\%$ and $518\Omega/N$ to strain and normal force, respectively.

W-148 A HIGH-Q ALL-FUSED SILICA SOLID-STEM WINEGLASS HEMISPHERICAL RESONATOR FORMED USING MICRO BLOW TORCHING AND WELDING 821
J.Y. Cho and K. Najafi
University of Michigan, USA

We report a new fabrication technology for making complete wineglass resonators through forming thin fused silica (FS) shell resonators integrated with arbitrarily sized FS solid stems through a simultaneous process of micro blow-torching and micro welding. The fabricated wineglass resonator operates at 22.6 kHz with long ring down time (35.9 s) and high quality factor (2.55 million). The ring down time and Q are the best reported values for micro FS devices.

M-149 GRAPHITE-ON-PAPER BASED TACTILE SENSORS USING PLASTIC LAMINATING TECHNIQUE 825
H.-P. Phan, D.V. Dao, T. Dinh, H. Brooke, A. Qamar, N.-T. Nguyen, and Y. Zhu
Griffith University, AUSTRALIA

We report for the first time a highly sensitive paper-based tactile sensor using laminated graphite drawn on paper. Due to a high gauge factor of 26.2, as well as its excellent humidity-resistance, plastic-laminated graphite-on-paper has a high potential for mechanical sensors. Additionally, the plastic lamination combined with the laser cutting technique proposed in this study will bring a step forward to the mass production of cleanroom-free fabrication and low-cost MEMS devices.

T-150 HIERARCHICAL WRINKLE STRUCTURING ON INSIDE WALL OF CLOSED MICRO CHANNEL 829
A. Takei and H. Fujita
University of Tokyo, JAPAN

This paper presents a surface structuring method for a closed micro channel. Because fine photolithography can only be made on a flat surface, it has been difficult to make complicated patterns on the inside walls of non-planar microfluidic channels. Here, we demonstrated that micro-scale hierarchical patterns can be formed on the inside walls of the micro channel simply by depositing a plastic thin film in the channel and stretching the micro channel.

W-151 MICROFABRICATION OF WIDE-MEASUREMENT-RANGE LOAD SENSOR USING QUARTZ CRYSTAL RESONATOR 833
Y. Murozaki, S. Sakuma, and F. Arai
Nagoya University, JAPAN

We present a wafer level fabrication process of the QCR load sensor that has three-layer structures; two Si-hold layers and a quartz layer. Using microfabrication and atomic diffusion bonding, the assembly process was simplified. The proposed sensor is easily integrated in outer package and design the measurement range. We succeeded in multi-biosignals (heartbeat, body motion) detection using fabricated QCR sensor and outer case.

M-152 SELF-CURVED DIAPHRAGMS BY STRESS ENGINEERING FOR HIGHLY RESPONSIVE PMUT 837
S. Akhbari[1], F. Sammoura[1,2], C. Yang[1], A. Heidari[3], D. Horsley[3], and L. Lin[1]
[1]*University of California, Berkeley, USA*, [2]*Masdar Institute of Science and Technology, UAE, and* [3]*University of California, Davis, USA*

A process to make self-curved diaphragms by engineering residual stress in thin films has been developed to construct highly responsive piezoelectric micromachined ultrasonic transducers (pMUT). This process enables high device fill-factor to achieve better than 95% wafer utilization with controlled formation of curved membranes.

T-153 THERMOCOUPLES ON TRENCH SIDEWALL IN CHANNEL FRONTING ON FLOWING MATERIAL 841
M. Shibata, T. Yamaguchi, S. Kumagai, and M. Sasaki
Toyota Technological Institute, JAPAN

Thermocouples on the trench sidewall fronting on the flowing material are fabricated by applying 3D photolithography. The first novelty is the fabrication technique to realize the device. And, the fabricated thermocouple is confirmed to work having the advantage sensing the flow temperature directly. The metals on the sidewall do not make the shadow allowing the observation inside the microchannel using the optical microscope.

W-154 VACUUM CAVITY FORMATION FOR HIGH THERMAL ISOLATION IN FLEXIBLE THERMAL SENSOR 845

P. Kim[1], S Shibata[1], and M. Shikida[2]
[1]Nagoya University, JAPAN and [2]Hiroshima City University, JAPAN

To realize MEMS sensors in the flexible fashion, we proposed to apply a Cu On Polyimide (COP) substrate as a starting material, and introduced a sacrificial etching for producing a cavity and an electrical feed through into the COP substrate. We also newly introduced a vacuum cavity realizing high thermal isolation in flexible thermal sensor, for the first time.

Materials for Physical Sensors

M-155 A STUDY OF ADHESION FORCES IN THICK EPITAXIAL POLYSILICON UNDER DYNAMIC IMPACT LOADING 849

S. Dellea[1], R. Ardito[1], B. De Masi[1], F. Rizzini[2], A. Tocchio[2], and G. Langfelder[1]
[1]Politecnico di Milano, ITALY and [2]ST Microelectronics, ITALY

We present a structure and a method for the characterization of impact and adhesion between MEMS moving and fixed parts: the focus is to monitor an inertial mass colliding with a stopper. From the measurements we evaluate the energy balance during impacts. The work analyzes the adhesion evolution after a number of collisions comparable to a 5-year operation. Results show growing and stabilizing adhesion forces of 170 nN. We also show the possibility to change and track the impact kinetic energy.

T-156 EXPLORING THE *Q*-FACTOR LIMIT OF TEMPERATURE COMPENSATED CMOS-MEMS RESONATORS 853

M.-H. Li, C.-S. Li, and S.-S. Li
National Tsing Hua University, TAIWAN

This work presents an in-depth study on the Q-factor of the passively temperature compensated CMOS-MEMS resonators through the collected material/experimental database and finite-element simulation. By adapting an anchor-loss-free double-ended tuning fork (DETF) resonator design, the intrinsic material loss is expected to be the major loss mechanism in CMOS-MEMS resonators that limits the maximum Q-factor below 3,400 at the frequency of interest (300 kHz–3 MHz).

W-157 NANOFIBER FORESTS AS A HUMIDITY-SENSITIVE MATERIAL 857

C. Lei[1,2], L. C. Tang[1,2], H.Y. Mao[1,3], Y. Wang[1,2], J.J. Xiong[2], W. Ou[1,3], Y. Ou[1,3],
A. J. Ming[1,3], D. Li[4], Q.L.Tan[2], W. B. Wang[1,3], D. P. Chen[1,3], and T. Liang[2]
[1]Chinese Academy of Sciences, CHINA, [2]North University of China, CHINA,
[3]Jiangsu R&D Center for Internet of Things, CHINA, and [4]Stanford University, USA

Nanofiber forests with high hydrophilicity are reported in this work. They are fabricated from polyimide(PI) by a plasma-stripping technique. In a relative humidity range of 50%-80%, nanofiber forest-based devices have a capacity ~50% larger than those of PI-based sensors. Besides, the absorption and desorption of moisture take less time. It is expected that the performance of such devices can be improved owing to the simple and fast processes for both nanofiber forests and the humidity sensors.

M-158 PIEZOELECTRIC PAPER FOR PHYSICAL SENSING APPLICATIONS 861

S.K. Mahadeva, K. Walus, and B. Stoeber
University of British Columbia, CANADA

We have developed robust and mechanically flexible piezoelectric paper and we have demonstrated its suitability for physical sensing at the example of a tactile sensor. This piezoelectric paper is mechanically strong and has the largest piezoelectric coefficient reported for paper to date (d_{33}=45.7$\Box$4.2 pC/N); this coefficient is comparable to that of commercially available piezoelectric polymers (polyvinylidene fluoride; PVDF d_{33}=30 pC/N).

Nanoscale Physical Sensors

T-159 A GRAPHENE ACCELEROMETER 865

A.M. Hurst[1,2], S. Lee[1], W. Cha[1], and J. Hone[1]
[1]Columbia University, USA and [2]Kulite® Semiconductor Products Inc., USA

This work presents an SU-8 clamped graphene nano-electro-mechanical-systems (GNEMS) accelerometer, with a SU-8 proof mass located at the center of the membrane. This GNEMS accelerometer is approximately three orders of magnitude smaller than state-of-the-art MEMS accelerometers with the graphene diameter of 3-5 μm and its proof mass diameter of 1-3 μm. The fabrication and experimental periodic calibration results show a repeatable response to a periodic input acceleration levels of ~40 gs.

W-160 A SOLID-GATED GRAPHENE FET SENSOR FOR PH MEASUREMENTS 869

Y. Zhu[1], C. Wang[1,2], N. Petrone[1], J. Yu[1], C. Nuckolls[1], J. Hone[1], and Q. Lin[1]
[1]Columbia University, USA and [2]Nankai University, CHINA

We develop and model a graphene field effect transistor (GFET) nanosensor that, with a back gate provided by a high-κ solid dielectric allows analyte detection in liquid media at low gate voltages (~1.5 V). On the basis of the experimental observations and quantitative analysis, we are able to propose that the charging of the electrical double layer capacitor, instead of the surface transfer doping, is the major mechanism responsible for the pH sensing.

M-161 MULTI-MODAL GRAPHENE POLYMER INTERFACE CHARACTERIZATION PLATFORM FOR VAPORIZABLE ELECTRONICS 873
V. Gund, A. Ruyack, K. Camera, S. Ardanuc, C. Ober, and A. Lal
Cornell University, USA

We report a novel graphene-based micromechanical resonant platform with resistive and mass-dependent frequency-sensing for thermal-response measurements of thin-film analytes. Resistance-temperature variation of atomically thin graphene, which also serves as the resistive heater, due to surface-interactions with spun-on analytes, and mass-sensing with silicon-nitride as structural layer provides unique dual-signal electrical and mechanical signatures of analytes.

T-162 TWO-DIMENSIONAL MoS_2 NANOMECHANICAL RESONATORS FREELY-SUSPENDED ON MICROTRENCHES IN FLEXIBLE SUBSTRATE 877
R. Yang[1], Z. Wang[1], P. Wang[1], R. Lujan[2], T.N. Ng[2], and P.X.-L. Feng[1]
[1]Case Western Reserve University, USA and [2]Palo Alto Research Center, USA

We demonstrate the first high-frequency ultrathin MoS2 nanomechanical resonators, freely-suspended on microtrenches (~13μm wide and 14μm deep) fabricated on flexible substrates, with bendability and stretchability. Through investigations of the device resonances via optical excitation and detection, we observe multimode resonances up to ~50MHz with the PDMS substrate under different bending and stretching conditions. This platform will enable studies of strain coupling effects in 2D crystals.

Other Physical Sensors

W-163 A SENSOR FOR STIFFNESS CHANGE SENSING BASED ON THREE WEAKLY COUPLED RESONATORS WITH ENHANCED SENSITIVITY 881
C. Zhao[1], G.S. Wood[1], J. Xie[2], H. Chang[2], S.H. Pu[1,3], H.M.H. Chong[1], and M. Kraft[4]
[1]University of Southampton, UK, [2]Northwestern Polytechnical University, CHINA, and
[3]University of Southampton Malaysia Camput, MALAYSIA, and [4]University of Liege, Montefiore Institute, BELGIUM

A novel MEMS resonant sensing device consisting of three weakly coupled resonators that is ultra-sensitive to stiffness change was designed, fabricated and electrically tested. By measuring amplitude ratio change of two resonators caused by mode localization, due to a change of spring stiffness of one resonator, a 49 times improvement in sensitivity compared to a previously reported 2DoF resonator sensor, and 4 orders magnitude enhancement compared to a 1DoF resonator sensor has been achieved.

M-164 CHIP-SCALE AEROSOL IMPACTOR WITH INTEGRATED RESONANT MASS BALANCES FOR REAL TIME MONITORING OF AIRBORNE PARTICULATE CONCENTRATIONS 885
M. Maldonado-Garcia[1], E. Mehdizadeh[1], V. Kumar[1], J.C. Wilson[2], and S. Pourkamali[1]
[1]University of Texas, Dallas, USA and [2]University of Denver, USA

This work presents chip-scale integration of a MEMS resonant mass balance along with an aerosol impactor on a single SOI. A three mask microfabrication process has been developed to produce the main components; mass balance, impactor nozzle, and impaction micro-chamber. In addition to extreme miniaturization of a conventionally bulky setup and allowing real-time particulate mass concentration data collection, this approach addresses misalignment issues between MEMS resonators and nozzle.

T-165 HARBOR SEAL WHISKER INSPIRED FLOW SENSORS TO REDUCE VORTEX-INDUCED VIBRATIONS 889
A.G.P. Kottapalli[1,2], M. Asadnia[1,2], J.M. Miao[1], and M. Triantafyllou[2,3]
[1]Nanyang Technological University, SINGAPORE, [2]Singapore-MIT Alliance for Research and Technology, SINGAPORE, and [3]Massachusetts Institute of Technology, USA

Harbor seals (Phoca vitulina) are able to track their prey underwater by detecting minute water movements using their whiskers. Through comparative experimental study conducted using two MEMS sensors -one possessing a circular cylindrical haircell and the other processing a haircell with whisker-like undulations, we validate the VIV reduction of in case of whisker geometry to be 50 times lower.

W-166 ISOTROPIC 3D SILICON HALL SENSOR 893
C. Sander, C. Leube, T. Aftab, P. Ruther, and O. Paul
University of Freiburg-IMTEK, GERMANY

This paper reports the first 3D Hall sensor with isotropic sensitivity for the three spatial components of the magnetic field. The silicon device has the shape of a hexagonal prism with symmetric sets of three contacts on its top and bottom surfaces. Sending currents obliquely across the device allows one to operate it as three mutually crossing, identical, and effectively orthogonal Hall sensors. We demonstrate a design achieving sensitivities of Sx=33.0 mV/VT, Sy=33.9 mV/VT and Sz=33.3 mV/VT.

M-167 MASS-FABRICATION COMPATIBLE MECHANISM FOR CONVERTING IN-PLANE TO OUT-OF-PLANE MOTION 897
I. Hotzen, O. Ternyak, S. Shmulevich, and D. Elata
Technion - Israel Institute of Technology, ISRAEL

We present a mechanism that converts in-plane to out-of-plane motion, which is fully compatible with mass-fabrication technology. The motion conversion ratio of the mechanism is constant over a wide range of motion, and this ratio can be easily tuned by adding or subtracting modular elements into the design of an otherwise unchanged planform. The mechanism enables harnessing well behaved in-plane comb-drive actuators to achieve a well behaved out-of-plane motion.

T-168 MONOLITHIC INTEGRATION OF MICRO MAGNETIC PILLAR ARRAY WITH ANISOTROPICMAGNETO-RESISTIVE (AMR) STRUCTURE FOR OUT-OF-PLANE MAGNETIC FIELD DETECTION 901

W.-M. Lai, F.-M. Hsu, W.-L. Sung, R. Chen, and W. Fang
National Tsing Hua University, TAIWAN

A novel integrate micro magnetic pillar array with anisotropic magneto-resistive (AMR) structure for out-of-plane magnetic field detection has been proposed and demonstrated. Through the Nickel pillar to be as magnetic concentrator, the out-of-plane magnetic field can be detected by AMR sensor, and this study propose the micro magnetic pillar array to enhance the magnetic conversion efficiency.

W-169 MULTI-COLOR IMAGING WITH SILICON-ON-INSULATOR DIODE UNCOOLED INFRARED FOCAL PLANE ARRAY USING THROUGH-HOLE PLASMONIC METAMATERIAL ABSORBERS 905

D. Fujisawa[1], S. Ogawa[1], H. Hata[1], M. Uetsuki[1], K. Misaki[1], Y. Takagawa[2], and M. Kimata[2]
[1]Mitsubishi Electric Corporation, JAPAN and [2]Ritsumeikan University, JAPAN

We report a silicon-on-insulator (SOI) diode uncooled infrared focal plane array (IRFPA) with through-hole plasmonic metamaterial absorbers (TH-PLMAs) for multi-color imaging with a 320x240 array format. Through-holes formed on the PLMA can reduce the thermal mass while maintaining both the single-mode and high absorption due to the plasmonic metamaterial structures, which realizes fast response and high responsivity.

M-170 PASSIVE WIRELESS TEMPERATURE SENSING WITH PIEZOELECTRIC MEMS RESONATORS 909

H. Fatemi, M.J. Modarres-Zadeh, and R. Abdolvand
University of Central Florida, USA

For the first time, a piezoelectric MEMS resonator is utilized for passive wireless temperature sensing with an accuracy of less than 0.1°C at 1m with a signal power of 500mW and 5dBi gain antennas. The high quality factor and low motional resistance of a 991MHz thin-film piezoelectric-on-silicon (TPoS) resonator are exploited to accurately determine the temperature from the change in the resonance frequency by taking the Fourier transform of the resonator's time-gated response.

W-171 SNR IMPROVEMENT IN AMPLITUDE MODULATED RESONANT MEMS SENSORS VIA THERMAL-PIEZORESISTIVE INTERNAL AMPLIFICATION 913

M. Mahdavi, A. Ramezany, V. Kumar, and S. Pourkamali
University of Texas, Dallas, USA

We studied, the effect of thermal-piezoresistive internal amplification on signal to noise ratio (SNR) of amplitude modulated resonant MEMS sensors showing the possibility to significantly improve the detection limit. It has been shown that as the thermal-piezoresistive amplification sets in, noise rms value increases with a slower rate than the boost in quality factor (Q) and output signal level, therefore the SNR value increases.

Sonic and Ultrasonic MEMS Transducers (Microphones, PMUTs, Etc.)

T-172 A RESONANT PIEZOELECTRIC MICROPHONE ARRAY FOR DETECTION OF ACOUSTIC SIGNATURES IN NOISY ENVIRONMENTS 917

A.A. Shkel, L. Baumgartel, and E.S. Kim
University of Southern California, USA

We report a MEMS acoustic resonator array with improved Automatic Speech Recognition (ASR) and signature detection characteristics in environments with high levels of acoustic interference. ASR experiments are performed, showing an increase of 62.7 percentage points in transcription accuracy with -15 dB Signal-to-Noise Ratio (SNR). The results of this study support the development of highly resonant acoustic sensors for a variety of pattern recognition applications.

W-173 AIR-COUPLED ALUMINUM NITRIDE PIEZOELECTRIC MICROMACHINED ULTRASONIC TRANSDUCERS AT 0.3 MHZ TO 0.9 MHZ 921

O. Rozen[1], S.T. Block[2], S.E. Shelton[2], R.J. Przybyla[2], and D.A. Horsley[1]
[1]University of California, Davis, USA and [2]Chirp Microsystems, Inc., USA

For the first time, air-coupled piezoelectric micromachined ultrasonic transducers (PMUTs) operating at frequencies ranging from 300 to 900 kHz were designed, fabricated and characterized. We also increased the fractional bandwidth by about 35% by patterning the diaphragm center into a ring or structural ribs, resulting in a reduction of the PMUT's mass. Fabrication was conducted using wafer-level bonding of a MEMS PMUT wafer to a CMOS wafer using a conductive metal eutectic bond. This process allows for close integration of PMUT arrays and signal processing circuitry and is used here to study the effects of wafer-level packaging on acoustic performance.

M-174 SOUND FOCUSING IN LIQUID USING A VARIFOCAL ACOUSTIC MIRROR 925

R. Aoki, N. Thanh-Vinh, K. Noda, T. Takahata, K. Matsumoto, and I. Shimoyama
University of Tokyo, JAPAN

This paper reports a method to concentrate sound in liquid in a desired location using a PDMS-based varifocal acoustic mirror. We used PDMS–air boundary as a parabolic sound reflector to concentrate sound. By adjusting the curvature radius of the acoustic mirror, we could change the position where sound was concentrated. We confirmed that our method was able to make the output of the acoustic sensor ten times larger than that without focusing in water.

T-175 BIMORPH PMUT WITH DUAL ELECTRODES 928

S. Akhbari[1], F. Sammoura[1], C. Yang[1], M. Mahmoud[2], N. Aqab[2], and L. Lin[1]

[1]*University of California, Berkeley, USA and [2]Masdar Institute of Science and Technology, UAE*

We have successfully demonstrated "bimorph" piezoelectric micromachined ultrasonic transducers (pMUT) with unique advantages, dramatically improving the device capabilities in the process. The bimorph pMUT utilizes two active AlN layers in a CMOS-compatible process. This innovative design is the first bimorph pMUT with two active piezoelectric layers separated by a common electrode.

MEMS for Electromagnetics
DC and Low Frequency Magnetic and Electromechanical Components and Systems

W-176 A LOW-NOISE SUB-500µW LORENTZ FORCE BASED INTEGRATED
MAGNETIC FIELD SENSING SYSTEM 932

S. Brenna, P. Minotti, A. Bonfanti, G. Laghi, G. Langfelder, A. Longoni, and A.L. Lacaita

Politecnico di Milano, ITALIA

A complete 1-D magnetic field sensing system including a z-axis Lorentz force MEMS sensor and an integrated circuit (ASIC) is presented. Measurement results show an achievable sensor resolution of 220 nT·mA/√Hz with an achievable bandwidth 100 Hz. The ASIC low-noise performance does not impair the minimum detectable magnetic field, mainly limited by the MEMS thermo-mechanical noise. Dissipating only 400 µW, the circuit satisfies the consumer electronics low-power requirements.

M-177 CHIP-SCALE ELECTRODYNAMIC SYNTHETIC JET ACTUATORS 936

S.G. Sawant[1], E.A. Deem[2], D.J. Agentis[3], L.N. Cattafesta[2], and D.P. Arnold[1]

[1]*University of Florida, USA, [2]Florida State University, USA, and [3]Virginia Tech, USA*

We report the first chip-scale electrodynamic synthetic jet actuator that integrates both a coil and permanent magnet via micro-fabrication. This is achieved by integrating bonded NdFeB powder magnets into standard silicon micro-machining processes. The device has a volume of 7.5 mm x 7.5 mm x 1.1 mm and generates a fluidic jet with a peak velocity of 2.1 m/s while operating at 180 Hz with 20 mW input power. The actuator has applications in flow control and active cooling of electronic devices.

T-178 FULLY-POLYMERIC NEM RELAY FOR FLEXIBLE, TRANSPARENT,
ULTRA-LOW POWER ELECTRONICS AND SENSORS 940

Y. Pan, F. Yu, and J. Jeon

Rutgers, The State University of New Jersey, USA

A fully-polymeric NEM relay based on a conductive polymer, Poly(3,4-Ethylenedioxythiophene):Polystyrene-Sulfonate (PEDOT:PSS), and dielectric polymers is proposed for the first time to enable flexible, transparent, ultralow-power electronics and sensors, and the first functional prototype fabricated using a five-mask low-thermal-budget process is demonstrated. Exploiting the water-absorption behavior of PEDOT:PSS, the potential use of the relay as a biochemical sensor is also demonstrated.

W-179 UHF PIEZOELECTRIC QUARTZ MEMS MAGNETOMETERS BASED ON
ACOUSTIC COUPLING OF FLEXURAL AND THICKNESS SHEAR MODES 944

H.D. Nguyen, J.A. Erbland, L.D. Sorenson, R. Perahia, L.X. Huang, R.J. Joyce, Y. Yoon,
D.J. Kirby, T.J. Boden, R.B. McElwain, and D.T. Chang

HRL Laboratories, USA

We report for the first time piezoelectric Quartz MEMS magnetometers based on acoustic coupling between resonance modes. The magnetic sensors employ a novel transduction scheme to upconvert the desired near-DC magnetic field signal (using the fundamental flexural mode) onto frequency modulated (FM) sidebands of the primary quartz thickness shear (TS) oscillation. First-generation devices exhibit flexural and TS resonances at 2.77kHz and 583.31MHz, respectively, and a magnetic sensitivity of 63.6V/T.

Free Space Optical Components and Systems (Displays, Lenses, Detectors)

M-180 A 45°-TILTED 2-AXIS SCANNING MICROMIRROR INTEGRATED
ON A SILICON OPTICAL BENCH FOR 3D ENDOSCOPIC OPTICAL IMAGING 948

C. Duan[1], W. Wang[1], X. Zhang[1], J. Ding[2], Q. Chen[2], A. Pozzi[1], and H. Xie[1]

[1]*University of Florida, USA and [2]WiO Technology Limited, CHINA*

This paper presents a 2-axis electrothermal single-crystal-silicon (SCS) micromirror that is tilted 45° out of plane on a silicon optical bench (SiOB). The tilt of the mirror is achieved with the bending of a set of stressed bimorph beams and the stop provided by the silicon sidewall. To the best of our knowledge, this is the first demonstration of an integrated SiOB with a 2-axis SCS mirror tilted at a fixed angle without assembly.

T-181 COMPACT NEAR-EYE DISPLAY SYSTEM USING A SUPERLENS-BASED
MICROLENS ARRAY MAGNIFIER 952

H.S. Park[1], R. Hoskinson[2], H. Abdollahi[2], and B. Stoeber[1]

[1]*University of British Columbia, CANADA and [2]Recon Instruments Inc., CANADA*

We present a new approach to make a very compact near-eye display (NED) using only two layers of microlens arrays (MLA) working in conjunction as a magnifying lens (MLA magnifier). The purpose of the MLA magnifier is to generate a virtual image of a display, positioned within several centimeters from the eye, at optical infinity to minimize the optical disparity between the surrounding scenery and the image on the display.

M-182 OPEN-STRUCTURE ELECTROWETTING DISPLAY
WITH CAPACITIVE SENSING FEEDBACK SYSTEM 956
S. Choi and J. Lee
Seoul National University, SOUTH KOREA

We report an open-structure electrowetting-based reflective display with capacitive sensing feedback that enables an effective self-dosing of ink, high contrast, and the precise control of color level. We introduce an display that can achieve such improvements via an open structure design and a capacitive feedback system including the effective ink dosing process, off color area being ~ 8% of viewable area, and precision control of color area even under a large variation of interfacial tension.

W-183 TUNABLE METAMATERIAL LENS ARRAY VIA METADROPLETS 960
Q.H. Song[1,2,] W.M. Zhu[2], W. Zhang[2], P.C. Wu[2], Z.X. Shen[2], Z.C. Yang[4], Y.F. Jin[4],
Y.L. Hao[4], T. Bourouina[3], Y. Leprince-Wang[1], and A.Q. Liu[2]
[1]*UPEM, Université Paris-Est, FRANCE,* [2]*Nanyang Technological University, SINGAPORE,*
[3]*ESIEE, Université Paris-Est, FRANCE, and* [4]*Peking University, CHINA*

This paper reports a liquid based tunable metamaterial which is using droplets as unit cell structures. It can function as a tunable lens array, whose focus spot can be continuously tuned, for the first time, from defocusing to sub-wavelength focusing in THz region. This work also develops a new tuning method to reconfigure the shapes of the liquid micro-droplets, which is using air pressure to expand the height of the micro channel so that the height of the droplets will be enlarged.

Manufacturing for Electromagnetic Transducers

M-184 FABRICATION OF PATTERNED MAGNETIC MICROSTRUCTURES
USING MAGNETICALLY ASSEMBLED NANOPARTICLES 964
C. Velez, I. Torres-Díaz, L. Maldonado-Camargo, C. Rinaldi, and D.P. Arnold
University of Florida, USA

Modeling and experimental characterization of a fabrication method for forming magnetic microstructures with complex shapes using self-assembled iron oxide (Fe3O4) magnetic nanoparticles. This method can potentially be used in roll-to-roll production of magnetic structures either patterned onto substrates or lifted off to create free-floating micromagnetic actuators.

Other Electromagnetic MEMS

T-185 ELECTRIC CONTACT STABILITY AND READOUT RESOLUTION
OF THE ANTIWEAR PROBE WITH A GROOVE AND OIL LUBRICATION
SLIDING SYSTEM FOR PROBE-BASED ARCHIVE MEMORIES 968
Y. Tomizawa, K. Toya, A. Oonishi, Y. Li, J. Hirota, M. Yabuki, I. Kunishima, and H. Shinomiya
Toshiba Corporation, JAPAN

The authors have proposed the novel concept of a sliding system called "AGO" (antiwear probe with a groove and oil lubrication) for probe-based archive memories. The system has been proven to have the ability to endure a meter-scale probe slide, not only in terms of electric contact stability but also in regards to the readout resolution degradation of the recorded pattern. This demonstrates the possibility of bringing probe-based memories into actual products used in data archiving, which requires limited time data access.

W-186 NEMS VARIABLE OPTICAL ATTENUATOR (VOA) DRIVEN BY OPTICAL FORCE 972
B. Dong[1,2], H. Cai[2], Y.D. Gu[2], Z.C. Yang[3], Y.F. Jin[3], Y.L. Hao[3], D.L. Kwong[2], and A.Q. Liu[1]
[1]*Nanyang Technological University, SINGAPORE,* [2]*Agency for Science, Technology and Research (A*STAR),*
SINGAPORE, and [3]*Peking University, CHINA*

We develop a NEMS optomechanical VOA driven by optical gradient force. The VOA is realized via waveguide based optical directional coupler. The gap between the directional coupler is controlled via optical force driven actuator. The doubly clamped silicon beam actuator is controlled by tuning the wavelength of control light. The NEMS VOAs have merits such as small dimension, low power consumption and good capability for all optical integration as compared with conventional MEMS based fiber VOA.

Photonic Components and Systems

M-187 A SUPER-REGENERATIVE OPTICAL RECEIVER BASED
ON AN OPTOMECHANICAL OSCILLATOR 976
T. Beyazoglu, T.O. Rocheleau, A.J. Grine, K.E. Grutter, M.C. Wu, and C.T.-C. Nguyen
University of California, Berkeley, USA

We present a super-regenerative optical receiver that detects on-off key modulated light input via the radiation-pressure gain of a self-sustained electro-opto-mechanical oscillator (EOMO). With oscillation amplitude a function of the intensity of light coupled into the oscillator, this device now allows data to be directly demodulated using only silicon-compatible materials, i.e., without the expensive III-V compound semiconductor materials often used in conventional optical receivers.

T-188 INTEGRATED PIEZOELECTRICALLY DRIVEN ACOUSTO-OPTIC MODULATOR 980
S. Ghosh and G. Piazza
Carnegie Mellon University, USA

This paper presents a new type of acousto-optic modulator based on the conjunction of a piezoelectric contour mode resonator with a photonic whispering gallery mode resonator. The monolithic aluminum nitride device exhibits coupling of piezoelectrically-generated lateral vibrations into a traveling-wave photonic ring resonator in a fully-integrated platform with electrodes directly patterned on the resonator body. We demonstrate the optical sensing of an actuated mechanical mode at 654 MHz.

**W-189 SPECTRALLY SELECTIVE INFRARED DETECTOR BASED
ON AN ULTRA-THIN PIEZOELECTRIC RESONANT METAMATERIAL** 984
Y. Hui and M. Rinaldi
Northeastern University, USA

We report on the first demonstration of a spectrally selective uncooled MEMS resonant IR detector based on an ultra-thin piezoelectric resonant metamaterial. High quality factor of 1407 and electromechanical coupling coefficient of 1.9%, and spectrally selective absorption (~40%) of long wavelength infrared radiation (8.8 μm with FWHM of 1.88 μm) in an ultra-low volume device were achieved, resulting in a fast (~650 μs) and high resolution (NEP ~7 nW/rt-Hz at 200 Hz bandwidth) MEMS IR detector.

RF MEMS Components and Systems

**M-190 A CMOS-MEMS ARRAYED RGFET OSCILLATOR USING
A BAND-TO-BAND TUNNELING BIAS SCHEME** 988
C.-H. Chin, C.-S. Li, M.-H. Li, and S.-S. Li
National Tsing Hua University, TAIWAN

This work reports a CMOS-MEMS Resonant-Gate Field Effect Transistor (RGFET) oscillator comprising only one single transistor. A band-to-band tunneling (BTBT) charging technique is implemented for the first time. Furthermore, this charging phenomenon on the floating gate can be well preserved for more than one day. Finally, a CMOS-MEMS RGFET self-sustained oscillator with only one active transistor is demonstrated with a decent far-from-carrier phase noise of -122 dBc/Hz.

**T-191 ACTIVE REFLECTORS FOR HIGH PERFORMANCE
LITHIUM NIOBATE ON SILICON DIOXIDE RESONATORS** 992
L. Shi and G. Piazza
Carnegie Mellon University, USA

We design, demonstrate and optimize active reflectors for enhancing the electromechanical coupling (k_t^2) and suppressing spurious modes in Laterally Vibrating Resonators (LVRs) based on X-cut ion-sliced Lithium Niobate (LN) thin film on silicon dioxide (SiO$_2$). Optimized active reflectors that resort to 100% metal coverage of the λ/4 extensions at the two ends of the resonant plate enable: (i) a considerable improvement of k_t^2 (up to 13%) (ii) spurious mode suppression, robustness to processing (iii) misalignment and (iv) over/under-etching.

**W-192 APPLICATION OF STATISTICAL ELEMENT SELECTION TO 3D INTEGRATED
ALN MEMS FILTERS FOR PERFORMANCE CORRECTION AND YIELD ENHANCEMENT** 996
A. Patterson[1], E. Calayir[1], G.K. Fedder[1], G. Piazza[1], B.W. Soon[2], and N. Singh[2]
[1]Carnegie Mellon University, USA and
[2]Agency for Science, Technology and Research (A*STAR), SINGAPORE

By 3D integration of an array of 12 nominally identical AlN MEMS sub-filters with a CMOS switching matrix and application of statistical element selection to the same system, we have built a self-healing filter offering 495 unique filter responses and a tuning range of 500 kHz for both center frequency and bandwidth. This system enables correction of intrinsic, fabrication-induced variation in filter performance that would otherwise severely limit the manufacturing yield of standalone filters.

M-193 DAMPING IN 1 GHZ LATERALLY-VIBRATING COMPOSITE PIEZOELECTRIC RESONATORS 1000
J. Segovia-Fernandez and G. Piazza
Carnegie Mellon University, USA

This work experimentally proves the physics of damping in this class of MEMS resonators. We first confute a previously developed theory of interfacial dissipation that assumed a stress (or Young modulus) jump between different materials and then find that damping is instead related to either interfacial dissipation due to a velocity jump or thermoelastic dissipation (TED) in the electrodes.

**T-194 DUAL-CLOCK WITH SINGLE AND MONOLITHICAL 0-LEVEL
VACUUM PACKAGED MEMS-on-CMOS RESONATOR** 1004
A. Uranga[1], G. Sobreviela[1], N. Barniol[1], E. Marigó[2], C. Tay-Wee-Song[2], M. Shunmugam[2], A.A. Zainuddin[2],
A. Kumar-Kantimahanti[2], V. Madhaven[2], and M. Soundara-Pandian[2]
[1]Universitat Autònoma de Barcelona, SPAIN and [2]Silterra, MALAYSIA

This paper demonstrates the feasibility of a novel fabrication approach of MEMS resonators above standard CMOS circuitry and with zero-level vacuum package. As a proof of concept a monolithical CMOS-MEMS-closed loop oscillator showing dual-clock capabilities (11.9 MHz and 24.5 MHz) is presented. These two frequencies correspond to two different resonator modes, specifically the torsional and vertical out of plane, of a paddle shaped MEMS resonator.

W-195 EXPERIMENTAL INVESTIGATION ON MODE COUPLING OF BULK MODE SILICON MEMS RESONATORS 1008

Y. Yang[1], E. Ng[1], P. Polunin[2], Y. Chen[1], S. Strachan[2], V. Hong[1],
C.H. Ahn[1], O. Shoshani[2], S. Shaw[2], M. Dykman[2], and T. Kenny[1]
[1]Stanford University, USA and [2]Michigan State University, USA

We present the effect of nonlinear elasticity on the coupling between different bulk modes of silicon MEMS resonators. From experimental data, the coupling has a strong dependence on the order and the shape of the coupled resonant modes, as well as the doping type/concentration, and crystal orientation, leading to a variety of complex and potential useful phenomena.

M-196 MEMS-BASED RF PROBES FOR ON-WAFER MICROWAVE CHARACTERIZATION OF MICRO/NANOELECTRONICS 1012

J. Marzouk, S. Arscott, A. El Fellahi, K. Haddadi, T. Lasri, L. Buchaillot, and G. Dambrine
University Lille 1, FRANCE

We demonstrate a radio frequency (RF) probe based on microelectromechanical systems (MEMS) design and processing technologies. The probe responds to the current needs of microelectronics requiring microwave characterization of nanoscale devices and systems having sub-micron pad sizes. The use of MEMS technologies enables the probe contact pad area dimensions to be reduced by a three orders of magnitude compared to existing commercial RF probes.

T-197 MICROMECHANICAL RING RESONATORS WITH A 2D PHONONIC CRYSTAL SUPPORT FOR MECHANICAL ROBUSTNESS AND PROVIDING MASK MISALIGNMENT TOLERANCE 1016

B. Figeys[1,2], B. Nauwelaers[2], H.A.C. Tilmans[1], and X. Rottenberg[1]
[1]imec, BELGIUM, [2]KU Leuven, BELGIUM

This paper reports on the design of ring-type electrostatically transduced bulk acoustic wave resonators designed for increased shock and vibration resistance. This was achieved through a 2D Phononic Crystal (PnC) support, simultaneously a mechanically strong and acoustically well-confined support. We manufactured SiGe-resonators at 137.8MHz with a Q-factor around 15k. Another feature to this design is the process tolerance of the Q-factor towards mask misalignment for the center support.

M-198 SILICON-MICROMACHINED SPACERS FOR UHF CAVITY RESONATORS 1020

D. Psychogiou, M.D. Sinani, and D. Peroulis
Purdue University, USA

This paper reports on a novel hybrid integration concept that enables the realization of high-quality (Q) factor, low-frequency cavity resonators with well-defined capacitive-loading and variable center frequency. It is based on a silicon-micromachined spacer that is mounted on top of a conventional CNC-machined metallic cavity to functionalize the resonator's capacitance. For the first time, it is demonstrated that low-frequency resonators with micrometer-scale gaps (10s of microns), relatively large Q-factor (459-505) and tunable response (18.5%) can be constructed without the need for post-fabrication tuning. To demonstrate these benefits, a resonator assembly was designed, built and experimentally tested at UHF band and for a frequency tuning range between 1424-1711 MHz.

W-199 SIMULTANEOUS MULTI-FREQUENCY SWITCHABLE OSCILLATOR AND FSK MODULATOR BASED ON A CAPACITIVE-GAP MEMS DISK ARRAY 1024

T.L. Naing, T.O. Rocheleau, and C.T.-C. Nguyen
University of California, Berkeley, USA

An array of capacitive-gap MEMS resonators with different frequencies combined with an ASIC amplifier, provides a first MEMS-based multi-frequency oscillator generating simultaneous oscillation outputs around 62MHz while employing only a single amplifier. Enabled via a softening non-linearity, amplitude is limited here for each MEMS resonator individually. Furthermore, electrical stiffness frequency tuning enables FSK modulation of the output waveform, offering a simple multichannel transmitter.

M-200 TAPERED PHONONIC CRYSTAL SAW RESONATOR IN GAN 1028

S. Wang, L.C. Popa, and D. Weinstein
Massachusetts Institute of Technology, USA

This paper presents a new Phononic Crystal (PnC) resonator design in which a tapered PnC is used to confine a 970 MHz SAW resonance in a GaN-on-Si platform. The use of a tapered PnC reflector in this work reduces the footprint of SAW resonators by >100X relative to the case of conventional metal grating reflectors while maintaining high Q. A 3.5X improvement in Q is experimentally demonstrated relative to uniform PnC reflectors of comparable dimensions.

THz MEMS Components and Systems

T-201 ENHANCED CONTROLLABILITY IN MEMS METAMATERIALS 1032

P. Pitchappa[1], C.P. Ho[1], Y. Qian[1], Y.-S. Lin[1], N. Singh[2], and C. Lee[1]
*[1]National University of Singapore, SINGAPORE and
[2]Agency for Science, Technology and Research (A*STAR), SINGAPORE*

We demonstrate a method to improve the controllabilty of the MEMS tunable metamaterials by individually actuating the alternate lines in the metamaterial array. This is the first step towards the realization of Programmable metamaterial, where each of the unit cell can be addressed independently.

PowerMEMS and Actuators
Actuator Components and Systems

W-202 BI-DIRECTIONAL EXTENDED RANGE PARALLEL PLATE ELECTROSTATIC ACTUATOR BASED ON FEEDBACK LINEARIZATION 1036

E.E. Moreira[1], F.S. Alves[1], R.A. Dias[2], M. Costa[2], H. Fonseca[2], J. Cabral[1], J. Gaspar[2], and L.A. Rocha[1]

[1]University of Minho, PORTUGAL and [2]International Iberian Nanotechnology Laboratory, PORTUGAL

A bi-directional extended range parallel-plate electrostatic actuator using feedback linearization control is presented in this paper. The actuator can have stable displacements up to 90% of the full-gap (limited by mechanical stoppers) on both directions, i.e, the device can move ±2um within a ±2.25um gap. The system has successfully tracked references until 1 kHz (limited by the dynamics of the device) and it presents a capacitor tuning rage of 17, using an actuation voltage from 0 to 10V.

M-203 BIOCOMPATIBLE CIRCUIT-BREAKER CHIP FOR TEMPERATURE REGULATION OF ELECTROTHERMALLY DRIVEN SMART IMPLANTS 1040

Y. Luo, M. Dahmardeh, and K. Takahata

University of British Columbia, CANADA

We present a thermoresponsive circuit breaker micromachined in a form of titanium-packaged chip for biomedical applications with a focus on electronic implants. This micro breaker has a temperature-sensitive cantilever actuator to serve as an absolute temperature limiter for the device of interest being protected from overheating, a critical safety feature for smart implants including those that are electrothermally active. Temperature regulation of a wireless heater powered by external RF field

T-204 CONTROLLABLE 'SOMERSAULT' MAGNETIC SOFT ROBOTICS 1044

T.S. Zhang, A. Kim, M. Ochoa, and B. Ziaie

Purdue University, USA

This paper reports controllable somersault magnetic soft robotics consisting of polymeric magnet embedded in a high friction silicone polymer. The soft structures are actuated by the rotation of a permanent magnet at fixed position and exhibit controllable linear movement in a flip-and-forward manner for extended distance on both horizontal and vertical non-magnetic surfaces. The control of direction of motion is also achieved.

W-205 CYBORG BEETLE: THRUST CONTROL OF FREE FLYING BEETLE VIA A MINIATURE WIRELESS NEUROMUSCULAR STIMULATOR 1048

T.T. Vo Doan, Y. Li, F. Cao, and H. Sato

Nanyang Technological University, SINGAPORE

We have developed a cyborg beetle, which is the hybrid of a miniature wireless communication system and a living beetle platform. We can remotely stimulate neuromuscular sites of the living beetle platform via the miniature system. In this study, we stimulated the subalar flight muscle, a major muscle directly inserted to the wing base of beetle, and demonstrated the thrust control of the cyborg beetle with graded response.

M-206 CYLINDRICAL HALBACH MAGNET ARRAY FOR ELECTROMAGNETIC VIBRATION ENERGY HARVESTERS 1051

I. Shahosseini and K. Najafi

University of Michigan, Ann Arbor (WIMS²), USA

This paper reports the design, optimization, and test results of a new magnetic structure for kinetic energy harvesters allowing seven-fold increase in power density compared to single-magnet configuration. For the first time, electromagnetic energy harvesters with "single cylindrical Halbach array" and "double-concentric Halbach array" magnetic structures are fabricated and tested.

T-207 FLUID SEPARATED VOLUMETRIC FLOW CONVERTER (FSVFC) FOR HIGH SPEED AND PRECISE CELL POSITION CONTROL 1055

T. Monzawa[1], S. Sakuma[2], F. Arai[2], and M. Kaneko[1]

[1]Osaka University, JAPAN and [2]Nagoya University, JAPAN

This paper proposes the on-chip Fluid Separated Volumetric Flow Converter (FSVFC) capable of high speed cell position control with high resolution, while the actuation fluid is physically separated from working fluid for biological considerations. By utilizing the newly developed on-chip comb shaped FSVFC, an online high speed vision sensor and a high speed PZT, we succeeded in controlling the position of a cell in microfluidic channel with the time constant of 12 ms and the resolution of 240 nm

W-208 ON-CHIP ENUCLEATION USING AN UNTETHERED MICROROBOT INCORPORATED WITH AN ACOUSTICALLY OSCILLATING BUBBLE 1059

I.S. Park, Y.R. Lee, S.J. Hong, K.Y. Lee, and S.K. Chung

Myongji University, SOUTH KOREA

This paper reports a novel on-chip enucleation method using an untethered microrobot incorporated with an acoustically excited microbubble, which will allow minimally invasive cell surgery for cloning techniques and biomedical applications. The proposed microrobot mainly consists of a compressible bubble for the manipulation of cells and twin permanent magnets for the manipulation of the microrobot in an aqueous medium.

M-209 OUT-OF-PLANE MICRO-FORCE FUNCTION GENERATOR WITH INHERENT SELF-FEEDBACK FOR MICRO-DEFORMATION MODIFYING 1063
X. Wang, D. Xiao, X. Wu, Z. Hou, Z. Chen, and H. He
National University of Defense Technology, CHINA

We propose a novel concept of out-of-plane micro-force function generator for micro-deformation modifying. The proposed generator is based on batch micro-fabricated polymer thermal actuators array and could actively modify micro-substrate warpage. This strategy constructively utilizes the inherent self-feedback for in-situ deformation control and has the potential for solving stress-induced problems of micro-fabricated devices.

Manufacturing for Actuators and PowerMEMS

T-210 A PAPER-LIKE MICRO-SUPERCAPACITOR WITH PATTERNED BUCKYPAPER ELECTRODES USING A NOVEL VACUUM FILTRATION TECHNIQUE 1067
C.-W. Ma, P.-C. Huang, and Y.-J. Yang   * Award Nominee
National Taiwan University, TAIWAN

This study reports a paper-like micro-supercapacitor with in-plane interdigital buckypaper electrodes on a filter membrane substrate. A vacuum filtration method assisted by lithography techniques is proposed for patterning buckypaper. The proposed micro-SC features advantages including a flexible structure, simple fabrication, easy chip integration, and high specific capacitance. The specific capacitance measured by cyclic voltammetry was 107.27 mF/cm2 at a scan rate of 20 mV/sec.

W-211 A POTASSIUM ELECTRET ENERGY HARVESTER FOR 3D-STACK ASSEMBLY 1071
K. Misawa[1], T. Sugiyama[2], G. Hashiguchi[2], H. Fujita[1], and H. Toshiyoshi[1]
[1]University of Tokyo, JAPAN and [2]Shizuoka University, JAPAN

We report an electrostatic energy harvester based on the potassium ion (K+) electret that could be stacked up into a 3D structure to multiply the output power. Vertical comb electrodes are implemented in a silicon-on-insulator (SOI) wafer with a relatively heavy mass in the handle layer to lower the resonance. A single substrate formation exhibited a 0.34 µW output at 310 Hz for a load resistance of 1 MOhm.

M-212 FABRICATION OF A THREE DIMENSIONAL CANTILEVERED VIBRATIONAL ENERGY HARVESTER USING SILVER INK 1075
S.-J. Chen, Y.-Y. Feng, and S.-Y. Liu
National Central University, TAIWAN

This paper reports a 3D micro electromagnetic energy harvester. Compared to state of the art, multiple layers of conductive coils are dispensed on the micro-machined cantilever diaphragm by an injecting machine, which will increase the potential power density of the harvester.

T-213 WAFER-LEVEL FABRICATION OF A TRIBOELECTRIC ENERGY HARVESTER 1078
M. Han, B. Yu, Z. Su, B. Meng, X. Cheng, X.-S. Zhang, and H. Zhang
Peking University, CHINA

We present a wafer-level fabrication method for triboelectric energy harvester (TEH), which, for the first time, fabricates the TEH completely in MEMS process, without any manually assembly. Compared to state of the art, the proposed TEH is batch fabricated in CMOS-compatible process and the reduced size allows it to be integrated with other electronic devices (e.g., keyboards). This device can produce 235 mV peak voltage at the frequency of 30 Hz, under the 100 MΩ external resistance.

Materials for Actuators and PowerMEMS

W-214 DISPLACEMENT MAGNIFICATION OF GEL ACTUATOR USING pNIPAAm-SU8 HYBRID STRUCTURE 1082
T. Kogure, R. Okada, S. Maeda, and S. Nagasawa
Shibaura Institute of Technology, JAPAN

A fabrication method for a hybrid structure of the poly-N-Isopropylacrylamide (pNIPAAm) gel as a soft material and the SU-8 as a solid material is proposed. This pNIPAAm-SU8 hybrid structure is utilized various applications such as gel actuators. Our hybrid structure was not broken through repeating swelling-shrinking states for 10 times. The SU-8 solid structure magnified a minute displacement of the gel which is occurred by the phase transition.

M-215 LIQUID-TOLERANT ELECTRET USING SUPER-LYOPHOBIC PILLAR SURFACE 1086
Y.-C. Chen, K.-Y. Song, K. Morimoto, and Y. Suzuki
University of Tokyo, JAPAN

For the first time, electret that can be used in the liquid environment has been realized toward higher energy density of electret generators and actuators. By using super-lyophobic overhanging pillar surface with SiO2 layer, stable Cassie-Baxter state for low-surface-tension liquid is sustained even with high surface potential. The pillar surface is successfully charged with soft X-ray photoionization. Surface potential after liquid contact has been significantly improved with the pillars.

T-216 GRAPHENE ONE-SHOT MICRO-VALVE: TOWARDS VAPORIZABLE ELECTRONICS 1090
V. Gund, A. Ruyack, S. Ardanuc, and A. Lal
Cornell University, USA

We report a micro-scale arrayable single-trigger valve of graphene transferred on silicon-nitride for vaporizable electronics. Graphene serves as a nanoscale barrier to oxygen diffusion and as a resistive heater for pulsed-power thermomechanical cleaving to expose sealed alkali metals for heat generation to vaporize polymer electronics. Our valve demonstrates long-storage lifetime, durability under pressure and low-power triggering.

W-217 PIEZOELECTRIC MICRO ENERGY HARVESTERS EMPLOYING ADVANCED (MG,ZR)-CODOPED ALN THIN FILM 1094

L.V. Minh[1], M. Hara[1], H. Kuwano[1], T. Yokoyama[2], T. Nishihara[2], and M. Ueda[3]
[1]*Tohoku University, JAPAN,* [2]*Taiyo Yuden Co., Ltd., JAPAN, and* [3]*Taiyo Yuden Mobile Technology Co., Ltd., JAPAN*

We report the new doped-AlN thin film, (Mg,Zr)AlN, based micro energy harvester. By co-doping Mg and Zr into AlN crystal, (Mg,Zr)AlN shows giant piezoelectricity and preserves low permittivity. (Mg,Zr)AlN has higher figure of merit (FOM=$e_{31}^2(\varepsilon_0\varepsilon)$)) than conventional PZT. The 13 at%-(Mg,Zr)AlN had the experimental FOM of up to 16.7 GPa. The micromachining harvester provided the high normalized power density of 3.72 mW.g^{-2}.cm^{-3}. This achievement was 1.5-fold increase compared to state of the art.

M-218 PULSE POLING WITHIN 1 SECOND ENHANCE THE PIEZOELECTRIC PROPERTY OF PZT THIN FILMS 1098

T. Kobayashi[1], Y. Suzuki[1,2], N. Makimoto[1], H. Funakubo[3], and R. Maeda[1]
[1]*National Institute of Advanced Industrial Science and Technology (AIST), JAPAN,*
[2]*Ibaraki University, JAPAN, and* [3]*Tokyo Institute of Technology, JAPAN*

We present simple but fast poling technique to enhance the piezoelectric property of PZT thin films. Application of pulse voltage to the PZT thin films on MEMS microcantilevers has resulted in large piezoelectric constant (d31) as high as 105 pm/V. It took only 1 second for poling the PZT thin films.

Nanoscale Actuators and PowerMEMS

T-219 A BETAVOLTAIC MICROCELL BASED ON SEMICONDUCTING SINGLE-WALLED CARBON NANOTUBE ARRAYS/SI HETEROJUNCTIONS 1102

M.G. Li and J. Zhang
Peking University, CHINA

This paper reports a novel betavoltaic microcell based on semiconducting single-walled carbon nanotubes (s-SWCNTs). The aligned arrays of p-type s-SWCNTs were prepared onto n-type silicon forming the p-n heterojunction as the energy conversion. This heterojunction displays good rectification characteristics with I_0=1.5pA and n=1.83. Under 7.8mCi/cm^2 ^{63}Ni irradiation, the microcell achieves higher performance of V_{OC}=62mV, J_{SC}=3.8µA/cm^2, FF=33.4% and η=9.8% compared with our previous devices.

Other Actuators and PowerMEMS

W-220 A MICROGRIPPER WITH A RATCHET SELF-LOCKING MECHANISM 1106

Y.C. Hao, W.Z. Yuan, H.M. Zhang, and H.L. Chang
Northwestern Polytechnical University, CHINA

This paper reports a novel electrostatic actuated microgripper with a ratchet self-locking mechanism which enables the longtime gripping without continuously applying the external driving signal such as electrical, thermal or magnetic fields. This greatly reduces the influence and damage on the gripped micro objects induced by the external driving signals.

PowerMEMS Components and Systems

M-221 A HIGH-EFFICIENT BROADBAND ENERGY HARVESTER BASED ON NON-CONTACT COUPLING TECHNIQUE FOR AMBIENT VIBRATIONS 1110

X. Wu and D.W. Lee
Chonnam National University, SOUTH KOREA

In this work, a high-efficient piezoelectric energy harvester based on non-contact coupling technique is proposed and characterized, which allows it, for the first time, to take advantage of multi-cantilevers and frequency-up conversion technique to enhance the power generation efficiency for ambient excitation. The unique energy harvester can effectively scavenge environmental vibration energy with a wide bandwidth. Aiming for high space efficiency, folded cantilevers are designed.

T-222 BIDIRECTIONAL THERMOELECTRIC ENERGY GENERATOR BASED ON A PHASE-CHANGE LENS FOR CONCENTRATING SOLAR POWER 1114

M.S. Kim, M.K. Kim, H.R. Ahn, and Y.J. Kim
Yonsei University, SOUTH KOREA

This paper reports a bidirectional thermoelectric energy generator (TEG) with double type lenses for concentrating solar power. When solar power was applied to the TEG, solar energy is concentrated by PMMA lens firstly. The concentrated energy is absorbed as heat energy through phase-change of PCM. And then, the liquid PCM lens focuses energy on the TEG. After removing energy source, the latent heat in PCM is released. Therefore, the proposed TEG generates energy steadily.

W-223 BIOTEMPLATED HIERARCHICAL NICKEL OXIDE SUPERCAPACITOR ELECTRODES 1118

S. Chu, K. Gerasopoulos, and R. Ghodssi
University of Maryland, College Park, USA

We present hierarchical Ni/NiO supercapacitor electrodes utilizing Tobacco mosaic virus (TMV) as bio-nanotemplates. The hierarchical electrodes were fabricated by integrating high aspect ratio silicon micropillars with thermally oxidized nickel-coated TMVs. An ultra-high areal capacitance of 585.9mF/cm2 was achieved with hierarchical Ni/NiO electrodes, exceeding the capacitance of nanostructured only and planar Ni/NiO by a factor of 3.4 and 29.7, respectively.

M-224 DOUBLY RE-ENTRANT CAVITIES TO SUSTAIN BOILING NUCLEATION IN FC-72 1122

T. Liu and C.-J. Kim

University of California, Los Angeles, USA

We report a micro/nano-machined surface cavity on which boiling nucleation resumes after ceasing in refrigerant FC-72 for a short time. Having the lowest surface tension of all liquids, FC-72 completely wets any existing material including Teflon so that all existing cavities get flooded once nucleation stops and could not restart boiling without excessive heat. We experimentally confirm the half-century old idea of doubly re-entrant cavities as a boiling site, encouraging further development.

T-225 EXPERIMENTALLY VERIFIED MODEL OF ELECTROSTATIC ENERGY HARVESTER WITH INTERNAL IMPACTS 1125

B.D. Truong, C.P. Le, and E. Halvorsen

Buskerud and Vestfold University College, NORWAY

We present experimentally verified progress on modeling of MEMS electrostatic energy harvesters with internal impacts on transducing end-stops. The model includes nonlinearities of the electromechanical transduction, the squeezed-film damping and the impact force. The comparison between simulation and measurement shows that these effects are crucial and gives good agreement for phenomenological parameters. This is a significant step towards accurate modeling of this complex system.

W-226 MONOLITHIC 2-AXIS IN-PLANE PZT LATERAL BIMORPH ENERGY HARVESTER WITH DIFFERENTIAL OUTPUT 1129

S. Nadig, S. Ardanuç, and A. Lal

Cornell University, USA

We report a novel 2-axis (X-Y) piezoelectric energy harvester, whose sensitive axis in-plane is rotationally invariant, a result achieved by spiral cascading of lateral bimorphs. This is different than conventional piezoelectric energy harvesters that are sensitive only along one axis or can realize multi-axis sensitivity through integration or assembly of multiple devices at different orientations.

M-227 SOLID-STATE FLEXIBLE MICRO SUPERCAPACITORS BY DIRECT-WRITE POROUS NANOFIBERS 1133

C. Shen[1], G. Luo[1], A. Kozinda[1], M. Sanghadasa[2], and L. Lin[1]

[1]University of California, Berkeley, USA and [2]US Army, USA

We report solid-state flexible micro supercapacitors based on direct-write porous polymer nanofibers. Compared with state-of-art supercapacitors, key innovations include: 1) porous 3D nanostructure of conductive nanofibers via the near-field electrospinning process; 2) flexible solid-state micro electrodes with high energy density using the pseudocapacitive effect; (3) simple and versatile process compatible with different substrates and surfaces.

T-228 STRETCHABLE WIRELESS POWER TRANSFER WITH A LIQUID ALLOY COIL 1137

S.H. Jeong[1] and Z.G. Wu[1,2]

[1]Uppsala University, SWEDEN and [2]Huazong University of Science and Technology, CHINA

A stretchable wireless power transfer (WPT) device was fabricated with a liquid alloy coil, which was integrated with rigid electronic chips in elastomer packaging. Tape transfer masking with spray deposition was applied for patterning a long coil of the liquid alloy. The WPT efficiency reached 10% at 140 kHz and worked with 25% strain. Different sizes of liquid alloy coils and soft magnetic composite cores were tested for a higher efficiency system.

W-229 THREE-AXIS PIEZOELECTRIC VIBRATION ENERGY HARVESTER 1141

E.E. Aktakka and K. Najafi

University of Michigan, USA

This paper reports for the first time a piezoelectric harvester for scavenging vibrational energy in all three-dimensions. The device is formed of optimized PZT/Si unimorph crab-legs such that the first three resonance modes are linear in-plane and out-of-plane vibrational modes with closely spaced frequencies. Partitioned electrodes collect vibrational energy in the transverse piezoelectric mode, and have different phases in their outputs according to the axis of the applied vibration.

M-230 VERTICAL CAPACITIVE ENERGY HARVESTER POSITIVELY USING CONTACT BETWEEN PROOF MASS AND ELECTRET PLATE - STIFFNESS MATCHING BY SPRING SUPPORT OF PLATE AND STICTION PREVENTION BY STOPPER MECHANISM 1145

T. Takahashi[1], M. Suzuki[1], T. Nishida[2], Y. Yoshikawa[2], and S. Aoyagi[1]

[1]Kansai University, JAPAN and [2]ROHM Co. Ltd., JAPAN

In a vertical capacitive energy harvester, two methods to effectively use the contact between proof mass and electret plate are proposed; one is to match the stiffness between plate and mass, which is effective to increase their contact duration. For this purpose, instead of a gel shock-absorber, a soft spring for supporting plate is employed. Another is to surely detach mass from plate after their contact, opposing electrostatic attraction. The output power was improved by 5 times up to 50 μW.

The 28th International Conference
on Micro Electro Mechanical Systems

MEMS PORTUGAL 2015
18-22 JANUARY 2015   ESTORIL

CONFERENCE CHAIRS
Jürgen Brugger
EPFL Lausanne, SWITZERLAND

Wouter van der Wijngaart
KTH Royal Institute of Technology, SWEDEN

Sponsored by: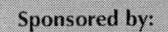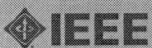

If you have questions regarding the installation, please contact:

The Printing House, Inc.
Phone: +1-608-873-4500   Fax: +1-608-873-4558
Hours: Monday through Friday, 8 am - 5 pm CST
E-mail: graphics@printinghouseinc.com

LETTER FROM THE CHAIRS

Welcome to the 27th IEEE International Conference on Micro Electro Mechanical Systems (MEMS 2015) in Estoril, Portugal!

The IEEE MEMS Conference series originated in 1987, and has been known as the IEEE International Conference on Micro Electro Mechanical Systems since 1999. Over the last decade, the MEMS community has experienced immense growth in science and technology of miniaturization, as well as commercialization, with a current global MEMS sales revenue of 10 Billion Euro.

This Conference was designed to continue building the international MEMS community by bringing together top academic researchers, students and key industrial players. We hope you will appreciate the cocktail of high profile plenary speakers and stringently selected scientific presentations in the glamorous, secluded, but yet easily reachable, location of Estoril, Portugal.

The two industrial invited plenary speakers represent the world's largest MEMS players in terms of sales volume and independent semiconductor manufacturing services in assembly and testing. Together with two academic plenary speakers they will provide insight into some of today's major trends in our field of MEMS research and development.

A total of 295 papers out of 719 submitted abstracts were carefully selected by the 47 experts of the Technical Program Committee (TPC). This year, for the first time, we implemented a double blind review process, to ensure scientific quality being the sole selection criterion. The presentations are arranged in a mixed single/parallel session format with 4 invited plenaries, 65 oral and 230 poster presentations. Twenty abstracts were nominated for the outstanding paper award program, based on their quality. The oral and poster nominees are marked in the programme with an *. The awards are to recognize excellence amongst the work presented by students. The outstanding paper awardees will be announced just prior to adjourning the Conference, late Thursday morning.

We would like to express our sincerest gratitude to all the authors who submitted their abstracts. Their high quality work serves as the foundation for the success of this Conference. The papers were selected by the TPC, made up of academic and industrial members, with equal representation from three regional divisions: Americas, Europe & Africa, and Asia & Oceania. Six sub-committees were formed in order to facilitate a careful review. Each abstract has been evaluated and rated by eight expert members of the TPC. We are grateful to all TPC members who volunteered their valuable time, including participation in a two-day on-site meeting in Berlin, Germany, for paper selection. We gratefully acknowledge the industrial support groups, exhibitors and benefactors for their involvement in this Conference, and the IEEE Robotics and Automation Society for their continued support of this meeting. The dedicated and relentless effort of Ms. Katharine Cline and her team at PMMI in managing this Conference is highly appreciated.

In closing, we hope you enjoy the mingling, technical presentations, exhibition booths and events of the Conference this week in Estoril!

Bem Vindos!

Jürgen Brugger
EPFL Lausanne, SWITZERLAND

Wouter van der Wijngaart
KTH Royal Institute of Technology, SWEDEN

ORGANIZING COMMITTEE

General Chairs

Jürgen Brugger, Ph.D.
EPFL, SWITZERLAND

Wouter van der Wijngaart, Ph.D.
KTH Royal Institute of Technology,
SWEDEN

STEERING COMMITTEE

Chairs

Farrokh Ayazi, Ph.D.
Georgia Institute of Technology, USA

Chang-Jin "CJ" Kim, Ph.D.
University of California, Los Angeles

Members

Farrokh Ayazi, Ph.D.
Georgia Institute of Technology, USA

Jürgen Brugger, Ph.D.
EPFL, SWITZERLAND

Lionel Buchaillot, Ph.D.
IEMN, FRANCE

Chang-Jin "CJ" Kim, Ph.D.
University of California, Los Angeles

Gwo-Bin "Vincent" Lee, Ph.D.
National Cheng Kung University, TAIWAN

Ellis Meng, Ph.D.
University of Southern California, USA

Clark Nguyen, Ph.D.
University of California, Berkeley, USA

Hiroshi Toshiyoshi, Ph.D.
University of Tokyo, JAPAN

Toshiyuki Tsuchiya, Ph.D.
Kyoto University, JAPAN

Xiaohong Wang, Ph.D.
Tsinghua University, CHINA

Wouter van der Wijngaart, Ph.D.
KTH Royal Institute of Technology,
SWEDEN

Hans Zappe, Ph.D.
University of Freiburg – IMTEK, GERMANY

TECHNICAL PROGRAM COMMITTEE

Reza Abdolvand, Ph.D.
University of Central Florida, USA

Farrokh Ayazi, Ph.D.
Georgia Institute of Technology, USA

Jürgen Brugger, Ph.D.
EPFL, SWITZERLAND

Junseok Chae, Ph.D.
Arizona State University, USA

Il-Joo Cho, Ph.D.
Korea Institute of Science and Technology,
SOUTH KOREA

Joao Pedro Conde, Ph.D.
Instituto Superior Tecnico, PORTUGAL

Andreas Dietzel, Ph.D.
Technische Universität Braunschweig,
GERMANY

Shih-Kang Fan, Ph.D.
National Taiwan University, TAIWAN

Philip Feng, Ph.D.
Case Western Reserve University, USA

Andreas Hierlemann, Ph.D.
ETH Zürich, SWITZERLAND

Eiji Iwase, Ph.D.
Waseda University, JAPAN

Jack Judy, Ph.D.
University of Florida, USA

Takeshi Kawano, Ph.D.
Toyohashi University, JAPAN

Chang-Jin "CJ" Kim, Ph.D.
University of California, Los Angeles, USA

Hanseup Kim, Ph.D.
University of Utah, USA

Yong-Kweon Kim, Ph.D.
Seoul National University, SOUTH KOREA

Manfred Kohl, Ph.D.
Karlsruhe Institute of Technology,
GERMANY

Haluk Kulah, Ph.D.
Middle East Technical University, TURKEY

Walter Lang, Ph.D.
University of Bremen, GERMANY

Chengkuo "Vincent" Lee, Ph.D.
National University of Singapore,
SINGAPORE

Bernard Legrand, Ph.D.
LAAS-CNRS, FRANCE

Wen Li, Ph.D.
Michigan State University, USA

Ellis Meng, Ph.D.
University of Southern California, USA

Norihisa Miki, Ph.D.
Keio University, JAPAN

Hyejin Moon, Ph.D.
University of Texas, Arlington, USA

Clark Nguyen, Ph.D.
University of California, Berkeley, USA

Takahito Ono, Ph.D.
Tohoku University, JAPAN

Jaeyeong Park, Ph.D.
Kwangwoon University, SOUTH KOREA

Siavash Pourkamali, Ph.D.
University of Texas, Dallas, USA

Mina Rais-Zadeh, Ph.D.
University of Michigan, USA

Niclas Roxhed, Ph.D.
KTH Royal Institute of Technology,
SWEDEN

Edin Sarajilic, Ph.D.
SmartTip B.V., THE NETHERLANDS

Minoru Sasaki, Ph.D.
Toyota Technological Institute, JAPAN

Tomonori Seki, Ph.D.
OMRON Corporation, JAPAN

Ashwin Seshia, Ph.D.
University of Cambridge, UK

Herbert Shea, Ph.D.
EPFL, SWITZERLAND

Logan Sorensen, Ph.D.
HRL Laboratories, USA

Boris Stoeber, Ph.D.
University of British Columbia, CANADA

Niels Tas, Ph.D.
University of Twente, THE NETHERLANDS

Hiroshi Toshiyoshi, Ph.D.
University of Tokyo, JAPAN

Fan-Gang Tseng, Ph.D.
National Tsing Hua University, TAIWAN

Evelyn Wang, Ph.D.
Massachusetts Institute of Technology, USA

Wei Wang, Ph.D.
Peking University, CHINA

Xiaohong Wang, Ph.D.
Tsinghua University, CHINA

Wouter van der Wijngaart, Ph.D.
KTH Royal Institute of Technology,
SWEDEN

Peter Woias, Ph.D.
University of Freiburg - IMTEK, GERMANY

Ryuji Yokokawa, Ph.D.
Kyoto University, JAPAN

TECHNICAL PROGRAM COMMITTEE

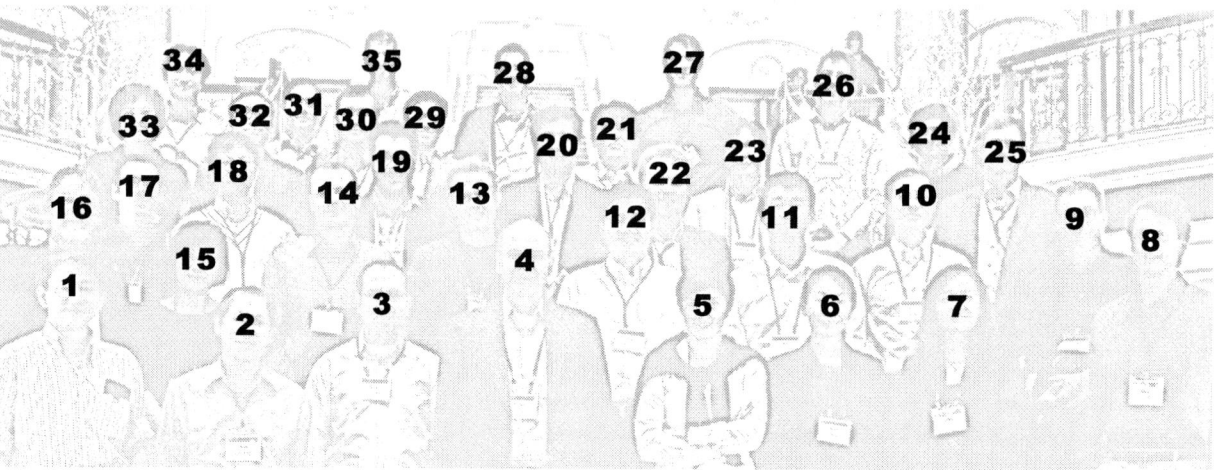

1	Hiroshi Toshiyoshi	22	Niels Tas
2	Clark Nguyen	23	Il-Joo Cho
3	Haluk Kulah	24	Ashwin Seshia
4	Peter Woias	25	Manfred Kohl
5	Mina Rais-Zadeh	26	Wouter van der Wijngaart
6	Wen Li	27	Edin Sarajilic
7	Xiaohong Wang	28	Farrokh Ayazi
8	Jack Judy	29	Junseok Chae
9	Joao Pedro Conde	30	Fan-Gang Tseng
10	Ryuji Yokokawa	31	Hanseup Kim
11	Wei Wang	32	Shih-Kang Fan
12	Chang-Jin "CJ" Kim	33	Ellis Meng
13	Andreas Dietzel	34	Norihisa Miki
14	Bernard Legrand	35	Andreas Hierlemann
15	Hyejin Moon		
16	Jürgen Brugger		
17	Boris Stoeber		
18	Eiji Iwase		
19	Siavash Pourkamali		
20	Takahito Ono		
21	Minoru Sasaki		

Not pictured: Reza Abdolvand, Philip Feng, Takeshi Kawano, Yong-Kweon Kim, Walter Lang, Chengkuo "Vincent" Lee, Jaeyeong Park, Niclas Roxhed, Tomonori Seki, Herbert Shea, Logan Sorensen, Evelyn Wang

ACKNOWLEDGEMENTS

We gratefully acknowledge the support and involvement of this Conference from the following companies and institutions as of the printing of 16 December, 2014:

Berkeley Sensors & Actuators Center (BSAC)

Coventor Sarl

DJ DevCorp

E&M

FemtoTools AG

Heidelberg Instruments Mikrotechnik GmbH

Institution of Engineering and Technology (IET)

ITmems s.r.l.

iX-factory GmbH

Journal of Micromechanics and Microengineering

MEMS and Nanotechnology Exchange

MEMS Industry Group

MEMS Journal

memsstar Limited

MicroChem Corp.

Micronarc

Muegge GmbH

Nanoscribe GmbH

OAI

Oxford Instruments

Plasma-Therm, LLC

POLYTEC GmbH

Silex Microsystems

SPTS Technologies

Tousimis

ULVAC

THE LONG PATH FROM MEMS RESONATORS TO TIMING PRODUCTS

E. Ng[1], Y. Yang[1], V.A. Hong[1], C.H. Ahn[1], D.B. Heinz[1], I. Flader[1], Y. Chen[1], C.L.M. Everhart[1], B. Kim[2], R. Melamud[2], R.N. Candler[2], M.A. Hopcroft[2], J.C. Salvia[2], S. Yoneoka[2], A.B. Graham[2], M.Agarwal[2], M.W. Messana[2], K.L. Chen[2], H.K. Lee[2], S. Wang[2], G. Bahl[2], V. Qu[2], C.F. Chiang[2], and T.W. Kenny[1] A. Partridge[3], M. Lutz[3], G. Yama[4] and G.J. O'Brien[4]

[1]Department of Mech.Eng., Stanford, [2]PhD Alumni of Stanford, [3]SiTime Inc, Mt. View, CA
[4]Robert Bosch Research and Technology Center, Palo Alto, CA

ABSTRACT

Research on MEMS Resonators began over 50 years ago. In just the last 10 years, there has been a series of important technological developments, and (finally!) success at commercialization. The presentation will highlight some key milestones along this path, describe some of the critical technology steps, and outline some of the important non-technological events within SiTime – all of these factors contributed to the successful outcome.

RESONATOR HISTORY

The history of MEMS and MEMS Resonators begins with the Resonant Gate Transistor [1], developed by Nathanson over 50 years ago. This is arguably the first device with all of the essential elements of MEMS. Nathanson described it as a time reference; and many challenging issues associated with fatigue, temperature coefficients, and reliability were already noted.

MEMS resonators were discussed as an important application of Silicon MEMS [2] by Petersen more than 30 years ago. In his paper, the challenge for success in MEMS Resonators is identified as (1) "...offering functions that cannot easily be duplicated by conventional..." approaches, (2) "...solve the inherent problems of mechanical reliability and reproducibility...", and (3) "...are fabricated by techniques totally compatible with standard IC processing...".

The first big step towards a manufacturable resonator technology was demonstrated 25 years ago by Tang and Howe [3], in the form of the surface micromachined comb drive resonator. While there were many challenges remaining to produce stable time references, the Tang and Howe comb drive became the basis for accelerometers and gyroscopes developed through the 1990s and is represented in all MEMS for consumer electronics today. Shortly thereafter, DeRooij demonstrated integration of capacitive, thermal, and piezoresistive resonators [4] and Cabuz and Esashi demonstrated resonant IR Sensors [5]. The field of Resonant MEMS was established!

To solve problems with stability, Esashi began focusing on packaging of MEMS resonators including getters [6], along with Guckel, Lee, and Chun [7-9]. In all of this work, researchers were focused on low pressure, hermeticity, and stability. In parallel with these Si-based activities, Ruby demonstrated the first FBARs [10]; this technology has led to the manufacture and sale of billions of MEMS resonators for RF filter applications.

Even with all this effort, success at meeting the goals outlined by Petersen was elusive. Packaged MEMS resonators were not stable enough for timing applications, and quartz technology was more than sufficient to satisfy market needs. To meet some particularly challenging goals for the US Department of Defense, Nguyen established a DARPA program on MEMS packaging [11], which focused on long-term stability and miniaturization.

THE *EPI-SEAL* PROCESS AND SITIME

In 1999, Robert Bosch opened its Palo Alto Research and Technology Center, adjacent to Stanford. Markus Lutz was among the first technologists at RTC, and came to Stanford seeking collaborators to help demonstrate a wafer-level encapsulation for inertial sensors using standard IC processes; Aaron Partridge was the first of many PhD students to get involved. The first trials utilized the Bosch epi-poly [12] for an initial cap layer and CVD oxide for sealing, and were presented at MEMS '01 [13]. These initial devices built showed insufficiently low pressure and poor long-term stability. To solve this problem, Partridge used epi-poly to form the seal, and it was immediately clear that a significant improvement had been achieved [14].

The main advantages of the epi-poly encapsulation process for MEMS resonators are : (1) Low pressure (near 1 millitorr), (2) Clean residual gas (H2 only), (3) No outgassing (from 1100C sealing step) and (4) Smooth sidewall surfaces (H2 annealing [15]).

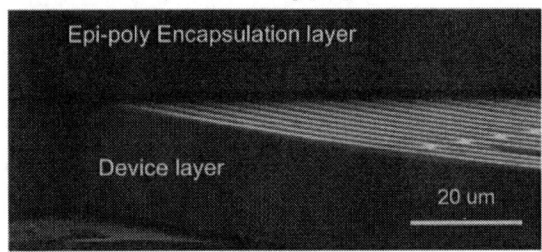

Figure 1 SEM Image of encapsulated MEMS resonator, sealed beneath epi-poly layer in encapsulation process.

With this process, we demonstrated ppm-level stability for extended periods [16], and began to consider formation of a company to commercialize the technology. In 2004, Kurt Petersen joined the team, and by late 2004, there was a technology license from Bosch, $5M in initial funding, and SiTime was "born" (www.sitime.com).

Around this time, there were other resonator startups, such as Discera (acquired by Micrel in 2013), Silicon Clocks (acquired by SiLabs in 2010), Harmonic Devices (acquired by Qualcomm in 2010), VTI (acquired by Murata in 2012), and Sand9 (independent). Of these, only SiTime has become a large-volume manufacturer of MEMS timing products, responsible for ~90% of MEMS timing products sold to date.

RECENT TECHNOLOGY EVENTS

From the beginning, it was known that the temperature dependence of the modulus of Silicon was too large, at -30 ppm/C, to meet precision timing applications. Early products from SiTime, and other companies, used calibration and digital compensation, which increased the power consumption and phase noise of the products, and made it very hard to compete with the established timing technologies. The most significant recent event in the development of MEMS-based timing was the demonstration of remedies for this problem, based on the use of materials with positive temperature coefficients of modulus (e.g., SiO_2 [16]), or near-degenerate doping [17]. Through these approaches, SiTime built temperature insensitive timing products, and used this breakthrough to greatly reduce power, noise and improve performance. A recent example product delivers a stable clock output with total power consumption less than 1 microWatt [SiTime SiT1532]. All of this led to the recent acquisition of SiTime by Megachips.

In parallel with these activities, Stanford has been active in demonstrating extensions to the basic timing technology. This includes heating the resonator within the encapsulation [18], complete models for the energy dissipation of resonators within this process [19], studies of fatigue [20] and adhesion [21], and suites of accelerometers, pressure sensors, gyroscopes, and other sensors [22] – all within a process similar to the SiTime *episeal* manufacturing process.

REFERENCES

[1] H.C. Nathanson and R.A. Wickstrom , "A Resonant Gate Silicon Surface Transistor with High-Q Bandpass Properties", *Appl. Phys. Lett*, 7, 84 (1965).

[2] K. Petersen, "Silicon as a Mechanical Material", *Proceedings of the IEEE* 70, 420 (1982).

[3] W.C. Tang, T.C. Nguyen, M.W. Judy and R.T Howe, "Electrostatic Comb Drive of Lateral Polysilicon Resonators", S&A A21-23, 328 (1990).

[4] C. Linder, M.Gretillat and N.F. de Rooij, "Realization of Different Polysilicon Resonators with Integrated Excitation and Detection Elements", Microelectronic Engineering 15, 411 (1991).

[5] C. Cabuz, S. Shoji, K. Fukatsu, E. Cabuz, K. Minami, and M. Esashi, "Fabrication and Packaging of a Resonant Infrared Sensor Integrated in Silicon", S&A A43, 92 (1994).

[6] H. Henmi, S. Shoji, Y. Shoji, K. Yoshimi and M. Esashi, "Vacuum Packaging for Microsensors by Glass-Silicon Anodic Bonding", S&A A43, 243 (1994).

[7] J.D. Zook, D.W. Burns, H. Guckel, J.J. Sniegowski, R.L. Engelstad and Z. Feng, "Characteristics of Polysilicon Resonant Microbeams", S&A A35 (1992).

[8] L. Lin, K. M. McNair, R. T. Howe, and **A**. P. Pisano, "Vacuum-Encapsulated Lateral Microresonators," Proceedings of Transducers 1993, pp. 270 (1993)

[9] B. Lee, S. Seok, and K Chun, "A Study on Wafer Level Vacuum Packaging for MEMS Devices", J.

Micromech. Microeng 13, 663 (2003).

[10] R. Ruby and P. Merchant, "Micromachined Thin Film Bulk Acoustic Resonators", IEEE IFC, 135 (1994).

[11] C.T.C. Nguyen, "The Harsh Environment Robust Micromechanical Technology (HERMiT) Program : Success and Some Unfinished Business", 2012 Microwave Symposium Digest, P. 1 (2012), and DARPA BAA 03-11 (2003).

[12] K. Funk, A. Schlip, M. Offenberg, B. Eisner, and F. Laermer, "Surface Micromachining of Resonant Silicon Structures," IEEE Transducers 1995.

[13] A. Partridge, A. Rice, T.W. Kenny and M. Lutz, "New Thin Film Epitaxial Polysilicon Encapsulation for Piezoresistive Accelerometers", IEEE MEMS, P. 54 (2001).

[14] R.N. Candler, M.A. Hopcroft, B. Kim, W.T. Park, R. Melamud, M. Agarwal, G. Yama, M. Lutz, and T.W. Kenny, "Long-Term and Accelerated Life-Testing of a Novel Single-Wafer Vacuum Encapsulation for MEMS Resonators", JMEMS 15, 1446 (2006).

[15] T. Sato, et.al., "Micro-structure Transformation of Silicon: A Newly Developed Transformation Technology for Patterning Silicon Surfaces using the Surface Migration of Silicon Atoms by Hydrogen Annealing", Jpn. J. Appl. Phys. 39, 5033. (2000).

[16] B.S. Berry and W.C. Pritchett, "Temperature Compensation for Constant Frequency Electromechanical Oscillators", IBM Tech. Discl. Bull. 14, (1971), and R. Melamud, et.al., "Temperature Compensated High-Stability Silicon Resonators", Appl. Phys. Lett 90, 244107 (2007).

[17] J.J. Hall, "Electronic Effects in Elastic Constants of N-Type Silicon", Phys. Rev 161, 756 (1967), and E.J. Ng, V.A. Hong, Y. Yang, C.H. Ahn, C.L.M. Everhart and T.W. Kenny, "Temperature Dependence of the Elastic Constants of Doped Silicon", JMEMS (2015).

[18] J.C Salvia, R. Melamud, S.A. Chandorkar, S.F. Lord and T.W. Kenny, "Real-Time Temperature Compensation of MEMS Oscillators Using an Integrated Micro-Oven and a Phase-Locked Loop", JMEMS 19, 192 (2010).

[19] R.Candler, A.Duwel, M.Varghese, S.Chandorkar, M.Hopcroft, W.Park, B. Kim, G. Yama, A. Partridge, M. Lutz and T.W. Kenny", Impact of Geometry on Thermoelastic Dissipation in Micromechanical Resonant Beams", JMEMS 15, 927 (2006).

[20] V. Hong, et.al., "Fatigue Experiments on Single Crystal Silicon in Oxygen-Free Environment", JMEMS (2015).

[21] D.B. Heinz, et.al., "Stiction Forces and Reduction by Dynamic Contact in Ultra-Clean Encapsulated MEMS Devices", IEEE MEMS (2015).

[22] C.F. Chiang, A.B. Graham, G.J. O'Brien, and T.W. Kenny, "A Single Process for Building Capacitive Pressure Sensors and Timing References with Precise Control of Released Area using lateral Etch Stop", IEEE MEMS, P. 519 (2012).

CONTACT

Thomas Kenny, kenny@cdr.stanford.edu

This page intentionally left blank.

GRAPHENE OXIDE MEMBRANES FOR PHASE-SELECTIVE MICROFLUIDIC FLOW CONTROL

Jennifer Gaughran[1], David Boyle[1,2], James Murphy[1] and Jens Ducrée[1,2*]*
[1]School of Physical Sciences, Dublin City University, Dublin, IRELAND
[2]Biomedical Diagnostics Institute, Dublin City University, Dublin, IRELAND

ABSTRACT

This paper investigates the unique properties of multilayer Graphene Oxide (GO) as a barrier selective to solvents, as well as the state of aggregation of fluids. To this end, we developed novel processes for the assembly of GO as membranes into polymeric microfluidic systems. We demonstrate that GO blocks pressurized air and organic solutions while it is permeable to water. It is planned to employ these GO membranes as a flow control element in a microfluidic system.

INTRODUCTION

Since its ground-breaking discovery in 2008 [1], a wide spectrum of fascinating characteristics of graphene and its compounds, such as the here considered graphene oxide (GO), have been extensively investigated by the scientific community. In recent years distinctive properties, such as its electrochemical responses [2], have also been measured by microfluidic systems. Ang *et al.* used a graphene transistor array in a microfluidic chip for the detection of malaria-infected red blood cells [3]. Lo *et al.* included glycidyl methacrylate functionalized graphene oxide within a hydrogel which showed a significant increase in size change when exposed to infrared radiation [4].

Flow control is of high importance for integrated Lab-on-a-Chip systems. Centrifugal microfluidics has emerged as a highly useful tool in the area of biomedical diagnostics as many of the necessary Laboratory Unit Operations (LUOs) in bioanalysis, such as plasma extraction, mixing and metering, can be fully automated and parallelised onto these "Lab-on-a-Disc" (LoaD) platforms. It also has the advantage over traditional microfluidic systems in that pumping is achieved without the use of modules on the peripheral instrument and also without fluidic / pneumatic interconnects. Instead, a simple spindle motor generates the force field to centrifugally propel liquids. Liquid handling steps such as plasma extraction from whole blood and LUOs can then be controlled by running a designated spin frequency protocol to modulate hydrodynamic interactions between on-disc liquids and microstructures [5]. The disc cartridge can be designed to store all required reagents and then safely encapsulate potentially biohazardous samples.

A particular challenge on integrated LoaD platforms is the omnipresence of the centrifugal field on the rotating cartridge. In order to coordinate sequential or parallel release of on-board stored liquid such as the sample and reagents, selective valving mechanisms have to be found. On the one hand, these can be hydrodynamic flow resistances or capillary barriers.

Noroozi *et al.* used pneumatic pumping as a method for mixing and accurate metering of reagents [6].

Grumann *et al.* reduced mixing times extensively by using a combination of rapid disc oscillation with the introduction of paramagnetic beads which are deflected by stationary magnets in the surface of the discs [7]. Siegrist *et al.* took steps towards a more integrated system by implementing a robust serial siphoning method of fluid handling [8]. However, these valving schemes can only act on the liquid phase, but not its vapour, thus making them unsuitable for long-term storage of liquid reagents which would be a common requirement of fully automated commercial point-of-use systems.

On the other hand, sacrificial materials have been introduced which must be removed on demand by physical or chemical stimuli. Incorporation of materials for functional enhancement of microfluidic systems has been of great interest for some time. Wax valves, for instance, can be actuated by exposure to a heat source [9]. Another example is hydrogels, which can act as valves actuated by size change in the presence of water [10]. Gorkin *et al.* used a dissolvable film material in conjunction with a pneumatic structure to develop a novel valving system on a centrifugal microfluidic platform [11]. Dimov *et al.* combined a hydrophobic membrane and dissolvable film to improve solid-phase purification of nucleic acids [12]. Kinahan *et al.* developed an event-triggered valving system using dissolvable films which allows for more dynamic control of fluids on a centrifugal microfluidic platform [13].

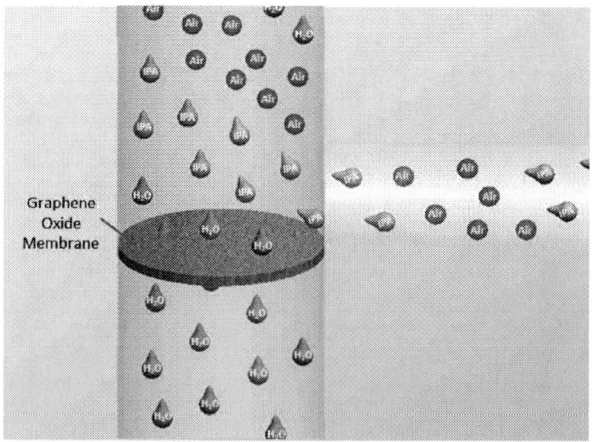

Figure 1: Illustration of the unique properties of GO. The GO membrane is entirely permeable to water, but it completely blocks air and organic solutions (IPA and EtOH).

A multitude of methods for graphene membrane fabrication have been developed with varying time scales and levels of repeatability. The first method used by Novoselov *et al.* was the scotch tape method, whereby layer upon layer of graphite was mechanically exfoliated

978-1-4799-7956-1/15 $31.00 © 2015 IEEE

away until a single layer (one atom thick) of graphene was achieved [1]. Other methods include Chemical Vapour Deposition (CVD). This method allows for very pure graphene substrates to be made but it is costly and lengthy.

However, while being explored by a broad scientific community, the wide spectrum of the often unique properties of graphene has so far not been implemented for microfluidic flow control; this is presumably due to the fact that it does not bind to polymer surfaces. To leverage our experiments, we first developed a scheme for the integration of GO membranes in common, polymeric microfluidic devices.

Once inclusion of graphene oxide in a microfluidic device can been achieved, however, its unique properties could enable a multitude of applications and advanced fluid flow control within a system. We investigated two such properties of the GO membranes; their solvent selectivity and air impermeability (Fig. 1).

GO TAB FABRICATION

GO membranes were synthesised from a suspension of GO flakes in water (Bluestone, Manchester, UK). The flake suspension was compressed into a membrane by vacuum filtration through a 0.45-μm pore size cellulose filter (Millipore, Cork, Ireland). The filtration process forces the flakes to distribute across the whole membrane surface as the flakes fill the pores of the cellulose forming a multilayer, free-standing stacked membrane. This created a 35-mm diameter membrane as defined by the size of the vacuum filtration process. The thickness of the membranes is governed by the volume and concentration of the GO flakes in suspension. The process ranges from four to eight hours, depending on the volume being filtered (Fig. 2a).

Once filtration was completed, the membranes were removed from the vacuum filtration set-up and left to dry for approximately four hours. During this time the membranes remained in contact with the cellulose filters. The completion of this drying step was crucial as prior handling of the membranes could lead to tearing. The membrane could then be pealed from the cellulose filters and be handled with ease (Fig. 2b).

In order to incorporate the GO into a microfluidic device a method for attaching the membrane to the polymer surface had to be devised. Using a method first shown by Gorkin et al., the membrane was adhered to double-sided Pressure Sensitive Adhesive (PSA), which featured a small, 1-mm diameter through hole. The contours and size of the GO and PSA tab was then flexibly defined using a precision knife cutter (Graphtec, Wrexham, UK). Next, the GO tab was integrated in a microfluidic system using the 'sticky' PSA backing (Fig. 2c, d).

Figure 2e shows an SEM image of a cross section of a GO membrane clearly displaying the multilayer stacked structure. This membrane possesses a thickness of approximately 10-μm. From this, the even distribution of the flakes can be observed. Also shown here is the relative clean cut of the tab edge, achieved using the precision knife cutter.

Figure 2: GO tab assembly and characterisation. a) Image of GO membrane after vacuum filtration through a cellulose filter. b) Free standing membrane after drying and a GO tab after fabrication. c, d) Assembly of GO tab. GO membrane is adhered to double-sided PSA with 1-mm through-hole cut out. e) SEM image of GO membrane showing multilayer stacked structure. f) Microscope image of GO tab showing uniform distribution of flakes and the acceptably clean cut from the knife cutter.

INVESTIGATION OF GO PROPERTIES

Two distinct properties of the GO were investigated; its solvent selectivity and air impermeability. These properties were tested over a given pressure range on two centrifugal microfluidic disc designs.

Solvent Selectivity

Figure 3 describes the disc used to test the solvent selectivity of the GO. The disc was made up of four layers of PMMA (green) and four layers of PSA (grey). It contained ten identical structures. Each structure consisted of a 3D architecture containing a Loading Chamber with a 1-mm hole residing at the base of the chamber. This vertical via connected to a channel which resided on the bottom layer of the disc. The hole was sealed by one of the GO tabs. The channel then connected to a Collection Chamber (Fig. 3b).

In order to test the solvent selective properties of the GO, three liquids were added to the chambers; deionized (DI) water, 2-propanol (IPA) and ethanol (EtOH). A 50-μL volume was added to the Loading Chamber and the disc was spun. Under centrifugation the liquid generated a hydrostatic pressure head acting on the GO tab. The rotational frequency was increased at regular intervals

978-1-4799-7956-1/15 $31.00 © 2015 IEEE

until the membrane yielded. In the case of DI water the plug passed through the GO at approximately 50 Hz (Fig. 3c). Notably, once the 'burst' pressure was exceeded the water flowed unhindered, with very low flow resistance. In contrast, the organic solutions IPA and EtOH were fully retained in the Loading Chamber, even at the highest spin frequency of 125 Hz that we could implement on our test stand.

This simple test demonstrates the complete impermeability of the GO membrane to organic solutions while offering very little flow resistance to water above a certain threshold frequency. Qualitative experiments further indicate that varying the thickness of the membranes can affect the burst pressure required for the passage of water.

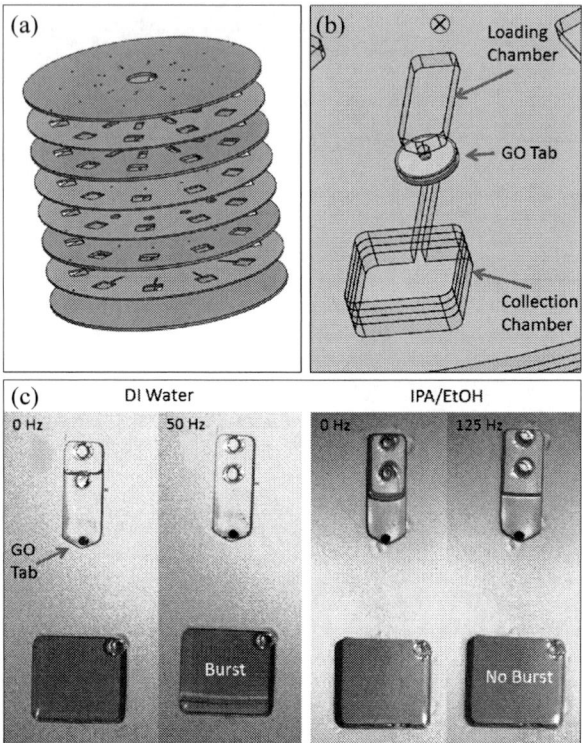

Figure 3: Design for testing the solvent selectivity of the GO. a) Multi-layer disc structure, with PMMA (green) and PSA (grey) layers. b) 3-dimensional architecture connecting loading chamber to collection chamber via a through-hole, sealed by a GO tab. c) Solvent selectivity of GO. DI water passes through at 50 Hz while the membrane is impermeable for the organic solvents IPA and EtOH, even at the much higher spin frequency of 125 Hz.

Air Impermeability

Figure 4 shows the design used to test the impermeability of the GO tabs to air. This disc consisted of three layers of PMMA and two layers of PSA. Each disc contained four identical structures; each of them consisted of a loading chamber and two side arms, which were connected via a single inlet channel. Both side arms exhibited a hole at the top which had been sealed by either a PSA or GO tab. For clarity in these tests, coloured food-dye in water was used, as the fluid would not make direct contact with the tabs.

A 180-μL volume was added to the loading chamber. This liquid trapped and enclosed the air inside the inlet channel and side arms. With the increase of the rotationally induced centrifugal field, liquid flows down the inlet channel to compress the air in the side arms. The liquid reached the interface between the channel and side arm, at a rotational frequency of 10 Hz. As the disc was spun faster, up to a rotational frequency of 100 Hz, the liquid entered the side arms and continued to rise (Fig. 4a). Figure 4b shows that there was no discernible difference of the liquid levels in both side arms at the upper spin frequency, thereby proving that the GO tab is air tight. If the GO tab had in fact been air permeable then the liquid level in the right side arm (the one containing the GO tab) would have been much higher.

Another interesting feature is the extremely high pressures that the tabs can withstand. When the liquid is initially loaded, sealing the air inside the channel and arms, the air inside is still under atmospheric pressure. As the spin frequency was increased to 10 Hz, the air has been compressed enough to generate an increase in pressure within the chamber, and thus begin exerted on the tabs, of approximately 25 mbar. Finally, at 100 Hz, there has been a significant increase in pressure of approximately 750 mbar. This demonstrates the high mechanical strength of the integrated GO membranes.

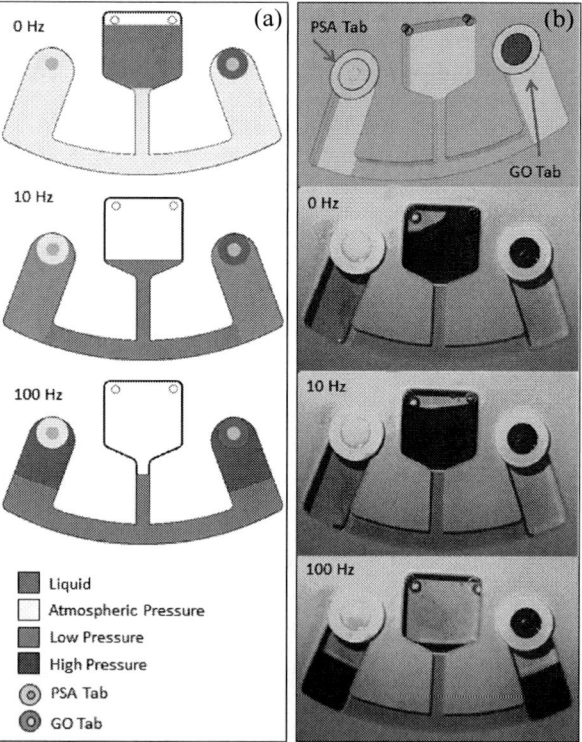

Figure 4: Design for measuring GO impermeability to air. The left and right arms are sealed by PSA and GO tabs, respectively. a) Operation under various air pressures. b) Image sequence showing that even at high centrifugally induced pressure-heads, hydrostatic equilibrium is maintained to demonstrate the impermeability of the GO membrane to air (as well as the integrity of the membrane and its seal with the disc substrate).

CONCLUSIONS

We have shown a new method for the integration of a GO tab within a polymeric microfluidic structure. Using centrifugal flow control, we have also investigated various unique features of GO, notably its solvent selectivity and air impermeability. We envision the engineering of a solvent-selective burst valve based on this surprising, selective permeability of GO membranes.

ACKNOWLEDGEMENTS

This work is funded under the Programme for Research in Third Level Institutions (PRTLI) Cycle 5. The PRTLI is co-funded through the European Regional Development Fund (ERDF), part of the European Union Structural Funds Programme 2007-2013.

REFERENCES

[1] K. S. Novoselov, A. K. Geim, S. V. Morozov, D. Jiang, Y. Zhang, S. V Dubonos, I. V Grigorieva, and A. A. Firsov, "Electric field effect in atomically thin carbon films," *Science*, vol. 306, no. 5696, pp. 666–669, Oct. 2004.

[2] M. Pumera, A. Ambrosi, A. Bonanni, E. L. K. Chng, and H. L. Poh, "Graphene for electrochemical sensing and biosensing," *TrAC Trends Anal. Chem.*, vol. 29, no. 9, pp. 954–965, Oct. 2010.

[3] P. K. Ang, A. Li, M. Jaiswal, Y. Wang, H. W. Hou, J. T. L. Thong, C. T. Lim, and K. P. Loh, "Flow sensing of single cell by graphene transistor in a microfluidic channel," *Nano Lett.*, vol. 11, no. 12, pp. 5240–5246, Dec. 2011.

[4] C.-W. Lo, D. Zhu, and H. Jiang, "An infrared-light responsive graphene-oxide incorporated poly(N-isopropylacrylamide) hydrogel nanocomposite," *Soft Matter*, vol. 7, no. 12, pp. 5604-5609, Jan 2011.

[5] R. Gorkin, J. Park, J. Siegrist, M. Amasia, B. S. Lee, J.-M. Park, J. Kim, H. Kim, M. Madou, and Y.-K. Cho, "Centrifugal microfluidics for biomedical applications," *Lab Chip*, vol. 10, no. 14, pp. 1758–1773, Jul. 2010.

[6] Z. Noroozi, H. Kido, M. Micic, H. Pan, C. Bartolome, M. Princevac, J. Zoval, and M. Madou, "Reciprocating flow-based centrifugal microfluidics mixer," *Rev. Sci. Instrum.*, vol. 80, no. 7, p. 075102, Jul. 2009.

[7] M. Grumann, A. Geipel, L. Riegger, R. Zengerle, and J. Ducrée, "Batch-mode mixing on centrifugal microfluidic platforms," *Lab Chip*, vol. 5, no. 5, pp. 560–565, May 2005.

[8] J. Siegrist, R. Gorkin, L. Clime, E. Roy, R. Peytavi, H. Kido, M. Bergeron, T. Veres, and M. Madou, "Serial siphon valving for centrifugal microfluidic platforms," *Microfluid. Nanofluidics*, vol. 9, no. 1, pp. 55–63, Nov. 2009.

[9] J. Park, Y. Cho, B. Lee, J. Lee, and C. Ko, "Multifunctional microvalves control by optical illumination on nanoheaters and its application in centrifugal microfluidic devices," *Lab Chip*, vol. 7, no. 5, pp. 557–564, May 2007.

[10] H. J. van der Linden, S. Herber, W. Olthuis, and P. Bergveld, "Stimulus-sensitive hydrogels and their applications in chemical (micro)analysis," *Analyst*, vol. 128, no. 4, pp. 325–331, Apr. 2003.

[11] R. Gorkin, C. E. Nwankire, J. Gaughran, X. Zhang, G. G. Donohoe, M. Rook, R. O'Kennedy, and J. Ducrée, "Centrifugo-pneumatic valving utilizing dissolvable films.," *Lab Chip*, vol. 12, no. 16, pp. 2894–2902, Aug. 2012.

[12] N. Dimov, E. Clancy, J. Gaughran, D. Boyle, D. Mc Auley, M. T. Glynn, R. M. Dwyer, H. Coughlan, T. Barry, L. M. Barrett, T. J. Smith, and J. Ducrée, "Solvent-selective routing for centrifugally automated solid-phase purification of RNA," *Microfluid. Nanofluidics*, Sep. 2014.

[13] D. J. Kinahan, S. M. Kearney, N. Dimov, M. T. Glynn, and J. Ducrée, "Event-triggered logical flow control for comprehensive process integration of multi-step assays on centrifugal microfluidic platforms.," *Lab Chip*, vol. 14, no. 13, pp. 2249–58, Jul. 2014.

CONTACT

*J. Gaughran, tel: +353 1 700 6012; jennifer.gaughran3@mail.dcu.ie
*J. Ducrée, tel: +353 1 700 5377; jens.ducree@dcu.ie

STRUCTURE-BASED SUPERHYDROPHOBICITY FOR SERUM DROPLETS

Tingyi "Leo" Liu[*] and Chang-Jin "CJ" Kim
Department of Mechanical and Aerospace Engineering
University of California, Los Angeles (UCLA), USA

ABSTRACT

We report that superhydrophobic (SHPo) surfaces based purely on surface structuring shows a robust super-repellency under a prolonged contact with serum droplet as an example of protein-rich biological fluids. In contrast, normal SHPo surfaces, which are based on surface chemistry and surface structuring, lose repellency and eventually get wetted in the same tests. This is the first report of a SHPo surface maintaining super-repellency to a biological fluid.

INTRODUCTION

Superhydrophobicity can be realized by combining surface roughness with a hydrophobic coating [1]. However, although superhydrophobic (SHPo) surfaces can super-repel water and most aqueous biological fluids upon initial contact, their long-term stability has been an issue in practice [2], especially for biological fluids [3]. Because proteins or biofilms accumulate on the hydrophobic material and turn it hydrophilic, the SHPo surface will eventually get wetted (i.e., Wenzel state) [3]. This report starts by proposing the following hypothesis. If SHPo surface is realized by roughness alone regardless of the material [4], not only (i) one can choose any low-fouling material (e.g., PEGylated) but also (ii) the repellency won't be affected anyway even if the biofluid changes the surface chemistry over time.

EXPERIMENT AND DESIGN

To test the above hypothesis, we compared the long-term liquid repellency of a regular and a proposed SHPo surface: a Teflon®-coated Si SHPo surface consisting of microposts (Fig. 1A) and a non-coated SiO$_2$ SHPo surface consisting of doubly re-entrant microposts (Fig. 1B) [4]. Advanced from the fabrication process reported in [5], the SiO$_2$ SHPo surface displayed a very large contact angle (~160°) similar to that on the Teflon®-coated SHPo surface (Fig. 1C&D) in spite of the highly hydrophilic nature of SiO$_2$ (i.e., <10° intrinsic contact angle). To assure a fair comparison, the two surfaces had an identical solid-liquid contact fraction (~5%) when a liquid is suspended on the posts. Figure 2 shows the experimental apparatus prepared for tests. A droplet of serum (40-50 μL) continuously slid back and forth horizontally (by moving the underlying surface) on a SHPo surface. Sheep sterile serum (from HemoStat Laboratories) was used as an example of protein-rich biological fluid since it contains all the components of blood except blood cells and the clotting factor. Clotting could not be allowed for tests involving droplets, which are surrounded with air and evaporate during the tests. The distance of horizontal oscillation was large enough to pass multiple posts but much smaller than the contact (wetting) radius of the droplet so that a large portion of the surface under the droplet is always in contact with the serum.

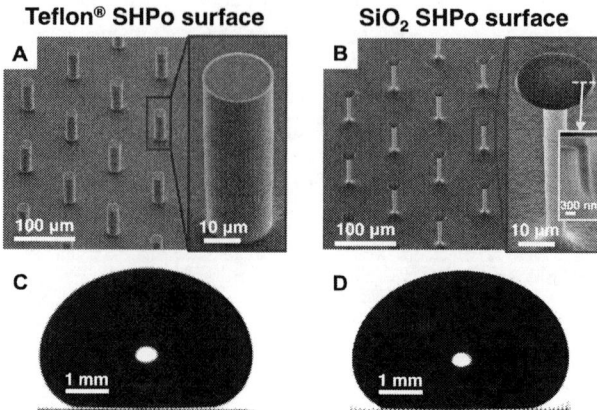

Figure 1: (A&B) SEM angled views of the SHPo surfaces fabricated for the current study. While the regular SHPo surface (A) consists of simple posts coated with Teflon®, the SiO$_2$ SHPo surface (B) consists of posts capped with a doubly re-entrant topology (cross section shown in the yellow inset) without any coating. The two surfaces have the same solid-liquid contact fraction when a liquid is suspended on the posts (i.e., in Cassie state). (C&D) A serum droplet forms the same large contact angle ($\theta_A \sim 160°$) on both surfaces at beginning. The light passing through the surface structures underneath the droplets further confirmed a successful Cassie state.

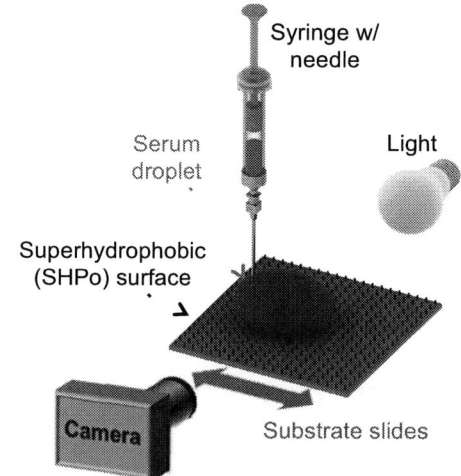

Figure 2: Experimental apparatus to study the longevity of SHPo surfaces in prolonged contact with a serum droplet. A serum droplet was created and held by a syringe on the test surface, which was placed on a motorized linear stage. An advancing and a receding meniscus were formed at the two ends of the droplet by sliding the linear stage periodically (500 μm/s for 4 seconds in each direction) through a computer control. The shape evolution of the droplet was recorded with the CCD camera at 20 fps for a prolonged time.

RESULTS AND DISCUSSION

Figure 3 shows the evolution of the droplet shapes during the fouling test. The receding contact angle of the serum droplet during the horizontal oscillation reflects the stickiness of the surface. As expected, both surfaces were non-sticky at the beginning and showed super-repellency to a serum droplet (surface tension ~50 mN/m [6]). However, on the Teflon® SHPo surface the right edge of the serum droplet showed a sign of pinning after ~30 min and got completely pinned at ~36 min. As the droplet shrunk by evaporation, the pinning caused the droplet to detach from the needle at ~65 min, after which the droplet stopped sliding. In contrast, on the SiO₂ SHPo surface the serum droplet showed no sign of pinning for at least ~141 min (terminated by the droplet encountering a surface defect), confirming its relative longevity. As shown in Fig. 4, when the tested surfaces were further inspected under SEM after the serum droplet was removed, an apparent biofilm residue was found on the Teflon® SHPo surface but not on the SiO₂ SHPo surface. Most of the residual biofilm found on the Teflon® SHPo surface was concentrated on the right edge, matching the location where pinning and loss of repellency were observed.

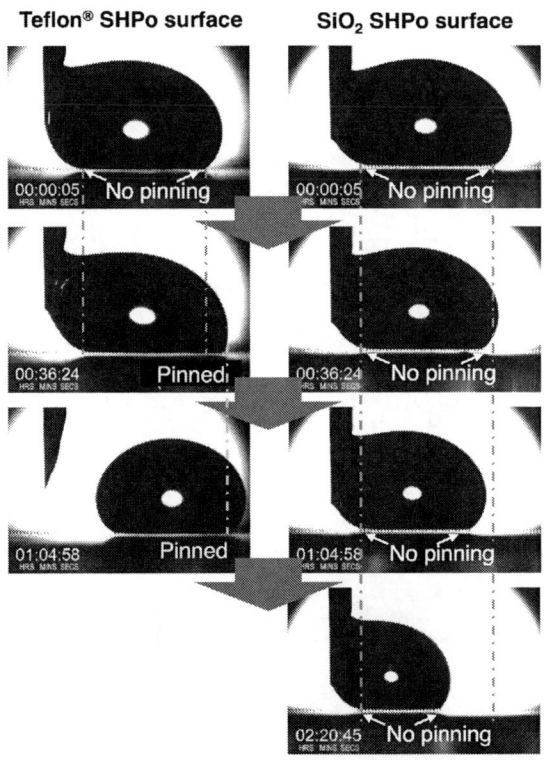

Teflon® SHPo surface   SiO₂ SHPo surface

Figure 3: Shape evolution of a serum droplet sliding on the two SHPo surfaces, demonstrating the long-term liquid-repellency of SiO₂ SHPo surface. (Left column) A serum droplet stuck to the Teflon® SHPo surface after ~36 minutes (manifested by the stretched droplet and increased contact diameter) and eventually detached from the needle due to the pinning and shrinkage. (Right column) In contrast, on the SiO₂ SHPo surface, the serum droplet continued to slide without pinning and stayed attached to the needle while shrinking. The time was shown at the lower left corner in the format of hh:mm:ss.

Teflon® SHPo surface   SiO₂ SHPo surface

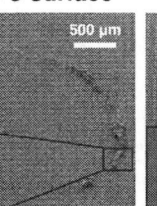

Figure 4: SEM angled views of SHPo surfaces after prolonged contact with serum by the above droplet-sliding test. (Left) Apparent residue found on the Teflon® SHPo surface indicates fouling of the surface. (Right) In contrast, no apparent residue was found on the SiO₂ SHPo surface. Note that on the Teflon® SHPo surface, the majority of the residue coincides with the pinned edge of the serum droplet in the droplet-sliding test.

CONCLUSIONS

We have investigated the long-term liquid repellency towards serum droplets on a structure-based SHPo surface compared with a usual SHPo surface based on material hydrophobicity and surface roughness. On the usual Teflon®-coated SHPo surface, the initial super-repellency was gradually lost and the surface became sticky, causing contact line pinning. In contrast, on the structure-based SHPo surface made of SiO₂, the initial super-repellency was retained much longer with no sign of degradation during the test.

ACKNOWLEDGEMENTS

The authors would like to thank Prof. Dino Di Carlo and Dr. Oladunni Adeyiga for their helpful discussion on selecting serum as an example of biological fluid.

REFERENCES

[1] D. Quéré, "Non-sticking drops," *Rep. Prog. Phys.*, vol. 68, pp. 2495–2532, 2005.

[2] P. Papadopoulos, L. Mammen, X. Deng, D. Vollmer, and H.-J. Butt, "How superhydrophobicity breaks down," *Proc. Natl. Acad. Sci.*, vol. 110, pp. 3254–3258, 2013.

[3] A. K. Epstein, T.-S. Wong, R. A. Belisle, E. M. Boggs, and J. Aizenberg, "Liquid-infused structured surfaces with exceptional anti-biofouling performance," *Proc. Natl. Acad. Sci.*, vol. 109, pp. 13182–13187, 2012.

[4] T. Liu and C.-J. Kim, "Turning a surface superrepellent even to completely wetting liquids," *Science*, vol. 346, 2014 (In Press).

[5] T. Liu and C.-J. Kim, "Microstructured SiO₂ surface repellant to liquids without coating," *Proc. Int. Conf. Solid State Sensors, Actuators and Microsystems (Transducers'13)*, Barcelona, Spain, June 2013, pp. 1609-1612.

[6] S. Lewin, "Blood Serum Surface Tension and its Potential," *Br. J. Haematol.*, vol. 22, pp. 561–566, 1972.

CONTACT

* T. Liu, tel: +1-310-825-3977; leolty@ucla.edu

MICROFABRICATED LIQUID CHAMBER UTILIZING SOLVENT-DRYING FOR IN-SITU TEM IMAGING OF NANOPARTICLE SELF-ASSEMBLY

Won Chul Lee[1,2]*, Jungwon Park[3,4,5]*, David A. Weitz[5], Shoji Takeuchi[1,2], and A. Paul Alivisatos[3,4]

[1]Institute of Industrial Science, The University of Tokyo, Tokyo, JAPAN
[2]ERATO Takeuchi Biohybrid Innovation Project, JST, Tokyo, JAPAN
[3]Department of Chemistry, University of California, Berkeley, USA
[4]Materials Sciences Division, Lawrence Berkeley National Laboratory, Berkeley, USA
[5]School of Engineering and Applied Sciences, Harvard University, Cambridge, USA
*These authors equally contributed to this work.

ABSTRACT

This work presents a microfabricated liquid-sample chamber for real-time TEM (Transmission Electron Microscopy) of nanoscale processes driven by liquid evaporation. Previous liquid-microchambers for TEM hold liquid samples in fully-closed (vacuum-tight) micro/nano-scale structures, thus evaporation-driven nanoscale phenomena couldn't be studied. The present work uses intended leakage/failure in bonding and e-beam induced heating in order to generate solvent-drying during TEM imaging, and the captured real-time nanometer-scale movies visualize critical steps in the self-assembly process of 2D nanoparticle arrays including the two-step crystallization process.

INTRODUCTION

Liquid-microchambers for TEM [1-3] (TEM liquid-cells, reviewed in [1]) have provided direct observation of nanoscale phenomena in a liquid phase, because the liquid-microchambers enable us to load/image liquid nano-samples in/with TEM. For these loading and imaging, the previous liquid-microchambers *1)* encapsulate liquid samples to prevent evaporation under TEM vacuum environment, *2)* provide ultra-thin (~200 nm) observation windows that electron-beams for imaging can pass through, and *3)* have the dimensions that fit in sample holders of conventional TEMs. However, their vacuum-tight encapsulation, which is the key for preventing liquid evaporation, also gives drawback: Nanoscale processes driven by liquid evaporation cannot be monitored in the previous microchambers, while evaporation-driven processes such as the self-assembly of 2D nanoparticle arrays are critically important in nanoscicence [4-7].

This work presents a simple method utilizing controlled liquid-evaporation in TEM liquid-microchambers (Fig. 1), and the in-situ TEM movies captured from the present microchamber deliver scientific findings in the solvent-drying mediated formation of 2D nanoparticle arrays. The present microchamber uses an indium spacer layer bonded at low temperature in order to generate leakage and/or failure on purpose, and this intended leakage/failure enables evaporation of liquid samples during TEM imaging. In detail, the present microchamber can hold liquid samples during vacuum pumping for TEM due to the very tiny leakage/failure, but then, during TEM imaging, local heating induced by electron-beam accelerates solvent evaporation from the imaging window through the leakage/failure (Fig. 1c).

Thus, the present liquid-microchamber provides excellent (self-triggered) conditions for the real-time nano-scale monitoring of evaporation-driven processes.

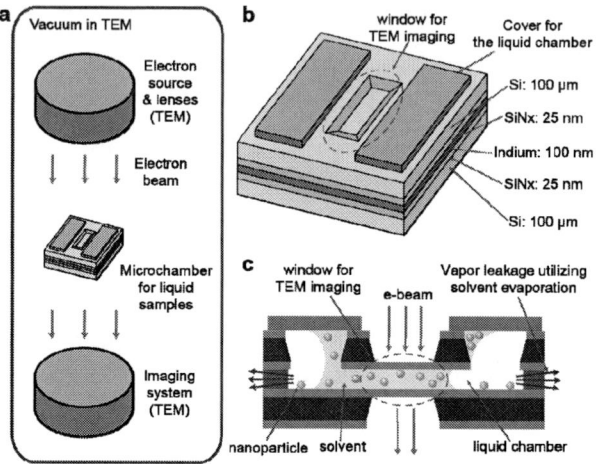

Figure 1: In-situ TEM imaging of the solvent drying mediated 2-D superlattice formation of nanoparticles. (a) Conceptual drawing of the experimental setup. (b,c) Schematic (b) and side (c) views of the present microchamber that utilizes real-time nanometer-scale imaging under solvent evaporation.

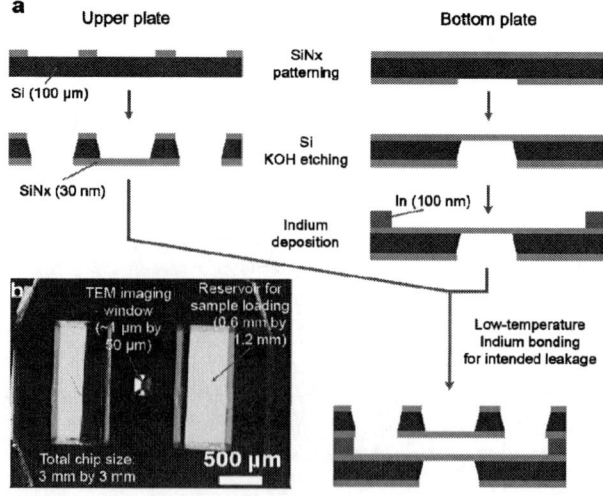

Figure 2: Fabrication of the present microchamber for in-situ TEM imaging. (a) Fabrication process. (b) Fabricated device.

EXPERIMENTAL RESULTS

The present microchamber is designed to hold liquid samples inside of thin (~150 nm thick) chamber structures. The key in the fabrication process (Fig. 2) is to use an indium spacer layer bonded at low temperature for generating leakage and/or failure on purpose. Pt nanoparticles with ~8 nm average diameter are synthesized by the reduction of ionic Pt precursors and dispersed in a 1:4 pentadecane: o-dichlorobenzene mixture with a small amount of oleylamine added in [3,8]. The nanoparticle solution is then loaded into the reservoirs, and the liquid microchamber is airtight for in situ TEM imaging.

Figure 3 shows the real-time nanometer-scale TEM movies captured with the fabricated microchamber. The movies visualize real-time formation processes of the two-dimensional nanoparticle arrays of Pt nanoparticles. In the movie, nanoparitcles are randomly dispersed in liquid (thus, move relatively freely) at the initial stage, and finally form two-dimensional arrays of stationary nanoparticles after solvent drying. The trajectories of nanoparticle motions (Fig.3a) show the formation process of nanoparticle arrays, and the formation process can be quantified using two variables: areal density (ρ) and crystallinity index (ψ_6, bond orientational order) as shown in Fig. 3b. The supperlattice formation shows the previously-known two-step crystallization process [3] (first, gradual increase of crystallinity index and areal density before 58 sec, and second, sudden jump of crystallinity index at 58 sec). The present method can visualize the formation process of 2D nanoparticle arrays for various conditions such as nanoparticle densities (volume fractions), nanoparticle materials, and substrate materials. New experimental findings can be discovered using the present method, which would improve theoretical understanding of the important nanoscale phenomenon, drying-mediated self-assembly at level of single nanoparticle dynamics [4-7].

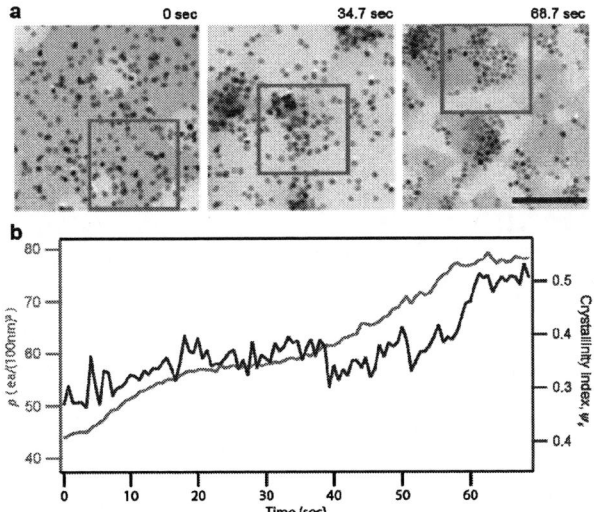

Figure 3: In-situ TEM imaging of the superlattice formation of Pt nanoparticles. (a) Still snapshots of the in-situ TEM movie [3]. (b) 2-D areal densities of particles (red) and bond-orientational order parameters (blue) as a function of time.

CONCLUSIONS

We present a simple method utilizing controlled liquid evaporation in microchambers for in-situ TEM imaging. The present microchamber uses an indium spacer layer bonded at low temperature in order to generate leakage and/or failure on purpose, and this intended leakage/failure enables evaporation of liquid samples during TEM imaging. The in-situ TEM movies captured from the present microchamber can deliver scientific findings in the solvent-drying mediated formation of 2D nanoparticle arrays.

ACKNOWLEDGEMENTS

This work was mainly supported by the Physical Chemistry of Semiconductor Nanocrystals Program, KC3105, Director, Office of Science, Office of Basic Energy Sciences, of the United States Department of Energy under contract DE-AC02-05CH11231. W.C.L and S.T. gratefully acknowledge support from the Takeuchi Biohybrid Innovation Project, Exploratory Research for Advanced Technology (ERATO), Japan Science and Technology (JST). J.P. and D.A.W gratefully acknowledge support from the Harvard MRSEC (DMR-0820484) and Amore-Pacific.

REFERENCES

[1] N. de Jonge, F.M. Ross, "Electron Microscopy of Specimens in Liquid", *Nat. Nanotechnol.*, vol. 6, pp. 695-704, 2011.

[2] H.M. Zheng, et al., "Observation of Single Colloidal Platinum Nanocrystal Growth Trajectories", *Science*, vol. 324, pp. 1309-1312, 2009.

[3] J. Park, et al., "Direct Observation of Nanoparticle Superlattice Formation by Using Liquid Cell Transmission Electron Microscopy", *ACS Nano*, vol. 6, pp. 2078-2085, 2012.

[4] K.J.M. Bishop, et al., "Nanoscale Forces and Their Uses in Self-Assembly", *Small*, vol. 5, pp. 1600-1630, 2009.

[5] J.L. Baker, et al., "Device-Scale Perpendicular Alignment of Colloidal Nanorods", *Nano Lett.*, vol. 10, pp. 195-201, 2010.

[6] F. Li, D.P. Josephson, A. Stein, "Colloidal Assembly: The Road from Particles to Colloidal Molecules and Crystals", *Angew. Chem., Int. Ed.*, vol. 50, pp. 360-388, 2011.

[7] T.P. Bigioni, et al., "Kinetically Driven Self Assembly of Highly Ordered Nanoparticle Monolayers", *Nat. Mater.*, vol. 5, pp. 265-270, 2006.

[8] C.K. Tsung, et al., "Sub-10 nm Platinum Nanocrystals with Size and Shape Control: Catalytic Study for Ethylene and Pyrrole Hydrogenation", *J. Am. Chem. Soc.*, vol. 131, pp. 5816-5822, 2009.

CONTACT

*W.C. Lee, tel: +81-3-5452-6616; wclee@iis.u-tokyo.ac.jp

SYNTHETIC MICROFLUIDIC PAPER

Jonas Hansson, Hiroki Yasuga, Tommy Haraldsson, and Wouter van der Wijngaart
Micro and Nanosystems, KTH Royal Institute of Technology, Stockholm, SWEDEN

ABSTRACT

We introduce a polymer synthetic microfluidic paper for lateral flow devices. The aim is to combine the high surface area of paper, or nitrocellulose, with the repeatability, controlled structure, and transparency of polymer micropillars. Our synthetic paper consists of a dense, high aspect ratio array of transparent pillars that are slanted and mechanically interlocked. We describe the manufacturing using multidirectional UV lithography and demonstrate successful capillary pumping of whole blood.

INTRODUCTION

Lateral flow devices (LFD), e.g. the pregnancy dipstick test, are traditionally manufactured using nitrocellulose substrates [1]. Nitrocellulose (Figure 1a), and the more recently introduced microfluidic paper substrates [2,3] (Figure 1b) consist of natural material based products and therefore suffer from batch-to-batch variations [4] in structure [3], surface properties, and optical transparency. This natural variability and the low material transparency [3] form a fundamental limit to the cut-off value (CV) that can be reached with such LFDs [1].

Micropillar arrays form a superior lateral flow substrate alternative [1] providing higher sensitivity, lower CVs, and higher resolution in quantitative measurements [5], at least in part due to the reproducibility in geometry and surface chemical characteristics [6]. Indeed, today's most sensitive commercial LFD [5] is polymer micropillar-based. Polymer micropillars for lateral flow devices have been manufactured in COC (Cyclic Olefin Copolymer) with an aspect ratio (a.r.) of 1 [7] (Figure 1c), in PMMA with a.r. 0.05 [8], in PDMS with a.r. 0.12 [9], and in SU-8 with a.r. 0.34 [9]. Downscaling and increasing the a.r. of pillars increases the reaction surface/volume but is limited by random pillar collapse due to capillary forces exerted between adjacent pillars during development or filling, The minimum required material stiffness required to avoid capillary collapse is

$$E_{crit} = \frac{32\sqrt{2}\gamma \cos^2\theta \cdot h^3 f(r)}{3d^4} \qquad (1)$$

where d, h and p are pillar diameter, height and pitch, respectively, θ is the equilibrium contact angle, γ is the liquid-vapor interfacial energy, and $f(r)$ is a function of $r = p/d$ [10]. Using silicon as a very high stiffness substrate material (Figure 1d) [11,12] addresses this issue. However, silicon LFDs are too costly for the majority of applications.

To overcome the problem of low polymer material stiffness we use multidirectional UV lithography that [13] has been used to create 3D inclined structures in photoresists previosly, and for lab-on-a-chip applications, mainly as an isoporous particle filter in SU-8 [14].

DESIGN

We introduce a microfluidic substrate that consists of a dense, high aspect ratio, stiff polymer micropillar array with thin slanted pillars that are mechanically interlocked (figure 1e).

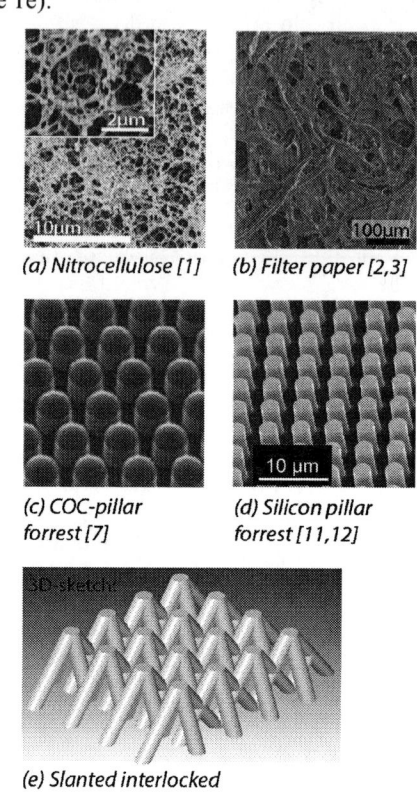

(a) Nitrocellulose [1]   *(b) Filter paper [2,3]*

(c) COC-pillar forrest [7]   *(d) Silicon pillar forrest [11,12]*

(e) Slanted interlocked polymer pillars

Figure 1. SEM images of previous lateral flow substrates (a-d), and a 3D-sketch of the synthetic microfluidic paper introduced in this work. Images a-d, reprinted with permissions (see References).

MANUFACTURING

The synthetic microfluidic paper is manufactured in an Off-Stoichiometry-Thiolene-Epoxy (OSTE+) polymer (OSTEMER Crystal Clear, Mercene Labs, Sweden). Photolithographic definition of the slanted pillars [15] is performed using a collimated, near-UV mercury lamp (OAI, Milpitas) (12 mW/cm² @ 365 nm) and the development process is done in acetone (VWR, USA) under ultrasonication (Model B220).

Fabrication of synthetic microfluidic paper is done in several steps (Figure 2). First, a flat solid OSTE+ layer is manufactured by casting a 1000 µm thick layer of OSTE+ prepolymer between a 100 µm thick transparency film (Nordic Office, Sweden) and a glass slide (VWR, USA), followed by illuminating the prepolymer for 60 s. On top of the polymerized OSTE+ flat layer, first 400 µm

(Device A) or 200 μm (Device B) spacers are placed, then OSTE+ prepolymer is added, after which the polymer is squeezed by a top lid consisting of a glass-chrome photomask (JD Photo-Tools, UK) that is protected by a polymer film ("Never Tear 53my Clear Cling", Antalis AB, Sweden). The entire stack is placed on a black paper (to avoid light reflection) and tilted ~45° under 30 s or 20 s of UV-exposure, for Device A and B, respectively. Subsequently, the stack is tilted ~45° in the opposite direction and illuminated using the same UV exposure conditions. In the directions perpendicular to the first two tiltings, exposure was performed with ~30° tilting for 20 s and 15 s, and for 15 s and 10 s, for Device A and B, respectively. The 4 exposures are followed by development in acetone for 5 min under ultrasonication, and a second thermal cure for 1 h at 110° C. Finally, the surface of the substrate is passivated by incubating the substrates in 1% w/w BSA solution for 10 min and subsequent drying at room temperature.

The interlocked pillar geometry is defined by the photomask pattern consisting of transparent circles with diameter (d) 100 μm and pitch (p) 300 μm (for Device A) or d = 50 μm and p = 100 μm (for Device B). The pillar arrays are fabricated as strips of 5 mm width and 50 mm length.

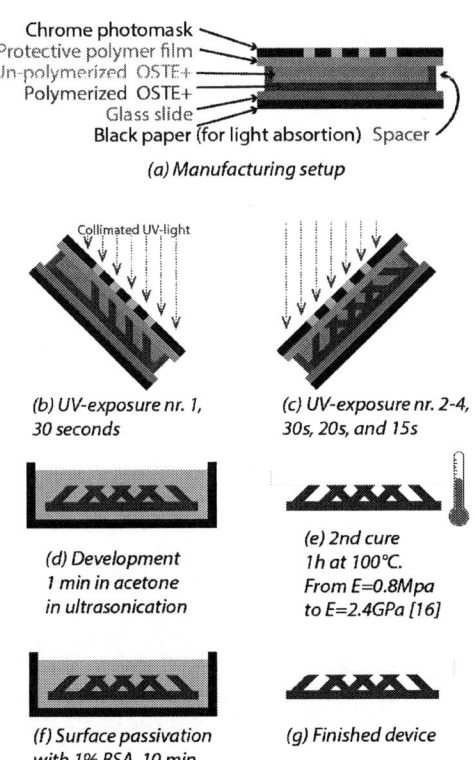

Figure 2: Schematic of the manufacturing of the synthetic microfluidic paper using multidirectional UV lithography.

For comparison, two devices were manufactured with vertical pillar arrays, using vertical UV-exposure, as "negative controls" using the masks with array geometries d = 100 μm, p = 300 μm (same mask as Device A) and d = 100, p = 200 μm (similar effective pillar density as Device A).

EXPERIMENTAL PROCEDURES

To evaluate the transparency of the slanted pillars, a human hair was inserted between the interlocked pillars along the bottom substrate of Device A, and imaged using brightfield microscopy (Leica M205C with Leica DFC290 camera).

The capillary filling properties of the synthetic microfluidic paper was tested by pipetting DI-water colored with red dye, or human whole blood at one end of the 5 mm x 50 mm slanted pillar array strips. The filling was filmed with a digital camera (Canon EOS 600D, with Canon Macro Lens EF 100 mm 1:2,8 USM) and still images at 7 time points (from t=0 s to t=150 s) were captured, cropped around the pillar strip and separated into different color channels. The filled area at each time point was calculated by subtracting the green channel value from the red channel value and classifying every pixel with an intensity >0 as filled. This method proved to provide accurate area measurement results. The volume per filled area was calculated by applying a known volume in the middle of an identical device, imaging the device after capillary filling was completed, and calculating the area with the method described above.

Finally, the synthetic microfluidic paper was tested as traditional paper by writing on it with a ball point pen (Pilot G2 07, red color). The word "Hello" was written on the slanted pillars (i.e. on the synthetic microfluidic paper), and the word "World" was written on the adjacent flat polymer substrate. Soon after writing a finger is swiped along both words to test the robustness of the writing on both types of surfaces.

RESULTS AND DISCUSSION

Device A and Device B (d = 50 μm, h = 200 μm, p = 100 μm) were both successfully manufactured with repeating units of slanted interlocked pillars (Figure 3).

The "negative control" devices with freestanding vertical pillars of similar geometry as Device A featured capillary pillar collapse during development in acetone (Figure 4). This is due to the E-modulus during acetone development $E_{dev} \approx 1$ MPa [16] $\ll E_{crit} \sim 2$ GPa.

Hence, the micropillar-interlocking configuration of the synthetic microfluidic paper introduced in this work results in a mechanical stability that allows miniaturization and increased aspect ratio of micropillars in capillary pumps beyond previously demonstrated vertical pillar geometries.

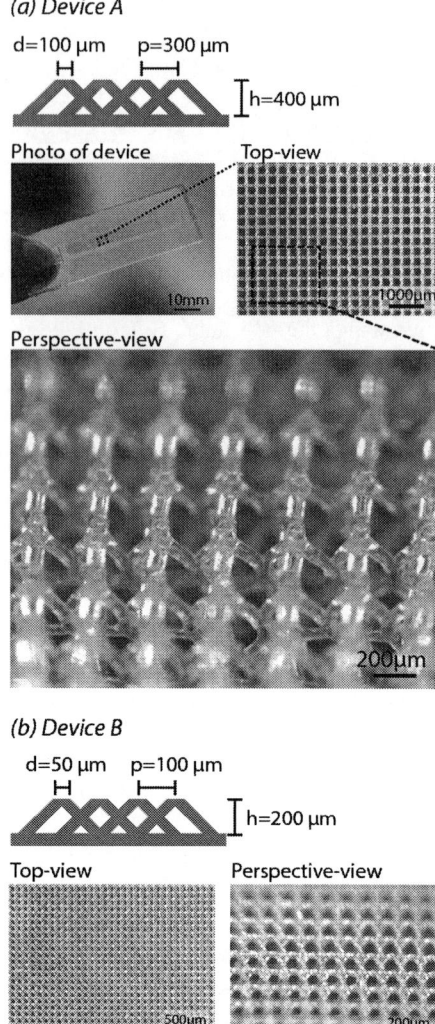

(a) Device A

d=100 µm   p=300 µm   h=400 µm

Photo of device   Top-view

Perspective-view

(b) Device B

d=50 µm   p=100 µm   h=200 µm

Top-view   Perspective-view

Figure 3: Manufactured devices

The human hair is very well visible through the slanted pillars, illustrating the good optical transparency of the devices (Figure 5). Unlike opaque materials such as nitrocellulose or paper, our synthetic paper transparency allows the simultaneous observation of several surface layers at several vertical positions, as illustrated in Figure 5a. In a LFD, the observation of reaction product on a multitude of vertically lined surface layers potentially enables measuring lower concentrations.

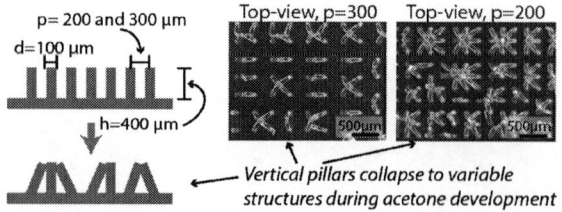

p= 200 and 300 µm

d=100 µm

h=400 µm

Top-view, p=300   Top-view, p=200

Vertical pillars collapse to variable structures during acetone development

Figure 4: Negative control collapse test of vertical freestanding pillars with similar size as device A

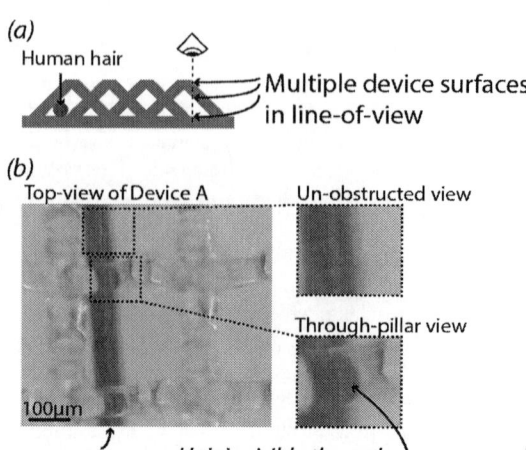

(a) Human hair — Multiple device surfaces in line-of-view

(b) Top-view of Device A   Un-obstructed view

Through-pillar view

Human hair   Hair is visible through the transparent pillars

Figure 5: Pillar transparency.

We successfully performed capillary pumping of whole blood at 2.4 µl/min in device A and pumping of aqueous food dye at 15 µl/min and 0.38 µl/min in devices A and B, respectively (Figure 6).

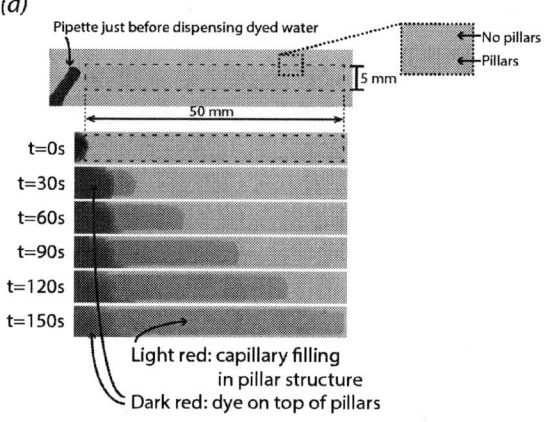

(a) Pipette just before dispensing dyed water   No pillars / Pillars   5 mm

50 mm

t=0s
t=30s
t=60s
t=90s
t=120s
t=150s

Light red: capillary filling in pillar structure
Dark red: dye on top of pillars

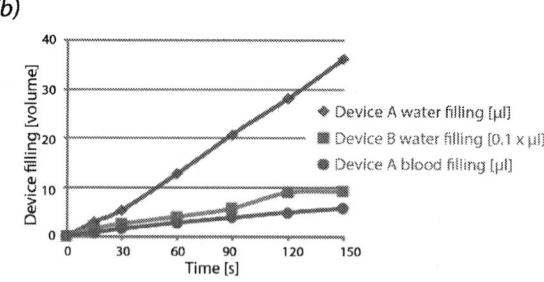

(b)

◆ Device A water filling [µl]
■ Device B water filling [0.1 x µl]
● Device A blood filling [µl]

Figure 6. Measurement of capillary pumping with device A & B. (a) Top-view images of device A capillary filling with dyed water. (b) Flow measurement of capillary pumping of water and whole blood.

Finally, we successfully tested our synthetic microfluidic paper as a traditional paper by writing on it with a ball-pen; without the pillars the ink easily smudges; with pillars it stays (Figure 7).

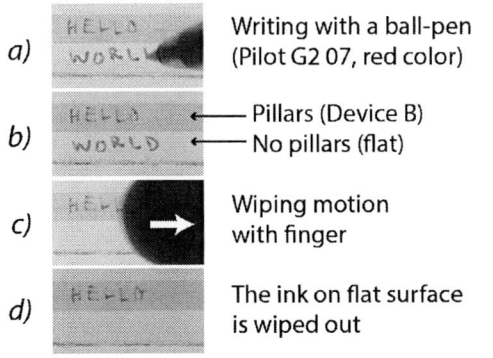

a) Writing with a ball-pen (Pilot G2 07, red color)

b) Pillars (Device B) — No pillars (flat)

c) Wiping motion with finger

d) The ink on flat surface is wiped out

Figure 7: Using Synthetic Microfluidic Paper as traditional paper

CONCLUSIONS

We have introduced a synthetic microfluidic paper that consists of a dense, high aspect ratio, array of transparent polymer pillars that are slanted and mechanically interlocked. The mechanical stability resulting from the pillar interlocking allows miniaturization and increased aspect ratio capillary pumps without pillar collapse as compared to freestanding vertical pillars that collapse randomly. The combination of pillar slanting and optical transparency allows observing several surface layers within the same line-of-view. Capillary pumping in synthetic microfluidic paper was demonstrated for whole blood at 2.4 µl/min and for aqueous food dye at 15 µl/min and 0.38 µl/min in Devices A and B, respectively. Finally, our synthetic microfluidic paper was demonstrated as a traditional paper by writing on it with a normal pen.

The results indicate that our synthetic microfluidic paper has the potential as substrate for highly sensitive lateral flow devices.

ACKNOWLEDGEMENTS

This work has been sponsored in part by the European Commission through the FP7 project ROUTINE. The human blood test was approved by [Swedish authority XX] in [approval number XX].

REFERENCES

[1] O'Farrell (2009). In R. Wong & H. Tse (Eds.), Lateral Flow Immunoassay (pp. 1–33). Humana Press.

[2] Martinez et al., (2007). Angewandte Chemie International Edition, 46(8), 1318–1320. doi:10.1002/anie.200603817

[3] Yetisen et al., (2013). Lab on a Chip, 13(12), 2210–2251. doi:10.1039/C3LC50169H

[4] Hayden et al., (2014). Clinical Chemistry, 60(6), 896–897. doi:10.1373/clinchem.2013.221028

[5] GlobeNewswire News Room, (Dublin, 2014, January 29). Retrieved May 12, 2014, from http://globenewswire.com/news-release/2014/01/29/605602/10066003/en/Trinity-Biotech-Announces-European-Approval-of-Guideline-Compliant-Point-of-Care-High-Sensitivity-Troponin-I-Product.html

[6] Retrieved May 12, 2014, from http://www.meritaspoc.com/Technology/

[7] Dudek et al., (2009). Langmuir, 25(18), 11155–11161. doi:10.1021/la901455g

[8] Mukhopadhyay et al., (2011). Nanoscale Research Letters, 6(1), 411. doi:10.1186/1556-276X-6-411

[9] Saha et al., (2009). Microfluidics and Nanofluidics, 7(4), 451–465. doi:10.1007/s10404-008-0395-0

[10] Chandra & Yang (2009). Langmuir, 25(18), 10430–10434. doi:10.1021/la901722g

[11] Zimmermann et al., (2006), Lab on a Chip, 7(1), 119–125. doi:10.1039/B609813D

[12] Xiao at al., (2010), Langmuir, 26(19), 15070–15075. doi:10.1021/la102645u

[13] Beuret, et al., (1994). IEEE MEMS '94, Proceedings (pp. 81–85). doi:10.1109/MEMSYS.1994.555602

[14] Prenen et al., (2011). Microfluidics and Nanofluidics, 10(6), 1299–1304. doi:10.1007/s10404-010-0763-4

[15] Vastesson et al., (2013). IEEE Transducers '13, Proceedings (pp. 408-411). doi:10.1109/Transducers.2013.6626789

[16] Carlborg et al., (2014). Journal of Polymer Science Part A: Polymer Chemistry, 52(18), 2604–2615. doi:10.1002/pola.27276

Image in Figure 1a, reprinted according to creative commons (CC) license from [17] Hirtz et al., (2013). Beilstein Journal of Nanotechnology, 4, 377–384. doi:10.3762/bjnano.4.44

Image in Figure 1b, reprinted with permission from from [18] Wang et al., (2014). Lab on a Chip, 14(4), 691–695. doi:10.1039/C3LC51313K Copyright (2014) The Royal Society of Chemistry

Image in Figure 1c, reprinted with permission from [7] Copyright (2009) American Chemical Society.

Image in Figure 1d, reprinted with permission from [12] Copyright (2009) American Chemical Society.

CONTACT

*T. Haraldsson, tel: +4687907794; tommyhar@kth.se

ASSEMBLED COMB-DRIVE XYZ-MICROSTAGE WITH LARGE DISPLACEMENTS FOR LOW TEMPERATURE MEASUREMENT SYSTEMS

Gaopeng Xue, Masaya Toda, and Takahito Ono
Graduate School of Engineering, Tohoku University, JAPAN

ABSTRACT

In this research, we report the novel design, fabrication and testing of an assembled comb-drive XYZ-microstage that produces highly decoupled motions into X-, Y-, and Z-directions for the three-dimensional (3D) scanning stage of magnetic resonance force microscopy. The XYZ-microstage based on assembling technology consists of three separated parts, i.e., a comb-drive XY-microstage, two comb-drive Z-microstages and a bottom silicon base substrate. The separated parts are assembled together by using micro manipulators and a guide block of stainless steel. It is demonstrated that the assembled XYZ-microstage can achieve large displacements of 25.2 μm in X direction, 20.4 μm in Y direction and 58.5 μm in Z direction.

INTRODUCTION

The concept of magnetic resonance force microscopy (MRFM) was firstly proposed by Sidles in 1991 [1]. Within several decades, MRFM has been applied to various fields including physics, chemistry, biology and material science as an effective characterization method [2]. As it can detect the densities of spin or radicals through a non-invasive method in nanometer scale, the MRFM is one of the promising approaches for a 3D imaging technique of biological samples [3]. However, the resolution is limited by the detectable minimum force originated in thermomechanical noise of a cantilevered force sensor [4]. To improve the detectable minimum force, usually, the measurements have been performed at low temperatures.

Microscanners for cryogenic measurements using MRFM require large stroke at low temperature with small affections to thermal variation. Piezoelectric actuators are widely used at low temperatures; however, the full range displacement of piezoelectric actuators decreases from ~40 μm at 300 K down to ~12 μm at 170 K and ~3 μm at 1.8 K [5]. The displacements are not enough for 3D imaging of biological species like cells.

Thermoelectric actuators are driven by thermal stress, but basically the actuator elements are heated up due to Joule heating; therefore, its use at cryogenic application is not basically suitable. Electromagnetic actuators cannot be applied to the MRFM system due to the leakage of magnetic field into the measurement system. In terms of magnetic force measurements for MRFM at low temperatures, electrostatic actuation would be the most applicable method for the micro stages. Various XYZ-microstages based on electrostatic actuators have been reported in literatures [6, 7]. However, the microscale displacements into out-of-plane direction are small due to limited planer structures [6] and space limitation of movements into vertical direction [7]. It is difficult to achieve large displacements into 3D directions using a monolithic XYZ-microstage.

This paper presents the novel design, fabrication and testing of a 3-axis comb-drive XYZ-microstage that produces highly decoupled motions into X-, Y-, and Z-directions with large displacements for the 3D scanning stage. The movements of XYZ-microstage are produced by a comb-drive XY-microstage into horizontal directions and by comb-drive Z-microstages into vertical direction, separately. The previous limitation of the displacement in Z-direction can be overcome by wafer-level assembling the XY-microstage with the two separated Z-axis microstages with capability of a large stroke.

DESIGN OF XYZ-MICROSTAGE

Figure 1 shows the schematic view of the designed comb-drive XYZ-microstage based on wafer-level assembling technology. The XYZ-microstage consists of three separated parts including a comb-drive XY-microstage, two comb-drive Z-microstages and bottom silicon base substrate. The Z-microstages are vertically mounted and fixed by a conductive glue on the silicon base substrate. Then, the XY-microstage is mounted on the Z-microstages and fixed by the conductive glue. Table 1 shows the main design parameters of the assembled XYZ-microstage.

Comb-drive Z-microstage

Figure 2 shows the schematic figure of the designed comb-drive Z-microstage. The comb-drive electrodes are symmetrically distributed to achieve a long actuated displacement. Two displacement sensors and two reference capacitances also based on the comb-drive configuration are integrated together. The movable silicon parts are supported by two conventional folded-flexure springs connected with fixed silicon parts. There are additional two conductive springs for electrical connection to outer driving source of the mounted XY-microstage.

Comb-drive XY-microstage

Figure 3 shows the schematic view of the designed comb-drive XY-microstage which refers to the previous article [8]. The XY-microstage is composed of two frames:

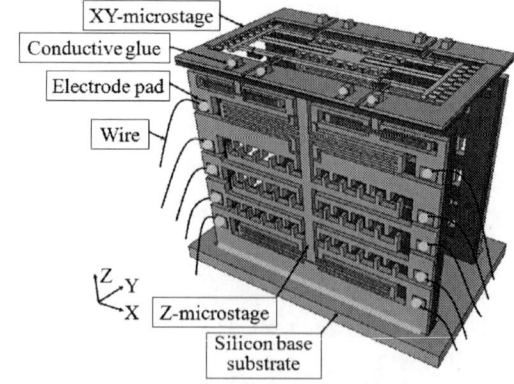

Fig. 1: Schematic view of the proposed comb-drive XYZ-microstage based on wafer-level assembling technology.

Table 1: Design parameters of the assembled Z-microstage and XY-microstage

Z-microstage			XY-microstage		
Support folded-flexure spring lever			**Serpentine spring lever for X-axis actuation**		
W=25 μm	L=3560 μm	T=200 μm	W=20 μm	L=3200 μm	T=200 μm,
			Number of levers	3	
Conductive folded-flexure spring lever			**Serpentine spring lever for Y-axis actuation**		
W=15 μm	L=2200 μm	T=200 μm	W=20 μm	L=3200 μm	T=200 μm,
Number of folds in one lever	2		Number of levers	4	
Comb fingers			**Comb fingers**		
W=15 μm	L=120 μm	T=200 μm	W=15 μm	L=135 μm	T=200 μm
Number of pairs per comb	180		Number of pairs for X-axis	160×2	
Number of combs	12		Number of pairs for Y-axis	150×2	
Initial overlap	15 μm		Initial overlap	60 μm	
Gap spacing	10 μm		Gap spacing	10 μm	
Mass of movable parts	$1.1×10^{-5}$ Kg		**Centre plate**	$1.0×1.0$ mm^2	
Resonant frequency	$2.9×10^2$ Hz		**Gap for insulation**	15 μm	
Dimension	$14.5 × 16.7$ mm^2		**Dimension**	$13.7 × 12.4$ mm^2	

Note. Each W, L, T represent width, length and thickness, respectively.
The structural material is based on silicon with Young's modulus of 130 GPa and density of 2.32 g.cm^{-3}.

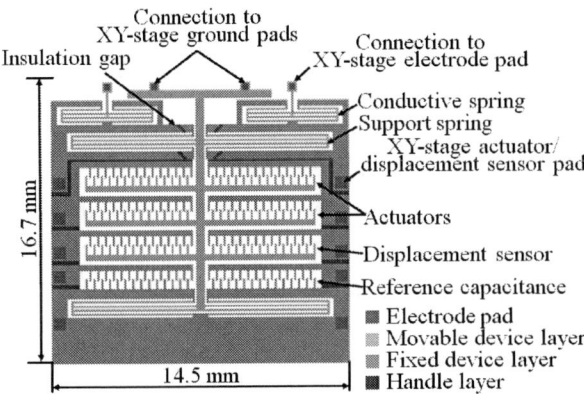

Fig. 2: Design of the Z-microstage fabricated from a SOI wafer.

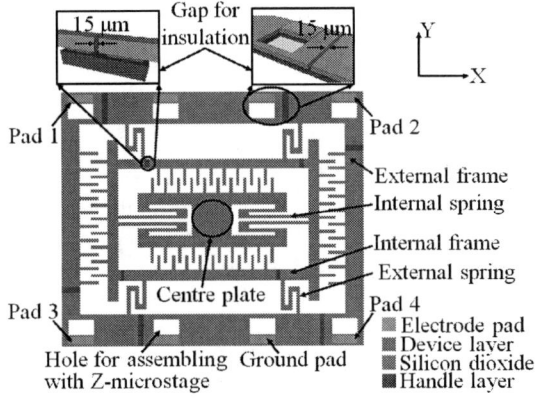

Fig. 3: Design of the XY-microstage fabricated from a SOI wafer.

an external stationary frame for X-axis actuation and an internal movable frame for Y-axis actuation. The centre stage can be actuated without crosstalk into X- and Y-directions. There are extra eight holes in the external stationary frame for assembling with Z-microstages. Because the movable elements of the XY-microstage are formed in the same device layer of SOI (silicon on insulator), some gaps are used to provide electrical insulation between the external frame and the internal frame.

Silicon Base Substrate

The silicon base substrate for assembling is fabricated from a silicon wafer with a thickness of 500 μm. In order to fix precisely the Z-microstages on the silicon base substrate in assembling process, two parallel grooves with a depth of approximately 300 μm and an area of 14.7 × 0.7 mm^2 are formed on the silicon base substrate. The groove size has margin (200 × 100 μm^2) for assembling. The distance between the two parallel grooves is designed as 10 mm, which is same as the distance (10.4 mm) between the two rows of parallel holes formed on the XY-microstage. The size of the silicon base substrate is designed as 12.4 × 15.6 mm^2.

MICROFABRICATION

Z-microstage

The Z-microstage is fabricated using a SOI wafer (200/1/400 μm in thickness) as the process chart is shown in Fig. 4. Cr-Au (30-300 nm) electric pads for the electrostatic actuators and the capacitive displacement sensors are formed by sputtering deposition and lift-off etching process (Fig. 4(b)). After the device structure on the top layer is completed using deep reactive ion etching (RIE) (Fig. 4(c)), the bottom layer is etched using deep RIE (Fig. 4(d)). The remained SiO$_2$ layer is removed by a BHF (HF+NH$_4$F) wet etching process (Fig. 4(e)). Since the Z-microstage and springs are supported by several narrow beams, the supporting beams are cut by the YAG-laser to release the movable part and the Z-microstage from the frame (Fig. 4(f)). Finally the released Z-axis microstage in liquid is dried by a supercritical CO$_2$ drier.

XY-microstage

The XY-microstage is fabricated also using a SOI wafer (200/1/400 μm in thickness) as shown in Fig. 5. Cr-Au (30-300 nm) electric pads for the electrostatic actuators and the capacitive displacement sensors are formed by sputtering deposition and lift-off processes (Fig.

978-1-4799-7956-1/15 $31.00 © 2015 IEEE

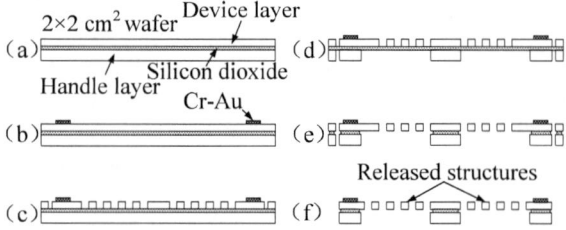

Fig. 4: Microfabrication process of Z-microstage.

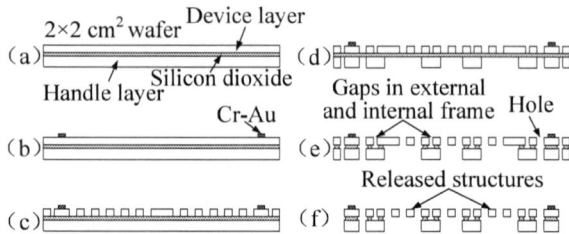

Fig. 5: Microfabrication process of XY-microstage.

5(b)). The first time deep RIE is performed for etching the device layer to form frames, comb fingers, springs, holes, and gaps for insulation in external frame (Fig. 5(c)). The second deep RIE process is performed for etching the handle layer to penetrate the holes and form the insulation gaps in internal frame (Fig. 5(d)). The remained SiO_2 layer is removed by BHF wet etching process (Fig. 5(e)). The support beams are cut by the YAG-laser to release the movable part and the XY-microstage from the frame (Fig. 5(f)). Finally the released XY-axis microstage in liquid is dried by supercritical CO_2 drier.

Fabrication Results

The Z-microstage and XY-microstage are fabricated respectively, as the optical micrographs are shown in Fig. 6 (a) and (b).

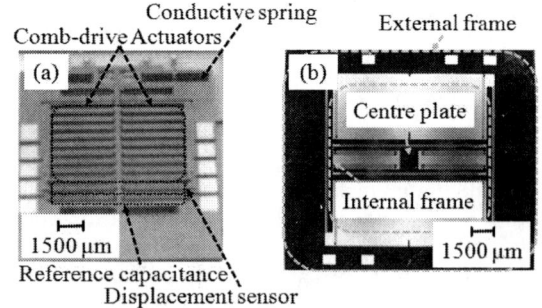

Fig. 6: Optical micrographes of the fabricated Z-microstage and XY-microstage.

ASSEMBLING PROCESS

As shown in Fig. 7, the base guide block of stainless steel with precisely cut is prepared to support the vertical positioning of the Z-microstages. The two clamping plates of Mo on the side walls of the base block are fixed with screws (Fig. 7(a)). A silicon base substrate is inserted into the bottom gap of the support base block. The wired Z-microstages are vertically mounted into the grooves of the silicon base substrate and fixed on the side walls of the block with the Mo-fixtures (Fig. 7(b)).

The XY-microstage is transferred with the thin Cu tray

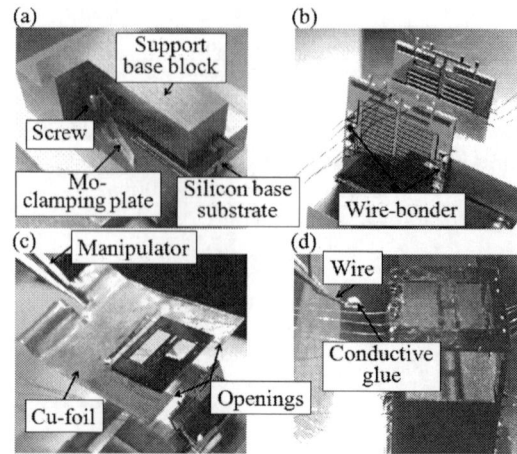

Fig. 7: Fabrication process of the XYZ-microstage based on assembling technology.

which is clamped by the tweezers on a manipulator (Fig. 7(c)). The Cu tray has rectangular openings for handling the XY-microstage. After the electrode pads of the Z-microstages are inserted into the holes of the XY-microstage, the Cu tray is removed away slowly.

The electrode pads of XY-microstage are electrically connected to the upper side of the electrode pads on Z-microstages by conductive glue. The conductive glue is putted on the tip of a wire and handled by a manipulator (Fig. 7(d)). Also the Z-microstages are electrically connected to the silicon base substrate by the conductive glue. After solidifying the conductive glue using a hot plate, the XYZ-microstage can be released from the support base block and Mo-fixtures.

EVALUATIONS

The fabricated XYZ-microstage is mounted on a printed board and the wires from XYZ-microstage are connected to the board for measurements, as shown in Fig. 8.

In order to evaluate the actuation performance of the fabricated XYZ-microstage, a DC power supply is used to provide actuation voltages of 0−120 V. Displacements into horizontal directions are measured by an optical

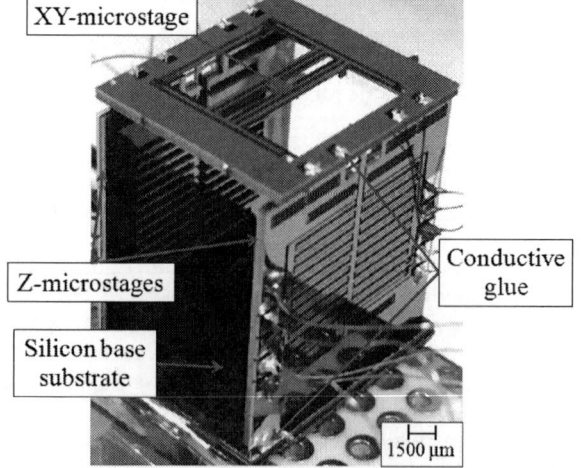

Fig. 8: Fabricated XYZ-microstage for evaluations.

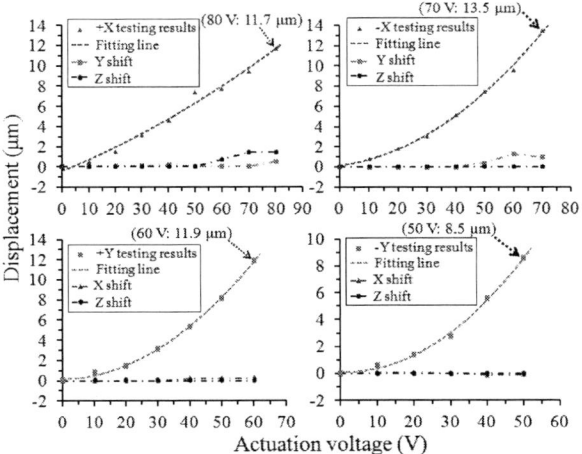

Fig. 9: *Optical testing results of displacements versus actuation voltages of the XYZ-microstage in parallel directions.*

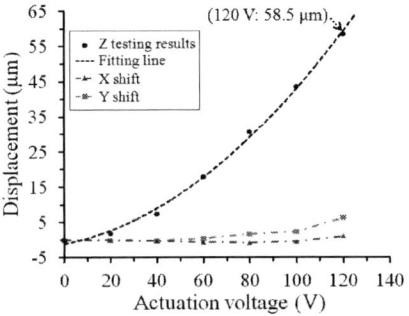

Fig. 10: *Optical testing results of displacements versus actuation voltages of XYZ-microstage in vertical direction.*

microscope with a digital camera. The vertical movements are also measured by an optical microscope which is tilted in horizontal direction. Figures 9 and 10 show the testing results of the actuation performance into horizontal directions and into vertical direction, respectively.

In Fig. 9, the positioning XYZ-microstage produces 11.7 μm displacement in +X direction at 80 V, 13.5 μm displacement in -X direction at 70 V, 11.9 μm displacement in +Y direction at 60 V and 8.5 μm displacement in -Y direction at 50 V, respectively. In Fig. 10, a displacement of 58.5 μm is observed with an actuation voltage of 120 V.

Simultaneously, cross-axis couplings are also measured, to verify the decoupled motions of the XYZ-microstage. The test results demonstrate that the designed structure provides highly decoupled motions in actuation voltage range of 0-50 V in ±X-, ±Y-axis and 0-60 V in Z-axis.

In order to investigate the performance of the fabricated device at low temperatures, the Z-microstage is installed into a cryostat equipment with a liquid N_2 tank. A displacement of 59.2 μm without any degradation is observed from displacement sensor at the low temperature of 104.5 K.

CONCLUSIONS

In this research, the comb-drive XYZ-microstage integrated with the capacitive displacement sensors has been constituted by three separate parts, including comb-drive XY-microstage, comb-drive Z-microstages and bottom silicon substrate, based on wafer-level assembly technology. The fabricated XYZ-microstage is capable of producing 11.7 μm displacement into +X direction, 13.5 μm displacement into -X direction, 11.9 μm displacement into +Y direction, 8.5 μm displacement into -Y direction and 58.5 μm displacement in vertical direction.

It is demonstrated that the comb-drive microstages with capacitive displacement sensors are promising applications with large displacement for low temperature. This novel XYZ-microstage based on assembling technology, can achieve large displacements into 3D directions.

ACKNOWLEDGEMENTS

This work was supported by Micro/Nano-machining Research and Education Center, Tohoku University, JAPAN.

REFERENCES

[1] J. A. Sidles, "Noninductive detection of singleproton magnetic resonance", Appl. Phys. Lett., vol. 58, pp. 2854-2856, 1991.

[2] S. Tsuji, T. Masumizu, Y. Yoshinari, "Magnetic resonance imaging of isolated single liposome by magnetic resonance force microscopy", J. Magn. Reson., vol. 167, pp. 211-220, 2004.

[3] C. L. Degen, M. Poggio, H. J. Mamin, C. T. Rettner, D. Rugar, "Nanoscale magnetic resonance imaging", PNAS, vol. 106, pp. 1313-1317, 2009.

[4] U. Gysin, S. Rast, P. Ruff, E. Meyer, "Temperature dependence of the force sensitivity of silicon cantilevers", PHYS. REV. B, vol. 69, pp. 045403, 2004.

[5] M. Fouaidy, G. Martinet, N. Hammoudi, F. Chatelet, S. Blivet, A. Olivier, H. Saugnac, "Full characterization at low temperature of piezoelectric actuators used for SRF cavities active tuning", in Digest Tech. Papers 2005 Particle Accelerator Conference, Knoxville, Tennessee, USA, May 16-20, 2005, pp. 728-730.

[6] X. Liu, K. Kim, Y. Sun, "A MEMS stage for 3-axis nanopositioning", J. Micromech. Microeng., vol. 17, pp. 1796-1802, 2007.

[7] K. Takahashi, M. Mita, H. Fujita, H. Toshiyoshi, "Switched-layer design for SOI bulk micromachined XYZ stage using stiction bar for interlayer electrical connection", J. Microelectromech. Syst., vol. 18, pp. 818-827, 2009.

[8] K. Laszczyk, S. Bargiel, C. Gorecki, J. Krężel, P. Dziuban, M. Kujawińska, D. Callet, S. Frank, "A two directional electrostatic comb-drive X–Y micro-stage for MOEMS applications", SENSOR ACTUAT A-PHYS., vol. 163, pp. 255-265, 2010.

CONTACT

*G. Xue, Tel: +81-022-7955810; Fax: +81-022-7955808; Email: setukoho@nme.mech.tohoku.ac.jp

PNEUMATIC BALLOON ACTUATOR WITH TUNABLE BENDING POINTS

Lanying Zheng[1], Shotaro Yoshida[1], Yuya Morimoto[1,2], Hiroaki Onoe[1,2] and Shoji Takeuchi[1,2]
[1] Institute of Industrial Science, The University of Tokyo, JAPAN
[2] Takeuchi Biohybrid Innovation Project, ERATO, JST, JAPAN

ABSTRACT

We propose a pneumatic balloon actuator capable of controlling its bending points. Local stiffness of the balloon actuator can be controlled by embedding low-melting-point-alloy that becomes solid by cooling. The bending points can be changed depending on the position of the melted low-melting-point-alloy since the pneumatic actuator bends at soft point. The advantage of our actuator is that its bending point can be simply controlled without changing the design of the whole device. We believe that the proposed actuation mechanism will be useful in designing highly flexible actuators for soft robotics.

INTRODUCTION

Pneumatic balloon actuator is mainly applied to micro-robot hand for medical use and soft robotics because it is small, flexible and safe [1-5]. Pneumatic balloon actuator is generally composed of two bonded layers with different stiffness, and the actuator bends by applied pressure because of the difference in stiffness. However, once the actuator is designed, the position of the bending point cannot be changed since it is determined by a structure of the actuator. The structure will be complex to achieve flexible and complicated actuation. To achieve such actuation effectively, flexible actuation mechanism that can bend at arbitrary point with simple structure is required. Here, we propose a double-layered pneumatic balloon actuator capable of controlling its bending point by changing local stiffness of the actuator (Fig.1).

Our actuator is composed of double-layered two different channels made of elastic silicon rubber (Ecoflex00-30®). The upper channel is for controlling the local stiffness of the actuator using bismuth-based low-melting-point-alloy (melting temperature: 47°C). The lower channel is used for actuation by air pressure. When the air pressure is supplied, the actuating channel inflates to bend the actuator. We can change local stiffness of the actuator by locally melting the alloy. The bending points can be changed depending on the position of the melted low-melting-point-alloy since the pneumatic actuator bends at soft points.

EXPERIMANTAL

Materials

The elastic silicone rubber Ecoflex00-30® was purchased from Smooth-On, Inc.. Low-melting-point-alloy (composition: Sn 8.3%, Bi 44.7%, Pb 22.6%, Cd 5.3%, In 19.1%) was purchased from Ematerial. Nichrome wire was purchased from The Nilaco Corporation.

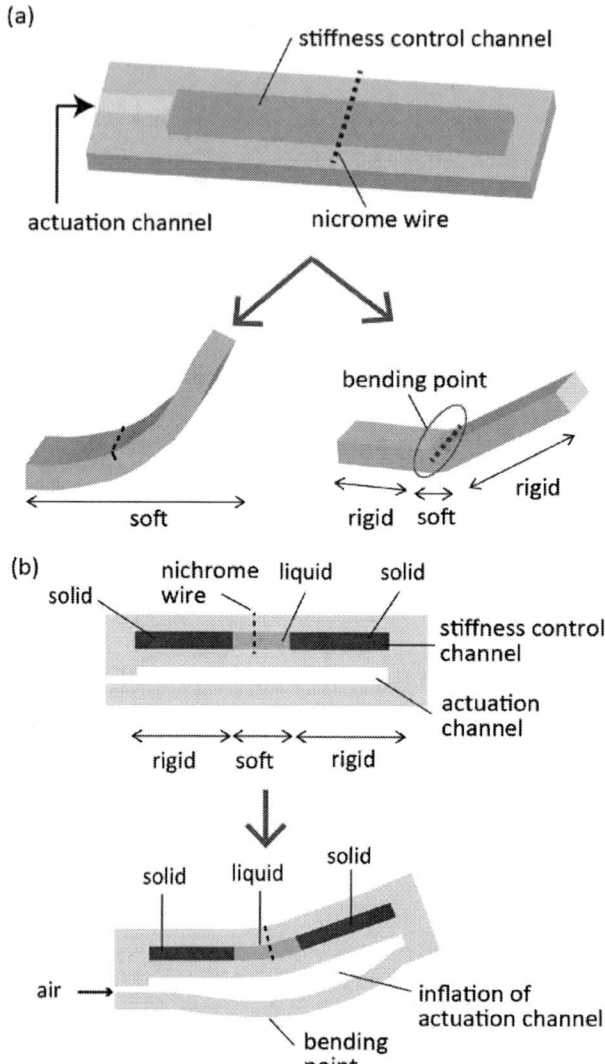

Figure 1. The concept of this study. (a) Proposed pneumatic balloon actuator is capable of controlling its bending points. (b) Local stiffness of the balloon actuator can be controlled by injected low-melting-point-alloy. The bending point can be changed depending on the position of melted low-melting-point-alloy since the pneumatic balloon actuator bends at soft points.

Fabrication of molds

We fabricated molds for formation of layers with channels made of Ecoflex. First, we designed the molds using a 3D modeling software (Rhinoceros, AppliCraft) and fabricated them by a 3D printer (AGILISTA-3100,

1. pour Ecoflex into the mold fabricated by 3D printer and cure Eccoflex 15 min at 75°C

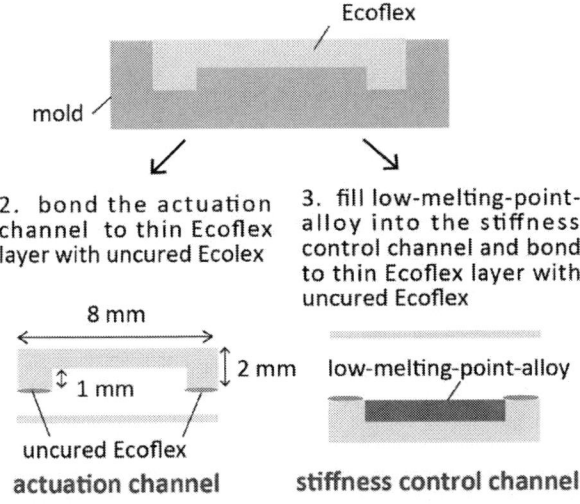

2. bond the actuation channel to thin Ecoflex layer with uncured Ecolex

3. fill low-melting-point-alloy into the stiffness control channel and bond to thin Ecoflex layer with uncured Ecoflex

4. bond actuation channel to stiffness control channel

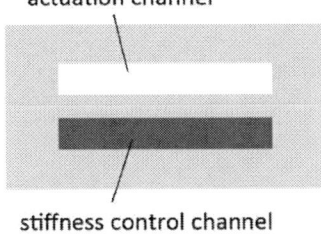

Figure 2. Fabrication process of the balloon actuator.

KEYENCE CORPORATION). The fabricated molds were sunk in water for half a day to remove sacrificing layers. Finally, we coated the molds with parylene to ease demolding of Ecoflex.

Fabrication of balloon actuator

We poured Ecoflex into the molds and cured it for 15 min at 75°C. After releasing Ecoflex layers from the molds, we first bonded the Ecoflex layer to a thin Ecoflex sheet fabricated on petri dishes with uncured Ecoflex as an adhesive agent. Subsequently, we cured them for 15 min at 75°C to form the actuation channel. Meanwhile, to form the stiffness control channel, we filled low-melting-point-alloy into the channel of the other Ecoflex layer and bonded to another thin Ecoflex sheet. Finally, we bonded both structures with uncured Ecoflex (Fig. 2).

Controlling stiffness of low-melting point-alloy

We first controlled the stiffness of low-melting-point-alloy using a heater and ice for preliminary experiments. To make the balloon actuator soft, we put the actuator on a hot plate set at 60°C and melt the alloy completely. To make the actuator hard, we left the actuator at room temperature to wait for the low-melting-point-alloy becoming rigid. When we make the actuator partially soft, we applied electricity controlled by a power supply machine (2400, Keithley Instruments) to the nichrome wire wound around the stiffness control channel. Joule heat of

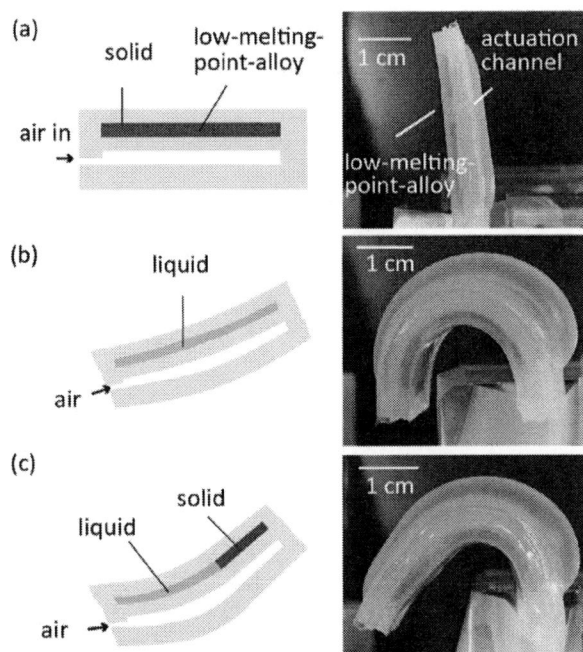

Figure 3. Change of bending points by locally melted low-melting-point-alloy. (a) When the whole alloy is solid. (b) When the whole alloy is melted. (c) When the alloy is locally melted.

the nichrome wire allow the low-melting-point-alloy partially melt. When we make the actuator partially hard, we put an ice on a part of the actuator after completely melting the alloy.

Inflation of the balloon actuator

We injected 3 ml air into the actuation channel using a syringe pump (KDS210, KD Scientific) to inflate the actuation channel. When we controlled air pressure in the actuation channel, we used a flow controller (MFCS100, FLUIGENT) and an air compressor (PC3-5.5T, YAEZAKI KUATSU CO., LTD.).

RESULTS AND DISCUSSION

Change of bending points by controlling stiffness of the low-melting-point-alloy

To confirm the change of bending points by controlling of the location of melted alloy, we compared drive status of the balloon actuator in three condition. First, we injected 3 ml air into the actuation channel when whole low-melting-point-alloy was solid (Fig. 3a). Second, we melted the whole alloy in the stiffness control channel and injected 3 ml air into the actuation channel (Fig. 3b). Third, we locally cooled the alloy using ice blocks after melting them, and injected 3 ml air into the actuation channel (Fig. 3c). When the whole alloy was solid, the balloon actuator did not bend. In addition, compared with Figure 3b and Figure 3c, we confirmed that the solid part in the balloon actuator caused the change of the bending point. From the results, we confirmed that the low-melting-point-alloy were useful material to control the bending points of the balloon actuator.

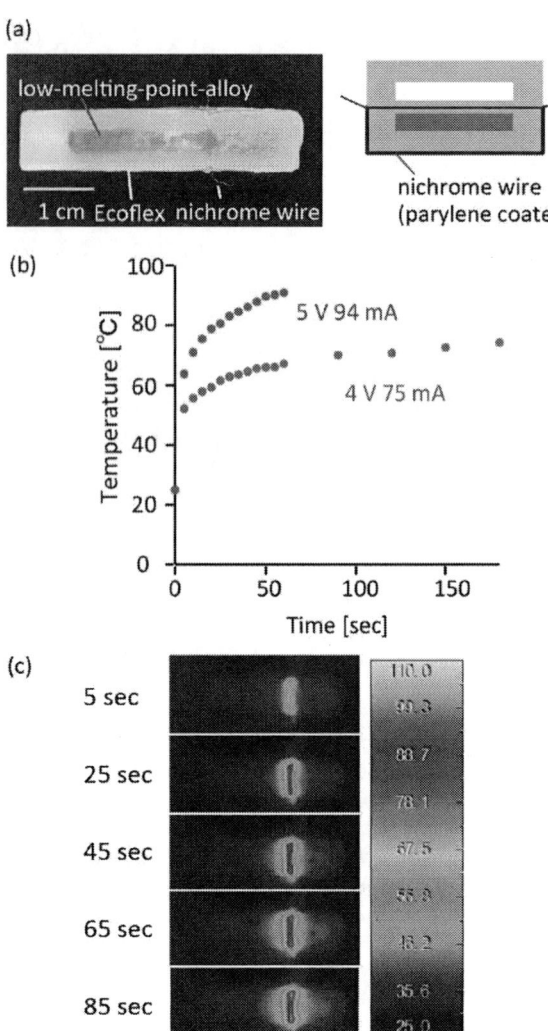

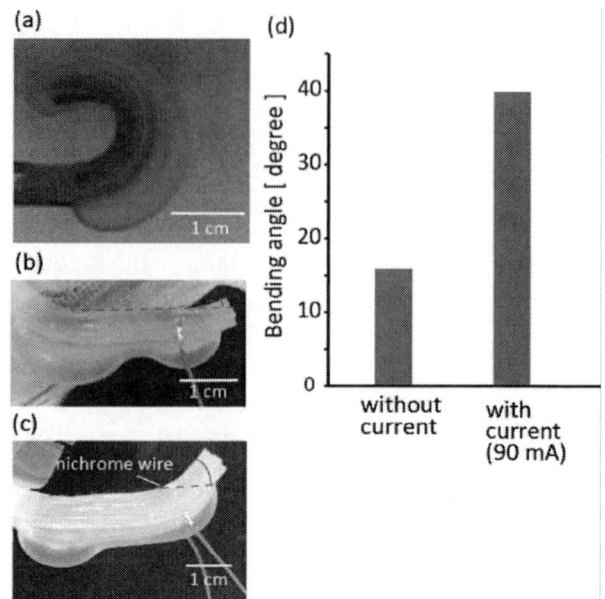

Figure 4. Transition of temperature by applied electricity to the nichrome wire. (a) Balloon actuator with a nichrome wire. (b) Transition of temperature by applied electricity to the nichrome wire. (c) Transition of surface temperature measured with a thermography. Applied current was 90 mA.

Control of the bending points using nichrome wire

We prepared the balloon actuator with the nichrome wire to control the local stiffness of the low-melting-point-alloy. The nichrome wire was wound around the stiffness control channel before bonding to the actuation channel. Figure 4a shows the balloon actuator with nichrome wire. Using the actuator, we measured transition of temperature around the nichrome wire when electricity was applied to nichrome wire. Temperature around the nichrome wire reached 50°C which is larger than the melting point of the alloy within 10 sec (Fig. 4b). Finally we could confirm that the alloy melted after 60 sec when applied 5 V (94 mA) and after 160 sec when applied 4 V (75 mA). The change of surface temperature caused by heated nichrome wire was local and did not reach to the whole device (Fig. 4c). From the results, nichrome wires were able to control local stiffness of the low-melting-point-alloy in the stiffness control channel.

Figure 5. Change of bending points by heated nichrome wire. Applied air pressure was 15 kPa. (a) When whole alloy is melted. (b) When whole alloy is solid. (c) When applying 70 mA to nichrome wire. (d) Comparison of bending angle between (b) and (c).

Demonstration of controlling the bending points

We demonstrated control of the bending points by locally melting the alloy with the nichrome wire (Figure 5). To confirm the change of bending points, we compared drive status of the balloon actuator in three condition. First, we applied 15 kPa to the actuation channel when the whole low-melting-point-alloy was melted (Fig. 5a). Second, we applied 15 kPa to the actuation channel when the whole low-melting-point-alloy was solid without applying electricity to the wire (Fig. 5b). Third, we applied 15 kPa to the actuation channel when the low-melting-point-alloy was partially melted with applying 90 mA to the wire (Fig. 5c). Figure 5d shows the comparison of bending angle between Fig. 5b and Fig 5c. The bending angle of the actuator increased due to locally melted alloy when we applied electricity to the nichrome wire. From the results, we confirmed that the bending points were controlled by heated nichrome wire.

CONCLUSION

We fabricated the pneumatic balloon actuator capable of controlling its bending points. The actuator was composed of the actuation channel and the stiffness control channel filled with low-melting-point-alloy. We confirmed that the bending points were able to be controlled by changing local stiffness of the low-melting-point-alloy by applying electricity to the nichrome wire. We believe that the balloon actuator will be a useful tool for soft robotics to realize flexible actuations. Furthermore, we expect the mechanism of the stiffness control will be applicable to functional materials for shape changing or shape keeping [6-8].

ACKNOWLEDGEMENTS

This work is partially supported by Takeuchi Biohybrid Project, ERATO, JST, Japan.

REFERENCES

[1] B. Gorissen, M. D. Volder, A. D. Greef and D. Reynaerts : "Theoretical and experimental analysis of pneumatic balloon microactuators", *Sensors and Actuators A: Physical*, vol.168, pp.58-65 (2011)

[2] O. C. Jeong and S.Konishi : "All PDMS Pneumatic Microfinger With Bidirectional Motion and Its Application", *Journal of Microelectromechaical Systems*, vol.15, pp.896-903 (2006)

[3] S. Konishi, F. Kawai and P. Cusin: "Thin flexible end-effetor using pneumatic balloon actuator", *Sensors and Actuators A: Physical*, vol.89, pp.28-35 (2001)

[4] B. Gorissen, W. Vincentie, F. Al-Bender, D. Reynaerts and M. D. Volder : "Modeling and bonding-free fabrication of flexible fluidic microactuators with a bending motion", *Journal of Micromechanics and Microengineering*, vol.23, pp.045012 (2013)

[5] B. Gorissen, R. Donose, D. Reynaerts, M. D. Volder : "Flexible pneumatic micro-actuators: analysis and prodution", *Procedia Engineering*, vol.25, pp.681-684 (2011)

[6] S. T. Chang, A. B. Ucar, G. R. Swindlehurst, R. O. Bradley IV, F. J. Renk and O. D. Velev : "Materials of Controlled Shape and Stiffness with Photocurable Microfluidic Endoskeleton", *Advanced Materials*, vol.21, pp.2803-2807 (2009)

[7] B. E. Schubert and D. Floreano : "Variable stiffness material based rigid low-melting-point-alloy microstructures embedded in soft poly(dimethylsiloxane) (PDMS)", *RSC Advances*, vol.3, pp.24671-24679 (2013)

[8] N. G. Cheng, A. Gopinath, L. Wang, K. Iagnemma and A. E. Hosoi : "Thermally Tunable, Self-Healing Composites for Soft Robotic Applications", *Macromolecular Materials and Engineering*, vol. 299, pp.1279-1284 (2014)

CONTACT

*L. Zheng, tel: +81-3-5452-6650;
r-tei@iis.u-tokyo.ac.jp

SELF-LIFTING ARTIFICIAL INSECT WINGS VIA ELECTROSTATIC FLAPPING ACTUATORS

Xiaojun Yan[1,2], Mingjing Qi[1,*], and Liwei Lin[3]

[1]School of Energy and Power Engineering, Beihang University, Beijing, China
[2]Collaborative Innovation Center of Advanced Aero-Engine, Beijing, China
[3]Mechanical Engineering Department, University of California at Berkeley, USA

ABSTRACT

We present self-lifting artificial insect wings by means of electrostatic actuation for the first time. Excited by a DC power source, biomimetic flapping motions have been generated to lift the artificial wings 5cm above ground (limited by the current experimental setup) under an operation frequency of 50-70Hz. Three achievements have been accomplished: (1) first successful demonstration of self-lifting electrostatic flying wings; (2) low power consumption as compared to other actuation schemes; and (3) self-adjustable rotating wing design to provide the lifting force. As such, this work can lead to a new class of electrostatic flapping actuators for artificial flying insects.

INTRODUCTION

Artificial flying insects can be considered as Micro Air Vehicles (MAVs) with wingspans of less than 3-5cm and potential applications in search, rescue, exploration, and reconnaissance [1]. These robots are designed to do maneuvers in limited spaces such as rooms, caves, and jungles that large flying robots won't be serviceable. To achieve such tasks, actuators of these robots are required to have biomimetic flapping motion (in terms of frequency, amplitude and trajectory etc.) with desirable simple design and low weight. The biomimetic flapping motion is the key to generate lift force, and the simple structural design is vital for autonomous flight. It has been a key challenge for current technologies to meet these requirements.

For the purpose of achieving fully autonomous flight, a few kinds of actuation schemes have been investigated to drive artificial flying insects, such as electrostatic actuator [2], piezoelectric actuator [3, 4], and DC motor [5, 6]. Among these schemes, the electrostatic actuator, which usually consists of fixed electrodes and moving electrodes (as flapping wings), is very attractive due to its low power consumption and ease of miniaturization. However, previous demonstrations have failed to generate effective lift force as the moving electrodes were not able to imitate the "rotational" flapping motions of real insect wings [2].

In 2013, an insect-scale flapping robot that achieves controlled and tethered flight has been demonstrated using piezoelectric flight muscles [3, 4]. Here, we demonstrate the first vertical lift motions of artificial insect wings via electrostatic flapping actuators [7]. The actuator can work at a resonance state under constant DC input without using any complex AC circuits, and directly drive the wings in a biomimetic way with measured lifting force up to more than 3mg. A prototype of the actuator has been designed and fabricated for the purpose of investigating its key characteristics in terms of wing motion and lift force.

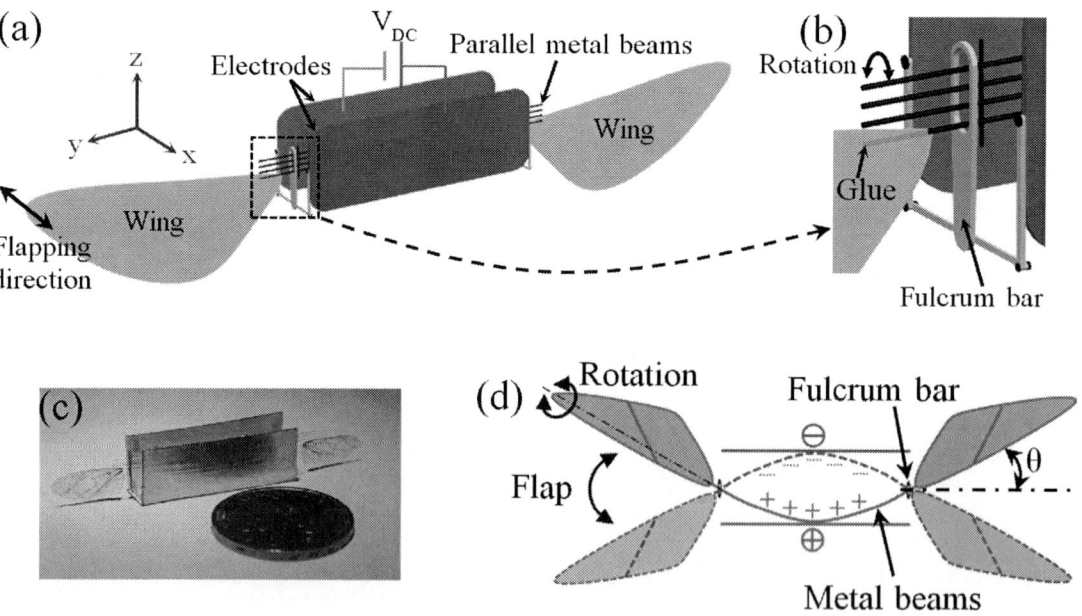

Figure 1: (a) Configuration of the actuator. (b) Detailed structure of the flapping wing assembly. (c) Photo of the actuator and a coin. (d) Top view of the flapping wing assembly in self-resonance state under the electrostatic actuation. Red and blue color configurations are two different states when beams are in contact with the negative and positive electrodes, respectively. In additional to the lateral flapping movements, the wings also have rotational motion to provide the lift force.

DESIGN

Figure 1a shows the configuration of the actuator, which consists of four parallel metal beams, two wings, two fulcrum bars and two electrodes. The parallel metal beams together with the two wings (as a flapping wing assembly) are placed in the middle of the two electrodes, and hold by the two fulcrum bars with oval holes. The electrodes are simply connected to a DC power source. Fig. 1b is the detailed assembly drawing, which shows that the two wings are glued at both ends of the lowest beam. A prototype device has been fabricated as shown in Fig. 1c, where the beams are made of metal alloy wires with length of 27mm and diameter of 60μm, and the wings are extracted from drone honey bees with the size of 12 mm in length and 5 mm in width.

It is noted that, the multiple beams, rather than a beam or a flat plate, are designed to enhance the electrostatic driving force and meanwhile reduce the air damping force. The oval holes on the fulcrum bars provide suitable support to the flapping wing assembly while allowing it to rotate around y-axis (the beam direction, Fig. 1b). The wings are only glued to the lowest beam for the purpose of inducing the wing rotation, which will be explained later in the next section.

The working principle of the actuator can be illustrated in Fig. 1d. When a DC voltage (V_{DC}) is applied, a steady electrostatic field will be generated between the two electrodes. Due to electrostatic induction effect, the electrostatic force will act on the parallel metal beams, and attract the beams toward one of the electrodes. With the increase of V_{DC}, the beams will move further until V_{DC} reaches the pull-in voltage, under which the beams can touch the positive electrode (red color) and get charged.

Afterwards, the charges with the same polarity repel each other and the electrostatic force will drive the beams to the opposite electrode - the negative electrode (blue color) and release the charges. Subsequently, the beams can be excited into a resonance state at their natural frequency through impacting the positive and negative electrodes alternately, and thus actuate the two wings into cyclic motions. During the resonance, the charge and discharge processes generate a series of pulse currents going through the circuit, which can be measured by connecting a resistor together with an oscilloscope in series. The pulse currents also can be used to obtain the resonant frequency and power consumption of the actuator [7, 8]. Under current prototype design, the typical resonant frequency is 50-70Hz, and the typical power consumption is 6-10mW.

It is noted that, the actuator is subjected to a static DC voltage, and its alternating driving force is generated and sustained by its self-resonance. From the view point of structural dynamics, the actuator actually undergoes a "self-exited vibration", which is similar to asynchronous motions of insect flight muscles [9]. While conventional actuators usually operate at a "force vibration" state, which needs complex AC circuits to generate, sense and feedback for the control of sustained resonance. The actuation scheme presented here is more aligned with the bionic actuation as compared with other conventional actuators.

PERFORMANCE TESTS

Both wing motion and lift force measurements have been conducted quantitatively in order to investigate the actuator's characteristics and explore the possibility for applications in artificial flying insects. A test system has been designed as shown in Fig. 2a, which consists of a

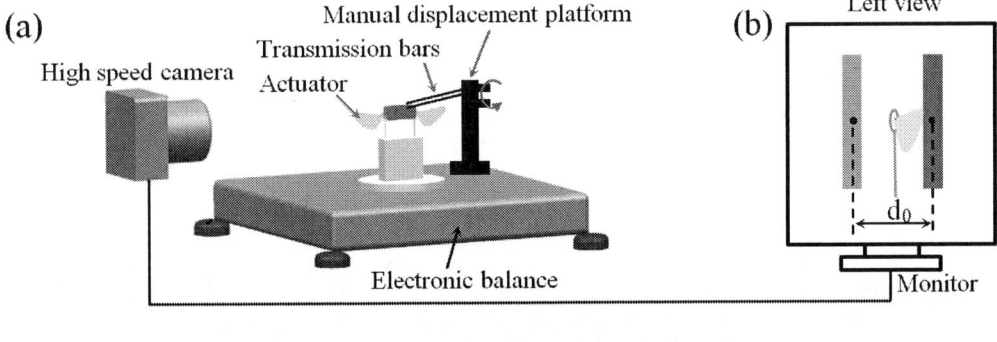

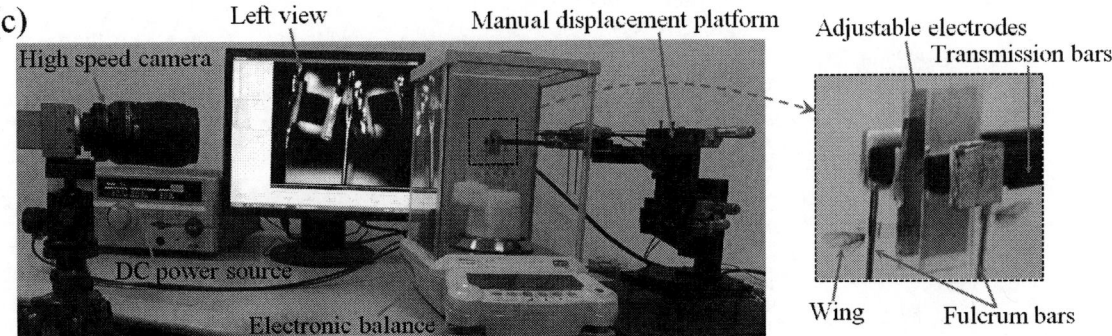

Figure 2: Test system of the actuator for wing motion and lift force measurements. (a) Measurement scheme. (b) A monitor showing left view of the actuator. (c) Photo of the whole test system.

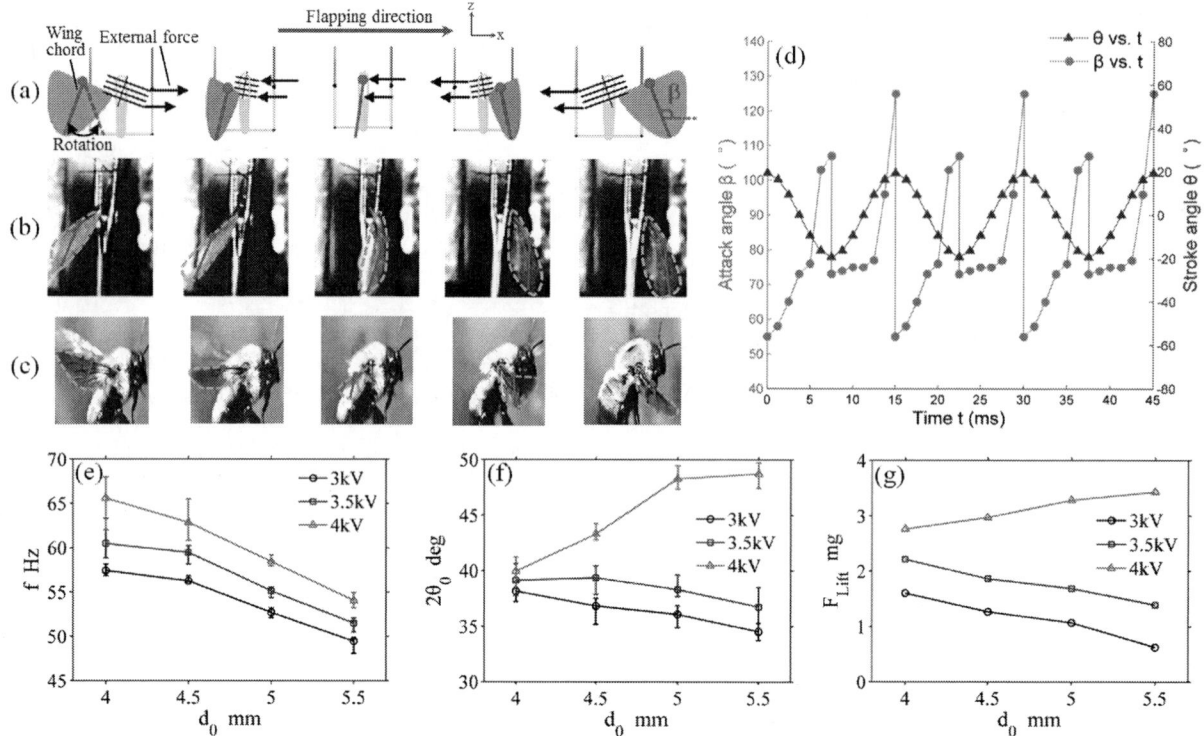

Figure 3: Step-by-step operations of the actuator in (a) illustrations and (b) high-speed experimental photos (d_0 = 4mm, V_{DC} = 4kV). (c) High-speed photos for wing motion of a real honey bee. (d) Stroke and attack angles of the wing during the vibration with frequency of 66.7Hz (d_0 = 4mm, V_{DC} = 4kV). Experimental measurements of (e) vibration frequency, (f) amplitude, and (g) lift force with respect to the input DC voltage (V_{DC}) and the gap distance (d_0).

manual displacement platform with two transmission bars, a high speed camera and an electronic balance. The actuator's electrode gap distance d_0 can be adjusted by the manual displacement platform through the transmission bars. Its wing motion can be captured by the high speed camera from the left view (Fig. 2b). And its average lift force can be measured based on the reduced weight readings on the electronic balance. Fig. 2c shows the photo of the whole test system.

Figure 3 shows the step-by-step operations of the flapping wing assembly in schematic illustrations (Fig. 3a) and photos taken by the high speed camera (Fig. 3b), which are directly compared with real insect wings in operation (Fig. 3c). In these images, the red solid lines represent the wing chord and β is the wing's attack angle. Since only the lowest beam is glued with the wings, the external force (resultant of electrostatic and air damping force) is different from other beams, which will make the lowest beam to vibrate out of phase and thus cause a rotation around y-axis up to 50 ° in the prototype device. The rotational motion has been proven to be key factor to generate high lift force in real insects [10]. While in the previously artificial flying insects, the rotational motion is realized through the passive deformation of an elastic wing hinge, which is usually subjected to short endurance due to fatigue problems [3, 5].

Fig. 3d plots the wing's stroke (θ, Fig. 1d) and attack angle (β, Fig. 3a) vs. time for several periods, which reveals that the wing is undergoing reciprocation at sinusoidal regime with stroke frequency of 66.7Hz, and

stroke amplitude of 38.49° (peak-to-peak). Figures 3e–3g record the stroke frequency (f), amplitude ($2\theta_0$), and lift force (F_{Lift}) under various electrode gap distances (d_0) and DC voltages (V_{DC}). It can be observed that, larger gap distance is beneficial to generating lift force in these tests, and higher DC voltages are required to sustain the higher flapping frequencies and amplitudes. These characteristics may be used for lift force optimization and flight attitude control in the future.

TAKEOFF TESTS

In order to demonstrate the self-lifting capability of the actuator, we have designed a vertical takeoff system based on the prototype device. As shown in Fig. 4a, the flapping wing assembly is supported by two long U-shape rails instead of the two fulcrum bars (Fig. 1b), and the electrodes are extended to give the wing assembly continuous electrostatic forces. These changes will allow the wing assembly to move upward along the U-shape rails if there is enough lift force generated by the flapping wings.

Test results show that, successful takeoffs have been achieved and recorded in video clips as shown in Fig. 4b, which illustrates that the flapping wing assembly (3.1mg in weight) is self-lifted 5cm above the ground once a V_{DC} of 5kV is applied. The lifting height is limited by the current experimental setup and can be further increased through extending the U-shape rails and electrodes.

As a comparison, same tests are also conducted for a metal beam assembly with no wings (no lift force). As

978-1-4799-7956-1/15 $31.00 © 2015 IEEE

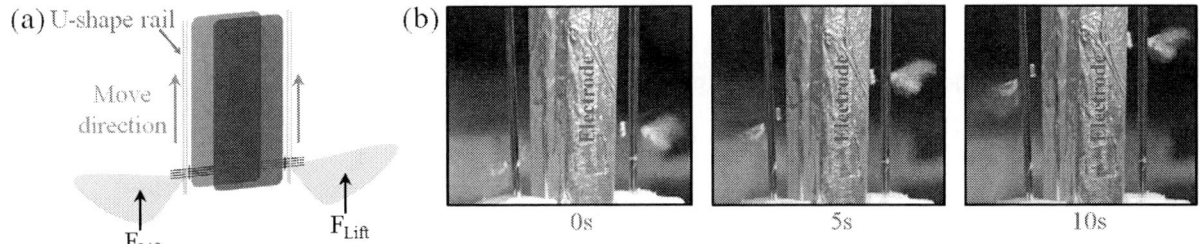

Figure 4: Vertical takeoff system for flapping wing assembly utilizing the electrostatic force. (a) Schematic view of the takeoff system. (b) Visual demonstration of the takeoff of the flapping wings with vibration frequency of 70Hz, and upward speed of 2.2mm/s. (Gap distance d = 5mm, V_{DC} = 5kV).

expected, without the wing structure, the assembly can't achieve vertical takeoff. The comparison tests demonstrate that it is the lift force generated by the flapping wings, not the climbing effect presented in previous references [11], results in the hovering of the flapping wing assembly. The successful takeoffs make the actuator promising in the area of micro-flying robots, where weight and size limits are major concerns.

CONCLUSIONS

This paper presents a self-excited electrostatic actuator with a simple construction and operation procedure, which can be excited into resonance under DC power source without using complex AC driving circuits. The actuator can drive two insect wings into reciprocation with rotational motion and thus generate effective lift force high enough to result in self-lifting of the flapping wing assembly. The wing motion and lift force measurements as well as takeoff tests in this paper demonstrate the feasibility of using this new actuator in flapping-wing robots. Our next work aims at further enhancing the lift force and reducing the total weight, for the purpose of realizing the takeoff of the whole actuator.

ACKNOWLEDGEMENTS

This work is supported by National Natural Science Foundation of China (Grant No. 11272025) and Defense Industrial Technology Development Program (Grant No. B2120132006).

REFERENCES

[1] M. Karpelson, G. Y. Wei, and R. J. Wood, "A review of actuation and power electronics options for flapping-wing robotic insects", *IEEE International Conference on Robotics and Automation,* Pasadena, CA, 2008.

[2] K. Suzuki, I. Shimoyama, and H. Miura, "Insect-model based microrobot with elastic hinges", *Journal of Microelectromechanical Systems,* vol. 3, pp. 4-9, 1994.

[3] K. Y. Ma, P. Chirarattananon, S. B. Fuller, and R. J. Wood, "Controlled Flight of a Biologically Inspired, Insect-Scale Robot", *Science,* vol. 340, pp. 603-607, 2013.

[4] R. J. Wood, "The first takeoff of a biologically inspired at-scale robotic insect," *IEEE Transactions on Robotics,* vol. 24, pp. 341-347, 2008.

[5] M. Azhar, D. Campolo, G.-K. Lau, L. Hines, and M. Sitti, "Flapping wings via direct-driving by DC motors", *International Conference on Robotics and Automation,* Karlsruhe, 2013, pp. 1397-1402.

[6] L. Hines, D. Campolo, and M. Sitti, "Liftoff of a Motor-Driven, Flapping-Wing Microaerial Vehicle Capable of Resonance", *IEEE Transactions on Robotics,* vol. 30, pp. 220-232, 2014.

[7] X. Yan, M. Qi, and L. Lin, "An autonomous impact resonator with metal beam between a pair of parallel-plate electrodes", *Sensors and Actuators A: Physical,* vol. 199, pp. 366-371, 2013.

[8] M. Qi, Z. Liu and X. Yan, "A low cycle fatigue test device for micro-cantilevers based on self-excited vibration principle", *Review of Scientific Instruments,* vol. 85, pp. 105005, 2014.

[9] R. Josephson, J. Malamud, and D. Stokes, "Asynchronous muscle: a primer", *J. Exp. Biol.,* vol. 203, pp. 2713-2722, 2000.

[10] M. H. Dickinson, F.-O. Lehmann, and S. P. Sane, "Wing Rotation and the Aerodynamic Basis of Insect Flight", *Science,* vol. 284, pp. 1954-1960, 1999.

[11] A. Degani, A. Shapiro, H. Choset, and M. T. Mason, "A dynamic single actuator vertical climbing robot", *IEEE/RSJ International Conference on Intelligent Robots and Systems,* San Diego, CA, 2007, pp. 2901-2906.

CONTACT

*Mingjing Qi, tel and fax: +86-10-82316356; qimingjing@buaa.edu.cn

SELF-ASSEMBLED HYDROGEL MICROSPRING FOR SOFT ACTUATOR

Koki Yoshida and Hiroaki Onoe

Department of Mechanical Engineering, Faculty of Science and Technology,
Keio University, Kanagawa, JAPAN

ABSTRACT

We found that a hydrogel microspring was formed by extruding sodium alginate pre-gel solution into calcium chloride solution using a bevel-tip microfluidic capillary. The formation of the microspring depends on the diameter of the capillary, flow rate and tip angle. As an example of soft actuator, the microspring including magnetic colloids was actuated by applying magnetic fields. We believe that the microspring will be used to various application including mechanical elements using soft materials.

INTRODUCTION

Soft and flexible gel actuators have attracted attentions for the realization of biomimetic soft-robots, artificial muscles and medical manipulators [1]. Although those gel actuators have highly-efficient energy conversion properties, one of the major issues to be solved for practical use is that gel actuators have relatively small magnitude of the displacement. On the other hand, in nature, microorganisms such as *vorticella* can move with large displacement using its stalk as a soft and flexible spring-shaped living actuator [2].

Thus, to overcome the problem, forming a gel actuator into a spring-shaped structure could be a promising strategy to expand the displacement. A couple of methods for making three-dimensional spring-shaped microstructures have previously been reported, for example the three-dimensional coil-shaped structures fabricated by two-photon stereolithography [3], and a spring-shaped microstructure made by material anisotropy [4]. However, there have been no reports that achieve the expansion and contraction of a hydrogel microspring actuator.

In this paper, we present a simple and easy method to spontaneously form spring-shaped microstructures for soft actuator (Figure 1). A calcium-alginate hydrogel microspring is continuously formed just by extruding sodium alginate pre-gel solution into calcium chloride solution with a bevel-tip capillary. Various materials such as colloids, polymer materials and microbe can be encapsulated in the calcium alginate hydrogel microspring. Thus it would be possible that the gel actuator with large displacement driven by various triggers such as magnetic field, light or chemical reaction could be achieved.

METHODS

BEVEL-TIP PREPARATION

We prepared two types of bevel-tip capillaries: a glass capillary and an injection needle. The glass capillary (outer diameter 1.5 mm, inner diameter 0.9 mm, length 900 mm, NARISHIGE, G-1.5) was sharpened by a tip-puller (NARISHIGE, PC-10). Then, the tip of the capillary was cut by a micro forge (NARISHIGE, MF-900) to adjust the tip of diameter, and beveled by a grinder (NARISHIGE, EG-44) (Figure 2). The half of the diameter of the glass

Self assembled hydrogel microspring

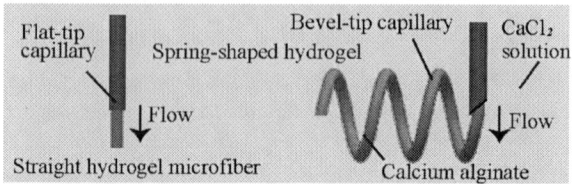

Magnetic actuation

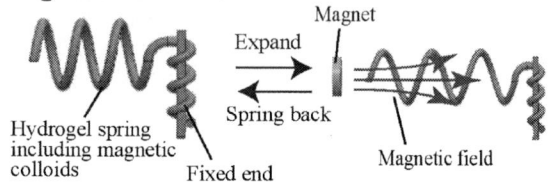

Figure 1: Concept of self-assembled hydrogel microspring

Glass capillary

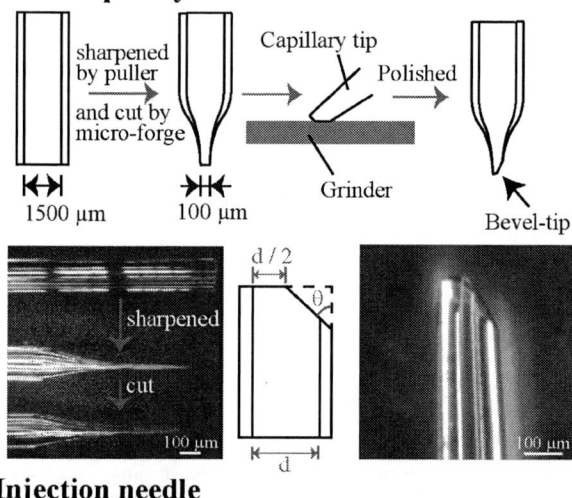

Injection needle

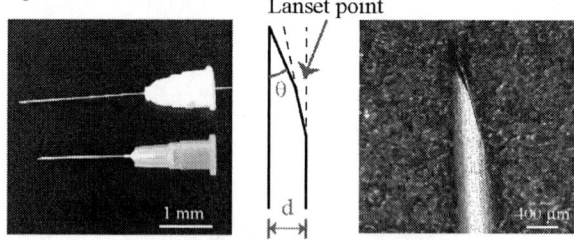

Figure 2: Bevel-tip preparation

capillary tip was grinded. The tip angle of the glass capillaries was adjusted to 20°-90°. As for injection needles, we used two different injection needles (TERUMO, NN-2719S, 27G, S.B and TERUMO, DN-3025K, 30G) that had the tip angle of 18°.

REAGENT PREPARATION

We prepared a 1.5% (v/v) sodium alginate solution (Wako, 194-13321) and 150 mM calcium chloride solution

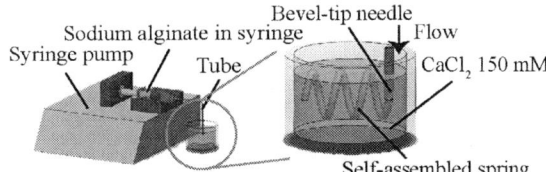

Figure 3: Fabrication setup for hydrogel microspring

(Wako, 090-00475) for the formation of hydrogel microspring using the glass capillary. 5% (v/v) fluorescent microbeads (life technologies, F8811, yellow-green fluorescent, 0.2 μm or F8810, red fluorescent, 0.2 μm) were encapsulated in the sodium alginate solution for visualization of the hydrogel. When we use the injection needle (TERUMO, DN-3025K, 30G), we used mixed solution of 0.5% (v/v) sodium alginate and 1.5% (v/v) propylene glycol alginate to adjust the viscosity.

For fabrication of the double-layered hydrogel microspring, two different sodium-alginate solutions labeled with fluorescent microbeads (life technologies,) were used.

For magnetic actuation of the hydrogel microspring, magnetic nanoparticles (Ferro Tec, EMG707) was mixed in the 1.5% (v/v) sodium alginate solution.

FABRICATION SETUP

A syringe (TERUMO, 1 mL) filled with the sodium alginate solution was set to a syringe pump (Kd Scientific, LEGATO 180). The syringe was connected to a tube (VICI, 1/16"×0.5 ETFE) by a connecter (ISIS, VPRF106). The bevel-tip capillary was connected to the tube by using silicon tube (inner diameter: 1 mm, outer diameter: 2 mm).

For the fabrication of the spring structure made of the calcium alginate hydrogel, we extruded the sodium alginate solution from the tip of the capillary into the calcium chloride solution at the constant flow rate, 0.5 μL/min (Figure 3). The calcium chloride solution was prepared in a petri dish. After the formation procedure, the fabricated hydrogel structure was moved to a glass plate with a drop of calcium chloride solution for the observation. We observed the fabricated hydrogel structure by an inverted fluorescence microscope (OYMPUS, IX73P1-22FL/PH).

RESULTS
RELATIONSHIP OF THE TIP ANGLE AND SHAPE OF HYDROGEL

Figure 4 shows the microscopic images of the beveled glass capillaries with the tip angle of 20°-90° (Figure 4 (a)-(e), the injection needle (Figure 4 (f)), and the calcium alginate hydrogel structures made by using the bevel-tip capillaries. Table 1 also summarized the result of what types of the hydrogel microstructures was formed. When the tip angle was around 50°-90°, the formed hydrogel structure was a straight string. On the other hand, when the tip angle was 20°, we could make hydrogel microspring. The boundary value of the tip angle whether the hydrogel microspring was formed or not was approximately 35°.

We could also succeeded in making hydrogel microsprings by using the injection needles. However, the microspring formation using the injection needles was

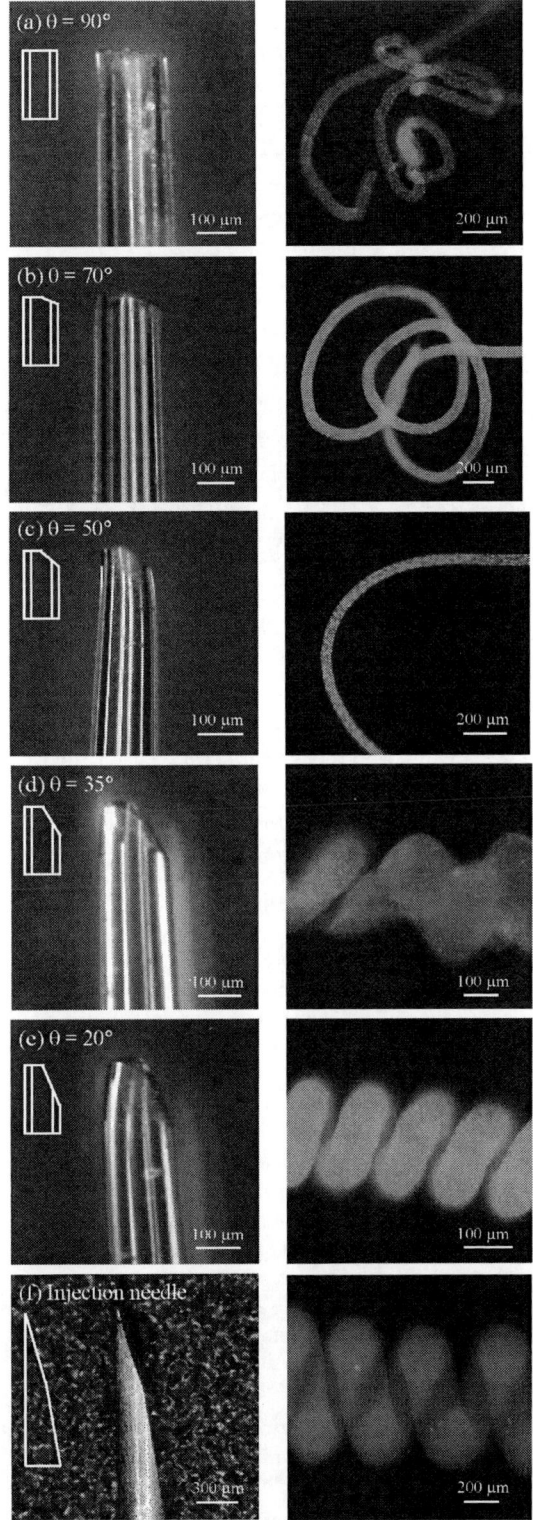

Figure 4: Beveled capillary tips and formed hydrogel structures

rather unstable compared to using the beveled glass capillaries. We considered that the one of the reasons was caused by the difference of the material between the glass capillaries and the injection needles: the difference may cause the different interfacial free energies between the sodium alginate solution and the inner surface of the capillaries/needles.

Table 1: The relationship between the tip angle of the glass capillaries and the shape of the hydrogel structure.

Tip angle (°)	Shape of hydrogel structure
90	straight
70	straight
50	straight
35	unstable
20	spring

Y-connecter setup   **Double-layered spring**

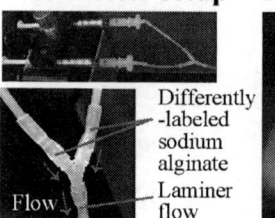
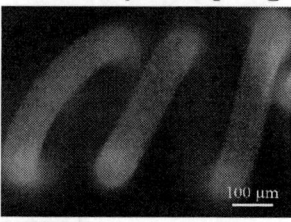

Figure 5: Set up using a Y-connecter (left) and an fluorescent image of the double-layered spring (right)

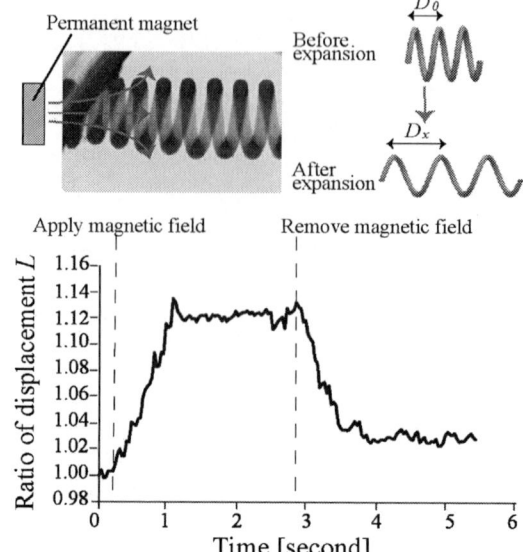

Figure 6: Magnetic actuation of the hydrogel microspring including magnetic colloids

DOUBLE-LAYERED SPRING

By using two-layer laminar flow created in the capillary with a Y-connecter (ISIS, VFY106Y), we fabricated double-layered hydrogel microsprings by using the glass capillary with tip angle of 20° (Figure 5). Both flow rates are equal. This demonstration indicates that an anisotropic spring composed of different materials could be formed in our fabrication method.

MAGNETIC ACTUATION

We demonstrated magnetic actuation of the hydrogel spring encapsulating magnetic colloids (Figure 6). The hydrogel microspring was made by using the injection needle (TERUMO, NN-2719S, 27G, S.B). By using a permanent magnet, magnetic field (~50 mT) was applied to the hydrogel maicrospring. We defined the pitch of the spring before applying magnetic field is D_0 and the one after applying magnetic field is D_x. The ratio of displacement ,L, is represented by

$$L = D_x / D_0 . \qquad (1)$$

The magnetic response of the hydrogel microspring was shown in Figure 6. We achieved the ratio of the displacement of the hydrogel microspring at ~10%.

CONCLUSION

We propose a formation method of forming calcium alginate hydrogel into microspring structures by using bevel-tip capillaries. The hydrogel microspring was fabricated continuously and spontaneously. The success of the spring formation depends on the tip angle of the capillaries. We also demonstrated the actuation of the hydrogel microspring encapsulating magnetic colloids by applying magnetic field. We believe that the microspring could be used to various application including mechanical elements using soft materials.

ACKNOWLEDGEMENT

This work was partly supported by Grant-in Aid for Young Scientist (A) (24686031), Japan Society for the Promotion of Science (JSPS), Japan.

REFERENCES

[1] M. Shahinpoor, Y. Bar-Cohen, J.O. Simpson, J. Smith, "Ionic polymer-metal composites (IPMCs) as biomimetic sensors, actuators and artificial muscles – a review", Smart Mater, Struct 7, 1998, R15-R30.

[2] S. Sareh, J. Rossiter, "Kirigami artificial muscles with complex biologically inspired morphologies", Smart Mater, Struct 22, 2013, pp 1-13.

[3] S. Tottori, L. Zhang, F. Qiu, K.K. Krawczyk, A. Franco-Obregón, B.J. Nelson, "Magnetic Helical Micromachines: Fabrication, Controlled Swimming, and Cargo Transport", Advanced Materials, Volume 24, 2012, pp 811-816.

[4] M.Yamada, S. Sugaya, M. Seki, "Microfludic synthesis of complex alginate fibers for the direction control of cell growth", 14th International conference on Miniaturized Systems for Chemistry and Life Science, October, 2010, Groningen, The Netherlands.

CONTACT

E-mail: K. Yoshida, yoshida-koki@a3.keio.jp, Tel: +81-45-566-1507.

PARAMETRIC DRIVE OF A TOROIDAL MEMS RATE INTEGRATING GYROSCOPE DEMONSTRATING < 20 PPM SCALE FACTOR STABILITY

D. Senkal[1], E.J. Ng[2], V. Hong[2], Y. Yang[2], C.H. Ahn[2], T.W. Kenny[2], and A.M. Shkel[1]
[1]University of California, Irvine, California, USA
[2]Stanford University, Palo Alto, California, USA

ABSTRACT

In this paper, we report parametric drive of a MEMS rate integrating gyroscope for reduction of drifts induced by drive electronics. A Toroidal Ring Gyroscope (TRG) with 100k Q-factor at 1760 μm diameter was fabricated in epitaxial silicon encapsulation (EpiSeal) process. Device was operated in rate integrating mode using a FPGA based control system. In contrast to the conventional amplitude control architecture, the central star electrode of the gyroscope, connected to a single drive channel was utilized for amplitude control of both modes. Due to the parametric pumping effect, energy added to each (x and y) mode is proportional to the existing amplitude of the respective mode. As a result, errors associated with finding the orientation of the standing wave and x-y drive gain drift are bypassed. Compared to conventional x-y drive architecture, as high as 14x improvement in scale factor stability was observed with parametric pumping, resulting in better than 20 ppm scale factor stability without any compensation or temperature stabilization.

INTRODUCTION

Degenerate mode Coriolis Vibratory Gyroscopes (CVGs) can be instrumented in two primary modes of operation: (1) rate gyroscope mechanization where the standing wave pattern is locked to a fixed orientation, (2) rate integrating (whole angle) gyroscope mechanization, where the standing wave is free to rotate under the effect of Coriolis forces. Rate integrating gyroscope offers a number of unique advantages compared to conventional vibratory rate gyroscopes, including mechanically unlimited dynamic range, low noise due to degenerate mode operation, and exceptional scale factor stability.

All axi-symmetric MEMS gyroscope architectures are capable of whole angle mechanization. Examples include ring gyros [1], disk gyros [2, 3], and 3-D wineglass gyroscopes [4, 5]. Typically, these architectures utilize two sets of drive electrodes for the two resonant modes of the gyroscope. As a result, amplitude control in whole angle mechanization would require finding the orientation of the standing wave and pumping energy along this direction using two sets of drive electrodes (vector drive). However, this method is susceptible to drift due to gain unbalance in drive electronics, errors in calculating the angle of the standing wave, as well as, the time delay between estimation of the standing wave and the actual amplitude command.

In this work, we explore parametric drive for amplitude control of MEMS rate integrating gyroscopes, which has previously been reserved for high performance macro-scale Vibratory Gyros [6]. In contrast to conventional amplitude control, a single drive channel connected to the central star electrode of the gyroscope was utilized for amplitude control of both modes (scalar

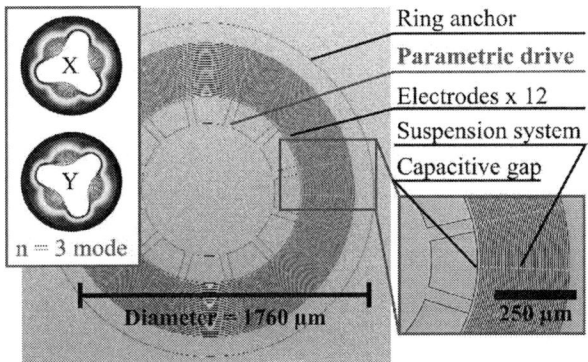

Figure 1: A 100k Q-factor, epitaxial silicon encapsulated Toroidal Ring Gyroscope was used in the experiments [7]. Device consists of an outer ring anchor, distributed suspension system and an inner electrode assembly.

drive). Even though a single drive channel is used, due to the parametric pumping effect, energy added to each (x and y) mode is proportional to the existing amplitude of the respective mode. This permits amplitude control of the standing wave at any arbitrary angle with minimal amount of perturbation. The scalar nature of the amplitude controller helps bypass errors associated with finding the orientation of the standing wave, time delay in the calculation, and x-y drive gain drift.

An additional benefit of parametric drive of MEMS gyroscopes is the minimization of the electrical feed-through between the actuation and pick-off channels [8, 9]. For a conventional MEMS gyroscope, actuation and pick-off signals occur at the same frequency. Any feed-through from the actuation signal will corrupt the pick-off channel, lowering overall performance of the system. Parametric drive mitigates this problem by separating the frequency of drive and pick-off channels. Since parametric drive frequency is a multiple of systems drive frequency, the electrostatic feed-through into the sense channel can be filtered out.

DESIGN
Gyroscope Architecture

A Toroidal Ring Gyroscope (TRG) with Q-factor of 100k at 70 kHz central frequency and 1.7 mm diameter was used for the experiments; the device consists of an outer ring anchor surrounding a distributed mass system and central electrode architecture [7], Fig. 1. As opposed to other axi-symmetric devices with central support structures vibration energy of TRG is concentrated at the innermost ring, Fig. 2. The distributed support structure decouples the vibrational motion from the substrate. This decoupling mitigates anchor losses into the substrate and prevents die/package stresses from propagating into the vibratory structure.

978-1-4799-7956-1/15 $31.00 © 2015 IEEE

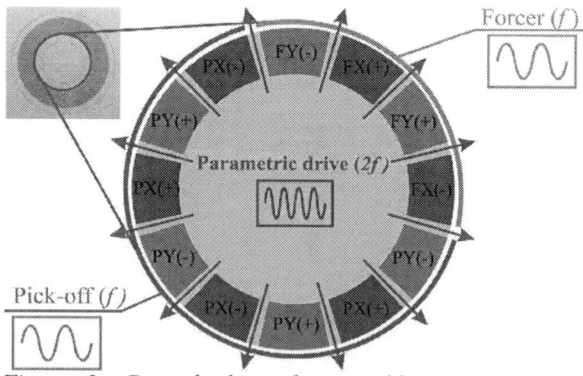

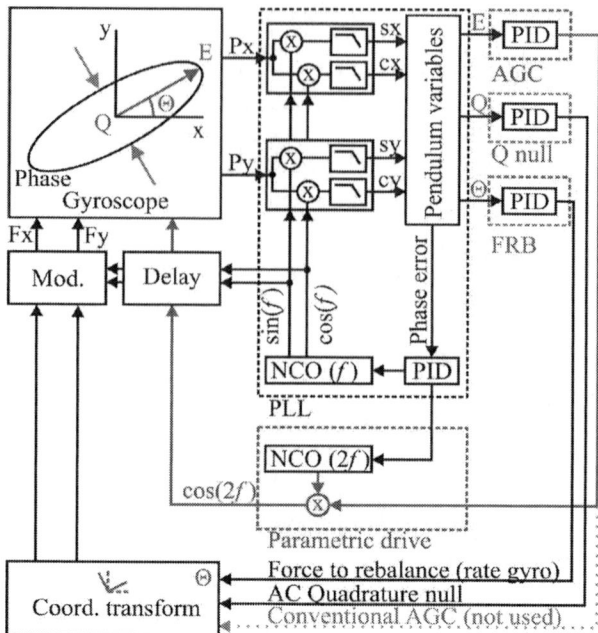

Figure 2: Central electrode assembly consists of 12 discrete electrodes, divided into 4 drive and 8 pick-off electrodes, and one star shaped parametric electrode.

Fabrication

The Toroidal Ring Gyroscope (TRG) was fabricated using a wafer-level epitaxial silicon encapsulation process (EpiSeal) [10]. EpiSeal process utilizes epitaxially grown silicon to seal the device layer at extremely high temperatures, which results in an ultra-clean wafer-level seal. This results in high vacuum levels (as low as 1 Pa) without the need for getter materials for absorption of sealing by-products. The epi-seal encapsulation process was proposed by researchers at the Robert Bosch Research and Technology Center in Palo Alto and then demonstrated in a close collaboration with Stanford University. This collaboration is continuing to develop improvements and extensions to this process for many applications, while the baseline process has been brought into commercial production by SiTime Inc.

Electrode Architecture

Device is operated in n = 3 mode, to help reduce the frequency splits associated with n = 2 mode on <100> silicon. Electrode assembly is located at the center of the gyroscope and consists of twelve discrete electrodes and one central star electrode, Fig 2. Discrete electrodes are distributed in groups of six onto the two degenerate wineglass modes. Out of six electrodes, two electrodes were used as a forcer and four as a pick-off for each mode, giving a total of four forcer and eight pick-off electrodes across the gyro.

In this work, the central star-shaped electrode is used for parametric pumping. Due to the twelve-pointed circular nature of this electrode, parametric pumping has equal contribution to the both degenerate n = 3 wineglass modes.

Control System

A Kintex 7 FPGA was used for controlling the gyro. The key component of this approach is a PLL loop that tracks the gyro motion at any arbitrary pattern angle as opposed to locking onto one of the primary gyro axis. Once the PLL lock is established, the FPGA extracts the slow moving variables: amplitude (E), quadrature error (Q), and pattern angle (θ) using the equations presented in

Figure 3: Rate integrating gyro controller with parametric drive, implemented on a Kintex 7 FPGA running at 1 MHz.

[11], Fig 3. A PID controller acts on each of these variables. These are Amplitude Gain Control (AGC) acting on E, quadrature null acting on Q and force-to-rebalance (FRB) that controls pattern angle (θ). For the whole angle mechanization, FRB is disabled so that the standing wave is free to precess. Once the correct command voltages for E, Q and θ are established, a coordinate transform around θ is performed to align these signals to the standing wave. This is followed by modulation of the command voltages at PLL frequency.

For the, parametric drive, a secondary numerically controlled oscillator (NCO) is used to generate a sine wave at twice the PLL frequency. This signal is applied to the central star-shaped electrode to parametrically pump energy in x-y plane at twice the resonance frequency, Fig 2. Due to the parametric pumping effect, energy added to each (x and y) mode is proportional to the existing amplitude of the respective mode [8]:

$$\ddot{x} + \frac{\omega_x}{Q_x}\dot{x} + \left(\omega_x^2 + \frac{F_p}{m_{eq}} sin(2\omega t + \phi_p)\right)x$$
$$= \frac{F_x}{m_{eq}} sin(\omega t + \phi_f) + 2\eta \dot{y}\Omega_z, \qquad (1)$$

$$\ddot{y} + \frac{\omega_y}{Q_y}\dot{y} + \left(\omega_y^2 + \frac{F_p}{m_{eq}} sin(2\omega t + \phi_p)\right)y$$
$$= \frac{F_y}{m_{eq}} sin(\omega t + \phi_f) - 2\eta \dot{x}\Omega_z, \qquad (2)$$

where ω_x, ω_y and Q_x, Q_y are the resonance frequencies and the Q-factors of the two degenerate modes, m_{eq} is the equivalent mass of the vibratory system, η is the angular gain factor, ω is the drive frequency and ϕ_p, ϕ_f are the phase of the parametric and vector drives respectively.

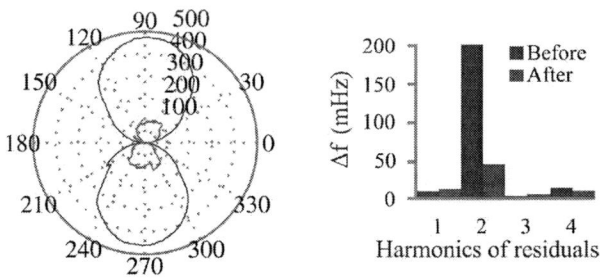

a) PLL frequency (Δf) with respect to pattern angle (mHz).

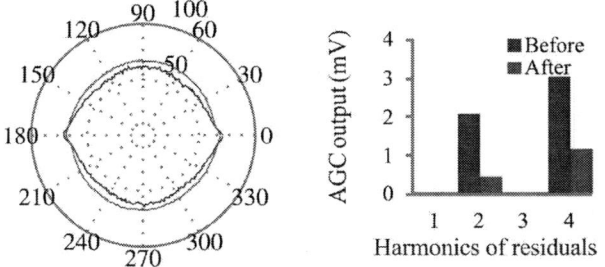

b) Parametric drive amplitude gain control (AGC) output (mV).

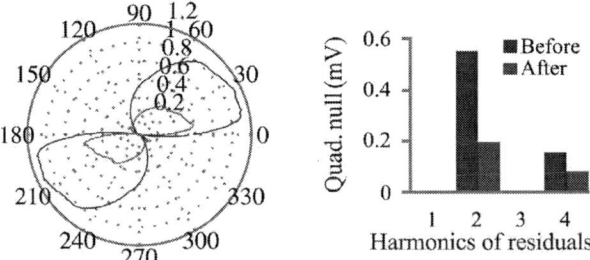

c) Quadrature null control output (mV).

Figure 4: Tuning of the gyro based on residuals of pattern angle data from PLL, parametric AGC and quadrature null loops.

This creates a preferential direction of pumping along the orientation of the standing wave without the need for any coordinate transformation around θ:

$$\theta = atan\left(\frac{y}{x}\right) \qquad (3)$$

Open loop parametric drive is typically unstable for nominal drive amplitudes [8, 9], which causes the gyro amplitude to increase exponentially for a fixed parametric drive signal. For this reason a secondary AGC controls the parametric drive voltage as to keep the gyro amplitude stable. This closed loop operation permits parametric drive of the gyro at a wide range of drive amplitudes, outside the stability boundary of open loop parametric drive.

Typical gyro startup procedure begins with driving the gyro to a preset amplitude using conventional (at resonance) drive. Once the PLL and AGC stabilize, drive signal is disabled and immediately parametric drive AGC is enabled. This switch occurs within one clock cycle of the FPGA and eliminates over-shoots in drive amplitude, which would otherwise occur while starting up the high-Q resonator.

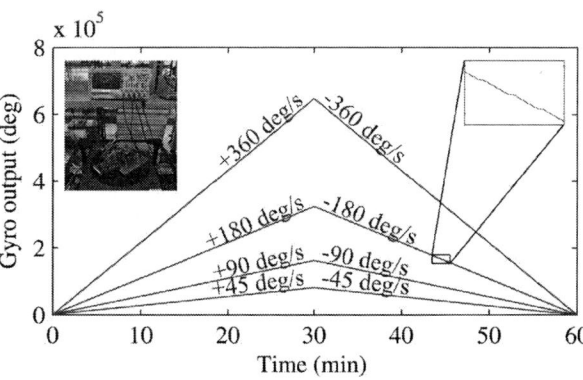

Figure 5: Experimental demonstration of rate integrating operation under parametric drive.

EXPERIMENTAL RESULTS

Pattern Angle Data

Pattern angle data for the gyroscope was obtained by changing the orientation of the standing wave (θ) using the force-to-rebalance loop and recording the gyro output with respect to pattern angle (θ). Fourier series expansion of pattern angle data from PLL, quadrature, and AGC loops was used for calibrating the gyro. Second harmonic of the PLL output provides run-time identification of the frequency split (Δf), which was used to tune the frequency split down to 50 mHz (700 ppb). Parametric AGC command signal displayed a 2 mV (~4%) variation on second harmonic, which was attributed to pick-off gain unbalance between x and y modes. Adjusting the pick-off gains in the FPGA, reduced this unbalance down to < 0.5 mV. The combined effect of tuning the frequency split and the pick-off gain unbalance resulted in an overall 2.5x reduction in the required quadrature null command signal, Fig 4.

Rate Integrating Gyro Response

In order to test the rate integrating operation, the gyro was driven using parametric drive and a constant rotation rate was applied for 1 hour, switching the direction at 30 minutes mark. Fig. 5 shows the unwrapped gyro response for four different speeds. This experiment was later repeated using conventional (vector) drive. A linear fit to the data revealed a combined electrical/mechanical angular gain factor of ~0.6. Comparison of residuals from both experiments is shown in, Fig 6. For all rate inputs the parametric drive resulted in better scale factor stability compared to conventional drive architecture. As predicted, the highest difference between conventional drive and parametric drive occurred at higher rotation rates. Time lag in calculation of the drive vector becomes more important at higher rotation rates as this lag can cause the drive vector to couple into the gyro output. As a result, any change in the drive vector amplitude either due to drive gain drifts or a Q-factor change in the resonator element can affect scale factor stability.

For 360°/s rate input over a 30 minute period standard deviation of accumulated error for conventional drive was 176° versus 13° for parametric drive. This resulted in 14x improvement for parametric drive and < 20 ppm scale factor stability overall, without any compensation or temperature stabilization.

978-1-4799-7956-1/15 $31.00 © 2015 IEEE   31

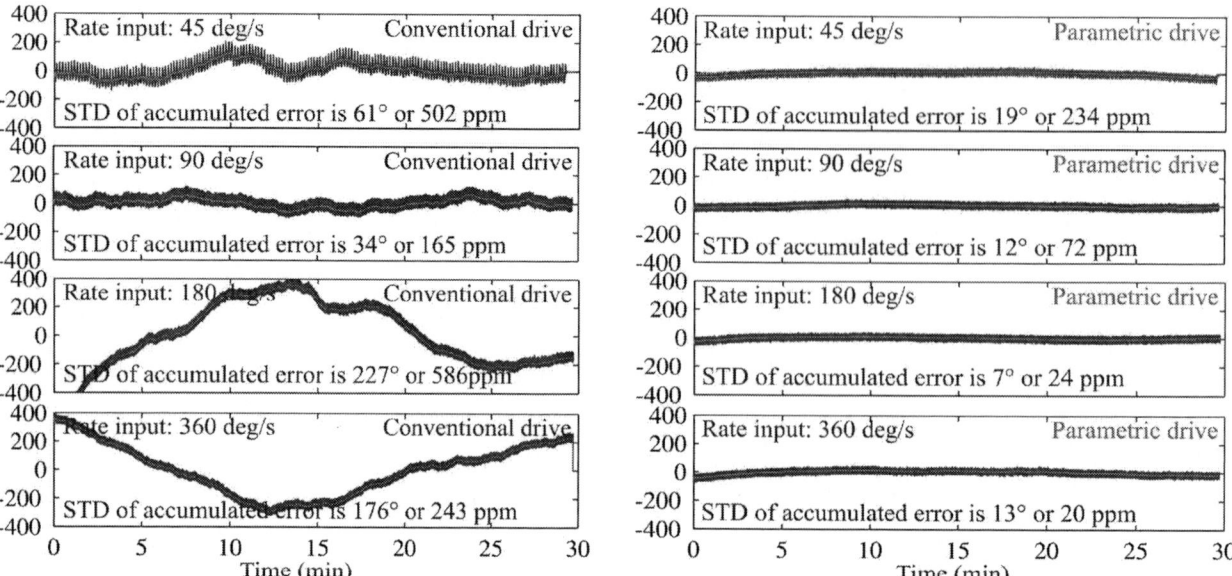

Figure 6: Comparison of residual errors of conventional drive and parametric drive for different rate inputs.

CONCLUSIONS

Parametric drive of a MEMS rate integrating gyroscope was presented for the first time. Parametric pumping was used to bypass the errors associated with finding the orientation of the standing wave, time delay in the calculation and x-y drive gain drift. This resulted in as high as 14x improvement in scale factor stability compared to conventional x-y drive and better than 20 ppm scale factor stability without any compensation or temperature stabilization. In addition, parametric drive reduces drive to pick-off electrical feed-through by creating a frequency separation between drive and pick-off channels.

Techniques presented in this paper can be used to improve performance of other axi-symmetric gyro architectures, such as ring, disk, and wineglass gyros.

ACKNOWLEDGEMENTS

Design and characterization of devices was done in UCI Microsystems Laboratory, fabrication was done at Stanford Nanofabrication Facility. This work was supported in part by the Defense Advanced Research Projects Agency Precision Navigation and Timing Program managed by Dr. A. Shkel and Dr. R. Lutwak under Contract N66001-12-1-4260, and in part by the National Science Foundation through the National Nanotechnology Infrastructure Network under Grant ECS-9731293

REFERENCES

[1] F. Ayazi and K. Najafi, "Design and fabrication of high-performance polysilicon vibrating ring gyroscope," in *IEEE MEMS*, Heidelberg, Germany, 1998, pp. 621-626.

[2] Z. Hao, S. Pourkamali, and F. Ayazi, "VHF single-crystal silicon elliptic bulk-mode capacitive disk resonators-part I: design and modeling," in *IEEE/ASME JMEMS*, 13, (6), pp. 1043–1053, 2004.

[3] T. H. Su, S. H. Nitzan, P. Taheri-Tehrani, M. H. Kline, B. E. Boser, and D. A. Horsley, "Silicon MEMS Disk Resonator Gyroscope with an Integrated CMOS

Analog Front-End," in *IEEE Sensors Journal*, 14, (10), pp. 3426–3432, 2013.

[4] D. Senkal, M. J. Ahamed, S. Askari, and A. M. Shkel, "1 million Q-factor demonstrated on micro-glassblown fused silica wineglass resonators with out-of-plane electrostatic transduction," in *Hilton Head*, Hilton Head Island, SC, USA, 2014, pp. 68–71.

[5] J. Cho, J.-K. Woo, J. Yan, R. L. Peterson, and K. Najafi, "A high-Q birdbath resonator gyroscope (BRG)," in *TRANSDUCERS*, Barcelona, Spain, 2013, pp. 1847–1850.

[6] D. M. Rozelle, "The hemispherical resonator gyro: From wineglass to the planets," in *AAS/AIAA Space Flight Mechanics Meeting*, 2009, pp. 1157–1178.

[7] D. Senkal, S. Askari, M. J. Ahamed, E. J. Ng, V. Hong, Y. Yang, C. H. Ahn, T. W. Kenny, and A. M. Shkel, "100k Q-factor toroidal ring gyroscope implemented in wafer-level epitaxial silicon encapsulation process," in *IEEE MEMS*, San Francisco, California, USA, 2014, pp. 9–12.

[8] K. M. Harish, B. J. Gallacher, J. S. Burdess, and J. A. Neasham, "Experimental investigation of parametric and externally forced motion in resonant MEMS sensors," in *IOP JMM*, 19, (1), pp. 15-21, Jan. 2009.

[9] L. A. Oropeza-Ramos, C. B. Burgner, and K. L. Turner, "Robust micro-rate sensor actuated by parametric resonance," *Sensors and Actuators A: Phyical.*, 152, (1), pp. 80–87, 2009.

[10] R. N. Candler, M. A. Hopcroft, B. Kim, W-T. Park, R. Melamud, M. Agarwal, G. Yama, A. Partridge, M. Lutz, T. W. Kenny, "Long-Term and Accelerated Life Testing of a Novel Single-Wafer Vacuum Encapsulation for MEMS Resonators," in *IEEE/ASME JMEMS*, 15, (6), pp. 1446-1456, 2006.

[11] D. D. Lynch, "Vibratory gyro analysis by the method of averaging," in *Intl. Conf. on Gyroscopic Tech. and Navigation*, St. Petersburg, Russia, 1995, pp. 26-34.

CONTACT

*D. Senkal, tel: +1-949-945-0858; dsenkal@uci.edu.

A 7PPM, 6°/HR FREQUENCY-OUTPUT MEMS GYROSCOPE

Igor I. Izyumin[1], Mitchell H. Kline[1], Yu-Ching Yeh[1], Burak Eminoglu[1], Chae Hyuck Ahn[2],
Vu A. Hong[2], Yushi Yang[2], Eldwin J. Ng[2], Thomas W. Kenny[2], and Bernhard E. Boser[1]
[1]University of California, Berkeley, California, USA
[2]Stanford University, Stanford, California, USA

ABSTRACT

We report the first frequency-output MEMS gyroscope to achieve < 7 ppm scale factor accuracy and $< 6°$/hr bias stability with a $3.24\,\text{mm}^2$ transducer. By implementing continuous-time mode reversal in an FM gyro, the rate signal is modulated away from DC, making the system insensitive to the resonant frequency of the transducer. The scale factor is almost entirely ratiometric, depending primarily on the mechanical angular gain factor of the transducer and the accuracy of the timing reference. Scale factor sensitivity to variations in quality factor, electro-mechanical coupling coefficients, and circuit drift is significantly reduced compared to conventional open-loop and force-rebalance operating modes.

INTRODUCTION

Achieving ppm-level scale factor stability in conventional amplitude-based MEMS gyroscopes has been a difficult challenge. As a typical example, [1] employs a sophisticated self-calibration technique to achieve an rms scale factor accuracy of 350 ppm over a $10\,°\text{C}$ temperature range.

Any sensor measures its input relative to a reference. For a rate gyro, the input has units of angular frequency; therefore, its reference must also be a frequency. However, amplitude-based rate gyros measure a force, either directly (in force-rebalance mode) or indirectly, by measuring sense axis displacement. The reference is constructed implicitly from a number of transducer and readout circuit variables, which include the effective mass, stiffness, various coupling factors (in turn determined by bias voltages), and readout circuit parameters, such as amplifier gains. It is very difficult to measure or control all of these parameters to the required level of accuracy, thus limiting the achievable scale factor stability.

Previously reported circular-orbit FM gyroscopes [2] potentially improve scale factor stability by sensing rate directly, without intermediate voltage-to-force conversions and associated errors. However, they are unable to distinguish rate from transducer resonant frequency drift, resulting in very poor bias stability (1550 °/hr in [2]). The bias drift is dominated by the temperature dependence of the resonant frequency on the Young's modulus of silicon, about 30 ppm/K. For a 30 kHz device, this translates to a temperature coefficient of 324 °/s/K, implying that $1\,\mu\text{K}$ accurate temperature regulation is needed for a 10 °/hr gyroscope. A second structure with opposite sensitivity reduces the coefficient to about 10 ppb/K, or 400 °/hr/K, but orders of magnitude improvement is still needed in order to reduce this error to < 10 °/hr.

This work presents an FM-based gyroscope operating mode that is insensitive to the resonant frequency of the transducer, while maintaining high scale factor stability.

LISSAJOUS FM OPERATION

Background

The proposed operating mode employs a mode mismatched but otherwise symmetric transducer. Both transducer axes are driven into oscillation at their natural frequency using sustaining oscillator loops. Because the axis frequencies are not equal, the proof mass follows a Lissajous trajectory. Figure 1 shows the block diagram of the gyroscope system, as well as expressions for the X and Y axis oscillation frequencies (for a Z-axis gyro) resulting from the analysis of the system differential equations in the complex baseband using the method of averaging [3, 4]. It can be seen that applied angular rate sinusoidally modulates the frequency of oscillation of both gyro axes. The depth of modulation is equal to the angular rate scaled by the angular gain factor α_z of the gyroscope, and the modulation period is determined by the frequency difference Δf.

The principle of operation may be understood by first considering the behavior of the circular orbit gyroscope as a function of the phase shift $\Delta\phi$ between the X and Y oscillations. When $\Delta\phi = \pm\pi/2$, the proof mass follows a circular orbit at the natural frequency of the transducer in an inertial frame of reference. Applied rotation causes the frame of the transducer to rotate relative to the orbit, changing the observed orbit frequency. When $\Delta\phi = 0$ or $\Delta\phi = \pm\pi$, the proof mass oscillates along a straight line, and the oscillation frequency is insensitive to angular rate. In the Lissajous FM (LFM) mode, $\Delta\phi$ is not a static value, but is instead a periodic ramp between $-\pi$ to $+\pi$ with a period $1/\Delta f$; as a result, the rate sensitivity of the circular orbit FM gyroscope is modulated sinusoidally at Δf.

Modulating the rate sensitivity causes the rate signal to move from DC to the frequency of modulation; this is iden-

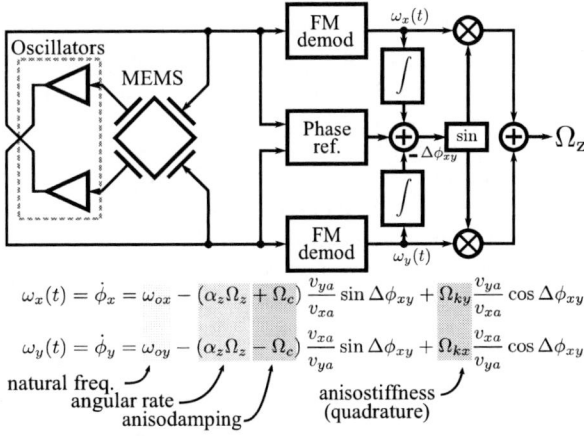

$$\omega_x(t) = \dot\phi_x = \omega_{ox} - (\alpha_z\Omega_z + \Omega_c)\frac{v_{ya}}{v_{xa}}\sin\Delta\phi_{xy} + \Omega_{ky}\frac{v_{ya}}{v_{xa}}\cos\Delta\phi_{xy}$$

$$\omega_y(t) = \dot\phi_y = \omega_{oy} - (\alpha_z\Omega_z - \Omega_c)\frac{v_{xa}}{v_{ya}}\sin\Delta\phi_{xy} + \Omega_{kx}\frac{v_{xa}}{v_{ya}}\cos\Delta\phi_{xy}$$

natural freq. · angular rate · anisodamping · anisostiffness (quadrature)

Figure 1: Block diagram of Lissajous FM gyroscope and expressions for X and Y axis instantaneous frequencies.

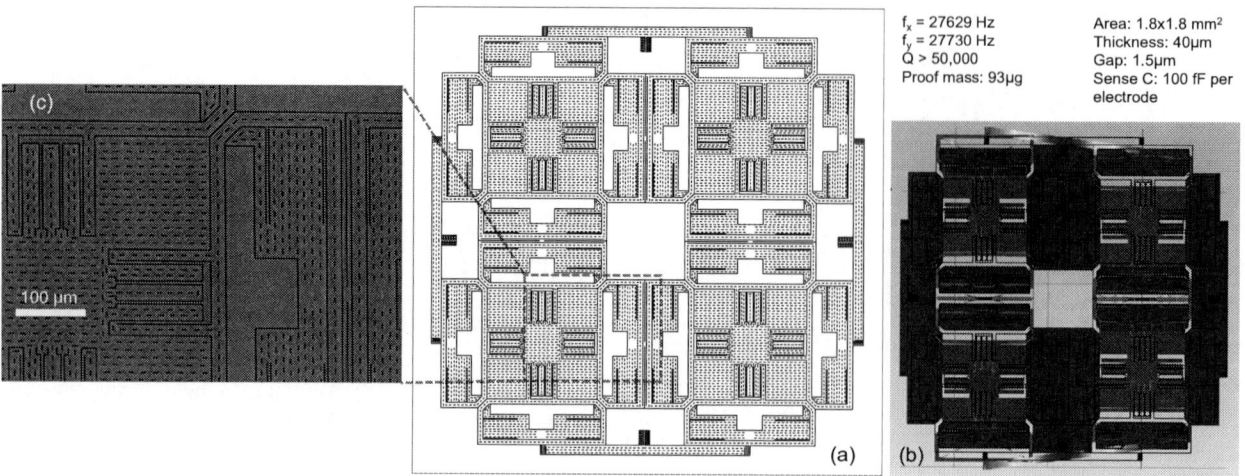

Figure 2: (a) layout of quad-mass MEMS transducer; (b) finite-element simulation of primary mode shape; (c) SEM photograph of fabricated structure.

tical to the chopper stabilization technique commonly used in precision amplifiers. As a result, the gyroscope is made insensitive to slow variations in the natural frequency of the transducer, as might result from temperature fluctuations and other drift sources. Furthermore, from the equations in fig. 1, it may be seen that the scale factor of the gyroscope depends only on the angular gain factor α_z (a stable, dimensionless parameter set by the transducer geometry) and the velocity amplitude ratios v_{xa}/v_{ya} and v_{ya}/v_{xa}. If the demodulated X and Y channel outputs are summed, the sensitivity to velocity amplitude mismatch is greatly reduced due to the reciprocal summation of the two amplitude ratios. For example, a relatively large mismatch of 0.5% between the X and Y axis velocities results in a scale factor error of only 12 ppm. Due to the periodic mode reversal inherent to LFM operation, summation of the X and Y outputs cancels reciprocal anisodamping errors, potentially improving bias stability. As in the ordinary AM mode, quadrature errors are rejected by phase-sensitive demodulation.

Signal processing

Angular rate measurement in the LFM operating mode requires high-resolution measurements of the frequencies of the X and Y axes. The FM demodulator must have sufficient bandwidth to not attenuate the tone at the split frequency. In order to recover the angular rate, the measured frequencies of the X and Y axes are integrated and then differenced, producing a phase ramp $\Delta\phi$. A digital phase detector is used to provide an initial condition for the integrators. Applying a sine operation to $\Delta\phi$ recovers the carrier; synchronous demodulation is then used to demodulate the angular rate signal to the baseband. The X and Y signals are then summed and filtered to produce the final rate output. The block diagram in fig. 1 illustrates this process.

Discussion

LFM operation presents several significant advantages over other operating modes. Unlike open-loop and force-rebalance operating modes, LFM gyros can achieve ppm-level scale factor stability without any additional calibration layers. Because the output signal is a frequency, rather

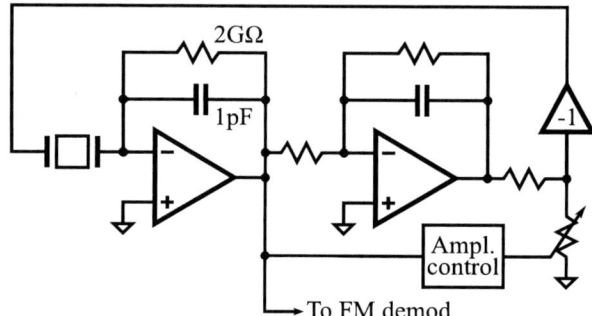

Figure 3: Sustaining oscillator differential half-circuit.

than an analog voltage or a current, a very high dynamic range can be accommodated without range switching. Unlike the circular-orbit FM mode, the LFM operating mode is not sensitive to natural frequency drift and does not impair bias stability. Because the required Δf is small, fully-symmetric transducers such as the quad-mass gyro can be employed, allowing the gyroscope to benefit from the high quality factor, vibration rejection, and other advantages inherent to such structures.

The choice of the split frequency Δf incurs some design trade-offs. The split frequency needs to be high enough to avoid injection locking and slow drift components near DC and accommodate the required gyroscope bandwidth. Unlike amplitude-based gyros, there is no trade-off with scale factor stability, allowing a significantly smaller split frequency to be used. Furthermore, because the natural frequency of each mode can be easily observed, active electrostatic tuning can be used to precisely control the split frequency, eliminating fabrication mismatch and drift. Since the noise contribution of the readout circuits depends on the frequency split, reducing the split frequency decreases circuit power quadratically. LFM can thus enable low-power operation while maintaining low ARW and high scale factor stability.

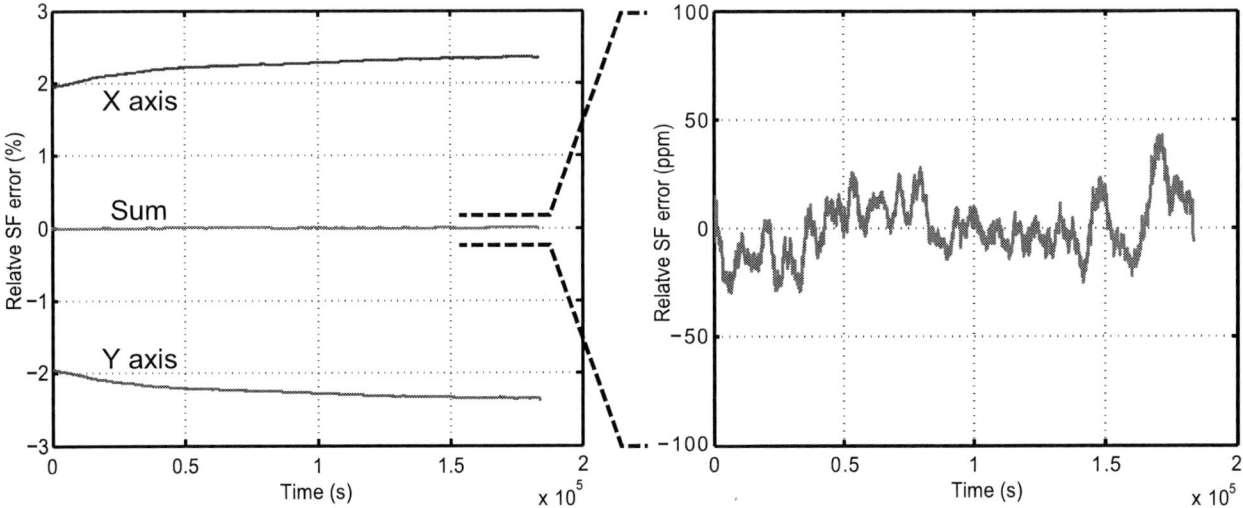

Figure 4: Time-domain scale factor measurement. A one-hour moving average was applied to the combined curve to reduce short-term ARW-related noise and show the long-term trend.

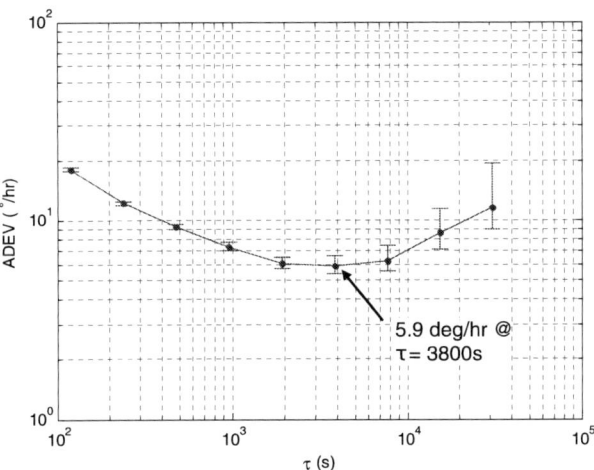

Figure 5: Bias Allan deviation.

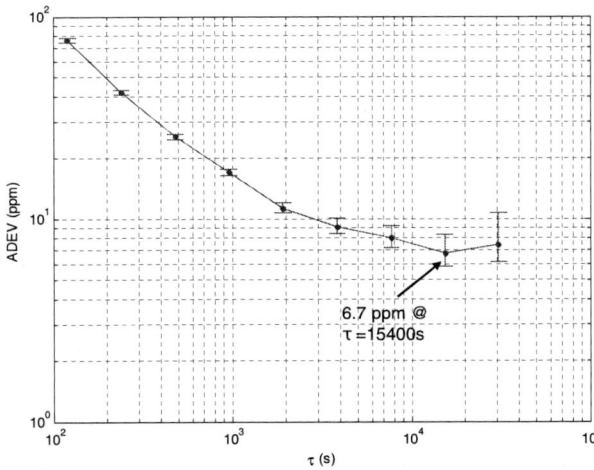

Figure 6: Scale factor Allan deviation.

EXPERIMENTAL CHARACTERIZATION

The LFM operating mode was characterized using a quad-mass structure similar to [5]. The layout, primary eigenmodes, design parameters, and measured resonant frequencies of the MEMS transducer are shown in fig. 2. The transducer was fabricated using the epi-seal vacuum encapsulation process described in [6]; fig. 2c shows an SEM micrograph of the fabricated structure. The structure uses differential parallel-plate actuation and sensing using 8 pairs of differential electrodes. The spring anchors are designed to minimize transfer of packaging stress to the springs. The transducer used for testing had an intrinsic Δf of 101 Hz (arising from fabrication mismatch); no electrostatic tuning was used.

The sustaining oscillators were implemented as discrete Pierce oscillators with opamp-based active integrators; the schematic of the differential half-circuit of a single oscillator channel is shown on fig. 3. FM demodulation was performed by a pair of low-power, high-resolution frequency-to-digital converters implemented on a custom ASIC [7]. The remaining demodulation operations were performed using digital signal processing.

The gyroscope bias and scale factor stability were characterized with a 50-hour test. The test consisted of repeated +90, 0, and $-90°/$s angular rate measurements, which allowed bias and scale factor stability to be measured simultaneously. Testing was performed at room temperature with no active temperature regulation.

Figure 4 shows the measured scale factor error drift over time. Due to velocity amplitude mismatch caused by circuit imperfections and drift, individual X and Y axis signals exhibited a 1.9% initial scale factor error and additional drift of 0.45% during the test; this is a typical performance level for a conventional AM gyro with no calibration circuits. Combining the X and Y signals almost entirely eliminates this error source: fig. 4 shows no significant drift trend for the summed output. The Allan variance of the scale factor reaches a minimum of 6.7 ppm at $\tau = 15400$s

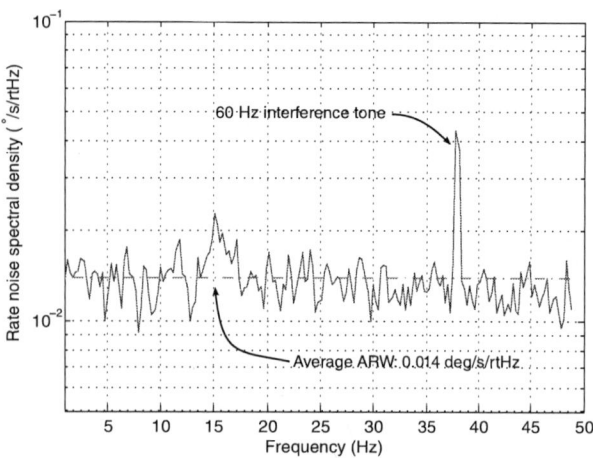

Figure 7: In-band rate noise power spectral density (angle random walk). The tone is caused by 60 Hz power line interference coupling to the sustaining circuits. The rate signal is at about 100 Hz; after demodulation the 60 Hz tone appears near 40 Hz.

(fig. 6), which is near the specified performance limit of the rate table used for this test.

Bias stability reaches a minimum of $5.9\,°/hr$ at $\tau = 3800\,s$ (fig. 5). This represents a $250\times$ improvement over the single-transducer circular orbit FM result in [2]. The ARW in fig. 5 is higher than actual due to noise folding resulting from the test protocol.

Figure 7 shows the measured rate noise density; the angle random walk is $0.014\,°/s/\sqrt{Hz}$, and is limited by the phase noise of the discrete oscillator circuits. The bandwidth of the gyroscope is 50 Hz; the full-scale rate exceeds $\pm1000\,°/s$.

CONCLUSIONS

This work has presented a frequency-readout operating mode for vibratory gyroscopes that dramatically improves bias stability over the circular-orbit FM mode while demonstrating state-of-the-art scale factor stability. The operating mode can take full advantage of high-Q fully-symmetric transducers to reduce cross-damping errors and power consumption. Unlike the whole-angle operating mode, Lissajous FM operation does not require a complex controller and is well-suited for low-cost low-power consumer-grade applications.

In addition to ensuring accurate scale factor, frequency readout permits a large dynamic range to be accommodated without encountering analog dynamic range limitations and without range switching. The mostly-digital implementation of the FM readout circuits is well-suited to implemen-

tation in deep-submicron, low-voltage CMOS processes and integration with sensor processors and SoCs.

ACKNOWLEDGEMENTS

The authors would like to thank TSMC and InvenSense for providing IC fabrication. This material is based upon work supported by the Defense Advanced Research Projects Agency (DARPA) under Contract No. W31P4Q-12-1-0001.

REFERENCES

[1] A. Trusov, I. Prikhodko, D. Rozelle, A. Meyer, and A. Shkel, "1 ppm precision self-calibration of scale factor in MEMS Coriolis vibratory gyroscopes," in *TRANSDUCERS 2013: 17th Int. Conf. Solid-State Sensors, Actuators and Microsystems*, Jun. 2013, pp. 2531–2534.

[2] M. Kline, Y. Yeh, B. Eminoglu, H. Najar, M. Daneman, D. Horsley, and B. Boser, "Quadrature FM gyroscope," in *2013 IEEE 26th Int. Conf. MEMS*, Jan. 2013, pp. 604–608.

[3] M. H. Kline, "Frequency modulated gyroscopes," Ph.D. dissertation, University of California, Berkeley, Dec. 2013.

[4] D. Lynch, "Vibratory gyro analysis by the method of averaging," in *2nd Saint Petersburg Int. Conf. Gyroscopic Technol. and Navigation I*, May 1995, pp. 26–34.

[5] A. Trusov, I. Prikhodko, S. Zotov, A. Schofield, and A. Shkel, "Ultra-high Q silicon gyroscopes with interchangeable rate and whole angle modes of operation," in *2010 IEEE Sensors*, Nov. 2010, pp. 864–867.

[6] R. Candler, M. Hopcroft, B. Kim, W.-T. Park, R. Melamud, M. Agarwal, G. Yama, A. Partridge, M. Lutz, and T. Kenny, "Long-term and accelerated life testing of a novel single-wafer vacuum encapsulation for MEMS resonators," *J. Microelectromech. Syst.*, vol. 15, no. 6, pp. 1446–1456, Dec. 2006.

[7] I. Izyumin, M. Kline, Y.-C. Yeh, B. Eminoglu, and B. Boser, "A 50 μW, 2.1 mdeg/s/$\sqrt{Hz}$ frequency-to-digital converter for frequency-output MEMS gyroscopes," in *ESSCIRC 2014 – 40th European Solid State Circuits Conf.*, Sep. 2014, pp. 399–402.

CONTACT

*I. I. Izyumin; izyumin@eecs.berkeley.edu

LARGE FULL SCALE, LINEARITY AND CROSS-AXIS REJECTION IN LOW-POWER 3-AXIS GYROSCOPES BASED ON NANOSCALE PIEZORESISTORS

Stefano Dellea[1], Federico Giacci[1], Antonio Longoni[1], Patrice Rey[2], Audrey Berthelot[2],
and Giacomo Langfelder[1]

[1]Politecnico di Milano, Italy
[2]CEA-Leti, Grenoble, France

ABSTRACT

This work presents in-plane and out-of-plane Coriolis rate gyroscopes based on nano-scale piezoresistive readout and using an eutectic bonding between the bottom wafer, where the sensor is formed, and the cap wafer, where routing and metal pads are fabricated. The gyroscopes feature a novel design with a central levered sense frame, to maximize the device symmetry and compactness. The position of the piezoresistive nano-gauges along the lever system optimizes the scale-factor. Operation on a ± 3000 dps full-scale-range (FSR) demonstrates quite competitive performance, with a linearity error lower than 0.25% and a cross-axis rejection 50x better than state-of-the art consumer gyroscopes.

INTRODUCTION

Most of Micro Electromechanical System (MEMS) gyroscopes in the consumer field rely on capacitive readout [1-2]. This approach is continuously challenged by the simultaneous needs for improved performance (to add new functions, like e.g. navigation), reduced power consumption (to allow e.g. the gyroscope to be always on) and miniaturization. Therefore, new technologies and new operation modes [3-4] have been recently given attention.

This work presents new Coriolis vibrating rate gyroscopes based on piezo-resistive sensing elements as the readout mean [5]. Both in-plane and out-of-plane structures are demonstrated. The interests in using nano-gauges as sensing elements are multiple: from the device point-of-view, miniaturization is achievable as the sensing element is simply formed by a few-μm-long, 250-nm-wide, crystalline Silicon beam, plus its anchor point. From the system point-of-view, the large output signal, together with the resistive readout (free of noise amplification issues given by parasitic capacitances) enables low power consumption from the readout electronics.

In addition, the devices of this work feature a new low-pressure wafer-level encapsulation. It is a gold-Silicon (Au-Si) eutectic bonding process, between the MEMS wafer, where the devices are fabricated, and the cap wafer, where two levels of metal interconnections and the pads are integrated. The Au-Si bonding provides both the sealing ring and the electrical contact between the device electrodes and the routing paths, wholly designed on the cap wafer. Again, this leads to an overall minimization of the occupied area.

From the design perspective, a new hinge-lever system, positioned at the device center, allows maximizing the device symmetry with respect to previous works [6] that used a lateral lever: improvements of in-plane devices performance and, for the first time, in-plane devices performance and, for the first time,

operation of out-of-plane devices are demonstrated through a complete set of experiments.

PROCESS AND DEVICE DESCRIPTION

The device fabrication on the MEMS wafer starts with a Silicon-on-insulator (SOI) substrate, with 2-μm buried oxide and 0.25-μm monocrystalline Silicon top layer. The nano-gauges are defined by deep ultraviolet (DUV) lithography and reactive ion etching (RIE). Then, a 1-μm-thick oxide is deposited to protect the gauges: following lithography and RIE etching maintain this layer only on top of the nano-gauges. Thick non-selective epitaxial growth is then done (both on crystalline Silicon and SiO_2) to form the 15-μm-thick MEMS part. A lithography step and a deep RIE (DRIE) define the MEMS structures and open the SiO_2 protective layer of the NEMS. Finally, the release of the sensor is achieved by hydrofluoric acid (HF) vapor etching. More details about the process can be found in [5]. The cap wafer is bonded on top of the MEMS wafer at a pressure of about 30 kN, leading to about 20 Ω resistance for an effective eutectic bonding contact area of $(14\ \mu m)^2$.

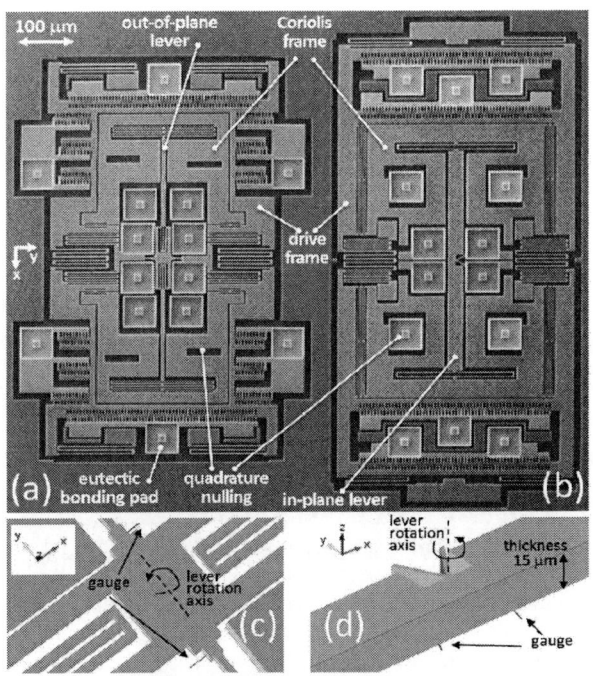

Figure 1: SEM of the y-axis (a) and z-axis (b) gyro-scopes: images are taken before bonding the sensor wafer to the cap wafer. In both the devices a central hinge-lever system (c-d) creates a differential stress on two suitably positioned, $5\mu m \times (250nm)^2$ piezo-resistive nano-gauges.

Fig. 1a-b is a scanning electron microscope (SEM) top view of an in-plane and an out-of-plane gyroscope. Note the absence of any routings on the MEMS wafer. The proof mass are perforated with $(2 \ \mu m)^2$ holes spaced 14 μm one another. The pads for the eutectic bonding are clearly visible, and have an overall area of about $(60 \ \mu m)^2$, to reach an effective contact area of $(14 \ \mu m)^2$: such mismatch was safely introduced to take into account possible misalignment between the two wafers during the Au-Si bonding process.

Both the shown devices rely on a tuning-fork (TF), doubly decoupled topology, with anchored drive frames, actuated along the x-axis, Coriolis frames, and levered sense frames. The double decoupling is needed to avoid too large, unwanted motion on the gauges. The optimum gauge position along the lever results from a compromise between scale-factor, linear range and maximum misalignment between the MEMS and NEMS masks.

The device (a) responds with out-of-plane, antiphase motion of the Coriolis frames to rates around the y-axis: a rotation of the lever is induced, causing a differential stress on two piezo-resistors, positioned as shown in the simulation of Fig. 1c. Similarly, the device (b) responds with in-plane differential motion of the Coriolis proof masses to z-axis rates: this induces a lever rotation and opposite stresses on the piezo-resistors (Fig. 1d).

In both the situations, cross-axis rates induce only very limited, common-mode stress on the gauges. Drive and sense modes are designed around 19 kHz, with about 1 kHz nominal frequency difference to operate in mode-split conditions [7]. Quadrature errors can be nulled through compensation electrodes designed within the Coriolis frames [2]. The overall device dimensions are 885x394 $(\mu m)^2$ for the z-axis and and 788x401 $(\mu m)^2$ for the y-axis. A 3-axis device would fit within a design area lower than 0.85x1.35 $(mm)^2$.

To infer the package pressure achievable through the used eutectic bonding process, an initial calibration of the quality factors of the z-axis structure was done in a vacuum probe station on bare wafers. Measurements were performed at different pressures in the range 0.1 mbar to 10 mbar, and at ambient pressure. The measurement technique is based on a *ITmems* MCP-G characterization platform [7, 8]. The same measurements were repeated once the devices are packaged, obtaining, on average, antiphase TF drive-mode quality factors in the order of 2000. From the calibration measurements, these values indicate an obtained pressure of about 2 mbar inside the sealed cavity.

INTERFACE ELECTRONICS

In operation, the proof mass is held to the ground potential. The drive mode is kept in antiphase oscillation through a trans-impedance amplifier (TIA) based primary loop, built around the drive resonator. A saturator (GAIN) and a variable gain stage (VGA) are used to set the AC and DC drive signal amplitude. A secondary loop, based on an integrator, a peak detector, and a comparator to a reference value, is used to control the variable gain, and in turn to set the drive motion to the desired amplitude of about 4.5 μm.

The sense-mode readout electronics is based on a Wheatstone bridge configuration, with 2 kΩ nominal value of each nano-gauge and of the dummy resistors of the bridge. The bridge output voltage is conditioned through an instrumentation amplifier (INA), demodulated, and digitized through a *Measurement Computing* USB acquisition board connected to a workstation with a *Labview* automated acquisition software. A schematic of the driving loops and of the sensing electronics of each gyroscope channel is shown in Fig. 2. Under these operating conditions, and assuming negligible aniso-elasticity and aniso-damping, the scale-factor $\Delta V_{out}/\Delta \Omega$ (output voltage change per unit angular rate) is written as:

$$\Delta V_{out}/\Delta \Omega = 1/180 \cdot x_d/\Delta f \cdot GF \cdot G_{des} \cdot G_{eln} \qquad (1)$$

where x_d is the drive mode displacement and Δf is the frequency split between the drive and sense modes.

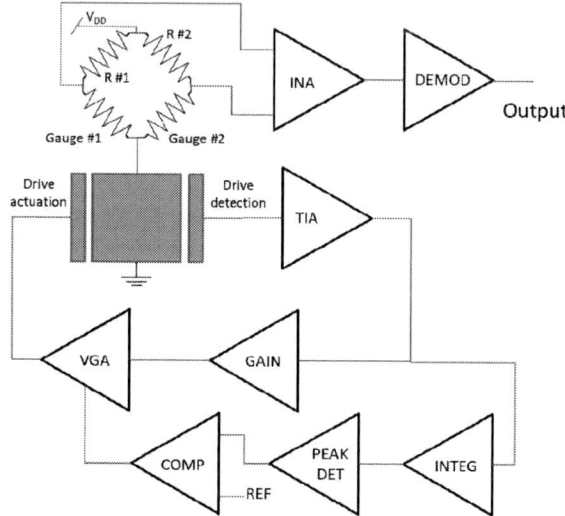

Figure 2: schematic representation of the driving loops and of the sensing electronics used to operate and readout the gyroscopes presented in this work.

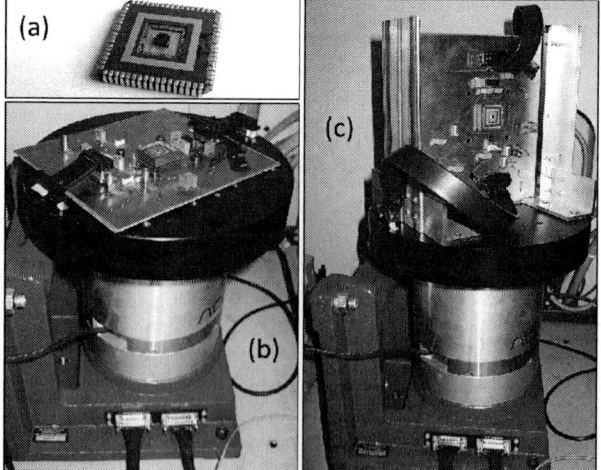

Figure 3: view of one device glued onto a ceramic carrier (a), which is then mounted on the board with the actuation/readout electronics. This is arranged on a rate table for z- (b) or y-axis (c) measurements.

GF is the piezoresistive gauge factor (relative resistance change per unit strain: the value is about 50 for the used process); finally, G_{des} is a geometry and process dependent constant, and G_{eln} is the readout electronics overall gain.

The bridge voltage causes a power dissipation by Joule effect in the nano gauges of 45 µW and 180 µW for the z- and y-axis devices. The y-axis bridge is indeed driven at a larger current to partially compensate the lower geometry factor G_{des} appearing in Eq. (1). A DC voltage of 15 V is summed to the AC drive signal by the VGA stage. The output low-pass filter is set to 500 Hz.

EXPERIMENTAL RESULTS

The devices are mounted on a ceramic carrier through adhesive tape, and positioned on the board that includes the described electronics. Fig. 3 shows the board aligned on an *Acutronic* AC1120-S rate table, arranged either for z- or y-axis sensing. This rotation table can achieve a maximum rate amplitude of ±2950 dps, and a -3 dB generated rate frequency of about 200 Hz at 12 dps.

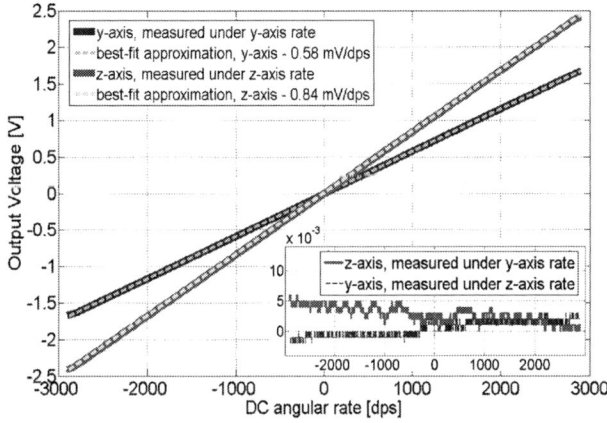

Figure 4: measured y- and z-axis scale-factor over ± 2950 dps. The linearity error, calculated as the deviation from the best linear fitting, is < 0.25% of the FSR. Cross-axis rejection (see the inset) is > 59 dB.

Fig. 4 reports the experimental devices response while ramping the angular rate from the minimum negative to the maximum positive values allowed by the rate table: in this range, the percentage deviation from the best linear fitting (indicated as well in the figure through dashed lines) is lower than 0.25 %, in line with the specifications required by consumer applications. The obtained scale-factors are 0.84 mV/dps and 0.58 mV/dps, on the two axes.

The response to cross-axis rates along the most critical direction (i.e. the one orthogonal to the drive motion) for each device is also shown in the inset of Fig. 4, with a measured rejection ratio larger than 59 dB (likely limited by manual alignment). This value is much better than typical performance found in off-the-shelf consumer gyroscopes.

Uncompensated quadrature errors on z-axis devices have an average value, measured over 29 samples, of 205 dps (6.8% of the 3000 dps FSR), with a standard deviation of 126 dps. The average quadrature error

measured before compensation on 22 y-axis devices is 3620 dps (121% of the 3000-dps full-scale), with a standard deviation of 2153 dps (worst case measurement is 5442 dps). Such low quadrature errors (compared e.g. to common findings, see e.g. [9]) are compensated both through electromechanical quadrature nulling electrodes and by in-phase demodulation.

The measured frequency response is partly affected by the rate table low-pass filter, resulting in a -3dB bandwidth of ~200 Hz, as shown by Fig. 5 for a z-axis device. The nominal mode-split is indeed in the order of 1 kHz, which should theoretically result in a larger flat frequency response.

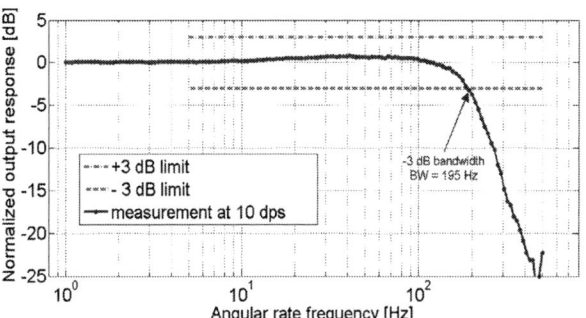

Figure 5: measured frequency response of a z-axis device under 12 dps sinusoidal rate stimuli. The graph is normalized to the DC value. The 200 Hz -3dB value is due to the rate table mechanical transfer function.

Allan variance measurements were captured in uncontrolled laboratory environment. Though the tiny dimensions of the suspended mass, Fig. 6 shows rate noise densities of 10 mdps/√Hz and 23 mdps/√Hz on the two axes. This white noise contributions is ascribed to the electronics. The bias stability is of about 10 mdps at 10 s of observation time. The worse result for the y-axis device than for the z-axis devices is caused by the lower scale-factor. In the left part of the figure the effects of the low-pass filter are visible. All the measured noise values are competitive with the state of the art (see e.g. [10-12]).

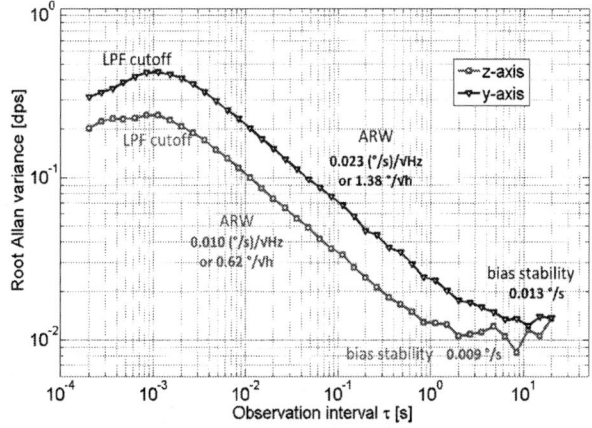

Figure 6: root Allan variance graphs measured on z- and y-axis devices as a function of the observation interval. Measurements are taken under electro-mechanical quadrature error nulling, in uncontrolled laboratory environment.

978-1-4799-7956-1/15 $31.00 © 2015 IEEE   39

Table 1 summarizes the obtained parameters for the two devices presented in this work.

CONCLUSIONS

The work proposed a novel design for miniaturized 3-axis gyroscopes based on piezoresistive nano-gauge detection. The achieved performance well compares against state-of-the-art consumer devices. Ongoing work is dedicated to the development of the drive and sense low-power integrated circuitry, as well as to the co-integration of the described 3-axis gyroscope with accelerometers [5] and magnetometers [13], to form 9-axis inertial measurement units.

Table 1: parameters measured on the gyroscopes, in uncontrolled laboratory conditions (T = 22°C ± 2°C).

z-axis parameter	Value	Unit
Area on MEMS wafer	885x394	$(\mu m)^2$
Full-scale range	> 3000	dps
Scale factor	0.84	mV/dps
Average quadrature*	205	dps
Bandwidth	> 500	Hz
Angle Random Walk	10	mdps/√Hz
Linearity error	< ±0.2	% FSR
Cross-axis rejection	> 59	dB
y-axis parameter	Value	Unit
Area on MEMS wafer	788x401	$(\mu m)^2$
Full-scale range	> 3000	dps
Scale factor	0.58	mV/dps
Average quadrature*	3620	dps
Bandwidth	> 500	Hz
Angle Random Walk	23	mdps/√Hz
Linearity error	< ±0.25	% FSR
Cross-axis rejection	> 65	dB

* value measured before electromechanical compensation

ACKNOWLEDGEMENTS

The authors thank Dr. G. Schierano and Dr. E. Brigo for helping with setup preparation. The work is supported by the European Union under the FP7-ICT program "Nine-axis inertial sensors based on piezoresistive nano-gauge Detection" (NIRVANA) under grant 288318. Experiments were done in the MEMS&3D lab facilities.

REFERENCES

[1] A. Sharma, M. F. Zaman, F. Ayazi, "A 104-dB Dynamic Range Transimpedance-Based CMOS ASIC for Tuning Fork Microgyroscopes", *IEEE Journ. of Solid-State Circuits*, Vol. 42, N. 8, 2007, pp. 1790-1802.

[2] S. Sonmezoglu, H. D. Gavcar, K. Azgin, S. E. Alper, T. Akin, "Simultaneous Detection of Linear and Coriolis Accelerations on a Mode-Matched MEMS Gyroscope", *Proc. MEMS 2014*, pp. 32-35.

[3] S. A. Zotov, I. P. Prikhodko, A. A. Trusov, A. M. Shkel, "Frequency Modulation Based Angular Rate Sensor", in *Proc. MEMS 2011, Cancun, MEXICO, January 23-27, 2011*, pp. 577-580.

[4] M. H. Kline, Y.-C. Yeh, B,k Eminoglu, H. Najar, M. Daneman, D. A. Horsley, B. E. Boser, "Quadrature FM Gyroscopes", in *Proc. MEMS 2013, Taipei, Taiwan, January 20 – 24, 2013*, pp. 604-608.

[5] P. Robert, V. Nguyen, S. Duraffourg, G. Jourdan, J. Arcamone, S. Harrison, "M&NEMS : A new approach for ultra-low cost 3D inertial sensor", in *Proc. IEEE Sensors 2009*, pp. 963-966.

[6] A. Walther, M. Savoye, G. Jourdan, P. Renaux, F. Souchon, P. Robert, C. Le Blanc, N. Delorme, O. Gigan, C. Lejuste, "3-Axis Gyroscope with Si Nanogage Piezo-Resistive Detection", in *Proc MEMS 2012, Paris, FRANCE, 29 January - 2 February 2012*, pp. 480-483.

[7] G. Langfelder, S. Dellea, A. Berthelot, P. Rey, A. Tocchio, A. F. Longoni, "Analysis of Mode-Split Operation in MEMS Based on Piezoresistive Nanogauges", *Journal of Microelectromechanical Systems*, accepted for publication, DOI: 10.1109/JMEMS.2014.2324032.

[8] ITmems s.r.l., MEMS Characterization Platform, MCP-G, *Product Datasheet*, [Online]. Available: http://www.itmems.it, accessed Nov. 2014.

[9] V. Kempe, "Gyroscopes," in *Inertial MEMS Principles and Practice*. Cambridge, U.K.: Cambridge Univ. Press, 2011, Ch. 8.

[10] Invensense ITG-3701, "Product Specification Revision 1.0", *product datasheet.* Available online at http://www.invensense.com/mems/gyro/documents/PS-ITG-3701.pdf. Accessed Nov. 2014.

[11] L3G3200D, "MEMS motion sensor: 3-axis digital output gyroscope", *product datasheet.* Available online at http://www.st.com/st-web-ui/static/active/en/resource/technical/document/datasheet/DM00043564.pdf. Accessed Nov. 2014.

[12] Maxim MAX21001, "Ultra-Accurate, Automotive, 3-Axis Digital Output Gyroscope", *product datasheet.* Available online at http://datasheets.maximintegrated.com/en/ds/MAX21001.pdf. Accessed Nov. 2014.

[13] D. Ettelt, P. Rey, M. Savoye, C. Coutier, M.Cartier, O. Redon, M. Audoin, A. Walther, P. Robert, Y. Zhang, F. Dumas-Bouchiat, N.M. Dempsey, J. Delamare, "A New Low Consumption 3d Compass Using Integrated Magnets and Piezoresistive Nano-Gauges", in Proc. *Transducers'11*, Beijing, China, June 5-9, 2011, pp. 40-43.

CONTACT

*Giacomo Langfelder, tel: +39-02-2399-3425; mail: giacomo.langfelder@polimi.it

THE ELECTROMECHANICAL RESPONSE OF A SELF-EXCITED MEMS FRANKLIN OSCILLATOR

Shai Shmulevich, Inbar Hotzen and David Elata

Faculty of Mechanical Engineering, Technion - Israel Institute of Technology, Haifa, ISRAEL

ABSTRACT

We present a self-excited MEMS Franklin oscillator, which responds in steady state vibrations when subjected to a sufficient dc voltage. The system is constructed from an electrostatically-floating rotor, which sequentially transfers charge between a source and drain electrodes. We present a comprehensive analysis of the electromechanical response of the system. Our analysis shows that current flows from the source to drain only when the rotor is in transition. Surprisingly, at contact of the rotor with either the source or the drain electrodes, there is no current in the system, and the charge transfer mechanism is essentially a recombination of opposite charges. This means that although we drive our system by 30 to 70 Volts, the contacts are essentially *cold switching*. Our experimental measurements of the dynamic response of the system are in good agreement with our model predictions.

INTRODUCTION

The Franklin bells-oscillator is named after Benjamin Franklin (1706-1790), but it was actually invented by Andrew Gordon (circa 1742) [1, 2]. The bell-oscillator is constructed from two conductive bells, each subjected to a different potential (Fig. 1) and a conductive bead that is suspended on an isolating string. The conductive bead sequentially transfers charge from one bell to the other. When the bead is charged, the electrostatic field between the bells always forces it towards the electrode with the opposite polarity. Franklin used the system to identify that *'electricity was in the air'* when the bells chimed continuously.

In recent years similar systems were implemented in nano-scale devices (e.g. [3]), and were shown to sustain self-oscillations.

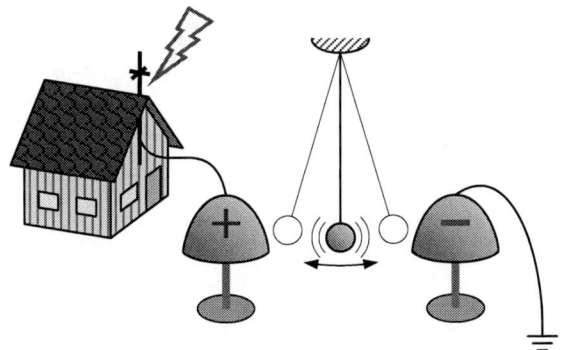

Figure 1: The Franklin bells-oscillator. A conducting bead sequentially transfers electrostatic charge between two bells when they are at different potentials.

Self-excited oscillations were also implemented with electrostatic MEMS switches, in which the rotor is directly subjected to the source voltage [4]. In this self-oscillating switch, the field between rotor and stator becomes infinite upon contact, and consequently the charge transfer mechanism is essentially of the *hot-switching* type. This hot-switching can be avoided (e.g. [5]), but at the price of increasing the system complexity.

In this work we present a comprehensive electromechanical model of the Franklin oscillator. We use our analysis to reveal several unexpected characteristics of the system. The model offers a rational method for designing such oscillators.

MODELING

A parallel-plates Franklin oscillator is schematically described in Fig. 2. The rotor of mass m is suspended on a linear flexure spring with stiffness k. In the undeformed state the rotor is symmetrically suspended between two fixed electrodes with a nominal gap g on both sides. The rotor is electrostatically floating and has a fixed parasitic capacitance C_p at its anchor. The source electrode is subjected to a constant dc voltage V, and the drain electrode is grounded. For simplicity we assume that $V > 0$, though the analysis applies for negative source voltages as well.

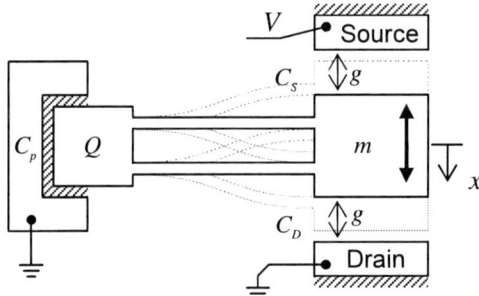

Figure 2: Schematic description of the MEMS parallel-plates Franklin oscillator.

We derive the system response using standard energy methods [6, 7]. The total potential energy of the system is given by

$$\psi = \frac{1}{2}kx^2 + \frac{1}{2}\left(C_D V_f{}^2 + C_S\left(V_f - V\right)^2 + C_p V_f{}^2\right) - (V - V_f)C_S V \tag{1}$$

Here x is the single degree-of-freedom of the system, $C_S = \varepsilon_0 A/(g+x)$ and $C_D = \varepsilon_0 A/(g-x)$ are the source and drain capacitance, ε_0 is the permittivity of free space, A is the area of the source and of the drain electrodes, and V_f is the (reactive) electrostatic potential of the floating rotor.

The first term on the left hand side of (1) is the potential energy of the elastic spring, the second term is the electrostatic potential energy of the three capacitors (C_S, C_D and C_p), and the last term is the electrostatic potential energy of the voltage source.

The voltage of the floating rotor and the capacitance of the system with respect to the voltage source are given by

$$V_f = \frac{Q + C_S V}{C_S + C_D + C_p}, \qquad \frac{1}{C} = \frac{1}{C_S} + \frac{1}{C_D + C_p} \qquad (2a,b)$$

Notice that with respect to the charge on the rotor, the two deformable capacitors and the parasitic capacitor are all connected in parallel (2a). In contrast, with respect to the source voltage, the parasitic capacitor C_p and the drain capacitor C_D are connected in parallel, and together they are connected in series with the source capacitor C_S (2b).

Using non-dimensional variables, the total potential of the system is given by

$$\tilde{\psi} = \frac{1}{2}\tilde{x}^2 + \frac{1}{2}\frac{\tilde{Q}^2(1-\tilde{x}^2) + 2\tilde{Q}\tilde{V}(1-\tilde{x}) - \left(1 + \tilde{C}_p(1-\tilde{x})\right)\tilde{V}^2}{2 + \tilde{C}_p(1-\tilde{x}^2)} \tag{3}$$

where

$$\tilde{x} = \frac{x}{g}, \quad \tilde{V}_i = \sqrt{\frac{\varepsilon_0 A}{kg^3}}V_i, \quad \tilde{C}_p = \frac{g}{\varepsilon_0 A}C_p, \quad \tilde{Q} = \frac{Q}{\sqrt{\varepsilon_0 A k g}} \tag{4}$$

In terms of these non-dimensional variables, the voltage of the floating rotor and the capacitance of the system with respect to the voltage source are given by

$$\tilde{V}_f = \frac{\tilde{Q} + \dfrac{1}{1+\tilde{x}}\tilde{V}}{\tilde{C}_p + \dfrac{2}{1-\tilde{x}^2}}, \qquad \tilde{C} = \frac{1 + \tilde{C}_p(1-\tilde{x})}{2 + \tilde{C}_p(1-\tilde{x}^2)} \tag{5a,b}$$

Finally, the force which is applied to the rotor in any state $(\tilde{V}, \tilde{Q}, \tilde{x})$ is given by

$$\tilde{f}_R = \tilde{x} - \frac{1}{2}\frac{\left[\begin{array}{c}4\tilde{x}\tilde{Q}^2 + 2\tilde{Q}\tilde{V}\left(2 + \tilde{C}_p(1-\tilde{x}^2)\right) \\ -\left(2\tilde{C}_p(1-\tilde{x}) + \tilde{C}_p^2(1-\tilde{x})^2\right)\tilde{V}^2\end{array}\right]}{\left(2 + \tilde{C}_p(1-\tilde{x}^2)\right)^2} \tag{6}$$

This expression for force can be used in the momentum equation to compute the dynamic response of the system.

Charge transfer mechanism

A schematic description of the charge transfer mechanism in one motion cycle is illustrated in Fig. 3. When the rotor is in contact with the source electrode (Fig. 3a), its voltage is equal to the source voltage and it carries a positive charge

$$\tilde{Q}_S = \left(\frac{1}{2} + \tilde{C}_p\right)\tilde{V} \tag{7}$$

If, as in our devices, $\tilde{C}_p \gg 1$ then most of this charge is loaded on the parasitic capacitor. When the rotor is in contact with the drain electrode (Fig. 3d), it is grounded and carries a negative charge

$$\tilde{Q}_D = -\frac{1}{2}\tilde{V} \tag{8}$$

where none of this charge is loaded on the parasitic capacitor.

It can be shown that as the rotor approaches the source electrode, i.e. $\tilde{x} \to -1 + \varepsilon$ with $0 \le \varepsilon \ll 1$ (Fig. 3f), the rotor voltage, the field in the source capacitor and the system capacitance, are given by

$$\begin{aligned}\tilde{V}_f &= \tilde{V} + [\tilde{Q} - \tilde{V}(\tilde{C}_p + \tfrac{1}{2})]\varepsilon + O(\varepsilon^2) \\ \tilde{E}_s &= -[\tilde{Q} - \tilde{V}(\tilde{C}_p + \tfrac{1}{2})] + O(\varepsilon) \\ \tilde{C} &= \tfrac{1}{2} + \tilde{C}_p + O(\varepsilon)\end{aligned} \tag{9a,b,c}$$

It follows from (9a) that irrespective of the charge $\tilde{Q}$ carried by the rotor, as the rotor approaches the source

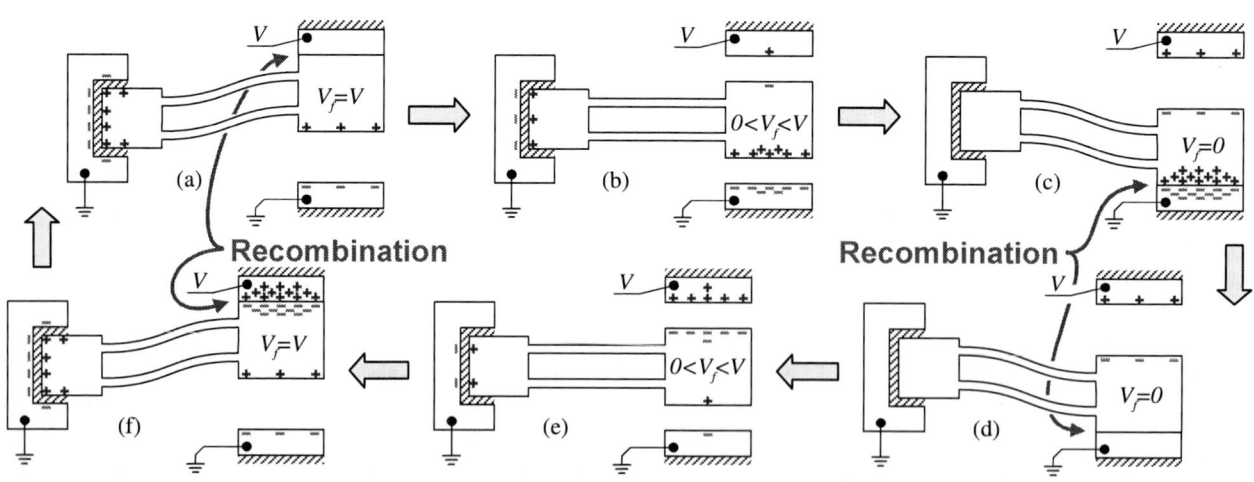

Figure 3: Charge transfer cycle. (a) Before detachment from the source, rotor charge is mostly loaded on the large parasitic capacitor at the anchor. (c) As it approaches the drain, the rotor voltage converges to the drain voltage (ground). At contact (c-d) opposite charges on the rotor and drain merely recombine - with no current flowing through the system. (f) As the rotor approaches the source, its voltage converges to the source voltage V. At contact (f-a) opposite charges on the rotor and source merely recombine, with no current flowing through the system.

electrode the rotor voltage approaches the source voltage $\widetilde{V}_f \to \widetilde{V}$.

From (9b) it follows that as the rotor approaches the source electrode, the field in the source capacitor converges to a finite value. This means that if $\widetilde{Q}$ and $\widetilde{V}$ are not excessively large, no electrostatic breakdown will occur before the rotor makes contact with the source electrode. In fact, when this contact occurs, positive charges on the source electrode and associated negative charges on the rotor simply *recombine* (i.e. are mutually annihilated as shown in Fig. 3f→a).

Likewise, when the rotor approaches the drain electrode, i.e. $\widetilde{x} = 1 - \varepsilon$ (Fig. 3c), the rotor voltage, the field in the drain capacitor and the system capacitance, are given by

$$\widetilde{V}_f = (\widetilde{Q} + \tfrac{1}{2}\widetilde{V})\varepsilon + O(\varepsilon^2)$$
$$\widetilde{E}_D = (\widetilde{Q} + \tfrac{1}{2}\widetilde{V}) + O(\varepsilon) \qquad (10\text{a,b,c})$$
$$\widetilde{C} = \tfrac{1}{2} + O(\varepsilon^2)$$

It follows from (10a) that irrespective of the charge $\widetilde{Q}$ carried by the rotor, as the rotor approaches the drain electrode (i.e. $\widetilde{x} \to 1$) the rotor voltage drops to zero.

From (10b) it follows that as the rotor approaches the drain electrode, the field in the drain capacitor is finite. This means that if $\widetilde{Q}$ and $\widetilde{V}$ are not excessively large, no electrostatic breakdown will occur before the rotor makes contact with the drain electrode. When this contact occurs, negative charges on the drain electrode and associated positive charges on the rotor, simply recombine, as shown in Fig. 3c→d.

All this means that though the system may be subjected to a substantial driving voltage, the contacts are essentially *cold switching*.

It is clear from (5b) that the system capacitance changes when the rotor is in transition. This means that when the rotor is in transition current flows through the system. However, from (9c) and (10c), it follows that as the rotor converges to any of the stators, the system capacitance converges to a finite value. This means that at contact of the rotor with either the source or the drain electrodes, there is no current through the system - which is compatible with the notion of cold switching.

Dynamic response

Instead of directly solving the momentum equation, we will next consider the fully-developed cyclic motion of the oscillator. We consider a cycle that begins as the electrode detaches from the source electrode (i.e. $\widetilde{x} = -1$) with charge $\widetilde{Q}_S$, given in (7). By the time the rotor has reached the drain electrode (i.e. $\widetilde{x} = 1$), the total potential energy in the system has changed by

$$\Delta \widetilde{\psi}_{S \to D} = \widetilde{\psi}\big|_{\widetilde{x}=1,\widetilde{Q}=\widetilde{Q}_S} - \widetilde{\psi}\big|_{\widetilde{x}=-1,\widetilde{Q}=\widetilde{Q}_S} = -\frac{1}{2}(1 + \widetilde{C}_p)\widetilde{V}^2 \quad (11)$$

As the rotor detaches from the drain electrode it carries a charge $\widetilde{Q}_D$, given in (8). By the time the rotor has reached back to the source electrode, the total potential energy in the system has changed by

$$\Delta \widetilde{\psi}_{D \to S} = \widetilde{\psi}\big|_{\widetilde{x}=-1,\widetilde{Q}=\widetilde{Q}_D} - \widetilde{\psi}\big|_{\widetilde{x}=1,\widetilde{Q}=\widetilde{Q}_D} = -\frac{1}{2}(1 + \widetilde{C}_p)\widetilde{V}^2 \quad (12)$$

In every half cycle the total energy in the system is reduced by the same amount, and that the total energy reduction per cycle is

$$\Delta \widetilde{\psi}_{cycle} = -(1 + \widetilde{C}_p)\widetilde{V}^2 \qquad (13)$$

Not surprisingly, this is identical to the product of the total charge extracted from the voltage source (per cycle) and the source voltage

$$-\left(\widetilde{Q}_S - \widetilde{Q}_D\right)\widetilde{V} = -\left(1 + \widetilde{C}_p\right)\widetilde{V}^2 \qquad (14)$$

In steady cyclic motion, this energy is dissipated by damping acting on the moving rotor, and by losses due to collisions of the rotor with the stators. We designed our oscillator devices with springs to make the collisions elastic. When the system is subjected to the source voltage, kinetic energy builds up until all the power supplied by the voltage source is dissipated by damping. Kinetic energy increases to a sufficiently high level where elastic forces and electrostatic forces hardly affect the velocity of the rotor, which consequently has a triangular wave-form trajectory. For such a trajectory, and assuming $\widetilde{C}_p = 100$, the rotor displacement and drain current are plotted in Fig. 4.

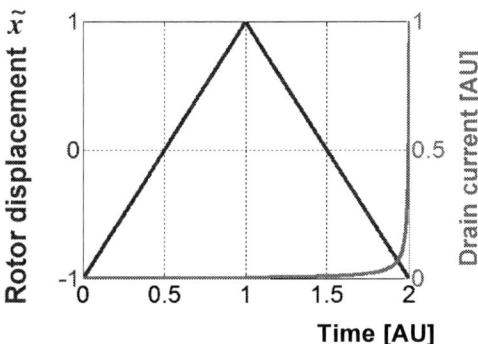

Figure 4: The predicted rotor trajectory and drain current in one cycle of the fully-developed periodic response.

Assuming that damping is linear (i.e. $f_{Damping} = -c\dot{x}$), it follows that the dissipated energy due to damping is proportional to velocity-squared (i.e. $W_{Damping} = -c\dot{x}^2$). Since the work input is proportional to voltage-squared (14), it follows that voltage squared is proportional to maximal velocity squared. This means that frequency is predicted to be linearly related to the absolute amplitude of the source voltage.

EXPERIMENTAL RESULTS

Test devices were fabricated using SOIMUMPs technology (run 47, [8]). A microphoto of a typical test device is shown in Fig. 5. The suspension flexures are designed to be $1000\mu m$ long and $5\mu m$ wide, and the device layer thickness is $25\mu m$. The gap between the fixed electrodes and the floating rotor is designed to be $6\mu m$. The elastic bumpers are shown in detail in the inset electrodes (inset in Fig. 5). Two small side-wall dimples were designed at the center of the bumpers on each side of the rotor. The purpose of these dimples was to prevent stiction (in fabrication) between the rotor and stators, and to ensure proper conduction when contact is made. To this end, a blanket gold layer was deposited (shadow masked) to improve conduction evident as the bright rectangle running across the device in Fig. 5.

The square ring structures fixed to the sides of the rotor were intended for handling and have no functional purpose. The parasitic capacitance in this device spans the entire device layer, except for the source and drain electrodes. The drain electrode was split to enable continuous attraction by using a lock-down voltage.

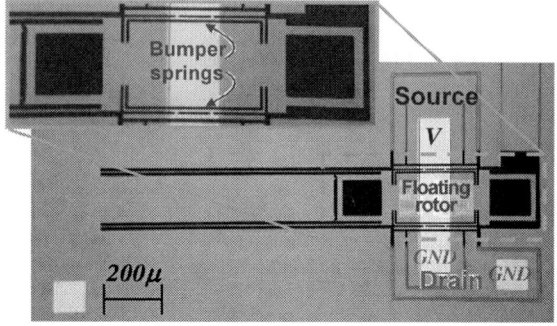

Figure 5: Microphoto of a typical device. The inset shows the elastic bumpers that cushion the mechanical impact between the rotor and the fixed electrodes.

A laser vibrometer was used to measure the rotor vibration motion. All measurements were performed at atmospheric ambient pressure. A typical response of the device is presented in Fig. 6. The source electrode was subjected to 48[V]. It is evident that the device sustains a self-oscillating response. In addition, measured motion response (dotted blue) fits a triangular wave-form (solid gray) as predicted.

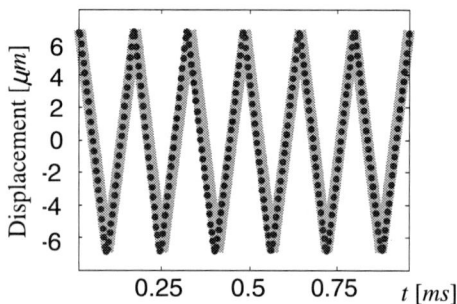

Figure 6: The rotor motion for V=48[V] measured at atmospheric pressure (dotted). The fit to the triangular wave-form (solid) confirms our model predictions.

Figure 7 presents the extracted oscillating frequency of the measured time response for different driving voltages (ranging 38 to 66[V]). The good fit to a linear curve confirms the predicted linear relation between applied voltage and oscillating frequency. Error bars in Fig. 7 indicate deviation of the measured oscillating

frequency which is attributed to non-ideal Ohmic contact between the rotor and fixed electrodes.

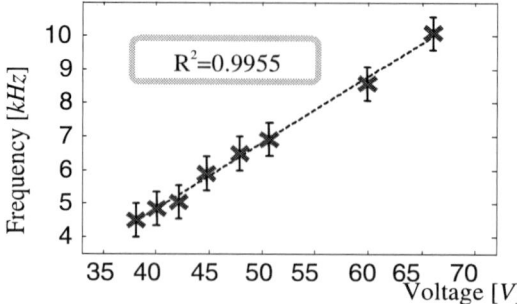

Figure 7: Oscillator frequency vs. source voltage measured at atmospheric pressure. The linear fit confirms our model predictions.

CONCLUSION

We have demonstrated a MEMS Franklin oscillator which responds in self-excited oscillations when subjected to a constant dc source voltage. The characterized test devices had an unnecessarily excessively large parasitic capacitance of 26 [pF] (the nominal capacitance was 11 [fF] which means that $\widetilde{C}_p = 2360$). Hence the average power consumed by the system was P=0.2÷1.2 [mW]. In future work the parasitic capacitance will be drastically reduced to achieve low power self-oscillations.

ACKNOWLEDGEMENT

This study was partially supported by the Russell Berrie Nanotechnology Institute (RBNI), Technion.

REFERENCES

[1] W. S. N. Trimmer and K. J. Gabriel, "Design Considerations for a Practical Electrostatic Micro-Motor," *Sensors and Actuators*, **11**, 189-206, 1987.

[2] M. B. Schiffer, K. L. Hollenback, and C. L. Bell, Draw the lightning down : Benjamin Franklin and electrical technology in the Age of Enlightenment. Berkeley: University of California Press, 2003.

[3] D. R. Koenig and E. M. Weig, "Voltage-sustained self-oscillation of a nano-mechanical electron shuttle," *Applied Physics Letters*, **101**, 213111, 2012.

[4] J. Bienstman, J. Vandewalle, and R. Puers, "The autonomous impact resonator: a new operating principle for a silicon resonant strain gauge," *Sensors and Actuators a-Physical*, **66**, 40-49, 1998.

[5] G. B. Torri, J. Bienstman, X. Rottenberg, H. A. C. Tilmans, C. Van Hoof, and R. Puers, "A MEMS autonomous switched oscillator," *IEEE-MEMS 2014*, San-Francisco, 2014.

[6] D. Elata and V. Leus, "Electromechanical modelling of electrostatic actuators," in Advanced RF Mems, S. Lucyszyn, Ed.: Cambridge U. Press, 2010, pp. 23-40.

[7] D. Elata, V. Leus, J. Provine, A. Hirshberg, and R. T. Howe, "Electromechanical Sensing of Charge Retention on Floating Electrodes," *IEEE-JMEMS*, **20**, 150-156, 2011.

A NANOMACHINED TUNABLE OSCILLATOR CONTROLLED BY ELECTROSTATIC AND OPTICAL FORCE

J. G. Huang[1,2,3], B. Dong[2,3], H. Cai[2], Y. D. Gu[2], J. H. Wu[1], T. N. Chen[1]
Z. C. Yang[4], Y. F. Jin[4], Y. L. Hao[4], D. L. Kwong[2] and A. Q. Liu[3†]

[1]School of Mechanical Engineering, Xian Jiaotong University, Xian 710049, China
[2]Institute of Microelectronics, A*STAR (Agency for Science, Technology and Research)
Singapore 117685
[3]School of Electrical & Electronic Engineering, Nanyang Technological University, Singapore 639798
[4]National Key Laboratory of Science and Technology on Micro/Nanofabrication
Institute of Microelectronics, Peking University, Beijing 100871, China

ABSTRACT

We develop a miniaturized electrostatically tunable optomechanical oscillator, whose frequencies can be electrostatically tuned by as much as 10%. By taking advantage of the optical and the electrical spring, the oscillator achieves a high tuning sensitivity without resorting to mechanical tension. Particularly, the high-Q optical cavity greatly enhances the system sensitivity, making it extremely sensitive to the motional signal, which is often overwhelmed by background noise.

INTRODUCTION

MEMS oscillators [1] typically have large footprint and high resonant frequency at the expense of large mechanical stiffness, which makes frequency tuning difficult. The optomechanical oscillator, instead, can achieve high resonant frequency with optical force utilized [2] while maintains the mechanical compliance. The high stability and low phase noise of the resonant optomechanical oscillator have been widely reported [3-4]. On the other hand, the resonant frequency tunability, another important potential property of the oscillator, remains unexplored. In fact, cavity optomechanical device is a good candidate for resonant frequency tuning because of its inherently nonlinear optical spring effect [5-6]. Here, we adapt a new method to tune the optical spring by making use of the electrostatic force. The electrostatic force changes the relative detuning of the input light, making the mechanical frequency of the optical resonator shift with optical spring changing. The small motional signal can be detected due to the high quality factor of the optical cavity, which greatly enhances the system sensitivity.

The proposed tunable nano mechanical oscillator are controlled by both the electrostatic force and the optical force. By taking advantage of the large nonlinearity of the optical force, the oscillator achieves a high tuning sensitivity without resorting to mechanical tension. The tunable optomechanical oscillator have merits such as small dimension, low power consumption and little requirement for complex feedback systems, which makes it a good candidate in future applications in silicon photonic circuits and on chip signal processing.

DESIGN AND WORKING PRINCIPLE

The proposed optomechanical oscillator consists of a cantilever, two metal pads, a ring resonator and a bus waveguide, as shown in Fig. 1(a). A high power drive light is pumped into the bus waveguide through the input port and coupled into the ring resonator. Due to the evanescent wave overlapping, the attractive optical force is generated between the ring resonator and the cantilever. By taking advantage of large optomechanical coupling coefficient, the blue-detuned pumping reduces the mechanical damping rate and leads to regenerative amplification of the thermal motion. As a result, the cantilever starts to oscillate. A low power signal light is then pumped into the bus waveguide which induces a small optical force on the cantilever. Meanwhile, an electric DC voltage is applied between the two metal pads and induces electrostatic force on the cantilever. The electrostatic force and the optical force by the signal light on the cantilever modulates the cantilever oscillation frequency. As the cantilever modulates the resonance frequency of the ring, the moved distance of the cantilever can be precisely detected by measuring the frequency shift at the output.

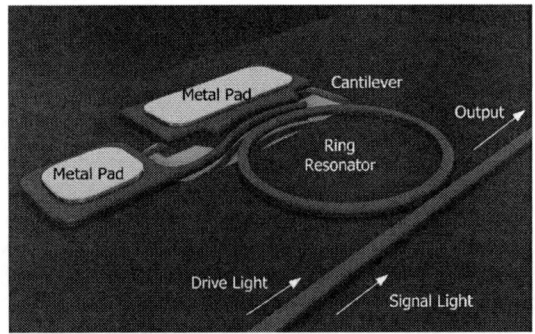

Figure 1: Schematic of the tunable nanomachined oscillator. The electrostatic force can drive the doped silicon cantilever to tune the frequency.

The strong optical spring effect can be easily tuned by changing the electrostatic force to reach the softening and hardening regimes. The finite element method (COMSOL Multiphysics) simulation on the cantilever deformation is shown in Fig. 2 (a). The ring has a diameter of 30 μm and the length of the cantilever is 7 μm. The electrostatic force and the optical force are in opposite direction. Compared with the optical force, the electrostatic force dominates in the deformation of the cantilever. The total structure can be reduced to a lumped theoretical model, which is shown in Fig. 2(b). Considering the nonlinearity of the optical and electrostatic force, the effective spring constant of the structure consists of three parts: the optical spring constant, the electrical spring constant and the original mechanical spring constant. The effective spring constant can either be increased or decreased when the optical spring constant is negative or positive, correspondingly, which depends on the relative detuning. It provides a good platform to realize a tunable oscillator.

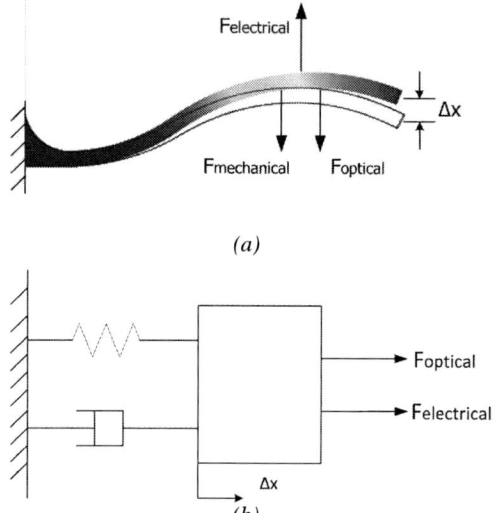

(a)

(b)

Figure 2: (a) Finite element method simulation of the deformation of the cantilever and (b) working principle of the effective spring model.

The optical energy stored in the ring resonator can be expressed as [7]

$$|a|^2 = \frac{k_e}{(k/2)^2 + (w_c - w_r - g_{om}x)^2} P_{in} \qquad (1)$$

where the power of the control light is P_{in}, k is the full-width at half-max linewidth of the optical resonance, k_e is the external coupling rate, w_c is the frequency of the control light, and w_r is the unperturbed resonance frequency of the ring, g_{om} is the optomechanical coupling coefficient and x is the mechanical deformation. The optical force can be expressed as [8]

$$F_0 = \frac{|a|^2 g_{om}}{w_l} \qquad (2)$$

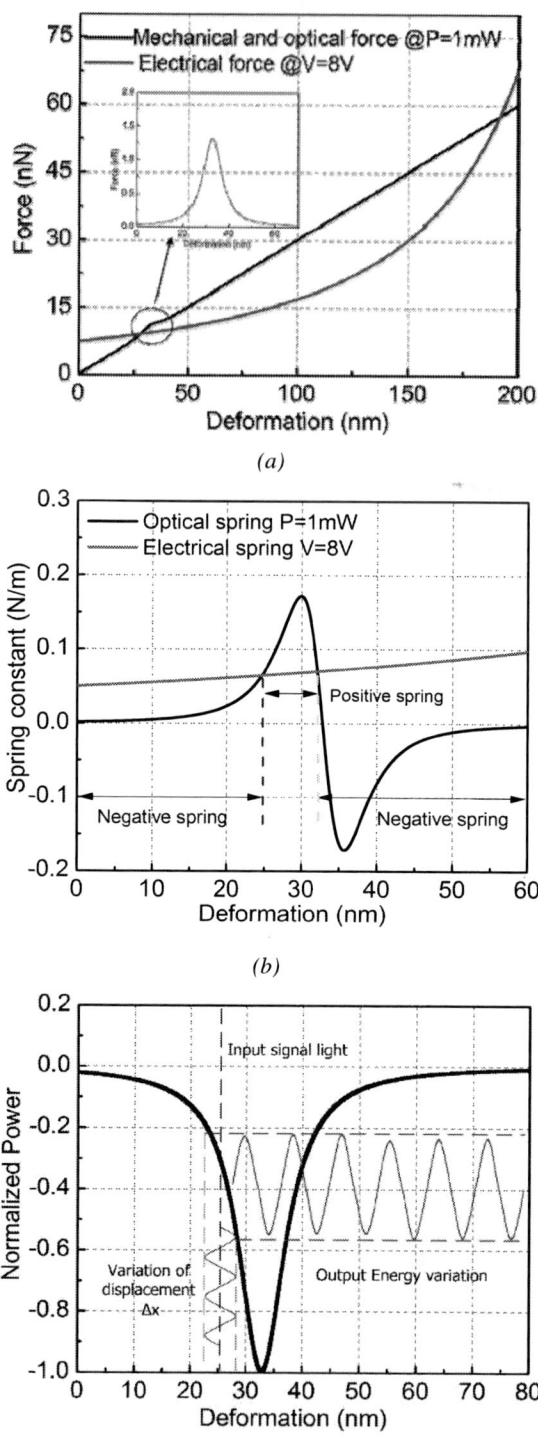

(a)

(b)

(c)

Figure 3: (a) Deformation of the cantilever at the

electrostatic and optical force. (b) The spring constant is function of the deformation and (b) the sensing principle by the ring cavity.

The strong light confinement in the ring makes the optical mode frequency highly sensitive to the gap between the ring and the beam with a theoretical optomechanical coupling coefficient g_{om} = 1.2 GHz/nm. The typical loaded optical quality factor for the fundamental TE mode is $Q = 1.8 \times 10^4$ (Finesse F = 171). The electrical actuators are formed by the metal pads and the highly doped silicon cantilever. The electrostatic force induced by a voltage V can be expressed as

$$F_e = \frac{\alpha A \varepsilon \varepsilon_0}{2(d-x)^2} V^2 \qquad (3)$$

where d is the capacitor gap and α is correction parameter for the fringing effect. Electrostatic force leads to the contraction of the capacitors, resulting in a wavelength shift of the cavity resonance.

To investigate the optical spring effect, we extend the optical force at the equilibrium point to second order expressed as

$$F_{optical} = F(x_0) + k_{o1}\delta x + k_{o2}\delta x^2 \qquad (4)$$

where k_{o1} is the spring constant and k_{o2} is the quadratic coefficient. In fact, the quadratic term can be neglected for the primary resonance. Therefore, the spring constant is dominant on the primary resonance of the cantilever beam. The electrical spring constant can also be derived through the Taylor extension. The effective spring constant of the structure can be expressed as

$$k_{effective} = k_{o1} + k_{e1} + k_m \qquad (5)$$

Figure 3(a) shows the deformation of the cantilever by the optical and electrostatic force. Compared with the optical force, the electrostatic force is much bigger. Therefore, the deformation of the cantilever is largely determined by the electrostatic force. However, the optical spring constant is much bigger than the electrical spring constant even at so much difference. The optical and electrical spring constant is shown in Fig. 3(b). The optical spring is not only bigger than electrical spring, but also shows negative or positive spring effect depending on the deformation. Figure 3(c) shows the sensing principle of the oscillator. A signal light is pumped into the bus waveguide to detect the small motion of the cantilever. Due to the high quality factor of the ring, the sensitivity of the system is very high.

FABRICATION PROCESSES

The scanning electron microscope (SEM) graphs of the tunable optomechanical oscillator are shown in Fig. 4. The oscillator with a footprint of 45 μm × 30 μm is fabricated on silicon-on-insulator (SOI) wafer with structure layer

thickness of 220 nm. The actuator is patterned by deep UV lithography and etched by plasma dry etching. The waveguide is covered by a layer of SiO_2 cladding (2 μm thick) which is deposited using plasma enhanced chemical vapor deposition (PECVD). In the release process, a 50-nm amorphous silicon layer is used as the hard mask to protect the structures not to be released. Then the buried-oxide layer is removed using HF-vapor with precise time control. The width of the suspension beam is 200 nm and total length is 7 μm, so that it can provide a spring constant comparable with the optical spring and be more sensitivity to the driven force. The suspension beam are highly doped to serves as a movable capacitor. The doping concentration is 2 ×10^{14} cm^{-3}. A pair of inversed mode converters was used to couple light into and out of the device on the waveguide alignment system platform. The device is tested in a vacuum chamber.

(a)

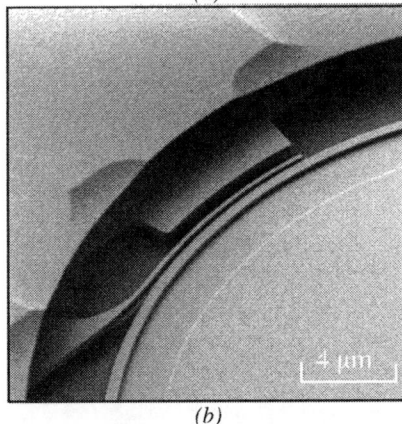

(b)

Figure 4: SEM images of the tunable oscillator.

EXPERIMENTS AND DISCUSSIONS

The measured transmission spectra of the waveguide at different electric voltage are shown in Fig. 5. When increasing the voltage from 0 V to 11 V, the cantilever is pulled against the ring, leading to the wavelength blue-shift with $\Delta\lambda$ = 0.58 nm. The shift of the resonant frequency is linear to the voltage in small range. At the same time, due to the small deformation, the tension in the cantilever can be neglected. The voltage can effectively control the deformation of the cantilever and tunes the optical spring constant. We use a high power drive light to stimulate the oscillation of the cantilever and then tune the frequency by applying different voltages. The measured RF spectra of the cantilever oscillation are shown in Figure 6. It can be seen that the oscillation frequency is increased from 4.2 MHz to 5 MHz when the voltage increases from 7.2 V to 9 V.

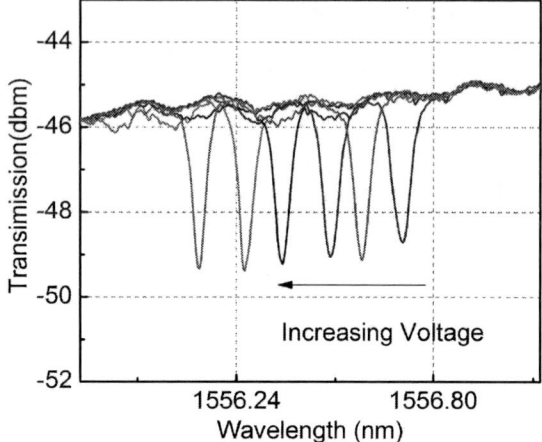

Figure 5: Transmission spectra of the cavity when increasing the voltage.

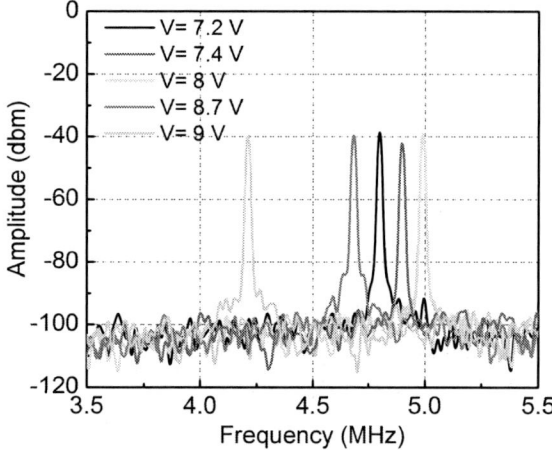

Figure 6: Resonance frequency of the oscillator tuned by the voltage.

CONCLUSIONS

In this work, an electrostatically tunable optomechanical oscillator is designed, fabricated and experimentally tested. The doped cantilever can be deformed by the electrostatic force, resulting in the shift of the ring resonance. The tuned optical spring constant changes the resonance frequency of the oscillator. The frequency is shifted from 4.2 MHz to 5 MHz in the experiment with the applied voltage increasing from 7.2 V to 9 V. The oscillator in the experiment shows a high tuning sensitivity without resorting to the mechanical tension. The tunable optomechanical oscillator have merits such as small dimension, low power consumption and little requirement for complex feedback systems, which makes it a good candidate for future applications in silicon photonic circuit and on chip signal processing.

ACKNOWLEDGMENTS

The work is supported by the Environmental and Water Industry Development Council of Singapore (EWI), RPC programme (Grant No: 1102-IRIS-05-01, 1102-IRIS-05-02, 1102-IRIS-05-04 and 1102-IRIS-05-05).

REFERENCES

[1] D. K. Agrawal, J. Woodhouse, and A. A. Seshia, "Modeling nonlinearities in MEMS oscillators," *Ultrasonic, Ferroelectrics and Frequency Control, IEEE Transactions on*, Vol. 60, pp. 1646-1659, 2013.

[2] W. M. Zhu, T. Zhong, A. Q. Liu, X. M. Zhang and M. Yu, "Micromachined optical well structure for thermooptic switching", *Appl. Phys. Lett*, Vol. 91, 261106,2007

[3] K. J. Vahala, "Back-action limit of linewidth in an optomechanical oscillator." *Physical Review A* 78, Vol. 2. 023832, 2008.

[4] B. Dong, H. Cai, G. I. Ng, P. Kropelnicki, J. M. Tsai, A. B. Randles, M. Tang, Y. D. Gu, Z. G. Suo and A. Q. Liu, "A nanoelectromechanical systems actuator driven and controlled by Q-factor attenuation of ring resonator," *Applied Physics Letters*, Vol 103, 181105, 2013.

[5] M. Ren, J. Huang, H. Cai, J. M. Tsai, J. Zhou, Z. Liu, Z. Suo, and A. Q. Liu, "Nano-optomechanical actuator and pull-back instability." *ACS Nano*, Vol 7, pp.1676–1681, 2013.

[6] H. Cai, B. Dong, J. F. Tao, L. Ding, J. M. Tsai, G. Q. Lo, A. Q. Liu and D. L. Kwong, "A nanoelectromechanical systems optical switch driven by optical gradient force," *Applied Physics Letters*, Vol 102, 023103, 2013.

[7] Q. Xu and M. Lipson, "All-optical logic based on silicon micro-ring resonators." *Optics Express* 15, Vol. 3, pp. 924-929, 2009.

[8] D. Van. Thourhout, J. Roels. "Optomechanical device actuation through the optical gradient force." *Nature Photonics* 4.4，pp. 211-217, 2010.

CONTACT

[†]A. Q. Liu, +65-67904336; eaqliu@ntu.edu.sg

NANO-OPTOMECHANICAL STATIC RANDOM ACCESS MEMORY (SRAM)

B. Dong[1,2], H. Cai[2], Y. D. Gu[2], Z. C. Yang[3], Y. F. Jin[3], Y. L. Hao[3], D. L. Kwong[2] and A. Q. Liu[1†]

[1]School of Electrical & Electronic Engineering, Nanyang Technological University, Singapore 639798
[2]Institute of Microelectronics, A*STAR (Agency for Science, Technology and Research)
Singapore 117685
[3]National Key Laboratory of Science and Technology on Micro/Nano Fabrication
Institute of Microelectronics, Peking University, Beijing 100871, China

ABSTRACT

This paper reports an on chip nano-optomechanical SRAM, which is integrated with light modulation system on a single silicon chip. In particular, a doubly-clamped silicon beam shows bistability due to the non-linear optical gradient force generated from a ring resonator. The memory states are assigned with two stable deformation positions, which can be switched by modulating the control light's power with the integrated optical modulator. The optical SRAM has write/read time around 120 ns, which is much faster as compared with traditional MEMS memory. Meanwhile, the write and read processes can happen concurrently without interference, which further reduces the time as compared with conventional electrical enabled SRAM.

INTRODUCTION

The exploration of information in modern society challenge the way we store and process information. There is always a strong demand for high speed, energy efficient and compact approaches to store signals in the current high speed and high capacity data processing system. Various approaches have been investigated such as magnetic storage, opto-electro based photonic memory and mechanical memory [1, 2]. MEMS/NEMS memory has been developed for more than 20 years. Most of the memories are based on either mechanical bistability [3] or material based bistability [4]. However, MEMS memory has drawbacks such as large scale and low speed. NEMS memory based on optomechanical effects shows great potential in optical storage on silicon photonic circuits due to its high speed, small size and good integrability [5-7].

This paper reports an on chip nano-optomechanical SRAM, which is integrated with light modulation system on a single silicon chip. The memory states are stored as deformation of doubly-clamped silicon beam, while the silicon beam shows bistability which is induced by the nonlinear optical gradient force. The optical SRAM can be utilized in all optical computing, with high speed (120 ns) and low power consumption. The CMOS compatible fabrication processes also make the optical SRAM cheap and can be easily integrated with other systems.

DESIGN AND THEORY

The schematic of the nano-optomechanical SRAM is shown in Fig. 1(a). Both "WRITE" and "READ" light from demultiplexer are coupled into the on-chip memory array. The "WRITE" light go through the integrated intensity modulator before entering the memory array. The intensity

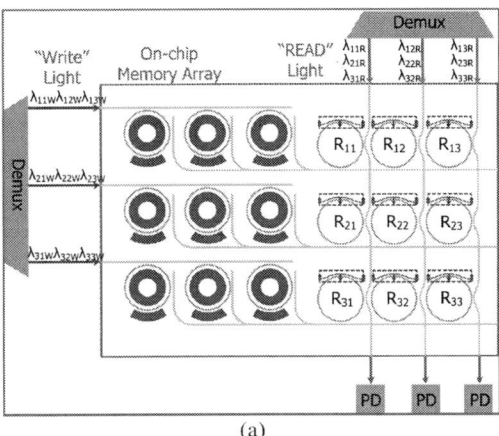

(a)

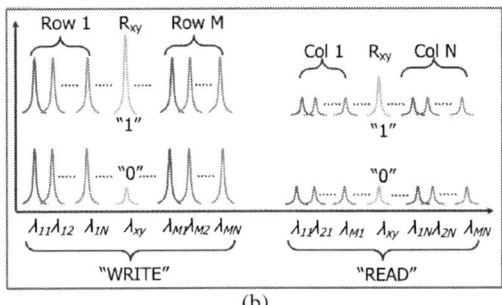

(b)

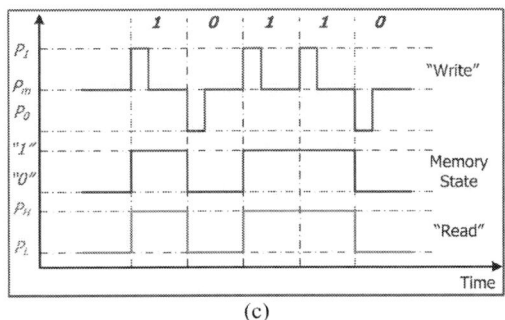

(c)

Figure 1: (a) Schematic of the nano-optomechanical memory, (b) layout of the optical memory and (c) spectrum illustration of the optical memory.

of "WRITE" light is modulated by the address controlled intensity modulator such that the intensity is increased when writing "1" and decrease when writing "0. Each wavelength can only interact with a single memory cell such that the deformation of a particular unit cell is modulated. Meanwhile, the "READ" light interacts with a single memory cell for memory states read out. (Fig 1(b)). The "WRITE" light and "READ" light are located in different spectra region to avoid interference. Fig. 1(c) shows the operation of the optical memory. By pumping in a pulsed high power (P_I) write light, the memory can be switched to state "1" while the transmission of read light is at high power level (P_H). The memory is switched to state "0" by further reducing the write light power to P_0. As a result, the transmission of the read light is low (P_L). Due to the bistability of the memory, the memory can maintain its state with a moderate power level (P_m) to save power consumption.

The schematic of the unit cell of the SRAM array is shown in Fig. (2). The unit cell consists of a bus waveguide, a ring resonator and a doubly-clamped silicon beam. Optical gradient force is generated by control light in the ring resonator over the side coupled silicon beam via evanescent wave. The ring resonator is 40 μm in diameter and 450 nm in width. The silicon beam is 200 nm in width and the coupling gap in-between is 200 nm.

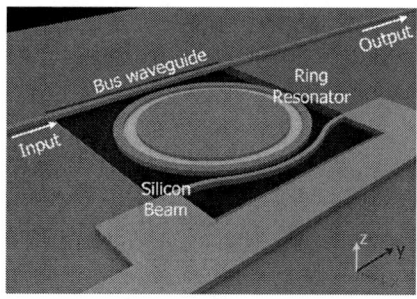

Figure 2: Schematic of nano-optomechanical memory unit cell

The operation of memory requires two lights to realize write and read processes. A red-detuned write light is coupled from the bus waveguide into the ring resonator. Thereafter, the optical gradient force is conducted between the ring resonator and the silicon beam, which pulls the silicon beam towards the ring resonator. The signal light, known as read light, is also coupled into the ring resonator and sensed at the output port. The wavelength of the read light overlaps with the resonance wavelength of the ring resonator. When the power of the write light is increased, the optical gradient force is increased and the mechanical arc moves towards the ring resonator along x direction. The displacement Δx of the silicon beam results in the change in the effective refractive index Δn_{eff} of the ring resonator, causing a red shift of the ring resonance wavelength $\Delta\lambda_r$. The read light can be transmitted through the waveguide since its wavelength is no longer overlapped with the

resonance wavelength now. The displacement of the silicon beam can affect the transmission of the read light by altering the resonance wavelength of the ring resonator. Therefore, by pumping a red-detuned write light with different power level, the transmission of the read light can be controlled and the statuses of the doubly-clamped silicon beam, which is known as the on-state and the off-sate of the switch, can be switched.

The deformation of the doubly-clamped silicon beam is determined by two forces: optical gradient force and mechanical spring force. The mechanical spring force, increases linearly as the deformation of the silicon beam increases. The generated optical force by the single wavelength laser λ_W can be expressed as

$$F_{optical}(x) = -\frac{2\gamma_e P_{optical}}{n_{eff}(x)} \frac{g_{om}(x)}{\Delta(x)^2 + \gamma^2} \qquad (1)$$

where γ_e is the external damping due to waveguide-ring coupling, $P_{optical}$ is the light power, n_{eff} is the effective refractive index, g_{om} is the optomechanical coupling coefficient, is the total damping coefficient and (x) is the laser detuning which is defined as

$$\Delta(x) = 2\pi c\left(\frac{1}{\lambda(x)} - \frac{1}{\lambda_W}\right) \qquad (2)$$

where (x) is the resonance wavelength of the ring resonator, which is related to the effective refractive index and therefore the deformation of the doubly-clamped silicon beam. The write light's wavelength detuning is therefore defined as

$$\delta = \lambda_W - \lambda(0) \qquad (3)$$

where $\lambda(0)$ is the resonance wavelength of the ring resonator at zero deformation.

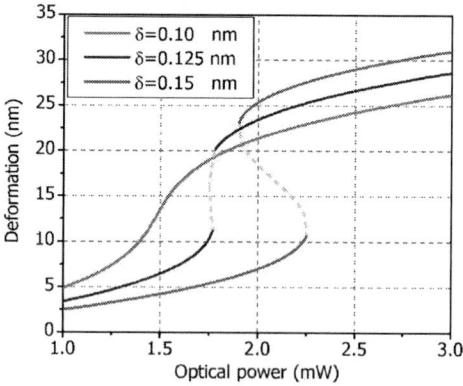

Figure 3: Bi-stability of the unit cell of SRAM at various laser detuning.

The doubly-clamped silicon deforms in different paths at different wavelength detuning as shown in Fig. 3. When the wavelength detuning is 0.1 nm, the doubly-clamped silicon beam shows only one stable position at all time. Therefore, the doubly-clamped silicon beam shows no bistability. When the wavelength detuning is increased to 0.125 nm, there is only a very narrow region that the doubly-clamped silicon beam shows two stable positions, and the transition curve is close to a vertical line. This is the critical condition where the doubly-clamped silicon beam starts to show bistability. When the wavelength detuning is increased to 0.175 nm, the bistable region is further broadened. Therefore, besides the write light power, the wavelength is critical to control the bistability of the doubly-clamped beam.

The doubly-clamped silicon beam shows bistability when it is deformed by the optical gradient force. But such bistability is not due to the mechanical properties of the doubly-clamped silicon beam, but the non-linear properties of the optical gradient force. The bistability of the doubly-clamped silicon beam can therefore be manipulated by controlling the light wavelength and power, providing more freedom in the operation of the optomechanical memory.

FABRICATION

Figure 4 shows the scanning electron microscope (SEM) images of fabricated nano-optomechanical SRAM with CMOS compatible nano-silicon-photonics technology.

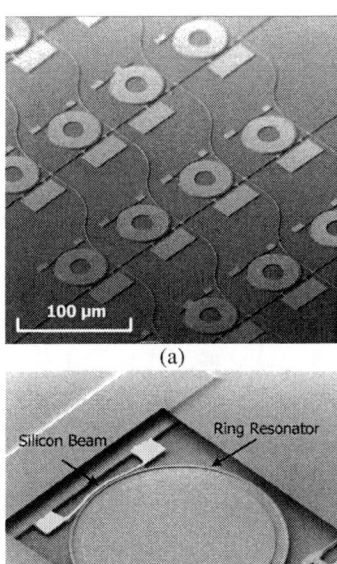

(a)

(b)

Figure 4: SEM images of the optical memory.

The waveguide structures have a width of 450 nm and a height of 220 nm. The doubly-clamped silicon beam is designed to be 200 nm in width while the coupling gap between the ring resonator and the silicon beam is 200 nm. The waveguides and ring resonators are patterned by a two-step deep UV lithography and RIE process. The rib structure has a 70-nm silicon slab layer to support the ring resonator. After etching, a 2-μm SiO_2 layer is deposited on the structure layers. A 40-nm Al_2O_3 is deposited and patterned, which is used as the protection film to protect the •xed structures and leave the window area opened for suspended structures. Finally, HF vapor selectively undercuts the buried oxide layer in the window area to release the movable structures.

EXPERIMENTS AND DISCUSSIONS

The intensity is experimentally tested by applying voltage across the p-i-n junction. Initially, the resonance wavelength of the ring resonator is 1587.94, which is the wavelength of "WRITE" light. By applying the voltage of the p-i-n junction, the resonance wavelength is red shifted due to the injection of the electrons. As a result, the transmission power of the "WRITE" light in reduced by 3 dB at 2.25 V and 2dB at 1.75 V. By controlling the voltage across the p-i-n junction, the intensity of the "WRITE" light can therefore be modulated.

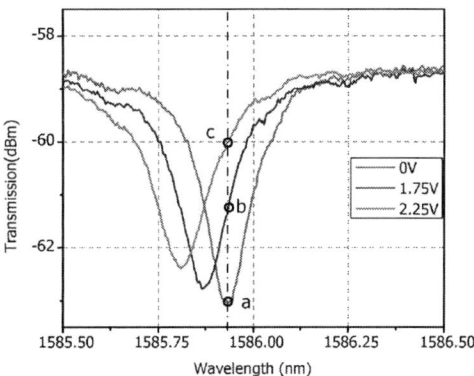

Figure 5: Tested transmission spectra of intensity modulator.

The power of the write light is modulated among P_1, P_m and P_0. The transmission spectra of the optical memory at various power of the red detuned (0.36 nm) write light is shown in Fig. 6. Initially, the power level is Pm (-8 dBm, the black line), the resonance wavelength is close to 1593 nm, which is the wavelength of the read light, the transmission of the read light is low, and the memory is at "0" state. When the power level is increased to P_1 (-6 dBm, the red line), the resonance wavelength shifts to 1591.6 nm, the transmission of the read light is high and the memory is at "1" state. After the power level is decreased to P_m again (the blue line), the resonance wavelength maintains its current position with negligible blue shift, and the memory remain at "1". When the power is further reduced to P_0 (-10 dBm, the pink line), the resonance wavelength reduces to

978-1-4799-7956-1/15 $31.00 © 2015 IEEE   51

the original position, and the memory state is switched to "0". The memory state remains at "0" when the power is changed to P_m (the green line) again. Therefore, the intensity difference of the read light between "0" and "1" is approximately 4.7 dB.

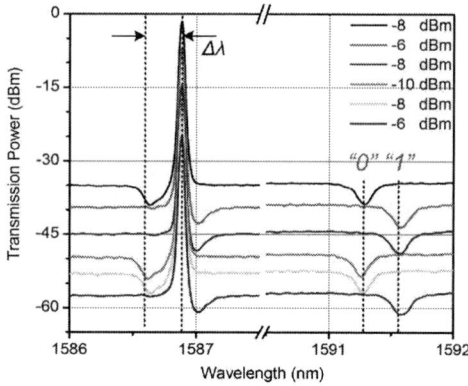

Figure 6: Transmission spectra of optical memory with various input powers.

The time domain measurement of the optomechanical memory is shown in Fig. 7. Owing to the small dimension, the optomechanical memory has faster response as compared with MEMS mechanical memory. To further increase the switching speed of the optical memory, the most direct way is to reduce the size of the doubly-clamped beam. However, smaller size of the doubly-clamped beam means a shorter interaction length for the optical gradient force and therefore smaller force. On the other hand, smaller ring resonator increases the bending loss of light circulating, which can further reduce the Q-factor and optical force.

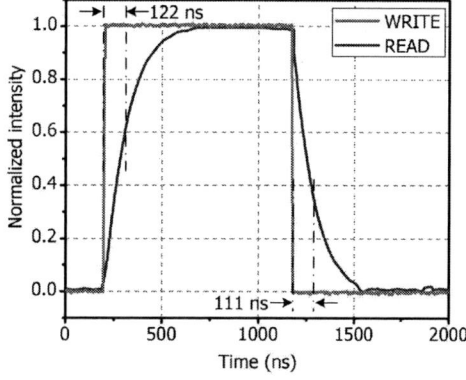

Figure 7: Time response of optomechanical memory.

CONCLUSIONS

In this work, a nano-optomechanical SRAM array is demonstrated. The memory states are stored as deformation of doubly-clamped silicon beam, while the silicon beam shows bistability which is induced by the nonlinear optical gradient force. On-chip p-i-n junction light intensity

modulator is integrated to tune the WRITE light's power. Photo detector can be integrated in future such the whole WRITE and READ process can be realized on a single chip. The optical SRAM can be utilized in all optical computing, with high speed (120 ns) and low power consumption. The CMOS compatible fabrication processes also make the optical SRAM cheap and can be easily integrated with other systems.

ACKNOWLEDGMENTS

The work is supported by the Environmental and Water Industry Development Council of Singapore (EWI), RPC programme (Grant No.: 1102-IRIS-05-01, 1102-IRIS-05-02, 1102-IRIS-05-04 and 1102-IRIS-05-05).

REFERENCES

[1] S. Zimmermann, A. Wixforth, J. P. Kotthaus, W. Wegscheider, and M. Bichler, "A Semiconductor-Based Photonic Memory Cell," *Science,* vol. 283, pp. 1292-1295, 1999.

[2] A. Uranga, J. Verd, E. Marigó, J. Giner, J. L. Muñóz-Gamarra, and N. Barniol, "Exploitation of non-linearities in CMOS-NEMS electrostatic resonators for mechanical memories," *Sensors and Actuators A: Physical,* vol. 197, pp. 88-95, 2013.

[3] M. Hoffmann, P. Kopka, and E. Voges, "Bistable micromechanical fiber-optic switches on silicon with thermal actuators," *Sensors and Actuators a-Physical,* vol. 78, pp. 28-35, 1999.

[4] G. Stegeman and E. Wright, "All-optical waveguide switching," *Optical and Quantum Electronics,* vol. 22, pp. 95-122, 1990.

[5] B. Dong, H. Cai, G. I. Ng, P. Kropelnicki, J. M. Tsai, A. B. Randles, M. Tang, Y. D. Gu, Z. G. Suo and A. Q. Liu, "A nanoelectromechanical systems actuator driven and controlled by Q-factor attenuation of ring resonator," Applied Physics Letters, Vol 103, 181105, 2013.

[6] M. Ren, J. Huang, H. Cai, J. M. Tsai, J. Zhou, Z. Liu, Z. Suo, and A. Q. Liu, "Nano-optomechanical actuator and pull-back instability," ACS Nano, Vol 7, pp.1676–1681, 2013.

[7] H. Cai, B. Dong, J. F. Tao, L. Ding, J. M. Tsai, G. Q. Lo, A. Q. Liu and D. L. Kwong, "A nanoelectromechanical systems optical switch driven by optical gradient force," Applied Physics Letters, Vol 102, 023103, 2013.

CONTACT

[†]A. Q. Liu, +65-67904336; eaqliu@ntu.edu.sg

A LOW-POWER MEMS TUNABLE PHOTONIC RING RESONATOR FOR RECONFIGURABLE OPTICAL NETWORKS

Carlos Errando-Herranz*, Frank Niklaus, Göran Stemme, and Kristinn B. Gylfason
KTH Royal Institute of Technology, Stockholm, Sweden

ABSTRACT

We experimentally demonstrate a low-power MEMS tunable photonic ring resonator with 10 selectable channels for wavelength selection in reconfigurable optical networks operating in the C band. The tuning is achieved by changing the geometry of the slot of a silicon slot-waveguide ring resonator, by means of vertical electrostatic parallel-plate actuation. Our device provides static power dissipation below 0.1 µW, a wavelength tuning range of 1 nm, and a narrow bandwidth of 0.1 nm, i.e. 10 nW static power dissipation per selectable channel for TE mode tuning.

INTRODUCTION

The current exponential increase in data communication, with annual growth rates up to 90% [1], follows from the ever-increasing demand for capacity in communication networks. This demand cannot be fulfilled by traditional copper networks due to fundamental limits in data transmission, making optical networks a requirement. A promising solution to further increase optical network capacity is to add spatial multiplexing on top of existing optical wavelength multiplexing [1].

Spatial multiplexing requires optical add/drop filters. Following the introduction of reconfigurable optical add/drop multiplexers (ROADMs) in 2003, the demand for flexibility and scalability in the communications network resulted in the progressive substitution of the traditional fixed optical add/drop multiplexers (FOADMs) with ROADMs. The currently available ROADMs systems are based on free space optics. Light is launched into free space from a fiber input towards a diffractive grating where the channels are spatially separated and directed towards an optical switching array. After switching and redirecting, the channels are recombined with a second grating and coupled back into the fiber system. Current switching arrays are formed by either MEMS mirrors or liquid crystals [2]. However, free-space-based ROADMs have a number of disadvantages such as large size, high insertion losses, and fabrication complexity, which translates directly to high costs. Key features for a future ROADM are good optical performance, high number of channels, low insertion loss, flexible channel plan, and low cost [2].

A promising approach to fulfill these requirements is the use of matrixes of single-wavelength reconfigurable add-drop filters (Fig.1). Silicon photonic ring resonators are excellent candidate building blocks for this scheme, due to their compactness and high optical quality factor (Q). Constant bandwidth filters are preferred, to maintain equal channel spacing. Furthermore, the number of available communication channels is determined by the filter bandwidth and its tuning range.

Resonance wavelength tuning in ring resonators can be achieved by using physical effects such as carrier injection, thermo-optic tuning, electro-optic materials, and MEMS actuation. Several approaches to tunable ring resonators have been presented. Wavelength tuning by free carrier injection enables high-speed modulation [3], but the optical loss induced by the interaction between photons and free carriers makes the method impractical for reconfigurable optical networks. Thermo-optic tuning yields long tuning ranges and constant bandwidth [4, 5]. However, for ROADM applications, power dissipation from thermally tuned rings causes thermal crosstalk, resulting in communication errors and thus limiting integration density. Integration of electro-optic materials can provide ring resonator tuning with low static power dissipation. However, with the electro-optic materials used so far, the devices show either low optical Q due to scattering from the tunable material [6], short tuning ranges [7], tuning instability due to resonance drift [8], or optical interference due to fabrication complexity [8, 9].

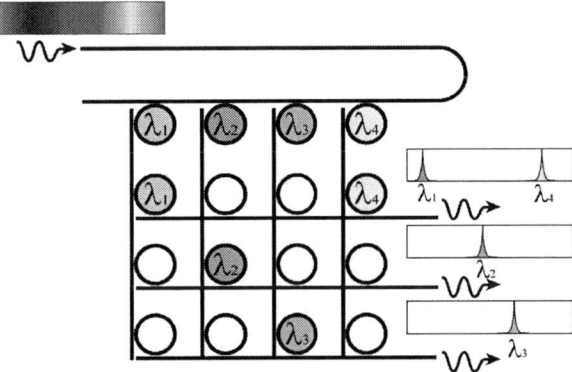

Figure 1: Matrix of reconfigurable ring resonators for optical wavelength band selection to several output ports. The use of such spatial multiplexing in reconfigurable optical networks has the potential to meet the currently increasing demand for data communication capacity.

MEMS-based optical tuning has been achieved by a suspended dielectric cantilever on top of a ring resonator. In this configuration, the small separation of cantilever and waveguide necessary for wavelength tuning sets the actuator operation close to pull-in, resulting in a controllable wavelength tuning range of only 120 pm [10]. A constant bandwidth and long tuning range were achieved by tuning the diameter of a silicon ring. The ring consisted of two suspended U-shaped waveguides optically connected by directional couplers. One of the waveguides is displaced by a comb drive actuator, changing the diameter of the ring and thus resulting in a shift in resonance wavelength [11]. However, the use of fully suspended waveguides makes anchoring necessary, which results in poor optical performance due to reflections caused by waveguide discontinuities at the anchors.

978-1-4799-7956-1/15 $31.00 © 2015 IEEE

In this work, we use vertical parallel-plate actuation of an SOI cantilever containing the inner part of a slot-waveguide ring resonator. This locally changes the effective refractive index of the resonant optical mode, resulting in a shift in resonance wavelength. The novel vertical actuation introduced in this work yields a discontinuity-free optical path, while setting the actuator operation point far from the pull-in state. A key advantage of electrostatic tuning of MEMS cantilevers is the low-power wavelength tuning, which results in negligible cross-talk between adjacent rings. Moreover, the perfect elasticity of the single-crystalline silicon cantilever provides a stable and hysteresis-free tuning [12]. A schematic of the device is shown in Fig. 2.

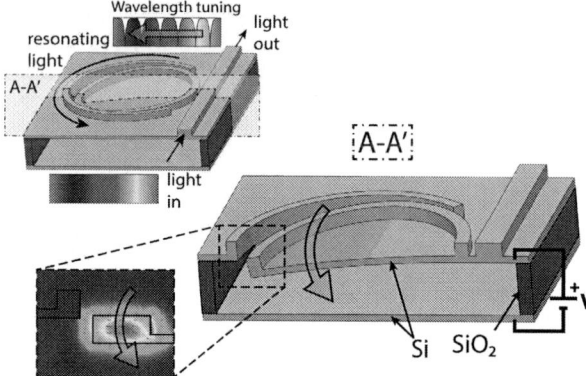

Figure 2: A schematic of the MEMS tunable ring resonator. Parallel-plate actuation (indicated with open arrow) of a free-standing cantilever (indicated in orange) that contains part of the inner rail of a slot-waveguide ring resonator changes the waveguide geometry. This locally affects the effective refractive index of the resonant optical mode, resulting in a shift in resonance wavelength. The inset shows the simulated optical mode.

FABRICATION

Our MEMS tunable ring resonator was fabricated by a very simple SOI-based process (Fig. 3) with two dry etch steps for the silicon device layer (resulting in two heights) and a wet SiO₂ under-etch. The first lithography step defines the ridge waveguides that form the ring resonator and the grating coupled bus waveguide. The second lithography step and the wet under-etch define the free standing cantilever. The cantilever is delimited by the fully etched slot waveguides, and its free suspended area is determined by the placement of etch holes.

The fabrication process starts with a clean SOI chip with a 220 nm crystalline silicon device layer and 2 μm buried oxide (Fig. 3A). This is a standard substrate specification used by the Epixfab silicon photonics foundries. Electron beam patterning of a 50 nm layer of a high-resolution negative electron beam resist (Hydrogen silsesquioxane, HSQ) defines the waveguide structures (Fig. 3B). The pattern is then transferred to the device layer by timed dry etch of silicon, resulting in ridge waveguide structures with 110 nm height on a 110 nm thick silicon slab (Fig. 3C). The patterned HSQ remains on the chip for the next lithography step.

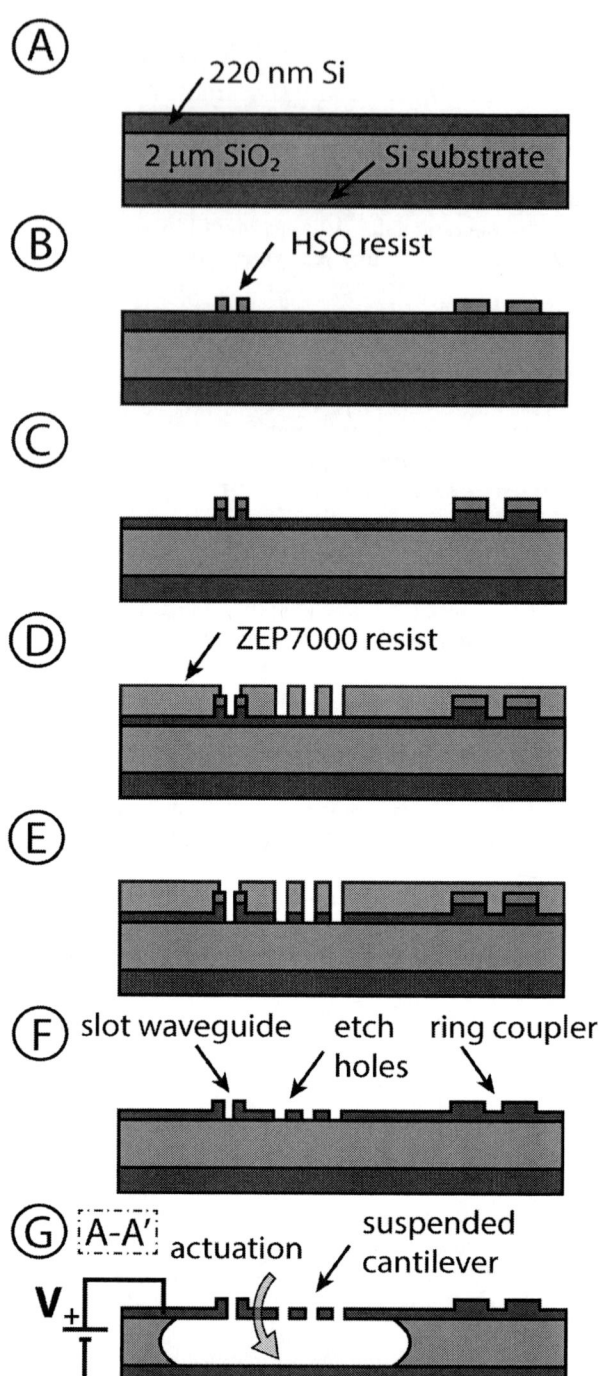

Figure 3: Cross-section (A-A' from Fig. 2) of the fabrication process. A clean SOI chip (A) was spun with HSQ resist and e-beam patterned (B) to define ridge waveguides by timed dry etch of silicon (C). The HSQ resist, in combination with a second e-beam patterned positive ZEP7000 resist (D) were used for a second through dry silicon etch (E) that defines the free-standing cantilever (F). After stripping the resists, the structures are released by HF wet oxide under-etch, followed by critical point drying (G). Applied voltage between the silicon substrate and the device layer deflects the cantilever, resulting in a resonance wavelength shift.

In order to define the free-standing cantilever, we e-beam patterned a 200 nm thick layer of positive e-beam resist (ZEP7000) after alignment to the existing structures (Fig. 3D). The subsequent through dry etch of silicon was then defined by the superposition of the two e-beam masks (Fig. 3E). This allows the creation of a self-aligned etched slot in the center of the waveguide, defined by the high resolution HSQ patterning of the first lithography step. After stripping the two resists (Fig. 3F), the patterned cantilever was then released via a 50% HF wet oxide etch, followed by critical point drying (Fig. 3G).

RESULTS AND DISCUSSION

Figure 4A is a top view of the design of our MEMS tunable ring resonator. Part of the etched slot, combined with etch-holes, defines a suspended cantilever. To ensure a discontinuity-free optical path, we designed smooth refractive index transitions in the silicon ring by using strip-to-slot couplers, and in the buried oxide layer by strategic etch hole positioning for tapering the transition between suspended and supported waveguides. An SEM image of our device is shown in Fig. 4B.

For optical characterization, TE polarized light is coupled in and out of the chip through optical fibers aligned to grating couplers fabricated on-chip. At resonance wavelengths, light couples from the bus waveguide into the ring. The transmitted light from the device is then measured by a wavelength domain component analyzer (Agilent Technologies 86082A). Our device shows a free spectral range (FSR) of 1.24 nm, a 0.1 nm -3 dB bandwidth, and a loaded Q of 16200 at a wavelength of 1573.8 nm (Fig. 5).

To characterize the tuning performance of the device, the silicon substrate was grounded. A voltage was then applied by direct contact of a compliant probe needle to the device layer of the chip. The resonator was actuated by a voltage of up to 56 V, inducing a wavelength shift of up to 1 nm (Fig. 5). Considering the 0.1 nm bandwidth of the ring resonator, the number of unique selectable channels obtained for this device is 10.

Below a 30 V applied voltage, our device showed a non-linear tuning rate (Fig. 6). This is most likely caused by the static deflection of the cantilever, due to relaxation of mechanical stress of the released silicon device layer. This initial deflection results in parts of the cantilever situated above and parts below the zero deflection plane. Actuation of the cantilever then results in combined effects that may counteract depending on the cantilever initial static deflection. When the cantilever bending has overpassed the zero deflection plane (i.e. at an actuation of 30 V), blueshift prevails. Consequently, from 30 to 56 V, we observe a linear -36 pm/V tuning rate, which compares favorably to the 13 pm/V shown with integrated LiNbO$_3$ [9]. The bandwidth variation for our device is below 20% in the linear tuning range and shows no apparent trend, which indicates that the optical losses in the ring are not related to the wavelength tuning (Fig.6).

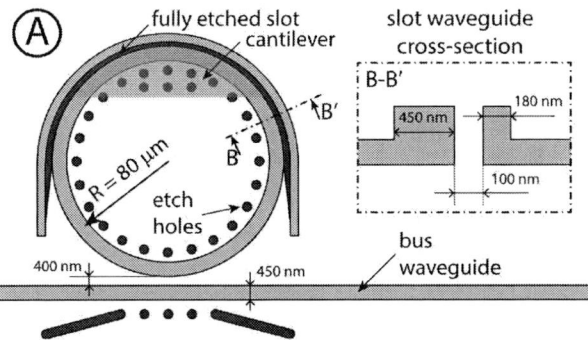

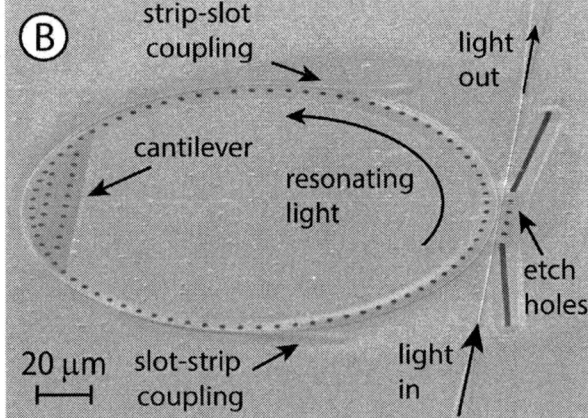

Figure 4: (A) Design of the MEMS tunable ring resonator including geometrical parameters. The inset (B-B') shows the slot-waveguide cross-section. The cantilever is indicated in orange. Light couples from the bus waveguide into the ring. Half of the ring is made of a slot-waveguide with a through etched slot. A suspended cantilever is formed by the through etched slot and the etch holes. Tuning is achieved by electrostatically bending down the cantilever. (B) SEM image of a MEMS tunable ring resonator. The under-etch pattern, indicated in green, ensures a discontinuity-free optical path.

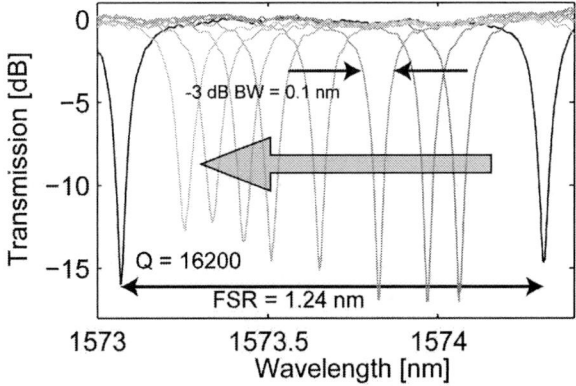

Figure 5: Transmission spectra for the device shown in Fig. 4 under a range of actuation voltages. The ring shows a Q of 16200 with an FSR of 1.24 nm. A bandwidth of 0.1 nm and a tuning range of 1 nm yield 10 selectable optical channels.

978-1-4799-7956-1/15 $31.00 © 2015 IEEE   55

Our MEMS tunable ring resonator has static power dissipation below 0.1 μW, which is orders of magnitude below the lowest reported value of 2.4 mW for thermally tuned ring resonators [5]. An optical tuning range of 1 nm and a bandwidth of 0.1 nm yield 10 selectable optical channels for our device. This results in a 10 nW static power dissipation per selectable channel.

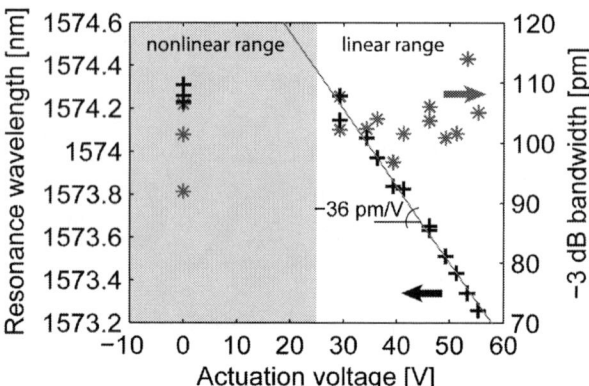

Figure 6: Our MEMS tunable ring resonator presents a linear tuning rate of -36 pm/V (+) with bandwidth variation below 20% (*) for actuation voltages between 30 and 56 V.

CONCLUSIONS

We have shown a MEMS tunable ring resonator with a sub 0.1 μW static power dissipation and stable tuning performance using a simple silicon fabrication process.

In particular, the presented device combines the extremely low power consumption of electrostatic MEMS, a very simple high-resolution SOI fabrication process, and 10 selectable channels, i.e. the highest number reported for MEMS tunable ring resonators.

Our results suggest that the presented MEMS tunable ring resonator technology has the potential to become an important building block for combined wavelength and spatial multiplexing in future reconfigurable optical networks.

ACKNOWLEDGEMENTS

This work was partially funded by the Swedish Research council and the European Research Council through the FP7-ERC-M&M's starting grant (No. 277879) and the FP7-ERC-xMEMS advanced grant (No. 267528).

REFERENCES

[1] P. J. Winzer, "Making spatial multiplexing a reality," *Nat Photon*, vol. 8, no. 5, pp. 345-348, May 2014.

[2] T. A. Strasser and J. L. Wagener, "Wavelength-Selective switches for ROADM applications," *Selected Topics in Quantum Electronics, IEEE Journal of*, vol. 16, no. 5, pp. 1150-1157, Sep. 2010.

[3] Q. Xu, B. Schmidt, S. Pradhan, and M. Lipson, "Micrometre-scale silicon electro-optic modulator," *Nature*, vol. 435, no. 7040, pp. 325-327, May 2005.

[4] X. Zheng, I. Shubin, G. Li, T. Pinguet, A. Mekis, J. Yao, H. Thacker, Y. Luo, J. Costa, K. Raj, J. E. Cunningham, and A. V. Krishnamoorthy, "A tunable 1x4 silicon CMOS photonic wavelength multiplexer/demultiplexer for dense optical interconnects," *Opt. Express*, vol. 18, no. 5, pp. 5151-5160, 2010.

[5] P. Dong, W. Qian, H. Liang, R. Shafiiha, D. Feng, G. Li, J. E. Cunningham, A. V. Krishnamoorthy, and M. Asghari, "Thermally tunable silicon racetrack resonators with ultralow tuning power," *Opt. Express*, vol. 18, no. 19, pp. 20 298-20 304, Sep. 2010.

[6] T.-J. Wang, W.-J. Li, and T.-J. Chen, "Radially realigning nematic liquid crystal for efficient tuning of microring resonators," *Opt. Express*, vol. 21, no. 23, pp. 28 974-28 979, 2013.

[7] C.-T. Wang, Y.-C. Li, J.-H. Yu, C. Y. Wang, C.-W. Tseng, H.-C. Jau, Y.-J. Chen, and T.-H. Lin, "Electrically tunable high q-factor micro-ring resonator based on blue phase liquid crystal cladding," *Opt. Express*, vol. 22, no. 15, pp. 17 776-17 781, 2014.

[8] W. De Cort, J. Beeckman, T. Claes, K. Neyts, and R. Baets, "Wide tuning of silicon-on-insulator ring resonators with a liquid crystal cladding," *Opt. Lett.*, vol. 36, no. 19, pp. 3876-3878, Oct. 2011.

[9] L. Chen, M. G. Wood, and R. M. Reano, "12.5 pm/V hybrid silicon and lithium niobate optical microring resonator with integrated electrodes," *Opt. Express*, vol. 21, no. 22, pp. 27 003-27 010, 2013.

[10] S. M. C. Abdulla, L. J. Kauppinen, M. Dijkstra, M. J. de Boer, E. Berenschot, H. V. Jansen, R. M. de Ridder, and G. J. M. Krijnen, "Tuning a racetrack ring resonator by an integrated dielectric MEMS cantilever," *Opt. Express*, vol. 19, no. 17, pp. 15 864-15 878, 2011.

[11] T. Ikeda and K. Hane, "A microelectromechanically tunable microring resonator composed of freestanding silicon photonic waveguide couplers," *Applied Physics Letters*, vol. 102, no. 22, pp. 221 113+, Jun. 2013.

[12] F. Zimmer, M. Lapisa, T. Bakke, M. Bring, G. Stemme, F. Niklaus, "One-Megapixel Mono-Crystalline Silicon Micro-Mirror Array on CMOS Driving Electronics Manufactured with Very Large Scale Heterogeneous Integration", *IEEE Journal of Microelectromechanical Systems*, Vol.20, No.3, pp.564-572, 2011.

CONTACT

*C. Errando-Herranz, tel: +46760692156; carloseh@kth.se

NEMS OPTICAL CROSS CONNECT (OXC) DRIVEN BY OPTICL FORCE

H. Cai[1], J. X. Lin[2], J. H. Wu[2], B. Dong[1,3], Y. D. Gu[1], Z. C. Yang[4], Y. F. Jin[4], Y. L. Hao[4], D. L. Kwong[1]
and A. Q. Liu[3]

[1]Institute of Microelectronics, A*STAR (Agency for Science, Technology and Research)
Singapore 117685
[2]School of Mechanical Engineering, Xi'an Jiaotong University, Xi'an 710049, China
[3]School of Electrical & Electronic Engineering, Nanyang Technological University, Singapore 639798
[4]National Key Laboratory of Science and Technology on Micro/Nano Fabrication
Institute of Microelectronics, Peking University, Beijing 100871, China

ABSTRACT

This paper presents a compact silicon-photonic based optical cross connect (OXC) driven by the optical gradient force. Each switch element consists of a waveguide-crossing-coupled micro-ring resonator and a suspended arc. The device is fabricated with a standard CMOS compatible process using deep-UV 248-nm lithography with a double-etch technique. A switching time of 0.24 μs is experimentally demonstrated. The proposed switch topology of the interconnections has potentials of employing a single wavelength channel or multiple wavelength channels, and provides channel selection from sets of input fibers and sets of output fibers.

INTRODUCTION

Over the past decade, efforts on silicon photonics are trying to revolutionize computing platforms by manufacturing optical communication devices using traditional complementary metal-oxide-semiconductor (CMOS) techniques, which support large-scale complicated circuits with high yield and high volume production. It will provide substantial size, cost and power savings over traditional optical communication solutions. A series of key components have been successfully demonstrated, including high-speed modulators [1], photo-detectors [2] and other various photonics devices [3-5]. With CMOS photonics technology developed and the device libraries enriched, building silicon photonic system with complex functionalities becomes possible.

With the increase of the complexity of integrated photonics circuits, routing topology will become a serious issue, similar to the case of electronics in its early days. The electronics industry provided a solution by adding more and more backend metal interconnect layers. However, such architectures will not apply to silicon photonics circuits, since it is extremely challenging to create multiple crystalline silicon layers (i.e. device layers). Furthermore, the electrical interconnection counterparts impose severe limitations including high-power consumption and electrical-limited data bandwidth [6]. The thermal issue associated with electronic interconnection has also been a daunting problem that prevents significant increase in the processor speed. Photonics alternatives may offer new possible solutions for the interconnections.

Meanwhile, using light to dynamically and stably redirect the flow of other beams has been a long-term goal in photonics integrated systems. In the future, optically transparent networks with high-complexity optical cross-connect features will be highly required. For example, it enables microprocessors to use light instead of electrical signal to communicate with transistors on a chip. And the key merits of the optical approach include low power consumption for signal transmissions and large data bandwidth carrier of optical channel. However, it is still challenging to realize a practical all-optical switching device in silicon owing to its weak optical nonlinearity.

In this paper, we propose an on-chip optical switch, comprising a 2 × 2 waveguide cross-grid array coupled with an array of photonic-mechanical based micro-ring resonator switches. Through resonance enhancement of both optical field and mechanical motion, all-optical switching operation is achieved. Such architecture could be a meaningful solution for the optical transparent networks.

THEORY AND DESIGN

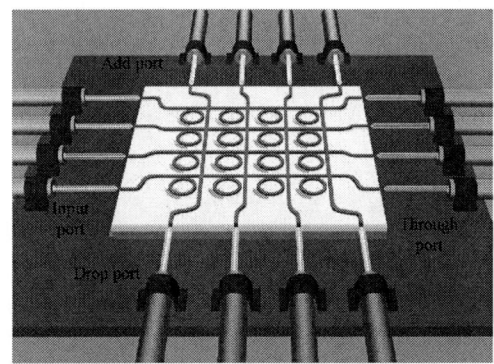

Figure 1: Schematic illustration of the n × n all-optical switch system. The tapered optical fiber array is aligned and bonded around the chip; both input lights and output lights are coupled into and out of the chip through the waveguide-based spot size converters, respectively.

A passive photonic-mechanical micro-ring resonator based switching element as building blocks for the n × n all-optical OXC system is proposed, as schematically illustrated in Fig. 1. The input-ports are connected with signal inputs, whereas the outputs are labeled as drop-port and through-port, respectively. The add-ports are mainly for the introduction of the control lights. By operating with only a single wavelength channel that is at an off-resonance wavelength of the micro-rings along the channel, a single transmission path between any input-port and direct through-port can be established. While by introducing the control light through any add-port, some corresponding micro-ring along the signal transmission path is switched on. As a result, the signal is directed to the corresponding drop-port.

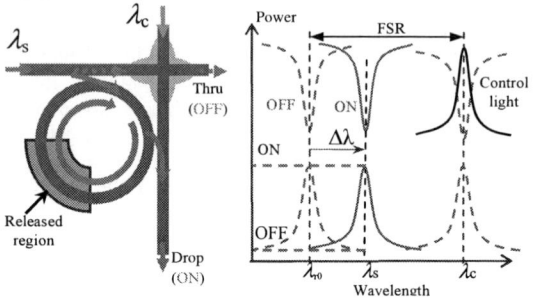

Figure 2: (a) Design of the switch element based on micro-ring resonator. A micro-ring resonator coupled to a waveguide crossing. Meanwhile the free standing arc as a part of the ring can be deflected by the optical gradient force. (b) Working principle of the switch, with the spectral response at different switching states.

The basic switch element is formed with a micro-ring resonator, coupled with two nano-waveguides, one for the drop port, and the other for the through port (i.e. 'Thru port'), as shown in Fig. 2(a). In particular, the part of the micro-ring is released from the substrate and becomes free-standing. The input signal (λ_s) is transmitted to the drop port only when the wavelength λ_s satisfies the micro-ring resonant condition (i.e. Drop port in the 'ON' state), which is expressed as

$$m\lambda_{r0} = n_{eff}L$$

where m is an integer, n_{eff} is the effective index of optical mode, and L is the length of the resonating cavity. Otherwise, the input signal light will be mostly guided to the through port (i.e. Drop port in the 'OFF' state), as shown in Fig. 2(a). In principle, with the absence of the control light, there is no transmission of the signal light (at λ_s) to the Drop port since most light passes through to the Thru port. On the other hand, when a strong control light with the wavelength of λ_c is introduced, a gradient optical force is formed through the evanescent wave coupling between the ring resonator and the substrate. The magnitude of the force is proportional to the optical power injected. It drives the free-standing arc to bend down, and changes the ring resonance condition. As a result, the wavelength of the signal light overlaps with the ring resonant wavelengths (i.e. the transmission 'dip'), leading to the signal light transmission to the Drop port.

Here, instead of using micro-disk geometry or fully suspended ring-spokes geometry [7] for the resonant cavity design, a free-standing arc is employed because of its lower mechanical stiffness and better stability. As the suspended arc is a part of the micro-ring, it can be bent down to the substrate by the optical gradient force. Meanwhile, due to its stiffness, a mechanical force pushes the arc upwards. When the two forces are balanced with each other, the suspended arc can be stabled at an intersection position.

The derivative of the optical energy with respect to the gap (between the arc and the substrate, as shown in the inset in Fig. 3) gives the gradient optical force [5-7]. Subject to a control light of a fixed wavelength and power, the optical force is nonlinearly related to the deflection. Such nonlinear function is illustrated in Fig. 3 for several levels of detuning, defined by the difference between the wavelength of the control light and the resonant wavelength of the ring (e.g. in the original state, with no arc deformation), $\Delta = \lambda_c - \lambda_{r0}$. The nonlinearity of the optical force is due to the interaction between the mechanical motion and the optical field. Owing to the cavity effect of the micro-ring resonator, the magnitude of the optical force depends on the resonant wavelength, while the power and the wavelength of the control light are constant. The deflection of the arc modifies the effective refractive index and thus tunes the resonant wavelength of the ring. Consequently, the optical force is nonmonotonic. In general, the deflection of the arc can be varied by changing either the wavelength λ_c or the power P_c of the control light. Here, we employ wavelength variation due to the following two reasons. Firstly, varying the wavelength of the control light provides an efficient positioning control approach, since the deformation of the arc x is inversely proportional to the square of Δ, or $x \propto (\lambda_c - \lambda_r)^{-2}$, while is proportional to the input power, or $x \propto P_{in}$. Secondly, wavelength variation helps to pose the limitation of the fluctuation of power variations. Once the arc deflects, the elasticity of the arc generates a mechanical force, F_{mech}, which tends to pull the arc back to its original position. In such two-force driving system, the mechanical force linearly increases with the deflection of the free-standing arc, while the optical force is in nonlinear manner. The two forces balance each other at the equilibrium points. The linear force becomes too large to be balanced by the nonlinear force when the deflection is above a certain value, which results in the pull-back instability. The mechanical spring constant of the arc is $k_{mech} = 0.14$ N/m, which is estimated by using the finite-element software COMSOL. The shape of the arc reaches a state of equilibrium when the optical force balances the mechanical force.

As the optical force is a nonlinear function of the deflection, for the control light of a fixed wavelength, the arc may deflect to multiple states of equilibrium. For example, as shown in Fig. 3, the two curves (F_{mech} and F_{opt}) intersect at three points (a, b, and c), when $\Delta = 0.19$ nm. As Δ increases, points b and c move closer, while point a approaches to zero-deflection position. Particularly, at a critical detuning value of $\Delta = 0.64$ nm, points b and c merge, and there is only one stable point s, corresponding to the arc largest deflection.

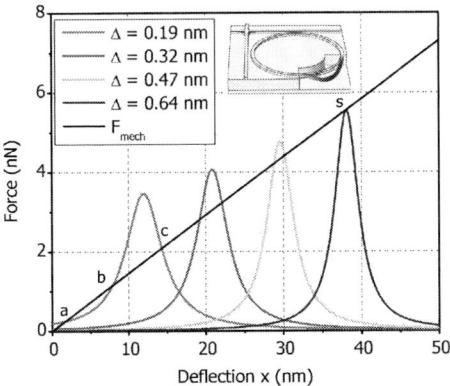

Figure 3: Simulations of the optical force and mechanical force under different wavelengths of control light.

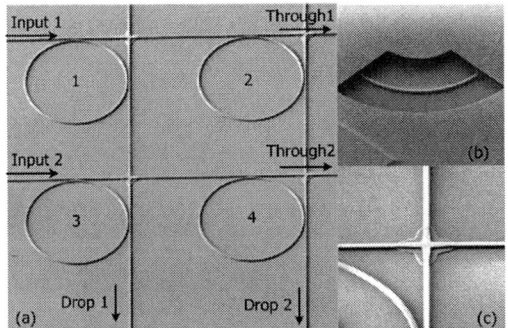

Figure 4: SEM images of the fabricated 2 × 2 array all-optical switch on the SOI wafer: (a) overview of the all-optical switch, (b) suspended arc, and (c) double-etched waveguide crossing.

RESULTS AND DISCUSSIONS

Figure 4 shows scanning electron microscope (SEM) images of a 2 × 2 all-optical switch device based on the proposed switching element, with a free-standing arc (as in Fig. 4(b)). It is fabricated on a standard silicon-on-insulator wafer, with a 220-nm thick (p-type) silicon structure layer and a 2-μm buried oxide (BOX) layer. The waveguides, including micro-rings and straight sections, have a width of 450 nm and a height of 220 nm. Except the suspending arc, all other waveguides have a silicon core surrounded by 2-μm SiO$_2$ claddings. The coupling gap between the ring and the straight waveguide is 200 nm. All the structures/devices are achieved through the technology of 248 nm deep-UV lithography patterning. The compact crossing is achieved using a double-etch technique. After the first deep-UV lithography patterning, the plasma dry etching is followed to transfer photoresist pattern into the silicon structure layer, leaving a 70-nm remaining silicon film, which is required by ridge crossing waveguide structure. The second

lithography and dry etching define the shallow-etch pattern and etch until the BOX layer. For low optical loss and symmetric waveguide geometry, a 2-μm PECVD SiO$_2$ layer is deposited as the upper cladding, so that the Si waveguide is surrounded by the SiO$_2$ cladding. Another film for the release window protection is then deposited, patterned and etched away, leaving the areas directly above the suspending arc features open for release process. Finally, HF-vapor selectively undercuts the buried oxide layer, leaving the free-standing structures.

Before the switch device characterization, the optical transmission measurement of the device is firstly conducted. A broadband light source is launched into the optical switch through the input port. The obtained transmission spectrum with typical resonance patterns is observed, with an average side-mode suppression ratio of more than 15 dB. In the switching experiments, both signal light and control light will be coupled through the input-port and add-port, respectively (with mode size converter at the tips) from taper fibers. A polarization controller is used to select only TE light injected into the waveguide. The lasing wavelength of the control light can be tuned with a step of 0.01 nm and its power is maintained to ensure a 2.8-mW optical power to be injected into the input waveguide. Compared with the control light, the power density of the signal light is weak and does not generate optical force. An optical spectrum analyzer with 0.001-nm resolution is used to measure the transmission spectrum of the ring resonator.

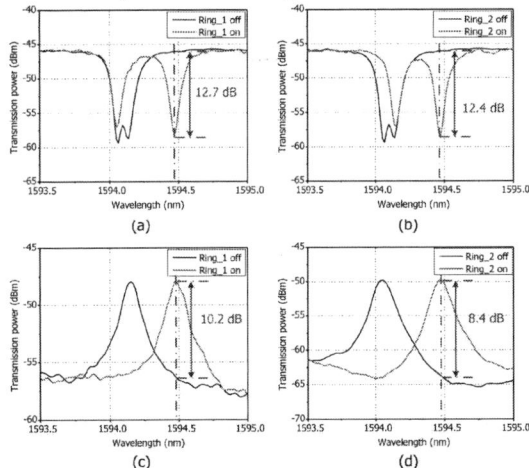

Figure 5: Transmission spectra at (a) Thru port 1 and (c) Drop port 1, without (black) and with (pink) control light introduced from Add port 1, respectively; transmission spectra at (b) Thru port 1 and (d) Drop port 2, without (black) and with (pink) control light from Add port 2, respectively.

In the OFF state, the input wavelength λ_s is different from the resonant wavelength of the micro-rings (Ring 1 and Ring 2 in Fig. 4(a)). Such input signal light at λ_s propagates directly from the input port to the through port

978-1-4799-7956-1/15 $31.00 © 2015 IEEE

(Thru port 1) when Ring 1 and Ring 2 are powered OFF. While in the ON state only Ring 1 is turned ON by the injection of the control light λ_c, and the resonant wavelength shifts to the same as the signal wavelength λ_s. In such resonance condition, the signal light is coupled to Ring 1 and directed to the Drop port 1. Hence, there is an obvious power drop (~12.7 dB) in the transmission spectrum observed at Thru port 1 as shown in Fig. 5(a). While for the above mentioned 'OFF' and 'ON' cases, the spectra monitoring at Drop port 1 are plotted in Fig. 5(c), corresponding to two different peaks detected. Similarly, the spectrum at Thru port 1 is shown in Fig. 5(b) when introduces the control light to turn 'ON' Ring 2 only, and the resonance wavelength of the Ring 2 is shifted. Hence, signal light is coupled to the Ring 2 and directed to the Drop port 2 (as shown in Fig. 5(d)), with an extinction ratio of 8.4 dB.

In the experiment of transient switching response measurement, a continuous wave (CW) from a tunable laser is modulated as a control light to be coupled into the switch device. 10% of the modulated light is tapped off with a coupler to be used as a reference for the switching response measurement. The other 90% of light is coupled into the control waveguide. The output light at the Drop port is coupled to the detection system through a tapered fiber, and the switching states are monitored with an oscilloscope for the switch response measurement. As expected, the applied modulated control signal causes a deflection of the arc. The normalized switching response is shown in Fig. 6. A 10% – 90% rise time and fall time are measured as 0.24 µs and 0.11 µs, respectively. It is also observed that there is no visible residual oscillation or overshoot in the switching response.

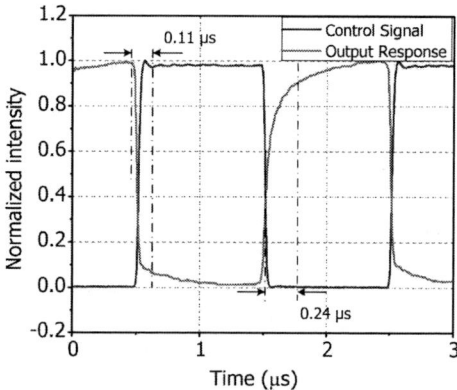

Figure 6: Dynamic response of the optical force driven optical switch, including the waveforms of the control and output response signals of the switch.

CONCLUSIONS

In conclusions, a 2 × 2 all-optical switch, achieved through optical force driving a suspended silicon arc, has been demonstrated. The experimental results measure an extinction ratio of more than 10 dB, and the switching time of about 0.24 µs. The initial study on the optical switch element, which is based on cascaded waveguide crossing-coupled micro-ring resonator, has shown its potential for optical cross-connects.

REFERENCES

[1] G. T. Reed, G. Mashanovish, F. Y. Gardes, and D. J. Thomson, "Silicon optical modulators", *Nature Photon.* vol. 4, pp.518-526, 2010.

[2] J. Michel, J. Liu, and L. C. Kimerling, "High-performance Ge-on-Si photodetectors", *Nature Photon.* vol. 4, pp. 527-534, 2010.

[3] W. Bogaerts, et al, "Silicon-on-insulator spectral filters fabricated with CMOS technology", *IEEE J. Sel. Top. Quantum Electron.*, vol. 16, pp. 33-44, Jan/Feb. 2010.

[4] L. Zhang, X. Tan, M. Yang, M. Qi, T. Hu, and J. Yang, "On-chip wavelength-routed photonics networks with comb switches", in *Proc. 9th IEEE int. Conf. GFP*, Aug. 2012, pp.279-281.

[5] Q. Fang, J. Song, X. Luo, L. Jia, M. Yu, G. Lo, and Y. Liu, "High efficiency ring-resonator filter with NiSi heater," *IEEE Photon. Technol. Lett.*, vol. 24, pp. 350–352, 2012.

[6] A. Scandurra, M. Lenzi, R. Guerra, F. G. Delia Corte, M. A. Nigro, "Optical interconnects for network on chip," *1st International Conference on Nano-Networks and Workshops (NanoNet '06)*, 2006, pp.1-5.

[7] H. Cai, B. Dong, J. F. Tao, L. Ding, J. M. Tsai, G. Q. Lo, et al., "A nanoelectromechanical systems optical switch driven by optical gradient force", *Appl. Phys. Lett.*, vol. 102, 023103, 2013.

[8] Povinelli, M. L.; Lonⲅcar, M.; Joannopoulos, J. D. "Evanescent-Wave Bonding between Optical Waveguide", *Opt. Lett.* vol. 30, pp. 3042–3044, 2005.

[9] M. Ren, J. Huang, H. Cai, J. M. Tsai, J. Zhou, Z. Liu, Z. Suo, and A.-Q. Liu, "Nano-optomechanical actuator and pull-back instability", *ACS Nano.* vol. 7, pp.1676-1681, 2013.

[10] J. F. Tao, J. Wu, H.Cai, Q. X. Zhang, J. M. Tsai, J. T. Lin, A. Q. Liu, A Nano-machined Optical Logic Gate Driven by Gradient Optical Force. Appl. Phys. Lett. 2012, 100, 113104.

[11] Y. F. Yu, J. B. Zhang, T. Bourouina, A. Q. Liu, "Optical-Force-Induced Bistability in Nanomachined Ring Resonator Systems", *Appl. Phys. Lett.* vol. 100, 093108. 2012.

CONTACT

H. Cai, caih@ime.a-star.edu.sg.

THE FUTURE OF MEMS SENSORS IN OUR CONNECTED WORLD

Gerhard Lammel

Bosch Sensortec GmbH, Reutlingen, GERMANY

ABSTRACT

The MEMS market is year after year growing faster than the average semiconductor industry. Over that time the largest technology driver for MEMS changed from automotive applications to consumer electronics dominated by smartphones. Beyond that, MEMS sensors become the heart of whole classes of new devices like fitness trackers, smart watches, virtual reality glasses and smart sensor nodes for the Internet of Things.

Silicon chips are only one part of the MEMS story, you need as well special mixed signal circuitry, low power data processing, smart algorithms and connectivity to transform raw signals into meaningful information. Multi-sensor applications & modules are playing an increasingly important role.

INTRODUCTION
Technological Background

The following 6 key technologies developed by Bosch are the basic micromachining processes used for most of today's sensor products:

- Growth of very thick layers of polysilicon, so-called "Epipoly" [1]
- High-precision and fast deep trench etching (the "Bosch DRIE-process") [2]
- Vapor phase etching for release of structures [3]
- Encapsulation for hermetical sealing [4], [5]
- APSM process – exact vacuum cavities in silicon (Advanced Porous Silicon Micromachining) [6]
- FlipCore geomagnetic field sensing in a thin-film semiconductor process

Bosch holds more than 1000 patent families on a large variety of MEMS technologies and sensor applications. With these technologies acceleration sensors, gyroscopes, pressure sensors, magnetometers and others were fabricated at Bosch with a cumulated volume of more than 4 000 000 000 parts until 2014.

Figure 1: Silicon structures created by the Bosch process for deep trenching. For size comparison, a hair with 90 μm diameter is shown on top of an acceleration sensor structure on the right.

THE THREE WAVES OF MEMS PROLIFERATION

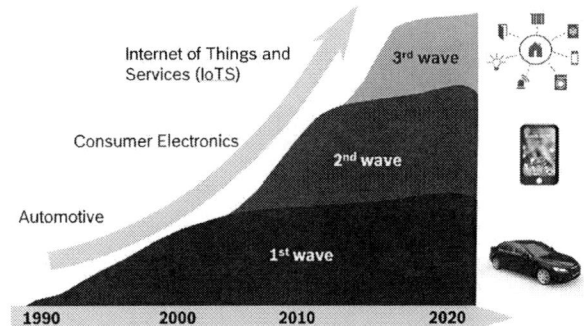

Figure 2: The three waves of MEMS proliferation

The first wave: Automotive

In the mid 90ies automotive MEMS started with airbag, yaw-rate and pressure sensors in high volume. In a typical car of today you find more than 50 MEMS sensors, which are used for [7]:

Engine Management
- Barometric Air Pressure (BAP)
- Diesel particulate filter
- Mass flow sensor…

Vehicle Dynamics Control
- Yaw rate sensor
- High pressure sensor
- IMU and low g sensors…

Safety Systems
- Rollover sensor
- Occupant weight sensor iBolt
- Pedestrian Contact Sensor (PCS)
- Upfront Sensor (UFS)
- Peripheral Pressure Sensor (PPS) …

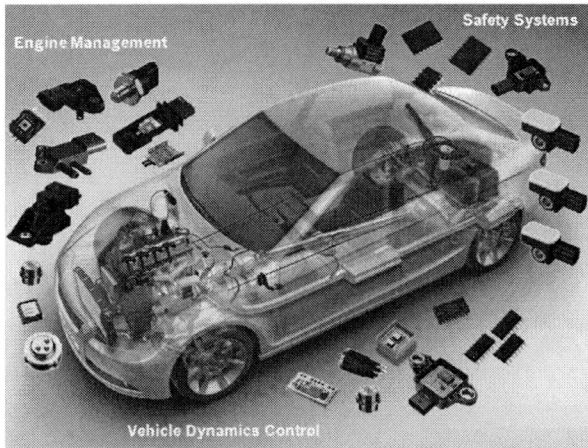

Figure 2: MEMS sensor applications in a car of today.

Figure 3: Consumer MEMS products shrink path.

With MEMS sensors the cars became safer, more fuel efficient and cleaner, which benefits our environment.

The second wave: Consumer electronics

In the mid of the last decade mobile phones started using MEMS sensors. Meanwhile they became the strongest driver of new MEMS developments, shrinking size and cost at a high pace while increasing the performance, see Fig. 3. Today's applications of MEMS sensors in smart phones are:

Orientation detection
- Portrait / landscape
- Upside down
- Free speech profile

User interface
- Gaming input
- Menu navigation
- Gesture recognition

Motion analysis
- Step counting
- Activity monitoring
- Power management

Pedestrian navigation
- Dead reckoning
- Floor level tracking
- Location based services

The third wave: Internet of Things and Services (IoTS)

If you want to increase comfort, security, health and productivity you have to create smart, interactive and sensitive systems. Beyond smartphones new classes of wearable devices emerge, like fitness trackers, smart watches and virtual reality glasses that are all loaded with MEMS sensor. They target the following applications:

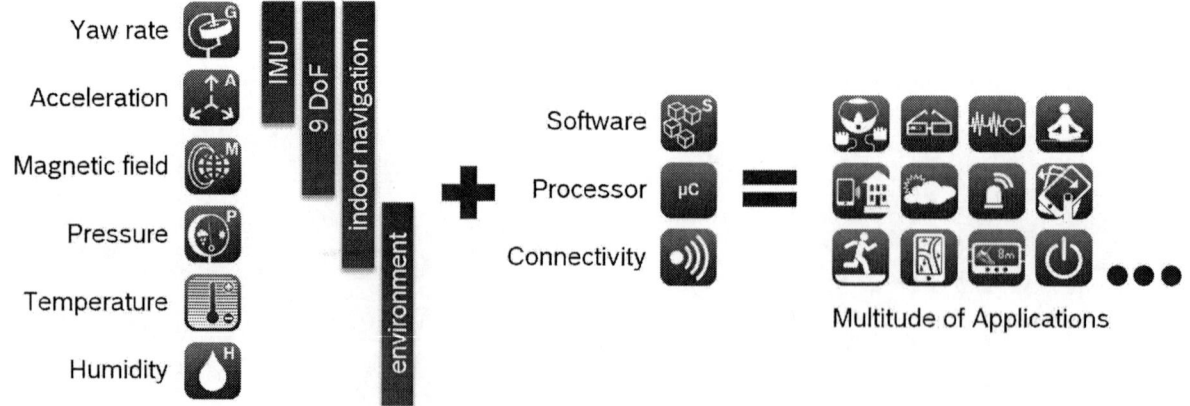

Figure 4: From measurands to sensor data fusion

978-1-4799-7956-1/15 $31.00 © 2015 IEEE

Wellbeing
- Sleep monitoring
- Activity classification
- Stress level detection

Fitness
- Speed and performance
- Fitness level
- Calorie monitoring
- Sports trainer

The Internet of Things also targets home automation and smart cities. Today we see that stand-alone devices like home appliances, heating, venting and air conditioning (HVAC) get connected by proprietary solutions within distributed networks. The future will shift from monitoring towards control of intelligently connected household devices and HVAC systems in function of persons present, their individual behavior, weather forecast, electricity price etc.

These sensor applications have in common that the sensor modules measure multiple physical quantities, are always on, have a processing unit with software and are connected to a communication network, see Fig. 4. The impacts on sensor requirements are:

- Reducing power consumption
- Reducing footprint
- Smart data fusion software
- Integration with micro controller and RF

FROM SENSOR SIGNALS TO USER EXPERIENCE

The cause-and-effect dependencies between use-case and technology are usually passing several steps. Three of these chains shall be shown here to explain the processing of different signals for their respective application.

From pressure sensors to faster navigation

An atmospheric pressure sensor can be used to measure the altitude above sea level. The measurement gets more precise if the weather dependent sea level pressure (average value 1013.25 hPa) is taken into account, which can be obtained online by service providers, e.g. from the closest airport.

A GPS can take up to several minutes from cold start to output the first position, also known as TTFF (time-to-first-fix). This time can be shortened by using the almanac data from mobile cell towers (A-GPS, Assisted GPS). The waiting time can be further decreased when the altitude above sea level is available in the system [8]. The reason is that the 3D location solution space for locking the satellite codes and pseudoranges is reduced. This feature was first used in high volume in the 2011 Google Nexus phone [9].

Sensor data fusion: more than the sum of its parts

A compass and a gyroscope both suffer from fundamental imperfections for finding the true north heading. The compass is distorted by steel constructions in buildings, electrical power lines and transformers,

electrical motors in printers, elevators etc., as well as by other components in the mobile device itself like loudspeakers and NFC antennas. The gyroscope has a temperature dependent yaw rate offset (bias) that leads to a heading drift when integrated.

With the help of Kalman filters [10], the errors of both sensors can be estimated continuously and cancelled mutually using the overdetermined properties of the equation systems. The computation power is considerable, since matrix inversion is needed. Additionally, the algorithms need to be optimized according to the sensors' characteristics.

This so called sensor data fusion to obtain the true orientation vector (roll, pitch, heading) in world coordinates has been successfully integrated in one component BNO055 (see Fig. 5). It contains acceleration sensor, gyroscope and magnetometer, each 3-axis, in total 9 DoF (degrees of freedom), together with a micro-processor and software to output the self-calibrated orientation in real time.

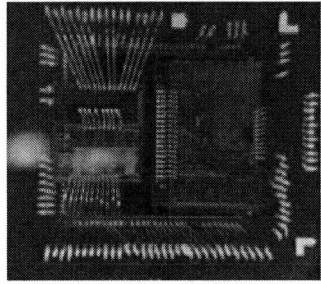

Figure 5: Multi chip package for true orientation sensor BNO055.

- Package size: 5.2mm x 3.8mm x 1.1mm
- Accuracy: 2° … 3° (static)
- Latency: 20 ms
- Robust to magnetic distortions
- Optimized HW/SW Codesign

This orientation vector can be combined with a step counter for dead reckoning, which is the extrapolation of the location from the last or to the first known fix from GPS or WiFi localization. The merging of MEMS sensor data with other sources improves the accuracy of localization especially indoors, e.g. in shopping malls and other public buildings.

As the power consumption of sensors decreases (e.g. less than 1 mA for Bosch Sensortec's BMI160 inertial measurement unit with a 3-axis gyroscope and 3-axis acceleration sensor) it is important not to waste power in signal processing. Therefore integrated sensor functions enable always-on applications, e.g. a step counter with less than 50 μA consumption for computation. This is orders of magnitudes less than the application processor of a smartphone would consume on average, even if it is woken up only for a very short time.

Beyond motion sensors, the pressure sensor used as altimeter can track indoor floor levels, which is not possible with GPS.

978-1-4799-7956-1/15 $31.00 © 2015 IEEE   63

Environmental sensors

We call the combination of pressure, humidity and ambient temperature sensor an environmental sensor, see Fig. 6. Such sensors are usually integrated in HVAC control systems, where the response time is not critical. Bosch Sensortec's BME280 shows a very fast response time of $\tau = 1$ sec. This enables completely new applications like detecting the proximity of human skin (which always evaporates some humidity) in contrast to other surfaces. Unlike optical proximity sensors, this environmental sensor can distinguish between a user and another object close to the device, which is helpful to save power by a more specific wake-up feature.

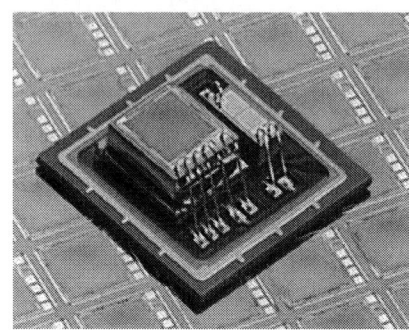

Figure 6: Environmental sensor BME280, 2.5 x 2.5 mm², here shown without metal cap on top.

CONCLUSION

Bosch sensors can be found in every second smartphone worldwide today.

Modules with a multitude of sensors, integrated low power signal processing e.g. for context awareness and specific applications, will change MEMS from the provision of raw data to meaningful data fusion output from sensor nodes for our connected world, see Fig 7.

ACKNOWLEDGEMENTS

I would like to thank all colleagues at Bosch Sensortec and at Robert Bosch GmbH for the excellent cooperation to achieve innovative products, pushing MEMS to the next level of user experience and customer benefit.

REFERENCES

[1] M. Offenberg, F. Laermer, B. Elsner, H. Muenzel, W. Riethmueller, "Novel Process For A Monolithic Integrated Accelerometer", in *Digest Tech. Papers Transducers '95 and Eurosensors IX Conference*, Stockholm, June 25-29, 1995.

[2] F. Laermer, A. Urban, "Milestones in deep reactive ion etching", *Digest Tech. Papers Transducers '05 Conference*, Seoul, Korea, June 5-9, 2005.

[3] V. Becker, F. Laermer, M. Offenberg, A. Schilp, patent US6558559 B1, DE19704454C2.

[4] A. Franke, A. Trautmann, A. Feyh, S. Knies, patent EP2197781B1.

[5] J. Gonska, R. Hausner, patent US8035209B2.

[6] G. Lammel, S. Armbruster, C. Schelling, H. Benzel, J. Brasas, M. Illing, R. Gampp, V. Senz, F. Schaefer, S. Finkbeiner, "Next generation pressure sensors in surface micromachining technology", *Digest Tech. Papers Transducers '05 Conference*, Seoul, Korea, June 5-9, 2005.

[7] J. Marek, U. Gómez, "MEMS (Micro-Electro-Mechanical Systems) for Automotive and Consumer Electronics", in Chips 2020, The Frontiers Collection, pp 293-314, 2012.

[8] http://electroiq.com/blog/2009/07/pressure-sensors-provide-indoor-competency-for-navigation/

[9] https://plus.google.com/+DanMorrill/posts/jVJhPyouWDP

[10] G. Welch, G. Bishop, "An Introduction to the Kalman Filter", July 24, 2006. http://www.cs.unc.edu/~welch/media/pdf/kalman_intro.pdf

CONTACT

Gerhard.Lammel@bosch-sensortec.com

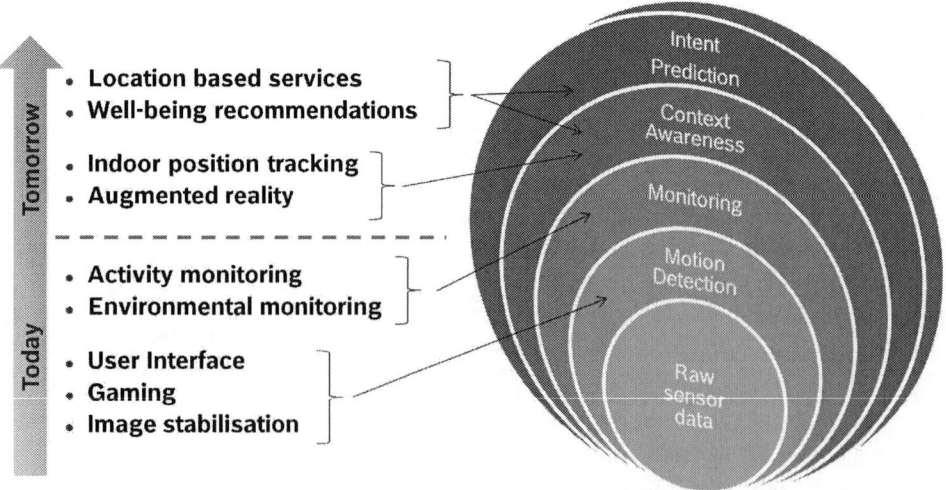

Figure 7: More advanced use-cases will drive systems.

978-1-4799-7956-1/15 $31.00 © 2015 IEEE

FABRICATION AND CHARACTERIZATION OF A NEW VARIFOCAL LIQUID LENS WITH EMBEDDED PZT ACTUATORS FOR HIGH OPTICAL PERFORMANCES

Stephane Nicolas[1], Marjolaine Allain[1], Claudine Bridoux[1], Stephane Fanget[1], Suzanne Lesecq[1], Mykhailo Zarudniev[1], Sebastien Bolis[2], Arnaud Pouydebasque[2] and Fabrice Jacquet[2]

[1]CEA, LETI, MINATEC Campus, Grenoble, FRANCE
[2]WAVELENS, Grenoble, FRANCE

ABSTRACT

This paper reports the fabrication and characterizations of a compact varifocal lens with an embedded PZT actuator. Optical aperture is typically up to 3 mm diameter with a total component thickness down to 400μm. High power efficiency (lower than 0.1 μW), high speed response time (down to 1 ms), high electro-optical performances (10 diopters optical power variation at 10V) and good optical quality (wavefront error lower than 50 nm) is reported through this paper.

INTRODUCTION

Today, there is a continuous trend in camera modules that are integrated in mobile phones or tablets to increase performances while still gaining on size and cost. Because of the size, power consumption and cost constraints in camera modules for mobile phones, standard functions such as optical auto-focus, zoom or optical image stabilization are more difficult to achieve in such devices compared to classical digital cameras.

Generally, the tuning of the focal length to perform auto-focus or zoom functions is obtained by mechanically adjusting the position of the lenses system using stepper motors. In camera phones, VCM (voice coil motor) based auto-focus devices are today mainly used. But VCM are now facing challenging size, power consumption and speed issues.

In order to address these challenges, developments of alternative solutions have been carried out which are, for the most part, illustrated by works on variable-focus (varifocal) liquid lenses [1,2]. These last devices can be identified by 2 main operating concepts. The first one is based on material properties modification as, for example, in liquid-crystal lenses [3-5]. In such devices, a non-uniform electric field is applied and the resulting spatial distribution of the refractive index across the liquid crystal layer leads to a change of the focus.

In the second type of device, the focal length modification is induced by a lens curvature change. This phenomenon can be obtained by changing the interface properties between two immiscible liquids using electrowetting [6].

Another way to modify the lens curvature is to bend a deformable membrane that encapsulates a liquid [7–16]. This can be obtained by applying a hydraulic pressure in a hermetic chamber which displaces the liquid towards the center of the lens and thus changing the curvature of the membrane. For example, thermal [7,8], electromagnetic [9], electrostatic [10-12], piezoelectric [13,14] or electro-active polymers [15,16] actuators were proposed to provide the desired pressure.

Based on this latest concept, an embedded electrostatic actuation has been firstly developed in our previous works in order to provide the liquid displacement necessary to bend the membrane [10-12].

In this paper, we present the integration of a thin film piezoelectric actuator that improves the varifocal lens performances. The use of a piezoelectric MEMS actuation instead of electrostatic actuation permits to greatly release some material and integration constraints. Firstly, the liquid is no more active in the actuation process and more references can be used with enhanced properties in terms of thermal stability, viscosity, etc. Moreover, instability phenomena inherent to electrostatic actuation such as the pull-in effect are no longer relevant with the piezoelectric actuation and a simple DC signal can be used instead of AC signal at high frequency to power the lens. The power consumption can also be strongly reduced. Finally, in terms of process integration, no needs to perform any process steps on the glass wafer with the piezoelectric actuation, which simplify the fabrication process and release the necessity of a precise alignment during the bonding between the silicon and the glass wafers.

ACTUATION PRINCIPLE

As an operating principle, when the PZT structure is actuated, the optical oil flows through the membrane center so that it modifies the membrane's curvature and introduces a focal length variation (Figure 1).

The varifocal lens consists of an optical flexible membrane released onto an optical oil-filled cavity, with integrated PZT actuators embedded at the membrane periphery (Figure 2).

Figure 1: Actuation principle of the varifocal liquid lens

978-1-4799-7956-1/15 $31.00 © 2015 IEEE   MEMS 2015, Estoril, PORTUGAL, 18 - 22 January, 2015

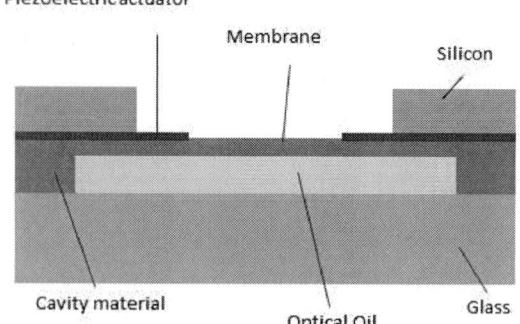

Figure 2: Cut-view of the varifocal liquid lens with embedded MEMS actuators.

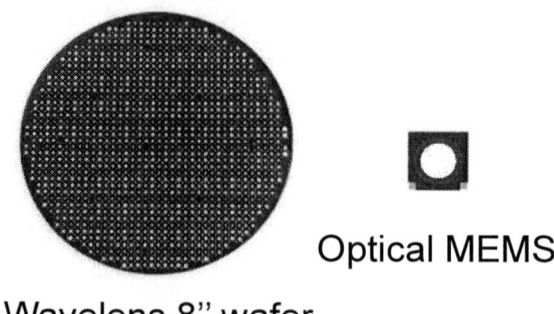

Figure 4: Pictures of the final optical MEMS wafer and device after singulation.

VARIFOCAL LENS FABRICATION

The device is achieved on 200 mm wafers and the process flow is briefly described in Figure 3. Note that this fabrication process is patented and only conventional clean room process steps are used.

First, a silicon bare wafer is used to realize the actuator parts, the electrical lines and contact pads. These functional parts are obtained by conventional deposition, lithography and etching steps. Then, a transparent flexible polymer membrane is deposited with a final thickness of several microns. Finally, a thick polymer layer is patterned in order to define the cavity. It has to be noted that this polymer has very good bonding properties toward silicon and glass materials.

The cavity is then filled by a liquid using an automatic dispensing tool that allows an accurate control of the volume for each cavity. The liquid is a clear, colorless oil which is very stable in a wide range of temperature and available commercially.

Then, the silicon base wafer is assembled by polymer bonding with a 200 mm unprocessed glass wafer. A specific bonding recipe has been developed in order to avoid any remaining bubbles inside the cavity after bonding.

Finally, the silicon wafer is thinned down to 100 µm and the remaining silicon is etched away by conventional Deep Reactive Ion Etching (DRIE) in the optical area and above the contact pads. The final membrane releasing is performed by the removal of an oxide etch stop layer using buffered oxide etch after DRIE such that the membrane surface quality is not altered. Pictures of the final wafer obtained are reported on Figure 4.

The wafer stack is then diced in order to singulate the components.

CHARACTERIZATIONS RESULTS

Optical bench description

The optical characterizations of the varifocal lenses are performed using a Shack-Hartmann wavefront sensor. Shack–Hartmann sensors are commonly used for wavefront characterization and measuring lens aberrations but it can also be used to determine a focal length with excellent accuracy. The set-up used for our characterizations has been already described in previous works [10].

Optical power measurements

On Figure 5, a typical optical power variation response of the varifocal liquid lens with an applied voltage ranging from 0 to 15 V is reported.

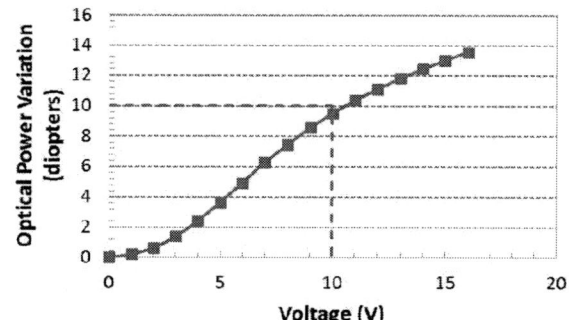

Figure 5: Typical optical power variation in diopters versus applied voltage obtained with our varifocal lens.

It is demonstrated that an optical power variation of 10 diopters can be obtained with a voltage as low as 10V. It has to be noted that due to the resulting small current

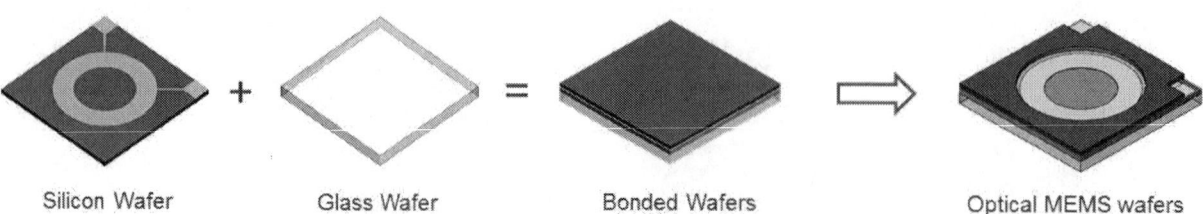

Figure 3: Fabrication of the varifocal liquid lens on 200mm wafer.

(few nA), the power consumption needed to actuate the varifocal lens is very low (< 0.1 µW). Note that the typical power consumption of VCM is 100 mW.

Wavefront error measurements

On Figure 6, a typical wave front error variation of the varifocal liquid lens with an optical power variation ranging from 0 to 14 diopters is reported.

A good optical quality is demonstrated since the wave front error is ranging from 20 to 40 nm which is about 20 times lower than the Shack-Hartmann laser's wavelength (λ= 632nm).

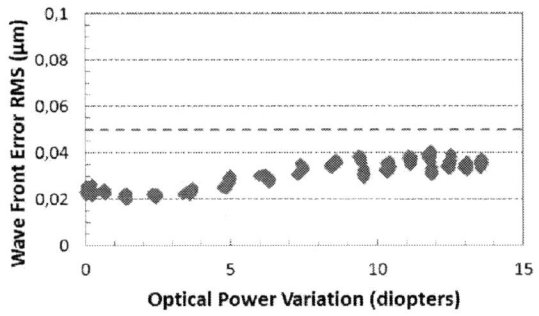

Figure 6: Typical wave front error RMS versus optical power variation obtained with our varifocal lens.

Response time measurements

For the response time evaluation the wavefront sensor measurement is triggered by the actuation voltage with a controlled delay time Δt (Figure 7).

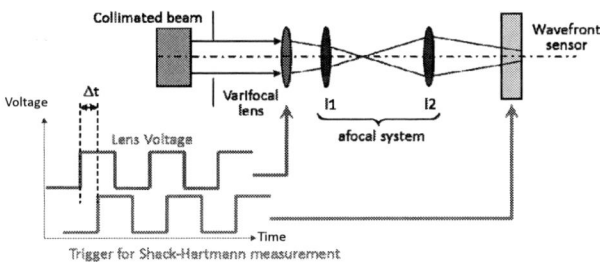

Figure 7: Schematic representation of the set-up used for response time measurement.

By changing the value of Δt below 1 ms up to 100 ms and measuring the optical power variation for each value, a response curve is obtained and an example is reported on Figure 8.

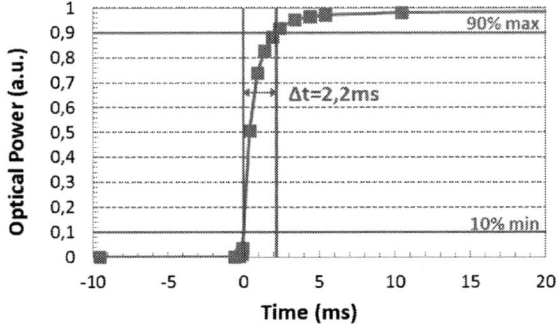

Figure 8: Typical response time obtained with our varifocal liquid lens.

Very fast response time can be achieved depending on the initial optical aperture of the varifocal lens. For small apertures, a response time of 1 ms can be obtained whereas, for larger apertures (3 mm), typical response time is 3 ms. This is one order of magnitude faster that the response time of a VCM.

Optical power variation after cycling test

In order to evaluate the reliability of the device and more specifically the actuation part, cycling test (switch from 0V to 10V) has been performed up to 10 million cycles at a frequency of 10 Hz. Regularly, the test is stopped in order to characterize the optical power at 0V and 10V. Then, it is possible to evaluate at each step the optical power variation using as a reference the initials values before cycling the varifocal lens.

The results after 10 million cycles are reported on Figure 9.

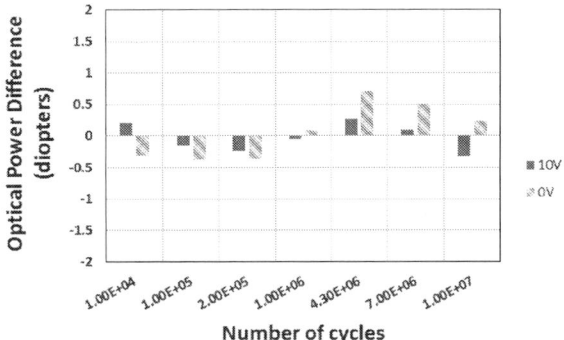

Figure 9: Optical power variation at 0V and 10V during 10 million cycles at 10 Hz.

An optical power variation within +/- 0.5 diopters is observed whatever the cycling steps chosen. This variation is typically within the measurement error of our optical characterization set-up. It can be concluded that there is no significant change in the optical power of the varifocal lens even after several millions of cycles.

Integration of the varifocal lens within a 5Mp camera module

Finally, the varifocal lens has been integrated on a 5 megapixel camera module and pictures were taken at different voltages as illustrated on Figure 10.

Without voltage, only the poster at 70 cm can be observed clearly. A significant change in the focus occurs at 6 V where the card at 17 cm becomes sharp. Finally, the bare code at 10 cm, blurred without voltage, can be resolved at 10 V.

CONCLUSION

In conclusion, the fabrication and characterizations of a compact varifocal lens with an embedded PZT actuator has been described.

The process used for the fabrication of the device is patented and based on standard MEMS processes that can guarantee high yield and reliability in view of a potential industrialization.

Thanks to the very good optical performance of the system, the optical power variation required for autofocus

978-1-4799-7956-1/15 $31.00 © 2015 IEEE

0 V – 70 cm   6 V – 17 cm   10 V – 10 cm

Figure 10: Images captured by the varifocal lens ahead of a 5 megapixel camera of objects situated at 70, 17 and 10 cm.

application (10 diopters) can be achieved for a wide range of optical aperture. Good optical quality, very low power consumption ($<0.1\mu W$) and fast response time (few ms) have been also demonstrated as well as the reliability of the actuation part (10 million cycles at 0 & 10 V).

These results demonstrate that this device is a promising candidate to provide compact, fast, low power and low cost varifocal lenses and could be an interesting alternative to VCM for camera phones and miniature cameras in a near future.

REFERENCES

[1] N.-T. Nguyen, "Micro-optofluidic Lenses: A review," Biomicrofluidics, vol. 4, no. 3, p. -, 2010.

[2] X. Zeng and H. Jiang, "Liquid tunable microlenses based on MEMS techniques," Journal of Physics D: Applied Physics, vol. 46, no. 32, p. 323001, 2013.

[3] K. Asatryan, V. Presnyakov, A. Tork, A. Zohrabyan, A. Bagramyan, and T. Galstian, "Optical lens with electrically variable focus using an optically hidden dielectric structure," Opt. Express, vol. 18, no. 13, pp. 13981–13992, Jun. 2010.

[4] H.-C. Lin and Y.-H. Lin, "An electrically tunable-focusing liquid crystal lens with a low voltage and simple electrodes," Opt. Express, vol. 20, no. 3, pp. 2045–2052, Jan. 2012.

[5] M. Ye, B. Wang, M. Uchida, S. Yanase, S. Takahashi, and S. Sato, "Liquid crystal lens with electrically controllable focal length," in Mechatronics and Automation (ICMA), 2011 International Conference on, 2011, pp. 635–639.

[6] B. Berge, "No Moving Parts, Liquid Lens Capability Realization Soon for Mass Production – Refraction Control using Voltage Change", NIKKEI ELECTRONICS, 2005.10.24. p.129-135.

[7] W. Wang and J. Fang, "Design, fabrication and testing of a micromachined integrated tunable microlens," Journal of Micromechanics and Microengineering, vol. 16, no. 7, p. 1221, 2006.

[8] W. Zhang, K. Aljasem, H. Zappe, and A. Seifert, "Completely integrated, thermo-pneumatically tunable microlens," Opt. Express, vol. 19, no. 3, pp. 2347–2362, Jan. 2011.

[9] H. Yu, G. Zhou, F. S. Chau, and S. K. Sinha, "Tunable electromagnetically actuated liquid-filled lens," Sensors and Actuators A: Physical, vol. 167, no. 2, pp. 602 – 607, 2011.

[10] A. Pouydebasque, S. Bolis, F. Jacquet, C. Bridoux, L. Zavattoni, S. Soulimane, S. Moreau, D. Saint-Patrice, C. Bouvier, C. Kopp, and S. Fanget, "Thin varifocal liquid lenses actuated below 10V for mobile phone cameras", Photonics West 2012, vol. 8252, p. 82520P–82520P–11.

[11] A. Pouydebasque, S. Bolis, C. Bridoux, F. Jacquet, S. Moreau, E. Sage, D. Saint-Patrice, C. Bouvier, C. Kopp, N. Sillon, S. Fanget, and E. Vigier-Blanc, "Process optimization and performance analysis of an electrostatically actuated varifocal liquid lens," in Solid-State Sensors, Actuators and Microsystems Conference (Transducer), 2011 16th International, 2011, pp. 578–581.

[12] A. Pouydebasque, C. Bridoux, F. Jacquet, S. Moreau, E. Sage, D. Saint-Patrice, C. Bouvier, C. Kopp, G. Marchand, S. Bolis, N. Sillon, and E. Vigier-Blanc, "Varifocal liquid lenses with integrated actuator, high focusing power and low operating voltage fabricated on 200mm wafers," Sensors and Actuators A: Physical, vol. 172, no. 1, pp. 280 – 286, 2011.

[13] J. Draheim, F. Schneider, T. Burger, R. Kamberger, and U. Wallrabe, "Single chamber adaptive membrane lens with integrated actuation," in Optical MEMS and Nanophotonics (OPT MEMS), 2010 International Conference on, 2010, pp. 15–16.

[14] H. Oku and M. Ishikawa, "High-speed liquid lens with 2 ms response and 80.3 nm root-mean-square wavefront error," Applied Physics Letters, vol. 94, no. 22, p. -, 2009.

[15] S. T. Choi, J. Y. Lee, J. O. Kwon, S. Lee, and W. Kim, "Varifocal liquid-filled microlens operated by an electroactive polymer actuator," Opt. Lett., vol. 36, no. 10, pp. 1920–1922, May 2011.

[16] K. Wei, N. W. Domicone, and Y. Zhao, "A tunable liquid lens driven by a concentric annular electroactive actuator," in Micro Electro Mechanical Systems (MEMS), 2014 IEEE 27th International Conference on, 2014, pp. 909–912.

CONTACT

*S. Nicolas, tel: +33(0)4 38789201; stephane.nicolas2@cea.fr

IMPLEMENTATION OF NANOPOROUS ANODIC ALUMINUM OXIDE LAYER WITH DIFFERENT POROSITIES FOR INTERFEROMETRIC RGB COLOR PIXELS AS HANDHELD DISPLAY APPLICATION

P.-H. Lo[1], G.-L. Luo[2], and W. Fang[1,3]

[1]NEMS Institute, [3]Power Mechanical Eng. Dept., National Tsing Hua Univ., Hsinchu, TAIWAN,
[2]Asia Pacific Microsystems Inc., Hsinchu, TAIWAN

ABSTRACT

This study presents the approach to implement the interferometric modulation color pixels application by using nanoporous anodic aluminum oxide (np-AAO) layer as the key suspended micro structure. In this work, the presented bi-stable color pixels have averaged reflectivity of 60%, and the central wavelengths are 651 nm, 554 nm and 452 for red, green and blue, respectively. Advantages of the presented np-AAO color-pixel are as follows, (1) porosity of np-AAO layer can be easily adjusted by low temperature fabrication process to modulate the refractive index of proposed device, (2) color-pixels of different porosities (refractive indices) can be implemented and integrated on single np-AAO layer by process for different colors modulation, and (3) anti-stiction coating is not required since contact adhesion force significantly reduced by porous surface. The study reports the design, fabrication, and experimental results of the color pixels to demonstrate the advantages of the presented device.

INTRODUCTION

In the always-connected world, consumer's requirement and usage of mobile devices grow rapidly. The display becomes the main interface between the user and the connected-world. While conventional display technologies perform well and conform to fundamental customer requirements, the growing usage of mobile devices in professional and personal life demands the development of new display technologies with efficaciously reduced power consumption and image quality independent of ambient light, especially under outdoors environment conditions [1].

The microelectromechanical systems (MEMS) interference modulation display concept is believed to be the promising idea for its simple layer stack, replaceable process, low power consumption and its operation simplicity. Therefore, MEMS interferometric color pixels modulated by film thickness have been launched in recent years [2-3]. However, the integration process of color pixel with different air gap is complicated. Furthermore, the liquid-crystal with voltage-controlled refraction index is employed to realize the wavelength tunable interference color pixels [4]. Nevertheless, low response time and polarization dependence of liquid-crystal are critical concerns [5]. Normally, the interferometric color pixels are capacitive MEMS switching mechanism for changing the pixel state. The suspended structure will contact with the substrate surface for the bi-stable modulation while in use. Hence stiction is frequently occurred in contact interfaces of interferometric color pixels. Several approaches have been discussed to reduce the stiction between contact interfaces, such as the surface with nano-texture patterned

has been exploited to reduce the surface adhesion [6-7]. However, complicated fabrication processes are required to prepare the pattern. Moreover, the ALD (atomic layer deposition) or SAM (self-assembled monolayer) anti-stiction coating is required to prevent stiction [8]. The above-mentioned design considerations for MEMS color pixel can be improved by utilizing the proper thin film materials with simple micro fabrication processes. Thus, this study implements the bi-stable MEMS color pixel using the promising nanoporous anodic aluminum oxide film.

DESIGN CONCEPT

Figure 1(a) exhibits the structure design of the proposed color pixel. The color pixel consists of three suspended np-AAO interference layers as sub-pixels respectively for R-G-B color. The cross-sectional schematic plot in Fig. 1(b) indicates the composition of multilayers stacking. The color pixel contains *Al* half-reflector, *Si* reflector, and *Al* driving electrode. The *Al* thin film (10nm) on top of the np-AAO layer is employed as the optical reflector. The suspended and deformable np-AAO structure (350nm) is employed to modulate the refractive index of proposed device. The output color of sub-pixel is adjusted by varying the porosity (i.e. refractive

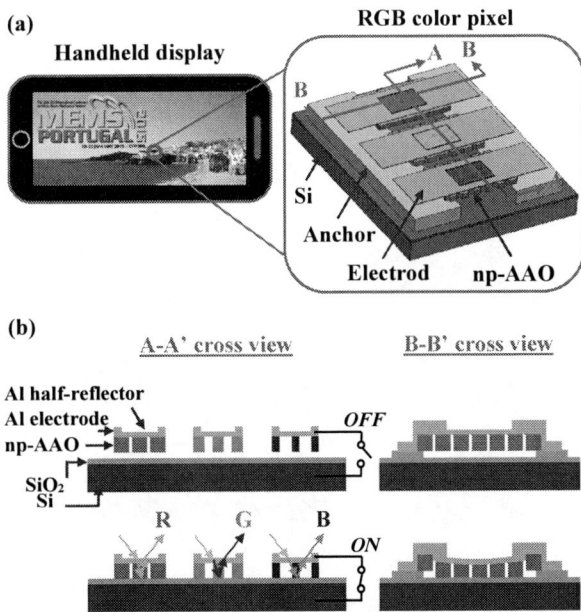

Figure 1: (a) Schematic plot of interferometric modulation color pixels, and (b) output wavelength can be modulated by varying the porosity (refractive index, n) of np-AAO layer.

index, n) of np-AAO layer. The *Ti* anchors are acted as the spacer to define the air gap between the np-AAO layer and substrate. The silicon substrate is exploited as the bottom electrode and the optical reflector as well. Note that the thermal oxide is an etching protection layer for *Si* substrate. The suspended np-AAO interference layer was pulled down by DC driving voltage to modulate output interference wavelength. Thus, the *Si* substrate reflector is fixed, while the *Al* film reflector is deformable for the wavelength modulation. In addition, the characteristic of reflected light beam can also be modulated by adjusting the air gap using electrostatic force.

FABRICATION PROCESS AND RESULTS

The schematic fabrication process steps of the present optical thin film filter are illustrated in Fig. 2. The substrate used for the fabrication of the presented device was a p-type (0.001–0.005 ohm-cm resistivity) silicon wafer. As indicated in Fig. 2(a), the thermal oxide and *poly-Si* were deposited to respectively act as the etch-stop and sacrificial layers. Thus, the air gap of the optical device was defined by the thickness of sacrificial layer. The *poly-Si* was etched for *Ti* anchors refill. Following, *Al* film was also deposited for the fabrication of np-AAO layer. After that, the np-AAO layer with nanopores array was grown via two-step anodization process [9], as shown in Fig. 2(b)–(d). The np-AAO of different porosities were then grown and

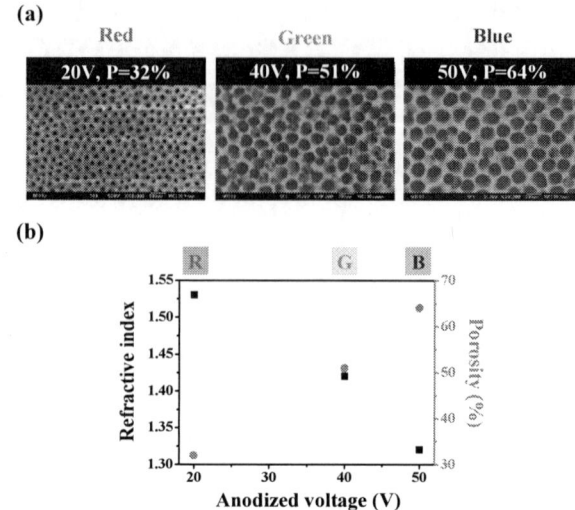

(a)

(b)

Figure 3: The refractive indices were successfully modulated by anodized voltage.

patterned as interference sub-pixels. During the anodization process, the wafer was firstly immersed into the 7°C aqueous solution with 0.3M oxalic acid and different porosities of np-AAO layer were achieved by varying the anodizing voltage. As illustrated in Fig. 2(e), the residual *Al* was removed by the *Al* etchant at 70°C. As depicted in Fig. 2(f), 100nm and 10nm *Al* films were respectively deposited and patterned for the top electrode and half-reflector. Moreover, the *Al* film with high reflectivity (>90%) over the entire visible spectral range [10] was employed to enhance the optical reflectivity of the device. Finally, the sacrificial *poly-Si* was isotropically etched using XeF$_2$ gas to suspend the np-AAO layer from substrate, as shown in Fig. 2(g).

FE-SEM images in Fig.3a show the np-AAO sub-pixels respectively anodized with 20~50V. Measurements in Fig.3b show the porosities P of three np-AAO sub-pixels were ranging P: 32~64%, and then the refractive indices n decreased from 1.53 to 1.32. Note that the wavelength of visible light is much larger than the pore size, hence the np-AAO can be considered as a uniform layer with an effective refractive index [11]. Thus red, green, and blue colors are respectively created by sub-pixels.

MEASUREMENT RESULTS

The micrographs in Fig.4 respectively show two actuation states of the color pixels. As shown in Fig.4a, no color was displayed by sub-pixels before applying driving voltage. In addition, as applying 30V DC voltages to the sub-pixels, the suspended membranes were pulled down to display red, green, and blue colors, as shown in Fig.4b. Thus, the bi-stable states of the suspended color pixels modulated by the driving voltages enable the change of displayed color. The test setup in Fig.5a was established to characterize optical spectral of reflected light of the devices. The setup is consisted of the microscope, spectrometer, source meter, and the test sample. The source meter was used to apply a DC voltage to actuate the

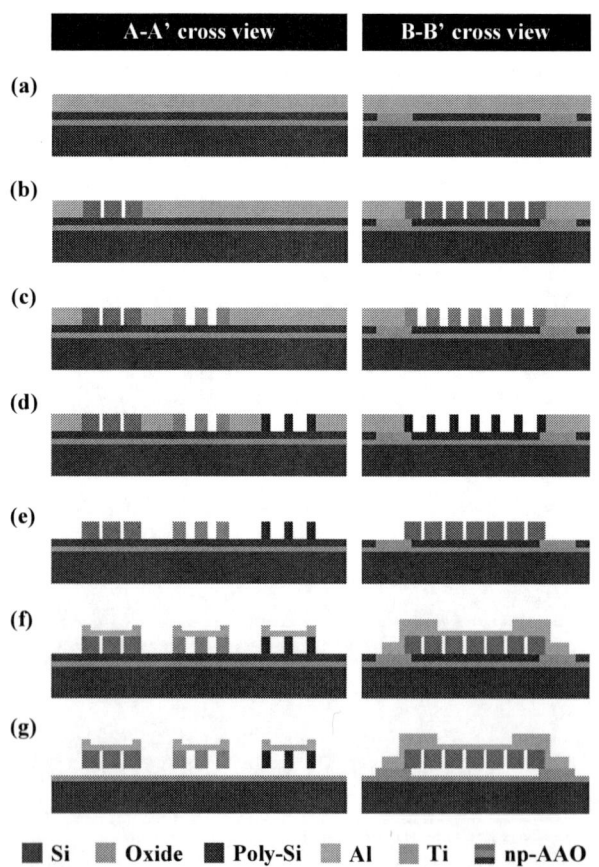

Figure 2: Fabrication process of the presented color pixels.

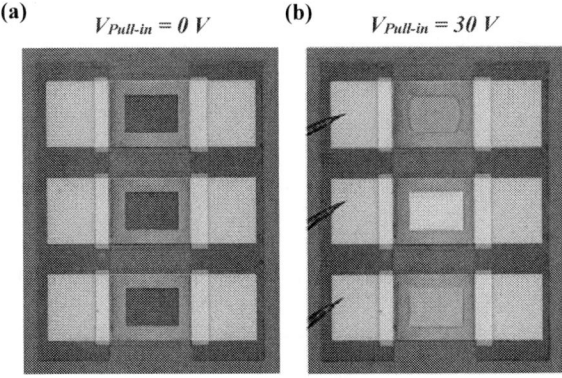

(a) $V_{Pull-in} = 0\ V$ **(b)** $V_{Pull-in} = 30\ V$

Figure 4: Demonstration for the presented color pixels.

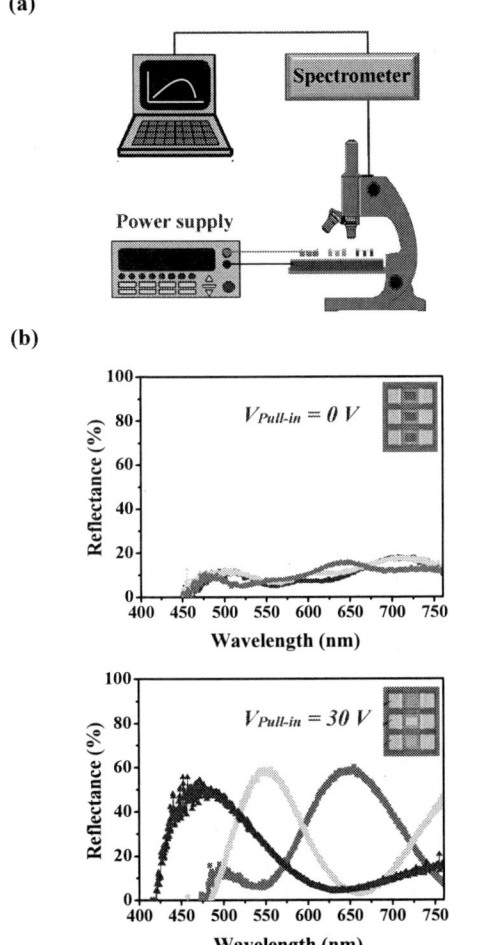

Figure 5: (a) Measurement setup, and (b) measured optical spectrum for primary RGB color pixel.

suspended thin film structure. The white light from microscope was incident perpendicularly onto the sample, and then the light reflected from device was measured by the spectrometer. In addition, the light reflected from the *Si* with 500nm *Al* coating film was used as a reference. Fig.5b shows the typical measured reflectance of the presented

color pixels. It indicates the relationship between the applying voltages and the light reflectance at different wavelength. The peak wavelength of light reflectance for red (651 nm), green (554 nm), and blue (452 nm) is 59%, 58%, and 55%, respectively. Moreover, this study employed a commercial AFM tip with spherical SiO_2 with a radius-of-curvature of 650nm fixed on the tip (Fig.6a) to investigate the real contact condition (np-AAO to SiO_2). Fig.6b illustrates the approach of AFM to measure the adhesion force between np-AAO film and oxide. The np-AAO layer was released from substrate. Thus, the AFM tip could touch the contact surface of np-AAO during the test. The AFM tip is initially placed on top of the sample (no contact), and then moves downward to approach and contact with the sample. An additional loading of the tip is specified to ensure a sufficient contact force between the tip and sample. After that, the AFM tip moves upward to retract from the sample. However, due to the existing of adhesion force, the cantilever of AFM will remain bending even retract after the initial contact point. Thus, the adhesion force is determined by the extra load to separate the AFM tip from sample. Measurement results in Fig.6c show the average adhesion force of samples on five different position and the error bar for these measurements. The adhesion forces of np-AAO layers were smaller than film with ALD coating [12]. In summary, the adhesion force was significantly reduced by increasing the porosity of np-AAO films to prevent the stiction.

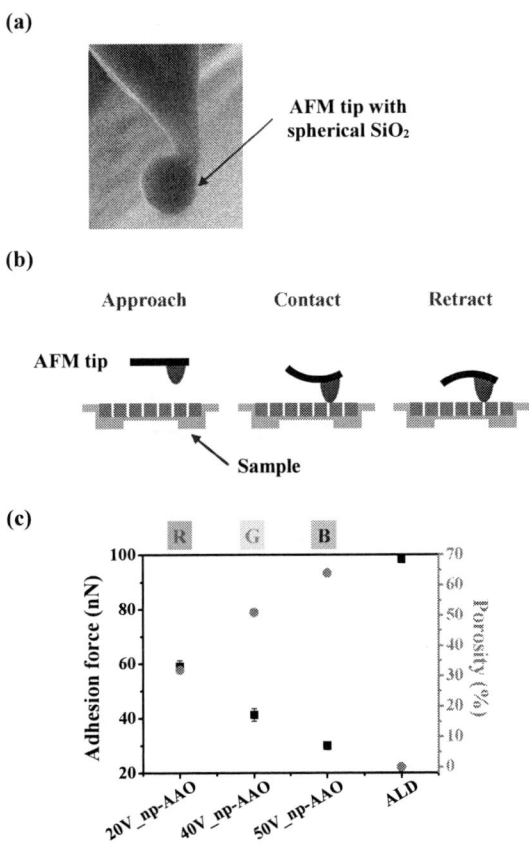

Figure 6: AFM tests for adhesion force.

978-1-4799-7956-1/15 $31.00 © 2015 IEEE

CONCLUSIONS

This study presents the approach to implement the color pixels by using the suspended np-AAO layer as the key deformable mechanical structure. The color pixels are driven by electrostatic force with bi-stable positions to modulate the display light of visible wavelength range. The porosity of np-AAO layer can be exploited to change the characteristics of color pixels. In this study, the np-AAO with porosity ranging from 32%–64% is demonstrated using the low temperature anodization process. Thus, the refractive index can be easily changed from 1.53 to 1.32. The peak wavelength of light reflectance for red (651 nm), green (554 nm), and blue (452 nm) is 59%, 58%, and 55%, respectively after the bi-stable modulation with a bias voltage of 30V. Furthermore, the experiments demonstrate that the adhesion force can be reduced by the porous surface of np-AAO layer to prevent the in-use stiction of color pixels. In summary, the np-AAO is found to be a promising material for color pixels application.

ACKNOWLEDGEMENTS

This paper was partially supported by Nation Science Council, Taiwan, under contract 103N2061E1, NSC 101-2221-E-007-069-MY3, NSC 102-2221-E-007-027 -MY3 and NSC 103-2811-E-007-008-MY3. The authors would also like to appreciate National Tsing Hua University. (Taiwan) in providing fabrication facilities.

REFERENCES

[1] C. Chui, P. D. Floyd, D. Heald, B. Arbuckle, A. Lewis, M. Kothari, B. Cummings, L. Palmateer, J. Bos, D. Chang, J. Chiang, L.-M. Wang, E. Pao, F. Su, V. Huang, W.-J. Lin, W.-C. Tang, J.-J. Yeh, C.-C. Chan, F.-A. Shu, Y.-D. Ju, "The iMoD™ display: considerations and challenges in fabricating MOEMS on large area glass substrates,"*Proc. of SPIE*, vol. 6466, pp 646609-1, 2007.

[2] D. Felnhofer, K. Khazeni, M. Mignard, Y. J. Tung, J. R. Webster, C. Chui, and E. P. Gusev, "Device physics of capacitive MEMS," *Microelectron. Eng.*, vol. 84, pp. 2158-2164, 2007.

[3] C.-Y. Lo, O. H. Huttunen, J. Hiitola-Keinanen, J. Petaja, H. Fujita, and H. Toshiyoshi, "MEMS-controlled paper-like transmissive flexible display," *Journal of Micro-electronic Systems*, vol. 19, pp. 410-418, 2010.

[4] C.-K. Liu, K.-T. Cheng, and A. Y.-G. Fuh., "Designs of High Color Purity RGB Color Filter for Liquid Crystal Displays Applications Using Fabry-Perot Etalons," *J. of Display Technology*, vol. 8, pp. 174-178, 2012.

[5] K. Hirabayashi, Y. Ohiso, and T. Kurokawa, "Polarization-independent tunable wavelength selective filter using a liquid crystal," *IEEE Photonic Technol. Lett.*, vol. 3, pp. 1091-1093, 1991.

[6] Min Zou, Hengyu Wang, P.R. Larson, K.L. Hobbs, M.B. Johnson and O. K. Awitor, "Ni nanodot-patterned surfaces for adhesion and friction reduction," *Tribology Letters*, vol. 24, pp. 137-142, 2006.

[7] L. Zhu, Y. Xiu, J. Xu, D. W. Hess, and C. P. Wong, "Optimizing geometrical design of superhydrophobic surfaces for prevention of microelectromechanical system (MEMS) stiction," *In Proceedings of the 56th Electronic Components and Technology Conference (ECTC)*, San Diego, USA, 30 May 2 June, 2006, pp. 1129-1135.

[8] R. Maboudian and R. T. Howe, "Critical Review: Adhesion in surface micromechanical structures," *J. Vac. Sci. Technol. B*, vol. 15, pp. 1-20, 1997.

[9] H. Masuda, H. Yamada, M. Satoh, H. Asoh, M. Nakao, and T. Tamamura, "Highly ordered nanochannel-array architecture in anodic alumina," *Applied Physics Letters*, vol. 71, pp.2770-2772, 1997.

[10] M. Bartek, J. H. Correia and R. F. Wolffenbuttel, "Silver-based reflective coatings for micromachined optical filters,"*J. Micromech. Microeng.*, vol. 9, pp 162-165, 1999.

[11] D. E. Aspnes, J. B. Theeten, and F. Hottier, "Investigation of effective-medium models of microscopic surface roughness by spectroscopic ellipsometry," *Phys. Rev. B*, 20, pp. 3292-3302, 1979.

[12] A. Londergan, E. Gousev and C. Chui, "Advanced Processes for MEMS-based Displays," *Proceedings of the Asia Display*, SID, vol. 1, pp. 107-112, 2007.

CONTACT

*W. Fang, tel: +886-3-574-2923; fang@pme.nthu.edu.tw

LOW-NOISE ALN-ON-SI RESONANT INFRARED DETECTORS USING A COMMERCIAL FOUNDRY MEMS FABRICATION PROCESS

Vikrant J. Gokhale[1], Cesar Figueroa[1], Julius Ming Lin Tsai[2], and Mina Rais-Zadeh[1,3]

[1] Department of Electrical Engineering and Computer Science, University of Michigan, Ann Arbor

[2]Invensense Inc., San Jose, CA, USA

[3] Department of Mechanical Engineering, University of Michigan, Ann Arbor, MI, USA

ABSTRACT

This work presents the first measured results for resonant AlN-based infrared (IR) detectors fabricated using a proprietary InvenSense AlN MEMS process. Resonators fabricated in the first fabrication run achieved high electromechanical performance with a Q of ~1400 at 115 MHz, insertion loss of 17.9 dB, and a motional impedance of 670 Ω. The detectors are coated with an IR absorber layer (SiN_x), and tested for response to IR radiation using a calibrated, traceable black body source. The estimated responsivity of the device is 210ppb/μW for the longwave-infrared (LWIR) spectrum. The detectors are expected to have low noise, with estimated NEDT and NEP of 51 mK and 52.7 pW/Hz$^{0.5}$, respectively. The resonators are fabricated in a hybrid MEMS/CMOS wafer level packaged die, allowing for CMOS-based routing and readout.

INTRODUCTION

Infrared (IR) detectors have been in use since the 1950s, with the primary uses being surveillance, security and defense, and astronomical imaging. While a number of technically advanced solutions exist for the high sensitivity and resolution requirements of these applications [1], the associated costs make them prohibitive for wide commercial use if high-end performance is to be maintained. The leading candidates for high performance and low cost include uncooled IR detector elements and small- to medium-format focal plane arrays (FPAs) fabricated using MEMS technology [2], [3]. Recent years have shown that piezoelectric electromechanical resonators have the potential to be used as very sensitive, low-noise, small footprint, and low-power uncooled IR detectors [4], [5]. Recent prototypes of single resonators [6], differential pairs [7], and IR sensing arrays [5], [8] have been successfully demonstrated using a variety of materials for the resonator themselves, as well as for the crucial high-efficiency IR absorber layer [9], [10]. The present work successfully transitions designs and processes developed in research labs to a repeatable commercial foundry MEMS/CMOS fabrication process using aluminum nitride (AlN) based resonators.

PROCESS AND DESIGN

Fabrication Process

The devices were fabricated as part of a multi-user wafer using commercial foundry processes at A*STAR IME (MEMS chip with low stress AlN-on-Si) and Global Foundries (CMOS chip) followed by wafer-level bonding at IME. The InvenSense AlN MEMS process begins by preparing an engineered silicon-on-insulator (SOI) wafer with cavities. A bare silicon (Si) wafer is first etched to form cavities on one side, followed by SiO_2 deposition. Next, the wafer with Si cavities is fusion bonded to another Si wafer and thinned down till the required device layer thickness is achieved. In this case, the goal was 5μm for the device layer. An AlN seed layer, bottom metal layer, and structural AlN layer are subsequently deposited. A standoff layer on top of the AlN structural layer is used to define the gap between the MEMS substrate and the CMOS substrate. The AlN structural layer is then patterned. The top metal and Ge layers are deposited and patterned to form top electrodes and AlGe bonding pads corresponding to CMOS top metal pads, respectively. After the contours of the MEMS structure are defined using deep reactive ion etch (DRIE), the MEMS substrate is bonded to a CMOS substrate to form electrical connections with CMOS circuitry. This step also achieves hermetic sealing. At the same time, it also provides extremely low electrical parasitics and a full utilization of the CMOS routing capability without the need for expensive and complicated through-silicon-via (TSV) processing. Finally, the MEMS substrate is thinned down and etched from the back with the port opening to allow direct contact with external environment.

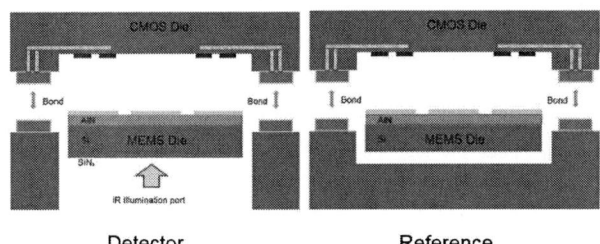

Figure 1: A schematic (not to scale) depicting the AlN-on-Si resonant IR detectors. The sense resonator (left) has an illumination port and is coated with silicon nitride, while the reference resonator (on right) is shielded from IR radiation. The MEMS resonator wafer is bonded to a CMOS readout IC at the wafer level.

Note that in the case of reference resonators used for differential measurements, the port etch (backside DRIE) is not used. The reference is shielded from IR illumination by the MEMS wafer and potentially by another metal shielding layer. Fig. 1 shows schematic cross-sections of both detector and reference resonators, with and without the IR illumination port, respectively. This process is suitable for MEMS devices or arrays of MEMS devices such as IR bolometers as it allows for high 'pixel' packing density without complicated multi-layer routing on the MEMS die. Finally, the bonded wafers are separated by

978-1-4799-7956-1/15 $31.00 © 2015 IEEE

dicing. Fig. 2 shows a microscope image of such a separated die, as seen from the bottom of the MEMS die. The die is a 4×3 array of AlN-on-Si resonators, with 11 detectors and one reference. Silicon nitride is coated on the device backside at the final step to act as the IR absorber layer (this step is done at the U. of Michigan cleanroom). The reference resonator is shielded from the deposition.

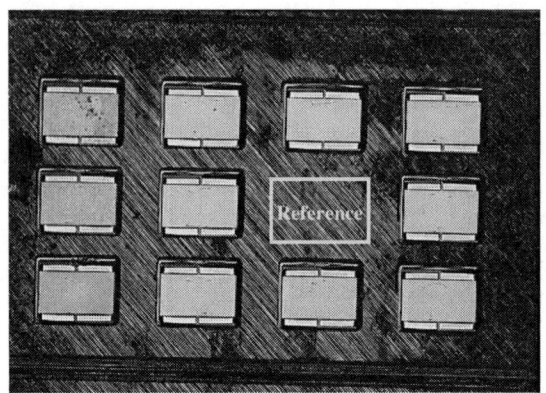

Figure 2: Optical microscope image showing an array of AlN-on-Si resonant IR detectors through their illumination ports (backside of the MEMS die). The reference resonator has no port, thus shielding it from IR radiation.

Design

The devices presented here are in-plane length-extensional mode resonators fabricated using AlN-on-Si as the structural materials. Multiple designs have been implemented in this batch, both as arrays and as single resonators. The resonators intended to work as IR detectors were designed to achieve low noise equivalent temperature difference (NETD) values (< 50 mK). One of the key factors in getting low NETD is to improve the thermal isolation of the resonators, specifically by decreasing the effective thermal conductance (G_{th}) of the tethers [4], [8]. One useful technique is to use crableg tethers instead of conventional straight tethers [10]. This increases the effective tether length (and thus decreases tether conductance) without decreasing the ratio of active absorbing area to the total etch cavity area which is the optical fill factor of the resonator pixel. As seen in Fig. 3, it is possible to get an effective tether length of 120 μm without sacrificing the fill factor. Simulations were used to determine the correct dimensions for the tethers to increase thermal isolation while at the same time not compromising the electromechanical performance of the resonator.

EXPERIMENTAL RESULTS

RF Performance of Resonators

The resonator demonstrates a high Q of 1380 at 115 MHz and a reasonably low insertion loss at -10 dBm of RF input (Fig. 4). Using an equivalent electrical model, we can calculate the motional impedance to be 670 Ω.

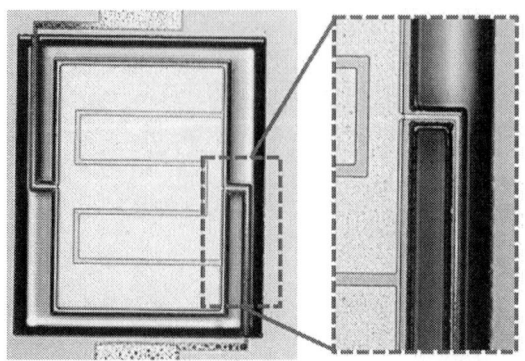

Figure 3: Optical microscope image of a sense resonator. This resonator is 190 μm × 128 μm in size. Magnified portion of the image highlights the use of a crableg tether that increases the thermal isolation of the device (desirable for high sensitivity and low noise), while retaining a compact total pixel size. The effective length of each tether is 120 μm. This image is taken from a deliberately de-bonded die to expose the resonator.

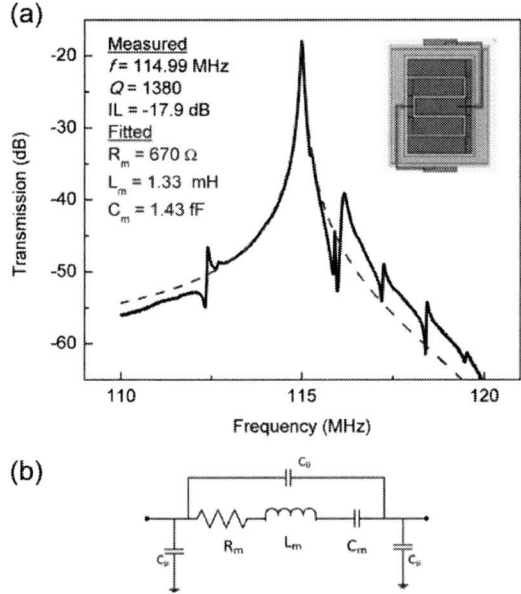

Figure 4: (a) Measured response of the AlN-on-Si resonator fabricated using a commercial foundry MEMS process, when actuated with -10 dBm of RF power. (b) BVD model of the two-port resonator.

Other devices from the same batch that had simple, straight conventional tethers (not crableg) were also tested. The RF performance of those devices is comparable to the crableg device measured here, with $f×Q$ values in the same range. As expected the devices with the simple tethers show markedly worse IR performance.

Infrared Response of the Prototype Resonators

The AlN-on-Si resonators are characterized for their IR performance in both near IR (NIR) and LWIR spectra. For the LWIR spectrum, we use a large area black body radiation source (Electro-Optical Industries' LES-100-4) with an emissivity of (0.97 ± 0.02), a temperature stability of ± 2 mK, and a planar emitting area of 102 mm × 102 mm. The source is freshly calibrated, and is traceable to NIST primary standards [11]. The high emissivity and temperature stability enable us to calculate the exact power distribution as a function of temperature and area of the source, according to the Stefan-Boltzmann Law. Given far-field radiation conditions and the known size of the resonator surface, we can calculate the power incident on the device. The resonant frequency of the device changes upon illumination with IR radiation, due to the absorption of radiation (in the IR absorber layer) and the subsequent conversion of that absorbed power into a temperature change in the resonator. Fig. 5 shows the IR measurement setup.

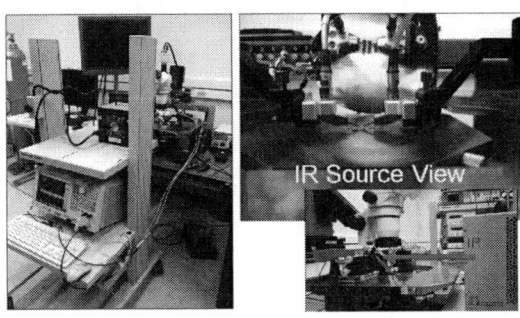

Figure 5: A photograph of the measurement setup. The device under test is mounted on a probe station, and its RF performance is measured using an Agilent VNA. For IR measurements, the device is illuminated using a wide-area black body source and a highly reflective gold mirror. All measurements are acquired remotely using a PC running custom-made LabVIEW code.

Fig. 6 shows a representative frequency response of a resonator to IR radiation when the source is at 80 °C (353 K). Based on the known temperature coefficient of frequency (TCF) and area of the devices, we can estimate the absorbed LWIR power to be ~3.1 µW (See Table 1).

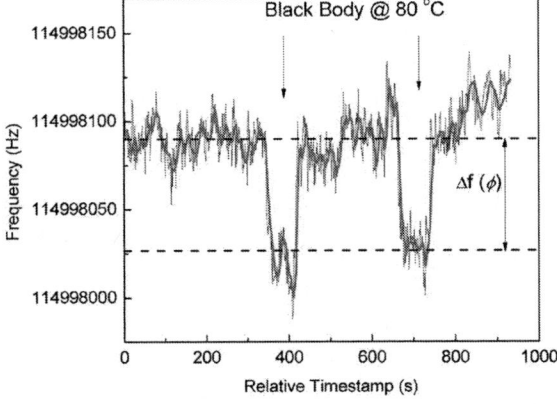

Figure 6: The frequency response of the device when irradiated by blackbody radiation from the calibrated source at 80 °C. The resonator absorbs ~ 3.1 µW of IR power in the LWIR spectra, leading to a perceptible frequency shift. This particular experiment was carried out in ambient air, with no temperature control.

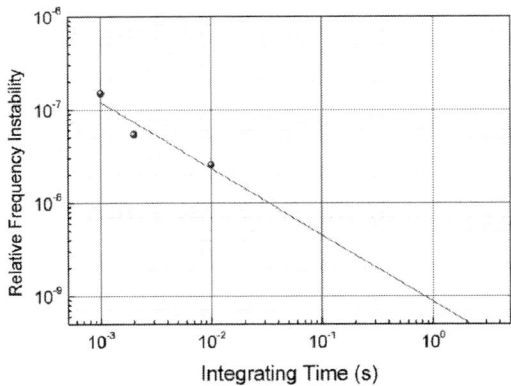

Figure 7: The relative frequency instability of the device as a function of integrating time. Measurements are acquired at ambient conditions, with no temperature or pressure control. For integrating time> 10 ms, the values are extrapolated. As expected, the device is more stable with larger integration times.

In order to characterize the differential performance of the resonators, we compare the IR response of two nearly identical detectors (with silicon nitride absorber) and one reference (without silicon nitride absorber). As expected, the reference is nearly invariant to stimulus. Using a near-IR radiation source (Ocean Optics IIL2000) and fiber-optic probes to couple light onto the specific resonators in questions we can compare the relative response (Fig. 8). Differential measurements can enable highly sensitive systems with common mode rejection of common mode effects. Details of the procedure and differential measurements are provided in our previous work [8]. For the NIR spectrum, the absorbed power is ~2.47 µW.

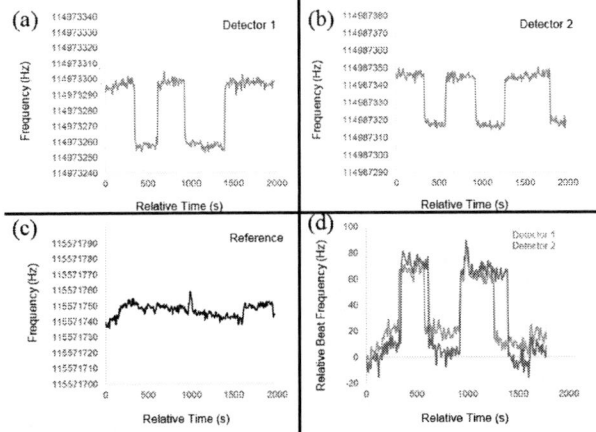

Figure 8: (a)-(c) The frequency response of two detectors and a reference when irradiated by NIR radiation. Note that the Y-axes have the same scales for these three graphs. The absorbed power is estimated to be on the order of ~ 2.47 µW of IR power leading to a perceptible frequency shifts in the detector but near-zero shift in the reference. (d) Using the relative beat frequency between detectors and reference we can achieve differential IR sensing.

Expected Infrared Metrics

Based on the design parameters and material properties of the resonators (Table 1), the AlN-on-Si resonant detectors are expected to have low-noise performance, with calculated NETD and NEP values of 51 mK and 52.7 pW/Hz$^{0.5}$, respectively [12]. Low NEDT is especially desirable, with commercial uncooled bolometer FPAs exhibiting NETD values on the order of 30 mK [13].

To an extent, the low NETD for these particular devices is due to the large surface area of the devices. For resonator pixels intended to be part of an IR imager, the area of the individual device must be reduced significantly. In order to keep NETD values at the same level, it is necessary to improve the thermal isolation of the devices (reduce G_{th}). For the current batch, the designs were primarily constrained by the process limitations of a multi-project fabrication process and the relatively large thickness of the Si device layer. With thinner and narrower tethers optimized for high thermal isolation, it is possible to get up to two orders of magnitude improvement in G_{th}, responsivity, and noise. Designs fabricated in the recent past [6], [10] using AlN-based resonant IR detectors have experimentally demonstrated thermal conductance as low as 10^{-5} W/K with the potential to go even lower.

Table 1. Design & Expected Performance Parameters for the Presented AlN-on-Si Resonant IR Detector.

Design Parameter	Value	Performance Metric	Value
Resonator Length	190 μm	Frequency	115 MHz
Resonator Width	128 μm	Q	1380
Tether Length	120 μm (Crableg)	Absorber	Silicon nitride
TCF	-30 ppm/K	Responsivity (LWIR)	210 ppb/μW
G_{th} (2 crableg tethers)	1.42×10^{-4} W/K	NETD (calculated)	51 mK
		NEP (calculated)	52.7 pW/Hz$^{0.5}$

CONCLUSION

This work lays the groundwork for high-performance uncooled, resonant IR detectors and detector arrays based on a commercially available and commercially viable technology. The success of this design and fabrication cycle is promising for better, more optimized processes and designs. The integration of resonant IR detectors with electronics and other MEMS based components on the same die would add significant functionality to commercial sensing and position control solutions. Similar resonators and arrays can also be used for a variety of other applications such as timing, mass sensing, flow-rate sensing, and biological/chemical sensing with only slight changes in the current designs.

ACKNOWLEDGEMENTS

The authors would like to thank Mr. Adam Peczalski

for assistance with the silicon nitride deposition. The work is supported by the MAST CTA under contract W911NF.

REFERENCES

[1] A. Rogalski, "Infrared detectors: status and trends," *Progress in Quantum Electronics*, vol. 27, pp. 59-210, 2003.

[2] M. Kimata, "Trends in small-format infrared array sensors," *IEEE SENSORS*, pp. 1-4, 2013.

[3] A. Rogalski, "Recent progress in infrared detector technologies," *Infrared Physics & Technology*, vol. 54, pp. 136-154, 2011.

[4] J. R. Vig, R. L. Filler, and Y. Kim, "Uncooled IR imaging array based on quartz microresonators," *IEEE/ASME Jounal of Microelectromechanical Systems (JMEMS)*, vol. 5, pp. 131-137, 1996.

[5] P. Kao and S. Tadigadapa, "Micromachined quartz resonator based infrared detector array," *Sensors and Actuators A: Physical*, vol. 149, pp. 189-192, 2009.

[6] Y. Hui, Z. Qian, G. Hummel, and M. Rinaldi, "Pico-watts range uncooled infrared detector based on a freestanding piezoelectric resonant microplate with nanoscale metal anchors," *Solid-State Sensors, Actuators and Microsystems Workshop*, Hilton Head Island, South Carolina, 2014.

[7] V. J. Gokhale, Y. Sui, and M. Rais-Zadeh, "Novel uncooled detector based on gallium nitride micromechanical resonators," *Proceedings of SPIE: Infrared Technology and Applications*, vol. 8353, p. 835319, 2012.

[8] V. J. Gokhale and M. Rais-Zadeh, "Uncooled infrared detectors using gallium nitride on silicon micromechanical resonators," *IEEE/ASME Journal of Microelectromechanical Systems (JMEMS)*, vol. 23, pp. 803 - 810, 2014.

[9] V. J. Gokhale, O. A. Shenderova, G. E. McGuire, and M. Rais-Zadeh, "Infrared absorption properties of carbon nanotube/nanodiamond based thin film coatings," *IEEE/ASME Journal of Microelectromechanical Systems (JMEMS)*, vol. 23, pp. 191-197, 2014.

[10] V. J. Gokhale, P. D. Myers, and M. Rais-Zadeh, "Subwavelength plasmonic absorbers for spectrally selective resonant infrared detectors," *IEEE Sensors*, Valencia, Spain, 2014.

[11] E.-O. Industries. (2014). Available: http://www.electro-optical.com/pdf/LES100-RevB.pdf

[12] P. W. Kruse, Uncooled Thermal Imaging: Arrays, Systems, and Applications. Bellingham, Washington, USA: SPIE Press, 2001.

[13] F. Niklaus, C. Vieider, and H. Jakobsen, "MEMS-based uncooled infrared bolometer arrays: a review," *Proc. SPIE*, vol. 6836, pp. 68360D-68360D, 2007.

CONTACT

*M. Rais-Zadeh, email: minar@umich.edu

HARD MASK FREE DRIE OF CRYSTALLINE SI NANOBARREL WITH 6.7NM WALL THICKNESS AND 50:1 ASPECT RATIO

Peng Liu, Fang Yang, Wei Wang, Wei Wang, Kui Luo, Ying Wang, and Dacheng Zhang*
Institute of Microelectronics, Peking University, Beijing, PR China

ABSTRACT

Because of the unique mechanical and electrical properties, silicon was widely used in IC/MEMS devices for decades. However, it is difficult to fabricate high aspect ratio structures when the feature size scaling down to 7nm. In this paper, single crystal Si nano barrel with wall thickness 6.7nm and aspect ratio 50:1 was achieved by top-down process. The fabrication was carried out by Electron Beam Lithography (EBL) for patterning and DRIE for pattern transfer, utilizing photoresist as etching mask directly. To predict the barrel profile, a model of ion transportation was established. This tactic provides a potential arsenal for the next generation of 3D-IC and NEMS devices.

INTRODUCTION

For the research of semiconductor in volume production, silicon had been widely studied for its good performance in electrical and mechanical properties. It is also convenient for the highly developed fabrication technique to construct Si nano devices, which was fully compatible with both IC and MEMS process.

According to the ITRS, the technology node name label will move to "7" in 2017, and the critical dimension (CD) of FinFET will shrinks to sub-7nm to pursuit the growing integration level. Limited by the fabrication process of high aspect ratio structures, the application of Si-based semiconductor in memory, processor, sensor and actuator faced grand challenge. Meanwhile, there was growing interest in the study of vertical FET performance theoretically [1], and several efforts have been carried out to realize 3D nano devices. For example, vertical Si nano wire was obtained for memory device [2]. The structure was etched using Si_3N_4 as hard mask and trimmed by the wet etching of thermal oxide. However, this fabrication method leads to low integration density and poor compatibility. In another fabrication process of 3D FET, the aspect ratio of critical structures were still limited below ~8:1 even using hard mask in dry etching [3]. Si nanowire array also studied for solar harvesting, which combine nanosphere lithography and wet etching [4]. It obtained relative high efficiency in energy conversion, whereas the metal assisted chemical etching leading to compatibility problems in mass production. Except these nanorods, which were usually solid cylinders, fabrication of hollow pillar was reported by redeposition method that involves annealing, sputtering and dry etching [5]. Yet, it was lack of uniformity and repeatability.

In this paper, conventional Si-based semiconductor equipment was utilized to realize vertical nano barrel array. Top-down fabrication strategy composed with EBL and DRIE was under taken. The etching process, which employed conventional photoresist as mask, was investigated theoretically and experimentally. High aspect ratio crystalline Si nano barrels with ultra-thin round wall were approach, which were representative and studied by SEM and TEM. This ability guaranties the critical structure for the next generation of nano devices with high repeatability and manufacturability.

FABRICATION

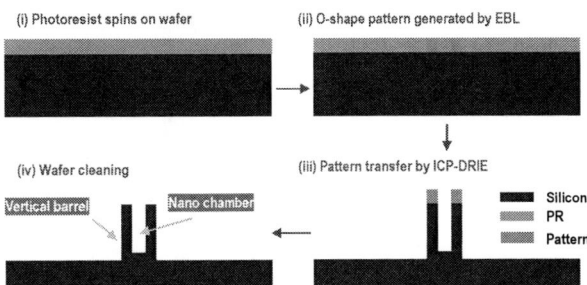

Figure 1: IC/MEMS compatible strategy to realize Si nano barrels, including (i) photoresist span on to wafer, (ii) EBL direct write of mask pattern, (iii) pattern transfer by DRIE and (iv) removing of photoresist and contamination.

As shown in Fig.1, novolak based negative tone photoresist AR-N7520.18 was span onto the wafer at rotation speed 5000 RPM, followed with a soft bake immediately on hotplate at 85 ℃ for 1 minute. The thickness of this organic film was measured as 80nm by interferometer. O-shaped patterns were generated by NanoBeam nB4 EBL system without proximity effect correction. The photoresist was exposed under acceleration voltage 80kV and the development was carried out under room temperature by immersing the wafer into developer. The developer was commercialized TMAH solution diluted with DI water.

After hard bake of the resist mask on hotplate, inductively coupled plasma (ICP) DRIE equipment was utilized to transfer the patterns into Si substrate. Varied RF power was applied on the wafer platen and constant 800W was applied on the source coil while the chamber pressure was fixed at 15mT. Fluorine based recipes (C_4F_8/SF_6) were studied and optimized extensively.

The mask and fabricated structure were observed and studied by scanning electron microscopy (SEM). After transferring the barrels from Si substrate onto copper mesh manually, transmission electron microscopy (TEM) was used to characterize the lattice structure.

IONS TRANSPORT AROUND BARREL

In ICP-DRIE process, negative DC bias on substrate leaded to directional movement of ions. These ions caused physical sputtering on substrate surface as well as passivation layer, and improved the chemical reactions between free radical and Si. Both physical and chemical enhancement contributed to the anisotropic etching [6]. In this work, a model of ions transportation was established to interpret the geometric morphology of nano barrels.

Affected by the electrostatic field (EF) due to DC bias, F^+ ions in plasma were accelerated to the substrate. As the height of barrels H was much shorter than the particle's mean free path, considering the ions approach the wafer surface without random scattering. Ions hit the substrate vertically in the open area, whereas deflected by EF laterally near the barrels. The trajectories of ions were discussed in 3 groups and illustrated as dash lines in Fig.2.

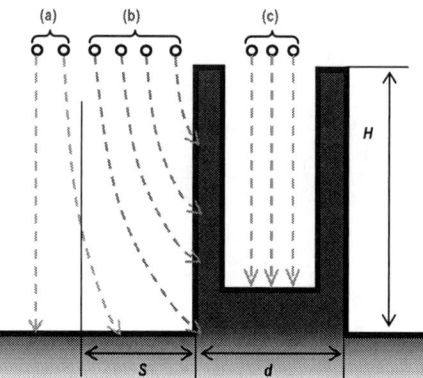

Figure 2: Ions' transportation nearby barrel was illustrated. Group (a) shows the trajectory partially deflected by lateral EF. Ions (b) crash into the wall. Considering the barrel as a uniform charged equipotential tube, EF was zero inside. Ions (c) hit the inner bottom directly without bending, which leading to high anisotropic etch ratio in the nano chamber.

In group (a), ions traveled through distance H in period t_z that satisfies

$$H = \int (V_z + a_z t_z) dt_z \qquad (1)$$

Where V_z is the vertical velocity of F^+ ions when they arriving at the structure. a_z is the acceleration of ions due to DC bias. For the sheath thickness is 1~2cm in ICP DRIE system, which is much longer than the height of the barrel, the increment of vertical velocity is negligible. So, equation (1) could be simplified as

$$H \approx \int V_z dt_z = V_z t_z \qquad (2)$$

As a conductor in plasma, the nano barrel was charged with electrons and kept equipotential with the wafer surface. Part of the ions in group (a) affected by lateral field and the trajectories were bended. Nearby the barrel, ions in group (b) bombarded the round wall before they hit the wafer surface. The travel time t_x satisfied $t_x < t_z$, where

$$S = \int_0^{t_x} a_{x(t)} t dt \qquad (3)$$

$a_{x(t)}$ is the acceleration toward the barrel which is a time dependent variable correlated with non-uniform EF around the barrel. S is the effective distance, within which the ions would be captured by the barrel.

However, considering the uniform charged nano chamber as a Faraday Cup, the EF was zero inside the barrel. Ions in group (c) fly into the nano chamber would neither be accelerated nor be deflected. The ions with directional movement tended to hit the chamber bottom directly instead of crashed into the round wall. More ions with kinetic energy removed the passivation layer on the inner bottom of the nano chamber, followed by physical

sputtering of Si substrate. Chemical reactions between radical and Si also be enhanced by the formation of lattice grid defection on the inner bottom.

Meanwhile, the conformance moving of ions and physical sputtering strengthen the wall angle selectivity. That is, the etching of inner wall face with positive tapper angle was boosted by ions bombardment, and the etching of inner wall face with negative tapper angle was declined by the insufficient passivation cleaning.

These transportation behaviors leaded to relative high anisotropic etching and kept the inner wall in vertical when varying the proportion of passivation.

RESULTS AND DISCUSSIONS
Barrel Arrays

Benefit from the high anisotropic etching in nano chamber, erect Si nano barrels with uniform wall thickness were successfully achieved using photoresist as etching mask. To characterize the nano structures, barrels array was staggering arranged and the nano barrel was sliced manually to obtain the axial cross-section view in Fig.3. The round wall thickness was measured as 6.7nm where the etching depth surpassed 330nm, and the aspect ratio was calculated more than 50:1. The barrels were also studied by TEM in Fig.4, which revealed <100> growth orientation and single crystal lattice grid.

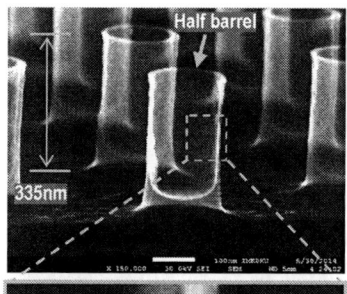

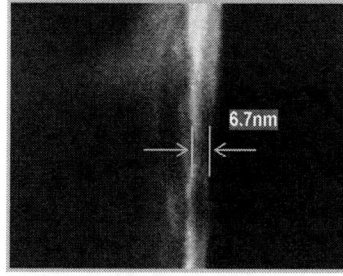

Figure 3: Barrels with aspect ratio 50:1 were obtained in Si (100) substrate. The barrels were sliced along axial direction manually. To ensure the target structure can be cutting through, this array was staggering arranged.

As etching mask in this work, O-shaped photoresist was shown in Fig.5 a. When scaling down to the resolution limitation of the photoresist [7], line edge roughness (LER) was comparable with line width and critical to realize the nano structure with feature size shrinks to sub-7nm. The SEM picture shown in Fig.5 b was taken from angle view, where the roughness was observed as particles that randomly spread. It can be predicted that the LER was mainly result from the molecule size of this organic

978-1-4799-7956-1/15 $31.00 © 2015 IEEE

material instead of the beam noise in EBL system.

Whereas for the Si nano barrels represented in Fig.3, round wall were etched with uniform thickness. It is indicated that the LER due to the resist material would not influence dry etching process evidently in this work.

Figure 4: TEM of nano barrel. Insert picture shows the growth orientation, and marked the image region by a dashed box. Lattice plane <100> can be observed perpendicular to the growth orientation, and the thickness of the round wall was measured as 6.76nm

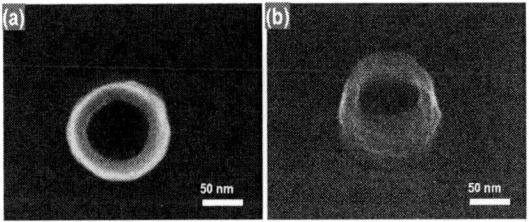

Figure 5: Top-view (a) of widely used organic photoresist ARN7520 as etching mask, which was generated by EBL on nB4 system. Roughness was observed as particles spread on the wall randomly (b), which mainly caused by molecule size limitation.

Figure 6: Barrel arrays with diameters ranging from 100 to 300nm were achieved in large area (0.5×1mm) with high uniformity. Each single barrel was spatial defined precisely with overlay error less than 10nm. Anisotropic etching of nano chamber was revealed universally.

To verify the uniformity and repeatability of this tactic, finely ordered barrels with diameters d ranging from 100 to 300nm were obtained in large area (0.5×1mm), and the SEM pictures were shown in Fig.6. These structures were transparent in SEM under acceleration voltage 30kV, and the inner profile could be investigated conveniently.

Polymer Residual

Before chemical cleaning of the wafer, organic contamination on the wall face was observed in Fig.7 a along with the remained photoresist on top of the barrels. This contamination resulted from polymerized free radicals and the insufficient removing of it. The de-passivation effect was partly accelerated by ions sputtering, which was shown as ions group (b) in Fig.2. With the increasing of etching depth H, the effective distance S expands. Growing number of ions bombard the barrel at the near bottom and leading to sufficient clean-up of contaminations. It resulted that the remained polymer was only observed on the barrels' top area, and was found scarcely near the footing area.

Profile Control

Dry etching process related physical damage on wafer surface had already been reported [8-9]. In this work, horizontal EF induced ions bombardment was observed on the round wall. Ions in group (b) were bended toward the wall. These ions caused lattice grid disorders and produced 1nm amorphous layer near top of the barrel (Fig.7 b). Meanwhile, more ions with higher energy and large incident angle hit the footing area, where the amorphous layer was measured thicker in Fig.7 c. Lattice defect was critical to mechanical and electrical performance of nano devices and should be eliminated in future works.

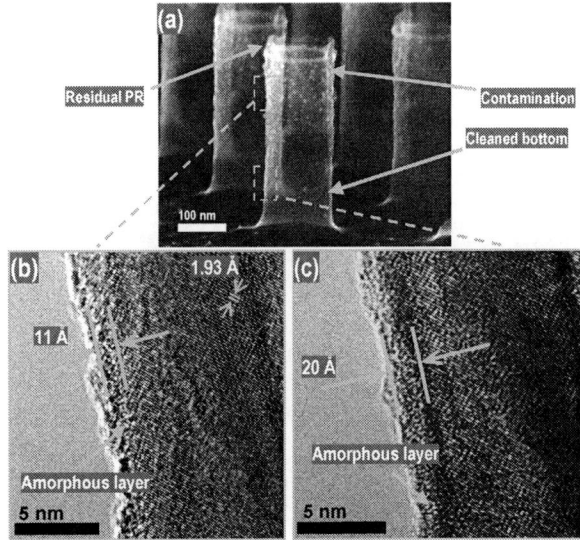

Figure 7: residual photoresist and polymer contaminations were shown in (a). Following the depassivation, the forming of amorphous layer near top of the barrel (b) was found thinner than bottom (c), which might cause by excessive physical bombardment laterally.

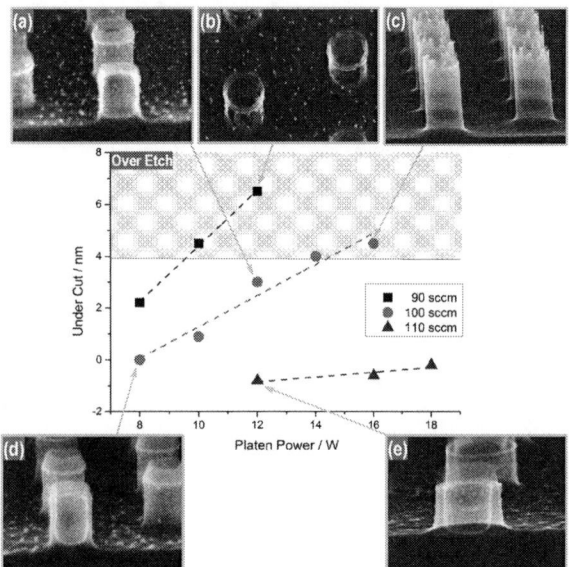

Figure 8: Outer face undercut vs. RF-power. Inserts shown the results of (a) optimized recipe (b) insufficient passivation (c) low selectivity (d) zero under-cut and (e) over passivation. All these barrels obtained erect inner chamber. Under-cut more than 4nm was marked as over etching and leaded to structure collapse.

To provide sufficient protection of resist mask in pattern transferring, passivation in DRIE was studied with RF power extensively. Undercut on outer face versus substrate RF power with different passivation gas flow rate was shown in Fig.8. As an optimized recipe, 100sccm C_4F_8 gas flow and 12W substrate RF power was applied during DRIE in this work, which gains high aspect ratio 3D structures shown as insert (a). Structures damage due to under-cut and etching selectivity were shown as insert (b) and (c) respectively. The recipes with under-cut more than 4nm were marked as over etching in the graph. With the variation of passivation gas flow rate, vertical inner wall face was gained universally. Tapper barrel with erect inner chamber was formed by over passivation in picture (e).

These evidences, including contamination clean up and profile control, proved the above ion transportation model.

CONCLUSIONS

In summary, nano barrels with wall thickness sub-7nm and aspect ratio 50:1 were successfully fabricated by microelectronic compatible top-down process. Lattice grid was revealed by TEM, which indicated single crystal Si nano barrels. An ions transportation model was established to interpret the forming of vertical nano chamber, and the physical damage on the wall face. This model was proved by the study of polymer contaminations, structure profile controlling and the thickness of amorphous layer.

Commercially used organic photoresist was utilized as etching mask directly. This construction method simplify the process design due to no hard mask was involved in. The hard mask free dry etching tactic avoided of compatible problems or additional CD loss. This high resolution 3D nano structure fabrication approach may find wide applications in 3D-IC and NEMS research.

ACKNOWLEDGEMENTS

This work was financed by the National Natural Science Foundation of China (NSFC: 91123026). The fabrication and characterization were supported by National Key Laboratory of Science and Technology on Micro/Nano Fabrication, Institute of Microelectronics, Peking University.

REFERENCES

[1] H. Liu, D. K. Mohata, A. Nidhi, V. Saripalli, V. Narayanan, S. Datta, "Exploration of vertical MOSFET and tunnel FET device architecture for sub l0nm node applications", in *70th Dev. Res. Conf.*, University Park TX, June 18-20, 2012, pp. 233-234

[2] Y. Sun, H. Y. Yu, N. Singh, K. C. Leong, E. Gnani, G. Baccarani, G. Q. Lo, D. L. Kwong, "Vertical Si nanowire based nonvolatile memory devices with improved performance and reduced process complexity", *IEEE Trans. on Electron. Dev.*, vol. 58, pp. 1329-1335, 2011

[3] A. Vandooren, D. Leonelli1, R. Rooyackers, K. Arstila, G. Groeseneken, C. Huyghebaert, "Electrical results of vertical Si n-tunnel FETs", in *Proc. of the Euro. Solid-State Dev. Res. Conf.*, Helsinki, September 12-16, 2011, pp. 255-258

[4] X. Wang, K. L. Pey, C. H. Yip, E. A. Fitzgerald, D. A. Antoniadis, "Vertically arrayed Si nanowire nanorod based core-shell p-n junction solar cells", *J. Appl. Phys.*, vol. 108, pp. 124303, 2010

[5] H.K. Jung, J. Choi, H. Na, D.S. Kwon, M.O. Kim, J.J Kang, J. Kim, "Lithography free fabrication of single crystalline silicon tubular nanostructures on large area", *Microelectron. Eng.*, vol. 98, pp. 325-328, 2012

[6] J. Bhardwaj, H. Ashraf, A. McQuarrie, "Dry silicon etching for MEMS", in *the Annual Meeting of ECS*, Montreal, May 4-9, 1997, pp. 1-13

[7] T. Borzenko, P. Fries, G. Schmidt, L.W. Molenkamp, M. Schirmer, "A process for the fabrication of large areas of high resolution, high aspect ratio silicon structures using a negative tone novolak based e-beam resist", *Microelectron. Eng.*, vol. 86, pp. 726-729, 2009

[8] Asahiko Matsuda, Yoshinori Nakakubo, Yoshinori Takao, Koji Eriguchi, Kouichi Ono, "Atomistic simulations of plasma process-induced Si substrate damage", in *Int. Conf. on IC Design & Tech.*, Pavia, May 29-31, 2013, pp. 191-194

[9] Koji Eriguchi, Yoshinori Takao, Kouichi Ono, "A new aspect of plasma-induced physical damage in three dimensional scaled structures", in *Int. Conf. on IC Design & Tech.*, Austin TX, May 28-30, 2014, pp. 1-5

CONTACT

*D.C. Zhang, dchzhang@ime.pku.edu.cn, tel: +86-10-62753130

SELF-HEALING METAL WIRE USING AN ELECTRIC FIELD TRAPPING OF GOLD NANOPARTICLES FOR FLEXIBLE DEVICES

Tomoya Koshi[1] and Eiji Iwase[1]
[1]Department of Applied Mechanics, Waseda University, Tokyo, JAPAN

ABSTRACT

We developed a self-healing metal wire using an electric field trapping of gold nanoparticles. A cracked metal wire on a stretchable substrate can get its conductivity again by the self-healing function. In this paper, first, we theoretically analyzed force acting on a nanoparticle and calculated a critical voltage which cause the electric field trapping. Next, we fabricated gold wires with artificially patterned cracks on a glass substrate and verified the self-healing function by experiments of a crack healing. Finally, we demonstrated the self-healing of a cracked metal wire on a stretchable substrate to show a usefulness of the self-healing for flexible devices.

INTRODUCTION

Recently, many researchers have reported flexible devices which have a bendability and/or a stretchability [1-2]. Especially, a stretchable wire is one of important components of flexible devices. To achieve the stretchability of wires, conductive elastomer wires [3] and curved metal wires [4-5] have been researched. However, the conductive elastomer wires do not have high conductivity in comparison with metal wires, and the curved metal wires are cracked when the wires are exposed to high or repeated stretching. Therefore, we focused on a self-healing metal wire to achieve both high conductivity and high stretchability.

The objective in this research is to provide a self-healing function to a metal wire for flexible devices. In this paper, "self-healing" means a selective healing of a crack by only applying a voltage to the cracked wire. To obtain a self-healing ability, we used an electric field trapping of gold nanoparticles by dielectrophoresis. Fig. 1 shows the schematic illustration of our self-healing. The metal wire is covered with a metal nanoparticle solution.

When a voltage is applied to a cracked wire, an electric field is generated only near the crack. The nanoparticles near the crack experience a dielectrophoresis force by the electric field. The dielectrophoresis force is an attractive force to the crack. When over a critical voltage is applied, the nanoparticles are trapped in the crack. This phenomenon is an electric field trapping, and we call the amplitude of this critical voltage a "healing voltage V_{heal}". As a result of applying the healing voltage, the crack is selectively healed by assembled nanoparticles bridging the crack.

ANALYSIS AND CALUCURATION

We theoretically analyzed force acting on a nanoparticle to calculate a healing voltage V_{heal}. In the case of applying a high frequency voltage, a total force acting on a nanoparticle near a crack is given by

$$F_{Total} = F_{VDW} + F_{ESR} + F_{DEP}, \quad (1)$$

where F_{VDW} and F_{ESR} are a Van der Waals force and an electrostatic repulsive force between the nanoparticle and the crack surfaces, and F_{DEP} is a dielectrophoresis force. F_{VDW} and F_{ESR} are determined by material properties of the nanoparticle and the solution [6]. On the hands, the time average of F_{DEP} is given by

$$\langle F_{DEP} \rangle = 2\pi\varepsilon \, r^3 \, \mathrm{Re}[\underline{K}(2\pi f)]\nabla E_{rms}^2, \quad (2)$$

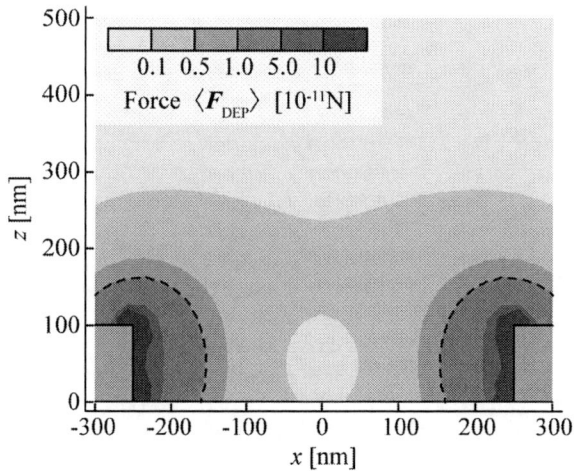

Figure 2. Calculation result of a $\langle F_{DEP} \rangle$ contour map near a crack in the case of 20-nm-radius gold nanoparticles in aqueous medium, 500 nm in crack width, 100 nm in wire thickness, 3.5 V in amplitude and 100 kHz in frequency of an applied voltage. The inside region of dotted line indicates a trapping region in which the strength of F_{DEP} is larger than the strength $F_{VDW}+F_{ESP}$ and gold nanoparticles are trapped to the crack.

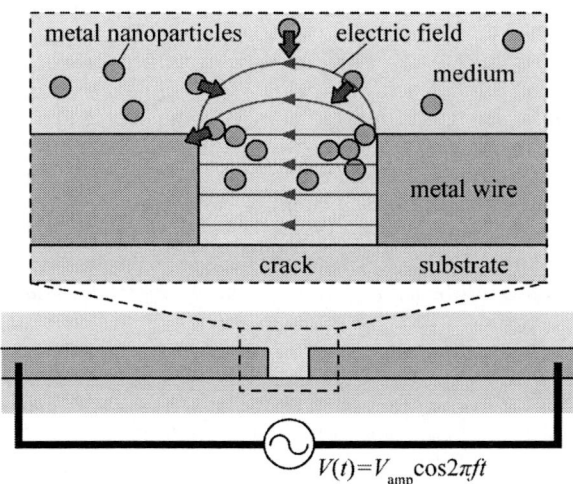

Figure 1. Schematic illustration of a self-healing of a cracked metal wire.

978-1-4799-7956-1/15 $31.00 © 2015 IEEE

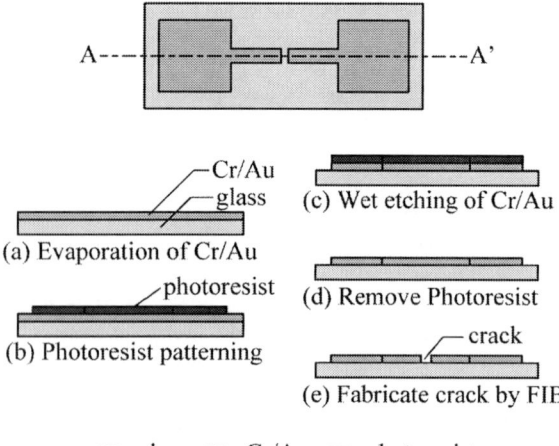

Figure 3. Fabrication process of a gold wire with a patterned crack on a glass substrate.

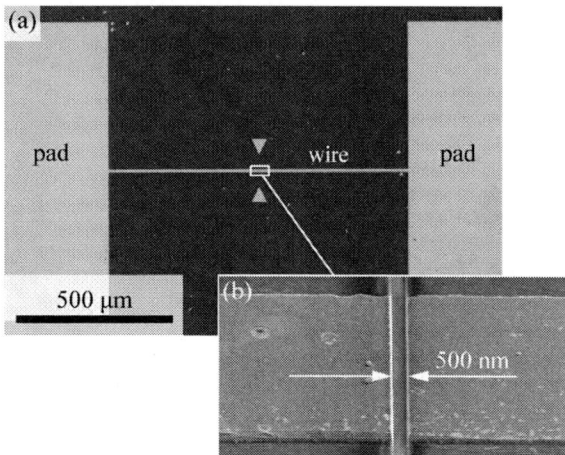

Figure 4. Optical and SEM images of a fabricated gold wire with a patterned crack on a glass substrate.

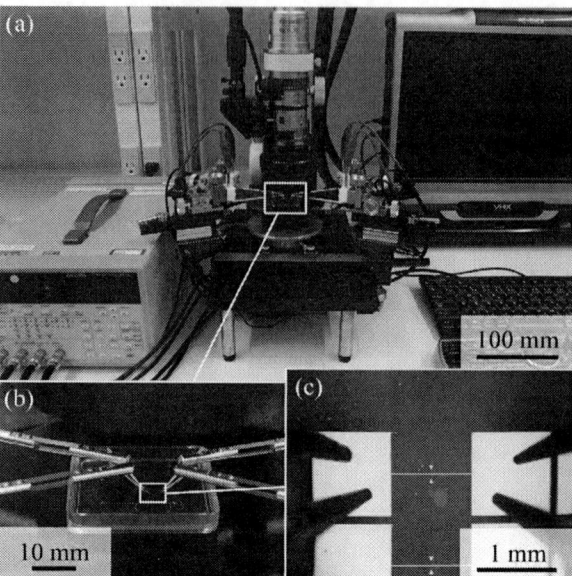

Figure 5. Experimental setup for a self-healing experiment.

medium, 500 nm in crack width, 100 nm in wire thickness and 100 kHz in frequency of the applied voltage.

EXPERIMENTS

We verified a self-healing function by experiments of a crack healing of a gold wire on a glass substrate. And then, we demonstrated a self-healing of a gold wire on a stretchable substrate.

We measured an impedance change of a cracked wire by self-healing and verified how wide a crack is healed. In the self-healing experiment, we used gold wires with an artificially patterned crack (500 to 1600 nm in width) on a glass substrate. Fig. 3 shows a fabrication process of a gold wire with a patterned crack. First, 10-nm-thick chromium (Cr) as an adhesive layer and 100-nm-thick gold (Au) were deposited on a 20 mm × 20 mm glass substrate by an electron beam evaporation. Next, the Cr and Au layers were patterned to wire shape by photolithography. In the end of the process, Focused Ion Beam (FIB) was used to fabricate 500- to 1600-nm-wide patterned cracks on a center of the gold wire (Fig. 3(e)). Fig. 4(a) shows a fabricated gold wire (10 μm in width and 100 nm in thickness) with a 500-nm-wide patterned crack on a glass substrate, and Fig. 4(b) is a Scanning Electron Microscope (SEM) image of the patterned crack. Fig. 5 shows the experimental setup for impedance measurement. We used 20-nm-radius gold nanoparticles in aqueous medium (Sigma-Aldrich, 741981) as a metal nanoparticle solution. The gold wires were covered with the gold nanoparticle aqueous suspension, and pads of the wire were connected to a LCR meter (NF Corporation, ZM2355) (left side of Fig. 5(a)) by micro-proves (Kyowariken, K-157MP) under a microscope (Keyence, VHX-2000) (right side of Fig. 5(a)). We applied AC voltage with 100 kHz in frequency to the gold wires and measured the impedance by a four probe method (Fig. 5(b), (c)). The applied voltage was increased at several tens of second intervals. Fig. 6 shows an

where r is a radius of the nanoparticle, ε is a permittivity of the medium, f is a frequency of the applied voltage, $\underline{K}(2\pi f)$ is a Clausius-Mosotti factor which represents a polarizability of the nanoparticle, and E_{rms} is an effective value of the electric field [7]. In this research, Re[$\underline{K}(2\pi f)$] is 1 because of using high conductive metal nanoparticles [8]. Eq. 2 shows that F_{DEP} can be changed by the applied voltage, therefore, we can control F_{DEP}. The electric field trapping is caused when F_{DEP} becomes large by large applying voltage and F_{Total} thereby becomes attractive force to a crack in Eq. 1. The amplitude of the critical voltage is V_{heal}.

Based on the above analysis, we calculated a V_{heal}. The nanoparticles will start to be trapped by electric field trapping when the strength of F_{DEP} becomes larger than the strength of $F_{VDW}+F_{ESR}$. The strength of $F_{VDW}+F_{ESR}$ near a crack surface is calculated to be about 1×10^{-11} to 1×10^{-10} N by using a DLVO theory [6]. Fig. 2 shows a calculated contour map of F_{DEP} near a crack by using Eq. 2. As a result of these calculations, we obtained that the electric field trapping is caused by applying about 1.3 to 4.0 V in the case of 20-nm-radius gold nanoparticles in aqueous

978-1-4799-7956-1/15 $31.00 © 2015 IEEE

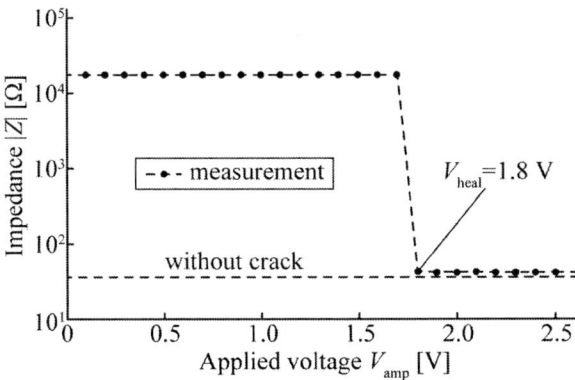

Figure 6. Impedance |Z| change according to applied voltage V_{amp} in the case of 500 nm in crack width.

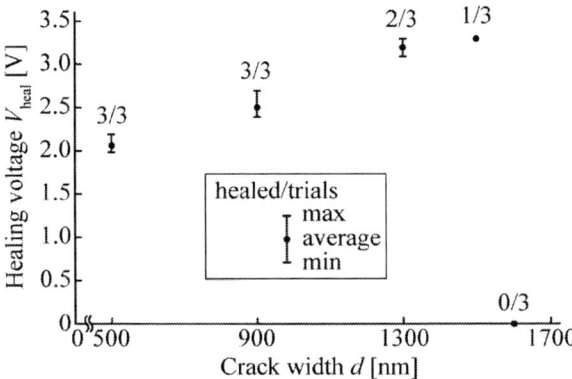

Figure 7. Relationship between healing voltage V_{heal} and crack width d.

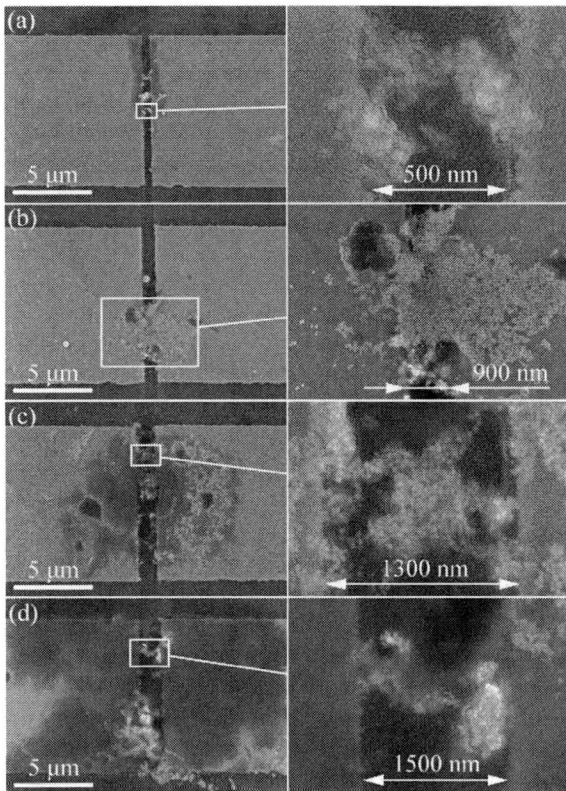

Figure 8. SEM images of cracks after self-healing experiments in the case of (a)500 nm, (b)900 nm, (c)1300 nm and (d)1500 nm in crack width respectively.

Figure 9. Optical and SEM images of a cracked gold wire on a PDMS substrate after self-healing experiment (V_{heal} was 1.6 V).

impedance change of a cracked wire with 500 nm in crack width between 0.1 to 2.5 V in amplitude of AC voltage. In Fig. 6, the impedance sharply dropped from about 10^4 to 10^1 Ω order at 1.8 V. This dropping of the impedance is attributed to a bridging of assembled gold nanoparticles in the crack by an electric field trapping, and the voltage of this dropping point indicates healing voltage V_{heal}. The impedance after self-healing was the same order as a gold wire with no cracks. This shows that our self-healing of a cracked metal wire has a high performance on healing ability.

Fig. 7 shows a relationship between healing voltage V_{heal} and a crack width. In the Fig. 7, numbers indicate the success rate of self-healings in 3 trails. The cracks with 500 to 1300 nm in width were healed over 2 times in 3 trails. The average values of V_{heal} were increased from 2.1 to 3.2 V. On the other hands, 1500-nm-wide cracks were healed only once in 3 trails, and 1600-nm–wide cracks were not healed in all trails. We suppose that gold wires were melted because of Joule heat caused by the applied voltage. In the case of success of self-healing, fluctuations of V_{heal} were about 0.3 V. As a reason of these fluctuations, the number of the particles in the trapping region (see Fig. 2) was stochastic. Therefore, the numbers of the assembled nanoparticles was changing every trials, and V_{heal} was changed. Fig. 8 shows SEM images of each crack width after self-healing process. We confirmed that assembled

nanoparticles were bridging the crack. In Fig. 8, assembled nanoparticles were bridging on a part of the crack. We considered that the nanoparticles might be preferentially trapped to a crack surface where a first nanoparticle was trapped, because F_{DEP} becomes stronger in a narrower crack.

We demonstrated a self-healing of a cracked wire on a stretchable substrate to show a usefulness of a self-healing for flexible devices. We used poly(dimethylsiloxane)

(PDMS) as a stretchable substrate. 10-nm-thick Cr and 100-nm-thick Au layers were deposited by using an electron beam evaporation. We used a ten times lower evaporation rate of 0.02 nm/sec to reduce wrinkles of metal layers on a PDMS substrate. A cracked gold wire with 10 μm in wire width, 100 nm in wire thickness and 270 nm in crack width was fabricated by a photolithography and an FIB milling. As a result of a crack healing, the gold wire was healed at 1.6 V, and we confirmed that assembled gold nanoparticles were bridging the crack (Fig. 9). Because this result is almost the same as the results on a glass substrate, it shows that our self-healing of the cracked wire is useful for flexible devices.

CONCLUSIONS

We proposed a self-healing metal wire using an electric field trapping of gold nanoparticles for flexible devices.

First, we theoretically analyzed forces acting on a nanoparticle: a Van der Waals force F_{VDW}, an electrostatic repulsive force F_{ESR} and a dielectrophoresis force F_{DEP}. In the case of 20-nm-radius gold nanoparticles in aqueous medium, 10 μm in wire width, 100 nm in wire thickness, and 100 kHz in frequency of AC voltage, we calculated that a cracked wire with 500 nm in crack width was healed by applying 1.3 to 4.0 V. Next, we fabricated gold wires which have a patterned crack with 500 to 1600 nm in crack width on a glass substrate by an FIB milling. As a result of experiments of a crack healing, a gold wire with up to 1300 nm in crack width was successfully healed by applying under 3.2 V. By SEM observation, we confirmed that gold nanoparticles were bridging the crack after the self-healing process. And the impedance of the self-healed gold wire was the same order as a gold wire with no cracks. Finally, we performed a self-healing on a PDMS substrate using a cracked gold wire with 10 μm in wire width, 100 nm in wire thickness and 270 nm in crack width, and showed the usefulness of a self-healing of the cracked wire for flexible devices.

ACKNOWLEDGEMENTS

This research was partially supported by TEPCO Memorial Foundation.

REFERENCES

[1] Dae-Hyeong Kim, Jianliang Xiao, Jizhou Song, Yonggang Huang, and John A. Rogers, "Stretchable, Curviliner Electronics Based on Inorganic Materials," *Adv. Mater.*, vol. 22, pp. 2108-2124, 2010.

[2] Mallory L. Hammock, Alex Chortos, Benjamin C.-K. Tee, Jeffrey B.-H. Tok, and Zhenan Bao, "25th Anniversary Article: The Evolution of Electronic Skin (E-Skin): A Brief History, Design Considerations, and Recent Progress," *Adv. Mater.*, vol. 25, pp. 5997-6038, 2013.

[3] Tsuyoshi Sekitani, Hiroyoshi Nakajima, Hiroki Maeda, Takanori Fukushima, Takuzo Aida, Kenji Hata, and Takao Someya, "Stretchable Active Matrix Organic Light-Emitting Diode Display using Printable Elastic Conductors," *Nat. Mater.*, vol. 8, pp.494-499, 2009.

[4] Darren S. Gray, Joe Tien, and Christopher S. Chen, "High-Conductivity Elastomeric Electronics," *Adv. Mater.*, vol. 16, pp. 393-397, 2004.

[5] Dae-Hyeong Kim, Nanshu Lu, Rui Ma, Yun-Soung Kim, Rak-Hwan Kim,Shuodao Wang, Jian Wu, Sang Min Won, Hu Tao, Ahmad Islam, Ki Jun Yu, Tae-il Kim, Raeed Chowdhury, Ming Ying, Lizhi Xu, Ming Li, Hyun-Joong Chung, Hohyun Keum, Martin McCormick, Ping Liu, Yong-Wei Zhang, Fiorenzo G. Omenetto, Yonggang Huang, Todd Coleman, John A. Rogers, "Epidermal Electronics," *Science*, vol. 333, pp. 838-843, 2011.

[6] Jacob N. Israelachvili, "Intermolecular and Surface Forces, Third Edition," *Academic Press*, 2011.

[7] Thomas. B. Jones, "Electromechanics of Particles," *Cambridge University Press*, 1995.

[8] Robert J. Barsotti, Jr., Michael D. Vahey, Ryan Wartena, Yet-Ming Chiang, Joel Voldman, and Francesco Stellacci, "Assembly of Metal Nanoparticles into Nanogaps," *Small*, vol. 3, pp. 488-499, 2007.

CONTACT

Tomoya Koshi, 3-4-1 Okubo, Shinjuku-ku, Tokyo 169-8555, JAPAN, Tel/Fax: +81-3-5286-2741, E-mail: koshi@akane.waseda.jp

CONTROLLED FABRICATION OF NANOSCALE GAPS USING STICTION

Farnaz Niroui, Ellen M. Sletten, Parag B. Deotare, Annie I. Wang,
Timothy M. Swager, Jeffrey H. Lang, and Vladimir Bulović
Massachusetts Institute of Technology, Cambridge, Massachusetts, USA

ABSTRACT

Utilizing stiction, a common failure mode in micro/ nano electromechanical systems (M/NEMS), we propose a method for the controlled fabrication of nanometer-thin gaps between electrodes. In this approach, a single lithography step is used to pattern cantilevers that undergo lateral motion towards opposing stationary electrodes separated by a defined gap. Upon wet developing of the pattern, capillary forces induce cantilever deflection and collapse leading to permanent adhesion between the tip and an opposing support structure. The deflection consequently reduces the separation gap between the cantilever and the electrodes neighboring the point of stiction to dimensions smaller than originally patterned. Through nanoscale force control achieved by altering device design, we demonstrate the fabrication of nanogaps having controlled widths smaller than 15 nm. We further discuss optimization of these nanoscale gaps for applications in NEM and molecular devices.

INTRODUCTION

As dimensions are continuously scaled down to achieve electronic, photonic and electromechanical devices with improved performance and novel principles, developing methods for the controlled fabrication of electrodes separated by nanometer-thin gaps is important to enabling reliably-functioning devices. Current methods of fabricating such nanogaps include oblique-angle shadow evaporation, electrochemical deposition, electromigration, mechanical break junctions, molecular junctions and etching of nanometer-scale sacrificial layers [1-5]. However, these approaches, mainly developed for two-terminal devices, commonly involve multiple processing steps and lack robustness and tunability, thus preventing effective incorporation into more complex multi-terminal designs. These limit their practical applications in integrated systems.

A common mode of failure in electromechanical systems is permanent adhesion between device components, referred to as stiction. Stiction arises due to the surface adhesion forces overcoming the elastic restoring force of a mechanically active structure, leading to its collapse and hindering its recovery. As surface adhesion forces increase with the decrease in gap dimensions, stiction becomes increasingly more challenging in NEM devices.

Here, we propose the use of stiction, typically considered an irreversible failure mode, to promote controlled fabrication of electrodes separated by nanoscale gaps of varying widths. The feasibility of this approach to form nanogaps with dimensions smaller than 15 nm, and their potential applications are investigated in this paper.

DESIGN PRINCIPLES

Capillary forces exerted on the mechanically-active structures of electromechanical systems by the drying of the liquid trapped in the small gaps of the devices after wet fabrication processing can readily lead to stiction [6]. Taking advantage of this stiction, our proposed method of fabricating nanogaps relies on control of the surface adhesion forces at the nanoscale to enable control of cantilever deflection that is used to tune the gap. The laterally-actuated cantilever and other device components are fabricated through a one-step lithography process. During the wet developing of the patterned structures, capillary forces induce deflection and eventual collapse of the cantilever. The stiction between the cantilever and a support structure enables formation of gaps smaller than originally patterned between the cantilever and additional electrodes located relative to the point of stiction. Through changes to the structural design and liquid phase processing step, surface adhesion forces caused by the capillary action can be adjusted to allow for precise control of the gap size.

FABRICATION

The fabrication scheme for stiction-induced formation of nanogaps is shown in Figure 1. Five layers of poly(methyl-methacrylate) (PMMA), a positive electron beam resist, are spun over a silicon (Si) substrate with 2 μm-thick thermal oxide (SiO_2). Each layer is spun at 2000 rpm for 45 s and baked at 180 °C for 90 s. The initial three layers of PMMA have molecular weight of 495 kg/mol (PMMA 495 A6) and the following two layers have molecular weight of 950 kg/mol (PMMA 950 A4). Next, the cantilever and other electrodes are defined by patterning the PMMA film using electron-beam (e-beam) lithography. The resist is developed in 1:3 dilution of methyl isobutyl ketone (MIBK) in isopropanol for 3 min, placed in an isopropanol bath to thoroughly rinse, and dried under a gentle stream of nitrogen normal to the surface. Finally, about 10 nm of chromium (Cr) and 100 nm of gold (Au) are deposited over the substrate using thermal evaporation to form the electrodes.

The five layers of PMMA of two different molecular weights with a total thickness of about 1.5 μm allow fabrication of large aspect ratio features with an undercut profile, a thinner base and a thicker top section. The undercut is achieved due to the differential dissolution rate of PMMA of varying molecular weights in MIBK; lower molecular weight PMMA has a faster dissolution rate than the higher molecular weight PMMA. The undercut prevents sidewall coverage during metal deposition and ensures electrical isolation between the electrodes. The undercut and the high aspect ratio of the PMMA cantilever also enable the structure to freely deflect due to an applied force.

During the wet-developing process, a capillary force is exerted on the cantilever (Electrode 1) in Figure 1. This force can cause deflection of the cantilever. If sufficiently large to overcome the spring restoring force, the cantilever collapses on to an opposing PMMA support structure (Electrode 2) and undergoes stiction. The deflection and stiction reduce the gap between the cantilever and other electrodes positioned along the length of the cantilever away from the point of stiction, leading to the formation of nanogaps smaller than originally patterned. The spring constant of the cantilever can be adjusted by altering its geometry and the relative positioning of counter electrodes. Then, by adjusting the surface adhesion forces, one can control the extent of deflection and its profile such that desired gap dimensions are achieved. The formed nanogaps can further be reduced in size by evaporating a thin-film of metal onto the PMMA structures to define the conductive electrodes. The reduction in gap size during this processing step is dependent on the thickness of the deposited film.

RESULTS AND DISCUSSION

The feasibility of utilizing stiction to develop nanogaps smaller than patterned in a single lithography step is shown in Figure 2. To promote stiction, the capillary forces must overcome the elastic restoring force of the cantilever, causing the irreversible collapse of the structure. By altering the design and processing conditions, including the dimensions of the cantilever, the patterned gap size, and the liquid used, the surface adhesion forces can be altered to ensure collapse of the active structure during the wet processing. The nanogap fabrication presented above is not limited to the material sets used in our proof-of-concept approach. Similar results are expected in cases where materials other than PMMA and Au are used. Furthermore, different materials selections provide an additional means of controlling the surface adhesion forces and the stiction process [8]. In the cantilever of Figure 2, fabricated using the scheme in Figure 1, stiction is promoted by decreasing the gap size between Electrode 1 and Electrodes 2 and 3 by approximately 30 nm. Once collapsed, an effective gap of ~50 nm between Electrodes 1 and 3 at the point closest to the stiction region is achieved compared to the 200 nm gap in a similar cantilever that has not undergone stiction.

Through engineering of the spring constant of the cantilever, and the geometry and relative placement of the opposing electrodes, the size of the nanogaps can be controlled. An example ~35 nm gap fabricated is shown in Figure 3. Figure 3b illustrates the reduction in the gap size to ~10 nm by positioning Electrode 3 closer to the point of stiction. The tunability of the gap dimension is further demonstrated in Figure 4 where gaps ranging from about 10 nm to 170 nm are formed. As shown, the process is not limited to two-terminal structures. Further, through the same lithography step, devices with multiple electrodes can be fabricated. These multi-terminal devices are more relevant to applications in integrated systems and allow

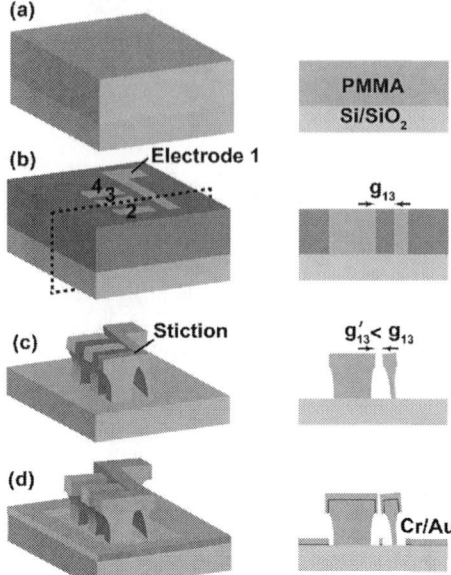

Figure 1: Controlled fabrication of nanoscale gaps using stiction; (a) multilayer PMMA e-beam resist is spun onto Si/SiO$_2$ substrate and baked after each spin, (b) resist is patterned using e-beam to define cantilever and opposing electrodes separated by a defined gap "g", (c) resist is developed in 1:3 solution of MIBK in isopropanol, rinsed in isopropanol bath and dried under a stream of nitrogen, (d) ~10 nm of Cr and ~100 nm of Au is evaporated onto the patterned structures defining the electrodes.

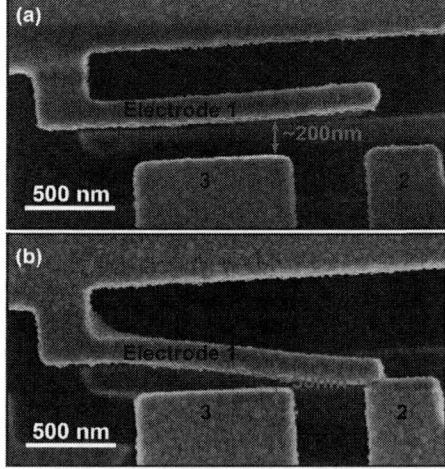

Figure 2: Stiction promotes formation of nanogaps smaller than patterned; (a) Scanning electron micrograph (SEM) of a cantilever with a gap of ~200 nm, (b) SEM of the same cantilever positioned closer to electrodes 2 and 3. Larger capillary force in (b) causes collapse of the cantilever, leading to stiction and reducing the gap between 1 and 3.

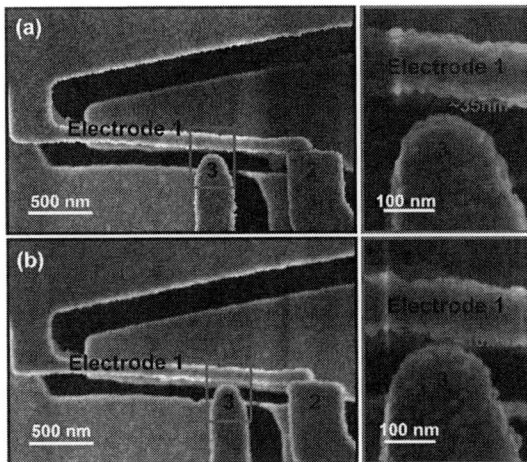

Figure 3: SEM of nanogaps fabricated using the scheme in Figure 1. Capillary forces during wet developing of the pattern cause stiction between Electrodes 1 and 2, while forming a gap of ~35 nm (a) and ~10 nm (b) between Electrodes 1 and 3. The smaller gap in (b) is achieved by positioning Electrode 3 closer to 2.

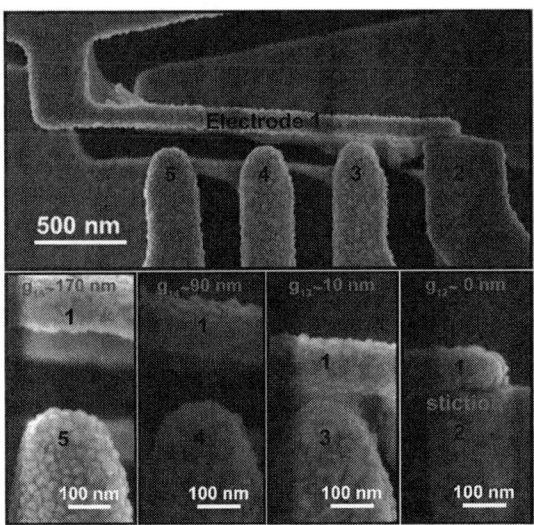

Figure 4: Optimizing device architecture and electrode placement relative to the point of stiction allow achieving nanogaps with controlled width, useful for fabrication of various multi-terminal devices.

exploration of more complex device designs and concepts.

A potential application of the proposed nanogap fabrication scheme is in the development of NEM switches, which, with their large on-to-off current ratios, abrupt switching and near-zero leakage currents have emerged as technology competitive with complementary metal-oxide-semiconductor transistors [8]. However, the need for large actuation voltages is among the main challenges preventing

integration of NEM switches into widely used systems. To reduce the operating voltage, decrease in the size of the switching gap is necessary, as it leads to an increase in the electrostatic force of actuation. The reduction of the nanogap size to the few-nanometer regime, however, has been a fundamental fabrication challenge. Typically, as the size of the switching gap decreases, the yield of functional devices lowers due to failure modes such as stiction during fabrication. In contrast, in our proposed approach the stiction is used to benefit nanogap formation. Depending on device design, the collapsed electrode can be made to remain stationary or undergo mechanical motion with an applied voltage. As shown in Figure 5, a smaller gap and a more flexible cantilever allow electromechanical modulation of the gap. In this two-terminal example a current modulation of 10^{11} is achieved within 3 V applied bias. The desired switching performance can be achieved through the optimization of the fabrication technique. This approach can also be extended to three terminal NEM switches that more closely resemble the conventional transistors. A prototype design is shown in Figure 5b.

Nanogaps have also been widely desired for applications in molecular electronics that rely on metal-molecule-metal junctions. These gaps are conventionally formed using techniques such as direct deposition of metallic contacts, which has the potential to damage the fragile organic molecular layer, leading to low yields of working devices and structures that lack robustness. Through the proposed stiction-induced nanogap fabrication, damage to the molecular layer can be avoided as the molecules are introduced into the gap after electrode formation through techniques such as vapor deposition or liquid-phase self-assembly. In self-assembly molecules are functionalized with terminal groups that enable selective attachment onto the electrode surfaces. For example

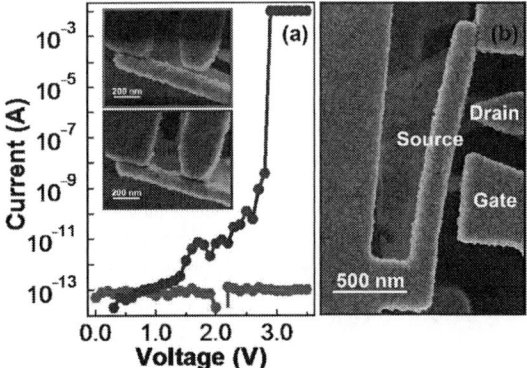

Figure 5: (a) Through altering device geometry to achieve smaller gaps and more flexible cantilevers, nanogaps can be fabricated to undergo electromechanical modulation (blue compared to red), useful for applications in NEM switches. Insets show corresponding devices with the bottom cantilever (blue) ~20 nm thinner than the top (red). (b) Prototype design of a three-terminal NEM switch.

978-1-4799-7956-1/15 $31.00 © 2015 IEEE

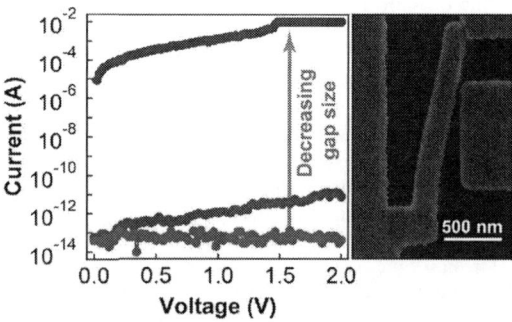

Figure 6: The current-voltage characteristics of three gaps of different sizes with molecular layers of fluorinated decanethiols self-assembled on the Au electrodes (left). Representative SEM image of the devices tested (right). Nanogaps of different sizes are achieved using the same design while changing the size of original gap patterned.

thiolated molecules self-assemble onto gold [9]. Figure 6 shows the current-voltage characteristics of nanogaps of different sizes with thin-films of fluorinated decanethiol self-assembled on the Au surface using vapor phase deposition. As expected, a general increase in current is observed with the decrease in gap size. Further interpretation of the data to extract information about the nature of the molecular junction requires more detailed experimentation. We speculate that as molecular layers are formed, further decrease in gap size can be achieved through changes in surface adhesion forces imposed by changes in surface properties. This approach is also valuable to form anti-stiction coatings in NEM devices to enhance operation reliability, and to form molecular switching gaps for quantum tunneling NEM switches [10].

Although providing a promising platform for fabrication of nanogaps, the proposed technique faces some challenges that must be addressed. As observed through the SEM images, the as-deposited metal electrodes have a large surface roughness, which introduces inhomogeneity in the gap size, makes the devices susceptible to electrical shorting and can alter device performance. This can be overcome by exploring materials alternative to Au and PMMA or through use of techniques such as atomic layer deposition to form atomically smooth surfaces. In addition, functional devices that require nanogaps with asymmetric metal electrodes of different materials can be envisioned. Enabling such structures requires modifications to the fabrication scheme that may increase its complexity.

CONCLUSION

A method for controlled fabrication of nanometer-scale gaps is proposed that utilizes stiction, a common source of failure in electromechanical systems. Capillary forces induced during liquid phase processing of the sample cause deflection of a movable electrode and consequent stiction. The deflection leads to reduction in the size of the gaps patterned. The experimental results support the feasibility

of the proposed method to fabricate nanogaps, with controlled widths smaller than 15 nm, through the engineering of surface adhesion forces. Further optimization of this versatile fabrication platform allows nanogap development for various applications including electromechanical and molecular devices.

ACKNOWLEDGEMENTS

This work is supported by the National Science Foundation (NSF) Center for Energy Efficient Electronics Science (E^3S) Award ECCS-0939514. F.N. acknowledges the support of Natural Sciences and Engineering Research Council of Canada (NSERC).

REFERENCES

[1] H. B. Akkerman, B. de Boer, "Electrical conduction through single molecules and self-assembled monolayers", *J. Phys.: Condens. Matter*, vol. 20, 013001, 2008.

[2] T. Li, W. Hu, D. Zhu, "Nanogap electrodes", *Adv. Mater.*, vol. 22, pp. 286-300, 2010.

[3] F. Streller, G. E. Wabiszewski, F. Mangolini, G. Feng, R. W. Carpick, "Tuanble, source-controlled formation of platinum silicides and nanogaps from thin precursor films", *Adv. Mater. Interfaces*, vol. 1, 1300120, 2014.

[4] J. O. Lee, Y. Song, M. Kim, M. Kang, J. Oh, H. Yang, J. Yoon, "A sub-1-volt nanoelectromechanical switching device", *Nature Nanotech.*, vol. 8, pp. 36-40, 2013.

[5] D. J. Beesley, J. Semple, L. K. Jagadamma, A. Amassian, M. A. McLachlan, T. D. Anthopoulos, J. C. deMello, "Sub-15-nm patterning of asymmetric metal electrodes and devices by adhesion lithography", *Nat. Commun.*, vol. 5, 3933, 2014.

[6] O. Raccurt, F. Tardif, F. Arnaud d'Avitaya, T. Varine, "Influence of liquid surface tension on stiction of SOI MEMS", *J. Micromech. Microeng.*, vol. 14, pp. 1083-1090, 2004.

[7] M. Toda, A. Yokoyama, N. V. Toan, N. Inomata, T. Ono, "Fabrication of nano-gap structures based on plastic deformation of strained Si springs by stiction effects", *Microsyst. Technol.*, pp. 1-6, 2014.

[8] O. Y. Loh, H. D. Espinosa, "Nanoelectromechanical contact switches", *Nature Nanotech.*, vol. 7, pp. 283-295, 2012.

[9] J. C. Love, L. A. Estroff, J. K. Kriebel, R. G. Nuzzo, G. M. Whitesides, "Self-assembled monolayers of thiolates on metals as a form of nanotechnology", *Chem. Rev.*, vol. 105, pp. 1103-1169, 2005.

[10] F. Niroui, P. B. Deotare, E. M. Sletten, A. I. Wang, E. Yablonovitch, T. M. Swager, J. H. Lang, V. Bulović, "Nanoelectromechanical tunneling switches based on self-assembled molecular layers", *IEEE 27th International Conference on Micro Electro Mechanical Systems*, pp. 1103-1106, 2014.

CONTACT

*F. Niroui, tel: +1-617-3248110; fniroui@mit.edu

DIFFERENTIAL MICRO-PIRANI GAUGE FOR MONITORING MEMS WAFER-LEVEL PACKAGE

Yang-Che Chen[1], Wei-Chu Lin[1], Hung-Sen Wang[1], Chen-Chih Fan[1], Keaton C.-H. Lin[1], Bruce C.S. Chou[1], and Mingo C.-M. Liu[1]

[1]Taiwan Semiconductor Manufacturing Company, Hsinchu, TAIWAN

ABSTRACT

The implementation of differential Pirani gauge for accurately measuring wafer-level package pressures was demonstrated. We proposed a multiple-sensor-solution, where two Pirani gauges were constructed under different pressures; one in sealed micro-cavity for measuring pressures and the other one in opened micro-cavity as a reference. Ambient pressure, structural dimension variations, and resistivity differences among wafers/lots, were captured through the differential scheme for error compensations, allowing accurate pressure determinations. Presented Pirani utilized small gaps (~2μm) between heater and dual heat sinks to obtain wide operation range (0.05~100 Torr) and high sensitivity (~10000 ppm/Torr). With 5X error reductions and high stabilities, the proposed device was successfully used in examining reliabilities and monitoring processes of wafer-level packages.

INTRODUCTION

Characterizing seal quality of wafer-level package is critical to successful MEMS manufactures. Moisture and gas penetrating into sealed micro-cavity easily happens in bad package, which may lead to the device offsets, device malfunctioning, and structural stictions. Detecting the final pressure of micro-cavity is the most direct method to check seal quality. Extracting Q-factors of vibrating micro-structures is widely used [1] but is limited by the long test time, and small operation range. Pirani gauge is an excellent solution for pressure measurement [2]. It has small size, and is highly sensitive to pressure changes, and easy to measure. However, implementing Pirani gauge in mass production, factors leading to pressure measurement errors, must be fully suppressed. Fast pressure determination method for testing the seal quality with high accuracy and stability is thus needed in a foundry.

In this work, we demonstrated a differential Pirani gauge based the multiple-sensor-scheme. An unsealed Pirani gauge was added and it captured the ambient pressure, manufacture errors, and resistivity variations for reducing the pressure measurement uncertainties. By extracting the differential signal from this scheme, a more robust pressure measurement method was achieved.

DESIGN

The electrical resistance of a bridge type micro-machined Pirani gauge can be expressed as [3]

$$R(T_{avg})=\rho_0 \cdot (1+\alpha \cdot T_{avg}) \cdot L/(e \cdot w) \qquad (1)$$

where T_{avg} is the average temperature across the beam and is a function of the pressure, bridge dimensions, and input current. ρ_0 is the electrical resistivity of the material. α is the material TCR. L, e and w represent the bridge's length, thickness, and width, respectively. Equation (1) indicates

that the output resistive signal is strongly dependent on the structural dimensions and material properties, which are easily affected by the process variations and thus vary among different wafers and lots (1 lot= 1 batch run= 25 wafers), resulting in wrong pressure measurements.

To minimize these measurement uncertainties caused by MEMS manufactures, a differential scheme was presented, where a bridge type Pirani gauge was integrated with an additional reference gauge. The differential scheme is illustrated in Figure 1(a), and the bridge type gauge design is demonstrated in Figure 1(b).

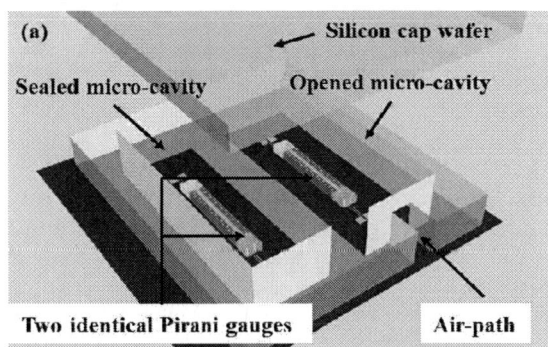

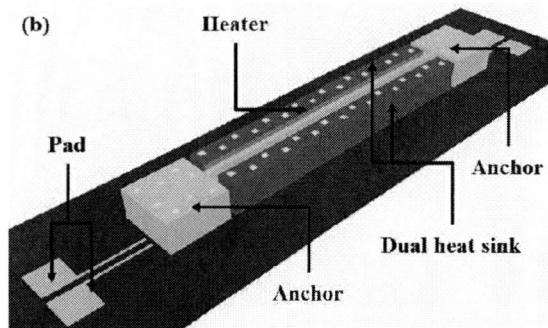

Figure 1: (a) Differential scheme. (b) Magnified view of one bridge type Pirani gauge with dual heat sinks.

In Figure 1(a), two identical gauges are encapsulated by silicon cap wafer in one chip. A wall is constructed between two gauges to separate them into two individual micro-cavities. Additionally, an air-path in the cap is constructed to open one of micro-cavities to ambient environment, making two different pressures in one chip (one is 1 atm and the other one is unknown). Pirani gauge under 1 atm is treated as reference gauge, from which the background noises from 1 atm ambient pressure, dimension variations, and doping differences are extracted. On the other hand, the sealed Pirani gauge is used to measure the unknown cavity pressure. Figure 1(b) shows the Pirani gauge having a free-standing bridge type heater in the center surrounded by dual heat sinks for lateral heat transferring. The area of heat loss from the heater to heat sink through gas is significantly increased to widen the gauge's dynamic range. In addition, the upper pressure measurement limit is enhanced due to the small air gaps between the heater and heat sinks, which is precisely defined by DRIE process in our platform.

In operation, each gauge is heated up and error compensation is achieved by reading out differential resistive signals. As a result, the output signal of the scheme can be written as

$$R= [R_s(T_{avg})- R_u(T_{avg})]/ R_u(T_{avg}) \qquad (2)$$

where R_s and R_u are the resistances of sealed and unsealed Pirani gauges, respectively.

EXPERIMET

Fabrication result

The proposed idea was realized in tsmc CMOS-MEMS monolithic platform [4]. This platform was developed for manufacturing motion sensors, such as accelerometers and gyroscopes, and thus their hemeticities need to be in-process monitored. The differential Pirani gauge was implemented in monitoring the seal quality of eutectic bond process for wafer-level package in this platform.

The fabrication results were demonstrated in Figure 2. Figure 2(a) shows the CMOS-MEMS wafer after the cap wafer was removed. Sealed and unsealed Pirani gauges were built in two cavities and separated by a wall. Cap wafer for sealing the CMOS-MEMS wafer with an air path made in its sidewall was shown in Figure 2(b). Still, air-path could be a weakness of the proposed device. After the wafer-level package, the wafer was treated with wet processes, such as the wet clean and die saw. Water and particles in these following processes easily flew into the opened micro-cavity and cause MEMS damages and stictions. IR image in Figure 2(c) was taken before the cap wafer was removed, and it shows no damages and stictions were found in the unsealed Pirani after the wet processes. However, particles were observed near the inner opening of air-path and these tiny particles (particle sizes are smaller than 0.1 um) do not greatly influence the performance of the device.

Device characterization

The device was calibrated inside the vacuum chamber as shown in Figure 3. A reference pressure gauge was used to monitor the chamber, while the leak valve allowed changing the pressure. Based on the standard four-point probe configuration, a current source with constant current was applied to differential Pirani gauge and the voltage drop across gauge was measured using Agilent B1500. Then, its resistive responses were extracted under different pressures ($P=10^{-4}$ Torr ~ P=760 Torr) to construct the calibration curve.

The normalized resistance variation versus pressures with respect to the resistance of the device under $P=10^{-4}$ Torr was plotted in Figure 4(a). It shows a sensitivity of 10000 ppm-resistance-change/Torr with the operation range of 0.05~100 Torr in constant current mode. Moreover, the concept of differential Pirani gauge reducing errors was proven in Figure 4(a) and (b). Two different devices exhibit almost identical calibration curves

with minimal resistance differences. Errors between two calibration curves obtained in Figure 4(a) were calculated and summarized in Figure 4(b) and only less than 5% deviations appeared.

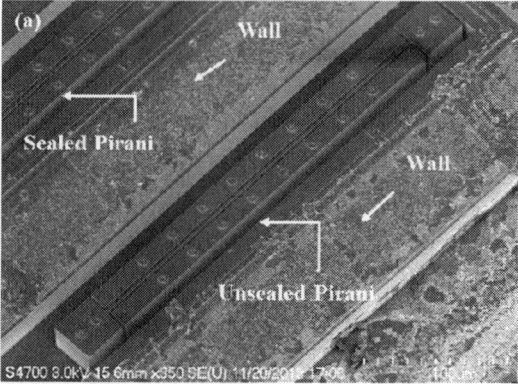

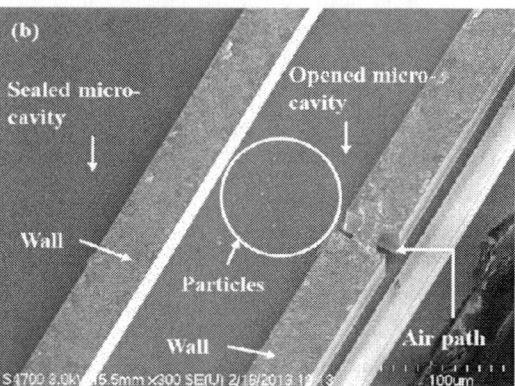

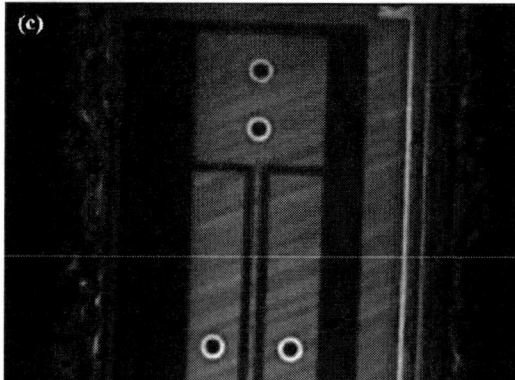

Figure 2: (a) De-capped CMOS-MEMS wafer. (b) Cap wafer, where air path was built for making two pressures in one chip. (c) IR image taken after fabrication processes completed, showing the unsealed Pirani gauge in a capped device. No structural damages or stiction were observed.

On the other hand, without the error compensation, calibration curves extracted from two single Pirani gauges were shown in Figure 5(a). More than 20% deviations between these two curves were observed in Figure 5(b). This deviation led to more than 20 Torr measurement errors among wafers. The measurement errors limited the application of conventional Pirani gauge for monitoring the gyroscope wafers, which require precise pressure controls.

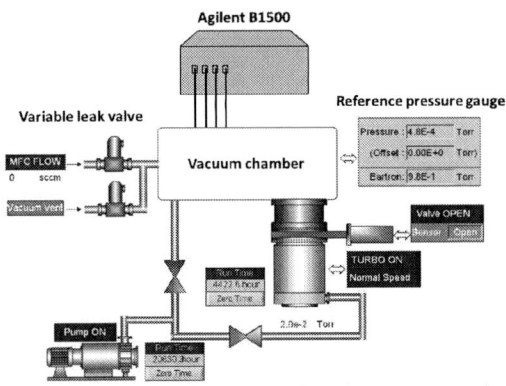

Figure 3: Experimental setup. The device was placed inside vacuum chamber with controllable pressure and temperature. The electrical signals were fed in and extracted by Agilent B1500.

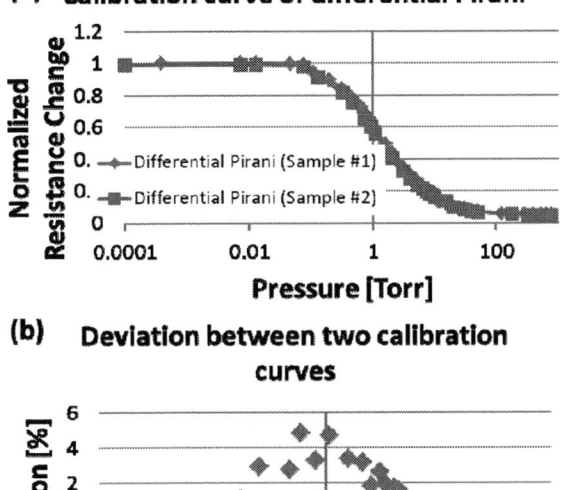

Figure 4: (a) Calibration curves of two differential Pirani gauges. (b) Deviations between two curves under different pressures are smaller than 5 %.

To implement the differential Pirani gauge in production lines, high reliability of the device is mandatory. This work analyzed the device's stabilities in Figure 6. In Figure 6(a), calibration curves were monitored for three weeks to demonstrate device's repeatability. This figure indicates the operation range and the sensitivity of the device do not change with time. Besides, specimens from different wafers and lots were tested in Figure 6(b) and 6(c). Deviations among samples were barely observed in Figure 6, implying that manufacture-induced errors were successfully suppressed and high accuracy was achieved. Small pressure measurement errors of monitoring wafer-level package could appear in mass production.

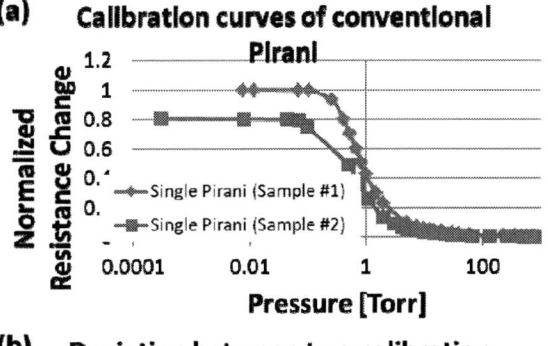

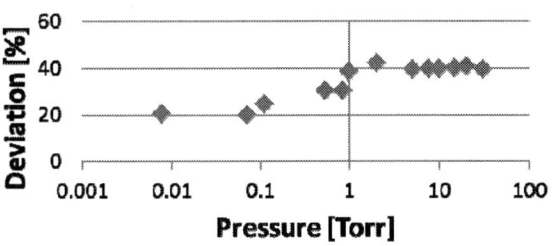

Figure 5: (a) Calibration curves of two conventional single Pirani gauges. (b) Deviations between two curves under different pressures are 20~ 40 %, which are too large for in-process monitoring and are not acceptable.

APPLICATION

Finally, this work implemented differential Pirani gauge for checking seal quality of tsmc CMOS-MEMS platform. The presented device was built in every reticle field, that is, there were 50 gauges in one single wafer. With these embedded monitor structures in wafers, the pressure distributions across the whole wafer were analyzed and used to examine the seal quality.

To test the robustness of package process, pressures of a vacuum-sealed wafer were measured before and after 96Hr wafer-level uHAST (un-biased High Accelerated Stress Test) tests. In Figure 7(a), most of the cavity pressures ranged from 10 to 40 Torr, which were the normal final pressures. But, three 760-Torr dies in the wafer center were detected, showing they were initially poor sealed. After the reliability test, Figure 7(b) shows bonding degradation happened in wafer center as more failed dies appeared and their pressures increased to higher than 100 Torr. This result suggested the bonding quality of this wafer was poor and its fabrication process needed to be improved.

Another application of presented device is for process recipe fine tuning and recipe selecting. Figure 8 illustrates the relationships between five bonding recipes to their final cavity pressures characterized by differential Pirani gauge. Constructing this relationship helps the development of high vacuum package process for MEMS gyroscopes and resonators. Figure 8 shows that this platform can achieve mid-vacuum levels (1~20 Torr) with uniform pressures across the wafer using bonding conditions of #1, #2, and #3.

(a)

Repeatability test (Same die)

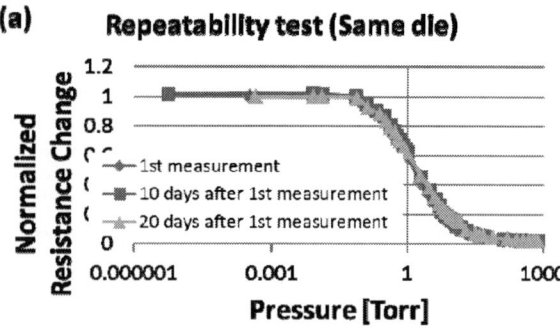

- 1st measurement
- 10 days after 1st measurement
- 20 days after 1st measurement

(b) Wafer-to-wafer variation (Same lot)

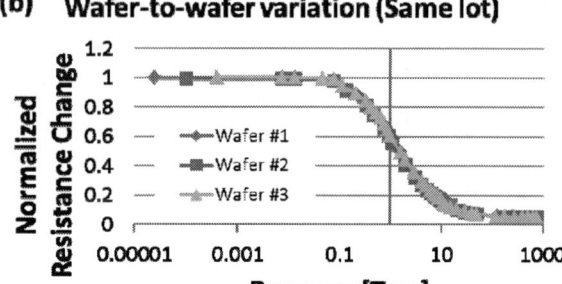

- Wafer #1
- Wafer #2
- Wafer #3

(c) Lot-to-lot variation

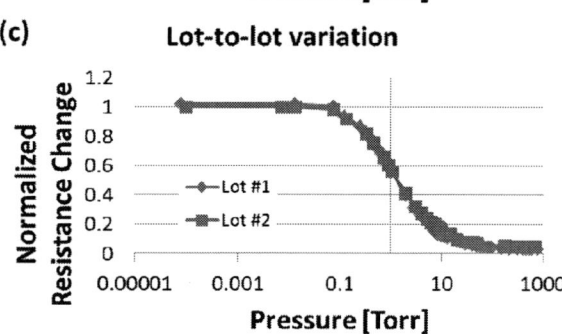

- Lot #1
- Lot #2

Figure 6: Stability test. (a) Repeatability test. (b) Wafer-to-wafer variation. (c) Lot-to-lot variation.

(a)

			38.99	38.31	33.21			
		29.02	22.05	18.39	15.91	21.70	35.98	
	28.99	18.61	14.70	14.36	14.47	13.89	21.55	35.03
36.57	23.24	15.52	18.89	43.30	18.85	12.21	14.91	26.05
	18.38	16.52	19.66	26.34	760.00	13.62	14.23	24.67
35.80	20.06	14.58	18.48	760.00	17.21	13.65	18.67	29.23
	34.57	20.92	16.21	760.00	16.12	18.05	26.20	
		33.60	24.92	21.10	22.97	28.06		

(b)

			31.06	30.64	26.45			
		22.21	15.81	14.03	14.06	19.01	27.93	
	21.41	13.27	11.70	41.56	11.64	12.27	16.79	28.17
27.83	16.50	12.49	13.79	281.81	111.25	204.87	13.70	20.97
	15.63	204.59	286.94	125.58	760.00	53.51	13.00	20.68
27.89	17.87	13.35	760.00	760.00	760.00	210.59	15.27	23.45
	35.31	17.67	14.36	760.00	13.58	14.54	19.97	
		27.20	20.00	17.43	17.64	22.30		

Figure 7: Implementation for wafer-level uHAST test (T=130°C, RH=85%, Time=96Hrs) (a) before and (b) after reliability tests. Package degradation in wafer center was detected. Unit in Figure 7 is Torr.

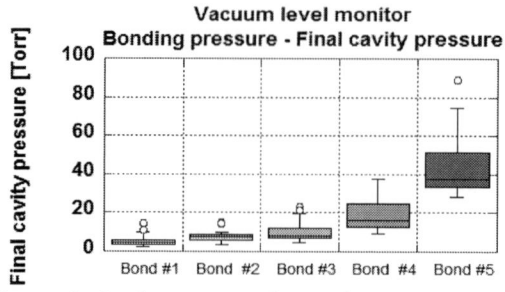

Figure 8: Implementation for package recipe fine tuning and selecting.

CONCLUSION

The idea of using differential Pirani gauge scheme for reducing the pressure measurement uncertainties was presented. The gauge was fabricated in tsmc CMOS-MEMS platform with small gaps (~2μm) between heater and dual heat sinks to obtain wide operation range (0.05~100 Torr) and high sensitivity (~10000 ppm/Torr). Experimental results proved the deviation was reduced to less than 5 % and the presented device had high stabilities. It was successfully implemented in the production line for monitoring the seal quality of MEMS wafer-level package process.

REFERENCES

[1] Chia-Fang Chiang, Andrew B. Graham, Brian J. Lee, Chae Hyuck Ahn, Eldwin J. Ng, Gary J. O'Brien, and Thomas W. Kenny, "Resonant pressure sensor with on-chip temperature and strain sensors for error correction", *in MEMS'13*, Taipei, Taiwan, 2013, pp. 45-48.

[2] Junseok Chae, Brain H. Stark, and Khalil Najafi, "A micromachined Pirani gauge with dual heat sinks," *IEEE Trans. Adv. Packag.*, vol. 28, pp. 619-625, 2005.

[3] Guillaume Schelcher, Filippo Fabbri, Elie Lefeuvre, Sebastien Brault, Philippe Coste, Elisabeth Dufour-Gergam, and Fabien Parrain, "Modeling and characterization of microPirani vacuum gauges manufactured by a low-temperature film transfer process," *J. Microelectromech. Syst.*, vol. 20, pp. 1184-1191, 2011.

[4] CM Liu, Bruce C.S. Chou, Robert Chin-Fu Tsai, Nick Y.M. Shen, Benior S.F. Chen, Emerson C.W. Cheng, Hsiao Chin Tuan, Alex Kalnitsky, Sean Cheng, Chung-Hsien Lin, Tien-Kan Chung, Kuei-Sung Chang, Yi-Shao Liu, "MEMS technology development and manufacturing in a CMOS foundry," *in Transducers'11*, Beijing, China, 2011, pp. 807-810.

CONTACT

*Y.C. Chen, tel: +886-3-5636688 Ext. 7076760; ychenv@tsmc.com

MICROFABRICATED ELECTROSTATIC PLANAR LENS ARRAY AND EXTRACTORS FOR MULTI-FOCUSED ION BEAM SYSTEM USING IONIC LIQUID ION SOURCE EMITTER ARRAY

Ryo Yoshida, Motoaki Hara, Hiroyuki Oguchi, and Hiroki Kuwano
Tohoku University, Sendai, Japan

ABSTRACT

This paper describes development of the electrostatic planar extractors and lenses integrated with the ionic liquid ion source (ILIS) array to realize multi focused ion beam system. The extractors and the lenses reduced ion beam divergence angle from 11° to 7.9°. Also, switching control of ion beam was demonstrated by operating separated extractors. Finally we performed Si substrate etching using various kinds of ionic liquid. As a result, 3.3 times larger sputtering yield of 7.3 atom/ion at 83% lower acceleration voltage than those of conventional Ga$^+$ focused ion beam (FIB) was achieved.

INTRODUCTION

Owing to today's market requirements, fabrication technologies of micro devices are gradually moving toward high-mix low-volume (HMLV) manufacturing from conventional low-mix high-volume (LMHV) one based on photolithography. FIB is promising for the HMLV manufacturing since it can offer various processes such as etching, deposition, imaging, and surface modification by controlling beam intensity and surrounding gas species [1]. However, throughput is extremely low.

Multiple beam processes using electron, ion, or X-ray is a key technology to improve the throughput, and was reported from many researchers [2-3]. So, we have proposed a micro-multi-FIB system [3]. Figure 1 shows a concept of that system. Emitters, extractors, and lenses are integrated to construct a multiple array. Ion beams can be emitted from each emitter concurrently or separately by on and off the extractor and the lens. These features enable us to obtain the various processes with high throughput.

In the prior works, ionic liquid ion source (ILIS) array exploiting ionic liquid (IL) of 1-ethyl-3-methylimidazolium tetrafluoroborate (EMIM-BF$_4$) was fabricated. Then, concurrent beam emission from the each emitter was confirmed, and reactive-ion-etching (RIE) of Si substrate was observed by using a quadrupole mass spectrometer (QMS) [3-4]. However, extractors and lenses were not discussed.

In this study, the electrostatic extractor and lens array was fabricated using deep RIE, and integrated with the ILIS array on wafer level. Performance of the device was evaluated experimentally. In addition, three kinds of ILs were applied to ion source and each etching characteristics was compared by microscope observation and mass spectrometry.

FABRICATION

An ILIS array consisting of micro needles and a reservoir for IL was fabricated by silicon bulk micromachining with single photo mask [3]. Then, IL of

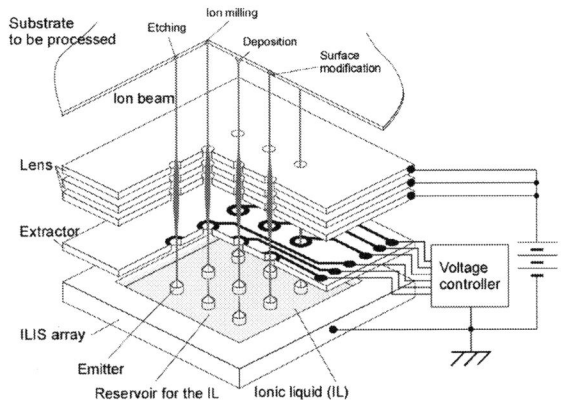

Figure 1: Concept image of the micro-multi-FIB system

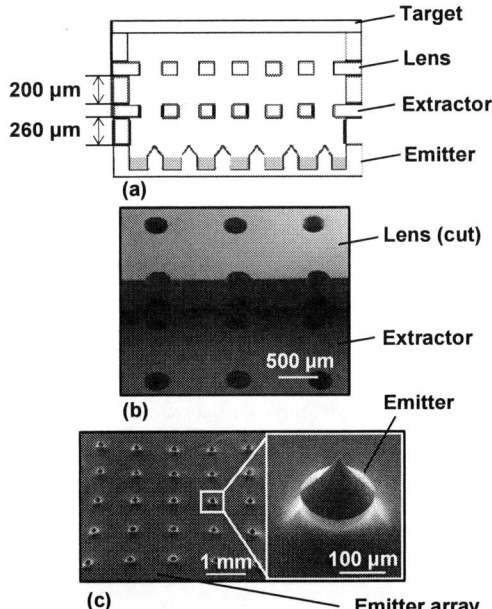

Figure 2: SEM images of a multi-FIB device: (a) cross sectional illustration of the device, (b) electrostatic lens and extractor array, and (c) ILIS emitter array

EMIM-BF$_4$ was poured into the reservoir by a micropipette. Arrays of extractors and lenses were fabricated by making through holes into Si substrates using deep RIE. Finally the fabricated arrays of ILIS, extractor, and lens were integrated by wafer-level-assembly. Figure 2 shows scanning electron microscopy (SEM) images of the fabricated device.

The diameter of the emitter was designed based on previous work and set to 200 μm [3]. The distance between each emitter was determined to 1 mm from finite element

978-1-4799-7956-1/15 $31.00 © 2015 IEEE

method (FEM) simulation (COMSOL). Figure 3 shows a contour image of the electrostatic field. Lens effect in each column was confirmed. Performance of the fabricated device was evaluated in a vacuum chamber with background pressure of less than 1×10^{-4} Pa.

ION BEAM FOCUSING

When emitting the ion beam toward a target, etched dimples as shown in the inset of Fig. 4 were formed. By plotting the diameter of these dimples as a parameter of the distance between the lens and the target, focusing effect of the lens can be evaluated. Figure 4 shows the relationship between the diameter and the distance when changing voltage between the extractor and the lens electrode V_{el}. In this evaluation, the voltage between the emitter and the extractor V_{ee} was fixed to 2.5 kV. As indicated by Fig. 4, beam diameter was 30% reduced by applying V_{el} of 5 kV, verifying beam-focusing ability of the device.

The emission ion current was observed to evaluate the beam stability as shown in Fig. 5. In this evaluation, V_{ee} was fixed to 2.2 kV. When V_{el} was 0 kV and 3 kV, average ion current was 550 nA and 356 nA, respectively. Also, deviation of the current was 40 nA and 27 nA with $V_{el} = 0$ kV and 3 kV. The possible reason of the current decrease is as follows: When the ions collided with the target (Si), secondary electrons were generated. The higher V_{el} generated stronger field to get out of the secondary electrons from the target and decreased the net amount of charge on the target. As a result the V_{el} decreased the current as shown in Fig. 5.

ION BEAM SWITCHING

Individual control of the emitter is necessary in the multi-FIB system. So, we newly fabricated an extractor plate as shown in Fig. 6. In this plate, each extractor was separated electrically, and integrated with an electrostatic lens. It was fabricated from silicon-on-insulator (SOI) wafer by silicon bulk micromachining. Figure 6 (b) shows a SEM image of the fabricated extractors.

Operation of the extractor array was demonstrated by applying voltage between the emitter and extractor arrays (V_{ee}). Figure 7 shows switching characteristics of ion current when V_{ee} was manually changed from 1.7 kV to 2.0 kV. Ion beam control by the extractor array was confirmed from this result.

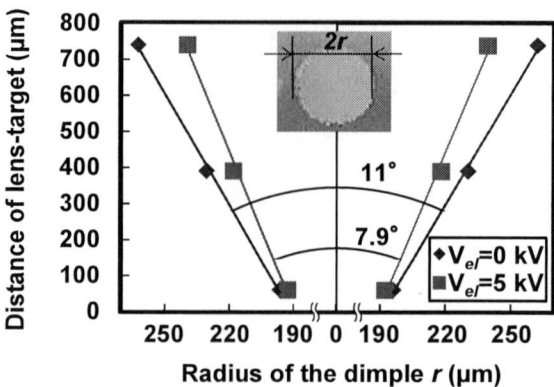

Figure 4: A dimple on the target formed by ion beam irradiation and relationship between distance from the planar lens array to the target and radius of the dimples

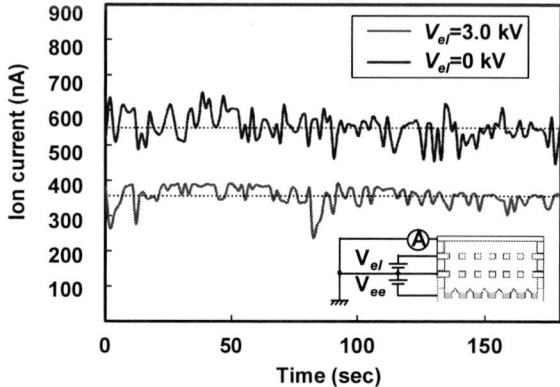

Figure 5: Beam emission characteristics of the fabricated device

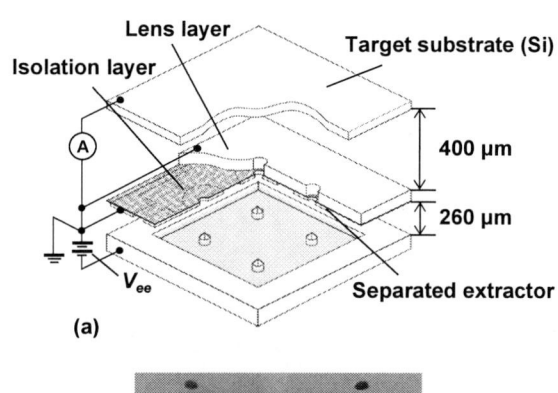

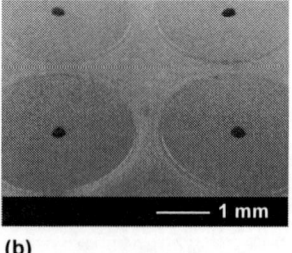

Figure 6: (a) Schematic illustration of the ILIS array with new extractor array, and (b) SEM image of the fabricated extractor

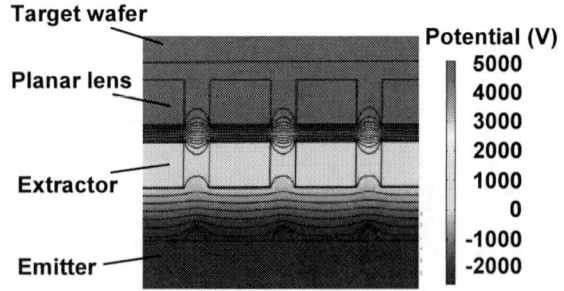

Figure 3: Contour image of the electrostatic field simulated by FEM analysis

978-1-4799-7956-1/15 $31.00 © 2015 IEEE   94

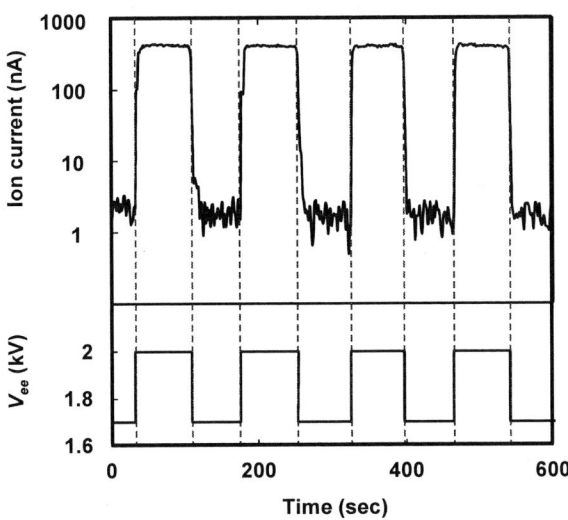

Figure 7: Ion current when 1.7 kV and 2.0 kV was applied to V_{ee} alternately

COMPARISON OF IONIC LIQUIDS

To establish the selection rule of the ILs for FIB, three kinds of ILs, namely EMIM-BF$_4$, 1-butyl-3-methylimidazolium hexafluorophosphate (BMIM-PF$_6$), and 1-Ethyl-3-methylimidazolium bis(trifluoromethylsulfonyl)imide (EMIM-TFSI), were compared with each other. For this experiment, the ILs containing fluorine ions were chosen among various ILs for the reactive etching of Si substrates.

Reactions between Si substrates and emitted ions were monitored during the beam emission by the QMS attached to the vacuum chamber. The extractor and the lens arrays were dismounted from the experimental setup to monitor the interaction clearly. In addition, prior to the measurements, the chamber was pumped down to the background pressure of from 1 to 2 ×10^{-5} Pa to suppress influence of residual gases. Figure 8 (a) shows the obtained mass spectra. Intensity of these spectra was normalized by the ion current because peaks of the spectra are proportional to ion current as shown in Fig. 8 (b). Peaks of SiF$^+$, SiF$_2^+$, SiF$_3^+$, and SiF$_4^+$ can be observed in Fig. 8 (a). These peaks indicate the reaction between fluorine atoms contained in the emitted ions and the Si substrate [5]. By comparing intensity of the peaks, we found that BMIM-PF$_6$ had the highest reactivity in those ILs. One of the reasons of this may be higher density of fluorine ions provided by BMIM-PF$_6$ than by other ILs (3.08×10^{-2} Å^{-3}, 2.33×10^{-2} Å^{-3}, 1.69×10^{-2} Å^{-3} for PF$_6^-$, BF$_4^-$, and TFSI$^-$ [6]).

Finally sputtering yield, that is the number of the sputtered atoms per emitted ion, was calculated from volume of the etched dimples and gross ion current [7]. Assuming that emitted ions have valence of -1, sputtering yield was calculated to be 7.3 atom/ion for BMIM-PF$_6$ when the acceleration voltage was 5 kV. This value was 3.3 times larger than that of conventional Ga$^+$ ion beam even though the applied voltage was 83% lower [8].

CONCLUSION

This study reported development of planar

electrostatic extractors and lenses integrated with the ionic liquid ion source (ILIS) emitter array. In the fabricated device, divergence angle of ion beam can be controlled by applying voltage to the integrated lens. As an experimental result, when the voltage between the extractor and the lens V_{el} was 5.0 kV, about 30% reduction of beam divergence angle was achieved. Also, an extractor plate, in which extractors were separated electrically each other and integrated with electrostatic lens, was newly developed. Using this plate, switching of ion beam emission was confirmed. In addition, in order to find suitable ionic liquid (IL) for our device, three kinds of ILs, which were 1-ethyl-3-methylimidazolium tetrafluoroborate (EMIM-BF$_4$), 1-butyl-3-methylimidazolium hexafluorophosphate (BMIM-PF$_6$), and 1-Ethyl-3-methylimidazolium bis(trifluoromethylsulfonyl)imide (EMIM-TFSI), were applied to ion source. Reactivity of these ions was compared each other by a quadrupole mass spectrometer (QMS). Consequently, BMIM-PF$_6$ indicated the highest reactivity for Si. Sputtering yield was calculated from volume of the etched dimples and gross ion current. It was 7.3 atom/ion when BMIM-PF$_6$ was used. This value was corresponding to 3.3 times larger than that of conventional Ga$^+$ ion beam although the applied voltage was 83% lower.

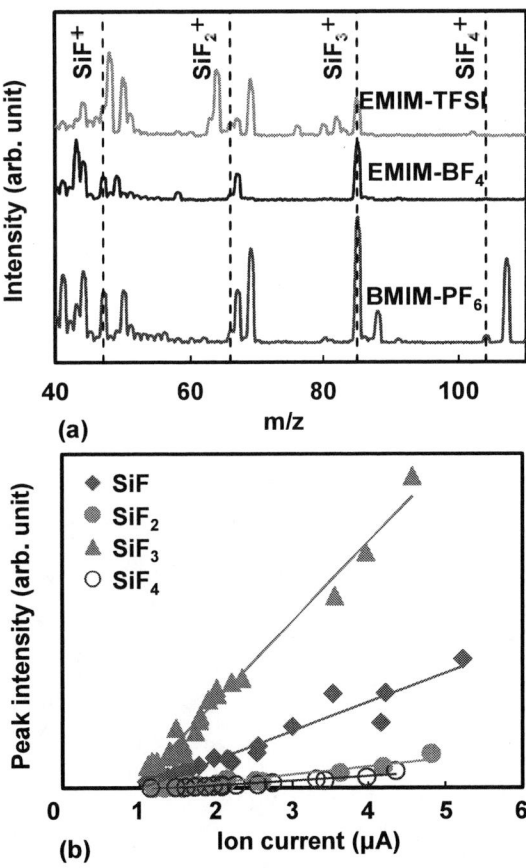

Figure 8: (a) Mass spectra of surrounding gas during ion emission, and (b) Relationship between peak intensity of mass spectra and ion current when BMIM-PF$_6$ was adopted as ion source

ACKNOWLEDGEMENTS

A part of this work was performed at micro/nano-machining research and education center, Tohoku University, Japan. This work was supported in part by the Challenging Exploratory Research from Ministry of Education, Culture, Sports, Science and Technology of Japan (MEXT) (Grant No. 12010289).

REFERENCES

[1] J. Gierak, "Focused Ion Beam Technology and Ultimate Applications", *Semicond. Sci Technol.*, Vol. 24, 043001, 2009.

[2] T. H. P. Chang, M. Mankos, K. Y. Lee, L. P. Muray, "Multiple Electron-Beam Lithography", *Microelectron. Eng.*, Vol. 57-58, pp. 117-135, 2001.

[3] T. Suzuki, M. Hara, H. Oguchi, H. Kuwano, "Arrayed Micro Ion Source with Ionic Liquid for Flexible and Concurrent MEMS Fabrication", *Sens. Actuators A: Phys.*, Vol. 215, pp. 161-166, 2014.

[4] R. Yoshida, M. Hara, H. Oguchi, T. Suzuki, H. Kuwano, "Concurrent Reactive Ion Etching Employing Micromachined Ionic Liquid Ion Source Array", *In Proc. MEMS2014*, San Francisco, Jan 26-30, 2014, pp. 463-466.

[5] H. F. Winters, I. C. Plumb, "Etching Reactions for Silicon with F Atoms: Product Distributions and Ion Enhancement Mechanism, *J. Vac. Sci. Technol. B*, Vol. 9, pp. 197-207, 1991.

[6] S. Zhang, N. Sun, X. He, X. Lu, X. Zhang, "Physical Properties of Ionic Liquids: Database and Evaluation", *J. Phys. Chem. Ref. Data*, Vol. 35, pp. 1475-1517, 2006.

[7] H. Yamaguchi, "Line Dose Dependence of Silicon and Gallium Arsenide Removal by A Focused Gallium Ion Beam", *J. Phys. Colloques*, Vol. 48, pp. 165-170, 1987

[8] J. Orloff, L. W. Swanson, M. Utlaut, "Fundamental Limits to Imaging Resolution for Focused Ion Beams", *Vac. Sci. Technol. B*, Vol. 14, pp. 3759-3763, 1996.

CONTACT

R. Yoshida, tel: +81-22-795-4771; yoshida@nanosys.mech.tohoku.ac.jp

NOVEL IONIC LIQUID – POLYMER COMPOSITE AND AN APPROACH FOR ITS PATTERNING BY CONVENTIONAL PHOTOLITHOGRAPHY

Natalia A. Bakhtina[1], Anja Voigt[2], Neil MacKinnon[1], Gisela Ahrens[2], Gabi Gruetzner[2], and Jan G. Korvink[1]

[1]IMTEK – Department of Microsystems Engineering, University of Freiburg, Germany
[2]micro resist technology GmbH, Berlin, Germany

ABSTRACT

A novel crosslinkable, conductive, highly transparent composite material based on a photoresist and an ionic liquid (the names of the composites are not announced here due to the current procedure of patenting) is presented. The composite possesses a good and stable ionic conductivity (up to 10 mS cm^{-1} at room temperature) over a wide frequency bandwidth (1 kHz – 1 MHz) and is optically transparent (transmission value of 90 % for a 170 µm thick film). In addition, an approach for the patterning of the composite material by conventional photolithography with a good spatial resolution (line width of 20 – 30 µm) is introduced. The unique properties of the material are utilized for time- and cost-saving direct manufacturing of electrically conductive, highly transparent microcomponents.

INTRODUCTION

Optically transparent, conductive materials have a wide range of applications. They find use in sensors, solar cells, displays, and other electronic components [1]. Commonly used materials for such applications are Transparent Conductive Oxides (TCOs), such as Indium Tin Oxide (ITO), Intrinsically Conductive Polymers (ICPs), such as poly(3,4-ethylenedioxythiophene) (PEDOT), and Electrically Conductive Polymer Composites (ECPCs).

Photoresists have a broad range of applications. The first main application of photoresists is their usage as resist masks or sacrificial layers for the generation of electrically conductive patterns in advanced semiconductor and micro-electrical-mechanical system (MEMS) devices via a pattern transfer process, i.e. lithography [2]. The second main application of photoresists, especially of negative-acting materials, is the direct manufacturing of permanent patterns, which can be used as a device material for the fabrication of precisely patterned and mechanically stable micro- and nano-structures in microfluidic systems [3], waveguides [4], and stamps [5]. There are a variety of crosslinkable materials for permanent applications available [6]. For some applications it is advantageous to have additional properties, like conductivity, already in the direct patternable materials to reduce process steps.

There are currently several known ECPC compositions in which a photoresist (an electronic insulator in its pure state) is mixed with various conductive filler particles in order to significantly increase the conductivity of the polymerized material. The examples of conductive fillers are terthiophene (3T) with copper (II) perchlorate [7], silver nanoparticles [8], graphite [9] or carbon black particles [10], protonically doped polyaniline (PANI) nanoparticles [11], and others [12,13]. However, these materials have several limitations: (1) the addition of filler particles requires control of material viscosity by addition of various solvents, influencing the photo-polymerization process; (2) film deposition on wafers when using microparticles may have a great effect on the surface morphology and, as a result, may lead to inhomogeneous conductive layers; (3) the resolution is constrained to 10 – 30 µm because of the light diffraction by filler particles [7, 8, 9]; (4) and finally, the cured material has low optical transparency over the visible range.

Ionic liquids (ILs) have appeared in recent years as novel compounds in materials research, and are already used in industrial processes due to several attractive characteristics [14,15]. ILs feature high ionic conductivity (up to 30 mS cm^{-1}), good solubility, low volatility, low flammability, and high thermal, chemical, and electrochemical stability. ILs have the immense advantage over ECPCs of being transparent and easy to produce, with a wide variety of anion/cation combinations which can be adjusted to tailor their physical properties. ILs seem to be promising materials for applications not only in flexible electronic devices, such as displays and photovoltaics [16], but also in electrochemical biosensors because of their good compatibility with biomolecules and enzymes, and even whole cells [17,18].

While ILs are conductive, the liquid nature of ionic liquids is an obstacle to applications where a predefined physical shape is required. For these applications, a process of solidification is necessary which constrains the liquid. It comprises the formation of a three-dimensional solid structure in the form of a polymer matrix, which entraps the ionic liquid in the porous network. In this material, liquid-like properties (e.g. charge transport) originate from the ionic liquid, whereas solid-like properties originate from the host polymer, which contributes to material flexibility while preventing the system from flowing.

Accordingly, this paper reports a new polymeric material based on a negative-acting photoresist (A, B, or C) and the ionic liquid (the names of the composites are not announced here due to the current procedure of patenting). The IL-polymer composites are ionically conductive (up to 10 mS cm^{-1} at room temperature) over a wide frequency bandwidth (1 kHz – 1 MHz) and are optically transparent (transmission value of 90 % for a 170 µm thick cast film). Conventional photolithography enables a single-step process for structuring of the material with a good spatial resolution (line width of 20 – 30 µm). Combining the advantages of both the material and the fabrication technique, time- and cost-saving direct manufacturing of electrically conductive microcomponents is demonstrated.

EXPERIMENTAL PART

Materials

Materials and preparation. The ionic liquid was purchased from IoLiTec, Ionic Liquids Technologies GmbH, Germany. A (Nanoscribe GmbH, Germany), B (micro resist technology GmbH, Germany), and C (MicroChem Corp., USA) liquid photoresists were used as purchased without further purification. The preparation of the developed composition was done by manual mixing in a glass vessel. The ionic liquid was added dropwise to the photoresist (A, B, or C). The mixture of components was stirred for a few minutes at room temperature under nitrogen to ensure a homogenous solution.

Methods

^{1}H NMR measurements. ^{1}H NMR spectra of the various components were measured on a Bruker AVANCE III 500 MHz spectrometer (Bruker BioSpin GmbH, Germany) operating at a frequency of 500.12 MHz. For each sample 8 averages each containing 16 K data points were collected over a spectral window of 12 ppm. Photoresist (A, B, or C) was diluted in $CDCl_3$ while IL was measured neat. The composite material (A, B, or C) in a 5 mm NMR tube was evacuated over night at 0.1 mbar to remove air bubbles. The samples was measured pre- and post-UV exposure (curing parameters: 365 nm, 1 minute interval, 2 – 5 cycles).

Spectral absorbance and optical transparency. For optical absorbance measurements, a droplet of the composite material was placed between two quartz glass slides (170 μm thick) and closed from the top. For optical transmission measurements, a 170 μm thick cast film was fabricated the same way and cured under UV-light (365 nm) for 2 min. For material C a step of soft baking at 80 °C for 1 hour took place before exposure and a step of post-exposure baking at 95 °C for 2 hours was required after exposure. The measurements were performed using a Cary 50 UV-vis spectrometer (Varian Inc., USA).

Electrochemical characterization. For the measurements of the ionic conductivity complex impedance spectroscopy was realized using an impedance/gain-phase analyzer (Solartron SI 1260 from Solartron Analytical Ltd., UK) and an electrochemical interface analyzer (Solartron Analytical SI 1287 from Solartron Analytical Ltd., UK). For the determination of the ionic conductivity of the bulk material, cast cylinders were prepared in UV-transparent PMMA pipe sections (outer diameter 16 mm, inner diameter 12 mm, distance between electrodes 5 mm), and fixed between two stainless steel electrodes. The UV curing was performed by irradiating the sample for different periods of time (3 – 11 minutes) at a wavelength of 365 nm through the UV-transparent sidewall. For material C a step of soft baking at 80 °C for 1 hour before exposure and a step of post-exposure baking at 95 °C for 2 hours took place before the measurements. A sinusoidal signal with an amplitude 10 mV was applied over the frequency range from 1 Hz to 1 MHz in normal atmosphere, and the impedance spectrum was recorded.

Conventional photolithography. A layer of material B was prepared by spin-coating, soft baked at 80 °C for 2 min and patternwise exposed with an exposure dose of

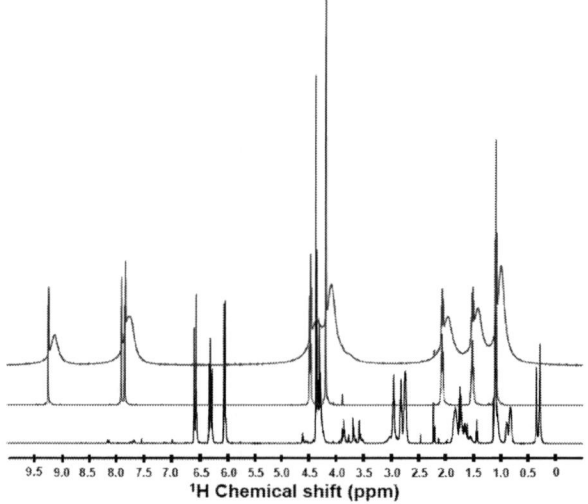

Figure 1: ^{1}H NMR of the cured composite material B (blue solid line). The spectra of the pristine IL (brown solid line) and 100 μL pristine uncured photoresist B in 500 μL chloroform (black solid line) are shown for comparison.

1000 mJ/cm^2 (MA6 Mask Aligner from Karl Süss, Germany). After application of a post-exposure bake at 130 °C for 10 min, a development for 20 min in a developer was followed by drying with compressed nitrogen.

RESULTS AND DISCUSSION

Identification of a transparent, conductive ionic liquid – polymer composite

It was first necessary to identify physically compatible combinations of an ionic liquid and a suitable negative-acting photoresist A, B, or C. The IL – polymer composite must be a homogeneous mixture that exhibits no sedimentation, crystallization, or phase separation before and after polymerization. Polymerized material was prepared by exposing a droplet (~ 10 μL) of the liquid mixture deposited on a glass slide to UV-light (365 nm) for different periods of time. Acceptable mechanical stability was achieved when the mixture of photoresist/IL was in the ratio 5:1. These samples were selected for further characterization.

For a better understanding of the IL – polymer network interactions, ^{1}H NMR measurements were done on the pure IL, pristine cured and uncured photoresists, and of the cured and uncured composite materials. The spectra suggest the absence of chemical reactivity of the ionic liquid components with the photoresist in materials A and C before and after the curing process. However, for material B, two populations of IL are observed after curing (Figure 1). The broad resonances suggests one population of cations is interacting with the cured resist (most likely physically, but potentially chemically), while the narrow resonances indicate a second population of cations that is free to diffuse.

Optical and electrical characterization

In the second step the properties (spectral absorbance,

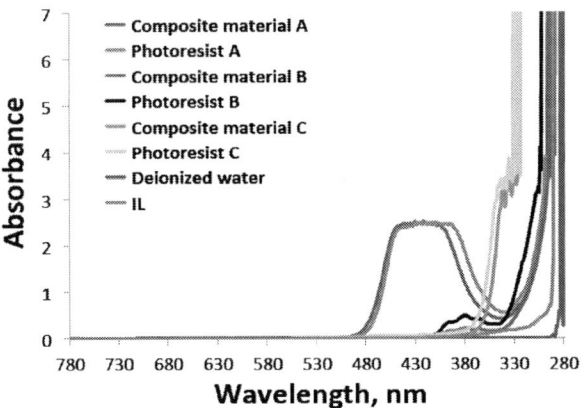

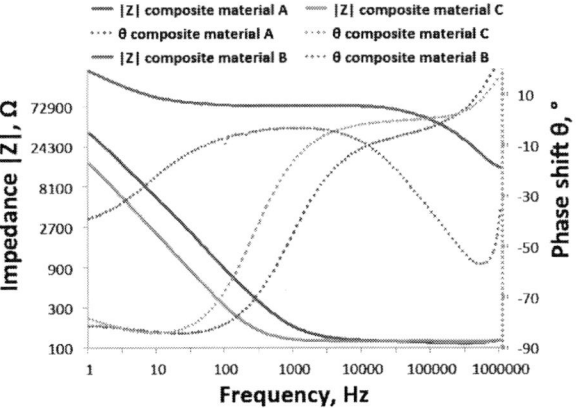

Figure 2: Spectral absorbance of the 170 µm thick layer of uncured composite material A (red solid line), composite material B (blue solid line), and composite material C (green solid line). The spectra for pure photoresists (A, B, and C), deionized water, and ionic liquid (IL) are shown for comparison.

Figure 4: Impedance magnitude |Z| of the composite material A (red solid line), composite materia B (blue solid line), and composite material C (green solid line) was recorded at the voltage amplitude V = 10 mV in normal atmosphere. The corresponding phase angles θ are shown in red, blue, and green dashed lines.

properties of the composite materials.

Secondly, we examined the optical transmission properties of the cured composite materials. All spectra showed high transmission values over nearly the entire visible range between 380 nm and 780 nm (red, blue, and green solid lines in Figure 3).

Finally, ionic conductivity values of the composite materials were determined from complex impedance measurements (Figure 4), in which a sinusoidal potential wave of amplitude 10 mV of varying frequency was applied to the tested sample. The impedance spectrum exhibited its maximum value at the lowest frequency and showed an impedance decrease towards higher frequencies until a plateau of the lowest impedance was reached at approximately 1 kHz. Low impedance values were measured for the composite materials A and C, while for type B higher values of the impedance were identified if compared to materials A and C (blue solid line in Figure 4). After the measurements the ionic conductivity (σ) was calculated according to Equation 1:

$$\sigma = \frac{1}{R} \cdot \frac{l}{h \cdot w} \qquad (1)$$

where R is an electrical resistance (Ω), l is a length (m), h is a height (m), and w is a width (m) of sample. If the phase angle θ (period difference between current and voltage) takes the value of 0 ° in Equation 2, the resistance of the electrical impedance is equivalent to the impedance magnitude itself:

$$R = |Z| \cdot \cos\theta \qquad (2)$$

The calculated conductivity of the composite materials, as well as other material characteristics, are summarized in Table 1.

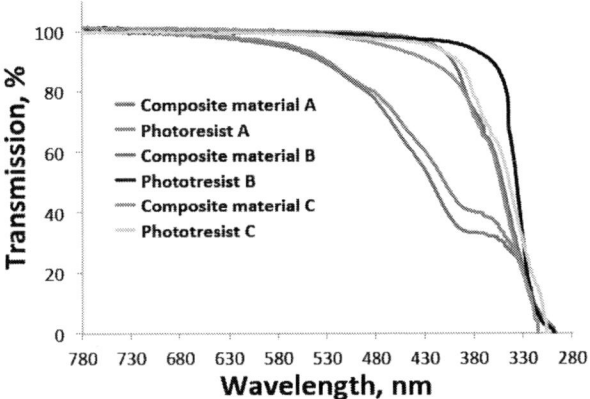

Figure 3: Optical transmission of the composite material A (red solid line), composite material B (blue solid line), and composite material C (green solid line). The 170 µm thick cast film was fabricated between two quartz glass slides, and cured by exposure to UV-light (365 nm). The spectra for pure photoresists (A, B, and C) are shown for comparison.

optical transparency, and electrical characteristics) of the composite materials were investigated. To garantee the full crosslinking of the material, the prefered photoresist/IL ratio in the mixtures was 5:1.

Optical properties of the material define the ability to transfer or absorb/scatter the light of a certain wavelength. Firstly, spectral sensitivity of the compositions to UV curing was examined by optical spectroscopy. While the absorbance spectrum of uncured material A (red solid line, Figure 2) showed high absorbance values (up to 3) between 400 nm and 450 nm, material B and C (blue and green solid lines, Figure 2) demonstrated a significant absorbance increase for wavelengths below 360 nm. A comparison of all three composite materials to the correspondent pure photoresists' absorbance showed that the addition of IL had no significant influence (i.e. only a slight shift towards shorter wavelengths) on the optical

Material patterning by conventional photolithography

Lithographic capabilities of the material facilitate a time- and cost-saving direct manufacturing of transparent, electrically conductive microcomponents by conventional photolithography. Figure 5 shows an example of the patterned material B (photoresist/IL is in the ratio 5:1)

978-1-4799-7956-1/15 $31.00 © 2015 IEEE

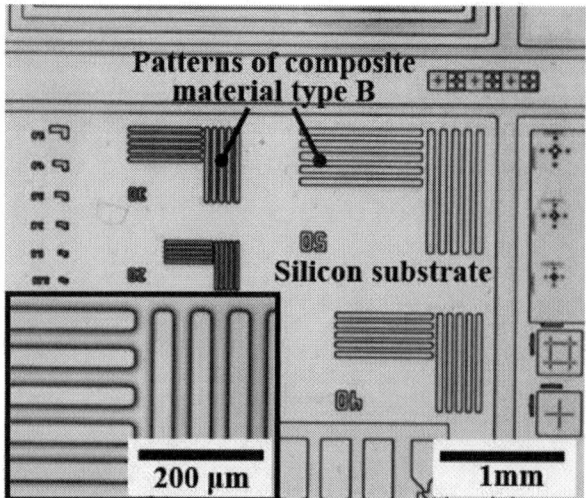

Figure 5: Photograph of the composite material B (15 μm height) patterned on a silicon wafer by conventional photolithography.

with the highest spatial resolution achieved to date.

Relatevely small droplets with diameters of 100 μm could be observed on the surface of the patterns of the material B one day after preparation. The shrinking process can cause internal stress in the IL-polymer matrix which might induce the release of IL at the material surface [19]. Consequently, the released ionic liquid might involve higher values of the impedance if compared to other composite materials (Figure 4). No droplets were observed on the surface of materials A and C.

CONCLUSIONS AND OUTLOOK

Three novel crosslinkable, conductive, highly transparent materials based on ionic liquid and photoresist have been identified that address the technical problems of current conductive photolithographic compositions. In contrast to commonly used materials consisting of crosslinkable materials and filler particles, composite materials of types A, B, and C have several advantages: (1) almost no effort with respect to optimizing the material handling has to be performed (e.g. no control of ambient conditions and no special pre- and post-processing), in particular, for the material A and B no care has to be taken to ensure evaporation of solvents before and after polymerization, which would have an effect on mechanical strength and electrochemical stability; (2) during deposition there is no need to control viscosity and sedimentation of filler particles; accordingly, uniform film deposition can be performed on quartz wafers by spin-coating and dropcasting; (3) material structuring with high resolution is possible due to homogeneity and high clarity of the mixture; (4) and finally, cured material has good optical transparency over the visible range. Moreover, the composite materials exhibit a good ionic conductivity [19]. They do not contain any plasticizers that are often used to obtain high ionic conductivity (up to 10^{-4} S cm^{-1}) [20]. The compositions also feature a wide range of material properties (i.e. tunable): the weight ratio of ionic liquid to the photoresist can be easily adjusted to tailor physical and chemical properties of the composition (up to 90 wt% of IL in the photoresist while remaining completely compatible).

The suitability of the materials for conventional photolithography has been proven. These achievements can be regarded as highly promising for a wide range of applications. Future work will focus on the further optimization of the fabrication parameters in order to address the observed challenges and on high resolution patterning of the compositions with the desired material properties by two-photon polymerization (2-PP) technique and nanoimprinting lithography.

ACKNOWLEDGEMENTS

We gratefully acknowledge financial support from the European Research Council (ERC) (contract number 290586), which funded this work. We also express our gratitude to the Laboratory for Biomedical

Table 1. Characteristics of the composite materials

Characteristics	Composite material		
	Type A	Type B	Type C
Solvent-free	Yes	Yes	No
Percentage of IL by weight	10 – 50 wt.%	10 – 50 wt.%	10 – 50 wt.%
Viscosity	0.1 – 0.5 Pa s	1 – 2 Pa s	0.05 – 2 Pa s
Film thickness upon spin coating at 1000 – 4000 rpm	5 – 40 μm	5 – 40 μm	3 – 40 μm
Spectral sensitivity for UV curing	300 – 480 nm	300 – 400 nm	300 – 370 nm
Optical transmission after curing at 400 nm	up to 60 %	up to 90 %	up to 85 %
Frequency window (f)	10 kHz – 1 MHz	100 Hz – 1 MHz	1 kHz – 1 MHz
Conductivity (σ)	1 – 10 mS cm^{-1}	0.001 – 0.01 mS cm^{-1}	1 – 10 mS cm^{-1}
Electrochemical window (EW)	up to 2.7 V	up to 2.7 V	up to 2.7 V

Microtechnology and the Laboratory for Chemistry and Physics of Interfaces (IMTEK, University of Freiburg), for providing access to equipment for impedance spectroscopy and UV-visible spectroscopy.

REFERENCES

[1] K. M. Ziadan, „Conducting Polymers Application" in *New Polymers for Special Applications* (Ed: A. de S. Gomes), Rijeka, Croatia 2012, Ch. 1.

[2] H. J. Levinson, *Principles of lithography*, Washington, USA, 2005.

[3] J. Zhang, K. L. Tan, G. D. Hong, L. J. Yang and H. Q. Gong, „Polymerization optimization of SU-8 photoresist and its applications in microfluidic systems and MEMS", *J. Micromech. Microeng.*, vol. 11, pp. 20–26, 2001.

[4] A. Borreman, S. Musa, A. A. M. Kok, M. B. J. Diemeer, and A. Driessen, „Fabrication of Polymeric Multimode Waveguides and Devices in SU-8 Photoresist Using Selective Polymerization", in *Proceeding of IEEE/LEOS Benelux Chapter*, Amsterdam, The Netherlands, 9 December, 2002, pp. 83–86.

[5] T. Koerner, L. Brown, R. Xie, and R. D. Oleschuk, „Epoxy resins as stamps for hot embossing of microstructures and microfluidic channels", *Sensor Actuat. B*, vol. 107, pp. 632–639, 2005.

[6] M. Töpper, T. Fischer, T. Baumgartner, and H. Reichl, „A comparison of thin film polymers for wafer level packaging", *Proceeding of IEEE/ECTC 2010*, Las Vegas, USA, 1–4 June, 2010, pp. 769–776.

[7] R. Abargues, P. J. Rodriguez-Canto, R. Garcia-Calzada, and J. Martinez-Pastor, „Patterning of Conducting Polymers Using UV Lithography: The in-Situ Polymerization Approach", *J. Phys. Chem. C*, vol. 116, pp. 17547–17553, 2012.

[8] S. Jiguet, A. Bertsch, H. Hofmann, and P. Renaud, „Conductive SU8 Photoresist for Microfabrication", *Adv. Funct. Mater.*, vol. 15, pp. 1511–1516, 2005.

[9] M. Benlarbi, L. J. Blum, C. A. Marquette, „SU-8-carbon composite as conductive photoresist for biochip applications", *Biosens. Bioelectron.*, vol. 38, pp. 220–225, 2012.

[10] N. Hauptman, M Zveglic, M. Macek, and M. K. Gunde, „Carbon based conductive photoresist", *J. Mater. Sci.*, vol. 44, pp. 4625–4632, 2009.

[11] U. M. Annaiyan, K. Kalantar-zadeh, Q. Fang, I. Cosic, "Development of a conductive photoresist with a mixture of SU-8 and HCL doped polyaniline," in *Proceeding of IEEE Tencon 2005*, Melbourne, Australia, 21–24 November, 2005, pp. 1B 07.2.

[12] GCM 3060 datasheet, hyperlink "http://www.gersteltec.ch/userfiles/1197841690.pdf", accessed: August, 2014.

[13] N. Damean, B. A. Parviz, J. N. Lee, T. Odom, G. M. Whitesides, „Composite ferromagnetic photoresist for the fabrication of microelectromechanical systems", *J. Micromech. Microeng.*, vol. 15, pp. 29–34, 2005.

[14] E. F. Borra, O. Seddiki, R. Angel, D. Eisenstein, P. Hickson, K. R. Seddon, and S. P. Worden, „Deposition of metal films on an ionic liquid as a basis for a lunar telescope", *Nature*, vol. 447, pp. 979–981, 2007.

[15] IoLiTec Ionic, Liquids Technologies GmbH, Germany, Ionic Liquids Today, hyperlink "http://www.iolitec.de/en/Download-document/665-Ionic_Liquids_Today_01-11.pdf.html", accessed: August, 2014.

[16] G. Scarpa, A. Idzko, A. Yadav, E. Martin, S. Thalhammer, "Toward Cheap Disposable Sensing Devices for Biological Assays", *IEEE Trans. Nanotechnol.*, vol. 9, pp. 527–532, 2010.

[17] J. Lewis, „Material challenge for flexible organic devices", *Mater. Today*, vol. 9, pp. 38–45, 2006.

[18] F. van Rantwijk and R. A. Sheldon, „Biocatalysis in ionic liquids", *Chem. Rev.*, vol. 107, pp. 2757–2785, 2007.

[19] U. Loeffelmann, N. Wang, D. Mager, P. J. Smith, and J. G. Korvink, „Solvent-Free Inkjet Printing Process for the Fabrication of Conductive, Transparent, and Flexible Ionic Liquid-Polymer Gel Structures", *J. Polym. Sci., Part B: Polym. Phys.*, vol. 1, pp. 38–46, 2012.

[20] C. A. Nguyen, S. Xiong, J. Ma, X. Lu, and P. S. Lee, "High ionic conductivity (PVDF-TrFE)/PEO blended polymer electrolytes for solid electrochromic devices", *Phys. Chem.*, vol. 13, pp. 13319–13326, 2011.

CONTACT

*N. A. Bakhtina, tel: +49 761 203 67477; natalia.bakhtina@imtek.uni-freiburg.de

TRIBOELECTRIFICATION BASED ACTIVE SENSOR FOR POLYMER DISTINGUISHING

B. Meng[1], X.L. Cheng[1], M.D. Han[1], H.T. Chen[1], F.Y. Zhu[1] and H.X. Zhang[1]

[1] Science and Technology on Micro/Nano Fabrication Lab, Institute of Microelectronics, Peking University, Beijing, CHINA

ABSTRACT

We present a novel sensor for polymer distinguishing among a group of known polymers based on the effects of triboelectrification and electrostatic induction. Multiple polymer-electrode cells are integrated on a flexible substrate, each cell produces an independent signal. The manufacture procedure of flexible printed circuit is employed to implement a low-cost and efficient fabrication of the device. According to the triboelectric serials, for different polymer groups, the friction layers can be well-selected. As an example, the distinguishing of polydimethylsiloxane, polyethylene and polyethylene terephthalate has been well demonstrated by employing polyimide and polystyrene as friction layers, showing potential applications in robotics and industrial fields.

INTRODUCTION

Triboelectrification (also called contact electrification) is a very traditional and familiar phenomenon. It has been 25 centuries since it was firstly observed by *Thales* using a piece of wool rubbing against an amber. Triboelectricfication occurs all around us but it's not always popular. Sometimes it even causes troubles and losses in our daily life, public affairs and industry manufactures. For centuries, it has been studied and applied in various fields [1-3].

Recently, using triboelectrification for energy harvesting was exploited [4-9]. Kinds of energy harvesting devices termed triboelectric generator were developed. The capturing of mechanical energy from human motions, ambient vibrations, and water/wind flow was investigated and applied to low-power devices [5-9], medical microsystems [7] and portable electronics [6]. These triboelectric generators employs simple structure design and facile fabrication procedure. It provides a much simpler approach for producing surface charge to realize electrostatic-based energy harvesting device beyond electret generator [10].

Our prior work has reported a so called STEG (single-friction-surface triboelectric generator), which incorporates a single fixed friction layer and a corresponding induction electrode [11-12], leading to a simplifying of the generator structure and an extending in the application range. Considering that the order of polymers in the triboelectric series is generally constant [3], such an open design of single-friction-surface can be also employed to produce active signal which reflects the property of two contacted surfaces as well. In this work, based on the single-friction-surface design, we developed an active sensor to distinguish different polymers from a group of known polymeric materials.

EXPERIMENTAL

Figure 1 outlines the schematic view of the polymer-distinguishing sensor. The device incorporates a three-layer structure, friction layer, electrode layer and substrate respectively. On a flexible polyimide (PI) substrate, multiple induction electrodes are integrated in a plane. Different polymer films are selected as the friction layers, and are located on top of the electrodes. Each induction electrode and the corresponding friction layer are patterned as an independent STEG cell. To reduce the interference among these cells, the induction electrodes were designed as sectors and located regularly within a circle of 10 mm in radius as the layout of a four-cell sensor shown in Figure 1b. Table 1 illustrates the thickness of each layer. This multiple-layer structure has a thickness of around 100 μm in total.

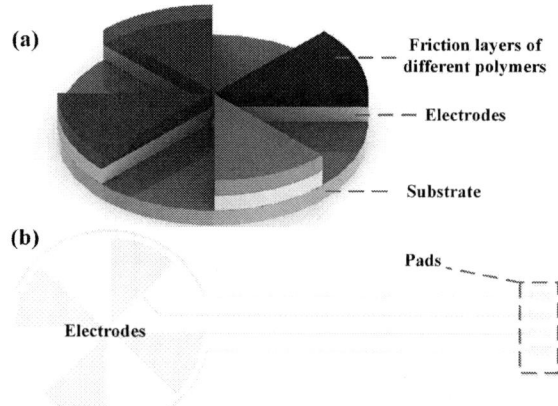

Figure 1: (a) Schematic of the polymer-distinguishing sensor with four STEG cells and (b) layout of the electrodes and pads.

Table 1: Parameters of the sensor.

Thickness of PI substrate	25 μm
Thickness of copper electrode	18 μm
Thickness of friction layer	50 μm

Figure 2 illustrates the fabrication process of the polymer-distinguishing sensor. A flexible-printed-circuit manufacture procedure was employed to realize the facile fabrication of the device. The fabrication started from a sheet of single-sided flexible copper-clad laminate, consisting of a PI substrate and a conductor copper film. The copper film was patterned by FeCl₃ etching after a lithography process. The patterned electrodes and pads were then coated with a thin film of gold using an electroless-nickel immersion-gold (ENIG) process, for a good connection with the external processing circuits. A PI thin film was then laminated on the electrodes as a package of the electrodes, preventing the electrodes from corrosion and short-connection with each other. The pads was exposed for signal read-out. Finally, the well-selected polymer films were tailored as the same with each electrode and laminated on the polyimide package film successively, acting as the friction layers. Each friction layer overlapped completely with the corresponding electrode.

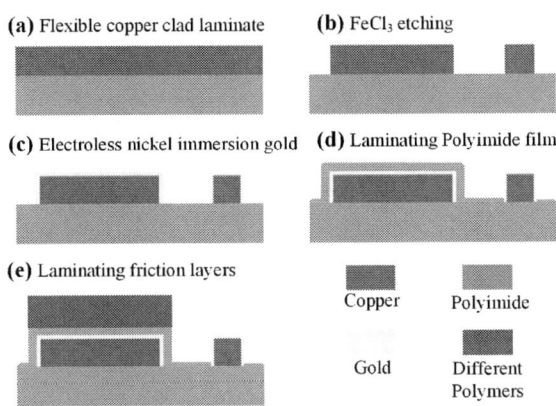

Figure 2: Fabrication process of the polymer-distinguishing sensor. (a) flexible copper-clad laminate, (b) FeCl₃ etching to pattern electrodes and pads, (c) electroless nickel and immersion gold on copper patterns, (d) laminating a PI thin film on the substrate with the pads exposed, (e) laminating different polymer films on the polyimide package.

Figure 3: Photographs of (a) a fabricated polymer distinguishing sensor with four electrodes and (b) a sensor fastened on a robotic arm.

Figure 3 shows the photographs of a fabricated polymer-distinguishing sensor with four STEG cells. Excellent flexibility and coin-sized dimension of the device makes it appropriate for multiple applications. As it's shown in Figure 3b, a polymer-distinguishing sensor was fastened on a robotic arm through an ethylene-vinyl-acetate (EVA) base.

RESULTS AND DISCUSSIONS

This polymer-distinguishing sensor works based on triboelectrification and electrostatic induction [11-14], as the schematic of working mechanism shown in Figure 4. When a single STEG cell is brought into contacting with a polymeric object, electrons will be transferred from one surface to another owing to contact electrification. On the condition that the polymeric material to be detected shows a tendency to attract electrons (Figure 4a) in triboelectrification, a negative voltage pulse followed by a positive one (which we define as a "0") will be observed in a single contact-separation cycle. For the case that the object to be detected donates electrons in triboelectrification (Figure 4b), the waveform of output voltage is a positive-negative one (which we define as a "1"). For a certain polymer, the value of output voltage may varies with the applied force, velocity and surface geometry, but the polarity will be only decided by their relative order in triboelectric serials. Thus, the polymers can be distinguished according to the output signals in turn with well-selected friction layers. Figure 5 illustrates the relative order of several polymers in triboelectric series. From negative to positive, the order is polytetrafluoroethylene (PTFE), polydimethylsiloxane (PDMS), PI, polyethylene (PE), polystyrene (PS) and polyethylene terephthalate (PET), respectively.

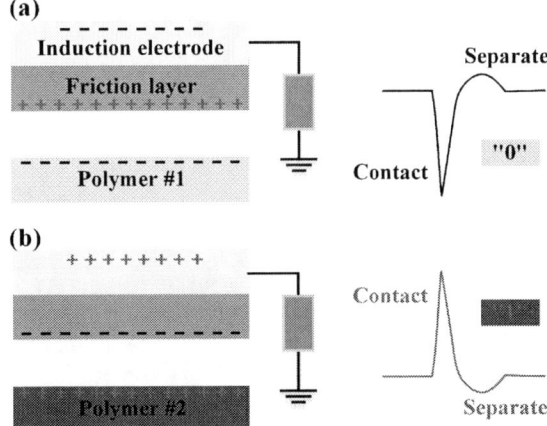

Figure 4: Working mechanism of the sensor. The polymer to be detected shows a tendency to (a) attract or (b) donate electrons against the fiction layer during contacting.

The performance of this sensor was characterized via repeated contact-separation cycles between the sensor and

978-1-4799-7956-1/15 $31.00 © 2015 IEEE   103

samples with a frequency of about 5 Hz. The output voltage was measured using a digital oscilloscope (Agilent DSO-X 2014A) with an impedance of 10 MΩ.

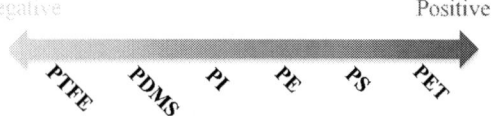

Figure 5: The relative order of PTFE, PDMS, PI, PE, PS and PET in triboelectric series.

As a primary example, we use PE as the friction layer to distinguish PET from PTFE. Figure 6 shows the time traces of the output voltage under repeated cycles. For a PET sample, continuous "1" waveform was observed (Figure 6a). In the case of a PTFE sample, obviously, the output switched to "0" ones (Figure 6b). Which keeps in line with the triboelectric serials. The measured absolute value of the peaks is around 1 V.

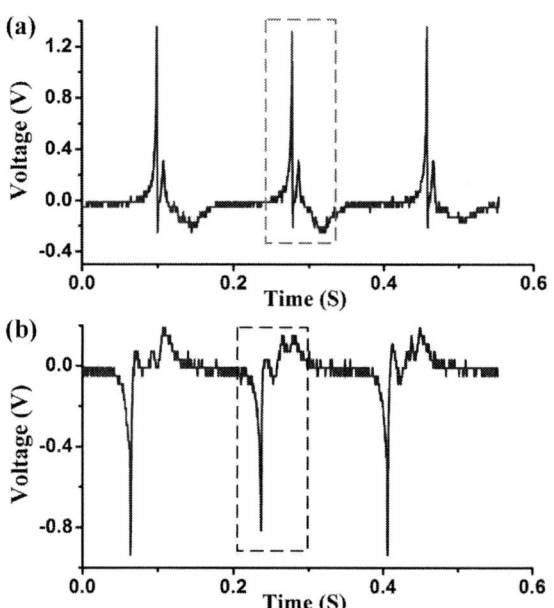

Figure 6: Time traces of the output voltage when using a single STEG cell with PE friction layer to distinguish (a) PET from (b) PTFE.

A further example was demonstrated to implement the distinguishing of PDMS, PE and PET from each other. Four STEG cells were integrated. Two PI friction layers and two PS friction layers were employed. Figure 7 shows the output waveforms of the 4 cells (PI, PI, PS, PS, respectively) during a single contacting-separation cycle when detecting sample #1 (PDMS, Figure 7a), sample #2 (PE, Figure 7b) and sample #3 (PET, Figure 7c). Owing to the asymmetry in applied force and the difference between the two kinds of friction

layer, the peak value of each voltage pulse in these waveforms differs. However, the polarities has shown good accordance. Outputs of "0000", "0011" and "1111" were observed respectively as we defined above. The results agreed well with the triboelectric serials (PET, PS, PE, PI, and PDMS, from positive to negative).

Usually, to distinguish N kinds of polymeric materials, at last N-1 kinds of selected friction layers are required. For the case that polymers are next to others in the triboelectric series (like PI and PE in the order shown in Figure 5), more kinds of friction layers with micro/nano surface patterns would be necessary.

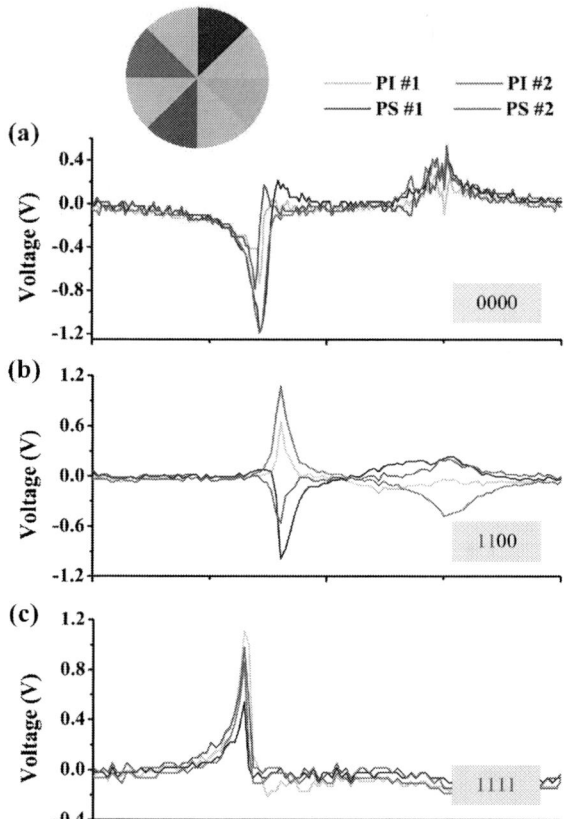

Figure 7: Waveforms of the output voltage when using four STEG cells with PI and PS as the friction layers to distinguish (a) sample #1 (PDMS), (b) sample #2 (PE) and (c) sample #3 (PET) from each other.

CONCLUSIONS

In summary, a novel active sensor was developed to realize a simple, fast and reliable material distinguishing among a group of known polymers. It works based on the coupling of triboelectrification and electrostatic induction effects. Multiple STEG cells are integrated on a flexible PI substrate as independent signal sources. The friction layers were selected according to the orders of the polymeric materials to be detected in triboelectric serials. Using flexible printed circuit manufacture procedure makes the fabrication low-cost and efficient and facilitates the integration of the sensor and external processing circuits. Using PI and PS as the two kinds of friction layers, the demonstration of distinguishing PDMS, PE and PET has been implemented successfully. This device is promising for multiple applications, such as robotics, artificial skins, industry and waste recycling, and so on.

ACKNOWLEDGEMENTS

This work is supported by the National Natural Science Foundation of China (Grant No. 61176103, 91023045 and 91323304), the National Hi-Tech Research and Development Program of China ("863" Project) (Grant No. 2013AA041102), the National Ph. D. Foundation Project (Grant No. 20110001110103) and the Beijing Natural Science Foundation of China (Grant No. 4141002).

REFERENCES

[1] L. B. Schein, "Recent progress and continuing puzzles in electrostatics", *Science*, vol. 316, pp. 1572-1573, 2006.

[2] H. T. Baytekin, A. Z. Patashinski, M. Branicki, B. Baytekin, S. Soh, B. A. Grzybowski, "The Mosaic of Surface Charge in Contact Electrification", *Science*, vol. 333, pp. 308-312, 2011.

[3] A. F. Diaz and R. M. Felix-Navarro, "A semi-quantitative tribo-electric series for polymeric materials: the influence of chemical structure and properties", *J. Electrostat.*, vol. 62, pp. 277-290, 2004.

[4] Z. L. Wang, "Triboelectric Nanogenerators as New Energy Technology for Self-Powered Systems and as Active Mechanical and Chemical Sensors", *ACS Nano*, vol. 7, pp. 9533–9557, 2013.

[5] F. R. Fan, Z. Q. Tian, Z. L. Wang, "Flexible triboelectric generator" *Nano Energy*, vol. 1, pp. 328-334, 2012.

[6] S. H. Wang, L. Lin, Z. L. Wang, "Nanoscale Triboelectric-Effect-Enabled Energy Conversion for Sustainably Powering Portable Electronics", *Nano Lett.*, vol. 12, pp. 6339-6346, 2012.

[7] X. S. Zhang, M. D. Han, R. X. Wang, F. Y. Zhu, Z. H. Li, W. Wang, H. X. Zhang, "Frequency-Multiplication High-Output Triboelectric Nanogenerator for Sustainably Powering Biomedical Microsystems", *Nano Lett.*, vol. 13, pp. 1168- 1172, 2013.

[8] B. Meng, W. Tang, X. S. Zhang, M. D. Han, W Liu, H. X. Zhang, "Self-Powered Flexible Printed Circuit Board with Integrated Triboelectric Generator", *Nano Energy,* vol. 2, pp. 1101-1106, 2013.

[9] B. Meng, W. Tang, X. S. Zhang, M. D. Han, X. M. Sun, W. Liu, H. X. Zhang, "A High Performance Triboelectric Generator for Harvesting Low Frequency Ambient Vibration Energy", in *Digest Tech. Papers MEMS'14 Conference*, San Francisco, January 26-30, 2014, pp. 343-349.

[10] Y. Suzuki, "Recent Progress in MEMS Electret Generator for Energy Harvesting", *IEEJ T Electr. Electr.*, vol. 6, pp. 101-111, 2011.

[11] B. Meng, W. Tang, Z. H. Too, X. S. Zhang, M. D. Han, W. Liu, H. X. Zhang, "A Transparent Single-Friction-Surface Triboelectric Generator and Self-Powered Touch Sensor", *Energy Environ. Sci.*, vol. 6, pp. 3235-3240, 2013.

[12] B. Meng, X. L. Cheng, X. S. Zhang, M. D. Han, W Liu, H. X. Zhang, "Single-friction-surface triboelectric generator with human body conduit", *Appl. Phys. Lett.*, vol. 104, pp. 103904, 2014.

[13] Y. Yang, H. Zhang, Z. H. Lin, Y. S. Zhou, Q. Jing, Y. Su, J. Yang, J. Chen, C. Hu, Z. L. Wang, "Human Skin Based Triboelectric Nanogenerators for Harvesting Biomechanical Energy and as Self-Powered Active Tactile Sensor System", *ACS Nano*, vol. 7, pp. 9213–9222, 2013.

[14] S. M. Niu, Y. Liu, S. H. Wang, L. Lin, Y. S. Zhou, Y. F. Hu, Z. L. Wang, "Theoretical Investigation and Structural Optimization of Single-Electrode Triboelectric Nanogenerators", *Adv. Funct. Mater.*, vol. 24, pp. 3332-3340, 2014.

CONTACT

*H. X. Zhang, tel: +86-10-62766570; zhang-alice @pku.edu.cn

SKIN BASED FLEXIBLE TRIBOELECTRIC NANOGENERATORS WITH MOTION SENSING CAPABILITY

Lokesh Dhakar[1,2], F. E. H. Tay[2,3], and Chengkuo Lee[1,]*

[1] Department of Electrical & Computer Engineering, National University of Singapore, Singapore
[2] NUS Graduate School for Integrative Sciences and Engineering, National University of Singapore, Singapore
[3] Department of Mechanical Engineering, National University of Singapore, Singapore

ABSTRACT

This paper presents a novel triboelectric nanogenerator (TENG) using outermost layer of human skin i.e. epidermis as an active triboelectric layer for device operation. The device is demonstrated to generate an open circuit voltage of ~90V with mild finger touch. The device uses PDMS nanopillar structures to improve the performance of contact electrification process which causes the charging of two surfaces. The device is demonstrated as a wearable self-powered device which can be used as a motion and activity sensor for a variety of applications.

INTRODUCTION

Looking for sustainable and green sources of energy generation is one of the important challenges which our society faces today. Part of this pursuit is designing miniaturized energy harvesting devices which can be used to provide sufficient power for autonomous operation of sensors and small-scale electronic devices. There are various sources of energy in the environment which can be converted in useful electrical energy including solar energy, temperature gradient, wind, pressure, vibrations etc. Out of these sources, mechanical energy based sources e.g. pressure and vibrations have an advantage of being unaffected by external environmental conditions like weather and temperature. To convert mechanical energy into useful electrical energy, piezoelectric [1], electromagnetic [2] and electrostatic [3] mechanism have been used conventionally. Recently, contact electrification or triboelectricity has been demonstrated as another possible mechanism to convert mechanical energy into useful electrical power in an efficient manner [4-6].

Contact electrification is the phenomenon of charging of two dissimilar surfaces when they are put in close contact. When these surfaces are pulled apart, a potential difference is generated between two surfaces. This potential difference can be used to generate power in an external load. Contact electrification phenomenon can also be used to sense force, pressure [7], motion [8], chemical properties [9] etc. which has led to development of many novel sensors. There is also a need for development of wearable sensors and energy harvesting devices for applications such as vital sign monitoring, electronic skin, soft-robotics and flexible electronics. In this paper, we present a wearable sensor for motion/activity sensing based on muscle movement. The device is also demonstrated as a triboelectric nanogenerator (TENG) which can generate energy to power small electronic devices or auxiliary functions of motion sensor. TENG uses epidermis which is the outermost layer of the human skin as a triboelectric layer.

The other triboelectric layer is a nano-structured PDMS layer. The contact between two layers results in contact electrification process and charging of skin and PDMS layers, which can then be used to generate energy.

DEVICE DESIGN AND OPERATING MECHANISM

Design and fabrication

For the fabrication of TENG (Figure 1a), an anodized aluminum oxide (AAO) substrate with nano-pores (Figure 1b) is spin-coated with a 10:1 mixture of polydimethylsiloxane (PDMS). The AAO template with PDMS layer is then kept for degassing for 2 hours to allow the PDMS enter the nano-pores. Thereafter, the samples are cured at 120°C for 2 hours to cure the PDMS layer. This PDMS layer is then carefully peeled off from the AAO substrate using tweezers so that not to break the nano-pillar like structure based ~500 μm thick PDMS layer. A Kapton layer is coated with a 50 nm gold using thermal evaporation process. The gold thin film acts as the electrode for TENG. The nano-patterned PDMS layer is then laid upon the Kapton layer which is already coated with gold. Kapton is chosen as the supporting layer for PDMS layer due to its mechanical properties and flexibility which helps to maintain device's overall strength and robustness. Figure 1d shows a photograph of the device depicting its flexibility. The as-fabricated device is then assembled on human skin and demonstrated as a motion sensor and energy harvesting device.

Operating mechanism

A schematic diagram of the TENG is shown in Figure 1c. This device is assembled on epidermis where nano-structured PDMS layer and epidermis acts as the triboelectric layers. As per triboelectric series, epidermis has a strong tendency to give up electrons as compared to PDMS which results in better performance of contact electrification and hence performance of TENG. As the two triboelectric layers come in close contact, the epidermis layer gets positively charges whereas the PDMS nano-pillar structures get charged with equal and opposite negative charges as shown in Figure 1a. As the PDMS layer is detached from epidermis layer, both the layers are charged and have a potential difference across them. As the gap between the PDMS and epidermis increases, there is an electric potential on the gold electrode due to the electric field created by triboelectrically generated charges. This electric potential can results into flow of electrons through a load resistor connected between the electrode and ground as shown in Figure 1c. Thereafter, as the gap decreases between the epidermis and PDMS layer, the electrons flow back to the

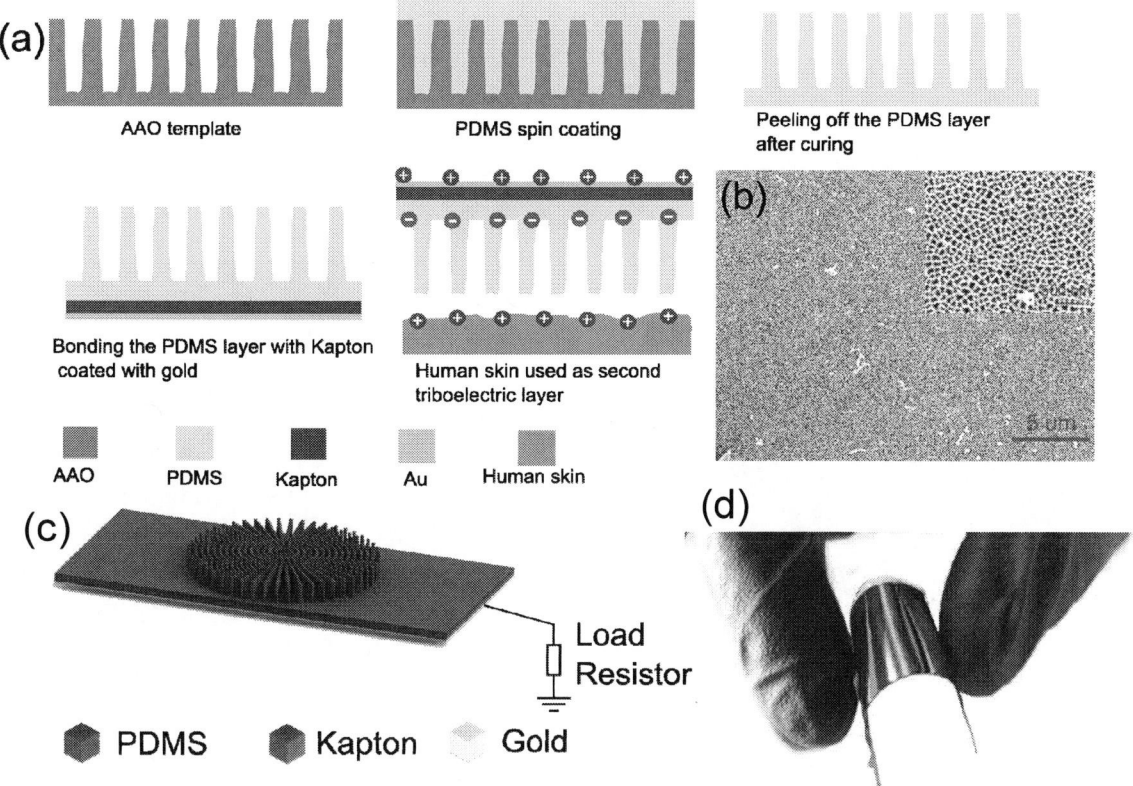

Figure 1: (a) Fabrication steps for the flexible TENG (b) Scanning electron microscopy (SEM) image of the AAO nanopores (c)schematic illustration of the TENG (d)a photograph of the actual fabricated device showing the flexibility.

ground from gold electrode. These charges flowing in the external circuit can be harvested to generate useful energy.

EXPERIMENTS AND RESULTS

Device Characterization

The fabricated PDMS samples were characterized using Atomic Force Microscopy (AFM) in tapping mode. Figure 2 shows an optical image of as-fabricated PDMS sample. The sample is divided into two parts, one part is fabricated using the AAO nanopores whereas other was peeled from area without any nanopore structures. As the device was characterized using AFM, the difference between the topography of two areas could clearly be seen. The part without AAO nanopores was flat as compared to the nano-pillar like structures on the part fabricated using AAO nanopores. This nano-structured topography results in the increased effective contact area between the epidermis and PDMS layers leading to the improved performance of device.

Testing as energy harvester

Flexible TENG was assembled on a human hand as shown in Figure 3(b). The assembly step was crucial as it has to be made sure that there is a minimal gap between the assembled TENG and skin (epidermis). As the hand is used to any activity, there is a movement in the muscle which translates into closure of gap and hence contact between epidermis and PDMS layer. The contact between the two surfaces results in generation of triboelectric

charges on both the surfaces. As the muscle device goes

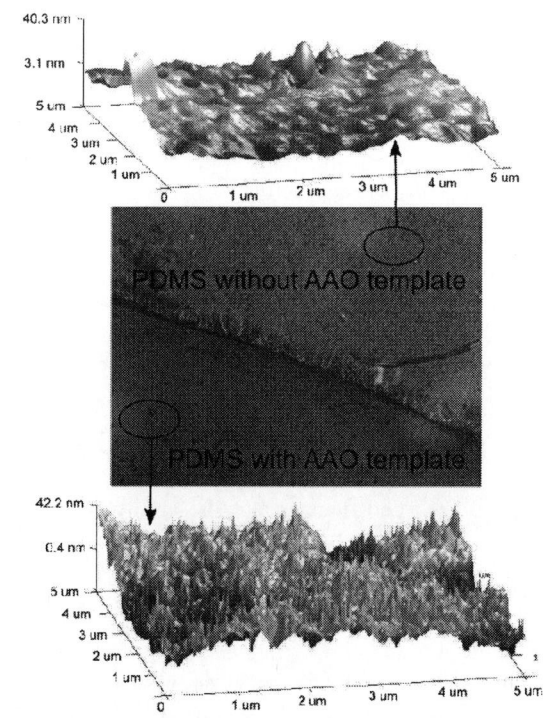

Figure 2: Optical image of as-fabricated PDMS sample with AFM images of topography of parts with and without nanopillar structures using AAO template.

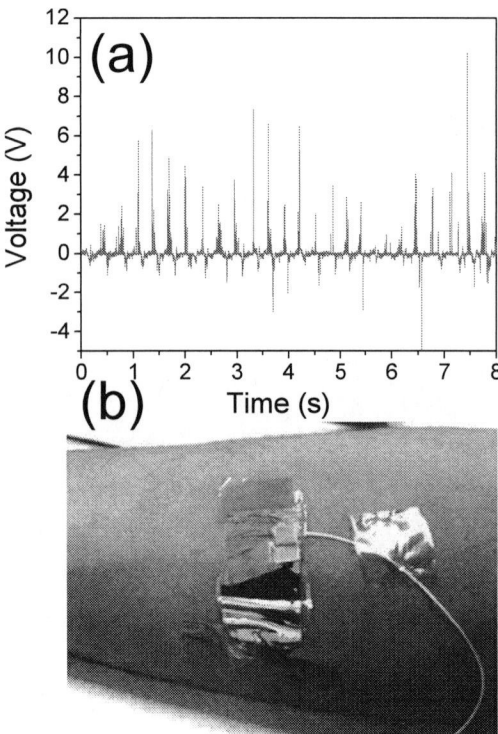

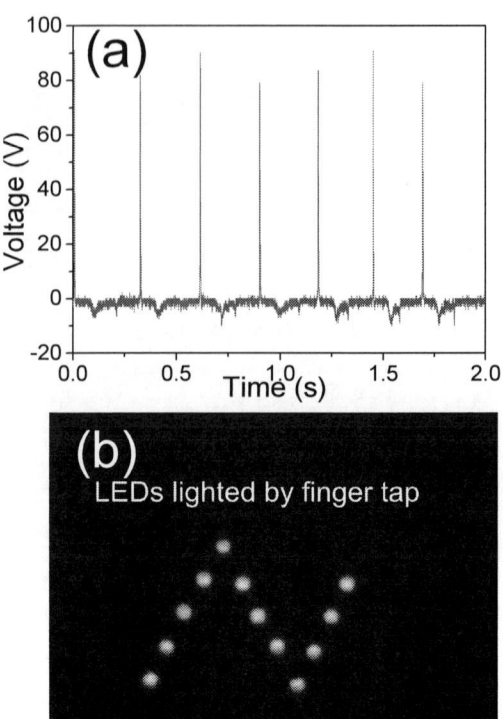

Figure 3: (a) Voltage output from fist movement of hand (b) Skin based nanogenerator assembled on human wrist

Figure 4: (a)Voltage output from finger tapping (b)LEDs lighted using mild finger touch on skin based TENG

back to the initial position, there is an electric potential developed on the gold electrode. This electric potential changes as the gap between epidermis and PDMS layer varies. This variation of electric potential results in inflow and outflow of electrons in the load resistor connected between the electrode and ground. This electron flow can be converted into useful electrical energy in the load resistor.

To measure the voltage output response of TENG, first set of experiments were conducted using with fist clenching and releasing activity of human hand. As the fist is clenched, there is a movement in the muscle resulting into the contact between epidermis and PDMS layer. As the muscle moves back to the initial position, the two triboelectric layers separate from each other. This contact-separation cycle leads to flow of electrons in the load resistor as discussed earlier. The device was able to generate electric potential using fist clenching and releasing motion. All these measurements were made using an oscilloscope with an internal impedance of 1 MΩ.

To check and demonstrate the ability of TENG to power electronic components as a power source, the device was connected to 12 commercially available LEDs. All these LEDs were connected in series. The device was again assembled on human skin and tapped mildly using finger to generate energy. As the device was touched and released with the finger, all the LEDs lighted up due to the electric potential created by TENG. Figure 4b shows an image of the lit up LEDs. The voltage generated by TENG was also measured using an oscilloscope with an internal impedance of 1 MΩ. The

highest peak voltage generated by the finger tapping on TENG assembled on epidermis was measured to be 90V as shown in Figure 4a.

These set of experiments demonstrated that TENG can potentially be used for powering small electronic devices by harvesting biomechanical energy. TENG can also be assembled on other body parts e.g. leg, arm etc. where the muscle movement due to daily activity like walking, running, lifting objects, can enable the contact and separation movement between the epidermis and nano-structured PDMS layer.

Motion/activity sensor testing

The flexible TENG device is also shown to be used as wearable sensor. The device was assembled was on hand and tested as a sensor for muscle movement due to any motion or activity of hand. As seen in Figure 5, a pulse is generated due to holding of an object using hand. The signal to noise (SNR) ratio for the TENG as a motion/activity sensor can be calculated using Equation 1.

$$SNR = \left(\frac{A_{signal}}{A_{noise}}\right)^2 \qquad (1)$$

where A_{signal} is the amplitude of required signal and A_{noise} is the amplitude of noise. As calculated from the acquired signal A_{signal} and A_{noise} were measured to be 6.34 V whereas and 0.19 V, respectively. Using these values in Equation 1, the SNR is calculated to be 1113.5. The calculated SNR value demonstrates the superior characteristics of TENG as a motion or activity sensor.

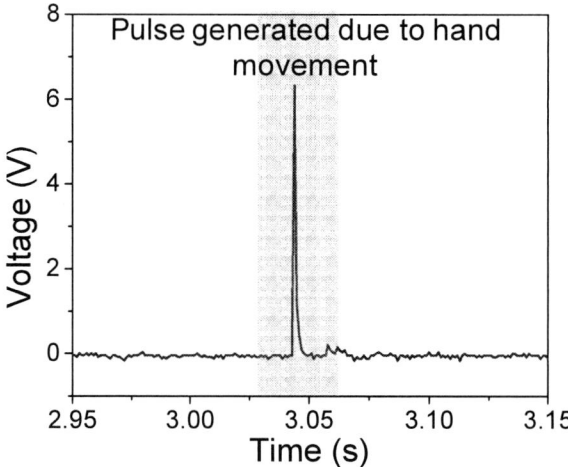

Figure 5: Testing of the flexible TENG as a skin based motion/activity sensor

This skin based wearable sensor can be used for fall detection in elderly, for advanced input devices using hand gestures, soft robotics and flexible electronics applications.

CONCLUSION

A contact electrification based flexible energy harvester has been developed. The TENG device used epidermis layer (skin) as an active triboelectric layer for the device operation. TENG is demonstrated to generate voltages up to 90V. Also it has been shown as a potential power source for small electronic devices or components. 12 commercially LEDs connecting in series were lit up using finger touching and releasing on TENG assembled on epidermis.

Moreover the device is also tested as a sensor which detects muscle movement due to holding of objects. The sensor was observed to demonstrate a signal to noise ratio of 1113.5. In summary, the presented flexible TENG device which can be used as a wearable self-powered sensor to track muscle movement due to any motion or activity.

ACKNOWLEDGEMENTS

This work is partially supported by the Faculty Research Committee (FRC) Grant (No. R-263-000-692-112) at the National University of Singapore, and the NRF2011 NRF-CRP001-057 Program 'Self-powered body sensor for disease management and prevention-orientated healthcare' (R-263-000-A27-281) from the National Research Foundation (NRF), Singapore.

We would also like to thank Prakash Pitchappa for his contribution in the testing of devices.

REFERENCES

[1] A. Erturk and D. J. Inman, "An experimentally validated bimorph cantilever model for piezoelectric energy harvesting from base excitations," *Smart Materials and Structures,* vol. 18, pp. 025009, 2009.

[2] S. P. Beeby, R. Torah, M. Tudor, P. Glynne-Jones, T. O'Donnell, C. Saha and S. Roy, "A micro electromagnetic generator for vibration energy harvesting," *Journal of Micromechanics and microengineering,* vol. 17, p. 1257, 2007.

[3] Y. Naruse, N. Matsubara, K. Mabuchi, M. Izumi, and S. Suzuki, "Electrostatic micro power generation from low-frequency vibration such as human motion," *Journal of Micromechanics and Microengineering,* vol. 19, p. 094002, 2009.

[4] F.-R. Fan, L. Lin, G. Zhu, W. Wu, R. Zhang, and Z. L. Wang, "Transparent triboelectric nanogenerators and self-powered pressure sensors based on micropatterned plastic films," *Nano letters,* vol. 12, pp. 3109-3114, 2012.

[5] L. Dhakar, F. E. H. Tay, and C. Lee, "Development of a Broadband Triboelectric Energy Harvester With SU-8 Micropillars," *J. Microelectromech. Syst,* vol. 99, , 2014.

[6] B. Meng, X. Cheng, X. Zhang, M. Han, W. Liu, and H. Zhang, "Single-friction-surface triboelectric generator with human body conduit," *Applied Physics Letters,* vol. 104, p. 103904, 2014.

[7] L. Lin, Y. Xie, S. Wang, W. Wu, S. Niu, X. Wen and Z. L. Wang., "Triboelectric active sensor array for self-powered static and dynamic pressure detection and tactile imaging," *ACS nano,* vol. 7, pp. 8266-8274, 2013.

[8] W. Yang, J. Chen, X. Wen, Q. Jing, J. Yang, Y. Su, G. Zhu, W. Wu and Z. L. Wang, "Triboelectrification Based Motion Sensor for Human-Machine Interfacing," *ACS Applied Materials & Interfaces,* vol. 6, pp. 7479-7484, 2014.

[9] Z. H. Lin, G. Zhu, Y. S. Zhou, Y. Yang, P. Bai, J. Chen and Z. L. Wang, "A Self-Powered Triboelectric Nanosensor for Mercury Ion Detection," *Angewandte Chemie International Edition,* vol. 52, pp. 5065-5069, 2013.

CONTACT

*Chengkuo Lee, tel: +65-65165865; elelc@nus.edu.sg

FLEXIBLE TRIBOELECTRIC AND PIEZOELECTRIC COUPLING NANOGENERATOR BASED ON ELECTROSPINNING P(VDF-TRFE) NANOWIRES

Xingzhao Wang[1,2], Bin Yang[1,], Jingquan Liu[1], Qing He[1], Honglei Guo[1], Chunsheng Yang[1], and Xiang Chen[1]*

[1]The National Key Laboratory of Science and Technology on Micro/Nano Fabrication Laboratory,
Department of Micro/Nano Electronics
[2]School of Biomedical Engineering, Shanghai Jiao Tong University, Shanghai, China

ABSTRACT

This paper studied a triboelectric and piezoelectric coupling nanogenerator based on MEMS technology. The electrospinning PVDF-TrFE nanofibers are not only used as a piezoelectric functional layer, but also as a friction layer of triboelectric generator. The other flexible friction layer is realized by PDMS films with doped multiwall carbon nanotubes (MWCNT). The sandwich-shaped nanogenerator's triboelectric output peak voltage, piezoelectric output peak voltage, triboelectric energy power, piezoelectric energy power, triboelectric energy volume power density and piezoelectric energy volume power density are 5V, 30V, 98.66µW, 9.74µW, 1.98mW/cm³ and 0.689mW/cm³ under the pressure force of 5 N, respectively. This device has some advantages such as flexibility, thickness controllability, double coupling mechanisms.

INTRODUCTION

Nanogenerators have been aroused tremendous interests due to their high effective efficiency and low-cost with organic nanomaterial [1]. The present micro/nanosystems face the problems of the compatibility and continuity from traditional batteries. Therefore, small scale energy harvester becomes a main trend to power systems in the applications such as medical devices, bio-sensors, health monitoring, wearable electronics, sensor networks, and so on. Conventional piezoelectric energy harvester is mostly used to transport wireless energy and thus widely applied in pacemaker [2-4].When the piezoelectric thin film is deformed under pressure force, it will produce electricity by piezoelectric effect. Piezoelectric transducer scavenges the energy from the environment, including vibration mechanics and ultrasound wireless transportation. Triboelectric energy harvester can easily occurred between two layers which have different dielectric constant. Triboelectric and piezoelectric devices can convert mechanical energy to electric one, thus provide an effective solution for the device's longstanding working. The nanogenerator based on MEMS technology can provide power source for series of micro devices and implantable chips. [5-6]

The piezoelectric and triboelectric effects have been utilized as extremely effective mechanisms to harvest mechanical energy by a new type of organic/polymer nanogenerators. The contacting between two surfaces can make the double side surfaces negative and positive charge. In the previous work, researches on two effects have been done and some structures have been studied [7]. However, the flexibility of the generator is limited by the industrialized production of PET/ITO film. Moreover, most of generators are based on a single mechanism for power generation [8]. Here, we proposed a flexible nanogenerator, which is a kind of triboelectric and piezoelectric coupling nanogenerator (NG) with excellent flexibility. A multi-wall carbon nanotube (MWCNT) doped PDMS film and electrospinning process fabricated P(VDF-TrFE) film are first used to be a friction layer in nanogenerator.

DESIGN & FABRICATION

Sandwich-shaped piezoelectric nanogenerators are widely used and usually includes top electrode, upper electrode and piezoelectric layer. The double-side electrodes of P(VDF-TrFE) nanofiber are fabricated by screen-printing conductive silver. And PDMS friction layer faces to silver electrode to form triboelectric NG. The nanogenerator schematic of two coupling mechanism is as shown in Fig 1(a). There are pillars patterns on the surface to enhance the friction performance between two friction surfaces. The patterns are fabricated by MEMS technology. PDMS film is also doped with MWCNT to enhance the triboelectric performance.

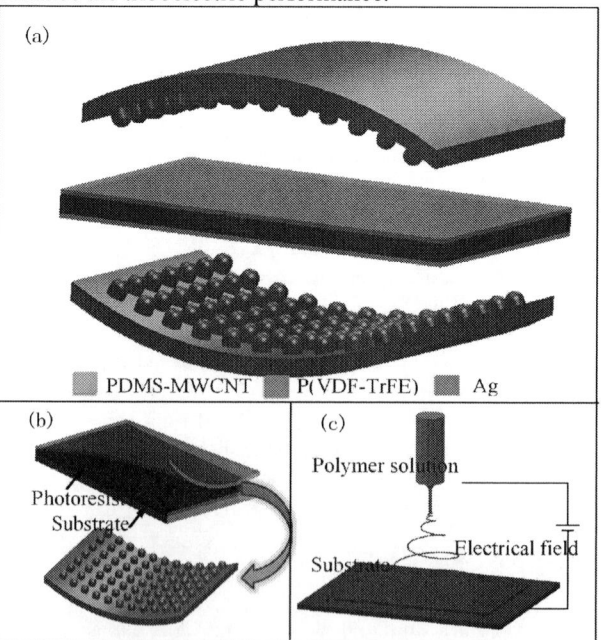

Fig 1: (a) The structure of the nanogenerator. (b) The process of fabrication the PDMS-MWCNT friction layer. (c) The electrospnning process of P(VDF-TrFE) nanofiber.

The dimension of each layer is 8mm×8mm. The

thickness of P(VDF-TrFE) nanofiber and PDMS-MWCNT films are 85μm and 45μm, respectively. It is worth mentioning that the thickness of each layer for nanogenerators is controllable, and could be determined for different applications.

In the steady equilibrium state, each layer of the generator is insulated and its packaging is to ensure the small gap between two friction layers. The value of surface area and volume is 0.64 cm^2 and 0.0128 cm^3. P(VDF-TrFE) is selected as the material for piezoelectric layer due to its high flexibility, excellent mechanical and chemical stability [9]. Electrospinning process has the advantage that the provided high bias electric field (usually from 10kV~80kV), which can directly array the electric dipole along the major axis direction of the fiber [10]. Therefore, part of α crystal phase inside of P(VDF-TrFE) nanofiber converses into β crystal phase. Then the PVDF-TrFE nanofibers can generate electric current under pressure outside.

PDMS is widely used for triboelectric generation causing due to its flexibility and easy deformation [11]. PDMS solution and MWCNT are mixed in the proportion of 1:1. Toluene is dissolved in the mixture and ultrasonic instrument is used to obtain uniform distribution for 24h. The dispersion of MWCNT is a key factor that influences the characterization of the PDMS film. In order to obtain microstructure, a silicon substrate is patterned with an array of cylinder by photolithography, as shown in Figure 1(b). Then mixture PDMS-MWCNT is coated on the substrate, pressed and peeled after a thermal curing process. The height of cylinder is realized by the thickness of photoresist.

During the electrospinning process, the syringe filled with P(VDF-TrFE) solution is connected to a high positive voltage of 10 kV. In the precursor solution, P(VDF-TrFE) (55/45) is dissolved in DMF solution, and the weight percentage is 15%. To get electrospinning solution, 55wt% acetone is dissolved which has good volatility. A grounded collector is placed 10cm away from the needle to gather the nanofibers. The injection speed of syringe pump is 0.4ml/h. Annealing process is needed and nanofibers is annealed under 85 ℃ in vacuum. In order to capture electrons from piezoelectric layer under deformation, conductive silver paste is covered on both sides by screen printing method. The thickness of electrode is about 10μm.

CHARACTERIZATION

The SEM images of P(VDF-TrFE) nanowires by electrospinning process are shown in Figure 2(a) and Figure 2(b). Electrospun nanowires show a disordered distribution. The microscope of the fabricated P(VDF-TrFE) white thin film is shown in the left corner at Fig 2(a).As shown in Figure 2(c) and Figure 2(d), the nature of the polymer crystalline phase presented in fibers can be identified by XRD and FTIR. XRD curve shows a typical β crystal phase peak with around $2\theta =19.48°$. And the vibration bands at 848 cm^{-1} and 1280 cm^{-1} are assigned to the absorption bands of the β crystal phase. The vibration brands at 1240 cm^{-1} is assigned to the absorption band of the γ crystal phase [12].

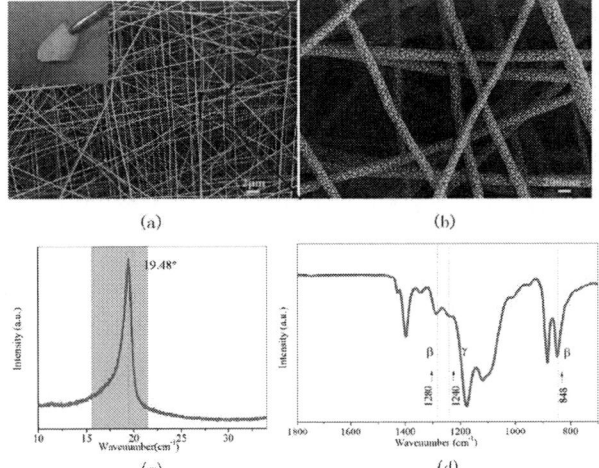

Fig 2: (a) - (b) SEM picture of P(VDF-TrFE) nanofibers. (c) XRD patterns and (D) FTIR spectra of the P(VDF-TrFE) nanofibers.

Figure 3(a) shows the characterization of the PDMS-MWCNT films. The fabricated film has tapered cylindrical structure as shown in Fig.3 (a). The diameter, height and gap of each tapered cylindrical structure's roots are 50μm, 50μm and 50μm, respectively. Alternatively, the exposed MWCNT surface of the film fiber can clearly be seen in the Fig 3(b). With doped MWCNT PDMS film, it can reduce the load resistance and thus change the conductivity at a certain extent.

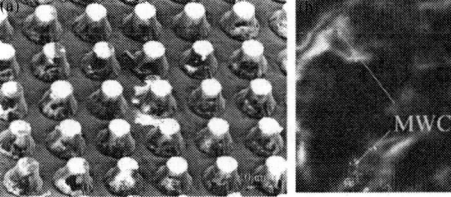

Fig 3: (a) SEM image of the PDMS-MWCNT flat surface. (b) Schematic diagram of the micro-patterned PDMS-MWCNT friction surface.

When an external force is applied on generator, PDMS-MWCNT film with microstructures and P(VDF-TrFE) nanofibers layer are brought into contact and the deformation of piezoelectric layer is occurred. The friction between PDMS-MWCNT film and silver electrode cause the separation of electrons. As the force is released, nanogenerator reverts back to the original position. Then the friction layers carry positive and negative charge and thus an electric field occurred. There's electricity generated through two effect respectively. Subsequently, the external force impact once again, the triboelectric and piezoelectric effects will be occurred during this process. The electric energy generation process is occurred by triboelectric and piezoelectric effects.

EXPERIMENT

The experiment process is shown in Fig 4. It is placed under the NG that a 3-axial force pressure sensor (ATI, NANO17) is to record the finger pressure applied to NG. It can monitor the value of x, y, z axial pressure in the computer. The relationship between the direction of press and the performance of NG will be studied in the next work. Several resistors are used to getting a variable resistance value, and installed in series with the oscilloscope. LTC3588 Nanopower Energy Harvesting Power Supply is placed and DC/AC modules inside convert AC to DC form from device since the mechanic force pressed to the surface of device. A DC type circuit is presented on the screen of oscilloscope. The NG compared with one eurocent is shown in Fig 4.

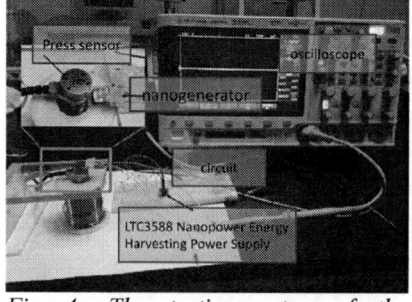

Fig 4: The testing setup of the performance of nanogenerator and the physical picture of NG.

The output voltage performance of the triboelectric nanogenerator under the frequencies varied from 1Hz to 4Hz when the pressure force of 5N by a bare finger covered in a PE glove is shown in Fig 5. The triboelectric open-circuit output voltage is shown in Fig 5(a) and the piezoelectric one is shown in Fig 5(b).It can be seen that with increasing frequency, the triboelectric peak output voltage of NG increased from 26V to 32V. Until the frequency increased upper to 3Hz, the growth rate is not obvious. The piezoelectric open-circuit output voltage is much lower than the triboelectric one. It is increased from 1.5V to 2.5V when the frequency is increased.

The test of triboelectric power performance is shown in Fig 6. The PDMS-MWCNT layer is connected with a silver layer so that generating a current flow of electrons by triboelectric effect when a downward force is applied to the NG.

As can be seen that the output voltage is increasing along with the increasing of resistance. Under the resistance of 20MΩ, it arrives at the maximum peak voltage of 30V, and the current of 1.53µA. After that, the voltage is decreasing along with the increasing of resistance, it can be explained by the following equation [13]:

$$R_{opt} = \frac{d}{\varepsilon WL}\frac{1}{f}$$

where the R_{opt} is the optimum resistance, ε is the permittivity of the NG, f is the frequency of the press force, L, d and W are the nanogenerator's length, thickness and width, respectively. Due to Ohmic loss, the decreasing of voltage along with the increasing of resistance occurred mostly.

Meanwhile, the current is decreasing along with the increasing of resistance from 6.4µA to 0.21µA. Thus 5MΩ is the optimum internal impedance and corresponding voltage and current are 22.2V and 4.44µA. And the maximum output power is 98.568µW.

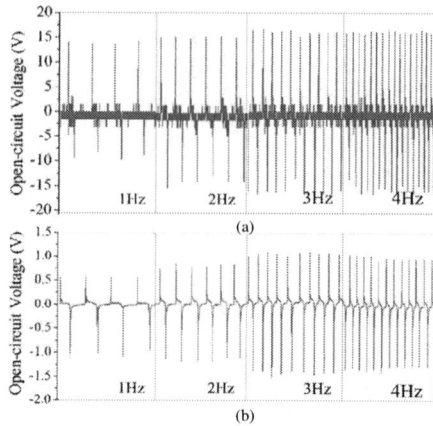

Fig 5: (a) The nanogenerator open-circuit voltage of the triboelectric effect under the force of 5N. (b) The nanogenerator open-circuit voltage of the piezoelectric effect under the force of 5N

As demonstrated, a piezoelectric layer is connected by two silver layers so that generating a current flow of electrons by piezoelectric effect when a downward force is applied to the NG.

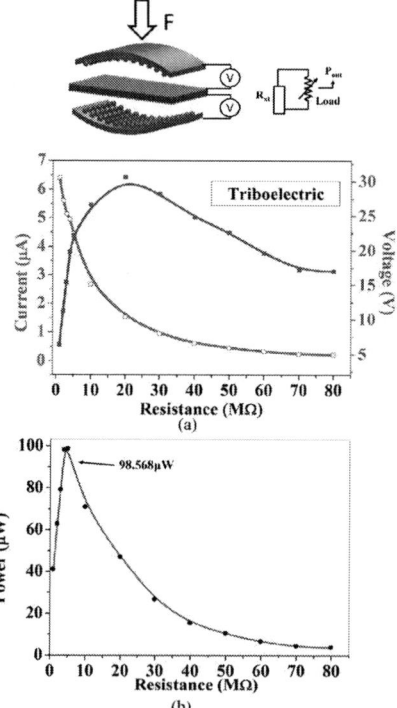

Fig 6: The triboelectric effect dependence of the output voltage and output power on external load resistance.

As can be seen in Fig 7, the maximum peak voltage is 21V at the resistance of 50MΩ. The piezoelectric maximum output power is 9.747µW at the resistance of 30MΩ. By connecting the two effects generated current to the circuit, it can realize the piezoelectric and triboelectric effects coupling NG and enhance its performance.

For this test data, it demonstrates that the optimum load resistance is different under triboelectric and piezoelectric effect. The contradictory leads to the determination of optimum load resistance when it is used in practical application. Thus triboelectric and piezoelectric effect electrodes need to be integrated in one circuit and test the performance under two mechanisms.

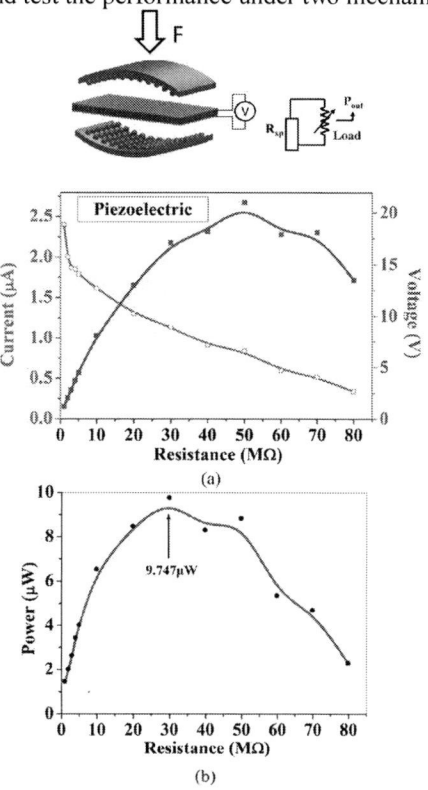

Fig 7: The piezoelectric effect dependence of the output voltage and output power on external load resistance.

CONCLUSION

In summary, a triboelectric and piezoelectric coupling nanogenerator based on MEMS technology is proposed and characterized. The dimension of nanogenerator is 8mm×8mm. It can be easily curled and bended, and thus it could be a promising power source for flexible electronic devices and portable electronic products. The output peak voltages of the nanogenerator based on the triboelectric and piezoelectric effect are 30.16V and 21V. The optimum output powers based on the triboelectric and piezoelectric effect are 98.568μW under 5MΩ and 9.747μW under 30MΩ, respectively. The volume power densities based on the triboelectric and piezoelectric effect are 1.98mW/cm^3 and 0.689mW/cm^3. Although the resistance of nanogenerator is bigger than others, its flexibility can't be ignored.

ACKNOWLEDGEMENTS

This work was supported in part by the National Natural Science Foundation of China under Grant No. 61204119, 51035005 and 61076107, 973 Program (2013CB329401), the Shanghai Pujiang Talent Program Sponsorship (No. 13PJ1405100), Innovation Program of Shanghai Municipal Education Commission under Grant No. 14ZZ019, the Science and Technology Department of Shanghai (No. 11JC1405700).

REFERENCE

[1] Wang S, Lin L, Wang Z L, "Nanoscale triboelectric-effect-enabled energy conversion for sustainably powering portable electronics", *Nano letters*, vol. 12, pp. 6339-6346, 2012.

[2] Wang, Z. L, "Self-powered nanotech". *Scientific American*, vol. 298, pp. 82-87, 2008.

[3] Karami M A, Inman D J, "Powering pacemakers from heartbeat vibrations using linear and nonlinear energy harvesters", *Applied Physics Letters*, vol.100, pp. 042901, 2012.

[4] Cooper D, WILKOFF B, MASTERSON M, et al, "Effects of extracorporeal shock wave lithotripsy on cardiac pacemakers and its safety in patients with implanted cardiac pacemakers", *Pacing and Clinical Electrophysiology*, vol.11, pp. 1607-1616, 1988,

[5] Panescu D, "MEMS in medicine and biology". *Engineering in Medicine and Biology Magazine, IEEE*, vol. 25, pp. 19-28, 2006.

[6] Wang Z L, "Towards Self‐Powered Nanosystems: From Nanogenerators to Nanopiezotronics", *Advanced Functional Materials*, vol. 18, pp. 3553-3567, 2008.

[7] Han M D, Zhang X S, Liu W, et al, "Low-frequency wide-band hybrid energy harvester based on piezoelectric and triboelectric mechanism", *Science China Technological Sciences*, vol. 56, pp. 1835-1841, 2013.

[8] Hou T C, Yang Y, et al, "Triboelectric nanogenerator built inside shoe insole for harvesting walking energy", *Nano Energy*, vol. 2, pp. 856-862, 2013.

[9] Mandal D, Yoon S, Kim K J, "Origin of Piezoelectricity in an Electrospun Poly (vinylidene fluoride‐trifluoroethylene) Nanofiber Web‐Based Nanogenerator and Nano‐Pressure Sensor", *Macromolecular rapid communications*, vol. 32, pp. 831-837, 2011.

[10] Baji, Avinash, et al, "Electrospinning of polymer nanofibers: effects on oriented morphology, structures and tensile properties", *Composites science and technology*, vol. 70, pp. 703-718, 2010.

[11] Fan, F. R., Lin, L., Zhu, G., Wu, W., et al "Transparent triboelectric nanogenerators and self-powered pressure sensors based on micropatterned plastic films", *Nano letters*, vol. 12, pp. 3109-3114, 2012.

[12] Cui, Z., Drioli, E., & Lee, Y. M, "Recent progress in fluoropolymers for membranes", *Progress in Polymer Science*, vol. 39, pp. 164-198, 2014.

[13] Wang Z L, "Self-powered nanosensors and nanosystems", *Adv Mater*, vol. 24, pp. 280−285, 2012

CONTACT

*B. Yang, tel: +86-021-34206683;
binyang@sjtu.edu.cn

HIGHLY RELIABLE MEMS RELAY WITH TWO-STEP SPRING SYSTEM AND HEAT SINK INSULATOR FOR POWER APPLICATIONS

Yong-Hoon Yoon, Yong-Ha Song, Seung-Deok Ko, Chang-Hoon Han, Geon-Sik Yun,
Min-Ho Seo and Jun-Bo Yoon
Department of Electrical Engineering, KAIST
291 Daehak-ro, Yuseong-gu, Daejeon 305-701, Republic of Korea

ABSTRACT

This paper reports remarkably reliable MEMS relays having a unique two-step spring system and a heat sink insulator. The two-step spring system is designed to reduce Joule-heating by lowering contact resistance. The heat sink insulator is proposed for efficiently removing heat generated in the contact area. These two features are firstly adopted in the MEMS relay for minimizing thermal damage in high current level, thus enhancing reliability significantly. The fabricated relay demonstrated contact resistance of 2 mΩ, which is the lowest value compared to the state-of-the-art [11]. In addition, the relay was operated up to 5.3×10^6 cycles at 1 V/200 mA in an air and hot switching condition with negligible contact resistance variation. The resulting lifetime is 500 times longer than that of the commercial MEMS relay measured in the same test setup.

INTRODUCTION

The electrostatically actuated MEMS relays for power applications have drawn considerable attention as a promising alternative to electromagnetic relays (EMRs) and solid state relays (SSRs) due to their effective power efficiency at on state, quasi-zero leakage current at off state and relatively small footprint [1], [2]. Various reports have been successfully published on this study with positive prospect [3-5].

In spite of the attractive performance and possibility for power application, the electrostatically actuated MEMS relays are still facing considerable difficulty for the commercialization. This is mainly due to poor reliability of the MEMS relays in high current level [6]. It is well known that the electrostatic force for operating the MEMS relays is relatively weak, which inevitably causes high contact resistance. The high contact resistance considerably increases Joule-heating, which results in thermal damage and in hence poor reliability of the MEMS relays. Furthermore, the high contact resistance affects badly to the reliability at high current level than low current level.

In order to overcome the critical reliability issue, new relay designs [7], [8] and new materials [9], [10] have been proposed. Among them, the relays with low contact resistance for minimizing Joule-heating have achieved remarkable improvement in the reliability [11], [12]. For example, Y. H. Song and et al. successfully demonstrated the reliable MEMS relays with world smallest contact resistance thanks to the stacked-electrode structure for achieving high contact force and low effective hardness of the contact material [11]. However, the removal of the generated heat has not been considered to the MEMS relays despite its positive effect speculated by simulation [13].

In this paper, we propose a new and efficient method to reduce the contact resistance and to remove the generated heat simultaneously. In order to meet this purpose, the two-step spring system and the heat sink insulator have been employed in the MEMS relays for the first time.

CONCEPT

Fig. 1 is the conceptual illustration of the proposed MEMS relays. In this structure, four crab legs which act as first spring support the source plate. The second spring is formed inside the source plate. Under the plate, meshed drain electrode is placed. The heat sink insulator is inserted under the drain to isolate drain and gate electrodes. The gate electrode is buried to attract the source plate.

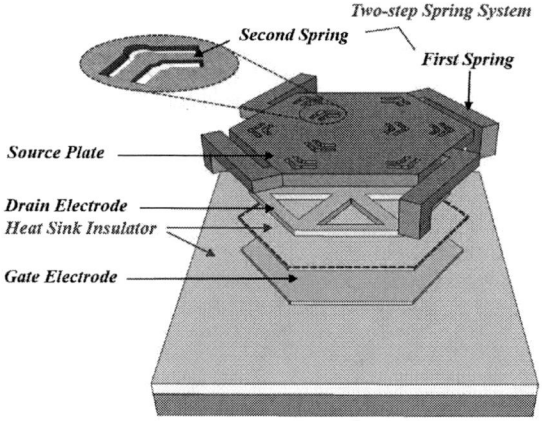

Fig.1 Conceptual illustration of the proposed MEMS relay with the two-step spring system (first spring and second spring) and the heat sink insulator.

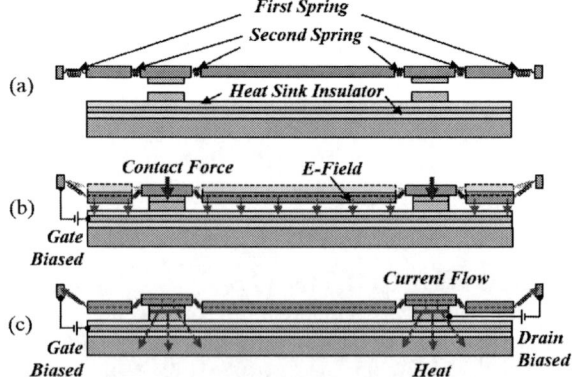

Fig.2 Actuation mechanism of the MEMS relay. (a) Initial state. (b) Contact state with increased contact force by the two-step spring system when the gate electrode is biased (c) Current flow state where the generated heat is effectively dissipated by the heat sink insulator.

978-1-4799-7956-1/15 $31.00 © 2015 IEEE   114   MEMS 2015, Estoril, PORTUGAL, 18 - 22 January, 2015

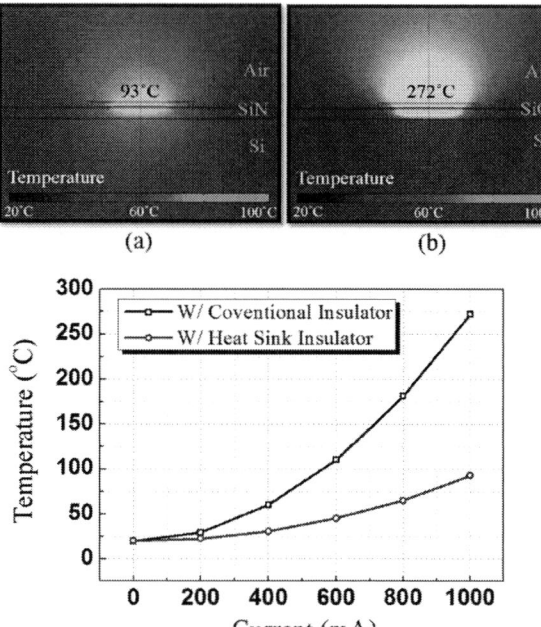

(a)　　　　　　　(b)

(c)

Fig.3 Finite element method (FEM) simulation results for thermal analysis. (a) Thermal profile with the heat sink insulator (SiN). (b) Thermal profile with the conventional insulator (SiO₂). (c) Temperature comparison as a function of the Joule-heating current.

Fig. 2(a) describes the cross sectional view of the proposed relays at initial state. As shown in Fig. 2(b), the grounded source plate moves down as a bias in the gate electrode increases until the plate contacts to the drain electrode (dot line). Conventional relay with a single spring system is stopped at this moment. Contrarily, the proposed relay's source plate additionally moves down by the second spring (solid line), which maximizes electrostatic force due to reduced air gap. As a result, the contact resistance is lowered, which induces the minimized Joule-heating at contact.

Fig. 2(c) shows that heat generated in the contact area by the drain-to-source current is efficiently dissipated to the substrate through a heat sink insulator. The heat sink insulator is realized by choosing a proper material of SiN having high thermal conductivity of 30.1 W/m·K which is 30 times higher than that of the generally used SiO₂ (1.04 W/m·K)

Fig. 3 represents finite element method simulation result for thermal analysis. Fig. 3(a) shows the uniform and lower temperature profile than Fig. 3(b) because the large amount of heat is dissipated down to the substrate via heat sink insulator (SiN). As a result, under the same Joule-heating condition, the temperature with the heat sink insulator (SiN) is considerably lower than their counter part (Fig. 3(c)). This simulation results can successfully anticipate that the heat sink insulator properly removes the generated heat in the contact area.

The two-step spring system and the heat sink insulator are proposed to relieve the thermal damage by minimizing the Joule-heating and removing the generated heat simultaneously.

FABRICATION

The proposed MEMS relays were fabricated with a six-mask process on the silicon substrate. First, the 1ˢᵗ SiN (0.3 μm) was deposited by low pressure chemical vapor deposition (LPCVD) for device isolation. Then, thermally evaporated Cr (1000 Å) was used as the gate electrode, where the Cr was selected for adhesion layer with SiN layer. Next, the 2ⁿᵈ SiN (1 μm) was deposited by plasma enhanced chemical vapor deposition (PECVD) for isolation between the gate and drain electrodes. A total of 5000-Å-thick Cr/Au layers were then deposited by evaporation and patterned by wet etching process to form the drain electrode.

In order to define initial air gap between the source plate and the drain electrode, the Ti/Cu/Ti sacrificial layers (totally 1.3 μm) were deposited. First, Ti (2000 Å) was deposited by sputter, which used as a diffusion barrier between Au and Cu. Subsequently, Cu (9000 Å) and Ti (2000 Å) were deposited as a sacrificial layer by sputter and thermal evaporator, respectively. Next, the sacrificial layers were patterned to form dimple and anchor as depicted in Fig. 4(d). A 1000-Å-thick Au was thermally evaporated as a contact material of the proposed MEMS relays. Next, thick photoresist (PR) was coated and patterned as a mold for electroplating. The PR mold was used to form an electroplated nickel film (10 μm) for the source plate. In the last step, the unnecessary PR mold, seed layer (Au) and, sacrificial layer (Ti/Cu/Ti) were removed. Then, critical point dryer was used for preventing initial stiction.

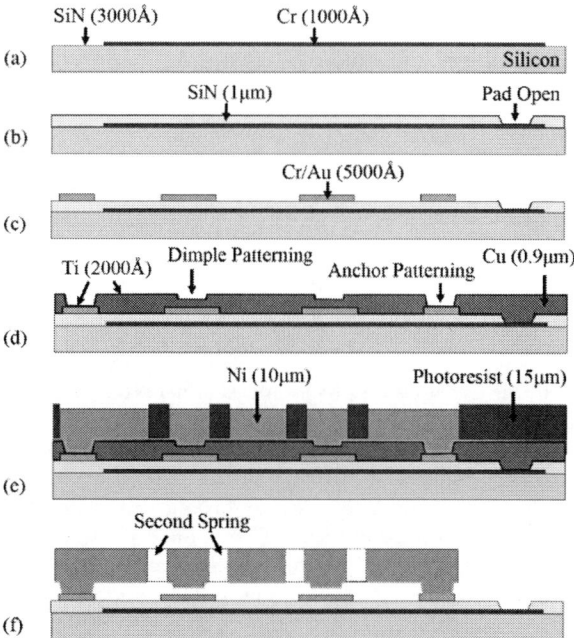

Fig.4 Overall fabrication process. (a) 1ˢᵗ SiN deposition and Gate electrode formation. (b) 2ⁿᵈ SiN deposition and pad open process for probing. (c) Drain electrode formation. (d) Sacrificial layer deposition with dimple and anchor patterning. (e) Seed layer deposition and nickel electroplating with a photoresist mold. (f) Releasing process by wet etching of the sacrificial layers.

978-1-4799-7956-1/15 $31.00 © 2015 IEEE

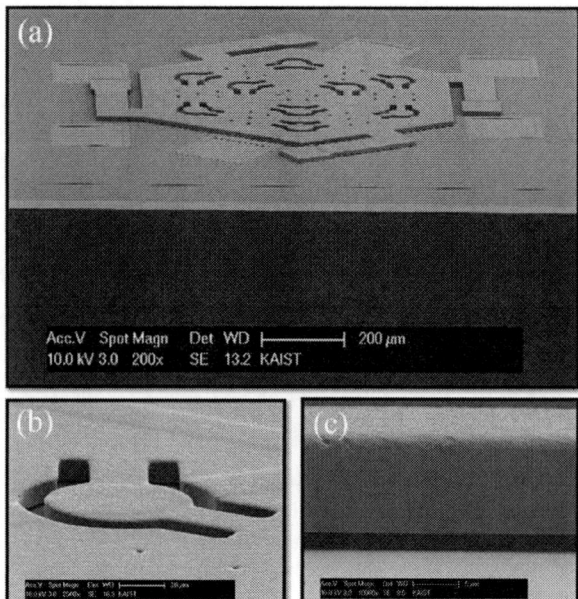

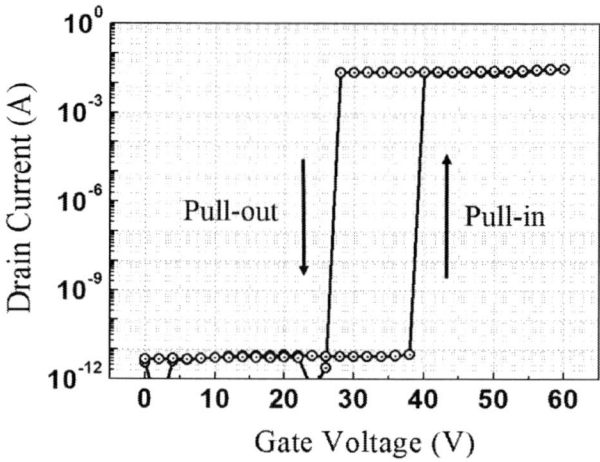

Fig.6 Measured drain current characteristic as a function of the gate voltage. Pull-in voltage was near 38V and pull-out voltage was near 25V

Fig.5 SEM photographs of the proposed MEMS relays. (a) Perspective overview. (b) Magnified view of the second spring. (c) Magnified view of the air gap.

Fig. 5(a) shows the scanning electron microscope image of the proposed MEMS relays with a source plate size of 600 μm x 600 μm. Fig. 5(b) and 5(c) represent the second spring and air gap respectively, which reveals that the proposed MEMS relays were successfully released.

RESULTS AND DISCUSSIONS

The measured actuation voltage (pull-in voltage) of the proposed MEMS relays was about 33 V ~ 42 V. The variation in pull-in voltage is originated from variation of plate's thickness during Ni electroplating. The calculated pull-in voltage was 35 V, which is comparable with the measured results. As shown in Fig. 6, the off current level was 10^{-12} A~10^{-11} A, which means that the proposed MEMS relays successfully disconnect a signal at off state.

Next, the contact resistance, the critical factor for the reliability, was measured. The Agilent 4156C parameter analyzer was used for precise measurement (the device can detect in the microvolt range) with a four-point probe method. A total of five devices were measured, and the testing current was 4 mA.

Fig. 7 represents the measured contact resistance with the aforementioned test setup. When the 40 V was biased to the gate electrode, the contact resistance of the MEMS relays with the two-step spring system exhibited about 50% lower value than their counterpart. This result implies that the contact force is successfully increased by the proposed two-step spring system. Furthermore, the contact resistance decreased further as the gate voltage was increased. The minimum value of the contact resistance was 2 mΩ at an applied voltage of 45 V, which breaks the previous world record to our best knowledge [11].

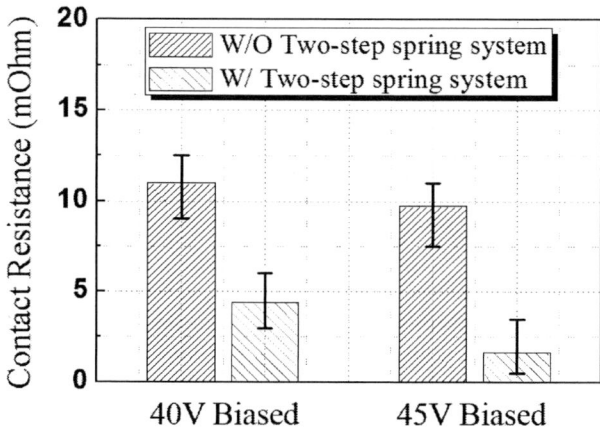

Fig.7 Measured contact resistance of the proposed MEMS relays with and without the two-step spring system. Minimum contact resistance was 2mΩ at 45V. The error bar means standard deviation of the five measured data.

The reliability test in high current level was also conducted. The Tabor-electronics function generator was used for deliberately making cyclic actuation signal, which has duty ratio of 20 %, frequency of 500 Hz, and the amplitude of 50 V. Then, the output signal was precisely detected by Keithley 2638B parameter analyzer. The testing was implemented in an air ambient (temperature: 23 °C, humidity: 35 %) and hot-switching condition.

Fig. 8 shows the reliability of the fabricated MEMS relays with the aforementioned test setup. The testing current of 200 mA was selected to validate the reliability at a high current level. As a result, The measured lifetime was 5.3×10^6 cycles, which is 500 times higher than that of the commercial MEMS Relay targeted for high power switching (RMSW 100HP from Radant Co.) in the same test setup. The failure mode of the both devices was due to the sudden increasing in the contact resistance.

978-1-4799-7956-1/15 $31.00 © 2015 IEEE

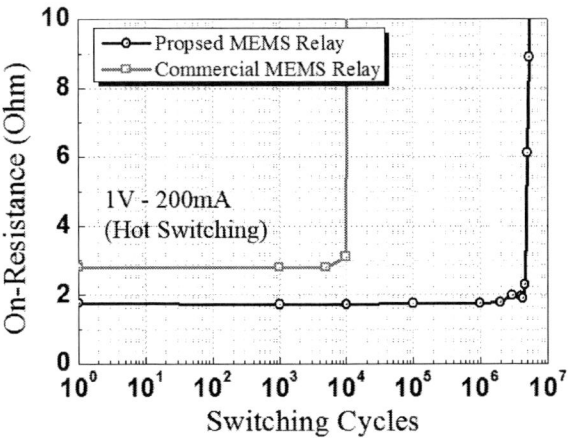

Fig.8 Measured lifetime of the proposed and commercial MEMS relays. The result was tested in an air and hot-switching condition.

CONCLUSION

An electrostatically actuated MEMS relays with the two-step spring system and the proper heat sink were suggested, fabricated and evaluated for the power applications. The proposed MEMS relays provided extremely low contact resistance of 2 mΩ, which is the smallest value among previously reported MEMS relays for power switching [11]. Also, they were able to operate up to 5.3×10^6 cycles at 200 mA current, which is 500 times higher than that of the commercial MEMS relays. The proposed MEMS relays with high reliability are expected to be considered as a promising device candidate for the power applications.

ACKNOWLEDGEMENT

This work was supported by TSE Co., Ltd.. The authors especially thanks to Chulho Kim, Chihun In, and Byungjun Seo, the supporting researchers in TSE Co., Ltd.. Also, this work was partially supported by the National Research Foundation of Korea (NRF) grant funded by the Korea government (MSIP) (CAFDC 5-4, NRF-2007-0056090)

REFERENCES

[1] H. S. Lee, C. H. Leung, J. Shi, and S. C. Chang, "Electrostatically actuated copper-blade microrelays," *Sens. Actuators A, Phys.*, vol. 100, no. 1,pp. 105–113, Aug. 2002.

[2] J. Qiu, J. H. Lang, A. H. Slocum, and A. C. Weber, "A bulk micromachined bistable relay with U-shaped thermal actuators," *J. Microelectromech. Syst.*, vol. 14, no. 5, pp. 1099–1109, Oct. 2005.

[3] J.-E. Wong, J. H. Lang, and M. A. Schmit, "An electrostatically-actuated MEMS switch for power applications," *in Proc. IEEE Int. Conf. Microelectromech. Syst.*, pp. 633–638, 2000.

[4] A. Cao, P. Yuen, and L. Lin, "Microrelays with bidirectional electrothermal electromagnetic actuators and liquid metal wetted contacts," *J. Microelectromech. Syst.*, vol. 16, no. 3, pp. 700–708, Jun. 2007.

[5] B. Li, C. Keimel, G .Claydon, J. Park, A. D. Corwin, and M. Aimi, "Power switch system based on microelectro mechanical switch", in *Digest Tech. Papers Transducers'11 Conference*, Beijing, pp. 675-678, June, 2011.

[6] G. M. Rebeiz, "RF MEMS Theory, Design and Technology", 1st ed. Hoboken, NJ: Wiley, 2003.

[7] Y. Shi and S. G. Kim, "A lateral, self-cleaning, direct-contact MEMS switch" in Proc. *18th IEEE Int. Conf. Microelectromech. Syst.*, pp. 195-198, 2005.

[8] L. L. W. Chow, J. L. Volakis, K. Saitou, and K. Kurabayashi, "Lifetime extension of RF MEMS direct contact switches in hot switching operations by ball grid array dimple design" *IEEE Electron Device Lett.*, vol. 28, no. 6, pp. 479-481, Jun. 1998.

[9] J. Schimkat, "Contact materials for microrelays" in *Proc. IEEE Int. Conf. Microelectromech. Syst.*, pp. 190-194, 1998.

[10] F. Ke, J. Miao, and J. Oberhammer, "A ruthenium-based multimetal-contact RF MEMS switch with a corrugated diaphragm", *J. Microelectromech. Syst.*, vol. 17, no. 6, pp.1447-1459, Dec. 2008.

[11] Y.-H. Song, D.-H. Choi, H.-H. Yang, and J.-B. Yoon, "An Extremely Low Contact-Resistance MEMS Relay Using Meshed Drain Structure and Soft Insulating Layer", *J. Microelectromech. Syst.*, vol. 20, no. 1, pp. 204–211, Feb. 2011.

[12] Y.-H. Song, C.-H. Han, M.-W. Kim, J. O. Lee, and J.-B. Yoon. "An electrostatically actuated stacked-electrode MEMS relay with a levering and torsional spring for power applications", *J. Microelectromech. Syst.*, vol. 21, no. 5, pp. 1208-1217, Oct. 2012.

[13] L. L. Mercado, T.-Y. T. L., S.-M. Kuo, V. Hause, and C. A., "Thermal solutions for discrete and wafer-level RF MEMS switch packages", *IEEE Tras. on Advanced Packaging*, vol. 26 , no. 3 , pp. 318-326, August, 2003.

CONTACT

*Y.-H. Yoon, tel: +82 42 350 5476, yhyoon@3dmems.kaist.ac.kr

SOLIDIFIED IONIC LIQUID FOR HIGH POWER-OUTPUT VIBRATIONAL ENERGY HARVESTERS

S. Yamada[1], H. Mitsuya [2], S. Ono[3], K. Miwa[3], and H. Fujita[1]
[1]CIRMM, IIS, The University of Tokyo, Tokyo, JAPAN
[2]SAGINOMIYA SEISAKUSHO. Inc., Tokyo, JAPAN
[3]Central Research Institute of Electric Power Industry, Tokyo, JAPAN

ABSTRACT

We propose a high power-output vibrational energy harvesting based on ionic liquid. Ionic liquid enables very large capacitance (1.0-10 μF cm^{-2}) on the electrode at bias voltage less than 1.9 V due to its extremely thin (~ 1 nm) electrical double layer. By mechanical squeezing and drawing the ionic liquid, that was solidified with a polymer additive, between a pair of electrodes at 15 Hz, we stably obtained the current output of 22 μA$_{p-p}$ cm^{-2} at 1.5 V.

INTRODUCTION

The concept of "Internet of Things[1]" is attracting much attention because it can provide abundant useful data from wireless sensors installed in all places to realize a safe and healthy society. Energy harvesting is a key for practical wireless sensor networks because it can eliminate periodical battery change. Energy harvesting is expected as the power source for sensor nodes because they work without maintenance and supply sensors with power until they get broken. Especially, the vibration type attracts much attention for its wide applicability, e.g. in rooms and tunnels or on the body.

To replace batteries of the wireless sensor nodes by energy harvesters, two major requirements are identified. First, energy harvesters should be small and inexpensive. MEMS based harvesters can fulfill the requirement. In addition they should convert low-frequency and wide-band vibrational energy efficiently into electrical energy. The ambient vibrations caused with the bridge swing or vehicle motion are composed of mainly low frequency under 20Hz [2]. Energy harvesters must generate power from such low frequency vibrations abundant in the environment.

Currently, researchers focus on 3 types of vibrational energy harvesters, a piezoelectric type, a capacitive type and an electromagnetic induction type. The capacitance type generates only low power. The electromagnetic induction type generates high power but it needs large magnets and bulky structures like coils. Therefore, this type is not compatible with MEMS technology. The piezoelectric type can be achieved by MEMS devices but has high spring constant and small mass due to its small size. The structure increases the resonant frequency of the piezoelectric type. To decrease the frequency, low-spring-constant suspensions and a large mass are favorable [3]. Thus, the structure becomes really fragile. Current devices cannot satisfy the requirement of generating large power from the low frequency source.

In our previous work, we fabricated a prototype energy harvester based on ionic liquid[4]. We prepared a pair of electrodes and put ionic liquid between them (Fig .1). The lower electrode was attached to a shaker. The ionic liquid was squeezed and drawn. As a result, the contact area became large and small periodically. Capacitance of the ionic liquid is proportional to the contact area between the ionic liquid and electrode. Therefore, the shaker motion drove stored charge to go out and in. Thus, we could retrieve output current (2.3μA cm^{-2} at 5Hz). However, the output is smaller than we expected. Due to wettability, ionic liquid tends to stick on the electrode when it is drawn. In a previous study, a hydrophobic layer was coated on electrodes to reduce wettability [5]. Additional insulation layer thickness decreased the capacitance of the electrical double layer and the output.

Here, we fabricated solidified ionic liquid (sIL) to solve the sticking liquid problem and increase the output

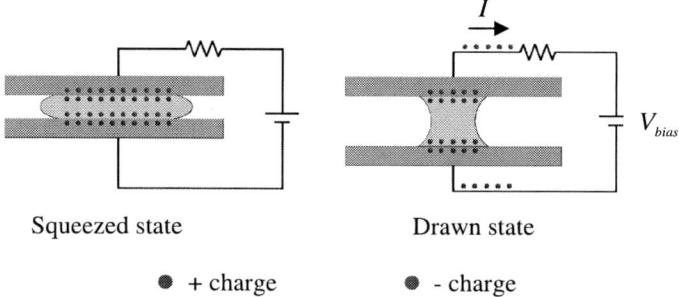

Figure 1. Schematic image of the power generation based on ionic liquid.

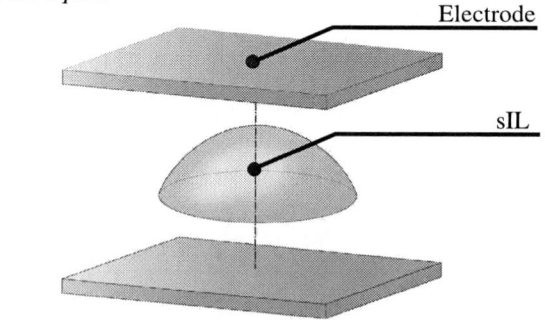

Figure 2. Schematic image of our setup.

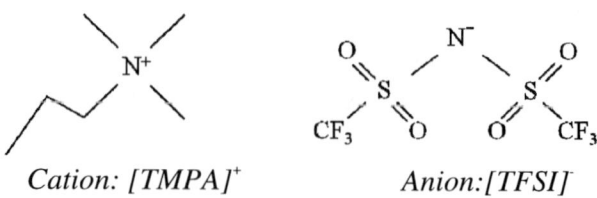

Figure 3. Chemical formula of sIL [TMPA]$^{+}$ [TFSI]$^{-}$.

978-1-4799-7956-1/15 $31.00 © 2015 IEEE   118   MEMS 2015, Estoril, PORTUGAL, 18 - 22 January, 2015

current (Fig.2). The characteristics of ionic liquid are maintained after solidification. The stability of gel shape increases the power generation. We are able to fabricate a high-power energy harvester with the low resonant frequency. The sIL doesn't require a spring-mass structure, making the device robust. Moreover, the sIL automatically forms electrical double layer of 1nm in thickness as ionic liquid does. On the contrary, the conventional capacitive type based on electrets has a gap of a few tens of μm [6], because it is difficult to keep the narrow gap while the mass vibrates. The electrical double layer makes the capacitance extremely high (~μF cm^{-2}) and we can obtain a large output. Wireless sensor nodes require 100μW to transmit data over 10m; this range of power generation is the target. In this paper, we propose a new power generation principle based on the solidified ionic liquid.

PRINCIPLE & CONCEPT
Ionic liquid

The ionic liquid has unique characteristics, such as the formation of the electrical double layer, the extremely low vapor pressure and the resistance to high temperature. For this reason, it has been utilized as lithium ion battery electrolytes [7]. The ionic liquid is non-flammable and able to decrease the risk of the fire. Recently, ionic liquid was used to improve solid electrochemical devices. Gerbaldi et al. [8] created polymer with ionic liquid. They merged a monomer (Bisphe-nol A ethoxylate dimethacrylate), ionic liquid (EMIPFSI) and photopolymerization starting agent (2-hydroxy-2-methyl-1-phenyl-1-propanone). The mixture was exposed to UV for approximately 3 minutes. The initiator absorbs UV and activates a serial chemical reaction to polymerize monomers. Finally they obtained solid electrolyte with ionic liquid. To fabricate sIL, we polymerized ionic liquid by UV exposure as well. We added a monomer to the ionic liquid ([TMPA]$^+$[TFSI]$^-$ Fig. 3) and solidified it by UV exposure.

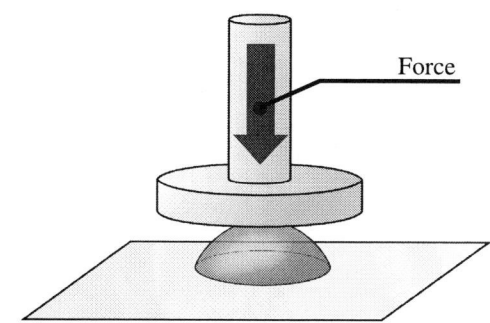

Figure 6. Schematic image of the force gauge.

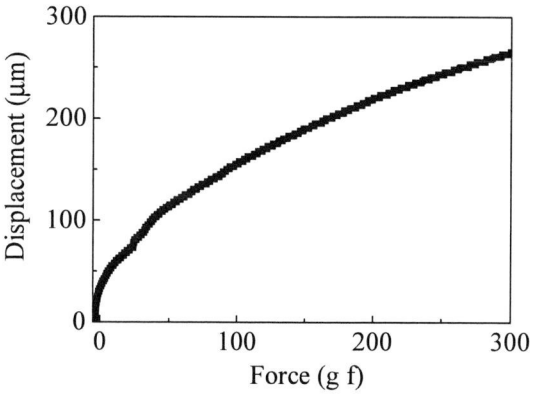

Figure 7. The displacement of the sIL vs force. The sIL was compressed 100μm with the force 38 gf.

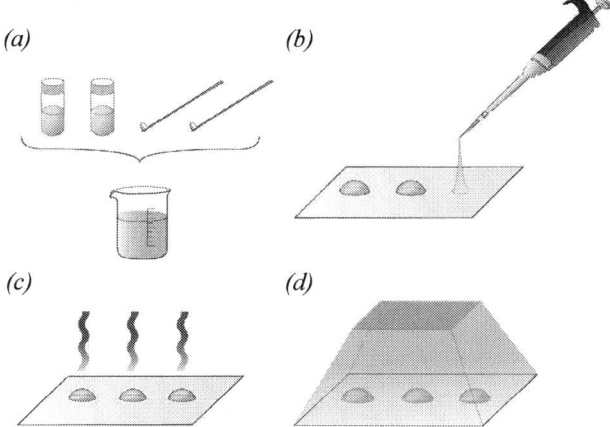

Figure 4. Fabrication process of the sIL.
(a)We measured and mixed 4 kinds of materials(ionic liquid, monomer, bismaleimide, benzophenone).(b)We put droplets of the mixture on a sheet with micropipette. (c) The sIL were heated for 7minutes at 100 degree. (d)We exposed UV on them for 20 minutes.

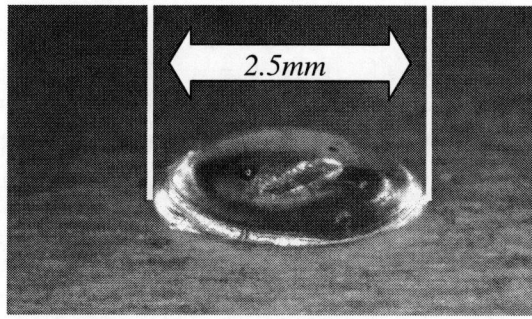

Figure 5. Photograph of the sIL.
Height: 0.67mm
Bottom area: 6.2mm^2

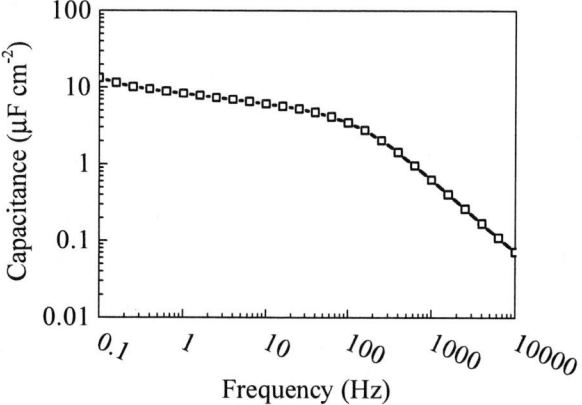

Figure 8. The capacitance of sIL vs frequency.
Capacitance: 8.3μF cm^{-2} at 1Hz.

Fabrication process

The fabrication process of the sIL is shown in Fig. 4. First of all, we prepared four kinds of materials, a monomer, the ionic liquid, bismaleimide and benzophenone. Those two chemicals work as crosslinking agent and polymerization agent, respectably, to get the ionic liquid polymerized by UV exposure. The materials were well mixed and dispensed on a sheet as droplets. We exposed UV for 20 minutes; photopolymerization started and the droplets got hardened, becoming the solidified ionic liquid. We show the solidified ionic gel in Fig. 5. It has a hemisphere like shape. Its height and diameter are 0.67 mm and 2.5 mm respectably. We squeeze and draw the sIL to generate the power. Before doing this, we measured the squeezing force by compressing the sIL with force gauge(Fig. 6). Figure 7 shows the deformation of the sIL vs force. According to the result, 38gf is needed to squeeze sIL for 100μm. We also measured the capacitance of sIL. The capacitance is 3-8μF cm^{-2}(Fig.8) between 1-100 Hz. From this measurement, it was confirmed the sIL kept the high capacitance of ionic liquid.

Principle

The power generating principle is as follows. We put sIL between electrodes and applied voltages between 0.5-2.0V which are within the potential window (Fig. 9); in this window the sIL remains as insulator. Please note in the real harvester the voltage source must be replaces by an electret. With the electrical double layer and insulation, we can obtain the high capacitance (8.3μF cm^{-2}). After the voltage application on sIL, vertical vibration is applied to the lower electrode. The contact area changes as the motion of the compression and pull. The capacitance is proportional to the contact area. The areal increase and decrease follow the sinusoidal motion of the electrode. Therefore, the stored charge goes in and out, causing the current flow in the circuit (Fig. 10).

Calculation

Before experiments, we estimated the output current with a simple model. Let us assume the capacitance of sIL in squeezed state is 10μF at V_{bias}= 1.0V. The external vibration force draws the electrode up and the contact area decreases from 1.0cm^2 to 0cm^2. Simultaneously, the

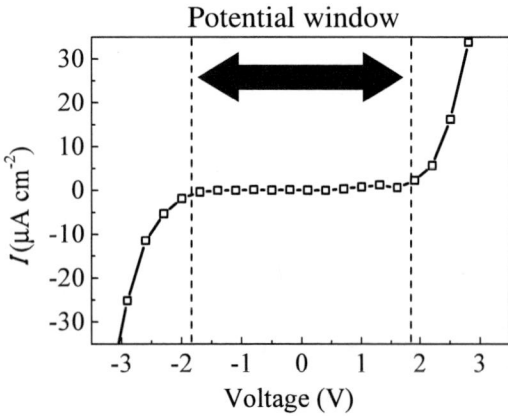

Potential window

Figure 9. The potential window of the solidified ionic gel, where electrical double layer works as insulator. Potential window: -1.9V to 1.9V

capacitance also decreases from 10μF to 0μF. The change of the stored charge is 10μC. If this change happens in 1s, the output current is 10μA.

EXPERIMENT

We prepared the setup and measured the output current to confirm the principle. The experimental setup is shown in Fig. 11. The upper electrode was fixed to the manipulator and lower one was attached to a shaker. We inserted the sIL between electrodes. We lowered the upper electrodes until it touched the gel. The shaker was connected to a function generator. We can control the shaker displacement and the frequency. The lower electrode went up and down for ± 50 μm by the shaker with the squeezing force of 38 gf. The contact areas between the electrode and sIL were 4.2mm^2 and 0mm^2 in the squeezed and drawn state respectably. We applied voltage on sIL and turned on the shaker. The sIL was periodically squeezed and drawn by the vertical sinusoidal vibration. The output current at load resistance R was I-V converted and monitored on an oscilloscope.

RESULT & DISCUSSION

The peak to peak value of the output current of 22 μAp-p cm^{-2} was detected in phase of shaker motion as shown in Fig. 12-(a). The leak current at 1.5 V was 0.63 μA cm^{-2} and smaller than the output. Hence, this output was

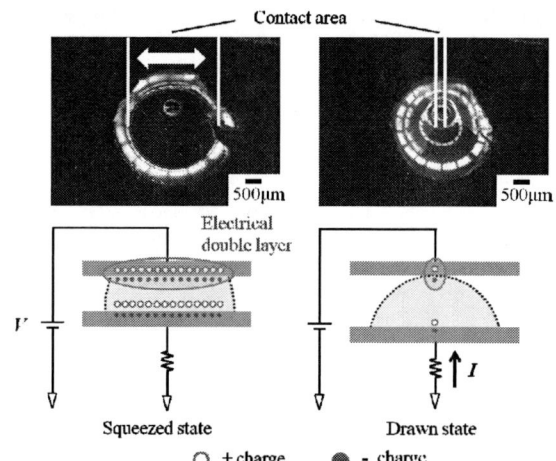

Figure 10. Photograph and schematic image of the power generation. By the compression of sIL, the contact area of the sIL and electrodes changed. Thus, the stored charge changes and current flows.

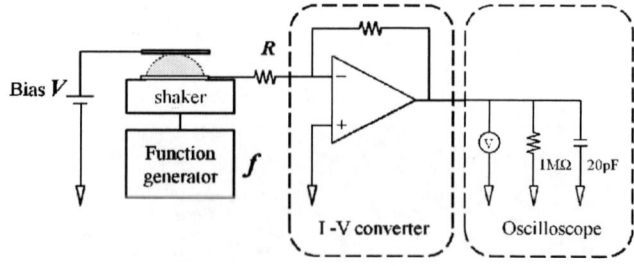

Figure 11. Schematic image of the experimental setup (R, V, f represent a loading resistance, bias voltage, frequency of the shaker respectively).

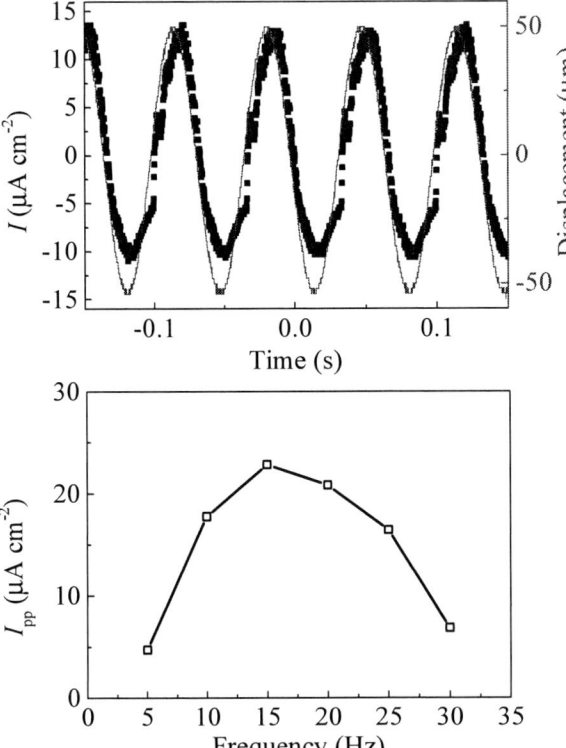

Figure 12. Result of the power generation. (a) Output current wave form at 15Hz, bias V= 1.5V, through load resistance =1kΩ (b) Frequency dependence of standardized output current at V_{bias}=1.5V, R=1kΩ. (Device contact area = 6.6 mm²)

owing to mechanical motion. We obtained larger current when the sIL expanded rather than contracted because the mechanical work done for contraction was converted not only to electrical energy but also to elastic energy. The output current was smaller than we expected. Several issues could influence on this power generation for example, dilution of ionic liquid, change of internal resistance and contamination. For now, we are investigating that.

The frequency response of the output current is shown in Fig. 12-(b) from 5 to 30Hz, at 1.5V and R = 1kΩ. The maximum output was 22μA$_{p-p}$ cm⁻² at 15 Hz for the device active area of 6.6mm². According to this result, the sIL generated the power in low frequency. This means energy harvesters based on sIL are compatible with ambient vibrations. Our sIL has a resonant frequency at 15Hz. We can optimize the resonant frequency varying the shape, density and stiffness. Depending on the target vibration, we can tune sILs with the optimal resonant frequency.

The output power P is expressed by $P=RI^2$ (R: resistance, I: current). The output is only 484nW$_{p-p}$ cm⁻² and it was smaller than the power which wireless sensors need. However, high output current (22μA cm⁻²) is generated at this point. There is a room for improvement of the device in its shape, stiffness, the electrical circuit, and by optimizing the load resistance. It will be possible to generate a few tens of μW

CONCLUSION

From the results, following conclusion can be drawn.

● We fabricated solidified ionic liquid from the mixture of ionic liquid and polymer. The stability in shape was improved. Furthermore, it kept the high capacitance (8.3μF cm⁻²) of ionic liquid.

● Just squeezed and drawn, the sIL generated high output current 22μA$_{p-p}$ cm⁻² at 15 Hz with 1.5 V bias. The simple structure enables us to eliminate complicated structures from energy harvesting devices.

● We obtained power between 1Hz and 30Hz. The output current is especially high from 10Hz to 20Hz where ambient vibration energy mainly exists. Devices based on sIL are able to retrieve power efficiently from such vibrations.

We are going to take following steps; (1) elimination of the external voltage, (2) expansion of frequency band, (3) design of devices optimized for ambient vibrations. Finally, we will fabricate μW energy harvesters based on sIL and replace the buttery of wireless sensor nodes.

ACKNOWLEDGEMENT

I would like to thank VLSI Design Education Center (VDEC) for mask fabrication. This work has been supported by the JSPS Core-to-Core Program.

REFERENCE

[1] L. Atzori, A. Iera, and G. Morabito, "The Internet of Things: A survey," Comput. Networks, vol. 54, no. 15, pp. 2787–2805, Oct. 2010.

[2] G. Lombaert, J. P. Conte, and M. Asce, "Random Vibration Analysis of Dynamic Vehicle-Bridge Interaction Due to Road Unevenness," J. Eng. Mech., vol. 138, no. 7, pp. 816–825, 2012.

[3] D. Shen, J.-H. Park, J. Ajitsaria, S.-Y. Choe, H. C. Wikle, and D.-J. Kim, "The design, fabrication and evaluation of a MEMS PZT cantilever with an integrated Si proof mass for vibration energy harvesting," J. Micromechanics Microengineering, vol. 18, no. 5, p. 055017, May 2008.

[4] S. Yamada, H. Mitsuya, and H. Fujita, "Vibrational Energy Harvester based on Electrical Double Layer of Ionic Liquid," to be published.

[5] W. Kong, P. Cao, X. He, L. Yu, X. Ma, Y. He, L. Lu, X. Zhang, and Y. Deng, "Ionic liquid based vibrational energy harvester by periodically squeezing the liquid bridge," RSC Adv., vol. 4, no. 37, p. 19356, 2014.

[6] Y. Suzuki, D. Miki, M. Edamoto, and M. Honzumi, "A MEMS electret generator with electrostatic levitation for vibration-driven energy-harvesting applications," J. Micromechanics Microengineering, vol. 20, no. 10, p. 104002, Oct. 2010.

[7] M. Armand, F. Endres, D. R. MacFarlane, H. Ohno, and B. Scrosati, "Ionic-liquid materials for the electrochemical challenges of the future.," Nat. Mater., vol. 8, no. 8, pp. 621–9, Aug. 2009.

[8] C. Gerbaldi, J. R. Nair, S. Ahmad, G. Meligrana, R. Bongiovanni, S. Bodoardo, and N. Penazzi, "UV-cured polymer electrolytes encompassing hydrophobic room temperature ionic liquid for lithium batteries," J. Power Sources, vol. 195, no. 6, pp. 1706–1713, Mar. 2010.

978-1-4799-7956-1/15 $31.00 © 2015 IEEE

FERROFLUID LIQUID SPRING FOR VIBRATION ENERGY HARVESTING

Yufeng Wang[*], Qian Zhang, Lurui Zhao, and Eun Sok Kim
Department of Electrical Engineering-Electrophysics
University of Southern California
Los Angeles, CA 900089-0271, USA

ABSTRACT

This paper reports ferrofluid liquid spring used to suspend a magnet array for harvesting vibration energy. A new idea of ferrofluid-based suspension is used for a low resonant frequency for a microfabricated vibration-energy harvester as well as high reliability. The ferrofluid liquid spring has reduced the resonant frequency of a microfabricated electromagnetic energy harvester to around 340 Hz, at which 36 nW is delivered into a matched load of 2.3 Ω from 7 g acceleration.

INTRODUCTION

With ever improving microelectronic and MEMS technologies coupled with Internet of Things (IoT), wearable devices are expected to be the next "big thing." However, their weight, volume and operating time are heavily limited by the power sources, as batteries add weight/volume and need to be recharged or replaced. Thus, an energy harvester which can supplement or replace battery is highly desirable.

Vibrations are in plenty of places and objects such as building walls, bridges, automobiles, airplanes, human body, etc. These ubiquitous vibration sources provide significant amount of renewable energy that can be harvested and used to power wearable devices as well as distributed sensors and actuators, wireless transceivers, etc. [1]. Unlike solar cell, vibration-energy harvester can provide power day and night, whether light is present or not. Common mechanical-to-electrical transduction mechanisms used in vibration-energy harvesters include capacitive [2], piezoelectric [3] and electromagnetic [4]. Among these techniques, electromagnetic energy harvester has the advantages of being able to drive low impedance load with high current level [5].

Vibration-energy harvesters are typically built on resonant structures with rigid suspension such as membrane [6], cantilever [7] or spring [8], and their fabrication processes are difficult, especially if the resonant frequency needs to be low and if they need to be fabricated with microfabrication process. Moreover, a rigid suspension is prone to breakage or failure under strong vibration or in long run. Thus, a sturdy suspension structure with low resonant frequency is highly desired. The ferrofluid-based liquid spring described in this paper is a promising solution. Ferrofluid as spring has small volume and good robustness under strong vibration. Additionally, as the magnets which are also used as proof mass are encapsulated by the ferrofluid and thus do not touch the frame, the friction between the magnets and the frame during vibration is greatly reduced. These advantages make the ferrofluid-spring-based energy harvester be more durable and efficient in producing electrical power from vibrating sources.

DESIGN

The schematic of the novel vibration energy harvester based on liquid spring is shown in Fig. 1. The ferrofluid-based liquid suspension is formed in an enclosed chamber made by bonding a micromachined silicon (with an electroplated copper coil) and a laser-machined acrylic frame. Due to its unique magnetic property, the ferrofluid automatically suspends the magnet array right in the middle of the chamber.

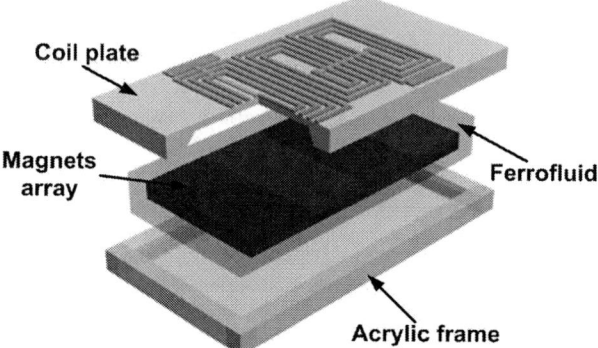

Figure 1: Schematic of the electromagnetic energy harvester based on liquid spring. The magnet array is suspended by ferrofluid in a chamber formed by a micromachined silicon (with electroplated copper coil) and a laser-machined acrylic frame.

The magnet array is formed by 4 NdFeB magnets with alternating north and south poles. This arrangement can provide a large magnetic field gradient. Electromotive force (EMF) can be produced when relative movement between the coil and magnet occurs due to the applied acceleration. To maximize the EMF, the center of the coil should be aligned with the boundary between two magnets where the magnetic flux gradient peaks. Thus, three coils in series are formed on a coil plate, since there are three boundaries in the magnet array.

Two designs of coil are compared: a square coil and a rectangular coil as shown in Fig. 2. According to Faraday's law, the induced EMF is

$$EMF = V.\sum_{k=1}^{n} B_k.L_k \qquad (1)$$

where V is the relative velocity between the coil and the magnet, while B_k and L_k are the magnetic flux density at the position of the kth wire and the length of the kth wire, respectively. The polarization of the magnet is along Z axis, thus the EMF mainly depends on the Z component of the magnetic flux density. Therefore, only the wires along the Y axis are going to cut the magnetic field line and contribute to EMF while the wires along the X axis only add to the resistance that needs to be minimized. The two rectangular coils occupy the same area at the one square coil, and thus the resistances for the two types are almost same as long as they have same wire width and spacing. However, only 50% of the wires in the square coil contribute to EMF, while 67% of the wires in the rectangular coil contribute. Thus, rectangular coils have been chosen in the fabricated devices.

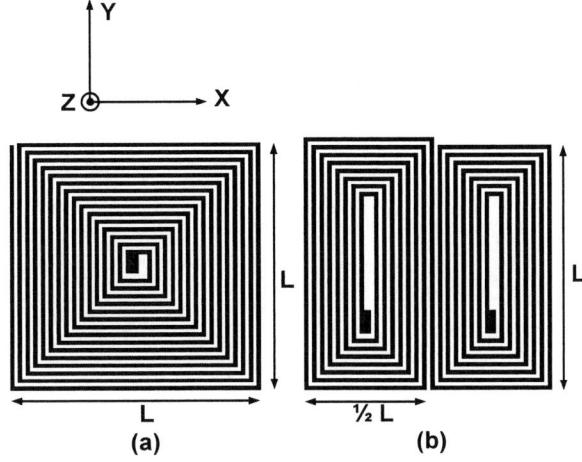

(a)   **(b)**

Figure 2: The polarization of the magnet is along Z axis, while the vibration is along X axis. Two designs of coil are compared: (a) square coil with side length of L and (b) rectangular coil with side lengths of L and 0.5 L.

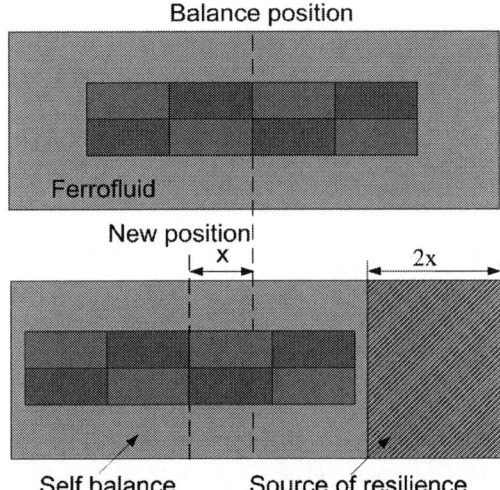

Figure 3: Conceptual illustration for liquid spring: (Top) Attractive force between the magnet array and ferrofluid makes the magnet stay in the middle automatically. (Bottom) When the magnets are displaced from its balanced position due to applied acceleration, the part of the ferrofluid which has no symmetric counterpart to neutralize the attraction draws the magnets back, with the force being proportional to the amount of the ferrofluid's unbalanced part.

As illustrated in Fig. 3, the ferrofluid works as a mechanical spring, since ferrofluid is a liquid that becomes strongly magnetized in the presence of a magnetic field, and is attracted by a magnet. The attractive forces counteract each other when the magnet array is in the middle with no applied acceleration. But once the magnet array is displaced from its balanced position (due to

applied acceleration), the part of the ferrofluid which has no symmetric counterpart will draw the magnets back, with the amount of the unbalanced part of the ferrofluid (or the force to pull back the magnets) being proportional to the deviated distance. The ferrofluid will also keep the magnet array right in the middle and align with the coil array automatically when there is no acceleration applied.

FABRICATION

Following the steps illustrated in Fig. 4, we etch a silicon wafer with KOH to form low pressure chemical vapor deposition (LPCVD) silicon-nitride micro-diaphragms for front-to-backside alignment [9], followed by a second KOH etching to form 200 μm deep trenches. Then Ti/Cu (20 nm/200 nm) is deposited by an evaporator, and patterned to form the connection between coils. After 1 μm-thick parylene is deposited for electrical insulation and patterned for electrical via, a second Ti/Cu (20 nm/200 nm) is deposited as a seed layer, followed by spin-coating and patterning of ~30 μm-thick photoresist (AZ4620) for a mold for the copper coil. After electroplating 30 μm thick copper, we remove the photoresist and then the seed layer, before bonding the silicon, with epoxy, to a laser-cut acrylic plate that contains NdFeB magnets in its recessed region. Then ferrofluid (Ferrotec, APG 1123) is filled through an inlet hole on the side of the acrylic plate. There exist some bubbles in the chamber after filling with ferrofluid. In order to fully fill the chamber, the device is pumped down to vacuum to remove the bubbles, and is filled with another dose of ferrofluid. Several repeated cycles of filling and vacuuming make the chamber be fully filled with ferrofluid without any bubble (Fig. 5). The inlet hole is sealed through depositing a 3 μm thick parylene layer. The fabricated device's dimensions and photos are shown in Table 1 and Fig. 6, respectively.

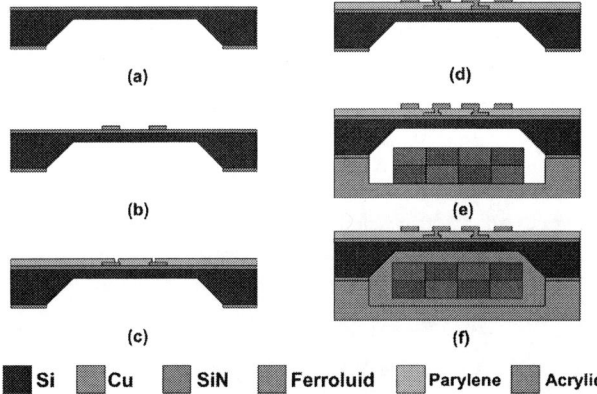

Figure 4: Brief microfabrication process of the energy harvester with ferrofluid as spring. (a) Etch Si from backside by KOH. (b) Form the connection electrode between coils to be deposited. (c) Deposit and pattern parylene isolation layer. (d) Evaporate Ti/Cu, followed by electroplating to form coils. (e) Assemble the magnets in a laser-cut acrylic chamber, and bond the chamber with the coil plate which has pre-etched trench. (f) Fill the chamber with ferrofluid and seal the inlet hole by depositing a 3 μm-thick parylene layer.

978-1-4799-7956-1/15 $31.00 © 2015 IEEE   123

Table 1: Physical dimension of the energy harvester.

Magnet size	6.4x3.2x0.8 mm³	Coil width	100 μm
Surface field	2,186 Gauss	Coil spacing	100 μm
Total weight	1 g	Coil thickness	~30 μm
Total volume	18.4x11x1.7 mm³	Resistance	2.3 Ω

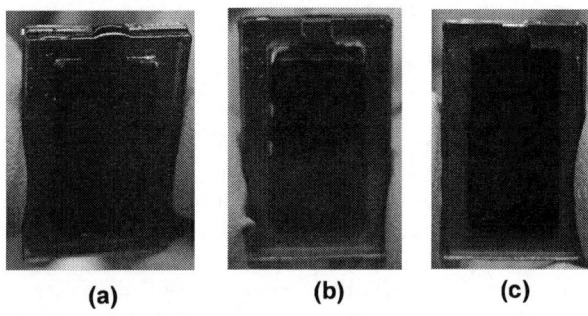

(a) **(b)** **(c)**

Figure 5: Photos of the chamber filled with ferrofluid: (a) before being pumped down to vacuum, (b) after being put into the vacuum (since the air bubbles have escaped from ferrofluid, there is large vacancy around the inlet), (c) after several cycles of filling and vacuuming (there is only small vacancy around the inlet, and the chamber is almost fully filled) .

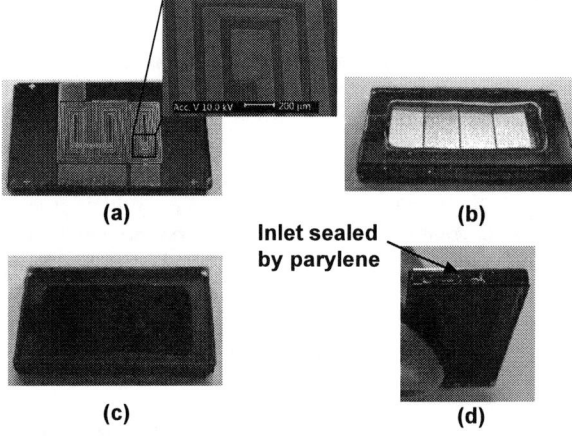

(a) **(b)**

Inlet sealed by parylene

(c) **(d)**

Figure 6: (a) Front-view photo of the energy harvester. Three electroplated coils are connected in series. (b) Back-view photo of the energy harvester before filling with ferrofluid. (c) Back-view photo of the energy harvester after filling with ferrofluid. (d) Photo showing the inlet hole sealed by depositing a 3 μm-thick parylene after the ferrofluid filling.

RESULTS AND DISCUSSION

The schematic of the testing platform is shown in Fig. 7. The fabricated energy harvester with ferrofluid spring is tested on a shaker that is driven by a function generator and calibrated by a laser Doppler displacement meter (LDDM). The output voltage of the harvester is amplified through a pre-amp and observed with an oscilloscope.

The measured output voltages as a function of frequency under various accelerations are shown in Fig. 8. At a fixed acceleration, the voltage depends on the vibration frequency and peaks at a resonant frequency. The measured resonant frequency is around 340 Hz, a relatively low value for a microfabricated harvester occupying 18.4×11×1.65 mm³ and weighing 1 gram, and the resonant

frequency decreases as the input acceleration increases, indicating that the liquid spring becomes softer as the vibrational amplitude increases.

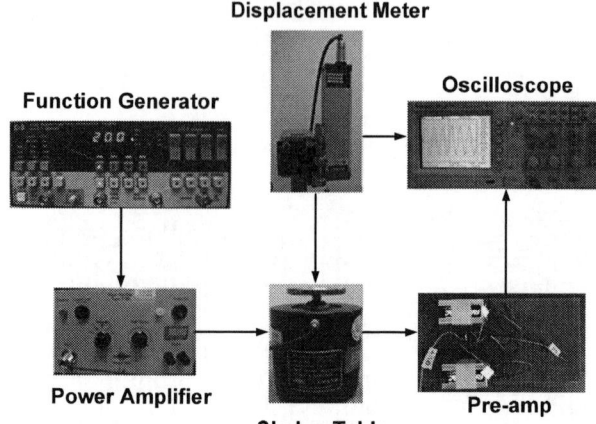

Figure 7: Schematic of the testing platform. A function generator provides the sinusoidal wave and drives the shaker table through a power amplifier. The shaker is calibrated with laser Doppler displacement meter. The signal from an energy harvester is amplified, and observed by and stored in an oscilloscope.

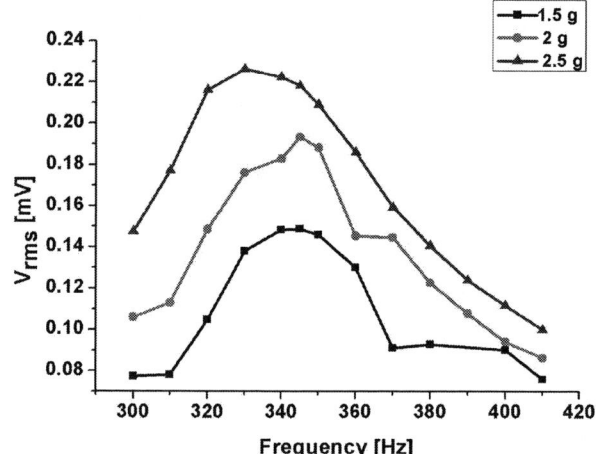

Figure 8: Frequency response of the energy harvester. The resonant frequency is around 340 Hz, and decreases when the input acceleration increases.

The root mean square (rms) of the output voltage and power that can be delivered to a matched load are shown in Fig. 9. The output voltage depends linearly on the applied acceleration, and 36 nW is delivered into a load of 2.3 Ω from 7 g acceleration (corresponding to 17 μm vibrational amplitude) at 320 Hz.

CONCLUSION

The new idea of using ferrofluid as liquid spring has been demonstrated. A magnet array with alternating north and south poles is levitated and aligned with a coil array automatically by ferrofluid. An energy harvester with ferrofluid spring has been fabricated with planar MEMS technology, and occupies a volume of 0.3 cc and weighs 1 gram. It produces an EMF of V_{rms}=0.58 mV and 36 nW power output (into 2.3 Ω load) at a resonant frequency of 320 Hz in response to a vibration amplitude of 17 μm (7 g at 320 Hz). The measured EMF vs. applied acceleration shows good linearity up to 7 g. These results show that the ferrofluid liquid spring is a sturdy suspension system that can achieve relative low resonant frequency while occupying small volume. Also, it can survive and maintain the performance under high input acceleration.

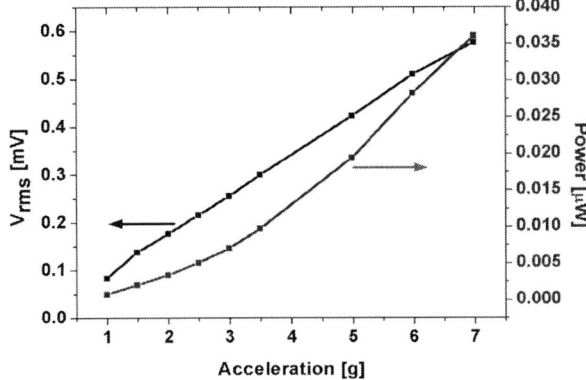

Figure 9: Measured power and V$_{rms}$ at the resonant frequency of 320 Hz vs. input acceleration: 36 nW power is delivered into a matched load of 2.3 Ω from 7 g acceleration (corresponding to 17 μm vibrational amplitude).

ACKNOWLEDGEMENTS
This article is based on the work supported by Defense Advanced Research Project Agency under grant no. N66001-13-1-4055.

REFERENCES
[1] S. Roundy, P. K. Wright, J. Rabaey, "A Study of Low Level Vibrations as a Power Source for Wireless Sensor Nodes," Computer Communications, vol. 26, pp. 1131-1144, 2003.
[2] T. Takahashi, M. Suzuki, T. Nishida, Y. Yoshikawa, S. Aoyagi, "Application of Paraelectric to a Miniature Capacitive Energy Harvester Realizing Several Tens Micro Watt - Relationship Between Polarization Hysteresis and Output Power," IEEE International Micro Electro Mechanical Systems Conference, Taipei, Taiwan, June 20-24, 2013, pp. 877-880.
[3] Y. Wang, T. Ren, Y. Yang, H. Chen, C. Zhou, L. Wang, L. Liu, "High-density PMUT Array for 3-D Ultrasonic Imaging Based on Reverse-bonding Structure", in IEEE International Micro Electro Mechanical Systems Conference, Cancun, Mexico, June 23-27, 2011, pp. 1035-1038.
[4] Q. Zhang, Y. Wang, E. S. Kim, "Power Generation from Human Body Motion through Magnet and Coil Arrays with Magnetic Spring", Journal of Applied Physics, vol. 115, pp. 064908-064908-5, 2014.
[5] S. P. Beeby, M. J. Tudor, N. M. White, "Energy harvesting vibration sources for Microsystems applications", Meas. Sci. Technol., vol. 17, pp. 175-195, 2006.
[6] Q. Zhang, S. J. Chen, L. Baumgartel, A. Lin, E.S.Kim, "Microelectromagnetic Energy Harvester with Integrated Magnets," Transducers '11, IEEE International Conference on Solid-State Sensors and Actuators, Beijing, China, June 5-9, 2011, pp. 1657-1660.
[7] H. Liu, C. J. Tay, C. Quan, T. Kobayashi, C. Lee, "Piezoelectric MEMS Energy Harvester for Low-Frequency Vibrations with Wideband Operation Range and Steadily Increased Output Power", IEEE.ASME J. Microelectromech. Syst., vol. 20, pp. 1131-1142, 2011.
[8] Q. Zhang, E. S. Kim, "Energy Harvesters with High Electromagnetic Conversion Efficiency through Magnet and Coil Arrays", in IEEE International Micro Electro Mechanical Systems Conference, Taipei, Taiwan, June 20-24, 2013, pp. 110-113.
[9] E. S. Kim, R. S. Muller, R. S. Hijab, "Front-to-Backside Alignment Using Resist-Patterned Etch Control and One Etching Step," IEEE/ASME J. Microelectromech. Syst., vol. 1, pp. 95-99, June 1992.

CONTACT
*Y. Wang, tel: +1-213-821-1611; yufengwa@usc.edu

AN ELECTROSTATIC ENERGY HARVESTER EXPLOITING VARIABLE-AREA WATER ELECTRODE BY RESPIRATION

Min-Ho Seo, Dong-Hoon Choi, Chang-Hoon Han, Jae-Young Yoo and Jun-Bo Yoon

Department of Electrical Engineering, Korea Advanced Institute of Science and Technology (KAIST), 291 Daehak-ro, Yuseong-gu, Daejeon, 305-701, Republic of Korea

ABSTRACT

This paper reports an electrostatic energy-harvester exploiting water-layer formed by respiration as a variable-area electrode. We discover that electrically conductive water-layer is instantly formed on a silicon-dioxide surface by exhaled-breath. We adopt this layer as a variable capacitive electrode for electrostatic energy-harvesting. The capacitance change was anticipated using a theoretical modeling and finite-element-method (FEM) simulation, and theoretical power-generation was estimated (~2 $\mu W/cm^2$ at 1 V). We then fabricated the prototype device and verified the capacitance change experimentally. Finally, the prototype showed charging and discharging characteristics by respiration successfully for being used as an energy-harvester driven by human-breath solely.

INTRODUCTION

In recent years, energy harvesters have gained much attention as portable power sources for self-sustainable wireless sensor networks [1-4]. Especially, a down-sized energy-harvesting device by means of micro-fabrication technologies provides self-powering capability for ubiquitous environmental- and bio-sensors without bulky energy storage components [5, 6]. For these advantages, various types of energy harvesters that harness different energy sources including natural environments, ambient physical and chemical stimulations, and even human-motions have been suggested [1-8]. Among these energy harvesters, researches employing human-motion for the energy source have been actively studied, because diverse and continuous human-motions such as physical motion, body-temperature, fluidic-blood, and breath allow the efficient and stable power generation.

As an energy-source, human-breath has significant advantages compared with other human-motions in terms of stability, sustainability and comfortability because human breathes with ease without break. Thus many researches have been focused on the power generation from the respiration. For example, piezo-electric and electromagnetic methods, which generate power by physically stimulating piezo- and magneto-materials, have been proposed [9-13]. In spite of great efforts, conventional methods, however, still need strong exhaled-breath or additional collecting-equipment to make an impact on target-materials, thus it restricts efficient and comfortable energy-harvesting. Here, we describe a simple and novel energy-harvester that uses the water vapor from respiration for the electrostatic power generation. The method reported here differs from previous works in a sense that a variable-area electrode by instantly condensed water-layer is used. The surface-wetting and evaporation occur naturally by respiration thus it enables, for the first time, us to generate electrostatic power from respiration.

We proved the proposed idea simply by means of measuring the capacitance change and current-flow on a fabricated prototype and demonstrated the feasibility of the proposed method for an energy-harvester.

CONCEPT

Schematics of the proposed device and its operational principle are shown in figure 1. The proposed device has a similar configuration with a typical metal-insulator-metal (MIM) structure, but it consists of a non-overlapped metal electrode on a silicon-dioxide (SiO_2) insulating layer (Fig. 1(a)). The principle of the proposed device is based on the capacitance change in an electrostatic energy harvester. The capacitance is formed between the top and the bottom electrodes, but the area of the top electrode is variable depending on a surface wetting which is natural phenomenon that generates water-layers on the SiO_2 surface by exhaled breath. The detailed surface-wetting can be explained by a typical water-vapor-pressure versus temperature diagram (Fig. 1b) [14]. On the diagram, the red-solid-line expresses the 'saturation-curve', equilibrium between evaporation and condensation of water. The orange area below the 'saturation-curve' presents an under-saturated state and the blue area above the 'saturation-curve' indicates a super-saturated state, respectively and the evaporation and condensation are

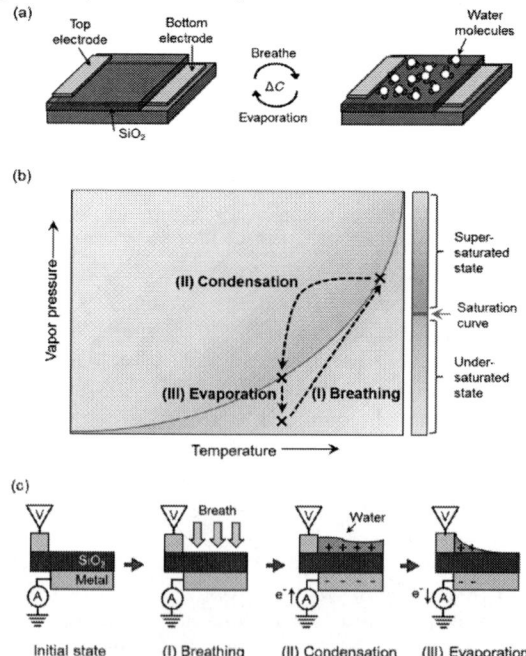

Figure 1: Schematic diagrams of (a) the proposed electrostatic energy harvester based on human respiration (b) a typical water-vapor-pressure versus temperature curve and (c) sequence of the water-layer formation and evaporation

dominant in the respective states. Under room temperature (~25 °C) and atmospheric pressure (~1 atm) conditions, the state of air around the device generally belongs to 'under-saturated-state'. In this state, the small capacitance is formed by fringe-field of the non-overlapped MIM structure (Initial state in Fig. 1(c)). When a human breathes onto the device (I. breath in Fig. 1(c)), the atmosphere around the device is abruptly replaced with exhaled-breath-molecule (~100 % relative humidity (RH) and ~36 °C temperature by human organ). At the moment, the humid exhaled breath air is instantly cooled by the ambient air and the device itself, thus water-molecule begins to be condensed on the SiO₂ surface, and that is similar to frost formed by exhaled breath on windows in winter. The water-condensation is continued until the air around the device becomes 'saturation', and that provides thick enough water-layer for the variable-area electrode (II. Condensation in Fig. 1(c)). Therefore, large capacitance can be formed by the enlarged water-electrode. Finally, the water on the electrode surface evaporates naturally due to its vapor-pressure (III. Evaporation in Fig. 1(c)), and the state of the device and its capacitance are returned to the original point (Initial state). In the above process, the variable capacitance allows charges to circulate from the ground or voltage-source to the device, thus it allows an electrostatic energy–harvesting [2, 7].

DEMONSTRATION

Figure 2(a) shows the fabrication flow of the proposed energy-harvester. A 0.7-μm-thick glass wafer was used for device isolation (I). The process begins with thermally evaporated chrome Cr (100 Å)/gold Au (1,000 Å) as the bottom electrode (II). Next, we deposited 1-μm-thick silicon-dioxide for the insulating layer using a plasma enhanced chemical vapor deposition (PECVD) at 350 °C (III). After the insulating layer deposition, the top electrode (Cr (100 Å)/Au (1,000 Å)) patterning was performed for the non-overlapped top metal electrode (IV) then some part of the insulating SiO₂ on the bottom electrode was removed by wet etching process (V). Finally, we defined the outer boundary of the variable-area electrode and the top and bottom contact pads using a naturally hydrophobic photoresist for a stable operation (VI). Figure 2(b) represents an optical image of the fabricated energy-harvesters. To prevent an electrical short between the top and bottom electrodes by respiration, their contact pads were formed far away from the outer boundary of the variable-area electrode.

RESULTS AND DISCUSSIONS

Before the energy-harvester experiments, we confirmed electrical properties of the proposed water electrode formed by exhaled breath. The properties were measured through a conventional current versus voltage characteristic and a transmission-line-measurement (TLM) on the devices consisting of two gold electrodes formed 100 μm-apart on a thermally oxidized silicon wafer using a parameter analyzer 4156C. Figure 3 shows the measured current-voltage curve. From -5 V to 5V range, the measured curve shows a similar shape with the curve of a

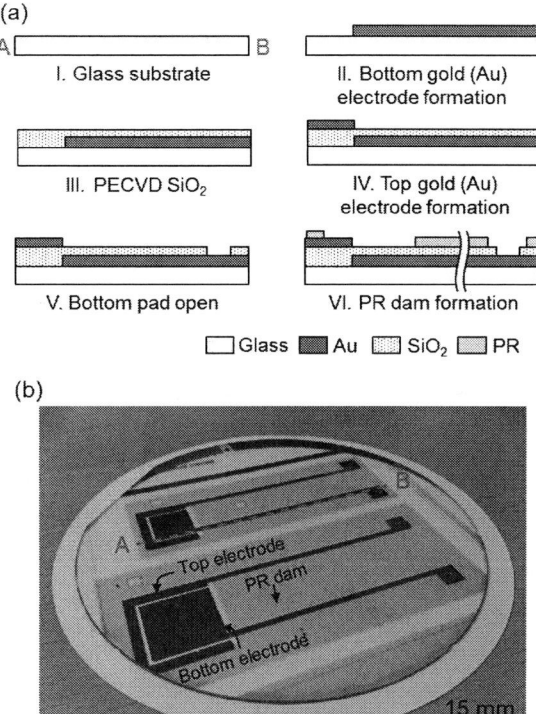

Figure 2: (a) Schematic of fabrication and (b) an optical image of the fabricated 1 cm × 1 cm and 1.5 cm × 1.5 cm sized devices, respectively.

typical water-electrolysis. This result enables us to convince that water-layer is successfully formed between the electrodes by respiration. To analyze sheet-resistance of the generated water-layer by respiration, we performed more I-V tests on various devices having two electrode distances from 50 to 300 μm, and we calculated the sheet-resistance by the process so called 'TLM' [15]. As shown in Fig. 4, the generated water-layer shows decreasing sheet-resistance as the applied voltage increases. We can explain this with accelerated electrolysis of the water-layer, which allows us to estimate sheet-resistance of the water-layer as ~45 Mohms/sq at 5 V. Note that, this is a suitable value for electrostatic discharging so that the generated water-layer can be used for the charge-transport as the variable-area electrode [16].

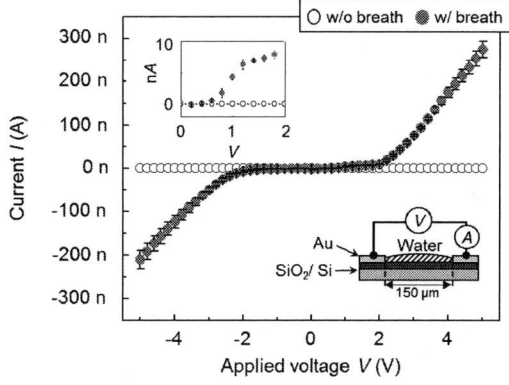

Figure 3: Current-Voltage curve of the water-layer formed on the 150 μm-separated electrodes by exhaled breath.

978-1-4799-7956-1/15 $31.00 © 2015 IEEE   127

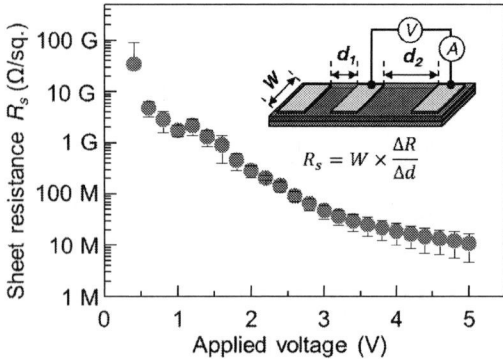

Figure 4: Sheet resistance of the water-layer as a function of the applied voltage. Inset shows schematics of the test set-up, and d and W present the distance between the electrodes and width of the electrode, respectively.

As mentioned before, the proposed water-layer formed by respiration plays a role as the variable-area electrode and it contributes to electrostatic energy-harvesting by means of changing capacitance. Thus we performed theoretical FEM simulation and calculation to confirm the feasibilities of its capacitive changing ability. Figure 5a shows the result of the electrical field simulation using Maxwell for the two situations: minimized-capacitance C_{min} (I. Initial state in Fig. 1(c)) and maximized-capacitance C_{max} (II. condensation). The FEM simulation result informs that the proposed water-electrode successfully generates uniform electric-field, resulting in appropriate capacitive change. Moreover, theoretically calculated capacitance values using Coventor-ware and the parallel-plate analytical model also indicate that capacitance can be changed by water-electrode (Fig. 5(b)). Note that the calculated capacitance were changed from ~1.6 pF to ~3.5 nF in a 1 cm × 1 cm device by the water-electrode, corresponding to the capacitance changing ratio of more than 2000, which can generate the electrostatic power of over 2 μW/cm^2 at 1 V theoretically [7].

Since we have confirmed the electrical properties and the feasibilities of the water-layer formed by human respiration for being used as the variable-area electrode for electrostatic energy-harvester, we could go further to characterize the experimental capacitance and current changes of the fabricated energy-harvester. Figure 6 represents the experimental result of the 1 cm × 1 cm device with real-time capacitance response. The

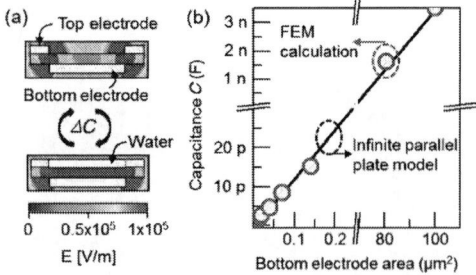

Figure 5: (a) FEM simulation of the capacitance change by water layer and (b) its theoretical calculation

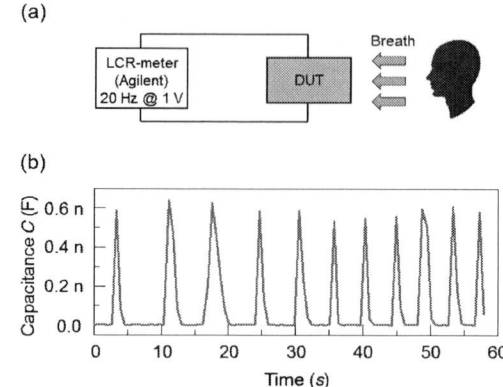

Figure 6: Real capacitive change test by respiration: (a) schematics of set-up and (b) measured result.

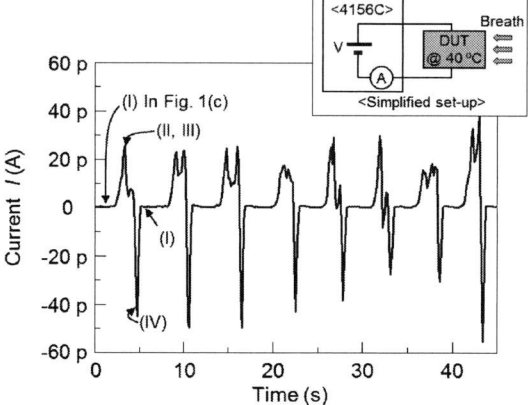

Figure 7: Measured current change due to human respiration. The charges move forward and backward directions by respiration under 0.1 V potential

capacitance was measured using LCR-meter (Agilent) by applying 20 Hz and 1 V AC-bias under ~20 °C temperature and ~50 % RH chamber condition, and a volunteer blew breath with ~30 cm apart from the device. The measured capacitance was changed drastically from ~2 pF to ~600 pF and response and relaxation times were fast enough (<0.65 sec) compared to normal breathing-interval. Note that the low C_{max} compared to the simulated value can be explained with an experimental limitation such as our LCR-meter measuring resolution (20 Hz as the lowest AC frequency to detect capacitance). Finally, we measured real charge-flow induced by respiration using a parameter analyzer 4156C under 40 °C temperature (Fig. 7). Even though this test does not allow us to detect short circuit current and open circuit voltage for power-estimation, it enables us to verify feasibilities of the proposed energy-harvester by measuring current-flow under applied static voltage (inset in Fig. 7). The measured current-flow represents the charging and discharging characteristics, which implies electrostatic energy-harvesting, by induced respiration and the maximum current value was 31 and -58 pA, respectively, at charging and discharging situations. The different current values can be explained with the dynamics of fluidic water-electrode causing the unstable charge-flow by the breath-pressure during the charging process. On the contrary, sharp and smooth current-flow

978-1-4799-7956-1/15 $31.00 © 2015 IEEE   128

was detected during the discharging, which is generated by natural evaporation without any mechanical disturbances. Thus we guess that the uneven current-flow was measured during the charging-discharging process. Note that even though the different shape of charging and discharging current-flow, the charge-conversion during the charging-discharging assure us that the proposed device can generate power from human-breath successfully.

CONCLUSION

In conclusion, we proposed and investigated a novel electrostatic energy-harvester using human-breath. The method exploits the natural phenomena of water condensation and evaporation by respiration and those play a role as forming variable-area electrode for a capacitive change, thus this approach enables us to generate power more efficiently and comfortably without any intentional or additional equipment. We theoretically and experimentally verified that the capacitance was dramatically changed by respiration and the capacitive change successfully generates current-flow for power generation. We anticipate that the proposed energy-harvesting method will provide many advantages including ease of power-generation and cost-effectiveness for wireless sensor networks.

ACKNOWLEDGEMENT

This research was supported by the Pioneer Research Center Program through the National Research Foundation of Korea funded by the Ministry of Science, ICT & Future Planning (2010-0019313).

REFERENCES

[1] S. Priya, D. J. Inman, Energy Harvesting Technologies, Springer, 2009.

[2] B. Yen and J. Lang, "A Variable-Capacitance Vibration-to-Electric Energy Harvester", *Circuits Syst. I Regul. Pap.*, vol. 53, no. 2, pp. 288–295, 2006.

[3] I. Robel and V. Subramanian, "Quantum dot solar cells. Harvesting light energy with CdSe nanocrystals molecularly linked to mesoscopic TiO2 films," *J. Am. Chem. Soc.,* no. 19, pp. 2385–2393, 2006.

[4] S. P. Beeby, M. J. Tudor, and N. M. White, "Energy harvesting vibration sources for microsystems applications," *Meas. Sci. Technol.*, vol. 17, no. 12, pp. R175–R195, 2006.

[5] X. Wang, J. Song, J. Liu, and Z. L. Wang, "Direct-Current Nanogenerator Driven by Ultrasonic Waves.," *Science*, vol. 316, no. 5821, pp. 102–5, 2007.

[6] X. Chen, S. Xu, N. Yao, and Y. Shi, "1.6 V Nanogenerator for Mechanical Energy Harvesting using PZT Nanofibers.," *Nano Lett.*, vol. 10, no. 6, pp. 2133–7, 2010.

[7] D.-H. Choi, C.-H. Han, H.-D. Kim, and J.-B. Yoon, "Liquid-based electrostatic energy harvester with high sensitivity to human physical motion," *Smart Mater. Struct.*, vol. 20, no. 12, p. 125012, Dec. 2011.

[8] M.-H. Seo, D.-H. Choi, I.-H. Kim, H.-J. Jung, and J.-B. Yoon, "Multi-resonant energy harvester exploiting high-mode resonances frequency down-shifted by a flexible body beam," *Appl. Phys. Lett.*, vol. 101, no. 12, p. 123903, 2012.

[9] Y. Qi and M. C. McAlpine, "Nanotechnology-enabled Flexible and Biocompatible Energy Harvesting," *Energy Environ. Sci.*, vol. 3, no. 9, p. 1275, 2010.

[10] C.-Y. Sue and N.-C. Tsai, "Human Powered MEMS-based Energy Harvest Devices," *Appl. Energy*, vol. 93, pp. 390–403, 2012.

[11] H.-I. Lin, D.-S. Wuu, K.-C. Shen, and R.-H. Horng, "Fabrication of an Ultra-Flexible ZnO Nanogenerator for Harvesting Energy from Respiration," *ECS J. Solid State Sci. Technol.*, vol. 2, no. 9, pp. P400–P404, 2013.

[12] Q. Zheng, B. Shi, F. Fan, X. Wang, L. Yan, W. Yuan, S. Wang, H. Liu, Z. Li, and Z. L. Wang, "In vivo Powering of Pacemaker by Breathing-driven Implanted Triboelectric Nanogenerator.," *Adv. Mater.*, vol. 26, no. 33, pp. 5851–6, 2014.

[13] C. Sun, J. Shi, D. J. Bayerl, and X. Wang, "PVDF Microbelts for Harvesting Energy from Respiration," *Energy Environ. Sci.*, vol. 4, no. 11, p. 4508, 2011.

[14] Y. Guissani and B. Guillot, "A Computer Simulation Study of the Liquid–Vapor Coexistence Curve of Water," *J. Chem. Phys.*, vol. 98, no. 10, p. 8221, 1993.

[15] A. Venugopal, L. Colombo, and E. M. Vogel, "Contact Eesistance in Few and Multilayer Graphene Devices," *Appl. Phys. Lett.*, vol. 96, no. 1, p. 013512, 2010.

[16] M. Narkis, G. Lidor, A. Vaxman, and L. Zuri, "New Injection Moldable Electrostatic Dissipative (ESD) Composites Based on Very Low Carbon Black Loadings," *J. Electrostat.*, vol. 47, no. 4, pp. 201–214, 1999.

CONTACT

Min-Ho Seo, tel:+82-350-5476, mhseo@3dmems.kaist.ac.kr

SILICON ANODE SUPPORTED BY CARBON SCAFFOLD FOR HIGH PERFORMANCE LITHIUM ION MICRO-BATTERY

Xiaozhao Li[1,2], Xiaohong Wang[1,2], and Siwei Li[1,2]*

[1]Tsinghua National Laboratory for Information Science and Technology, PR China
[2]Institute of Microelectronics, Tsinghua University, Beijing, PR China

ABSTRACT

We have developed a novel silicon/void/carbon anode for lithium ion batteries, with high specific capacity (2200 mAh/g in initial cycle) and superior cyclability (1200 mAh/g after 80 cycles). The high capacity is enabled by the sufficient silicon nanoparticles (SiNPs) at anode. And long cycle life is ensured by the nano void space between SiNPs and carbon scaffold, providing enough space for expansion and contraction of SiNPs during the process of lithium ion intercalation and deintercalation. By using templating method from photoresist (SU-8) with nanoparticles, it's possible to implement direct prototyping of three dimensional (3D) micro-battery on chip.

INTRODUCTION

Electrochemical energy storage has become a critical technology for a variety of applications, including electric vehicles, portable electronic devices and renewable energy storage. Rechargeable lithium-ion batteries have been widely considered as the most useful devices for electrochemical energy storage over the past decades because of their relatively high gravimetric and volumetric capacity [1]. To meet the ever increasing high-performance requirements and for applications, Li-ion batteries are in desperate need of novel electrode materials with high energy density and long cycle life. For high energy density, the electrode materials must possess higher specific capacity than current commercial electrode materials. And much research has been devoted to improving the performance and cycle life of high-capacity electrode materials. Recently, silicon has emerged as one of the most attractive high-capacity anode materials for next-generation Li-ion batteries because of its natural abundance, low discharge potential, and most importantly, high theoretical capacity (4200 mAh/g in a composite of $Li_{4.4}Si$), which is an order of magnitude beyond that of conventional graphite anodes (theoretical capacity of 372 mAh/g) [2]. Unfortunately, silicon undergoes large volume change (over 300%) during lithium-ion intercalation and deintercalation [3]. The volume change can lead to anode structure failure and electrical contact loss, and also causes continuous consumption of the active material and Li-ion for reformation of solid electrolyte interphase (SEI) layers [4]. As a consequence, the electrode suffers rapid capacity decay and poor reversibility.

Recently, a number of works have been focused on the reduction of the dimensions of the silicon phase to reduce the limitations. Switching from bulk to nanoscale morphologies, several strategies have been proposed to reduce the structural degradation due to the volume change during cycling and therefore to overcome the mechanical instability in bulk silicon. These nanostructure strategies such as thin films [5, 6], nanowires/nanotubes [7, 8] and nanoparticles [9, 10] facilitate relaxation of the stress

associated to the intercalation/deintercalation process and thus demonstrating high capacities in initial cycles. However, most of these all-silicon anodes still demonstrate obvious capacity fade due to the intrinsic low electrical conductivity of silicon, which may lead to silicon's separation from current collector after long cycles. Many materials with good conductivity have been tested to stabilize Si-collector contact. Among these efforts, carbon demonstrates to be effective not only in enhancing the electronic conductivity but also in stress relief due to its lightweight and ductile nature. The composite of silicon and carbon is gaining more and more attentions in the search of new high-performance Li-ion battery anodes. Much progress in terms of cycle stability has been made by designed silicon/carbon nanocomposite anodes from various carbon sources, such as graphene, PVDF, pitch, acetylene and polydopamine [11]. Some of these anodes also introduce suitable networks of interconnected pores in order to further buffer the volume change of the silicon.

However, none of the above-mentioned silicon-based techniques are compatible with the microfabrication protocols. As such, it's difficult to implement micro-batteries and hard to fulfill the high energy and power density requirements for powering of particular Micro-electromechanical Systems (MEMS) such as smart dusts, which are generally fabricated on a limited footprint surface.

In this work, a novel Si/void/C nanocomposite anode for high-energy Li-ion micro-batteries is designed and fabricated. SU-8 photoresist is used as the carbon source to construct a scaffold to support SiNPs for the first time, aiming at the opportunity to combine with MEMS technology. The SiNPs make the main contribution to the high specific capacity, and the carbon scaffold from SU-8 builds an efficient conductive pathway. With rationally designed void space to accommodate silicon expansion and contraction during cycling, the anode maintains a stable structure and achieves a long cycle life.

DESIGN AND FABRICATION

With the help of nanoscale fabrication techniques, the limitations of carbon-based or all-silicon-based anodes could be addressed by designing and controlling the microstructure of Si/C composite anode. The principle is based on the fact that graphitic carbon is able to release the stress generated by expansion and contraction of silicon, meanwhile the high conductivity of carbon can make up the electrical connection failure caused by silicon volume change.

In this work, we use SiNPs as the main active materials and SU-8 photoresists as the carbon source, which consists of epoxy resin SU-8, triaryl sulfonium salt and organic solvent. When exposed to ultraviolet light, the resin will form a highly structured cross-linked matrix.

978-1-4799-7956-1/15 $31.00 © 2015 IEEE

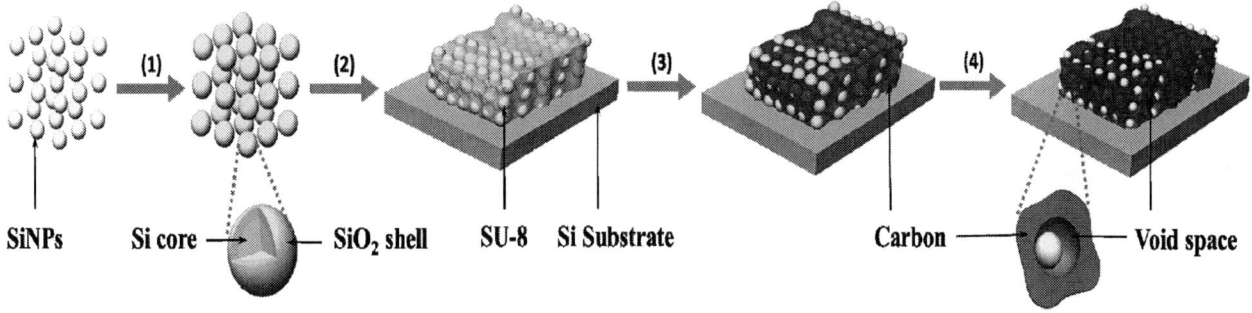

SiNPs   Si core   SiO₂ shell   SU-8   Si Substrate   Carbon   Void space

Figure 1: Schematic of the design and fabrication of the Si/void/C nanocomposite anode. 1) TEOS oxidation process to generate SiO₂ sacrificial layer; 2) Mixing Si/SiO₂ nanocomposite particles with SU-8 photoresist and spin-coating on Si substrate; 3) Pyrolysis process to carbonize SU-8; 4) Etching away SiO₂ layer, forming void space between carbon scaffold and SiNPs.

Moreover, SU-8 can retain micro patterns after high-temperature carbonization, which has been widely utilized in carbon Micro-electromechanical Systems (C-MEMS) and micro power devices, such as micro supercapacitors [12] and three-dimensional micro batteries [13]. The good thermal stability and chemical resistance of SU-8 shown in these researches inspire a brand new way to fabricate on-chip batteries with Si/C composite anode, using a Si/SiO₂-templated synthetic procedure.

The scheme of the overall design and fabrication of the Si/void/C nanocomposite anode is shown in Fig.1. In the 1st step, SiNPs (30~50 nm in diameter) coated by SiO₂ layer are obtained by typical Stöber method, with hydrolysis and condensation of tetraethoxysilane (TEOS) in a water-alcohol-ammonia medium [14]. The thickness of the SiO₂ sacrificial layer and therefore the void space size can be easily controlled by controlling the coating time, PH, and TEOS concentration. In order to accommodate the maximum volume expansion (300%) of each SiNPs, the thickness of sacrificial coating layer should be over 10nm. In following experiments, the thickness of the SiO₂ layer is controlled to be about 10~15 nm, considering size variation. In step 2, the Si/SiO₂ particles are dispersed in SU-8 photoresist to form Si/SiO₂-templated SU-8 composite, where SU-8 serves as the precursor of subsequent carbon scaffold. Considering both specific capacity and stability of the electrode, the ratio of each component in this step is 20wt% for Si/SiO₂ particles and 80wt% for SU-8. After spin-coated on silicon substrate, the Si/SiO₂-templated SU-8 composite is exposed under ultraviolet light, forming highly structured cross-linked polymer matrix. In step 3, the polymer contained in the materials is pyrolyzed in nitrogen atmosphere at 900 °C for 2h. And the SU-8 is turned into carbon scaffold to provide a considerably strong mechanical support for the particles. At last, to remove the SiO₂ sacrificial layer, the Si/SiO₂/C composite is etched in HF aqueous solution and the space once occupied by the nano SiO₂ layer becomes void space.

The SEM image of the cross-section of the Si/void/C nanocomposite anode before cycling is shown in Fig.2 (a). We can see that SiNPs are embedded in carbon scaffold with some nano void space between them clearly, which provides enough space for expansion of individual SiNPs

during lithium-ion intercalation.

The working electrodes are fabricated using a typical aqueous slurry method. Si/void/C nanocomposite electrode is prepared by homogeneously mixing appropriate amount of Si/void/C nanocomposite, carbon black and carboxymethyl cellulose sodium (CMC) binder to form a slurry, with a mass ratio of 8:1:1. In this step, to obtain the optimal proportion of silicon and carbon in following tests, four samples with different silicon content in weight percentage (35%, 50%, 60%, and 75%) are designed and synthesized. The viscous slurry is then cast onto a copper foil and dried at 80 °C under vacuum. For electrochemical tests, coin-type cells (CR2025) are fabricated inside a N₂-filled glove box, using circular Si/void/C nanocomposite electrode as the working electrode and Li metal foil as the counter electrode. As shown in Fig.2 (b), two electrodes are separated by a Celgard 2400 separator, which is soaked in the electrolyte. 1 mol/L LiPF₆ in a mixture of ethylene carbonate, diethyl carbonate and dimethyl carbonate (EC : DEC : DMC, 1: 1 : 1 by volume) is used as the electrolyte, and with 10% fluoroethylene carbonate (FEC) as the electrolyte additive to increase the

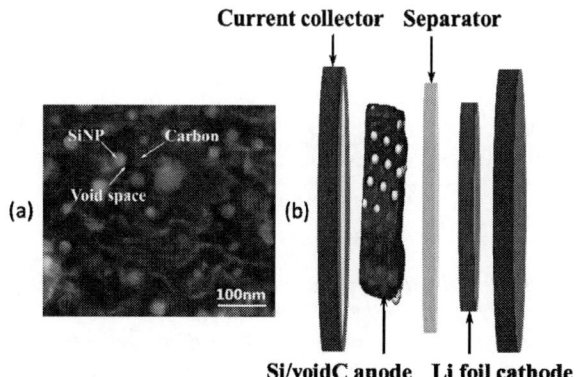

Figure 2: a) SEM image of the cross-section of Si/void/C nanocomposite anode. There is void space between SiNPs and carbon scaffold; b) Packaging diagram of Li-ion battery: Si/void/C nanocomposite as the anode; Li foil as the cathode.

cycling efficiency and improve cycling stability.

RESULTS AND DISCUSSION

The galvanostatic charge-discharge tests are performed using a battery testing system (BTS-3000, Neware) at room temperature, with voltage cutoffs set at 0.01V and 1.2V vs. Li/Li$^+$. Specific capacity is calculated based on the total mass of Si/void/C composite. The charge/discharge rate is calculated with respect to the theoretical capacity of silicon (4000 mAh/g). Therefore, a rate of 1C corresponds to a current density of 4000 mA/g.

For preliminary test, the SiNPs account for about only 35% of the anode composite by weight, thus the anode has a theoretical capacity of 1470 mAh/g, on the premise that the capacity contributed by carbon is negligible when compared with silicon. From the discharge/charge voltage profiles for different cycles in Fig.3, the actual initial capacity is about 1200 mAh/g. In the first few discharge processes, there is a long discharge plateau below 0.3V due to the SiNPs crystalline structure. The discharge/charge curves remain similar in shape from 1st to 10th cycle with a small decrease in the capacity, indicating a stable electrochemical behavior. This excellent cycling stability is attributed to the presence of void space and the structural stability of carbon scaffold from SU-8, therefore individual SiNPs can expand and contract inside the carbon scaffold without rupturing the outer shape of the structure. The capacity retention is 71% after 30 cycles. The irreversible capacity loss is a result of the irreversible insertion of Li into silicon and amorphous carbon. Besides, the continuous formation of SEI layers could make a significantly influence, because the carbon accounts for over 60% of the composite and directly contact the electrolyte, with a high surface area structure. The inset in Fig.3 shows the composite anode on a copper foil and the

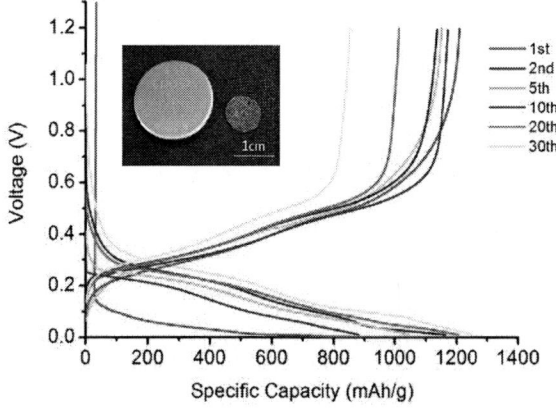

Figure 3: Galvanostatic discharge and charge profiles of the Si/Void/C nanocomposite anode (35wt% Si) for different cycles: 1st, 2nd, 5th, 10th, 20th and 30th. The inset is the photograph of Si/void/C composite anode (right) and packaged battery device (left).

packaged cell for testing.

To optimize the proportion of silicon and carbon, four samples with different silicon content in weight percentage (35%, 50%, 60%, and 75%) are designed and tested. As shown in Fig.4, the initial capacity increases with the silicon content. This indicates that the presence of SiNPs

indeed makes a significant contribution to the specific

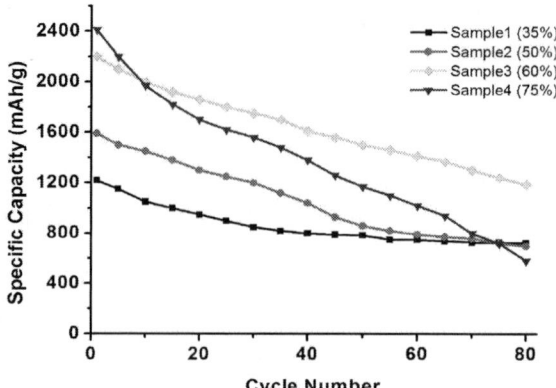

Figure 4: Galvanostatic cycling of Si/void/C with different Si content (wt%), from sample 1 to 4:35%, 50%, 60%, 75%.

capacity increase. While in long term cycles, the sample 3 (60wt% Si) shows the best performance, which is 1200 mAh/g after 80 cycles. As for the sample 4 (75wt% Si) which has the highest occupancy of silicon, its capacity declines fast from 2400 mAh/g in the initial cycle to 580 mAh/g after 80 cycles, ultimately lower than that of any other samples far from our expectation. This is mainly because of the low electrical conductivity and structure instability of too much silicon, similar to all-silicon anodes. From these results, we can learn that the proportion of silicon and carbon is one of the crucial factors to achieve the excellent performance of the Si/void/C anode. Silicon contributes to the capacity increase while carbon plays a vital role in maintaining stable structure and electrical contact. There should be a "trade-off" between high specific capacity and long cycle life, by controlling mass fraction.

The Si/void/C nanocomposite demonstrates much higher capacity and better capacity retention when compared to pure SiNPs and Si/C composite without void space (with the same wt% of silicon), as shown in Fig.5. Our Si/void/C nanocomposite anode has an initial specific capacity of 1200 mAh/g and a capacity of 850 mAh/g remaining after 30 cycles. Under the same conditions, the

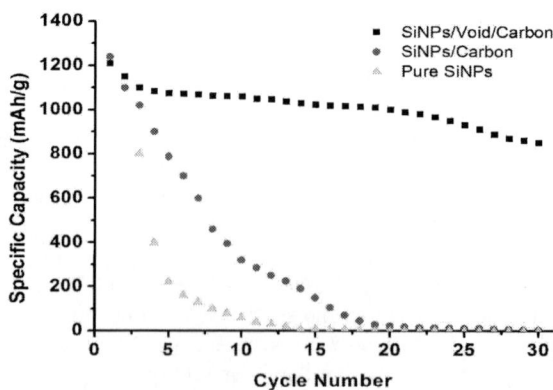

Figure 5: Galvanostatic cycling of different structures: a) pure SiNPs; b) Si/C composite (without void space) and c) Si/C composite (with void space).

*The Si content in b) and c) are both 35%

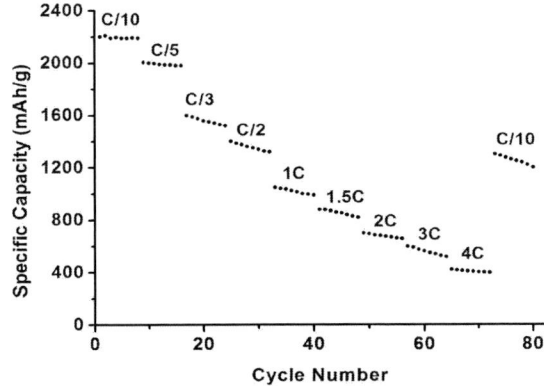

Figure 6: Charge capacity of Si/void/C (60wt% Si) anode under various rates from C/10 to 4C. At the high rate of 4C, the anode achieves a specific capacity of 395mAh/g, higher than the theoretical capacity of graphite anode.

pure SiNPs anode has a high initial capacity but decays quickly to almost zero after only several cycles, due to its structure failure, SEI formation and poor electrical conductivity. Si/C composite, which is without void space inside, shows a similar initial capacity to Si/void/C structure. However, poor capacity retention of less than 10% is observed after only 15 cycles. The above comparison has strongly proved that the Si/void/C composite anode shows outstanding performance, mainly because of the advantage of allowing the silicon particles to expand freely, with a stable and conductive carbon scaffold to support silicon.

The anode's cycle performance under various C rates from C/10 to 4C is also tested to verify the effect of amorphous carbon scaffold on electrochemical performance. Even at the high rate of 4C, the anode still achieves a specific capacity of 395mAh/g, which is higher than that of the conventional graphite anode (372mAh/g) in theory, as shown in Fig.6. This result suggests a stable structure and efficient conductive path with the help of carbon scaffold from SU-8.

CONCLUSIONS

In summary, a novel Si/void/C nanocomposite anode from Si/SiO$_2$-templated SU-8 are designed and fabricated. It demonstrates a high capacity (over 2000 mAh/g), stable cycling behavior (55% capacity retention after 80 cycles) and outstanding performance under high discharge/charge rate. Carbon scaffold serves to release the stress and provide an efficient conductive pathway, with void space inside to accommodate volume expansion during cycling. Hence the structural stability and cycling performance of the silicon-based anode are greatly improved. Furthermore, this technology has a great potential to implement patterned silicon anode and direct prototyping of high performance on-chip 3D micro-battery in the future.

ACKNOWLEDGMENTS

This work is supported by the National Natural Science Foundation of China (No. 61474071), 973 program (No. 2009CB320304), and 863 program (No. 2009AA04Z319) of China. The authors also thank the members in the laboratory of Department of Chemistry, Tsinghua University, for providing necessary experimental equipments and helpful instructions.

REFERENCES

[1] J. M. Tarascon, M. Armand, "Issues and Challenges Facing Rechargeable Lithium Batteries," *Nature*, vol. 414, pp. 359-367, 2001.

[2] A. S. Arico, et al., "Nanostructured Materials for Advanced Energy Conversion and Storage Devices," *Nature Materials*, vol. 4, pp. 366-377, 2005.

[3] B. A. Boukamp, G. C. Lesh, R. A. Huggins, "All-solid Lithium Electrodes with Mixed-conductor Matrix," *J. Electrochem. Soc.*, vol. 128, pp. 725-729, 1981.

[4] S. Nadimpalli, et al., "Quantifying Capacity Loss due to Solid-electrolyte-interphase Layer Formation on Silicon Negative Electrodes in Lithium-ion Batteries," *J. Power Sources*, vol. 215, pp. 145-151, 2012.

[5] T. Takamura, et al., "A Vacuum Deposited Si Film Having a Li Extraction Capacity over 2000 mAh/g with a Long Cycle Life," *J. Power Sources*, vol. 129, pp. 96-100, 2004.

[6] M. S. Park, et al., "Electrochemical Properties of Si Thin Film Prepared by Pulsed Laser Deposition for Lithium Ion Micro-batteries," *ELECTROCHIMICA ACTA*, vol. 51, pp. 5246-5249, 2006.

[7] C. K. Chan, et al., R. A. Huggins, and Y. Cui, "High-performance Lithium Battery Anodes Using Silicon Nanowires," *Nature Nanotechnology*, vol. 3, pp. 31-35, 2008.

[8] H. Kim, J. Cho, "Superior Lithium Electroactive Mesoporous Si@Carbon Core-Shell Nanowires for Lithium Battery Anode Material," *Nano Letters*, vol. 8, pp. 3688-3691, 2008.

[9] X. S. Zhou, et al., "Facile Synthesis of Silicon Nanoparticles Inserted into Graphene Sheets as Improved Anode Materials for Lithium Ion Batteries," *Chem. Commun.*, vol. 48, pp. 2198-2200, 2012.

[10] A. Magasinski, et al., "High-performance Lithium-ion Anodes Using a Hierarchical Bottom-Up Approach," *Nature Materials*, vol. 9, pp. 353-358, 2010.

[11] M. L. Terranova, et al., "Si/C Hybrid Nanostructures for Li-ion Anodes: an Overview," *J. Power Sources*, vol. 246, pp. 167-177, 2013.

[12] C. W. Shen, et al., "Direct Prototyping of Patterned Nanoporous Carbon: A Route from Materials to On-chip Devices," *Scientific Reports*, vol. 3, 2013.

[13] H. S. Min, et al., "Fabrication and properties of a carbon/polypyrrole three-dimensional microbattery," *J. Power Sources*, vol. 178, pp. 795-800, 2008.

[14] W. Stober, A. Fink, E. Bohn, "Controlled Growth of Monodisperse Silica Spheres in Micron Size Range," *Journal Of Colloid And Interface Science*, vol. 26, p. 62-&, 1968.

CONTACT

*X. Wang, Phone: +86-10-62798432, Fax: +86-10-62771130, Email: wxh-ime@mail.tsinghua.edu.cn

STATUS AND FUTURE TRENDS
OF THE MINIATURIZATION OF MASS SPECTROMETRY

Richard R.A. Syms[*1]

[1]EEE Dept., Imperial College London, London, UK

ABSTRACT

An overview of mass spectrometers incorporating miniaturization technology is presented. A brief history of conventional mass spectrometry is given, followed by a summary of the status of miniaturized systems. Applications for miniature/portable systems are reviewed, and opportunities and challenges are discussed.

INTRODUCTION

The development of microelectromechanical systems over the last 30 years has been dramatic, and MEMS now impact on most industrial sectors. Specific materials, device configurations and packaging have been developed for each application domain, and MEMS are available as commodity items such as accelerometers or laboratory instruments such as atomic force microscopes. Until recently, one area that stubbornly resisted miniaturization was mass spectrometry. The aim of this paper is to describe the historical difficulties faced by this sector, progress to date, and the potential for new applications.

HISTORICAL BACKGROUND

Mass spectrometry - the analysis of ions in vacuum by their mass-to-charge ratio m/z - is over a century old. Its beginnings can be traced to Eugen Goldstein, who discovered positive ions in 1886, and Wilhelm Wien, who analyze them in 1898 using crossed electric and magnetic fields. Joseph Thomson, who measured m/z for electrons in 1898, had by 1913 obtained mass spectra for a range of ions and separated the isotopes of neon. From then on, magnetic separation progressed rapidly. In 1918, Arthur Dempster developed the first sector magnet with direction focusing. In 1919, Thomson's student, Francis Aston, constructed the first mass spectrometer with velocity focusing. In 1932, Kenneth Bainbridge proved Einstein's mass-energy equivalence, and by 1933 had raised the resolution m/Δm to 600. In 1934, Joseph Mattauch and Richard Herzog introduced the double-focusing spectrograph. Ernest Lawrence, inventor of the cyclotron, developed 'calutrons' for separating uranium isotopes in 1942 during the Manhattan project, while Alfred Nier constructed portable mass spectrometers as leak detectors.

Because magnetic separators were large and heavy, different physical principles were investigated after the Second World War. In 1946, William Stephens proposed the use of time-of-flight. Angus Cameron demonstrated an early TOF-MS in 1948, and William Wiley and Ian McLaren of Bendix developed a TOF-MS with space and velocity focusing in 1955. Boris Mamyrin developed an improved method of compensation – the reflectron – in 1973. By 1987, Koichi Tanaka was able to use soft ionisation and TOF-MS to analyze intact proteins with m/z up to 100,000. In 1953, Wolfgang Paul and Helmut Steinwedel proposed the use of inhomogeneous time-varying electric fields for separation, in two different

structures: the quadrupole filter and the quadrupole ion trap. RF quadrupoles are workhorse instruments, with applications ranging from residual gas analysis to space exploration. Related components such as RF ion guides are used for ion transport, while collision cells are used for ion cooling and fragmentation. In 1978, Richard Yost and Chris Enke introduced the triple quadrupole, in which the first quadrupole is used for initial mass analysis, a second for collision-induced dissociation, and the third for analysis of the resulting fragments, enabling so-called tandem mass spectrometry. The quadrupole ion trap was developed commercially for Finnegan by Raymond March in 1983, and the linear quadrupole trap (which has an increased trapping volume) by Jae Schwartz in 2002. Other milestones include the development of the Fourier transform ion cyclotron resonance mass spectrometer by Alan Marshal and Melvin Comisarow in 1976, and the Orbitrap by Alexander Makarov in 1999, based on Kingdon's ring-and-wire ion trap of 1923.

Accompanying the mass filtering techniques above is a set of different ionization methods. Among the earliest is electron ionization (developed for solids by Arthur Dempster in 1918 and for gases by Walter Bleakney in 1929). Later methods include flame ionization (Allan Hayhurst, 1966), glow discharge and inductively-coupled plasma ionization (Alan Gray, 1975), atmospheric pressure (chemical) ionization, photoionization, laser desorption ionization, matrix-assisted laser desorption ionization (Franz Hillenkamp and Michael Karas, 1985) and electrospray ionization (investigated by Malcolm Dole in 1968). More recently, desorption electrospray ionization (R. Graham Cooks, 2004), has made a significant impact. Some sources involve ionization in vacuum, others at atmospheric pressure. In each case methods for transferring an analyte (which could be a solid, a liquid or a gas) or an ion stream into vacuum are required. These techniques include solid phase microextraction (SPME; Janus Pawliszyn, 1990), the membrane inlet interface and the jet expansion interface (developed by John Fenn in 1984, and the key to electrospray mass spectrometers). Other components such as air amplifiers and ion funnels are used to increase the coupling of ions into the mass spectrometer inlet, and nanospray sources are often used in place of electrospray.

In addition, complex mixtures are often separated before analysis. The first 'hyphenated' technique, gas chromatography-mass spectrometry (GC-MS) was introduced by Roland Gohlke and Fred McLafferty at Dow Chemical in 1956. Later techniques include liquid chromatography-mass spectrometry (LC-MS), capillary electrophoresis-mass spectrometry (CE-MS) and ion mobility spectrometry-mass spectrometry (IMS-MS).

MINIATURE MASS SPECTROMETERS

Miniaturization of mass spectrometers began in the

978-1-4799-7956-1/15 $31.00 © 2015 IEEE

1990s, around a decade after the initial surge of interest in MEMS. Many reviews are now available, see e.g. [1-3]. However, while progress on other sensor types was rapid, progress on mass spectrometers was crushingly slow.

Several factors were responsible. The main users were the chemical and pharmaceutical industries and chemical and biomedical researchers. Big pharma was developing high-throughput screening, a method of drug discovery that involves preparing many near-identical compounds for serial analysis using a small number of high-performance instruments. Instrument manufacturers were therefore exclusively focused on improving performance, using large, complex systems. Where there was an interest in miniaturization, it was overwhelmingly concentrated on combining chip-based separation with electrospray ionization, following the observation that integration reduced peak broadening in chromatography. An additional factor was that microfluidic technology was accessible to most university chemistry departments, leading to hundreds of publications on similar devices [4-9]. This led to excellent technical results, but did nothing to address the problem that the instrument with the largest cost and footprint was the mass spectrometer.

To begin with, miniaturization of mass spectrometers was mainly driven by alternative needs and budgets in space exploration and defense. However, many of those with appropriate knowledge of mass spectrometry lacked access to microfabrication technology, while those with access to the technology lacked design insight. In any case, MEMS processes were in their infancy and struggled to realize the three-dimensional structures required, and the materials most suited to planar processing - silicon and thin films of insulators and metals - generally did not have appropriate characteristics. As a result, early attempts at miniaturization involved two approaches. The first involved taking conventional technology such as CNC machining or PCB fabrication to its limits, and generally resulted in useful, but relatively large, systems. The second involved the use of MEMS fabrication, and yielded much smaller systems that generally worked poorly, if at all. As a result, few miniaturized systems have reached market.

In addition, if miniaturization were to be attempted, it was not clear which of the bewilderingly large number of sample introduction, ionization or mass filtering techniques to use. It was also not clear how to partition a mass spectrometer system (which requires many other additional components (such as an ion detector, a vacuum chamber, valves, a pressure gauge and a two-stage vacuum pumping system) into viable sub-units, or how to package them. Thus, most teams initially attempted to develop miniaturized sub-components contained in significantly larger conventional vacuum chambers. Only later were monolithically integrated chambers attempted. Fortunately, one manufacturer (Pfeiffer) had developed a compact turbo-pump that could be battery powered and hence was ideally suited to small or portable mass spectrometers. Vacuum ionization was used in all early systems, based on either electron impact or glow discharge sources. Gaseous analytes were introduced via a needle valve and liquids using SPME. Detection was generally carried out using a microchannel plate or channeltron electron multiplier, although less sensitive integrated systems used Faraday cups.

Within this paradigm, attention could be concentrated on the mass filter. Although significant efforts were made to develop powerful miniature magnetic sectors and detector arrays [10-14], and operating Wien filters were developed [15-17], it was quickly realized that magnetic separators had poor size scaling laws. Of the non-magnetic approaches, the most popular have been the time-of-flight filter, and RF-driven structures such as the quadrupole, the linear ion trap, the quadrupole ion trap, and a variant known as the toroidal trap. Most have been constructed as arrays, in an attempt to recover some of the sensitivity lost by miniaturization. Some other approaches (Fourier transform ion cyclotron resonance) have proved less popular, while others (the Orbitrap, which requires very complex 3D electrodes) have yet to be attempted.

For time-of flight filters, the necessary electrode structures are planar. Here the key difficulty has been to obtain sufficient mass separation in a short distance. However, some impressive results have been obtained with short, chip-scale flight paths. Electron impact ionization and photo-ionization have both been used to generate a short initial pulse of ions, and microfabricated Wiley-McLaren and plug-assembled reflectron electrode structures have both been developed to compensate for uncertainty in position and velocity [18-21].

RF-driven filters and traps can use continuous ion sources. In this case, the difficulty has been to realize the 3D electric fields needed. Two approaches have been used. The first is dimensional optimization of simpler surfaces such as planes and cylinders to approximate the fields generated by hyperbolic electrodes. Because it can be constructed as a simple three-layer stack, with the central electrode a cylindrical hole, the cylindrical ion trap, which approximates a hyperbolic quadrupole trap, has proved extremely popular; however, the quadrupole mass filter, quadrupole ion guides and the rectilinear ion trap have all been successfully miniaturized.

The second is the use of planes carrying multiple electrodes, each held at a different potential, to approximate the desired field. Most RF structures have suffered from the poor electrical isolation offered by thin layers of SiO_2, which typically results in heating of the substrate, if it is a semiconductor. More advanced MEMS technologies based on glass or ceramic insulators are therefore needed to achieve a high enough mass range.

Using semi-conventional technologies, NASA and JPL scientists and contractors have developed an impressive range of miniature mass spectrometers, including systems based on magnetic separators, time-of-flight filters, rotating fields and quadrupoles [22, 23]. However, the most successful use of such technologies has been the work of R. Graham Cooks and co-workers at Purdue University, which led to the spin-out company Griffin, subsequently ICx Technologies and now part of FLIR Systems (current product: Griffin 460, linear ion trap). Cooks has developed many miniaturized filters, including cylindrical and linear ion traps and ion trap arrays [24-28]. A large number of cylindrical [29-32] and linear [33-35] ion trap variants have subsequently been demonstrated, based on planar, arc-shaped, triangular and

978-1-4799-7956-1/15 $31.00 © 2015 IEEE

asymmetric electrodes. Cooks has also introduced many important techniques for portable systems such as paper spray, desorption electrospray ionization and discontinuous ion introduction.

Planar multi-electrode systems, which allow three-dimensional fields to be realised without the need for shaped electrodes, have been extensively investigated at Brigham Young University. Examples include multi-electrode cylindrical and linear ion traps [36, 37], and the so-called halo ion trap, which approximates a toroidal trap [38, 39]. Torodial traps have been commercialised at the Brigham Young spin-out Torion, now partnered with Smiths Detection. Both have products for homeland security applications (Tridion-9 and Guardion, respectively; both toroidal ion traps). Other manufacturers of portable mass spectrometers include 908 Devices (M908; handheld cylindrical ion trap).

Due to the expense and cycle times associated with microfabrication, MEMS mass spectrometers have received less attention; however, this approach has the greatest potential to drive down cost in production. The most successful application of MEMS technology has been the work of Imperial College on quadrupoles (Fig. 1), originally in collaboration with Liverpool University, that led to the spin-out Microsaic Systems [40-42].

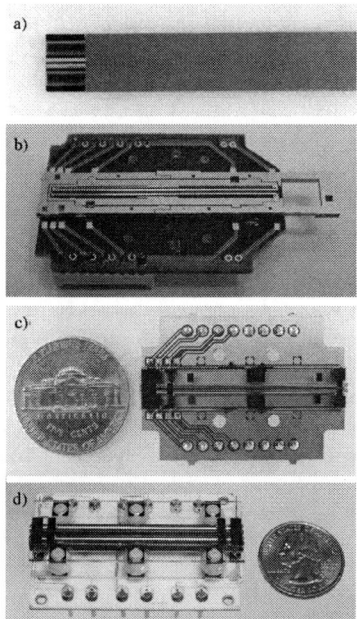

Figure 1: MEMS quadrupoles in a) 1996, b) 2004, c) 2009 and d) 2014 (b – c courtesy Microsaic Systems).

Microsaic manufactures a benchtop quadrupole-based ESI-MS (4000 MiD) using separately packaged MEMS components for ionization, ion transmission into vacuum, ion guidance and ion filtering [43-45]. A key advantage is that components are easily cleaned and replaced, and ultimately may be considered consumables. The system is small enough to allow all components (including pumps) to be contained in an enclosure smaller than a PC and mounted in a fume hood. The mass filter is based on cylindrical electrodes held in silicon-on-glass mountings, and all aspects of performance (sensitivity, resolution and mass range) have steadily improved as experience has

been gained. Other benchtop MS systems are available from 1st Detect (MMS-100; cylindrical trap) and Advion (Expression; quadrupole). Functionality is increasing; Microsaic has recently demonstrated a MEMS triple quadrupole [46], and tandem MS has been demonstrated using linear ion traps [47]. It is likely that hyphenated systems will be developed in the near future.

Attempts have been made to increase the degree of integration, by combining a mass filter with a source and/or a detector. Efforts have also been made to create an entire analysis system inside a microfabricated vacuum chamber. Performance and progress have both generally been slower, because integration prevents the separate optimization of individual components. Notable efforts on Wien filters include those at Northrop Grumman and CEA LETI. Similar advances in time-of-flight filters have been achieved at Ajou University, the University of Hamburg-Harburg and CEA LETI/DAM. Quadrupole filters based on square electrodes (which can be fabricated using patterning and etching) have also been developed at MIT following work on assembled devices [48, 49].

Advances in ion sources include improvements to cold-cathode electron impact ionization sources [50-54] and plasma sources [55-59]. Components designed to improve ion coupling, such as air amplifiers [60], ion funnels and ion carpets [61], have also been realized using planar technology. However, there appears to have been little progress on integrated ion detectors, due to the nature of the materials required (which must have a high secondary electron yield in multiplying detectors such as channeltrons and microchannel plates). As a result, mass detection has often tended to rely on the availability of components developed for conventional systems or under military night vision programs.

Gas pumps are now well developed [62, 63] and have been integrated with planar micromachined gas chromatography systems [64]. Although the poor scaling of dead volumes and clearances limits performance, MEMS vacuum pumps have also recently made considerable progress, with advances in many of the major types (vapor, orbitron, turbo, sputter ion and Knudsen pumps [65-70]). Vacuum pumps have also been integrated with MEMS vacuum gauges [71]. It is now possible to envisage MEMS mass spectrometers equipped with microfabricated pumping systems, although many years of development are likely to be needed to achieve sufficient base pressure and mechanical reliability.

APPLICATIONS

Applications of miniature mass spectrometry are increasing rapidly as instrument capabilities improve. In the military and homeland security sectors, global increases in terrorism are accelerating development of systems for the detection of explosives [72] and chemical and biochemical weapons [73]. In more general security, applications are developing in forensics [74] and in detection of illegal drugs [75] or performance enhancing drugs in sport. Environmental applications include volcanology [76], and monitoring of pollution in the air, soil, sea and inland waterways [77]. Related applications include the detection of adulterants in food [78]. Industrial applications include drug discovery and on-line monitoring of synthesis [79]. Medical applications include direct diagnosis of infection [80] and cancer diagnosis by analysis of vapors generated during electrosurgery [81]. General scientific applications include atmospheric science and monitoring of cabin air quality during space exploration [82]. Most recently, there has been considerable interest in the use of ion trapping in quantum technology. In this case, the emphasis is on creating, transporting and confining ion packets rather than chemical analysis [83-84].

CONCLUSIONS

Miniature mass spectrometers have made considerable advances in recent years. Performance is now good enough to enable applications that require portability, small size or low cost. The increasing availability of such systems – which allow the instrument to be taken to the sample, rather than the other way round - implies many new opportunities for *in situ* analysis [85]. Mainframe instruments are evolving into desktop systems, backpack-mounted systems into hand-held units, and vehicle-mounted systems into airborne systems. The new paradigm will generate changes broadly analogous to those driven by the development of personal computers.

ACKNOWLEDGEMENTS

The Author grateful acknowledges the major contributions of colleagues at Liverpool University, Imperial College and Microsaic Systems: Munir Ahmad, Max Bardwell, William Boxford, Ed Chrichton, Neil Dash, Peter Edwards, Martin Geear, Andrew Holmes, Guodong Hong, Andrew Malcolm, Richard Moseley, Alex Onishenko, Shane O'Prey, Marc-André Schwab, Steve Taylor and Steve Wright.

REFERENCES

[1] Z. Ouyang, R.G. Cooks, "Miniature mass spectrometers", *Ann. Rev. Anal. Chem.*, vol. 2, pp. 10.1-10.28, 2009.

[2] T. Sikanen, S. Franssila, T.J. Kauppila et al., "Microchip technology in mass spectrometry", *Mass Spect. Rev.*, vol. 29, pp. 351-391, 2009.

[3] S. Arscott, "SU-8 as a material for lab-on-a-chip-based mass spectrometry", *Lab Chip*, vol. 14, pp. 3668-3689, 2014.

[4] R. Ramsey, J. Ramsey, "Generating electrospray from microchip devices using electroosmotic pumping", *Anal. Chem.* vol. 69, pp. 1174-1178, 1997.

[5] L. Licklider, X.Q. Wang, et al., "A micromachined chip-based electrospray source for mass spectrometry", *Anal Chem.*, vol. 72, pp. 367-75, 2000.

[6] G.A. Schultz, T.N. Corso, S.J. Prosser, et al., "A fully integrated monolithic microchip electrospray device for mass spectrometry", *Anal. Chem.*, vol. 72, pp. 4058-4063, 2000.

[7] P. Griss, J. Melin, J.Sjödahl et al., "Development of micromachined hollow tips for protein analysis based on nanoelectrospray ionization mass spectrometry", *J. Micromech. Microeng.*, vol. 12, pp. 682-687, 2002.

[8] J.M. Lazar, J, Grym, F. Foret, "Microfabricated devices: a new sample introduction approach to mass spectrometry", *Mass Spect. Rev.*, vol. 25, pp. 573-594, 2006.

[9] P.R. Chiarot P.R., R.B. Mrad, "An overview of electrospray applications in MEMS and microfluidic systems", *J. Microelectromech. Syst.*, vol. 20, pp. 1241-1249, 2011.

[10] M.P. Sinha, A.D. Tomassian, "Development of a miniaturized, lightweight magnetic-sector for a field-portable mass spectrograph", *Rev. Sci. Inst.*, vol. 62, pp. 2618-2620, 1991.

[11] R.B. Darling, A.A. Scheidemann, K.N. Bhat, et al., "Micromachined Faraday cup array using deep reactive ion etching", *Sensors and Actuators A*, vol. 95, pp. 84-93, 2002.

[12] C.A. Bower, K.H. Gilchrist, M.R. Lueck et al., "Microfabrication of fine-pitch high aspect ratio Faraday cup arrays in silicon", *Sensors and Actuators A*, vol. 137, pp. 296-301, 2007.

[13] M.P. Sinha, M. Wadsworth, "Miniature focal plane mass spectrometer with 1000-pixel modified-CCD detector array for direct ion measurement", *Rev. Sci. Instrum.*, vol. 76, 025103, 2005.

[14] O. Hadjar, W.K. Fowler, G. Kibelka, et al., "Preliminary demonstration of an IonCCD as an alternative pixelated anode for direct MCP readout in a compact MS-based detector", *J. Am. Soc. Mass Spectrom.*, vol. 23, pp. 418-424, 2012.

[15] C. Freidhoff, "Mass spectrograph on a chip", Proc. 1997 IEEE Aerospace Conference, Aspen, Colorado, Feb 1 - 8, vol. 3, pp. 32, 1997.

[16] J.A. Diaz, C.F. Giese, W.R. Gentry, "Sub-miniature E x B sector-field mass spectrometer", *J. Am. Soc. Mass Spectrom.*, vol. 12, pp. 619-632, 2001.

[17] N. Sillon, R. Baptist, "Micromachined mass spectrometer", *Sensors and Actuators B*, vol. 83, pp. 129-137, 2002.

[18] H.J. Yoon, J.H. Kim, E.S. Choi, S.S. Yang, K.W. Jung, "Fabrication of a novel micro time-of-flight mass spectrometer", *Sensors and Actuators*, vol. A97-8, pp. 441-447, 2002.

[19] E. Wapelhorst, J.P. Hauschild, J. Muller, "Complex MEMS: a fully integrated TOF micro mass spectrometer", *Sensors and Actuators A*, vol. 138, pp. 22-27, 2007.

[20] J. Fox, R. Saini, K. Tsui, G. Verbeck, "Microelectromechanical system assembled ion optics: an advance to miniaturization and assembly of electron and ion optics", *Rev. Sci. Inst.*, vol. 80, 093302, 2009.

[21] C.-M. Tassetti, R. Mahieu, J.-S Danel et al., "A MEMS electron impact ion source integrated in a micro time-of-flight mass spectrometer", *Sensors and Actuators B*, vol. 189, pp. 173-178, 2013.

[22] O.J. Orient, A. Chutjian, V. Garkanian, "Miniature, high-resolution, quadrupole mass-spectrometer array", *Rev. Sci. Instrum.*, vol. 68, pp. 1392-1397, 1997.

[23] S. Boumsellek, R.J. Ferran, "Tradeoffs in miniature quadrupole designs", *J. Am. Soc. Mass Spectrom.*, vol. 12, pp. 633-640, 2001.

[24] E.R. Badman, R.G. Cooks, "A parallel miniature

cylindrical ion trap array", *Anal. Chem.*, vol. 72, pp. 3291-3297, 2000.

[25] E.R. Badman, R.G. Cooks, "Cylindrical ion trap array with mass selection by variation in trap dimensions", *Anal. Chem.*, vol. 72, pp. 5079-5086, 2000.

[26] Z. Ouyang, G.X. Wu, Y.S. Song et al., "Rectilinear ion trap: Concepts, calculations, and analytical performance of a new mass analyzer", *Anal. Chem.*, vol. 76, pp. 4595-4605, 2004.

[27] M. Fico, M. Yu, Z. Ouyang, R.G. Cooks, W.J. Chappell, "Miniaturization and geometry optimization of a polymer-based rectilinear ion trap", *Anal. Chem.*, vol. 79, pp. 8076-8-82, 2007.

[28] M. Fico, J. Maas, S.A. Smith, A.B. Costa, Z. Ouyang, W.J. Chappell, R.G. Cooks, "Circular arrays of polymer-based miniature rectilinear ion traps", *Analyst*, vol. 134, pp. 1338-1347, 2009.

[29] O. Kornienko, P.T.A. Reilly, W.B. Whitten, J.M. Ramsey, "Micro ion trap mass spectrometry", *Rapid Comm. in Mass Spect.*, vol. 13, pp. 50-53, 1999.

[30] S. Pau, C.S. Pai, Y.L. Low, J. Moxom, P.T.A. Reilly, W.B. Whitten, J.M. Ramsey, "Microfabricated quadrupole ion trap for mass spectrometer applications", *Phys. Rev. Lett.*, vol. 96, 120801, 2006.

[31] F.H.W. van Amerom, A. Chaudhary et al., "Micro-fabrication of cylindrical ion trap mass spectrometer arrays for handheld chemical analyzers", *Chem. Eng. Comm.*, vol. 195, pp. 98-114, 2008.

[32] A. Chaudhary, F.H.W. van Amerom, R.T. Short, "Development of microfabricated cylindrical ion trap mass spectrometer arrays", *J. Microelectromech. Syst.*, vol. 18, pp. 442-448, 2009.

[33] A.T. Clare, L. Gao, B. Brkic, et al., "Linear ion trap fabricated using rapid manufacturing technology", *J. Am. Soc. Mass Spectrom.*, vol. 21, pp. 317-322, 2010.

[34] Y. Xiao, Z. Ding, C. Xu et al., "Novel linear ion trap mass analyzer built with triangular electrodes", *Anal. Chem.*, vol. 86, pp. 5733-5739, 2014.

[35] Z.-Y. Zhang, C. Li, C.-F. Ding, et al., "A novel asymmetric arc-shaped electrode ion trap for improving the performance of a miniature mass spectrometer", *Rapid Comm. in Mass Spectrom.*, vol. 28, pp. 1764-1768, 2014.

[36] Z. Zhang, Y. Peng, B.J. Hansen et al. "Paul trap mass analyzer consisting of opposing microfabricated electrode plates", *Anal. Chem.*, vol. 81, pp. 5241-4248, 2009.

[37] A. Li, B.J. Hansen, A.T. Powell, A.R. Hawkins, D.E. Austin, "Miniaturization of a planar-electrode linear ion trap mass spectrometer", *Rapid Comm. in Mass Spectrom.*, vol. 28, pp. 1338-1344, 2014.

[38] D.E. Austin, M. Wang, S.E. Tolley et al., "Halo ion trap mass spectrometer", *Anal. Chem.*, vol. 79, pp. 2927-2932, 2007.

[39] M. Wang, H.E. Quist, B.J. Hansen et al., "Performance of a halo ion trap mass analyzer with exit slits for axial ejection", *J. Am. Soc. Mass Spectrom.*, vol. 22, pp. 369-378, 2011.

[40] R.R.A Syms, T.J. Tate, M.M. Ahmad, S. Taylor, "Fabrication of a microengineered quadrupole electrostatic lens", *Elect. Lett.*, vol. 32, pp. 2094-

2095, 1996.

[41] R.R.A. Syms, T.J. Tate, M.M. Ahmad, S. Taylor, "Design of a microengineered quadrupole electrostatic lens", *IEEE Trans. Electron Dev.*, vol. 45, pp. 2304-2311, 1998.

[42] M. Geear, R.R.A. Syms, S. Wright, A.S. Holmes, "Monolithic MEMS quadrupole mass spectrometers by deep silicon etching", *J. Microelectromech. Syst.*, vol. 14, pp. 1156-1166, 2005.

[43] S. Wright, S. O'Prey, R.R.A. Syms, G. Hong, A.S. Holmes, "Microfabricated quadrupole mass spectrometer with a Brubaker prefilter", *J. Microelectromech. Syst.*, vol. 19, pp. 325-337, 2010.

[44] S. Wright, R.R.A. Syms, R.W. Moseley et al., "MEMS-based nanospray ionisation mass spectrometer", *J. Microelectromech. Syst.*, vol. 19, 1430-1443, 2010.

[45] A. Malcolm, S. Wright, R.R.A. Syms et al. "A miniature mass spectrometer for liquid chromatography applications", *Rapid Comm. in Mass Spectrom.*, vol. 25, pp. 3281-3288, 2011.

[46] S. Wright, A. Malcolm, C. Wright, S. O'Prey, N. Dash, E. Chrichton, R.W. Moseley, W. Zaczek, R.R.A. Syms, P. Edwards, "Design and performance of a prototype miniature triple quadrupole mass spectrometer", *Proc. 61st ASMS Conf.*, Minneapolis, Minnesota, June 6-13, paper MP 297, 2013.

[47] L. Li, X. Zhou, J.W. Hager, Z. Ouyang, "High efficiency tandem mass spectrometry analysis using dual linear ion traps", *Analyst*, vol. 104, pp. 4779-4784, 2014.

[48] L.F. Velasquez-Garcia, A.I. Akinwande, M. Martinez-Sanchez, "Precision hand assembly of MEMS subsystems using DRIE-patterned deflection spring structures: an example of an out-of-plane substrate assembly", *J. Microelectromech. Syst.*, vol. 16, pp. 598-612, 2007.

[49] K. Cheung, L.F. Velasquez-Garcia, A.I. Akinwande, "Chip-scale quadrupole mass filters for portable mass spectrometry", *J. Microelectromech. Syst.*, vol. 19, pp. 469-483, 2010.

[50] T.E. Felter, "Cold cathode field emitter array on a quadrupole mass spectrometer: route to miniaturization", *J. Vac. Sci Tech. B*, vol. 17, pp. 1993-1996, 1999.

[51] O. Kornienko, P.T.A. Reilly, W.B. Whitten, et al. "Field-emission cold-cathode EI source for a microscale ion trap mass spectrometer", *Anal. Chem.*, vol. 72, pp. 559-562, 2000.

[52] H.J. Yoon, S.H. Song, et al., "Fabrication of an ion source using carbon nanoparticle field emitters for a micro time-of-flight mass spectrometer", *Proc. Transducers '07*, pp. 1067-1070, 2007.

[53] C.A. Bower, K.H. Gilchrist et al., "On-chip electron-impact ion source using carbon nanotube field emitters", *Appl. Phys. Lett.*, vol. 90, 124102, 2007.

[54] K. Han, Y. Lee, D. Jun, et al., "Field emission ion source using a carbon nanotube array for micro time-of-flight mass spectrometer", *Japn. J. Appl. Phys.*, vol 50, 06GM04, 2011.

[55] J.W. Frame, "Microdischarge devices fabricated on silicon", *Appl. Phys. Lett.*, vol. 71, pp. 1165-1167,

1997.

[56] J.A. Hopwood, "A microfabricated inductively coupled plasma", *J. Microelectromech. Syst.*, vol. 9, pp. 309-313, 2000.

[57] A. Bass, C. Chevalier, M.W. Blades, "A capacitatively coupled microplasma (CCµP) formed in a channel in a quartz wafer", *J. Anal. At. Spectrom.*, vol. 16, 9190921, 2001.

[58] J. Hopwood, F. Iza, S. Coy, et al., "A microfabricated atmospheric pressure microplasma source operating in air", *J. Phys. D. Appl. Phys.*, vol. 38, pp. 1698-1703, 2005.

[59] J.-P. Hauschild, E. Wapelhorst, J. Müller, "Mass spectra measured by a fully integrated MEMS mass spectrometer", *Int. J. Mass Spectrom.*, vol. 264, pp. 53-60, 2007.

[60] P. Jurcicek, H. Zou, S. Gao, "Design, simulation and fabrication of a MEMS-based air amplifier for electrospray ionization", *J. Micro/Nanolithography, MEMS and MOEMS*, vol. 12, 023006, 2013.

[61] S.N. Anthony, D.L. Shinholt, M.F. Jarrold, "A simple electrospray interface based on a DC ion carpet", Int. J. Mass Spectrom., vol. 371, pp. 1-7, 2014.

[62] H. Kim, A. Astle, K. Najafi et al., "A fully integrated high-efficiency peristaltic 18-stage gas micropump with active microvalves", Proc. 20th IEEE MEMS Conf., Kobe, Japan, pp. 131-134, 2007.

[63] A. Besharatian, K. Kumar, R.L. Peterson et al., "A scalable, modular, multistage, peristaltic, electrostatic gas micropump", Proc. 25th IEEE MEMS Conf., Paris, Jan 29-Feb 2, pp. 1001-1004, 2012.

[64] Y. Qin, Y.B. Gianchandani, "iGC2: an architecture for micro gas chromatographs utilizing integrated bi-directional pumps and multi-stage preconcentrators", *J. Micromech. Microeng.*, vol. 24, 065011, 2014.

[65] Doms M., Müller J., "A micromachined vapor jet pump", *Sensors and Actuators A*, vol. 199, pp. 462-467, 2005.

[66] H.W.P. Koops, "Proposal of a miniature orbitron vacuum pump for MEMS applications", *Proc. SPIE*, vol. 5838, pp. 38-42, 2005.

[67] C.M. Waits, M. McCarthy, R. Ghodssi, "A microfabricated spiral-groove turbopump supported on microball bearings", *J. Microelectromech. Syst.*, vol. 19, pp. 99-108, 2010.

[68] A.J. Gannon, G.V. Hobson, M.J. Shea et al., "MEMS-scale turbomachinery based vacuum roughing pump", *J. Turbomachinery*, vol. 136, 101002, 2014.

[69] S.R. Green, R. Malhotra, Y.B. Gianchandani, "Sub-Torr chip-scale sputter-ion pump based on a Penning cell array architecture", *J. Microelectromech. Syst.*, vol. 22, pp. 309-317, 2013.

[70] S. An, N.K. Gupta, Y.B. Gianchandani, "A Si-micromachined 162-stage two-part Knudsen pump for on-chip vacuum", *J. Microelectromech. Syst.*, vol. 23, pp. 406-416, 2014.

[71] T. Grzebyk, A. Gorecka-Drzazga, J.A. Dziuban, K. Maamari, "Integration of a MEMS-type vacuum pump with a MEMS-type Pirani gauge", Proc. 27th Int. Vacuum Nanoelectronics Conf., July 6-10, pp. 206-297, 2014.

[72] N.L. Sanders, S. Kothari, G. Huang, et al., "Detection of explosives as negative ions directly from surfaces using a miniature mass spectrometer", *Anal. Chem.*, vol. 82, pp. 5313-5316, 2010.

[73] Y. Seto, K.-M. Meiko K. Tsude et al., "Sensitive monitoring of volatile chemical warfare agents in air by atmospheric pressure chemical ionization mass spectrometry with counter-flow introduction" Anal. Chem., vol. 85, pp. 2659-2666, 2013.

[74] L.T. Demoranville, T.M. Brewer, "Ambient pressure thermal desorption ionization mass spectrometry for the analysis of substances of forensic interest", *Analyst*, vol. 138, pp. 5332-5337, 2013.

[75] Kirby A.E., Lafreniere N.M., Seale B., et al., "Analysis on the go: quantitation of drugs of abuse in dried urine with digital microfluidics and miniature mass spectrometry", *Anal. Chem.*, vol. 86, pp. 6121-6129, 2014.

[76] J.A. Diaz, Pieri D., Arkin C.R. et al., "Utilization of in situ airborne MS-based instrumentation for the study of gaseous emissions at active volcanoes" *Int. J. Mass Spectrom.*, vol. 295, pp. 105-112, 2010.

[77] J.N. Smith, A. Keil, J. Likens et al., "Facility monitoring of toxic industrial compounds in air using an automated, fieldable, miniature mass spectrometer" *Analyst*, vol. 135, pp. 994-1003, 2010.

[78] P. Li, P. Wei, H.-C. Hsu, R.G. Cooks, "Direct analysis of 4-methylimidazole in foods using paper spray mass spectrometry" *Analyst*, vol. 138, pp. 4624-4630 (2013)

[79] D.L. Browne, S. Wright, B.J. Deadman, et al., "Continuous flow reaction monitoring using an on-line miniature mass spectrometer", *Rapid Comm. in Mass Spectrom.*, vol. 26, pp. 1999-2010, 2012.

[80] A.K. Jarmusch, V. Pirro, K.S. Kerian, R.G. Cooks, "Detection of strep throat causing bacterium directly from medical swabs by touch spray-mass spectrometry", Analyst, vol. 139, pp. 4785-4789, 2014.

[81] J. Balog, L. Sasi-Szabo, J. Kinross et al., "Intraoperative tissue identification using rapid evaporative ionization mass spectrometry", *Sci. Transl. Med.*, vol. 5, 194ra93, 2013.

[82] M.P. Sinha, E.L. Neidholdt, J. Hurowitz, et al. "Laser ablation miniature mass spectrometer for elemental and isotopic analysis of rocks" *Rev. Sci. Inst.*, vol. 82, 094102, 2011.

[83] P. See, G. Wilpers, P. Gill, et al., "Fabrication of a monolithic array of three dimensional Si-based ion traps", *J. Microelectromech. Syst.*, vol. 22, pp. 1180-1189, 2013.

[84] R.C. Sterling, H. Rattanasonti, S. Weidt S., et al., "Fabrication and operation of a two-dimensional ion-trap lattice on a high-voltage microchip", *Nature Communications*, vol. 5, 4637, 2014.

[85] J.M. Perkel, "Miniaturizing mass spectrometry", *Science Magazine*, Feb. Issue, pp. 928-930, 2014.

CONTACT

*R.Syms, tel: +44-207-5946203; r.syms@imperial.ac.uk

SHORT-RANGE AND HIGH-RESOLUTION ULTRASOUND IMAGING USING AN 8 MHZ ALUMINUM NITRIDE PMUT ARRAY

Yipeng Lu[1], Hao-Yen Tang[2], Stephanie Fung[1], Bernhard E. Boser[2] and David A. Horsley[1]
Berkeley Sensor and Actuator Center
[1]University of California, Davis, USA
[2]University of California, Berkeley, USA

ABSTRACT

Ultrasound imaging uses costly bulk piezoelectric transducers and high voltage (200V+) electronics. Low-cost and low-voltage ultrasound transducers would enable many new applications in healthcare, biometrics, and personal health-monitoring. Here, we demonstrated short-range (~mm) and high-resolution (<100 μm) imaging based on piezoelectric micromachined ultrasonic transducers (PMUTs) and a 1.8 V interface ASIC. The PMUTs use piezoelectric Aluminum Nitride (AlN), which has the advantages of low-temperature (<400 °C) deposition and compatibility with CMOS fabrication but has a relatively low piezoelectric constant (e_{31}=-0.5 C/m^2), making detection of ultrasound signals from tiny (50 μm) PMUTs a challenging task. To solve this problem, we developed an ASIC with a low-noise analog front-end pre-amplifier that is impedance matched to the PMUT. Furthermore, a novel beam-forming and scanning method was demonstrated to achieve a sub-100μm focus size and 70 μm scanning step. Pressure map measurement from phased PMUT array and pulse echo imaging results were demonstrated using 1-D and 2-D phantoms.

INTRODUCTION

Ultrasonic transducers have been used in many applications and famous for nondestructive testing (NDT) and medical imaging. Conventional ultrasonic transducers are largely based on bulk piezoelectric ceramic with poor acoustic coupling to air or liquids, and additional matching layer is required. In contrast, micromachined ultrasonic transducers (MUTs) have a compliant membrane structure with low acoustic impedance for good coupling to air and liquids. Compared with well-developed capacitive MUTs (CMUTs), PMUTs [1] do not require a high polarization voltage and small gap [2], and thereby reduce circuit and fabrication complexity.

Previous research on PMUT pulse-echo imaging was based on lead zirconium titanate (PZT) [3-4], a material with high piezoelectric coefficients and high relative permittivity, which make the receiving amplifier easier to design and remote electronics feasible. Relative to PZT, AlN is lead-free and is compatible with CMOS fabrication, making it attractive for highly integrated, low-cost PMUT arrays. However, AlN has lower piezoelectric coefficients and low relative permittivity, which result in PMUTs with lower pressure sensitivity in transmitting and lower charge output in receiving. Therefore AlN PMUTs make ultrasound pulse-echo detection more challenging and require a low-noise and impedance-matched local pre-amplifier.

In addition, high fill-factor PMUT arrays are needed to minimize grating lobes and increase acoustic efficiency per unit area. However, reported PMUTs have large dimensions and pitch and therefore low fill-factor; this results from fabrication using through-wafer etching [5]. Front-side etching using a sacrificial layer and etch holes has been used to make a high fill-factor PMUT array, demonstrated in MEMS-2014 [6], but required a complicated multi-layer fabrication process and an additional layer to seal the etch holes after the release etch. PMUTs based on cavity SOI wafers have the advantages of a simple fabrication process and a high fill-factor, with device characterization first demonstrated in Hilton-Head-2014 [7], and integrated circuit details and tissue-phantom imaging demonstration to appear [8]. Here, we show for the first time short-range and high-resolution imaging using fluid-immersed AlN PMUTs.

DESIGN

A focused, narrow acoustic beam is essential for high resolution, pulse-echo ultrasound imaging. When using a single ultrasound transducer, two methods can be used to obtain a narrow acoustic beam: higher working frequency and larger PMUT dimensions [9]. Calculated acoustic beam patterns at a distance 1.5 mm from a single 50 μm diameter PMUT with various working frequencies are shown in Figure 1. The results show that ~400 MHz working frequency is required to achieve sub-100 μm beam-width. Achieving such a high frequency requires a PMUT working in thickness mode rather than flexural mode, resulting in poor acoustic impedance matching to fluid and tissue. In addition, higher working frequency will result in much greater acoustic attenuation in these media.

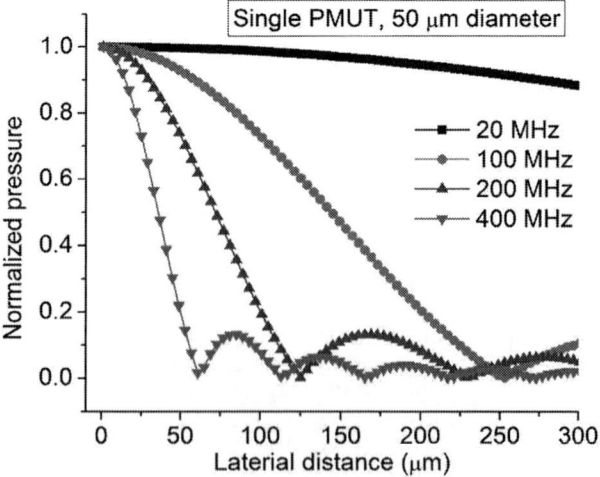

Figure 1: Calculated beam patterns at 1.5 mm away from a single PMUT. A 100 μm -6dB beam-width occurs at 400 MHz.

Figure 2: Optical images of a 72×9 PMUT array.

Here, rather than using a single high-frequency transducer, we use an 8 MHz phased array to achieve narrow acoustic beam-width. Optical images of the 72×9 PMUT array are shown in Figure 2. The array, composed of 50 μm diameter PMUTs with 70 μm pitch, was fabricated using cavity SOI wafers [7]. The 9 PMUTs in each column are connected together. Phased-array beam-forming is conducted using groups of transducers that are fewer than the whole 72-column array. Figure 3 shows calculated acoustic beam patterns for a 15-column group with various beam-forming pitches. The array's acoustic beam-width at 1.5 mm away from the array narrows with increasing PMUT pitch (70 μm, 140 μm, 210 μm) because the aperture of the array is increased. Conversely, a small pixel is required for high resolution ultrasonic imaging, and achieving high fill-factor requires minimizing the PMUT pitch. To satisfy both of these objectives, a novel beam-forming and scanning method was developed, as shown in Figure 4. High fill-factor is achieved through a small 70 μm PMUT pitch, and a narrow beam-width is achieved by using a beam-forming

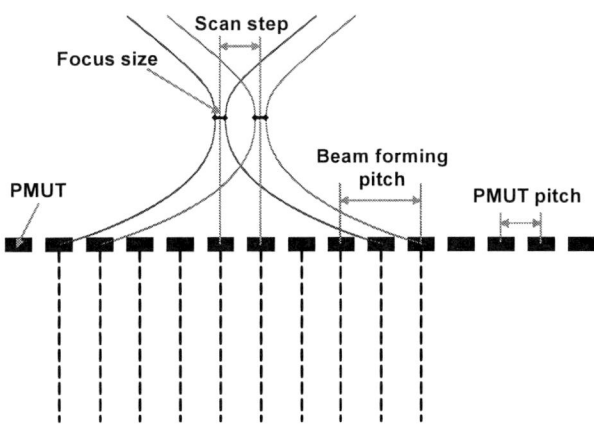

Figure 4: Beamforming using groups. The beam-forming pitch is shown as twice the PMUT pitch but may be any integer multiple. The focused spot is translated by switching from the 1st group of odd-numbered PMUTs (blue) to the 2nd group of even numbered PMUTs (red). The scan step is equal to the PMUT pitch.

pitch that is an integer multiple of the PMUT pitch, thereby achieving a larger aperture from a small group of PMUTs. The focused beam can be scanned by sequentially switching between groups with a small step size that is defined by the PMUT pitch. As illustrated in Figure 4, the phase delays within a group are symmetric; this allows an N-channel amplifier to drive a group containing $2N$-1 columns of PMUTs. In experiments, amplifiers with $N = 7$ and $N = 8$ channels were used, corresponding to 13-column and 15-column PMUT groups.

EXPERIMENT RESULTS

Ultrasound experiments were conducted with the array immersed in a fluid (Fluorinert FC-70, 3M) that has similar acoustic impedance with that of human tissue (Z = ~1.5 MRayls) and high electrical resistance to eliminate the need to insulate the wire-bonds and pads. A needle hydrophone with 40 μm effective diameter (Precision

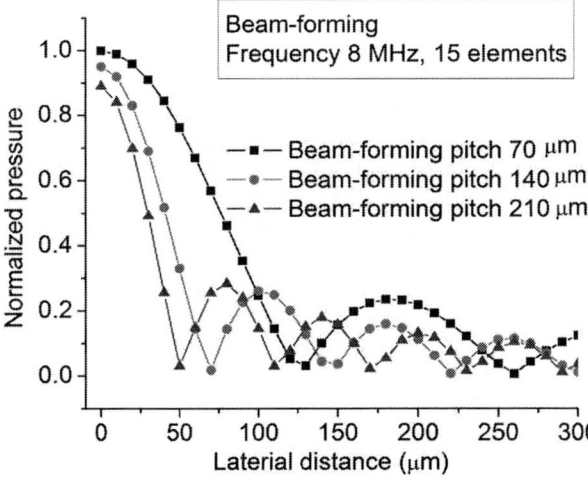

Figure 3: Beam-forming PMUT array is used to achieve narrow acoustic beam at lower frequency, 8 MHz, where low acoustic attenuation makes the beam-forming array favorable.

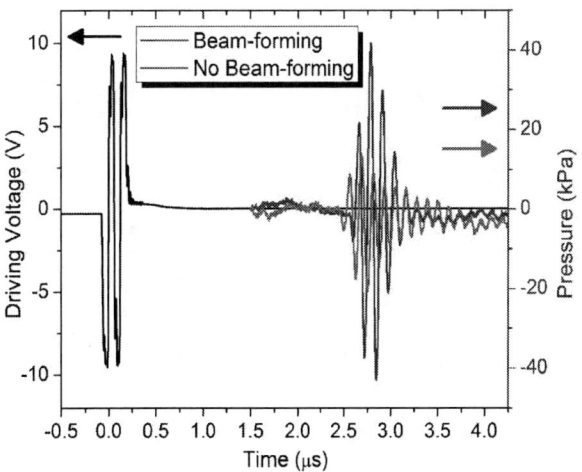

Figure 5: Measured pressure from a 15-column PMUT group (70 μm beam-forming pitch) with and without beam-forming control.

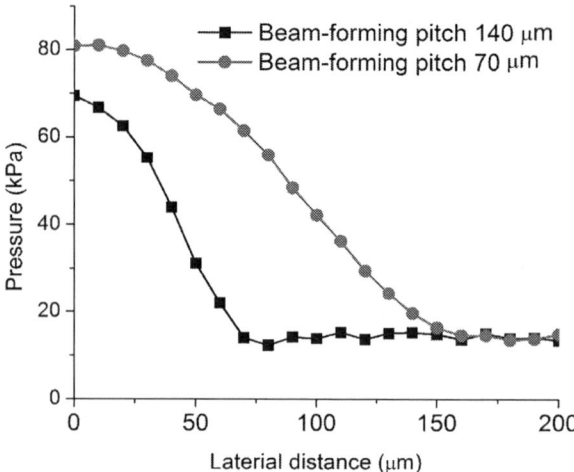

Figure 6: 8 MHz transmit pressure measured using a hydrophone laterally scanned at a distance 1.5 mm from the array surface.

Acoustics) was used to measure the acoustic pressure from a 15-column group of PMUTs driven with 2-cycles of 8-MHz 18 V_{pp} pulses. Shown in Figure 5, the measurement results demonstrate that beam-forming increases the pressure amplitude to 80 kPa peak-to-peak, ~2.5× the pressure produced from the same PMUT group without beam-forming (all PMUTs driven with the same phase delay). Furthermore, beam-forming results in a short acoustic pulse, <0.5 μs, corresponding to ~200 μm axial pulse-echo imaging resolution.

The acoustic pressure patterns of a 15-column group with 70 μm and 140 μm beam-forming pitch were measured using the needle hydrophone, as shown in Figure 6. The beam-forming phase increments were selected to focus the beam 1.5 mm away from the array. In agreement with simulation results shown in Figure 3, larger beam-forming pitch, 140 μm, results in reduced peak pressure from 80 kPa to 70 kPa, but narrower beam-width, from 200 μm to 90 μm diameter. Using 140 μm beam-forming pitch, the 2-D acoustic pressure field

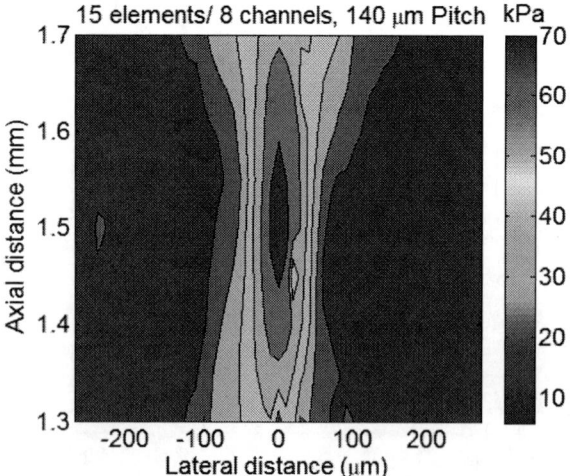

Figure 7: 8 MHz transmit pressure measured using a hydrophone; 2-D acoustic pressure pattern in x-z plane (PMUT array in x-y plane).

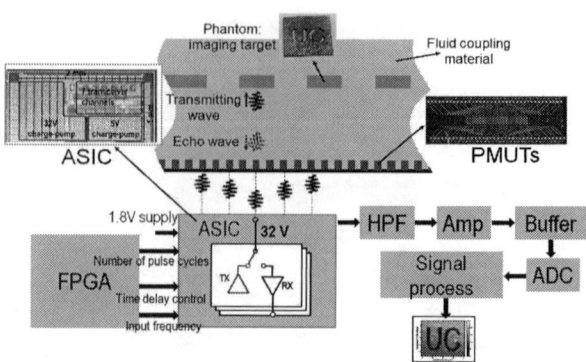

Figure 8: System diagram of pulse-echo imaging using AlN cavity SOI PMUT array and 1.8V 180 nm CMOS ASIC interface [8].

measured in the x-z plane (x is the lateral dimension and z is the axial dimension) is shown in Figure 7. The focused beam demonstrates ~0.4 mm depth-of-focus, which is suitable for many imaging applications.

Pulse-echo imaging was conducted using a custom 1.8V 7-channel interface ASIC [8]. Figure 8 shows the system diagram. Because the capacitance of the 9 AlN PMUTs in each column is low (~pF, including bond pads), a low-power on-chip 32V charge pump is capable

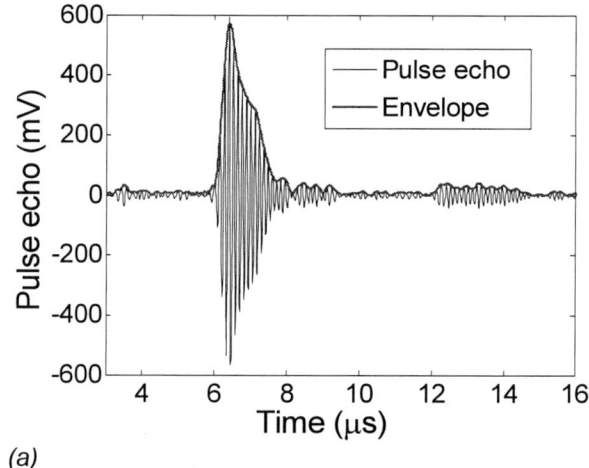

(a)

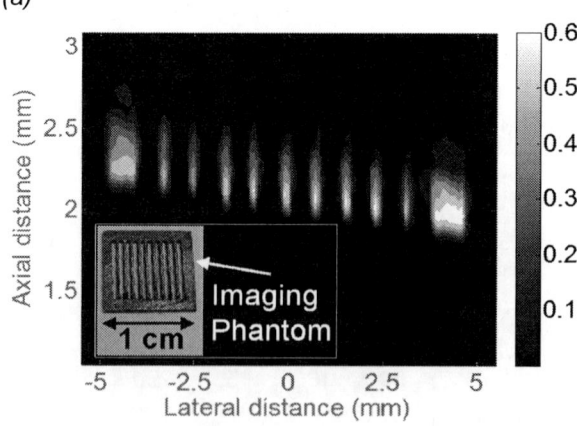

(b)

Figure 9: 1D B-scan pulse-echo imaging of a tilted steel phantom: (a) time-domain plot of received echo signal; (b) imaging result.

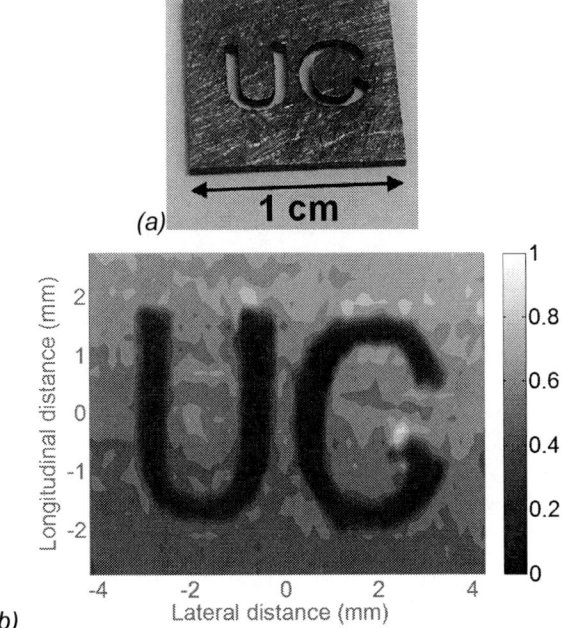

(a)

(b)

Figure 10: (a) Steel phantom and (b) C-scan pulse echo image.

of providing sufficient output current to drive a 15-column PMUT group. Imaging experiments were conducted using fluid-immersed steel phantom targets patterned via laser cutting. While 15-columns are used to transmit a focused beam, the echoes were received using a single column of PMUTs in the middle of the transmitting group for simpler signal process. If all received signals from the 15 columns of PMUTs are collected and processed, receiving beam-forming can be used to achieve higher SNR and imaging resolution. Since the 72×9 PMUT array is only 5 mm × 0.6 mm big, it is too small to image a large 10 mm × 10 mm phantom using electronic scanning. Instead, mechanical scanning was used to demonstrate the imaging performance. Figure 9(a) shows the pulse-echo time response resulting from a steel phantom ~2 mm away from the array. The second acoustic echo is also clearly observed in the time response. Figure 9(b) shows the B-scan image constructed from a lateral scan of this phantom. Figure 10 shows a high-resolution (<100 μm) C-scan image of a 2-D steel phantom, which demonstrates the feasibility for short-range and high-resolution imaging using AlN PMUTs.

CONCLUSION

This work shows that AlN PMUT arrays are suitable for ultrasound imaging. A method was demonstrated to produce a small 90 μm diameter focused beam using a group of PMUTs with a beam-forming pitch that is twice the PMUT pitch. This method was used to produce high-resolution pulse-echo images. Because AlN MEMS devices are already in mass production, AlN PMUTs can be easily commercialized. However, until now, most work has focused on PZT PMUTs, since they produce much higher signal levels than AlN PMUTs. Here, we demonstrated that an AlN PMUT array, combined with a

local custom interface ASIC, can produce high resolution ultrasound images. Because the ASIC is close to the PMUT array, rather than remotely connected through cables, parasitic capacitance is reduced, resulting in higher signal-to-noise ratio. Additionally, the fact that AlN has low dielectric constant means that lower input currents are needed to drive the array, and a 1.8V to 32V on-chip charge-pump provided sufficient power for this purpose.

ACKNOWLEDGEMENTS

The authors thank the UC Berkeley Marvell Nanofabrication Laboratory for device fabrication, and Berkeley Sensor and Actuator Center (BSAC) Industrial Members for financial support.

REFERENCES

[1] P. Muralt and J. Baborowski, "Micromachined ultrasonic transducers and acoustic sensors based on piezoelectric thin films," Journal of Electroceramics, vol. 12, pp. 101-108, Jan-Mar 2004.

[2] Y. Lu, A. Heidari, and D. A. Horsley, "A High Fill-Factor Annular Array of High Frequency Piezoelectric Micromachined Ultrasonic Transducers," *J. Microelectromechanical Systems*, p. 1, 2014.

[3] D.F. Dausch, J.B. Castellucci, D.R. Chou,and O.T. Von Ramm, "Theory and Operation of 2-D Array Piezoelectric Micromachined Ultrasound Transducers", *IEEE Transactions on Ultrasonics Ferroelectrics and Frequency Control*, vol. 55, no. 11, 2008, pp. 2482-2492.

[4] Y. Lu, R. G. Polcawich, O. Rozen, H.-Y. Tang, S. Fung, B. E. Boser, G. L. Smith, and D. A. Horsley, "Broadband piezoelectric micromachined ultrasonic transducers based on dual resonance modes," *Proc. IEEE MEMS*, Estoril, Portugal, 2015.

[5] W. Liao, W. Liu, J. E. Rogers, F. Usmani, Y. Tang, B. Wang, H. Jiang and H. Xie, "Piezoelectric Micromachined Ultrasound Transducer Array for Photoacoustic Imaging", *Proc. Transducers*, Barcelona, 16-20 June 2013, pp. 1831-1834.

[6] Y. Lu, A. Heidari, S. Shelton, A. Guedes and D.A. Horsley, "High frequency piezoelectric micromachined ultrasonic transducer array for intravascular ultrasound imaging", *Proc. IEEE MEMS*, San Francisco, 2014, pp. 745-748.

[7] Y. Lu, S. Shelton, and D. A. Horsley, "High frequency and high fill factor piezoelectric micromachined ultrasonic transducers based on cavity SOI wafers," *Proc. Solid-State Sensors, Actuators and Microsystems Workshop*, Hilton Head, SC, 2014, pp. 131-134.

[8] H. Tang, Y. Lu, S. Fung, D.A. Horsley, B.E. Boser, "Integrated Ultrasonic System for Measuring Body-Fat Composition", *Proc. ISSCC*, San Francisco, Feb 2015.

[9] D. Blackstock, Fundamentals of Physical Acoustics. : John Wiley & Sons, 2000.

CONTACT

*Yipeng Lu, tel: +1-530-752-5180; yplu@ucdavis.edu.

MEASUREMENT OF SURFACE ACOUSTIC WAVES PROPAGATION USING A PIEZORESISTIVE CANTILEVER ARRAY

Nguyen Minh-Dung, Pham Quang-Khang, Nguyen Thanh-Vinh,
Kiyoshi Matsumoto and Isao Shimoyama
The University of Tokyo, Tokyo, JAPAN

ABSTRACT

This paper presents an approach to measure the propagation of surface acoustic waves (SAW), using a piezoresistive cantilever array. SAW was measured by cantilevers designed at the liquid-substrate interaction area by utilizing the structure of liquid-cantilever-air, which is highly sensitive and is able to measure acoustic wave at high frequency. Experiment results demonstrate that the measurable range was confirmed from 0.1 MHz up to 100 MHz. Phase shift and propagation decay were also observed along the propagating direction of SAW.

INTRODUCTION

SAW devices can be found in literature incorporating various applications such as radio frequency communication [1] or biochemical sensors [2] or optical modulators [3]. Up to present, the basic structure of SAW devices consists of a piezoelectric substrate and a comb-shaped interdigitated transducer (IDT). When the IDT is applied with an electrical field, the piezoelectric material is deformed and produces surface acoustic waves. On the other hand, when SAWs arrive at the IDT with a matched frequency, an electrical signal can be measured between the electrodes. Generally, for piezoelectric materials, the signal to noise ratio is supposed to decrease when the device is miniaturized. Moreover, the electric signal would become much smaller in case of mismatch between the pitches of IDT and SAWs.

In this study, we propose a different approach to measure SAW signals using piezoresistive cantilevers covered with a liquid droplet, which was used to convey the vibration caused by SAWs to the cantilevers (Figure 1). The advantage of piezoresistive material is its high sensitivity even in miniaturization. Indeed, in the fabricated device, the longest dimension of the piezoresistive cantilever was approximately 150 µm, and the thickness was 300 nm. The key point here is that the ultra-thin cantilever was placed at the interface of liquid and air. The gap surrounding the cantilever was downscaled to prevent the liquid leak and to provide a multilayer structure of liquid – cantilever – air, which is proved to be high sensitive to pressure change [4] and ultrasonic vibration [5]. In this paper, we propose that this structure can also be utilized to measure SAWs for a wide frequency range. Additionally the propagation of SAWs can be observed with piezoresistive cantilevers that were fabricated along the propagating direction.

Regarding the miniaturized structure and simple fabrication of a cantilever on an SOI wafer, our proposed SAWs sensing device can be expected to lead new research fields and applications related to SAWs.

METHOD AND DESIGN

We measure SAW propagation by utilizing the high

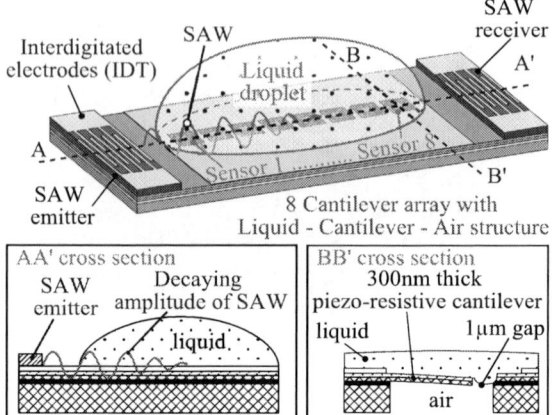

Figure 1: Skeptical diagram of the SAW device. A piezo-resistive cantilever array was used to measure the propagation of SAW on an SOI wafer.

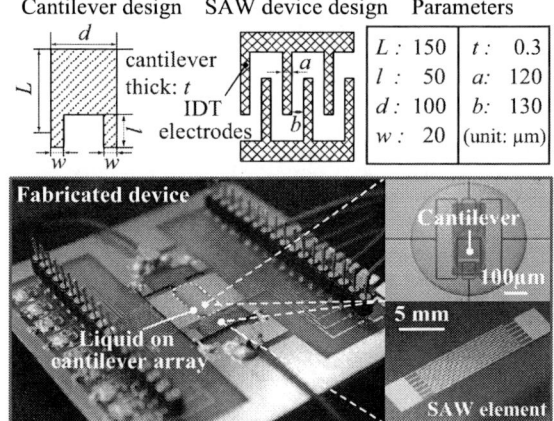

Figure 2: Design and images of the proposed device.

sensitivity and wide dynamic range of the liquid-cantilever-air multilayer structure. The gap between the piezoresistive cantilever and surrounding substrate wall was downscaled to 1 µm to prevent liquid leak. Indeed, a single cantilever can measure acoustic vibration up to a limited narrow frequency range in air. But when it is designed at the interface of liquid and air, the measurable range becomes much wider. We took advantage of this effect and designed a cantilever array lined along the propagating direction to measure SAW in this study. The cantilever array was covered by a liquid droplet, which was silicone oil (HIVAC-F4, Shinetsu Silicone, Japan). The design and images of the device are shown in Figure 2. Fabrication process of the piezoresistive cantilever on an SOI wafer is briefly provided in previous research of our group [6]. SAW device consisted of an IDT on LiNO3 substrate and then was glued on the SOI substrate, where

978-1-4799-7956-1/15 $31.00 © 2015 IEEE 144 MEMS 2015, Estoril, PORTUGAL, 18 - 22 January, 2015

the 8 cantilever array was fabricated, using epoxy. Epoxy was used since it could transfer SAWs from piezoelectric substrate to SOI wafer efficiently.

EXPERIMENTS AND RESULTS

Figure 3 describes the frequency characteristic of the liquid-cantilever-air structure in receiving the waves, which was swept from 0.1 MHz to 100 MHz, from SAW emitter. The output voltages of the sensors (i.e. piezoresistive cantilever) were divided by a reference signal, which was the output voltage of the SAW receiver. The result shows the wide dynamic range of the liquid-cantilever-air structure. The experiment measuring SAW propagation with 8 cantilever array is shown in Figure 4. The responses to SAW emitter from 0.1 MHz to 1 MHz were evaluated. The phenomenon of phase shift can be clearly observed from this result. Additionally, the amplitude measured at each sensor decreased when the distance to the SAW emitter became farer. In Figure 5, we normalized the amplitude of all sensors' response and investigated its dependence on the space position and on emitted SAW frequency. This result indicates that the decay factor was larger for higher frequency.

CONCLUSION

In this study, we proposed a method to measure SAW and its propagation on an SOI substrate using a piezoresistive cantilever array covered with a liquid droplet. The structure of liquid-cantilever-air enables the cantilevers to measure vibration at high frequency up to at least 100 MHz. The phase shift and amplitude decay were confirmed in the experiments. Decay factor was bigger for higher frequency.

ACKNOWLEDGEMENTS

This work is partly supported by New Energy and Industrial Technology Development Organization (NEDO). The photolithography masks were designated and created by using VLSI Design and Education Center (VDEC)'s 8 inch EB writer F5112 + VD01, which is donated by ADVANTEST Corporation.

REFERENCES

[1] C.K. Campbell, *Surface Acoustic Wave Devices for Mobile and Wireless Communications*, Academic Press, Inc.: Orlando, FL, USA, 1998.

[2] R. Dessy, "Surface acoustic wave probe for chemical analysis. I. Introduction and instrument description", *Analytical Chemistry.*, 51(9), pp. 1458-1464, 1979.

[3] A. Yariv, et al., *Optical Waves in Crystals: Propagation And Control of Laser Radiation*, John Wiley and Sons: NewYork, 1984.

[4] N. Minh-Dung, et al., "A Sensitive Liquid-Cantilever Diaphragm for Pressure Sensor", *Proceedings of IEEE MEMS2013*, pp. 617-620, 2013.

[5] P. Quang-Khang, et al., "Multi-axis force sensor with dynamic covering ultrasonic range", *Proceedings of IEEE MEMS2014*, pp. 769 - 772, 2014.

[6] M. Gel, et al., "Force sensing sub-micrometer thick cantilevers with ultra-thin piezoresistors by rapid thermal diffusion", *Journal of Micromechanics and Microengineering*, vol. 14, pp. 423-428, 2004.

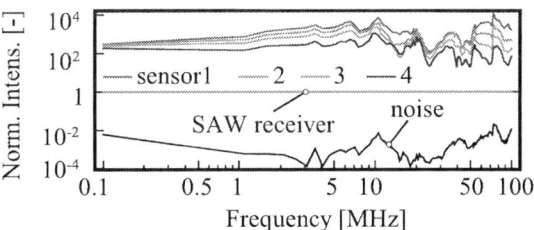

Figure 3: Frequency characteristic of the sensors (piezoresistive cantilevers). The sensor outputs were divided by the SAW receiver's output.

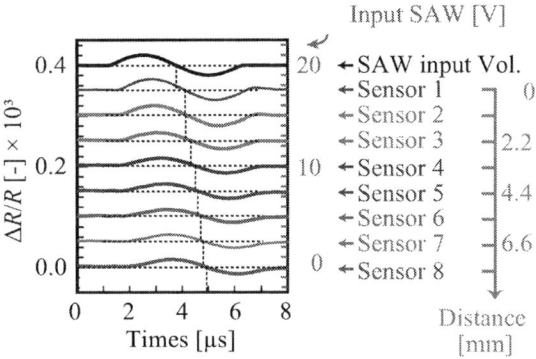

Figure 4: Measurement of SAW propagation with the fabricated piezoresistive cantilever (sensor) array at 0.2 MHz. Phase shift was confirmed between the SAW input and sensors' responses.

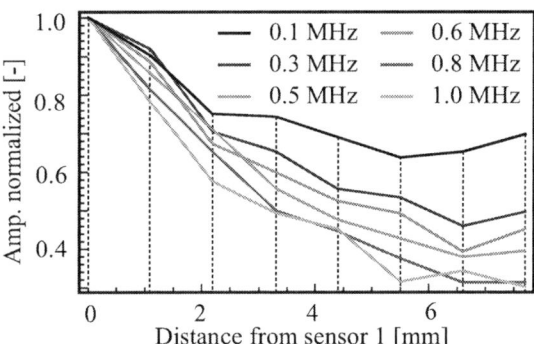

Figure 5: Amplitude decay of the propagating SAW with distance from the sensor 1 for several frequencies.

CONTACT

***Nguyen Minh-Dung, Shimoyama lab, Graduate school of information science and technology, the University of Tokyo, Tokyo, Japan.
Tel: +81-3-5841-6318, Fax: +81-3-3818-0835
E-mail: minhdung@leopard.t.u-tokyo.ac.jp
Lab website: http://www.leopard.t.u-tokyo.ac.jp
Address: Eng. Bldg. 2, Rm. 81B (8th Floor), 7-3-1 Hongo, Bunnkyo-ku, Tokyo, Japan (Postal code: 113-8656)

BROADBAND PIEZOELECTRIC MICROMACHINED ULTRASONIC TRANSDUCERS BASED ON DUAL RESONANCE MODES

Yipeng Lu[1], Ofer Rozen[1], Hao-Yen Tang[2], Gabriel L. Smith[3], Stephanie Fung[1], Bernhard E. Boser[2], Ronald G. Polcawich[3] and David A. Horsley[1]

[1]University of California, Davis, USA
[2]University of California, Berkeley, USA
[3]US Army Research Laboratory, USA

ABSTRACT

Piezoelectric micromachined ultrasonic transducers (PMUTs) have the potential for broad bandwidth, thus enabling high resolution imaging. However, previous PMUTs had fractional bandwidths of only ~50% or smaller because of the thick multilayer PMUT structure. Here, we demonstrate broadband PZT PMUTs that achieve a 97% fractional bandwidth by utilizing a thinner structure excited at two adjacent mechanical vibration modes. PMUTs were fabricated and characterized in the mechanical, electrical and acoustic domains, and a 30 μm × 200 μm ribbon PMUT demonstrates a large displacement sensitivity of 500 nm/V in air and pressure response of 0.3 kPa/V in fluid, equivalent to 13.6 kPa/V average pressure on the PMUT surface, measured 1.4 mm away from the PMUT.

INTRODUCTION

Micromachined ultrasonic transducers (MUTs) have numerous advantages over conventional ultrasound transducers. The most well-known MUTs are capacitive MUTs (CMUTs) [1,2]. Relative to CMUTs, piezoelectric MUTs (PMUTs) [3] have the advantages that they do not require a high-voltage bias source, reducing circuit complexity, and they do not require small capacitive gaps, reducing fabrication complexity [4]. Until recently, PMUTs were less well-developed than CMUTs because thin-film piezoelectric materials technology was immature, resulting in low and poor reproducibility of piezoelectric coefficients. Here, we present PMUTs fabricated using chemical solution deposited (CSD) lead zirconium titanate (PZT) thin films using a fabrication process that previously has demonstrated mechanical relays, resonator/filters, and microwing actuators [5].

A secondary issue suffered by PMUTs is that they are multilayer structures, often resulting in lower fractional bandwidth than monolayer CMUT devices. Broadband transducers are desirable in imaging applications because the bandwidth determines the axial imaging resolution. The bandwidth of a MUT is proportional to damping ($\propto$ membrane area, A) and inversely proportional to mass ($\propto$ $A \times$ thickness). Therefore, the bandwidth is determined in part by the thickness of the membrane, and for the same damping, thicker membranes have greater mass and therefore lower fractional bandwidth. Prior PMUTs had low fractional bandwidth due to their thick membranes: 26% in [4] and 43% in [6]. Here, we demonstrate a PMUT with a thin membrane (0.5μm PZT/ 0.5μm SiO₂) that achieves a broad fractional bandwidth through excitation at two adjacent mechanical vibration modes. Because the two modes have overlapping bandwidths, the final bandwidth is increased by nearly a factor of two.

Finally, it is desirable to minimize the space between PMUTs in an array in order to minimize grating lobes and increase acoustic efficiency. However, PMUTs fabricated using through-wafer etching have large dimensions and therefore low fill-factor [7,8]. While PMUTs based on cavity SOI wafers [9,10] can achieve fine pitch, cavity SOI wafers have relatively poor thickness control (on the order of ±0.5 μm), necessitating thicker PMUT membranes and resulting in narrow bandwidth. Front-side etching enables thinner PMUT membranes, but makes it more difficult to precisely define the released PMUT's dimensions, a parameter which has a strong effect on the resonant frequency. While the PMUT boundaries can be precisely defined using a buried sacrificial layer [11], this buried layer increases fabrication complexity and cost. Here, we proposed a novel process, where front-side etching is used to release PMUT membrane and patterned thick metal layers are used to define the effective boundary of each PMUT.

DESIGN

Figure 1 shows a laser confocal microscope image of a 50 μm diameter PMUT fabricated based on the process flow [5] shown in Figure 2. The circular PMUT is composed of a unimorph membrane (0.05 μm Pt/0.5μm PZT (52/48)/0.1μm Pt/0.036μm TiO₂/0.5μm SiO₂), actuated via an annular top electrode, and released through a central etching hole. The thin PMUT membrane was designed to obtain broad fractional bandwidth of each resonance mode.

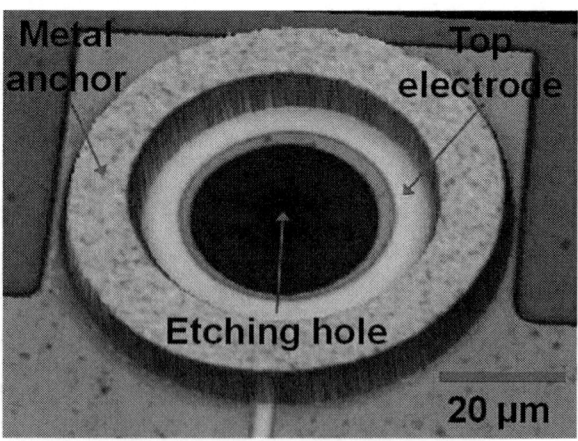

Figure 1: A laser confocal microscope image of a 50 μm circular PMUT anchored with a thick metal ring; the PMUT is actuated via an annular top electrode and released through a central etching hole.

As shown in Figure 2, a timed front-side XeF_2 Si etch is used to release PMUT membrane. However, it is challenging to precisely define the released PMUT's dimensions via timed etching. Therefore, a 4 μm thick gold layer is designed at the boundary of the PMUT membrane to provide additional mechanical support. This gold layer defines the effective PMUT anchor, reducing the sensitivity of the resonance frequencies to overetch during the XeF_2 release step.

Furthermore, it is desirable to reduce parasitic capacitance by removing the high dielectric constant PZT ($\varepsilon = 1200$) from beneath the bond pads and interconnect in order to improve the PMUT's receive sensitivity. This also prevents damage to the PZT layer from wire-bonding. The connection between the PMUT's Pt top electrode and the Au interconnect layer is made via a plated gold air bridge that prevents shorting of the top and bottom electrode layers. The air bridge is formed using a sacrificial photoresist (PR) layer, Figure 3(c), that is removed at the completion of the process using an oxygen plasma, Figure 3(d). As a result of removing the PZT layer from beneath the bond pads and interconnect, the capacitance of a 50 μm diameter PMUT is reduced from 128 pF to 60 pF, doubling the PMUT's receive sensitivity.

RESULTS

The measured displacement frequency response of a

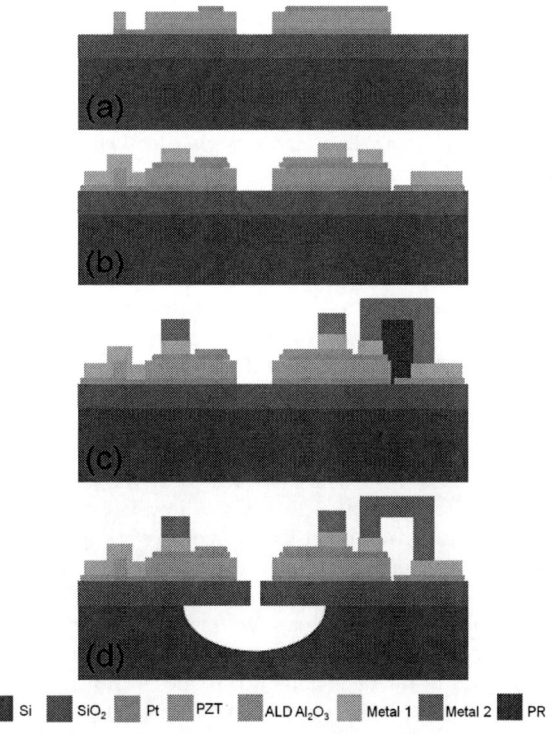

Figure 2: Fabrication process flow: deposition and patterning of (a) passive layer of SiO_2, bottom Pt / TiO_2, PZT(52/48) and top Pt; (b) ALD barrier Al_2O_3 or HfO_2 / Al_2O_3 and Metal 1 (Au/Pt/Cr); and (c) sacrificial PR and Metal 2 (Au); (d) PR removal, patterning passive SiO_2 and XeF_2 etching Si to release the membrane.

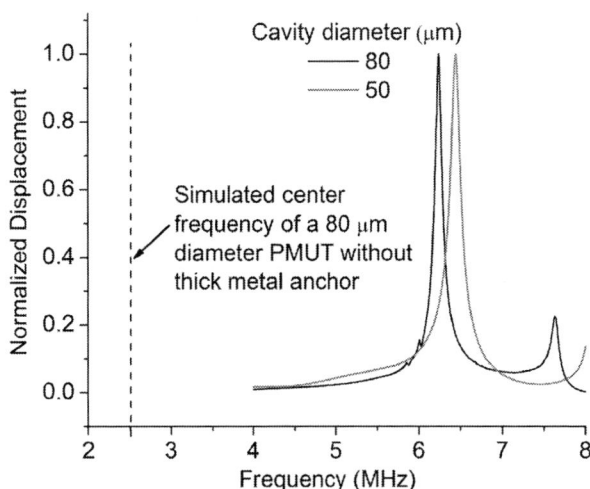

Figure 3: Measured frequency response in air showing the 7% frequency variation that occurs with 60% overetch.

50 μm PMUT with the thick metal anchor is shown in Figure 3. The measurement results show that a 60% (80 μm diameter) overetch produced only a 0.5 MHz shift in the 6.5 MHz center frequency. Finite element method (FEM) simulation results (COMSOL Multiphysics) show that this overetch would result in a 4MHz shift for PMUTs without the thick metal anchor. Figure 4 shows the measured and simulated resonant frequency in air versus overetch percentage (8% to 65%) for PMUTs with and without the thick metal anchor. PMUTs with thick metal anchors show ~10× lower frequency variation when compared with the simulated results for PMUTs without additional metal anchors.

Laser confocal microscope images of a single 30 μm × 200 μm rectangular ribbon PMUT are shown in Figure 5. The ribbon PMUT is designed to have broad bandwidth by selecting dimensions to create closely spaced resonance frequencies of the (1,1) and (1,3) vibration

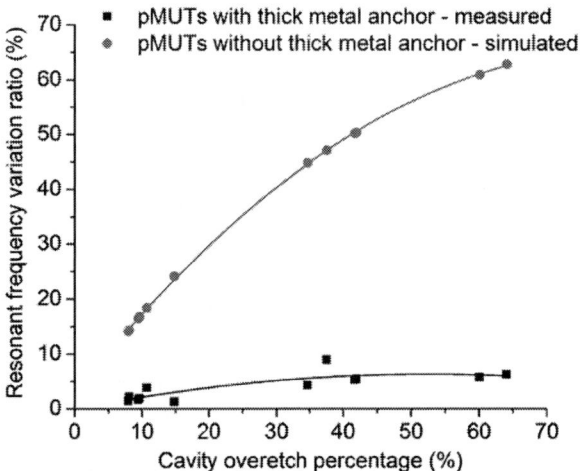

Figure 4: Measured frequency variation versus overetch percentage. PMUTs with the thick metal anchor show ~10× lower frequency variation when compared with simulated results for PMUTs without additional metal anchor.

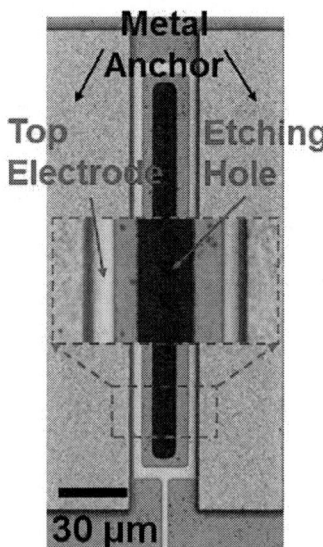

Figure 5: Image of a single 30μm×200μm PMUT. Inset: close-up showing the top electrode and 3 μm etching holes.

modes. PMUTs were characterized in air using a laser Doppler vibrometer (LDV). Frequency response measurements, Figure 6, show that the odd harmonic modes ((1,1), (1,3), …) are excited and demonstrate that the (1,1) and (1,3) modes are separated by approximately 0.4 MHz. The peak displacement sensitivity, 500 nm/V, occurs at the (1,1) mode's 5.5 MHz resonance frequency.

Displacement frequency response measurements were conducted at various dc bias voltages between -10 V to +10 V to characterize the dependence of the resonant frequency and the displacement sensitivity (nm/V) on the bias voltage. The measurement results, Figure 7, show that both parameters are both controlled by dc bias voltage and the PZT hysteresis loop and polarization switching are observed. Polarization switching is indicated by the minima of the measured resonant frequency in Figure 7(a), and by the sign reversal of the

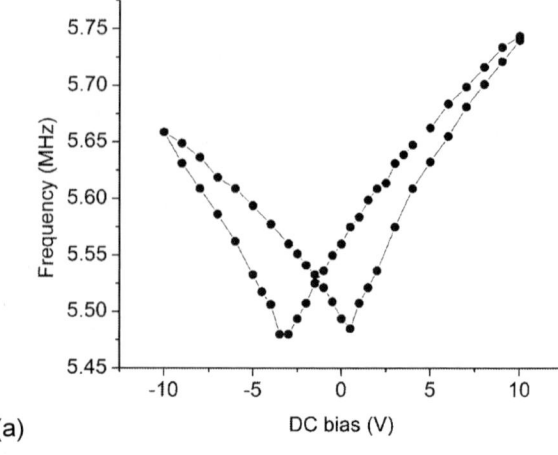

(a)

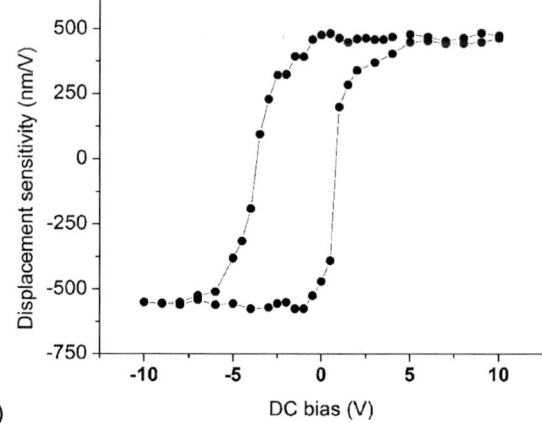

(b)

Figure 7: PMUT resonant frequency (a) and displacement sensitivity at resonance (b) measured in air versus dc bias.

measured displacement sensitivity in Figure 7(b). Both plots show that polarization switching occurs at +0.5 V and –4 V, corresponding to coercive fields $E_c = 1$ V/μm and –8 V/μm. The asymmetric hysteresis and butterfly loop may be caused by preferential alignment along the c-axis of the film, due to defect dipoles or residual stress within the film [12].

Ultrasound experiments were conducted with a single 30μm×200μm ribbon PMUT immersed in fluid (Fluorinert FC-70). The PMUT was driven with a 20V, 4 MHz pulse using a custom 1.8 V interface ASIC [13]. The excitation was unipolar to avoid repoling the PZT. The acoustic pressure was measured using a needle hydrophone 1.4 mm away from the PMUT. The time-domain acoustic response and frequency-domain response (computed via FFT) are shown in Figure 8. The frequencies of the (1,1) and (1,3) modes in fluid are at 3.7 MHz and 5.3 MHz, lower than those in air (shown in Figure 4) due to the mass loading of the fluid. Finally, the ribbon PMUT demonstrates a large 97% fractional bandwidth at 3.7 MHz center frequency due to the thin PMUT membrane and the excitation of the (1,1) and (1,3) modes, which have overlapping bandwidth.

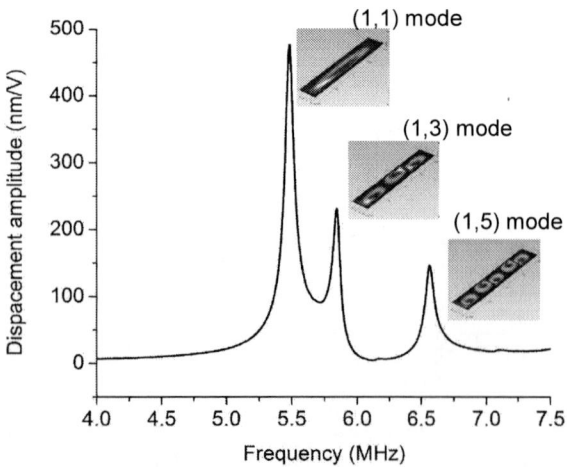

Figure 6: Measured displacement frequency response in air shows that the odd harmonics ((1,1), (1,3), …) modes are excited.

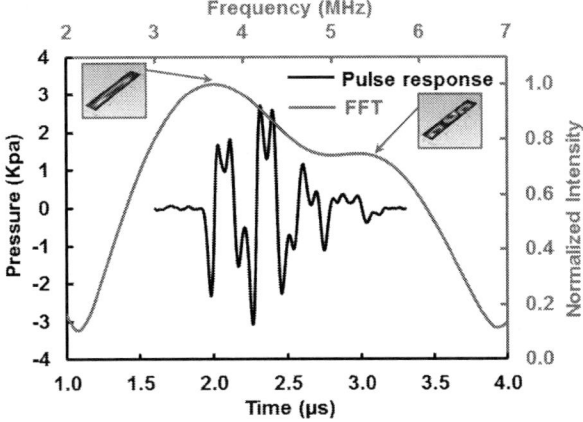

Figure 8: Measured acoustic pressure in fluid (FC-70) from a single $30\mu m \times 200\mu m$ PMUT demonstrates a large 97% fractional bandwidth at 3.7 MHz center frequency due to the thin pMUT membrane and the excitation of the (1,1) and (1,3) modes, which have overlapping bandwidths.

CONCLUSION

PMUTs were fabricated having a thin membrane composed of 0.5 µm piezoelectric PZT layer and 0.5 µm passive SiO_2 layer. To reduce the effect of overetching on the PMUT's resonant frequencies, a plated thick metal layer at the PMUT boundary was demonstrated, which proved to lower the frequency variation ~10× when compared with FEM simulations of PMUTs without the metal anchor. Instead of studying the PZT hysteresis loop using a specific instrument, we directly measured the displacement frequency response of PMUTs at small (-10 dBm) ac voltage and various dc bias voltages, and the results show that polarization switching occurred at coercive fields of E_c = 1 V/µm and –8 V/µm. Finally, due to the PMUT's thin membrane, two vibration modes were designed with overlapping bandwidths. The resulting ribbon PMUT has a broad 97% fractional bandwidth at 3.7 MHz center frequency, indicating its potential for high resolution imaging.

ACKNOWLEDGEMENTS

The authors thank Joel Martin, Brian Power, Luz Sanchez of the US Army Research Laboratory (ARL), Steven Isaacson of General Technical Services, and the UC Berkeley Marvell Nanofabrication Laboratory for device fabrication, and Berkeley Sensor and Actuator Center (BSAC) Industrial Members for financial support.

REFERENCES

[1] F. L. Degertekin, R. O. Guldiken, and M. Karaman, "Annular-ring CMUT arrays for forward-looking IVUS: Transducer characterization and imaging," *IEEE Transactions on Ultrasonics Ferroelectrics and Frequency Control*, vol. 53, pp. 474-482, 2006.

[2] I. O. Wygant, M. Kupnik, and B. T. Khuri-Yakub, "Analytically Calculating Membrane Displacement and the Equivalent Circuit Model of a Circular CMUT Cell," *Proc. IEEE International Ultrasonics Symposium*, 2008 pp. 2111-2114.

[3] P. Muralt and J. Baborowski, "Micromachined ultrasonic transducers and acoustic sensors based on piezoelectric thin films," *Journal of Electroceramics*, vol. 12, pp. 101-108, 2004.

[4] Y. Lu, A. Heidari, S. Shelton, A. Guedes and D.A. Horsley, "High frequency piezoelectric micromachined ultrasonic transducer array for intravascular ultrasound imaging", *Proc. IEEE MEMS*, San Francisco, 2014, pp. 745-748.

[5] G.L. Smith, J.S. Pulskamp, L.M. Sanchez, D.M. Potrepka, R.M. Proie, T.G. Ivanov, R.Q. Rudy, W.D. Nothwang, S.S. Bedair, C.D. Meyer, and R.G. Polcawich, "PZT-Based Piezoelectric MEMS Technology", *J. Am. Ceram. Soc.*, 95 [6] 1777–1792, 2012.

[6] J. Jung, S. Kim, W. Lee, and H. Choi, "Fabrication of a two-dimensional piezoelectric micromachined ultrasonic transducer array using a top-crossover-to-bottom structure and metal bridge connections," *Journal of Micromechanics and Microengineering*, vol. 23, Dec 2013, pp. 1-9.

[7] D.E. Dausch, J.B. Castellucci, D.R. Chou, and O.T. Von Ramm, "Theory and Operation of 2-D Array Piezoelectric Micromachined Ultrasound Transducers", *IEEE Transactions on Ultrasonics Ferroelectrics and Frequency Control*, vol. 55, no. 11, 2008, pp. 2482-2492.

[8] W. Liao, W. Liu, J. E. Rogers, F. Usmani, Y. Tang, B. Wang, H. Jiang and H. Xie, "Piezoelectric Micromachined Ultrasound Transducer Array for Photoacoustic Imaging", *Proc. Transducers*, Barcelona, 2013, pp. 1831-1834.

[9] Y. Lu, S. Shelton, and D. A. Horsley, "High frequency and high fill factor piezoelectric micromachined ultrasonic transducers based on cavity SOI wafers," *Proc. Solid-State Sensors, Actuators and Microsystems Workshop*, Hilton Head, SC, 2014, pp. 131-134.

[10] Y. Lu, H. Tang, S. Fung, B.E. Boser and D.A. Horsley, "Short-Range and High-Resolution Ultrasound Imaging Using an 8 MHz Aluminum Nitride PMUT Array", *Proc. IEEE MEMS*, Estoril, 2015.

[11] Y. Lu, A. Heidari, and D. A. Horsley, "A High Fill-Factor Annular Array of High Frequency Piezoelectric Micromachined Ultrasonic Transducers," *J. Microelectromechanical Systems*, p. 1, 2014.

[12] A. Gruverman, B.J. Rodriguez, A.I. Kingon, R.J. Nemanich, A.K. Tagantsev, J.S. Cross and M. Tsukada, "Mechanical stress effect on imprint behavior of integrated ferroelectric capacitors", *Applied Physics Letters*, 83 [4] 728-730, 2003.

[13] H. Tang, Y. Lu, S. Fung, D.A. Horsley, B.E. Boser, "Integrated Ultrasonic System for Measuring Body-Fat Composition", *Proc. ISSCC*, San Francisco, Feb 2015.

CONTACT

*Yipeng Lu, tel: +1-530-752-5180; yplu@ucdavis.edu.

WHEN CAPACITIVE TRANSDUCTION MEETS THE THERMOMECHANICAL LIMIT: TOWARDS FEMTO-NEWTON FORCE SENSORS AT VERY HIGH FREQUENCY

S. Houmadi[1,2], B. Legrand[3,4], J.P. Salvetat[2], B. Walter[5], E. Mairiaux[5], J.P. Aimé[1], D. Ducatteau[5], P. Merzeau[2], L. Buisson[2], J. Elezgaray[1], D. Théron[5], and M. Faucher[5]

[1]CNRS, CBMN, allée Geoffroy Saint Hilaire, Bât. B14, F-33600 Pessac, FRANCE
[2]CNRS, CRPP, 15 avenue Schweitzer, F-33600 Pessac, FRANCE
[3]CNRS, LAAS, 7 avenue du colonel Roche, F-31400 Toulouse, FRANCE
[4]Univ de Toulouse, LAAS, F-31400 Toulouse, FRANCE
[5]IEMN, CNRS UMR 8520, avenue Poincaré, F-59652 Villeneuve d'Ascq, FRANCE

ABSTRACT

We show that the capacitive transduction of a MEMS device using a setup based on a microwave detection scheme achieves the measurement of the thermomechanical noise spectrum of a high-frequency (>10 MHz) high-stiffness (>10^5 N/m) resonator, reaching the outstanding displacement resolution of 1 fm/√Hz. This result paves the way for vibrating sensors with exquisite force resolution in the fN/√Hz range, enabling large-bandwidth measurements of mechanical interactions at small scale and rheology of fluids at very high frequency. An example of application is given and concerns atomic force microscopy images of bio-molecular assemblies.

INTRODUCTION

MEMS resonators are the basis of multiple devices ranging from time-frequency applications to a variety of sensors [1-3]. These devices are able to detect minute changes in external interactions through variations of amplitude, phase, or frequency. Whatever the physical data that have to be measured (time, force gradient, mass, acceleration, magnetic field, pressure…), the resolution (R) and limit of detection (LOD) are key figures of such sensors. They depend not only on the micromechanical design but also on the signal detection scheme and the transduction efficiency. Indeed, they are strongly related to signal-to-noise and signal-to-background ratios (SNR and SBR). The field of atomic force microscopy (AFM) makes a significant use of vibrating sensors in order to detect tip-surface interactions [4]. Apart from currently available cantilevers or tuning fork probes, mostly operating in the range 30 kHz-300 kHz, significant advances have been made in order to propose high-speed AFM (HSAFM). An example is the use of an optimized AFM setup and small cantilevers with low spring constants [5,6]. In term of an approach renewed by the use of MEMS, high-stiffness resonators offer the benefits of higher working frequencies and larger measurement bandwidths. Thus, they allow the investigation of fast dynamics and high-frequency properties of many phenomena occurring in surface sciences. Moreover, small amplitude operation is achieved with no risk of jump-to-contact, leading to new insights for AFM where the tip vibrates with sub-100fm amplitude in constant interaction with the surface forces [7]. However, high stiffness imposes strong constraints on the measurement resolution: to benefit from the best resolution, femtometer displacements have to be sensed at the thermomechanical limit, which is tremendously demanding in terms of electromechanical transduction. Capacitive sensing is easily integrated in MEMS devices but the low efficiency and parasitic crosstalk at high frequencies are detrimental to SNR and SBR. Recent advances in detection using microwaves have shown that the capacitive transduction can achieve exquisite displacement resolution and compete with optical methods [8]. Here we show that it makes possible to measure the thermomechanical vibration of high-stiffness resonators, paving the way for ultimate sensitivities.

DEVICE DESIGN AND FABRICATION

The MEMS resonator for AFM was designed in order to work above 10 MHz, with a well-defined mechanical mode. The device exploits an in-plane vibration mode of a silicon ring and the chosen mode is the elliptical one. The frequency can be adjusted by varying the lateral dimensions, and the mode provides a true normal motion with respect to the surface to be imaged. This implies an AFM setup where the resonator chip will be mounted vertically [9].

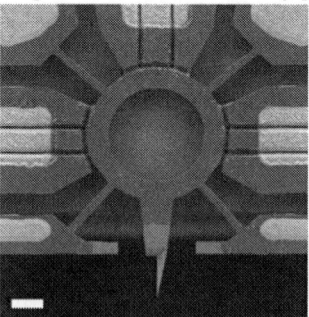

Figure 1: Scanning electron microscope image of the MEMS ring resonator used as the DUT (scale bar 10 μm). Ring inner/outer radii: 20/30 μm, thickness: 5 μm. The capacitive airgap is 50 nm wide and the tip radius is 10 nm.

A 60 μm long tip was placed at one of the vibration antinodes of the ring, and four beams anchor the resonator to the substrate thanks to unreleased large areas of the SOI device layer. The ring inner/outer radii are 20/30 μm, respectively. Given the wafer specifications, we evaluated by finite elements modeling the effective stiffness at the tip location to 200±30 kN/m at the resonance frequency of 12.5 MHz. The fabrication process was very similar to the one described in ref. [10] and made use of silicon-on-insulator (SOI) wafers. The resonator body was first etched in the whole 5 μm thick device SOI layer, followed by several thermal oxidation and chemical de-oxidation cycles. This regenerates low roughness sidewalls, which is a prerequisite for a reliable fabrication of narrow capacitive airgaps using sacrificial oxide spacers. Then, polycrystalline silicon deposition and patterning were performed in order to define electrodes both for detection and actuation. The resonator process specificity also included steps for nanotip fabrication. To do so, TMAH etching was performed thanks to a hard mask previously deposited by LPCVD (Low temperature oxide). Then, back-end processing steps were performed and consisted of n+ ion implantation, ohmic contact formation combined to pads patterning. This defines transmission lines that are necessary for subsequent RF-based detection. Finally, the resonator holder was etched using DRIE starting from the wafer backside. As a final step, a HF wet etching was performed in order to release the resonator. All the steps made use of UV lithography and collective processing. An example of a released device is shown in figure 1.

MICROWAVE DETECTION SCHEME

Experiments were performed using a detection scheme built on prior works [11] and presented in figure 2. An incident wave is reflected by the transducer capacitor and is thus phase-modulated by the time variation of the resonator displacement. The circulator directs the reflected wave to the frequency mixer. It operates a homodyne detection using a reference wave that is phase-shifted in order to down-convert the motional component of the signal; this being in quadrature with respect to the main part of the reflected signal. The converted signal, proportional to the resonator's body displacement, is further amplified and filtered before being analyzed. The microwave signal has been adjusted to optimize the sensitivity as described in [11] and the circuit parts have been carefully selected for very low noise/low loss operation. Figure 3 shows the spectral response of the MEMS resonator in air at room temperature, clearly evidencing the resonance close to 12.591 MHz driven by the thermomechanical noise with a Q factor of 556. Taking into account an effective stiffness value of k=200 kN/m at the tip location, the thermomechanical amplitude was estimated at the resonance frequency, by using $A_{TM} = (2k_B TQ/\pi f_{res}k)^{1/2}$, as 0.8 fm/√Hz, leading to a LOD of 0.65 fm/√Hz. Forced oscillations were obtained by applying a dc + ac voltage to a second capacitive transducer (V_{dc} = 4 V, V_{ac} = 10 mV$_{peak}$). Figure 4 shows the frequency sweep close to the mechanical

resonance. The Lorentzian behavior of the magnitude signal and the large phase rotation at the resonance confirm negligible effects of any parasitic input/output coupling signal floor.

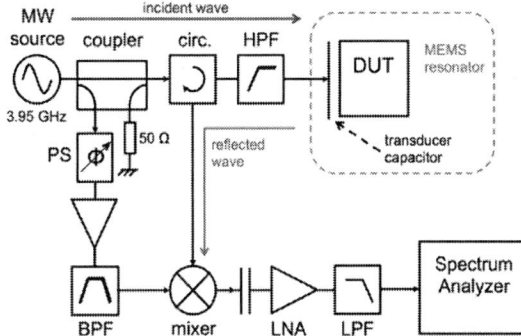

Figure 2: Schematic view of the microwave detection set-up of the MEMS vibration. MW: microwave, circ.: circulator, HPF/LPF/BPF: high/low/band pass filter, PS: phase shifter, LNA: low noise amplifier, DUT: device under test.

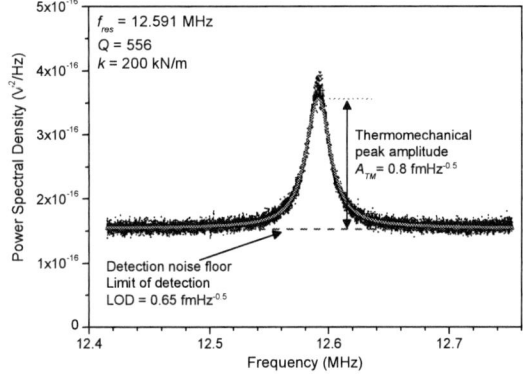

Figure 3: Vibrational noise spectrum of the DUT in air at room temperature. Red line corresponds to a Lorentzian fit. The detection was limited by the set-up noise floor. LOD is determined knowing that the power of uncorrelated noise sources adds up.

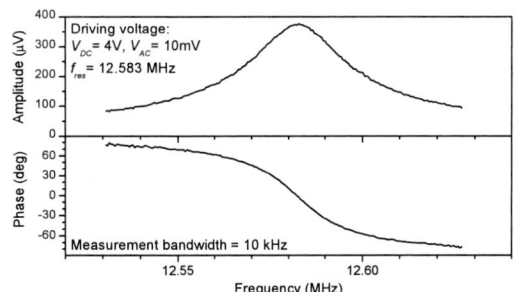

Figure 4: Frequency response of the DUT driven in forced oscillations by a dc + ac voltage applied to a second capacitive transducer. V_{dc} = 4 V, V_{ac} = 10 mV.

The shift of the resonance (- 8 kHz) towards the low frequencies observed in figure 4 with respect to figure 3 is attributed to the spring softening effect caused by the bias voltage $V_{dc} = 4$ V. The vibrational spectrum at a fixed driving frequency ($f = 12.573$ MHz) was measured. It is shown in figure 5. The signal peak corresponding to the forced oscillations is clearly detected above the vibrational noise spectrum, leading to an optimal SNR. The mechanical displacement was determined from the measurement using a calibration based on the thermomechanical peak amplitude, leading to a forced oscillation amplitude of 16 pm. Allan deviation was then computed, showing in figure 6 that white noise dominated up to $\tau = 1$s and that a 1.4 fm amplitude resolution ($1 \times \sigma$) was achieved for $\tau = 4$s.

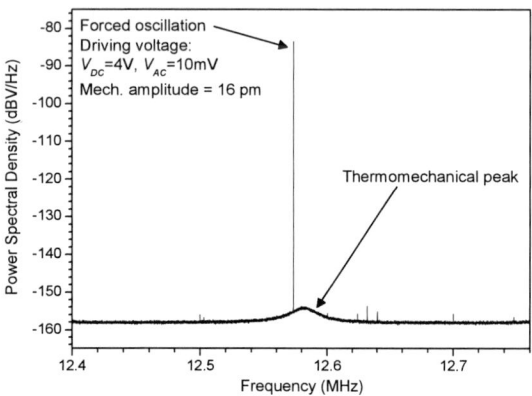

Figure 5: Vibrational spectrum of the DUT driven by a dc + ac voltage at 12.573 MHz. The resolution bandwidth used for the measurement is 14 Hz.

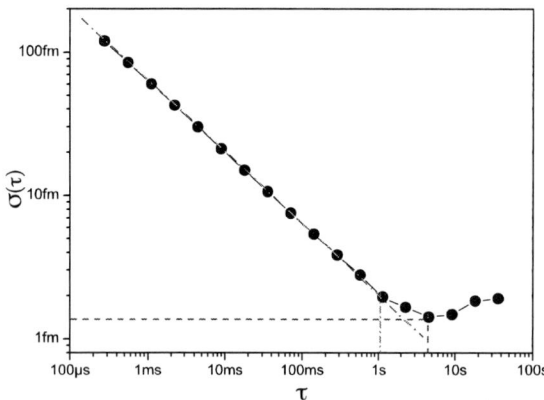

Figure 6: Allan deviation $\sigma(\tau)$ calculated from the vibration amplitude of the MEMS resonator. For an integration time less than 1s, the measurement resolution was dominated by white noise evidenced by a slope of -0.5.

APPLICATION TO AFM IMAGING

The MEMS device was integrated as a probe in a specifically designed AFM microscope shown in figure 7. The silicon chip supporting the microfabricated probe (fig. 1) was glued on the tip of a V-shaped mini-PCB (inset fig. 7) with silver epoxy, and connected by ball bonding to gold HF-compatible transmission lines through gold wires. The mini-PCB was equipped with two HF connectors, one for MEMS excitation and another for microwave detection. The probe holder was screwed to a L-shaped aluminum block that was fixed to a XYZ precision stage (Newport 562-XYZ). The Z-axis displacement was controlled by a stepper motor actuator (Newport TRA12PPD) dedicated to surface approach. Excitation of the resonator and signal acquisition from the microwave detection circuit were performed by a lock-in amplifier (HF2LI from Zurich Instrument). The AFM scanner was a 5 µm × 5 µm × 5 µm Picocube^{TM} piezo scanner (P-363, PI) driven by the E-536 controller. A Nanonis AFM controller was used for driving the instrument, with a Labview-based modular software that allows automatic surface approach, feedback loop control and image acquisition. The 5 kHz PID controller of HF2LI was also used for fast scans. Samples consisted of DNA origamis deposited on freshly cleaved mica from buffer solution. DNA origamis were obtained following Rothemund's procedure by annealing a buffer solution containing thermally denatured long viral DNA strands (extracted from M13mp18 virus) and an excess of 200 short DNA 'staple' strands [12,13]. Staples, being complementary to selected portions of viral DNA sequence, bind to and enforce a particular folding of the viral DNA scaffold upon cooling. In our case, the final shape of the folded structure is a rectangle, of lateral size 60 nm × 100 nm. Besides being a convenient ruler at the nanoscale, DNA origamis offer the possibility to graft additional DNA strands with 2 nm resolution. In our case, we added 9 double stranded protrusions, linked to the origami by 3 thymidines, in order to ensure proper flexibility.

Figure 7: Image of the AFM using MEMS ring-resonators. The setup enables to mount the devices vertically with respect to the sample surface. Inset: close view of the probe holder.

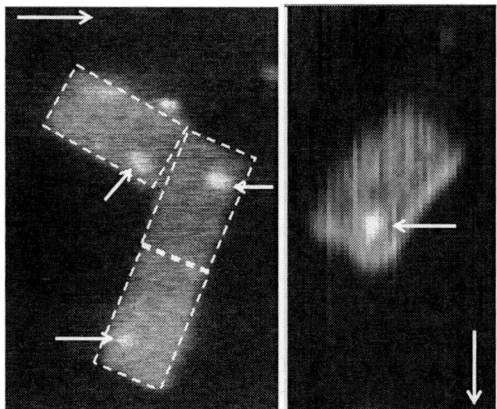

Figure 8: AFM images: left, DNA origami obtained at frame rate of 1 image/87s. Image size: 210×300 nm², 512×512 px. Right, single DNA origami obtained at a frame rate of 1.25 image/s. Image size: 125×250nm², 96×48 px. Yellow arrows indicate the scanning direction; white arrows highlight the DNA motif on the top of the origami.

Imaging was performed in air using amplitude modulation mode with free vibration amplitude $A_0 = 16$ pm and setpoint A_{sp}=14.5 pm. The MEMS was driven at the frequency $f = 12.573$ MHz, in the same conditions as in figure 4, which corresponds to a 45° phase shift. The tip-surface interaction given by $<f_{ts}> = k/Q(A_0^2 - A_{sp}^2)^{1/2}$ was estimated to be 2.4 nN. Figure 8 shows AFM images of the sample acquired with the MEMS AFM probe. Origamis were clearly defined and an overhanging DNA fuzzy blob corresponding to the 9 ds DNA motif was visible and not damaged even at frame rate up to 1.25 fps. From the LOD, it is estimated that a force resolution of 280 fN/√Hz could be achieved. Low-frequency fluctuations in the images are attributed to a residual coupling of the microscope with external mechanical vibrations.

CONCLUSION

Detection of a MEMS resonator motion at the femtometer scale was achieved using capacitive transduction, allowing observation of the thermomechanical spectrum of microresonators with high stiffness constant and calibration of the displacement. Based on these results, Atomic Force Microscopy can be envisioned with femtoNewton force resolution.

ACKNOWLEDGEMENTS

We acknowledge Renatech Network and in particular the IEMN clean room staff. This work has been supported by the French National Agency through project PIA VIBBnano ANR-10-NANO-04-01, and project VIBRATIONS ANR-12-EMMA-0039-01. We also acknowledge financial support from the European Community's Seventh Framework Program FP7/2007-2013

under Grant 210078 and Grant 297518.

REFERENCES

[1] C.T.-C. Nguyen, "Microelectromechanical devices for wireless communications", *in proc. IEEE MEMS'98 conference*, Heidelberg, Jan. 25-29, 1998, pp. 1-7.

[2] R.T. Howe and R.S. Muller, "Integrated resonant-microbridge vapor sensor", in *Proc. Int. Electron Devices Meeting*, San Francisco CA, Dec. 1984. pp. 213-216.

[3] C.C.-F Chiang, A.B. Graham, B.J. Lee, A. Chae Hyuck, E.J. Ng, G.J. O'Brien, and T.W. Kenny, "Resonant pressure sensor with on-chip temperature and strain sensors for error correction", *in proc. IEEE MEMS'13 conference*, Taipei, Jan. 20-24, 2013, pp. 45-48.

[4] F. J. Giessibl, "Forces and frequency shifts in atomic-resolution dynamic-force microscopy", *Phys. Rev. B* 56, 16010 (1997).

[5] T. Ando, T. Uchihashi, and N. Kodera, "High-speed atomic force microscopy", *Jpn. J. Appl. Phys.* 51, pp. 08KA02-1-15 (2012).

[6] T. Ando, T. Uchihashi, and T. Fukuma, "High-speed atomic force microscopy for nano-visualization of dynamic biomolecular processes", *Prog. Surf. Sci.* 83, pp. 337-437 (2008).

[7] D.S. Wastl, A.J. Weymouth, and F.J. Giessibl, "Optimizing atomic resolution of force microscopy in ambient conditions", *Phys. Rev. B* 87, 245415 (2013).

[8] B. Legrand, D. Ducatteau, D. Théron, B. Walter, and H. Tanbakuchi, "Microwave reflectometry: A high-resolution technique for measuring vibration of capacitive microresonators", *in proc IEEE Transducers*, Barcelona, June 16-20, 2013, pp. 566-569.

[9] M.Faucher, B. Walter, A.S. Rollier, K. Seguini, B. Legrand, G. Couturier, J.P. Aime, R. Boisgard, and L. Buchaillot, "Proposition of Atomic Force Probes Based on Silicon Ring-Resonators" *in proc. IEEE TRANSDUCERS'07*, Lyon, June 10-14, 2007, pp. 1529-1532.

[10] B. Walter, M. Faucher, E. Mairiaux, Z. Xiong, L. Buchaillot, and B. Legrand, "4.8 MHz AFM nanoprobes with capacitive transducers and batch-fabricated nanotips", *in proc. IEEE MEMS 2011*, Cancun, Jan 23-27, 2011, pp. 517-520.

[11] B. Legrand, D. Ducatteau, D. Théron, B. Walter, and H. Tanbakuchi, "Detecting response of microelectro-mechanical resonators by microwave reflectometry", *Appl. Phys. Lett.*, 103, 5, 053124 (2013).

[12] P.W.K. Rothemund, N. Papadakis, and E. Winfree, "Algorithmic self-assembly of DNA Sierpinski triangles", *PLoS Biology* 2, 12, pp. 2041-2053 (2004).

[13] P.W.K. Rothemund, "Folding DNA to create nanoscale shapes and patterns", *Nature*, 440, pp. 297-302 (2006).

978-1-4799-7956-1/15 $31.00 © 2015 IEEE

MEMS OXYGEN TRANSPORTER TO TREAT RETINAL ISCHEMIA

Dongyang Kang[1], Karthik Murali[2], Nicholas Scianmarello[1], Jungwook Park[1], Jay Han-Chieh Chang[1], Yang Liu[1], Kai-Tang Chang[1], Yu-Chong Tai[1], and Mark S. Humayun[2]

[1]California Institute of Technology, Pasadena, USA
[2]University of Southern California, Los Angeles, USA

ABSTRACT

For the first time, a paradigm shift in the treatment of retinal ischemia is proposed: providing localized supplemental oxygen to the ischemic tissue via an implanted MEMS device. A passive MEMS oxygen transporter is designed, built and tested in both artificial eye models and porcine cadaver eyes to confirm various hypotheses. The finite element modeling results predict that the proposed approach can treat complete retinal ischemia.

INTRODUCTION

In the United States, the leading cause of blindness is diabetic retinopathy [1], which results in retinal hypoxia (i.e. lack of oxygen), secondary to retinal ischemia (i.e. inadequate blood flow), ending in retinal nerve cell death. Unfortunately, current treatments of diabetic retinopathy have significant drawbacks. Laser photocoagulation results in a constricted peripheral visual field as well as delayed dark adaptation [2], while pars plana vitrectomy increases the risk of iris neovascularization as well as elevated intraocular pressure [3]. However, localized, supplemental intravitreal oxygen therapy has been proposed as a therapy for retinal ischemia [4]. This is done by delivering supplemental oxygen locally to hypoxic tissue and preventing progression of the ischemic cascade. This paper reports the first MEMS device (MEMS oxygen transporter) to transport oxygen from the oxygen-rich sub-conjunctival space to oxygen-deficient vitreous humor and then the ischemic retina.

First, the oxygen permeability of silicone is measured using a custom setup. Its high oxygen permeability allows fast passage of oxygen, making it an ideal material for the device. Then artificial eye models are used to simulate the diffusive behavior of oxygen in the static vitreous gel and the saccade-induced convective transport in the vitreous chamber following vitrectomy. The latest research has led to a finding that vitreous consumes oxygen due to the antioxidative capacity of ascorbate acid in the vitreous humor [5]. The reaction kinetics of the ascorbate-dependent oxygen consumption is studied via measuring the decay of oxygen tension in vitreous samples of porcine cadaver eyes, and confirmed using our device as an oxygen source in porcine cadaver eyes. Finally, finite element modeling is used to map three-dimensional oxygen transport in human eyes, predicting that the proposed approach can restore the retinal oxygen consumption rate (i.e. oxidative metabolic rate) to healthy level even for complete retinal ischemia (i.e. zero oxygen supply from retinal capillaries).

DEVICE DESIGN AND FABRICATION

The device has three major components (Figure 1): a bag (placed underneath the permeable conjunctiva and resting on the impermeable sclera), a cannula (penetrating the sclera at the pars plana), and a diffuser (placed in the posterior vitreous). The oxygen partial pressure (pO_2) underneath the conjunctiva is measured to be ~100mmHg for a rabbit when the eye is open, and around ~15mmHg in the vitreous humor. The pO_2 gradient will drive sub-conjunctival oxygen to permeate into the bag, transport along the cannula and diffuse into the vitreous and then the retina. The diffuser is designed to enter through a 3mm-long surgical incision by folding it. The cannula encapsulates a type 304 stainless steel tube (OD = 0.02in, ID = 0.016in) that can be bent to hold the device on and inside the eye as in Figure 1a. A photo of a finished device is shown in Figure 1b.

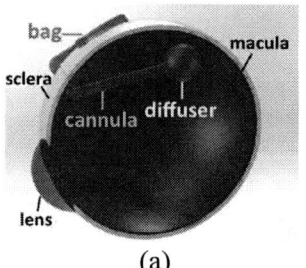 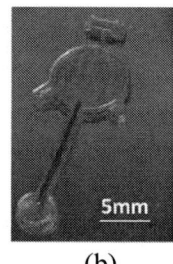

(a) (b)

Figure 1: (a) Device placement in the eye. (b) A photo of a finished device.

The device fabrication (Figure 2a) has two main steps: mold fabrication and silicone casting. For the mold fabrication, three layers of negative dry film photoresist (WBR2120, single-layer thickness 120μm, DuPont™) are laminated on a fresh silicon wafer, followed by post-lamination baking (65°C, 20min), UV light exposure (~400mJ/cm²) and development (AZ340 developer:DI water=1:4, ~40min), defining the side shells of the device. Then a second three-layered lamination followed by the same lithography process defines the top and bottom flat shells of the device. The resulting mold is coated with Parylene C for the easy release of subsequent silicone cast due to week adhesion between them. For the casting of silicone (MED4-4210, two-part, medical-grade, NuSil Technology LLC), base and curing agent are mixed at a 10:1 ratio by weight, degassed under vacuum, and applied onto the patterned mold. Excess silicone is removed, leaving silicone only inside the mold. This is followed by a partial curing at 65°C for 30min. The resulting partially cured silicone cast is only one half of the final device. A type 304 stainless steel tube of desired length (e.g., 12mm) is cut and assembled together with two silicone casts such that it is inside the cannula of the final device. The assembly utilizes uncured silicone followed by fully curing (100°C, 8 hours) as the glue to produce an enclosed chamber, as the final device. The geometry and the dimensions of the device (Figure 2b) remain the same throughout the following experiments and computational modeling. This device will work as an implant, and its biocompatibility is ensured by containing only medical grade silicone and type 304 stainless steel, well known for good biocompatibility.

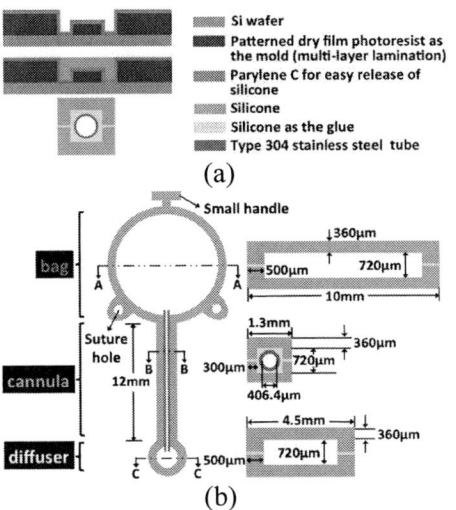

(a)

Si wafer
Patterned dry film photoresist as the mold (multi-layer lamination)
Parylene C for easy release of silicone
Silicone
Silicone as the glue
Type 304 stainless steel tube

(b)

Figure 2: (a) The fabrication process of the device. (b) The geometry and the dimensions of the device.

EXPERIMENTS, CHARACTERIZATION AND COMPUTATIONAL MODELING

The oxygen permeability of silicone is measured using a custom setup as in Figure 3a, according to the equation $\ln\left[(pO_2^t - pO_2^{ambient})/(pO_2^0 - pO_2^{ambient})\right] = -RTA\tilde{p}/(V_{tot}l)\,t$, where pO_2^t is pO_2 in the gas chamber at time t, pO_2^0 is pO_2^t at $t=0$, $pO_2^{ambient}$ is pO_2 in the air, $\sim$160mmHg, R is the universal gas constant, T is the room temperature, A is the silicone membrane area, l is the silicone membrane thickness, V_{tot} is the gas chamber volume, and $\tilde{p}$ is the oxygen permeability. Here, A, l, and V_{tot} are the estimated average values over the deflation process. The differential pressure $pO_2^t - pO_2^{ambient}$ is measured by a pressure gauge. This equation can be derived from the definition of the gas permeability $J = \tilde{p}(\Delta p/l)$, where J is the gas permeation molar flux and $\Delta p/l$ is the applied pressure gradient across the membrane thickness l. Then the oxygen permeability $\tilde{p}$ can be determined from the slope of the plot $\ln\left[(pO_2^t - pO_2^{ambient})/(pO_2^0 - pO_2^{ambient})\right]$ vs. time t. The result of a test is shown in Figure 3b. The mean oxygen permeability of silicone over three tests is determined to be $(3.49\pm0.78)\times10^4\ \mu L\cdot\mu m/(mm^2\cdot day\cdot atm)$.

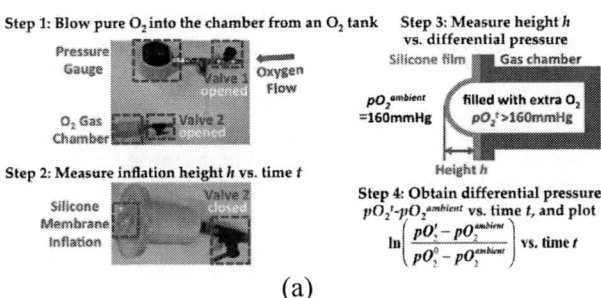

Step 1: Blow pure O_2 into the chamber from an O_2 tank
Pressure Gauge
Oxygen Flow
Valve 1 opened
O_2 Gas Chamber
Valve 2 opened

Step 2: Measure inflation height h vs. time t
Silicone Membrane Inflation

Step 3: Measure height h vs. differential pressure
Silicone film
Gas chamber
Valve 2 closed
$pO_2^{ambient}$ =160mmHg
filled with extra O_2
pO_2^t >160mmHg
Height h

Step 4: Obtain differential pressure pO_2^t-$pO_2^{ambient}$ vs. time t, and plot $\ln\left(\dfrac{pO_2^t - pO_2^{ambient}}{pO_2^0 - pO_2^{ambient}}\right)$ vs. time t

(a)

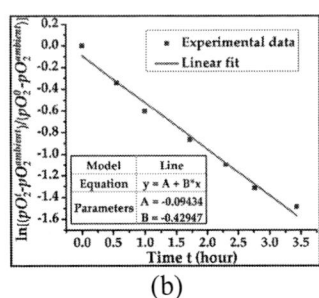

(b)

Figure 3: (a) Experimental procedure to measure the oxygen permeability of silicone. (b) A plot of $\ln\left[(pO_2^t - pO_2^{ambient})/(pO_2^0 - pO_2^{ambient})\right]$ vs. time t from a test, in which the oxygen permeability of silicone $\tilde{p}$ is calculated to be $4.04\times10^4\ \mu L\cdot\mu m/(mm^2\cdot day\cdot atm)$.

There are four main oxygen transport processes for the device. We define the oxygen flow resistance as the ratio of the difference between the upstream and downstream pO_2, ΔpO_2, to the oxygen molar flow rate. The flow resistance R_1 comes from sub-conjunctival oxygen permeation into the bag, R_2 from oxygen diffusion along the cannula, R_3 from oxygen permeation out of the diffuser and R_4 from oxygen diffusion through the vitreous to the retina. R_1 and R_3 are calculated from the measured oxygen permeability of silicone and the definition of the gas permeability. R_2 is calculated assuming one-dimensional Fick's law and the oxygen diffusivity inside the cannula equal to the oxygen-air binary diffusivity at 25°C and 1atm, $0.2cm^2/s$. R_4 is calculated by COMSOL Multiphysics 4.4 simulation assuming $pO_2=0$ at the retina (i.e. the worst case) and dissolved oxygen diffusivity in the vitreous equal to that in water at 25°C, $2\times10^{-5}cm^2/s$. Henry's law of solubility $pO_2 = K_H c$ is used to convert the molar concentration c of dissolved oxygen in the vitreous to pO_2 and the Henry constant K_H is assumed equal to that for water at 25°C, $769.2L\cdot atm/mol$. The calculated values of R_1–R_4 and a circuit model for the oxygen transport processes as a series combination of R_1–R_4 are shown in Figure 4. R_4 is significantly larger than all the other three resistances, implying that the oxygen diffusion in the vitreous is the limiting process. Assuming $pO_2=100$mmHg in the sub-conjunctival space, pO_2 in the diffuser is calculated to be 98mmHg.

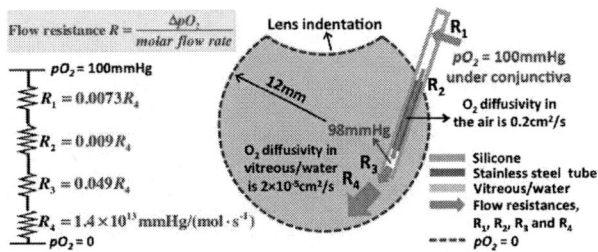

Figure 4: Flow resistances calculated by theory for R_1, R_2 and R_3 and simulation for R_4 and a circuit model for the oxygen transport processes. Oxygen diffusion in the vitreous is the limiting process and pO_2 is high in the diffuser even for the worst case, i.e. $pO_2=0$ at the retina.

An artificial human eye model of the same dimension

978-1-4799-7956-1/15 $31.00 © 2015 IEEE

as in [6] is created in hard silicone, and outside coated with epoxy to prevent any oxygen permeation through the shell, hence imposing a no-flux boundary condition (BC). It is then filled up with deoxygenated water (pO_2= 15mmHg initially). With a fabricated device mounted on it and exposed to the air in which pO_2=160mmHg, the pO_2 value in the water at a point 3mm away from the diffuser is measured by an oxygen sensor (NeoFox-GT, Ocean Optics, Inc.) and the four oxygen transport processes are modeled by a single three-dimensional COMSOL simulation (Figure 5). Here, the oxygen permeation process consists of two steps: oxygen dissolution into the silicone and diffusion inside the silicone. Another form of Henry's law of solubility $c= S \times pO_2$ is used to convert the molar concentration c of dissolved oxygen in silicone to pO_2. The solubility coefficient S satisfies $\tilde{p} = SD$, which can be derived from the definition of the gas permeability and Fick's law. Here, $\tilde{p}$ is the measured oxygen permeability of silicone, and D is the oxygen diffusion coefficient in silicone, assuming D =7.88×10^{-5}cm^2/s [7]. Both pO_2 and oxygen molar fluxes across the silicone-air and silicone-water interfaces are set to be continuous. The stainless steel tube wall is modeled as a zero-thickness impermeable material.

To simulate saccade-induced convective transport of oxygen, the eye model is shaken with sinusoidal rotation at a frequency of 5Hz and amplitudes of 5° and 20°. The pO_2 measurement is done at the same point as the static case and the plots in Figure 5c show a significantly faster pO_2 increase, demonstrating saccade-enhanced transport effect. The 3D unsteady saccade-induced laminar flow is modeled by COMSOL to obtain average flow streamlines on the equatorial and vertical planes of the eye model for the 20° rotation over one period when the flow field becomes periodic with time (Figure 5e).

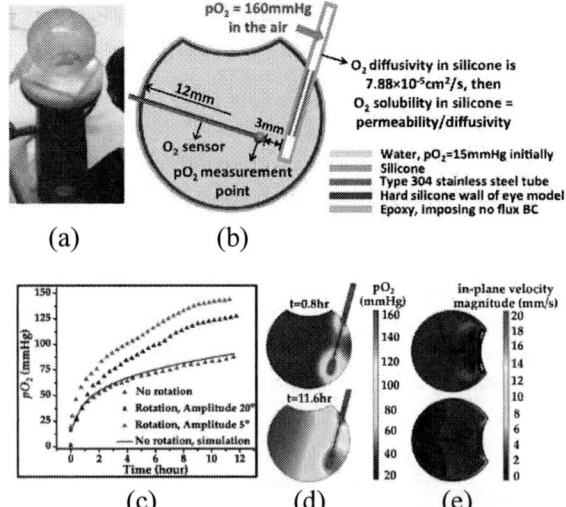

Figure 5: *Artificial eye model experiment. (a) A photo of the setup. (b) An illustration. (c) The experimental and simulation results of pO_2 increase with time. (d) Simulation of pO_2 profiles at t=0.8hr (upper) and 11.6hr (lower) for the static case. (e) Simulation of average flow streamlines on the equatorial (upper) and vertical planes (lower) of the eye model for the 20° rotation.*

Porcine cadaver eyes (Sierra Medical Inc.) are used to

verify our results *in vitro* (Figure 6a). First, the ascorbate-mediated oxygen consumption rate versus pO_2 is measured in a vitreous sample collected with a vitrector. The reaction kinetics [8] can be best described by a pseudo-first-order rate equation since the vitreous ascorbate is supplied in great excess and its concentration remains constant, which is valid in healthy physiological conditions. The vitreous oxygen consumption rate is then approximately linearly proportional to pO_2 (Figure 6b), and thus the oxygen molar concentration.

With a fabricated device mounted on a porcine cadaver eye, steady pO_2 values in the vitreous at points ranging from 0 to 5mm away from the diffuser are measured (Figure 6c). A similar 3D COMSOL simulation for oxygen transport processes is done, however, the chamber radius is changed to 9mm to mimic the porcine cadaver eye, and the vitreous oxygen consumption is included using a first-order reaction kinetics. The simulation results are in accordance with the measurement data (Figure 6d).

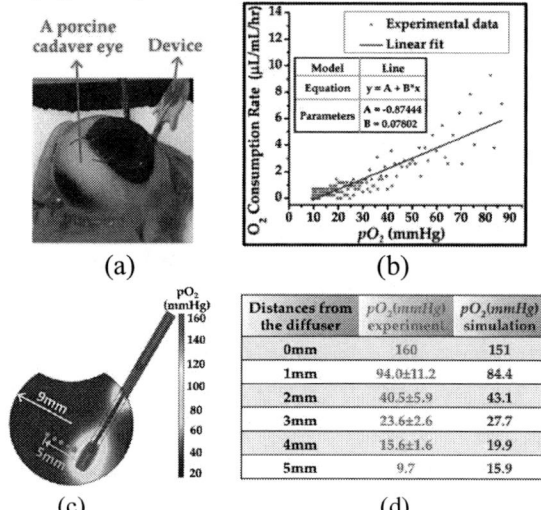

Figure 6: *In-vitro porcine cadaver eye experiments. (a) A photo of the setup. (b) Vitreous O_2 consumption rate vs. pO_2 fit by a first-order reaction rate model. (c) Simulation of the steady pO_2 profile. (d) Measured pO_2 at points from 0 to 5mm away from the diffuser, in accordance with the simulation results.*

Finally, oxygen transport processes in a human eye are modeled by a 3D COMSOL simulation, taking into account first-order lens oxygen consumption [9], ascorbate-mediated vitreous oxygen consumption [9] and oxygen diffusion, supply and consumption in the retina using a four-layered model [10], and all the relevant equations and parameters which can be found in [9, 10]. The convection-enhanced transport effect is simulated by increasing the absolute diffusivity of oxygen in the vitreous body [9]. The simulation results focus on the oxygen consumption rate (i.e. oxidative metabolic rate) in the macula, an oval-shaped area of ~6mm in diameter at the center of the retina, which is responsible for central, high acuity vision. Both pO_2 and oxygen molar fluxes across the silicone-vitreous and vitreous-retina interfaces are set to be continuous. The stainless steel tube wall is again modeled as a zero-thickness impermeable material. The results predict that even for a complete retinal ischemia (i.e. zero oxygen supply from retinal capillaries), an o-ring-shaped diffuser if placed near the macula and combined with partial vitrectomy such that saccade-induced convection will occur, can provide sufficient oxygen flux to the retina, maintaining it at healthy level.

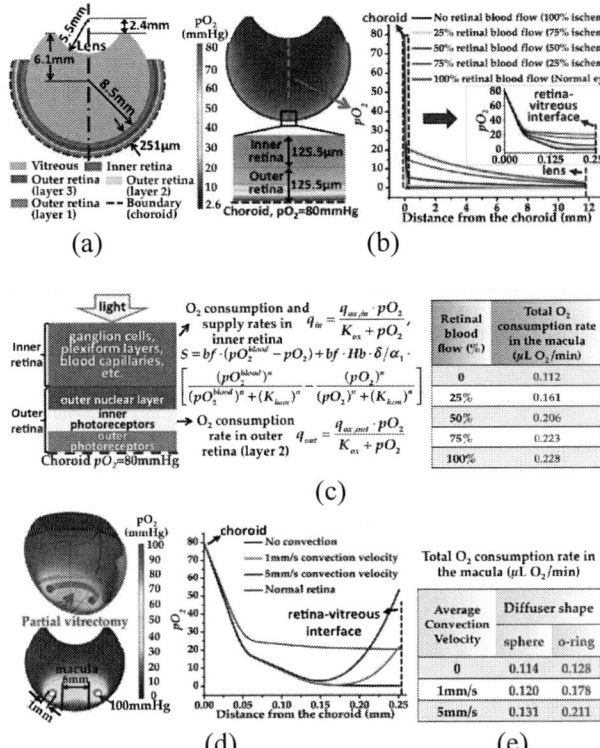

Figure 7: 3D simulation of O_2 transport processes in human eyes. (a) Computation domain [9-11]. (b) pO_2 profiles for various degrees of ischemia. (c) Four-layered model for the retina [10] and total O_2 consumption rates in the macula for various degrees of ischemia. (d) pO_2 profiles for complete ischemica treated by an o-ring-shaped diffuser in which $pO_2=100mmHg$. (e) Comparison between spherical (2mm in diameter, having the same surface area as our disk-shaped diffuser) and o-ring-shaped diffusers.

CONCLUSIONS

Both our experimental and modeling work demonstrate the feasibility of using our MEMS oxygen transporter to treat retinal ischemia that occurs in diseases such as diabetic retinopathy. Even for complete retinal ischemia, an o-ring-shaped diffuser can maintain the health of retinal cells if it is placed near the macula and combined with partial vitrectomy such that saccade-induced convection will occur.

ACKNOWLEDGEMENTS

This work is funded by National Institute of Health. The authors gratefully acknowledge the help of all the members from the Caltech Micromachining Lab and the USC Eye Institute.

REFERENCES

[1] N. Congdon, B. O'Colmain, C. C. W. Klaver, R. Klein, B. Munoz, D. S. Friedman, *et al.*, "Causes and prevalence of visual impairment among adults in the United States," *Archives of Ophthalmology,* vol. 122, pp. 477-485, Apr 2004.

[2] M. B. Landers, E. Stefansson, and M. L. Wolbarsht, "PANRETINAL PHOTO-COAGULATION AND RETINAL OXYGENATION," *Retina-the Journal of Retinal and Vitreous Diseases,* vol. 2, pp. 167-175, 1982.

[3] A. Goto, M. Inatani, T. Inoue, N. Awai-Kasaoka, Y. Takihara, Y. Ito, *et al.*, "Frequency and Risk Factors for Neovascular Glaucoma After Vitrectomy in Eyes With Proliferative Diabetic Retinopathy," *Journal of Glaucoma,* vol. 22, pp. 572-576, Sep 2013.

[4] W. Abdallah, H. Ameri, E. Barron, G. J. Chader, E. Greenbaum, D. R. Hinton, *et al.*, "Vitreal Oxygenation in Retinal Ischemia Reperfusion," *Investigative Ophthalmology & Visual Science,* vol. 52, pp. 1035-1042, Feb 2011.

[5] Y.-B. Shui, N. M. Holekamp, B. C. Kramer, J. R. Crowley, M. A. Wilkins, F. Chu, *et al.*, "The Gel State of the Vitreous and Ascorbate-Dependent Oxygen Consumption," *Archives of Ophthalmology,* vol. 127, pp. 475-482, Apr 2009.

[6] A. Stocchino, R. Repetto, and C. Cafferata, "Eye rotation induced dynamics of a Newtonian fluid within the vitreous cavity: the effect of the chamber shape," *Physics in Medicine and Biology,* vol. 52, pp. 2021-2034, Apr 2007.

[7] M. C. Kim, R. H. W. Lam, T. Thorsen, and H. H. Asada, "Mathematical analysis of oxygen transfer through polydimethylsiloxane membrane between double layers of cell culture channel and gas chamber in microfluidic oxygenator," *Microfluidics and Nanofluidics,* vol. 15, pp. 285-296, Sep 2013.

[8] M. H. Eisonperchonok and T. W. Downes, "KINETICS OF ASCORBIC-ACID AUTOXIDATION AS A FUNCTION OF DISSOLVED-OXYGEN CONCENTRATION AND TEMPERATURE," *Journal of Food Science,* vol. 47, pp. 765-&, 1982 1982.

[9] B. A. Filas, Y. B. Shui, and D. C. Beebe, "Computational Model for Oxygen Transport and Consumption in Human Vitreous," *Investigative Ophthalmology & Visual Science,* vol. 54, pp. 6549-6559, Oct 2013.

[10] M. W. Roos, "Theoretical estimation of retinal oxygenation during retinal artery occlusion," *Physiological Measurement,* vol. 25, pp. 1523-1532, Dec 2004.

[11] R. K. Balachandran and V. H. Barocas, "Contribution of Saccadic Motion to Intravitreal Drug Transport: Theoretical Analysis," *Pharmaceutical Research,* vol. 28, pp. 1049-1064, May 2011.

CONTACT

*Dongyang Kang, tel: +1-626-395-3885; email: dkang@caltech.edu

A NEW MONOLITHICALLY INTEGRATED MULTI-FUNCTIONAL MEMS NEURAL PROBE FOR OPTICAL STIMULATION AND DRUG DELIVERY

Yoojin Son[1, 2], Hyunjoo Jenny Lee[1], Jeongyeon Kim[1], C. Justin Lee[1], Eui-Sung Yoon[1], Tae Geun Kim[2] and Il-Joo Cho[1]

[1] KIST (Korea Institute of Science and Technology), Seoul, South Korea
[2] Korea University, Seoul, South Korea

ABSTRACT

We present a monolithically integrated multi-functional MEMS neural probe by integrating both an embedded microfluidic channel for drug delivery and a SU-8 optical waveguide for optical stimulation. In this work, we used a controlled glass reflow process to form an embedded glass layer which serves as both the cover for the microfluidic channel and the cladding layer for the optical waveguide. Using this simple fabrication process, we achieved a compact structure integrated with multiple stimulation modalities. Using our multifunctional neural probe, we demonstrated successful *in vivo* experiments by optically and chemically activating neurons and recording neural spike signals from individual neurons with a high SNR.

INTRODUCTION

Recently, there has been an increasing interest in neuroscience to understand our brain and to investigate the underlying mechanisms of neurological disorders such as Parkinson's and Alzheimer's diseases by examining neural networks in brain [1,2]. To clarify the neurological diseases, it is important to measure and analyze the signals of cerebral nerves - the smallest unit of brains. For directly measuring these neural signals from individual neurons, MEMS neural probes are promising candidates because of their distinctive advantages: small size, capability to apply deep-brain stimulation, and integration of various types of stimulation modality such as electrical, optical, and chemical stimulation with minimal damage [3]. In specific, the capability to apply various types of stimuli in a single platform allows us to control the brain circuits with high spatial and temporal resolution. For instance, injecting a virus at local target using a neural probe with drug delivery capability facilitates neuroscientists to generate a new disease mouse model [4]. Also, neural probes developed for optogenetic applications are capable of activating and inhibiting neurons with high spatial resolution by targeting only certain types of cells.

However, most of the previously reported neural probes were integrated with only a single functionality [5,6]. A multifunctional neural probe that integrated both optical and chemical stimulation modality was previously reported but the probe dimension was not suitable for *in vivo* experiments for mice and a complicated fabrication process was required [7]. To accurately control the brain circuits, precise spatial control with a high resolution is important. In addition, minimization of damage incurred on brain cells during insertion is critical in obtaining reliable neural signals *in vivo*. Our previous work achieved a thin neural probe with an integrated optical waveguide on

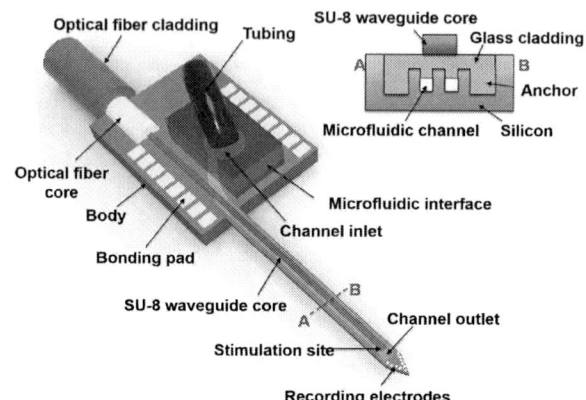

Figure 1: Conceptual diagram of the proposed multi-functional neural probe integrated with an embedded microfluidic channel and an optical waveguide.

multiple shanks, which offered four stimulation sites, and another silicon probe that was integrated with a microfluidic channel for a selective delivery of chemicals in mice [8,9].

In this work, we present a neural probe that integrates two stimulation modalities in a single platform. We developed a new multifunctional MEMS neural probe that is integrated with an embedded microfluidic channel for drug delivery and a SU-8 optical waveguide for an optical stimulation. We used a controlled glass reflow process [10] to form a glass layer in the silicon substrate that serves as a transparent cover for the microfluidic channel and as a particle-free thick cladding layer of the optical waveguide. This layer not only simplifies the fabrication process but also allows us to achieve a compact integrated structure (Figure 1).

FABRICATION PROCESS

The fabrication process of the proposed multi-functional neural probe is shown in Figure 2. The fabrication process starts from forming the glass layer in the silicon substrate. First, on a silicon-on-insulator (SOI) wafer, 25-μm high and 5-μm wide cavities are formed by a deep reactive ion etching (DRIE) using a photoresist mask and an oxide mask. After removing the soft mask, we etch another 10 μm of silicon. Then, a 100-μm-thick borosilicate glass wafer is anodically bonded to a patterned SOI wafer in a vacuum. We reflowed glass at 750°C for two hours using rapid thermal annealing (RTA). The channel height after glass sealing is 10 μm, which is readily controlled by adjusting reflow condition. After the glass reflowing process, unnecessary glass is removed by chemical mechanical polishing (CMP). While the planarized surface is available for next fabrication processes, the substrate now consists of the embedded glass layer which serves as a cover for microchannels and a cladding layer for optical waveguides (Figure 1 inset).

Next, a 300-nm-thick SiO_2 insulation layer is deposited using plasma-enhanced chemical vapor deposition (PECVD) and a 300-nm-thick gold and 20-nm-thick chrome adhesion layers are deposited and patterned using lift-off process. The signal lines are protected by another 400-nm-thick SiO_2 insulation layers and the contact for the microelectrodes are opened followed by microelectrode patterning and 150-nm-thick iridium (Ir) sputter deposition. A 15-μm-thick SU-8 core layer for light transmission is patterned on top of the cladding layer. The refractive index of SU-8 is 1.58, which is larger than the refractive index of glass cladding layer to reduce the propagation loss. Then, a U-groove to position a 125-μm-diameter optical fiber is patterned by DRIE. Then, a 60-μm-wide outlet and a 1-mm-mm wide inlet are patterned and etched using DRIE and RIE. Finally, the shape of the probe shank is defined and released using backside DRIE.

Figure 3 shows scanning electron microscopy (SEM) images of the successfully fabricated multi-functional neural probe integrated with an optical waveguide and

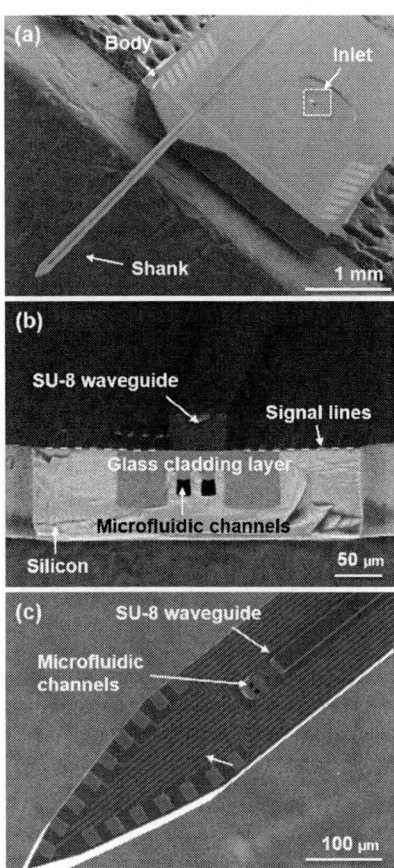

Figure 3: SEM pictures of the fabricated multi-functional neural probes showing (a) probe shank and body, (b) cross section of the probe shank, and (c) tip of the shank with the waveguide, microfluidic channel, and electrode arrays.

microfluidic channels. The cross-sectional SEM image illustrates the compact structure; the SU-8 core is patterned on top of the thick glass cladding layer that serves as a cover for the embedded microfluidic channel (Figure 3(a)). The recording sites are located near the optical and chemical stimulation sites to record neural signals from stimulated neurons (Figure 3(c)).

EXPERIMENTAL RESULTS
Packaging and characterization

We successfully fabricated 40-μm-thick MEMS neural probes integrated with optical waveguides and microfluidic channels. The fabricated multifunctional neural probe was attached on a customized printed circuit board (PCB) and electrically connected through wire bonding. An average impedance of iridium microelectrodes was 0.8 MΩ at 1 kHz, which is low enough to detect neural spikes with a high signal-to-noise ratio (SNR). Also, we packaged the probe with a multimode optical fiber (GIF50, Thorlabs, Newton, NJ) using an index-matching UV-epoxy and a black-epoxy for an optical interface. A custom-designed polydimethylsiloxane (PDMS) interface was bonded to the probe body for fluidic interface resulting in a compact

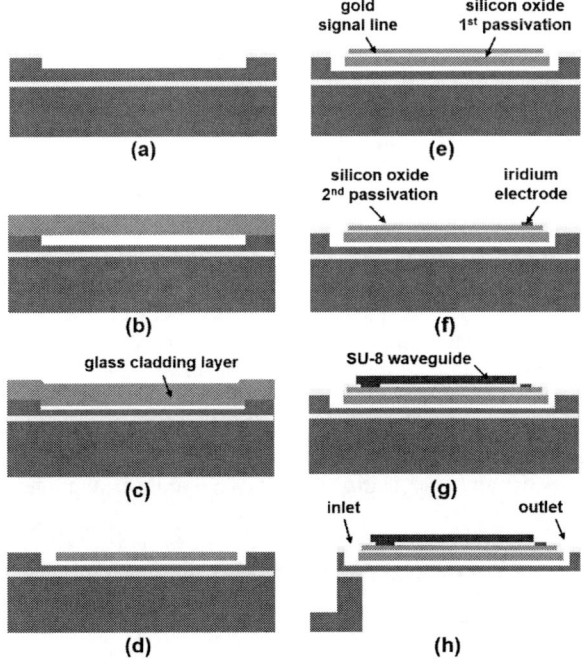

Figure 2: Fabrication process flow: (a) cavity formation, (b) glass anodic bonding, (c) glass cladding and embedded channel formation using reflow, (d) CMP, (e) passivation layer deposition and signal line patterning, (f) passivation layer deposition and electrode patterning, (g) SU-8 waveguide patterning, and (h) groove patterning, inlet and outlet definition, and release of the structure.

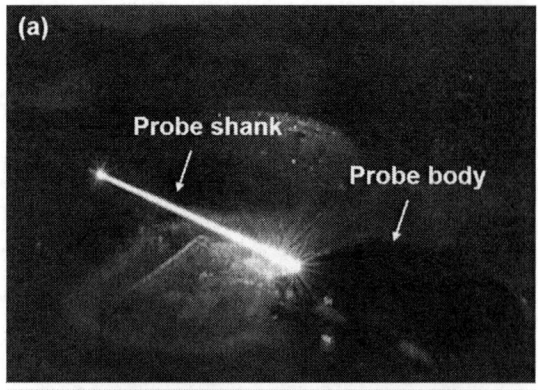

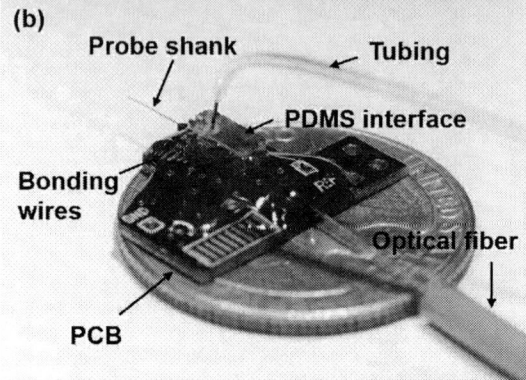

Figure 4: Optical pictures of (a) transmitted light exiting from the end of the packaged neural probe and (b) the packaged probe coupled with an optical fiber and fluidic tubing.

We characterized both the optical property of the integrated waveguides and flow rates of microfluidic channels. First, we measured optical output power of the packaged multi-functional neural probe. By using SU-8 waveguide, we successfully transmitted blue light (λ=473 nm) from the light source (ADR-2301, RGBlase LLC, CA) and measured the maximum output optical power of 0.9 mW, which is enough to stimulate the neurons. Also, we measured a flow rate of water through the microfluidic channels with air pressure ranging from 50 kPa to 120 kPa in air and agarose gel. The agarose gel was used because it has a higher density than the brain [11]. The flow rate of 0.1 µl/min was obtained at 200 kPa of applied pressure. The linear flow rate characteristic confirms that we can

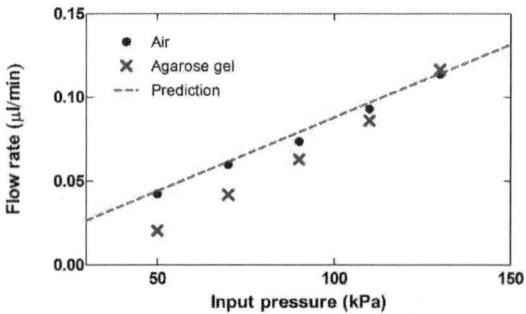

Figure 5: Plot of flow rates measured at different applied pressure values in a pressure-driven injection system.

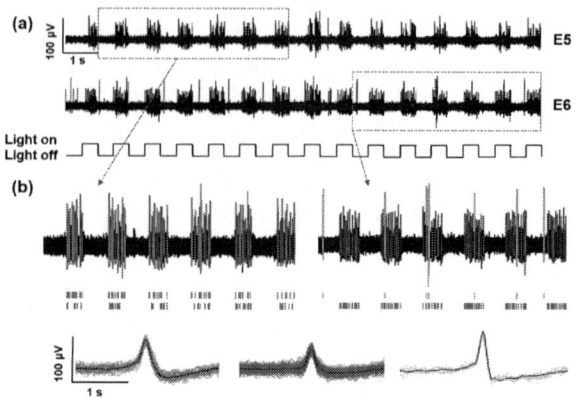

Figure 6: Recorded neural signals with optical stimulation: (a) 2-representative neural signals which is synchronized with light pulses, (b) sorted spike signals (blue, red, green) from activated neurons with raster plots.

precisely control the small quantity of drugs that are suitable for small animal experiments (Figure 5).

In vivo experiment

We also demonstrated successful *in vivo* experiments using our fabricated multifunctional neural probes. During the *in vivo* experiments, our probe penetrated the mouse brain without bending or fracturing. First, we conducted the *in vivo* experiment with an optical stimulation using a transgenic ChR2-YFP mouse. We inserted the fabricated probe into the hippocampus region of an anesthetized mouse brain. Recorded neural signals were synchronized with a train of blue light (λ = 473 nm) pulses. The frequency of applied square waveform was 1 Hz with a duty cycle of 50%. The neural activities from the activated neurons on the multi-electrodes were recorded using Neuralynx system. As shown in Figure 6(a), recorded neural signals were synchronized with the stimulation light pulses and the neural activities were dramatically increased by the light stimulation. This confirms that the neurons in the hippocampus region of a transgenic mouse were successfully stimulated (Figure 6(c)).

Also, the functionality of the fabricated multifunctional neural probe with a drug delivery capability was demonstrated by using a wild type mouse. We chose

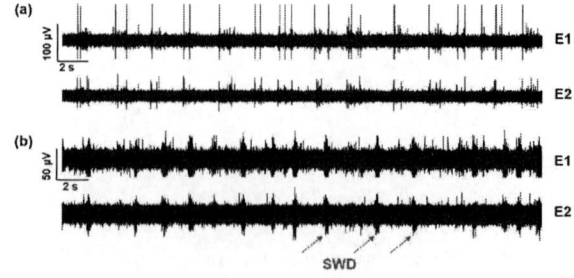

Figure 7: Recorded neural signals with drug delivery: (a) neural signals before drug injection, and (b) after injection of 1 µl of Baclofen showing distinctive spike-wave discharge (SWD).

978-1-4799-7956-1/15 $31.00 © 2015 IEEE

seizure-inducing drugs to observe dramatic changes in neural signals. Before delivering the drugs, we detected distinctive neural spike signals from individual neurons on multiple channels (Figure 7(a)). By injecting 1 μl of Baclofen using the embedded microfluidic channels to the thalamus region of mouse brain, we successfully induced an absence seizure. We observed distinct changes in the neural signals after the injection of baclofen; spike-wave-discharge (SWD), a unique pattern of the neural signals that indicates absence seizure, was observed (Figure 7(b)) [12].

CONCLUSION

In this paper, we proposed and presented a multi-functional MEMS neural probe that is integrated with an optical waveguide for optical stimulation and an embedded microfluidic channel for drug delivery. By embedding the microchannels in the silicon substrate, we achieved a thin probe with a thickness of 40 μm which is suitable for mouse experiments with minimal damage. The presented thin multi-functional MEMS neural probe allows altering the status of certain regions in a brain by injecting drugs and controlling the brain circuits through light stimulation on the same region. In *in vivo* experiments, recorded neural signals were synchronized with blue light pulses and the neural activities were dramatically increased by the blue light stimulation. Also, we successfully delivered 1 μl of Baclofen using the microfluidic channel to the thalamus region of a brain and induced absence seizure.

This multi-functional neural probe will be an important neuroscience tool with a wide range of applications in neuroscience including investigation of brain functionalities, neural circuits, and underlying mechanisms of brain diseases. By applying optical and chemical stimulation, this probe would greatly support developing a new disease model, investigating various cerebral nerves and brain diseases, and researching causes and cures of brain diseases.

REFERENCES

[1] J. G. Bernstein and E. S. Boyden, "Optogenetic Tools for Analyzing the Neural Circuits of Behavior", *Trends in Cogn. Sci.* Vol. 15, pp. 592-600, 2011.

[2] A.B. Schwartz, "Cortical Neural Prosthetics", *Ann. Rev. Neurosci.,* Vol. 27, pp. 487-507, 2004.

[3] T. -I. Kim, J. G. McCall, Y. H. Jung, X. Huang, E. R. Siuda, Y. Li, J. Song, Y. M. Song, H. A. Pao, R. H. Kim, C. Lu, S. D. Lee, I. –S. Song, G. C. Shin, R. Hasani, S. Kim, M. P. Tan, Y. Huang, F. G. Omentto, J. A. Rogers and M. R. Bruchas, "Injectable, Cellular-Scale Optoelectronics with Applications for Wireless Optogenetics", *Science*, Vol. 340, pp. 211-216, 2013.

[4] A. R. Adamantidis, F. Zhang, A. M. Aravanis, K. Deisseroth and L.d. Lecea, "Neural Substrates of Awakening Probed with Optogenetic Control of Hypocretin Neurons", *Nature*, Vol. 450, pp. 420-424, 2007.

[5] S. Royer, B. V. Zemelman, M. Barbic, A. Losonczy, G.Buzsaki and J. C. Magee, "Multi-array Silicon

Probes with Integrated Optical Fibers: Light-assisted Perturbation and Recording of Local Neural Circuits in the Behaving Animal", *J. of Neurosci.*, Vol. 31, pp. 2279-2291, 2010.

[6] A. Pongrácz, Z. Fekete, G. Márton, Z. Bérces, I. Ulbert and P. Fürjes, "Deep-brain silicon multielectrodes for simulataneous in vivo neural recording and drug delivery", *Sens. Actuators B-Chem.*, Vol. 189, pp. 97-105, 2013.

[7] B. Rubehn, S. B. E. Wolff, P. Tovote, A. Lüthi and T. Stieglitz, "A polymer-based neural microimplant for optogenetic applications: design and first in vivo study", *Lab on a Chip*, 13, pp. 579-588, 2013.

[8] Y. Son, H. J. Lee, D. Kim, Y. K. Kim, E. –S. Yoon, J. Y. Kang, N. Choi, T. G. Kim and I. J. Cho, "MEMS neural probe array for multiple-site optical stimulation with low-low optical waveguide by using thick glass cladding layer", *Proc. 27th IEEE MEMS Conference, San Francisco*, pp. 853-856, 2014.

[9] Y. Kim, H. J. Lee, D. Kim, Y. K. Kim, S. H. Lee, E. –S. Yoon and I. J. Cho, "A new MEMS neural probe integrated with embedded microfluidic channel for drug delivery and electrode array for recording neural signal", *Proc. 17th IEEE Transducers, Barcelona*, pp. 876-879, 2013.

[10] H. J. Lee, Y. Son, D. Kim, Y. K. Kim, N. Choi, E.-S. Yoon, and I.-J. Cho, "A new thin silicon microneedle with an embedded microchannel for deep brain drug infusion", *Sens. Actuators B-Chem.*, 2014 (accepted).

[11] Z. –J. Chen, G. T. Gillies, W. C. Broaddus, S. S. Prabhu, H. Fillmore, R. M. Mitchell, F. D. Corwin and P. P. Fatouros, "A realistic brain tissue phantom for intraparenchymal infusion studies", *J. Neurosurgery*, Vol. 101, pp. 314-322, 2004.

[12] H. Nersesyan, F. Hyder, D. L. Rothman and H. Blumenfeld, "Dynamic fMRI and EEG recordings during Spike-Wave Seizures and Generalized Tonic-Clonic Seizures in WAG/Rij Rats", *J. of Cerebral Blood Flow & Metabolism*, Vol. 24, pp. 589-599, 2004.

MINIATURIZED 3×3 OPTICAL FIBER ARRAY FOR OPTOGENETICS WITH INTEGRATED 460 NM LIGHT SOURCES AND FLEXIBLE ELECTRICAL INTERCONNECTION

Michael Schwaerzle, Philipp Elmlinger, Oliver Paul, and Patrick Ruther

Department of Microsystems Engineering (IMTEK), University of Freiburg, GERMANY

ABSTRACT

We report on the design, fabrication, assembly, as well as optical and thermal characterization of a novel MEMS-based optical probe array for optogenetic research. The system allows the optical stimulation of neural brain tissue at 3×3 independently controllable spots. It comprises nine high-efficiency light emitting diodes (LED) integrated in a micromachined silicon (Si) housing that ensures the mechanical stability and precise alignment of 5-mm-long optical glass fibers with a diameter of 125 μm. The fibers transmit the light emitted by the LEDs (center wavelength 456 nm) into the brain tissue. A highly flexible polyimide (PI) ribbon cable provides the electrical interconnection of the LEDs to a control unit. The compact housing (volume less than 2.2 mm³) is beneficial for chronic system implantations in rodents. Furthermore, the electrical-only interconnection of this array represents a distinct advantage over conventional systems using external light sources interfaced by optical fibers. The optical characterization of the array reveals an optical output power of up to 1.28 mW/mm² at a drive current and duty cycle of 30 mA and 10%, respectively. The corresponding temperature increase of the silicon housing is 2.2 K only.

INTRODUCTION

Research in the field of neuroscience deals with various aspects of brain function and dysfunction. Common to all efforts is the request for a basic understanding of complex neural networks. This is addressed by, among other approaches, the observation and modification of neural brain activity, ideally with high spatial and temporal resolutions in combination with cell type specificity enabling a precise analysis of the underlying brain function.

A wide variety of neural probes, from simple wire electrodes [1] via micromachined electrode arrays [2] to CMOS-integrated high-density electrode arrays [3], have been developed for the purpose of extracellular electrical recordings. These probes can be applied for electrical tissue stimulation as well. However, high current densities, insufficient spatial resolution and cell type unspecificity prohibit a more detailed analysis.

Optogenetics represents a relatively new experimental approach in neuroscience to interact with neural tissue and control the activity of individual types of neurons with light [4,5]. So far, optical fibers connected to external lasers have mostly served the purpose of delivering light to targeted brain areas. They have done so either directly or connected to silicon-based neural probes equipped with waveguides [6]. The stiffness of the optical fibers impedes, however, in vivo experiments with freely behaving animals. Integrated light sources [7-9] circumvent this drawback by

providing an electrical-only system control. As an example, bare laser diode chips were integrated by our group on a relatively large Si probe base [7]. On the other hand, the integration of LED chips on a flexible substrate [8] to be inserted into the brain tissue bears the risk of highly localized heat dissipation causing an unacceptable temperature increase.

In this study we extend our recent approach [9] to a first implantable 3×3 array of such high-efficiency LED chips for considerably more sophisticated optogenetic studies.

ARRAY DESIGN

The probe design relies on the four components shown in Fig. 1. Optical fibers on the one hand guide the light into deeper brain structures. They represent those parts of the array to be implanted into the brain tissue while all other components remain outside the brain. Bare LED chips are used as miniaturized light sources. The alignment, protection, and stability of these two optical components is ensured by a miniaturized Si housing. The electrical interconnection of the LEDs is realized with a custom-made highly flexible ribbon cable. The cross-sectional view in Fig. 1(c) highlights the three functions of the housing. First, it hosts the LED chip in a recess above an aperture. Second, it ensures that the optical fibers assume position and align their

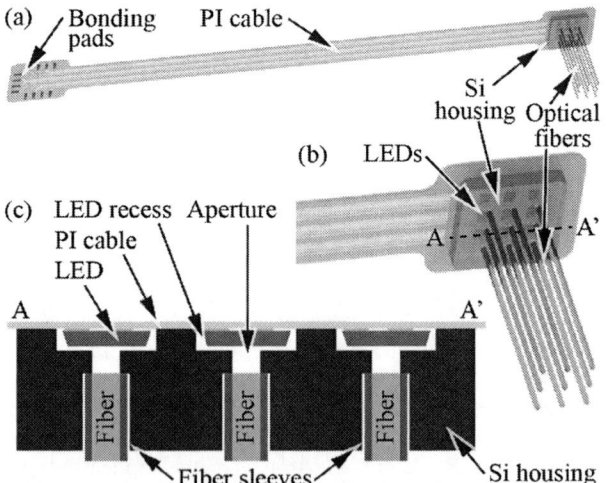

Figure 1: (a) Overall schematic of the novel 3×3 optogenetic probe array with bare LED chips, optical fibers, Si housing, and PI cable; (b) detailed schematic of the Si housing aligning nine optical glass fibers with individually controllable LEDs; (c) cross-section of assembled 3×3 probe array with three LEDs aligned to optical fibers and protected by the Si housing.

direction to the LED chips. Finally, it allows the light emitted by the chip to couple into the fiber through the aperture between the fiber and LED recesses. The PI cable shown in yellow comprises electroplated gold pads and constitutes the electrical interface to the LED chips.

FABRICATION AND ASSEMBLY

The fabrication processes of the custom-made parts, i.e. the Si housing and PI cable, are summarized in Fig. 2. The Si housing is processed using a three-step deep reactive ion etching (DRIE) process (Fig. 2, left side). The process is based on standard clean room fabrication and starts with a 380-μm-thick double side polished Si substrate. In a first step, both sides are covered by a 4.6-μm-thick stress-compensated SiO_x/Si_xN_y layer stack using plasma enhanced chemical vapor deposition (PECVD). This masking layer is then patterned on both wafer sides by reactive ion etching (RIE). A first 70-μm-deep DRIE step into the substrate defining the aperture relies on a photoresist (PR) as the etch mask. After removal of the PR, a second 70-μm-deep DRIE step produces the recesses for the LED and extends the apertures to a depth of 140 μm. Finally, a third 240-μm-deep DRIE from the wafer rear results in the guiding sleeves for the optical fibers. At the same time it opens the apertures and releases the silicon housing chips from the silicon substrate.

The fabrication process of the highly flexible polyimide ribbon cables adapts a previously published process [10]. It is illustrated on the right side of Fig. 2. The cable is composed of two 5-μm-thick PI layers spin-coated onto four-inch silicon wafer. Sandwiched between these PI layers is a 250-nm-thick Pt layer patterned by a lift-off process using AZ5214 image reversal resist. The interconnection lines running along the cable are thickened by a 1-μm-thick gold (Au) layer electroplated in a mask-less process step {Fig. 2(c2)}. A single RIE process using a 30-μm-thick

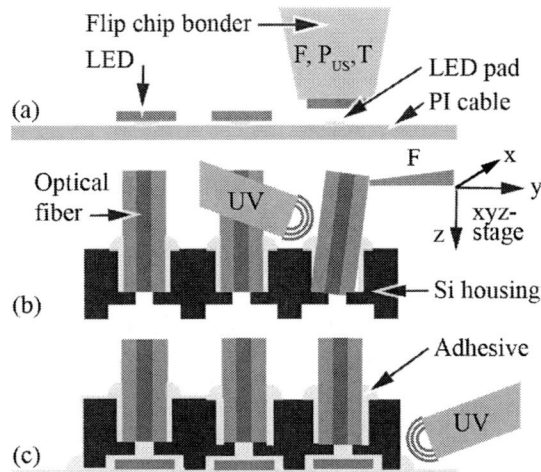

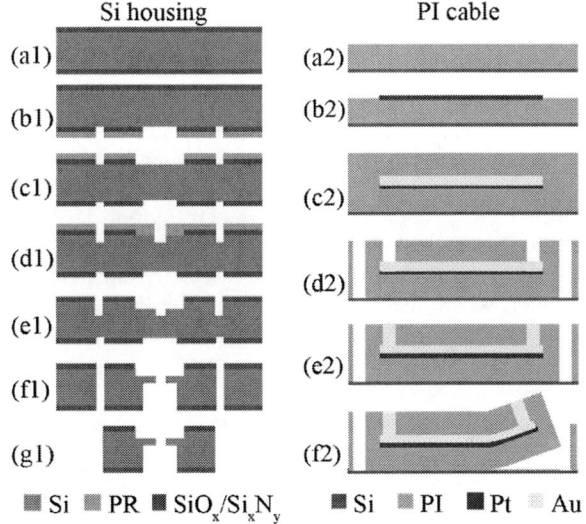

Figure 2: Fabrication process of Si housing using (a1) PECVD, (b1, c1) RIE, and two-sided DRIE to define (d1) the aperture, (e1) LED recess, and (f1) fiber sleeve. The polyimide (PI) cable process applies (b2) lift-off patterning of the metal lines, (c2, e2) electro-plating, and (d2) RIE to define vias and cable geometry.

Figure 3: Assembly and packaging of the probe array by (a) flip-chip bonding of bare LED chips and (b) UV-curable adhesive fixation of fibers after alignment using an xyz stage, and (c) fixation to the PI cable.

AZ 9260 masking layer is used to pattern the cable and to simultaneously open the bonding pads {Fig. 2(d2)} which are further strengthened with an additional 6-μm-thick electro-plated Au layer {Fig. 2(e2)}. Finally, the cables are delicately peeled off the substrate.

The used optical glass fibers have a high numerical aperture of 0.22 ensuring a high coupling efficiency (Fiber-Tech VIS-IR, Leoni Fiber Optics GmbH, Neuhaus-Schierschnitz, Germany), a core diameter of 105 μm and a cladding thickness of 10 μm. The fibers are cut at 90° to an application specific length of a few millimeters using a conventional wafer saw. The preparation of the fiber dicing starts with the removal of the fiber coating. Then several centimeter long bare glass fibers are fixed in 200-μm-wide trenches in a silicon handle wafer using a PR. Single fibers can be separately removed by dissolving the PR. Finally, the fiber facets are polished to improve the coupling efficiency.

The array assembly is illustrated in Fig. 3 and starts with the fixation of the bare LED chips onto the PI cable. A flip-chip bonder (Fineplacer 96λ, Finetech GmbH, Berlin, Germany) is used to align and assemble the bare LED chips (C460TR2227-S2100, Cree Europe GmbH, Unterschleißheim, Germany) on the cable. The chips have a size of 220×270×50 μm³ and emit with a specified center wavelength of 460 nm. The successful attachment of the LED chips to two 80-μm-wide Au bond pads on the PI cable requires an appropriate bonding tool and optimized bonding parameters. A bond force of 1.5 N was applied at a temperature of 120°C and an ultrasonic energy of 300 mW. The force and temperature were kept for 2 minutes, whereas the ultrasonic power was applied for 1 second. The glass fibers are manually inserted into their sleeves in the Si housing and vertically aligned using a Si needle controlled by an xyz-stage. The fibers are fixed with an UV-curable adhesive (NOA164, Norland, Cranbury, NJ, USA). In a final assembly step the Si housing with the glass fibers is aligned with respect to the bonded LED chips on the PI cable. An additional drop of the UV-curable adhesive fixes the housing onto the PI cable.

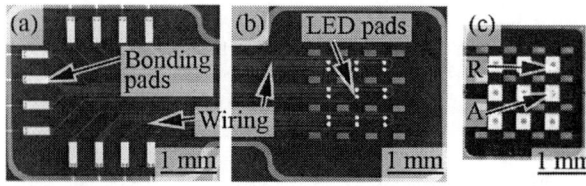

Figure 4: Photograph of (a) bonding area of PI cable and (b) LED pads for flip-chip bonding; (c) Si housing for nine LEDs and optical fibers (A: aperture, R: LED recess).

Results of the custom-made component fabrication and system assembly are shown in Figs. 4 and 5, respectively. Figure 4(a) shows the bonding pad area of the PI cable to interface with the external instrumentation as well as the individual wiring for the nine LED chips. Figure 4(b) details the LED section of the PI cable with 18 electroplated circular Au bond pads (diameter 80 μm, pitch 135 μm) matching the LED pad layout. The Si housing with a footprint of 2.3×2.5 mm² is shown in Fig. 4(c) with its 110-μm-diameter apertures (A) and its LED recesses (R) of 250×290×70 μm³ matching the fiber core and LED dimensions, respectively. The fiber sleeves with a diameter of 130 μm on the rear of the housing are not visible.

Figure 5(a) shows a PI cable carrying nine flip-chip bonded LEDs on their respective contact pads. The LED pitch is 550 μm in both directions. Each LED has a separate p-wire while three LEDs share a common n-contact. The result of the fiber assembly is illustrated in Fig. 5(b). The Si housing carries nine 5-mm-long vertically aligned glass fibers which are inserted into the sleeves and fixed in them.

ARRAY CHARACTERIZATION

The optical functionality of the assembled LEDs is demonstrated in Fig. 6(a-c) indicating various illumination patterns of individually controlled LEDs. Figure 6(d) shows a fully assembled system in operation. All nine LEDs are connected to the PI cable wiring and driven by a pulsed current source. Light emitted by each LED towards the corresponding fiber through the aperture of the silicon housing is coupled into the fiber core and guided by the fiber to the stimulation site. Undesired stray light is efficiently blocked by the housing. Nevertheless, some light is observed to leak through the gaps of less than 5 μm between fibers and housing.

The optical characterization includes the determination of the spectral light distribution and the optical output intensity emitted from the stimulation sites (Fig. 7). Both measurements are performed using an integrating sphere (ISP-50-I-USB, Ocean Optics, Ostfildern, Germany). For these measurements, the assembled array is aligned in front

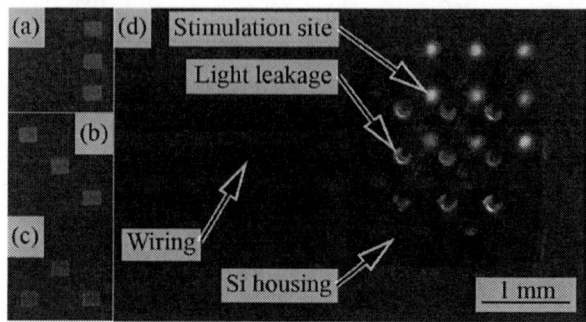

Figure 6: (a)–(c) Illumination patterns by individual control of LEDs. (d) Assembled system with nine active LEDs aligned to optical fibers indicated by bright stimulation sites. Stray light is observed at the fiber bases.

of the opening aperture of the integrating sphere. The optical characterization of the wavelength shows a peak emission at 456 nm {solid curve in Fig. 7(a)}. It overlaps with the normalized response curve of ChR2 indicated in Fig. 7(a) by the dashed curve [11]. When characterizing the output power P_{opt} per fiber, the duty cycle was fixed to 10% at a frequency of 100kHz and variable drive currents of up to 45 mA. Representative optical output power densities are shown in Fig. 7(b) for nine fibers of a 3×3 array as a function of the LED current. The observed variations in output power among different fibers are attributed mainly to the variability in the alignment of the fibers and in the quality of their coupling surfaces. For comparison, the optical output power of LEDs mounted on a PI cable without Si housing and fibers was determined to be 98.9±2.6 mW/mm² at a driving current of 30 mA and a duty cycle of 10%. For the fiber-based system an optical output power from 0.89 to 1.28 mW/mm² is achieved using the same operation parameters {cf. Fig. 7(b)}. This is equivalent to a coupling efficiency of 0.88 to 1.27 %. Despite the high losses, the stimulation of optogenetically modified neurons seems possible in view of published power densities between 1 and 5 mW/mm² required for optogenetic experiments [5].

To analyze the self-heating of the LEDs and thus the Si housing, in particular at high currents and duty cycle values, we measured the surface temperature of the Si housing with a high resolution IR camera (PI450, Optris GmbH, Berlin, Germany) providing a thermal resolution of 40 mK. For this measurement, a worst case scenario with the array floating in air is compared with a thermally more realistic situation where the array is lying on agarose gel, a well-established brain tissue model. In both cases, the surface

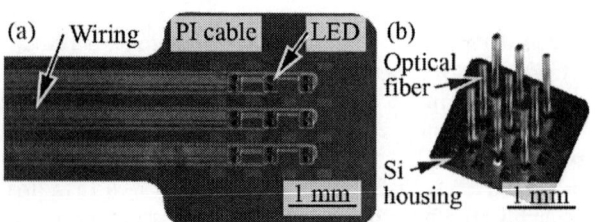

Figure 5: (a) Nine LED chips flip-chip bonded to a PI cable with respective wiring leads. (b) Nine 5-mm-long optical fibers adhesively fixed in the sleeves of a silicon housing.

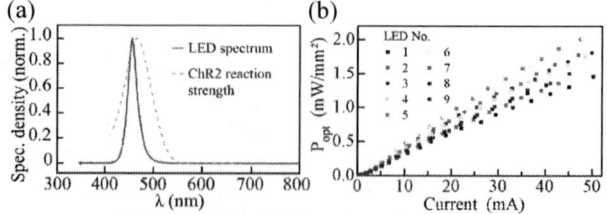

Figure 7: (a) Spectral light distribution with peak emission at 456 nm (solid line) and ChR2 reaction strength (dashed line) [11]. (b) Optical output power density of nine fibers from a 3×3 array operated at a constant duty cycle of 10% as a function of the drive current.

978-1-4799-7956-1/15 $31.00 © 2015 IEEE

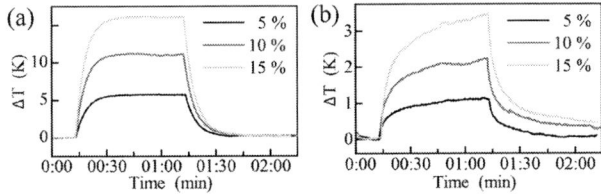

Figure 8: (a) Thermal response of a 3×3 array Si housing heated by driving an LED with 30 mA and duty cycles of 5%, 10 % and 15%. (b) Thermal response of an array placed on agar gel.

temperature was measured as a function of time once the driving current of 30 mA was switched on and off. In the case of the floating array, a temperature increase of 5.8 K, 11.2 K, and 16.1 K is achieved for duty cycles of 5 %, 10 %, and 15 % {Fig. 8(a)}, respectively. As shown in Fig. 8(b), the temperature increase for the array on agar gel is strongly reduced to moderate values in the range of the limit of 2 K requested by the ISO standard 14708-1 for active implants. The measured values after one minute of operation are 1.1 K, 2.2 K, and 3.5 K for duty cycles of 5 %, 10 %, and 15 %, respectively. When the array is operated under any of these conditions, the 2 K limit is not reached within the first 18 seconds.

CONCLUSIONS

This paper presented the design, processing, assembly, and characterization of a 3×3 array with optical functionality. It demonstrated the successful combination of bare LED chips and optical glass fibers interfaced by a custom made Si housing which aligns and protects the optical components. The LEDs are electrically interconnected with a highly flexible ribbon cable allowing the array and its connector to be positioned freely with respect to each other. With a housing volume of less than 2.2 mm³ the array is suitable for in-vivo experiments. The system has been characterized optically and thermally. Insertion tests into an agar brain tissue model still need to show how the fiber array interacts with its intended object of study. The currently flat diced fiber ends are definitely sub-optimal. Tapered or sharpened fiber tips may mechanically be better suited. However, it should be kept in mind that any modification of the fiber tip shape will influence the light emission as well. Finally, there seems room for further improvement regarding optical coupling efficiency and, thus, achievable optical output power density.

ACKNOWLEDGMENTS

This work was (partly) supported by BrainLinks-Brain-Tools Cluster of Excellence funded by the German Research Foundation (DFG, grant number EXC 1086). The authors gratefully acknowledge Michael Reichel and Armin Baur (both IMTEK-RSC) for support during the cleanroom fabrication, and Michael Kroener (IMTEK-Design of Microsystems) for providing the infrared camera.

REFERENCES

[1] M. A. L. Nicolelis, "Methods for Neural Ensemble Recordings", 2nd ed., *CRC Press*, 2007.

[2] K. D. Wise, A. M. Sodagar, Y. Yao, M. N. Gulari, G. E. Perlin and, K. Najafi, "Microelectrodes, microelectronics, and implantable neural microsystems", *Proc. IEEE*, vol. 96, no. 7, pp. 1184-1202, 2008.

[3] K. Seidl, M. Schwaerzle, I. Ulbert, H.P. Neves, O. Paul, and P. Ruther, "CMOS-based high-density silicon microprobe arrays for electronic depth control in intracortical neural recording-characterization and application", *IEEE J. Microelectromech. Syst.*, vol. 21, no. 6, pp. 1426-1435, 2012.

[4] E. S. Boyden, F. Zhang, E. Bamberg, G. Nagel, and K. Deisseroth, „Millisecond-timescale, genetically targeted optical control of neural activity", *Nat. Neurosci.*, vol. 8, 2005, pp. 1263-1268.

[5] O. Yizhar, L. E. Fenno, T. L. Davidson, M. Moguri, and K. Deisseroth, "Integrated device for optical stimulation and spatiotemperal electrical recording of neural activity in light-sensitized brain tissue", *Neuron*, vol. 71, pp. 9-34, 2011.

[6] S. Royer, B.V. Zemelman, M. Barbic, A. Losonczy, G. Buzsáki, and J.C. Magee, "Multi-array silicon probes with integrated optical fibers: light-assisted perturbation and recording of local neural circuits in the behaving animal", *Europ. J. Neurosci.*, vol. 31, pp. 2279-2291, 2010.

[7] M. Schwaerzle, K. Seidl, U.T. Schwarz, O. Paul, and P. Ruther, "Ultracompact optrode with integrated laser diode chips and SU-8 waveguides for optogenetic applications", *IEEE MEMS Conf. Proc.*, 2013, pp. 1029-1032.

[8] T.-I. Kim, J.G. McCall, Y.H. Jung, X. Huang, E.R. Siuda, Y. Li, J. Song, Y.M. Song, H.A. Pao, R.-H. Kim, C. Lu, S.D. Lee, I.-S. Song, G. Shin, R. Al-Hasani, S. Kim, M.P. Tan, Y. Huang, F.G. Omenetto, J.A. Rogers, and M.R. Bruchas, "Injectable, cellular-scale optoelectronics with applications for wireless optogenetics", *Science*, vol. 340, pp. 211-216, 2013.

[9] M. Schwaerzle, P. Elmlinger, O. Paul, and P. Ruther, „Miniaturized Tool for Optogenetics Based on an LED and an Optical Fiber Interfaced by a Silicon Housing", *Proc. IEEE EMBS Conf.*, 2014, pp. 5252-5255.

[10] S. Kisban, J. Kenntner, P. Janssen, R.v. Metzen, S. Herwik, U. Bartsch, T. Stieglitz, O. Paul, and P. Ruther, "A novel assembly method for silicon-based neural devices", *IFMBE Conf. Proc.*, 2009, pp. 107-110.

[11] K. Deisseroth, "Controlling the Brain with Light", *Scientific American*, vol. 303, pp. 48-55, 2010.

CONTACT

M. Schwaerzle, tel: +49-761-203 67913;
michael.schwaerzle@imtek.uni-freiburg.de

FLEXIBLE END-EFFECTOR INTEGRATED WITH SCANNING ACTUATOR AND OPTICAL WAVEGUIDE FOR ENDOSCOPIC FLUORESCENCE IMAGING DIAGNOSIS

Yoko Muramatsu, Taizo Kobayashi, Satoshi Konishi
Ritsumeikan University, Shiga, JAPAN

ABSTRACT

This paper presents a flexible end-effector integrated with a scanning actuator and optical waveguide for endoscopic fluorescence imaging diagnosis. The tiny and functional end-effector can also be introduced into the abdominal cavity or the digestive tract. Pneumatic balloon actuator (PBA) is used as the scanning actuator for the end-effector in consideration of its soft and safe features. SU-8 optical waveguides are integrated onto the PBA structure made of polydimethylsiloxane (PDMS). The excitation light (404 nm) is transmitted through an optical fiber to the waveguides. The outgoing light is scanned by the bending motion of PBA. The waveguide works as a detector as well. Our developed device has successfully scanned, excited, and detected fluorescence beads (515 nm) distributed in a pseudo tissue.

INTRODUCTION

Optical micro devices for an endoscope have been reported by many groups [1-6]; magnastatic mirror for rotational scanning, electromagnetic micromotor for rotary mirror [For example [5]]. Recently, MEMS neural probe integrated with SU-8 optical waveguide was reported [6]. Optical waveguides have been used for the light transmission in various medical applications such as laser surgery, dental treatment and imaging diagnosis [1-4]. On the other hand, we have developed all polydimethylsiloxane (PDMS) pneumatic balloon actuator (PBA) which has small, soft and safe features. PBA has been applied to various medical applications such as a retractor for an endscope and transplantation tool of cell sheet in an eye [8].

In this work, we intend to utilize PBA to drive optical components for endoscopic imaging diagnosis in consideration of its soft and safe features. This work integrates PBA with SU-8 waveguide for both excitation and detection toward fluorescence imaging. Scanning and detecting characteristics of the developed device are examined and reported in this paper.

DESIGN AND FABRICATION

Figure 1 shows a schematic of the flexible end-effector integrated with a PBA and SU-8 optical waveguides. We designed the width of end-effector narrow enough (less than 10 mm) to be attached through the inner slot of endoscope. One of the waveguides is used for excitation light, and the other is used for fluorescent light. The proposed device was fabricated through a series of PDMS-molding for PBA, SU-8 (3050, Microchem Corp., MA, USA) lithography for waveguide, and integration of SU-8 waveguide with PBA by vacuum ultraviolet (VUV) -assisted bonding process. PDMS (Sylpot184, Dow Corning Toray Co., Ltd., Tokyo, Japan) structure for pneumatic channels is formed by 50 μm high SU-8 mold and bonded by 150 μm thick PDMS film to form balloon and channels. 30-μm-thick SU-8 patterns were employed as optical core on a PDMS cladding layer.

RESULTS & DISCUSSIONS

Evaluation of Actuator integrated with Waveguide

Bending angle (θ; defined in insertion of Fig. 2(a)) of the developed devices with and without SU-8 waveguide was characterized in Fig. 2(a). Both devices could show movable angles from 0 to 120° at the pressure ranging from 50 to 120 kPa without serious restriction by SU-8 waveguide. Photographs of bending along the deformation of millimeter-scaled PBA are shown in Fig. 2 (b).

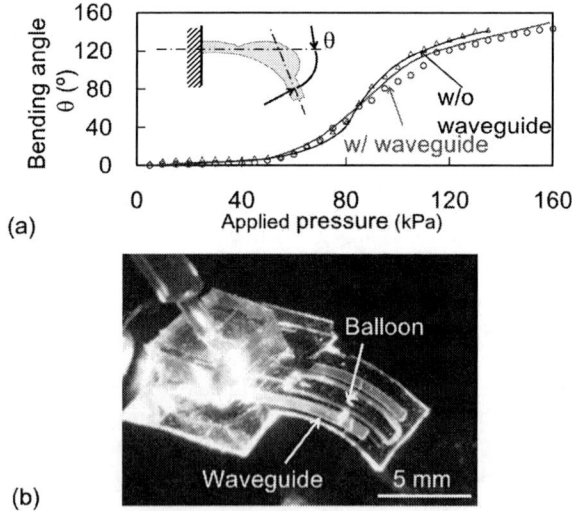

(a)

(b)

Figure 2: Characterization of bending angle of the developed device, (a) bending angle as a function of applied pressure into PBA, photographs of (a) the integrated device.

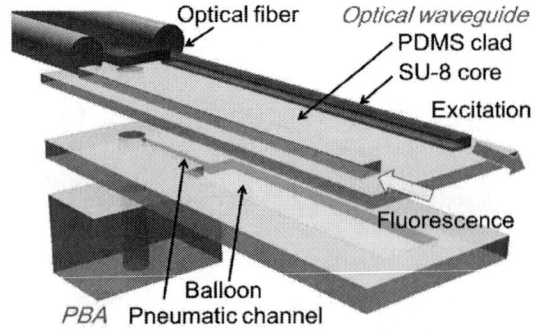

Figure 1: A schematic view of the proposed flexible end effector.

Optical Characterization as Scanner for Endoscopic Fluorescence Imaging

Figure 3 shows lighting characterization using the fabricated device with varying the applied pressure into PBA from 0 to 85 kPa. The angles were calculated from the central position of bright spot at pseudo tissue coated with fluorescent beads (Fig. 3(a)). The device could control lighting angle from 0 to 75° by changing the applied pressure into PBA (Fig. 3(b)). It was observed in the preliminary experiment that the beam divergence of SU-8 waveguide showed approximately 35° of FWHM (Data is not shown here). Figure 4 shows the light-receiving characterization. The light intensities reflected from pseudo tissue with 1 mm-width fluorescent line were measured at both excitation and fluorescent wavelengths of 404 and 515 nm with varying bending angles from 0 to 120° (Fig. 4(a)).

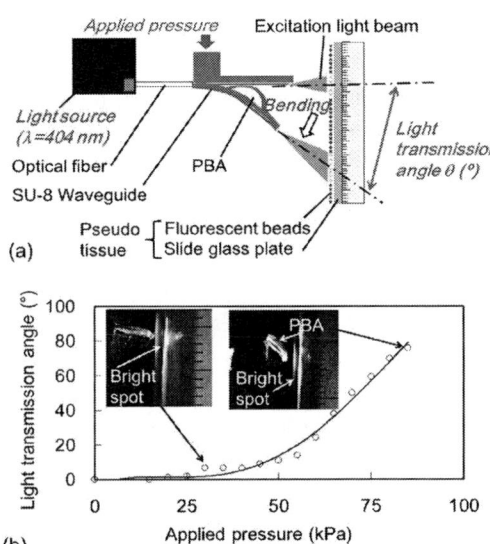

(a)

(b)

Figure 3: Lighting characterization, (a) experimental configuration, and (b) lighting angle vs applied pressure into PBA.

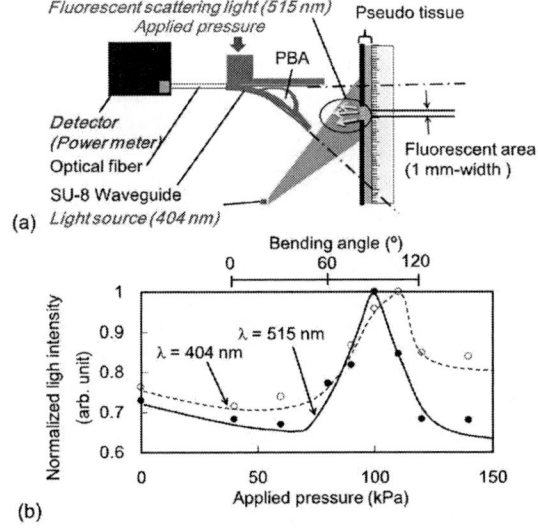

(a)

(b)

Figure 4: Light-receiving characterization, (a) experimental configuration, and (b) distribution of detected light intensity vs applied pressure.

The height of the fluorescent line was set at under 5 mm from the position of SU-8 core. The distributions of light intensities at both wavelengths were measured by changing the applied pressure into PBA (Fig. 4(b)). The light intensity shown in Figs. 4 was measured using an optical power meter.

CONCLUSION

A flexible end-effector integrated with a scanning actuator and optical waveguide was developed. The fabricated device could control lighting angle transmitted through SU-8 waveguide from 0 to 75° by changing the applied pressure into PBA. The light intensities reflected from pseudo tissue were also successfully measured at both excitation and fluorescent wavelengths of 404 and 515 nm with varying bending angles. Our developed device would be useful as the tiny and functional end-effector for endoscopic fluorescence imaging diagnostics in the abdominal cavity or the digestive tract.

REFERENCES

[1] C. M. Lee, C. J. Engelbrecht, T. D. Soper, F. Helmuchen, and E. J. Seibel, " Scanning Fiber Endoscopy with Highly Flexible, 1 mm Catheterscopes for Wide-field, Full-color Imaging", *J. Biophotonics*, vol. 3, pp. 385-407, 2010.

[2] D. C. Abeysinghe, S. Dasgupta, J. T. Boyd, and H. E. Jackson, " A Novel MEMS Pressure Sensor Fabricated on an Optical Fiber", *IEEE Photonics Technology Letters*, vol. 13, pp. 993-995, 2001.

[3] J. Su, J. Zhang, L. Yu, and Z. Chen, " In vivo Three-dimensional Microelectromechanical Endoscopic Swept Source Optical Coherence Tomography" *Optics Express*, vol. 15, pp. 10390-10396, 2007.

[4] G. N. Merberg, " Current Status of Infrared Fiber Optics for Medical Laser Power Delivery ", *Lasers in Surgery and Medicine*, vol. 13, pp. 572-576, 1993.

[5] N. Weber, H. Zappe, and A. Seifert, " Endoscopic Optical Probes for Linear and Rotational Scanning", *Proc. MEMS2013*, Taipei, January 20-24, 2013, pp. 1065-1068.

[6] Y. Son, H. J. Lee, D. Kim, Y. K. Kim, E-S. Yoon, J. Y. Kang, N. Choi, T. G. Kim, and I-J. Cho, " MEMS Neural Probe Array For Multiple-Site Optical Stimulation with Low-loss Optical Waveguide by Using Thick Glass Cladding Layer, *Proc. MEMS2014*, San Francisco, January 26-30, 2014, pp. 853-856.

[7] O. C. Jeong and S. Konishi, "All PDMS Pneumatic Microfinger With Bidirectional Motion and Its Application", *ASME/IEEE JMEMS*, vol. 15, pp. 896-902, 2006.

[8] M. Tokida, T. Obara, M. Takahashi, and S. Konishi, " Integration of Cell Sheet Sucking and Tactile Sensing Functions to Retinal Pigment Epithelium Transplantation Tool, *Proc. MEMS2010*, Hong Kong, January 24-28, 2010, pp. 316-319.

CONTACT

*S. Konishi; konishi@se.ritsumei.ac.jp

FULLY IMPLANTABLE AND RESORBABLE WIRELESS MEDICAL DEVICES FOR POSTSURGICAL INFECTION ABATEMENT

H. Tao[1,&], S.W. Hwang[2,&], B. Marelli[3], B. An[3], J.E. Moreau[3], M. Yang[3], M.A. Brenckle[3], S. Kim[2], D.L. Kaplan[3], J.A. Rogers[2], and F.G. Omenetto[3,4]

[1]State Key Laboratory of Transducer Technology, Shanghai Institute of Microsystem and Information Technology, CAS, Shanghai, China
[2]Department of Materials Science and Engineering, Beckman Institute for Advanced Science and Technology, and Frederick Seitz Materials Research Laboratory, University of Illinois at Urbana-Champaign, Urbana, USA
[3]Department of Biomedical Engineering, Tufts University, Medford, USA
[4]Department of Physics, Tufts University, Medford, USA

ABSTRACT

We present a therapeutic application of a microfabricated implantable and resorbable medical device made out of fully degradable materials by demonstrating in vivo elimination of bacterial infection by wireless activation of the device after implantation. The device disappears upon its completion, requiring no retrieval.

INTRODUCTION

For many years a principal objective of developing implantable devices was to guarantee that such devices would resist degradation within the human body and do not induce immune responses[1]. However, recently the demand for materials that degrade with precisely controlled characteristics (i.e. degradation rate and products) has increased dramatically along with a rapid surge of interest in degradable devices[2]. Especially, implanted devices and materials for thermal therapy to manipulate the body or tissue temperature for the treatment of disease can be traced back to the earliest medical practices and flourishes more than ever in modern research thanks to the advances in both materials (especially organic and inorganic nanoparticles)[3-5] and energy sources (ultrasound, radiofrequency waves, microwaves, and lasers)[6-9]. One fundamental challenge/requirement for non-invasive "on demand" heat delivery is to achieve spatial precision, i.e. conforming a barely working dose of heat in small tissue volumes, vertically and horizontally, leaving minimum degradation (sub)products and little damage to surrounding tissues[3].

We present here an electronically addressable, biologically degradable medical device that combines resorbable electronics and thermo-therapy through wireless control. A potential application of such devices for infection abatement (for example, surgical site infection treatment) is demonstrated.

METHOD AND RESULTS

The device is manufactured by using degradable and bioresorbable components. Silk - known and used as a high quality textile material for thousands of years - has extended its splendour to the field of biomedical engineering, as a "versatile" biomaterial, for its remarkable biocompatibility and unique mechanical and optical properties[10]. Here we report a silk-based integrated medical device for wirelessly controllable localized thermo-therapeutic treatments with all components degradable and resorbable, including a serpentine magnesium (Mg) resistor serving as the heating element that is connected with a inductively coupling Mg coil for remote heating, fabricated on a flexible silk substrate and further encapsulated in a silk "pocket" (not shown) to control the lifetime of the device (Fig. 1).

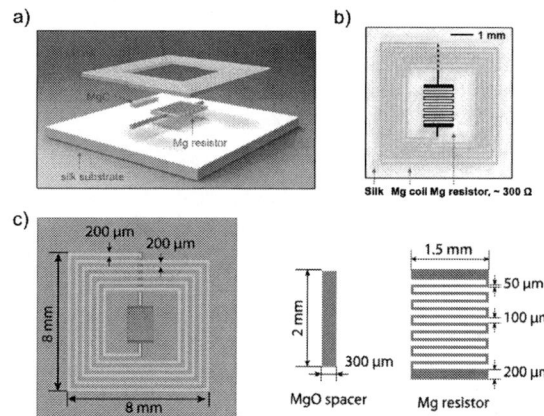

Figure 1: Geometries and dimensions of the device used in the experiment.

Silk fibroin aqueous solutions were prepared as previously described. Briefly, *Bombyx mori* cocoons were boiled for certain period of time ranging from 15 minutes to 60 minutes (varying with different applications and lifetime of devices), in an aqueous solution of 0.02 M sodium carbonate, followed by a throughout rinse using deionized water. After 2 days drying in a chemical hood, the silk fibroin was dissolved in an aqueous solution containing 9.3 M lithium bromide at 60 °C for 4 hours. The solution was then injected in Slide-a Lyzer dialysis cassettes (MWCO 3500, Pierce, Rockford, IL) and was dialyzed against deionized water for 48 hours (8 water changing in an interval of 6 hours).

The silk solution was cast on a flat surface (i.e. the bottom of polystyrene petri dishes) and was left drying at ambient conditions for 12 hours, resulting in silk fibroin films. The thickness of silk film can be precisely controlled by adjusting the volume and concentration of the silk solution and the casting area. For example, a dose of 0.2 mL/cm^2 of 6 wt% silk solution produces films of ~ 100 μm thick.

The fabrication involves a series of chemical-free processes including shadow mask deposition for the resistor and coils, thin film casting for the silk substrates, and low temperature embossing (T~85°C, for the formation of the encapsulating silk pocket). The Mg serpentine resistor was fabricated on a silk substrate of ~ 50 μm thickness and had a resistance of ~ 300 ohms, determined by the thickness (i.e. ~ 200 nm) of the 1st Mg deposition. After the deposition of a passivation layer of MgO, a 6 turn receiving coil (~ 2 μm) was deposited through a 2nd deposition step in order to connect and power the serpentine resistor. This was accomplished by inductive coupling to an external coil of ~ 5mm in diameter.

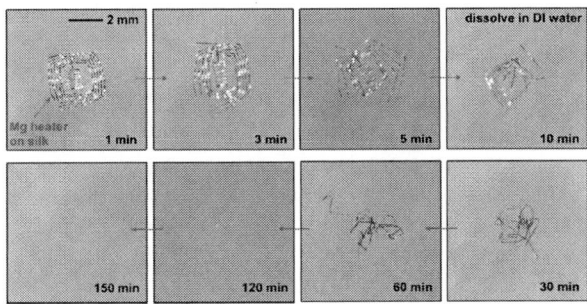

Figure 2: Optical images of full dissolution of the device (without any encapsulation) in DI water at room temperature within 2.5 h.

The device (prior to encapsulation) rapidly disintegrates (in ~ 5 mins) and fully dissolves (in ~ 150 mins) when immersed in DI water (Fig.2). The lifetime of the same device could be potentially prolonged by several orders of magnitude with suitable encapsulation strategies. Thermal embossing/lamination technique was used in this work for thermal control of silk fibroin film crystallinity. Moreover, with reflow upon heating, silk can act as a glue by controlling its thermal state. Briefly, the device to be encapsulated was placed in between two pieces of pre-treated (i.e. annealed to be water-insoluble) silk films. A few tiny drop of silk solution were applied around the edges to help sealing the silk "pocket". A detailed description of this technique (including embossing parameter optimization and the relationship of life time and embossing conditions) will be given elsewhere.

A series of *in vitro* and *in vivo* experiments were conducted to evaluate the performance of as-fabricated devices with a commercial IR camera (FLIR SC645). This device can act as an option for embedded infection management by thermal treatment when systemic antibiotic treatment alone is insufficient due to the rapid emergence of antibiotic-resistant infectious strains. The performances of the device were first evaluated *in vitro*, followed by *in vivo* studies in mice. An *in vitro* setup was used to explore the parameter space related to the therapeutic function of the device, specifically the effect of temperature and duration of heat treatment on bactericidal performance.

This was carried out by placing the devices underneath bacterial cultures of *Staphylococcus aureus* grown on agar plates and wirelessly powering them for

heating through inductive coupling via a primary coil (Fig.3a).

Lyophilized *S. aureus* cultures were reconstituted and expanded according to instructions provided by ATCC. To test susceptibility, bacteria cultures were grown in liquid Tryptic Soy Broth for 18–24 h to an optical density (OD_{600}) between 0.8 and 1 (corresponding to a viable count of approx. 10^7–10^8 CFU/mL).

Antibacterial effect *in vitro* was estimated based on the principle of the Kirby-Bauer Susceptibility Test where antibacterial effect is assessed by comparing zones of clearance in bacterial lawns. Briefly, 50 μl of the S. aureus culture were plated on Tryptic Soy Agar plates. The devices were placed on a primary coil for wireless powering/heating. The heating temperature was controlled by adjusting the input power of the primary coil using an IFI Sccx100 RF amplifier and a commercial infrared camera.

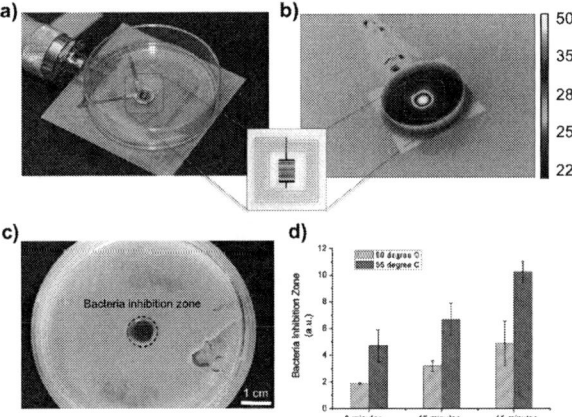

Figure 3: (a) The devices were placed underneath bacterial culture of Staphylococcus aureus grown on an agar plates. (b) The device was wirelessly powered to achieve desired temperature monitored by an IR camera. (c) A clear zone of inhibition, after heat treatment and overnight incubation, was found to correspond to the area of heat treatment application. (d) Increases in power (thus temperature) and duration can both enhance the overall bacterial inhibition effects.

The infrared heat map of the powered device shows a central region with a sharp temperature differential of ~ 28 °C between the core (resistor) and the untreated areas (Fig.3b). Higher temperatures are attainable by adjusting the input RF power in the primary coil accordingly. Once exposed, the bacteria plates (N=6) were immediately placed in a 37 °C incubator and then examined for bactericidal effects the next day. Following incubation and bacterial lawn formation, zones of inhibition were found to correspond to the areas of heat treatment application (Fig.3c). Increases in both power (thus temperature) and duration are feasible and can enhance the overall bacterial inhibition effects (Fig. 3d).

This resorbable device has the potential to be used as an implantable infection mitigation device. To evaluate this, in vivo studies were performed by implanting the devices in BALB/c mice and then infected with a subcutaneous injection of S. aureus at the device implantation site for surgical site infection mimicking purpose.

All animal experiments were conducted in accordance with Institutional Animal Care and use Committee protocols. Two sets of 10-minute heat treatments were carried out after bacteria injection with an input power of 100 mW and 500 mW at 80 MHz. The corresponding skin temperatures of 42 °C (labelled as low temp) and 49 °C (labelled as high temp) were respectively observed. After 24 hours, infected wounds formed at the site of injection.

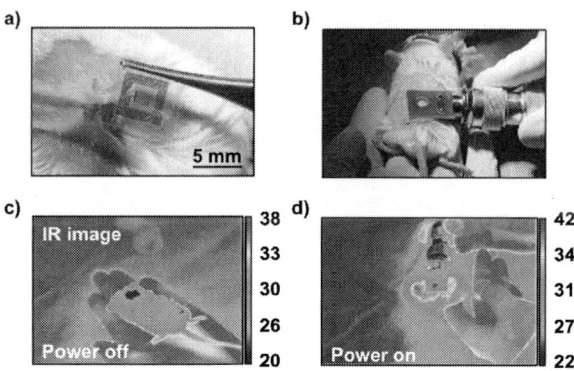

Figure 4: (a)&(b) Optical images of the device before and after implantation. (c)&(d) IR pictures of the implanted device before and after applying the rf power.

The effectiveness of the device was evaluated by excising the infected tissue site and assessing the normalized number of colony forming units (CFU) in the homogenate (n=3) using standard plate counting methods. The tissue shows a clear bacterial count reduction for thermally treated mice (Fig. 5).

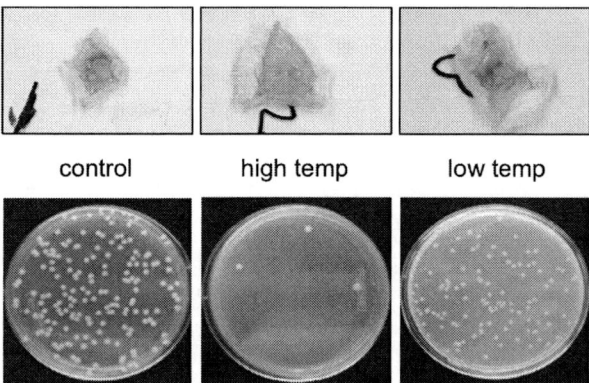

Figure 5: The infected tissues were collected 24 hours after thermal treatments and were assessed by counting the normalized number of colony forming units (CFU) in the homogenates (n=3) using standard plate counting methods.

To further evaluate the degradation process of the device *in vivo*, devices prepared in the same fashion as described previously were implanted and post-operatively wirelessly activated in the sub-dermal region of BALB/c mice. It is noted that the encapsulation was implemented in a way that the devices were able to survive the surgery process and the initial function checking point and started to degrade within a few hours to better access the device's degradation behaviors after finishing its function. Examination showed that the entire device fully degraded after 15 days (Fig. 6).

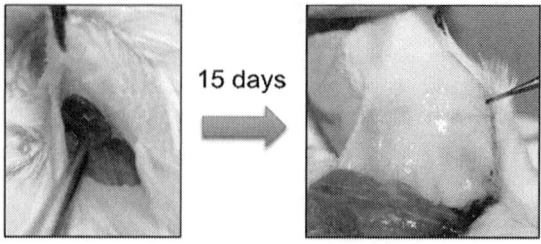

Figure 6: Devices were implanted and examined after 15 days showing no traces of the implanted device.

DISCUSSION

Recent studies demonstrate a class of complete water-soluble and fully resorbable silicon-based components, shedding light on new classes of biodegradable devices with integrated functionalities beyond the sole and more specific function offered by currently available devices (e.g. resorbable sutures, degradable intravascular stents, and matrices for drug release). A disadvantage of the original embodiment of this class of resorbable electronics is the relatively slow dissolution rates of silicon and silicon oxide being used as active elements and the passivation layer respectively, which range from weeks to months, depending strongly on temperature, pH, and thickness[11]. With only Mg and MgO used, devices reported in this work dissolve much faster via hydrolysis and get consumed in several hours (as opposed to weeks) once exposed to water and PBS solution. The rate of dissolution of silk (serving as the material for encapsulating and also mechanical supporting material) can be controlled over a wide range of time (from minutes to months, if not longer). Both lifetime (i.e. device staying functional) and existence time (i.e. device being fully degraded) of the device can both be specifically adjusted, chosen via the crystallinity of the silk.

One of greatest successes in localized thermal therapies in the past decade has been achieved by using biocompatible gold nanoparticles/nanoshells as the near infrared (NIR) absorber (i.e. the thermal coupling agent) for thermal ablative therapy for cancer. The size of those inert particles used is crucial in terms of getting enough high optical activities and precise delivery *in vivo*. Bio-distribution studies revealed that gold nanoparticles of smaller sizes (e.g. less than 50 nm in diameter) showed widespread distribution in nearly all the tissues including blood, liver, lung, heart and even brain, while larger particles tended to get accumulated locally in organs. Detailed studies of chronic toxicity of gold nanoparticles of different sizes, configurations and concentrations, especially long term ones, are greatly needed. The devices reported here operate at radio frequencies with broader operating window (for potentially multi-band selective triggering) and considerably greater penetration depth, which opens up new avenues to novel implantable thermo-therapeutic medical devices that deliver applicable thermal therapy on demand and then get fully resorbed by metabolic pathways of the organism in the absence of residual side effects.

ACKNOWLEDGEMENTS

The authors gratefully acknowledge support from NSF-INSPIRE Grant (DMR-1242240), and from the NIH P41 (EB002520). H.T. and F.G.O. would like to thank the Science and Technology Commission of Shanghai Municipality for the support under the International Collaboration Project (14520720400). M.A.B. would like to thank the ASEE for their support under the NDSEG fellowship.

REFERENCES

[1] J. M. Anderson, "Biological Responses to Materials", *Annu Rev Mater Res*, 2001, 31, pp. 81-110.

[2] S. P. Lyu, D. Untereker, "Degradability of Polymers for Implantable Biomedical Devices", *Int J Mol Sci*, 2009, 10, pp. 4033-4065.

[3] L. R. Hirsch, R. J. Stafford, J. A. Bankson, S. R. Sershen, B. Rivera, R. E. Price, J. D. Hazle, N. J. Halas, J. L. West, "Nanoshell-mediated Near-infrared Thermal Therapy of Tumors under Magnetic Resonance Guidance", *P Natl Acad Sci USA*, 2003, 100, pp. 13549-13554.

[4] I. II. El-Sayed, X. H. Huang, M. A. El-Sayed, "Selective Laser Photo-thermal Therapy of Epithelial Carcinoma Using Anti-EFGR Antibody Conjugated Gold Nanoparticles", *Cancer Lett*, 2006, 239, pp. 129-135.

[5] D. P. O'neal, L. R. Hirsch, N. J. Halas, J. D. Payne, J. L. West, "Photo-thermal Tumor Ablation in Mice Using Near Infrared-absorbing Nanoparticles", *Cancer Lett*, 2004, 209, pp. 171-176.

[6] M. G. Skinner, M. N. Iizuka, M. C. Kolios, M. D. Sherar, "A Theoretical Comparison of Energy Sources – Microwave, Ultrasound and Laser – for Interstitial Thermal Therapy", *Phys Med Biol*, 1998, 43, pp. 3535-3547.

[7] C. J. Simon, D. E. Dupuy, W. W. Mayo-Smith, "Microwave Ablation: Principles and Applications", *Radiographics*, 2005, 25, pp. S69-83.

[8] A. S. Wright, L. A. Sampson, T. F. Warner, D. M. Mahvi, F. T. Lee, "Radiofrequency versus Microwave Ablation in a Hepatic Porcine Model", *Radiology*, 2005, 236, pp. 132-139.

[9] M. Nikfarjam, V. Muralidharan, C. Christophi, "Mechanism of Focal Heat Destruction of Liver Tumors", *J Surg*, Res 2005, 127, pp. 208-223.

[11] H. Tao, D. L. Kaplan, F. G. Omenetto, "Silk Materials – A Road to Sustainable High Technology", *Adv Mater*, 2012, 24, pp. 2824-2837.

[12] S. W. Hwang, H. Tao, D. H. Kim, H. Y. Cheng, J. K. Song, E. Rill, M. A. Brenckle, B. Panilaitis, S. M. Won, Y. S. Kim, Y. M. Song, K. J. Yu, A. Ameen, R. Li, Y. W. Su, M. M. Yang, D. L. Kaplan, M. R. Zakin, M. J. Slepian, Y. G. Huang, F. G. Omenetto, J. A. Rogers, "A Physically Transient Form of Silicon Electronics", *Science*, 2012, 337, pp. 1640-1644.

CONTACT

*H. Tao, tel: +86-21-62511070; tiger@mail.sim.ac.cn or J.A. Rogers, tel: +1-217 244 4979; jrogers@uiuc.edu or F.G. Omenetto, tel: +1-617 627 4972; fiorenzo.omenetto@tufts.edu

&H.T. and S.W.H. contributed equally to this work.

TUNING THE FIRST INSTABILITY WINDOW OF A MEMS MEISSNER PARAMETRIC RESONATOR USING A LINEAR ELECTROSTATIC ANTI-SPRING

Shai Shmulevich, Inbar Hotzen and David Elata

Faculty of Mechanical Engineering, Technion - Israel Institute of Technology, Haifa, ISRAEL

ABSTRACT

We demonstrate frequency tuning of the first instability window of a MEMS Meissner parametric resonator. Our parametric resonator includes a transducer which provides a negative electrostatic stiffness that is not affected by motion. We therefore refer to this transducer as a *linear anti-spring*. We achieve parametric excitation by time-modulation of this negative electrostatic stiffness. Our design is rather robust to fabrication tolerances. In contrast, most state-of-the-art MEMS parametric resonators are either detrimentally affected by a nonlinear electrostatic stiffness, or are far more sensitive to fabrication tolerances.

INTRODUCTION

Electrostatic MEMS resonators have been proposed in the pioneering work of Nathanson [1]. Since then much effort has been invested to improve their performance to enable their use in filtering, clocking and sensing applications [2]. In the last two decades, significant work was devoted to exploit the advantages of nonlinear dynamics in micro-systems. Specifically, much attention was given to the possibility of designing and utilizing parametric resonators [3].

Parametric resonators are dynamic systems in which at least one physical parameter (mass, damping or stiffness) varies periodically in time. For specific cases, the periodic modulation of the system parameter results in a harmonic vibration, even when no additional forces are applied. Often, the system parameter which is modulated in time is the stiffness, as in the case of the Mathieu [4] and Meissner [5] equations. The periodic response of a parametric system may be stable or unstable. The system response is commonly presented by a stability map (Fig. 1). The stable and unstable regions are indicated over the domain spanned by the stiffness modulation amplitude (vertical axis) and the modulation period squared (horizontal axis).

One interesting property of parametric resonators is the sharp transition between stable and unstable responses. These sharp transitions enable sensitive frequency-shift detection for sensing applications [6]. Moreover, the exponentially unbounded response is also appealing for large displacement applications [7]. The efficient energy pumping of parametric systems allows enhanced dynamic response referred to as parametric amplification [8]. Another appealing property is that the applied driving frequency may differ from the response frequency. This allows to use parametric resonators in filtering [9]. In general, parametric actuation opens new possibilities and offers advantages for MEMS transducers.

An important feature that will facilitate application of parametric resonators in MEMS devices is the possibility

to tune the frequency of an instability window. Previous investigations [10, 11] presented the possibility of modifying the boundaries of an instability window, which also resulted in rotation and distortion of that window. In most MEMS parametric resonators the modulated stiffness is not purely time-dependent but is also affected by motion itself [3, 6, 10-13]. Due to this nonlinear effect, tuning of the instability window is often associated with its rotation and distortion.

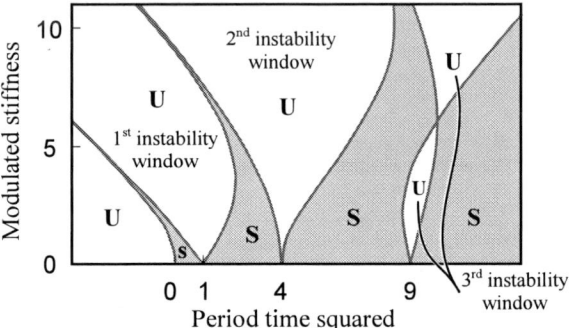

Figure 1: Stability map of the classic Meissner parametric resonator. The response of the system may be stable (shaded regions marked by 'S') or unstable (clear regions marked by 'U'). This map presents the first three instability windows. The first instability window is associated with the $2\omega_n$ excitation frequency.

Recently, we demonstrated a parametric resonator which is based on a linear anti-spring [14]. The stiffness of the anti-spring transducer may be modulated in time by modulating its tuning voltage, but for a given tuning voltage this stiffness is *unaffected by motion*. Hence, this device may be operated as a classic parametric resonator and specifically may be driven as a Meissner or Mathieu resonator - depending on the modulation scheme.

In the present work, we show analytically and experimentally, that this type of device may perform as a *tunable* parametric resonator. Specifically, we present the tuning of the first instability window of a Meissner parametric resonator, and demonstrate that this tuning does not distort the instability window.

A TUNABLE PARAMETRIC RESONATOR

The parametric resonator used in this work is schematically illustrated in Fig. 2. The system is constructed from two comb-drive transducers connected in series, such that their rotors are mechanically coupled. The lower transducer (Fig.2) is a standard double-sided comb-drive actuator which can be operated as a linear resonator [15]. The upper transducer is an electrostatic anti-spring with linearly tapered length of the rotor fingers

[16]. For a given voltage, the anti-spring induces a diverging electrostatic force that is linearly proportional to motion. This means that the transducer induces a constant negative stiffness which is only affected by the applied voltage.

When voltages are only applied to the anti-spring (i.e. no electrostatic driving forces are applied by the lower standard transducer), the equation of motion of the system is given by

$$m\ddot{x} + c\dot{x} + \left(k - \kappa_{el}V_{tune}^{2}\right)x = 0 \qquad (1)$$

where

$$\kappa_{el} = n\frac{\varepsilon_0 w}{gOL}$$

Here m is the rotor mass, c is the linear damping coefficient, k is the linear stiffness of the elastic suspension and κ_{el} is the electrostatic stiffness coefficient, where n is the number of tapered fingers of the rotor on each side, ε_0 is the permittivity of free-space, w is the device thickness, g is the gap between rotor and stator fingers, and OL is the initial overlap, where motion is in the range $-OL < x < OL$.

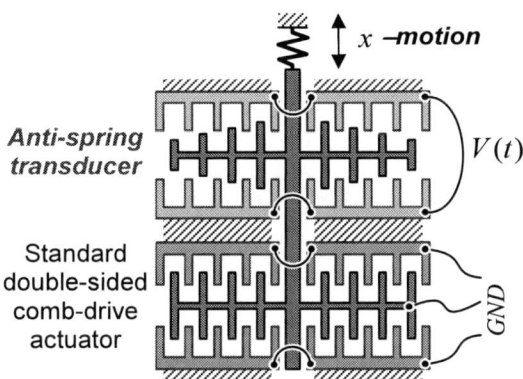

Figure 2: Schematic view of the device. The top anti-spring transducer is used for parametric excitation and frequency tuning. The bottom standard actuator is used to drive the system as a linear resonator, for measuring the natural frequency, the damping coefficient, and the frequency tuning properties of the anti-spring. The shown electrical connections are relevant for operation of the system as a parametric resonator.

It is clear from Eq. (1) that the system stiffness may be tuned-down by applying a dc voltage V_{tune}, and that for any given V_{tune} the stiffness is unaffected by motion x.

Applying a time-modulated voltage to the anti-spring will result in parametric excitation. This work focuses on the classic parametric Meissner resonator, which is parametrically excited by a *square wave-form* modulation of stiffness [5]. Since the modulated stiffness in (1) is proportional to voltage-squared, it is preferable that the voltage is of the form [14],

$$V(t) = \sqrt{V_{dc}^{2} + V_{tune}^{2}} + V_{ac}\,\mathrm{sgn}\left(\cos(\omega t)\right) \qquad (2)$$

Whenever V_{ac} is modified, V_{dc} is adapted to maintain a constant value of $V_{eff}^{2} = V_{dc}^{2} + V_{ac}^{2}$. Applying this voltage scheme to (1) yields,

$$m\ddot{x} + c\dot{x} + \left[\begin{array}{c} k - \kappa_{el}\left(V_{eff}^{2} + V_{tune}^{2}\right) \\ -2\kappa_{el}V_{ac}\sqrt{V_{eff}^{2} - V_{ac}^{2} + V_{tune}^{2}}\,\mathrm{sgn}(\cos(\omega t)) \end{array}\right]x = 0$$
$$(3)$$

The first two terms in the square brackets constitute the constant stiffness, and the third term constitutes the time-modulated stiffness of the system. The form of Eq. (3) follows the Meissner equation [5], and includes a linear damping term. It is clear that the constant stiffness of the system may be tuned by applying different values of V_{tune} on top of a square wave-form. This enables to drive the system as a classic parametric Meissner resonator, and in addition, shift the instability window with no expected rotation.

EXPERIMENTAL VERIFICATION

Test devices were fabricated using SOIMUMPs technology (Run 47, [17]). A microphoto of a typical device is presented in Fig. 3. Two independent anti-spring transducers (upper) and a standard double-side comb drive (lower), are connected in series sharing the same rotor.

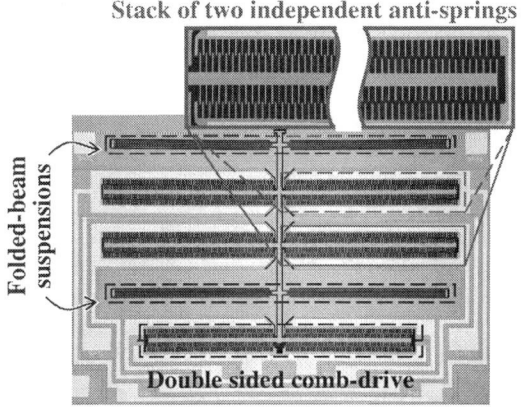

Figure 3: Microphoto of the test device. The two top transducers are of the anti-spring type, where each may be operated independently. The inset shows the finger length tapering of the anti-spring transducer. The standard double-sided comb-drive actuator is at the bottom.

The rotor is suspended on four folded-beam springs and its nominal natural frequency is close to $500Hz$. The finger width and minimal trench were both designed to be $4\mu m$, and the device layer thickness is $25\mu m$. The anti-spring rotor has 162 fingers on each side, and their length is tapered by $0.25\mu m/finger$. The maximal travel distance was designed to be $\pm10\mu m$. The standard actuator enables to operate the system as a linear resonator. This allows to independently measure: the natural frequency of the system; the damping coefficient of the system; and more importantly, the frequency tuning properties of the anti-spring.

Tuning the linear resonance with the anti-spring

The anti-spring property was characterized at atmospheric pressure. The rotor was subjected to a constant dc bias voltage of V_{dc}=5V. A low V_{ac} signal (typically 0.1$VRMS$) was applied to the standard comb-drive stator, and a dc tuning voltage, $V_{dc}+V_{tune}$, was applied to both stators of the anti-spring. A laser vibrometer was used to measure the vibration velocity. Reasonable SNR was achieved by a lock-in amplifier, which provided the reference ac signal to the standard comb-drive actuator, and acquired the signal from the vibrometer with respect to the same ac frequency. Figure 4 presents the fundamental frequency-squared as function of the tuning voltage-squared, showing an excellent linear fit. This validates the predicted linear tuning functionality of the anti-spring.

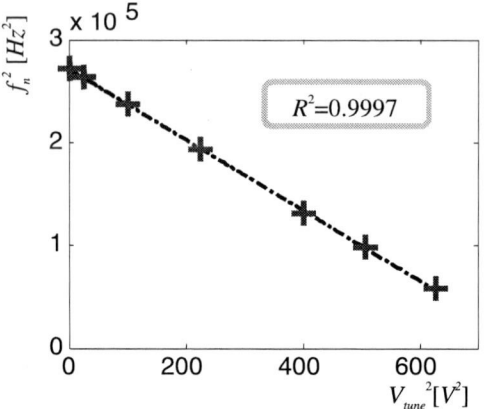

Figure 4: *Fundamental resonance frequency-squared as function of tuning voltage-squared, when the system is driven as a linear resonator, and the anti-spring is subjected to different settings of the tuning voltage V_{tune}. The excellent linear fit validates the predicted tuning property of the anti-spring. These results are compatible with the findings presented in [18].*

Measured response of a tuned Meissner resonator

The test device was next operated as a Meissner resonator, by applying time-modulated voltage to one of the single anti-springs, according to Eq. (2). All other pads were grounded. The experiments were performed at ambient vacuum of 6$mTorr$. Modulation frequency was slowly swept in the vicinity of double the natural frequency of the system. This frequency is associated with the first instability window of the Meissner equation.

At the detection of an unbounded motion increase, the system was brought back to rest. Then, the parametric actuation was repeated in that very same frequency, to verify that the unbounded response indeed results from parametric excitation. Figure 5 presents a typical measured developed response of the device excited by modulation of 0-6 [V]. The time-modulated signal has twice the frequency of the response signal, as expected in parametric excitation of the first instability window. The motion (green line in Fig. 5) is eventually bounded by limiter bumps at ±10μm.

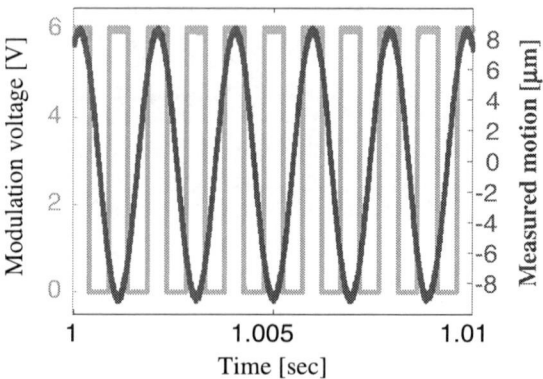

Figure 5: *Measured motion (right scale) for a modulated square wave signal of 6[V] (left scale). The modulated signal has twice the frequency of the response signal, as expected for a parametric response of the first instability window.*

To characterize the first instability window of the Meissner resonator, different voltage modulation amplitudes were applied according to Eq. (2) for each setting of the tuning voltage, V_{tune}. For every modulation amplitude, the modulation frequency was swept - back and forth - to identify the boundaries of the instability window where the response switches from stable to unstable motion. Figure 6 presents the first instability window shifted down along in frequency. It is evident that tuning is not associated with rotation of the instability window.

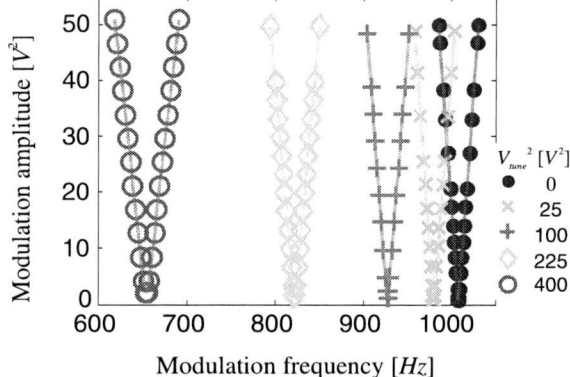

Figure 6: *Measurements of the first instability window of the Meissner resonator, for different settings of the tuning voltage (V_{tune}^2). It is evident that the instability window is shifted down in frequency, with no rotation. Frequency of the window tip was down-tuned by ~35%.*

Figure 7 presents the same measurements, where for each tuning voltage, frequency is normalized by the tip frequency, and modulation is similarly normalized to maintain consistency of Eq. (3). It is evident that all the shifted windows correlate well with the predicted boundaries of the first instability window of a Meissner resonator (solid black). Moreover, tuning does not introduce any distortion of the window boundaries.

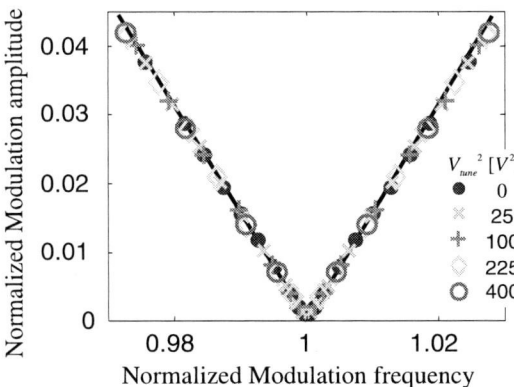

Figure 7: When each of the results in Fig. 6 is normalized by its respective tip frequency, they all overlay the predicted boundaries of the first instability window (black solid line). It is evident that the shifted windows are not distorted by the tuning.

CONCLUSION

In this work we demonstrated a parametric resonator in which the first instability window may be shifted with no rotation or distortion of its boundaries. Specifically, a classic Meissner parametric resonator was realized, and its parametric response in the $2\omega_t$ window was shifted by 35%, without rotating this instability window.

The excellent fit of the parametric response to the classic Meissner equation, and the clean frequency shifting of the instability window, are enabled by the ideal response of the anti-spring transducer. This anti-spring is superior to previous schemes of electrostatic stiffness modulators, because it does not induce motion depended nonlinearities.

In the analysis of the electrostatic forces (Eq. (1)), we ignored electrostatic fringing fields. The out-of-plane fringing fields are fixed because the gap between rotor and stator fingers is constant. Therefore the effect of the out-of-plane fringing field is a constant factor. Another fringing field occurs (in the plane of the device) in front of the fingers which are close to being intertwined. However, if the number of rotor fingers n is sufficiently large, these fringing fields are also constant, though their location 'hops' between adjacent fingers as the rotor engages or disengages with each stator. These two fringing-field factors, affect the amplitude of the electrostatic field, but because they are constant, they do not affect the functional relation in Eq.(1). This means that (uniform) fabrication inaccuracies such as larger or narrower comb-fingers, will not affect the linearity of the anti-spring transducer.

ACKNOWLEDGEMENT

This study was partially supported by the Farkas Family Fund for Research, and by the Russell Berrie Nanotechnology Institute (RBNI), Technion, Israel Institute of Technology.

REFERENCES

[1] H. C. Nathanson et al., "The Resonant Gate Transistor," *IEEE Trans. on Elect. Dev.*, **14**, 117-133, 1967.

[2] C. T. C. Nguyen, "MEMS technology for timing and frequency control," *IEEE Trans. Ultrasonics Ferroelectrics Freq. Control*, **54**, 251-270, 2007.

[3] K. Moran, C. Burgner, S. Shaw, and K. Turner, "A review of parametric resonance in microelectro-mechanical systems," *NOLTA, IEICE*, **4**, 198-224, 2013.

[4] E. Mathieu, "Mémoire sur Le Mouvement Vibratoire d'une Membrane de forme Elliptique" *Journal de Mathématiques Pures et Appliquées*, **13**, 137-203, 1868.

[5] E. Meissner, "Ueber Schüttelerscheinungen in Systemen mit periodisch veränderlicher Elastizität" *Schweizer Bauzeitung*, **72**, 95-98, 1918.

[6] K. L. Turner, S. A. Miller, P. G. Hartwell, N. C. MacDonald, S. H. Strogatz, and S. G. Adams, "Five parametric resonances in a microelectromechanical system," *Nature*, **396**, 149–152, 1998.

[7] C. Guo and G. K. Fedder, "A quadratic-shaped-finger comb parametric resonator," *JMM*, **23**, 095007, 2013.

[8] R. B. Karabalin, X. L. Feng, and M. L. Roukes, "Parametric Nanomechanical Amplification at Very High Frequency," *Nano Letters*, **9**, 3116-3123, 2009.

[9] S. Shaw, K. Turner, J. Rhoads, and R. Baskaran, "Parametrically Excited MEMS-Based Filters," Proc. *IUTAM*, Rome, 2003.

[10] S. G. Adams, F. M. Bertsch, K. A. Shaw, and N. C. MacDonald, "Independent tuning of linear and nonlinear stiffness coefficients," *JMEMS*, **7**, 172-180, 1998.

[11] J. F. Rhoads, S. W. Shaw, K. L. Turner, and R. Baskaran, "Tunable Microelectromechanical Filters that Exploit Parametric Resonance," *J Vibration and Acoustics*, **127**, 423-430, 2005.

[12] J. F. Rhoads et al. "Generalized parametric resonance in electrostatically actuated microelectro-mechanical oscillators," *JASA*, **296**, 797-829, 2006.

[13] B. E. DeMartini, J. F. Rhoads, K. L. Turner, S. W. Shaw, and J. Moehlis, "Linear and nonlinear tuning of parametrically excited MEMS oscillators," *JMEMS*, **16**, pp. 310-318, 2007.

[14] S. Shmulevich, I. Hotzen, and D. Elata, "An ideal MEMS parametric resonator using a tapered comb drive," *Eurosensors 2014*, Brescia, 2014.

[15] C. Marxer et al. "An Electrostatic actuator with large dynamic range and linear displacement-voltage behavior for a miniature spectrometer," *Transducers '99*, Sendai, Japan, 1999.

[16] K. B. Lee and Y. H. Cho, "A triangular electrostatic comb array for micromechanical resonant frequency tuning," *S&A*-A, **70**, 112-117, 1998.

[17] http://www.memscap.com/products/mumps/soimumps.

[18] D. Scheibner et al., "Characterization and self-test of electrostatically tunable resonators for frequency selective vibration measurements," *S&A-A*, **111**, 93-99, 2004.

A VISCOMETER BASED ON VIBRATION OF DROPLETS ON A PIEZORESISTIVE CANTILEVER ARRAY

Nguyen Thanh-Vinh, Kiyoshi Matsumoto, and Isao Shimoyama
The University of Tokyo, Tokyo, JAPAN

ABSTRACT

This paper reports a method to measure viscosity based on the resonant vibration of droplets on a piezoresistive cantilever array. We demonstrate that viscosity of small droplets (3μL) can be estimated from the attenuation rate of the cantilever output during free-decay of the droplet vibration. The estimation for liquid sample with different viscosity using water-glycerol solutions were carried out.

INTRODUCTION

It has been known that a liquid droplet can perform the vibration in multiple modes under the application of the periodical external force or displacement. Moreover, it is also shown that viscosity of a droplet can be extracted from the attenuation rate β (s^{-1}) of the droplet vibration using following relationship [1]:

$$\mu \approx \beta m^{2/3} \rho^{1/3} \qquad (1)$$

where μ is the liquid viscosity; m and ρ are droplet mass and liquid density, respectively. This method to estimate viscosity offers a benefit of small sample volume which is highly desired in point-of-care testing and bio-sensing. However, conventional methods for measuring droplet vibration rely on high-speed image inquiry or optical detection [1-4]. This drawback of such sophisticated observation schemes makes it difficult to realize simple viscometers based on droplet vibration. Besides, for droplets with high viscosity, the vibration amplitude becomes small and cannot be easily captured by a high-speed camera.

In this paper, we use a piezoresistive cantilever array to detect the droplet vibration as shown in Figure 1 (a). The cantilever array is fabricated on a pillar-typed superhydrophobic surface. Here, the superhydrophobic surface plays a role of droplet levitation to enhance the vibration amplitude. When the droplet vibrates on the cantilever array, the normal force acting on each cantilever changes periodically causing the change in the cantilever resistance. Hence, the vibration of the droplet can be detected by simply measuring the resistance changes of the cantilevers and the viscosity of the droplet can be estimated from the attenuation rate of the cantilever output as shown in Figure 1 (b). Moreover, the proposed MEMS-based cantilevers also have an advantage of high-sensitivity to detect a delicate motion of the droplet which may be impossible for high-speed image inquiry.

We investigate the relationship between the attenuation rate of the cantilever output and the viscosity of the droplet and verify the sensing principle. We also measure the normal force distribution along the diameter of the contact area of a

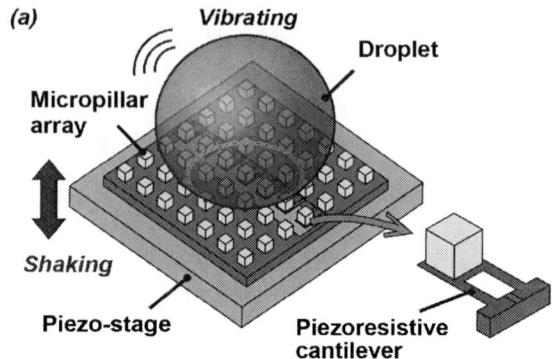

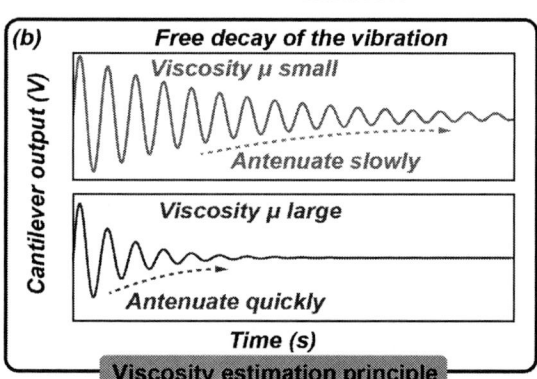

Figure 1: Conceptual illustration (a) and sensing principle (b) of the proposed viscometer. The vibration of a droplet is measured by an array of piezoresistive cantilevers. First the resonance vibration of the droplet is applied by shaking the piezo-stage. The viscosity of the droplet is estimated from the attenuation rate of the cantilever output after the piezo-stage is switched off.

vibrating droplet in order to find out the optimized location of the cantilever.

SENSOR FABRICATION & CALIBRATION

Sensor fabrication

The fabrication process of the cantilever array is shown in Figure 2, which is similar to that reported previously [5, 6]. Here, we briefly introduce the process. An SOI wafer was first doped by ion implantation. Next, the top layer Si was deposited with an Au/Cr layer functioning as electrical wire. The cantilevers were formed by patterning the Au/Cr layer and then etching the Si by using inductively coupled plasma RIE (ICP-RIE). After that, piezoresistors at at the hinges of the cantilevers were revealed by wet-etching the Ar/Cr layer.

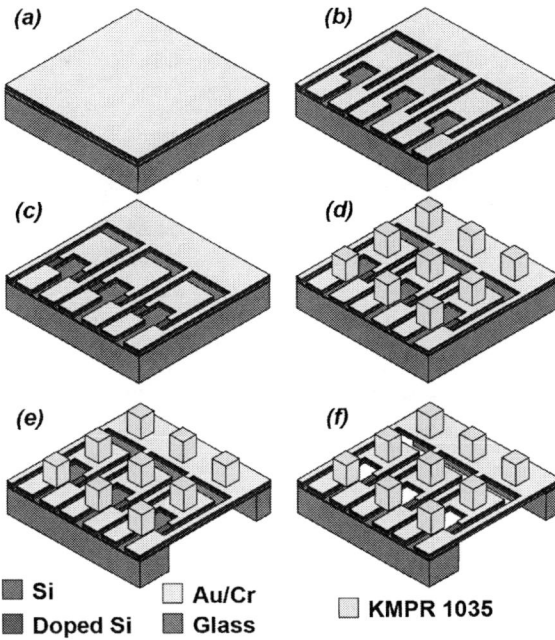

Si / Au/Cr / Doped Si / Glass / KMPR 1035

Figure 2: Fabrication process of the piezoresistive cantilever-micropillar array.

Pillar size: 30µm × 30µm × 35µm, Pillar pitch: 75µm
Cantilever thickness: 300 nm

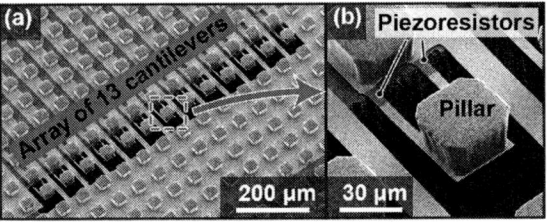

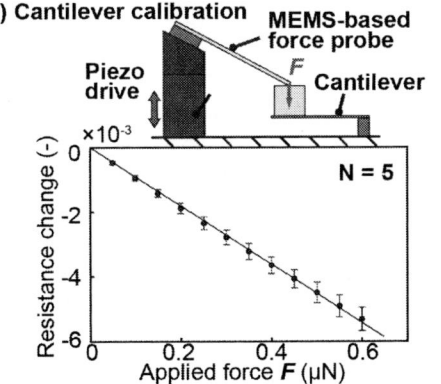

Figure 3: Fabrication result and calibration of a cantilever.

The micropillar array was then fabricated by patterning a KMPR 1035 photoresist layer. In the next step, the handle Si layer was etched using ICP-RIE with Al mask to create a through-hole underneath the cantilevers. The box layer was finally etched by vapor HF to reveal the cantilevers. The fabricated sensor was then attached and electrically connected to a printed circuit board using ultrasonic wire-bonder.

The fabricated cantilevers are shown in Figure 3 (a) and (b).The thickness of the cantilever was 300 nm. The size of each micropillar and pitch of the micropillar array were $30\times30\times35$ µm and 75 µm, respectively. The micropillar array was cleaned by O_2 plasma etching and then coated with a thin film of C_4F_8 to increase the hydrophobicity.

Sensor calibration

The calibration of the fabricated cantilever is shown in Figure 3 (c). Due to the small size and high flexibility of the cantilever, we use a MEMS-based force probe to calibrate the cantilever. The fabrication and calibration of the force probe can be found in our previous paper [6]. First, the force probe was attached to a piezo-stage by which the vertical displacement of the probe can be controlled. The normal force F was applied by pressing the force probe on the micropillar of the cantilever. The relationship between the applied force and the resistance change of the cantilever was shown in the graph. The result shows that the resistance change of the cantilever was proportional to the applied force, which means that the applied force can be measured by monitoring the resistance of the sensor.

EXPERIMENT AND RESULTS

Experimental setup

The experimental setup to investigate the relationship between the vibration of the droplet and its viscosity is shown in Figure 4. The fabricated cantilever array was attached to a piezo-stage (P-753.11C, PI Polytech). Liquid droplets were deposited on the cantilever array using a glass-tube affixed to a syringe. The position of the droplets were adjusted by dragging them by the glass-tube while observing by a microscope (Keyence, VH-5910) from above. The vibration of the droplet was induced by shaking the piezo-stage in the vertical direction at the resonant frequency of the droplet. During the droplet vibration, the resistance change of the sensor were monitored using a Wheatstone-bridge circuit and an amplifier at the sampling rate of 2 kHz. Moreover, the motion of the droplet was captured by a high speed camera (PHOTRON Inc., Japan, SA-XTKY03) at 2000 fps.

Experimental results

When detecting the droplet vibration by a cantilever, it is necessary to determine the optimized location of the cantilever. To maximize the cantilever output, the cantilever should be positioned at the location where the force fluctuation is largest. Thus, in our first experiment we investigated the normal force change distribution during the droplet vibration using the fabricated array of thirteen cantilevers and confirmed that the normal force change was

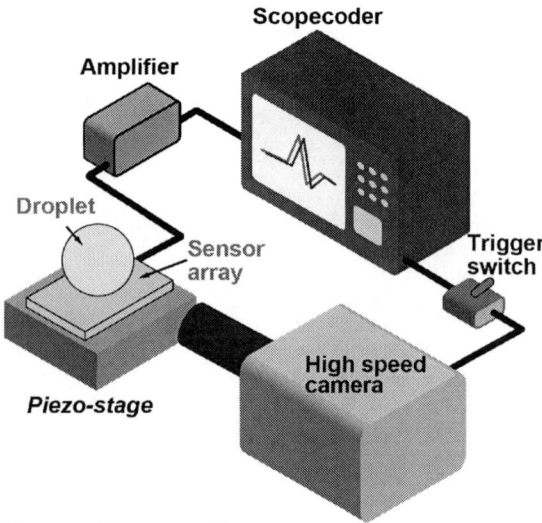

Figure 4: Experimental setup.

Table 1. Viscosity of the prepared liquid samples.

Water - glycerol solution	Viscosity (mPa·s)
Water	0.97
Glycerol 17%	1.9
Glycerol 25%	2.4
Glycerol 33%	3.0
Glycerol 50%	6.8
Glycerol 75%	42.7

largest at the cantilever located at the edge of the contact area.

In the next experiment, the relationship between liquid viscosity and the output of the cantilevers were investigated. In this study, liquid samples with different viscosity were prepared using water-glycerol mixture with different mixing ratio. Because glycerol and water have similar surface tension (63 and 72 mN/m, respectively), but very different viscosity (1000 and 1 mPa·s); the prepared mixtures with various viscosity but the ignorable surface tension difference. The viscosity of the prepared samples was first measured using EMS viscometer (Kyoto Electronics Manufacturing Co., EMS-1000) at 20 degree C which was the same with the temperature of the experiment. The measurement result shown in Table 1 indicates that the viscosity of liquid sample increase as the glycerol/water ratio increases.

Figure 5 shows the viscosity estimation by the cantilever array. When the piezo-stage is switched-off, the vibration of the droplet will attenuate reflecting in the attenuation of the cantilevers' outputs. We focused on the cantilever on the The droplets volume was 3 µL for all liquid sample. As shown in Figure 5 (a), vibration of more viscous droplet attenuates faster. Moreover, as shown in Figure 5 (b), the attenuation rate of the cantilever output is almost proportional to viscosity, which agrees well with the relationship in Equation (1). Therefore, the result indicates the ability to measure the

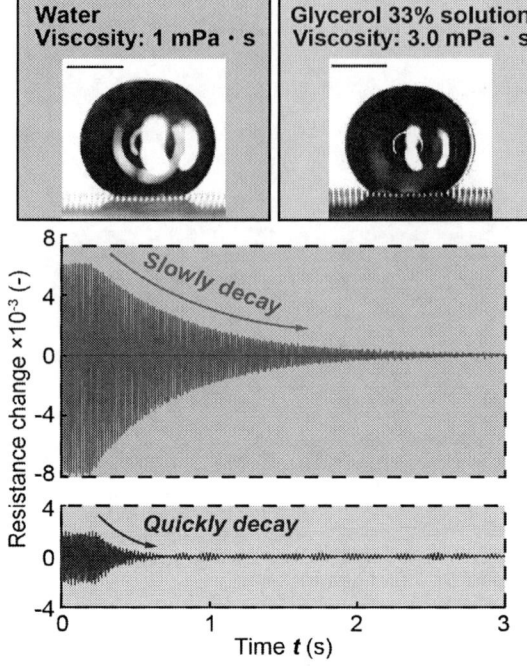

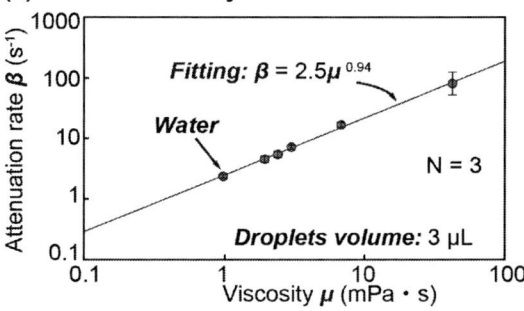

Figure 5: Estimation of droplet viscosity based on the attenuation rate of the vibration.
(a) The more viscous droplet attenuates faster due to a larger damping factor.
(b) Relationship between viscosity of the droplets and the attenuation rate of the cantilever output.

viscosity of small droplets by our proposed method.

CONCLUSIONS

This paper reported the method to measure the viscosity using an array of piezoresistive cantilever. The proposed method provides a benefit of small amount of required liquid sample. We demonstrated that viscosity of small droplets (3 µL) could be estimated from the attenuation rate of the cantilever output during the free-decay of the droplet vibration.

ACKNOWLEDGEMENT

The photolithography masks were made using the University of Tokyo VLSI Design and Education Center (VDEC)'s 8 inch EB writer F5112 + VD01 donated by ADVANTEST Corporation. This work was partially supported by JSPS KAKENHI Grant Numbers 25000010, 24656162. N. Thanh-Vinh was supported by Research Fellowship of the Japan Society for the Promotion of Science (JSPS) for Young Scientists, Japan.

The authors would like to thank Professor Keiji Sakai, Dr. Taichi Hirano and Dr. Yuji Shimokawa for the viscosity measurement of the liquid samples using EMS Viscometer.

REFERENCES

[1] S. S James, "Resonant properties of sessile droplets; contact angle dependence of the resonant frequency and width in glycerol/water mixtures", *Soft Matter*, 8, pp. 399-407, 2012.

[2] S. Mettu, M. K. Chaudhury, "Vibration Spectroscopy of a Sessile Drop and Its Contact Line", *Langmuir*, 28(39), pp. 14100-14106, 2012.

[3] P. M. McGuiggan, D. A. Grave , J. S. Wallace, S. Cheng, A. Prosperetti, M. O. Robbins, "Dynamics of a Disturbed Sessile Drop Measured by Atomic Force Microscopy (AFM)", *Langmuir*, 27(19), pp. 11966-11972, 2011.

[4] G. McHale, S. J. Elliott, M. I. Newton, D. L. Herbertson, K. Esmer, "Levitation-Free Vibrated Droplets: Resonant Oscillations of Liquid Marbles", *Langmuir*, 25(1), pp. 529-533, 2009.

[5] N. Thanh-Vinh, H. Takahashi, K. Matsumoto, I. Shimoyama, "Interaction forces during the sliding of a water droplet on a textured surface," *The 27th IEEE International Conference on Micro Electro Mechanical Systems (MEMS2014)*, San Francisco, USA, pp. 979-982, 26-30 January, 2014.

[6] N. Thanh-Vinh, H. Takahashi, K. Matsumoto, I. Shimoyama, "Two-axis MEMS-based Force Sensor for Measuring the Interaction Forces during the Sliding of a Droplet on a Micropillar Array," *Sensors and Actuators A: Physical*, 2014. DOI: 10.1016/j.sna.2014.09.015.

CONTACT

*Nguyen Thanh-Vinh,
Tel: +81-3-5841-6318; vinh@leopard.t.u-tokyo.ac.jp

HOLLOW MEMS MASS SENSORS FOR REAL-TIME PARTICLES WEIGHING AND SIZING FROM A FEW 10 NM TO THE µM SCALE

Céline Hadji[1,2], Clément Berthet[1,2], François Baléras[1,2], Martine Cochet[1,2], Béatrice Icard[1,2], and Vincent Agache[1,2]

[1]Univ. Grenoble Alpes, F-38000 Grenoble, FRANCE
[2]CEA, LETI, MINATEC Campus, 17 rue des Martyrs, F-38054 Grenoble, Cedex9, FRANCE

ABSTRACT

This paper reports hollow MEMS plate oscillators for mass sensing in liquid with an extracted mass resolution of 4 femtograms. The sensors performances – 10,000-range Q factor and ppb-range frequency stability – make them amenable to individual particles metrology from a few 10 nm up to the micrometer diameter range. These devices are operating in air, connected through a customized *plug and play* test platform and do not need to work in a vacuum-sealed package.

INTRODUCTION

MEMS technologies have a tremendous impact in the field of mass sensing and offer constantly improved mass resolution enabled by microfabrication and sensor downscaling. They have become a relevant alternative to gold standard techniques for particle metrology in fluid (dynamic light scattering, flow cytometry), with applications in pharmaceutical, biological and environmental fields. A relevant paradigm introduced by E. Stemme [1] consists in confining a fluidic channel inside a mechanical oscillator. The oscillator vibrates in a dry environment, while fluid is delivered inside the embedded channel to significantly reduce the viscous damping and signal losses. This technique enables the weighing and accurate counting of particles as small as 10 nm; the frequency excursions can be measured individually for a large number of particles, allowing precise mass distribution measurement with an unequaled sensitivity (down to the attogram scale). S. R. Manalis and coworkers were then able to measure cell growth in fluid [2] and weight nanoparticles [3] using vacuum-packaged hollow flexural cantilevers. To reach such mass resolution, sensors must be optimized in terms of frequency stability (sub-ppm frequency fluctuations), revealed by their Allan deviation σ parameter, which relies on the resonator quality factor Q. A. Heidari and coworkers found Q factor of 1,800 using filled Lamé mode mass sensor in air [4], while Manalis' team found Q factor around 15,000 in vacuum and Allan deviation ranging from 4 to 8 ppb at an averaging time of 1 ms [3].

This paper reports square or disk plate resonators with an embedded microchannel operating in Lamé mode configuration. Both actuation and readout transduction rely on capacitive electrodes. Our sensors demonstrated in atmospheric conditions Q factors up to 26,000 and Allan deviation down to 3 ppb at an averaging time of 20 ms.

DESCRIPTION OF OUR DEVICES

The sensors vibrating structure consists in a disk or square silicon plate from 50 to 200 µm-wide, similarly to [5], with an embedded microchannel etched along its edges, through which the fluid and particles can transit. Two capacitive electrodes drive an alternative voltage to actuate the sensor at its resonant frequency f_0; two sensing electrodes collect the motional current arising from the plate motion. When particles are traveling through the device, the frequency shift δf between the output signal and the plate's natural frequency allows accurate weighing of these particles by measuring their buoyant mass m_B:

$$\delta m = m_B = -2\, m_{eff} \frac{\delta f}{f_0} \qquad (1)$$

where m_{eff} is the plate's effective mass.

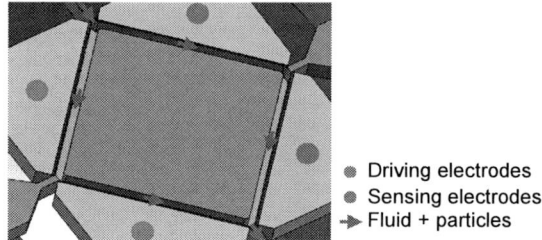

Figure 1: Schematic view of a typical hollow square plate sensor with capacitive electrodes for driving and readout.

- Driving electrodes
- Sensing electrodes
- Fluid + particles

Overview of the geometry

Figure 1 and Figure 2 illustrate the geometry of typical sensors.

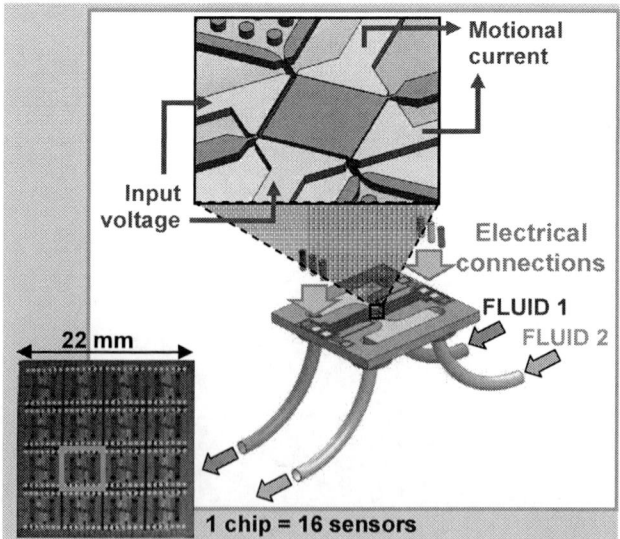

Figure 2: Photograph of one chip and schematic view of one square plate sensor.

Figure 1 exhibits a close-up schematic view of the vibrating plate; in the center, the square or disk resonator

978-1-4799-7956-1/15 $31.00 © 2015 IEEE

plate; in yellow, the four (two driving and two sensing) facing electrodes separated by a 1 µm-wide air gap from the resonator edges; in blue, the embedded microchannel in a bypass configuration. Figure 2 shows the geometry of a whole sensor, with electrical connections enabled on top of the chip and backside fluidic supply.

Figure 3 shows SEM pictures of two sensors (square and disk hollow plates), as well as mode shapes (extracted from COMSOL®) corresponding to the plates' Lamé resonant frequency.

New features

In contrast to previous generation of hollow MEMS plates [6], these new devices feature specific improvements allowing faster operation, comparable mass resolution with larger dynamic range. In addition to previous square resonators, disk hollow plate sensors were designed. The plate thickness was increased from 3 to 15 µm, providing 3 µm-wide microchannel section which allows metrology of larger particles as compared to previous study and offers a greater dynamic range. Moreover, larger facing surfaces increase the capacitive transduction efficiency and signal-to-noise ratio. The embedded microchannel is now in a bypass configuration between upstream and downstream channels to enable faster filling fluidic procedure and increase the analysis throughput. This bypass channel network is supplied with fluid via four backside inlet and outlet ports.

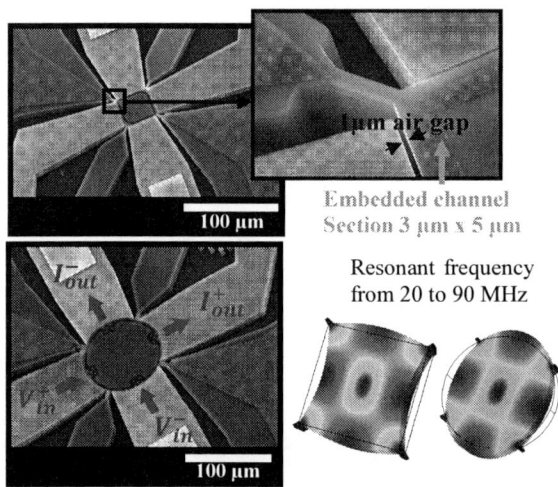

Figure 3: SEM pictures of a 50 µm-wide square plate and a 100 µm-wide disk plate. COMSOL® simulations are given to illustrate the plates' Lamé mode shape in Lamé.

The constraints of wire bonding and capillaries gluing have been also eliminated for these devices, by physically decoupling electrical connections (front side of the device) and fluidic supply (backside), which enables fast connection with a dedicated platform (hosting electrical pogo pins contacts and µfluidic tubing).

Fabrication process

Figure 4 describes the devices fabrication main steps. Two silicon-on-insulator (SOI) wafers are processed to elaborate the sensors (a); the whole fabrication relies on 200 mm wafer MEMS technology. A first SOI bottom wafer is etched to pattern the embedded microchannel, followed by another photolithography process combined with deep reactive-ion etching (DRIE) to define the bypass channels (b). This "bottom" SOI wafer is then assembled with a "top" SOI wafer using silicon direct fusion bonding procedure. Then, the top SOI handle substrate and BOX layer are removed successively by coarse then fine grinding (c) and buffered HF etching. Metallic pads and tracks are patterned by sputtering deposition, photolithography and wet etching of a 2 µm-thick AlSi layer (d). The sensor geometry and its four transduction electrodes are then patterned by photolithography and RIE technique (e). Finally, a backside etching process is performed to define both the releasing cavity under the plate and the fluidic inlets and outlets of the system (f).

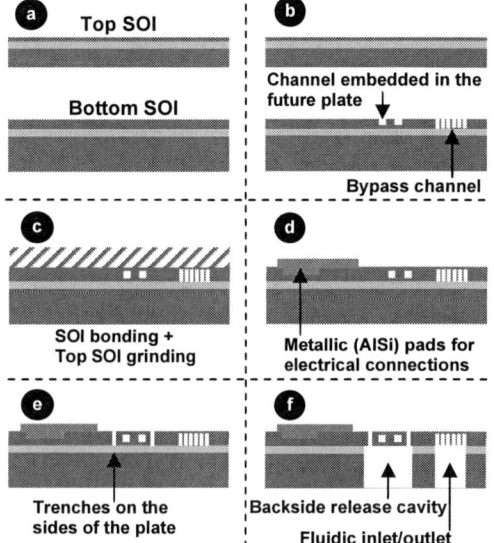

Figure 4: Description of the technological processes involved in the sensors' fabrication.

DEVICE CHARACTERIZATION
Connection with the setup

The chip interfacing with the whole setup is performed by a customized *plug and play* platform (Figure 5), which is hosting pogo pins for electrical contact, microfluidic PTFE-coated tubings and o-rings for hermetic fluidic connection. This platform allows a fast connection with no need for wire bonding nor capillaries gluing, thus highly limiting the risks of pollution and channel clogging.

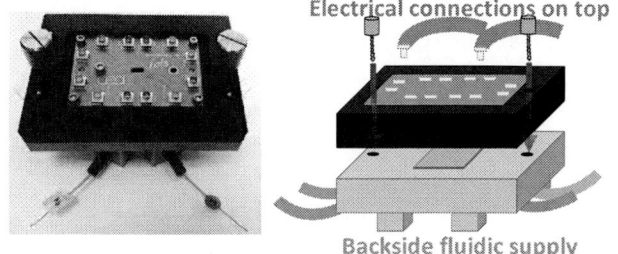

Figure 5: Photograph and schematic view of the customized plug and play *platform.*

978-1-4799-7956-1/15 $31.00 © 2015 IEEE   181

The electrical setup is given in Figure 6. It is based on a Lock-in amplifier, mixers and AC sources to perform heterodyne detection.

One AC source delivers a signal at frequency ω (20 to 90 MHz) which is used to drive both the sensor and a power splitter to be mixed with a signal from a second AC source at a frequency ω − Δω. The frequency of the resulting signal is then down mixed to audio frequency Δω (100 kHz) similarly to what is described in [7]. Same mixing process is applied to the sensor output signal before being sent to a low noise amplifier. Finally, the Lock-in amplifier performs demodulation in order to extract the phase difference between a reference signal and the signal collected from the sensor output electrodes.

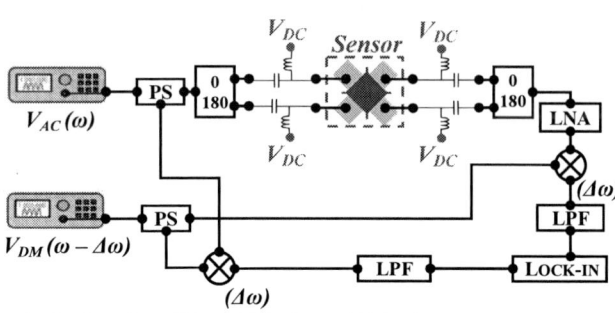

LPF = Low-Pass Filter | DM = Down-Mixing |
LNA = Low-Noise Amplifier | PS = Power Splitter

Figure 6: Experimental setup used for the electrical characterization of the sensors.

Four Tee Bias allow the DC polarization of the electrodes while grounding the sensor plate to avoid electrolysis disturbance which would occur by polarizing the channel walls in contact with conductive liquids. This setup does not degrade the Q factor or the frequency stability at short observation times.

Experimental results

Typical transmission curves were obtained by actuating these new sensors at different frequencies around their natural resonant modes. From these curves the values of the sensors' quality factors Q were extracted. Typically, 10,000 range Q factors were obtained in air, with a maximum value around 26,000.

Figure 7 shows the transmission amplitude and phase response from open-loop experiments, carried out on two 150 µm-wide square plate sensors with the same channel geometry. The two devices are referred to as sensor C and sensor D. We applied a 1.26 V_{rms} AC voltage and a 80 V DC voltage; the Lock-in time constant was set at 30 ms. The sensors' natural resonant frequency at $V_{DC} = 80$ V is about 27.56 MHz; the relative difference between the resonant frequencies of the two devices is less than 0.01% which indicates the good reproducibility of our fabrication process.

Allan deviation computation was also performed from phase-locked loop (PLL) measurements. During this experiment, the sensor is locked at its resonant frequency using a feedback loop, while the frequency fluctuations are acquired. The Allan deviation σ is obtained by averaging these fluctuations for different times τ and was

plotted for averaging times from 20 ms to 2 s, with a total acquisition time of 1 hour. The Lock-in integrating time was chosen equal to 10 ms; higher filtering time constants may artificially decrease the Allan deviation [8].

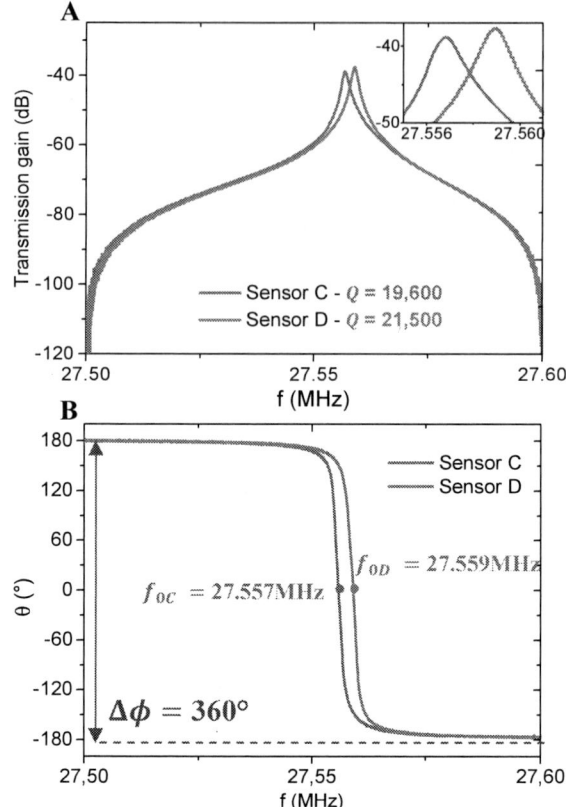

Figure 7: (A) *Transmission curves extracted from open-loop measurements carried out on two 150 µm-wide square plate sensors (sensors C and D). (B) Phase variations measured during this experiment.*

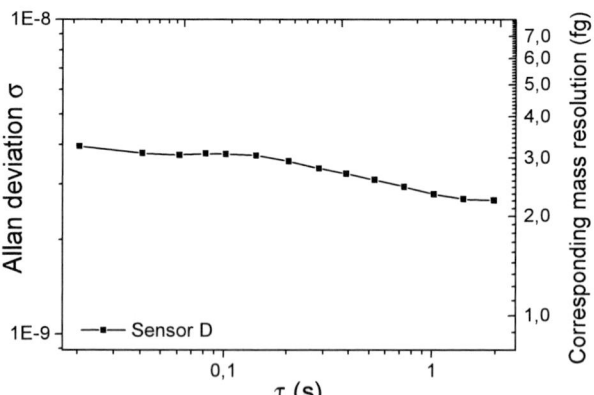

Figure 8: Allan deviation curve obtained for a 150 µm-wide square-plate sensor (1 h acquisition time).

Expected performances

The mass resolution of the sensors can be estimated from the Allan deviation computation. Indeed, the mass resolution corresponds to the smallest variation that cannot be attributable to frequency fluctuations, i.e.:

$$\delta m = 2 \cdot \sigma \cdot m_{eff} \qquad (2)$$

where $m_{eff} = \alpha \cdot \rho \cdot t \cdot W^2$ is the effective mass of the plate; ρ, t and W are respectively the plate's density, thickness and width; $\alpha = 0.5$ for a square plate and $\alpha = 0.28$ for a disk plate.

A mass resolution of 4 fg was then highlighted for sensor D with an averaging time of 20 ms. This corresponds for instance to the buoyant mass of a 75 nm-diameter gold nanoparticle in water. This expected value has to be confirmed by later fluidic experiments, though similar Q factors are expected in presence of fluid as detailed in [9].

Using Equation (2) one can estimate the buoyant mass of a particle depending on the observed frequency shift. A real-time monitoring of the sensors' resonant frequency was performed by applying DC pulses to simulate successive particles traveling across the chip; varying the applied bias voltage leads to an increased electrostatic stiffness k_{el}, resulting in smaller effective stiffness k_{eff} and resonant frequency. Figure 9 shows the curve extracted from this computation, with examples of particles which could be weighted from such frequency shifts. For instance, a 8 Hz frequency shift corresponds to 230 fg which is the buoyant mass of a 290 nm-diameter gold nanoparticle.

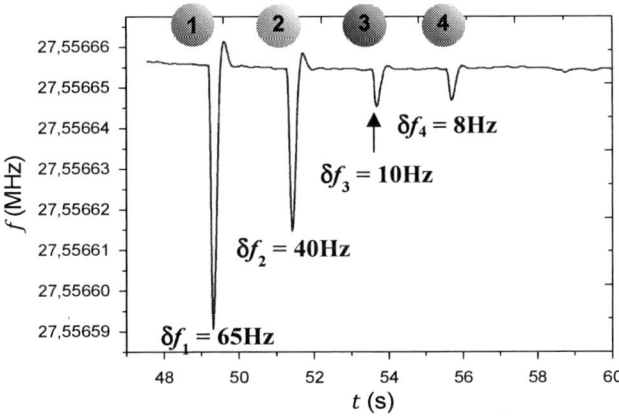

Figure 9: Real-time monitoring of frequency fluctuations by applying V_{DC} increments to mimic particles transiting across the sensor. 1: 580 nm-diameter gold nanoparticle (NP) / 2: 900 nm-diameter TiO₂ NP / 3: 350 nm-diameter mercury droplet / 4: 290 nm-diameter gold NP.

CONCLUSION

This paper has reported hollow MEMS mass sensors for particle monitoring in fluid, with ppb-range frequency resolution at ambient conditions. These sensors are fabricated via typical CMOS technologies and do not require additional packaging steps, which makes them easy to be connected. They should allow fast, precise and non-destructive characterization of liquid samples, which makes them well suited for particle metrology.

The devices also offer a larger dynamic range (gold nanoparticles with a diameter from 100 nm to 2 μm) than previous similar studies, enabled by the micrometric dimensions of the embedded channel together with the fg-range sensor mass resolution; this makes these sensors

more versatile and compliant with wider application spectrum.

REFERENCES

[1] E. Stemme, J. Ekelöf, and L. Nordin, "Measuring liquid density with a tuning-fork transducer," *IEEE Trans. Instrum. Meas.*, vol. 32, no. 3, pp. 434–437.

[2] T. P. Burg, M. Godin, S. M. Knudsen, W. Shen, G. Carlson, J. S. Foster, K. Babcock, and S. R. Manalis, "Weighing of biomolecules, single cells and single nanoparticles in fluid.," *Nature*, vol. 446, no. 7139, pp. 1066–9, Apr. 2007.

[3] S. Olcum, N. Cermak, S. C. Wasserman, K. S. Christine, H. Atsumi, K. R. Payer, W. Shen, J. Lee, a. M. Belcher, S. N. Bhatia, and S. R. Manalis, "Weighing nanoparticles in solution at the attogram scale," *Proc. Natl. Acad. Sci.*, no. Table 1, pp. 2–7, Jan. 2014.

[4] A. Heidari, Y.-J. Yoon, M. K. Park, W.-T. Park, and J. M.-L. Tsai, "High sensitive dielectric filled Lamé mode mass sensor," *Sensors Actuators A Phys.*, vol. 188, pp. 82–88, Dec. 2012.

[5] J. E.-Y. Lee, J. Yan, and A. a Seshia, "Low loss HF band SOI wine glass bulk mode capacitive square-plate resonator," *J. Micromechanics Microengineering*, vol. 19, no. 7, p. 074003, Jul. 2009.

[6] G. Blanco-Gomez and V. Agache, "Experimental Study of Energy Dissipation in High Quality Factor Hollow Square Plate MEMS Resonators for Liquid Mass Sensing," *J. Microelectromechanical Syst.*, vol. 21, no. 1, pp. 224–234, Feb. 2012.

[7] I. Bargatin, E. B. Myers, J. Arlett, B. Gudlewski, and M. L. Roukes, "Sensitive detection of nanomechanical motion using piezoresistive signal downmixing," *Appl. Phys. Lett.*, vol. 86, no. 13, p. 133109, 2005.

[8] E. Mile, G. Jourdan, I. Bargatin, S. Labarthe, C. Marcoux, P. Andreucci, S. Hentz, C. Kharrat, E. Colinet, and L. Duraffourg, "In-plane nanoelectromechanical resonators based on silicon nanowire piezoresistive detection," *Nanotechnology*, vol. 21, no. 165504, 2010.

[9] V. Agache, G. Blanco-Gomez, M. Cochet, and P. Caillat, "Suspended nanochannel in MEMS plate resonator for mass sensing in liquid.," *Proc. MEMS 2011*, pp. 157–160.

CONTACT

C. Hadji | +33(0)4 38 78 14 43 | celine.hadji@cea.fr
V. Agache | +33(0)4 38 78 26 53 | vincent.agache@cea.fr

DEVELOPMENT OF MICROFLUIDIC RESONATORS VIA SILICON-ON-NOTHING TECHNIQUE

Joohyun Kim and Jungchul Lee*

Department of Mechanical Engineering, Sogang University, Seoul, Korea

ABSTRACT

This paper reports wafer-level batch fabrication of microfluidic resonators by employing high temperature annealing of a pre-structured silicon wafer with periodic cylindrical pits for the first time. Upon high temperature annealing of pre-structured silicon, the surface silicon atoms around cylindrical pits moved and merged into long channel-shaped cavities with closed lids which are known as silicon-on-nothing (SON) structures [1]. Then, the wafer was planarized, cavities were oxidized, and resulting oxide microtubes were released to make free-standing resonators. Finally, metal reflectors were deposited for optical readout of resonance frequency. We have confirmed the functionality and integrity of the fabricated microtubes by injecting dyed solutions and measured resonance frequency of dry microtube resonators at various partial vacuums. The proposed fabrication processes are not only cost-effective but also potentially enable wafer-scale batch fabrication of nanofluidic resonators.

INTRODUCTION

Fluidic channel integrated resonators (fluidic resonators) have drawn continuous attention due to their unprecedented mass resolution for particles suspended in liquid. Mass, density, volume, surface charge, and deformability of single cells, bacteria, and single micro-/nanoparticles [2-7] have been quantitatively measured at the single particle level with the help of such novel devices. To further improve the performance of fluidic resonators, there is continuous effort to reduce their physical sizes and now it is possible to weigh a single virus in liquid [8-9]. The desire to push the limit of this technique will be continuing to weigh and detect particle matters lighter than a single virus.

Fabrication processes for fluidic resonators reported so far rely on silicon fusion bonding [2] or sacrificial processes [10] both of which become problematic upon downsizing of fluidic channels to be integrated. To further scale down fluidic resonators, we should rely on dedicated nanofabrication techniques that are typically slow, expensive, available to limited researchers, thus impractical. Besides the downsizing limit of two aforementioned fabrication methods, they are not capable of switching resonator structural materials due to the limited process compatibility and their fabrication cost and time are relatively high and long, respectively. Since we are now facing the resolution limit of batch microelectromechanical systems (MEMS) fabrication techniques, we need to develop a totally new way to batch fabricate fluidic resonators with nanochannel integrated.

While we were heavily searching for alternatives to fabricate fluidic resonators which potentially address all issues with currently available methods, we found an interesting way to make embedded cavities called silicon-on-nothing (SON) [1]. SON structures have been mainly proposed to improve dielectric barrier strength for miniaturized semiconductor devices [11] and have been recently employed fabricate MEMS devices such as biological patch clamps [12] and pressure sensors [13] but not yet explored to fabricate fluidic channel integrated resonators.

This paper reports a novel approach to fabricate fluidic resonators which potentially overcomes aforementioned issues in fusion bonding and sacrificial processes by employing SON structures for the first time. Fluidic resonators were successfully fabricated and their resonant behaviors were thoroughly characterized under various partial vacuums.

PROCESS DEVELOPMENT AND DEVICE FABRICATION

Fabrication processes started with a 6-inch double sided polished (100) silicon wafer (see Fig. 1). The silicon wafer has been anisotropically etched using a deep reactive ion etcher (DRIE) to define U-shaped arrays of cylindrical

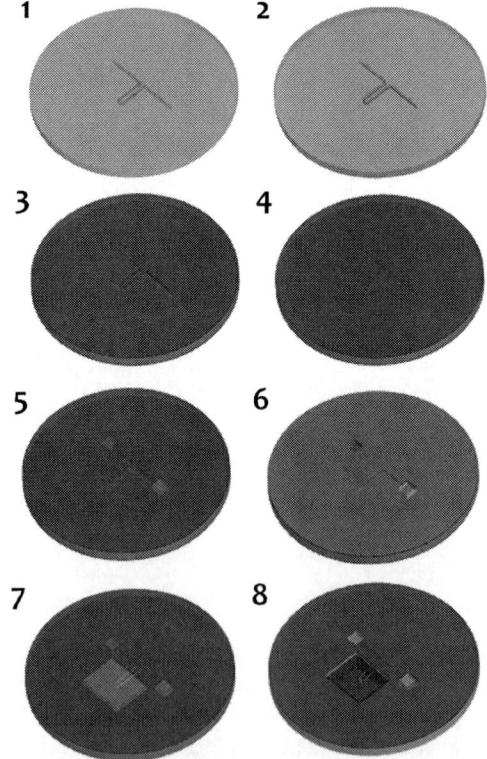

Figure 1: Fabrication processes for the silicon wafer. 1-Hole array patterning using stepper and DRIE, 2-High Temperature Argon annealing, 3-Polysilicon deposition, 4-Chemical mechanical polishing, 5-Fluidic ports opening using DRIE, 6-Oxidation, 7-Removal of surface oxide and silicon surrounding the oxide tube using DRIE, 8-Silicon dioxide tube release and metal deposition.

pits (process 1 in Fig. 1) then annealed at 1150 °C for 10~20 minutes in a rapid thermal processing (RTP) chamber purged with argon. Due to the limitation of

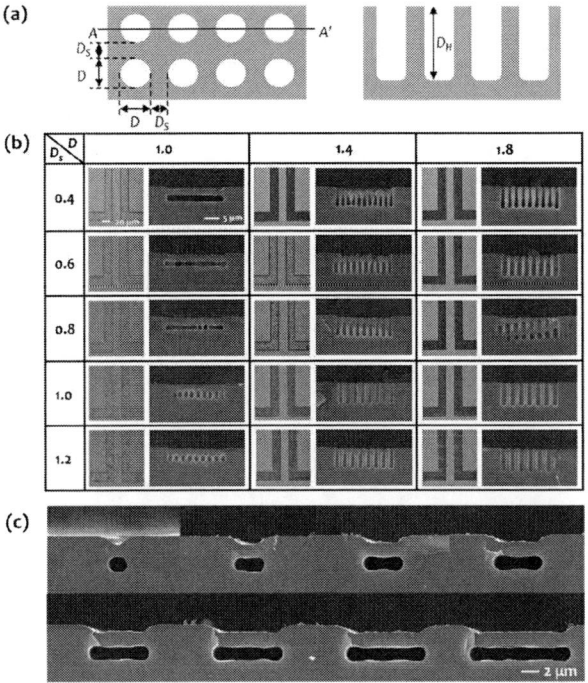

Figure 2: (a) Array configuration of cylindrical pits. D and D_H: diameter and depth of each cylindrical pit, D_S: distance between adjacent pits. (b) Optical (top view) and scanning electron (cross sectional view) micrographs of the annealed silicon wafer showing optimal geometric conditions for SON structures. All units are μm. (c) Scanning electron micrographs of SON structures with varying widths fabricated by different array configuration of cylindrical pits.

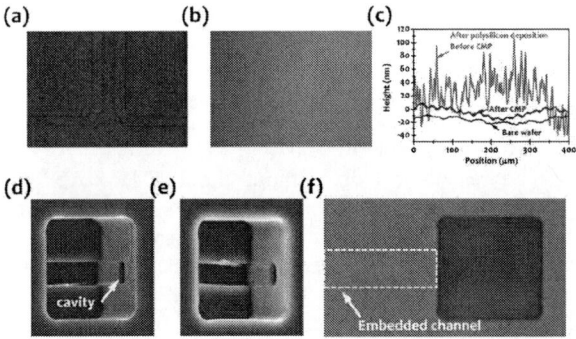

Figure 3: Optical micrographs of the silicon wafer around the SON structures (a) after LPCVD polysilicon deposition and (b) after CMP, (c) atomic force microscope topographic measurements of a bare silicon wafer, the wafer after polysilicon deposition, and the wafer after CMP, (d) scanning electron micrograph of the fluidic access port before oxidation, (e) scanning electron and (f) optical micrographs of the fluidic access port after oxidation.

annealing time with RTP at such a high temperature, 60~90-second annealing was repeated several times. High temperature annealing makes silicon atoms around cylindrical pits move and merge into long channel-shaped cavities with closed lids. Simultaneously, about 1.5-μm deep concave grooves above all cavities were formed (process 2 in Fig. 1).

The initial spacing between neighboring cylindrical pits (center-to-center distance, $D+D_S$) and the aspect ratio (a ratio of height to diameter of individual pits, D_H/D) were experimentally optimized to result in a nicely connected cavity (see Fig. 2). For a given D_S, decreasing D is favorable to form connected SON structures as shown in the first row of Fig. 2(b). For a given D, small and large D_S result in connected and separated SON structures, respectively. After optimizing the spacing and aspect ratio, the number of pit array in the width direction was varied to make fluidic resonators with various widths as shown in scanning electron micrographs (SEMs) for diced SON structures in Fig. 2(c).

To remove the concave groove above each cavity which hinders the vacuum packaging to be followed, 3-μm thick low pressure chemical vapor deposition (LPCVD) polysilicon was deposited first to mitigate the groove (process 3 in Fig. 1) and the wafer underwent chemical mechanical polishing (CMP) (process 4 in Fig. 1). Without the deposited polysilicon, many cavities are possibly exposed thus devices fail during the CMP process. Figures 3(a) and (b) show optical micrographs of polysilicon deposited SON structures before and after the CMP process. Their RMS roughnesses measured with an atomic force microscope were 20.7 and 5.3 nm, respectively (see Fig. 3(c)). The roughness of the polished wafer similar to that of a brand new wafer, 3.9 nm, is enough for anodic bonding to be developed later on.

After both ends of embedded cavities were anisotropically etched to expose interiors of cavities to atmosphere (process 5 in Fig. 1), the wafer was dry oxidized in a tube furnace (process 6 in Fig. 1). As a result, 300-nm thick oxide layers were formed at wafer surfaces and interiors of cavities. Figures 3 (d)~(f) show scanning electron and optical micrographs of deep reactive ion etched opening at one end before and after the oxidation. Either charging or color change provide evidence for the formation of thin oxide shell due to the internal oxidation.

Once the surface oxide was removed via reactive ion etching (RIE) (process 7 in Fig. 1), oxide microtube structures were released from the front side to leave U-shaped microtube resonators by combination of DRIE and SF_6 etch to minimize undercuts which generally increase the length and effective mass of each resonator, decrease the resonance frequency, thus deteriorates mass resolution. Figure 4(a) shows SEMs of hollow U-shaped microtube resonators with different cavity widths. Resonators were focused ion beam milled to inspect the cavity cross-section and confirm tube structures (see two SEMs on the right in Fig. 4(a)). Finally, 20-nm thick titanium and 130-nm thick gold were sequentially deposited on top of free standing microtube resonators (process 8 in Fig. 1). Figure 4(b) shows three U-shape microtube resonators with different lengths and widths.

978-1-4799-7956-1/15 $31.00 © 2015 IEEE

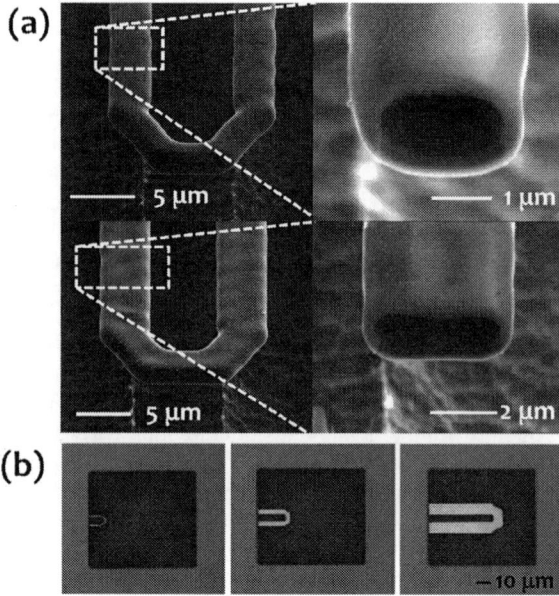

Figure 4: (a) Scanning electron and (b) optical micrographs of hollow U-shaped microtube resonators. Resonators are focused ion beam milled prior to SEM imaging and metallized (Ti/Au:20 nm/130 nm) prior to bright field microscope imaging via lift-off.

RESULTS AND DISCUSSION

Amorphous silicon was deposited on both sides of a cleaned 6-inch borosilicate wafer and the amorphous silicon layer on the front side was patterned using the RIE with a patterned photoresist as an etch mask. Then, the borosilicate wafer was isotropically etched to mainly create microfluidic channels for sample delivery and waste collection. Currently, the process optimization for vacuum anodic bonding between the processed silicon and

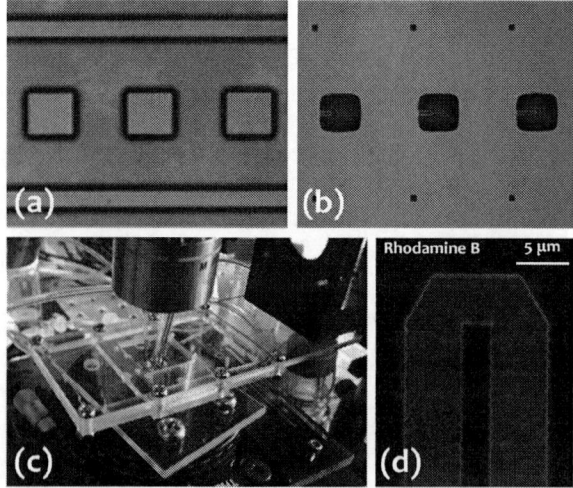

Figure 5:(a) Replica molded PDMS cover, (b) finalized silicon chip with an array of silicon dioxide tube resonators, (c) custom alignment and packaging setup, (d) a U-shaped tube resonator dyed with rhodamine B.

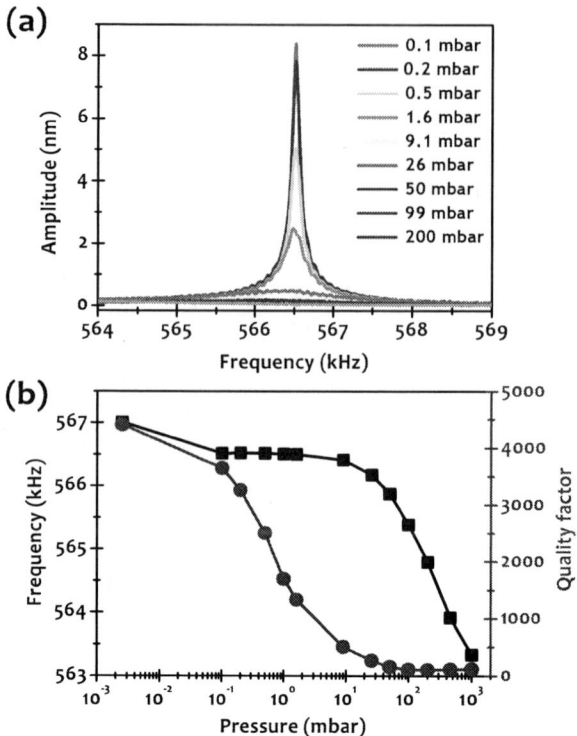

Figure 6: (a) Amplitude spectra at various levels of partial vacuum, (b) Resonance frequency and quality factor as functions of the pressure.

borosilicate wafers is ongoing. In the meantime, polydimethylsiloxane (PDMS) covers were prepared by double replica molding onto the etched borosilicate wafer and packaged over diced resonator chips by using a custom alignment setup (see Figs. 5(a)~(c)). After the alignment and packaging, tubings were inserted on top of the PDMS covers to deliver samples and collect wastes. Deionized water mixed with rhodamine B was injected into microtube resonators and their interior walls were successfully dyed even after microtube resonators were dried (see Fig. 5(d)). The chip used was not metallized intentionally for fluorescent observation. No leakage was found between the PDMS cover and the resonator chip with the applied pressure regulated below 100 kPa.

The resonator chip without the PDMS cover was placed inside a vacuum chamber and resonant behaviors of each resonator were optically investigated. Figure 6(a) shows the amplitude spectra of a specific microtube resonator measured at various partial vacuums. The vibration amplitude greatly increased with decreasing pressure. Figure 6(b) shows the resonance frequency and quality factor of the microtube resonator as functions of the pressure. The quality factor increased more than 30 times when the chamber pressure decreased from 1 atm to 0.1 mbar. Both resonance frequency and quality factor tended to saturate with the pressure level below 0.1 mbar. Currently, we are focusing on on-chip vacuum packaging by anodic bonding with a getter deposited borosilicate wafer for fast and reliable operation.

CONCLUSIONS

This paper presents a novel approach to fabricate fluidic resonators by employing high temperature annealing of pre-structured single crystalline silicon wafers. We experimentally optimized the initial spacing between neighboring cylindrical pits and the aspect ratio to obtain nicely connected SON cavity structures. Internal oxidation of cavities resulted in silicon dioxide shells which became microtube resonators after device release. Liquid samples could be introduced into microtube resonators via a pressure-driven flow. Dry silicon dioxide microtube resonators exhibited relatively high quality factor of ~4,000 at moderate vacuum around 0.1 mbar. The proposed method for fabricating fluidic resonators is not only cost and time effective but also has the potential to batch fabricate nanofluidic resonators.

ACKNOWLEDGEMENTS

This research was supported in part by Basic Science Research Program through the National Research Foundation of Korea (NRF) funded by the Ministry of Science, Information, and Communication Technology, and Future Planning (NRF-2013R1A1A1076080) and in part by National Research Foundation Grant funded by the Korean Government (NRF-2011-220-D0014).

REFERENCES

[1] K. Sudoh, H. Iwasaki, R. Hiruta, H. Kuribayashi, R. Shimizu, "Void Shape Evolution and Formation of Silicon-on-Nothing Structures During Hydrogen Annealing of Hole Arrays on Si (001)", *J. Appl. Phys.*, vol. 105, 083536, 2009.

[2] T. P. Burg, M. Godin, S. M. Knudsen, W. Shen, G. Carlson, J. S. Foster, K. Babcock, S. R. Manalis, "Weighing of Biomolecules, Single Cells, and Single Nanoparticles in Fluid", *Nature*, vol. 446, pp. 1066-1069, 2007.

[3] P. Dextras, T. P. Burg, S. R. Manalis, "Integrated Measurement of the Mass and Surface Charge of Discrete Microparticles Using a Suspended Microchannel Resonator", *Anal. Chem.*, vol. 81, pp. 4517-4523, 2009.

[4] A. K. Bryan, A. Goranov, A. Amon, S. R. Manalis, "Measurement of Mass, Density, and Volume During the Cell Cycle of Yeast", *Proc. Nat. Acad. Sci.,* vol. 107, pp. 999-1004, 2009.

[5] M. Godin, F. F. Delgado, S. Son, W. H. Grover, A. K. Bryan, A. Tzur, P. Jorgensen, K. Payer, A. D. Grossman, M. W. Kirschner, S. R. Manalis, "Using Buoyant Mass to Measure the Growth of Single Cells", *Nat. Methods,* vol., 7, pp. 387-390, 2010.

[6] W. H. Grover, A. K Bryan, M. Diez-Silva, S. Suresh, J. M. Higgins, S. R. Manalis, "Measuring Single-Cell Density", *Proc. Nat. Acad. Sci.* vol. 108, pp. 10992-10996, 2011.

[7] S. Son, A. Tzur, Y. Weng, P. Jorgensen, J. Kim, M. W. Kirschner, S. R. Manalis, "Direct Observation of Mammalian Cell Growth and Size Regulation", *Nat. Methods*, vol. 9, pp. 910-912, 2012.

[8] J. Lee, W. Shen, K. Payer, T. P. Burg, S. R. Manalis, "Toward Attogram Mass Measurements in Solution with Suspended Nanochannel Resonators", *Nano Lett.*, vol. 10, pp. 2537-2542, 2010.

[9] S. Olcum, N. Cermak, S. C. Wasserman, K. S. Christine, H. Atsumi, K. R. Payer, W. Shen, J. Lee, A. M. Belcher, S. N. Bhatia, S. R. Manalis, "Weighing Nanoparticles in Solution at the Attogram Scale", *Proc. Nat. Acad. Sci.,* vol. 111, pp. 1310-1315, 2014.

[10] M. F. Khan, S. Schmid, Z. J. Davic, S. Dohn, A. Boisen, "Fabrication of Resonant Micro Cantilevers with Integrated Transparent Fluidic Channel", *Microelec. Eng.*, vol. 88, pp. 2300-2303, 2011.

[11] M. Jurczak, T. Skotnicki, M. Paoli, B. Tormen, J. Martins, J. L. Regolini, D. Dutarte, P. Ribot, D. Lenoble, R. Pantel, S. Monfray, "Silicon-on-Nothing (SON)-an Innovative Process for Advanced CMOS", *IEEE Trans. Electron. Dev.*, vol. 47, pp. 2179-2187, 2000.

[12] F. Zeng, Y. Luo, L. Yobas, M. Wong, "Self-Formed Cylindrical Microcapillaries Through Surface Migration of Silicon and Their Application to Single-Cell Analysis", *J. Micromech. Microeng.* vol. 23, 055001, 2013.

[13] X. Hao, S. Tanaka, A. Masuda, J. Nakamura, K. Sudoh, K. Maenaka, H. Takao, K. Higuchi, "Application of Silicon on Nothing Structure for Developing a Novel Capacitive Absolute Pressure Sensor", *IEEE Sens. J.*, vol. 14, pp. 808-815, 2014.

CONTACT

*Jungchul Lee, Dept. of Mechanical Engineering, Sogang University, Tel: +82-2-705-7973; Fax: +82-2-712-0799; E-mail: jayclee@sogang.ac.kr

A MICROFABRICATED, FLOW DRIVEN MILL FOR THE MECHANICAL LYSIS OF ALGAE

Justin Millis[1], Laurie Connell[2], Scott D. Collins[1], and Rosemary L. Smith[1]
[1]The MicroInstruments and Systems Laboratory, University of Maine, Orono, USA
[2]School of Marine Science, University of Maine, Orono, USA

ABSTRACT

This paper reports a novel, new means of mechanically lysing algae using a microfabricated, flow-driven grinding mill. The mill is demonstrated to be effective in the mechanical lysis of the dinoflagellate genus *Alexandrium*, a neurotoxin producing algae, responsible for Red Tide and paralytic shellfish poisoning. Pre and post milling cell counts revealed lysing efficiencies as high as 96%. Selective amplification of lysate samples was successfully performed, using Polymerase Chain Reaction (PCR) and primers specific to *Alexandrium*, demonstrating that the released DNA is suitable for downstream analysis.

INTRODUCTION

Harmful algal blooms (HABs), sometimes called "red tides", are a worldwide health issue. Some algal species produce high potency toxins (e.g. Paralytic Shellfish Toxin, PST) that concentrate in the shellfish that feed on them, consequently poisoning the human consumer. Since some harmful algal blooms, e.g. *Alexandrium*, do not generally involve sufficiently high cell densities to produce discoloration of the water, the "gold standard" detection method for a HAB is manual cell counting. This method is slow, and due to the morphological similarity among species, requires a high level of training for identification. With HABs increasing worldwide, and further increase anticipated with global climate change [1] the need is growing for rapid and inexpensive onsite analysis methods in order to ensure and maximize safe shellfish harvest areas.

The microfabricated mill presented in this paper was investigated as part of an on-going project to develop low cost, portable instruments [2] for the detection of PST producing species of the dinoflagellate genus *Alexandrium*. The motivation for mechanical cell lysing over a chemical means was to produce a lysate compatible with an immunoassay detection of released neurotoxin, in addition to nucleic acid identification, immediately downstream. Several micromechanical lysing methods have been previously reported to be effective with mammalian blood cells, including the use of shredding projections [3,4]. This method was tried but found to be inefficient in lysing armored dinoflagellates, such as *Alexandrium*. The projections generally just impaled the cells, which quickly collected and obstructed flow.

DESIGN AND FABRICATION

The grinding mill, presented here, is inspired by the horizontal-wheel or Norse mill, which is one of the earliest known, water-powered type of grinding mills. The microfabricated mill features two microfluidic channels: one that supplies the fluidic driving force to the vanes and turns the milling "stone", and one that supplies the algae sample to the grinding surface on the opposite side. A cross-section illustration of the device is shown in Figure 1. With this design, the relative flow rates of the drive and that of the sample can be optimized to levitate the millstone and eliminate cellular debris from clogging the drive mechanism. To the authors' knowledge, this is the first reported demonstration of a microfluidic grinding mill and its application to cell lysis.

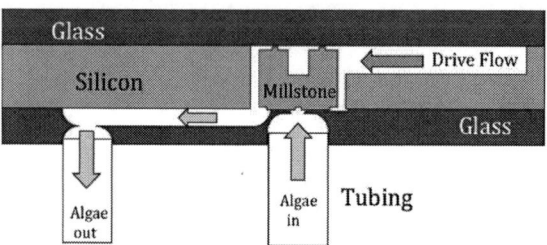

Figure 1: Cross-section of assembled device. Algae sample enters at base of the millstone, and exits into adjacent channel. Drive flow is on top.

Figure 2: Autodesk CFD 2D simulation of flow velocity.

The vane geometry, and positioning of the inlet and outlet for the drive flow channel were informed using Autodesk Computational Fluid Dynamic simulation software. An example 2D plot of flow velocity is shown in Figure 2. The simulations revealed behavior observed with completed devices, most notably instability due to build up of equal pressure between two adjacent fins, directing one in the forward and one in the reverse direction. A split in the inlet flow path was made in an attempt to eliminate such instability by applying force at two, out of phase fins (presenting different angles). It was later found that providing a fluid flow path across the center of the stone was the most reliable means for eliminating the pressure build-up and resulted in highly reproducible start-up behavior on all similarly made devices.

The microfluidic channels and the millstone are fabricated in a single silicon wafer, using double-sided photolithography and Deep Reactive Ion Etching (DRIE) of silicon. The fabrication starts with oxidation of a double-side polished, four-inch diameter, silicon wafer. The oxide layer is first patterned on the front side to delineate bumps that will sit on top of each vane, reducing the area in contact with the upper glass and friction. Next, oxide on the reverse side of the wafer is patterned, using front to back alignment, to define the grinding surface pattern of the millstone. The bottom surface is next patterned with a thick photoresist that defines the millstone and its housing. The photoresist serves as etch mask for a DRIE etch halfway through the wafer (≈250μm). After stripping the resist, the backside of the wafer is DRIE etched again for another 10μm, this time using the patterned oxide mask, producing the millstone grinding pattern (Figure 3).

Figure 3: Millstone (4 mm diameter) grinding surface patterns: Left- Cross, Right- Curve. Ridge height = 10 μm

Next, the topside (drive) fluidic channels and the vanes are patterned in photoresist and the silicon is DRIE etched ≈200 μm. At this point, the vanes and drive channel are defined and the mill wheel remains attached to the silicon substrate by a membrane of silicon, 15-25 μm in thickness. The photoresist is stripped, and the wafer is bonded to a carrier wafer using thermally conductive paste (COOL-GREASE™ (CGR7016). The wafer is again DRIE etched to release the moving wheel and to form the oxide masked bumps on top of the vanes. The wafer is separated from the carrier, cleaned and the millstones are collected. The topside of the silicon wafer is anodic bonded to a Pyrex 7740 glass wafer. The bonded pair is then diced into individual devices 35 mm long by 15 mm wide. Holes are made using a diamond drill to access the supply and exit of the driving fluidic channel. The mill wheels are inserted into the housing using a vacuum pick up tool. Photographs of a completed device (top view) and an isolated millstone (perspective) are shown in Figure 4.

The flow channels that direct the algae sample to the millstone surface are etched into a Pyrex 7740 glass wafer. A Graphtec Craft Robo Pro knife plotter was used to cut the algae sample flow channel and access hole patterns into high-tack UV curable tape. Patterned tape was adhered to the front and back of the glass substrate and placed in a vacuum oven set at 70 C and left over night (15-18 hrs). Both sides of the glass substrate were

then etched in concentrated HF acid (49 wt%) for 40 min, using the wafer tape as mask. The access holes are etched all the way through the glass, and the channel is etched to a depth of ≈ 200 μm. A portion of the tape is removed around the inlet access hole on the channel side, and the exposed glass surface is etched another 8 min to set a 15-20 μm gap between the bottom of the millstone and the glass surface after assembly. The rest of the tape is then removed, the glass cleaned, diced and anodic bonded to the silicon parts. The fluidic drive and sample channels are connected to syringe pumps via short lengths of tubing that are inserted and glued to the access holes.

Figure 4. Photograph of fabricated device, viewed through glass (top- drive side). Insert, on right shows the millstone removed, in perspective. Mill diameter •4 mm.

EXPERIMENTAL RESULTS

The millstone angular velocity versus time, obtained using Autodesk CFD for an input fluidic pressure of 30 psi is shown in Figure 5. The angular velocities of microfabricated mills were measured using a stereoscope and a high-speed camera (Motion Scope M3) running at 16000 fps. For these measurements, the driving fluid flow was delivered by pressurizing a bottle of deionized water to the desired input pressure. The grinding side was supplied with a water flow rate of 0.5mL/min, which lifts the stone above the bottom glass surface and reduces friction. Results from two measurements, made with 30 psi drive pressure, are superimposed on the simulation plot in Figure 5.

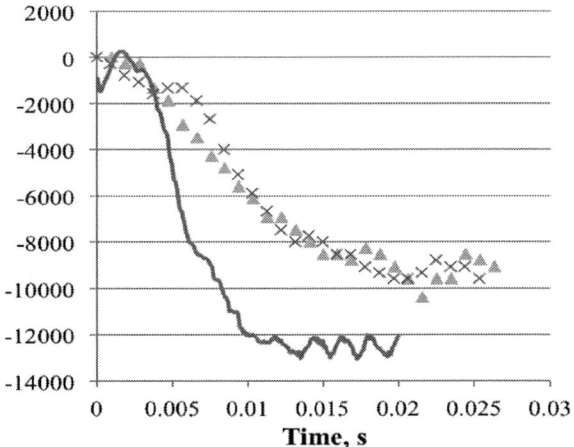

Figure 5: Plot of simulated angular velocity (solid-blue) in RPM, versus time in seconds, with a driving fluidic pressure drop of 30 psi. The measured angular velocities versus time of 2 fabricated mills are superimposed (×, Δ).

The overall behavior of the fabricated devices was quite similar to that produced by simulations. Typical values for measured steady state angular velocities were 10000 RPM, which is about 20% less than the simulated result. Differences are attributed to measurement error and frictional forces that were not included in this model, i.e. in the simulations, the millstone is assumed to spin freely.

The efficacy of algae lysing using the mill was assessed using a traditional microscope cell counting chamber. Samples of algae were taken directly from culture flasks. A 1 mL sample was introduced into the counting chamber to determine the initial cell density. A 5 mL sample was then run through the mill, collected, and a comparative cell count performed on 1 mL of lysate. Comparative tests were performed using different millstone patterns, sample cell densities and sample flow rates. The cell count results, given in Table 1, yielded lysing efficiencies as high as 96% when using a textured millstone and a sample flow rate of 0.5 mL/min. These optimized conditions were repeated on additional samples, producing similar results (95-98% efficiency). Interestingly, a surprisingly high lysing efficiency (>70%) was also obtained using an unpatterned millstone. This indicates that cells experience a high shear stress as they traverse the rapidly spinning surface, producing lysis.

Table 1:Results of Cell Counts, Pre and Post Milling

Mill no.	Mill-stone Pattern	Sample Flow Rate ml/min	Drive Flow Rate ml/min	Density PRE Mill cells/ml	Density POST Mill cells/ml	Lysing Efficiency %
1B	Smooth	1	20	2125	632	70.26
3B	Curve	1	20	2125	586	72.42
5B	Cross	1	20	2125	243	88.56
5B	Cross	1	20	5015	630	87.44
5B	Cross	0.5	20	5015	205	95.91

Phytoplankton contain chloroplasts that fluoresce strongly when released from the cells. Fluorescence spectra were obtained with pre- and post-milled samples of algae using a benchtop fluorescence spectrometer (Quantamaster T-format spectrofluorimeter, Photon Technology International). Figure 6 shows a large increase in fluorescence of samples after passage through the microfabricated mill, another indication of cell lysing.

The suitability of the released DNA for downstream applications, such as detection and analysis, was demonstrated by PCR amplification using primers specific to *Alexandrium*. Post-milled samples were desalted using Illustra Microspin G-25 columns (GE Healthcare, Buckinghamshire, UK) prior to PCR according to the manufacturers' instructions. PCR was carried out using 100ng of genomic DNA and primer pair D1R/D2C as previously described [5]. PCR products were visualized on a 1.8% TAE I.D. NA agarose gel (Lonza, Rockland, ME, USA) and SYBR Safe stain (Life Technologies, Carlsbad, CA, USA) under UV light. The fluorescent image of four separately lysed samples, after amplification and gel electrophoresis is shown in Figure 7. Lanes 3,4, 6, and 7 show positive for amplified DNA.

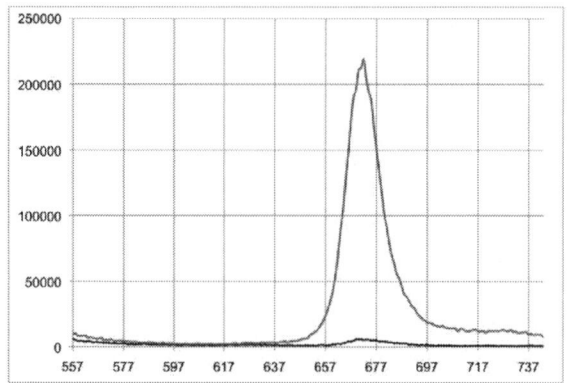

Figure 6. Fluorescence spectra of pre-milled (black) and post-milled (red) algae samples. The x-axis is optical wavelength in nm ; y-axis is counts, arbitrary units.

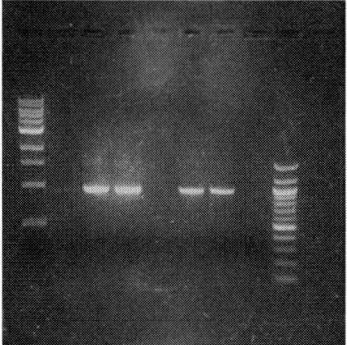

Figure 7: Gel electrophoresis of PCR amplified lysate. 1.8% agarose gel visualized using SYBR Safe. Lane 1 is 1kb size ladder, lanes 2 and 10 are blanks, lanes 5 and 8 are negative controls. Lanes 3,4,6 and 7 contain 100ng genomic DNA as PCR template. Lane 9 is a 100bp size ladder.

ACKNOWLEDGEMENTS

The authors acknowledge partial financial support for this project from the University of Maine's Institute for Molecular Biophysics and NSF award CBET-0854020 . The authors thank Dr. Paul Millard, Dept. of Chemical and Biological Engineering, University of Maine, for his assistance with fluorescence spectroscopy.

REFERENCES

[1] S. K. Moore, V. L. Trainer, N. J. Mantua, et al, "Impacts of Climate Variability and Future Climate Change on Harmful Algal Blooms and Human Health", *Environmental Health*, **7 Suppl 2**, S4, 2008.

[2] J. Duy, R. L. Smith, S. D. Collins and L. B. Connell, "A Field-deployable Colorimetric Bioassay for the Rapid and Specific Detection of Ribosomal RNA", *Biosensors and Bioelectronics*, **52**, 433-437, 2014.

[3] S-S. Yun, S. Y. Yoon, M-K. Song, et al, "Handheld Mechanical Cell Lysis Chip with Ultra-sharp, Nano-blade Arrays for Rapid Intracellular Protein Extraction", *Lab on a Chip*, **10**, 1442-1446, 2010.

[4] D. Di Carlo, K. H. Jeong and L. P. Lee, "Reagentless Mechanical Cell Lysis by Nanoscale Barbs in Microchannels for Sample Preparation", Lab on a Chip, **3(4)**, 287-291, 2003.

[5] C. A. Scholin, M. Herzog, M. Sogin and D M Anderson, "Identification of Group- and Strain-Specific Genetic Markers for Globally Distributed Alexandrium (Dinophyceae). 2. Sequence Analysis of a Fragment of the LSU rRNA Gene." *Journal of Phycology,* **30(6),** 999-1011, 1994.

CONTACT

R. L. Smith,+1-207-5813361; rosemary.smith@maine.edu

CLUSTER SIZING OF CANCER CELLS BY RAIL-BASED SERIAL GAP FILTRATION IN STOPPED-FLOW, CONTINUOUS SEDIMENTATION MODE

Macdara Glynn[1], Charles Nwankire[1], David Kinahan[1] and Jens Ducrée[1]

[1]Biomedical Diagnostics Institute, School of Physical Sciences, Dublin City University, IRELAND

ABSTRACT

In addition detecting circulating tumour cells (CTCs) in blood, the presence of multi-cellular clusters has recently been identified to carry further information pertaining to patient outcome. We present a label-free method of measuring the range and load of clusters in a blood sample using a size-exclusion rail operated by centrifugal microfluidics. A negative selection strategy first enriches clusters from a whole blood sample; these clusters are then processed along the rail where they resolve to a series of collection bins according to size. Analysis of the occupancy of these bins then provides metrics on the cancer-cell load carried in the blood.

INTRODUCTION

There is increasing evidence that in addition to their presence, the propensity of CTCs to form multi-cellular clusters bears significant information of cellular resistance to chemotherapy and overall prognosis [1]. Both, individual CTCs and clusters thereof, occur when cellular events detach from the site of a primary tumour and enter the blood stream. We here define an event as either a single cell or a cluster. These CTCs may then promote metastasis.

A number of microfluidic systems isolate candidate CTC events by using a positive detection strategy where cells with selected epitopes are targeted and manipulated *via* the binding of reagents specific to the epitopes of interest [2, 3]. While such strategies can be very sensitive to such cells of interest, the inherent heterogeneity within CTC populations can limit the use of the strategy as there are no known epitopes that are common to all CTCs identified. For example, EpCAM expressing cells in blood are commonly used in commercial positive isolation strategies for the identification of CTCs. However, some CTCs do not express EpCAM; and indeed, in many carcinomas EpCAM is under-expressed or even absent [4]. Similarly, enrichment strategies by flow-based filtration are susceptible to clogging [5]; also sizing of co-localized clusters on a filter as well as subsequent removal of target cells tends to be challenging. The technique presented here alleviates these caveats by combining surface biomarker purification and differential size filtration to directly characterize the clinically relevant cluster load in blood using a two-stage centrifugo-microfluidic strategy (Fig. 1).

The first stage involves an off-chip blood processing step. When a whole blood sample is to be tested, the red blood cells (RBCs) are first lysed using standard hypotonic lysis protocol. Following this, a negative isolation strategy is used to overcome the inherent phenotypic heterogeneity of CTCs. Here, the white blood

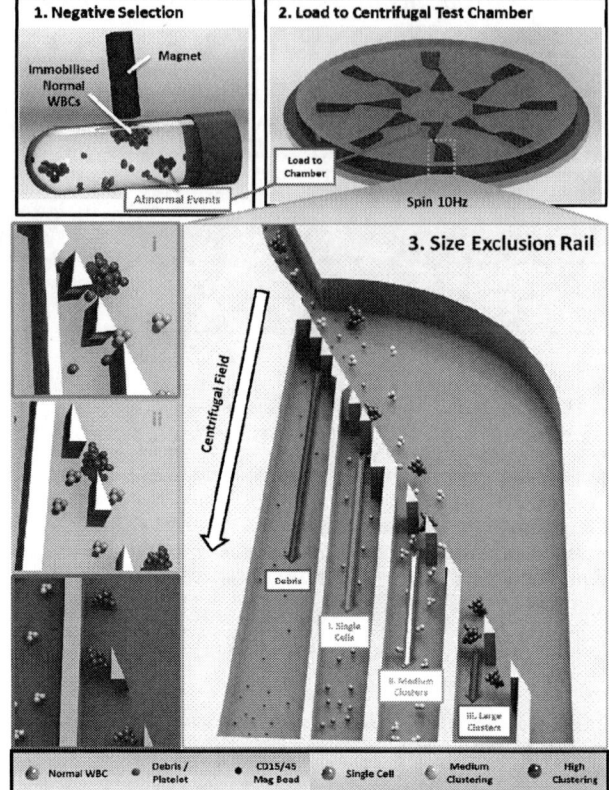

Figure 1: Full Protocol *(Top) Negative Selection and 3D representation of a full disc equipped with 8 test chambers. (Lower Right) The size based segregation section of the test chamber showing binning of single, medium and large clusters. (Lower Left) Zoom of bin gates. For clarity, a chamber with only four bins is represented.*

cells (WBCs) are incubated with a mixture of CD15 and CD45 super-paramagnetic beads which bind to all naturally occurring WBCs while abnormal events (candidates for CTCs or CTC-based clusters) lack expression of CD15 or CD45 and therefore stay unbound. Healthy WBCs are then gently removed from the solution by placing the tube in proximity to a magnet (Fig. 1, top left). The cellular events that remain in solution are considered to be abnormal events. These events are then placed into the centrifugal test chamber (Fig. 1, top right) where they are sorted according to size. The sorting is carried out by a rail that spans the top of eight discrete collection bins. The apertures in the rail that gates the top of each bin increase in size the further along the rail an event progresses, with the larger events resolving to later bins. Examination of the filling of the bins then provides an indication of the range and load of abnormal cell clusters harboured in the original whole blood sample.

SYSTEM DESIGN

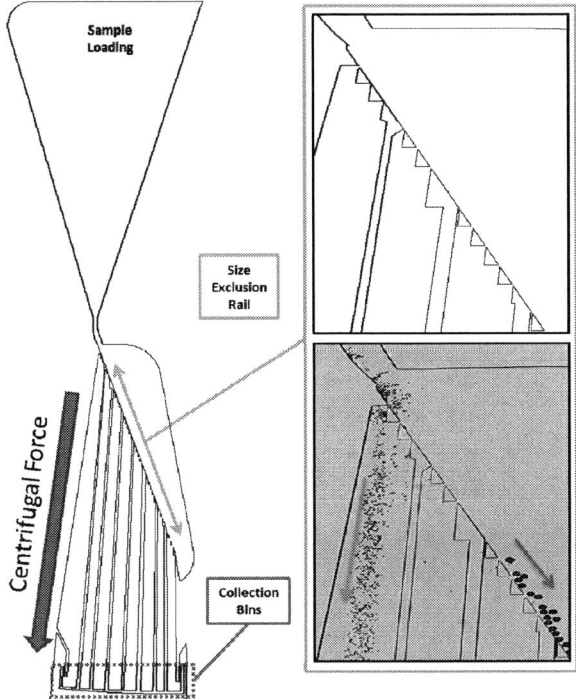

Figure 2: Schematic of the Size Exclusion Rail. (Left) Schematic of a full test chamber. (Orange border) The Critical Size Exclusion Rail during processing of 5 μm (green arrow) and 40 μm (red arrow) beads. For clarity, beads are shown rather than cells.

The entire, dead-end microfluidic chamber is first fluidically primed so that the system operates in stopped-flow sedimentation mode. The isolated abnormal events are loaded to the Sample Loading chamber *via* the loading port and the disc is spun at 10 Hz. On exiting the loading chamber, events are guided along a Size Exclusion Rail which is slightly inclined against the direction of the centrifugal field (Fig. 2, orange box). This rail is composed of flat-edged pillars, with the interspersed gaps that gate the entry to the underlying bin progressively increasing in size. Bins are oriented in the radial direction so events penetrating the rail centrifugally sediment into the bin. The gate size increases from initially 4.7 μm (isolates debris and highly deformable cells) to 130 μm in the last of the eight bins. The events then resolve by the centrifugal force to the base of the collection bins where they can be observed by microscopy.

METHODOLOGY
CHIP MANUFACTURE
The microfluidic discs used in this paper were formed from polydimethylsiloxane (PDMS; Dow Corning, MI) mixed at a ratio of 10:1 base and curing agent. The procedures for making a master and for securing the co-rotating magnet in the PDMS have been described in detail elsewhere [6-67]. Loading holes and vents were defined in the PDMS at appropriate locations using a dot punch. The PDMS slab containing the microfluidic features was placed on a 100-mm glass base disc and allowed to bond for 1 min. Finally, the glass / PDMS disc was mounted to a PMMA base. To prime the microchannels and structures, the disc was placed under vacuum for at least 1 hour, following which a large drop of priming buffer (phosphate-buffered saline [PBS] pH 7.4, 0.1% w/v bovine serum albumin [BSA], 1 mM EDTA) was immediately placed on the surface of the PDMS, covering the sample ports of the loading chamber. Degas-driven flow then primed the channels.

BLOOD PROCESSING AND CELL CULTURE
Blood was extracted directly from healthy donors *via* intravenous extraction. To prevent coagulation samples were drawn into tubes containing EDTA. Blood was isolated and prepared fresh, directly before experimental use.

HL60, colo794 and sk-mel28 cells (DSMZ, Braunschweig, Germany) were cultured in 75 cm^2 flasks in RPMI 1640 media, with 10% un-inactivated foetal bovine serum, 100 U mL^{-1} penicillin, and 100 μg mL^{-1} streptomycin. Cultures were maintained at 37 °C with 5% CO_2. Where indicated, cells were fluorescently stained with NucBlue Live Cell Stain (Life Technologies) according to the instructions of the manufacturer.

Experimental samples were prepared by spiking 1000 cellular events into 1 ml of whole blood. Un-spiked whole blood was used as a control sample. The RBCs were lysed by adding 1 ml whole blood to 9 ml of hypotonic lysis buffer (BD Biosciences, Franklin Lakes, USA) and incubated for 15 minutes. WBCs were then isolated by centrifugation at 200 X g for 5 minutes and resuspension in 1 ml of priming buffer. CD45 (80 μl) and CD15 (16 μl) Dynabeads (Life Technologies) were added and incubation was carried out in a 2 mL Eppendorf tube with rotation for 10 minutes. Negative isolation was performed by slowly bringing a permanent NdFeB magnet into proximity to the incubation tube while rocking the tube horizontally. Cells expressing CD15 or CD45 were immobilized to the side of the tube and the remaining cellular events were placed into a new tube and centrifuged at 500 x g for 5 minutes to pellet. The final pellet was resuspended in 40 μl and 5 μl of this sample was loaded to each of the chambers in the disc. The disc was spun counter-clockwise at 10 Hz for 30 minutes. The resolution of material to the collection bins was observed by both bright-field and fluorescent microscopy.

RESULTS
To test the ability of the size exclusion rail to show the range of cluster sizes inherent to a spiked cell line in blood, we first showed that the cell lines had a propensity to cluster; and to measure the range and extent of the clustering. All three cell lines (HL60, colo794, and sk-mel28) were observed under light microscopy and each event observed was scored according to the number of cells within the event (Fig. 3).

In Figure 3, HL60 cells occur primarily as single celled events, with little or no propensity to form clusters. A low level of clustering was observed for colo794 cells, where 78% of events existed as single cells but clusters composed of up to 4 cells were observed. Finally, sk-mel28 cells showed a high range of cluster sizes from one to more than eight cells, with less than 22% contribution of single celled events.

We then spiked cancer cells into whole blood, processed as described, and recorded the distribution of cellular events in the collection bins using fluorescence microscopy (Fig. 4). The occupancy distribution of the bins closely correlated with the range of cluster sizes intrinsic to the specific cell line. For example, HL60 cells localized almost exclusively to the first bin, while sk-mel28 events were distributed across the eight bins (Fig. 3), correlating with the tendency of these populations to form events ranging from 1 to more than 8 cells (Figs. 4 and 5). Loss of larger clusters through steric restriction was minimal, underpinning an advantage over flow-based filtration methods which may be clogged by larger clusters. Also of note, HL60 and sk-mel28 cells are known to be EpCAM+ and EpCAM-, respectively. Positive isolation systems that target EpCAM as the primary marker of CTCs would hence have failed to identify and enrich the sk-mel28 cells. The negative isolation protocol used in the presented strategy had no difficulty isolating all spiked cell lines regardless of their phenotype.

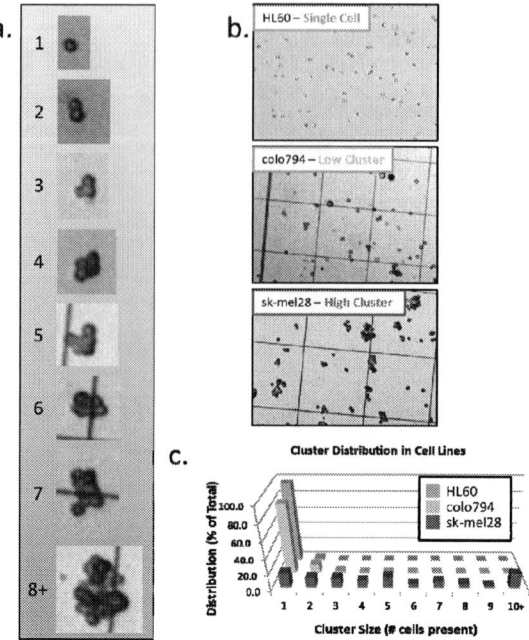

Figure 3: Cell cluster distribution. a) Examples of cell clusters of indicated sizes. b) Images of cell lines that present as single, low clustering, and high clustering configurations. c) Distribution of clusters in all three lines is shown in the 3D histogram.

Normal Blood

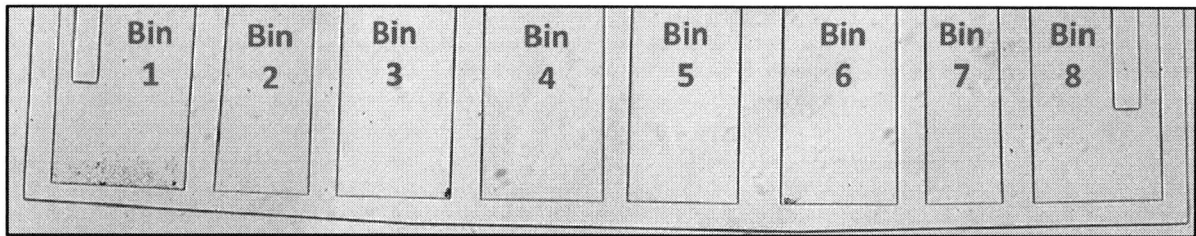

Spiked Blood (sk-mel28)

Figure 4: Image of size bins following processing of normal blood and blood spiked with high clustering sk-mel28 cells. In normal blood, only the first bin is occupied with debris and platelets. Spiked blood occupies bins according to the distribution of cluster sizes inherent to the spiked cell line. Cells have been stained using the NucBlue nuclear fluorescent stain for visual clarity.

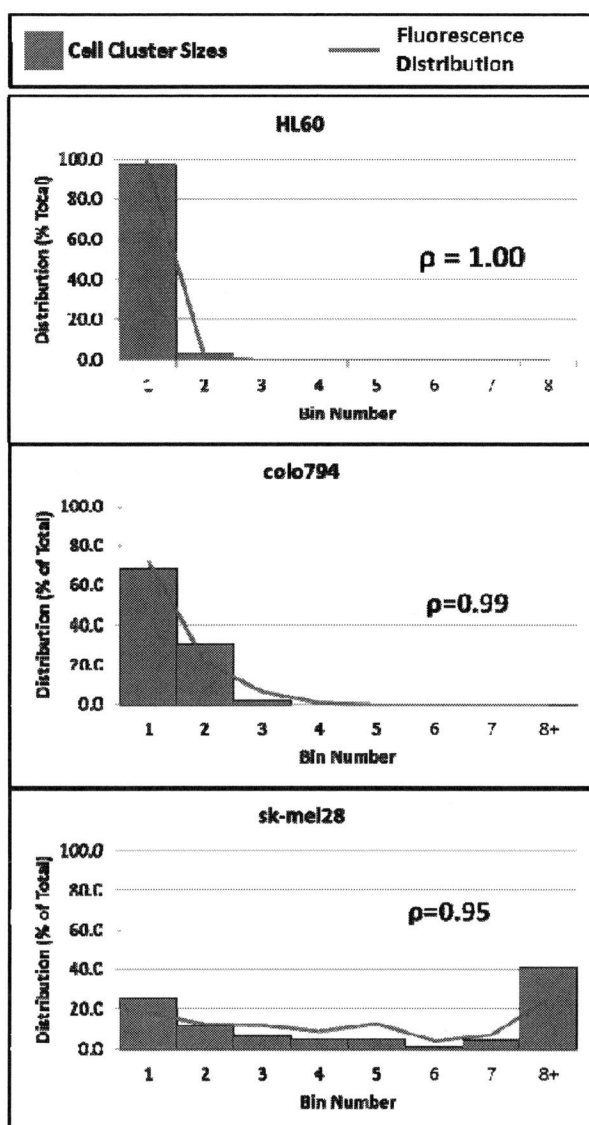

Figure 5: Data analysis. *Correlation analysis of cell clustering (bar charts, data same as the bar chart data in figure 3c) and fluorescent signal distribution (red line graph) in the bins after sample processing. Correlation values between expected cell cluster sizes and measured distribution across the bins are shown.*

CONCLUSIONS

In conclusion, we have quantified the differential propensities of cancer cell lines for clustering by a novel 2-stage, continuous centrifugal sedimentation strategy in stopped-flow mode. The chip-based results were verified against microscope-based read-out. Current work is directed towards on-disc integration of the negative isolation purification protocol on the upstream side in a sample-to-answer point-of-care device.

ACKNOWLEDGEMENTS

This work was supported by Enterprise Ireland under grant No. CF 2011 1317, the ERDF and the Science Foundation Ireland under grant No 10/CE/B1821.

REFERENCES

[1] Jian-Mei Hou, Matthew G. Krebs, Lee Lancashire, Robert Sloane, Alison Backen, Rajeeb K. Swain, Lynsey J.C. Priest, Alastair Greystoke, Cong Zhou, Karen Morris, Tim Ward, Fiona H. Blackhall and Caroline Dive. Clinical Significance and Molecular Characteristics of Circulating Tumor Cells and Circulating Tumor Microemboli in Patients With Small-Cell Lung Cancer *Journal of Clinical Oncology*, 2012, 30, 525-532

[2] Daniel Kirby, Macdara Glynn, Gregor Kijanka, and Jens Ducrée. Rapid and cost-efficient enumeration of rare cancer cells from whole blood by low-loss centrifugo-magnetophoretic purification under stopped-flow conditions. *Cytometry Part A*, 2014, DOI:10.1002/cyto.a.22588. status: in print.

[3] Fredrik I. Thege, Timothy B. Lannin, Trisha N. Saha, Shannon Tsai, Michael L. Kochman, Michael A. Hollingsworth, Andrew D. Rhim, and Brian J. Kirby. Microfluidic immunocapture of circulating pancreatic cells using parallel EpCAM and MUC1 capture: characterization, optimization and downstream analysis. *Lab on a Chip*, 2014, 14, 1775-1784

[4] Min-Ji Kim, Na Young Choi, Eun Kyung Lee, Myung-Soo Kang, Identification of novel markers that outperform EpCAM in quantifying circulating tumor cells. *Cellular Oncology*, 2014 ePub, DOI 10.1007/s13402-014-0178-4

[5] Oscar Lara, Xiaodong Tong, Maciej Zborowski, Jeffrey J. Chalmers. Enrichment of rare cancer cells through depletion of normal cells using density and flow-through, immunomagnetic cell separation. *Experimental Hematology*, 2004, 32, 891-904

[6] Daniel Kirby, Jonathan Siegrist, Laëtitia Zavattoni, Robert Burger, and Jens Ducrée. Centrifugo-magnetophoretic particle separation. *Microfluidics & Nanofluidics*, 13(6):899–908, 2012

[7] Macdara Glynn, Daniel Kirby, Danielle Chung, David J. Kinahan, Gregor Kijanka, and Jens Ducrée. Centrifugo-magnetophoretic purification of CD4+ cells from whole blood towards future HIV/AIDS point-of-care applications. *Journal of Laboratory Automation*, 19(3):285–296, 2014

CONTACT

1. Macdara Glynn, tel: +353-1-700-7899; macdara.glynn@dcu.ie
2. Jens Ducrée, tel: +353-1-700-5377; jens.ducree@dcu.ie

TUNABLE DIELECTROPHORESIS FOR SHEATHLESS 3D FOCUSING

Yu-Chun Kung[1], Daniel L. Clemens[2], Bai-Yu Lee[2] and Pei-Yu Chiou[1]

[1]University of California, Los Angeles, Department of Mechanical and Aerospace Engineering, USA
[2]University of California, Los Angeles, Department of Medicine, USA

ABSTRACT

We report on a novel tunable insulator-based dielectrophoresis (TiDEP) for three-dimensional, sheathless, single-stream cell and bacteria focusing. For the first time, objects as small as sub-micron sized bacteria are continuously focused in the center of a channel without sheath flows. TiDEP is realized by sandwiching a 3D microfluidic channel with high aspect ratio sidewalls between two conductive, featureless indium tin oxide (ITO) glass substrates. The tunable electric field pattern is realized by deforming the channel's sidewalls to change its cross-sectional geometry. Compared to prior DEP works, this new TiDEP can provide a long DEP interaction distance to focus small objects into a single stream in the center of a channel in high-speed flows.

INTRODUCTION

The ability to manipulate cells is crucial to many biomedical applications, including isolation and detection of rare cells, concentration of cells from dilute suspensions, trapping, and positioning of individual cells for characterization. Commonly used methods include dielectrophoresis [1], mechanical [2], magnetophoresis [3], acoustics [4], optical [5], and hydrodynamics [6]. Among these methods, dielectrophoresis (DEP) has been widely used for manipulating cells, DNA molecules, virus, and colloid particles. In convention, DEP manipulation is achieved by creating non-uniform electric fields using patterned electrodes. With the advancement of microfabrication technologies, different patterns of microelectrodes can be created to satisfy the needs of versatile DEP applications [7]. In addition to patterned electrodes, insulating structures (obstacles) have also been demonstrated to be able to produce non-uniform electric fields for DEP manipulation, termed as insulator-based DEP (iDEP) [8]. In iDEP, insulating constriction was employed to squeeze the electric field in a conductive medium, hence producing a local maximum of high-electric-field gradient. The structures of iDEP devices are mechanically robust and chemically inert, and the gas evolution caused by electrolysis at metallic DEP electrodes can be avoided since there is no need for metal electrodes. However, a high driving voltage is required for iDEP devices, and the current throughput is low.

Most importantly, none of the demonstrated DEP platforms can achieve sheathless sample focusing in high speed flows, especially for small objects. This is an important function for lab-on-a-chip systems, especially for applications in flow cytometry and cell sorting. Prior works on inertial focusing, acoustic focusing, and electrokinetic focusing have been demonstrated. However, they cannot provide effective focusing of small objects due to rapidly decreasing forces as particle size shrinks. DEP forces have been shown to be able to trap submicron-sized particles. However, prior electrode or insulator based DEP devices can only provide trapping under low flow speed (100's μm/s).

Here, we demonstrate a novel TiDEP device that can achieve tunable 3D cell focusing in the center of a channel without sheath flows. TiDEP aims to provide the first microfluidic device capable of providing effective continuous 3D sheathless focusing of submicron objects such as bacteria, as well as micron-sized mammalian cells, in high speed flows.

TUNABLE INSULATOR-BASED DIELECTROPHORESIS (TIDEP)

Dielectrophoresis (DEP) is a phenomenon in which a particle in a non-uniform electric field can experience an electrostatic force moving the particle towards stronger electric field regions if the particle is more polarizable than the medium, which is called positive DEP. A particle moves to weaker electric field regions if the particle is less polarizable than the medium, which is called negative DEP. To migrate a particle in an electric field using dielectrophoresis, there must be electric field gradient. In a uniform electric field region, although particles are polarized, no net DEP forces can be induced to move particles.

Figure 1(a,b) shows the schematic of a TiDEP device. A single layer PDMS thin film with high aspect ratio (HAR 4) walls is bonded between two featureless ITO glass substrates. Figure 1(c) shows the picture of a completed microfluidic device. By applying pressure to the two side channels, the middle channel can be deformed as shown in Figure 1(b). Application of an electrical bias to the two featureless ITO electrodes, electric field streamlines are focused in the middle neck of the deformed channel. This creates highly non-uniform electric field with a maximum field enhancement occurring in the center (Figure 1(d)).

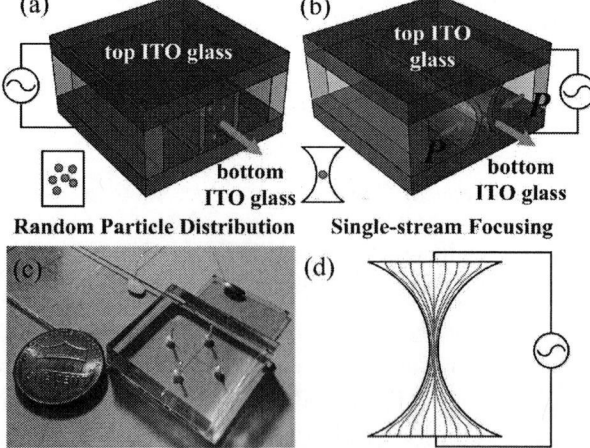

Figure 1: Schematic of a tunable insulator-based DEP (TiDEP) focusing device. (a) An open PDMS microchannel with thin, high-aspect-ratio sidewalls is sandwiched between two conductive indium tin oxide (ITO) coated

glass substrates. An a.c. signal is applied between the top and bottom substrate. (b) When the channel sidewalls are pressurized, the channel deforms to form a narrow neck in the middle and focus the electric field lines. Positive DEP responding particles would migrate to the center of the channel where the maximum electric field occurs. (c) Photograph of a fabricated device. (d) A schematic showing crowding electric field streamlines in the neck of a deformed channel.

DEVICE FABRICATION

Figure 2 shows the schematic of the microfabrication process flow of TiDEP devices, a process previously demonstrated in [9].

Step1: *Fabrication of master molds.* SU-8 mold masters on silicon wafers is fabricated using photolithography (Figure 2(a)). All masters need to be surface treated with trichloro (1H,1H,2H,2H-perfluorooctyl) silane (97%, Sigma-Aldrich, USA), also called PFOCTS, to facilitate later demolding.

Step2: Fabrication of hybrid stamps. It starts from preparing the Sylgard 184 silicone elastomer mixture (Dow Corning Corporation, Miland, USA) The weight ratio of Base : Curing agent is 10 : 1. Few drops of this mixture are poured into a petri dish. A suitable size of polystyrene plastic plate is cut and pressed against the bottom of the petri dish under a pressure of 3 psi. A thin layer of polydimethylsiloxane (PDMS) with a thickness of roughly 30μm is formed between the petri dish and the plastic plate. Additional uncured PDMS is poured to fill up the petri dish, and followed by a curing step at 60^0C in an oven for 12 hours. A hybrid stamp is formed when the plastic plate together with a thin PDMS layer on its surface is peeled off from the petri dish (Figure 2(b)). The hybrid stamp is also surface treated with PFOCTS as in Step 1 for 6 hours. To fabricate PDMS thin film with through-layer structures, uncured PDMS is poured onto the master mold, pressed by the hybrid stamp under a pressure of 4psi, and cured at 50^0C in an oven for an hour.

Step 3: Demolding PDMS films from master mold. During the demolding process, the cured PDMS thin film has stronger adhesion to the hybrid stamp rather than the master mold since more PFOCTS is coated on the master mold due to a longer treatment time (Figure 2 (c)).

Step 4: Transfer the PDMS thin film. Oxygen plasma treatment is performed on both the PDMS thin film on the hybrid stamp and the substrate to be bonded. No alignment is needed. (Figure 2(d)). The bonded set is baked in an oven at 60^0C for 2 hours.

Step 5: Removing hybrid stamp. It starts from peeling off the bulk PDMS part on the plastic plate (Figure 2(e)), and followed by dissolving the polystyrene plastic plate in an acetone bath for 4 hours (Figure 2(f)). This leaves a thin residual PDMS film on the substrate that can be easily peeled off from the device due to prior PFOCTS treatment (Figure 2(g)) to finish the transferring. This mechanically gentle releasing technique allows us to transfer PDMS thin

film with fragile substrates, such as a high aspect ratio vertical wall.

Step 6: Cover the device with a top ITO glass coverslip with oxygen plasma bonding to finish the device fabrication. (Figure 2(h)).

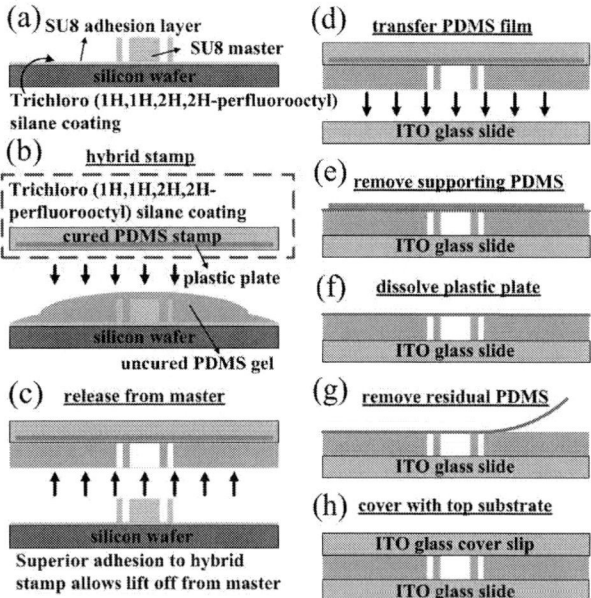

Figure 2: Schematic of fabrication process flow using a plastic plate embedded hybrid stamp. (a) A SU8 master is treated with PFOCTS to facilitate later demolding. (b) Uncured PDMS mixture is poured on the master, and pressed against the hybrid stamp. (c) Due to less PFOCTS treatment on the hybrid stamp compared to the master, the casted PDMS film tends to adhere to the hybrid stamp and allows to be peeled off from the master. (d) Film transferred and bonding is achieved through oxygen plasma treatment. (e) Remove the support PDMS on the hybrid stamp. (f) Dissolve the polystyrene plastic plate in acetone. (g) Remove the residual PDMS thin film to complete the removal of a hybrid stamp. (h) Cover the device with an ITO glass coverslip to complete the fabrication process by oxygen plasma bonding.

Figure 3(c)(d) compares the simulated electric field distribution in a channel with and without deformation, respectively. Since the sidewalls can be continuously deformed to eventually contact each other and seal the gap, this device allows the tuning of electric field strength in the middle of the channel by controlling the applied pressure in the side channels. The narrower the neck, the larger the electric field enhancement is generated in the neck. This creates a strong electric field gradient for focusing small objects such as bacteria. Cells or particles experiencing positive DEP forces in the channel are attracted to the center of the channel, providing a continuous 3D sheathless cell focusing function for potential applications in microfluidic flow cytometers and cell sorters.

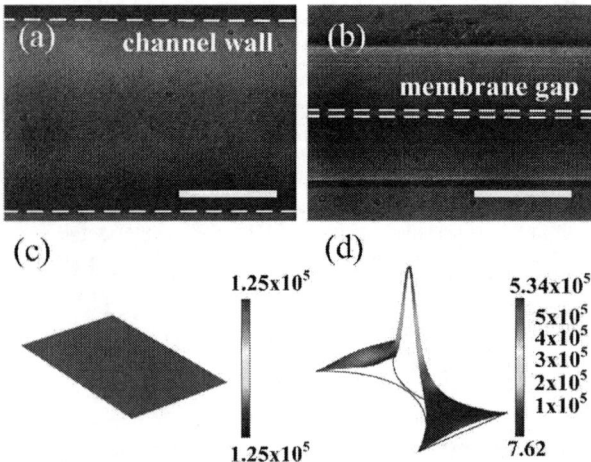

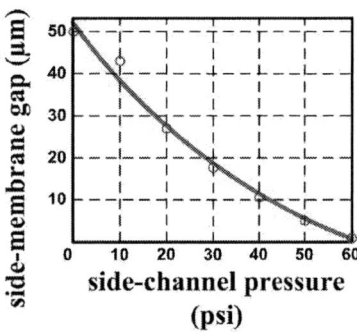

Figure 3: (a) Microscopy images of a deformable microchannel with a dimension of 50μm in width, 80μm in height, and a sidewall thickness of 20 μm. (b) The channel sidewalls are under of a pressure of 60psi. (c) Electric field distribution in a microchannel without deformation under an ac bias of 10V peak-to-peak at a frequency of 5MHz. (d) When the channel is deformed to form a neck of only 0.5μm wide, strong field enhancement is generated in the middle of this channel. Scale bar: 50μm

Based on the size of particles been focused, the gap spacing between sidewalls can be tuned accordingly by changing the applied pressure in the side channels. The measured gap spacing between continuously deformed sidewalls under different pressure is shown in Figure 4.

Figure 4: Measured deformed gap spacing under different pressure.

EXPERIMENTAL RESULTS

The performance of 3D DEP focusing on this deformable channel is tested using three types of cells, HeLa-GFP with a diameter of 15μm, GFP-E. Coli that is 2 μm long and 0.7 μm in diameter, and submicron GFP-Francisella Live Vaccine Strain (LVS) bacteria that is 1 μm long and 0.25 μm in diameter. Before pumping the sample into the device, Hela-GFP cells was suspended in an isotonic buffer consisting of 8.5% sucrose and 0.3% dextrose and with a conductivity of 30 mS/m. GFP-E. Coli and GFP-LVS were suspended in a buffer solution of 10% sucrose in 20 mM HEPES and with a conductivity of 50 mS/m. The sample was injected continuously into the microchannel by a syringe pump (kdScientific, 780100). A function generator (Agilent, 33220A) and a power amplifier (ENI, Model 2100L) was used to provide the a.c. voltage source. The DEP response of the cells were monitored under an inverted fluorescence microscope (Carl Zeiss, Axio Observer.A1), and recorded by a CCD camera (Carl Zeiss, AxioCam MRm). For the HeLa-GFP focusing, side channel was pressured at 30psi to create a 18μm wide gap, slightly bigger than the size of HeLa cells. When an a.c. signal (300kHz, 56.610V peak-to-peak) was applied, HeLa-GFP cells are focused to the center of the channel (Figure 5(b)), and remained focused in the downstream even at regions without deformed channels due to the laminar flow nature in microfluidics. To focus E. Coli and GFP-LVS, the sidewalls of the channel is further deformed to a smaller gap of only 2μm and 1μm accordingly to create stronger electric field strength and gradient for focusing such smaller objects. When there was no a.c. signal applied, GFP-E. Coli and GFP-LVS were randomly distributed in the channel (Figure 5(c)(e)). Under the application of an a.c. signal (5MHz, 56.610V peak-to-peak), positive DEP forces focus GFP-E. Coli and GFP-LVS to the center of the channel to form a single stream fluorescence trace as shown in Figure 5(d)(f).

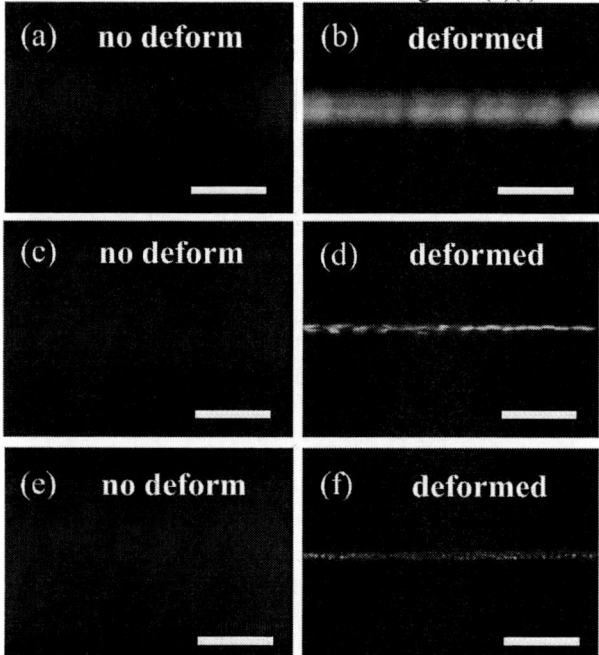

Figure 5: The focused fluorescence traces of different sizes of biological objects. All samples are suspended in low conductivity medium ranging from 0.03 S/m to 0.05 S/m. An a.c. signal of 56.610V peak-to-peak with frequencies varying from 300 kHz to 5 MHz is applied. (a,b) show the focused stream of GFP-Hela (spherical shape, 15μm in diameter) at an average flow speed of 1 cm/sec with a gap spacing of 16 μm; (b,c) show that GFP- Escherichia coli (rod-shape, 2μm long and 500nm wide) are focused at an average speed of 5 mm/sec with a gap spacing of 3 μm; and (e,f) show that GFP-Francisella Live Vaccine Strain bacteria (rod-shape, 1μm long and 250nm wide) are focused at an average speed of 1 mm/sec with a gap spacing of 1 μm. Scale bar: 25μm

978-1-4799-7956-1/15 $31.00 © 2015 IEEE

ACKNOWLEDGEMENTS

This work is supported by NSF DBI 1256178, NSF ECCS1232279, and Cal. Cap. LLC through an industry sponsored research agreement

REFERENCES

[1] P.-Y. Chiou, A. T. Ohta, and M.-C. Wu, "Massively parallel manipulation of single cells and microparticles using optical images," *Nature,* vol. 436, pp. 370 - 372, 2005.

[2] D. D. Carlo, D. Irimia, R. G. Tompkins, and M. Toner, "Continuous Inertial Focusing, Ordering and Separation of Particles in Microchannels," *Proceedings of the National Academy of Sciences of the United States of America,* vol. 104, pp. 18892 - 18897, 2007.

[3] N. M. Karabacak, P. S. Spuhler, F. Fachin, E. J. Lim, V. Pai, E. Ozkumur, *et al.*, "Microfluidic, marker-free isolation of circulating tumor cells from blood samples," *Nature Protocols,* vol. 9, pp. 694-710, 2014.

[4] X. Dinga, Z. Peng, S.-C. S. Lin, M. Gerid, S. Lie, P. Lia, *et al.*, "Cell separation using tilted-angle standing surface acoustic waves," *Proceedings of the National Academy of Sciences of the United States of America,* vol. 111, pp. 12992-12997, 2014.

[5] M. M. Wang, E. Tu, D. E. Raymond, J. M. Yang, H. Zhang, N. Hagen, *et al.*, "Microfluidic sorting of mammalian cells by optical force switching," *Nature Biotechnology,* vol. 23, pp. 83-87, 2005.

[6] M. Yoshida, K. Tohda, and M. Gratzl, "Hydrodynamic micromanipulation of individual cells into," *Analytical Chemistry,* vol. 75, pp. 4686-4690, 2003.

[7] J. Voldman, "Electrical forces for microscale cell manipulation," *Annual Review of Biomedical Engineering,* vol. 8, pp. 425-454, 2006.

[8] C.-F. Chou, J. O. Tegenfeldt, O. Bakajin, S. S. Chan, E. C. Cox, N. Darnton, *et al.*, "Electrodeless dielectrophoresis of single- and double-stranded DNA," *Biophysical Journal,* vol. 83, pp. 2170-2179, 2002.

[9] Y.-C. Kung, K.-W. Huang, Y. Yang, Y.-J. Fan, and P.-Y. Chiou, "Fabrication of 3D Microfluidic Networks with a Hybrid Stamp," in *IEEE 26th International Conference on Micro Electro Mechanical Systems (MEMS),* Taipei, Taiwan, 2013, pp. 915-918.

CONTACT

*Y.-C. Kung, tel: +1-310-825-7457;
yuchunkung@ucla.edu

CELL MANIPULATION METHOD BASED ON VIBRATION-INDUCED LOCAL FLOW CONTROL IN OPEN CHIP ENVIRONMENT

Takeshi Hayakawa, Shinya Sakuma, and Fumihito Arai

Department of Micro-Nano Systems Engineering, Nagoya University, Nagoya, JAPAN

ABSTRACT

We present a novel cell manipulation method using vibration-induced local flow in open chip environment. By applying circular vibration to micropillars on a chip, local whirling flow is induced around the micropillars. From the observation of this unique phenomenon, we propose the concept of cell manipulation in open chip environment. We analyze this phenomenon theoretically, and evaluate the effect of the frequency and the amplitude of applied vibration. We design the micropillar array according to the analysis for transportation for oocytes. We apply the proposed method to transportation of mouse oocytes and confirm that the velocity of transportation is approximately 25 μm/s.

INTRODUCTION

Recent progress in microfabrication and micromanipulation technologies enables cell manipulation, and it plays important roles in the biotechnology. One of the most powerful candidates for future cell manipulation technique is on-chip cell manipulation. It makes possible to high-accuracy and high-throughput cell manipulation with low initial costs.

Environment on a microfluidic chip for cell manipulation can be categorized in two types: (i) close chip environment and (ii) open chip environment. The close chip environment is constructed by substrate and cover, such as glass substrate and PDMS cover. It enables high-throughput cell manipulation by using flow control in a closed microchannel by an external pump [1]. However, the closed structure must have interface to external environment, such as tube connection part. Target cells are often lost at these interface part and it is difficult to apply the close chips for rare cells such as oocytes. Therefore, the interface part of close chip to external environment is serious problems for rare cell application.

On the other hands, the open chip environment is constructed without any cover and can be easily accessed from external environment. Since this feature is suitable for rare cell applications, flow control for cell manipulation on the open chip environment without pump control is highly required. Conventionally, cell manipulation methods such as micromanipulator [2,3], optical tweezers [4], dielectrophoresis [5,6], or surface acoustic wave [7,8] are used on the open chips. However, micromanipulator requires complex system settings and operator skills. Manipulation force of optical tweezers are very small ($\approx$ pN) and difficult to apply to large cells such as oocytes ($\approx$ 100 μm). To use dielectrophoresis or surface acoustic wave, we have to fabricate multi-electrodes or piezoelectric actuators on a chip and it leads to complex chip structure and control system.

In this study, we propose a novel cell manipulation method on the open chip environment based on vibration-induced flow. This method enables cell manipulation on open chip environment with simple chip structure and system setup.

CONCEPT

Flow inductions by vibrating objects are well known phenomenon and it can be observed even in microscale environment [9]. Therefore, applications of this phenomenon for micromanipulations on a chip have been attracting attentions. In our previous research, we proposed flow induction by using magnetically driven on-chip microrobot and application to cell manipulation [10]. However, by using the on-chip robot, manipulation area is restricted to working area of the robot and difficult to apply manipulations in large area. In this study, we utilize this phenomenon for cell manipulations by different approach.

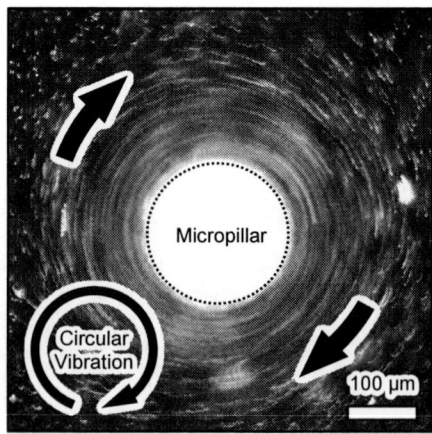

Figure 1: Microscopic image of vibration-induced flow around a micropillar fabricated on a chip

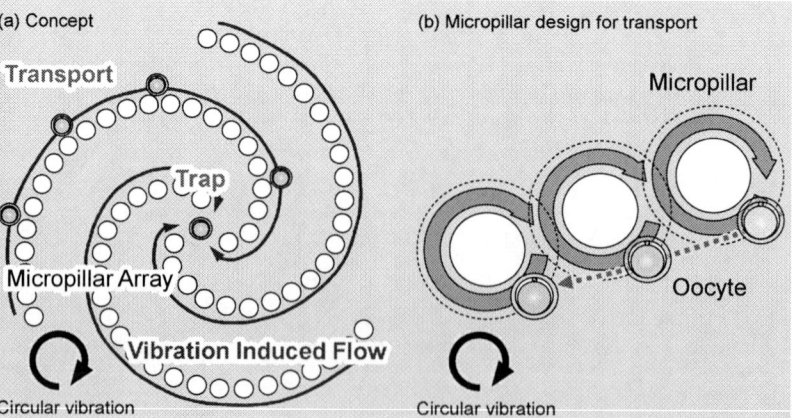

Figure 2: Concept of oocyte transportation manipulation by vibration-induced flow: (a) conceptual image of oocyte transportation, (b) micropillar array design for oocyte transportation

We confirm that the similar flow induction can be achieved when applying a vibration to microchip that has micropillar pattern as shown in Fig. 1. By applying circular vibration to the chip, local whirling flow can be induced around ($\approx$ 10-100 μm) a micropillar even in an open chip environment. Furthermore, the direction and velocity of the flow can be changed by changing direction, amplitude and frequency of applied vibration. By considering these features of the phenomenon, we can realize cell manipulation on the open chip environment by using micropillar patterns and circular vibrations. We previously applied this phenomenon to transportation of small somatic cells ($\approx$20 μm) [11].

As a new application of this manipulation method, we propose an oocyte transportation and trapping as shown in Fig. 2(a). The transportation is achieved by fabricating micropillar array and inducing flow along the array by applying vibration to the chip, as shown in Fig. 2(b). By patterning spiral array, we can transport target oocyte from all area of the patterning to the center of the spiral. At the center of the spiral, we can trap the transported oocytes by patterning three micropillars as a triangle shape. When we apply circular vibration to the chip, the vibration-induced flow around each micropillars are interfered at the center of the triangle, and trapping flow can be induced. Therefore, we can achieve oocyte transportation and trap by using micropillar array and vibration induced flow.

THEORETICAL ANALYSIS

To design the micropillar array for oocyte manipulation, we firstly analyze this phenomenon theoretically. Details of analytical calculations are explained in reference [11] and we omit the details in this paper. The acquired stream function $\psi_{st}^{(1)}$ with 1st order approximation which is written as following Eq. (1) through (3).

$$
\psi_{st}^{(1)} = \pm \left[r^4 \left(\frac{1}{48} \int_a^r \frac{1}{x} \rho(x)dx + c_1 \right) \right.
$$
$$
+ r^2 \left(-\frac{1}{16} \int_a^r x\rho(x)dx + c_2 \right)
$$
$$
+ \left(\frac{1}{16} \int_a^r x^3 \rho(x)dx + c_3 \right)
$$
$$
\left. + \frac{1}{r^2} \left(-\frac{1}{48} \int_a^r x^5 \rho(x)dx + c_4 \right) \right]
$$
(1)

$$
\rho(r) = \frac{2\pi^3 f^3 A_0^2}{\eta^2} \left[2X + 2X^* - 2\frac{a^2}{r^2}CZ^* - 2\frac{a^2}{r^2}C^*Z \right.
$$
$$
\left. -4XX^* + 4ZZ^* \right]
$$
(2)

$$
C_1 = -\frac{1}{48}\int_a^\infty \frac{1}{x}\rho(x)dx, \quad c_2 = \int_a^\infty \frac{1}{16}x\rho(x)dx,
$$
$$
C_3 = \frac{a^4}{16}\int_a^\infty \frac{1}{x}\rho(x)dx - \frac{a^2}{8}\int_a^\infty x\rho(x)dx,
$$
(3)
$$
c_4 = -\frac{a^6}{24}\int_a^\infty \frac{1}{x}\rho(x)dx + \frac{a^4}{16}\int_a^\infty x\rho(x)dx
$$

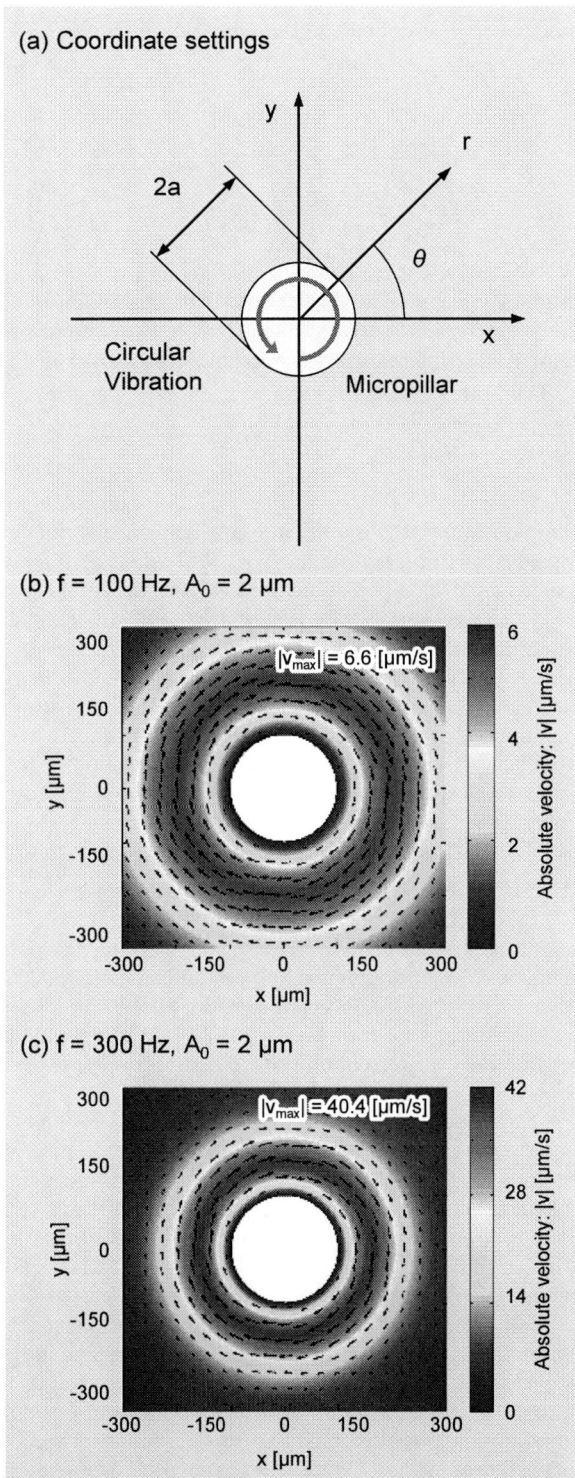

(a) Coordinate settings

(b) f = 100 Hz, A_0 = 2 μm

(c) f = 300 Hz, A_0 = 2 μm

Figure 3: Theoretical analysis of vibration-induced flow: (a) Coordinate settings of analysis model, numerical calculation results of 2D velocity distribution under the conditions of (b) f = 300 Hz, A_0 = 1μm, (c) f = 300 Hz, A_0 = 2μm

Where, the coordinate settings are indicated in Fig. 3(a), f and A_0 are frequency and amplitude of applied vibration, respectively. X, Z and C are defined by 1st kind Hankel function $H_n^{(1)}$ as defined in following Eq. (4) and (5).

$$X = \frac{H_0^{(1)}(\epsilon r)}{H_0^{(1)}(\epsilon a)}, Y = \frac{H_1^{(1)}(\epsilon r)}{H_0^{(1)}(\epsilon a)}, Z = \frac{H_2^{(1)}(\epsilon r)}{H_0^{(1)}(\epsilon a)} \qquad (4)$$

$$C = \frac{H_0^{(1)}(\epsilon a)}{H_0^{(1)}(\epsilon a)}, \epsilon = \left(i\frac{\omega}{\eta}\right)^{1/2} \qquad (5)$$

Where, ϵ is a scale parameter having a dimension of length. By using relationship between stream function and flow velocity in Eq. (6), we can calculate two dimensional velocity distribution around micropillar as shown in Fig. 3(b) and (c).

$$v_r = -\frac{1}{r}\frac{\partial \psi}{\partial \theta}, v_\theta = \frac{\partial \psi}{\partial r}, \qquad (6)$$

From Eq. (1) and (2), it is expected that vibration-induced flow is strongly dependent on the frequency of applied vibration f. We calculate the two dimensional velocity distribution as shown in Fig. 3(b) and (c) when we set the frequency $f = 100, 300$ Hz, and amplitude $A_0 = 2$ μm. Here, please notice that the color scale is different in each figures because absolute values of the velocity are quite different. From Fig. 3, we can see the maximum value of flow velocity is widely changed when we change the frequency of applied vibration. These results indicate that the flow velocity can be changed by changing the frequency of applied vibration.

DESIGN OF MICROPILLAR ARRAY

Design factors of the micropillar array are micropillar height, radius and patterns. Firstly, we determine the pillar height as 200 μm so that the micropillar can cover oocytes whose size is ranged in approximately 80-120 μm. Then, we determine the micropillar radius as 100 μm by considering a resolution of patterning process by photoresist.

For transportation of oocytes, pitch of micropillars is determined as 50 μm so that the target oocytes do not pass through between the pillars. And also, the channel width between each spiral pattern is determined as 240 μm so that the target oocyte with maximum size ($\approx$ 120 μm) is not affected by vibration-induced flow around next spiral pattern.

For trapping of oocytes, we determine the distance of micropillars patterned in triangle geometry. The target oocyte with maximum size ($\approx$ 120 μm) should be introduced at the center of the triangle without affected by vibration-induced flow around next micropillar. The flow induced area width ($\delta \approx 50\sim60$ μm) can be estimated by previous numerical plots. Therefore we set the pillar distance as 180 μm.

EXPERIMENTS

We construct proposed cell manipulation system on an inverted microscope (Nikon Corporation, Eclipse Ti). Images of target oocytes is acquired by CCD camera (Point Grey Research KK, Flea 3) attached to the microscope.

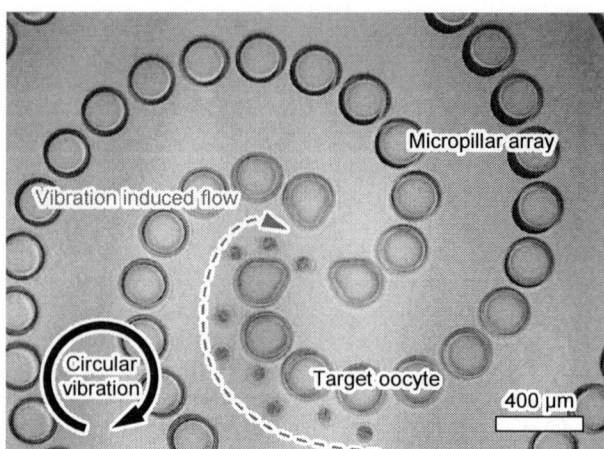

Figure 4: Demonstration of oocyte transportation

Micropillar array on the microfluidic chip is fabricated by SU-8 (Nippon Kayaku Co. Ltd., SU-8 3050) on glass substrates by standard photolithography process. Circular vibration is applied to the chip by piezoelectric actuators (Nihon Ceratec Co. Ltd., PAC166J). Output signal of DA board (Interface, PCI-3340) on a control PC is applied to the piezoelectric actuators after amplified by an amplifier ((TOYO Corporation, 9400, gain: x50).

RESULTS AND DISCUSSION

Successive photograph of oocyte transportation is shown in Fig. 4. We succeed in transportation of oocytes along fabricated micropillar array as shown in concept Fig. 2(a). We evaluate the transportation velocity by image analysis and the velocity is 25 ± 2.5 μm/s under the condition of frequency $f = 360$ Hz, input voltage is 2V (equivalent to $A_0 = 3.5$ μm). In our current design of micropillar array, diameter of the spiral pattern is 3 mm and maximum length of transportation route is 3.5 mm. Thus, we can transport target oocytes to the center of spiral pattern within approximately 3 min wherever oocytes is dropped on the pattern. Furthermore, Fig. 4 shows that the target oocyte is trapped at the center of the spiral, as proposed by the concept in Fig. 2(a). These results indicate that we can transport target oocytes from outside of the observation area on open-chip, and trap the target oocyte at the center of spiral patterning.

CONCLUSIONS

In this paper, we propose a novel cell manipulation method based on vibration-induced flow in open chip environment. By patterning micropillars on a chip and applying circular vibration to the chip, local whirling flow is induced around micropillars. By utilizing this local flow and patterned micropillar array on a chip, we demonstrate the transportation of oocyte. We succeed in oocyte transportation with 25± 2.5 μm/s velocity. The proposed method enables cell manipulation in open chip environment and provides easy to handle and low costs cell manipulation system.

ACKNOWLEDGMENT

This study was financially supported by Grant-in-Aid for JSPS Fellows Number 13J03580 and Grant-in-Aid for Challenging Exploratory Research Number 26630094.

REFERENCES

[1] S. Sakuma, K. Kuroda, C. H. D. Tsai, W. Fukui, F. Arai, and M. Kaneko, "Red blood cell fatigue evaluation based on the close-encountering point between extensibility and recoverability", *Lab on a Chip*, vol. 14, pp. 1135-1141, 2014.

[2] K.Yanagida, H.Katayose, H.Yazawa, Y.Kimura, K.Konnai, and A.Sato, "The usefulness of a piezo-micromanipulator in intracytoplasmic sperm injection in humans", *Human Reproduction*, vol. 14, pp. 448-453, 1998.

[3] A. Ramadan, K. Inoue, T. Arai and T. Takubo, "New Architecture of a Hybrid Two-Fingered Micro–Nano Manipulator Hand: Optimization and Design", *Advanced Robotics,* vol. 22, pp. 235-260, 2008.

[4] K. Onda and F. Arai, "Multi-beam bilateral teleoperation of holographic optical tweezers", *Optics Express*, vol. 20, pp. 3633-3641, 2012.

[5] J. Voldman, M. L. Gray, M. Toner, and M. A. Schmidt, "A Microfabrication-Based Dynamic Array Cytometer", *Analytical Chemistry*, vol. 72, pp. 3984-3990, 2002.

[6] P. Benhal, J. G. Chase, P. Gaynor, B. Obackc and W. Wang, "AC electric field induced dipole-based on-chip 3D cell rotation", *Lab on a Chip*, vol. 14, pp. 2717-2727, 2014.

[7] J. Shi, D. Ahmed, X. Mao, S. S. Lin, A. Lawita and T. J. Huang, "Acoustic tweezers: patterning cells and microparticles using standing surface acoustic waves (SSAW)", vol. 9, pp. 2890-2895, 2009.

[8] X. Ding, S. S. Lin, B. Kiraly, H. Yue, S. Li, I. Chiang, J. Shi, S. J. Benkovic, and T. J. Huang, "On-chip manipulation of single microparticles, cells, and organisms using surface acoustic waves", *PNAS*, vol. 109, pp.11105-11109, 2012.

[9] J. Holtsmark, I. Johnsen, T. Sikkeland, and S. Skavlem, "Boundary layer flow near a cylindrical obstacle in an oscillating incompressible fluid", *The Journal of the Acoustical Society of America*, vol. 26, pp. 26-39, 1954.

[10] M. Hagiwara, T. Kawahara, and F. arai, "Local streamline generation by mechanical oscillation in a microfluidic chip for noncontact cell manipulations", *Applied Physics Letters*, vol. 101, p. 074102, 2012.

[11] T. Hayakawa, S. Shinya, and F. Arai, "A Single Cell Extraction Chip Using Vibration-Induced Whirling Flow and a Thermo-Responsive Gel Pattern", *Micromachines*, vol. 5, pp. 681-696, 2014.

CONTACT

*T. Hayakawa, tel: +81-52-789-5026;
e-mail:t-hayakawa@biorobotics.mech.nagoya-u.ac.jp

ULTRA SENSITIVE LORENTZ FORCE MEMS MAGNETOMETER WITH PICO-TESLA LIMIT OF DETECTION

Varun Kumar, Mohammad Mahdavi, Xiaobo Guo, Emad Mehdizadeh and Siavash Pourkamali

Electrical Engineering Department, University of Texas at Dallas, Richardson, USA

ABSTRACT

This work presents ultra-high sensitivities for Lorentz Force resonant MEMS magnetometers enabled by internal thermal-piezoresistive vibration amplification. Up to 2400X increase in sensitivity has been demonstrated by tuning the resonator bias current to maximize its internal amplification factor boosting the effective Quality Factor (Q) from its intrinsic value of 680 to 1.14×10^6 (1675X amplification). For a bias current of 7.245mA, where the sensitivity of the device is maximum (2.107mV/nT), the noise floor is measured to be as low as 2.8 pT/√Hz. This is by far the most sensitive MEMS Lorentz force magnetometer demonstrated to date.

INTRODUCTION

Magnetic sensors have found their way into a variety of applications such as Magneto-encephalography [1], magnetic anomaly detection, mineral prospecting [2], munitions fusing, magnetic compass [3], automotive sensors, respiratory measurements [4], and magnetic memory readout. Depending on the magnitude of the magnetic field to be detected, a number of techniques currently exist. Devices based on existing techniques include Hall Effect Sensors (μT-T), anisotropic magnetoresistance sensors (AMR- nT-mT), Optical Fiber [5] & Fluxgate sensors (nT-mT). Search Coils [6] and SQUIDS can detect extremely small fields (down to femto-Tesla) however, Search coils are quite bulky and unable to detect static magnetic fields, and SQUIDS on the other hand require cryogenic cooling and have high sensitivity to electromagnetic interference, thus requiring a sophisticated infrastructure (liquid helium supply, glass-fiber-reinforced epoxy Dewar vessels, and electromagnetic shielding). MEMS Magnetometers have an edge over the abovementioned conventional counterparts due to their small size, low cost, lower power consumption and operation simplicity. Such properties offer unrivalled advantages, especially when it comes to medical applications, such as magneto-encephalography, where compact arrays of ultra-sensitive sensors are desirable. Resonant Lorentz force magnetometers are one of the most common categories of MEMS magnetometers that can be implemented in silicon without the need for any special magnetic materials, and unlike magneto-resistive and fluxgate sensors they are free from hysteresis. Such devices take advantage of high Q microscale resonant structures to turn small Lorentz forces into measurable vibration amplitudes. Resonant Lorentz force magnetometers are typically operated at frequencies in the tens to hundreds of kHz, which helps significantly suppress the low frequency noise.

Limit of detection (LOD) for Lorentz force resonant MEMS magnetometers are typically in the lower μT to higher nT range, which is many orders of magnitude higher than the required LOD for most medical applications (pT to aT). A number of approaches have been reported to increase the sensitivity of such sensors. This includes using novel topologies [7], electronic tunneling [8] and parametric amplification [9]. In [9], the force-to-displacement transduction of a resonant sensor was increased via artificially increasing the resonator quality factor through modulation of the spring constant of the device at twice its natural frequency. Sensitivity was parametrically amplified by 50 folds using this approach to 39nT/√Hz. However, operation of parametrically amplified devices as practical sensors is quite challenging due to sophistication of supporting electronics. In this work, the previously demonstrated internal thermal-piezoresistive amplification [12] within the micromechanical silicon structure is used to reach much larger vibration amplitudes for the same magnetic actuation, consequently achieving much higher sensitivity [10] [11].

Internal Amplification

Internal thermal-piezoresistive amplification in resonant structures comprising of longitudinally stressed beams has been previously demonstrated [10] [11] [12]. As the resonator vibrates, the alternating tensile and compressive stress in the longitudinally stressed beams modulates their electrical resistance due to the piezoresistive effect. Modulation of electrical resistance, while maintaining a constant bias current or voltage across the beam, modulates the ohmic loss and therefore Joule's heating in the beams. This turns the beam to a thermal actuator generating a thermal actuation force component which can either amplify or attenuate the resonator vibration amplitude at resonance frequency depending on the polarity of the structural piezoresistive coefficient and thermal delays in the system. Previous results have proven that in case of constant current biased beams, in order for the extra thermal actuation force component to be in phase with the vibration of the resonator and hence amplify the vibration amplitude, longitudinal piezoresistive coefficient of the structural material should be negative [12]. Thus, in n-type single-crystalline silicon structures (negative piezoresistive coefficient), with adequate DC bias current flowing through an extensional mode beam, large vibration amplitudes can result from minute Lorentz actuation forces due to internal amplification of vibration amplitude. Figure 1 illustrates the interactions between the four physical domains (Magnetic, Mechanical, Electrical and Thermal) involved in the thermal-piezoresistive amplification process. The Lorentz force from an external magnetic field acting on a current carrying part of the resonator leads to an actuation force and therefore structural vibrations that are then internally amplified by the thermal-piezoresistive interactions within the extensional beam embedded in the structure.

978-1-4799-7956-1/15 $31.00 © 2015 IEEE

In effect, the resonator absorbs energy from the DC bias source and uses it to partially compensate the mechanical losses that limit the vibration amplitude at resonance. Increasing the piezoresistor bias current increases the absorbed energy leading to further amplification (higher effective Q) and eventually even self-sustained oscillation of the resonator (absorbed energy larger or equal to mechanical losses). In this work, the above-mentioned principle has been utilized to amplify the displacement resulting from Lorentz actuation force and demonstrate ultra-high sensitivities to magnetic fields for such sensors.

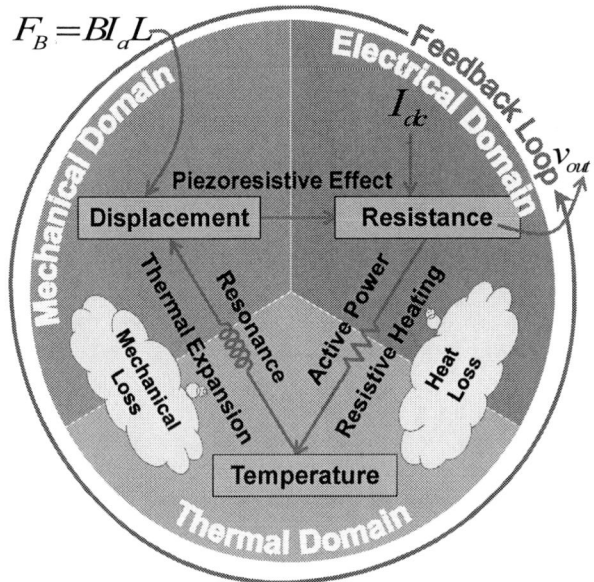

Figure 1: Schematic diagram showing interactions between different domains (Magnetic, Mechanical, Electrical and Thermal) in a Lorentz force magnetometer with thermal-piezoresistive internal amplification.

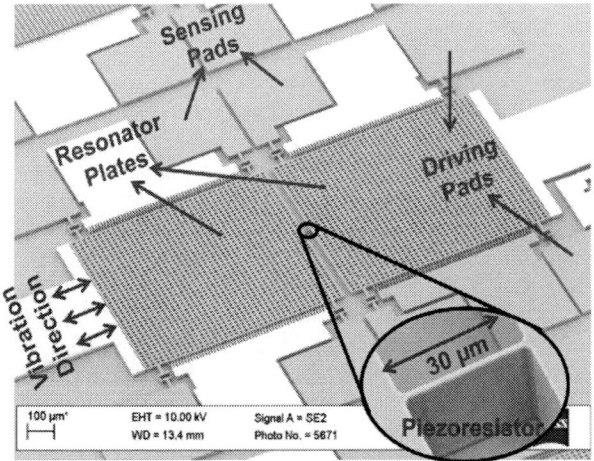

Figure 2: SEM view of the 400 kHz dual plate in-plane resonator. The inset shows a zoomed in view of the amplifying beam (piezoresistor-30 μm ×1.5 μm × 15 μm).

DEVICE DESCRIPTION

The dual plate monocrystalline silicon resonant structure of Figure 2 was fabricated on an SOI substrate (15μm thick device layer) using a single-mask micro-machining process and operated as a Lorentz Force Magnetometer. The resonator body was patterned and defined using deep reactive ion etching (DRIE) of silicon device layer and devices were released by removing the SOI buried oxide layer in hydrofluoric acid (HF). Holes on the resonator plates are provided to facilitate and accelerate undercutting of the large resonant plates.

The 30 μm long, 1.5 μm wide and 15 μm thick beam in the middle of the resonator connecting the two 400 μm × 400 μm resonator plates acts as the amplifying longitudinal beam. When the resonator resonates in its in-plane mode, this piezoresitor acts as a strain gauge that undergoes periodic tensile and compressive stress. The driving pads located on the two sides of the resonator plates are to generate the actuating Lorentz Force. Passing an AC current at the natural frequency of the resonator between the two driving pads results in a Lorentz Force that can actuate the resonator in its in-plane extensional mode. In this resonant mode, the resonator plates move back and forth as shown in Figure 3. The resulting vibration amplitude of the resonator due to the magnetically induced Lorentz Force is amplified by the mechanical Q of the resonator. The alternating stress in the piezoresistor beam leads to fluctuations in its electrical resistance. The sense pads connected to the two resonator plates are biased with a DC bias current which results in the internal amplification as discussed above. The same DC current also results in a modulated output voltage across the sense terminals which is proportional to the device vibration amplitude.

MEASUREMENT RESULTS

Measurement Setup

To test the resonator of Figure: 2 as a Lorentz Force MEMS magnetometer, a relatively long current carrying wire was placed next to the device which would act as the source of the magnetic field. The wire was kept at a specified distance from the device and in a direction such that the Lorentz Force from the applied magnetic field would actuate the resonator in its in-plane extensional mode. To avoid interference between the Lorentz force driving current and output signal of the device, a DC current was applied to the device for Lorentz force generation, and an AC magnetic field at the device resonant frequency was used to actuate the device. To generate the AC field, the RF output of the network analyzer was connected to the wire (P1 and P2- see Figure 3). The other advantage of using an AC field is that the device can operate without the interference from the Earth's magnetic field. A DC current I_{dc} passing through the silicon beam (P3 and P4- see Figure 3) allows piezoresistive readout of device resonance response (the same current is also responsible for internal amplification). Figure 3 shows the measurement setup and the electrical connections for device testing. The resonator frequency responses were obtained for different current amplitudes in the wire and for different the distances between the wire and the device which would in-turn change the magnitude of the magnetic field.

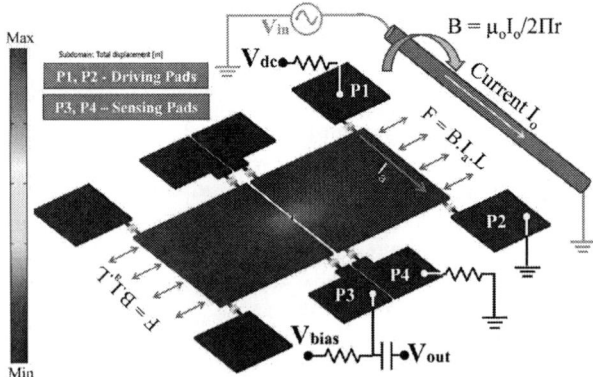

Figure 3: Finite element modal analysis of the resonator of Figure 2 showing the in-plane resonance mode and the measurement setup and its electrical connections.

Results

As the DC bias current between the sense pads increases, the onset of the internal thermal-piezoresistive amplification increases the vibration amplitude for the same actuating magnetic field. This effect shows up as an increase in the measured Q factor of the resonance peak in the device frequency response. Figure 4 shows the measured effective Quality Factor versus the bias current applied to the device demonstrating the amplification effect. The effective Q of the resonator increases from its intrinsic value of 680 (at 5.164 mA) to 1.14×10^6 (at 7.245 mA) under atmospheric pressure. The inset in Figure 4 shows the frequency response from the network analyzer for the magnetometer operating at 7.245 mA bias current with a magnetic field of 3.5 nT. An effective Q of 1.14×10^6 is obtained at this bias current and the measured output voltage is 7.548 mV. The maximum sensitivity (2.107mV/nT) of the magnetometer is achieved at this bias current. Figure 5(a) illustrates the frequency responses at field intensity of 3.5 nT for different piezoresistor bias currents showing how the output signal amplitude increases by increasing the bias current. Figure 5(b) illustrates the measured output voltage amplitudes (left y-axis) at resonance versus the magnitude of magnetic field (nT) for different bias currents. There is a ~2400X improvement in sensitivity (from 0.9 μV/nT to 2.107 mV/nT) when the bias current is increased from 5.164 mA to 7.245mA. The increase in output amplitudes (and thus sensitivity) at higher currents is partly due to higher piezoresistive sensitivity (higher piezoresistor bias current) and partly due to internal vibration amplification. To demonstrate the effect of internal amplification alone, sensitivity figure of merit (FOMS), defined as sensitivity divided by the bias current, is illustrated in Fig. 5 (b) (right y-axis) by the slope of the lines at different bias currents, showing a ~1620X improvement as a result of internal amplification alone. Figure 5(a) and 5(b) clearly shows that FOMS increases proportional to the resonator effective Q-factor as the bias current increases.

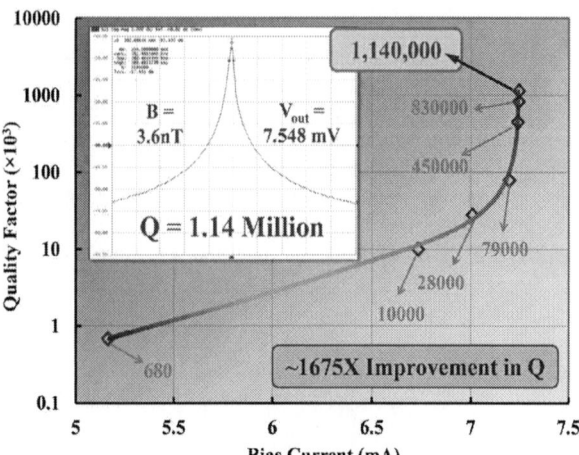

Figure 4: Graph showing measured effective Quality Factor versus the bias current demonstrating the Q and vibration amplification effect. Inset- Network Analyzer response for piezoresistor bias current of 7.245 mA.

Noise and Limit of Detection

It should be noted that although the thermo-mechanical noise will also be amplified due to internal amplification, only noise components at close vicinity of the resonance frequency will be amplified by a factor close to the sensor output signal. Therefore, overall signal to noise ratio of the sensor is expected to improve. The minimum detectable magnetic field in such sensors is limited by different sources of noise including thermo-mechanical noise which is the most dominant component in micromechanical systems. The thermo-mechanical noise magnitude depends on the temperature and mechanical damping. To compare the amplification in noise due to the effect of thermal-piezoresistive amplification with the rate of the increase in signal, the noise behavior of the sensor was studied using a lock-in amplifier. For a bias current of 7.245mA, where the sensitivity of the device is maximum (2.1mV/nT), the noise floor is measured to be as low as 5.83 μV/√Hz, which translates to 2.8 pT/√Hz. The measurement results are summarized in Table 1, which highlights the Sensitivity, FOMS, effective Q and the Noise Floor values for different bias currents. Although it is evident from the Table that the noise amplitude will increase along with the signal amplification, overall signal to noise ratio (SNR) of the sensor improves with increase in bias current. Increasing the bias current leads to a temperature increase within the piezoresistor, thus raising the noise level, an increase in the signal level and the reduction in bandwidth at a higher pace will however result in the improvement in SNR.

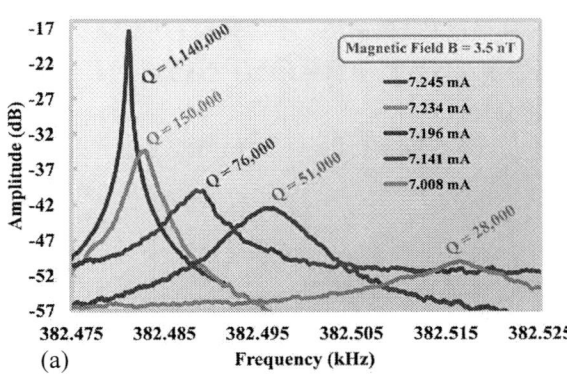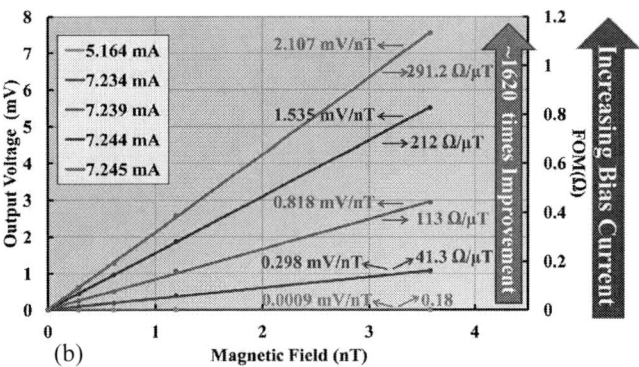

Figure 5: (a) Resonant responses of the device with different bias currents under constant magnetic field intensity of 3.5 nT. (b) Graph showing the output voltage amplitude and the FOMS values versus the magnetic field intensity for different bias currents.

Table 1: Sensitivity, FOMS Quality Factor and noise floor values for the magnetometer at different bias currents.

Bias Current (mA)	5.164	6.733	7.141	7.196	7.236	7.239	7.243	7.244	7.245
Sensitivity (µv/nT)	0.89	17.87	90.73	145.44	542.05	818.23	1188	1535.7	2107.8
FOMS (Ω/µT)	0.18	2.7	12.7	20.2	74.9	113	164	212	291.2
Quality Factor	680	1×10^4	2.8×10^4	7.9×10^4	28.5×10^4	45×10^4	6.3×10^5	8.3×10^5	1.1×10^6
Noise (pT/√Hz)	2340.3	264.64	61.71	39.34	10.72	7.11	4.90	3.79	2.76

CONCLUSION

Internal self amplification of a micro-scale resonant Lorentz Force magnetometer with piezoresistive readout was demonstrated. The sensitivity of the device made up of n-type silicon was improved by ~2400X. Close to ~1620X improvement in the magnetometer sensitivity figure of merit was validated. It is expected that by thinning down the piezoresistive amplifying beam and design optimizations, much higher sensitivities can be obtained potentially allowing compact low power sensor arrays for biomedical applications and brain mapping.

ACKNOWLEDGEMENTS

This work was supported by the US National Science Foundation under ECSS award #1345161.

REFERENCES

[1] D. Niarchos, "Magnetic MEMS: key issues and some applications," Sensors & Actuators A,Phys., Vol 106, no. ½, pp 255-262, Sep 2003.

[2] E. K. Ralph, "Comparison of a Proton and a Rubidium Magnetometer for Archaeological Prospecting," Archaeometry, vol. 7, no. 1, pp. 20–27, Jun. 1964.

[3] M. Li, V. T. Rouf, M. J. Thompson, and D. . Horsley, "Three-Axis Lorentz-Force Magnetic Sensor for Electronic Compass Applications," J. Microelectromechanical Syst., vol. 21, no. 4, pp. 1002–1010, Aug 2012.

[4] S. Levine, D. Silage, D. Henson, J. Y. Wang, J. Krieg, J. LaManca, and S. Levy, "Use of a triaxial magnetometer for respiratory measurements," J. Appl. Physiol. Bethesda Md 1985, vol. 70, no. 5, pp. 2311–2321, May 1991.

[5] J. Lenz and A. S. Edelstein, "Magnetic sensors and their applications," IEEE Sens. J., vol. 6, no. 3, pp. 631–649, Jun. 2006.

[6] A. Pérez Galván, B. Plaster, J. Boissevain, R. Carr, B. W. Filippone, M. P. Mendenhall, R. Schmid, R. Alarcon, and S. Balascuta, "High uniformity magnetic coil for search of neutron electric dipole moment," Nucl. Instrum. Methods Phys. Res. Sect. Accel. Spectrometers Detect. Assoc. Equip., vol. 660, no. 1, pp. 147–153, Dec. 2011.

[7] W. Zhang and J. E. Lee, "A horseshoe micromachined resonant magnetic field sensor with high quality factor," IEEE Electron Device Lett., vol. 34, no. 10, pp. 1310–1312, Oct. 2013.

[8] L.M. Miller, J.A. Podosek, E. Klruglick, "A µ-Magnetometer based on Electron Tunneling, Micro Electro Mechanical Systems, pp 467-472, Feb 1996.

[9] M. Thompson & D. Horsley, "Parametrically Amplified MEMS Magnetometer," IEEE Transducers, pp. 1194-1197, 2009.

[10] E. Mehdizadeh, V. Kumar, and S. Pourkamali, "Sensitivity Enhancement of Lorentz Force MEMS Resonant Magnetometers via Internal Thermal-Piezoresistive Amplification," IEEE Electron Device Lett., vol. 35, no. 2, pp. 268–270, Feb. 2014.

[11] E. Mehdizadeh, V. Kumar, and S. Pourkamali, "High Q Lorentz Force MEMS Magnetometer with Internal Self-Amplification," IEEE Sensors 2014, pp. 706–709, October 2014.

[12] A. Rahafrooz, and S. Pourkamali, "Thermal-piezoresistive energy pumps in micromechanical resonant structures," IEEE Trans. Electron Devices, vol. 59, no. 12, pp. 3587–3593, Dec. 2012.

CONTACT

V. Kumar, vxk120630@utdallas.edu
S. Pourkamali, siavash.pourkamali@utdallas.edu

HIGH FREQUENCY MICROWAVE ON-CHIP INDUCTORS USING INCREASED FERROMAGNETIC RESONANCE FREQUENCY OF MAGNETIC FILMS

Kisik Koh[1]*, Donald S. Gardner[2], Chen Yang[1], Kevin P. O'brien[2], Noureddine Tayebi[2] and Liwei Lin[1]
[1]University of California, Berkeley, USA
[2]Intel Corporation, USA

ABSTRACT

The fabrication and characterization of high frequency on-chip inductors using sputtered magnetic films with an improved frequency range is presented. Reducing the sputtering power in the deposition process was found to result in smoother film surfaces and stronger uniaxial magnetic anisotropy and increased the FMR of CoZrTaB from 1.48 GHz to 2.13 GHz. A magnetic-core, on-chip inductor was fabricated using the CoZrTaB films. Results have shown 150% higher inductance and a larger Q-factor up to 1.2 GHz as compared to an air-core inductor.

INTRODUCTION

Magnetic thin films with superior properties such as high permeability and high ferromagnetic resonance (FMR) frequency are desirable in various applications including integrated on-chip inductors and transformers. For example, in the area of radio and microwave frequency components and systems such as GHz-range transceivers, the FMR limits of magnetic materials have played a critical role in designing devices for high frequency operation [1]. Maintaining high magnetic permeability for magnetic thin films at high frequency is also important for magnetic recording devices as their operating frequencies have been continuously increasing [2,3].

FMR occurs when a ferromagnetic material absorbs energy from an oscillating magnetic field at a sharply defined frequency, which correspond to transitions between the energy levels split by a uniform magnetic field. Classically, this phenomenon is related to the Larmor precession of the magnetic moment along the static field that includes the anisotropy field and demagnetizing field in a magnetic material and resonance occurs when an alternating field is applied at the Larmor frequency [4]. The oscillating magnetic field causes the magnetic moments in the sample to precess with the precession frequency of the magnetization depending on the orientation of the material, the strength of the magnetic field, as well as the macroscopic magnetization of the sample [5]. The energy losses occur at FMR frequencies as the oscillatory electron spins absorb it. Under higher frequencies, the magnetic moments fail to follow the direction of the alternating field resulting in rapid decrease or relaxation of the magnetic permeability.

Achieving alignment of the magnetic moments for proper domain structures is critical for their dynamic behavior in high frequency devices. Generally, magnetization by coherent rotation of magnetic moments has faster response (which leads to higher magnetic resonance frequencies) and lower loss than magnetization by domain wall motion [6]. In addition, the controlled domain structures are used to obtain high frequency

permeability; the magnetization in the domains oriented perpendicular to the magnetic flux path change magnetization by rotating and not by domain wall movements [7]. By applying external magnetic fields during magnetic material deposition (e.g., sputtering and electroplating), magnetic films compatible with standard complementary metal-oxide semiconductor (CMOS) processing can be prepared [6, 8].

In this paper, we investigate the FMR of CoZrTaB thin films by changing the sputtering power during the deposition processes under an external magnetic field. As the sputtering power is lowered from 1,000 to 150 Watts, smoother surfaces [9] and larger uniform uniaxial magnetic anisotropy are obtained. Magnetic films with smooth surfaces have been previously reported as being favorable for strong uniaxial magnetic anisotropy of cobalt [10]. On the other hand, high sputtering power can result in rough surfaces and weak uniaxial magnetic anisotropy as shown in Fig. 1. The FMR frequency of the magnetic thin film was found to increase by lowering the sputtering power. In addition, these films are integrated in the fabrication of on-chip inductors for microwave (>1 GHz) applications using the magnetic thin films with high FMR as the magnetic-core material.

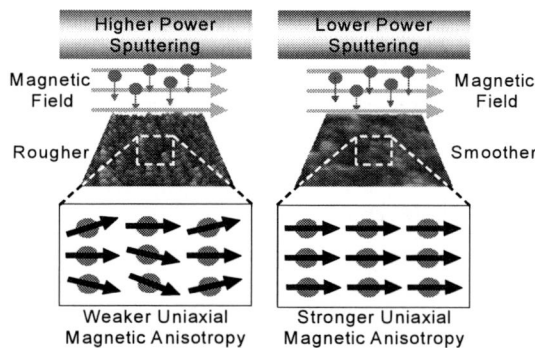

Figure 1: Schematics of magnetic anisotropy results from different sputtering power under an external magnetic field during the deposition process. Lower sputtering power results in smoother surface and stronger uniaxial magnetic anisotropy and the FMR is increased.

SPUTTERED MAGNETIC THIN FILMS USING DIFFERENT SPUTTERING POWERS

Magnetic thin films of CoZrTaB are sputtered under an external magnetic field during the deposition process to generate the in-plane uniaxial anisotropy. This anisotropy helps to minimize energy loss from the magnetic hysteresis behaviors under alternating magnetic fields. When the sputtering power is reduced from 1,000 Watts to 150 Watts, the RMS surface roughness of the deposited thin films is reduced from 4.352 to 3.831 nm as shown in the Atomic Force Microscopy (AFM) measurements in Figs. 2a and 2b. The scanning area is $3\times3\mu m^2$. Both films were amorphous based on Electron Microprobe and X-ray crystallography measurements (not shown in this paper).

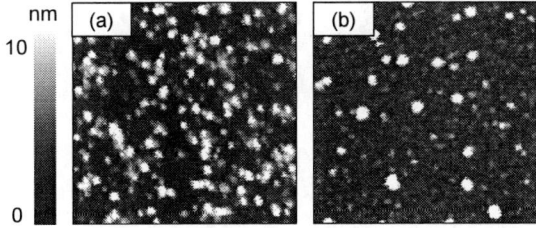

Figure 2: AFM images of CoZrTaB films deposited with a sputtering power of: (a) 1000W and (b) 150W, respectively.

The deposited magnetic thin films are tested in a Vibrating Sample Magnetometer (VSM) for magnetic property characterization (see Fig. 3). The anisotropy field (H_k, applied magnetic field at the crossing of the saturation lines between easy and hard axes) increases from 25 to 50 Oe when the sputtering power is reduced from 1,000 to 150 Watts. This result validates Fig. 1 that depositions under lower sputtering power results in stronger anisotropy.

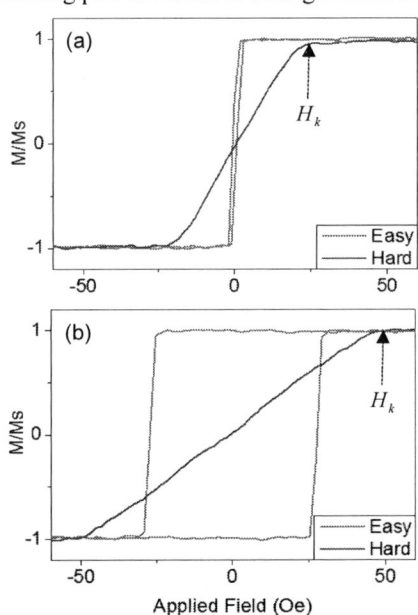

Figure 3: M-H loops from CoZrTaB thin films sputter deposited using (a) 1,000 W and (b) 150 W power.

MICROWAVE PROPERTY OF MAGNETIC THIN FILMS

When an alternating magnetic field is applied to the hard axis of the film (in-plane and perpendicular to the axis of the anisotropy), the magnetic permeability is expressed in the complex form ($\mu = \mu' - j\mu''$) which accounts for the phase angle from the magnetization delay. In general, the real part, μ' represents the capability of amplifying the

magnetic flux which enhances the inductance of inductors, while the imaginary part, μ'' describes the magnetic losses.

The dynamic behavior of the magnetization in thin films is widely described by the Landau-Lifshitz model which assumes a coherent magnetization rotation with the evolution of the magnetization in the model expressed as [11]:

$$\frac{d\vec{M}}{dt} = -|\gamma|\vec{M}\times\vec{H} - \frac{\alpha\gamma}{|\vec{M}|}\vec{M}\times\left[\vec{M}\times\vec{H}\right] \quad (1)$$

where M is the saturation magnetization, H is the effective field including external fields and demagnetizing fields, γ is the gyromagnetic ratio, and α is the damping constant. The Landau-Lifshitz model can be solved for the complex permeability, μ, along the hard axis as a function of frequency, f [12], and the FMR frequency, f_r, is defined as that frequency at which the real part of the permeability is zero as [3]:

$$\mu(f) = 1 - \frac{M_s\gamma^2\left(j2\pi f\frac{\alpha}{\gamma} + (H_k + 4\pi M_s)\right)}{(2\pi f)^2 - j2\pi f\alpha\gamma(2H_k + M_s) - H_k(H_k + 4\pi M_s)(1+\alpha^2)\gamma^2} \quad (2)$$

$$f_r = \frac{\gamma}{2\pi}\sqrt{4\pi H_k M_s} \quad (3)$$

where M_s is the saturation magnetization, H_k is the anisotropy field, γ is the gyromagnetic ratio, and α is the damping constant.

The increase of the uniaxial in-plane anisotropy by lowering the sputtering power during the deposition process as shown in Fig. 3 can leads to an increase in the FMR frequency. Permeability measurements versus frequency show a gradual increase in the FMR from 1.48 GHz to 2.13 GHz with decreasing deposition power (see Fig. 4), using a permeameter (Ryowa Electronics Co.). Table 1 further summarizes the FMR frequencies and the real part values of permeability at 100 MHz for thin films deposited using different sputter deposition powers. The anisotropy was found to increase as the sputtering power is reduced with a corresponding increase in the FMR frequency. Increased anisotropy however leads to reduced real values of the permeability along the hard axis as the enhanced stiffness in magnetization results in more difficult rotations for the magnetic moments to move from the easy to the hard axis consistent with Snoek's law [13].

Table 1: Summary of FMR frequencies and μ' at 100 MHz for CoZrTaB films made using different sputtering powers

Sputtering Power (W)	FMR (GHz)	μ' @ 100 MHz
1000	1.48	1077
700	1.55	901
300	1.78	570
150	2.13	380

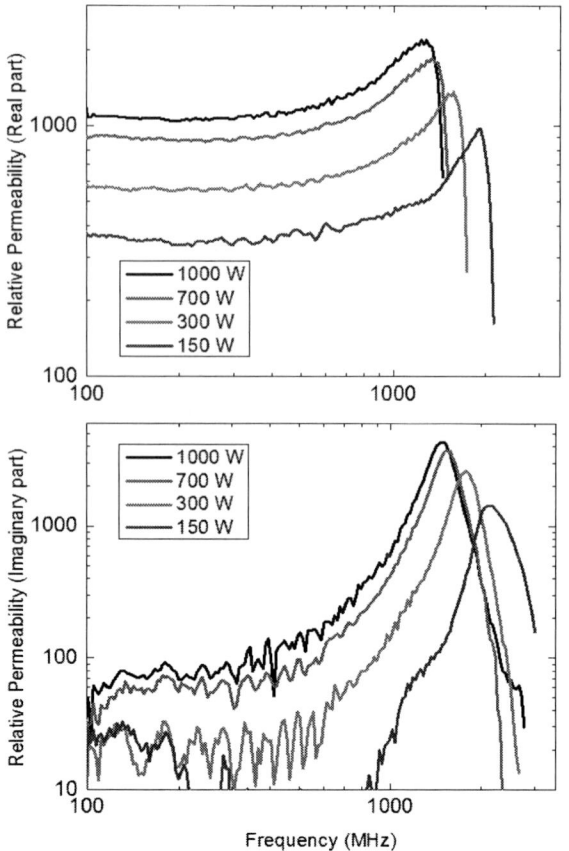

Figure 4: Microwave permeability spectra from thin films fabricated using four different sputtering powers: real part (top) and imaginary part (bottom).

FABRICATION & CHARACTERIZATION OF ON-CHIP INDUCTORS

Magnetic thin films with enhanced FMR frequencies are then used in the monolithic fabrication of on-chip inductors. Figure 5 illustrates the fabrication process flow. Stripline shape inductors 1000 μm-long, 10 μm-wide, and 1 μm-thick made of copper are constructed. The process starts with a silicon wafer as the substrate with a 6 μm-thick LPCVD silicon dioxide film on top to reduce substrate losses from eddy currents in the bulk silicon at high frequencies. A 100 nm-thick CoZrTaB film is then deposited using the lift-off process in Fig. 5a. The process follows with a 100 nm-thick silicon nitride thin film deposition and a 1 μm-thick copper thin film deposition and then patterning by a second lift-off process as shown in Fig. 5b. A 100 nm-thick silicon nitride film is then deposited and vias are etched using a dry etching process. The second 100 nm-thick CoZrTaB thin film is deposited and then patterned using a third lift-off process as shown in Fig. 5c. The final structure consists of a 1000 μm-long stripline surrounded by magnetic films with a closed-loop path for the magnetic flux.

Figure 6a shows an optical image of the fabricated device (top view), where a 1000 μm-long stripline inductor is covered by CoZrTaB with an area of 900 × 80 μm².

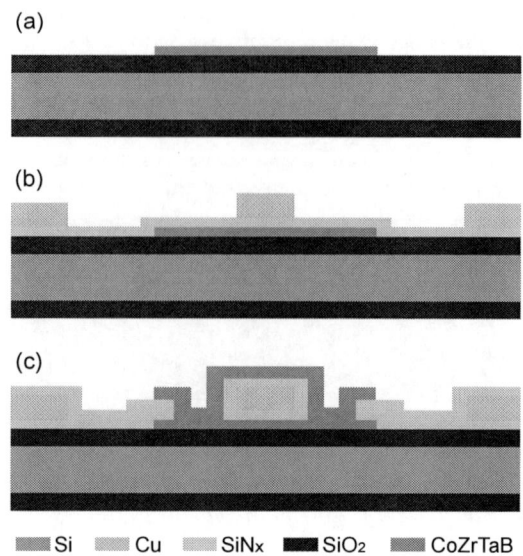

Figure 5: Schematic diagram of fabrication process flow: three lift-off processes for (a) bottom magnetic layer, (b) copper metallization, and (c) top magnetic layer.

Measurements of the inductor samples are performed by using GSG probes and the Vector Network Analyzer (VNA). To remove the external effects from the device under test (DUT), de-embedding is conducted by using the OPEN structure in Fig. 6c. The de-embedded scattering parameters from the VNA are converted to inductance, L, and quality factor, Q, at each frequency, f [8, 14].

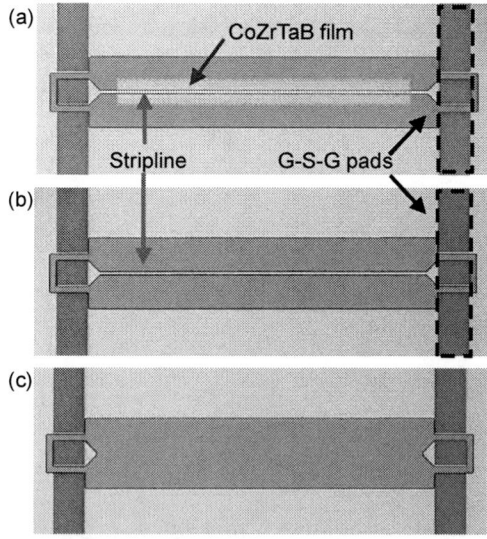

Figure 6: (a) A fabricated stripline inductor with magnetic material using CoZrTaB; (b) An air-core stripline inductor without CoZrTaB; (c) OPEN structure for the de-embedding.

Stripline inductors with and without the CoZrTaB films deposited at 300W are tested. Experimental results on the inductance and quality factor with respect to frequency are plotted in Figs. 7a and 7b, respectively. The magnetic-core inductors are compared with the air-core inductor (a control inductor without the integration of magnetic material) with the same dimensions. It is found that the magnetic core inductor has 2.5 times higher inductance than that of the air-core inductor with a cut-off frequency at 2.1 GHz. Furthermore, the magnetic-core inductor has a larger Q-factor than that of the air-core inductor up to 1.2 GHz.

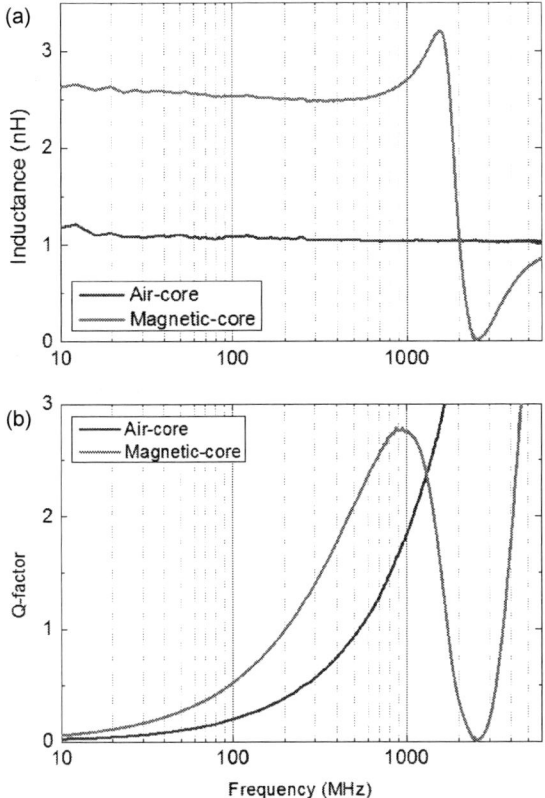

Figure 7: Measurements of (a) inductance and (b) Q-factor versus frequency of an inductor with CoZrTaB (magnetic-core) and without CoZrTaB (air-core).

CONCLUSION

During the sputtering deposition process under external magnetic fields, the change of surface roughness of the sputtered film due to the different sputtering powers can modify the in-plane anisotropy and consequentially increase the FMR frequency of CoZrTaB films. With the increased FMR, the magnetic material has been integrated with a stripline inductor fabrication process to achieve larger inductance and Q-factor than those of an air-core inductor. These experimental efforts toward improving the FMR frequency by modifying the sputtering process will result in magnetic-core microwave (>1 GHz) on-chip inductors.

ACKNOWLEDGEMENT

This work is supported in part by SRC (Semiconductor Research Corporation). These devices are fabricated in the UC Berkeley Marvell Nanofabrication Laboratory. The authors would like to thank to Dr. Ouk Jae Lee, Dr. Jeffrey P Clarkson and Casey Glick for valuable help and discussion.

REFERENCES

[1] V. Korenivski, "GHz magnetic film inductors," *J. Magn. Magn. Mater.*, vol. 216, pp. 800–806, 2000.

[2] X. Chen, Y. G. Ma, and C. K. Ong, "Magnetic anisotropy and resonance frequency of patterned soft magnetic strips," *J. Appl. Phys.*, vol. 104, no. 1, p. 013921, 2008.

[3] E. Van de Riet and F. Roozeboom, "Ferromagnetic resonance and eddy currents in high-permeable thin films," *J. Appl. Phys.*, vol. 81, pp. 350–354, 1997.

[4] J. M. D. Coey, *Magnetism and Magnetic Materials*, Cambridge, 2010.

[5] S. V. Vonsovskii, *Ferromagnetic Resonance: The Phenomenon of Resonant Absorption of a High-Frequency Magnetic Field in Ferromagnetic Substances*, Oxford: Pergamon, 1966

[6] N. Wang, E. O'Sullivan, P. Herget, B. Rajendran, L. E. Krupp, L. T. Romankiw, B. C. Webb, R. Fontana, E. A. Duch, E. A. Joseph, S. L. Brown, X. Hu, G. M. Decad, N. Sturcken, K. L. Shepard, and W. J. Gallagher, "Integrated on-chip inductors with electroplated magnetic yokes," *J. Appl. Phys.*, vol. 111, no. 7, p. 07E732, 2012.

[7] S. Chikazumi, *Physics of Magnetism*, Krieger, 1978.

[8] D. S. Gardner, G. Schrom, P. Hazucha, F. Paillet, T. Karnik, S. Borkar, R. Hallstein, T. Dambrauskas, C. Hill, C. Linde, W. Worwag, R. Baresel, and S. Muthukumar, "Integrated on-chip inductors using magnetic material," *J. Appl. Phys.*, vol. 103, no. 7, p. 07E927, 2008.

[9] M.-T. Le, Y.-U. Sohn, J.-W. Lim, and G.-S. Choi, "Effect of Sputtering Power on the Nucleation and Growth of Cu Films Deposited by Magnetron Sputtering," *Mater. Trans.*, vol. 51, no. 1, pp. 116–120, 2010.

[10] M. Li, G.-C. Wang, and H.-G. Min, "Effect of surface roughness on magnetic properties of Co films on plasma-etched Si(100) substrates," *J. Appl. Phys.*, vol. 83, no. 10, p. 5313, 1998.

[11] L. Landau and E. Lifshitz, "On the Theory of the Dispersion of Magnetic Permeabilityin Ferromagnetic Bodies," *Phys. Z. der Sow.*, vol. 8, pp. 153-169, 1935.

[12] O. Geradin, J. Ben Youssef, H. Le Gall, N. Vukadinovic, P. M. Jacquart, and M. J. Donahue, "Micromagnetics of the dynamic susceptibility for coupled Permalloy stripes, " *J. Appl. Phys.*, vol. 88, no. 10, pp. 5899–5903, 2000.

[13] J. L. Snoek, "Dispersion and absorption in magnetic ferrites at frequencies above one Mc/s," *Physica*, vol. 14, pp. 207-217, 1948.

[14] K. Büyüktas, K. Koller, K.-H. Müller, and A. Geiselbrechtinger, "A New Process for On-Chip Inductors with High Q-Factor Performance," *Int. J. Microw. Sci. Technol.*, vol. 2010, pp. 1–9, 2010.

CONTACT

*K. Koh, tel: +1-510-507-2111; kskoh@berkeley.edu

FUSION OF CANTILEVER AND DIAPHRAGM PRESSURE SENSORS ACCORDING TO FREQUENCY CHARACTERISTICS

Ryo Watanabe, Nguyen Minh-Dung, Hidetoshi Takahashi, Tomoyuki Takahata,
Kiyoshi Matsumoto and Isao Shimoyama
The University of Tokyo, Tokyo, JAPAN

ABSTRACT

This paper reports on an approach to measure sensitive barometric pressure by fusing a cantilever-based differential pressure sensor (DPS) and a commercially available diaphragm-based absolute pressure sensor (APS). At high frequency, the DPS can detect smaller pressure change than the APS. At low frequency, the APS shows less drift absolute pressure than the DPS. By utilizing the DPS and APS at each advantageous frequency, we propose a highly sensitive and drift-free measurement of barometric pressure.

INTRODUCTION

Many types of absolute pressure sensors (APS) have been developed for conventional applications such as personal mobile systems and automotive navigation systems. In these days, the APSs are especially utilized for activity logging application to measure altitude converted from absolute pressure. Generally, these types of APSs are based on a diaphragm with a sealed cavity [1-2]. However, the resolution is not good enough to measure small change.

In contrast, a DPS has been proposed to measure change in barometric pressure using a piezo-resistive cantilever with an unsealed cavity [3]. When barometric pressure changes, the cantilever of DPS bends. Although the DPS can obtain more accurate measurement than the APS for a few seconds, a response of the DPS has drifts caused by integration error for a long time. Using the DPS, it is difficult to measure the absolute pressure at low frequency.

We propose a sensor fusion method of a cantilever-based DPS and a diaphragm-based APS to obtain sensitive and drift-free absolute pressure measurement as shown in Figure 1. By applying high-pass filter and low-pass filter for the DPS and APS, respectively, we can take the advantage of them as shown in Figure 2. The barometric pressure can be utilized in the application of the personal mobile systems requiring high sensitive altitude measurement.

PRINCIPLE AND METHOD

Differential Pressure Sensor

Figure 3 shows the principle to measure the barometric pressure by a DPS [3]. It consists of an opened cavity and a micro piezoresistive cantilever. When barometric pressure around the DPS increase, difference of pressure between inside and outside cavity leads to the piezoresistive cantilever deformation. The deformation is able to be detected by the resistance change. Therefore, monitoring the resistance change enable us to measure barometric pressure change.

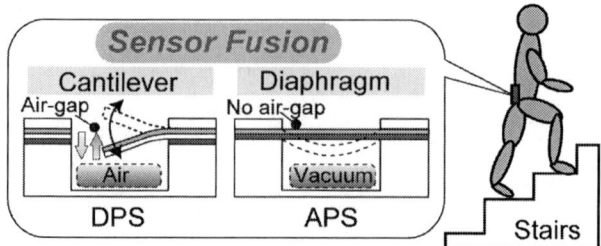

Figure 1: The cantilever based differential pressure sensor (DPS) and diaphragm based absolute pressure sensor (APS) measure the barometric pressure change which can be converted into change in altitude.

	DPS	APS	Proposed method
Sensitivity	*High*	*Low*	*High*
Absolute Pressure	*Impossible*	*Possible*	*Possible*

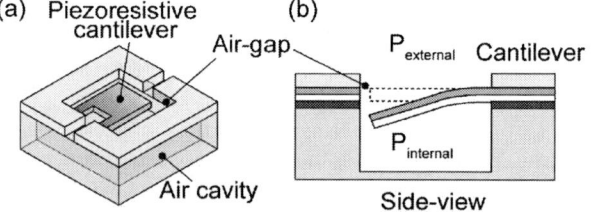

Figure 2: The advantage and disadvantage of the differential pressure sensor (DPS) and absolute pressure sensor (APS)

Figure3: (a) Principle of barometric pressure measurement by the DPS with an opened cavity. (b) Deformation of the cantilever in the condition of $P_{external} > P_{internal}$

Sensor Fusion Method

Our method to obtain sensitive and drift-free absolute pressure is fusing the DPS and APS on a frequency domain. As shown in Figure 4, a transfer function of the DPS is characterized by the calibration experiment with a rotating wheel. By multiplying the response of the resistance change from the DPS, $\Delta R/R_{diff}$ by the transfer function, a barometric pressure change, P_{DPS}, is calculated. To obtain P_{high} with drift-free, a high pass filter is applied to for P_{DPS}, in order to cut the integration error.

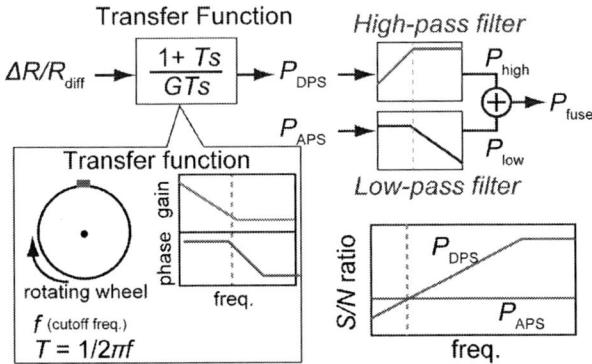

Figure 4: Digital signal processing flow to obtain the sensitive and drift-free barometric pressure

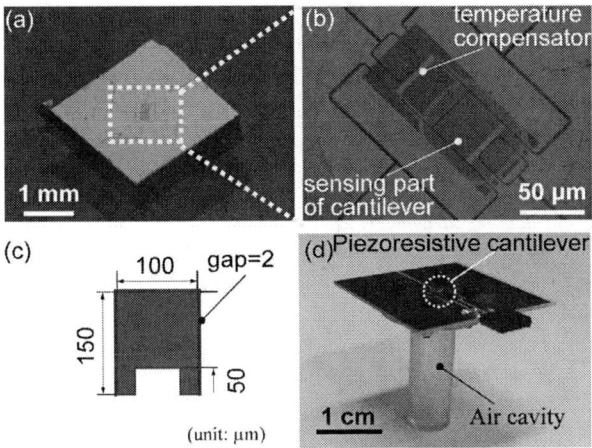

Figure 5: (a) (b) The fabricated DPS. (c) The size of the designed DPS. (d) The DPS placed on the opened cavity.

In order to derive P_{low} with drift-free, the low pass filter is applied to P_{APS}. Cut-off frequency for the low-pass filter and the high-pass filter are defined by a crossing point of the S/N ratio between the DPS and APS. Finally, highly sensitive and drift-free absolute pressure measurement is achieved by combining the filtered barometric pressure of the DPS and APS.

SENSOR AND DEVICE DESIGN

Figure 5(a) shows two types of peizoresistive cantilever fabricated at the center of a chip. It was formed by a silicon on insulator wafer. As shown in Figure 5(b), one of the cantilever was sensing part. The other was fabricated to compensate the effect of temperature. The fabrication process of the sensor was described in the previous research [4]. As shown in Figure 5(c), the size of sensing cantilever was 150 μm × 100 μm × 0.3 μm. Since three end of the cantilever was free, it was able to detect smaller barometric pressure change than the diaphragm based APS. The size of the gap between the sensing part of the cantilever and its surrounding frame was designed on 2.0 μm. Figure 5 (d) shows that the DPS was placed on an opened air cavity which was made by 3D Printer. An opened air cavity contained 2.0 ml.

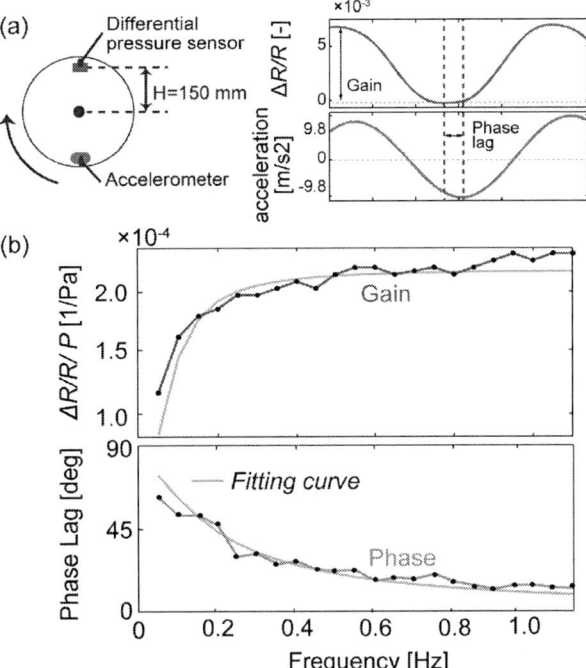

Figure 6: (a) Schematic image of calibration experimental setup the resistance change by the DPS. (b) Bode diagram consisting of Gain and Phase lag of the DPS

CALIBRATION

Figure 6(a) shows an experiment to investigate the frequency characteristic of the DPS by using a rotating wheel of 300 mm in diameter. The frequency range of the rotating wheel was performed from 0.05 Hz to 1.1 Hz. The red line in the graph describes the resistance change of the DPS at 0.4 Hz. Figure 6(b) shows bode diagram including gain and phase delay of the DPS. The gain was calculated by peak-to-peak of resistance change from the DPS divided by input pressure calculated by altitude change. The phase lag at a certain frequency was defined as differences between an angular calculated by an acceleration sensor and response of the angular of the DPS. The transfer function was characterized by the bode diagram.

Noise level of the APS was defined as 2.0 Pa in a data sheet of MEMS absolute pressure sensor from Omron Corporation [5]. On the other hand, S/N ratio of the DPS was calculated from the gain and noise level of the DPS. In the end, a crossing frequency point between the DPS and the APS in S/N ratio was calculated to be 0.075 Hz.

EXPERIMENT AND RESULTS

Figure 7 shows experiment with the sensor fusion of the DPS and APS. A person with the DPS and APS walked down stairs of 15 steps. The height of each step was 180 mm. It took approximately two seconds to step down each step of the stairs.

Figure 7(a) shows the resistance change of the DPS. By utilizing the calculated transfer function, the resistance change of the DPS was converted into the barometric pressure change (Figure 7(b)). Although every steps could be observed, there were a large drift due to integration error. To cut the integration error, a high pass filter of 0.075 Hz cutoff frequency was applied (Figure 7(c)).

978-1-4799-7956-1/15 $31.00 © 2015 IEEE

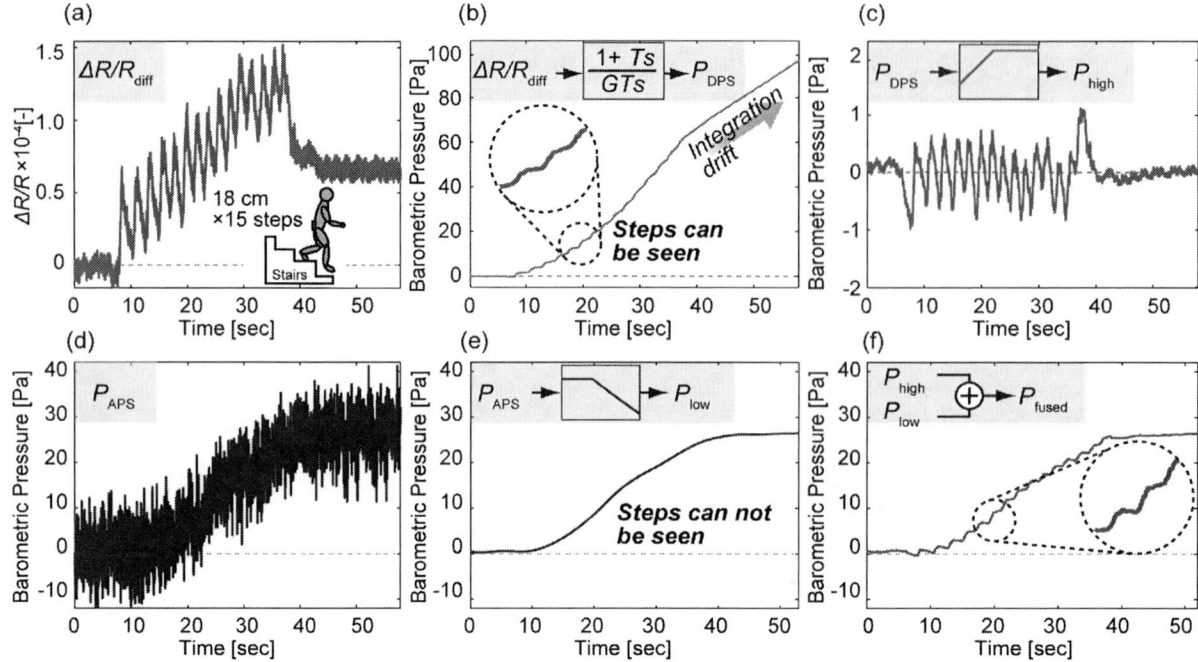

Figure 7: Experimental result of walking down on the stairs by measuring barometric pressure

On the other hand, Figure 7(d) shows the barometric pressure change obtained by the APS, Omron Corporation's MEMS absolute pressure sensor (2SMPB-01-01) [5]. The result indicates that the height of each step was too small for the APS sensor to detect clearly due to low sensitivity. The APS signal was passed through a low pass filter with 0.075 Hz cut-off frequency to reduce the noise (Figure 7(e)). By combining the processed signals of both DPS and APS, we were able to obtain a sensitive and drift-free measurement enough to observe every steps of the stairs during whole walk (Figure 7(f)).

CONCLUSION

In this paper, we reported fusion of the cantilever based DPS and diaphragm based APS on frequency domain. The transfer function of the DPS was characterized by the calibration experiment. By taking advantage of the DPS and APS on frequency domain, barometric pressure from both sensors went through high-pass filter and low-pass filter, respectively. Finally, we were able to obtain sensitive and drift-free barometric pressure on the experiment of walking down the stairs with them.

ACKNOWLEDGEMENT

The photolithography masks were made using the University of Tokyo VLSI Design and Education Center (VDEC)'s 8 inch EB writer F5112 + VD01, which was donated by ADVANTEST Corporation. This work was partly supported by JSPS KAKENHI Grant Number 25000010.

REFERENCES

[1] J. N. Palasagaram and R. Ramadoss, "MEMS capacitive pressure sensor fabricated using printed circuit processing techniques," *Sensors Journal*, vol. 6, article no.6, 2006.

[2] Y. Zhang, et al, "An ultra-sensitive, high-vacuum absolute capacitive pressure sensor," in *Proc. IEEE. MEMS '01 Conference*, Interlaken, January 21-25, 2001, pp. 166–169.

[3] N. Minh-Dung, H. Takahashi, T. Uchiyama, K. Matsumoto, I. Shimoyama, "A barometric pressure sensor based on the air based on the air-gap scale effect," *Applied Physics Letters*, vol.103, no. 14, article no. 143505, 2013.

[4] H. Takahashi, N. Minh-Dung, K. Matsumoto and I. Shimoyama, "Differential pressure sensor using a piezoresistive cantilever," *Measurement Science and Technology*, vol. 24, no.5, article no. 055304, 2013.

[5] Omron,2SMPB-01-01(datasheet), http://www.omron.com/ecb/products/sensor/21/2smpb-01-01.html

CONTACT

*R. Watanabe, Tel: +81-3-5841-6318
E-mail: r_watanabe@leopard.t.u-tokyo.ac.jp

DYNAMICALLY-BALANCED FOLDED-BEAM SUSPENSIONS

Shai Shmulevich, Inbar Hotzen and David Elata

Faculty of Mechanical Engineering, Technion - Israel Institute of Technology, Haifa, ISRAEL

ABSTRACT

We present a complete methodology for designing a new folded-beam suspension which responds as a linear spring at the fundamental resonance. This is in sharp contrast to the response of standard folded-beam suspensions. The *static* response of the standard folded-beam suspension is linear over a wide range of motions. But, surprisingly, the *dynamic response* of the standard folded-beam suspension is *strongly nonlinear* for small motion amplitudes that are larger than the *width* of the flexure beams. We have previously shown experimental evidence of this problem with the standard suspension. In contrast, the stiffness of the new dynamically balanced folded-beam suspension is not affected by motion amplitude. In the present work we show new experimental evidence demonstrating that the new design solves this problem.

INTRODUCTION

Electrostatic comb-drives are prevalent in MEMS because they enable two features, large motions [1] and a linear relation between voltage and motion [2]. By suspending such comb-drives on linear elastic springs, linear electrostatic resonators can be constructed [1]. Linear resonators with high quality-factors find many applications as filters, sensors, and oscillators. The suspension of choice in many comb-drive resonators is the folded-beam suspension [1-4]. Standard folded-beam suspensions (Fig. 1a) promise a linear mechanical response in the primary axis of motion for motions up to 20% of the length of the flexure. In addition, this suspension also provides sufficient transverse rigidity to suppress the side pull-in instability [5, 6].

In recent work [7], we demonstrated that the *dynamic response* of the standard folded-beam suspension is *inherently nonlinear*. Though this suspension is designed to respond as a linear mechanical spring in *static* applications [6], in *periodic dynamic* motion, the inertia of the flying-bar induces axial stresses in the flexure-beams. These axial stresses are similar to those induced in clamped-clamped beams which are subjected to bending. In clamped-clamped beams this effect results in a nonlinear response known as strain stiffening or as membrane stiffening [8]. This nonlinear response becomes dominant for motions proportional to half the beam width instead of the beam length [9]. Consequently, the standard folded-beam suspension exhibits strong nonlinear response for periodic motions beyond the beam width. Furthermore, simulations in [7] show that when the inertia of the flexures and flying-bars are negligible, the dynamic response of the standard folded-beam becomes linear. Since the inertia of flexures and flying-bars cannot be ignored, practical design must consider these effects.

In this work we present the notion of dynamically balanced folded-beam suspensions, which responds as a linear spring at the fundamental resonance. In the new

suspension the anchored beams are shorter by a specific factor, such that at the fundamental resonance no axial stresses are induced in the flexures, and the dynamic response is therefore linear. We present a methodology for designing practical dynamically-balanced folded-beam suspensions, which fully accounts for the inertia of the flying-bars and flexures. We also present experimental validation of the new suspension.

FOLDED-BEAM SUSPENSION - REVISITED

The standard folded beam suspension is constructed from eight identical flexure beams, connecting the shuttle to the anchors through two stiff flying-bars (Fig 1a). In static loadings, it follows from symmetry considerations that when the shuttle moves by 2Δ, the flying-bars move by Δ. Consequently, all beams deform to the same specific shape or its mirror image. Therefore, all forces induced by the flexures to each flying-bar, cancel-out identically (detailed at the right of Fig. 1a).

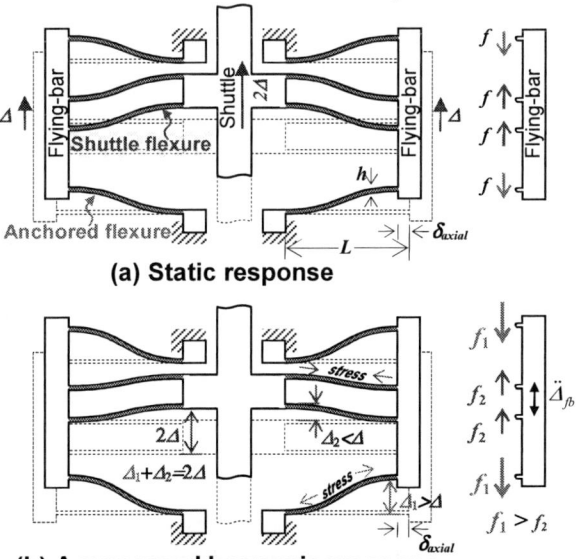

Standard Folded-Beam Suspension

(a) Static response

(b) A presumed harmonic response

Figure 1: Deformed shape of the standard folded-beam suspension: (a) Static response. All flexures deflect by Δ, which is half the shuttle motion, and they all have an identical deformation shape and axial contraction. All flexures induce equal forces to the flying-bar, such that they cancel-out identically in a static state. (b) A presumed harmonic response. The anchored flexures must induce a larger restoring force to affect the acceleration of the flying-bars. But this induces membrane stiffening: axial tension in the anchored flexures (solid red) and axial compression in the shuttle flexures (solid green). It follows that due to this membrane stiffening, the dynamic response of the standard folded-beam suspension cannot be linear.

978-1-4799-7956-1/15 $31.00 © 2015 IEEE 215 MEMS 2015, Estoril, PORTUGAL, 18 - 22 January, 2015

It follows that each flying-bar is in static equilibrium, and that each flexure deforms as an elastic clamped-rotationally constrained Euler-Bernoulli beam. It can be shown that the effective axial contraction, δ_{axial}, of such a beam is given by

$$\delta_{axial} = \frac{3}{5}\frac{\Delta^2}{L} + O(\Delta^4 / L^3) \tag{1}$$

If the flexure cannot contract axially (as in clamped–clamped flexures) then axial stress will be induced, resulting in nonlinear membrane-stiffening. It can be shown that this nonlinearity becomes dominant for displacements of the order of half the beam width [9].

The standard folded-beam suspension is intentionally designed with eight *identical* flexures, to ensure that all undergo the same axial contraction, and that no axial stress is induced.

However, at the fundamental resonance of the system, the flying-bar must be accelerated and decelerated in its harmonic motion. This means that the net force applied to the flying-bar by the flexures *cannot* cancel-out. Specifically, the anchored flexures must exert a larger restoring force than the shuttle flexures (Fig. 1b), to affect such accelerations.

Seemingly, the only way the anchored flexures may induce a larger restoring force, is if their relative edge displacement Δ_1 is larger than that of the shuttle flexures, Δ_2 (Fig. 1b). This means that the flying-bars will travel more than half the distance of the shuttle. Consequently, since the axial contraction of the anchored and shuttle flexures is constrained by the stiff flying-bar to be the same, the anchored flexure will be in axial tension while the shuttle flexure will be in axial compression (Fig. 1b).

This effect will be negligible if the displacements Δ_i are very small and therefore, the associated axial contraction will be insignificant. However, in [7] we have shown that for displacements on the order of the width h of the beams, this nonlinear membrane-stiffening becomes dominant resulting in a nonlinear response of the suspension.

DYNAMICALLY BALANCED SUSPENSION

To achieve a harmonic motion of the flying-bars, the restoring forces induced by the anchored flexures must be larger than those of the shuttle flexures. In our new design, this is achieved by shortening the length of the anchored flexures (Fig 2a). This results in a nonzero resultant force on the flying-bar, which is necessary for its acceleration. A simplified model of a quarter of the new suspension is presented in Fig. 2b.

Next, we present the equations of motion which govern the fundamental resonance response of the system. A full derivation and more explanations will be provided in a future publication.

The motion equations are given by

$$-\lambda^4 \frac{EI}{\rho A L_2^4} m_{sh}(\Delta_1 + \Delta_2) = -f_{2sh}$$

$$-\lambda^4 \frac{EI}{\rho A L_2^4} m_{fb}\Delta_1 = -f_{2fb} - f_{1fb} \tag{2}$$

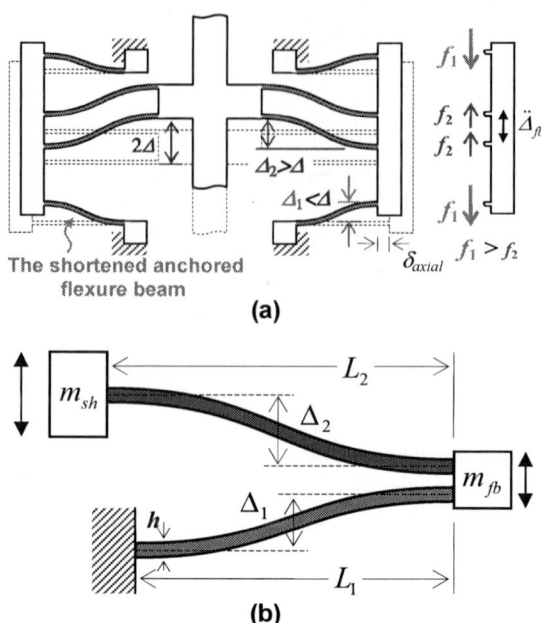

Dynamically Balanced Folded-Beam Suspension

Figure 2: The dynamically balanced folded-beam suspension. (a) The suspension has flexures beams of different lengths. The anchored flexure is shortened by a specific ratio such that it may induce a larger force to change the momentum of the flying-bar, but simultaneously its axial contraction is equal to that of the shuttle flexure. (b) A simplified model of the dynamically balanced folded-beam suspension. Due to the symmetries of the full suspension, the simplified model considers only a quarter of the system.

Here the eigenvalue is $\lambda = \omega^{\frac{1}{2}}(\rho A / EI)^{\frac{1}{4}}L_2$ where ω is the angular resonance frequency, ρ and E are the density and elasticity modulus of the flexure material. A and I are the area and second-moment of the flexures cross-section. m_{sh} and m_{fb} are the masses of the shuttle and flying-bar, respectively (Fig. 2b). The edge forces on the flying-bar and shuttle are given by,

$$f_{2sh} = \lambda^3 \frac{EI}{L_2^3}\left[\frac{(\Delta_1 + \Delta_2)\Gamma_a(\lambda) - \Delta_1\Gamma_b(\lambda)}{1 - \cos(\lambda)\cosh(\lambda)}\right]$$

$$f_{2fb} = -\lambda^3 \frac{EI}{L_2^3}\left[\frac{(\Delta_1 + \Delta_2)\Gamma_b(\lambda) - \Delta_1\Gamma_a(\lambda)}{1 - \cos(\lambda)\cosh(\lambda)}\right] \tag{3}$$

$$f_{1fb} = \lambda^3 \frac{EI\Delta_1}{L_2^3}\frac{\Gamma_a(\lambda L_1 / L_2)}{1 - \cos(\lambda L_1 / L_2)\cosh(\lambda L_1 / L_2)}$$

where $\Gamma_a(\lambda) = \sin(\lambda)\cosh(\lambda) + \cos(\lambda)\sinh(\lambda)$ and $\Gamma_b(\lambda) = \sin(\lambda) + \sinh(\lambda)$.

The set of motion equations (2) may be solved for the resonance frequency (i.e. eigenvalue) and the ratio $\alpha = \Delta_1/\Delta_2$ (eigenmode). As in any eigenvalue problem, the amplitude of motion (e.g. Δ_2) is arbitrary. However, for a given L_2 and an arbitrary L_1, this linear analysis may result in incompatible axial contractions $\delta_1 = 0.6\Delta_1^2/L_1$ and $\delta_2 = 0.6\Delta_2^2/L_2$, according to (1).

978-1-4799-7956-1/15 $31.00 © 2015 IEEE   216

To solve this problem, we augment (2) with the constraint $\delta_1 = \delta_2$ such that the axial contractions of all beams are equal. This constraint can be rewritten as

$$L_1 / L_2 = \Delta_1^2 / \Delta_2^2 = \alpha^2 \qquad (4)$$

Simultaneous solution of (2) and (4) will result in the resonance frequency (eigenvalue), the ratio $\alpha = \Delta_1 / \Delta_2$ (eigenmode), and the correct length $L_1 = \alpha^2 L_2$ which ensures no axial stress is induced in the flexure beams - *regardless of motion amplitude*. The motion amplitude (e.g. Δ_2) is still arbitrary, because we are still solving a linear eigenvalue problem.

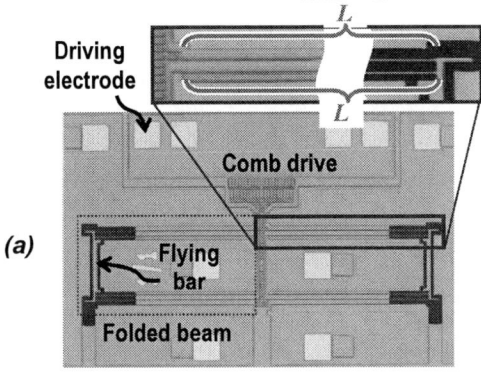

Standard folded-beam suspension
Two flexures of *equal* lengths

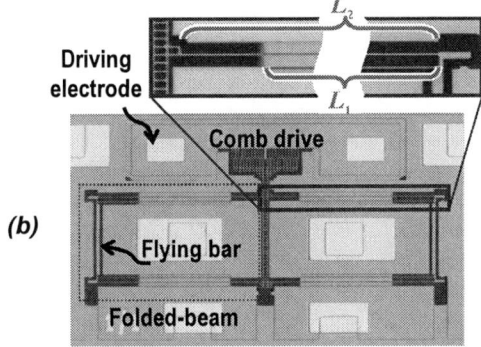

The new Dynamically-Balanced suspension
Two flexures of *different* lengths

*Figure 3: Microphotos of the two types of fabricated test devices (a) An electrostatic comb-drive resonator with a **standard** folded-beam suspension. All eight flexure beams have equal lengths. (b) An electrostatic comb-drive resonator with the **dynamically-balanced** folded-beam suspension. All anchored flexures are shortened by a specific factor.*

EXPERIMENTAL VALIDATION

Test devices were fabricated using SOIMUMPs technology (runs 41 and 47, [10]). The test devices are electrostatic comb-drive resonators suspended on folded-beam suspensions. Two types of test devices were fabricated: one device with a standard folded-beam suspension with beams of equal length (Fig. 3a), and the other with a dynamically-balanced suspension with a shorter anchored beam (Fig. 3b). The devices were designed with an arbitrary mass ratio of $m_{sh} = m_{fb}$. The

flexure beams were designed to be $h=3\mu m$ wide, $t=25\mu m$ thick, and $L_2=600\mu m$ long, except for the shorter beam in the dynamically-balanced suspension. For these devices the shorter beam was designed to be $L_1=497\mu m$ long. This length was determined by solving (2) and (4), with the appropriate masses of shuttle and flying bars, material properties, and the geometric parameters h, t, and L_2 detailed above.

The dynamic response of each test device was characterized in ambient pressure of $\sim 1Torr$. The device was subjected to a driving voltage with a constant component of $V_{dc}=5V$, and a harmonic component V_{ac} in the range of $0.02VRMS$ to $1.2VRMS$. A laser vibrometer measured vibration displacement amplitudes. Reasonable SNR was achieved by using a lock-in amplifier which provided the reference ac signal to the comb-drive actuator, and acquired the signal from the vibrometer with respect to the same ac frequency.

Figure 4 presents measured time response at resonance frequency for small ($\sim 0.5\mu m$) and large ($\sim 6\mu m$) motions relative to the flexure width ($3\mu m$). It is evident that the device with the standard suspension exhibits a nonlinear response in large motion amplitudes (Fig. 4a), whereas the device with the dynamically-balanced suspension maintains its linear response in large motion amplitudes (Fig. 4b).

Standard folded-beam suspension

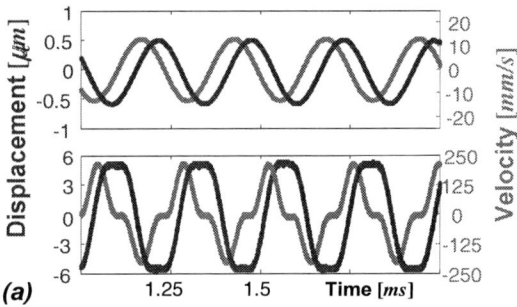

The new Dynamically-Balanced suspension

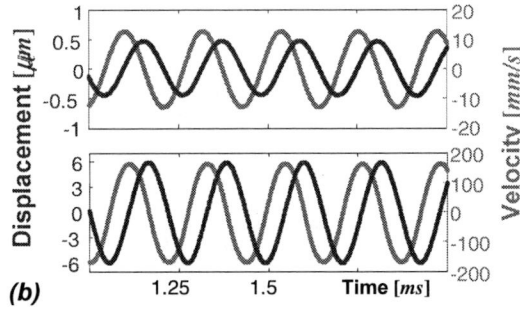

*Figure 4: Measured displacement and velocity at resonance. (a) With the **standard** folded-beam suspension, for small motion amplitudes ($0.5\mu m$) the response is linear, but for larger motion amplitudes ($6\mu m$) the response is clearly nonlinear. (b) With the **new dynamically-balanced** folded-beam suspension, the measured response is linear, for both small ($0.5\mu m$) and large ($6\mu m$) motion amplitudes.*

Figure 5 presents measured displacement amplitudes for frequency sweeps, with different *ac* voltage settings. For comparison, the frequencies in each figure are normalized by the natural frequency of the related device (measured for the smallest V_{ac}). The frequency sweeps of the standard device (Fig. 5a) exhibit considerable nonlinear stiffening, which increases with increasing V_{ac}. It is clear that when motion amplitude increases above $\sim 1.5 \mu m$ (i.e. about half of the width h of the flexure beams), the nonlinear stiffening becomes more apparent. In contrast, the device with the dynamically-balanced folded-beam suspension (Fig. 5b) shows only marginal stiffening. We attribute this to fabrication inaccuracies which affect both the width of flexures (measured to be $2.5 \mu m$ instead of $3 \mu m$), and the masses of the shuttle (including the comb-drives) and flying-bars.

Standard folded-beam suspension

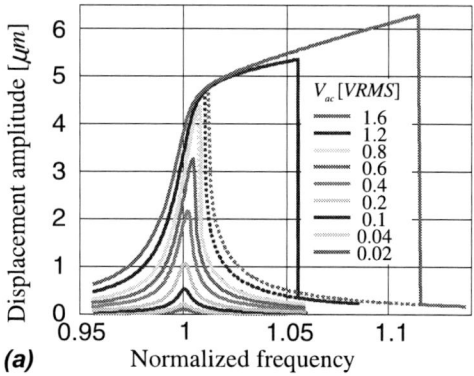

(a)

The new Dynamically-Balanced suspension

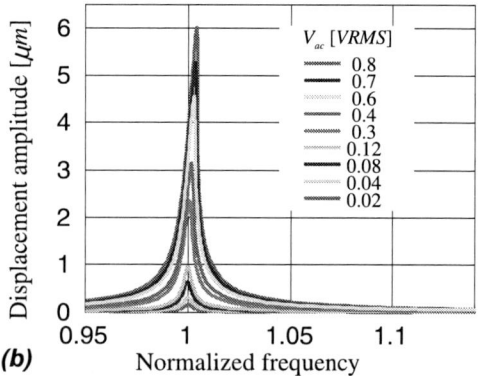

(b)

Figure 5: Measured frequency sweeps of displacement amplitude for different ac settings. The legend indicates the required ac voltage to achieve the desired displacement. For each device, frequency is normalized by the nominal natural frequency, which is measured at small amplitudes (i.e. V_{ac}=0.02VRMS). (a) Measurements of the resonator with the standard folded-beam suspension. For motion amplitudes larger than half the width of the $3 \mu m$ wide flexures, the response becomes strongly nonlinear. (b) Measurements of the resonator with the new dynamically-balanced folded-beam suspension. Clearly the new design drastically improves the linearity of the resonance response.

Though the balanced device shows marginal stiffening it does not exhibit any nonlinear bifurcation, which is very dominant in the standard device.

CONCLUSION

In previous work we demonstrated that the dynamic response of the standard folded-beam suspension is essentially nonlinear at the fundamental resonance of the system. We have shown that this nonlinear response becomes dominant when the motion amplitude is on the order of the *width* of flexure beams.

In this work we presented the methodology for designing a dynamically-balanced folded-beam suspension, which has a linear response at the fundamental resonance. We experimentally validated the design by comparing the response of two similar comb-drive resonators: one with the standard folded-beam suspension, and the other with the new design.

Since the new suspension is linear at the fundamental resonance, it is clear that its response at any other frequency - and specifically under static loading - may be expected to exhibit nonlinearities. However, this is insignificant since so many resonators are optimized to perform at a specific frequency.

ACKNOWLEDGEMENT

This study was partially supported by the Farkas Family Fund for Research, and by the Russell Berrie Nanotechnology Institute (RBNI), Technion.

REFERENCES

[1] W. C. Tang, T. C. H. Nguyen, and R. T. Howe, "Laterally Driven Polysilicon Resonant Microstructures," *Sensors and Actuators*, **20**, 25-32, 1989.

[2] C. Marxer, O. Manzardo, H. P. Herzig, R. Dandliker, and N. De-Rooij, "An Electrostatic actuator with large dynamic range and linear displacement-voltage behavior for a miniature spectrometer," *IEEE-Transducers '99*, Sendai, Japan, 1999.

[3] S. Shmulevich, M. Lerman, and D. Elata, "On the quality of quality-factor in gap-closing electrostatic resonators," *JMM*, **23**, 115010, 2013.

[4] C. T. C. Nguyen and R. T. Howe, "An integrated CMOS micromechanical resonator high-Q oscillator," *J. Solid-State Circuits*, **34**, 440-455, 1999.

[5] C. Guo and G. K. Fedder, "A quadratic-shaped-finger comb parametric resonator," *JMM*, **23**, 095007, 2013.

[6] V. Jaecklin, C. Linder, N. De Rooij, and J. Moret, "Micromechanical comb actuators with low driving voltage," *JMM*, **2**, 250, 1992.

[7] R. Legtenberg, A. W. Groeneveld, and M. Elwenspoek, "Comb-drive actuators for large displacements," *JMM*, **6**, 320-329, 1996.

[8] S. Shmulevich, A. Joffe, I. Hotzen, and D. Elata, "Are folded-beam suspensions really linear?," *Eurosensors 2014*, Brescia, 2014.

[9] E. S. Hung and S. D. Senturia, "Extending the travel range of analog-tuned electrostatic actuators," *IEEE-JMEMS*, **8**, 497-505, 1999.

[10] http://www.memscap.com/products/mumps/soimumps.

MEMBRANE-BASED CHEMOMECHANICAL TRANSDUCER FOR THE DETECTION OF APTAMER-PROTEIN BINDING

Jun-Kyu Choi[1] and Junghoon Lee[2]
[1]Small machines incorporation, Seoul, South Korea
[2]Seoul National University, Seoul, South Korea

ABSTRACT

We report a membrane-based chemomechanical transducer for the sensitive detection of surface molecular reaction through a highly reliable common mode rejection (CMR) technique. Chemomechanical transduction, originally based on the micro-cantilever, offers potential benefits: label-free assay, and real-time monitoring of molecular interaction via mechanical deformation [1, 2]. Membrane-based approaches have been proposed to overcome the inherent limitations of the micro-cantilever system, but most results were either inconclusive or far from practical standards. Here we show clear-cut detection of molecular binding using a membrane transducer fabricated with conventional MEMS technology. This goal is achieved through the implementation of CMR that rejects physical effects such as pressure and temperature, leaving only specific chemical binding responsible for resulting signal. We demonstrate highly specific recognition of thrombin protein by using DNA aptamer immobilized on the membrane surface with the limit of detection down to ~3 pM, and the wide dynamic range > 5×10^4.

INTRODUCTION

The cost of health care has been exponentially increased due to aging of the world`s population; molecular diagnostics market has become important for early prevention of the disease before costly medical treatment. Chemomechanical transduction for biological and chemical sensing has generated wide interest in the expanding applications of molecular diagnostics and drug discovery because of fast, efficient, and high-throughput detection technology of molecular interaction without exogenous labels. Micro- and nanometer scale cantilevers were mainly suggested and used for these purposes. The micro-cantilever and some polymer devices demonstrated the detection of surface stress change caused by specific bio molecular interactions. These approaches were, however, hardly realized into a compact device for point-of-care test (POCT) due to the optical detection equipment and environmental noises [3].

For sensing surface stress, thin membrane transducer on the other hand has several advantageous characteristics over the cantilever approaches. First, shell structures are more robust and stable than cantilevers, but is still very sensitive to surface reaction [4], therefore, can be easily functionalized and probed by using commercially available fabrication techniques [5]. Secondly, the reaction surface is physically isolated from the electrical sensing surface which can be easily implemented for a precise low-noise capacitance measurement technology [6]. The isolated reaction surface provides a diverse platform which is accessible by both liquid and gas samples. Finally, electric measurement of sensing structure can be readily scaled and multiplexed with different reaction chamber [7, 8].

In the previous research [9, 10], a polymer based thin membrane transducer was introduced to show the possibility of biomolecular detection. To improve the lower sensitivity of the membrane transducer compared with cantilever type transducer, polymer material which has lower rigidity was used. The advantage of the low rigidity was offset by the decrease of reliability in wet environments and fabrication uniformity compared with micro machined materials; silicon oxide, poly silicon, and silicon nitride. In addition to difficulties in fabricating a completely integrated device with polymer membrane, polymer based transducer has issues in the packaging of fluidic channel and measurement circuits because of high flexibility, low melting point, and sensitivity to water.

The reliability of chemical and biological diagnosis is crucial for commercial products, as reliability often dominates the device designs. In essence, the membrane transducer will respond with much higher signal to physical disturbances such as pressure and thermal deformation. No previous approach clearly addressed these issues, resulting in unreliable polymeric membrane devices [9], unreasonable signal range [11], and premature results with less practical significance [12]. Silicon-based MEMS fabrication provides core technologies to realize integrated systems with high reliability and advanced functionalities. A key to overcoming the aforementioned challenges is the rigorous implementation of common mode rejection (CMR) with highly reliable fabrication, and precision noise control.

Here we report a highly reliable chemomechanical transducer system, consisting of a MEMS fabricated membrane transducer, custom-immobilized receptor molecule, and high precision capacitive measurement system. A fully integrated thin membrane transducer is developed by using micromachining with silicon nitride membrane material that is inert to chemical interaction and commonly used in MEMS sensors. We demonstrate a high performance signal detection with clear CMR process.

EXPERIMENTAL

Membrane-based chemomechanical transducer system

Our membrane transducer has a sensing surface, analyte reaction chamber, and read out structure, which is composed of seven layers fabricated through mixed surface and bulk micromachining. Figure 1 shows our transducer and fabrication process. Two devices are next to each other, sharing the analyte solution through a microfluidic well. Receptor molecule is immobilized only on one of the device surfaces. Molecular binding on the 0.5-μm thick silicon nitride membrane produces a small deflection in addition to the bending by all other physical effects such as hydraulic pressure, heat fluctuation and even non-specifically chemical adsorption. This deformation can be detected by changing the distance between membrane and an opposite metal structure which offers a precise capacitance gap.

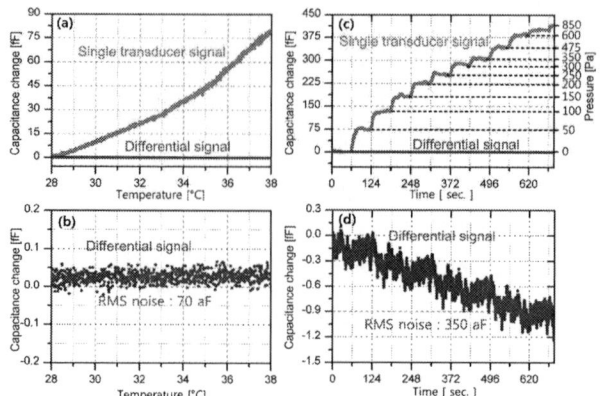

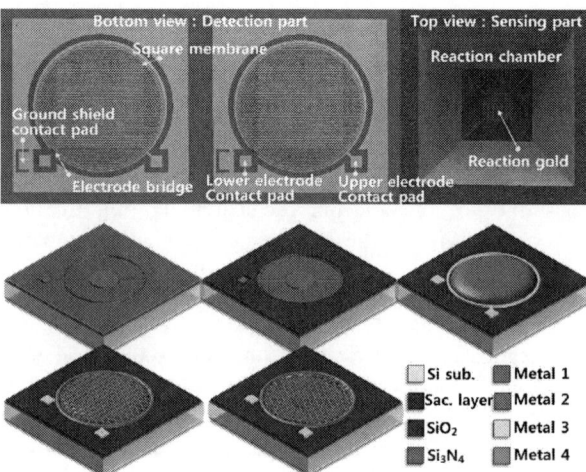

Figure 1: Fabricated sensors with capacitive sensing configuration. Major fabrication process consists of film deposition/patterning, electroplating, sacrificial etching, and bulk etching.

When measured signals from the two devices are subtracted only the signal by specific molecular binding remains. It is of crucial importance to maintain the two devices physically identical for this CMR. We developed a production quality fabrication process, a packaging method, and a low noise signal processing to guarantee reasonable resemblance between the two adjacent devices. We first demonstrate the physical signal rejection and noise characteristics of our device.

One of the key challenges of the membrane-based chemomechanical transducers are regarding how the external disturbances can be overcome. Low noise signal obtained by subtracting the measured capacitance values between the two devices shows the excellent rejection performance of the thermal and pressure effects as described in figure 2. Thermal characterization of the sensor was performed to evaluate sensor performance and test if the membranes were fully released from the substrate. The plot of normalized sensor capacitance versus temperature for a fully released membrane is shown in figure 2(a). The large thermal response of capacitance change is 9.0 fF/K which results from the composite membrane, coated with the sandwich of metal layers of different area and shape to enhance the sensitivity of surface stress. However, the response resulting from the CMR of the array sensor to the change of temperature from 28°C to 38°C was less than ~ 100 aF as shown in figure 2(b). Figure 2(c) also indicates that the capacitance change of a single membrane transducer is vulnerable to dynamic pressure chance when the flow velocity in micro-channel over the membrane was increased. The flow velocity was translated into the pressure variation with Hagen-Poiseuille equation, indicated on the graph of figure 2(c). Again, the measured signal with the CMR of full pressure on both sensors was only ~ 1fF which is insignificant compared with the individual signal of ~ 400 fF.

Figure 2: (a) Cap by thermal deformation (b) CMR thermal signal (noise ~70 aF) (c) Cap by hydraulic pressure (d) CMR pressure signal (noise ~350 aF)

Chemomechanical transduction

The high specificity and affinity of aptamer make it possible to achieve higher detection sensitivity and selectivity in many sensing formats. The fundamental principle of aptamer-based bioassays can also enable the best configuration for the chemomechanical transduction of bio-recognition events. In this paper, we used the protein thrombin and thrombin binding aptamer (TBA) pair for the demonstration of biosensing. Since thrombin is a key protein in the regulation of thrombosis and haemostasis, the aptamer that recognizes thrombin has been extensively investigated. Special assay for stable and sensitive receptor structure was developed for our membrane-based transducer system by adjusting molecular binding density and blocking protocol. For the densely self-assembled monolayer SAM formation, the completely fabricated membrane sensor was coated with 11-mercapto urethane acid (MUA), to prohibit non-specific molecular adsorption [13, 14]. The dense SAM also prevents the surface stress fluctuation due to adsorption of ion charge on the defect sites which include Au binding sites unoccupied with SAM. On top of the SAM layer, we immobilized a thiol-modified TBA (HTQ: 5'-GGTTGGTGTGGTTGG-3') by using 4-amino phenyl maleimide with linker. This immobilization process will result in the surface density of ~ 9.1×10^{12}/cm^2 [15].

When target protein thrombin (100 nM) was introduced in our receptor-coated transducer, all three signals from sensor, reference, and difference clearly manifested the extraction process of key molecular binding as shown in figure 3. Thrombin protein was freely diffused without external mixing when injected. Therefore, the capacitance value of both sensor and reference transducer slowly increased with about 15 fF increase due to non-specific adsorption (left axis). However, the distinct signal from the sensor transducer was higher by about 10 fF because of the specific molecular interaction (right axis). The TBA recognized thrombin proteins, and the reaction induced the increase of intermolecular repulsion forces which caused the compressive stress on the membrane [10]. Figure 3 clearly shows the rejection process of other physical events.

978-1-4799-7956-1/15 $31.00 © 2015 IEEE

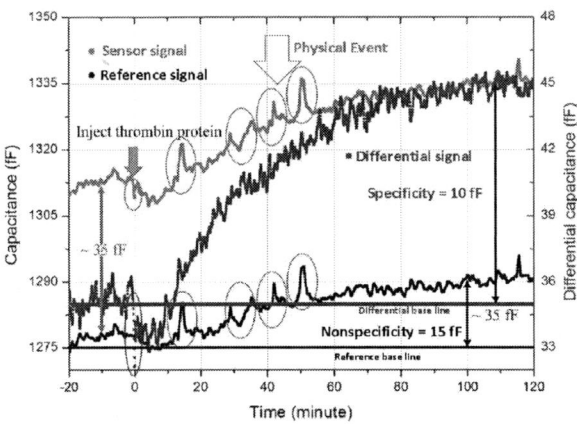

Figure 3: Time-course response of TBA binding with thrombin (100 nM). Left axis is for sensor and reference signals, and right axis is for differential signal. Physical event occurred by a short temperature fluctuation when the reaction chamber was opened.

RESULTS AND DISCUSSION

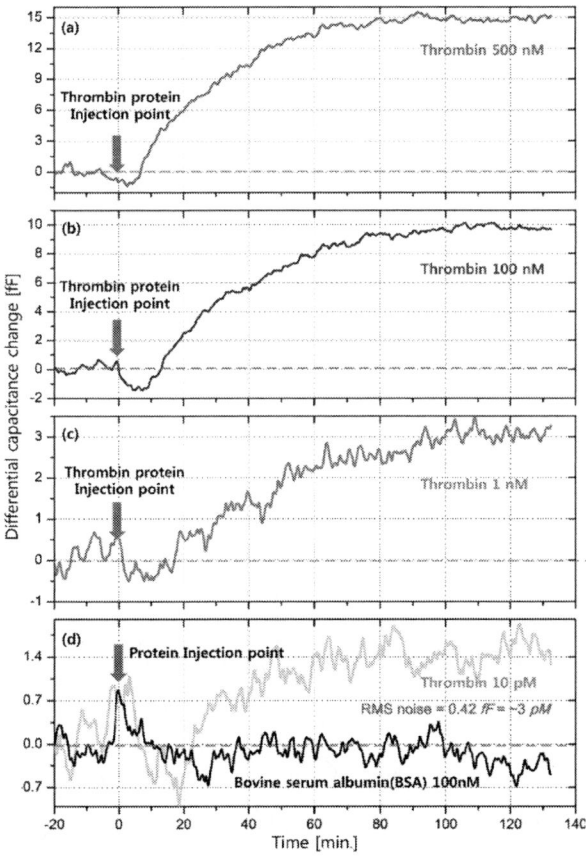

Figure 4: Figure 4: CMR signals with various target thrombin concentration. Every signal has time constant of ~ 60 min which can be improved by introducing a mixing technique. LOD estimated by min noise is ~ 3 pM. Dynamic range > 5×10⁴. BSA control has no significant response.

Figure 4 shows example results of testing various target concentrations. Low-pass filtered at 0.1 Hz, the resulting signal has a very low noise with the limit of detection at ~3 pM. The central deflection of membrane according to the thrombin protein interaction of 500 nM was 24 nm according to the calculation with our model to produce the measured signal of 15 fF. The distance of electrodes depending on the thickness of sacrificial layer was 2 μm which have to be subtracted with the deflection of 1.3 μm due to hydraulic pressure. According to finite element simulation shown figure 5, the amount of deflection that corresponds to this deformation was the compressive surface stress of 165 mN/m. The surface stress of 165 mN/m was quite higher than that of usual aptamer interaction at 500 nM target protein. It needs to be pointed out that the surface stress of 51 mN/m was obtained with the signal of cocane-target aptamer-based cantilever sensor at the concentration of 500 μM [16]. Additional work is needed to explain other income-prehensible effects.

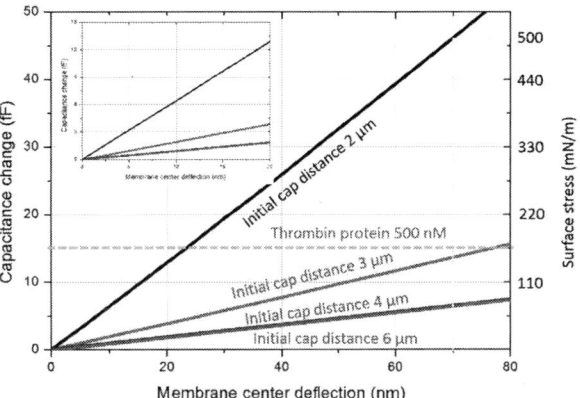

Figure 5: Capacitance calculation according to membrane deflection.

The dynamic range of our chemomechanical detection was investigated by performing binding experiments at a total of seven different thrombin concentrations (over 2 times for each concentration). Figure 6 shows the differential capacitance change due to membrane deflection as a function of thrombin concentration (logarithmic scale). The dynamic range of our sensor was > 5×10^4.

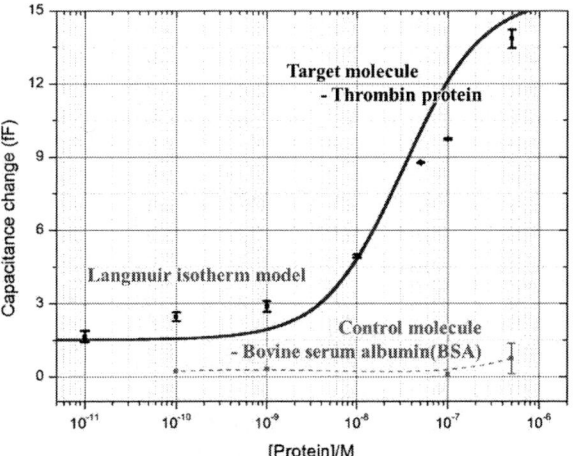

Figure 6: Collection of repeated test results for obtaining the affinity and dynamic range.

We further investigated the use of this platform for finding the dissociation constant (K_D) of TBA-thrombin binding as shown in figure 7. Curve fitting with the langmuir isotherm model results in the K_D of 26 nM which is close to reported value [17]. This task of obtaining K_D would need a high-end equipment such as SPR setup, but our system could accomplish it with a single chip. Additional salient features include quantitative, time course signal, and array-ready system, all on a single chip platform, leading to high performance POCT.

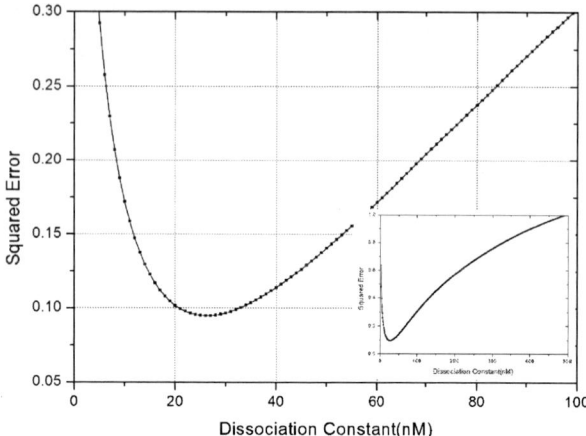

Figure 7: The curve fitting of square error vs. KD value.

CONCLUSIONS

This research showed the development of membrane-based chemomechanical transducer, for the sensitive detection through a highly reliable common mode rejection technique. Our work is a showcase where core MEMS technology for the new transducer has successfully overcome the challenges encountered by previous approaches. A POCT that challenges the performance indices of central testing equipment is on the horizon.

ACKNOWLEDGEMENTS

This work was supported by the Small and Medium Business Administration as Start-up Commercialization for Global market (S2230578). And the fabrication was performed at the Interuniversity Semiconductor Research Center (ISRC) in Seoul National University.

REPERANCES

[1] M. K. B. J. Fritz, H. P. Lang, H. Rothuizen, P. Vettiger, E. Meyer, H.-J. Gu¬ntherodt, Ch. Gerber, J. K. Gimzewski, "Translating Biomolecular Recognition into Nanomechanics," *Science*, vol. 288, pp. 316-318, 2000.

[2] E. B. M. a. M. L. R. J.L. Arlett, "Comparative advantages of mechanical biosensors," *Nature nanotechnology*, vol. 6, pp. 203-215, 2011.

[3] S. K. Ram Datar, Sangmin Jeon, Peter Hesketh, Scott Manalis, Anja Boisen, and Thomas Thundat, "Cantilever Sensors: Nanomechanical Tools for Diagnostics," presented at the MRS BULLETIN, 2009.

[4] E. T. Carlen, M. S. Weinberg, C. E. Dubé, A. M. Zapata, and J. T. Borenstein, "Micromachined silicon plates for sensing molecular interactions," *Applied Physics Letters*, vol. 89, p. 173123, 2006.

[5] Z. C. Wu, K. Griffiths, H. R. Xu, J. Ma, X., "A novel silicon membrane-based biosensing platform using distributive sensing strategy and artificial neural networks for feature analysis," *Biomed Microdevices*, vol. 14, pp. 83-93, Feb 2012.

[6] E. T. C. A.M. Zapata , E. S. Kim1, J. Hsiao1, D. Traviglia1, M.S. Weinberg, "BIOMOLECULAR SENSING USING SURFACE MICROMACHINED SILICON PLATES," presented at the The 14th International Conference on Solid-State Sensors, Actuators and Microsystems, Lyon, France, 2007.

[7] S.-H. S. Lim, D. Raorane, S. Satyanarayana, and A. Majumdar, "Nano-chemo-mechanical sensor array platform for high-throughput chemical analysis," *Sensors and Actuators B: Chemical*, vol. 119, pp. 466-474, 2006.

[8] T. Xu, Z. Wang, J. Miao, L. Yu, and C. M. Li, "Micro-machined piezoelectric membrane-based immunosensor array," *Biosens Bioelectron*, vol. 24, pp. 638-43, Dec 1 2008.

[9] S. Satyanarayana, McCormick, Daniel T. and Majumdar, Arun, "Parylene micro membrane capacitive sensor array for chemical and biological sensing," *Sensors and Actuators B: Chemical*, vol. 115, pp. 494-502, 2006.

[10] M. Cha, J. Shin, J. H. Kim, I. Kim, J. Choi, N. Lee*, et al.*, "Biomolecular detection with a thin membrane transducer," *Lab Chip*, vol. 8, pp. 932-7, Jun 2008.

[11] V. Tsouti, M. Filippidou, C. Boutopoulos, P. Broutas, I. Zergioti, and S. Chatzandroulis, "Self-Aligned Process for the Development of Surface Stress Capacitive Biosensor Arrays," *Procedia Engineering*, vol. 25, pp. 835-838, // 2011.

[12] G. Yoshikawa, T. Akiyama, S. Gautsch, P. Vettiger, and H. Rohrer, "Nanomechanical membrane-type surface stress sensor," *Nano Lett*, vol. 11, pp. 1044-8, Mar 9 2011.

[13] R. J. White, N. Phares, A. A. Lubin, Y. Xiao, and K. W. Plaxco, "Optimization of electrochemical aptamer-based sensors via optimization of probe packing density and surface chemistry," *Langmuir*, vol. 24, pp. 10513-8, Sep 16 2008.

[14] M. Godin, V. Tabard-Cossa, Y. Miyahara, T. Monga, P. J. Williams, L. Y. Beaulieu*, et al.*, "Cantilever-based sensing: the origin of surface stress and optimization strategies," *Nanotechnology*, vol. 21, p. 075501, 2010.

[15] P. L. Simon D. Keighley , Pedro Estrela, Piero Migliorato, "Optimization of DNA immobilization on gold electrodes for label-free detection by electrochemical impedance spectroscopy," *Biosens and Bioelectronics*, vol. 23, pp. 1291-1297, 2008.

[16] A. S. Kyungho Kang, Marit Nilsen-Hamilton, and Pranav Shrotriya, "Aptamer Functionalized Microcantilever Sensors for Cocaine Detection," *Langmuir*, vol. 27, pp. 14696–14702, 2011.

[17] H. Hasegawa, K.-i. Taira, K. Sode, and K. Ikebukuro, "Improvement of Aptamer Affinity by Dimerization," *Sensors*, vol. 8, pp. 1090-1098, 2008.

CONTACT

*J. Lee, Tel: +82-2-880-9104; jleenano@snu.ac.kr

HIGHLY SENSITIVE SERS DIAGNOSIS FOR BACTERIA BY THREE DIMENSIONAL NANO-MUSHROOMS AND NANO-STARS-ARRAY SANDWICHED ON BACTERIAL AGGREGATION

Chun-Wei Lee[1], Jen-Kuei Wu[1], and Fan-Gang Tseng[1,2]

[1] Department of Engineering and System Science, National Tsing Hua University, Hsinchu 300, Taiwan R.O.C

[2] Research Center for Applied Sciences, Academia Sinica, Taipei 115, Taiwan R.O.C

ABSTRACT

This paper reports a highly sensitive SERS Diagnosis system by incorporating three dimensional Nano-Mushrooms and Nano-Stars-Array sandwiched on Bacterial Aggregation. Through the action of ACEOF and nano-mushroom/bacteria/nano-stars-array self-aggregation process, the signal can be much enhanced by 5 orders of magnitude in 5 minutes. Detection limit can approach 1 bacterium/ml from the analysis result.

INTRODUCTION

Typical researches in SERS were focusing on fabricating nano-structures providing large Raman amplification[1][2]. However, these structures were mostly fixed on a substrate, as a result, highly concentrated analyte is required to filling into the hot spot region for obtaining enough RS signals. On the other hand, several other researches combined analyte capturing methods to increase analyte local concentration for enhancing SERS signal [3][4] to improve the detection limit. However, for detecting bacteria, due to steric mismatching issue between bacteria and nano SERS structures, the SERS hot spots may either easily miss the detectable regions of bacterial or providing not enough encountering points for signal enhancement. As a result, a 3D combination of surface nanostructures with freely mobile SERS particles may resolve the aforementioned problem for fully extracting bacterial surface information. In this study, a surface nano-stars-array incorporated with mobile nano-mushrooms are employed together. Through ACEO flow, they can form self-aggregated 3D structures with bacteria, thus signals from all over the bacterial surface can be fully obtained, as shown in Fig. 1.

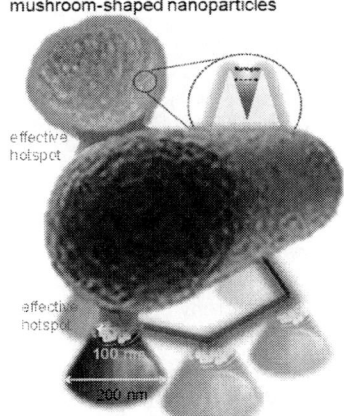

mushroom-shaped nanoparticles

Figure 1 : mushroom-shaped nanoparticles and nanostars Au nanopillar arrays produce hot spot on the top side and bottom side of a bacterium, Respectively

THEORY

AC Electroosmosis(ACEO)

ACEO is a kind of electrodynamics of manipulate the flow field to achieve transmission, hybrid or collect particulates in solution.[5] The positive electrode surface attracts ions form an electrical double layer, the electric field drive the ions on the electrode move to the -X direction and then stimulate the flow field. when the signal is switched to the negative electrode, positive ions attracted to form an electrical double layer. The flow field will stimulate by electric field to the -X direction. In other words, the flow field always has net flow forward to the electrode. Fig. 2

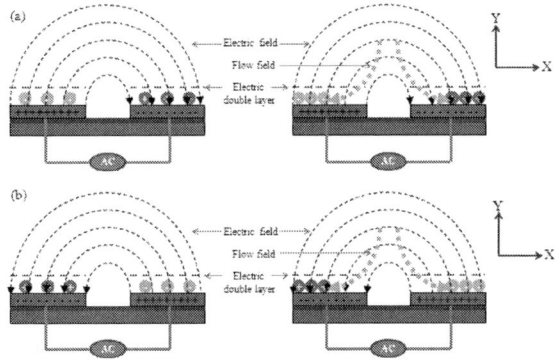

Fig 2: Schematic of the ACEO

Raman Scattering

When the incident light beam to a sample (molecules or crystals), the photons scattered by collisions with molecules, with a change of momentum or energy can be divided into elastic collision and inelastic collisions. Elastic collision: incident light only change in direct due to the change in momentum, and the scattering photon energy is the same with incident photon, it named Reyleigh scattering. inelastic collisions: the scattered photon involved energy change and makes the photon frequency (wavelength or wave number) up or down in the collision, it named Raman scattering. The amount of photon frequency change is related the sample. It means we can distinguish the sample species via detect the Raman scattering.[6] Fig. 3

Although Raman spectroscopy is a powerful analytical method for distinguish the structures of materials, it's low signal intensity limit the development and applications (I=Io x 10-9). Until the surface enhanced Raman scattering(SERS) is reported by Fleischm. We can enhance the intensity of Raman signal several orders by roughened surfaces of properly metal substrates (Au or Ag).[6][7]

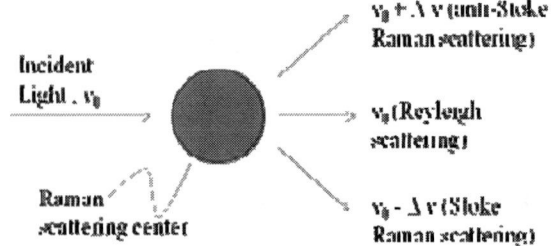

Fig 3: Schematic of the Raman scattering

EXPERIMENTAL

Principle

The 3D SERS aggregation integrated three important components, including a nanao-mushroom with corrugated surface nanostructures, which can enhance 20-100 folds RS signal [7], and a nano-stars-array by anisentropic O_2 plasma etching on fused silica substrate with nano-gold particles as masks, which can provide 5 times signal enhancement by SERS and array resonance. Finally, Alternative current electroosmosis (ACEO) flow was employed to preconcentrate nano-mushrooms and analytes, such as E.coli, on the top of nano-stars-array at the detection region for forming 3D aggregation structure, as schematically shown in Fig. 4.

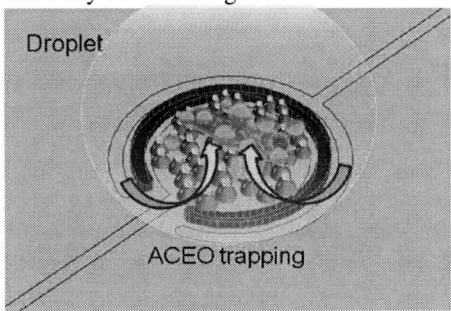

Figure 4: Schematic of the ACEO enhanced SERS sensing on concentrated bio-molecules and nano-particles to the nanopillar arrays.

Fabrication

As shown in Fig. 5, the Au nano-mushrooms nanoparticles were fabricated on silicon substrate. Polystyrene beads(PSBs) self-assembed a monolayer in PDMS well.[6] After the completion of PSB arrangement, an oxygen plasma treatment was utilized to create rough surface Then, 30-nm gold was deposited onto the corrugated PSBs. Fig. 5(A) shows the SEM image of the Au nano-mushrooms. Nano-star array chip fabrication are as follows. PSBs self-assembed a momolayer on fused silica wafer, then deposit 30nm Au on the substrate. After that, remove the PSBs ,and the triangular shape array of Au will stay at the gap of PSBs. Annealing this chip to form nano gold array and as nano mask. Finally, anisentropic O_2 plasma etching will form nanostar-array chip as shown in Fig.5(B).

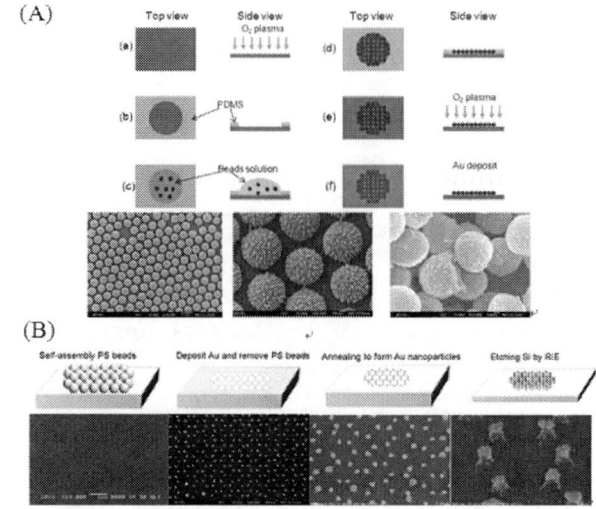

Figure 5 : Fabrication process and the SEM images of fabricated structures of Au nano-mushrooms(A) and Nanostars-array (B)

RESULTS AND DISCUSSION

In this configuration, analytes and Au nano-mushrooms can be brought into the center stagnation point by ACEO flow, and the flow drag force at the center will be reduced by the assembled/stacked Au nanoparticles and thus enhance the concentration ability of the analyte from 10^3(ACEOF's concentration) into 10^5. At the same time, SERS signal can be increased by 20-100 times due to the close packing of Au nanoparticles on the top side and Au/Si nanopillars at bottom side of bacteria. This configuration can rapidly detect chemical or bio-molecules without location-finding and molecule labeling by directly obtaining 3-D SERS signal from the aggregated analytes in the central.

The aggregation of analytes and Au nano-mushroooms is shown in Figure 6, in which all the analystes can be brought into the center stagnation point by ACEO flow, and the concnetration increased with time. The localy concnetration is increase 1000 times in 5 minutes.

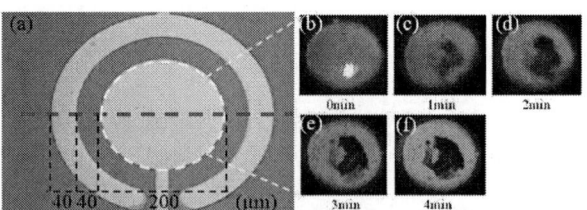

Figure 6.(a)OM image of ACEOF chip.(b)~(f) Analytes and Au nanoparticles can be brought into the center stagnation point by ACEO flow with time increase and the chip provides concentration ability about 10^3 folds .

The Raman spectrum and SEM image of E. coli (100 μl, 100-1/μl) aggregated together with Au nano-mushrooms on nano-stars-array is shown in Figure 7, demonstrating a 5 folds signal enhancement by 3D integration of Nano-mushrooms and roughness nano-stars-array for E. coli detection when compared to that of only surface nano-star arrays. In Fig. 8, the detection limit of this 3D configuration can be found at 1

978-1-4799-7956-1/15 $31.00 © 2015 IEEE   224

E.coli in 1ml (red line), which is potentially useful for rapid bacterial diagnosis in food or blood samples.

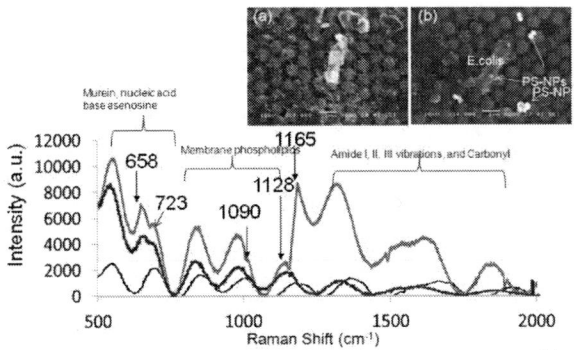

Figure 7 : The Raman spectrum of E. coli (100 μl, 10⁰·¹/μl) dropped with nano-mushrooms on roughness (a, red line) and smooth (b, blue line) nano-star-array. Roughness nano-satr-array provided a signal enhancement of 5 times higher than smooth nano-star-array. The signal of E. coli on fused Silica was employed as a reference (black line)

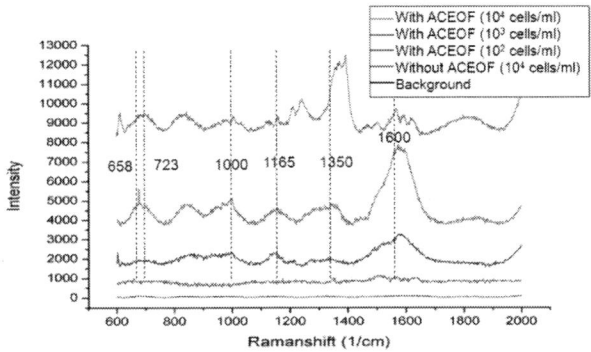

Figure 8:The Raman spectrum of E. coli with different concentration Sandwiched with Au-Coated nano-mushrooms and roughness Au nano-stars nano-cones-array. The signal of E. coli on fused Silica was employed as a reference (black line).

CONCLUSION

In this paper, we have demonstrated a novel method to enhance the intensity of Raman signal. This method incorporate three dimensional Nano-Mushrooms and Nano-Stars-Array sandwiched on Bacterial Aggregation.
We provide 3D SERS singnal of analyte that avoid information lost from the analyte surface. And we can find one E.coli in 1ml in 5 minutes. Due to the above reasons, we believe this novel method has large potential in rapid bacterial diagnosis in food or blood samples.

ACKNOWLEDGEMENTS

THIS WORK WAS FINANCIALLY SUPPORTED BY THE NATIONAL SCIENCE COUNCIL (NSC) OF TAIWAN UNDER GRANT NUMBER
MOST 103-2321-B-007 -004

REFERENCE

[1]. P. K. Jain, W. Huang, M. A. El-Sayed, *Nano Letters 2007*, pp. 2080-2088

[2]. J. Xie, Q. Zhang, J. Y. Lee, Daniel I. C. Wang, *ACS. Nano 2008*, pp. 2473-2480

[3]. S.J. Ye, Y.N. Mao, Y.Y. Guo, S.S. Zhang*, *Trends in Analytical Chemistry 2014*, pp. 43- 54

[4]. L. Lesser-Rojas, P. Ebbinghaus, G. Vasan, M.L. Chu, A. Erbe*, C.F. Chou*, *Nano Letters 2014*, pp. 2242−2250

[5]. M. R. Bown A C. D. Meinhart, Microfluid Nanofluid 2006, pp. 513–523

[6]. K. Kneipp, H. Kneipp, I. Itzkan, R.R. Dasari and M.S. Feld, *J. Phys.: Condens. Matter*, 2002, pp. 597-624

[7]. H.-Y. Hsieh, J.-L. Xiao, C.-H. Lee, T.-W. Huang, C.-S. Yang, P.-C. Wang, and F.-G. Tseng, The Journal of Physical Chemistry C, 2011, pp. 16258-16267

CONTACT

*F.G.Tseng, tel: + 886-3-5715131 ext. 34270 ; fangang@ess.nthu.edu.tw

SIMULTANEOUS IMPEDANCE SPECTROSCOPY AND STIMULATION OF HUMAN IPS-DERIVED CARDIAC 3D SPHEROIDS IN HANGING-DROP NETWORKS

Sebastian C. Bürgel[1], Yannick Schmid[1], Irina Agarkova[2], David A. Fluri[2], Jens M. Kelm[2],
Andreas Hierlemann[1] and Olivier Frey[1]
[1]ETH Zurich, Basel, Switzerland
[2]InSphero AG, Schlieren, Switzerland

ABSTRACT

Here, we present electrical impedance spectroscopy (EIS) data of human iPS-derived cardiac 3D spheroids with electric stimulation integrated in a hanging drop network. Microscopy videos of the beating spheroids were correlated with synchronously obtained EIS recordings. For stimulation, the spheroid was exposed to a continuous sinusoidal electric field – in contrast to traditional pulse trains. This stimulating field was supplied via the same electrodes that were used for the EIS recordings. Our measurements revealed a beating frequency modulation upon tuning the stimulation signal amplitude.

INTRODUCTION

Cardiac dysrhythmia and cardiotoxicity are the most common causes for drug withdrawals from the market and termination of drug development at late stages [1]. Earlier identification of these potentially fatal side effects requires new in-vitro toxicity screening methods. It has been shown that cells arranged in 3D spheroids mimic in-vivo conditions more closely than traditional 2D cell cultures; spheroids, however, require novel systems for culturing and analysis [2]. Reconfigurable hanging drop networks can be used to combine spheroid formation and microfluidic culturing in a single platform [3]. Here, we add an exchangeable EIS plug-in to a similar hanging drop network, we characterize the beating of human iPS-derived cardiac spheroids by EIS as well as optical means, and we monitor the beating frequency upon electrical stimulation via the same integrated microelectrodes.

MATERIALS AND METHODS

Hanging Drop Network

The hanging drop network depicted in Figure 1 accommodates up to eight spheroids in separate hanging drops that are interconnected by microchannels. The perimeter of the hanging drops is confined by hydrophobic PDMS rims. The PDMS piece is bonded onto a glass substrate, and the entire chip is then flipped with the PDMS structure facing downwards. The chip is therefore open at the bottom, so that drops are suspended in air hanging from the chip. The substrate can then be placed on top of an inverted microscope. The EIS plug-in comprising the platinum electrodes is inserted into a rectangular recess in the PDMS matrix that has been realized for one of the drops. By using this plug-in approach, most of the droplet system consists of PDMS, which is bio-compatible, and the surface and wetting properties of which can be easily modified through O_2-plasma treatment.

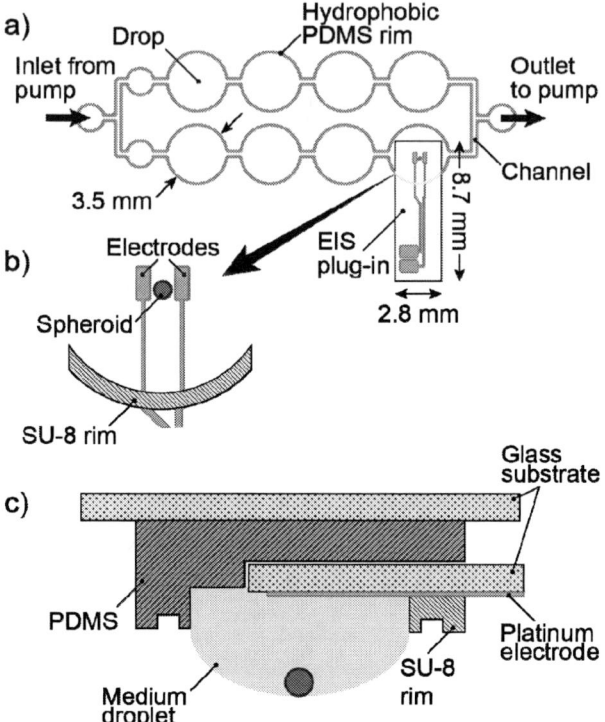

Figure 1: **Schematic** overview of the device in a), close-up of the measurement region in b) and side view of the drop with the EIS plug-in in c). The removable EIS unit features a set of coplanar electrodes for readout and stimulation of the spheroid. Inlet and outlet are connected to a syringe pump, the electrode pads are connected to an impedance spectroscope and a transimpedance amplifier for EIS readout and stimulation.

The hanging drop network was fabricated by using standard soft lithography methods as described previously [3]. The EIS plug-in was fabricated on a glass substrate with integrated platinum microelectrodes patterned by lift-off. The plug-in also featured an SU-8 wall, which sealed tightly to the sidewalls of the drop rim to yield a leakage-free overall droplet compartment and overall microfluidic network. After SU-8 development, the glass wafer was diced into separate plug-ins.

Electrical Setup

The EIS plug-in of the chip was wire-bonded to a custom printed circuit board facilitating the electrical interfacing of the platform. An HF2 impedance spectroscope was used to provide the input voltages for the EIS measurements, as well as delivering the stimulation signals to the spheroid. The output current was transformed into a voltage by using an HF2TA trans-impedance amplifier and then fed back to the impedance spectroscope (both from Zurich Instruments AG, Switzerland). The applied signal amplitudes for EIS measurements were 100 mV per frequency while simultaneously using up to eight individual frequencies between 10 kHz and 15 MHz. For stimulation, a 1 kHz sine-wave of 4.5 V - 8 V amplitude was applied in parallel to the EIS measurement via the same electrode pair. Data were recorded on a PC and later analyzed using a custom Matlab tool (Mathworks Inc, USA). For visualization of the signal magnitude and phase spectra, the baseline has been subtracted.

Microscopy

The microchip was placed in a holder frame on the stage of an inverted microscope (Leica DMI 6000B), which was inside a custom heating chamber and stage top incubator operating at 37°C in 5% CO_2 atmosphere at 95% humidity. Videos were acquired with a Leica DFR 320 camera (Leica Microsystems, Switzerland) at 30 fps, an exposure time of 1 ms, 2 x 2 binning and pixel depth of 8 bit to accommodate sufficiently high video frame rates, which can resolve temporal dynamics of the beating process. The inlet and outlet of the chip were connected to Nemesys syringe pumps (Cetoni GmbH, Germany) to perfuse the network with medium.

Image Analysis

The recorded videos were analyzed by using a custom C# tool. For each frame, the preceding frame was subtracted, so that the motion of the spheroid could be extracted from this differential video. The number of pixels in the differential image with an intensity value above a defined threshold was then determined. Therefore, the optical signals have low values during periods of little or no spheroid motion and peaks during the beating of the spheroid. Visualization of the differential image sequence shows white pixels in areas with large motion of the spheroid tissue and black areas elsewhere.

Spheroid Formation

Human iPS-derived cardiac microtissue spheroids and cardiac maintenance medium were obtained ready-to-use from InSphero AG, Zurich, Switzerland.

For experiments, a pre-formed spheroid was inserted into the hanging drop containing the EIS plug-in by manual pipetting.

RESULTS AND DISCUSSION

Optical and EIS Beating Analysis

The cardiac spheroids were beating spontaneously, as was observed optically and by EIS in parallel, and the results are shown in Figures 2 and 3. The differential optical signal in Figure 2 (bottom curve, left axis) showed a spike, whenever the spheroid was beating. The smaller secondary peak is an artefact of the differential image analysis method and pertains to the same beat as the larger primary peak. Each peak in the optical signal was coinciding with a peak in the 5.3 MHz impedance magnitude signal (upper curve, right axis). The spheroids were beating regularly, when fresh medium was continuously supplied. The arrhythmic beating shown in Figure 2 was evoked upon stopping perfusion through subsequent nutrient depletion (about 10 minutes).

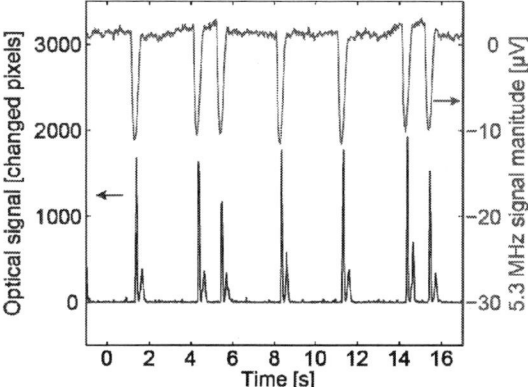

*Figure 2: **Beating sequences** obtained by microscopic video recordings (blue, left axis) and EIS-recording (green, right axis) show a clear correlation, which indicates that EIS is a suitable tool to extract the beating frequency of the cardiac spheroid in the hanging droplet.*

The change in the impedance signal due to the beating of the spheroid is analyzed in more detail in Figure 3: The image sequence (Figure 3a) and impedance magnitude and phase spectra (Figure 3b and c) show the very same beating cycle. The spectra were obtained from the time-domain signals, which were similar to the one shown in Figure 2, by subtracting the baseline level from the beating-induced spike. These peak-to-baseline signals were evaluated at each of the eight frequencies in terms of magnitude and phase component to obtain the spectra in Figures 3b and 3c.

During the maximal contraction of the spheroid (t=0, green frame and green line), the magnitude over the entire spectral range decreased, while the phase spectrum showed a change in slope. Furthermore, the EIS data showed – in agreement with optical analysis – that spheroid contraction is about 5 times faster than full relaxation: The differential images as well as the EIS results revealed that the contraction of the spheroid, accompanied by a corresponding decrease in the signal magnitude spectrum, took approximately 0.2 s. The full relaxation of the spheroid to its initial state, and the corresponding increase of the signal magnitude spectrum back to the original values, however, took more than 1 s.

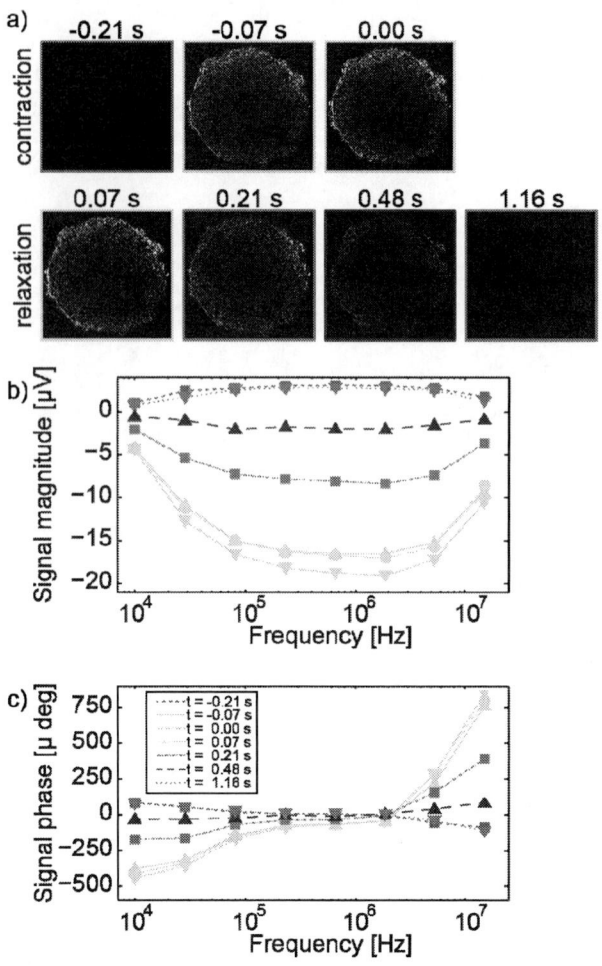

Figure 3: **Single beat characteristics** as observed optically in a) and through EIS in terms of signal magnitude in b) and phase spectra in c.). The timing of all three recorded signal is identical, and the times are displayed in a) and c).

Stimulated Beating

Upon applying a 1 kHz sine wave of amplitudes between 4.5 V – 8 V to a regularly beating spheroid, the spheroid beating frequency could be tuned (Figure 4a, b). As observed optically and through EIS, the 1 kHz stimulation signal influenced the beating frequency of the spheroid that could be modulated in dependence of the stimulus amplitude. Larger stimulus amplitudes led to higher beating frequencies of up to 3 Hz at 8 V. The observed relationship between the evoked beating frequency and applied stimulation signal amplitude was almost linear and had a slope of 0.7 Hz/V (Figure 4c). Amplitudes above 8 V were not accessible due to hardware limitations; lower frequencies (below 4.5 V stimulus amplitude) produced arrhythmic beating. The amplitude-frequency modulation did not show hysteresis behavior upon ramping the stimulus amplitude up and down.

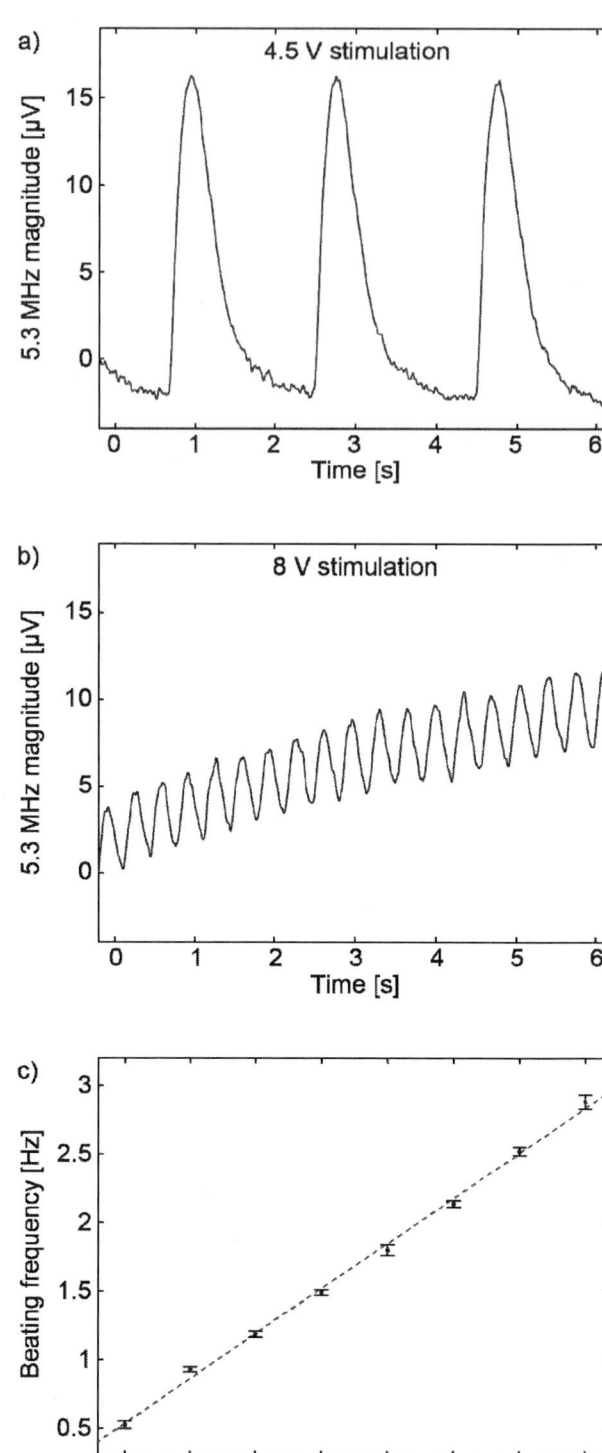

Figure 4: **Stimulated beating** with amplitude-frequency modulation and parallel EIS readout. Slow beating at an applied stimulation signal amplitude of 4.5 V is shown in a), fast beating upon applying 8 V in b), a stimulation voltage sweep and the resulting spheroid beating frequencies in c). Each data point in c) corresponds to at least 9 beating cycles from which the mean and standard deviations were obtained and displayed.

CONCLUSION

We presented the successful integration of an EIS unit into a hanging drop network, and were able to simultaneously characterize the beating of human cardiac spheroids optically and by means of EIS. The detection of arrhythmias as shown in Figure 2 for the case of medium depletion or the detection of irregularities in contraction/relaxation duration are fundamentally important. Such features may also occur upon drug administration and may then be directly correlated to the nature of the applied different drugs and the respective dosages. We further observed an amplitude/frequency modulation upon varying the electrical stimulation signal that was applied to the cardiac spheroids. Advantages of EIS over optical detection methods include the possibility for device integration, the associated higher temporal resolution (typically better than 1 kHz) and label-free non-invasive multi-parameter analysis.

The system presented here holds great potential for addressing key questions of cardiotoxicity in the emerging field of 3D microtissues.

ACKNOWLEDGEMENTS

This work was financially supported by the FP7 of the EU through the project "Body on a chip", ICT-FET-296257 and an individual Ambizione Grant 142440 of the Swiss National Science Foundation for Olivier Frey

REFERENCES

[1] R. R. Shah, "Drugs, QT interval prolongation and ICH E14: the need to get it right", *Drug Safety*, vol. 28(2), pp. 115-25, 2005.

[2] J. M. Kelm, M. Fussenegger, "Microscale tissue engineering using gravity-enforced cell assembly", *Trends in Biotechnology*, vol. 22(4), pp. 195–202, 2004.

[3] O. Frey, P. M. Misun, D. A. Fluri, J. G. Hengstler, A. Hierlemann, "Reconfigurable microfluidic hanging drop network for multi-tissue interaction and analysis", *Nature Communications*, vol. 5, pp. 4250, 2014.

CONTACT

*S.C. Bürgel, tel: +41774186541; sbuergel@ethz.ch

DROPLET FLOW FOCUSING FOR MOLECULAR BINDING DETECTION

Sunggu Kim and Junghoon Lee

Seoul National University, Seoul, South Korea

ABSTRACT

We report the detection of interfacial molecular binding via droplet generation in a flow focusing device. DNA hybridization at the phospholipid causes interfacial tension shift at the oil-water boundary which changes the mode of droplet production in the device. Droplet size decreased ~30% due to a complementary binding compared with non-complementary and control cases. Our report includes a molecular protocol to functionalize the interface with single-strand (ss) DNA as receptor and flow condition tuning for an unambiguous distinction of complementary binding.

INTRODUCTION

Droplet-based microfluidics finds diverse applications in emulsion forming [1], a gas bubble sensor [2] and synthesis of polymer particles [3], but never been used for the detection of a specific molecular interaction. Droplets of various utilities were "fabricated" by dispersing immiscible fluids, predominantly oil and water, using devices such as T-junction and flow focusing microchannels [4,5]. Researchers tested various conditions including flow rate, viscosity, and interfacial tension that affect the droplet shape and size [6]. Researchers have investigated to understand physical characteristics underlying flow-focusing. Especially, break up and formation mechanism of droplets [7] and transition of droplet generation scheme such as jetting and dripping [8]. Some researchers have found out electrical or temperature effect on generation patterns of droplets [9, 10].

Even though there was plenty of knowledge regarding flow-focusing droplet generations, no approaches have been carried out yet to use the flow-focusing system as a molecular sensor. The interfacial tension shift by molecular binding can be strong enough to cause the droplet size change at a fixed flow rate [11]. Unlike solid based molecular detection, droplet-based soft interface detection has many benefits available from previous resources such as easy change of receptors and no requirement for additional electrical devices with everything installed in a soft chip of, e.g., 2 cm × 1 cm in size.

Here we report the lipid binding protocol of single-strand (ss) DNA receptor-lipid which is adjusted for an optimized flow condition with a specific design to effectively catch the pattern change of droplet generation. Molecular detection was accomplished by observing the pattern of droplet formation with an appropriate optical system. Figure 1 shows the PDMS-based "standard" flow focusing device as our choice for droplet generation to implement our idea.

SAMPLE PREPARATION

Microchannel Structure and Fabrication

The flow-focusing device consists of a droplet forming orifice, PBS buffer, and oil inlet and their corresponding flow channels. All structures were fabricated using patterned PDMS molding process (Fig. 2). Two immiscible liquid flows were merged, and formed droplets through the orifice. The widths of the PBS, oil, droplet generation channels and orifice were 70 μm, 70 μm, 150 μm, and 20 μm respectively.

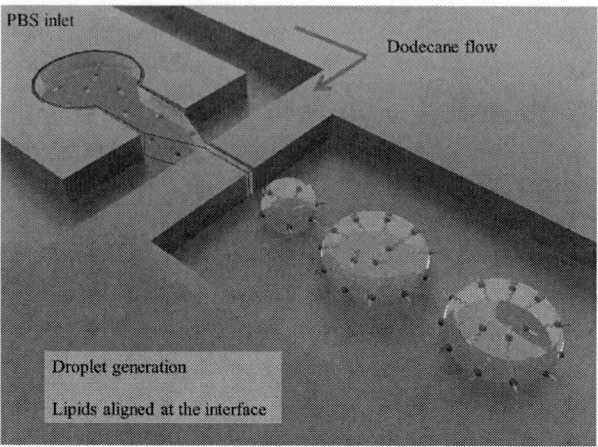

Figure 1: Droplet generation in a flow-focusing device. Lipids are aligned at the oil-water interface.

We fabricated PDMS-based microfluidic flow-focusing device. First, the SU-8 (SU-8 3050) mold was constructed on a silicon wafer using standard photo-lithography technique, whose uniform height was ~80μm. Second, 10:1 PDMS-solvent mixture was poured on top of the SU-8 mold and cured at 65 °C for one hour. Then, the patterned PDMS was peeled from the mold, and placed on top of flat PDMS layer which was already prepared and then cured overnight at 65 °C.

Based on the hydrophobic nature of material, oil (dodecane) was used as a continuous phase, and aqueous buffer (PBS) was used as a dispersed phase. To ensure the stable droplet generation and reduce internal flow resistance in the microchannel, dodecane was flowed first through all channel paths before droplet generation.

Testing Method

To validate the concept of flow-focusing molecular detection, microfluidic set-up was established to provide liquid flows into the PDMS microchannels with precise control of flow rate, and droplet image was optically monitored. The schematic of flow-focusing device was describe in figure 1. Dodecane and lipid-PBS were prepared in two syringes which were connected to inlets of PDMS microchannels. Both syringes were driven by syringe pump (KDS101, KD scientific), to give desired flow rates of both liquids into the microchannels. We used dodecane as a continuous phase while lipid-PBS was used as dispersed phase. There was no surfactant added in both liquids which generally used for the prevention of droplet coalescence in flow-focusing device. This could prevent the competition between lipids and surfactant at droplet interface which would reduce the effect of interfacial tension shift by DNA binding. We analyzed only several droplets right out of the orifice before coalescence occurred. Even if numerous droplets were recorded at a single time, merged droplets were not counted.

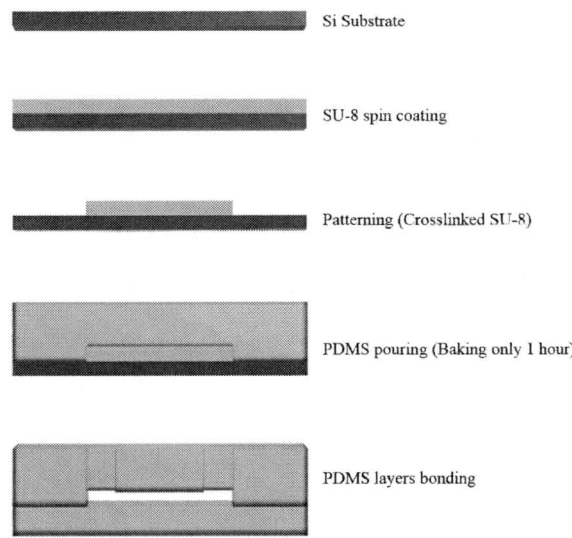

Figure 2: Fabrication flow of flow-focusing microchannel

Optical measurement system consists of a microscope, a high-speed video camera and the image processing tool. The microscope (ECLIPSE LV150N, Nikon) was connected to a high-speed video camera (Phantom Miro, Nikon) to record the droplet generation process. In every droplet generation condition, video was recorded after 5 minutes of stabilization time to obtain a constant droplet generation. After recording was finished, the video converted to the frames of images to be analyzed in a custom MATLAB code. At least 50 successive droplets were collected for statistical analysis.

Aptamer-conjugated Lipid Preparation

Phospolipids were used as a surfactant as well as ssDNA binding receptor in all cases. All lipids were dissolved in PBS buffer and introduced to microchannel through by the syringe pump. Two lipids were combined to use. First, maleimide functionalized methoxypoly(ethylene glycol)distearoyl phosphatidyl ethanolamine (DSPE-PEG(2k)-maleimide, Avanti Polar Lipid) was used as modified lipid receptor which occupied 2% of whole population for optimized binding affinity. Second, 2-distearoyl-sn-gycero-3-phosphocholine (DSPC, Avanti Polar Lipid) was merely used as spacer molecule for stable droplet generation which occupied remaining 98%.

DSPC and DSPE-PEG(2k)-maleimide mixture was prepared to 2.5 mg to covalently bond with ssDNA. 1 mL of PBS with 2 wt% glycerol was added to prepare aqueous lipid solution. Brief sonication and agitation were then applied. The thiol-end of thiol modified ssDNA (SH-5'-GGTTGGTGTGGTTGG-3', SH-DNA) was initially protected by disulfide bond. 10 μL of 1.0 N dithiothreitol (DTT) was added to SH-DNA (10 μL, 100 μM) to break the disulfide bond. After 15 min, washing

solution was applied to remove excess DTT and unwanted thiol fragments from SH-DNA by extracting with ethyl acetate three times. The activated SH-DNA was mixed with the lipid solution by sonication. The conjugation steps were accomplished consecutively.

For the tests of DNA bindings, same amount (10 μL of 100 μM) of complementary DNA (5'-CCAACCACA CCAACC-3') and non-complementary DNA (5'-ATCT AACTGCTGCGC-3') were prepared and applied to ssDNA-lipid solution in separate samples. Note that all lipid solutions were diluted 10 times and injected to syringes to avoid any sediment. Figure 3 summarizes the protocol for an effective regulation of interfacial tension associated with the molecular binding.

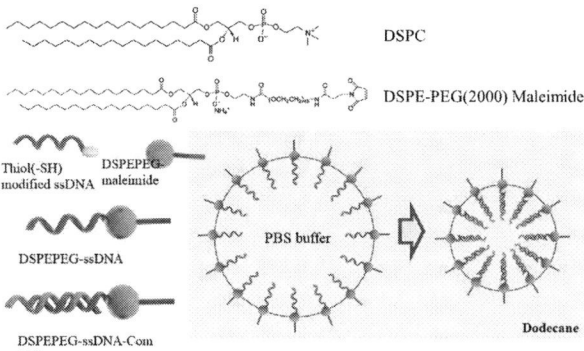

Figure 3: (a) Molecular structure of phospholipids, (b) Thiol-maleimide coupling and decrease of droplet size after DNA hybridization.

RESULTS AND DISCUSSION

Droplet Size Tuning

We tested various combination of flow speed for both phases in search of the condition for optimized droplet generation. We achieved a stable droplet generation that sharply discriminates the binding event when the flow rates of oil and lipid-PBS were 30 μL/min and 3 μL/min, respectively (Ca=0.00161 without binding). As the differences between both flow rate increases with high flow rate of continuous phase, the droplet size decreases and the binding event effectively reflect to the droplet size change (Fig. 4).

Verification of Lipid Conjugation and DNA Hybridization

To check the covalent conjugation of DSPE-PEG(2k)-maleimide and SH-ssDNA as well as complementary binding of DNA-DNA, we observed fluorescence image at stationary liquid droplets. First, we made emulsions using PBS and lipids mixture with same conditions. Lipids dissolved in the PBS and dodecane were prepared in 1.5 μL viral with 10:1 volume ratio. Brief sonication and 10 min agitation were applied to obtain white emulsion layer on top of the solution. Emulsion layer was transferred to a new viral and washed three times with PBS. Complementary DNA was then applied to the emulsion. After 30 minutes, we washed the emulsion three times again, and check the fluorescence through the microscope. On the microscope, the emulsion droplet was ~50 μm similar to our tuned-droplet size in the flow-focusing device. In Figure 5, we confirmed the binding at the interface via the fluorescence when the labeled complementary DNA was hybridized with the ssDNA.

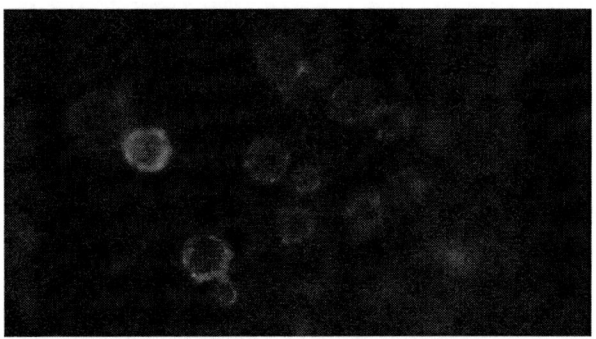

Figure 4: Flow rate conditions and droplet generation patterns. Concentration of lipids is 0.025wt% in PBS buffer. Stable droplet generation occurred @ flow condition: Q_{PBS}= 3 μL/min, Q_{oil}=30 μL/min, Ca=0.00161. Droplet frequency = 200~700 Hz.

Figure 5: Fluorescence image of droplets. Aptamer of lipid hybridizes fluorescence labeled target DNA.

DNA Hybridization and Droplet Size Shift

Figure 6 shows droplet pattern that was dependent on the molecular binding. Large droplet (~50 μm) was produced in the absence of the specific binding. Hybridization caused the dramatic reduction in droplet size (~35 μm), and the radical shift in flow pattern represented by higher droplet frequency, and less merging downstream (Figure 6 (a)). Control test with a non-complementary ssDNA shows that this size reduction was only caused by the specific binding (Figure 6 (b)). Droplet size maintained the same with non-complementary case when there was no added additional DNA and even there was no ssDNA-receptor modification for the lipids. Further testing with various target concentration clearly demonstrates that our approach can be used for the quantification of binding event (Figure 6 (c)). Droplet size gradually decreased as target DNA concentration increased and droplet generation frequency also increased as a result. These consequences

matched to previous report that the DNA hybridization induces free energy changes at the surface [12] as the droplet size changes due to the interfacial energy shift.

(a)

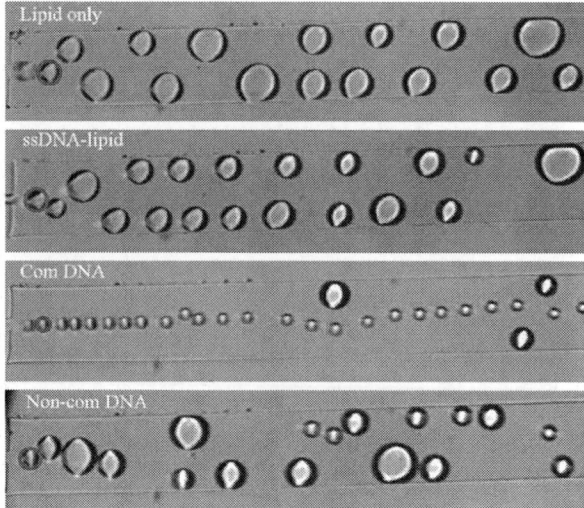

(b)

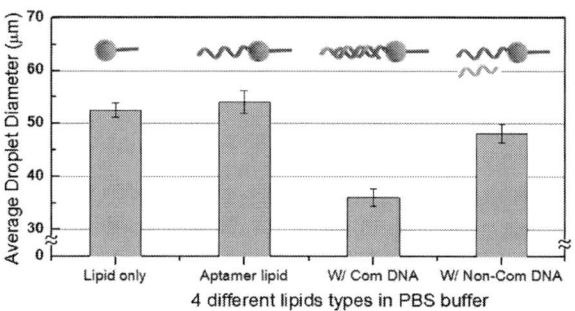

(c)

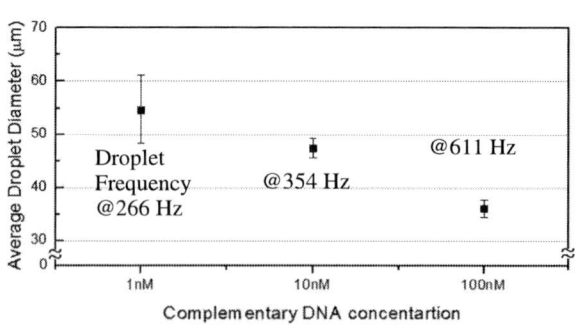

Figure 6: (a) Droplet generation pattern dependent on complementary binding, (b) Diameter of droplets, (c) Concentration dependence (50 droplets measured each case, 4 tests)

CONCLUSION

We confirmed the DNA hybridization could be detected by droplet size change in flow-focusing device, and also show the target concentration of DNA. With the possibility for integration with sensor platform, our first demonstration shows a new approach for a sensing using a soft interface in a dynamic situation. With the inherent benefits of label free assay, simple operation, and potential for incorporating other contents, our "platform" protocol opens up new applications in variety of molecular detection.

ACKNOWLEDGEMENTS

This work was supported by the ICT & Future Planning as Global Frontier Project (CISS-2012M3A6A 6054193) and the Brain Korea 21 Plus Project in 2014. And the fabrication was performed at the Interuniversity Semiconductor Research Center (ISRC) in Seoul National University.

REFERENCES

[1] S-H. Huang, W-H. Tan, F-G, Tseng, S. Takeuchi, J. "A Monolithically Three-dimensional Flow-focusing Device for Formation of Single/double Emulsions in Closed/open Microfluidic Systems", *J. Micromech. Microeng,* 16, p. 2336, 2006.

[2] A. Bulbul, A.S. Basu, H. Kim, "Characterization of Microbubbles of Multiple Gases in Microfluidic Channels", *µTAS '13*, Freiburg, Germany, Oct. 27-31, 2013.

[3] M. Seo, Z. Nie, S. Xu, M. Mok, P. C. Lewis, R. Graham, E. Kumacheva, "Continuous Microfluidic Reactors for Polymer Particles", *Langmuir,* 21, pp. 11614-11622, 2005.

[4] T. Torsen, R.W. Roberts, F.H. Arnold, S.R. Quake, "Dynamic Pattern Formation in a Vesicle-Generating Microfluidic Device", *Phys. Rev. Lett.,* 86, pp. 4163-4166, 2001.

[5] S.L. Anna, N. Bontoux, H.A. Stone, "Formation of Dispersions using "Flow Focusing" in Microchannels ", *App. Phy. Lett.* 82, p. 364, 2003.

[6] T. Ward, M. Faivre, M. Abkarian, H.A. Stone, "Microfluidic Flow Focusing: Drop size and Scaling in Pressure versus Flow-rate-driven Pumping", *Electrophoresis*, 26, pp. 3716-3724, 2005.

[7] P. Garstecki, M.J. Fuerstman, H.A. Stone, G.M. Whitesides, "Formation of Droplets and Bubbles in a Microfluidic T-junction — Scaling and Mechanism of Break-up", *Lab Chip*, 6, pp. 437-446, 2006.

[8] A.S. Utada, A. Fernandez-Nieves, H.A. Stone, D.A. Weitz, "Dripping to Jetting Transitions in Coflowing Liquid Streams", *Phys. Rev. Lett.,* 99, p. 094502, 2007.

[9] S.H. Tan, B. Semin, J-C. Baret, "Microfluidic Flow-focusing in AC Electric Fields", *Lab. Chip*, 14, p. 1099, 2014.

[10] C.A. Stan, S.K.Y. Tang, G.M. Whitesides, "Independent Control of Drop Size and Velocity in Microfluidic Flow-Focusing Generators Using Variables Temperature and Flow Rate", *Anal. Chem.*, 81, pp. 2399-2402, 2009.

[11] C.-Y. Chao, D. Carvajal, I. Szleifer, K.R. Shull, "Drop-Shape Analysis of Receptor-Ligand Binding at the Oil/Water Interface", *Langmuir,* 24, pp. 2472-2478, 2008.

[12] J. Fritz, M.K. Baller, H.P. Lang, H. Rothuizen, P.

Vettiger, E. Meyer, H.-J. Guntherodt, Ch. Gerber, J.K. Gimzewski, "Translating Biomolecular Recognition into Nanomechanics", *Science*, 288, pp. 316–318, 2000.

CONTACT

*Junghoon Lee, tel: +82-2-880-9104; jleenano@snu.ac.kr

SEMICONDUCTOR IC PACKAGING, THE NEXT WAVE

CP Hung

Corporate R&D, Advanced Semiconductor Engineering, Inc., TAIWAN

ABSTRACT

MEMS sensor growth in the future is expected with inevitable trend on miniaturization for smartphone and wearable devices. In this presentation, the next wave architecture of 3D IC packaging solution overview versus traditional packaging technologies will be demonstrated and discussed, for applications including MEMS.

INTRODUCTION

Recently, micro electromechanical systems (MEMS) sensors are widely used in smartphone and wearable products. In the miniaturization requirements, the traditional MEMS packages are normally quad flat no-leads (QFN) or land grid array (LGA) with bonding wire, have reaching limit on size and performance. Wafer level package by implementing 3D IC structure with TSV (Through Silicon Via) technology can thus be applied as enhancing solution.

In this presentation, advantages of 3D IC technology with TSV releasing the limit of SoC (System on Chip), and enabling comprehensive SiP (System in Package) will be discussed. Latest reliability results will be presented, as well as re-positioning semiconductor value chain and collaborating model.

SEMI INDUSTRY AND MEMS

The new development of electronic products is always accompanied by the evolution of semiconductor technology. From the computer products and consumer electronics of the paradigms shift, gradually changing to or integrating with communications are seen in the mainstream market. The SoC node also progresses from 0.18um, 90nm to 28nm, and today's 20nm. In the following years, the development direction of smartphone / wearable devices will be not only the pursuit of 'right' performance, minimized size, effective power consumption, but also have all sorts of sensing features.

The history of motion MEMS sensor is began around 1990 and applied in some specific industrial application. From 2004 to 2007, consumer products including game console, adopt three-axis accelerometers with high performance, size shrinking, and cost advantages. By the rising of the smartphone, it has started the important wave of MEMS sensors, including accelerometer, magnetometer and gyroscope, pressure sensor, recently ambient light sensor, temperature sensor on the system to realize the new hybrid sensing functionality. The trend of implementing more or new type of sensors will continue [1].

IC PACKAGE TECHNICAL TREND AND TSV ADVANTAGE

Modern trend of the semiconductor IC packaging as showed in Fig. 1 has two major directions: one is with large scale size such as 3D IC / 2.5D flip-chip packages for high I/O (Input/Output), another one is with miniaturized size such as wafer level packages. One of the common features of these two directions is implementing TSV technology.

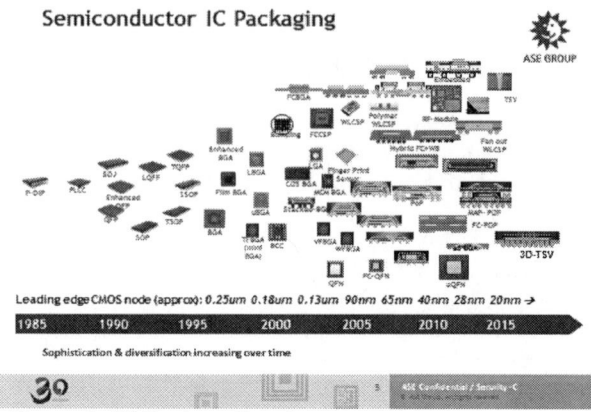

Figure 1: IC Package Roadmap

In the foreseeable future, with more growth of MEMS sensors in smartphone, size will be more critical. Fig. 2 shows the typical evolution of compact multi-axis accelerometers. In the 4mm^2 migrating to 2mm^2 new MEMS package, TSV is one of the enabling technologies. Through direct chip-to-chip stacking, TSV has not only simplified interconnection from traditional side-by-side chips structure, it also integrates multi-chip in vertical way with significant size reduction. The device could also benefit higher frequency bandwidth with TSV [2]. For MEMS components integrated with controller, the miniaturization has being discussed with feasible design shrunk from 2mm by 2mm to 1.5mm by 1.2mm, with 16% of thickness reduction [3].

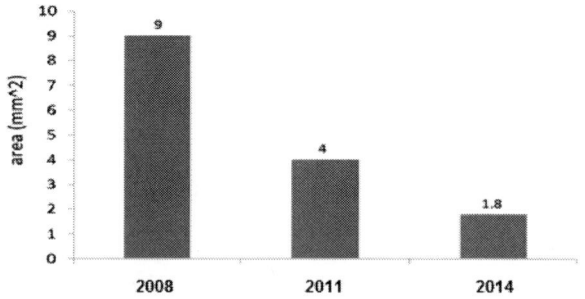

Figure 2: The trend of MEMS sensor device size.

REFERENCE

[1] A. Oouchi, "Plastic molded package technology for MEMS sensor evolution of MEMS sensor package", *International Conference on Electronics Packaging (ICEP)*, Japan, April 23-25, 2014, pp. 371-375.

[2] C. H. Chen, K. C. Lu, C. Y. Hung, P. N. Lee, M. J. Wang, C. P. Hung, H. M. Tong and T. S. Horng, "GHz High Frequency TSV for 2.5D IC Packaging", *International Microelectronics assembly and packaging society*, USA, 2012.

[3] D. E. Serrano, "Design and analysis of accelerometer", in *SENSORS*, USA, 2013.

ROTATIONAL CHAMBERS ON FLUIDIC CHANNELS FOR THE REPETITIVE FORMATION OF OPTICALLY OBSERVABLE LIPID-BILAYER MEMBRANES

Fumiaki Tomoike[1], Taishi Tonooka[1], Toshihisa Osaki[2], and Shoji Takeuchi[1]

[1]CIBiS-IIS, the University of Tokyo, Tokyo, Japan
[2]Kanagawa Academy of Science and Technology, Kanagawa, Japan

ABSTRACT

We develop a device adapted for repetitive formation of horizontal lipid bilayer membranes. This device allows simultaneous optical and electrophysiological measurements of the formed membranes. We integrated a rotational chamber on a fluidic channel separated by a parylene micropore. The rotational motion is designed to form a bilayer repeatedly. This rotational process emulates the conventional painting method, in which a thick lipid –oil layer at a micropore was made thinner by hand work to obtain a bilayer. The bilayer formation with our device was examined optically and electrically. The simultaneous measurement device will be useful for better understanding of bilayer features and membrane protein incorporation.

INTRODUCTION

Lipid bilayer is a major component of the boundary of individual cells, and compartmentalizes organelles within the cells. These compartments restrict diffusion of various molecules and render specific reaction space. For transport and communication between the compartments, membrane proteins in the lipid bilayer play important roles. Because of their functions, the membrane proteins are a major target for drug discovery [1]. However, the analysis of membrane protein remains difficult. One of the difficulties in the membrane protein analysis is that most membrane proteins form proper structure and exhibit the function only in a lipid bilayer. Therefore, the membrane protein analysis requires the techniques for forming lipid bilayer. The artificial lipid-bilayer is one of the most common models for the cellular membrane. Originally, Muellar *et al.* suggested the method to form an artificial lipid bilayer between two chambers as the painting method [2]. In the painting method, lipid-oil layer, containing lipid molecules, is formed at a micropore between two chambers by a paintbrush. By hydrodynamic pressure from aqueous solutions in chambers and van der Waals force between lipid molecules, the lipid-oil layer becomes thinner, and a lipid bilayer is formed. Although the painting method is simple, considerable experience and skill are required. The advances in MEMS technology have opened new methods for bilayer formation [3, 4], yet most of those methods remained difficult. The droplets contact method is one of the easiest and simplest methods for artificial lipid bilayer formation [5] and used for the analysis of various ion channels [6]. Recent advances in the contact method have shown that lipid bilayers can be formed repeatedly by shifting the location of the droplets [7]. This method enables the reformation of lipid bilayer without reloading solutions and the rapid acquisition of extensive electrophysiological data. However, the method is hardly compatible with microscope observation of the bilayer, because the bilayer is formed in a vertical manner to a focal plane of a microscope. Although there were some methods for optical observable bilayer [8-14], it is impossible to simply reform the lipid bilayer with the methods. We recently reported a device that formed optically observable lipid bilayer based on the droplets contact method [15]. In this method, however, multiple manipulation was necessary to form a lipid bilayer repeatedly.

In this study, we proposed a device for optically-observable lipid bilayer formation (Figure 1). Our device emulates the painting method, and the lipid bilayer is able to be repeatedly formed just by the rotational motion of chambers. In this proceedings, we demonstrate repetitive formations of the formed lipid bilayer as a confirmation of bilayer formation. Finally, we observed the activity of nanopore-forming proteins after the formation of lipid bilayer by use of our device.

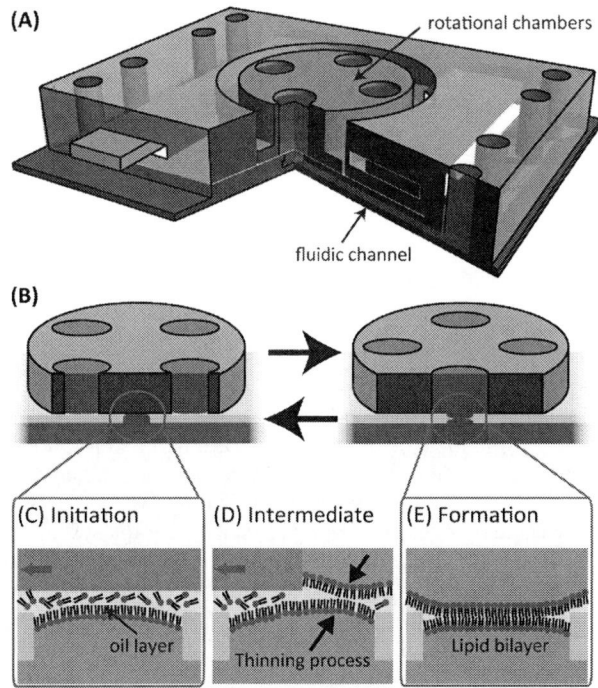

Figure 1: The concept of our device. (A) Overall structure of our device. Our device contains chambers and fluidic channels. As shown in panel B, the state of the pore between the chamber and the micro-fluidic channel can be set by rotating the chambers. (C) At initial position, PMMA plate forms a thick lipid-oil layer. (D) In intermediate, the lipid-oil layer was sandwiched between aqueous solutions in the chamber and the fluidic channel. (E)Finally, lipid bilayer was formed.

978-1-4799-7956-1/15 $31.00 © 2015 IEEE

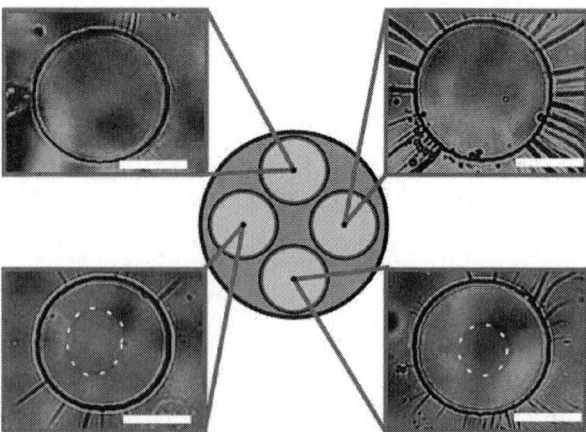

Figure 2: The repetitive formation of BLMs. White dot lines indicate the edges of BLMs. Scale bars indicate 50 μm. White and black numbers indicate the position of the chambers and the pores.

Experimental Procedures

Chemicals

In all experiments, diphytanoylphosphatidylcholine (DPhPC) (Avanti Polar Lipids, USA) was used as lipid material and dissolved in n-decane (Kanto Chemical, Japan) at a concentration of 15 mg mL^{-1}. Buffer components, KCl, K_2HPO_4, and KH_2PO_4 were purchased from Wako Pure chemical Industries, Japan. Alpha-hemolysin (αHL) was purchased from Sigma Aldrich, USA. Buffer solutions were prepared with ultrapure water from MilliQ. Main body and thin film with micropores in our device were made of PMMA substrate (Nitto Jushi Kogyo, Japan) and parylene C (Specialty Coating Systems, USA), respectively. A positive photoresist, S1818 (Shipley, USA), mask materials, aluminum wire (Nilaco, Japan), S1818 developer, NMD-3 (Tokyo Ohka Kogyo, Japan), and aluminum etchant (Wako Pure chemical Industries, Japan) were used in the photolithography process.

Fabrication

Our device is composed of two parts, the basal part and the rotational chambers. The basal part and rotational chambers contain fluidic channels and chambers, respectively. The main bodies of these two parts were fabricated by micromachining (MM-100, Modia systemes, Japan). To stabilize the formed bilayer, the region of lipid bilayer formation was limited using micropores with a diameter of 100 μm, fabricated by combining a patterned parylene sheet and PMMA plate [16]. The pores in the parylene sheet were located at the center of the contact regions of aqueous solutions in the rotational chambers and the fluidic channels. The patterned parylene sheet was fabricated by photolithography process. First, 2 μm-parylene C was deposited on No.1 microscope glass cover slides (Matsunami, Japan) by Parylene deposition system 2010 (Specialty Coating Systems), and aluminum was deposited as an etching mask. Aluminum was patterned by photolithography. Parylene was etched by oxygen plasma, using RIE10-NR (Samco, Japan). The

etched parylene sheet was bound onto basal part by a solvent adhesive, acrysunday (Acrysunday, Japan). After the attachment of the parylene sheet to the PMMA plate, the glass substrate was removed from the parylene sheet.

Bilayer formation

To form the lipid-bilayers, the fluidic channels were firstly filled with the aqueous solution, and the lipid solution was applied onto the contact surface between the basal parts and rotational chambers. The rotational chambers was then placed on the basal part and fixed by the insertion of the stoppers. Finally, aqueous solutions were filled into the rotational chambers. The cyclic motion of the rotational chambers moved aqueous solutions of the chambers on the aqueous solutions in the fluidic channel. As shown Figure 1, PMMA region of the rotational chambers formed a thick lipid-oil layer, like the paintbrush in painting method at the initial state (Figure 1C). At the bilayer formation state, the aqueous solutions in the chambers and fluidic channels were brought into contact at the pores of the parylene sheet. Finally, the lipid-oil layer became thinner (Figure 1D), and the lipid bilayer was formed (Figure 1E). The formed membranes were observed by an inverted light microscope (Olympus, Japan).

Membrane thickness determination

In this experiment, Ag/AgCl electrodes were inserted into both the chamber and fluidic channel. The aqueous solution was composed of 1 M KCl and 10 mM phosphate buffer adjusted at pH 7.2. The area of the formed bilayer was analyzed by ImageJ software (NIH, USA). The capacitance value was measured by using the amplifier, Patch/Whole Cell Clamp Amplifier CEZ2400 (NIHON KOHDEN, Japan) and digitizer, Digidata 1440A (Molecular devices, USA). The thickness of lipid bilayer was estimated based on the correlation between the capacitance value and the thickness of thin layer estimates the lipid bilayer thickness.

αHL signal measurement

Incorporation of αHL into the bilayer membrane was confirmed by signal current recordings. The same buffered electrolyte was used but additionally mixed 100 nM αHL monomer. The current recordings were performed with a patch-clamp amplifier with digitizer integrated, PICO (TECELLA, USA). A potential of 100 mV was applied.

RESULTS

At first, we observed the formation of lipid bilayer by an inverted light microscope. In the formation state, black-lipid-membranes (BLM) were formed and expanded (Figure2). Next, repetitive lipid bilayer formation was demonstrated. BLM was broken and reformed repeatedly just by cycling the rotational chambers. This character improves the efficiency of lipid bilayer formation. When a lipid bilayer formation is failed, another bilayer formation can be restarted without washing and reloading the solutions.

To confirm that the observed BLM was indeed a lipid-bilayer, we determined the thickness of the formed membrane from the simultaneous measurement of optical and electrophysiological measurement. The thickness of lipid bilayer is estimated from the capacitance value. The thickness of BLM was estimated as 8.7 nm, which was comparable to the thickness of a lipid bilayer previously reported, approximately 5 nm [17].

Using our device, we further observed the incorporation of αHL into the lipid bilayer. αHL is one of the toxic and soluble protein. Under the existence of a lipid bilayer, αHL monomers incorporate into the lipid bilayer spontaneously and form heptameric cylindrical nanopores [18, 19]. Because its pore-size is constant, the conductance value via αHL pore depends on the number of incorporated αHL pores. When the aqueous solution contained αHL, an electric current step was observed after the formation of the BLM. The conductance value was consistent with the literature data [20]. In addition, we observed a time delay in the electric current step compared with the lipid bilayer formation.

CONCLUSION

We developed the device for repetitive optically observable lipid bilayer by emulating the conventional painting method. Because this device is simple to form the bilayer repeatedly, it is suitable for the membrane protein analysis.

ACKNOWLEDGEMENTS

This research was partly supported by Grant-in-Aid for Exploratory Research (Project No. 26560431) from Japan Society for the Promotion of Science, and the Regional Innovation Strategy Support Program of MEXT, Japan.

REFERENCES

[1] M. A. Yildirim, K. I. Goh, M. E. Cusick, A. L. Barabasi, M. Vidal, "Drug-target network" *Nature biotechnology*, 25, 1119-1126, 2007.

[2] P. Mueller, D. O. Rudin, H. T. Tien, W. C. Wescott, "Reconstitution of cell membrane structure in vitro and its transformation into an excitable system" *Nature*, 194, 979-980, 1962.

[3] P. Kongsuphol, K. B. Fang, Z. P. Ding, "Lipid bilayer technologies in ion channel recordings and their potential in drug screening assay" *Sensor Actuat B-Chem*, 185, 530-542, 2013.

[4] M. Zagnoni, "Miniaturised technologies for the development of artificial lipid bilayer systems" *Lab Chip*, 12, 1026-1039, 2012.

[5] K. Funakoshi, H. Suzuki, S. Takeuchi, "Lipid bilayer formation by contacting monolayers in a microfluidic device for membrane protein analysis" *Anal Chem*, 78, 8169-8174, 2006.

[6] R. Kawano, Y. Tsuji, K. Sato, T. Osaki, K. Kamiya, M. Hirano, T. Ide, N. Miki, S. Takeuchi, "Automated parallel recordings of topologically identified single ion channels" *Sci Rep*, 3, 1995, 2013.

[7] Y. Tsuji, R. Kawano, T. Osaki, K. Kamiya, N. Miki, S. Takeuchi, "Droplet Split-and-Contact Method for High-Throughput Transmembrane Electrical Recording" *Anal Chem*, 85, 10913-10919, 2013.

[8] E. E. Weatherill, M. I. Wallace, "Combining Single-Molecule Imaging and Single-Channel Electrophysiology" *Journal of molecular biology*, DOI: 10.1016/j.jmb.2014.07.007, 2014.

[9] V. C. Stimberg, J. G. Bomer, I. van Uitert, A. van den Berg, S. Le Gac, "High yield, reproducible and quasi-automated bilayer formation in a microfluidic format" *Small*, 9, 1076-1085, 2013.

[10] C. Shao, E. L. Kendall, D. L. DeVoe, "Electro-optical BLM chips enabling dynamic imaging of ordered lipid domains" *Lab Chip*, 12, 3142-3149, 2012.

[11] E. L. Kendall, C. Shao, D. L. DeVoe, "Visualizing the growth and dynamics of liquid-ordered domains during lipid bilayer folding in a microfluidic chip" *Small*, 8, 3613-3619, 2012.

[12] T. Tonooka, K. Sato, T. Osaki, R. Kawano, S. Takeuchi, "Lipid Bilayers on a Picoliter Microdroplet Array for Rapid Fluorescence Detection of Membrane Transport" *Small*, DOI: 10.1002/smll.201303332, 2014.

[13] T. Ide, T. Ichikawa, "A novel method for artificial lipid-bilayer formation" *Biosens Bioelectron*, 21, 672-677, 2005.

[14] H. Suzuki, B. Le Pioufle, S. Takeuchi, "Ninety-six-well planar lipid bilayer chip for ion channel recording fabricated by hybrid stereolithography" *Biomed Microdevices*, 11, 17-22, 2009.

[15] F. Tomoike, T. Tonooka, S. Takeuchi, "Formation of optically-observable lipid bilayer membrane by sliding chambers on a fluidic channel" in SAN ANTONIO, TEXAS U.S.A., Oct. 26-30, 2014, pp. 1793-1794.

[16] M. Mayer, J. K. Kriebel, M. T. Tosteson, G. M. Whitesides, "Microfabricated Teflon Membranes for Low-Noise Recordings of Ion Channels in Planar Lipid Bilayers" *Biophysical journal*, 85, 2684-2695, 2003.

[17] H. Fujiwara, M. Fujihara, T. Ishiwata, "Dynamics of the spontaneous formation of a planar phospholipid bilayer: A new approach by simultaneous electrical and optical measurements" *J Chem Phys*, 119, 6768-6775, 2003.

[18] S. Bhakdi, J. Tranum-Jensen, "Alpha-toxin of Staphylococcus aureus" *Microbiological reviews*, 55, 733-751, 1991.

[19] L. Song, M. R. Hobaugh, C. Shustak, S. Cheley, H. Bayley, J. E. Gouaux, "Structure of staphylococcal alpha-hemolysin, a heptameric transmembrane pore" *Science*, 274, 1859-1866, 1996.

[20] J. J. Kasianowicz, E. Brandin, D. Branton, D. W. Deamer, "Characterization of individual polynucleotide molecules using a membrane channel" *P Natl Acad Sci USA*, 93, 13770-13773, 1996.

CONTACT

*F. Tomoike, tel: +81-3-5452-6650; tomoike@iis.u-tokyo.ac.jp

MICROTUBULE SORTIING WITHIN A GIVEN ELECTRIC FIELD BY DESIGNING FLEXURAL RIGIDITY

Naoto Isozaki*[1], Hirofumi Shintaku[1], Hidetoshi Kotera[1], Edgar Meyhöfer[2], and Ryuji Yokokawa[1]
[1]Kyoto University, Kyoto, JAPAN
[2]University of Michigan, Ann Arbor, Michigan, USA

ABSTRACT

We propose a method to control gliding directions of kinesin-propelled microtubules (MTs) corresponding to their flexural rigidity (*EI*). We prepared two kinds of MTs having different *EI* and their trajectories within an electric field were clearly separated. Therefore, this study demonstrated the *EI*-altered MTs can be workhorses to sort/concentrate various combinations of target molecules loaded on MTs.

INTRODUCTION

Biomolecular motors have been extensively studied for utilizing as high-performance nanoscale actuators *in vitro*. Kinesin is one class of biomolecular motors that function in intracellular transport by converting a chemical energy of adenosine triphosphate (ATP) to a mechanical motion along MT filaments [1]. The kinesin-MT motility can be recapitulated *in vitro* with the inverted molecular configuration where MTs glide on a kinesin-coated substrate (Fig. 1) [2]. With developments of techniques for labeling MTs with target molecules through avidin-biotin [3, 4] and antigen-antibody binding [5, 6] and DNA hybridization [7, 8], the kinesin-propelled MTs have been regarded as molecular shuttles. The shuttles are expected to be workhorses to sort/concentrate target molecules from a solution including multiple molecules [4, 9].

Since random directions of MT gliding hamper to realize the molecular sorter/concentrator using the kinesin-MT shuttles, the main challenge in the field has been developing a method to define MT gliding directions. Although some methods have been reported with microfabricated tracks [4, 10, 11] or external fields [9, 12-16], they guided MTs in only one designated direction because all MTs in the tracks or fields behaved in the same manner without active control of a field [9]. Our group reported a method to design MT surface charge density through labeled-molecules that allowed them to glide in different directions depending on their electrophoretic mobilities in a given electric field [17]. However, the method necessitates different charges of target molecules, which will be sorted by MT gliding. To establish a method independent of the intrinsic characteristic of labeled-molecules to define MT gliding directions is important for future applications.

To address this issue, we focused on the intrinsic characteristic of MTs, *EI*, as a tuning parameter to define MT trajectories. Considered that the free leading tip of an MT is a micro-scale cantilever beam that is anchored at the frontmost kinesin, when an electric field was applied in the negative x direction of the xy-plane and MTs entered the field at the origin in the positive y direction, the MT trajectory was defined as follows:

$$y(x) = A \cdot arccos\left(e^{-\frac{x}{A}}\right) \quad (1)$$

$$A = \frac{3EI}{c_\perp(\mu_{e,\perp}+\mu_{EOF})E_e\langle d\rangle^2} \quad (2)$$

where $c_\perp$ is the perpendicular Stokes drag coefficient per unit length of an MT tethered to the surface via kinesin, E_e is the field intensity, μ_{EOF} is the electro-osmotic mobility, and $\langle d\rangle$ is the distance between active kinesins [13]. As shown in equation (2), the A value defines the curvature of MT trajectory and depends on the value of *EI*, which can be altered by changing MT-polymerizing conditions [18, 19] or MT-associated proteins [20, 21]. Here, we prepared two MT groups having different *EI*; one is polymerized in the presence of guanosine 5′-triphosphate (GTP, GTP-MTs) and the other is guanylyl-(α,β)-methylene-diphosphonate (GMPCPP, GMPCPP-MTs). Since GMPCPP-MT have the 2.1-fold higher *EI* than that of GTP-MT [18], GMPCPP-MT is expected to show larger A value than that of GTP-MT. They were assayed in a polydimethylsiloxane (PDMS) device shown in Fig. 2, which consisted of four channels; channels 1–4 were for MT injection before introducing an ATP solution, MT alignment along the wall, MT sorting according to their *EI*, and MT capturing, respectively. In this study, MT landing and gliding in channels 1–3 were performed. We confirmed that the *EI*-altered MT trajectories were in accordance with the cantilever model and the MTs can be sorted with more than 80% precision according to their *EI*.

MATERIALS AND METHODS

Protein Preparation

Kinesin, consisted of human kinesin (amino acid residues 1–573) with an N-terminal histidine tag, was purified as described [22]. Phosphocellulose (PC) tubulin was purified from porcine brains by two assembly-disassembly cycles and PC chromatography, and

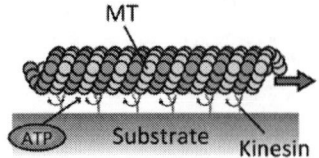

Figure 1: Schematic view of the inverted motility assay. MT glides on a kinesin-coated substrate.

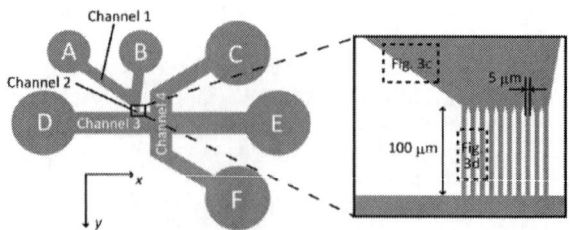

Figure 2: Schematic view of the PDMS device. The device consisted of channels 1–4 and reservoirs A–F.

stored in liquid nitrogen [21]. Tetramethylrhodamine (TAMRA, red)-tagged tubulin (T-Tu) and AlexaFluor 488 (green)-tagged tubulin (A-Tu) were prepared by adding a 10-fold molar excess of amine-reactive TAMRA (C-1171; Invitrogen) and AlexaFluor 488 (A-20000; Invitrogen), respectively. Resulted labeling stoichiometry was 0.45–0.70 for T-Tu and 0.27–0.30 for A-Tu [23]. Recycled-tubulin (R-Tu) was purified from PC tubulin after an assembly-disassembly cycle to eliminate non-polymerizable tubulin. R- and A-Tu were mixed at 2:1 in the presence of 1 mM dithiothreitol (DTT) and 1 mM GMPCPP and polymerized into GMPCPP-MT at 37°C for 30 min. T-Tu and GTP were used for GTP-MT instead of A-Tu and GMPCPP, respectively. Non-polymerizable tubulin was removed by MT pelleting in an ultracentrifuge (himac CX100GX2; Hitachi, 163000 × g, 15 min, 27°C) and resuspension in BRB80 buffer (80 mM PIPES, 1 mM MgCl$_2$, and 1 mM EGTA, pH 6.8) with 20 μM paclitaxel.

Device Fabrication

The PDMS device shown in Fig. 2 was fabricated by a standard PDMS molding process and bonding to a coverslip. The entire channel was 10 μm in height. Channels 1–4 were 500 μm, 5 μm, 1 mm, and 1 mm in width, respectively. Channels 2–3 were 100 μm and 8 mm in length, respectively. Channel 2 consisted of 3–10 channels. Following spin coating of negative photoresist (SU-8 3010; Microchem) onto a silicon wafer, it was exposed to UV light at 150 mJ cm^{-2} using a mask aligner (PEM-800; Union Optical) through a photomask having the channel patterns. The photoresist was developed by SU-8 developer (MicroChem) for 3 min at 40°C, resulting in a mother mold for the channels. After silanizing the mold with trichloro-(1H,1H,2H,2H-perfluorooctyl)silane (448931-10G; Sigma-Aldrich) for 2 h, PDMS having channels on their surface was fabricated by pouring PDMS prepolymer (Silpot 184; Dow Corning Toray) on the mold and cured for 2 h at 80°C. Reservoirs A–F were made by punching the PDMS, and then it was hydrophilized by plasma treating for 40 s at 40 W (Covance; Femto Science) with a coverslip (24 × 36 × 0.17 mm^3) and placed on the coverslip. Reservoirs A–B and C–F were 2 mm and 3 mm in diameter, respectively.

MT Gliding Assay

BRB80 buffer including 0.1 mg ml^{-1} kinesin and 1.2 mg ml^{-1} casein was introduced into the device. It was incubated for 6 min for nonspecific adsorption of kinesin to glass substrate and washed out. After filling reservoirs C–F with buffer solution, reservoir B was filled with a GMPCPP-MT solution or a mixture of GTP- and GMPCPP-MT solution. Reservoir A was empty to make differences in water pressure among reservoirs. The MT solution was incubated for 6 min to immobilize MTs on the kinesin-coated substrate and washed out. MTs started gliding after injection of an ATP solution consisting of 0.5 mM ATP in BRB80 with 0.30 mg ml^{-1} casein and an O$_2$-scavenging system composed of 8.0 μg ml^{-1} catalase, 25 mM D-glucose, 20 μg ml^{-1} glucose oxidase, 1% 2-mercaptoethanol, 20 mM DTT, 3.0 mM 1,1′-ferrocenedimethanol (322-49071; Wako Pure Chemical Industries, Ltd.), 10% glycerol, and 20 μM paclitaxel, followed by application of an electric field with an average intensity of 4 kV m^{-1} (6487; Keithley Instruments).

Optical Imaging and Analysis

MTs were observed under an IX73 inverted epifluorescence microscope (Olympus) with an excitation filter (GFP/DsRed-A-OMF; Opto-Line International, Inc.), absorbance filter (A11400-04; Hamamatsu Photonics), charge-coupled device camera (ORCA-D2; Hamamatsu Photonics), 10× (NA 0.3) objective, and 60× (NA 1.35) and 100× (NA 1.4) oil-immersion objectives. Exposure time was 500 ms (1 frame/s) using a ND3 or ND6 filter with a shutter (VMM-D3; Uniblitz). Optical images were stored as sequential image files in TIFF format using HCImage software (Hamamatsu Photonics). Trajectories were defined by tracking MT leading tips with the Mark2 image analysis software (provided by Dr. Kenya Furuta, National Institute of Information and Communications Technology) at the sampling rate of 0.5 frame/s. Here, we tracked only MTs which entered channel 3 after gliding in channel 2. Coordinate data were exported to MATLAB software (MathWorks); A and R^2 values and MT orientation angles respect to the x-axis for each trajectory were exported through a custom-written MATLAB algorithm. A and R^2 values were calculated by fitting MT trajectories with equation (1) by a least squares method.

RESULTS AND DISCUSSION

We confirmed the PDMS channel was fabricated without leakage of solution by introducing red ink (Fig. 3a, b). GMPCPP-MTs were landed only within channel 1 just before introducing the ATP solution owing to the applied flow fields (Fig. 3c). After introducing the ATP solution, GMPCPP-MTs started gliding and their gliding directions were aligned along channel 2 because the narrow channels

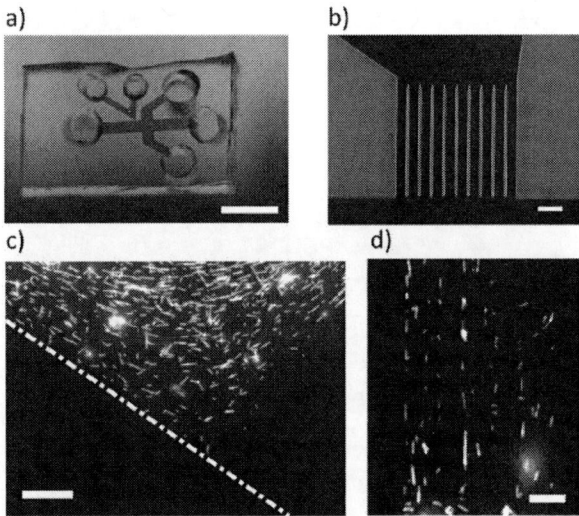

Figure 3: a) Overview of the fabricated PDMS device. Red ink visualized the channels. Scale bar = 5 mm. b) Bright field image of channel 2 filled with red ink. Scale bar = 20 μm. Fluorescence images of GMPCPP-MTs c) in channel 1 before introducing the ATP solution and d) in channel 2 while their gliding. Scale bars = 20 μm. Images c) and d) correspond to the areas shown in Fig. 2.

physically prevented a GMPCPP-MT from making a U-turn (Fig. 3d). GMPCPP-MT orientation angles in channel 1 and channel 2 were $79.4 \pm 48.5°$ (mean $\pm$ SD, $n = 70$) and $87.6 \pm 6.5°$ ($n = 35$), respectively, which shows the SD values were significantly decreased while gliding in channel 2 with a width of 5 μm (Fig. 4).

Once GTP-MT and GMPCPP-MT entered channel 3 applied an electric field of 4 kV m^{-1}, their gliding directions were changed toward the anode due to their negative charges. Sequential fluorescence images for the MTs are shown in Fig. 5a. There was difference in trajectories between a GTP-MT and a GMPCPP-MT. Some MTs existed in channel 3 just after introducing the ATP solution because they were detached from kinesin in channel 1 and attached to kinesin in channel 3

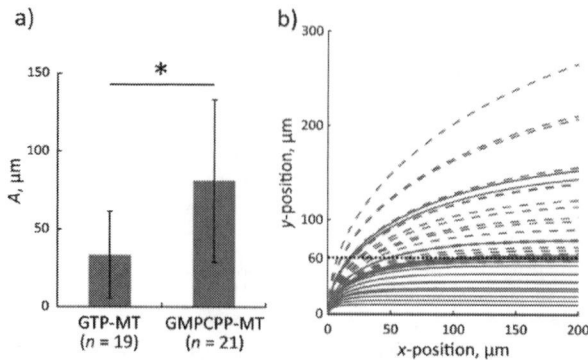

Figure 6: a) A values for GTP- and GMPCPP-MT. There were significant differences (*p < 0.01, t-test). b) MT trajectories plotted by the calculated A values for GTP-MT (red lines) and GMPCPP-MT (green dashed lines). Their trajectories were categorized into two groups. The black dotted line represents y = 60 μm.

unintentionally by flow fields while the solution introduction. As mentioned above, we omitted the existed MTs from evaluation procedures. Trajectories of the MTs shown in Fig. 5a were plotted and fitted with equation (1) (Fig. 5b). A value for each MT was calculated and the GMPCPP-MT showed higher A than that of the GTP-MT corresponding to the difference in EI.

As a result of analyzing, A values for GTP- and GMPCPP-MT were 33.4 ± 28.1 μm ($n = 19$) and 81.3 ± 52.2 μm ($n = 21$), respectively, and there were significant differences by t-test at the level of $p < 0.01$ (Fig. 6a). The ratio of mean A value for GMPCPP-MT to that for GTP-MT was close to that of EI (2.4 vs. 2.1). It was consistent with the fact that the A in equation (1) is in proportion to the EI of an MT. Furthermore, a mean R^2 value for each MT was more than 0.98. Therefore, the cantilever model fitted trajectories of our EI-altered MTs and known values of other parameters enable to predict MT trajectories without performing assay experiments.

By substituting A values for equation (1), trajectories of GTP- and GMPCPP-MTs were plotted (Fig. 6b). It showed that there were 84% of GTP-MTs at $y < 60$ μm and 86% of GMPCPP-MTs at $y > 60$ μm at $x = 100$ μm, that is, a partition placed at the coordinate $(x, y) = (100$ μm, 60 μm) enable to sort the two kinds of MTs with more than 80% precision. Optimizing parameters $\mu_{e,\perp}$, E_e, and $\langle d \rangle$ will enhance the difference in trajectories and improve the precision.

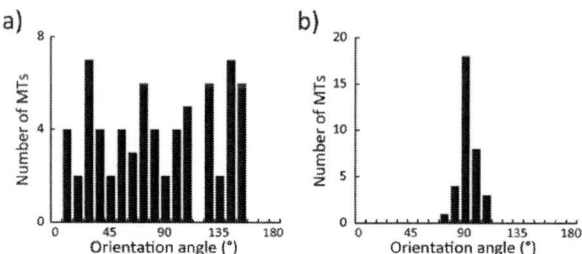

Figure 4: Distribution of MT orientation angles with respect to x-axis a) in channel 1 and b) in channel 2. The angle distribution was significantly reduced while their gliding in channel 2. Bin = 9°.

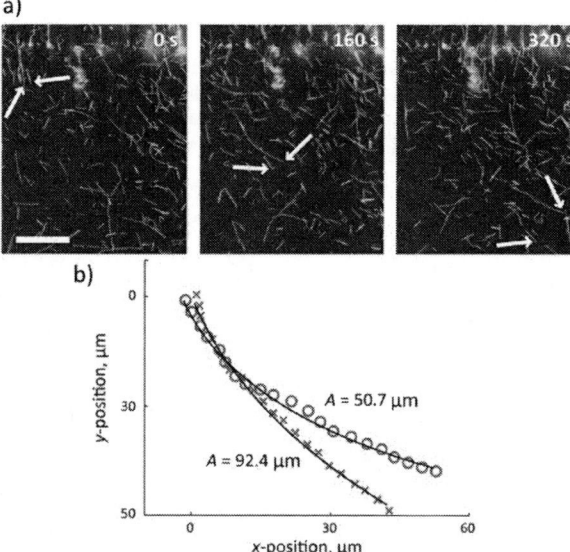

Figure 5: a) Sequential fluorescence images of gliding GTP-MTs (red) and GMPCPP-MTs (green) with 160-s intervals in channel 3 applying an electric field of 4 kV m^{-1}. The field direction was from right- to left-hand side. White arrows point leading heads of a GTP-MT and a GMPCPP-MT. Scale Bar = 20 μm. b) Trajectories of MTs shown in a). Red circles and green crosses indicate the coordinates of leading heads of the GTP- and GMPCPP-MT, respectively. Black lines indicate curves fitted with equation (1). A values for the GTP- and GMPCPP-MT were 50.7 μm and 92.4 μm, respectively ($R^2 > 0.99$).

CONCLUSION

In summary, we proposed a method to sort two groups of EI-altered MTs under an electric field. Gliding directions of GTP- and GMPCPP-MTs landed only within channel 1 were aligned in channel 2 and changed toward the anode in channel 3. Their trajectories depended on the EI of the MTs, which were consistent with the cantilever model, and were sorted with more than 80% precision. The method has the advantage of simplicity and broad applicability; solutions were introduced without syringe pumps, MTs were prepared just by changing polymerization conditions, and entire MT surfaces were left free to be used as a cargo carrier. Therefore, the

978-1-4799-7956-1/15 $31.00 © 2015 IEEE

EI-altered MTs can be workhorses for molecular sorter/concentrator by optimizing assay conditions, labeling MTs with different target molecules such as cancer markers, and capturing the sorted MTs in channel 4.

ACKNOWLEDGEMENTS

This research was supported by PRESTO, JST, JSPS KAKENHI Grant Number 25709018 to R.Y. and Grant-in-Aid for JSPS Fellows Grant Number 26•2439 to N.I.

REFERENCES

[1] M. J. Schnitzer and S. M. Block, "Kinesin hydrolyses one ATP per 8-nm step," *Nature,* vol. 388, pp. 386-390, 1997.

[2] M. K. A. Rahim, T. Fukaminato, T. Kamei, and N. Tamaoki, "Dynamic Photocontrol of the Gliding Motility of a Microtubule Driven by Kinesin on a Photoisomerizable Monolayer Surface," *Langmuir,* vol. 27, pp. 10347-10350, 2011.

[3] G. D. Bachand, S. B. Rivera, A. K. Boal, J. Gaudioso, J. Liu, and B. C. Bunker, "Assembly and transport of nanocrystal CdSe quantum dot nanocomposites using microtubules and kinesin motor proteins," *Nano Lett.,* vol. 4, pp. 817-821, 2004.

[4] C. T. Lin, M. T. Kao, K. Kurabayashi, and E. Meyhofer, "Self-contained, biomolecular motor-driven protein sorting and concentrating in an ultrasensitive microfluidic chip," *Nano Lett.,* vol. 8, pp. 1041-1046, 2008.

[5] S. Ramachandran, K. H. Ernst, G. D. Bachand, V. Vogel, and H. Hess, "Selective loading of kinesin-powered molecular shuttles with protein cargo and its application to biosensing," *Small,* vol. 2, pp. 330-334, 2006.

[6] T. Fischer, A. Agarwal, and H. Hess, "A smart dust biosensor powered by kinesin motors," *Nat. Nanotechnol.,* vol. 4, pp. 162-166, 2009.

[7] S. Taira, Y. Z. Du, Y. Hiratsuka, K. Konishi, T. Kubo, T. Q. Uyeda, *et al.*, "Selective detection and transport of fully matched DNA by DNA-loaded microtubule and kinesin motor protein," *Biotechnol. Bioeng.,* vol. 95, pp. 533-538, 2006.

[8] S. Hiyama, Y. Moritani, R. Gojo, S. Takeuchi, and K. Sutoh, "Biomolecular-motor-based autonomous delivery of lipid vesicles as nano- or microscale reactors on a chip," *Lab Chip,* vol. 10, pp. 2741-2748, 2010.

[9] M. G. L. van den Heuvel, M. P. De Graaff, and C. Dekker, "Molecular sorting by electrical steering of microtubules in kinesin-coated channels," *Science,* vol. 312, pp. 910-914, 2006.

[10] Y. Hiratsuka, T. Tada, K. Oiwa, T. Kanayama, and T. Q. Uyeda, "Controlling the direction of kinesin-driven microtubule movements along microlithographic tracks," *Biophys. J.,* vol. 81, pp. 1555-1561, 2001.

[11] J. Clemmens, H. Hess, J. Howard, and V. Vogel, "Analysis of microtubule guidance in open microfabricated channels coated with the motor protein kinesin," *Langmuir,* vol. 19, pp. 1738-1744, 2003.

[12] B. M. Hutchins, M. Platt, W. O. Hancock, and M. E. Williams, "Directing transport of CoFe2O4-functionalized microtubules with magnetic fields," *Small,* vol. 3, pp. 126-131, 2007.

[13] M. G. van den Heuvel, M. P. de Graaff, and C. Dekker, "Microtubule curvatures under perpendicular electric forces reveal a low persistence length," *Proc. Natl. Acad. Sci. U.S.A.,* vol. 105, pp. 7941-7946, 2008.

[14] I. Dujovne, M. van den Heuvel, Y. Shen, M. de Graaff, and C. Dekker, "Velocity modulation of microtubules in electric fields," *Nano Lett.,* vol. 8, pp. 4217-4220, 2008.

[15] R. R. Agayan, R. Tucker, T. Nitta, F. Ruhnow, W. J. Walter, S. Diez, *et al.*, "Optimization of isopolar microtubule arrays," *Langmuir,* vol. 29, pp. 2265-2272, 2013.

[16] T. Kim, E. Meyhofer, and E. F. Hasselbrink, "Biomolecular motor-driven microtubule translocation in the presence of shear flow: modeling microtubule deflection due to shear," *Biomed. Microdevices,* vol. 9, pp. 501-511, 2007.

[17] N. Isozaki, S. Ando, T. Nakahara, H. Shintaku, H. Kotera, E. Meyhofer, *et al.*, "Control of microtubule trajectory within an electric field by altering surface charge density," under revision.

[18] B. Mickey and J. Howard, "Rigidity of Microtubules Is Increased by Stabilizing Agents," *J. Cell Biol.,* vol. 130, pp. 909-917, 1995.

[19] T. L. Hawkins, D. Sept, B. Mogessie, A. Straube, and J. L. Ross, "Mechanical properties of doubly stabilized microtubule filaments," *Biophys. J.,* vol. 104, pp. 1517-1528, 2013.

[20] H. Felgner, R. Frank, and M. Schliwa, "Flexural rigidity of microtubules measured with the use of optical tweezers," *J. Cell Sci.,* vol. 109, pp. 509-516, 1996.

[21] H. Felgner, R. Frank, J. Biernat, E. M. Mandelkow, E. Mandelkow, B. Ludin, *et al.*, "Domains of neuronal microtubule-associated proteins and flexural rigidity of microtubules," *J. Cell Biol.,* vol. 138, pp. 1067-1075, 1997.

[22] A. Hyman, D. Drechsel, D. Kellogg, S. Salser, K. Sawin, P. Steffen, *et al.*, "Preparation of modified tubulins," *Methods Enzymol.,* vol. 196, pp. 478-485, 1991.

[23] R. C. Williams, Jr. and J. C. Lee, "Preparation of tubulin from brain," *Methods Enzymol.,* vol. 85 Pt B, pp. 376-385, 1982.

CONTACT

*N. Isozaki, tel: +81-75-383-3687, fax: +81-75-383-3681; isozaki.naoto.72s@st.kyoto-u.ac.jp

CARVING OF PROTEIN CRYSTAL BY HIGH-SPEED MICRO-BUBBLE JET USING MICRO-FLUIDIC PLATFORM

S. Takasawa[1], T. Syu[1], and Y. Yamanishi[1, 2]
[1]Shibaura Institute of Technology, Tokyo, Japan
[2]Japan Science and Technology Agency (JST) PRESTO, Japan

ABSTRACT

This paper reports a novel micro-fluidic platform for carving protein crystal with electrically driven high-speed mono-dispersed micro-bubble jet. This minimally invasive micro-processing method overcomes the difficulties of processing, holding and positioning fragile material such as protein crystal underwater. The combination of using electrically-induced micro-bubble knife and microfluidic channel provide effective carving of protein crystal by ablation crystal and draining of chips by free vortex flow in microchannel. Three-dimensional positioning of crystal was sufficiently achieved by the configuration of effective micro-fluidic channels. The protein crystal can be carved to the desired shape to fit to X-ray analysis effectively. It seemed that it has potential to contribute to more precise protein analysis

INTRODUCTION

The study of proteins is emerging in biochemical and biomedical research with the aim of designing new therapeutic drugs [1]. It is important to obtain protein structure for understanding protein functions. And it is promising to produce a protein crystal and to analyze using X-ray in order to identify the structure of protein. Generally, the combination of solution component such as neutral salt and organic solvent and employment of crystallization method is very important to make protein crystal. For example, one of the conventional protein crystallization methods is hanging drop method and sitting drop method which use supersaturation of the protein solution and classified as vapor diffusion [2][3]. In case of the method, a drop composed of protein and regent is placed in the arrayed well chamber with reservoir solution. Typically, the drop contains a lower reagent concentration than the reservoir. And water in the drop vaporizes and transfers to the reservoir. Then the concentration increases to an optimal level for crystallization. Finally, the protein molecular is crystallized in the protein drop. However obtaining diffraction-quality single crystals of membrane crystals is extremely difficult [4]. At present, there exists no well-established technique for producing high-quality membrane crystals, resulting in a small number of reports on membrane protein crystals. Manual operation with tiny knife, which is the only method currently employed, frequently failed to accomplish the desired treatment. The lack of effective processing tools retards the progress of membrane protein crystallography [5]. Recently, authors have invented the electrically-induced bubble knife [6, 7] as minimally invasive cell surgery tool without any thermal damage. And simultaneous ablation and injection was successfully operated by using electrically-induced micro-bubble knife for biomedical applications. The present work employed the bubble-knife for processing protein crystal to remove impurity of protein and carving crystal to be an ideal shape for X-ray analysis.

CONCEPT

The concept of the processing of protein crystal is shown in Figure 1. The glass coated copper wire whose tip has a space was discharged underwater. First, a bubble was produced in the space by electrolysis or by local heat. Next, a bubble grew up until it fit to the glass wall of the space by applying electric pulses. Then, after the continuous electric pulses were applied, the high speed mono-dispersed micro-bubbles were continually generated eventually. Figure 2 shows a photo of high speed mono-dispersed micro-bubbles underwater produced by electrically induced bubble knife taken by high-speed camera. This mono-dispersed bubble flow can make an ablation of the crystal surface due to by the cavitation when the bubbles are hit to the surface of it and are collapsed. Figure 1(a) shows the concept of processing the surface of protein. As soon as after the high speed bubble is processing protein crystal, the processed protein chips, which has a risk of deteriorate the protein crystal, are collected by glass capillary using negative pressure or draining of free-vortex in the microchannel automatically. Fig. 1 (b) shows a protocol of removing impurity part of protein and Fig. 1 (c) shows the carving to an ideal shape for X-ray analysis which is typically either spherical or cubic shape. Fig. 3 shows the concept view of microfluidic platform to have three-dimensional accurate positioning of protein crystal with x-y and x-z rotating fluidic motion. Fig. 4 shows the overall integrated experimental setting of micro-fluidic platform to process protein crystals effectively.

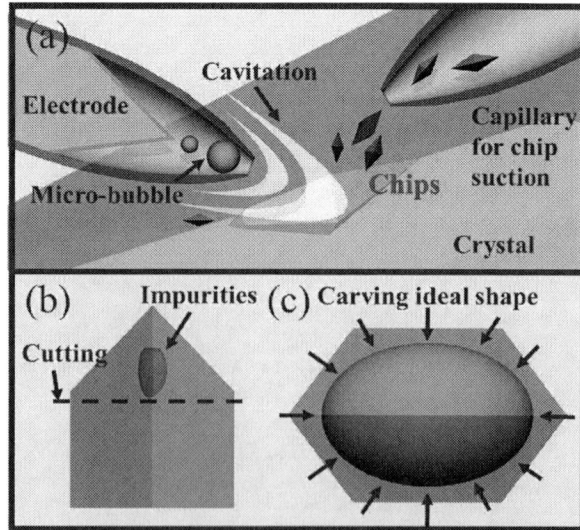

Figure 1: Schematics of carving of protein crystal using electrically-induced bubble knife and concept of carving to diffraction-quality crystals. The micro-bubble can process the surface of the crystal with high-speed.

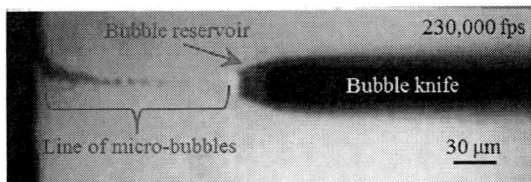

Figure2: High-speed mono-dispersed micro-bubbles generated by electrically induced bubble knife underwater discharging. The size of micro-bubble is about 5μm with the speed of about 5 m/s.

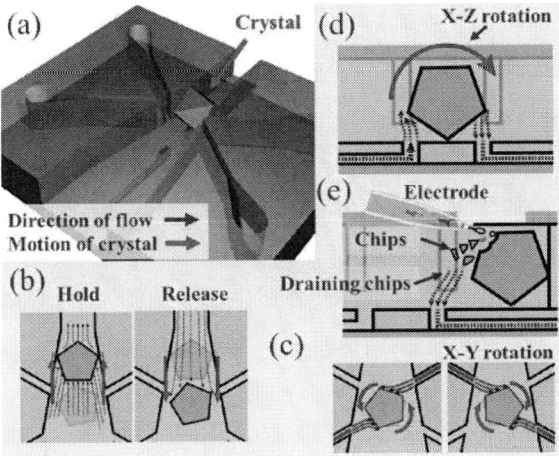

Figure 3: Microfluidic platform to manipulate protein for the accurate three-dimensional positioning. Inlet of devise is connected to the peristaltic pump. Outlets were arrayed to the fluid flows flow along the surface of channel to make the flow rotating.

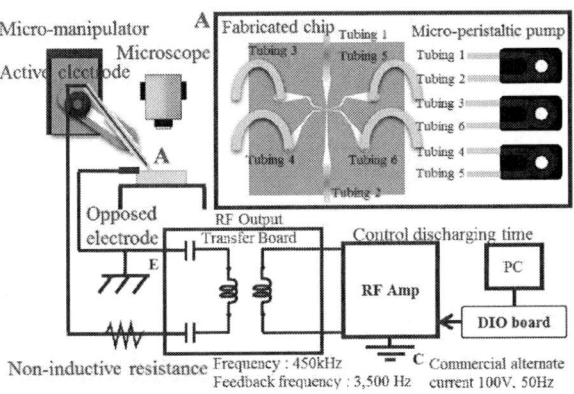

Figure 4: Overall experimental setup of integrated micro-fluidic platform to process protein crystal. Output voltage was amplified in the general electric knife circuit. Non-inductive resistance (10.82 k Ω) was installed to the circuit and the chip is connected to the peristaltic pump (RP-TX, Aquatech Co.Ltd.)

The ablation and suction probe are installed from the top of the intersection of micro-channels. Electrical circuit is also shown to produce electrically-induced high-speed bubble jet to process protein crystal. Discharging time and timing of electrically-induced bubbles were controlled by digital

input output board. In order to demonstrate to process the protein crystal, the lysozyme is employed as a sample of protein crystal. The lysozyme is capable of easily performing the treatment and the crystallization. The composition and volume of solution to produce lysozyme crystal for the present study is shown in table 1. The microfluidic chip is produced by general photo lithographic technology by using thick negative photoresist of SU-8. Two or three layer of PDMS sheet which are transcribed by SU-8 pattern were stacked and bounded by plasma treatment.

Table 1. Experiment condition of reservoir solution and protein solution to produce Lysozyme crystal

	Compositions	Volume (μl)
Reservoir solution	5M NaCl	50
	1M Na acetate buffer pH4.5	5
	80% glycerol	31.2
	Pure water	13.8
Protein solution	Lysozyme(100mg/ml)	1
	Reservoir solution	2

EXPERIMENT

The demonstration of processing crystal is shown in Fig. 5, and which shows that the striking of micro-bubble whose size is less than 100 μm processed the surface of protein successfully by using cavitation phenomenon. Fig. 6 shows the demonstration of carving of protein by high-speed bubbles jet to carve one corner of a protein crystal to be a round shape successfully. Fig. 7 shows the profiles of depth of ablation area and size of ablation area on the surface of protein as a function of number of applied electric pulses which linked to that dispensed number of bubbles. Monotonically increase of depth and size of ablation region is observed which is particularly observed for the condition of applied power of 3 W. Because the increase of depth is much higher than size of ablation, the aspect ratio of the processed hole becomes increasing. High aspect ratio results in effective carving of protein. The electrically-induced bubble knife injects the bubbles and suction probe is collecting the ablation chip under microfluidic control. Finally, Fig. 8 shows the maneuverability of protein crystal on a chip which capable to apply three dimensional microfluidic force to crystals. Combination of the shear stress of fluid and the sudden impact of the high-speed of micro-bubbles provide effective carving of crystal to round shape.

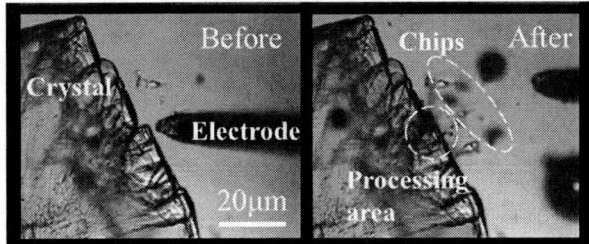

Figure 5: Photos of processing of protein crystal (Lysozyme) by using bubble knife.

978-1-4799-7956-1/15 $31.00 © 2015 IEEE

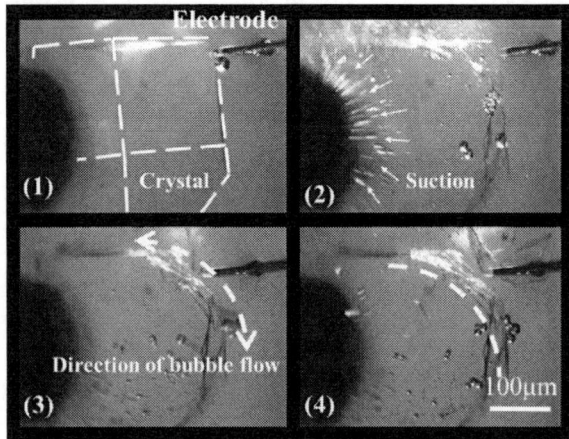

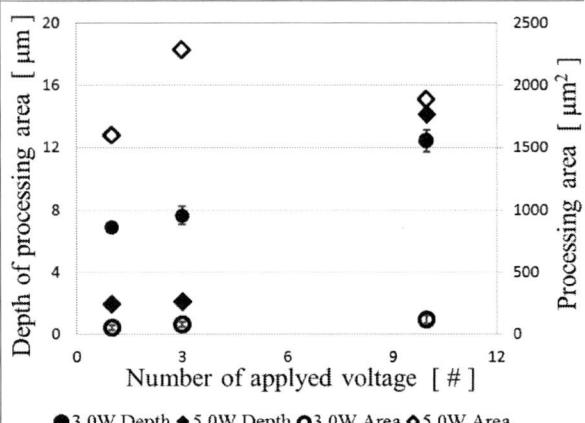

Figure 6: Sequence photos of demonstration of carving of protein crystal (Lysozyme) to round shape.

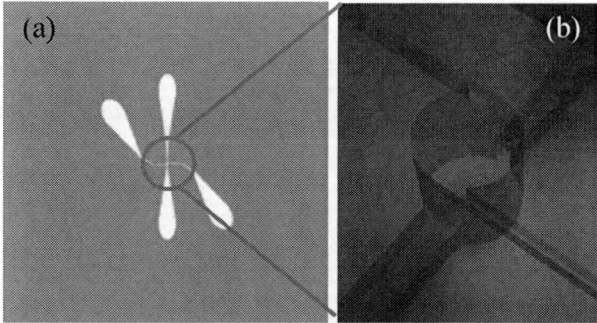

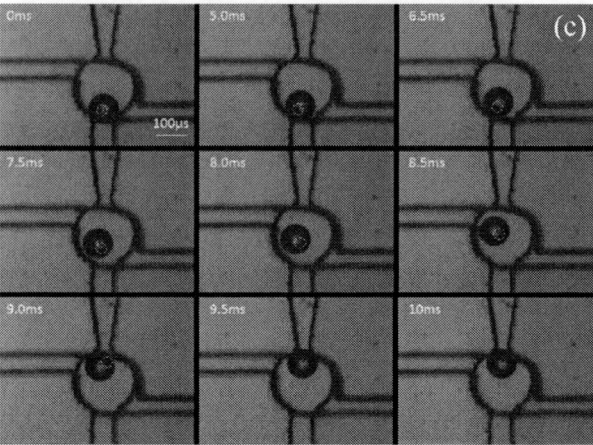

Figure 8: Three dimensional microfluidic channel to manipulate crystal positioning. (a) and (b) show overview the fabricated channels which has the main-microchannel (lower side) to hold the protein crystal and the sub-microchannel (upper side) to rotate the protein crystal (c) shows fabricated micro-fluidic chip for positioning micro-beads.

Figure 7: Size and depth of ablation region as a function of the number of applied pulses.

CONCLUSIONS

A novel micro-fluidic platform for processing protein crystal with high-speed micro-bubble jet was successfully operated. This minimally invasive micro-processing method overcomes the difficulties of processing fragile material under water. This micro-fluidic platform can be powerful tool for protein crystallography and contribute to development of bio-medical researches.

ACKNOWLEDGEMENTS

This work was partly financed by the Ministry of Education, Culture, Sports, Science and Technology (25289059, 25630091) and Japan Science and Technology Agency (JST) PRESTO program. The authors thanked to Prof. G. Kurisu and Prof. H. Tanaka for valuable advices related to protein crystals.

REFERENCES

[1] M. Congreve, C.W. Murray, T.L. Blundell, "Structural biology and drug discovery", Drug Discovery Today, vol. 10, pp.895-907, 2005.

[2] N.E. Chayen and E. Saridakis, " Protein crystallization for genomics: towards high-throughput optimization tech-niques", Acta Cryst, D58, 921, 2002.

[3] N.E. Chayen, "Comparative Studies of Protein Crystallization by Vapour-Diffusion and Microbatch Techniques", Acta Cryst, D54, 8, 1998.

[3] E. P. Carpenter, et al., Curr Opin Struct Biol. 18(5): pp.581–586, (2008).

[4] H. Kitano, et al., Journal of Bioscience and Bioengineering, pp. 50-53, (2005).

[5] H. Kuriki, Y. Yamanishi, S. Sakuma, S. Akagi, and Fumihito Arai, "Local Ablation of a Single Cell Using Micro/Nano Bubbles", Journal of Robotics and Mechatronics, Vol.25, No.3, pp.476-483, 2013

[6] H. Kuriki, et al., Proc. MEMS 2013, pp. 209-211.

CONTACT

*Y.Yamanishi, tel: +81-3-5859-8013;
yoko@shibaura-it.ac.jp

978-1-4799-7956-1/15 $31.00 © 2015 IEEE

A NOVEL CONFIGURATION OF TACTILE SENSOR TO ACQUIRE THE CORRELATION BETWEEN SURFACE ROUGHNESS AND FRICTIONAL FORCE

Ryogo Kozai, Kyohei Terao, Takaaki Suzuki, Fusao Shimokawa, and Hidekuni Takao
Faculty of Engineering, Kagawa University, JAPAN

ABSTRACT

In this paper, a novel configuration of (horizontal type) MEMS tactile sensor that can interact with micro surface roughness at a high resolution are proposed and reported for quantification of the fingertip sense. Two-axis movements of the contactor-tip are independently detected by the two independent suspensions. The contactor is sticking out of the side surface of the silicon chip, and its over-range motion is protected by the embedded stopper structure during the sweeping motion for tactile sensing on the object. In the evaluation experiment, Correlation between the surface shape and the local frictional force were successfully obtained at the same time and at the same point for many samples (paper, plastic, rubber etc.) for the first time.

INTRODUCTION

Various tactile sensors that have human-like tactile sense have been developed and reported. However, it is difficult to realize quantification of human hand feeling using previous devices. Many tactile sensors have diaphragm structure which contacts on the surface vertical to the object [1-4]. The stroke of the structure is not usually long enough, and detection performance of the micro roughness of the contact of the object is poor. On the other hand, tactile sensor using polymer films such as PVDF [5-6] is difficult to use for independent sensing of the sheer forces and contact force. So, multi-axis sensor which can detect two-axis force independently with high special resolution is required for detailed analysis and quantification of fingertip sense.

In this paper, a new sensing principle and structure of the horizontal type MEMS tactile sensor are reported. It has a novel configuration of tactile sensor which can react to micro roughness at a high resolution. Two axis forces are independently detected by the two independent suspension structures. Piezoresistive detection circuits are integrated on them. In the following sections, device configuration, sensor design, device fabrication, and evaluation are discussed.

CONFIGURATION OF THE SENSOR

Figure 1 shows the configuration of horizontal type MEMS tactile sensor. The sensor consists of contactor which contacts to the measurement object, the suspension to the support, and the chip frame including function of the over-range stopper of the contactor. Contactor tip is sticking out of the frame side by 50μm and it is controllable by CAD layout. The over-range stopper prevents destruction of the structure such as suspension by excessive displacement of the contactor. The reference plane of the chip side works as the stopper for x-axis movement, and the side frame of the contactor works as the y-axis over-range stopper. Contactor tip protrudes from the plane of the tip edge. When the measuring object is pressed by the contactor, the chip is swept in parallel to the chip side surface, and stress caused by the displacement is generated on the suspensions leading to contactor. In the device, two suspension units are integrated for independent detection of two-axis motion. Integrated piezoresistors are formed by diffusion process on the suspension fabricated by single crystalline silicon. Individual axis detection circuits detect both the x-axis direction depth displacement and y-axis displacement corresponding to the local frictional force at the contactor-tip. Since the components of the sensor structure is flexibly designed simply by CAD layout, mechanical properties and specifications of the tactile sensor can be optimized of various materials.

Figure 2 shows the sensing operation of the sensor. Contactor tip and the reference plane of the chip side contact is touching the sample surface. Sensing of the tactile sensor is performed by scanning the sensor in the same manner with human finger's sweeping motion. During the measurement, contactor follows the surface roughness close to the diameter of the contactor tip. Since the sensor can detect both surface roughness and the local frictional force depending on the surface morphology, correlation between the surface shape and the local frictional force are obtained at the same time and at the same point. Analyzing the correlations, the material property and the features can be extracted and quantized.

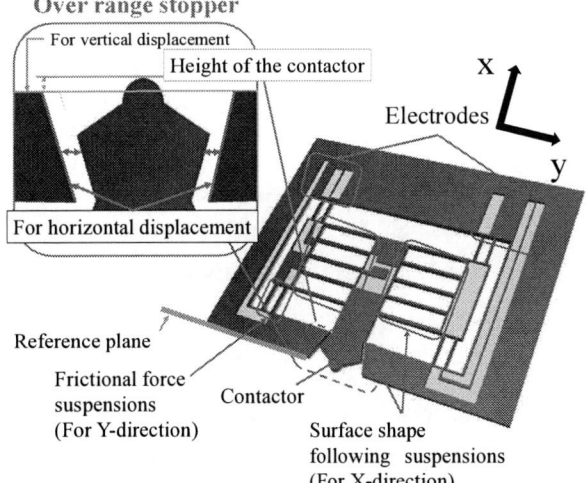

Figure 1: Configuration of the horizontal type MEMS tactile sensor device

The sensor can be fabricated in 1-D array that have different diameters of the contactor tips. The tactile sensor can realize additional functions as well as increase the accuracy of measurement by taking advantage of this feature. If two different contactor tips are arrayed in the same chip, it is possible to extract micro scale roughness of the target surface removing the effect of large undulation, which is obtained by subtraction of two signals.

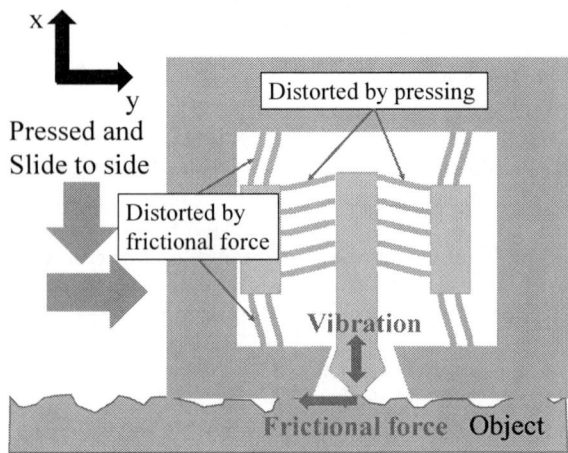

Figure 2: Sensing operation image of the sensor

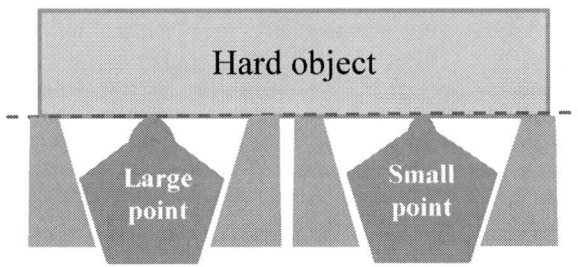

Pushing depth Difference is small

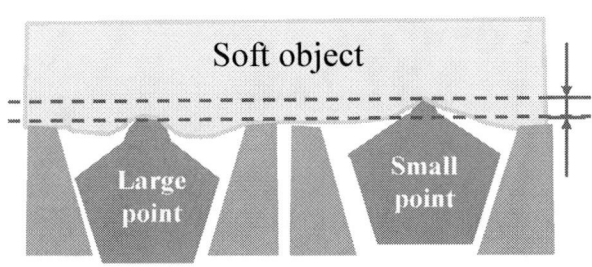

Pushing depth Difference is large

Figure 3: The concept of determines the hardness

Hardness detection is also easy using the new tactile sensor as explained as following. Figure 3 shows two different contactors touching on two different materials. If the tips are touching on a hard object, there is no difference in the amount of indentation between the two contactors. However, if the touching material is soft enough, smaller diameter of the contactor tip increases the indentation value. It is possible to integrate the ability of hardness detection based on the different design of the contactors. Since the important dimensions of the components are flexibly designed only by the CAD layout, mechanical vibration properties and measurement force range of the device can be flexibly controlled.

LAYOUT OF THE SENSOR

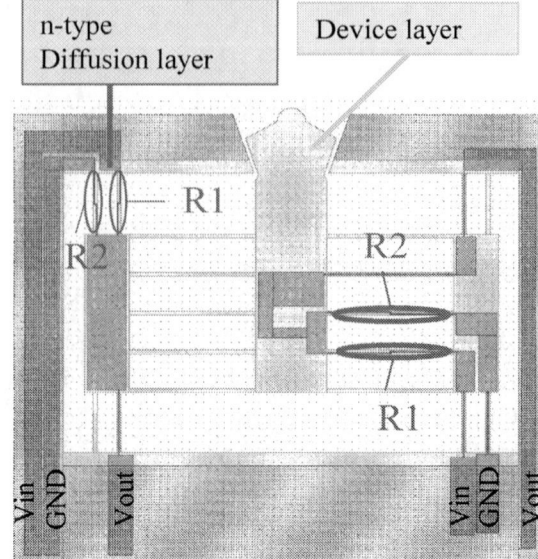

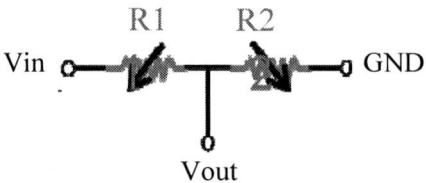

Figure 4: Stress detector circuit diagram using piezoresistors

The horizontal tactile sensor is fabricated by the active layer of SOI wafers. Suspensions and contactors as the movable structure are formed in the 50µm thickness device layer. Figure 4 shows the voltage divider of stress detection circuit fabricated by impurity diffusion on suspension beams. Two stress detection circuits are integrated to detect the displacement of the contact depth and the frictional force independently. So, it is possible to acquire independent signals of x-direction (i.e. object surface roughness) and y-direction (i.e. frictional force).

The projection of the tip is designed as 50µm from the reference plane. Movable range of the contactor is limited by the over-range stopper structure to be below 50µm for x-direction and 35µm for y-direction. The width and lenght of the x-axis suspensions are 30µm and 900µm, respectively. On the other hand, the y-axis suspensions have width of 30µm and length of 700µm. Based on an estimation of the noise level and the sensitivity, the detectable minimum displacement of the contactor is calculated as around 1µm. In the layout design, two kinds of sensor structures are integrated on a chip. The diameters of the contactors are 100µm and 200µm, respectively.

FABRICATION OF THE SENSOR

Figure 5 shows the fabrication process flow of the tactile sensor. Patterning of the diffusion layer for circuit formation is performed first on the device layer of SOI wafers using photolithography step (a, b). The diffusion layer is formed by a phosphorus diffusion process. Resist pattern of the movable structure is formed on device layer

978-1-4799-7956-1/15 $31.00 © 2015 IEEE   246

by photolithography (c). A Cr film is sputtered on the backside, and it is patterned for the following Deep-RIE (d). Deep-RIE of the device layer is performed to fabricate the movable structure (e). Remove the oxide layer around the movable structure (f). Deep-RIE is performed to the handle layer from the back side to release the sensor structure (g). Finally supporting photoresist is removed and the device is completed (h).

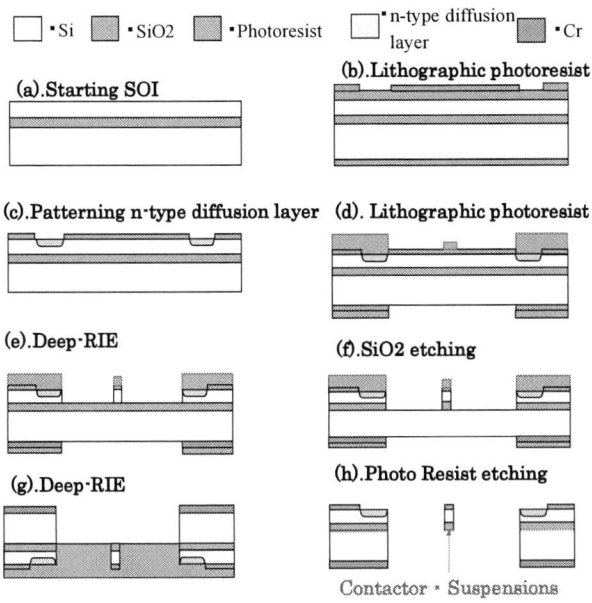

Figure 5: Process flow of the tactile sensor device

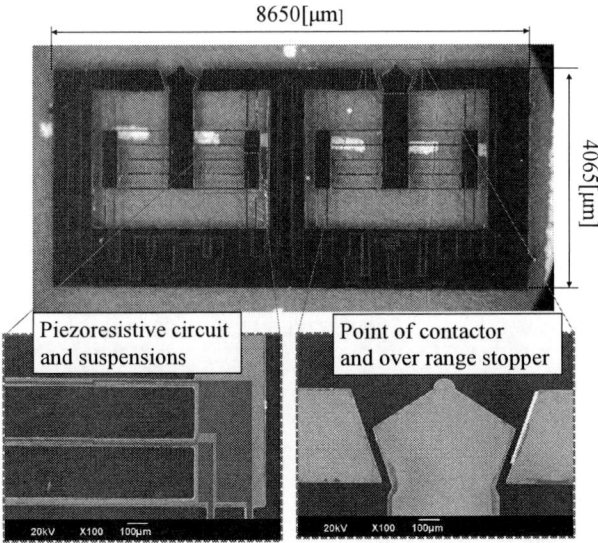

Figure 6: Photograph and SEM images of the produced device

Figure 6 shows a photograph and SEM images of the completed tactile sensor based on the processes. The contactor tips, over-range stoppers, and piezoresistors circuit formed on the suspensions are fabricated in a designed sizes. The two kinds of tactile sensors with different sizes of contactors (100μm and 200μm in diameter) are successfully fabricated as shown.

ELECTRICAL CHARACTERIZATION

Fabricated sensor chip was mounted on a universal circuit board, and it is used as the device package. It was confirmed that the output signals from the stress detection circuit of the n-type diffusion layer changes with the displacement of the contactor motion. Figure 7 shows the output characteristic of the piezoresistor circuit integrated on the x-axis suspensions (surface shape detection) for applied normal forces to the contactor. The applied force to the contactor in x-axis direction was controlled by a precise micro-force gauge and stepping motor. The output signal is in proportion to the normal force in this range, and the force sensitivity is about 39.0μV/mN/V.

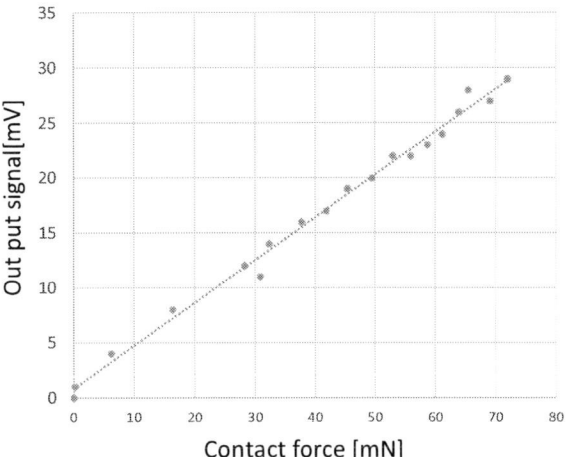

Figure 7: Relationship of the contact force and output signal of the circuit on surface shape following suspensions

Figure 8 shows the evaluation experiment of the fabricated tactile sensor using a sample plastic ruler as the measurement object. It is well recognized that the contactor tip is pushed to the left side by the touching sample. Under the situation, if the sample is swept on the sensor side surface, the sample slips on the reference plane, and the tip of the contactor moves on the two directions. As explained in the previous section, the frame around contactor has the function of the over-range stopper. In the experiments, the range of motion of the contactor was well limited by the stopper structure, and the suspensions were not destroyed by scanning on various sample surfaces. Even if the frictional force of the rubber surface is quite large, the sensor device was able to be scanned without any problem in the device structures. The sensor structure is quite robust, and many samples can be measured with this high sensitivity silicon sensor because it has functional over-range protection. Based on the obtained sensitivity and the noise floor of the fabricated sensor, it can detect micro roughness of below 1μm range.

Figure 9 shows the output voltage waveform when a plastic sample is swept on the sensor reference plane in parallel to the y-axis direction. The surface shape signal (x-axis, Orange line) changes according to the surface micro roughness. On the other hand, the y-axis signal changes depending on the friction on the tip at the moment. Normal force and frictional force are well correlated. However, the correlation coefficient is not 1.0, or -1.0, and it may strongly dependent on the material of sample. The frictional force signal includes information of averaged frictional coefficient and "stick and slip motion" of the contactor tip. Analyzing the correlation between the two-axis signals, important information to quantize the material properties will be obtained using the tactile sensor device.

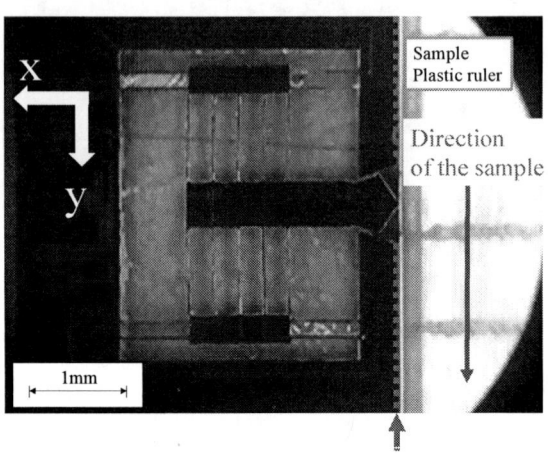

Figure 8: Photograph of the tactile sensing operation.

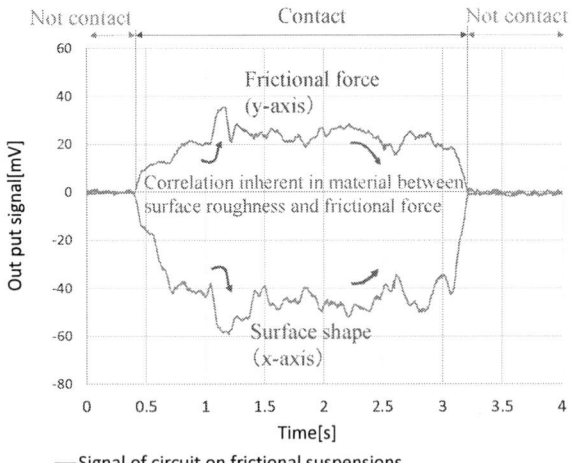

Figure 9: Output voltage waveform when the sample is swept on the sensor

CONCLUSION

In this paper, a new sensing principle and structure of the horizontal type MEMS tactile sensor has been realized and evaluated. It has a contactor structure formed on the side surface of a silicon chip for sweeping motion of tactile sensing. Integrated piezoresistive detection circuits and mechanical structures including suspension springs were designed and fabricated to realize highly sensitive two-axis tactile sensor. In the evaluation experiment, signal outputs were obtained in vertical direction and horizontal direction of the contactor motion by the sweeping on various samples. Analyzing the correlation of x and y axis motions and stick and slip motion obtained at the small tip area, micro-scale surface properties of materials will be explicated for quantification of the fingertip sense.

ACKNOWLEDGEMENTS

This research has been partially supported by the Ministry of Education, Culture, Sports, Science, and Technology through a Grant-in-Aid for Scientific Research (B),25289104,2014.

REFERENCES

[1] H. Takao, M. Yawata, K. Sawada, M. Ishida, "A multifunctional integrated silicon tactile imager with arrays of strain and temperature sensors on single crystal silicon diaphragm", *Sensors and Actuators A Physical*, vol. 160, pp. 69-77, 2010.

[2] Y. Maeda, K. Terao, T. Suzuki, F. Shimokawa, and H. Takao, "A Novel Integrated Tactile Image Sensor for Detection of Surface Friction and Hardness Using the Reference Plane Structure", in *Digest Tech. Papers IEEE Sensors 2012 Conference*, Taipei, October, 2012, pp. 40-43

[3] C.C. Wen , W. Fang, "Tuning the sensing range and sensitivity of three axes tactile sensors using the polymer composite membrane" *Sensors and Actuators, A Physical* vol.145–146, pp. 14–22, 2008

[4] N. Thanh-Vinh, N. Binh-Khiem, H. Takahashi, K. Matsumoto, I. Shimoyama, "High-sensitivity triaxial tactile sensor with elastic microstructures pressing on piezoresistive cantilevers" *Sensors and Actuators, A Physical*, vol. 215, pp. 167-175, 2014

[5] H.K. Kim, S. Lee, K.S. Yun, "Capacitive tactile sensor array for touch screen application" *Sensors and Actuators, A Physical*, vol. 165, pp. 2–7, 2011

[6] T. Okuyama, M. Hariu, T. Kawasoe, M. Kakizawa, H. Shimozu, and M. Tanaka, "Development of tactile sensor for measuring haia touch feeling" *Microsystem technologies,* vol.17(5-7) pp.1153-1160, 2011

CONTACT

Prof. Hidekuni Takao,
Faculty of Engineering, Kagawa University,
2217-20, Hayashi, Takamatsu, Japan.
Tel/Fax: +81-87-864-2331;
E-mail: takao@eng.kagawa-u.ac.jp

TUNNELING PIEZORESISTIVE TACTILE SENSING ARRAY FABRICATED BY A NOVEL FABRICATION PROCESS WITH MEMBRANE FILTERS

Cheng-Wen Ma, Ting-Hao Lin, and Yao-Joe Yang
National Taiwan University, Taipei, TAIWAN

ABSTRACT

In this work, a highly-sensitive tactile sensor array using the tunneling piezoresistive effect is presented. The sensing element, which is made of multi-wall carbon nanotubes and polydimethylsiloxane (MWCNT and PDMS) conductive polymer, was patterned with microdome structures by a novel fabrication process on a membrane filter substrate. The fabricated sensing device features advantages such as ultra-high sensitivity, flexibility, and simple fabrication process. The tunneling piezoresistive effects of the interlocked microdome structures with different MWCNT concentrations are demonstrated. The resistance change of the sensor array due to different elbow-bending motion was measured. Force images were also obtained by using an 8×8 sensing array with different patterns.

INTRODUCTION

Flexible tactile sensors are essential components in various applications, including electronic skin [1, 2], flexible touch screen [3], wearable electronics [4], and bio-monitoring [5]. In [6], an ultrathin, low-modulus, lightweight, stretchable "skin-like" membranes, on which electrodes, electronics, sensors, power supply, and communication components were fully integrated, was proposed. The membrane could conformally laminate onto the human skin surface, much like a tattoo. Lai et al. presented a resistive sensing array capable of retaining and erasing tactile images [7]. The sensing material was prepared by dispersing multiwalled carbon nanotubes and silver nanoparticles through polydimethylsiloxane (PDMS) with the assistance of the dielectrophoresis technique. In [8], a self-powered triboelectric sensor based on flexible thin-film materials was reported. Voltage signals were generated by contact electrification in response to a physical contact. Gong et al. reported a sensitive flexible pressure sensor by sandwiching thin gold nanowire-impregnated tissue paper between two thin PDMS sheets [9]. The proposed fabrication process is low-cost and scalable. Nie et al. proposed a flexible tactile sensing device which consists of an iontronic microdroplet array device by employing an ultra-large interfacial capacitance at the elastic droplet–electrode contact [10]. In [11], a highly twistable tactile sensing array was reported. The proposed device, which employs extendable spiral electrodes, is highly flexible so that it can conform to complex surfaces without damaging the skin structure and the metal interconnects on the sensing array.

In recent years, ultra-high-sensitive tactile sensors with embedded microstructure arrays were proposed. Schwartz et al. reported capacitive flexible tactile organic thin film transistors with microstructured PDMS dielectrics [12]. Wang et al. presented the fabrication of large-area patterned PDMS conducting thin films with uniformly microstructured patterns which was realized by using a high quality silk textile as a mould [13]. Park et al. proposed a flexible tactile sensing film based on composite elastomer layers that contain interlocked microdome arrays with giant tunneling piezoresistance [14]. Zhu et al. reported a flexible resistive tactile sensor with PDMS microstructured arrays deposited with graphene [15]. The aforementioned works exhibited high sensitivities as well as rapid responses. However, special molds were usually required for creating microstructures which are essential for increasing sensitivities.

In this work, we present a tactile sensing array with the tunneling piezoelectric effect with microdome structures. By patterning conductive polymer on a membrane filter using the soft-lithography process, sensing elements with microdome structures can be easily realized.

DEVICE DESIGN

Fig. 1 shows the schematic of the proposed tactile sensing array, which consist of a PDMS substrate, patterned conductive polymer layer, and Cr/Au layer. Each tactile sensing cell includes interlocked microdome structures which were formed by patterning conductive polymer using the soft-lithography technique on a filter membrane substrate. The sensing device features advantages such as ultra-high sensitivity, flexibility, low cost, and simple fabrication process.

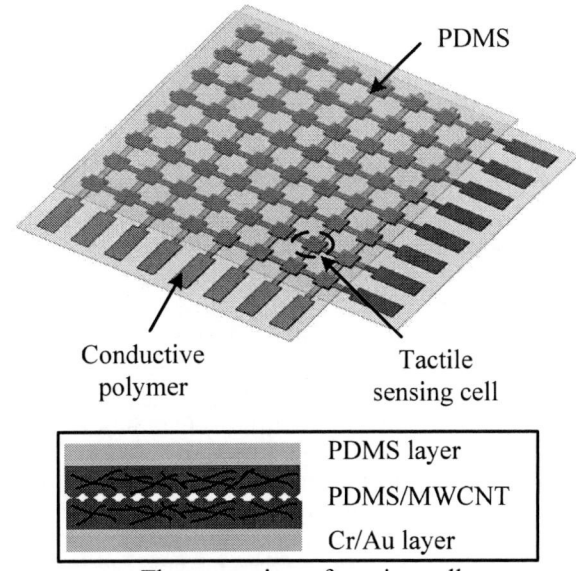

Figure 1: Schematic of the proposed tunneling piezoresistive tactile sensor array.

SENSING PRINCIPLE

Fig.2 shows the operational principle of this tactile sensor array. As an external pressure is applied, the contact area between the two contacting microdome structures increases significantly, which in turn changes the tunneling resistance at the contact area. The tunneling piezoresistive device is much more sensitive than the typical conductive polymer materials which were governed by the percolation theory [16]. The design of interlocked microdome structure features the advantage of fast response as compared with the typical flexible planar conductive composite films, due to the fact that immediate pressure-induced surface deformation of the microdomes in the interlocked geometry gives rise to the rapid variation of contact area, results in fast resistance change.

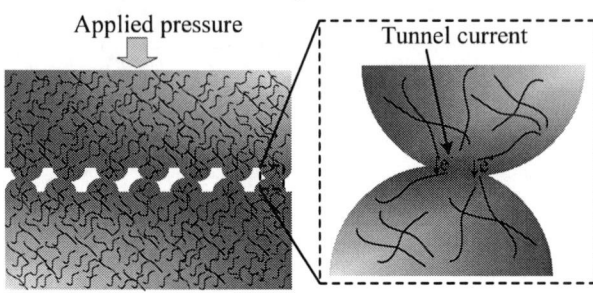

Figure 2: Schematic showing the working principle of this tactile sensor.

FABRICATION

For the preparation of the conductive polymer composite mixtures, multiwalled nanotubes (MWCNTs, Golden Innovation Business Co., Ltd.) with diameters of 20 nm and lengths ranging from 30 to 50 μm were dispersed in hexane and PDMS (Sylgard 184 A, Dow Corning Corporation) prepolymer at the ratio of 4:1 in a glass bottle. The prepolymer was fully stirred by using a magnetic stirrer for 6 h. Then, the curing agent (Sylgard 184 B, Dow Corning Corporation) was mixed at a 10:1 ratio, and was degassed in a vacuum chamber. The MWCNT concentration is 6 wt% in the PDMS/MWCNT prepolymer.

Fig. 3 describes the fabrication process. Firstly, SU-8 photoresist (SU-8 2050, MicroChem Corporation) is spin-coated on a handling wafer and a nylon membrane filter (nylon membrane filter, pore size: 1 μm, MS®) is placed on the top of the SU-8 film (Fig. 3(a) and (3b)). Then, SU-8 photoresist was spin-coated (Fig. 3(c)) and patterned (Fig. 3(d) and 3(e)). PDMS/MWCNT conductive polymer was filled in the trenches between SU-8 walls (Fig. 3(f)). Microdome structures were transferred to the surface of the conductive polymer contacting the filter membrane. Then, a Cr/Au layer is deposited (Fig. 3(g)). In Figure 3(h), the patterned conductive polymer was peeled off from the filter membrane. The film was then place on a thin PDMS film (Fig. (3i)). Finally, two patterned conductive polymer films were combined, and the sides with microdome structures faced each other. The films were fixed with silicone gel (Fig. (3j)).

Fig. 4(a) shows the scanning electron microscopy (SEM) image of the nylon membrane filter. Numerous small pores on the nylon membrane filter structure were observed. Fig. 4(b) shows the SEM image of the conductive polymer film with microdome structures transferred from the filter membrane. The maximum height and base-width of the microdome structure are about 5 μm. The assembled devices are shown in Fig. 5(a). The thickness of the film is about 400 μm. Fig. 5(b) shows the fabricated 8 × 8 flexible tactile sensing array. The size of each tactile sensing cell is 1 mm × 1 mm. The pitch between tactile sensing cells is 2 mm.

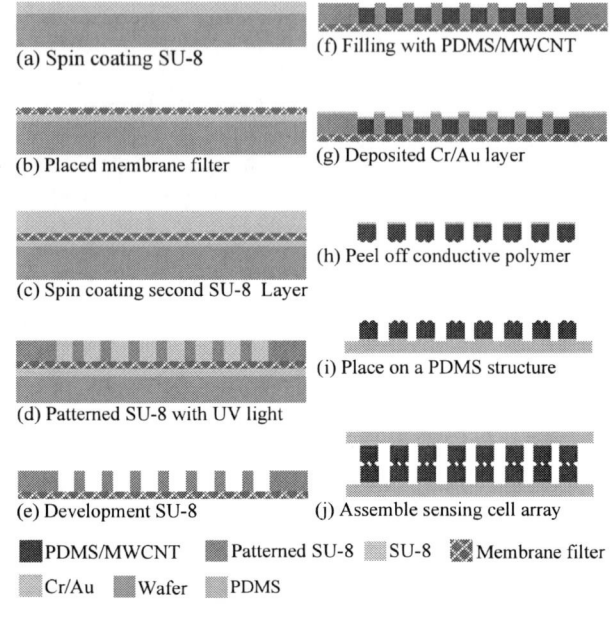

Figure 3: The fabrication process of the tunneling piezoresistive tactile sensing array.

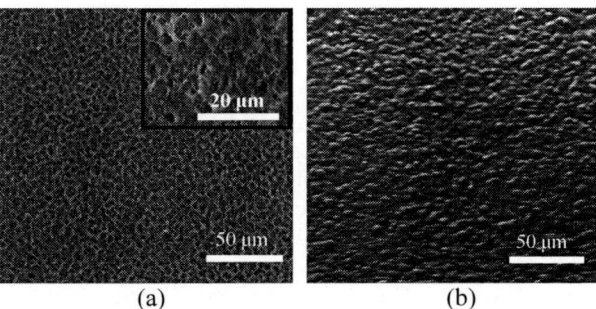

Figure 4: SEM pictures. (a) The surface of nylon membrane filter. (b) Composite elastomer with micro-domes on the surface.

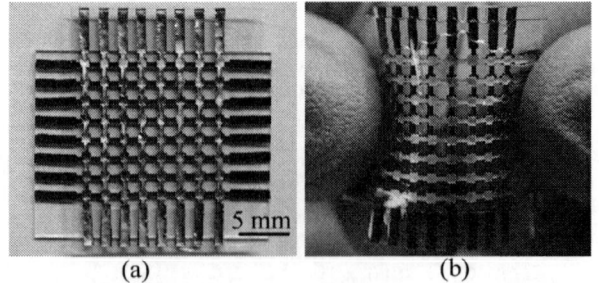

Figure 5: (a) The assembled device. (b) The fabricated flexible 8×8 tactile sensing array.

MEASUREMENT AND DISCUSSION

Figure 6 illustrates the experimental setups for testing the sensing arrays by applying normal forces. A force gauge (HF-1, ALGOL Engineering Co.) with maximal resolution of 1 mN was used to measure the applied force. The force gauge was fixed at a vertical (z-axis) translational stage that had a displacement resolution of 1 μm. A poly-methylmethacrylate (PMMA) rod with a circular cross-section was connected to sensing end of the force gauge. The resistance change of each tactile sensing cell was measured using a source meter (Model 2400, Keithley Instruments).

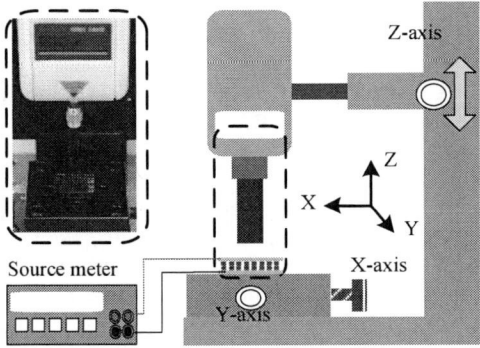

Figure 6: Experimental setup for testing the sensing arrays by applying normal forces..

Figure 7 shows the measured responses with CNTs concentrations of 6 and 8 wt% in the MWCNT/PDMS polymer composites. The piezoresistive effects are primarily caused by the variation in tunneling resistance when the microdome arrays deformed under external stress. As shown in the figure, the resistance changes sharply under small pressure force. The measure results also showed that the device with higher CNT concentration gives higher change in relative resistance under the same pressure change, due to the larger electrical contact region in the interlock area for the films with higher CNT concentrations. These measured results indicate excellent repeatability and highly sensitivity (-0.5/kPa).

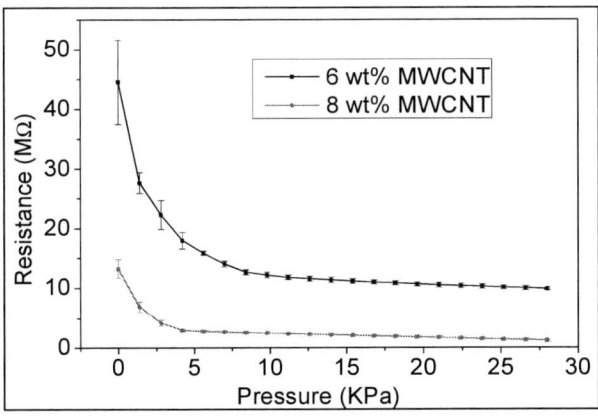

Figure 7: Measured response with different PDMS/MWCNT concentration.

Fig. 8 shows the measured resistance vs. human elbow bending motions. The experiment exhibit the potential application that our sensors can be integrated onto a wearable device for monitoring human motions. A fabricated tactile sensing cell was attached onto the bottom of the elbow of a volunteer for detect the bending movements. When the volunteer's elbow was gradually folded, as shown in the insets A to E, the resistance of the sensing cell decreased. As the elbow unfolded, the resistance increases to the initial value (E to F).

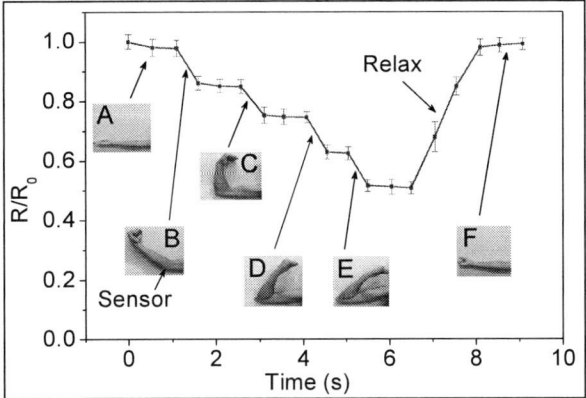

Figure 8: The measured resistance vs. human elbow bending motions.

Fig. 9 shows the normal force images captured by an 8 × 8 tactile sensing array. The force images were induced by using different acrylic stamps ('2', '0', '1' and '5'). The pictures of these stamps are also shown in the figure. Due to sharp resistance changes of sensing cells when pressed, the crosstalk between each sensing cell is quite small when compared with typical planar conductive composite films. The shapes of the stamps are clearly resolved by the sensing array. Note that these results are measured by the scanning readout circuit.

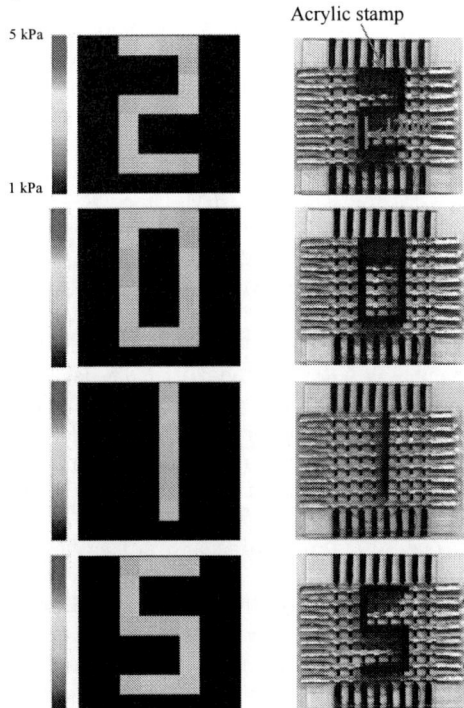

Figure 9: The normal force images captured by a 8 × 8

978-1-4799-7956-1/15 $31.00 © 2015 IEEE

tactile sensing array.

CONCLUSION

This work presented a highly-sensitive tactile sensor array using the tunneling piezoresistive effect. A novel method was proposed to fabricate microdome structures which are essential for the tactile sensing cells. An experimental setup for normal force measurement was implemented. The measured tunneling piezoresistive effects of the interlocked microdome structures with different MWCNT concentrations were also demonstrated. In addition, we demonstrated a data curve by integrating a tactile sensor onto a human elbow to detect the bending movements. We also demonstrated tactile images induced by different stamps using an 8 × 8 tactile sensing array.

ACKNOWLEDGEMENTS

This work was supported in part by the National Science Council, Taiwan, R.O.C. (Contract No: NSC 100-2221-E-002-075-MY3).

REFERENCES

[1] B. C. K. Tee, C. Wang, R. Allen, and Z. Bao, "An electrically and mechanically self-healing composite with pressure- and flexion-sensitive properties for electronic skin applications", *Nat Nano,* vol. 7, no. 12, pp. 825-832, 2012.

[2] K. Takei, T. Takahashi, J. C. Ho, H. Ko, A. G. Gillies, P. W. Leu, R. S. Fearing, and A. Javey, "Nanowire active-matrix circuitry for low-voltage macroscale artificial skin", *Nat Mater,* vol. 9, no. 10, pp. 821-826, 2010.

[3] M. Y. Cheng, X. H. Huang, C. W. Ma, and Y. J. Yang, "A flexible capacitive tactile sensing array with floating electrodes", *Journal of Micromechanics and Microengineering,* vol. 19, no. 11, p. 115001, 2009.

[4] K. Ig Mo, J. Kwangmok, K. Ja-Choon, J.-D. Nam, L. Young Kwan, and C. Hyouk Ryeol, "Development of soft-actuator-based wearable tactile display", *Robotics, IEEE Transactions on,* vol. 24, no. 3, pp. 549-558, 2008.

[5] M. Kaltenbrunner, T. Sekitani, J. Reeder, T. Yokota, K. Kuribara, T. Tokuhara, M. Drack, R. Schwodiauer, I. Graz, S. Bauer-Gogonea, S. Bauer, and T. Someya, "An ultra-lightweight design for imperceptible plastic electronics," *Nature,* vol. 499, no. 7459, pp. 458-463, 2013.

[6] D.-H. Kim, N. Lu, R. Ma, Y.-S. Kim, R.-H. Kim, S. Wang, J. Wu, S. M. Won, H. Tao, A. Islam, K. J. Yu, T.-i. Kim, R. Chowdhury, M. Ying, L. Xu, M. Li, H.-J. Chung, H. Keum, M. McCormick, P. Liu, Y.-W. Zhang, F. G. Omenetto, Y. Huang, T. Coleman, and J. A. Rogers, "Epidermal electronics", *Science,* vol. 333, no. 6044, pp. 838-843, 2011.

[7] L. Yu-Tse, C. Yung-Ming, and Y. J. J. Yang, "A novel cnt-pdms-based tactile sensing array with resistivity retaining and recovering by using dielectrophoresis effect", *Journal of Microelectromechanical Systems,* vol. 21, no. 1, pp. 217-223, 2012.

[8] G. Zhu, W. Q. Yang, T. Zhang, Q. Jing, J. Chen, Y. S. Zhou, P. Bai, and Z. L. Wang, "Self-powered, ultrasensitive, flexible tactile sensors based on contact electrification", *Nano Letters,* vol. 14, no. 6, pp. 3208-3213, 2014.

[9] S. Gong, W. Schwalb, Y. Wang, Y. Chen, Y. Tang, J. Si, B. Shirinzadeh, and W. Cheng, "A wearable and highly sensitive pressure sensor with ultrathin gold nanowires", *Nat Commun,* vol. 5, 2014.

[10] B. Nie, R. Li, J. D. Brandt, and T. Pan, "Iontronic microdroplet array for flexible ultrasensitive tactile sensing", *Lab on a Chip,* vol. 14, no. 6, pp. 1107-1116, 2014.

[11] M. Y. Cheng, C. M. Tsao, Y. Z. Lai, and Y. J. Yang, "The development of a highly twistable tactile sensing array with stretchable helical electrodes", *Sensors and Actuators A: Physical,* vol. 166, no. 2, pp. 226-233, 2011.

[12] G. Schwartz, B. C. K. Tee, J. Mei, A. L. Appleton, D. H. Kim, H. Wang, and Z. Bao, "Flexible polymer transistors with high pressure sensitivity for application in electronic skin and health monitoring", *Nat Commun,* vol. 4, p. 1859, 2013.

[13] X. Wang, Y. Gu, Z. Xiong, Z. Cui, and T. Zhang, "Silk-molded flexible, ultrasensitive, and highly stable electronic skin for monitoring human physiological signals", *Advanced Materials,* vol. 26, no. 9, pp. 1336-1342, 2014.

[14] J. Park, Y. Lee, J. Hong, M. Ha, Y.-D. Jung, H. Lim, S. Y. Kim, and H. Ko, "Giant tunneling piezoresistance of composite elastomers with interlocked microdome arrays for ultrasensitive and multimodal electronic skins", *ACS Nano,* vol. 8, no. 5, pp. 4689-4697, 2014.

[15] B. Zhu, Z. Niu, H. Wang, W. R. Leow, H. Wang, Y. Li, L. Zheng, J. Wei, F. Huo, and X. Chen, "Microstructured graphene arrays for highly sensitive flexible tactile sensors", *Small,* vol. 10, no. 18, pp. 3625-3631, 2014.

[16] Stauffer, D. and Aharony, "A Introduction to Percolation Theory", 2nd ed. London: Taylor & Francis, 1992.

CONTACT

*Cheng-Wen Ma, tel: +886-2-33664941#802, E-mail: jeson@mems.me.ntu.edu.tw

A TACTILE SENSOR WITH THE REFERENCE PLANE FOR DETECTION ABILITIES OF FRICTIONAL FORCE AND HUMAN BODY HARDNESS AIMED TO MEDICAL APPLICATIONS

Yusaku Maeda, Kyohei Terao, Takaaki Suzuki, Fusao Shimokawa, and Hidekuni Takao
Faculty of Engineering, Kagawa University, Kagawa, JAPAN

ABSTRACT

In this study, a novel tactile sensor with detection abilities of human body hardness and frictional force is reported. A new device configuration of back-side contact is proposed to realize a higher sensitivity of hardness, and low sensitivity to normal force. Miniaturization and sensitivity improvement of tactile sensor are important to apply the sensor to palpation inside of the body. Surface frictional force was successfully measured with rubber block in real time using our tactile image sensor. A completed device was evaluated, and shore A hardness was measured in the range from 1 HS to 54 HS. This performance corresponds to the hardness detection ability of adiposus in human body.

INTRODUCTION

Recently, endoscopic surgery and arteriosclerosis treatment using a catheter have been developed. There has been a growing interest in these treatment methods because they are possible to decrease risk of the complications from the wound. However, determine of lesion site by skin sensation is difficult in the operation, since a doctor cannot touch to organs of patient. There remains an ever-increasing interest and challenge to develop highly sensitive tactile sensor with below 3 mm size for detection of organ hardness. Especially, hardness is very important information to identify the lesion site like organs in medical practice. Recently, highly sensitive tactile sensors have been developed for realization of "human's skin like" sensor device [1-2]. Also, various tactile sensors with detection ability of hardness have been developed and reported [3-4], and they have hardness of 10HS have been recognized by the smallest resolution. However, many organs and skin hardness is distributing in the range from 4HS to 50 HS [5-7], and improvement of resolution is necessary in hardness detection of human organs. Previously, we have developed integrated multifunctional tactile imagers with detection abilities of 3-axis force and temperature distribution [8]. Furthermore, hardness detection proposed by our group using the "reference plane" structure showed a detection ability of hardness [9]. However, hardness detection of below 23 HS has not been achieved in the device because of the noise floor, and this value is not enough in the detection of human organs hardness (4-50HS). In addition, detection of frictional force is necessary for stable hardness detection, since in order to guarantee precise indentation to the organ.

In this study, a new configuration of the backside-contact type tactile sensor has been fabricated and evaluated toward hardness detection of more flexible target like humans' organs in size of below 3mm. Also, frictional force detection has been successfully demonstrated using our tactile image sensor.

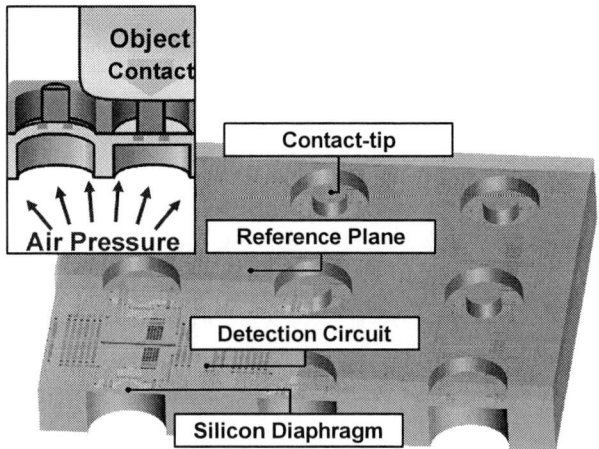

Figure 1: Conceptual diagram of the tactile sensor for detection of frictional force and body hardness by the reference plane structure.

Figure 1 shows a conceptual diagram of our tactile image sensor with the "reference plane". The sensor device structure can detect hardness and frictional force from the surface of the object. The sensor structure consists of integrated detection circuits, contact-tips fabricated on silicon diaphragm array, and the reference plane structure around the moving parts (i.e. the diaphragms). The contact-tips on the diaphragm transmit the contact-force from the measuring object to silicon diaphragms. The reference plane has important role for detection principle, and has another role of over-range stopper to prevent over-range displacement of the diaphragm. The tips and the reference plane are fabricated with an SU-8 layer originally. However, the contact-tip and reference plane are fabricated handle layer of SOI wafer as mention in following. An air pressure is applied from the backside of the device to make the contact-tips be higher than the reference plane surface.

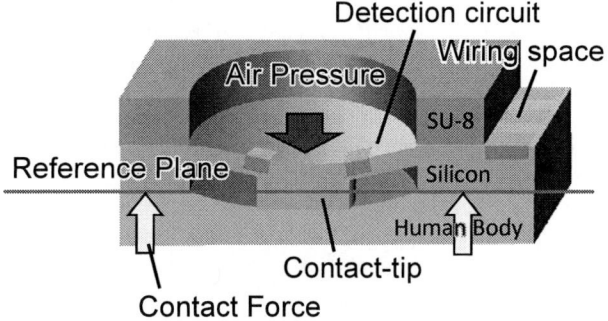

Figure 2: Cross sectional diagram of the back-side contact type tactile sensor.

CONFIGURATION AND PRINCIPLE

The new configuration of the tactile sensor with backside-contact type improves the sensitivity and reduce the device size. Figure 2 shows cross section of the device in this study. The reference plane structure and contact-tip is formed by using a handle layer of the SOI wafer. Silicon contact-tip is moved by the normal force applied from the measuring object surface. Even if the object is very soft like human organs, it is possible to prevent direct contact of the silicon diaphragm with the touching object, since contact-tip formed by the handle layer is thicker than our previous device with SU-8 layer. In addition, this configuration is advantageous to reduce the device size, since contact-tip is placed on opposite side to the circuit wiring space.

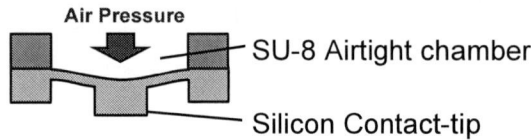

(a) Initial state of the one pixel tactile sensor with silicon contact-tip.

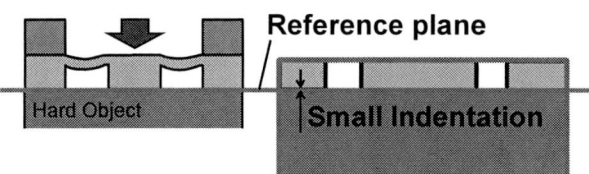

(b) Hardness measurement with hard material.

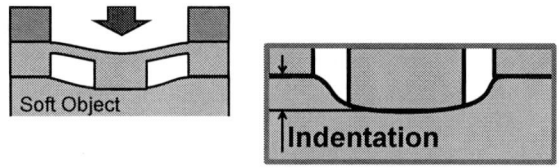

(c) Hardness measurement with soft material.

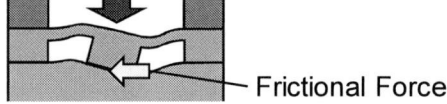

(d) Frictional force measurement from object.

Figure 3: Measurement principles of hardness and frictional force using the silicon reference plane structure.

Detection of normal-force to the contact-tip and frictional force is performed by the integrated piezoresistors around the contact-rip and the diaphragm. Signal conversion is performed by bridge circuits. Detection of hardness is realized by contact-tip and reference plane basically independent from the contact force to the measuring object. Measuring the indentation-height from the reference plane to the object, the hardness such as shore A index can be quantified for soft objects like human organs. Figure 3 (a) is initial state of the device when air pressure is applied to diaphragm. If

the touching object is very hard, the indentation-height from the reference plane to the object becomes very small as shown in Figure 3 (b). On the other hand, indentation-height increases as the touching material gets softer as illustrated in Figure 3 (c). The reference plane is very helpful to detect the indentation-height to the object. Surplus contact force for hardness measurement is supported by the reference plane. In addition, if the applying pressure from the diaphragm is reduced, the range of hardness measurement can be lower for measurement of softer materials. Detection of frictional force is realized by detecting the contact-tip rotation as shown in figure 3 (d).

DEVICE FABRICATION

The tactile sensor was fabricated with our integration process of silicon circuit and MEMS. Figure 4 shows the fabrication flow of the tactile sensor. The starting material of the tactile sensor is SOI substrate with 5µm-thick device layer and 475µm-thick handle layer in p-type. First of all, piezoresistors and wiring of the circuit is formed by n-type diffusion layer for detection of indentation-height and frictional force (Figure 4 (a)). Mechanical movable structures are formed after the circuit formation process. The silicon diaphragm for contact-tip motion sensing is fabricated by deep etching of the handle layer using ICP-RIE. At this point, the contact-tip and the reference plane structure are formed at the same time (Figure 4 (b)). Finally, air pressure chamber is fabricated with an SU-8 layer (Figure 4 (c)).

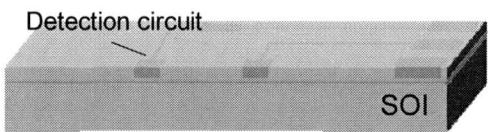

(a) Formation of n-type diffusion layer on 5 µm thick SOI wafer.

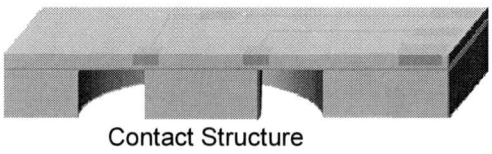

(b) Formation of contact structure reference plane and contact-tip structure necessary for the measurement principle by ICP-RIE.

(c) Formation of air pressure chamber and load support structure using SU-8.

Figure 4: Process flow of the tactile sensor device with detection abilities of hardness and frictional force in flexible object.

978-1-4799-7956-1/15 $31.00 © 2015 IEEE

Figure 5 shows a photograph of the fabricated sensor chip designed in 2.3mm × 2.5mm size. Left photograph of the figure 5 is contact-side of the fabricated device. The diaphragms diameter is 1000µm, and the contact-tip diameter is 400µm. Right photograph of the figure 5 is detection circuit side of the fabricated device.

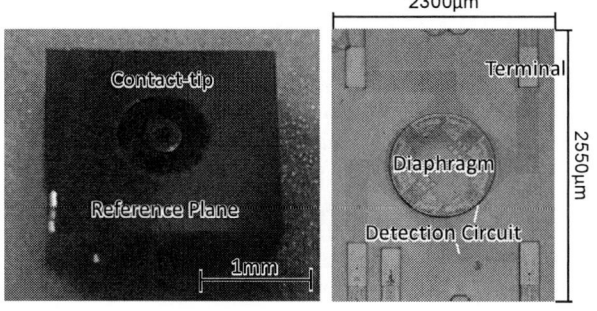

Figure 5: Photograph of a fabricated backside contact type tactile sensor with silicon contact-tip and reference plane.

DEVICE EVALUATION

Figure 6 shows a measured output signal from a pixel in the tactile image sensor for various frictional force input in time domain. In this experiment, frictional forces in the positive direction and the negative direction are applied by a rubber block to the contact-tip of a sensor pixel. The signal is responding in real time to the input frictional force. This result suggests that reduces the contact force dependence on hardness detection.

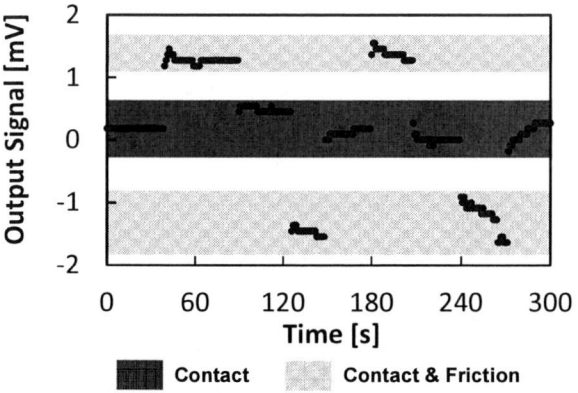

Figure 6: Detection result of real time frictional force by the tactile image sensor.

Figure 7 shows the evaluation system of hardness detection for the fabricated sensor. This system consist of force gauge on z-axis position controller package and an air pump. Air pressure was applied sensor diaphragm by the pump to generate to the initial displacement with contact-tip. The position controller has 50nm resolution in Z-axis direction. Sample object was placed on the same axis of the contact-tip, when sensor package was driven along to Z-axis by the position controller. The applied air pressure is 10kPa, and drive voltage of the sensor is 2.2V.

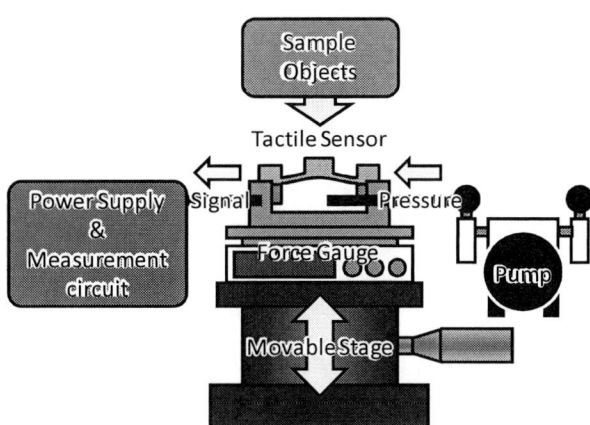

Figure 7: An evaluation system in hardness detection.

Figure 8 shows obtaining linear relationship between output signal and normal force from object to contact-tip. Enough high sensitivity for at least 1mN normal input force was obtained. The force sensitivity of the sensor is 485µV/mN/V.

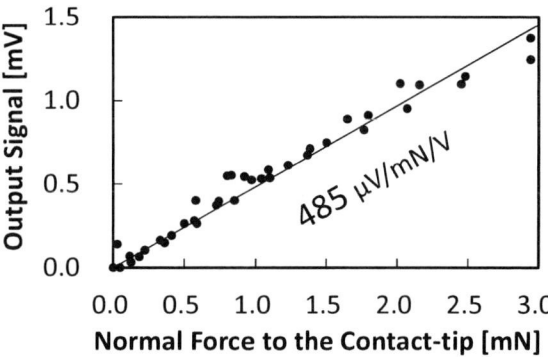

Figure 8: Measured relationship between output voltage and input normal force to the contact-tip.

Evaluation of hardness detection was performed with the same system. Before the evaluation, "Shore A hardness" of 5 objects were measured by a durometer (JIS K 6301 A) to use them as the references of hardness. The reference object was attached to the tip of the force gauge. Hardness measurement with the fabricated device was performed by pressing the reference objects against to the sensor surface. The indentation-height of the contact-tip from reference plane to the object corresponds to the object hardness (softness). Figure 9 shows measured dependencies of the fabricated sensor signals on the contact force for various hardness objects. Since the sensor has the reference plane structure around contact-tip, their output signals are saturated in a force range of contact force from 0.2 N to 5.0 N. The saturated output signal is increasing along with the hardness of objects. This result is demonstrating the advantage of the "reference plane" that reduces the contact force dependence on hardness detection. Figure 10 shows the relationship between the output signal at a 0.2N contact force and Shore-A hardness of the touching object.

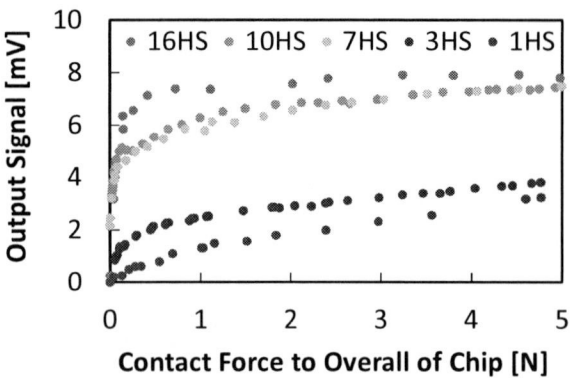

Figure 9: Measured relationship between the output signal and contact force from the measured objects.

As shown in the figure, a linear relationship between the output and sample hardness is obtained in the range from 1 HS to 54 HS. The minimum detectable hardness in this experiment is less than adiposus (4HS). This result corresponds to the ability to identify the lesion site of human organs, since measurable hardness range of this sensor includes hardness of organs like the liver and the stomach. The hardness sensitivity of the measured device is 462μV/HS/V. Through the experiment, detection of hardness corresponding to the human organs has been successfully demonstrated.

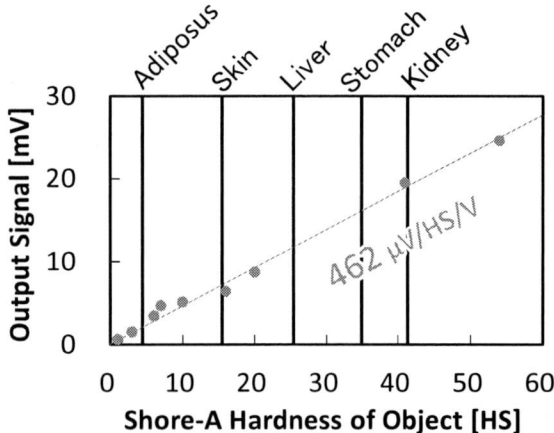

Figure 10: Measured relationship between the output signal and Shore A hardness of the measured samples. Annotation on plot is hardness of each human organ.

CONCLUSIONS

In this paper, a new configuration of tactile sensor with hardness detection abilities of flexible object surface have been proposed and reported. Detection of frictional force and hardness are realized with the reference plane structure around the contact-tip. The tactile sensor device was designed and fabricated with combination of a conventional LSI/MEMS integration process and SU-8 polymer structure formation. A real-time response of the frictional force measurement was demonstrated with our tactile image sensor. Also, a linear relationship between the output and hardness is obtained in the range from 1 HS to

54 HS, which is important to apply the device to organ hardness measurement. Under the situation that contact force is not precisely measured. The minimum detectable of hardness was below 3HS, and the proposed detection principles of hardness have been successfully demonstrated through the evaluation experiments.

ACKNOWLEDGEMENTS

This research has been partially supported by the Ministry of Education, Culture, Sports, Science, and Technology through a Grant-in-Aid for Scientific Research (B), 25289104, 2014.

REFERENCES

[1] M. Sohgawa, D. Hirashima, Y. Moriguchi, T. Uematsu, W. Mito, T. Kanashima, M. Okuyama, H. Noma, "Tactile sensor array using microcantilever with nickel–chromium alloy thin film of low temperature coefficient of resistance and its application to slippage detection", *Sensors and Actuators A.*, vol. 186, pp. 32-37, 2012.

[2] H. Kim, S. Lee, K. Yun, "Capacitive tactile sensor array for touch screen application", *Sensors and Actuators A.*, vol. 165, pp. 2-7, 2011.

[3] Y. Hasegawa, M. Shikida, T. Shimizu, T. Miyaji, H. Sasaki, K. Sato, K. Itoigawa, "Amicromachined active tactile sensor for hardness detection", *Sensors and Actuators A,.* vol. 114, pp. 141-146, 2004.

[4] J. Engel, J. Chen, Z. Fan, C. Liu, "Polymer micromachined multimodal tactile sensors", *Sensors and Actuators A.*, vol. 117, pp. 50-61, 2005.

[5] V. I. Egorov, I. V. Schastlivtsev, E. V. Prut, A. O. Baranov, R. A. Turusov., "Mechanical properties of the human gastrointestinal tract", *J Biomech.*, vol. 35(10), pp. 1417-1425, 2002.

[6] H. Yamada, F. G. Evans, *Strength of Biological Materials*, The Williams & Wilkins Company, 1970

[7] P. G. Agache, C. Monneur, J. L. Leveque, and J. De Rigal, "Mechanical properties and Young's modulus of human skin in vivo", *Arch Dermatol Res.*, vol. 269(3), pp. 221-232, 1980.

[8] H. Takao, M. Yawata, K. Sawada, M. Ishida, "A multifunctional integrated silicon tactile imager with arrays of strain and temperature sensors on single crystal silicon diaphragm", *Sensors and Actuators A Physical*, vol. 160, pp. 69-77, 2010.

[9] Y. Maeda, K. Terao, T. Suzuki, F. Shimokawa, and H. Takao, "A Novel Integrated Tactile Image Sensor for Detection of Surface Friction and Hardness Using the Reference Plane Structure", in *Digest Tech. Papers IEEE Sensors 2012 Conference*, Taipei, October, 2012, pp. 40-43

CONTACT

Prof. Hidekuni Takao,
Faculty of Engineering, Kagawa University,
2217-20, Hayashi, Takamatsu, Japan.
Tel/Fax: +81-87-864-2331;
E-mail: takao@eng.kagawa-u.ac.jp

978-1-4799-7956-1/15 $31.00 © 2015 IEEE

6-AXIS FORCE/TORQUE SENSOR
FOR SPIKE PINS OF SPORTS SHOES

H. Ishido, H. Takahashi, A. Nakai, T. Takahata, K. Matsumoto, and I. Shimoyama
The University of Tokyo, Tokyo, JAPAN

ABSTRACT

This paper reports on a method to measure forces and torques acting on spike pins of sports shoes. We fabricated a 2 mm × 2 mm × 0.3 mm 6-axis force/torque sensor chip, which consists of 6 piezoresistive beams. From the resistance changes of the 6 beams, 6 components of forces and torques can be detected. The sensor chip was connected to a flexible printed circuit board (PCB) and covered with thin polydimethylsiloxane (PDMS). Then, the sensor chip and flexible PCB were embedded in spike-pin-shaped epoxy resin. We calibrated the spike-pin-shaped sensor, and confirmed that 6-axis forces and torques were able to be detected by the proposed sensor.

INTRODUCTION

Recently, many researches related to measurement of walking and running have been reported to analyze human gait [1-6]. In these researches, it is important to measure the Ground Reaction Force (GRF) because the GRF is the only external force acting on human during walking or running except for the gravity [3-6].

In previous researches, the GRF has been generally measured by a force plate [3-4]. Using a force plate, the total force acting on shoe soles can be measured at one step of walking or running. As another approach to measure the GRF, multi-axis force sensors have been developed and utilized to be attached on shoe soles or insoles [5-6]. The measurement method with force sensors is suitable to measure distribution of the GRF. Additionally, the GRF can be measured at every step with force sensors attached on shoe soles.

In athletic sports, athletes wear spike shoes of which soles are not flat, but have many pins (Figure 1). Because the GRF of a runner wearing spike shoes is larger than that of a runner wearing other shoes [4], it is predicted that the pins contribute to the increase of the GRF and faster running. Additionally, it is also predicted that most part of the GRF acts on the spike pins biting an athletic field (Figure 2). Therefore, it is important to evaluate forces and torques acting on spike pins to analyze the function of shoes or running of athletes.

In this paper, we reports on a method to measure forces and torques acting on spike pins. Forces and torques can be detected by one sensor chip embedded in the root of a spike pin as shown in Figure 2. Previous force/torque sensors cannot be embedded in spike pins which are 2–3 mm in radius, because of the size or shape of the sensors [7-8]. Therefore, we fabricated a 6-axis force/torque sensor chip which was able to be embedded in spike pins. This compact sensor can be achieved by forming sensing elements only from 6 simple straight beams, which is the least to detect 6 components of forces and torques. The size of the sensor chip was 2 mm × 2 mm × 0.3 mm. The sensor chip was covered with thin polydimethylsiloxane

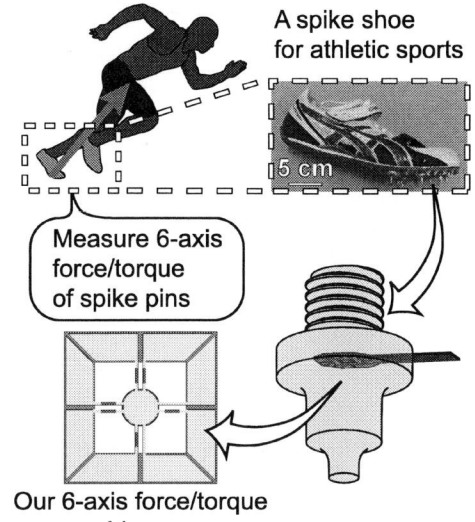

Figure 1: Concept sketch of the 6-axis force/torque sensor for spike pins.

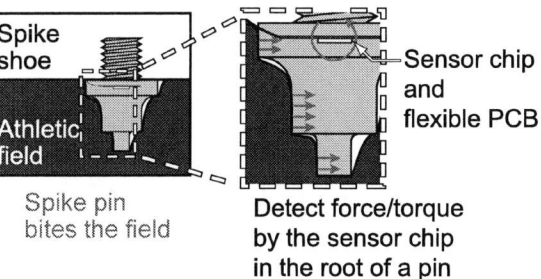

Figure 2: Schematic image of the spike pin biting the athletic field.

(PDMS) and embedded in spike-pin-shaped epoxy resin. Then, the spike-pin-shaped sensor was calibrated to forces and torques.

DESIGN

The design of the sensor chip is shown in Figure 3. The sensor chip is composed of 8 beams extending from the center part to four directions. Among the 8 beams, 6 beams have piezoresistive layers in the middle of the beams. From resistance changes of the 6 beams, 6 components of forces and torques can be detected. Among the 6 beams, 4 beams have piezoresistive layers on their top surfaces and sidewalls, and other 2 beams have piezoresistive layers only on their top surfaces. The other 2 beams are connected to the ground. The length, width and thickness of all beams are 360 μm, 20 μm and 20 μm, respectively.

The principle of the 6-axis force/torque detection is shown in Figure 4. As shown in Figure 4 (a), F_z, T_x and T_y

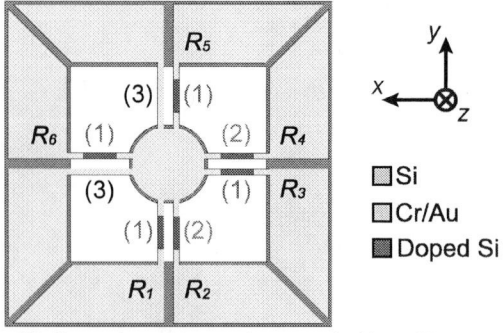

(1) Doped on top surface and sidewall
(2) Doped on top surface
(3) Ground beams

Figure 3: Design of the sensor chip.

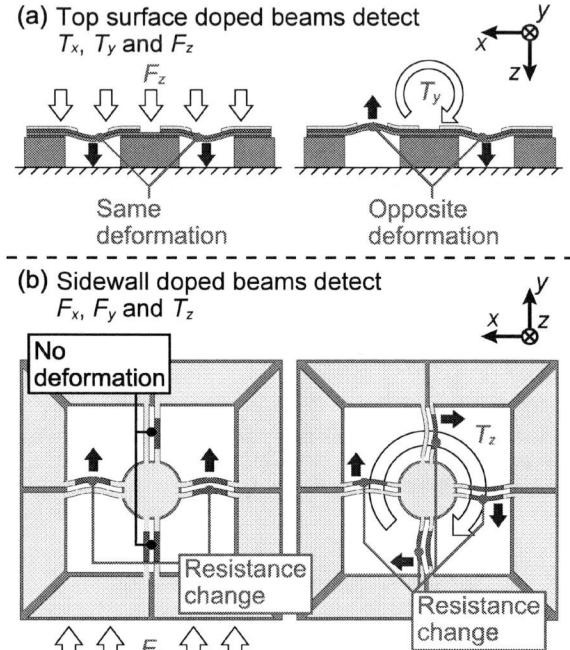

Figure 4: Principles of the 6-axis force/torque detection.

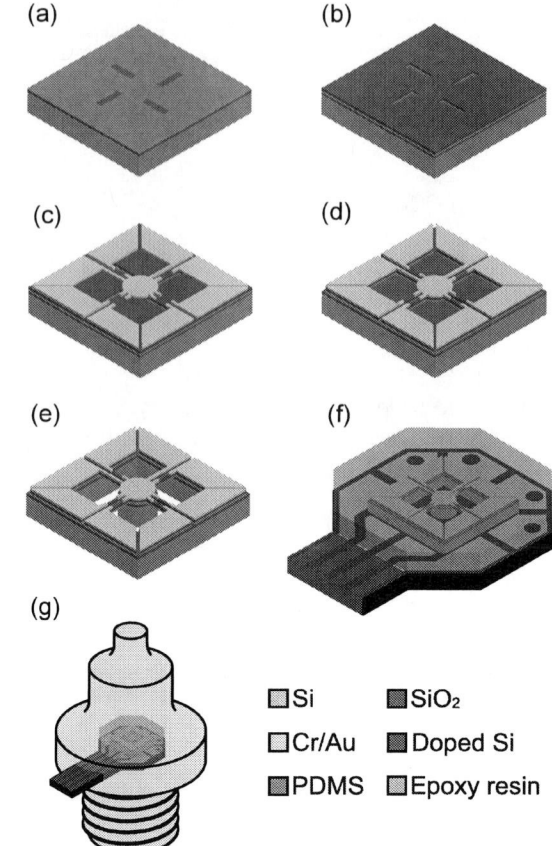

Figure 5: Fabrication process of the spike-pin-shaped sensor.

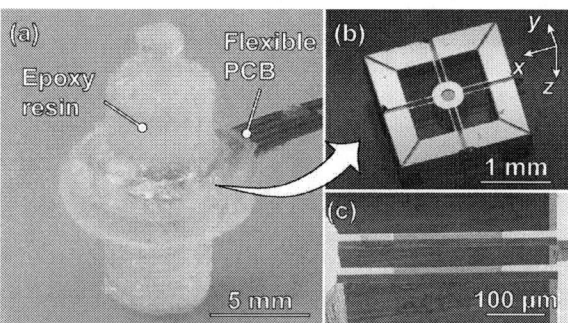

Figure 6: Photographs of (a) the spike-pin-shaped sensor and (b) the 6-axis force/torque sensor chip. (c) SEM image of the beams.

can be detected by all 6 beams. When F_z is applied to the chip, all beams deform to the same direction. On the other hand, when T_x or T_y is applied to the chip, 4 beams perpendicular to the axis of rotation deform to up or down, and the other 4 beams have no vertical deformation. From these differences of the deformation of the beams, F_z, T_x and T_y can be detected separately.

F_x, F_y and T_z can be detected by 4 beams of which the sidewall is doped as shown in Figure 4. When shear force like F_x or F_y is applied to the chip, 4 beams perpendicular to the direction of the force deform, and the other 4 beams have no deformation. Therefore, the resistances of 2 beams of which sidewalls are doped change. On the other hand, when T_z is applied to the chip, all of the beams deform. Thus, the resistances of the 4 beams of which sidewalls are doped change. From these difference of the resistance changes, F_x, F_y and M_z can be detected separately.

FABRICATION

Figure 5 shows the fabrication process of the spike-pin-shaped sensor. The detailed information of the fabrication process of the sensor chip was described in the previous work [9]. In the process, we used a 20 μm / 1 μm / 300 μm thick Silicon on Insulator wafer. First, the device Si layer was etched to form doping holes (Figure 5 (a)). Second, a doped layer was formed on both top surface of the device Si layer and sidewalls of the etched holes by rapid thermal diffusion (Figure 5 (b)). Third, a Cu and an Au layer were deposited on the doped layer and patterned by liftoff process (Figure 5 (c)). Next, the device Si layer was etched to form the beam structures (Figure 5 (d)).

 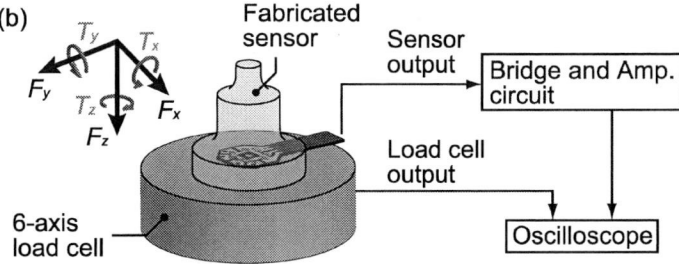

Figure 7: (a) Photographs of experimental setup. (b) Schematic image of experimental setup.

Then, the handle Si layer was etched from the backside and the SiO_2 layer was etched to release the beams (Figure 5 (e)). In this way, the 6-axis force/torque sensor chip was fabricated. After that, the sensor chip was connected to the flexible printed circuit board (PCB) with the electrically conductive paste. Then, the sensor chip and PCB was covered with thin PDMS (Figure 5 (f)). Finally, the sensor chip covered with PDMS was embedded in spike-pin-shaped epoxy resin (Figure 5 (g)). The spike-pin-shaped resin was formed in a mold built by a 3D printer (Objet Eden 260V, Stratasys). Figure 6 shows photographs of the spike-pin-shaped sensor and fabricated sensor chip, and a scanning electron microscope (SEM) image of the beams.

EXPERIMENTS AND RESULTS

We measured the response of the fabricated sensor to 6-axis force/torque. The experimental setup is shown in Figure 7. The fabricated sensor was fixed on the center of a commercially available 6-axis load cell (SI-130-10, ATI Industrial Automation) for calibration. The forces and torques measured by the load cell are corresponding to those acting on a shoe sole on which the spike pin is fixed. The piezoresistive beams of the sensor chip were used as resistors in the Wheatstone-bridge circuits. The outputs of the bridge circuits were connected to an oscilloscope through instrumentation amplifiers. The outputs of the 6-axis load cell were also connected to the oscilloscope synchronizing to the outputs of the bridge circuits. Calibration matrix from the resistance changes to force/torque can be calculated from the outputs of the fabricated sensor and those of the 6-axis load cell.

Forces and torques were applied to the fabricated sensor by pushing at the 9 different positions. Figure 8 shows the experimental results. We show the 3 sets of the results as examples. The pushed positions are shown in the Figure 8 (a). The upper graphs show the forces and torques applied to the spike-pin (Figure 8 (b)). The middle graphs show the resistance changes of the 6 piezoresistive beams according to each force/torque (Figure 8 (c)). Calibration matrix was calculated from the applied forces and torques and the resistance changes by least squares method. The calibration matrix was expressed in an equation shown in Figure 9. The calibrated forces and torques were calculated from the resistance changes of the beams and the calibration matrix (Figure 8 (d)).

When the calibrated force/torque (Figure 8 (d)) is compared to the applied one (Figure 8 (b)), it is observed that the directions of each calibrated force/torque correspond to that of the applied one. It is also observed that the errors of F_z (blue line) and T_z (blue dotted line) between the calibrated force/torque (Figure 8 (d)) and the applied force/torque (Figure 8 (b)) are larger than the errors of other components. As shown in Figure 8 (c), the sensitivity to F_z is lower than other directions of force/torque. It is indicated that this lower sensitivity to F_z contribute to the larger error. As shown in Figure 8 (b), the applied T_z is smaller than torques around other axes because the forces were applied to the sensor only by pushing, not twisting around z-axis. It is possible that this contributed to the larger error between the calibrated T_z and the applied T_z. By improving these points, the sensor would be more sensitive or have smaller errors.

These experimental results suggest that the proposed sensor structure can be used to measure forces and torques acting on spike pins.

CONCLUSIONS

In conclusions, we proposed the spike-pin-shaped sensor in which the 6-axis force/torque sensor chip was embedded. The sensor chip consisted of 6 piezoresistive beams and the size of the chip was 2 mm × 2 mm × 0.3 mm. The chip was covered with thin PDMS and embedded in epoxy resin. The fabricated sensor was calibrated to forces and torques. It is confirmed that the spike-pin-shaped sensor was able to detect 6-axis force/torque.

ACKNOWLEDGEMENTS

The photolithography masks were made using the University of Tokyo VLSI Design and Education Center (VDEC)'s 8 inch EB writer F5112 + VD01 donated by ADVANTEST Corporation. This work was supported by JSPS KAKENHI Grant Number 25000010.

REFERENCES

[1] S. Studenski, S. Perera, K. Patel, *et al.*, "Gait Speed and Survival in Older Adults," *Journal of American Medical Association*, Vol. 305, No. 1, pp. 50–58, 2011.

[2] T. G. Supuk, A. K. Skelin and M. Cic, "Design, Development and Testing of a Low-Cost sEMG System and Its Use in Recording Muscle Activity in Human Gait," *Sensors*, Vol. 14, pp. 8235–8258, 2014.

[3] C. F. Munro, D. I. Miller and A. J. Fuglevand, "Ground reaction forces in running: a reexamination," *Journal of Biomechanics*, Vol. 20, No. 2, pp. 147–155, 1987.

[4] S. Logan, I. Hunter, J. T. Hopkins, J. T., J. B. Feland and A. C. Parcell, "Ground reaction force differences between running shoes, racing flats, and distance

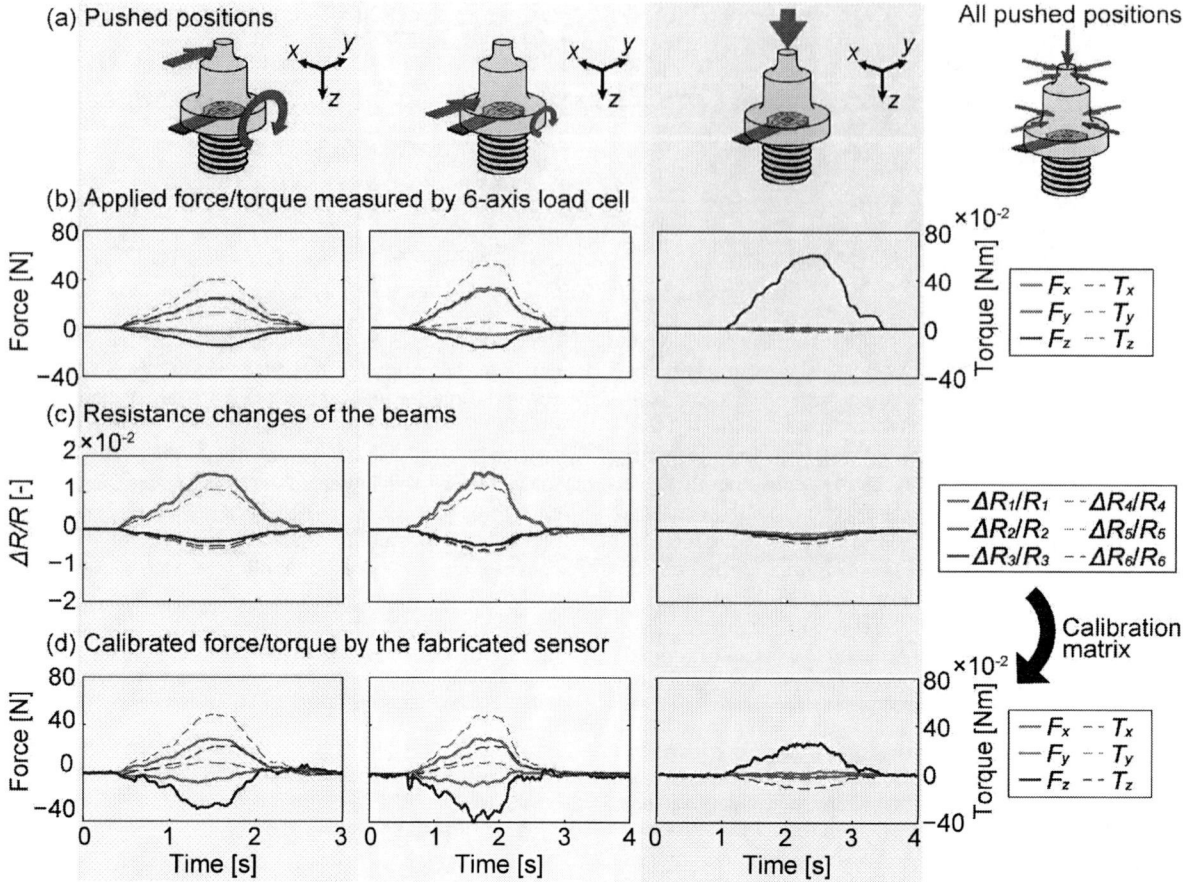

Figure 8: Experimental results. (a) Pushed positions. Graphs of (b) Forces and torques applied to the spike pin, (c) Resistance changes of the 6 beams, (d) Calibrated forces.

$$
\begin{array}{ccc}
F\,[\mathrm{N}] & T\,[10^{-2}\mathrm{Nm}] & \Delta R/R\,[-]
\end{array}
$$

$$
\begin{pmatrix} F_x \\ F_y \\ F_z \\ T_x \\ T_y \\ T_z \end{pmatrix}
=
\begin{pmatrix}
-29.3 & -17.2 & 71.6 & -93.1 & 65.1 & 67.6 \\
-2.0 & 21.1 & -16.1 & 13.3 & 0 & -1.8 \\
134.4 & -222.8 & 61.2 & -320.1 & 76.1 & 287.7 \\
6.2 & 9.0 & -21.4 & 21.2 & -14.0 & -10.5 \\
26.3 & -18.6 & 27.1 & -58.1 & 35.2 & 50.5 \\
-49.8 & 22.6 & 6.5 & 18.7 & 70.0 & 5.2
\end{pmatrix}
\begin{pmatrix} \Delta R_1/R_1 \\ \Delta R_2/R_2 \\ \Delta R_3/R_3 \\ \Delta R_4/R_4 \\ \Delta R_5/R_5 \\ \Delta R_6/R_6 \end{pmatrix}
\times 10^2
$$

Figure 9: Calibration matrix.

spikes in runners," *Journal of Sports Science and Medicine*, Vol. 9, pp. 147–153, 2010.

[5] P. H. Veltink, C. Liedtke, E. Droog and H. Kooij, "Ambulatory Measurement of Ground Reaction Forces," *IEEE Transactions on Neural Systems and Rehabilitation Engineering*, Vol. 13, No. 3, pp. 423–427, 2005.

[6] A. Nakai, A. Nagano, H. Takahashi, K. Matsumoto and I. Shimoyama, "Measurement of 3-axis Stress Distribution at Human Sole by MEMS 3-axis Force Sensors Placed on Insole," *No. 13-2 Proceedings of the 2013 JSME Conference on Robotics and Mechatronics, Japan*, 2013.

[7] F. Beyeler, S. Muntwyler and B. J. Nelson, "A Six-Axis MEMS Force-Torque Sensor With Micro-Newton and Nano-Newtonmeter Resolution," *Journal of Micro-electromechanical Systems*, Vol. 18,

No. 2, pp. 433–441, 2009.

[8] R. A. Brookhuis, H. Droogendijk, M. J. de Boer, R. G. P. Sanders, T. S. J. Lammerink, R. J. Wiegerink and G. J. M. Krijnen, "Six-axis force-torque sensor with a large range for biomechanical applications," *Journal of Micromechanics and Microengineering*, Vol. 24, No. 3, 2014.

[9] H. Takahashi, A. Nakai, N. Thanh-Vinh, K. Matsumoto and I. Shimoyama, "A triaxial tactile sensor without crosstalk using pairs of piezoresistive beams with sidewall doping," *Sensors and Actuators A: Physical*, Vol. 199, pp. 43–48, 2013.

CONTACT

*H. Ishido, tel: +81-3-5841-6318;
Email: ishidou@leopard.t.u-tokyo.ac.jp

3D PRINTED RF PASSIVE COMPONENTS
BY LIQUID METAL FILLING

Chen Yang[1], Sung-Yueh Wu[1,2], Casey Glick[1], Yun Seok Choi[3], Wensyang Hsu[2], and Liwei Lin[1]
[1] University of California, Berkeley, USA
[2] National Chiao Tung University, Hsinchu, Taiwan
[3] Samsung Electronics Co., Ltd., Hwaseong-si, Korea

ABSTRACT

We present three-dimensional (3D) micro-scale electrical components and systems by means of 3D printing and a liquid-metal-filling technique. The 3D supporting polymer structures with hollow channels and cavities are fabricated from inkjet printing. Liquid metals made of silver particles suspension in this demonstration are then injected into the hollow paths and solidified to form metallic elements and interconnects with high electrical conductivity. In the proof-of-concept demonstrations, various radio-frequency (RF) passive components, including 3D-shaped inductors, capacitors and resistors are fabricated and characterized. High-Q inductors and capacitors up to 1 GHz have been demonstrated. This work establishes an innovative way to construct arbitrary 3D electrical systems with efficient and labor-saving processes.

INTRODUCTION

3D printing is a polymer direct-write technology which has attracted great interests for various applications in rapid prototyping due to its geometric flexibility in design and manufacturing [1]. Recently, researchers have started to use 3D printing and other technologies to make 3D micro devices such as microfluidic systems [2-8] but few have studied in the rapid metallization for 3D micro systems. Ideally, metal-embedded 3D micro systems could enable various new applications. For example, a solenoid coil can serve as the inductor in *LC*-resonant circuits, as an actuator in solenoid actuation devices, or an inductive coupling component in electrical transformers. Currently, most published works still have the lithography-defined planar electrodes in these micro systems using rather complicated fabrication and bonding processes.

Several efforts have been explored to use bonding wires or direct-written liquid metals to form 3D conductive structures [9-12]. These metallic structures are suspended in air and difficult to fabricate. Here we propose a novel method to generate arbitrary 3D microstructures with embedded metallic elements by the liquid-metal-filling technique for a variety of possible applications. In this work, we demonstrate several 3D RF passive devices and systems as the proof-of-concept.

DESIGN

Figure 1 illustrates the proposed 3D design and fabrication steps. First, functional 3D structures are designed and constructed by the 3D printing technique. The hollow micro channels and cavities are embedded in the 3D structures to be filled with liquid metal pastes. As an example, a hollow solenoid-shaped channel is formed as shown in Figure 1. Second, in order to facilitate the liquid metal filling step in the next fabrication process, injecting

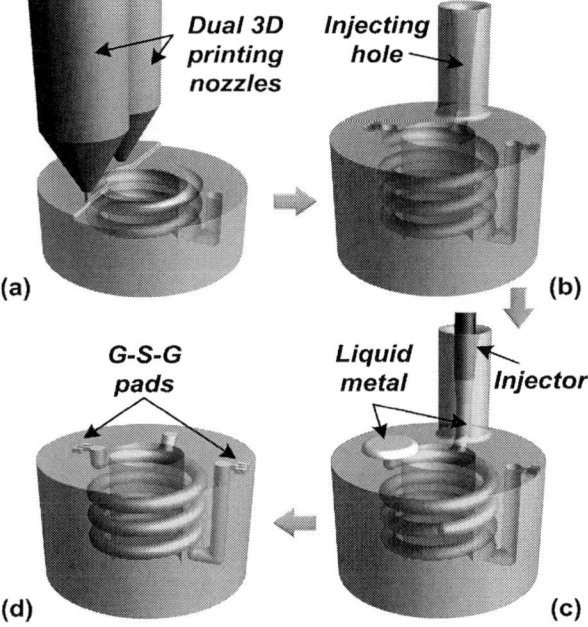

Figure 1: Schematic diagram of the proposed 3D fabrication method including the liquid metal filling technique for RF passive components. (a) 3D printing to form structures with hollow channels (a solenoid-type inductor as an example); (b) a finished 3D structure with the injection hole; (c) liquid metal filling; and (d) surface planarization to remove the injection hole and extra metal.

holes are designed and fabricated with solenoid channels. For example, the solenoid-inductor structure as shown has designated cavities as the ground-signal-ground (G-S-G) pads on the top surface of the device. Third, after the 3D printing process, liquid metal paste is injected via the injecting hole into the designed channels and cavities to form electrical structures. The overflow of the liquid metal at the outlets on the top surface can be used as the contact pads. Forth, the solidification process cures the liquid metal to form solid structures and the top surface of the device is planarized to remove the injecting hole and overflowed liquid metal.

An array of RF passive components has been designed, as shown in Figure 2, including resistors, inductors and capacitors with various geometries. The goal is to demonstrate and test the system-level performance of 3D micro devices fabricated by the proposed methodology and filled with liquid metal. Specific issues to be investigated include the precision and accuracy of the physical structure, the electrical conductivity of the liquid filled metal, and the characteristics of the fabricated components and systems.

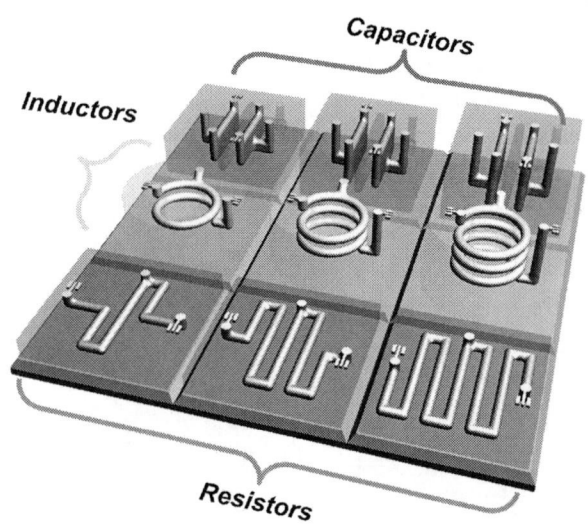

Figure 2: The schematic diagram of the 3D RF passive component array, including parallel-plate capacitors, spiral-shape inductors, and meandering-shape resistors. These RF components are utilized to demonstrate the feasibility of the proposed process.

FABRICATION AND EXPERIMENTS

The fabrication process uses the 3D printing machine, ProJet™ HD 3000, based on the fused deposition modeling (FDM) technology [13] with a printing resolution of 30 µm. During the printing process, polymer materials are heated and ejected from the nozzles of the inkjet printer. Building (VisiJet® EX 2000 [14]) and sacrificial materials (VisiJet® S100 [15]) are deposited alternatively from the dual nozzles to form the printed samples, in which the building material defines the molding structure while the sacrificial material occupies the hollow channels.

A post-printing process is conducted to remove the sacrificial materials. First, the whole 3D printed sample is immersed in a mineral oil bath at 80 °C to dissolve the sacrificial material. Second, the residual mineral oil is removed by detergent and water in sequence thoroughly. Afterwards, the liquid metal, silver suspension (Pelco® 16040-30) is injected into the micro channels and cavities. The as-filled sample is kept at room temperature for 2 hours for the solidification process.

The fabricated passive components are shown in Figure 3. It is noted that the volume of silver suspension shrank after the solidification process, and therefore leads to voids inside the metal traces. By optimizing the silver suspension concentration and repeating the filling operations, the voids could be minimized for better electrical conductivity. The cross-section view in Figure 3c indicates the metal filling inside the solenoid coil.

The electrical performances of the fabricated passive components are characterized. The DC *I-V* curves of resistors are measured by a semiconductor parameter analyzer (HP 4145B). The RF *S*-parameter spectra of the inductors and capacitors are measured by a Cascade G-S-G probe station and a network analyzer (Agilent E5071A). The parasitic effects of G-S-G pads are de-embedded accordingly.

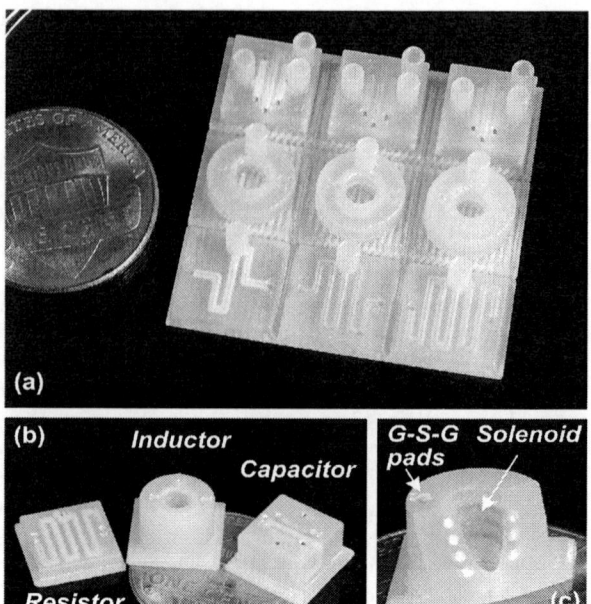

Figure 3: (a) An optical photo of the 3D printed structure as compared with a US 1-cent coin before the liquid metal filling process. (b) Fabricated 3D RF passive components after the liquid filling and curing process. (c) The cross-section view of a 3.5-turn solenoid-type inductor showing the cured liquid metal.

RESULTS

Resistors

The DC *I-V* curves of two resistors with different designs are shown in Figure 4. The equivalent conductivity σ of the filled metal is calculated as:

$$\sigma = \frac{1}{R} \cdot \frac{L}{S} \qquad (1)$$

where R is the total resistance, L is the length of the conductor, and S is the cross-sectional area by considering the void effect. The cross-sectional shape of the metal

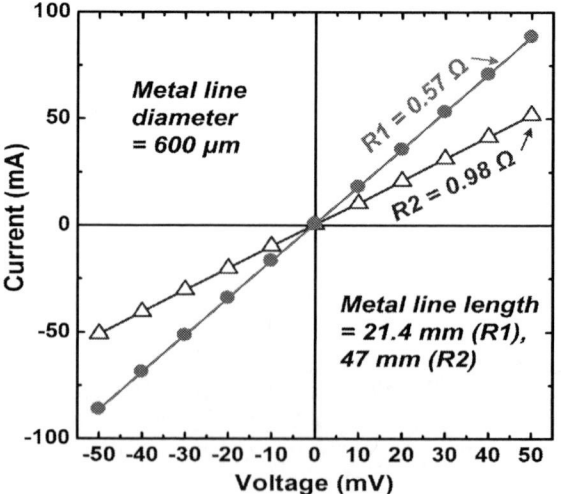

Figure 4: Measured DC *I-V* curves of two 3D printed resistors.

traces is circular with a diameter of 600 μm and the length is 21.4 mm and 47 mm, for resistor R1 and R2, respectively. The calculated average σ is 2.30×10^6 S·m^{-1}, which is about 86% of the ideal conductivity of silver paste at 2.67×10^6 S·m^{-1} [16]. This difference may result from the remaining solvent of the injected metal, which could be further improved by optimizing the solidification process in the future.

Inductors

The measured S-parameters of the fabricated inductors are converted to Y-parameters and then the inductor performances are extracted as [17]:

$$L = \frac{\mathrm{Im}\left(\dfrac{1}{Y_{11}}\right)}{2\pi f}, \qquad (2)$$

$$Q = \frac{\mathrm{Im}\left(\dfrac{1}{Y_{11}}\right)}{\mathrm{Re}\left(\dfrac{1}{Y_{11}}\right)}. \qquad (3)$$

where L is the total inductance, Q is the quality factor, and f is the frequency.

Figure 5 shows the measured inductance and quality factor of inductors with different numbers of coil turns, N. These solenoid-shaped inductors have a designed diameter of 4 mm. The cross-sectional shape of the metal traces is circular with a diameter of 600 μm. Line spacing between adjacent winding is 400 μm. In Figure 5a, the measured total inductance L increases as N increases. For example, the inductances at 0.8 GHz are 19 nH, 36 nH and 66 nH for inductors with 1.5, 2.5 and 3.5 turns, respectively. For each inductor, the L increases first as the frequency increases, and reaches up to maximum due to self resonance. For example, the inductance of the 2.5-turn inductor increases from 36 nH at 0.8 GHz to over 1500 nH around 1.14 GHz. Then the inductance drops quickly. The frequency at which the L drops to zero, i.e., the self-resonance frequency f_0 is 1.24 GHz, 1.15 GHz and 1.04 GHz for these inductor samples, respectively. It is noted that larger N corresponds to smaller f_0 due to larger inductance and parasitic capacitance. Figure 5b shows the measured quality factors. The Q-factor first increases as the frequency increases and then drops down to zero due to the high loss at self-resonance frequency. It is noted that higher inductance leads to lower quality factor due to higher energy losses. For example, the 1.5-turn inductor has the smallest inductance and highest Q about 31 at 0.95 GHz, while the 3.5-turn inductor shows maximum Q of 11 at 0.72 GHz. During the magnetic energy storage cycles in the inductors, the energy loss mechanisms mainly include the skin-effect induced ohmic losses in the conductor and the electric field energy losses due to parasitic capacitance.

Capacitors

The measured S-parameters of the capacitors are converted to Y-parameters and then the capacitor performances are extracted as [18]:

$$C = \frac{1}{2\pi f} \cdot \mathrm{Im}(Y_{11}). \qquad (4)$$

where C is the total capacitance.

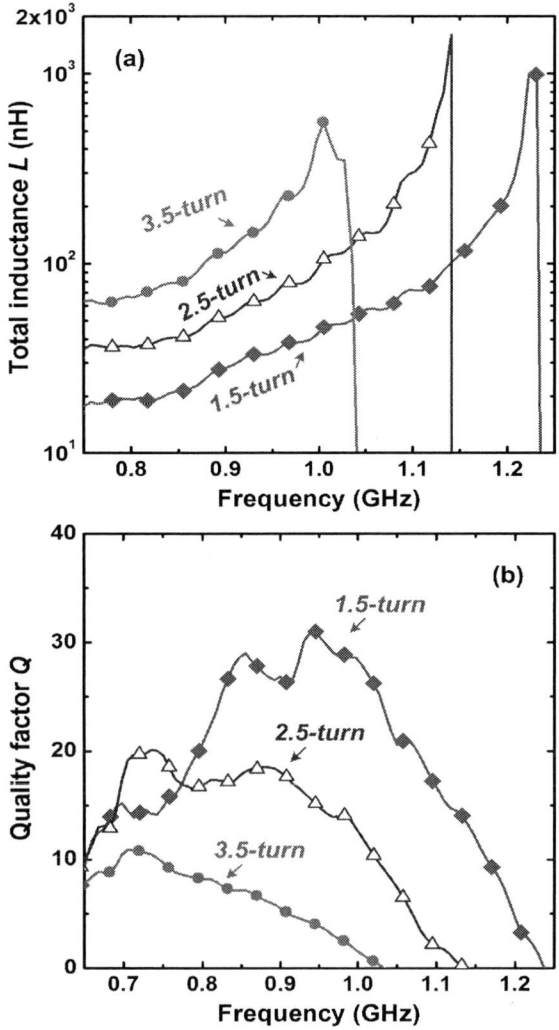

Figure 5: Measurement results of 3D printed RF inductors with different number of turns. (a) Total inductance L, which varies from 20nH to 40nH at 1GHz. All the devices' self-resonance frequencies are over 1GHz. (b) Quality factor Q.

The measured total capacitance of a parallel capacitor is shown in Figure 6. The rectangular-shaped parallel-plate type capacitor has an area of 10.1 mm^2 with a gap of 1400 μm. The capacitance increases first as the frequency increases, and reaches a maximum value about 2.7 pF at 0.96 GHz as the frequency is close to the self-resonance. Afterwards, it drops down to zero at the self-resonance frequency, which is about 1.12 GHz.

CONCLUSION

The design and fabrication steps and processes to construct three-dimensional (3D) micro-scale components and structures by the combination of 3D printing and liquid metal filling technique have been developed. The 3D structures with hollow channels and cavities are first built using the polymer inkjet printing process. Silver particles in the form of suspension pastes are injected into the hollow channels and solidified to form metallic elements

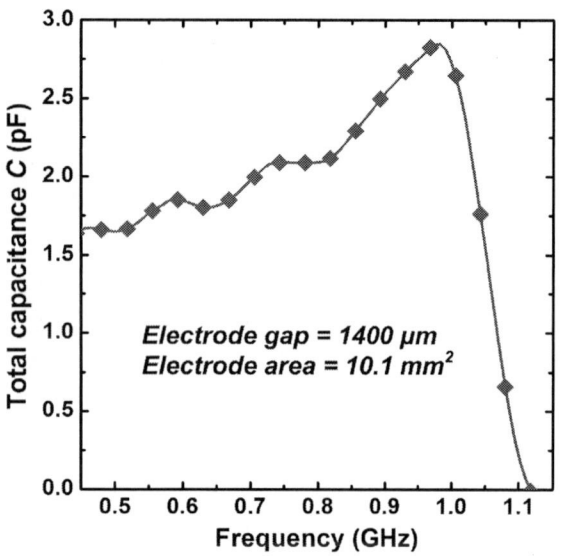

Figure 6: Measured total capacitance C of 3D printed RF capacitor. The device's self-resonance frequency is 1.1 GHz.

and interconnects inside the 3D structures. By optimizing the silver suspension concentration and repeating the filling operation, the voids inside the cured silver paste could be minimized for good conductivity. Various RF passive components (including inductors, capacitors and resistors) are fabricated and characterized. As such, this paper has developed and demonstrated a new class of manufacturing process to construct arbitrary 3D structures with the metallization process. Possible applications could extend to different kinds of 3D micro electromechanical systems with embedded metallic components, such as 3D packaging and lab-on-a-chip to name a few.

ACKNOWLEDGMENT

This work is partially supported by Samsung Electronics, Inc. Mr. Sung-Yueh Wu is supported by the Ministry of Science and Technology of Taiwan under Grant 103-2917-I-009-192. The authors would also like to thank the help from Prof. Albert P. Pisano for the measurement equipment.

REFERENCES

[1] H. Lipson and M. Kurman, *Fabricated: The New World of 3D printing*, John Wiley & Sons, 2013.

[2] D. Therriault, S. R. White, and J. A. Lewis, "Chaotic Mixing in Three-Dimensional Microvascular Networks Fabricated by Direct-Write Assembly," *Nature Materials*, vol. 2, pp. 265-271, 2003.

[3] P. J. Kitson, M. H. Rosnes, V. Sans, V. Dragone, and L. Cronin, "Configurable 3D-Printed Millifluidic and Microfluidic 'Lab on a Chip' Reactionware Devices," *Lab on a Chip*, vol. 12, pp. 3267-3271, 2012.

[4] O. H. Paydar, C. N. Paredes, Y. Hwang, J. Paz, N. B. Shah, R. N. Candler, "Characterization of 3D-Printed Microfluidic Chip Interconnects with Integrated O-Rings," *Sensors and Actuators A: Physical*, vol.

205, pp. 199-203, 2014.

[5] G. Comina, A. Suska, and D. Filippini, "PDMS Lab-on-a-Chip Fabrication Using 3D Printed Templates," *Lab on a Chip*, vol. 14, pp. 424-430, 2014.

[6] B. C. Gross, J. L. Erkal, S. Y. Lockwood, C. Chen, and D. M. Spence, "Evaluation of 3D Printing and Its Potential Impact on Biotechnology and the Chemical Sciences," *Analytical Chemistry*, vol. 86, pp. 3240-3253, 2014.

[7] J. L. Erkal, A. Selimovic, B. C. Gross, S. Y. Lockwood, E. L. Walton, S. McNamara, R. S. Martin, and D. M. Spence, "3D Printed Microfluidic Devices with Integrated Versatile and Reusable Electrodes," *Lab on a Chip*, vol. 14, pp. 2023-2032, 2014.

[8] K. G. Lee, K. J. Park, S. Seok, S. Shin, D. H. Kim, J. Y. Park, Y. S. Heo, S. J. Lee, and T. J. Lee, "3D printed modules for integrated microfluidic devices," *RSC Advances*, vol. 4, pp. 32876-32880, 2014.

[9] J. Hu and M. -F. Yu, "Meniscus-Confined Three-Dimensional Electrodeposition for Direct Writing of Wire Bonds," *Science*, vol. 329, pp. 313-316, 2010.

[10] J. J. Adams, E. B. Duoss, T. F. Malkowski, M. J. Motala, B. Y. Ahn, R. G. Nuzzo, J. T. Bernhard, and J. A. Lewis, "Conformal Printing of Electrically Small Antennas on Three-Dimensional Surfaces," *Advanced Materials*, vol. 23, pp. 1335-1340, 2011.

[11] C. Ladd, J. -H. So, J. Muth, and M. D. Dickey, "3D Printing of Free Standing Liquid Metal Microstructures," *Advanced Materials*, vol. 25, pp. 5081-5085, 2013.

[12] B. Y. Ahn, E. B. Duoss, M. J. Motala, X. Guo, S. -I. Park, Y. Xiong, J. Yoon, R. G. Nuzzo, J. A. Rogers, and J. A. Lewis, "Omnidirectional Printing of Flexible, Stretchable, and Spanning Silver Microelectrodes," *Science*, vol. 323, pp. 1590-1593, 2009.

[13] http://www.3dcreationlab.co.uk/pdfs/projet-hd-3000.pdf

[14] http://www.3dsystems.com/products/datafiles/visijet/msds/visijet-crystal/24184-S02-00-B-MSDS-US-English-EX-200-Crystal.pdf

[15] http://www.shapeways.com/rrstatic/material_docs/msds-frosted.pdf

[16] http://www.tedpella.com/technote_html/16040-30%20TN.pdf

[17] C. Yang, F. Liu, T. -L. Ren, L. -T. Liu, G. Chen, X. -K. Guan, A. Wang, and H. -G. Feng, "Ferrite-Integrated On-Chip Inductors for RF ICs," *IEEE Electron Device Letters*, vol. 28, pp. 652-655, 2007.

[18] I. Kwon, M. Je, K. Lee, Senior Member, IEEE, and H. Shin, "A Simple and Analytical Parameter-Extraction Method of a Microwave MOSFET," *IEEE Transactions on Microwave Theory and Techniques*, vol. 50, pp. 1503-1509, 2002.

CONTACT

*C. Yang, tel: +1-510-642-8983; chenyang@berkeley.edu.
C. Yang and S.-Y. Wu contributed equally to this work.

A CONVENIENT METHOD TO FABRICATE MULTILAYER INTERCONNECTIONS FOR MICRODEVICES

Jia Li, Supin Chen, and Chang-Jin "CJ" Kim
Mechanical and Aerospace Engineering Department
University of California, Los Angeles (UCLA), California 90095, U.S.A.

ABSTRACT

We report a new method to fabricate multilayer interconnections without requiring wet or dry etching or deposition of insulating layers. Three levels of electrical connections are obtained by merely repeating deposition, photolithography, and anodization of a metal layer. Without the need to etch metal layers or deposit and etch insulation layers, the overall process is simple, cheap, safe, and of low temperature. While the utility is general for a wide variety of microdevices and electronics, in this paper we demonstrate one application by developing a low-cost fabrication of a large-array electrowetting-on-dielectric (EWOD) chip that requires three metal layers.

INTRODUCTION

Microdevices that have a large number of electrical connections often require multilayer interconnections. However, there are tradeoffs for the typical approaches for multilayer interconnections. The IC fabrication technique used for electrical interconnections of many MEMS devices [1] is not practical for applications that require large areas such as microfluidics. PCB is often utilized for its low-cost multilayer routing [2] but limits the selection of substrate material and integration with MEMS fabrication. It also adds significant bulk. Other unique interconnection techniques, such as glass-in-silicon wafer processing [3], require expensive processing. A convenient fabrication method is desired to meet general interconnection requirements.

Anodized tantalum (i.e., tantalum pentoxide), an electrically stable (dielectric strength = 6-7 MV/cm) material with high corrosion resistance that is used in high performance capacitors [4] and microelectronics [5], is a low-cost alternative for dielectric materials. In this method, conduction lines are made of tantalum, and insulation is tantalum pentoxide electrochemically grown from the tantalum. Other metals that can be anodized, such as aluminum, may also be used instead of tantalum. In this report, we explore successive repetition of patterned anodization to obtain multilayer interconnections.

FABRICATION METHOD

Our multi-level interconnection is achieved by repeating the following three steps for each level: (1) deposition of a valve metal layer, (2) selective through-metal oxidation using photolithography and anodization, and (3) selective partial oxidation of the metal using photolithography and anodization. In this paper, we develop a three-level interconnection, following the process flow presented in Figure 1.

In Figure 1(a), a 250 nm of tantalum was sputtered on glass substrate. In Figure 1(b), a 12 µm-thick photoresist

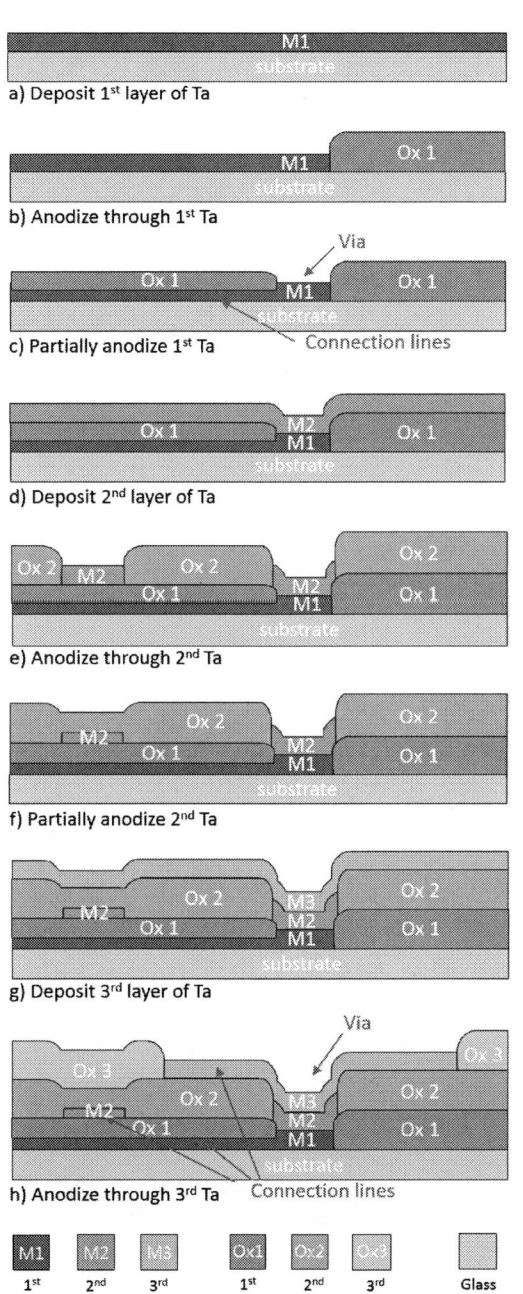

a) Deposit 1st layer of Ta

b) Anodize through 1st Ta

c) Partially anodize 1st Ta

d) Deposit 2nd layer of Ta

e) Anodize through 2nd Ta

f) Partially anodize 2nd Ta

g) Deposit 3rd layer of Ta

h) Anodize through 3rd Ta

M1	M2	M3	Ox1	Ox2	Ox3		Glass
1st	2nd	3rd	1st	2nd	3rd		
Metal (Ta) layers			Oxide (Ta₂O₅) layers				Substrate

Figure 1: Process flow to fabricate the proposed multi-level interconnections, drawn for three layers. Each anodization step was performed while masked by patterned PR. Tantalum was used in the current report for the metal layers.

(PR) (AZ4620) was patterned on tantalum, followed by anodization in 0.01% KI solution using a tantalum sheet as the cathode. The bias voltage applied between cathode and anode was gradually increased to 300-350 V to ensure full anodization. The full anodization through the entire thickness of a metal layer was important because this anodized material electrically separates all the connection lines within the same layer. We confirmed the completion of the through-layer anodization by measuring either the voltage drop between cathode and KI solution or the decrease in current density [6] during the anodization step. In Figure 1(c), the existing PR was removed, and another 12 µm-thick PR was spin-coated and patterned, followed by a partial anodization using 100-150 V. The PR-masked regions were protected from anodization to function as conductive vias. The remaining regions not masked by the PR were partially anodized at the tantalum surface to grow an insulation layer on top. The PR is, then, removed. In Figures 1(d)-1(f), the same three steps were repeated to fabricate the second layer of interconnection. In Figures 1(g) and 1(h), the final, third layer interconnection was made by tantalum deposition and full anodization.

During anodization process in Figures 1(b), 1(e) and 1(h), PR needs to protect the underneath tantalum from being anodized at a bias up to 300-350 V. The electrical strength of PR is a key factor. Our test has shown that thick and fully hard-baked PR significantly improved the quality of the tantalum electrode lines defined by the full anodization. Partially anodized areas showed up at the electrode edges when the patterned PR failed to withstand the voltage. Although we used a general-purpose PR not designed to withstand high voltage for convenience, a new protection material with higher electrical strength would be desired in future research.

In Figure 1(b), the maximum thickness of tantalum we could fully anodize was around 250 nm with 300-350 V. However, the process was stackable by repeating the metal deposition and full anodization multiple times. In this report, we successfully repeated deposition and anodization of 250 nm tantalum three times to obtain a total of 750 nm tantalum and 1.5 µm tantalum pentoxide. In Figure 1(h), the height difference between the lowest and the highest point was around 750 nm. A planarization process may be needed if thicker layers or more layers of interconnections are necessary.

TEST RESULT

Because the resistance of conduction line is determined by the thickness of the remaining tantalum after partial anodization and affected by various local process parameters, such as current density, ionic concentration, temperature, etc., we performed a series of characterization tests. We exposed the two ends of test conduction lines to facilitate resistance measurement using a probe station. As shown in Figures 2(a) and 2(b), conduction lines of layer 1 and layer 2 were formed by partial anodization of tantalum intended to anodize a half of the original tantalum thickness. The conduction line in layer 3 was expected to be twice as thick as those in layer 1 and layer 2 because no partial anodization process was needed for the top layer. In theory the resistance in layer 3 should be half of that in layer 1 or layer 2. The resistances

of connection lines were measured to range 2.5-16 kOhm for the 1st (bottom) layer, 2.7-13.8 kOhm for the 2nd (middle) layer, and 0.6-7.3 kOhm for the 3rd (top) layer, confirming the connectivity. With the same width and thickness across the device, resistances of connection lines in the same layer varied by their lengths.

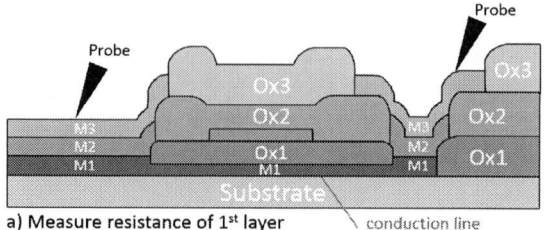

a) Measure resistance of 1st layer

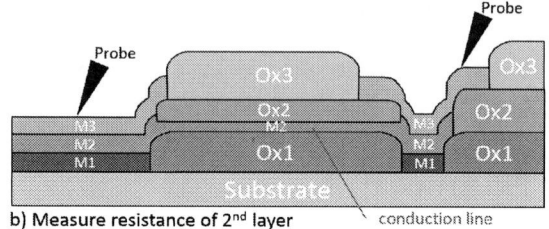

b) Measure resistance of 2nd layer

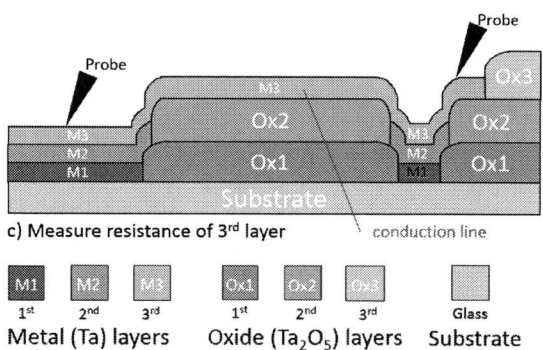

c) Measure resistance of 3rd layer

Metal (Ta) layers   Oxide (Ta₂O₅) layers   Substrate

Figure 2: Resistance measurement for each layer. The 3rd conductive layer was thicker than the 1st and 2nd because it was not anodized.

Due to the possibility of imperfect full anodization leaving enough residual tantalum beneath tantalum pentoxide to short adjacent conduction lines, we also measured resistance between conduction lines designed to be isolated from each other for each of the three conduction layers. Resistance between adjacent connection lines was measured to be infinite, confirming electrical isolation.

APPLICATION DEMONSTRATION

Utility of the proposed interconnections was demonstrated by using the technique in fabricating an electrowetting-on-dielectric (EWOD) [7] device containing 100 independently controlled electrodes spaced as an array of 10x10 EWOD pads that required a multilayer electrical connection. On a typical EWOD chip a relatively high voltage (50-100 V) is applied on each electrode to manipulate droplets. Defect-free insulation between adjacent electrodes and their connection lines is essential for successful high voltage operation.

978-1-4799-7956-1/15 $31.00 © 2015 IEEE   266

As shown in Figure 3, we developed fabrication of EWOD device with multilayer interconnection. For the EWOD dielectric layer, we deposited a 1 μm silicon nitride by PECVD rather than partially anodizing the 3rd tantalum layer because of the polarity and frequencies dependencies [8]. The fabrication was completed by spin-coating Teflon to form the hydrophobic topcoat. On the device, 10x10 EWOD electrode pads were made with the 3rd tantalum layer, and each EWOD pad was exclusively connected to one of the 100 contact pads at one edge of the chip through the 3 layers of connection lines. The transparent areas in Figures 3(a) and 3(b) indicate opaque tantalum (750 nm thick) has been fully anodized to transparent tantalum pentoxide (1.5 μm-thick) on transparent glass substrate. A sine wave of 70 V_{rms} and 1 kHz was used to perform all four basic droplet-manipulation functions (i.e., generation, moving, splitting, and merging [7]) with DI water. Figure 4 shows sequential movement of a droplet on EWOD chip. The demonstration was performed using a handheld EWOD control system developed in our lab.

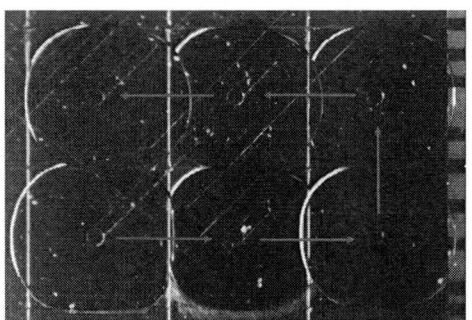

Figure 4: Overlapping images of sequential droplet movement on a demonstration EWOD device with a cover plate consisting of Teflon on a transparent indium tin oxide electrode on glass. Arrows indicate the moving direction of the droplet.

CONCLUSION

We have developed a simple fabrication method to obtain multilayer interconnections without wet or dry etching or deposition of insulating layers. The process is cheap, safe, and of low temperature. As an example, a 3-level interconnection was fabricated on a glass substrate and tested for electrical conductivity and insulation. Its utility has been demonstrated for EWOD microfluidics by sequentially actuating a droplet with a voltage of 70 V_{rms}, but we foresee other types of microdevices benefiting from the proposed interconnection method in the future.

The critical length (i.e., the feature size) in this preliminary development was 10 μm, much larger than the vertical dimension of each layer (200 nm to 500 nm). Further development is needed to decrease the feature size and understand suitable application areas.

ACKWLEDGEMENTS

This work was supported in part by the Department of Energy [DE-SC0005056]. We thank Dr. Yao-Wen Hsu for his assist in Teflon coating and EWOD chip testing.

REFERENCES

[1] P. F. van Kessel and L. J. Hornbeck, *Proc. IEEE*, vol. 86, pp. 1687-1704, 1998.

[2] J. Gong and C.-J. Kim, *J. Microelectromech. Syst.*, vol. 17, pp. 257-264, 2008.

[3] R. M. Haque and K. D. Wise, *Proc. IEEE Int. Conf. MEMS*, Jan. 2011, Cancun, Mexico, pp. 995-998.

[4] S. G. Byeon and Y. Tzeng, *IEEE Trans. Electron Dev.*, vol. 37, pp. 972-979, 1990.

[5] C. Chaneliere, J. L. Autran, R. A. B. Devine and B. Balland, *Materials Science and Engineering R*, vol. 22, pp. 269-322, 1998.

[6] K. Ueno, S. Abe, R. Onoki and K. Saiki, *Journal of Applied Physics*, vol. 98, 114503, 2005.

[7] S. K. Cho, H. Moon and C.-J. Kim, *J. Microelectromech. Syst*, vol. 12, pp. 70-80, 2003.

[8] L. X. Huang, B. Koo, and C.-J. Kim, *J. Microelectromech. Syst*, vol. 22, pp. 253-255, 2013.

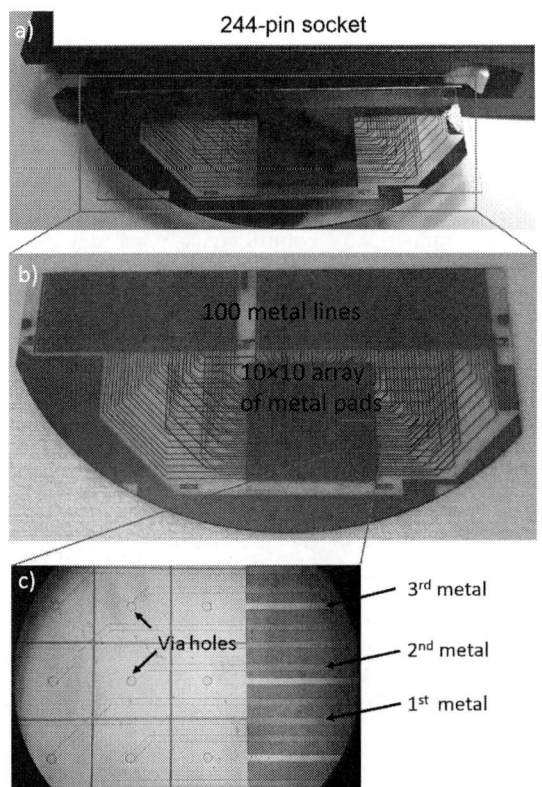

Figure 3: The proposed multi-layer interconnection technology applied to EWOD devices as a demonstration. (a) A fabricated EWOD chip with 100 independently controllable EWOD pads is inserted into a custom-developed handheld operating system through 244 pin socket. (b) The EWOD chip shows the 10×10 array of EWOD pads individually connected to the 100 contact pads at the top edge of the chip. (c) An optical micrograph showing 9 EWOD pads with vias connected to 9 conduction lines of 3 different layers. The 1st and 2nd metals are visible through the overlaying tantalum pentoxide that is transparent.

CONTACT

Jia Li; killeyzib@gmail.com

DEVELOPMENT OF AN ATMOSPHERIC PRESSURE AIR MICROPLASMA JET FOR THE SELECTIVE ETCHING OF PARYLENE-C FILM

Honglei Guo, Jingquan Liu, Zhaoyu Wang, Xingzhao Wang,
Xiang Chen, Bin Yang and Chunsheng Yang

National Key Laboratory of Science and Technology on Micro/Nano Fabrication Laboratory, Key Laboratory for Thin Film and Micro fabrication of Ministry of Education, Department of Micro/Nano Electronics, Shanghai Jiao Tong University, People's Republic of China

ABSTRACT

This paper develops a novel and simple process device based on an atmospheric pressure air microplasma jet for the selective etching of parylene-C film. In order to realize the selective etching, a quartz glass microtube (100 μm, inner diameter) is employed to generate the air microplasma jet. Experimental results demonstrated Micro-holes, micro-trenches on parylene-C film were successfully fabricated by the air microplasma jet without causing any heat damage to films and using any masks, and the etching rate reached 5.14μm/min. Due to its operating at ambient conditions, this process device can be easily integrated with roll-to-roll systems for large-scale manufacturing of flexible electronic devices in the future.

INTRODUCTION

Much attention has been paid to parylene-C film by MEMS researchers in the field of applications ranging from microstructures to BioMEMS[1] and microfluidics systems[2] because it has particularly attractive properties including low defect density, transparency, chemical inertness, highly biocompatibility and biostability[3]. The use of parylene-C as a structural material requires the development of micropatterning techniques for its selective removal[4, 5]. Nowadays, conventionally applied techniques to pattern parylene-C film include optical lithography combined with plasma-chemical etching and microcontact printing[6]. However, the non-automatic micromachining and the requirement of mask and vacuum system during processing not only restrict the flexibility in microfabrication but also increase the processing time and fabrication cost.

To address the above concerns, some alternative microfabrication techniques need to be developed for the selective removal of parylene-C film. Nowadays, atmospheric pressure plasma jets have recently been reported to etch some materials because they can obtain reactive ion/radical species with high concentration at atmospheric pressure without an expensive pumping system. For instance, various kinds of atmospheric plasma jet devices have been developed for the removal of silicon[7, 8], Kapton[9] and Biofilms[10]. However, the line width of these plasma jets is usually in the order of millimeters in size and cannot realize the microfabrication of materials. Furthermore, in order to produce homogenous plasma jets at atmospheric pressure, a large volume of argon or helium gas (5-20L/min) mixed with a small fraction of reactive gas (CF_4, O_2) are required and thus to increase the operation cost and process complexity. Additionally, previous researches of plasma induced polymer removal at atmospheric pressure showed a relatively low etch rate[9, 11]. For instance, the etching

rate of Kapton with a helium/oxygen plasma jet was 42-83 nm/s. Consequently, it's not an effective way to selectively etch parylene-C film through the existing plasma jet sources.

In this paper, an atmospheric pressure air microplasma jet (AμPJ) source was developed for the selective etching of parylene-C film. In order to reduce the line width of plasma jet, a quartz glass microtube with a fine nozzle of 100μm in diameter was employed to fabricate the plasma jet source, and a stainless steel needle electrode (Ø100μm) was used as high voltage electrode. Furthermore, compared to the existing plasma jet sources used to remove polymer films, the air plasma jet operated with air rather than helium, argon or Ar/He-O_2 gas mixtures. These enable the air microplasma jet to have two prominent advantages: 1) the line width of this plasma jet was much smaller than that of traditional plasma jet; 2) the plasma jet contained a great deal of oxygen reactive species with lower cost, which would increase the etching etch of parylene-C film. At last, through a numerically controlled X-Y motion platform integrated with the substrate or the air microplasma jet, several different microstructures on parylene-C film were fabricated by this air microplasma jet, and the morphologies of microstructures had been characterized.

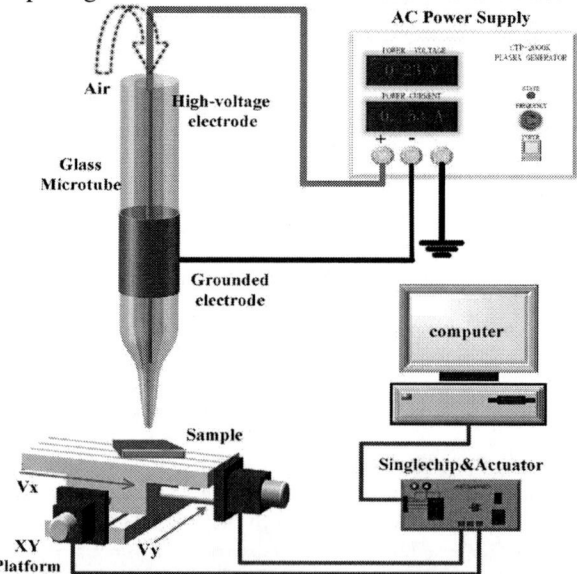

Figure 1: Schematic diagram of an air microplasma jet device for the selective etching parylene-C film

EXPERIMENTAL METHOD

A simple and novel etch device based on an air microplasma jet was schematically shown in Figure 1. This device mainly included plasma discharge device, AC power supply and a numerically controlled X-Y motion platform. The plasma discharge device was composed of a stainless steel needle electrode, a grounded electrode and a quartz glass microtube. The stainless steel needle electrode was located at the center of quartz glass microtube with the inner diameter 0.5mm at origin end and 100μm at the micromachining fine end. SEM and optical microscope images of the microtube were shown in Figure 2a and 2b. The glass microtube was wrapped by a copper tape (10mm width) acted as the grounded electrode. The distance between this tape and the tip of the stainless steel needle electrode was 5mm. As shown in Figure 1, the sample on the X-Y motion platform was placed on the downstream of the plasma jet.

Since the air microplasma jet was generated by air discharge, the microplasma would contain a mass of oxygen reactive species which could react with polymer films, thus resulting in the formation of volatile carbon oxides (CO, CO_2) and H_2O and then remove polymer films.[4] In order to characterize the etching behavior of the air microplasma jet, three types of etching experiment by the plasma jet were carried out to fabricate micro-holes, linear and circular micro-trenches on parylene-C film. During the experiment, the distance between the nozzle tip and the sample was 1mm and the applied plasma jet was the same. After finishing the etching process, depth profiles and morphologies of microstructures were characterized by optical microscope and Scanning Electron Microscope (SEM), Wyko NT1100 optical profiler.

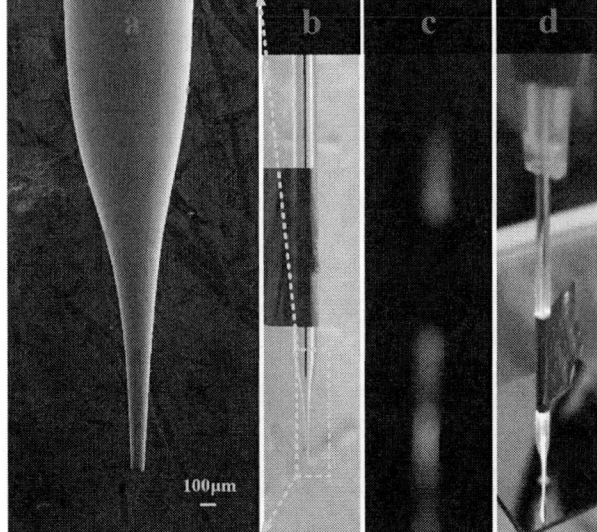

Figure 2: (a) SEM image of the microtube; (b) Digital image of the plasma discharge device; (c) Digital image of the microplasma jet in the dark condition; (d) The air microplasma jet was applied to etch the sample.

RESULTS AND DISCUSSION

The microplasma jet was initiated by an AC sinusoidal power supply with adjustable peak voltage from 0 to 30 kV and a frequency of 20.5 kHz. When a voltage of 9 kV (peak-to-peak) was applied on the centric electrode, homogeneous and stable plasma would be generated within the microtube. This plasma would reduce the breakdown voltage of the corona discharge at the tip of the inner electrode by increasing the seed

electrons number.[12] Therefore, a homogenous discharge could be prevented from transiting to arc discharge which could result in the formation of a high-temperature discharge.[13] With increasing the applied voltage to 11.2 kV (peak-to-peak), a long plasma plume eject from the nozzle of the microtube into the ambient air. As shown in Figure 2c, the plasma jet generated by air discharge looked like purple homogenous plasmas. Additionally, since the silicon wafer played the role of floating electrode as the plasma plume approached it, the plasma jet would focus between the inner electrode and the silicon wafer. Figure 2d indicated that the plasma plume became much longer and brighter when the air microplasma jet was deployed to etch parylene-C film on the silicon wafer.

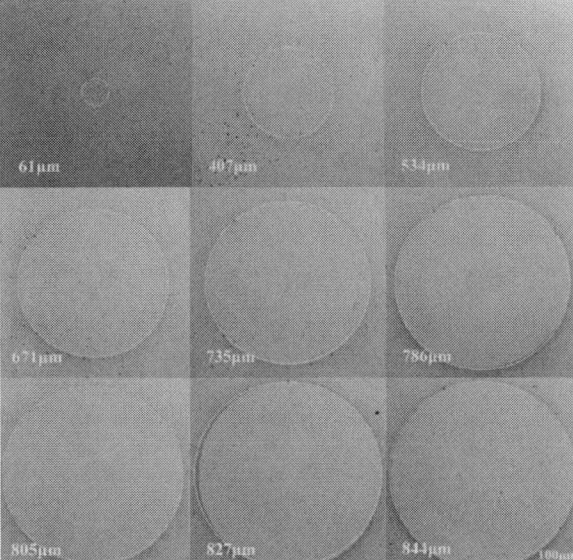

Figure 3: Optical microscopy images of micro-holes on the parylene-C film etched by the AμPJ. (The diameter of the hole was noted on the bottom left, and the etch parameters were: the etch time is 30s, 60s, 90s, 120s, 150s, 180s, 240s, 300s and 360s, respectively; the distance L was 1mm; the applied voltage was 11.2 kV.)

As shown in Figure 3, the size of micro-hole became larger with the increasing of etch time, but remained unchanged after reaching the maximum degree. The green circle in Figure 3 was fitted out based on the theorem that three points along the edge of the hole determinate a circle. As shown in Figure 3, the edge of the hole matched well with the corresponding green circle, which indicated that the intensity of the plasma jet was uniform. Furthermore, neither polymer residues nor carbon residues appeared within micro-holes. As illustrated in Figure 4a and 4b, parylene-C films around the edge of micro-holes fabricated by the air microplasma jet didn't behave carbonization phenomenon, and parylene-C film was etched layer by layer. Therefore, this air microplasma jet wouldn't cause any heat damage to parylene-C film. In addition, Figure 4c and 4d showed the 3-D morphologies of micro-holes on parylene-C film. It was observed that micro-holes were obviously symmetrical and the surface of etched areas was smooth, which indicated that the intensities of plasmas in the same level surface were almost no difference. Figure 5 demonstrated the relationship between the depth profiles of micro-holes and the etch time. The profile curves were almost symmetric. When the etch time was 30s, the V-shaped profile was appeared. Based on this profile curve as a function of etching time, it's calculated that the etch rate of parylene-C film reached 5.14μm/min. With further increasing of the etching time, the shape of depth profiles changed to be inverted trapezoid. The reason for this phenomenon was that all parylene-C film in the treatment area had been removed by the air microplasma jet. When the etching time arrived 120s, the sidewall of the hole was almost vertical, which was consistent with the 3-D morphology of micro-hole in Figure 4d. Therefore, it was obvious that this air microplasma jet exhibited excellent selectivity for the etching of parylene-C film.

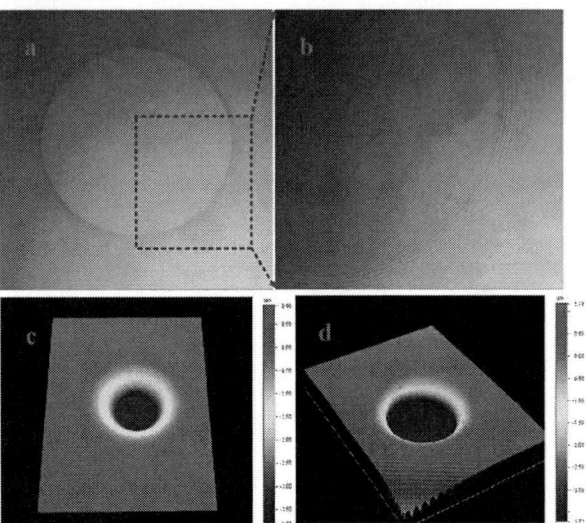

Figure 4: (a) Optical microscope image of micro-hole fabricated by the air microplasma jet with the etch time of 60s; (b) Magnified details of small areas of micro-hole; (c) 3-D morphology of micro-hole, the etching time is 60s; (d) 3-D morphology of micro-hole, the etching time is 120s.

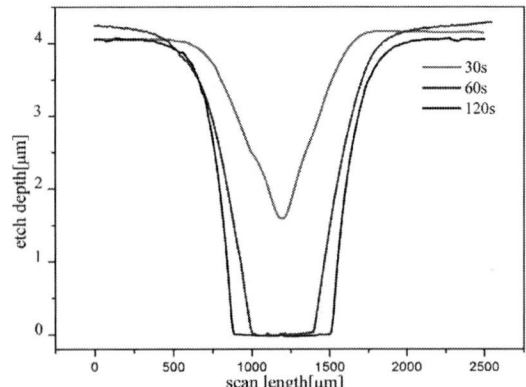

Figure 5: Depth profiles of micro-holes fabricated by the air microplasma jet

To obtain relatively complex microstructures on the parylene-C film, the X-Y motion system was controlled to move along a straight line at the speed of 2 mm/min and 1 mm/min, respectively. With the same operating conditions, straight micro-trenches were obtained on

parylene-C film after the etching of 240s. As shown in Figure 6a, 6b, 6c and 6d, the line width of straight micro-trenches were about 200μm and 500μm, respectively. Since the relative scan speed determined the etch time, the line width of micro-trench fabricated by the air microplasma jet would change with the relative scan speed. Additionally, it could be concluded from Figure 6b and 6d that parylene-C film on the etching area was uniformly removed in the line direction and the etched areas showed smooth surface. Furthermore, a circular micro-trench on parylene-C film was realized by the air microplasma jet. Figure 6e and 6f illustrated SEM image and 3-D morphology of the circular micro-trench. The sidewall of the circular micro-trench showed nearly vertical and a smooth etched surface. Therefore, these experimental results demonstrated that this developed atmospheric pressure air microplasma jet was capable of being used to fabricate complex microstructure on parylene-C film without masks.

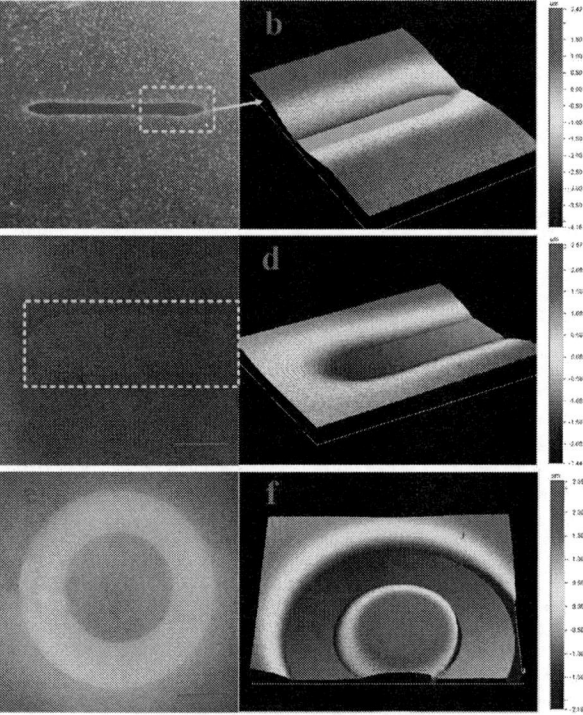

Figure 7: (a)Optical microscope image of a linear micro-trench, the scan speed is 2 mm/min; (b)3-D morphology of the straight micro-trench with the line width of 200μm; (c) SEM image of a linear micro-trench, the scan speed is 2 mm/min; (d) 3-D morphology of the straight micro-trench with the line width of 500μm; (e)SEM image of circular micro-trench created by the air microplasma jet on parylene-C film with the etching time of 270s; (f)3-D morphology of the circular micro-trench. (Scale bar: 500μm)

For practical applications of the atmospheric pressure air microplasma jet to large area *manufacturing of microstructures*, an integrated assembly of microplasma jets such as shown in figure 8 would be designed and prepared, which would be more efficient and can be operated at high pressure ranges including atmospheric pressure. Additionally, this plasma jet process device can be easily integrated with roll-to-roll systems for large-scale manufacturing of flexible electronic devices in the future.

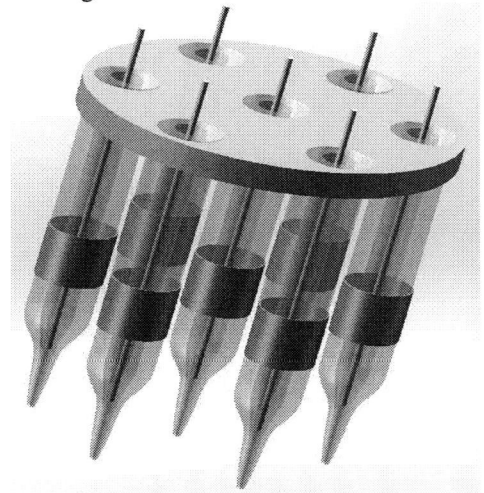

Figure 8: Schematic diagram of microplasma jet array for large-scale manufacturing of microstructures

CONCLUSIONS

The selective etching of parylene-C film has been achieved by a developed atmospheric pressure air microplasma jet device. By combining the air microplasma jet with a numerically controlled X-Y motion platform, micro-holes, straight and circular micro-trenches on parylene-C film was successfully fabricated without using any masks and causing heat damage to film. The etch rate of parylene-C film was as high as 5.1μm/min, and the minimum liner width of the micro-trench reached 200μm. In summary, the simple and effective etch process technique based on an atmospheric pressure air plasma jet would have potential application in polymer microfabrication and might promote the development of flexible MEMS device.

ACKNOWLEDGEMENTS

This work is partly supported by the National Natural Science Foundation of China (No. 51035005, 61176104), Shanghai Municipal Science and Technology Commission (No.11JC1405700, 13511500200). The authors are also grateful to the colleagues for their essential contribution to this work.

REFERENCES

[1] J. Charmet, J. Bitterli, O. Sereda, M. Liley, P. Renaud, and H. Keppner, "Optimizing Parylene C Adhesion for MEMS Processes: Potassium Hydroxide Wet Etching," *Microelectromechanical Systems, Journal of,* vol. 22, pp. 855-864, 2013.

[2] C.-Y. Shih, Y. Chen, and Y.-C. Tai, "Parylene-strengthened thermal isolation technology for microfluidic system-on-chip applications," *Sensors and Actuators A: Physical,* vol. 126, pp. 270-276, 2006.

[3] Y. Li, Q. Xie, W. Wang, M. Zheng, H. Zhang, Y. Lei, *et al.,* "Parylene C-on-photoresist (POP): a low temperature spacer scheme for polymer/metal nanowire fabrication," *Journal of Micromechanics and Microengineering,* vol. 21, p. 067001, 2011.

[4] E. Meng, P.-Y. Li, and Y.-C. Tai, "Plasma removal of Parylene C," *Journal of Micromechanics and Microengineering,* vol. 18, p. 045004, 2008.

[5] C. Liu, "Recent Developments in Polymer MEMS," *Advanced Materials,* vol. 19, pp. 3783-3790, 2007.

[6] M. Thomas, J. Borris, A. Dohse, M. Eichler, A. Hinze, K. Lachmann, *et al.,* "Plasma Printing and Related Techniques – Patterning of Surfaces Using Microplasmas at Atmospheric Pressure," *Plasma Processes and Polymers,* vol. 9, pp. 1086-1103, 2012.

[7] T. Ichiki, R. Taura, and Y. Horiike, "Localized and ultrahigh-rate etching of silicon wafers using atmospheric-pressure microplasma jets," *Journal of applied physics,* vol. 95, pp. 35-39, 2003.

[8] H. Paetzelt, T. Arnold, G. Böhm, F. Pietag, and A. Schindler, "Surface Patterning by Local Plasma Jet Sacrificial Oxidation of Silicon," *Plasma Processes and Polymers,* vol. 10, pp. 416-421, 2013.

[9] J. Jeong, S. Babayan, V. Tu, J. Park, I. Henins, R. Hicks, *et al.,* "Etching materials with an atmospheric-pressure plasma jet," *Plasma Sources Science and Technology,* vol. 7, p. 282, 1998.

[10] K. Fricke, I. Koban, H. Tresp, L. Jablonowski, K. Schröder, A. Kramer, *et al.,* "Atmospheric Pressure Plasma: A High-Performance Tool for the Efficient Removal of Biofilms," *PLoS ONE,* vol. 7, p. e42539, 2012.

[11] M. H. Jung and H. S. Choi, "Photoresist etching using Ar/O$_2$ and He/O$_2$ atmospheric pressure plasma," *Thin Solid Films,* vol. 515, pp. 2295-2302, 2006.

[12] X. Li, N. Yuan, P. Jia, and J. Chen, "A plasma needle for generating homogeneous discharge in atmospheric pressure air," *Physics of Plasmas (1994-present),* vol. 17, p. 093504, 2010.

[13] X. Li, P. Jia, N. Yuan, T. Fang, and L. Wang, "One atmospheric pressure plasma jet with two modes at a frequency of several tens kHz," *Physics of Plasmas (1994-present),* vol. 18, pp. -, 2011.

CONTACT

*J.Q. Liu, Tel: +86-21-34207209; jqliu@sjtu.edu.cn

ELECTROPLATED STENCIL REINFORCED WITH ARCH STRUCTURES FOR PRINTING FINE AND LONG CONDUCTIVE PASTE

Pi-Hsun Chen and Che-Hsin Lin [*]

[1]National Sun Yat-sen University, Kaohsiung, 804, TAIWAN

ABSTRACT

This study presents an MEMS-based stencil reinforced with arch structures and a surrounding buffer reservoir for printing conductive paste of fine and long lines. The developed reinforced stencil successfully solves the problems came with the conventional stencil structure including limited printable line width and ease of fracture. A novel process was developed to fabricate a thin yet robust electroplated stencil by using two AZ4620 layers and one SU-8 layer as the electroplating molds. A precise stencil with a long and high-density line structure can be produced with the developed method. The printing results show that the developed stencil is capable of printing parallel lines of 20 μm in pitch. The printable length of the fine parallel lines is longer than 10 mm with the arch structure reinforced stencil. In addition, the developed stencil is capable of printing closed ring patterns with small pitch, which is not possible to be printed using conventional stencil or screen printing technologies. The MEMS-based stencil developed in the present study will give substantial impact on the paste printing technologies.

INTRODUCTION

Printing technology is an emerging technique in daily life because it is a straight forward method to rapid producing patterns and documents in various industry including publisher, housewares and many others. Screen printing is the most commonly adopted method in printing techniques since this method is capable of transferring various "inks" on various substrates including metal, polymer, ceramic and dielectric materials. The screening printing techniques are also associated with a number of advantages including low-cost and energy saving, large-area fabrication. Recently, this techniques technology has also been used in a varieties of advanced manufacturing fields such as solar cell electrodes [1], bio-sensing sensor electrodes [2], chemical sensing electrodes[3], telecommunication devices [4] and even drug delivery systems[5]. Over the developed industries relating printing techniques, printed electronics have the biggest market share and have attracted a number of interests.

There trend for developing the printing technology in printed electronics is to continuously print patterns of large area and to print high-density lines of small feature size. Several roll-to-roll or sheet-to-sheet processes such as gravure printing and rotary screen printing, have been developed for continuously printing electronic patterns. However, the printable feature size for typical gravure printing is limited at around 50–100 μm.[6] Rotary screening printing is an alternative approach for continuously printing

patterns of large area. Nevertheless, the stainless mesh of the screen can limit the printable feature size of the patterns. The quality of printing is affected by several variables such as substrate properties, ink properties, and other mechanical parameters including the printing speed, blade pressure, printing direction and mesh count [6, 7]. In general, the coarse resolution of screen printing was around 75 μm since the diameter for the SS fabric to produce the woven mesh is usually greater than 40 μm. The screen fabrics might block the paste injection path and cause incomplete printed patterns. Although the recent improvements suggested that features as small as around 20 μm was reported, a delicate stainless mesh and surface preparation processes were required for achieving this fine patterns [8, 9].

On the other hand, the stencil printing is an alternative way to produce consistence patterns without the physical limitations of the woven mesh. The stencil printing is composed of a metal mask which rests directly in contact with the surface of the board during printing. Stencil printing has been widely used for patterning conductive wires or solder bumps for flip-chip packaging in the semiconductor industry [10]. Stencils printed paste bumps are also usually adopted in printed circuit board (PCB) industry for surface mount assembly applications. The thickness of the printed paste is relatively thicker for this application since a high-temperature reflow process is required to attach the surface mount components onto the solder paste printed on the PCB. Therefore, commercial available stencils are made in stainless sheet of the thickness greater than 150 μm due to the strength and the thick paste requirements. In practices, the patterns on stencils are typically generated with the wet chemical etching, laser-ablation or electroforming processes [11]. Stencils fabricated with the isotropic etching of metal and laser ablation usually have the critical dimension bigger than 150 μm [12]. This big structure dimension limits the printable feature size of the printed patterns. In addition, both the chemical etched stencil and the laser ablated stencil exhibited a rough surface on the side walls which might be another issue during printing. Although stencil produced with electroplating provides a smaller structure dimension, the printable feature size and the line width is typically limited at around 100 μm. Recently, the micro-machined stencil fabricated with electroplating technique for printing fine patterns of the size smaller than 50 μm for flip-chip bonding was reported [13]. Although the printable feature size was greatly improved by the electroplating process, the minimum feature size for the stencil printed patterned still could not meet the requirement for pattering ultra-fine conductive wire for touch panel applications. Therefore, there is still a need to develop a new method to produce high quality stencils for fine pattern printing.

CONCEPT AND DESIGN

The conventional stencils are usually produced by etching the thin stainless sheet such that the thickness is usually thicker than 150 μm.(Fig. 1a) Reducing the thickness and the designed patterns of the stencil is the most straight-forward method to print fine patterns with stencil printing. An electroplated thin stencil was also developed for printing the patterns of the dimension smaller than 30 μm was reported previously [14]. However, the structure was not robust enough such that the stencil was easy to be broken during printing (Fig. 1b). In addition, stiction happened around the region composed of dense and long line patterns. Fig. 1c shows the concept of the proposed stencil reinforced with arch structures and the buffer reservoir. The conductive paste flows underneath the arch structures and forms a successive line during squeezing the paste (Fig. 1d).

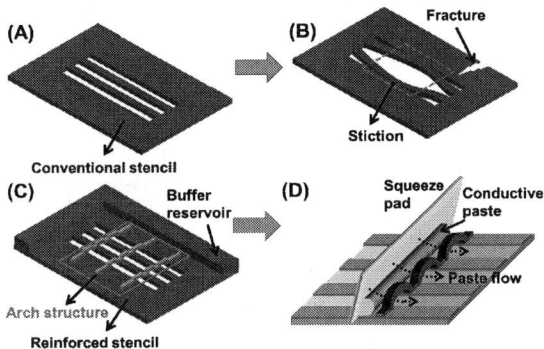

Figure 1: (A) Conventional stencil structure (B) Stiction and torn fracture in conventional stencil (C) Reinforced stencil structure (D) stencil printing of conductive paste using the reinforced stencil.

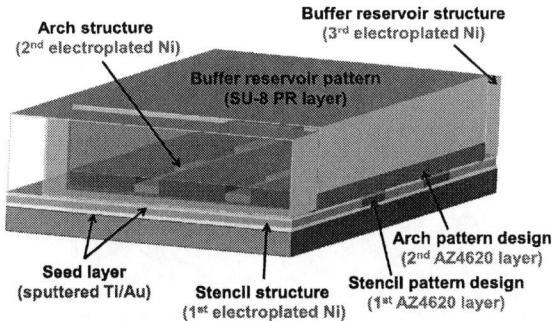

Figure 2: Structural design for producing the reinforced stencil using the combination of AZ4620 and SU-8 as the electroplating molds.

Fig. 2 presents the structural design for producing the reinforced stencil using the combination of AZ4620 and SU-8 as the electroplating molds. In order to produce the thin

stencil with arch reinforced structures and the buffer reservoir, this work develops a three-step electroplating process utilizing the combination of two layers of AZ4620 positive photoresist (PR) and one layer of SU-8 negative PR for nickel electroplating. With this approach, the produced stencil layer is thin and robust for printing fine patterns without stiction. Moreover, the thick metal layer to form the reservoir provides the mechanical strength to prevent the thin stencil from breaking. Therefore, the developed stencil reinforced with arch structure exhibits small feature size and enough mechanical strength for fine pattern printings. The developed stencil is capable of printing parallel fine lines of longer length, which is difficult to be printed using conventional stencil printing method.

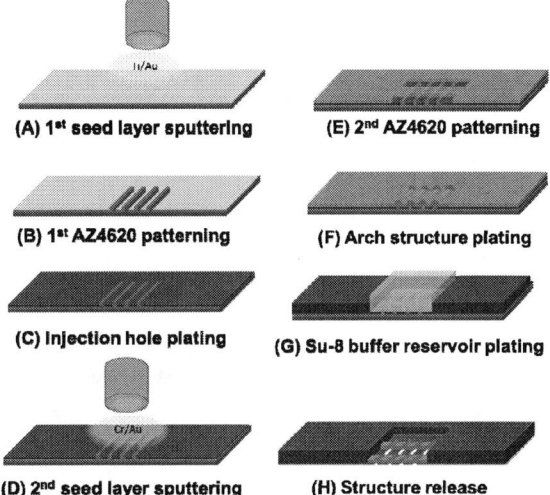

Figure 3: The simplified fabrication process for producing the reinforced stencil with the arch structure and the buffer reservoir.

FABRICATION

Figure 3 shows the simplified fabrication process for producing the stencil reinforced with the arch structures and buffer reservoir. Low-cost microscope slides were used as the substrates for producing the stencils. In order to meet the requirement for metal electroplating and structure releasing, the Ti/Au layers of 50 nm in thickness were sputter coated on the substrate. (Fig. 3a) The thicker Ti layer was used as the adhesion and sacrificial layers in the process since Ti strongly reacts with HF-based solution and is able to serve as the sacrificial layer for structure releasing. With this simple method, the nickel plated stencil could be released without adhering on the glass substrate. A 5-μm AZ4620 layer was patterned using a standard photolithography process (Fig. 3b) to for the pattern for printing. The patterned substrate was then immersed into a nickel sulfamate bath for 1st nickel plating. A 2-μm nickel layer was plated to define the printing patterns on the stencil with a current density of 1.0 ASD in a nickel chloride bath of 50°C and pH 3.8 (Fig. 3c). Note that

978-1-4799-7956-1/15 $31.00 © 2015 IEEE   273

prior to the nickel plating process, the substrate was treated with O_2 plasma to remove the photoresist residuals and enhance the surface wettability of the patterned substrate. Once the 1st metal layer was formed, another 50-nm Cr/Au layer was sputtered on the nickel plated substrate (Fig. 3d). An 8-μm AZ4620 PR layer was again patterned and then a 6-μm nickel was plated to define the arch structures to reinforce the thin stencil (Fig. 3e and 3f). These arch structures greatly enhanced the robust of the stencil which also prevented the dense line structures from stiction. Once the 2nd nickel layer was plated, a SU-8 layer of 40 μm in thickness was then spun on the electroplated substrate for the formation of the thicker nickel layer. The high transparency and the negative-tone properties of SU-8 PR made it easy to align the patterned structures. The substrate was again plated with the 3rd nickel layer of 30 μm in thickness after patterning the SU-8 layer. Note that the current density for this thick nickel plating was ramped from 0.5 ASD to 3 ASD (Fig. 3g). The 3rd plated substrate was immersed in a 75°C PG remover (Micro-chem Inc. USA) for 24 h to remove both AZ4620 and SU-8 photoresists. The produced stencil structure was finally released by immersing the substrate into a diluted HF solution (1.0 M). The HF solution strongly reacted with Ti without rapid attacking the nickel such that the plated stencil could be released in a short period of time. The released structure was cleaned and mounted on a laser-cut PMMA template using UV glue for stencil fixation.

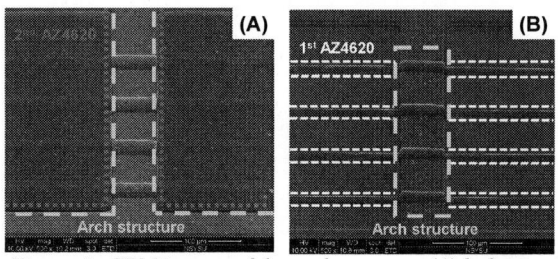

Figure 4: SEM images of the arch structure (A) before and (B) after removing the AZ4620 PR layers.

RESULTS AND DISCUSSION

Figure 4 shows the SEM images of the arch structures before (Fig. 4a) and after (Fig. 4b) removing the AZ4620 PR layer. Results showed that the stencil patterns and arch structures were well defined with the developed fabrication procedures. In addition, the arch structure exhibited good adhesion onto the 1st nickel layer without pealing. Since the arch structures were oriented perpendicular to the printing patterns of the stencil, the 6-μm thick arch structure could efficiently prevent the thin neighboring structures from stiction after releasing. Figure 5 presents the eagle view and the closed-up views for the developed stencils after releasing from the glass substrate. It is noted that the continuous length for the stencil pattern was 100 mm which was challenging for releasing the structure without stiction and fracture. The SEM image clearly showed that the surface of the stencil was smooth and there was no significant bending or stretching caused by the electroplating induced residual stress. (Fig. 5a)

Figure 5b shows the close-up view for the produced arch structure across the stencil injection lines. The aperture for the arch structures was around 30 μm in width and 6 μm in height. The printing paste could flow through the aperture of the arch structure and transfer onto the substrate to form complete line patterns.

Figure 5: (A) Eagle view and (B)(C) closed-up views for the produced arch structures.

Commercial available silver paste (TEC-PA-060, Inktec, Korea) was used to evaluate the printing performance of the developed stencil reinforced with the arch structures. The specification of the paste indicated that the viscosity of the paste was with 65000 cp. The typical film thickness for the printed pattern was 3-4 μm while printing with a screen of 10-μm thick emulsion. The substrate for printing the patterns was a 120 μm thick polyethylene terephthalate (PET) film. The stencil printing was achieved using a home-built printing module with the conditions that scratching speed of 150 mm/s and the normal applied force of 15 N. Figure 6a shows the optical images of the printed conductive pastes on PET substrate using the developed stencil. The length of the printed lines were around 500 μm such that no arch structure was required to prevent from the stiction. Results showed that the silver paste lines with the width and pitch of around 20 μm and 30 μm were well defined on the PET substrate. Figure 6b presents the printed parallel long lines using the stencil with the arch structures. Results showed that dense and parallel lines with the length longer than 100 mm could be clearly printed. The capability of the developed stencil will meet the requirement for the applications of modern touch panel devices.

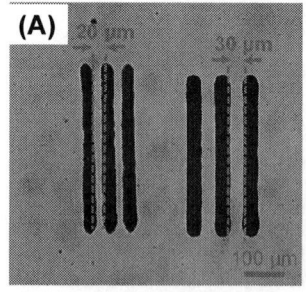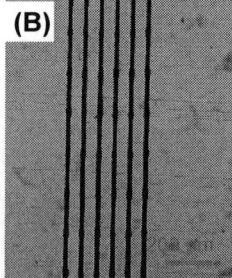

Figure 6: Optical images for the stencil printed conductive paste (A) fine pitch lines (B) long lines of 100 mm in length.

Figure 7a presents the SEM images showing the printed silver paste on the PET substrate. Results showed that the printed patterned paste was well defined. Although there were node patterns corresponding to the arch structures, the measured width variation for the node was less than 35%. In addition, no smear was observed for the printed patterns with the developed stencil. Figure 7b shows the closed-up view for the printed conductive paste. Results shows that the printed paste exhibited well-confined edge and uniform surface morphology. The measured height of the printed paste was around 2.0 μm corresponding to the thickness of the stencil layer and surface energy of PET substrate. The reinforced stencil developed in the present study will greatly enhance the capability of stencil printing technology.

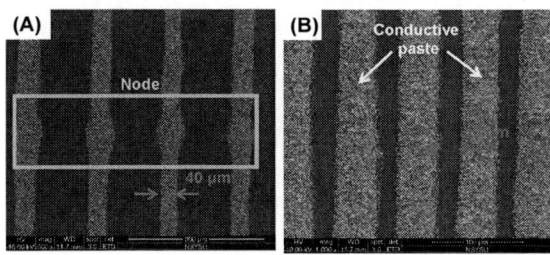

Figure 7: SEM images of the node patterns beneath the arch structure (B) fine pitch conductive paste.

CONCLUSION

A novel MEMS-based stencil reinforced with arch structures and a thick buffer reservoir for printing conductive paste of fine and long lines was developed. The arch-structures greatly improved the robust of the thin stencil layer such that the stiction issue could be prevented. Results showed that the silver paste lines with 20 μm in width and 100 mm in length was successfully printed with the developed stencil. The SEM views confirmed that the conductive paste exhibited good morphology and pattern confinements. The MEMS-based stencil developed in the present study will give substantial impact on the modern printing technology.

ACKNOWLEDGEMENTS

The authors would like to thank the financial supports from Ministry of Science and Technology (MOST) and Metal Industries Research and Development Center (MIRDC) of Taiwan.

REFERENCE

[1]Park, C., Kwon, T., Kim, B., Lee, J., Ahn, S., Ju, M., Balaji, N., Lee, H., and Yi, J.: 'Front-side metal electrode optimization using fine line double screen printing and nickel plating for large area crystalline silicon solar cells', Mater Res Bull, 2012, 47, (10), pp. 3027-3031

[2]Taleat, Z., Khoshroo, A., and Mazloum-Ardakani, M.: 'Screen-printed electrodes for biosensing: a review (2008-2013)', Microchim Acta, 2014, 181, (9-10), pp. 865-891

[3]Li, M., Li, Y.T., Li, D.W., and Long, Y.T.: 'Recent developments and applications of screen-printed electrodes in environmental assays-A review', Anal Chim Acta, 2012, 734, pp. 31-44

[4]Pranonsatit, S., and Lucyszyn, S.: 'Micromachined screen printing (MaSPrint) technology for RF MEMS applications', in Editor (Ed.)^(Eds.): 'Book Micromachined screen printing (MaSPrint) technology for RF MEMS applications' (IEEE, 2005, edn.), pp. 3-6

[5]Kolakovic, R., Viitala, T., Ihalainen, P., Genina, N., Peltonen, J., and Sandler, N.: 'Printing technologies in fabrication of drug delivery systems', Expert Opin Drug Del, 2013, 10, (12), pp. 1711-1723

[6]Hrehorova, E., Rebros, M., Pekarovicova, A., Bazuin, B., Ranganathan, A., Garner, S., Merz, G., Tosch, J., and Boudreau, R.: 'Gravure Printing of Conductive Inks on Glass Substrates for Applications in Printed Electronics', J Disp Technol, 2011, 7, (6), pp. 318-324

[7]Manessis, D., Patzelt, R., Ostmann, A., Aschenbrenner, R., and Reichl, H.: 'Technical challenges of stencil printing technology for ultra fine pitch flip chip bumping', Microelectron Reliab, 2004, 44, (5), pp. 797-803

[8]Menard, E., Meitl, M.A., Sun, Y.G., Park, J.U., Shir, D.J.L., Nam, Y.S., Jeon, S., and Rogers, J.A.: 'Micro- and nanopatterning techniques for organic electronic and optoelectronic systems', Chem Rev, 2007, 107, (4), pp. 1117-1160

[9]Schwanke, D., Pohlner, J., Wonisch, A., Kraft, T., and Geng, J.: 'Enhancement of Fine Line Print Resolution due to Coating of Screen Fabrics', Journal of microelectronics and electronic packaging, 2009, 6, (1), pp. 13

[10]Bhat, S.N., Rao, G.A., Dinesh, N., and Baliga, B.: 'Photo-Defined Electrically Assisted Etching Method for Metal Stencil Fabrication', Components, Packaging and Manufacturing Technology, IEEE Transactions on, 2011, 1, (7), pp. 1116-1121

[11]Bhat, S.N., Rao, G.A., Dinesh, N.S., and Baliga, B.N.: 'Photo-Defined Electrically Assisted Etching Method for Metal Stencil Fabrication', Ieee T Comp Pack Man, 2011, 1, (7), pp. 1116-1121

[12]Kay, R., and Desmulliez, M.: 'A review of stencil printing for microelectronic packaging', Solder Surf Mt Tech, 2012, 24, (1), pp. 38-50

[13]Kay, R.W., Stoyanov, S., Glinski, G.P., Bailey, C., and Desmulliez, M.P.Y.: 'Ultra-fine pitch stencil printing for a low cost and low temperature flip-chip assembly process', Ieee T Compon Pack T, 2007, 30, (1), pp. 129-136

[14]Chen, P.H., and Lin, C.H.: 'Fabrication of ultra-thin stencil with buffer reservoir utilizing the combination of AZ4620 and SU-8 electroplating molds', Proceedings of the IEEE International Conference on Micro Electro Mechanical Systems (IEEE MEMS 2014), San Francisco, USA Jan. 26-30, 2014.

FABRICATION AND CHARACTERIZATION OF SINGLE CRYSTALLINE 4H-SIC MEMS DEVICES WITH N-P-N HOMOEPITAXIAL STRUCTURE

Feng Zhao and Allen V. Lim

Micro/Nanoelectronics and Energy Laboratory, Department of Electrical Engineering, School of Engineering and Computer Science, Washington State University, Vancouver, WA, USA

ABSTRACT

This paper reports single crystalline 4H-SiC MEMS with homoepitaxial n-p-n structure and its resonant characteristics under electrostatic actuation. Single crystalline fully exploits the superior material properties of SiC for operations in harsh environments. Compared to previously report p-n structure, the n-p-n structure makes electrostatic actuation applicable which is essentially important for applications of resonators and actuators to sensor devices. Such n-p-n structure, complementing the p-n structure, also further extends the capability of monolithic integration between SiC MEMS and electronic devices and circuits with not only p-n configurations such as diodes, but also n-p-n configurations such as BJTs and MOSFETs, etc.

INTRODUCTION

As a wide bandgap semiconductor, silicon carbide (SiC) has superior material properties such as its large bandgap energy, chemical resistance, mechanical and high temperature robustness, radiation hardness, and biocompatibility. Among the different polytypes (3C, 4H, 6H, etc.), single crystalline 4H-SiC attracts research interest in both electronic device [1-3] and MEMS device [4-6] applications due to their potential to fully exploit the superior material properties of SiC for operations in harsh environments, such as high temperature, chemical and biomedical, radiation, etc. 4H-SiC has homoepitaxial single crystalline SiC film grown on single crystalline SiC substrate, which is different from other polytypes such as 3C-SiC, with single or polycrystalline SiC on Si or silicon-on-insulator (SOI) substrate. The 4H-SiC substrate is more promising for harsh environment and the lower interfacial defect density between SiC film and substrate, and high quality material by homoepitaxial growth of 4H-SiC provide potential to further improve the properties of devices, for example, the resonant characteristics of MEMS. However, in order to fabricate MEMS devices, isotropic etching process is needed to undercut single crystalline SiC to release free-standing structures. This is challenging due to the extreme chemical stability of SiC, as it can practically only be etched in molten KOH above 600°C, which is not applicable for MEMS fabrication.

In recent years, various techniques have been developed to address this challenge, including bulk micromachining such as etching from the backside of SiC substrate and through the wafer by inductively coupled plasma (ICP) or deep reactive ion etching (DRIE) [7, 8], ultrasonic drilling [9], and laser milling [10], etc. Recently another promising technique using dopant-selective photoelectrochemical (PEC) etching was developed to fabricate 4H-SiC MEMS with a p-n structure [4, 5]. Such p-n structure has high quality homoepitaxial p-SiC suspended microstructures on single crystalline n-SiC substrate. The advantages of this technique include process simplicity, well control, low cost, and capability to fabricate complex MEMS structures. In this paper, we report the process development of 4H-SiC MEMS with n-p-n homoepitaxial structure by PEC etching. Characterization results from force-displacement measurement and electrostatically-actuated resonance of the MEMS structures will be presented. With the middle p-SiC layer separating the top n-SiC free-standing structure from the bottom n-SiC substrate, n-p-n structure allows electrostatic actuation, which is important to drive resonators and actuators, and also extends monolithic integration of MEMS with electronic devices and circuits in n-p-n configurations.

DEVICE FABRICATION

A 4H-SiC wafer with an n-type substrate, followed by a p-type layer (thickness: 2 μm) and another n-type layer on the top (thickness: 1 μm) was chosen as the starting material. The thickness of the top n-SiC layer determines the thickness of the final suspended free-standing microstructures. Wafer was first cleaned by standard RCA solution 1 and 2. A SiO_2 layer was deposited by PECVD on the SiC wafer surface which is later used as the hard etch mask. Photolithography was performed to pattern the lateral layout of the MEMS structures. Reactive ion etching (RIE) with CHF_3/Ar mixed gas was applied to etch SiO_2, and SF_6 to etch SiC. The n-SiC in the etch windows must be removed completely to expose the underneath p-SiC for later PEC etching. After removing the SiO_2 layer, the wafer was cleaned again by RCA, and photoresist mask was patterned for lift-off of metal contact electrodes used later for electrostatic actuation of MEMS devices. Nickel (Ni) was deposited on the front and backside of the wafer, followed by lift-off. A rapid thermal annealing process was carried out at 950°C for 60 sec in Ar ambient to transform Ni to Ohmic contacts. The process flow is shown in Figure 1 (a~f).

Dopant-selective PEC etching was applied to release the free-standing n-SiC MEMS structures. The chemical reaction of n-SiC and p-SiC in KOH and principle of dopant-selective PEC etching 4H-SiC has been explained elsewhere [5]. The wafer was immersed in a dilute (5% by weight) potassium hydroxide (KOH) solution which was connected through a platinum wire to the cathode of a DC power supply. The backside of the wafer was isolated from KOH and connected to the anode of the DC power supply. A UV light source from a 100 W mercury arc lamp with a 250nm~400nm bandpass filter was focused through the objective of a florescence microscope to expose the wafer surface. A digital multimeter was connected in the circuit to monitor the current follow and adjust the DC bias voltage. Current density is a very critical parameter in controlling the etching rate of SiC and roughness of the wafer surface after PEC etching. The PEC etching setup is shown in Figure 1(g). When a proper DC bias voltage was selected, the top n-SiC is intact while the underneath both p-SiC layer and n-SiC substrate is etched by employing the different flat-band potentials of n-SiC and p-SiC in the KOH solution, from which the top n-SiC layer was released to form the free-standing microstructures.

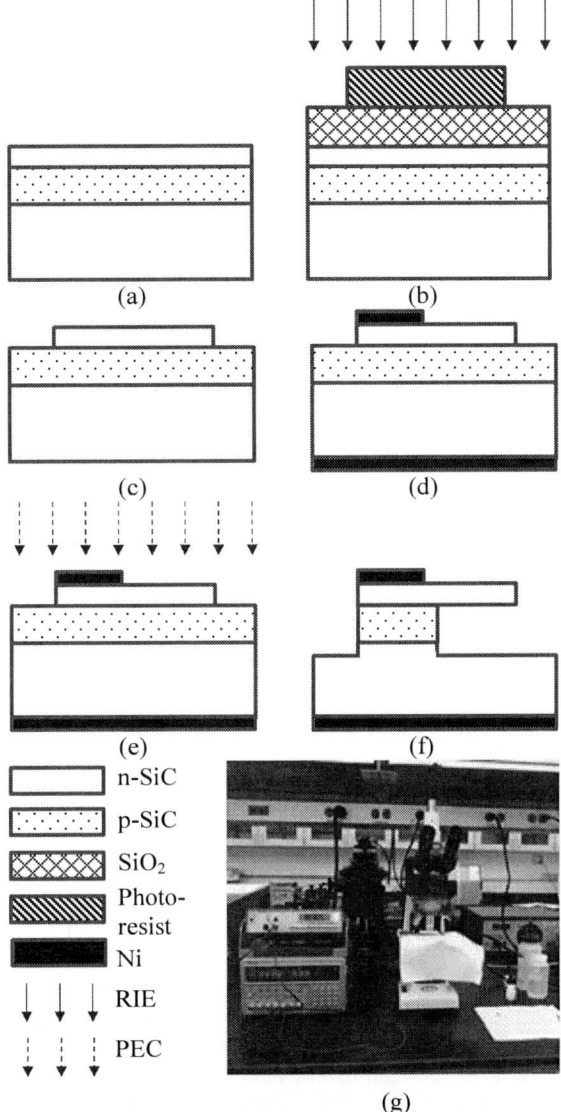

Figure 1: Process flow: (a) starting SiC wafer with n-p-n structure. (b) SiO₂ hard mask deposition, photoresist patterning, and RIE etching. (c) SiO₂ and photoresist removal. (d) Ni deposition and lift-off. (e) PEC etching in KOH. (f) final released structure. (g) PEC etching setup.

RESULTS AND DISCUSSION

The lateral undercutting process during PEC etching was monitored and shown in Figure 2 (a~d). The lateral etching gradually removed both p-SiC layer and underneath n-SiC substrate, eventually released completely the top n-SiC microstructures. The released SiC thin film is transparent under optical microscope due

to the large bandgap energy ($E_{G(SiC)}$=3.2 eV) of 4H-SiC and thin film thickness (1 μm as shown in the SEM image in Figure 2 (e)). The final released microstructure with a 50 μm-diameter platform suspended by four 80 μm-long beams is shown in Figure 2 (e). The straightness and flatness of the platform and beams indicate the absence of internal stress or strain gradient. Figure 2(e) clearly shows that the top 1 μm-thick n-SiC layer was retained in etching, while the p-SiC layer acted as the sacrificial layer and was removed by PEC etching. In addition, since the thickness of p-SiC layer is 2 μm, but the cap between the top n-SiC film and n-SiC substrate is more than 3 μm as shown in the SEM image, it indicates that the n-SiC substrate was also partially etched.

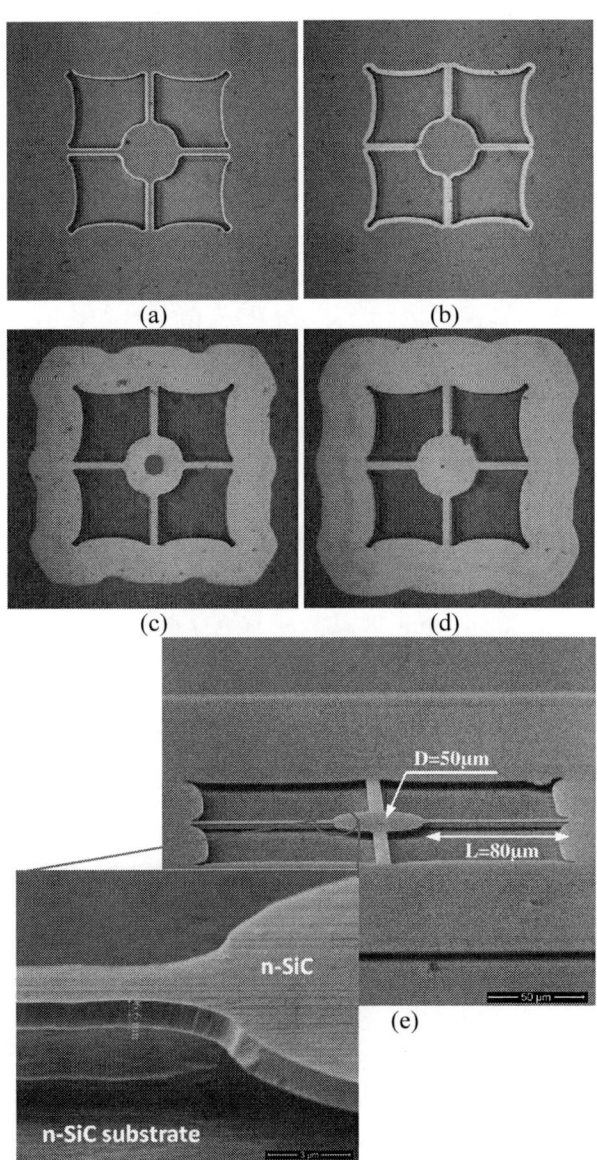

Figure 2: Undercutting 4H-SiC by PEC after (a) 5 min, (b) 10 min, (c) 60 min, and (d) 75 min. The SEM image of the released structure with a close-look is shown in (e).

A force-displacement measurement was carried out on the microstructure shown in Figure 2 using a

978-1-4799-7956-1/15 $31.00 © 2015 IEEE

nanomanipulator system. The total displacement z which is the sum of the deflection of the probe tip and the bending of the suspended microstructure is measured by a high precision stage with nanometer resolution. The force F applied on the microstructure is sensed with a resolution of 5 nN. The test setup and resulted curve are shown in Figure 3. A clear movement of the microstructure under the force by probe tip was observed through a high-definition camera with a microscope objective, which proves the suspended microstructure is free-standing and undercutting of this 50 μm-diameter n-SiC film by PEC etching was successful. The slope F/z gives the total spring constant $1/k_{tot}=1/k_p+1/k_m$ [11], where k_p and k_m are the spring constant of the probe tip and microstructure. Since the tungsten probe tip of the nanomanipulator is very rigid with a high spring constant, k_{tot} is approximately equal to k_m, which is 43.48 N/m derived from Figure 3(c).

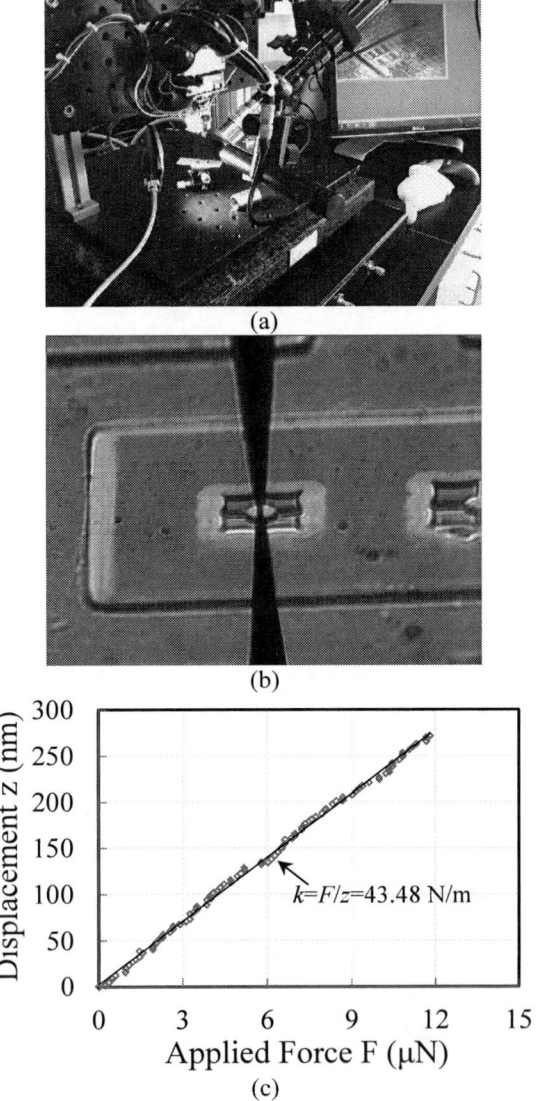

Figure 3: Force-displacement measurement (a) test setup, (b) high-definition camera image showing the probe tip contact and push down the suspended microstructure, and (c) resulted F-z curve.

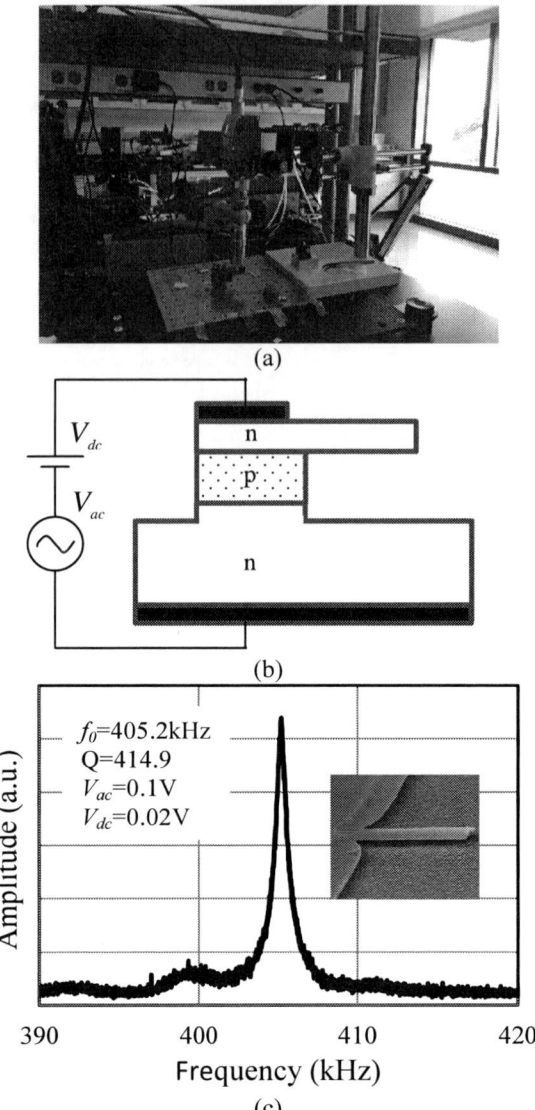

Figure 4: (a) The laser Doppler setup used for dynamic measurement, (b) schematic illustration of electrostatic actuation of a 4H-SiC microcantilever (75×10×1 μm), and (c) frequency response from the cantilever.

Electrostatic actuation was also applied successfully on a microcantilever (75×10×1 μm) and measurement was performed in air and at room temperature. The dynamic response of the cantilever under actuation was collected by a laser and a Doppler vibrometer shown in Figure 4 (a). The cantilever was excited by applying ac voltage with dc bias between the n-SiC cantilever and n-SiC substrate through the Ohmic Ni contact electrodes as illustrated in Figure 4(b). With the top n-SiC cantilever separated from the bottom n-SiC substrate by the p-SiC layer in the middle, the electrostatic force was applicable. The amplitude of vibration as a function of frequency was recorded and shown in Figure 4(c) with the first mode resonant frequency f=405.2 kHz. The quality factor was calculated as $Q=f/\Delta f$, where Δf is the full-width at half-maximum (FWHM). A 0.98 kHz FWHM was obtained from Figure 4(c), which gives the quality factor Q=414.9 of this electrostatic excited 4H-SiC cantilever.

CONCLUSION

Single crystal 4H-SiC MEMS was fabricated by standard microfabrication combined with dopant-selective PEC etching process on a 4H-SiC wafer with n-p-n homoepitaxial structure. The p-SiC acted as the sacrificial layer during PEC etching and also the electrical isolation layer for electrostatic actuation of the final free-standing microstructure. An n-SiC film with a diameter of 50 μm was successfully released and proven by force-displacement measurement. The resonant property of a 75 μm-long cantilever under electrostatic actuation was characterized, and a resonant frequency of 405.2 kHz and a quality factor of 414.9 were obtained in air and at room temperature. The n-p-n structure provides the capability of electrostatic excitation, and also for MEMS to integrate with other electronic devices and control circuits to form integrated systems on the 4H-SiC platform, which are desirable for harsh environment applications.

ACKNOWLEDGEMENTS

Feng Zhao acknowledges the support from National Science Foundation (ECCS-1307237 monitored by Dr. Anupama B. Kaul).

REFERENCES

[1] C. A. Fisher, M. R. Jennings, Y. K. Sharma, D. P. Hamilton, P. M. Gammon, A. P. Tomás, S. M. Thomas, S. E. Burrows, P. A. Mawby, "Improved Performance of 4H-SiC PiN Diodes Using a Novel Combined High Temperature Oxidation and Annealing Process", *IEEE Trans Semiconductor Manufacturing*, vol. 27, pp. 443-451, 2014.

[2] K. W. Chu, W. S. Lee, C. Y. Cheng, C. F. Huang, F. Zhao, L. S. Lee, Y. S. Chen, C. Y. Lee, M. J. Tsai, "Demonstration of Lateral IGBTs in 4H-SiC," *IEEE Electron Device Letters*, vol. 34, pp. 286-288, 2013.

[3] A. Salemi, B. Buono, A. Hallén, J. Ul Hassan, P. Bergman, C. M. Zetterling, M. Östling, "Fabrication and Design of 10 kV PiN Diodes Using On-Axis 4H-SiC", *Materials Science Forum*, vol. 778-780, pp. 836-840, 2014.

[4] N. Watanabe, T. Kimoto, J. Suda, "Fabrication of Electrostatic-Actuated Single-Crystalline 4H-SiC Bridge Structures by Photoelectrochemical Etching", *Proc. of SPIE*, pp. 7926-11, 2011.

[5] F. Zhao, M. M. Islam, "Fabrication of Single-Crystal Silicon Carbide MEMS/NEMS for Bio-sensing and Harsh Environments", in *Proc. IEEE MEMS*, Cancun, January 23-27, 2011, pp. 261-263.

[6] F. Zhao, M. M. Islam, C. F. Huang, "Photoelectrochemical Etching to Fabricate Single-Crystal SiC MEMS for Harsh Environments", *Materials Letters*, vol. 65, pp. 409-412, 2011.

[7] F. A. Khan I. Adesida, "High Rate Etching of SiC Using Inductively Coupled Plasma Reactive Ion Etching in SF6-based Gas Mixtures", *Applied Physics Letters*, vol. 75, pp. 2268-2270, 1999.

[8] G. Beheim, C. S. Salupo, "Deep RIE Process for Silicon Carbide Power Electronics and MEMS", in *Proc. MRS Spring Meeting*, San Francisco, May 24-28, 2000, pp. T8.9.

[9] J. P. Desbiens, P. Masson, "ArF Excimer Laser Micromachining of Pyrex, SiC and PZT for Rapid Prototyping of MEMS Components", *Sensors and Actuators A*, vol. 136, pp. 554-563, 2007.

[10] W. Kang, M. T. A. Saif, "A novel SiC MEMS Apparatus for in situ Uniaxial Testing of Micro/nanomaterials at High Temperature", *Journal of Micromechanics and Microengineering*, vol. 21, pp. 105017, 2011.

[11] C. Serre, P. Gorostiza, A. Perez-Rodriguez, F. Sanz, J. R. Morante, "Measurement of Micromechanical Properties of Polysilicon Microstructures with an Atomic Force Microscope", *Sensors and Actuators A*, vol. 67, pp. 215-219, 1998.

CONTACT

*F. Zhao, tel: +1-360-5469187; feng.zhao@wsu.edu

INTEGRATION OF DISTRIBUTED GE ISLANDS ONTO SI WAFERS BY ADHESIVE WAFER BONDING AND LOW-TEMPERATURE GE EXFOLIATION

F. Forsberg[1], N. Roxhed[1], C. Colinge[2], G. Stemme[1] and F. Niklaus[1]

[1] Micro and Nanosystems, KTH Royal Institute of Technology, Stockholm, Sweden
[2] California State University, Sacramento, USA

ABSTRACT

We present a novel and highly efficient wafer-level batch transfer process for populating silicon (Si) wafers with distributed islands of thin single-crystalline germanium (Ge) layers. This is achieved by transferring Ge from a Si wafer containing thick Ge dies to a Si target wafer by adhesive wafer-bonding and subsequent low-temperature Ge exfoliation.

INTRODUCTION

The combination of electronic, MEMS and photonic functions on a common Si substrate enable high-performance heterogeneous microsystems such as infrared detector arrays, optical gyroscopes or components for optical communication systems [1-7]. However due to the lattice mismatch between Si and typical photonic materials such as Ge, gallium-arsenide (GaAs) or indium phosphide (InP), it is in most cases not possible to directly deposit or grow these type of high-quality photonic materials on top of a Si substrate. An attractive approach to overcome this problem is to transfer a layer of the high-quality photonic material from its original substrate to the Si substrate. This has been implemented mainly by using chip-level processes based on adhesive bonding [8, 9] or direct bonding [10]. In these approaches the photonic material donor substrate typically is sacrificially etched after the bonding step to leave a thin layer of the photonic material on top of the Si substrate. Thereafter the photonic devices on the Si substrate can be defined and formed from the photonic material. However, these processes are very resource demanding since the comparably expensive photonic material substrates are sacrificially removed by grinding and/or etching. Furthermore, due to the difference in coefficient of thermal expansion (CTE) between Si and e.g. Ge or GaAs, it is extremely challenging to utilize large-scale and high-yield wafer-to-wafer bonding at elevated temperatures. An innovative approach to address this problem is to populate the Si wafer with pre-dices dies of the photonic material using pick-and-place positioning of the dies on the wafer along with a subsequent adhesive wafer bonding step. All dies on the Si wafer can then be thinned and processes in parallel fashion using standard wafer-scale processes [3, 6]. Another related process that has been proposed is based on batch-transfer of radially expanded die arrays to achieve efficient layer transfer [11]. However, both these approaches rely on sacrificial removal of the excess photonic donor substrate. In this work we present a wafer-level batch transfer process that is based on transferring Ge from a Si wafer that is containing Ge dies to a Si target wafer by adhesive wafer-bonding and subsequent low-temperature Ge exfoliation from the Ge dies. The bulk of

the Ge dies is remaining on the Si wafer and can in principle be reused to transfer subsequent layers from the remaining bulk Ge dies. Thus, the proposed approach circumvents problems caused by the CTE mismatch between Si and photonic material wafers and avoids the sacrificial removal of the comparably expensive photonic base substrate.

CONCEPT OF THE INTEGRATION PROCESS AND EXPERIMENTS

The conceptual idea for the transfer of thin Ge layers from Ge dies onto a Si wafer consists of starting with a Si wafer containing distributed island of Ge dies. This is used as the Ge donor-wafer which is then adhesively bonded to a receiving Si substrate. Adhesive bonding is attractive for heterogeneous integration processes due to its insensitivity to topographies or particles at the surfaces to be bonded and the resulting high-yield bond interfaces [1]. Since both the target wafer and the donor wafer containing the Ge dies are made of Si, there is no significant CTE mismatch between the two wafers, which avoids problems related to a CTE mismatch during bonding. The hydrogen-implanted Ge dies are exfoliated at a comparably low temperature of 300°C [5], which leaves transferred 1 µm thick Ge layers on the Si target wafer. Figure 1 shows the detailed process scheme that was implemented in this work. In this process, hydrogen is implanted in a Ge wafer surface to a depth of about 1 µm as indicated in Figure 1a. Thereafter a 1 µm thick layer of silicon nitride is deposited on the surface of the Ge wafer using plasma-enhanced chemical vapor deposition as depicted in Figure 1b. Next, the wafer is attached to an expandable UV-release tape and diced into mm-sized dies as shown in Figure 1c. The array with dies is then expanded as depicted in Figure 1d using a matrix expander [4]. A Si wafer is spin-coated with AP3000 adhesion promoter (Dow Chemical Company) at 3000 rpm and spun dry. This is followed by spin-coating a layer of BCB (Cyclotene 3022-46, Dow Chemical Company) that is mesitylene-diluted 1:1 by weight at 5000 rpm onto the Si wafer as shown in Figure 1e. The dilution of the BCB is done to obtain a very thin BCB coating with a thickness of about 550 nm. The spin-coated BCB on the Si wafer is soft-baked at 150°C for 3 min on a hotplate and subsequently slightly crosslink for 5 min at 180°C in an oven. Thereafter, the tape with the expanded Ge die array is pressed onto the BCB-covered wafer and baked for 10 min at 80°C as shown in Figure 1f. The dicing tape is removed by exposing the tape for 10 min to UV-radiation as shown in Figure 1g. A second BCB-coated Si wafer is prepared by spin-coating the wafer with the adhesion promoter AP3000 at 3000 rpm until the wafer is dry and then spin-coating an undiluted layer of BCB (Cyclotene 3022-46) at 5000 rpm on the Si wafer. This results in a 2.4 µm thick layer of BCB-layer, as depicted in Figure 1h. The BCB-coated Si wafer is then bonded to the Si wafer containing the Ge dies as illustrated in Figure 1i. The wafer bonding is performed in a Suss Microtec SB8 wafer bonder. The combined wafer bonding and Ge exfoliation process consists of 24 h baking at 130°C, followed by a temperature ramp of 1°C per minute to 300°C. The bonding temperature is kept stable for 10 min followed by cooling the wafer stack to room temperature. The temperature cycling steps crack the thin hydrogen-implanted layer from the surface of the Ge dies at the depth of the hydrogen-implantation [12]. This leaves thin, exfoliated Ge layers bonded with BCB on the target wafer. The bulk Ge dies remain bonded to the Si wafer as illustrated in Figure 1j.

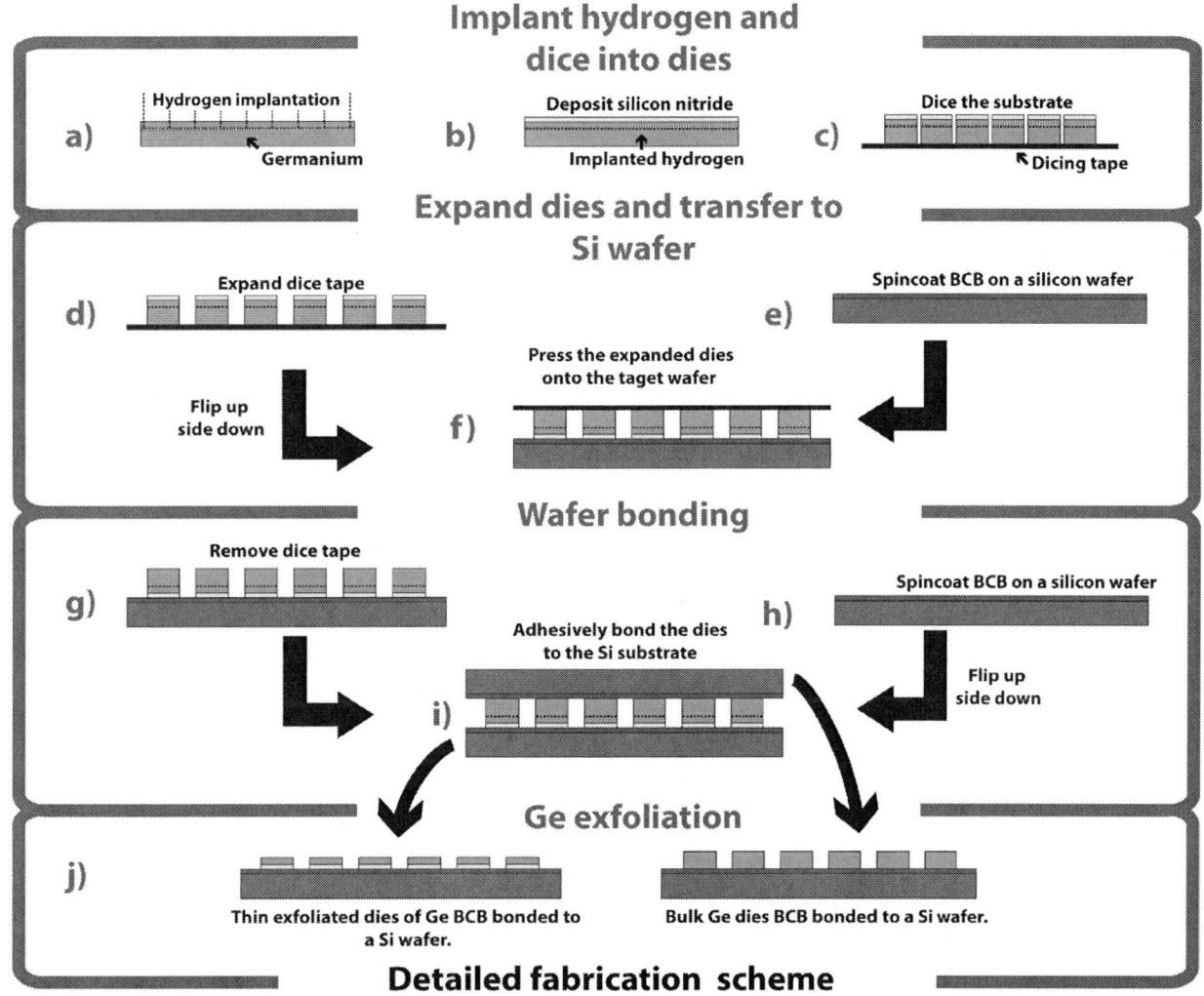

Figure 1: (a) Hydrogen is implanted into a Ge wafer. (b) Deposit 1 µm PECVD silicon nitride onto the Ge wafer. (c) Dice the Ge wafer. (d) Expand the dicing tape to separate the dies. (e) Spin coat BCB onto a Si wafer. (f) Press the expanded die array onto the Si wafer. (g) Remove the dicing tape after the transfer. (h) Spin-coat BCB on a second Si wafer. (i) Press the Si wafer onto the Si wafer containing the Ge dies. Bond the two wafers. (j) Exfoliation of the thin hydrogen-implanted Ge layer by temperature ramping to 300 °C. This process causes the cracking of a thin layer of Ge from the surface of the Ge dies. Thus, 800 nm thick exfoliated Ge layers are left on the Si target wafer. The bulk Ge dies remain bonded on the Si handle wafer.

RESULTS

Figure 2 depicts experimental results from studying the fabrication process as described in Figure 1. Figure 2a shows a 15 mm square Ge die that is attached to an expandable dicing tape. Figure 2b shows the same Ge die after dicing and tape expansion to separate the Ge dies before their transfer to a Si wafer using adhesive bonding.

Figure 2c shows four arrays of Ge dies bonded to a Si wafer as depicted in Figure 1g. The dies are squares and have sizes of 2 mm, 2.5 mm, 3 mm and 5 mm respectively. Figures 3 and 4 illustrate the results following the wafer bonding and Ge exfoliation steps as depicted in Figure 1i and 1j. Figure 3a and 3b are images of Si wafers with bonded and exfoliated Ge dies.

978-1-4799-7956-1/15 $31.00 © 2015 IEEE   281

Figure 3a shows a Si handle wafer with adhesively bonded bulk Ge dies after the Ge exfoliation step. Figure 3b shows the respective Si target wafer containing the transferred 1 μm thick Ge layer islands. Figure 4a shows the results from an EDX material analysis to confirm that Ge has been transferred. Figure 4b shows one bulk Ge die on the donor wafer after exfoliation. Figure 4c shows a transferred and exfoliated Ge layer island on the target Si wafer.

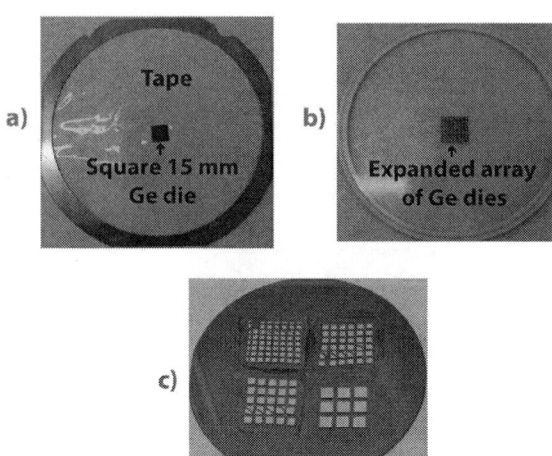

Figure 2: (a) Square 15 mm Ge die attached to an expandable UV-release dicing tape. Results following the step shown in Figure 1c. (b) Expanded die array. Results following the step shown in Figure 1d. (c) Four arrays of dies transferred and BCB-bonded to a 100 mm diameter Si wafer. Results following step shown in Figure 1g.

Figure 3: (a) Bulk Ge dies BCB bonded to a 100 mm diameter Si wafer. Results following step shown in Figure 1j. (b) Thin exfoliated Ge layer islands that are BCB bonded to a 100 mm diameter Si wafer. Results following step shown in Figure 1j.

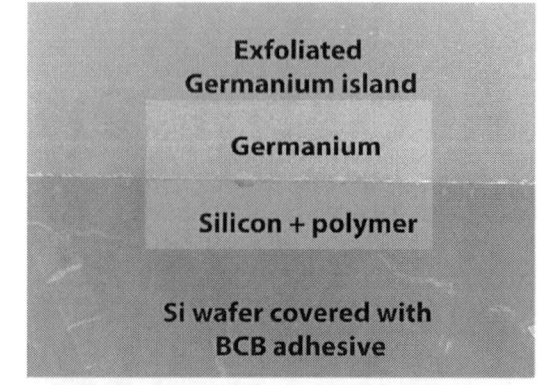

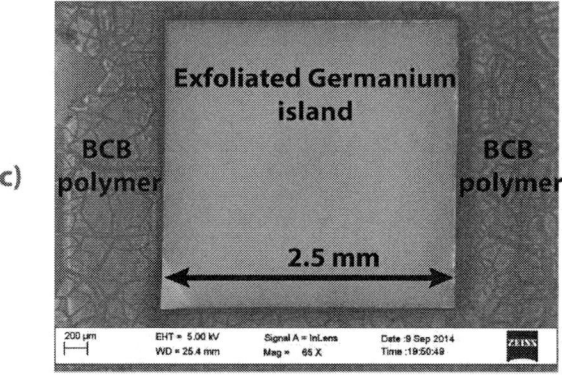

Figure 4: (a) EDX scan confirming that Ge was transferred. The color coding corresponds to concentration of Ge and Si. (b) SEM image of a bulk Ge die on the Si donor wafer after exfoliation. (c) Transferred and exfoliated Ge thin layer island that is bonded to the Si target wafer.

The results from the experiments depicted in Figures 2, 3 and 4 conclusively demonstrate that it is possible to use adhesive bonding and subsequent exfoliation of thin Ge layers to transfer Ge from a donor wafer and populate a Si target wafer with thin-film Ge islands in a controlled way. This approach could be a step towards cost-competitive and resource efficient heterogeneous integration of active photonic materials in Si-based platforms. Potential applications could be advanced heterogeneous Si photonics for optical communication systems and high-performance infrared bolometer arrays.

978-1-4799-7956-1/15 $31.00 © 2015 IEEE   282

CONCLUSIONS

In summary, we have demonstrated the viability of a new process scheme for resource efficient and wafer-level transfer of high-quality photonic materials, in this case Ge, on top of a silicon wafer. The proposed heterogeneous integration process is based on adhesive wafer bonding with BCB as the intermediate bonding layer and subsequent exfoliation of a thin layer from the photonic material donor substrate. The process can be implemented on wafer-level and is potentially more resource efficient than conventional schemes using chip-to-wafer bonding together with sacrificial substrate removal.

ACKNOWLEDGEMENTS

This work was supported by the European Commission through the Grant No.277879 and Grant No.267528.

REFERENCES

[1] F. Forsberg, et al., "CMOS-Integrated Si/SiGe Quantum-Well Infrared Microbolometer Focal Plane Arrays Manufactured With Very Large-Scale Heterogeneous 3D Integration", *IEEE Journal of Selected Topics in Quantum Electronics*, Vol.21, 2014, DOI: 10.1109/JSTQE.2014.2358198.

[2] C. Sorrentino, J.R. Toland, "Ultra-Sensitive Chip Scale Sagnac Gyroscope Based on Periodically Modulated Coupling of a Coupled Resonator Optical Waveguide", *Optics Express*, Vol.20, No.1, pp.354-363, 2012.

[3] G. Roelkens, et al., "III-V/Si Photonics by Die-to-Wafer Bonding", *Materials Today*, Vol.10, No.7, pp.36-43, 2007.

[4] M. Smit, J. Brouckaert, D. Van Thourhout, R. Baets, R. Nötzel, G. Roelkens, "Adhesive Bonding of InP/InGaAsP Dies to Processed Silicon-On-Insulator Wafers using DVS-bis-Benzocyclobutene", *Journal of The Electrochemical Society*, Vol.153, pp.1015-1019, 2006.

[5] A. Gassenq, F. Gencarelli, J. Van Campenhout, Y. Shimura, R. Loo, G. Narcy, B. Vincent, G. Roelkens, "GeSn/Ge Heterostructure Short-Wave Infrared Photodetectors on Silicon", *Optics Express*, Vol.20, No.25, pp.27297-27303, 2012.

[6] M. Lapisa, G. Stemme, F. Niklaus, "Wafer-Level Heterogeneous Integration for MOEMS, MEMS and NEMS", *IEEE Journal of Selected Topics in Quantum Electronics*, Vol.17, No.3, pp.629-644, 2011.

[7] J. Brouckaert, G. Roelkens, D. Van Thourhout, R. Baets, "Thin-film III–V Photodetectors Integrated on Silicon-on-Insulator Photonic ICs", *Journal of Lightwave Technology*, Vol.25, No.4, pp.1053-1060, 2007.

[8] F. Niklaus, G. Stemme, J.-Q. Lu, R.J Gutmann, "Adhesive Wafer Bonding", *Journal of Applied Physics*, Vol.99, No.1, pp.031101.1-031101.28, 2006.

[9] F. Niklaus, P. Enoksson, E. Kälvesten, G. Stemme, "Low Temperature Full Wafer Adhesive Bonding", *Journal of Micromechanics and Microengineering*, Vol.11, No.2, pp.100-107, 2001.

[10] D. Liang, J.E. Bowers, D.C. Oakley, A. Napoleone, D.C. Chapman, C.-L. Chen, P.W. Juodawlkis, O. Raday, "High-quality 150 mm InP-to-silicon epitaxial transfer for silicon photonic integrated circuits", *Electrochemical and Solid-State Letters*, Vol.12, No.4, pp.H101-H104, 2009.

[11] F. Forsberg, N. Roxhed, T. Haraldsson, Y. Liu, G. Stemme and F. Niklaus, "Batch Transfer of Radially Expanded Die Arrays for Heterogeneous Integration Using Different Wafer Sizes", *IEEE Journal of Microelectromechanical Systems*, Vol. 21, No. 5, pp. 1077-1083, Oct. 2012.

[12] I.P. Ferain, K.Y. Byun, C.A. Colinge, S. Brightup, M.S. Goorsky, "Low Temperature Exfoliation Process in Hydrogen-Implanted Germanium Layers", *Journal of Applied Physics*, Vol.107, No.5, pp.054315, 2010.

CONTACT

* F. Forsberg; tel: +46-73 3944176; ffors@kth.se

MULTILAYER ETCH MASKS FOR 3-DIMENSIONAL FABRICATION OF ROBUST SILICON CARBIDE MICROSTRUCTURES

Karen M. Dowling[1], Ateeq J. Suria[2], Ashwin Shankar[1],*
Caitlin A. Chapin[2] and Debbie G. Senesky[1,3]

[1]Electrical Engineering Department, Stanford University, Stanford, CA
[2]Mechanical Engineering Department, Stanford University, Stanford, CA
[3]Aeronautics and Astronautics Department, Stanford University, Stanford, CA

ABSTRACT

This paper details the creation of 3-dimensional (3-D) microstructures in 4H-silicon carbide (4H-SiC) substrates with a plasma etch process that utilizes multilayer etch masks. An inductively coupled plasma (ICP) etch process (SF_6/O_2) for SiC was developed and etch rates as high as ~1 μm/min, a selectivity of 60:1 (SiC to Ni), and aspect ratio dependent etch characteristics were demonstrated. In addition, the selectivity of atomic layer deposited (ALD) Al_2O_3 etch masks to 4H-SiC is reported for the first time. Using this unique process, the microfabrication of complex microstructures (mechanical gears, Lego®-like bricks, and poker chips) is presented. The use of 4H-SiC as the structural material enables such microstructures to be utilized under high cycles of wear, within elevated temperatures, and within chemically corrosive environments.

INTRODUCTION

Complex 3-D microstructures (sensor, actuator, and microfluidic devices) made from Si, Ni, and polymer materials (e.g., PMMA) have been fabricated using deep reactive-ion etching (DRIE) [1], [2], lithography, electroplating, and molding (LIGA) [3]. This is enabled through the high selectivity (above 300:1 for Si to SiO_2) and fast etch rates (up to 10 μm/min) of DRIE, as well as the high aspect ratio (1.5 μm x 350 μm) of LIGA processes [2], [3]. However, Si, Ni and polymer materials are limited to benign operating conditions due to their low melting points, fatigue under high cycles of wear, and susceptibility to chemical corrosion. Thus, SiC has been investigated for microstructure design due to its inherent thermal stability, chemical inertness, and wear-resistance [4].

3-D features have been previously made in polycrystalline SiC using a 3-D patterned Si mold and high rate atomic pressure chemical vapor deposition (AP CVD) [5]. However, it is desired to bulk micromachine single-crystalline SiC to leverage its electrical properties and high thermal conductivity. 3-D fabrication techniques for single-crystalline SiC, have yet to be demonstrated and will aid in the realization of robust microstructures that can operate within extreme harsh environments such as micro-scale combustion systems, microchannels for hot fluids, and wear-prone mechanical structures (Figure 1).

Previous efforts in bulk SiC micromachining have focused on development of inductively coupled plasma (ICP) based etches using fluorine, chlorine, and bromine chemistries [6]–[9]. In addition, photoelectrochemical etching has been used to etch n-type 3C-SiC and has

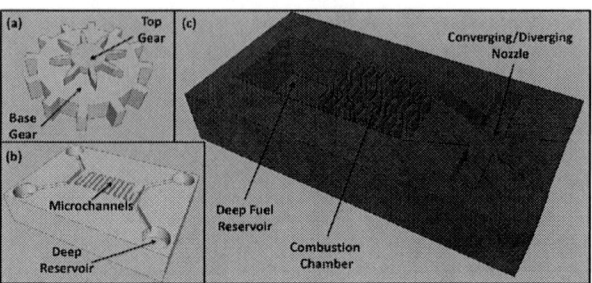

Figure 1: Schematic image of robust 3-D SiC microstructures that are enabled through a multilayer mask process: (a) stacked mechanical gears, (b) bio-compatible, heat-tolerant, and transparent microchannels, and (c) robust microthrusters.

reached etch rates as high as 100 μm/min [10] but shows poor directionality, which is required for many MEMS components [8]. Femtosecond laser irradiation has been shown as a maskless technique to pattern deep (up to 350 μm) via holes [11]. In addition, there has been tremendous work in improving the via hole plasma etching of SiC, but these efforts have been for sub 1:1 aspect ratio features [12]. G. Beheim et al. have carefully characterized the effects of etch rates with respect to many plasma etch parameters with an indium-tin-oxide mask to obtain etch depths over 100 μm [13]. Large depths (> 200 μm) and high aspect ratio (> 10:1) features have also been demonstrated using an electroplated Ni mask, and opening widths of 10 μm and 50 μm up to 250 μm [14]. Recently, new mask candidates have been considered such as AlN [15].

This work demonstrates a multilayer mask etch process to create 3-D microstructures in 4H-SiC substrates. The fabrication process for three experiments is described and the results are presented. First, the SiC etch rate and mask selectivity for three materials (Ni, SiO_2, and Al_2O_3) are studied at varying plasma bias powers. Additionally, aspect ratio dependent etch characteristics are presented with 5 μm to 110 μm opening width variation. Finally, a gallery of high-resolution images of fabricated 3-D structures (mechanical gears, Lego®-like bricks, and poker chips) on a transparent and robust 4H-SiC substrates is shown for the first time. The use of 4H-SiC as a structural material enables such microstructures to be utilized under high cycles of wear, within elevated temperatures, and within chemically corrosive environments unlike common MEMS structural materials [4].

978-1-4799-7956-1/15 $31.00 © 2015 IEEE

FABRICATION

A 4H-SiC substrate (Cree Inc.) was used for this experiment. Due to the high cost of the SiC substrates, dies of 1 cm by 1 cm were used for the experiments. Before processing, SiC samples underwent a piranha clean (H_2SO_4:H_2O_2, 9:1). Next, the samples were rinsed sequentially in deionized water, acetone, methanol and isopropanol. The samples were processed with one or two stages of standard 1:1 contact photolithography to perform the etch selectivity study, observe the aspect ratio dependent etch rate, and create the 3-D structures. The SiC plasma etch (PlasmaTherm's LL-ICP Metal Etch System) baseline recipe selected for this work used 1000 W ICP power, 50 W of RF bias power, 5 mTorr of pressure, 9:1 ratio of SF_6/O_2 gas chemistry, and a total flow rate of 60 sccm.

Etch Rates and Mask Selectivity

Three types of etch masks were used for this study: evaporated Ni, plasma enhanced chemical vapor deposition (PECVD) SiO_2 and plasma atomic layer deposited (ALD) Al_2O_3.

Ni (200 nm) was evaporated and lifted-off from photoresist patterned 4H-SiC. ALD Al_2O_3 (60 nm) was deposited at 250°C on 4H-SiC using a plasma ALD process (Ultratech/Cambridge Nanotech Fiji System). It was then patterned with photoresist and plasma etched (PlasmaTherm's LL-ICP Metal Etch System) with BCl_3 chemistry comprising a total flow rate of 50 sccm, at an etch rate of 45 nm/min. PECVD SiO_2 (2 µm) was deposited at 350°C on 4H-SiC, photoresist was patterned, followed by an etch in an ICP etch system (PlasmaTherm's LL-ICP Oxide Etch system) with an O_2/CHF_3 chemistry at an etch rate of 0.2 µm/min. Once all the etch masks were patterned, the 4H-SiC was etched with the selected SiC etch baseline recipe and then etch rates for RF bias power values of 25, 50, 100, 150 and 200 W was investigated.

The selectivity is obtained by calculating the ratio of etch rates of 4H-SiC to the etch masks for the same parameters. Using a stylus-based profilometer (Tencor Alphastep 500), the SiC etch depth was determined for a given time with the mask. The Ni mask was removed and the sample re-measured, to calculate the final thickness of the mask. The difference of this provides the etch depth of the Ni mask from which the Ni etch rate is determined. This technique was previously used by K. Williams et al. in 1996 [16]. The SiO_2 selectivity was measured by processing a Si wafer with 2 µm deposited in the plasma etch for 1 minute, and the SiO_2 film thickness was measured before and after using an optical profilometer (Nanospec 010-180). The ratio of the SiC etch rate to SiO_2 etch rate gives the SiC:SiO_2 selectivity. The thin Al_2O_3 film was completely removed within 1 minute of plasma etching and as a result, the selectivity was estimated using the ratio of SiC etch depth to the starting thickness of Al_2O_3 [17]. Etch rate and selectivity were measured in three to five consistent locations across the 1 cm^2 SiC die, and data is presented with the average and standard deviation of those measurements.

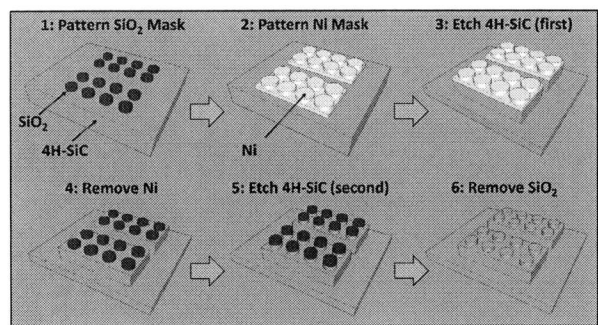

Figure 2: Schematic image of the plasma etch process with multilayer masks (Ni and SiO_2) used to create 3-D microstructures in 4H-SiC

Aspect Ratio Dependent Etch Rate

To investigate the effect of opening dimensions on the etch rate of 4H-SiC, the dies were patterned with a mask design that had several opening dimensions using standard lithography. Ni (200 nm) was evaporated as described earlier with a 10 nm Ti adhesion layer. Next the same sample was evaporated with an additional 500 nm of Ni and 10 nm Ti adhesion layer, making a total mask thickness of approximately 750 nm. This metal was lifted off to serve as the etch mask for this investigation. PlasmaTherm's ICP Metal etch system was utilized to etch the 4H-SiC 3-D structures. The recipe used a 1000 W ICP power, 50 W of RF bias power, 7 mTorr of pressure, 9:1 ratio of SF_6/O_2 and a total flow rate of 60 sccm. The sample was etched for 90 minutes with an average large opening etch rate of approximately 0.6 µm/min. The sample was then imaged using variable pressure scanning electron microscopy (Hitachi S-3400N SEM) to characterize the aspect ratio dependent etch rate.

3-D Etching using Multilayer Etch Masks

The fabrication process used for the development of 3-D microstructures is presented in Figure 2. Post dehydration (hot plate at 115°C for 10 minutes), the sample was coated with a 2 µm thick PECVD SiO_2 film at 350°C. Following this, the sample underwent a standard lithography process. The SiO_2 thin layer was patterned using photoresist as an etch mask. ICP etching was used to etch the 2 µm PECVD SiO_2 layer. Another iteration of standard lithography was performed onto which 200 nm of Ni is evaporated and lifted-off.

To perform the first level of the complex SiC etch patterning, SF_6/O_2 chemistry is used. The baseline ICP etch recipe was used to etch the 4H-SiC 3-D structures to 10 µm. The sample was then dipped in a nickel etchant (Transene Company, Inc.) to remove the leftover Ni mask. Next the sample was etched for a second time to 1 µm using the patterned SiO_2 as the mask. The remaining SiO_2 was stripped using a wet etch with hydrofluoric acid (49% HF). The samples were then imaged using SEM.

RESULTS

Etch Rates and Mask Selectivity

As mentioned previously, three etch masks were used in this investigation: evaporated Ni, PECVD SiO_2, and

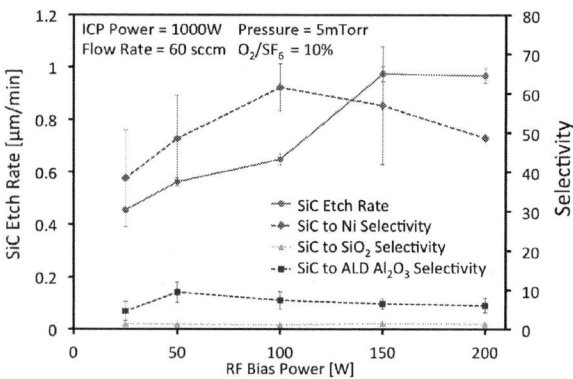

Figure 3: Measured 4H-SiC etch rates and mask selectivity with respect to RF bias power for three etch masks (evaporated Ni, PECVD SiO_2 and ALD Al_2O_3).

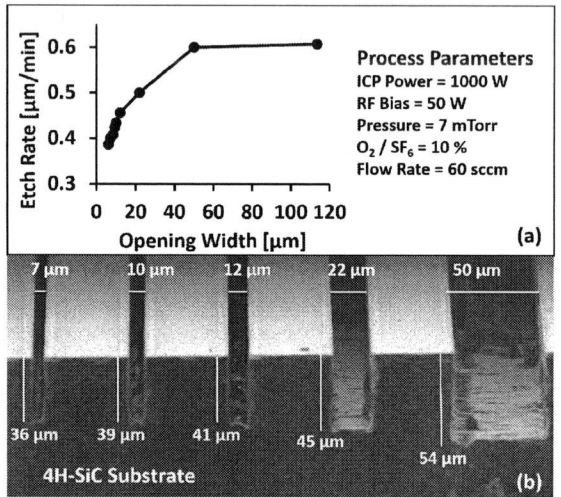

Figure 4: (a) Measured etch rate for various mask opening dimensions showing aspect ratio dependent etch rate. (b) Cross-sectional SEM image of 4H-SiC microtrenches of varying widths.

ALD Al_2O_3, and were chosen due to their high etch selectivity. Figure 3 shows the etch rate of 4H-SiC and the SiC to hard mask selectivity values for varying plasma bias power. A SiC etch rate as high as approximately 1μm/min was observed at a bias power of 150 W and a lower end etch rate of about 0.45 μm/min at a bias power of 25 W. In addition, a SiC to Ni etch selectivity of 60:1 and an etch rate of 1 μm/min were obtained at a RF bias power of approximately 100 W and approximately 150 W, respectively. Furthermore, a decrease in selectivity is observed at higher bias power These results agree with previous SiC etch studies [13], [18]. A SiC:SiO_2 selectivity ratio of 1:1 was observed, which is near the expected selectivity [13]. In addition, the etch characterization showed SiC to ALD Al_2O_3 selectivity values between 5:1 and 10:1, surpassing typical selectivity values for SiO_2 (< 3:1). These preliminary results show that ALD Al_2O_3 is promising as a mask for multilayer SiC etching. The moderate selectivity can be leveraged when a diversity of masks are needed for complex microstructures.

Aspect Ratio Dependent Etch Rate

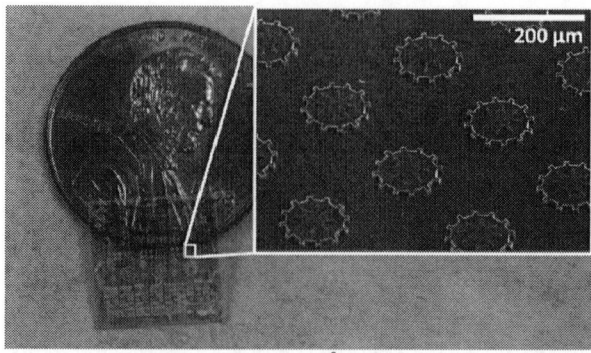

Figure 5: Image of a 1 cm^2 4H-SiC die (optically transparent) that was used to microfabricate a variety of microstructures with inset showing SEM of gear array with minimal micromasking.

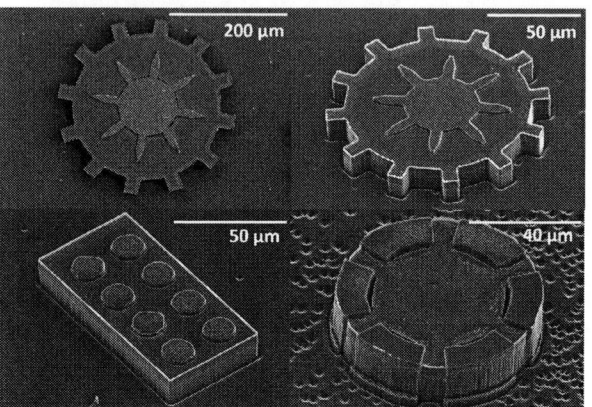

Figure 6: SEM gallery of various 4H-SiC 3-D microstructures created using this multilayer mask technique: (a) top-view of 300-μm-diameter gear stack, (b) 100-μm-diameter gear stack, (c) Lego®-like brick, and (d) poker chip.

An aspect ratio dependent etch was observed for feature openings between 5 μm and 110 μm. Figure 4 shows the SiC etch rate dependence on opening width, and a SEM of trenches with opening widths varying from 7 μm to 50 μm. As expected, there is a decrease in etch rate (and overall etch depth) with a decrease in starting opening feature. This trend has been widely reported in Si etching literature [2] and SiC etching literature [14]. Additionally, there is an "open area" saturation value of etch rate, where larger openings will etch nominally at the same rate. To obtain high aspect ratio trenches, the modified etch rate must be considered for starting mask feature size. Additionally, there are observable pointed features in the corners of each channel due to microtrenching [13], and these features converge in narrow trenches over longer etch times.

3-D Etching using Multilayer Etch Masks

The examination of the selectivity and aspect ratio dependent etch rates enable a set of etch recipes to realize 3-D etch structures using multilayer masks. An array of structures (stacked gears) microfabricated in a transparent, 1 cm^2 4H-SiC substrate is shown in Figure 5. Figure 6 presents an image gallery of the first complex microstructures (mechanical gears, a Lego®-like brick, and a poker chip) created in 4H-SiC using multilayer mask etching. The base features of these structures have a height of 10 μm and the top features are 1 μm tall. The structures in Figure 6a to Figure 6c were etched using the baseline SiC plasma etch recipe shown in Figure 3 with a bias power of 50 W. Figure 6d, the poker chip, was etched using a 100 W bias power, and shows rough artifacts on the etch surface due to micromasking [9]. The nickel mask is sputtered and redeposited on the etch surface creating non-desired roughness. This roughness can be avoided by using reduced bias powers. Overall, these results demonstrate the potential to create 3-D structures in SiC from use of multilayer masks.

CONCLUSIONS

This work shows the effect of plasma RF bias power on the 4H-SiC etch rate and selectivity to various etch masks including evaporated Ni, PECVD SiO_2, and ALD Al_2O_3. A SiC to Ni etch selectivity of 60:1 and an etch rate of 1 μm/min were obtained at a RF bias power of ~100 W and ~150 W, respectively. These results agree with previous SiC etch studies [13], [18]. In addition, we observed an aspect ratio dependent etch, with a maximum depth achieved of 54 μm. Finally, we presented the first 3-D structures made in SiC using this multilayer mask etching technique. It is suggested that future work be performed to create 3-D features using multilayer etch masks including ALD Al_2O_3 and to demonstrate ultra-deep (above 100 μm) 3-D features in SiC.

ACKNOWLEDGEMENTS

This material is based upon work supported by the United States Air Force and DARPA under Contract No. FA-8650-13-C-7374. The authors would like to thank Mr. Jonathan Allison (The Boeing Company) and Prof. Goodson's group at Stanford for insightful discussions, as well as the Stanford Nanofabrication Facility (SNF) and Cell Sciences Imaging Facility (CSIF).

REFERENCES

[1] J. Yeom, Y. Wu, J. C. Selby, and M. a. Shannon, "Maximum achievable aspect ratio in deep reactive ion etching of silicon due to aspect ratio dependent transport and the microloading effect," *J. Vac. Sci. Technol. B Microelectron. Nanom. Struct.*, vol. 23, no. 6, pp. 2319–2329, 2005.

[2] Y. Mita, M. Mita, A. Tixier, J.-P. Gouy, and H. Fujita, "Embedded-mask-methods for mm-scale multi-layer vertical/slanted Si structures," in *Proceedings IEEE Thirteenth Annual International Conference on Micro Electro Mechanical Systems*, 2000, pp. 300–305.

[3] D. T. Haluzan and D. M. Klymyshyn, "High-Q LIGA-MEMS vertical cantilever variable capacitors for upper microwave frequencies," *Microw. Opt. Technol. Lett.*, vol. 42, no. 6, pp. 507–511, Sep. 2004.

[4] R. Maboudian, C. Carraro, D. G. Senesky, and C. S. Roper, "Advances in silicon carbide science and technology at the micro- and nanoscales," *J. Vac. Sci. Technol. A Vacuum, Surfaces, Film.*, vol. 31, no. 5, p. 050805, 2013.

[5] N. Rajan, M. Mehregany, C. A. Zorman, S. Stefanescu, and T. P. Kicher, "Fabrication and testing of micromachined silicon carbide and nickel fuel atomizers for gas turbine engines," *J. Microelectromechanical Syst.*, vol. 8, no. 3, pp. 251–257, 1999.

[6] P. Leerungnawarat, K. P. Lee, S. J. Pearton, F. Ren, and S. N. G. Chu, "Comparison of F_2 plasma chemistries for deep etching of SiC," *J. Electron. Mater.*, vol. 30, no. 3, pp. 202–206, Mar. 2001.

[7] D. Gao, R. T. Howe, and R. Maboudian, "High-selectivity etching of polycrystalline 3C-SiC films using HBr-based transformer coupled plasma," *Appl. Phys. Lett.*, vol. 82, no. 11, pp. 1742–1744, 2003.

[8] M. Mehregany, C. a. Zorman, and N. Rajan, "Silicon carbide MEMS for harsh environments," *Proc. IEEE*, vol. 86, no. 8, pp. 1594–1609, 1998.

[9] M. Gad-el-Hak, Ed., *Design and Fabrication*, 2nd ed. London: CRC Press, 2005.

[10] J. S. Shor, X. G. Zhang, and R. M. Osgood, "Laser-Assisted Photoelectrochemical Etching of n-type Beta-SiC," *J. Electrochem. Soc.*, vol. 139, no. 4, pp. 1213–1216, 1992.

[11] V. Khuat, Y. Ma, J. Si, T. Chen, F. Chen, and X. Hou, "Fabrication of through holes in silicon carbide using femtosecond laser irradiation and acid etching," *Appl. Surf. Sci.*, vol. 289, pp. 529–532, Jan. 2014.

[12] P. Chabert, "Deep etching of silicon carbide for micromachining applications: Etch rates and etch mechanisms," *J. Vac. Sci. Technol. B Microelectron. Nanom. Struct.*, vol. 19, no. 4, p. 1339, 2001.

[13] G. Beheim and C. S. Salupo, "Deep RIE Process for Silicon Carbide Power Electronics and MEMS," *MRS Proc.*, vol. 622, p. T8.9.1, Mar. 2011.

[14] S. Tanaka, K. Rajanna, T. Abe, and M. Esashi, "Deep reactive ion etching of silicon carbide," *J. Vac. Sci. Technol. B Microelectron. Nanom. Struct.*, vol. 19, no. 6, p. 2173, 2001.

[15] D. G. Senesky and A. P. Pisano, "Aluminum nitride as a masking material for the plasma etching of silicon carbide structures," in *Proceedings IEEE Thirteenth Annual International Conference on Micro Electro Mechanical Systems*, 2010, pp. 4–7.

[16] K. Williams and R. Muller, "Etch rates for micromachining processing," *J. Microelectromechanical Syst.*, vol. 5, no. 4, pp. 256–269, 1996.

[17] K. R. Williams, K. Gupta, and M. Wasilik, "Etch rates for micromachining processing-part II," *J. Microelectromechanical Syst.*, vol. 12, no. 6, pp. 761–778, Dec. 2003.

[18] J. Hopkins, G. Nicholls, and L. Lea, "Plasma sources for high-rate etching of SiC," *Solid State Technology*, pp. 61–64, May-2005.

CONTACT

*K. Dowling, tel: +1-650-7256973; kdow13@stanford.edu

NEMS BY MULTILAYER SIDEWALL TRANSFER LITHOGRAPHY
Dixi Liu[1], Richard R.A. Syms*[1] and Munir M. Ahmad[1]
[1]EEE Dept., Imperial College London, London, UK

ABSTRACT

This paper reports an extension of a recently demonstrated technique to fabricate nano-electro-mechanical systems (NEMS) using sidewall transfer lithography (STL). The process uses three pattern transfer steps. Each step only requires optical lithography, making the method suitable for low-cost, wafer scale fabrication. The first two involve STL and are used to form nanoscale features such as suspension beams. These may now intersect, breaking an important restriction of single-layer STL NEMS. The third involves conventional lithography and is used to form microscale features such as anchors. Current nanoscale features have a width of 100 nm and an aspect ratio of 50 : 1. The new process should allow mass parallel fabrication of complex NEMS.

INTRODUCTION

Nanoelectromechanical systems have many potential advantages over conventional MEMS, including improved sensitivity in mechanical sensors based on elastic suspensions. However, current fabrication techniques involve either an expensive direct-writing system such as electron-beam or AFM lithography to form all nanoscale features [1, 2], or a wafer-scale replication process such as soft lithography or nanoimprint lithography that itself requires nanoscale mastering [3, 4]. In each case the cost and complexity of the patterning step may prevent the translation of NEMS into practical use. Recently, a fabrication process based on sidewall transfer lithography [5-11] has been demonstrated that allows ultrathin suspensions to be combined with microscale anchors and hence form NEMS such as the electrothermal actuator in Fig. 1 [12].

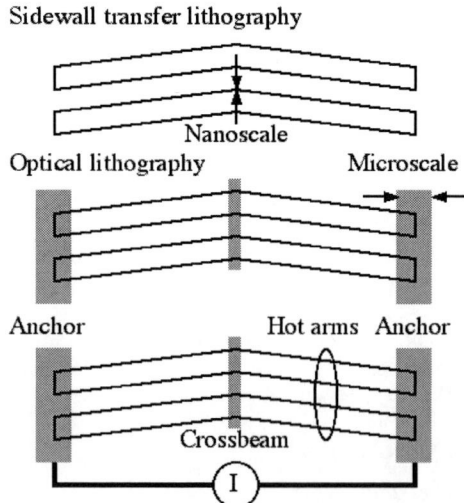

Figure 1: Single-layer STL NEMS process, for an example single-axis electrothermal stage.

Optical lithography is used to form an initial microscale pattern, which is transferred into a shallow silicon mesa. The mesa is coated with a conformal material, and etched to leave only the vertical parts of the coating (which follow the mesa perimeter, and have a width equal to the coating thickness) as a nanoscale mask. Conventional lithography is used to add any necessary micron-scale features, and the combined pattern is transferred into silicon using deep reactive ion etching (DRIE). Suspended parts are then freed to allow motion, by etching of a sacrificial oxide interlayer.

MULTILAYER STL NEMS

Sidewall transfer lithography is potentially attractive for mass production of NEMS, since it allows wafer-scale fabrication of nanoscale features using widely available equipment. However, it suffers from key topological constraints. Since the nanoscale patterns follow closed polygonal perimeters, they must have constant width, be continuous and not be self-intersecting. It is also difficult to combine nanoscale features and separations; the features currently have microscale separations. Overcoming these limitations should extend the range of possible applications. Here we demonstrate a three-layer process involving two STL steps and one additional conventional lithography step that can allow overcome one constraint, namely to allow intersecting nanoscale features, for example as required in a two-axis electro-thermal stage with intersecting suspensions (Fig. 2).

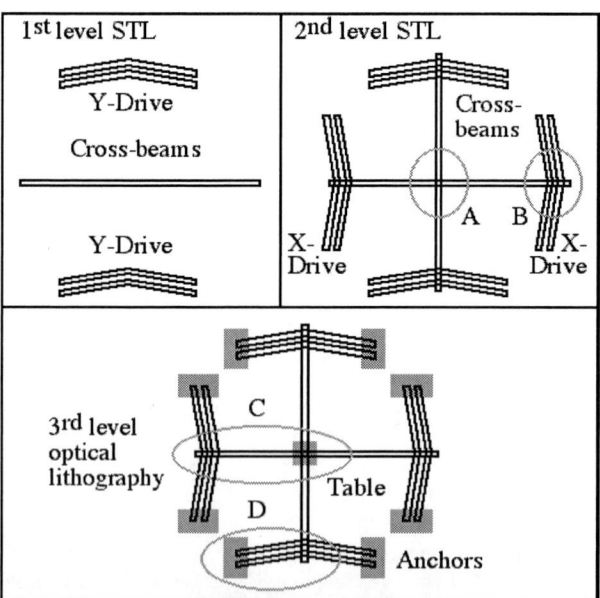

Figure 2: Multi-layer STL NEMS process, for an example two-axis electrothermal stage.

Intersecting nanoscale features can be formed using repetitions of a comparable process based on conventional lithography and reactive ion etching to form a shallow mesa, followed by common deposition of a conformal coating whose horizontal surfaces are then removed. The remaining vertical surfaces form a sidewall mask that now follows the combined perimeter of the overlaid mesa patterns (Fig. 3). A final conventional lithography step can again be used to add microscale features such as anchors, and the whole pattern can be transferred into the silicon as before, using deep reactive ion etching.

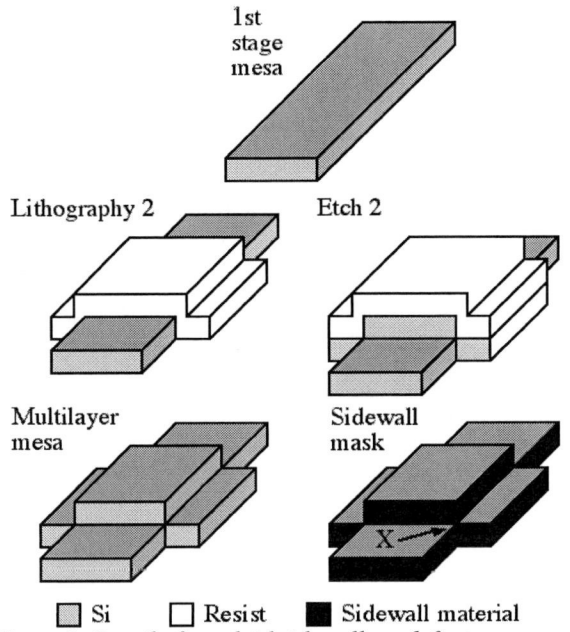

Figure 3: Detail of overlaid sidewall mask features.

The modified process still suffers from several limitations. Firstly, additional photomasks are required, since the nanoscale features must be spread over multiple layers. However, the masks contain relatively large features, and may therefore be low cost. Secondly, overlaid nanoscale features can suffer from microscale alignment errors, and it is important that these do not impact on device operation. Thirdly, overlaid features suffer a gradual degradation in separation, since a thicker resist is required to planarise mesa features formed in earlier stages. Finally, intersecting features must have shallow steps at the surface derived from the original mesa pattern. However, despite these limitations, surprisingly good results can still be obtained.

EXPERIMENTAL RESULTS

NEMS were fabricated in 100 mm diameter bare Si and bonded silicon-on-insulator (BSOI) wafers, using the process flow shown in Fig. 4. Here, for simplicity, overlaid nanoscale features are shown side by side; however, they may clearly intersect one another. The two STL steps were carried out by patterning and etching a first set of mesas (steps 1-3), and then patterning and etching a second set (steps 4, 5). A low-stress conformal metal coating was then deposited, and horizontal surfaces of this metal were removed by directional etching (steps 6, 7). The result was a surface mask defining overlaid nanoscale features. A third conventional lithography step was then carried out to add any microscale parts (step 8). The combined pattern was transferred into the silicon

substrate by deep reactive ion etching (steps 9, 10). Suspended mechanical parts were freed by etching of sacrificial oxide (step 11) and metal was deposited over the entire structure to provide electrical contact (step 12).

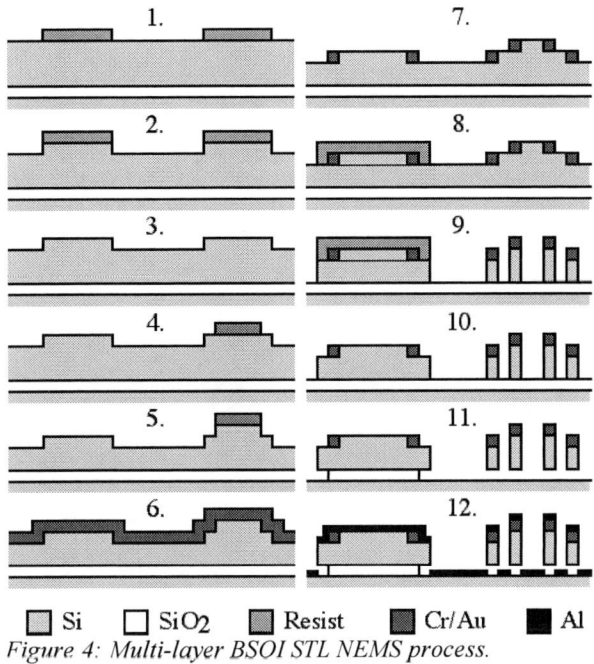

Figure 4: Multi-layer BSOI STL NEMS process.

Patterning was carried out using a Quintel Q7000 mid-UV contact mask aligner. 0.4 μm thickness of Shipley S1805 photoresist was used for patterning the first STL layer; 1.5 μm of Shipley S1813 resist was used for the second, and the same resist was used to define the anchors. Development was carried out using Microposit MF-319, followed by an O_2 plasma descum.

Mesa etching was carried out using a STS single chamber multiplex inductively coupled plasma (ICP) etcher, using a cyclic process based on SF_6 and C_4F_8. The DRIE parameters were 6.8 mTorr pressure, 350 W coil RF power, 11 W platen power and 150 V DC bias, with a 10 sec etch cycle using 50 sccm SF_6 and 5 sccm O_2 followed by a 5 sec passivation cycle using 80 sccm C_4F_8.

Sputter deposition and etching were carried out using a Nordiko RF sputter coater. A conformal layer of 10 nm Cr and 100 nm Au was first deposited, and horizontal layers of this material were then etched away by RF sputtering in Ar gas at 2×10^{-3} mbar pressure. The Au layer was used to form a low-stress metal sidewall mask, while the higher-stress Cr was used to ensure adhesion.

Deep silicon etching was again carried out using the ICP DRIE. However, the cycle time of the etch step was reduced to 4 sec to prevent erosion of the nanoscale beam by scalloping, while the cycle time of the passivation step was similarly reduced to maintain the width of the beam at approximately 100 nm for the full etch depth.

Fig. 5 shows scanning electron microscope photographs of structures at different stages during processing. Fig. 5a shows intersecting mesas formed by two consecutive cycles of patterning a layer of photoresist and then transferring the resulting features into the Si to a depth of 500 nm by DRIE. The terraced nature of the

978-1-4799-7956-1/15 $31.00 © 2015 IEEE

compound mesa may clearly be seen. Figs. 5b and 5c show details of the mesas after deposition of Cr/Au and sputter etching. The sidewall has a stepped structure that follows the original mesa pattern. It has been eroded to roughly half the original mesa height by sputter etching, as has the edge of the Si mesa. However, continuous joints are clearly formed between the two layers of sidewall metal. Figure 5d shows high aspect ratio (HAR) nanoscale features formed after the combined pattern has been transferred into the silicon to a total depth of 5 µm by deep etching. The nanoscale features are again continuous from layer-to-layer. The remaining sidewall mask can be seen at the top of the beam. Despite this, the HAR features are not distorted by residual stress.

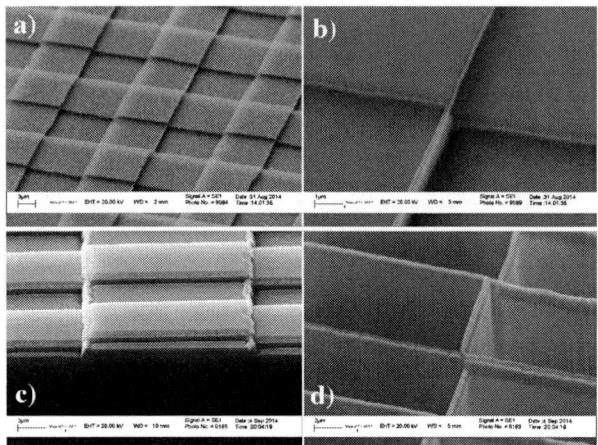

Figure 5: SEM images of a) etched multilayer mesa, b) and c) sidewall mask, d) HAR Si nanostructure.

Using this process, intersecting features such as the crossbeams at position A in the two-axis stage shown in Fig. 2 and the chevron electrothermal drive at B can easily be fabricated. Fig. 6a shows images obtained at A in a part-completed device using a SEM and a Veeco optical interferometer, while Fig. 6b shows the corresponding images obtained at B. Although the interferometer lacks sufficient resolution to visualize the nanoscale structures completely, the 3D views highlight the residual terracing at the top of the beams.

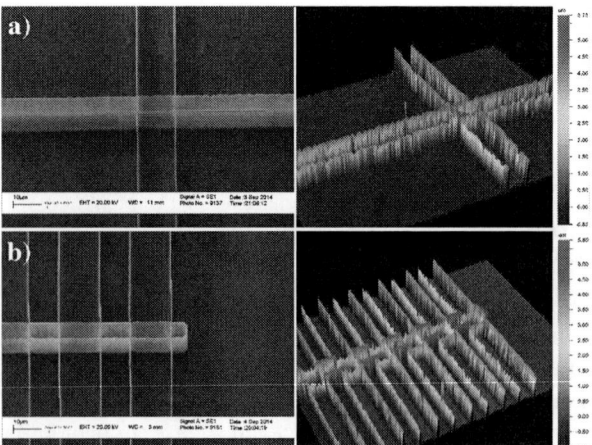

Figure 6: SEM and interferometric images of a) cross-beams (A in Fig. 2) and b) electrothermal drive (B).

We have verified that the results of multilayer sidewall processing are effectively independent of the assignment of features to the two different STL layers. For example, Fig. 7a shows the layout of a set of dies containing nested and stacked nanoscale features. In each case, two die variants are shown, with the features assigned to the two STL layers in reverse order. Figs. 7b and 7c show fabricated structures; the same result is clearly achieved in each case.

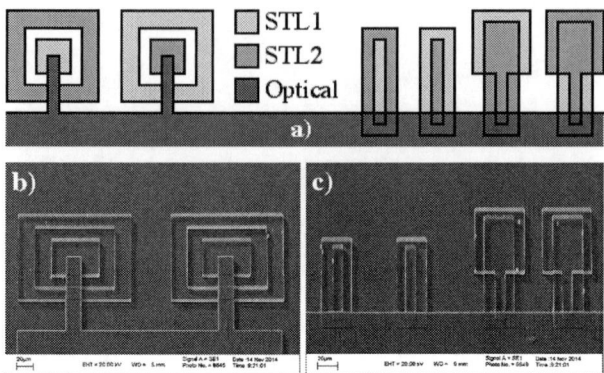

Figure 7: a) CAD layouts for nested (LH) and stacked (RH) nanoscale structures; b) and c) SEM images.

With the addition of microscale anchors, complete device structures can be realized. Fig. 8a shows details of the two-axis stage in Fig. 2 in position C, showing one chevron electrothermal actuator together with the crossbeams and central table, and Fig. 8b shows a close-up of the attachment of the actuator beams to their anchor points. The ends of the polygons defining the beam array are buried in the anchors, so these common features have no effect on the final device.

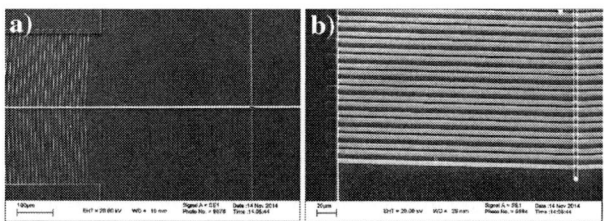

Figure 8: SEM images of a) actuator and table (region C in Fig. 2) and b) anchor (D).

We have also verified that devices may be fabricated on the same wafer using single- and double-layer STL patterning simultaneously. For example, the intersections between the crossbeams and the actuators in the x-y stage in Fig. 2 may instead be achieved using additional link-bars, as shown in Fig. 9a. Similarly, the intersections between the crossbeams and the central table may be achieved by sub-dividing the beams into non-intersecting segments that are linked by the table itself. Figs. 9b and 9c show SEM views of the resulting structures at F and G respectively. Similar static arrangements have clearly been achieved using this alternative layout; however, one disadvantage may be a reduction in dynamical performance due to the inertial mass of the additional microscale components.

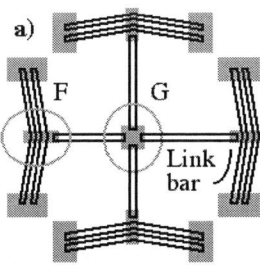

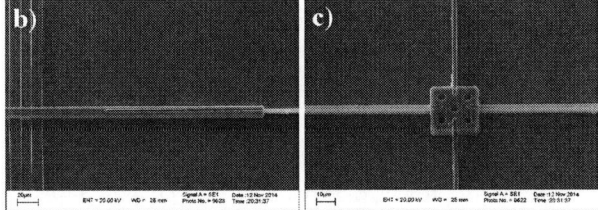

Figure 9: a) Single-layer STL NEMS X-Y stage; b) and c) SEM views at F and G.

PROCESSING ISSUES

The process suffers from several minor difficulties associated with lithography, etching and stress.

Lithography - care is needed to ensure complete planarisation of all mesas during spin coating of resist; failure to do so inevitably results in voids in the structure during DRIE. Similarly care is needed to ensure complete exposure and development of the resulting resist layers, since they must have a varying thickness.

Etching - the edge quality of first set of mesas is slightly degraded by the RIE step used to form the second set. In addition, the physical sputtering used to strip the horizontal metal surfaces does also remove some of the vertical surfaces of the sidewall mask. Generally, this is unimportant, but care is needed to avoid erosion of links between the two metal layers, for example at the point X in Fig. 3. Sputter etching can lead to re-deposition of chromium spots; however, these can be removed using a wet etch step in cerium and dilute nitric acid.

Stress - the metal sidewall mask currently remains in situ after completion of processing, and residual stress can cause some distortion of the underlying Si nanostructure. We have verified that some of these difficulties can be avoided using a SiO_2 sidewall mask, since this can be removed more easily from horizontal layers by RIE and then stripped completely by vapor-phase HF etching, for example during undercut of suspended structures.

Finally, we note that the final metal layer may be localized to the anchors by depositing and patterning the metal after step 7 in Fig. 4, before the final DRIE step.

CONCLUSIONS

A multi-layer sidewall transfer lithography process has been developed for mass parallel fabrication of NEMS. Nanoscale features are spread over multiple photomasks, and repetitive cycles of optical lithography and shallow etching are used to transfer these features into a terraced structure. The sidewalls are coated with a common layer of low stress metal whose vertical surface provides a mask for etching. The combined pattern is transferred into the silicon as a high aspect ratio nanostructure by deep reactive ion etching, together with other microscale features. Sacrificial layer processing then allows formation of movable suspended parts. Using this process, multilayer nanoscale structuring of silicon has been demonstrated with a width of 100 nm and an aspect ratio of 50 : 1.

REFERENCES

[1] H.G. Craighead, "Nanoelectromechanical systems", *Science*, vol. 390, pp. 1532-1535, 2002.

[2] Z.J. Davies, G. Abadal, O. Hansen and X. Borisé, "AFM lithography of aluminium for nano-electromechanical systems", *Ultramicroscopy*, vol. 97, pp. 467-472, 2003.

[3] S.R. Quake and A. Scherer, "From micro- to nano-fabrication with soft materials", *Science*, vol 290, pp. 1536-1540, 2000.

[4] P.B. Grabiec, M. Zaborowski, K. Domanski, T. Gotszalk and I.W. Rangelow, "Nano-width lines using lateral pattern definition for nano-imprint template fabrication", *Microelectr. Engng.*, vol. 73-74, pp. 599-603, 2004.

[5] W.R. Hunter, T.C. Holloway., P.K. Chatterjee and A.F. Tasch, "A new edge-defined approach for submicrometer MOSFET fabrication", *IEEE Electron. Dev. Letts.*, vol. 2, pp. 4-6, 1981.

[6] P. Vettiger, P. Buchmann K. Dätwyler, G. Sasso and B.J. Van Zeghbroeck, "Nanometer sidewall lithography by resist silylation", *J. Vac. Sci. Tech. B*, vol. 7, pp. 1756-1759, 1989.

[7] D.S.Y. Hsu, N.H. Turner, K.W. Pierson and V.A. Shamamian, "20 nm linewidth platinum pattern fabrication using conformal effusive-source molecular precursor deposition and sidewall lithography", *J. Vac. Sci. Tech. B*, vol. 10, pp. 2251-2258, 1992.

[8] U. Hilleringmann, T. Vierigge and J.T. Horstmann, "A structure definition technique for 25 nm lines of silicon and related material", *Microelectr. Engng.*, vol. 53, pp. 569-572, 2000.

[9] K.H. Chung, S.K. Sung, D.H. Kim, W.Y. Choi, C.A. Lee, J.D. Lee and B.G. Park, "Nanoscale multi-line patterning using sidewall structure", *Jpn. J. Appl. Phys.*, vol. 41, pp. 4410-4414, 2002.

[10] Y.-K. Choi, T.-J. King and C. Hu, "A spacer patterning technolgy for nanoscale CMOS", *IEEE Trans. Electron Devices*, vol. 49, pp. 436-441, 2002.

[11] J.T. Horstmann and K.F. Goser, "New fabrication technique for nano-MOS transistors with W = 25 nm and L = 75 nm using only conventional optical lithography", *Microelectr. Engng.*, vol. 61-62, pp. 601-605, 2002.

[12] D.Liu and R.R.A. Syms, "NEMS by sidewall transfer lithography", *J. Microelectromech. Syst.*, DOI 10.1109/JMEMS.2014.2313462, 2014.

CONTACT

*R.Syms, tel: +44-2075946203; r.syms@imperial.ac.uk

NEW SCALABLE MICROFABRICATION METHOD FOR SURFACE ION TRAPS AND EXPERIMENTAL RESULTS WITH TRAPPED IONS

S. Hong[1], M. Lee[1], H. Cheon[1], J. Ahn[2], M. Kim[2], T. Kim[2] and D. D. Cho[1]

[1]ISRC/ASRI, Department of Electrical and Computer Engineering, Seoul National University, Republic of Korea

[2]Quantum Tech. Lab, SK Telecom, Republic of Korea

ABSTRACT

This paper presents a new microfabrication method for surface ion traps and experimental results with trapped ions. Fabricating ion trap chips is a very formidable task because the top electrodes are vertically spaced more than 10 μm from the bottom electrodes with an indented dielectric layer in the middle. Previous ion traps were fabricated using TEOS timed etch or tungsten sacrificial etch techniques. This paper presents a new microfabrication method, using copper as a sacrificial material for an aluminum-oxide-aluminum ion trap structure. Using the developed method the overhang dimensions of the top aluminum electrodes can be accurately controlled. Fabricated ion trap chips are installed in a 1×10^{-11} Torr vacuum environment for ion trapping experiments. Successful results in trapping strings of $^{171}Yb^{+}$ and $^{174}Yb^{+}$ ions as well as manipulating $^{171}Yb^{+}$ ions for qubit operation are demonstrated.

INTRODUCTION

Ion traps are one of the promising physical implementations of quantum information processing (QIP), because of the long coherence time and the capability of individual qubit manipulation [1]. Earlier ion traps were built by conventional machining and manual assembling. However, to implement advanced quantum algorithms, arrays of ion traps need to be integrated [2]. For this purpose, various MEMS fabrication technologies are being applied. Several designs of microfabricated ion trap have been reported, until settling to the current "surface ion trap" design that utilizes aluminum surface electrodes with a thick insulating dielectric layer [3, 4].

One of the practical challenges in the microfabrication of the surface ion trap is the realization of overhang structures at the edges of electrodes. The overhang structures shadow the upper parts of the dielectric pillars to reduce the disturbance from the built-up charges on the sidewall of the dielectric pillars. The dielectric pillars mechanically support the top electrodes and must be sufficiently thick to prevent electrical breakdowns when large RF voltages are applied between the top and bottom electrodes. Surface ion traps with the overhung electrodes have been used by two groups [5, 6]. One approach seems to have fabricated overhung electrodes by depositing four layers of plasma enhanced chemical vapor deposited (PECVD) tetraethyl orthosilicate (TEOS) films, followed by wet etching to define the overhang [5]. This method results in jagged and slanted sidewalls. In addition, the variations of overhang dimensions are inevitable. In the other approach, tungsten is used as a sacrificial layer [6]. The first surface ion trap developed by our group is similar to the first approach but uses four layers of PECVD SiO$_2$ films to construct 14-μm thick dielectric pillars. The PECVD SiO$_2$ films are measured to have substantially less residual stresses than PECVD TEOS films, which can allow easier handling during ensuing process steps. However, our results also showed jagged sidewalls as well as variations in the overhang length [7].

In this paper, a new fabrication method for surface ion trap chips is presented. We use PECVD SiO$_2$ films as dielectric pillars and copper as a sacrificial layer to fabricate straight-line dielectric pillars and accurately-patterned electrode overhangs. Copper has been used in various MEMS applications as a sacrificial material to release polymer [8] and nickel [9], but our work is the first time copper is used to release aluminum-oxide-aluminum sandwich structures fabricated on a silicon wafer. Using the microfabricated surface trap chips, we performed experiments on trapping ions in an ultra-high vacuum (UHV) environment. This paper shows detailed fabrication method and ion-trapping experimental results.

DESIGN OF ION TRAP

In the surface ion trap, all electrodes are fabricated in the same plane as shown in the schematic (Fig. 1). The radial confinement of the ion is achieved using the two planar RF electrodes, and the position of the RF null where the ion is trapped is placed above the two RF electrodes. Multiple DC electrodes are also fabricated outside the RF electrodes. These outer DC electrode function as the RF ground. Also, by applying different control voltages to these outer DC electrodes, the trapped ion can be axially

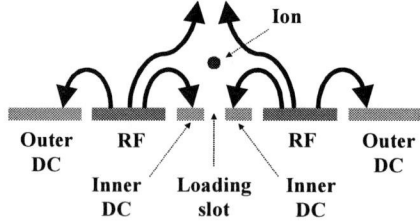

Figure 1: Schematic of a surface ion trap. The red and blue rectangles indicate RF and DC electrodes respectively. The curved arrows denote the direction of electric field when RF voltage is positive.

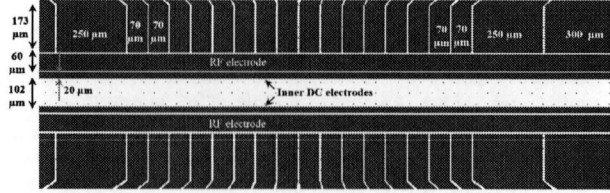

Figure 2. Layout and dimensions of the top electrodes. All the gaps between electrodes are 8 μm.

confined. Furthermore, by applying time-varying control voltages to the outer DC electrodes, the trapped ion can be moved or shuttled along the axis. The inner DC electrodes also function as the RF ground. However, the main function of the inner DC electrode is to keep the DC potential at the RF null point minimum, when the voltages applied to the outer DC electrodes are varied. In the schematic neutral atoms which are photoionized by diode lasers are loaded through the loading slot between the two inner DC rails.

The dimensions of the electrodes for our trap chip are shown in Figure 2. This design is based on that of Sandia's "Thunderbird" chip [6] which has been used successfully by many different research groups through their collaborative research program. We further optimized electrode dimensions to increase the trap depth and decrease the q parameter. We also designed the electrodes with rounded corners. The details of design optimization are beyond the scope of this paper and are omitted for brevity.

FABRICATION
Fabrication Process

The detailed fabrication process of the developed copper sacrificial layer method is shown in Figure 3. First, dielectric layers which consist of 0.5-μm thermally grown silicon dioxide and 0.2-μm low pressure chemical vapor deposited (LPCVD) silicon nitride are formed. These dielectric layers isolate the bonding pads from the silicon substrate. Above the dielectric films, a 1.5-μm thick aluminum ground layer is deposited. The aluminum ground layer and the dielectric layers are photolithographically patterned in sequence, to provide bonding pads for DC electrodes and a silicon opening for penetrating the loading slot at the end of the fabrication process (Fig. 3(a)). A 14-μm thick SiO_2 film pillar is fabricated by a PECVD process in four layers to control the residual stress. The SiO_2 film is dry etched to form dielectric pillars (Fig. 3(b)). Generally, an RF voltage of several hundred volts is applied to the RF rail, and to prevent electrical breakdowns thick dielectric layers are required. A 300-Å titanium film and a 1000-Å copper film are deposited as the seed layer for the ensuing copper electroplating process. The electroplated copper and the PECVD SiO_2 are planarized by a chemical mechanical polishing (CMP) process (Fig. 3(c)). A 1.5-μm thick aluminum electrode layer is deposited and dry etched, and the boundaries of the electrodes are defined (Fig. 3(d)). In this step, the overhang length is determined from the lateral distance between the electrode patterns and oxide pillar patterns. After patterning the electrodes, the silicon substrate is etched by a deep reactive ion etching (DRIE) from the backside. The backside DRIE step is essential to prohibit an excessive process time of the frontside DRIE process to penetrate the loading slot. The overhang structures are released by the removal of the copper sacrificial materials and the seed layers, using a wet etching process (Fig. 3(e)). A copper etchant (AP-100, Trensene) and a titanium etchant (TiW-30, Trensene) are chosen because they remove the target materials without damages to the aluminum and SiO_2 parts. Finally, the

fabrication process is completed by penetrating the loading slot through the frontside DRIE process (Fig. 3(f)).

Fabrication Results

A top view of the mircorfabricated surface ion trap is shown in Figure 4(a), which identifies a pair of RF electrodes, a pair of inner DC rails, segmented DC control electrodes, and a loading slot. Figure 4(b) shows the cross-section of the microfabricated surface ion trap. The vertical sidewalls of oxide pillars are straight and the overhang length is approximately 2 μm. Note that if the overhang length is too large, the top aluminum electrode can bend under large applied voltages.

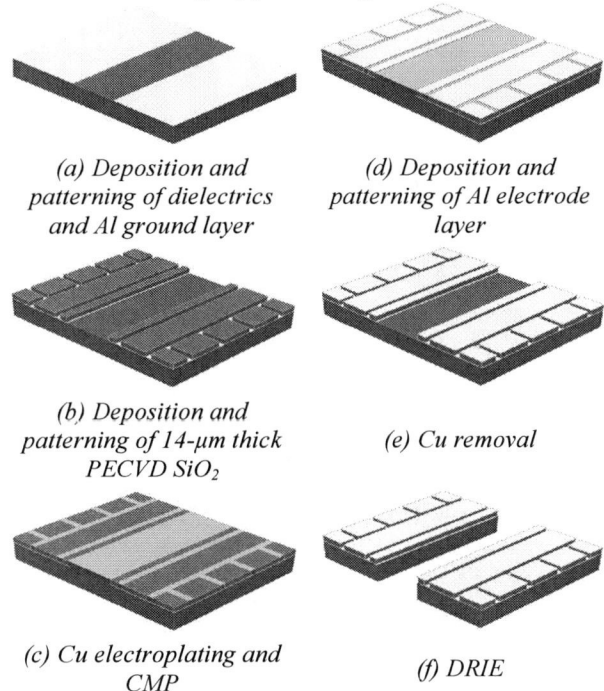

(a) Deposition and patterning of dielectrics and Al ground layer

(d) Deposition and patterning of Al electrode layer

(b) Deposition and patterning of 14-μm thick PECVD SiO_2

(e) Cu removal

(c) Cu electroplating and CMP

(f) DRIE

Figure 3: Fabrication process flow of the copper sacrificial layer method.

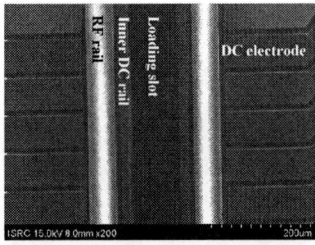

(a) Top view of the fabricated trap chip.

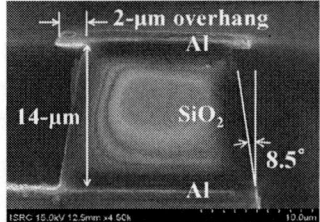

(b) Cross-section of fabricated trap chip.

Figure 4: Scanning electron micrograph (SEM) images of the fabricated surface ion trap chips.

978-1-4799-7956-1/15 $31.00 © 2015 IEEE   293

EXPERIMENT

Experimental Setup

Figure 5 shows a schematic of our experimental setup. A microfabricated chip is mounted on the commercial ceramic chip carrier (IPKX0F1-8180BA, NTK) using an epoxy compound, and gold wires are used to connect the bonding pads to the chip carrier. The ion trap package is installed in an ultra-high vacuum (UHV) chamber, and the pressure of the UHV chamber is lowered to 1×10^{-11} Torr. A metallic ytterbium source with an oven is located at backside of the ion trap package to provide neutral atoms. The ion trap package is electrically connected to the outside of the UHV chamber by feed-throughs. The RF signal generated by a function generator is amplified by a helical resonator and applied to the RF electrodes of the surface trap chip. The control DC voltages are generated using a digital-to-analog converter (DAC) which is controlled by a field-programmable gate array (FPGA) board, and applied to the DC electrodes via the electrical feed-throughs. To image and count the photons emitted from the trapped ions, an electron multiplying charge coupled device (EMCCD) and photon multiplier tube (PMT) are used, respectively. To trap and manipulate the ytterbium ion four diode lasers are used, which is a well-defined procedure [10]. A 399-nm laser is used to photoionize neutral ytterbium atoms. A 935-nm and a 638-nm laser are used for the Doppler cooling. A 369.5-nm laser is used in the photoionization step, the Doppler cooling step, and the qubit operation step including state initializations and detections. A microwave horn is equipped outside the UHV chamber to generate a resonant microwave field which induces Rabi oscillation of the ^{171}Yb$^+$ ion qubit.

Experimental Results

By applying an RF signal with a peak-to-peak voltage of approximately 320 V at 25.0 MHz, up to three ^{171}Yb$^+$ ions and up to six ^{174}Yb$^+$ ions have been trapped. Figure 6 shows the EMCCD image of the trapped ion string which consists of three ^{171}Yb$^+$ ions. The lifetime of the trapped ions measured to be over 24 hours with laser cooling. As shown in Fig. 7, the internal state of the ion qubit can be manipulated using the diode lasers and the microwave field. After Doppler cooling (Fig. 7(a)), all the ions are initialized to the $|0\rangle$ state (Fig. 7(b)). The probabilities of $|0\rangle$ and $|1\rangle$ state are determined by the applying time of the microwave field (Fig. 7(c)), and statistically estimated through the detection step (Fig. 7(d)). Figure 8 shows the experimental

results of the Rabi oscillation of the qubit internal state. Each probability is estimated by iterative experiments over 10,000 times. In Figure 8, it is also shown that if a detuning of the microwave exists, the maximum probability of $|1\rangle$ cannot reach one hundred percent.

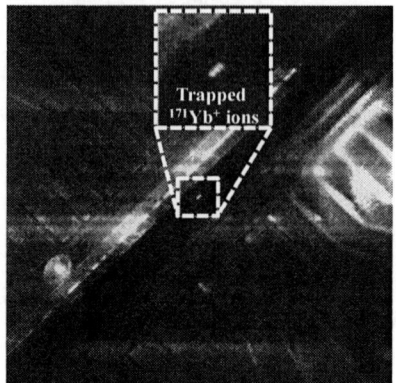

Figure 6. EMCCD image of trapped ^{171}Yb$^+$ ions (image of the electrode was overlapped for better comprehension of the picture)

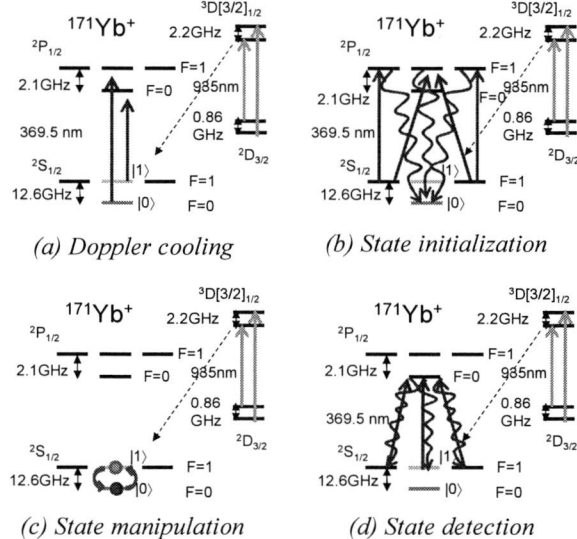

(a) Doppler cooling   (b) State initialization

(c) State manipulation   (d) State detection

Figure 7. Experimental procedure of single qubit state manipulation.

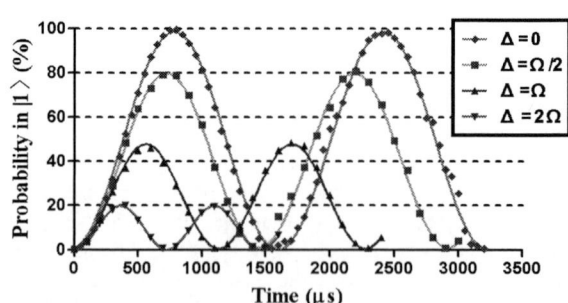

Figure 8. Experimental results of Rabi oscillations induced by 12.6428 GHz microwave with various detuning (Δ), where Ω is Rabi frequency in the resonance case (625 kHz in our experiments)

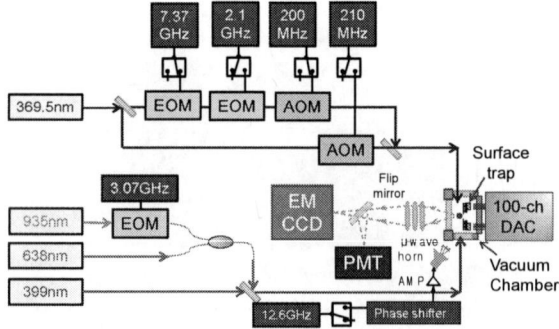

Figure 5. Schematic of experimental setup

CONCLUSION

In this paper, a detailed fabrication method for surface ion trap chip using copper sacrificial layer is presented. Dielectric pillars of 14-μm thick PECVD SiO_2 films are patterned, and copper is electroplated as sacrificial layer. Then, an aluminum layer is deposited and patterned to provide top electrodes. By removing the copper sacrificial layer through a wet-etching process, the overhang structures of the electrodes are fabricated. The trap chip fabricated by this method has a well-controlled straight-line sidewall profiles. The length of the electrode overhang for reducing the disturbance from the built-up charges on the dielectric layers is also controlled to approximately 2 μm. Ion strings with up to three $^{171}Yb^+$ ions and up to six $^{174}Yb^+$ ions have been trapped above our microfabricated surface ion traps, and the lifetime of the trapped ions are over 24 hours with the cooling lasers on. In addition, single qubit operations using a $^{171}Yb^+$ ion are achieved. Rabi oscillations of the qubit state are induced by a microwave, and the qubit state can be clearly distinguished from the photon statistics.

REFERENCES

[1] C. Monroe and J. Kim, "Scaling the ion trap quantum processor," Science, vol. 339, no. 6124, pp. 1164-1169, 2013.

[2] D. Kielpinski, C. Monroe, and D. J. Wineland, "Architecture for a large-scale ion-trap quantum computer," Nature, vol. 417, no. 6890, pp. 709-711, 2002.

[3] S. Seidelin et al., "Microfabricated surface-electrode ion trap for scalable quantum information processing," Phys. Rev. Lett., vol. 96, no. 25, pp. 253003, 2006.

[4] K. Wright et al., "Reliable transport through a microfabricated X-junction surface-electrode ion trap," New J. of Phys., vol. 15, no. 3, pp. 033004, 2013.

[5] D. R. Leibrandt et al., "Demonstration of a scalable, multiplexed ion trap for quantum information processing," Quant. Inform. Comput., vol. 9, no. 11, pp. 901-919, 2009.

[6] D. Stick et al., "Demonstration of a microfabricated surface electrode ion trap," arXiv:1008.0990v2, 2010.

[7] T. Kim et al., "Development of quantum repeater based on ion trap," In proceeding of the 4th International Quantum Optics Workshop, 2013.

[8] J. Park and M. Allen, "Development of magnetic materials and processing techniques applicable to integrated micromagnetic devices," J. of Micromech. and Microeng., vol. 8, no. 4, pp. 307, 1998.

[9] P. Zavracky et al., "Micromechanical switches fabricated using nickel surface micromachining," J. of Microelectromechanical Systems, vol. 6, no. 1, pp. 3-9, 1997.

[10] S. Olmschenk et al., "Manipulation and detection of a trapped Yb^+ hyperfine qubit," Physical Review A, vol. 76, no. 5, pp. 052314, 2007.

CONTACT

*D. Cho, tel: +82-2-880-8371; dicho@snu.ac.kr

POST-RELEASE STRESS-ENGINEERING OF SURFACE-MICROMACHINED MEMS STRUCTURES USING EVAPORATED CHROMIUM AND IN-SITU FABRICATED RECONFIGURABLE SHADOW MASKS

Ratul Majumdar[1], Vahid Foroutan[1], and Igor Paprotny[1]

[1]Department of Electrical and Computer Engineering,
University of Illinois, Chicago, IL, USA

ABSTRACT

We present a novel *post-release* stress-engineering process that can provide out-of-plane curvature to initially planar MEMS structures. This process uses a layer of evaporated Chromium (Cr) with intrinsic compressive stress to provide the required curvature. The stressor layer is applied *post-release*, and is patterned using shadow masks fabricated *in-situ* as part of the surface micromachining fabrication process. The masks can be reconfigured to provide variable stressor layer coverage of the underlying structures without the use of photolithography. A model is presented that encompasses the increase in deflection, and thus non-uniform chromium coverage, during the stress-engineering of released structures.

INTRODUCTION

Stress-engineering has been previously used to generate out-of-plane curvatures of initially planar microstructures to construct microturbines [1], stress-engineered MEMS microrobots [2], or catalytic microengines. In all these applications, a stressor layer is deposited onto initially planar microfabricated structures. As deposited, the stressor layer has an intrinsic stress (either tensile or compressive), which causes the structures to bend. The stressor layer can be patterned to precisely define the coverage geometry, and thus the deflection of the stress-engineered structure [2]. Traditionally this is done prior to release using photolithography. However, such approach has two potential drawbacks. It limits the application of the stress-engineering to post-processing devices received from a third part foundry [3] and may results in large amount of metal being exposed during the release etch, potentially promoting galvanic attack if a wet process is used.

This paper describes a novel post-release stress engineering process that is based on blanket deposition of the stressor layer on released surface micromachined structures. A model for post-release stress-engineering is presented, along with the theoretical and experimental results verifying the new model, and the use of reconfigurable *in-situ* fabricated shadow masks which are used to pattern the stressor layer.

POST-RELEASE STRESS-ENGINEERING

Out-of-plane deflection of planar structures such as the steering arm of a MicroStressBot to vary the pull-down and release voltages to enable independent control [2], or to promote static stability of stress-engineered MEMS microfliers [4].

The stress-engineering process described in this work involves the deposition of a 50-150 nm thick layer of evaporated Chromium (Cr) on initially planar structures. After deposition, Cr has an intrinsic compressive residual stress, which induces out-of-plane curvature.

The stress-engineering model presented in [1] assumes that stress-engineering occurs prior to structural release. Consequently, the stress-engineering model was extended to account for the gradual bending of fabricated microstructures during blanket Cr-deposition onto released structures. This model uses a numerical finite element approximation, and applies to a planar cantilever. Assume the analysis is done on a cantilever of length l which is divided into b segments of equal length. Within each segment the chromium thickness after deposition is assumed to be constant. The total deposition time is divided into equal time intervals of duration c. The initial stress (σ_{ini}) is calculated with the flat cantilever. For each time step j and for each segment i the deflection $d(j,i)$, radius of curvature $\rho(j,i)$ and Cr thickness $h_2(j,i)$ is calculated as:

$$d(j,i) = \rho(j,i-1) - \rho(j,i-1)\cos\left(iL/b\rho(j,i-1)\right), \tag{1}$$

$$\rho(j,i) = \frac{6E_1h_1^2E_2h_2^2(j,i) + (4E_1h_1^3E_2 + 3E_1E_2h_1^3)h_2(j,i) + E_1^2h_1^4}{6h_1h_2^2(j,i)E_1\sigma_{ini} + 6h_1^2E_1h_2(j,i)\sigma_{ini}}, \tag{2}$$

$$h_2(j+1,i) = h_2(j,i) + \left(cr\cos(\theta(j,i))\right). \tag{3}$$

For the next segment $i+1$ in the same time interval j the Cr thickness $h_2(j,i+1)$ can be calculated from the angle of deflection $\theta(j,i)$. This iterative process is continued for consecutive time steps until total time of deposition is reached. Calculated cantilever profiles based on the model from [1] (pre-release) and eqs. (1) − (3) (post-release) is shown in Fig. 1.

The post-release stress engineering model matches well with cantilever profiles extracted experimentally. Fig. 2 shows the profiles from two cantilevers pairs (one cantilever was partially shadowed using a shadow mask) subject to 75 nm (a and b) and 150 nm (c and d) Cr stressor layer. The dashed line show the measured profile using interferometric micrography.

IN-SITU FABRICATED RECONFIGUBLE SHADOW MASKS

In order to achieve precise deflection of the stress-engineered structures, the stressor layer must be precisely patterned to define the desired out-of-plane deflection. In-situ fabricated shadow masks (shadow masks fabricated on die using the same structural material) are used to

978-1-4799-7956-1/15 $31.00 © 2015 IEEE

define the deposition of the stressor layer in absence of lithography. Two types of shadow masks are used: (a) *sacrificial shadow masks* where mask segments are removed and discarded to expose the desired structure to the stressor layer and (b) *hinged shadow masks*, where the masks can be folded over the desired structure to shield it from Cr deposition.

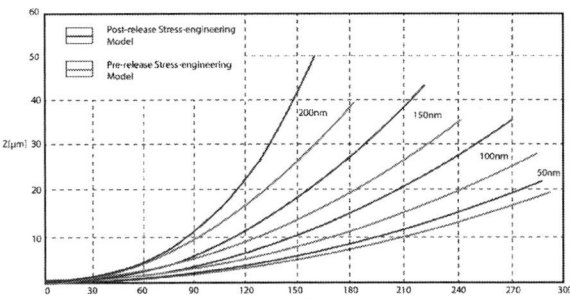

Figure 1: *Calculated variation in profiles for the pre-released [1] (red) and post-released (blue) stress engineering model for different Cr thickness(50 nm,100 nm,150 nm and 200nm) measured for constant stress of 680 MPa.*

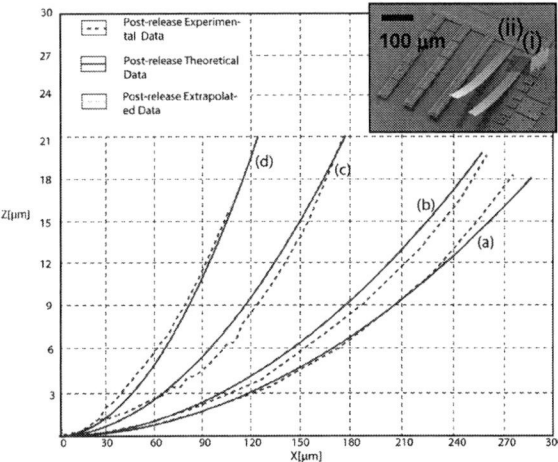

Figure 2: *Profiles (a-d) showing a comparison between the experimental (dotted) and theoretical data (continuous) for post-release stress engineering model (Eq. (1) - (3)) for cantilevers (i) and (ii) (inset). Profiles (a) and (c) are for cantilever (ii) for Cr thickness of 75 nm and 150 nm, respectively, while (b)and (d) are for cantilever (ii) for Cr thickness of 75 nm and 150 nm, respectively. (Cantilever (i) was partially shadowed during Cr deposition.)*

Fig. 3 shows the arrangement of the sacrificial masks on the test structures and the subsequent deflection of the test structures after their removal and deposition of Cr-layer. The areas covered by the shadow masks (which were removed prior to imaging) are clearly seen as darker shadows (less conductive) areas on the image. The shadow masks were removed manually using micro probes.

Shadow masks with hinges at the base allow them to be fabricated without the need for additional polysilicon layers on top of the structures that are patterned, however these mask might allow for less precise patterning of the underlying Cr stressor layer. The hinged masks are simply rotated in place to provide adjustable coverage of the fabricated structures. Fig. 4 shows several types of hinged shadow mask structures used for protecting parts of the structures from blanket deposition of the stressor layer.

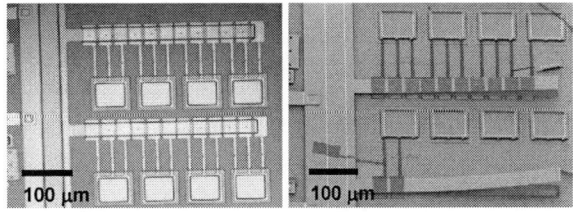

Figure 3: *Sacrificial shadow masks: (left) Optical micrograph of polysilicon cantilevers with sacrificial modular shadow masks. (right) SEM of the same cantilevers with their curvature defined by the removal of elements from the sacrificial shadow masks.*

Figure 4: *Hinged shadow masks: (left) Segmented hinged shadow masks providing partial coverage for cantilever beams. (right) Large hinged shadow mask structures for patterning the arms of MicroStressBots.*

The shadow masks shown in this work require an external force for either removal or flipping before the Cr layer deposition process. This force can be provided manually with help of microprobes. To facilitate automatic reconfiguration of the shadow masks, scratch drive actuators (SDA) can perhaps be utilized which generates enough force to break the sacrificial shadow masks or flip the hinges. Such automatic reconfiguration is currently under development in our laboratory.

Applications of this novel process includes fabrication of stress engineered MEMS microrobots and MEMS microfliers. Fig. 5.left shows precise patterning of two MicroStressbots without affecting the backplate of SDA. Fig. 5.right shows a chassis of a stress-engineered MEMS microflier after the deposition and patterning of the stressor Cr layer. Convex chassis of the microfliers are necessary to maintain in-flight stability [4]. The sacrificial shadow masks is designed in such a way that the main body (i) curls up leaving the wings (ii) perfectly flat.

978-1-4799-7956-1/15 $31.00 © 2015 IEEE

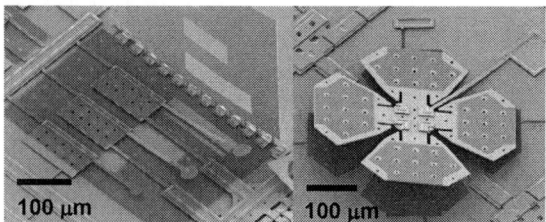

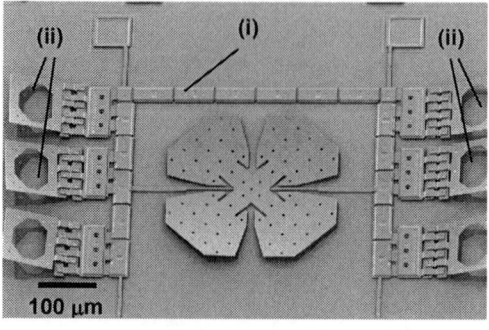

Figure 5: SEM of two applications of in-situ masked post-release stress engineering. (left) A hinged shadow mask is used to define the curvature of the steering arms for stress-engineered MEMS microrobots. (right) Sacrificial masks are used to define the curvature of an untethered MEMS microflier [4].

STRESSOR-RESITING STRUCTURES

During a blanket deposition of a stressor layer using physical vapor deposition, all structures within line-of-sight of the source are covered with the stressor material. However, it may be desirable to have structures that are immune to stress-engineering, i.e. structures that do not curve, or curve very little during the application of the stressor layer. For example, batched transfer structures called the transfer frames are fabricated attached to the microrobots (Fig. 6) or microfliers (Fig. 7) and are used to transport devices over to their operating environment. It is desirable that the body of such transfer frames do not bend after the application of the stressor layer.

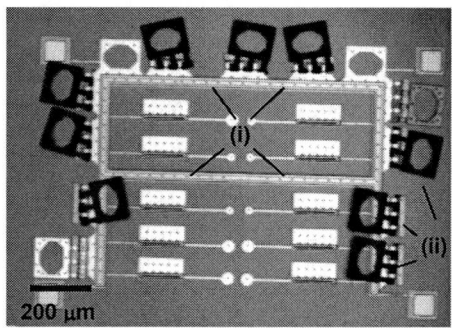

Figure 6: Optical Micrograph of a transfer frame used to transport the MicroStressBots to the operating environment. The frame consists of the body (i) and hinges (ii) used to secure the microprobe tips prior to transport.

In order to minimize the effect of the stressor layer on the exposed frame chassis, we have developed structures which are designed to resist curving after blanket deposition of the stressor layer. These structures are in essence covered with permanently attached shadow masks. These shadow masks prevent the reducing its undesired curvature. Table 1 shows the configuration of permanent shadow masks used on the transfer frames and their respective cross sections. The values of radius of curvature for the three cross sections demonstrate the effectiveness of each structure in withstanding the stress from Cr layer.

Figure 7: Transfer frame attached to a microflier after CR deposition. The frame consists of the body (i) and hinges (ii) used to secure the microprobe tips prior to transport.

In order to minimize the effect of the stressor layer on the exposed frame chassis, we have developed structures which are designed to mitigate the effects of the stressor layer on their curvature. These structures have various I-beam type cross sections, and are in essence covered with permanently attached shadow masks using an additional structural layer. These shadow masks prevent accumulation of stress the body of the mask, reducing the undesired curvature. Table 1 shows the configuration of permanent shadow masks used on the transfer frames and their respective cross sections. The values of radius of curvature for the three cross sections after the deposition of 75 nm of Cr illustrate the effectiveness of each structure in withstanding the stress from the blanket deposited stressor layer.

Table 1: Cross-sections of cantilevers (i)-(iii) from Fig. 2 above, with their respective radii of curvature (r) after the deposition of 75 nm of Cr.

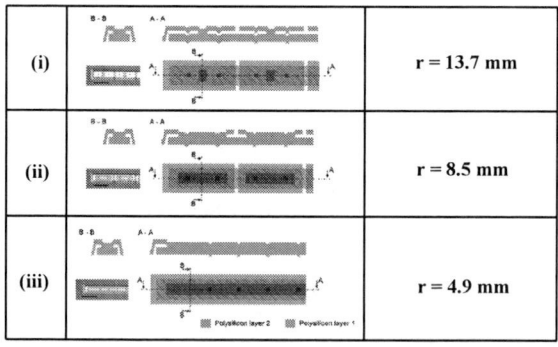

(i)		r = 13.7 mm
(ii)		r = 8.5 mm
(iii)		r = 4.9 mm

The recorded data showing the profiles for cantilevers (i) - (iii) after blanket deposition of 75 nm (red) and 150 nm (blue) Cr is displayed in Fig. 8. The apparent resistance to the stressor layer deposition of cantilever (i) is obvious. The variable radius of curvature is also visible on the SEM image on Fig. 9.

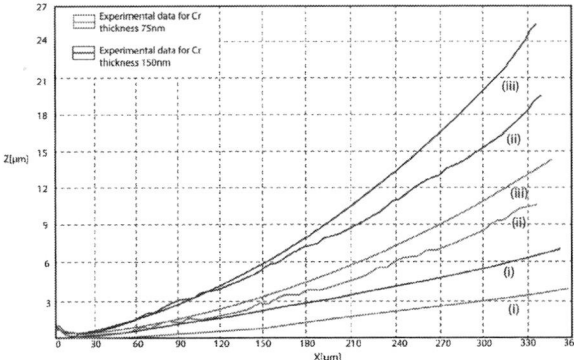

Figure 8: Difference in radius of curvature during blanket Cr deposition of 75 nm (red) and 150nm (blue) for cantilevers (i) – (iii) The resistance of cantilever with profile (i) to blanket stressor layer deposition is clear.

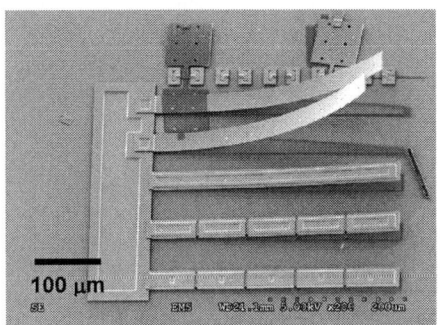

Figure 9: SEM image showing the difference in radius of curvature after blanket Cr deposition for cantilevers (i) – (iii) with varying cross sections. The resistance of cantilever with profile (i) to blanket stressor layer deposition is clear.

Non-Uniform Cantilever Cross-Section

In order to model the non-uniform cross-sections (i) – (iii) in table 1 the test structure is assumed to be made of two segments (1) composite I-beam and (2) composite rectangular beam as shown in Fig. 10. For simplification each segment is considered as a symmetric I-beam with uniform Cr deposition on the surface. The rectangular beams between two I-beams are relatively of small length compared to the length of I-beams and length of the test structure. Hence, stress engineering model assuming uniform Cr deposition can be used. The thickness of the stressor layer is considered to be much less than the actual test structure so that the neutral axis is approximately taken along the middle of the I-beam. The radius of curvature $\rho^{'}$ and ρ for two segments (1) and (2) can be calculated as:

$$\rho^{'} = \frac{E_2 h_2^3 w + E_1[b_1(2d_s + d_s^{'})^3 - b_1^{'} d_s^{'3}]}{6\sigma[2d_s b_1 + d_s^{'}(b_1 - 2b_1^{'})][h_2 + d_s + d_s^{'}]} \quad , \quad (4)$$

$$\rho = \frac{6E_1 d_s^2 E_2 h_2^2 + (4E_1 d_s^3 E_2 + 3E_1 E_2 d_s^3)h_2 + E_1^2 d_s^4}{6d_s h_2^2 E_1 \sigma + 6d_s^2 E_1 h_2 \sigma} \quad (5)$$

The radius of curvature in both the cross sections (i) and (ii) is varied by varying parameter $b_1^{'}$ (whereas cross section (iii) is achieved by setting $b_1^{'}$ and b_2 to zero.) Increasing $b_1^{'}$ parameter leads to increasing radius of curvature which can be seen clearly in eq. (5) while keeping all the other parameters constant. By inserting values of all the parameters and computing $\rho^{'}$ and ρ, it can show that radius of curvature of segment (1) is greater than (2). This leads to flattening of the test structure after Chromium deposition thus nullifying the intrinsic stress of the stressor layer. The radius of curvature shown in table 1 confirms the theory presented in this section.

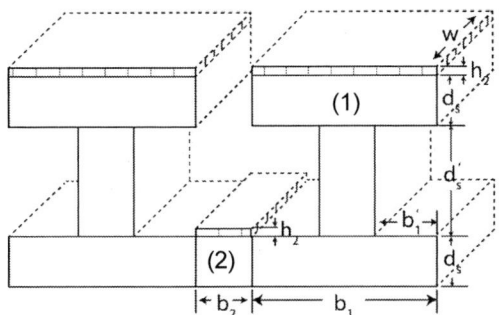

Figure 10: A part of the test structure with two composite I-beams (1) separated by a composite rectangular beam (2) along with the parameters used to calculate the combined radius of curvature.

CONCLUSION

In this work, we presented a novel post-release stress-engineering process for surface-micromachined MEMS structures. The stressor layer is patterned using in-situ fabricated reconfigurable shadow masks. We have shown a model that takes into account the continuous change in curvature as the stressor layer is deposited onto a released structure. Future work includes embedded actuators that automatically reconfigure (adjust) the shadow masks for a desired planar deflection either prior to, or during, the stressor layer deposition process.

REFERENCES

[1] C.-L. Tsai and A. K. Henning, "Out-of-plane microstructures using stress engineering of thin films," Proc. Microlithogr. Metrol. Micromach., vol. 2639, pp. 124–132, 1995

[2] B. R.Donald, , C. Levey, C. McGray, I. Paprotny, and D. Rus. An Untethered, Electrostatic, Globally-Controllable MEMS Micro-Robot. Journal of Microelectromechanical Systems (J.MEMS), 2006; 15(1):1-15

[3] D. Koester, A. Cowen, R. Mahadevan, M. Stonefield, and B. Hardy, PolyMUMPs Design Handbook, A MUMPs Process, 10th ed. Grenoble, France: MEMSCAP, 2003.

[4] Foroutan V., Majumdar R., Mahdavipour O., Ward S. P., and Paprotny I., "Levitation of Untethered Stress-Engineered Microflyers using Thermophoretic(Knudsen) Force," in the Technical Digest of the Hilton Head Workshop 2014: A Solid-State Sensors, Actuators and Microsystems Workshop, (2014), pp: 105-106

CONTACT

*I.Paprotny, tel: +1(802)996-1924;paprotny@uic.edu

978-1-4799-7956-1/15 $31.00 © 2015 IEEE

SELF-ASSEMBLY OF MICROCOMPONENTS USING THE ENTROPIC EFFECT

U. Okabe, T. Okano, and H. Suzuki
Faculty of Science and Engineering, Chuo University, JAPAN

ABSTRACT

We propose the use of the entropic effect (the depletion volume effect), which is at work in the assembly of biomolecules, for the self-assembly of artificially engineered microcomponents. When the solution contains macromolecule at relatively high concentration, microcomponents formed assembled structures. The bonding energy is not originated from the surface; it is generated by increasing the translational entropy of macromolecules in the solution. We expect that use of the depletion volume effect promotes the search of the global free-energy minimum in the system by avoiding being trapped to the local minima in the self-assembly process.

INTRODUCTION

Engineered self-assembly has been one of the major challenges in MEMS societies since it is expected to realize massively parallel manufacturing of heterogeneous parts in milli, micro, and nano-scale [1]. Most of the MEMS self-assembly studies utilize the capillary force (*e.g.*, hydrophilic/hydrophobic patterning, adhesives, and low melting-point metals) [2] [3] and electrostatic force [4]. Although many successful results are reported, the self-assembled structures achieved to date are relatively simple; *e.g.*, the number of component types has been only one to several. Since the capillary and electrostatic forces are relatively strong especially in small scales, misaligned components cannot escape from the local free-energy minima. Due to this reason it has been difficult to design the self-assembly system that has specific bonding patterns. In turn, biological molecules (*e.g.*, proteins) and cellular components self-assemble into very specific structures. One of the reasons is because they utilize weak bonds (e.g., van der Waals and hydrogen bonds) as well as the entropic force, enabling the search of the global minimum.

THEORY

We here propose the use of weak bonding force generated by the entropic effect. The classical hypothesis suggests that, when macromolecules (or any small particles with overwhelming abundance) and other relatively larger objects are both present in solution (Fig. 1a), larger objects aggregate together to reduce the volume around them, which is limited by the gyration radius of macromolecules (depletion volume, V_{dep}) (Fig. 1b) [5, 6]. This transition is favored because it increases the volume where macromolecules are able to move freely ($V_{free}= V_{system} - V_{dep}$), thereby increasing the translational entropy of the system. In terms of the thermodynamics, the entropic gain is expressed as

$$\Delta S = N_m k_B \ln\left(\frac{V_{free} + \Delta V_{dep}}{V_{free}}\right), \qquad (1)$$

where N_m is the number of macromolecules in the system,

k_B the Boltzmann constant, V_{free} the initial free volume, and ΔV_{dep} the change of the depletion volume due to overlapping. Accordingly the free energy of the system decreases as $-T\Delta S$, where T is the temperature. Decrease of entropy due to the clustering of larger object is neglected here since they are few in numbers compared to N_m.

When the microcomponents with defined shapes are used as larger objects, they should accumulate and bind to the position in which the contact area becomes the maximum, decreasing greater extent of V_{dep}. Since this bonding energy is weak (not far from the energy of thermal fluctuation), misaligned components readily dissociate to search for the global minima (more stable bonding). The extent of the bonding energy can be readily modulated by adjusting the size and concentration of macromolecules without any surface treatment.

EXPERIMENTS AND RESULTS
Self-assembly of micro polystyrene beads

First we examined the condition in which the depletion volume effect became apparent using the 3 μm polystyrene beads as a large particle and polyethylene glycol (PEG) with various molecular weight (MW) as macromolecules. Mixture of beads (Polybead® #17134, Spherotec, 0.54 %v/v at final concentration), PEG, and sodium lauryl sulfate (SDS, 0.1%w/v) was introduced between the cover glass slides separated by the double-sided tape as spacers. SDS, the ionic surfactant, was included to avoid the hydrophobic interaction between beads, but the electric double layer is shorter than the effective distance of the depletion volume effect (~

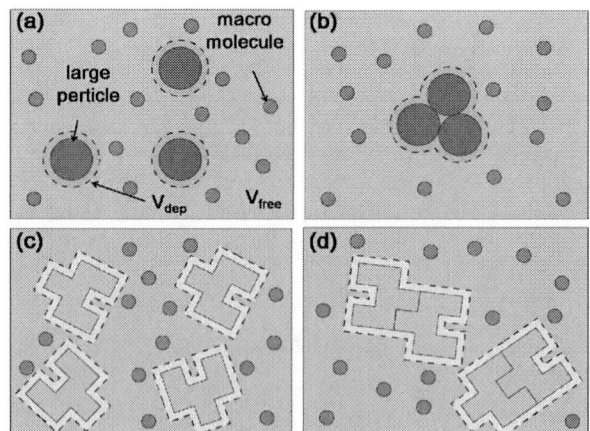

Figure 1: Schematics of the depletion volume effect. In (a), large particles were suspended in the macromolecule solution. Macromolecules cannot approach to the large particles closer than its gyration radius, forming the depletion volume. Aggregation of large particles is entropically favored since it increases the volume where the macromolecules can move freely (b). When engineered microcomponents with defined shape are used as large particles, they form ordered clusters (c, d).

978-1-4799-7956-1/15 $31.00 © 2015 IEEE   300   MEMS 2015, Estoril, PORTUGAL, 18 - 22 January, 2015

gyration radius of macromolecules). The motion of the beads was captured via the inverted microscope (Nikon Ecripse TS100) with the time-lapse recording.

Since the specific density of polystyrene (~ 1.05) is slightly heavier than water, all beads settled on the coverglass slide within ~ 1 h, and they jiggled in the 2D plane due to the Brownian motion. When PEG was not present in the solution, beads remained dispersed even after 3.5 h (Fig. 2, a-b). When the PEG with MW of 6 kDa and 50 kDa was included, the result was the same until 1 mM (6 mg/mL) and 25 μM (1.25 mg/mL), respectively. PEG molecules were insoluble to water above these weight concentrations. However, when the PEG with 2,000 kDa was included at above 0.5 μM (1 mg/mL), beads gradually formed clusters (Fig. 2 c-d, Table 1). Note that, although the weight concentrations are on the same order, clustering was observed only with the large PEG molecule.

The observed tendency can be understood from the thermodynamics point of view. Under the assumption of dilute solution, decrease in the free energy per a pair of two larger spherical particles in contact is written as

$$\Delta G = -c_m T k_B \Delta V_{dep}$$
$$\approx -2\pi \, c_m T k_B r_m^2 R_p, \quad (2)$$

where c_m, r_m, and R_p respectively represents the number density of macromolecule, the gyration radius of the macromolecule, and the radius of the larger particle,

respectively. Note that this simplification is valid when $r_m \ll R_p$. Since the free energy change is proportional to the square of r_m, it is reasonable to understand that the clustering was dependent on the MW of PEG, not on the weight concentration (the estimated r_m for 2,000 Da, 50 kDa, and 6 kDa is 91, 41, and 0.38 nm, respectively [7]). Furthermore, the working range of the entropic force is on the order of r_m, most likely contributing to the effective clustering. In addition, we have to mention that the once clustered beads were dispersed after 4.5 h.

Self-assembly of microcomponents

Next we applied this method for the assembly of microfabricated structures made of SU-8 photoresist. Fabrication and experimental procedure is shown in Fig. 3. A thin layer of gelatin (10 %w/v), as a sacrificial layer, was spin-coated on the 30 × 30 mm² coverglass slide (~ 1 μm thick after spin-coating at 3,000 RPM). Then, SU-8 2 was spin-coated at 3,000 RPM. Pattern of the microcomponents was directly exposed using the maskless exposure system (DL-1000, Nanosystem Solutions). The square-shaped microcomponents, with the sides of 5 μm and thickness of 2 μm, are shown in Fig. 3(e). Similarly, the arrow-head shape with square corners with 6 μm width, consisting of 2 μm short edges, was fabricated as shown in Fig. 3(f).

SU-8 microcomponents were released as follows. First, the glass plate was heated to 70 °C, and 5 mL of PEG solution pre-heated to 70 °C was poured as depicted in Fig. 3(c). With this procedure, nearly one-million microcomponents were dispersed into the φ50 mm plastic dish. Similarly to the beads assembly experiment, 0.1%w/v SDS was included in the solution.

The dish containing the microcomponents was agitated with the rotary shaker with orbital motion,

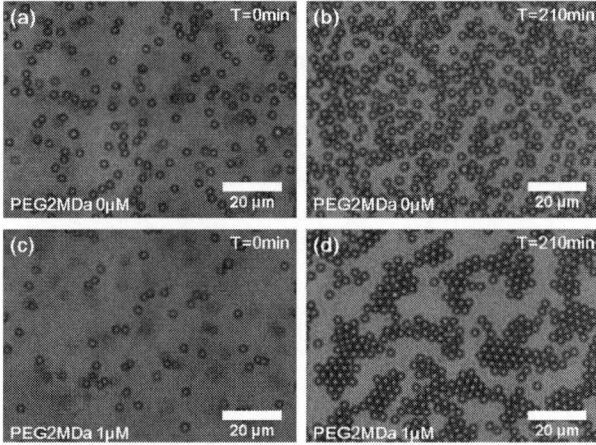

Figure 2 Behavior of 3 μm polystyrene beads on the coverglass surface. In (a), immediately after the beads was introduced, they were dispersed in the 3D space but gradually settled on the bottom. (b) However, they remained dispersed without macromolecules in the solution. In (c), 2,000 kDa PEG with 1 μM concentration was included in the solution. In this case, they gradually formed clusters on the glass surface. Image at 3.5h was shown in (d).

Table1 Tested conditions. Circles represent the condition in which clustering of beads was observed, and cross symbols represent conditions in which beads remained dispersed.

MW of PEG	1 mM	25 μM	1 μM	0.5 μM	0.1 μM	0.05μM
2,000 kDa	-	-	○	○	×	×
50 kDa	-	×	×	×	×	×
6 kDa	×	×	×	×	×	×

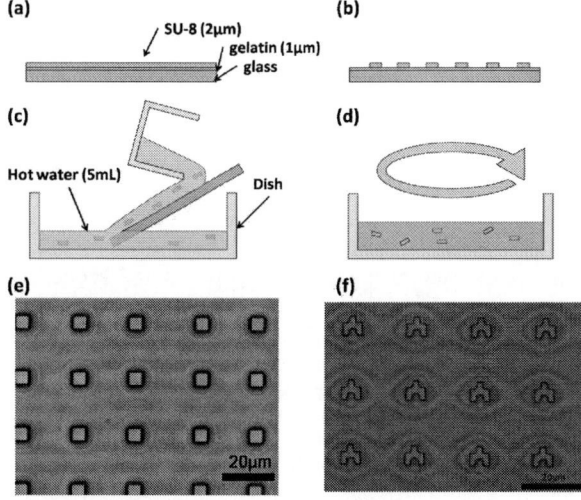

Figure 3: Fabrication of microstructures. SU-8 photoresist was patterned on the gelatin sacrificial layer (a, b). The structures were released into the petri dish by pouring the hot water (c). The dish containing structures was agitated on the rotary shaker at various rotating speed (d).

because the SU-8 microcomponents did not jiggle with the Brownian motion due to the high specific density (~ 2.0). The experimental result performed with rectangular components is shown in Fig. 4(a~d). Upon agitation at 120 rpm, components gradually migrated toward the center of the dish, but they formed clusters only when the solution contains PEG. Although this was the dissipative system (external energy was introduced), we found that clustering due to the depletion volume effect took place similarly to the beads system which was driven only with the thermal fluctuation.

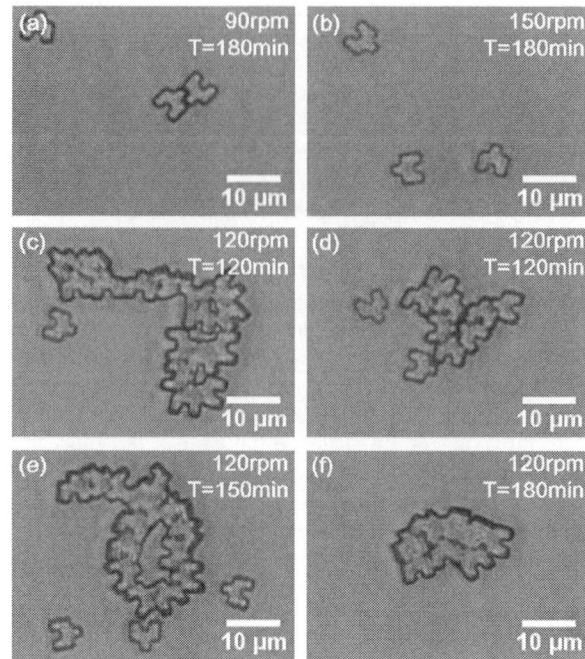

Figure 5: Results with microcomponents having complementary convex and concave sides. They bound with their corners fitted complementary. At 120 rpm, large clusters were often observed.

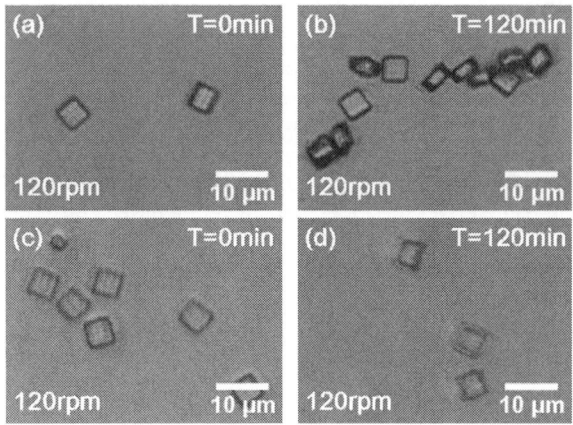

Figure4 : Assembly of square microcomponents. (a, b) When solution contained 2 MDa PEG, microcomponents formed some clusters after 2h. (c, d) They remained dispersed without PEG in the solution.

Next we attempted the assembly of microcomponents with complementary shape. The arrow-head shape shown in Fig. 3(f) has one convex and one concave sides, which are complementary. In this case, we tested the effect of various rpm of the rotary shaker, which should correspond to the temperature in the thermodynamic systems. Images were taken at 0.5 h interval until 5 h after the start of agitation.

Typical microscope images are shown in Fig. 5. At 90 and 150 rpm, large clusters did not appear, although there were a few components connected together (Fig. 5 a,b). In the meantime, at 120 rpm, clusters consisting of more than 10 components were found frequently at 2 to 3 h agitation (Fig. 5 c-f). We assume that, at 90 rpm agitation, the chance of components contacting was small because the random motion of the components was not active. Meanwhile, at 150 rpm, agitation might be too strong so that bonding due to the depletion volume effect was torn apart. Clustering due to the depletion attraction was apparent at 120 rpm.

Finally we analyzed the bonding patterns in the results with 120 rpm agitation. By looking at every pair of components within the clusters exampled in Fig. 5(c-f), we categorized each bonding according to the number of 2 μm short edges in contact. For example, the perfect matching of the complementary concave and convex edges has five short edges in contact, which are denoted as category A. Accordingly, bonding patterns with 3, 2, and 1 short edges were denoted as categories B, C, and D, respectively. Bonding with 45° angle (error alignment) was also found frequently, which was denoted as category E. Note that only one example in each category is shown in Fig. 6(a). There are some other bonding patterns in each category, and they are summed in the bar graphs in Fig. 6(b-f).

The relative frequency of bonding in each category, obtained at 1 to 3 h agitation is summarized in Fig. 6 (b-f). At 1 h agitation (Fig. 6b), the frequencies of A to D bonding was relatively flat, although bonding with 3 and 2 edges was slightly more frequent. As the time passed, bonding with larger number of short edges gradually increased; at 1.5 (Fig. 6c), B bonding became dominant, and at 2 h (Fig. 6d), frequency of A bonding increased drastically. Afterward, the frequency of A bonding gradually decreased (Fig. 6 e-f). The frequency of E bonding was kept high (20 to 40%) throughout the experiment, indicating this bonding pattern is robust in the present experimental design.

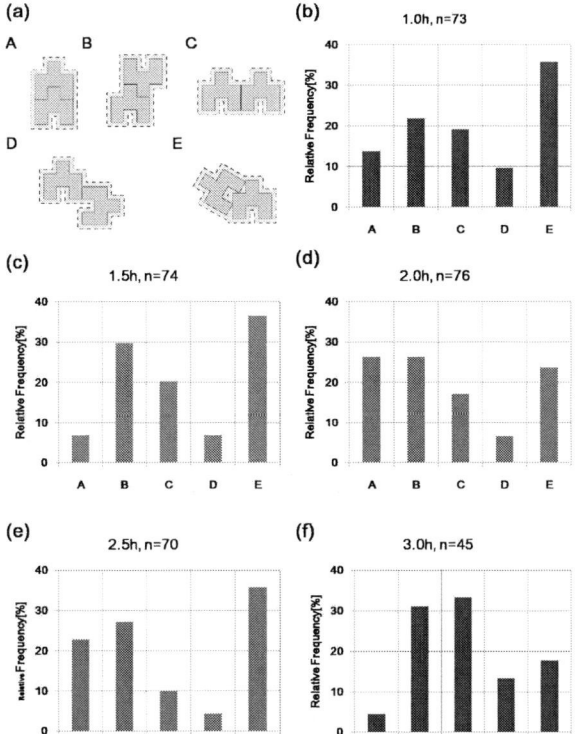

Figure 6: (a) Examples of bonding patters having various number of short edges in contact. (b-f) Relative frequency of each bonding category obtained at various agitation time. (b) 1 h, (c) 1.5 h, (d) 2 h, (e) 2.5 h, (f)3 h.

DISCUSSION

In the result shown in Fig. 6, the system moved toward the state with smaller depletion volume by increasing the number of edges in contact until 2 h agitation. In turn, the contact area of component became less after 2 h. This tendency is similar to the results obtained with the polystyrene beads. This reverse effect cannot be explained with the hard-sphere approximation of macromolecules. We speculate that PEG molecules were gradually adsorbed on the surface of microcomponents, and the hard-sphere approximation became not valid. Further study is needed to clarify the cause of the reverse effect. It will also be possible to use the spherical nano-beads instead of the macromolecules [6], although use of macromolecules is advantageous in ease and cost of the experiment.

We also found that the unintended bonding with 45° angle (Fig. 6a, E) was found at the highest frequency in the present component design. This result indicates that it is difficult to escape from these local minima, due to the steric barrier. We think that this problem can be alleviated by optimizing the component design. As the biomolecules in living systems assembles with extreme specificity, there are a lot to learn from biology [8].

CONCLUSION

By using the entropic effect (the depletion volume effect) of macromolecule in the solution, self-assembly of mesoscopic components are shown to be plausible. The new improvements in this work are twofold. (1) Assembly of artificial components was firstly shown to be possible with dissolved background macromolecules, which had been tested only with the nano spheres [6]. (2) The depletion-volume attraction can work in the dissipative system, with external energy is introduced to the system. Recently the hypothesis of statistical mechanics can be applied to the dissipative system consisting of the mllimeter scale objects [9]. Our result suggests that we can utilize the theory of statistical mechanics and thermodynamics to the self-assembly of microcomponents.

REFERENCES

[1]   M. Mastrangeli W. Ruythooren1, C. V. Hoof, J-P. Celis, "Conformal Dip-coating of Patterned Surfaces for Capillary Die-to-Substrate Self-Assembly," *J. Micromech. Microeng.*, 19, 083001, 2009.

[2]   R. R. A. Syms, E. M. Yeatman, V. M. Bright, G. M. Whitesides, "Surface Tension-Powered Self-Assembly of Microstructures — The State-of-the-Art," *J. MEMS*, 12(4), pp.387-417, 2003.

[3]   T. D. Clark, J. Tien, D. C. Duffy, K. E. Paul, G. M. Whitesides, "Self-Assembly of 10-μm-Sized Objects into Ordered Three-Dimensional Arrays," *J. Am. Chem. Soc.*, 123, pp.7677-7682, 2001.

[4]   H. Onoe, K. Matsumoto, I. Shimoyama, "Three-Dimensional Sequential Self-Assembly of Microscale Objects," *Small*, 8, pp.1383-1389, 2007.

[5]   S. Asakura and F Oosawa, "Interaction between Particles Suspended in Solutions of Macromolecules," *J. Polym. Sci.*, 33, pp.183-192, 1958.

[6]   A. D. Dinsmore, A. G. Yodh, D. J. Pine, "Entropic Control of Particle Motion Using Passive Surface Microstructures," *Nature*, 383, pp.239-242, 1996.

[7]   S. Kawaguchi, G. Imai, J. Suzuki, A. Miyahara, T. Kitano, K. Ito "Aqueous Solution Properties of Oligo- and Poly(Ethylene Oxide) by Static Light Scattering and Intrinsic Viscosity," *Polymer* Vol. 38 No. 12, pp. 2885-2891, 1997.

[8]   G. M. Whitesides and M. Boncheva, "Beyond Molecules: Self-Assembly of Mesocopic and Macroscopic Components," *PNAS*, 99, pp.4769-4774, 2002.

[9]   S. Tricard, C. A. Stan, E. I. Shakhnovich, G. M. Whitesides, "A Macroscopic Device Described by a Boltzmann-like Distribution," *Soft Matter*, 9, pp.4480-4488, 2013.

CONTACT

*H. Suzuki, tel: +81-3-3817-1827; suzuki@mech.chuo-u.ac.jp

THREE-DIMENSIONAL INTEGRATION OF SUSPENDED SINGLE-CRYSTALLINE SILICON MEMS ARRAYS WITH CMOS

Zhen Song[1], Yuxin Du[1], Miao Liu[1], Shujie Yang[1], Dong Wu[1], and Zheyao Wang[1,2,]*

[1]Insitute of Microelectronics, and Tsinghua National Laboratory for Information Science and Technology, Tsinghua University, Beijing, China

[2]Innovation Center for MicroNanoelectronics and Integrated System, Beijing, China.

ABSTRACT

We present a generic three-dimensional (3-D) integration method to fabricate suspended single crystalline silicon (SCS) MEMS arrays on CMOS. The method is based on transferring the SCS device layer of SOI on the top of CMOS. For general purpose, thin MEMS arrays with long supporting arms are demonstrated. The fabrication process is clarified in details. Mechanical reliability of the suspended structure is addressed with drop tests and is improved by optimizing the Cu plating parameters. The residue stress is compensated by tailoring the thickness of the oxide layer and flat free-standing structures are achieved. This method is applicable to a large variety of SCS MEMS including accelerometers, gyroscopes, micro-mirrors, RF MEMS switches, and resonators.

INTRODUCTION

Integration of MEMS with CMOS circuits is an important technology which provides several merits, including enhanced signal transduction, reduced chip pinout, improved immunity from electromagnetic interference, and potentially lower cost compared with multichip implementations [1]. MEMS structure typically contains movable or suspended membrane structures which sense or control optical, physical, or chemical quantities. Meanwhile the CMOS circuits have functions like amplification, addressing, electric excitation, filtering and analog-to-digital conversion. It is crucial to use high-performance materials for the MEMS devices. Single-crystalline silicon (SCS) is always preferred for its excellent mechanical and electrical properties. For example, the mechanical stability of micromirrors made of SCS is superior over that made of metals as there are no recrystallization effects in SCS hinges. Also the achievable optical quality, the surface roughness and the uniformity of SCS surfaces is superior compared to most other surfaces [2].

Monolithic integration is a straightforward method which means that the MEMS devices are directly processed on the CMOS substrate. It has been used in inertial sensors [3], micromirrors [4], RF switches [5] and resonators [6]. The main constraint of monolithic integration is the CMOS thermal budget (<450 ℃) which excludes many high performance materials like single crystalline materials.

Flip-chip bonding is a possible way to overcome the material and process incompatibilities in the integration of MEMS and CMOS [7]. However the relatively large bumps and pads limit the integration densities and device miniaturization. Moreover, the bonding process may damage the fragile free-standing MEMS structures.

Transfer bonding based three-dimensional integration, which transfers suitable device layer (e.g. SCS) from a support wafer to a CMOS wafer and connects them with through-silicon-vias (TSV), is a promising approach to realize the integration of SCS MEMS with CMOS [8]. It combines the advantages of monolithic integration and flip-chip bonding as MEMS and CMOS can be processed and optimized independently and the electrical interconnections between them can be miniaturized and of high density. Integrated devices based on this technology have been reported e.g. micromirrors [2] , RF filters [9], RF switch [10], atomic force microscope (AFM) tips [11].

This paper presents a generic method to realize 3-D integration of SCS MEMS with CMOS. Key challenges including fabrication process, mechanical reliability, and residue stress induced deflection are addressed. For general purpose, thin MEMS arrays with long supporting arms are demonstrated with improved mechanical reliability and compensated residue stress.

PROCESS FLOW

The method is based on transferring the SCS device layer of SOI on top of CMOS. As shown in Fig.1 (a), SOI and CMOS wafers are fully processed and optimized independently. Next, the SOI wafer is flipped and bonded with the CMOS wafer using polyimide as the adhesive layer, as shown in Fig.1 (b). After removing the SOI substrate using mechanical grinding and etching, the SOI device layer is transferred on top of CMOS, as shown in Fig.1 (c). Then contact holes are etched through the device layer and the bonding polyimide until the contact pads on

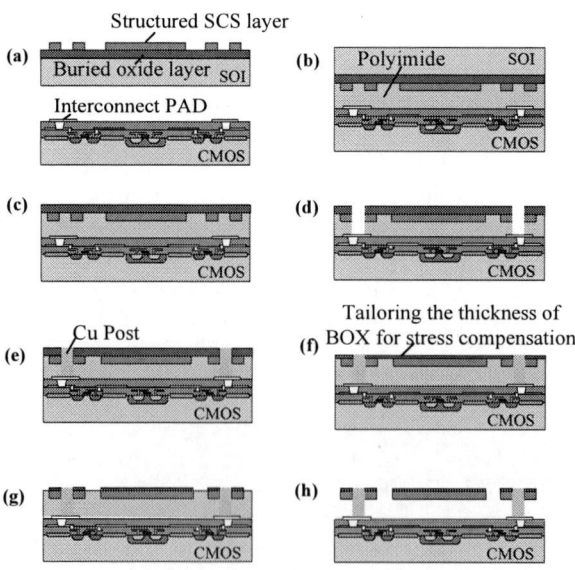

Figure 1: Schematic of process flow

the CMOS layer, and Cu posts are electroplated to form conductive TSVs as both the mechanical support and the electrical interconnect to the MEMS structures, as shown in Fig.1 (d) and Fig.1 (e). To compensate residual stress, the thickness of the buried oxide (BOX) is tailored by etching or deposition, as shown in Fig.1 (f). Finally the oxide layer is patterned to expose etching windows, through which the bonding polyimide is etched away using isotropic dry etching to suspend free-standing MEMS structures, as shown in Fig.1 (g) and Fig.1 (h).

EXPERIMENTAL RESULTS

SOI wafers with a 400nm device layer and a 500 nm BOX are used to demonstrate the 3-D integration method. The CMOS wafers are fabricated with 0.35µm CMOS technology. Wafers are bonded using polyimide as the adhesive polymer. Before spin-coating of polyimide, the wafers are pre-cleaned with deionized water rinse and O_2 plasma to remove inorganic particles and organic residues. The particles or organic residues may cause to bonding voids, which would lead to crack or delamination of the transferred SCS after the SOI substrate is removed. Thus pre-cleaning is a stringent requirement for reliable bonding.

As the SCS layer was patterned and structured before bonding, alignment between the MEMS layer and the CMOS circuits is required during bonding. After optimization of bonding parameters, void-free bonding is achieved using aligned wafer bonders with post-bonding misalignment error less than 3 µm [12]. Then the substrate of SOI wafers is thinned to 70 µm using mechanical grinding and then completely removed by silicon etching. Different silicon etching methods are evaluated to realize intact film transfer, as shown in table 1. Although wet etching methods using KOH and HNA are more economical, they attack the polyimide bonding interface severely, causing delamination of transferred SCS from CMOS wafer. SF_6 ICP etching is employed to remove the remaining substrate silicon to finalize intact film transfer.

Table 1: Different methods to remove the silicon substrate to realize film transferring.

Etching method	KOH*	HNA**	SF_6 ICP
Etching rate of silicon (µm/min)	0.8	0.8	1.1
Etching rate of Si-polyimide interface (µm/min)	19	1500	--
Etching rate of SiO_2-polyimide interface (µm/min)	2.6	1000	--
Etching rate of CMOS passivation-polyimide interface (µm/min)	100	110	--
Result	Film delaminated	Film delaminated	**Film intact**

*: 33wt %.   **: HF: Nitric: Acetic Acid=1:3:8

The mechanical reliability of the suspended structures is investigated by drop tests. The naked dies with suspended MEMS structures fall to the concrete ground freely and randomly, and it is founded that the transferred thin MEMS structures are delaminated from the Cu posts (Fig.2) if using conventional 3-D processes, causing fatal mechanical reliability problem. This problem may be mainly caused by the weak strength of the Cu post. This problem is completely solved by optimizing the Cu plating parameters to enhance the mechanical strength through tailoring the plated Cu grains.

Figure 2: Mechanical reliability test: Without optimization of Cu plating, transferred MEMS structures delaminated from the Cu plugs.

Distortion of the ultra-thin suspended MEMS structures caused by residue stress is a common problem that is encountered in thin structure transfer. Without balance, the residual stress causes significant stress gradient and lead to deflection of suspended MEMS structures, as shown in Figure 3. This problem is solved by compensating the residue stress through precisely tailoring the thickness of the oxide layer (Fig.1 (f)). It is based on the fact that symmetrical materials and thickness across the supporting legs can balance the residual stresses and thus eliminate the stress gradient. The thickness of the compensating oxide layer depends on the specific applications, and can be optimized by experimental trial. In this demonstration, the residue stress is compensated and flat free-standing structures are achieved by controlling the BOX to 400 nm. For MEMS structures with large thickness, the residual stress caused distortion can be negligible.

Figure 3: Residue stress induced distortion of the free-standing films. (Oxide thickness is 500nm)

Figure 4: Compensated residue stress. (Oxide thickness is 400nm, after the mechanical reliability test).

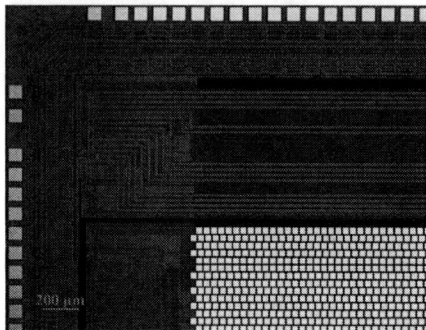

Figure 5: CMOS circuits including a switched capacitor integrator and a 6-bit cyclic analog-to-digital converter.

After solving the problems of mechanical strength and residual stress, thin MEMS structures are well integrated on the surface of CMOS chips, as shown in Figure 4. Although the width of the supporting legs is only 2 μm, it is strong enough to support the MEMS structures.

The CMOS circuits include a switched capacitor integrator and an analog-to-digital converter (ADC). Fig.5 shows the optical photo of the CMOS circuits. Electrical test before 3-D integration is performed and the results show that the CMOS circuits work well. After 3-D integration, the CMOS circuits are tested again and the outputs of the integrator and the ADC are shown in Figure 6. The results show that they work well and demonstrate that the 3-D integration processes have no negative impacts on the CMOS circuits.

CONCLUSIONS

Three-dimensional integration of suspended SCS MEMS array with CMOS is demonstrated with improved mechanical reliability and compensated residue stress. Void free bonding interface is a key challenge for the transfer bonding. A combination of mechanical grinding and SF_6 ICP etching is a reliable way to remove the SOI substrate to realize intact film transfer. Free-standing structures may suffer mechanical reliability problem and Cu plating is optimized in terms of mechanical strength. Residue stress and stress gradient lead to distortion of the ultra-thin suspended MEMS structures and it can be compensated by tailoring the thickness of the oxide layer. The CMOS circuits functions well after 3-D integration and MEMS releasing. It is a generic method and can be used to a large variety of MEMS devices including optical MEMS devices, inertial sensors, and RF MEMS devices.

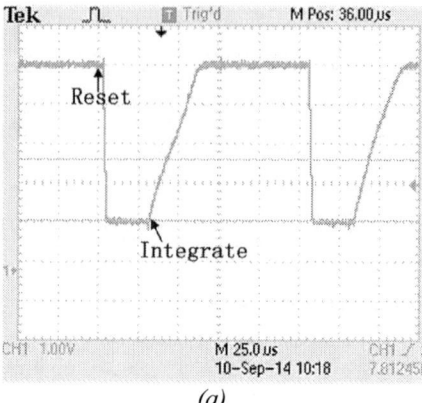

(a)

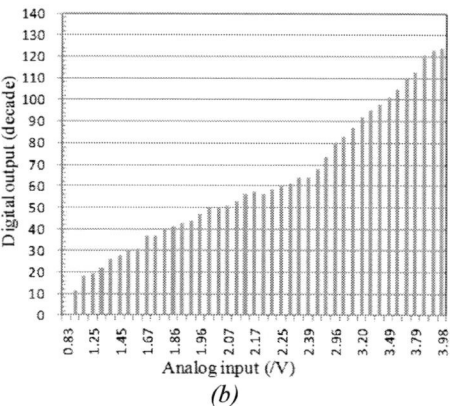

(b)

Figure 6: CMOS circuit test results after 3-D integration: (a) Switched capacitor integrator; (b) 6-bit cyclic analog-to-digital converter.

ACKNOWLEDGEMENTS

This work was supported in part by 973 Program under Grant 2011CBA00603 and NSFC under Grant 61271130.

REFERENCES

[1] G. K. Fedder, et al., "Technologies for cofabricating MEMS and electronics", *Proc. IEEE*, vol. 96, pp. 306-322, 2008.

[2] F. Niklaus, et al., "Arrays of monocrystalline silicon micromirrors fabricated using CMOS compatible transfer bonding", *J. Microelectromech. Syst.*, vol. 12, pp. 465-469, 2003.

[3] M. W. Judy, "Evolution of integrated inertial MEMS technology", *in Digest Tech. Solid-State Sensor, Actuator and Microsystems Workshop*, Hilton Head Island, June 6-10, 2004, pp. 27-32

[4] P. F. Van, et al., "A MEMS-based projection display", *Proc. IEEE*, vol. 86, pp. 1687-704, 1998.

[5] J. D. Brazzle, et al., "A hysteresis-free platinum alloy flexure material for improved performance and reliability of MEMS devices", in *Digest Tech. Papers Transducers'03 Conference*, Boston, June 8-12,

2003, pp. 1152-1155.

[6] W. L. Huang, Z. Ren, and C. T. Nguyen, "Nickel vibrating micromechanical disk resonator with solid dielectric capacitive-transducer gap", in *Proc. IEEE Int. Freq. Contr. Symp.*, Jun. 2006, pp. 839–847.

[7] F. F. Faheem, et al., "Flip-chip assembly and liquid crystal polymer encapsulation for variable MEMS capacitors", *IEEE Trans. Microw. Theory Tech.*, vol. 51, pp. 2562-2567, 2003.

[8] M. Lapisa, et al., "Wafer-Level Heterogeneous Integration for MOEMS, MEMS, and NEMS", *IEEE J. Select. Topics Quantum Electron.*, vol. 17, pp. 629-644, 2011.

[9] T. Matsumura, et al., "Multi-band radio-frequency filters fabricated using polyimide-based membrane transfer bonding technology", *J. Micromech. Microeng.*, vol. 20, pp. 095027, 2010.

[10] R. Guerre, et al., "Wafer-level transfer technologies for PZT-based RF MEMS switches", *J. Microelectromech. Syst.*, vol. 19, pp. 548-560, 2010.

[11] M. Despont, et al., "Wafer-scale microdevice transfer/interconnect: its application in an AFM-based data-storage system", *J. Microelectromech. Syst.*, vol. 13, pp. 895-901, 2004.

[12] Z. Song, et al., "Void-free BCB adhesive wafer bonding with high alignment accuracy", *Microsystem Technologies*, accepted.

CONTACT

*Z Wang, tel: +86-10-62772748;
fax: +86-10-62771130; z.wang@tsinghua.edu.cn

TRAJECTORY CONTROL OF MEMS FALLING OBJECT FABRICATED BY SU-8 MULTILAYER STRUCTURE

Hokuto Yamane[1] and Sumito Nagasawa[1]

[1]Dept. of Eng. Sci. and Mech., Shibaura Institute of Technology, Tokyo, JAPAN

ABSTRACT

In this paper we propose a trajectory control method for a MEMS falling object as shown in Figure 1. The MEMS falling object is consisted of two units, an autorotation part and a non-rotation part. The autorotation part keeps its attitude stable with the gyro-effect of the autorotation phenomenon. The non-rotation part keeps a non-rotation state by using the air breaking boards. This non-rotation part controls its falling trajectory and the scattering region. By using large falling objects, aerodynamics of the falling object was characterized, e.g. falling speed, rotational speed, etc. Then the MEMS falling object was designed considering with this aerodynamics. The MEMS falling object was fabricated with a method of the SU-8 multi-layer structure. A MEMS autorotation part whose wing length is 6mm in diameter rotates at 4,800 rpm in the wind-tunnel successfully.

INTRODUCTION

In the small world, flight is good transfer strategy due to the scaling effects. Recently a small flying robot that can take dynamic balance has been developed [1-3]. However, keeping the dynamic balance for flight is now still a major problem. On the other hand some seeds (called "flying seed", e.g. maple and hornbeam seed) fall slowly and stably by the autorotation phenomenon. This autorotation phenomenon can be seen when the flying seed has a wing. The wing rotation generates leading-edge vortices (LEV) at the leading edge of the seed's wing. The LEV generates pressure difference between the upper surface and the lower surface of the wing. This is the mechanism that a lift force is generated. The LEV and associated characteristics of the autorotation are investigated actively [4]. Since plant seeds have various ways to disperse themselves, these dispersal methods can be utilized in the engineering systems by analyzing its essentials [5, 6].

In this paper we propose a trajectory control method for a MEMS falling object. Figure 2 indicates applications of the falling objects. The first useful goal is in agricultural applications. By attaching this proposed object to real plant

seeds, a sowing seeds task can be achieved to the wide area. Although these objects with the seeds are released at the same location, the falling objects fall in the wide area by their designed aerodynamics. For these types of tasks, this falling object has to be made of the biodegradable materials for reducing the environmental load.

Another goal is for environmental applications. By attaching a monitoring sensor and an electric battery, the environmental data (e.g. weather data) in wide area can be obtained. In addition, by attaching a control unit with microcomputers, the falling object can control its falling trajectory more aggressively for various advanced applications. Objectives of this study are to design a falling object modeled after the flying seeds and to control the falling region of the falling object.

FUNCTIONS OF FALLING OBJECT

Wider falling area can be achieved by controlling each aerodynamic characteristics of the object. Our proposed falling object is consists of the two parts; first one is an autorotation part and the other one is a non-rotation part as shown in Figure 3. The autorotation part keeps its attitude stable by the gyro effect, and the non-rotation part controls the falling trajectory by using the two air-breaking boards.

Since the air-breaking board is mounted with an angle of attack, the air flow generated by falling gives a lateral direction force as shown in Figure 4. Another physical forces, a friction at the axis, a counterforce of the rotational flow, a counterforce of the falling flow, are working

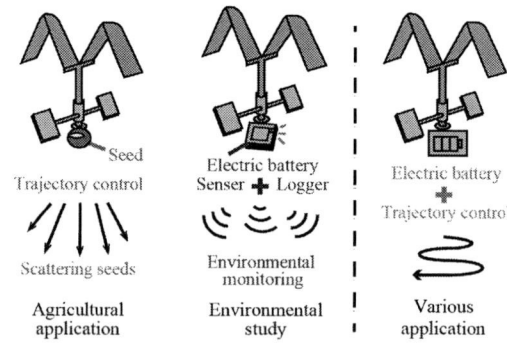

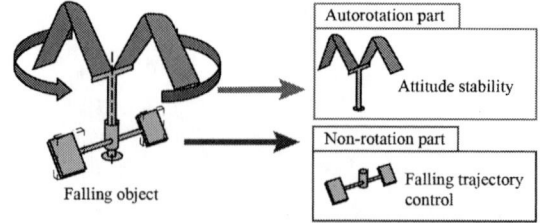

Figure 2: Useful applications using the falling objects.

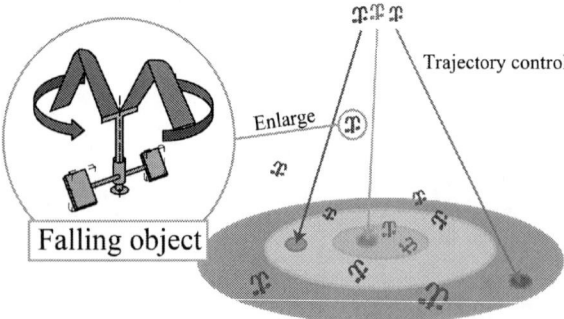

Figure 1: Schematic diagram of this study. We design a falling object modeled after the plant seeds and control the falling regions of the objects.

Figure 3: Explanation of functions of the falling object. The autorotation part keeps its attitude stable by the gyro-effect. The non-rotation part controls the falling trajectory by using the two air-breaking boards.

978-1-4799-7956-1/15 $31.00 © 2015 IEEE   308   MEMS 2015, Estoril, PORTUGAL, 18 - 22 January, 2015

around the axis as shown in Figure 5. Non-rotating state of the non-rotation part is achieved by balancing these forces (torques) around the axis, by adjusting the angle of attack or the areas of the air-breaking boards. In contrast to the rotational direction, by unbalancing of these forces in the lateral direction, a lateral force works to the whole object. As the result, by the effect of this lateral force, the diagonal falling of the falling object is achieved.

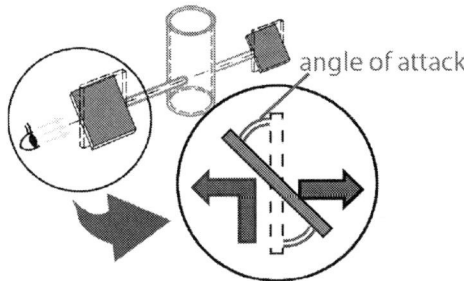

Figure 4: Mechanism of the falling object which falls diagonally. The falling object gets lateral reaction forces from the two air-breaking boards with the angle of attack.

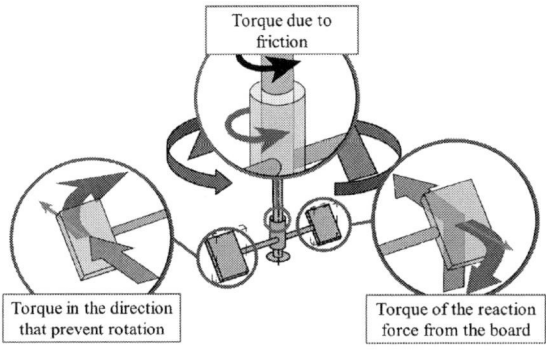

Figure 5: Physical forces concerning to the non-rotation part. These forces are a friction at the axis, a counterforce of the rotational flow, a counterforce of the falling flow, respectively.

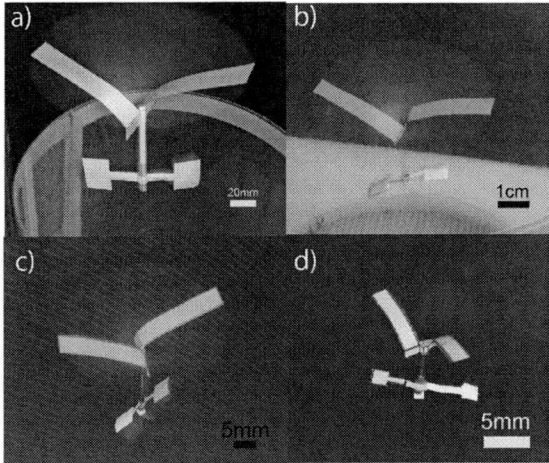

Figure 6: Photo of the paper falling object which is rotating. They have different wing area respectively. (a)3200mm², (b)800mm², (c)200mm²,(d)50mm² wing area.

FABRICATION OF FALLING OBJECT

Paper falling objects

We fabricated four types of the paper falling objects. Figure 6 depicts photos of the paper falling object that is floating in a vertical wind-tunnel. They have 3000, 800, 200, 50 mm² in wing area, respectively. The paper falling object consists of the paper parts and a plastic straw or a brass wire. The autorotation part was fabricated by bonding the paper wing to the brass wire or the plastic straw. We also fabricated the non-rotation part by attaching two paper plates on the end of the brass pipe or the plastic straw. These plates are called as air-breaking boards. The falling object gets lateral reaction forces from the boards with the angle of attack.

MEMS autorotation part

Figure 7 shows the process diagram to fabricate a MEMS wing using SU-8 3050. SU-8 3050 is one of the photosensitive negative resists. First of all, we form a thin film of LOR 30B on the surface of the SiO$_2$ substrate using a spin coater. LOR 30B is not a photosensitive resist. This resist was used as a sacrificing layer. Then we baked the sample at 180 °C for 10 minutes in an oven. Secondly, we formed a thin film of SU-8 on the surface of the LOR using a spin coater. Then we baked the sample at 95 °C for 15minutes on a hot plate (pre-baking). After that, SU-8 was exposed via a photo mask by the h-line, and then SU-8 was baked at 95 °C for 5 minutes on the hot plate (P.E.B.). The SU-8 multilayer structure was fabricated by repeating from (2) to (4) for 4 times [7]. Finally, we developed SU-8 for 15 minutes using the SU-8 developer, and then the SU-8 multilayer structure was released from the SiO$_2$ substrate by dipping in NMD-3 for 24 hours.

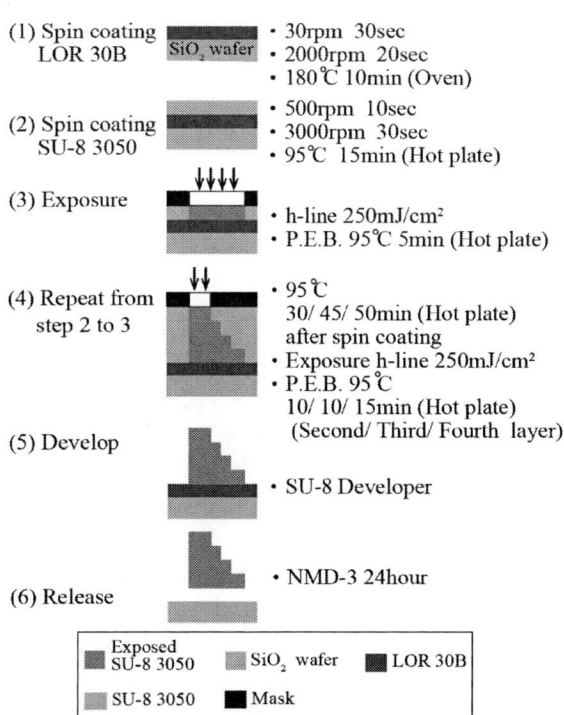

Figure 7: Fabrication process of the multilayer SU-8 structure. The SU-8 MEMS wing was fabricated by this process.

We fabricated three types of the MEMS wings which are different in the shape of the wing and wing area as shown in Figure 8. They have 12.5, 6.25, and 6.25 mm^2 in wing area, respectively. We fabricated a MEMS autorotation part by bonding the MEMS wing to the end of the thin brass wire.

The SU-8 multilayer structure was employed to make the angle of attack of the small wing. In this paper, we fabricated only the MEMS autorotation part. Therefore, the MEMS non-rotation part is need to fabricate in future work. Figure 9 illustrates a fabrication design of the MEMS non-rotation part using the locking mechanism and hinges.

EXPERIMENTAL RESULTS

Experimental setup

Figure 10 shows our experimental setup using a vertical wind-tunnel. When the paper falling object or the MEMS autorotation is floating in the vertical wind-tunnel, its rotational speed and wind velocity are measured. The flow speed of the wind tunnel was controlled by changing the fan speed. The paper falling objects are floating and rotating in the vertical wind-tunnel. The rotational speed and the wind velocity were measured using a strobe system and a hot-wire anemometer. A rotational state of the MEMS autorotation part was recorded by using a high-speed camera system as shown in Figure 11.

Rotational speed of the falling object

Figure 12 shows the relationship between the rotational speed and the wing area. Since the vertical wind-tunnel ability was not enough, the MEMS autorotation parts did not reach to the steady-state. Therefore rotational speed of the MEMS autorotation parts was measured by using a High-speed camera. The rotational speed is faster as the wing area becomes smaller. The maximum rotational speeds of the paper falling objects and the MEMS autorotation part were 5022 rpm and 4800 rpm, respectively.

Wind velocity of the falling object

Figure 13 shows the relationship between the wind velocity and the wing area. The wind velocity is faster as the wing area becomes smaller. Since the gyro-effect becomes stronger when the rotational speed is faster, the attitude of the falling object becomes stable as its size becomes smaller.

These results mean that a falling velocity of the falling object becomes faster as its size becomes smaller. The wing area of the MEMS autorotation part was smaller than the wing area of the paper falling object. Despite that, the rotational speed of the MEMS autorotation parts was slower than the rotational speed of the paper falling object. This problem can be caused by the material difference of the falling objects. The wing made of SU-8 is difficult to deform elastically because SU-8 is a hard material. However, because the wing made of paper is flexible, the wing made of paper is easy to deform into the optimum pitch angle to rotate faster. Therefore, in the future, for the MEMS SU-8 wing the optimum pitch angle has to be designed.

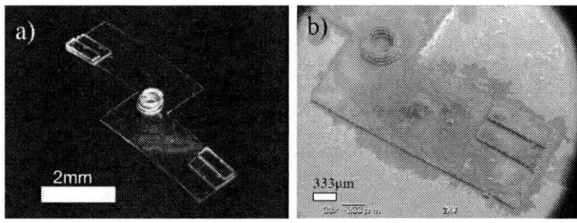

Figure 8: (a) Photo of a wing made of SU-8. (b) SEM Photo of a MEMS wing. The wing length is 6 mm in this MEMS SU-8 wing. The thickness of the SU-8 single layer is 5µm. The MEMS SU-8 wing has an angle of attack.

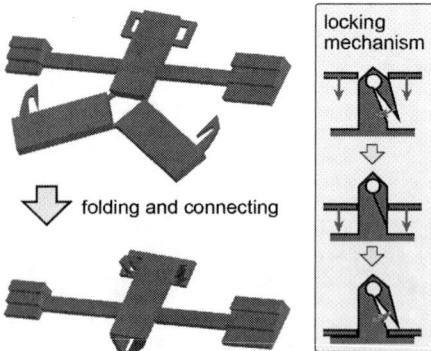

Figure 9: Micro structure of non-rotation part we suggests. It is required the design by using the locking mechanism and hinge.

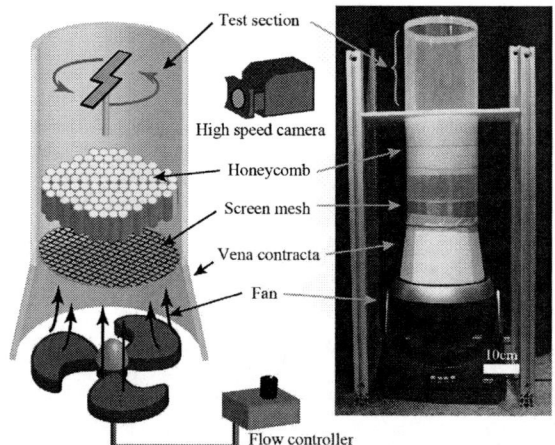

Figure 10: Experimental setup. The paper falling object and the MEMS SU-8 autorotation part are floating in this vertical wind-tunnel. Rotational speed of the objects is measured by using a strobe system and a high-speed camera system.

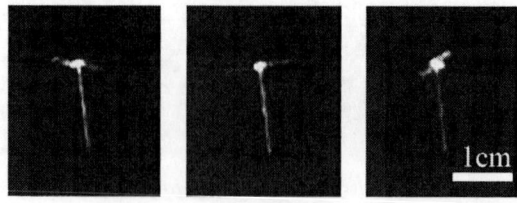

Figure 11: A rotational state of the MEMS autorotation part was recorded by using a high-speed camera system at 1000 fps.

978-1-4799-7956-1/15 $31.00 © 2015 IEEE

Evaluation of falling region

Figure 14 shows the falling areas using the paper falling objects and the only autorotation part. The only autorotation part fell straightly with a little dispersal. The paper falling objects fell spirally with different curvatures. The two paper falling objects with different sizes were used in this experiment. By the effect of the non-rotation part the falling region of the paper falling object becomes wider than the only autorotation part.

CONCLUSIONS

In this paper we propose a trajectory control method for a MEMS falling object. The falling object consists of the autorotation part and the non-rotation part. The first useful goal is in agricultural applications. By attaching this proposed object to real plant seeds, a sowing seeds task can be achieved to the wide area.

A MEMS falling object was successfully fabricated using the method of the SU-8 multilayer structure. The paper falling objects with different wing areas were also fabricated. Then the rotational experiment of the MEMS autorotation part and the paper falling objects were demonstrated in the vertical wind-tunnel. The characteristics of the rotational speed and the wind velocity were measured and discussed. The maximum rotational speeds of the paper falling objects and the MEMS

autorotation part were 5022 rpm and 4800 rpm, respectively. The wind velocity is faster as the wing area becomes smaller. Since the gyro-effect becomes stronger when the rotational speed is faster, the attitude of the falling object becomes stable as its size becomes smaller.

The falling region was evaluated using the only autorotation part and the paper falling objects. The effect of the non-rotation part which makes the falling region wider was clearly confirmed.

REFERENCES

[1] Kevin Y. Ma, Pakpong Chirarattananon, Sawyer B. Fuller, Robert J. Wood, "Controlled Flight of a Biologically Inspired, Insect-Scale Robot," *Science* 3 May 2013, Vol. 340 no. 6132, pp. 603-607, DOI: 10.1126/science.1231806.

[2] Orlowski, Christopher T., and Anouck R. Girard. "Dynamics, stability, and control analyses of flapping wing micro-air vehicles." *Progress in Aerospace Sciences* 51 (2012): 18-30.

[3] Wood, R. J., *et al.* "Microrobot design using fiber reinforced composites." *Journal of Mechanical Design* 130.5 (2008): 052304.

[4] D. Lentink, *et al*, "Leading-Edge Vortices Elevate Lift of Autorotating Plant Seeds", *Transactions of the Science, AAAS, Journal of Applied Mechanics*, Vol. 324, pp. 1438-1440, (2009).

[5] Nathan, Ran, et al. "Mechanisms of long-distance dispersal of seeds by wind." *Nature* 418.6896 (2002): 409-413.

[6] Nathan, Ran, et al. "Mechanisms of long-distance seed dispersal." *Trends in Ecology & Evolution* 23.11 (2008): 638-647.

[7] MATA, Alvaro; FLEISCHMAN, Aaron J.; ROY, Shuvo. "Fabrication of multi-layer SU-8 microstructures," *Journal of micromechanics and microengineering*, 2006, 16.2: 276.

CONTACT

*S. Nagasawa, tel: +81-3-5859-8063; nagasawa@shibaura-it.ac.jp

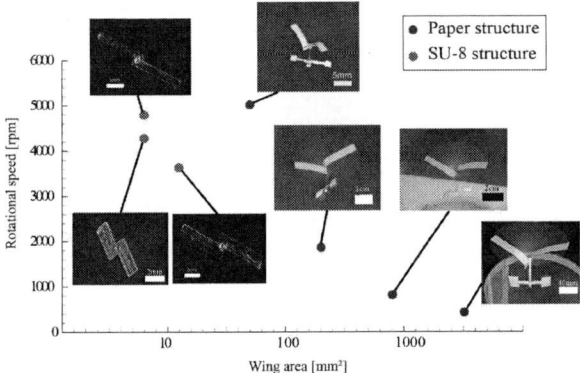

Figure 12: Relationship between the rotational speed and the wing area. The rotational speed is faster as the wing area becomes smaller.

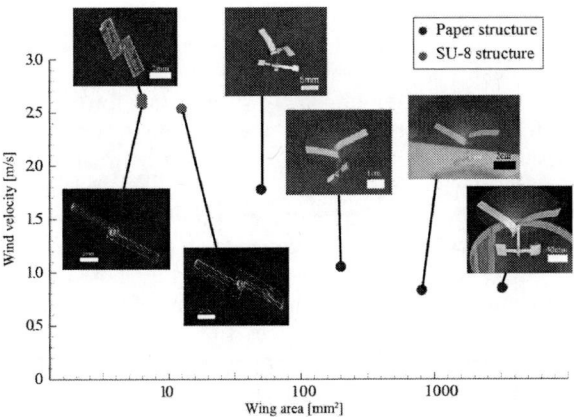

Figure 13: Relationship between the wind velocity and the wing area. The wind velocity is faster as the wing area becomes smaller.

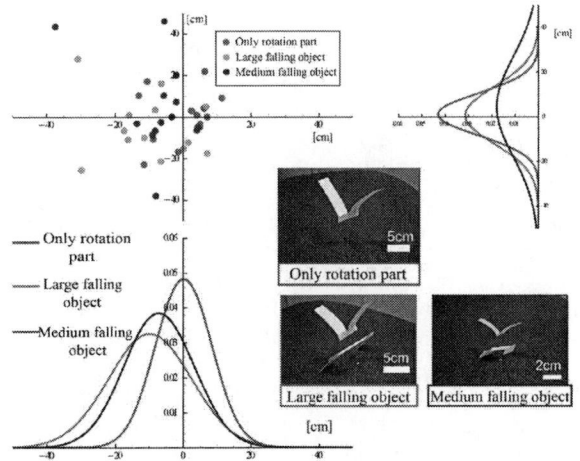

Figure 14: Standard distribution fitting to the scattering distributions of the falling positions.

WAFER-SCALE INTEGRATION OF CARBON NANOTUBE TRANSISTORS AS PROCESS MONITORS FOR SENSING APPLICATIONS

Kiran Chikkadi[1], Wei Liu[1], Cosmin Roman[1], Miroslav Haluska[1], and Christofer Hierold[1]

[1]Department of Mechanical and Process Engineering, ETH Zurich, Zurich, Switzerland

ABSTRACT

We report on the fabrication of carbon nanotube transistors designed as process control monitors for applications such as gas and pressure sensing. We demonstrate the concept for an integration process used for gas sensor fabrication on a 100 mm wafer. From a maximum possible of 9348 devices, 4463 working devices were fabricated, of which 2702 semiconducting were identified, allowing the extraction of distributions of threshold voltage, minimum device resistance, process yield and wafer uniformity data. We show a fabrication yield of 29% for transistors (for semiconducting tubes only), a median minimum-resistance of 122 kΩ (interquartile dispersion: 179 kΩ), and a threshold voltage of -0.75 V (interquartile dispersion: 1.46 V). Distribution profiles of transistor parameters over the wafer are also quantified.

INTRODUCTION

Advances in carbon nanotube electronics over the last few years have generated interest in the manufacturability of transistors, for logic as well as for sensing applications [1-3]. Devices based on individual single-walled carbon nanotubes (SWNTs) can be utilized as piezoresistive transducers in pressure sensors [4] due to their impressive gauge factors.

The sensitivity of SWNTs to gases such as NO_2 and SO_2 also suggested their use in gas sensors [5] with ultra-low power capabilities, exploiting the particular advantage of individual nanotubes [6].

Nonetheless, the fabrication of carbon nanotube devices on a large scale is currently hindered by the poor control of the electrical properties of the fabricated devices [7-10]. CNFETs often show widely distributed resistance, threshold voltage and non-idealities such as hysteresis. Therefore, test structures designed to obtain distributions of the relevant device parameters are essential for process control and feedback.

Test structure design

The test structures proposed in this study are designed for nanotubes which are not aligned, i.e., their orientation is not actively controlled during nanotube growth. Currently, although aligned growth of low-diameter nanotubes on quartz and sapphire substrates [11] have been shown, the growth of high-quality, larger-diameter carbon nanotubes with a defined orientation is still challenging, due to which techniques to integrate unaligned nanotubes into devices are highly relevant.

Since about 33% of the nanotubes are statistically metallic or quasi-metallic, devices with such nanotubes in the channel do not function as transistors (i.e., they do not turn OFF at any gate voltage). As the number of nanotubes increases, the likelihood of having at least one metallic nanotube increases. In order to maximize the number of functional transistors on the wafer, we focus on maximizing the yield of single-nanotube devices.

We have previously shown that even with non-aligned nanotubes, moderate yields of single-tube devices can be achieved by designing the electrode geometry with respect to the length and density distributions of the nanotubes. The devices discussed in this study are back-gated transistors, where the substrate acts as the gate, and the nanotubes are contacted by 10 pairs of opposing electrodes, as visible in Figure 1c. A total of 60 electrodes are present per chip. The electrode gaps were designed to be 2 µm. Further discussion on optimization of electrode designs can be found in [10].

Device fabrication

For fabricating the devices, we use a 100 mm Si wafer with 70 nm SiO_2 as the gate dielectric. Deep-UV (220 nm wavelength) photolithography was used for catalyst island definition and metal electrode patterning. Ferritin was applied as the catalyst precursor, for nanotube growth. SWNT growth was performed at 850°C in a CH_4/H_2 atmosphere. 60 nm of Au (with 2 nm Cr as adhesion layer) was used as the contact metal. The devices were finally passivated using atomic layer deposited Al_2O_3 (40 nm) in order to avoid degradation. Figure 1a shows an image of the final wafer with fabricated devices, while Figure 1b-d show successive magnifications of one set of electrodes on the wafer. In Figure 1d, two devices where SWNTs have successfully bridged the electrodes are visible, as marked by arrows.

RESULTS AND DISCUSSION

After fabrication, the wafer was probed in an automated prober (Cascade Microtech Summit 12000) using a semiconductor parameter analyzer (Agilent B1500A). A total of 9840 electrodes were present on the wafer in pairs of 20 electrodes (60 electrodes per chip, 164 chips). For each pair of 20 electrodes a maximum of 19 devices can be obtained without the formation of electrical loops. Therefore, the maximum number of possible devices is 164 × 19 × 3 = 9348. A total of 4463 working devices were found on the wafer. The devices were classified using an automated routine in MATLAB, with devices being characterized as semiconducting if they showed an OFF-state (current below 1 nA) and as quasi-metallic if they did not show an OFF state. With this classification, 2702 devices showed semiconducting behavior (resulting in a fabrication yield of 29%), with the remaining devices showing quasi-metallic behavior.

To reveal the distribution of devices over the wafer, the number of devices per chip (three 20-electrode pad rows) was extracted, as shown in Figure 2.

978-1-4799-7956-1/15 $31.00 © 2015 IEEE

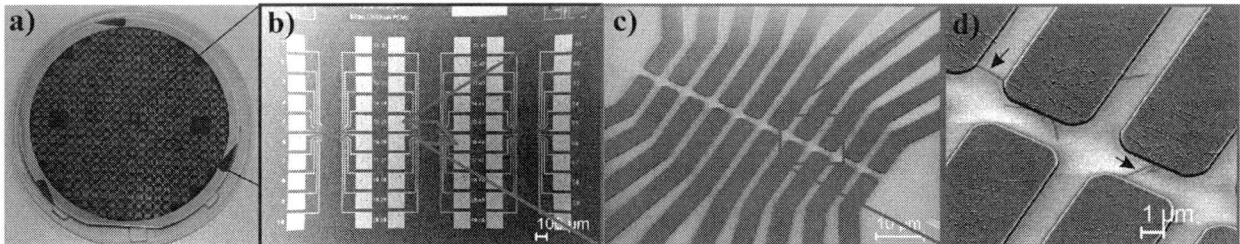

Figure 1: Single-walled carbon nanotube devices fabricated on a 100 mm wafer. The wafer consists of 4463 devices out of a maximum of 9348. Of these devices, 2702 were semiconducting. b)-d) Close-up SEM images of devices on the wafer. Two devices are visible in d), marked by arrows.

Figure 2a shows a histogram of the number of devices per chip (bins normalized such that area under the histogram is 1), while Figure 2b shows a wafer map indicating the spatial distribution of the devices. The black areas between the relevant chips show areas where other test structures for monitoring the length and density distributions (see [12] for more information) are present. The histogram in Figure 2a fits a normal distribution with a mean of 26.5 and standard deviation of 10.8 devices. In order to understand if this variation over the wafer was a result of purely random fluctuations from the growth process, we performed Monte Carlo simulations of the growth processes and estimated the variation in the number of devices obtained per chip (i.e., each set of 60 electrodes). The results from the simulations suggest an average of 26±8.5 devices per chip. The variations on the wafer can therefore be assumed to be largely a result of natural variations in the carbon nanotube length and density over the wafer. Nevertheless, the standard deviation of the measurements is marginally higher, indicating that other spatial variations may also be present. Figure 2b also suggests that there are clear patterns in the wafer map (a significant drop in the device yield at the bottom left of the wafer is visible, which was found to be due to a scratch on this part of the wafer. These chips were excluded from the remainder of the analysis).

For the analysis of device parameter distributions, only the semiconducting devices are considered. For all the analysis presented here, the gate voltage used was -10 V to +10 V; both forward and reverse sweeps were performed. The source-drain voltage was fixed at 100 mV. Figure 3 shows the transfer characteristics of 100 randomly selected nanotube transistors from the ensemble of 2702 devices. Only the forward sweep is plotted for clarity, gate hysteresis was nevertheless observed. The transistors show ambipolar characteristics with a dominant n-branch.

To estimate parameter distributions, the minimum device resistance was estimated for the n-branch at V_g = +10 V. The threshold voltage was calculated using the V_g-axis intercept of the maximum-slope point on the transfer characteristics.

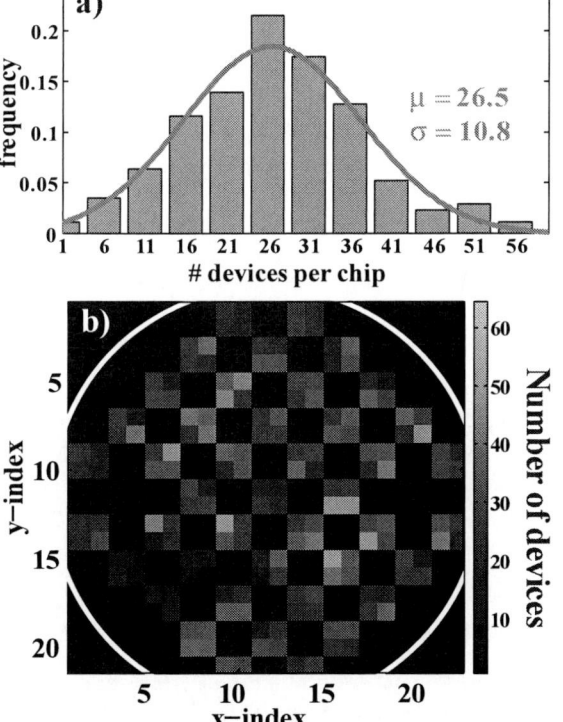

Figure 2: a) The number of devices per chip is normally distributed with a mean μ = 26.5 devices per chip, which is in close agreement with simulations, which predict 26 devices per chip. b) Wafer map of devices on the wafer. The black areas on the wafer consist of other structures. Each pixel represents one chip.

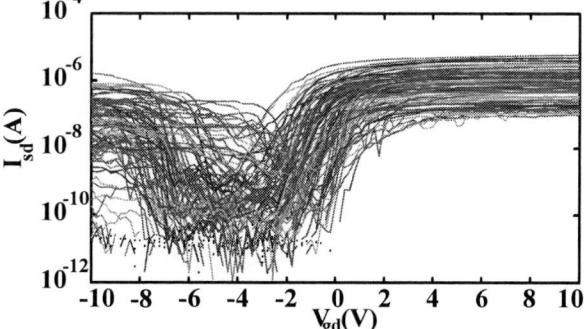

Figure 3: Randomly selected sample of transfer characteristics from 100 SWNT transistors from the wafer. The transistors show ambipolar characteristics, but with the n-branch current higher on average than the p-branch. Gate sweeps are plotted only in one direction for clarity, but a median hysteresis of 2.5 V is recorded from the sweeps. V_{sd} = 100 mV.

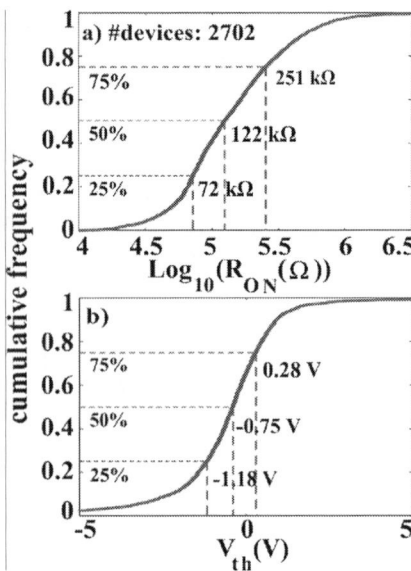

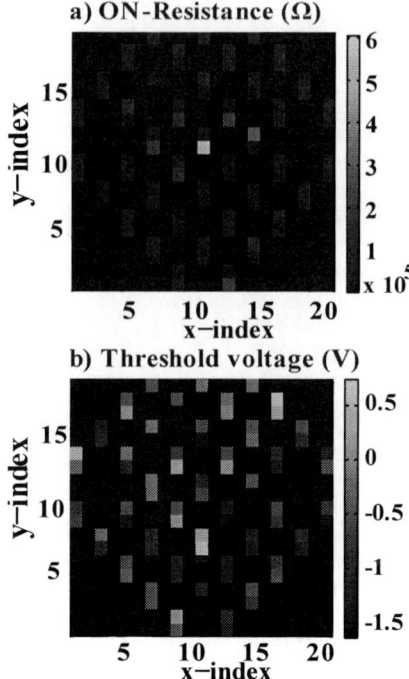

Figure 4: a) Cumulative frequency of minimum device resistance, measured on the n-branch at $V_g = +10$ V. The x-axis is in log-scale. The 25 and 75 percentile values are shown, indicating a spread of 179 kΩ. b) Cumulative frequency for the threshold voltage, indicating the interquartile spread of 1.46 V and a median threshold voltage of -0.75 V.

The median value of the minimum device resistance is found to be about 122 kΩ, with a spread of about 179 kΩ between the 25 and 75 percentiles. The median resistance value is lower than the values reported by Khamis et al [9]. and Martin-Fernandez et al [13]; Furthermore, 95% of the devices fall below 684 kΩ. This is comparable to the resistance fluctuations on the same single nanotube as reported by Zhang et al. [14], who reported that a majority of their devices were below 800 kΩ. This highlights the role of contact resistance contributions to the total device resistance. Large variations in contact resistance mask the effect of resistance variations originating from nanotubes of different chirality. It has been noted that process residues such as photoresists, which cannot be efficiently removed from the nanotube surface, may be a major cause for these large fluctuations in resistance [9]. In another separate study, we have shown that application of a protective layer can reduce the overall resistance as well as its dispersion by avoiding contact between the nanotube and the photoresist [15].

The threshold voltage median is at 0.75 V, with an interquartile dispersion of 1.46 V. The fluctuations in threshold voltage may be a result of nanotubes with different chirality, as well as the presence of charge traps in the substrate which could lead to significant fluctuations in threshold voltage, as we have shown earlier [16]. Nonetheless, the threshold voltage distribution, when normalized with respect to the gate voltage sweep range, is comparable to those reported by Park et al. [8] and Franklin et al. [7].

Further information on the spatial variations of these parameters on the wafer may be obtained through wafer maps shown in Figure 5. For plotting these wafer maps,

Figure 5: Wafer maps for a) Minimum resistance and b) threshold voltage, indicating fluctuations over the wafer. The wafer map is plotted combining devices from two adjacent chips together.

the devices from two adjacent chips were combined in order to have a sufficient number of devices. Each chip on average has about 16 n-type devices, so combining two adjacent chips results in about 30 devices on average for each of the plotted points in Figure 5.

Figure 6 shows a scatter plot between the minimum device resistance and the threshold voltage, where each point corresponds to the threshold voltage and ON-resistance values from Figure 5. The ON-resistance scale is plotted in log-scale for ease of visualization. The Pearson coefficient of this dataset is 0.27, which suggests only a weak correlation for the sample set at a 10% significance level. The calculated p-value assuming the presence of no correlation as the null-hypothesis is 0.03, suggesting that the presence of correlation cannot be ruled out at a significance level of 0.05 [17].

This analysis cannot exclude the possibility that the spatial variations in device parameter distributions across the wafer are the result of a systematic variation in the process. For instance, non-uniform exposure of the photoresist could lead to repeatable patterns of areas on the wafer with different amounts of process residues. Such problems can be monitored over different runs, and run-to-run consistencies in patterns on the wafer can be tracked. Pattern matching algorithms can be employed, for instance, to track such variations and thereby provide useful information for process development.

CONCLUSIONS

We have presented a framework that can extract device parameter distributions of individual CNFETs for monitoring complex fabrication processes which employ

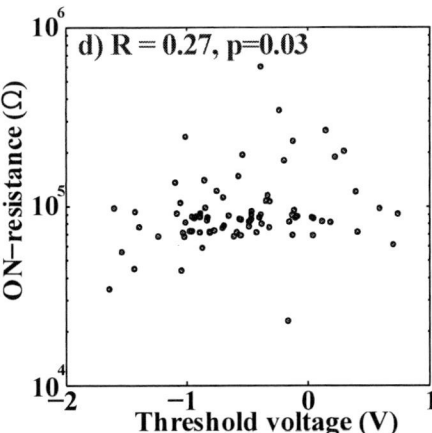

Figure 6: Scatter-plot of the mean ON-resistance and threshold voltage values shown in Figure 5. The Pearson correlation coefficient is estimated to be 0.27, and the p-value as 0.03.

CNFETs as their basic elements, such as pressure sensors and gas sensors. The devices were fabricated on a 100 mm wafer using a deep-UV (220 nm) lithography process. Electrical measurements on the wafer yielded data from 4463 devices, of which 2702 were semiconducting. Device parameters such as threshold voltage, device resistance, device type and their variations over the wafer were quantified. A median device resistance of 122 kΩ and a median threshold voltage of -0.75 V were obtained. The versatility of the framework is also expected to make it widely applicable for use in sensor fabrication processes using nanotube transistors as their functional elements.

ACKNOWLEDGEMENTS

We thank Shih-Wei Lee, Valentin Döring and Matthias Muoth for their support. We are grateful to the COST action MP0901 for financial support. The support from the staff at the FIRST and BRNC clean rooms at ETH Zurich are also acknowledged.

REFERENCES

[1] B. R. Burg, T. Helbling, C. Hierold, and D. Poulikakos, "Piezoresistive pressure sensors with parallel integration of individual single-walled carbon nanotubes," *Journal of Applied Physics,* vol. 109, 2011.

[2] K. Chikkadi, C. Roman, L. Durrer, T. Suess, R. Pohle, and C. Hierold, "Scalable Fabrication of Individual SWNT Chem-FETs for Gas Sensing," *Procedia Engineering,* vol. 47, pp. 1374-1377, 2012.

[3] A. Vijayaraghavan, S. Blatt, D. Weissenberger, M. Oron-Carl, F. Hennrich, D. Gerthsen, *et al.*, "Ultra-Large-Scale Directed Assembly of Single-Walled Carbon Nanotube Devices," *Nano Letters,* vol. 7, pp. 1556-1560, 2007.

[4] T. Helbling, C. Roman, and C. Hierold, "Signal-to-Noise Ratio in Carbon Nanotube Electromechanical Piezoresistive Sensors," *Nano Letters,* vol. 10, pp. 3350-3354, 2010.

[5] T. Zhang, S. Mubeen, N. V. Myung, and M. A. Deshusses, "Recent progress in carbon nanotube-based gas sensors," *Nanotechnology,* vol. 19, p. 332001, 2008.

[6] K. Chikkadi, M. Muoth, V. Maiwald, C. Roman, and C. Hierold, "Ultra-low power operation of self-heated, suspended carbon nanotube gas sensors," *Applied Physics Letters,* vol. 103, p. 223109, 2013.

[7] A. D. Franklin, G. S. Tulevski, S.-J. Han, D. Shahrjerdi, Q. Cao, H.-Y. Chen, *et al.*, "Variability in Carbon Nanotube Transistors: Improving Device-to-Device Consistency," *ACS Nano,* vol. 6, pp. 1109-1115, 2012.

[8] H. Park, A. Afzali, S. J. Han, G. S. Tulevski, A. D. Franklin, J. Tersoff, *et al.*, "High-density integration of carbon nanotubes via chemical self-assembly," *Nature Nanotechnology,* vol. 7, pp. 787-791, 2012.

[9] S. M. Khamis, R. A. Jones, and A. T. C. Johnson, "Optimized photolithographic fabrication process for carbon nanotube devices," *AIP Advances,* vol. 1, p. 022106, 2011.

[10] K. Chikkadi, M. Haluska, C. Hierold, and C. Roman, "Process Control Monitors for Individual Single-walled Carbon Nanotube Transistor Fabrication Processes," in *IEEE International Conference on Microelectronic Test Structures (ICMTS), 2013* pp. 173-177.

[11] J. Xiao, S. Dunham, P. Liu, Y. Zhang, C. Kocabas, L. Moh, *et al.*, "Alignment Controlled Growth of Single-Walled Carbon Nanotubes on Quartz Substrates," *Nano Letters,* vol. 9, pp. 4311-4319, 2009.

[12] K. Chikkadi, C. Roman, and C. Hierold, "Process control monitors for single-walled carbon nanotube based sensor fabrication processes," in *Micro Electro Mechanical Systems (MEMS), 2013 IEEE 26th International Conference on,* 2013, pp. 275-278.

[13] I. Martin-Fernandez, M. Sansa, M. J. Esplandiu, E. Lora-Tamayo, F. Perez-Murano, and P. Godignon, "Massive manufacture and characterization of single-walled carbon nanotube field effect transistors," *Microelectronic Engineering,* vol. 87, pp. 1554-1556, 2010.

[14] X. Zhang, D. Chenet, B. Kim, J. Yu, J. Tang, C. Nuckolls, *et al.*, "Fabrication of hundreds of field effect transistors on a single carbon nanotube for basic studies and molecular devices," *Journal of Vacuum Science & Technology B,* vol. 31, p. 06FI01, 2013.

[15] W. Liu, K. Chikkadi, S.-W. Lee, C. Hierold, and M. Haluska, "Improving non-suspended carbon nanotube FET performance by using an alumina protective layer," *Sensors and Actuators B: Chemical,* vol. 198, pp. 479-486, 2014.

[16] K. Chikkadi, M. Muoth, W. Liu, V. Maiwald, and C. Hierold, "Enhanced Signal-To-Noise Ratio in Pristine, Suspended Carbon Nanotube Gas Sensors," *Sensors and Actuators B: Chemical,* vol. 196, pp. 682-690, 2014.

[17] S. Goodman, "A Dirty Dozen: Twelve P-Value Misconceptions," *Seminars in Hematology,* vol. 45, pp. 135-140, 2008.

CONTACT:

*K. Chikkadi, tel: +41446322538, kiranc@ethz.ch

FABRICATION OF A MONOLITHIC CARBON MOLD FOR PRODUCING A MIXED-SCALE PDMS CHANNEL NETWORK USING A SINGLE MOLDING PROCESS

Yunjeong Lee[1], Yeongjin Lim[1], and Heungjoo Shin[1]

[1]Ulsan National Institute of Science and Technology (UNIST), ULSAN, KOREA

ABSTRACT

We introduce a novel batch fabrication technique of a monolithic carbon mold for producing a mixed-scale polydimethylsiloxane (PDMS) channel network consisting of nanochannels and microchannels with micro-pillars. Nano- fluidics has attracted much attention because of their distinguishing properties. However, the research on the nanofluidics is limited by complex nanofabrication techniques. In this paper, a mixed-scale monolithic carbon mold was simply fabricated using a batch carbon-MEMS process consisting of two successive UV-lithography processes and a single pyrolysis. Then, PDMS channel networks was completed by soft molding process. By modulating pyrolysis conditions, the surface energy of the pyrolyzed carbon mold could be optimized for efficient PDMS channel demolding.

INTRODUCTION

The nanofluidic channel is the nano-conduit including at least one-dimension below 100-nm [1]. Because of special phenomena found in nanofluidc devices such as ion rectification effect [2], ion concentration polarization [3], nanocapillarity [4] that are not found in micro/macro scale channels, nanofluidic devices are used in various applications including nanofluidic diode [5], water desalination [6], biosensor [7] and nanoelectroporation [8]. The microfluidic channel is essential to supply fluid into the nanofluidic channel. Various fabrication technologies for the mixed-scale fluidic devices have been reported such as e-beam lithography (EBL) [9], focused ion-beam (FIB) milling [10] and nano-imprint lithography (NIL) [11]. However, complex, time-consuming processes and high production cost as well as difficulty in alignment between microstructure and nanostructure limit their applications. Soft lithography using a polymer replica overcame the low yield problem even though it required FIB milling processes for nanostructures [12]. Although soft lithography enables cost effectiveness and mass production, low modulus of the soft materials such as soft PDMS limits broad application because of the limitations such as roof collapse, pairing, sagging and shrinking. These limitations could be overcome using hard PDMS owing to its high modulus [13]. Some research groups introduced simple fabrication of PDMS nanochannels using the roof collapse of the soft PDMS microchannels [14]. However, the roof collapse method also requires separate microchannel fabrication.

In this research, carbon MEMS technology and soft molding using hard PDMS were used for direct molding of mixed-scale PDMS channel network without using a replica. The pyrolysis process in carbon MEMS converts micro-sized polymer structures into carbon nanostructures because of a dramatic volume reduction [15, 16] so that

batch fabrication of nano-sized molds (thickness = ~ 50 nm, width = ~ 500 nm) could be fabricated using conventional UV lithography. Therefore, two UV lithography processes using a thin photoresist layer and a thick photoresist layer respectively enables the formation of a semi-mixed-scale polymer structure, and the polymer structures can be converted into a monolithic carbon mixed-scale channel mold. The surface energy of the pyrolyzed carbon was changed depending on the pyrolysis temperature. Therefore the pyrolysis temperature was controlled for efficient PDMS channel demolding and thus the monolithic carbon mold could be reused more than 40 times without any anti-adhesion layer coating process. Because of anisotropic volume reduction in pyrolysis, originally vertical polymer side walls were converted into smoothly tapered carbon side walls. Therefore, the PDMS channel network can guide fluids and particles from microchannels to nanochannels more efficiently. This enables effective entrapment of particles such as cells at the junction of microchannel and nanochannel for the application of nanoelectroporation. As the mixed-scale channel material, hard PDMS was used and micro-pillar structures were integrated to the microchannel to prevent channels collapse in glass bonding process.

EXPERIMENTAL

Fabrication of mixed-scale PDMS channel networks

Mixed-scale carbon structure was fabricated using a single pyrolysis process subsequent to two-step UV-lithography processes on a 6-inch Si substrate as shown in Figure 1. First, the polymer structure was fabricated using UV lithography. The bare silicon wafer was spin-coated with a thin SU-8 negative photoresist layer (thickness ~ 300 nm) for a nanoscale carbon channel mold and the substrate was soft baked on the hot plate. After UV exposure, the Si substrate was post-exposure-baked and developed. Then, a thick SU-8 photoresist layer (thickness ~ 10 μm) for a microscale carbon structure was spin-coated on the substrate and baked. After the Si substrate was cooled down slowly for minimizing thermal stress, the rest of photolithography steps were completed. To complete carbon mold fabrication, the semi-mixed-scale polymer structure was heated up to 900°C in the furnace in vacuum and cooled down naturally.

Finally, a hard PDMS layer was spin-coated on the mixed-scale carbon mold without any anti adhesion layer coating and was cured in a convection oven. Then, soft PDMS was poured on top of the carbon mold and placed in a vacuum chamber for degassing bubbles, and cured in the convection oven. The cured PDMS layer was peeled off from the Si substrate and bonded to a glass slide after oxygen plasma treatment.

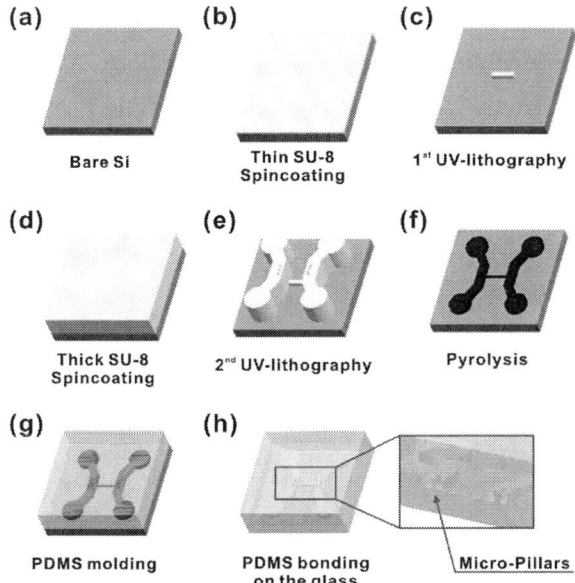

Figure 1: Fabrication steps for a mixed-scale PDMS channel networks.

Surface free energy measurement

Carbon samples for measuring surface energy were prepared by pyrolyzing SU-8 pads on Si substrate at various temperature conditions ranging from 400°C to 900°C. The contact angle was measured using deionized water and diiodomethane (CH_2I_2). The values of contact angles from 10 sample were averaged for the calculation of surface free energy. Surface free energy was calculated based on the Owens-Wendt geometric mean equation [17].

Fluorescence imaging

The functionality of the mixed-scale PDMS channel system was confirmed by filling 0.1mM fluorescein isothiocyanate (FITC) solution into the channel network and taking fluorescent images of the channel network using an inverted microscope. The 0.1mM FITC solution was prepared in the deionized water.

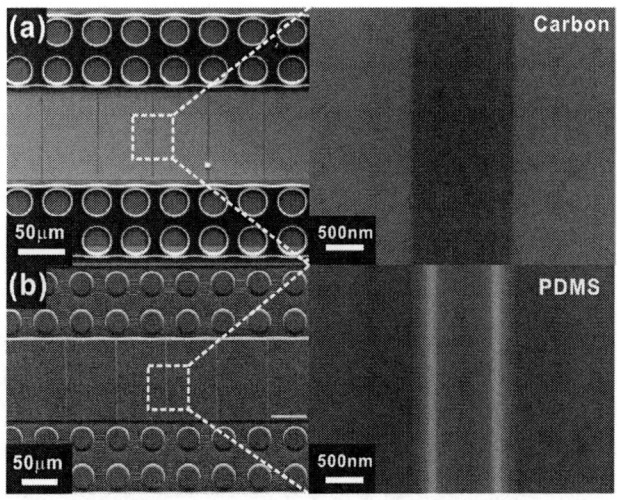

Figure 2: SEM images of (a) pyrolyzed carbon molds and (c) corresponding mixed-scale PDMS channel networks.

RESULTS

Morphology of mixed-scale channel networks

Scanning electron microscopy (SEM) images of mixed-scale carbon molds and PDMS channel networks are shown in Figure 2. A 1-μm-wide, 300-nm-thick, and 100-μm-long SU-8 photoresist structure was successively converted into a 500-nm-wide, 50-nm-thick, and 120-μm-long carbon nanochannel mold as shown in Figure 2 (a). The PDMS nanochannel matched well with the cabon nanochannel mold as shown in Figure 2 (b). Microchannels and micro-pillars were also well replicated from the carbon mold.

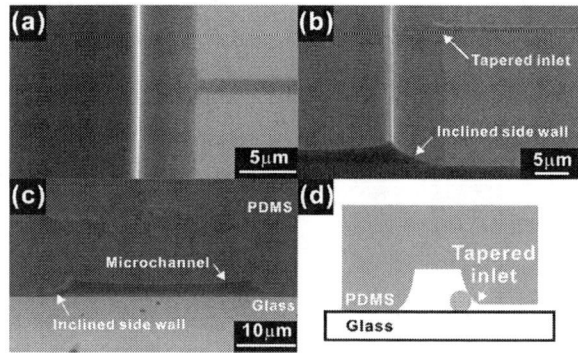

Figure 3: SEM images of a mixed-scale carbon mold; (a) top-view, (b) tilted-view. (c) Section-view of PDMS microchannel. (d) Schematic of an efficient entrapment of a micro-particle at the tapered inlet of a nanochannel.

The side walls of the SU-8 photoresist microstructure was converted into gradually tapered side walls after pyrolysis because of good adhesion of the photoresist structure on the substrate and relatively free volume shrinkage at the top and side walls of the photoresist structure as shown in Figure 3 (a) and (b).. This inclined side wall enables efficient PDMS demolding process and prevents mold wear that can be found frequently at the edges of the mold with vertical side walls. Figure 3 (c) shows a PDMS microchannel with inclined side walls that was bonded on a glass slide.

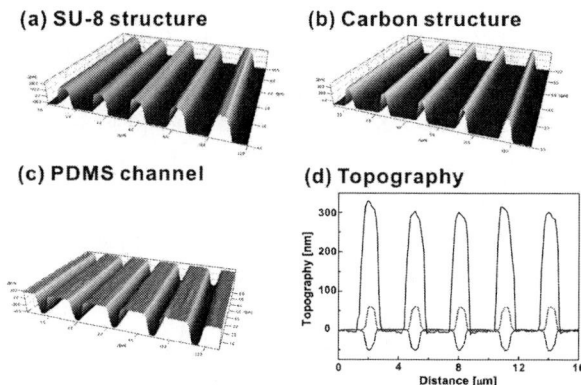

Figure 4: 3D AFM images of (a) SU-8 structures, (b) carbon nanochannel molds, and (c) PDMS nano-channels. (d) Sectioned views of (d) SU-8 structures (black), carbon molds (red) and PDMS channels (blue).

978-1-4799-7956-1/15 $31.00 © 2015 IEEE

The atomic force microscopy (AFM) surface scanning data confirmed the large volume reduction (approximately 83% in thickness reduction) of the nanoscale carbon mold after pyrolysis and efficient PDMS molding process as shown in Figure 4.

Surface free energy of the carbon surface

Figure 5 shows surface free energies of the pyrolyzed carbons at various temperatures (400 °C ~ 900 °C) calculated using contact angle measurement. The surface free energy was increased drastically between 400 °C and 500 °C because of the formation of solid carbon (C_∞) at the temperature region.

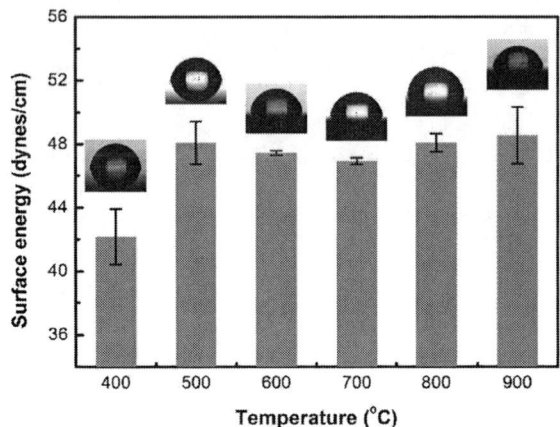

Figure 5: Surface free energy data of carbon structures pyrolyzed at various temperature conditions

Fluorescence imaging

The fluorescent image of the mixed-scale channel network filled with 0.1mM FITC solution confirms uniform nanochannel foramtion without leakage and channel collapse as shown in Figure 6.

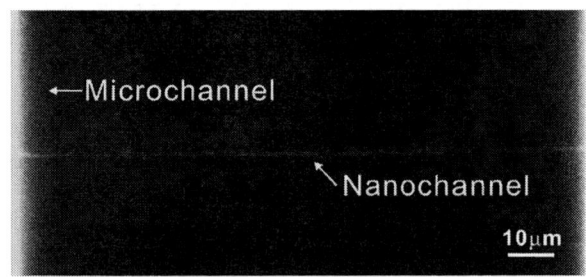

Figure 6: Fluorescence image of the PDMS nanochannel filled with 1mM FITC solution.

CONCLUSION

We introduced a simple batch-fabrication process for the mixed-scale PDMS channel networks using two-step UV-lithography processes, pyrolysis and a single molding process. The dramatic volume reduction in pyrolysis enabled the formation of convex mixed-scale monolithic carbon mold. The inclined side walls and high surface free energy of the carbon molds enhanced the efficiency of the demolding process. The mixed-scale carbon molds are expected to be used in nanoelectroporation for cell transfection with improved efficiency owing to the novel architecture of gradually tapered side walls as shown in Figure 3 (d).

ACKNOWLEDGEMENTS

This work was supported by Basic Science Research Program through the National Research Foundation of Korea (NRF) funded by the Ministry of Education (NRF-2013R1A1A2009711). We are grateful for technical assistance to the staff members at UCRF (UNIST Central Research Facilities) and UOBC (UNIST Olympus Biomed imaging Center) in UNIST.

REFERENCES

[1] C. Duan, W. Wang, Q. Xie, "Fabrication of Nanofluidic Devices", *Biomicrofluidics*, vol. 7, pp. 026501, 2013.

[2] J. M. Perry, K. Zhou, Z. D. Harms, S. C. Jacobson, "Ion Transport in Nanofluidic Funnels", *ACS Nano*, vol. 4, pp. 3897-3902, 2010.

[3] T. A. Zangle, A. Mani, J. G. Santiago, "Theory and Experiments of Concentration Polarization and Ion Focusing at Microchannel and Nanochannel Interfaces", *Chem. Soc. Rev.*, vol. 39, pp. 1014-1035, 2010.

[4] J. W. van Honschoten, N. Brunets, N. R. Tas, "Capillarity at the Nanoscale", *Chem. Soc. Rev.*, vol. 39, pp. 1096-1114, 2010.

[5] L. Cheng, L. J. Guo, "Nanofluidic Diodes", *Chem. Soc. Rev.*, vol. 39, pp. 923-938, 2010.

[6] S. J. Kim, S. H. Ko, K. H. Kang, J. Han, "Direct Seawater Desalination by Ion Concentration Polarization", *Nat. Nanotechnol.*, vol. 5. pp. 297-301, 2010.

[7] A. E. Muñiz, A. Merkoçi, "A Nanochannel/ Nanoparticle-Based Filtering and Sensing Platform for Direct Detection of a Cancer Biomarker in Blood", *Small*, vol. 7, pp. 675-682, 2011.

[8] P. E. Boukany, A. Morss, W. Liao, B. Henslee, H. Jung, X. Zhang, B. Yu, X. Wang, Y. Wu, L. Li, K. Gao, X. Hu, X. Zhao, O. Hemminger, W. Lu, G. P. Lafyatis, L. J. Lee, "Nanochannel Electroporation Delivers Precise Amounts of Biomolecules into Living Cells", *Nat. Nanotechnol.*, vol. 6, pp. 747-754, 2011.

[9] R. Yokokawa, Y. Yoshida, S. Takeuchi, T. Kon, H. Fujita, "Undirectional Transport of a Bead on a Single Microtubule Immobilized in a Submicrometre Channel", *Nanotechnology*, vol. 17, pp. 289-294, 2006.

[10] L. D. Menard, J. M. Ramsey, "Fabrication of Sub-5nm Nanochannels in Insulating Substrates Using Focused Ion Beam Milling", *Nano Lett.*, vol. 11, pp. 512-517, 2010.

[11] J. Wu, R. Chantiwas, A. Amirsadeghi, S. A. Soper, S. Park, "Complete Plastic Nanofluidic Devices for DNA Analysis via Direct Imprinting with Polymer Stamps", *Lab chip*, vol. 11, pp. 2984-2989, 2011.

[12] F.I. Uba, S.R. Pullaguria, N. Sirasunthorn, J. Wu, S.

Park, R. Chantiwas, Y.K. Cho, H. Shin, S. A. Soper, "Surface Chanrge, Electroosmotic Flow and DNA Extension in Chemically Modified Thermoplastic Nanoslits and Nanochannels", *Analyst*, DOI: 10.1039/c4an01439a.

[13] H. Schmid, B. Michel, "Siloxane Polymers for High-Resolution, High-Accuracy Soft Lithography", *Macromolecules*, vol. 33, pp. 3042-3049, 2000.

[14] J. Heo, H. J. Kwon, H. Jeon, B. Kim, S. Kim, G. Lim, "Ultra-high-aspect-orthogonal and Tunable Three Dimensional Polymeric Nanochannel Stack Array for BioMEMS Applications", *Nanoscale*, vol. 6, pp. 9681-9688, 2014.

[15] Y. Lim, J. Heo, M. Madou, H. Shin, "Monolithic Carbon Structures Including Suspended Single Nanowires and Nanomeshes as a Sensor Platform", *Nanoscale Res. Lett.,* vol. 8, pp. 492-450, 2013.

[16] Y. Lim, J. Heo, H. Shin, "Fabrication and Application of a Stacked Carbon Electrode Set Including a Suspended Mesh Made of Nanowires and a Substrate-Bound Planar Electrode toward for an Electrochemical/ Biosensor Platform", *Sens. Actuator B-Chem.,* vol. 192, pp. 796-803, 2014.

[17] D.K. Owens, R.C. Wendt, "Estimation of the Surface Free Energy of Polymers", *J. Appl. Polym. Sci.*, vol. 13, pp. 1741-1747, 1969.

CONTACT

*H. Shin, tel: +82-52-217-2315, hjshin@unist.ac.kr

FABRICATION OF PATTERNABLE NANOPILLARS FOR MICROFLUIDIC SERS DEVICES BASED ON GAP-INDUCED UNEVEN ETCHING

Y. Wang[1,2], L. C. Tang[1,3], H. Y. Mao*[1,2], C. Lei[1,3], W. Ou[1,2], J. J. Xiong[3], Y. Ou[1,2], A. J. Ming[1,2],
D. Li[4] and D. P. Chen[1,2]

[1]Key Laboratory of Microelectronics Devices & Integrated Technology, Institute of Microelectronics,
Chinese Academy of Sciences, Beijing 100029, P. R. China
[2]Smart Sensor Engineering Center, Jiangsu R&D Center for Internet of Things,
Wuxi 214135, P. R. China
[3]National Key Laboratory for Electronic Measurement Technology, North University of China,
Taiyuan 030051, P. R. China
[4]Department of Electrical Engineering, Stanford University, California, 94305, United States

ABSTRACT

In this work, a lithography-free approach for fabricating patternable nanopillars is reported. The key technique of the approach is to introduce a gap over the substrate by covering it with a cap, which contains through holes and the material on its lower surface has similar etching rate with the substrate. By this means, uneven etching of the substrate is induced under the through holes, consequently, nanopillars are fabricated into patterns corresponding to the holes. By sputtering a thin layer of Ag on the nanopillar patterns, obvious Raman enhancement can be observed. Since the large areas around the patterns are protected from anisotropic etching and metal sputtering, they are flat enough to be bonded with PDMS caps thus to form microfluidic Surface-enhanced Raman Scattering devices, and the whole fabrication process for the devices is simplified.

INTRODUCTION

Since Surface-enhanced Raman Scattering (SERS) effect was discovered, a way of trace detection, even at single molecule level [1], has been paved due to the continuous innovations and breakthroughs in research of SERS-active substrates. So far, the SERS substrates have been increasingly applied in various fields, such as biomedical, DNA sequencing [2, 3], agricultural and environmental analysis [4], in which the capability of quantitative and real-time detection is of great importance [5]. Similar to the SERS-active substrates, microfluidic SERS devices have advantages of high sensitivity, while in the mean time, they are capable of reaching a real-time detection and distributing analytes uniformly by avoiding coffee-ring effect [6] and thus to achieve relatively low measurement errors [5].

As have been reported, microfluidic SERS devices usually contain SERS-active substrates, which are covered by caps with microchannels [5, 7]. To achieve surface plasma resonance thus to realize SERS effect on the substrates as well as in the devices, nanoscale roughness or structures are required. Recently, various nanopillars are reported to function as the nanostructures, and meanwhile, preparation of the nanopillars become crucially important for the realization of high-performance microfluidic SERS devices. So far, different methods for fabricating nanopillars have been studied and applied extensively, including an oxygen-plasma-stripping-of-photoresist

(OPSOP) technique [8] and other approaches which can fabricate highly repeatable nanostructures, such as electron beam (EB) lithography, focused ion beam (FIB) etching [9], nanosphere lithography (NSL) [10] and nanoimprint lithography [11]. Both EB and FIB can realize extremely tiny structures in arrays, however, the nature of serial processes demands a comparatively lengthy exposure time, thus their broader range of applications are restricted. NSL combining the advantages of both top-down and bottom-up approaches needs monolayer nanoparticles in large areas, which is of difficulty and increases complexity of the process [12]. OPSOP technique has a good controllability over dimensions and densities of the nanopillars, but the process requires several material-deposition and etching steps, which are time-consuming and also involve regions around the nanopillar patterns. As a result, bonding the substrates to PDMS caps would become difficult because of the surface roughness.

In this work, a novel, simple and lithography-free method for fabricating desirable nanopillars patterns is proposed, and the approach can provide a fairly smooth surface around the patterns. Based on this method, the bonding between PDMS caps and SERS-active substrates becomes much easier. As a consequence, a new way for fabricating microfluidic SERS devices is further developed.

FABRICATION

In single-step experiments to figure out etching parameters, small pieces of wafers are usually attached to larger substrates and then suffered to etching processes. As taken for granted, the areas on the substrates revealed to etching gases would be etched evenly, thus a relatively smooth surface could be obtained. However, in our experiments, nanopillar patterns with certain widths were observed on the substrates around the margins of the small wafers. Based on such observations, a novel lithography-free approach for fabricating patternable nanopillars is proposed.

The fabrication process for patternable nanopillars is schematically depicted in Fig. 1. At the beginning of the process, through holes are perforated in a cap, and the lower surface of the cap is covered with a substrate-alike material layer, which has similar etching rate with substrate, as illustrated in Fig. 1(a). Then, the cap is attached to the substrate using adhesive tapes, which are of certain thicknesses, therefore, a gap is introduced between the cap and the substrate (shown in Fig. 1(b)). Afterwards, reactive ion etching (RIE) is adopted to etch the substrate. By this means, nanopillars are patterned within regions corresponding to the through holes (shown in Fig. 1(c)).

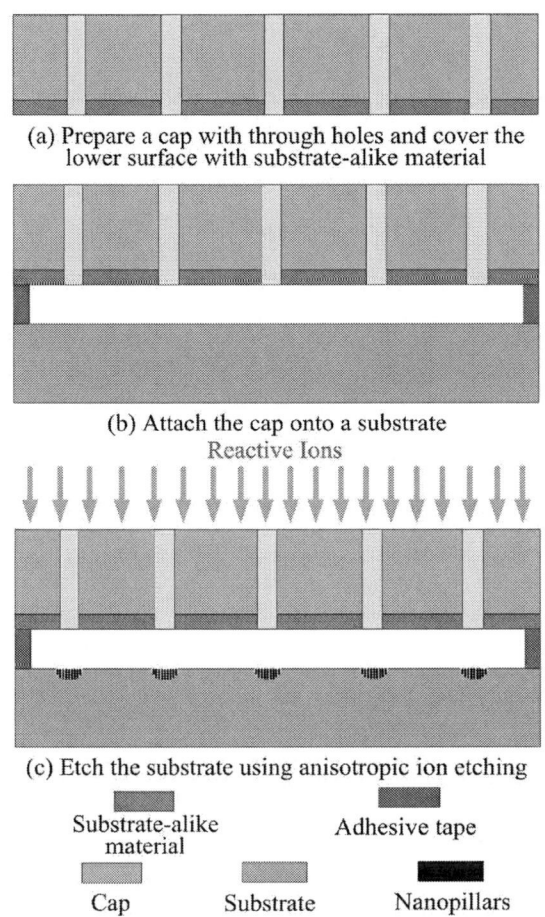

(a) Prepare a cap with through holes and cover the lower surface with substrate-alike material

(b) Attach the cap onto a substrate

Reactive Ions

(c) Etch the substrate using anisotropic ion etching

Substrate-alike material   Adhesive tape

Cap   Substrate   Nanopillars

Figure 1: Fabrication process of nanopillar patterns

More specifically, in the experiment to obtain silicon nanopillar patterns, a glass wafer with through holes was used as a cap, α-Si used as the substrate-alike material and Si wafer as the substrate. When the cap was attached to the substrate, a ~200 μm gap was introduced between the two layers, and then RIE was adopted to etch the Si substrate through the holes. In the etching, RF power was 350 W, Cl_2/He flow rates were 180/400 sccm. As a result of 180 seconds of etching, patternable nanopillars were formed. Figure 2 demonstrates the scanning electron microscopy (SEM) images of the obtained nanopillars under a circular through-hole (with a diameter of ~400 μm), where, nanopillars are distributed in patterns with a width of ~60 μm along the border of a large circle (Fig. 2(a)). Figure 2(b) exhibits a top-down view of Region B in (a), where nanopillars with different densities are displayed. The margin of a pattern that has a nanopillar region inside and a slightly etched area outside is displayed in Fig. 2(c), and this figure is also a cross-section view of Region C in (a). Magnified images of the nanopillars distributed in different regions, e.g. Region E, F, G, and H, are shown in Fig. 2(d)~(h).

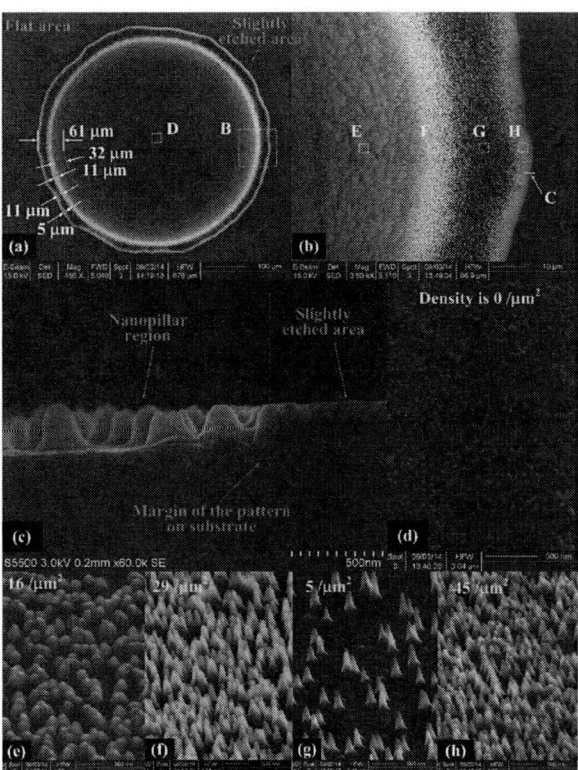

Figure 2: SEM images of a nanpillar pattern and magnified images at different positions of the pattern. (a) large circular pattern with nanopillars; (b) top-down view of Region B in (a); (c) cross-section view of Region C in (b); (d)-(h) magnified images of nanopillars in regions from D to H in (a) and (b), in which the bar is 500 nm.

METHODOLOGY

The mechanism for forming the nanopillars can be ascribed to the varying and non-uniform ion distributions caused by the gap and the substrate-alike material, the details are illustrated in Fig. 3. On account of the reason that almost every ion going directly through the center of the holes reacts with the substrate, the concentration of the etching ions that reach the hole centers is maintained at a high level, thus at these positions, the etching rate reaches its highest. Near the hole-edges, nevertheless, the substrate-alike material and the substrate surface are revealed to etching gases, thus part of the ions would transfer laterally into the gap, then the ion concentration at these positions is decreased, which, as a result, gives rise to a slower etching rate. Simultaneously, ion concentration near the hole-edges is non-uniformly distributed for the same reason. Consequently, an uneven etching of the substrate leading to nanopillars is achieved. Figure 3 also shows a magnified graph of nanopillars distributed within a region corresponding to a perforation. As shown in Fig. 3, Region 1 presents the very short and sparse nanopillars around the central areas of the circle where the material is almost fully removed. Conversely, Region 2 demonstrates an inadequate reaction between the Si substrate and the ions, therefore, obvious nanopillars are observed in this region where the ions are non-uniformly distributed. Within the gap, on the other hand, the anisotropic features of the ions is reduced, thus the etching rate in this region decreases rapidly, subsequently, nanopillars are generated only within small regions near the hole-edges. This is also the reason for the generation of nanopillars in Region 3. In this way, by adjusting sizes and profiles of the perforations, nanopillars with desirable patterns can be obtained.

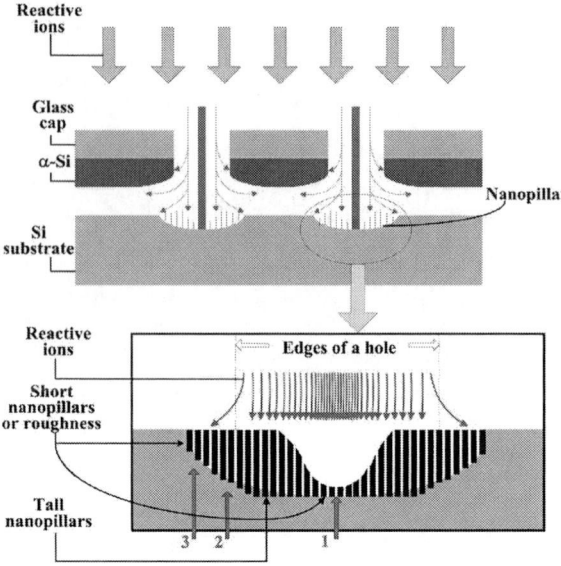

1: Central areas on the Si substrate are etched by reactive ions with high concentration
2: Marginal areas on the Si substrate are etched by reactive ions with lower concentration
3: Part of reactive ions transfer into the gap and react with α-Si and Si substrate with a low vertical speed, the reaction only takes place near the edges of the hole

Figure 3: Schematic graph for the mechanism of the gap-induced-uneven-etching approach, by using glass with through holes as a cap, α-Si as the substrate-alike material, Si as the substrate. Nanopillar morphologies are different at different positions, but can be controlled by adjusting width of the holes in the cap.

CHARACTERIZATION AND APPLICATION

SERS features of the nanopillars formed in this work were tested by sputtering a thin Ag layer (50 nm) on them and using R6G solutions with different concentrations as analytes. A 1 μL droplet of R6G analyte with concentration of 1 μg/L (Analyte A) was dropped on a nanopillar-based SERS substrate, and the substrate was with a surface area of 1.0×1.0 cm^2. The SERS spectrum of Analyte A at the position of Region B in Fig. 2 (a) is illustrated as Curve 1 in Fig. 4. As for comparisons, a 1 μL droplet of R6G analyte with concentration of 2 mg/L (Analyte B) was dropped on a glass slide with the same surface area. Curve 2 in Fig. 4 represents the spectrum of Analyte B. By comparing these two curves, an enhancement factor around 10^7 can be obtained, demonstrating potential applications of nanopillars in microfluidic SERS devices.

In conventional fabrication processes for nanopillar-based SERS-active substrates, it is difficult to get a very smooth surface around the patterns, as these areas are also suffered to anisotropic etching processes.

Therefore, bonding the substrates to particular caps seems to be uneasy.

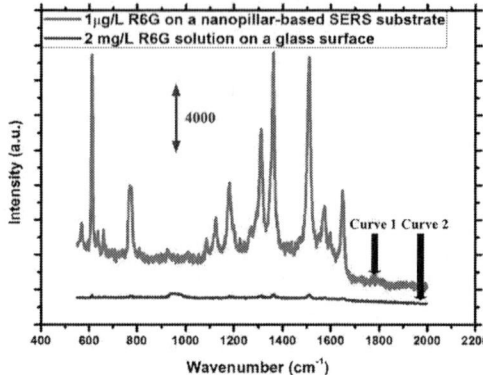

Figure 4: SERS measurement results, with an enhancement factor around 10^7, demonstrating potential applications of nanopillars in microfluidic SERS devices.

Fortunately, the method proposed in this work provides a way to obtain a very flat surface around the nanopillar patterns, thus it has potential to overcome the problem. To testify the smooth features of the areas beside the nanopillar patterns, atomic force microscopy (AFM) was adopted to analyze the roughness, for comparison, a polished Si wafer was also detected. The AFM images are shown in Fig. 5, which demonstrates that the surfaces of the areas are as smooth as the polished Si wafer. Therefore, the bonding in fabrication of microfluidic SERS devices would be easily achieved.

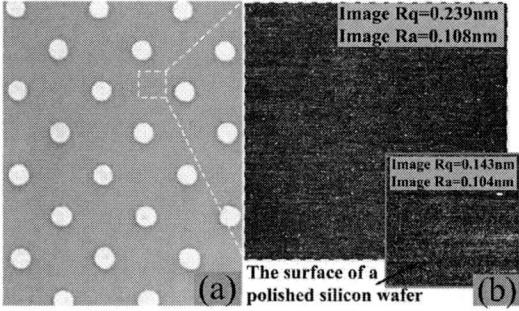

Figure 5: (a) Ag patterned in nanopillar regions realized by the gap-induced-uneven-etching approach; (b) roughness comparison between areas around the Ag patterns and surfaces on a polished Si wafer.

With all these features, a process for fabricating SERS devices based on our approach is proposed, which is quite simple and can be easily managed, as is schematically depicted in Fig. 6. Firstly, the method described in Fig. 1 is utilized to generate nanopillar patterns (shown in Fig. 6(a)). Subsequently, the cap with perforations is further used as masks in Ag sputtering. For the substrate with nanopillars, only regions corresponding to the perforations are revealed, thus only these regions are covered with Ag, and the surface beside them are well protected (Fig. 6(b)). In this way, lithography and lift-off technique for patterning the Ag layer become unnecessary. Finally, the cap is taken off from the substrate, and a PDMS cap with microchannels is bonded to the substrate. Consequently, a microfluidic SERS device is achieved (Fig. 6(c)).

978-1-4799-7956-1/15 $31.00 © 2015 IEEE   322

Moreover, such a gap-induced-uneven-etching approach can also be applied to produce nanopillars of other materials, including SiO_2, Si_3N_4, metal, glass and others. Figure 7 shows SiO_2 nanopillars generated by using a glass cap over a Si substrate covered with SiO_2. The average diameter of the SiO_2 nanopillars is about 100 nm, and the density is about 30 /μm^2. Since the features of glass are similar to those of SiO_2, thus it is possible that nanopillar patterns can be obtained on glass substrates. Therefore, transparent SERS devices become realizable.

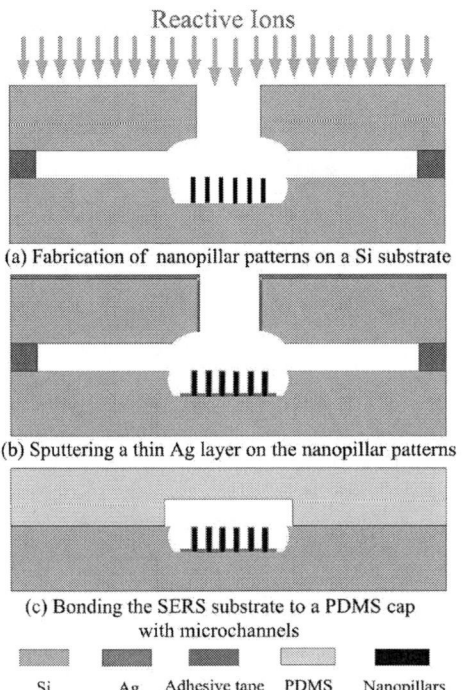

(a) Fabrication of nanopillar patterns on a Si substrate

(b) Sputtering a thin Ag layer on the nanopillar patterns

(c) Bonding the SERS substrate to a PDMS cap with microchannels

| Si | Ag | Adhesive tape | PDMS | Nanopillars |

Figure 6: Fabrication process of microfluidic SERS devices.

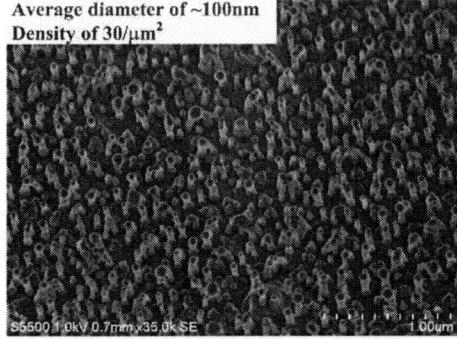

Figure 7: A large area of SiO_2 nanopillars achieved by the gap-induced-uneven-etching approach. In the fabrication, a SiO_2 layer covered by a glass cap with through holes is etched.

CONCLUSIONS

A gap-induced-uneven-etching technique for fabricating patternable nanopillars is proposed, which is simple, fast and lithography-free. Based on this approach, a simplified fabrication process for microfluidic SERS devices is further developed. Similarly, such an approach can be applied to other materials, including SiO_2, which

makes transparent SERS devices possible. Meanwhile, based on this approach, a new route for nanopillar patterns to be applied in other fields is opened up.

ACKNOWLEDGMENT

This work was supported in part by Jiangsu Natural Science Foundation (Grant No. BK20131098), the National Natural Science Foundation of China (Grant No. 61401458, 61335008 & 61176114). Special acknowledgements are due to Prof. W.G. Wu and Mr. Y. F. Mao from Institute of Microelectronics, Peking University for their help with the experiment and preparation of SEM images.

REFERENCES

[1] M. Fleischman, P. J. Hendra, A. J. McQuillan, "Raman-Spectra of Pyridine Adsorbed at a Silver Electrode," *Chem. Phys. Lett.*, vol.26, pp. 163-166, 1974.

[2] S. J. Park, T. A.Taton, C. A. Mirkin, "Array-Based Electrical Detection of DNA using Nanoparticle Probes," *Science,* vol. 295, pp. 1503-1506, 2002

[3] L. A. Gearheart, H. J. Ploehn, C. J. Murphy, "Oligonucleotide Adsorption to Gold Nanoparticles: A Surface-Enhanced Raman Spectroscopy Study of Intrinsically Bent DNA," *J. Phys. Chem.* B, vol. 105, pp. 12609-12615, 2001.

[4] D. A. Stuart *et al.*, "In Vivo Glucose Measurement by Surface-Enhanced Raman Spectroscopy," *Anal. Chem.,* vol. 78, pp. 7211, 2006.

[5] H. Y. Mao *et al.*, "Microfluidic Surface-Enhanced Raman Scattering Sensors Based on Nanopillar Forests Realized by an Oxygen-Plasma-Stripping-of-Photoresist Technique," *Small*, vol. 10, pp. 127-134, 2014.

[6] X. Y. Shen, C. M. Ho, T. S. Wong, "Minimal Size of Coffee Ring Structure," *J. Phys. Chem.* B, vol. 114, pp. 5269-5274, 2010.

[7] H. Y. Mao *et al.*, "Silicon Nanopillar-Forest Based Microfluidic Surface-Enhanced Raman Scattering Devices," *IEEE MEMS 2011*, pp. 968-971, Jan. 2011.

[8] H. Y. Mao *et al.*, "The Fabrication of Diversiform Nanostructure Forests Based on Residue Nanomasks Synthesized by Oxygen Plasma Removal of Photoresist." *Nanotechnology*, vol. 20, pp. 445304, 2009.

[9] L. Xia, W. Wu, J. Xu, Y. Hao, Y. Wang, "3D Nanohelix Fabrication and 3D Nanometer Assembly by Focused Ion Beam Stress-Introducing Technique," *IEEE MEMS 2006*, pp. 118-121, Jan. 2006.

[10] K. Kempa, *et al.*, "Photonic Crystals Based on Periodic Arrays of Aligned Carbon Nanotubes," *Nano Lett.*, vol. 3, pp.13-18, 2003.

[11] R. Alvarez-Puebla *et al.*, "Nanoimprinted SERS-active Substrates with Tunable Surface Plasmon Resonances," *J. Phys. Chem.* C, vol. 111, 6720, 2007.

[12] P. Colson, C. Henrist, R. Cloots, "A Powerful Method for the Controlled Manufacturing of Nanomaterials," *J. Nano Mat.,* vol. 1, pp. 948510, 2013.

CONTACT

*H.Y. Mao, tel: 86-10-82995934; maohaiyang@ime.ac.cn

A LOW-COST AND LABEL-FREE ALPHA-FETOPROTEIN SENSOR BASED ON SELF-ASSEMBLED GRAPHENE ON SHRINK POLYMER

Shota Sando, Bo Zhang, and Tianhong Cui
University of Minnesota, Minneapolis, USA

ABSTRACT

Self-assembled graphene by layer-by-layer (LbL) self-assembly on a shrink polymer substrate demonstrates nano wrinkles after heating. As a result, graphene with nano wrinkles shows different wettability and surface roughness, depending on shrink temperatures. The graphene on shrink polymer performs sensing applications for pH and alpha-fetoprotein (AFP) detection with advantages of low cost and label free, due to self-assembly technique, ease of modification, and antigen-antibody surface reaction.

INTRODUCTION

Graphene has exhibited intriguing properties on an electron scale since its discovery [1]. There have been tremendous applications of graphene reported in the last decade. As graphene has gained increased exposure to the scientific community, researchers continue to seek further applications [2]. While getting easier to produce graphene by today's technology such as chemical vapor deposition (CVD) [3], it still relies on complicated systems which cost much funds and take long time.

LbL self-assembly methods presented in this work promises to be simple and widely applicable to fabrication process. We report experimental results in controlling surface morphology with both shrink-induced polymer structures and self-assembled graphene nanoplatelets. Through our experiments we produced tunable shrink induced graphene on shape-memory polymers adjusted by shrink temperatures, promising to widen the range of graphene applications to molecular detections and high-surface-area conductors [4]. In addition, we were able to change the feature size of nanowrinkles, roughness, and wettability. Compared with conventional fabrication methods of graphene nanostructures and devices, the combination of LbL self-assembly and shape memory polymers as the substrate demonstrates a relatively low-cost and simple process, because there are no requirements for sophisticated systems such as CVD, photolithography, and e-beam lithography.

This work shows that sensors based on shrink-induced graphene are applicable to label-free molecule detections. While Enzyme-linked immunosorbent assay (ELISA) is still the most popular method to detect specific proteins in today's medicine, label-free biosensors have attracted great attention due to their simplicity and ease of use [5]. The sensor described in this paper demonstrates the detection of alpha-fetoproteins (AFP), one of the most important tumor biomarkers associated with liver cancer and ovarian cancer. The results indicate shrink-induced graphene is practically applicable to high-performance biosensors.

SENSOR FABRICATION

LbL self-assembly in this work requires three types of charged suspensions to form graphene sheets, including poly(diallyldiamine chloride) (PDDA, positively charged), poly(styrene sulfonate) (PSS, negatively charged), and pristine graphene platelets solution. The concentrations of PDDA and PSS solutions were 1.5 and 0.3wt%, respectively. Pure Sheets MONO (1 or 2 layers of graphene, 0.25 mg/mL) from Nanointegris, Inc., was used as the graphene suspension solution (negatively charged).

Polystyrene (PS) was selected as a shrink polymer substrate. The sequence of immersion of the PS substrate is: [PDDA (10 min) + PSS (10 min)]$_2$ + [PDDA (10 min) + graphene suspension (20 min)]$_5$. Graphene sheets were formed on the PS substrate (Fig. 1 Bottom).

The PS substrate can be deformed and shrunk by heating over 100°C. On the other hand, the graphene sheets are non-shrunk materials. Hence, heating the PS substrate with graphene shrinks in size, inducing graphene nanowrinkles (Fig. 1 Middle).

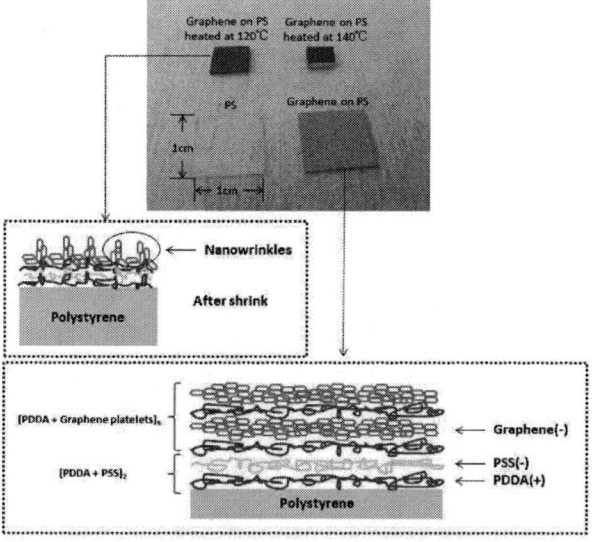

Figure 1: (Top) Image of 1 cm×1 cm polystyrene (PS) diced with and without LbL self-assemnbled graphene, and graphene on PS heated at 120°C and 140°C, showing shrinkage and losing transparency after heating; Structure of LbL self-assembled graphene on PS after shrinkage (Middle) and before shrinkage (Bottom)

Conductive epoxy from Chemtronics, Inc. was used to form electrodes for graphene on a PS substrate after heating (Fig. 2 a). For AFP sensors, poly-L-lysine (PLL) solution (0.1%), bovine serum albumin (BSA) solution (7.5%), dulbecco's phosphate buffer solution (DPBS), and anti-AFP (rabbit, 0.05 mg/ml) were purchased from Sigma-Aldrich. The sensing area was immersed in PLL solution for 1h at room temperature, and then was washed three times by DI water. Next, it was immersed in anti-AFP solution diluted down to 1 µg/ml in a refrigerator for overnight. After washing by DPBS, the sensing area was soaked in BSA solution (2%) at room temperature for 2h, and was washed by DPBS again. The sensor was immediately used after antibody modification.

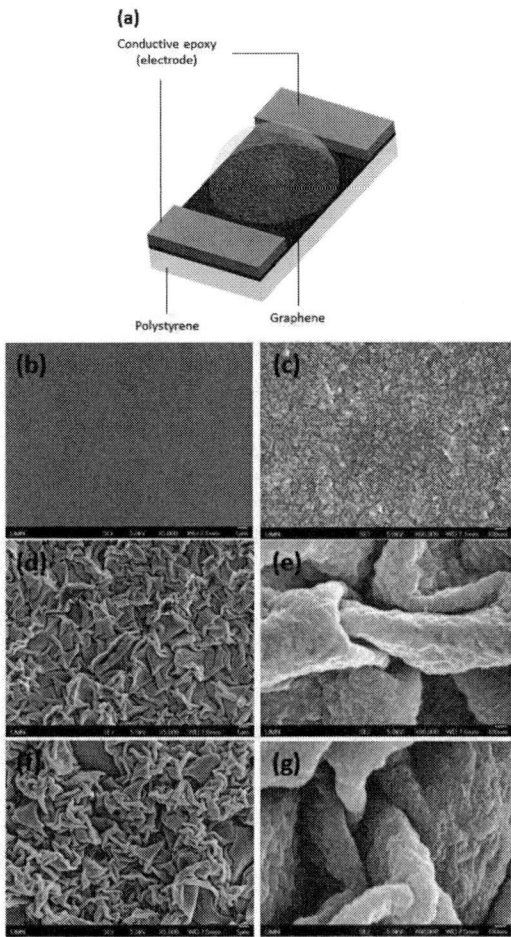

Figure 2: (a) Schematic image of AFP sensor; (b), (c) SEM image of graphene on PS; (d), (e) SEM images of graphene on PS heated at 120°C; (f), (g) SEM images of graphene on PS heated at 140°C. Magnitude for (b),(d), and (f) is ×5,000. Magnitude for (c), (e), and (g) is ×60,000.

RESULTS AND DISCUSSION

Since graphene sheets on a PS substrate are thin layers, the surface shows transparency. When heated mismatch of interface between shrink material (PS) and non-shrunk material (graphene sheets) causes nanowrinkles. As shown in Fig.1 Top, graphene sheets on a PS substrate obviously lost its transparency after heated at 120 and 140 °C, compared with one before shrinkage. SEM images of graphene surface on the PS substrate were taken before and after heating. The surface of graphene sheets before heating was flat, indicating no nanowrinkles (Fig. 2 b, c). The surface after heating showed tremendous amount of nanowrinkles (Fig. 2 d-g). Higher shrink temperatures induced relatively rougher and larger nanowrinkles. To

derive this, we measured the shrink rate of PS substrates. The PS substrate 1 cm by 2 cm were prepared for this experiment, and heated in an oven for 10 min at different temperatures. As shown in Fig. 3, higher shrink temperature causes higher shrinkage rate, and the PS substrate start to shrink at 100°C. Herein, the shrink rate is calculated by following equation.

$$\text{Shrink Rate} = \frac{(S_0 - S_T)}{S_0} \times 100 \qquad (1)$$

where S_0 is the initial area of the PS substrates (1 cm×2 cm = 2 cm^2) and S_T is the area after heated at T °C. The area of the PS substrate gets relatively smaller with higher temperatures, resulting in a rougher surface.

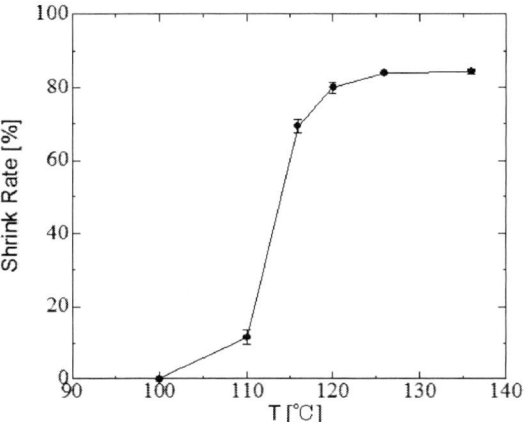

Figure 3: Shrink rate of PS substrates with different temperatures

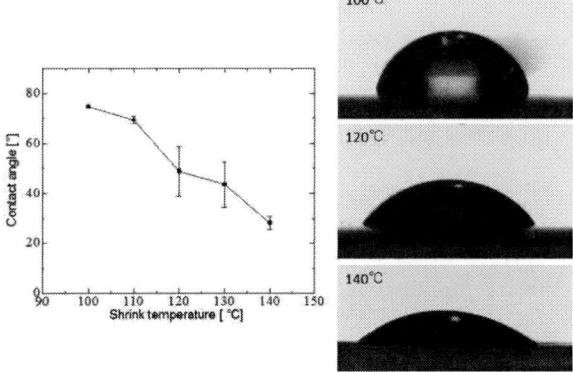

Figure 4: (Left) Contact angle of graphene on PS vs shrink temperatures; (Right) image of contact angles with varying shrink temperatures

The contact angle was measured for graphene on PS substrates with various shrink temperatures (Fig. 4). The contact angle decreased with the increase of shrink temperatures as increase of roughness, indicating that adjustment of shrink temperature leads to tunable surface profile on roughness and area. It can be explained by Wenzel's model. The model explains a relation between contact angle and surface roughness by the following equation:

978-1-4799-7956-1/15 $31.00 © 2015 IEEE   325

$$\cos\theta^* = r\cos\theta \qquad (2)$$

where θ^* is the apparent (measured) contact angle, r is the ratio of actual and projected surface area (roughness ratio), and θ is the Young contact angle. According to this equation, hydrophilic surface (r < 90°) becomes more hydrophilic with roughness, and in contrast, hydrophobic surface (r > 90°) becomes more hydrophobic with roughness. This certainly corresponds to our results shown in Fig. 4. Graphene sheets on shrink polymer with tunable roughness and wettability may have a great potential in microfluidics applications.

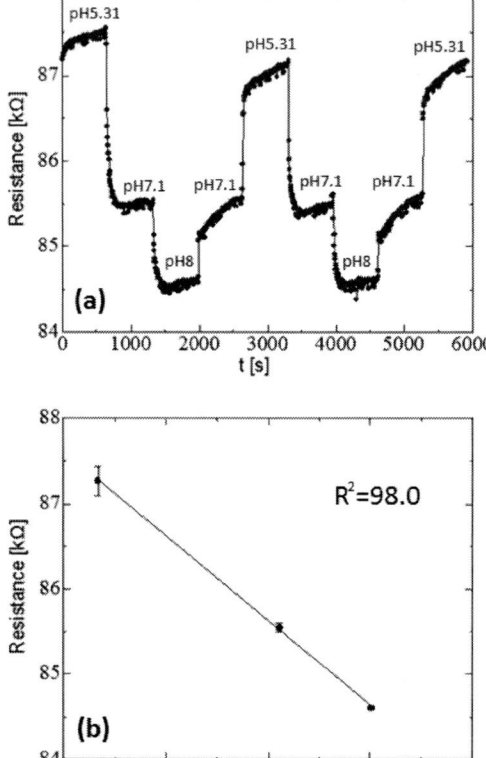

Figure 5: (a) Resistance shift of AFP sensor with different pH of PBS (pH: 5.31, 7.1, 8.0); (b) Resistance vs pH

Demonstrating self-assembled graphene sheets on a shrink polymer as a sensor is of importance to wide range of applications. We investigated pH response without any modification to the graphene surface after heating. Three different pH solutions in Phosphate buffer saline (PBS) were prepared for this experiment. The solutions were introduced on the sensor, and the resistance shift was measured by Agilent Data logger (34970A, Agilent, Inc.). As shown in Fig. 5, the resistance increased with decrease of pH, though as-prepared sensors do not have selectivity to both hydroxonium ions (H_3O^+) and hydroxyl ions (OH⁻). However, since both of the ions working as acceptors and donors are able to lead the modulation of the graphene channel conductance due to its ambipolar characteristic,

and the segregation of ions at the graphene/ electrolyte interface was reported [6], it is anticipated that the sensor can respond to pH change with constant shift of resistance.

The self-assembled graphene sheets after heating showed p-type, which means the main carriers are holes. This has good agreement with the results of pH measurement by the sensor. Lower pH results from higher density of H_3O^+ ion, and therefore holes in the graphene repulsively recede from interface of the solution, causing increase of resistance. The resistance shift can be also explained by Nernst equation in terms of potential shift at the graphene/ electrolyte interface [7]. The general expression for the sensitivity of the electrostatic potential to changes in the bulk pH is:

$$\frac{\delta\psi_0}{\delta pH} = -2.3\frac{kT}{q}\alpha \qquad (3)$$

$$\alpha = \frac{1}{\left(2.3kTC_{diff}/q^2\beta_{int}\right) + 1} \qquad (4)$$

where ψ_0 is the change of the insulator-electrolyte potential, k is the Boltzmann constant, T is the absolute temperature, q is the elementary charge, α is a dimensionless sensitivity parameter, C_{diff} is the differential capacitance, and β_{int} is the intrinsic buffer capacity. α takes a value between 0 and 1 depending C_{diff} and β_{int}. According to the equation, the surface potential shift is -59.2 mV/pH in case of $\alpha=1$. This potential corresponds to gate voltage shift on a graphene field-effect transistor. The electrical characteristics of p-type graphene reveals on the left half from a vertex of its I_d-V_g curve, called Dirac point. The characteristics show that the resistance increases with increase of gate voltage (decrease of pH). Thus, the sensor can respond to its environment change well.

For a biosensor application, the sensor was tested in AFP detections. After immobilization of AFP antibody onto the surface of graphene according to aforementioned method, the sensor was characterized by applying DPBS containing different concentrations of AFP antigen.

When monitoring antigen-antibody interactions with change of total charges on the sensing surface, whole molecules must be within the Debye length. However, conventional biosensors were unable to target molecules due to drawbacks in sensitivity. The Debye length exponentially decreases with increase of ionic strength, and as a result, the screening length of human serum of blood is less than 1 nm, much shorter than length of an antibody (~15 nm). To obtain unambiguous response caused by antigen-antibody reaction, testing solution should be diluted or desalted, so that the Debye length is well above the length of antigen-antibody complexes [8]. However, taking a step of dilution is not practical in terms of actual diagnosis due to either an extra requirement to higher detection limit or a risk of losing target molecules in the sample. To overcome the problems, we used a technique which takes a procedure of changing ionic strength after antigen-antibody reaction [9].

We prepared two testing solutions, DPBS containing AFP antigen and diluted DPBS. The ionic strength of as-prepared DPBS was 154.4 mM, corresponding to a Debye length of 0.78 nm. On the other hand, diluted DPBS was adjusted for its ionic strength down to 0.15 mM (×1000), a Debye length of 25 nm, as a reference buffer solution. First, the diluted DPBS was introduced on the sensor, and after stabilization of the output (resistance), we exchanged the solution to DPBS with AFP antigen. After 30 min, enough time for equilibrium of antigen-antibody reactions, we changed the solution back to the diluted DPBS. This procedure enabled to adjust the Debye length to longer enough to detect captured AFP antigens by AFP antibodies while monitoring. We repeated this procedure several times for different concentrations of AFP antigen.

As shown in Fig. 6, resistance shift corresponding to different concentrations of AFP antigen shows good linearity. A normalized resistance was used to obtain clear readout and to cancel individual difference of sensors. A resistance of the diluted solution before introducing any AFP antigen was used as an initial resistance, R_0, and other resistance of the diluted DPBS was used as R. ΔR is the difference between R and R_0, and therefore, the y axis in Fig. 6 ($\Delta R/R_0$) shows percentage of resistance shift before and after antigen-antibody reactions with the reference buffer solution.

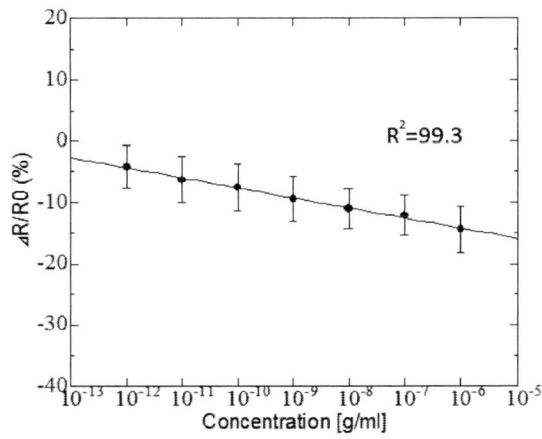

Figure 6: Result of AFP biomarker detection. Normalized resistance represented as $\Delta R/R_0$ (%). Estimated detection limit is 1 pg/mL.

AFP antigen is negatively charged in DPBS (pH ~7.3), for its isoelectric point (pI) is ~4.7. Therefore, regarding the resistance shift, the decrease of resistance with increasing concentrations of AFP antigen, was reasonable, and can be explained by the same discussion as the resistance shift in the pH testing section. The sensor demonstrates a lower detection limit down to 1 pg/mL. We estimate that the graphene surface structures with nanowrinkles played an important role of giving antigens and antibodies more chances to meet together due to rougher surface with larger area. This result promises that the graphene sensor is capable of being used as a label-free AFP sensor.

CONCLUSION

Shrink-induced graphene with LbL self-assembly exhibited tunable surface properties for bio-sensing applications. Utilization of shape memory polymers enables facile tuning of graphene surface by adjusting

shrink temperatures, resulting in tunnable surface roughness and wettability. Detection limits of graphene sheets with nanowrinkles was also investigated, showing a good response to detect very low concentrations of captured antigens based on the antigen-antibody reactions. In addition, we successfully achieved a relatively low-cost and label-free sensor based on its simple fabrication process and ease of surface modification.

The sensor holds enormous potential for applications to microfluidics and biosensors. We will investigate further fundamental principles and applications of tunable shrink-induced graphene sheets by LbL self-assembly.

ACKNOWLEDGEMENTS

The authors thank the assistance of fabrication and characterization from the Minnesota Nano Center and the Characterization Facility at the University of Minnesota.

REFERENCES

[1] F. Schwlerz, "Graphene Transistors", *Nature Nanotecnology*, vol. 5, pp. 487-496, 2010.

[2] D. Chen, et al., "Graphene-based Materials in Electrochemistry", *Chemical Society Reviews*, vol. 39, pp. 3157-3180, 2010.

[3] Q. Yu, et al., "Control and Characterization of Individual Grains and Grain Boundaries in Graphene Grown by Chemical Vapour Deposition", *Nature Materials*, vol. 10, pp. 443-449, 2011.

[4] C. Fu, et al., "Tunable Nanowrinkles on Shape Memory Polymer Sheets", *Advanced Materials*, vol. 21, pp.1-5, 2009.

[5] F. Patolsky, et al., "Fabrication of Silicon Nanowire Devices for Ultrasensitive, Label-free, Real-time Detection of Biological and Chemical Species", *Nature Protocols*, vol. 1, no. 4, pp. 1711-1724, 2006.

[6] D. J. Cole, et al., "Ion Adsorption at the Graphene/Electrolyte Interface", *The Journal of Physical Chemistry Letters*, vol. 2, pp. 1799-1803, 2011.

[7] R.E.G. V. Hal, et al., "A Novel Description of ISFET Sensitivity with the Buffer Capacity and Double-Layer Capacitance as Key Parameters", *Sensors and Actuators B*, vol. 24, pp. 201-205, 1995.

[8] A. Kim, et al., "Direct Label-Free Electrical Immunodetection in Human Serum Using a Flow-Through-Apparatus Approach with Integrated Field-Effect Transistors", *Biosensors and Bioelectronics*, vol. 25, pp. 1767-1773, 2010.

[9] E. Stern, et al., "Importance of the Debye Screening Length on Nanowire Field Effect Transistor Sensors", *Nano Letters*, vol. 7, no. 11, pp. 3405-3409, 2007.

CONTACT

* T. Cui, tel: +1-612-626-1636; tcui@me.umn.edu

978-1-4799-7956-1/15 $31.00 © 2015 IEEE

HIGH-TOPOGRAPHY SURFACE FUNCTIONALIZATION BASED ON PARYLENE-C PEEL-OFF FOR PATTERNED CELL GROWTH

Florian Larramendy[1,2], Daniela Serien[2], Shotaro Yoshida[2], Laurent Jalabert[2], Shoji Takeuchi[2] and Oliver Paul[1]

[1]Department of Microsystems Engineering – IMTEK, University of Freiburg, Freiburg, GERMANY
[2]Institute of Industrial Science, The University of Tokyo, Tokyo, JAPAN

ABSTRACT

This paper introduces a new technique for patterning functionalization layers on substrates with high-topography. The method is based on a parylene-C template shaped by a structured, sacrificial photoresist layer and attached to the substrate where functionalization is not intended. After photoresist removal and surface functionalization, the parylene layer is peeled off, leaving all areas initially covered by the sacrificial polymer functionalized. The technique has several advantages: (i) In contrast to microcontact printing, it allows surfaces with complex topographies to be functionalized; (ii) complex functionalization patterns are possible; (iii) the parylene structure can be reutilized. We successfully demonstrate the technique with the guided growth of neuron-like PC12 cells on honeycomb-shaped protein patterns on micropillars and microwells. The range and limits of the technique are analyzed and discussed in detail.

INTRODUCTION

The analysis of neuronal cells embedded in networks of controlled geometry has grown rapidly over the past decade and stays important for biological studies and medical research. The positioning of individual cells has become a key technique for cell engineering applications such as cell therapy and brain regeneration. Various techniques such as chemical treatment, microfluidic guiding or microfluidic trapping, negative dielectrophoresis, mechanical constraining and surface functionalization have been used to position and fix cells in desired locations.

The state of the art in surface functionalization is enabled by microcontact/nanocontact printing [1] which consists of transferring surface-functionalizing molecules in well-defined patterns onto substrates using micro/nano-structured PDMS stamps. Inherently, this technique is unsuitable for surfaces with pronounced topographies. Alternatively, it is envisionable to use parylene peel-

off [2, 3] on more complex topographies. However this technique has only served to realize simple geometries with isolated functionalized patches without interconnections. For more advanced studies of cell interactions, the functionalization of surfaces with complex topographies is desirable [4]. This paper addresses this goal.

FABRICATION

To demonstrate the advantages of the novel technique, we created high-topography surfaces on silicon substrates. This section introduces the substrate fabrication, parylene process and cell culture method.

Substrate fabrication

High-topography surfaces with microwells and micropillar arrays [5] were realized on two-inch standard {100} silicon substrates. Silicon microstructures were fabricated using photolithography (with Shipley resist S1805) followed by deep reactive ion etching. Scalloping was minimized by lowering the chemical contribution of the ion species with a radio frequency (RF) coil power of about 300 W and by enhancing their physical contribution with a platen RF power of 30 W. The resulting etch rate of about 1 µm/min allowed the precise control of the etch depth. Remaining photoresist was removed by oxygen plasma.

Different geometries were realized to study the effect of topography on the functionalization process. Microwells were etched with a width of 40 µm and a depth of 5

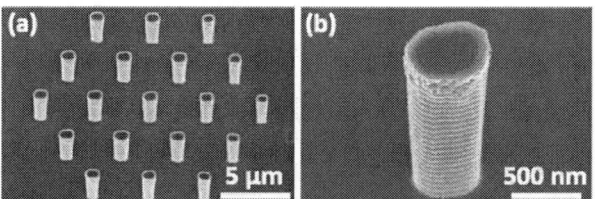

Figure 1: (a) Scanning electron micrographs of micropillar array and (b) single micropillar.

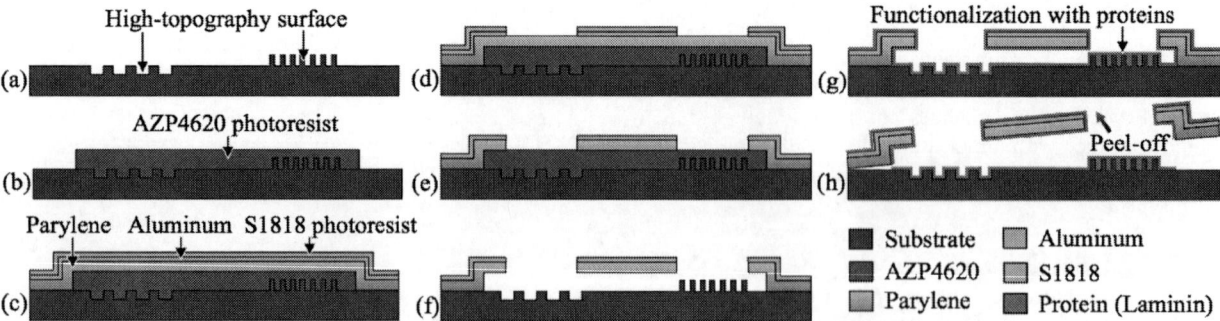

Figure 2: Sacrificial-photoresist and parylene-template based fabrication process for patterning surface-functionalizing proteins on substrates with high surface topography, i.e., protruding and recessed microstructures.

or 10 μm. The micropillar arrays are composed of 19 pillars {Fig. 1 (a)} spaced by 5 μm. Micropillars have a diameters of 300 to 700 nm, depending of the photoresist mask size, and are 3, 5, or 7 μm high {Fig. 1 (b)}.

Parylene-C template fabrication

The process steps leading to the parylene-C template are summarized in Fig. 2. Starting with a substrate that comprises protruding and recessed microstructures {Fig. 2 (a)}, the areas to be functionalized are first protected using a sufficiently thick sacrificial photoresist layer {Fig. 2 (b)}. Here, we worked with 10-μm-thick AZP 4620 photoresist. A 10-μm-thick parylene-C layer (from Specialty Coating Systems, Indianapolis, USA) is then deposited. At this thickness, the layer including its more fragile sidewalls proved to be thick enough to easily survive the mechanical strain due to the peel-off. The parylene layer is then structured using a 50-nm-thick aluminum (Al) mask thickened by a 3-μm-thick Shipley S1818 photoresist {Fig. 2 (c)}. The opening across the photoresist and the Al layer is realized in one step using NMD-3 developer {Fig. 2 (d)}. Next, anisotropic RIE cuts through the parylene, cf. Fig. 2 (e). Then, the sacrificial photoresist is dissolved in acetone {Fig. 2 (f)}. The Al layer is left on the parylene, without any utility however for the rest of the process. It could as well be removed, which would however represent an additional process step. An O_2 plasma removes residual solvent and photoresist and makes the parylene hydrophilic. Surface-functionalizing proteins in phosphate-buffered saline (PBS) solution are then dispensed. They densely cover all exposed parylene-C and substrate surfaces. After rinsing, the parylene-C layer is ready to be peeled off the substrate using tweezers, leaving only the originally photoresist-covered surfaces on the substrate functionalized.

Cell culture

The experiments were performed using cells that had undergone fewer than ten passages. As described previously [6], neuron-like PC12 cells (Riken BRC Cell Bank, Japan) were seeded onto laminin-coated chips with Dulbecco's Modified Eagle's Medium (DMEM) (D5796, Sigma) supplemented with 1% penicillin-streptomycin (P4333, Sigma) and 10 ng/ml nerve growth factor (N6009, Sigma). 30,000 cells/cm^2 are deposited onto the chips. Incubation was carried out under standard conditions for neuron-like PC12 cells, i.e., 5% CO_2 at 37°C. For seven consecutive days, the medium was left unchanged. This duration is required for the development of cell extensions. In this study, we did not distinguish between axons and dendrites and subsequently term all extensions neurites.

EXPERIMENTAL RESULTS

Process range

First, we explored the limits of new process. Of interest are the minimum diameter of functionalized islands and the minimum width and maximum length of functionalized lines. We obtained well-controlled protein patches with diameters down to 600 nm. At this

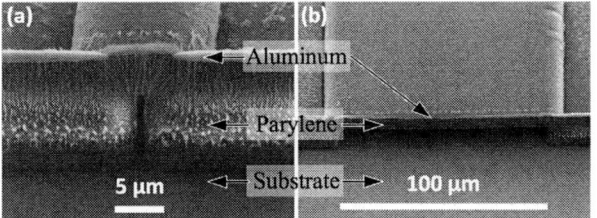

Figure 3: Scanning electron micrograph of (a) 1-μm-wide and (b) 100-μm-wide microchannels realized in parylene.

dimension, squares and disks on the mask both result in protein disks on the substrate due to the small ratio of the design width to the etch depth (1 μm : 10 μm). For the microchannels, we successfully functionalized 1-μm-wide to 100-μm-wide {Fig. 3} microchannels. By equipping long microchannels with openings spaced by 1 mm for efficient photoresist removal, we succeeded in demonstrating a maximum functionalized line length of 2.5 cm, after O_2 plasma treatment for higher hydrophobicity of the parylene layer. Filling the microchannels with aqueous solutions does not detach the parylene layer from the substrate.

Planar substrate

More complex functionalization geometries were realized as well, as shown with the honeycomb-shaped parylene-C structure in Fig. 4, consisting of open dome-shaped reservoirs with a diameter of 50 μm connected by microchannels of 10 μm width. The result is a hexagonal lattice of protein disks connected by protein lines, dimensioned to host one neuron-like PC12 cell per disk,

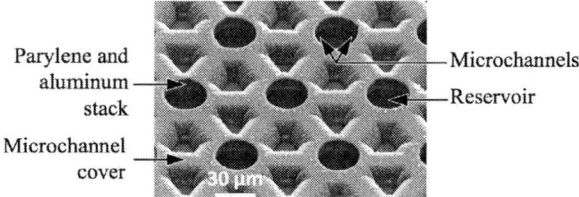

Figure 4: Scanning electron micrograph of honeycomb arrangement of parylene containers connected by microchannels.

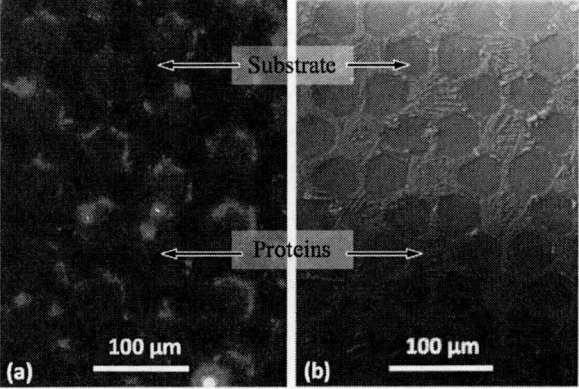

Figure 5: (a) Fluorescence and (b) scanning electron micrographs of laminin layer patterned by parylene peel-off technique.

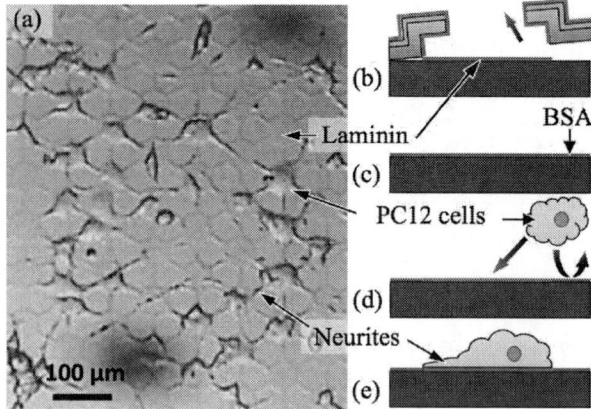

Figure 6: (a) Optical micrograph of neuron-like PC12 cells after 5 days of culture on laminin honeycomb pattern; (b)-(e) cell culture protocol.

with pairwise interconnection by neurites. Our technique was first tested with a mixture of rhodamine-tagged laminin (LMN01-A Cytoskeleton, Inc.) and laminin (L2020, Sigma, total 50 µg/mL, 25% rhodamine tag) and observed with fluorescence microscopy (Olympus IX, Olympus) to prove that the expected patterns were produced. Rhodamine-tagged laminin can be detected using excitation and detection wavelengths of 535 nm and 585 nm, respectively. Fig. 5 shows the resulting fluorescence and scanning electron micrographs of a laminin structure. Before cell growth, an incubation of the substrate with a 5%-solution of bovine serum albumin (BSA) in PBS for 2 hours is necessary. Indeed, although cells prefer to grow on laminin, they grow on silicon and glass as well. By covering all substrate areas free of laminin, BSA effectively suppresses cell adhesion in those areas [7]. After five days of cell culture, the majority of cells were found to attach to the intended areas and neurites to grow on the laminin lines in-between, as evidenced by Fig. 6.

Surface with complex topographies

One of the advantages of the novel technique is that it offers the possibility to pattern substrates with high-topography. The choice of micropillar arrays and micro-wells was motivated by the fact that with microcontact printing it is difficult to coat the entire surface of a micropillar without damaging it; it is similarly challenging to functionalize the bottom of a microwell

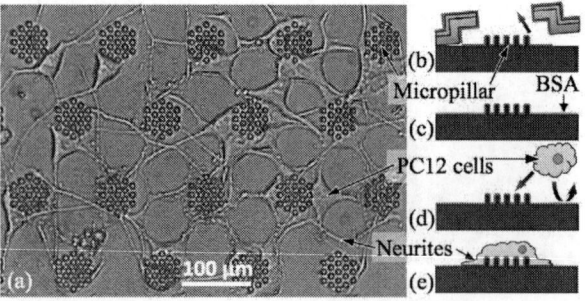

Figure 7: (a) Optical micrograph of neuron-like PC-12 cells grown on micropillar arrays with neurites guided by laminin lines; (b)-(e) cell culture protocol.

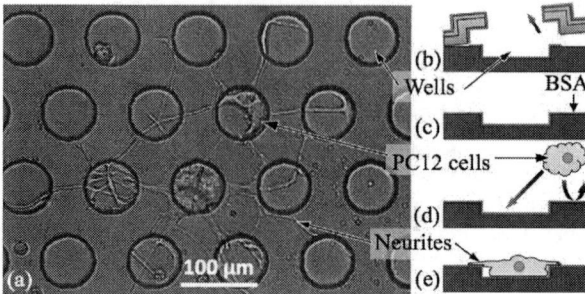

Figure 8: (a) Optical micrographs of neuron-like PC-12 cells grown in microwells with neurites guided by laminin lines; (b)-(e) cell-culture protocol.

without perfectly aligned microcontact printing stamp.

The same honeycomb template as above is used for this test; micropillar arrays and microwells were designed to be positioned in the middle of the honeycomb reservoirs. The design intends to favor the growth of neuron-like PC12 cells on micropillar arrays and in microwells and to guide the development of neurites between the cells. The new functionalization technique was then applied to these high-topography substrates. Results of equivalent quality were obtained for identical functionalization patterns on micropillars with heights up to 7 µm and diameters down to 1 µm {Fig. 7} and in microwells with depths down to 10 µm {Fig. 8}.

Several designs with different geometries were tested including the self-explaining pattern shown in Fig. 9 (a). After observation by optical microscopy (Olympus IX, Olympus), neuron-like cells were visualized with the LIVE/DEAD® Viability/Cytotoxicity Kit for mammalian cells (L3224, Molecular Probes). In this test viable cells are labeled with calcein AM detectable as green fluorescence, whereas dead cells are labeled with ethidium homodimer (EthD-1) producing red fluorescence. Excitation of the live stain occurs at 494 nm with fluorescent emission at 530 nm. For the dead stain, the excitation is at 528 nm, with fluorescent emission at 617 nm. The same pattern as in optical microscopy was observed with the live/dead test {Fig. 9 (b)}. Pictures taken after the death of individual cells prove the validity

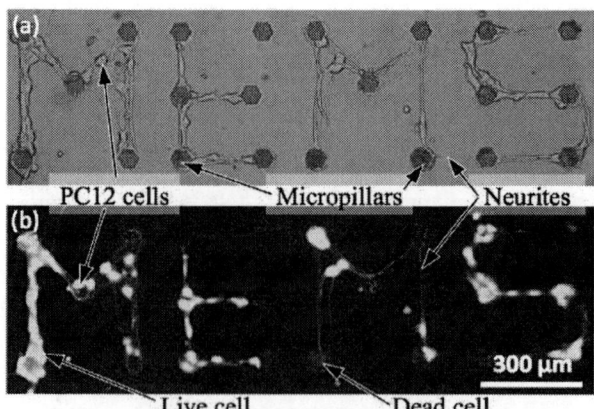

Figure 9: (a) Optical micrographs of neuron-like PC-12 cells grown on micropillar arrays with neurites guided by laminin lines to form 'MEMS' and (b) its equivalent in fluorescence microscopy after live-dead test.

of the test. Similarly successful results were obtained with all tested designs (alphabet, honeycomb matrix, square matrix).

Moreover, after peeling off the parylene layer, with the alignment marks still present on the parylene mask, this technique offers the possibility to re-use the same mask with a new substrate [8]. For this purpose, the parylene layer needs to be wetted with ethanol, positioned on the new substrate with respect to alignment mark, dried, and treated in a vacuum chamber to remove air bubbles. We successfully functionalized three different substrates consecutively with the same parylene layer.

DISCUSSION

Our designs have demonstrated the validation of the novel concept. However, against expectation some turned out not to be optimal for neuron-like PC12 cell culture as intended. In fact, the soma of neuron-like PC12 cells measures between 10 and 40 µm and their neurites are narrower than 1 µm. As we observed on a square matrix design in Fig. 10, neuron-like PC12 cells are located not only on micropillar arrays, but also on the connection lines between neighboring micropillar arrays due to the width of the connection lines of 10 µm. Thinner connection lines would therefore be preferable. Also parallel growth of neurites distant by 10 µm was observed, which results from the higher concentration of laminin along the edges of the parylene mask. This effect was reduced by 1-µm-wide microchannels serving as connection lines. Finally, we also observed some non-specific cell adhesion on BSA-protected areas close to the laminin coating. Although BSA reduces the non-specific cell adhesion on the substrate, it does not prevent it completely. Likely the non-specific adhesion is promoted by the vicinity of large numbers of healthy cells on the laminin patterns.

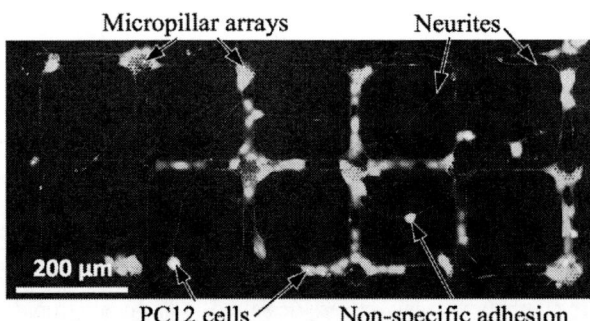

Figure 10: Fluorescent micrograph of neuron-like PC-12 cells grown on micropillar arrays with neurites guided by laminin lines on square matrix design after live/dead test.

CONCLUSION

A new technique for patterning functionalization layers on substrates with high topography has been reported in this paper. The technique is based on the peel-off of a parylene layer deposited on a structured sacrificial photoresist. It was successfully demonstrated with the guided growth of neuron-like PC12 cells on different patterns of cell-growth promoting proteins on micropillars, in microwells, and in-between. This work provides a technology that promises to be useful for a number of applications including cell biology, microfluidics, and lab-on-a-chip systems.

ACKNOWLEDGEMENTS

Financial support by project EUJO-LIMMS (no. 295089) funded by the EU 7th Framework Program is gratefully acknowledged.

REFERENCES

[1] D. Qin, Y. Xia, G. M. Whitesides, "Soft lithography for micro- and nanoscale patterning", *Nature Protocols*, 2010, pp. 491-502.

[2] C. P. Tan, B. R. Cipriany, D. M. Lin, and H. G Craighead, "Nanoscale Resolution, Multicomponent Biomolecular Arrays Generated By Aligned Printing With Parylene Peel-Off", *Nano Letters*, 2010, pp. 719-725.

[3] C. P. Tan, B. R. Seo, D. J. Brooks, E. M. Chandler, H. G. Craighead, and C. Fischbach, "Parylene peel-off arrays to probe the role of cell–cell interactions in tumour angiogenesis", *Integr. Biol.*, 2009, 1, pp. 587-59.

[4] J. Held, J. Gaspar, P. J. Koester, C. Tautorat, A. Cismak, A. Heilmann, W. Baumann, A. Trautmann, P. Ruther, and O. Paul, "Microneedle arrays for intracellular recording applications" *MEMS 2008*, Tucson, AZ, USA, January 13-17, 2008, pp. 268-271.

[5] A. M. P. Turner, N. Dowell, S. W. P. Turner, L. Kam, M. Isaacson, J. N. Turner, H. G. Craighead, and W. Shain, "Attachment of astroglial cells to microfabricated pillar arrays of different geometries", *Journal of Biomedical*, 2000, pp. 430-441.

[6] K. J. Tomaselli, C. H. Damsky, and L. F. Reichardt "Interactions of a neuronal cell line (PC12) with laminin, collagen IV, and fibronectin: identification of integrin-related glycoproteins involved in attachment and process outgrowth", *Journal of Cell Biology*, 1987, vol. 105, 5, pp. 2347-2358.

[7] K. Shimizu, H. Fujita, E. Nagamori "Micropatterning of single myotubes on a thermoresponsive culture surface using elastic stencil membranes for single-cell analysis", *Journal of Bioscience and Bioengineering*, 2010, vol. 109, 2, pp. 174-178.

[8] D. Wright, B. Rajalingam, J. M. Karp, S. Selvarasah, Y. Ling, J. Yeh, R. Langer, M. R. Dokmeci, and A. Khademhosseini, "Reusable, reversibly sealable parylene membranes for cell and protein patterning", *J. of Biomed. Mater. Res. 85A*, 2007, pp. 530-538.

CONTACT

F. Larramendy, Tel: +49-761-203-7191;
Email: florian.larramendy@imtek.uni-freiburg.de

LASER TREATED GLASS PLATFORM WITH RAPID WICKING-DRIVEN TRANSPORT AND PARTICLE SEPARATION CAPABILITIES

Manuel Ochoa, Hongjie Jiang, Rahim Rahimi, and Babak Ziaie
Purdue University, West Lafayette, IN, USA

ABSTRACT

Wicking and particle separation are two required capabilities for many microfluidics and lab-on-a-chip devices, but they often require multiple materials and structures (e.g., paper, polymer filters) which are difficult to integrate with established microfabrication techniques and materials. In this work, we combine both properties into a single glass platform with a straightforward and economical fabrication process. By laser machining soda lime glass with a specific power and laser speed, we create channels defined by an array of micro cracks (3–4 μm) which provide particle separation properties and simultaneously enable rapid liquid transport (up to 24.2 mm/s) as a result of capillary forces from the crevices and laser-induced surface hydrophilization.

INTRODUCTION

Glass has traditionally been one of the workhorse materials in the fabrication of microfluidic and other biomedical lab-on-a-chip devices [1]–[3]. This is due to its optical transparency, rigidity, bio-compatibility, and ease of surface modification/ functionalization. Recently, however, researchers have investigated lower-cost, flexible substrates such as functionalized polymers [2], [4], [5] and paper [6], [7], whose remarkable inherent wicking property and filtration capabilities have allowed the realization of a variety of passive analytical microsystem [8]–[10]. Wicking, in particular, offers the advantage of passive liquid transport [11], [12] without the need for a micropump, thus significantly reducing the system complexity and cost. Similarly, passive filtration via a mesh structure provides an inexpensive particle separation method [13], [14] which can be easily extended (e.g., by using a larger filtration region) for high throughput operations. Together, wicking and particle separation are two principal necessities for many microfluidics and lab-on-a-chip applications, but they often require multiple distinct materials (e.g., paper, polymer filters) which are difficult to integrate with established microfabrication techniques and materials. Imparting such capabilities (i.e., wicking and filtering) to glass would allow for a unique platform that combines the reliability and established surface chemistry of glass with the passive fluidic transport and filtration properties of paper. As a first step towards this goal, we developed an economical laser-surface-treatment method to fabricate glass with wicking properties, rapid liquid transport, and particle separation capabilities.

DESIGN

The platform consists of a network of open microfluidic channels defined by a multitude of inter-connected surface cracks created in a controlled fashion. Figure 1a-b illustrates the cross-sectional structure and working principle of a channel. The cracks comprising each channel feature diameters no larger than 3–4 μm, allowing for rapid wicking action due to capillary forces; the liquid flow is further aided by the addition of hydrophilic functional groups created by laser treatment. When an aqueous suspension of particles is deposited on one end of the channel, the liquid seeps into the cracks and is drawn across the length of the channel. Any particles larger than the surface cracks are then filtered from the liquid and remain at the beginning of the channel. Multiple liquids can be rapidly transported throughout a network of channels in this manner, and they can optionally be mixed at junctions. In such case, the cracked glass structure aids in rapid, proper mixture; since the cracks split the liquid into a series of streams, the junctions enable intertwining of the various fluid streams, allowing for enhanced diffusion and fast mixing. Such extraordinary capabilities are achieved on glass by simply laser-machining a soda-lime glass substrate using a CO_2 laser at specific settings.

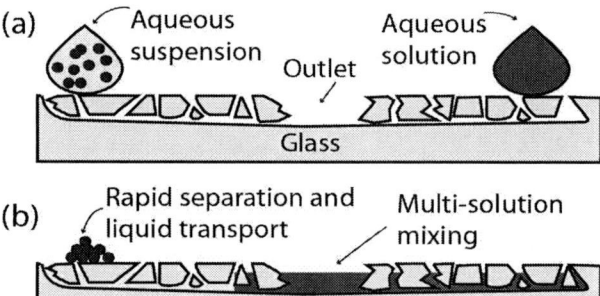

Figure 1: Cross-sectional schematic of the laser-machined glass structure showing its wicking mechanism. (a) An aqueous solution prior to contacting the glass. (b) The solution wicks into the fractured glass and quickly propagates across the channel.

FABRICATION

The fabrication process of the wicking glass platform is straightforward and economical, Figure 2. First, a microchannel network is designed in vector graphics software and imported onto a commercial CO_2 laser engraver system (PLS6MW, Universal Laser Systems, Inc., Scottsdale, AZ). The system then inscribes the pattern on a standard soda lime glass slide (GOLD SEAL® Micro Slide), producing cracked glass microchannels on the surface. The channel geometries and resolution are limited only by the spot size of the laser (as small as 30 μm for our system). This process imparts high energy bursts onto the glass surface, causing localized thermal shock in the regions of lasing. As a result, many micro-crevices/cracks are generated. This effect is unique to soda lime glass and has been previously used to selectively remove the surface of soda lime glass

978-1-4799-7956-1/15 $31.00 © 2015 IEEE   332   MEMS 2015, Estoril, PORTUGAL, 18 - 22 January, 2015

substrates to create concave patterns that behave as microchannels [15]. However, we extend the technique to more than simply material removal. By controlling the laser parameters (power and processing speed), it is possible to control the morphology of the channels such that the cracked regions are not released from the rest of the substrate. Instead, they provide an interconnected network of channels that enable capillary action for wicking while simultaneously acting as a mesh for filtering particles.

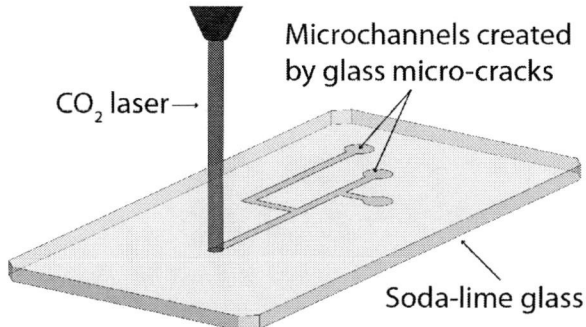

Figure 2: Fabrication process of the wicking glass. Create design in CAD software and laser-ablate design onto a soda-lime glass slide. The resulting micro-cracks and glass segments remain attached to the substrate to create wicking traces.

EXPERIMENTAL PROCEDURE

Qualitative characterization

The wicking and filtration properties of the platform were characterized quantitatively and qualitatively. For the qualitative assessments, T-junction channels (two 1 mm wide × 5 mm long channel segments feeding into a 1 mm × 15 mm channel) were fabricated. Two aqueous dyes, a yellow dye and a blue dye containing silica spheres (9–13 μm), were then deposited at the inlet of the short channel segments; their propagation and separation properties were recorded by photographs under a microscope. The micro-structure of the glass channels was also observed by scanning electron microscopy (SEM) to confirm the filtration properties. For this, each channel was loaded with an aqueous suspension containing a mixture of particles (i.e., glass microspheres, iron oxide microparticles, silver nanoparticles) of various sizes (200 nm–13 μm). The liquid was allowed to wick and subsequently dry prior to imaging the beginning and end regions of the channel with SEM.

Quantitative characterization

The liquid transport properties of the platform were quantitatively investigated by measuring the average wicking speed of an aqueous dye. For this, we created 1 mm wide × 30 mm long channels using various parameter values for the laser system (i.e. laser power and laser processing speed). The channels were tested immediately after processing by depositing 1 μL of an Evans Blue solution at the beginning of the channel and measuring the average time elapsed until the wicking solution reached the end of the channel using a hand-held stopwatch; the channel length was sufficiently large to allow for reliable precision when starting and stopping the stopwatch.

To better understand influence of the laser upon the surface chemistry of the glass, we performed spectroscopic analysis on untreated and laser-treated glass samples. The data were processed using CasaXPS software to separate closely over-lapping peaks.

RESULTS AND DISCUSSION

Figure 3a shows an overview photograph of a T-junction channel created on a glass slide. The laser-machined region is clearly distinguishable from the rest of the substrate. After depositing two aqueous dyes (with particle suspensions) into the right-most circular regions, the dyes advance towards each other along their respective channel until they meet and begin to mix, Figure 3b. This magnified image elucidates the efficient manner in which mixing occurs. Essentially, the crack pattern splits each incoming dye into a set of micro-streams which are then inter-twined with those of the other dye, resulting in improved diffusion. The image also shows the straight edge created by laser machining; the cracks do not propagate beyond the intended channel region, thus allowing full containment of the liquid. Figure 3c shows the channels after the two dyes have propagated the length of the entire length; mixing is identified by the green highlights on the left-hand side of the horizontal channel. A closer look at the inlet of the blue channel (Figure 2d) reveals the glass spheres originally suspended in the blue dye; thus, the channel also behaves as a sieve while simultaneously providing wicking action.

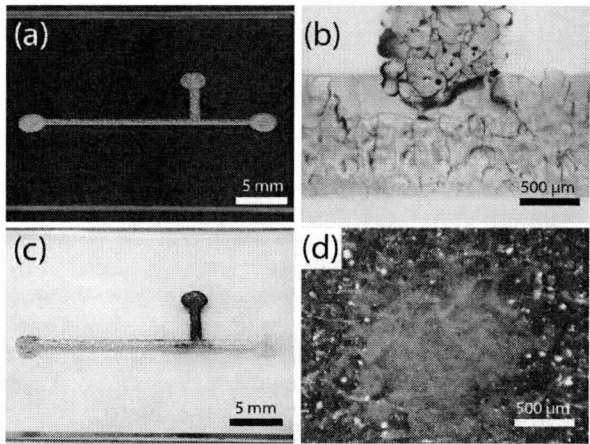

Figure 3: (a) Empty pair of intersecting channels on a glass slide; (b) magnified view of a junction, showing the beginning of mixing; (c) filled channel; (d) magnified view of glass micro spheres remaining at the inlet of the (blue) channel.

Figure 4 shows the images captured by scanning electron microscopy. Figure 4a highlights typical large crevices on the channels; with a width of about 3–4 μm, they are sufficiently small to filter out suspension components larger than 4 μm. As Figure 4b shows, the glass structure retains the large particles (> 4 μm) at the beginning, allowing only those smaller than 3–4 μm to reach the end of the channel, Figure 4c. Thus, the cracks comprising the channels are an effective and easy-to-integrate method for size-controlled particle separation in microfluidics; in particular this mesh size renders them suitable for biological or biomedical applications requiring separation of cells from a liquid medium (e.g., extracting blood plasma from whole blood for further analysis at the end of the channel).

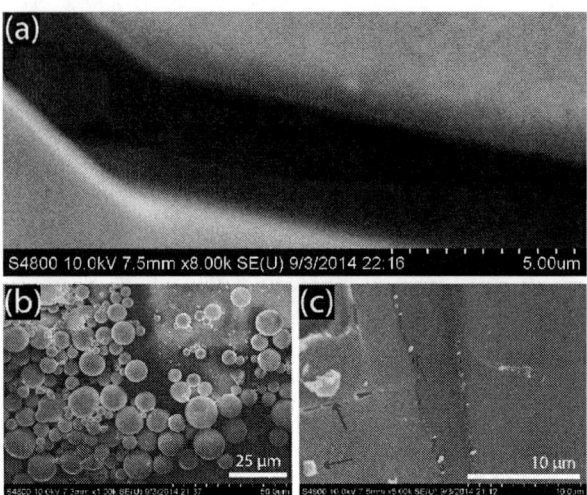

Figure 4: SEM images. (a) Example of the 3–4 μm crevices comprising channels; (b) various particles of sizes 200 nm–13 μm at the inlet of the channel; (c) end of the channel showing that the largest remaining particles (indicated by red arrows) which can traverse the glass cracks are in the size range < 3–4 μm.

The results of the transport speed experiments are plotted in Figure 5 and represent the average of three samples. The data show a strong dependence of wicking velocity on both, the power and processing speed of the laser system. In particular the average wicking velocity increases with increasing laser power and decreasing speed. These relationships can be understood in terms of the energy imparted on the glass by the laser. With high power or low speeds (slow movement of the laser head), the system imparts more energy (and hence more thermal-shock/cracking) compared to with low power and high speeds. By using a power of 75 W and processing speed of 0.8 mm/ms, we achieved an average wicking velocity of 24.2 mm/s. This value is significantly faster than the typical 2 mm/s reported for filter paper [16], allowing for liquid transport and mixing as well as particle separation at speeds superior to those of paper microfluidics. Additionally, unlike with paper, the wicking velocity of the glass platform is controllable along a continuous spectrum of values down to 0.8 mm/s by varying the laser parameters. This broad range of possible wicking velocities makes the platform suitable for many applications requiring controlled liquid transport.

The high flow rates observed in Figure 5 are attributed to a combination of both physical and chemical phenomena in the channels, namely, capillary forces and surface hydrophilization as a result of laser processing. The influence of capillary forces can be understood based on the abundance and long, narrow geometry of the crevices comprising the channels, as described above. The chemical effects can be observed by spectroscopic analysis, as shown in Figure 6.

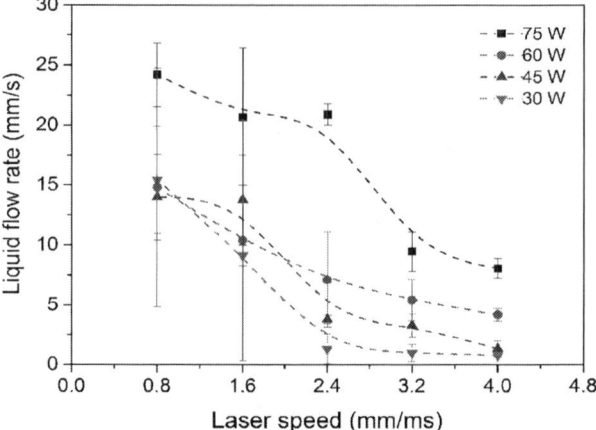

Figure 5: Resulting liquid flow rate as a function of laser fabrication parameters (power and speed). The flow rate can be controlled in the range of 0.8–24.2 mm/s.

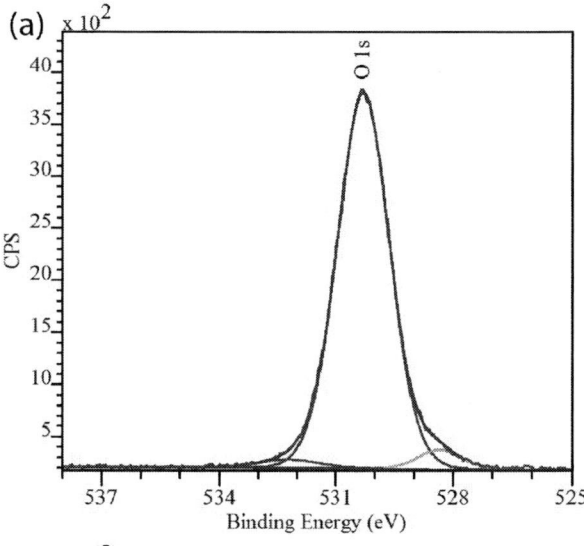

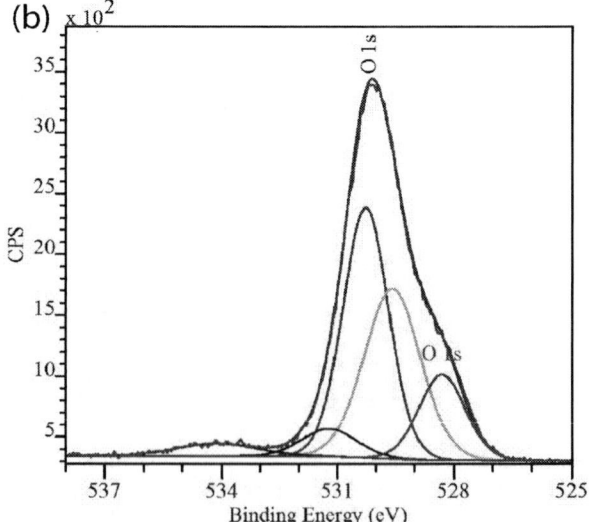

Figure 6: XPS data of the O 1s spectra for soda-lime glass (a) before, and (b) after laser treatment. The data suggest an increase in the amount of hydrophilic functional groups (e.g.,–OH) on the glass after laser treatment.

Figures 6a and 6b show the O 1s spectra for untreated and laser-treated glass, respectively. A comparison between the two reveals an additional oxidation peak in the laser-treated sample, corresponding to Si–$(OH)_x$ groups (at 531 eV) [17]. Such increased concentration of hydrophilic species at the surface are expected to contribute to the increased hydrophilicity of the glass after laser treatment. Hence, the combination of capillary forces together with chemical surface modification enable the observed rapid liquid transport.

CONCLUSIONS

We have developed a glass-based microfluidic platform that combines high-speed wicking-driven liquid transport with size-based particle separation. Its fabrication is economical and is based on a low-cost CO_2 laser machining technique, which allows for rapid prototyping and is scalable to large-volume manufacturing. The platform offers a passive particle separation mechanism through its series of glass crevices comprising the channels, with a sieve size of about 3–4 μm. Additionally, it is possible to transport aqueous liquids at a velocity of up to 24.2 mm/s, which is tunable down to 0.8 mm/s by adjusting the operation parameters of the laser system (i.e., laser power and processing speed). Such high velocity is attributed to capillary forces created by the crevices, as well as hydrophilization of the channels during fabrication; the latter point is confirmed by XPS analysis. The cracked glass structure is a versatile platform that combines the wicking and filtering capabilities of paper with the mechanical and chemical reliability of glass.

ACKNOWLEDGEMENTS

The XPS data was obtained at the Surface Analysis Facility of the Birck Nanotechnology Center, Purdue University. The authors thank the staff of the Birck Nanotechnology Center for their support. Funding for this project was provided in part by the National Science Foundation under grant EFRI-BioFlex #1240443.

REFERENCES

[1] P. N. Nge, C. I. Rogers, and A. T. Woolley, "Advances in microfluidic materials, functions, integration, and applications.," *Chem. Rev.*, vol. 113, no. 4, pp. 2550–83, May 2013.

[2] E. K. Sackmann, A. L. Fulton, and D. J. Beebe, "The present and future role of microfluidics in biomedical research.," *Nature*, vol. 507, no. 7491, pp. 181–9, Mar. 2014.

[3] B. Ziaie, A. Baldi, M. Lei, Y. Gu, and R. A. Siegel, "Hard and soft micromachining for BioMEMS: review of techniques and examples of applications in microfluidics and drug delivery," *Adv. Drug Deliv.*

Rev., vol. 56, no. 2, pp. 145–172, Feb. 2004.

[4] M. Ochoa, R. Rahimi, and B. Ziaie, "Flexible sensors for chronic wound management.," *IEEE Rev. Biomed. Eng.*, vol. 7, pp. 73–86, Jan. 2014.

[5] M. Kitsara and J. Ducrée, "Integration of functional materials and surface modification for polymeric microfluidic systems," *J. Micromechanics Microengineering*, vol. 23, no. 3, p. 033001, Mar. 2013.

[6] G. Chitnis, Z. Ding, C. Chang, C. A. Savran, and B. Ziaie, "Laser-treated hydrophobic paper: an inexpensive microfluidic platform.," *Lab Chip*, vol. 11, no. 6, pp. 1161–1165, Mar. 2011.

[7] W. K. Tomazelli Coltro, C.-M. Cheng, E. Carrilho, and D. P. de Jesus, "Recent advances in low-cost microfluidic platforms for diagnostic applications.," *Electrophoresis*, pp. 2309–2324, Mar. 2014.

[8] E. W. Nery and L. T. Kubota, "Sensing approaches on paper-based devices: A review," *Anal. Bioanal. Chem.*, vol. 405, pp. 7573–7595, 2013.

[9] A. K. Yetisen, M. S. Akram, and C. R. Lowe, "Paper-based microfluidic point-of-care diagnostic devices.," *Lab Chip*, vol. 13, no. 12, pp. 2210–51, Jun. 2013.

[10] A. W. Martinez, S. T. Phillips, G. M. Whitesides, and E. Carrilho, "Diagnostics for the developing world: microfluidic paper-based analytical devices.," *Anal. Chem.*, vol. 82, no. 1, pp. 3–10, Jan. 2010.

[11] S. Ravi, D. Horner, and S. Moghaddam, "A novel method for characterization of liquid transport through micro-wicking arrays," *Microfluid. Nanofluidics*, pp. 1–9, 2013.

[12] E. Fu, S. A. Ramsey, P. Kauffman, B. Lutz, and P. Yager, "Transport in two-dimensional paper networks," *Microfluid. Nanofluidics*, vol. 10, pp. 29–35, 2011.

[13] E. Y. Kenig, Y. Su, A. Lautenschleger, P. Chasanis, and M. Grünewald, "Micro-separation of fluid systems: A state-of-the-art review," *Sep. Purif. Technol.*, vol. 120, pp. 245–264, Dec. 2013.

[14] P. Sajeesh and A. K. Sen, "Particle separation and sorting in microfluidic devices: a review," *Microfluid. Nanofluidics*, vol. 17, no. 1, pp. 1–52, Nov. 2013.

[15] Z. K. Wang and H. Y. Zheng, "Investigation on CO(2) laser irradiation inducing glass strip peeling for microchannel formation.," *Biomicrofluidics*, vol. 6, no. 1, pp. 12820–1282012, Mar. 2012.

[16] K. T. Hodgson and J. C. Berg, "The effect of surfactants on wicking flow in fiber networks," *J. Colloid Interface Sci.*, vol. 121, no. 13, pp. 22–31, 1988.

[17] A. U. Alam, M. M. R. Howlader, and M. J. Deen, "Oxygen Plasma and Humidity Dependent Surface Analysis of Silicon, Silicon Dioxide and Glass for Direct Wafer Bonding," *ECS J. Solid State Sci. Technol.*, vol. 2, no. 12, pp. P515–P523, Oct. 2013.

CONTACT

* M. Ochoa, tel: +1-562-546-2669; ochoam@purdue.edu

LIPOSOME ARRANGEMENT CONNECTED WITH AVIDIN-BIOTIN COMPLEX FOR CONSTRUCTING FUNCTIONAL SYNTHETIC TISSUE

Hiroshige Hamano, Toshihisa Osaki, and Shoji Takeuchi
Institute of Industrial Science, The University of Tokyo, JAPAN
Kanagawa Academy of Science and Technology, JAPAN

ABSTRACT

This paper describes a method to arrange liposomes into organized structures with biochemical binding of streptavidin-biotin complex, inspired by the biological anchoring junction. This approach enhances the stability of the liposome structure, which would provide an improved model for a liposome-based tissue in synthetic biology.

INTRODUCTION

Cells *in vivo* are basic components of living organism. Various specialized functions emerge by complicated interactions of biological molecules inside of the cell [1]. Recently, *in vitro* reconstruction of those functions within a cellular size has been challenged in science and engineering fields, and as a part of them, DNA replication and protein synthesis have been successfully mimicked. On the one hand, artificial cell studies using aqueous micro-droplets or liposomes mainly focus on reproducing the functions in a confined system. There have been numbers of studies reported such model systems presenting metabolic reactions, growth of a model cell, division cycles, and even Darwinian evolution [2-4]. A next challenge beyond the mimicking of properties of single cells will be the reconstruction of living-tissue structures and functions. In recent years, water-in-oil droplets have been applied as a unit of tissue-like synthetic structure [5]. This droplet network realized molecular transportation via nanopore-forming protein as well as bending motion triggered by osmotic change [5]. However, the connection manners between the model cells in those works did not follow the realistic tissue systems. The model cells were contacted with a single lipid bilayer, while living cells are faced one another with lipid

bilayer–lipid bilayer contacts with membrane proteins or extracellular matrices [1].

We previously reported an array of giant liposomes as the prerequisite platform to develop a tissue-like liposome arrangement consisting of lipid bilayer–lipid bilayer contacts [6]. Commonly, giant liposomes are produced by a gentle hydration or an electroformation method, yet these methods have difficulty in controlling the size, density and position of the formed liposomes; therefore, it is hardly feasible to form a liposome array by these methods. In MEMS 2011, we developed a method that enabled a precise lipid patterning by the integration of an electrospray deposition (ESD) technique and a micro-fabrication process [7, 8]: By the ESD of lipids on a conductive/non-conductive patterned substrate, the sprayed lipids were electrically led only to the conductive regions. With a simple hydration process of these dried lipid patterns, we succeeded in the formation of giant liposomes in an array format with a narrow range of the size distribution, and moreover obtained the desired sizes of liposomes by changing the sizes of lipid patterns, close to common cellular sizes. However, even though the distance between the liposomes was set to adhere each other, the platform was not able to maintain an organized liposome structure [6].

The cell membrane-cell membrane connection formed by a specialized protein is called as anchoring junction. An example is transmembrane cadherin protein that tightly connects and enables to transmit stress through cytoskeleton [1]. In this work, inspired by the structure of the anchoring junction, we integrated an adhesive connection between the liposomes in a biological manner using the avidin-biotin complex into the lipid patterning method as shown in Figure 1 [3]. We also measured the

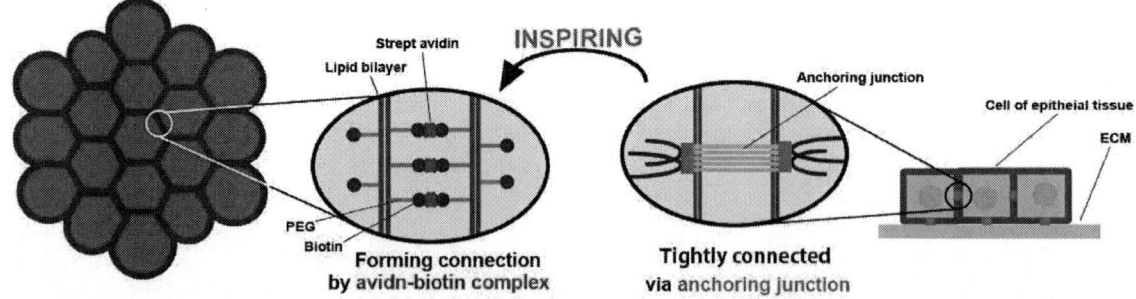

Figure 1: Conceptual diagram of an anchoring junction in vivo and its synthetic model using the avidin-biotin complex. Membrane protein complex called anchoring junction expressed on the epithelial tissue realizes maintenance of organized shape of tissue, and transmission stress. For these functions, the membrane proteins tightly adhere cells to cells. Here, inspired by this structure, we are motivated to construct arrangement of liposomes adhered by avidin-biotin complexes for model cell research.

avidin-biotin effects on the stability of the organized liposome structure by using fluorescent observation.

EXPERIMENTAL
Fabrication of Patterned Substrates

We first fabricated the conductive/non-conductive patterns on a substrate using a common photolithography technique. As shown in Figure 2-A, a polymer film of poly(chloro-p-xylylene) (parylene C) was deposited on an indium-tin-oxide (ITO)-coated glass slide. Then, an Al-layer and a photoresist (S1818) were coated on the substrate by using a vacuum vapor deposition and a spin-coating, respectively. The photoresist was patterned by UV lithography, and the Al-layer was then patterned by an Al etchant. Finally, the polymer layer was patterned by O_2 plasma, and the residual Al and photoresist were removed. The ITO area was exposed after the fabrication process. A microscopic image of the fabricated substrate is shown in Figure 2-a.

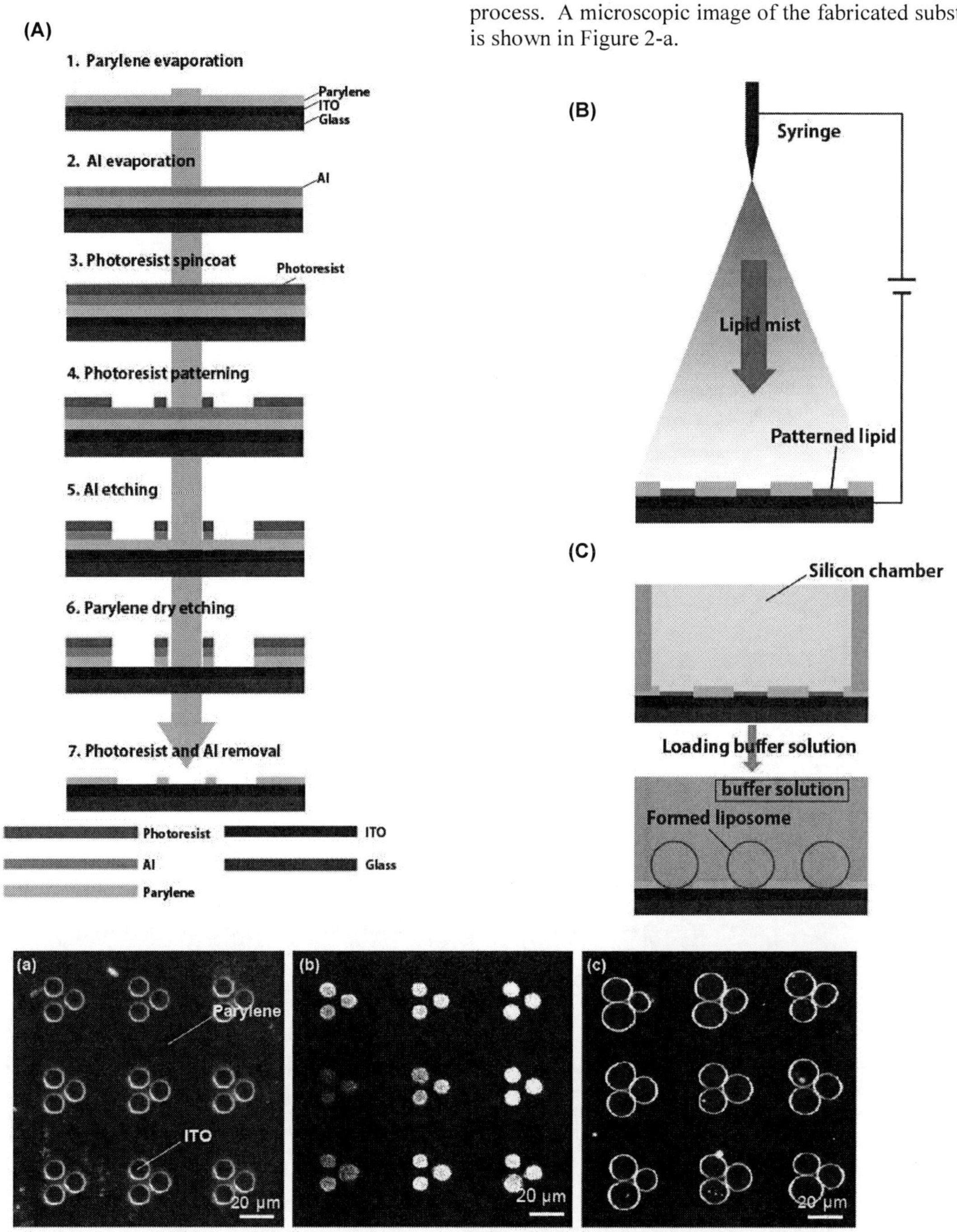

Figure 2: Preparation of the adhered liposome arrangement. (A) Fabrication process of conductive/non-conductive patterned substrate. (a) Microscopic image of fabricated substrate with 10 µm-diameter patterns. (B) Conceptual diagram of the electrospray setup for lipid deposition on the patterned substrate. (b) Confocal image of patterned lipid on the substrate. (C) Illustration of the hydration process of the patterned lipid. (c) Confocal image of the liposome assemblies.

Electrospray Deposition of Lipids

Schematic illustration of the ESD setup was shown in Figure 2-B. Due to the electrification of the lipid solution, the sprayed lipid was deposited on the conductive pattern on the substrate (Figure 2-b). In this work, we used a mixture of DOPC and DOPG lipids as the basic components. A biotin-conjugated lipid was also mixed to construct biochemical binding via streptavidin-biotin complex (Biotin-DSPE, 5wt%). A rhodamine-labeled lipid was added for fluorescence observation (Rhod-DPPE, 1wt%). The tip of the glass capillary was between 10 and 15 μm in diameter. After the deposition, the samples were kept in vacuum before use. Lipids were purchased from Avanti Polar Lipids, AL, USA. All chemicals of this experiment were used without further purification. After the desiccation, a buffer solution was added on the deposited micro-pattern to trigger hydration of the lipid pattern that forms an assembled liposome structure consisting of lipid bilayer through a self-assembled process of lipid molecules (2-C, 2-c).

RESULTS AND DISCUSSION

Formation of Adhered Liposomes

After forming liposome assemblies, we added streptavidin solution into the silicon chamber to adhere the liposomes one another through the avidin-biotin complex (Figure 3-A). We confirmed whether the streptavidin effectively alters the shape of the contact surface between the liposomes. As shown in Figure 3-B, the contact area increased by the loading of streptavidin. Before the addition, we observed a void area at the middle of three liposomes (see Figure 3-B left); after the addition, however, the void disappeared and the contact area appeared instead (Figure 3-B right). We also examined the line profile of the fluorescence intensity between the start and end points shown in Figure 3-C. The intensity before loading showed the maximum value at around 5 μm, which indicates that the two liposomes are contacted at that point. On the other hand, the intensity after loading was rather constant between 0 and 10 μm and closed to the value at the contact point before streptavidin loading. We consider that the loading increased the contact surface area; the adhesion of liposome membranes can be observed as transition of fluorescence intensity of contact area. We also took time-lapse images of a larger-size liposome arrangement after the addition of the streptavidin solution (Figure 4). As shown in the images, liposomes gradually adhered after the streptavidin loading. We confirmed that the adhesion process required about 20 min. In the center area of the liposome arrangement, any adhesion areas did not exist. This is probably because the liposomes in the area were smaller than the surrounding liposomes in the arrangement. This result indicates that avidin-biotin comples required a certain extent of contact area for the adhesion of liposomes.

Liposome Arrangement Stability

We also investigated the stability of the form of the liposome arrangement. According to the time-course observation of the formed arrangements, the contact area between the liposomes was almost constant, and at least its change during 1h observation was extremely small. This complex is able to maintain the organized structure during continuous observation, and clearly improved the stability compared to our previous technique.

CONCLUSIONS

In this work, we presented a stable organized liposome arrangement by the addition of a streptavidin solution into liposome assemblies previously reported [6]. We observed that the streptavidin changed the form of liposomes at the contact interface. The time-lapse observation of the liposomes exhibited gradual changes of such liposome

(A)

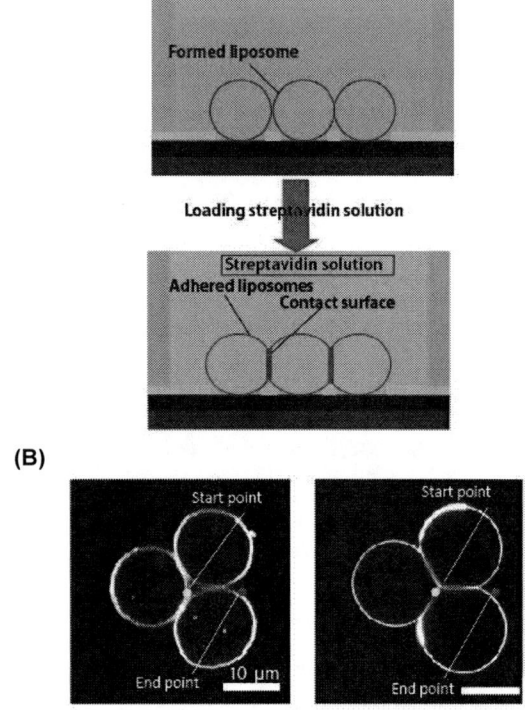

(B)

(C)

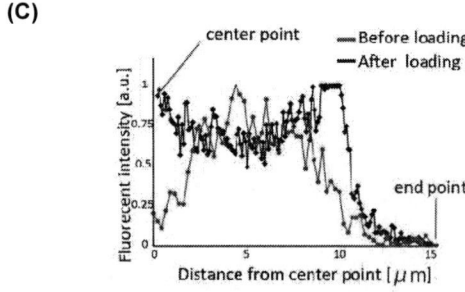

Figure 3: (A) Schematic illustration of the liposome adhesion process triggered by a streptavidin solution. (B) Confocal images of liposome arrangements with (right) and without (left) the streptavidin. Yellow and blue points in the images are the start and end points of the line profiles shown in 3-C. (C) Line profiles of the fluorescence intensity before and after the streptavidin addition.

shapes after the addition of streptavidin. The size of the organized structure was stable during 1h observation. This result indicated that the adhesion using the streptavidin-biotin complex maintained the organized structure. For future studies, we propose the introduction of channel-forming membrane proteins between the liposomes to realize continuous molecule transportation.

ACKNOWLEDGEMENTS

The authors deeply acknowledge the technical supports provided by Ms. Nose and Ms. Uchida (KAST). This works was partly supported by JSPS (Grant-in Aid for Challenging Exploratory Research; 26600066), and MEXT (Platform for Dynamic Approaches to Living System), Japan.

REFERENCES

[1] B. Alberts, A. Johnson, J. Lewis, M. Raff, K. Roberts, P. Walter, *Molecular Biology of the Cell 5E: Reference Edition*, Garland Science, 2008.

[2] K. Nishimura, T. Matsuura, T. Sunami, H. Suzuki, T. Yomo, "Cell-free protein synthesis inside giant unilamellar vesicles analyzed by flow cytometry", *Langmuir,* vol. 28, pp. 8426-8432, 28, 2012.

[3] K. Kurihara, M. Tamura, K. Shohda, T. Toyota, K. Suzuki, T. Sugawara, "Self-reproduction of supramolecular giant vesicles combined with the amplification of encapsulated DNA", *Nat. Chem.,* vol. 3, pp. 775-781, 2011.

[4] N. Ichihashi, K. Usui, Y. Kazuta, T. Sunami, T. Matsuura, T. Yomo, "Darwinian evolution in a translation-coupled RNA replication system within a cell-like compartment", *Nat. Commun.,* vol. 4, p. 2494, 2013.

[5] G. Villar, A. D. Graham, H. Bayley, "A tissue-like printed material", *Science*, vol. 340, pp. 48-52, 2013.

[6] H. Hamano, T, Tonooka, T. Toshihisa, S. Takeuchi, "Highly Packed Liposome Assemblies Toward Synthetic Tisssue", *Proc. IEEE MEMS 2014,* Sanfrancisco, pp. 17-19, 2014.

[7] T. Osaki, K. Kuribayashi-Shigetomi, R. Kawano, H. Sasaki, S. Takeuchi, "Uniform-sized liposome array formation with gentle hydration", *Proc. IEEE MEMS 2011*, Cancun, pp. 103-106, 2011.

[8] T. Osaki, K. Kamiya, R. Kawano, H. Sasaki, S. Takeuchi, "Towards artificial cell array system: Encapsulation and hydration technologies integrated in liposome array", *Proc. IEEE MEMS 2012*, Paris, pp. 333-336, 2012.

CONTACT

*H. Hamano, Institute of Industrial Science, The University of Tokyo, 4-6-1 Komaba, Meguro, Tokyo 153-8505, Japan; Tel: +81-3-5452-6650; Fax: +81-3-5452-6649; Email: hamano@iis.u-tokyo.ac.jp

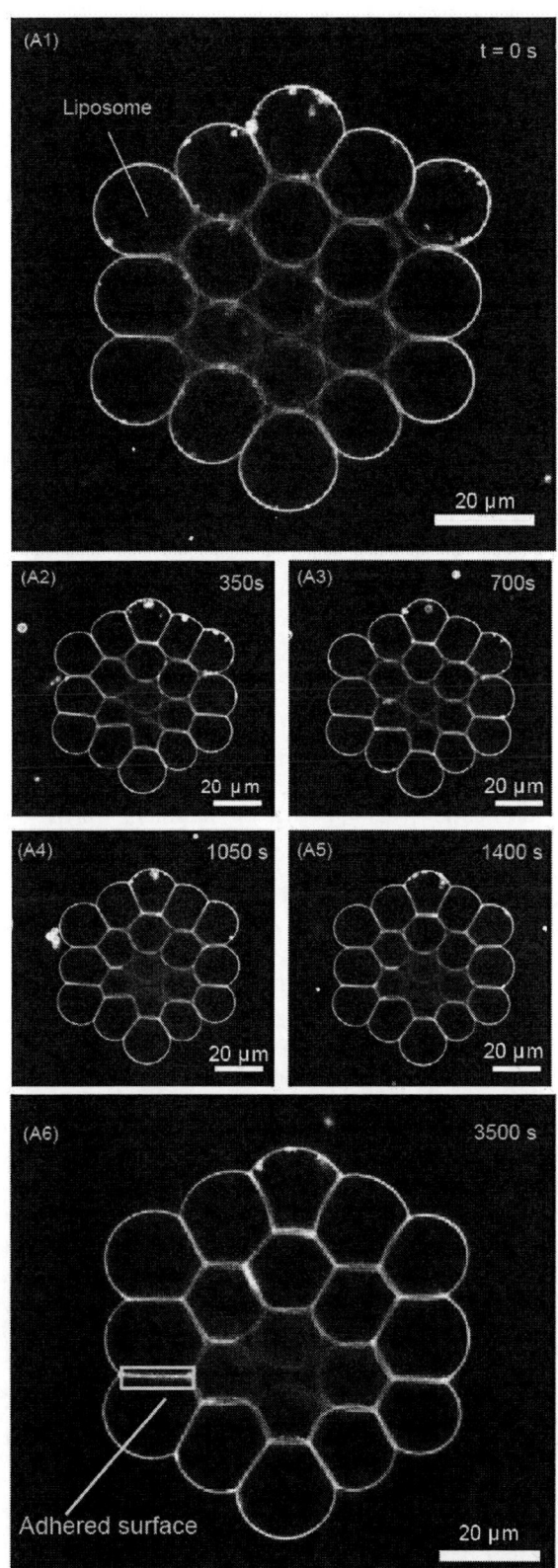

Figure 4: Time-lapse confocal images of liposome arrangements after the loading of streptavidin.

MAGNESIUM-EMBEDDED LIVE CELL FILTER FOR CTC ISOLATION

Yang Liu[1], Jungwook Park[1], Tong Xu[2], Yucheng Xu[2], Jay Han-Chieh Chang[2], Dongyang Kang[1], Xiaoxiao Zhang[1], Amir Goldkorn[2] and Yu-Chong Tai[1]

[1]California Institute of Technology, Pasadena, USA
[2]University of Southern California, Los Angeles, USA

ABSTRACT

This paper reports a novel Magnesium-embedded cell filter for Circulating Tumor Cell (CTC) capture, release and isolation. The new and novel feature is the use of thin-film Mg to release the captured CTCs based on the fact that any Cl⁻ containing culture medium can readily etch Mg away [1]. The releasing and the isolation of each individual CTC are demonstrated here. After filtration process, the filter is submerged in PBS to facilitate Mg etching. The top PA-C filter pieces break apart from the bottom after Mg completely dissolves, enabling captured CTC cells to detach from the filter. The released CTC can then be easily aspirated into a micropipette, and then for further, such as, DNA mutation analysis.

INTRODUCTION

Live CTC capture from whole blood has been identified to be an unmet need for cancer research [2]. Two types of MEMS devices have been proposed. One is to use AdCAM-antibody capture, and the other uses MEMS filters that target the inherently larger CTCs by size [3, 4]. Furthermore, however, what the researchers really need is isolated live CTC cells for following crucial analyses such as DNA mutation. The AdCAM-antibody capture method will require some biochemical ways to break lose the CTCs and a safe method that doesn't harm the CTCs is yet to be proven. Culturing of captured CTCs has been performed on slot filter to prove CTC viability [4]. On the other hand, the size-based mechanical filtering always has some CTCs stuck on the filter holes so they cannot be isolated without damage. This work then reports the first time-delayed releasing of captured live CTCs using a novel buried sacrificial Mg layer underneath the capturing filters. As the thin-film Mg can readily be dissolved by the CTC culture medium (hence biodegradable), there's no need for any additional chemical to release the CTCs so that filtered CTCs can be identified, released and picked up by automated pipette. This work reports the design, fabrication, test and verification of such a filter.

Magnesium (Mg) and magnesium alloys have drawn great attention as biodegradable materials because of their degradability by Cl⁻-containing solution such as saline [1]. In the meantime, Mg is totally compatible with MEMS fabrication techniques. High purity Mg pellets for E-beam evaporation are commercially available. This indicates that magnesium is an interesting dual "sacrificial and biodegradable MEMS material". It also has a great potential to be incorporated with many parylene-based devices for implant applications [5-14] besides the Mg-embedded CTC filter reported in this paper.

An illustration of the CTC capture, release and isolation is shown in Fig.1, following the order that (a) the CTC gets captured during filtration, (b) the CTC deforms and get stuck in the filter hole, (c) the Mg is etched, the filter pieces separate and the CTC is released, (d) the pipette comes and picks up the CTC to complete the isolation.

FABRICATION

Fig.2 shows the fabrication process of the magnesium-embedded cell filter for CTC capture. A 4-inch wafer is prepared for fabrication after piranha plus buffered hydrofluoric acid (BHF) cleaning, and then hexamethyldisilazane (HMDS) treatment respectively. Parylene-C (PA-C) film is then deposited over the wafer. 0.1µm-thick aluminum (Al) is thermally evaporated over the PA-C and the photoresist AZ1518 is spin-coated, exposed and developed to pattern Al as a plasma etching mask. Oxygen plasma (400W, 300mT) is used to etch through the openings on PA-C film all the way down to silicon surface (Fig.2a). Next, Mg is evaporated over the PA-C using E-beam according to the recipe reported previously [1] (Fig.2b). Thick photoresist AZ4620 is then patterned through lithography as a sacrificial layer (Fig.2c). Another run of oxygen plasma (50W, 200mT, 1 minute) is used to roughen the surface of PA-C, followed

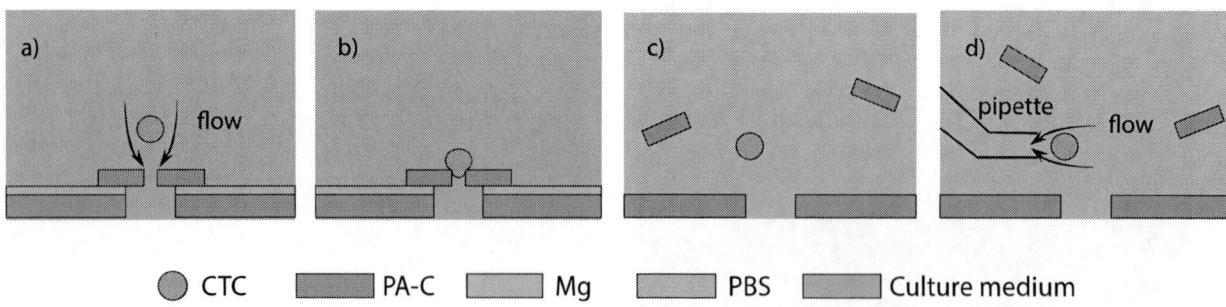

Figure 1: (a) the CTC gets captured during filtering, (b) the CTC cell deforms and get stuck in the filter hole, (c) the Mg is etched and the filter pieces separate and the CTC cell is released, (d) the pipette comes and picks up the CTC to complete the isolation.

by a 30-second BHF cleaning to make the surface hydrophobic. A final 10-μm-thick PA-C layer is deposited over the device (Fig.2d). Thick photoresist AZ4620 is patterned as etching mask, and the openings on the top PA-C layer is then etched through by oxygen plasma (400W, 300mT) (Fig.2e). Finally, the sacrificial photoresist is taken away by acetone, and the magnesium-embedded cell filters are peeled off from silicon in DI water (Fig.2f). The SEM photo is shown in Fig.3.

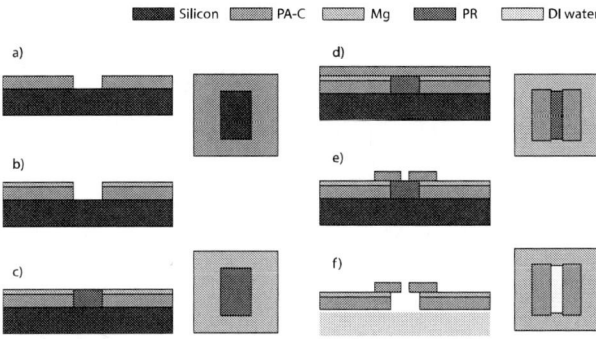

Figure 2: Fabrication process of the filter: a) PA-C deposition/patterning on Si, b) Mg deposition, c) Photoresist spin-coating/patterning, d) PA-C deposition, e) PA-C patterning, f) Photoresist dissolving and peeling off device from Si substrate.

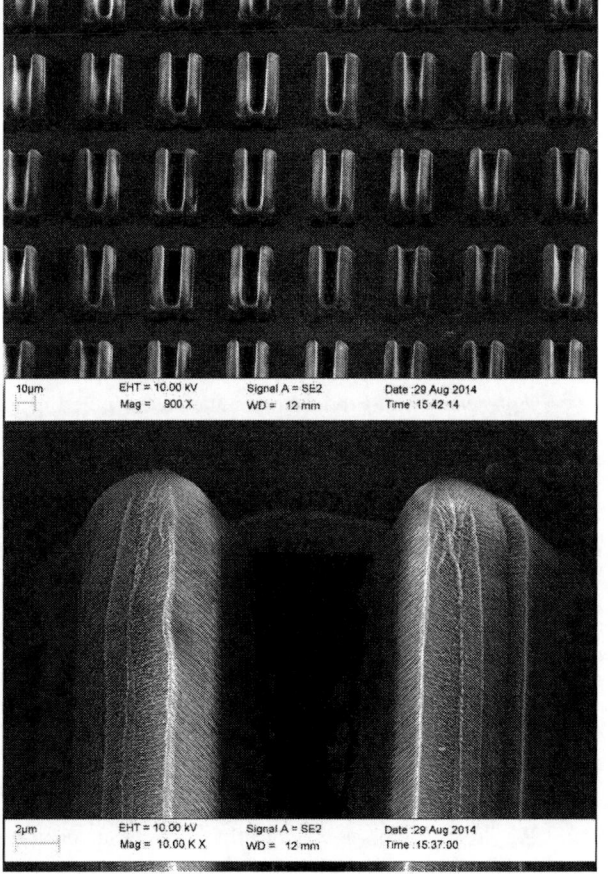

Figure 3: Top, SEM of Magnesium-embedded CTC filter array; Bottom, single Magnesium-embedded CTC filter slot.

MODELING

In order to achieve the goal that Mg stays during filtration while etched away sometime after filtration, the etching properties are studied ahead of the experiment. Experimental data are collected at different chlorine concentrations, namely 0.02, 0.04, 0.08, 0.2, 0.5mol/L. Etching rates are calculated by taking derivatives with respect to etching lengths over time. Modeling is based on the "combined first-and-second order" principle, which has been reported previously in [1]. First-order only model indicates chemical reaction only while second-order only model indicates diffusion only. Etching data fit the model (Fig.4). At low concentration the etching rate is dominated by the chemical reaction rate while at high concentration diffusion mechanism dominates. The Mg etching length, which is half of the width of the overlapping area between the top and bottom PA-C, is determined to be 3 micron according to the biodegradable etching results in saline reported in [1], which provides enough time for the top PA-C to detach from bottom PA-C film.

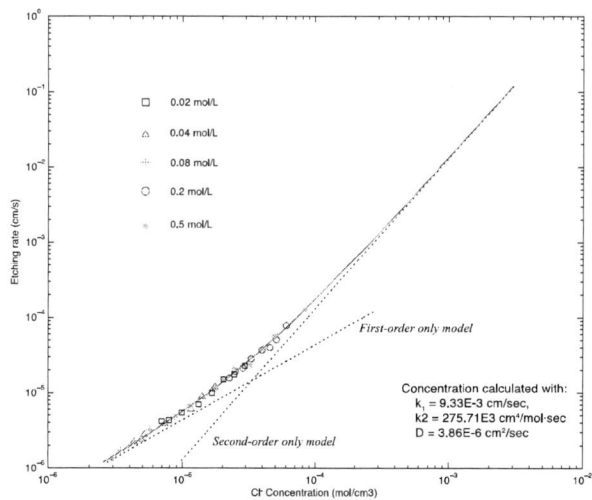

Figure 4: Etching rate of Mg channel dependence on concentration of Cl⁻ is studied. Data fit the model well.

RESULTS

Filtration and soaking

MDA-MB-231 breast cancer cells (15-16 μm in diameter) were used for filtration demonstration. 2 μL/s filtration rate was applied. After filtration, the filter was soaked in PBS for the dissolution of the Mg sacrificial layer, as shown in Fig.1c. Fig.5a shows that Mg hasn't been etched away 1 minute after filtration, which is desirable. The top PA-C is detached from the bottom when Mg is totally etched away (Fig.5b), which verifies that no additional chemical is needed to release the captured CTCs.

Aspiration system

Fig.6 shows the setup of CTC cell isolation. A manipulator, which is connected to a syringe, controls the micropipette and the syringe pump is used to create the suction force to aspirate released CTCs into the micropipette. Fig.7 shows the CTCs are safely aspirated into the micropipette, after Mg completely releasing from the filter.

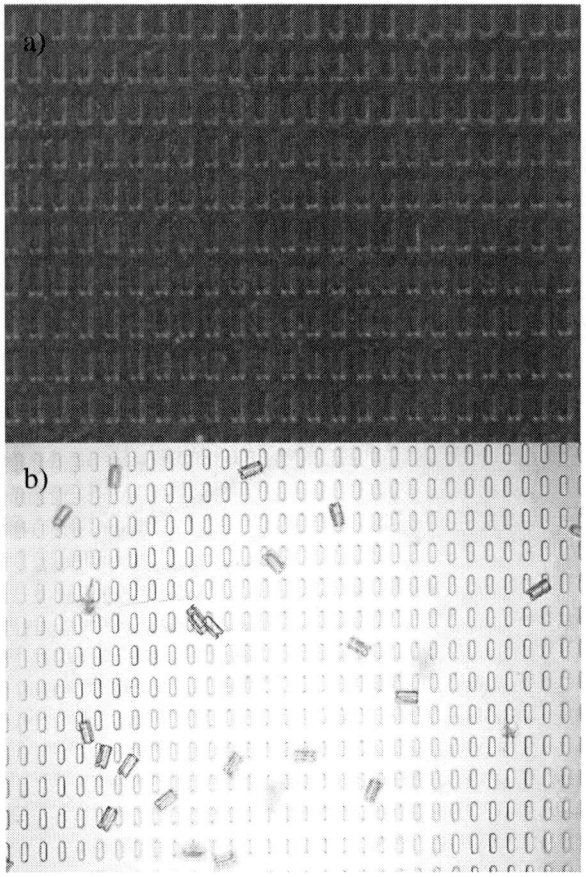

Figure 5: a) Mg dissolving process in saline: After 1 min. Mg remains during filtration process. b) After 150 min. Mg is totally etched away. Top parylene free-released from the bottom parylene.

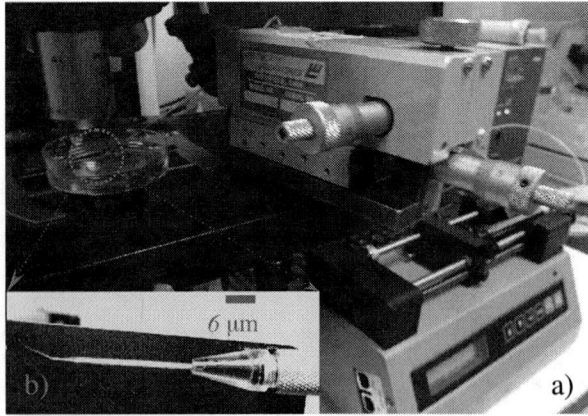

Figure 6: a) Setup for CTC aspiration into micropipette, b) close-up of the micropipette with 20-μm ID tip.

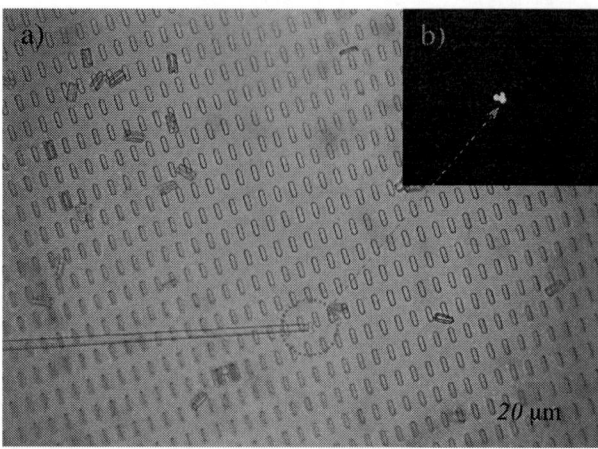

Figure 7: a) CTCs are aspirated into the micropipette, after Mg-releasing from the filter. b) Fluorescence photo of CTCs at the tip of micropipette.

CONCLUSION

This paper reports the design and fabrication of a novel Magnesium-embedded cell filter for CTC capture. Based on the fact that any Cl⁻ containing culture medium can readily etch Mg away, thin-film Mg sacrificial layer is micro-machined to achieve that the captured CTCs can be released without adding any additional chemical. The feasibility of this Mg-embedded CTC cell filter has been successfully demonstrated.

ACKNOWLEDGEMENTS

The authors gratefully acknowledge the help of all the members in California Institute of Technology (Caltech) Micromachining Lab, and University of Southern California (USC) Keck School of Medicine Norris Comprehensive Cancer Center, as well as experimental assistance of Mr. Trevor Roper.

REFERENCES

[1] Y. Liu, J. Park, J.H. Chang, Y.C. Tai, "Thin-film magnesium as a sacrificial and biodegradable material", in *Digest Tech. Papers MEMS'14 Conference*, San Francisco, January 26-30, 2014, pp. 656-659.

[2] S. Nagrath, L.V. Sequist, S. Maheswaran, D.W. Bell, D. Irimia, L. Ulkus, M.R. Smith, E.L. Kwak, S. Digumarthy, A. Muzikansky, P. Ryan, U.J. Balis, R.G. Tompkins, D.A. Haber, M. Toner, "Isolation of rare circulating tumour cells in cancer patients by microchip technology", *Nature*, vol. 450, pp. 1235-1239, 2007.

[3] S. Zheng, H. Lin, R.J. Cote and Y.C. Tai, "A novel 3D micro membrane filtration device for capture viable rare circulating tumor cells from whole blood", in *Digest Tech. Papers Hilton Head '08 Conference*, Hilton Head Island, June 1-5, 2008, pp. 134-137.

[4] B. Lu, T. Xu, S. Zheng, A. Goldkorn, Y.C. Tai, "Parylene membrane slot filter for the capture, analysis and culture of viable circulating tumor cells", in *Digest Tech. Papers MEMS'10 Conference*, Hong Kong, January 24-28, 2010, pp. 935-938.

[5] Y. Liu, J. Park, R.J. Lang, A. Emami-Neyestanak, S. Pellegrino, M.S. Humayun, and Y.C. Tai, "Parylene origami structure for intraocular implantation", in *Digest Tech. Papers Transducers'13 Conference*, Barcelona, June 16-20, 2013, pp. 1549-1552.

[6] J.H. Chang, Y. Liu, D. Kang, M. Monge, Y. Zhao, C.C. Yu, A. Emami, J. Weiland, M. Humayun, and Y.C. Tai, "Packaging study for a 512 channel intraocular epiretinal implant", in *Digest Tech. Papers MEMS'13 Conference*, Taipei, January 29-Feburary 2, 2012, pp. 353-356.

[7] J.H. Chang, Y. Liu, D. Kang, and Y.C. Tai, "Reliable packaging for parylene-based flexible retinal implant", in *Digest Tech. Papers Transducers'13 Conference*, Barcelona, June 16-20, 2013, pp. 2612-2615.

[8] J.H. Chang, Y. Liu, and Y.C. Tai, "A low-temperature parylene-C-to-silicon bonding using photo-patternable adhesives and its application", in *Digest Tech. Papers Transducers'13 Conference*, Barcelona, June 16-20, 2013, pp. 2217-2220.

[9] J.H. Chang, Y. Liu, and Y.C. Tai, "Long term glass-encapsulated packaging for implant electronics", in *Digest Tech. Papers MEMS'14 Conference*, San Francisco, January 26-30, 2014, pp. 1127-1130.

[10]J.H. Chang, "Wireless parylene-based retinal implant", *Dissertation (Ph.D.), California Institute of Technology*, 2014.

[11]J.H. Chang, R. Huang, and Y.C. Tai, "High density 256-channel chip integration with flexible parylene pocket", in *Digest Tech. Papers Transducers'11 Conference*, Peking, June 5-9, 2011, pp. 378-381.

[12] J.Y.H. Kim, Y. Liu, N. Scianmarello, and Y.C. Tai, "Piezoelectric Parylene-C MEMS microphone", in *Digest Tech. Papers Transducers'13 Conference*, Barcelona, June 16-20, 2013, pp. 39-42.

[13] D. Kang, A. Standley, J.H. Chang, Y. Liu, and Y.C. Tai, "Effects of deposition temperature on Parylene-C properties", in *Digest Tech. Papers MEMS'13 Conference*, Barcelona, January 20-24, 2013, pp. 389-390.

[14] J.Y.H. Kim, Y. Liu, N. Scianmarello, P. Satsanarukkit and Y.C. Tai, "Ice Fishing Micro channels with sub-micron pores", in *Digest Tech. Papers NEMS'13 Conference*, Suzhou, April 7-10, 2013, pp. 1080 - 1083.

CONTACT
*Y. Liu, tel: +1-626-4374445; ylliu@caltech.edu

MICRO FLUIDIC CHAMBER WITH THIN SI WINDOWS FOR OBSERVATION OF BIOLOGICAL SAMPLES IN VACUUM

Hideki Hayashi, Masaya Toda, and Takahito Ono

Graduate school of engineering, Tohoku University, Sendai, Japan

ABSTRACT

A micro fluidic chamber with 178 nm-thick single crystal Si windows on a micro channel has been developed. Because of these thin windows, the aquatic sample inside of the channel can be observed by scanning electron microscopy. Secondary electrons from a sample in the channel are able to be detected in vacuum with an acceleration voltage of 15 kV, where the emission current is 75 µA. The micro fluidic chamber is possibly applied to cell imaging via the Si thin window in vacuum using magnetic resonance force microscopy.

INTRODUCTION

The observation of the inside phenomenon of living cells at a high resolution is a demand for investigating the mechanism of apoptosis and canceration process in detail. The surface structure of cancer cells is observed in nano-scale using scanning electron microscopy (SEM) and transmission electron microscopy (TEM). Magnetic resonance force microscopy (MRFM) has been developed for observing the inner structure of materials in nano-scale [1]. However, those technologies require a high vacuum condition for high resolution. In the previous work, an adherent cell was observed using special setup which is the combination of the inverted SEM and optical microscope [2]. A floating particle in liquid has been observed with a common TEM system using micro fluidic chamber [3]. In these systems, a thin SiN window is used for separating vacuum from liquid because of its high mechanical strength.

A nuclear magnetic resonance imaging (MRI) and an electron spin resonance (ESR) imaging are used to analyze inner structures of biological samples as non-invasive observation techniques. The resolution of MRI has been limited by the spatial resolution of electromagnet field. The resolution of ESR imaging was achieved into sub-micron [4]. On the other hand, MRFM is the imaging technique based on high sensitive mechanical cantilever sensing with nano-meter resolution. ESR-MRFM has been developed for 3D imaging of biological samples [5]. ESR-MRFM is a candidate for imaging tools of living cells based on radicals. Since crystal defects in SiN induce background noise for ESR-MRFM measurement, SiN might not be the suitable material as the window in the system.

In this work, the micro fluidic chamber with thin single crystal Si windows is developed for observing biological samples with vacuum analytical system such as SEM and MRFM. The micro chamber consists of micro channels and cell trapping slits to observe cells under the window.

DESIGN

A single crystal Si is selected as the material of the thin window on the micro chamber for cell observation in vacuum. Figure 1 shows the schematics of the micro fluidic chamber. The size of the chamber chip is 10 mm × 10 mm × 0.3 mm.

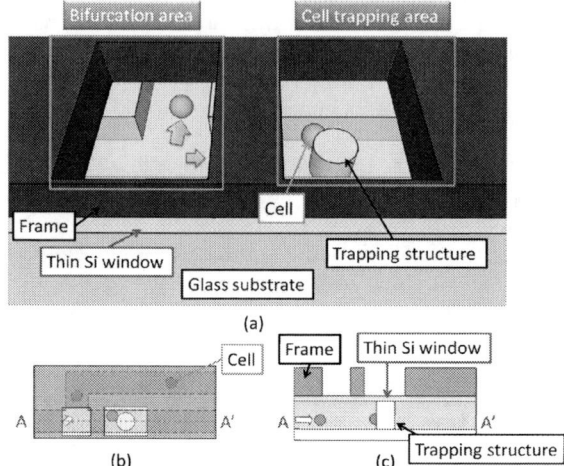

Figure 1: Schematic view of the micro fluidic chamber. (a) 3D view, (b) Top view, (c) Side view. The areas of bifurcation (left) and cell trapping (right) are fabricated under thin Si windows. Once a cell is trapped by the glass pole, the other cells go to the bypass channel.

Window Material

For SEM observation, the detected electron needs to pass through the window from the inner channel to vacuum. Electron penetration depth R_G in solid material with atomic number ranging from 10 to 15 is given by

$$R_G = \frac{3.98 \times 10^{-5} E_B^{1.75}}{\rho}, \qquad (1)$$

where E_B is the electron energy in kilovolts, ρ is density of the solid material [6]. From this equation, electrons with an acceleration voltage of 5.7 kV can travel 400 nm distance in Si. At least, less than half of the traveling distance is required for Si window thickness. To observe the secondary electrons from the channel, the higher acceleration voltage than 5.7 kV should be required because the kinetic energy of electrons diminishes by inelastic scattering process in the channel and the window. After electrons pass through the window and go into inside of the chamber, they are also scattered by water or sample. The most of the electron are estimated to be scattered within 11 µm traveling distance in water from Monte Carlo simulation at an acceleration voltage of 30 kV [7].

If a 200 nm-thick Si window is considered, the maximum window size is designed to be 25 µm × 50 µm from finite element method simulation using COMSOL as shown in figure 2. The applicable maximum pressure of 711 MPa to the window is calculated with assuming atmospheric pressure. The diaphragm is expected to be robust enough because the allowable shear stress of single crystal Si is approximately 1 ~ 4 GPa.

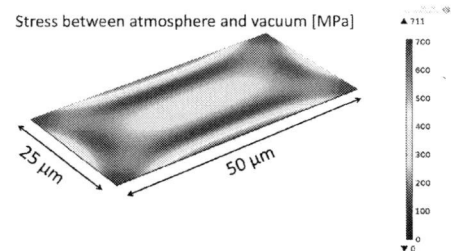

Figure 2: Simulated image of the single crystal Si diaphragm. Maximum stress is 711 MPa.

Channel design

The trapping system of the designed micro channel is based on hydrodynamics. There are two kinds of channels, i.e. trapping main channel and bypass channels, as shown in figure 3 (a). The main channel has trapping structure. The bypass channel is the connection pass to the next main channel with a trapping structure. The flow resistances of the bypass channel is designed due to the length of the channel. When the trapping area in the main channel is free, the main channel has a lower flow resistance than the bypass channel. Once a cell is trapped, the flow resistance increases in the main channel. Then following cells flow to the next trapping structure via the bypass channel.

To determine the dimension of the micro channel, the equivalent electrical circuit [8-9] as shown in figure 3 (b) is considered. R_{M1} and R_{M3} are the front and rear flow resistances of the main channel, respectively. R_{M2} is the flow resistance at the trapping area. R_B is the flow resistance of the bypass channel. Once a cell trapped, R_{M2} increase drastically, then other cells flow into the bypass channel. In our estimation, the boundary condition of the same pressure drop value across both channels is used because the end points are connected. Pressure drop at main channel ΔP_M and bypass channel ΔP_B are same and the flow rate Q_M and Q_B is estimated from the equation $\Delta P_M = \Delta P_B$. If Q_M / Q_B is calculated to be more than 1, cells enter the main channel when they are not trapped. In this work, both channels are designed as $3 < Q_M/Q_B < 4$ for effective trapping.

The height and width of the micro channel are designed to be 20 μm and 25 μm, respectively. There are serial trapping structures for flowing cell below the windows in the micro channel. To trap a single cell, the width of the trapping structure is less than 5 μm for various cells. The aspect ratio of the trapping channel is required to be over 4

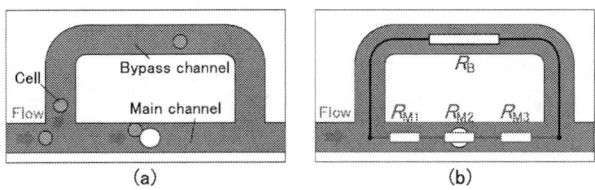

Figure 3: Schematic view of micro channel structure. (a) Overview of the main channel and the bypass channel. (b) Estimation of the flow resistances using the equivalent electrical circuit.

FABRICATION

Micro channel structure is fabricated using a 300 μm-thick Tempax glass. Cr and Au are deposited on the glass using sputtering [Figure 4 (a)]. A 2 μm-thick Ni layer is formed by electroplating on the Au layer patterned with a photolithography [Figure 4 (b)]. After wet etching of Au and Cr layers using metal etchants, the glass is etched by SF₆ reactive ion etching (RIE) using the Cr-Au-Ni layer as a mask [Figure 4 (c)] [10]. After the Ni, Au and Cr are removed by wet etching, two 1 mm diameter holes for sample inlet and outlet are made by sandblast [Figure 4 (d)].

The device layer of the SOI (Silicon On Insulator) wafer (200 nm / 400 nm / 300 μm) is used as the window layer of the chamber. 1st Deep RIE is performed on the bottom side of the SOI wafer for making window part [Figure 4 (e)]. This SOI wafer is bonded to the prepared Tempax glass using anodic bonding [Figure 4 (f)]. The handling layer of the bonded SOI wafer is etched by the 2nd Deep RIE till observing the opened insulation layer at the window part. Finally the insulation layer is etched by vapor HF [Figure 4 (g)]. For liquid loading, small fitting eyelets (Inside Φ 1 mm, Outside Φ 1.5 mm, Flange Φ 3.4 mm) to tubes are glued with araldite on the backside of the glass. Silicone tube (Inside Φ 1 mm, Outside Φ 1.5 mm) and Teflon tube (Inside Φ 1 mm, Outside Φ 1.64 mm) are connected to the eyelets.

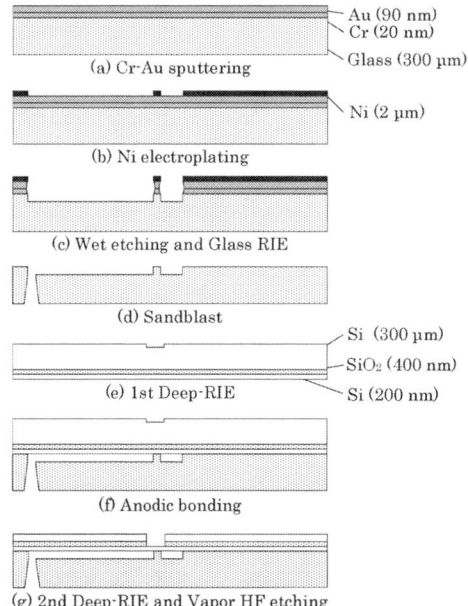

Figure 4: Fabrication process of the micro fluidic chamber which is started from silicon on insulator wafer and Tempax glass.

RESULTS AND DISCUSSIONS

SEM image of the fabricated trapping structure in the micro channel at process step of figure 4 (c) is shown in figure 5. The taper angle of 75° at the side wall and 63° around the pole can be observed. The gaps between the glass pole and the micro channel wall serve as the trapping slit; however, the slit width at the bottom of the channel consequently became narrower than the designed width.

978-1-4799-7956-1/15 $31.00 © 2015 IEEE   345

The optical image of the completed micro fluidic chamber is shown in figure 6. The flat window without any residual stress during the anodic bonding was seemingly observed.

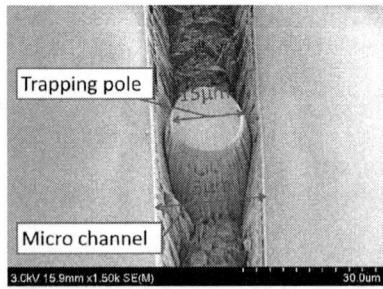

Figure 5: SEM image of the fabricated micro channel. The width and depth are 25 μm and 20 μm, respectively. The 15 μm diameter trapping pole is fabricated in the channel.

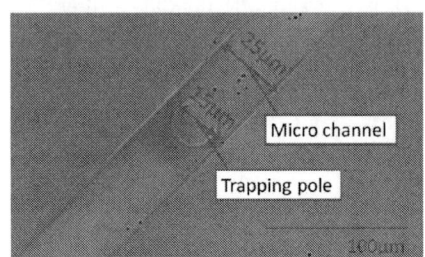

Figure 6: Optical image of the completed micro fluidic chamber.

Fabricated thin Si windows

The thickness of the fabricated thin Si window after bonded with the glass is 178 nm, which is measured using a microscopic optical interferometer (DF-1037R1, Techno Synergy, Japan). As the first evaluation of the thin Si window, the window is observed in the SEM chamber under conditions of electron acceleration voltages of 5 kV, 10 kV, 15 kV, 20 kV, 25 kV, and 30 kV. The observed SEM images at 5 kV and 30 kV acceleration voltages on the window part are shown in figure 7. At the acceleration voltage of 5 kV, the surface of the window was observed because the penetration depth of electrons is not enough to observe inside the window. On the other hand, the inside view of the chamber was observed at an acceleration voltage of 30 kV.

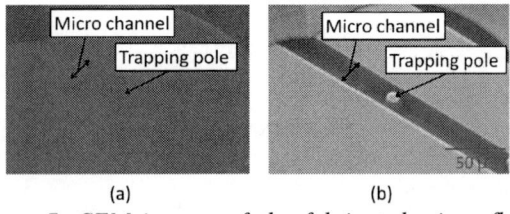

Figure 7: SEM images of the fabricated micro fluidic chamber at acceleration voltages of (a) 5 kV, (b) 30 kV.

Liquid loading in vacuum

The schematic image of the liquid loading system in the vacuum chamber is shown in figure 8. Using the tube connection with the eyelet fitting, DI water is loaded into the channel with negative pressure. The tubes are fed through a ICF 70 flange sealed with epoxy resin (Torr seal).

After the vacuum chamber is vacuumed with a turbo molecular pump, the vacuum level is achieved to 7.6×10^{-4} Pa after 24 hour with filled water in the channel. Here, it is confirmed that 178 nm thin Si window can separate between ultra-high vacuum and liquid of atmospheric pressure.

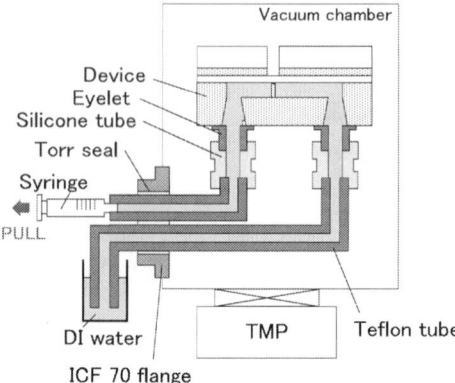

Figure 8: Schematic image of the liquid loading system to the vacuum chamber.

SEM observation of liquid

After the micro fluidic chamber is installed into a SEM chamber, DI water is loaded into the micro channel. After the SEM chamber is evacuated by an oil diffusion pump, the pressure is below 0.1 Pa. Then, the micro chamber is observed at an acceleration voltage of 5 kV, as shown in figure 9. Only the surface of the window was observed at 5 kV as shown in figure 7(a). The water in the channel is impervious to the SEM observation at low electron energy. The window on the channel with filled water is observed to be seemingly flat in vacuum condition by SEM image.

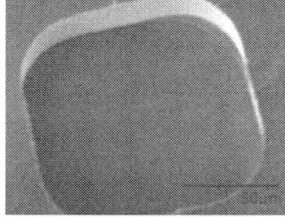

Figure 9: SEM image of the micro fluidic chamber at an acceleration voltage of 5 kV after loading liquid.

With increasing the acceleration voltage of the electron, the surface of the inserted water in the chamber is able to be observed at 15 kV, where the emission current is 90 μA. The scanning speed of the electron beam is 15.6 kHz per 100 μm and the frame rate of SEM monitor is 29 fps. The motion of the inserted water surface is observed as shown in figure 10.

Figure 11 shows the difference of the SEM images at acceleration voltages of 20 kV and 25 kV. At the acceleration voltage of 25 kV, the side wall of the micro channel was clearly observed than that at 20 kV. The difference of the observed depth is approximately several micron. Calculated electron penetration depth in water is 10.5 μm and 10.1 μm at acceleration voltage of 25 kV and 20 kV, respectively. For the cell observation by SEM, the

978-1-4799-7956-1/15 $31.00 © 2015 IEEE   346

target sample should be close to the window.

At acceleration voltages of 20 kV and 25 kV, sometimes bubbles appear in water as shown in figure 12. When the bubble happens continuously, the volume of the bubble is increased to reach to the bifurcation area. The reached bubble at the bifurcation area flows away to the bypass channel immediately. After the generated bubble goes away to the bypass channel, another bubble appears under the monitoring window. It seems to require the optimizations of the condition of SEM observation. Also the chamber can be applied to MRFM because it has the thin windows of single crystal Si without defection.

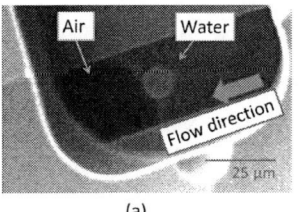 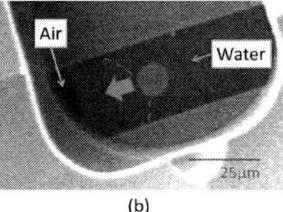

(a)   (b)

Figure 10: SEM image of the liquid surface moving in the micro fluidic chamber at an acceleration voltage of 15 kV and an emission current of 75 µA.

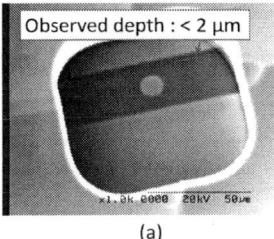 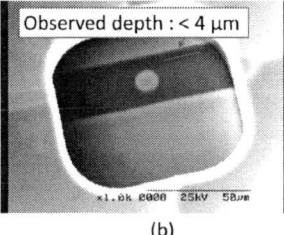

(a)   (b)

Figure 11: SEM image of the water in micro fluidic chamber at acceleration voltages of (a) 20 kV and (b) 25 kV.

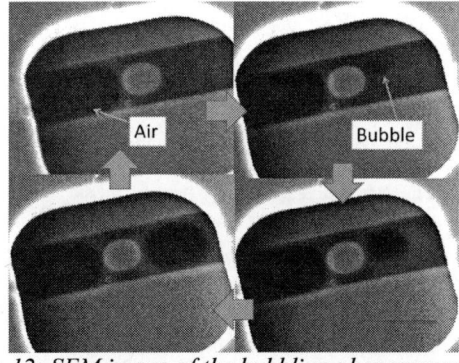

Figure 12: SEM image of the bubbling phenomenon in the micro fluidic chamber at the acceleration voltage of 25 kV.

CONCLUSIONS

The micro fluidic chamber with 178 nm-thick Si window for the observation of biological samples in vacuum has been developed. Increasing the acceleration voltage of the electron, the surface of the inserted DI water filled in the chamber is able to be observed at 15 kV, where the emission current is 90 µA. Since the observed depth is a few micron even if the acceleration voltage was 25 kV, the target sample is required to be close to the window for

observation. Sometimes bubbles appear in water at acceleration higher than 20 kV. The optimization of the condition of SEM observation would be required for stable measurement.

ACKNOWLEDGEMENTS

Parts of this work were performed in the Micro/Nano machining Research Education Center (MNC) and Micro System Integration Center (µSIC) of Tohoku University.

REFERENCES

[1] Y. J. Seo, M. Toda, Y. Kawai, and T. Ono, "Ultrasensitive Si nanowire probe for magnetic resonance detection", *Proceedings of 27th IEEE International Conference on Micro Electro Mechanical Systems*, San Francisco, January 26-30, 2014, pp. 151-154.

[2] H. Nishiyama, M. Suga, T. Ogura, Y. Maruyama, M. Koizumi, K. Mio, S. Kitamura, and C. Sato, "Atmospheric scanning electron microscope observes cells and tissues in open medium through silicon nitride film", *Journal of Structural Biology.*, vol. 172, pp. 191-202, 2010.

[3] J E. Evans, K L. Jungjohann, P C. K. Wong, G H. Dutrow, I. Arslan, and N D. Browning, "Visualizing macromolecular complexes with in situ liquid scanning transmission electron microscopy", *Micron*, vol. 43, pp. 1085-1090, 2012.

[4] A. Blank, E. Suhovov, R. Halevy, L. Shtirberg, and W. Harneit, "ESR imaging in solid phase down to sub-micron resolution", *Physical Chemistry Chemical Physics*, Vol.11, pp. 6689-6699, 2009.

[5] S. Tsuji, Y. Yoshinari, H. S. Park, and D. Shindo, "Three dimensional magnetic resonance imaging by magnetic resonance force microscopy with a sharp magnetic needle", *Journal of Magnetic Resonance*, vol. 178, pp. 325-328, 2006.

[6] T. E. Everhart, and P. H. Hoff, "Determination of Kilovolt Electron Energy Dissipation vs Penetration Distance in Solid Materials", *Journal of Applied Physics*, vol. 42, pp. 5837, 1971.

[7] A. Bogner, P. H. Jouneau, G. Thollet, D. Basset, and C. Gauthier, "Monte Carlo Simulation of Water Radiolysis for Low-energy Charged Particles", *Journal of Radiation Research*, vol.47, pp. 69-81, 2006.

[8] W. Tan, and S. Takeuchi, "A trap-and-release integrated microfluidic system for dynamic microarray applications", *Proceedings of National Academy of Sciences*, vol. 104, pp.1146-1151, 2007.

[9] A. C. Rowat, J. C. Bird, J. J. Agresti, O. J. Rando, and D. A. Weitz, "Tracking lineages of single cells in lines using a microfluidic device", *Proceedings of National Academy of Sciences*, vol. 106, pp.18149-18154, 2009.

[10] X. Li, T. Abe, and M. Esashi, "Deep reactive ion etching of pyrex glass", *Proceedings of 13th IEEE International Conference on Micro Electro Mechanical Systems*, Miyazaki, January 23-27, 2000, pp.271-276.

CONTACT

*Masaya Toda, tel: +81-22-795-5810;
mtoda@nme.mech.tohoku.ac.jp

RAPID, LOW COST FABRICATION OF CIRCULAR CROSS-SECTION MICROCHANNELS BY THERMAL AIR MOLDING

Thanh- Qua Nguyen[1], and Woo-Tae Park[1, 2,*]

[1]Department of Mechanical and Automotive Engineering, Seoul National University of Science and Technology, SOUTH KOREA

[2]Convergence Institute of Biomedical and Biomaterial Engineering, Seoul National University of Science and Technology, SOUTH KOREA

ABSTRACT

This paper demonstrates a simple fabrication process of polydimethylsiloxane (PDMS) circular cross section microfluidic channels by using a PDMS master mold and thermal air molding. Based on this technique, circular cross section microchannel can be easily produced in a wide range of dimensions from 10 μm to 500 μm with simple bench top equipment. This technique can create perfect circular channels without any plasma activated bonding and alignment process. We can also apply this technique to fabricate circular shape microchannel network for mimicking the vascular system, micro concaves for spheroid culture, micro nozzles for droplet generation, and micro patch clamps for cell immobilization.

INTRODUCTION

The field of microfluidics has grown rapidly since the introduction of rapid prototyping methods, especially the widely accepted soft lithography technique. However, this method only allowed the fabrication of rectangular cross section microchannels [1]. Since many tissue engineering systems, such as human blood vessel mimicry require circular cross section channels for uniform mechanical stress under culture conditions (Figure 1 a, d), there has been a strong interest in developing technologies to fabricate circular cross section micro channels. Recently, several fabrication methods were introduced by replicating from the mold of semi-circular shape cross section channel and then aligning and bonding two half PDMS semi-circular shape microchannels [2-5]. The semi-circular molds were fabricated from reflowed positive photoresist [2], laser writing or micro milling [3, 4], and isotropic etched silicon wafers [5]. The huge disadvantage of bonding two pieces of device method is malalignment during bonding due to elasticity of PDMS and manual handling process. The malalignment causes discontinues in final profile of the channel, especially when the channel diameter is down to tens of micron. Therefore, other methods to directly fabricate cylinder microchannel have been introduced by coating solvent diluted PDMS to modify rectangular channels [6], using sucrose as a sacrificial template material [7], absorbing polymethylmethacrylate (PMMA) solution into PDMS molds [8], and expanding of degradable poly lactic acid [9]. However, these methods require expensive mold masters or complex, and time consuming fabrication steps. The novel method reported here is simpler with a few fabrication steps, and requires minimum equipment.

DESIGN

To fabricate the device with rounded cross-section channel (Figure 1b), we design the master device in (Figure 1c) with the width of 50 μm in honey cone shape and 200 μm in main channels to mimic the capillary networks. For other applications, the height of SU-8 channel was 75 μm, 50 μm, 25 μm for microvascular device, micro concaves, micro nozzles, multi-width channels respectively. The Silicon molds were casted with PDMS (Sylgard-184) with a mixing ratio of 10: 1 (base: curing agent).

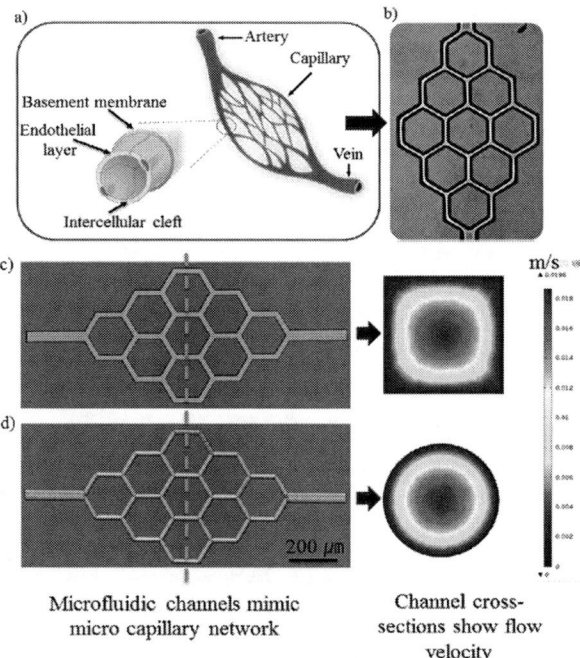

Figure 1: (a) Human capillary network with cylinder shape lumen (b) Simulation model with micro-channels mimic the capillary network in the microfluidic chip with round shape cross-section (c) Comsol simulation of rectangular channels with nonuniform flow profile at same distance from center (d) Comsol simulation of circular uniform flow profile at same distance from center. This flow profile resembles the natural flow of blood vessels.

FABRICATION

The fabrication process is described in Figure 2. The process starts with fabricating a pattern using SU-8. Then a PDMS stamp was created by molding the SU-8. The PDMS stamp was then treated with an acid buffer solution of hydroxyl propylmethyl cellulose (HPMC) to prevent adhesion between PDMS stamp and subsequent PDMS molding. The treated PDMS stamp was then bonded to a partially cured PDMS prepolymer. The PDMS prepolymer was partially cured at 75°C for 15 minutes to increase the viscosity and surface tension. This step creates such a thin solid membrane on the surface of prepolymer to prevent the mold channel collapsing during mold contact. After the PDMS stamp was contacted with the partially cured PDMS, the stack was heated on a hot plate with the condition of 90°C for 20 minutes. The air trapped inside the mold expands and creates the semi-circular microchannel. The profile of the microchannel depends on the air trapped cavity, the curing temperature, the width of channel, and the standing time for increasing the pressure inside the channel. The PDMS stamp was then detached from the circular microchannel without any damage (due to HPMC surface treatment), and the semi-circular microchannel was bonded to another partially cured PDMS layer and the stack was heated again on a hot plate to complete the circular cross section microchannel (Figure 2 b).

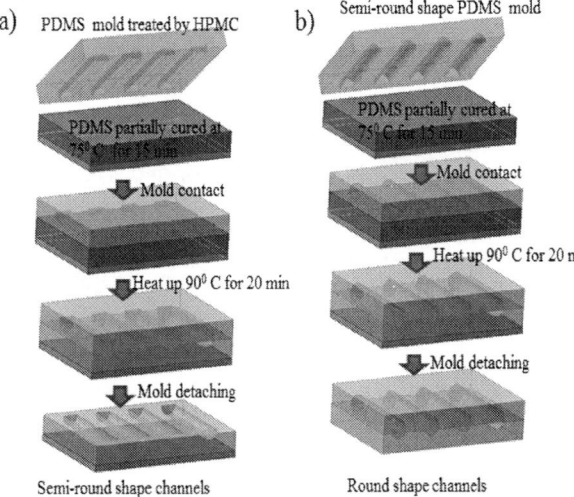

Figure 2: Schematic diagrams of the fabrication procedure for circular cross-section channel: (a) Semi-round shape cross-section microchannel fabrication; PDMS mold treated with HPMC is temporarily bonded to a partially cured PDMS wafer. The stack is then heated up and the trapped air expands to form semi-circular channels. The original mold is detached without damage (b) Fabrication of circular cross-section microchannel; the semi-circular PDMS channel wafer is bonded to another partially cured PDMS wafer. The stack is heated up and the trapped air expands to form circular cross section channels without any alignment or plasma bonding.

RESULTS

This method produced various channel sizes from 10 μm to 500 μm (Figure 3 d) with multiple depth channels. Moreover, round shape channels could be obtained without any alignment and plasma bonding process (Figure 3 c). Depending on the desired width of channel, the fabrication parameters should be optimized for fully rounded shape or elliptical cross-section shape. The SEM images in Figure 5

shows the many applications our method can be used to form micro concave wells to complex channel networks.

We characterized the semi-round shape of channel based on thermal pressure- assisted expansion of partial cured PDMS. The deformation of the channel depends on the width of mold channel, surface tension of PDMS prepolymer, and temperature of the heater. Figure 4 shows the relationship between molding temperature from 70°C to 95°C in hot plate and aspect ratio of PDMS expanded cavities. The increasing of incubation temperature increased the aspect ratio of cavities due to thermal effect. Thus, we can create many replicas with different geometries with the same mold, reducing the cost of SU-8 mold fabrication.

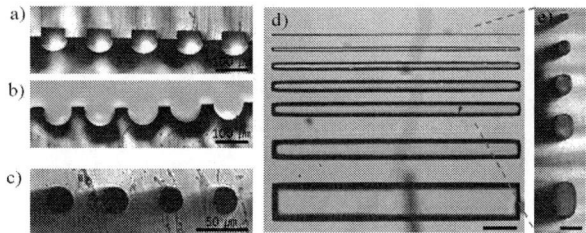

Figure 3: Bright-field images of PDMS microchannel: (a) PDMS rectangle cross-section master on top and semi-round shape stamp (b) Semi-round shape cross-section channels after detaching the master (c) Round shape channels (d) Various microchannels width (scale bar 500 μm) (e) Cross-section of different size of channels (scale bar 100 μm).

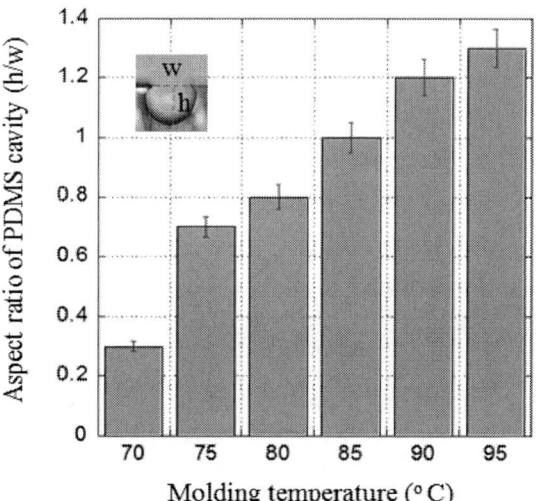

Figure 4: Characterization semi-round shape channels between aspect ratio of height and width of microchannels under different molding conditions.

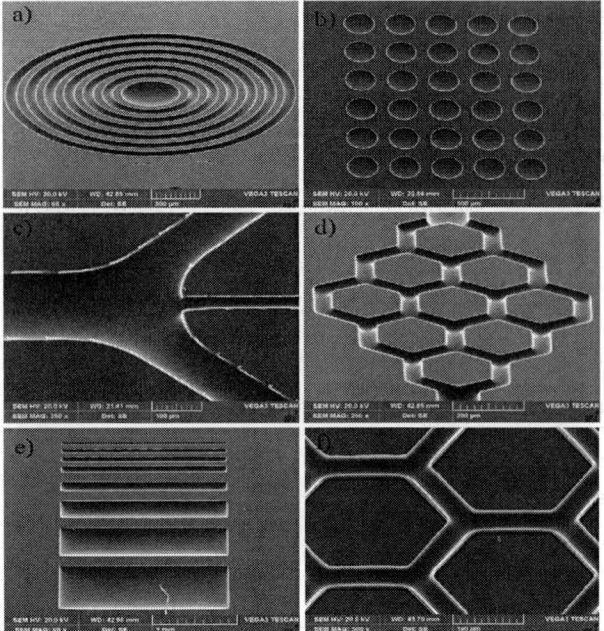

Figure 5: SEM images showing the micro features with round shape channels such as (a) concentric micro channel rings (b) micro concave wells (c) micro nozzle (d) micro capillary network (e) microchannel with different widths (f) micro channel intersections with 50 µm in width.

CONCLUSIONS

We have developed a novel PDMS molding method to rapidly fabricate circular cross-section microfluidic channels with many advantages over previous approaches. This is a straightforward and low-cost fabrication process that can create complex channel topology. We successfully fabricated concentric micro channel rings, micro concaves, micro nozzles for droplet generation device, and interconnected microcapillary networks. We think the microcapillary network application can particularly benefit from the cell culture favored round cross section channels and the capability of complicated network patterning. Our technique may also be used for other lab-on-a-chip applications that need the round shape channels such as a patch clamp device that can measure multiple cells simultaneously on a single chip.

ACKNOWLEDGEMENTS

This work was partly supported by Basic Science Research Program through the National Research Foundation of Korea (NRF) funded by the Ministry of Science, ICT and Future Planning (NRF-2013R1A1A1012616) and under the framework of international cooperation program managed by National Research Foundation of Korea (NRF-2013K2A1B8054280). This work was also supported by the Radiation Technology R&D program through the National Research Foundation of Korea funded by the Ministry of Science, ICT & Future Planning (NRF-2013M2A2A9043274).

REFERENCES

[1] Xia Y, et al., "Soft Lithography," Annu. Rev. Mater. Sci, vol 28,pp. 153–84, 1998.

[2] Jong Seob Choi, et al., "Fabrication of a circular PDMS microchannel for constructing a three-dimensional endothelial cell layer," Bioprocess Biosyst Eng, vol. 36, pp. 1871–1878, 2013.

[3] M.E. Wilson, et al., "Fabrication of Circular Microfluidic Channels by Combining Mechanical Micromilling and Soft Lithography," Lab Chip, vol. 11, pp. 1550-1555, 2011.

[4] Dong Hyuck Kam, et al., "Three-dimensional biomimetic microchannel network by laser direct writing," J. Laser Appl , vol. 20, pp.185-192, 2008.

[5] Jong Seob Choi, et al., "Fabrication of various cross-sectional shaped polymer microchannels by a simple PDMS mold based stamping method," BioChip J. vol 6, pp. 240-246, 2012.

[6] L.K. Fiddes, et al., "A circular cross-section PDMS microfluidics system for replication of cardiovascular flow conditions," Biomaterials, vol 31, pp. 3459–3464, 2010.

[7] Jiwon Lee, et al., "Sucrose-based fabrication of 3D-networked, cylindrical microfluidic channels for rapid prototyping of lab-on-a-chip and vaso-mimetic devices," Lab Chip, vol 12, pp. 2638–3642, 2012.

[8] Sung Hoon Lee, et al.," Use of directly molded poly(methyl methacrylate) channels for microfluidic Applications," Lab Chip, vol 10, pp. 3300-3306, 2010.

[9] Jen-Huang Huang, et al., "Embedding Synthetic Microvascular Networks in Poly(Lactic Acid) Substrates with Rounded Cross- Sections for Cell Culture Applications," PLoS One., vol 8, pp. e73188, 2013.

CONTACT

[*]Woo-Tae Park, Tel: +82-2-970-6354.
Email: wtpark@seoultech.ac.kr

SKIN-EQUIVALENT INTEGRATED WITH PERFUSABLE CHANNELS ON CURVED SURFACE

Nobuhito Mori[1], Yuya Morimoto[1,2], and Shoji Takeuchi[1,2]
[1]Institute of Industrial Science, the University of Tokyo, JAPAN
[2]Takeuchi Biohybrid Innovation Project, ERATO, JST, JAPAN

ABSTRACT

This conference proceeding describes a method to construct a skin-equivalent cultured on a curved surface. The skin-equivalent consists of not only the dermis/epidermis but also perfusable channels. We embedded an anchoring structure in a culture device to prevent the horizontal contraction of the tissue during cultivation. Owing to perfusion of culture medium via the channels, we can culture the epidermis at the air-liquid interface for cornification. This method enables the skin-equivalent construction on a curved surface; the curved surface is necessary for the construction of 3D complex skin surface and the covering of living skin on the 3D surface of biohybrid robots.

INTRODUCTION

Skin-equivalent is an *in vitro* skin model composed of the dermis layer made of collagen gel containing dermal fibroblasts and the epidermis layer made of epidermal keratinocytes [1]. The skin-equivalent can be used as a tool for the dermatological investigation, a skin graft in the treatment of patients with burns or scars and the creation of cosmetics and drugs [2]–[4]. In addition, the skin-equivalent may be utilized as a skin of biohybrid robots to mimic appearance and texture of human skin.

For the applications described above, many studies have been conducted to construct various elements in the skin-equivalent or either of dermis/epidermis layers such as hair follicle [5], melanocyte [6], immune cell [7], nerve [8], sweat gland [9], dermal papilla [10], [11] and blood capillary [12]–[14]. However, although perfusable channels and curved surfaces are important elements for the applications including cosmetics test, skin grafting and the skin of biohybrid robots, skin-equivalent with both elements has not been reported. This is because the contraction of the dermis [15] eventually collapsed the channels in culture and it was difficult to culture the epidermis at the air-liquid interface that is necessary for cornification on a curved surface.

In this study, we show the method to prevent the dermis from horizontal contraction and fabricate the channels to enable culturing the epidermis at the air-liquid interface for cornification (Figure 1). Since our culture device has contraction prevention structures (i.e. anchoring structures) at connections, the dermis cultured in the device does not contract horizontally but contracts only vertically. Thus, we can form channels in the dermis without collapse by removing wires after the dermis has contracted adequately in a vertical direction.

In addition, we can integrate the contraction prevention structures with a finger-shaped mold as a demonstration of a curved surface and form channels in the dermis cultured on the finger-shaped mold. Owing to the channels, even on the curved surface, we can culture the

skin-equivalent at the air-liquid interface that is a mandatory condition for cornification of the epidermis.

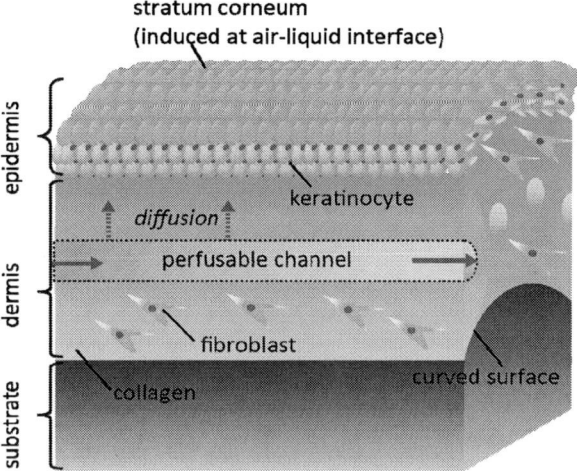

Figure 1: Concept of our skin-equivalent composed of dermis and epidermis layers. The dermis consists of collagen gel containing dermal fibroblasts. The epidermis is made of epidermal keratinocytes. The dermis also has channels for perfusing medium. By perfusing medium, keratinocyte can be cultured at the air-liquid interface for cornification even on the curved

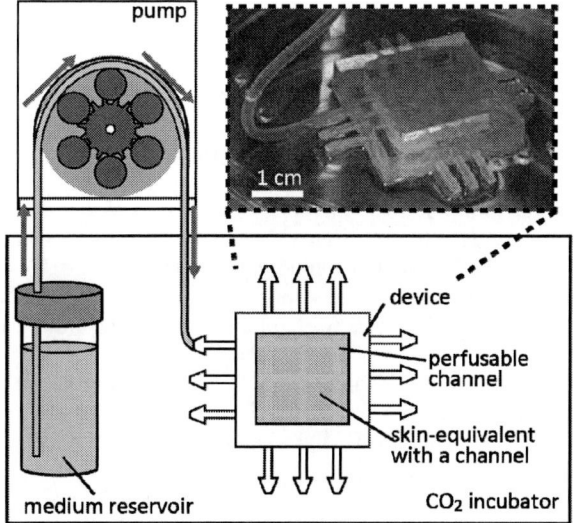

Figure 2: Description of the device. The device and medium reservoir were in a CO_2 incubator. One of the connection was connected to a peristaltic pump and the medium reservoir via a silicone rubber tube. Since the skin-equivalent in the device had perfusable channels, the medium was able to flow in it. The other connections were open to drain the medium. The drained medium was discarded via the holes of a dish where the device

978-1-4799-7956-1/15 $31.00 © 2015 IEEE   351   MEMS 2015, Estoril, PORTUGAL, 18 - 22 January, 2015

MATERIALS AND METHODS
Materials

Fibroblast Growth Medium-2 (FGM), growth factors for FGM (0.5 mL of human fibroblastic growth factor (hFGF-B), 0.5 mL of insulin, 10 mL of feral bovine serum (FBS), 0.5 mL of gentamicin and amphotericin (GA-1000)), KGM-Gold Keratinocyte Growth Medium (KGM) and growth factors for KGM (2 mL of bovine pituitary extract (BPE), 0.5 mL of human epidermal growth factor (hEGF), 0.5 mL of insulin, 0.5 mL of hydrocortisone, 0.5 mL of transferrin, 0.25 mL of epinephrine, 0.5 mL of GA-1000) were purchased from Lonza. Dullbecco's modified eagle's medium (DMEM) and penicillin-streptomycin solution were purchased from SIGMA-Aldrich. FBS for supplementing DMEM was purchased from Biosera Ltd. Phosphate buffered saline (PBS) was purchased from Cell Science & Technology Institute, Inc. Type-I collagen solution (IAC-50) was purchased from KOKEN Co., Ltd. L-ascorbic acid phosphate magnesium salt n-hydrate (ascorbic acid) was purchased from Wako Pure Chemical Industries, Ltd.

FGM and KGM were supplemented with growth factors. DMEM was supplemented with 10% (v/v) FBS, 100 U/mL penicillin, 100 µg/mL streptomycin and 250 µM ascorbic acid.

Cell culture

We cultured normal human dermal fibroblasts and normal human epidermal keratinocytes purchased from Lonza by FGM and KGM, respectively. We changed the medium once every 2 or 3 days and subcultured them when they were 70 to 90% confluence. We used them for constructing the skin-equivalent immediately after they had reached 100% confluence. We discarded them after passage number 15.

Device design and perfusion system

We fabricated the culture device by 3D printer (AGILISTA-3100, KEYENCE Corporation). The device had a contraction prevention structure at the edge of the connection and nylon wires (diameter 520 µm, 2400007084092, TOA-STRINGS Co., Ltd.) in the connection. The inner diameter of the connection was 700 µm. The device was a square 24 mm on a side at the inner bottom and 8.8 mm in inner height. We exposed the device to UV light overnight for sterilization.

As shown in Figure 2, one of the connection was connected to a peristaltic pump (AC-2110II, ATTO Corporation) and the medium reservoir via a silicone rubber tube. The silicone rubber tube had been sterilized by 70% ethanol and UV light in advance. The device and medium reservoir were in a CO_2 incubator (MCO-5AC, SANYO Electric Co., Ltd). The other connections were open to drain the medium. The drained medium was discarded via the holes of a polystyrene dish where the device was placed.

Culture of skin-equivalent with perfusable channels

At first, we mixed 10×PBS, collagen solution, and FGM containing 5×10^5 cells/mL of fibroblasts at the ratio of 1:9:18 and filled the device with the mixed solution to make the dermis. The mixed solution in the device was incubated in the incubator at 37°C for 1 hour for the gelation of the collagen. After the gelation, we immersed the device in DMEM and cultured it for more than 5 days in order to contract the dermis adequately in a vertical direction. Subsequently, we changed the medium surrounding the device from DMEM to the mixed medium composed of DMEM and KGM at the ratio of 1:1. We seeded 1×10^6 of keratinocytes suspended in the mixed medium on the dermis. After culturing for 1 day, we removed the surrounding medium and wires to fabricate the channels. We perfused the medium via the channels at approximately 2 mL/hour and cultured the keratinocytes at the air-liquid interface for more than 5 days to induce cornification.

Skin-equivalent on finger-shaped mold

The finger-shaped mold was fabricated by 3D printer based on the data made by scanning a real finger using 3D scanner (HEW-100U, Hamano Engineering Co., Ltd.). The finger-shaped mold was integrated with the medium reservoir for the cultivation of the dermis layer and the same connections with the device described in the previous subsection. We constructed the skin-equivalent on the finger-shaped mold in the same procedure with the previous subsection.

RESULTS AND DISCUSSION
Effectiveness of contraction prevention structure

As a result of culturing the dermis for 5 days, the size of the dermis in the device without the contraction prevention structure became 45% in area, while the dermis of the device with the structure became 64% in area. According to this result, the structure is effective in preventing contraction. It appears that the dermis is anchored to the contraction prevention structure since the collagen gel containing fibroblasts wraps around the structure.

Perfusion via channels

We checked the channels were perfusable by perfusing fluorescent micro-beads via the channels. As a result, even after 7 days of perfusion culture, the channel could still be perfused. It seems that the channels does not collapse since the dermis might have adequately contracted when the wires are removed.

Morphology and function of skin-equivalent

By haematoxylin-eosin staining, we confirmed that the skin-equivalent had the dermis/epidermis layers were formed when cultured by perfusion of the medium via the channels, while the epidermis layer was not formed when cultured without the perfusion of the medium. This results indicates that the nutrition might be supplied to the epidermis by diffusion from the channels and the air-liquid interface might be formed.

We also observed that the medium dropped on the skin-equivalent was repelled after 9 days rather than 1 day of perfusion culture. This result indicates that cornification was induced since it is known that the stratum corneum repels liquid.

978-1-4799-7956-1/15 $31.00 © 2015 IEEE

Skin-equivalent on finger mold

We successfully constructed the dermis with perfusable channels on the finger-shaped mold and cultured the keratinocytes on the dermis by perfusion via the channels. We confirmed that the skin-equivalent constructed on the mold had the dermis/epidermis layers by haematoxylin-eosin staining. This result indicates that the nutrition diffusion from the channels appropriately occurs and the cultivation at the air-liquid interface is possible even on the curved surface.

CONCLUSION

We fabricated a device to construct the skin-equivalent composed of the dermis/epidermis layers and perfusable channels. Using the device, we successfully constructed the skin-equivalent with the stratum corneum by perfusion of the culture medium via the channels. We also applied this method to a finger-shaped mold and successfully constructed the skin-equivalent that had the dermis/epidermis layers and perfusable channels on the mold. We believe that our skin-equivalent can be utilized for the applications such as cosmetics test, skin grafting and the skin of humanoid robots.

ACKNOWLEDGEMENT

Y. Morimoto is supported by Supporting of the Matsuda Foundation. N. Mori is supported by Grant-in-Aid for Japan Society for the Promotion of Science (JSPS) Fellows (26-7064).

REFERENCES

[1] E. Bell, H. P. Ehrlich, D. J. Buttle, and T. Nakatsuji, "Living tissue formed in vitro and accepted as skin-equivalent tissue of full thickness.," *Science*, vol. 211, no. 4486, pp. 1052–1054, Mar. 1981.

[2] C. A. Brohem, L. B. D. Cardeal, M. Tiago, M. S. Soengas, S. B. D. Barros, and S. S. Maria-Engler, "Artificial skin in perspective: concepts and applications," *Pigment Cell & Melanoma Research*, vol. 24, no. 1, pp. 35–50, 2011.

[3] Y. Morimoto, R. Tanaka, and S. Takeuchi, "Construction of 3D, Layered Skin, Microsized Tissues by Using Cell Beads for Cellular Function Analysis," *Advanced Healthcare Materials*, vol. 2, no. 2, pp. 261–265, 2013.

[4] S. T. Boyce, R. J. Kagan, D. G. Greenhalgh, P. Warner, K. P. Yakuboff, T. Palmieri, and G. D. Warden, "Cultured skin substitutes reduce requirements for harvesting of skin autograft for closure of excised, full-thickness burns," *Journal of Trauma-Injury Infection and Critical Care*, vol. 60, no. 4, pp. 821–829, 2006.

[5] M. Michel, N. L'Heureux, R. Pouliot, W. Xu, F. A. Auger, and L. Germain, "Characterization of a new tissue-engineered human skin equivalent with hair," *In Vitro Cellular & Developmental Biology-Animal*, vol. 35, no. 6, pp. 318–326, 1999.

[6] M. Regnier, C. Duval, J. B. Galey, M. Philippe, A. Lagrange, R. Tuloup, and R. Schmidt, "Keratinocyte-melanocyte co-cultures and pigmented reconstructed human epidermis: Models to study modulation of melanogenesis," *Cellular and Molecular Biology*, vol. 45, no. 7, pp. 969–980, 1999.

[7] V. Facy, V. Flouret, M. Regnier, and R. Schmidt, "Langerhans cells integrated into human reconstructed epidermis respond to known sensitizers and ultraviolet exposure," *Journal of Investigative Dermatology*, vol. 122, no. 2, pp. 552–553, 2004.

[8] K. B. English, N. Stayner, G. Krueger, and R. P. Tuckett, "Tactile Function in Skin-Equivalent Grafts," *Experimental Neurology*, vol. 115, no. 1, pp. 104–108, 1992.

[9] S. Huang, Y. A. Xu, C. H. Wu, D. Q. Sha, and X. B. Fu, "In vitro constitution and in vivo implantation of engineered skin constructs with sweat glands," *Biomaterials*, vol. 31, no. 21, pp. 5520–5525, 2010.

[10] G. Lammers, G. Roth, M. Heck, R. Zengerle, G. S. Tjabringa, E. M. Versteeg, T. Hafmans, R. Wismans, D. P. Reinhardt, E. T. P. Verwiel, P. L. J. M. Zeeuwen, J. Schalkwijk, R. Brock, W. F. Daamen, and T. H. van Kuppevelt, "Construction of a Microstructured Collagen Membrane Mimicking the Papillary Dermis Architecture and Guiding Keratinocyte Morphology and Gene Expression," *Macromolecular Bioscience*, vol. 12, no. 5, pp. 675–691, 2012.

[11] B. R. Downing, K. Cornwell, M. Toner, and G. D. Pins, "The influence of microtextured basal lamina analog topography on keratinocyte function and epidermal organization," *Journal of Biomedical Materials Research Part A*, vol. 72A, no. 1, pp. 47–56, 2005.

[12] A. S. Klar, S. Guven, T. Biedermann, J. Luginbuhl, S. Bottcher-Haberzeth, C. Meuli-Simmen, M. Meuli, I. Martin, A. Scherberich, and E. Reichmann, "Tissue-engineered dermo-epidermal skin grafts prevascularized with adipose-derived cells," *Biomaterials*, vol. 35, no. 19, pp. 5065–5078, 2014.

[13] A. F. Black, F. Berthod, N. L'heureux, L. Germain, and F. A. Auger, "In vitro reconstruction of a human capillary-like network in a tissue-engineered skin equivalent.," *FASEB journal: official publication of the Federation of American Societies for Experimental Biology*, vol. 12, no. 13, pp. 1331–40, Oct. 1998.

[14] M. Ponec, A. El Ghalbzouri, R. Dijkman, J. Kempenaar, G. van der Pluijm, and P. Koolwijk, "Endothelial network formed with human dermal microvascular endothelial cells in autologous multicellular skin substitutes," *Angiogenesis*, vol. 7, no. 4, pp. 295–305, 2004.

[15] T. Nishiyama, N. Tominaga, K. Nakajima, and T. Hayashi, "Quantitative-Evaluation of the Factors Affecting the Process of Fibroblast-Mediated Collagen Gel Contraction by Separating the Process into 3 Phases," *Collagen and Related Research*, vol. 8, no. 3, pp. 259–273, 1988.

CONTACT

* N. Mori; tel: +81-3-5452-6650; mori1985@iis.u-tokyo.ac.jp

TFT DISPLAY PANEL TECHNOLOGY AS A BASE FOR BIOLOGICAL CELLS ELECTRICAL MANIPULATION – APPLICATION TO DIELECTROPHORESIS –

Agnès Tixier-Mita[1], Bertrand-David Ségard[2], Young-Jin Kim[3], Yukiko Matsunaga[3], Hiroyuki Fujita[3] and Hiroshi Toshiyoshi[1]*

[1]RCAST, The University of Tokyo, Tokyo, JAPAN
[2]LIMMS/CNRS-IIS UMI 2820, The University of Tokyo, Tokyo, JAPAN
[3]IIS, The University of Tokyo, Tokyo, JAPAN

ABSTRACT

This paper reports for the first time the use of TFT (Thin Film Transistor) technology of display panels for biological cells electrical manipulation. This technology allows to have high density distributed transparent micro-electrodes, independently controllable, covering centimeter-size glass substrates. This technology is much superior to usual micro-technology used for Multielectrode Arrays (MEAs), which allows only millimeter size surface with micro-electrodes, and with a limited number of 64 micro-electrodes maximum. The chosen application, to demonstrate the capability of such technology, is dielectrophoresis on micro-beads and yeast cells.

INTRODUCTION

Analyze and the in-vitro characterization of biological entities (tissues, cells, proteins or DNA) is essential to understand their role, their possible disease, their interactions and their viability in their original environment. The most widely used technique uses optical microscopy and fluorescent dyes to target the aspect to be observed. However, staining cells or tissues requires heavy preparation and dyes are recognized to be potentially toxic to the biological entity. Moreover, this technique is not portable.

Another approach is to develop electrical devices, made by micro-fabrication technology, for analyze, characterization and manipulation of the biological materials. The advantages of electrical devices is that they are portable, the electrical structures (electrodes for instance) have micro-scale dimensions, same as the ones of biological cells, and no fluorescent staining is necessaryfor the detection.

In addition, they might offer complementary data or confirm some studied phenomena with the optical approach. Comparison of both approaches can be illustrated in the case of the study of electroactive cells. The first usage of micro-electrodes for experiments on cells was applied to neurons, as early as the 80's [1]. In such studies, excitation of neurons, as well as registration of action potential can be performed. Activity of neurons can also be visualized using standard optical microscope and fluorescence technique, but the value of the potential can't be obtained, and local excitation of one particular neuron can't be performed, as with the electrical approach.

This electrical approach, combined with microfluidic devices, helps in the development of bio-MEMS, to develop smarter, distributed systems for local sensing or manipulation of the biological material.

In-vitro biology deals with mainly 2D material: 2D array of electrodes is then an important development in bio-MEMS devices. Three main technologies for 2D-microelectrodes array can be reported. The first one is known as Multi-Electrode-Arrays (MEAs) [2]. Standard MEAs consist into 8x8 arrays of metallic electrodes made by micro-fabrication. They allow to stimulate or register the activity of electroactive cells cultured on their surface [3]. Usually, the MEAs are integrated in a whole electronic system able to control one by one the set of 64 electrodes, proposed by companies like MED64. The second one uses CMOS technology to fabricate a very dense array of independent micro-electrodes, which are multiplexed [4,5]. The number of micro-electrodes can reach more than 10000. Finally, Organic-TFT technology proposes a large transparent array of sensing sites [6]. The transparency is an advantage as optical characterization can then be performed at the same time to confirm or complete the electrical data obtained electrically.

However the MEAs and CMOS technologies have some limits due to the low density (64 electrodes for the MEAs) or the small surface occupied by the electrodes: about $1mm^2$ to $4mm^2$. In addition, the Organic-TFT is mainly used for sensing, and no voltage can be applied with it.

The technology proposed here uses the TFT technology developed for display panels. Its unique feature is that it gathers all the advantages of the previous technologies: transparent electronics, centimeter-scale electrode array, 2-D high density independent electrodes. In addition, according to the design, it can be used as well to apply a potential, as to perform measurements (sensing). Many applications can be investigated: electrical manipulation by dielectrophoresis, electrical stimulations or sensing.

This article presents for the first time TFT display panel technology used for biological applications. In particular, the substrates have been used for manipulation of micro-beads and yeast cells using dielectrophoresis, as a demonstration.

DESCRIPTION OF THE DEVICE

Liquid Crystal Displays panels (LCD) based on TFT technology are composed of an upper glass, with polarizer and color filter, and a lower glass, with pixels electrodes which ON/OFF status is individually controlled by individual TFTs. In between the two glasses, there is the liquid crystal and a spacer. Figure 1 describes the typical structure of these panels. In this research, the basic idea is to use the so called "lower glass" of LCD panels to take advantage of the high density distributed electrode array.

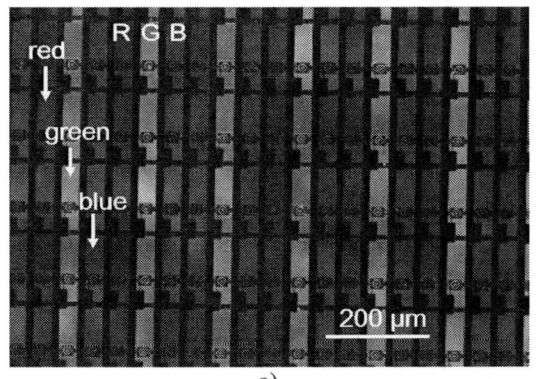

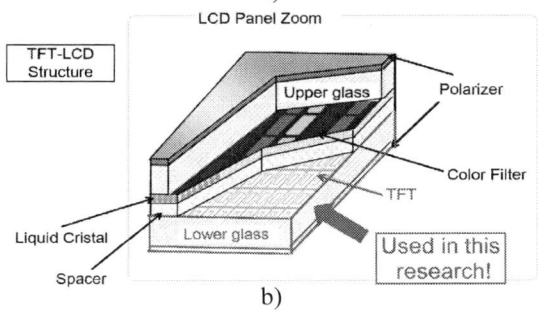

Figure 1: Structure of a TFT-LCD panel. a) Top-view of the panel observed with a microscope. b) TFT-LCD panel structure. This is the lower glass which has been used in this research.

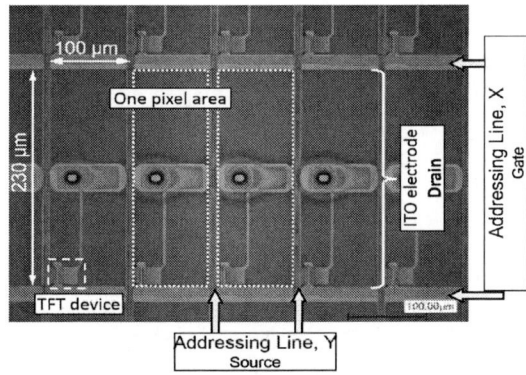

Figure 2: Close view of the lower glass of the panel with some pixels with TFT. When a potential is applied to one line gate and to one line source, the TFT at their intersection is switched ON. It polarizes the ITO electrode connected to its drain.

The "lower glass" is made of a glass substrate with transparent Indium Tin Oxide (ITO) electrodes in a dense array. The electrodes have dimensions of some ten micrometers in width and 100 to few hundreds microns in height, separated by few microns. These dimensions are well compatible with biological material like biological cells which have dimensions from 15 micrometers in diameter. Each ITO electrode is connected to the drain of

one TFT and is controlled by it. To adjust the drain potential of that TFT and its ITO electrode, the corresponding source and gate have to be polarized.

Figure 2 shows a closer view of some ITO electrodes, with their TFT. When a source line and a gate line are polarized, only the TFT placed at their intersection can be controlled.

Experiments were performed to show that these devices are well transparent and can be used for optical analyses using fluorescence, with an inverted microscope, usually used in biological experiments. The substrate can also welcome microfluidic device like microchannel. Figure 3 presents the results of fluorescent micro-beads flowing inside a PDMS microchannel placed on the TFT substrate and observed with an inverted microscope.

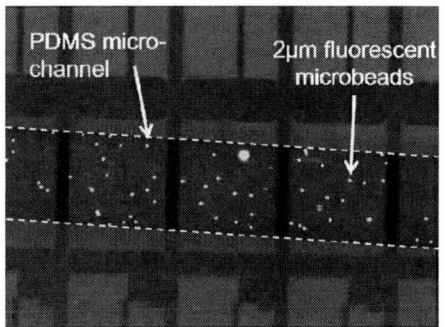

Figure 3: Observation with an inverted microscope of fluorescent beads flowing in a PDMS channel.

EXPERIMENTS
Generalities about Dielectrophoresis
The first type of experiments which is conducted with this device is dielectrophoresis (DEP), for demonstration on biological material. In DEP phenomena, dielectric particles (micro-beads or biological cells) can be electrically moved and manipulated when it is subjected to a non-uniform electric field. The non-uniform electric field creates a dielectrophoretic force which provokes the motion of the particles. In the case of the TFT substrate, non-uniform electric field can be easily "patterned" by applying a potential to the electrodes of interest. The strength of the force and its repulsive of attractive aspects are strongly dependent on: the particles and medium electrical properties, the particles size and shape and the frequency of the electric field [7].

The question of the frequency is the first point which has been investigated. Actually, standard TFT-LCD panels are working with DC voltage at the level of the electrodes (drain). However to obtain dielectrophoresis, an AC voltage is necessary. Some measurements of the potential coming out at one ITO electrode have been performed, when both the gate and the source are polarized. During these measurements, the whole system was in a dry environment (no micro-fluidic), a probe was placed on one ITO electrode, a 25V DC voltage was applied to the gate and an AC voltage was applied to the source. The AC voltage at the source was applied with a frequency ranging

978-1-4799-7956-1/15 $31.00 © 2015 IEEE   355

between 100 Hz and 1 MHz.

Figure 4 shows the normalized output voltage at the drain according to the frequency. Similar results were obtained for voltage at the source ranging between 1V and 5V. An attenuation of almost 20dB of the output signal is obtained when the frequency reaches 100 MHz. This attenuation is due to the characteristics of the transistor.

Despite the 20dB attenuation for higher frequencies, the TFT substrates have been used to try DEP experiments.

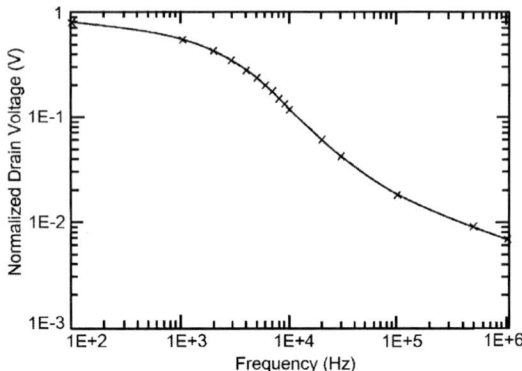

Figure 4: Characterization of the TFT device: the drain voltage at one ITO electrode has been measured and normalized. The gate is polarized by a 25V DC voltage, while an AC voltage is applied at the source. An attenuation of the drain voltage is observed when the frequency increases.

Dielectrophoresis Experiments

For the experiments of dielectrophoresis, 10 µm polystyrene micro-beads as well as yeast cells have been used. The medium used is de-ionized water (DIW), which has a low conductivity of more than 10 MΩ cm.

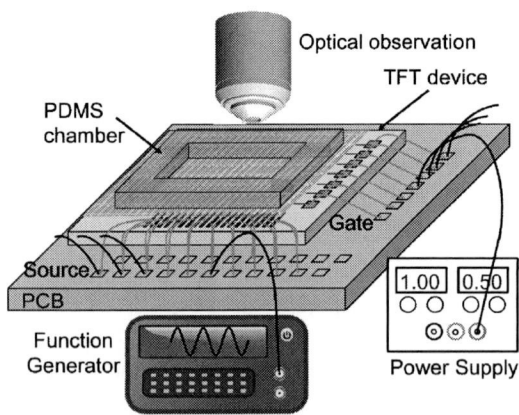

Figure 5: Experimental set-up used for the experiments. The gate and source lines terminals are connected by wire bonding to a PCB. The gate lines used during the experiments are connected to a power supply. The source lines are connected to a function generator.

A PDMS chamber is placed on the substrate to welcome the solution of micro-beads or yeast. The terminal pads of the different gate lines and the different source lines are bonded to a Printed Circuit Board (PCB) for easier application of the potentials.

The gate pads which have been chosen are connected to a power supply, while the chosen source pads are connected to a function generator. Optical observation is performed, during the experiments, with a standard microscope. Figure 5 shows the set-up used during the experiments.

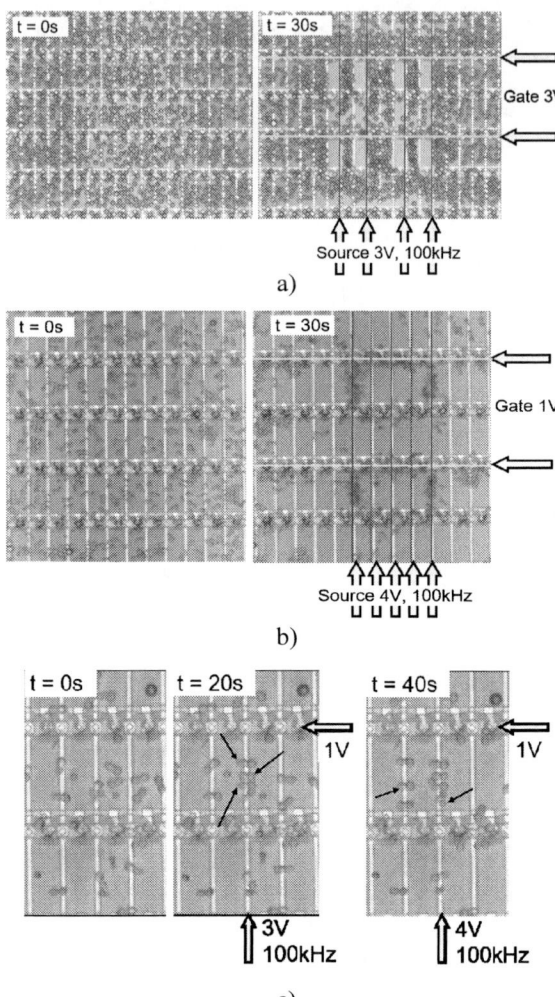

Figure 6: Experimental results of dielectrophoresis. a) Beads present a negative dielectrophoresis. b) and c) Cells present a positive dielectrophoresis.

The micro-beads or the yeast cells solution is then introduced inside the PDMS chamber. The authors had to wait about 5 to 10 min minutes, before beginning DEP experiments, in order that the particles fall on the substrate. Then, potential is applied to the chosen gate and source. As soon as the potential is applied the micro-beads or the yeast cells begin to move. Several experiments were performed with a potential applied on the gate of 0.5 V, 1 V, 2V, 3V or 4 V. At the source, the potential applied, according to the experiments, was: 0.5 V, 1V, 2V, 3V, 4V or 5V. Several frequencies were also investigated: 1 kHz, 10 kHz, 100 kHz, 1MHz and 10 MHz. When higher potential value was applied, the electrodes began to be damaged. DEP was observed in an efficient way for a frequency of 100 kHz. For other frequencies, almost no motion could be observed. By increasing the potential, the displacement of the particles happens more quickly. Figure 6 shows the different results obtained for a set of value of potential which was optimized: a) for micro-beads; b) and c) for yeast cells.

For the micro-beads, it can be clearly observed that they tend to escape from the areas with high potential. This phenomena is called negative dielectrophoresis: the micro-beads escape from the polarized electrodes and go towards the minimum of potential. For yeast cells, this is the contrary. They experience positive dielectrophoresis: the yeast cells are attracted towards the maximum of potential, which means the edges of the polarized electrodes.

DISCUSSION

The results presented on Figure 6 clearly demonstrate that it is possible to use TFT display panels for electrical manipulation of biological cells in a distributed way. Even if DC voltage is a standard applied potential to these panels, the measurements show that AC voltage can also be used. The measurements, Figure 4, show a clear attenuation of about 20dB of the output potential at the drain (ITO electrode), for the frequencies used for DEP. But, despite this attenuation, the amplitude of the potential seems to be high enough to allow DEP phenomena.

The experiments presented here show first steps in the development of TFT substrates for biological applications. With regard to the results, further applications can be investigated in addition to DEP: electro-stimulation of cells, electroporation or electro-fusion are among the application which should be possible with such technology.

CONCLUSIONS

In this article, one part of TFT display panel, the "lower glass" actually, has been used as a substrate for biological application. The unique advantages of such substrates, compared to more standard Multi-Electrode Arrays (MEAs), are that they combine a large (centimeters dimension) and dense 2D array of micro-electrodes, controlled individually by TFTs, with transparent substrate and micro-electrodes. In addition such substrates are micro-fluidic compatible.

With such substrates as well electrical as optical experiments, using fluorescence and inverted microscopy observation, can be performed on biological material, confirming or completing each other.

For demonstration of the possibilities of such technology, the devices have been applied to electrical manipulation of micro-beads and yeast cells using dielectrophoresis technique. The experiments performed showed that as well DC as AC voltage can be used and that as well micro-beads as yeast cells can experience DEP phenomena for a frequency of 100 kHz. Micro-beads experienced negative DEP, while yeast cells experienced positive DEP.

Based on these results, further experiments will be investigated: electro-stimulation of cells, electroporation or electrofusion.

ACKNOWLEDGEMENTS

The authors would like to thank Takuya Takahashi for his good advices and help for the connection of the TFT substrate with the PCB.

REFERENCES

[1] J. Pine J, "Recording action potentials from cultured neurons with extracellular microcircuit electrodes", *Journal of Neuroscience Methods*, vol. 2(1), pp. 19-31, 1980.

[2] K.H. Boven, M. Fejtl, A. Möller, W. Nisch, A. Stett, "On Micro-Electrode Array Revival." In: Baudry M, Taketani M, eds. *Advances in Network Electrophysiology Using Multi-Electrode Arrays.* New York: Springer Press; pp. 24-37, 2006.

[3] F. Morin, N. Nishimura, L. Griscom, B. Le Pioufle, H. Fujita, Y. Takamura, E. Tamiya, "Constraining the connectivity of neuronal networks cultured on microelectrode arrays with microfluidic techniques: A step towards neuron-based functional chips", *Biosensors and Bioelectronics*, vol. 21(7), pp. 1093-1100, 2006.

[4] A. Hierlemann, U. Frey, S. Hafizovic, F. Heer, "Growing Cells atop Microelectronic Chips: Interfacing Electrogenic Cells in Vitro with CMOS-based Microelectrode Arrays", *Proceedings of the IEEE*, vol. 99, No. 2, pp. 252-284A, 2011.

[5] D.J. Bakkum, U. Frey, M. Radivojevic, T.L. Russell, J. Müller, M. Fiscella, H. Takahashi, A. Hierlemann. "Tracking axonal action potential propagation on a high-density microelectrode array across hundreds of sites", *Nature Communications*, vol. 4: 2181, 2013.

[6] P. Lin, F. Yan, "Organic Thin Film Transistors for Chemical and Biological Sensing", *Advanced Materials*, vol. 24, pp. 34-51, 2012.

[7] R. Pethig, G.H. Markx, "Applications of dielectrophoresis in biotechnology", *Trends in Biotechnology*, vol. 15(10), pp. 426-432, 1997.

CONTACT

*A. Tixier-Mita, tel: +81-3-5452-5131; agnes@iis.u-tokyo.ac.jp

3D MORPHOLOGY RECONSTRUCTION OF HIGH ASPECT RATIO MEMS STRUCTURE BY USING AUTOFLUORESCENCE OF PARYLENE C

Lingqian Zhang[1], Yaoping Liu[1], Fang Yang[1, 2, 3], Wei Wang[1, 2, 3],*
Dacheng Zhang[1, 2, 3], and Zhihong Li[1, 2, 3]

[1]Institute of Microelectronics, Peking University, Beijing, 100871, China
[2]National Key Lab of Micro/Nano Fabrication Technology, Beijing, 100871, China
[3]Innovation Center for Micro-Nano-electronics and Integrated System, Beijing, 100871, China

ABSTRACT

This paper reported a MEMS fabrication compatible, damage free method for in-process 3D morphology reconstruction of high aspect ratio microstructure. As a novel morphology tracer, Parylene C thin film was conformally deposited onto the structure and annealed at high temperature under N_2. The autofluorescence of Parylene C was considerably enhanced by the annealing, which made it possible to image the microstructure. By scanning with a confocal microscopy, 3D morphology of the microstructure was reconstructed. The preliminary result indicated that microstructure with width of 8 μm and depth of 34 μm (34.1 μm actual depth by SEM) was successfully and accurately measured by this method.

INTRODUCTION

Morphology measurement during the process is always of great significance to micromachining. Various methods and instruments have been developed, including surface profiler, AFM and SEM. However, for the limitation of probe size and DOP (depth-of-field), it is still a great challenge to reconstruct 3D morphology of high aspect ratio microstructure without breaking it. For example, cross section of the high aspect ratio microstructure is often needed for accurate measurement of depth, which requires a destructive dicing process. After the destructive measurement, it is always not appropriate for the structure to continue the posterior processes.

As a material compatible with the traditional MEMS technique, Parylene C has many advantages and unique properties. It can be chemical vapor deposited into nanometers to tens of microns with good coating effect. Even for high aspect ratio microstructures, a conformal coating can also be achieved [1]. The removal and patterning of Parylene C can be realized by oxgen plasma etching [2].

In recent years, the autofluorescence of Parylene C has been reported for its biomedical impacts and applications [3-6]. Some methods, such as UV-illumination and Green-illumination, to change the intensity and characteristics of Parylene C fluorescence were proposed. Fluorescence mechanisms were discussed for the variation of Parylene C autofluorescence intensity [3].

This work utilized the autofluorescence of Parylene C, as a tracer to visualize the microstructure under confocal microscopy. Because of the high conformal deposition capability of the Parylene C and the powerful 3D fluorescence detection ability of the confocal laser scanning system, a nondestructive 3D morphology reconstruction of high aspect ratio MEMS structure is achieved.

DESIGN

The operation process to achieve 3D morphology reconstruction was schematically illustrated in Figure1.

Firstly, a thin Parylene C film (5000 Å) was deposited onto the substrate with fabricated high aspect ratio structures. Then the structure with Parylene C film was annealed at 270 °C for 1 hour under N_2 to improve the autofluorescence intensity. Confocal Laser Scanning System (Leica TCS SP5) was used to obtain fluorescence signals in different depths. Because of the high position precision in the vertical direction, the obtained 3D fluorescence information can be used to reconstruct 3D morphology of the MEMS structure. Finally, Parylene C film was etched off by oxygen plasma to recover the initial surface for posterior micro fabrication.

The whole process was fabrication compatible, and structure damaging was avoided during the morphology measurement.

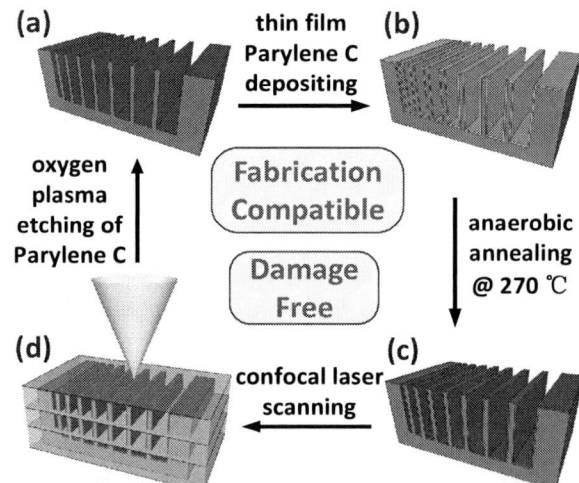

Figure 1: Operation process of the present 3D morphology reconstruction. (a) high aspect ratio microstructure; (b) structure coated with thin Parylene C; (c) annealing for fluorescence enhancement; (d) 3D morphology reconstruction with fluorescence signals in different depths.

EXPERIMENTS

Property of Parylene C autofluorescence

Generally, autofluorescence of Parylene C is weak for high resolution measurement, although it was reported to be strong enough to interfere with the biological fluorescence detection. To enable the present morphology

---

978-1-4799-7956-1/15 $31.00 © 2015 IEEE   358   MEMS 2015, Estoril, PORTUGAL, 18 - 22 January, 2015

measurement, the Parylene C coated substrate was annealed at high temperature under N_2 to enhance the fluorescence signals. Relationship between the annealing temperature and autofluorescence intensity was studied. Considering the glass transition temperature of Parylene C is 290 °C, the upper bound of treatment temperature was set as 270 °C to avoid the reflow.

Autofluorescence intensity was observed by the Confocal Laser Scanning System. Fluorescence spectrum was taken by FLS920 spectrometer (Edinburgh Instruments Ltd.).

3D morphology reconstruction

By vertical stepping of focal plane, optical slices of Parylene C autofluorescence in different depth were obtained by the Confocal Laser Scanning System, as illustrated in Figure 1(d). The vertical step of confocal scanning was set as 1μm. Afterwards, a 3D fluorescence distribution model was established from these optical slices and 3D structure morphology was reconstructed. 3D fluorescence distribution under different views were also obtained by data processing.

Fluorescence information of the 3D model can be used for extraction of various parameters. In this work, structure depth was extracted as a symbol of the 3D high aspect ratio morphology.

Meanwhile, the actual structure depth was measured as a comparison with the extracted structure depth. Cross section observation method mentioned in the introduction was taken by SEM.

Verification of process compatibility

By oxygen plasma etching of Parylene C, the measured microstructure was expected to recover to the initial state for posterior fabrication processes. To verify the process compatibility of this 3D morphology reconstruction method, cross section observations under SEM was conducted. Parylene C coating status on the initial structure, Parylene C deposited structure and Parylene C etched structure were compared from the cross section photos.

RESULTS

Effect of annealing temperature

Relationship between the annealing temperature and autofluorescence was shown in Figure 2(a). Three groups of annealed Parylene C autofluorescence images by Confocal Laser Scanning System were listed. It was presented that the fluorescence intensity of annealed Parylene C was stronger than the control, and the fluorescence was enhanced significantly with the increasing of annealing temperature, especially from 190 °C to 250 °C. When annealed at 270 °C, the fluorescence was the brightest, which meant that 270 °C was the best annealing temperature for the 3D morphology reconstruction.

Fluorescence spectrum of annealed Parylene C was shown in Figure 2(b). The spectrum peak value showed the same variation trend with the Parylene C fluorescence intensity. Since long conjugated structure has more $\pi \rightarrow \pi^*$ electronic transition and leads to stronger fluorescence

intensity, it was believed that the significant enhancement of Parylene C autofluorescence was caused by conjugate chain length increment during the annealing.

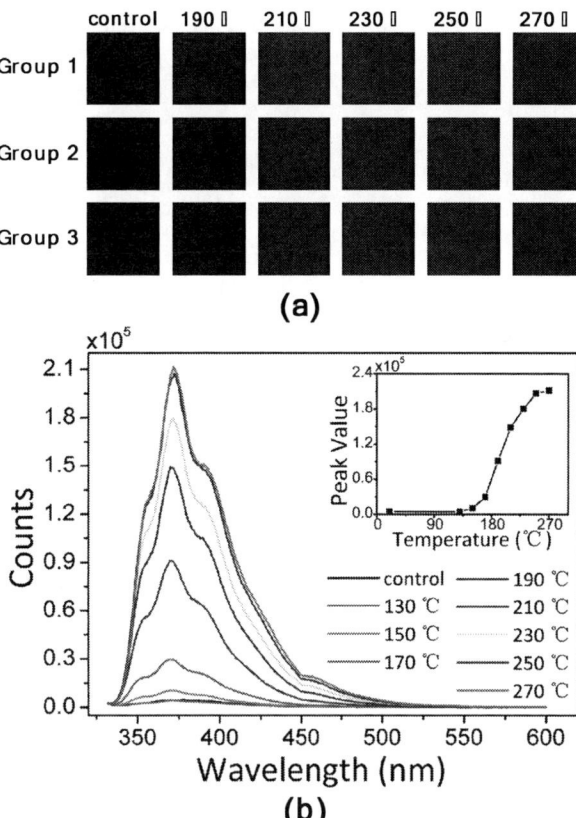

Figure 2: Effect of annealing temperature on Parylene C autofluorescence. (a) three groups of fluorescence images by Confocal Laser Scanning System; (b) fluorescence spectrum of annealed Parylene C.

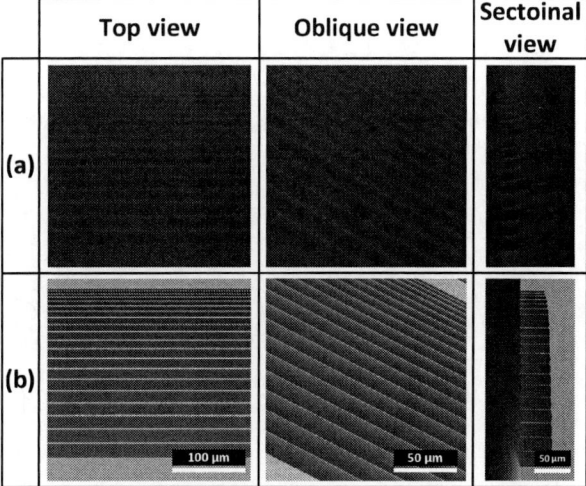

Figure 3: Reconstructed 3D morphology by confocal laser scanning, with SEM photos shown as comparison. (a) 3D fluorescence image; (b) SEM observation.

3D morphology reconstruction

Figure 3 showed the reconstructed 3D morphology of

the high aspect ratio structure, with SEM images as comparisons. 3D fluorescence distribution of the high aspect ratio microstructure under different views were illustrated. The images of top view, oblique view and sectional view of the 3D fluorescence morphology were basically in accordance with the SEM photos, while the sectional view by SEM was taken after breaking the structure.

Depth extracting of the microstructure

Depths of the high aspect ratio structures were extracted from the 3D morphology, as shown in Figure 4. As the vertical step was 1μm in the confocal scanning, there was a certain error in the extracted depth. Smaller scanning step would reduce the error, and somehow increase data calculation amount as well. The results indicated that this 3D model was able to display microstructure morphology properly, with featured width down to 8 μm and depth up to 34 μm.

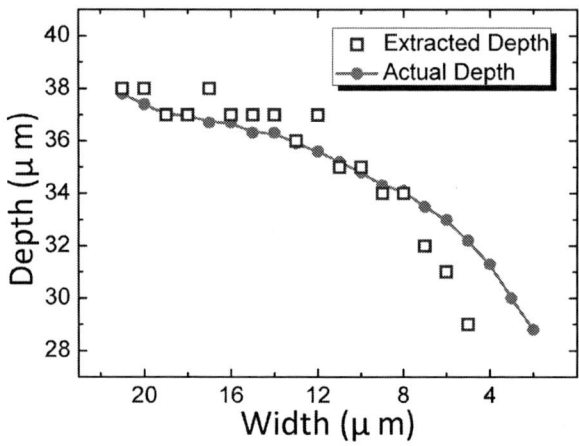

Figure 4: Extracted depths of the high aspect ratio structures, with actual depths obtained by SEM measurements.

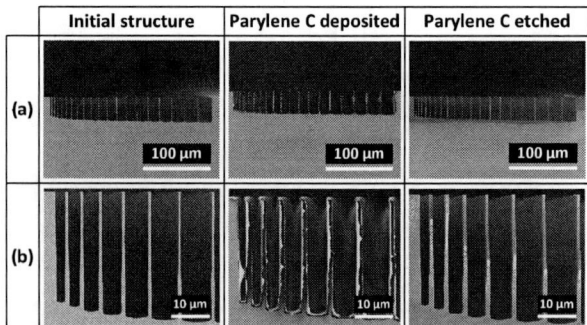

Figure 5: Verification of fabrication compatibility. (a) overall view; (b) detail view. Conformal coating and completely removal of Parylene C were achieved on the microstructure. (Parylene C crimping was caused by the stress when dicing)

Fabrication process compatibility

Fabrication process compatibility was verified by SEM cross section observation, as shown in Figure 5. Compared with the initial structure, conformal deposition and completely removal of 5000 Å Parylene C were

achieved in the following process.

CONCLUSIONS

In conclusion, this work introduced a MEMS fabrication compatible, damage free method for in-process 3D morphology reconstruction of high aspect ratio MEMS structure.

By conformal depositing of Parylene C on the microstructure, 3D morphology information of the high aspect ratio MEMS structure was obtained. After annealing at 270 °C under N_2, the autofluorescence of Parylene C was considerably enhanced, which made it possible for confocal scanning and 3D fluorescence information imaging. Finally, reconstruction of 3D morphology and data extracting of structure depth were achieved. The preliminary result indicated that microstructure with width of 8 μm and depth of 34 μm was successfully and accurately measured by this method. It was believed that 3D morphology reconstruction with higher precision can be achieved by improving horizontal resolution, reducing vertical scanning step and post-processing of fluorescence distribution data.

ACKNOWLEDGEMENTS

This work was financially supported by the Major State Basic Research Development Program (973 Program) (Grant No. 2011CB309502 and 2015CB352103) and the National Natural Science Foundation of China (Grant Nos. 81471750, 91023045 and 91323304).

REFERENCES

[1] Y.H. Lei, W. Wang, H.Q. Yu, Y.C. Luo, T. Li, Y.F. Jin, H.X. Zhang and Z.H. Li, "A parylene-filled-trench technique for thermal isolation in silicon-based microdevices", *Journal of Micromechanics and Microengineering*, 2009, Vol. 19, pp. 1-7.

[2] E. Meng, P.Y. Li, and Y.C. Tai, "Plasma removal of Parylene-C", *Journal of Micromechanics and Microengineering*, 2008, Vol. 18, pp. 1-13.

[3] B. Lu, S. Zheng, and Y.C. Tai, "Parylene background fluorescence study for biomems applications", *Transducers 2009*, Denver, Colorado, USA, pp. 176-179, June, 2009.

[4] B. Lu, S. Zheng, B. Q. Quach, and Y.C. Tai, "A study of the autofluorescence of parylene materials for μTAS applications", *Lab on a Chip*, 2010, Vol. 10, pp. 1826-1834.

[5] A. T. Ciftlik, D. G. Dupouy, and M. A. M. Gijs, "Programmable Parylene-C bonding layer fluorescence for storing information on microfluidic chip", *Lab on a Chip*, 2013, Vol. 13, pp. 1482-1488.

[6] D. G. Dupouy, A. T. Ciftlik, and M. A. M. Gijs, "Use of Parylene-C bonding layer fluorescence as reference for on-chip imaging and detection applications", *MicroTAS 2013*, Freiburg, Germany, pp. 260-262, October, 2013.

CONTACT

*Wei Wang, Peking University, Beijing, 100871, China; tel: +86-10-62769183; w.wang@pku.edu.cn

A NANOWIRE GAUGE FACTOR EXTRACTION METHOD FOR MATERIAL COMPARISON AND IN-LINE MONITORING

Issam Ouerghi, Julien Philippe, Carine Ladner, Pascal Scheiblin,
Laurent Duraffourg, Sébastien Hentz and Thomas Ernst
CEA, LETI, MINATEC Campus, 17 rue des Martyrs - 38054 GRENOBLE Cedex 9, France

ABSTRACT

We propose a new non-destructive extraction method of gauge factor (GF) of nanowires (NW) for in-line monitoring of this parameter and piezoresistive material properties comparisons. Unlike destructive conventional techniques which also suffer from reproducibility issues, this method allows a direct measurement of the GF locally at the nanoscale and at the wafer level. GFs have been reliably measured on a wide range of silicon-based NEMS resonators with different designs, crystalline structures and doping levels. For monocrystalline devices, the extracted values are in good agreement with typical values obtained for NWs fabricated with well-controlled top-down processes. These values are also compared with polysilicon (polySi) NEMS, which look promising for low cost solutions.

INTRODUCTION

Since the fifties, monocrystalline silicon is well known and used for its strong piezoresistive properties. Silicon-based piezoresistive devices have been introduced in several sensing applications, such as pressure sensors, accelerometers, cantilever force sensors, inertial sensors and strain gauges [1]. Moreover, piezoresistivity is now commonly used in CMOS, known as substrate [2] induced- or process [3] induced-strain for mobility enhancement. The piezoresistive effect is quantified by a change in the electrical resistivity when a mechanical stress is applied. The relative resistance variation is proportional to the axial strain and is simply expressed as $\Delta R / R = \gamma \times \varepsilon$ where γ is the piezoresistive GF. This resistance variation is mainly modeled by the piezoresistive tensor which accounts for the modulation of holes and electrons effective masses under strain [4]. Commonly, a high GF value for p-type silicon is associated to a longitudinal stress following the [110] direction. In this case, equal opposite values are obtained for longitudinal and transverse piezoresistive coefficients [1].

Piezoresistive coefficients and associated GFs of a bulk material are usually extracted by the four-point bending measurement [5-6]. The sample should be cleaved ($2 \times 8 \, \text{cm}^2$ in our experimental apparatus for example) and placed on a support rod to be submitted to a tensile or a compressive strain. This technique is slow, destructive since the samples must be cleaved, and suffers from repeatability issues. These drawbacks make it impossible to process GF data statistically at the wafer level. Moreover, at the nanoscale (NEMS devices) the piezoresistance is subject to surface and or dimensional effect that may affect its value as recently evidenced by several authors, which make relevant a simple, non-destructive method to monitor this property. For instance, a giant piezoresistance effect demonstrated by [7] was attributed by interface effects [5] and not reproduced with well controlled top-down processing, as used in the following study [8].

In this study, we demonstrate a new method to extract the local longitudinal GF directly at the wafer level. To this end, NWs are embedded into a simple nanomechanical device which can be used for in-line testing during fabrication or co-integration of NEMS devices. The results are then compared with the conventional method and some reference values to validate this approach.

NEMS STRCUTURE AND FABRICATION

The GF-testing device design is inspired from a NEMS sensor for chemical or mass sensing applications [9-10], see Figure 1. In this work, the device is only used in DC mode to properly study NW gauges properties. Its fabrication process is summarized in Figure 2. The NEMS is structured from a Silicon On Insulator (SOI) wafer by etching the top silicon. The top silicon and the buried oxide layers thicknesses are respectively 160 and 400nm. Electrical wires and pads are made with aluminum silicide (AlSi) metallic layer. The structure is then released by vapor hydrofluoric acid (HF) etching. The NEMS structure consists of two suspended piezoresistive p-doped silicon NWs connected to a rigid-enough lever (a cantilever) in a symmetric bridge configuration. The lever is electrostatically actuated, which induces high axial stress in the two piezoresistive gauges thanks to a large lever arm.

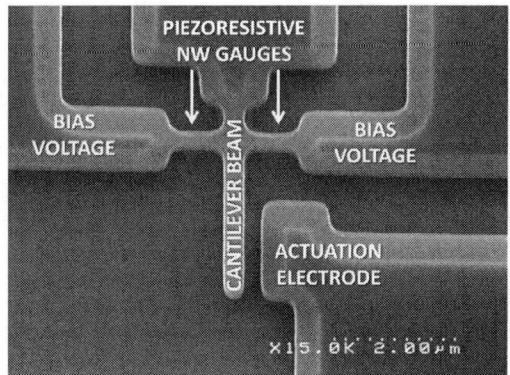

Figure 1: Top view SEM of the NEMS structure. Piezoresistive gauges are connected to a lever arm.

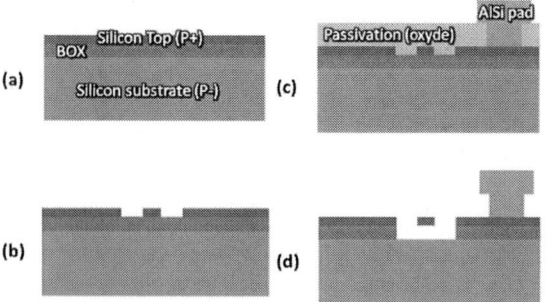

Figure 2: NEMS fabrication sequences: (a) Boron atoms implantation and activation, (b) Etching of the silicon top layer to define the NEMS (c) Passivation, metal deposition and etching for defining the metallic pads and (d) Release of the structure with a vapor HF to obtain suspended structure.

EXPERIMENT

With a van der Pauw method, we measured the resistivity of the layer before the NEMS fabrication step, therefore a nominal gauge resistance R_g of 1.5 kΩ was extracted. Considering the access resistances, this result is consistent with the electrical resistance measurement on the gauges. The maximum actuation voltage was fixed at 25 V before the collapse or the breakdown of the beam (experimentally we found a pull in voltage superior to 30 V). A low detection current through the gauges, below 100 µA, prevents from any actuation perturbation and avoids the NW gauges self-heating. The corresponding bias voltage is 250 mV. The gauge relative resistance is expected to vary with the DC actuation voltage applied to the electrode V_{act}. Indeed, the stress within one gauge can be expressed as $\sigma = \alpha_s \times E \times V_{act}^2$ where E is the Young's modulus of silicon and α_s is some transduction factor (in V^{-2}). Electromechanical Finite Element simulations were performed to obtain precise values of α_s (see Figure 3).

On the other hand, the relative piezoresistive gauge resistance variation is $\Delta R_g / R_g = \gamma \times \sigma / E$. Experimentally, resistance of a single gauge was monitored while varying V_{act}. Figure 4 shows the electrical current through the gauge and relative resistance variation with respect to the actuation voltage, a high signal-to-noise ratio is obtained thanks to the significant lever arm of this design. As expected, the relative resistance variation follows a quadratic law with the actuation voltage: $(\Delta R_g / R_g)_{exp} = \alpha_e \times V_{act}^2$, where α_e is an experimental factor (in V^{-2}). Nevertheless, the proximity of the actuation electrode induces a linear field effect within the NWs, which adds up to the purely piezoresistive effect in the I-V curves shown in Figure 4(a)-(b). Poisson equations and continuity transport equations were solved numerically using an Arora mobility model [11] with a commercial TCAD tool (Silvaco's ATLASTM) to model the contribution of this effect (insets in Figure 4 (a)-(b)), yielding the simple equation: $I \approx 10^{-9} V_{act}$. This linear effect does not impact piezoresistive properties and can be subtracted from the I-V curves for GF extraction (Figure. 4 (c) and (d)). The device corresponds to a bridge configuration and the gauges feel an opposite stress for the same actuation voltage (applied on the same actuation electrode): the gauge resistance increases with the actuation voltage due to a tensile stress while the other gauge resistance decreases due to a compressive stress. These observations are consistent with others experiments for P-type silicon [5]. Finally, $\alpha_e = \gamma \times \alpha_s$ is plotted over α_s in Figure 5 for different device designs. The linear fit yields a GF value of 84 ± 2 for both tensile and compressive stresses.

DISCUSSIONS

In order to compare the experimental results with theoretical predictions, the theoretical GF value from the works of Richter et al [12] were analyzed. They used both a 6×6 k·p Hamiltonian and a tight binding model to predict the shear piezoresistive coefficient of p-type silicon according to both the dopant concentration and the temperature. At room temperature and for a doping level of 1×10^{19} cm^{-3}, they find a percentage of about 70 ± 5 % of the shear piezoresistive coefficient maximum value. That corresponds to other references [13] [14]. In the [110] direction, the maximum value (corresponding to a doping level $<1 \times 10^{16}$ cm^{-3}) of the longitudinal piezoresistive coefficient π_l measured is 71.8×10^{-11} Pa^{-1} [15].

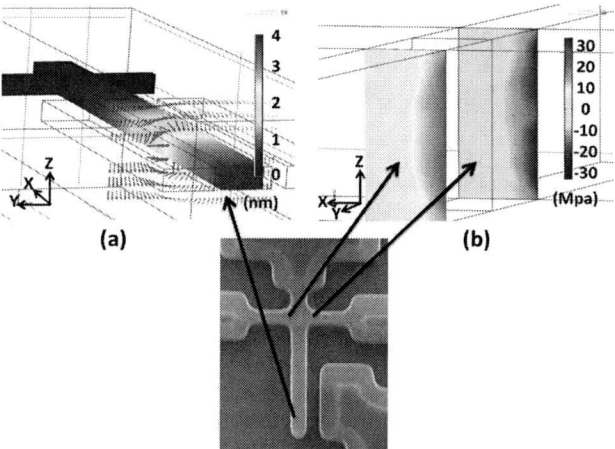

Figure 3: (Color) Electromechanical simulation of NEMS for an actuation voltage of 15 V. (a) The color legend and red arrows respectively represent the beam deflection and the electric field lines. (b) Simulated gauge stress on the surfaces between the gauges and the beam. Negative and positive values respectively represent a tensile and compressive stress. The average pressure on the surfaces is 8.0 MPa.

978-1-4799-7956-1/15 $31.00 © 2015 IEEE   362

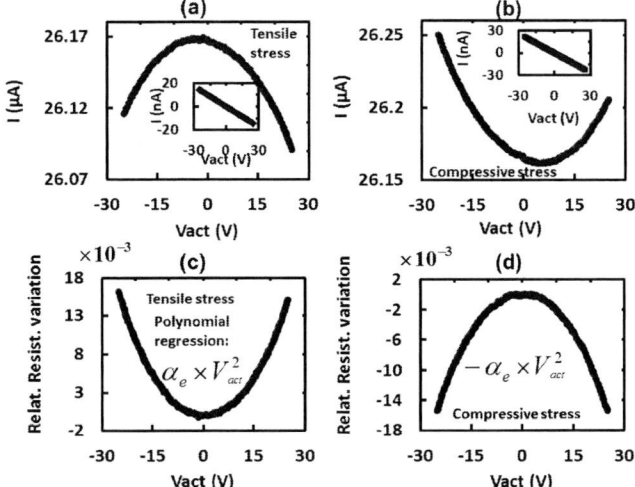

Figure 4: (a-b) I-V measurements of the gauge current versus the actuation voltage. The beam deflection involves a tensile stress on one gauge (a) and a compressive stress on the other (b). (c-d) Relative gauge resistance variations as a function of actuation voltage. A linear component of the current as a function of the actuation voltage (insets of (a) and (b)) due to a field effect between the gauges and the actuation electrode is suppressed.

Therefore the doping levels used in our experiments involves a theoretical GF value of $\gamma \approx \pi_l \times E \times 0.70 \pm 0.05 \approx 85 \pm 6$. Comparison of the GF values between our method and the four-point bending measurement was performed on rectangular-box-shaped devices ($3 \times 0.25 \times 0.25\,\mu m^3$). I-V measurements for each applied stress were performed to extract the resistance variation. As shown in Figure. 6, the GF is extracted thanks to the linear fit of the relative resistance variation as a function of strain curve and equal to 88 ± 4. This points out that our method is consistent with the conventional method and the expected theoretical value, i.e. 84 ± 2, 88 ± 4 and 85 ± 6 respectively. Giant piezoresistance effects [7] in suspended Si NWs have been quite widely studied these last years. This phenomenon does not concern highly doped top-down NW [8]. Moreover, the NW cross-section in this work is one order of magnitude larger than the NWs concerned by the giant piezoresistance effect. This characterization method could be extended on other piezoresistive materials providing the signal to noise ratio is sufficiently high. As example our method allowed for easy measurement of three different polySi NEMS GFs, briefly presented in Table 1.

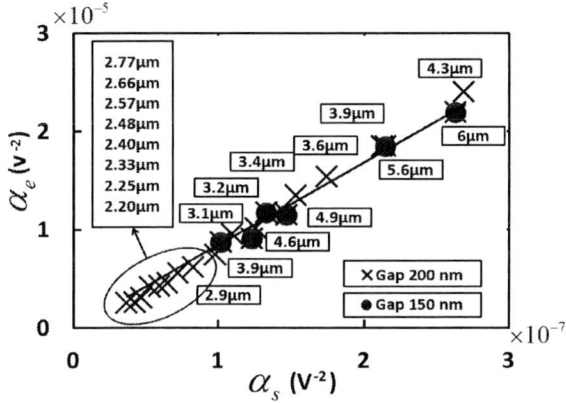

Figure 5: Comparison between simulation and experimental coefficients for different actuation-beam gaps and beam lengths. The linear fit gives directly the GF value.

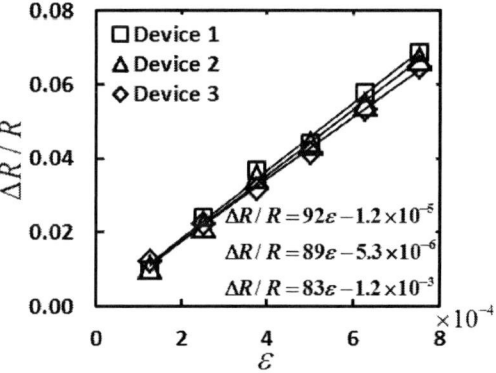

Figure 6: Four-point bending measurement performed on three devices ($3 \times 0.25 \times 0.16\ \mu m3$) The linear fit gives directly the GF value.

Table 1. Comparison between extracted GFs with conventional and new method. Results are in good agreement. Unlike monocrystalline silicon, GF value for PolyB material increases with doping level.

	PolyA	PolyB		PolyC	Single crystal silicon	
Deposition conditions	0.175 Torr, 580° C	0.2 Torr, 620° C (columnar)		0.375 Torr, 580° C	SOI wafers	
Doping level	$2.9 \times 10^{20} cm^{-3}$	$10^{19}\ cm^{-3}$	$10^{20}\ cm^{-3}$	$7 \times 10^{19} cm^{-3}$	$10^{19}\ cm^{-3}$	$10^{20}\ cm^{-3}$
GF from conventional method	22	15	33	19	88	68
GF from the new method	**18**	**13**	**32**	**15**	**84**	**53**

CONCLUSIONS

In conclusion, a new characterization and extraction method of GF was presented. This design can be added as a test structure to monitor this parameter in parallel with other NEMS devices. Therefore it is fast enough to characterize a large number of NEMS at the wafer level and is hence compatible with NEMS VLSI. The results are consistent with theoretical data and with the conventional method and can be extended to other piezoresistive materials.

ACKNOWLEDGEMENTS

This work was supported by the European Research Council, Grant No. 240382 – DELPHINS project.

REFERENCES

[1] A. A. Barlian, W.-T. Park, A. J. R. Mallon, A. J. Rastegar, and B. L. Pruitt, "Review: Semiconductor Piezoresistance for Microsystems," *Proc. IEEE*, vol. 97, no. 3 Mar. 2009.

[2] F. Andrieu, O. Weber, T. Ernst, O. Faynot and S. Deleonibus, "Strain and channel engineering for fully depleted SOI MOSFETs towards the 32 nm technology node," *Microelectronic Engineering*, vol. 84, pp. 2047–2053, Sept.–Oct. 2007.

[3] F. Andrieu, T. Ernst, C. Ravit, M. Jurczak, G. Ghibaudo and S. Deleonibus, "In-Depth Characterization of the Hole Mobility in 50-nm Process-Induced Strained MOSFETs," *IEEE Electron Dev. Lett.*, vol. 26, no. 10, Oct. 2005.

[4] J. Bardeen and W. Shockley, "Deformation Potentials and Mobilities in Non-Polar Crystals," *Phys. Rev.*, vol. 80, no. 1, Oct. 1950.

[5] A. Koumela, D. Mercier, C. Dupré, G. Jourdan, C. Marcoux, E. Ollier, S. T. Purcell, and L. Duraffourg, "Piezoresistance of top-down suspended Si nanowires," *J. Nanotechnology*, vol. 22, 395701, Sept. 2011.

[6] W.L Wang, X. Jiang, K. Taube and C. Klages "Piezoresistivity of polycrystalline p-type diamond films of various doping levels at different temperatures", *J. Appl. Phy.*, vol. 82, 2, 1997.

[7] R. He and P. Yang, "Giant piezoresistance effect in silicon nanowires," *nature nanotechnology*, vol. 1, pp. 42-46, Oct. 2006.

[8] J.S Milne and A.C.H. Rowe, "Giant Piezoresistance Effects in Silicon Nanowires and Microwires," *Phy. Rev. Lett.*, vol. 105, no. 22, Nov. 2010.

[9] M. S. Hanay, S. Kelber, A. K. Naik, D. Chi, S. Hentz, E. C. Bullard, E. Colinet, L. Duraffourg and M. L. Roukes, "Single-protein nanomechanical mass spectrometry in real time", *Nature Nanotechnology*, 7, pp. 602–608, 2012.

[10] E. Mile, G. Jourdan, I. Bargatin, S. Labarthe, C. Marcoux, P. Andreucci, S. Hentz, C. Kharrat, E. Colinet, and L. Duraffourg, "In-plane nanoelectromechanical resonators based on silicon nanowire piezoresistive detection," *Nanotechnology*, vol. 21, 165504, Mar. 2010.

[11] ND. Arora, JR. Hauser and DJ. Roulston, "Electron and Hole Mobilities in Silicon as a Function of Concentration and Temperature," *IEEE Trans. Electron Devices*, vol. 29, no. 2, Feb. 1982.

[12] J. Richter, J. Pedersen, M. Brandbyge, E. V. Thomsen, and O. Hansen, "Piezoresistance in p-type silicon revisited," *J. Appl. Phys.*, vol. 104, 023715, Jul. 2008.

[13] O. N. Tufte and E. L. Stelzer, "Piezoresistive Properties of Silicon Diffused Layers," *J. Appl. Phys.*, vol. 34, pp. 313-318, Feb. 1963.

[14] Y. Kanda, "A Graphical Representation of the Piezoresistance Coefficients in Silicon," *IEEE Trans. Electron Devices*, vol. 29, pp. 64-70, Jan. 1982.

[15] C. S. Smith, "Piezoresistance Effect in Geruianium and Silicon," *Phys. Rev.*, vol. 94, no. 1, pp. 42-49, Apr. 1954.

CONTACT

Thomas Ernst, Thomas.ernst@cea.fr

A THREE-STEP MODEL OF BLACK SILICON FORMATION IN DEEP REACTIVE ION ETCHING PROCESS

Fuyun Zhu[1], Chen Wang[2], Xiaosheng Zhang[1], Xin Zhao[2], and Haixia Zhang[2]*

[1] Science and Technology on Micro/Nano Fabrication Lab, Institute of Microelectronics, Peking University, Beijing, CHINA
[2] Insitute of Robotics and Automatic Information System, Nankai University, Tianjin, CHINA

ABSTRACT

A three-step model used for modeling and simulation of black silicon formation in DRIE (Deep Reactive Ion Etching) process is presented. It divides the plasma etching system into plasma layer, sheath layer and sample surface layer. At the same time, it combines quantum mechanics, sheath dynamics and diffusion theory together based on plasma environment to predict the probability distribution of etching particles so as to simulate the final etching results. The simulation results show very good coincidence with experimental images, proving the applicability of this theory and it's promising to make black silicon formation in DRIE process to be controllable and repeatable.

INTRODUCTION

As microdevices scaling down to nanoscale, controllable formation of nanostructures by traditional MEMS process becomes a challenging technique. Plasma-based processes have attracted major attention owing to their outstanding simplicity, efficacy and cost-efficiency. Black silicon, named due to its impressive black surface, is a typical example of plasma-made nanostructures [1-2]. Taking advantage of its outstanding performance of low-reflectivity and super hydrophobicity, black silicon is experiencing quantum leaps in many application fields, such as solar cells, optical sensors, and microfluidics and so on [3-4].

The black silicon surface can be fabricated in several methods, which are always combinations of nanolithography and mask-assisted etching, like electron beam, focused ion beam (FIB) and normal Bosch DRIE (Deep Reactive Ion Etching) [5-6]. However, they usually have inherent limits, such as large structure size, low-efficient, low aspect ratio and uncontrollability. Unlike those mask-based fabrication methods, an optimized DRIE process which is maskless, low-cost, high-efficient, single-step and controllable, was employed to achieve black silicon [7-8].

In order to control the plasma-based black silicon process, the research on black silicon formation mechanism is becoming more and more significant. There have been many efforts from different research groups, such as investigating relationship between experimental process parameters and structure shape of black silicon, or analyzing chemical reactions related to the formation process [8-9]. However, there is little research on presenting the practical formation process of black silicon theoretically.

In this study, the etching process is described as probability event. A three-step model, combining quantum mechanics, sheath dynamics and diffusion theory, is developed to model the motion of etching particles, estimate the energy distribution on substrate and simulate the formation process of needle-like nanostructures.

OPTIMIZED DRIE PROCESS

According to Figure 1, DRIE process is usually carried out using SF_6/C_4F_8 in a mode of cyclic etching-passivation steps. Compared to normal DRIE process, there appear nanomasks after a few cycles in the optimized DRIE process, and they will work as the nanomasks during the following DRIE process, thus nanostructures are formed. It is widely believed that the generation of nanomasks is caused by the residual of polymeric layer deposited during passivation step [8]. Therefore, the generation of nanomasks should be the key point of investigating the formation mechanism of black silicon in plasma-based process.

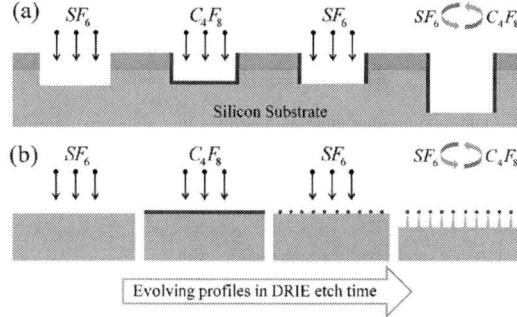

Figure 1: Schematic illustration of black silicon formation: (a) normal DRIE process, (b) optimized DRIE process.

Figure 2 shows the schematic of a typical dual source DRIE system which consist of an ICP (inductively coupled plasma) module and a CCP (capacitively coupled plasma) module [9]. High-density plasma glow is created by ICP source. Then the ions from plasma glow region are leaded by a RF (Radio Frequency) power towards the wafer surface with increasing ion energy. In fact, where RF power works is an electric field called sheath, which is between the plasma and the sample electrode [10].Thus, the plasma etching system can be divided into three parts, i.e. plasma layer, sheath layer and sample surface layer. The etching particles enter from above, go through the plasma layer and sheath layer, then impinge the sample surface. There is distinct particle motion state in each part.

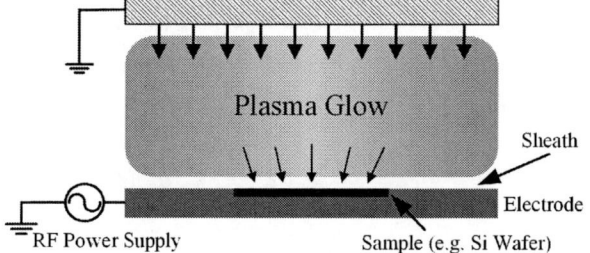

Figure 2: Schematic of a typical dual source DRIE system which consists of both coin module and plate module.

THREE-STEP MODEL FOR SIMULATION OF BLACK SILICON FORMATION

Figure 3 is a schematic illustration of the proposed three-step model. As mentioned above, the plasma etching system is divided into plasma layer, sheath layer and sample surface layer. Correspondingly, quantum mechanics, sheath dynamics and diffusion theory are adopted in each layer according to the movement characteristics of etching particles. A 6μm x 6μm area in Figure 3 is the selected simulation area. These three layers are in sequence. The simulation result of last layer will become the initial value for the next layer. As passivation process and radical etching can be regarded as uniform chemical deposition or reaction, only the motion of ions during etching period should be taken into consideration.

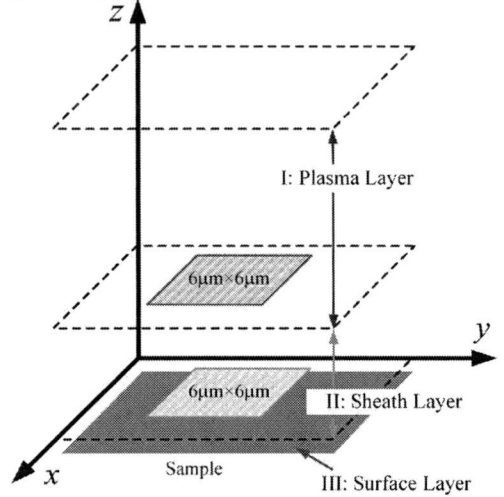

Figure 3: Schematic of three-step model: (I) plasma layer, (II) sheath layer and (III) sample surface layer.

Plasma Layer

Everything possesses wave property and it's particularly notable for microscopic particles. There exists uncertainty in the motion of object and the matter wave is also called probability wave. Analyzing shows the etching particles' motion during DRIE quite conforms to the typical motion elaborated in quantum mechanics [11-12]. So quantum mechanics is introduced here to estimate the probability that etching particles reach each point. In quantum mechanics, motion of particles can be described by Schrodinger equation:

$$ i\hbar \frac{\partial \varphi_p}{\partial t} = -\left[\frac{\hbar^2}{2M} \nabla^2 + V(\vec{r},t) \right] \varphi_p \qquad (1) $$

Here, $V(\vec{r},t)$ is field potential at position $\vec{r}$, M is mass of particles and $\hbar$ is Dirac constant. φ_p is function of matter wave, as shown in Equation (2):

$$ \varphi_p = \varphi_p(\vec{r},t) = A\exp(\frac{i(\vec{p} \cdot \vec{r} - Et)}{\hbar}) \qquad (2) $$

where $\vec{p}$ and E are momentum and kinetic energy of particles, respectively.

According to theory of plasma sheath [10-11], $V(\vec{r},t)$ is considered as zero in plasma layer. It's assumed

that there exist two principal momentum magnitudes for the incident particles of the order $10^{-20} \sim 10^{-21}$kg·m/s. After taking them into Equation (2) and (1), the probability distribution of etching particles on the simulation area (red area in Figure 3) can be achieved, as shown in Figure 4. It can be taken approximately as energy distribution because the probability and etching capability are of positive correlations. Meanwhile, this result also serves as initial value for the model in sheath layer. Thus, the difference in etch rate is caused by the uneven probability distribution of etching particles, moreover, the difference in etch rate is the principal contributor to self-masking effect.

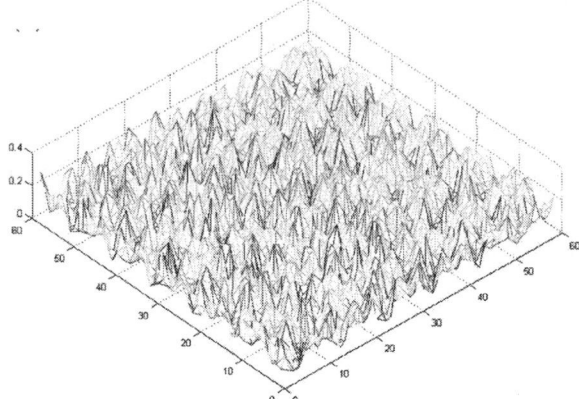

Figure 4: Simulated probability distribution on bottom surface of plasma layer

Sheath Layer

After leaving plasma layer, etching particles will go through the sheath layer before impinging on the sample. The electrode that the substrate is placed upon is often independently RF-biased to provide the electric fields within sheath layer and to incease the ions energy because the potential drop caused by RF bias voltage mainly happens in sheath layer. The quantum mechanics can also be used to describe the motion of etching particles in sheath layer. However, $V(\vec{r},t)$ in Equation (1) is changed. It is nonlinear for this layer due to the sheath potential.

To calculate the sheath potential expression, a self-consistent model of sheath dynamics was used [10, 13]. It consists of equations describing the charge transport in sheath layer coupled to an equivalent circuit model to predict thickness of sheath layer, sheath potential and spatiotemporal variation of potential in sheath layer. All the physical quantities of sheath layer are considered to vary only in the vertical direction towards plate because it is much larger than that in horizontal direction.

Fig.5 shows spatiotemporal variation of potential in sheath layer. The curve oscillates over time in the direction as the blue arrows both in x-axis direction and y-axis direction indicate. The thickness of sheath layer equals to the distance from electrode to the point where potential value equals zero, i.e. the thickness of sheath layer equals to the x-value of point A. So the oscillation curve A shows time dependent sheath thickness calculated from the sheath model. Similarly, the potential on electrode equals to the y-value of point B. So the oscillation curve B shows time dependent potential on bottom electrode.

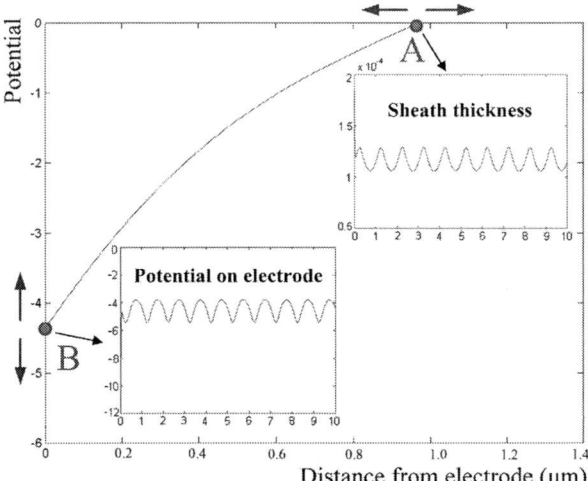

Figure 5: Simulated oscillation curves in sheath model: spatiotemporal variation of potential in sheath layer.

With the sheath potential derived from the above, Equation (1) can be solved by numerical iteration and the probability distribution on the simulation area in sheath layer（orange area in Figure 3）is shown in Figure 6. Compared to Figure 4, it's easy to notice that the sheath potential not only increases the maximum probability but also enhances concentration, making it easier for formation of high-aspect-ratio structures.

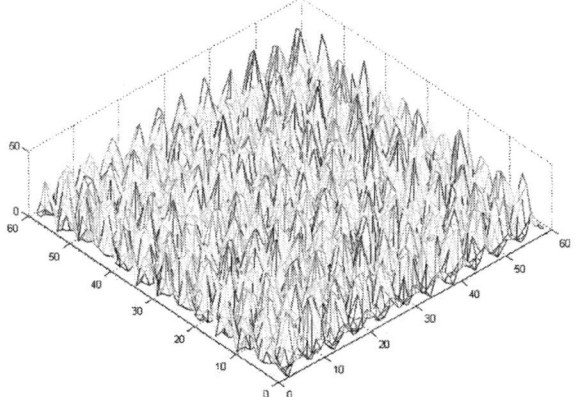

Figure 6: Simulated probability distribution on bottom surface of sheath layer.

Sample Surface Layer

When etching particles impinge on the sample surface, they are supposed to be affected by electric field generated by stored charge in the substrate. Two reasons are considered here: the particles are quite close to the charge on surface; the main factor that influences the final result in this layer is the plate-parallel motion of particles because sheath potential is considered to vary only in the vertical direction towards plate. A diffusion equation [14] shown below is used to model the process.

$$\frac{\partial h}{\partial t} = -D_h \nabla^2 h \qquad (3)$$

Here, h is a variable of height which can be obtained after sheath layer simulation. D_h is diffusion coefficient.

According to above, the particles will diffuse from convex part to concave part on the surface. As a result, the concave part will receive more particles and be etched more.

RESULTS AND DISCUSSION

Finally, the simulated surface morphology can be obtained by rendering them in OpenDX. Furthermore, a matrix of simulated 3D images of nanostructures can be mapped out through varying momentum magnitude within reasonable range as is shown in Fig.7. Here, it is assumed that there exist two kinds of momentum magnitude in the experimental environment, the x-axis stands for one momentum magnitude while y-axis stands for the other. Under different combination of momentum magnitude, there will form nanostructures with different morphology. In this matrix, there are 45 simulated nanostructure morphologies which differ from one another. We believe for each of them, there should be a corresponding experimental sample with the same nanostructures, though we cannot realize all of them now subject to the experimental limitation.

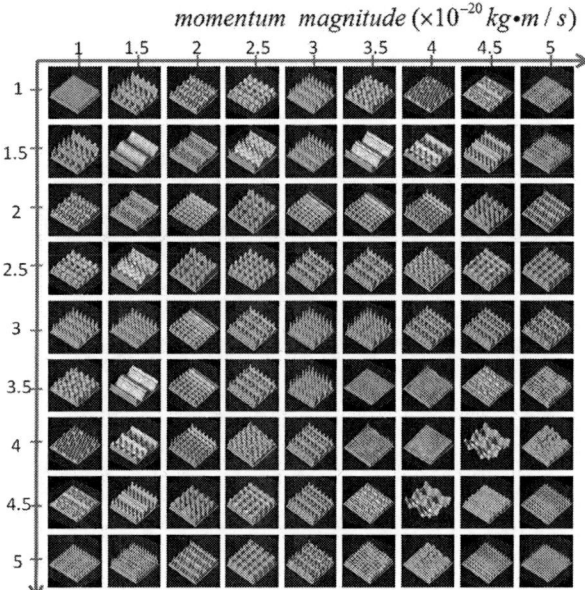

Figure 7: A matrix of 3D simulated nanostructures with reasonable range of momentum magnitude of etching particles (the matrix is symmetric).

In order to observe more clearly, four groups of comparison between simulation results and experimental SEM images of black silicon are listed in Fig.8. As we can see, they are consistent with each other both in profile and structure scale. Besides, they have the same distribution characteristic, such as some coterminous structures and the spacing between nanostructures. If we try to vary the momentum magnitude of etching particles in a larger range, it is promising to obtain more distinct nanostructure morphologies.

CONCLUSIONS

This paper focuses on the exploration of black silicon formation mechanism in DRIE process. It proposes a brand new simulation opinion in this field. Quantum mechanics, sheath dynamics and diffusion theory are combined together to simulate the probability distribution of etching particles. This three-step model gives out very inspiring results to control and predict the fabrication process. It's worth further research to improve the model and establish a better theory system for plasma etching process.

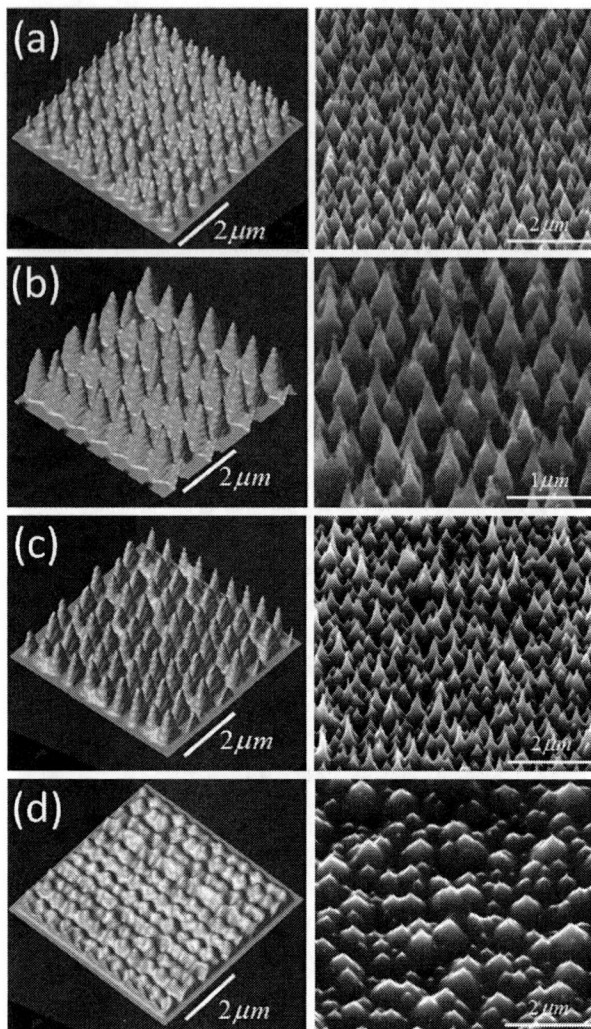

Figure 7: Comparisons between simulation result and SEM image of black silicon.

(a) $\left|\overrightarrow{p_1}\right| = 4e-20 kg \cdot m/s, \left|\overrightarrow{p_2}\right| = 2e-20 kg \cdot m/s$;

(b) $\left|\overrightarrow{p_1}\right| = 3e-20 kg \cdot m/s, \left|\overrightarrow{p_2}\right| = 1e-20 kg \cdot m/s$;

(c) $\left|\overrightarrow{p_1}\right| = 4e-20 kg \cdot m/s, \left|\overrightarrow{p_2}\right| = 1e-20 kg \cdot m/s$;

(d) $\left|\overrightarrow{p_1}\right| = 4.5e-20 kg \cdot m/s, \left|\overrightarrow{p_2}\right| = 3,5e-20 kg \cdot m/s$ ·

ACKNOWLEDGEMENTS

This work is supported by the National Natural Science Foundation of China (Grant No. 61176103, 91023045 and 91323304), the National Hi-Tech Research and Development Program of China ("863" Project) (Grant No. 2013AA041102), and the National Ph. D. Foundation Project (Grant No. 20110001110103) and the Beijing Natural Science Foundation of China (Grant No. 4141002).

REFERENCES

[1] Jiann Shieh, Srikanth Ravipati, Fu-Hsiang Ko and Kostya Ostrikov, "Plasma-made silicon nanograss and related nanostructures", *J. Phys. D: Appl. Phys.*, 44 174010, 2011.

[2] Fu-Yun Zhu, Qi-Qi Wang, Xiao-Sheng Zhang, Wei Hu, Xin Zhao, Hai-Xia Zhang, "3D nanostructures reconstruction based on SEM imaging principle and Applications", *Nanotechnology*, 25 (2014) 185705.

[3] Junshuai Li, HongYu Yu, She Mein Wong, Gang Zhang, Xiaowei Sun, Patrick Guo-Qiang Lo, and Dim-Lee Kwong, "Si nanopillar array optimization on Si thin films for solar energy harvesting", *Appl. Phys. Lett.*, 95: 033102, 2009.

[4] N. J. Peter, X.S. Zhang, S G. Chu, F.Y. Zhu, H. Seidel and H.X. Zhang, "Tunable wetting behavior of nanostructured poly(dimethylsiloxane) by plasma combination treatments", *Appl. Phys. Lett.*, 101: 221601, 2012.

[5] Manfrinato V R, Zhang L H, Su D, et al. "Resolution Limits of Electron-Beam Lithography toward the Atomic Scale". Nano Lett., 13: 1555-1558, 2013.

[6] Gierak J, "Focused ion beam technology and ultimate applications", *Semicond. Sci. Technol.*, 24: 043001, 2009.

[7] K. Lilienthal, M. Fischer, M. Stubenrauch, A. Schober, "Self-organized nanostructures in silicon and glass for MEMS, MOEMS and BioMEMS", *Materials Science and Engineering B*, 169, 78-84, 2010.

[8] F.Y. Zhu, X.S. Zhang, W. Hu, H.X. Zhang. "Controllable formation and optical characterization of silicon nanocone-forest using SF_6/C_4F_8 in cyclic etching-passivation process". *Proc. IEEE Conf. NEMS*, 2013, 1034-1037.

[9] H V Jansen, M J de Boer, S Unnikrishnan, M C Louwerse and M C Elwenspoek, "Black silicon method X: a review on high speed and selective plasma etching of silicon with profile control: an in-depth comparison between Bosch and cryostat DRIE processes as a roadmap to next generation equipment", *J. Micromech. Microeng.* 19 (2009) 033001 (41pp).

[10] E. A. Edelberg and E. S. Aydil, "Modeling of the sheath and the energy distribution of ions bombarding rfbiased substrates in high density plasma reactors and comparison to experimental measurements", *J. Appl. Phys.* 86 (1999) 4799 (14pp).

[11] P K Shukla and B Eliasson, "Recent developments in quantum plasma physics", *Plasma Phys. Control. Fusion*, 52 (2010) 124040 (15pp).

[12] R. P. Feynman and A.R. Hibbs, Quantam Mechanics and Path Integrals (Dover Publications, New York) pp. 2-10, 2010.

[13] Becker F, Rangelow I W, and Kassing R. "Ion energy distributions in SF_6 plasmas at a radiofrequency powered electrode". *J. Appl. Phys.*, 1996, 56: 80.

[14] R. L. Mace and W. H. Matthaeus, "Velocity space diffusion of charged particles in weak magnetostatic fields: Nonlinear effects, model constraints, and implications for simulations", *Phys. Plasmas*, 19 (2012) 032309 (12pp).

CONTACT

* H.X. Zhang, Tel: +86-10-62766570;
zhang-alice@pku.edu.cn

ANOMALOUS RESISTANCE CHANGE OF ULTRASTRAINED INDIVIDUAL MWCNT USING MEMS-BASED STRAIN ENGINEERING

K. Yamauchi[1], T. Kuno[1], K. Sugano[1], and Y. Isono[1]
[1]Graduate School of Engineering, Kobe University, Kobe, JAPAN

ABSTRACT

This research clarified the anomalous electric resistance change of ultrastrained multi-walled carbon nanotube (MWCNT), as well as its mechanical properties, using the Electrostatically Actuated NAnotensile Testing device (*EANAT*) mounted on the *in-situ* SEM nanomanipulation system. The Young's modulus of MWCNT and its shear stress during interlayer sliding deformation were estimated from the load-displacement curve. The electrical resistance of the MWCNT was 215 kΩ without strain, which was similar to the previously reported value, however the anomalous resistance change was observed under enormous strain. Although the resistance change ratio was almost constant during interlayer sliding of the MWCNT, it specifically showed a sharp raise at the end of the sliding in spite of the MWCNT not breaking mechanically. The molecular dynamics (MD) simulation provided a good understanding that the atomic reconfiguration due to the hard sticking at the edge of extracted outer layer of MWCNT might induce the sharp raise of resistance without its mechanically breaking. This result reported here is extremely important for reliability of MWCNT interconnects.

INTRODUCTION

Multi-walled carbon nanotubes (MWCNTs) are one of promising materials as ultralow-resistance interconnect for via structures of the half-pitch 32 nm fabrication node [1], [2]. However, via structures formed by MWCNTs have to withstand thermal stress generated by the device operation, which will lead to interlayer sliding deformation of MWCNTs resulting in changing electrical resistance. Thus, basic information of the electrical properties for mechanically strained individual MWCNT is required to reliable design of MWCNT interconnects.

We have so far developed Electrostatically Actuated NAnotensile Testing device (*EANAT*) and established MEMS-based elastic strain engineering technique to characterize mechano-electric properties for individual nanowires [3]-[5]. The latest designed *EANAT* includes two functions: electrostatically driven actuators to produce a tensile force and a capacitive displacement sensor to measure tensile displacement. The design details of *EANAT* and mechanical properties of the individual MWCNT synthesized by atmospheric pressure-chemical vapor deposition have been reported in the previous paper [5]. A resolution of 0.28 nm in the displacement has been achieved with the cantilever motion amplification system incorporated into the capacitive sensor. The deformation accompanied by the repeated stick-slip and hard sticking events like telescopic motion was observed in individual MWCNTs under uniaxial loading, and the shear strength at a stick-slip event during interlayer sliding was directly derived from the shear interaction force.

The present research focuses on clarifying a complete mechano-electric characterization for MWCNT during an interlayer sliding deformation under uniaxial tensile strain, and also reports on the MD tensile simulation of MWCNT model using the bond-order potential function in order to provide a good understanding the electrical resistance change during interlayer sliding such as a stick-slip event between layers on an atomic scale.

EXPERIMENTAL AND ANALYTICAL PROCEDURES

MEMS-based Elastic Strain Engineering

The structure and dimensions of the *EANAT* are basically same as those of the previously developed device [4], [5]. Fig. 1 shows a schematic of the *EANAT* and its lumped mechanical model. An individual MWCNT is made to bridge the 5-μm-gap in the specimen area using the *in-situ* SEM nanomanipulation system. Thermally driven actuators are installed on the left- and right-sides of the movable frame in the specimen area to avoid vibrations of the frame when an individual MWCNT is implanted in the *EANAT*. An electrostatically driven comb actuator is separately arranged at an interval of 4 μm (= g_1) from the movable portion of the specimen area. Capacitive displacement sensors are set at an interval of 3 μm (= g_2) from the lower part of the comb actuator employed for the tensile loading system. The movable portion of the specimen, the comb actuator and the displacement sensors are suspended by each supporting beam from fixed anchor parts.

In the lumped mechanical model, the force generated by the comb actuator deflects the supporting beams of the comb actuator with a stiffness of K_a at the beginning of device operation. This force is continually applied for elastic deformation of the sensor structures with a stiffness of K_{ds}, the supporting beams of the specimen area with a stiffness of K_s, and an individual MWCNT bridged at the specimen area. Therefore, the tensile force of MWCNT can be obtained from a total stiffness of K_a, K_{ds} and K_s, and a difference in the actuator displacements between before and after the failure of sample. Here, the doubly-clamped beams used as the supporting beams at the actuator and the specimen areas show the nonlinearity of stiffness caused by tensile stress induced in the longitudinal direction of the beams under a large deformation [5].

The developed differential capacitor with arc shaped combs incorporated into the cantilever motion amplification system achieved a scale factor larger than 1 fF/nm, and it carried an advantage of a large capacitance change and an accurate calibration of the capacitive displacement sensor [5].

Fig. 2 shows SEM images of the *EANAT* used in this research. The device has been fabricated by a conventional bulk micromachining process using an SOI wafer with 35-μm thick active layer on the 4-□m thick SiO2 and 400-μm thick Si substrate [4], [5].

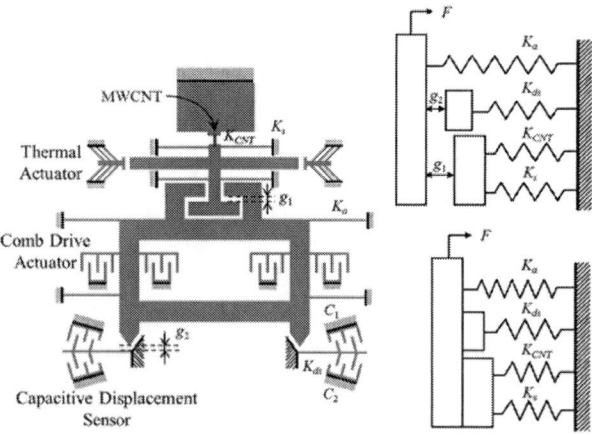

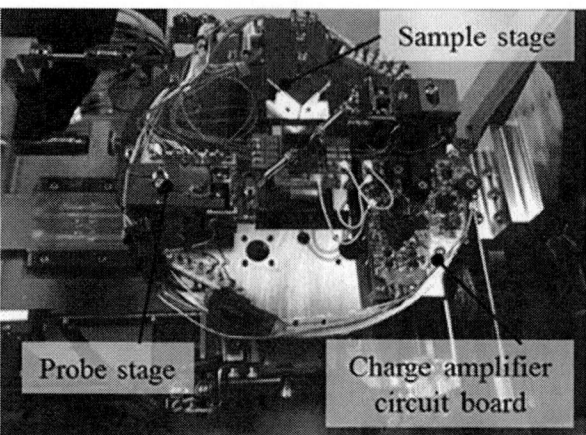

Fig. 1 Schematics of EANAT and its lumped mechanical model.

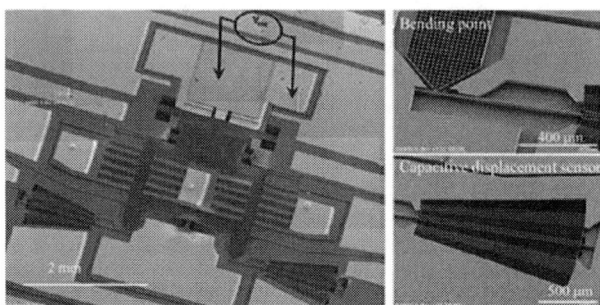

Fig. 2 SEM images of the EANAT.

Fig. 3 Photographs of the in-situ SEM nanomanipulation system; the outside view of the system, and the EANAT mounted on the sample stage.

The *in-situ* SEM nanomanipulation system, as shown in Fig. 3, was employed to pick up an individual MWCNT from a substrate and to fix it to an *EANAT*. The nanomanipulation system includes two probe stages and one sample stage. Each of the probe stages has 3 degrees of freedom (DOF) and the sample stage has 2 DOF. A PZT actuator installed in each stage for fine displacement control allows for a positioning resolution of 0.1 nm, and the successive stepping motion of the PZT for long stroke movement uses a working space of the probe stages of 10 mm (±5 mm) in one direction. The charge amplifier circuit board for the differential capacitor of *EANAT* is also set on the system on the side of the sample stage in order to reduce electrical noise as much as possible [5].

The SEM images of nanomanipulation for an individual MWCNT are shown in Figs. 4. The MWCNT is fixed on the probe by electron beam induced contamination in the SEM with a higher emission current, i.e. the deposition of carbonaceous material over the sample surface. After cutting the vicinity of the fixed end of the MWCNT by a focused electron beam, the MWCNT is moved to the test position previously marked by photolithography on the specimen area of the *EANAT* and is attached to the device by the electron beam induced carbon-based-contamination.

Molecular dynamic simulation

Fig. 5 shows the MWCNT model with five cylindrical carbon layers used in tensile MD simulation. The carbon layers are alternatively placed like a telescope. Both ends

of even-numbered layers are not fixed at the end of the models for realizing their interlayer sliding deformations. The unit ring of carbon cylinder model with the minimum diameter of 1.59 nm is also shown in the figure, which is composed of zigzag structures in the longitudinal direction. The single CNT layer corresponds to 30 unit rings, and its length is 13.05 nm (=L). Thus, the total length of the telescopic MWCNT models is 19.6 nm (=1.5L) since the overlap region between the outer and inner layers is defined as 0.25L.

The classical MD simulation was carried out using the bond-order empirical potential function proposed by D.W. Brenner [6], as the following equations:

$$E = \sum_i \sum_{j>i} \left[V_R(r_{ij}) - \overline{B_{ij}} V_A(r_{ij}) \right]. \tag{1}$$

E is the binding energy for the carbon potential, which is the sum of the repulsive and attractive terms given by

$$V_R(r_{ij}) = f_c(r_{ij}) \frac{D_{ij}^{(e)}}{S_{ij}-1} \exp\left(\sqrt{2S_{ij}} \beta_{ij}\left(R_{ij}^{(e)} - r\right)\right), \tag{2}$$

and

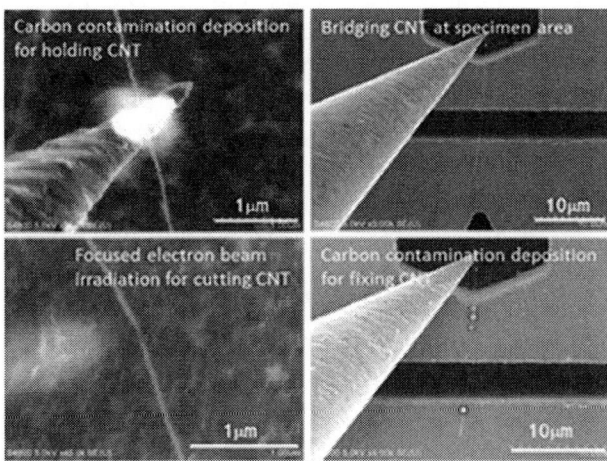

Figs. 4 Typical manipulating operations for individual MWCNT; (a) holding, (b) cutting, (c) bridging, and (d) fixing MWCNT.

$$V_A(r_{ij}) = f_c(r_{ij}) \frac{S_{ij} D_{ij}^{(e)}}{S_{ij} - 1} \exp\left(\sqrt{\frac{2}{S_{ij}}} \beta_{ij} \left(R_{ij}^{(e)} - r\right)\right), \quad (3)$$

respectively.

$$f_c(r_{ij}) = \begin{cases} 1, & r < R_{ij}^{(1)} \\ \left[1 + \cos\frac{\pi\left(r - R_{ij}^{(1)}\right)}{R_{ij}^{(2)} - R_{ij}^{(1)}}\right]\Big/2, & R_{ij}^{(1)} < r < R_{ij}^{(2)} \quad (4) \\ 0, & r > R_{ij}^{(2)} \end{cases}$$

$f_c(r_{ij})$ is the function that restricts the pair potential to nearest neighbors, and $\overline{B_{ij}}$ is the empirical bond-order function.

The periodic boundary condition was employed at the fixed ends of the MWCNT model in the x-direction. For applying uniaxial strain to the model, the periodic boundary length was expanded at the strain of 1.0D-6 per every time step after thermal energies of atoms stabilized. MD simulation was run at 0.1 K in the microcanonical ensemble. Time series of the position for atoms was calculated by Newton's equation of motion using the Verlet algorithm by the time step of 0.1 fs.

RESULTS AND DISCUSSIONS

Resistance change of ultrastrained individual MWCNT

Fig. 6 shows the load-displacement curve for the MWCNT with a mean diameter of 60 nm and a length of 15 μm. The deformation behavior in the figure is similar to that obtained previously reported MWCNTs [5]. Although the tensile load increases with an increase in tensile displacement, regions of decrease and saturation exist in the curves before failure of the sample. Linear increases in load are observed at the initial tensile deformation stage owing to elastic deformation. After the elastic region, the load reduces and fluctuates above an elongation of 0.85 μm with an increase in displacement, and it linearly increase again with the similar slope as the elastic region. This trend in the load-displacement curve is very close to the result by M-F. Yu *et al.* [7]. They clarified that individual MWCNTs

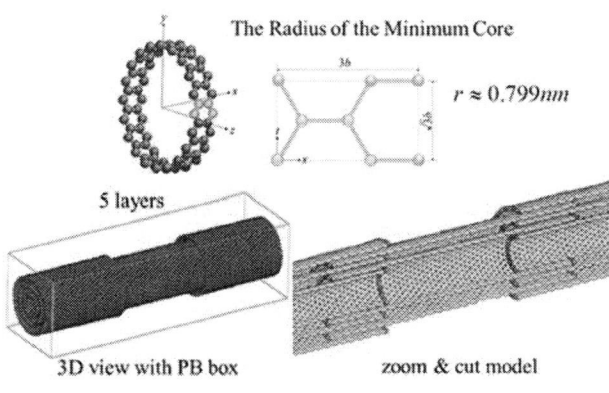

Fig. 5 The telescopic MWCNT model with five cylindrical layers for tensile MD simulation.

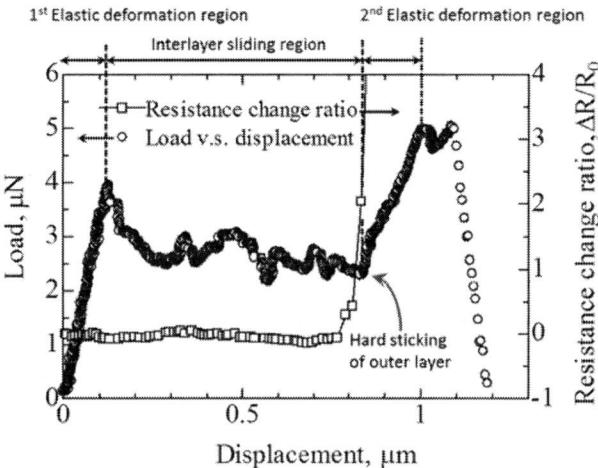

Fig. 6 The load-displacement curve and the variation of electrical resistance change ratio with increasing the displacement for MWCNT.

showed interlayer sliding deformation behavior with stick-slip events between interlayers when the stress saturated or reduced during tensile loading.

The Young's modulus and the shear stress during the interlayer sliding showed 418 GPa, and 60.1 MPa, respectively. The electrical resistance of MWCNT was 215 kΩ without strain. However, the anomalous resistance change was observed under enormous strain in the figure in spite of the MWCNT not breaking mechanically. The resistance change ratio was almost constant during the interlayer sliding, but it showed a sharp raise at the end of the sliding without breaking.

Figs. 7 show SEM images of MWCNT after its breaking. The fractured edge of the outer layer has kept the straight external form without plastic deformation, while the opposite edge at the right side of the image shows tiered external form that is caused by the inside of MWCNT has been pulled up. The interlayer sliding of the MWCNT like a telescopic motion of CNT layer had been produced. The interlayer sliding probably corresponds to the saturated region with the stick-slip event observed in the load-displacement curve.

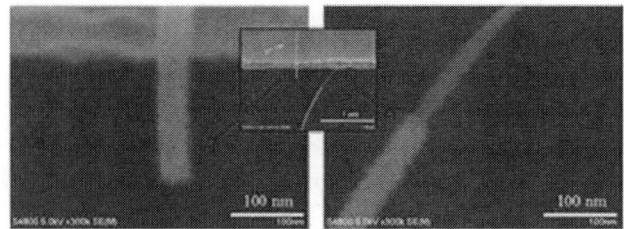

Fig. 7 SEM images of MWCNT after its breaking.

Interlayer sliding in MD simulations

Fig. 8 shows the force-displacement curve and the snapshot of MWCNT model obtained from the tensile MD simulation using the bond-order empirical potential function. After several stick-slip events during the interlay sliding, a linear increase of tensile force was observed; at this time, edge atoms of the outer layer was strongly stuck to those of the inner layer and the reconfiguration of edge atoms was produced in the model. The tensile force was kept at the large value for a little while, and it gradually decreased with an increase of displacement. Therefore, it was expected that the anomalous resistance change of MWCNT without mechanically breaking in the experiment was caused by the atomic reconfiguration due to hard sticking at the edge of the outer layer.

REFERENCES

[1] F. Kreupl, *et al.*, *Microelectronic Eng.*, Vol. 64, Issues 1-4, pp. 399-408, 2002.

[2] Jun Li, *et al.*, *Appl. Phys. Lett.* Vol. 82, No. 15, pp. 2491-2493, 2003.

[3] M. Kiuchi, *et al.*, *J. Micromech. and Microeng.*, IOP, Vol. 18, No. 6, 065011(10 pages), 2008.

[4] H. Ohmori, *et al.*, *Proceedings of MEMS2012*, Paris, 29 January - 2 February, 2012, pp. 412-415.

[5] Hyun-Jin, Oh, *et al.*, *J. Microelectromech. Syst.*, Vol. 23, No. 4, pp. 944-954.

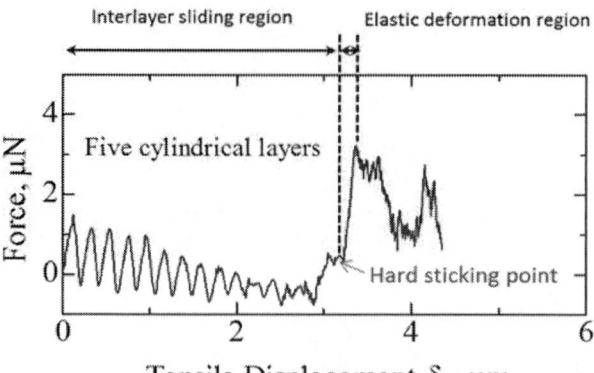

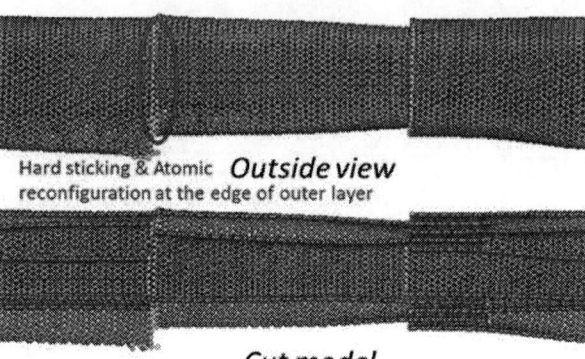

Fig. 8 The force-displacement curve and the deformed model in the tensile test simulation.

[6] D. W. Brenner, *Phys. Rev. B* **42** (15), pp. 9458-9471, 1990.

[7] M-F. Yu, *et al.*, *Science*, Vol. 287, pp. 637-640, 2000.

CONTACT

*Y. Isono, tel:+81-78-803-6145;isono@mech.kobe-u.ac.jp

FABRICATION OF TETRAPOD-SHAPED AL/NI MICROPARTICLES WITH TUNABLE SELF-PROPAGATING EXOTHERMIC FUNCTION

Keita Inoue[1], Toshihisa Fujito[1], Kazuhiro Fujita[2], Yoshikazu Kuroda[2], Katsuhisa Takane[2], and Takahiro Namazu[1]*
[1]University of Hyogo, Himeji, JAPAN
[2]Gauss Co., Ltd., Aioi, JAPAN

ABSTRACT

This paper reports micron-sized tetrapod shape Al/Ni particles fabricated by injection molding and electroless plating. We realize fabricating the 3D tetrapod micro particles with self-propagating exothermic function that is able to tunable by changing Al powder's diameter and simultaneously by keeping porous Al tetrapod's void fraction constant. The maximum surface temperature and high temperature duration during the reaction differ from sputtered Al/Ni multilayer film's exothermic performances. The tetrapod's exothermic performances can be freely controlled in response to the applications.

INTRODUCTION

By applying a small energy, self-propagating exothermic reaction in Al/Ni multilayer films can be seen during producing NiAl compound as reported in MEMS 2006 [1]. There are excellent features such as easy fabrication, easy activation, fast reaction propagation (10m/s), controllable heat energy, low cost, and zero emission, which enable us to use the films as a local heat source for solder bonding for MEMS as reported in Transducers 2009 and 2011 [2, 3] and for power devices [4]. The reactive films can be easily fabricated by sputtering or vapor evaporation, but it is difficult to form them to arbitrary 3D shapes because the films consist of two different metals with different etching characteristics. If the exothermic function can be applied to arbitrary 3D shaped particles, the application will expand [5, 6]. However no self-propagating exothermic particles have been realized so far.

The purpose of this study is to develop a new technique for fabricating 3D shaped Al/Ni micro particles with self-propagating exothermic function. We propose an experimental technique that is the combination of powder injection molding for tetrapod-shaped porous Al micro particles and electroless plating for Ni electroplating into pores [6]. To control the exothermic performance, such as maximum temperature and reaction duration, during the reaction, Al powder diameter ranging from 3 to 30μm is varied under the condition of constant void fraction of 33%. The performances are evaluated by means of differential scanning calorimeter (DSC) analysis, infrared radiation thermometer measurement, and high-speed charge coupled device (CCD) camera observation.

FABRICATION

A process flow for porous Al tetrapod-shaped micro particles using injection molding technique is illustrated in Fig. 1. Al powders with the diameter of 3, 10, and 30μm were used. First an organic binder and the powders were mixed. The void fraction of the product after degreasing is

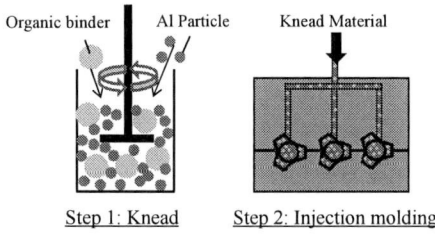

Figure 1: Fabrication process of micron-sized porous Al tetrapod particles by injection molding.

Figure 2: SEM photograph of produced porous Al tetrapod particles using 10μm-diameter Al powders.

determined by the volume ratio of the powder to the binder. In this study, to produce Al/Ni tetrapod particles with the atomic ratio of 3:2 in Al:Ni, the volume fraction of the binder was controlled to be approximately 33%. After the mixed materials were kneed, injection molding was carried out. We have used micron-sized tetrapod shape mold to demonstrate 3D shape Al/Ni exothermic particles. Degreasing was then conducted at 450°C for 1 hour for fabrication of porous structures, followed by sintering at 580°C for 2 hours to increase the rigidity of porous Al tetrapod particles.

Fig. 2 shows produced porous Al tetrapod particles with 10μm Al powders. All the particles could be finely shaped, although the photographs are not included in this paper. The whole size of the particles was approximately 900μm, and the diameter of each leg was about 300μm. These values were slightly smaller than the mold size because of volume shrinkage during sintering. The void fraction was kept constant at 33~34% in all the tetrapod particles.

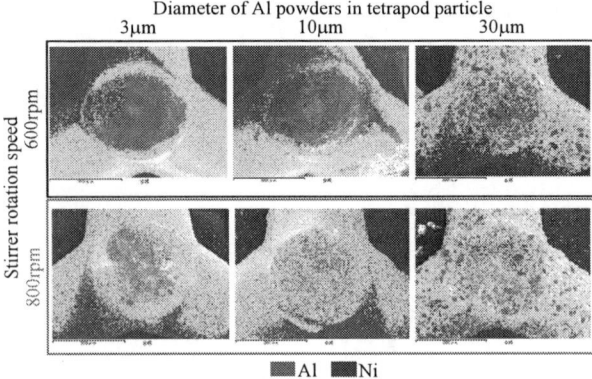

(a) EDX maps of the particles' cross-section after plating.

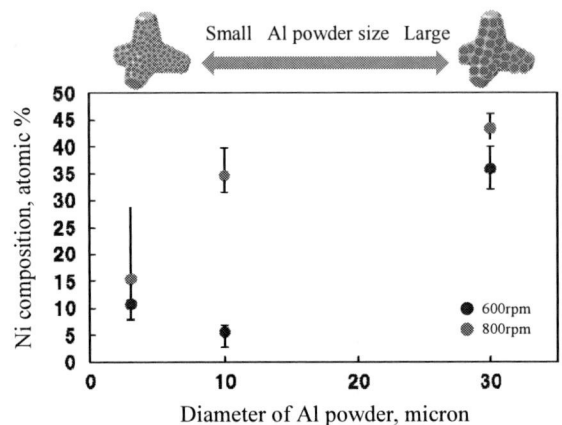

(b) Relation between Al powder size and Ni composition.
Figure 3: EDX analysis result of the cross-section of Al/Ni tetrapod particles after electroless Ni plating.

After fabrication of porous Al tetrapod particles, Ni electroless plating was conducted to provide the particles with self-propagating exothermic function. First surfactant is applied to porous Al particles for 60min to enhance wettability of the surface. Then zincate treatment is conducted for 60min to enhance adhesion of electroless Ni coating. Finally electroless Ni plating is conducted at 65°C for 48hours to infill Ni into pores in particles. Fig. 3 (a) shows representative EDX maps of the cross-section of the micro particles after electroless plating. The red and green-colored portions indicate Al and Ni, respectively. After the plating with 600rpm, Ni could not grow inside the particles with 3 and 10μm Al powders because the cavity size was very small. Ni could grow inside the particles with 30μm Al powders. By increasing rotation speed in plating to 800rpm, Al particles with 10 and 30μm Al powders could be filled with Ni.

Fig. 3 (b) shows the relationship between Al powder size of Al/Ni tetrapod particles and Ni composition. The blue and red-colored plots indicate stirrer rotation speed of 600 and 800rpm during plating, respectively. Ni could be grown inside the particles with only 30μm at 600rpm and with 10 and 30μm at 800rpm. The particles finely fabricated by injection molding and plating show Ni composition of around 40%. This indicates that faster flow rate in plating is necessary for finely fabricating

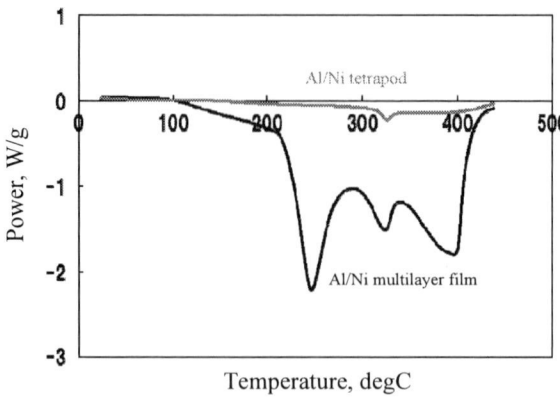

Figure 4: Representative DSC analysis result of Al/Ni tetrapod particles along with the result of Al/Ni film.

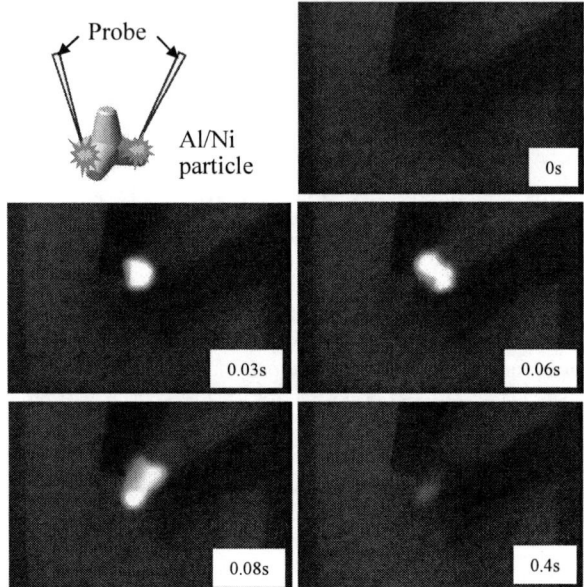

Figure 5: Snapshots of the reaction test for an Al/Ni tetrapod particle. By supplying an electric power, the particle was able to react.

Al-Ni-iteration-structured micro particles. In the case of incompletely-plated Al/Ni particles, Ni composition ranges from 5 to 30at.%.

CHARACTERIZATION

Fig. 4 shows representative DSC curve for Al/Ni tetrapod particles. The red and black-colored lines indicate DSC curves for the particles and sputtered Al/Ni film, respectively. Since the DSC curve for Al/Ni particle has one exothermic peak at around 320°C, the particles is found to have exothermic reaction function. However, the reaction performance is possibly different from the film's performance because the particles' peak intensity was weak compared with the film's peak.

After the confirmation that the particles showed exothermic reaction, we have tried to ignite the particles by supplying an electric power via two micro probes made of tungsten carbide. The voltage and current were set to 0.1V and 0.5A, respectively. Fig. 5 shows snapshots of the reaction test for an Al/Ni particle. A movie during the

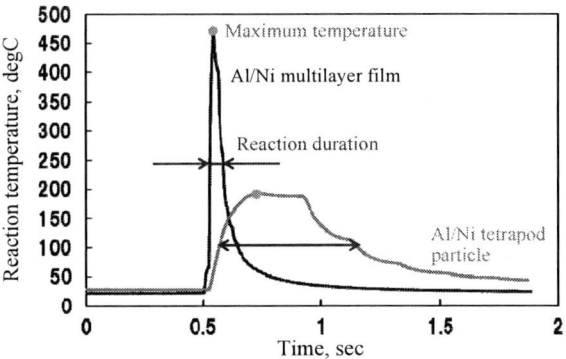

Figure 6: Representative temperature measurement result of Al/Ni tetrapod particle along with the result of sputtered Al/Ni film.

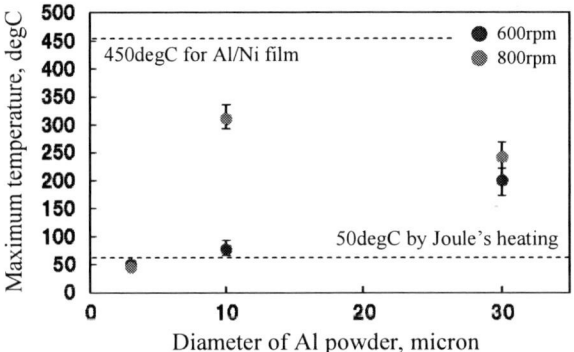

(a) Al powder size vs. maximum temperature

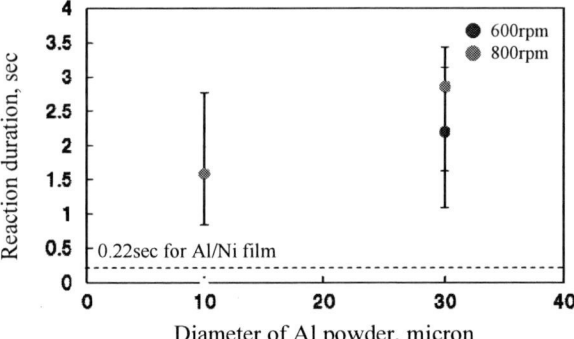

(b) Al powder size vs. reaction duration

Figure 7: Influence of Al powder size on heat performance of Al/Ni tetrapod particles.

reaction was shot with a high-speed CCD camera. By supplying the electric power, the tetrapod-shaped particle could react. The reaction gradually propagated along the tetrapod legs one by one. The speed was very slow compared with the film's speed, approximately 10m/s.

Fig. 6 shows representative result of reaction temperature measurement. The red and black-colored lines indicate Al/Ni particle's reaction and the film's reaction, respectively. It is found that there is a big difference between the particle's reaction and film's reaction. Compared with sputtered Al/Ni film, the maximum temperature of the particle was very low, but the reaction duration was very long. In this paper, reaction duration was defined as the time of full width at half maximum.

The relationship between Al powder size used for Al/Ni particles and maximum temperature during the reaction is shown in Fig. 7 (a). The red and blue-colored plots indicate stirrer rotation speed in plating, 600 and 800rpm, respectively. The incomplete-plating particles were heated to 50°C because of Joule's heating during the test. The maximum reaction temperature of 10μm-powder particles at 800rpm was approximately 300°C, lower than the film's temperature. The temperature decreased to 200°C with increasing Al powder size to 30μm. In the powder size of 30μm, the difference in stirrer rotation speed between 600 and 800rpm did not influence the reaction temperature. Fig. 7 (b) shows the relationship between Al powder size and reaction duration. The mean reaction duration of 10μm-powder Al/Ni particles' reaction was 1.6sec, which is 7.3 times longer than that of the film's reaction. The duration increased to 3.0sec with increasing the powder size to 30μm. The reason is thought to be that, in the case of larger Al powders, it takes longer time for atomic diffusion during producing NiAl compound. Therefore, by changing Al powder size under the constant porosity, that is, the constant atomic ratio, the exothermic reaction performance can be controlled.

CONCLUSIONS

We have successfully fabricated tetrapod-shaped Al/Ni micro particles with self-propagating exothermic reaction function by using a combined technique of powder injection molding and electroless plating. The injection molding technique provided the fabrication of porous Al tetrapod micro particles using three-different-diameter Al powders. The electroless molding technique provided Ni growth the inside of the pores. The produced Al/Ni particles could react by supplying an electric power. By changing the Al powder size in the particles, the maximum temperature and reaction duration could be controlled.

ACKNOWLEDGEMENTS

The authors express their gratitude to Dr. Takano, The New Industry Research Organization, Kobe, Japan, for fruitful discussion on the fabrication of porous Al/Ni particles and electroless plating into micro pores.

This study was partly supported by Grant-in-Aid of The Canon Foundation, Japan, and also partially supported by Grant-in-Aid for Challenging Exploratory Research, Japan.

REFERENCES

[1] T. Namazu, H. Takemoto, H. Fujita, Y. Nagai, and S. Inoue, "Self-Propagating Explosive Reactions in Nanostructured Al/Ni Multilayer Films as A Localized Heat Process Technique for MEMS", *Proc. of the 19th IEEE Int. Conf. on Microelectromech. Syst., MEMS 2006, Istanbul*, pp. 286-289, 2006.

[2] K. Ohtani, Y. Yamano, T. Namazu, and S. Inoue, "Strength Evaluation of Lead-Free-Solder Joint Fabricated by Exothermic Film Local Heating", *Proc. of the 15th Int. Conf. on Solid-State Sensors, Actuators and Microsystems, Transducers 2009, Denver*, pp. 172-175, 2009.

[3] T. Namazu, K. Ohtani, K. Yoshiki, and S. Inoue, "Crack Propagation Direction Control of Crack-Less Solder Bonding Using AlNi Flash Heating Technique", *Proc. of the 16th Int. Conf. on Solid-State Sensors, Actuators and Microsystems, Transducers 2011, Beijing*, pp. 1368-1371, 2011.

[4] S. Miyake, S. Kanetsuki, K. Morino, J. Kuroishi, and T. Namazu, "Thermal Resistance Analysis of Solder Joints Fabricated by Self-propagating Exothermic Reaction", *Proc. of 27th Int. Microprocesses and Nanotechnology Conf., MNC2014, Fukuoka*, 6P-7-93, 2014.

[5] T. Matsuda, S. Inoue, and T. Namazu, "Micro-sized Exothermic Reactive Particles Fabricated by Sputtering to Mesh Substrates", *Proc. of 26th Int. Microprocesses and Nanotechnology Conf., MNC2013, Hokkaido*, 8B-8-4, 2013.

[6] K. Inoue, T. Fujito, K. Fujita, Y. Kuroda, K. Takane, and T. Namazu, "Fabrication of Self-propagating Exothermic Microparticles Using Injection Molding and Electroless Plating Techniques", *Proc. of 27th Int. Microprocesses and Nanotechnology Conf., MNC2014, Fukuoka*, 6C-4-2, 2014.

CONTACT

T. Namazu, Tel: +81-79-267-4962; Fax: +81-79-267-4975
E-mail: namazu@eng.u-hyogo.ac.jp
http://www.eng.u-hyogo.ac.jp/mse/mse12/index.html

HIGH-PRODUCTIVE FABRICATION METHOD
OF FLEXIBLE PIEZOELECTRIC SUBSTRATE
Hirotaka Hida[1], Shun Yagami[1], Akira Sakurai[1], and Isaku Kanno[1]*
[1]Department of Mechanical Engineering, Kobe University, Kobe, JAPAN

ABSTRACT

This paper reports a simple and high-productive fabrication method of flexible piezoelectric substrate. To achieve a low-cost fabrication process of the substrate, we developed a novel transfer technique of piezoelectric thin films which were formed on temporary stainless-steel substrate to target PDMS (Poly-dimethylsiloxane) substrate by using wet etching process. We experimentally clarified that $Pb(Zr,Ti)O_3$ (PZT) thin films transferred on PDMS substrate, which is developed flexible piezoelectric substrate, have piezoelectric properties by evaluating crystal structures and an inverse piezoelectric effect. This fabrication method might allow us to efficiently develop novel MEMS (micro-electro-mechanical systems) devices such as wearable sensors and artificial muscle in large quantities at a low cost.

INTRODUCTION

Piezoelectric thin films have been widely used in various types of MEMS (micro-electro-mechanical systems) applications such as microphone [1], energy harvesters [2, 3] and actuators for optical devices [4, 5] because of their advantages; a low driving voltage, high mechanical responsibility and self-generation characteristic [6, 7]. During film deposition process of piezoelectric materials, high temperature treatment at 500 °C or above is generally required for crystallization. Thus, heat resistant materials such as silicon or metals have been used as deposition substrates. However, as a result, mechanical characteristic of piezoelectric MEMS devices strongly depend on mechanical properties of substrate materials, for example, high stiffness, high mass-density and brittleness of silicon. Therefore, to expand MEMS device applications such as wearable sensors and artificial muscles, high flexibility of the deposition substrate is one of the essential properties.

To improve selectivity of substrate material for forming functional thin-films including piezoelectric

materials, several thin-film transfer methods have been reported by using laser lift-off process [8, 9], physical peel-off process with adhesive and release layer [10]. However, there are some technical issues, for example, low productivity and high cost fabrication process. Meanwhile, thin-film transfer method using wet etching process [11] has several advantages; high productivity and low cost fabrication process. However, there are still technical issues with conventional wet-etching based transfer method. One of the issues is that it is difficult to directly form film electrodes on piezoelectric thin-films for using as sensors and actuators with piezoelectric effect. In addition, a photolithography process was preliminarily required for transferring the piezoelectric thin-films.

Here, we developed a simple and efficient wet-etching based transfer method of piezoelectric thin-films without using preliminary processes in clean room to fabricate flexible piezoelectric substrate at a low cost. For expanding applications using the piezoelectric substrate, we introduced a PDMS (Poly-dimethylsiloxane) film containing carbon which works as both conductive- and adhesion-layer for piezoelectric thin films during transfer process.

METHOD AND EXPERIMENTS
Flexible piezoelectric substrate

Figure 1 shows a schematic view of developed flexible piezoelectric substrate. The developed substrate consists of PDMS film as a main substrate, piezoelectric thin film, upper/lower film electrodes and black PDMS layer which contains carbon black. The black PDMS layer has two roles; to act as an adhesion layer between the piezoelectric thin film and bottom electrode and to improve electrical conductivity between the top/bottom electrodes. The upper electrode is implanted into PDMS substrate and electrically connected to external devices via metal bumps.

Fabrication process

The fabrication process of flexible piezoelectric substrate is as shown in figure 2. We used a stainless steel (SS304) having two centimeters square and thickness of 30-200 mm as a temporary substrate for transfer method based on wet etching process, and lead zirconate titanate (Pb(Zr,Ti)O3, PZT) which has superior piezoelectric properties as piezoelectric thin-film material.

The detailed fabrication process is as mentioned below.
(a) Preparation of temporary substrate: Titanium (< 20 nm in thickness) and platinum (approximately 100 nm in thickness) thin films are deposited on stainless steel substrate by spattering method at room temperature to form bottom electrode.
(b) PZT deposition process: PZT thin film was derived from sol-gel precursor (Pb:Zr:Ti=1.1:0.52:0.48 in 10 %(W/W), Mitsubishi Material Corporation, Japan) on Pt/Ti/SS304 substrate. The sol-gel precursor was applied

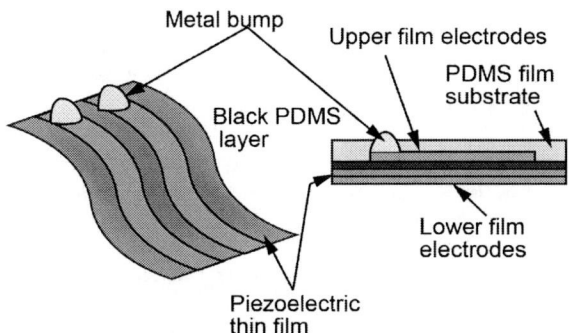

Figure 1: Schematic view of developed flexible piezoelectric substrate.

978-1-4799-7956-1/15 $31.00 © 2015 IEEE

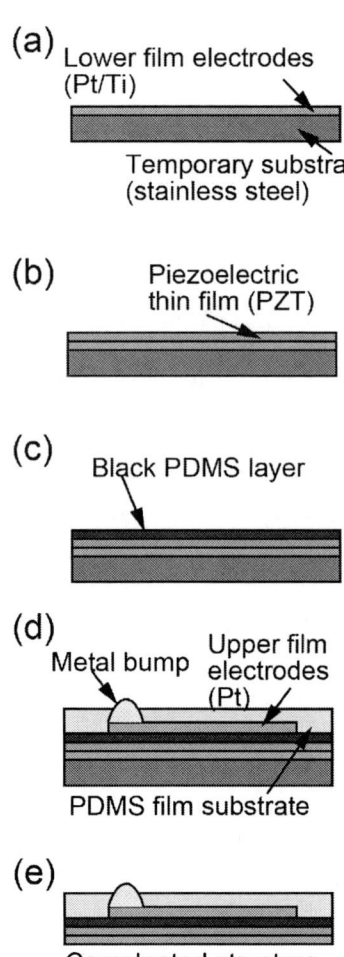

(a) Lower film electrodes (Pt/Ti)
Temporary substrate (stainless steel)

(b) Piezoelectric thin film (PZT)

(c) Black PDMS layer

(d) Metal bump   Upper film electrodes (Pt)
PDMS film substrate

(e) Compleated structure

Figure 2: Fabrication process of flexible piezoelectric substrate. (a) Deposition of Pt/Ti films, (b) PZT deposition by sol-gel method, (c) Black PDMS deposition, (d) Deposition of PDMS and Pt films and forming metal bumps, and (e) Transferring PZT thin-film to black PDMS by using wet etching process

on the substrate by a spin-coating method (first: 1000 rpm for 10 seconds, second: 3000 rpm for 30 seconds) and then dried at 300 °C for five minutes on hot plate. After repeating coating and heat process five times, the PZT layers were annealed in oven at 600 °C (10 °C/min) for 10 minutes to obtain perovskite crystal structures without film cracks caused by residual stress. The thickness of PZT thin film was finally up to 2 μm by repeating a series process consisting of coating, heating and annealing six times.
(c) Black PDMS deposition: PDMS precursor was mixed with carbon black suspending solution in 1.1:0.5 by mass by using a vortex mixer for four minutes. Then, the mixture was defoamed in vacuum to improve uniformity for one hour. After defoaming, the mixture was applied on the PZT thin film by spin-coating method (first: 500 rpm for 5 seconds, second: 6000 rpm for 120 seconds) and cured at 90 °C for five minutes on hot plate. The thickness of black PDMS was around 5 μm.
(d) Forming upper electrode and metal bump: Platinum thin-film was patterned on black PDMS with shadow mask

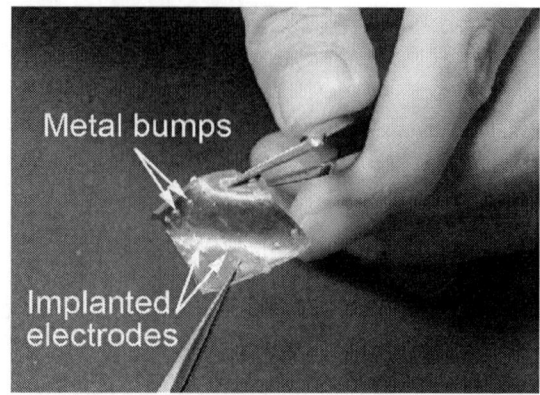

Figure 3: Photograph of developed flexible piezoelectric substrate.

as upper film electrode by sputtering method at RT, and the metal bump was manually formed with conductive silver paste. Then, PDMS substrate (> 50 μm-thick) was formed by spin coating method for physically protecting the upper electrode during wet etching process of stainless-steel substrate.
(e) Stainless-steel wet etching process: The temporary stainless-steel substrate was removed by wet etching process. An etchant, which composed of 20 % (V/V) aqueous ferric chloride and 2.5 % (W/V) citric acid, was heated at 60 °C and the whole substrate was dipped in the etchant. A measurement value of etching rate was about 2-3 μm/min.

RESULTS AND DISCUSSION

Figure 3 shows a photograph of fabricated flexible piezoelectric substrate by using our developed method. We experimentally confirmed that PZT thin films were efficiently transferred onto black PDMS layer using wet etching process. The fabricated substrate can be freely deformed without breaking implanted PZT and electrode thin-films. In contrast, PZT thin films were not transferred by just contacting PDMS surface physically. These results show that non-covalent force may be generated in an interface between PZT films and PDMS during wet etching process.

Crystal structure analysis

XRD (X-ray diffraction) patterns of PZT thin film and platinum before and after transfer process are as shown in Fig. 4. From these results, crystal structure of transferred PZT thin films have preferentially oriented on a (110)/(101) plane.

Characterization of piezoelectric actuator

For demonstrating the fabricated flexible piezoelectric substrate as an actuator, we evaluated deformation behaviors of the substrate by applying varied voltage conditions. Figure 5 shows a schematic view of experimental setup for measuring deformation of the fabricated substrate by applying voltage. This time, the fabricated substrate was fixed on a metal support frame with conductive glue and thus lower electrode was electrically connected to a ground via metal support frame.

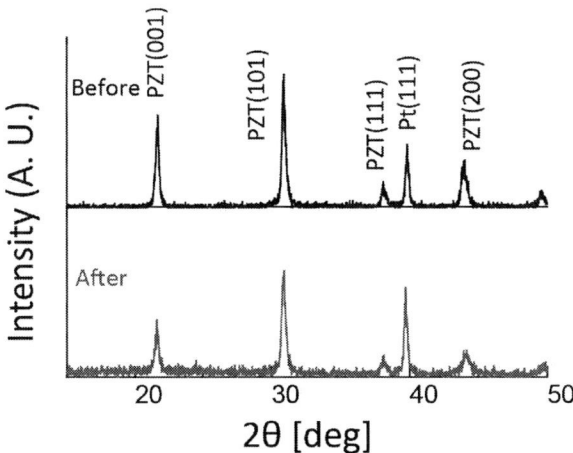

Figure 4: XRD pattern of PZT thin films before and after transfer process.

We measured the deformation at the center of the substrate by using laser interferometer as shown in Fig. 5.

First, we evaluated the relationship between the substrate deformation and polarity of applied voltage. We observed that the fabricated substrate can be symmetrically actuated when applying negative unipolar voltage as shown in fig. 6 (a). On the other hand, the substrate was asymmetrically deformed by applying bipolar voltage (fig. 6 (b)). It is assumed that the asymmetrical deformation of the substrate was occurred by polarization characteristic of PZT thin film and/or mechanical characteristic of fabricated substrate including initial shape condition.

Figure 7 shows relationship between mechanical vibration amplitude of substrate and applied voltage value. In this measurement, we applied sine-wave voltage ranging from zero to negative value at one hertz frequency. The vibration amplitude was linearly increased by increasing applied voltage. From these results, we confirmed that the fabricated substrate was deformed by piezoelectric effect.

We finally characterized a frequency response of the fabricated substrate as actuators. Figure 8 shows measurement results of frequency response curve of actuated flexible piezoelectric substrate. We confirmed that the developed substrate can be actuated by applying voltage with frequency less than 100 Hz. The mechanical vibration amplitude of the substrate was gradually decreased by increasing frequency of applied voltage. This amplitude decrease may be caused by mechanical characteristic of PDMS substrate, in addition, electrical effects due to the black PDMS layer which has a measurement resistance value less than several tens of ohms

CONCLUSIONS

We have demonstrated a simple and low-cost fabrication method of flexible piezoelectric substrate by using wet-etching based transfer technique of PZT thin film. By evaluating crystal structure and driving characteristics, we experimentally clarified the fabricated piezoelectric substrate have piezoelectric properties. This method might contribute for developing novel MEMS devices such as wearable sensors and artificial muscle of

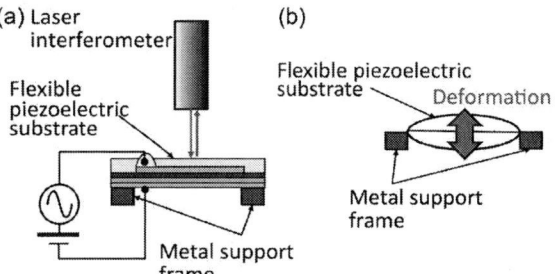

Figure 5: Demonstration of developed flexible piezoelectric substrate as an actuator. (a) Schematics of experimental setup and (b) Deformation behavior of the substrate.

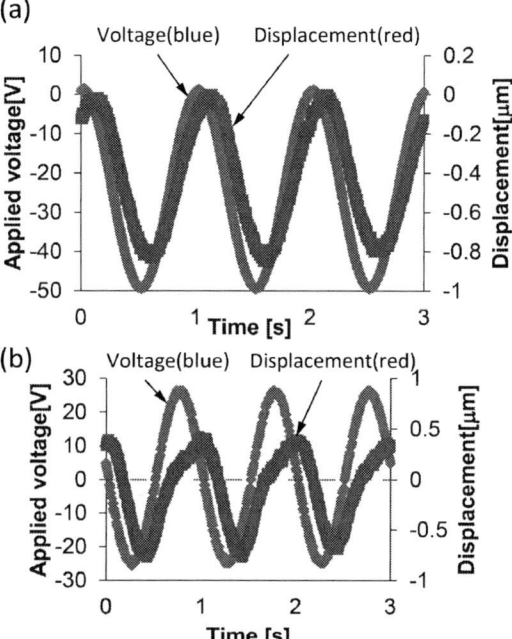

Figure 6: Measurement of substrate displacement by applying voltage at 1 Hz. Applied voltage condition: (a) negative unipolar voltage (0 to -50 V) and (b) bipolar voltage (+25 to -25 V).

high mass productivity at low cost in the future.

ACKNOWLEDGEMENTS

The authors thank Mitsubishi Material Corporation, Japan, for providing PZT sol-gel precursor.

REFERENCES

[1] M. L. Kuntzman, J. Gloria Lee, N. N. Hewa-Kasakarage, D. Kim, and N. a Hall, "Micromachined piezoelectric microphones with in-plane directivity", *Appl. Phys. Lett.*, vol. 102, no. 5, 054109 (4p), 2013.

[2] S.-G. Kim, S. Priya, and I. Kanno, "Piezoelectric MEMS for energy harvesting", *MRS Bull.*, vol. 37, no. 11, pp. 1039–1050, 2012.

[3] N. E. Dutoit, B. L. Wardle, and S.-G. Kim, "Design Considerations for Mems-Scale Piezoelectric

978-1-4799-7956-1/15 $31.00 © 2015 IEEE   379

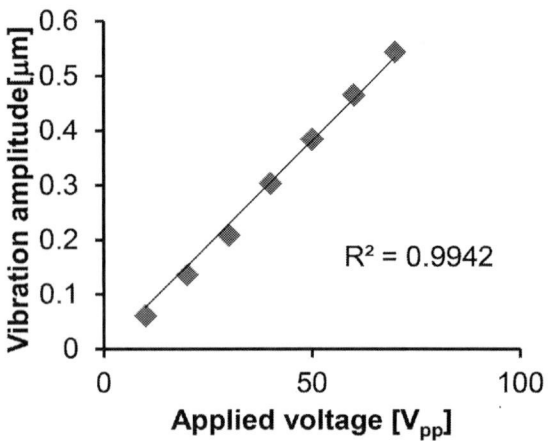

Figure 7: Relationship between the vibration amplitude of the substrate and applied voltage value (negative unipolar, 1 Hz)

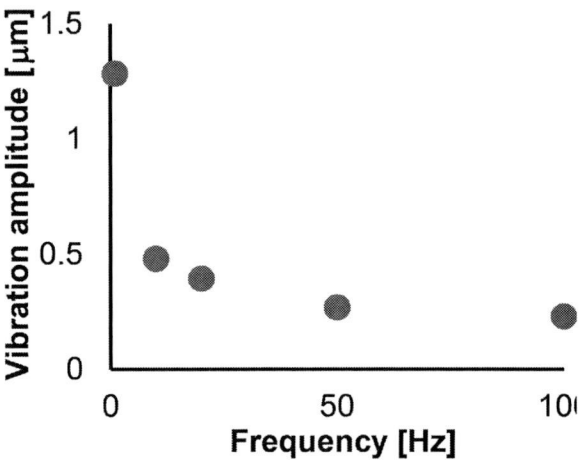

Figure 8: Frequency response curve of vibration amplitude of developed flexible piezoelectric substrate. Voltage condition: 0 to -50 V.

Mechanical Vibration Energy Harvesters", *Integr. Ferroelectr.*, vol. 71, no. 1, pp. 121–160, 2005.

[4] F. Filhol, E. Defaÿ, C. Divoux, C. Zinck, and M.-T. Delaye, "Resonant micro-mirror excited by a thin-film piezoelectric actuator for fast optical beam scanning", *Sensors Actuators A Phys.*, vol. 123–124, pp. 483–489, 2005.

[5] S. Matsushita, I. Kanno, K. Adachi, R. Yokokawa, and H. Kotera, "Metal-based piezoelectric microelectromechanical systems scanner composed of Pb(Zr, Ti)O3 thin film on titanium substrate", *Microsyst. Technol.*, vol. 18, no. 6, pp. 765–771, 2012.

[6] D. Polla, "Microelectromechanical systems based on ferroelectric thin films", *Microelectron. Eng.*, vol. 29, pp. 51–58, 1995.

[7] S. Trolier-McKinstry and P. Muralt, "Thin Film Piezoelectrics for MEMS", *J. Electroceramics*, vol. 12, no. 1/2, pp. 7–17, Jan. 2004.

[8] Bansal, R. Hergert, G. Dou, R. V. Wright, D. Bhattacharyya, P. B. Kirby, E. M. Yeatman, and a. S. Holmes, "Laser transfer of sol–gel ferroelectric thin films using an ITO release layer", *Microelectron. Eng.*, vol. 88, no. 2, pp. 145–149, 2011.

[9] R. Delmdahl, R. Pätzel, and J. Brune, "Large-Area Laser-Lift-Off Processing in Microelectronics", *Phys. Procedia*, vol. 41, pp. 241–248, 2013.

[10] H. Kozuka, A. Yamano, T. Fukui, H. Uchiyama, M. Takahashi, M. Yoki, and T. Akase, "Large area ceramic thin films on plastics: A versatile route via solution processing", *J. Appl. Phys.*, vol. 111, no. 1, p. 016106, 2012.

[11] Y. Qi, N. T. Jafferis, K. Lyons, C. M. Lee, H. Ahmad, and M. C. McAlpine, "Piezoelectric ribbons printed onto rubber for flexible energy conversion", *Nano Lett.*, vol. 10, no. 2, pp. 524–8, Feb. 2010.

CONTACT

*H. Hida, tel: +81-78-803-6058; hida@mech.kobe-u.ac.jp

MECHANICAL CHARACTERIZATION OF THIN FILMS USING A MEMS DEVICE INSIDE SEM

Changhong Cao, Brandon Chen, Tobin Filleter*, and Yu Sun*
University of Toronto, Canada

ABSTRACT

A MEMS device was developedfor mechanical characterization of 2D ultra-thin films.The device utilizes electrothermal actuators to apply uniaxial tension. The robust design makes the device capable of withstanding both dry and wet transfer of 2D ultra-thin film materials onto the suspended structures of the device. Fracture stress of thin graphene oxide (GO) films was measured.

INTRODUCTION

Two-dimensionalthin films such as graphene and graphene oxide (GO)have been shown to possess outstanding mechanical behavior[1, 2], promising applications in composites, batteries, and electronics [3-5]. For mechanical characterization, conventional tensile stagesare used to test macroscopic films (micrometers thick and above), and nanoindentationwas previouslyused to characterize mechanical properties of monolayer films [6, 7]. These existing experimental techniques are not able to characterize mesoscale films (tens of nanometers thick) whichare important for applications in energy storage [8] and electronic devices [9].

Strength characterization of multilayer thin films by indentation requires anin-direct analysis including the knowledge of interfacial properties between layers; however, such mechanisms are not currently well understood. Indentation methods also have the limitation of probing a smalllocal film area. Moreover, 2D films are in general too small in size to be tested on conventional tensile testers due to both geometric as well as force resolution limitations. Therefore, a MEMS device for tensile testing nanometer thick thin film nanomaterials is required for strength characterization.

Unlike MEMS testing of 1D nanomaterial (nanotubes and nanowires), which generally relies on nanomanipulation to place a sample onto the testing platform[10, 11], MEMS for mechanical characterization of 2D thin films requiredifferent transfer approaches. Take the transfer of graphene as an example. Graphene transfertechniquescan be classified into two major categories: dry hard contact transfer and multi-stage chemical wet etching transfer. Lee et al.[2] transferred graphene onto holey transmission electron microscopy (TEM) grids by hard pressing the graphene/PDMS onto the target substrate to form an intimate contact between graphene and the substrate. Suk et al.[12] introduced PMMA as a handle layer to protect graphene film during the transferring process, and then etched PMMA away when graphene is securely adhered to the substrate, during which the target substrate needs to be fully immersed in analcohol solution. In order to survive these transferring processes, the device must be robust.

In addition, larger forcesare also required during tension to displace meso-scaled2D films(vs. displacing 1D nanostructures) due to larger cross-sectional areas. A single nanowire/nanotube typically has a diameter of several nanometers and tens of nanometers, while 2D films can easily cover hundreds of nanometers or a few microns. The significantly larger force requirement demands the actuator beams to be much stiffer than the ones for tensile testing of 1D nanomaterials. Based on these requirements, V-beam electrothermal actuators were chosen in this work.

In order to monitor the evolution of failure of materials during tension, the MEMS device must be made SEM and/or TEM compatible. In this case, heat released from the electrothermal actuator needs to be well dissipated to minimize heat-induced drift of SEM images. High temperatures can also introduce thermal stress into the nanomaterial under test. Finally, the alignment of the two ends of the actuation shuttles where the edges of 2D films are anchored must be well controlled. A slight misalignment can cause failure of material transferand/or artifacts in measurement results. Device design should be symmetrical on the two sides of the sample under test, and hence, any unwanted stress from one side during fabrication can be counteracted to keep both actuation shuttles on the same plane.

DEVICE DESIGN, FABRICATION, AND CALIBRATION

The MEMS device consists of two symmetrical sets of V-beam electrothermal actuators, as shown in Fig. 1. The2μm gap between the two actuation shuttles enable SEM imaging in a confined region at high magnifications. Each electrothermal actuator contains eight pairs of V-beams (500μm long, 10μm wideand 10μm thick) and seven pairs of heat sink beams (100μm long, 10μm wideand 10μm thick). Two groups of heat sink beams were used to dissipate heat generated evenly and holdthe shuttle straight in the sample testing area. Features on the device were intentionally made wide and thick to tolerate the transferof different types of thin films.Finite element analysis was conducted to guide the device design.

The device was fabricated using the MicralyneMicraGEM-Si fabrication process schematically described in Fig. 2. Two SOI wafers were DRIE etched separately and then bonded using conductive adhesives. After patterning and metalizing, devices were released from the main wafer. In order to achieve potential TEM compatibility, the top surface area of the whole device was intended to be maximized. Thus, when back etching is conducted to form a through window, stress exerted by its own weight can be minimized.

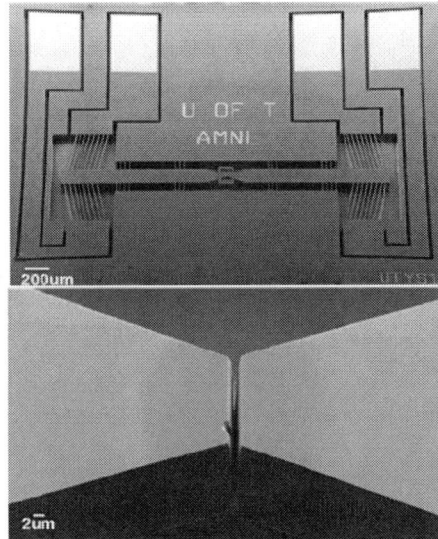

Figure 2(Top) SEM image of the MEMS device. (Bottom) Zoom-in image of actuation shuttles with a 2μm gap in between as red boxed in the top image.

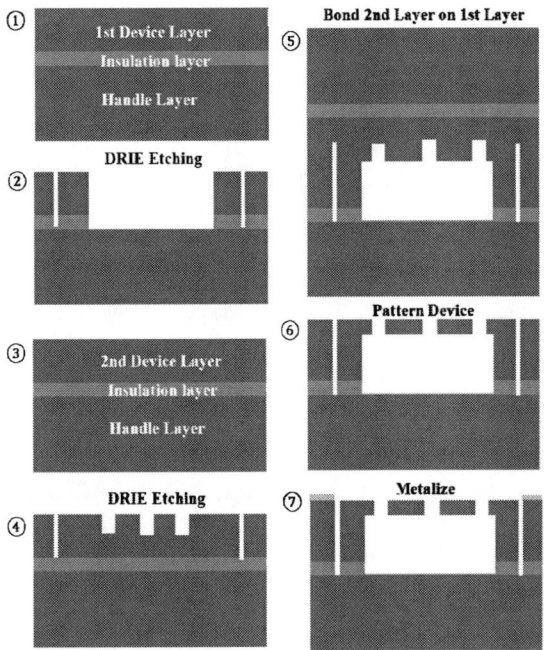

Figure 1 Fabrication process for constructing the MEMS tensile tester.

After fabrication, devices were calibrated under SEM imaging before and after graphene transfer. Transfer of graphene was used as a process to verify whether the device was able to tolerate the transfer steps. From the calibration results summarized in Fig. 3, it can be seen that the transfer processes did not cause significant influence to the performance of the device.The displacement resolution is better than 2 nm. This allows fine control of tensile strain, which is critical for testing thin films because of their low ductility. The calibration results also match well with multiphysics simulations shown as a polynomial line fit in Fig. 3. When a voltage of 4V is applied, a 25% strain can

be achieved, which is above the strain required to load thin films to failure[2, 7]. At a givenapplied voltage, the displacement difference with and without the thin film on the device is used to calculate applied force to the thin film using the calculated stiffness of the actuator based on Hooke's Law [13].

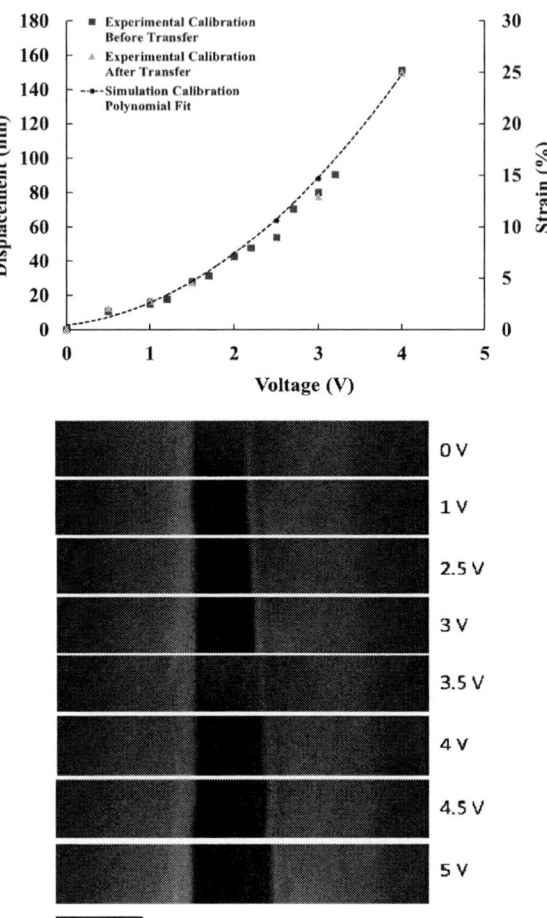

Figure 3 (Top) Calibration results. Displacement corresponds to the distance change between two edges of the actuation shuttles. (Bottom) SEM images showing the gap distance change as a function of applied voltage. (Scale Bar 1μm)

MATERIAL PREPARATION

Graphene oxide (GO) thin film was prepared by a similar method we reported previously[1]but with a larger thickness. A water solution with GO flakes , with a carbon to oxygen ration of 4 to 1 as measured by XPS, was drop casted onto the center of the two actuation shuttles, using a custom-built robotic micropipette system[14]. A water drop containing GO flakes was formed on top of the gap region. After air drying and baking at 90°C, a thin GO film was suspended over the two sides of the actuation shuttles. Film thickness for each sample tested was measured usingatomic force microscopy (AFM) in tapping mode. Fig. 4(a) shows a representative topography image of a GO filmsuspended on the MEMS device. Fig. 4(b) shows the height profile of the film across the red dash line labelled on Fig. 4(a).

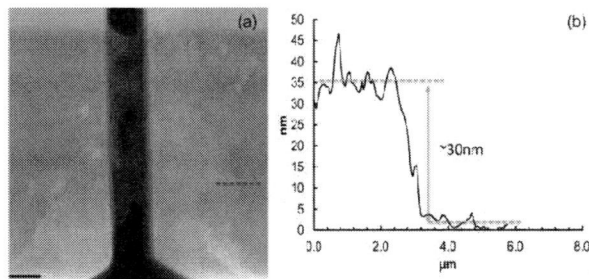

Figure 4 (a) AFM tapping mode topography scan of suspended GO suspended on MEMS device. Red dashed line represents where height profile was taken (scale bar: 5μm). (b) Height profile corresponding to the red dash line in (a). Thickness of the GO film was measured to be 30nm.

RESULTS

Fig. 5(a) shows a suspended GO film with a pre-existing crack(less than 10% of the sample length)that was tested under uniaxial tension. When actuation voltages were increased (0.5V step size), the GO film was stretched until brittle failure suddenly occurred. The crack initiated at the pre-cracked tip and subsequently propagated across the entire length of the sample. Zhang et al.[15]found similar behavior for two layer graphene films. From Fig. 5(b), tensile stress was found to increase with respect to the increase of applied voltage, and drop significantly when the crack started propagating.Stress was calculated by dividing applied force by the cross section area of the sample measured via AFM (thickness) and SEM (width).

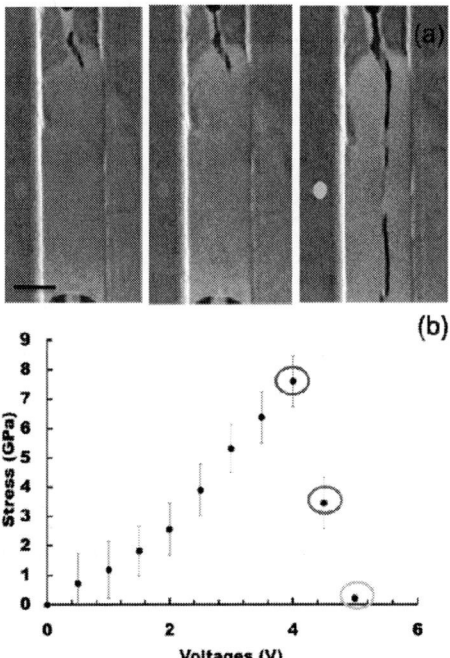

Figure 5 (a) SEM images showing crack propagation in a GO thin film during tensile testing. The arrow indicates pre-existing crack. Scale bar: 1μm. (b) Stress versus applied voltage data. Blue, red and yellow correspond to images with same color in (a).

As expected the meso-scaled thin GO films (tens of nanometer thick) were found to have a fracture stress lower than the strength previously measured for monolayer GO of 27.3GPa [1] due to the pre-existing crack, but interestingly, the fracture stress is similar to that of two layer CVD graphene of 2-9 GPa[15]. Our present work focuses on modeling to quantify the fracture behavior of these GO films.

CONCLUSION

This paper reported an electrothermal actuator-based MEMS device for 2D thin film mechanical characterization in situ SEM. The device is robust enough to tolerate dry and wet transfer of 2D thin films such as graphene and graphene oxide (GO). The MEMS device has a displacement resolution better than 2 nm, enabling fine control of tensile strain. The measured GO films had thickness of ~30 nm, and the measured fracture stress was 7-8 GPa.

ACKNOWLEDGEMENTS

Authors would like to acknowledge Canada Foundation of Innovation (CFI) and Natural Sciences and Engineering Research Council of Canada (NSERC) for funding this project and CMC Microsystems for fabrication assistance.

CONTACT

*Tobin Filleter, Tel: +1416-978-5877;
Filleter@mie.utoronto.ca;
*Yu Sun, Tel: +1416-946-0549;
Sun@mie.utoronto.ca;

REFERENCES

[1]   C. Cao, M. Daly, C. V. Singh, Y. Sun, and T. Filleter, "High strength measurement of monolayer graphene oxide," *Carbon (In Press)*.

[2]   G.-H. Lee, R. C. Cooper, S. J. An, S. Lee, A. van der Zande, N. Petrone, *et al.*, "High-Strength Chemical-Vapor–Deposited Graphene and Grain Boundaries," *Science,* vol. 340, pp. 1073-1076, 2013.

[3]   K. S. Kim, Y. Zhao, H. Jang, S. Y. Lee, J. M. Kim, J. H. Ahn, *et al.*, "Large-scale pattern growth of graphene films for stretchable transparent electrodes," *Nature,* vol. 457, pp. 706-710, Feb 5 2009.

[4]   J. K. Lee, K. B. Smith, C. M. Hayner, and H. H. Kung, "Silicon nanoparticles-graphene paper composites for Li ion battery anodes," *Chem Commun (Camb),* vol. 46, pp. 2025-7, Mar 28 2010.

[5]   G. Eda and M. Chhowalla, "Graphene-based composite thin films for electronics," *Nano Letters,* vol. 9, pp. 814-818, 2009.

[6]   C. Lee, X. D. Wei, J. W. Kysar, and J. Hone, "Measurement of the elastic properties and

intrinsic strength of monolayer graphene," *Science,* vol. 321, pp. 385-388, Jul 18 2008.

[7] D. A. Dikin, S. Stankovich, E. J. Zimney, R. D. Piner, G. H. B. Dommett, G. Evmenenko, *et al.,* "Preparation and characterization of graphene oxide paper," *Nature,* vol. 448, pp. 457-460, Jul 2007.

[8] N. Li, Z. P. Chen, W. C. Ren, F. Li, and H. M. Cheng, "Flexible graphene-based lithium ion batteries with ultrafast charge and discharge rates," *Proceedings of the National Academy of Sciences of the United States of America,* vol. 109, pp. 17360-17365, Oct 2012.

[9] S. Bae, H. Kim, Y. Lee, X. F. Xu, J. S. Park, Y. Zheng, *et al.,* "Roll-to-roll production of 30-inch graphene films for transparent electrodes," *Nature Nanotechnology,* vol. 5, pp. 574-578, Aug 2010.

[10] Y. Zhang, X. Y. Liu, C. H. Ru, Y. L. Zhang, L. X. Dong, and Y. Sun, "Piezoresistivity Characterization of Synthetic Silicon Nanowires Using a MEMS Device," *Journal of Microelectromechanical Systems,* vol. 20, pp. 959-967, Aug 2011.

[11] H. D. Espinosa, R. A. Bernal, and T. Filleter, "In Situ TEM Electromechanical Testing of Nanowires and Nanotubes," *Small,* vol. 8, pp. 3233-3252, 2012.

[12] J. W. Suk, A. Kitt, C. W. Magnuson, Y. Hao, S. Ahmed, J. An, *et al.,* "Transfer of CVD-Grown Monolayer Graphene onto Arbitrary Substrates," *ACS Nano,* vol. 5, pp. 6916-6924, 2011/09/27 2011.

[13] J. J. Brown, J. W. Suk, G. Singh, A. I. Baca, D. A. Dikin, R. S. Ruoff, *et al.,* "Microsystem for nanofiber electromechanical measurements," *Sensors and Actuators A: Physical,* vol. 155, pp. 1-7, 10// 2009.

[14] E. Shojaei-Baghini, Y. Zheng, and Y. Sun, "Automated micropipette aspiration of single cells," *Ann Biomed Eng,* vol. 41, pp. 1208-16, Jun 2013.

[15] P. Zhang, L. Ma, F. Fan, Z. Zeng, C. Peng, P. E. Loya, *et al.,* "Fracture toughness of graphene," *Nat Commun,* vol. 5, 04/29/online 2014.

ROOM TEMPERATURE SYNTHESIS OF SILICON DIOXIDE THIN FILMS FOR MEMS AND SILICON SURFACE TEXTURING

Akarapu Ashok, and Prem Pal

Department of Physics, Indian Institute of Technology Hyderabad, Telangana, India

ABSTRACT

In the present work, the room temperature deposited silicon dioxide thin films are explored for the fabrication of microelectromechanical systems (MEMS) components and the surface texturing for crystalline silicon solar cell applications. The etch rates of as-grown oxide films are investigated in different concentration tetramethyl-ammonium hydroxide (TMAH) and potassium hydroxide (KOH) solutions at different temperatures. In 25 wt% TAMH, the as-grown oxide is demonstrated as structural and masking layers for the fabrication of various kinds of MEMS components. Furthermore, the as-grown oxide is exploited as etch mask in KOH to texturize silicon wafer surface without using lithography.

INTRODUCTION

Numerous applications of SiO_2 thin films in several fields have stimulated extensive research on its synthesis and characterization. In MEMS, SiO_2 is commonly used as etch mask, sacrificial and structural layers. The key advantage of SiO_2 as structural material for the fabrication of freestanding structures is its smaller Young's modulus compared to silicon and silicon nitride [1, 2]. In addition, SiO_2 is one of the widely used etch mask materials in wet anisotropic etching using alkaline etchants [3]. Besides wet anisotropic etching for silicon bulk micromachining, alkaline solutions are commonly employed in surface texturing of crystalline silicon (c-Si) to reduce the light reflectance, enhance light trapping, and thereby improve conversion efficiency of silicon based solar cells [4, 5].

In various deposition techniques, thermal oxidation and chemical vapor deposition (CVD) methods are most widely used for the synthesis of oxide layer on silicon substrate. However, these processes involve high thermal budget, heavy equipment, expensive and toxic gases, etc. Among several techniques, anodic oxidation of silicon is a method which can be performed even below room temperature. Besides, it offers several advantages such as simple experimental set-up, low cost, ease of modifying the oxide properties, non-involvement of toxic/expensive gases, etc. [6, 7]. Moreover, the oxide properties are easily tailored by varying the electrolyte composition and/or the growth parameters.

In the present work, SiO_2 thin films are synthesized using anodic oxidation technique at room temperature. The etch rates of as-grown oxide films are investigated in different concentrations of TMAH and KOH to explore them as etch mask and structural layer for MEMS and to texturize the silicon surface for solar cell applications.

EXPERIMENTAL DETAILS

Czochralski (Cz) grown one side polished three inch P-type Si{100} wafers (resistivity 1-10 Ω cm) are used for the deposition of SiO_2 thin films using anodic oxidation method. Aluminum is deposited on the rough surface side of the wafer using DC sputtering for ohmic contact. A two-electrode electrochemical set-up as schematically shown in Fig. 1 is employed for oxide deposition. The wafer is fixed in a customarily designed wafer holder which provides gold contact on its back-side. In the experimental set-up, the silicon wafer is fixed as anode and the platinum gauge mesh (90% Pt, 10% Ir) as cathode. The electrodes are separated by fixed distance of 1.5 cm. KNO_3 of 0.04 M and a fixed volume percentage H_2O are added in the ethylene glycol solvent. Here, KNO_3 serves as electrolyte in the solvent. The *pH* of the solution is maintained at 4.

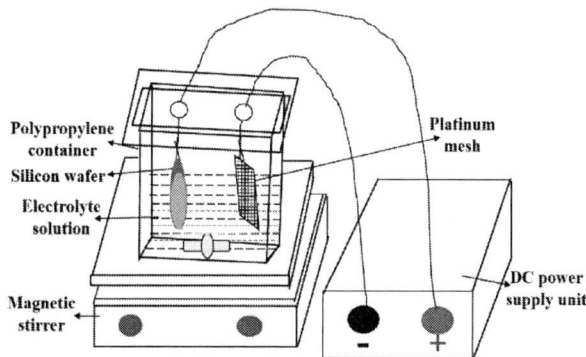

Figure 1: Schematic diagram of experimental setup for the anodic oxidation of silicon.

In the present work, oxide is deposited using potentiodynamic regime in which the oxide growth is started in constant current mode at 8 mA/cm^2 and the oxidation is continued until the forming voltage reaches the predetermined voltage. Thereafter, the process is continued in constant voltage (i.e. potentiostatic) mode for 15 minutes. Prior to oxide growth, the silicon wafers are cleaned sequentially in acetone and deionized water (DI) for 5 minutes by ultrasonic cleaning method. In order to remove the native oxide layer before oxidation the samples are given a 2% hydrofluoric acid (HF) dip followed by thorough rinse in DI water. After oxidation, the samples are thoroughly cleaned in DI water to get rid of the adsorbed glycol solvents. The etch rates of as-grown SiO_2 films are determined in different concentrations of TMAH and KOH at 60, 68 and 76 °C using a constant temperature etch bath. The desired TMAH concentration is prepared by diluting the commercially available 25 wt% TMAH (99.999%, Alfa Aesar) solution using DI water. KOH pellets (99.99%, Alfa Aesar) are dissolved in DI water to prepare the wanted concentration of KOH solution. Chip size samples diced from the full size oxidized wafer are used for etch rate study. Etching is performed in a Teflon made container equipped with reflux condenser to prevent the evaporation of etchant in order to maintain the constant concentration of the etchant solution during etching. Samples are held in a PFA made chip holder which can accommodate several samples at a time. All experiments are carried out at constant temperature with an accuracy of ±1 °C. Ellipsometry (J.A. Woolam, model: M-2000D) is employed to determine the thickness of as-grown oxides. Scanning electron microscope (SEM) (Zeiss, model: SUPRA40) is used to observe the surface morphology of etched samples and the fabricated MEMS structures. Atomic force microscopy (AFM) (Bruker, model: DIMENSION icon with Scan Asyst) is used to examine the surface topography of the textured silicon samples.

RESULTS AND DISCUSSION

Ellipsometric measurements: Variable angle spectroscopic ellipsometry is used to measure the thickness of the as-grown oxide films. The thickness of oxide films is measured at 65°, 70° and 75° angle of incidence. Fig. 2 shows the effect of forming voltage on oxide thickness developed at constant current density of 8 mA/cm^2 in the electrolyte containing 2.7 vol% H_2O.

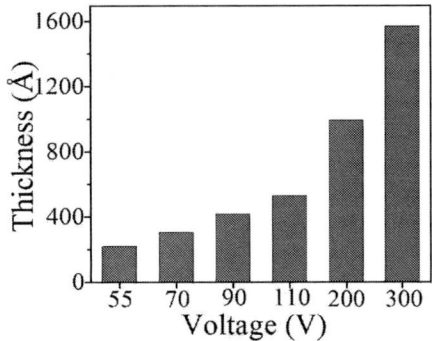

Figure 2: Effect of forming voltage on oxide thickness for the films deposited in the electrolyte containing 2.7 vol% H_2O at 8 mA/cm^2.

It can be observed from Fig. 2 that the oxide thickness increases with the forming voltage, which indicates the strong dependence of oxide thickness on forming voltage. In the present study, maximum thickness of 156 nm at 300 V and minimum thickness of 22 nm at 55 V are obtained. The thickness of the oxide developed in the electrolyte containing 3.7 vol% H_2O at 110 V is measured to be 53 nm. The same trend is observed for the films deposited in the electrolyte containing 2.7 vol% H_2O. Moreover, it is noticed that the film thickness and growth rate are almost same for both water concentrations (i.e. 2.7 and 3.7 vol%) used in the present work.

Etch rate study in TMAH: In alkaline solutions, TMAH provides very high Si/SiO$_2$ etch selectivity and therefore an appropriate choice for the fabrication of oxide microstructures as well as the rectangular cavities using oxide as mask layer [8]. In order to exploit the anodically grown oxide film as etch mask and structural material for MEMS, the etch rate of the oxide deposited in the electrolyte containing 2.7 vol% H_2O at 300 V is investigated in 5 wt% and 25 wt% TMAH at 60, 68 and

76 °C. Fig. 3 presents the effect of TMAH concentration and temperature on the etch rate of oxide. It can be noticed from Fig. 3 that the oxide etch rate increases with temperature and decreases with increase in TMAH concentration. As 25 wt% TMAH exhibits lower oxide etch rate, it is a preferable choice for the fabrication of suspended microstructures and the rectangular cavities using as-grown oxide as structural and etch mask layers, respectively. Moreover, 25 wt% TMAH provides high undercutting at the convex corners of the mask patterns in comparison to low concentration TMAH solution, which is highly desirable for the fast release of suspended structures [3].

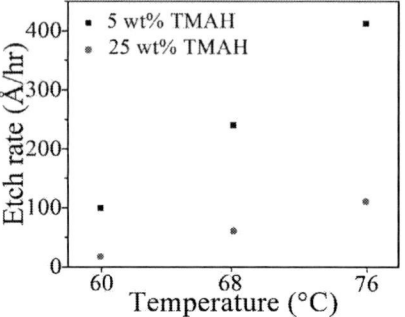

Figure 3: Effect of TMAH concentration and temperature on oxide etch rate.

Fabrication of MEMS structures: The as-grown anodic oxide films on silicon wafers are patterned using photolithography process in which the mask edges are aligned parallel to <110> in order to avoid any undercutting due to misalignment [9]. The primary flat of the wafer is used as reference for the alignment of mask edges along <110> direction. As discussed in previous section, 25 wt% TMAH provides high undercutting and low oxide etch rate, it is employed as an etchant for the fabrication of freestanding structures and rectangular cavities using as-grown anodic oxide as structural and mask layers, respectively. In both cases, etching is done at 60 °C. Fig. 4 presents the SEM images of fabricated overhanging structures. Different size cavities fabricated in {100} silicon wafer using anodic oxide as mask are shown in Fig. 5. It can be easily noticed in Fig. 5 that the top silicon surface is smooth and pits free, indicating that the silicon surface is well protected by the as-grown oxide during etching. The successful fabrication of different types of structures (i.e. freestanding and fixed) confirms that the anodic oxide grown at room temperature can be employed as structural as well as etch mask layer in TMAH solution.

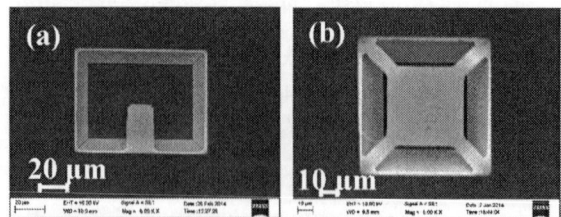

Figure 4: Suspended structures (a) cantilever and (b) diaphragm of anodic SiO$_2$ fabricated in 25 wt% TAMH at 60 °C.

978-1-4799-7956-1/15 $31.00 © 2015 IEEE

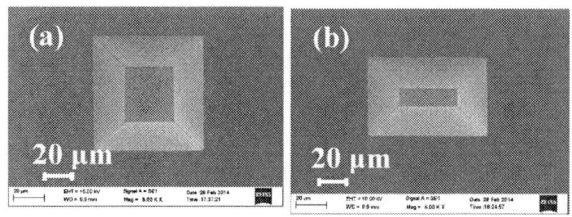

Figure 5: Different size cavities in Si{100} fabricated in 25 wt% TAMH at 60 °C by employing anodic SiO₂ as etch mask. SEM has been taken after oxide removal.

Etch rate study in KOH: KOH is preferred over TMAH when high anisotropy between {111} and {100} is required [10]. In this paper, KOH is explored to investigate the feasibility of as-grown anodic SiO₂ as etch mask. The etch rates of as-grown oxides are measured in different concentration KOH solutions at 60, 68 and 76 °C and are presented in Fig. 6. It can be seen that the etch rate increases with increase in KOH concentration as well as temperature. These results suggest that only low concentration KOH is a suitable choice for employing as-grown oxide as mask. On comparison of results presented in Figs. 3 and 6, it can be concluded that the etch rate of as-grown oxide in KOH is much higher than that in TMAH.

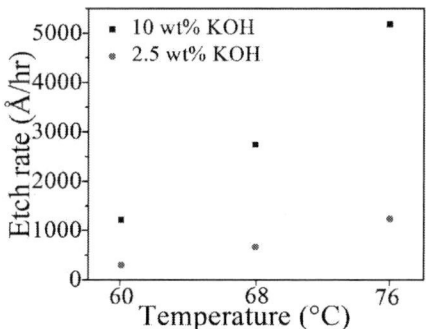

Figure 6: Effect of KOH concentration and temperature on oxide etch rate.

Surface texturing of Si{100}: In order to reduce the light reflectance to improve the efficiency, surface texturing is one of the common practices in silicon based solar cells. The texturing of silicon surface is realized by forming either random upright pyramids or inverted pyramids using alkaline etchants such as KOH, TMAH, etc. [4, 5, 11]. The surface textured with inverted pyramids yields high efficiency compared to the surface textured with normal pyramids [11, 12]. However, the involvement of lithography process in the formation of inverted pyramids is a major drawback as it adds extra cost to solar cell production. Therefore, the lithography free processes are to be investigated for the texturization of the surface with inverted pyramids. In this work, in order to exploit the as-grown oxide film for surface texturing of silicon surface without using lithography, silicon samples with 53 nm thick oxide deposited in 3.7 vol% H₂O added electrolyte are etched in KOH solution.

In order to investigate the effect of KOH concentration on surface texturing, etching is performed in 2.5 wt% and 10 wt% KOH at 60 °C. In each concentration, etching time is optimized for the maximum surface coverage of inverted pyramids. It is done by etching the large number of samples under same condition for different periods of time, ranging from 1 to 45 minutes in one-minute intervals. The maximum coverage of inverted pyramids is observed before all etch pits coalesces. Figs. 7a and 7b exhibit the surface morphology of samples etched in 2.5 wt% and 10 wt% KOH at 60 °C for the optimized etching times, respectively.

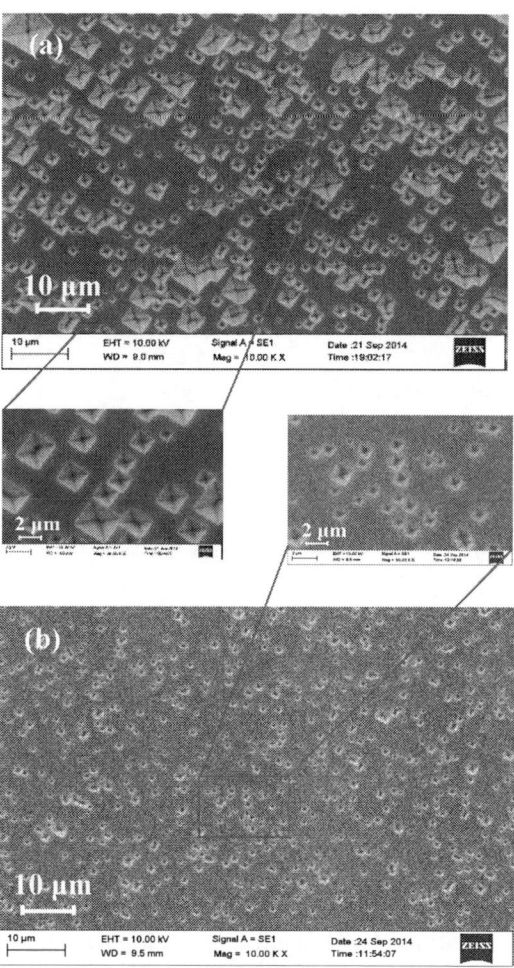

Figure 7: Surface morphology of Si{100} after etching in (a) 2.5 wt% KOH for 39 min (b) 10 wt% KOH for 20 min at 60 °C using anodic oxide as etch mask.

It is easily visible in Fig. 7 that the size of the pits decreases as the KOH concentration increases. The reason is that the etch rate of the oxide increases with the increase of KOH concentration as presented in Fig. 6. Therefore, the sample in 10 wt% KOH is exposed for less time in comparison to 2.5 wt% KOH, resulting in smaller size cavities/pyramids. It can be seen in SEM pictures that the size of pyramids are not identical. Moreover they are randomly distributed. It happens because the pinholes in oxide layer during etching, which are responsible for the formation of cavities, develops randomly and increases with time.

978-1-4799-7956-1/15 $31.00 © 2015 IEEE   387

Atomic force microscopy (AFM) in tapping mode is used to examine the surface topography of the samples. Figs. 8a and 8b depict the 2D AFM images along with the cross sectional profiles of the samples etched in 2.5 wt% and 10 wt% KOH, respectively. Similar to SEM results presented in Fig. 7, AFM images confirm that the size of inverted pyramids decreases with increase in KOH concentration. The size of the inverted pyramid ranges from 1 to 6 µm and depth from 200 to 600 nm for the sample etched in 2.5 wt% KOH for 39 minutes, whereas for the sample etched in 10 wt% KOH, the width and depth vary from 1 to 3 µm and 200 to 500 nm, respectively. It can be understood from this basic study that the low concentration KOH is preferable for the better surface coverage of the large size inverted pyramids. There is still more room to improve the coverage of inverted pyramids by optimizing the oxide properties (e.g. porosity), KOH concentration and etching temperature.

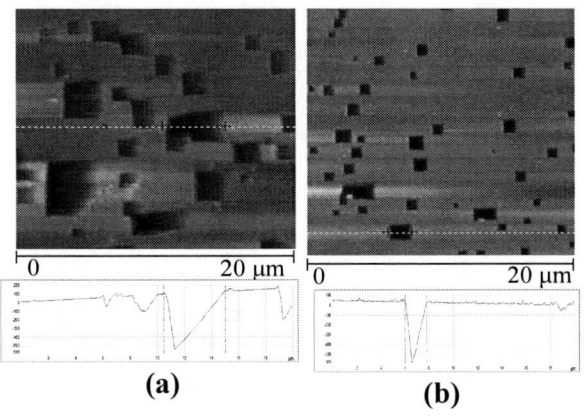

(a) **(b)**

Figure 8: AFM image of Si{100} after etching in (a) 2.5 wt% KOH for 39 min (b)10 wt% KOH for 20 min at 60 °C using anodic oxide as etch mask.

CONCLUSIONS

Silicon dioxide thin films are developed using anodic oxidation technique at room temperature and explored for MEMS as well as surface texturing of crystalline {100} silicon surface. The as-grown oxide layer is successfully demonstrated as structural and masking layers for the realization of various shapes overhanging structures and the cavities of different sizes in Si{100} using bulk micromachining in 25 wt% TMAH solution. High concentration TMAH (i.e. 25 wt%) and low concentration KOH (i.e. 2.5 wt%) are found most appropriate to use as-grown anodic oxide for MEMS and surface texturing applications, respectively. The surface texturing of silicon demonstrated in this paper is an economical process as it does not involve lithography and the oxide is deposited at room temperature using a simple technique. In solar cell fabrication, texturized surfaces are employed to reduce the light reflectance to increase the efficiency and the process developed in this paper can be adapted for the same purpose.

REFERENCES

[1] Y. Tang, J. Fang, X. Yan, H-F. Ji, "Fabrication and characterization of SiO₂ microcantilever for microsensor application", *Sens. Actuators B*, vol. 97, pp. 109–113, 2004.

[2] P. Li, X. Li, G. Zuo, J. Liu, Y. Wang, M. Liu, D. Jin, "Silicon dioxide microcantilever with piezoresistive element integrated for portable ultraresoluble gaseous detection", *Appl. Phys. Lett.*, vol. 89, pp. 074104 (3pp), 2006.

[3] P. Pal, K. Sato, "Fabrication methods based on wet etching process for the realization of silicon MEMS structures with new shapes," *Microsystem Technologies*, vol. 16, pp. 1165-1174, 2010.

[4] A. K. Chu, J. S. Wang, Z. Y. Tsai, C. K. Lee, "A simple and cost-effective approach for fabricating pyramids on crystalline silicon wafers", *Solar Energy Materials & Solar Cells*, vol. 93, pp. 1276–1280, 2009.

[5] D. Iencinella, E. Centurioni, R. Rizzoli, F. Zignani, "An optimized texturing process for silicon solar cell substrates using TMAH", *Solar Energy Materials & Solar Cells*, vol. 87, pp. 725–732, 2005.

[6] A. Ashok, P. Pal, "Growth and etch rate study of low temperature anodic silicon dioxide thin films", *The Scientific World Journal*, vol. 2014, pp. 106029 (9pp), 2014.

[7] G. Mende, H. Flietner, M. Deutscher, "Optimization of anodic silicon dioxide films for low temperature passivation of silicon surfaces", *J. Electrochem. Soc.*, vol. 140, pp. 188-194, 1993.

[8] P. Pal, K. Sato, M. A. Gosalvez, B. Tang, H. Hida, M. Shikida, "Fabrication of novel microstructures based on orientation dependent adsorption of surfactant molecules in TMAH solution," *J. Micromech. Microeng.*, vol. 21, pp. 015008 (11pp), 2011.

[9] W. H. Chang, Y. C. Huang, "A new pre-etching pattern to determine <110> crystallographic orientation on both (100) and (110) silicon wafers", *Microsystem technologies*, vol. 11, pp. 117-128, 2005.

[10] M. A. Gosalvez, P. Pal, N. Ferrando, H. Hida, K. Sato, "Experimental procurement of the complete 3D etch rate distribution of Si in anisotropic etchants based on vertically micromachined wagon wheel samples", *J. Micromech. Microeng.*, vol. 21, pp. 125007 (14pp), 2011.

[11] J. Kim, D. Inns, K. Fogel, D. K. Sadana, "Surface texturing of single-crystalline silicon solar cells using low density SiO₂ films as an anisotropic etch mask", *Solar Energy Materials & Solar Cells*, vol. 94, pp. 2091–2093, 2010.

[12] M. Morenon, D. Daineka, P. Roca i Cabarrocas, "Plasma texturing for silicon solar cells: from pyramids to inverted pyramids-like structures", *Solar Energy Materials & Solar Cells*, vol. 94, pp. 733–737, 2010.

CONTACT

*P. Pal, prem@iith.ac.in

SIZE EFFECT ON BRITTLE-DUCTILE TRANSITION TEMPERATURE OF SILICON BY MEANS OF TENSILE TESTING

A. Uesugi[*], Y. Hirai, T. Tsuchiya, and O. Tabata
Department of Micro Engineering, Kyoto University, Kyoto, JAPAN

ABSTRACT

This paper reports the size effect on brittle-ductile transition temperature (BDTT) of single crystal silicon (SCS) using different width specimens (120-μm long, 5-μm thick and 4 or 9-μm wide). The BDTT was characterized using tensile testing in a vacuum from room temperature to 600 °C. The fractured specimens showed that the slips were occurred at 500 °C and above, which indicated that the BDTT of micrometer-sized silicon decreased compared to millimeter-sized structures. By comparing the temperature ranges of the slip occurrences among the different sized specimens including other researchers' reports, we found that the length along slips-propagation direction, i.e. thickness, might dominate the BDTT.

INTRODUCTION

Single crystal silicon (SCS) is one of the standard materials in MEMS devices, which need mechanical deformation of structural materials for their operations, so evaluation of its mechanical property with high accuracy is important for the reliability improvement.

Understanding of high-temperature mechanical property is necessary for the reliability of MEMS in harsh environments, e.g. engine room sensors, gas sensors, and aerospace sensors. The temperature effect causes various mechanical property changes; decrease in the elastic constants at high temperature [1], change in fracture modes and strengths [2], and transition of fracture behaviors from brittle to ductile [3][4][5]. In this research, brittle ductile transition (BDT) was investigated as an important factor for the MEMS reliability which may bring drift on output caused by plastic deformations.

The BDT temperature (BDTT) of SCS bulk structure was reported as about 600 °C. Recently, the decrease in BDTT of nano- or micro-scale specimens has been reported; the smaller structure has the lower BDTT [3][5][6]. However, due to difficulties in measurements, the variety of measured structure sizes has been insufficient to investigate the size effect on BDTT in detail. Moreover, structures of few micrometers wide and thick which are standard dimensions in MEMS, has never been reported.

For further investigation of the size effect, a high-temperature tensile testing method adaptable to wide range of specimen dimensions is required. In this paper, we report development of a high-temperature tensile testing machine, its improvements for better measurement accuracy at high temperature, and testing results of two different-width <110> SCS specimens. The discussion is focused on observation of failure behavior in the relation to testing temperature, especially on the size effect on BDTT by observing slips.

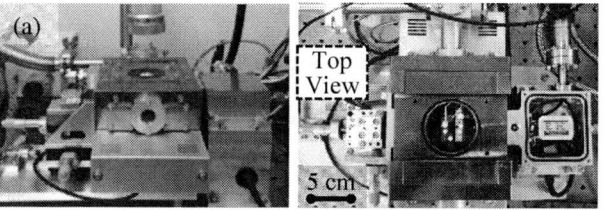

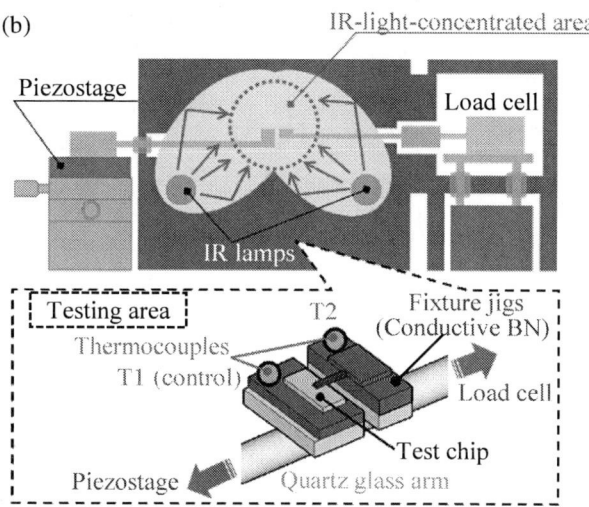

Figure 1: High-temperature tensile testing machine in a vacuum. (a) Outlook of machine. (b)Cut-view of machine using electrostatic-force grip and IR heating.

EXPERIMENTS

High Temperature Testing Machine

The mechanical properties of SCS microstructure at high-temperature was investigated using tensile testing with an electrostatic-force grip and an IR light heating [1]. Figure 1 shows the testing machine in a vacuum which suppresses oxidation of specimens. In the vacuum chamber, lights from two IR lamps are concentrated at the testing area. Since the fixture jigs at testing area, shown in fig.1b, are supported with quartz arms, heat conduction from the testing area to the load cell and the piezostage is suppressed.

The specimen chucking method helps to surmount difficulties in high-temperature measurements. On chucking a specimen, a DC voltage is applied between the specimen and the chucking probe to generate electrostatic force. The fixture jigs as well as the test chips and the probe in the testing area are small, so due to the small heat capacity, it is possible to heat the testing area quickly. Another feature is that the specimen and the probe can be aligned after the testing area heated to a testing temperature. Using the electrostatic grip, the test parts of specimens are not subjected to bending or compression forces caused by thermal expansion of equipment.

Temperature of the specimen is monitored and

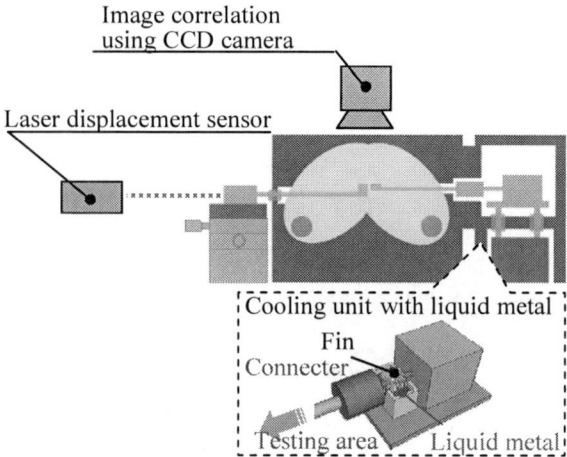

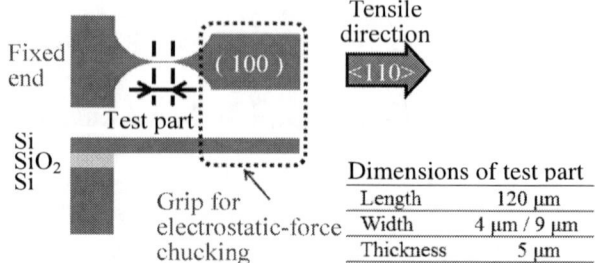

Figure 3: Specimens fabricated on (100) SOI wafer with <110> tensile direction.

Figure 2: Set-up improvements. Stage displacement was monitored using laser displacement sensor and heat conduction to load-cell was suppressed using liquid metal.

controlled by using two thermocouples on the both-sides fixture jigs. The temperature drift of the controlled specimen side (T1) is about 1 °C at 100 °C and about ±0.1 °C at higher than 200 °C. The temperature uniformity of the testing area was confirmed by measuring temperature difference (T1-T2) between the two sides, which was smaller than 20 °C at 600 °C.

Measurement Stability Improvements

For analysis of plastic or ductile failures from measured stress-stage displacement curves, higher measurement accuracy is required. Since measurement accuracy problems were found from the measured stress-stage displacement curves in the initial testing at high temperature (as shown in Figure 4), the measurement set-up was improved as shown in Figure 2.

One of the problems was the fluctuation of measured stress at high temperature. The stress fluctuation occurred even after specimen fracture, which indicates that this problem would be caused by the load cell temperature drift. In order to suppress temperature rise of the load cell, a liquid-metal cooling unit was inserted at the loading axis. The cooling unit provides a bypath of heat conduction from the axis to the load cell basement whose temperature is kept near room temperature (RT) by a water-cooling of the vacuum chamber. With testing temperature of 600 °C, the temperature at the cooling unit decreased from 60 °C to 35 °C. The improvement realized force measurement with negligible small fluctuation, as shown in Figure 6.

The other problem was low credibility of stage position measurement; as shown in Figure 4, the measured curves had large temperature dependence in the slope at their initial stages. The dependence was much larger than that calculated from the theoretical stiffness changes of the equipment and the specimen which was estimated in several percent from RT to 600 °C. In order to check temperature effect on measured displacement by the embedded displacement sensor, the piezostage displacement was measured using both a laser displacement sensor and image correlation using CCD camera. Both are not affected by temperature at the testing area. At 600 °C, as a result, the stage position measured by

the stage was larger by 15 %, compared with both the other two methods, while there was no difference in conducted three methods at RT. Adopting the laser displacement sensor, the measured slope change was reasonable in stress-stage displacement curves, as shown in Figure 6.

Specimen

The specimens were fabricated on (100) SOI wafers with tensile direction for <110>. Dimensions of the test part are 120 μm in length, 5 μm in thickness, 4 μm or 9 μm in width. The two different-width specimens were prepared for investigation of size effect on BDTT.

RESULTS
4-μm-wide specimens

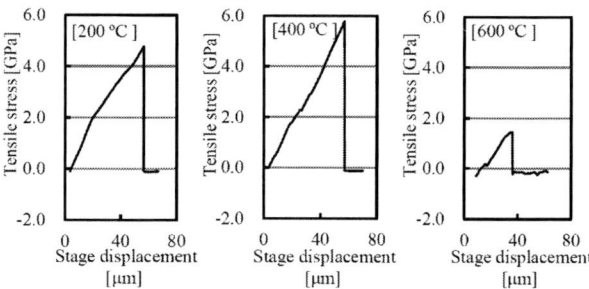

Figure 4: Tensile stress-stage displacement curves of 4-um -wide specimens at 200 °C, 400 °C, and 600 °C.

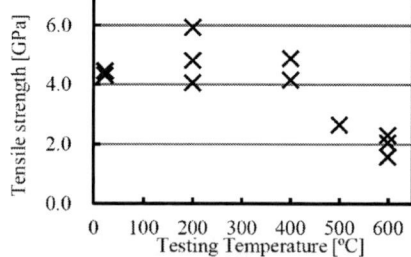

Figure 5: Tensile strength as a function of temperature.

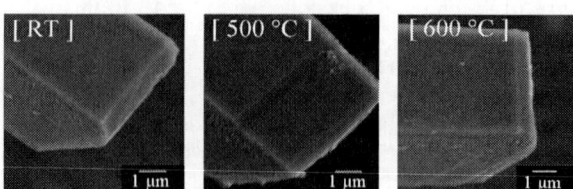

Figure 6: 4-μm-wide specimens fractured at RT, 500 °C, and 600 °C. Specimens tested at 500 °C and 600 °C had slips.

The tensile testing of two types of specimens was conducted at the temperature range from RT to 600 °C in a vacuum. Figure 4 shows the stress-stage displacement curves of 4-μm-wide specimens. Since the tests were conducted before the set-up improvements, the measured curves had the fluctuations. However, this fluctuation did not affect tensile strength evaluation, since it was small and slow compared to the stress change at fracture. Figure 5 shows the tensile strength as a function of the testing temperature. The tensile strength showed significant decrease by more than 50 % over 400 °C.

The decrease is thought to be caused by slips. The fractured specimens were observed using scanning electron microscope (SEM), as shown in Figure 6. Only specimens tested at 500 °C and 600 °C showed slips on the surfaces. The results ensured that BDTT of the micrometer-sized beam specimens is below 500 °C. The temperature range of the slips occurrences seems to correspond to that of strength decrease. Moreover, coexistence of fracture surfaces and dislocation slips indicates a possibility that the steps of dislocation slips caused stress concentration or cross-sectional area reduction to cause the decrease of nominal strength over 400 °C.

9-μm-wide specimens

9-μm-wide specimens were tested after the set-up improvement. Figure 7 shows measured stress-stage displacement curves. Compared with the results of 4-μm-wide specimens, the obtained curves had small fluctuation. The slopes of the curves at 500 °C and 600 °C decrease with tensile stress increase, while slips occurrences could not be distinguished from the curves.

Specimens with 9-μm width also showed the slips on (111) crystal planes at 500 °C and 600 °C, as shown in Figures 8 to 10. Fracture shapes difference was also observed depending on temperature. From RT to 400 °C, similar fracture surfaces were observed. The fractures started at the sidewall damages and propagated mainly on (111) crystal planes. On the other hand, fracture mainly on (110) crystal plane was observed at 500 °C. Furthermore, the fracture surfaces at 600 °C consisted of a cluster of facets. Severe necking was also observed on the specimens fractured at 600 °C, which indicates that the fracture was affected more by ductile behavior.

The fracture surfaces showed the temperature effect on fracture propagation; change from fracture on (111) crystal planes to fractures on (110) crystal planes between 400 °C and 500 °C. The change seems different from reported fracture mode change [2]; change from fracture on (110) crystal planes to fractures on (111) crystal planes between 60 °C and 80 °C. This deference might be caused by difference in surface roughness of specimens. Since brittle fracture is affected by shape and location of fracture origin, the rough sidewall-surface in this research caused different fracture shapes. In addition, the surface steps caused by slips at high temperature might also contribute to the fracture mode change.

DISCUSSIONS

The size effect on BDTT, focusing on slips

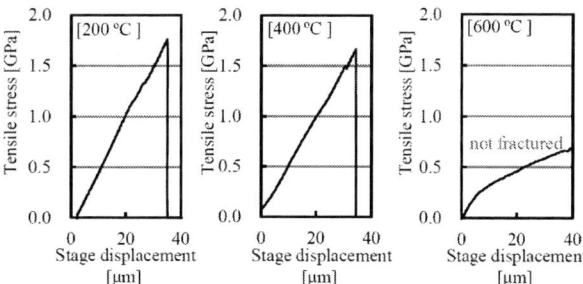

Figure 7: Tensile stress-stage displacement curves of 9-μm wide specimens at 200 °C, 400 °C, and 600 °C.

Figure 8: 9-μm-wide specimens fractured at RT, 200 °C, 400 °C. All specimens had fracture surfaces on (111) starting at their sidewall damage.

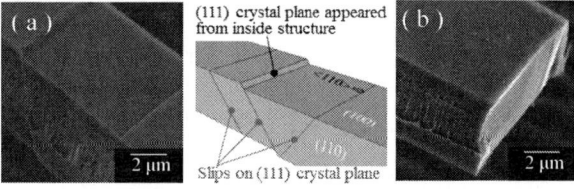

Figure 9: 9-μm-wide specimens tested at 500 °C. (a) Slips on (111) of un-fractured specimen. (b) Specimen with fractured on (110).

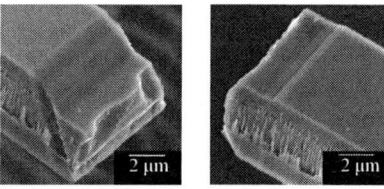

Figure 10: Paired ends of fractured 9-μm-wide specimen at 600 °C with slips and necking deformation.

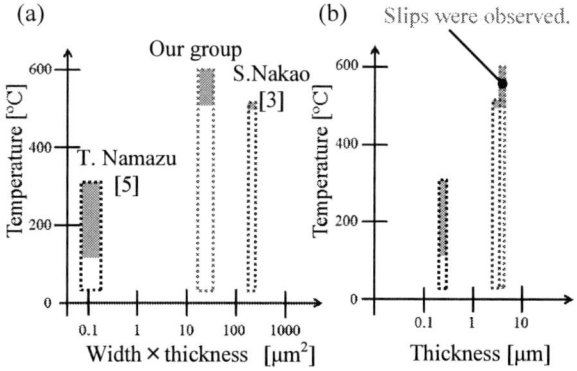

Figure 11: Size effect on BDTT. BDTT were plotted (a) against cross-sectional area and (b) against thickness. Dashed frames indicated specimen dimension and temperature range. Orange-colored indicated occurrence of surface slips lines.

occurrences temperature is discussed here, by comparing these results with the reported results where similar surface slips on (111) crystal planes were observed on micrometer-sized structures at 500 °C by *Nakao et al.* [3] and on sub-micrometer–sized structures above 100 °C by *Namazu et al.* [5]. Since the measured specimen shape in each research is different, the temperature range of slips occurrences were compared on the cross sectional area and on thickness of specimen, as shown in Figure 11.

The cross sectional area of this report (ca. 20 ~ 45 μm^2) is in between two previous reports (ca. 0.07 ~ 0.16 μm^2 and ca. 200 μm^2, respectively). However, the lowest temperature of slip occurrence of this research is same as that of the larger structure, and that of the sub–micrometer -sized structures decreased significantly. Considering the BDTT of bulk structures: 600 °C, the size effect based on cross sectional area seems not consistent. On the other hand, the thicknesses of the micrometer-sized structures are similar (4 μm and 5 μm, respectively), which corresponds to the similarity of the lowest temperature of slip occurrence.

From the above comparisons, thickness would describe well the size effect on the BDTT. We thought that the length along slips-propagation direction might dominate the BDTT. In order to confirm the hypothesis, we plan to measure smaller structures with different crystal orientations to change length along slips-propagation.

ACKNOWLEDGEMENTS

This work was supported by Grant-in-Aid for JSPS Fellows. A part of this work was supported by Kyoto University Nano Technology Hub in "Nanotechnology Platform Project" sponsored by the Ministry of Education, Culture, Sports, Science and Technology (MEXT), Japan.

REFERENCES

[1] T. Tsuchiya, T. Ikeda, A. Tsunematsu, K. Sugano and O. Tabata, Sensors and Materials, 22, 1 (2010), pp. 1-10.

[2] S. Nakao, T. Ando, M. Shikida, and K. Sato, J. Micromech. Microeng., 18 (2008), pp.1-7.

[3] S. Nakao, T. Ando, M. Shikida, K. Sato, J. Micromech. and Microeng., 16 (2006), pp. 715-720.

[4] T. A. Taylor and C. R. Barrett, Mater. Sci., 10 (1972), pp.93-102.

[5] T. Namazu, Y. Isono and T. Tanaka, J. Microelectromech. Syst., 11, 2 (2002), pp. 125-135.

[6] X. Han, K. Zheng, Y. Zhang, X. Zhang, Z. Zhang and Z. Wang, Adv. Mater., 19, 16 (2007), pp. 2112-2118.

CONTACT

* A. Uesugi, tel: +81-75-383-3693;
a_uesugi@nms.me.kyoto-u.ac.jp

STICTION FORCES AND REDUCTION BY DYNAMIC CONTACT IN ULTRA-CLEAN ENCAPSULATED MEMS DEVICES

D.B. Heinz[1], V.A. Hong[1], T.S. Kimbrell[1], J. Stehle[2], C.H. Ahn[1], E.J. Ng[1],
Y. Yang[1], G. Yama[2], G.J. O'Brien[2] and T.W. Kenny[1]
[1]Department of Mechanical Engineering, Stanford University, USA
[2]Robert Bosch RTC, Palo Alto, CA, USA

ABSTRACT

We demonstrate the consistent and manageable nature of surface adhesion and stiction forces in MEMS devices fabricated using the high-temperature epitaxial encapsulation process. In this encapsulation process (commercialized by SiTime), there are no chemical anti-stiction films or getters. Data from more than 2000 test structures with more than 80 design variations from three different fabrication runs were gathered in this study. Surprisingly, the adhesion force is shown to be independent of design geometry. The measured adhesion forces (18-25uN) are small enough for inertial sensors. In addition, we demonstrate anti-stiction bump stops with springs for a sliding contact, which reduce the probability of stiction by over 50%.

INTRODUCTION

Epitaxial encapsulation is a unique high-temperature, wafer-scale, pure silicon packaging technique. The encapsulation occurs at 1100 °C in near vacuum, resulting in oxide-free encapsulation and extremely pure silicon surfaces. In addition the process offers narrow gaps, smooth sidewalls and an environment free of impurities [1]. This allows for the fabrication of stable and high quality resonators. These features also suggest that stiction might be problematic, but much of prior work on stiction [2] does not apply because of the unique characteristics of this encapsulation process..

Efforts to fabricate integrated MEMS sensors that combine high frequency resonators and high displacement sensors have made it essential to understand the stiction that exists in this process. Previously it has been shown that release from stiction can occur repeatedly in test structures fabricated in this process [3].

Successful high reliability MEMS devices often do not simply rely upon the native stiction force, and instead incorporate additional anti-stiction measures. Unfortunately, the high-temperature encapsulation precludes the use of many of the typical anti-stiction techniques such as self- assembled monolayers [4]. These organic films have shown to significantly reduce stiction, but will not survive the high-temperature silicon deposition. Another approach to reducing stiction in unencapsulated devices relies upon contact springs to create a dynamic contact and reuse collision energy to increase likelihood of release. This has proven very effective in eliminating stiction, but once again requires materials and complex 3-D structures that are unavailable in many MEMS processes, including our pure silicon encapsulation [5].

DESIGN

In order to study the stiction properties of devices fabricated in the epitaxially encapsulated process, a series of test structures were designed. Two distinct types of structures were designed to study both process (fabrication) stiction and worst case in-use stiction.

In-process stiction device designs

To better understand in-process stiction, a series of cantilevers were designed. The length, thickness, gap size, effective mass and bump stop configuration were varied, resulting in a series of designs like those shown in figure 1.

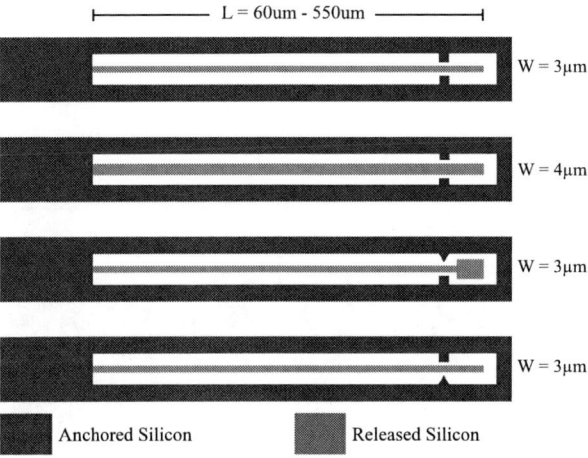

Figure 1 - Schematic of cantilever variations for studying process stiction

This type of design allows us to differentiate the effects of resonant frequency, spring stiffness, and released length, among other parameters. The contact areas were varied between point/tangential contact and 20um contact length. The devices had resonant frequencies between 20kHz and 500kHz.

In-use stiction device design

The in-use stiction test structures are meant to mimic the basic configuration of a high displacement inertial sensor, like an accelerometer, as these types of sensors are most likely to experience stiction problems. A series of electrostatically sensed and actuated "pull-in" structures allow us to apply a large force and measure the release or stiction failure of the device. These devices are shown in figure 2. A series of different bump stop designs were fabricated and tested. The bump stop shown in figure 2 mounts the contact point on the tip of a spring. This has the dual purpose of storing some potential energy to aid in the release, and causing a sliding contact. The variations of the spring are designed with different spring stiffness, to store different amounts of energy per unit displacement, and different spring angles, to vary the amount of sliding. Additionally, some reference designs were included that had traditional fixed bump stops.

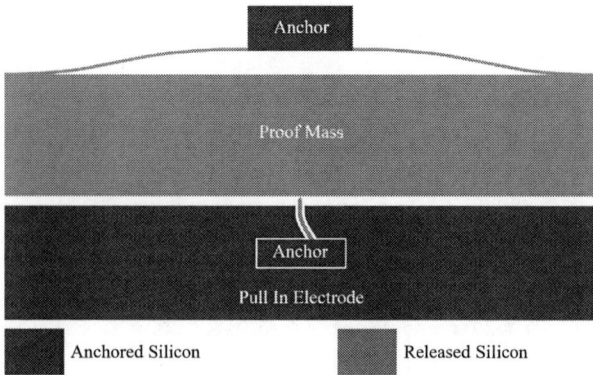

Figure 2 - "Pull-in" test structures for studying in-use stiction

FABRICATION

The devices were fabricated in several different runs of the epitaxial encapsulation process. This process, developed at Stanford University, in cooperation with Robert Bosch GmbH, is an ultra high-temperature, wafer-scale, oxide free process, the details of which have been reported extensively elsewhere [1].

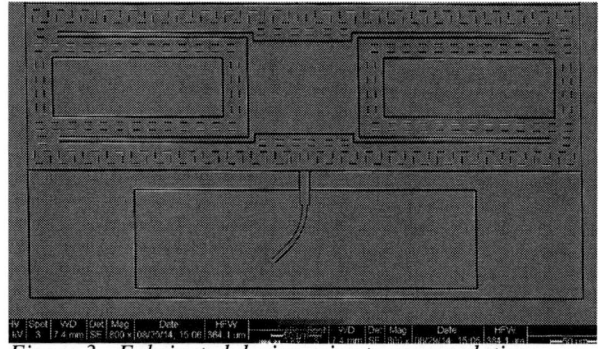

Figure 3 - Fabricated device prior to encapsulation

It is of particular relevance to note, however, that the devices considered in this study were fabricated in several separate runs of this process, and included notably different process variations. The earliest devices were fabricated in 2012, using the basic process on a 40um device layer. A second and third set of devices were fabricated in 2013, again on a 40um device layer, but with additional process steps for the inclusion of electrodes mounted to the cap, and different doping levels. Finally, another fabrication run was conducted in 2014 on a 20um device layer, with additional nitride layers included as etch stops [6]. Other small variations exist between each run, such as tools used, anneal times, etc.

TEST METHODOLOGY

The testing procedure for the in-process stiction test structures is quite straightforward. During processing the released devices are subjected to several large magnitude inertial forces. During spin coating, and even just wafer handling, the devices can experience accelerations as large as 5000g. These inertial forces cause many of the lower stiffness devices to contact the sidewalls. Subsequently, the forces are removed, and if the spring restoring force is greater than the surface adhesion (stiction) force, the device will release. If the stiction force is larger, however, the device will remain adhered. Upon the conclusion of processing, the devices are probed to detect conductive contact with the protruding bump stops. The minimum spring constant required to release the device is then an indication of the stiction force.

In use stiction requires additional actuation and sensing to mimic the conditions of sidewall contact occurring post-fabrication. To force the contact, a voltage was applied between the proof mass and anchored proof mass electrode (see figure 1). The position of the proof mass was determined by measuring the capacitance between these same electrodes (Agilent E4980A). An example of this measurement is shown in figure 4.

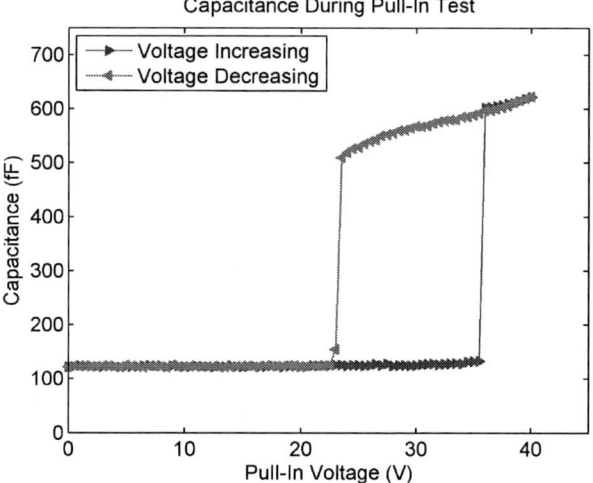

Figure 4 – Typical capacitance measurement during electrostatic pull in

RESULTS

In this work, the results from testing over 1,800 cantilevers for process stiction were compiled and examined against several independent variables. The successful release from stiction was best predicted by the stiffness of the cantilever beam, while resonant frequency and released length displayed much weaker correlations. The correlation between spring stiffness and release is shown in figure 5.

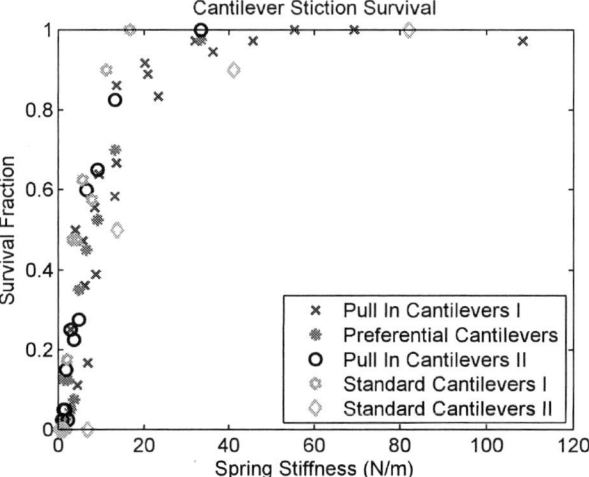

Figure 5 - Observed probability of successful release from stiction forces by cantilever equivalent spring stiffness. Each point represents 8-40 devices

Two interesting observations may be made from this result. First, we see that despite non-trivial differences in the processing steps, the devices from separate fabrication runs obey the same trend and are very tightly grouped. Furthermore, they appear to remain fairly consistent over time, as some devices were tested immediately after fabrication, while others were fabricated over a year before testing. The second observation is that devices with spring stiffness greater than 13.5 N/m have a high likelihood of survival, indicating a relatively modest stiction force of 18uN.

The in-use stiction test is meant to cause a worst case scenario impact. Electrostatic actuation causes a high acceleration of the proof mass as soon as it passes the unstable pull in point, at 1/3 of the initial gap, and additionally, the force continues to be applied after the initial impact. Once again, the important metric is whether the device survives pull-in, without developing fatal stiction forces. The data for this test is shown in figure 6.

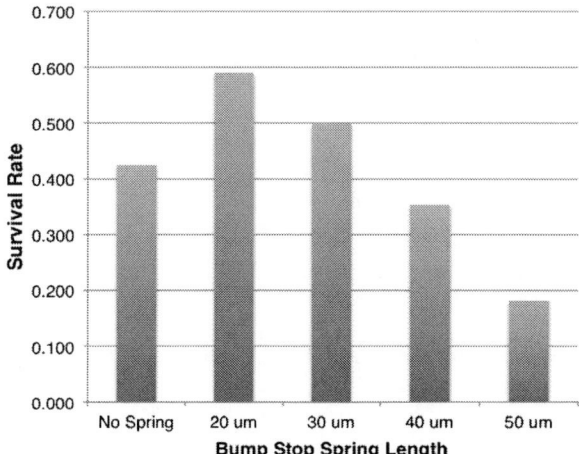

Figure 6 - Results of electrostatic pull in testing, categorized by length of the bump stop spring

The survival rates indicated from this test are cumulative. In order to be considered a survival, the test device must be successfully fabricated and released, must correctly indicated an electrostatic pull-in and finally must release to the original resting position. If the device fails at any of these steps, it is considered a failure, even in the cases where this is not directly attributable to stiction. We use this method to guarantee a conservative result that can be used by designers with confidence.

The bump stops mounted on the longer springs are not particularly successful, and those at 40um or longer actually increase stiction related failures. Additional test show that these springs are insufficiently stiff to arrest the motion of the proof mass during pull-in, causing the proof mass to electrically short to the pull-in electrodes in some cases. The shorter springs, however, improve performance relative to the reference case. The 20um spring bump stops display a marked improvement. The survival rate for these devices is over 40% higher than the base case.

CONCLUSIONS

The large number of devices tested and strong trends demonstrated by the data allow us to form a number of conclusions and design guidelines for sensors integrated in this fabrication process.

Process Stiction

The consistent nature of the stiction forces across different fabrication runs and with different contact geometries is unexpected. Traditional descriptions of stiction suggest that it should be strongly dependent upon contact area. We expect the devices to have very different contact areas either due to the different designs, or the non-deterministic sidewall profiles (see figure 7).

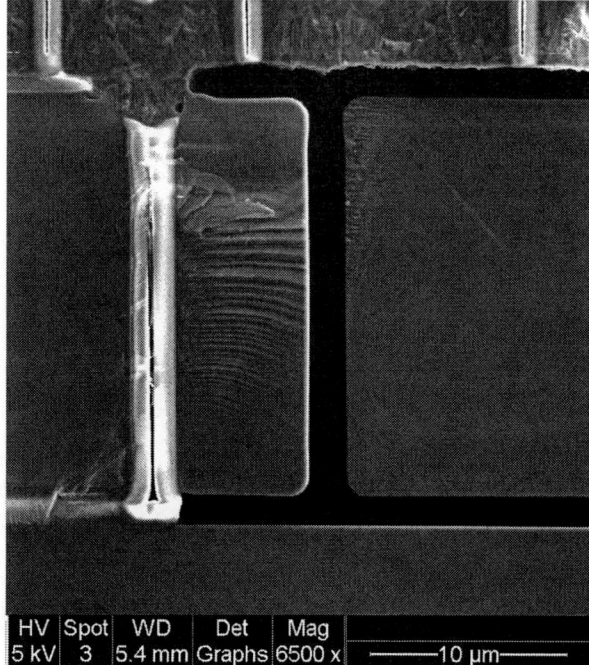

Figure 7 - Uneven sidewall profiles are dependent upon etch and annealing processes

Despite this, however, the data from different devices, wafers and fabrication runs overlaps and is consistent with the hypothesis of a contact that is dominated exclusively by asperities. The area and amount of asperity deformation is more determined by the forces and material properties than by the micron-scale geometry. This is a strong affirmation of expectations formed by our previous work [3]. One consequence of this result is that we may use this data as a design guideline for future devices, and be reasonably confident that small perturbations will not invalidate the results.

In-Use Stiction

Despite the concerns about stiction in epitaxially encapsulated MEMS devices, we see that devices do not necessarily experience fatal stiction. This is true even for high mass, high displacement devices in the style of an inertial sensor, such as those used for this test. In addition, we can see that even a simple compliant contact can substantially reduce the likelihood of a device becoming permanently adhered. The spring bump stops used in this study were not optimized, and the results are clearly very sensitive to parameter variations. This implies that further refinements should be able to improve greatly upon this initial improvement.

Both the modest amount of process stiction, and the improvement upon the native stiction properties are highly encouraging for the possibility of integrated MEMS inertial sensors in this process. The high temperature epitaxial encapsulation process proves to have good resistance to stiction, and despite incompatibility with some traditional anti-stiction methods, there exist straightforward mechanical means to reduce problems caused by stiction.

ACKNOWLEDGEMENTS

This work was supported by DARPA grant N66001-12- 1-4260, "Precision Navigation and Timing program (PNT)," managed by Dr. Robert Lutwak, and DARPA grant FA8650-13-1-7301, "Mesodynamic Architectures (MESO)," managed by Dr. Jeff Rogers. Student support was provided to D.B. Heinz from the Stanford Graduate Fellowship.This work was performed in part at the Stanford Nanofabrication Facility (SNF), which is supported by the National Science Foundation through the NNIN under Grant ECS-9731293

REFERENCES

[1] R. N. Candler et al., "Single wafer encapsulation of MEMS devices," IEEE Trans. Adv. Packag., vol. 26, no. 3, pp. 227–232, Aug. 2003.

[2] R. Maboudian et al., "Tribological Challenges in Micromechanical Systems," Tribology Letters, Vol. 26 2 (2002)

[3] D. B. Heinz et al., "Characterization of stiction forces in ultra-clean encapsulated MEMS devices," *IEEE MEMS*, pp. 588-591, Jan 2014.

[4] R. Maboudian et al., "Stiction reduction processes for surface micromachines," *Tribology Letters,* vol. 3, pp. 215-221 (1997)

[5] M.W. Miles, "MEMS devices with stiction bumps," US Patent US 7554711 B2 (2006)

[6] C.F. Chiang et al., "A novel, high-resolution resonant thermometer used for temperature compensation of a cofabricated pressure sensor," *Solid-State Sensors, Actuators, and Microsystems Workshop, Hilton Head* 2012, pp. 54-57, Jun 2012.

CONTACT

D.B. Heinz, +1-650-736-0044, dheinz@stanford.edu

STUDY OF THE HYBRID PARYLENE/PDMS MATERIAL

Dongyang Kang[1], Sanae Matsuki[2], and Yu-Chong Tai[1]
[1]California Institute of Technology, Pasadena, USA
[2]Northeastern University, Boston, USA

ABSTRACT

This paper reports the mechanical behavior and barrier property of the hybrid parylene/PDMS material. The repetitive uniaxial tensile tests are done to characterize its mechanical behavior and the water vapor transmission rate is measured to evaluate its barrier property. The experimental data are in accordance with the composite material theory. A novel approach of facilitating the diffusion and penetration of parylene coatings into PDMS using in-situ heated deposition is presented. The parylene depth profiling in PDMS and 180° peel tests demonstrate that parylene deposition at elevated temperatures shows enhanced pore sealing capability. A theoretical model is proposed, featuring an infinitely long cylindrical PDMS pore model, free molecular flow and time-varying pore geometry during the deposition. There is only one unknown parameter in the model: the PDMS pore diameter. By fitting the numerical solutions of the theoretical model to the parylene depth profiling curves, the PDMS pore diameter is estimated to be ~6.02nm.

INTRODUCTION

Parylene is widely used in various industries because of its many excellent properties, such as superior barrier property used to protect the electronic devices against damages from moisture and corrosive etchants [1-3]. The hybrid parylene/PDMS material has also been investigated, such as parylene-caulked PDMS [4, 5] and parylene coatings inside PDMS microchannels [6], attempting to take advantage of the excellent barrier property of parylene. However, the mechanical behavior such as the uniaxial tensile stress-strain relation and the barrier property such as the water vapor transmission rate (WVTR) of the hybrid parylene/PDMS material have not been well documented. The understanding of these properties will be very important to the design and the application of this hybrid parylene/PDMS material.

The pore-sealing feature of parylene conformal coating has been utilized to reduce gas or moisture permeation of various porous materials. For examples, parylene-caulked PDMS is promising for long-term pneumatic balloon actuator [4], and parylene coatings onto porous ultralow-*k* interlayer dielectrics are used to prevent precursor penetration during subsequent metallorganic deposition [7]. However, these demonstrations only use parylene deposition at room temperature (RT). In-situ heated deposition has led to the thorough investigation of deposition temperature effects on thermal, structural and mechanical properties of Parylene C [8]. This paper reports another application of in-situ heated parylene deposition: improving the diffusion and penetration of parylene into PDMS. The hypothesis is that the surface mobility of parylene monomer is a strong increasing function of temperature so the gaseous monomer tends to diffuse further inside the PDMS pore at elevated deposition

temperatures. The enhanced pore sealing capability of parylene at elevated deposition temperatures is verified by experiments.

MECHANICAL BEHAVIOR AND BARRIER PROPERTY

In order to investigate the mechanical behavior of the hybrid parylene/PDMS material, we first prepare 92μm-thick PDMS (Sylgard 184, Dow Corning, base: curing agent=10:1, cured at 70°C for 75min) samples coated with 0.64μm-thick Parylene C (PA-C). Reactive ion etching (RIE) is used to etch away 0.3μm surface PA-C of some samples. Reiterated uniaxial tension tests are performed using a commercial machine (Q800, TA Instruments). In each iteration, samples are loaded till the strain $_1$ is reached, then relaxed, preparing for the subsequent loading till the strain $_2 > _1$ is reached. Stress-strain relations for the parylene/PDMS samples with and without RIE etching are shown in Figure 1a and b, respectively.

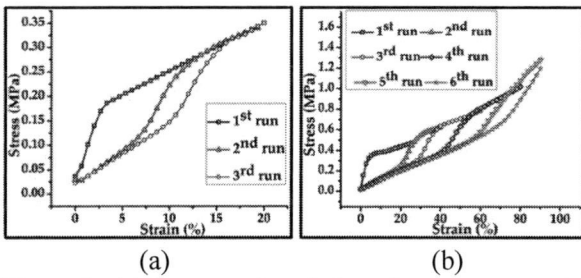

(a)   (b)

Figure 1: (a) Stress-strain relations for three successive uniaxial tensile tests of a parylene/PDMS sample with RIE etching. 0.3μm surface PA-C is etched away. (b) Stress-strain relations for six successive uniaxial tensile tests of a parylene/PDMS sample without RIE etching.

The WVTR is measured using a commercial water vapor permeability tester (TSY-W3, Labthink). Samples include pure PDMS, and PDMS coated with 0.64μm PA-C with and without RIE etching. In the case of RIE etching, either 0.3μm or 0.64μm surface PA-C is etched away. The thickness of PDMS is 92μm. Thermal annealing treatments at two different temperatures, 80°C in a convection oven and 180°C in a vacuum oven, for 8 hours are done to study the thermal annealing effect on the WVTR of the samples. The WVTR experimental data (Figure 2) show that the thermal annealing treatment can decrease the WVTR of the hybrid parylene/PDMS. The reason is likely that the high temperature annealing can decrease the WVTR of parylene by increasing its crystallinity and decrease the WVTR of PDMS by increasing its degree of cross-linking.

978-1-4799-7956-1/15 $31.00 © 2015 IEEE   397   MEMS 2015, Estoril, PORTUGAL, 18 - 22 January, 2015

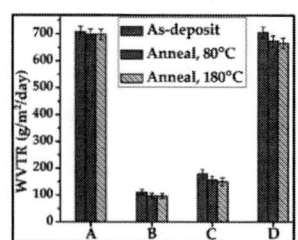

Figure 2: The WVTR experimental data for various parylene/PDMS samples with and without annealing.

Based on the composite material theory, the Young's modulus of the hybrid E_{hybrid} is given by the "rule of mixtures",

$$E_{hybrid} = \frac{E_{PAC}t_{PAC}}{t_{PAC}+t_{PDMS}} + \frac{E_{PDMS}t_{PDMS}}{t_{PAC}+t_{PDMS}}. \quad (1)$$

The WVTR of the hybrid $WVTR_{hybrid}$ is given by,

$$\frac{1}{WVTR_{hybrid}} = \frac{1}{WVTR_{PAC}} + \frac{1}{WVTR_{PDMS}}. \quad (2)$$

Here, E_{PAC}, $WVTR_{PAC}$ and t_{PAC} are the Young's modulus, the WVTR and the thickness of PA-C, respectively; E_{PDMS}, $WVTR_{PDMS}$ and t_{PDMS} are the Young's modulus, the WVTR and the thickness of PDMS, respectively.

The experimental data of the Young's modulus and the WVTR of parylene, PDMS and the hybrid are summarized in Table 1, and the theoretical results of the Young's modulus and the WVTR of the hybrid are summarized in Table 2. Here, the experimental result of the Young's modulus of the hybrid parylene/PDMS is calculated from the slope of the stress-strain curve for the first uniaxial tensile test run.

Table 1: The experimental data of the Young's modulus and the WVTR of parylene, PDMS, and the hybrid.

	92μm PDMS	0.64μm PA-C	0.34μm PA-C	92μm PDMS +0.64μm PA-C	92μm PDMS +0.64μmPA-C +RIE etch away 0.3μm PA-C
Young's modulus (MPa)	0.82	1902	1780	13.52	7.15
WVTR (g/m²/day)	708.1	132.7	254.8	102.8	180.1

Table 2: The theoretical results of the Young's modulus and the WVTR of the hybrid parylene/PDMS.

	92μm PDMS + 0.64μm PA-C	92μm PDMS +0.64μmPA-C +RIE etch away 0.3μm PA-C
Young's modulus (MPa)	13.95	7.37
WVTR (g/m²/day)	111.8	184.7

The experimental data of the mechanical and barrier properties of the hybrid parylene/PDMS material are in accordance with the theoretical results, implying that the parylene coatings inside PDMS pores almost do not contribute to the macroscopic mechanical or barrier properties of the hybrid parylene/PDMS material.

ENHANCED PARYLENE PENETRATION INTO PDMS: EXPERIMENTAL RESULTS AND THEORETICAL MODELING

A closed-loop temperature control system is built and put inside a parylene coater (PDS 2035CR, Specialty Coating Systems, Inc) as in Figure 3.

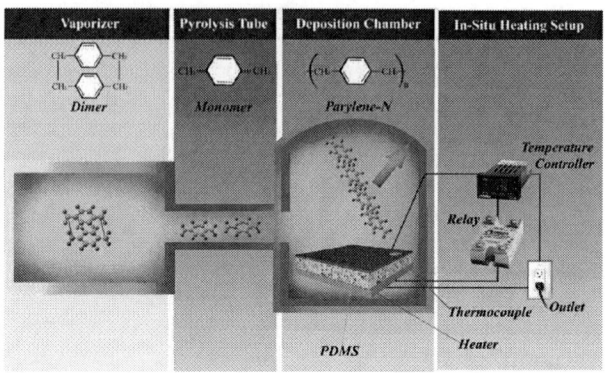

Figure 3: Parylene deposition process with an in-situ heating setup. Parylene N heated deposition process is shown as one example.

PDMS samples are molded in a small Petri dish and cured at 70°C for 75 minutes. Parylene N (PA-N), C, D (PA-D) and HT (PA-HT) (SCS products) are then coated on PDMS for studies. First, each type of parylene is deposited at RT onto PDMS. Some samples are then thermally annealed at 80°C in a convection oven for 8 hours, and 180°C in a vacuum oven (to prevent oxidation) for 8 hours, respectively. Next, each type of parylene is deposited at 80°C onto PDMS. The parylene penetration profile into PDMS is obtained using a secondary-ion-mass spectrometry (SIMS) (IMS 7f-GEO, CAMECA).

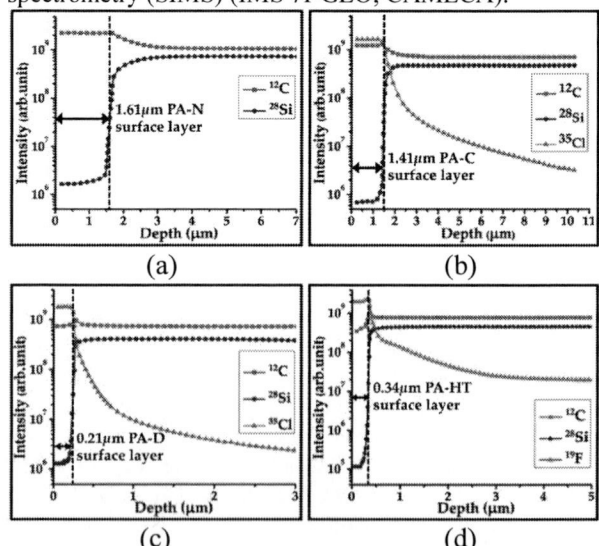

Figure 4: SIMS data of the depth profiling of Parylene N, C, D and HT inside PDMS for the deposition at 80°C. ^{12}C profile represents PA-N (a), ^{35}Cl profiles represent PA-C (b) and PA-D (c), and ^{19}F profile represents PA-HT (d). ^{28}Si represents PDMS.

SIMS sputter-etches the surface into the depth and measures specific atomic species distribution simultaneously, then parylene depth profiling curves are obtained (Figures 4 and 5).

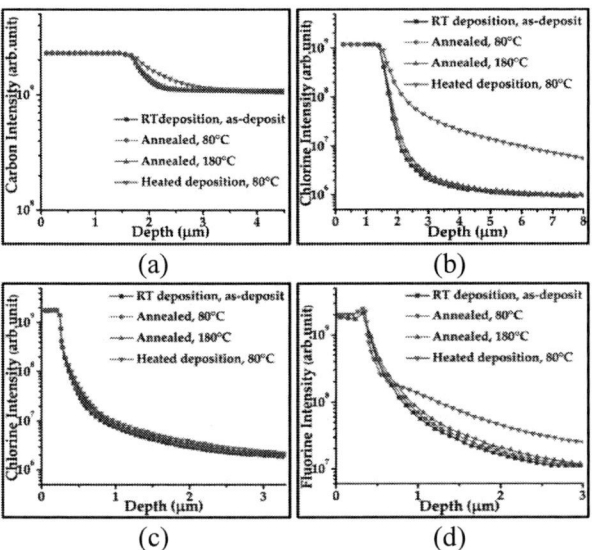

(a)　　　　　　　(b)

(c)　　　　　　　(d)

Figure 5: SIMS depth profiling curves for conventional RT deposition, post-deposition thermal annealing at 80°C and 180°C for 8 hours, and in-situ heated deposition at 80°C for each type of parylene. (a) PA-N. (b) PA-C. (c) PA-D. (d) PA-HT.

For mechanical adhesion tests, 90μm-thick PDMS samples are spin-coated on silicon wafers, cured and coated with PA-C at RT and 80°C, respectively. Then the hybrid samples are cut into 8mm-wide strips and mounted on a tensile test machine (Q800, TA Instruments) to perform the 180° peel tests (Figure 6).

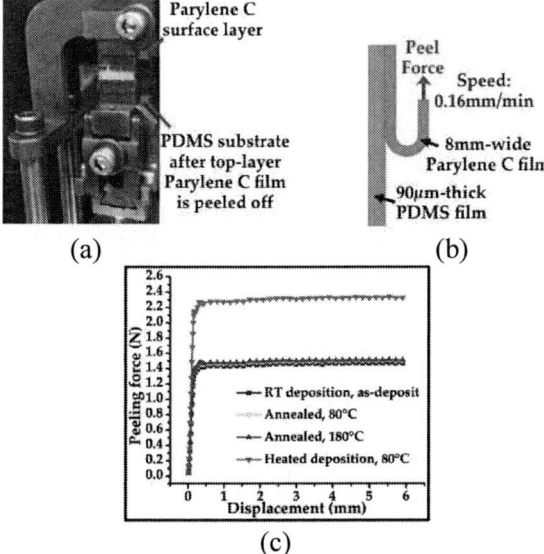

(a)　　　　　　　(b)

(c)

Figure 6: (a) A photo of the setup. (b) An illustration of the 180° peel test. (b) Experimental data of 180° peel tests for PDMS coated with Parylene C, prepared by conventional RT deposition, 80°C and 180°C annealing and 80°C deposition.

The depth profiling curves from Figures 4 and 5 demonstrate that in-situ heating significantly increases the penetration depths for almost all parylene types. In addition, post-deposition thermal annealing affects the penetration depths and the WVTR in a very limited way. The adhesion strength can be evaluated using the equation $R = \frac{F}{w}\left[2 + \frac{F/wt}{2E}\right]$, where R is the adhesion strength, F the peel force, w the width of the testing strips, and t and E the thickness and Young's modulus of parylene layer [9]. The adhesion strength ratio of the hybrid parylene/PDMS prepared by deposition at 80°C to RT is found almost equal to the ratio of their peeling forces, i.e. ~1.6. The data support the hypothesis that in-situ heated deposition is a very effective means to enhance the pore sealing capability of parylene coatings.

During the parylene deposition, the parylene monomer gas flow inside PDMS pores is in the free molecular flow regime, for which intermolecular collisions rarely happen. The Knudsen diffusion coefficient for the free molecular flow is only dependent on the molecular weight, the deposition temperature and the PDMS pore diameter [10]. A parylene deposition rate model [11] is used and becomes the reaction term in the mass balance equation. The pore shape and geometry vary with time during the deposition. Then a theoretical model is proposed, in which each PDMS pore is approximated as an infinitely long cylindrical tube as in Figure 7a. The deposition pressure for parylene usually ranges from 10mT to 100mT, under which the mean free math of the parylene monomer is of the order of 0.5mm. Since PDMS is known to be a nanoporous material, its average pore size is expected to be of the order of 10nm. Therefore, the mean free path λ of the parylene monomer in the vacuum deposition chamber is far larger than the characteristic length L of the PDMS pore, i.e. the pore diameter. In another word, the Knudsen number $\lambda/L \simeq 0.5\text{mm}/10\text{nm}$ is much larger than 10, hence the free molecular flow condition is satisfied.

The PA-N deposition rate [11] $R_d(x,t)$ depends on both the molar concentration $c(x,t)$ of the PA-N monomer and the deposition temperature T,

$$R_d(x,t) = \frac{(1-\theta)(MRT)^{1/2}}{(2\pi)^{1/2}\rho(1 + Ae^{-(E_d - E_a)/RT})}c(x,t). \quad (3)$$

The pore radius $h(x,t)$ evolution with time t due to the parylene deposition onto the wall of the pore is,

$$\frac{\partial}{\partial t}h(x,t) = -R_d(x,t). \quad (4)$$

As discussed above, the PA-N monomer gas flow inside PDMS pores is a Knudsen diffusion process, hence the molar flux $j(x,t)$ of the PA-N monomer is given by,

$$j(x,t) = -\frac{2h(x,t)}{3}\sqrt{\frac{8RT}{\pi M}}\frac{\partial c(x,t)}{\partial x}. \quad (5)$$

The mass balance equation for this one-dimensional flow model is given by,

$$\frac{\partial\left[\pi h^2(x,t)c(x,t)\right]}{\partial t} = -\frac{\partial\left[\pi h^2(x,t)j(x,t)\right]}{\partial x} \tag{6}$$

$$-2\pi h(x,t)R_d(x,t)\frac{\rho}{M}.$$

The boundary and initial conditions are,

$$c(0,t) = \frac{45\mathrm{mT}}{RT}, \ \ c(\infty,t) = 0, \ \ c(x,0) = 0, \tag{7}$$

$$\text{and } h(x,0) = a.$$

Here, M is the molar mass of the PA-N monomer, ρ is the PA-N film density, R is the gas constant, E_d and E_a are activation energies, A is a constant, θ is the coverage constant, and a is the initial pore radius, which is the only unknown and fitting parameter. The equations (3-7) are solved numerically, and the computed deposition depth profiles at different times for different temperatures are shown in Figure 7b and c. The values of the parameters in the deposition rate model can be found in [11].

The numerical solutions of the theoretical model is fitted to PA-N depth profiling curves, and the PDMS pore diameter $2a$ is determined to be ~6.02nm.

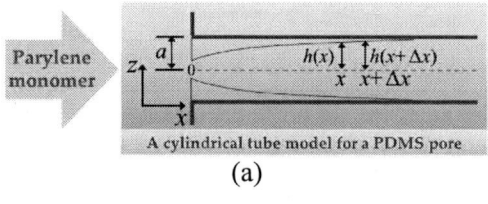

(a)

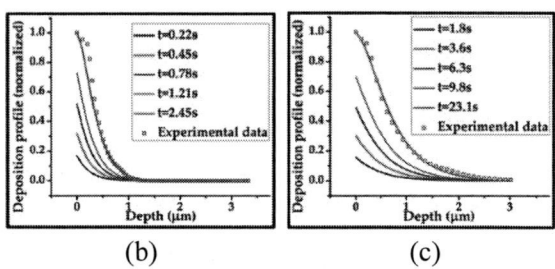

(b)   (c)

Figure 7: (a) A one-dimensional flow model for parylene deposition into pores. (b) and (c) Numerical solutions for the depth profiles of Parylene N deposited at room temperature (b) and 80°C (c) into PDMS at different times, compared with the experiment data. The pore diameter of PDMS is determined to be ~6.02nm.

ACKNOWLEDGEMENTS

The authors gratefully thank all the members from the Caltech Micromachining Lab.

REFERENCES

[1] J. H. C. Chang, L. Yang, K. Dongyang, and T. Yu-Chong, "Reliable packaging for parylene-based flexible retinal implant," in *Solid-State Sensors, Actuators and Microsystems (TRANSDUCERS & EUROSENSORS XXVII), 2013 Transducers & Eurosensors XXVII: The 17th International Conference on*, 2013, pp. 2612-2615.

[2] J. H. C. Chang, Y. Liu, D. Y. Kang, M. Monge, Y. Zhao, C. C. Yu, *et al.*, "PACKAGING STUDY FOR

A 512-CHANNEL INTRAOCULAR EPIRETINAL IMPLANT," in *26th Ieee International Conference on Micro Electro Mechanical Systems*, ed, 2013, pp. 1045-1048.

[3] J. H. C. Chang, D. Y. Kang, Y. C. Tai, and Ieee, "HIGH YIELD PACKAGING FOR HIGH DENSITY MULTI-CHANNEL CHIP INTEGRATION ON FLEXIBLE PARYLENE SUBSTRATE," in *2012 Ieee 25th International Conference on Micro Electro Mechanical Systems*, ed, 2012.

[4] S. Sawano, K. Naka, A. Werber, H. Zappe, S. Konishi, and Ieee, "Sealing method of PDMS as elastic material for MEMS," in *Mems 2008: 21st Ieee International Conference on Micro Electro Mechanical Systems, Technical Digest*, ed New York: Ieee, 2008, pp. 419-422.

[5] Y. Lei, Y. Liu, W. Wang, W. Wu, and Z. Li, "Studies on Parylene C-caulked PDMS (pcPDMS) for low permeability required microfluidics applications," *Lab on a Chip*, vol. 11, pp. 1385-1388, 2011 2011.

[6] J. Flueckiger, V. Bazargan, B. Stoeber, and K. C. Cheung, "Characterization of postfabricated parylene C coatings inside PDMS microdevices," *Sensors and Actuators B-Chemical*, vol. 160, pp. 864-874, Dec 15 2011.

[7] C. Jezewski, C. J. Wiegand, D. X. Ye, A. Mallikarjunan, D. L. Liu, C. M. Jin, *et al.*, "Molecular caulking - A pore sealing CVD polymer for ultralow k dielectrics," *Journal of the Electrochemical Society*, vol. 151, pp. F157-F161, 2004.

[8] D. Kang, S. Andrew, J. H. Chang, Y. Liu, and Y.-C. Tai, "Effects of deposition temperature on Parylene-C properties," in *Micro Electro Mechanical Systems (MEMS), 2013 IEEE 26th International Conference on*, 2013, pp. 389-390.

[9] K. Kendall, "THIN-FILM PEELING - ELASTIC TERM," *Journal of Physics D-Applied Physics*, vol. 8, pp. 1449-1452, 1975.

[10] E. M. Tolstopyatov, "Thickness uniformity of gas-phase coatings in narrow channels: I. Long channels," *Journal of Physics D-Applied Physics*, vol. 35, pp. 1516-1525, Jul 2002.

[11] J. B. Fortin and T. M. Lu, "A model for the chemical vapor deposition of poly(para-xylylene) (parylene) thin films," *Chemistry of materials*, vol. 14, pp. 1945-1949, May 2002.

CONTACT

*Dongyang Kang, tel: +1-626-395-3885, email: dkang@caltech.edu

SYNTHESIS OF CARBON NANOTUBES-NI COMPOSITE FOR MICROMECHANICAL ELEMENTS APPLICATION

Zhonglie An[1,2], Masaya Toda[1], Go Yamamoto[3], Toshiyuki Hashida[3], and Takahito Ono[1]

[1]Graduate School of Engineering, Tohoku University, Sendai, Japan
[2]Micro System Integration Center, Tohoku University, Sendai, Japan
[3]Fracture and Reliability Research Institute, Tohoku University, Sendai, Japan

ABSTRACT

We present the fabrication and characterization of a silicon micromirror with carbon nanotubes (CNTs)-nickel (Ni) composite beams, and evaluate the mechanical stability of the micromirror in terms of resonant frequency. A novel electroplating method is developed for the synthesis of the CNTs-Ni composite. The weight fraction of the CNTs in the electroplated composite is 2.6 wt%, and the ultramicroindentation hardness of the composite is 18.6 GPa. The maximum variation of the resonant frequency of the fabricated micromirror during a long term stability test is approximately 0.25%, and its scanning angle is approximately 20°. It shows the potential ability of the CNTs-Ni composite for micromechanical elements application.

INTRODUCTION

A CNT is one of the most typical one-dimensional nanomaterials in an order of micrometers in length and nanometers in diameter. The Young's modulus and bending strength of multi-walled CNTs (MWCNTs) have been reported to be approximately 1 TPa, as much as 14 GPa, respectively [1]. Therefore, researches for the practical applications of CNTs, especially CNT composites, have been actively pursued [2]. As a type of CNTs-metal composites, a CNTs-Ni composite can substantially improve Young's modulus, hardness, tensile strength and fracture strength in comparison with a Ni film due to the hardness improvement of the CNTs and the good interfacial bonding between the CNTs and the Ni matrix [3-4].

On the other hand, resonant scanning micromirrors based on microfabrication technologies have been conceptualized for a wide range of image display applications [5]. It requires high resonant frequency, wide scanning range and mechanical stability during operation, which highly depends on the mechanical elements of the micromirror. The CNTs-Ni composite can be one of candidates for some applications of the scanning micromirror in which the micromechanical elements of the micromirror are difficult to satisfy critical mechanical requirements [6].

In this paper, we present the novel electroplating method of a CNTs-Ni composite with a high content of CNTs and evaluate the mechanical property of the composite. We also present the fabrication and characterization of a micromirror with the CNTs-Ni composite beams and evaluate the shear modulus of the composite beams compared with that of pure Ni. To confirm the ability of the composite beams, mechanical stability of the micromirror is evaluated in terms of the resonant frequency, and scanning angle of the micromirror is demonstrated.

SYNTHESIS AND FABRICATION

Synthesis of CNTs-Ni Composite

MWCNTs used for an electroplating of the CNTs-Ni composite were synthesized by a catalytic chemical vapor deposition (CCVD) method followed by high temperature annealing [7]. The typical diameter and length of the pristine MWCNTs from scanning electron microscopy (SEM) and transmission electron microscopy (TEM) measurements ranged from 30 to 120 nm (average: 70 nm) and 1 to 23 μm (average: 9 μm), respectively. The CNTs-Ni electroplating suspension is prepared as following procedure.

A subset of the pristine MWCNTs was refluxed in a 3:1 (volume ratio) concentrated $H_2SO_4:HNO_3$ mixture at 70°C for 2 h, washed thoroughly with distilled water to be acid-free [8]. To increase the CNTs content in CNTs-Ni composite, the MWCNTs were suspended into a 300 mL (5 wt%) aqueous PDDA (diallyldimethylammonium chloride) solution and sonicated for 30 min. PDDA was used as a polyelectrolyte to introduce positive charges on the CNTs surface [11]. The suspension was then followed by filtration, which is accompanied with distilled water to remove excess PDDA. The pretreated CNTs are added into a sulfamate Ni electroplating solution, and the solution is sonicated for 1 h to produce a homogenous suspension. Finally, the CNTs-Ni composite is deposited on a Cr-Au conducting layer by dispersion electroplating using the CNT-Ni suspension as shown in Fig. 1. A pulse-reverse controlled condition is utilized to ensure the homogenous dispersion of CNTs in the composite. The parameters of the pulse-reverse electroplating are listed in table. 1.

The electroplated CNTs-Ni nanocomposite thin film is polished by chemomechanical polishing (CMP) to form a flat surface (roughness Ra < 1 μm) for mechanical property evaluation. The thickness of the composite is approximately 8 μm. As a reference sample, a Ni film with

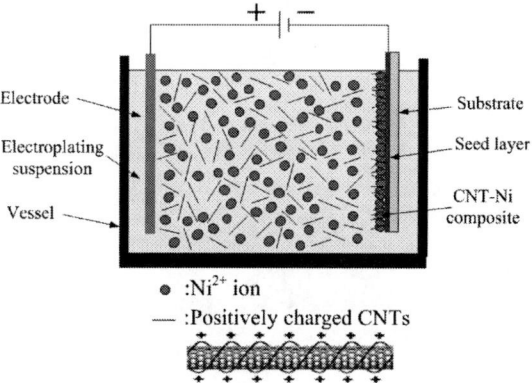

Figure 1: Schematic diagram of CNTs-Ni composite electroplating.

Table 1: Deposition parameters of electroplating for the CNTs-Ni composite.

$Ni(NH_2SO_3)_2$	350	g/l
$NiCl_2$	5	g/l
H_3BO_3	40	mg/l
CNT	1.8	g/l
Temperature	45	°C
Current ($I_{forward}$, $I_{reverse}$, I_{off})	40, -30, 0	mA/cm^2
Period ($t_{forward}$, $t_{reverse}$, t_{off})	100, 8, 100	ms
Time duration	60	min

a thickness of 8 μm was also deposited by electroplating using a sulfamate Ni electroplating solution.

Design and Fabrication of Micromirror

A micromirror generally consists of a mirror plate with a high reflectivity metal on the plate and two torsional beams made by micromachining. Working frequency and scanning angle mostly depends on the micromechanical property of the torsional beams [5]. The CNTs-Ni nanocomposite is applied to micromechanical elements including the torsional beams, which needs high mechanical strength. Figure 2 shows a schematic diagram of the micromirror with the electroplated composite beams. The composite beams adhere to the Si mirror and Si support frame. The micromirror was made from a Si wafer with a thickness of 300 μm. The fabrication process is schematically shown in Fig. 3. (a) Cr-Au (50 nm- and 150 nm-thick) thin films were deposited by sputtering on the Si substrate as an adhesion and seed layer. (b) Torsional beams of the micromirror were patterned on the substrate by photolithography. (c) The CNTs-Ni composite was deposited on the patterned substrate by electroplating with a pulse-reverse current for uniform deposition. (d) After removing resist, the Cr-Au thin films were etched by wet etching. (e) Then, the micromirror pattern was formed on the back side of the substrate by photolithography. (f) Finally, the Si substrate underneath the micromirror was etched by deep reactive ion etching (DRIE) to release the micromirror structure and the Cr-Au thin films were removed by etching.

CHARACTERIZATION
Composition Analysis of the CNTs-Ni Composite

The CNTs-Ni composite was evaluated using filed-emission scanning electron microscopy (FE-SEM).

A surface image of the electroplated CNTs-Ni composite is shown in Fig. 4 (a). It can be seen that CNTs

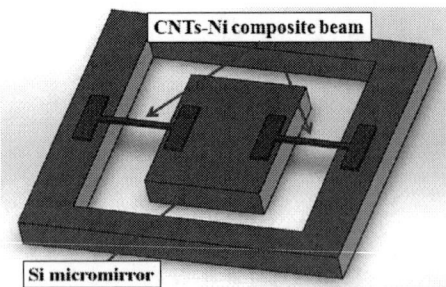

Figure 2: Schematic diagram of the micromirror with the electroplated CNTs-Ni composite beams.

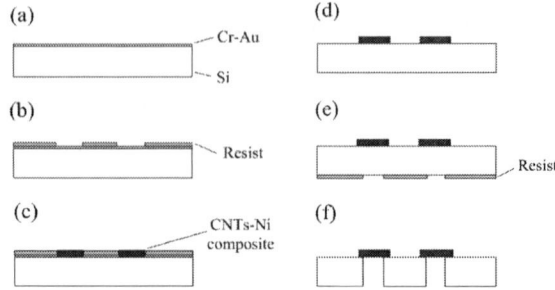

Figure 3: Fabrication process of a micromirror. (a) Cr-Au seed layer sputtering on a Si wafer (b) Photolithography for patterning of the micromirror beams (c) CNTs-Ni composite electroplating (d) Resist removing and Cr-Au etching (e) Backside photolithography for patterning of the micromirror (f) Deep RIE, resist removing and Cr-Au etching.

are incorporated into the Ni matrix and Ni thin films are deposited on the surface of CNTs. The CNTs provide a template for the deposition of the Ni thin film on the carbon surface. CNTs remained on the surface of the composite was polished, which form a flat surface as shown in Fig. 4 (b). CNTs are embedded into the Ni matrix and a surface roughness R_a of approximately 10 nm is obtained for the evaluation of the mechanical property of the composite.

X-ray photoelectron spectroscopy (XPS) analysis was carried out to obtain a weight fraction of the CNTs in the composite. After Ar$^+$ sputtering to remove contaminants and oxide on the surface, atomic concentrations of Ni, carbon and oxygen are analyzed as shown in Fig. 5. The oxygen originates in a small amount of remained slurry on the surface. From the atomic concentration of the carbon, the weight fraction of the CNTs is calculated to be 2.6 wt%, which is higher than a maximum weight fraction of 1.2 wt% reported by other researches [3, 10].

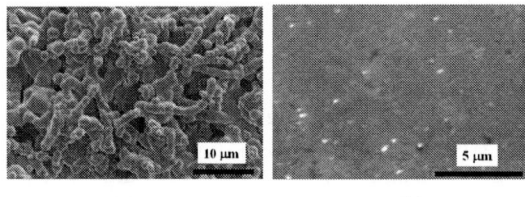

(a)   *(b)*

Figure 4: SEM images of the CNTs-Ni composite (a) surface image after electroplating (b) surface image after CMP.

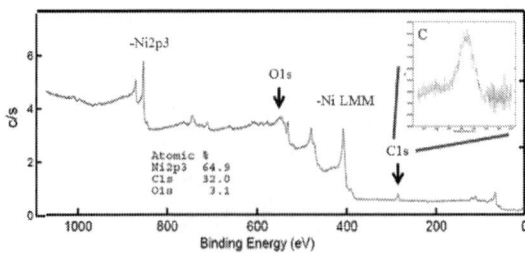

Figure 5: XPS spectrum of CNTs-Ni composite after Ar+ sputtering to remove contaminants.

Mechanical Properties of the CNTs-Ni Composite

An ultramicroindentation test was utilized to measure the hardness of the CNTs-Ni composite using a thin film mechanical property tester (MH4000, NEC Corporation) [11]. Figure 6 illustrates the typical load-indentation depth curves of the CNTs-Ni composite and the Ni film. We estimate the hardness with the penetration depth range of $0.1 \sim 0.2$ μm, and over 10 times of indentation were performed for each sample.

The measured hardness of the CNTs-Ni composite and the Ni film are 18.6 ± 2.0 GPa and 13.8 ± 1.8 GPa, respectively. Both samples have elastic recovery and show plastic deformation after removing the load. The hardness of the CNTs-Ni composite is larger than that of the Ni film, it should be rationalized by the addition of CNTs. The result shows that higher mechanical strengthening is expected in the CNTs-Ni composite than that of Ni film. It indicates that the composite would be applicable to micromirror beams.

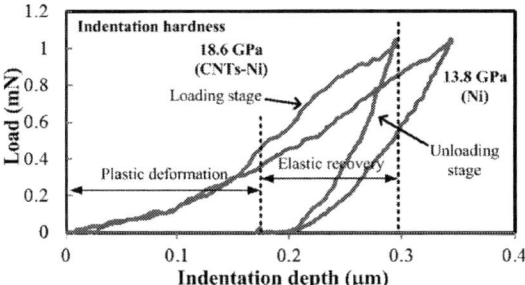

Figure 6: Load-indentation depth curves for CNTs-Ni composite and Ni film.

RESULTS AND DISCUSSION

Figure 7 shows the SEM image of a fabricated micromirror with electroplated CNTs-Ni composite beams. The Si mirror was 1000 μm in width and 2000 μm in length. The electroplated composite beams were 100 μm in width, 500 μm in length and 12 μm in thickness. Micromirror was successfully fabricated with the proposed fabrication process.

The resonant frequency of the fundamental torsional vibration mode of the micromirror was measured using a laser Doppler vibrometer. The fundamental resonant frequency f is given by

$$f = \frac{1}{2\pi}\sqrt{\frac{k}{J_p}}, \qquad (1)$$

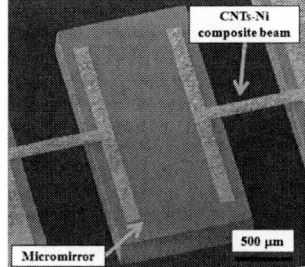

Figure 7: SEM image of a fabricated micromirror with the CNTs-Ni composite beams.

where J_p is mirror inertia depending on the mass and size of the mirror, k is torsional spring constant. The spring constant k is given by

$$k = \frac{2GJ_t}{l_s}, \qquad (2)$$

where G and l_s are sheer modulus and the length of the beam, J_t is the polar moment of area of a non-circular beam.

From the two equations described above, shear modulus of the beam is given by

$$G = \frac{2l_s\pi^2 J_p}{J_t}f^2, \qquad (3)$$

The shear modulus of the composite beam of the micromirror shown in Fig. 7 is calculated to be 105 GPa. If the Young's modulus of CNT and Ni are supposed to be 1000 GPa and 200 GPa, also both of Poisson's ratio of Ni and the composite are supposed to be 0.31, the theoretical shear modulus of the CNTs-Ni composite with 2.6 wt% of CNT should be $97 \sim 100$ GPa, and the theoretical shear modulus of the pure Ni should be 76 GPa. Therefore, it is thought that the experimental shear modulus of the composite beam is comparable with the theoretical value of the composite beam. Also, it indicates that the addition of CNTs in the Ni matrix effectively improved the shear modulus of the CNTs-Ni composite compared with the pure Ni.

In order to investigate the mechanical strengthening effect and the dynamic performance of the CNTs-Ni composite, a long term stability of the resonant frequency at the fundamental torsional vibration mode was measured. The micromirror was vibrated at the resonant frequency with a high constant driving power using a piezo ceramic actuator. The experiment was performed in a vacuum chamber with a pressure of $\sim 10^{-3}$ Pa. A reflected laser beam from a laser to the surface of the micromirror was recorded on a screen as shown in Fig. 8. The reflected beam forms a vertical line on the screen with a longest length of h, which is tuned by the driving power of the actuator, and the driving frequency was adjusted for tracing the resonant frequency during this experiment.

Figure 9 shows the long term stability measurement result of the micromirror with the composite beams. The resonant frequency was varied at the beginning and kept at stable resonance during about 13 days with a maximum variation of approximately 2.7 Hz, which corresponds to 0.25% of the stable resonant frequency. It can be

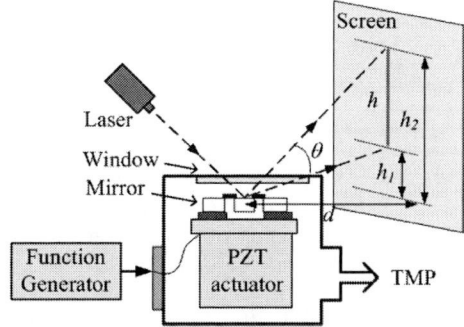

Figure 8: Experimental setup for tracing resonant frequency and scanning angle measurement.

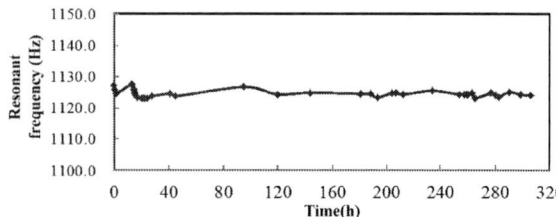

Figure 9: Long term stability measurement result of resonant frequency of the micromirror.

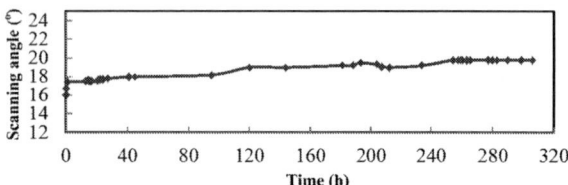

Figure 10: Long term measurement result of scanning angle of the micromirror.

rationalized that the resonant frequency was changed due to the plastic deformation of the beams and reached a stable condition. After the variation at the beginning, the resonant frequency is slightly fluctuated around the stable resonant frequency value during the measurement period. Considering the relatively small amount of frequency variation, we believe that the composite microelement of the micromirror is robust and reliable.

The scanning angle of the micromirror was also measured during the long term stability measurement of the resonant frequency as shown in Fig. 8. The upper point h_2 and lower point h_1 of the reflected beam were recorded. The scanning angle θ is calculated using following equation.

$$\theta = \arctan(\frac{h_2}{d}) - \arctan(\frac{h_1}{d}), \qquad (4)$$

where d is the distance from a reflecting point on the micromirror to the screen.

The scanning angle of the mirror with the CNTs-Ni composite beams gradually increased at the beginning and reached a stable value of approximately 20° as shown in Fig. 10. Due to the addition of CNTs, the spring constant of the composite beam was increased by equation (2). The composite beam is hard to rotate because the increase of the spring constant is found in the composite. Considering the size of mirror and the dimension of the composite beam expressed in equations (1) and (2), the mirror design and dimension of the composite beam should be further considered to show wider scanning angle without the decreasing of the resonant frequency.

CONCLUSIONS

Dispersion electroplating of a CNTs-Ni composite thin film with positively charged CNTs was developed and their CNTs content in the electroplated composite was 2.6 wt%. The measured ultramicroindentation hardness of the composite was 18.6 GPa, it was enhanced from 13.8 GPa of the hardness of the Ni film by the content of the CNTs.

A silicon micromirror with the CNTs-Ni composite beams was designed and fabricated. The addition of CNTs in the composite effectively improved the shear modulus compared with the pure Ni. The resonant frequency of the micromirror at the fundamental vibration mode was kept at a stable value with a maximum variation of 2.7 Hz during the long term stability test, which corresponds to 0.25% of the resonant frequency. A scanning angle of the micromirror with the composite beams was kept at a stable condition of 20° during the stability test. The optimal design of the Si micromirror and the CNTs-Ni composite should be considered for appropriate applications.

REFERENCES

[1] E. W. Wong, P. E. Sheehan, and C. M. Lieber, "Nanobeam mechanics:Elasticity, strength, and toughness of nanorods and nanotubes", Science, Vol. 277, pp. 1971–1975, 1997.

[2] P. Calvert, "Nanotube composites-A recipe for strength", Nature, Vol.399, pp.210-211, 1999.

[3] G. Shen, Y. Cheng, and L. Tsai, "Synthesis and characterization of Ni-P-CNT's nanocomposite film for MEMS application", IEEE transactions on Nanotechnology, Vol.4, No.5, 2005.

[4] Y. Sun, J. Sun, M. Liu, and Q. Chen, "Mechanical strength of carbon nanotube–nickel nanocomposites", Nanotechnology, Vol. 18, No. 50, 505704, 2007.

[5] S. T. S. Holmstrom, U. Baran, and H. Urey, "MEMS Laser Scanners: A Review", Journal of Microelectromechanical Systems, Vol. 23, No. 2, pp. 259-75, 2014.

[6] T. Tsuchiya, "Evaluation of mechanical properties of MEMS materials and their standardization", Reliability of MEMS: Testing of Materials and Devices, pp. 303, 2008.

[7] Y. A. Kim, T. Hayashi, M. Endo, Y. Kaburagi, T. Tsukada, J. Shan, K. Osato, and S. Tsuruoka, "Synthesis and structural characterization of thin multi-walled carbon nanotubes with a partially facetted cross section by a floating reactant method", Carbon, Vol. 43, No. 11, pp.2243-50, 2005.

[8] G. Yamamoto, M. Omori, T. Hashida, and H. Kimura, "A novel structure for carbon nanotube reinforced alumina composites with improved mechanical properties", Nanotechnology, Vol. 19, 315708, 2008.

[9] P. Kanaujia, D. Pardasani, A. Purohit, V. Tak, and D. Dubey, "Polyelectrolyte functionalized multi-walled carbon nanotubes as strong anion-exchange material for the extraction of acidic degradation products of nerve agents", Journal of Chromatography A, Vol. 1218, pp. 9307-13, 2011.

[10] Y. Suzuki, S. Arai and M. Endo, "Electrodeposition of Ni-P alloy-multiwalled carbon nanotube composite films", Journal of The Electrochemical Society, Vol. 157, No. 1, D50-D53, 2010.

[11] Y. Tsukamoto, H. Yamaguchi, and M. Yanagisawa, "Mechanical properties of thin films: measurements of ultramicroindentation hardness, Young's modulus and internal stress", Thin Solid Films, Vol. 154, pp.171-181, 1987.

CONTACT

*Zhonglie An, tel: +81-22-795-5810;
zhonglie@nme.mech.tohoku.ac.jp

3D STRUCTURAL FORAMATION UTILIZING GLASS TRANSITION OF A PARYLENE FILM

Tetsuo Kan, Akihiro Isozaki, Hidetoshi Takahashi, Kiyoshi Matsumoto and Isao Shimoyama
The University of Tokyo, Tokyo, JAPAN

ABSTRACT

This paper reports on a fabrication method of three-dimensional micro structures supported by a Parylene thin film. The three-dimensional structure formation procedure starts with an out-of-plane actuation of the Si surface micromachined structure coated with a thin Parylene film. At the same time of the actuation, the environmental temperature was elevated above the glass transition temperature T_g of the Parylene, 80-100 °C, and then cooled down below T_g. In this process, the rearrangement of the Parylene film happens above T_g, and the consolidated Parylene after the cooling down maintains three-dimensional micro structures as actuated. In a proof-of-principle experiment, the three-dimensional spiral structures formed by 300-nm-thick silicon were fixed by 1-µm-thick Parylene film.

INTRODUCTION

It is a fundamental issue to fabricate a three-dimensional structure with the MEMS surface micromachining technology. It has been reported that three-dimensional MEMS deformable structures work as functional elements in fields such as optics and so on [1]. For a practical use, it is important to keep three-dimensional structures as actuated.

Several methods were proposed to fix three-dimensional structures formed by actuated flat MEMS structures. Our group proposed a magnetic-field assisted rising up of MEMS structures and fixed them with latch elements [2, 3]. In addition, we also reported a magnetic-field assisted standing-up of micro cantilevers, fixing them with a Parylene coating [4]. The latching way, however, requires a special structure for latching. The Parylene fixing way requires the actuation having affinity to the vacuum Parylene deposition process when the actuation is being performed. It thus restricts choice of the force for a MEMS structure actuation.

In this paper, we propose another way of three-dimensional structure fixation which does require neither a complicated structure for fixing nor a vacuum process at the time of actuation.

FIXATION METHOD

The three-dimensional structure formation procedures are summarized in a heat diagram in figure 1. The main point lies in utilization of T_g transition of Parylene film. First, a Si microstructure (a cantilver in this figure) is coated with Parylene. Then, the structure is actuated by an external force. At the same time, the structure was heated above T_g of Parylene film. Because the rearrangement of the Parylene film occurs at this step, the polymer memorizes the shape of the deformed structure. After a sufficient heating duration, the structure is cooled down below T_g and the external force application is also finished. Because the Parylene keeps its deformed-shape and is consolidated, three-dimensional structure can be fabricated after the release of the external force.

As a proof-of-principle experiment, we applied this scheme to make a three-dimensional spiral structure from a planar Si spiral, which is similar to a metamaterial structure we previously reported [5]. Design parameters and SEM photographs are shown in figure 2. The planar Si spirals were fabricated using an SOI (Silicon-On-Insulator) wafer which had a 300-nm-thick top layer (BOX (Buried Oxide) layer thickness: 400 nm, handle layer thickness: 200 µm). First, the 300-nm-thick Si layer was etched to make planar spirals. Then, the handling Si layer was etched by DRIE. The BOX layer was removed by vapor HF etching to make the planar spirals free-standing. An outer diameter of the planar spiral was 150 µm, a spiral beam width was 6 µm, and number of turns was 5. Then, a 1-µm-thick Parylene film was coated on the spiral surface (figure 2).

The three-dimensional structure formation procedure is performed using a setup shown in figure 3. A programmable electric furnace was used to control a heating profile. Pneumatic force by N_2 gas was employed as an external force to actuate the spirals in the

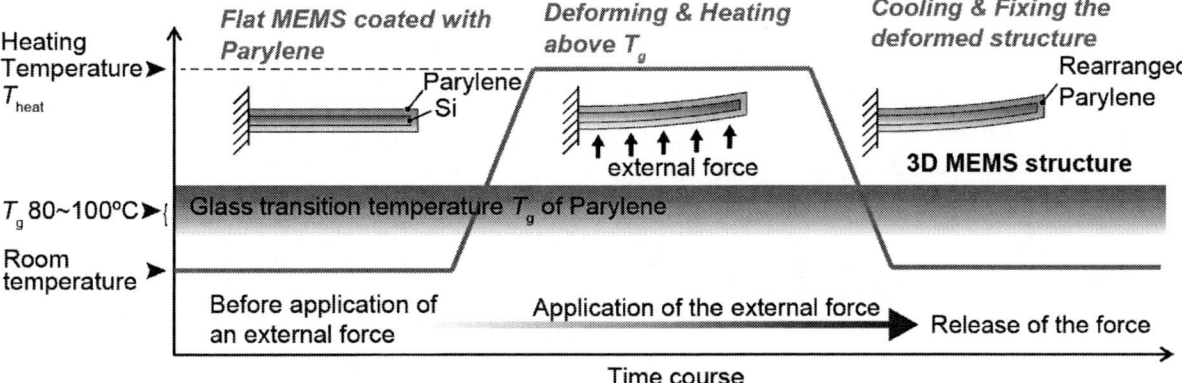

Figure 1: Thermal steps to fix 3-dimensional MEMS structure using Parylene glass transition.

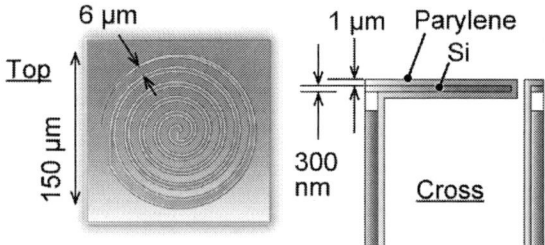

Figure 2: Device designs and configurations of top view and cross sectional view.

out-of-plane direction using an air supply chamber (figure 3). Because the furnace had an exhaust port at its top, an N_2 supply tube was introduced inside the furnace through the port as shown in figure 3. Length of the N_2 supply tube was about 2 m, and N_2 pressure was monitored at the start of the tube by a sensor. A spiral array chip was mounted on an aluminum jig we fabricated to apply pressure from the bottom side of the spiral to make a three dimensional spiral inside the furnace. Then, the heating and cooling experiment was performed.

EXPERIMENTS AND RESULTS

Three-dimensional MEMS structures were fixed following the proposing method with the furnace setup. As a parameter, the heating temperatures T_{heat} were changed above T_g, that is 100, 125, 150, 170, and 200 °C during 1.7 kPa application for actuation. The temperature of the furnace was increased from room temperature to T_{heat} at a rate of 20 °C/min. Duration of the heating at T_{heat} was 15 min. Then, the furnace was left for natural cooling until the furnace temperature reached 80 °C where the structure was picked out of the furnace. It was found that fixation of three-dimensional structures functioned. Scanning electron micrographs revealed that the deformed three-dimensional

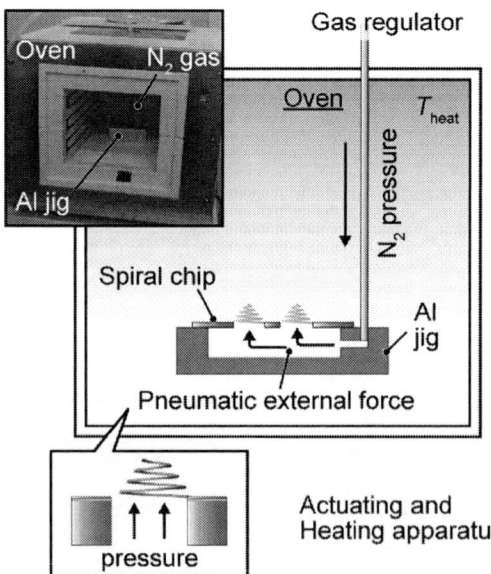

Figure 3: Experimental setup to perform heating and three-dimensional structure fixation during pneumatic actuation.

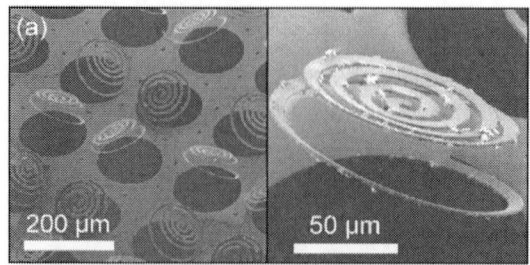

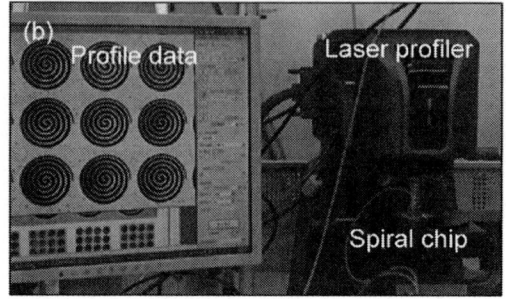

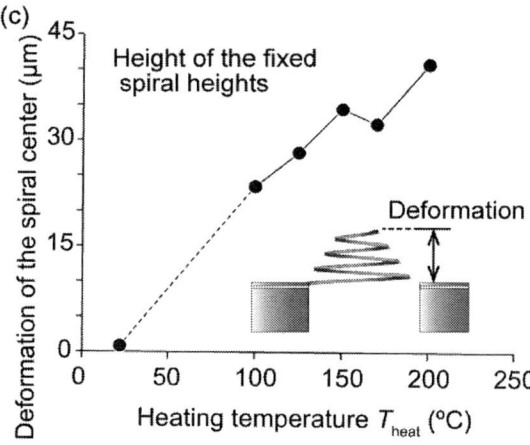

Figure 4: (a) SEM images of the fixed three-dimensional spiral structures, (b) An experimental setup to measure the deformation of the structure using a laser scanning optical microscopy, (c) The relationship between the heating temperature T_{heat} and the height of the center of the spiral.

structures were maintained after releasing of the pneumatic force as shown in figure 4(a). The relationships between the heating temperature T_{heat} and the out-of-plane deformation were measured. The out-of-plane deformation is defined as the height difference between the center of the spiral and the substrate surface, which was measured using a laser scanning optical microscopy shown in figure 4(b). The definition of the deformation is also given as an inset in figure 4(c). An increase in T_{heat} tended to provide higher deformation, indicating that rearrangement of the Parylene polymer is promoted as the increase in T_{heat}.

In a second experiment, the mechanical resonant frequencies were obtained for the spirals of before-Parylene-deposition and after-Parylene-deposition (figure 5). These properties were measured using laser Doppler vibrometer (figure 5(a)). For the measurement, the chip where spirals were formed was mounted on a PZT

978-1-4799-7956-1/15 $31.00 © 2015 IEEE   406

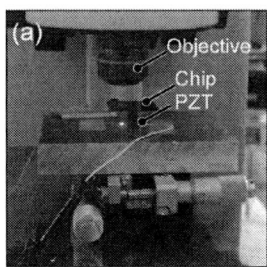

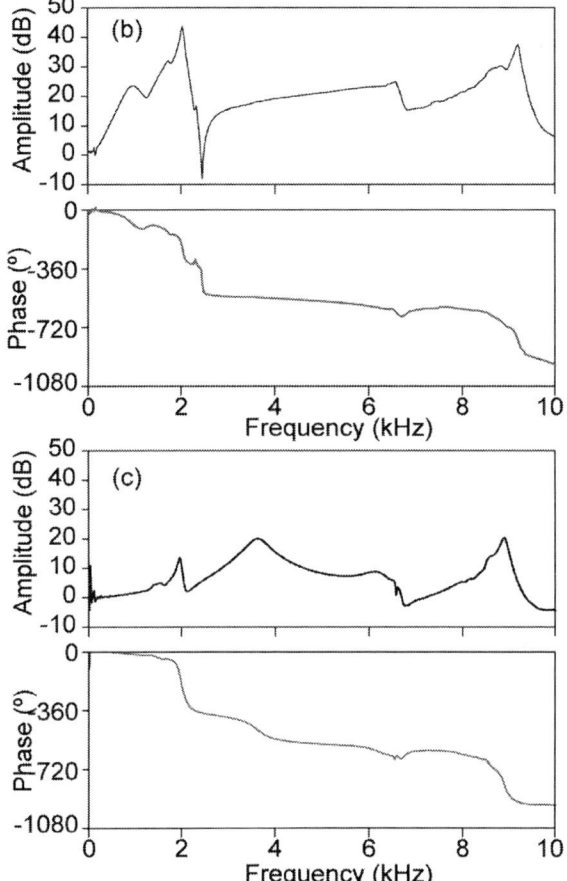

Figure 5: (a) An experimental setup for measuring mechanical resonant properties of the structure (left side) and an image of the device area to be measured. (b) Mecanical resonant properties before the Parylene deposition, and (c) Mecanical resonant properties after the Parylene deposition.

disc for vibration application. The laser spot was focused on the spiral beam, and the vibrations of the spiral beam were measured. The deformation amplitude was shown as dB, which is defined as logarithm of a ratio of deformations at the spiral beam and the substrate surface. Comparison of the mechanical resonant characteristics between figure 5(b) and (c) indicated stiffening of the beams by the Parylene deposition.

In a third experiment, an ability of shape reconfiguration of the spiral structure was evaluated. The spiral device used in the first experiment with 200 °C heating was flipped and mounted on the air supply jig, and

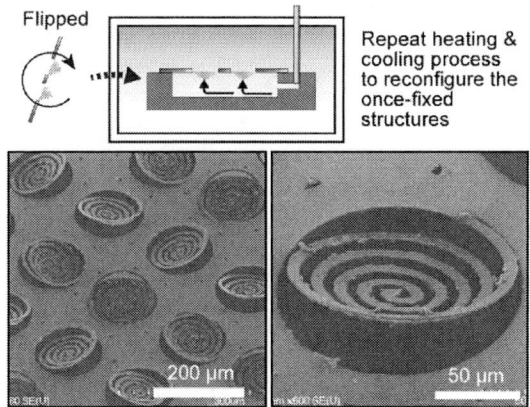

Figure 6: Reconfiguration of spiral structure after the first fixation.

2.5 kPa pressure was applied to induce an anti-directional out-of-plane deformation. The heating and cooling procedures were also performed using the same experimental furnace setup in the first proof-of-principle experiment. After the fixation steps, the spirals exhibited downward convex shapes by an SEM observation as shown in figure 6. Although it was impossible for the three-dimensional spiral structures to return to the initial flat spiral, ability of the reconfiguration after the fixation by the same heating and cooling procedure was confirmed.

CONCLUSION

Fixation method of three-dimensional MEMS structures using Parylene T_g transition was proposed. As a proof-of-principle experiment, a flat spiral structure with 1-μm-thick Parylene coating was actuated by N2 pressure, and simultaneously heated and cooled down. The experimental results indicated that the proposing method worked to obtain three-dimensional spiral. The reconfiguration of the spiral structure by repeating of the heating steps was possible. This method is widely applicable for a fixation of MEMS deformable structure in a non-vacuum environment.

ACKNOWLEDGEMENTS

This research was partly supported by Toyota Motor Corporation. This research was also partially supported by JSPS KAKENHI (26706008). The photolithography masks were made using the University of Tokyo VLSI Design and Education Center (VDEC)'s 8 inch EB writer F5112 + VD01 donated by ADVANTEST Corporation.

REFERENCES

[1] P.F. van Kessel, L.J. Hornbeck, R.E. Meier, M.R. Douglass, "A MEMS-based projection display," Proceedings of the IEEE , vol.86, no.8, pp.1687-1704, 1998.

[2] Eiji Iwase, Isao Shimoyama, "Multi-Step Sequential Batch Assembly of Three-Dimensional Ferromagnetic Microstructures with Elastic Hinges," Journal of Microelectromechanical Systems, vol. 14, no. 6, pp. 1265-1271, 2005.

978-1-4799-7956-1/15 $31.00 © 2015 IEEE

[3] Eiji Iwase, Isao Shimoyama, "A Design Method for Out-of-Plane Structures by Multi-Step Magnetic Self-Assembly," Sensors and Actuators A-Physical, vol. 127, no. 2, pp. 310-315, 2006.

[4] Kentaro Noda, Kazunori Hoshino, Kiyoshi Matsumoto, Isao Shimoyama, "A Shear Stress Sensor for Tactile Sensing with the Piezoresistive Cantilever Standing in Elastic Material," Sensors and Actuators A: Physical, vol. 127, no. 2, pp. 295-301, 2006.

[5] Tetsuo Kan, Akihiro Isozaki, Natsuki Kanda, Natsuki Nemoto, Kuniaki Konishi, Makoto Kuwata-Gonokami, Kiyoshi Matsumoto, and Isao Shimoyama, "Spiral metamaterial for active tuning of optical activity," Applied Physics Letters, vol. 102, no. 22, article no. 221906, 2013.

CONTACT

*T. Kan, tel: +81-3-5841-0461;
kan@leopard.t.u-tokoy.ac.jp

A NOVEL FABRICATION AND WAFER LEVEL HERMETIC SEALING METHOD FOR SOI-MEMS DEVICES USING SOI CAP WAFERS

Mustafa Mert Torunbalci[1], Said Emre Alper[1], and Tayfun Akin[1,2]

[1]MEMS Research and Application Center, Middle East Technical University, Ankara, TURKEY
[2]Dept. of Electrical and Electronics Eng., Middle East Technical University, Ankara, TURKEY

ABSTRACT

This paper presents a novel and inherently simple all-silicon fabrication and hermetic packaging method developed for SOI-MEMS devices, enabling lead transfer using vertical feedthroughs formed on an SOI cap wafer. The processes of the SOI cap wafer and the SOI-MEMS wafer require a total of five inherently-simple mask steps, providing a combined process and packaging yield as high as 95%. The hermetic encapsulation is achieved by Au-Si eutectic bonding at 400°C. The package pressure is measured as 1 Torr without any getter activation, and the package is proved to remain hermetic even after various temperature cycling tests. The shear strength of the fabricated chips is measured to be above 15 MPa, indicating a mechanically strong bonding.

INTRODUCTION

In the recent years, SOI-based MEMS fabrication processes are widely used for the development of high performance MEMS devices for a variety of applications [1, 2]. The use of SOI substrates allows the selection of stress-free and high quality single crystal silicon layers with optimum structural thickness, electrical properties, and desired crystallographic orientation. The structure of the SOI substrates also makes them attractive to be used for the packaging of MEMS devices, where hermetic encapsulation and lead transfer are the key packaging requirements. There are already well-established hermetic encapsulation methods applied to the MEMS devices with lateral feedthroughs, where glass frit [3] and metal-based [4, 5] bonding techniques are used. However, these techniques suffer from either high bonding temperatures and large bond rings or the need for additional dielectric for isolation and insufficient step coverage due to the limited material thickness. These limitations can be eliminated by using vertical feedthroughs, which are typically achieved with complex via-refill and trench-refill processes. METU-MEMS Research and Application Center has already reported a method for fabricating SOI-based and simple vertical feedthroughs that does not require any complex via or trench-refill process steps, called as the advanced-MEMS (*aMEMS*) process [6, 7]. Up to now the aMEMS process has been applied to the MEMS structures that are fabricated on glass substrates [6, 7]. This paper demonstrates for the first time the adaptation of the aMEMS process to an all-silicon (glass-free) SOI-MEMS device. The processes of the SOI cap wafer and the SOI-MEMS wafer require a total of five inherently-simple mask steps, providing a combined process and packaging yield as high as 95%. Moreover, the pitch of the vertical feedthroughs are shown to be reduced to half-size compared to [7], yielding a smaller size for the packaged chip.

PACKAGE DESIGN and FABRICATION

Package Design

Figure 1 shows the three-dimensional (3D) view of the proposed method in which *aMEMS* process is adapted for the wafer level hermetic packaging of SOI-MEMS devices. Vertical feedthroughs and the sealing walls are formed by highly doped silicon and simultaneously structured by etching the device layer of an SOI cap wafer. The via openings are formed by etching the handle layer of the SOI cap wafer, while the non-etched part of the handle layer acts as a roof for the encapsulated sensor. The signal transfer is achieved by getting simple wire bonds from the exposed faces of the vertical feedthroughs, eliminating any need for via-refill, as in [8].

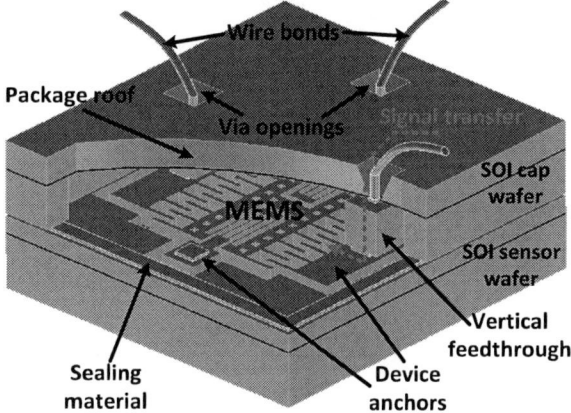

Figure 1: 3D view of the proposed package design.

Fabrication

Figure 2 shows the main process steps of the proposed method requiring and SOI sensor wafer and an SOI cap wafer. The SOI sensor wafer with a 35 μm-thick <111> device silicon layer is patterned and etched using DRIE, followed by a vapor HF process for releasing the structures (Fig. 2-a). The process of the SOI cap wafer starts by forming the via openings on 100μm-thick <100> handle layer of the SOI wafer with KOH etching. The thickness of the handle layer determines the vertical feedthrough pitch, and it is selected as 350 μm with the use of a 100 μm thick handle layer in this study. Next, Cr/Au pad metals are formed inside the via openings for wire bonding purposes by using the lift-off technique (Fig. 2-c). A second Cr/Au layer, to serve as the sealing material, is patterned on the device silicon layer of the SOI cap wafer with lift-off as shown in Fig. 2-d. The last step of the cap wafer preparation is the simultaneous formation of cavity and vertical feedthroughs by etching the device silicon layer of the SOI cap wafer by DRIE until reaching to the buried oxide layer as shown in Fig. 2-e. The SOI sensor and cap wafers are then bonded with the Au-Si eutectic bonding technique at 400°C by using a bond pressure around 3 MPa, during which the hermetic sealing and lead transfer are achieved simultaneously. After the wafer-level hermetic packaging process, the signal transfer between the hermetically sealed sensor and the outer world can simply be achieved by conventional wire bonds (Fig. 2-f).

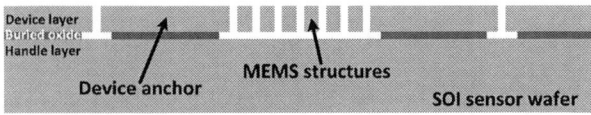

a) Structure etch and dry release with DRIE and vapor HF

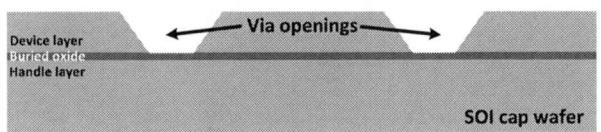

b) Formation of via openings on the handle layer of the SOI cap wafer with KOH

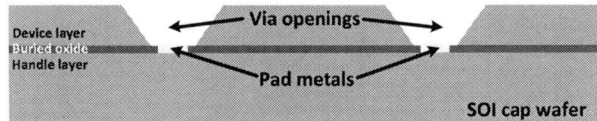

c) Formation of Cr/Au pad metals inside the via openings with lift off

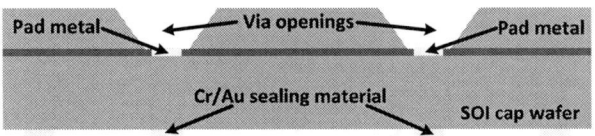

d) Formation of Cr/Au sealing material on the device layer

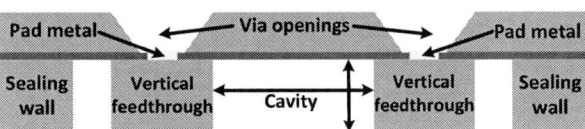

e) Formation of cavity and vertical feedthroughs simultaneously with DRIE

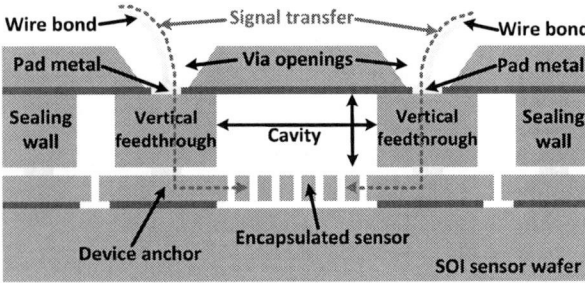

f) Wafer level hermetic packaging with Au-Si eutectic bonding and wire bonds for signal transfer

Figure 2: Main process steps of the proposed method.

Fabrication Results

Figure 3 shows MEMS chips fabricated, whereas Figure 4 shows the close-up SEM pictures of an encapsulated MEMS resonator, including the via regions. The via pitch size has been reduced from 700 µm in [7] to 350 µm by using a thinner handle layer for the SOI cap wafer, allowing to reduce the die size by 42% compared to [7] that contains the identical MEMS resonator inside. The die size in this study is 5.0 mm x 4.5 mm.

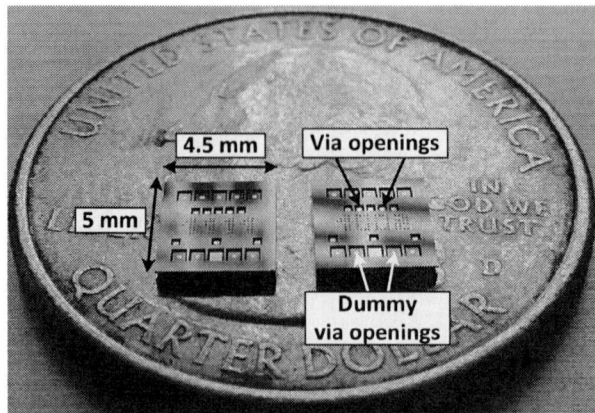

Figure 3: Photograph of the packaged dies after dicing.

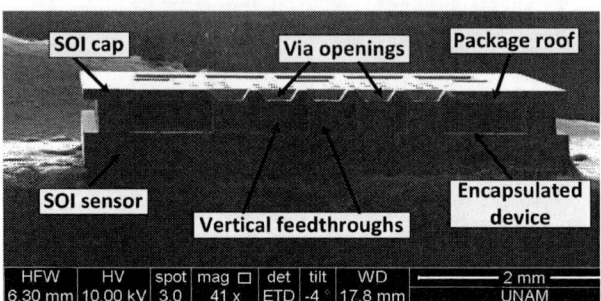

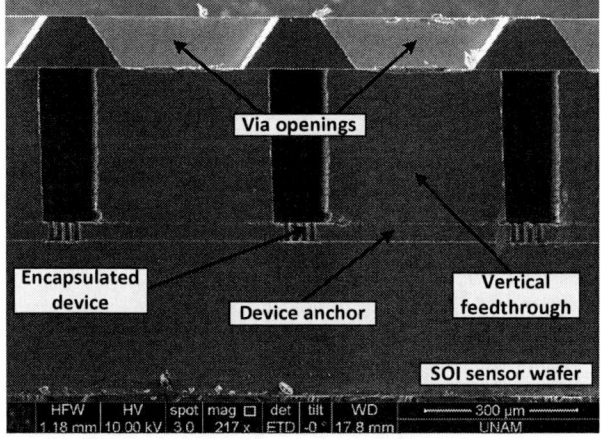

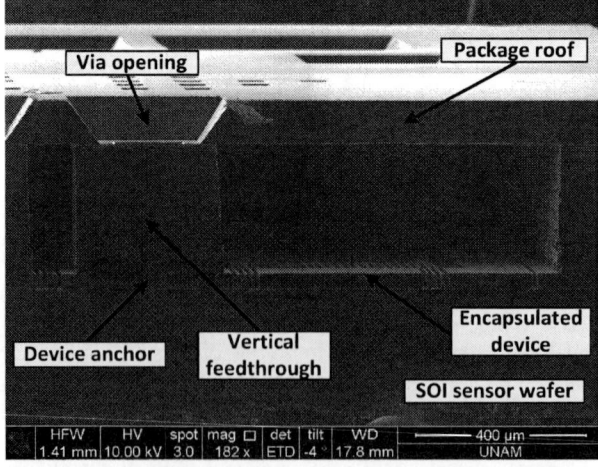

Figure 4: SEM pictures of a MEMS structure encapsulated using the proposed method.

978-1-4799-7956-1/15 $31.00 © 2015 IEEE

CHARACTERIZATION RESULTS

Various tests are performed on the packages fabricated with the proposed method in order to measure the package pressure, production yield, bonding strength, and thermal robustness.

Package Pressure and Production Yield

The package pressure is indirectly measured by tracking the quality factors of the MEMS resonators encapsulated with the proposed method. The quality factor of the MEMS resonators directly depends on the air damping that provides a direct measure of the pressure change inside the encapsulated cavity.

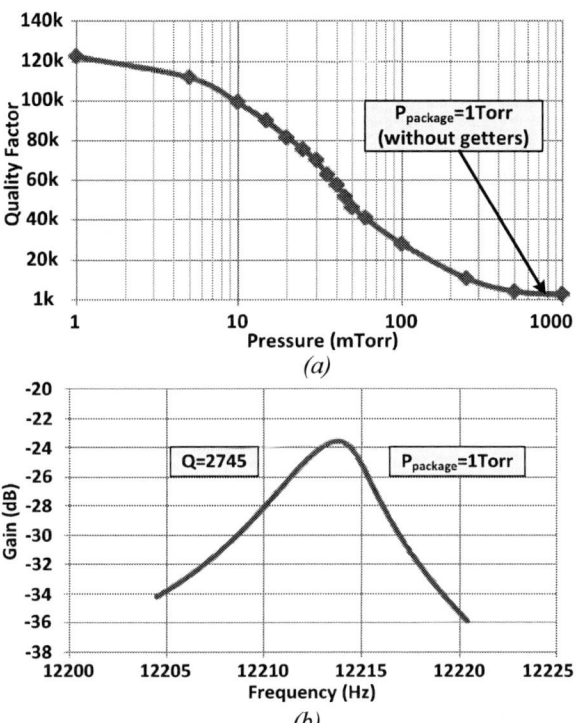

(a)

(b)

Figure 5: Test results of the fabricated prototypes: (a) Quality factor versus pressure characteristics of a tuning fork resonator obtained in a controlled-pressure chamber before wafer-level packaging process. (b) Resonance characteristics of the same resonator after wafer-level packaging, showing a measured quality factor of 2745 corresponding to a cavity pressure of 1 Torr, limited by the absence of a thin-film getter inside the cavity.

The characterization of the resonators on the MEMS sensor wafer is performed before and after the wafer-level encapsulation process. First, the MEMS sensor wafer is placed inside a controlled vacuum chamber in which the pressure can be adjusted in between 10 μTorr and 10 Torr. The resonance characteristics of the resonators located on this wafer are measured at different pressures using an Agilent 35670A Dynamic Signal Analyzer, Agilent E3631A Power Supply and an external readout circuit. The quality factors of the resonators are then extracted from these measurements. Finally, the quality factor versus pressure graph is obtained and the pressure inside the packages is predicted by using the graphs. Figure 5 presents the results of the tests performed on a benchmark

MEMS resonator prior to and after the wafer-level hermetic packaging process. This approach helps verifying the hermeticity of the wafer-level sealed cavities as well as detecting the pressure inside them. Test results show that the wafer-level packaged MEMS resonators have quality factors around 2700 which corresponds to a pressure around 1Torr, limited by the absence of a thin-film getter inside the cavities. The cavity pressures can be reduced down to 1-10 mTorr with the use of a proper getter, as already verified experimentally in [7].

The production yield is another critical parameter in order to demonstrate the success of the proposed method. In order to check to the production yield of the proposed method, wafer level tests are performed on randomly selected 50 MEMS resonators on a wafer level vacuum packaged wafer. According to these test results, the proposed method shows a combined fabrication and packaging yield of 94% (47/50).

Bonding Strength

The mechanical robustness of the proposed method is evaluated by shear strength measurements. These measurements are performed on several fabricated chips by using a conventional shear test tool. Figure 6 shows that 10 of the 11 randomly-selected packaged chips have shear strengths above 15MPa with an average strength of 28 MPa. Figure 7 shows the photographs and SEM picture of cap and sensor chips separated from each other during the shear test. The Au-Si bond interface is observed to withstand the shear test, as the handle layer of the SOI cap is broken from the silicon-oxide fusion bond interface of the SOI, but not the Au-Si eutectic bonding interface, demonstrating the bonding strength.

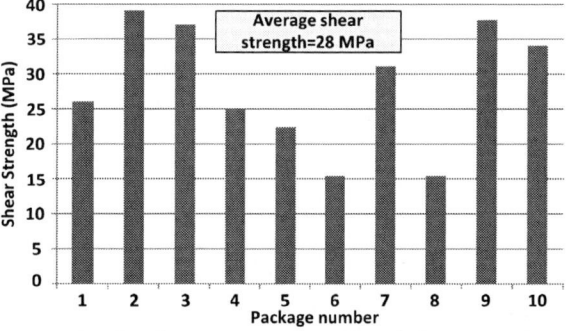

Figure 6: The shear test results of 10 different packaged chips. The measured results show that all of the packaged chips have shear strength above 15 MPa.

Thermal Robustness

Thermal robustness of the packaged chips is verified by thermal cycling tests. The packaged chips are subjected to thermal cycling tests between 100°C and 25°C for 5 cycles with 10 minute duration at each temperature and instant movements between hot and cold states. The quality factor of a sample MEMS resonator is initially measured as 2745 whereas it is reduced to 2325 after the thermal cycling test. The packaged chips remained hermetic even after the thermal cycling tests, but there is a slight reduction in the quality factor, which corresponds to an increase in the package pressure. This is an expected result since there is no active getter inside the encapsulated cavity, which otherwise could prevent the effects of outgassing occurred during the thermal cycling tests. Overall, the packaged chips are still hermetic and functional at the end of this thermal cycling test.

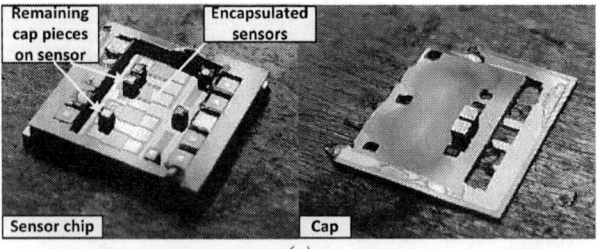

(b)

Figure 7: Photographs (a) and SEM picture (b) of cap and sensor chips separated from each other during the shear test. The Au-Si bond interface is observed to withstand the shear test, as the handle layer of the SOI cap is broken from the silicon-oxide fusion bond interface, but not from the Au-Si eutectic bonding interface, demonstrating the bonding strength.

CONCLUSIONS

This paper reports an all-silicon, inherently-simple, high-yield, and mechanically-strong fabrication and wafer-level hermetic encapsulation process adapted from the original aMEMS process of the METU-MEMS Research and Application Center. The hermetic packaging is achieved by Au-Si eutectic bonding at 400°C using a bond pressure of 3 MPa. The fabricated prototypes are verified to be operational with cavity pressures around 1Torr, which can be further reduced by using a proper getter inside the cavity. The process of the SOI sensor and cap wafers require a total of five inherently-simple mask steps, providing a combined process and packaging yield as high as 94%. The mechanical strength of the packages are evaluated by conventional shear tests, providing a minimum and average shear strengths of 15 MPa and 28 MPa, respectively, indicating a mechanically-strong bonding. The robustness of the packages has also been verified with thermal cycling tests performed between 100°C and 25°C with 10 minute duration at each temperature and instant movements between hot and cold states. The hermeticity of the

packages is verified to be preserved after this test, although the package pressure increased slightly due to the absence of a getter that could buffer the outgassing occurred during the tests. In summary, the proposed fabrication method provides an inherently-simple, high-yield, and mechanically-strong solution for the manufacturing and wafer-level hermetic packaging of a wide-range of MEMS devices.

ACKNOWLEDGEMENTS

This project is supported by The State Planning Organization (DPT) of Turkey (currently Ministry of Development) with a project titled Industrial Micro-Electro- Mechanical Systems (MEMS). The authors also would like to thank to Eyüp Can Demir for the shear tests of the chips.

REFERENCES

[1] A. A. Trusov, A. R. Schofield, and A. M. Shkel, "Micromachined Rate Gyroscope Architecture with Ultra High Quality Factor and Improved Mode Ordering," *Sensors and Actuators A,* Vol. 165, 2011, pp. 26-34.

[2] C. Guo, E. Tatar, and G. Fedder, "Large Displacement Parametric Resonance using a Shaped Comb Drive," *in Digest Tech. Papers MEMS'13 Conference,* Taipei, Taiwan, January 20-24, 2013, pp. 173-176.

[3] R. Knechtel, "Glass Frit Bonding: An Universal Technology for Wafer Level Encapsulation and Packaging," *J. Microsys. Technol.,* Vol. 12, No.1, 2005, pp. 63-68.

[4] W. C. Welch and K. Najafi, "Gold-Indium Transient Liquid Phase (TLP) Wafer Bonding for MEMS Vacuum Packaging," *in Digest Tech. Papers MEMS'08 Conference,* Tucson, Arizona, USA, January 13-17, 2008, pp. 807-809.

[5] J. S. Mitchell and K. Najafi, "A detailed study of yield and reliability for vacuum packages fabricated in a wafer level Au-Si eutectic bonding process," *in Digest Tech. Paper Transducers'09 Conference,* Denver, Colorado, USA, June 21-25, 2009, pp.841-844.

[6] S. E. Alper, M. M. Torunbalci, and T. Akin, "Method of Wafer Level Hermetic Packaging with Vertical Feedthroughs," *PCT/TR2013/000298,* September 2013.

[7] M. M. Torunbalci, S. E. Alper, and T. Akin, "Wafer Level Hermetic Encapsulation of MEMS Inertial Sensors using SOI Cap Wafers with Vertical Feedthroughs," *in Digest Tech. Papers ISISS'14 Conference,* Laguna Beach, California, USA, March 25-26, 2014, pp. 1-2.

[8] A. C. Fischer, J. G. Korvink, N. Roxhed, G. Stemme, U. Wallrabe, and F. Niklaus, "Unconventional Applications of Wire Bonding Create Opportunities for Microsystem Integration," *J. Micromech. Microeng.,* Vol. 23 No. 8, 2013, pp. 1-18.

CONTACT

*M. M. Torunbalci, tel: +90-543-2251423; mtorunbalci@mems.metu.edu.tr

BONDING MECHANISM IN THE VELCRO CONCEPT SI-SI LOW TEMPERATURE DIRECT BONDING TECHNIQUE

Shervin Keshavarzi[1,2], Ulrich Mescheder[1], and Holger Reinecke[2]

[1]Institute of Applied Research (IAF), Furtwangen University, Furtwangen, Germany
[2]Institute of Microsystems Engineering (IMTEK), Freiburg University, Freiburg, Germany

ABSTRACT

This work presents a bonding mechanism between needle-like surfaces for room temperature Si-Si direct bonding similar to the Velcro-principle, a fully CMOS compatible approach suitable for system integration using Si-motherboard concept. The proposed bonding model is superior to other presented models since it considers humidity effect and the deformation mechanism of the needles during the bonding.

INTRODUCTION

Ongoing miniaturization of microsystems devices, smarts systems, and used packages "smart dust" necessitates a new bonding technique with self-alignment capability at low temperature to swap the conventional pick and place approaches. Porous Silicon based technology has been exhibited strong permanent bonds between needle-like surfaces similar to Velcro concept at room temperature [1]. Although the proposed bonding model based on Van der Walls force approach [2] indicates agreement between measurement and simulation results, it does not cogitate the existence of condensed vapor or capillary bridges due to humidity which might make the capillary force dominant over the Van der Waals forces [3] as well as the deformation mechanism of needles due to applied bonding load.

SURFACE GENERATION

The needle-like surfaces can be simply generated in the region between porous silicon generation and electro-polishing through electrochemical dissolution of Si wafer in HF/Water solution. Choosing proper anodization and material parameters such as current density, substrate type, doping concentration, backside resistivity, and electrolyte concentration, the pore morphology of porous silicon can be reformed from sponge type to needle type.

The Velcro-like surfaces (Fig. 1) are generated via anodization of the surface of a low doped p-Si wafer in a double tank cell with a constant current density of 70mA/cm² in 7wt. % HF/Water solution for 40 minutes, and dried through Ethanol and Pentane for 15 minutes; respectively.

The single needles are bundled through a self-bonding mechanism due to induced capillary force during rinsing and drying processes. However, the surfaces with single needles can be obtained by some additional treatments after needle formation [1].

The bundled needle density, diameter of curvatures (tip diameter of bundled needles) and heights of bundled needles are measured (Table. 1) via optical technique using SEM pictures since the surface structures could not be investigated through AFM due to large interaction between AFM tip and needles and height of needles itself.

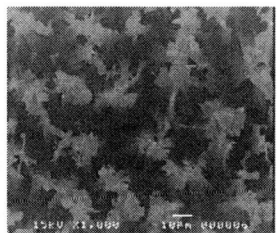

Figure 1: SEM pictures of generated bundled Velcro-like surfaces. Left: The needle like surface in 45° tilted view in respect to x-axis. Right: The needle like surface in top view.

Table 1: Surface properties of the specific fabricated Velcro-like surface.

Wafer Resistivity (Ωcm)	10-20
Bundled needle density ($No./cm^2$)	2.134×10^5
Diameter of curvature of the bundled needles (μm)	Max = 16.2 Min = 4.4 Ave =10.4
Height of bundled needles (μm)	Max = 56.8 Min = 37.5 Ave = 46.9

BONDING AND FORCE MEASURMENT

The generated needle-like surfaces are cut with a laser to small ($0.64cm \times 0.64cm$) chips and bonded together by pressing them with different applied bonding load at the room temperature. The bond strength (pull-off force) is measured at the point where the bonded chips are detached via a special designed force measurement unit, and the results are shown in Figure. 2. A maximum bonding force of 11.2N is obtained for a 6.33kg bonding load which corresponds to bond strength of 0.273MPa.

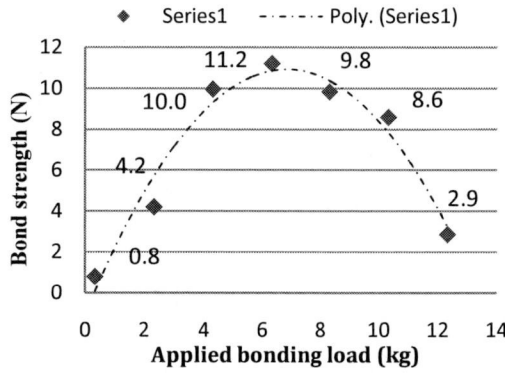

Figure 2: Measured bond strength at various applied bonding load for active surface area of 0.41 cm².

The reduction of distance between the two surfaces as function of applied bonding load is measured by a laser triangulation device (Keyence LC2430) at four different points, and the average values for three different measurements (M1-M3) are shown in Fig.3.

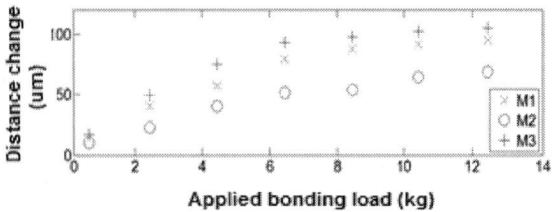

Figure 3. The measured change of distance between two surfaces as function of applied bonding load.

BONDING MECHANISM

Surface forces and nanoscale surface interactions play significant role in bonding mechanism between two surfaces. Various interactive forces such as Van der Waals forces, electrostatic forces, capillary forces, and hydrogen bridging can be observed within two solid surfaces [4]. In the case of hydrophilic surfaces or existence of humidity, the capillary force is always dominant over other forces [3]. The Van der Waals forces can have significant influence and be dominant only in the case of dry environment [5].

When two objects are brought into contact, they essentially make a slight slit around the contact area. If the involved surfaces are lyophilic in respect to the surrounding vapor, some vapor condense into a liquid phase and form a meniscus [6]. The meniscus causes an attractive force due to direct action of surfaces tension of the liquid around the periphery of the meniscus and pulls the objects towards each other.

In general, the bonding mechanism of two Velcro-like surfaces can be divided in two parts. i) Deformation mechanism where the needles compress, bend, interlace and break ii) Adhering mechanism where the needles from opposite surfaces adhere to each other or to substrates.

DEFORMATION MECHANISM

Compressing, bending, followed by interlacing and for certain conditions breaking are the main parts of deformation mechanism of needles during bonding. All four or three can occur to a needle/needles depending on the applied bonding load and the geometry of the involved needles.

A simple model for the deformation mechanism of the needles during the bonding can be described as follow: when two needle-like surfaces are brought into contact without external force, needles from opposing surfaces touch each other mostly from tips and aside from their centerlines and system stays at first equilibrium state. A slight compression on tips of needles followed by buckling or lateral bending, sliding and maybe breaking of needles occur when a bonding load is applied. At this stage, the tallest needles which have high aspect ratios and touch the counter surfaces first are compressed and bent laterally sooner and more than the shorter ones. The laterally bent needles slide asides and interlace with opposing needles. Normally, bent needles have much

lower stability than straight needles, and are less resistive to bonding loads; hence, they cannot generate sufficient counter force and break and shorten easily. Once the tall needles with high aspect ratio break and shorten, other needles endure more force: the force which is already applied on them plus the force which had caused the tall needles to break. This additional force causes lower aspect ratio needles to bend or break as well. This process continues until the system reaches a new equilibrium state where the shorter needles / larger number of needles are involved in counter force mechanism, and generate sufficient counter force to stand the applied bonding load.

Compression

A simple model can be assumed for occurrence of compression during bonding. Normally, compression may occur to a needle at two stages during the bonding. First, at the beginning when the bonding load is applied to the needle. At this stage the compression is fully elastic and followed by sliding; second, at the last stage when the needles are interlaced and the system is reaching the new equilibrium state. At this point, the compression can either elastic or fracture.

Since a needle is presented as a cylinder with a hemisphere head, its total compression can be simply expressed as summation of compression of a hemisphere and a cylinder under the same amount of a load.

The compression length ω of a hemisphere can be obtained through a model proposed in [7] as:

$$F = \frac{4}{3} E R^{1/2} \omega^{3/2} \tag{1}$$

Whereas the compression length of a cylindrical part can be calculated by [8]:

$$F = AE\omega L^{-1} \tag{2}$$

Where F is the compressive force, E is Young's Modulus, R is the radius of hemisphere and cylinder, L is the length of the cylinder, and A is the cross sectional area of the cylinder.

Bending

The lateral bending behavior of a needle can be described as a cylindrical cantilever beam under a shear point load. By using the strain stored energy of the cantilever beam and taking its first derivative with respect to applied load, the displacement δ of a beam in the direction of applied load can be obtained through [8]:

$$\delta = \frac{1}{3} FL^2 E^{-1} I^{-1} \tag{3}$$

where $I = \frac{\pi R^4}{4}$ is the moment of inertia of the cylinder with radius R.

Breaking

If the maximum stress σ applied to a needle is larger than its ultimate yield strength σ_{yield}, the needle will fail or break. As the needles are out of brittle c-Si, they will break directly after leaving the elastic range, without undergoing plastic deformation. Using cylindrical cantilever beams, the maximum stress on a needle (pint B in Fig. 5) can be obtained by combining stress induced by shear load P, and axial load F (Fig. 5) component of the bonding load as [8]:

$$\sigma_x = FA^{-1} + PRaI^{-1} \qquad (4)$$

Here, a is the distance between the applied load and the point where the stress needs to be calculated.

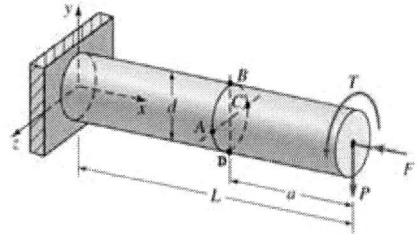

Figure 5:A cantilever beam under various point loads [8].

ADHERING MECHANISM

The adhering mechanism between needle-like surfaces can be modelled based on enhanced uncoupled multi-asperity model [9] where the influence of each needle is considered separately from the others, and total bonding force is obtained by summing up all contributing needles by taking both tip and side interaction of needles into account. In this way, the interaction between two needle-like surfaces is simplified to interaction between one needle-like surface and a flat surface [10], a needle is represented as a cylinder with hemisphere head (tip), and total interaction is divided into two parts (Fig.5): a) interaction between tip and substrate (F_{T-S}) b) interaction between neighbored or interlaced needles (F_N).

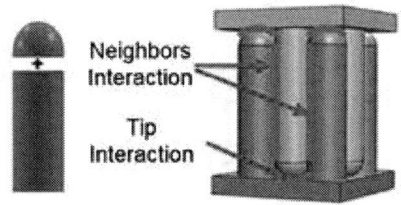

Figure 5: Schematic representation of a needle as a cylinder with a hemisphere head and their interactions in bonding mechanism. Blue lines: possible capillary bridges

The total adhering force at any distance (d) between two surfaces is then obtained through:

$$F_{adhering} = 2\sum_0^m \left((F_{T-S}) + \sum_0^n (F_N) \right) - \sum_0^n (F_N) \qquad (5)$$

where m is the number of needles in one surface which has $\sigma_x > \sigma_{yield}$, and n is the number of neighbors for each needle.

Capillary forces between the tip and the substrate

The model presented in [11] is used to describe the capillary forces between the tip of a needle and the substrate. A sphere with radius R stays at the distance D away from a plane (Fig. 6). The liquid enclosed has a filling angle of β and forms meniscus with radius r, and contact angles θ_1 and θ_2 with the sphere and the plane, respectively. According to Butt and Kappl, the capillary force can be described by Laplace pressure difference as:

$$F_{T-S} = 2\pi\gamma R \left(\cos(\theta_1 + \beta) + \cos\theta_2 - \frac{D}{r} \right) \qquad (6)$$

At constant vapor pressure, r can be calculated through Kelvin equation as:

$$r = -\frac{\lambda_K}{\ln\left(\frac{P}{P_o}\right)} \qquad (7)$$

where λ_K is constant and called Kelvin length, P is actual vapor pressure, and P_o is the saturation vapor pressure.

The capillary force decreases linearly by increasing distance between two objects until the meniscus breaks, and the range in which the capillary force are present follows the subsequent relation:

$$\cos(\theta_1 + \beta) + \cos(\theta_2) \geq \frac{D}{r} \qquad (8)$$

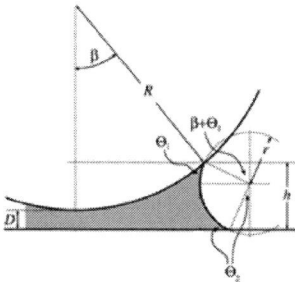

Figure 6: 2D schematic of sphere and plan interaction and its geometry parameters (Retrieved from[11, 12]).

Capillary force between interlaced needles

To describe the capillary force between interlaced needles, the cylinder-plane model proposed in [12] is modified and employed. For a cylinder with radius R at the distance D away from a plane (Fig. 6), the capillary force can be expressed as:

$$F_N = 2L\gamma \left(\frac{R\sin\beta}{r} + \sin(\theta_2 + \beta) \right) \qquad (9)$$

where L is the length of the cylinder, and γ is the surface tension of the liquid.

In the case of two parallel cylinders, $R_{eq} = \frac{R_1 R_2}{R_1 + R_2}$ should be considered instead of R [13, 14], and common length between cylinders should be used instead of L. R_1 and R_2 are the radiuses of the interacting cylinders.

SIMULATION

The measured parameters (Table 1) are fed into the model programmed in MATLAB. For an active bonding area of 0.41cm², 874000 needles are generated with normal distributions in height and radius of curvatures (s. Fig. 1). The overall capillary force in respect to the distance between two surfaces for various filling angle (β) considering only one neighbor for each needle is calculated and shown in Fig. 7. A relative humidity of

45% is considered to compute radius of curvatures of the meniscus and the contact angles θ_1 and θ_2. The 70% of SiO_2 yield strength is assumed as breaking point of needles while they were under a load force which acted on them with angle of 70° in respect to x-axis.

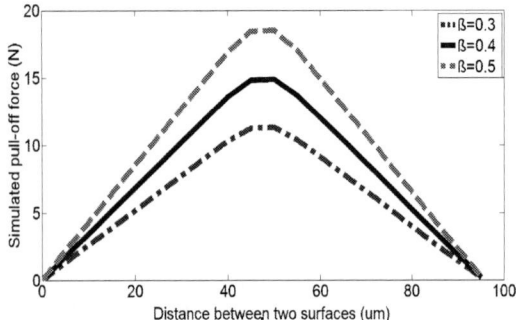

Figure 7: The total simulated capillary force in respect to distance between surfaces for various filling angles (ß).

The change in distance between two surfaces in respect to applied bonding load (Fig. 8) is calculated by finding the
corresponding distance between two surfaces for each measured bonding strength in the simulation curves, and subtracted it from twice the maximum height of needles (distance between two surface at equilibrium state when no bonding load applied).

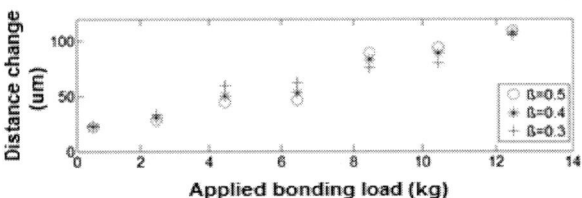

Figure 8: The calculated change of distance between two surfaces in respect to various applied bonding loads.

COCLUSION

A simple Silicon based technology to generate needle-like surfaces which can result in strong permanent bonds at room temperature similar to Velcro principle is presented. A sufficient bonding force to attach and hold a 27mg chip (~ weight of 23mm², 525µm Si wafer) with active bonding area of 1mm² against 1000g acceleration can be obtained using this technique. The bonding mechanism between such surfaces is modelled by taking both interaction and deformation mechanism of needles into account. The interaction mechanism is expressed by means of capillary force approach whereas the deformation mechanism is described mainly through cantilever beam approach. Both measurement and simulation results follow the same shape of the curve and are very close in term of amplitude (simulation curve with filling angle ß=0.3°) which illustrates the validity and accuracy of the proposed bonding mechanism.

ACKNOWLEDGEMENTS

The authors would like to thank the Ministry of Science, Research and Art at the state of Baden Württemberg, Germany for providing the financial support in the framework program "Kooperatives Promotionskolleg Generierungsmechnismen für Mikrostrukturen".

REFERENCES

[1] P. Jonnalagadda, U. Mescheder, A. Kovacs, and A. Nimoe, "Nanoneedles based on porous silicon for chip bonding with self-assembly capability," *Phys. Status Solidi C*, vol. 8, no. 6, pp. 1841–1846, 2011.

[2] S. Keshavarzi, U. Mescheder, and H. Reinecke, "Characterization and simulation of low temperature Si-Si direct bonding through Velcro-like surfaces based on porous silicon," in *2014 IEEE 27th International Conference on Micro Electro Mechanical Systems (MEMS)*, pp. 1119–1122.

[3] Y. Rabinovich, J. Adler, M. Esayanur, A. Ata, R. Singh, and B. Moudgill, "Capillary forces between surfaces with nanoscale roughness," *Adv. in Col. and Inter. Sci.*, vol. 96, no. 1-3, pp. 213–230, 2002.

[4] P. Prokopovich and V. Starov, "Adhesion models: from single to multiple asperity contacts," *Adv. in Col. and Inter. Sci.*, vol. 168, no. 1-2, pp. 210–222, 2011.

[5] T. S. Chow, "Nanoscale surface roughness and particle adhesion on structured substrates," *Nanotechnology*, vol. 18, no. 11, p. 115713, 2007.

[6] H.-J. Butt and M. Kappl, *Surface and Interfacial Forces*. Wiley-VCH Verlag GmbH & Co., 2010.

[7] L. Kogut and I. Etsion, "Elastic-Plastic Contact Analysis of a Sphere and a Rigid Flat," *J. Appl. Mech*, vol. 69, no. 5, p. 657, 2002.

[8] J. M. Gere and S. Timoshenko, *Mechanics of materials*, 4th ed. Boston: PWS Pub Co, 1997.

[9] G. G. Adams and M. Nosonovsky, "Contact modeling- forces," *Tribology Inter.*, vol. 33, pp. 431–442, 2000.

[10] P. Prokopovich and S. Perni, "Comparison of JKR- and DMT-based multi-asperity adhesion model: Theory and experiment," *Colloids and Surfaces A: Physicochemical and Engineering Aspects*, vol. 383, no. 1-3, pp. 95–101, 2011.

[11] H.-J. Butt and M. Kappl, "Normal capillary forces," *Adv. in Col. and Inter. Sci.*, vol. 146, no. 1-2, pp. 48–60, 2009.

[12] J. Vitard, S. Regnier, and P. Lambert, "Study of cylinder/plan capillary force near millimeter scale and experimental validation," in *2007 IEEE International Symp. on Assembly and Manufacturing*, pp. 221–226.

[13] S. Chen and T. Wang, "General solution to two-dimensional nonslipping JKR model with a pulling force in an arbitrary direction," *Journal of colloid and interface science*, vol. 302, no. 1, pp. 363–369, 2006.

[14] Q. Li, V. Rudolph, and W. Peukert, "London-van der Waals adhesiveness of rough particles," *Powder Technology*, vol. 161, no. 3, pp. 248–255, 2006.

CONTACT

*S. Keshavarzi, tel: +49-7723-9202809; kesh@hs-furtwangen.de.

FET PROPERTIES OF SINGLE-WALLED CARBON NANOTUBES INDIVIDUALLY ASSEMBLED UTILIZING SINGLE STRAND DNA

Kosuke Hokazono, Yoshikazu Hirai, Toshiyuki Tsuchiya, and Osamu Tabata
Kyoto University Graduate School of Engineering, Kyoto, JAPAN

ABSTRACT

This paper reports a new assembly process for isolated single-walled carbon nanotubes (SWCNTs) on MEMS structures utilizing single strand DNA (ssDNA) and the electrical properties of SWCNT field effect transistors (FETs). Mono-dispersed SWCNT solution was prepared by wrapping biotin modified ssDNA around SWCNT and the SWCNTs were assembled onto gold electrodes using biotin-avidin bindings. The isolated SWCNT bridges between electrode gaps of 100 to 300 nm wide were successfully demonstrated. To improve electrical contacts, electroless gold deposition was employed. The I_d-V_g curves in back-gate FET configurations showed either metallic or semiconducting properties and we confirmed that the isolated SWCNTs were individually assembled between the electrodes.

INTRODUCTION

SWCNT is a nano-carbon material with a structure of rolled up graphene sheet. It has many attractive properties such as high electron mobility, high Young's modulus and low mass density. That is why it is expected to be applied to elements of nano devices. Especially, a lot of researches on SWCNT-FET have been done [1, 2], which uses its electronic properties. Moreover, gas sensors [3] and biosensors [4], which utilize SWCNT's large specific surface area, have been proposed. For these kinds of applications, a batch integration of SWCNTs onto the device is required. Dielectrophresis [5], localized growth [6] and transfer from a single crystal quartz wafer [7] have been developed by other researchers. However, it is difficult to avoid bundling in the dielectrophoresis method, and it is difficult to control chirality in the growth method and the transfer method. We are developing a parallel integration of single SWCNTs onto MEMS device structures by the newly proposed method using ssDNA.

ASSEMBLY METHOD

The proposed assembly method is schematically drawn in Figures 1. At first, SWCNTs are mono-dispersed by wrapping biotin modified ssDNA [8]. Biotinylated thiol is assembled onto gold electrodes. The dispersed SWCNTs are assembled onto gold electrodes using biotin-avidin bindings. Finally, electroless gold deposition is employed in order to improve electrical contacts.

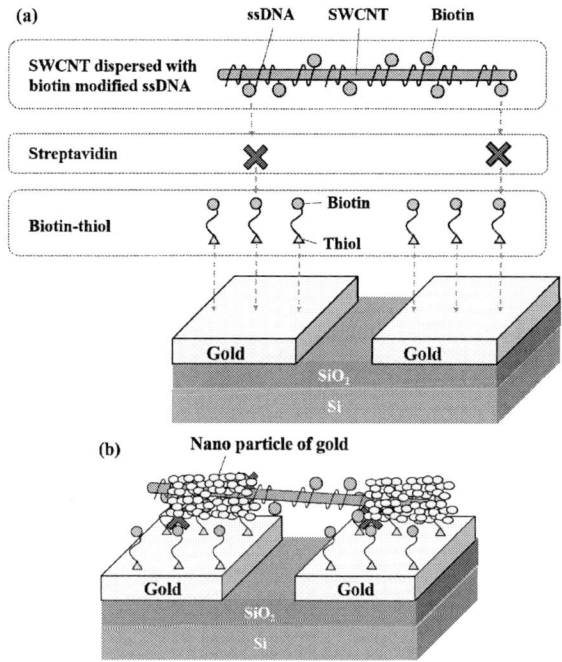

Figure 1: SWCNT assembly method. (a) binding using biotin modified ssDNA (b) electroless gold deposition.

EXPERIMENTS

Dispersion of SWCNTs using biotin modified ssDNA

CoMoCAT SWCNT in powder style (Sigma Aldrich Japan, 704113) was used in experiment. This product contains both metallic and semiconducting SWCNTs. The sequence of ssDNA (Operon Biotechnology) was designed to be 40 thymines according to the report that poly-thymine has the highest dispersion efficiency among homopolymers [8]. 0.01 mg of SWCNT was suspended with 60 μL aqueous solution (32 μM ssDNA and 0.1 M NaCl) and ultrasonically agitated for 90 min. Obtained solution was separated by centrifugation for 90 min at 16000 g, 4 °C. After centrifugation, the supernatant was collected, which is called as the sample solution below.

The dispersion of SWCNT was confirmed using an atomic force microscope (AFM; Research Institute of Biomolecule Metrology, Nano Live Vision) in a liquid environment, a transmission electron microscope (TEM; JEOL, JEM2100) and near-infrared fluorescence spectroscopy. The sample for TEM observation was prepared by dropping the sample solution on a TEM grid with 20 μm holes and drying in a vacuum chamber.

SWCNT assembly onto gap electrodes

Gold gap electrodes were fabricated on a Si substrate with 100 nm thick thermal oxide. The designs of the electrodes are shown in Figure 2. The electrodes of triangular and rectangular shapes of 100 to 300 nm gap were fabricated with the processes using electron beam

(EB) lithography, EB deposition of chromium and gold and lift-off. One chip with 1 cm square has 10 pairs of electrodes of each design.

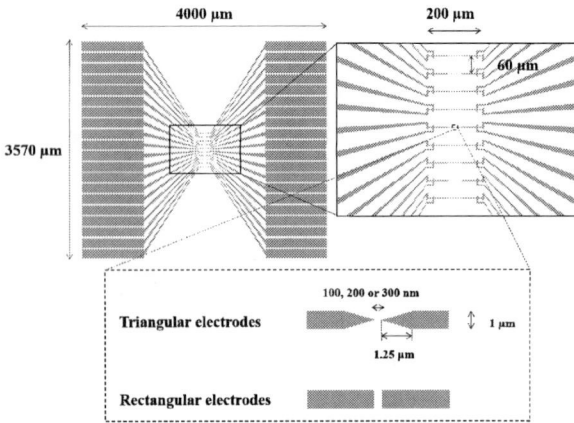

Figure 2: Electrode design.

Fabricated chips were cleaned with piranha solution and soaked into 20 μM biotinylated thiol (ProChimia Surfaces, FT 005) ethanol solution for 12 hours and rinsed with ethanol and distilled water. The chips were soaked into 5 μg/mL streptavidin (Aldrich Japan, 85878) aqueous solution for 1 hour and were rinsed with distilled water. 2 uL of the sample solution was dropped on the electrode patterns and the chips were incubated for 1 hour and were rinsed with distilled water.

Electroless gold deposition (Nanoprobes, GoldEnhance LM/Blot) was performed on the samples after SWCNT assembly for the improvement of electrical contacts.

Measurement of FET properties

The electrical properties of the samples were measured as back-gate FET configurations, as schematically shown in Figure 3. The drain current I_d as a function of the gate voltage V_g was measured at the constant source-drain voltage of V_{sd} = 30 mV.

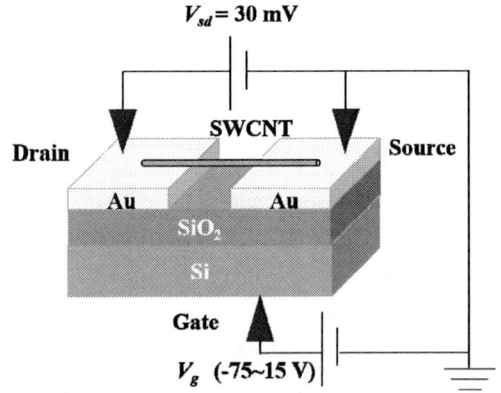

Figure 3: Measurement setup of SWCNT-FET.

RESULTS

Dispersion of SWCNTs using biotin modified ssDNA

AFM images of the sample solution are shown in Figures 4. There are a lot of tube-like objects which have

about 2 nm height and length ranging from 200 nm to 700 nm as shown in Figure 4a. Figure 4b shows a magnified image. A tube at the center looks wrapped by thin lines with 0.8 nm height indicated by the yellow arrows. Figure 5 shows the TEM image of suspended tube-like objects which have the average width of 2.6 nm between islands around the hole of the TEM grid. The islands were unstable during the observation and the image was unclear. The islands are thought to be crystals of NaCl and the charge up seems to be the reason for instability. Considering that SWCNTs which have the diameter of about 1 nm are wrapped with biotin modified ssDNA, those in the AFM and TEM images should be isolated SWCNTs.

The result of fluorescence spectroscopy exhibited some peaks as shown in Figure 6, which also supports the dispersion of SWCNTs in the solution. Each peak corresponds to the chirality of (6, 5), (7, 5) and (8, 3) [9].

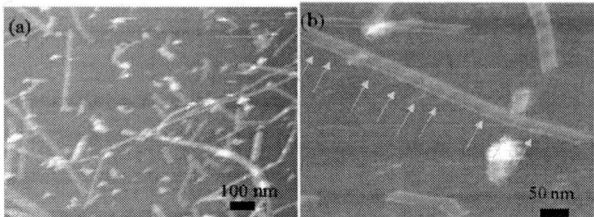

Figure 4: AFM images of ssDNA modified SWCNT.

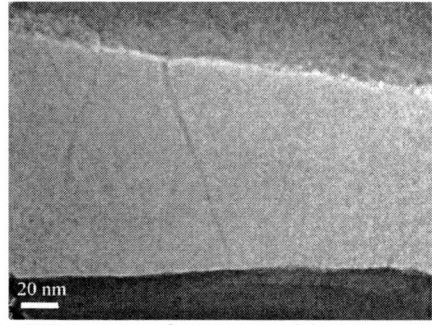

Figure 5: TEM image of ssDNA modified SWCNT.

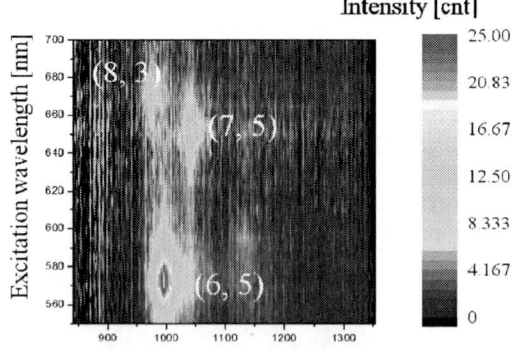

Figure 6: Fluorescence spectroscopy of ultrasonicated solution. Described (n, m) indicates the chirality of each peak according to the reference [9].

SWCNT assembly onto gap electrodes

The electrode gaps after SWCNT assembly were observed using a field emission scanning electron microscope (FE-SEM; Hitachi High-Technologies, SU8000). Figure 7 shows SWCNTs bridged over the electrode gap. There was no tube on Si substrate surface. From these results, specific assembly of SWCNTs onto gold electrodes was confirmed.

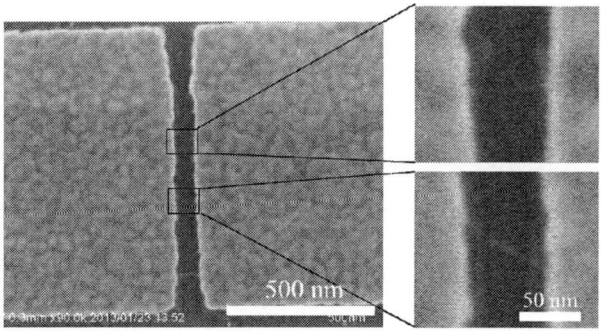

Figure 7: SWCNTs bridged on the electrode gap.

FET properties

The I_d-V_g curves for the samples before and after electroless gold deposition are shown in Figures 8 and 9, respectively. Before gold deposition, all samples exhibit a low drain current of less than 1 nA showing a hysteresis as shown in Figure 8. The wrapping ssDNA and biotin-avidin bonding may inhibit electrical contacts. After gold deposition, however, some exhibit a large and constant current as shown in Figure 9a, and the others exhibit p-type FET properties as shown in Figure 9b, which indicate on-state I_d at negative V_g. The existence of the two different properties is consistent because the sample solution contained both metallic and semiconducting SWCNTs. In other words, properties like Figures 9a and 9b are from metallic and semiconducting SWCNTs, respectively. Figure 9c is a logarithmic plot of Figure 9b which confirms that the FET properties are from SWCNTs assembled on the electrode gaps because it is known that FETs exhibit exponential curves at the sub-threshold region [10], which is appeared as a linear line in the logarithmic plot as indicated in the figure.

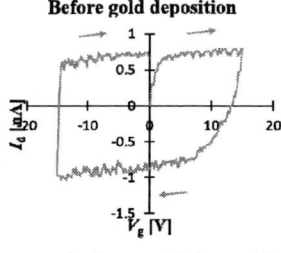

Figure 8: I_d-V_g curve before gold deposition.

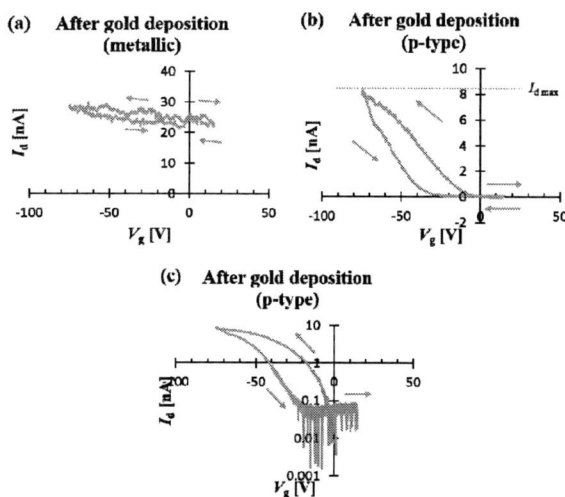

Figure 9: I_d-V_g curves from the sample after gold deposition. (c) is logarithmic plot of (b); the orange line indicates linear fitting between -40 V and -20 V.

The percentiles of p-type, metal and mixed (p-type and metal) among 10 gaps for each design are shown in Table 1. Figures 10 show the maximum drain current $I_{d\,max}$ of p-type FETs extracted from the results. $I_{d\,max}$ should be proportional to the number of SWCNTs between the electrodes [11]. The triangular electrodes with 200 nm and 300 nm gap may have only one SWCNT because $I_{d\,max}$ has similar values and small deviations. The other electrodes may have multiple tubes, which agree with the FE-SEM image shown in Figure 7. Supposing the triangular electrodes with 200 nm and 300 nm gap have only one SWCNT, the mean number of SWCNTs assembled between the triangular electrodes with 100 nm gap is 4.5. The mean value of $I_{d\,max}$ of the electrodes with 200 nm gap is 0.67 nA, which is smaller than the on-state current of 30 nA reported in the reference [1], which uses a SWCNT prepared by chemical vapor deposition directly on a silicon substrate with 225 nm thermal oxide and gold for the contacts on the SWCNT. The all samples did not saturate in spite of high negative voltage. This fact indicates the weak gate coupling in our samples. The slopes d(log I_d)/dV_g of the exponential parts of the curves are shown in Figures 11. All slopes take the similar values irrespective of the numbers of tubes estimated from $I_{d\,max}$, and these values also look small compared with the reference [1], which also demonstrate the weak coupling of our samples.

Table 1: Measured conductor types of assembled SWCNTs

Electrode shape	Gap distance [nm]	p-type [%]	Mixed [%]	Metallic [%]
Triangular	100	100	0	0
	200	80	0	0
	300	70	0	0
Rectangular	100	50	30	20
	200	40	40	20
	300	60	40	0

The difference of slope comes from the configuration difference. The samples in this research have SWCNTs located on the gold contacts, which differ from the gold contacts deposited on the SWCNT in the reference [1].

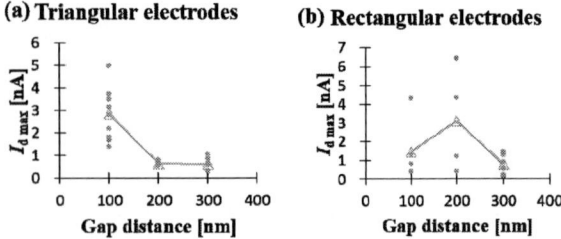

Figure 10: $I_{d\ max}$ of samples with triangular electrodes (a) and rectangular electrodes (b) which exhibit p-type FET properties after gold deposition; blue plots indicate measured values and orange triangles indicate average values.

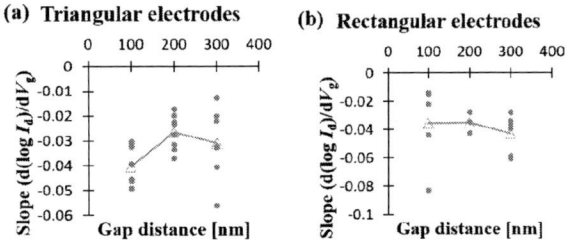

Figure 11: Slope at sub-threshold region of p-type SWCNT after gold deposition; blue plots indicate measured values and orange triangles indicate average values.

CONCLUSIONS

A novel assembly process for making isolated SWCNTs on MEMS structures by wrapping ssDNA has been proposed in this paper. The pair electrodes with gaps of 100 to 300 nm wide were fabricated using the EB lithography and lift-off process. The isolated SWCNTs were successfully assembled over the gold electrode gaps using biotin-avidin bindings. Electroless gold deposition was performed for improvement of electrical contacts. The I_d-V_g curves in back-gate FET configurations showed either metallic or semiconducting properties. From the maximum drain current $I_{d\ max}$, we confirmed that the isolated SWCNTs were individually assembled at the gap of the triangular electrodes with 200 nm and 300 nm in the distance.

Our work has a great significance because our novel assembly method may open the door to fabricate hundreds of SWCNT-FETs or SWCNT sensors in parallel.

ACKNOWLEDGEMENTS

We would like to express my gratitude to Prof. Hierold and Mr. V. Döring on the electrical measurement at ETH Zurich and Prof. Sumigawa for TEM observation.

A part of this work was supported by Kyoto University Nano Technology Hub in "Nanotechnology Platform Project" sponsored by MEXT, Japan.

REFERENCES

[1] M. Radosavljevic et al., "Nonvolatile Molecular Memory Elements Based on Ambipolar Nanotube Field Effect Transistors", *Nano Letters*, Vol. 2, No. 7, pp. 761-764, 2002.

[2] S. Rosenblatt et al., "High Performance Electrolyte Gated Carbon Nanotube Transistors", *Nano Letters*, Vol. 2, No. 8, pp. 869-872, 2002.

[3] J. Kong et al., "Nanotube Molecular Wires as Chemical Sensors", *Science*, Vol. 287, pp. 622-625, 2000.

[4] Y. Lin et al., "Glucose Biosensors Based on Carbon Nanotube Nanoelectrode Ensembles", *Nano Letters*, Vol. 4, No. 2, pp. 191-195, 2004.

[5] R. Krupke et al., "Simultaneous Deposition of Metallic Bundles of Single-walled Carbon Nanotubes Using Ac-dielectrophoresis", *Nano Letters*, Vol. 3, No. 8, pp. 1019-1023, 2003.

[6] M. Muoth et al., "Hysteresis-free Operation of Suspended Carbon Nanotube Transistors", *Nature Nanotechnology*, Vol. 5, pp. 589-592, 2010.

[7] N. Patil et al., "Wafer-scale Growth and Transfer of Aligned Single-walled Carbon Nanotubes", *IEEE Transactions on Nanotechnology*, Vol. 8, No. 4, pp. 498-504, 2009.

[8] M. Zheng et al., "DNA-assisted Dispersion and Separation of Carbon Nanotubes", *Nature Materials*, Vol. 2, pp. 338-342, 2003.

[9] S. M. Bachilo et al., "Structure-assigned Optical Spectra of Single-walled Carbon Nanotubes", *Science*, Vol. 298, pp. 2361-2366, 2002.

[10] S. M. Sze, *Semiconductor Devices, Physics and Technology*, John Wiley and Sons, 1985.

[11] A. D. Franklin et al., "Current Scaling in Aligned Carbon Nanotube Array Transistors With Local Bottom Gating", *IEEE Electron Device Letters*, Vol. 31, No. 7, pp. 644-646, 2010.

CONTACT

* K. Hokazono, tel: +81-75-383-3693; k_hokazono@nms.me.kyoto-u.ac.jp

IMPACT-INDUCED HARDENING PACKAGE
FOR TACTILE SENSORS USING DILATANT FLUID
Tomoyuki Takahata, Kiyoshi Matsumoto, and Isao Shimoyama
The University of Tokyo, Tokyo, JAPAN

ABSTRACT

To realize high-sensitive and shock-resistant tactile sensor, the sensing element was surrounded by dilatant fluid, which is soft against a static force and hard against an impact force. The applied static force was concentrated to the sensor, whereas the impact force was dispersed to the substrate. We have experimentally shown that the shock-resistant nature of the sensor with dilatant fluid package was 4 to 16 times as large as that of without the fluid.

INTRODUCTION

We propose that tactile sensors' shock-resistant nature is enhanced by surrounding the sensing element by dilatant fluid (Figure 1). Dilatant fluid is solidified by impact normal stress [1], as well as strong shear stress [2], that means the fluid is normally soft and gets hard only when it receives an impact force.

High-sensitive tactile sensors are needed to give robots tactile sensation. At the same time, the sensor should be shock-resistant when the robots work in the real world, in which the sensor was attacked by something. Tactile sensors have generally soft structure to enhance the sensitivity. However the soft structure may be broken by an impact force.

Since the dilatant fluid package utilize the applied stress to change its hardness, it does not need any additional energy. This is the key point different from other hardness-tunable fluid like electrorheological fluid. One can make a tactile sensor shock-resistant only by embedding the sensor into dilatant fluid.

DESIGN

Figure 2 shows cross sections of devices used in experiments on the shock-resistant behavior of the dilatant fluid package. As a representative example of tactile sensors, a strain sensor embedded in an elastic body (polydimethylsiloxane, PDMS) was used. Device A had the dilatant fluid package (Figure 2(a)). Device B and C were used for control experiments. Device B is an example of sensitive tactile sensors (Figure 2(b)). The applied force was concentrated to the sensing element owing to the small-area elastic body. Device C is an example of shock-resistant tactile sensors (Figure 2(c)). The force was dispersed by the large-area elastic body, sacrificing the sensitivity.

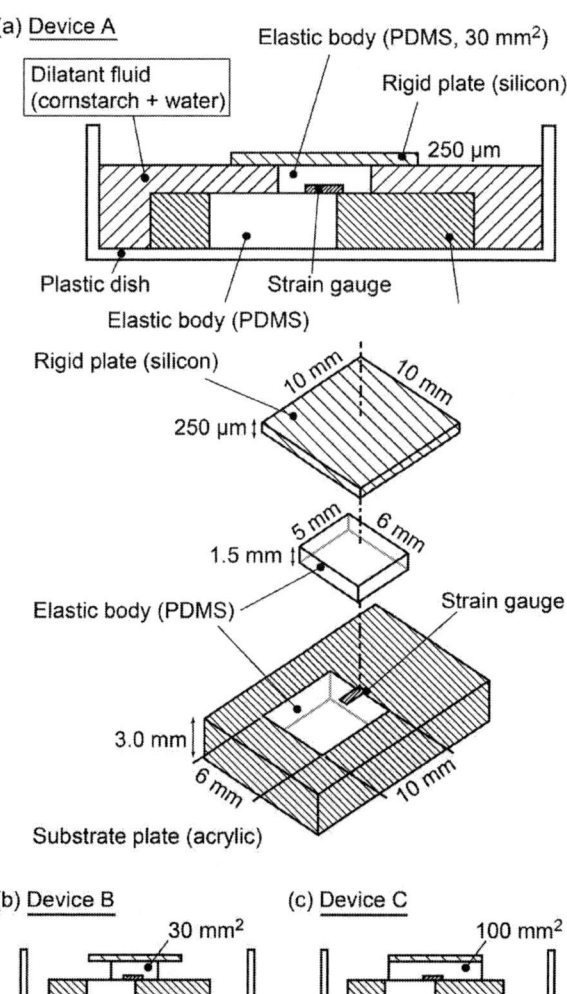

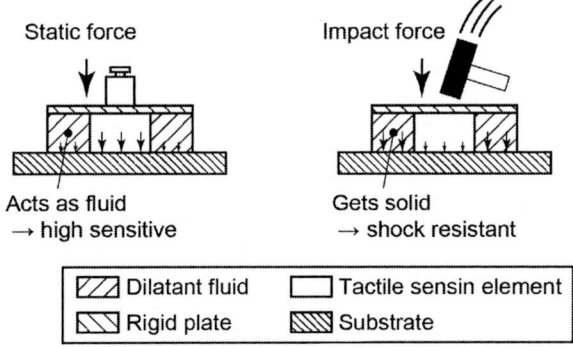

Figure 1: A concept sketch of the shock-induced hardening package using dilatant fluid.

Figure 2: Cross sections of devices used for experiments on the shock-resistant behavior of dilatant fluid package. In all devices, the same rigid plate (10 mm×10 mm silicon wafer) and substrate plate, and the strain sensor was used. (a) The upper elastic body (6 mm×5 mm) was surrounded by dilatant fluid. (b) The device with smaller elastic body (6 mm×5 mm). (c) The device with wider elastic body (10 mm×10 mm, same as the cover plate).

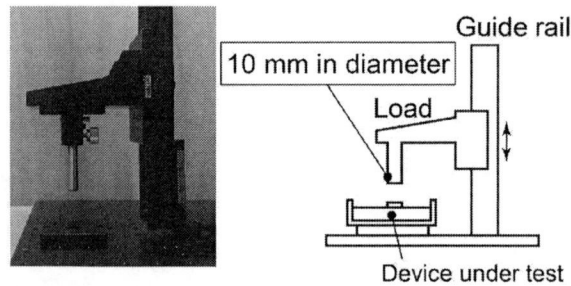

Figure 3: Experimental setup for applying static load and impact.

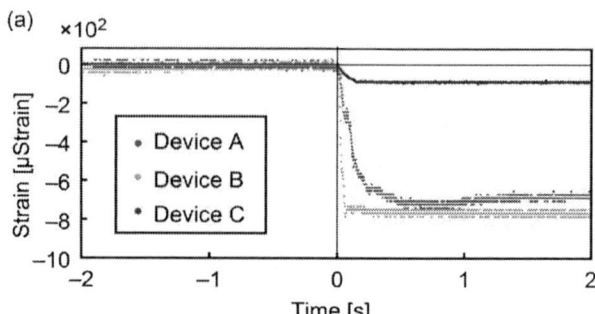

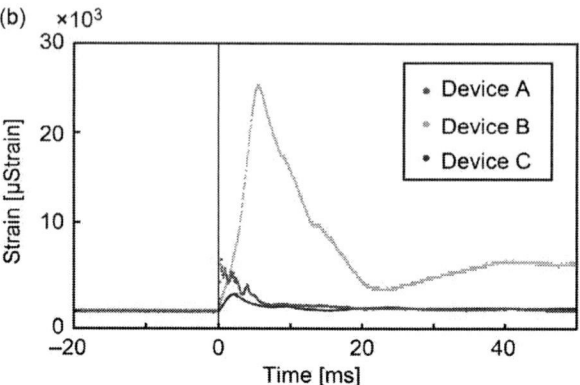

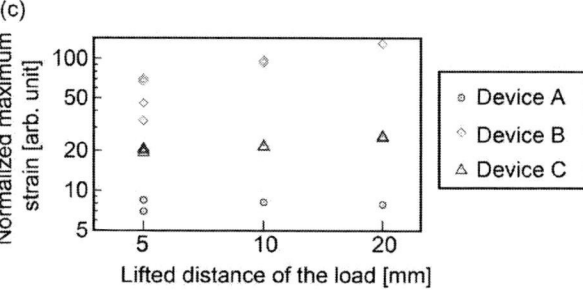

Figure 4: Results of experiments. (a) Measured strain of static force. The load was taken off at 0 s. (b) Measured strain of impact force when the lifted distance was 5 mm. The load was contact with the device at 0 ms. (c) Double logarithmic plot of normalized maximum strain versus lifted of the load.

The diameters of cornstarch granules were about 40μm, ranging from 10 to 80 μm. The granules of the surface of the dilatant fluid were be able to be seen.

A load of 0.5 kg in mass was placed on a guide rail in order to be let fall vertically (Figure 4). In experiment (1) of applying static force, the load was initially put on the rigid plate then taken off. In experiment (2) of applying impact force, the load was lifted by 5, 10, and 20 mm, and was let fall to the device.

RESULTS

As results of experiment (1) (Figure 5(a)), the strain due to the load of the device A, B, and C was 6.9×10^2, 7.6×10^2, and 9.0×10 micro strains, respectively. By comparing A with B, the sensitivity for static force was reduced by only 9.2% using the dilatant fluid package. By comparing C with B, the sensitivity of the device C was reduced by 88% because of the large-area elastic body.

The results of experiment (2) with 5 mm lifting distance was shown in Figure 5(b). The maximum strain of the device A, B, and C were 5.8×10^3, 2.5×10^4, and 1.9×10^3 micro strains, respectively. Comparing A with B, the maximum strain was reduced by 77%. From the perspective of shock-resistance, the maximum strain is important because a tactile sensor is broken when the applied strain exceeds the limit of the soft structure in the sensor.

Figure 5(c) shows maximum strains with the impact forces normalized by the strains with the static force. The smaller normalized maximum strain indicates that the device is more shock-resistant keeping high-sensitivity. The value of the device A was ranging from 7.0 to 8.5, whereas the values of the device B and C exceed 30 and 20, respectively. The sensor surrounded by dilatant fluid was 4 to 16 times as shock-resistant as that without fluid.

CONCLUSION

We have proposed the method to enhance the shock-resistant nature of tactile sensors by surrounding them with dilatant fluid. In static force experiments, the sensitivity of the device with dilatant fluid packaging reduced by only 9.2% than that of fluid. In impact experiments, the maximum strain was reduced by 77%. Considering shock-resistance index as the maximum strain with the impact forces divided by the strain with the static force, the device got 4 to 16 times as shock-resistant owing to the fluid.

ACKNOWLEDGEMENTS

This work was supported by JSPS KAKENHI Grant Number 25630086.

REFERENCES

[1] S. R. Waitukaitis and H. M. Jaeger, "Impact-activated solidification of dense suspensions via dynamic jamming fronts," *Nature*, vol. 487, pp. 205–209, 2012.

[2] X. Cheng, J. H. McCoy, J. N. Israelachvili, and I. Cohen, "Imaging the Microscopic Structure of Shear Thinning and Thickening Colloidal Suspensions," *Science*, vol. 333, pp. 1276–1279, 2011.

CONTACT

*T. Takahata, takahata@leopard.t.u-tokyo.ac.jp

MICRO DEVICES INTEGRATION WITH STRETCHABLE SPRING EMBEDDED IN LONG PDMS-FIBER FOR FLEXIBLE ELECTRONICS

Wei-Lun Sung[1], Chao-Lin Cheng[1], and Weileun Fang[12]

[1]Department of Power Mechanical Engineering, and [2]Institute of NanoEngineering and MicroSystems, National Tsing Hua University, Hsinchu, Taiwan

ABSTRACT

This study presents a PDMS (Polydimethylsiloxane) fiber integrated with multi devices scheme using stretchable electroplating copper spring. Each device was located on the node and embedded in PDMS-fiber. Thus, devices are mechanically connected by PDMS-fiber and electrically connected by inner stretchable spring. Thus, large-area and flexible applications can be achieved. Advantages of this approach: (1) length magnification by stretchable spring; (2) thicker stretchable spring embedded in PDMS provides well mechanical and electrical characteristics; (3) node acts as a hub for devices implementation and integration; (4) partially stretched spring could reduce the resistance variation by external loads. A 6.2cm PDMS-fiber with sensors and LED is implemented using a 2.4cm node-spring components fabricated on 4-inch wafer. PDMS-fiber longer than 30cm can be achieved using different spring design.

INTRODUCTION

Large area flexible electronic devices have attracted attentions and also been extensively investigated in past decades. These devices were implemented to meet requirements of various innovative applications, such as flexible display [1], active antennas [2], and wearable devices [3-9]. The wearable devices especially show their potential for the integrating and interfacing of micro-processors, micro-sensors, and other functional devices with the human body. Various approaches have been reported to implement wearable, large-area, and flexible devices for different applications, such as the device with detected bio-signal information [4-5], record environment data [6], and energy storage [7]. The wearable devices with sensors and actuators can be attached to clothes or directly wrapped around non-planar biological surfaces. Moreover, the micro-devices can also be embedded in textile fibers [8]. Thus, the characteristics of flexible, stretchable, and stable in different loading conditions are required for the wearable devices.

Among the aforementioned applications, conductive fiber [4] and screen printed electrodes [5-7] technologies are developed to meet the design requirements. The conductive fiber and yarn can be coated with a very thin layer of silver or copper on polyamide. Conductive fiber can also be made of silver or copper threads twisted with natural or artificial fibers [9]. Moreover, the screen printing electrodes are adopted in textile technology. The ink is passed through and patterned by a screen through the squeegee process. Thus, the pattern can be printed on the textile [7]. However, it is challenge to integrate these technologies with traditional semiconductor/MEMS processes/devices.

Thus, this study extends the concept in [10] to design a flexible PDMS-fiber integrated with multi devices

scheme. The PDMS-fiber provides solid nodes for the implementation and integration of micro devices. Moreover, stretchable springs are fabricated by mature microelectronics fabrication and spread out for length/area magnification. Electroplated copper as stretchable springs provides well electrically connection between different devices. The discrete micro-devices are connected by stretchable springs and embedded in PDMS to form a long flexible PDMS-fiber. Thus, the stretchable spring embedded in PDMS provides the required electrical and mechanical characteristics. This study demonstrates a PDMS-fiber of 8 interconnection micro-devices to integrate with different functions devices, and further implemented a 5×5 PDMS-fiber array by weaving to form a large-area, flexible electronics application.

DESIGN CONCEPT

The proposed design of PDMS-fiber consists of solid nodes and stretched springs. Fig.1 shows the MEMS devices and stretched spring embedded in PDMS forming a long flexible PDMS-fiber. Each device can be implemented or integrated on the solid node of the 1D PDMS-fiber. The discrete micro-devices can be directly implemented by fabrication process on nodes. Moreover, the heterogeneous micro-devices will be integrated on solid nodes by bonding process. The copper stretchable

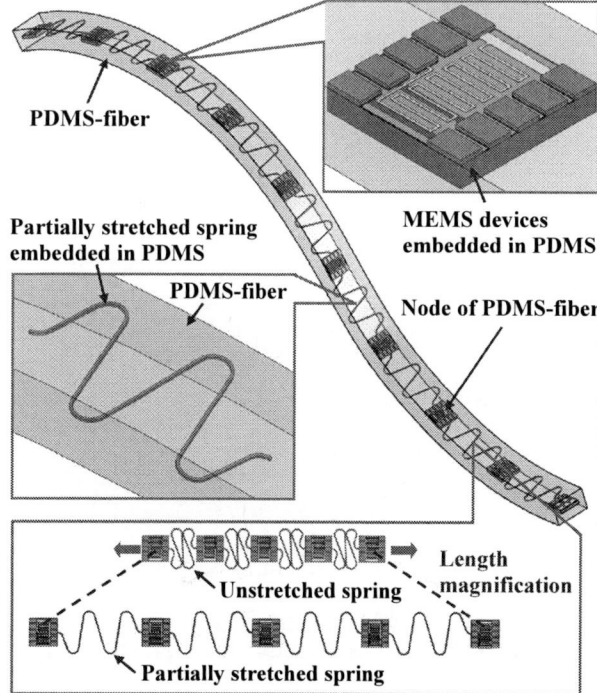

Figure 1: Schematic of PDMS-fiber with stretchable spring.

spring embedded in the PDMS is designed to provide the required electrical and mechanical properties.

In applications, the discrete MEMS devices can be expanded to increase the length by stretchable spring. This study firstly demonstrated a partially stretched spring as in Fig.1 to eliminate resistance change under different loading conditions. Note that the length magnification can be defined by spring design. This study also showed the integration of 5×5 PDMS-fiber array by weaving for 2D applications.

FABRICATION PROCESS

This study employs the existing process modules and materials to realize the stretchable electroplated Cu springs with Si/Glass sensing nodes, and further embedded the devices into PDMS. The heterogeneous integration of semiconductor/MEMS devices on sensing nodes is easily achievable. In Fig.2a, the 1st metal layers (Cr/Cu) and dielectric film (Si_3N_4) were deposited and patterned on substrate. As shown in Fig.2b, the 2nd metal layers (Cr/Au) were then deposited and patterned. The 1st and 2nd metal layers and intermediate dielectric films were mainly served as electrical routings in this study. These metal layers could also be employed as sensing element. In Fig.2c, the Ti/Cu films were deposited to respectively act as the seed layer for electroplating and as the sacrificial layer for spring suspension. The thick Cu structure layer was then electroplated and patterned. After that, dicing saw was used to define splitting trenches at backside of wafer, as in Fig.2d. A 100μm thick Si was left after the dicing process and the Ti/Cu sacrificial layers under the Cu structure is removed. As in Fig.2e-f, the splitting trenches were broken to allow the stretch of springs, and then placed inside the acrylic mold for the following PDMS molding. The liquid PDMS was filled into the acrylic mold and then cured at 100°C for 90minutes. Vacuum chamber was employed to remove the bubbles trapped in PDMS during the curing process. Finally, PDMS-fiber with embedded stretchable spring and functional devices on nodes can be obtained after removing acrylic mold as in Fig.2g. Note that various micro devices can be further implemented or integrated on nodes using bonding, bulk micromachining, etc.

RESULTS AND DISCUSSIONS

Typical fabrication results are shown in Fig.3. The unstretched spring was shown in Fig.3a. The electroplated Cu springs are suspended on the substrate, and the thickness and width of Cu spring are 20μm and 10μm, respectively. The results indicate the Cu spring has good yield and small initial deformation (by residual stress). The Cu spring will deform and stretch as the external load applied. Thus, the Cu spring was stretched for only 60% of its full extension (full extension means the springs stretched from flexures-springs to straight lines), when embedded into the PDMS fiber (in Fig.2f). Fig.3b shows the partially stretched springs embedded in PDMS. The spring has a 5mm extension in its fixed ends (the fully extensions of the spring is 8.1mm). Fig.3c shows the PDMS-fiber with embedded spring of 5mm extension is further stretch for 0.5mm. The PDMS-fiber necking can be obtained. Moreover, the Cu spring embedded in PDMS-fiber remains unbroken after stretching. Fig.3d

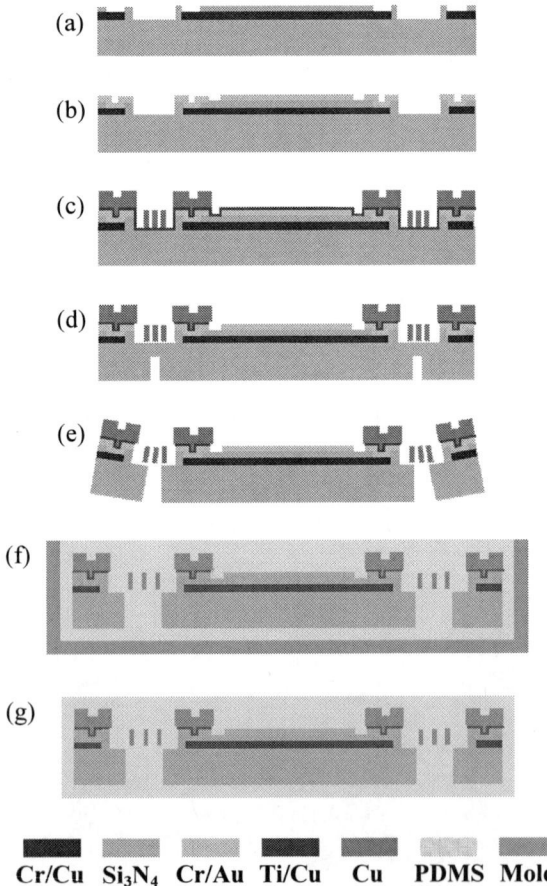

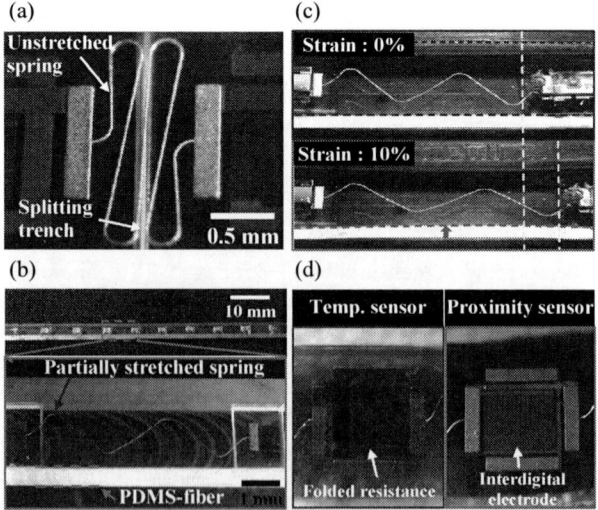

Figure 2: Fabrication process steps of long PDMS-fiber with stretchable spring base on embedded polymer architecture.

Figure 3: Typical fabrication results, (a) unstretched spring, (b) partially stretched spring embedded in PDMS, (c) springs with 0% and 10% strains and (d) sensors integration.

shows resistive-temperature and capacitive-proximity sensors integrated on nodes and embedded in PDMS. These two sensors are directly fabricated on rigid Si nodes. Nevertheless, the flexible electronics can be achieved as the discrete rigid Si nodes are connected by the deformable Cu springs.

Fig.4 shows LED chips mounting on nodes of PDMS-fiber to demonstrate the copper springs embedded in PDMS are electrically connected. In Fig.4a, the LED chips are mounted on the nodes of the PDMS-fiber. The chip dimensions of commercially available LED are 1.6mm (l)×0.8mm (w)×1.0mm (h). Solder paste is coated in between the electrodes of LED chip and nodes using the dispensing systems. After mounting the LED chip, the chip-network is heated to 230°C for 10 minutes. Finally, the PDMS-fiber with LED array is achieved through processes in Fig.2f-g. Fig.4b shows the 8 LED chips integrated and embedded in PDMS-fiber. Lighting test with 1mA demonstrates the spring for electrical routing. Fig.4c shows the preliminary bending test. The result display the PDMS-fiber could tolerate a bending radius-of-curvature of 15mm.

Fig.5 shows electrical tests under tensile, twisting, and bending forces. Fig.5a shows the setup for tensile test. The source meter is used to apply input current and measure the resistance change simultaneously. The free-ends of PDMS-Fiber are fixed on the position stages. Thus, the displacement of PDMS-fiber can be precisely controlled. The measurement results indicate the variation of resistance-change before 15% strain is less than 0.05%. As strain reach 15%, resistance is significantly changed due to spring broken. Further increasing strain to 22%, PDMS-fiber will be broken from the interface of node and PDMS. The micrographs show the LED chips are still lighting with 12% tensile strain. Moreover, the PDMS-fiber can revert to original position after unload the tensile strain. Fig.5b shows applying twisting force on PDMS-fiber for electrical test. As applied rotation angles ranging 0~1080°, variation of resistance-change is less than 0.09%. The micrographs show the twisting of PDMS-fiber with 1080° rotation angle. Moreover, the PDMS-fiber also attach on different curved surface for bending tests as shown in Fig.5c. As radius-of-curvature ranging 5mm~50mm, the variation of resistance-change is less than 0.29%. The micrograph shows the PDMS-fiber is wound around the 5mm radius-of-curvature rod, and the LED chips are still lighting. These tests demonstrate the partially stretched spring embedded in PDMS-fiber has the flexibility to tolerate different loading conditions, and further reduce the resistance variation by external load.

Micrograph in Fig.6a shows lighting test in water for LED chips embedded in PDMS-fiber. Result demonstrates LED chips are effectively sealed and protected by flexible PDMS-fiber during operation. Fig.6b shows the schematic of 5×5 PDMS-fiber manufactured by weaving. Fig.6c-d shows the LED lighting tests for weaved PDMS-fiber array. Fig.6c shows the PDMS-fiber is manually weaved to construct a large-area 2D flexible electronics. The area of PDMS-fiber array is about 30mm×30mm. Fig.6d further shows the result of LED lighting test. The LED embedded in weaved PDMS-fiber array have been successfully

lighted with 1mA input current. Note that weaved PDMS-fiber array can further integrate with MEMS devices for flexible and larger-area applications. Output signals of each node can be detected by adding springs for more electrical-routing or switch sensing-circuit.

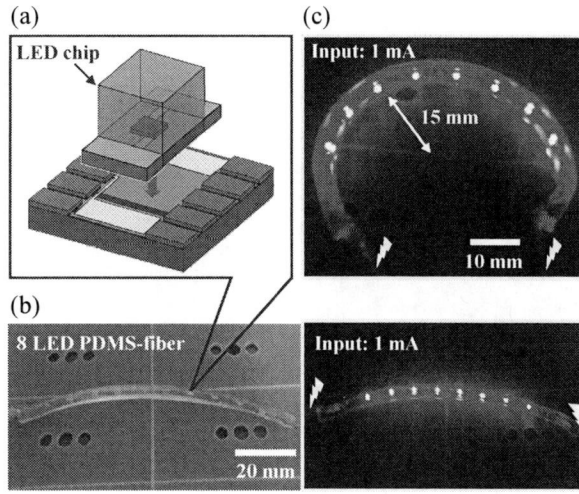

Figure 4: LED chips integration and result, (a) LED chip integration (b) LED PDMS-fiber lighting test and (c) preliminary bending test.

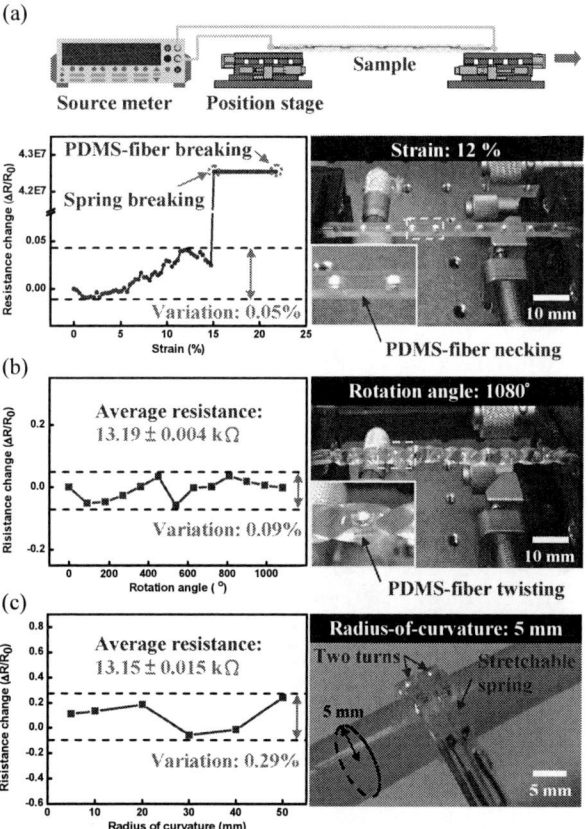

Figure 5: PDMS-fiber electrical tests setup and results for applying (a) tensile force, (b) twisting force, and (c) bending force.

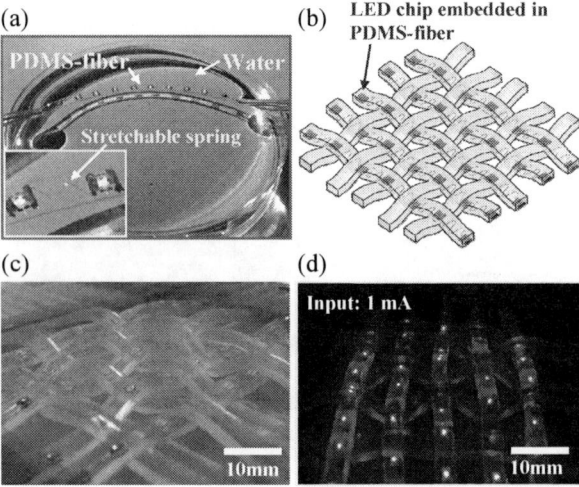

Figure 6: (a) water protection test, (b) schematic of 5x5 PDMS- fiber weaving test and (c-d) results of weaving test before and after LED lighting.

CONCLUSIONS

In this study, the PDMS-fiber integrated with multi devices scheme using stretchable electroplating copper spring has been proposed and demonstrated. The nodes house the device for functional PDMS-fiber, and provide a solid substrate for implemented and integrated devices easily. The stretchable spring can be stretched and expands the length before PDMS molding process. It can be reduce the limitation of fabricated area for large length application. Moreover, the stretchable spring embedded in PDMS provides well electrical and mechanical characteristics to tolerate different loading conditions. The measurement results demonstrate the capability to tolerate the tensile, twist, and bend loading. The weaving test also demonstrate the feasibility of PDMS-fiber for large-area, flexible electronics application. Note that the cross-section area of PDMS-fiber is currently limited by the LED chip size. It can be further integrated other micro-devices to achieve smaller cross-section area. Other micro-devices (temperature sensor and proximity sensor) integrated with PDMS-Fiber have been fabricated, and the performance of these devices will be characterized.

ACKNOWLEDGEMENTS

This research is based on the work supported by National Science Council of Taiwan under grant number NSC 102-2221-E-007 -027 -MY3 and NSC 101-2221-E-007 -069 -MY3. The author would like to express his appreciation to CNMM of the National Tsing Hua U. and the Nano Facility Center of National Chiao Tung U. in providing fabrication facilities.

REFERENCES

[1] C. Hong, T. T. Tang, R. P. Pan, and W. Fang, "Nanoporous anodic aluminum oxide (np-aao) alignment layer on pet/ito substrate for flexible liquid crystal display application," *in Proc. IEEE MEMS*, Cancun, Mexico, January 23-27, 2011

[2] F. Declercq and H. Rogier, "Active integrated wearable textile antenna with optimized noise characteristics," *IEEE Trans. Antennas Propag.*, vol. 58, no. 9, pp. 3050-3054, 2010

[3] A. Lymperis and R. Paradiso, "Smart and interactive textile enabling wearable personal applications: R&D state of the art and future challenges," *Proc. 30th Ann. Int. IEEE EMBS Conf.*, Vancouver, Canada, August 20-24, 2008

[4] M. D. Rienzo, F. Rizzo, G. Parati, G. Brambilla, M. Ferratini and P. Castiglioni, "MagIC system: a new textile-basedwearable device for biological signal monitoring. Applicability in daily life and clinical setting," *in Proc. 27th Ann. Int. IEEE EMBS Conf.*, Shanghai, China, September 1-4, 2005

[5] S. Yao and Y. Zhu, "Wearable multifunctional sensors using printed stretchable conductors made of silver nanowires," *Nanoscale*, 6, pp. 2345-2352, 2014

[6] K. Malzahn, J. R. Windmiller, G. Valdes-Ramirez, M. J. Schoning and J. Wang, "Wearable electrochemical sensors for in situ analysis in marine environments," *Analyst*, 136, pp. 2912-2917, 2011

[7] K. Jost, D. Stenger, C. R. Perez, J. K. McDonough, K. Lian, Y. Gogotsi and G. Dion, "Knitted and screen printed carbon-fiber supercapacitors for applications in wearable electronics," *Energy Environ. Sci.*, 6, pp. 2698-2705, 2013

[8] F. Carpi and D. D. Rossi, "Electroactive polymer-based devices for e-textiles in biomedicine," *IEEE Trans. Inf. Technol. Biomed.*, vol. 9, no. 3, pp. 295-318, 2005

[9] F. Axisa, P. M. Schmitt, C. Gehin, G. Delhomme, E. Mcadams, and A. Dittmar, "Flexible technologies and smart clothes for citizen medicine, home healthcare, and disease prevention," *IEEE Trans. Inf. Technol. Biomed.*, vol. 9, no. 3, pp. 325-336, 2005

[10] W. L. Sung, W. C. Lai, C. C. Chen, K. Huang and W. Fang, "Micro devices integration with large-area 2D chip-network using stretchable electroplating copper spring," *in Proc. IEEE MEMS*, San Francisco, USA, January 26-30, 2014

CONTACT

* W. Fang, tel: +886-3-5742923; fang@pme.nthu.edu.tw

PREPARATION OF WAFER LEVEL GLASS-EMBEDDED HIGH-ASPECT-RATIO PASSIVES USING A GLASS REFLOW PROCESS

Mengying Ma, Jintang Shang, and Bin Luo

Key Lab of MEMS of Ministry of Education, Southeast University, Nanjing, CHINA

ABSTRACT

This paper presents an innovative, uncomplicated and inexpensive method based on a glass reflow process to fabricate glass-embedded passives for RF MEMS packaging. Experimental results show that various glass-embedded passives, including annular cylindrical and coaxial cylindrical conductive through-holes, plate and coaxial torus trench capacitors, square spiral, circular spiral and meander trench inductors, and filters, can be manufactured void-free with vertical and smooth sidewall and large signal pathways. This glass-embedded design implements passives with large thickness and provides much surface space for 3D MEMS integration. The HFSS simulation results show that a minor thickness increase contributes to a large increase in Q and a relatively small decrease in L. The best Q (38.9-83.9) and L (19.9-15 nH) were achieved by using square spiral inductors.

INTRODUCTION

Traditional surface manufacturing technology produces planar RF components [1], further developing to 3D system by a relatively complicated stack-based process utilizing interposers with conductive through-holes [2].

Improved design of trench RF passive components embedded in silicon substrate [3] provides larger signal pathways and releases more surface space for 3D integration. However, signal coupling and electric leakage is a thorny problem for conductive silicon substrates [4, 5]. Although an additional dielectric layer is a solution, the compactness and thickness of this layer are restricted, which aggravates the electrical isolation failure, especially in RF MEMS [6, 7]. Moreover, simple optical detection method cannot be used to detect the inner mechanical failure of Si substrate due to its opacity.

The glass has high electrical resistivity and low loss tangent even at high-frequency ranges [8], which indicates its potential for RF components substrates fabrication and interposer application. It also has the superiorities of insulation, chemically stability, low coefficient of thermal expansion (CTE), and transparency for internal reliability monitoring. The research orientation on an alternative for silicon substrates with glass substrates is promising and significant.

Traditional glass substrate processing technologies consist of hollow structure formation by direct glass processing technology and hollow structure filling by electroplating. Fabricating void-free embedded passives with a high aspect ratio is a challenge for this design. DRIE of glass not only requires complicated processing steps to form masks but also suffers from a slow etching rate [9]. Isotropic wet etching of glass with BHF solution is unable to form vertical structures with high-aspect ratio [10]. Sandblasting results in conical shape holes and rough surface [11]. Direct laser drilling contributes to residual stress and defects [12]. In addition, precise and complicated technology is needed to deposit compact seed layer before Cu electroplating process.

In this work, it is the first time to prepare embedded high-aspect-ratio passives in glass substrates by a glass reflow process for MEMS/IC packaging. This simplified technology exclusives traditional hollow structure formation by direct glass processing technology and seed layer deposition process. This paper will present a complete set of design, manufacturing and characterization of glass-embedded high-aspect-ratio passives.

DESIGN

To investigate the feasibility of glass reflow process and the effects of thickness-dependent RF characteristics, different kinds of glass-embedded passives were fabricated, including cylindrical, annular cylindrical and coaxial cylindrical conductive through-holes, plate and coaxial torus trench capacitors, square spiral, circular spiral and meander trench inductors, and filters. The material of these passives is highly-doped silicon or copper.

Considering the biocompatibility, insulativity and transparency of glass, the conductive through-holes can be used for 3D stacked interconnection of biomedical devices, RF system and optical communication system. As the thermal conductivity of Si ($\sim$150W/m·K) and Cu ($\sim$400W/m·K) is much higher than that of glass ($\sim$1.16 W/m·K for Pyrex 7740), the through-holes can also be parts of a cooling module for high power heat dissipation in glass substrate. The glass-embedded capacitors can either be used as decoupling capacitor or bypass capacitor to eliminate noise and maintain RF system stability. The combination of inductance and capacitance creates diverse filters, resonators, etc.

FABRICATION

The manufacturing technology is based on a glass reflow process and seedless electroplating process [13]. A schematic view of the fabrication process of passives with different materials (Highly-doped Si or Cu) is depicted in Figure 1.

Glass Reflow Process

The glass reflow process is shown in Figure 1(1). In this process, high-aspect-ratio passive structures were firstly patterned in a highly-doped silicon wafer by a Deep Reactive Ion Etching (DRIE) technology. Here a 4', 500 um thick Si wafer (p-type, 0.002 Ω-cm) was etched into 180 -350um deep.

During the DRIE process, the carbon in the photoresist, the etching gas (mostly C_2F_4) and the silicon substrate are easy to react to form invisible polymer firmly stick on to the etched surface. The following high temperature treatment would further transform the invisible polymer to large visible polymer particles. Therefore, the etched wafer should undergo a strict

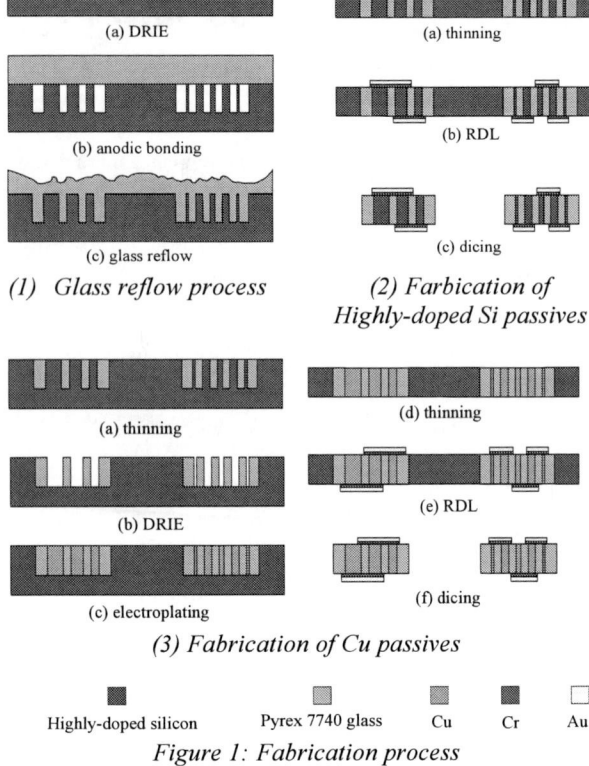

(a) DRIE	(a) thinning
(b) anodic bonding	(b) RDL
(c) glass reflow	(c) dicing

(1) Glass reflow process | *(2) Farbication of Highly-doped Si passives*

(a) thinning	(d) thinning
(b) DRIE	(e) RDL
(c) electroplating	(f) dicing

(3) Fabrication of Cu passives

Highly-doped silicon   Pyrex 7740 glass   Cu   Cr   Au

Figure 1: Fabrication process

photoresist stripping process. An oxygen plasma asher was used here to dry etch the photoresist. Then the Si wafer was cleaned in H_2SO_4:H_2O_2 (3:1) solution at 100℃ for 15 min. After that, the Si wafer was cleaned following the standard cleaning procedure.

The etching thickness is about 180 um to 350 um for 500 um thick Si wafer to ensure that the wafer does not fracture during subsequent processing steps. The etching thickness defines the final thickness of the glass substrate. If it is too thin, it is not easy to perform the following grinding and polishing process. If it is too thick, the remaining silicon layer might crack easily under vacuum (10-3Torr) during the reflow process.

Secondly, a 4 inch, 500 um thick Pyrex 7740 glass was anodic bonded onto the etched highly-doped Si wafer at 400℃, 800V and a vacuum atmosphere ($•10^{-3}$ Torr) by a bonding equipment (eg. EVG-501).

Thirdly, the bonded wafers were heated in atmosphere at 900℃ for 3 h, so that the molten glass slumped into the cavities driven by the pressure difference between the inside vacuum and outside atmosphere pressure . A low vacuum pressure inside the sealed cavities is crucial important to ensure that the difference of the pressure inside and outside the cavity is sufficient to make the molten glass fill the cavities completely and seal the etched silicon structures without voids.

Glass-embedded Highly-doped Si Passives Preparation

As shown in Figure 1(2), to fabricate highly-doped Si passives, the glass-Si wafers were firstly thinned at both sides to expose the intermediate structures after the upper glass reflow process. Usually traditional grinding and polishing technology was used to perform the wafer

thinning process. But we found that the grinding process easily led to residual stress, defects and other failures, which made the thin composite wafer crack easily. Hence, in Figure 1(2)-(a), a wet etching followed by a polishing process was preferred. The glass-silicon wafers were immerged in HNA solution (HF: HNO_3: CH_3COOH=1:2:3) at room temperature for 2 hours. Thus, the Si wafer was etched ~240 um with an etch rate of ~4 um/min by agitation. Then the glass wafer was wet etched in 40% HF solution at room temperature for 2 hours with an etch rate of ~3.6 um/min. Finally, a chemical mechanical polishing (CMP) process was performed to expose the embedded passive structures and through holes on both sides for next step interconnection. The fabricated composite wafer was ~160 um. Side etch happened in the whole wet etch process, but it did not influence the main structures or the following process.

The redistribution layers (RDL) were then patterned on the surface of the wafer for next step interconnection. The metal line was composed of two layers from bottom to top: Cr with a thickness of 500 nm patterned by Physical Vapor Deposition (PVD) process and gold with a thickness of 2 um patterned by electroplating process. Finally, the wafer was diced into single units for interconnection as required.

Glass-embedded Cu Passives Preparation

As shown in Figure 1(3), to obtain glass-embedded Cu structures, only the glass wafer was wet etched in 40% HF solution and polished to expose the embedded silicon passive structures after glass reflow process, which were then removed by DRIE.

Copper was electroplated to fill in the vacancy of the etched embedded structures with remaining highly-doped Si substrate as the seed layer. Compare to the common seed layer deposited by Chemical Vapor Deposition (CVD) or electroplating process, the highly-doped Si layer was more uniform and compacter. Copper sulfate electrolyte acted as the electroplating solution here. The electroplating rate of Cu was ~0.5 um/min with a current density of 27mA/dm².

After the seedless electroplating process, the remaining highly-doped Si layer was grinded and polished by Buehler Beta Twin Variable Speed Grinder-Polisher. The wheel speed and force were small enough to reduce the residual stress and defects. The following RDL and dicing process was as same as that performed in highly-doped Si passives process.

Experimental Results

The experimental results show that various passives were fabricated successfully, including cylindrical, annular cylindrical and coaxial cylindrical conductive through-holes, plate and coaxial torus trench capacitors, square spiral, circular spiral and folding type trench inductors, and filters. The close-up details of some fabricated samples are shown in Figure 2.

The thickness of the final glass substrate was 150 um. The cylindrical conductive TGV had a radius from 25 um to 50 um. The inner radius/outer radius of annual TGV were 50 um/100 um, 100 um/200 um, and 150 um/300 um, respectively. The coaxial cylindrical conductive through-holes were the combination of the above two types

978-1-4799-7956-1/15 $31.00 © 2015 IEEE

through-holes. The width/space of capacitors were respectively 100 um/30 um and 100 um/100 um. The meander Cu inductor had a turns number of 3.5, width of 100 um and space of 150 um. The square spiral inductor had a turns number of 3, width of 100 um and space of 100 um. The circular spiral inductor had a turns number of 3, width of 100 um and space of 100 um. The filters consisted of four twin spiral inductors with each had a width of 100 um and space of 100 um.

By theoretical calculation, the DC resistance is respectively 1.34 mΩ, 11.5 mΩ and 9.8 mΩ for cylindrical Cu TGV (radius=25 um, thickness=150 um), meander Cu inductor (turns number=3.5, width=100 um, space=150 um, thickness=150 um) and square spiral inductor (turns number=3, width=100 um, space=100 um). At 5 GHz, considering the skin effect, the above inductor has an R of 1.85Ω, Q of 56.4 and L of 3.03 nH, and the above square spiral inductor has an R of 1.58Ω.

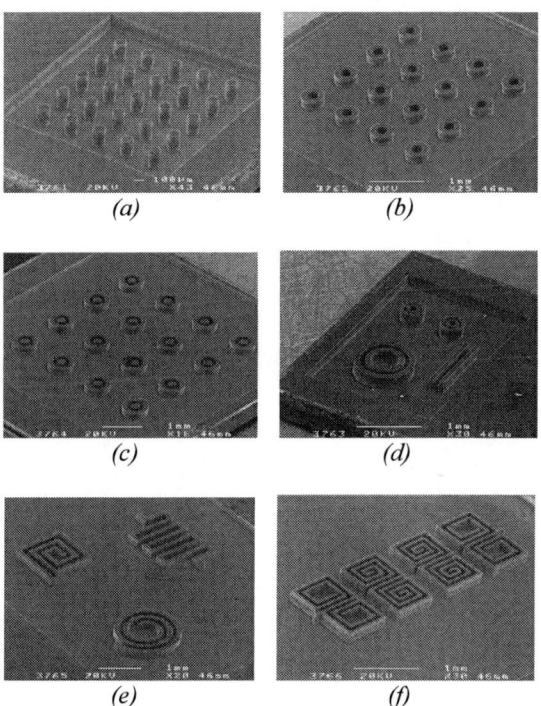

(a)   *(b)*

(c)   *(d)*

(e)   *(f)*

Figure 2: The trench RF passives structures fabricated by DRIE. (a) Cylindrical through-holes; (b) annular through-holes; (c) coaxial cylindrical through-holes; (d) plate and coaxial torus trench capacitors; (e) square spiral, circular spiral and meander trench inductors; (f) filter.

SIMULATION

HFSS was used to simulate the electrical parameters (inductance, Q-factor) of four type inductors: glass-embedded Si meander inductor, glass-embedded Cu meander inductor, planar Cu meander inductor and glass-embedded Cu square spiral inductor. Only the planar Cu meander inductor is on the surface of the glass substrate while others are glass-embedded. Figure 3 shows the geometric design parameters of the meander inductors and the spiral inductor, typically 50 um in width and 100 um in

space. The inner length of square spiral inductor is 500 um. In consideration of bulk micro-fabrication process and surface micro-fabrication process, the designed thickness of the glass-embedded inductors ranges from 20 um to 100 um at an interval of 5um while that of the planar inductor ranges from 1 um to 20 um at an interval of 1 um.

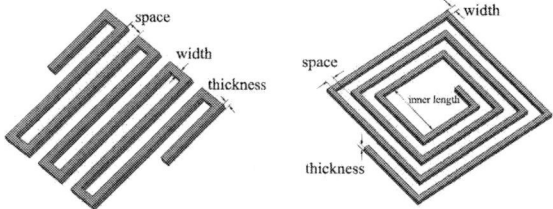

Figure 3: Models of meander inductor and spiral inductor

The combination of FEM quasi-static analysis and a lumped-element equivalent circuit model analysis with a sweep frequency varying from 0-20 GHz was adopted here. The parameters compared for the field simulation include: (1) materials (highly-doped Si and Cu), (2) structures (meander and square spiral), (3) thickness (0-100 um), and (4) packaging method (glass-embedded and planar). Figure 4 summarizes the numerous simulation results.

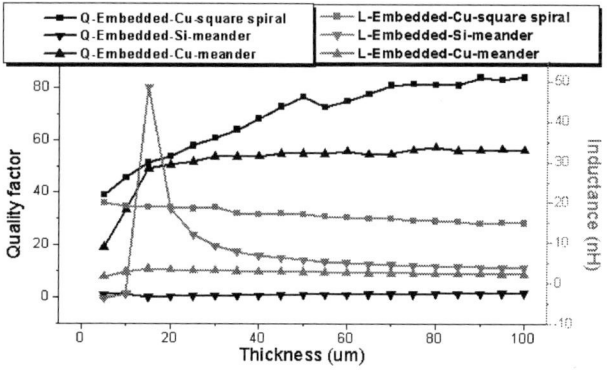

(a) glass-embedded inductors

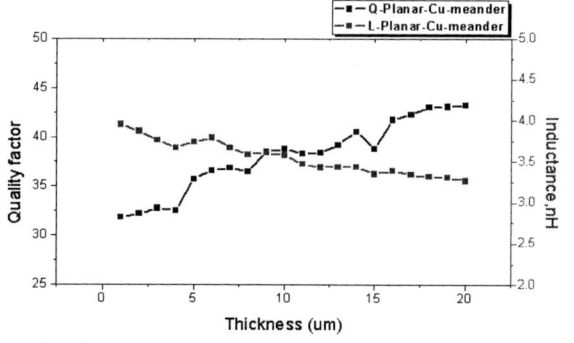

(b) planar Cu meander inductor

Figure 4: Simulation results of maximum Q and L versus thickness

From Figure 4 we can draw that, for embedded inductors, when the thickness is increased from 5 um to 100 um, the maximum Q increases by 63.6% (1.07-1.75), 191.5% (19.23-56.05), 115.7% (38.93-83.96) for glass-embedded Si meander inductor, glass-embedded Cu meander inductor and glass-embedded Cu square spiral inductor, while L decreases by 44.1%(6.9-3.86 nH), 37.2% (3.55-2.23 nH), 24.6% (19.9-15 nH) for each type. The analysis of L above excludes unstable inductance of glass-embedded Si meander inductors with the six smallest thickness. For the planar Cu meander inductor, when the thickness is increased from 1 um to 20 um, the maximum Q increases by 35.7% (31.86-43.24) while L decreases by 17.2% (3.95-3.27 nH). In fact, the thickness of glass-embedded passives can be much larger to acquire a higher Q.

Though heavy doping ($\sim 8 \times 10^{18} cm^{-3}$) improves the conductivity of Si, its dielectric loss is still larger than that of metal, which results in a too low Q-factor (0.16-1.75) for glass-embedded Si meander inductor. As a result, doped Si is not an appropriate material for RF passives. Considering the simple glass reflow process and the matched Coefficient of Thermal Expansion (CTE) of Si to that of Pyrex 7740 glass, Si can still be used as the TGV material in other fields[14].

For meander inductors, simulation shows that L can be significantly increased by increasing the space, and Q can be improved by increasing width. However, it may also increase the packaging area. Glass-embedded Cu square spiral inductor acquired the optimized Q (38.9-83.9) and L (19.9 nH-15 nH) when the thickness is 5-100 um, and it occupies less area. Nevertheless, it has a smaller resonant frequency (f) from 1.5-1.8 GHz, compared to 6.8-17.2 GHz for glass-embedded Si meander inductor, 4.5-6.2 GHz for glass-embedded Cu meander inductor, and 3.7 to 3.9 GHz for planar Cu meander inductor.

CONCLUSION

This paper demonstrates a novel fabrication technology based on a glass reflow process for high aspect ratio passives. High aspect ratio passives were prepared successfully by a glass reflow process. Highly doped silicon and copper were used as conductive materials for passives embedded in glass substrates respectively. HFSS simulation shows that, when the passives thickness is the only variable, Q has a large increase as the thickness increases while L has a smaller decrease at the same time. Thus, we can take the passives thickness as an extra design parameter for glass-embedded passives to obtain the required Q while the corresponding L is not changed much. The fabrication process and performance parameters of glass-embedded filters etc. will be presented in the conference.

ACKNOWLEDGEMENTS

This work is supported by the National Natural Science Foundation of China (No. 51275091). The authors would like to thank Kai Zheng and Li Zhang from Jiangyin Changdian Advanced Packaging Co. Ltd, Junwen Liu from Jiangsu Intellisense Technology Co., Ltd. The authors also acknowledge the support of National Science and Technology Major Project (Project number: 2013ZX02501).

REFERENCES

[1] M. G. Allen, "MEMS Technology for the Fabrication of RF Magnetic Components", *J. IEEE Transactions on Magnetics*, vol. 39, no. 5, pp. 3073-3078, 2003.

[2] J. H. Lau , "TSV Manufacturing Yield and Hidden Costs for 3D IC Integration", in *60th ECTC*, Las Vegas, June 1-4, 2010, pp. 1031-1042.

[3] T. Pan, A. Baldi, E. D. Venn, R. F Drayton, "Fabrication and Modeling of Silicon-Embedded High-Q Inductors", *J. Micromechanics and Microengineering. 15*, pp. 849–854, 2005.

[4] M. Pfost, H. M. Rein, "Modeling and Measurement of Substrate Coupling in Si-Bipolar IC's up to 40 GHz", *J. Solid-state Circuits*, vol. 33, no. 4, pp. 582-591, 1998.

[5] M. Lee, J. Cho, J. Kim, J. Kim, J. Kim, "Noise Coupling of Through-Via in Silicon and Glass Interposer", in *63th ECTC*, Las Vegas, May 28-31, 2013, pp. 1806-1810.

[6] X. Liu, Q. Chen, P. Dixit, R. Chatterjee, R. R. Tummala, S. K. Sitaraman, "Failure Mechanisms and Optimum Design for Electroplated Copper Through-Silicon Vias (TSV)", in *59th ECTC*, San Diego, May 26-29, pp. 624-629, 2009.

[7] A. C. Reyes, S. M. El-Ghazaly, S. Dorn, M. Dydyk, D. K. Schroder, H. Patterson, "High Resistivity Si as a Microwave Substrate", in *46th ECTC*, Orlando, May 28-31, pp. 382-391, 1996.

[8] V. Sukumaran, T. Bandyopadhyay, V. Sundaram, R. Tummala, "Low-cost thin glass interposers as a superior alternative to silicon and organic interposers for packaging of 3-D ICs", *J. Transactions Components, Packaging and Manufacturing Technology*, vol. 2, no. 9, pp. 1426–1433, 2012.

[9] X. H. Li, T. Abe, M. Esashi, "Deep Reactive Ion Etching of Pyrex Glass Using SF6 Plasma", *J. Sensors and Actuators A 87*, pp. 139-145, 2001.

[10] C. Iliescu, B. Chen, J Miao, "Deep Wet Etching-Through 1mm Pyrex Glass Wafer For Microfluidic Applications", in *Proceedings of the IEEE MEMS*, Hyogo, January 21-25, 2007, pp. 393-396.

[11] H. Wensink, J. W. Berenschot, H. V. Jansen, M. C. Elwenspoek, "High Resolution Powder Blast Micromachining", in *13th International Workshop on MEMS*, Miyazaki, pp. 769–774, 2000.

[12] V. Sukumaran, Q. Chen, F. Liu, N. Kumbhat, T. Bandyopadhyay, H. Chan, S. Min, C. Nopper, V. Sundaram, R. Tummala, "Through-Package-Via Formation and Metallization of Glass Interposers", in *60th ECTC*, Las Vegas, pp.557-563, 2010.

[13] J. Y. Lee, S. W. Lee, S. K. Lee, J. H. Park, "Wafer Level Packaging For RF MEMS Devices Using Void Free Copper Filled Through Glass Via", in *26th MEMS*, Taipei, Jan 20-24, 2013, pp. 773-776.

[14] R. M.Haque, K. D. Wise, "An Intraocular Pressure Sensor Based On A Glass Reflow Process", in *Solid-State, Actuators, and Microsystems Workshop*, Hilton Head Island, June 6-10, 2010, pp. 49-52.

CONTACT

*J.T. Shang, tel: +86 13913869603; jshang@seu.edu.cn; shangjintang@hotmail.com

"CELL PINBALL": WHAT IS THE PHYSICS?
Ryo Murakami[1], Makoto Kaneko[1], Shinya Sakuma[2], and Fumihito Arai[2]
[1]Osaka University, Suita, JAPAN
[2]Nagoya University, Nagoya, JAPAN

ABSTRACT

During the deformability test of Red Blood Cell (RBC) by utilizing a micro fluidic channel, we found an interesting phenomenon where some RBCs behave just like elastic pinball with the motion in the perpendicular direction with respect to the main flow line, while most of RBCs simply move along the main flow line. This phenomenon is called "Cell Pinball". Through visualization, we found that the RBC being in Cell Pinball mode rotates around the perpendicular axis to the flow line and the rotating direction has one-to-one relationship with the moving direction without any exceptions. We also found that the rotating axis exists slightly behind the center of gravity with respect to the flow direction. This geometrical configuration makes an unstable condition for the cell under the fluid force, which eventually produces the motion perpendicular to the main flow line.

WHAT IS "CELL PINBALL"?

It is well known that there is a close relationship between the deformability of RBC and various diseases. Due to this, there have been many groups discussing the deformability of RBC by using micro fluidic channel [1, 2,3,4,5]. During the deformability test of RBC, just by chance, we discovered "Cell Pinball" phenomenon where the RBC being in Cell Pinball mode moves a different direction compared with those of other RBC as shown in Figure 1(b), (c) and (d). Figure 1(e) shows a RBC that goes straight although it is in Cell Pinball mode. No report has been done on "Cell Pinball" so far. This paper will show that under what condition "Cell Pinball" occurs and what relation does exist between the motion and the moving direction of RBC, through visualization technique and modeling.

EXPERIMENTAL SYSTEM

Figure 2 shows an overview of the experimental system where it is composed of a microfluidic chip with a microchannel in it, a microscope for observing the cells in the microfluidic channel, a high-speed vision system for capturing motion, a pressure sensor for measuring the internal pressure, and a syringe pump for generating a driving force to the diluted blood.

MICRO FLUIDIC CHIP

Figure 3 shows the schematic view of the microfluidic chip. The width and the depth of the microchannel are 50 micrometers and 4 micrometers, respectively. This microfluidic chip is made by using the PDMS molding method. We make pattern SU-8 on the Si substrate by using laser lithography technique and transcribe of channels by PDMS molding. Finally, we attach the molded PDMS to the glass substrate after O_2 plasma treatment[6,7,8,9,10].

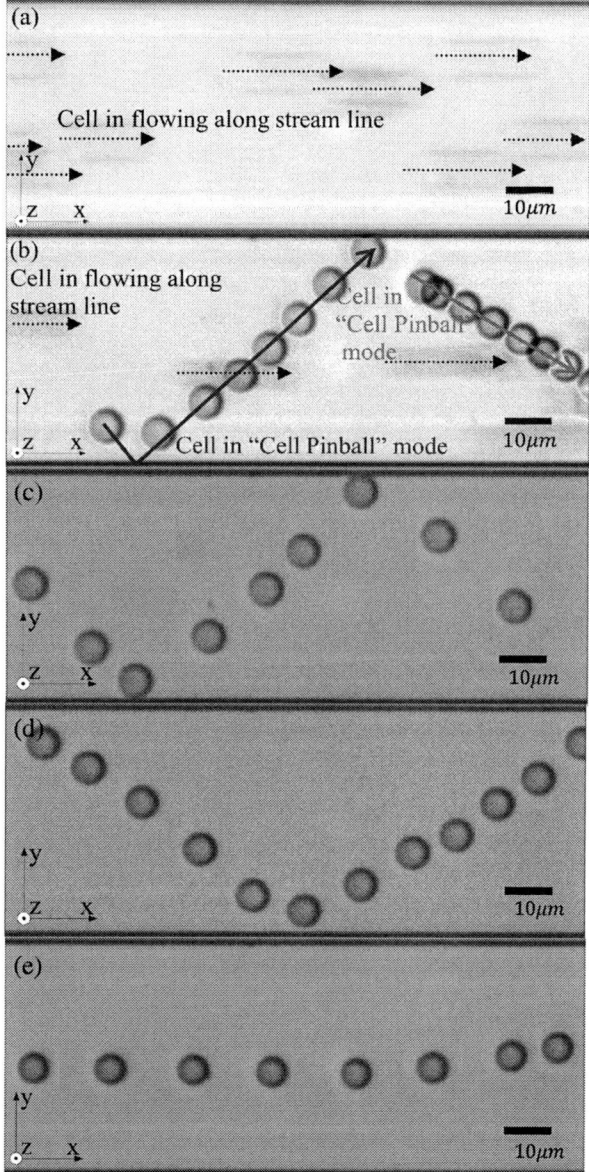

Figure 1: Various examples of Cell Pinball

EXPERIMENT

Cell Pinball is supposed to occur under the condition that RBC makes contact with both the top and the bottom surfaces of the microchannel. Based on this consideration, we change the size and the shape of RBC intentionally by adjusting the NaCl percentage of saline. Under such a condition, the RBC can swell due to the osmotic pressure[11,12,13].

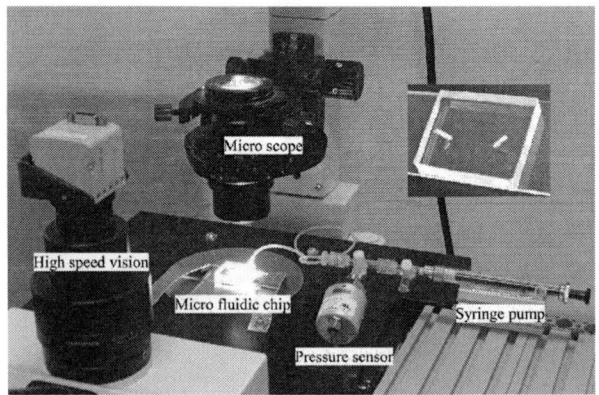

Figure 2: An overview of the experimental system

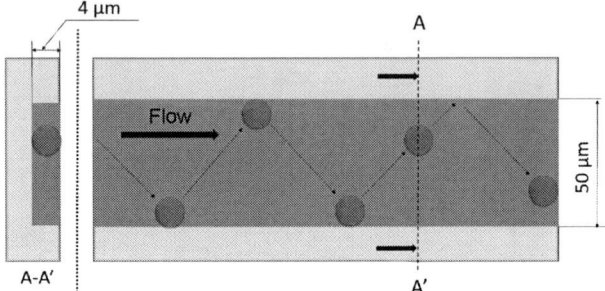

Figure 3: The schematic view of the microfluidic channel

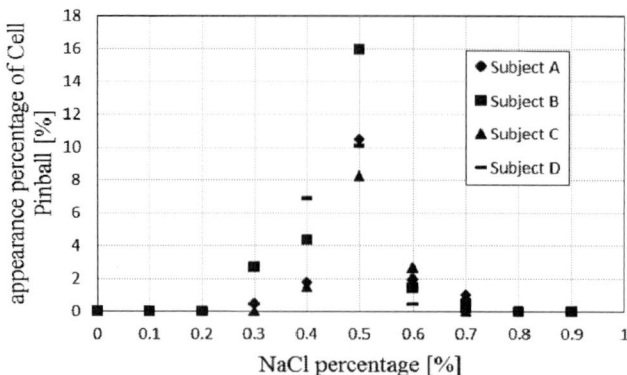

Figure 4: The relationship between NaCl percentage and appearance percentage of Cell Pinball

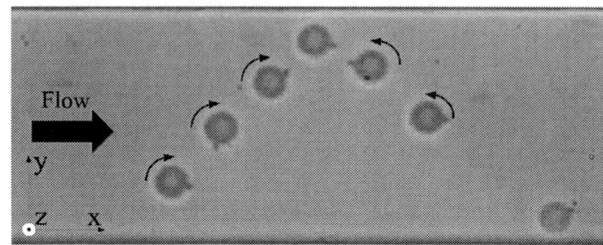

Figure 5: A result of the visualization

RESULT

Figure 4 shows the relationship between the appearance percentage of Cell Pinball and the NaCl percentage. We find that the appearance percentage of Cell Pinball increases as NaCl percentage decreases until it is about 0.5%, and then it decreases to zero as NaCl percentage further decreases due to the collapse of the RBCs. This tendency is observed for all blood samples taken from the subjects.

VISUALIZATION

To know what is happening for the RBC being in Cell Pinball mode, we visualize the motion of the RBC by attaching a micro bead (carboxyl latex, 0.8 μm; Invitrogen) on the RBC membrane as a marker[14,15,16]. Figure 5 shows the result of the visualization. In Figure 5, the cell rotates the clockwise direction until it hits to the upper wall and it changes the rotating direction in the counter clockwise direction after changing the moving direction. We can find one-to-one relationship between the rotating direction and the moving direction. If the RBC rotates clockwise, it goes upward and vice versa. Without any exceptions, this tendency is observed in other all results of the visualization. From this experimental result, we believe that it is necessary for the RBCs being in Cell Pinball mode to rotate in order to deviate from streamlines.

MODELING

The fact in the Figure 5 suggests the model as shown in Figure 6 where there exists the representative contact point a bit behind the centroid line[17,18,19,20,21]. Such cell is unstable under the main flow and will eventually

start to rotate around the contact point easily. Figure 7 is the top view of Figure 6. For example, as shown in Figure 7(a), a RBC rotates counter-clockwise and moves as shown in Figure 7(a) and Figure 7(b). Since the stream is in x-direction, RBC deforms with its movement and the position of the contact point is shifted downward as shown in Figure 7(c). We believe that this sequence eventually makes a movement perpendicular to the direction of the flow. This consideration of motion based on the model very nicely coincides with the real motion shown in Figure 5. By using this model, we can explain why the rotating motion is generated in the horizontal flow.

GEAR MECHANISM
Equation of Cell Pinball

Given the fact that RBCs being in "Cell Pinball" are rotating around the perpendicular axis, it is very natural that we suspect there is some relationship between the rotating motion and the moving direction. If we suppose non translational slip happens for with respect to rotation of the cell during rotating motion, the motion of the RBC being in Cell Pinball mode should happen based on the gear mechanism and there should be one-to-one relationship between the rotating direction and the moving motion. Now, let L, n and S be the distance between two walls, the rotational times of cell and the distance between the centroid line and the contact line, respectively. Then, if we take the gear motion into consideration, we can find this kind of relationship as shown in the following equation.

$$S = \frac{L}{2\pi n} \qquad (1)$$

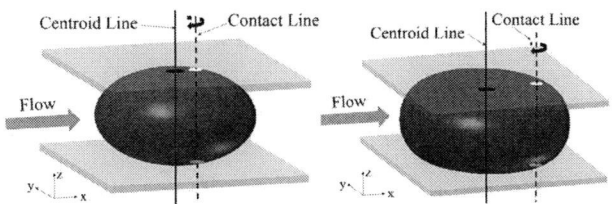

Figure 6: Contact models between red blood cell and two plates

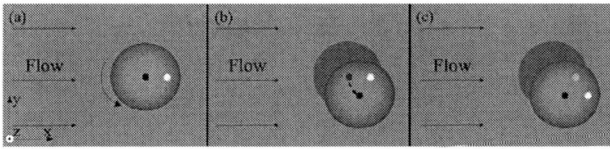

Figure 7: A possible mechanism of Cell Pinball phenomenon

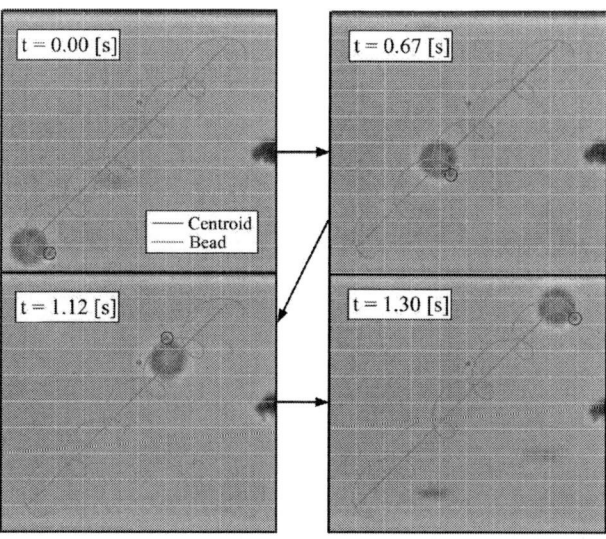

Figure 8: The comparison with the experimental data

Since L and n can be obtained through visualization of cell motion, we can compute S by using the above equation. In one experiment, we obtain L and n as 38.64 μm and 5.06, respectively. By using this result, we compute S as 1.22 μm. Since, the radius of the RBC is 3.59 μm, S is located just between the centroid line and the boundary of the RBC, witch is physically possible.

Locus of Cell Pinball

By using the computed contact point, we can obtain expected the locus of the motion of the bead based on the gear mechanism as shown in Figure 8. In order to obtain the locus, the initial and the terminal positions of the RBC and the bead are used as the boundary condition. The blue line and the red line show the locus of the centroid of the RBC and the position of the bead, respectively. In Figure 8, we compare the expected locus with the real motion of the RBC and the bead. There is a very nice coincidence between the real bead motion and the estimated locus. This coincidence means that Equation (1) can estimate the basic motion of Cell Pinball and certainly there is the gear mechanism as we expected between the RBC and the wall of the channel.

CONCLUDING REMARKS

We newly discovered a zigzag motion of RBC in microchannel under the condition that the RBC makes contact with both the top and the bottom plates of the microchannel. By attaching a micro bead on the surface of RBC, we showed that the RBC rotates around the contact line and by using the model, we succeeded in explaining the basic physics of Cell Pinball, especially why the rotating motion is generated in the horizontal laminar flow and why there is one-to-one relationship between the rotation direction and the moving direction. Also based on the gear mechanism, we found an equation that can estimate the basic motion of the RBC being in Cell Pinball mode. From the result, we could confirm the reliability of the model and the equation. In our future work, first we will confirm the motion of RBC that goes straight although it looks like "Cell Pinball" as shown in Figure 1(e) by using the visualization technique. We suspect that such RBCs are not rotating. Next, we are going to try to explain why RBCs

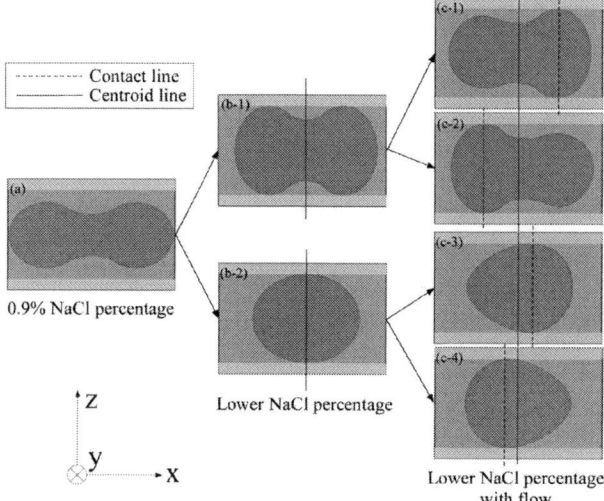

Figure 9: Possible contact ways between red blood cell and channel walls

being in Cell Pinball mode bound as if they were just like elastic pinballs by using the models we suggest. We will also visualize the contact area between RBC and the plates, so that we can make sure the contact line in more details and support our models. In order to get more information about contact area, we are planning to use a confocal laser scanning microscope[22,23,24]. By using a confocal laser scanning microscope, we can obtain a full three dimensional shape of RBC in the microchannel and then, by considering the relationship between the environment and the obtained shape of RBC, we will be able to estimate where the exact contact point is. Figure 9 shows various possible cases where the representative contact point is located and we will confirm that which model is correct. We also think that the diameter of the RBC being in Cell Pinball mode correlates closely with its movement direction and we will ascertain this by using image analysis. There might be some relationship between the deformability of RBC and the numerical value of "S" and as a result *S* may have a correlation with some diseases.

ACKNOWLEDGEMENTS

This research is supported by Grant-in-Aid for Scientific Research on innovative Areas (Hyper Bio Assembler for 3D Cellular Innovation) #23106003, Grant-in-Aid for Challenging Exploratory Research #26630098 and "Creating Hybrid Organs of the future" of the Ministry of Education, Culture, Sports, Science and Technology(MEXT) of Japan.

REFERENCES

[1] S. Sakuma, K. Kuroda, C. Tsai, W. Fukui, F. Arai and M. Kaneko: "Red blood cell fatigue evaluation based on the close-encountering point between extensibility and recoverability", *Lab on a chip*, 14, pp.1135-1141, 2014.

[2] Y. C. Chou, G. Y. Chen, Y. C. Lin and G. J. Wang: "A lab-on-a-chip capillary network for red blood cell hydrodynamics", *Microfluid Nanofluid*, 9, 585-591, 2010.

[3] M. S. Amin, Y. K. Park, N. Lue, R. R. Dasari, K. Badizadegan, M. S. Feld and G. Popescu: "Microrheology of red blood cell membranes using dynamic scattering microscopy", *OPTICAL EXPRESS*, Vol. 15, No. 25, 17001-17009, 2007.

[4] M. Dao, C. T. Lim and S. Suresh: "Mechanics of the human red blood cell deformed by optical tweezers", *Journal of the Mechanics and Physics of Solids*, 57, 2259-2280, 2003.

[5] O. K. Baskurt, D. Gelmont, and H. J. Meiselman: "Red Blood Cell Deformability in Sepsis", *American Journal of Respiratory and Critical Care Medicine*, Vol. 157, 421-427, 1998.

[6] T. Fujii: "PDMS-based microfluidic devices for biomedical applications", *Microelectronic Engineering*, Volume61-62, 907-914, 2001.

[7] T. Vilkner, D. Janasek and A. Manz: "Micro Total Analysis Systems. Recent Developments", *analytical chemistry*, 76(12), 3373-3386, 2004.

[8] C. H. Wang, G. B. Lee: "Automatic bio-sampling chips integrated with micro-pumps and micro-valves for disease detection", *Biosensors and Bioelectronics*, Volume21, Issue 3, 419-425, 2005.

[9] P. S. Dittrich, K. Tachikawa, and A. Manz: "Micro Total Analysis Systems. Latest Advancement and Trends", *analytical chemistry*, 78(12), 3887-3908, 2006.

[10] P. S. Dittrich, and A. Manz: "Lab-on-a-chip: microfluidics in drug discovery", *Nature Reviews Drug Discovery*, 5, 210-218, 2006.

[11] C .Tsai, M. Kaneko, S. Sakuma and F. Arai: "Distinct patterns of Cell Motion Inside a Micro-Channel under Different Osmotic Conditions", *International Conference of the IEEE EMBS*, 5525-5528, 2013.

[12] D. A. Goldstein and A. K. Solomon: "Determination of Equivalent Pore Radius for Human Red Cells by Osmotic Pressure Measurement", *The Journal of General Physiology*, Vol. 44, no. 1, 1-18, 1960.

[13] Y. Tan, D. Sun, J. Wang and W. Huang: "Mechanical Characterization of Human Red Blood Cells Under Different Osmotic Conditions by Robotic Manipulation With Optical Tweezers", *IEEE Transactions on Biomedical Engineering*, Vol. 57, No. 7, 1816-1825, 2010.

[14] J. Dupire, M. Socol and A. Viallat: "Full dynamics of a red blood cell in shear flow", *PNAS*, vol. 109, no.51, 20808-20813, 2012.

[15] S. Choi, S. Song, C. Choi and J. K. Park: "Sheathless Focusing of Microbeads and Blood Cells Based on Hydrophoresis", *Small*, 4, no.5, 634-641, 2008.

[16] L. Bitsch, L. H. Olesen, C. H. Westergaard, H. Bruus, H. Klank and J. P. Kutter: "Micro particle-image velocimetry of bead suspensions and blood flows", *Experiments in Fluids*, 29, 505-511, 2005.

[17] T. L. Steck: "THE ORGANIZATION OF PROTEINS IN THE HUMAN RED BLOOD CELL MEMBRANE", *The Journal of Cell Biology*, Vol. 62, 1-19, 1974.

[18] Y. Liu, W. K. Lui: "Rheology of red blood cell aggregation by computer simulation", *Journal of Computational Physics*, 220, 139-154, 2006.

[19] D. A. Fedsov, B. Caswell and G. E. Karniadakis: "A Multiscale Red Blood Cell Model with Accurate Mechanics, Rheology, and Dynamics", *Biophysical Journal*, Volume 98, 2215-2225, 2010.

[20] M. P. M. Marinkovic, K. T. Turner, J. P. Butler, J. J. Fredberg and S. Suresh: "Viscoelasticity of the human red blood cell", *American Journal of Physiology*, Vol. 293, no. 2, C597-C605, 2007.

[21] R. Waugh and E. A. Evans: "THEMOELASTICITY OF RED BLOOD CELL MEMBRANE", *Biophysical Society*, Vol. 26, 115-132, 1979.

[22] J. S. Park, C. K. Choi and K. D. Lihm: "Optically sliced micro-PIV using confocal laser scanning microscopy(CLSM)", *Experiments in Fluids*, Volume 37, Issue 1, 105-119, 2004.

[23] A. Ovsianikov, M. Malinauskas, S. Schlie, B. Chichkov, S. Gittard, R. Narayan, M. Löbler, K. Sternberg, K. –P. Schmitz and A. Haverich: "Three-dimensional laser micro-and nano-structuring of acrylated poly(ethylene glycol) materials and evaluation of their cytoxicity for tissue engineering applications", *Acta Biomaterialia*, Volume 7, Issue 3, 967-974, 2011.

[24] K. Karlsson, P. E. Danielsson, R. Lenz, A. Lijeborg, L. Majlöf, and N. Åslund: "Three-dimensinal microscopy using a confocal laser scanning microscope", *Optics Letters*, Vol. 10, Issue 2, 52-55, 1985.

CONTACT

*Ryo Murakami, tel: +816-6879-7333;
murakami@hh.mech.eng.osaka-u.ac.jp

A FULLY MONOLITHIC MICROFLUIDIC DEVICE FOR COUNTING BLOOD CELLS FROM RAW BLOOD

John Nguyen[1], Yuan Wei[1], Yi Zheng[1], Chen Wang[2], and Yu Sun[1]

[1]University of Toronto, Toronto, CANADA

[2]Mount Sinai Hospital, Toronto, CANADA

ABSTRACT

This paper reports a monolithic microfluidic device capable of complete blood constituent enumeration from raw blood. On-chip sample processing (e.g. dilution, lysis, and filtration) and downstream single cell measurement were fully integrated to enable sample preparation and single cell analysis from whole blood on a single device. The device consists of two parallel sub-systems that perform sample processing and electrical impedance measurements for measuring RBC and WBC parameters. Experimental characterization of patient blood samples validated the system's capability for performing on-chip raw blood processing and measurement.

INTRODUCTION

Blood constituents consist of proteins, glucose, mineral ions, hormones and blood cells including red blood cells (RBCs or erythrocytes), white blood cells (WBCs or leukocytes) and platelets. The enumeration and characterization of these blood cell constituents can provide important clinical insight. For example, an abnormally low RBC concentration is a characteristic of insufficient oxygen delivery and can be indicative of potential disorders such as anemia and leukemia [1]. WBCs play a vital role in immune response to infection and foreign bodies. Measuring WBCs can be a preliminary step in the diagnosis of a number of conditions such as inflammatory processes, bone marrow alterations, and immune disorders.

In whole blood, RBCs outnumber WBCs by a ratio of 1000:1, making it challenging to distinguish between target cell groups and remaining blood cell population. Hence, sample preparation to either sort, concentrate, isolate or separate target blood cells is necessary to facilitate downstream measurement. Dilution for RBC analysis requires large sample volumes, and/or serial dilution steps which contribute to an accumulation of non-systematic errors making approximation of the native RBC concentration nontrivial.

Similarly, the isolation of WBCs and the depletion of contaminating RBCs are critical for analysis to eliminate RBCs' masking effect. In laboratory settings conventional methods for WBC isolation are size based differential centrifugation and selective bulk lysis targeting RBCs. Overall, these methods for raw blood dilution and isolation are macroscale in nature and are limited by the required blood sample volume, cell quality, processing time, operation efficiency and variability. Current microfluidic devices enumerating blood cell sizes provide incomplete CBC information [2],[3], lack on-chip processing [4], and require multiple specialized devices [5] that need stringent flow rate control necessitating bulky pumping mechanisms.

This paper reports a microfluidic device capable of performing blood sample preparation for downstream on-chip RBC and WBC analysis. As a demonstration of our processing capabilities, the device integrates on-chip sample preparation and single cell electrical impedance measurement on a compact platform producing complete blood count information from whole blood (Fig. 1).

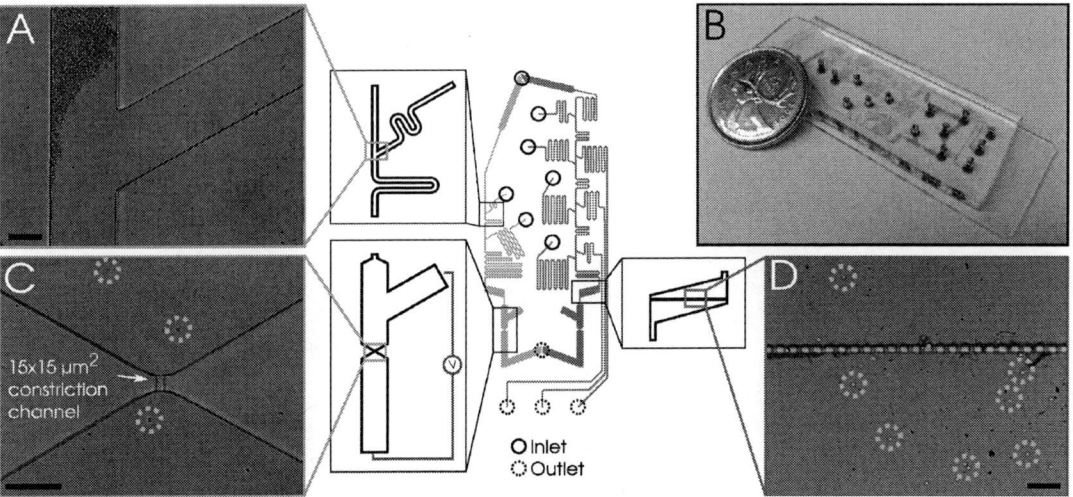

Figure 1: Schematic highlighting key modules for RBC (red) and WBC (blue) processing. (A) Module used to lyse RBCs with lysis solution. (B) Fully monolithic microfluidic device for blood cell sample preparation on 18 × 48 mm² footprint. (C) Downstream electrical measurement. Electrode placement and electrical measurement module denoted in schematic figure by 'V'. WBCs are circled. (D) Inline filtration structures used to eliminate debris. RBCs are circled. All scale bars represent 100 μm.

Robust fluidic handling of whole blood is enabled by integrating dilution, lysis, quench, filtration and electrical analysis. The use of a single pneumatic pressure source in our device reduces bulk and operational complexity. Integrated filtration permits the measurement of over 10,000 cells per test.

SYSTEM OVERVIEW

The device uses coulter counters for electrical enumeration and analysis in continuous flow after sample pre-processing on chip. The device operates in a parallel fashion, siphoning raw/whole blood from a single sample inlet into two separate streams for concurrent RBC and WBC processing and measurement. Parallel sample processing is necessary to achieve required dilution ratios as dictated by measurement requirements (1:10,000 for RBCs) and lysis protocols (1:23 for WBCs).

In Coulter counter measurement, 'coincident events' also known as 'coincidence' is the consequence of multiple objects simultaneously present in the detection region, making accurate enumeration difficult. Dilution can reduce the number of coincident events from the overwhelmingly large concentration of RBCs (3,500-5,000 cells/µL) in raw blood. To achieve optimal dilution, four separate 1:10 dilution modules are serially employed before the sample reaches in-line filtration modules (Fig. 1(D)). A modified serial dilution scheme employing multiple inlets of similar solutions is used in our device design (right side of Fig. 1). Each dilution module connects to its own individual buffer and waste outlet allowing flexible modifications to overall operation. An example of a modification is to introduce different buffer solutions at various inlets for multiple solutions mixing.

To properly analyze WBCs, sample lysis (Fig. 1(A)) and quench is necessary for removing RBCs due to the large density discrepancy between RBCs and WBCs. Following lysis, flow is filtered before finally undergoing electrical measurement (10-990 kHz) in the constriction region (Fig. 1(C)). In cases where lysis is not 100% complete, the minute remaining RBC population can be easily distinguished from WBCs.

Coulter counter measurement requires channel dimensions to be comparable to cell size in order to obtain a high cell to detection volume ratio. Debris can plague the use of constriction channels, and debris can vary in size, deformability and material make-up. The lysis of RBCs is rarely 100% debris free and produces highly deformable RBC ghosts which are remnant RBC membranes without internal components. Individually RBC ghosts can freely travel but can cluster and collectively impede electrical measurements by clogging the constriction channel. Thus, filtration is a necessary module in sample preparation for single cell analysis. The use of an in-line filtration module can drastically reduce the occurrences of constriction channel clogging and extend experiment duration by forcing the sample flow through numerous small and parallel orifices having sizes comparable to the downstream constriction channel.

Implementing filtration structures can increase device fluidic resistance and can limit the device's ability to operate with low pneumatic pressures. The overall

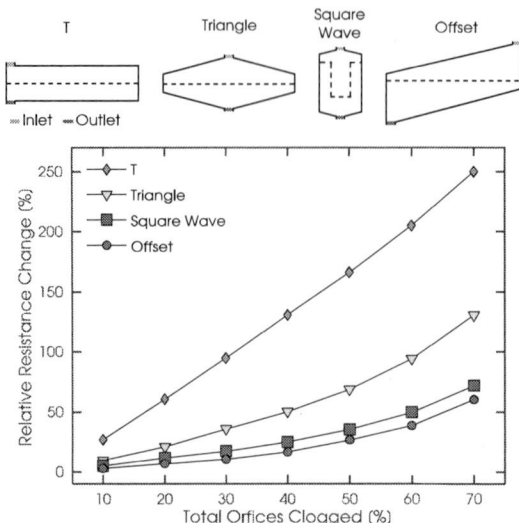

Fig. 2: Comparisons of existing and our new offset filter configurations. Filter orifice array denoted by dashed lines. At 50% of filter orifices become clogged, our offset filter increases fluidic resistance only by 26.8%.

fluidic resistance of a filtration module can be reduced by a factor of n by maximizing the number of parallel orifices/filter channels (n). Additionally, as more filter orifices are clogged, there is an undesirable increase in fluidic resistance which should be minimized. Fig. 2 summarizes finite element simulation results that theoretically compare filter performance of several existing filtration configurations with our new 'offset' filter configuration. This improvement allows for more accurate fluidic handling for longer durations and increases total sample throughput.

MATERIALS AND METHODS

Blood samples were obtained from healthy donors via venipuncture (Mount Sinai Hospital, Toronto, Canada). Hematological parameters of whole blood samples varied within healthy physiological ranges (Lymph %: 8-32, WBC concentration: 10-13 cells/nL, MCV: 83-96 fL, RBC concentration: 3-5 cells/pL). Blood samples were anticoagulated with EDTA anticoagulant (ethylenediaminetetraacetic acid 1.5 mg ml⁻¹) and stored at room temperature prior to use within 12 hours of withdrawal. For each patient sample, an aliquot of the sample was also measured by a standard hematology analyzer (Sysmex XN-9000, Sysmex America, Illinois) for reference.

PBS with 1% w/v BSA was used for device incubation and as RBC diluent. Lysis and quench solutions were prepared based on previously established bulk lysis protocols[6]. Briefly, ~10 µL of whole blood is mixed on-chip in a 1:12 ratio with lysis solution (0.12% v/v formic acid and 0.05% saponin in DI) and subsequently neutralized by 1:5.3 with a quench solution (0.6% w/v sodium carbonate in 1X PBS).

The device consists of a single layer of polydimethylsiloxane (PDMS), fabricated using standard soft lithography. Overall, the device has three separate areas each with distinct cross sectional dimensions, including the sample preparation and reagent handling channels (200 μm×40 μm), intermediate channels between electrodes (1,000 μm×40 μm), and constriction region for single cell measurement (15 μm×15 μm).

Prior to experiments, devices were incubated with PBS, all inlets and outlets were sealed, and the device was pressurized to remove trapped gas pockets via diffusion through the PDMS channel wall. Whole blood, lysis and quench solutions were pipetted into their respective inlet ports/reservoirs. Two sets of Ag/AgCl electrodes, one for each measurement module, were plugged in the device. Sample and dilution outlets were then connected externally to a custom pneumatic pressure source that drives fluidic flow.

RESULTS AND DISCUSSION
RBC enumeration and characterization

A high dilution ratio of 10,000× best mitigated the effect of RBC coincidence and produced results closest to reference RBC concentration benchmarked by the Sysmex commercial hematology analyzer (Fig.3). At 1,000× and 100× dilutions, the mistaken identification of multiple cells as a single cell resulted in much lower concentration measurement results. The stochastic nature of coincidence makes it difficult to adjust simply using a correction factor. The combined average standard deviation for dilution ratios of 1,000× and 100× was 1.076 cells/pL. A total of 104,735 RBCs from 10 samples were subsequently measured, and our devices produced concentration results with strong correlation with reference results (R^2=0.83) and an average difference of +0.1 cells/pL.

The measured electrical volume is difficult to directly equate with its single cell volume when considering its resting biconcave shape and its deformed parachute shape in microfluidic channels. However, the electrical volume exhibited by a population of RBCs can be correlated to its mean cell volume, which is similar to the procedures performed in commercial hematology analyzers. Fig. 4 is an experimentally generated RBC size histogram showing various RBC indices that were

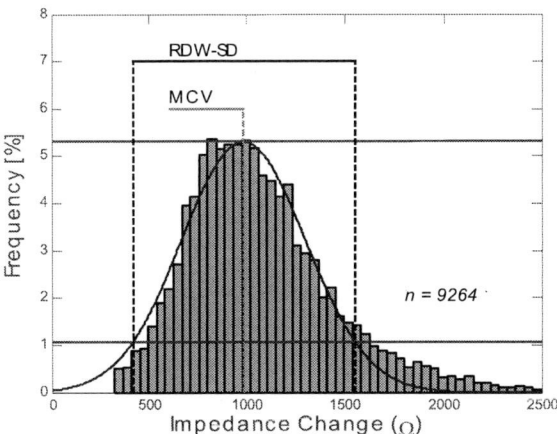

Fig. 4 Size histogram for RBC analysis highlighting specific measured RBC indices: mean corpuscular volume (MCV) and distribution width (RDW-SD).

WBC enumeration and differential

There are two major WBC cell types, including lymphocytes and a grouped category of granulocytes and monocytes. We first confirmed the inherent cell size difference between lymphocytes and non-lymphocytes in a post-lysis sample using microscopy imaging. There is an apparent size difference which generates a bimodal distribution of WBCs, centered around 70 fL for lymphocytes and 210 fL for non-lymphocytes, which concurs with size characteristics determined by hematology analyzers.

Individually, RBC ghosts produce minimal electrical signals, but a relatively high concentration of RBC ghosts can collectively cause false identification with lymphocytes. Hence, our measurement used low and high frequency impedance data in tandem to selectively distinguish WBCs from RBC ghosts. Moderately high frequency impedance data (100-400 kHz) was used to identify WBC peaks and synchronized with both lower frequencies (10 kHz) and higher frequencies (990 kHz) to generate multi-frequency information.

Fig. 5 shows scatter graphs of low frequency (i.e., cell electrical volume) and opacity data for various lymphocyte ratios within and out of normal physiological ranges, measured on different samples. Opacity data here cannot conclusively differentiate between granulocytes and monocytes, but low frequency data in combination with high frequency data for cell identification was sufficient to generate a predictable bimodal size histogram and provided 2-WBC differentials. Minimums in the size histograms were identified as lymphocyte flags, and cells were categorized according to their relative position to these generated gating flags. The microfluidic device performed on-chip raw blood processing and then conducted enumeration and size characterization of WBCs (97,305 total from 10 samples). Electrical measurements were compared to reference results from a standard hematology analyzer.

Cell loss from sedimentation and unwanted chemical lysis can contribute to undercounting. The relatively short constriction channel limits the maximum device flow rate which can promote cell sedimentation if

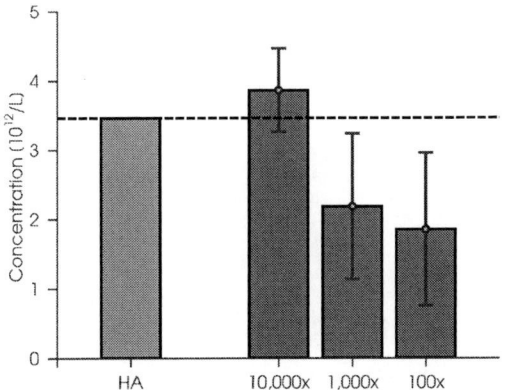

Fig. 3: RBC enumeration results at different dilution ratios. 10,000x dilution produced results most close to reference RBC concentration benchmarked by commercial hematology analyzer (HA).

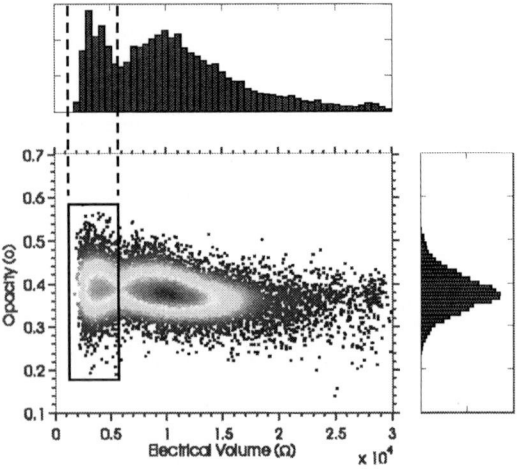

Fig. 5: Scatter graph for WBC measurement showing opacity and electrical volume from a healthy patient sample. Automated gating algorithms identify lymphocyte population (Lymph %: 23.9% vs. Reference: 22.5%).

cell velocities are too low. Strong linear correlations between device and reference results indicate the use of a correction factor or internal calibration can be used to account for cell loss from both sedimentation and chemical procedures. This is particularly applicable in a microfluidic environment due to its repeatable, uniform lysis and dilution conditions.

CONCLUSION

This paper reported a fully monolithic microfluidic system for simultaneously performing dilution and lysis of whole blood before conducting on-chip electrical enumeration and white blood cell differential. The filter and channel designs can be broadly applicable and can extend constriction channel lifetimes to enable the measurement of a high number of cells (e.g., over 10,000 cells). Raw blood samples from 10 patients were processed on chip and electrically analyzed on the device to measure several blood cell concentrations and parameters. Comparative results between reference hematology analyzers and the microfluidic device for specific RBC indices, RBC and WBC concentrations, and WBC differentials show strong correlations, proving the microfluidic device's capability to perform whole blood pre-processing and measurement. More generally, the work demonstrates that this platform can be used to handle solutions of varied viscosities and mixing ratios and can provide sample preparation for single blood cell analysis as a monolithic device.

ACKNOWLEDGEMENTS

Authors also acknowledge financial support support from the Government of Canada through Grand Challenges Canada and Natural Sciences and Engineering Research Council of Canada (NSERC) via an NSERC E.W.R. Steacie Fellowship.

REFERENCES

[1] B. George-Gay, "Understanding the complete blood count with differential.," *J. Perianesthesia Nurs.*, vol. 18, no. 2, pp. 96–117, 2003.

[2] N. N. Watkins, U. Hassan, G. Damhorst, H. Ni, A. Vaid, W. Rodriguez, R. Bashir, "Microfluidic CD4+ and CD8+ T Lymphocyte Counters for Point-of-Care HIV Diagnostics Using Whole Blood," *Sci. Transl. Med.*, vol. 5, no. 214, pp. 214ra170–214ra170, 2013.

[3] U. Hassan, N. N. Watkins, C. Edwards, and R. Bashir, "Flow metering characterization within an electrical cell counting microfluidic device," Lab Chip, vol. 14, no. 8, p. 1469, 2014.

[4] D. Holmes, D. Pettigrew, C. H. Reccius, J. D. Gwyer, C. van Berkel, J. Holloway, D. E. Davies, and H. Morgan, "Leukocyte analysis and differentiation using high speed microfluidic single cell impedance cytometry.," Lab Chip, vol. 9, no. 20, pp. 2881–9, Oct. 2009.

[5] C. van Berkel, J. D. Gwyer, S. Deane, N. G. Green, N. Green, J. Holloway, V. Hollis, and H. Morgan, "Integrated systems for rapid point of care (PoC) blood cell analysis.," Lab Chip, vol. 11, no. 7, pp. 1249–55, Apr. 2011.

CONTACT

*Y. Sun, tel: +1-416-946-0549; sun@mie.utoronto.ca

978-1-4799-7956-1/15 $31.00 © 2015 IEEE

A MICROWELL DEVICE FOR MEASUREMENT OF MEMBRANE TRANSPORT OF ADHERENT CELLS

Y. Okada, M. Tsugane, and H. Suzuki

Faculty of Science and Engineering, Chuo University, JAPAN

ABSTRACT

We developed the microwell device for measurement of membrane transport of single adherent cells. As the cells in a population (*e.g.*, tumor) is inevitably heterogeneous, a technique to measure the transport activities at a single-cell level is needed. When adherent cells were cultured on the microwells with ~10 µm diameter, they spread over the opening to form the closed picoliter space. Thus, molecules exported from cells accumulate in such a space and be detected by fluorescence imaging. In this report, we show that, by employing horizontal microwell design, materials exported from the cell membrane can be visualized without overlapping with the cell, increasing the S/N ratio of the fluorescence signal. Efflux of the cancer drug transported by the multidrug resistance protein was detected.

INTRODUCTION

Import and export of materials through the plasma membrane are vital for cells, and are also important for medical diagnosis. For instance, tumor cells gain resistance to cancer drugs by expressing the multidrug transporter proteins. Conventionally such activities are assessed by analytical methods as the average property of the cell population. However, it has been widely recognized that the cells in a single population are heterogeneous in various aspects (*e.g.*, chromosome number and gene expression) [1]. A method to measure the membrane transport activities at single cell level will provide the invaluable information not only for basic biology but for future diagnosis. Previously measurement of the efflux of drug from single bacterial cells were demonstrated by encapsulating *E. coli* expressing the AcrB transporter in microwells [2]. However, this approach is not applicable for adherent cells, which forms cell-to-cell junctions. We here attempt to utilize the ability of cells to adhere to the substrate covered with the extracellular matrix to enclose the microwell for detection of exported materials.

MATERIAL AND METHODS

Microwell Design

The microwell architecture for the single-cell transport assay is shown in Fig. 1. When the adherent cells are cultured on a substrate with the microwell smaller than the cell size, they stretch over the opening to close its space. If some materials or molecules are exported through the plasma membrane, they accumulate in this tiny space. However, with the conventional microwells having the vertical axis, cells overlap with the microwells to hinder the fluorescence imaging on the inverted microscope (Fig. 1a). Thus, we fabricated the microwell that stems from the side of the mainchannel to the horizontal direction (Fig. 1b). In this design, cells do not intervene the imaging of the microwells.

Fabrication Process

The microwell device was fabricated by laminating the three PDMS layers formed via standard soft-lithography. First, 10-µm thick SU-8 photoresist was patterned into the channel shape with horizontally protruding microwells (10 to 15 µm width and 30 µm length). The 40-µm thick second layer of SU-8 was formed only in the shape of the channel. From this mold two-layered PDMS structure was obtained. Another PDMS structure with 40-µm thick channel was fabricated separately. These two PDMS slabs were aligned and bonded. After exposing to the oxygen plasma, one of the

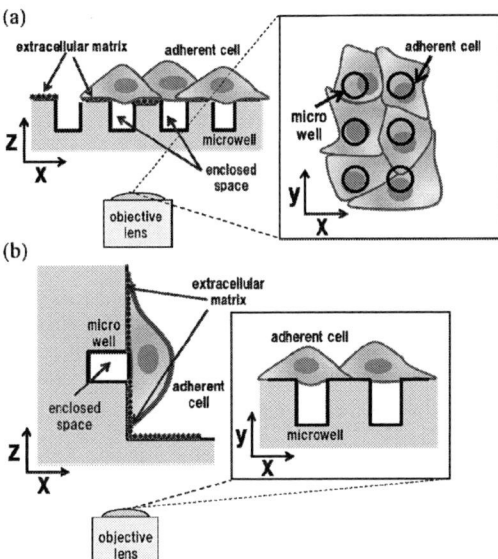

Figure 1: Conceptual diagram of cell adhesion over the microwell device. (a) Cell adhesion to vertical microwell device. (b) Cell adhesion to the horizontal microwell device.

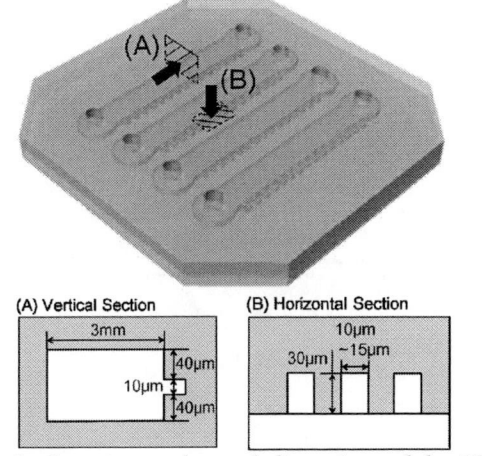

Figure 2: Overview and actual dimensions of the PDMS device with horizontal micro-wells.

PDMS structure was placed over another with ethanol as a lubricant. After alignment under the microscope, the device was settled for ~ 3 h at 50 °C and for ~ 1 h at 70 °C. As ethanol evaporated, two slabs were covalently bonded. The final structure with actual dimensions is illustrated in Fig. 2.

Cell culture

We used HeLa cell as a model adherent cell. The cells were incubated in Dulbecco's modified Eagle's medium (DMEM) (Sigma-Aldrich) supplemented with 10% fetal bovine serum (Invitrogen) and penicillin/streptomycin (50 U/ml; Sigma-Aldrich) at 37 °C in 5% CO_2. The cells were removed from culture dishes with 0.25% trypsin in EDTA solution and resuspended in buffer consisting of 130 mM NaCl, 5.4 mM KCl, 1.8 mM $CaCl_2$, 0.8 mM $MgCl_2$, 5.5 mM glucose and 10 mM Hepes-NaOH (pH 7.3). For checking the hermetic sealing, dextran conjugated with cascade blue (M.W. of 3 or 10 kDa, Invitrogen) at 4 mM concentration was included in the buffer. The cells were plated at a density of 3.0×10^7 cells/mL in microwell devices.

For the transport experiment, HeLa cells over-expressing the ATP-binding Cassette (ABC) transporter protein MDR1 (Multidrug resistant 1) was prepared as follows. GFP-tagged MDR1 plasmid vector (OriGene) was transfected into cells by using Lipofectamine®2000 (Invitrogen). After 1-2 days of transfection, cells were used for the transport experiment. Cells expressing MDR1 can be spotted by the green fluorescence from GFP. For monitoring the anti-cancer drug with fluorescence, Paclitaxel conjugated with BODIPY 564/570 (Invitrogen, P7501) was used.

Experimental Procedure

Prior to the experiment, microchannel wall was coated with the extracellular matrix by incubating 50 µg/mL fibronectin for 30 min. Sample preparation before microscope observation was conducted as follows (Fig. 3). First the device was placed under vacuum for 10 min, to avoid the air bubbles trapped in the microwells, and the main channel was filled with the cell suspension (1). After being sandwiched by the glass slides as a support (b), the device was placed vertically with the microwell side located downward (3). During 1.5 h incubation, cells sediment to the side of the channel and adhere over the microwells. Then, the cells in excess were washed out (4), and reagents for assays (e.g., drugs) were introduced. Finally the device was placed on the microscope stage for imaging (5). The epi-fluorescence microscopy (IX 51, Olympus) was used in most of the experiments, with one exception (Fig. 5) obtained using the laser scanning microscopy (LSM-700, Carl Zeiss).

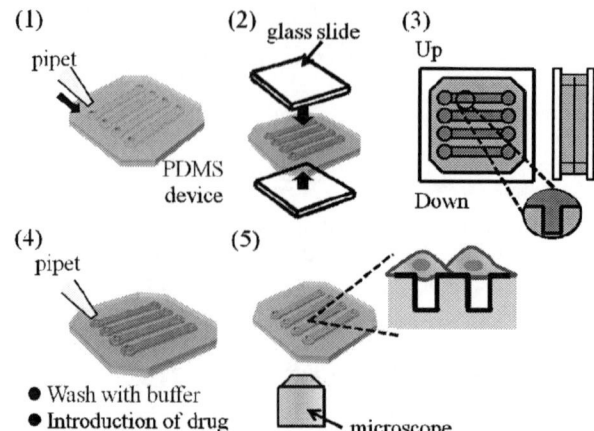

Figure 3: Experimental procedure. (1) Introduction of the cell suspension. (2) Cover with the glass lid. (3) Place the chip in the vertical direction during incubation. (4) Washing excess cells and introduction of drugs. (5) Microscope imaging.

EXPERIMENTS

Visualization of Adherent Cells

First we examined the morphology of HeLa cells on the microwells. The typical image obtained using the vertical microwell device (Fig. 1a), obtained for comparison, is shown in Fig. 4(a). After 1.5 h incubation, cells formed confluent culture over the microwell structures. However, it was not clear if cells penetrated into the wells. Another problem was that, as the cells overlap with the wells, background fluorescence from cells deteriorated the signal from the microwell space.

With the horizontal microwell device, images exampled in Fig. 4(b) was obtained. Vertical morphology of the HeLa cell on the floor of the main channel was clearly visible. It was also clear that most of the cells did not completely penetrate into the microwell. It was confirmed that images of cells and microwells were spatially uncoupled using the horizontal mirowell device.

Next we tested the ability of this device for the unconventional imaging. As a model of cell function occurring across the plasma membrane, swelling of the membrane was induced by exposing to ethanol. In practice, after cells adhesion, 0.5% ethanol in the buffer was introduced into the channel. This solution was supplemented with 5 µg/mL DiI for fluorescence labelling of the lipid membrane. After 5 min incubation, the channel was washed with the buffer, and subjected to microscope observation. As seen in Fig. 5, plasma membrane extruded into the microwells was clearly observed. In some cases, membrane was expelled as vesicles and settled in bottom of the microwell. Although this demonstration may not have any physiological meaning, such direct visualization of the membrane phenomena in the vertical direction will provide the new tool to the cell biology [3].

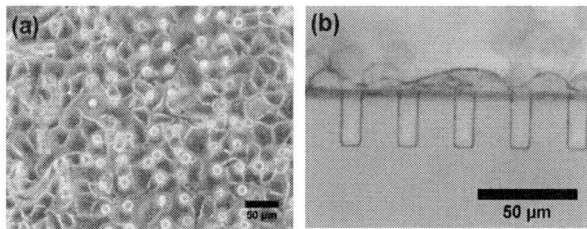

Figure 4: Comparison of the two different microwell architecture. (a)Phase-contrast image of HeLa cell incubated over the vertical (conventional) microwell device. (b) Image of the cell incubated over the horizontal microwell device. Note that the same inverted microscope setup was used for imaging.

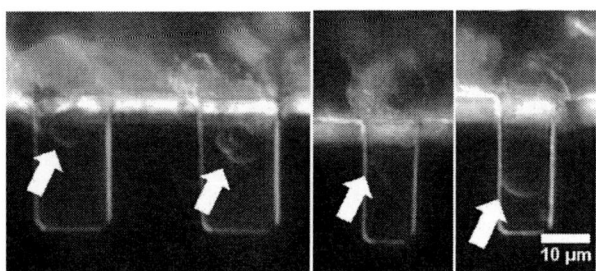

Figure 5: Extrusion of the plasma membrane stimulated by ethanol exposure. Membrane deformation of the adherent cells can be visualized from the direction horizontal to the culture plane.

Sealing Test

For detecting the molecules accumulated in microwells, sealing between the cell and the opening is impoartant. We tested the sealing of the enclosed space by monitoring the fluorescently labelled dextran. The dextran, included in the initial cell suspension, was washed after incubation steps. Under the fluorescence imaging with UV excitation, we sought the wells emitting the blue fluorescence, and obtained the time-lapsed images. Typical image of the microwell with blue fluorescence is shown in Fig. 6(a). From the time-lapsed images, we extracted the time course of the blue fluorescence intensity averaged in some of the microwells using the ImageJ software (Fig. 6b). In these wells, dextrans with both M.W. were kept above the background level (intensity in nearby dark wells) for at least 30 min, showing that hermetic sealing was achieved. Typically 10 to 20 % of the microwells resulted in good sealing in the present procedure, which was often realized by the slight invasion of the plasma membrane into the well (Fig.6a).

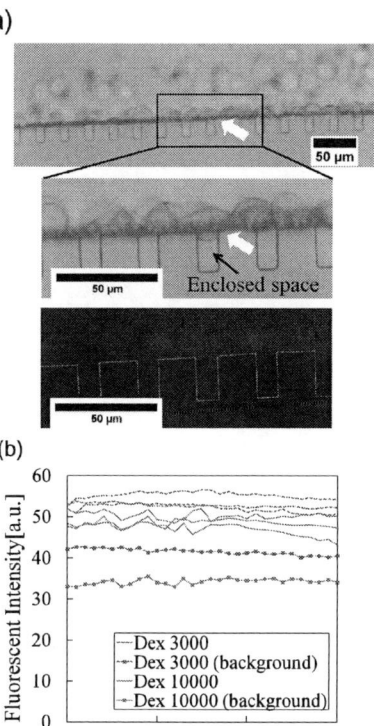

Figure 6: Sealing test. Fluorescence-conjugated dextran was enclosed in the microwell. (a) After washing, the fluorescence remained especially in the microwell into which plasma membrane slightly intruded (white arrow). (b) Time course of the fluorescence intensity from dextran in microwells. The bright fluorescence remained for at least 30 min, showing hermatic sealing.

Detection of Transport Activity of MDR1

Finally we measured the transport activity of MDR1, which was over-expressed in the HeLa cell. The bright-field image and two-colored fluorescence image, showing dextran in blue and GFP-tagged MDR1 in green, are shown in Fig. 7 (a) and (b), respectively. Such microwell with remaining dextran covered with MDR1 expressing cell was selected for imaging of paclitaxel. Typical time-lapsed images of red fluorescence from Paclitaxel-BODIPY is shown in Fig. 7(c). From a set of microwells, red fluorescence intensities were quantified. The time-course of the intensity relative to the initial value is shown in Fig. 7(d). We confirmed that, in the microwells with hermetic sealing (those containing dextran covered with MDR1 expressing cell), the concentration of Paclitaxel-BODIPY was gradually increased (red lines). In contrast, in uncovered microwells and/or those without dextran, intensity level remained constant (blue lines).

This result shows that, by using the horizontal microwell architecture, transport of the drug can be detected using the ordinal epi-fluorescence microscope setup. Since the size of the well is smaller than the typical cell size, the transported materials in a well should be attributed to a single cell.

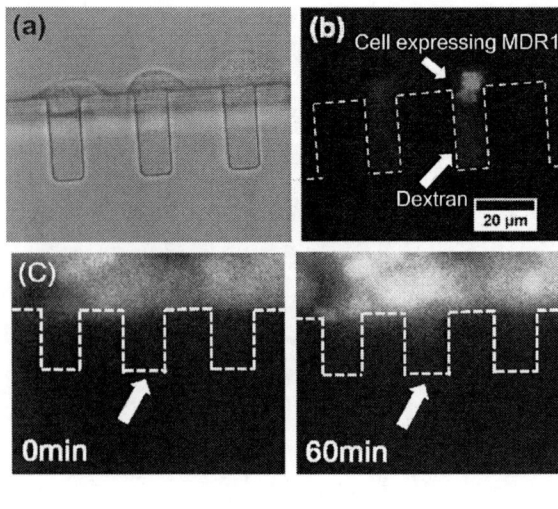

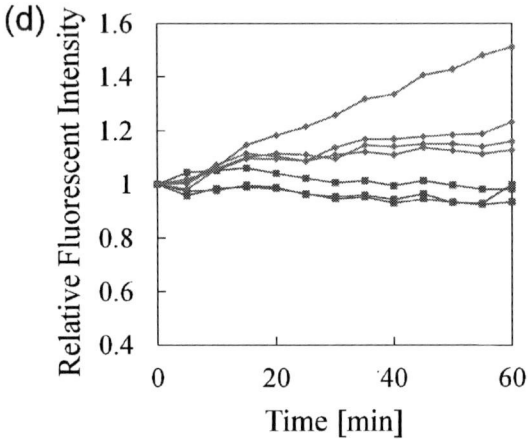

Figure 7: Transport experiment of fluorescence-labelled drug (paclitaxel) with the MDR1-expressed cells. (a) Bright-field image, showing cells lying over the microwells. (b) Fluorescence image. Blue-fluorescence in the well shows the localization of dextran marker, and green-fluorescence shows the expression of GFP-tagged MDR1. (c) Red-channel fluorescence images showing the transport of paclitaxel (white arrows). (d) Time courses of red fluorescence in microwells, in which the increase of fluorescence intensity was observed in some wells.

DISCUSSION & CONCLUSION

With the aid of horizontal microwell architecture, the sectional image of the adherent cells covering the microwell was clearly obtained. We found that, although extracellular matrix was not selectively patterned only on the floor, cells did not fully penetrate into the microwell. This adhesion property resulted in formation of the closed space between cell and well, as we originally expected. In some reports, patterning of the extracellular matrix was necessary to avoid cell structure penetrating into the recessed surface of micrometric structures [4]. This aspect might be dependent on the characteristics of cell species, which requires further investigation.

Furthermore, the horizontal configuration was shown to be effective in spatially uncoupling the cell and well in the inverted microscope imaging. Presence of the fluorescent molecules in microwells can be readily detected with the ordinal epi-fluorescence microscope. As a result, transport of fluorescence-conjugated anti-cancer drug via multidrug resistance protein was detected. We believe the present microwell device is also applicable for imaging various cellular phenomena involving the influx and efflux of materials across the plasma membrane of the cultured cells.

REFERENCES

[1] R. A. Burrell et al, "The causes and consequences of genetic heterogeneity in cancer evolution", *Nature* 501, pp. 338–345, 2013.

[2] R. Iino et al., "Rapid detection of drug efflux from single bacterial cell enclosed in femtoliter chamber array," *Proc. 11th μTAS*, pp. 757-759, 2007.

[3] Teshima et al., "Magnetically responsive microflaps reveal cell membrane boundaries from multiple angles", *Adv. Mat.*, vol. 26, pp. 2850-2856, 2014.

[4] J. L. Tan, "Cells lying on a bed of microneedles: An approach to isolate mechanical force", PNAS, vol. 100, No.4, pp. 1485-1489, 2003.

ALGINATE HYDROGEL BASED 3-DIMENSIONAL CELL CULTURE AND CHEMICAL SCREENING PLATFORM USING DIGITAL MICROFLUIDICS

Subin M. George and Hyejin Moon
The University of Texas at Arlington, Arlington, TX, USA

ABSTRACT

We report the development of a microfluidic device that is the first of its kind to generate uniform arrays of individually addressable cell seeded calcium alginate gels for 3D cell culture using electrowetting on dielectric (EWOD) digital microfluidics (DMF). This is combined with EWOD DMF's multiplexing abilities to demonstrate how a single device is capable of forming cell seeded alginate hydrogels, generating different concentrations of chemicals and delivering these generated chemical concentrations to different alginate gels to observe the effect of chemicals on 3D cultured cells. This lays the foundation for a more efficient and versatile 3D cell culture and chemical screening platform.

INTRODUCTION

The application of alginate hydrogels for 3D cell culture has already been well documented [1]. Incorporating alginate hydrogels into continuous microfluidic devices for cell culture and screening purposes, however, has been found to be problematic due to the difficulty in forming well controlled gel shapes. As such, methods have been limited to continuous microfluidic devices that formed a single large hydrogel exposed to a chemical gradient [2]. Robotic spotters have been used to form multiple alginate spots on a single device, but could not isolate alginate gels and exposed them all to the same chemical concentrations [3]. Current approaches mostly rely on droplet microfluidics since other methods fail to generate alginate hydrogels of regular and uniform shapes. However, droplet microfluidic methods fail to allow for integrated operations and require separate devices to carry out different steps [4].

In order to allow for 3D cell culture and chemical screening using alginate hydrogels, an electrowetting on dielectric (EWOD) digital microfluidics (DMF) approach is proposed. EWOD DMF allows for discrete nanoscale volumes of drops to be generated and transported on chip [5]. This can be utilized to accurately direct the motion of sodium alginate and calcium chloride droplets and control their method of merging to yield uniformly shaped alginate gels. This can then be used for cell culture and chemical screening applications.

DESIGN

The formation of alginate gels on EWOD DMF has never been studied in detail before. Conventionally, drops of sodium alginate are immersed in calcium chloride solution to form 3D alginate gel environment. Therefore, the most obvious immediate method to form an alginate gel on EWOD DMF would be to merge drops of sodium alginate and calcium chloride. The schematic of operations required to carry out such a mode of gel formation on EWOD DMF is shown in Figure 1.

However, this approach was found to yield very irregular shaped gels with alginate hydrogels.

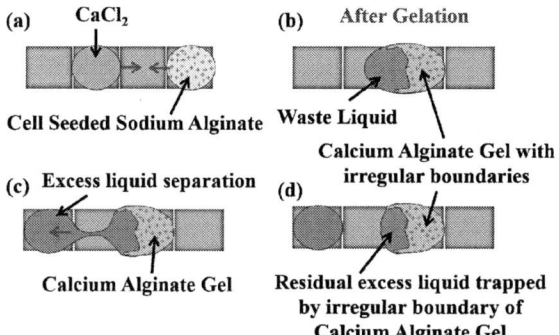

Figure 1: Schematic showing how direct merging of sodium alginate with calcium chloride (a) would result in the formation of irregularly shaped gels (b) which would trap liquid during removal of excess liquid (c)-(d).

Irregular shaped gels were found to retain pockets of liquid that hamper complete removal of excess liquid. These pockets of liquid would dilute any incoming liquid, making it impossible to accurately deliver specific concentrations of chemicals to the cells in the gel for screening purposes and studies. Without the ability to form gels with regular shapes and size, a reliable 3D cell culture and chemical screening would be impossible.

Gels with irregular shape also provided additional resistance to the separation of excess liquid. Applying greater force to remove excess liquid would result in the gel being displaced from its gel formation site and being carried away with the excess liquid. Such events would lead to the failure of the DMF system.

Thus in order to ensure uniform gel formation, a new encapsulation design was developed. This method relies on delivering sodium alginate to a circular gel post followed by encapsulation with calcium chloride. Once the sodium alginate is completely encapsulated by calcium chloride, the two liquids are allowed to merge to initiate gel formation. This method would immediately define the outside boundary of the alginate gel resulting in a uniformly circular gel.

The ability to separate the encapsulation step from the merging step relies on the adoption of a separator ring as seen in Figure 2. This separator ring allows the sodium alginate to be surrounded by the calcium chloride without accidental merging occurring. Merging is then carried out on demand by actuation of the separator ring electrode.

In order to ensure that premature merging of sodium alginate and calcium chloride does not occur, precision dispensing of cell seeded sodium alginate is carried out at the gel formation sites. Thus merging of the two drops only occurs by deliberate actuation of the separator ring electrode only after complete encapsulation.

978-1-4799-7956-1/15 $31.00 © 2015 IEEE   443   MEMS 2015, Estoril, PORTUGAL, 18 - 22 January, 2015

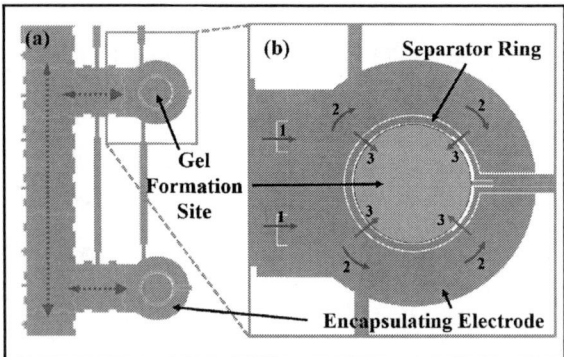

Figure 2: (a) shows a magnified mask design around the gel formation sites of a EWOD platform. Blue dotted arrows show the pathways for fluid. Inset magnified image (b) shows the separator ring electrode (green) that allows for 'on-demand' merging of the calcium chloride with sodium alginate. Blue arrows with numbers show how calcium chloride enters the encapsulating electrode (1) and then encapsulate the gel formation site (2). Actuation of the separator ring electrode causes calcium chloride to flow into and merge with sodium alginate at the gel formation site as seen by red arrows (3).

METHODS AND MATERIALS
Cell Culture and Alginate Hydrogel Preparation

Human breast cancer cells (MCF-7) obtained from American Type Culture Collection (ATCC) were maintained in culture medium (D-MEM/F12, Life Technologies, Carlsbad, CA), and supplemented with 4% fetal bovine serum, 2 mM l-glutamine, 100 μg/mL penicillin/streptomycin, and 0.01 mg/mL insulin. Cells were incubated in 25 cm^2 T-flasks at 37°C and 5% CO2 while being sub-cultured every 3-4 days at ~ 80% confluency.

In order to prepare 0.5% wt/vol sodium alginate solutions, first stock solutions of sodium alginate were formed by dissolving low viscosity sodium alginate powder (Sigma Aldrich, St. Louis, MO – CAS 9005-38-3) in deionized water to obtain a concentrated 4% wt/volume sodium alginate solution. This concentrated alginate solution was diluted in a 1:4 ratio with MCF-7 cells suspended in DMEM/F12 to obtain a 0.5% wt/vol of cell seeded sodium alginate solution with a concentration of 0.5-1 x 10^6 cells/mL. 100 mM calcium chloride solutions were prepared by dissolving calcium chloride crystals (Sigma Aldrich) in DMEM/F12 culture media. In order to prevent biofouling [6], pluronic F-68 was added to all culture media containing reagents at 0.04% wt/vol concentration.

EWOD Device Fabrication and Setup

EWOD devices were fabricated at the University of Texas at Arlington's Nanofabrication facility in a manner similar to that used by Wijethunga et al [7]. Briefly, a bottom EWOD chip consists of patterned ITO electrodes on a glass substrate, 5μm-thick dielectric layer of Su8-2005, and 150 nm-thick topmost hydrophobic layer of Teflon whereas a top EWOD chip is formed with an ITO and a Teflon layers only. EWOD bottom and top chips were assembled by means of 100 μm thick double sided

kapton tape and mounted into a custom holder that interfaced with a DAQ through which AC voltages could be transmitted from an AC signal generator and amplifier to the EWOD electrodes. Signals transmitted to the EWOD were programmed through scripts that were fed into a LABView program that controlled the DAQ.

Fluorescent Dye Preparation

Fluorescent dyes Hoechst 33342 (Life Technologies) was used to stain all the cells present in the hydrogels while Propidium Iodide (PI) (Life Technologies) was used to stain all the dead cells. H-33342 and PI were dissolved in sterile culture grade water to obtain final dye concentrations of 5ug/ml of H-33342 and 8ug/mL of PI.

DMSO Preparation and Dilution Protocol

A stock solution of DMSO was prepared by mixing sterile culture grade DMSO (Sigma Aldrich), deionized water, Hoechst-33342 and Propidium Iodide to obtain a 50% v/v DMSO solution with the same final dye concentration of H-33342 and PI as used to prepare the fluorescent dye solution described earlier. This solution was loaded into the reservoir and on chip dilution was carried out using a serial dilution protocol similar to that used by Park et al [8]. In this manner, concentrations of 50%, 25% and 12.5% DMSO were obtained while the 0% concentration was a control drop of fluorescent dye.

Targeted Chemical Delivery Experiment Protocol

In order to carry out the targeted drug delivery experiments, MCF-7 cell seeded alginate hydrogels were created at designated tissue post sites. This was done by dispensing cell seeded sodium alginate drops and delivered to designated tissue post sites. Calcium chloride drops was similarly dispensed from its parent reservoir and brought to the tissue post sites where they were merged with sodium alginate drops and gelation was allowed to occur. During this gelation process, simultaneous preparation of different concentrations of DMSO (0%, 12.5%, 25%, and 50%) was carried out in another area on the same chip using the DMSO dilution protocol as described earlier. After allowing sufficient time for gelation (~7 minutes), excess liquid from the tissue posts was extracted and dispensed to waste. The 4 different DMSO solutions were then delivered to the target tissue posts and the whole chip was incubated for 30 minutes in a 37°C, 5% CO2 humidified incubator following which fluorescence images were then taken using the protocol described below.

Fluorescence Microscopy and Cell Counting

Fluorescence images were taken using fluorescence microscope (Olympus BX-51) to visualize the cells in the gel posts stained by the fluorescent dyes. H-33342 dye stained all the cells present blue when viewed under DAPI filters while dead cells were labelled red by PI dye when viewed under TRITC filters.

Fluorescent images thus obtained were combined using ImageJ (NIH) so that the dead cells labelled red were overlaid with all the cells present in the sample that were stained blue. The resulting image clearly allowed viable cells to be distinguished from dead cells.

For experiments where viability was measured, cell counting was carried out manually using combined fluorescent images and the cell counter plugin in ImageJ software. This yielded the total number of live and dead cells, which were then used to calculate the viability percentage. The targeted chemical delivery experiment was repeated 3 times and results were presented as viability percentage ±1 standard deviation.

RESULTS

The proposed design for uniform gel formation was tested by delivering extra-large drops of sodium alginate to the gel formation site. Precision dispensing of sodium alginate was carried out directly at the gel formation site as seen in Fig 3 (a)-(c) resulting in sodium alginate drops that stayed within the boundary of the gel formation site.

Calcium chloride was then delivered to the encapsulating electrode and allowed to surround the sodium alginate as seen in Fig 3 (d)-(e). The separator ring was found to be successful in preventing merging of the two liquids as clearly seen in Fig 3 (e) where encapsulation has taken place but the drops are still separated. In order to initiate gel formation, the separator ring electrode is then actuated as seen in Fig 3 (f), which causes merging of the sodium alginate and calcium chloride to occur. After allowing for 7 minutes of gelation time, excess liquid removal was successfully carried out as seen in Fig 3 (g)-(i). This method resulted in uniformly circular gels that were of consistent size and shape and could be repeatedly formed.

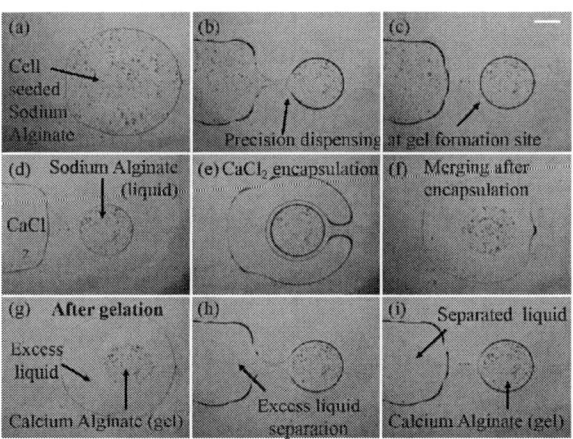

Figure 3: Sequence of alginate gel formation starting with precision sodium alginate dispensing at gel formation sites (a)-(c). This is followed by encapsulation with calcium chloride (d)-(e) and subsequent merging (f) through actuation of the separator ring electrode. After gelation, calcium alginate hydrogels are formed and excess liquid is removed from the calcium hydrogel as shown in (g)-(i). Scale bar in (c) represents 0.25 mm

To demonstrate the targeted chemical delivery abilities of this proposed screening platform, four MCF-7 seeded calcium alginate hydrogels were formed on chip as per the encapsulation design. While gelation was taking place, different concentrations of DMSO (0, 12.5, 25 and 50 %) were formed on chip as per the method described in the methods and material section. Once the calcium alginate hydrogel gelation was complete, excess liquid was removed and discarded to waste and the 4 different DMSO concentration drops were delivered to 4 gels and incubated for 30 minutes. Fig. 4 shows one set of results from such an experiment where cells were seeded at a high density to better allow visualization of the variation in cell. Based on the fluorescence images, a clear difference in viability of cells is observed with viability decreasing as DMSO concentration increases. Due to the high cell density used, this particular data set was not quantified and was only used for visualization purposes.

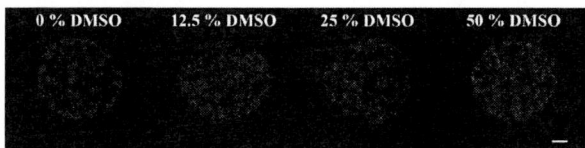

Figure 4: Fluorescent images showing effects of on chip DMSO delivery at 4 different concentrations (0%, 12.5%, 25% and 50%) to MCF-7 cell seeded calcium alginate hydrogel post. Blue dots represent live cells while red dots represent dead cells. Scale bar = 100 micron.

In order to determine the repeatability of this experimental procedure, 3 sets of the same experiment were carried out with cells seeded at lower density (in order to facilitate counting of cells). The results of the replicate experiments are shown in Fig. 5. The 0% DMSO concentration serves as the control gel to show the viability of cells formed on chip. The 12.5% DMSO gels showed a very slight decrease in viability compared to the control gels and this is to be expected since it is well known that exposure to low concentrations ~10% DMSO for short durations of time have a negligible effect on viability.

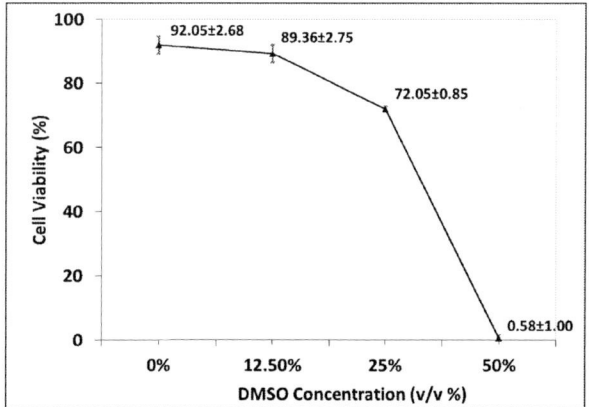

Figure 5: Plot showing quantified data for the effects of DMSO delivery at 0%, 12.5%, 25% and 50% concentrations on cell viability after delivery and incubation for 30 minutes to separate gel posts. Error bars (in red) represent ±1 standard deviation, n=3.

The gels exposed to 25% and 50% concentrations of DMSO show a marked decrease in viability with 25% DMSO concentrations showing a viability of 72.05% ± 0.85 while the 50% DMSO concentrations killing off all the cells in the 30 minute exposure time.

In order to demonstrate the ability to note changes in response of cells to DMSO over time, cell seeded alginate gels were formed on chip. While gelation was allowed to happen, different concentrations of DMSO (0%, 12.5% and 25%) were prepared on the same chip through serial dilution. Once gelation was complete and excess liquid was separated, the 3 different DMSO concentration drops were delivered to the tissue posts and the device was incubated. Fluorescent images were taken at 30 minutes intervals to measure the effect of DMSO over a span of two hours on cell laden hydrogels.

Fig. 6 shows the resultant viabilities obtained at each time step for each gel post exposed to DMSO concentration. From the graph, it can clearly be seen that there is minimal decrease in viability of the cells in the control gel exposed to 0% DMSO. However a decrease in cell viability could be seen over the first 60 minutes for the 12.5% DMSO treated gel followed by no major increase in cell death at 90 minutes and 120 minutes. The 25% DMSO treated gel on the other hand showed a consistent increase in cell death at every 30 minute interval for the first 90 minutes with the cell death rate starting to taper off at 120 minutes.

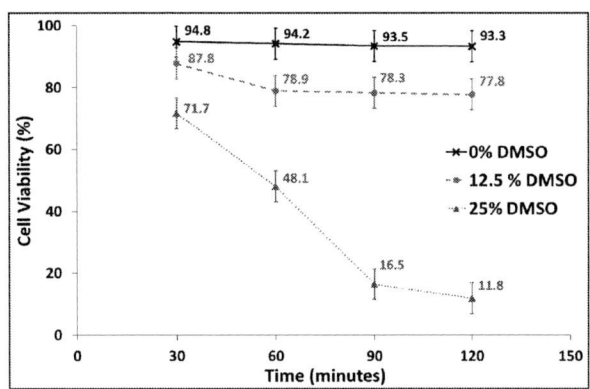

Figure 6:   Plot showing time lapse variation in cell viability % during a single experiment (n=1) as hydrogels are exposed to varying DMSO concentration (0%,12.5% and 25%) for 30,60,90 and 120 minutes.   Error bars represent ±5% error to account for errors in cell counting.

Such time-lapse experiments provide valuable insight into the effect of chemicals over time on cells without having to carry out multiple experiments. Being able to monitor the variation in cell death during the same experiment over time reveals that cell death in the 12.5% DMSO post increases only over the first 60 minutes and after that remains more or less constant. This kind of observation relating the nature of cell death with both time and concentration would be hard to make using conventional experimental setups without having to carry out multiple experiments. The EWOD DMF platform provides a simple and convenient means to take rapid images of multiple gel posts over multiple time frames with ease allowing investigators the ability to observe the transient effect of chemicals on cells cultured in a 3D environment.

CONCLUSION

An encapsulation design to allow the formation of uniform alginate hydrogels on EWOD DMF was developed. This encapsulation design was tested for use as a 3D cell culture platform for chemical screening and the differential effects of DMSO delivery on cell viability in alginate hydrogels were demonstrated. A time-lapse experiment was also demonstrated showing how the effects of different chemical concentrations on cells could be tracked temporally. This showcases the potential for EWOD DMF to be used for hydrogel based cell culture and chemical screening.

ACKNOWLEDGEMENTS

Authors acknowledge the support by National Science Foundation CAREER award (grant no. ECCS-1254602).

REFERENCES

[1] K. Y. Lee and D. J. Mooney, "Alginate: properties and biomedical applications.," *Prog. Polym. Sci.*, vol. 37, no. 1, pp. 106–126, Jan. 2012.

[2] A. P. Wong, R. Perez-Castillejos, J. Christopher Love, and G. M. Whitesides, "Partitioning microfluidic channels with hydrogel to construct tunable 3-D cellular microenvironments.," *Biomaterials*, vol. 29, no. 12, pp. 1853–61, Apr. 2008.

[3] M.-Y. Lee, R. A. Kumar, S. M. Sukumaran, M. G. Hogg, D. S. Clark, and J. S. Dordick, "Three-dimensional cellular microarray for high-throughput toxicology assays.," *Proc. Natl. Acad. Sci. U. S. A.*, vol. 105, no. 1, pp. 59–63, Jan. 2008.

[4] L. Yu, M. C. W. Chen, and K. C. Cheung, "Droplet-based microfluidic system for multicellular tumor spheroid formation and anticancer drug testing.," *Lab Chip*, vol. 10, no. 18, pp. 2424–32, Sep. 2010.

[5] S. K. Cho, H. Moon, and C.-J. Kim, "Creating, transporting, cutting, and merging liquid droplets by electrowetting-based actuation for digital microfluidic circuits," *J. Microelectromechanical Syst.*, vol. 12, no. 1, pp. 70–80, Feb. 2003.

[6] V. N. Luk, G. C. Mo, and A. R. Wheeler, "Pluronic additives: a solution to sticky problems in digital microfluidics.," *Langmuir*, vol. 24, no. 12, pp. 6382–9, Jun. 2008.

[7] P. A. L. Wijethunga, Y. S. Nanayakkara, P. Kunchala, D. W. Armstrong, and H. Moon, "On-chip drop-to-drop liquid microextraction coupled with real-time concentration monitoring technique.," *Anal. Chem.*, vol. 83, no. 5, pp. 1658–64, Mar. 2011.

[8] S. Park, P. A. L. Wijethunga, H. Moon, and B. Han, "On-chip characterization of cryoprotective agent mixtures using an EWOD-based digital microfluidic device.," *Lab Chip*, vol. 11, no. 13, pp. 2212–21, Jul. 2011.

CONTACT

*Hyejin Moon, tel: +1-817-272-2017;
hyejin.moon@uta.edu

AN ELECTROCHEMICAL MICROFLUIDIC PAPER-BASED GLUCOSE SENSOR INTEGRATING ZINC OXIDE NANOWIRES

*Xiao Li, Chen Zhao, and Xinyu Liu**
McGill University, Montreal, CANADA

ABSTRACT

This paper reports, for the first time, an electrochemical microfluidic paper-based analytical device (EµPAD), featuring a highly-sensitive working electrode (WE) decorated with zinc oxide nanowires (ZnO NWs), for glucose detection in human serum. Besides common features of µPADs such as low cost, high portability/disposability, and ease of operation, the reported EµPAD has three additional advantages. (i) It provides higher sensitivity and lower limit of detection (LOD) than previously reported µPADs, owing to the high electron-transfer efficiency and high surface-to-volume ratio of the ZnO NWs. (ii) It does not need light-sensitive electron mediator (usually required in enzymatic glucose sensing), leading to enhanced biosensing stability. (iii) The ZnO NWs are directly synthesized on the paper substrate via low-temperature hydrothermal growth, representing a simple, low-cost, highly-consistent, and mass-producible process.

INTRODUCTION

Microfluidic paper-based analytical devices (µPADs) represent a newly emerging platform for such applications as diagnostic biosensing and environmental monitoring [1]. MicroPADs can be fabricated at low cost, and used with ease in resource-limited settings by end users without professional skills [2]. These merits make them particularly promising for promoting public health in developing countries, where costs of diagnoses are a major concern, and resources such as electricity and professional equipment are not always accessible. They may also contribute to health care in the developed world, as they may eventually enable users to readily monitor their health conditions in a point-of-care fashion. Many µPAD designs adopted electrochemical sensing mechanisms (thus called electrochemical µPADs – EµPADs) because of its high accuracy and sensitivity [3-5].

With rapid advancements in the field of paper-based microfluidics, the research focus has shifted to creating fully-functional µPADs with superior analytical performance for practical diagnoses [6], for which the introduction of biosensing nanomaterials to µPADs has shown remarkable potential. The benefits of using nanomaterials for biosensing mainly stem from their unique physical/chemical characteristics [7]. For instance, researchers have integrated multi-walled carbon nanotubes (MWCNTs) into µPADs to covalently immobilize probe antibodies and enhance the electrical output of electrochemical immunoassays by utilizing MWCNT's high surface-to-volume ratio and superior electronic conductivity [4]. Graphene has been also used for similar purpose [8]. Another nanomaterial that has been utilized on µPADs is gold nanorod for optical

biosensing [9]. Nevertheless, the preparation of those nanomaterials is usually complex and costly, which may hinder the wide use of µPADs. In addition, their integration into µPADs after synthesis may also raise concerns about the nanomaterials' attachment stability and distribution uniformity on paper substrates.

Zinc oxide nanowires (ZnO NWs) have shown their outstanding capacities in biosensing [10]. They possess not only high surface-to-volume ratio, but also high biocompatibility, non-toxicity, and fast electron transfer. Even more, ZnO NWs have isoelectric point (IEP) around 9.5, which is much higher than many enzymes and therefore allows them to immobilize enzymes with high efficiency via electrostatic attractions. Therefore, they are a popular candidate in enzymatic electrochemical sensing of targets such as glucose [11], uric acid [12], and phenol [13]. Another highly desirable property of ZnO NWs is that they can be synthesized on various substrates (e.g., plastics [14] and cellulose paper [15]) through a low-cost hydrothermal process. To date, there is no previous work on developing µPADs using ZnO nanomaterials as the biosensing component.

Here, we present the first EµPAD integrating ZnO NWs for electrochemical enzymatic detection of glucose in serum. An array of circular paper channels (as reaction areas) are formed on a paper substrate by wax patterning, and electrodes are patterned on top of the reaction areas by stencil printing of conductive inks. The EµPAD features a separate layer of working electrode (WE), where the paper piece is covered by carbon ink and ZnO NWs are directly grown on the carbon ink through a low-cost hydrothermal process. Glucose oxidase (GOx) is then immobilized on the ZnO NWs by simple addition. The paper channel layer and the WE layer are stacked together to form a complete EµPAD. We characterize the device via a series of electrochemical measurements, and demonstrate improved response of the device to glucose. Glucose detections in both phosphate buffered saline (PBS) and human serum are performed, and the device's sensing performance is analyzed. The reported ZnO-NW EµPAD covers the major clinically-relevant range for glucose sensing, and provides sensitivity and limit of detection (LOD) both superior to existing EµPADs and commercial meters.

EXPERIMENTAL DESIGN
Fabrication of ZnO-NW EµPAD

An explored schematic of the ZnO-NW EµPAD is shown in Figure 1A. The device consists of two layers of chromatography paper (Whatman® CHR #1). A reaction zone is patterned in the top layer via the wax printing technique (wax was printed by a Xerox ColorQube 8570 printer and heated on a 150 °C hot plate to form hydrophobic wax barriers in paper). A counter electrode (CE) and a reference electrode (RE) were then patterned

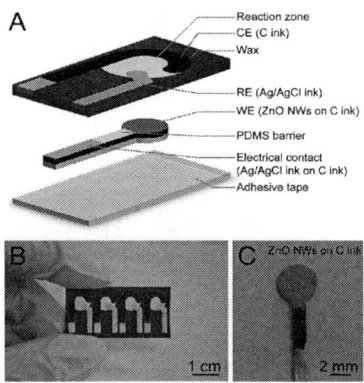

Figure 1: (A) Schematic of the ZnO-NW EµPAD. (B)(C) Photographs of (B) an array of EµPADs and (C) a WE grown with ZnO NWs in its circular area (the gray color is from ZnO NWs).

on top of the reaction zone by stencil-printing carbon ink (E3456, Ercon) and Ag/AgCl ink (E2414, Ercon), respectively. Underneath the CE, an Ag/AgCl strip was printed in advance, to enhance the electrical conductivity between the CE and the measurement setup. Inks were dried on a 65 °C hot plate for 20 min. The bottom layer of paper was cut into a shape shown in Figure 1(A) using a CO_2 laser cutter (VLS 2.30, Universal Laser Systems), where the top surface of the circular area (2 mm in radius) was then covered by carbon ink as the working electrode (CE) via stencil printing. An Ag/AgCl strip was patterned on the paper beam of the WE layer for electrical connection. After the inks were dried, polydimethyl-siloxane (PDMS) mixed with it curing agent (w/w ratio: 10:1) was dropped to both sides of the CE layer to form a hydrophobic barrier close the circular area, so that liquid added to the WE could be confined in the circular area. The circular WE area was finally decorated with ZnO NWs through a hydrothermal process, which will be described in details in the next section.

For enzymatic glucose sensing, 2 µL of 2.5 mg/mL GOx solution (Aspergillus niger, 147.9 U·mg⁻¹, Sigma-Aldrich) was dropped to the circular area of the ZnO-NW WE, dried at room temperature for 5 min. 4 µL of 0.5% (weight) Nafion® solution was then dropped to the circular area and dried at room temperature for 5 min. The Nafion® coating of the WE could stabilize GOx on ZnO NWs. The two layers of paper were finally attached together using a tape to form a complete device.

Hydrothermal Growth of ZnO NWs on WE

The synthesis of ZnO NWs includes two steps: (i) seeding of ZnO nanoparticles (NPs) on WE, and (ii) hydrothermal growth of ZnO NWs on the seeded ZnO-NPs. In the first step, 20 mL of 4 mM zinc acetate dihydrate (ZAD) and 20 mL of 4 mM sodium hydroxide (SH) solutions were prepared in 200 proof ethanol, and the solutions were stirred in a 70 °C oven to completely dissolve the chemicals. 20 mL of 200 proof ethanol was then added to the ZAD solution, and the mixture was then heated in a 70 °C oven for 30 min. After cooling down to room temperature, the SH solution was slowly added into the ZAD solution with constant stirring, and the mixture was placed in a 60 °C oven for 2 h to obtain a colloidal

solution of ZnO NPs. The paper beam of the WE layer (coated with carbon ink and PDMS barrier) was clapped between two 1 mm thick glass slides, with the circular WE areas exposed. 4 µL of ZnO-NP seeding solution was then added to the circular WE area, and the WE layer was dried on a 100 °C hot plate for 3 min. For each WE, this seeding and drying process was repeated 6 times. After seeding, 25 mL of aqueous growth solution, with 50 mM zinc nitrate hexahydrate, 25 mM hexamethylenetetramine and, 0.372 M ammonium hydroxide, was prepared in a 25 mL glass bottle, and two pieces of four-WE-arrays were immerged into the growth solution. The glass bottle was then sealed and heated in an 86 °C oven for 8 h hydrothermal growth. After growth, the WE layer was thoroughly washed in deionized water and ethanol, and dried for final device assembly.

Electrochemical Measurements

In glucose detection, the three electrodes of an EµPAD were connected by metal clips to a precision potentiostat (PGSTAT302N, Metrohm). 5 µL of PBS or human serum spiked with D-(+)-glucose (Sigma-Aldrich) was added to the reaction zone on the top of the sensor. After 50 s incubation, the potentiostat started the measurements.

RESULTS AND DISCUSSIONS
Fabrication of ZnO-NW EµPAD

Figures 1B shows an array of four ZnO-NW EµPADs after assembly, which was readily fabricated from a single run. The top paper layer of four reaction zones and four sets of CEs and REs were conveniently attached with a bottom paper layer of four WEs. The distance between centers of adjacent reaction zones was made following the standard tip-to-tip pitch (9 mm) of multi-channel pipette, so that multiple reactions can be performed in parallel on a sensor array by simultaneously adding sample solution(s) into different reaction zones with a multi-channel pipette.

Figure 1C shows a photograph of the WE layer after ZnO-NW growth. The circular area of carbon ink turned from black to gray, indicating uniform coverage of ZnO NWs. During ZnO-NW growth, the PDMS barrier successfully confined the growth solution in the circular area and thus no ZnO NWs were observed outside the circular area (on the paper beam). The PDMS barrier also held the sample solution which vertically wicked from the reaction zone in top paper layer, so that the target biomarker in the sample solution could react with the corresponding enzyme on ZnO NWs more efficiently.

Characterization of ZnO NWs on WE

To the best of our knowledge, we demonstrated the first hydrothermal growth of ZnO NWs on carbon ink printed on paper. Carbon ink printed on paper is relatively hydrophobic but affinitive to ethanol. The seeding solution made in ethanol was thus well spread over the whole circular area of WE. Scanning electron microscopy (SEM) imaging of the ZnO-NW WE shows that high-density ZnO NWs uniformly covered the whole carbon-ink-coated WE area (Figure 2B). Because paper is highly

978-1-4799-7956-1/15 $31.00 © 2015 IEEE

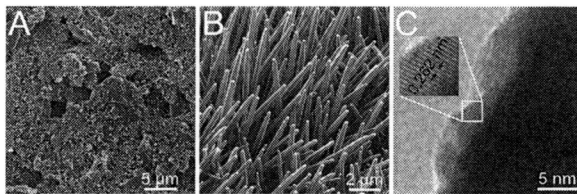

Figure 2: (A) SEM image of carbon-ink WE surface before ZnO-NW growth. (B) Zoomed-in view of ZnO NWs (width: 298.7 ± 54.1 nm, N=10) grown on carbon-ink WE. (C) TEM image of ZnO-NW crystal lattice.

porous and its cellulose microfibers were not smooth, the carbon ink coating did not form a smooth substrate for ZnO-NW growth (Figure 2A). The non-smoothness of the carbon ink surface made the synthesized ZnO NWs not uniformly oriented (Figure 2B). Transmission electron microscopy (TEM) imaging of single ZnO NWs reveals the lattice edges along the ZnO NW (Figure 2C). We measured the distance between two edges to be 0.262 nm, which corresponds to the distance between adjacent (0002) planes in wurtzite ZnO crystal units.

Characterization of ZnO-NW EµPAD

To demonstrate the electrochemical performance of the ZnO-NW EµPAD, we compared the cyclic voltammetry (CV) responses of EµPADs with and without ZnO NWs (both with pre-stored GOx), with 5 mL of pure PBS and 5 mM glucose solution added to the devices. As shown in Figure 3, for both types of EµPADs, the presence of glucose in the solution led to an increase in peak magnitude of the measured CV current, compared with the current output from pure PBS. Clear current responses were observed in voltage ranges of: (i) >0.3 V for ZnO-NW EµPAD; and (ii) >0.8 V for Non-NW EµPAD. In addition, the peak of current response of ZnO-NW EµPAD was 3.5 times higher than that of Non-NW EµPAD. Thus, we confirmed that the ZnO NWs can improve the performance of glucose sensing, through both lowering the voltage required to generate electrochemical response and elevating the current response. In many other electrochemical glucose sensors (including paper-based glucose sensors) using GOx, electron mediators are used to lower the working voltage and increase the current response [16]. However, electron-mediators are usually light sensitive, which could cause biosensing instability.

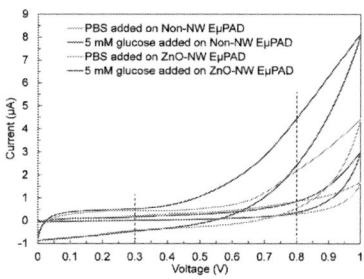

Figure 3: Cyclic voltammetry responses of EµPADs (with and without ZnO NWs) with PBS or glucose added. Response of ZnO-NW EµPADs to glucose is 3.5 times higher than that of Non-NW EµPAD (measured at 0.8 V).

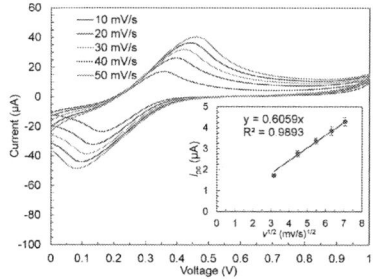

Figure 4: CV of the EµPAD using 10 mM $K_3[Fe(CN)_6]$ in 1 M KCl aqueous solution at various scan rates. Inset: The linear relationship between the anodic peak current and the square root of the scan rate, measured from the CV.

The integration of ZnO NWs into EµPADs eliminates the need of electron mediators and can thus potentially improve the device stability.

We conducted CV characterization of the ZnO-NW EµPAD using aqueous solution of 10 mM $K_3[Fe(CN)_6]$ in 1 M KCl. The CV response of the device shows typical redox peaks (Figure 4). The anodic current peak of the cyclic voltammogram was proportional to the square root of the CV scan rate. These results indicate a good electrochemical characteristic of the developed EµPAD.

Calibration of ZnO-NW EµPAD

The CV characterization results (Figures 3 and 4) implies the feasibility of operating the ZnO-NW EµPAD at a voltage as low as 0.3 V. However, for better comparison of glucose sensing results to previous work, we chose 0.8 V as the working voltage for chrono-amperometry (CA), which was often used in ZnO-NW glucose sensors [11, 17]. As shown in typical CA curves (Figure 5A), the capacitive current quickly dropped after the application of a 0.8 V voltage, and the Faradaic current eventually dominated. After the current was stabilized, we took the average current value in 79–80 s as

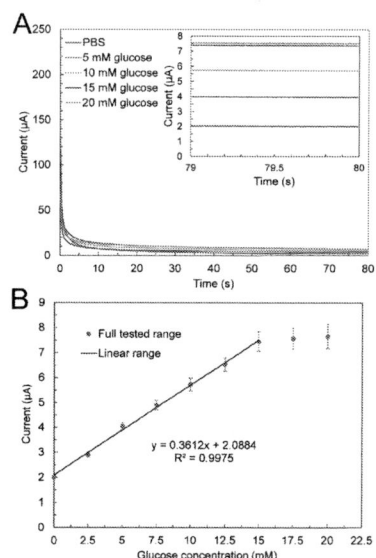

Figure 5: (A) Typical CA curves (0.8 V applied) and (B) calibration plot (N=5) of glucose detection in spiked PBS.

978-1-4799-7956-1/15 $31.00 © 2015 IEEE 449

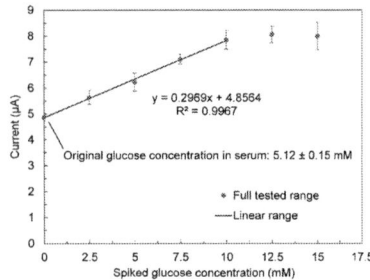

Figure 6: Calibration plot of glucose detection in spiked human serum (N=5).

the readout for calibration.

The calibration results using glucose-spiked PBS samples reveal a linear range of 0–15 mM (Figure 5B). The calibration was also performed using spiked human serum, which is a more practical and complex sensing environment with the existence of various large molecules. The calibration results revealed a linear range of 5.12–15.12 mM (Figure 6). The lowest glucose concentration of 5.12 mM reflects the original glucose level of the human serum sample, which was measured using a commercial glucose meter. The linear range of the ZnO-NW EµPAD covers the major clinically relevant range of glucose in human blood. Based on the calibration results from PBS samples, the sensitivity and LOD of the ZnO-NW EµPAD is summarized in Table 1. Both parameters obtained from our device are superior to the commercial glucose meters and a previously reported EµPAD [8] with similar test conditions. Although we used human serum rather than whole blood in this work, it is totally feasible to integrate a layer of filtering paper/membrane on top of the device's reaction zone for blood cell filtration [18]. The above results demonstrate the feasibility of using our ZnO-NW EµPAD for glucose detection in real blood samples.

CONCLUSIONS

We developed a novel EµPAD with ZnO NWs directly grown on the paper-based carbon-ink electrode, for glucose sensing. The synthesis of ZnO NWs was completed through a low-cost, facile hydrothermal process. The integration of ZnO NWs on EµPAD lowered the working voltage and boosted the current response, and the ZnO-NW EµPAD retained proper characteristic as an electrochemical system. Calibration results on both PBS and human serum revealed a linear range up to 15 mM, covering the major clinically relevant range. The ZnO-NW EµPAD provided a sensitivity of 2.88 µA·mM^{-1}·cm^{-2}, and a LOD of 94.7 µM, both of which are better than

Table 1. Analytical performance comparison.

Glucose sensor	Linear range (mM)	Sensitivity (µA·mM^{-1}·cm^{-2})	LOD (µM)
This work	0-15	2.88	94.7
Ref [8]	0-20	1.025	350
Commercial meter [9]	0-30	N/A	830

these of commercial glucose meters and previously reported EµPADs.

REFERENCES

[1] A. W. Martinez, et al., "Three-dimensional microfluidic devices fabricated in layered paper and tape", *P. Natl. Acad. Sci. USA,* vol. 105, pp. 19606-19611, 2008.

[2] A. W. Martinez, et al., "Diagnostics for the developing world: microfluidic paper-based analytical devices", *Anal. Chem.,* vol. 82, pp. 3-10, 2010.

[3] W. Dungchai, et al., "Electrochemical detection for paper-based microfluidics", *Anal. Chem.,* vol. 81, pp. 5821-5826, 2009.

[4] D. J. Zang, et al., "Electrochemical immunoassay on a 3D microfluidic paper-based device", *Chem. Commun.,* vol. 48, pp. 4683-4685, 2012.

[5] Z. Nie, et al., "Integration of paper-based microfluidic devices with commercial electrochemical readers", *Lab Chip,* vol. 10, pp. 3163-3169, 2010.

[6] A. K. Yetisen, et al., "Paper-based microfluidic point-of-care diagnostic devices", *Lab Chip,* vol. 13, pp. 2210-2251, 2013.

[7] F. Patolsky, et al., "Nanowire sensors for medicine and the life sciences", *Nanomedicine,* vol. 1, pp. 51-65, 2006.

[8] Y. F. Wu, et al., "Paper-pased microfluidic electrochemical immunodevice integrated with nanobioprobes onto graphene film for ultrasensitive multiplexed detection of cancer biomarkers", *Anal. Chem.,* vol. 85, pp. 8661-8668, 2013.

[9] A. Abbas, et al., "Multifunctional analytical platform on a paper strip: separation, preconcentration, and subattomolar detection", *Anal. Chem.,* vol. 85, pp. 3977-3983, 2013.

[10] Z. W. Zhao, et al., "ZnO-based amperometric enzyme biosensors", *Sensors,* vol. 10, pp. 1216-1231, 2010.

[11] J. F. Zang, et al., "Tailoring zinc oxide nanowires for high performance amperometric glucose sensor", *Electroanalysis,* vol. 19, pp. 1008-1014, 2007.

[12] F. F. Zhang, et al., "Immobilization of uricase on ZnO nanorods for a reagentless uric acid biosensor", *Anal. Chim. Acta,* vol. 519, pp. 155-160, 2004.

[13] B. X. Gu, et al., "Tyrosinase Immobilization on ZnO Nanorods for Phenol Detection", *J. Phys. Chem. B,* vol. 113, pp. 377-381, 2009.

[14] S. H. Ko, et al., "Digital selective growth of ZnO nanowire arrays from inkjet-printed nanoparticle seeds on a flexible substrate", *Langmuir,* vol. 28, pp. 4787-4792, 2012.

[15] Y. H. Wang, et al., "A paper-based piezoelectric touch pad integrating zinc oxide nanowires", in *Micro Electro Mechanical Systems (MEMS), 2014 IEEE 27th International Conference on,* 2014, pp. 781-784.

[16] C. Zhao, et al., "A microfluidic paper-based electrochemical biosensor array for multiplexed detection of metabolic biomarkers", *Sci. Technol. Adv. Mat.,* vol. 14, pp. 054402-054408, 2013.

[17] D. Pradhan, et al., "High-Performance, flexible enzymatic glucose biosensor based on ZnO nanowires supported on a gold-coated polyester substrate", *ACS Appl. Mater. Inter.,* vol. 2, pp. 2409-2412, 2010.

[18] T. Songjaroen, et al., "Blood separation on microfluidic paper-based analytical devices", *Lab Chip,* vol. 12, pp. 3392-3398, 2012.

CONTACT

*X. Liu, tel: +1 514 398 1526,
fax : +1 514 398 7365; xinyu.liu@mcgill.ca

978-1-4799-7956-1/15 $31.00 © 2015 IEEE

DRY REAGENT STORAGE IN DISSOLVABLE FILMS AND LIQUID TRIGGERED RELEASE FOR PROGRAMMED MULTI-STEP LAB-ON-CHIP DIAGNOSTICS

Gabriel Lenk, Göran Stemme, and Niclas Roxhed
KTH Royal Institute of Technology, Stockholm, SWEDEN

ABSTRACT

Multi-step lab-on-chip assays typically require multiple manual pipetting steps of different reagents and a sophisticated design to run the assay, which is generally unfavorable for point-of-care applications. This work presents the use of dissolvable polyvinylalcohol films for (1) reagent storage and release together with (2) timed valving in capillary driven microfluidics. This allows four different volumes to be split up from a single liquid which is applied to the inlet of the chip. PVA captured reagents are released to each volume forming four different solutions which are separately released in a timed sequence to a common target zone. The presented chip thereby enables a single liquid-triggered multi-reagent sequence with the potential to be used for advanced point of care diagnostics.

INTRODUCTION

Lab on a chip (LOC) systems are a powerful tool for bioanalysis and point of care (POC) diagnostics. However, many of the LOC microfluidic systems require extensive peripheral equipment for fluid handling such as pumps and valves. More self-sustainable operable systems are based on capillary microfluidics. Here the liquid flow is self-powered, meaning driven by the capillary forces in a microchannel, or in other microstructures such as pillar forests or paper substrates. The self-powered liquid flow is one of the reasons why capillary driven devices for diagnostics, or lateral flow assays (LFA), have commercially become a successful microfluidic platform with millions of devices fabricated each year at a very low cost [1]. The most prominent example for LFAs is the pregnancy stick which only requires a small amount of urine to detect pregnancy within 5 to 10 minutes. Although LFAs have been used for a long time for diagnostic applications, the function of LFAs has been limited to single or dual step chemistry. Recently, LFAs based on paper microfluidics have gained renewed interest and LFAs with more advanced functions based on capillary paper substrates have been demonstrated during the last few years. Basic novel functions such as dissolvable sugar bridges to interrupt capillary flow after a certain time [2] and dissolvable sugar delays in a paper substrate to control the wicking speed of liquid in the paper [3] have been presented. This allows for more complex fluid handling of several liquids in parallel which is needed for automated multi-step assays on paper, where different liquids are delivered sequentially after an "initialization" step of an assay [3]. In another approach, newly introduced retention burst valves and low aspect ratio trigger valves for capillary microfluidics in PDMS channels were used to realize a similar functionality [4]. By accurately controlling the microfluidic properties of these circuits, several reagents can be preloaded to their inlet ports before initiating the system by adding the sample. The addition of the sample automatically triggers a sequential flow of the sample and the preloaded reagents one after another over a common detection zone. However both presented concepts require the initial application of the different reagents to their inlet ports and hence several manual handling steps.

Substance release from dissolvable films has been shown previously, mostly for drug release from oral drug strips [5]. For microfluidic devices, reagent storage can be realized in a liquid form, usually requiring a composite-foil pouch which is mechanically actuated [6] or in dried form immobilized in a substrate [7]. Also water soluble electro spun fibers have been used for enzyme storage and on-chip release [8]. However, release of several reagents in a timed manner is often difficult, requiring a complex microfluidic system or several manual pipetting or actuation steps.

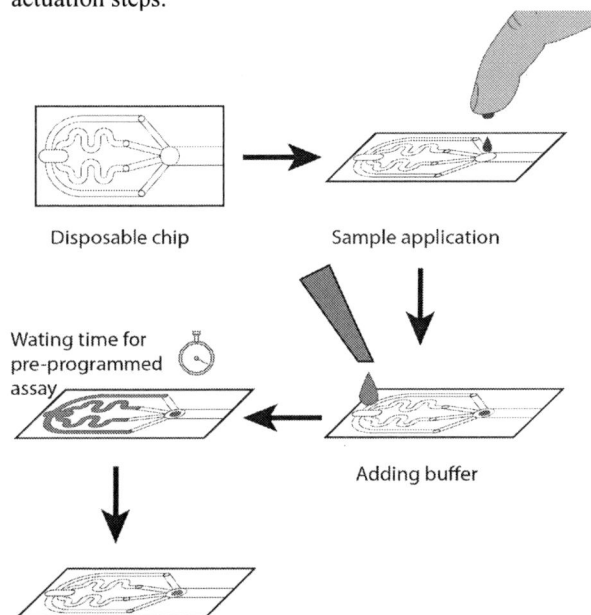

Figure 1: Concept for POC multi-reagent assay. After adding a sample, for example a droplet of blood from a fingertip, a buffer fluid for dissolving the different reagents is added which automatically initiates a sequence of different events on the chip. The result can be read, as for example, by a colorimetric read out.

In this work we build on previous works where we demonstrated volume metering in a capillary driven device [9] and timed reagent release with pre-programmed dissolvable valves made of thin dissolvable films to achieve a multi-step timing in capillary

---

978-1-4799-7956-1/15 $31.00 © 2015 IEEE   451   MEMS 2015, Estoril, PORTUGAL, 18 - 22 January, 2015

microfluidics [10]. Here these features are combined with dissolvable reagent release films in capillary microchannels. This enables a microfluidic system that reduces manual handling and allows for on-chip reagent storage and an easy-to-handle point-of-care assay with multiple reagents. Furthermore, it becomes possible to develop assays which only require two manual steps, the application of the sample to the chip, and the application of a buffer solution for dissolving the different reagents (see Figure 1).

METHOD

To demonstrate the concept of reagent release from water soluble PVA layers two different designs have been developed and tested. The first design consists of a single capillary channel with metering function as previously presented in [9]. These test channels were equipped with PVA layers for substance release at the channel bottom, and the same layer at the channel outlet for timing of the outlet valve. The second design is a demonstrator device with four parallel metering channels with release layers and additional functionalities such as a common inlet, a structure that separates the four channel volumes from each other and four timed release valves for sequential release of the volumes.

Materials

For the fabrication of all chips, sheets of different materials were laminated together. For the definition of the microfluidic structures, a 100 µm capillary spacer tape (IVD 090448PV1.001/09, Tesa GmbH, Germany), with a 50 µm PET core and a 25 µm acrylic adhesive on both sides was used. As channel cover and bottom 100 µm thick Xerox transparency foils (003R96002 Type C, Xerox Co. Ltd., USA) with a hydrophilic surface treatment to assist the capillary suction were used. All integrated paper parts consists of Whatman 903® DBS paper (Whatman plc, UK). The dissolvable films consist of polyvinylalcohol (PVA) (Mowiol® 8-88 Mw ~67,000 and Mowiol® 4-88 Mw ~31,000, Sigma Aldrich Inc., USA).

Fabrication

The fabrication of all devices was based on a lamination process. For structuring the different material sheets, a GRAPHTEC CE 5000 cutting plotter (Graphtec America Inc., USA) was used. The pattern of all layers was designed in Adobe Illustrator and then each layer was separately plotted to the transparencies using Cutting Master 2 (Graphtec America Inc., USA), a plug-in for Adobe Illustrator. All paper parts were cut with a CO2 laser. The different components were assembled by lamination in a GBC® HeatSeal® H600 Pro laminator (GBC Inc., USA). Dissolvable films were fabricated prior to the lamination process by preparing different solutions of the PVA mixed with either food dye for visualization or with 10 vol.-% of microparticles. The solutions were then spin coated on silicon wafers and dried at 60° C and then peeled of and transferred to the lamination process.

REAGENT RELEASE CHANNELS

To demonstrate and evaluate reagent storage and release in dissolvable films, different thicknesses of dye-colored PVA films and PVA films loaded with microparticles were fabricated and applied to the channel bottom as shown by the schematic cross section in Figure 2.

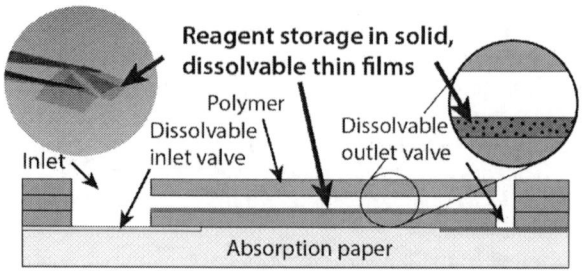

Figure 2: A metering channel with a reagent release layer at the channel bottom, an inlet absorption valve and an outlet emptying valve. The small inlay shows dissolvable films in two different colors on a tweezers tip. The inset illustrates the particle loaded film at the channel bottom.

Successful film dissolving and hence reagent release has been evaluated by analyzing the channel surface before and after testing of the chips. The full schematic sequence during reagent release from PVA layers can be seen in Figure 3.

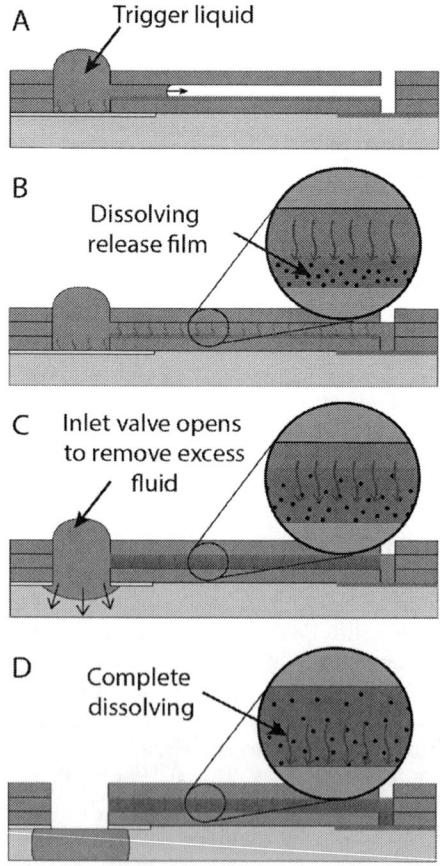

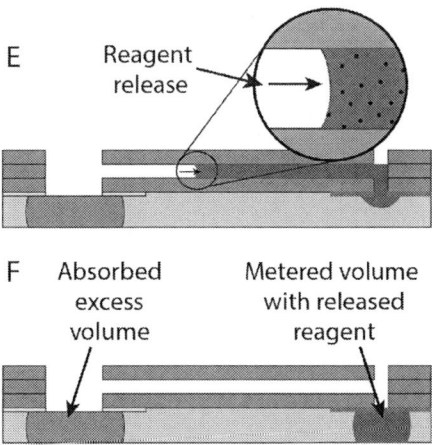

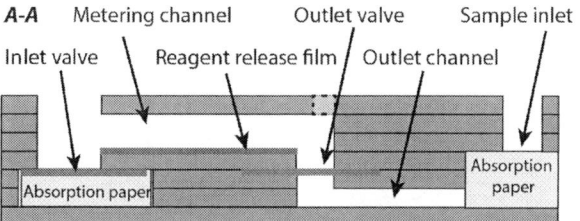

Figure 3: Sequence of reagent release in a metering channel. A) Applying a liquid to the inlet initiates the process on the chip by simultaneously activating the inlet valve dissolving and capillary channel filling. B) Reagent release layer starts dissolving. C) Complete inlet valve dissolving results in excess-volume absorption and channel inlet pinch-off. D) Reagent release layer completely dissolved. E) Complete outlet dissolving after the release layer has been completely dissolved. The liquid mixed with the particles from the release layer is absorbed by the paper under the outlet. F) The metered channel volume with the reagent from the dissolved layer is completely absorbed by the paper at the outlet.

Figure 4: Cross section of the demonstrator device. Compared to Figure 2, a second microchannel was added at the outlet which is in contact with the target zone. The metering channel is 200 μm high allowing the 100 μm high outlet channel to empty the release/metering channel by capillary forces after the outlet valve has been completely dissolved. This cross section is an unfolded cross section along the dashed line (A-A) in Figure 5.

DEMONSTRATOR DEVICE

The demonstrator device (see Figure 4 and 6) uses dissolvable films for fluidic timing and reagent storage medium at the same time. Four channels are connected to the same inlet (see Figure 5A) so that when a liquid is applied all four channels fill simultaneously by capillary action. Once the channels are completely filled, a thin dissolvable film at the inlet area is dissolved and a paper matrix absorbs any excess volume still present at the inlet. This results in a liquid pinch-off at each channel inlet, separating the four channels into discrete liquid volumes. In each channel a dissolvable film, laminated to the channel bottom starts to dissolve once it comes in contact with the liquid. The film releases the reagent/particles stored in the film homogenously along the channel length to the defined liquid volume. At the channel outlets, the thickness of each of the outlet valves controls the timing of the sequence and also the duration the liquid is in contact with the reagent-carrying films. Once an outlet valve is open the liquid is capillary sucked out from the channel and the liquid is flushed over the target and sample zone.

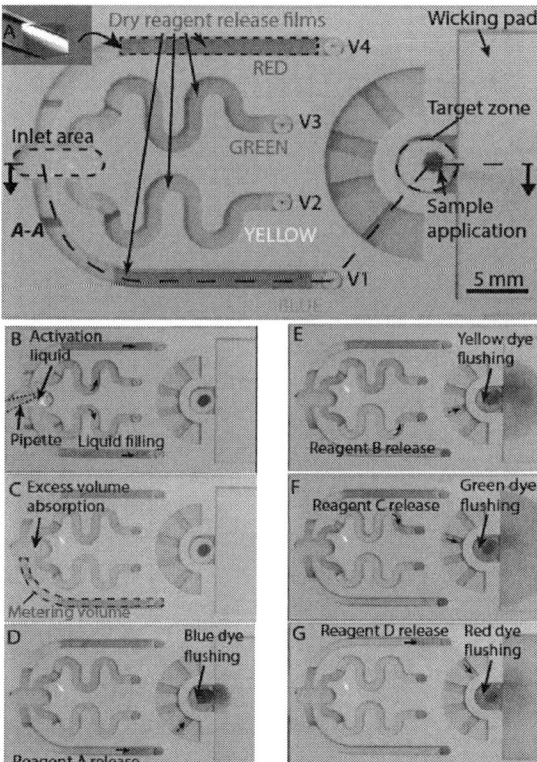

Figure 5: Top view of the demonstrator chip. A) The four different channels with the common inlet area. The sample is applied directly to the target zone. B) Activation with the buffer liquid and simultaneous channel filling of all channels. C) Metering by excess volume absorption. D)-F) Sequential release of the different reagents to the common target zone.

RESULTS

Dissolvable layers up to 30 μm could successfully be dissolved and absorbed at the outlet by an absorbent paper matrix. Particle-carrying layers up to 26 μm thickness containing 10 vol.-% of micro-particles could be dissolved and sucked out of the channel likewise (Figure 7). For the tested channels this equals a volumetric particle concentration of 1.2 vol.-%. The

978-1-4799-7956-1/15 $31.00 © 2015 IEEE   453

thicker the PVA storage layer in the channel, the higher the viscosity of the liquid becomes after dissolving which slows the liquid flow down during emptying.

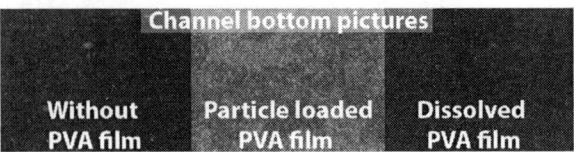

Figure 6: Microscope pictures of the channel bottom: A) Before application of the dissolvable film. B) After laminating a 26 μm thick bright dissolvable film with 10 vol.-% microspheres to it. C) After dissolving the particle-loaded film to the metering channel and emptying of the channel. No residues of the particles and the dissolvable film are visible.

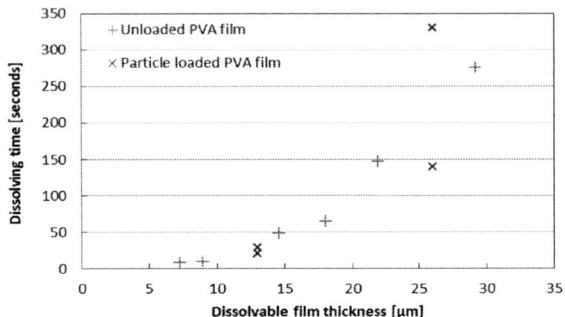

Figure 7: Dissolving times for layers with different thicknesses with 10 vol.-% particle load (red plus) and without particle load (blue cross). The dissolving times were measured between wetting of the outlet valve (Figure 3B) and opening of the outlet valve (Figure 3F).

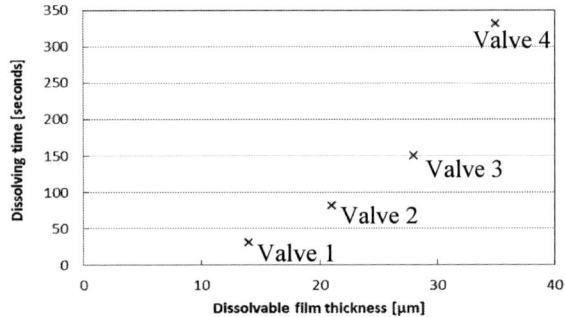

Figure 8: Timing of the different steps of the multi-release mechanism.

For the demonstrator device, 10 μm thick release layers were successfully dissolved after undyed DI water was applied to the inlet. The release valves sequentially opened after 31, 82, 150, 332 seconds, respectively (Figure 7), and released the now dye colored liquid. Thus, the target zone with the sample was sequentially flushed with blue, yellow, green and red dyed colored liquid in an automatic fashion. After liquid filling and emptying, the metering channels, previously containing colored reagent films, became uncolored, demonstrating that the films were successfully dissolved and carried away by the liquid in the channel.

CONCLUSION

The demonstrated concept with integrated reagent/particle storage in PVA films has been developed and shown to be feasible. Activation with a single liquid initiates a pre-programmed sequence resulting in four different solutions being flushed over a common target zone. This allows realizing a multi reagent assay with the use of only one triggering buffer liquid. The time scale needed for each assay and the storage and stability of reagents in PVA films has to be evaluated assay specifically.

ACKNOWLEDGEMENTS

This work was supported by the Stockholm county council and the European Research Council (ERC) through the Advanced Grant No: 267528.

REFRENCES

[1] S. Haeberle et al. "Microfluidic platforms for lab-on-a-chip applications. Lab Chip, vol. 7, pp. 1094-1110, 2007.

[2] J. Houghtaling et al. "Dissolvable Bridges for Manipulating fluid Volumes in Paper". Anal. Chem., vol. 85, pp. 11201-11204, 2013.

[3] B. Lutz et al. „Dissolvable fluidic time delays for programming multi-step assays in intrument-free paper diagnostics." Lab Chip, vol. 13, pp. 2840-2847, 2013.

[4] R. Safavieh et al. "Capillarics: pre-programmed, self-powered microfluidic circuits built from capillary elements." Lab Chip, vol. 13, pp. 4180-4189, 2013.

[5] B.P. Panda et al. "Development of innovative orally fast disintegrating film dosage forms: A review." Int. J. Pharm, vol. 5 (2), pp. 1666-1674, 2012.

[6] T. v. Oordt et al. „Miniature stick-packaging-an industrial technology for pre-storage and release of reagents in lab-on-a-chip systems." Lab Chip, vol. 13, pp. 2888-2892, 2013.

[7] S. Ramchandran et al. "Long-term dry storage of an enzyme-based reagent system for ELISA in point-of-care devices." Analyst vol. 139, pp. 1456-62, 2014.

[8] M. Dai, A. Senecal, S.R. Nugen. Electrospun water-soluble polymer nanofibers for the dehydration and storage of sensitive reagents." Nanotechnology, vol. 25, 225101 (8pp), 2014.

[9] G. Lenk et al. "A disposable chip enabling metering in dried blood spot sampling." μTAS 2013, pp. 281-283.

[10] G. Lenk et al. "Delay valving in capillary driven devices based on dissolvable thin films." μTAS 2014, pp. 216-219.

CONTACT

*G.Lenk, tel: +46762921966; glenk@kth.se

ON-CHIP DETECTION OF WILD 3R, 4R AND MUTANT 4R TAU THROUGH KINESIN-MICROTUBULE BINDING

S. P. Subramaniyan[1], M. C. Tarhan[2], S. L. Karsten[3], H. Fujita[2],
H. Shintaku[1], H. Kotera[1], and R. Yokokawa[1]
[1]Department of Micro Engineering, Kyoto University, JAPAN
[2]LIMMS, Institute of Industrial Science, The University of Tokyo, JAPAN
[3]NeuroInDx Inc., Signal Hill, CA, USA

ABSTRACT

We report the successful demonstration of an on-chip tau detection system based on the difference in landing rate and binding density of microtubules (MTs) on a kinesin surface. Tau detection device comprises of a MT reservoir, channel and collector region with an overhang structure. We assayed MTs decorated with three tau types in the kinesin coated device. Since the increase in fluorescence intensity (FI) at the collector regions reflected the type of tau decorated on MTs, thus by measuring the FI we were able to distinguish wild 3R, 4R and P301L mutant tau.

INTRODUCTION

Recent claims on tau facilitating the spread of neurodegeneration among the brain cells [1], has underlined the importance of tau as a biomarker. Elaborate study relating to the distribution of tau isoforms and mutants in the brain and CSF could not be consolidated into developing an early detection method for various neurodegenerative conditions; one reason is that a standardized tau detection method which does not employ antibodies has not been established. We developed a non-immunological tau detection method and reported two new detection parameters: landing rate and MT binding density (MEMS 2014 [2]). In this paper we show the device design and experimental results of tau detection assay on a chip. The microfluidic device is integrated such that it has retained the assay sensitivity, to differentiate wild 3R vs 4R and 4R vs P301L. Here, P301L is a biomarker for progressive neurodegeneration involving Corticobasal region.

With six different tau isoforms in human brain, tau proteins play a key role in maintaining the structural integrity of neuronal transportation. Disruption of normal tau isoform ratio (3R: 4R) itself serves as a prelude to neurodegeneration [3]. Apart from over and/or down regulation of specific tau isoforms, many tau mutants especially those confined to tau-MT binding repeats (MTBR) have been reported to cause rapid neurodegeneration [4]; one such is P301L mutation that occurs only in tau with four binding repeats.

Previously, by recreating the intercellular transport system in a microfluidic device incorporating the MT-kinesin system, many applications for sorting [5-7], concentrating, transporting [8] and detecting [9,10] various biomolecules have been reported. Herein, when MTs decorated with tau isoform/mutant were assayed on a kinesin coated surface, the number of MTs binding to the kinesin surface was strictly regulated by the type of tau that decorated it. Therefore along with the velocity of MT, the landing rate and binding density also varied significant

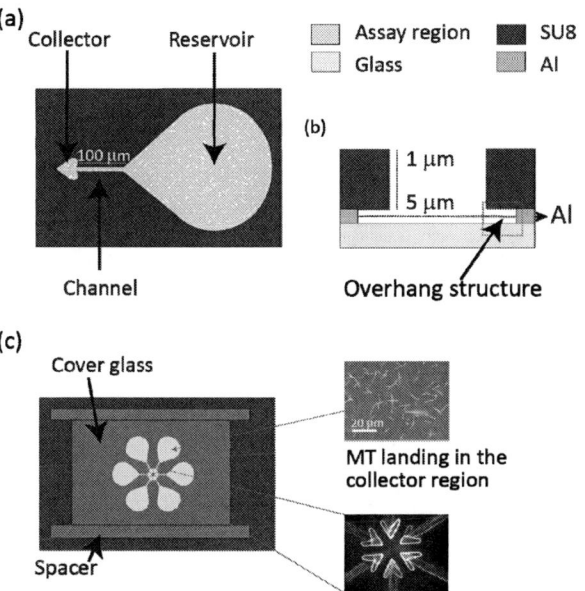

Figure 1. Schematic representation of tau detection device. a) Top view of the device, yellow color denotes the assay region b) Cross sectional view of the device channel showing the overhang structure, and c) Construct of the microfluidic device i) MTs binding at the reservoir and ii) MT accumulated at the collector region (scale bar, 20 μm)

enough that we could eventually differentiate among tau isoforms and mutants; with respect to the number of binding repeats (3R and 4R) and position of point mutation respectively.

Following this, we designed and fabricated a microfluidic device, which can rapidly differentiate tau isoform and mutants. The details are discussed below.

DESIGN

We constructed a microfluidic device comprising of a MT reservoir, a channel ($l\sim100$ μm and $w\sim5$ μm) and a collector region (Figure 1). Our prime focus to retain the assay sensitivity in the microfluidic device was achieved by focusing on the following points: (1) to detain the MTs that glide into the collector region, (2) to increase the probability of MTs landing in the reservoir region, (3) to prevent random entry of MTs into the assay region (yellow part in Figure 1a and c) and (4) the confinement of MTs within the assay region. These conditions were attained by sensible designing and fabrication steps described below.

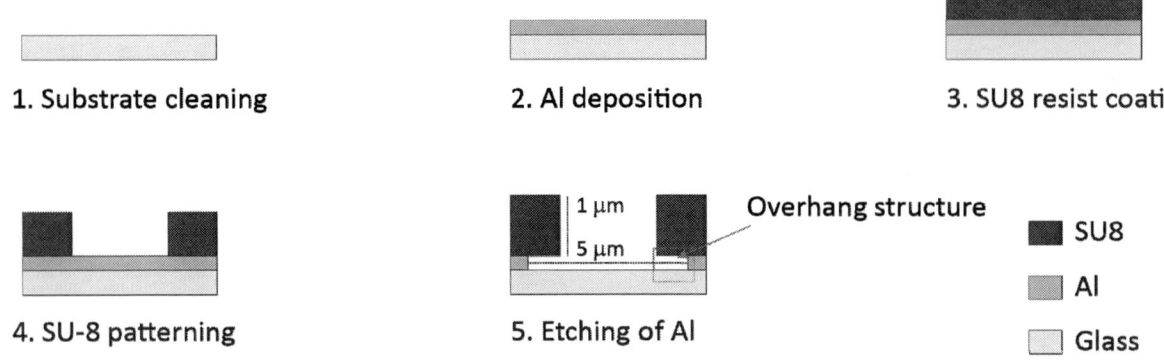

1. Substrate cleaning   2. Al deposition   3. SU8 resist coating

4. SU-8 patterning   5. Etching of Al

1 μm   5 μm   Overhang structure

■ SU8
▨ Al
□ Glass

Figure 2. Schematic representation of the fabrication process of tau detection device

We designed the collector region in the shape of an arrow head, so that once the MT enters the collector it is repeatedly circulated within the collector region. The reservoir region was designed many times larger than the collector region in order increase the probability of MTs landing in the reservoir region. The fabrication technique and device topology modification employed further aided in retaining the assay sensitivity in the device.

EXPERIMENTAL

Microfluidic device fabrication

The glass substrate (24 mm × 36 mm; Matsunami Glass) was cleaned using piranha solution (H_2SO_4:H_2O_2 = 3:1) at 80°C for 20 min. On the cleaned substrate Aluminum (~150 nm thickness) was coated by thermal deposition (VPC-260F, ULVAC). A UV sensitive negative photoresist SU8 3005 was spun (6000 rpm, 30 sec) on Al coated glass substrate resulting in ~1 μm thick film. The resist was exposed through a chrome mask, to UV-light at 51 mJ cm^{-2}. The photolithographically patterned resist surface was developed in SU8-developer and rinsed in IPA

Development of overhang structure

In order to prevent the MTs from gliding out of the assay region an overhang structure was developed. Al underlying the SU8 pattern was over-etched at room temperature by an Al etchant containing a mixture of HNO_3, H_3PO_4, CH_3COOH and H_2O. As a result of the over-etching, following the removal of the sacrificial Al layer from the assay region, Al underlying the SU8 along the periphery of assay region was gradually etched out; this result in a SU8 overhang structure along the periphery of the assay region (Figure 1a and 2). Al etching was carried out just prior to the tau detection assay in order to preserve the clean assay surface.

Topological patterning of assay region in the microfluidic device

The fabricated microfluidic device was sandwiched with a cover glass (12 mm × 18 mm; Matsunami Glass), using a paraffin tape (spacers) (0.0127 mm thickness) as shown in Figure 1. Before introducing the protein systems into the device, nonspecific kinesin binding was eliminated by pre-treating the device with a PEO (Poly ethylene oxide) triblock

(PEO−PPO−PEO) copolymer named Pluronic surfactant (2 mg ml^{-1}) which blocks the undesirable kinesin binding to the hydrophobic SU8 region leaving only the glass surface kinesin-adhesive [11]; thereby further preventing the MTs from randomly entering/leaving the assay region. The pluronic coated surface was thoroughly rinsed with DIW twice and once with BRB80 buffer before the assay.

Preparation of proteins

The protein preparation is described elsewhere [12,13]. In brief; Recombinant Homo sapiens kinesin (residues 1–573) was diluted to 30 μg ml^{-1} in dilution buffer (2.5 mg ml^{-1} casein, 1 mM ATP, and 1 mM $MgCl_2$). MTs were prepared by polymerizing labeled (TAMRA-labled) and unlabeled tubulin (1:10 molar ratio) at 37°C for 30 min in BRB80 buffer containing

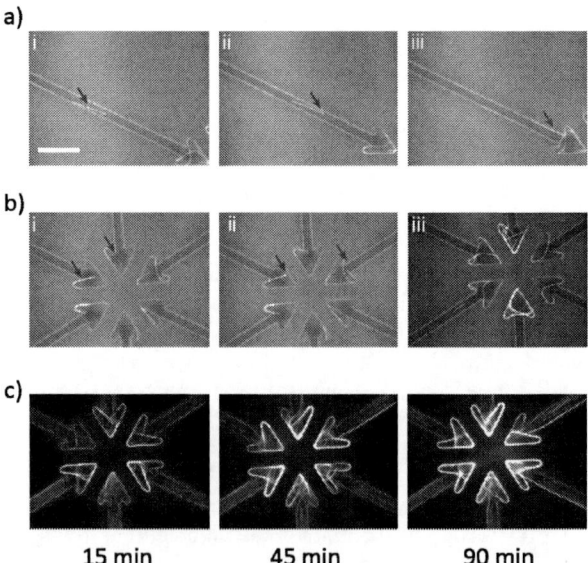

15 min   45 min   90 min

Figure 3. Evaluation of tau detection device. a) MT gliding from the reservoir region through the narrow channel, b) MTs redirected into the arrow head shaped collector region and c) accumulate of MTs in the collector region at different time intervals. The red arrow head in a) (i)-(iii) point at a MT gliding towards the collector regions, in b) (i)-(iii) arrow heads point at MTs being redirected and circulated within the collector region (scale bar 20 μm).

978-1-4799-7956-1/15 $31.00 © 2015 IEEE

and 1 mM GTP. Polymerized MTs were stabilized by (40 µM) paclitaxel. Recombinant taus were purchased from rPeptide (Bogart, GA, USA) diluted in DIW and stocked at -80°C.

Protein assay in the microfluidic device

For the tau detection assay 5 µM MT solution in BRB80 containing 10 µM paclitaxel was made and the MTs were sheared by passing them 30-35 times through the needle attached to a syringe of gauge 22S and length 51 mm in order to obtain MTs of uniform length. Tau protein was incubated at 37°C for 30 min at a final concentration of 1 µM in a solution containing 0.5 µM MTs. Then the tau-MT solution was diluted with motility solution just before introducing into the microfluidic device (1:4 ratio). Casein 10 µl, (0.5 mg ml^{-1}) was flushed into pluronic surfactant (2 mg ml^{-1}) treated flow cell and held within for 5 min at room temperature. Next, 10 µl of the kinesin solution was introduced into the flow cell and held for 5 min. Finally, 10 µl of a tau-decorated MT or undecorated-MT solution was added into the flow cell that was then immediately sealed with grease. After incubation for 5 min, images were acquired. The MTs were visualized using a fluorescent microscope (IX71, Olympus) equipped with a 100× oil objective and a charge-coupled device camera (ORCA-R2, Hamamatsu, Japan). Fluorescent images were stored using HDR-35 recording software (Hamamatsu).

RESULTS

Tau detection assay in the microfluidic device

Tau detection assay was carried out in the fabricated microfluidic device with an overhang structure. The protein solutions were sequentially introduced into the microfluidic device. The MTs landed in the reservoir region and glide through the narrow channel (Figure 3a) towards the MT collector. The arrow head shaped structure channelized and circulated the MTs within the collector region as shown in Figure 3b. Thus on reaching the arrow head shaped collector, MTs were successfully retained for 25 min and beyond (Figure 3c).

In addition, the SU8 overhang structure circumscribe the assay region efficiently retained the MTs from gliding out of the assay region Especially, those MTs entering the arrow head shaped collector could not glide out of the collector region throughout the observation period (> 90 min). Further, the kinesin non-adhesive SU8 surface prevented MT gliding on the non-assay SU8 region thus blocking random entry or exit of the MTs across the assay region.

Rapidity of the assay in the microfluidic device

Firstly, the 2N4R tau decorated and undecorated MTs were assayed in the microfluidic device. The MT collectors were observed for 25 min and recorded at every 5 min time interval. FI profiles of the collectors are plotted in Figure 4. The FI of 2N4R tau decorated MT collectors were significantly lower (p<0.001) compared to undecorated MT collectors at 5 min of incubation (Figure 4). The FI at the collectors gradually increased with time, depending on whether MTs were 2N4R tau decorated or undecorated.

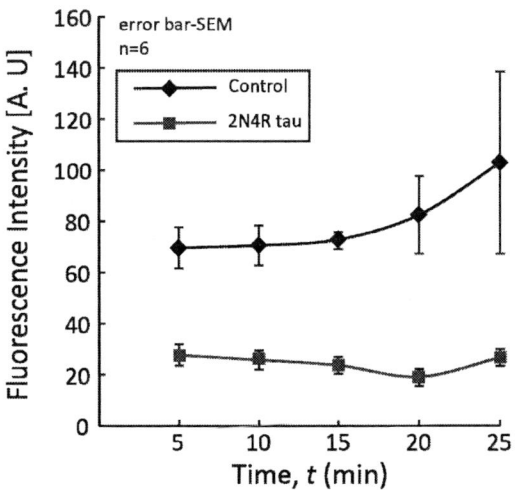

Figure 4. The Fluorescence intensity (FI) profile of MT collector region, black line indicates the FI at the undecorated MT collectors (control) and the blue line indicates the FI at the 2N4R tau decorated MT collectors (n= number of MT collector analyzed).

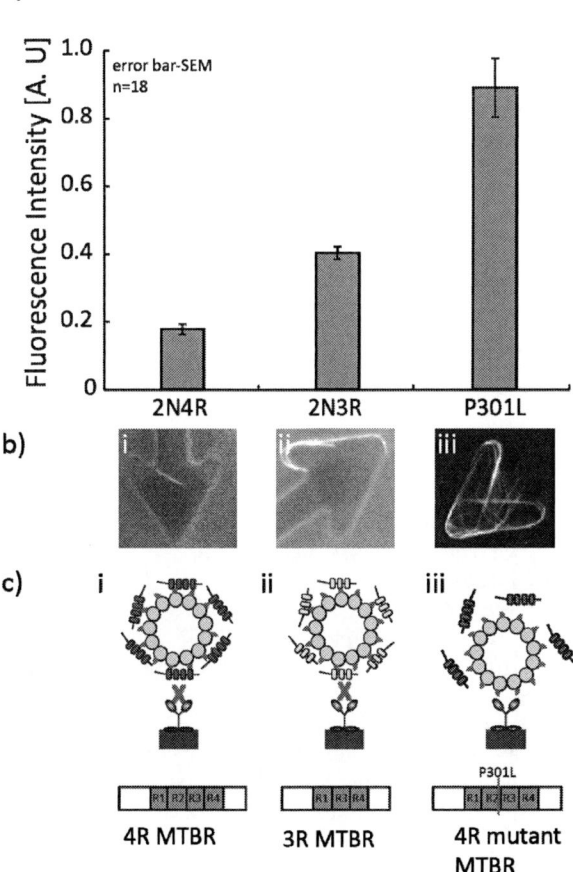

Figure 5. a) FI profile of MT collector of devices assayed with MT decorated by 2N4R, 2N3R and P301L. b) (i)-(iii) collector regions of 2N4R, 2N3R and P301L assayed device respectively. c) (i)-(iii) schematic representation of C. S of MT decorated with 2N4R, 2N3R and P301L tau landing on kinesin; MTBR of the corresponding tau. "X" denotes tau hampering with kinesin MT binding.

978-1-4799-7956-1/15 $31.00 © 2015 IEEE   457

Tau detection method incorporating knesin-MT system successfully detected 2N4R tau decorated MTs from undecorated MTs on-chip in 5 min. Next, we assayed two wild tau isoforms with identical projection domain that differ only in the number of MTBR: 2N3R (3 MTBR) and 2N4R (4 MTBR); along with P301L (a point mutated 2N4R tau). Schematics of the corresponding MTBR are shown in the Figure 5c. The FI at the corresponding MT collectors were observed after 5 min of MT incubation.

Differentiation of tau isoforms

The 2N4R tau decorated MT collectors had significantly ($p<0.001$) lesser FI than 2N3R tau decorated MT collectors (Figure 5a and5b). The MT density of 2N4R tau decorated MT was lesser than the 2N3R tau decorated MT in the respective reservoirs. This is due to the relatively greater interference offered by 2N4R-MTBR to MT-kinesin interaction when compared with that of 2N3R-MTBR, which can be explained by the presence of R2 MTBR in 2N4R tau (Figure 5c).

Differentiation of wild and mutant taus

The FI at the collectors collecting the MT decorated with P301L tau in were significantly greater ($p < 0.001$) than the wild tau decorated MT (Figure 5b, iii) collectors. This is due to the non-binding of P301L tau to the MT surface, which allows the MT to bind readily to the kinesin coated reservoir region(as illustrated in Figure 5c, iii), unlike the wild tau decorated MTs where tau interferes with the MT-kinesin interaction in the assay region. Therefore we were able to differentiate P301L tau from the wild taus on-chip.

The density of MT at collector region is determined by the number of MTs in the reservoir region which in turn depends on the type of tau decoration on MT. Thus we were able to identify the type of tau decorating the MT from the FI measurement at the corresponding MT collectors in the tau detection device.

CONCLUSION

Tau detection method incorporating kinesin-MT system was successfully integrated into a microfluidic device with an overhang structure, comprising of reservoir, channel and an arrow head shaped collector region, the MTs were retained in the assay region/collectors for more than 90 min. Wild and mutant tau , along with tau isoforms with different number of MTBR were successfully differentiated on-chip. The tau detection assay can also be extended into identifying other MAPs and MAP mutants.

The highlight of this detection method is that, in addition to identifying the type of tau, one can also determine the biological functionality of the protein which cannot be achieved through conventional protein detection methods.

ACKNOWLEDGEMENTS

This work is supported by Nakatani foundation; JSPS and NSF under the Japan U.S. Cooperative Science Program and SPIRITS-Kyoto University as part of the Program for Promoting the Enhancement of Research Universities MEXT, Japan.

REFERENCES

[1] Guo, J.L. and V.M.Y. Lee, "Cell-to-cell transmission of pathogenic proteins in neurodegenerative diseases", *Nature Medicine*, Vol. 20, No. 2, pp. 130-138, 2014.

[2] Subramaniyan, S.P., et al. "Detection of mutations in the binding domain of tau protein by kinesin-microtubule gliding assay". in *Micro Electro Mechanical Systems (MEMS)*, 2014 IEEE 27th International Conference on 2014. pp. 314-317.

[3] Chen, S.F., et al., MAPT Isoforms: "Differential Transcriptional Profiles Related to 3R and 4R Splice Variants". *Journal of Alzheimers Disease*, Vol. 22, No. 4, pp. 1313-1329, 2010.

[4] Wolfe, M.S., "The role of tau in neurodegenerative diseases and its potential as a therapeutic target". *Scientifica*, Vol. 2012, Article ID. 796024, 2012.

[5] van den Heuvel, M.G., M.P. de Graaff, and C. Dekker, "Molecular sorting by electrical steering of microtubules in kinesin-coated channels". *Science*, Vol. 312, No. 5775, pp. 910-914, 2006.

[6] Lin, C.T., et al., "Self-contained, biomolecular motor-driven protein sorting and concentrating in an ultrasensitive microfluidic chip". *Nano Lett*, Vol. 8, No. 4, pp. 1041-1046, 2008.

[7] Fischer, T., A. Agarwal, and H. Hess, "A smart dust biosensor powered by kinesin motors". *Nat Nanotechnol*, Vol. 4, No. 3, pp. 162-166, 2009.

[8] Carroll-Portillo, A., et al., "In vitro capture, transport, and detection of protein analytes using kinesin-based nanoharvesters". *Small*, Vol. 5, No. 16, pp. 1835-1840, 2009.

[9] Ramachandran, S., et al., "Selective loading of kinesin-powered molecular shuttles with protein cargo and its application to biosensing". *Small*, Vol. 2, No. 3, pp. 330-334, 2006.

[10] Soto, C.M., et al., "Toward single molecule detection of staphylococcal enterotoxin B: mobile sandwich immunoassay on gliding microtubules". *Anal Chem*, Vol 80, No. 14, pp. 5433-5440, 2008.

[11] Clemmens, J., et al., "Mechanisms of microtubule guiding on microfabricated kinesin-coated surfaces: Chemical and topographic surface patterns". *Langmuir*, Vol. 19, No. 26 pp. 10967-10974. 2003.

[12] Yokokawa, R., et al., "Simultaneous and bidirectional transport of kinesin-coated microspheres and dynein-coated microspheres on polarity-oriented microtubules". *Biotechnol Bioeng*, Vol 101, No. 1, pp 1-8, 2008.

[13] Williams RC, Jr., Lee JC. "Preparation of tubulin from brain". *Methods Enzymol* Vol. 85 Pt B, pp. 376-385, 1982.

CONTACT

*S. P. Subramaniyan, tel: +81-75-383-3687
Subramaniyan.subhathirai.76x@st.kyoto-u.ac.jp

THE CAPILLARY NUMBER EFFECT ON THE CAPTURE EFFICIENCY OF CANCER CELLS ON COMPOSITE MICROFLUIDIC FILTRATION CHIPS

Cong Zhao[1], Rui Xu[1,3], Kui Song[2], Dayu Liu[4], Shuo Ma[2], Chen Tang[2], Chun Liang[3,6], Yitshak Zohar[5], and Yi-Kuen Lee[1,2]

[1]Division of Biomedical Engineering, HKUST, Hong Kong, China
[2]Department of Mechanical and Aerospace Engineering, HKUST, Hong Kong, China
[3]Division of Life Science, HKUST, Hong Kong, China
[4]Guangzhou First Municipal People's Hospital, Guangzhou, China
[5]Department of Aerospace and Mechanical Engineering, University of Arizona, Tucson, AZ, USA
[6]Biomedical Research Institute, Shenzhen PKU-HKUST Medical Center, Shenzhen, China

ABSTRACT

We present a systematic study of the Capillary number (*Ca*) effect on the capture efficiency of cancer cells on a composite microfluidic filtration chip. By altering the *Ca* in microchip experiments, the balance between the viscous force and the cell cortical tension affecting the capture efficiency has been investigated experimentally and analyzed theoretically. A 'Phase Diagram' for the capture efficiency of microfiltration chips is presented, for the first time, as a function of the normalized cell diameter and *Ca*. A critical value of *Ca*, around 0.03~0.04, has been identified for enhancing the capture efficiency of cancer cells. The phase diagram is found to be consistent with the results of cancer-cell capture in microfiltration systems reported previously by others. The diagram can be a useful tool for designing the next generation microfiltration devices for isolating circulating tumor cells.

INTRODUCTION

Circulating tumor cells (CTCs) are cancer cells that shed from original solid tumors and entered the peripheral blood. Detection of CTCs in blood samples can serve as liquid biopsy, which is a minimally invasive method for cancer diagnosis and therapy monitoring.

Based on the differences in size and deformability between CTCs and hematologic cells, various microfiltration devices have been developed for capturing CTCs from the peripheral blood [1]. The features of these devices usually include: high throughput, simplicity and low cost. In comparison with biochemical approaches, targeting epithelial markers expressed by CTCs, such label-free microfiltration devices can potentially capture metastatic tumor cells undergoing epithelial to mesenchymal transition (EMT), which is associated with a loss of expression of epithelial markers [2].

The capture efficiency, also termed as recovery, is defined as the ratio between the number of captured cells and the initial number of target cancer cells in the test sample. The capture efficiency is a key performance criterion for the evaluation of the microfiltration devices. Both numerical simulations [3] and experimental studies [4] have been carried out in effort to improve the capture efficiency of such devices. However, due to the complexity of cell deformation during the cell-pore interaction in microfilters, there is still a lack of quantitative guidelines for selecting design parameters such as: pore size, applied pressure, flow rate, or cell-suspension viscosity for the fabrication in these microfiltration devices.

Unlike solid particles, a cell under hydrodynamic load behaves like a liquid drop with certain surface tension. Thus, to investigate the cell-pore interaction during filtration, the cell is modeled in this study as a Newtonian liquid droplet enclosed by a cortical layer with constant tension [5]. By introducing the Capillary number (*Ca*) as the dominant control parameter, the capture efficiency can be explored as a result of the balance between viscous force due to fluid flow and the cell cortical tension.

Here we fabricated composite microfluidic filtration chip (CMFC) integrated with Polycarbonate (PC) membrane filters varying in pore size. Using membrane filters with 4 different pore sizes, 3 different cell lines and 4 filtration buffers of different viscosities, the *Ca* was adjusted to systematically study its effect on the capture efficiency of cancer cells. The results were summarized in a 'Phase Diagram' to characterize the capture efficiency of microfiltration chips as a function of normalized cell diameter and *Ca*. For more in depth understanding of cell-pore interaction in microfilters under hydrodynamic loading, a 2-DOF model for the deformation and capture of single cell in a pore has also been proposed.

MATERIALS AND METHODS

Fabrication of CMFC

The CMFC was fabricated using standard soft photolithography. Following the one mask process shown in Figure 1, CMFC devices were fabricated as a sandwich structure of two Polydimethylsiloxane (PDMS) layers with a Polycarbonate (PC) filter membrane (Whatman, UK) in-between. The PC membrane was treated with (3-Aminopropyl) triethoxysilane (APTES; Sigma-Aldrich, USA) to form a strong bonding with the PDMS layers.

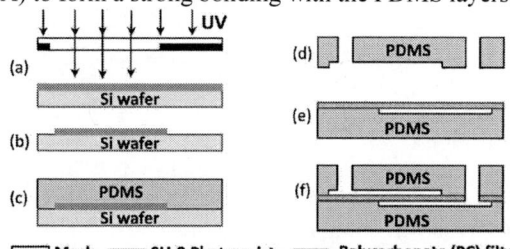

Figure 1: Fabrication process of a composite microfluidic filtration chip (CMFC). (a) UV exposure; (b) baking and development; (c) molding of PDMS slab; (d) punching the PDMS layer for the inlet and outlet; (e) and (f) packaging of CMFC using a chemical assisted method.

Cell culture and mechanical properties

Three cancer cell lines: HeLa, HEK293 and MCF7 cells having different cortical tensions were used in the experiments. The cells were cultured using the Dulbecco's Modified Eagle Medium (DMEM; Sigma-Aldrich, USA) with 10% fetal bovine serum (FBS) in an incubator at 5% CO_2 and 37°C. The mechanical properties of the three cell lines, listed in Table 1, were collected from previous reports obtained experimentally by either micropipette aspiration or AFM indentation of live cell samples.

Table 1. Mechanical properties of the cell lines.

Parameter	Cell line		
	HeLa	HEK293	MCF7
Cell mass m (ng)	5.6	1.4	3.1
Cell damping coefficient c (g/s)	9.7 [6]	1.3 [7]	5.5 [8]
Cell cortical tension k (mN/m)	2.5 [9]	0.6 [10]	0.15 [11]
Initial cell diameter d_c (μm)	22	14	18

Manipulation of Capillary number

The Capillary number is defined as:

$$Ca = \mu V / \sigma \qquad (1)$$

where μ is the viscosity of the cell suspension, V is the flow velocity, and σ is the cell cortical tension. The Ca was manipulated by altering all three variables. The cortical tension was modified by conducting experiments with different cell types. The flow velocity in the CMFC was determined as a result of the opening factors of the membrane filters with pore sizes of 5, 8, 10 and 12 μm. Under a constant flow rate of 5 mL/hr (Reynolds number < 0.1), the average velocity in the CMFC devices with the four different filters ranged from 1.6 to 3.2 mm/s.

To adjust the viscosity of the cell suspension, Polyvinylpyrrolidone (PVP) was added into phosphate buffered saline (PBS) since PVP has been used to alter the plasma viscosity in many biological studies and found to be biocompatible [12]. Prior to on-chip experimentation, fluorescence microscopy confirmed that the cell size and viability had not changed due to the addition of PVP at concentrations of 0, 5, 10 and 20 (w/v %). The corresponding cell-suspension viscosity with these PVP concentrations ranged from 0.9 to 23.6 mPa·s at 25°C.

In the present study, Ca varies over three orders of magnitude between 0.0005 and 0.5; the square root of Ca, sqrt(Ca), is then introduced ranging from 0.02 to 0.7 to simplify the numerical computations.

Experimental setup

A cell suspension with a volume of 2 mL (~2,000 cells) was injected into each CMFC using a digital syringe pump, as shown in Figure 2, at a flow rate of 5 mL/hr. The CMFC device was placed under a fluorescence microscope (BX41, Olympus, Japan) equipped with a CCD camera. The captured cells were labeled with Acridine Orange/Ethidium Bromide (AO/EB; Sigma-Aldrich, USA) fluorescence dyes to distinguish between dead cells in red and live cells in green. The captured cells were enumerated using digital image processing. The capture efficiency was then determined as the ratio between the number of captured cells and the total number of injected cells.

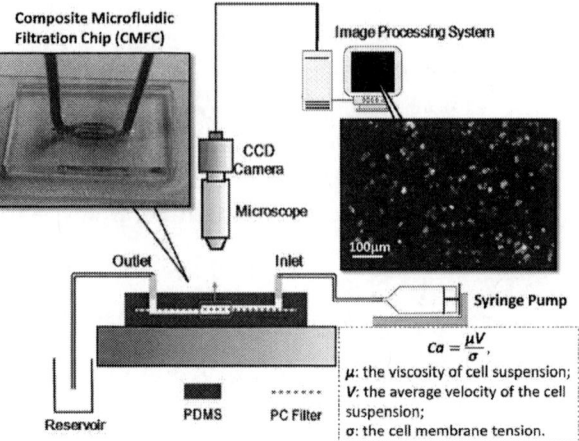

Figure 2: The schematic diagram of the experiment setup for testing the CMFC (left inset). The right inset shows the HeLa cells captured on the filter after filtration experiment.

RESULTS AND DISSCUSSION

Viscous effect on capture efficiency

As shown in Figure 3, the capture efficiency of HeLa cells increased with increasing cell-suspension viscosity. However, the capture efficiency of HEK293 and MCF7 cells initially increased but then decreased with increasing filter pore size (8~12 μm). This clearly indicates that viscosity is not the only mechanism affecting the capture efficiency. Considering the cell-pore interaction in microfilters as a balance between the viscous force due to fluid flow and the cell cortical tension, the different cortical tensions among the three cell lines should be taken into account in the form of Capillary number.

For the CMFC device with a 5 μm filter, the capture efficiency did not change much even in the experiment with the highest viscosity due to the upper limit of cell deformation. Although the cell deformed significantly under a high viscous force, it was still captured by the 5 μm filter. Furthermore, in experiments with similar Ca and the same CMFC device but different cell types, the capture efficiencies were also different. Therefore, the sizes of the filtration pores and the cells should be included as parameters influencing the overall capture efficiency.

'Phase Diagram' of capture efficiency in the CMFC

Following dimensional analysis, all of the experimental results are summarized in a 'Phase Diagram' as depicted in Figure 4. The phase diagram presents the capture efficiency as a function of the initial cell size normalized by the pore size (d_c/d_p) and sqrt(Ca). By setting a cut-off capture efficiency at 70%, a clear boundary can be identified separating between the high and low capture efficiency regimes. This phase diagram can provide guidelines for designing optimized microfiltration systems to capture certain cancer cells, including the design of filter, filtration throughput and sample preparation.

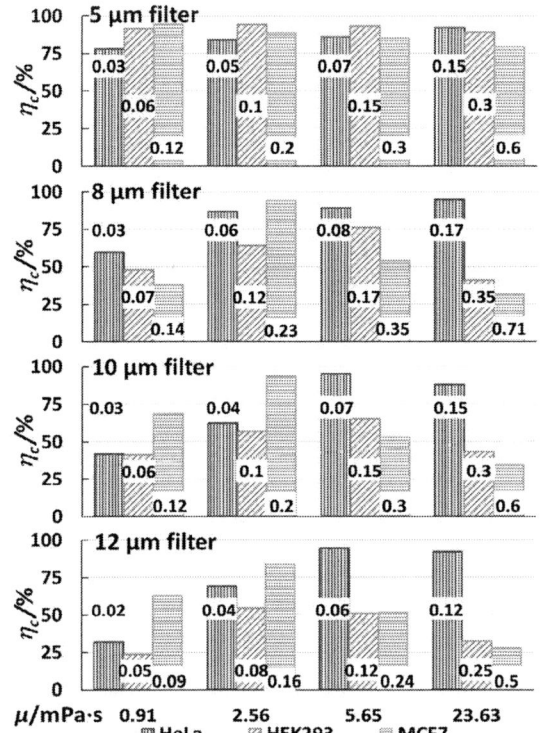

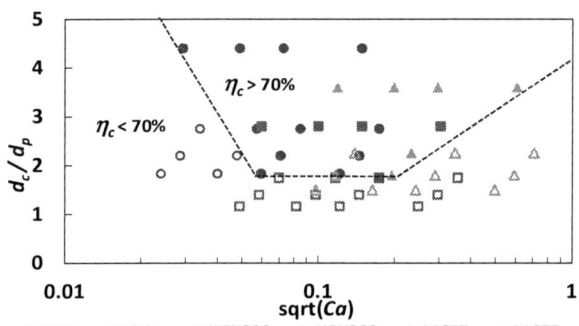

Figure 3: The capture efficiency (η_c) of HeLa, HEK293 and MCF7 cells using CMFC with different pore sizes and cell suspensions at different viscosities μ. The square root of Capillary number, sqrt(Ca), in every experiment is marked on the corresponding column.

Figure 4: The 'Phase Diagram' of the CMFC's capture efficiency (η_c) (70% as the cut-off point) as a function of the normalized cell size with respect to pore size (d_c/d_p) and the square root of Capillary number sqrt(Ca).

Verification of the 'Phase Diagram'

In order to verify the proposed guideline, we included in the phase diagram with the results from several relevant reports on CTC capture in microfiltration systems. Zheng *et al.* developed a membrane filter using parylene-C with a pore size of 7~8 µm. The device has been tested by spiking LNCap cells into PBS and whole human blood, resulting in capture efficiency > 85% in both tests [13]. Lim *et al.* fabricated a microsieve device on a silicon-on-insulator (SOI) wafer with a pore size around 10 µm yielding a capture efficiency > 80% in experiments with MCF7 and HepG2 cells spiked into PBS and whole human blood [14]. Hosokawa *et al.* fabricated microcavity array with different pore sizes of 8.4, 9.1, 10 and 10.6 µm by electroforming of nickel. In experiments using MCF7 and SW620 cells spiked into whole human blood, they obtained a capture efficiency of 50%~90% using microcavity array devices with different pore sizes. When driving NCL-H358 cells spiked into whole human blood through the 9.1 µm microcavity array under flow rates of 6, 30, 60, 90 and 120 mL/hr, the capture efficiency decreased from > 90% to around 50% [15].

The proposed guideline is therefore consistent with the experimental results reported elsewhere for microfiltration systems (Figure 5). This suggests that the phase diagram of capture efficiency as a function of d_c/d_p and sqrt(Ca) is a reliable guideline for designing microfiltration systems regardless of differences in filter-construction materials.

The d_c/d_p value consistently shows to be about 2 in previous reports summarized above. Theoretically, a larger d_c/d_p would lead to a higher capture efficiency. When dealing with human whole blood samples, however, the purity of captured CTCs should be considered as well. In order to achieve a high purity, the hematologic cells, especially leukocytes (~8 µm in diameter), should be completely removed from the filter. Thus, the general guideline for achieving both satisfactory capture efficiency and purity is to fix the d_c/d_p around 2 and sqrt(Ca) around 0.1 in microfiltration systems designed for CTC capture.

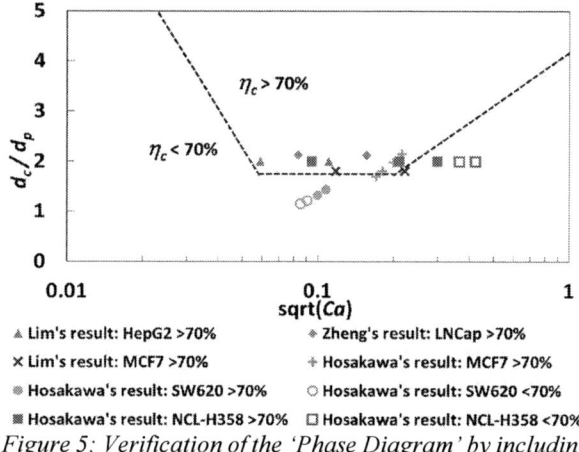

Figure 5: Verification of the 'Phase Diagram' by including in the experiment results from various microfiltration systems in the literature. The dashed line is the same boundary defined by the Phase Diagram in Figure 4.

Simplified two-DOF model

By representing a cell as a mass-spring-damper system, a 2-DOF model for the deformation and capture of a single cell in a pore is proposed. As shown in Figures 6(a) and 6(b), the cell deforms under the shear force resulting in increased size in the x direction, Δx; while in the y direction, the cell deformation Δy is also increased to enter the filtration pore under the pressure force. Once the center of the cell mass reached half of the membrane height (h), $\Delta y/(R_c+0.5h) > 1$), where R_c is the initial radius of the cell, the cell is considered to pass through the pore. By solving the ODEs, the cell deformation in x and y directions can be plotted as a function of Ca, shown in Figure 6 (c) and (d) for HeLa and HEK293 cells as examples. The critical value of Ca for enhanced capture of HeLa, HEK293 and MCF7 has been identified to be 0.03, 0.04 and 0.038, respectively.

In comparison with the modeling results, the values of Ca in the HeLa-cell experiments are within the range of enhanced cell capture. However, the Ca values for HEK293 and MCF7 cells are only partially within that range. This explains the results shown in Figure 3, in which the capture efficiency of HeLa cells increased with increasing Ca, while the capture efficiency of HEK293 and MCF7 cells initially increased but then decreased when the Ca exceeded the critical value.

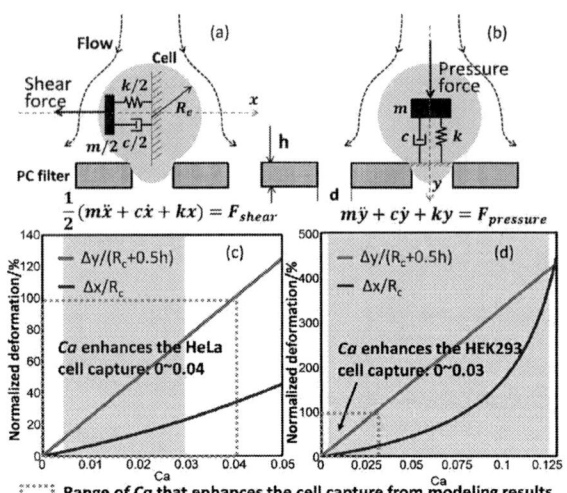

Figure 6: Simplified 2-DOF model for capturing mechanics of a single cell in x (a) and y (b) directions to study the Ca effect on the cell deformation in a pore. (c) and (d) show the comparison of the Ca effect on the capture efficiency of HeLa and HEK293 cells.

CONCLUSIONS

In this work, CMFC devices integrated with PC membranes with different pore sizes were utilized to study the Ca effect on the capture efficiency of cancer cells. A phase diagram characterizing the capture efficiency of microfiltration chips as a function of normalized cell diameter and Ca is proposed and verified. Utilizing a 2-DOF model, derived for predicting the deformation and capture of a single cell in a pore, a critical value of Ca is identified for enhancing capture efficiency of cancer cells in the microfiltration system.

ACKNOWLEDGEMENTS

This research was partially supported by Hong Kong RGC GRF grant (No. 16205314), a research grant from NSFC, China (No. 81171418), and a research grant from Shenzhen Dept. of Science and Information, China (JCYJ20130329110752138).

REFERENCES

[1] L. Yu, S. R. Ng, Y. Xu, H. Dong, Y. J. Wang, C. M. Li, "Advances of lab-on-a-chip in isolation, detection and post-processing of circulating tumour cells," *Lab Chip,* vol. 13, pp. 3163-3182, 2013.

[2] K. Polyak, R. A. Weinberg, "Transitions between epithelial and mesenchymal states: acquisition of malignant and stem cell traits," *Nat Rev Cancer,* vol. 9, pp. 265-273, 2009.

[3] Z. Zhang, J. Xu, B. Hong, X. Chen, "The effects of 3D channel geometry on CTC passing pressure - towards deformability-based cancer cell separation, *Lab Chip,* vol. 14, pp. 2576-2584, 2014.

[4] F. A. W. Coumans, G. van Dalum, M. Beck, L. W. M. M. Terstappen, "Filter Characteristics Influencing Circulating Tumor Cell Enrichment from Whole Blood," *PLoS ONE,* vol. 8, p. e61770, 2013.

[5] E. Evans, A. Yeung, "Apparent viscosity and cortical tension of blood granulocytes determined by micropipet aspiration. " *Biophys J,* vol. 56, pp.151–160, 1989.

[6] E. H. Zhou, F. D. Martinez, J. J. Fredberg, "Cell rheology: Mush rather than machine," *Nat Mater,* vol. 12, pp. 184-185, 2013.

[7] N. Khatibzadeh, A. A. Spector, W. E. Brownell, B. Anvari, "Effects of Plasma Membrane Cholesterol Level and Cytoskeleton F-Actin on Cell Protrusion Mechanics," *PLoS ONE,* vol. 8, p. e57147, 2013.

[8] S. Moreno-Flores, R. Benitez, M. dM Vivanco, J. L. Toca-Herrera, "Stress relaxation and creep on live cells with the atomic force microscope: a means to calculate elastic moduli and viscosities of cell components," *Nanotechnology,* vol. 21, p. 445101, 2010.

[9] J.-Y. Tinevez, U. Schulze, G. Salbreux, J. Roensch, J.-F. Joanny, E. Paluch, "Role of cortical tension in bleb growth," *PNAS,* vol. 106, pp. 18581–18586, 2009.

[10] A. Beyder, F. Sachs, "Electromechanical coupling in the membranes of Shaker-transfected HEK cells," *PNAS,* vol. 106, pp. 6626-6631, 2009.

[11] J.-D. Kim, M. Waleed, Y.-G. Lee, "Stiffness measurement of a biomaterial by optical manipulation of microparticle," *SPIE BiOS,* p. 85951G, 2013.

[12] L. Zhao, B. Wang, G. You, Z. Wang, H. Zhou, "Effects of different resuscitation fluids on the rheologic behavior of red blood cells, blood viscosity and plasma viscosity in experimental hemorrhagic shock," *Resuscitation,* vol. 80, pp. 253-258, 2009.

[13] S. Zheng, H. Lin, J. Q. Liu, M. Balic, R. Datar, R. J. Cote, Y. C. Tai, "Membrane microfilter device for selective capture, electrolysis and genomic analysis of human circulating tumor cells," *J Chromatogr A,* vol. 1162, pp. 154-161, 2007.

[14] L. S. Lim, M. Hu, M. C. Huang, W. C. Cheong, A. T. L. Gan, X. L. Looi, S. M. Leong, E. S. -C. Koay, M. -H. Li., "Microsieve lab-chip device for rapid enumeration and fluorescence in situ hybridization of circulating tumor cells, *Lab Chip,* vol. 12, pp. 4388-4396, 2012.

[15] M. Hosokawa, T. Hayata, Y. Fukuda, A. Arakaki, T. Yoshino, T. Tanaka, T. Matsunaga, "Size-Selective Microcavity Array for Rapid and Efficient Detection of Circulating Tumor Cells," *Anal Chem,* vol. 82, pp. 6629-6635, 2010.

CONTACT

*Y. -K. Lee, tel: +852-2358-8663; meyklee@ust.hk

3D CULTURE OF MOUSE IPSCS IN HYDROGEL CORE-SHELL MICROFIBERS

Kazuhiro Ikeda[1,2,3], Teru Okitsu[1,2], Hiroaki Onoe[1,2] and Shoji Takeuchi[1,2]
[1]Institute of Industrial Science, The University of Tokyo, JAPAN
[2]ERATO Takeuchi Biohybrid Innovation Project, JST, JAPAN
[3]Graduate School of Life and Environmental Sciences, University of Tsukuba, JAPAN

ABSTRACT

This paper reports the culturing and expansion of mouse induced pluripotent stem cells (iPSCs) in hydrogel core-shell microfibers; the core consists of iPSCs with or without extracellular matrix (ECM) proteins, and the shell is composed of calcium alginate. We revealed that mouse iPSCs cultured in the micro-scale space with ECM proteins sustain their pluriotency efficiently. This 3D culture system may be a useful tool to expand iPSCs for clinical use.

INTRODUCTION

iPSCs are expected to be used for regenerative medicine as a source of cells because iPSCs are generated from somatic cells by transgenesis of four genes and have abilities of self-renewal and differentiation into three germ layers, called pluripotency [1]. Figure 1a shows the conventional culture system of iPSCs. Generally iPSC maintenance is difficult due to spontaneous differentiation of iPSCs. In addition, this culture system is not xeno-free and is unsuitable for obtaining large number of cells enough to be used clinically because 2D culture system is less efficient than that of 3D in terms of industrial mass culture. However 3D suspension culture systems of iPSCs tend to make uncontrollable huge cell aggregation. This phenomenon causes cell death and spontaneous differentiation by poor diffusion of gas and nutrition in the cell aggregation. These obstacles prevent iPSCs from medical applications, so new feeder-free 3D culture systems of iPSCs are necessary.

EXPERIMENTAL

Conventional 2D culture

We used mouse iPSC line, iPS-MEF-Ng-20D-17, expressing GFP by Nanog promoter [2]. As conventional 2D culture system, the iPSCs were cultured on feeder cells. The feeder cells were mitomycin C treated or X-irradiated mouse embryonic fibroblasts. We prepared feeder layer by seeding feeder cells on gelatin coated plastic dishes and culturing overnight. The iPSCs were seeded on the feeder layer, maintained for three days and subcultured by seeding on other feeder layer. In this culture system, we used mouse ES medium (DMEM high glucose with 15% FBS, 1% 100x NEAA, 1% 100x L-Glutamin, 110 uM 2-mercaptoethanol and 1000 Units/mL Leukemia inhibitory factor).

3D culture in core-shell microfibers

We proposed a feeder-free 3D culture system to expand mouse iPSCs by using a microfluidic device [3]. Figure 1b shows the schematic diagram. We prepared feeder-free mouse iPSCs by culturing on gelatin coated

plastic dishes. Next, we fabricated hydrogel core-shell microfibers encapsulating the mouse iPSCs suspended in neutralized native collagen without feeder cells. Diameter of the core-shell microfibers is 200 μm and the core diameter is 100 μm. The diameter is small enough to allow for diffusion of gas and nutrition [4]. After iPSCs in the microfibers had been cultured for three days, we evaluated our 3D culture system of iPSCs in terms of expansion property and pluripotency by comparing with the standard 2D culture on feeder cells. Furthermore, we changed the core ECM proteins from native collagen to 1) mixture of native collagen and fibrinogen and 2) culture medium

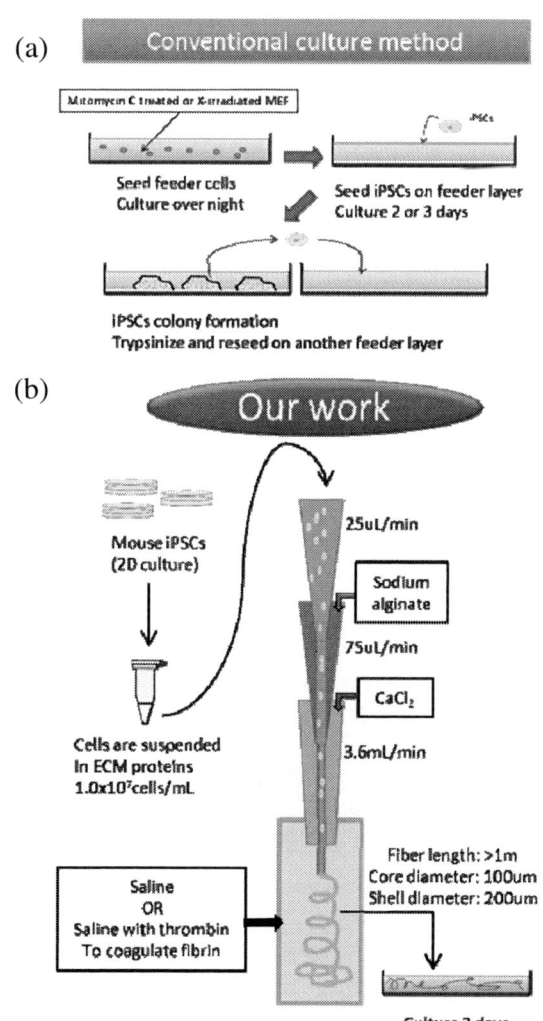

Figure 1: (a) Conventional mouse iPSC culture system on feeder cells. (b) Schematic diagram of fabrication of hydrogel core-shell microfiber encapsulating mouse iPSCs by the microfluidic co-axial device.

978-1-4799-7956-1/15 $31.00 © 2015 IEEE   463   MEMS 2015, Estoril, PORTUGAL, 18 - 22 January, 2015

(ECM-free). To coagulate fibrinogen and make fibrin, thrombin was added in the saline that receives the fabricated core-shell microfiber and incubated for ten minutes in water bath. All cell-laden core-shell microfibers were cultured by mouse ES medium. After three days, cells in the core-shell microfibers are examined by the same ways.

RESULTS

iPS-MEF-Ng-20D-17 expresses GFP by Nanog promoter. Because Nanog is a pluripotency marker, pluripotent cells fluoresce by GFP but differentiated cells do not [2]. Figure 2 shows images of mouse iPSCs in hydrogel core-shell microfibers containing native collagen as the ECM protein. Mouse iPSCs in the microfibers were initially dispersed, but finally constructed fiber-shaped cell aggregations and expressed Nanog-GFP. FACS analysis revealed that Nanog-GFP expression ratio of mouse iPSCs cultured in the core-shell microfibers are as high as that of conventional 2D culture system (Figure 3). Next, we changed the ECM proteins from native collagen to a mixture of native collagen and fibrin. After culturing, almost the same microfibers were formed. Lastly, we encapsulated mouse iPSCs suspended in medium (ECM-free) in the core-shell microfiber. The cells aggregated in a few hours after fabrication. After culturing, the cells in the ECM-free core-shell microfibers made fiber-shaped cell aggregation too, but these microfibers contain a fair amount of space without cells. In terms of pluripotency ratio, there was no difference between the microfiber with a mixture of native collagen and fibrin and that with native collagen only. However, without ECM proteins, Nanog-GFP expression ratio was lower (Figure 4). It may indicate the importance of ECM proteins to expand iPSCs in 3D culture systems.

CONCLUSION

This work presented the novel 3D culture system for maintaining mouse iPSCs. The cells in the core-shell microfiber with or without ECM proteins could sustain their pluripotency, but the cells with ECM proteins sustain it more efficiently. In future works, we will optimize this 3D culture system and human iPSCs culture system by the same way.

ACKNOWLEDGEMENTS

This work was supported by Takeuchi Biohybrid Innovation Project, ERATO, Japan Science and Technology Agency (JST). We thank Hoshimi Aoyagi and other lab members for their kind technical support.

REFERENCES

[1] K. Takahashi, S. Yamanaka "Induction of pluripotent stem cells from mouse embryonic and adult fibroblast cultures by defined factors."*Cell*, 126(4), 663-76 (2006)

[2] K. Okita, T. Ichisaka, S. Yamanaka "Generation of germline-competent induced pluripotent stem cells." *Nature* 448 313-317 (2007)

[3] H. Onoe, T. Okitsu, A. Itou, M. Kato-Negishi, R. Gojo, D. Kiriya, K. Sato, S. Miura, S. Iwanaga, K. Kuribayashi-Shigetomi, YT. Matsunaga, Y. Shimoyama, S. Takeuchi. "Meter-long cell-laden microfibers exhibit tissue morphologies and functions." *Nature materials*, 12(6), 584-90 (2013)

[4] RH. Li, DH. Altreuter, FT. Gentile. "Transport characterization of hydrogel matrices for cell encapsulation." *Biotechnol Bioeng.* 1996 May 20;50(4):365-73.

CONTACT

*Kazuhiro Ikeda, tel +81-3-5452-6545; ikeda-k@iis.u-tokyo.ac.jp

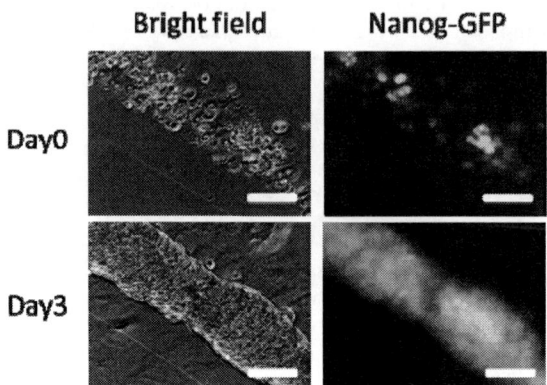

Figure 2: Photograph of mouse iPSCs cultured in the hydrogel core-shell microfiber. (Scale bar = 100um)

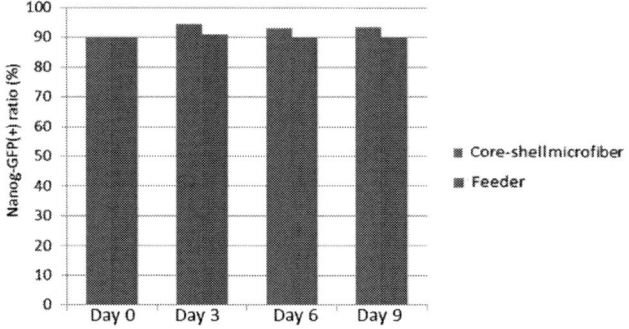

Figure 3: iPSCs in core-shell microfiber sustain Nanog-GFP(+) ratio as high as iPSCs on feeder cells even after two passages.

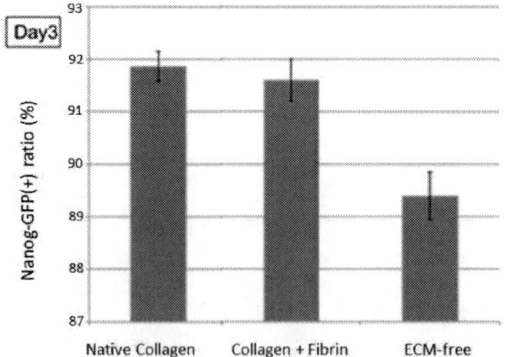

Figure 4: Core-shell microfibers with ECM proteins sustain higher iPSC's Nanog-GFP(+) ratio than ECM-free microfibers.

ALIGNMENT OF COLLAGEN NANOFIBERS IN 2D SUBSTRATES USING CYCLIC STRETCH

Eun Ryel Nam[1,2], Won Chul Lee[1,2], and Shoji Takeuchi[1,2]
[1]Institute of Industrial Science, The University of Tokyo, JAPAN
[2] ERATO Takeuchi Biohybrid Innovation Project, JST, JAPAN

ABSTRACT

This paper presents a new method to fabricate 2-dimensional collagen sheets whose internal molecular structures form highly-aligned nanofibers in one direction. In this work, collagen nanofibers are self-aligned in fully 2-dimensional substrates by applying cyclic stretch during the gelation process of collagen solution, and the fabricated collagen sheets induce the alignment of cells without any mechanical force within the cultivation period.

INTRODUCTION

Collagen fiber orientation plays a critical role in maintaining cell structures and functions for many tissues in vivo, but the alignment of collagen fibers has been difficult to realize in vitro [2-4]. Previously, studies involving aligned collagen fiber substrates have relied on animal explants, magnetic alignment [5], and rotational seeding [6], but the specialized preparation required for these methods prevents widespread application in cell biology (reviewed in [2-4]). Recently, the alignment methods based on microfluidic channels [1] and microtransfer molding [7] were suggested, but they also have limitations because they could only either fabricate 1-dimensional lines (microchannel shap) of aligned-collagen substrates or align outer shapes of collagen structures (not collagen molecules). Thus, in vitro fabrication of aligned-collagen sheets, which can advance 2D substrates for cell culture, has not been fully achieved yet.

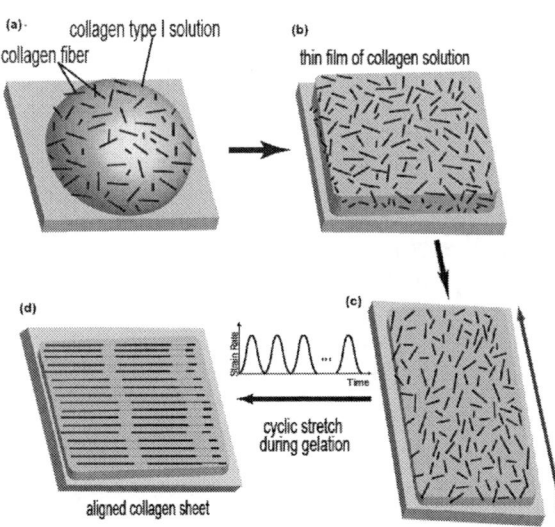

Figure 1: Schematic diagram of the construction of the aligned collagen sheet. (a) Randomly arranged collagen fibers in collagen type I solution. (b) Formation of thin film of collagen solution to increase geometric constriction. (c) Cyclic stretch and releasing of the thin film continued during gelation. (d) Aligned collagen sheet after the cyclic stretch.

MATERIALS AND METHODS

This paper presents an 'aligned collagen sheet' using spin coat and cyclic stretch as shown in Fig.2. We increased geometric constraint by decreasing thickness of the collagen sheet. And we successfully constructed thin and highly aligned collagen film, which had fibrils of vertical axis to the stretch direction by providing continuous mechanical loading to fibrils in the sheet. In addition, we found that this sheet induced the alignment of cells and did not need any mechanical force within the cultivation period.

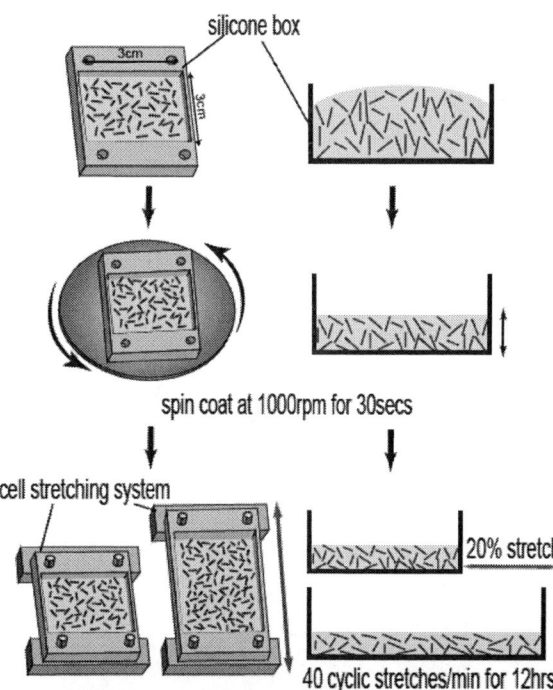

Figure 2: Process flow of producing the aligned collagen sheet. Collagen type I solution in a silicone box. Thin collagen film on the bottom of the box by spin-coating and discretion of the height of collagen solution. The silicone box with the thin collagen film onto a device of cell stretching system in the cell culture incubator.

RESULTS AND DISSCUSSION

Applying cyclic mechanical force to collagen sheets induced an alignment of collagen fibers. Interestingly, the arrangement of collagen fibers was perpendicular to the stretch direction (Fig.3b). However, fibers in collagen gel, the control, were observed no particular alignment under a light and confocal laser microscope (Fig.3a). Therefore, the geometric constrain and the cyclic stress would align

978-1-4799-7956-1/15 $31.00 © 2015 IEEE   465   MEMS 2015, Estoril, PORTUGAL, 18 - 22 January, 2015

the fibrils. 3T3 cell was seeded on the aligned collagen sheet, and a cell culture dish at a density of 3 x10^4 cells / cm^2, and cultivated for 3 days. The cells were adhered to and proliferated on the collagen film (Fig. 4b). In addition, the cells were aligned and expanded along the direction of collagen fibers with immunostaining (Fig. 4d). On the contrary, no specific direction of cell arrangement was detected on the cell in the cultivated dish (Fig. 4a and 4c). This result supposed that highly aligned scaffold might indicate cell alignment without mechanical stress during cultivation.

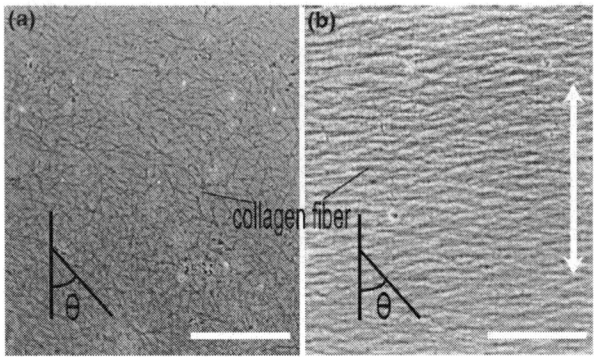

Figure 3: Collagen sheet with the cyclic stretch under a light microscope. The arrangement of collagen fibers in the collage sheet with the cyclic stretch was perpendicular to the stretch direction (b). The collagen gel without the cyclic stretch under a light microscope (a) displayed no particular fiber alignment. The arrow indicates the direction of the cycle stretch. (scale bar : 50um)

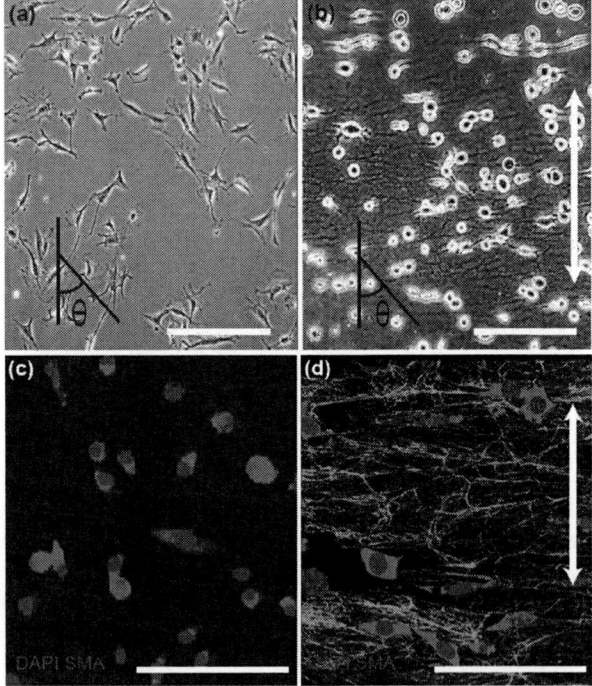

Figure 4: 3T3 cells on the culture dish under a light (a) and confocal laser microscope (c). No specific direction of cell arrangement was detected on the cells in the cultivated dish after 3 days cultivation. 3T3 cells on the aligned collagen sheet under a light (b) and a confocal laser

microscope (d). The cells on the collagen sheet were aligned and expanded along the direction of collagen fibers. The arrow indicates the direction of the cycle stretch. (scale bar : 50um)

CONCLUSION

In this work, the 2-dimensional collagen sheets with the highly-aligned nanofibers were fabricated by the cyclic stretch. Cells cultivated on the aligned collagen sheet were also arranged in one direction without the cyclic stretch during cultivation.

ACKNOWLEDGEMENT

This research wassupported by "Grant-in-Aid for Research Activity Start-up" of Japan Society for the Promotion of Science. (145500000410) and the Takeuchi Biohybrid Innovation Project, Exploratory Research for Advanced Technology (ERATO), Japan Science and Technology (JST), Japan.

REFERENCES

[1] P. Lee, R. Lin, J. Moon, and L. P. Lee, "Microfluidic alignment of collagen fibers for in vitro cell culture," *Biomedical Microdevices*, vol. 8, pp. 35–41, 2006.

[2] C. H. Lee, A. Singla, and Y. Lee, "Biomedical applications of collagen," *Int J Pharm Pharm Sci*, vol. 221, pp. 1–22, 2001.

[3] J. W. Ruberti and J. D. Zieske, "Prelude to corneal tissue engineering—Gaining control of collagen organization," *Prog Retin Eye Res*, vol. 27, pp. 549–577, 2008.

[4] V. H. Barocas, and R. T. Tranquillo, "An anisotropic biphasic theory of tissue–equivalent mechanics: The interplay among cell traction, fibrillar network deformation, fibril alignment, and cell contact guidance." *J. Biomech. Eng.* vol. 119, pp. 137–145, 1997

[5] C. Guo, and L. J. Kaufman, "Flow and magnetic field induced collagen alignment." *Biomaterials*, vol. 28, pp. 1105-1114, 2007

[6] N. K. Weidenhamer, and R. T. Tranquillo, "Influence of cyclic mechanical stretch and tissue constraints on cellular and collagen alignment in fibroblast-derived cell sheets." *Tissue Eng Part C Methods*, vol. 19, pp. 386-395, 2013

[7] N. Naik, J. Caves, E.L. Chaikof, and M.G. Allen, "Generation of spatially aligned collagen fiber networks through microtransfer molding." *Adv Healthc Mater*, Vol. 3, pp.367-374, 2014

CONTACT

*E.R. Nam,
tel: +81-3-5452-6616; namel@iis.u-tokyo.ac.jp

BONDING-FRIENDLY PCPDMS: DEPOSITING PARYLENE C INTO PDMS MATRIX AT AN ELEVATED TEMPERATURE

Yaoping Liu[1], Lingqian Zhang[1], Wei Wang[1, 2, 3] and Wengang Wu[1, 2, 3]*

[1]Institute of Microelectronics, Peking University, Beijing, 100871, CHINA
[2]National Key Laboratory of Science and Technology on Micro/Nano Fabrication, 100871, CHINA
[3]Innovation Center for Micro-Nano-electronics and Integrated System, Beijing, 100871, CHINA

ABSTRACT

This paper reported a simple and effective process of bonding-friendly Parylene C-caulked PDMS (pcPDMS) for low-permeability required microfluidics. Parylene C was deposited into PDMS matrix at an elevated temperature (higher than 135°C) to caulk the permeable sites. The so-prepared pcPDMS can be directly bonded with oxygen plasma treatment just as pristine PDMS. SEM EDAX and Laser scanning confocal microscopy (LSCM) were introduced to characterize the Parylene C caulked status in the PDMS matrix based on the specific Cl element component and the firstly-found temperature-sensitive autofluorescence of Parylene C. The preliminary results indicated that the present bonding-friendly pcPDMS can successfully suppress the diffusion of small molecules into the PDMS matrix.

INTRODUCTION

PDMS is a commonly-used material in microfluidics due to its good characteristics: cheap, transparent, biocompatible and easy to fabricate. However, there come some issues when it is applied in biomedical researches. Among which the diffusion of small molecules caused by porosities in PDMS is the most serious and still being problem. Many researchers have attempted to solve this problem through different approaches. The developed methods are mainly two types, one is to change of curing time and base/curing agent ratios, and the other is surface coating. The straightforward variations of curing time and base/curing agent ratios really work to change porosity in PDMS bulk, but the performance is very limited. The surface coating relied on various surface modification and heavy manual operation. Besides, the shape and size of microchips could be altered after coating. These methods are time-consuming and also exits a questionable long-time reliability because there is a risk that coating layer would partially lift off from the surface during high-temperature reaction or under other extreme condition.

Parylene C, a well-known biocompatible polymer, is often deposited as passivation coating to achieve low adhesion and good barrier properties to moisture, inorganic and organic molecules. Young et al., proposed PDMS-based micro PCR chip with Parylene C coating and identified the possibility of preventing adsorption of *Taq* DNA polymerase and DNA. But the size and shape of this chip were limited because Parylene C deposition onto the inner surface of chips was carried out after PDMS bonding due to the bonding problem caused by Parylene C coating on PDMS surface. [1] Sawano and our previous work demonstrated a method for low-permeable PDMS fabrication by depositing Parylene C onto PDMS substrate.

[2, 3, 4] The so-called pcPDMS achieved a good suppression of small molecule diffusivity by effectively sealing the permeable sites in PDMS matrix. However, their process suffers a long time plasma etching to remove the over-deposited Parylene C layer and a hazardous BHF treatment to reactivate the surface for the posterior bonding. Moreover, it is very tricky to control the etching time length to only remove the Parylene C layer on the top surface of PDMS but not influence Parylene C in the PDMS matrix.

Here we developed an easy and performant one-step process to prepare a bonding-friendly pcPDMS via depositing Parylene C into PDMS matrix at temperature higher than 135°C. The plasma boding strength was tested via manual breaking from the bonding interface of pcPDMS. To prove the suppression effect of small molecules for the so-prepared bonding-friendly pcPDMS, the diffusion of Rhodamine B was also characterized in the pristine PDMS and the fabricated pcPDMS microfluidic chips.

EXPERIMENTAL RESULTS AND DISCUSSIONS

Device Fabrication

Fig.1 schematically shows the fabrication process. First, 10:1 PDMS was used to make the flat PDMS and PDMS microchannel with the standard soft lithography. Second, the prepared PDMS substrate was heated to an elevated temperature (higher than 135°C) with a homemade heating system installed in the Parylene C deposition chamber (PDS 2010, SCS, USA.) to prepare the so-called pcPDMS. Third, the pcPDMS microfluidic chip could be prepared after 5 s oxygen plasma treatment (oxygen flow rate 1500 sccm, chamber pressure 35 Pa, and etching power 100 W) induced irreversible bonding. The main structure of the chip is the multi tortuous microchannel and the dimension of the single microchannel is 60 μm × 50 μm.

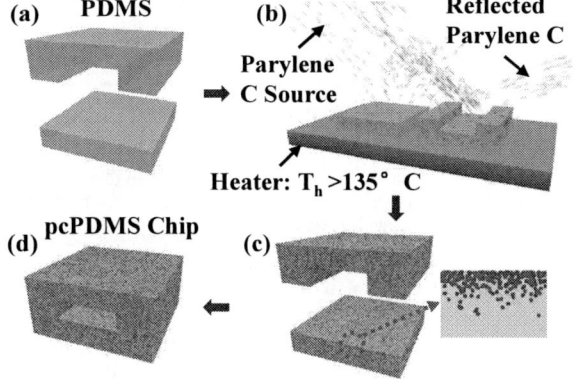

978-1-4799-7956-1/15 $31.00 © 2015 IEEE

Figure 1: Schematic illustration of fabrication process, (a) preparation of flat PDMS and PDMS microchannel with the standard soft lithography, (b) Parylene C deposition into PDMS at 135°C with a home-made heating system, (c) pcPDMS bonding by oxygen plasma treatment for 5s.

Fig.2 (a) shows the same transparency of pcPDMS with the pristine PDMS. The broken points from the bulk pcPDMS in the manual peeling test reveals that the bonding strength of pcPDMS is also compatible to that of PDMS, shown in Fig. 2 (b). And the cross-sectional SEM pictures of the so-prepared pcPDMS microfluidic chip was shown in Fig.2 (c).

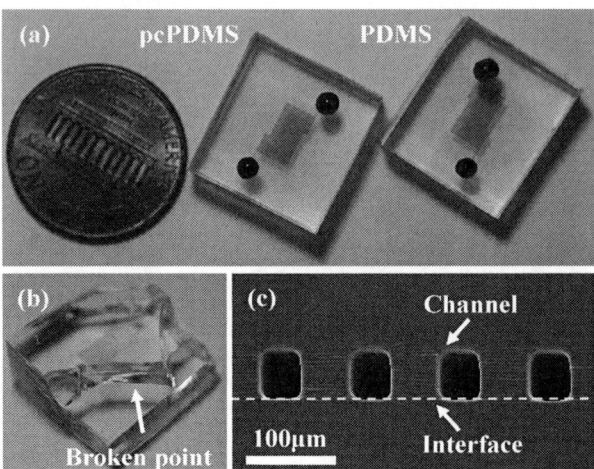

Figure 2: Transparency of the pristine PDMS and the present pcPDMS (a), manual peeling test of bonded pcPDMS microfluidic chip (b) and cross-sectional SEM view of the bonded pcPDMS.

As the sticking coefficient of Parylene C deposition decreases dramatically with the temperature increment, the deposition rate on the PDMS surface with temperature above 135°C was small enough to avoid forming an intact film. [5] This enables Parylene C monomers migrate deeper into the PDMS matrix and caulk the nano-scale sites there. Therefore the so-prepared pcPDMS could be bonded directly through oxygen plasma treatment with no continuous Parylene C film forming on top of surface.

Characterizations

The SEM EDAX line scan and the LSCM slice scan were used to characterize the Parylene C caulking status in this paper.

For the SEM EDAX, there is no Cl element in PDMS molecule while no Si element in Parylene C molecule, so the ratio of Cl/Si could suggest the depth of Parylene C in the PDMS matrix, as indicated in Fig.3 (a). For the sample preparation of SEM EDAX, a silicone rubber (704 RTV Silicone Rubber, Nanda, China) was manually coated to the pcPDMS surface, forming a 704-pcPDMS-PDMS sandwich structure, and deposited at room temperature overnight. A longitudinal cut was placed from the top of the so-called sandwich to get a thin sliced SEM EDAX sample. Under the SEM EDAX system, a spot detection was firstly swept near the interface of 704 and PDMS to define the range of line scan of Cl and Si element. The

percentage of Cl/Si indicates the diffusion depth of Parylene C into the PDMS matrix, as shown in Fig. 3(a).

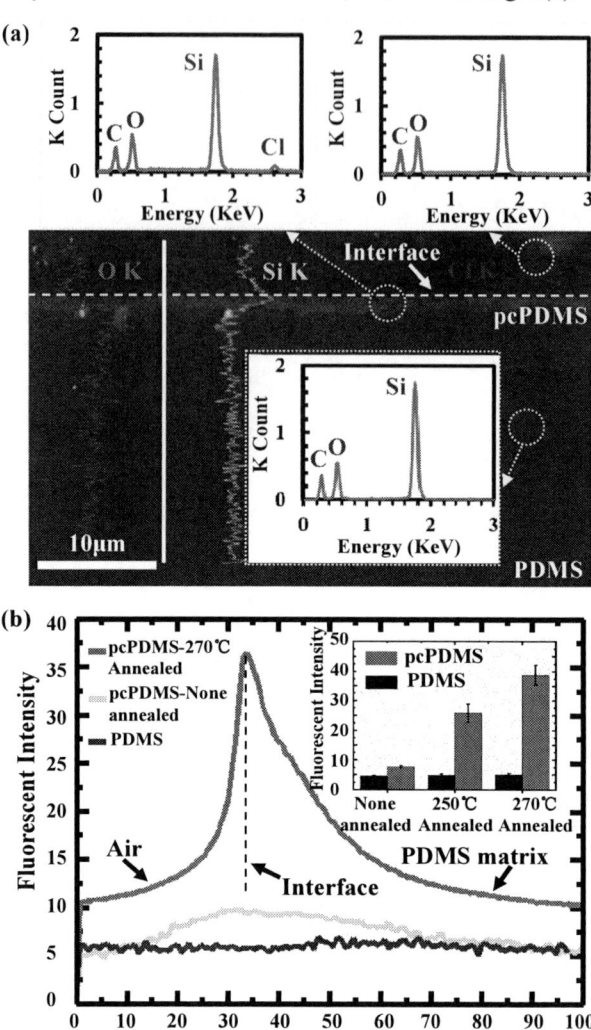

Figure 3: The SEM EDAX results (a): there is a characteristic spectra peak of Cl element in Parylene C spot analysis, while no for PDMS, and LSCM results (b): the distribution in the depth direction of the Parylene C blue autofluorescence, the inserted one shows the differences for different annealing treatments.

The LSCM characterization is based on the autofluorescence of Parylene C. The autofluorescence of Parylene has been reported by Kochi et al., [6] and Lu et al., [7] found that blue autofluorescence intensity of Parylene C was much higher than green or red fluorescence and accordingly the blue autofluorescence was selected to identify Parylene C from PDMS. Moreover, we firstly found that annealing in vacuum at 270°C or above can considerably increase the autofluorescence contrast between Parylene-C and PDMS, as shown in the inserted pictures of Fig.3 (b), which provide an alternative way to characterize the caulking status. In our LSCM characterization, the pcPDMS samples were thermally treated by annealing in vacuum at 270°C and then scanned layer by layer at a specified step size (0.5 μm) with the 340-380 nm excitations and 435–485 nm emissions (Gate voltage: 845.8 V, Smart offset: -1.2%, Pinhole: 60.63 μm). The intensity change of scanned fluorescent photos was analyzed with MATLAB processing to characterize the distribution of Parylene C in the depth of PDMS matrix.

Both the SEM EDAX and LSCM results indicated the caulking depth of Parylene C in PDMS is more than 5μm.

Rhodamine B Diffusion Test

To identify the blocking effect of small molecules of the so-prepared bonding-friendly pcPDMS, we introduced Rhodamine B to characterize the diffusion coefficient in the pcPDMS and PDMS microfluidic chips. 150μM Rhodamine B (Sigma, S9012) in water was pumped through PDMS and pcPDMS microfluidic channels and where the fluorescent signals were recorded at different time points after sample loading with a CCD camera under a fluorescent microscope (Leica DMI 6000B, Germany). The filled large reservoirs guaranteed the microfluidic channels filled with enough Rhodamine B molecules during the recording period, eliminating the influence of evaporation. The fluorescent intensities in images were analyzed by MATLAB to define the diffusion width of Rhodamine B.

Fluorescent intensities and the dynamic diffusion length of Rhodamine B in the so-prepared pcPDMS and pristine PDMS microfluidic chips at 5 min, 10 min, 20 min, 30 min, and 40 min after the sample loading are shown in Fig. 4 (a). Compared to the pristine PDMS, the diffusion of Rhodamin-B was significantly suppressed in the pcPDMS device, as shown in Fig.4 (a). The diffusivity of Rhodamine B in the native PDMS and pcPDMS was fitted based on the experimental measurements according to Fick's first law with the diffusion data at 5 min under a constant-source diffusion assumption. The diffusivities of Rhodamin-B in the pristine PDMS and the pcPDMS were $2.43\pm0.019\times10^{-13}$ m^2/s and $1.21\pm0.023\times10^{-13}$ m^2/s, respectively, shown in Fig. 4(b).

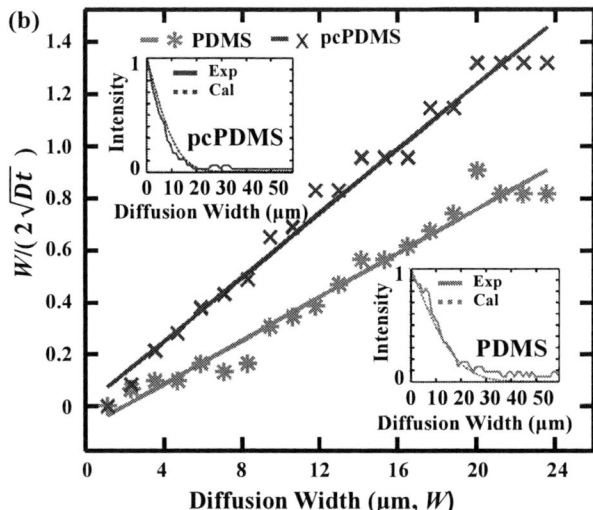

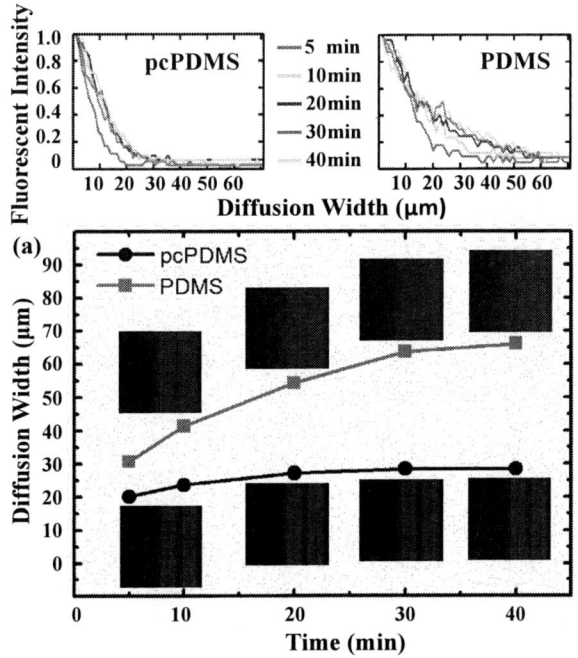

Figure 4: Diffusion of Rhodamine B and its fluorescent intensity distribution in pcPDMS and pristine PDMS microchannels (a) and the fitted diffusivities of Rhodamine B (b).

ACKNOWLEDGEMENTS

This work was financially supported by the National Natural Science Foundation of China (Grant No. 81471750 and 91023045), the Major State Basic Research Development Program (973 Program) (Grant Nos. 2009CB320300 and 2011CB309502).

REFERENCES

[1] Y. Shin, K. Cho, S. Lim, S. Chung, S. Park, C. Chung, D. Han and J. Chang, "PDMS-based micro PCR chip with Parylene coating", J. Microemech. Microeng., 13 (5), pp. 768-774, 2003.

[2] S. Sawano, K. Naka, A. Werber, H. Zappe and S. Konishi, "Sealing method of PDMS as elastic material for MEMS", IEEE MEMS 2008, Tucson, AZ, USA, January13–17, 2008.

[3] Y. Lei, Y. Liu, W. Wang, W. Wu and Z. Li, "Studies on Parylene C-caulked PDMS (pcPDMS) for low permeability required microfluidic applications", Lab. Chip., 11 (7), pp. 1385-1388, 2011.

[4] Y. Lei, Y. Liu, W. Wang, W. Wu and Z. Li, "Fabrication and characterization of Parylene C-caulked PDMS for low-permeable microfluidics", IEEE MEMS 2011, Cancun, MX, USA, January 23–27, 2011.

[5] J. B. Fortin and T. M. Lu, "A model for the chemical vapor deposition of Poly(para-xylylene) (Parylene) thin films", Chem. Mater., 14 (5), pp. 1945-1949, 2002.

[6] M. Kochi, K. Oguro and T. Mita, "Photoluminescence of solid aromatic polymers-I. Poly (*p*-Xylylene)", Eur. Polym. J., 24 (10), pp. 917-927, 1988.

[7] B. Lu, S. Zheng, B. Q. Quach and Y. Tai, "A study of the autofluorescence of parylene materials for μTAS applications", Lab Chip., 14 (11), pp. 1826-1834, 2010.

CONTACT

*Wei Wang, Tel: +86-010-62769183

E-mail: w.wang@pku.edu.cn

CHEMICALLY RESPONSIVE PROTEIN-PHOTORESIST HYBRID ACTUATOR

Daniela Serien[1], and Shoji Takeuchi[1,2]
[1] Center for International Research on Integrative Biomedical Systems,
Institute of Industrial Science (CIBiS-IIS), The University of Tokyo
[2] Takeuchi Biohybrid Innovation Project, ERATO, JST

ABSTRACT

We report the multiphoton fabrication of hybrid microstructures of photoresist and chemically responsive protein hydrogel for microactuation, such as a lever and a rotary stepper. By two-step direct laser writing (DLW) technology, we combine chemically responsive protein hydrogel with mechanical robust photoresist into pH-responsive hybrid actuators that contain only biocompatible materials. The fabrication can be performed separately, without adding to the complexity of device fabrication. We observe micrometer-range motion of the photoresist components. These microactuators may also serve as a pH- or salt-concentration-sensor that measure and interact with their environment by their motion as immediate feedback.

INTRODUCTION

Chemically responsive gels are considered smart materials because of their ability to sense and response to a chemical stimulus [1]. Chemical actuation is interesting for various purposes: chemical robots [2], drug delivery [3], cell culture [4] and microactuation [1, 4-6]. Amongst those applications, chemical actuators are interesting because they can act as sensors as well as manipulation tools. Chemical actuators need to fulfill the following key features [1]: (i) mechanical robustness, (ii) capability of force transfer, and (iii) chemical scalability to adjust response time as desired. Since response time is often diffusion based [1] and force transfer and mechanical robustness are challenging at macro-scale, hydrogel actuation and regulation seems advantageous at micro-scale. Microfabrication of chemically responsive hydrogels has been proposed for conventional lithography processing in layers [5, 6] as well as for mask-free fabrication process [4,7-11].

This mask-free fabrication process enables not only arbitrary 3-dimensinal fabrication, but also leads to protein hydrogel based on a protein cross-linking mechanism [12]. This pure protein hydrogel is biocompatible [4, 8], can be composed of various proteins [4, 9] and maintain the protein function [9].

The chemical responsiveness of protein hydrogel can be scaled by adjusting lithography parameters or chemical parameters such as concentration and mixture of proteins [4, 7]. We recently showed that it can be patterned with nm-scale feature sizes by direct laser writing (DLW) [10]. However, protein hydrogel is mechanically not robust. Its softness also leads to poor force transfer. In this work, we combine the scalable and biocompatible protein hydrogel with a mechanically robust photoresist to increase its feasibility for microactuation (Fig. 1).

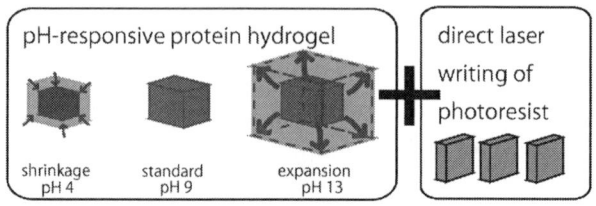

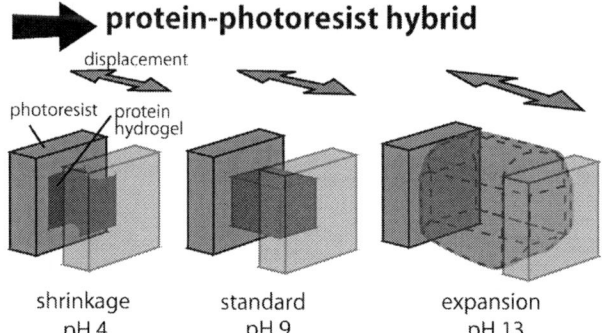

Figure 1: Concept figure of combination of chemically responsive protein hydrogel and conventional mechanically robust DLW photoresist into a hybrid fabricated by a two-step DLW process.

MATERIALS AND METHODS

Preparing Protein Hydrogel

We modified the recipe of the working solution for protein hydrogel by adding rhodamine B (Wako Pure Chemical Industries, Ltd.) dissolved in DMSO (dimethyl sulfoxide, Wako Pure Chemical Industries, Ltd.) to a working solution of 1 mM of the photosensitizer flavin adenine dinucleotide (FAD, F-6625, Sigma-Aldrich) with 200 mg/mL bovine serum albumin (BSA, A-7906, Sigma-Aldrich) in 0.02 mM HEPES (Sigma-Aldrich, H-3375) buffer at pH 7.4 with 50% DMSO.

Fabrication Process

For the fabrication of two-component hybrid microstructures, we first fabricate components with the negative photoresist IPL-780 (Nanoscribe, Germany). After development, we expose the sample to air-born O2-plasma of 18 W for 15 min to increase hydrophilicity, enabling proper and permanent contact between the two components. Then, we continue with a second fabrication step, utilizing tools for alignment provided by Nanowrite software (Nanoscribe, Germany). Eventually, the two-component hybrid structure is released and exposed to different pH conditions.

RESULTS

By laser scanning microscopy, we confirmed the details of the two-component hybrid structure at different pH-values (Fig. 2): Two photoresist plates connected by a protein hydrogel block change their spacing due to pH change; pH=13 leads to expansion of the protein hydrogel, hence increasing the spacing, when the pH=4 causes shrinkage of the hydrogel, hence reducing the spacing.

This displacement within hybrid structures repetitively occurred when starting from pH=9 and then cycling between pH=13 and pH=4. Comparison between the deformation of pure protein microstructures and protein-photoresists microstructures illustrates the advantage of a hybrid microstructure: When the displacement change due to expansion is half of the pure protein hydrogel expansion, the displacement change due to shrinkage is more than double of the pure protein hydrogel expansion. Overall, the hybrid structure demonstrates increased gain of the hydrogel response. Additionally, the response of strong expansion and little shrinkage was translated to a response balanced between expansion and shrinkage.

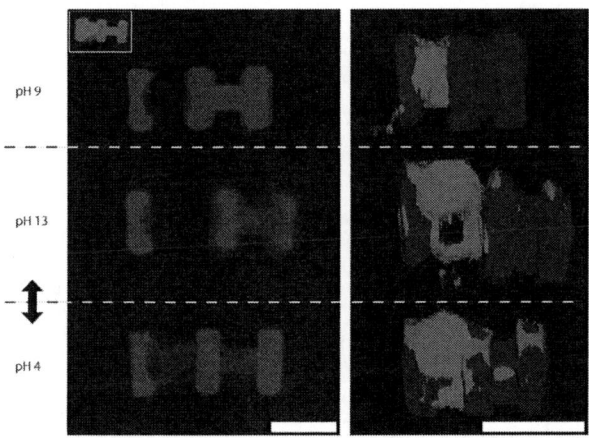

Figure 2: Protein-photoresist hybrid actuation by pH. Top view and side view of fluorescence confocal microscopy results for different pH-values. Photoresist autofluorescence is depicted in blue for both top and side view, and protein autofluorescence in red for side view only. Inset shows top view with both fluoresecence imaging channels activated. Scale bars represent 10 μm.

CONCLUSION

We introduced protein-photoresist hybrid structures fabricated as two-component microstructures by DLW. We demonstrate the displacement of photoresist components by the pH-response of protein hydrogel components. In the future, chemically-actuated hybrid structures might complement microfluidic devices, and function as pH-sensors possibly with motion as feedback.

ACKNOWLEDGEMENTS

This work was partly supported by JSPS Core-to-Core Program (C2C), A. Advanced Research Networks. D.S. receives a scholarship from the Ministry of Education, Culture, Sports, Science and Technology of Japan.

REFERENCES

[1] J. M.G. Swann, and A. J. Ryan, "Chemical actuation in responsive hydrogels," *Polymer International*, vol. 58, pp. 285-289, 2009

[2] S. Maeda, Y. Hara, S. Nakamaru, H. Nakagawa3 and S. Hashimoto, "Chemical Robots," *On Biomimetics (InTech) Capt. 12*, pp. 253-272, 2011

[3] A. V. Reis, M. R. Guilherme, O. A. Cavalcanti, A. F. Rubira, and E. C. Muniz, "Synthesis and characterization of pH-responsive hydrogels based on chemically modified Arabic gum polysaccharide," *Polymer*, vol. 47, pp. 2023–2029, 2006

[4] B. Kaehr, and J. B. Shear, "Multiphoton fabrication of chemically responsive protein hydrogels for microactuation," *PNAS*, vol. 26, pp. 8850-8854, 2008

[5] J. Guan, H. He, D. J. Hansford, and L. J. Lee, "Self-Folding of Three-Dimensional Hydrogel Microstructures," *Journal of Physical Chemistry B*, vol. 109, pp. 23134-23137, 2005

[6] D. H Gracias, "Stimuli responsive self-folding using thin polymer films," *Current Opinion in Chemical Engineering*, vol. 2, pp. 112–119, 2013

[7] J. D. Pitts, P. J. Campagnola, G. A. Epling, and S. L. Goodman, "Submicron Multiphoton Free-Form Fabrication of Proteins and Polymers: Studies of Reaction Efficiencies and Applications in Sustained Release," *Macromolecule*, vol. 33, pp. 1514-1523, 2000

[8] Bryan Kaehr, Richard Allen, David J. Javier, John Currie, and Jason B. Shear, "Guiding neuronal development with in situ microfabrication", *PNAS*, vol. 101 (46), pp. 16104–16108, 2004

[9] E. C. Spivey, E. T. Ritschdorff, J. L. Connell, C. A. McLennon, C. E. Schmidt, and J. B. Shear, "Multiphoton Lithography of Unconstrained Three-Dimensional Protein Microstructures," *Advanced Functional Materials*, vol. 23, pp. 333-339, 2013

[10] D. Serien and S. Takeuchi, "Direct laser writing of 3D protein structures with nanoscale feature sizes," *The 27th IEEE International Conference on Micro Electro Mechanical Systems (MEMS 2014)*, pp.471-473, 26-30 Jan. 2014

[11] D. Serien and S. Takeuchi, "Elastic Wire-Frame Microparticles of Cross-Linked Bovine Serum Albumin," *The 18th International Conference on Miniaturized Systems for Chemistry and Life Sciences (MicroTAS 2014)*, pp. 2489-2490, 26-30 Oct. 2014

[12] J.D. Spikes, H.-R. Shen, P. Kopečková, and J. Kopeček, "Photodynamic Crosslinking of Proteins. III. Kinetics of the FMN- and Rose Bengal-sensitized Photooxidation and Intermolecular Crosslinking of Model Tyrosine- containing N-(2-Hydroxypropyl)-methacrylamide Copolymers," *Photochemistry and Photobiology*, vol. 70, pp.130–137, 1999

CONTACT

*D. Serien; serien@iis.u-tokyo.ac.jp

CORE-SHELL MICROPARTICLE SYNTHESIS IN DROPLET MICROFLUIDICS USING A SINGLE STEP POLYMERIZATION

Xiamo Zhou, Yang Sun, Anna Finne-Wistrand, Wouter van der Wijngaart* and Tommy Haraldsson
KTH Royal Institute of Technology, Stockholm, SWEDEN

ABSTRACT

We present, for the first time, a method for the synthesis of core-shell microparticles in a single polymerization step using two-phase droplet microfluidics. We verify the successful generation of core-shell microparticles using the novel synthesis approach. To enable (future) cell encapsulation, this novel method is specifically designed to allow: i) uncomplicated synthesis of core-shell microparticles and ii) well-controlled and reproducible shell thickness with a perfectly centred core.

INTRODUCTION

Core-shell polymer structures have been proposed and investigated since 1961 [1], and have been used in a large scale in applications such as impact modifiers, surface coating, catalysis, sensing and drug delivery [2]-[4]. A core-shell structure usually has at least two different components, constituting the core and the shell of the particles respectively, in which the core-shell composite expresses superior properties not possessed by any of the single component [5].

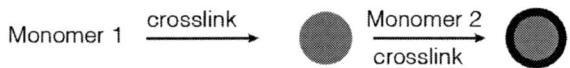
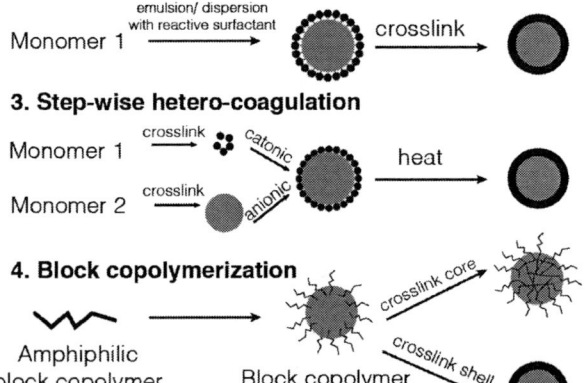

Figure 1: Illustration of previous methods for core-shell structure preparation.

Traditional core-shell micro/nano particle synthesis methods include dispersion, suspension, emulsion and combined polymerization ranging from a single step to multiple steps [5], [6]. The most common methods are two-stage emulsion polymerization, reactive surfactant emulsion polymerization, step-wise hetero-coagulation and block copolymerization (Fig. 1). Previous core-shell particle manufacturing techniques using capillary microfluidic based emulsion can be classified in two approaches (Fig. 2). In the two-step polymerization approach, the core is first created and a shell is added subsequently, which imparts complexity and cost. The three-phase double emulsion approach creates a good size uniformity but results in poorly controlled core-shell morphology where the core is easily displaced from the centre, the shell has varying thickness, and/or the core-shell morphology varies from batch to batch due to difficulty in controlling three phase systems [7], [8].

 Two step or double emulsion polymerization gives a "moon-like" structure

 External stimuli can cause core swelling thus break the shell coating

 Thin shell structure is impossible to achieve: shear force in emulsion polymerization is too high to maintain the thin shell formation.

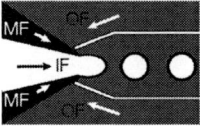 Three-phase double emulsion is hard to control, resulting in a shell structure with ill-defined core placement and shell thickness.

Figure 2: Illustration of problems inherent to previous droplet microfluidic core-shell microparticle synthesis methods.

NOVEL CONCEPT

Here, we present a novel and uncomplicated, microfluidics based particle synthesis platform as a promising solution for generating precisely controlled core-shell geometries (Fig. 3). Our platform features microdroplet generation of a disperse droplet phase into a continuous phase, followed by a UV exposure that *simultaneously* crosslinks the droplets and grafts a surrounding shell. The dispersed phase consists of water dissolved hydrogel monomer (PEGDA) and photoinitiator (Irgacure 2959). The continuous phase consists of shell components (GPTA) and surfactant (Span 80) dissolved in toluene. The UV exposure creates radicals from initiators and polymerizes the PEGDA droplet, during which the core radicals diffuse to the droplet interface, where they trigger the polymerization of the GPTA shell in a grafting from fashion. The result of this simultaneous PEGDA core and GPTA shell polymerization is particles with an evenly distributed, well-defined shell, generated at a fairly high synthesis speed of up to 400 particles/min.

978-1-4799-7956-1/15 $31.00 © 2015 IEEE

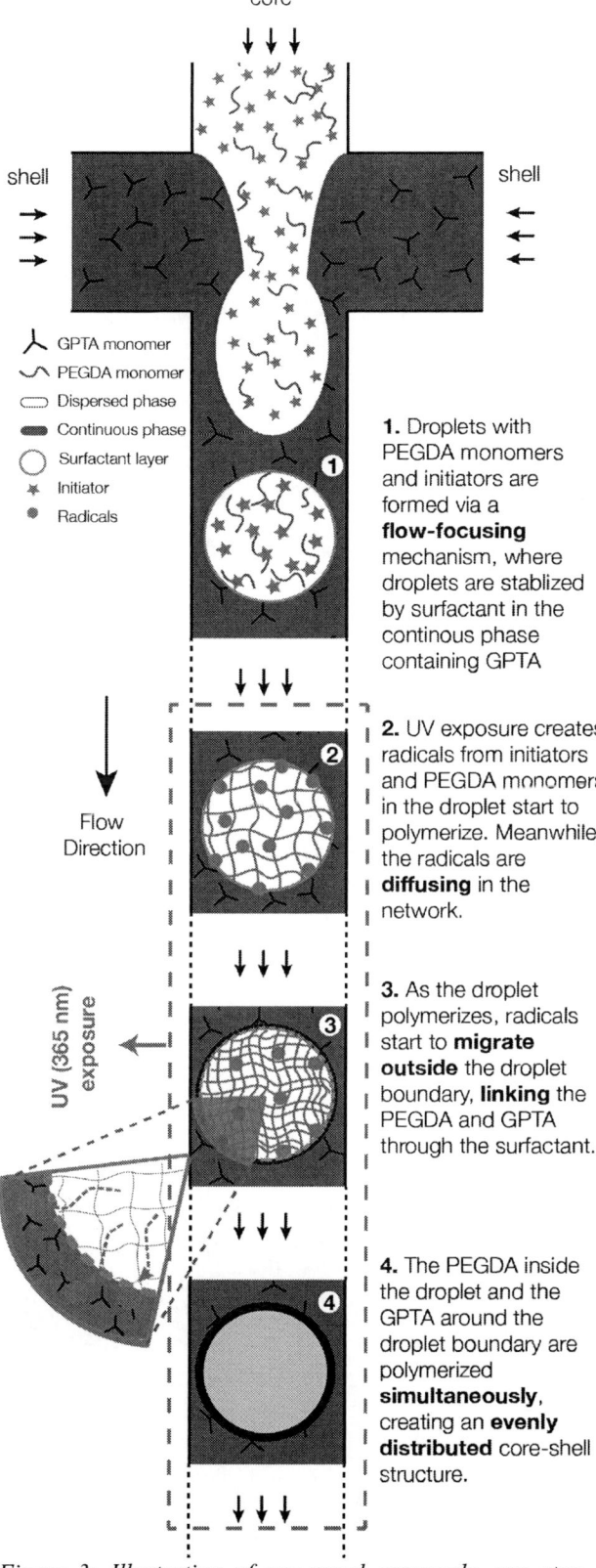

1. Droplets with PEGDA monomers and initiators are formed via a **flow-focusing** mechanism, where droplets are stablized by surfactant in the continous phase containing GPTA

2. UV exposure creates radicals from initiators and PEGDA monomers in the droplet start to polymerize. Meanwhile, the radicals are **diffusing** in the network.

3. As the droplet polymerizes, radicals start to **migrate outside** the droplet boundary, **linking** the PEGDA and GPTA through the surfactant.

4. The PEGDA inside the droplet and the GPTA around the droplet boundary are polymerized **simultaneously**, creating an **evenly distributed** core-shell structure.

Figure 3: Illustration of our novel approach: one step core-shell polymer synthesis mechanism.

EXPERIMENTAL

The microfluidic synthesis platform consists of a 75 x 25 mm^2 microfluidic chip made of off-stoichiometry thiol-ene-epoxy polymer (OSTE+ 50%, Mercene Labs AB, Sweden, cured by OAI Model 30 Collimated UV Light Source, USA) placed underneath a UV station (Fig. 4; UV LED 365nm, 10W, KT-electronic, Germany). The two inlets of the chip are connected to syringe pumps (NE-1000, New Era Pump Systems, Inc. NY, USA) using PEEK tubing (1/32'' OD, Mengel Engineering, USA) and corresponding NanoPort™ Assemblies (N-126S, Upchurch Scientific, IDEX Health & Science LLC). The syringe for the continuous phase solution is loaded with 30% w/w glycerol propoxylate triacrylate (GPTA, Sigma-Aldrich, Sweden) and 0.5% w/w Span 80 (Sigma-Aldrich, Sweden) dissolved in toluene (Sigma-Aldrich, Sweden), and the syringe for dispersed phase is loaded with water solution containing 25% w/w poly(ethylene glycol) diacrylate (PEGDA, MW = 6 kDa, Sigma-Aldrich, Sweden) as hydrogel monomer, and 0.5% w/w 2-Hydroxy-4'-(2-hydroxyethoxy)-2-methylpropiophenone (Irgacure 2959, Sigma-Aldrich, Sweden) as photoinitiator. A collecting vessel is connected to the outlet of the chip by PTFE tubing.

The microfluidic device concatenates a droplet generating channel junction with a polymerization section, which consists of a 600 mm long meandering channel (Fig. 4).

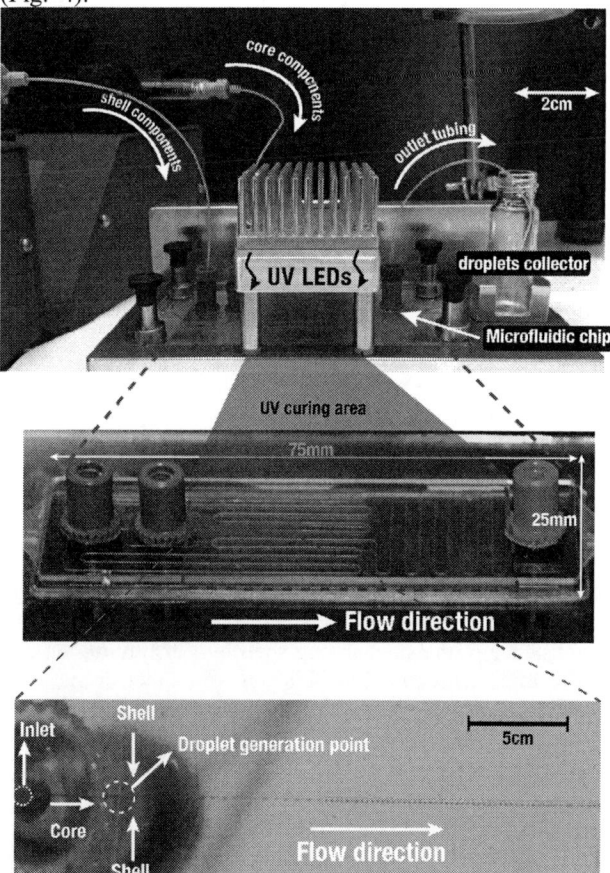

Figure 4: UV station setup for core-shell synthesis (upper); the microfluidic chip with long meander channels for core-shell particle preparation (middle); the droplets (red dyed) generated from flow-focusing microfluidic chip at the speed of ~400 particles/min. The droplets diameter is ~100µm (lower).

We performed three experiments. Experiment A was designed to generate core-shell microparticles with the above described synthesis method. Control experiments B & C were designed to investigate the diffusion barrier function of the polymerized shell. In both experiments B and C, the PEGDA in the dispersed phase was substituted with polyethylene glycol (PEG, Sigma-Aldrich, Sweden), which has no functional groups for photo crosslinking and which is thus expected to remain in its liquid state during UV exposure. The photo-initiator (Irgacure 2959) is absent in the dispersed phase in experiment B, while present in experiment C (see Table 1). Hence, we expect shell polymerization only in experiment C. Red food dye was added to all dispersed phases for visualisation.

RESULTS

Experiment A successfully generated core-shell particles (Table 1, A). A polymerized core-shell polymer structure collected from the outlet tubing shows a homogenous whitish shell around a red dyed PEGDA core, indicating a well-polymerized structure. The elongated shape is attributed to the hydrodynamic forces during polymerization.

Control experiment B resulted in a stream of dissolved red-dyed PEG emanating from the droplet. The lack of photoinitiator in the PEG core prevents formation of a polymerized protective shell. Once in contact with water in the collection vessel, the droplet rapidly disintegrates.

Control experiment C resulted in a batch of coloured beads in the collection vessel. This indicates the presence of a polymerized shell which protects the un-corsslinked liquid core from the water in the collection vessel.

The difference between B and C can only be attributed to the successful formation of a shell in C and the absence of such shell in B.

CONCLUSIONS

We have demonstrated a single step polymerization method to prepare size-controlled and uniformly distributed core-shell microparticles in a microfluidic device. The polymerization mechanism is initiated by radical diffusion from the core components towards the core-shell boundary, thus polymerizing a homogenous thin shell around the core as well as crosslinking the core and shell polymers simultaneously. We also verify the existence of the surrounding shell by substituting the core monomer with a non-crosslinkable monomer, from which the experimental group with the presence of photoinitiator formed a shell and maintain the droplet shape while the one with the absence of photoinitiator lysis to water.

ACKNOWLEDGEMENTS

We gratefully acknowledge the financial support by Barncancerfonden, Sweden. We also express our gratitude to Mercene Labs AB for providing the OSTE+ polymers.

Table 1: Experimental results of core-shell particle formation (A) and control experiments (B&C) to verify the existence of the shell.

EXPERIMENT -A- Fully polymerized core-shell particle	
Dispersed phase components	DI water PEGDA (25% w/w) Irgacure 2959 initiator (0.1% w/w) Red dye
Continuous phase components	Toluene GPTA (30% w/w) Span 80 surfactant (0.5% w/w)
Phenomenon observed	A homogenous white shell around a red core, the elliptical shape is created by hydrodynamic force and well preserved by polymerization. (Picture taken from collection tubing)
EXPERIMENT -B- Unpolymerized PEG without shell results in PEG droplets that dissolve in water	
Dispersed phase components	DI water PEG monomer (25% w/w) Red dye

Continuous phase components	Toluene GPTA (30% w/w) Span 80 surfactant (0.5% w/w)
Phenomenon observed	Unpolymerized hydrophilic core with no shell due to the absence of initiator in the core component results in PEG lysis into water instantly.
EXPERIMENT -C- **Un-crosslinked PEG with shell results in stably encapsulated PEG droplets**	 2mm **Red particles**
Dispersed phase components	DI water PEG monomer (25% w/w) Irgacure 2959 initiator (0.1% w/w) Red dye
Continuous phase components	Toluene GPTA (30% w/w) Span 80 surfactant (0.5% w/w)
Phenomenon observed	Un-crosslinked hydrophilic core with a polymerized shell due to the migration of initiator in the core component maintains its shape in water for a while. Hence, a shell exists.

REFERENCES

[1] L. J. Hughes and G. L. Brown, "Heterogeneous polymer systems. I. Torsional modulus studies," Journal of Applied Polymer Science, vol. 5, no. 17, pp. 580–588, Sep. 1961.

[2] A. K. Khan, B. C. Ray, J. Maiti, and S. K. Dolui, "Preparation of core-shell latex from co-polymer of styrene-butyl acrylate-methyl methacrylate and their paint properties: Pigment & Resin Technology: Vol 38, No 3," Pigment & Resin Technology, vol. 38, no. 3, pp. 159–164, 2009.

[3] C. D. J. and and L. A. Lyon, "Shell-Restricted Swelling and Core Compression in Poly(N-isopropylacrylamide) Core−Shell Microgels," Macromolecules, vol. 36, no. 6, pp. 1988–1993, Feb. 2003.

[4] W.-H. L. and H. D. H. Stöver, "Monodisperse Cross-Linked Core−Shell Polymer Microspheres by Precipitation Polymerization," Macromolecules, vol. 33, no. 12, pp. 4354–4360, Jun. 2000.

[5] H. JW, P. IJ, L. SB, and K. DK, "Preparation and characterization of core-shell particles containing perfluoroalkyl acrylate in the shell," Macromolecules, vol. 35, no. 18, pp. 6811–6818, 2002.

[6] M. Jonsson, O. Nordin, E. Malmström, and C. Hammer, "Suspension polymerization of thermally expandable core/shell particles," *Polymer*, vol. 47, no. 10, pp. 3315–3324, May 2006.

[7] Y.-L. Yu, R. Xie, M.-J. Zhang, P.-F. Li, L. Yang, X.-J. Ju, and L.-Y. Chu, "Monodisperse microspheres with poly(N-isopropylacrylamide) core and poly(2-hydroxyethyl methacrylate) shell," Journal of Colloid and Interface Science, vol. 346, no. 2, pp. 361–369, Jun. 2010.

[8] J.-W. Kim, A. S. Utada, A. Fernández-Nieves, Z. Hu, and D. A. Weitz, "Fabrication of Monodisperse Gel Shells and Functional Microgels in Microfluidic Devices," Angewandte Chemie International Edition, vol. 46, no. 11, pp. 1819–1822, Mar. 2007.

CONTACT
*Wouter van der Wijngaart
Tel:+4687906613
E-mail: wouter@kth.se

VASCULAR NETWORK FORMATION FOR A LONG-TERM SPHEROID CULTURE BY CO-CULTURING ENDOTHELIAL CELLS AND FIBROBLASTS

Tomoya Hayashi[1], Hisako Takigawa-Imamura[2], Koichi Nishiyama[3], Hirofumi Shintaku[1], Hidetoshi Kotera[1], Takashi Miura[2] and Ryuji Yokokawa[1]

[1]Kyoto University, Kyoto, JAPAN
[2]Kyusyu University, Fukuoka, JAPAN
[3]Kumamoto University, Kumamoto, JAPAN

ABSTRACT

In this paper, we present a poly-dimethylsiloxane (PDMS) microfluidic device to create a vascular network for a long-term spheroid culture, of which network and spheroid are consist of human umbilical vein endothelial cells (HUVEC) and normal human lung fibroblasts (LF), respectively. Following device design, fabrication, and fundamental evaluation of HUVEC sprouting conditions, we visualized that HUVEC networks were successfully formed by the co-culture with LFs and reached a LF-based spheroid. Moreover, perfusability of the network was evaluated by injecting fluorescent microbeads. This platform will be applicable for long-term tissue cultures to understand morphogenesis and modeling of blood vessel functions.

INTRODUCTION

Morphogenesis is a process that controls the spatial distribution of cells during embryonic development. This process has been studied for revealing which physiological cues have significant effect on the development. It is important basic knowledge for organ regeneration. Although it has been studied by transgenic mice [1] and tissue culture [2], the culture duration is limited to a few weeks, resulting in the observation limited to the short period of morphogenesis. This is caused by the important difference between *in vitro* and *in vivo* tissue culture, where vascular network *in vivo* tissue is actively connected but not *in vitro*. Due to the lack of vascular network *in vitro*, oxygen and nutrients are not sufficiently supplied to inner cells of a tissue and metabolites are not removed. It results in hypoxia and/or necrosis, especially when the tissue is over 100 μm in diameter [3]. *In vivo* vascular networks, mainly composed of endothelial cells (ECs), supply them to almost all tissues. They sprout toward growth factors such as vascular endothelial growth factors (VEGF) [4]. They are formed by two steps: vasculogenesis, a formation occurring by a *de novo* production of vascular meshwork and angiogenesis, a new blood vessel formation by the growth and sprouting of existing blood vessels. Angiogenesis is an important process found in development, wound healing and formation of tumor. To extend the culture duration *in vitro* for further understanding of morphogenesis in the later stage, long-term tissue culture by integrating vascular networks and tissue culture *in vitro* is in high demand.

Toward this end, some specifically designed culture methods have been reported: perfusion bioreactor[5][6], air-medium interface culture [7], and rotation culture [8]. However, externally applied oxygen and nutrients are not fully delivered to the center of spheroids and tissues. Therefore, it is desirable that tissues are cultured with vascular networks as seen *in vivo*.

In vitro studies of vascular networks, conventional planar culture are used for the investigation of fundamental behavior of ECs: meshwork formation [9], barrier function and mechanosensitive response [10]. However, such 2D system lacks the 3D context. To reproduce *in vivo* like 3D environment, culture with extracellular matrix (ECM) as a scaffold or using EC coated beads in ECM are studied to recreate vascular networks [11]. However, these are close-ended networks. They are not able to be used as a flow channel to supply oxygen and nutrients to a tissue. In previous research about reproduction of vascularized tissue, EC coated spheroids [12] or EC co-cultured spheroids [13] were used. They showed that ECs can form lumen networks between spheroids or in the co-cultured spheroid. However, these networks are also close-ended and cannot be externally accessed. Long-term tissue culture requires open lumen vascular networks to supply

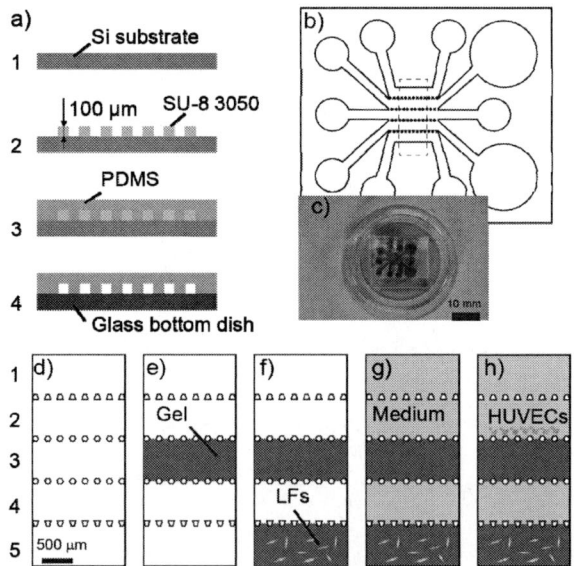

Figure 1: a) Fabrication process b) Schematic illustration of device 1. Five channels (w = 700 mm, d = 100 mm) are separated by pillar array (dia. = 100 mm), enabling selective gel injections due to hydrophobicity of PDMS. c) Device 1 filled with red ink. d-e) Experimental procedure to observe HUVEC sprouting. d) Initial dry condition. e) Fibrin gel injection to channel 3. f) Injection of LF suspended in fibrin gel to channel 5. g) LF culture with EGM-2 medium. h) HUVEC injection to channel 2 and sprouting into fibrin gel in channel 3.

oxygen and nutrients.

Recently, microfluidic devices have improved the cell culture microenvironment. Vascular networks were formed in microfluidic devices by arranging ECM and EC locations selectively [14][15] and by using EC-coated microfibers [16]. However, previous studies targeted to study a cancer metastasis model through the formed vascular networks, and to our knowledge utilization for tissue culture has not been reported.

Here, we suppose an on-chip HUVEC angiogenesis to a LF spheroid as a tissue model using a microfluidic device

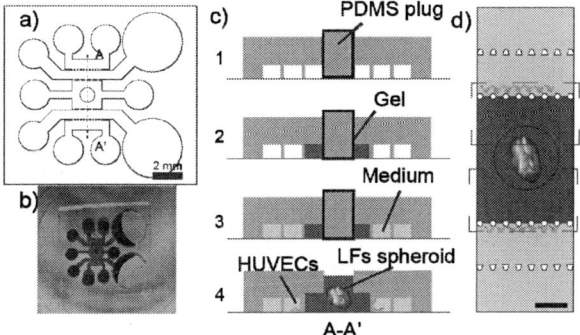

Figure 2: a) Schematic illustration of device 2. Channel 3 is widened to create a spheroid culture well. b) Device 2 filled with red ink. c) Cross-sectional illustration at A-A'. c1) PDMS plug was inserted to the well and contacted to the bottom glass. c2) Fibrin gel injection to channel 3. c3) Other four channels were filled with EGM-2 medium. c4) LF spheroid and HUVECs were introduced for sprouting assay. d) Schematic of HUVEC sprouting to the spheroid (topview). Scale bar is 500 µm.

referring to the report of Kim *et al* in 2013 [14]. LFs, which have a role to promote angiogenesis in the process of wound healing, are often used for angiogenesis. At first, the effect of two different fibrinogen concentrations and three LFs concentrations on angiogenesis were examined. To the result, we optimized the culture condition and co-culture HUVECs and a LF spheroid in a microfluidic device. And we confirmed the networks were connected with the spheroid and it had an open lumen structure.

EXPERIMENTAL PROCEDURES

Device fabrication process

Microfluidic device was constructed by the patterned poly-dimethylsiloxane (PDMS) and glass bottom dish. SU-8 3050 (MicroChem) photoresist was spin coated on a silicon substrate cleaned with Piranha solution ($H_2SO_4:H_2O_2 = 3:1$) for 10 min at 70°C. The photoresist was baked for 45 min at 95°C, and exposed to UV light at 250 mJ/cm^2 (PEM 800, Union Optical), followed by post-exposure bake for 5 min at 95°C and development by SU-8 developer (MicroChem) for 15 min. PDMS was cast (elastomer:curing agent = 10:1) (Dow Corning) and heat-cured for 2 h at 80°C. The PDMS slab and glass bottom dish were bonded after exposing to O$_2$ plasma (COVANCE, Femto Sicence) and heated at 80°C for 24 h (Fig.1a).

Evaluation of sprouting growth

It is known that HUVECs sprouts are promoted by secretary factors from fibroblasts [17]. To co-culture LFs and HUVECs at designated locations at micrometer-scale, device 1 was designed to have five channels (1: medium channel, 2:

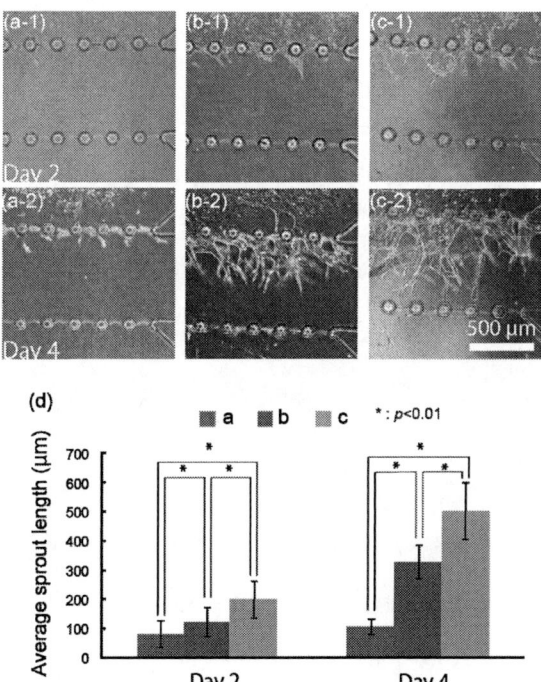

Figure 3: a-c) HUVEC sprout length in three different LF concentrations: 0, 0.3 and 1 × 10⁶ cells/ml. d) Quantified sprout length. Higher LF concentration induces longer sprouting.

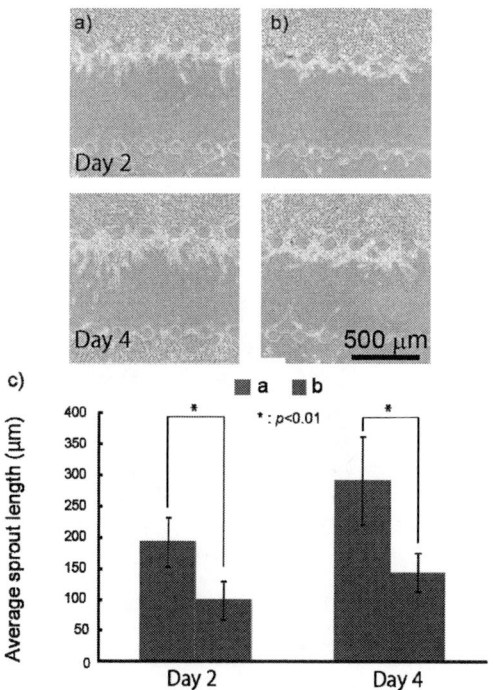

Figure4: a-b) HUVEC sprout length in two different fibrinogen concentrations: 2.5 and 5.0 mg/ml. c) Quantified sprout length. Lower (softer) fibrinogen concentration is preferable for longer sprouting.

978-1-4799-7956-1/15 $31.00 © 2015 IEEE

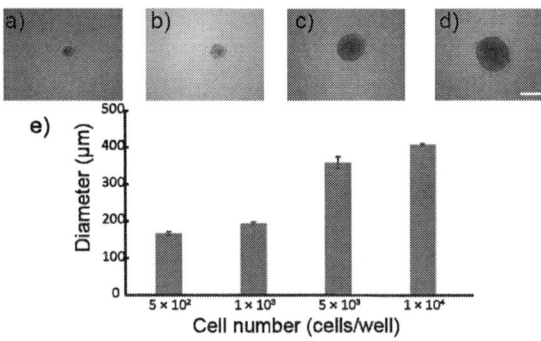

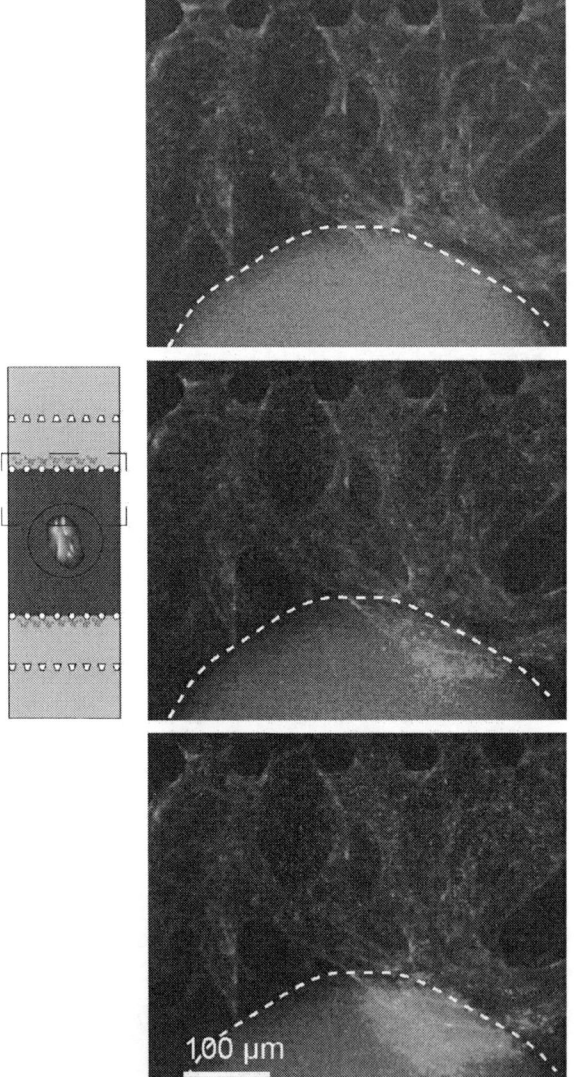

Figure 5: a-d) Phase-contrast images of LF spheroids at day 2. Initial cell concentrations: 5×10^2, 1×10^3, 5×10^3, 1×10^4 cells/well. Scale bar, 300 µm. e) Quantification of the spheroids diameter. The results are mean $\pm$ s.d., N =5.

HUVEC channel, 3: angiogenesis channel, 4: medium channel, and 5: LF channel). These channels were separated by hexagonal pillars with 100 µm in diameter and 100 µm spacing (Fig. 1b). This device configuration enables to apply growth factors from LFs to HUVECs. We first examined how two fibrinogen concentrations (2.5 and 5.0 mg/ml) and three LF seeding concentrations (0, 0.3, and 1.0×10^7 cells/ml) affect the sprout length.

Cells and gel were introduced as follows: 2.5 mg/ml or 5.0 mg/ml fibrinogen and 50 U/ml thrombin both in PBS were mixed at a 100:1 ratio to prepare fibrin gel solution. The mixture was immediately injected into the angiogenesis channel (channel 3 in Fig. 1d). LF suspension was prepared in the fibrinogen solution at three target concentrations after removing cells from culture dishes using 0.25% Trypsin-EDTA. The suspension and thrombin solution were mixed at a 100:1 ratio and immediately injected into the LF channel (channel 5 in Fig. 1d). These gels were allowed to be polymerized for 5 min at room temperature. Then EGM-2 medium (Lonza) was introduced into the other channels and the device was incubated for 24 h in an incubator (37°C, 5% CO_2) to remove air bubbles from the gel-medium interface. Next, HUVECs suspended in EGM-2 at the concentration of 5×10^7 cells/ml were introduced into the HUVEC channel and the device was tilted 90° for 20 min in the incubator to attach HUVECs to the gel-medium interface between

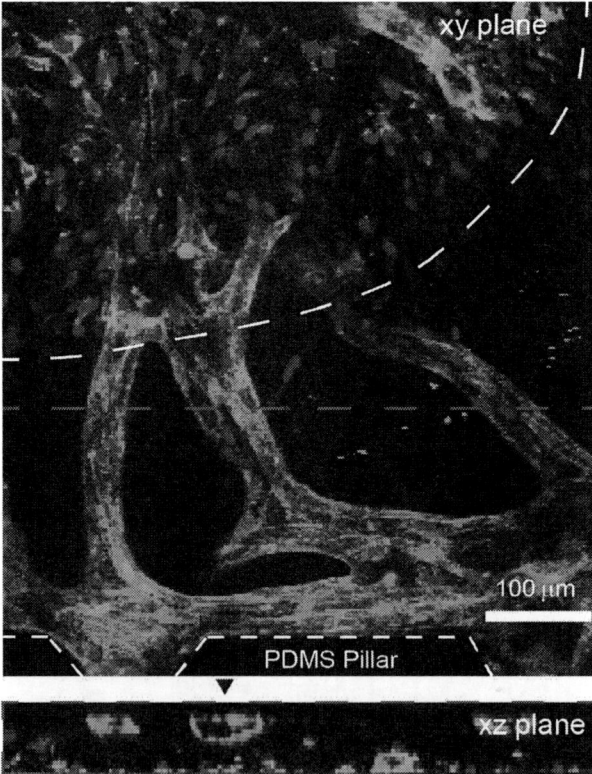

Figure 6: Fluorescence time-lapse images of microbeads flowing through HUVECs network to the spheroid culture well. Nuclei (Hoechst 33342, blue), polymer microspheres red fluorescing (R0100, Duke Scientific) and HUVECs (FITC-conjugated Lectin from Ulex europaeus, green).

Figure 7: Confocal fluorescence microscopy image. White dotted lines show the spheroid chamber area. b) XZ plane shows the cross-sectional image of red dotted line in XY plane. Black triangle shows the lumen structure. Nuclei (Hoechst 33342, blue) and HUVECs (FITC-conjugated Lectin from Ulex europaeus, green).

channel 1 and 2. Medium was changed every 3 days. The devices were put in a culture dish with soaked Kimwipes to keep moisture.

Co-culture of HUVECs and a LF spheroid

Device 2 was designed to co-culture a LF spheroid and HUVEC networks (Fig. 2a). Referring to the fact that LFs growth factors reaches up to 1 mm, a spheroid culture well (1 mm in diameter) was designed to locate at the center of angiogenesis channel with 2 mm width (channel 3 in Fig. 2a). To adjust a spheroid size to the well, LFs at four different concentrations were seeded on a 96-well plate (PrimeSurface, Sumitomo Bakelite) and their diameter were measured at day 2.

Adopting optimal concentrations obtained by the previous experiments, LF spheroid (1.0×10^4 cells/well), HUVECs (5×10^6 cells/ml) and fibrin gel (2.5 mg/ml) were used in device 2 as follows. A PDMS plug was inserted to the spheroid culture well and contacted to the bottom dish. Fibrin gel was injected into the channel 3 and incubated for 5 min at room temperature. After pulling the plug out of the well, the spheroid in fibrin gel was introduced into the well and the device was incubated for 5 min at room temperature again. HUVECs suspended in EGM-2 was introduced into channel 2 and the device was tilted 90° for 20 min at 37°C. Then HUVECs was also introduced into the other side channel and tilted again. Medium was changed every 3 days.

Immunostaining

HUVECs were labeled with FITC-conjugated lectin from *Ulex europaeus* (Sigma, L9006). Hoechst 33342 (Life technology, H1399) was used to stain cell nuclei. After applying these regents, the device was incubated for 2 h at 37°C.

RESULT AND DISUCUSSION
Evaluation of sprout length

HUVECs sprouting into fibrin gel was observed in any condition in the device 1 at day 2 and day4 (Fig. 3a-c). The sprout length was longer at higher LFs concentrations and lower fibrinogen concentration (Fig. 4a,b). These results indicate that factors secreted from LFs could reach to HUVECs through the 700 µm wide channel and it has an enough quantity to promote angiogenesis. The sprout length was longer at lower fibrinogen concentration. However, in case of lower than 2.5 mg/ml fibrinogen, the gel could not solidify.

Co-culture of HUVECs and a LF spheroid

2.5 mg/ml fibrinogen was selected in device 2 experiment. LFs formed a spheroid at any concentration within 2 days and their diameter increased with the increase of cell density (Fig. 5). To introduce the spheroid into the 1-mm chamber, a spheroid made of 1×10^4 cells/well was chosen. Until day 4-6, HUVECs sprouted toward LF spheroid. Fig. 6 and Fig. 7 show HUVEC sprouting toward LF spheroid visualized by epi-fluorescence microscopy (at Day 4) and confocal fluorescence microscopy (at Day 6), respectively. The lumen structure of the sprout was observed (*xz* plane in Fig. 7). Red fluorescent microbeads (Duke Scientific,

R0100) injected into the network from channel 2 flowed into the spheroid chamber through the vascular network without leakage, which demonstrated perfusability of the network (Fig. 6). Figure 7 also shows that HUVECs (green) sprouted into the inside of the LF spheroid (blue surrounded by white dotted line). Our new device enables to co-culture HUVECs and LF spheroid and connect HUVECs network to LF spheroid.

CONCLUSION

We proposed a PDMS device to create vascular network for a long-term spheroid culture. We successfully visualized that HUVECs networks reached LF spheroid in 2.5 mg/ml fibrin gel. This is the first report of success in connecting open lumen vascular networks with a spheroid.

ACKNOWLEDGEMENTS

This research was partially supported by Japan Society for the Promotion of Science (JSPS) KAKENHI Grant Number 25600060; CREST, Japan Science and Technology Agency (JST); Toyota Foundation. T.H. is supported by Wakisaka Foundation for his travel.

REFERENCES

[1] K. Sekine, *et al.*, *Nature*, 21, 138–141, 1999.
[2] D. Hartmann and T. Miura, *J. Theor. Biol.*, 1–20, 2006.
[3] H. S. Bell, *et al.*, *Neuropathol Appl. Neurobiol.*, 27, 291–304, 2001.
[4] H. Gerhardt, *et al.*, *J. Cell Biol.*, 161, 1163–77, 2003.
[5] R. Pörtner, *et al.*, *J. Biosci. Bioeng.*, 100, 235–45, 2005.
[6] F. W. Janssen, *et al.*, *Biomaterials*, 27, 315–23, 2006.
[7] H. Nogawa and T. Ito, Development, 121, 1015–22, 1995.
[8] B. Unsworth and P. Lelkes, *Nat. Med.*, 1998.
[9] Y. Kubota, *et al.*, *J. Cell Biol.*, 107, 1589–98, 1988.
[10] D. Ingber and J. Folkman, *J. Cell Biol.*, 109, 1989.
[11] M. N. Nakatsu, *et al.*, *Microvasc. Res.*, 66, 102–112, 2003.
[12] M. Inamori, H. Mizumoto, and T. Kajiwara, *Tissue Eng.*, 15, 2009.
[13] J. Rouwkema, J. de Boer, and C. a Van Blitterswijk, Tissue Eng., 12, 2685–93, 2006.
[14] S. Kim, *et al.*, *Lab Chip,* 13, 1489–500, 2013.
[15] J. J. H. Yeon, *et al.*, *La*b Chip, 12, 2815–22, 2012.
[16] H. Onoe, *et al.*, *Nat. Mater.*, 12, 584–90, 2013.
[17] L. A. Kunz-Schughart, *et al.*, Am. J. Physiol. *Cell Physiol.*, 290, C1385–98, 2006.

CONTACT

*T. Hayashi, tel: +81-075-383-3687;
hayashi.tomoya.87u@st.kyoto-u.ac.jp

HIGHLY CONTROLLABLE THREE-DIMENSIONAL SHEATH FLOW DEVICE FOR FABRICATION OF ARTIFICIAL CAPILLARY VESSELS

J. Ito[1], R. Sekine[1], D.H. Yoon[1], Y. Nakamura[2], H. Oku[2], H. Nansai[2], T. Chikasawa[2], T. Goto[2], T. Sekiguchi[3], N. Takeda[2], and S. Shoji[1]

[1]Major in Nanoscience and Nanoengineering, Waseda University, Tokyo, JAPAN
[2]Major in Life Science and Medical Bioscinece, Waseda University, Tokyo, JAPAN
[3]Institute for Nanoscience and Nanotechnology, Waseda University, Tokyo, JAPAN

ABSTRACT

This paper reports a highly controllable three-dimensional (3D) sheath flow device for fabrication of artificial capillary vessels. Three step sheath injection type 3D flow device which realizes wide core and sheath structure variations by simply flow rate control was applied to fabricate double-layer coaxial Core-Sheath microfibers applicable for long micro capillary vessels by aligning and cultivating vascular endothelial cells. As a result, about a 3-centimeter-long microfiber which has the vascular endothelial cells fused mutually in the center was successfully formed after three days culture.

INTRODUCTION

Fabricating 3D engineered tissues by assembling individual cells attracts attention for applications of regenerative medicine. Although the conventional cell culture has been performed on dishes with functional and/or patterned surfaces, such two-dimensional culture restricts the cellular construct to thin and short structure [1]. However, most of tissues and organs in the human body are thick and long, so fabricating 3D body tissues is a challenging target at the current study of tissue engineering.

For achieving this target, it is necessary to supply sufficient nutrient and oxygen into and remove wastes from deep interior of engineered tissues to keep them viable. Therefore, providing blood vessels is crucial, and fabricating artificial blood vessels is required in tissue engineering. Artificial blood vessels of medium to large caliber have already been realized, and are put in practical use in the bypass surgery [2]. On the other hand, a study of fabricating tubular microfibers by cultivating vascular endothelial cells with collagen gel in microchannels was reported as an example of artificial blood vessels of the small caliber [3]. Since vascular endothelial cells form tubular structure spontaneously in 3D culture, millimeter long capillary vessels were realized along with microchannels in this study. However, practical artificial capillary vessels are preferably much longer for networking with other living body tissues. Therefore, we aim at longer capillary vessel fabrication by a microfluidic device.

In order to fabricate various hydrogel microfibers, microfluidic devices using microfabricated nozzle array [4], polydimethylsiloxane (PDMS)-based cylindrical channels [5], roller-assisted microchannels [6], and capillary combined microfluidic channels [7-9] have been employed. We also applied 3D sheath flow device to fabricate centimeter long nerve bundles using nerve cells as a core flow [10]. In this study, we focused on fabrication of the layered gel fibers wherein PC12 cells, model cells of neuron. As a result, the layered gel fibers which have the PC12 cells in the center were successfully fabricated, and greater than 80% of their cells in the fibers were viable in three days culture. In addition, the PC12 cells in the gel fiber elongated neurites in parallel to core layer, and engineered nerve bundles could be effectively fabricated.

The technology for this nerve bundles is applicable to construct capillary vessels since sizes of both tissues are μm in width and cm/m in length. This paper focuses on fabrication of long capillary vessels through cultivating vascular endothelial cells by collagen gel using 3D sheath flow device.

CONCEPT

Vascular endothelial cells can make tubular structures from being cultivated by collagen gel with them aligned in microchannels within 24–48 hours of seeding as shown in Figure 1. These tubes exhibit cell–cell junction formation characteristic of early stage capillary vessels, and the diameter of them could be controlled by varying collagen concentrations or channel width [3]. Therefore, long capillary vessels whose diameters are controllable can be fabricated by cultivating vascular endothelial cells within microfibers generated with microfluidic devices.

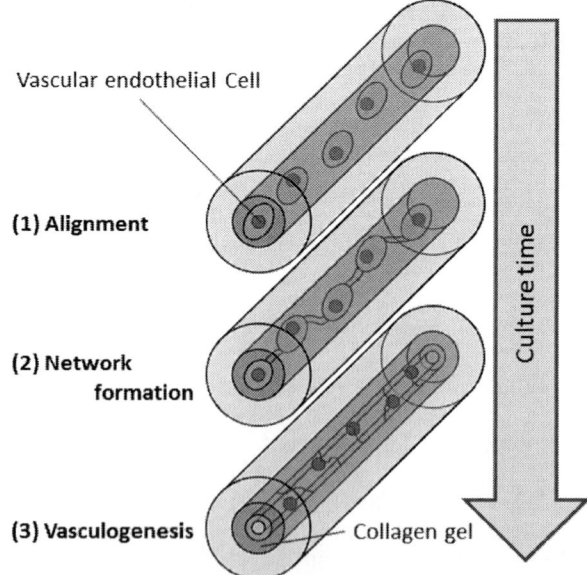

Figure 1: Conceptual diagram of tube-like structure formation by cultivating vascular endothelial cells with collagen gel

978-1-4799-7956-1/15 $31.00 © 2015 IEEE

Figure 2 shows a design of 3D sheath flow device for fabrication of artificial capillary vessels. Since the device has sheath flow microchannels with three kinds of height, double-coaxial laminar flow can be achieved by narrowing core flow from four directions of the left, right, top and bottom. The detailed principle of double-coaxial laminar flow formation and two-layer coaxial Core-Sheath microfiber fabrication is shown in Figure 3. At first, the core flow of cell suspension is focused on the bottom by the first sheath flow of alginate solution. Then, the core flow is lifted up and centered by the second sheath flow. Finally, a gel fiber is fabricated in the pool by gelation of alginate solution with calcium chloride aqueous solution supplied from downstream channel.

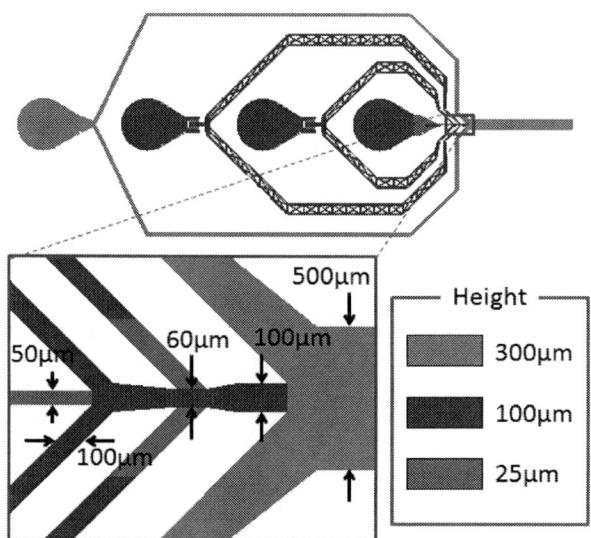

Figure 2: Design of 3D sheath flow device for fabrication of artificial capillary vessels

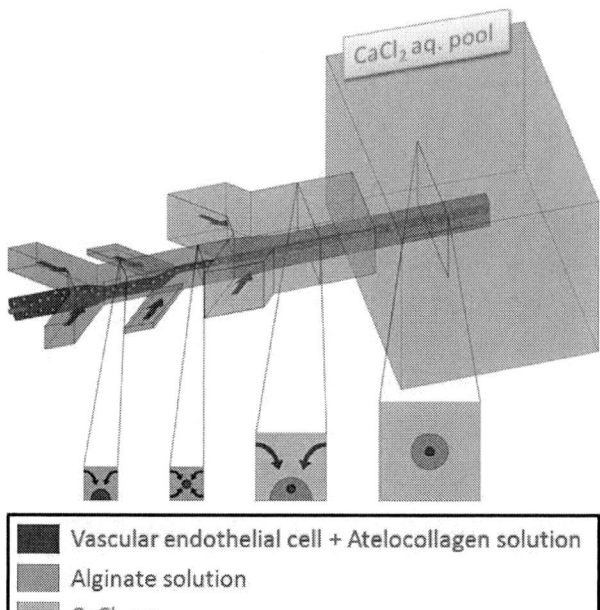

Figure 3: Principle of double-coaxial laminar flow formation and double-coaxial Core-Sheath microfiber fabrication

EXPERIMENT

Device Fabrication

Device fabrication process is shown in Figure 4. In order to fabricate microchannels with three kinds of height, mold of the device was fabricated by three steps of photolithography using SU-8 (SU-8 3050, MicroChem). At first, SU-8 is spin-coated on a silicon substrate, and exposure and development are performed (Figure 4(a)). Then, SU-8 spin coat of the second layer and exposure are carried out (Figure 4(b)), and the process is performed again for the third layer (Figure 4(c)). After the mold was fabricated by development (Figure 4(d)), the PDMS base resin and curing agent are mixed in 10:1 ratio and poured onto the SU-8 mold. Following degassing process, they are cured at 75 degrees for 45 minutes (Figure 4(e)), and the structure of PDMS is bonded with a PDMS coated glass substrate by O_2 plasma treatment after PDMS replicating from the mold (Figure 4(f)).

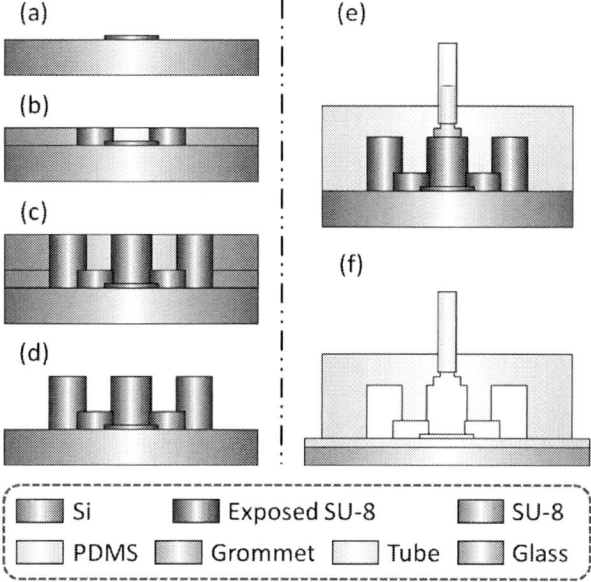

Figure 4: Device fabrication process; (a) First layer patterning (25 μm heights), (b) Second layer coat and exposure (100 μm heights), (c) Third layer coat and exposure (300 μm heights), (d) Development, (e) PDMS molding, and (f) plasma bonding

Experimental Setup

For fluidic experiments to check whether 3D sheath flow can be formed, syringe pumps (KDS210, KD Scientific) were used to control volumetric flow rates of core flows and sheath flows. In order to evaluate 3D sheath flow formation, a confocal laser scanning microscope (TCS SL, Leica) which can observe the cross-section of microchannels was employed.

In microfiber fabrication experiment, syringe pumps (Pump 11 Elite, Harvard Apparatus) were used as instruments for controlling the diameter of microfibers by adjusting flow rates. For observation of the fabricated gel microfiber, optical and fluorescence microscopes (ECLIPSE TE300 and Ti-E, Nikon) were employed.

Materials

In 3D sheath flow formation experiment, we used 75% glycerin solution and 100% glycerin as core and sheath flows, respectively. 75% glycerin solution has almost same viscosity as atelocollagen solution used as the core in actual microfiber fabrication. Similarly, the viscosity of 100% glycerin is almost equal to that of alginate solution employed as the first/second sheath. Additionally, 75% glycerin solution of the core was dyed by the Rhodamine B for cross-sectional flow observation of microchannels with a confocal laser scanning microscope.

On the other hand, in microfiber fabrication experiments, atelocollagen, fibrinogen, bovine aortic endothelial cells (BAECs), NIH-3T3 cells (mouse embryonic fibroblast cell lines), alginate solution, calcium chloride solution, and thrombin solution containing Hank's Balanced Salt Solution (HBSS) were employed. For the core, atelocollagen and fibrinogen with BAECs were used. Since fibrinogen changes to fibrin by chemical reaction with thrombin and facilitate vascularization [11], it was also chosen with alginate solution as materials of the first/second sheath. In order to gel alginate, calcium chloride solution was employed as the third sheath.

RESULT AND DISCUSSION
3D Sheath Flow Formation

The cross-sectional flow observation result to check whether 3D sheath flow can be formed using the proposed device is shown in Figure 5. 75% glycerin solution dyed by Rhodamine B and 100% glycerin were chosen as core and sheath flows considering the viscosity of actual application. White square of the confocal microscopic image shows the microchannel. As expected, it turned out that the core is narrowed down toward the center by the sheath, and the core about 20 μm in diameter was able to be obtained. Since the core flow was located in the center of microchannel, it was confirmed that expected 3D sheath flow was formed in the proposed device.

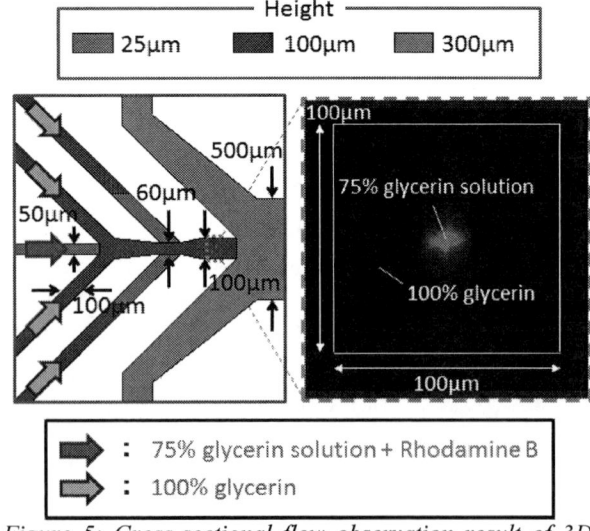

Figure 5: Cross-sectional flow observation result of 3D sheath flow formation experiment (White square; microchannel)

Microfiber Fabrication for Artificial Capillary Vessels

We applied 3D sheath flow device to form long single layer cell tubular structure microfiber using vascular endothelial cells with atelocollagen as the core, alginate solution as the first/second sheath, and calcium chloride solution as the third sheath. Figure 6 shows the microscopic images of cell microfiber after 1 day (a) and 3 days (b). About a 3-centimeter-long microfiber was fabricated, and the embedded BAECs were fused mutually in a line. Then, the proposed device was applied to form double layer cell tubular structures. Since an original capillary vessel has pericytes around vascular endothelial cells, NIH-3T3 cells instead of pericytes were introduced to sheath flow as well as BAECs to core flow. Hoechst stained BAECs with atelocollagen solution were used as the core while NIH-3T3 cells with alginate solution and calcium chloride solution were employed as the first/second sheath and the third sheath, respectively. Figure 7 shows fluorescence images of the fabricated double layer cell structure. As a result, embedding rate of BAECs to the core layer was 70.4%, and localization of each cell was successfully confirmed.

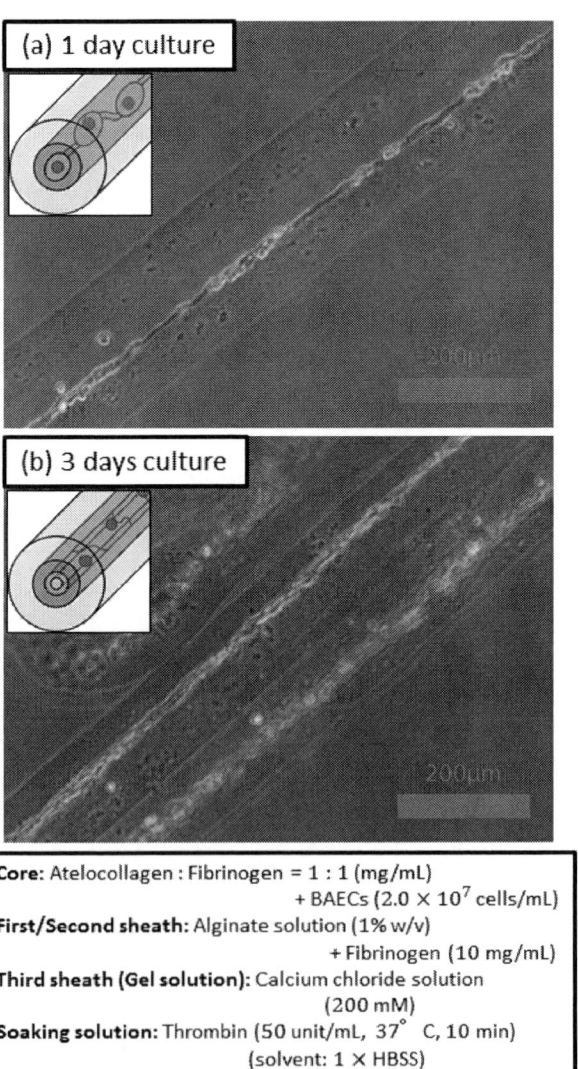

Core: Atelocollagen : Fibrinogen = 1 : 1 (mg/mL)
+ BAECs (2.0 × 10⁷ cells/mL)
First/Second sheath: Alginate solution (1% w/v)
+ Fibrinogen (10 mg/mL)
Third sheath (Gel solution): Calcium chloride solution (200 mM)
Soaking solution: Thrombin (50 unit/mL, 37° C, 10 min)
(solvent: 1 × HBSS)

Figure 6: Microscopic images of the fabricated cell microfiber after 1 day (a) and 3 days (b)

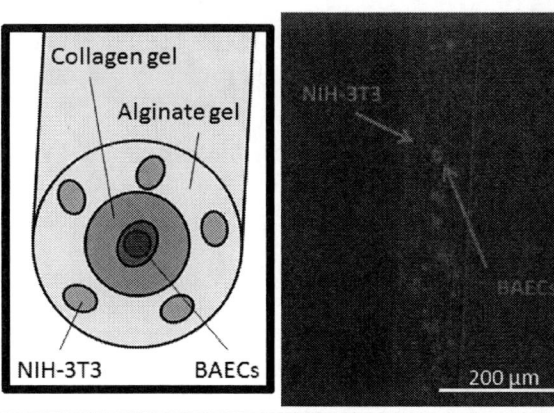

Core: Atelocollagen (2 mg/mL)
+ Hoechst stained BAECs (2.0×10^7 cells/mL)
First/Second sheath: Alginate solution (1% w/v)
+ NIH-3T3 (1.0×10^7 cells/mL)
Third sheath (Gel solution): Calcium chloride solution
(200 mM)

Figure 7: Fluorescence images of the fabricated double layer cell structure

CONCLUSION

We designed and fabricated a functional 3D sheath flow device for fabrication of artificial capillary vessels. In order to fabricate double-layer coaxial Core-Sheath microfibers applicable for long micro capillary vessels, the PDMS-fabricated device which can control core and sheath flows in three dimensions by three sheath injections was used. Through observation of the cross-section of microchannels with a confocal laser scanning microscope, it was confirmed that 3D sheath flow was formed. Consistently, in microfiber fabrication, double-coaxial laminar flow can be achieved by optimized two step alginate solution injection and successive calcium chloride solution injection. Furthermore, about a 3-centimeter-long microfiber with BAECs fused mutually in the center was fabricated through three days culture. Then, in order to apply proposed method for double layer cell tubular structures construction, first step verification toward the fabrication of artificial capillary vessels was also performed. As a result, 3D sheath flow including cells in each layer was maintained stably, and double layer cell structures of core and sheath were formed successfully.

ACKNOWLEDGEMENTS

This work is partly supported by Japan Ministry of Education, Culture, Sports, Science and Technology Grant-in-Aid for Scientific Basic Research (S) No. 23226010. The authors thanks for Nanotechnology Platform and Center for Advanced Biomedical Sciences of Waseda University for their technical assistances.

REFERENCES

[1] K. Shimizu, H. Fujita, and E. Nagamori, "Alignment of skeletal muscle myoblasts and myotubes using linear micropatterned surfaces ground with abrasives", *Biotechnology and Bioengineering*, Vol. 103, pp. 631-638, 2009.

[2] Michel R. Hoenig, Gordon R. Campbell, Barbara E. Rolfe, and Julie H. Campbell, "Tissue-engineered blood vessels: Alternative to autologous grafts?", *Arteriosclerosis, Thrombosis, and Vascular Biology*, Vol. 25, pp. 1128-1134, 2005.

[3] Srivatsan Raghavan, Celeste M. Nelson, Jan D. Baranski, Emerson Lim, and Christopher S. Chen, "Geometrically Controlled Endothelial Tubulogenesis in Micropatterned Gels", *Tissue Engineering Part A*, Vol. 16, pp. 2255-2263, 2010.

[4] S. Sugiura, T. Oda, Y. Aoyagi, M. Satake, N. Ohkohchi, and M. Nakajima, "Tubular gel fabrication and cell encapsulation in laminar flow stream formed by microfabricated nozzle array", *Lab Chip*, Vol. 8, pp. 1255-1257, 2008.

[5] E. Kang, S. Shin, K. Lee, and S. Lee, "Novel PDMS cylindrical channels that generate coaxial flow, and application to fabrication of microfibers and particles", *Lab Chip*, Vol. 10, pp. 1856-1861, 2010.

[6] J. Su, Y. Zheng, and H. Wu, "Generation of alginate microfibers with a roller-assisted microfluidic system", *Lab Chip*, Vol. 9, pp. 996-1001, 2009.

[7] K. Hirayama, D. Kiriya, H. Onoe, and S. Takeuchi, "Biofilms in Hydrogel Core-shell Fibers" *Proc. MEMS 2011*, pp. 845-848, 2011.

[8] W. Lan, S. Li, Y. Lu, J. Xu, and G. Luo, "Controllable preparation of microscale tubes with multiphase co-laminar flow in a double co-axial microdevice", *Lab Chip*, Vol. 9, pp. 3282-3288, 2009.

[9] T. Takei, S. Sakai, H. Ijima, and K. Kawakami, "Development of mammalian cell-enclosing calciumalginate hydrogel fibers in a co-flowing stream", *Biotechnology Journal*, Vol. 1, pp. 1014-1017, 2006.

[10] Hitomi Oku, Yutaro Nakamura, Rui Sekine, Junichi Ito, Yoon Donghyun, Tetsushi Sekiguchi, Shuichi Shoji and Naoya Takeda, "Controllable Three-Dimensional Sheath Flow Microfluidic Device Focusing and Embedding Cells in Center of Double Layered Hydrogel Fiber for Tissue Engineering Applications", *International Conference on BioSensors, BioElectronics, BioMedical Devices, BioMEMS/NEMS and Applications 2013 (Bio4Apps 2013)*, 2013.

[11] Ayelet Lesman, Jacob Koffler, Roee Atlas, Yaron J. Blinder, Zvi Kam, and Shulamit Levenberg, "Engineering vessel-like networks within multicellular fibrin-based constructs", *Biomaterials*, Vol. 32, pp. 7856-7869, 2011.

CONTACT

*J. Ito, tel: +81-3-5286-3384;
j_ito@shoji.comm.waseda.ac.jp

MODELING NON-WETTING PERFORMANCES OF SUPERLYOPHOBIC SURFACES BASED ON LOCAL CONTACT LINE

Zhiwei Wang, and Tianzhun Wu[*]

Shenzhen Institutes of Advanced Technology, Chinese Academy of Sciences, Shenzhen, CHINA

ABSTRACT

Superlyophobic surfaces (SLS) are promising as the novel universal platform for microfluidics due to the unique super-repellency and extremely low adhesion for almost all liquids, however, there are no adequate models which can predict well their non-wetting performances, especially the pressure stability and the contact angle hysteresis (CAH). This paper proposes a pressure stability criterion and a CAH analytic expression based on the local contact line analysis, which are in excellent agreement with the experimental results. These achievements enable precise non-wettability design for both water and oil in digital microfluidics.

INTRODUCTION

Biologically inspired superhydrophobic surfaces (SHS), which show high apparent contact angles (CAs) of typically above 150° and low contact angle hysteresis (CAH) for water and other aqueous solutions, have been elaborately investigated in the past decades [1]. However, usually SHS do not repel low surface tension liquids, such as oil, organic solutions, and aqueous solutions with surfactants, and they are apt to be contaminated by oil and organic solutions, leading to the loss of water-repellent performance [2]. As a extension and enhancement of the traditional SHS, superlyophobic surfaces (SLS) can repel almost any liquid and provide high apparent CA and low CAH, hence they can effectively minimize liquid waste, prevent contamination and reduce flow resistance, for which they have attracted much interest in recent years [3]. Due to their excellent non-wetting performances, SLS provide a promising versatile platform for the manipulation of various fluids [4], and hence enable many potential applications, such as self-cleaning, anti-bacteria, chemical shielding, precise control of droplet shape and motion, and generation of droplet arrays [3].

Although there are many attractive characteristics of SLS and great progresses have been achieved on the design and fabrication of SLS, engineering and practical application of SLS are still highly limited nowadays. One bottleneck is new fabrication or synthesis techniques have to be developed for particular materials and well controlled geometrics is necessary for SLS, which was unblocked recently in our group using a facile and inexpensive microfabrication approach for high-performance and mass-production of SLS on various curable materials [5]. Another bottleneck is the non-wetting performances, especially the pressure stability (the robustness of the Cassie-Baxter state) and CAH. There have been several preliminary studies in the evaluation or prediction of the non-wetting performances, for example, Nguyen et al. [6] proposed a simple model by assuming the curvature radius is a linear function of pillar space and pillar length to predict the critical impalement pressure; Our group also [4] introduced a normal stress to represent the solid-liquid interaction and proposed two dimensionless factors to predict the droplet wetting state on SLS. However, the non-wetting performances are still difficult to predict exactly, which hindered the design and fabrication of SLS with controllable non-wetting performances. Hence herein we modeled the non-wetting performances of SLS, including the pressure stability and CAH, based on the analysis of local contact line. We first derive a uniform expression to predict the maximum Laplace pressure on SLS with various overhang surface structures. Then based on this, we present a criterion to predict the pressure stability of SLS by introducing two dimensionless parameters. At last, we propose a new approach to model the CAH on SLS by introducing contact line fraction. These contributions may be helpful to understand the wetting mechanism and also shed new light for design and fabrication of high-performance SLS for various applications.

PRESSURE STABILITY ANALYSIS

Two Failure Modes for SLS

Nowadays, almost any random or uniform micro/nano structures, such as "T" structure, inversed trapezoid structure, sphere structure, etc., can be fabricated through various MEMS based processes [3]. According to the design criteria of SLS, several requirements, such as overhang structure, minimum suspending height, pinning condition, curvature requirement, and so on, should be satisfied [7]. For the suspending droplets on SLS with overhang micro/nano surface structures, the hydrostatic pressure and downward Laplace pressure induced by the convex droplet top are balanced by the upward Laplace pressure caused by surface geometries [7]. Liquid-air interface should have convex meniscus in micro/nano cavities to provide an upward Laplace pressure, hence overhang or re-entrant structures are necessary for SLS to support droplets. Nevertheless, overhang or re-entrant structure is not a sufficient condition for SLS and there will be transition from Cassie-Baxter state to Wenzel state on some surfaces with overhang or re-entrant structure. There are two fundamental failure modes for SLS [8], the T^* failure mode, in which the local contact angle gets too large for equilibrium and the liquids slide down along the side wall of the structure, and the H^* failure mode, in which the bending liquid-air interface touches the bottom of structure and leading to full wetting. Essentially, H^* failure mode can be avoided given sufficient height and stiffness of micro/nano structures. Hence, the key issue in improving the pressure stability is to prevent T^* failure. According to Laplace pressure equation, $P_{La} = 2\gamma/R$, for a specific liquid, the only way to increase P_{La} is to reduce curvature radius R of the meniscus interface on surface cavity structures. Hence it is crucial for predicting Laplace pressure to obtain the expression of curvature radius R. For a specific surface, intrinsic contact angle is constant, therefore ideally R can be determined by geometric dimensions of structures and the pinning angle, which is constrained by the Gibbs inequalities [7].

Unified Pressure Expression

Since the concave profile for overhang structures is essentially unstable [7], there are three different types of side profiles for overhang micro/nano structures: vertical, incline, and convex, as shown in Figure 1. By analyzing the pinning of liquid on the "T" structure, inversed trapezoid structure, and sphere structure separately, based on the Laplace pressure formulation and the pinning condition on these typical overhang microstructures, we can obtain a uniform expression for curvature radius:

$$R = \frac{L/2 - D/2}{\sin(\beta/2)} = \frac{L-D}{2\sin(\theta - \varphi)} \quad (1)$$

Hence the maximum Laplace pressure can be expressed as:

$$P_{La}^{max} = 2\gamma \left(\frac{\sin(\theta_x - \varphi_x)}{(L_x - D_x)} + \frac{\sin(\theta_y - \varphi_y)}{(L_y - D_y)} \right) \quad (2)$$

Where θ is the pinning angle, φ is the slope angle, L and D are respectively the pitch distance and cap size of the surface structures on the contact horizontal plane. The subscript x and y represents two orthogonal directions on the surface. According to the Gibbs inequalities [7], the pinning angle θ in Figure 1 is confined by the Young's contact angle, $\theta \leq \theta_Y$. While on the other hand, Laplace pressure is the only upward force that supports the Cassie-Baxter state droplet, hence the pinning angle θ should be larger than the slope angle φ, $\theta > \varphi$, to keep convex meniscus liquid-air interface in the micro/nano cavity to provide upward Laplace pressure. So we have:

$$\varphi < \theta \leq \theta_Y \quad (3)$$

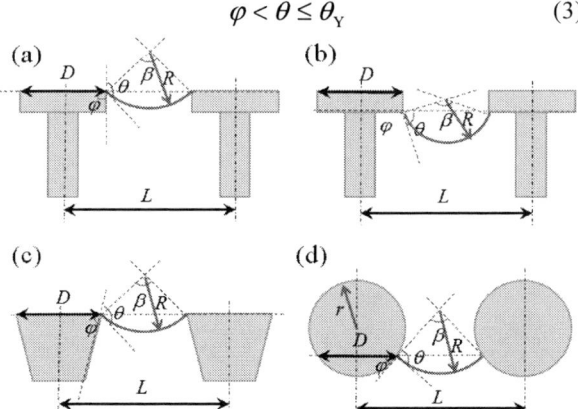

Figure 1: Pinning of contact line on typical overhang surface structures. (a) upper edge of "T" structure, (b) lower edge of "T" structure, (c) inversed trapezoid structure, (d)sphere structure.

Criteria for Pressure Stability

Since Laplace pressure is the only upward force that supports the Cassie-Baxter state droplet, the uniform expression Eq. (2) can be used for the prediction of the critical impalement pressure. Hence for the two failure modes namely the T^* and H^* failure modes, we proposed two dimensionless parameters as criteria for the Cassie-Baxter state:

$$P^* = \ln(P_{La}/P_{down}) \quad (4)$$

for the T^* failure mode, $P_{down} = \rho g h_{drop}$ is the downward hydrostatics pressure [4], where h_{drop} is the height of droplet, and

$$H^* = \ln(H_D/H_S) \quad (5)$$

for the H^* failure mode, where H_D and H_S are the actual height of micro/nano structures and the maximum suspension height of liquid. Clearly the necessary but insufficient condition for droplets to be in Cassie-Baxter state should be $P^* > 0$ and $H^* > 0$. From the uniform expression of curvature radius R in Eq. (1), the maximum suspension height of liquid H_s can be expressed as:

$$H_S = R(1 - \cos(\beta/2)) = (L-D)\frac{1 - \cos(\theta - \varphi)}{2\sin(\theta - \varphi)} \quad (6)$$

Experimental Verification

In order to validate the pressure stability analysis above, we fabricated T-shape micro structures SLS on Si substrate using deep reactive ion etching (DRIE), as shown in Figure 2(a). By optimizing DRIE process conditions, we can obtain highly uniform overhang micro structures with small undercuts (~300 nm). Linear, diamond, as well as square patterns with a series of different cap sizes and pitch distances is fabricated as illustrated in Figure 2(b-d). The cap sizes ranged from 3 μm to 30 μm and pitch distances from 30 μm to 360 μm with etching height varied from 16 μm to 25 μm.

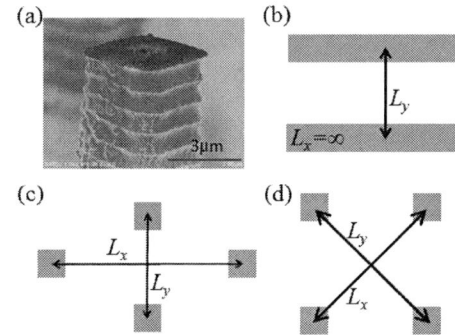

Figure 2: Geometric layout fabricated by deep reactive ion etching on Si used for pressure stability analysis. (a) SEM images of the 3D SLS microstructures with slight undercuts. (b) linear, (c) diamond and (d) square layouts of the SLS microstructures.

Both deionized water ($\gamma = 73$ mN/m, $\rho = 1000$ kg/m^3) with a volume of about 5 μL and hexadecane ($\gamma = 27$ mN/m, $\rho = 770$ kg/m^3) with a volume of about 2 μL was used as probe liquids. We assume the fabricated micro surface structures are perfectly T-shape, hence deionized water will be pinned on the upper edge of the T-shape with $\varphi = \pi/2$ and hexadecane will be pinned on the lower edge of the T-shape with $\varphi = 0$, as seen in Figure 1(a-b). The wetting states of water and hexadecane droplets on various SLS are plotted on Figure 3. For SLS with $H^* < 0$, all the droplet are in Wenzel state, because $H^* < 0$ means the etching height of micro/nano structures is smaller than the required height for liquid pinning, and results in H^* failure. All droplets should be in Cassie-Baxter state for $H^* > 0$ and $P^* > 0$ in an ideal situation, while experimental results indicated that droplet can be on the Cassie-Baxter state only when $H^* > 0$ and $P^* > 3$, which is because SLS with a small P^* is still fragile to disturbances acted on the droplet and effective pressure brought by the droplet impact. Hence the droplets on such SLS tends to irreversible transition from the Cassie state to the Wenzel state. Therefore, in our experiments P^* should be larger than about 3 for SLS.

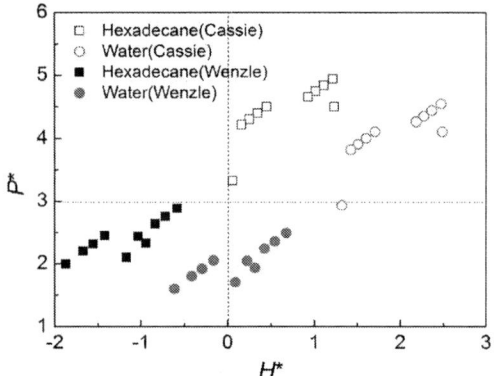

Figure 3: The wetting state of droplets on SLS of linear, diamond and square layouts at different P^* and H^*. Experimental results indicate droplet can be on the Cassie state only when $H^*>0$ and $P^*>3$.

CONTACT ANGLE HYSTERESIS ANALYSIS
Previous CAH Models Based on Solid Fraction

CAH, defined as the differences between the minimum receding and the maximum advancing contact angles, corresponding to the force that draws back droplet motion. It is very important for the practical application of SLS and has attracted plenty of research interest. Most existing CAH models are based on the solid fraction f_s, in which a widely used CAH model is proposed by Reyssat and Quéré [9]. They investigated the CAH generated by strong dilute defects and introduced a spring stiffness to quantitative characterize the energy stored in the distorted TCL, thus CAH is expressed as:

$$\Delta \cos \theta = a / 4 f_s \ln(\pi / f_s) \qquad (7)$$

where a is a coefficient depending on the detail of the line distortion and the experimental results of Reyssat and Quéré on strong dilute cylindrical pillar arrays suggests $a=3.8$ [9].

Experimental fabrication

Considering the fact that surfaces with same f_s and different TCL exhibit different advancing and receding contact angles [10], we suppose the existing CAH models may be defective for non-uniform surfaces. Hence we fabricated SLS of uniformly distributed and non-uniformly distributed discrete-ring layouts with different δ and $\Delta\delta$ (see Figure 4). In present experiments, $\Delta\delta$ ranges from $2°$ to $3°$ and δ ranges from $6°$ to $30°$.

The discrete-ring layouts depicted in Figure 4 are ideal choices for the TCL investigation since contact areas of droplet on the surface is circular and therefore it is easy to obtain the contact length proportion of TCL along the rings. In present experiments, for each sample, contact angles were measured at least 10 times with a drop shape analysis system (KRUSS, DSA 25) in the laboratory environment

(23±2°C, humidity 50±5%) with a standard deviation of less than 2°. We found that Eq. (7) cannot fit our experimental results and cannot give a meaningful fit in Figure 5, which indicates the failure of the model by Reyssat and Quéré [9] for the discrete-ring layouts.

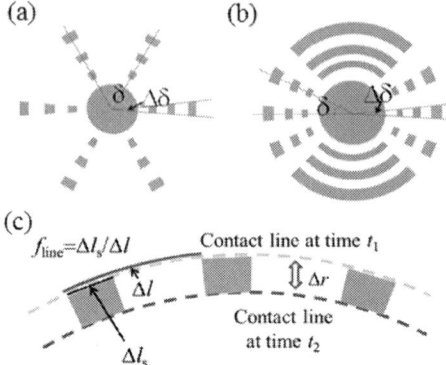

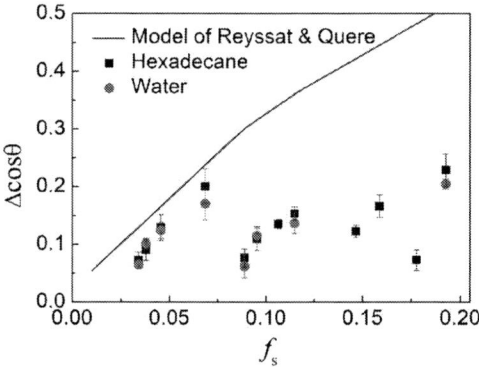

Figure 4: Geometric layouts used for CAH analysis. (a) uniformly distributed and (b) non-uniformly distributed discrete-ring layout. (c) definition of contact line fraction f_{line} and the contact line motion.

Figure 5: Relation between the commonly used f_s and measured CAH on the SLS of discrete ring layouts [9]. There is no meaningful function between CAH and f_s.

Present CAH Models Based on Contact Line Fraction

Since CA is merely determined by the TCL [11], we expect to clarify the relationship between CAH and the TCL details. For the unit micro structure with length Δl shown in Figure 4(c), the flow resistance caused by CAH acted on the solid part Δl_s only, hence we have:

$$F_r = \gamma \Delta \cos \theta \Delta l_s \qquad (8)$$

Denoting ε as the maximum energy stored in unit surface micro/nano structure with length Δl, hence the force represented by the energy stored in the distorted TCL can be expressed as:

$$F_s = \varepsilon / \Delta l \qquad (9)$$

At critical point just before droplet motion, the flow resistance should be equal to the force represented by the energy stored in the distorted TCL, $F_r = F_s$, so we have:

$$\gamma \Delta \cos \theta \Delta l_s = \varepsilon / \Delta l \qquad (10)$$

For dilute and small micro/ nano structures, $\Delta l_s << \Delta l$ and $\Delta l << R_{\text{CP}}$, where R_{CP} is radius of the contact perimeter, so refer to the analysis of Reyssat and Quéré [9], the energy stored in unit surface micro/nano structure can be expressed as:

$$\varepsilon = 1/2Ku^2 = 1/4a\gamma\Delta l_s^2 \ln(2\Delta l/\Delta l_s) \quad (11)$$

Considering CAH should be dependent on TCL other than f_s. We define the contact line fraction f_{line} as

$$f_{\text{line}} = \Delta l/\Delta l_s \quad (12)$$

Finally, using Eq. (10-12), we can obtain the following expression to predict the CAH of SLS:

$$\Delta\cos\theta = a/4f_{\text{line}}\ln(2/f_{\text{line}}) \quad (13)$$

The relation between contact line fraction f_{line} and CAH are plotted on Figure 6, in which the prediction from Eq. (13) fits the measured CAH results fairly well when $a = 1.2$. The good agreement with experimental measurement indicates the applicability of Eq. (13) to predict the CAH on SLS with micro-structured surfaces. From the results we can conclude the CAH on SLS is depended on the local contact line fraction and detailed line distortion only, and the liquid surface tension has little influence on the CAH, at least in present discrete-ring layouts. The detailed line distortion in other layouts is still not clear and calls for further investigations.

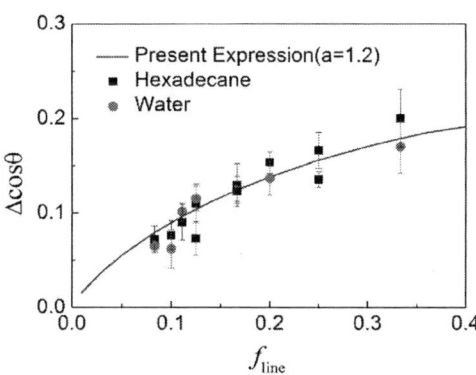

Figure 6: Relation between f_{line} and measured CAH. Prediction from present expression fits the results fairly well when a=1.2.

CONCLUSIONS

In this paper we modeled the pressure stability and CAH from the aspect of local contact line to exactly predict the non-wetting performance of SLS. We first derivate a uniform expression to predict the maximum Laplace pressure on various typical overhang surface structures and we further investigated the pressure stability on SLS based on the expression. By introducing two dimensionless parameters, P^* and H^*, a criterion for SLS was proposed and validated by experiment of DRIE on Si substrate.

Considering contact angle is determined by surface property in the vicinity of the contact line and not exactly on the global characteristics under contact area, CAH should be dependent on the contact line fraction f_{line} other than the commonly used f_s. Hence we proposed an analytic formulation for CAH of SLS by introducing f_{line} and found the line distortion coefficient of 1.2 fits experimental results fairly well for both water and hexadecane.

The contributions are helpful in overcoming the difficulties of quantitative prediction and evaluation the performance of SLS, as well as understanding the underlying wetting mechanisms. The results also shed new light for design and fabrication of high-performance SLS for microfluidics and other fields.

ACKNOWLEDGEMENTS

This research was supported by National Science Foundation of China (No. 51105388, 51406221, 51475451) and China Postdoctoral Science Foundation funded project (No. 2014M552253).

REFERENCES

[1] E. Celia, T. Darmanin, E. T. de Givenchy, S. Amigoni, F. Guittard, "Recent advances in designing superhydrophobic surfaces", *Journal of Colloid and Interface Science*, vol. 402, pp.1-18, 2013.

[2] A. Tuteja, W. Choi, M. Ma, J. M. Mabry, S. A. Mazzella, G. C. Rutledge, G. H. McKinley, R. E. Cohen, "Designing Superoleophobic Surfaces", *Science*, vol. 318(5856), pp. 1618-1622, 2007.

[3] K. Liu, Y. Tian, L. Jiang, "Bio-inspired superoleophobic and smart materials: Design, fabrication, and application", *Progress in Materials Science*, vol. 58(4), pp. 503-564, 2013.

[4] T. Wu, Y. Suzuki, "Engineering superlyophobic surfaces as the microfluidic platform for droplet manipulation", *Lab on a Chip*, vol. 11(18), pp. 3121-3129, 2011.

[5] L. Yuan, T. Wu, W. Zhang, S. Ling, R. Xiang, X. Gui, Y. Zhu, Z. Tang, "Engineering superlyophobic surfaces on curable materials based on facile and inexpensive microfabrication", *Journal of Materials Chemistry A*, vol. 2(19), pp. 6952-6959, 2014.

[6] T. P. N. Nguyen, P. Brunet, Y. Coffinier, and R. Boukherroub, "Quantitative Testing of Robustness on Superomniphobic Surfaces by Drop Impact", *Langmuir*, vol. 26(23), pp. 18369-18373, 2010.

[7] T. Wu, Y. Suzuki, "Design, microfabrication and evaluation of robust high-performance superlyophobic surfaces", *Sensors and Actuators B: Chemical*, vol. 156(1), pp. 401-409, 2011.

[8] A. Cavalli, P. Bøggild, F. Okkels, "Parametric Optimization of Inverse Trapezoid Oleophobic Surfaces", *Langmuir*, vol. 28(50), pp. 17545-17551, 2012.

[9] M. Reyssat, D. Quéré, "Contact Angle Hysteresis Generated by Strong Dilute Defects", *The Journal of Physical Chemistry B*, vol. 113(12), pp. 3906-3909, 2009.

[10] L. Gao, T. J. McCarthy, "Reply to 'comment on how Wenzel and Cassie were wrong by Gao and McCarthy' ", *Langmuir*, vol. 23(26), pp. 13243-13243, 2007.

[11] L. Gao, T. J. McCarthy, "How Wenzel and Cassie Were Wrong", *Langmuir*, vol. 23(7), pp. 3762-3765, 2007.

CONTACT

*Tianzhun Wu, tel: +86-755-86392339; tz.wu@siat.ac.cn

MICROFLUIDIC SELECTION OF APTAMERS USING COMBINED ELECTROKINETIC AND HYDRODYNAMIC MANIPULATION

Tim Olsen[1], Jing Zhu[1], Jinho Kim[1], Renjun Pei[2], Milan N. Stojanovic[3], and Qiao Lin[1]

[1]Department of Mechanical Engineering, Columbia University, New York, NY, USA
[2]Suzhou Institute of Nano-Tech and Nano-Bionics, Chinese Academy of Sciences, Suzhou, China
[3]Department of Medicine, Columbia University, New York, NY, USA

ABSTRACT

This paper presents a microfluidic device that integrates the entire process of aptamer selection *via* bead-based biochemical in which affinity selected target-binding oligomers are electrokinetically transferred for amplification, while the amplification product is transferred back for affinity-selection via pressure-driven fluid flow. This hybrid oligomer manipulation leverages the advantages of both electrokinetic and pressure-driven oligomer manipulation by reducing the amount of flow control elements (e.g. valves) while avoiding exposure of the target to electric field induced, potentially structure altering, environmental conditions (e.g. pH). Efficient manipulation of reagents and reaction products is achieved through bead-based affinity selection and amplification, which, combined with hybrid transfer, allows integration of all steps of the SELEX process on a single chip. The microfluidic device is thus capable of closed-loop, multi-round aptamer enrichment without manual intervention or use of off-chip instruments, as demonstrated by selection of DNA aptamers against the protein IgE with high affinity ($K_D = 12$ nM) in a rapid manner (4 rounds in 10 hours).

INTRODUCTION

Aptamers are affinity receptors isolated from large randomized libraries of oligonucleotides (oligomers) through an *in-vitro* process of iterative affinity-selection and amplification, termed systematic evolution of ligands by exponential enrichment (SELEX). Aptamers have been developed for a variety of targets such as small molecules, peptides, amino acids, proteins, cells, viruses, and bacteria, with broad applications to basic biological sciences, drug discovery, and clinical diagnostics and therapeutics Recently microfluidics has been applied to improve efficiency and minimize assay time for the SELEX process. However, existing devices have limitations related to the use of multiple buffers and off-chip amplification of oligomers [1, 2]. Our own recent work addressed these issues using solid-phase capture and amplification of oligomers, which were manipulated using electrophoresis [3] exposing targets to potentially target compromising electric fields or pressure-driven fluid flow [4] that required a significant number of flow-control components. This paper presents a device that uses a hybrid approach, in which affinity selected target-binding oligomers are electrokinetically transferred to another chamber where they are amplified by polymerase chain reaction (PCR), and the amplified oligomers are transferred back using pressure-driven flow for further affinity selection. This hybrid approach leverages the advantage of both electrokinetic and pressure-driven flow methods while limiting the drawbacks of using one alone by avoiding the exposure of target proteins to unfavorable electric fields

while reducing the number of flow components, thereby simplifying the device design and improving the operation efficiency.

PRINCIPLE AND DESIGN

The iterative, closed-loop isolation and amplification process for the isolation of target-binding oligomers is achieved in two microchambers coupled by electrokinetic and hydrodynamic transfer methods (**Fig. 2**).

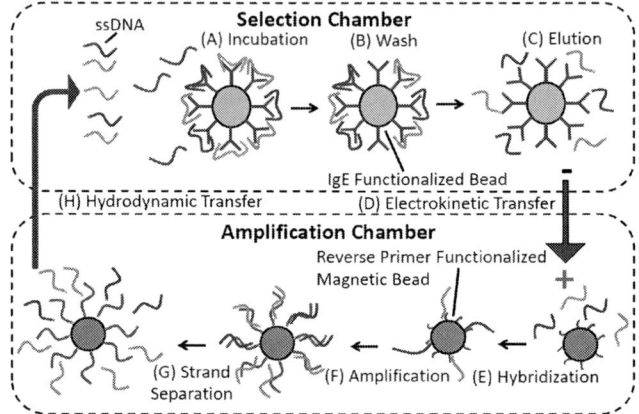

Figure 1: Principle of microfluidic aptamer selection: (A) ssDNA with random sequence binds to IgE-functionalized beads in the selection chamber. (B) Weak binders are removed by washing; (C) strong binders are thermally eluted and (D) transferred to the amplification chamber through electrokinetics. (E) The strands are captured by magnetic beads with surface-immobilized reverse primers and (F) amplified through PCR. (G) The amplified single strands are thermally released from bead surfaces and (H) transported back to the selection chamber through pressure-driven flow for the next round of affinity-selection.

Briefly, target-functionalized beads are introduced into the selection chamber of the device and immobilized on a weir structure. Single-stranded DNA (ssDNA) oligomers are then infused into the device where they bind to the target-immobilized beads. Weakly binding oligomers are removed by multiple buffer washes. Next, primer-functionalized magnetic beads are introduced into the amplification chamber and held via an external magnet. The remaining oligomers strongly bound to the target-functionalized beads are then thermally eluted and transferred to the amplification chamber through electrokinetics where they hybridize to the reverse primers on the magnetic bead surfaces. PCR reagents are then introduced and the magnetic bead bound oligomers are amplified through PCR. Next, the target-functionalized beads in the selection chamber are replaced with new beads. The amplified oligomer product is thermally

978-1-4799-7956-1/15 $31.00 © 2015 IEEE

released from the beads and transferred through modest pressure-driven flow for further affinity-selection using the replenished bead-immobilized target molecules which have not been subjected to any potential electrolysis-induced damage.

The device consists of two (selection and amplification) microchambers each of 1.7 uL volume. The microchambers are equipped with electrode ports for the insertion of platinum wires which generate an electric field for electrokinetics. The selection microchamber features a weir structure for capturing microbeads. The selection and amplification microchambers are connected *via* two microchannels (Fig. 1a): one filled with agarose gel that allows electrokinetically driven ssDNA migration while preventing bulk flow (Fig. 1b), the other equipped with microvalves actuated by pressurized oil from another layer of channels above (Fig. 1c).

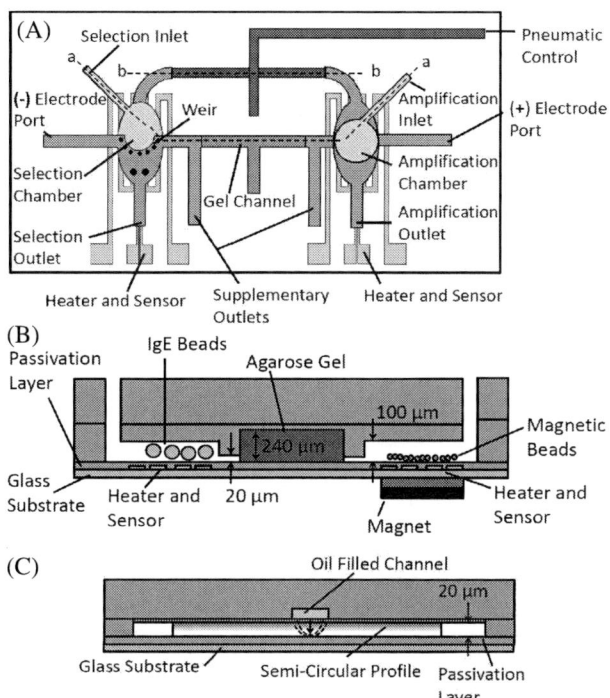

Figure 2: Schematic of the microfluidic aptamer selection device: top view (a), and cross-sectional views along line a-a (B) and line line b-b (C). The selection chamber (1.7 µL) contains IgE-functionalized microbeads retained by a microweir. The amplification chamber (1.7 µL) contains primer-functionalized magnetic beads retained by a magnet. ssDNA strands are transferred by electrophoresis from the selection chamber to the amplification chamber, and reversely by pressure-driven flow.

EXPERIMENTAL

Materials

All chemicals were purchased from Sigma-Aldrich (St. Louis, MO) unless otherwise indicated. Deoxyribonucleotide triphosphates (dNTPs) and GoTaq Flexi DNA polymerase were obtained from Promega Corp. (Madison, WI). Randomized oligomer library (5' – GCC TGT TGT GAG CCT CCT GTC GAA – 45N – TTG AGC GTT TAT TCT TGT CTC CC – 3') and primers (Forward

Primer: 5' – FAM – GCC TGT TGT GAG CCT CCT GTC GAA -3', and Reverse Primer: 5' – dual biotin – GG GAG ACA AGA ATA AAC GCT CAA – 3') were synthesized and purified by Integrated DNA Technologies (Coralville, IA). Human Myeloma Immunoglubulin E (IgE) was purchased from Athens Research and Technology (Athens, GA), and NHS-activated microbeads were purchased from GE Healthcare (Little Chalfont, Buckinghamshire, United Kingdom). Dulbecco's phosphate-buffered saline (D-PBS), and streptavidin coupled magnetic beads (Dynabeads® M-270 Streptavidin) were purchased from Invitrogen (Carlsbad, CA).

Microchip Fabrication

The microfluidic device was fabricated using conventional multi-layer soft-lithography techniques. First, a layer of AZ-4620 positive photoresist (Clariant Corp. Somerville, NJ) was spin-coated on a silicon wafer (Silicon Quest International, Inc., San Jose, CA), exposed to ultraviolet light through photomasks, developed, and baked to form the round-shaped flow channel that can be sealed completely. Then, using the same silicon wafer, SU-8 (MicroChem, Newton, MA) layers were spin-coated and developed to define the flow layer. In parallel, a layer of SU-8 photoresist was patterned on another silicon wafer to establish the control. Meanwhile, chrome (10 nm) and gold (100 nm) thin films were thermally evaporated on a glass slide, patterned through photolithograph, and wet etched to form heaters and temperature sensors. The heater and sensor were passivated by spin-coating PDMS prepolymer solution and curing at 72 °C for 30 minutes.

Subsequently, PDMS prepolymer solution (base and curing agent mixed in a 10:1 ratio) was spin-coated onto the silicon wafer bearing the flow layer, and cured on a hotplate at 72 °C for 15 minutes (Fig. 3C). Another PDMS prepolymer solution was cast onto the control layer silicon wafer and cured on a hotplate at 72 °C for 30 minutes. The resulting control layer PDMS slab was peeled off from the mold, punched to form a pneumatic inlet, and bonded to the PDMS membrane on the silicon mold bearing the flow layer features. The bonded slab was then peeled from the flow layer wafer. After punching inlets and outlets, the slab was bonded to a glass slide bearing the heater and temperature sensor. Finally, molten agarose gel was injected in the gel inlet of the device to and cured at room temperature to form the electrokinetic transfer channel. A fabricated device can be seen in Figure 2.

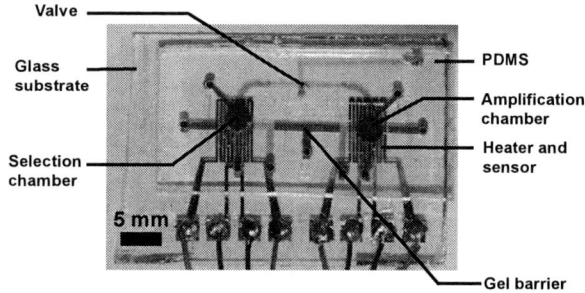

Figure 3: Photograph of a fabricated device.

Experimental Procedure

NHS-activated microbeads are functionalized with protein by incubation with IgE. The functionalized microbeads are then introduced into the selection chamber of the device until approximately 40% of the selection chamber volume was occupied by beads. Selection of oligomers is then performed by infusing randomized library (1 uM) into the device (10 uL/min) for 10 minutes, followed by multiple washes with PBS buffer (20 uL/min) to remove weakly binding oligomers for 15 minutes. Next, primer functionalized magnetic beads are introduced into the amplification chamber of the device and held by an external magnet. Tris-boric acid electrolyte buffer is then injected into the device and platinum wires are inserted into the electrode inlets of each chamber with a 50 V potential difference applied between them for 35 minutes. Meanwhile, strongly bound oligomers remaining in the selection chamber are thermally eluted (50 °C) using the integrated heater and temperature sensor. The 25 V/cm electric field induced by the platinum wires electrokinetically transfers the thermally eluted oligomers to the positive electrode in the amplification chamber where the oligomers then hybridize to the reverse primers immobilized on the magnetic bead surfaces. The platinum wires are removed from the device eliminating the electric field, PCR reagents are introduced into the amplification chamber and bead-based PCR progresses utilizing the heater and temperature sensor located beneath the amplification chamber. A PCR process of 95 °C for 10 seconds, 59 °C for 30 seconds, and 72 °C for 10 seconds is used. After 20 cycles of PCR thermocycling, the IgE-functionalized microbeads are removed from the selection chamber and replaced with new IgE-functionalized microbeads. The valve is then opened and oligomers are released from the bead surfaces by heating to 95 °C. The released oligomers are transported back to the selection chamber through the opened valve via pressure-driven flow (20 uL/min) for further affinity selection with the replenished microbeads. This closed-loop process is repeated for a total of four affinity selections and four 20-cycle PCR amplifications.

RESULTS AND DISCUSSION

To demonstrate multi-round, closed-loop affinity selection and amplification, washing waste from four rounds of selection and the strongly bound thermally eluted ssDNA from the fourth round were collected, amplified (16 cycles PCR) and imaged with gel electrophoresis (Fig. 4). Since the brightness of bands in a gel image represents the amount of oligomers in the eluent loaded in the lane, comparison of the band intensities allowed investigation of the selection process. In the first round some oligomers were in the washing waste after the completion of the washing process as indicated by the presence of a band in lane W19. However, the increase in band intensity from W19 (selection 1, wash 9) from W21 (selection 2, wash 1) suggests that oligomers were successfully eluted from the bead surfaces, electrokinetically transferred and amplified by PCR. The lack of a band in W29 (selection 2, wash 9) shows that weakly bound oligomers were removed from the washing process. The increase in band intensity in from W29 to W31 (selection 3, wash 1) and W39 (section 3, wash 9) to W41 (selection 4, wash 1) suggests that, again, oligomers were successfully thermally eluted, electrokinetically transferred to the amplification chamber, amplified by PCR, and transferred back to the selection

chamber via pressure driven flow. The presence of a band in the elution lane (E), while W49 (selection 4, wash 9) lacks a band, suggests oligomers were strongly bound after four rounds of selection and amplification and were thermally eluted from the bead surfaces.

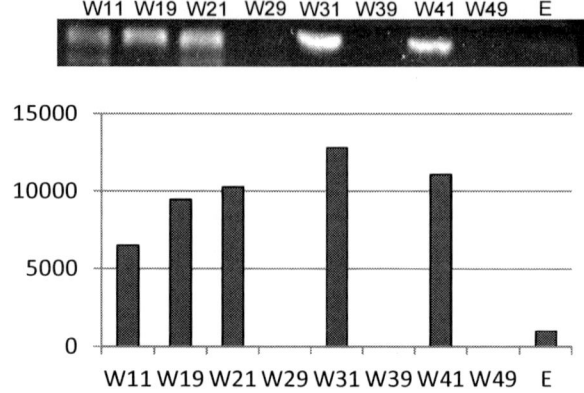

Figure 4: (A) Gel electropherogram of amplified eluents obtained during closed loop selection and amplification. (B) Bar graph depicting intensities, representing the amount ssDNA in the eluent, of lanes W11- W29: Lane W11: selection 1, wash 1; Lane W19: selection 1, wash 9; Lane W21: selection 2, wash 1; Lane W29: selection 2, wash 9; Lane 31: selection 3, wash 1; Lane 39: selection 3, wash 9; Lane 41: selection 4, wash 1; Lane 49: selection 4, wash 9; Lane E: thermally eluted ssDNA.

The enriched aptamer pool collected from the thermal elution of the fourth selection round was further investigated for its affinity and specificity using a fluorescence binding assay [1]. Six different concentrations (100 nM, 50 nM, 25 nM, 12.5 nM, 6.25 nM and 3.125 nM) of fluorescently tagged oligomers (enriched aptamer pool or randomized library) were incubated with IgE-functionalized beads in triplicate 100 μL volumes. After incubating the oligomers with the beads for 30 minutes, the beads were washed and the bound oligomers were thermally eluted (95 °C). The eluted oligomers were collected and their relative amounts were determined with a Wallac EnVision Multilabel Reader fluorescent spectrometer.

When the enriched pool was incubated with IgE-immobilized beads, washed, and bound oligomers were thermally eluted and measured, the fluorescent intensity rapidly increased until reaching an asymptote (Fig. 5). This indicated that the affinity of the enriched oligomer pool considerably improved after the microfluidic SELEX process. Assuming monovalent binding, the dissociation constant (K_D) of the enriched pool was determined to be approximately 12 nM, which is consistent with that of existing IgE aptamers.

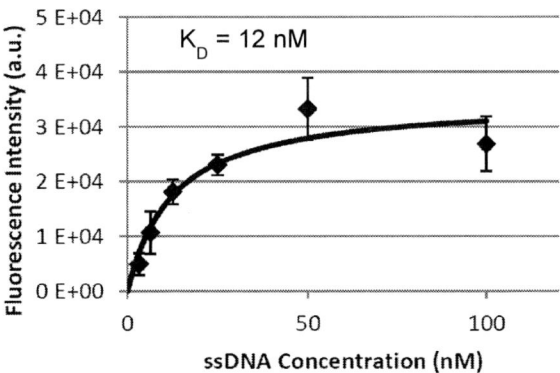

Figure 5: Fluorescence based binding affinity measurements of enriched pool towards IgE-functionalized microbeads. Error bars represent standard deviations from triplicate measurements.

When the fluorescently tagged randomized library used to initiate aptamer selection was incubated with IgE-immobilized beads representing the amount of target-bound oligomers, increased without reaching an asymptote (Fig. 6). A nominal disassociation constant, also based on the assumption of monovalent binding, was estimated to be on the order of 620 nM. This is over 50 times the value of K_D estimated for the enriched pool above, which suggests that there was negligible binding between the randomized library and IgE and demonstrates the ability of the microfluidic device to isolate aptamers against protein targets.

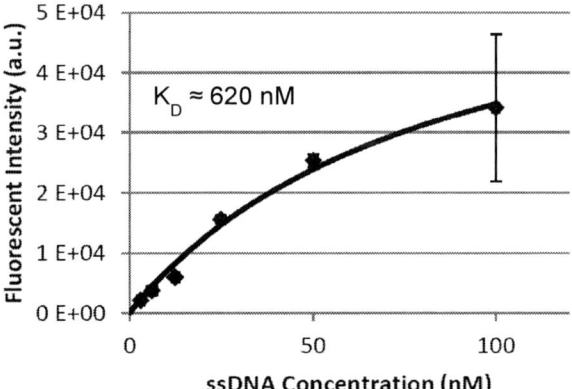

Figure 6: Fluorescence based binding affinity measurements of randomized library used to initiate SELEX towards IgE-functionalized microbeads. Error bars represent standard deviations from triplicate measurements.

CONCLUSION

We have demonstrated a microfluidic–device for the isolation and amplification of human IgE-specific oligomers using combined electrokinetic and pressure driven transfer methods. In the device IgE-binding oligomers are affinity selected against target functionalized beads, electrophoretically transferred to an amplification chamber, hybridized to microbeads, amplified by PCR, and transferred back to the selection chamber by pressure driven flow for further affinity selection. This hybrid approach reduces the number of pressure-driven flow components and avoids unfavorable electrolysis-induced affinity-selection environmental conditions to simplify the device design and improve the device operation efficiency. Experimental results show successful multi-round closed-loop enrichment of IgE-targeting oligomers with high affinity ($K_D = 12$ nM). These results demonstrate the utility of the device to rapidly isolate aptamers with high affinity to protein targets.

ACKNOWLEDGEMENTS

We gratefully acknowledge financial support from the National Institutes of Health (grant numbers: 8R21GM104204 and U19 AI067773).

REFERENCES

[1] M. Cho, Y. Xiao, et al., "Quantitative Selection of DNA Aptamers through Microfluidic Selection and High-Throughput Sequencing," *Proc Natl Acad Sci U S A* 107: 15373-15378, 2010.

[2] C.J. Huang, H. I. Lin, et al., "An integrated microfluidic system for rapid screening of alpha-fetoprotein-sepcific aptamers". *Biosensors and Bioeletronics*, 28-45, 2012.

[3] J. Kim, J. P. Hilton, et al., "Nucleic acid isolation and enrichment on a microchip." *Sensors and Actuators A: Physical*, 183-190, 2013.

[4] J. Zhu, T. Olsen, et al., "A microfluidic device for isolation of cell-targetting aptamers." *Proc. MEMS 2013*, 242-245, 2013.

[5] T. Olsen, J. Zhu, et al., "A microfluidic device for isolation of affinity oligonucleotides using combined electrokinetic and hydrodynamic manipulation." *Proc. MicroTAS 2014*, 1521-1524, 2014.

CONTACT

*T. Olsen, tel: +1-212-854-4981; tro2104@columbia.edu

A NOVEL DENSITY-BASED DIELECTROPHORETIC PARTICLE FOCUSING TECHNIQUE FOR DIGITAL MICROFLUIDICS

Ehsan Samiei[1], Hojatollah Rezaei Nejad[1], and Mina Hoorfar[1]

[1]University of British Columbia, Kelowna, Canada

ABSTRACT

In the present study a particle focusing technique has been developed for digital microfluidics (DMF) based on cumulative effects of negative dielectrophoretic (nDEP) and gravitational forces. The mechanism of the proposed technique has been studied for different particles, and different magnitudes and frequencies of the applied voltage. Particle focusing using the proposed technique was found to be more efficient for particles with a large diameter and high density. To show a potential application of this method, it was applied to collect particles on one side of a large droplet which was split on a DMF platform into two daughter droplets with very low and very high particle concentrations.

INTRODUCTION

Developing microfluidic systems to miniaturize the conventional macroscopic bioassays received great attention in the past decade. These systems can be categorized into continuous and digital microfluidics [1, 2]. In the continuous microfluidics, liquid is pumped through microchannels designed for a specific purpose. This can limits the reconfigurability of continuous microfluidic devices [1]. Digital microfluidics (DMF), on the other hand, operates based on handling discrete liquid droplets without requiring peripheral devices (such as syringe pumps) [1]. In addition, DMF systems are reconfigurable, programmable, and easy to fabricate.

Particle manipulation and focusing are primary steps in many biological and chemical assays [3, 4]. Despite the advances in developing techniques for manipulating small volumes of droplets on the DMF platform [5], the technology is still premature for processing particles carried by the liquid droplets. Therefore, one crucial step to extend the application of the DMF systems is to develop particle manipulation and focusing techniques that can be implemented on the DMF systems.

In recent years, several techniques have been developed for particle manipulation, which have mostly been implemented into continuous microfluidics. These techniques include electrophoresis [5], magnetic [6], optics [7], acoustics [8], electrothermal [9], hydrodynamics [10] and dielectrophoresis [11]. Majority of these techniques, however, are difficult, if not impossible, to implement into the DMF platform. Among these techniques, dielectrophoresis has shown promising results in particle positioning and focusing on the DMF platform [11, 12]. Dielectrophoresis (DEP) techniques are based on manipulation of polarizable particles in a non-uniform electric field. There are two types of DEP: negative dielectrophoresis (nDEP) and positive dielectrophoresis (pDEP) for which the particle located in a non-uniform electric field moves toward regions with low and high gradients of the electric field, respectively. In our recent articles, we have shown that by altering the shape of the DMF electrodes, the particles can be patterned [11] or concentrated [12] at a desired position in the droplet using nDEP.

In the present study the particles of different sizes and densities are focused using simultaneous effect of nDEP and gravity. As a result, the presented technique does not require any alteration in the shape of the conventional DMF square electrodes and is capable of concentrating particles with a relative density (i.e., the density of the particle over the density of the surrounding medium) as low as 1.06. Increasing the amplitude or frequency of the applied voltage within a certain range increases the focusing efficiency, while beyond that range they hinder particle focusing. It is shown that this technique can be used to concentrate particles in one side a droplet manipulated on a DMF platform. Then the droplet is split into two daughter droplets with very low and very high particle concentrations.

EXPERIMENTAL SETUP

The two plate (sandwiched) DMF configuration is used for this study. The actuating electrodes are designed on the bottom plate while the ground electrode covers the entire top plate.

Fabrication

The bottom plate was fabricated using an Au and Cr coated glass slide. The substrate was made by sputtering 25 nm of Cr and 65 nm of Au on a glass slide consecutively. The electrodes were patterned on the substrate by photolithography. As the dielectric layer, a 4-μm layer of parylene C was deposited on the electrodes using Parylene Deposition System 2010 Labcoater. Finally, a Teflon (3% wt) layer was spun and post baked (140 °C for 30 min) on the substrate as the hydrophobic layer. The top plate was an ITO-coated glass slide (Delta Technologies LTD) coated with a Teflon layer with the same procedure as the bottom plate.

Experimental design

The experiments were divided into two parts. In the first part, a 1.5-mm square electrode was used to study particle concentration. The gap height between the two plates was set as 300 μm for all experiments. Two different frequencies of f =1 and 10 kHz and amplitudes of U = 50 and 100 V were used for the applied voltage. To study the effect of the particle density, two different particle types were used including iron core silica (Si) particles (ρ = 3-4 g/cm^3) and polystyrene (PS) particles (ρ = 1.06 g/cm^3). All the samples were prepared with DI water and the particles with a volume fraction of 0.0017 (27000 particle/μL for 5-μm diameter and 1000 particle/μL for 15-μm diameter particles). For all the experiments, a 1.3 μL droplet of the sample was dispensed on the electrode and squeezed with the top plate to cover an area larger than the electrode. For each experiment the electrode was actuated for 90 seconds.

In the second part of the experiments, the feasibility of the proposed technique for particle focusing in a droplet was studied. An array of 1.5-mm electrodes is used for both purposes of nDEP particle manipulation, and electrowetting on dielectric (EWOD) droplet actuation.

For all experiments, the camera is focused on the particles on the surface of the bottom plate. Therefore, the blurry image of particles shows they are in the bulk of the liquid rather than on the surface.

RESULTS AND DISCUSSION

This study includes two parts. The first part is designed to study the mechanism of the proposed particle focusing technique. The second part is designed to show the application of this technique for particle concentration in droplets.

Mechanism

Particles (Si and PS) with a diameter of 5 μm were used for the first part of the experiments. Figure 1 shows the results in which the yellow particles are Si and red ones are PS. In the first row of the images presented in Figure 1 the amplitude and frequency of the applied voltage are $U = 50$V and $f = 1$ kHz, respectively. In the time frames of 45s and 90s, the particles are slightly concentrated. This means that the particles on the electrode which are close to the edge are pushed toward the center of the electrode. For both cases the image of the particles is relatively focused, showing the particles are on the surface of the bottom plate. The gradient of the electric field has its maximum value on the edge of the electrode and it decreases as the distance from the edge increases. Therefore, the center of the electrode has the lowest gradient on the electrode and for the areas around the electrode the gradient of the electric field decreases continuously as the distance from the edge is increased. The motion of the particles from the edge of the electrode

toward its center shows that the force applied to the particles is nDEP [12, 13]. As it can be observed in the images in the first row of Figure 1, the area close to the edge of the electrode is clearer for the Si particles than for the PS particles; however, the Si particles are more concentrated in the central part of the electrode compared to the PS particles.

To increase the force applied to the particles, the amplitude of the applied voltage is increased [12, 13] to 100V. The results are shown in the second row of Figure 1. It can be observed that the Si particles, which have a significantly higher density than the PS particles, are concentrated much faster than the 50V case; while the PS particles with a density close to that of the droplet are not concentrated. The image of the Si particles is also focused; whereas the image of the PS particles is blurry. This shows that the PS particles are lifted from the bottom surface. In essence, an increase in the magnitude of the applied voltage hinders particle focusing for the PS particles while improves it for the Si particles. This indicates that the upward vertical component of the nDEP force affects the PS particles (which have low density) by lifting them away from the bottom plate and moving them towards the bulk of the liquid.

In the experiments shown in the third row of Figure 1, the amplitude of the voltage is kept at $U = 100$V while the frequency is increased to $f = 10$ kHz. This increase in the frequency increases the magnitude of the nDEP force [12, 13]. For this case, the Si particles still remain on the surface (as they are very dense) and concentrated on a very small area at the center of the electrode. This indicates that the particle weight dominates the vertical component of the nDEP force. On the other hand, the PS particles are completely lifted from the surface as the nDEP dominates the gravitational force. An increase in the nDEP force (due to the increase in the frequency) results in the motion of the PS particles toward the areas around the electrode with significantly lower gradient and intensity of the electric field.

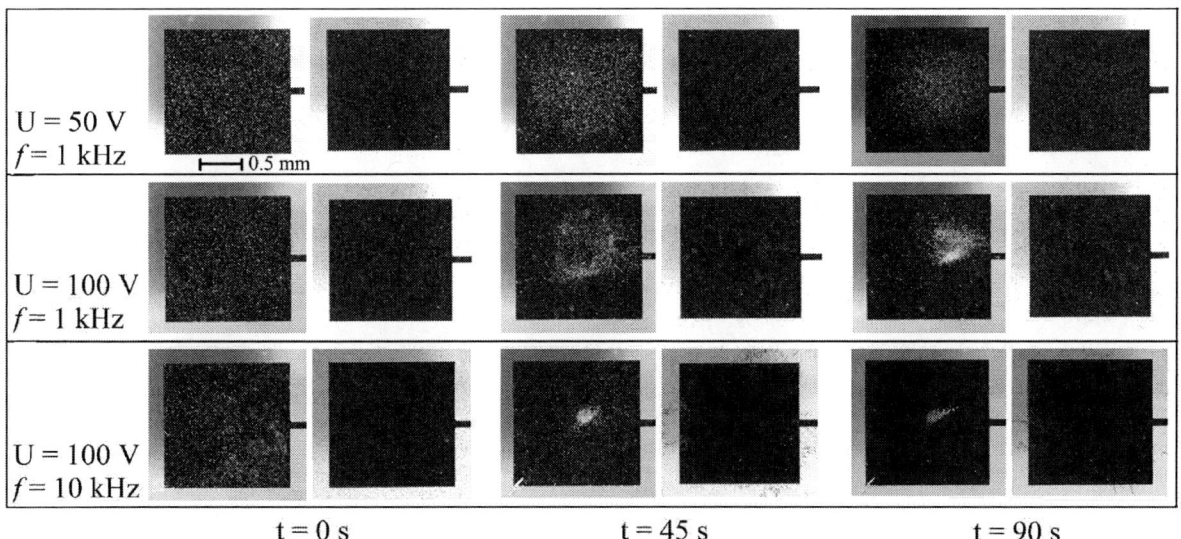

Figure 1: Sequence images of particle focusing using nDEP on a 1.5-mm square electrode in a DMF platform. The Si and PS particles are yellow and red, respectively. Both types of particles have a diameter of 5 μm.

In addition to density, the effect of the size of the particles was studied by repeating the experiments with the PS particles with an average diameter of 15µm. The results are shown in Figure 2 in which the first row of the images were acquired when the voltage of U = 50V with the frequency of f = 1 kHz was applied. For this case, the PS particles are pushed slightly away from the edge in a way that the particles initially positioned on the electrode are pushed toward its center and the ones outside are pushed away from the electrode. All particles are still on the surface, however focusing is not efficient enough.

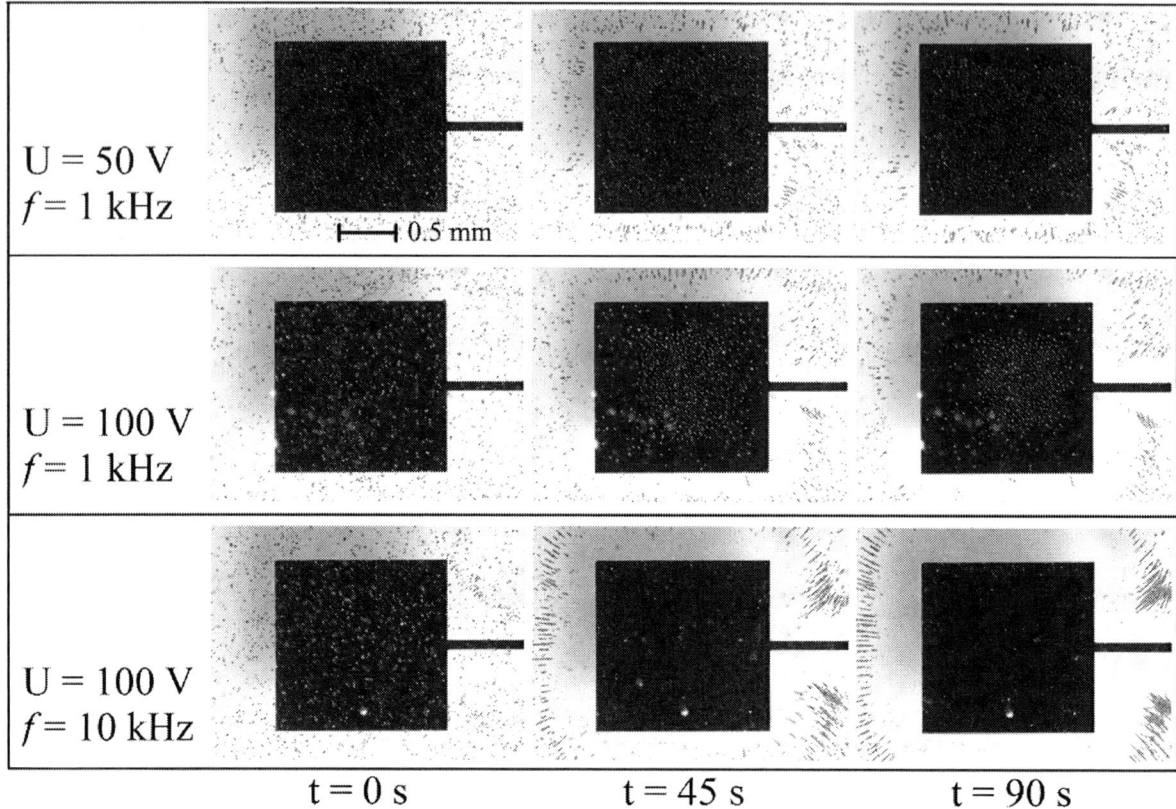

Figure 2: Sequence images of particle focusing using nDEP on a 1.5-mm square electrode in a DMF platform. Particles are PS with a diameter of 15 µm.

In the second row of images presented in Figure 2, the frequency is kept at f = 1 kHz while the amplitude is increased to U = 100V. It can be observed that the particles located initially on the electrode are focused very well, and the concentration on the central region is significantly increased over time. The images of the particles are focused, illustrating that they are still on the surface, and hence, the vertical component of the nDEP force has not overcome the gravitational force.

In the third row of the images, the voltage is kept at 100V while the frequency is increased to 10 kHz. Therefore, the magnitude of the nDEP force has increased significantly. It is observed that the particles become blurry in the image acquired at t = 45s. This means that they are lifted from the surface as the vertical component of the nDEP force overcomes the gravitational force. The decrease in the concentration shows the particles have been pushed away from the electrode. The image acquired at t = 90s shows the electrode is almost clear, and majority of the particles have been pushed away from the electrode.

The comparison between the results obtained for 5-µm (Figure 1) and 15-µm (Figure 2) PS particles indicates that the particle size is another important factor for particles focusing using the proposed technique. The 5µm-PS particles are barely focused at a low voltage (at 50 V for which the vertical component of the nDEP force is still not large); while the 15µm-PS are focused even at a higher voltage (100 V).

Application

In the previous section it was shown that by controlling the amplitude and frequency of the applied voltage it is possible to focus particles on the central region of an electrode using nDEP. In this part, the proposed technique will be used to concentrate the particles in a droplet. As it is shown in Figure 3, an array of 1.5-mm square electrodes is used for this part of the experiments. A 1.3-µL droplet containing the Si particles is dispensed on the chip covering an area twice that of the electrode. The first and second images in Figure 3 show that by actuating one of the electrodes the particles on the same electrode are focused at the center. Then, the droplet is moved downward to concentrate the particles in

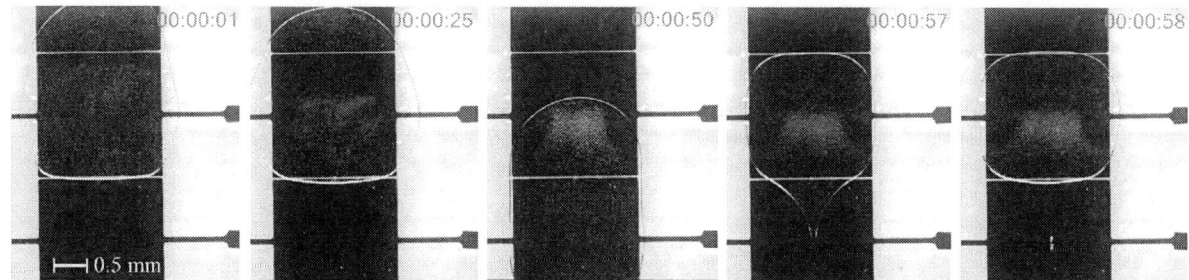

Figure 3: Particle concentration in a droplet in a DMF platform is shown. The first and the second frames show particle focusing on one electrode using nDEP. The other frames show splitting of the mother droplet into two daughter droplets with high and low particle concentrations on an array of electrodes. Particles are Si with a diameter of 5 μm.

the remaining portion of the droplet. Following this step, the droplet is pulled more downward to contact three electrodes simultaneously (see the third image in Figure 3). This way the particles are concentrated on one side of the droplet; while the other side has a very low concentration. Finally, the two far electrodes are actuated to pull the droplet from the two sides while the electrode in between is deactivated (see the fourth image in Figure 3). At this moment, a neck forms in the middle. Immediately the neck breaks up and the mother droplet is split into two daughter droplets. One daughter droplet contains the majority of the particles (the fifth image in Figure 3); while the other droplet has a very low concentration of the particles.

CONCLUSIONS

This study presents a new technique for focusing particles in digital microfluidic devices using the cumulative effects of negative dielectrophoresis (nDEP) and gravity. It is shown that the particles resting on an electrode can be focused in the central part of the electrode using nDEP. However, if the magnitude of the nDEP force increases (by increasing the magnitude or frequency of the voltage) the vertical component of this force can overcome the gravitational force and lift the particles. This will result in either dispersion of the particles over the entire electrode area or moving them towards the outside of the electrode. Therefore, the amplitude and frequency of the applied voltage should be controlled to achieve a high particle focusing efficiency. This technique is more efficient for particles with higher densities and larger diameters.

The proposed technique was finally used to concentrate the particles on one side of the droplet which was then split into two daughter droplets, one containing majority of the particles and the other with a very low concentration of the particles. It was shown that the proposed density-based dielectrophoretic technique has a potential to be used for droplet purification in DMF platforms.

REFERENCES

[1] R. B. Fair, "Digital microfluidics: is a true lab-on-a-chip possible?" Microfluid. Nanofluid. vol. 3, pp. 245–281, 2007.

[2] E. K. Sackmann, A. L. Fulton and D. J. Beebe, "The present and future role of microfluidics in biomedical research", Nature, vol. 507, pp. 181–189, 2014.

[3] J. Kim, M. Johnson, P. Hill and B. K. Gale, "Microfluidic sample preparation: cell lysis and nucleic acid purification", Integr. Biol., vol. 1, pp. 574–586, 2009.

[4] D. R. Gossett, W. M.Weaver, A. J. Mach, S. C. Hur, H. T. Tse, W. Lee, H. Amini and D. Di Carlo, "Label-free cell separation and sorting in microfluidic systems", Anal. Bioanal. Chem., vol. 397, pp. 3249–3267, 2010.

[5] S. K. Cho, Y. Zhao and CJ Kim, "Concentration and binary separation of micro particles for droplet-based digital microfluidics", Lab Chip vol. 7, pp. 490–498, 2007.

[6] R. Sista, A. Eckhardt, V. Srinivasan, M. Pollack, S. Palanki and V. Pamula, "Heterogeneous immunoassays using magnetic beads on a digital microfluidic platform", Lab Chip, vol. 8, pp. 2188–2196, 2008.

[7] D. G. Grier "A revolution in optical manipulation" Nature, vol. 424, pp. 810–816, 2003.

[8] T. Laurell, F. Petersson and A. Nilsson, "Chip integrated strategies for acoustic separation and manipulation of cells and particles", Chemical Society Reviews, vol. 36, pp. 492–506, 2007.

[9] M. Lian , N. Islam and J. Wu, "AC electrothermal manipulation of conductive fluids and particles for lab-chip applications", IET Nanobiotechnology, vol. 1, pp. 36–43, 2007.

[10] D. Di Carlo, L. Y. Wu and L. P. Lee, "Dynamic single cell culture array", Lab Chip, vol. 6, pp. 1445–1449, 2006.

[11] H. Rezaei Nejad, O. Z. Chowdhury, M. D. Buat and M. Hoorfar, "Characterization of the geometry of negative dielectrophoresis traps for particle immobilization in digital microfluidic platforms", Lab. Chip., vol. 13, pp. 1823–1830, 2013.

[12] H. Rezaei Nejad and Mina Hoorfar, "Purification of a droplet using negative dielectrophoresis traps in digital microfluidics", Microfluid. Nanofluid., DOI: 10.1007/s10404-014-1446-3, 2014.

CONTACT

*M. Hoorfar, tel: +1-250- 8078804;
  mina.hoorfar@ubc.ca

ACTIVE CONTROL OF CHEERIOS EFFECT FOR DIELECTRIC FLUID

Junqi Yuan and Sung Kwon Cho*
University of Pittsburgh, Pittsburgh, USA

ABSTRACT

This paper presents a novel method of manipulating objects floating on the free surface of dielectric fluids by combining "Cheerios effect" with di-electrowetting. Cheerios effect is a common phenomenon in which small floating objects are either attracted or repelled by the sidewall via capillary effects. The wettability of the sidewall surface is known to be a key parameter for this phenomenon. By controlling the contact angle of the sidewalls via di-electrowetting and thus distorting the adjacent interfaces, Cheerios effect can be controlled to manipulate floating objects. In this control, the titling angle of the sidewall is found to be critical. Theoretical calculation and experimental results are compared for various titling angles and applied voltages. The theory qualitatively agrees with the experiments. Using this method, a linear transportation of millimeter-sized floating object in a channel is achieved. By sequentially applying a voltage to the arrayed di-electrowetting electrodes on the titled sidewalls and by carefully controlling the time interval between switching shifts, the floating object can be propelled continuously along the channel.

INTRODUCTION

'Cheerios effect' is a common phenomenon named after observations that breakfast cereal flakes floating in milk tend to be attracted to or repelled from the sidewall of bowl[1, 2]. This attraction/repulsion is driven by lateral capillary forces generated between floating objects and sidewall. In fact, not only cereal flakes but also any floating objects that distort the air-liquid interface can be attracted to or repelled from the sidewall. For example, air bubbles are aggregated and attracted by a glass sidewall (Figure 1(a)). In nature, some water-walkers utilize the capillary forces to climb an inclined meniscus, otherwise

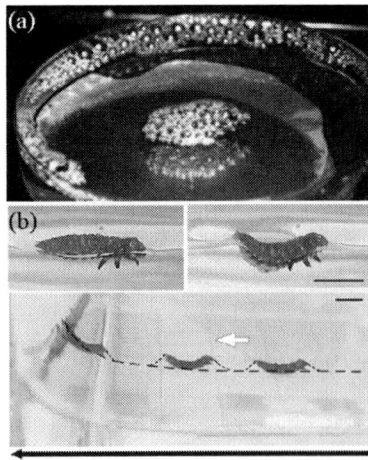

Figure 1: Cheerios Effect. (a) Bubbles migrate to the wall[2]; (b) Meniscus climbing behavior of the water lily leaf beetle[4].

they would not be able to do so by using traditional propulsion methods[3]. Normally these small animals deform the water surface and maintain a certain body posture, such as arcing the back, to generate capillary forces and to climb the meniscus (Figure 1(b)) [4].

A common perception of Cheerios effect is that the direction of the lateral capillary force (i.e., attraction or repulsion) at the air-liquid interface solely depends on the surface wetting properties (wettability) of the floating objects and wall[5-7]. For example, a hydrophobic (lyophobic) floating object is attracted by another hydrophobic (lyophobic) floating object or hydrophobic (lyophobic) sidewall, but it is repelled by any hydrophilic (lyophilic) ones. Vella and Mahadevan [2] challenged this theory with a simple experiment. By maintaining the similar surface wettability of the two floating objects but changing the weight of one object, it was shown that the force between the floating objects could be reversed. That is, the density of the floating object is another important parameter. A theoretical analysis reveals that interface slopes near the floating particle and wall are most important in determining the capillary force[2, 8, 9]. For a vertical plate, the interface slope is mainly predicted by the wettability of the surface. For floating objects, however, it is more complicated because the surface wettability, density and geometry of the floating object all affect the interface slopes.

Recently, some attempts have been made to fabricate and investigate meniscus-climbing devices by utilizing the Cheerios effect. Hu et al. [10] and Yu et al. [11] used a bent plastic or metal sheet to distort the air-liquid interface to mimic the meniscus-climbing behavior of the natural insect. However, it is not easy to precisely adjust the shape of the sheet to manipulate the air-liquid interface. Yuan and Cho[12] reported an active control method of Cheerios effect by using electrowetting to change the wettability (contact angle) of floating objects and sidewalls. By installing electrowetting electrodes on the floating objects and sidewalls and applying electric potential to the electrodes, the air-water interface can be distorted to generate lateral capillary forces between the object and sidewall. However, this electrowetting method works well with conductive liquids but is not directly applicable for non-conductive (dielectric) liquids since electrowetting is mainly effective for conductive liquids.

In this paper, we introduce di-electrowetting to change the wettability of dielectric liquids and thus to control Cheerios effect. Combining di-electrowetting with the Cheerios effect, we control attraction and repulsion between a small object floating on dielectric liquid and the sidewall. In this control, we theoretically and experimentally discovered that the slope angle is another critical parameter in addition to the wettability and density of the floating objects. Finally, we demonstrate that Cheerios effect can continuously propel small floating

978-1-4799-7956-1/15 $31.00 © 2015 IEEE
496
MEMS 2015, Estoril, PORTUGAL, 18 - 22 January, 2015

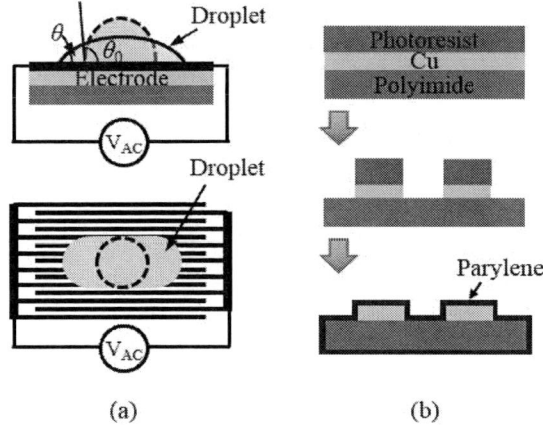

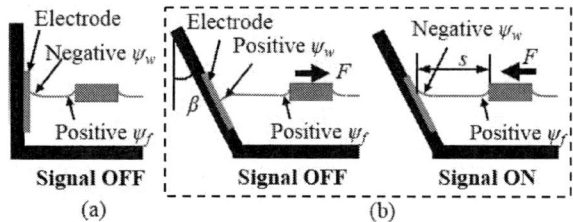

Figure 4: (a) Di-electrowetting electrode vertically submerges in dielectric fluid. (b) Control concept of Cheerios Effect by di-electrowetting.

Figure 2: (a) Configuration of di-electrowetting (side view and top view). Dashed lines show initial shape of droplet. Solid lines show droplet shape after actuation; (b) Microfabrication flow for di-electrowetting electrodes.

objects along a linear path when an array of electrodes is energized sequentially in a controlled manner. Detailed fabrication, testing and results are described in the following sections.

WORKING PRICIPLE

While electrowetting works well with conductive liquid under DC as well as AC inputs at low frequency (typically < 1 kHz), di-electrowetting allows us to actuate non-conductive liquids with high frequency AC inputs. The contact angle of dielectric liquid can be controlled in a wide range with a low degree of contact angle hysteresis[13, 14]. Di-electrowetting uses interdigitated electrodes and AC input of high frequency (Figure 2(a)). When an AC input is applied to the electrodes, the contact angle θ_0 will decrease to θ and the droplet will spread on the electrode surface along the electrode direction. The droplet will return to its initial shape when the input signal is removed.

Figure 2(b) shows a microfabrication flow for di-electrowetting electrodes. A thin, flexible sheet with a Cu layer of 9 μm in thickness (DuPont Pyralux® flexible Cu product) is patterned by standard lithography and wet etching process and is then coated with 3 μm of Parylene

layer for the dielectric layer. Both width of and spacing between interdigitated electrodes are 75 μm. On the top of surface, a thin Teflon layer is dip-coated for lyophobic purpose. Because dielectric fluid with low viscosity is preferable for reduced flow resistance, propylene carbonate is selected for the dielectric testing fluid (its viscosity is about 20 times lower than that of 1,2 propylene glycol used in Refs. 13 and 14).

Figures 3(a) and 3(b) show reversible spreading and contracting of the propylene carbonate droplet under 300 V_{RMS} and 10 kHz AC signal. The change in the contact angle is large (over 50°, even initially lyophilic), which is higher than that by typical electrowetting (about 40°). When the interdigitated di-electrowetting electrode vertically and partially submerges in the dielectric liquid, the contact line quickly and reversibly climbs up upon di-electrowetting activation (Figure 3(c)). Because the change in the contact angle is larger, the climbing height of the contact line is also higher than that of electrowetting.

The di-electrowetting electrode is installed on the sidewall to control Cheerios effect (Figure 4). The attraction and repulsion behavior between the wall and floating object is mainly determined by interface slopes near the wall (ψ_w) and floating object (ψ_f). Generally speaking, when the signs of ψ_w and ψ_f are the same, the

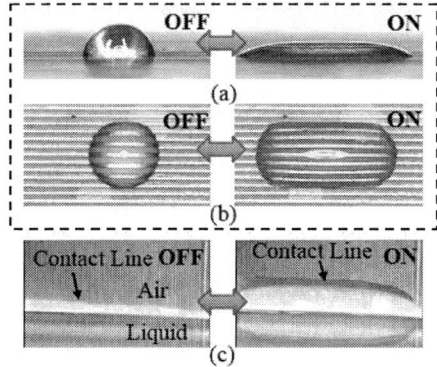

Figure 3: (a) side view & (b) top view: Droplet actuated by di-electrowetting; (c) Contact line climbs vertically submerged di-electrowetting electrodes. (V = 300 V_{RMS}, f = 10 kHz)

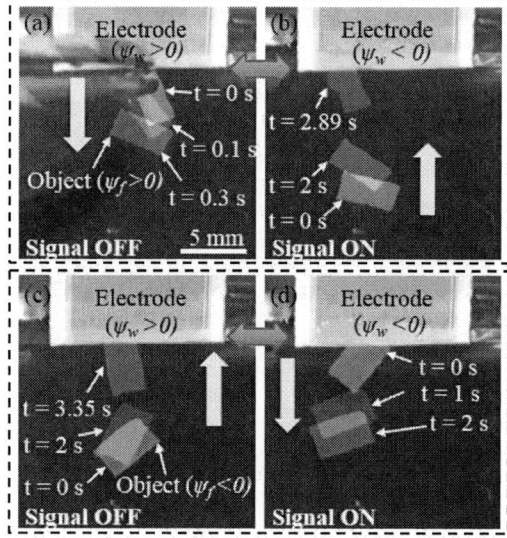

Figure 5: Initial Repulsion (a) of object (low-density, positive ψ_f) is switched to attraction (b) by di-electrowetting. Initial attraction (c) of object (high-density, negative ψ_f) is switched to repulsion (d) by di-electrowetting. (top views, β = 30°, V = 300 V_{RMS}, f = 10 kHz)

978-1-4799-7956-1/15 $31.00 © 2015 IEEE   497

wall and the object repel each other; otherwise, they attract. For water, ψ_w near the hydrophobic vertical wall is initially positive and becomes negative upon electrowetting signal activation. For dielectric liquid, however, ψ_w near the vertical wall is always negative since most of dielectric fluids have contact angles of less than 90° (Figure 4(a)) due to the low surface tension. In this case, the sign of ψ_w cannot be reversed by di-electrowetting, so the capillary force cannot switch between attraction and repulsion. This problem is resolved by tilting the wall by angle β. With a proper value of β, ψ_w can be initially positive, so repulsion initially occurs between the wall and object ($\psi_w > 0$, $\psi_f > 0$, Figure 4(b) left). Upon di-electrowetting activation, ψ_w changes to be negative because the contact angle decreases. As a result, attraction occurs between the wall and object ($\psi_w < 0$, $\psi_f > 0$, Figure 4(b) right).

RESULTS

Experimental verifications of Figure 4(b) are shown in Figure 5. When di-electrowetting signal is off, a low-density floating object ($\rho \approx 0.1$ g/cm^3) with positive meniscus slope ψ_f is initially repelled by the tilted sidewall ($\beta = 30°$) that has positive ψ_w (Figure 5(a)). When the di-electrowetting signal ($V = 300$ V$_{RMS}$, $f = 10$ kHz) is turned on (Figure 5(b)), the contact angle reduces and the meniscus slope ψ_w becomes negative. As a result, the capillary force is changed to attraction. On the contrary, when a high-density floating object ($\rho \approx 1.5$ g/cm^3) is used, the initial slope ψ_f is negative. The initial attraction (Figure 5(c)) is switched to repulsion (Figure 5(d)) by di-electrowetting ($V = 300$ V$_{RMS}$, $f = 10$ kHz). All these switching operations are reversible.

By solving the Young-Laplace interface differential

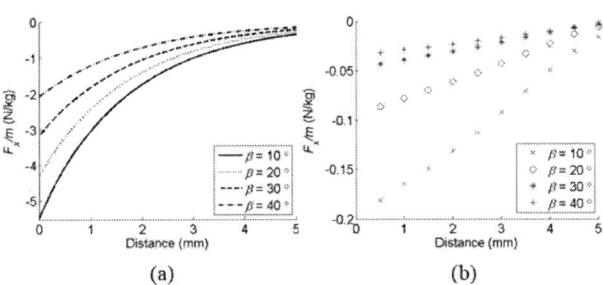

Figure 6: Force/object mass vs. tilting angle β. (a) Theoretical prediction; (b) Experimental result. ($V = 300$ V$_{RMS}$, $f = 10$ kHz).

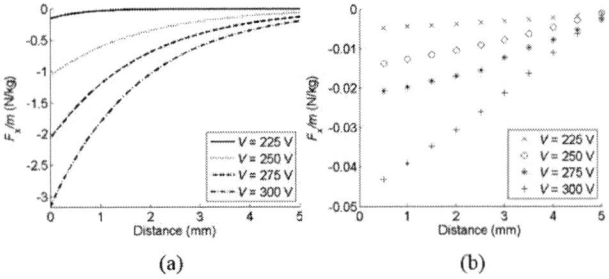

Figure 7: Force/object mass vs. applied voltage. (a) Theoretical prediction; (b) Experimental result. ($f = 10$ kHz, $\beta = 30°$)

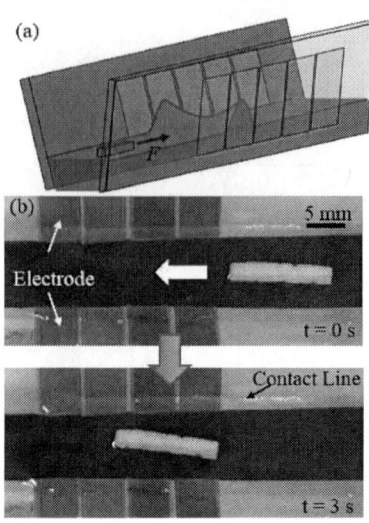

Figure 8: (a) Continuous propulsion concept; (b) Top view of continuous transporting in channel. Pairs of electrodes are activated sequentially from right to left with duration of 0.7 sec. ($V = 300$ V$_{RMS}$, $f = 10$ kHz, $\beta = 30°$)

equation, the lateral capillary force between the vertical plate and spherical floating particle with radius R can be theoretically obtained.[8]. When the plate submerges into liquid with a tilting angle β, the ratio between the lateral capillary force F_x and mass of spherical object m (that is, acceleration) in Cheerios effect can be obtained as:

$$F_x/m = -\pi\sigma/m \left[-2q\left(r\sin\psi_f\right)^2 K_1(2qs)\right.$$
$$+2r\sin\psi_f\sqrt{1-\sin(\theta+\beta)}e^{-qs}$$
$$\left.+q\left(r\sqrt{1-\sin(\theta+\beta)}e^{-qs}\right)^2\right], \qquad (1)$$

where σ is the air-liquid surface tension, $q^{-1} = L_c \approx \sqrt{\sigma/\rho_l g}$ the capillary length, ρ_l the density of liquid, g the gravitational acceleration, r the radius of the contact line on the spherical floating object, s the horizontal distance between the floating object and sidewall, and K_1 the modified Bessel function. Here, r is obtained from the vertical force balance of the floating particle.

Figures 6 and 7 show comparisons between experimental results and theoretical calculations according to Equation (1). A rectangular floating object with low density ($\rho \approx 0.1$ g/cm^3) is used in the experiments. It is initially repelled by the sidewall. When di-electrowetting is turned on, the movement is recorded by a high-speed camera. Then, the acceleration of the object is obtained from the second time-derivative of the displacement field. The width of the floating object is measured and equivalently converted to the particle radius R in the theoretical calculation. The overall trend shown in Figures 6 and 7 is that the capillary force increases when the distance decreases. As shown in Figure 6, as the tilting angle β increases, the attraction force on the object of positive ψ_f decreases. If the tilting angle is fixed, for example $\beta = 30°$ in Figure 7, as the applied voltage increases, the lateral capillary force increases. It is easy to understand these two effects from Equation (1) that the sum of θ and β is important to determine the direction of the lateral capillary force. The effect of increasing β is equivalent to the effect of increasing θ (which corresponds to lowering the applied voltage). The reason for quantitative discrepancy is that the theory does not take viscous drag into account. However, during the experiment, the object experiences a large viscous force due to the large contact area and high viscosity of the propylene carbonate (~2.5 times larger than that of water).

When simply turning di-electrowetting on or off with a single electrode, the motion of objects is limited to the space between the wall and objects (one-time movement toward or far away from the sidewall). Figure 8(a) illustrates how to continuously transport a floating object in channel by di-electrowetting. An array of electrodes is attached to each of two tilted sidewalls. ψ_w is initially positive, so the object with positive ψ_f automatically stays in the middle of channel due to symmetric repulsion by both sidewalls. When a pair of electrodes (one on each side wall) is activated together, the meniscus slopes ψ_w switch from positive to negative. As a result, the distorted interface simulates the wall effect in Figure 4(b) and generates a pulling force on the floating object toward the activated electrodes. Due to symmetry, the lateral forces on the object are cancelled out. Only a net force on the object is toward the elevated interface (along the channel), so the object is propelled along the channel path. By shifting the activation to the next pair, a continuous movement is achieved.

Figure 8(b) shows a low-density object ($\psi_f > 0$) is propelled in the channel with two arrays of di-electrowetting electrodes (5 mm width). Upon shifting activations of electrode pairs to the left ($V = 300$ V$_{RMS}$, $f = 10$ kHz) with duration of 1 sec, the object is step-by-step propelled to the left. A sequential signal is provided by a microcontroller (ATMEL ATtiny24A). Relays (Panasonic AQW614EH) are used for transmitting a high AC voltage from the amplifier to the electrodes. The present control of Cheerios effect would provide an efficient tool in many micro/nano particle manipulation and processes on phase interfaces.

CONCLUSION

This work presents a novel propulsion method combining Cheerios effect and dielectrowetting. This method works with dielectric fluids. By electrically controlling the wettability of dielectric fluids and then reversing the sign of the slope angle near the side wall, a 1-D motion for the floating object toward or away from the wall is controlled. Here, the tilting angle of the sidewall is found to be critical for switching between attraction and repulsion. The acceleration of floating particle for the various tilting angles and applied voltages are measured and compared with theory. The experimental results qualitatively agree with the theoretical calculation. Finally, a continuous propulsion method is achieved using the Cheerios effect. By sequentially activating microfabricated electrode arrays, linear translations of floating objects in the small channel are demonstrated.

ACKNOWLEDGEMENTS

This research is supported by NSF Grant No. ECCS-1029318.

CONTACT

* Sung Kwon Cho, Tel: +1-412-624-9798; skcho@pitt.edu

REFERENCES

[1]   J. Walker, "The Flying Circus of Physics, 2nd ed.," ed. New York: Wiley, 2007.

[2]   D. Vella and L. Mahadevan, "The "Cheerios effect"," *Am. J. Phys.,* vol. 73, pp. 817-825, 2005.

[3]   R. Baudoin, "La physico-chimie des surfaces dans la vie des Arthropodes aeriens des miroirs d'eau, des rivages marins et lacustres et de la zone intercotidale," *Bull. Biol. France Belg.,* vol. 89, pp. 16-164, 1955.

[4]   D. L. Hu and J. W. Bush, "Meniscus-climbing insects," *Nature,* vol. 437, pp. 733-736, 2005.

[5]   G. K. Batchelor, *An Introduction to Fluid Dynamics.* Cambridge U.K.: Cambridge University Press, 1967.

[6]   D. J. Campbell, E. R. Freidinger, J. M. Hastings, and M. K. Querns, "Spontaneous assembly of soda straws," *Journal of Chemical Education,* vol. 79, pp. 201-202, 2002.

[7]   J. C. Berg, *An Introduction to Interfaces & Colloids: The Bridge to Nanoscience.* Singapore World Scientific, 2009.

[8]   P. A. Kralchevsky, V. N. Paunov, N. D. Denkov, and K. Nagayama, "Capillary Image Forces I. Theory," *J. Colloid Interface Sci.,* vol. 167, pp. 47-65, 1994.

[9]   V. N. Paunov, P. A. Kralchevsky, N. D. Denkov, and K. Nagayama, "Lateral Capillary Forces between Floating Submillimeter Particles," *Journal of Colloid and Interface Science,* vol. 157, pp. 100-112, 1993.

[10]   D. L. Hu, M. Prakash, B. Chan, and J. W. M. Bush, "Water-walking devices," *Exp Fluids,* vol. 43, pp. 769-778, 2007.

[11]   Y. Yu, M. Guo, X. Li, and Q. S. Zheng, "Meniscus-climbing behavior and its minimum free-energy mechanism," *Langmuir,* vol. 23, pp. 10546-50, Oct 9 2007.

[12]   J. Q. Yuan and S. K. Cho, "Free surface propulsion by electrowetting-assisted 'Cheerios effect'," presented at the Micro Electro Mechanical Systems (MEMS), 2014 IEEE 27th International Conference on, San Francisco, CA, 2014.

[13]   G. McHale, C. V. Brown, M. I. Newton, G. G. Wells, and N. Sampara, "Dielectrowetting driven spreading of droplets," *Phys Rev Lett,* vol. 107, p. 186101, Oct 28 2011.

[14]   G. McHale, C. V. Brown, and N. Sampara, "Voltage-induced spreading and superspreading of liquids," *Nat Commun,* vol. 4, p. 1605, 2013.

BIOMECHANICAL CANAL SENSORS INSPIRED BY CANAL NEUROMASTS FOR ULTRASENSITIVE FLOW SENSING

A.G.P. Kottapalli[1,2*], M. Asadnia[1,2], J.M. Miao[1] and M.S. Triantafyllou[2,3]

[1]School of Mechanical and Aerospace Engineering, Nanyang Technological University, SINGAPORE

[2]Center for Environmental Sensing and Modeling (CENSAM), Singapore-MIT Alliance for Research and Technology (SMART), SINGAPORE

[3]School of Mechanical Engineering, Massachusetts Institute of Technology, USA

ABSTRACT

Fishes use their mechanosensory lateral-line system to detect minute disturbances underwater. The lateral-lines consist of superficial and canal neuromast (SN and CN) sensory sub-systems. Unlike SNs which are exposed to external flow directly, the CNs, due to the presence of canals, have a higher immunity to noise and greater signal selectivity. In this paper, we present the design, fabrication and experimental characterization of arrays of zero-powered and ultrasensitive MEMS piezoelectric hair cell sensors encapsulated into biomimetic canals. The experimental characterization of the MEMS canal encapsulated sensors in the presence of steady and unsteady flow conditions validates the biomechanical high-pass filtering function of the canals.

INTRODUCTION: BIOINSPIRATION

Although being blind, *Astyanax mexicanus fasciatus* also known as the blind cavefish demonstrates an uncanny ability to swim at high-speeds without collision with its surrounding objects. Due to impaired sight, the mechanosensory lateral-line cues are certainly of great significance to the blind cavefish as compared to other eyed fish [1]. The lateral-line is a primary mechanosensory organ found in most fishes which enables them to perform various behavioral abilities like object tracking, energy efficient maneuvering, rheotaxis (orientation to currents) and schooling (match their self-swimming speed and direction with the neighbors in the school) [1-3]. Figure 1(a) shows a photograph of the blind cavefish. Figure 1(b) shows the scanning electron microscopic (SEM) image of the CN system showing a series of pores.

Each neuromast sensor consists of hair cells that are embedded into a soft gelatinous material called cupula. The hair cells are connected to the afferent fibers at the base and form the principal sensing elements while the cupula couples the motion of the surrounding water to the embedded hair cells and increases the viscous drag and pressure force on the hair cells. The sensory elements called neuromasts are mainly divided into two types of sub-modalities called as the superficial neuromasts (SN) and the canal neuromasts (CN) each performing a unique function which is vital for the fish in order to perform complete underwater sensing [3]. Both the SNs and the CNs consist of a similar structural organization of the hair cells and the cupula with minor variations but major

differences in cupula shape and dimensions. As the name suggests, the superficial neuromasts are present superficially on the surface of the fish's body and are directly exposed to external flow. SNs and the CNs are often called as velocity and acceleration sensors respectively [1].

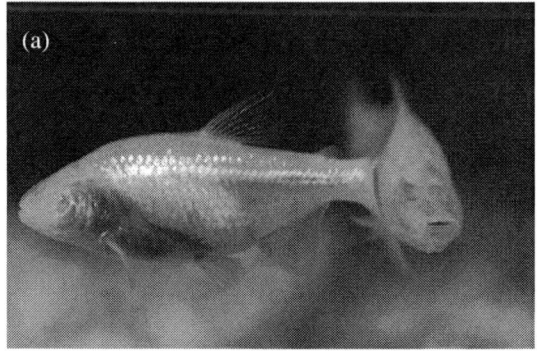

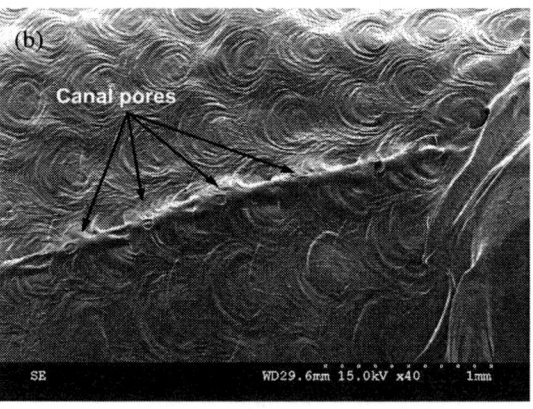

Figure 1: Mechanosensory lateral-line system in the blind cavefish (a) A photograph of the blind cavefish. The visibly observable pale pink skin and the atrophied eyes are the features believed to have developed due to survival in dark caves for long periods of time (b) An SEM image of the canal neuromast system of the lateral-line. The canal sensory system consists of a series of pores that leads the external flow to reach the embedded neuromasts.

The major difference between the canal and the superficial neuromasts is that the CNs are embedded in sub-dermal channels and expose to external flow through series of pores on the skin of the fish that lead to the channel. A single neuromast exists embedded between two pore openings. In the simplest case, the canal can be considered as a straight-walled tube with pores through which disturbances in water reach the neuromast. The CN is actuated only when there is a pressure difference between the consecutive pores within which the neuromast is located. The motion of water inside the canal is impeded by the inertia of water and the friction offered by the canal walls. At low frequencies, the friction generated by the walls of the canal is dominant and the water movements inside the canal are significantly smaller than the movements outside the canal [1]. Therefore, the canal acts as a bio-mechanical high-pass filter cutting off the steady-state laminar flow (dc flow) and the low frequency oscillatory flows (ac flows). The design of the height of the canal significantly impacts the cut-off frequencies of the bio-mechanical high-pass filter that the canal forms.

MEMS ARTIFICIAL CN SENSORS

In the past, few research groups have devoted their attention towards developing MEMS flow sensors inspired by the biological hair cell sensors on various species [4-13]. In the research involving artificial lateral-line sensor development, most work was confined to developing flow sensors mimicking the superficial neuromast, whereas, the importance of a biomimetic canal structure in enhanced sensing of relevant flows and bio-mechanical filtering of background low-frequency flows has not received much attention. In our work in the past [4-10], we developed SN inspired hair cell sensors encapsulated in biomimetic artificial hydrogel cupula. In this paper, we present the development of artificial MEMS CN sensors. The biomechanical filtering ability of the canal is illustrated through experiments.

Artificial CN sensor fabrication

The artificial MEMS CNs consist of three major components- the piezoelectric MEMS sensing membrane, high-aspect ratio hair cell and a biomimetic canal that houses the sensors. The piezoelectric sensing membrane is fabricated by a combination of MEMS micromachining and sol-gel methods for the growth of thin-film $Pb(Zr_{0.52}Ti_{0.48})O_3$. The hair cell fabricated separately through μ-stereolithography (μ-SLA), is mounted at the center of the PZT membrane. A microscopic image of the piezoelectric hair cell sensors is shown in figure 2(b).

An array of ten piezoelectric hair cell sensors is developed by mounting the sensors on a flexible liquid crystal polymer (LCP) film patterned with gold interconnects. A flexible canal is developed by PDMS molding process. A negative mold of the required canal structure is fabricated by rapid prototyping process. The distance between two successive pillars in the 3D-printed mold is same as the distance between successive hair cells in the array. Then a thoroughly mixed and degassed PDMS solution is poured and cast inside the mold and allowed to dry on its own at room temperature. The

PDMS structure is then peeled out of the mold. The PDMS canal is carefully mounted on top of the array of hair cell sensors by viewing through a microscope to ensure accurate location of the canal. A schematic of the organization of the hair cells within the array is shown in figure 3.

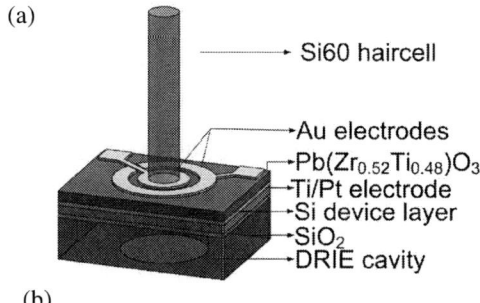

(a)

(b)

Figure 2: Piezoelectric hair cells sensor (a) schematic of the sensor showing various materials employed in the device fabrication. (b) Photograph of the MEMS PZT hair cell sensor. The high-aspect ratio hair cells are fabricated through stereolithography process.

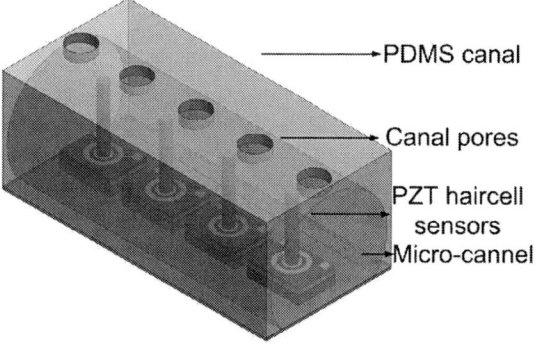

Figure 3: A schematic showing the organization of the hair cell sensors within the PDMS canal encapsulation.

EXPERIMENTAL RESULTS

The artificial canal sensors are characterized under various fluid conditions. Steady (dc) flows and unsteady (ac) flows of desired velocities are generated by using a water tunnel and dipole stimulus (vibrating sphere) respectively. A comparison of the performance of the hair cell sensors without the canal (superficial sensors) and the hair cell sensors with the canal (canal sensors) is conducted by introducing both the sensors into similar experimental flow conditions and analyzing their outputs. Therefore, the experiments illustrated clearly demonstrate the utility of the canal in eliminating dc and low-frequency flows. A steady-state laminar flow of known velocity is generated in the water tunnel by adjusting the motor frequency (The motor frequency is calibrated with respect to the flow velocity). A vibrating sphere (dipole) driven by a function generator through a power amplifier acts as an ac flow source. An ac flow of known frequency and amplitude can be generated using the dipole set-up. The goal of this experiment is to evaluate the ability of the canal encapsulated sensors in effectively detecting the ac flow (signal) in the presence of strong dc flow (noise) and compare the results with those of the superficial sensor. In addition to dc flow, the noise in this system also consists of random and irregular hydrodynamic signatures that demonstrate the broadband spectra with no dominant frequencies. The artificial SN and the CN sensors are mounted at the center of the test-section of the water tunnel in the vicinity of the dipole source as shown in figure 4. Signals from the superficial sensor and one sensor of the canal array are acquired though two channels of the NI-DAQ and recorded in LABVIEW.

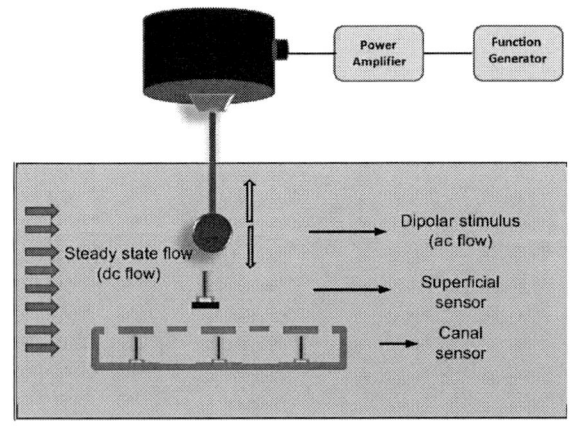

Figure 4: A schematic of the experimental set-up showing a steady (dc) flow generated in a water tunnel and unsteady (ac) flow generated using a dipole stimulus.

The dipole is set to vibration at a frequency of 35Hz and an amplitude of 250mV$_{rms}$. The data acquired from both the artificial canal and the superficial sensors is as shown in figure 5. The data shown in plots in figure 5 shows the fast Fourier transform (FFT) of the signal acquired from both the sensors showing the frequency content of the data acquired. The data shown in red is the signal collected from the superficial sensors and the signal from the canal encapsulated sensors is shown in blue. As it can be seen from the results, the superficial sensors pick up a intense low frequency noise as compared to the canal sensors. The canal sensors not only filter away all the low

frequency noises, but also show a clear peak corresponding to the signal at 35 Hz. However the signal (at 35 Hz) can also be observed in the output of the superficial sensors as well, a very low frequency ac signal (< 15 Hz as observed from figure 5) could have been completely buried under the noise and would be indistinguishable by the superficial sensors.

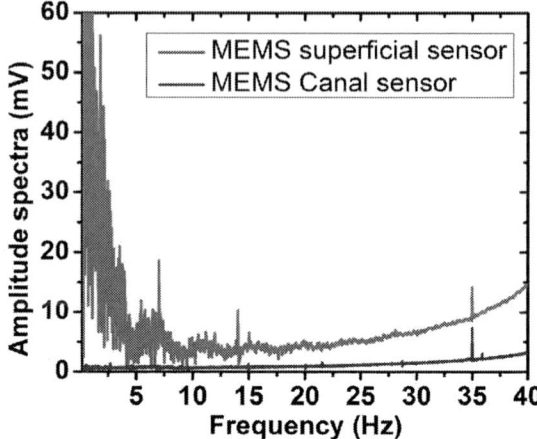

Figure 5: Experimental results showing that MEMS CN sensors reject the noises generated by dc and low-frequency ac flows. CN sensors demonstrate high-sensitivity in sensing the stimulus generated oscillatory flows.

CONCLUSION

Blind cave fishes that survive in deep waters are bestowed with the finest set of sensors that are designed through evolution. Studies conducted by the biologists reveal that the fish uses the SN sensors perform flow velocity sensing and the CN sensors perform flow acceleration sensing. The canals function as biomechanical filters with a frequency-based signal selection capability. This work presents the development of bio-inspired flexible canal encapsulated arrays of piezoelectric hair cell sensors for underwater acceleration sensing. A comparison between the performance of the artificial superficial and canal sensors clearly elucidates the ability of the canal structure in eliminating low frequency noises.

ACKNOWLEDGEMENTS

This research was funded by the Singapore National Research Foundation (NRF) through the Singapore-MIT Alliance for Research and Technology (SMART) Centre, Center for Environmental Sensing and Modeling (CENSAM) IRG.

REFERENCES

[1] J. C. Montgomery, S. Coombs, and M Halstead, "Biology of the mechanosensory lateral line in fishes," *Rev. Fish Biol. Fisher.*, vol. 5, pp. 399-416, 1995.

[2] G. S. Triantafyllou, M. S. Triantafyllou, and M. A. Grosenbaugh, "Optimal thrust development in

978-1-4799-7956-1/15 $31.00 © 2015 IEEE

oscillating foils with application to fish propulsion," *J. Fluids. Struct.*, vol. 7, pp. 205-224, 1993.

[3] J. C. Montgomery, S. Coombs, and C. F. Baker, "The mechanosensory lateral line system of the hypogean form of *Astyanax fasciatus*," *Evol. Biol. Fish.*, vol. 62, pp. 87-96, 2001.

[4] A. G. P. Kottapalli, M. Asadnia, J. M. Miao, G. Barbastathis and M. Triantafyllou, " A flexible liquid crystal polymer MEMS pressure sensor array for fish-like underwater sensing," *Smart. Mater. Struct.*, vol. 21, p. 115030, 2012.

[5] M. Asadnia, A.G.P. Kottapalli, Z. Shen, J. M. Miao, and M. Triantafyllou, "Flexible, and surface-mountable piezoelectric sensor arrays for underwater sensing in marine vehicles," *IEEE Sensors J.* vol. 13, pp. 3918-3925, 2013.

[6] A. G. P. Kottapalli, M. Asadnia, J. M. Miao, and M. Triantafyllou, "Electrospun nanofibrils encapsulated in hydrogel cupula for biomimetic MEMS flow sensor development," in *Proc. IEEE MEMS'13 Conference,* Taipei, January 20-24, 2013, pp. 25-28.

[7] A. G. P. Kottapalli, M. Asadnia, J. M. Miao, G. Barbastathis, and M. S. Triantafyllou, "Polymer MEMS pressure sensor arrays for fish-like underwater sensing applications," Micro. Nano. Lett., vol. 7, pp. 1189-1192, 2013.

[8] A. G. P. Kottapalli, M. Asadnia, J. M. Miao, and M. Triantafyllou, "Touch at a distance sensing: Lateral-line inspired MEMS flow sensors," *Bioinspir. Biomim.* vol. 9, p. 046011, 2014.

[9] A. G. P. Kottapalli, C. W. Tan, M. Olfatnia, J. M. Miao, G. Barbastathis and M. Triantafyllou, "A liquid crystal polymer membrane MEMS sensor for flow rate and flow direction sensing applications," *J. Micromech. Microeng.*, vol. 21, p. 085006 , 2011.

[10] A. G. P. Kottapalli, M. Asadnia, J. M. Miao, and M. Triantafyllou, "Harbor seal inspired MEMS artificial micro whisker sensors," in *Proc. IEEE MEMS'14 Conference,* San Francisco, January 26-30, 2014, pp. 741-744

[11] M. E. MConney. et. al., "Bioinspired material approaches to sensing," *Adv. Funct. Mater.*, vol. 19, pp. 2527-2544, 2009.

[12] S. Peleshanko et.al., "Hydrogel-encapsulated microfabricated hair cells mimicking fish cupula neuromast," *Adv. Mater.*, vol. 19, pp. 2903-2909, 2007.

[13] N. Chen, C. Tucker, J. M. Enge, Y. C. Yang, S. Pandya and C. liu "Design and characterization of artificial haircel sensor for flow sensing with ultrahigh velocity and angular sensitivity,", *J. Microelectromech. Syst.* Vol. 16, pp. 999-1014, 2007.

CONTACT

*A.G.P. Kottapalli, tel: +65-93722843; Center for Environmental Sensing and Modeling (CENSAM), Singapore-MIT Alliance for Research and Technology (SMART), ajay@smart.mit.edu

BAKING-POWDER DRIVEN CENTRIPETAL PUMPING CONTROLLED BY EVENT-TRIGGERING OF FUNCTIONAL LIQUIDS

David J. Kinahan[1], Robert Burger[2], Abhishek Vembadi[1], Niamh A. Kilcawley[1], Daryl Lawlor[1], Macdara T. Glynn[1] and Jens Ducrée[1]

[1]Biomedical Diagnostics Institute, National Centre for Sensor Research, Dublin City University, IRELAND

[2]Department of Micro- and Nanotechnology, Technical University of Denmark (DTU), Kongens Lyngby, DENMARK

ABSTRACT

This paper reports radially inbound pumping by the event-triggered addition of water to on-board stored baking powder in combination with valving by an immiscible, high-specific weight liquid on a centrifugal microfluidic platform. This technology allows making efficient use of precious real estate near the center of rotation by enabling the placement of early sample preparation steps as well as reagent reservoirs at the spacious, high-field region on the perimeter of the disc-shaped rotor. This way the number of process steps and assays that can be integrated on these of this "Lab-on-a-Disc" (LoaD) cartridge can be significantly enhanced while maintaining minimum requirements on the intrinsically simple, spindle-motor based instrumentation.

INTRODUCTION

By now LoaD platforms have shown great benefit for sample-to-automation of bioanalytical assays for a manifold of applications such as point-of-care diagnostics [1-3]. The underlying centrifugal microfluidic liquid handling scheme allows actuation by a simple, low-cost spindle motor. An inherent feature of this paradigm is the unidirectional nature of the radially outbound centrifugal field which severely constricts the number of assay steps such as metering and mixing that can be integrated on the limited real estate of a typically CD-sized LoaD cartridge. Furthermore, upstream sample preparation processes, such as blood centrifugation, and reagent reservoirs must be placed near the centre of the disc where space is most precious and the centrifugal field lowest.

To mitigate these limitations, a number of methods for pumping against the centrifugal field have been developed. These can include the addition of energy from external sources, transfer of energy to the liquid through the spindle motor and the storage of energy on the disc. External energy sources include connection to air bottles [4], the use of a thermal heat source to expand trapped gas (and thus displace liquid radially inwards) [5] and the use of external electrical energy to electrolytically displace liquid [6].

These methods tend to increase the complexity of instrumentation with respect to the system-innate spindle motor. Centrifugo-pneumatic pumping has been used to transiently story centrifugal energy in a compressed gas volume on the disc; at a reduced spin rate, the gas expands to centripetally pump liquid without need of any additional instrumentation [7]. While conceptually simple and elegant, this mechanism relies on highly 'dynamic' pumping implemented by powerful spindle motors for rapid changes of the spin rate during compression and expansion of the enclosed gas volume. Similarly, this pumping mechanism is largely enabled through use of high-resistance microchannels and, similarly, its efficiency is largely based on ratio of the flow resistances between inlet and outlet microchannels.

Potential energy has also been stored on-disc. Positive [8] displacement pumping uses an ancillary liquid to push a sample radially inwards. This can be implemented using an intermediary air pocket to ensure that the liquids do not come in contact. In a similar approach, negative [9] liquid displacement-based pumping, an ancillary liquid is generates an underpressure which draws a sample radially inwards. While promising, these pneumatic methods are linked to rather sensitive sample loading procedures. Recently, variations on this scheme have been introduced where relatively heavy immiscible ancillary liquids, such as oils [10] and fluorocarbons 11], have been used to displace samples radially inwards based through directly contacting the samples. This approach can also be used to accurately meter the samples during the pumping operations. However, a drawback to each 'potential energy' storage mechanism is that the ancillary liquid consumes valuable on-disc real-estate; while the most efficient pumping will be enabled by the ancillary liquid being located radially inwards.

In this work we present a pumping mechanism based on storage of chemical energy in ubiquitous, low-cost baking powder [12]. This pumping mechanism is governed by event-triggered control flow [13] and thus making it broadly independent of spin rate. Therefore the mechanism does not impact upstream and downstream Laboratory Unit Operation (LUOs) such as mixing and metering. Additionally, the pumping structure occupies minimal space and, connected via pneumatic channels, can be located at arbitrary locations on the disc.

OPERATION AND METHODS

Operating Principle

In our pumping concept (Fig. 1), the release of CO_2 from on-board stored baking powder [10] pressurizes a chamber to drive centripetal pumping. The pumping chamber is composed of four compartments; one containing the baking powder, one the ancillary liquid (water), the third contains a heavy immiscible liquid (Fluorocarbon FC-40; specific gravity ~1.85) while the fourth chamber transiently stores the sample. The FC-40 compartment is sealed by a water dissolvable film (DF). The other chambers are in pneumatic communication without allowing the interchange of liquids. When the sample enters the pumping chamber, the DF (called the VF)

978-1-4799-7956-1/15 $31.00 © 2015 IEEE

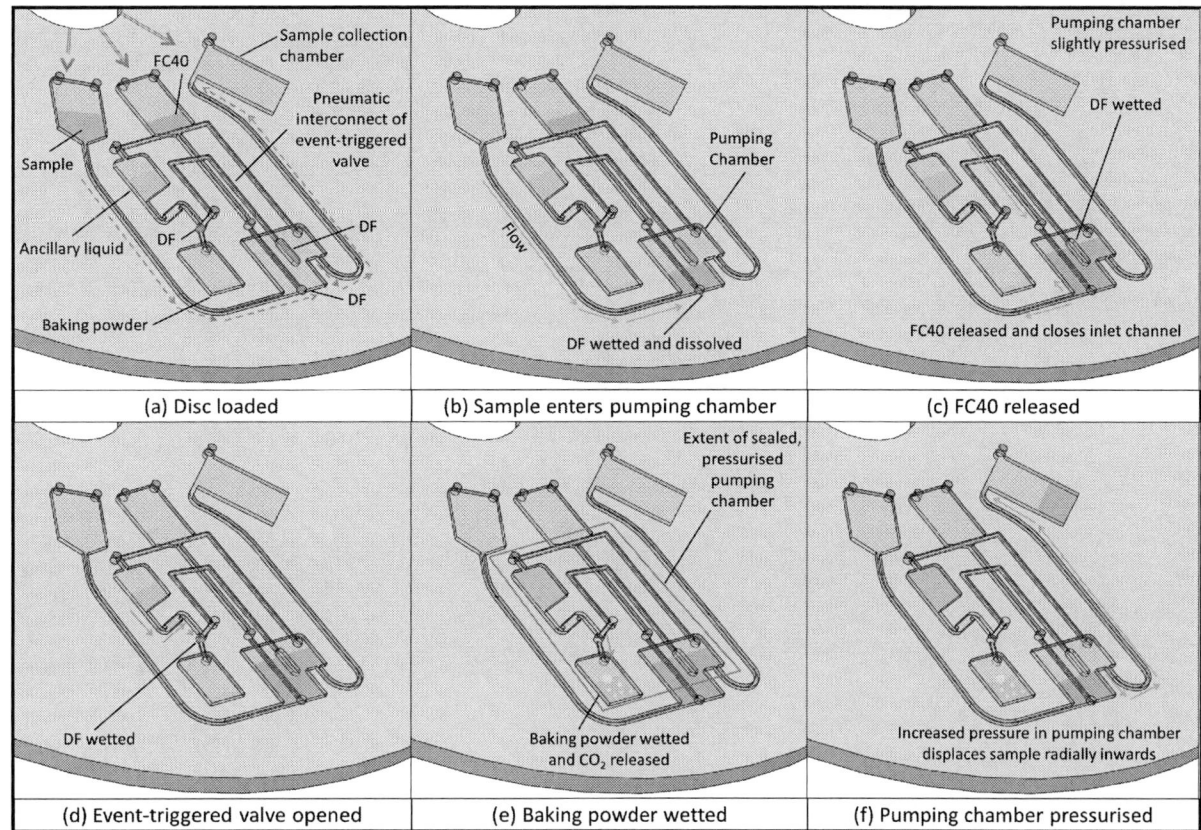

Figure 1 - Centripetal pumping structure. Note in the multilayer disc shown in Figure 2 some connecting channels are hidden as opaque materials are used during manufacture. (a) Disc-stored reagents. The orange arrows indicate the direction of centrifugal force and the dashed red line shows the nominal path of the sample. (b) Upon spinning the sample is centrifugally driven into the pumping chamber. Note the loading chamber is open to atmosphere and the pumping mechanism is triggered by the presence of the sample. Thus the centripetal pump is independent of spin-protocol. (c) FC-40 is released and closes the inlet channel. (d) The pumping liquid is displaced upwards and the event triggered valve is actuated. (e) The release the ancillary liquid (DI Water) to wet the baking powder and trigger the release of CO_2. Note the solid orange triangle represents the extent of the pressurised pumping chamber. (f) The arrival and centrifugal stabilization of the high-density FC-40 effectively seals the inlet channel so the emerging gas expansion displaces the sample through the only outlet radially inwards.

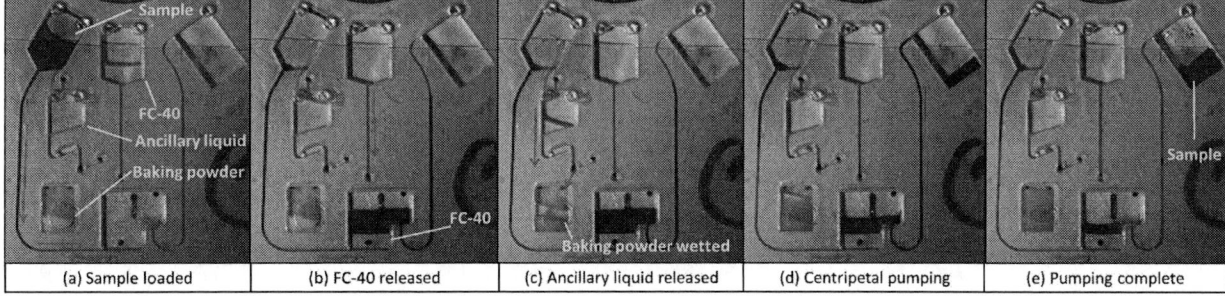

Figure 2: Centripetal Pumping. (a) Sample loaded and flowing radially outwards. (b) Clear liquid at the base of the pumping chamber is FC-40 has a higher specific weight than water so it is layered by the centrifugal field at the bottom where it thus effectively seal the inlet channel. (c) Ancillary liquid is released, activates the baking powder to generate CO_2. (d-e) Gas expands to centripetally pump the sample through the radially inbound channel.

dissolves and releases the FC-40. This liquid flows underneath the sample to fill the inlet channel. Through dissolving the control film (CF), the sample then opens the event-triggered valve and liquid is released to activate the baking powder. The subsequent release of CO_2 increases the gas pressure in the pumping chamber. However, due to the differing densities between the FC-40 (filling the inlet chamber) and sample (filling the outlet chamber), only the sample is pumped radially inwards. Thus the FC-40 acts akin to a check valve, preventing backflow of the sample during pumping.

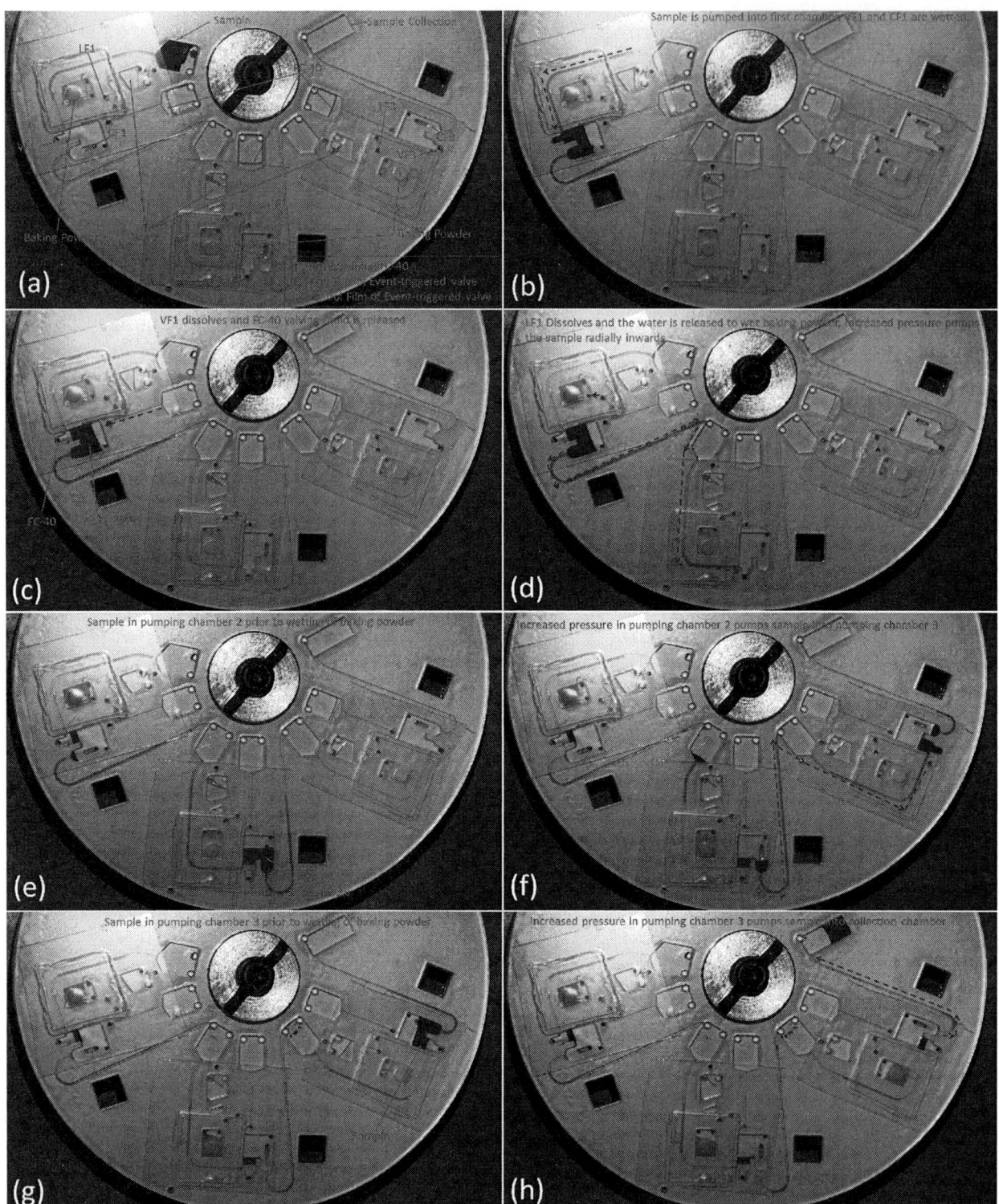

Figure 3: Repeated Centripetal Pumping of a Sample (a) Sample is loaded and flowing radially outwards. (b) Sample enters the first pumping chamber and wets DFs (VF restraining the FC-40 valving liquid and CF of an event-triggered valve). (c) the FC-40 (Clear liquid) is released and blocks the inlet channel. (d) Ancillary liquid (water) is released, activates the baking powder to generate CO_2 and pumps the sample radially inwards and into pumping chamber 2 (e-h) Process is repeated through pumping chamber 2 and pumping chamber 3 until the sample enters a collection chamber located radially inwards. Note this pumping occurs at a constant disc spin rate and is triggered only by the entry of the liquid into the pumping chambers.

Fabrication

The disc is prototyped by stacking multiple, specifically designed adhesive and structural layers as previously described [11]. Briefly, the disc is manufactured from laminates of PMMA bonded with pressure sensitive adhesive (PSA). DFs are mounted to block vertical vias where required. Note also that 'lower level' microchannels (not visible in the images here) provide liquid and pneumatic vias.

CONCLUSIONS AND OUTLOOK

As shown in Figure 2, this pumping mechanism can be used to pump a liquid, initially located at the centre of the disc, to the periphery and back to a radially central location. Figure 3 demonstrates the implementation of this mechanism in a series of three pumping chambers. Here, with the disc rotating at a constant spin rate, the sample is pumped inwards and outwards three times.

The pumping mechanism presented here expands the capabilities of the centrifugal platform in a number of ways. As the pump is only triggered by the presence of the sample, the structure operates widely independent of the spin rate. Thus, and unlike most previous implementations of centripetal pumping, we represent a module which can readily be inserted at any point of an on-disc workflow. Additionally, this pump takes up a comparatively little of space and, due to pneumatic connecting channels, the ancillary liquid and baking powder can be located at remote, arbitrary locations on the disc cartridge. Similarly, unlike displacement based (potential energy) pumping methods, the module saves valuable real-estate near the centre of the disc. While some sample is lost during pumping (representing a lack of efficiency), this is primarily owed to our present system design and prototyping methods.

These pumps have further application towards storage of reagents on the periphery of the disc. Here, the ancillary liquid could be released to wet baking powder at a pre-determined rotational frequency. This variant significantly increases the on-disc real-estate available for reagent storage and permits the spin rate controlled release of reagents.

REFERENCES

[1] Gorkin, R., Park, J., Siegrist, J., Amasia, M., Lee, B. S., *et al.,* "Centrifugal microfluidics for biomedical applications" *Lab on a Chip*, vol. 10, pp. 1758-1773, 2011.

[2] Ducrée, J., Haeberle, S., Lutz, S., Pausch, S., von Stetten, F. *et al* "The centrifugal microfluidic Bio-Disk platform", *Journal of Micromechanics and Microengineering*, vol. 17 pp 103, 2007.

[3] Madou, M., Zoval, J., Jia, G., Kido, H., Kim, J *et al.,* "Lab on a CD", *Annu. Rev. Biomed. Eng.*, vol. 8, pp 601-628, 2006.

[4] Kong, M. C., and Salin, E. D. "Pneumatically pumping fluids radially inward on centrifugal microfluidic platforms in motion", *Analytical chemistry*, vol. 82, pp. 8039-8041, 2010.

[5] Abi-Samra, K., Clime, L., Kong, L., Gorkin III, R., Kim, *et al.,* "Thermo-pneumatic pumping in centrifugal microfluidic platforms", *Microfluidics and nanofluidics*, vol. 11, pp. 643-652, 2011.

[6] Noroozi, Z., Kido, H., and Madou, M. J. "Electrolysis-induced pneumatic pressure for control of liquids in a centrifugal system", *Journal of The Electrochemical Society*, vol. 158, pp. 130-P135, 2011.

[7] Zehnle, S., Schwemmer, F., Roth, G., von Stetten, F., Zengerle, R. *et al.,* "Centrifugo-dynamic inward pumping of liquids on a centrifugal microfluidic platform", *Lab on a Chip*, vol. 12, pp. 5142-5145, 2012.

[8] Kong, M. C., Bouchard, A. P., & Salin, E. D. "Displacement pumping of liquids radially inward on centrifugal microfluidic platforms in motion", *Micromachines*, vol. 3, pp.1-9, 2011.

[9] Soroori, S., Kulinsky, L., Kido, H., and Madou, M. "Design and implementation of fluidic micro-pulleys for flow control on centrifugal microfluidic platforms" *Microfluid Nanofluid*, vol. 16, pp. 1117-1129, 2014.

[10] Kilcawley, N., Kinahan, D., Nwankire, C., Glynn, M., and Ducrée, J. "Buoyancy-driven centripetal pumping for nested sample preparation in bioassays" In *Proceedings of the 18th International Conference on Miniaturized Systems for Chemistry and Life Sciences,* October 26-30, 2014, San Antonio, Texas, USA.

[11] Kim, T.H, Park, J. and Cho, Y.K., "Fully Integrated Molecular Diagnostics of Pathogenic Microorganisms on a Disc" in *Proceedings of the 18th International Conference on Miniaturized Systems for Chemistry and Life Sciences,* October 26-30, 2014, San Antonio, Texas, USA.

[12] Ahn, Chong H. "Disposable Polymer "Smart" Lab-on-a-Chip for Point-of-Care Testing (PocT) in Clinical Diagnostics", in *Proceedings of The 13th International Conference on Solid-State Sensors, Actuators and Microsystems*, Seoul, Korea, June 5-9, 2005.

[13] Kinahan, D. J., Kearney, S. M., Dimov, N., Glynn, M. T., and Ducrée, J. "Event-triggered logical flow control for comprehensive process integration of multi-step assays on centrifugal microfluidic platforms", *Lab on a Chip*, vol. 14, pp. 2249-2258, 2014.

CONTACT

*J. Ducrée, tel: +353-1-700-5377; jens.ducree@dcu.ie

CONTINUOUS-FLOW DIELECTROPHORETIC SORTING OF PARTICLES VIA 3D SILICON ELECTRODES FEATURING CASTELLATED SIDEWALLS

Xiaoxing Xing, and Levent Yobas

The Hong Kong University of Science and Technology, Hong Kong, P. R. of CHINA

ABSTRACT

Continuous-flow dielectrophoretic sorting of particles has been demonstrated using a simple microfluidic design incorporating 3D electrodes with castellated sidewalls. Two variations of the design have been fabricated, slightly differing in their sidewall and separation junction profiles, through a single-mask process on silicon-based platforms. These 3D silicon electrodes have shown the capacity to segregate polystyrene beads into distinct flow layers along the channel depth while delivering them to separate outlets through a downstream junction of a specific design. The utility of either structure has been showcased by sorting a mixture of 1 and 10 µm beads based on their size continuously at a velocity of 1.5 mm/s.

INTRODUCTION

Over the years, dielectrophoresis (DEP) has attracted tremendous attention as a powerful technique for selective manipulation of particles in lab-on-a-chip (LOC) systems. DEP-activated cell sorting (DACS), in particular, has shown potential for medical diagnosis, cell therapeutics, and drug discovery [1-2]. DACS is advantageous over the commercialized particle separation techniques as these techniques typically rely on foreign labels such as fluorophores (fluorescence activated cell sorting, FACS) and functionalized beads (magnetic activated cell sorting, MACS) all targeting specific surface ligands. Being label free, DACS discriminates particles even in the absence of specific surface ligands based on their induced dipole moment in a non-uniform electric field, reducing the complexity and cost of the separation procedure that may arise from external labels [3].

Electrodes and their configurations play a pivotal role in setting a non-uniform electric field across the sample. Conventional designs utilize thin-film coplanar electrodes albeit known to be effective for those particles near the electrodes whereas for those at a distance, the induced force diminishes quickly [4]. Placing thin-film electrodes on the chamber top and bottom addresses this concern to some extent and yet necessitates face-to-face precise alignment during fabrication [1]. Volumetric electrodes instead extend the electric field presence along the entire channel depth and thus exert an effective dielectric force field on particles regardless of their positions within the chamber. Such electrodes have been typically made of a thick metal, pyrolyzed thick resist, or heavily doped silicon [2, 4, 5]. Fabrication of thick metal electrodes such as high aspect ratio metal posts requires complex and costly electroplating. Conducting posts can also be made of SU8 photoresist pillars pyrolyzed into carbon posts albeit often with dimensional inconsistencies as a result of variable shrinkage at high temperatures. These electrodes require additional photomasking steps to pattern their electrical

leads and to further integrate them with fluidic channels. The single crystal silicon electrodes structured through deep reactive ion etching (DRIE) offer a fairly compact way of constructing DACS systems where they serve as the electrodes as well as chamber sidewalls [4]. Nevertheless, none of these volumetric electrodes displays a genuine 3D profile; they project a quasi-2D profile without structural variation along the chamber depth and thus unable to exert effective forces in that direction.

We recently introduced a unique design featuring 3D silicon electrodes with castellated sidewalls [6-8]. A benefit of castellated sidewalls apparent in our design is that they reduce the complexity of overall electrofluidic integration to a single photomasking step. Such sidewalls allow fluidic paths to pass through the 3D electrodes without interrupting their electrical continuity; this has led to a 3D interdigitated comb array where each silicon digit is perforated with an array of lateral pores. We showed the utility of the array on the isolation and enumeration of circulating tumor cells (CTCs) from leukocytes based on their dielectric signature at a high cell-loading density [6-8]. This design, however, falls short of taking full advantage of castellated sidewalls for continuous-flow separation. Such feature is rather introduced here using two separate designs that slightly differ in their sidewall profile as well as downstream junction layout as illustrated in Figure 1. In either design, a straight main channel emerges from a gap 100 µm wide and 20 mm long between the 3D silicon electrodes, connecting a single inlet to a downstream outlet after branching out daughter channels at a nearby junction. In one design, streams branching out to either side have to flow beneath a bridge structure that is simply monolithic extension of the siding silicon electrode. This particular design also features a distinct sidewall profile; the curved sidewalls directly join the channel floor.

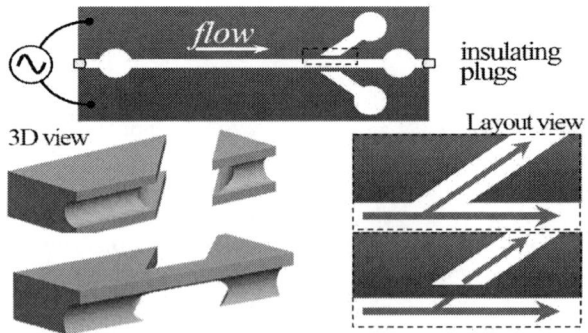

Figure 1: Illustrations of the two designs using 3D silicon electrodes for continuous-flow DEP sorting of particles. The overall layout with their junction regions (the dashed rectangle) detailed in lower panels per each design and shown next to their 3D rendering.

978-1-4799-7956-1/15 $31.00 © 2015 IEEE

MATERIAL AND METHODS
Microfabrication

The devices without bridges were directly fabricated on a silicon-on-insulator (SOI) substrate whereas those featuring bridges were realized on a composite substrate formed by silicon and glass anodic bonding. Both the substrates offered silicon crystals (100) oriented and doped heavily (0.005 Ω-cm) for forming the electrodes. The fabrication process borrowed from our earlier work [6-8]. First, a SiO$_2$ hard mask was patterned on silicon as per the layouts and the exposed regions were etched using DRIE to form the upper planar sidewalls. The structure was then deposited with low-temperature oxide (LTO) upon which the LTO was removed from the channel floor through reactive ion etching (RIE), leaving the top surface and sidewalls passivated. The bulk silicon was next removed from the exposed floor in isotropic dry etch, forming the curved sidewalls. In the devices featuring bridges, the curved sidewalls reached the glass substrate while pinching and undercutting the bridges at the junction. In the remaining devices, a second DRIE step followed the isotropic etch and formed the lower planar sidewalls down to the buried oxide layer.

Experimental

The devices were characterized with fluorescent polystyrene beads in various sizes (Bangs Laboratories, Ex/Em 480/520nm). The beads were washed at 5000 rpm for 2 min and then resuspended in deionized (DI) water. Homogeneous suspensions of 10 and 15 μm beads were obtained at a density of 5×10^6 mL^{-1}. For the sorting experiments, samples were prepared by mixing 10 and 1 μm beads at a density of $0.5 \times$ and 5×10^7 mL^{-1}, respectively. The devices were permanently enclosed with a cover featuring inlet/outlet fluidic ports and electrical vias in polydimethylsiloxane (PDMS). Samples were introduced at a controlled rate using a syringe pump (Harvard Apparatus). A sine-wave voltage excitation was delivered through a transformer connected to a power amplifier (AL-50HFA, Amp-Line Corp., NY) supplied by a function generator (Tektronix CFG250). The voltage was monitored on an oscilloscope (Tektronix TDS 2012C) and an epi-fluorescence microscope (Nikon ECLIPSE FN1) equipped with a colored CCD camera (SPOT, RT3 Mono) was used to acquire the device images.

Numerical Modeling

The 3D device simulations were performed using COMSOL Multiphysics Software v3.5 (Comsol Inc., MA). The electric field maps were derived based on the spatial derivative of the potential field obtained by numerically solving the Laplace equation. The boundary condition was set to a sine-wave excitation of 7 V-rms at 400 kHz. The streamlines were obtained from the Navier–Stokes equation numerically solved under the assumptions of incompressible laminar flow and no slippage. A constant flow speed of 3 mm/sec was set at the inlet, corresponding to the highest flow rate used in the experiments whereas no viscous stress was assigned to the outlet as the fluidic boundary conditions. The fluidic and electrical properties of the materials were set as per our previous work [8].

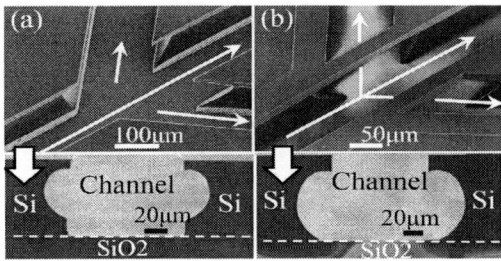

Figure 2: SEM images of the junctions and sidewall profiles from the two fabricated designs shown in oblique and cross-sectional (lower panel) views.

RESULTS AND DISCUSSION
Fabrication Results

Figure 2 shows scanning electron microscopy (SEM) images of the completed designs revealing their junction and channel (electrode) profiles. In the images, electrode or channel profiles can be directly compared between the two designs. The design without the bridge electrodes has smaller features than the latter design across the channel depth including both the upper planar sidewalls and the curved sidewalls. In the former, each curved sidewall surrounds a space that spans about 25 μm wide and 50 μm deep between the upper and lower planar sidewalls 10 and 15 μm deep, respectively. In the design featuring the bridge electrodes, the curved sidewalls marks a volume that extends 45 μm wide and 80 μm deep beneath the planar sidewalls 20 μm deep. These curved sidewalls also rest in direct contact with the channel floor. The bridge electrodes remain intact and show no sign of buckling.

Simulation and Experimental Results

For the design featuring no bridge at the junction, the simulated electric field intensity across the main channel is shown in Figure 3. The arrows indicate the direction of the dielectric force. The dashed lines AA' and BB' mark the focal planes for which the particle trajectories predicted based on numerically simulated streamlines are shown in subsequent panels and compared against those images experimentally obtained from such particles delivered at 1.67 μL/min under 7 V-rms sine-wave excitation at either 1 or 400 kHz. The electric field profile across the channel depth recalls that of the traditional planar castellated electrodes. This profile however persists along the entire channel length except for the junction. The field gets intensified near the sites facing the planar sidewalls, in particular around the pointing corners neighboring the curved sidewalls. These are the sites where the beads under positive DEP (pDEP) would be drawn to whereas those under negative DEP (nDEP) would be repelled as illustrated in Figure 3a. The beads under nDEP are likely to align with the channel center where a local field minimum exists or with the curved sidewalls where the field intensity is of the lowest. This picture is supported by the predicted trajectories and validated by the experimental results in Figure 3b. The beads under 1 kHz excitation are shown railed by the planar sidewalls in the main channel and then diverted to the side channels. Further into the side channels, the beads start to redistribute freely as the field dies off.

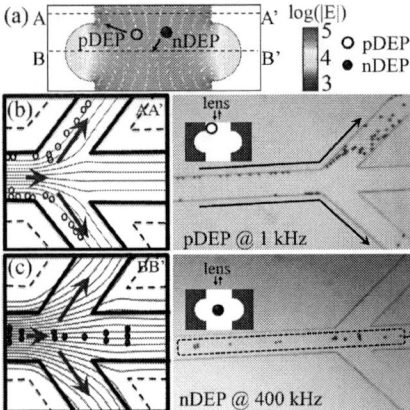

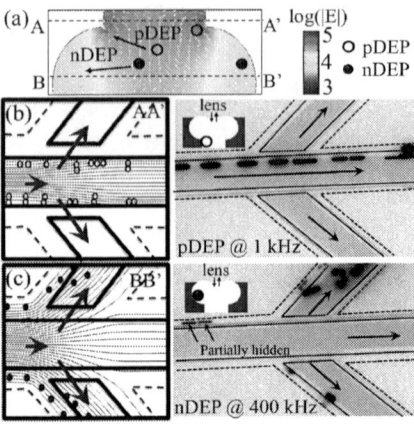

Figure 3: (a) Electric field intensity simulated across the design without the bridges. (b,c) Streamlines and particle trajectories in AA' and BB' focal planes as per the dashed lines in (a) simulated under (b) pDEP and (c) nDEP. The solid and dashed boundaries are the planar and curved sidewalls, respectively. Micrographs: 10 µm beads in DI water introduced at 1.67 µL/min and with a sine-wave excitation of 7 V-rms at (b) 1 and (c) 400 kHz. Insets: illustrations of bead positions along the channel depth.

Figure 5: (a) Electric field intensity simulated across the design featuring bridges. (b,c) Streamlines and particle trajectories in AA' and BB' focal planes as per the dashed lines in (a) simulated under (b) pDEP and (c) nDEP. The solid and dashed boundaries are the planar and curved sidewalls, respectively. Fluorescence micrographs: 15 µm beads in DI water applied at 1.67 µL/min with a sine-wave excitation of 7 V-rms at (b) 1 and (c) 400 kHz. Insets: illustrations of bead positions along the channel depth.

With the frequency increased to 400 kHz, the beads are depicted experiencing nDEP with nearly all aligned to the channel center migrating to the main outlet bypassing the side channels. Those that happen to reposition next to the curved sidewalls, albeit hidden under the overhangs should reemerge again in the side channels as the field is diminished. Yet, the experiments have shown no evidence of beads diverted to the either side channel at 400 kHz. Additional experiments have further shown that the beads under nDEP could be found in one of the two equilibrium positions depending on the field frequency (Figure 4). While the beads experience pDEP are drawn to the planar sidewalls at low frequencies (1 kHz) as in Figure 4a, they are repelled from such regions with the increased frequencies. At intermediate frequencies (e.g., 100 kHz), the beads are found to prefer positions near the curved sidewalls completely hidden from view, Figure 4b, whereas at high frequencies (e.g., 400 kHz), they prefer to move to the channel center. This trend is believed to be due to the four corners of the curved sidewalls forming a strong field cage at the channel center that can prevent the beads migrating toward the sidewalls at high frequencies. This is not noticeable in the other design, as the particles cannot be strongly confined in space between a pair of corners (field maxima).

For the design featuring junction bridges, the corresponding results are presented in a similar way in Figure 5. The beads under pDEP at low frequencies continue to align with the planar sidewalls as the field gets intensified (Figure 5a). However, unlike in the former design, those beads are prevented from entering the side channels and instead railed to the main outlet by the active bridges (Figure 5b). At high frequencies, the beads are pushed against the curved sidewalls and yet stay partially visible (Figure 5c). The reason for the partial visibility is that the design was observed upside down during the experiments as illustrated by the sketches inserted on the respective images. The bridges on the either side being at a different level than the beads cannot stop those beads from entering the side channels.

Sorting of Beads

The two designs have been further demonstrated for the task of size-based separation of the polystyrene beads. Figure 6 depicts representative images obtained during the continuous-flow separation of a binary mixture of 10- and 1-µm beads in DI water at a rate of 0.83 µL/min and under a field excitation of 14 V-rms. For the design devoid of bridges, the field was applied intentionally at a high frequency (500 kHz) so as to mobilize 10-µm beads to the channel center under nDEP activation so that they can be collected at the main outlet. At this frequency, 1-µm beads are seen experiencing pDEP albeit too weak to be directly railed on the planar sidewalls to downstream and yet still enough to keep the beads at a close distance to be diverted to the side channels. Almost all those beads ended up in the side channels except for very few which managed to escape along with 10-µm beads. Contrarily, none of the 10-µm beads has been found in the side channels.

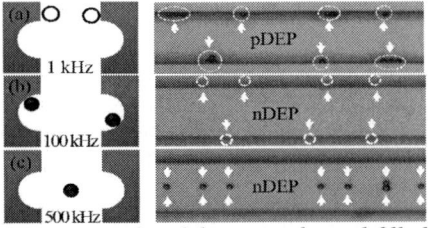

Figure 4: Micrographs of the main channel filled with 10 µm beads in DI water under (a) pDEP at 1 kHz, and (b,c) nDEP at (b) 100 and (c) 500 kHz. Illustrations: bead positions along the channel depth.

978-1-4799-7956-1/15 $31.00 © 2015 IEEE   510

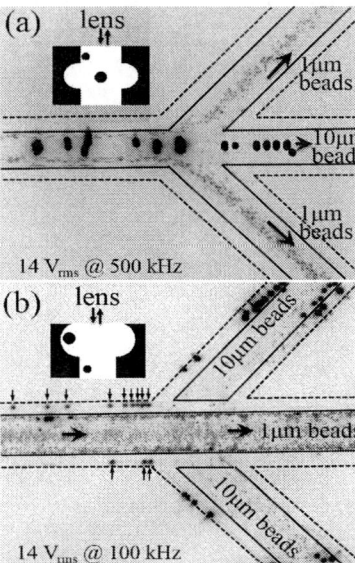

Figure 6: Fluorescence micrographs of the designs (a) without and (b) with bridges during the separation of 10 and 1 μm beads in DI water introduced at 0.83 μL/min with a sine-wave excitation of 14 V-rms at (a) 500 and (b) 100 kHz. The solid and dashed boundaries refer to the planar and curved sidewalls, respectively. The bold arrows indicate the flow direction. Insets: illustrations of bead positions along the channel depth.

In the design featuring junction bridges, the field was applied at an intermediate frequency (100 kHz) since the effect of pDEP became noticeably weak on 1-μm beads at higher frequencies. A cursory comparison between the two images confirms this as the 1-μm beads are seen traveling being closer to the planar sidewalls in the design with a lower activation frequency. However, a significant fraction of the 1-μm beads within that design are also seen dispersed in the channel center. This is believed to be due to buoyancy force acting on these beads and the location of the planar sidewalls with respect to the channel depth. It should be noted that the design was imaged again upside down during the experiments to spot those beads near the curved sidewalls. All the dispersed beads in the channel center can be seen bypassing the side channels along with those 1-μm beads near the planar sidewalls. Some of those beads under insufficient pDEP, however, escaped towards the curved sidewalls and ended up in the side channels contaminating the 10-μm beads. Almost all the 10-μm beads entered the side channels.

CONCLUSION

The 3D silicon electrodes have been integrated into microfluidics featuring castellated sidewalls and utilized for continuous sorting of polystyrene beads based on size. The DEP-based separation has been performed under a pressure-driven flow with a velocity up to 1.5 mm/s, an order of magnitude higher than our previous design where the separation was kept in batch mode [7, 8]. Integrating these miniaturized 3D electrodes tangential to the laminar flow streams allowed for the continuous-mode separation of particles by diverting them into distinct equilibrium positions along the channel width as well as the channel

depth. The integration has been realized through a single-step photomasking process in which the sidewalls were structured by tailoring silicon etch profile. The specific profile has also allowed for the creation of suspended active structures within the same process steps. These active bridges when placed at a junction have been shown to rail the particles under pDEP to a downstream outlet while allowing those in lower layers to freely flow underneath and get diverted to a separate outlet. The 3D electrodes with a tailored sidewall profile should not be limited to the specific design or fabrication process shown here and can be explored for further beneficial ways of coupling the electric field profile and flow streamlines toward an effective sorting of particles and cells. The continuous sorting of particles at a high speed makes such devices attractive for high-throughput sorting applications, and they can be further integrated as a front-end module in LOC systems for biomedical analysis. We are currently developing such devices for the enumeration of CTCs from blood.

ACKNOWLEDGEMENTS

This work was supported by the Research Grant Council of Hong Kong under Grant 621711.

REFERENCES

[1] J. Park, B. Kim, S. K. Choi, S. Hong, S. H. Lee, and K. I. Lee, "An efficient cell separation system using 3D-asymmetric microelectrodes," *Lab Chip*, vol. 5, pp. 1264-1270, 2005.

[2] R. Martinez-Duarte, R. A. Gorkin, K. Abi-Samra, and M. J. Madou, "The integration of 3D carbon-electrode dielectrophoresis on a CD-like centrifugal microfluidic platform," *Lab Chip*, vol. 10, pp. 1030-1043, 2010.

[3] T. B. Jones, *Electromechanics of Particles:* CAMBRIDGE University Press, 1995.

[4] F. E. H. Tay, L. M. Yu, A. J. Pang, and C. Iliescu, "Electrical and thermal characterization of a dielectrophoretic chip with 3D electrodes for cells manipulation," *Electrochim Acta*, vol. 52, pp. 2862-2868, Feb 10 2007.

[5] L. Wang, L. A. Flanagan, N. L. Jeon, E. Monuki, and A. P. Lee, "Dielectrophoresis switching with vertical sidewall electrodes for microfluidic flow cytometry," *Lab Chip*, vol. 7, pp. 1114-1120, 2007.

[6] X. X. Xing, M. Y. Zhang, and L. Yobas, "Interdigitated 3-D Silicon Ring Microelectrodes for DEP-Based Particle Manipulation," *J Microelectromech S*, vol. 22, pp. 363-371, Apr 2013.

[7] X. Xing, L. Yobas, *MEMS2014 conference*, San Francisco, Jan 26-30, 2014, pp. 951-954.

[8] X. X. Xing, R. Y. C. Poon, C. S. C. Wong, and L. Yobas, "Label-free enumeration of colorectal cancer cells from lymphocytes performed at a high cell-loading density by using interdigitated ring-array microelectrodes," *Biosens Bioelectron*, vol. 61, pp. 434-442, Nov 15 2014.

CONTACT

*L. Yobas, tel: +852-23587068; eelyobas@ust.hk

ELECTRON BEAM SWITCHED TRAPPING AND RELEASE OF NANOPARTICLES ON NANOPORE ARRAY

Takayuki Hoshino[1] and Kunihiko Mabuchi[1]
[1]The University of Tokyo, JAPAN

ABSTRACT

We demonstrated switching of trap and release of nanoparticles using an inverted-electron beam lithography (I-EBL). 240-nm nanobeads suspended in pure water were trapped and released on nanopore array. The nanopores in a silicon nitride membrane generated trapping flows for the beads by infinitesimal leakages toward directly connected high vacuum via the nanopores. The incident electron beam selectively induced release of the trapping beads into the solution by Coulomb force.

INTRODUCTION

Chemical reactions could be modulated by the concentrations of the chemical species dispersed in the solution. Thus in-situ modulation of solute localization and concentration as in spatio-temporal pattern would contribute the precise stimulus to biomolecule analysis. Laminar flows in microchannels had been applied to modulate the concentration gradation of the solutes. [3] Mixing and dilution with some solutions were utilized for the single cell analysis and biomolecular analysis. [1-3] Optical trapping [4] and laser induced dielectrophoresis [5,6] could manipulate a living cell and small devices. Although these devices very useful for the analysis of the extremity small dynamic systems, these were limited the manipulation resolution because of light diffraction and fabrication techniques.

Thus, we demonstrated here the switching method for the controlling localization and concentration of nanoparticles in solution using nanopore array and the inverted-electron beam lithography (I-EBL). The nanopore array that interconnected a solution and vacuum locally generated infinitesimal flows in the solution by infinitesimal water leakages, and they trapped the nanoparticles on each nanopores. The I-EBL could generate electrostatic repulsive force to the trapped nanoparticle to disperse the nanoparticles into the solution. We demonstrated trapping the nanobeads and dispersion of them using four nanopores in a silicon nitride (SiN) membrane.

EXPERIMENTS

Experimental setup

Figure 1 shows the experimental setup for the nanobeads trapping and release. The inverted-electron beam lithography (I-EBL) [7-12] was utilized for this study. The I-EBL consisted of an inverted EB and a fluorescent microscope in co-axially to observe simultaneously. A 100-nm thick SiN was work as a barrier for the vacuum of Schottky type EB system from wet atmospheric samples. [7] 2.5 keV of accelerating voltage was fine focused on the SiN. Vacuum was kept to below 10^{-3}Pa at main vacuum chamber and kept ultra-high vacuum at beam emitter using

differential pumping.

240-nm nanobeads (fluorescent beads) were dispersed in deionized water (DI water) on the SiN. The displacements of the nanoparticles were measured and recorded using sCMOS camera at 50 fps using high NA water immersion type objective lens (OL). Fluorescent images were observed via FITC-filter (Ex: 488 nm, BA: 530 nm).

Porations in the SiN and nanobeads manipulations were sequentially operated on this setup.

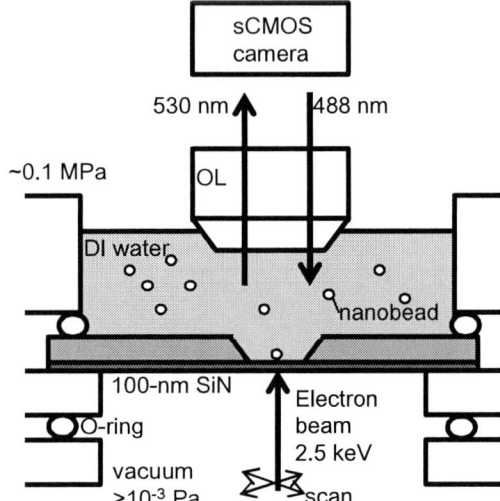

Figure 1: Experimental setup for nanobeads trapping and release using an inverted-electron beam lithography. 100-nm thick SiN was set over the I-EBL optics, which was work as a barrier for the vacuum of EB system from wet atmospheric samples. 240-nm nanobeads were dispersed in deionized water on the SiN. Porations in the SiN and nanobeads manipulations were sequentially operated on this setup. The displacements of the nanoparticles were measured and recorded using sCMOS camera.

In-situ poration in SiN

Porations in the SiN were operated on this experimental setup. The four nannopores (P1-P4) were fabricated in-situ in the 100-nm thick SiN membrane by using this I-EBL with high current dose. Since 100-nm thick free-standing SiN had enough strength to separate the pressure conditions of atmospheric solution (~0.1 MPa) and high vacuum ($<10^{-3}$ Pa), the perforated nanopores were fabricated in the SiN membrane via I-EBL to generate nanoflow of water in the target solution, as shown in figure 2.

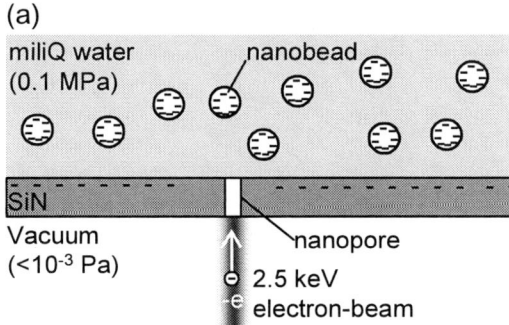

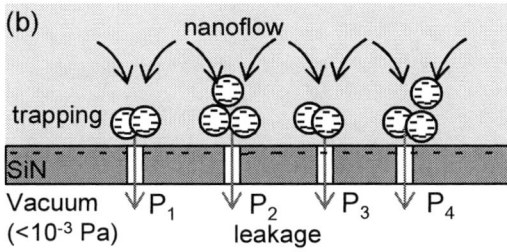

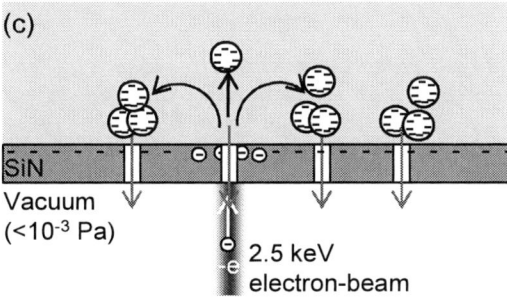

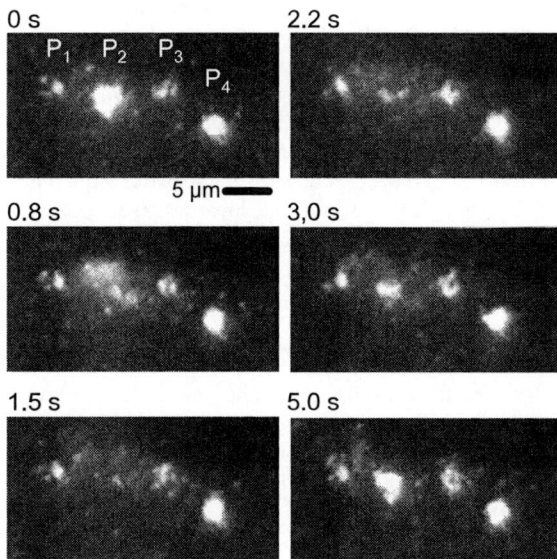

Figure 3: Fluorescent video images of nanobeads trapping and release. The nanobeads were trapped and their clusters were appeared on each nanopores P_1-P_4. EB was incident on P2 at 0.8 s. At 1.5 s, EB irradiation has been finished. After 2.2 s, the floating nanobeads were trapped on P_2 again and the concentration at P_2 was increasing.

Figure 2: Nanobeads trapping and release using nanopores and I-EBL. The free-standing SiN membrane separated water solution in atmospheric pressure and high vacuum chamber which had EB optics. (a) High dose EB perforated nanopores at arbitrary positions on a free-standing SiN membrane. (b) The nanopores connected water solution and high vacuum to leak water via nanopores. The leakage flow collected floating nanobeads (c) EB induced nanobeads dispersion due to electrostatic repulsive force. The dispersed nanobeads were released from the incident pore and were caught next pores such as side-by-side delivery.

Nanobeads trapping and release

Figure 2 shows the procedure of nanobeads trapping and release. Water was constantly leaked toward vacuum and evacuated with rapidly because of infinitesimal water leakage. Smaller diameter of the nanopores than that of the dispersed nanobeads could contribute separation of the nanobeads and water leakage at the nanopores, so the nanopores normally sunk infinitesimal flow of water. From the vacuum chamber, an electron beam was selectively incident to the particular nanopores to generate

negative electrostatic repulsive force to the trapping nanobeads, then the nanobeads were dispersed into the solution. The 240-nm fluorescent beads were dispersed in pure water with pluronic F-127 on the SiN membrane. The responses were simultaneously observed using the fluorescent microscope with x100 objective, and recorded using sCMOS camera with 50 fps at fluorescent imaging.

RESULTS AND DISSCUSSIONS

The water leakage flow generated by the perforated nanopores could collect floating nanobeads in the solution and trapped their clusters on all nanopores (at 0 s in figure 3). When the spot EB was incident to particular nanopore P2, the responses observed by the fluorescent imaging showed that the nanobeads were selectively dispersed into the solution and concentration of the nanobeads at P2 was rapidly decreased during EB irradiation (at 0.8-1.5 in figure 3). The spread nanobeads from P2 were trapped on immediate neighbor nanopores P1 and P3 as shown in figure 4. After EB irradiation, beads concentration at P2 was slightly recover.

The electron beam had several tens nanometers resolutions, therefore the poration could formed with small diameter below the nanobeads diameter. Our previous study of the I-EBL resolution shows 120 nm on the upper surface of the SiN at line-and-space patterns. [11] Furthermore we reported electrokinetic force could be generated to the solution on the SiN with tens micrometer region around the beam spot in pure water, and micrometers in electrolyte solution. [9-11] These electrokinetic repulsion force could contribute the selected release on single pore.

The water leakage from nanopores were extremity few

compared to evacuation capacity, therefore the vacuum condition was kept well for emitting and fine focusing the beam. Introducing such rare water and gas flows into vacuum chamber were generally utilized for EB/focused ion beam chemical vapor depositions at nanotechnology, and vacuum down and contamination in beam column was also solved at those systems using the differential pumping. Thus, water leakage induced damages to the EB system and performance decrement of the EB drawing would expect to be limited in our system.

CONCLUTION

Here, we demonstrated switching of trapping and releasing the nanobeads using the I-EBL, and also confirmed applicability for a horizontal side-by-side manipulation in the nanopore array in water solution. This technique would contribute nanomanipulation of nanomixing and handling for biomolecular analysis.

ACKNOWLEDGEMENTS

This work was partly supported by The Precise Measurement Technology Promotion Foundation.

REFERENCES

[1] Kazuki Iijima, Noritada Kaji, Yukihiro Okamoto, Manabu Tokeshi, and Yoshinobu Baba, "SINGLE MOLECULE ENZYMATIC KINETICS IN SUBCELLULAR-SIZED NANOSPACES USING PNEUMATIC VALVE-ASSISTED ATTO-LITER CHAMBER ARRAY DEVICES," *Proc. Micro Total Analysis Systems 2011*, pp. 1092–1094.

[2] C. Y. Lee, C. L. Chang, Y. N. Wang, and L. M. Fu, "Microfluidic Mixing: A Review," International *Journal of Molecular Sciences*, vol. 12, no. 5, pp. 3263–3287, 2011.

[3] S. Takayama, E. Ostuni, P. LeDuc, K. Naruse, D. E. Ingber, and G. M. Whitesides, "Laminar flows: Subcellular positioning of small molecules," *Nature*, vol. 411, no. 6841, p. 1016, Jun. 2001.

[4] A. Ashkin, "Optical trapping and manipulation of neutral particles using□lasers," Proceedings of the National Academy of Sciences, vol. 94, no. 10, pp. 4853 –4860, May 1997.

[5] V. Velasco, A. H. Work, and S. J. Williams, "Electrokinetic concentration and patterning of colloids with a scanning laser," *ELECTROPHORESIS*, vol. 33, no. 13, pp. 1931–1937, Jul. 2012.

[6] S. Maruo and N. Yoshimura, "Polymeric micromachines driven by laser-induced negative dielectrophoresis," in *2011 International Symposium on Micro-NanoMechatronics and Human Science (MHS)*, 2011, pp. 327–332.

[7] Takayuki Hoshino, Kunihiko Mabuchi, "Closed-looped in situ nano processing on a culturing cell using an inverted electron beam lithography system," *Biochemical and Biophysical Research Communications*, vol.**432** (2), pp. 345-349, 2013.

[8] Takayuki Hoshino, Keisuke Morishima, "Electron-Beam Direct Processing on Living Cell Membrane", *Applied Physics Letters*, **99**, pp.174102, 2011.

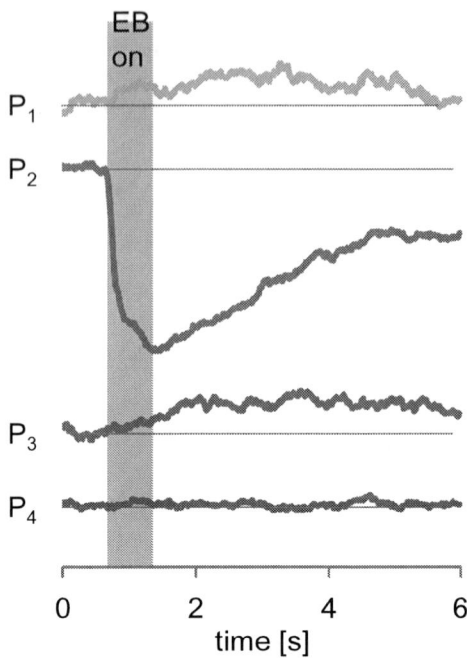

Figure 4: Fluorescent intensity (ΔF/F) at the nanopore regions. Time sequence of EB incident was indicated as dark pattern. P_2 was reduced and then neighbor P_1 and P_3 were slightly increased.

[9] Hiroki Miyazako, Kunihiko Mabuchi, Takayuki Hoshino, "2-D Electrokinetic Nano-manipulation for Aqueous Solution by Using a Simple Scanning Electron Beam," *Proceedings of The 18th International Conference on Miniaturized Systems for Chemistry and Life Sciences (Micro TAS 2014)*, M.442e, San Antonio (USA), 26-30 Oct. 2014, pp. 1677-1679 (M.442e).

[10] Takayuki Hoshino, Akira Wagatsuma, Kunihiko Mabuchi, "Direct Chemical-Computer Interface for Living Cell Analysis," *Proceedings of The 17th International Conference on Miniaturized Systems for Chemistry and Life Science (µTAS 2013)*, W.020b, Freiburg (Germany), 27-31 October, 2013.

[11] Takayuki Hoshino, H. Miyazako, A. Wagatsuma and K. Mabuch, "Measurement of Intracellular Strain Energy Distributions by Using Inverted-Electron Beam Lithography," *Proceedings of The 26th International Microprocesses and Nanotechnology Conference (MNC 2014)*, 7P-11-93, Fukuioka, (JAPAN), 5-7 November 2014.

[12] Takayuki Hoshino, Keisuke. Morishima, "Closed-looped Nano Stimulation Microscope for Living Cell Membrane", *Proceedings of 2012 7th IEEE International Conference on Nano/MicroEngineering and Molecular Systems (NEMS 2012)*, T4C-6, Kyoto (Japan), March 5-8, 2012, pp. 117-120.

[13] T. Hoshino, K. Watanabe, R. Kometani, T. Morita, K.

Kanda, and Y. Haruyama, T. Kaito, J. Fujita, M. Ishida, Y. Ochiai, and S. Matsui: "Development of three-dimensional pattern-generating system for focused-ion-beam chemical-vapor deposition", *Journal of Vacuum Science & Technology B*, Vol.21, No.6, pp.2732-2736, 2003.

[14] H. W. P. Koops, R. Weiel, D. P. Kern, and T. H. Baum, "High resolution electron beam induced deposition," *Journal of Vacuum Science Technology B: Microelectronics and Nanometer Structures*, vol. 6, no. 1, pp. 477 –481, Jan. 1988.

CONTACT
*T. Hoshino, tel: +81-3-5841-6880; takayuki_hoshino@ipc.i.u-tokyo.ac.jp

ENCODING AND MANIPULATING MICROCOMPONENT ON ELECTROMICROFLUIDIC PLATFORM

Min-Yu Chiang[1], San-Yuan Chen[1], and Shih-Kang Fan[2]

[1]Department of Materials Science and Engineering, National Chiao Tung University, Hsinchu, TAIWAN

[2]Department of Mechanical Engineering, National Taiwan University, Taipei, TAIWAN

ABSTRACT

We report an electromicrofluidic (EMF) platform for (1) encoding programmable microcomponents (building blocks) using the embedded colloidal particles or mammalian cells and (2) manipulating and assembling microcomponents in two or three dimensions (2D or 3D). This platform is not only accessible to encapsulated particles with various properties, e.g., conductive/dielectric or synthetic/biological, but also to multiple hydrogel materials synthesized differently for further assembly of 2D or 3D complex constructs.

INTRODUCTION

Engineering microcomponents or architectures on microscale and with ordered and complex patterns are of importance in various fields, such as microelectro-mechanical systems (MEMS) [1], photonics [2], metamaterials [3], and tissue engineering [4], because of the controllable and tunable properties in optics, electricity, magnetism, and biology. Previous studies have demonstrated the microcomponent synthesis, such as emulsification [5], photolithography [6], microfluidic-assisted synthesis [7], stop-flow lithography [8]. Varied microcomponents with a wide range of dimension from nanometer to millimeter necessary depend on the applications. For example, in the photonics applications, microcomponents are usually made on micrometer or sub-nanometer scale, while tissue engineering requires micrometer to millimeter constructs. However, the synthesis of the microcomponents with ordered structure and the assembly of constructs with combined and varied properties, such as diverse stiffness, are still challenging.

With the abilities of manipulating fluids and particles, an electromicrofluidic (EMF) platform creates, transports, splits, and merges liquid droplets and drives suspended particles, which facilitates a portable device to perform on-chip chemical and biological procedures that conduct enzyme [9] and cells [10] assays. On the EMF platform, droplets (e.g., conductive, dielectric, and aqueous droplets) and particles (e.g. polystyrene beads and mammalian cells) are controlled by electromechanical manners including electrostatic forces, dielectrophoresis (DEP) and electrowetting-on-dielectric (EWOD).

Our previous study has demonstrated that liquid hydrogels were manipulated and assembled to form a heterogeneous architecture with fluorescent particles encoded by DEP [11]. In this study, we further manipulate multiple hydrogel precursors with different solidification methods, including chemical and irradiation crosslinking, to synthesize microcomponents driven by EWOD or DEP force. Additionally, each microcomponent can be encoded

with suspended particles using DEP force before cross-linking. The particle-encoded microcomponents are further manipulated, assembled, or stacked in liquid hydrogel precursor solution by DEP force.

PRINCIPLE

EWOD and DEP

EWOD and DEP are investigated for manipulations of multiple hydrogel precursor droplets that are crosslinked with various manners. Chemical crosslinked polyacrylamide, and irradiation crosslinked PEG-DA (poly(ethylene glycol) diacrylate) were successfully driven in air medium or in oil shells along the electrodes between two parallel plates. The conductive aqueous hydrogel precursor droplets including Matrigel and polyacrylamide are driven by EWOD through the contact angles change on the two sides of the actuated droplet. When applying an AC signal between one of the patterned driving electrodes on the bottom plate and the unpatterned electrode on the top plate, contact angle of the droplet on the powered electrode decreases and generates pumping pressure to move the droplet towards the energized electrode. The EWOD driving force (F_{EWOD}) in the parallel plate device is described as:

$$F_{EWOD} = \frac{\varepsilon_0 \varepsilon_D W}{2t} V_D^2 \qquad (1)$$

where ε_0 is the permittivity of vacuum, ε_D is the relative permittivity of the dielectric layer, W is the width of the patterned driving electrodes, t is the thickness of the dielectric layer, and V_D is the applied voltage across the dielectric layer.

In contrast, dielectric hydrogel precursor droplets are driven by DEP with none or limited contact angle change with the applied signals. When a sufficient electric field is applied between the electrodes on the two plates, a surface force exerts on the liquid/medium interface (e.g., oil/air boundary). DEP drives the dielectric PEG-DA precursor droplet with higher permittivity towards high electric field regions between two parallel plates. The DEP force (F_{LDEP}) is expressed as:

$$F_{LDEP} = \frac{\varepsilon_0 (\varepsilon_{Oil} - \varepsilon_{Air}) W}{2d} V_{LDEP}^2 \qquad (2)$$

where ε_{Oil} and ε_{Air} are the relative permittivities of oil and air, respectively, V_{LDEP} is the voltage between the plates, and d the distance between the parallel plates.

By using EWOD and DEP forces, conductive and dielectric hydrogel precursor droplets are driven on the EMF platform. The droplets are future crosslinkable to generate microcomponents on the platform. Moreover, the suspended particles and cells are actuated by particle DEP to encode the microcomponents as shown in Fig. 1.

(a)

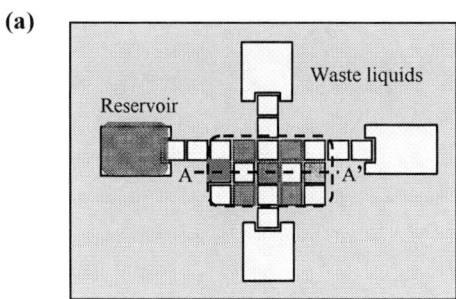

(b) Particles arrangements and microcomponents formation

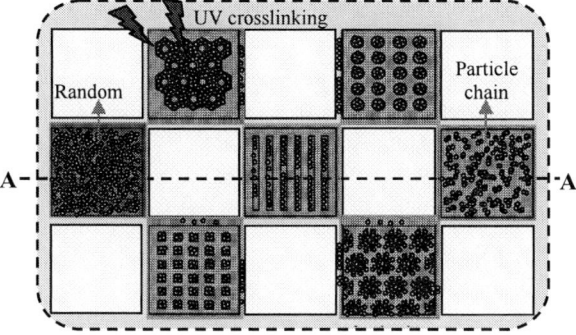

(c) Assembly of microcomponents in hydrogel precursor solution

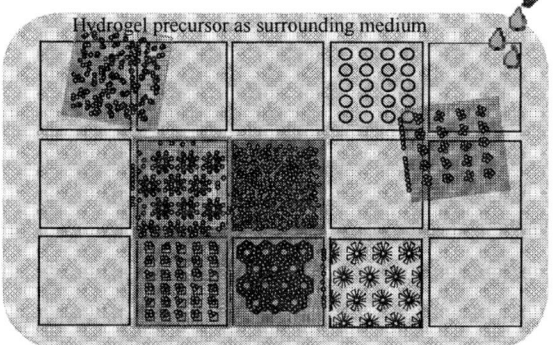

(d) A-A' cross-sectional view

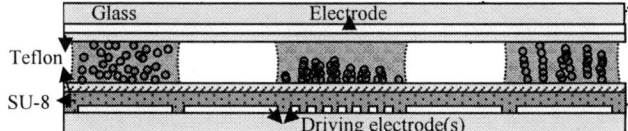

Figure 1: The schematic illustration of an electromicrofluidic (EMF) platform. (a) Top view of the electrode design. (b) With various pattern designs, hydrogel precursor droplets and particles are manipulated and microcomponents are solidified after crosslinking. (c) Microcomponents assembly in hydrogel precursor on the EMF platform. (d) A-A' cross-sectional view of (b) and (c).

DESIGN

The EMF device consists of top and bottom plates. The bottom plate is composed of 200-nm-thick ITO electrodes, 1.7-µm-thick SU-8 dielectric layer and 55-nm-thick Teflon layer (Fig. 1). The top glass plate consists of a blank ITO and a Teflon layer. The conductive and dielectric hydrogel precursor droplets are pumped by EWOD and DEP forces, respectively; while the particles/cells encapsulated in the hydrogel precursor are driven by DEP using patterned electrodes that generate desired non-uniform electric fields. The microcomponents are then crosslinked, for example, by UV light as shown in Fig. 1b. After the microcomponents are synthesized, liquid hydrogel precursor is pipetting to the EMF platform again for suspending the microcomponents (Fig. 1c). The microcomponents are subsequently actuated and assembled by DEP force to form a 2D construct. The cross-sectional view of the microcomponents with encapsulated and ordered particles is shown in Fig. 1d.

EXPERIMENT

Multiple Hydrogel Precursor Droplets Manipulation

As shown in Fig. 2a and 2b, multiple hydrogel precursors (i.e. polyacrylamide and PEG-DA) containing different fluorescent particles (7 µm red particles and 5 µm green particles) were actuated on the EMF platform). PEG-DA hydrogel droplets were manipulated by applying a square wave AC signal with 1 kHz and 100 V_{RMS}. After irradiating 320-500 nm and 70 mW/cm^2 UV light for 8 s, as shown in Fig. 2 (b)-(d), microcomponents with particles with different concentration and arrangements were obtained.

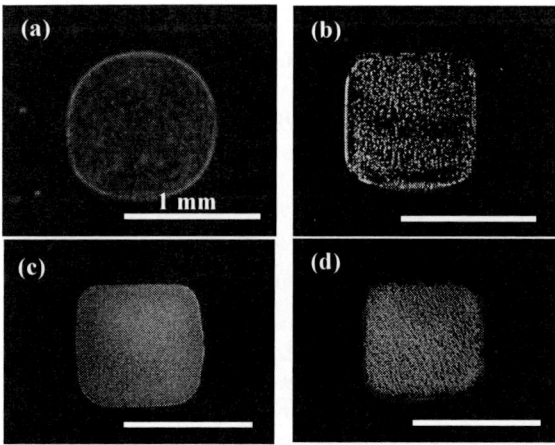

Figure 2: Synthesis of microcomponents composed of various hydrogel materials and particles. (a) Chemically crosslinked polyacrylamide. (b) UV crosslinked PEG-DA. (c) Microcomponent containing random particles. (d) Microcomponent with polarized particle chains.

978-1-4799-7956-1/15 $31.00 © 2015 IEEE

Microcomponents Manipulations

The microcomponents can be further stacked (Fig. 3(a)-3(c)) or assembled (Fig. 3(d)-3(f)) in hydrogel precursor by DEP to form a 2D or 3D construct on the EMF platform. After the 100 μm-high microcomponents were synthesized, we increased the spacers from 100 μm to 300 μm between two parallel plates and pipetted another liquid phase hydrogel precursor to resuspend these microcomponents. Then we applied an AC signal (100 Hz, 115 V_{RMS}, square wave) to manipulate the microcomponents. The microcomponents were repelled from the strong electric field regions to the weak electric field area. Stacking and assembly of the microcomponents were demonstrated on the EMF platform by DEP force.

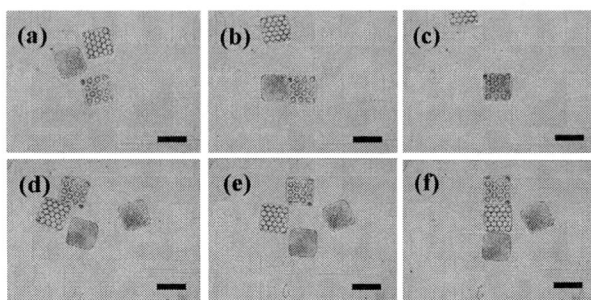

Figure 3: Microcomponents manipulations on an EMF platform. (a)-(c) Stacking of multiple microcomponents with different patterns. (d)-(f) Assembling microcomponents. Scale bar: 1 mm.

Cells Arrangements in Microcomponents

The biological particles encapsulated and arranged in biocompatible microcomponents (gelatin methacrylate, GelMA) were also demonstrated as shown in Fig. 4(a)-(d). In this study, we utilized 5% (w/v) GelMA hydrogel precursor with 0.5% (w/v) photoinitiator I2959 and 0.04% (w/v) pluronic F127 dissolved in low conductivity buffer consisting of 10 mM HEPES, 0.1 mM CaCl$_2$, 59 mM D-glucose and 236 mM sucrose, at pH of 7.35, as the material of microcomponent and NIH 3T3 fibrobalsts as the biological particles. By applying a high frequency AC signal (1 MHz, 2.2 V_{RMS}, sine wave), fibroblasts were attracted to the strong electric field area and reorganized in the microcomponent. Then the cell-ordered microcomponent was further crosslinked by UV light (320-500 nm, 70 mW/cm^2) for 8s.

From the cytoimmunostaining results, NIH/3T3 fibroblasts encapsulated in microcomponents with random or ordered pattern all remained good viability and proliferated well after 5 days (Fig. 4a-4b) or 2 days (Fig. 4c-4d) in culture.

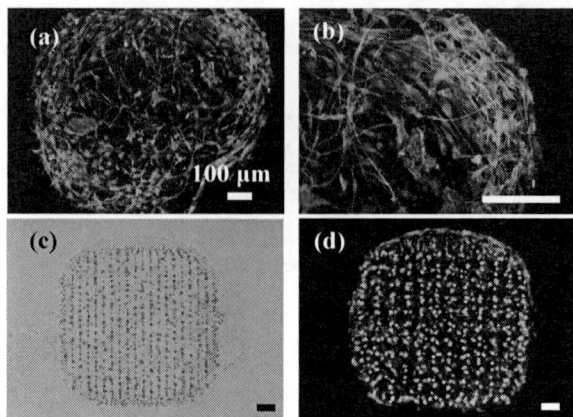

Figure 4: Microcomponents with random or ordered cells. (a)-(b) NIH3T3 fibroblasts randomly encapsulated in GelMA microcomponent. (c)-(d) NIH3T3 fibroblasts encapsulated in GelMA microcomponent with ordered patterns.

CONCLUSIONS

On the EMF platform, we demonstrate the preparation of microcomponent that are synthesized from varied materials that crosslinked differently and are encoded with accessible particles. The microcomponents were further stacked and assembled on the platform. Mammalian cells were actuated to form encoded microcomponents made of biocompatible hydrogel for 3D cell culture.

REFERENCES

[1] R. J. Knuesel, H. O. Jacobs, Proc. Natl. Acad. Sci. U.S.A. 2010, 107, 993-998.

[2] Y. Lu, Y. Yin, Y. Xia, Adv. Mater. 2001, 13, 34-37.

[3] N. I. Zheludev and Y. S. Kivshar, Nat. Mater. 2012, 11, 917-924.

[4] U. A. Gurkan, S. Tasoglu, D. Kavaz, M. C. Demirel, U. Demirci, Adv. Healthcare Mater. 2012, 1, 149-158.

[5] M. Han, X. Gao, J. Z. Su, and S. Nie, Nat. Biotechnol. 2001, 19, 631-635.

[6] S. Tasoglu, E. Diller, S. Guven, M. Sitti, and U. Demirci, Nat. Commun. 2014, 5, 3124.

[7] D. Dendukuri, and P. S. Doyle, Adv. Mater. 2009, 21, 4071-4086.

[8] Y. K. Cheung, B. M. Gillette, M. Zhong, S. Ramcharan, and S. K. Sia, Lab Chip 2007, 7, 574-579.

[9] E. M. Miller and A. R. Wheeler, Analytical Chemistry 2008, 80, 1614-1619.

[10] S. Srigunapalan, I. A. Eydelnant, C. A. Simmons, and A. R. Wheeler, Lab Chip 2008, 12, 369-375.

[11] M.-Y. Chiang, Y.-W. Hsu, H.-Y. Hsieh, S.-Y. Chen, and S.-K. Fan, MicroTAS 2014, 96-98.

CONTACT

*S.-K. Fan, Tel: +886-2-33664515; skfan@fan-tasy.org

IN VITRO DYNAMIC FERTILIZATION BY USING EWOD DEVICE

Lung-Yuan Chung[1], Hsien-Hua Shen[1], Yu-Hsiang Chung[1], Chih-Chen Chen[1], Chia-Hsien Hsu[2],
Hong-Yuan Huang[3] and Da-Jeng Yao[1]*

[1]Institute of NanoEngineering and Microsystems, National Tsing Hua University, Hsinchu, Taiwan.
[2]Institute of Biomedical Engineering and Nanomedicine, National Health Research Institutes, Miaoli, Taiwan.
[3]Department of Obstetrics and Gynecology, Chang Gung Memorial Hospital, Taoyuan, Taiwan.
*E-mail: djyao@mx.nthu.edu.tw

ABSTRACT

This research demonstrates the in vitro fertilization (IVF) technique by using the digital microfluidic (DMF) system. The DMF device has been proved with biocompatibility and used in this research for mouse gametes in vitro fertilization and embryos culture based on the dispersed droplet form. Dynamically moving, cell culture and observation can be achieved by the time-lapse microscopy in an incubator. The fertilization rate of IVF on DMF is 34.8%, and about 25% inseminated embryos dynamically cultured on DMF chip could develop into 8-cell stage. The result indicated that the DMF system has the potential used for assisted reproductive technology.

INTRODUCTION

A digital microfluidic (DMF) platform is a different method for discrete droplets processing instead of using the conventional pump to create continuous-flows in the channel. Currently, electrowetting-on-dielectric (EWOD) is the most widely accepted actuation mechanism for the DMF system, micro and nanoliter-sized droplets can be easily created, transported, separated, and merged. The attraction force of EWOD is induced by applying voltage between electrodes, which is commonly quantified by the amount of contact-angle reduction. The relation between applied voltages and contact-angle reduction can be formulated by combining Lippmann's equation and Young's equation, as shown in Equation (1), where $\theta(V)$ and θ_0 are corresponding contact angles at applied voltage V and 0 V, individually, ε and ε_0 are the dielectric constant of the dielectric and the air, respectively, γ_{LG} is the liquid–gas interfacial energy, and t is the thickness of the dielectric layer[1]. See Nelson and Kim[2] for the related equations in more general forms and the definition of the non-dimensional electrowetting number Ew.

$$\cos\theta(V) - \cos\theta = \frac{\varepsilon\varepsilon_0}{2\gamma_{LG}t}V^2 = Ew \qquad (1)$$

Today, DMF device has been widely applied for biomedical applications[3] because the advantages of little consumption of reagents, rapid reactions and modest cost. Also, there are researches discuss the feasibility and advantages of cell culture with the DMF platform. When cells are manipulated or cultured in a continuously flowing channel, the contamination problem might be existed; moreover, it is difficult to manipulate and observe an individual or a few amount cells in a continuously flowing channel. Therefore the DMF system is considered as a high potential tool for single-cell droplets manipulation. In 2011, Lammertyn's research team used lift-off fabrication to define the cell culture regions on the electrodes[4], the

cells could attach on the patches and remain viable for up to three days. In 2012, Wheeler's research team also provided an upside-down cell culture method by creating regions of cell culture on the top cap of a parallel-plate DMF chip[5]. Derek G. et al. in 2014 has presented the development of a DMF for the vitrification of mammalian embryos for clinical in vitro fertilization (IVF) applications[6].

With the increasing clinical utilization of assisted reproductive technology (ART), scientists make effort to improve the rates of success following assisted reproduction. IVF technique removes ovum from the woman's ovaries and manually fertilizes with sperm in the tube. Successfully inseminated embryos would then be cultured in an oil-covered static medium microdroplet on a petri dish for 2–6 days and finally implanted back into the female's uterus. The repeating manual pipetting and washing procedures might cause the loss and prospective stress of temperature, pH, osmolality, light, and mechanical stress on the embryos [7]. Continuously flowing microchannel has been used in IVF to increase the sample quality and culture environment[8].

The research team tried to bring the DMF into IVF. Based on the advantages of DMF system, the expectation could be an applicable tool for IVF to increase the rates of fertilization in vitro and provide a stable culture environment for the blastocysts formation. The chip biocompatibility problem was discussed in the beginning of this research. 2-cell mouse embryos were statically cultured on chip until the hatching status to prove that the DMF chip exhibits well biocompatibility. Another important problem in this research is that if the embryos would be damaged or injured during electrowetting droplets manipulation. To solve this problem, the 2-cell status embryos were dynamically cultured on the DMF system to blastocyst or hatching status and compared with the control group. After the confirmation of DMF chip has well biocompatibility and is harmless to embryos, the chip was used for mouse oocytes and sperm droplets manipulation and on chip fertilization. Female and male gametes could be well mixed by DMF electrodes without pipetting, and then the inseminated embryos were dynamically cultured between the electrodes from 2-cell to 8-cell status. Finally, the success rate of IVF by DMF was compared with the control group.

978-1-4799-7956-1/15 $31.00 © 2015 IEEE

MATERIAL AND METHOD

Pluronic F127 and cell culture oil

Contamination of the DMF chip surface remains a serious problem when the biomedical reagents are used. Biomolecules would stick on the chip surface and decrease the hydrophobicity of hydrophobic layer which increases the operating voltage of manipulation of sample droplets and could be easily result in the dielectric breakdown. Pluronic in a small dose is a proven non-toxic and safe surfactant, which can be applied in cell culturing or tissue engineering [9]. In this research, pluronic F127 additive (0.08 % mass/volume) was added to the human tubal fluid (HTF) medium to decrease biofouling and also to enhance protein stability.

The cell culture oil is a sterile, light paraffin produced by OVOIL™ that is used for covering of medium during in vitro cell culture procedures to prevent evaporation of medium. Due to the volume of the microdroplets used in this research are quite small, less than 10µl, the culture oil must be used for to prevent medium evaporation.

The DMF system and the microscope incubator

The hardware of the DMF system used in this research includes a digital I/O controller (NI USB-6509), a control board, an AC power amplifier (AA303), a signal generator (Agilent 33220A) and a clamp (Yokowo). For real-time embryos observation during cell culture, a microscope incubator system was used, this contains: a gas flow controller, temperature controller and a time-lapse microscope in an incubator. The gas flow controller provides stable CO_2 together with N_2 flow inside the incubator, and the time-lapse microscope (Leica DMI 6000) is able to capture real-time image at multi-points. The image of the DMF system and the microscope incubator is shown in Figure 1. During the experiment, the DMF chip would be connected with the clamp via a cable make it able to deliver the electrical signal from the system to the clamp inside the cell culture incubator.

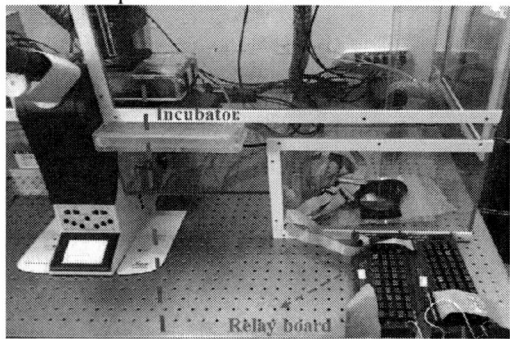

Figure 1: DMF system and the microscope incubator. The left picture shows a DMF clamp inside a cell culture incubator under the time-lapse microscopy. The setup in the right picture is the DMF controlling system.

Chip fabrication process & packaging improvement

The fabrication process of the coplanar type electrode EWOD chip used for IVF is show in Figure 2. After indium tin oxide (ITO) electrodes were patterned, 450 nm plasma-enhanced chemical vapor deposited (PECVD) Si_3N_4 and a 100 nm spin-on-glass (SOG) were spin coated on an ITO electrode as a compound dielectrics. Then 150 nm Teflon was spin coated on the dielectric layer as a hydrophobic layer. Finally, a Polydimethylsiloxan (PDMS) ring was attached on the chip surface.

In the previous design, the chip used for embryo culture manipulated droplets between a top cap and the bottom chip. Oil must be injected in the empty space to prevent sample evaporation as shown in Figure 3 (a). The thick oil layer would block the gas exchange of the medium that leads to the decreasing of the embryo survival rate. The solution for this can be done that discarded the top cap and used a PDMS ring as an oil pool for embryo culture as shown in Figure 3 (c) & (d). The thin oil layer not only avoids sample evaporation but also enables the medium gas exchange as shown in Figure 3(b).

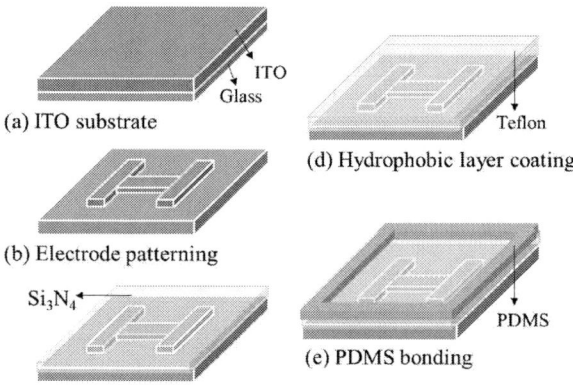

Figure 2. EWOD chip fabrication process. (a) Starten from ITO glass substrate. (b) Electrode patterning by photo lithography. (c) Dielectrics deposition. (d) Hydrophobic layer coating. (e) PDMS bonding.

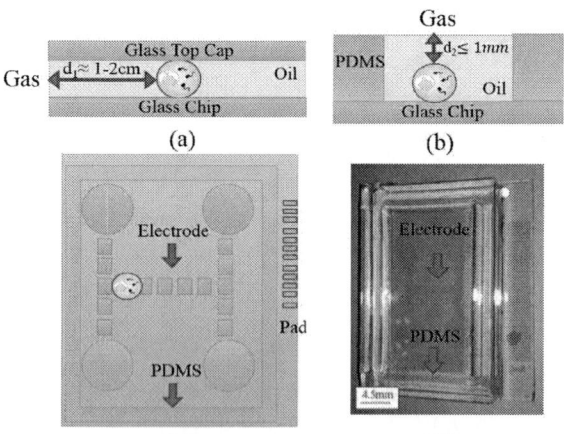

Figure 3: (a) Cross-sectional view of the traditional DMF chip package composed by a top cap and bottom chip. (b) Improved chip package by using a PDMS ring as the sidewall. (c) Top view of the improved chip package. (d) Picture of the DMF chip.

978-1-4799-7956-1/15 $31.00 © 2015 IEEE

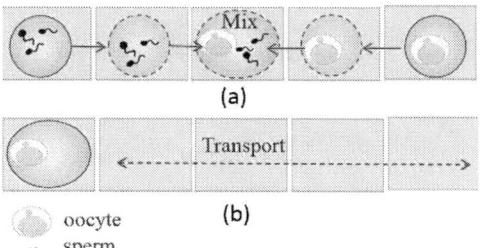

(a)

(b)

oocyte

sperm

Figure 4: The automatic programming DMF droplets manipulation. (a) Manipulation of sperm and oocyte droplets for on chip IVF process. (b) Dynamic culture after fertilization.

Embryo collection

Imprinting Control Region (ICR) mouse was used in this research. Female mice were injected intraperitoneally with Pregnant Mares Serum Gonadotrophin (PMSG), 5 IU (Sigma, U.S.A.). 48 hours after injection, the female mice were further injected intraperitoneally with Human Chonionic Gonadotrophin (hCG) 5 IU (Sigma, U.S.A.). Immediately after hCG injection, each female mouse was caged with a male mouse for mating (E_0 days). 36 hours post hCG injection, the female mice were sacrificed and the oviducts were dissected then placed in a petri dish containing HTF medium, with zygotes for an embryonic development culture were released on tearing the oviducts.

In vitro fertilization and dynamic culture on chip

Male and female mouse gametes droplets were prepared for DMF manipulation. A 5µl HTF microdroplet containing sperms (2×10^6 mL^{-1}) and a 5µl HTF microdroplet containing 10 oocytes were dropped on the electrodes and covered with culture oil. The DMF electrodes were operated with 60V at 500Hz square signal to attract both medium droplets and mixed them together for fertilization as shown in Figure 4 (a).

The embryos after insemination were dynamic cultured on the chip by automatic program, droplets were moved back and forth between 3-5 electrode pairs for 30 seconds, having the interval between each movement is 30 minutes as shown in Figure 4 (b). The embryos after inseminated were continuously and dynamically cultured for 3 to 4 days until blastocyst status. The development of the embryo was individually observed on Day 1 ($E_{2.5}$), Day 2 ($E_{3.5}$), Day 3 ($E_{4.5}$), and Day 4 ($E_{5.5}$) with the time-lapse microscope.

RESULT

Chip biocompatibility testing

In the biocompatibility testing, the 2-cell status embryos were cultured on the DMF chip without any voltage applied. The culture result on $E_{5.5}$ is shown in Table 1. About 96.6% (N=32) embryos could finally developed into blastocyst (B.C.) status cultured on chip, and about 3.4% embryos were dead and stay on 4-cell stage during the 4 days culture process. The embryo surviving rates is quite high, and the result can be comparing with the control group cultured on the traditional petri dish. The chance for embryos developing into B.C. for both cultured on chip and on dish are nearly the same or more than 95%.

Table 1: Embryo development for biocompatibility testing.

	4-cell	Morula	B.C.
On DMF chip	1/32	0/32	31/32
On petri dish (control)	0/32	1/32	31/32

Table2. Fertilization rates of the on chip and on dish IVF.

	Not fertilized	Fertilized
On DMF chip	15/23	8/23
On petri dish (control)	46/118	72/118

Table3. Embryo development after dynamic culture.

	2-cell	4-cell	8-cell
On DMF chip	1/8	1/8	2/8
On petri dish (control)	28/72	32/72	12/72

Gametes droplets manipulation and fertilization by DMF platform

Male and female mouse gametes were prepared in the 5µl HTF microdroplets and dropped on the chip as shown in Figure 5 (a). The left HTF droplet contained about 1×10^4 sperms, and there were 10 oocytes in the right HTF droplet. Both of the droplets were covered with culture oil. The DMF electrodes were operated with 60V at 500Hz square signal to move both droplets and finally mixed with each other as shown in Figure 5 (b)-(c). Then the mixed droplet was moved between electrode pairs to make the droplet well mixed so the oocytes were surrounded by sperms as shown in Figure 5 (d). The fertilization result can be seen from Table 2, the fertilization rate by using DMF chip and petri dish are 34.8% (N=23) and 61% (N=118), respectively. After fertilized, the embryos were statically cultured on chip and continuously observed under the time-lapse microscope in the incubator until $E_{4.5}$. Figure 6 shows the dynamic culturing process, the HTF medium with embryos inside was transported between 3 electrode pairs. The culture result is shown on Table 3. About 25% (N=8) embryos cultured on DMF chip developed into 8-cell stage, and about 16.7% (N=72) embryos cultured on petri dish developed into 8-cell stage.

The result shows that the fertilization rate of IVF through DMF system is lower than traditional control group. One of the possible factors is the cleanness of the medium droplet after fertilization. Without the top cap, the DMF chip provides not only droplet mixing and transporting functions but also separation. It is difficult to change the medium after fertilization on DMF chip by pipetting because the medium change process could easily result in the loss of embryos. Therefore, the medium exchange problem must be solved in the future IVF work.

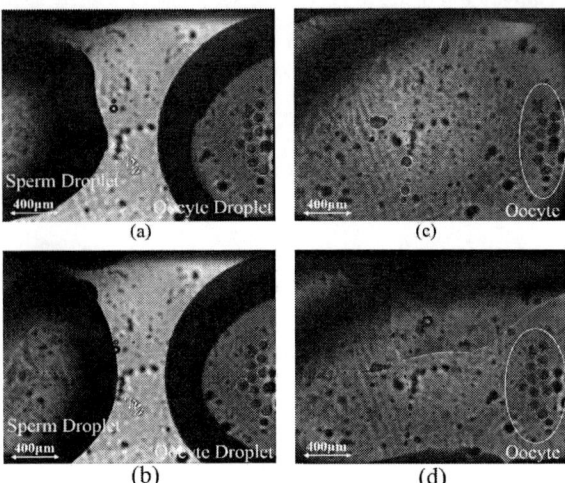

Figure 5: Male and female mouse gametes in HTF microdroplets were manipulated by DMF. (a) Before manipulation. (b) Turned on the electrode between two droplets. (c) Droplets were mixed for IVF. (d) Turned off the electrode.

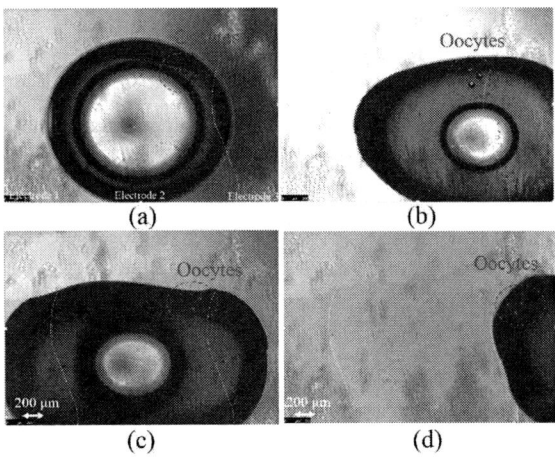

Figure 6: Embryos were dynamically cultured on chip by automatic program. The droplets were transported between electrode pairs.

CONCLUSION

This study successfully applied the DMF system for in vitro fertilization. A PDMS ring bonded DMF chip enables gas exchange of the HTF medium droplet covered by culture oil. Static culture of the embryos on the chip has high probability develop from 2-cell status into blastocyst cell, which has proven the biocompatibility of the DMF chip. HTF microdroplets containing male and female mouse gametes can be mixed for IVF by DMF droplet manipulation. The inseminated embryos can be dynamically cultured, transported between electrode pairs at regular intervals. All the droplet movements and cell culture images can be captured by the time-lapse microscopy in the incubator system. The fertilization rate of IVF on DMF platform and traditional petri dish are 34.8% and 61%, separately. Currently, the medium exchanging problem needs to be fixed so that the increase of fertilization rate of IVF on DMF. In dynamic culture, about 25% embryos could developed into 8-cell stage, and about 16.7% embryos cultured on petri dish developed into 8-cell stage. The DMF culture result is similar, even a little higher than the control group, which suggested that the DMF droplet manipulation method is harmless to the embryos. In the future, culture environment enhancement and system stability are needed to improve in order to make the DMF system more applicable to assisted reproductive technology.

ACKNOWLEDGEMENTS

The authors would like to acknowledge Solar Applied Materials Technology Corp. for providing technical support and the material (ITO glass) for this research. This work was partially supported by the Ministry of Science and Technology, Taiwan, under grant no. NSC 100-2628-E-007-014-MY3.

CONTACT

* D.J. Yao, tel: +886-3-5715131~42850; djyao@mx.nthu.edu.tw

REFERENCES

[1] B. Berge, "Electrocapillarity and wetting of insulator films by water," *Comptes Rendus de l'Academie des Sciences Serie II,* vol. 317, pp. 157–163, 1993.

[2] W. C. Nelson and C.-J. C. Kim, "Droplet Actuation by Electrowetting-on-Dielectric (EWOD): A Review," *Journal of Adhesion Science and Technology,* vol. 26, pp. 1747-1771, 2012 2012.

[3] H.-H. Shen, S.-K. Fan, C.-J. Kim, and D.-J. Yao, "EWOD microfluidic systems for biomedical applications," *Microfluidics and Nanofluidics,* vol. 16, pp. 965-987, 2014/05/01 2014.

[4] D. Witters, N. Vergauwe, S. Vermeir, F. Ceyssens, S. Liekens, R. Puers*, et al.*, "Biofunctionalization of electrowetting-on-dielectric digital microfluidic chips for miniaturized cell-based applications," *Lab on a Chip,* vol. 11, pp. 2790-2794, 2011.

[5] S. Srigunapalan, I. A. Eydelnant, C. A. Simmons, and A. R. Wheeler, "A digital microfluidic platform for primary cell culture and analysis," *Lab on a Chip,* vol. 12, pp. 369-375, 2012.

[6] D. G. Pyne, L. Jun, M. Abdelgawad, and S. Yu, "Automated vitrification of mammalian embryos on a digital microfluidic device," in *Micro Electro Mechanical Systems (MEMS), 2014 IEEE 27th International Conference on,* 2014, pp. 829-832.

[7] Y. Xie, F. Wang, E. E. Puscheck, and D. A. Rappolee, "Pipetting causes shear stress and elevation of phosphorylated stress-activated protein kinase/jun kinase in preimplantation embryos," *Molecular Reproduction and Development,* vol. 74, pp. 1287-1294, Oct 2007.

[8] R. H. Hester P, Clark S, Walters E, Beebe D, Weeler MB, "Enhanced cleavage rates following in vitro maturation of pig oocytes within polydimehtylsiloxane-borosilcate microchannels," *Theriogenology,* vol. 57:723, 2002.

[9] V. N. Luk, G. C. H. Mo, and A. R. Wheeler, "Pluronic Additives: A Solution to Sticky Problems in Digital Microfluidics," *Langmuir,* vol. 24, pp. 6382-6389, 2008/06/01 2008.

LIPOPHILIC-MEMBRANE BASED ROUTING FOR CENTRIFUGAL AUTOMATION OF HETEROGENEOUS IMMUNOASSAYS

Rohit Mishra[1], Rizwan Alam[2], David J. Kinahan[1], Karen Anderson[2] and Jens Ducrée[1]
[1]Biomedical Diagnostics Institute, Dublin City University, Dublin, IRELAND
[2]The Biodesign Institute, Arizona State University, Tempe, USA

ABSTRACT

We demonstrate centrifugal [liquid handling] automation of an Enzyme-Linked Immuno-Sorbent Assay (ELISA) for the detection of anti-p53 antibodies in whole blood. On this "Lab-on-a-Disc" (LoaD) platform, all unit operations were implemented by event-triggered rotational flow control. In order to avoid interference during absorbance measurement from the solid phase in this heterogeneous assay format, it is pivotal that the intermediate reaction product is eventually forwarded from the incubation chamber to a distinct optical measurement chamber. To this end we have devised routing of flows by lipophilic film valves (LFVs) which remain intact in aqueous and selectively dissolve when exposed to an ancillary, oleophilic solvent.

INTRODUCTION

In this work we specifically consider centrifugal microfluidic systems towards the development of conceptually simple and cost-efficient bioanalytical point-of-care (POC) applications [1]. The full integration and automation of underlying liquid handling protocols while keeping the instrumentation compact and cost-efficient is a common design goal of these "Lab-on-a-Disc" (LoaD) systems [2]. As during spinning all liquids are subject to the same radial force field, valves have to be opened selectively in time and space for implementing sequential and parallel process steps. Many early-stage LoaD systems have introduced capillary barriers or siphons to orchestrate flow control. Recent developments showed the integration of functional materials into microfluidics [3], for instance disc-based ferrowax plugs were molten by an external laser [4], thus increasing the complexity of the instrument. Also centrifugo-pneumatic dissolvable films (DF) valves were presented which are opened by rotationally induced contact with aqueous solutions such as most biosamples and reagents [5].

We introduce here a novel, sacrificial membrane which is dissolvable in lipophilic liquids. This flow control element initially blocks aqueous solutions until an ancillary, oleophilic liquid is introduced from its back side to route the next aqueous solution, e.g. an elution buffer, to designated outlet. This event-triggered [5] router is implemented to demonstrate the on-disc integration of a bead-based, i.e. heterogeneous immunoassay.

In comparison to planar surfaces in bioassays, microparticle based assays provide much higher surface area thus substantially improving performance and have hence found direct applications in various areas like immunoassay formats (e.g.: ELISA) and bioprocessing steps like purification and extraction [6]. Typically such applications feature several steps of washing and incubation. Bead-based ELISAs are an essential tool for the detection of antigens and antibodies for clinical diagnostics. The final step in such assays is commonly the reaction of the enzymatic substrate in the bead column (immobilized capture antibodies on bead surface in the reaction chamber) and then eventual quenching with an acid before measuring the absorbance values. In an automated point-of-care ELISA, the absorbance measurement is ideally performed on disc, thus requiring a clear pathway for optical detection. The presence of a solid phase hence makes it mandatory to route the eventual enzymatic substrate away from the reaction chamber to a final quenching / measurement chamber to avoid any disturbance in absorbance measurements of the end product.

In LoaD systems, this has been accomplished previously by using laser-actuated valves [4]. We address this issue by introducing a routing mechanism based on lipophilic membranes that are selectively dissolved on demand by an immiscible liquid. This "transistor mode" operation with an ancillary liquid makes them particularly interesting since the routing does not require any additional instrumentation other than the simple spindle motor while allowing purely rotational flow control over the fluid flow.

Using lipophilic membrane routing, we present here the integration of a multi-step ELISA assay encompassing plasma separation from whole blood, incubation with capture antibodies on beads, multiple buffer washes and eventual routing of post-reaction enzymatic substrate to a measurement chamber performed on the disc in its entirety based on event-triggered rotational flow control [5].

SYSTEM DESIGN
Lipophilic valve based routing

Lipophilic membranes are biocompatible materials [7] that are hydrophobic in nature and selectively dissolve in bio-reagent grade oleophilic solvents such as light oil [8]. When in contact with an aqueous solutions, the hydrophobic nature of the membrane keeps the liquid from wetting and dissolving it. On the contrary, oleophilic solvents wet and dissolve the membranes while being immiscible with aqueous solutions. This implies that the membranes can act as a solvent-selective, intermediate barrier between two channels. Additionally, due to their softness and plasticity, they can be easily assembled into multi-layer valve composites cut into shapes.

We have devised a 3D structure to integrate such membranes into our microfluidic flow system (Fig.1). The lipophilic membrane is integrated under a microchannel in order to restrict the flow of aqueous solutions, acting as a transient barrier for fluid communication between two separate channels. When an aqueous solution is required to be routed from the top channel away from the normal exit, the lipophilic membrane is dissolved by mineral oil to open up the second channel and direct the subsequent aqueous solution into a designated chamber. (Fig.2)

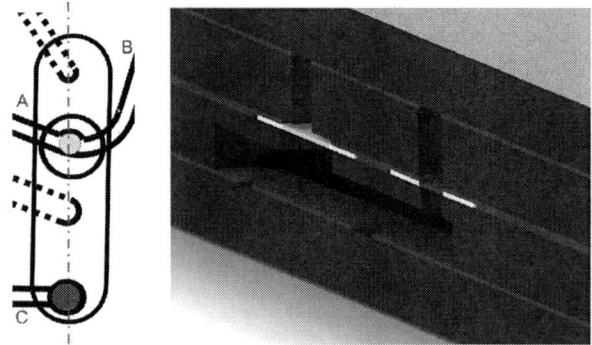

Figure 1: Lipophilic membrane based router. Cross section shows the placement of the valves inside the router. (Yellow tab- Lipophilic membrane, Blue tab- DF, Red layers-PSA; Black layers- PMMA). When the lipophilic valve is triggered, the aqueous solution from the inlet channel A routes to channel C instead of the usual channel B.

Figure 3 shows a fully integrated system for a disc-based ELISA for the detection of anti-p53 antibodies in whole blood. Individual components are: A) plasma extraction, B) reagent storage, C) bead incubation with surface-immobilized p53 capture antibody, D) waste, E) pneumatic chamber and F) the solvent-selective router. The vector of dissolvable-film valves in the waste (D) orchestrates the serial release of wash and ELISA reagents to the incubation chamber (C) as well as the ancillary liquid originally stored in chamber F to the back of the lipophilic membrane. Mixing in (C) is boosted by its centrifugo-pneumatic coupling to the compression chamber E while swiftly varying the spin rate. The incubation chamber (C) features two outlets, one of them leading via a siphon to the waste (D). The second outlet is initially sealed by a lipophilic membrane and opened to the detection chamber G by introducing the oleophilic ancillary liquid (F) from the reverse side.

In the first steps of the ELISA, sample, wash and reagents are routed from their respective reservoirs (A, B1-4) through the incubation chamber (C) where they are rotationally stirred and subsequently guided through the siphon into the waste (D). The final trigger concurrently releases the colour-generating enzymatic substrate (TMB) into C as well as the ancillary liquid to wet the first lipophilic valve. The concurrent dissolution from the backside of the second lipophilic membrane in the routing chamber with the ancillary liquid then routes the TMB from the reaction chamber through the centrifugally preferred outlet to G where the preloaded acid quenches the reaction before absorbance detection (Fig.2).

METHODOLOGY
Disc fabrication and assembly

The LoaD is an eight layered structure composed of four layers of poly methyl-methacrylate (PMMA, 1.5 mm thick) bonded by intermediate pressure sensitive adhesive sheets (PSA, 86 μm) as shown in Fig.4. All features on the PMMA layers (Disc OD 130 mm, vents, reservoirs, router components) were laser ablated (30-W CO_2 laser, Epilog, USA). The PSA layers were structured with a knife cutter (Graphtec, Japan). The topmost PSA layer consists of the bulk of the microchannels connecting the reservoirs containing the sequence of unit operations of the ELISA. Appropriate air exits are engraved in the PMMA layer above this in the top Vents layer.

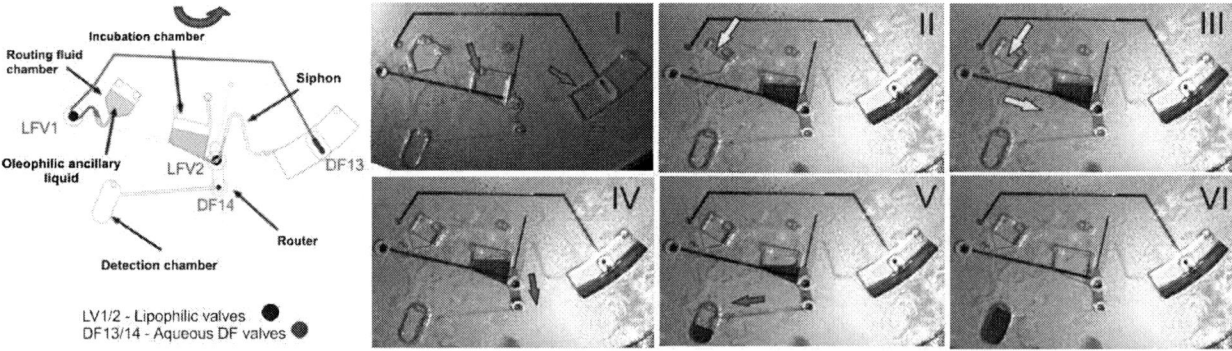

Figure 2: Trigger and release mechanism for the lipophilic-film valves (LFV) for routing of aqueous solutions. The aim is to route the final liquid from the incubation chamber (containing beads) to the detection chamber rather than across the siphon to waste. Red and yellow arrows indicate flow of aqueous and oleophilic ancillary liquid, respectively. Please note that steps II through IV are performed at spin frequency higher than the critical frequency of the siphon. (I) Aqueous solution from incubation chamber flows over to waste through the siphon triggering the release of the ancillary liquid in the routing fluid chamber. (II) DF13 valve triggers the ancillary liquid to wet the LFV1. (III) Once LFV1 is dissolved, the oleophilic liquid flows to the Router (engraved in a lower layer) and wets LFV2 but not the DF14. (IV) LFV2 is dissolved in the ancillary liquid releasing the aqueous solution from the incubation into the routing chamber where it phase separates below because of higher density enabled by centrifugally induced stratification of aqueous and oleophilic phases. (V) The aqueous solution wets DF14 thus opening up the channel to the detection chamber as in Section VI.

978-1-4799-7956-1/15 $31.00 © 2015 IEEE   524

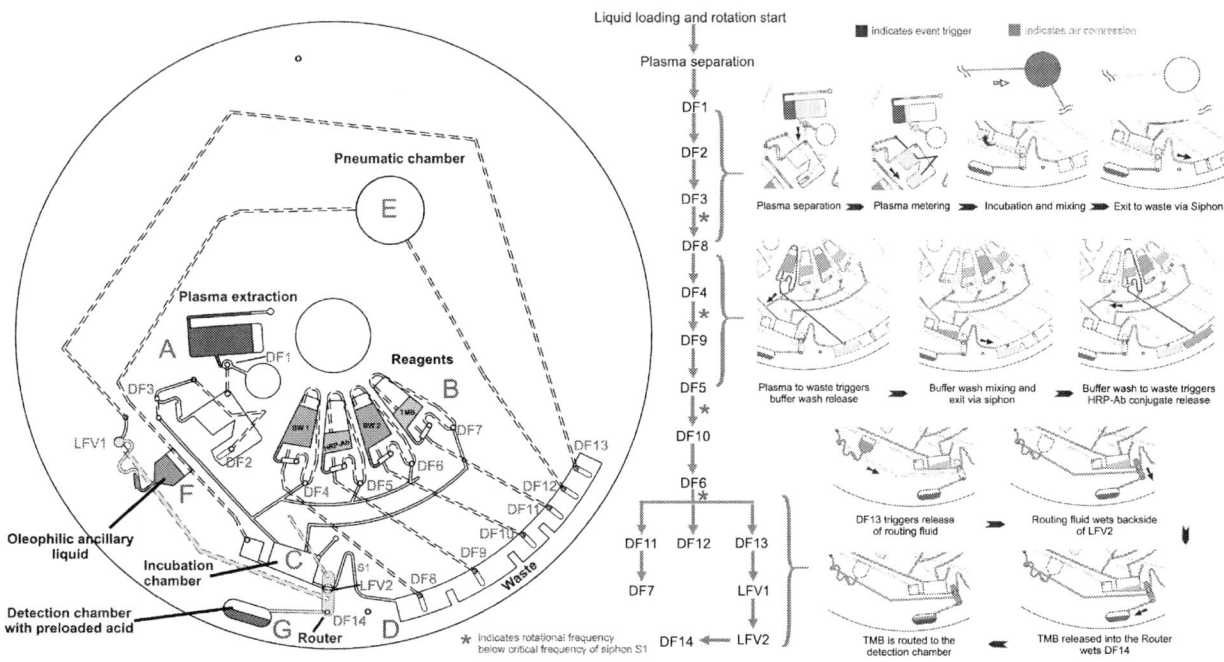

Figure 3: ELISA miniaturization on a centrifugal microfluidic Lab-on-a-Disc (LoaD) platform with event flow. The LoaD integrates all assay steps ranging from plasma extraction to the eventual absorbance measurement for quantification. The disc architecture is constituted by chambers for: A) plasma extraction, B) reagent storage (BW-buffer wash, HRP-Ab – bioconjugate of HRP with secondary antibody, TMB- substrate), C) incubation, D) waste, F) router (highlighted in yellow are the elements of the lipophilic routing mechanism) and G) detection chamber. Dissolvable-film valves (DFs) in the segments of the waste chamber (D) event-trigger the successive release of liquids along the ELISA. The final dissolution of the lipophilic film valve (LFV2) at the other outlet of (C), routes the final TMB solution to detection chamber (Fig.2).

The reservoir PMMA layer includes storage, incubation chambers and waste chambers. The DF and lipophilic membrane valves are sandwiched in between the DF support and DF cover PSA layers. The lower air channels PSA layer includes the pneumatic connections that are the core of the event-triggered structure allowing the release of reagents, buffer washes and the routing fluid on the disc. The lowest Base PMMA layer contains the vent for the router and seals the lower channels. All microchannels are 500 μm wide except for the oleophilic liquid (which measures 800 μm). All features of the device were designed using SolidWorks 2013 (Dassault systems, USA). Routing of the fluids using DF valves and event-triggered formation was accomplished as demonstrated before [5].

Protocol for anti-p53 antibody ELISA

A 140-μl blood sample is centrifuged for 5 minutes in the separation chamber to extract a plasma volume of 40 μl. This metered volume is incubated with the preloaded capture antibody + p53 protein coated beads in the incubation chamber for 10 minutes. During these steps, the mixing is induced by pulsed compression / decompression of a pneumatic chamber connected to the main incubation chamber (Fig. 3). Thereafter the beads are washed with the first buffer wash of 80 μl for 30 seconds. This is followed by incubation of the beads with the HRP tagged Secondary Antibody (40 μl) for 10 minutes. Then the final buffer wash (80 μl) is released and it incubates with the beads for 30 seconds. The last trigger after this wash eventually releases the enzymatic substrate

TMB (40 μl) and also the ancillary oleophilic liquid. The TMB incubates for 7 minutes with the beads to undergo color change to blue. The LFV router sets the point in time for the release of the TMB from the incubation chamber to the final detection chamber. The reaction is then quenched by the pre-loaded 40 μl of acid in the final measurement chamber. The resulting solution is measured for absorbance at 425 nm.

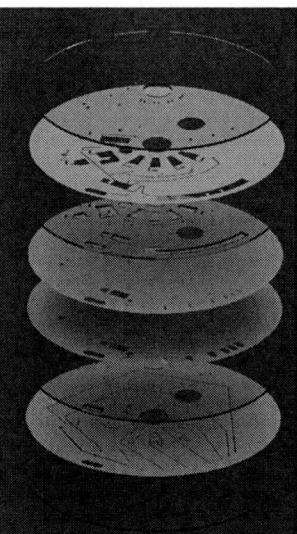

Figure 4: 3D render of the exploded view of the eight layered ELISA disc showing transparent PMMA layers with intermediate PSA layers (red) for bonding. In descending order from the top: 1. Vents, 2. Micro-channel, 3. Reservoirs, 4. DF Cover, 5. DF support, 6. Valving routes layer, 7. Lower air channels, 8. Base with vents.

RESULTS

Microfluidic testing

A fully operational microfluidic routing of a fluid using a lipophilic membrane is demonstrated in Fig. 2. The purpose is to route an aqueous fluid from the main incubation chamber to the detection chamber as opposed to the normal route that is taken by the aqueous solution across the siphon. In a multi-step operation the last liquid to enter the waste event triggers the release of the oleophilic solution into the router. The sequential dissolution of the two lipophilic membranes in the router arrangement (LFV1 and LFV2) by the ancillary liquid eventually triggers the release of the aqueous solution from the incubation chamber into the router. Here the wetting of the DF14 valve by the aqueous solution (assisted by the centrifugal stratification due to lower density ancillary liquid) opens up the channel leading to the detection chamber.

Effect of reduced reagent/buffer volumes on ELISA sensitivity

One of the most important aspects of moving to a microfluidic platform from the standard bench-top ELISA protocol is the significant reduction in the volumes of the reagents and the buffer washes. For on-disc implementation, standard volumes used on 96 well plates (100 µl) were reduced to 40 µl for the TMB and Sec-Ab conjugate, and two single intermediate 80 µl for the buffer washes (instead of two 100 µl 3X washes i.e. 300 µl each wash). Figure 5 summarizes the results for the detection of the anti-p53 antibody on a well plate using the reduced on-disc volumes; the limit of detection (LOD) was determined as 0.7 ng ml^{-1}. In comparison to data from bench-top protocol, we found no significant reduction in the sensitivity of the ELISA (data not shown).

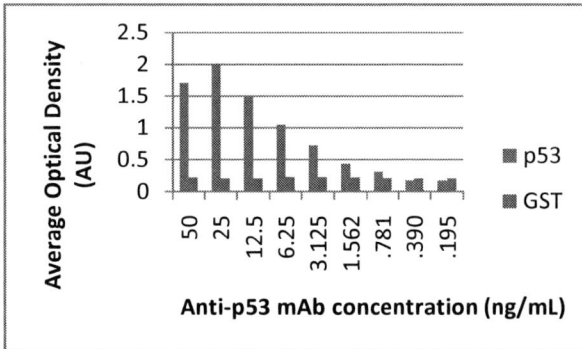

Figure 5: Detection of anti-p53 antibodies using ELISA with on-disc 40-µl volume (2.5 times lower than the bench top protocol) for reagents and a single 80-µl wash buffer. Average optical density of WT p53 proteins probed with monoclonal anti-p53 antibodies and detected using HRP secondary IgG is plotted against concentration (realizing an LOD of 0.7 ng ml^{-1}.)

DISCUSSION AND CONCLUSION

We have integrated selectively dissolvable lipophilic membranes into a centrifugal microfluidic platform for routing of aqueous solutions without the need for external actuation modules on the instrument other than a simple, frequency-controlled spindle motor. We have shown its direct applicability for bead-based ELISAs where further downstream sample handling is required before eventual detection. This is a significant improvement on current systems that require complex instrumentation approach for such routing. We also have demonstrated a fully functional microfluidic system for a bead-based-ELISA to detect anti-p53 antibodies in blood.

Further steps will include the optimization of the on-disc protocol of the ELISA. We also plan to extend the platform to further clinically relevant antibodies towards a flexible point-of-care device. Such a lab-on-a-disc system holds significant potential where intensive instrumentation and sample preparation are least desired.

ACKNOWLEDGEMENTS

This work was supported by the Science Foundation Ireland under TIDA grant Nos. 13/TIDA/B2692 and 10/CE/B1821, and the ASU-DCU Catalyst Fund.

REFERENCES

[1]   R. Gorkin *et al.* "Centrifugal microfluidics for biomedical applications" Lab Chip, 2010, 10, 1758-1773

[2]   C. D. Chin *et al.* "Commercialization of microfluidic point-of-care diagnostic devices" Lab Chip, 2012, 12, 2118-2134.

[3]   M. Kitsara and J. Ducrée "Integration of functional materials and surface modification for polymeric microfluidic systems" J. Micromech. Microeng. 2013. 23, 033001.

[4]   J.-M. Park *et. al.* "Multifunctional microvalves control by optical illumination on nanoheaters and its application in centrifugal microfluidic devices" Lab Chip, 2007,7, 557-564.

[5]   D. J. Kinahan *et al.* "Event-triggered logical flow control for comprehensive process integration of multi-step assays on centrifugal microfluidic platforms" Lab Chip, 2014,14, 2249-2258.

[6]   C.T. Lim and Y. Zhang "Bead-based microfluidic immunoassays: The next generation" Biosens. Bioelectron., 2007, 22, 1197–1204.

[7]   Hyun J *et al.* "Patterning cells in highly deformable microstructures: effect of plastic deformation of substrate on cellular phenotype and gene expression" Biomaterials, 2006, 27 (8), pp. 1444-1451.

[8]   R. Kodzius *et al.* "Inhibitory effect of common microfluidic materials on PCR outcome" Sensors and Actuators, B: Chemical, 2012, 161 (1), pp. 349-358.

CONTACT

*Jens Ducrée, jens.ducree@dcu.ie

LOW RESISTANCE LIQUID MOTION FOR ENERGY HARVESTING

Ankur Goswami, Shashank Gowda, Abinash Tripathy, Diptanu Roy,
Venkatesh Bharadwaja and Prosenjit Sen

Centre for Nano Science and Engineering, Indian Institute of Science, Bangalore, India

ABSTRACT

Low resistance motion of liquids on a well-defined path is beneficial for several MEMS based applications including energy harvesting and switching. By eliminating the contact line we demonstrate low resistance motion of a liquid bulge on pre-wetted strips. The bulge appears on wetted strips due to a morphological instability. The wetted strip confines the mercury bulge and defines its path of motion. Resistance to initiate motion of the bulge was studied experimentally and compared to other cases. An electret based energy harvesting device using bulge motion has been fabricated and tested.

INTRODUCTION

A multitude of remotely placed electronic and micro-electro-mechanical devices require self-sustaining power sources. In particular there is an increasing demand for powering portable personal computing devices, wireless distributed sensor networks [1] and implantable / wearable biomedical devices [2]. Energy harvesting is considered as a solution to obtain the miniaturized and theoretically inexhaustible energy source for these devices. Energy harvesting refers to the conversion of ambient waste energy (from sources like wind, heat, vibrations, etc.) into useful electrical energy.

Vibration based energy harvesters are primarily piezoelectric [3], electromagnetic [4] or electrostatic [5]. Most of these devices are based on resonant structures and operate at relatively high frequencies [6]. Small deviations from these frequencies results in a substantial loss of obtained power. For harvesting environmental vibration a broadband large displacement device capable of achieving high conversion efficiency in non-resonant mode is preferable. A sliding mass has been demonstrated as a solution in the form of a variable capacitance electrostatic energy harvester [7]. The use of a fluid instead of a solid as the sliding mass has advantages in no wear and reduced losses from friction [5].

Electrostatic energy harvesters need to be biased to operate. Some of the ways of biasing are providing external voltage, using materials with different work functions or by permanently embedding charges in a dielectric to form an electret [8]. Boland et al. [5] were the first to introduce liquid rotor electret power generator as a form of energy harvesting device that converts mechanical energy to electrical energy based on the principle of a variable permittivity capacitor. Motion of a liquid droplet in between the electrodes of a pre-charged capacitor leads to a change in its permittivity. This change in capacitance leads to a changing open circuit voltage of the capacitor which is given by

$$V(t) = \frac{Q}{C(t)} \qquad (1)$$

where V is the voltage, Q is the electret charge and C is the capacitance. Mercury was used as the liquid because it has a high surface tension and hence not heavily affected by electrowetting. The conductive liquid metal droplet acted as a dielectric of infinite permittivity. This device was very apt for harnessing random environmental vibrations because it did not require a sinusoidal driving motion and it could operate at non resonant state.

Yang et al. [8, 9] have reported a similar droplet based energy harvesting device with interdigitated electrodes. The devices were able to achieve a better efficiency in comparison to the previous demonstration [5]. A Polytetrafluoroethylene (PTFE) electret used was charged during the sputter deposition of the polymer itself. A 1.2 mm mercury droplet required a tilt angle of about 15° to overcome the contact angle hysteresis and set the liquid droplet in motion. The motion of the droplet was confined using PDMS baffles.

Both the devices required to overcome a significant contact angle hysteresis (a tilt angle greater than 15°) before droplet motion was initiated. This meant a significant amount of vibrational force was required to overcome hysteresis before any harvesting was possible. This leads to reduced energy harvesting efficiency. Moreover extra energy would be wasted in detaching the droplet from the external baffles (side walls) used in these designs. It is also important to realize that for optimized energy conversion for electrostatic harvesting the droplet should move in a well-controlled optimum path with respect to the electrodes. All the previous designs have no control over the path that the droplet follows.

Superhydrophobic surfaces are known to reduce hysteresis to negligible limits. It is however not trivial to confine droplets on superhydrophobic surfaces as they tend to escape and stick to more wettable regions and sidewalls. There is also very little control on the path a droplet takes on superhydrophobic surfaces, making them unsuitable for energy harvesting and switching applications. Here we demonstrate low resistance motion of a liquid bulge on a pre-wetted line. In addition to reduced resistance the proposed configuration confines the bulge and restricts its motion to a pre-defined path.

DEVICE FABRICATION

Device Design

The electret based energy harvesting device consists of two plates which will be referred to as the electret and the base plate. The base plate consists of a "mercury-philic" strip in a "mercury-phobic" domain (see Figure 1). The morphology of the liquid on the strip depends on its volume relative to the strip width [10, 11]. At small volumes of mercury the rectangular strip is wetted to form a liquid cylindrical cap. Above a critical volume, once the contact angle is above 90°, a liquid bulge appears due to a morphological instability. This bulge does not have any contact line and can move along the strip with viscous losses only. At even higher volumes, above a second critical volume, the bulge spreads to form a contact line with the "mercury-phobic" region. In this regime the bulge starts to experience additional contact angle hysteresis losses. Hence, for a given strip width an optimal volume range exists which leads to minimum contact line losses and allows bulge motion at smaller tilt angles. The wetted strip also confines the motion of the liquid bulge to a well-defined path. In this configuration the liquid bulge does not contact the sidewalls and hence reduces losses in energy conversion.

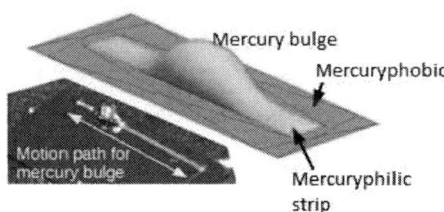

Figure 1 : Morphology of the mercury bulge on the gold strip as shown by surface evolver and by experiment.

The electret plate consists of a spin coated dielectric (Teflon) under which a set of rectangular electrodes are spaced periodically. The dielectric is charged by means of high voltage corona discharge. The charges trapped in the dielectric ensure that the top plate behaves as a permanent source of electric field - an electret. The liquid bulge is in contact with both the top and bottom plate, with the top plate oriented in such a way that the electrodes are perpendicular to the path of the liquid on the bottom plate (see Figure 2).

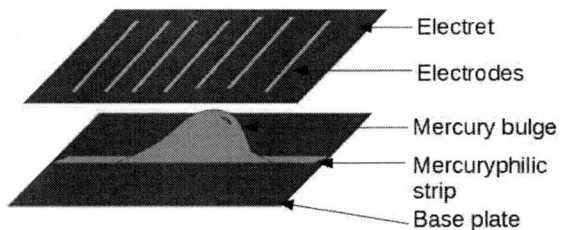

Figure 2 : Schematic of the orientation of electrodes on the electret plate with respect to liquid bulge path.

Base Plate Fabrication

Si wafers were cleaned with RCA1 solution (5:1:1 $H_2O:H_2O_2:NH_4OH$ at 75 °C) for 10 min and RCA2 solution (6:1:1 $H_2O:H_2O_2:HCl$ at 75 °C) for 10 min to remove organic and metallic contaminants and prepare the wafers for thermal oxidation. 200 nm of SiO_2 was grown using thermal oxidation. The wafers were then cleaned with acetone, IPA and DI water. Dehydration bake was performed at 250 °C for 10 min. Lift off process was used to pattern the Cr/Ni/Au metal lines as "mercury-philic"

regions. For this lift off resist LOR10A was spin coated at 4000 rpm for 40 s. The wafers were then heated at 160 °C for 2 min and then left at room temperature for 15 min. S1813 was then spin coated at 4000 rpm for 40 s. Following a soft bake at 110°C for 1 min photolithography was performed at 45 mJ/cm^2 using a mask aligner (EVG 620). The wafers were developed in AZ351B solution. After baking for 3 min at 110 °C development of LOR10A was done using MF26A. Cr/Ni/Au of thicknesses 20/200/40 nm were deposited by Tecport E-beam evaporator. The lift off process was completed by agitating the wafers in PG remover solution.

Electret Plate Fabrication

The process is the same as the base plate fabrication but for a few modifications. 500 nm of SiO_2 was grown after cleaning with RCA. Following the complete lift off process, SiO_2 of 500 nm was deposited on the wafer using plasma enhanced chemical vapour deposition (PECVD). S1813 was spin coated at 4000 rpm for 40 s. The wafers were patterned to expose the contact pads. BHF was used to etch away the SiO_2 over the gold contact pads. The S1813 was then removed using acetone and IPA. 0.5 wt% silica nanoparticles (50 - 70 nm) were then dissolved in ethanol and spin coated on the wafer at 7500 rpm for 40 s. The samples were baked at 250°C for 15 min. Teflon (AF1600 from Dupont) was used as received and spin coated onto the wafer at 6000 rpm for 40 s. The wafers were baked at 110 °C for 2 min and then at 250 °C for 40 min to remove all the solvents (Figure 3).

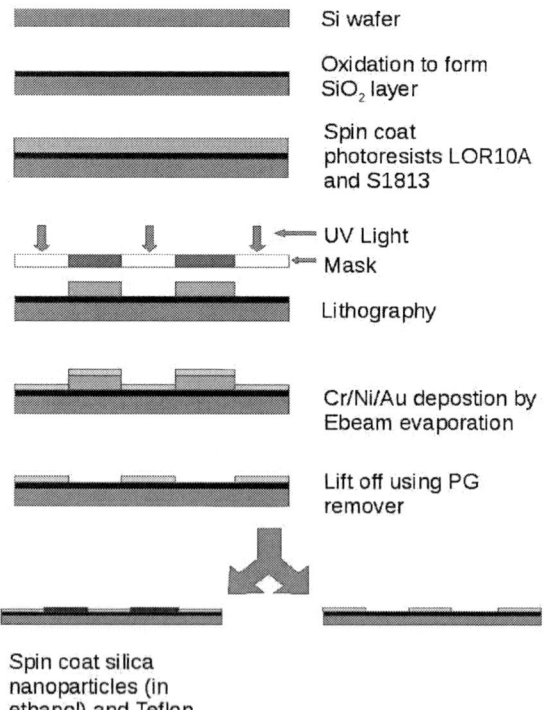

Figure 3 : Fabrication procedure for electret and base plate.

Corona Discharge

The electret plate was subjected to corona discharge (Figure 4) to embed charges into the dielectric. 10 kV voltage source is connected to an Aluminium rod from which a Cu wire of 0.5 mm diameter is suspended. 1 kV voltage source is connected to a mesh which is placed between the discharge tip and the sample [12]. The electret plate is heated to 200 °C and the voltage is increased until the tip of the wire glows and a current starts flowing, indicating the ionization of air around the tip. The charged ions/electrons bombarding the Teflon layer gets trapped in the dielectric. The mesh is used to distribute the charges homogenously throughout the top plate.

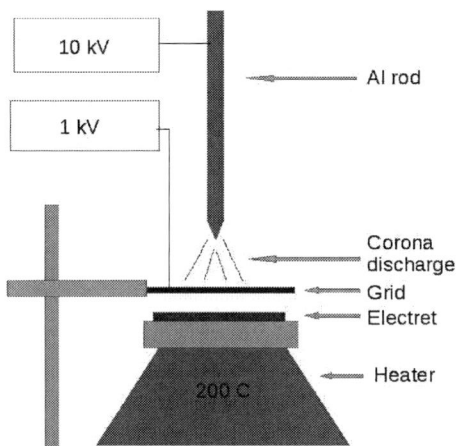

Figure 4 : Schematic of the corona discharge setup.

EXPERIMENT

Contact angle hysteresis

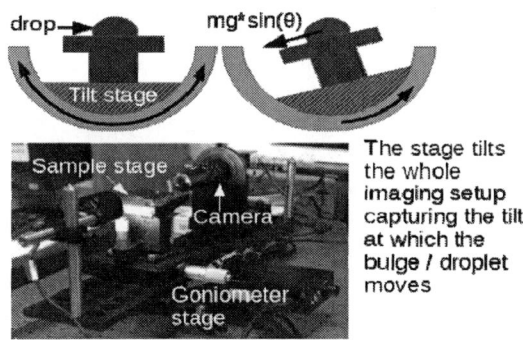

Figure 5 : Goniometer schematic and setup.

As discussed above an optimum volume range exists for which the contact line losses are minimized. For our configuration of 700 μm line width the volume was maintained between 3.5 - 5 μL to ensure losses are minimum. The tilt angle at which the liquid bulge is displaced from its initial position was obtained using a goniometer (Newport BGS80CC) (Figure 5) and a camera (Edmund Optics). For our strips tilt angles as low as 0.8° was sufficient to initiate bulge motion (Figure 6). For

comparison we experimentally measured the tilt required to initiate motion of similar volume droplets on different surfaces. The force required to initiate liquid bulge motion is about 4x less than required to initiate droplet motion on a Teflon coated surface and about 20x lesser than droplet motion on a Si substrate (Figure 7).

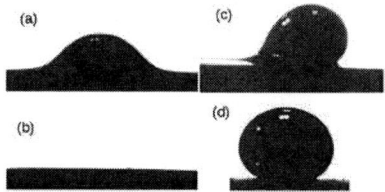

Figure 6 : Mercury bulge on the stripped surface. Image taken from the goniometer camera. (a) The bulge at 0.5° tilt and (b) The bulge moved at 0.8° tilt angle. (c) The bulge does not escape the wetting film even when the tilt is 40°. (d) Resistance to droplet motion where droplet is sitting on the mercury-phobic portion.

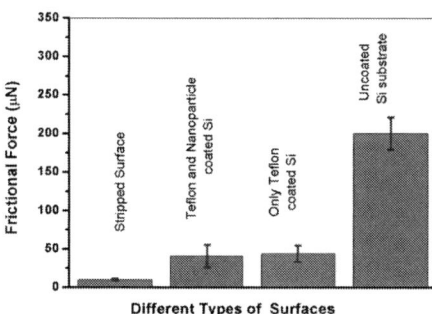

Figure 7 : Force to initiate liquid bulge motion on stripped surface is 4x lesser than a droplet on Teflon coated low hysteresis surface and 20x lesser than a normal Si substrate.

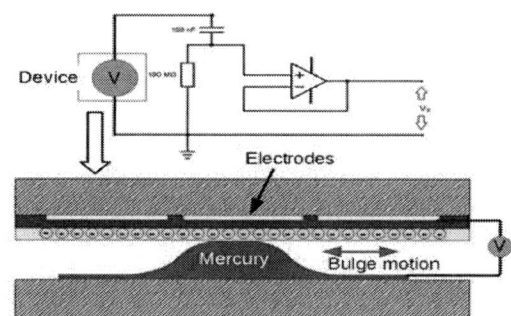

Figure 8 : Schematic of energy harvesting device and readout circuit.

Output measurement

The liquid bulge is formed on the wetting line of the base plate. The electret plate is then placed such that it is in contact with the liquid bulge and the orientation is as shown in Figure 4. Wires are soldered onto the contact pads of both the electret and base plates. The electret plate leads are connected to an input amplifier, while the base plate electrode is connected to the ground. The output of the amplifier is connected to an oscilloscope

(Agilent DSO1102B) to detect the output of the energy harvesting device. Voltage peaks of up to 2V (Figure 9) were obtained when the liquid bulge comes in contact with the electrode on the electret plate from which the output was being read.

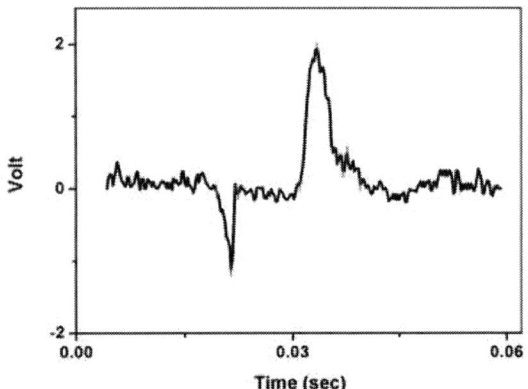

Figure 9 : Voltage signal from the readout circuit as a result of mercury bulge motion due to tilting.

CONCLUSIONS

In this work we have demonstrated that mercury bulge on a pre-wetted strip serves two primary purposes in this particular application of an energy harvesting device: a) low resistance to liquid motion and; b) achieving a well-controlled path for the liquid motion with respect to the electrodes. The force required to initiate bulge motion is extremely small. The force was quantified by measuring the tilt angle to initiate bulge motion using a goniometer. Quantitatively the force required is approximately 4x and 20x lesser than the conventionally used surfaces - Teflon coated silicon and uncoated silicon respectively for a given volume of the mercury. Minimal losses ensure higher harvesting efficiency of the device. Voltages up to 2V were obtained. The approximate peak power from a single electrode calculated as V^2/R_L is 40 nW._This type of energy harvesting device is suitable for low frequency applications.

ACKNOWLEDGEMENTS

The authors would like to thank Indian Institute of Science, Bangalore and Unilever R&D Bangalore for financial support. The authors would also like to thank Ms. Ayushi Patel from IIT Gandhinagar for surface evolver support and Prof. Rudra Pratap from CeNSE, IISc for useful discussions.

REFERENCES

[1]. R. J. M Vullers, R. van Schaijk, I. Doms I, C. Van Hoof, R.Mertens, "Micropower energy harvesting", *Sol. Stat. Ele.*, vol. 53, no. 7, pp. 684–693, 2009

[2]. E. Romero, R. O. Warrington, and M. R. Neuman, "Energy scavenging sources for biomedical sensors", *Physiol. Meas.*, vol. 30, no. 9, pp. R35–R62, 2009.

[3]. H. Shen, J. Qiu, M. Balsi, "Vibration damping as a result of piezoelectric energy harvesting", *Sens. Actuators A, Phys.*, vol. 169, no. 1, pp. 178–186, 2011.

[4]. T. Galchev, K. Hanseup, K. Najafi, "Micro power generator for harvesting low-frequency and nonperiodic vibrations", *J. Micromech. Microeng.*, vol. 20, no. 4, pp. 852–866, 2011.

[5]. J. S. Boland, J. D. M. Messenger, K.W. Lo, Y.C. Tai, "Arrayed liquid rotor electret power generator systems", *Proc. 18th IEEE Int. Conf. Micro Electro Mech. Syst.*, Feb. 2005, pp. 618–621.

[6]. P. D. Mitcheson, P. Miao, B. H. Stark, E. M. Yeatman, A. S. Holmes, T. C. Green, "MEMS electrostatic micropower generator for low frequency operation", *Sens. Actuators A, Phys.*, vol. 115, no. 2–3, pp. 523–529, 2004.

[7]. H. Lo and Y. C. Tai, "Parylene-based electret power generators", *J. Micromech. Microeng.*, vol. 18, no. 10, pp. 104006-1-104006-8, 2008.

[8]. Z. Yang, E. Halvorsen, T. Dong, "Electrostatic Energy Harvester Employing Conductive Droplet and Thin-Film Electret", *J. Micromech. Microeng.*, vol. 23, no. 2, pp. 315-323, 2013.

[9]. Z. Yang, E. Halvorsen, T. Dong, "Power generation from conductive droplet sliding on electret film", *Appl. Phys. Lett.*, vol. 100, no. 21, pp. 213905-1-213905-4, 2012.

[10]. H. Gau, S. Herminghaus, P. Lenz, and R. Lipowsky, "Liquid Morphologies on Structured Surfaces: From Microchannels to Microchips", Science, vol. 283, no. 5398, pp. 46-49, 1999.

[11]. M. Brinkmann and R. Lipowsky, "Wetting morphologies on substrates with striped surface domains", *J. Appl. Phys.*, vol. 92, no. 8, pp. 4296-4306, 2002.

[12]. M. Kranz, M. G. Allen, and T. Hudson, "In situ wafer-level polarization of electret films in MEMS acoustic sensor arrays", *Sens. Actuators A, Phys.*, vol. 188, pp. 181-189, 2012.

CONTACT

Dr.Prosenjit Sen

Centre for Nano Science and Engineering, Indian Institute of Science, Bangalore 560012, India.

Email: prosenjits@cense.iisc.ernet.in

Ph No: +91-80-2293-3516

MICROFLUIDIC SWITCHING DEVICES SHOWING CONTROLLABLE HYSTERESIS

Salvador Ortiz[1] and Julián A. Lell[2]

[1]Departamento de Micro y Nanotecnología, Comisión Nacional de Energía Atómica
Av. Gral Paz 1499, San Martín, CP 1650, Prov. de Buenos Aires, Argentina

ABSTRACT

We present a microfluidic MEMS device that is capable of switching among different states, thus behaving as an effective flip-flop. The devices consist of linear microfabricated deLaval nozzles with three exit channels, and we analyze the response as a function of input and output pressures, feed gas, and dimensions. In all cases, we have seen the appearance of a vortex, whose direction of swirl changes sign according the input pressure, and showing hysteresis. This type of response can be used to realize history-aware fluidic devices.

INTRODUCTION

Vortex dynamics describing this type of behavior date back to the 1920 's and 30's, relating to the Coanda effect and its applications to aeronautics[1]. The study of fluidics, as an option to vacuum and solid-state electronics, became popular in the 60's and 70's, but were abandoned with the advent of the p-n-p transistor[2]. Recently, the widespread development of lab-on-chip systems recenters the attention on microdevices that can be useful for pumping and/or fluid control, applications where the behavior shown here can be useful[3].

The devices under study consist of linear converging/diverging DeLaval nozzles, with three different output channels: one in the center and two lateral ones. Critical design parameters are throat size, ranging from 200mic to 1mm, and position of the skimmers that separate the output. Devices was made out of a 530mic thick silicon wafer, through a Deep Reactive Ion Etching (RIE, Oxford PlasmaLab 80) process up to 270mic. After etching, the nozzle was sealed by means of Anodic Bonding (EVG) with a Borofloat Glass wafer. Inlet and oulet holes were performed at the back side of silicon wafer, using the same technique of silicon dry etching described before. These wafers, after dicing, were inserted into a custom-made chamber for controlled circulation of gases through inlet and outlets (see Fig.1).

All processes related to design, fabrication, characterization and measurements were carried out at MEMS microfabrication facilities located at CNEA's Centro Atómico Constituyentes, in Buenos Aires, Argentina.

Previous to their fabrication, the devices were simulated numerically using Ansys Fluent[4], predicting this type of behavior at certain output pressure ranges. This was further corroborated with a simple analytical model for a vortex subject to an external flow, giving rise to switches in the vortex swirl direction according its vertical position.

At the time of designing and fabricating these devices, this effect was not expected, and not desired for the application that they were designed for. However, they show a non trivial complex behavior, that can be interesting from a microfluidical point of view, and can lead to applications profiting from the hysteresis in their response.

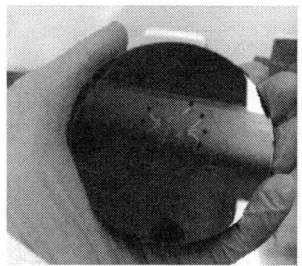 

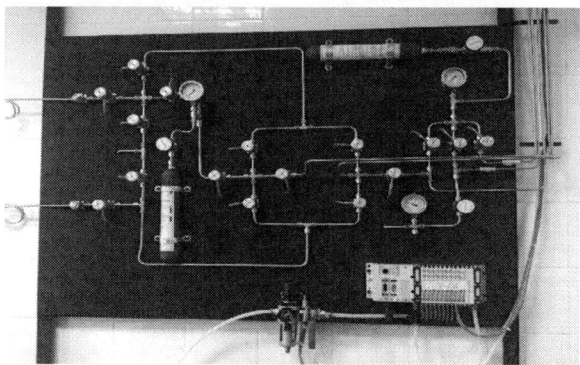

Figure 1: Fabricated microfluidic devices (top left); measurement chamber (top right); and manifold for displacement of gases during experiments (bottom).

RESULTS AND ANALYSIS

Measurement of gas flow through different channels was carried out using mass flow controllers (Brooks GFD SolidSense II, 20 slpm) and pressure regulators (Linde). Experiments were carried out with several gases: Argon, Nitrogen, and Sulphur Hexafluoride, all showing the behavior described above.

Simulations in Fluent were carried out under the following conditions: Density-based solver, using the Advection Upstream Splitting Method (AUSM), with Least-squares cell based gradient computation. Quadrilateral meshes were generated by a Delaunay approach under Gmsh[5], with cell count on the order of 40k. Convergence with residuals under 10^{-10} was reached in all cases, within less than 50k iterations.

Supersonic flow was attained in all of our simulations, and the total flow through the nozzles corresponds accurately with the 1-D models given by the standard Rankine-Hugoniot relations[6].

Concerning the distribution of flow among the three possible exits, we have observed that, in certain regimes (outlet pressure ~1 atm, flows lower than ~10 sccm), gas

978-1-4799-7956-1/15 $31.00 © 2015 IEEE   531   MEMS 2015, Estoril, PORTUGAL, 18 - 22 January, 2015

can be suctioned from one of the outlets, even though it is held at a lower pressure than the inlet. This rather counter-intuitive result is explained by the presence of a vortex in the fluid, which is the subject of this work. In Figs 2-3, we show experimental and simulated response for a typical device, with throat separation of 200mic, and skimmer placed at 1/3 and 2/3 of the total exit dimensions. Output atmospheric pressure was used, whereas input pressure ranges from 1.5 to 5 bar.

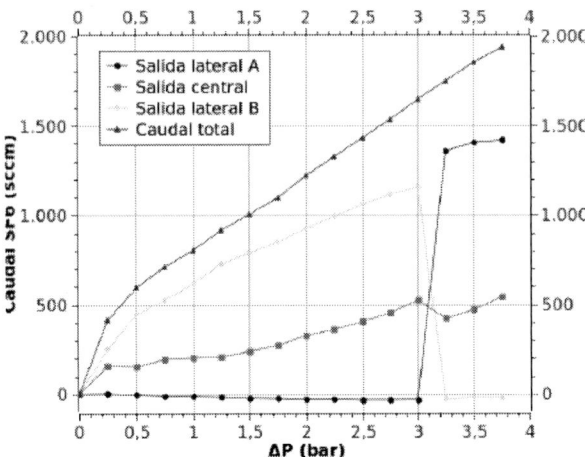

Figure 2: Experimental measurement of flow for a nozzle with 200mic throat, as a function of the pressure difference (outlet pressure at 1atm, gas is Ar)

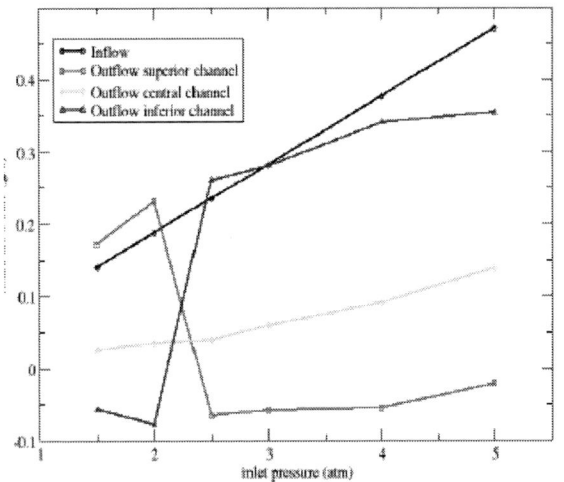

Figure 3: Fluent simulation for the measurement shown in Fig. 2.

Clearly, the device shows a vortex that causes suction from one of the exits, evidenced by a negative mass flow. Furthermore, this negative flow switches from one channel to another, as inlet pressure varies, owing to a variation in the vortex's position and direction of swirl.

This can be visualized in detail for the simulations (Fig. 4), where we have included two velocity fields corresponding to the same simulation parameters (i.e.: Ar, pin=3atm, pout=1atm), the only difference being that one

case was initiated from a solution with a clockwise vortex (lower inlet pressure) whereas the other one was initiated from a counter-clockwise vortex (higher inlet pressure). Concerning the total mass flux, they are exactly the same for both cases, even though their distribution varies radically.

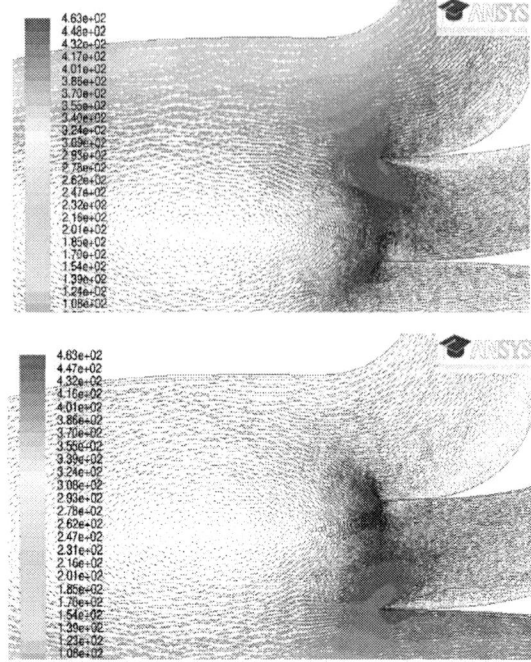

Figure 4: Coexistence of solutions for the simulated flow: clockwise vortex located below the center of the device (top), and counter-clockwise vortex located above center (bottom), for the same case (see text).

Coexistence and hysteresis of solutions was also observed in the experimental results. Similar results were obtained for different nozzles and operating gases.

MATHEMATICAL MODEL

The dynamics of this device can be emulated by a mathematical model consisting of a "rigid body" subject to an effective potential, which represents the interaction between the vortex and the background flow. The body is allowed to drift in one dimention (x axis) and to have an angular momentum perpendicular to the lateral drift (z axis). Full explanation of this model and its correlations to our device is beyond the scope; here we will briefly describe its main aspects.

At central position (x=0), no torque is applied to the body, but as soon as the body has been moved in any direction (positive or negative), a proportional torque will be applied to emulate the action of gas flow entering into the device. The angular momentum gained by the rigid body will drag it to the left if the sign of angular momentum is positive, and to the right if the angular momentum is negative, thus giving positive feedback to the original perturbation. Once the rigid body is far away from the center of the device, the angular momentum gained is big enough to sustain the chosen state (left or right) even in the presence of strong random perturbation.

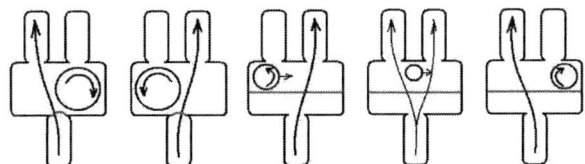

Figure 5: Graphical depiction of dynamical behavior of the body representing the vortex in the mathematical model; flow rate increases from left to right. The red line shows the shock wave position.

If the flow rate is increased, a shock wave will invade the area where the "rigid body" is allowed to exist, thus limiting the "size" of the body (i.e. the maximum magnitude of the angular momentum). If the flow rate is too large, the angular momentum of the body will be too small to keep the state of the device despite random perturbations, so sudden state flip may (and will) occur. Refer to Fig. 5 or a graphical picture of these situations. For the program to work as realistic as possible, several constraints were added:

1) the flow rate is affected by vortex "size" (L) and position (x), so flow rate is minimized when the vortex is at x=0 position, and maximized when it is at either extremes of the device. The higher the size of the vortex, the higher the influence it has on the flow rate.

$$Q = 1 - \frac{0.08 |L|}{e^{\left(\frac{100 . x^2}{L^2}\right)}}$$

2) drag forces tend to "decelerate" the spin and translation speed of the vortex. This is achieved by the addition of dissipative terms in the numerical integration equations.

$$L = L + dL . dt$$
$$x = x + dx . dt$$
$$t = t + dt$$

$$dx = dx + \left(- L + F_{wall} + noise(t)\right) . dt - 0.02 dx$$
$$dL = Q . sign(x) . x^4 - 2L$$

3) a white noise perturbation is applied to the vortex position, in order to take into account the small turbulence that can be present in the system, and is responsible for switching between the different metastable solutions, when coexistence occurs.

4) an exponential "potential wall" keeps the vortex inside the device for big displacements, and permits the bumping of the vortex against the wall.

$$F_{wall} = \frac{10}{\left(1 + e^{(20 . (x - gap))}\right)} - \frac{10}{\left(1 + e^{(- 20 . (x + gap))}\right)}$$

With these simple rules, it is possible to reproduce qualitatively the behavior shown by our devices, as shown in Fig. 6: on one side, we plot the position and size of the vortex for stationary state (after 10^4 iterations), as a function of the forcing flow. This plot shows a region of

undefined behavior, wherein the state of the device does not stabilize, and is strongly perturbed by the noise; for higher values of flow rate, the dynamics stabilize asymptotically to a well-defined state (in this case a positive L). In the lower plot, we show further detail, by plotting the time evolution for two of the flows referenced in the previous plot: one stable case, and one unstable, showing a strong fluctuation that can be related to coexistence of solutions.

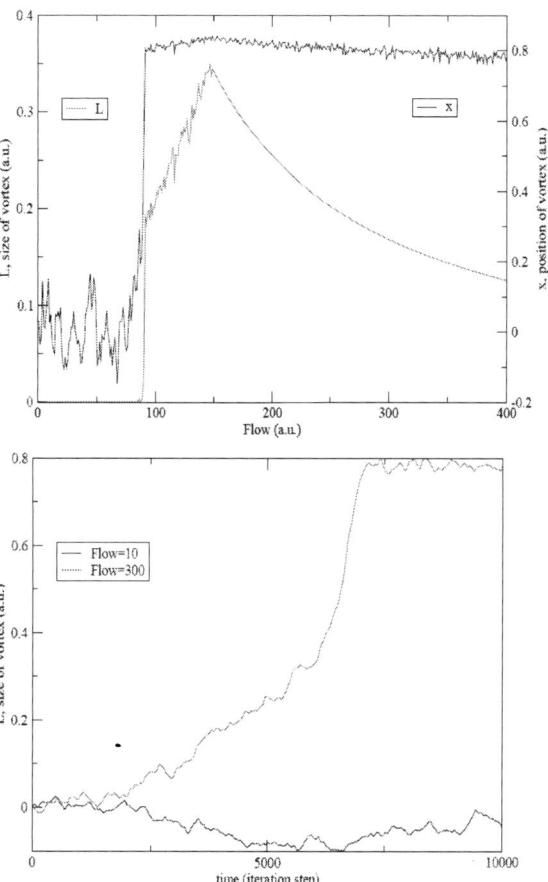

Figure 6: Numerical output for the mathematical model described in the text; stationary response as a function of flow rate, showing a an unstable regime associated to coexistence of solutions (top); and time evolution for two exemplary flows (bottom).

CONCLUSIONS AND FURTHER WORK

We have shown, experimentally and numerically, the appearance of an interesting, non-trivial behavior in very simple microfluidic devices, with possible applications to with possible applications to control logic and flags memory in extremely high EM field, radiation and temperature environments. The physical reasons for this behavior can be found in the highly non-linear interaction between a vortex and a constant-velocity field, as shown by our simple 1-D model. In order to fully understand this mechanism, future work needs to be carried out, specifying precisely the ranges and conditions that give rise to this behavior.

ACKNOWLEDGEMENTS

The authors thank all employees and researches at MEMS facilities in CNEA, for their efforts in fabricating the devices shown in this article.

REFERENCES

[1] Tritton, D.J., Physical Fluid Dynamics, Van Nostrand Reinhold, 1977

[2] A. Groisman, M. Enzelberger, S.R. Quake, Microfluidic Memory and Control Devices, Science 300, 955 (2003);

[3] B. Mosadegh et al, Integrated elastomeric components for autonomous regulation of sequential and oscillatory flow switching in microfluidic devices, Nature Physics, Vol. 6 (2010), pp 433-437 doi: 10.1038/nphys1637

[4] Ansys (r) Fluent: www.ansys.com;

[5] C. Geuzaine and J.-F. Remacle. Gmsh: a three-dimensional finite element mesh generator with built-in pre- and post-processing facilities. International Journal for Numerical Methods in Engineering 79(11), pp. 1309-1331, 2009, available online at http://geuz.org/gmsh/

[6] A. Shapiro, The Dynamics and Thermodynamics of Compressible Fluid Flow, The Ronald Press Company, 1953.

CONTACT

*Salvador Ortiz tel: +54-11-6772-7047; salvador,ortiz@cnea.gov.ar

MICROFLUIDIC HANGING-DROP PLATFORM FOR PARALLEL CLOSED-LOOP MULTI-TISSUE EXPERIMENTS

Saeed Rismani Yazdi[1,2], Amir Shadmani[1], Andreas Hierlemann[1], and Olivier Frey[1]

[1]ETH Zurich, Dept. Biosystems Science and Engineering, Basel, Switzerland
[2]Politecnico di Milano, Dep. Mechanical Engineering, Milan, Italy

ABSTRACT

We present a new on-chip pumping approach for microfluidic hanging-drop networks that are used for experiments with 3D microtissue spheroids. The pump includes a pneumatic chamber located directly above one of the hanging drops and uses the liquid-air-interface for flow-control. With this approach several independent hanging drop networks (HDN) can be operated in parallel with only one single pneumatic actuation line. The pump concept enables closed-loop medium circulation between different organ models for body-on-a-chip applications and allows for multiple simultaneous assays in parallel.

INTRODUCTION

The design of multi-organ devices is the next step towards more biomimetic in-vitro models. Configurable hanging-drop networks have been used to combine microtissue spheroids, representing 3D organ models, with microfluidic technology in order to realize continuous tissue-tissue interaction and study complex multi-organs systems [1]. Up to date, medium flow in such systems has been controlled by using external pumps and tubing. Tubing increases medium volume, bears the risk of increased compound adsorption at surfaces and compromises ease-of-use and the potential for highly parallelized experiments. Several on-chip micropump designs have been reported for different applications [2-3]. However, these micropumps were designed for driving liquid flow in closed microfluidic systems and are, therefore, of limited use in open microfluidic systems, such as the hanging-drop networks. Here, we present a novel integrated pump system for microfluidic hanging drop networks that helps to overcome these limitations.

DESIGN AND OPERATION PRINCIPLE

Figure 1a shows the design of one microfluidic hanging-drop loop. The loop consists of 10 hanging drops, each with 3.5 mm base diameter (circular areas); three of them form the pneumatically driven hanging-drop pump, and seven can be used to host the respective microtissues under investigation. The hanging drops are connected through 200-μm-wide bottom-open microchannels.

The pump is operated by pressurizing a pneumatic chamber located above the center pump-drop through a pressure control line (indicated in red in Figure 1a). The pressure increase deflects the membrane between pneumatic chamber and drop ceiling (500 μm thick) and closes, at the same time, an integrated valve at the inlet channel of the drop (Figure 1b). The valve is used to prevent back-flow and to obtain unidirectional flow.

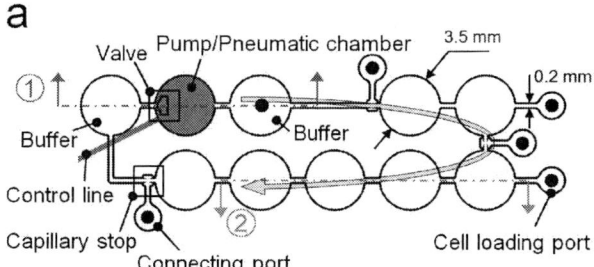

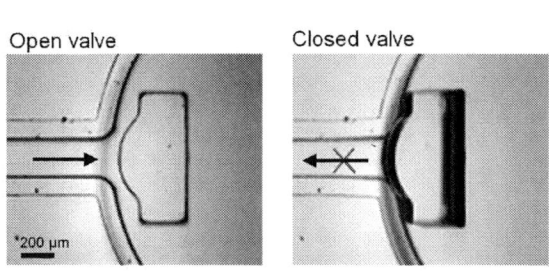

Figure 1: (a) Architecture and concept of the integrated micropump in a microfluidic loop of hanging drops. Two buffer drops are located before and after the pump-drop. A pneumatic chamber is integrated above the pump-drop (indicated in red, cross-section 1 in Figure 2). (b) Open and closed state of the valve. During pump actuation, the valve blocks the channel inlet and prevents reverse flow.

The membrane deflection virtually increases the volume of the drop. Based on Young-Laplace's equation, the decrease of the radius of the air-liquid-interface results in a pressure increase (P1>P2, Figure 2-1A) inducing a directional flow from the center pump-drop to the right neighboring drop (Figure 2-1B). When the pump membrane returns to resting state, liquid from both neighboring drops flows into the center drop thereby restoring equilibrium conditions (Figure 2-1C). This sequence is repeated with each pump stroke, and a circular and uniform flow through the hanging-drop loop can be achieved by applying an optimized actuation protocol. The two drops up- and downstream of the center pump-drop act as low-pass filters to reduce flow pulsation.

The other seven drops can be used to host spherical microtissues and have been arranged in groups of 2 and 5 drops that can be loaded with microtissues derived from different cell types (Figure 2-2). After cell medium has been filled into the microfluidic network, all three sub-networks (pump-unit, 2-drop-unit, 5-drop-unit) are fluidically interconnected upon opening of the capillary stop valves by supplying a small amount of liquid through the connecting ports. (Please refer to Ref. 1 for more details).

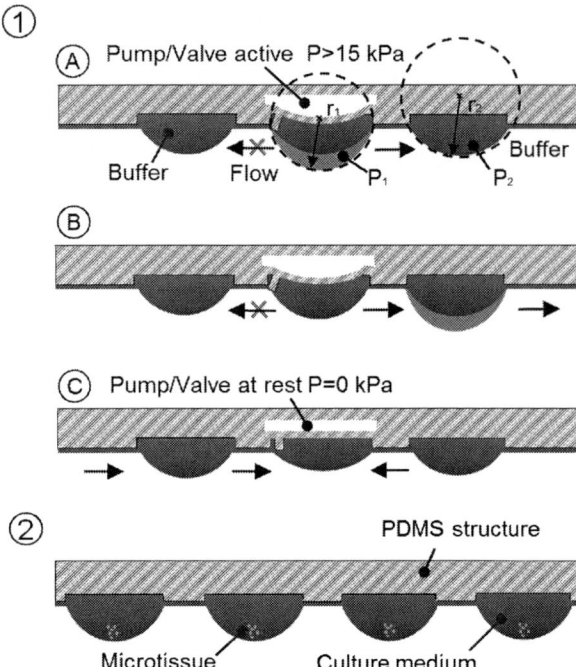

Figure 2: Cross-sections (1) and (2) as indicated in Figure 1, showing the schematic of pump actuation in a sequence. Upon pressurizing the pneumatic chamber above the pump-drop (indicated in red in Figure 1), the membrane deforms and closes the valve. At the same time, the volume change in the hanging drop below induces a pressure difference between pump drop and neighboring drop to the right and generates a directional fluid flow. The pump drop is then refilled after releasing the pressure from both neighboring drops. The two neighboring drops up- and downstream act as buffers to reduce pulsation. (2) Seven drops can be used to host different tissue types (e.g., 2 and 5 of a certain type) for tissue-tissue interaction studies.

MATERIALS AND METHODS

The microfluidic device with 8 independent loops was fabricated by using multilayer soft lithography. For both layers, pump and hanging-drop network (HDN), SU-8 100 negative photoresist (MicroChem, USA) was spin coated on thoroughly cleaned 4-inch silicon wafers. A rotation speed of 1500 rpm for 30 sec was selected to achieve a photoresist layer of 200 µm thickness. Two soft-bake steps at 65 °C for 30 min and at 95 °C for 90 min were applied. The wafer was then exposed to ultraviolet light to initiate crosslinking followed by two successive post-exposure baking steps at 65 °C for 5 min and at 95 °C for 30 min. The aforementioned steps were repeated for a second layer to reach final patterns of 450 µm height. The wafer was subsequently developed for 30 min in SU-8 developer (mrDev 600), rinsed with isopropyl alcohol, de-ionized water and dried with a nitrogen gun.

Two different transparency masks were used for UV exposure of the SU-8 layers of the HDN, whereas only one was used for the pump layer. Before replica molding, wafers were coated with trichlorosilane (Sigma-Aldrich,

Switzerland) to reduce adhesion between SU-8 and the PDMS elastomer. Microfluidic devices were casted from polydimethylsiloxane (PDMS, Sylgard 184, Dow Corning Corp., USA). PDMS base and curing agent were mixed in a ratio of 10:1 and degassed before pouring onto the SU-8 mold. Next, the PDMS was cured at 80 °C for 1 hour. The micropatterned replicas were cut and peeled off from the SU-8 mold. An inlet hole for the control line was punched into the pump layer. After rinsing with acetone and isopropyl alcohol, air drying and oxygen plasma activation, the pump layer was precisely aligned onto the HDN layer. In order to improve irreversible oxygen plasma bonding and prevent compressed air leakage, both layers were manually pressed against each other and kept in an oven at 80 °C for 1 hour. Then, inlet ports were punched through the bonded layers to be able to supply medium and/or cell suspensions (Figure 3). A pump layer of 6 mm thickness was used in order to provide the required mechanical stability for the device. The micropatterned side of device was covered with Scotch tape after a cleaning step for storage. Prior to device usage, the surface of the HDN layer was covered with a PDMS layer with openings in the drop areas and then activated with oxygen plasma in order to increase wettability of the PDMS. In this way, only the surfaces that come in contact with liquid were rendered hydrophilic, whereas rim structures were kept hydrophobic and prevented liquid overflow. Particularly, the wetting of the valve surface turned out to significantly affect pumping performance, as it changed the pump drop shape and caused additional fluid flow over the activated surface region.

For characterization experiments each individual hanging drop loop was filled with 80 µl suspension of micro-beads (5 µm diameter) in de-ionized water through the inlet holes by using a multichannel pipette.

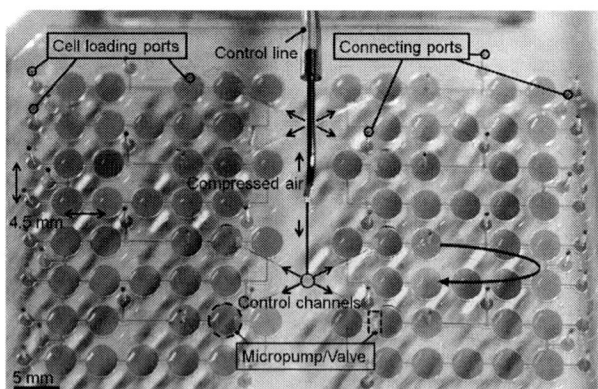

Figure 3: Photograph of the microfluidic hanging drop platform consisting of 8 identical hanging drop loops (4 loops on the left side are mirrored with respect to Fig. 1). All micropumps are connected to a single control line for parallel actuation. Alternating colored liquid show the 10 drops in each of the 8 independent loops. The drops are arranged in a 384-well plate grid (i.e. 4.5-mm pitch) for automated read out.

Characterization tests were performed on an inverted microscope (Olympus IX81) using a 5X objective in bright field mode. To minimize evaporation of the liquid from the hanging drops during experiments, microfluidic devices were placed into a closed humidified chamber. A single tube connected the pneumatic control line over a 21-gauge 90-degree-bent needle to all pneumatic chambers. The control line was connected to a 3/2-way miniature solenoid valve (Festo, Germany) controlled via a data acquisition (DAQ) device (National Instruments) operated by LabVIEW programming software. A custom-designed printed-circuit board was used to connect a DAQ card to the solenoid switch. A pressure controller was used to regulate the applied positive pressure. Compressed air (typically 30 kPa) was routed to the 8 pneumatic chambers through identical and symmetric control channels. This layout assures identical actuation frequency and membrane deflection heights and, consequently, identical flow-rates for all eight loops. The device was designed to be compatible with 384-well plate formats and multichannel pipettes (4.5 mm pitch between inlet ports of adjacent loops) to facilitate medium and cell loading as well as automated imaging.

Particles were traced along the length of the channel after the pump drops by using a digital camera with an image acquisition rate of 10 frames per second. Average particles speeds were measured manually by using ImageJ image analysis software so that the respective flow-rates could then be calculated.

The pump actuation protocol (shown as a square wave in Figure 5) consists of two states maintained for two different durations. T-on refers to the time when the pressure is applied to the pneumatic chamber. In this state the membrane is deflected, and the valve is closed. T-off stands for the time at resting state, when no pressure is applied, the membrane is at its initial flat position, and the valve is open. The rather large thickness of the PDMS membrane ensures short transition times.

RESULTS

The performance of the pumping system was investigated for different pump protocols. Figure 4 shows time-lapse images of ink circulating in the hanging-drop loop, which indicates that the integrated pumping system was able to successfully generate gradual unidirectional fluid flow. Before the pump was activated (t = 0 min), 4 µl of blue ink were added to one of the hanging drops. Some of the liquid penetrated into both neighboring drops. Upon actuation of the pump, the fluid flows clockwise, and the blue link is transported along the drops and channels through the system. Complete circulation and mixture of ink and water was observed after about 15 min of continuous pumping.

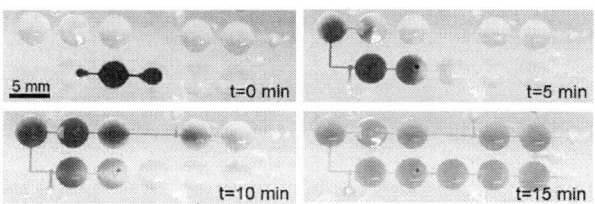

Figure 4: Time series of images showing the circulation of a small volume of blue ink within one loop of drops as a result of the integrated pumping. (Protocol: t-off = 0.025 s, t-on = 0.050 s).

Figure 5 presents experimental results of the achieved flow-rates in dependence of the different pump actuation protocols.

First, the valve was closed by applying a positive pressure in excess of 15 kPa. It has been observed that increasing the applied pressure generates higher flow-rates. Here, 30 kPa relative pressure were applied, which deformed the membrane by approximately 400 µm in the center. During each stroke, the membrane expansion adds a virtual volume of 2 µl to the hanging drop.

The t-off time was set to 0.025 s, whereas the t-on time was varied between 0.025 and 0.3 s. A constant flow-rate was achieved after 20 pump strokes. As can be seen in Figure 5, the flow-rate decreased with increasing the t-on time. Further, the pump operation generated pulsatile fluid flow, while pulsation frequency and amplitude were strongly depending on the pump actuation protocol. Flow-pulsations were larger at lower actuation frequencies (114% relative flow variation with respect to the average flow-rate for t-on = 0.2 s versus 49% relative flow variation with respect to the average flow-rate for t-on = 0.025 s).

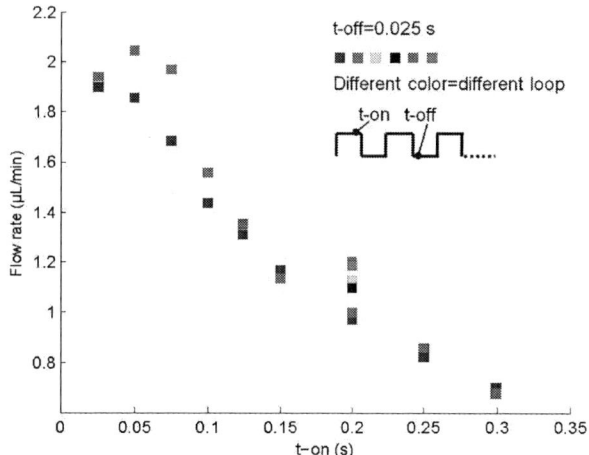

Figure 5: Measured flow-rates in dependence of the pump actuation protocol (indicated as a square wave, t-on stands for the time when the membrane is deflected, t-off when it is in the resting position). The different colors of the dots are measurements in different hanging-drop loops and demonstrate the low variability.

978-1-4799-7956-1/15 $31.00 © 2015 IEEE   537

For t-on = 0.2 s and t-off = 0.025 s the flow-rate variation between the different loops on the same platform has been measured. The relative variation was below 15%. These results demonstrate that the generated flow-rates can be precisely modulated by adjusting pump actuation protocol and operation conditions.

CONCLUSIONS

The design, fabrication, and characterization of a novel micropump, integrated into completely open microfluidic systems, have been presented in this paper. Unidirectional simultaneous fluid flow for 8 independent hanging-drop networks has been generated via an on-chip pumping system actuated through one single pressurized control line. The developed technique offers unique opportunities for parallel medium circulation in hanging drop networks, and is of great importance for multi-tissue experiments.

ACKNOWLEDGEMENTS

This work was supported by the European Union FP7 grant under the project 'Body on a chip' ICT-FET-296257 and an individual Ambizione Grant 142440 of the Swiss National Science Foundation for Olivier Frey.

REFERENCES

[1] O. Frey, P. M. Misun, D. A. Fluri, J. G. Hengstler, A. Hierlemann, "Reconfigurable microfluidic hanging drop network for multi-tissue interaction and analysis", *Nat. Commun.*, 5:4250, 2014.

[2] Brian D. Iverson, Suresh V. Garimella, "Recent advances in microscale pumping technologies: a review and evaluation", *Microfluid and Nanofluid*, 5:145-174, 2008.

[3] C.K. Byun, K. Abi-Samra, Y-K Cho, S. Takayama, "Pumps for Microfluidic Cell Culture", *Electrophoresis*, 35:245-257, 2014.

CONTACT

*Olivier Frey, tel: +41-61-3873344; olivier.frey@bsse.ethz.ch

ONE CORE-FIVE SHEATHS COAXIAL FLOW FORMATION USING MULTILAYER STACKED FLOW FOCUSING STRUCTURE

D.H. Yoon[1], J. Ito[1], N. Takeda[2], T. Sekiguchi[3], and S. Shoji[1]

[1]Faculty of Science and Engineering, Waseda University, Tokyo, JAPAN
[2]Graduate School of Advanced Science and Engineering, Waseda University, Tokyo, JAPAN
[3]Nanotechnology Research Center, Waseda University, Tokyo, JAPAN

ABSTRACT

This research presents efficient coaxial sheath flow formation using a three-dimensional PDMS device. In the device, one core and five sheaths are simply formed with low diffusion between different samples. Stacking and alignment of six PDMS layers of same structure allowed for fabrication of the proposed 3D device. Only one point alignment of center channel for the sheath provides free from misalignment that was a critical problem of multi-stacked PDMS device. Furthermore, the number of samples is infinitely expandable by increase in the number of stacking layers. The coaxial sheath flow is useful for biological fiber formation which requires multilayer of different materials such as artificial blood vessels or muscles.

INTRODUCTION

Three-dimensional flow focusing is one of interesting technology in a microfluidics field. For example, the 3D focusing allows for sample alignment for more efficient sorting or separation [1], or formation of mixing flow [2] or coaxial flow [3]. These functional flows have been realized by sophisticated channel design, flow control, and fabrication process.

Especially, with development of tissue engineering in recent years, the focusing and sheath flows were applied to the formation of artificial tissues having fiber shape, such as blood vessels or muscles.

Kiriya et al. reported the cell embedded hydrogel fiber using glass capillary tubes [4], Hu et al. utilized the 3D structure for hydrogel fiber formation [5], Lee et al. used cylindrical PDMS channels for the 3D focusing and fiber formation [6]. However, many researches are still remaining on a cell alignment or single layer fiber formation only.

Moreover, to realize the multiple layers of the real tissues, researchers have being asked more complicated devices which can inject large number of different sheath materials. However, conventional researches required highly precise device fabrication and long focusing area [7]. Thus, diffusion between the samples and reagents in the area for the sheaths affects the fiber structure. Sophisticated techniques for the fabrication were also an obstacle for practical applications.

To prevent the diffusion, efficient focusing in a short channel length is essential. Coaxial flow from a 3D structure provides improved focusing performance however, complicated structures are required. In this paper, we realized highly efficient multilayer sheath and focusing flow using 3D PDMS structure and practical fabrication method of the 3D device.

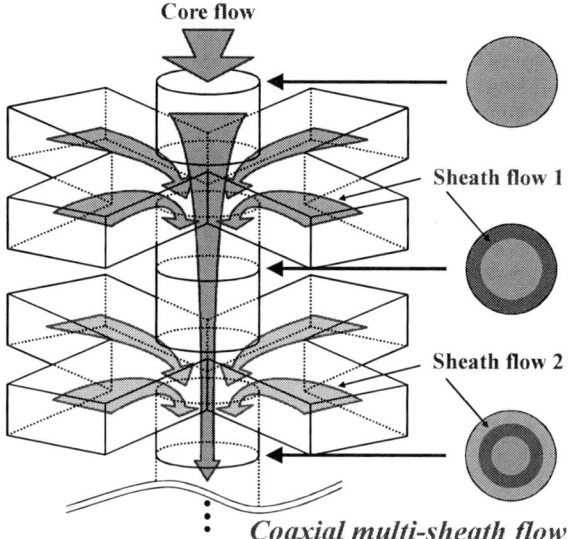

Figure 1: Formation principle of the core and multi-sheath flow in the 3D fluidic structure.

PRINCIPLE

Figure 1 shows a scheme of the 3D sheath flows of multiple layers in a proposed 3D device. A core fluid is introduced from center of the 3D device initially and the flow is focused by first sheath flow introduced from four directions which are separated from one inlet channel. By this 3D flow, core and sheath flow of cylindrical shape are formed. Then, the cylindrical core and sheath are focused again by second sheath flow as the same principle of the first focusing. The multilayer coaxial flow can be formed and the number of layers can be increased by repetition of this focusing. Furthermore, efficient focusing within a short distance utilizing multi-directional injection provides low diffusion between each interface.

DESIGN AND FABRICATION

For the complex 3D coaxial sheath flow, relatively simple designs were used, as shown Figure 2 (a). One sheath flow was obtained only a single 3D structure. Inlet channel of first layer is divided into four branch channels for uniform focusing and the flow is introduced into cylindrical channel for the coaxial flow. The 3D structure was fabricated by two step SU-8 patterning and thickness of first and second layer was 60 μm and 80 μm, respectively. SEM image of focusing area of the SU-8 mold and PDMS structure is shown in Figure 2 (b).

978-1-4799-7956-1/15 $31.00 © 2015 IEEE

(a) Design and dimension of fluidic channel

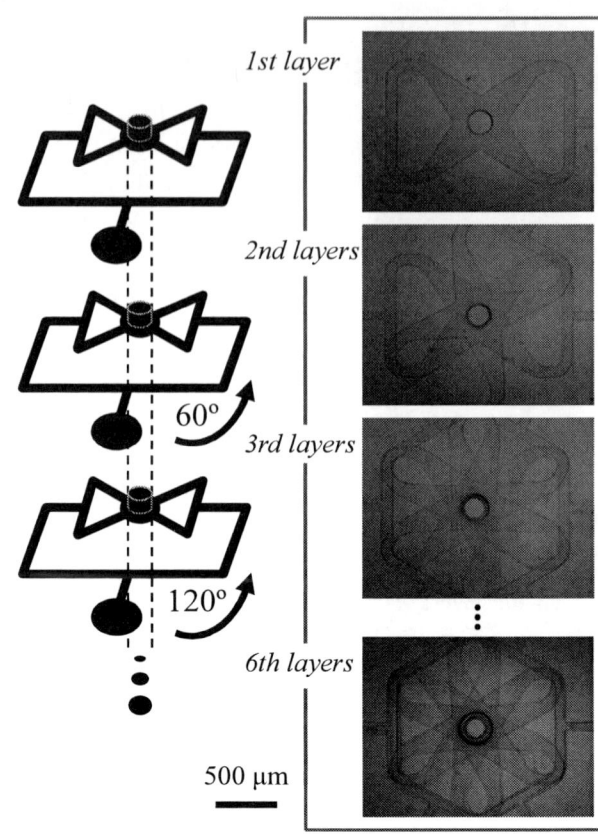

Second layer
(thickness: 80 μm)
(height: 140 μm)

First layer
(thickness: 60 μm)

Flow direction

(b) SEM images of focusing area

SU-8 mold   *PDMS layer*

Figure 2: Design and detailed size of the focusing area (a), and SEM images of focusing area.

Figure 4: Stacking process of the layers and visualized the results.

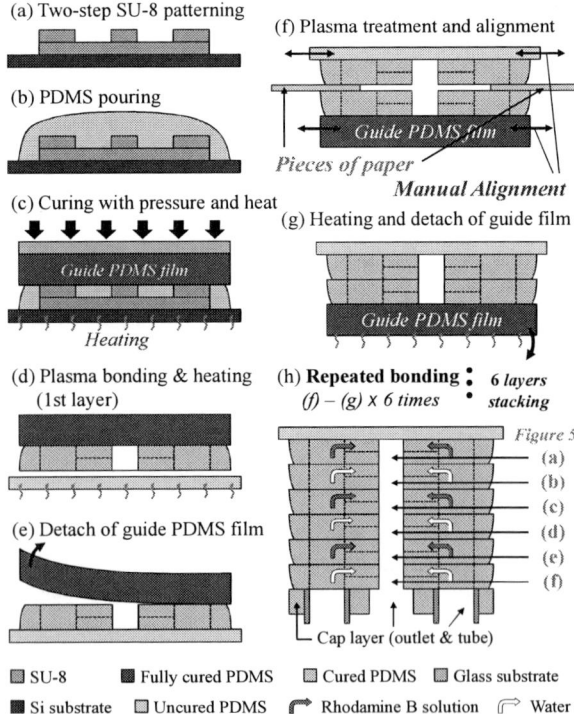

(a) Two-step SU-8 patterning
(b) PDMS pouring
(c) Curing with pressure and heat
Guide PDMS film
Heating
(d) Plasma bonding & heating (1st layer)
(e) Detach of guide PDMS film
(f) Plasma treatment and alignment
Guide PDMS film
Pieces of paper
Manual Alignment
(g) Heating and detach of guide film
Guide PDMS film
(h) **Repeated bonding :** **6 layers**
(f) – (g) x 6 times **stacking**
Figure 5
Cap layer (outlet & tube)

SU-8   Fully cured PDMS   Cured PDMS   Glass substrate
Si substrate   Uncured PDMS   Rhodamine B solution   Water

Figure 3: Fabrication process of the 6 layers PDMS structure (12 layers fluidic channels).

The 3D microfluidic structure was fabricated by stacking of the 3D PDMS structure. Especially, fully cured PDMS guide film was used to form through holes and to handle the thin patterned PDMS layers [8]. The guide PDMS film was formed on a non-patterned Si wafer at 80°C for 1 hour and the film was baked one more time after detaching from the substrate to reduce shrinkage of patterned structures. After pouring the uncured PDMS on the SU-8 mold, pressure of 100kPa was applied to the film on the structure and PDMS, while the PDMS was cured at 80°C for 30 min (Figure 3 (c)). The cured thin PDMS structure can be simply detached with the thick guide film from the SU-8 structure due to strong adhesion force [9], and first layer was bonded on the glass substrate after O_2 plasma treatment (Aiplasma, Matsushita Electric Works) as shown in Figure 3 (d). Heating at 70°C for 10 min assists fast and strong bonding of the attached layers.

Then the thick guide film was detached alone from the 3D PDMS layer (Figure 3 (e)). Second layer was formed by same process as Figure 3 (c), and detached from the SU-8 mold. As shown in Figure 3 (f), the layer was aligned under microscope using pieces of paper as spacer after plasma treatment. The paper was practically very useful for the precise alignment of PDMS layers. After the alignment and attachment of the layers, they were heated and the guide film was removed. By six times repeating the process, plasma treatment, alignment, heating, and detaching, the full structure was obtained (Figure 3 (h)).

978-1-4799-7956-1/15 $31.00 © 2015 IEEE

The microscopic images from the top side of the fabricated device are shown in Figure 4. In order to introduce six different liquids layer by layer, a PDMS structure was rotated and aligned about 60 degree from top layer. As shown Figure 4, center circle of cylindrical channel was the most important alignment point. Precise alignment is not required except center channel. Consequently, the paper spacer and one point alignment leads to successful fabrication of the precise multilayer fluidic device. Practical misalignment was less than 20 μm in the fully stacked device.

EXPERIMENTS

The fabricated fluidic channels were treated by Pluronic F-127, 10% solution in water for 1 hour to prevent bubble remaining in the channels. To visualize the focusing flow, Rhodamine B aqueous solution was introduced to layer 1, 3, and 5, while water was injected to layer 2, 4, and 6. The flows were controlled by syringe pumps (KDS 210, KD Scientific), and injection rate of all fluids was 5 μl/min. The results were visualized by confocal microscopy (Leica DMIRE2, Leica), as shown in Figure 5.

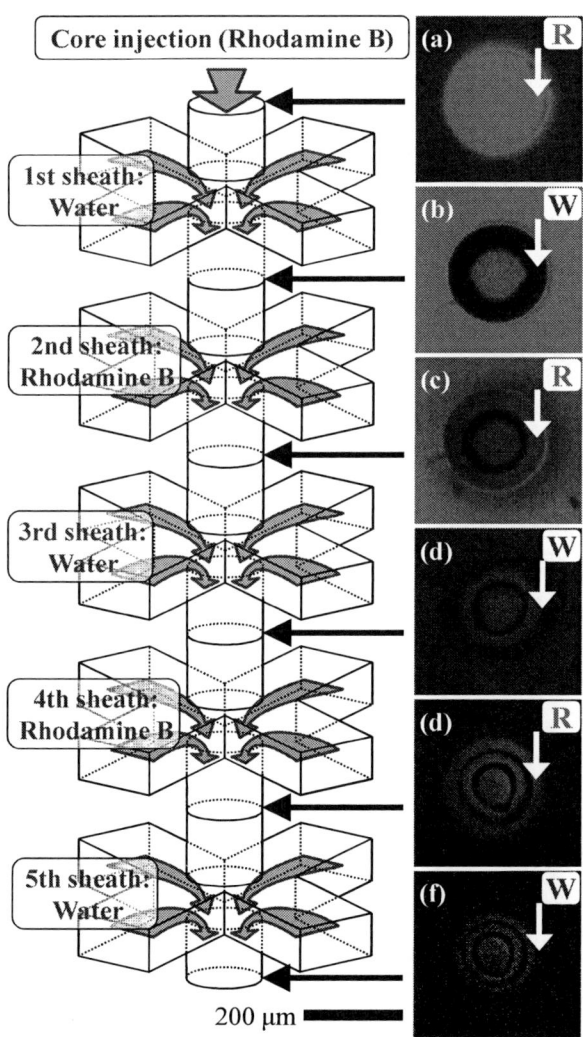

Figure 5: Images of one core-five sheath coaxial flow in each focusing layer visualized by confocal microscopy.

Initially introduced core (Figure 5 (a)) was focused by sheath flow (Figure 5 (b)) and they were focused repeatedly by next sheaths (Figure 5 (c)-(f)). The multilayer coaxial flow was obtained successfully by the proposed focusing method. One core and five sheaths were clearly observed and each sheath was thinner than 20 μm when the six fluids were fully introduced. The clear coaxial circles and thin layers confirm low diffusion between each interface. It is under investigation that various formation control of core and sheath under different flow conditions.

Furthermore, this device is now applied to the formation of artificial blood vessels. Cells, Alginate solution, and Calcium chloride solution are used for the fiber formation, and improvement and optimization of the device are in progress.

DISCUSSION

This 3D focusing flow is also being studied using a device formed by different fabrication method such as 3D printer. In recent years, precision of the 3D printer has been rapidly improved and many devices fabricated by this methods are already reported [10, 11]. Since this focusing scheme requires only 3D structure, it seems that the 3D printer will be a useful tool for fabrication of flow focusing devices.

As a future work, multi-core and multi-sheath flow is under investigation by considering our proposed device. Multi inlets of first layer will allow for multiple injections of different materials and the samples will be focused by flows from following layers for the formation of more functional biological fiber.

CONCLUSION

Single core and multi sheath coaxial flow was successfully formed by proposed 3D multilayer PDMS device. The 3D structure allowed formation of efficient 3D focusing flow within short distance and the flow provided low diffusion between the core and sheath interfaces. The 3D structure was fabricated by multi-stacking process of PDMS structure which requires only center channel's precise alignment. Moreover, the number of samples and sheaths are infinitely expanded by increase in the number of stacking layers.

This structure will be applied to formation of artificial blood vessels and formation of more functional biological fiber using improved structure is under investigation.

ACKNOWLEDGEMENTS

This work is partly supported by Japan Ministry of Education, Culture, Sports Science & Technology (MEXT) Grant-in-Aid for Scientific Basic Research (S) No. 23226010. And the authors thank for MEXT Nanotechnology Platform Support Project of Waseda University.

REFERENCES

[1] T. Arakawa, Y. Shirasaki, T. Aoki, T. Funatsu, and S. Shoji, "Three-dimensional sheath flow sorting microsystem using thermosensitive hydrogel", *Sensors and Actuators A*, vol. 135, pp. 99-105, 2007.

[2] Y. Gambin, C. Simonnet, V. VanDelinder, A. Deniz, and A. Groisman, "Ultrafast microfluidic mixer with three-dimensional flow focusing for studies of biochemical kinetics", *Lab Chip*, vol. 10, pp. 598-609, 2010.

[3] E. Kang, S. J. Shin, K H. Lee, and S. H. Lee, "Novel PDMS cylindrical channels that generate coaxial flow, and application to fabrication of microfibers and particles", *Lab Chip*, vol. 10, pp. 1856-1861, 2010.

[4] D. Kiriya, R Kawano, H. Onoe, and S. Takeuchi, "Microfluidic control of the internal morphology in nanofiber-based macroscopic cables", *Angew. Chem. Int. Ed.*, vol. 51, pp. 7942-7947, 2012.

[5] M. Hu, R. Deng, K. M. Schumacher, M. Kurisawa, H. Ye, K. Purnamawati, J. Y. Ying, "Hydrodynamic spinning of hydrogel fiber", *Biomaterials*, vol. 31, pp. 863-869, 2010.

[6] K. H. Lee, S. J. Shin, Y. Park, and S. H. Lee, "Synthesis of cell-laden alginate hollow fibers using microfluidic chips and microvascularized tissue-engineering applications", *Small*, vol. 5, pp. 1264-1268, 2009

[7] Y. Jun, E. Kang, S. Chae, and S. H. Lee, "Microfluidic spinning of micro- and nano-scale fibers for tissue engineering", *Lab Chip*, vol. 14, pp. 1245-2160, 2014

[8] D. H. Yoon, T. Sekiguchi, J. S. Go, and S. Shoji, "Continuous size-selective separation using three dimensional flow realized by multilayer PDMS structure", *MEMS 2012*, Paris, Jan. 29-Feb. 2, 2012, pp. 1025-1028

[9] M. Zhang, J. Wu, L. Wang, K. Xiao, and W. Wen, "A simple method for fabricating multi-layer PDMS structures for 3D microfluidic chips", *Lab chip*, vol. 10, pp. 1199-1203, 2010.

[10] G. Comina, A, Suska, and D. Filippini, "Low cost lab-on-a-chip prototyping with a consumer grade 3D printer", *Lab Chip*, vol. 14, pp. 2978-2982, 2014.

[11] G. Comina, A, Suska, and D. Filippini, "PDMS lab-on-a-chip fabrication using 3D printed templates", *Lab Chip*, vol. 14, pp. 424-430, 2014.

CONTACT

*D.H. Yoon, tel: +81-3-5286-3384; yoon@shoji.comm.waseda.ac.jp

978-1-4799-7956-1/15 $31.00 © 2015 IEEE

RECONFIGURABLE MICROFLUIDIC DILUTION GENERATION FOR QUANTITATIVE ASSAY

Jinzhen Fan[1,], Baoqing Li[1,2], and Tingrui Pan[1]*

[1]Micro-Nano Innovations (MiNI) Laboratory, University of California, Davis, USA.
[2]Department of Precision Machinery and Precision Instrumentation, University of Science and Technology of China, Hefei, China

ABSTRACT

In this paper, we present a reconfigurable microfluidic dilution device for high-throughput quantitative assays, without the assistance of continuous fluidic pump or vacuum source. This device is capable to produce discrete logarithmic concentration profiles ranging from 1 to 100 fold dilution in parallel from a fixed sample volume (e.g., 10 μL). To facilitate mixing and reactions in the chambers, two acoustic microstreaming actuation mechanisms have been investigated for easy integratability and accessibility. We characterized the dilution performance by both colorimetric and fluorescent means.

INTRODUCTION

In biological and biochemical studies, dilution generation has been a critical preparatory step and the gold standard to generate quantitative readings for a wide variety of assays [1-3]. The purpose of serial dilution is to create a series of discrete concentration profiles from an original analyte solution. Common applications of dilution generation include sample preparation for quantitative detection[4], calibration curve generation in immunoassays[5], and IC_{50} determination of drug response[6]. In the serial dilution protocols, similar steps are repeated in a sequence by either repetitive manual operations[7] or an expensive robotic machine[8]. To automate this laborious serial dilution step in quantitative assaying, extensive research and development have been contributed to establishing concentration dilution gradients using microfluidic devices pioneered by Whiteside group [9-14]. However, current dilution designs require a continuous flow along with a pumping source, often unavailable in clinics or point-of-care settings. More importantly, the clinical samples are typically collected in a fixed volume, and thus, the continuous flow assays either are incompatible with the sample size or result in decreased sensitivity and accuracy[15].

To tackle these aforementioned challenges, we report the first reconfigurable microfluidic dilution generator, producing discrete dilution profiles from a fixed sample volume of 10 μL, without needing any complicated equipment or pumping source. This dilution generator is an integrated platform which consists of switchable channels, metering reservoirs, and pressure-activatable Laplace valves [16, 17]. Simple manual pipetting steps that activate the Laplace valves are involved in the operation. Following the sequential loading of a sample, a diluent, and a detection reagent into the individual metering chambers, the top microfluidic layer will be reconfigured, and metered chemicals will be collected into the reaction chambers in parallel by pipetting, where the detection will be conducted. Acoustic mixing strategies have been integrated to facilitate mixing process at the end of the procedure and ensure solution uniformity across the device. In this paper, we demonstrated the dilution function by proportional combination of two colorimetric dyes and logarithmic dilution of bovine serum albumin (BSA)-fluorescent solution (FITC) conjugate. Overall, this portable chip is capable to serve as a facile tool for running generic quantitative assays for daily monitoring tasks in fields and biochemical laboratories.

DESIGN

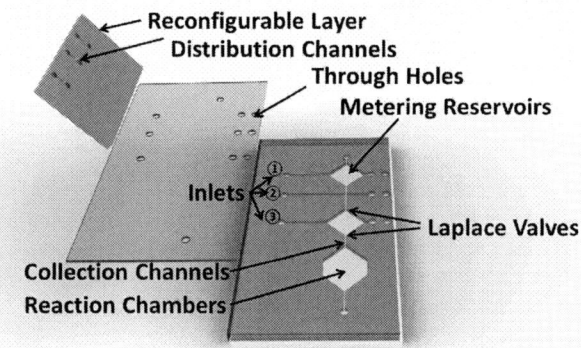

Figure 1: Structure and assembly of one of the eight dilution units on the reconfigurable microfluidic dilution generation device. ①, ② and ③ indicate the inlets of diluent channel, sample channel, and detection reagent channel, respectively.

This dilution generation device consists of three micropatterned layers, the reconfigurable top layer, the through-hole membrane, and the bottom layer. Structure of one of the eight dilution units is shown in Fig. 1. The top layer with horizontal distribution channels connects metering reservoirs in the bottom layer through the through-hole membrane. Sample, diluent, and detection reagents injected through three inlets will separate in metering chambers. In our current design, volumes of sample metering reservoirs are set to increase in a logarithmic fashion at base 2 (e.g., the first one is 18 nL, the second one is 36 nL, etc.), while that of the diluent chambers decrease in a complementary way which keeps total volume constant (e.g., the first one is 2268 nL, the second one is 2250 nL, etc.), leading to formation of a standard binary dilution profile.

Fig. 2 illustrates a generic protocol to perform a quantitative assay on this device. Initially, a sample, a diluent, and a detection reagent are loaded in a sequence by a micropipette through the inlets into the horizontal distribution channels to be metered (Fig. 2b). In the next step, the top layer with horizontal distribution channels can

a.

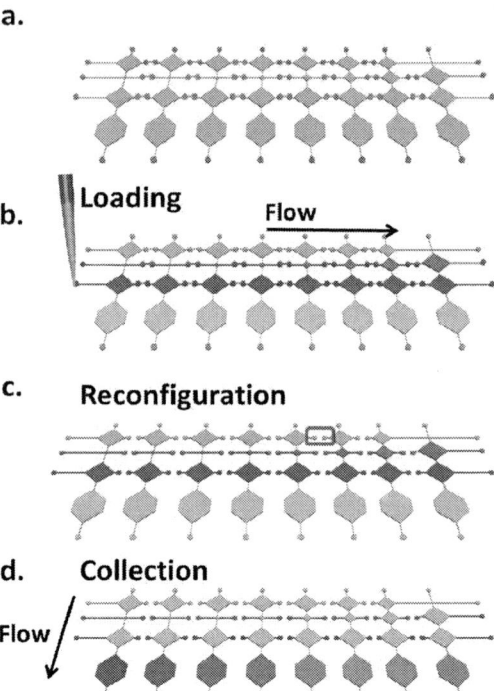

b. Loading

Flow

c. Reconfiguration

d. Collection

Flow

Figure 2: Illustration of dilution operation procedure. a) Empty device. b) Device after injection of diluent, sample, and detection reagent. c) Device after reconfiguration. d) All liquids are retracted to the final chambers and mix with each other to form a log 2 dilution profile.

be manually peeled off from the reconfigurable network and replaced with a solid piece of silicone, blocking the though holes. To activate fluidic passage in the vertical direction (Fig. 2c), a negative pressure will be applied by a micropipette at the outlet to draw all chemical contents from each vertical passage into corresponding reaction chamber, where active acoustic mixing strategies can be utilized to facilitate the targeted assay (Fig. 2d). In summary, the microfluidic dilution generator enables three simple steps of operation: loading, reconfiguration, and collection/reaction, which can be applicable to a wide variety of biological and biochemical assays with high-throughput and quantitative outcomes.

METHODS

Direct laser micromachining was used to fabricate all aforementioned three layers of the microfluidic dilutor in one simple engraving step. In detail, pre-designed fluidic channel patterns for each layer were directly engraved on a thin silicone sheet (254 μm thickness, Rogers Corporation Bisco™ HT-6240 Transparent Solid Silicone) by a desktop CO_2 pulsed laser engraver (VersaLaser, Universal Laser) with controlled parameters (e.g. power and raster speed). Subsequently, the engraved silicone sheets were rinsed and cleaned by ethanol and DI water. In the following step, the middle and bottom layers of the fluidic network was bonded together and fixed to a glass slide through an oxygen plasma treatment (Harrick Plasma, 1 minute at a high power setting). At last, the reconfigurable top layer

was placed and adhered on top of the middle layer without any permanent bonding.

For acoustic microstreaming mixing setup, a circular piezoelectric transducer with radius of 15 mm was purchased from RadioShack (Model 273-073) and fixed onto a polystyrene substrate by epoxy glue. Subsequently, the dilution generator chip was bonded to the same substrate by an oligomer transfer method reported previously [18]. To activate the mixing process, 1.28 kHz square waves signal (a peak-to-peak voltage of 5 Volts and 50% duty ratio) generated by a function generator (Agilent 33220A) was applied to the transducer. Alternatively, an ultrasonic water bath (Branson 200) was used to implement the active mixing process during which the microfluidic dilution generator was directly immersed into the water tank of the cleaner with the top surface sealed by a flat silicon layer.

RESULTS

Colorimetric Dilution Experiment

In order to evaluate the parallel dilution generation performance of our chip, colorimetric dilution experiments were conducted. Prior the test, 1.5 g /L aqueous solutions of soluble dyes (Jacquard iDye Natural Fabric Dye No. 451 Blue and No. 449 Red) were prepared by dissolving dry powder in water. In the colour dye dilution experiment, a distinct colour gradient was formed in the bottom reaction chambers after performing the parallel dilution process directly on the chip by mixing blue and red dye solutions. As can be seen in Fig. 3, the gradient gradually changed from a pure blue colour on the left to pure red on the right with intermediate colours in between due to the increasing ratio of red dye.

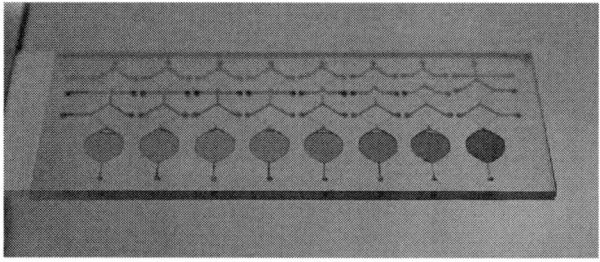

Figure 3: Dilution profiles on chip characterized by colorimetric red and blue dyes.

Mixing Characterization

The performance of the bubble-actuated acoustic mixing has been evaluated by dynamically capturing photos of blue and red dye mixing assisted by acoustic microstreaming. As illustrated in Fig. 4, a major vortex was immediately formed in the centre of the chamber, and the liquid mixture rotated clockwise to blend into separate coloured regions. After 3-minite actuation, the coloured boundaries were gradually blurred and colour homogeneity has been reached. To achieve the optimal mixing efficiency, the piezoelectric driving frequency was set at 1.28 kHz, the resonance frequency of the air-liquid interface, which can be estimated by Rayleigh–Plesset equation[19]. In comparison, the control group without active mixing showed marginal change in the coloured mixing in 3 minutes. In brief, the acoustic bubble actuation

978-1-4799-7956-1/15 $31.00 © 2015 IEEE

achieves an efficient mixing of the dilution solutions provided by the micro vortexes generated by vibrations of the trapped air bubbles. The 5 Volts driving voltage requirement is compatible with smartphone interface, enabling on-site use and point-of-care use of this device.

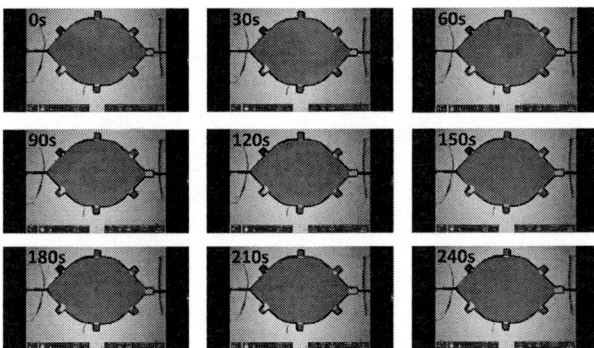

Figure 4: Dynamic snapshots of bubble-array acoustic mixing in a final chamber every 30 seconds.

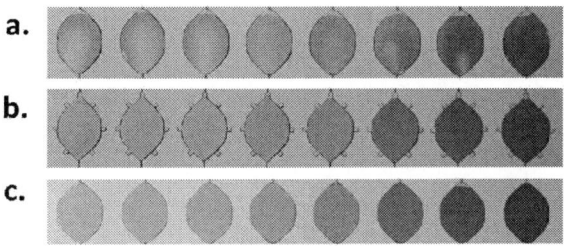

Figure 5: Comparison of mixing performance. a) Control, without external mixing, b) after bubble-array microstreaming mixing, and c) after ultrasonic water bath mixing.

Fig. 5 shows the comparison of mixing performance of two acoustic mixing methods, bubble-induced microstreaming mixing, and ultrasonic water bath mixing. Fig. 5a shows the dilution profile generated by the chip in reaction chambers without external acoustic mixing, as a control. Fig. 5b shows the picture of dilution profile after bubble-array microstreaming mixing with six air pockets, and Fig. 5c shows the dilution profile after ultrasonic water bath mixing, without any peripheral air pockets. As it can be seen, ultrasonic water bath mixing could reach the same degree of uniformity as bubble-array microstreaming mixing, which suggests that we can produce dilution device without electronic parts at a lower cost and provide it to sites where basic ultrasonic cleaner is available.

Quantitative Fluorescent Dilution Experiment

To further evaluate the accuracy of this dilution generation, a quantitative fluorescent dilution experiment was conducted using the same chip design. Fig. 6 shows the normalized fluorescent intensities vs normalized concentrations by performing the dilution on the chip and by manual serial dilution. As shown in manual dilution group, the fluorescent intensities had a linear relationship with FITC concentrations in this range. A fitting curve was generated from manual dilution group as a reference. For on-chip dilution group, the eight data points represented results in eight chambers, while the concentrations were calculated values by chip design. Fluorescent intensities of on-chip dilution group followed a clear binary declination relationship in these eight chambers, demonstrating a log 2-based dilution profile. Deviation of on-chip diluted group from manual diluted group in the lowest concentrations can be attributed to dead volumes and the physical absorption of fluorescent molecules by the silicone[20]. In summary, quantitative fluorescent results demonstrated that this dilution generator is capable of generating discrete parallel dilution profile across multiple reaction chambers with a logarithmic concentration distribution.

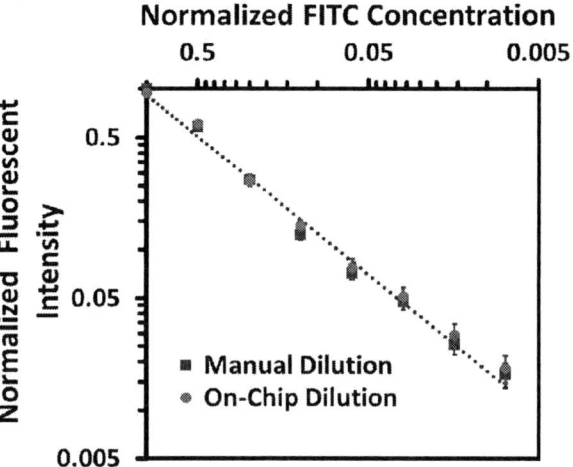

Figure 6: Fluorescent intensities generated by on-chip dilution (red round dots), and by manual dilution (blue rectangular boxes). Mean values with standard derivation bars of 4 repeats are plotted. The blue dotted line is the fitting curve of manual dilution group as a reference curve.

CONCLUSION

This paper reports a reconfigurable microfluidic dilution device for high-throughput quantitative assay, which can produce discrete logarithmic/binary concentration profiles ranging from 1 to 100 fold dilution in parallel from a fixed sample volume (of 10 µL) without assistance of continuous fluidic pump or vacuum source. The microfluidic dilution generator has been characterized by both qualitative colorimetric and quantitative fluorescent means. Two plug-and-play acoustic mixing strategies have been implemented specifically for regular chemical or biological lab settings. Mixing performance has been characterized with at least 20 times enhanced in the mixing rate. The operation procedure of simply pipetting in a "sample-to-answer" fashion makes it applicable for non-trained personnel to perform generic quantitative biomolecular assays. In conclusion, permitting easy-to-operate protocols, quantitative readouts, and low chemical consumptions, the reconfigurable microfluidic dilution generator can potentially offer a generic assay platform for homogeneous biochemical and biological analyses in regular laboratories, point-of-care and low-resource environments.

ACKNOWLEDGEMENT

This work has been in partly supported by National Science Foundation Awards ECCS-0846502 and DBI-1256193 to TP. JF acknowledges the fellowship support from the Howard Hughes Medical Institute Integrating Medicine into Basic Science (HHMI-IMBS) Training Program at UC Davis. Authors would also like to thank Dr. Shirley J. Gee, Dr. Candace S. Bever, and Dr. Bruce D. Hammock (the Department of Entomology) for generously providing the BSA-FITC reagents and BCA kits. In addition, we would like to thank Siyuan Xing and Yijun Zhang for assistance on chip design and graphs.

REFERENCES

[1] Peterson, G.L., *Determination of Total Protein.* Methods in Enzymology, 1983. 91: p. 95-119.

[2] Sissons, J.G.P., L.K. Borysiewicz, and J. Cohen, *Immunology of infection.* Immunology and medicine series. 1994, Dordrecht ; Boston: Kluwer Academic Publishers. x, 245 p.

[3] Larsen, K., *Creatinine Assay by a Reaction-Kinetic Principle.* Clinica Chimica Acta, 1972. 41(1): p. 209-&.

[4] Stanley, J., *Essentials of immunology & serology.* 2002, Albany, NY: Delmar Thomson Learning. xxi, 538 p.

[5] Minor, L., *Handbook of assay development in drug discovery.* Drug discovery series. 2006, Boca Raton: CRC/Taylor & Francis. 464 p.

[6] Chen, C.Y., A.M. Wo, and D.S. Jong, *A microfluidic concentration generator for dose-response assays on ion channel pharmacology.* Lab on a Chip, 2012. 12(4): p. 794-801.

[7] Harmening, D., *Clinical hematology and fundamentals of hemostasis.* 3rd ed. 1997, Philadelphia: F.A. Davis Co. xv, 743 p.

[8] Mack, D.R., *Method and apparatus for liquid addition and aspiration in automated immunoassay techniques.* 1989, Google Patents.

[9] Jiang, X., et al., *A miniaturized, parallel, serially diluted immunoassay for analyzing multiple antigens.* J Am Chem Soc, 2003. 125(18): p. 5294-5.

[10] Dertinger, S.K.W., et al., *Generation of gradients having complex shapes using microfluidic networks.* Analytical Chemistry, 2001. 73(6): p. 1240-1246.

[11] Folch i Folch, A., *Introduction to bioMEMS.* 2013, Boca Raton: CRC Press. xxxv, 492 p.

[12] Lee, K., et al., *Generalized serial dilution module for monotonic and arbitrary microfluidic gradient generators.* Lab Chip, 2009. 9(5): p. 709-17.

[13] Yamada, M., et al., *A microfluidic flow distributor generating stepwise concentrations for high-throughput biochemical processing.* Lab Chip, 2006. 6(2): p. 179-84.

[14] Jensen, E.C., et al., *Digitally programmable microfluidic automaton for multiscale combinatorial mixing and sample processing.*

Lab on a Chip, 2013. 13(2): p. 288-296.

[15] Lenth, R.V., *Some practical guidelines for effective sample size determination.* American Statistician, 2001. 55(3): p. 187-193.

[16] Zimmermann, M., P. Hunziker, and E. Delamarche, *Valves for autonomous capillary systems.* Microfluidics and Nanofluidics, 2008. 5(3): p. 395-402.

[17] Man, P.F., et al., *Microfabricated capillarity-driven stop valve and sample injector.* Micro Electro Mechanical Systems - Ieee Eleventh Annual International Workshop Proceedings, 1998: p. 45-50.

[18] Ding, Y.Z., et al., *Universal Nanopatternable Interfacial Bonding.* Advanced Materials, 2011. 23(46): p. 5551-+.

[19] Leighton, T., *The acoustic bubble.* 1994: Academic press.

[20] Sasaki, H., et al., *Parylene-coating in PDMS microfluidic channels prevents the absorption of fluorescent dyes.* Sensors and Actuators B: Chemical, 2010. 150(1): p. 478-482.

CONTACT

*J. Fan, tel: +1-530-5746640; jinfan@ucdavis.edu

SELF-MIXING BY ON-CHIP PREPARATION OF AQUEOUS TWO PHASE SYSTEMS AND ITS INFLUENCE ON EXTRACTION KINETICS

Pavithra A. L. Wijethunga and Hyejin Moon

The Mechanical and Aerospace Engineering, The University of Texas at Arlington, Texas, USA

ABSTRACT

This paper introduces an advantageous self-mixing phenomenon created during the formation of aqueous two-phase systems in a digital microfluidic device, for the first time. The influence of such self-mixing on enhancement of on-chip extraction kinetics is quantified and reported, highlighting a significant mixing capability in the absence of forced mixing. Results suggest potential applications of the self-mixing concept in portable microfluidics, where implementing complicated mixing sequences or device geometries are problematic.

INTRODUCTION

An aqueous two-phase system (ATPS) consists of two distinct liquid phases, yet both phases are aqueous. An ATPS is formed by mixing two structurally different polymers (or one polymer and one salt) in water at high concentrations. The resulting mixture settles into two immiscible phases. ATPSs generally provide gentle environments for biological samples. They also provide easy mass transfer environments, because both phases are aqueous. Due to these advantages, they are popularly used as solvents for separations and extractions of biological molecules in large scale. Recent interest on microscale ATPS applications enabled several ATPS characterization studies in micro channels[1]. We reported the first such study on an electrowetting on dielectric (EWOD) digital microfluidic (DMF) in [2], where EWOD DMF's capability to prepare ATPS was investigated despite the confined environment of nL volume droplets inside a chip. Building upon that study, this paper reports the interesting self-mixing phenomenon created during the on-chip ATPS preparation.

Furthermore, in order to quantify the effect of self-mixing phenomena, its influence on the kinetics of an ATPS-based on-chip extraction was evaluated. This drop-to-drop extraction process resembles the process reported in [3]. The qualitative and quantitative data of the improved drop-to-drop aqueous two-phase extraction (ATPE) kinetics are presented in this paper. To our best knowledge, a self-mixing condition reported here and an investigation of its potential use as an alternative to forced mixing condition were never reported elsewhere.

EXPERIMENTAL

Materials

Following reagents were used to prepare samples; polyethelene glycol (PEG)($C_{2n}H_{4n+2}O_{n+1}$, 8000 g/mol), dextran (DEX) ($H(C_6H_{10}O_5)_xOH$, 500,000 g/mol), and ammonium sulphate (($NH_4)_2SO_4$, 132.14 g/mol). To obtain a particular ATPS configuration, two pure solutions at designated concentrations were prepared such that mixing them in 1:1 volume ratio ended up with the desired ATPS configuration. Altogether, 12 pure solutions of polymer and salt were prepared to form six ATPSs.

Three of them were polymer/polymer (P/P) ATPSs and the other three were polymer/salt (P/S) ATPSs. They were selected by observing the biphasic curves (Figure 1). Biphasic curves indicate the threshold concentrations, beyond which the resultant solution will separate into two phases. The selected ATPSs are shown in Figure 1 superimposed on their respective biphasic curve.

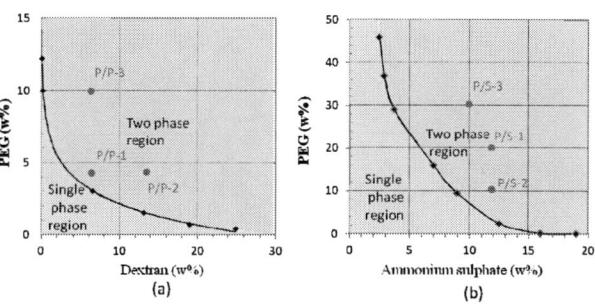

Figure 1: Selected ATPSs for this study superimposed on relevant biphasic curves. (a) Polymer/polymer (P/P) systems, and (b) Polymer/salt (P/S) systems. Data for biphasic curves for P/P and P/S systems were extracted from literature[4-5]

Device and experimental setup

Figure 2(a) shows the fabricated layers of EWOD DMF device. A device consisted of arrays of 2 mm X 2 mm planar electrodes on which the polymer and salt solution droplets were dispensed and translated by applying voltages to these electrodes sequentially. The fabrication procedure is the same as previously reported [3].

As shown in Figure 2(b), a function generator and a voltage amplifier were used to produce the AC voltage (90-100 V_{rms} at 4 kHz) required to activate electrodes. A transmitting light source together with a digital microscopic camera system (HIROX KH 1300) was used for visualization of two-phase formation. Microscope images were recorded with a 5 fps and 640 x 480 resolution. The automated setup consists of data acquisition hardware, a custom built control circuit with switches and a custom built LABVIEW software (www.ni.com) program to automatically drive droplets in a desired pattern.

On-chip ATPS formation

Experiments were carried out by initially dispensing droplets of two pure solutions. For P/P systems, one droplet from pure PEG solution and the other from pure DEX solution were created. For P/S systems, droplets of pure PEG solution and pure ammonium sulphate solution were dispensed. Then, these two equal volume droplets were brought into contact. At this point, ATPS formation begins and interface dynamics appears for some systems, creating a self-mixing phenomenon.

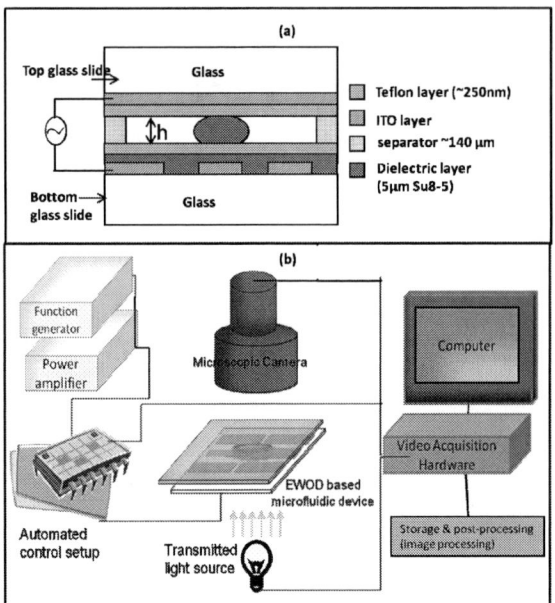

Figure 2: (a) Cross sectional view of the EWOD digital microfluidic device and (b) Experimental setup.

Off-chip ATPS formation

The same configurations of P/P and P/S systems were prepared in macro scale using vials. Pure solutions of PEG and DEX (or PEG and ammonium sulphate) were vigorously mixed in 1:1 volume ratio for 10 minutes using a vortex. It was confirmed that 10 minutes of vortex mixing is sufficient and longer mixing times do not change the final volume ratio of two phases. Once the mixture was settled into two clean phases, samples from each phase were carefully dispensed.

Drop-to-drop ATPE experiment

A food dye (Assorted food color containing FD&C Yellow No. 5 in liquid form) was selected as a solute for extraction experiments. To prepare samples, 2 μL dye was dissolved into 125 μL donor solution. Upon merging the two droplets, the transfer of dye molecules from donor droplet to extractant droplet was observed and recorded. The recorded data were post-processed using an image processing technique to measure dye concentration in the donor droplets. This technique allowed convenient real time data collection. For each experiment, calibration curves which relate dye concentration with a color parameter, particularly the *b* color parameter defined in *CIE Lab* color model, were prepared and used.

Three different modes were tested for comparison. Mode 1 is dye extraction in off-chip prepared ATPS with forced mixing. Forced mixing was provided by driving the merged droplet along 5 electrodes in a cross junction (moving up to down and left to right in an alternating sequence) to provide sufficient mixing. Mode 2 is dye extraction during on-chip ATPS formation with self-mixing aids, but without forced mixing. Mode 3 is extraction in off-chip prepared ATPS without forced mixing.

For Mode 2 test, pure solutions of DEX (for P/P system) and ammonium sulphate (for P/S system) were used as donors and pure solution of PEG was used as extractant. Then one drop of dye containing donor and one drop of extractant were brought to contact to initiate ATPS formation on chip, but no forced mixing was provided externally. For Mode 1 and 3 tests, DEX-rich (for P/P system) and salt-rich (for P/S system) phases were used as donors and PEG-rich phase was used as extractant.

RESULT AND DISCUSSION

Interface dynamics: self-mixing phenomena

For some concentration combinations, interesting interface dynamics were observed as soon as the pure PEG and pure DEX droplets merged to form ATPS. In order to isolate the effect, all electrodes on the device were supplied with ground voltage soon after merging. Unique interface dynamics for each system were observed as shown in Figure 3. The maximum extent of interface dynamics in each system was observed at different elapsed time after merging as per the visual observations. Note that the dye color in P/P-2 did not alter the interface dynamics, but the dye was only used for better visualization purpose. Based on visual observations, some systems showed strong interface dynamics (Figure 3 (d) and (e)), while the others showed either moderate dynamics (Figure 3 (a) and (b)), or no dynamics at all (Figure 3 (c) and (f)).

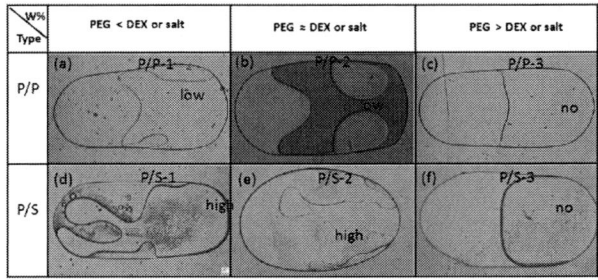

Figure 3: Interface dynamics during the formation of ATPS ('low', 'no', or 'high' is the qualitative classification of the interface dynamics based on visual observations), (a)-(c) PEG/DEX systems (P/P), (d)-(f) PEG/(NH$_4$)$_2$SO$_4$ systems (P/S).

It is reasonable to assume that the initial momentum of the two moving droplets acting against each other are cancelled off. Thus the observed interface dynamics can be considered solely by surface tension driven flow. Two causes of surface tension driven flow are speculated. First, the excess surface energy (the difference in total interface energies of the two droplets before and after merging) could be dissipated to create the fluidic motion at the interface [6]. Second, marangoni effect due to variations in concentration at the interface might cause the fluidic motion at the interface[7]–[9]. Polymer molecules and salt ions at high concentration at each side of the interface tend to transfer across the interface. Then this transfer can create local variations of concentration, thus, interfacial tension which leads to a marangoni effect. According to

978-1-4799-7956-1/15 $31.00 © 2015 IEEE

the in-depth review by Stone [9], the velocity profile induced by marangoni effect has a magnitude, $u=\Delta\sigma/\mu(1+\lambda)$, where u, $\Delta\sigma$, μ, and λ are velocity, variations in interfacial tension, viscosity of the continuous phase and viscosity ratio of the two phases, respectively. This signifies the importance of system properties in determining velocity at the interface (hence the amount of self-mixing), as all the variables, $\Delta\sigma$, μ, and λ, are directly influenced by the concentration of polymer and/or salt. The induced local velocity then acts as viscous force against local interface tension, creating deformations and droplet breakups at the interface. Such deformations are characterized by the capillary number ($Ca = u\mu/\sigma$), a dimensionless number given by the ratio of viscous force to interface tension force for a given flow type [9].

A large Ca denotes a dominant viscous force over interfacial tension indicating higher chance of deforming the interface. Since here both u and Ca are dependent on concentration, clearly, different interface dynamics in the absence of external force can be expected for different concentration combinations as observed in Figure 3. In other words, the results imply that the interface dynamics can be tuned largely by changing the concentrations even when the same P/P or P/S combination is used. Creating strong interface dynamics within the system could be considered as self-mixing phenomena. Engineering such interface dynamics to provide efficient mass transfer is beneficial in confined microfluidics, especially without providing any external force.

Enhanced on-chip drop-to-drop ATPE by self mixing

Quantitative and qualitative data that compare the kinetics of on-chip drop-to-drop ATPE using P/S-1 and P/P-2 are shown in Figure 4. To quantify extraction kinetics, fractional conversion (FA) was plotted in Figure 4(a) and (c). FA was calculated as the decrease in dye concentration in donor droplet with respect to its initial concentration. The time-laps images in Figure 4 (b) and (d) show the interface dynamics (e.g. self-mixing) developments in a merged donor/extractant droplet.

Extraction with mode 1 reaches approximately a steady FA value within 100 s indicating that the extraction process has reached at the equilibrium. This relatively quick set of equilibrium was achieved at the expense of exerting EWOD actuation to create forced mixing in the merged drop. As shown in the first rows of Figure 4 (b) and (d), EWOD actuation create strong fluidic motion within a merged drop. The extraction process is significantly facilitated by this forced mixing and reaches to equilibrium quickly.

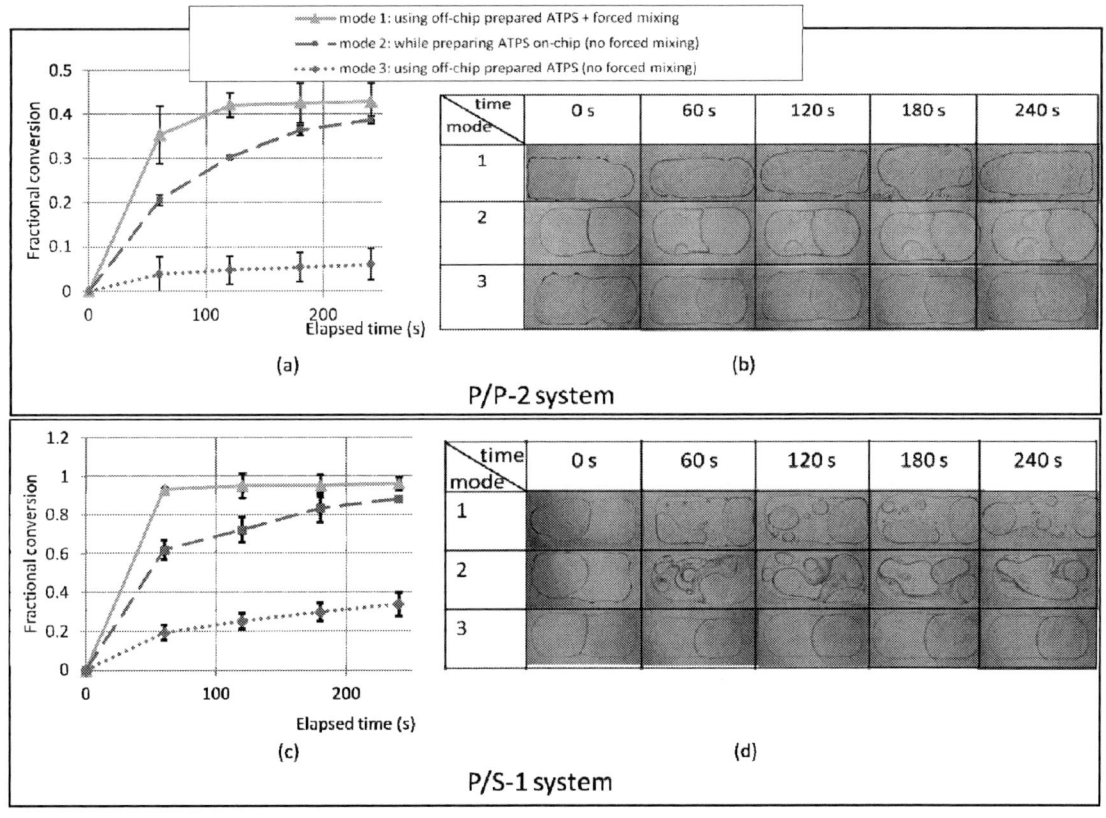

Figure 4: On-chip drop-to-drop ATPE kinetics for P/P-2 and P/S-1. (a, c) Quantitative comparison of extraction kinetics between the three cases shown in legend. Fractional conversion= (Donor drop's initial concentration - concentration at time t)/initial concentration. (b, d) Visual comparison of extraction kinetics over time for the three cases.

Mixing effect is well demonstrated when comparing extraction kinetics in mode 1 and 3. Without mixing (mode 3), extraction cannot reach to the equilibrium in the same time duration and the FA is much smaller than that of mode 1. Note that the third rows of Figure 4(b) and (d) show negligible fluidic motion, therefore, extraction is solely done by diffusion. However, in the extraction with mode 2 that is in the absence of forced mixing, FA value is reaching closer to the equilibrium. This clearly demonstrates that interfacial dynamics during ATPS formation (shown in second rows of Figure 4 (b) and (d)) provides self-mixing and enhances extraction process.

Another observation to note is that even though the interface dynamics were relatively stronger in P/S-1 (mode 2 in Figure 4(d)) than in P/P-2 (mode 2 in Figure 4(b)), both systems showed similar enhancement in extraction kinetics and reached to equilibrium in similar time duration. In both the systems, dye is extracted to PEG phase. However the concentration of PEG in P/S-1 is about four times the concentration of PEG in P/P-2, which makes the extractant in P/P-2 to be less viscous than the extractant in P/S-1. Therefore the diffusivity of dye and the counter phase molecule or ion into the extractant in P/P-2 is higher because of the higher contribution from diffusive mass transfer. This difference in diffusivity could be a reason for enhanced extraction kinetics in P/P-2 similar to P/S-1, even though the less interface dynamics are present.

Overall results point out that extraction kinetics have significantly improved when performed in mode 2 when compared with mode 3. Since both modes were not influenced by a stirring or any externally provided factor to enhance the process, the difference in extraction kinetics purely signify the enhancement due to interface dynamics. Hence the results highlight the successful application of self-mixing condition in performing or enhancing ATPE process without needing a forced mixing. Although the application of forced mixing shows the best results, implementing complicated electrode actuation sequences could be a bottleneck in some applications, such as in portable microfluidics. For example, the basic electrowetting operations on an EWOD DMF using finger actuation[10] shows the potential of developing portable DMF devices that does not require power input in complicated sequences. For such situations, utilizing a self-mixing as in mode 2 will be a simple solution that does not require any modification to the typical device design.

CONCLUSION

When the ATPS droplets are formed on EWOD DMF, interesting interface dynamics were observed. Each system showed unique deformations and propagation patterns, which indicates the tunable dynamics are achievable through changing concentrations. As a potential application of interface dynamics, on-chip ATPE processes were performed. Interface dynamics provided a self-mixing condition and the effect of which is comparable to the forced mixing on EWOD DMF. As the dynamics were initiated at the interface, it is intuitive to predict that minor changes in electrode design to increase the interfacial area between the two phases may further elevate the interface dynamics. For example, the electrode designs to have one pure solution droplet encapsulated by the other pure solution droplet soon after merging, is such a design to increase interfacial area. It is worth investigating the effect of such modifications on the interface dynamics and self-mixing capabilities. This concept of self-mixing introduces a simple option to perform ATPE in the absence of forced mixing, such as in a portable micofluidics where implementing complicated mixing sequences or device geometries are problematic.

ACKNOWLEDGEMENTS

Authors acknowledge the support by National Science Foundation CAREER award (grant no. ECCS-1254602).

REFERENCES

[1] S. Hardt and T. Hahn, "Microfluidics with aqueous two-phase systems," Lab Chip, vol. 12, no. 3, pp. 434–442, 2012.

[2] P. A. L. Wijethunga and H. Moon, "A Study of On-Chip Aqueous Two Phase System Formation and its Applications," in ASME 2012 Third International Conference on Micro/Nanoscale Heat and Mass Transfer, 2012, pp. 851–856.

[3] P. A. L. Wijethunga, Y. S. Nanayakkara, P. Kunchala, D. W. Armstrong, and H. Moon, "On-chip drop-to-drop liquid microextraction coupled with real-time concentration monitoring technique," Anal. Chem., vol. 83, no. 5, pp. 1658–1664, 2011.

[4] S. C. Silverio, P. P. Madeira, O. RodrÃguez, J. A. Teixeira, and E. A. Macedo, "deltaG (CH2) in PEG-Salt and Ucon-Salt Aqueous Two-Phase Systems," J. Chem. Eng. Data, vol. 53, no. 7, pp. 1622–1625, 2008.

[5] H. Walter and G. Johansson, "Methods in Enzymology," Aqueous Two-Phase Systems, vol. 228. 1994.

[6] A. V Anilkumar, C. P. Lee, and T. G. Wang, "Surface tension induced mixing following coalescence of initially stationary drops," Phys. Fluids A Fluid Dyn., vol. 3, no. 11, pp. 2587–2591, 1991.

[7] L. E. Scriven Sterling, C. V., "The marangoni effects," Nature, vol. 187, p. 186, 1960.

[8] Z.-S. Mao and J. Chen, "Numerical simulation of the Marangoni effect on mass transfer to single slowly moving drops in the liquidâ€"liquid system," Chem. Eng. Sci., vol. 59, no. 8, pp. 1815–1828, 2004.

[9] H. A. Stone, "Dynamics of drop deformation and breakup in viscous fluids," Annu. Rev. fluid Mech., vol. 26, pp. 65–102, 1994.

[10] C. Peng, Z. Zhang, C.-J. C. J. Kim, and Y. S. Ju, "EWOD (electrowetting on dielectric) digital microfluidics powered by finger actuation.," Lab Chip, vol. 14, no. 6, pp. 1117–22, Mar. 2014.

CONTACT

*Hyejin Moon, tel: +1 817 272 2017,
fax: +1 817 272 2952; Hyejin.moon@uta.edu

SINGLE-LAYER MICROFLUIDIC CURRENT SOURCE
VIA OPTOFLUIDIC LITHOGRAPHY

C.C. Glick, S. Peng, M. Chung, K. Korner, M. Veale, C. Liu, J. Moore,
A. Chu, A. Buckley, K. Iwai, R.D. Sochol, and L. Lin
University of California, Berkeley, USA

ABSTRACT

This work marks the first use of *in-situ* photopolymerization to create single-layer microfluidic devices which serve as ultra-low *Reynolds Number* (*Re*) *current sources* to regulate fluid flow rate independent of operating pressures. Autonomous fluidic components are an emerging aspect of micro/nanofluidic circuits and applications; however, many existing fluidic applications require specific pressure and/or flow rate conditions to perform optimally, and many require complex and expensive fabrication procedures. Here we introduce single-layer microfluidic system which utilize a spring and piston system – fabricated *in situ via* optofluidic lithography – to passively constrain fluid flow rate to a value independent of operating pressure. Experimental results revealed controlled flow rates of 29.2 ± 0.8 µl/min (from P = 50-100 mbar) and a maximum small-signal resistivity of 141.1 mbar-min/µl, which represents the highest performance for a low-pressure microfluidic current source.

INTRODUCTION
Microfluidic Current Control Mechanisms

Microfluidic systems offer numerous benefits for biochemical applications, such as low cost, rapid reaction times, low reagent volumes, and the ability to mix, transport, and/or array chemical processes involving beads, cells, or other biomolecules [1-4]. The ability to dynamically control microfluidic systems in a self-regulating manner is therefore critical for the long-term development of lab-on-a-chip (LOC) technology; at present, the majority of integrated microfluidic systems require substantial external regulation during device operation [5-8]. In recent years, research has shifted toward developing new, self-regulating microfluidic technologies [9-15]. However, producing dynamic microfluidic control systems in an efficient, cost-effective manner has remained elusive and the focus of widespread research efforts.

For example, optofluidic "domino diodes" are capable of passive flow rectification only [16]. While many biological experiments require precise reagent flow rates unavailable through such passive controls, building devices that function as hydraulic analogs to electronic "current sources" has proved a considerable challenge. Previously microfluidic current sources have required three PDMS layers and pressure > 1 ATM for current regulation [17], or have been single-layer devices PDMS devices that only function at high *Reynolds number* (*i.e., Re* > 80) [18].

Optofluidic Lithography for Microfluidic Circuitry

Previous works have demonstrated optofluidic lithography in Teflon-based systems do fabricate freely moving pistons and check-valves in two- to four-layer microfluidic systems. However, the efficacy of such components remains limited due to the high pressures required for operation [19-20]. Maskless optofluidic lithography has been used to create railed systems which allow free movement at low pressure [21-23]. Single-layer optofluidic regulatory devices have been developed which require changes in fluid properties, such as Temperature and pH, but do not respond to flow parameters (*e.g.,* pressure and flow velocity) [24].

In earlier works, we presented microfluidic "gain" valves – which can use low pressures to close valves at higher pressures – using free-floating microstructures (fabricated *via* optofluidic lithography) [25]. However, these valves exhibit motion based only on flow pressure and do not respond directly to fluidic current, and thus lack some of the flexibility found in analogous electronic devices. To expand the capabilities of these optofluidic circuit elements, we introduce a hybrid current source utilizing optofluidic lithography to fabricate a single-layer microfluidic system with a moving piston element which can passively control fluid flow rates independent of operating pressures.

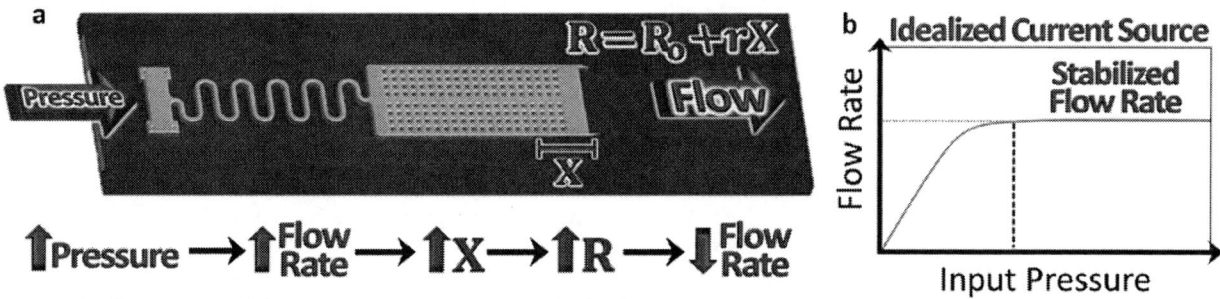

Figure 1: Illustrations of the current source concept. (a) The device consists of a spring-mounted piston which slides within a microchannel according to the fluid forces exerted on it. As pressure increases, the increased flow rate forces the piston into the high-resistivity portion of the microchannel, increasing the hydrodynamic resistance of the system. The increased resistance then lowers the flow rate in a negative feedback loop, causing the flow rate to auto-stabilize. (b) Behavior of an ideal current source. Once the pressure passes a critical threshold value, the flow rate remains constant regardless of the pressure applied to the device.

CONCEPT

Figure 1 a-c highlights the concepts behind the single-layer microfluidic current source. The device consists of a freely-moving piston mounted to a spring which moves in and out of a narrow microchannel according to pressure and shear hydrodynamic forces exerted on it. The motion of the piston enables a *negative feedback* system within the current source; as pressure and flow increase, the piston's motion increase the resistance of the system, which then to stabilizes the flow rate. According to Poiseuille's Law for low Re systems, flow rate (Q) is proportional to pressure drop (ΔP) divided by hydrodynamic resistance (R). In general, R is proportional to the length of a channel and its resistivity (r), a factor which is highest where the piston is located in the narrow part of the channel. For this system, resistance is a roughly linear function the piston's position within the narrow channel, approximated by

$$R(x) = R_0 + rx \tag{1}$$

where R_0 is a constant parasitic resistance which does not depend on channel position, r is piston resistivity, and x is the length of the piston contained within the narrow channel. By balancing spring restoring force ($k\,x$) with pressure force (ΔP) and shear force (σ), we can determine the pressure and flow rate as parametric functions of x:

$$\left(P(x), Q(x)\right) = \left(\frac{k\,x\,(R_0 + rx)}{\sigma + rx}, \frac{k\,x}{\sigma + rx}\right) \tag{2}$$

These equations demonstrate that Q tends toward a flat value k/r while the spring is free to extend, and resistance reaches a constant value $R_{max} = R_0 + rL$ when the piston reaches its maximum deformation length L.

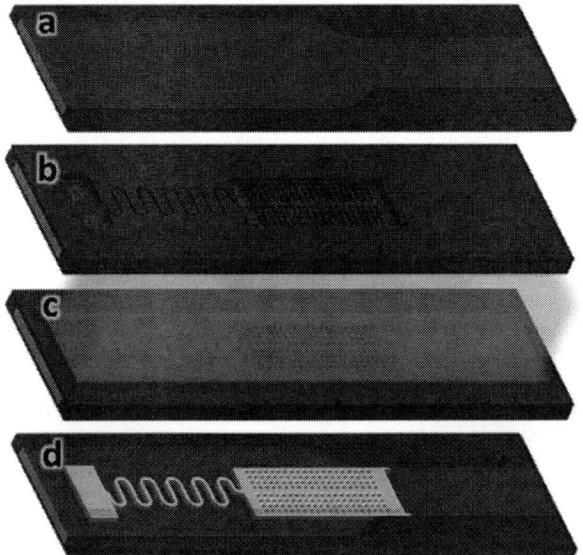

Figure 2: The optofluidic fabrication process. (a) A photocurable liquid – PEGDA with 1% photoinitiator – is loaded into a PDMS microchannel. (b) A photomask is placed in contact with the device which is then (c) exposed to UV light, causing PEGDA to photopolymerize. The piston does not stick to the PDMS due to contraction of the cured PEGDA and oxygen layers in to the PDMS. (d) The mask is removed and the remaining photocurable liquid is drained and replaced by non-photocurable PEGDA.

FABRICATION

The bulk microfluidic system was fabricated using standard soft lithography processes, as described in previous work [16,25]. Briefly, micromolded poly (dimethyl-siloxane) (PDMS), with microchannel heights of 110 μm, was thermally bonded to glass slides coated with 80 μm of PDMS(*via* PDMS-PDMS bonding techniques) to prevent the optofluidic components from sticking to the glass slides.

Figure 2 shows conceptual illustrations of the optofluidic lithography process to create freely moving pistons within a PDMS microchannel. First, a solution of poly(ethylene glycol) diacrylate (PEGDA) with 1% photoinitiator (2,2-dimethoxy-2-phenylaceto-phenone) was loaded into the device (Fig. 2a). A photomask was placed in contact with the device and aligned (Fig. 2b). The device was then exposed to UV, causing the PEGDA to photopolymerize. Due to an oxygen layer in the PDMS, the cured PEGDA does not bond to the PDMS, and due to contraction of the PEGDA during exposure, a thin gap forms, which allows for lubricated piston motion. Figure 3 shows optical images of the fabricated gain device, illustrating the piston in both 'open' and 'closed' positions.

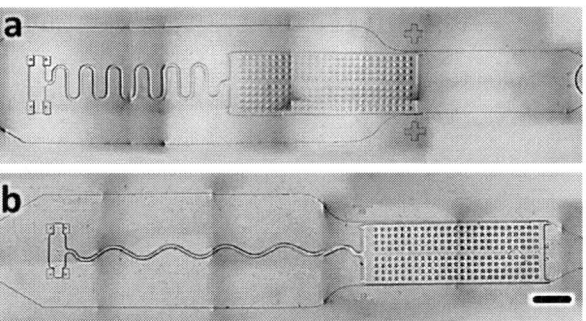

Figure 3: Images of fabricated current source device in (a) relaxed spring configuration (under no flow conditions) and (b) extended spring configuration (under forward fluid flow). Micrographs depict images stitched together with photo-editing software. Scale bar = 500 μm

RESULTS AND DISCUSSION
Simulations

Figures 4 shows the results of three-dimensional COMSOL Multiphysics simulations of the microfluidic current source. In Figure 4a, fluid pressure drop across the piston occurs primarily when the piston is located within the narrow channel. This is due to the much higher resistivity in this region, as shown by the numerical results in Figure 4b. Note that the resistivity of the narrow channel plus piston is more than 60x greater than the resistivities of the other sections of the current source, demonstrating that the resistance of this device becomes dominated by this region as the piston moves farther into the channel.

Figure 4c shows an important limitation for current sources using an asymmetric spring design: lateral distortion forces arise even when the spring is stretched in a purely axial direction. These lateral forces in turn push the piston against the channel wall, causing drag forces that prevent the piston from withdrawing smoothly from the channel creating the hysteresis effect seen in Figure 5.

978-1-4799-7956-1/15 $31.00 © 2015 IEEE

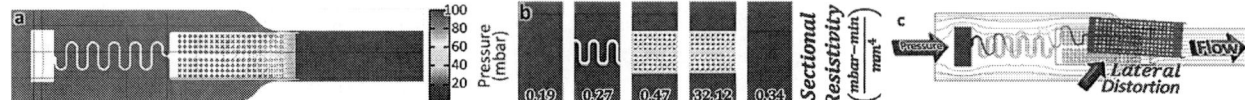

Figure 4: Three-dimensional COMSOL Multiphysics simulations of a microfluidic current source. (a) Simulation results depicting pressure field around the piston in the extended position. (b) Numerical results showing the resistivity of the individual sections of the current source channel. Note that the fourth resistivity (i.e., the narrow channel plus piston) is > 60x higher the other parts of the current source, demonstrating that the resistance of this region dominates as the piston moves into the channel. (c) Simulation with lateral constraints removed. Asymmetries in the spring cause lateral distortion forces which push the piston against the wall. The resultant drag causes device performance hysteresis (seen in Figure 5).

Device hysteresis can be limited in several ways. First, the optofluidic exposure time can be reduced, which prevents overexposure of the piston sides, reducing overall friction. Second, the process can be controlled to prevent movement of the microfluidic device during exposure. Third, a symmetric spring design can be used to prevent the lateral distortions entirely – this method is sub-optimal because *ceteris paribus*, a symmetric spring has a *16x* larger spring constant than does an asymmetric spring, resulting in devices that are too stiff to move. Finally, micropillars within the channels can act to constrain the piston to axial motion within the channel.

Flow Rate-Pressure Experiments

To investigate the overall performance of the current source, experiments were performed where the current source was subjected to a varying pressure sweep at its input and output terminals. Pressure differential applied and resultant flow rates were measured (*via Fluigent MFCS* and *Flowell* sensors) for the duration of the sweeps and the resultant flow rate-pressure (QP) curves were fitted using the aforementioned theoretical model and plotted.

In Figure 5, the *QP* curve for a single current source is compared with the linear response of an equivalent empty channel. During the forward sweep (red), flow rate levels off at a maximum value of 29.2 ± 0.8 µl/min (from $P = 50$-100 mbar) due to the action of the current source. When the piston reaches maximum extension, the device begins to act as an ordinary resistor, thereby limiting the device's dynamic pressure range. This range can be extended primarily by increasing the length of the optofluidic piston.

An extreme hysteresis curve (green) shows that during the reverse pressure sweep, the piston remained partially extended. An empty channel (blue) yields a linear response, as is expected from the Poiseuille-flow resistive model.

In 5, the *QP* curves are plotted for current sources with differing spring amplitudes and spring stiffness. Note that as the spring constant increases, the current stabilization is not reached until higher pressures.

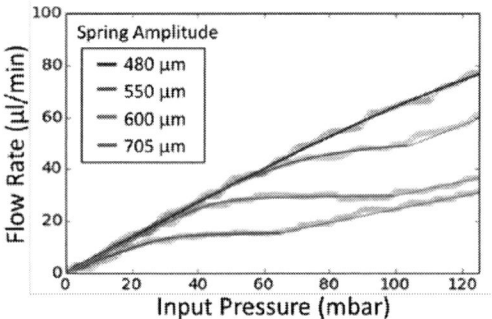

Figure 6: Data comparing the QP response of four microfluidic current source devices. Device response is determined by spring amplitude; shorter springs have stiffer spring constants and larger response times. All curves have been fitted with the same analytical model.

CONCLUSION

Autonomous, low-cost microfluidic circuit components offer significant potential to improve the functionality of lab-on-a-chip applications. In this work, we presented a single-layer system for implementing microfluidic current control, with fabrication *via in situ* optofluidic lithography. The methodology presented has several adaptations over previous methods; for example it implements the current source using a fabrication method that requires only a single layer of PDMS substrate and two photolithographic processing steps, has enabled controlled flow rates as low as 16 µl/min (*i.e.* for *Re < 0.15*). The high pressure stability of these results suggest that the presented methodology could numerous Lab-on-a-Chip applications in situations where precise reagent delivery is essential.

Several adaptations of this method could be used to improve regulate fluid flows in a variety of conditions. For example, decreases piston width lowers the steady-state flow rate, and increasing piston length increases the operating range. For future work, we plan to investigate hysteresis-reduction techniques to make the device behave more like an ideal fluidic current source. Due to its ease of manufacture and low flow rates available, this microfluidic current source can be adapted to offer a simple, yet versatile, component for microfluidic and lab-on-a-chip circuits and systems.

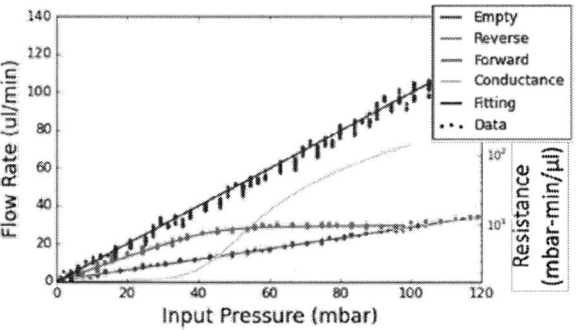

Figure 5: Data collected from a single microfluidic current source (spring amplitude = 600 µm). During the forward sweep (red), flow rate levels off at a maximum value of 30.3 µl/min due to the action of the current source. After 100mbar, the piston reaches maximum extension and resistance (pink) becomes constant. When the sweep is reversed (green), piston drag against the channel walls leads to the hysteresis effect shown. An empty channel (blue) yields a linear response.

ACKNOWLEDGEMENTS

The authors thank all the members of the Liwei Lin Laboratory, the Luke P. Lee Laboratory, and the *Micro Mechanical Methods for Biology* (*M³B*) Laboratory Program. This research is supported in part by the DARPA N/MEMS program under the Micro/Nano Fluidics Fundamentals Focus (MF3) center and by the National Science Foundation.

REFERENCES

[1] W. H. Tan and S. Takeuchi, "A trap-and-release integrated microfluidic system for dynamic microarray applications," Proceedings of the National Academy of Sciences of the United States of America, vol. 104, pp. 1146-1151, Jan 2007.

[2] R. D. Sochol, B. P. Casavant, M. E. Dueck, L. P. Lee, and L. Lin, "A dynamic bead-based microarray for parallel DNA detection," Journal of Micromechanics and Microengineering, vol. 21, p. 054019, 2011.

[3] R. D. Sochol, S. Li, L. P. Lee, and L. Lin, "Continuous flow multi-stage microfluidic reactors via hydrodynamic microparticle railing," Lab on a Chip, vol. 12, pp. 4168-4177, 2012.

[4] R. D. Sochol, M. E. Dueck, S. Li, L. P. Lee, and L. Lin, "Hydrodynamic resettability for a microparticle arraying system," Lab on a Chip, vol. 12, pp. 5051-5056, 2012.

[5] M. A. Unger, H. P. Chou, T. Thorsen, A. Scherer, and S. R. Quake, "Monolithic microfabricated valves and pumps by multilayer soft lithography," Science, vol. 288, pp. 113-116, Apr 7 2000.

[6] T. Thorsen, S. J. Maerkl, and S. R. Quake, "Microfluidic large-scale integration," Science, vol. 298, pp. 580-584, Oct 18 2002.

[7] J. Wang, H. C. Fan, B. Behr, and Stephen R. Quake, "Genome-wide single-cell analysis of recombination activity and de novo mutation rates in human sperm," Cell, vol. 150, pp. 402-412, 2012.

[8] D. C. Leslie, C. J. Easley, E. Seker, J. M. Karlinsey, M. Utz, M. R. Begley, et al., "Frequency-specific flow control in microfluidic circuits with passive elastomeric features," Nature Physics, vol. 5, pp. 231-235, Mar 2009.

[9] J. A. Weaver, J. Melin, D. Stark, S. R. Quake, and M. A. Horowitz, "Static control logic for microfluidic devices using pressure-gain valves," Nature Physics, vol. 6, pp. 218-223, Mar 2010.

[10] B. Mosadegh, C.-H. Kuo, Y.-C. Tung, Y.-s. Torisawa, T. Bersano-Begey, H. Tavana, et al., "Integrated elastomeric components for autonomous regulation of sequential and oscillatory flow switching in microfluidic devices," Nature Physics, vol. 6, pp. 433-437, Jun 2010.

[11] C. C. Glick, M. Zhang, L. Wang, and W. Wen, "Electrorheological fluid-based microfluidic logic gates," in Caltech SURF Seminar Day Proceedings, California Institute of Technology, California, Oct 19 2010.

[12] N. S. G. K. Devaraju and M. A. Unger, "Pressure driven digital logic in pdms based microfluidic devices fabricated by multilayer soft lithography," Lab on a Chip, vol. 12, pp. 4809-4815, 2012.

[13] B. Mosadegh, T. Bersano-Begey, J. Y. Park, M. A. Burns, and S. Takayama, "Next-generation integrated microfluidic circuits," Lab on a Chip, vol. 11, pp. 2813-2818, 2011 2011.

[14] E. F. Hasselbrink, T. J. Shepodd, and J. E. Rehm, "High-pressure microfluidic control in lab-on-a-chip devices using mobile polymer monoliths," Analytical Chemistry, vol. 74, pp. 4913-4918, Oct 1 2002.

[15] B. J. Kirby, T. J. Shepodd, and E. F. Hasselbrink Jr, "Voltage-addressable on/off microvalves for high-pressure microchip separations," Journal of Chromatography A, vol. 979, pp. 147-154, 2002.

[16] R. D. Sochol, C. C. Glick, K. Y. Lee, T. Brubaker, A. Lu, M. Wah, et al., "Single-layer "domino" diodes via optofluidic lithography for ultra-low reynolds number applications," in Micro Electro Mechanical Systems (MEMS), 2013 IEEE 26th International Conference on, 2013, pp. 153-156.

[17] E. P. Kartalov, C. Walker, C. R. Taylor, W.F. Anderson, and A. Scherer. "Microfluidic vias enable nested bioarrays and autoregulatory devices in Newtonian fluids," PNAS, vol. 130 pp. 12280-12284, 2006.

[18] I. Doh, Y. H. Cho "Passive flow-rate regulators using pressure-dependent autonomous deflection of parallel membrane valves," Lab Chip pp. 2070-2075, 2009.

[19] D. J. Beebe, J. S. Moore, J. M. Bauer, Q. Yu, R. H. Liu, C. Devadoss, et al., "Functional hydrogel structures for autonomous flow control inside microfluidic channels," Nature, vol. 404, pp. 588-+, Apr 6 2000.

[20] D. Kim and D. J. Beebe, "A bi-polymer micro one-way valve," Sensors and Actuators a-Physical, vol. 136, pp. 426-433, May 1 2007.

[21] S. E. Chung, Y. Jung, and S. Kwon, "Three-dimensional fluidic self-assembly by axis translation of two-dimensionally fabricated microcomponents in railed microfluidics," Small, vol. 7, pp. 796-803, Mar 21 2011.

[22] S. E. Chung, W. Park, H. Park, K. Yu, N. Park, and S. Kwon, "Optofluidic maskless lithography system for real-time synthesis of photopolymerized microstructures in microfluidic channels," Applied Physics Letters, vol. 91, 23 July 2007.

[23] S. E. Chung, W. Park, S. Shin, S. A. Lee, and S. Kwon, "Guided and fluidic self-assembly of microstructures using railed microfluidic channels," Nature Materials, vol. 7, pp. 581-587, 2008.

[24] M. Vazquez and B. Paull, "Review on recent and advanced applications of monoliths and related porous polymer gels in micro-fluidic devices," Analytica Chimica Acta, vol. 668, pp. 100-113, Jun 4 2010.

[25] C. C. Glick, R. D. Sochol, K. T. Wolf, N. Shahmohhamadi, S. Miller-Hack, V. Jayaprakash, K. Iwai, L.P. Lee, L. Lin. "Pressure gain in single-layer microfluidic devices via optofluidic lithography," in 2013 Transducers and Eurosensors: 17th International Conference on, pp. 404-407, 2013.

CONTACT

*C. C. Glick, tel: +1-530-5192324; cglick@berkeley.edu

THIN–FILM EDGE ELECTRODE LITHOGRAPHY ENABLING LOW-COST COLLECTIVE TRANSFER OF NANOPATTERNS

Yongfang Li[1,2], Akihiro Goryu[1], Kunhan Chen[2], Hiroshi Toshiyoshi[2] and Hiroyuki Fujita[2]
[1]Corporate Research & Development Center, Toshiba Corporation, Kawasaki, JAPAN
[2]The University of Tokyo, Tokyo, JAPAN

ABSTRACT

This paper reports a new lithography method using thin-film edge electrodes (TEEs) to collectively transfer nanopatterns by generating oxide on the substrate surface via an electrochemical reaction (ECR). Nanometric thick TEEs are formed on the sidewall of insulating stamping structures. ECR-based oxide patterns have the same width and shape as the TEEs because ECR is induced only between the conductor and the substrate. Oxide nanopatterns of 300 nm and 70 nm in width were collectively transferred on Si substrate in a millimeter-scale area.

INTRODUCTION

Lithography has been one of the key drivers for the semiconductor industry for the past several decades. However, the physical limits of mainstream optical lithography are coming and being replaced in time with the next-generation lithography (NGL) technologies such as extreme ultraviolet lithography (EUVL), electron beam lithography (EBL) and nanoimprint lithography (NIL) [1, 2, 3]. EBL [4] and EUVL [5] are limited mainly by the throughput and implementation costs. NIL is positioned as the most popular choice due to its inherent simplicity and low cost of operation [6, 7]. However, resolution of NIL depends on the pattern size of the mold limited by the resolution of lithography. In this study, we proposed a novel lithography method of thin-film edge electrode lithography (TEEL) as a solution.

CONCEPT AND PRINCIPLE

As shown in Fig.1, TEEL combines a thin-film edge electrode mold (TEEM) with an ECR. The TEEM consists of a base and insulating patterns with nanometric-thick TEEs and nanometric-thick insulating layer alternately formed on those of sidewall (Fig.1 a, b). The principle of TEEL is quite similar to local anodic oxidation lithography [8, 9] and nanoelectrode lithography [10]. The TEEM and the surface of target material are usually covered by a thin film of absorbed water in air. When the TEEM thus TEEs proximally approaches the target material, these absorbed layers come in contact, and a water meniscus is produced because of the capillary effect. With the application of a corresponding electric field, an electrochemical reaction is initiated in the water interface between the electrodes and the substrate through the water bridge. If the target surface is positively charged and the TEEs of TEEM are negatively charged, an ECR is induced to generate oxide patterns in the water meniscus between the TEEs and the surface of target material. Therefore, nanopatterns corresponding to the TEEs rather than the insulating patterns can be transferred because ECR takes place only between the conducting portion of the mold and the substrate (Fig.1 c).

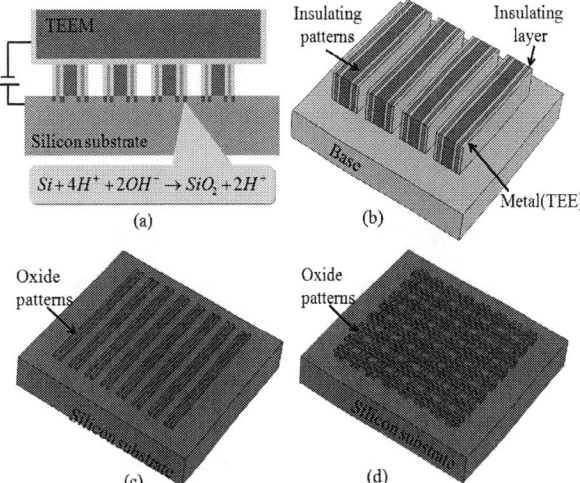

$$Si + 4H^+ + 2OH^- \rightarrow SiO_2 + 2H^-$$

Figure 1: Schematic of the proposed thin-film edge electrode lithography (TEEL). (a) Principle of TEEL. (b) Schematic of the proposed thin-film edge mold (TEEM). (c) Schematic of the transferred oxide patterns using single TEEL. (d) Schematic of the transferred oxide patterns using multiple TEEL.

This approach allows us to select the area of target material in which ECR occurs and thus generate oxide patterns by applying the bias voltage to part of TEEs. Furthermore, several kinds of patterns more complicated than TEEs of TEEM can be transferred by conducting TEEL in different directions for multiple times (Fig. 1 d) using one TEEM.

After TEEL, an etching process can be performed to transfer the shape of the oxide patterns to the target material. When the target material for patterning cannot be oxidized, a hard mask such as a Si or a metal layer can be formed on the target surface before performing TEEL. Therefore, TEEL can be used for pattering of any materials [11].

Figures 2 (a) and (b) show the transfer process of conventional UV-type NIL and the proposed TEEL. The main difference between the two methods is whether resist film exists or not and whether the transfer resolution depends on the pattern size of the mold or not. Conventional NIL uses the mold to physically transfer pattern on resist film, which can cause defective patterns when releasing the mold. Compared to NIL, TEEL is able not only to reduce the number of process steps thus fabrication cost but also to avoid defect of pattern, thereby improving the accuracy because of the resistless process. The resolution of conventional NIL is dependent on the pattern size of the mold, which is limited by the resolution of the conventional lithography. On the other hand, the width of the transferred oxide pattern based on TEEL is dependent on the lateral thickness of TEEs, and the super-narrow space of the transferred patterns is decided by the thickness of the insulating layer. Therefore, the resolution of TEEL is only dependent the lateral thickness of TEEs and the insulating layer but not on pattern size, which can be easily thinned for higher resolution.

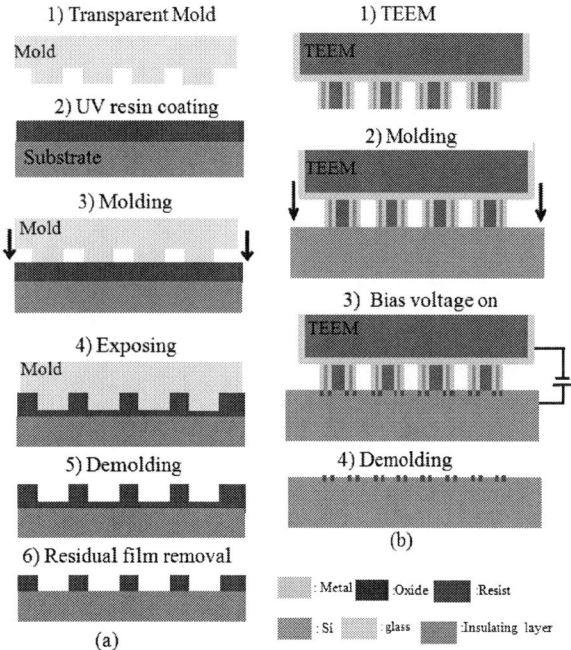

Figure 2: (a) Process chart of UV-type NIL. (b) Process chart of the proposed TEEL.

STRUCTURE OF PROTOTYPE TEEM

A prototype of TEEM shown in Fig.3 is used to verify the concept of TEEL. As shown in the figure, the prototype TEEM is composed from three parts, insulating stamping structures, nanometric-thick single-layer TEEs formed on the sidewall of the stamping structures, and a base. The TEEs are mechanically and electrically connected to the base via the insulating patterns on the base. The stamping structures made of silicon (Si) act as insulating patterns during ECR of TEEL. The TEEs enables not only to generate nano-patterns smaller than the insulating patterns on TEEM via an ECR but it also enables collective transfer and thus high throughput. In addition, gold (Au) is used as the TEEs material of TEEM for its soft contact, good electrical conductivity, and chemical stability.

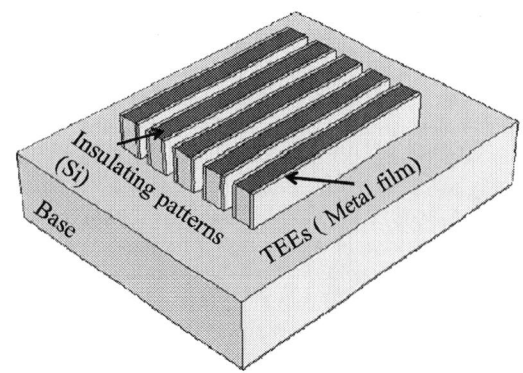

Figure 3: Schematic of prototype TEEM.

FABRICATION

The schematic of fabrication process flow is shown in Fig. 4. The fabrication process begins with a p-type (100)-oriented Si substrate with a thickness of 525 μm and an electrical resistivity of 1-10 Ω · cm. First, lithography process was conducted to form a resist pattern on Si substrate for insulating patterns of TEEM. Then, insulating patterns were shaped on the mold by DRIE (Deep Reactive Ion Etching). After that, a chromium film as an adhesion layer for gold and silicon, and a gold film for creating the TEEs were deposited through a sputtering process. At last, TEEs were shaped by lift-off process. Two kinds of TEEM with 300-nm-thick and 70-nm-thick TEEs were fabricated, which are shown in Fig. 5 and Fig.6 respectively. Both the two TEEMs had a pattern area size of 10 mm×10 mm, and the insulating patterns on the mold had a height of 4 μm and a half pitch of 5 μm, 10 μm, 15 μm and 20 μm.

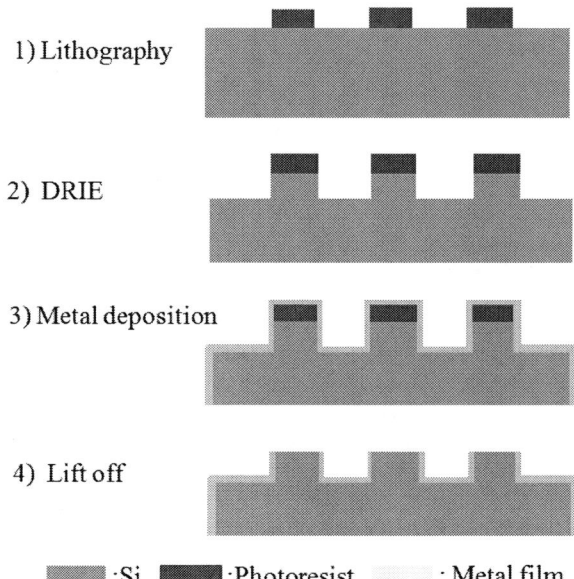

Figure 4: Fabrication Process flow.

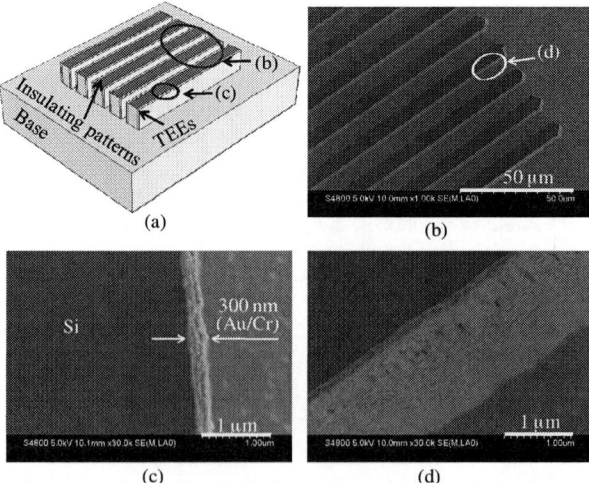

Figure 5: (a) Schematic of the prototype TEEM. (b) SEM image of the fabricated TEEM with 300-nm-thick TEEs. (c) Tow view of the fabricated TEE. (d) Close-up SEM image of the developed TEE.

978-1-4799-7956-1/15 $31.00 © 2015 IEEE   556

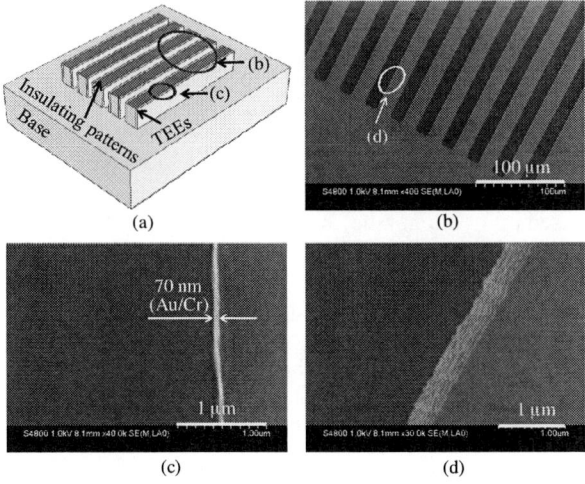

Figure 6: (a) Schematic of the prototype TEEM. (b) SEM image of the fabricated TEEM with 70-nm-thick TEEs. (c) Tow view of the fabricated TEE. (d) Close-up SEM image of the developed TEE.

TEEL

As shown in Fig. 2 (b), the transfer process of TEEL consists of 4 steps: 1) set the TEEM and Si substrate on the pressing unit; 2) applied a force to ensure the contact between the TEEM and substrate; 3) applied a bias voltage between the TEEM and the substrate to induce electrochemical reaction for generating oxide pattern corresponding to TEEs; 4) released the TEEM from the substrate by unloading the force. The setup for performing TEEL shown in Fig.7 was composed of a pressing unit and an electrical unit. The pressing unit enabled application of force ranging from 0 to 500 N, while the electrical unit could apply a constant bias voltage from 0 to 40 V in increments of time from 10μs to 1s. A p-type (100) oriented Si substrate with an electrical resistivity of 5-10 Ω · cm was used for target material of TEEL.

At first, the TEEM shown in Fig.5 with 300-nm-thick TEEs was used for pattern transfer. TEEL was conducted with 150 N contacting force and pulsed bias voltage of 20 V applied between the TEEM and the substrate in air with a humidity of 58% and a temperature of 23.7 °C. To prevent the target surface from Joule heating, the bias voltage with a pulse duration of 1 s and a period of 1 s was applied. TEEL was carried out twice and the total time of bias application was 4 s and 4 min. The transferred oxide patterns with a total time of bias application of 4s and 4 min were shown in Fig. 8 (a)-(b) and Fig.8(c)-(d) respectively. From Fig.8, it can be seen that oxide patterns with a line width of about 300 nm were collectively transferred on the Si substrate in an area of hundreds of micrometer. The total area size in which pattern transfer was conformed was in a millimeter scale. Furthermore, the transferred oxide pattern shown in Fig. 8(c) became more uniform than that in Fig. 8 (a). These results indicated that ECR would proceed sufficiently as the time of bias application increases, resulting in uniform transfer by absorbing the contact variation between the substrate and the TEEs of TEEM.

Moreover, both the oxide patterns shown in Fig.8 (a) and (c) were thought to be transferred by the TEEs formed

on the insulating patterns with a half pith of 15 μm on TEEM. However, the transferred pattern shown in Fig. 8(a) has a half pitch of about 30 μm, while the transferred pattern shown in Fig. 8(c) has a half pitch of about 15 μm. If the TEEs of the fabricated prototype of TEEM have the same height and the TEEM are exactly parallel to the Si substrate, only the oxide lines of 15 μm interval could have been transferred. However, the shape of the prototype TEEM especially the TEEs was not perfect and the parallelism between the TEEM and the substrate was not controlled during TEEL, which might have resulted in a partial contact of the TEEs with the substrate. Therefore, only one side of the TEEs formed on the sidewall of insulating patterns came in contact with the substrate to generate oxide patterns. When the total time of bias application became large enough, ECR could proceed sufficiently to generate patterns even when there was a small gap between the TEEs and the substrate. It will explain why the oxide patterns with a 15 μm interval corresponding to the interval of the insulating patterns were transferred when increasing the time of bias application from 4 s to 4 min.

Figure 7: Picture of the setup for performing TEEL

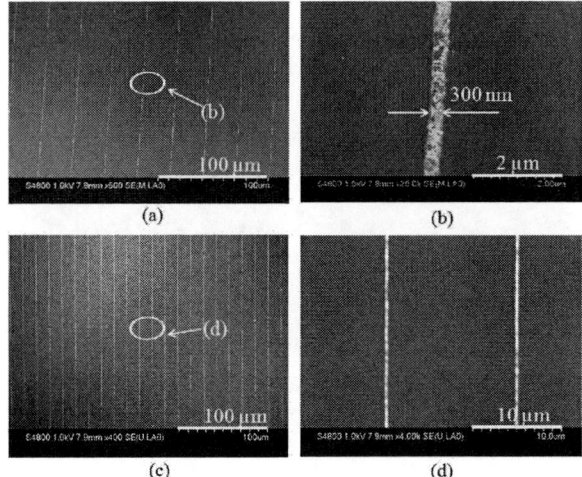

Figure 8: (a) SEM image of the transferred oxide patterns when applying the bias voltage of 20 V for 4s. (b)Close-up SEM image of the transferred oxide patterns in (a). (c) SEM image of the transferred oxide patterns when applying the bias voltage for 4min. (d) Close-up SEM image of the transferred oxide patterns in (c).

978-1-4799-7956-1/15 $31.00 © 2015 IEEE   557

In order to transfer the oxide pattern to Si substrate, dry etching using a SAMCO RIE-10 NR system was performed with transferred oxide pattern served as the mask shown in Fig.8(c). RIE was conducted under the conditions as follows: an etching gas of SF_6 at a flow rate of 50 sccm, a pressure of 5 Pa, a power of 50 W and an etching time of 10 s. Oxide patterns after 10 s etching are shown in Fig.9. From Fig.9, it can be seen that oxide patterns still remained and that Si was successfully etched. The etched oxide pattern was measured by AFM (Atomic Force Microscope) and the etching thickness was found to be about 40 nm. Furthermore, the width of the oxide pattern was reduced from 300 nm to 200 nm after a 10 s RIE etching, which was thought to have been caused by the thickness difference in the center and the edge of the transferred oxide patterns. These results indicated that Si patterns with a finer line width could be achieved by transferred oxide patterns generated by TEEL.

Next, TEEL was performed using the fabricated TEEM with 70-nm-thick TEEs shown in Fig.6. Conditions of TEEL were the same as described above, and the total time of bias application was 4 min. The transferred oxide patterns shown in Fig.10 had a line width of about 70 nm and an interval of 15 μm, which were consistent with the width of the TEEs and half pitch of the insulating patterns. These results indicated that the line width of the transferred oxide pattern was reduced correspondingly by reducing the thickness of TEEs.

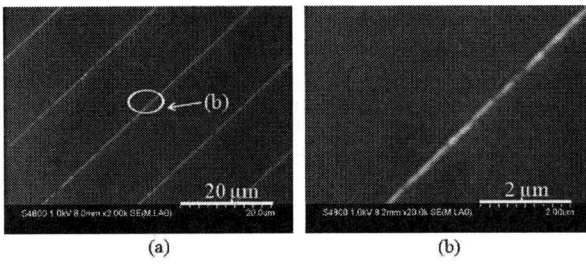

(a)   (b)

Figure 9: (a) SEM image of the transferred oxide pattern shown in Fig.8(c) after 10 s RIE etching. (d) Close-up SEM image of the transferred oxide pattern after 10 s RIE etching.

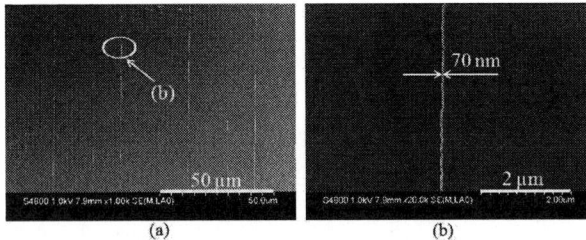

(a)   (b)

Figure 10: (a) SEM image of the transferred oxide patterns by TEEM with 70-nm-thick TEEs. (b) Close-up SEM image of the transferred oxide patterns.

CONCLUSION

A novel lithography method of TEEL using TEEs to collectively transfer nanopatterns by generating oxide on the substrate surface via an ECR was proposed as a candidate of NGL for realizing high resolution and high through put while maintaining low implementation cost. A

prototype of TEEM with a nanometric thick single-layer TEEs formed on the sidewall of the micro-scale insulating patterns was used to verify the concept of TEEL. Two kind of TEEMs with 300-nm-thick and 70-nm-thick TEEs were fabricated using microelectromechanical systems techniques. Oxide nanopatterns with a line with of 300 nm and 70 nm were collectively transferred on Si substrate based on ECR, and the concept of TEEL was found to be effective.

We aimed to develop TEEL to realize transfer resolution of sub-20-nm in high throughput and low cost so that this technique would be available for industrial use. We are currently reducing the thickness of TEEs for high resolution and improving the TEEM and the transfer equipment for uniform transfer of patterns in large scale.

ACKNOWLEDGEMENT

The authors would like to thank R. Inanami at Center for Semiconductor Research & and Development, Toshiba for valuable discussions and helpful suggestions.

REFERENCES

[1] M. Rothschild, et al., "Recent Trends in Optical Lithography", The Lincoln laboratory Journal, Vol. 14, No.2, pp. 221-236, 2003.

[2] L.R. Harriott, "Limits of Lithography", Proc. Of the IEEE, vol. 89, No. 3, pp. 336-374, 2001.

[3] J. Bendik, D. Brandt, "Next Generation Lithography: As we push down to 11 nm lines and spaces", NIST TIP White Paper, TIP Critical National Needs Ideas: February 15, 2011.

[4] M. Saito, et al., "The Performances and Challenges of Today's EB Lithography and EB-resist Materials", J. Photopolym. Sci. Technol., Vol. 27, No. 4, 2014.

[5] S. Kyoh, et al., "EUVL challenges toward 1x nm Generation", J. Photopolym. Sci. Technol., vol.24, No.1, 2011.

[6] Stephen Y. Chou, et al., "Nanoimprint Lithography", J. Vac. Sci. Technol. B 14, 4129, 1996.

[7] S.W. Pang, et al., "Direct nano-printing on Al substrate using SiC mold", J. Vac. Technol. B 16, pp. 1145-1149, 1998.

[8] J. A. Dagata, et al., "Role of space charge in scanned probe oxidation," J. Appl. Phys., vol. 69, pp. 6891–6900, Dec. 1998.

[9] P. M. Campbell, et al., "AFM-based fabrication of Si nanostructures," J. Appl. Phys. B, vol. 227, no. 1, pp. 315–317, 1996.

[10] A. Yokoo, "Nanoelectrode Lithography", Jpn. J. Appl. Vol. 42, pp.L92-L94, 2003.

[11] H. Namatsu, et al., "chemical nanoimprint lithography for step-and-repeat Si etching", J. Vac. Sci. Technol. B 25(6), pp.2321-2324, 2007.

CONTACT

*Y.F. Li, tel: +81-44-549-2361; yongfang.li@toshiba.co.jp

ULTRAFINE PARTICLE COUNTER USING A MEMS-BASED PARTICLE PROCESSING CHIP

Hong-Lae Kim[1], Jang Seop Han[1], Sang-Myun Lee[1], Hong-Bum Kwon[1], Jungho Hwang[1], and Yong-Jun Kim[1]
[1]School of Mechanical Engineering, Yonsei University, KOREA

ABSTRACT

This paper reports on the full realization of an ultrafine particle monitoring system, including a MEMS-based particle processing chip and signal processing circuits. Unlike a conventional liquid-based microfluidic chip, the proposed particle processing chip handles a mixture of gas and particles. The proposed particle monitoring system is suitable for routine ambient air monitoring due to its small size, ease of use and low cost. The detection performance of the proposed system was evaluated through measurements of particle number concentration and compared with that of commercial instrument.

INTRODUCTION

Recently, the demands for routine monitoring of airborne particles have increased due to their adverse health effects. In particular, ultrafine particles (UFPS; diameter less than 100nm) have a greater potential for adverse health effects compared with larger-sized particles [1,2]. These particles are easily inhaled and reach the deepest parts of respiratory system [1]. Moreover, most of the ultrafine particles are mainly formed by gas to particle conversion of harmful compounds [2]. However, it is difficult to monitor the concentration of UFPs by conventional optical detectors, because the light scattering intensity of the UFPs is very low due to their small size. Also, commercially available particle detection instruments are not suitable for routine exposure monitoring because they are large and expensive [3]. In this paper, a low-cost and compact particle detection system using a MEMS-based particle processing chip is proposed for routine monitoring of UFPs. Because of their negligible mass compared with larger-sized particles, the detection performance was evaluated through measurements of particle number concentration. Unlike the previous particle detection chip [4], the proposed particle processing chip was designed to measure the UFPS, and integrated system containing signal processing circuits was realized. Also, the detection performance of the proposed particle counter was compared with that of commercial instrument.

DESIGN AND FABRICATION

As shown in Fig. 1, the particle processing chip is composed of a micro virtual impactor and a tip-to-nozzle micro corona charger. The micro virtual impactor separates the airborne particles into a straight low-velocity flow (minor flow) and a perpendicular high-velocity flow (major flow) based on the inertial of particles. The inlet flow containing large and small particles is accelerated through an injection nozzle. The particles with a large mass follow a straight channel known as a minor channel. On the other hand, particles with a small mass move to the side channels, which are major channels normal to the injection nozzle (Fig. 1). The virtual impactor was designed to have cut-off diameter of 200nm, which means that particles less than 200nm move to major channels. The tip-to-nozzle corona charger is used for electrical charging of classified particles. If a high voltage is applied between a discharging electrode and a ground electrode, corona discharge occurs, and gaseous ions are generated. The particles are charged due to collision of the particles with migrated ions. The nozzle type ground electrode was designed to reduce the electrostatic precipitation of UFPs caused by applied voltage. And then, the charged UFPs are captured in the filter connected to an electrometer and generate an induced current (Fig. 1).

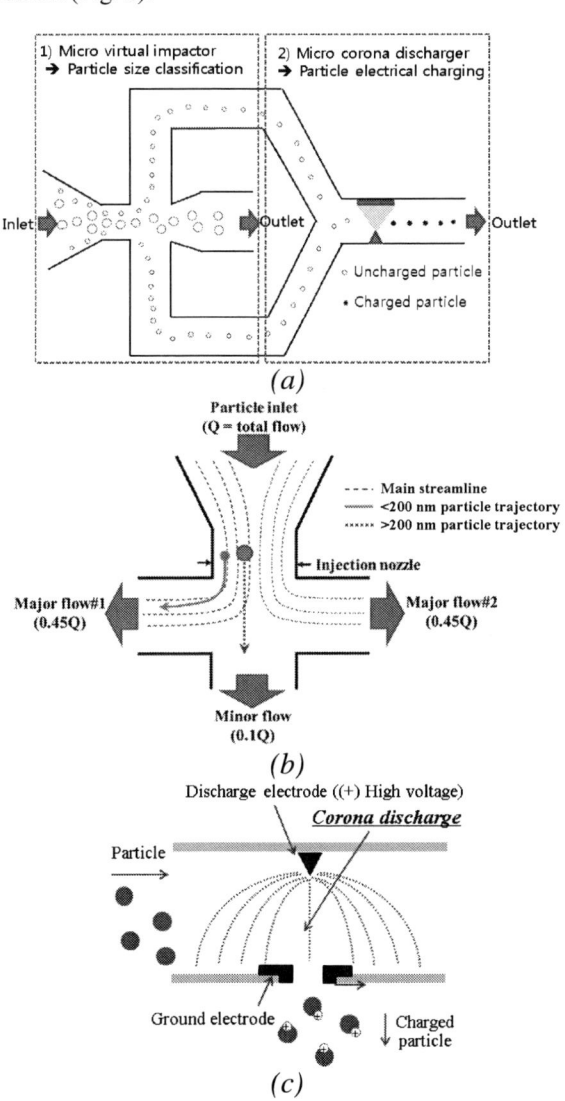

Figure 1: (a) Microfluidic layout of the particle detection chip, (b) particle classification and (c) charging process

The induced current is proportional to the flow rate and the number concentration of particles. Therefore, under constant flow condition, it is possible to measure the number of particles in real time by measuring the induced current.

Fig. 2 shows a simplified fabrication process of the particle processing chip. The fabrication of top electrode began with patterning the silicon dioxide layer on a four inch silicon wafer. The sharp silicon tip was realized by anisotropic wet etching of the silicon wafer using a 20%wt potassium hydroxide and water (KOH, J.T. Baker, USA) aqueous solution. The fabricated silicon tip was 35 μm in height and 50 μm in width on average. Subsequently, a fresh 1 μm-thick silicon dioxide layer for insulation was thermally grown on the sharp silicon tip. After this step, a 3 μm-thick titanium-copper layer was deposited by DC sputtering and patterned. The fabrication of bottom plate also began with patterning the silicon dioxide layer on a silicon wafer. The nozzle type ground electrode was realized by anisotropic wet etching of the silicon wafer. Subsequently, a 3 μm-thick titanium-copper layer was deposited and patterned. Then, the microchannel was realized using a 200 μm-thick Su-8 layer.

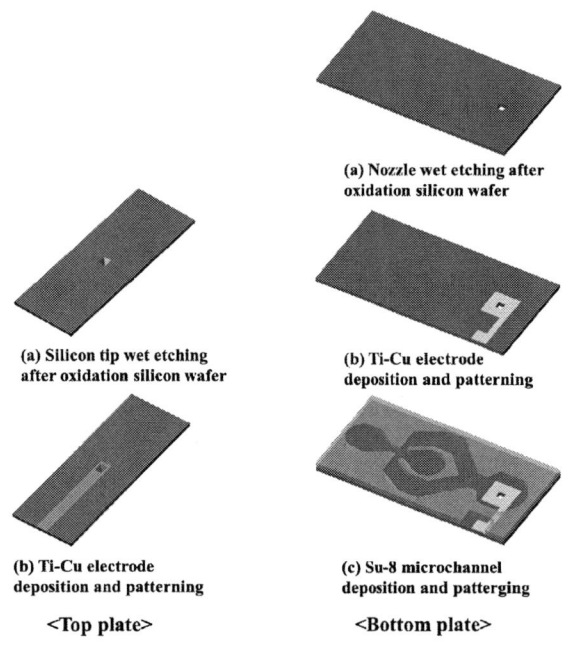

Figure 2: Simplified fabrication process of the MEMS-based particle processing chip

Fig. 3 shows schematic view of signal processing circuits for measuring UFPs. Signal processing circuits are composed of a small-sized high voltage converter, a low current sensing circuit, and a MCU (micro controller unit). The high voltage converter supplies several kV level voltages for particle charging. The low current sensing circuit transforms fA level currents, carried by charged particles, into voltages. In order to minimize an effect of the outside electrical noise, a high integrity PCB design, shielding, and electronic packaging method was utilized. The MCU converts analog signal into digital signal, and

controls the all functions of the system such as high voltage converting circuit and air pumps. The manufactured ultrafine particle counter was shown in Fig. 4.

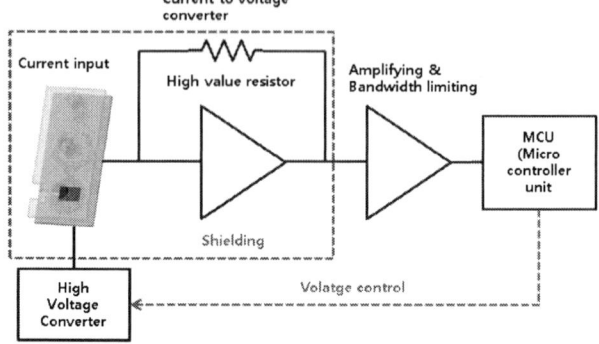

Figure 3: Schematic view of signal processing circuits

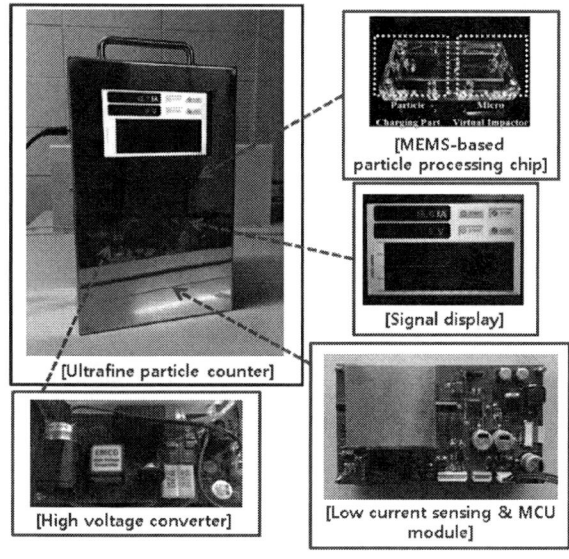

Figure 4: Optical photograph of the manufactured ultrafine particle counter

EXPERIMENT AND RESULTS

Fig. 5 shows a schematic diagram of the experimental setup for classification and detection of UFPs. It is divided in four parts as (a) UFPs generator (b) UFPs size classifier (c) particle classification efficiency measurement and (d) particle concentration measurement. Compressed air was used as a carrier gas, after oil droplets, moisture, and contamination particles were removed by a clean air supply. Sodium chloride (NaCl) particles were used as test particles and were generated from electrically heated tube furnace (Lenton Furnaces, model GTF12/25/364). Once NaCl particles were generated in the aerosol generator, the desired NaCl particle concentration after the tube furnace was controlled using an aerosol conditioner. The NaCl particles ranging from 40 to 300nm were supplied to the proposed particle processing chip and the particle classification efficiency was measured using a condensation particle counter (CPC, TSI model 3022A). The flow rate is controlled by a vacuum pump and CPC, to a major flow of 0.3 *lpm* and minor flow of 0.03 *lpm*. After then, the particles were electrically charged by corona discharge and the induced current was measured by electrometer.

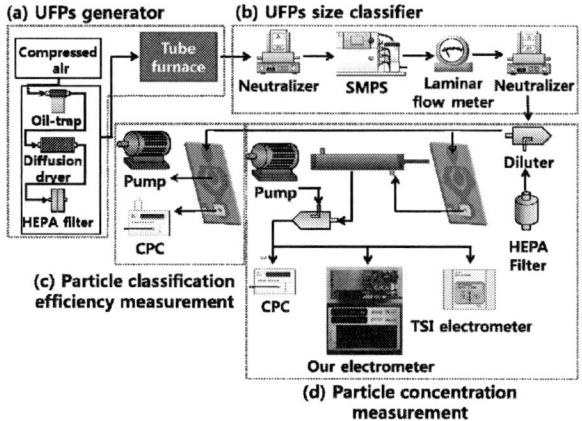

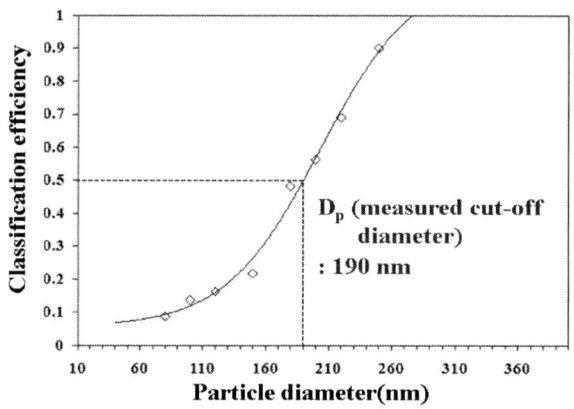

Figure 5: Schematic diagram of the experimental setup for classification and detection of ultrafine particles

Fig. 6 shows the particle classification efficiency of the proposed ultrafine particle counter. The collection efficiency on y-axis is defined as the ratio of the particle number concentration in minor flow to that of in inlet flow. The measured cut-off-diameter was 190 nm, and the deviation from the designed value was approximately 5%.

Figure 6: The particle classification efficiency of the proposed ultrafine particle counter

Fig. 7 shows the measurement of the induced current carried by charged NaCl particles as function of the particle number concentration. The 4 kV is applied to the particle processing chip for corona discharge which charges UFPs. The induced current carried by charged particles was measured by a low current sensing electrometer. The low current sensing electrometer was compared with the commercial aerosol electrometer (TSI model 3068A). By using the measured induced current, the number concentration of particles can be calculated.

$$N = \frac{I}{PneQ} \qquad (1)$$

where, N is the number concentration of particles, I the induced current by charged particles, P the particle penetration, n the number of charges, e the elementary charge, Q the flow rate. Fig. 8 shows the comparison result of the proposed ultrafine particle counter and commercial condensation particle counter. Overall, the difference was below 15%.

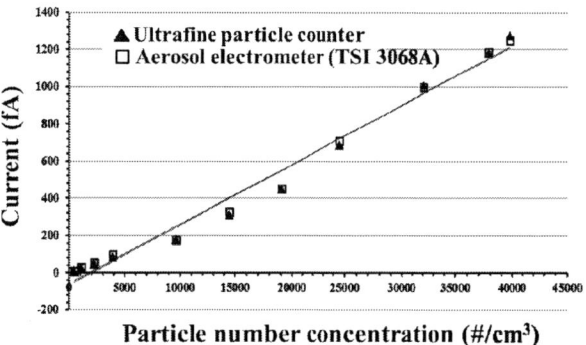

Figure 7: The measured currents as a function of the number concentration of the NaCl particles

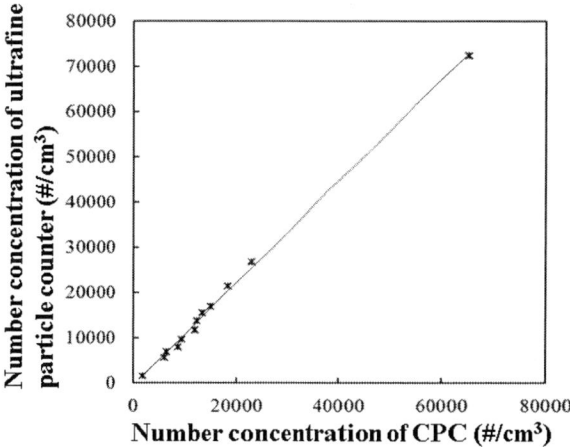

Figure 8: Comparison between proposed ultrafine particle counter and CPC in terms of number concentration

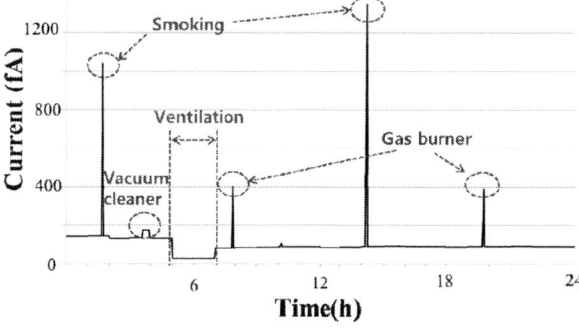

Figure 9: Experimental result of the 24-hour particle monitoring in laboratory

Fig. 9 shows the result of the 24-hour particle monitoring in our laboratory. There was no degradation in performance during long-term operation of proposed particle counter.

CONCLUSION

The miniaturized ultrafine particle counter using MEMS-based particle processing chip was designed and fabricated for real-time and on-site monitoring of UFPs. The particle counter is composed of a mems-based particle processing chip and signal processing circuits. Unlike a conventional liquid-based microfluidic chip, the proposed particle processing chip handles a mixture of gas and particles. The fabricated particle processing chip demonstrates successful classification of particles according to their sizes and charging of particles. The fabricated signal processing circuits also demonstrate successful measurement of induced current carried by charged particles for calculating particle number concentration. The proposed particle monitoring system is suitable for routine ambient air monitoring due to its small size, ease of use and low cost.

ACKNOWLEDGEMENTS

This subject is supported by Korea Ministry of Environment as "The Conversing Technology Program"

REFERENCES

[1] J. H. Lee, Y. S. Kim, K. S. Song, H. R. Ryu, J. H. Sung, J. D. Park, H. M. Park, N. W. Song, B. S. Shin, D. Marshak, K. H. Ahn, J. E. Lee, I. J. Yu, "Biopersistence of silver nanoparticles in tissues from Sprague-Dawley rats", *Part. Fibre. Toxicol.*, vol. 10:36, pp.1-14, 2013

[2] Andrea D'Anna, "Combustion-formed nanoparticles", *P. Combust. Inst.*, vol. 32, pp. 593-613, 2009.

[3] M. Marjamaki, J. Keskinen, D.-R, Chen and D.Y.H. Pui, "Performance evaluation of the Electricla Low-Pressure Impactor (ELPI)", *J. Aerosol Sci.,* vol. 31, pp. 249-261, 2000.

[4] Y. H. Kim, D. H. Park, J. H. Hwang and Y. J. Kim, "Integrated particle detection chip for environmental monitoring", *Lab on a chip*, vol. 8, pp. 1950-1956, 2008

CONTACT

*Y.J. Kim, tel: +82-2-2123-2844; yjk@yonsei.ac.kr

VERTICAL MEMBRANE MICROVALVES IN PDMS

Jonas Hansson, Mikael Hillmering, Tommy Haraldsson, and Wouter van der Wijngaart
Micro and Nanosystems, KTH Royal Institute of Technology, Stockholm, SWEDEN

ABSTRACT

We present the design, realization and evaluation of the first leak-tight vertical membrane pneumatic microvalve. The design freedom in the vertical valve configuration allows for a flow throughput per footprint area that is increased two orders of magnitude compared to horizontal membrane microvalves.

INTRODUCTION

Normally open pneumatic valves, consisting of one flow channel layer and one pneumatic control channel layer separated by a horizontal membrane [1] (Figure 1a), constitute one of the most common microvalve types due to their ease of manufacturing in PDMS [2]. Sundararajan et al. [3] introduced vertical membrane microfluidic actuators (Figure 1b). Despite several efforts [3-5], vertical membranes have been unable to fully close microchannels due to restrictions in their mechanical deformability at the membrane edges [5], hence, vertical membranes microfluidic actuators are instead used as variable flow resistors [6], mixers [3], droplet generators [6], focusing microlenses [8], or particle traps [9].

(a) Horizontal membrane valves [1]

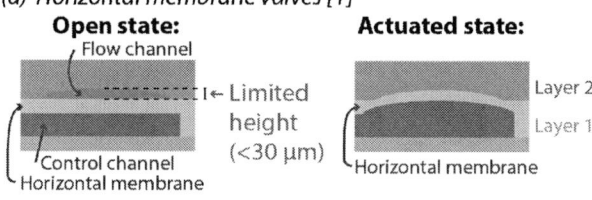

(b) Vertical membrane actuators [3-5]

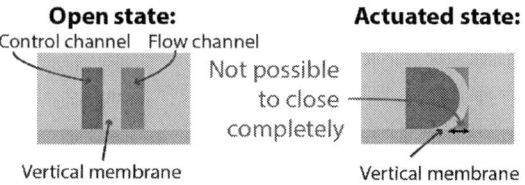

(c) Vertical membrane valves

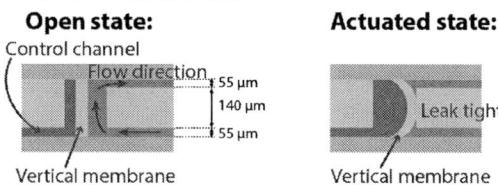

Figure 1: Cross-sectional side-views of pneumatic microvalves. (a) Horizontal membrane valves are manufactured using 2 microstructured layers, while (b) Vertical membrane actuators and (c) Vertical membrane valves are manufactured using 1 microstructured layer.

DESIGN

In comparison to horizontal membrane valves, our novel design (Figure 1c and Figure 2) features 3D, instead of 2D, flow microchannels, in which a vertical membrane actuates a vertical flow channel section. The high degree of freedom in designing vertical channel cross-sections allows for:

1) adjusting the shape of the flow channel to the local deformability of the membrane, allowing for the first time leak tight closure of the flow channel using a vertical membrane; and

2) increasing the flow channel cross-section, allowing a reduced pressure drop, alternatively an increased flow throughput, per footprint area, thus potentially enabling a higher degree of miniaturization and a denser integration of microvalves than allowed by horizontal membrane designs.

The flow channels connecting the vertical valve are both 55 μm high, and 140 μm vertically separated, hence the total thickness of the microstructured fluidic layers is 250 μm (see Figure 1c). The shape of the 56 μm x 408 μm cross-section of the vertical flow channel is depicted in Figure 2a.

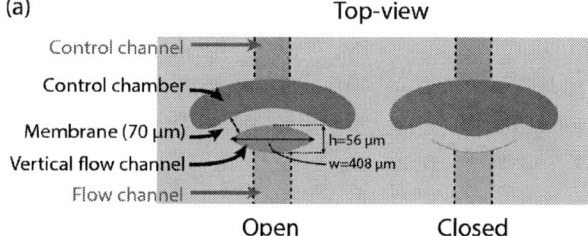

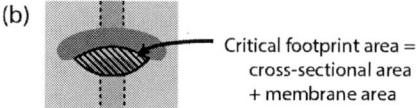

Figure 2. Top-view of the valve design. (a) Top-view schematic of actuation. (b) Critical footprint area calculation

MANUFACTURING

We realized the valves as a glass-PDMS-PDMS layered structure using an automatically aligning double-sided molding process introduced by Karlsson et al. [10] (Figure 2). The bottom glass layer and the top PDMS layer both being unstructured removes the need for layer alignment.

The dual layer micromolds were fabricated using two-step spin-coating and photolithographic patterning of SU-8 2025 (MicroChem Corp, USA) on 4" silicon wafers using an emulsion film photomask, with subsequent development of the SU-8 according to manufacturer protocols. The height of the first SU-8 layer defines in-plane channels and microfluidic structures, and the height

---

978-1-4799-7956-1/15 $31.00 © 2015 IEEE   563   MEMS 2015, Estoril, PORTUGAL, 18 - 22 January, 2015

of the second layer defines both guiding structures for alignment of the pairing molds and vertical interconnects. Thereafter, molds were saw-diced into 8 mm x 12 mm microchips to facilitate handling and to enable pre-alignment of the molds during the molding process.

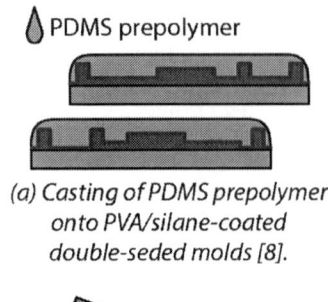

(a) Casting of PDMS prepolymer onto PVA/silane-coated double-seded molds [8].

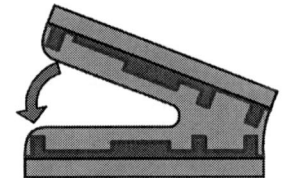

(b) Folding of one mold on top of the other.

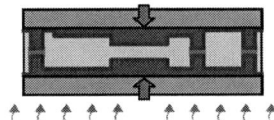

(c) Automatic mold alignment using guiding structures. Molds are pressed together. Curing of PDMS (70 °C, 40 min).

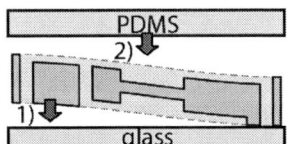

(d) Bonding of the structured PDMS layer to (1) a glass bottom substrate, and (2) a PDMS lid.

Figure 3. Manufacturing process.

PDMS prepolymer (Sylgard 184, Dow Corning, USA) was mixed (1:10, curing agent:base) and casted onto PVA/silane-coated micromolds and degassed to remove any bubbles introduced during the casting process. The mold halves were folded onto each other with the pattern facing inwards and manually pre-aligned by ensuring that the external mold edges were aligned. Pressure was applied towards the guiding structures for a good fit and thereafter held together with clamps during 40 min curing at 70°C.

The stack composed of two mold halves enclosing a cured PDMS layer was placed in a water bath in which ultrasonication was performed 20-60 minutes to separate the molds by dissolving the PVA which lines the patterned side of the mold halves.

Floatation transfer [10] of the PDMS layer was then performed to a carrier substrate consisting of a rough polycarbonate (PC) surface for plasma treatment (40 W, 15 s; FEMTO A, Diener electronic GmbH, Germany) onto the destination glass substrate. This was followed by a second plasma treatment to bond the non-structured PDMS top layer onto the microstructured PDMS middle layer.

MATERIALS AND METHODS

The flow channel inlet was connected to a pressure-controlled container with deionized (DI) water. The pneumatic control port was coupled to a pneumatic source, consisting of a piston, that can be manually regulated to obtain specific actuation pressures. The flow channel was coupled via a flow sensor (ASL 1600-10, Sensirion AG, CH) to a waste container at atmospheric pressure. Pressure sensors (ELFA, Sweden) were used to monitor the pressure in the DI water container and at the pneumatic control source.

The pressure-flow characteristics for the valve were measured as follows. After priming the valves with DI water, the DI water container was pressurized at 50 kPa. The pneumatic channels were pressurized between 0 and 300 kPa in increments of 25 kPa. The valve was imaged using a top-view microscope and photographs were taken at pneumatic pressures of 0, 100, and 300 kPa. The perimeter of the channel cross-section and membrane wall was traced using imaging software, from which the valve flow channel cross-sectional area was calculated (see figure 2b).

Fluid conductance was measured by pressurizing the flow channel at 50 kPa while keeping the valve at ambient pressure, measuring the output flow, and calculating the conductance for the system from the measured flowrate and applied flow pressure.

RESULTS AND DISCUSSION

Figure 4a shows the microscope images of the valve in open and closed states. At 0 kPa pneumatic pressure, the valve is open; at 100 kPa, the membrane is deflected but not entirely closing the channel; and at 300 kPa, the valve is entirely closed. Pressure-flow characteristics are plotted in Figure 4b. At 50 kPa liquid pressure, it closes at a pneumatic pressure between 125 and 150 kPa. These results are comparable to those for horizontal membrane microvalves, which close at 40 kPa pneumatic pressure for 0 kPa of liquid pressure [1] (Table1).

978-1-4799-7956-1/15 $31.00 © 2015 IEEE

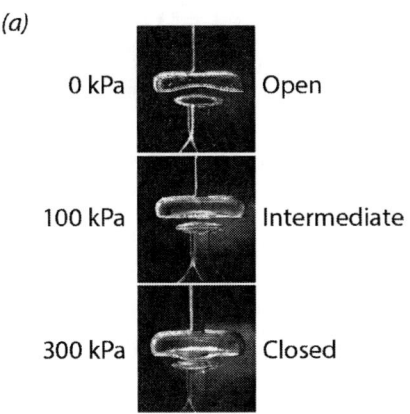

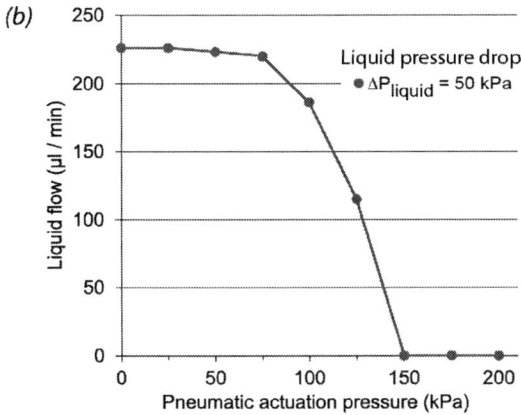

Figure 4: Valve actuation results. (a) Top view of the actuated valve. (b) Pneumatic performance

The footprint area normalized volumetric flow throughput of the vertical membrane microvalve in the open state is 120 $\mu l/(kPa\cdot s\cdot mm^2)$, i.e. two orders of magnitude higher than that of horizontal membrane microvalves (Table 1).

Table 1: Comparison of valve performance

	Horizontal membrane valves [1]	Vertical membrane valves
Fluid conductance per footprint area [$\mu l/(kPa\cdot s\cdot mm^2)$]	1.3	120
Closing pressure [kPa]	40	125

CONCLUSIONS

We have demonstrated the first leak-tight vertical membrane pneumatic microvalve. The valve closes under pneumatic control pressures in the range 100-150 kPa, i.e. similar to those for horizontal membrane microvalves. The footprint area normalized volumetric flow throughput of the vertical membrane microvalve in the open state is 120 $\mu l/(kPa\cdot s\cdot mm^2)$, i.e. two orders of magnitude higher than that of horizontal membrane microvalves

The uncomplicated 3D manufacturing and high fluidic conductance of this valve makes it particularly suited to high flowrate applications and large-scale integration.

ACKNOWLEDGEMENTS

This work has been sponsored in part by the European Commission through the FP7 project ROUTINE.

REFERENCES

[1] Unger et al., Science, vol. 288, no. 5463, pp. 113–116, Apr. 2000.
[2] Au et al., Micromachines, vol. 2, no. 4, pp. 179–220, May 2011.
[3] Sundararajan et al., Lab Chip, vol. 5, no. 3, pp. 350–354, Feb. 2005.
[4] Abate et al., Applied Physics Letters, vol. 92, no. 24, p. 243509, Jun. 2008.
[5] Lee et al., J. Micromech. Microeng., vol. 17, no. 5, p. 843, May 2007.
[6] Doh et al., Lab Chip, vol. 9, no. 14, pp. 2070–2075, Jul. 2009.
[7] Abate et al., Applied Physics Letters, vol. 94, no. 2, p. 023503, Jan. 2009.
[8] Shao et al., Microsyst Technol, vol. 19, no. 11, pp. 1823–1828, Nov. 2013.
[9] Kim et al., Microfluid Nanofluid, vol. 13, no. 5, pp. 835–844, Nov. 2012.
[10] Karlsson et al., Proc. MicroTAS 2012, Okinawa, Japan, pp. 659-661, Oct. 2012.

CONTACT

*T. Haraldsson, tel: +4687907794; tommyhar@kth.se

3D HUMAN CARDIAC MUSCLE ON A CHIP: QUANTIFICATION OF CONTRACTILE FORCE OF HUMAN IPS-DERIVED CARDIOMYOCYTES

Yuya Morimoto[1, 2], Saori Mori[1, 2], and Shoji Takeuchi[1, 2]

[1]Institute of Industrial Science, The University of Tokyo, Tokyo, JAPAN
[2]Takeuchi Biohybrid Innovation Project, ERATO, JST, Tokyo, JAPAN

ABSTRACT

We propose a method for constructing fiber-type three-dimensional (3D) tissue of human iPS-derived cardiomyocytes and quantifying its contractile force in response to the addition of drug. By culturing the cardiomyocytes in micropatterned hydrogel with anchors, we succeeded in fabrication of the fibers with aligned cardiomyocytes and fixation of the fiber edges to the anchors. Since the fiber generated contractile force in a single direction due to alignment of cardiomyocytes, we can measure the contractile force accurately. Furthermore, as a demonstration of drug testing, we quantified contractile frequency and force in accordance with concentrations of pilsicainide. We believed that the fiber of human iPS-derived cardiomyocytes will be used in pharmacokinetic applications for drug development.

INTRODUCTION

Since heart is one of the most important organs for sustaining life, avoidance of adverse effect to cardiomyocytes is required in drug development. Recently, instead of animal testing, systems using human iPS-derived cardiomyocytes are demanded for assay of the adverse effect. The conventional assay systems are generally based on two-dimensional (2D) culture or 3D culture using spheroids or hydrogel substrates [1, 2]. However, measurement of contractile force is difficult in both culture systems because adhesion of cardiomyocytes on 2D substrates restricts their contractions and embedding measurable parts is difficult for conventional 3D culture systems (Fig. 1(a)). Here, we propose the fiber of human iPS-derived cardiomyocytes whose edges are fixed on anchors. Due to non-adhesion of the fiber on any substrates except its edges, the fiber can contract freely in 3D environment. In addition, the cardiomyocytes are aligned to the fiber direction by determining dimensions of the fiber. Since the contractions of the aligned cardiomyocyte fiber are in a single direction, we can measure the contractile force from the deformation of a cantilever (Fig. 1(b)).

MATERIAL AND METHOD

Materials

Plating medium and maintenance medium were purchased from Cellular Dynamics International, Inc.. Matrigel was purchased as hydrogel for cardiomyocytes from Corning, Inc.. MPC polymer (phosphorylcholine (PC)-based polymers) was purchased from NOF Corporation. Phosphate buffered saline (PBS) was purchased from SIGMA-Aldrich. TrypLE was purchased from life technologies. Fibronectin was purchased from Biomedical Technologies Inc.. Pilsicainide was purchased

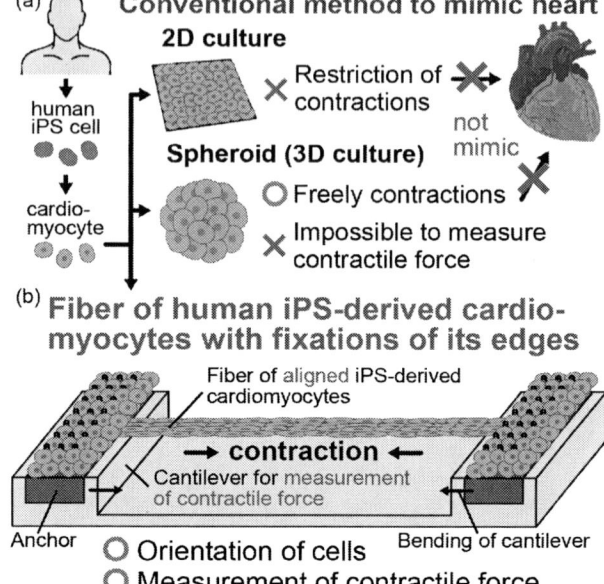

Figure 1: (a) Illustration of conventional method to culture human iPS-derived cardiomyocytes. Culture on 2D substrates and in spheroids is not appropriate for cardiomyocytes. (b) Concept of a fiber of human iPS-derived cardiomyocytes with fixations of its edges. Since cardiomyocytes are oriented to the fiber direction, the fiber contracts to the fiber directions. From deformation of a PDMS cantilever by contractions of the fiber, we can estimate contractile force.

from SIGMA-Aldrich. Type-I collagen (Cell Matrix) was purchased from Nitta Gelatin Inc.. Other chemicals were purchased from Kanto Chemical Co., Inc and Wako Pure Chemical Industries. Unless otherwise specified, all water used in the experiments refers to ultra-pure water obtained from a Millipore system having a specific resistance of 18MΩ•cm and sterilized by an autoclave treatment.

Cell preparation

Single-cell suspensions of iPS-derived cardiomyocytes (iCell® Cardiomyocytes) were purchased from Cellular dynamics International, Inc.. At first, we cultured the iPS-derived cardiomyocytes on culture dishes with Plating medium. After exchange of Plating medium according to manufacturer's protocol, we washed cardiomyocytes with PBS and collected them by a treatment of TrypLE.

Formation of PDMS devices

For formation of the fiber of human iPS-derived cardiomyocytes, we used polydimethylsiloxane (PDMS) devices comprised of a PDMS stamp and a PDMS substrate. The PDMS devices were replicated from resin molds. We designed the resin molds using a 3D modeling software (Rhinoceros, AppliCraft) and, for fabrication of them, used a commercial stereolithography modeling machine (Perfactory, Envision Tec, Germany) with photoreactive acrylates resin ("R11, 25-50 μm layers", Envision Tec), consisting of acrylic oligomer, dipentaerythritol pentaacrylate, propoxylated trimethylolpropane triacrylate, photoinitiator and stabilizers. After the formation of the resin molds using stereolithography and curing them by exposure of ultraviolet (UV) light to them, we coated a 2 μm thick parylene layer to accurately transcribe the shapes of the resin molds to PDMS devices [3]. The PDMS device was fabricated by casting the liquid prepolymer composed of a mixture of 10:1 silicone elastomer and curing agent (Sylgard 184, Toray). The mixture was cured at 75 °C for 1.5 h. Subsequently, the PDMS device was peeled from the resin mold and cleaned with ethanol.

To sterilize the PDMS devices, we exposed UV lights to them for 1 night. After sterilization, the PDMS devices were placed on a hot plate at 60 °C for drying. Finally, we coated MPC polymer on the surface of PDMS devices for blocking cell adhesion after an O_2 plasma treatment.

Fabrication of anchors

To produce anchors with poles, we designed them with the 3D modeling software and formed them using the commercial stereolithography modeling machine similar to the formation of the resin molds for PDMS devices. After the formation of the anchors, we coated a 2 μm thick parylene layer and fibronectin on a surface of the anchors to enable cell adhesion to the anchors.

Formation of iPS-derived cardiomyocyte fiber

We formed the fiber of aligned free-standing cardiomyocytes using the fabrication method (Fig. 2(a-d)). In this process, we prepared the line patterned PDMS stamp and the PDMS substrate since line patterning achieves the alignment of the cells [4, 5]. After putting anchors with poles coated with fibronectin at the edges of the PDMS substrate and hydrogel solution with iPS-derived cardiomyocytes on the substrate (Fig. 2(a)), we sandwiched the solution by the PDMS stamp and substrate (Fig. 2(b)). Subsequently, we located the PDMS device including the solution in an incubator under 37 °C during 10 minutes for gelating hydrogel solution. After peeling off the PDMS stamp (Fig. 2(c)), we cultured iPS-derived cardiomyocytes in the hydrogel using the maintenance medium. As a result, the iPS-derived cardiomyocytes connected with each other so that the cardiomyocyte fiber was constructed (Fig. 2(d)).

Measurement of contractile force

When we measured contractile force of the iPS-derived cardiomyocyte fiber, we placed the fiber to a PDMS device with thin cantilevers (Fig. 2(e)); the cantilever contacted with the anchor fixed to the edges of the fiber and was deformed by contractile force of the fiber. Finally, we observed the cantilever deformation when the

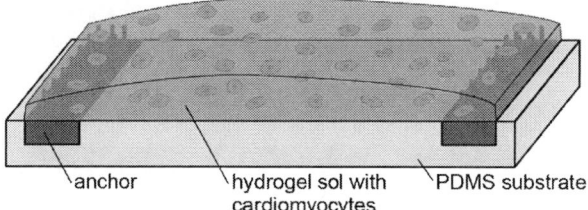

(a) Placing hydrogel with cardiomyocytes

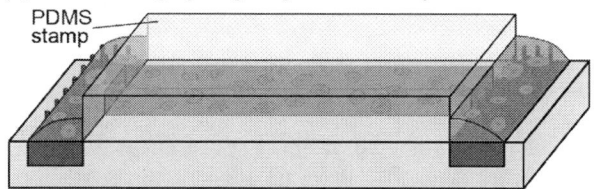

(b) Sandwiching hydrogel by PDMS stamp

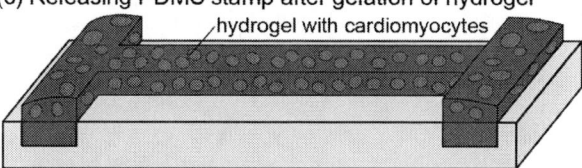

(c) Releasing PDMS stamp after gelation of hydrogel

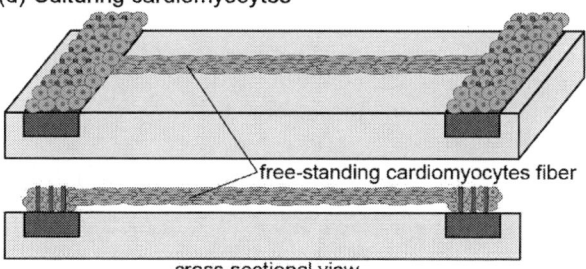

(d) Culturing cardiomyocytes

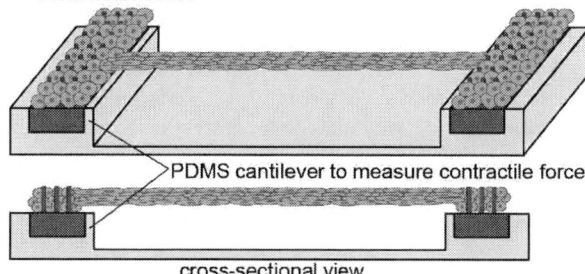

(e) Placing the fiber on PDMS device to measure contractile force

Figure 2: Process flow of producing a human iPS-derived cardiomyocyte fiber and measuring the contractile force of the fiber. (a-c) Micropatterning hydrogel (mixture with type-I collagen and matrigel) with human iPS-derived cardiomyocytes using a PDMS stamp with a line pattern. (d) Constructing free-standing cardiomyocyte fiber after culture. The fiber edges are fixed on anchors. (e) Placing the fiber to a PDMS device for measurement of contractile force.

fiber contracted and calculated contractile force using the formula for the cantilever deformation.

RESULTS AND DISCUSSION
Morphology or the iPS-derived cardiomyocyte fiber

Using our proposed method, we were able to construct

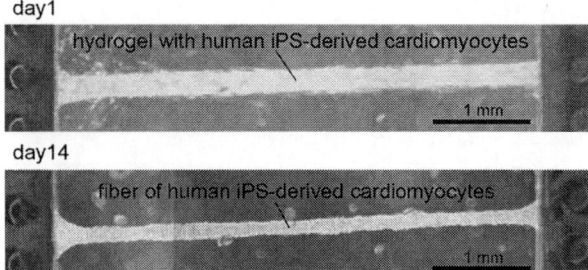

day1

hydrogel with human iPS-derived cardiomyocytes

1 mm

day14

fiber of human iPS-derived cardiomyocytes

1 mm

Figure 3: Time lapse images of the cardiomyocyte fiber. Cardiomyocytes adhered to each other and keep the fiber shape with the fixation of the fiber edges.

iPS-derived cardiomyocyte fiber which edges were fixed to anchors (Fig. 3). For validation of the iPS-derived cardiomyocyte fiber, we evaluated its morphology firstly. From observation of immunostained fiber using antibody of α-actinin, we found that cardiomyocytes in the fiber had orientated sarcomeres to same direction, indicating that the cardiomyocytes were matured and aligned. Since contractions of the cardiomyocytes were synchronized after 1day of culture, we evaluated interval and contractile distance of spontaneous contractions. Although the intervals of the fiber was similar to that of spheroid, contractile distance of the fiber was longer than that of the spheroid. The results indicate that contractility of the fiber is better than that of the spheroid.

Contractility of the iPS-derived cardiomyocyte fiber

For quantification of contractility of the iPS-derived cardiomyocyte fiber, we measured contractile force of the fiber using the PDMS device with cantilevers (Fig. 4). Because electrical stimulations were triggers for contractions of cardiomyocytes [6, 7], we compared contractile force of spontaneous contractions and contractions triggered by electrical stimulations. Using our system, we can quantitatively confirm the difference of contractile force in several conditions. As a demonstration of drug testing, we analyzed contractile force of the cardiomyocyte fiber after addition of pilsicainide. Although electrical stimulations worked as triggers for contractions, frequency of spontaneous contractions and contractile force decreased when we added pilsicainide over 10μM. The result shows that pilsicainide prevented depolarization of cardiomyocytes as Na^+ channel blockers, indicating that the fiber of human iPS-derived cardiomyocytes has potentials to become pharmacokinetic models for drug development.

CONCLUSION

Using the PDMS stamp, the PDMS substrate and the anchors with poles, we can fabricate the fiber of aligned free-standing iPS-derived cardiomyocyte. Furthermore, we can estimate contractile force of the fiber from the deformation of the PDMS cantilever. Thus, we believe that the iPS-derived cardiomyocyte fiber will be a useful tool to analyze functions such as contractility of cardiomyocyte as a tissue model for drug testing.

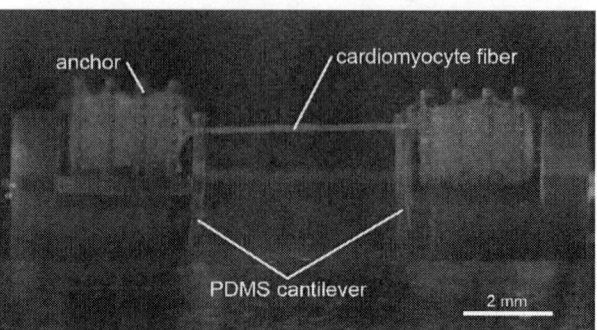

Figure 4: The fiber of human iPS-derived cardiomyocytes on the PDMS device to measure its contractile force.

ACKNOWLEDGEMENTS

The authors thank M. Onuki and F. Sakai for their technical assistance. Y. Morimoto is supported by Grant-in-Aid for Young Scientists (Start-up) 26889016 of the Japan Society for the Promotion of Science (JSPS), Supporting of the Matsuda Foundation, Supporting of Casio Science Promotion Foundation and Supporting for assistant professor of Institute of industrial science, the university of Tokyo.

REFERENCES

[1] J. Zhang, G. F. Wilson, A. G. Soerens, C. H. Koonce, J. Yu, S. P. Palecek, J. A. Tomson and T. J. Kamp, "Functional Cardiomyocytes Derived From Human Induced Pluripotent Stem Cells", *Circ. Res.*, vol 104, pp. e30-e41, 2009.

[2] H. J. Evans, J. K. Sweet, R. L. Price, M. Yost and R. L. Goodwin, "Novel 3D culture system for study of cardiac myocyte development", *Am. J. Physiol. Heart Circ. Physiol.*, vol. 285, pp. H570-H578, 2003.

[3] Y. T. Matsunaga*, Y. Morimoto* and S. Takeuchi, "Molding cell Beads for Rapid Construction of Macroscopic 3D Tissue Architecture", *Adv. Mater.*, vol. 23, pp. H90- H94, 2011. (*: equal contribution)

[4] Y. Morimoto, M. Kato-Negishi, H. Onoe and S. Takeuchi, "Three-dimensional neuron-muscle constructs with neuromuscular junctions", *Biomaterials*, vol. 34, pp. Seattle, October 2-6, 2011, pp. 789-791.

[5] K. Nagamine, T. Kawashima, T. Ishibashi, H. Kaji, M. Kanzaki and M. Nishizawa, "Micropatterning contractile C2C12 myotubes embedded in a fibrin gel", *Biotechnol. Bioeng.*, vol. 105, pp. 1161- 1167, 2010.

[6] A.W. Feinberg, A. Feigel, S.S. Shevkoplyas, S. Sheehy, G.M. Whitesides and K.K. Parker, "Muscular Thin Films for Building Actuators and Powering Devices", *Science*, vol. 317, pp. 1366-1370, 2007.

[7] J. C. Nawroth, H. Lee, A. W. Feinberg, C. M. Ripplinger, M. L. McCain, A. Grosberg, J. O. Dabiri and K. K. Parker, "A tissue-engineered jellyfish with biomimetic propulsion", *Nature Biotech.*, vol. 30, pp. 792-797, 2012.

CONTACT

*Y. Morimoto, +81-3-5452-6650;
y-morimo@iis.u-tokyo.ac.jp

MEMS 2015 KEYWORD INDEX

2D Crystal 877
2D Phononic Crystal 1016

3D 284
3D Cell Culture 443, 643
3D Fabrication 1071
3D Hall Sensor 893
3D Integration 304, 996
3D Microfluidics 196
3D Morphology Reconstruction 358
3D Photolithography 841
3D Printing 261
3D Structure 405, 539

4H-SiC 276

6-Axis Sensor 730

A

Accelerometer (s) 849, 865
Acoustics 921
Acousto-Optic Modulators 980
Actuator 114
Adherent Cells 439
Adhesion 393, 849
Adhesive Bonding 280
Adipocyte 643
Affinity Binding 488
AFM (Atomic Force Microscope) 150, 732, 752
Air Flow 710
Air-Molding 348
ALD (Atomic Layer Deposition) 771, 805
Alginate 26
Alignment 465
Allan Deviation 180
Allan Variance 37
Aluminum Nitride - AlN 73, 140, 373, 837, 921, 928, 984, 1094
Amplitude-Modulated Sensor 913
AMR Sensor 901
Amyloid 604
Anchor Loss 797
Anode 130
Anodic Silicon Dioxide 385
Anodized Metal 265
Aptamer 488, 569, 581
Aqueous Two Phase Extraction 547
Array (s) 917, 952

Assembling	14
Autofluorescence	358
Autorotation	308

B

Baking Powder	504
Band-to-Band Tunneling	988
BAW Gyroscope	789
Bead (s)	431, 604
Betavoltaic Microcell	1102
Bimorph	928
Bio-Dissolvable	636
Bio-Inspired	500, 706, 889
Bio-MEMS	336, 354
Biochemical Sensor	612
Biodegradable	340
Biomarker	455
Biomimetics	500, 889
Biosensor	447, 573
Biosignal Detection	833
Biotemplate	1118
Black Silicon	365
Blood Cell Measurement	435
Body-on-a-Chip	535
Boiling	1122
Bovine Serum Albumin (BSA)	470
Breast Cancer Cell	698
Breath	126
Broadband	146
Bubble	655
Buckypaper	1067

C

C-Reactive Protein	581
Cantilever	144, 718
Capacitive Feedback	956
Capacitive Humidity Sensor	767
Capacitive Transduction	150, 219
Capillary Filling	10
Capillary Force	413
Capillary Number	459
Capillary Vessel	480
Carbon Nanotubes & Devices	249, 312, 369, 401, 417, 744, 1102
Carbonization	817
Castellated Electrode	508
Cavitational Microstreaming	1059
Cavity	1122
Cell Fiber	643
Cell Filter	340
Cell Lysing	188
Cell Manipulation	200, 1059
Cell Migration	624, 628
Cell Origami	589
Centrifugal	192, 504, 523
Chemical Analysis	608

Chemomechanical Transducer	219
Circuit-Breaker Chip	1040
Circulating Tumor Cell	192, 340, 459
CMOS-MEMS	732, 736, 767, 853, 988
Cold Switching	41
Collagen	465
Comb-Drive	14, 215
Commercialization	1
Communication	53
Complete Blood Count	435
Consumer Electronics	61
Contact Angle Hysteresis	484
Contact Electrification	102
Continuous Particle Sorting	508
Core-Shell	472
Core-Shell Microfiber	463
Cryogenic Measurement	14
Crystal	242
Crystal Silicon	389
Current Source	551
Curved PMUT	837
Curved Surface	351
Cyborg Insect	1048
Cylindrical	801

D

Deep Reactive Ion Etching	288, 365
Deep Silicon Etching	885
Deposition	467
Di-Electrowetting	496
Diabetic Retinopathy	154
Diagnosis	447
Diamond	801
Dielectrophoresis	196, 354, 508
Digital Microfluidics	443, 492, 519
Dilatant Fluid	421
Dilution	543
Direct Laser Writing	470
Displacement Magnification	1082
Dissolvable Film	451
Doppler Effect	640
Double Re-Entrant	6, 1122
Droplet Detection	785
Droplet Vibration	176
Drug Cocktail Optimization	658
Drug Delivery	158
Drug Screening	566
Dry Etching	77
Dual Resonance Modes	146
Durotaxis	628

E

Elastomer Packaging	1137
Electret	527, 1071, 1086, 1145
Electric Contact	968
Electric Field	238
Electric Field Trapping	81
Electrical Double Layer	118
Electrical Impedance Measurement	435
Electrical Stimulation	649
Electro Tactile Display	649
Electrochemical Etching	649
Electrochemical Impedance	620, 662
Electrochemical Reaction	555
Electrodynamics	936
Electrokinetic Force	512
Electroless Plating	373
Electromagnetic	122
Electromagnetic	1051, 1075
Electromechanical Coupling	992
Electromicrofluidic	516
Electron Beam	512
Electronic Skin	760
Electrophoresis	488
Electroplating	272, 401
Electrospray	134
Electrostatic	45, 1125
Electrostatic Actuator (s)	22, 276, 897, 1036
Electrostatic Lens	93
Electrowetting	496, 956
Electrowetting-on-Dielectric (EWOD)	265, 519, 547
Embeded Passives	427
Encapsulated	393
End-Stop Transducer	1125
Endoscopic Optical Imaging	948
Endothelial Cell	480
Energy Harvesting	122, 126, 118, 527, 1051, 1071, 1075, 1078, 1094, 1110, 1129, 1145
Energy Pump	913
Energy Storage	1133
Enhanced Controllability	1032
Entropic Effect	300
Enzyme-Linked Immunosorbent Assay	523
Epitaxial	393
Epi-Seal	1
Etching	284
Eutectic Bonding	409
Exfoliation	280
Extracellular Matrix	463

F

Fabrication	1078
Ferrofluid	122
Ferromagnetic Resonance	208
Fiber Array	162
Fiber Sensor	775
Field Effect Transistor	417, 581

Fixation 405
Flapping Wing 22
Flexible 110, 636, 726, 845
Flexible Device (s) 81, 106
Flexible Electronics 1133
Flexible Sensors/Actuators 377
Flexible Substrate 877
Flexion Sensing 760
Flexural Rigidity 238
Flow 620
Flow Control 2
Flow Focusing 539
Flow Lithography 666
Flow Sensor 500, 889
Folded-Beam Suspensions 215
Food Texture 646
Force Measurement 702
Force/Moment Transducer 736
Force/Torque Sensor 257
Free Molecular Flow, Knudsen Diffusion 397
Free-Standing Structures 304
Frequency Characteristic 212
Frequency Modulation 33
Frequency Tuning 789
Frictional Force 245, 253
FSK Modulator 1024
Functional Paper 861
Functional Surface 829
Fused Silica 821

G

Gallium Nitride 1028
Gas Chromatography 771
Gas Sensor (s) 312, 779
Gauge Factor 361
Glass 332
Glass Blowing 805
Glass Substrate 427
Glass Transition 405
Gold Nanoparticles 81
Graphene 324, 569, 600, 865, 869, 873, 1090
Graphene Oxide 2, 381
Graphite on Paper 825
Gyroscope (s) 33, 37, 801, 813, 821

H

H_2S 779
Halbach Array 1051
Hall-Effect 893
Hanging Drop 226, 535
Hard Mask Free 77
Heat Spreader 616
Hemolysin 235
Hierarchical Electrode 1118
High Aspect Ratio 77, 358, 427, 805

High Frequency 150
High Power 118
High Speed Flow 196
High Topography 328
Human Unbilical Vein Endothelial Cell 476
Humidity Sensing 783, 857
Hybrid Parylene 397
Hybrid Silicon 682
Hybrid Structure 1082
Hydrocephalus 620, 662
Hydrogel (s) 26, 585
Hysteresis 531

I

Impactor 885
Impedance 604
Impedance Spectroscopy 226
Implants 620
In Vitro Fertilization (IVF) 519
Induced Pluripotent Stem Cell 463
Inductor 208
Inductors 261
Inertial Harvester 1141
Infrared Sensing 73, 905, 984
Injection 655
Injection Molding 373
Inkjet Printing 632
Instrumentation 849
Integrated Light Source 162
Integration 417
Interface 496
Interface Dynamics 547
Interfacial Dissipation 1000
Interferometric 69
Internal Impact 1125
Intracellular Calcium Level 651
Ion Sensor 785
Ion Source 93
Ion Trap 292
Ionic Liquid 93, 97, 118, 783
IPL Photoresist 470

K

Keyboard 1078
Kinesin 694

L

Lab on a Disc 504, 523
Lab-on-a-Chip 10, 451, 563
Large Displacements 14
Laser 817
Laser Doppler Velocimeter 748
Laser Machined 332
Lateral Flow Device 10
Leading-Edge Vortex 308

Length Extensional Mode 785
Lens Array 960
Li-Ion Battery 130
Linear Springs 215
Lipophilic Membranes 523
Liquid 144
Liquid Alloy Patterning 1137
Liquid Metal 261
Liquid Phase Detection 612
Liquid Sensing 718
Liquid Spring 122
Lithium Niobate 992
Live Cancer 324
Load Sensor 833
Lorentz Force 204, 944
Low-Melting-Point-Alloy 18
Low-Noise Low-Power Electronics 932
Low-Temperature Bonding 413
Lung Cancer 600
Lung Fibroblast 476

M

Magnesium 340
Magnetic 604
Magnetic Actuator 1044
Magnetic Devices 936
Magnetic Field Sensor 893
Magnetic Flux Concentrator 901
Magnetic Forces 964
Magnetic Micropillars 624
Magnetic Particles 964
Magnetic Resonance Force Microscopy 344
Magnetic Resonance Imaging 674
Magnetic Structure 1051
Magnetic Thin Films 208
Magnetometer 204, 932, 944
Magnetophoresis 192
Magnifier 952
Mass Sensing 180
Mass Spectrometer 134
Material Reduction 775
Material Transfer 280
MATRIX 467
Mechanotransduction 624
Membrane Sensor 219
Membrane Transport 439
MEMS on CMOS 1004
Metamaterial 960, 984
Metastability 531
Micro Airflow Sensor 714
Micro Electro Mechanical System 37, 180, 292, 296, 381, 714, 718, 932,
.... 992, 1012, 1016, 1032, 1036, 1098
Microactuators 936
Microbial Fuel Cell 573, 577
Microcantilever 752

Microchamber 344
Microchannel 666, 702, 841
Microcoil 674
Microcomponent (s) 300, 516
Microdisk 698
Microdispenser 658
Microdroplet 960
Microelectro Mechanical System Micromirrors 948
Microelectro Mechanical System Technology 110
Microelectrode 686
Microfabrication 268, 292
Microfiltration 459
Microfluidic (s) 2, 180, 188, 332, 348, 431, 472, 480, 504, 531, 543, 551, 581
Microfluidic Chip 200
Microgenerator 1094
Microgripper 1106
Microlens 952
Micromachining 284
Micromanipulator 1059
Micromirror 401
Micromixer 658
Micropatterning 336, 829
Microphones 917
Micropillar 628
Microplasma 268
Micropore 235
Micropump 535
Microrobotics 296
Microstructure 964
Microsupercapacitor 1067, 1133
Microtissue Spheroid 226
Microtube 184
Microtubule 238, 694
Microvalve 563
Microwave Characterization 1012
Microwell Device 439
Mode Coupling 1008
Mode Localization 881
Mode Matching 789
Molecular Detection 230
Molecular Sorter 238
Molybdenum Disulfide (MoS2) 877
Monolithic Integration 771
MOSFET 686
Motion Conversion 897
Motion Sensor 106
Multi-Axis 1141
Multi-Frequency 1024
Multi-Roof Tile-Shaped Vibration Modes 718
Multi-Step-Assay 451
Multilayer Interconnection 265
Multilayer Sidewall Transfer Lithography 288
Myoelectric 678

N

n-p-n 276
Nanoactuator 694
Nanocomposite 130, 744
Nanofiber Forests 857
Nanofluidic (s) 512
Nanofluidic Crystal 593
Nanomaterial 447
Nanoparticle 8, 608
Nanopatterns 555
Nanopillars... 320
Nanopore 512
Nanoporous Anodic Aluminum Oxide 69
Nanoscale Gap 85
Nanosensor 869
Nanotribology 968
Nanowire 361
Negative Dielectrophoresis 492
NEMS 57, 288, 686, 865, 940, 972
Neural Probe 158, 616, 636, 682
Neurodegeneration 455
Neuromodulation 651
Neuromuscular Stimulation 1048
Neurons 328
Nickel Oxide Supercapacitor 1118
Non-Immuno Protein Assay 455
Nonlinear Damping 1125

O

One-Shot Valve 1090
One-Step Polymerization 472
Open Microfluidics 535
Optical Cross Connect 57
Optical Force 57
Optical Microsensor 748
Optical Receivers 976
Optical Resonators 976
Optical Stimulation 162
Optical Waveguide 166
Optofluidic Lithography 551
Optogenetic (s) 158, 162, 616
Optomechanical 45, 49
Optrode 682
Organ Hardness 253
Organ on a Chip 566
Oscillating Bubble 1059
Oscillator (s) 41, 793, 976, 988, 1004, 1024
Out-of-Plane 1063
Out-of-Plane Motion 897
Ovenization 793, 809
Overhanging Pillar 1086
Oxygen Transporter 154

P

P-N Heterojunction	1102
Package	421
Packaging	1
Paper-Based Devices	825
Paper-Based Microfluidics	10, 447
Paper-Based Sensor	577
Parametric Drive	29
Parametric Resonators	172
Particle Focusing	492
Particle Processing Chip	559
Particulate Matter	885
Parylene	268, 328, 358, 397, 405, 467, 612, 662, 682
Parylene Photonics	682
Passive	909
Patency Sensor	662
PDMS	316, 397, 431, 467, 476, 539, 563, 624, 817
PDMS Fiber	423
Perfluorohexane	1122
Perfusable Channel	351
Phase Change Lens	1114
Phononic Crystal	797, 1028
Photolithography	97
Photonics	53, 980
Photoresist	97
Piezo Electric Micromachined Ultrasound Transducer (pMUT)	140, 921, 928
Piezoelectric	944, 1129, 1141
Piezoelectric Actuators	65
Piezoelectric Coefficient	861
Piezoelectric Gyroscope	789
Piezoelectric Material	377
Piezoelectric Paper	861
Piezoelectric Resonators	1000
Piezoelectric Thin Film	1094
Piezoelectric Transducer	651
Piezoresistive Cantilever	176, 670, 706
Piezoresistor	144, 245, 361, 730, 825
Pirani	89
Pixel	69
Planer Bilayer	235
Plant-on-a-Chip	702
PLL	793
Pneumatic Balloon Actuator	166
Pneumatic Pump	666
Point-of-Care-Test	219
Poling	1098
Poly(3,4-Ethylenedioxythiophene): Polystyrene-Sulfonate	940
Poly(vinylidenefluoride-co-trifluoroethylene)	110
Polycrystalline Diamond	616
Polyethylene Glycol	636
Polymer Characterization	873
Polymer Distinguishing	102
Polymerase Chain Reaction	488
Polysilicon	361
Position Control	1055

Potassium Ion 1071
Pressure Sensing 722
Pressure Sensor 89
Pressure Stability 484
Printed Sensor 756
Probe-Based Memory 968
Process Monitors 312
Processing 435
Protein 242
Protein Cross-Linking 470
Protein Pattern 328
Pseudo Inverse 736
Pulse Wave 670
PVDF Nanofiber 678
PZT 146, 1098
PZT Thin Film 377

Q

Q Enhancement 204
Quadrature Error 37
Quadruple Filter 134
Quality Factor 797, 821, 853, 1000
Quantitative Assay 543
Quantum Information Processing 292
Quartz 944
Quartz Crystal Resonator 833

R

Ratchet 1106
Rate Integrating Gyroscope 29
Reactive Ion Etching 93
Red Blood Cell 431, 1055
Reference Plane 253
Reflective Display 956
Reflective Micro Objective 632
Reflow Process 427
Relay 114, 940
Reliability 114
Resonator (s) 1, 41, 73, 172, 184, 204, 215, 612, 698, 785, 797, 801, 809,
.... 821, 853, 877, 881, 885, 909, 913, 984, 992, 1020, 1008
Resorbable Electronics 168
Resorcinol-Formaldehyde Aerogel 767
Respiration 126
Retinal Ischemia 154
RF Ground-Signal-Ground (GSG) Probes 1012
RF MEMS 976, 1020
RF Passives 261
RF Resonator 1020
RGFET 988
ROADM 53
Root Growth Characterization 702
Running 257

S

SAW	144, 1028
Scale Factor Stability	29, 33
Scanning Confocal Fluorescence Microscopy	632
Scanning Electron Microscopy	344
Scanning Probe Microscopy	732, 968
Screen Printing	272
Selectivity	756
Self-Assembly	8, 300, 600, 608, 964
Self-Excited	41
Self-Excited Vibration	22
Self-Healing	81, 996
Self-Locking Mechanism	1106
Self-Mixing	547
Self-Powered Sensor	106
Self-Propagating Exothermic Reaction	373
Sensitivity	881
Sensor (s)	324, 593, 600, 752, 917
Sensor Fusion	212
SERS	223, 608
Serum	6
Shape-Memory-Alloy Actuator	1040
Shear Stress Sensor	710
Sheathless Focusing	196
Sheet	465
Shrink Polymer	324
Side-Wall Doping	710
SiGe Ring Resonator	1016
Signal Processing	61
Silicon	385
Silicon Carbide (SiC)	284, 698
Silicon Microelectrode Array	636
Silicon Nanoparticles	130
Silicon on Nothing	184
Silicon Optical Bench	948
Silk	168
Silver Nanoparticles	775
Single Cell Analysis	589
Single Crystalline	276
Single Molecule Sensing	596
Skin-Equivalent	351
Smart Glove	760
Smart Implant	1040
SNR	913
Soft Actuator	18, 26
Soft Material	1082
Soft Robotics	1044
Soft X-Ray Charging	1086
SOI MEMS Devices	409
SOI Technology	1012
Solid-Gate	869
Sound Focusing	925
Spherical Resonator	805
Spike Pin	257
Spring	26

SRAM 49
Statistical Element Selection 996
Stencil 272
Stencil Lithography 632
Stiction 85, 393
Stiffness Gradient 628
Strain Sensor (s) 678, 744, 756
Stress Effects on MEMS 813
Stress Engineering 296, 837
Stress Mapping 736
Stretchable 744, 817
Stretchable Electronics 1137
Stretchable Spring 423
SU-8 308
Superhydrophobic 6
Superlens 952
Superlyophobic 484, 1086
Surface Enhanced Raman Scattering 320
Surface Roughness 245
Surface Tension 496
Surface Texturing 385
Switch 114
Switchable Sensitivity 674
Synthetic Jet Actuators 936

T

Tactile Sensor (s) 245, 249, 253, 421, 726, 740, 825
Tantalum Pentoxide 265
Tau Protein 455
Teeth Vibration 646
TEM Liquid Cell 8
Temperature 1063
Temperature Compensation 809
Temperature Effect 389
Temperature Regulation 1040
Temperature Sensor 756, 909
Temperature Stable 793
Tensile Test 381, 369, 389
Terahertz 960
TFT Susbtrate 354
Thermal Sensor 845
Thermal-Piezoresistive 732, 885
Thermocouple 841
Thermoelastic Dissipation 1000
Thermoelectric Energy Generator 1114
Thin Film Edge Electrode Lithography 555
Thin Film Encapsulation 996
Thin Si Film 344
Tissue Engineering 348, 351, 480
Torque 730
Touch Sensor 861
Transducers 917
Transient Electronics 873, 1090
Transmitter 1024
Trench Sidewall 841

Triboelectric 1078
Triboelectric Energy Harvester 106
Triboelectric Generator 102
Tunable Metamaterial 1032
Tunable Oscillator 45
Tunable Resonator (s) 172, 1020
Tunable Ring Resonator 53

U

U-Shaped 775
Ultrafine Particle Counter 559
Ultrasonic Receiver 640
Ultrasonic Transducer 140, 146
Ultrasound 837, 928
Ultrasound Imaging 140
Ultrasound Stimulation 651
Underwater 925
User Experience 61

V

Vacuum Cavity 845
Vacuum Interface 134
Valve 666
Vapor-Liquid-Solid Growth 686
Variable Capacitors 1036
Varifocal Acoustic Mirror 925
Varifocal Liquid Lens 65
Velcro-Like Surfaces 413
Velocity Control 694
Vertical-Parallel-Plate Array 767
Vibration Energy Harvesting 1141
Vibration-Induced Flow 200
Viscosity 176

W

Wafer Level Packaging 409, 833
Wafer-Scale Synthesis 312
Water-Electrode 126
Wet Anisotropic Etching 385
Wet Etching Process 377
Whole Angle Mechanization 29
Wicking 332
Wireless 909
Wireless Power Transfer 1137
Wrinkle 829

X

XYZ Microstage 14

Z

Zebrafish ECG 690
ZnO Nanowire 764, 779

MEMS 2015 Author Index

A

Abdollahi, H.	952
Abdolvand, R.	909
Aftab, T.	893
Agache, V.	180
Agah, M.	771
Agarkova, I.	226
Agarwal, M.	1
Agentis, D.J.	936
Ahmad, M.M.	288
Ahn, C.H.	1, 29, 33, 393, 809, 1008
Ahn, H.R.	1114
Ahn, J.	292
Ahrens, G.	97
Aimé, J.P.	150
Akbar, M.	771
Akhbari, S.	837, 928
Akin, T.	409
Akita, I.	686
Akita, S.	756
Aktakka, E.E.	1141
Alam, R.	523
Alivisatos, A.P.	8
Allain, M.	65
Alper, S.E.	409
Alves, F.S.	1036
Amjadi, M.	744
An, B.	168
An, Q.L.	857
An, Z.	401
Anderson, K.	523
Aoki, R.	925
Aoyagi, S.	1145
Aqab, N.	928
Arai, F.	200, 431, 833, 1055
Ardanuç, S.	873, 1090, 1129
Ardito, R.	849
Arie, T.	756
Arnold, D.P.	936, 964
Arscott, S.	1012
Asadnia, M.	500, 678, 889
Asai, K.	686
Ashok, A.	385
Ayazi, F.	789

B

Bahl, G.	1
Bakhtina, N.A.	97
Baléras, F.	180
Barniol, N.	1004
Barz, F.	636
Baumgartel, L.	917
Becker, F.	736
Becker, M.F.	616
Beebe, T.	690
Bergbreiter, S.	726
Bergman, L.A.	752
Berthelot, A.	37
Berthet, C.	180
Beyazoglu, T.	976
Bharadwaja, V.	527
Bittner, A.	718
Blanche, T.J.	682
Block, S.T.	921
Boden, T.J.	944
Bolis, S.	65
Bonfanti, A.	932
Boser, B.E.	33, 140, 146
Bourouina, T.	960
Boyle, D.	2
Brenckle, M.A.	168
Brenna, S.	932
Bridoux, C.	65
Bright, V.M.	805
Brooke, H.	825
Brugger, J.	632
Buchaillot, L.	1012
Buckley, A.	551
Buisson, L.	150
Bulović, V.	85
Bürgel, S.C.	226
Burger, R.	504

C

Cabral, J.	1036
Cai, H.	45, 49, 57, 972
Calayir, E.	996
Camera, K.	873
Candler, R.N.	1
Cao, C.	381
Cao, F.	1048
Cattafesta, L.N.	936
Cha, W.	865
Chamanzar, M.	682
Chang, D.T.	944
Chang, H.L.	881, 1106

Chang, J.H.-C.	154, 340
Chang, K.-T.	154
Chang, K.-W.	581
Chang, W.-H.	581
Chapin, C.A.	284
Charalambides, A.	726
Chen, B.	381
Chen, C.-C.	519
Chen, D.P.	320, 857
Chen, H.T.	102
Chen, K.	555
Chen, K.L.	1
Chen, P.-H.	272
Chen, Q.	948
Chen, R.	740
Chen, R.S.	901
Chen, S.	265
Chen, S.-J.	1075
Chen, S.-Y.	516
Chen, T.N.	45
Chen, T.-Y.	740
Chen, X.	110, 268
Chen, Y.	612
Chen, Y.	1, 809, 1008
Chen, Y.-C.	89
Chen, Y.-C.	1086
Chen, Z.	1063
Cheng, C.-L.	423, 767
Cheng, J.	726
Cheng, X.L.	102, 1078
Cheon, H.	292
Chiang, C.F.	1
Chiang, M.-Y.	516
Chiao, M.	624
Chikasawa, T.	480
Chikkadi, K.	312
Chin, C.-H.	988
Chiou, P.-Y.	196
Cho, D.D.	292
Cho, H.	752
Cho, I.-J.	158, 651
Cho, J.Y.	821
Cho, S.K.	496
Choi, D.-H.	126
Choi, G.	577
Choi, J.-K.	219
Choi, S.	573, 577
Choi, S.	956
Choi, Y.S.	261
Chong, H.M.H.	881
Chou, B.C.S.	89
Chu, A.	551

Chu, C.-H.	581
Chu, S.	1118
Chuang, S.-T.	740
Chun, Y.E.	651
Chung, L.Y.	519
Chung, M.	551
Chung, S.K.	1059
Chung, V.P.J.	767
Chung, Y.-C.	740
Chung, Y.-H.	519
Clemens, D.L.	196
Cochet, M.	180
Colinge, C.	280
Collins, S.D.	188
Connell, L.	188
Costa, M.	1036
Cui, T.	324, 600

D

Dahmardeh, M.	1040
Dambrine, G.	1012
Dao, D.V.	825
De Masi, B.	849
Deem, E.A.	936
Dellea, S.	37, 849
Denman, D.J.	682
Deotare, P.B.	85
Dhakar, L.	106
Dias, R.A.	1036
Ding, J.	948
Dinh, T.	825
Dohi, T.	674
Dong, B.	45, 49, 57, 972
Dowling, K.M.	284
Du, Y.	304
Duan, C.	948
Ducatteau, D.	150
Ducrée, J.	2, 192, 504, 523
Duraffourg, L.	361
Dykman, M.	1008

E

El Fellahi, A.	1012
Elata, D.	41, 172, 215, 897
Elezgaray, J.	150
Elmlinger, P.	162
Eminoglu, B.	33
Erbland, J.A.	944
Ernst, T.	361
Errando-Herranz, C.	53
Everhart, C.L.M.	1

F

Fan, B. ... 616
Fan, C.-C. ... 89
Fan, J. ... 543
Fan, S.-K. ... 516
Fang, W. ... 69, 423, 767, 901
Fanget, S. ... 65
Fatemi, H. ... 909
Faucher, M. ... 150
Fedder, G.K. ... 813, 996
Feng, P.X.-L. ... 698, 877
Feng, Y.-Y. ... 1075
Figeys, B. ... 1016
Figueroa, C. ... 73
Filleter, T. ... 381
Finne-Wistrand, A. ... 472
Flader, I. ... 1, 809
Fluri, D.A. ... 226
Fonseca, H. ... 1036
Foroutan, V. ... 296
Forsberg, F. ... 280
Fraiwan, A. ... 573, 577
Frey, O. ... 226, 535
Fujisawa, D. ... 905
Fujita, H. ... 118, 354, 455, 555, 829, 1071
Fujita, K. ... 373
Fujito, T. ... 373
Funakubo, H. ... 1098
Fung, S. ... 140, 146

G

Gardner, D.S. ... 208
Gaspar, J. ... 1036
Gaughran, J. ... 2
George, S.M. ... 443
Gerasopoulos, K. ... 1118
Gertsch, J. ... 805
Ghodssi, R. ... 1118
Ghosh, S. ... 980
Giacci, F. ... 37
Giner, J. ... 805
Glick, C.C. ... 261, 551
Glynn, M. ... 192
Glynn, M.T. ... 504
Gokhale, V.J. ... 73
Goldkorn, A. ... 340
Goryu, A. ... 555
Goswami, A. ... 527
Goto, T. ... 480
Gowda, S. ... 527
Graham, A.B. ... 1

Gray, J.M.	805
Grine, A.J.	976
Gruetzner, G.	97
Grutter, K.E.	976
Gu, Y.D.	45, 49, 57, 972
Gund, V.	873, 1090
Guo, H.	110, 268
Guo, X.	204
Gylfason, K.B.	53

H

Habasaki, S.	666
Haddadi, K.	1012
Hadji, C.	180
Haluska, M.	312
Halvorsen, E.	1125
Hamano, H.	336
Han, C.-H.	114, 126
Han, J.S.	559
Han, M.D.	102, 1078
Hansson, J.	10, 563
Hao, Y.C.	1106
Hao, Y.L.	45, 49, 57, 960, 972
Haq, M.	616
Hara, M.	93, 1094
Harada, S.	756
Haraldsson, T.	10, 472, 563
Hashida, T.	401
Hashiguchi, G.	1071
Hassett, D.J.	577
Hata, H.	905
Hayakawa, T.	200
Hayashi, H.	344
Hayashi, T.	476
Hayashida, Y.	748
He, H.	1063
He, Q.	110
Heidari, A.	801, 837
Heinz, D.B.	1, 393
Hentz, S.	361
Hida, H.	377, 702
Hierlemann, A.	226, 535
Hierold, C.	312
Higashiyama, T.	702
Higurashi, E.	748
Hillmering, J.	563
Hirai, Y.	389, 417
Hirota, J.	968
Ho, C.P.	1032
Hodjat-Shamami, M.	789
Hokazono, K.	417
Honda, W.	756

Hone, J.	865, 869
Hong, S.	292
Hong, S.J.	1059
Hong, V.A.	1, 29, 33, 393, 1008
Hoorfar, M.	492
Hopcroft, M.A.	1
Horsley, D.A.	140, 146, 801, 837, 921
Hoshino, T.	512
Hoskinson, R.	952
Hotzen, I.	41, 172, 215, 897
Hou, Z.	1063
Houmadi, S.	150
Hsiai, T.	690
Hsiao, A.Y.	643
Hsu, C.-H.	519
Hsu, F.-M.	901
Hsu, W.	261
Huang, H.	779
Huang, H.-Y.	519
Huang, J.G.	45
Huang, L.X.	944
Huang, P.-C.	1067
Huang, W.-Y.	658
Hui, Y.	984
Humayun, M.S.	154
Hung, CP	234
Hurst, A.M.	865
Hwang, J.	559
Hwang, S.W.	168

I

Icard, B.	180
Ikeda, K.	463
Inoue, K.	373
Ishida, M.	686
Ishido, H.	257
Isono, Y.	369
Isozaki, A.	405, 706
Isozaki, N.	238
Ito, J.	480, 539
Ito, T.	748
Iwai, K.	551
Iwase, E.	81
Izyumin, I.I.	33

J

Jacot-Descombes, L.	632
Jacquet, F.	65
Jalabert, L.	328
Jeon, J.	940
Jeong, B.	752
Jeong, S.H.	1137

Jia, H.	698
Jiang, H.	332
Jin, W.	662
Jin, Y.F.	45, 49, 57, 960, 972
Joss, D.	632
Joyce, R.J.	944

K

Kan, T.	405
Kanao, K.	756
Kaneko, M.	431, 722, 1055
Kaneko, T.	670
Kang, D.	154, 340, 397
Kang, J.Y.	604
Kanno, I.	377, 702
Kao, W.-J.	581
Kaplan, D.	168
Karsten, S.L.	455
Kawano, T.	686
Kazama, R.	710
Kelm, J.M.	226
Kenny, T.W.	1, 29, 33, 393, 809, 1008
Keshavarzi, S.	413
Keum, H.	752
Khademolhosseini, F.	624
Khomenko, A.	616
Kilcawley, N.A.	504
Kim, A.	585, 1044
Kim, B.	1
Kim, B.J.	620, 662
Kim, C.-J.	6, 265, 1122
Kim, E.S.	122, 917
Kim, H.-L.	559
Kim, J.	158
Kim, J.	184
Kim, J.	488
Kim, J.	752
Kim, M.	292
Kim, M.J.	604
Kim, M.K.	1114
Kim, M.S.	1114
Kim, P.	845
Kim, S.	168, 752
Kim, S.	230
Kim, T.	292
Kim, T.G.	158
Kim, T.S.	651
Kim, Y.J.	1114
Kim, Y.-J.	354
Kim, Y.-J.	559
Kimata, M.	905
Kimbrell, T.S.	393

Kinahan, D.J. 192, 504, 523
Kirby, D.J. 944
Kitamura, N. 649
Kline, M.H. 33
Ko, K. 651
Ko, S.-D. 114
Kobayashi, T. 166, 1098
Kogure, T. 1082
Koh, K. 208
Konishi, S. 166
Korner, K. 551
Korvink, J.G. 97
Kosanic, D. 632
Koshi, T. 81
Kotera, H. 238, 455, 476, 694
Kottapalli, A.G.P. 500, 678, 889
Kozai, R. 245
Kozinda, A. 1133
Kraft, M. 881
Kucera, M. 718
Kumagai, S. 841
Kumar, V. 204, 885, 913
Kumar-Kantimahanti, A. 1004
Kung, Y.-C. 196
Kunishima, I. 968
Kuno, T. 369
Kuroda, Y. 373
Kuwano, H. 93, 1094
Kwon, H.-B. 559
Kwon, K.-Y. 616
Kwong, D.L. 45, 49, 57, 972

L

Lacaita, A.L. 932
Lacour, S.P. 760
Ladner, C. 361
Laghi, G. 932
Lai, W.-M. 901
Lal, A. 873, 1090, 1129
Lammel, G. 61
Lang, J.H. 85
Langfelder, G. 37, 849, 932
Lapatki, B. 736
Larramendy, F. 328
Lasri, T. 1012
Lawlor, D. 504
Le, C.P. 1125
Lee, B.-Y. 196
Lee, C. 106, 1032
Lee, C.J. 158, 651
Lee, C.-W. 223
Lee, D.W. 1110

Lee, G. 732
Lee, G.-B. 581, 658
Lee, H. 573
Lee, H.J. 158, 651
Lee, H.K. 1
Lee, J. 184
Lee, J. 219, 230, 628, 956
Lee, J. 752
Lee, J. 764
Lee, J.-H. 651
Lee, J.E.-Y. 797
Lee, K.Y. 1059
Lee, M. 292
Lee, S. 628
Lee, S. 865
Lee, S.-M. 559
Lee, S.H. 604
Lee, W.C. 8, 465
Lee, Y. 316, 764
Lee, Y.-K. 459
Lee, Y.R. 1059
Legrand, B. 150
Lei, C. 320, 857
Lell, J.A. 531
Lenk, G. 451
Leprince-Wang, Y. 960
Lesecq, S. 65
Leube, C. 893
Li, B. 543
Li, C.-S. 853, 988
Li, D. 320, 857
Li, D. 775
Li, J. 265
Li, M.-H. 853, 988
Li, M.G. 1102
Li, S. 130
Li, S.-S. 853, 988
Li, T. 726
Li, W. 616
Li, X. 130
Li, X. 447
Li, X. 612, 779, 785
Li, Y. 555, 968
Li, Y. 1048
Li, Z. 358
Liang, C. 459
Liang, T. 857
Lim, A.V. 276
Lim, C.J. 624
Lim, Y. 316, 764
Lin, C.-H. 272
Lin, C.-L. 581

Lin, J.X. 57
Lin, K.C.-H. 89
Lin, L. 22, 208, 261, 551, 801, 837, 928, 1133
Lin, Q. 488, 569, 869
Lin, T.-H. 249
Lin, W.-C. 89
Lin, Y.-S. 1032
Liu, A.Q. 45, 49, 57, 960, 972
Liu, C. 551
Liu, C.C. 624
Liu, D. 288
Liu, D. 459
Liu, G. 569
Liu, J. 110, 268
Liu, M. 304
Liu, M.C.-M. 89
Liu, P. 77
Liu, S. 714
Liu, S.-Y. 1075
Liu, T. 6, 1122
Liu, W. 312
Liu, X. 447
Liu, Y. 154, 340, 690
Liu, Y. 358, 467
Lo, C.-Y. 740
Lo, P.-H. 69
Longoni, A. 37, 932
Lu, Y. 140, 146
Lu, Z.-R. 698
Lujan, R. 877
Luo, B. 427
Luo, G. 1133
Luo, G.-L. 69
Luo, K. 77
Luo, Y. 1040
Lutz, M. 1

M

Ma, C.-W. 249, 1067
Ma, M.Y 427
Ma, S. 459
Mabuchi, K. 512
MacKinnon, N. 97
Madhaven, V. 1004
Maeda, R. 1098
Maeda, S. 1082
Maeda, Y. 253
Mahadeva, S.K. 861
Maharbiz, M.M. 682
Mahdavi, M. 204, 913
Mahmoud, M. 928
Mairiaux, E. 150

Majumdar, R.	296
Makimoto, N.	1098
Makino, H.	686
Maldonado-Camargo, L.	964
Maldonado-Garcia, M.	885
Mansour, R.R.	732
Mao, H.Y.	320, 857
Marelli, B.	168
Marigó, E.	1004
Marzouk, J.	1012
Matsui, D.	608
Matsui, R.	640
Matsuki, S.	397
Matsumoto, H.	589
Matsumoto, K.	144, 176, 212, 257, 405, 421, 640, 646, 670, 706, 710, 730, 783, 925
Matsunaga, Y.	354
McElwain, R.B.	944
McFarland, D.M.	752
Mehdizadeh, E.	204, 885
Melamud, R.	1
Meng, B.	102, 1078
Meng, E.	620, 662
Merzeau, P.	150
Mescheder, U.	413
Messana, M.W.	1
Meyhöfer, E.	238
Miao, J.M.	500, 678, 714, 889
Michaud, H.O.	760
Miki, N.	649
Millis, J.	188
Ming, A.J.	320, 857
Minh, L.V.	1094
Minh-Dung, N.	144, 212, 640, 670
Minotti, P.	932
Misaki, K.	905
Misawa, K.	1071
Mishra, R.	523
Mitsuya, H.	118
Miura, T.	476
Miwa, K.	118
Modarres-Zadeh, M.J.	909
Monzawa, T.	1055
Moon, H.	443, 547
Moore, J.	551
Moreau, J.E.	168
Moreira, E.E.	1036
Mori, N.	351, 589
Mori, S.	566
Morimoto, K.	1086
Morimoto, Y.	18, 351, 566, 589, 596, 666
Morishita, Y.	730
Morita, N.	748

Mukherjee, T.	813
Mukouyama, Y.	666
Murakami, R.	431
Murali, K.	154
Muramatsu, Y.	166
Murashige, K.	674
Murozaki, Y.	833
Murphy, J.	2

N

Nadig, S.	1129
Nagasawa, S.	308, 1082
Naing, T.L.	1024
Najafi, K.	821, 1051, 1141
Najar, H.	801
Nakahara, T.	694
Nakai, A.	257, 730
Nakamura, Y.	480
Nam, E.R.	465
Namazu, T.	373
Nansai, H.	480
Nauwelaers, B.	1016
Ng, E.J.	1, 29, 33, 393, 809, 1008
Ng, T.N.	877
Nguyen, C.T.-C.	976, 1024
Nguyen, H.D.	944
Nguyen, J.	435
Nguyen, N.-T.	825
Nguyen, T.-Q.	348
Nicolas, S.	65
Niklaus, F.	53, 280
Niroui, F.	85
Nishida, T.	1145
Nishihara, T.	1094
Nishiyama, K.	476
Nobukawa, A.	596
Noda, K.	925
Nogami, H.	748
Norford, L.K.	714
Norouzpour-Shirazi, A.	789
Notaguchi, M.	702
Nuckolls, C.	869
Nwankire, C.	192

O

O'Brien, G.J.	1, 393
O'Brien, K.P.	208
Ober, C.	873
Ochoa, M.	332, 817, 1044
Ogawa, S.	905
Oguchi, H.	93
Oh, S.-J.	651

Okabe, U. 300
Okada, R. 1082
Okada, Y. 439
Okano, T. 300
Okitsu, T. 463, 643, 666
Oku, H. 480
Olfat, M. 732
Olsen, T. 488
Omenetto, F.G. 168
Omi, S. 655
Ono, S. 118
Ono, T. 14, 344, 401
Onoe, H. 18, 26, 463
Oonishi, A. 968
Ortiz, S. 531
Osaki, T. 235, 336, 596
Ou, W. 320, 857
Ou, Y. 320, 857
Ouerghi, I. 361
Ozoe, K. 702

P

Pal, P. 385
Pan, S. 714
Pan, T. 543
Pan, Y. 940
Paprotny, I. 296
Park, H.S. 952
Park, I. 744
Park, I.S. 1059
Park, J. 8, 154
Park, J. 340, 690
Park, J.H. 585
Park, W.-T. 348
Partridge, A. 1
Patterson, A. 996
Paul, O. 162, 328, 636, 736, 893
Pei, R. 488
Peng, S. 551
Perahia, R. 944
Peroulis, D. 1020
Petrone, N. 869
Pettit, C. 752
Pfusterschmied, G. 718
Phan, H.-P. 825
Philippe, J. 361
Piazza, G. 980, 992, 996, 1000
Pitchappa, P. 1032
Polcawich, R.G. 146
Polunin, P. 1008
Popa, L.C. 1028
Pourkamali, S. 204, 885, 913

Pouydebasque, A.	65
Pozzi, A.	948
Przybyla, R.J.	921
Psychogiou, D.	1020
Pu, S.H.	881

Q

Qamar, A.	825
Qi, M.	22
Qian, Y.	1032
Qu, V.	1
Quang-Khang, P.	144, 670

R

Rachet, B.	632
Rahimi, R.	332, 817
Rais-Zadeh, M.	73, 793
Ramezany, A.	913
Rechenberg, R.	616
Reinecke, H.	413
Rey, P.	37
Rezaei Nejad, H.	492
Rinaldi, C.	964
Rinaldi, M.	984
Rismani Yazdi, S.	535
Rizzini, F.	849
Rocha, L.A.	1036
Rocheleau, T.O.	976, 1024
Rogers, J.A.	168
Roman, C.	312
Rottenberg, X.	1016
Roxhed, N.	280, 451
Roy, D.	527
Rozen, O.	146, 921
Ruiz-Díez, V.	718
Ruther, P.	162, 636, 893
Ruyack, A.	873, 1090

S

Saha, B.	628
Saito, D.	801
Sakuma, S.	200, 431, 833, 1055
Sakurai, Y.	377
Salvetat, J.P.	150
Salvia, J.C.	1
Samiei, E.	492
Sammoura, F.	837, 928
Sánchez-Rojas, J.L.	718
Sander, C.	736, 893
Sando, S.	324
Sanghadasa, M.	1133
Sarkar, N.	732

Sasaki, M.	841
Sato, H.	1048
Sawada, R.	748
Sawahata, H.	686
Sawant, S.G.	936
Scheiblin, P.	361
Schmid, U.	718
Schmid, Y.	226
Schmidt, F.	736
Schwaerzle, M.	162
Scianmarello, N.	154
Ségard, B.-D.	354
Segovia-Fernandez, J.	1000
Sekiguchi, T.	480, 539
Sekine, R.	480
Sen, P.	527
Senesky, D.G.	284
Senkal, D.	29
Seo, M.-H.	114, 126
Serien, D.	328, 470, 589
Shadmani, A.	535
Shaffer, E.	632
Shahosseini, I.	1051
Shakeel, H.	771
Shang, J.	427
Shankar, A.	284
Shaw, S.	1008
Shelton, S.E.	921
Shen, C.	1133
Shen, H.-H.	519
Shen, Z.X.	960
Shi, L.	992
Shibata, M.	841
Shibata, S.	845
Shikida, M.	845
Shimokawa, F.	245, 253
Shimoyama, I.	144, 176, 212, 257, 405, 421, 640, 646, 670, 706, 710, 730, 783, 925
Shin, H.	316, 764
Shin, K.-S.	604
Shinomiya, H.	968
Shintaku, H.	238, 455, 476, 694
Shkel, A.A.	917
Shkel, A.M.	29, 805
Shmulevich, S.	41, 172, 215, 897
Shoji, S.	480, 539
Shoshani, O.	1008
Shunmugam, M.	1004
Sinani, M.D.	1020
Singh, N.	996, 1032
Sletten, E.M.	85
Smith, G.L.	146
Smith, R.L.	188

Sobreviela, G.	1004
Sochol, R.D.	551
Son, Y.	158
Song, K.	459
Song, K.-Y.	1086
Song, Q.H.	960
Song, Y.-H.	114
Song, Z.	304
Soon, B.W.	996
Sorenson, L.D.	944
Soundara-Pandian, M.	1004
Stehle, J.	393
Stemme, G.	53, 280, 451
Stoeber, B.	861, 952
Stojanovic, M.N.	488
Strachan, S.	1008
Strathearn, D.	732
Su, Z.	1078
Subramaniyan, S.P.	455
Sugano, K.	369, 608
Sugiyama, T.	1071
Sun, C.	775
Sun, Y.	381, 435
Sun, Y.	472
Sung, W.-L.	423, 901
Suria, A.J.	284
Suzuki, C.	646
Suzuki, H.	300, 439
Suzuki, M.	1145
Suzuki, T.	245, 253
Suzuki, Y.	1086, 1098
Swager, T.M.	85
Syms, R.R.A.	134, 288
Syu, T.	242

T

Tabata, O.	389, 417, 608
Tabrizian, R.	789
Tai, Y.-C.	154, 340, 397, 690
Takagawa, Y.	905
Takahashi, H.	212, 257, 405, 706, 710
Takahashi, K.	655
Takahashi, T.	1145
Takahata, K.	1040
Takahata, T.	212, 257, 421, 640, 646, 670, 710, 925
Takane, K.	373
Takao, H.	245, 253
Takasawa, S.	242
Takeda, N.	480, 539
Takei, A.	829
Takei, K.	756
Takei, Y.	640, 646, 670, 783

Takeuchi, S. 8, 18, 235, 328, 336, 351, 463, 465, 470, 566, 589, 596, 636, 643, 666
Takigawa-Imamura, H. 476
Tan, Q.L. 857
Tanaka, M. 686
Tang, C. 459
Tang, H. 698
Tang, H.-Y. 140, 146
Tang, L.C. 320, 857
Tao, H. 168
Tarhan, M.C. 455
Tatar, E. 813
Tay, F.E.H. 106
Tay-Wee-Song, C. 1004
Tayebi, N. 208
Teixidor, J. 760
Terao, K. 245, 253
Ternyak, O. 897
Thanh-Vinh, N. 144, 176, 925
Théron, D. 150
Tilmans, H.A.C. 1016
Tixier-Mita, A. 354
Tocchio, A. 849
Toda, M. 14, 344, 401
Tomizawa, Y. 968
Tomoike, F. 235
Tonooka, T. 235, 596
Torres-Díaz, I. 964
Torunbalci, M.M. 409
Toshiyoshi, H. 354, 555, 1071
Toya, K. 968
Triantafyllou, M.S. 500, 678, 889
Tripathy, A. 527
Truong, B.D. 1125
Tsai, C.-H.D. 722
Tsai, J.M.L. 73
Tseng, F.-G. 223
Tsuchiya, T. 389, 417, 608
Tsugane, M. 439

U
Ueda, M. 1094
Uesugi, A. 389
Uetsuki, M. 905
Uranga, A. 1004

V
Vakakis, A.F. 752
Veale, M. 551
Velez, C. 964
Vembadi, A. 504
Vo Doan, T.T. 1048
Voigt, A. 97

W

Walter, B.	150
Walus, K.	861
Wang, A.I.	85
Wang, B.	593
Wang, C.	365, 569, 869
Wang, C.	435
Wang, H.-S.	89
Wang, J.	785
Wang, K.	658
Wang, P.	877
Wang, S.	1
Wang, S.	1028
Wang, W.	77
Wang, W.	77, 358, 467, 593
Wang, W.	948
Wang, W.B.	857
Wang, X.	110, 268, 1063
Wang, X.	130
Wang, Y.	77
Wang, Y.	122
Wang, Y.	320, 857
Wang, Y.-L.	581
Wang, Z.	268
Wang, Z.	304
Wang, Z.	484
Wang, Z.	877
Watanabe, R.	212
Weber, A.J.	616
Wei, X.	573
Wei, Y.	435
Weinstein, D.	1028
Weitz, D.A.	8
Wijethunga, P.A.L.	547
Wijngaart, van der, W.	10, 472, 563
Wilson, J.C.	885
Wood, G.S.	881
Wu, D.	304
Wu, J.H.	45, 57
Wu, J.-K.	223
Wu, M.C.	976
Wu, P.C.	960
Wu, S.-Y.	261
Wu, T.	484
Wu, W.	467
Wu, X.	698
Wu, X.	1063
Wu, X.	1110
Wu, Z.	793
Wu, Z.	1137

X

Xiao, D. 1063
Xie, H. 948
Xie, J. 881
Xie, S. 632
Xing, X. 508
Xiong, J.J. 320, 857
Xu, K. 775
Xu, P. 612, 779, 785
Xu, R. 459
Xu, T. 340
Xu, Y. 340
Xue, F. 714
Xue, G. 14

Y

Yabuki, M. 968
Yagami, S. 377
Yama, G. 1, 393
Yamada, S. 118
Yamagiwa, S. 686
Yamaguchi, T. 841
Yamamoto, G. 401
Yamamoto, Y. 756
Yamane, H. 308
Yamanishi, Y. 242, 655
Yamauchi, K. 369
Yan, X. 22
Yang, B. 110, 268
Yang, C. 208, 261, 801, 837, 928
Yang, C.S. 110, 268
Yang, F. 77, 358
Yang, J. 569
Yang, M. 168
Yang, R. 877
Yang, S. 304
Yang, W. 573
Yang, Y. 1, 29, 33, 393, 809, 1008
Yang, Y.-J. 249, 1067
Yang, Z. 714
Yang, Z.C. 45, 49, 57, 960, 972
Yao, D.-J. 519
Yasuga, H. 10
Yeh, Y.-C. 33
Yip, M.-C. 767
Yobas, L. 508
Yokokawa, R. 238, 455, 476, 694
Yokoyama, T. 1094
Yoneoka, S. 1
Yoo, J.-Y. 126
Yoon, D.H. 480, 539

Yoon, E.-S. 158, 651
Yoon, J.-B. 114, 126
Yoon, Y. 944
Yoon, Y.-H. 114
Yoshida, K. 26
Yoshida, R. 93
Yoshida, S. 18, 328, 589
Yoshikawa, Y. 1145
Yu, B. 1078
Yu, F. 612, 785
Yu, F. 940
Yu, H. 612, 775
Yu, J. 869
Yu, L. 620, 662
Yu, S. 775
Yu, W. 817
Yuan, J. 496
Yuan, W.Z. 1106
Yun, G.-S. 114
Yun, K.-S. 651

Z

Zainuddin, A.A. 1004
Zarudniev, M. 65
Zhang, B. 324, 600
Zhang, D. 77, 358
Zhang, H. 365, 1078
Zhang, H.M. 1106
Zhang, H.X. 102
Zhang, J. 1102
Zhang, L. 358, 467
Zhang, Q. 122
Zhang, R. 593
Zhang, T.S. 1044
Zhang, W. 960
Zhang, X. 340, 690
Zhang, X. 948
Zhang, X.-S. 365, 1078
Zhang, Y. 714
Zhao, C. 447
Zhao, C. 459
Zhao, C. 881
Zhao, F. 276
Zhao, L. 122
Zhao, W. 593
Zhao, X. 365
Zheng, D. 779
Zheng, L. 18
Zheng, Y. 435
Zhou, X. 472
Zhu, F.Y. 102, 365
Zhu, H. 797

Zhu, J.	488
Zhu, W.M.	960
Zhu, Y.	569, 869
Zhu, Y.	825
Ziaie, B.	332, 585, 817, 1044
Zohar, Y.	459

9781479979561

2015 28th IEEE International Conference on Micro Electro Mechanical Systems (MEMS 2015)

Estoril, Portugal
18-22 January 2015

IEEE Catalog Number: CFP15MEM-POD
ISBN: 978-1-47997-956-1

2015 28th IEEE International Conference on Micro Electro Mechanical Systems

(MEMS 2015)

Estoril, Portugal
18-22 January 2015

Pages 569-1148

IEEE Catalog Number: CFP15MEM-POD
ISBN: 978-1-4799-7956-1

Copyright © 2015 by the Institute of Electrical and Electronic Engineers, Inc
All Rights Reserved

Copyright and Reprint Permissions: Abstracting is permitted with credit to the source. Libraries are permitted to photocopy beyond the limit of U.S. copyright law for private use of patrons those articles in this volume that carry a code at the bottom of the first page, provided the per-copy fee indicated in the code is paid through Copyright Clearance Center, 222 Rosewood Drive, Danvers, MA 01923.

For other copying, reprint or republication permission, write to IEEE Copyrights Manager, IEEE Service Center, 445 Hoes Lane, Piscataway, NJ 08854. All rights reserved.

This publication is a representation of what appears in the IEEE Digital Libraries. Some format issues inherent in the e-media version may also appear in this print version.

IEEE Catalog Number: CFP15MEM-POD
ISBN 13:   978-1-4799-7956-1

Additional Copies of This Publication Are Available From:

Curran Associates, Inc
57 Morehouse Lane
Red Hook, NY 12571 USA
Phone: (845) 758-0400
Fax: (845) 758-2633
E-mail: curran@proceedings.com
Web: www.proceedings.com

MEMS 2015 Program Schedule

Monday, 19 January

07:55 **Opening and Welcome Address**
Jürgen Brugger, *EPFL Lausanne, SWITZERLAND*
Wouter van der Wijngaart, *KTH Royal Institute of Technology, SWEDEN*

Invited Plenary Speaker I
Session Chairs:
J. Brugger, *EPFL Lausanne, SWITZERLAND*
H. Toshiyoshi, *University of Tokyo, JAPAN*

08:15 **THE LONG PATH FROM MEMS RESONATORS TO TIMING PRODUCTS** 1
E. Ng[1], Y. Yang[1], V.A. Hong[1], C.H. Ahn[1], D.B. Heinz[1], I. Flader[1], Y. Chen[1], C.L.M. Everhart[1], B. Kim[2],
R. Melamud[2], R.N. Candler[2], M.A. Hopcroft[2], J.C. Salvia[2], S. Yoneoka[2], A.B. Graham[2], M. Agarwal[2],
M.W. Messana[2], K.L. Chen[2], H.K. Lee[2], S. Wang[2], G. Bahl[2], V. Qu[2], C.F. Chiang[2], **Thomas W. Kenny, Ph.D.**[1],
A. Partridge[3], M. Lutz[3], G. Yama[4] and G.J. O'Brien[4]
[1]Stanford University, USA, [2]PhD Alumni of Stanford University, USA, [3]SiTime Inc, USA, and [4]Robert Bosch RTC, USA

Session I - Micro and Nanofluidics
Session Chairs:
B. Stoeber, *University of British Columbia, CANADA*
F.G. Tseng, *National Tsing Hua University, TAIWAN*

09:30 **GRAPHENE OXIDE MEMBRANES FOR PHASE-SELECTIVE MICROFLUIDIC FLOW CONTROL** 2
J. Gaughran, D. Boyle, J. Murphy, and J. Ducrée   * Award Nominee
Dublin City University, IRELAND

We investigate the unique properties of Graphene Oxide (GO) as a barrier selective to the solvent and state of aggregation of the fluid. To this end, we developed novel processes for the assembly of GO as membranes into polymeric microfluidic systems. We show that GO completely blocks pressurized air and organic solutions while it is permeable to water. These GO membranes are then employed as a flow control element in a microfluidic system.

09:45 **STRUCTURE-BASED SUPERHYDROPHOBICITY FOR SERUM DROPLETS** 6
T. Liu and C.-J. Kim
University of California, Los Angeles, USA

We report that superhydrophobic (SHPo) surfaces based purely on surface structuring shows a robust super-repellency under a prolonged contact with serum droplet as an example of protein-rich biological fluids. In contrast, normal SHPo surfaces, which are based on surface chemistry and surface structuring, lose repellency and eventually get wetted by the same tests. This is the first report of a SHPo surface not degraded by a biological fluid.

10:00 **MICROFABRICATED LIQUID CHAMBER UTILIZING SOLVENT-DRYING
FOR IN-SITU TEM IMAGING OF NANOPARTICLE SELF-ASSEMBLY** 8
W.C. Lee[1,2], J. Park[3,4,5], D.A. Weitz[5], S. Takeuchi[1,2], and A.P. Alivisatos[3,4]
*[1]University of Tokyo, JAPAN, [2]Japan Science and Technology Agency (JST), JAPAN
[3]University of California, Berkeley, USA, [4]Lawrence Berkeley National Laboratory, USA, and [5]Harvard University, USA*

We present a microfabricated liquid-sample chamber for real-time TEM (Transmission Electron Microscopy) of nanoscale processes driven by liquid evaporation. The present chamber (TEM liquid-cell) uses intended leakage/failure in its bonding process in order to generate solvent-drying during in-situ TEM imaging. The captured real-time nanometer-scale TEM movies visualize critical steps in the self-assembly process of 2 dimensional nanoparticle arrays.

10:15 **SYNTHETIC MICROFLUIDIC PAPER** 10
J. Hansson, H. Yasuga, T. Haraldsson, and W. van der Wijngaart
KTH Royal Institute of Technology, SWEDEN

We demonstrate a polymer synthetic microfluidic paper with the aim to combine the high surface area of paper or nitrocellulose with the repeatability, controlled structure, and transparency of polymer micropillars for lateral flow devices. It consists of a dense, high aspect ratio, stiff polymer micropillar array with thin slanted pillars that are interlocked, and is manufactured with multidirectional UV lithography in Off-Stochiometry-Thiol-Ene-Epoxy polymer.

10:30 **Break & Exhibit Inspection**

Session II - Actuators
Session Chairs:
M. Kohl, *Karlsruhe Institute of Technology, GERMANY*
F. Niklaus, *KTH - Royal Institute of Technology, SWEDEN*

11:00 **ASSEMBLED COMB-DRIVE XYZ-MICROSTAGE WITH LARGE DISPLACEMENTS FOR LOW TEMPERATURE MEASUREMENT SYSTEMS** 14
G. Xue, M. Toda, and T. Ono
Tohoku University, JAPAN

In this research, we report the novel design, fabrication and testing of an assembled comb-drive XYZ-microstage that produces highly decoupled motions into X-, Y-, and Z-directions for the 3D scanning stage of magnetic resonance force microscopy. It is demonstrated that the assembled XYZ-microstage can achieve large displacements of 20.4 micrometre in X direction, 25.2 micrometre in Y direction and 58.5 micrometre in Z direction.

11:15 **PNEUMATIC BALLOON ACTUATOR WITH TUNABLE BENDING POINTS** 18
L. Zheng[1], S. Yoshida[1], Y. Morimoto[1,2], H. Onoe[1,2], and S. Takeuchi[1,2]
[1]University of Tokyo, JAPAN and [2]Japan Science and Technology Agency (JST), JAPAN

We propose a pneumatic balloon actuator capable of controlling its bending points. Local stiffness of the balloon actuator can be controlled by injected low-melting-point-alloy. The bending point can be changed depending on the position of the rigid low-melting-point-alloy. We believe that the proposed actuation mechanism will be useful in designing highly flexible actuators for soft robotics.

11:30 **SELF-LIFTING ARTIFICIAL INSECT WINGS VIA ELECTROSTATIC FLAPPING ACTUATORS** 22
X. Yan[1,2], M. Qi[1], and L. Lin[3]
[1]Beihang University, CHINA, [2]Collaborative Innovation Center of Advanced Aero-Engine, CHINA, and [3]University of California, Berkeley, USA

We present a self-lifting artificial insect wings using electrostatic actuation for the first time. Excited by a DC power source, biomimetic flapping motions have been generated to lift the artificial wings under an operation frequency of 50-70Hz. Three achievements have been accomplished: (1) first successful demonstration of electrostatic flying wings; (2) low power consumption as compared to other actuation schemes; and (3) self-adjustable rotating wing design to provide the lifting force.

11:45 **SELF-ASSEMBLED HYDROGEL MICROSPRING FOR SOFT ACTUATOR** 26
K. Yoshida and H. Onoe
Keio University, JAPAN

We found that the hydrogel microspring was formed by extruding sodium alginate pre-gel solution into calcium chloride using a bevel-tip microfluidic capillary. The formation of the microspring depends on the diameter of the capillary, flow rate and tip angle. As an example of soft actuator, the microspring including magnetic colloids was actuated by applying magnetic fields. We believe that the microspring will be used to various application including mechanical elements using soft materials.

12:00 **Lunch & Exhibit Inspection**

13:00 **Poster/Oral Session I**
Poster presentations are listed by topic category with their assigned number starting on page 14.

15:00 **Break & Exhibit Inspection**

Session III - Gyroscopes
Session Chairs:
F. Ayazi, *Georgia Institute of Technology, USA*
A. Seshia, *University of Cambridge, UK*

15:30   PARAMETRIC DRIVE OF A TOROIDAL MEMS RATE INTEGRATING GYROSCOPE DEMONSTRATING < 20 PPM SCALE FACTOR STABILITY 29
D. Senkal[1], E.J. Ng[2], V. Hong[2], Y. Yang[2], C.H. Ahn[2], T.W. Kenny[2], and A.M. Shkel[1]
[1]University of California, Irvine, USA and [2]Stanford University, USA

In this paper, we report parametric drive of a MEMS rate integrating gyroscope for reduction of drifts induced by drive electronics, resulting in < 20 ppm scale factor stability. Due to the parametric pumping effect, energy added to each (x & y) mode is proportional to the existing amplitude of the respective mode. As a result, errors associated with finding the orientation of the standing wave and x-y drive gain drift can be bypassed, demonstrating > 20x improvement in scale factor stability.

15:45   A 7PPM, 6°/HR FREQUENCY-OUTPUT MEMS GYROSCOPE 33
I.I. Izyumin[1], M.H. Kline[1], Y.C. Yeh[1], B. Eminoglu[1], C.H. Ahn[2],
V.A. Hong[2], Y. Yang[2], E.J. Ng[2], T.W. Kenny[2], and B.E. Boser[1]
[1]University of California, Berkeley, USA and [2]Stanford University, USA

We report the first frequency-output MEMS gyroscope to achieve <7 ppm scale factor accuracy and <6 deg/hr bias stability with a 3.24mm^2 transducer. By employing continuous-time mode reversal, the rate measurement is made insensitive to the resonant frequency of the transducer. The scale factor is almost entirely ratiometric; scale factor sensitivity to transducer and circuit parameters is significantly reduced compared to conventional open-loop and force-rebalance operating modes.

16:00   LARGE FULL SCALE, LINEARITY AND CROSS-AXIS REJECTION IN LOW-POWER 3-AXIS GYROSCOPES BASED ON NANOSCALE PIEZORESISTORS 37
S. Dellea[1], F. Giacci[1], A. Longoni[1], P. Rey[2], A. Berthelot[2], and G. Langfelder[1]
[1]Politecnico di Milano, ITALY and [2]CEA-Leti, FRANCE

The work presents 3-axis rate gyroscopes based on nanoscale piezoresistive readout and eutectic bonding between a bottom wafer, where the sensor is formed, and a cap wafer, where routing and pads are fabricated. The design features a central levered sense frame, to maximize symmetry, compactness and vibration rejection. Operation on a ±3000 dps full-scale shows competitive performance, with linearity errors <0.2% and cross-axis rejections 50x better than state-of-the-art consumer gyroscopes.

16:15   THE ELECTROMECHANICAL RESPONSE OF A SELF-EXCITED MEMS FRANKLIN OSCILLATOR 41
S. Shmulevich, I. Hotzen, and D. Elata
Technion - Israel Institute of Technology, ISRAEL

We present a self-excited MEMS Franklin oscillator, which responds in steady state vibrations when subjected to a dc voltage. The system is constructed from a floating rotor, which transfers charge between a source and drain electrodes. Current flows through the system only when the rotor is in transition. Surprisingly, at contact of the rotor with either the source or the drain electrodes, there is no current, and the charge transfer mechanism is essentially a recombination of opposite charges.

Session IV - Photonics
Session Chairs:
J. Conde, *Instituto Superior Tecnico, PORTUGAL*
C. Lee, *National University of Singapore, SINGAPORE*

16:30   A NANOMACHINED TUNABLE OSCILLATOR CONTROLLED BY ELECTROSTATIC AND OPTICAL FORCE 45
J. G. Huang[1,2,3], B. Dong[2,3], H. Cai[2], Y. D. Gu[3], J. H. Wu[1], T. N. Chen[1],
Z. C. Yang[4], Y. F. Jin[4], Y. L. Hao[4], D. L. Kwong[3], and A. Q. Liu[3]
[1]Xi'an Jiaotong University, CHINA, [2]Agency for Science, Technology and Research (A*STAR), SINGAPORE,
[3]Nanyang Technological University, SINGAPORE, and [4]Peking University, CHINA

We develop a miniaturized electrically tunable optomechanical oscillator, whose frequencies can be electrostatically tuned by as much as 10%. By taking advantage of the optical and the electrical spring, the oscillator achieves a high tuning sensitivity without resorting to mechanical tension. Particularly, the high-Q optical cavity greatly enhances the system sensitivity, making it extremely sensitive to the motional signal, which is often overwhelmed by spurious coupling or background.

16:45   NANO-OPTOMECHANICAL STATIC RANDOM ACCESS MEMORY (SRAM) 49
B. Dong[1,2], H. Cai[2], Y. D. Gu[2], Z. C. Yang[3], Y. F. Jin[3], Y. L. Hao[3], D. L. Kwong[2], and A. Q. Liu[1]
[1]Nanyang Technological University, SINGAPORE, [2]Agency for Science, Technology and Research (A*STAR),
SINGAPORE, and [3]Peking University, CHINA

We develop an on chip NEMS optomechanical SRAM, which is integrated with light modulation system on a single silicon chip. In particular, a doubly-clamped silicon beam shows bistability due to the non-linear optical gradient force generated from a ring resonator. The memory states are assigned with two stable deformation positions, which can be switched by modulating the control light's power with the integrated optical modulator.

**17:00 A LOW-POWER MEMS TUNABLE PHOTONIC RING RESONATOR
FOR RECONFIGURABLE OPTICAL NETWORKS** **53**

C. Errando-Herranz, F. Niklaus, G. Stemme, and K.B. Gylfason
KTH Royal Institute of Technology, SWEDEN

* Award Nominee

We experimentally demonstrate a low-power MEMS tunable photonic ring resonator with 10 selectable channels for wavelength selection in reconfigurable optical networks. The tuning is achieved by changing the geometry of a silicon slot-waveguide ring resonator by vertical electrostatic parallel-plate actuation. Our device provides static power dissipation below 0.1 μW, a wavelength tuning range of 1 nm, and a bandwidth of 0.1 nm, i.e. 10-8 watts per selectable channel, the lowest number reported.

17:15 NEMS OPTICAL CROSS CONNECT (OXC) DRIVEN BY OPTICL FORCE **57**

H. Cai[1], J. X. Lin[2], J. H. Wu[2], B. Dong[1,3], Y. D. Gu[1], Z. C. Yang[4], Y. F. Jin[4],
Y. L. Hao[4], D. L. Kwong[1], and A. Q. Liu[3]
[1]Agency for Science, Technology and Research (A*STAR), SINGAPORE,
[2]Xi'an Jiaotong University, CHINA, [3]Nanyang Technological University, SINGAPORE, and [4]Peking University, CHINA

We report a nano-silicon-photonic optical cross connect driven by optical gradient force, which demonstrates the all-optical OXC system on silicon. A switching time of 170 ns is experimentally demonstrated, which is much faster than that of conventional optical switches. In addition, the proposed switch system has the advantages of compact size (35 μm x 35 μm for switching element), high extinction ratio and low power consumption.

17:30 Adjourn for the Day

Tuesday 20, January

08:25 **Announcements**

Invited Plenary Speaker II
Session Chairs:
H. Toshiyoshi, *University of Tokyo, JAPAN*
J. Brugger, *EPFL Lausanne, SWITZERLAND*

08:30 **THE FUTURE OF MEMS SENSORS IN OUR CONNECTED WORLD** 61
Gerhard Lammel, Ph.D.
Bosch Sensortec GmbH, GERMANY

Session V - Micro-Optics
Session Chairs:
I.-J. Cho, *Korea Institute of Science and Technology, SOUTH KOREA*
B. Legrand, *LAAS-CNRS, FRANCE*

09:15 **FABRICATION AND CHARACTERIZATION OF A NEW VARIFOCAL LIQUID LENS WITH EMBEDDED PZT ACTUATORS FOR HIGH OPTICAL PERFORMANCES** 65
S. Nicolas[1], M. Allain[1], C. Bridoux[1], S. Fanget[1], S. Lesecq[1], M. Zarudniev[1], S. Bolis[2], A. Pouydebasque[2], and F. Jacquet[2]
[1]CEA, LETI, MINATEC Campus, FRANCE and [2]WAVELENS, FRANCE

This paper reports the fabrication and characterizations of a compact Varifocal microlens with an embedded MEMS actuator. Optical aperture is typically between 1.5 mm to 3 mm diameter with a total component thickness down to 400µm. High power efficiency (< 0.1 µW), high speed response time (down to 1 msec), high electro-optical performances (10 diopters optical power variation at 10V) and good optical quality (wavefront error lower than 50 nm) are reported through this paper.

09:30 **IMPLEMENTATION OF NANOPOROUS ANODIC ALUMINUM OXIDE LAYER WITH DIFFERENT POROSITIES FOR INTERFEROMETRIC RGB COLOR PIXELS AS HANDHELD DISPLAY APPLICATION** 69
P.-H. Lo[1], G.-L. Luo[2], and W. Fang[1]
[1]National Tsing Hua University, TAIWAN and [2]Asia Pacific Microsystems Inc., TAIWAN

This study employs the nanoporous anodic aluminum oxide (np-AAO) for interferometric modulation color pixels. Advantages of the proposed color-pixel are, (1) porosity of np-AAO layer can be adjusted to modulate therefractive index, (2) color-pixels of different porosities can be implemented on single np-AAO layer for different colors modulation, (3) the morphology of np-AAO with Al half-reflector could scatter reflect light to enhance view-angle, and (4) anti-stiction coating is not required.

09:45 **LOW-NOISE ALN-ON-SI RESONANT INFRARED DETECTORS USING A COMMERCIAL FOUNDRY MEMS FABRICATION PROCESS** 73
V.J. Gokhale[1], C. Figueroa[1], J.M.L. Tsai[2], and M. Rais-Zadeh[1]
[1]University of Michigan, USA and [2]Invensense Inc,, USA

This work presents the first measured results for resonant AlN-based IR detectors fabricated in a proprietary AlN MEMS process. Resonators fabricated in the first fabrication run achieved high electromechanical performance (Q of ~1400 at 115 MHz), an infrared responsivity of 10.7%/W, and low noise, as evidenced by an NEDT of 51 mK and an NEP of 52.7 pW/Hz^0.5. The resonators are fabricated in a hybrid MEMS/CMOS wafer level packaged die, allowing for CMOS-based routing and readout.

10:00 **Break & Exhibit Inspection**

Session VI - Novel Fabrication
Session Chairs:
A. Dietzel, *Technische Universität Braunschweig, GERMANY*
H. Moon, *University of Texas, Arlington, USA*

10:30 **HARD MASK FREE DRIE OF CRYSTALLINE SI NANOBARREL WITH 6.7NM WALL THICKNESS AND 50:1 ASPECT RATIO** 77
P. Liu, F. Yang, W. Wang, W. Wang, K. Luo, Y. Wang, and D. Zhang   *** Award Nominee**
Peking University, CHINA

Crystal Si barrel with wall thickness of 6.7nm and aspect ratio of 50:1 were fabricated using IC/MEMS compatible process. PR mask was generated by e-beam lithography then etched into Si by fluorine based DRIE directly, without hard mask transfer. To achieve high anisotropic etching, a model of ions transportation around barrels in DRIE was established and studied. This tactic with high repeatability and manufacturability provides an arsenal for the next generation of 3D nano devices fabrication.

10:45 **SELF-HEALING METAL WIRE USING AN ELECTRIC FIELD TRAPPING OF GOLD NANOPARTICLES FOR FLEXIBLE DEVICES** 81
T. Koshi and E. Iwase
Waseda University, JAPAN

* Award Nominee

We developed a self-healing metal wire for flexible devices. A cracked metal wire on a stretchable substrate can get its conductivity again by the self-healing ability using an electric field trapping of gold nanoparticles. First, we analyzed the electric field trapping. Next, we fabricated cracked wires on a glass substrate and verified the self-healing by experiments. Finally, we demonstrated the self-healing on the stretchable substrate to show a usefulness for flexible devices.

11:00 **CONTROLLED FABRICATION OF NANOSCALE GAPS USING STICTION** 85
F. Niroui, E.M. Sletten, P.B. Deotare, A.I. Wang, T.M. Swager, J.H. Lang, and V. Bulović
Massachusetts Institute of Technology, USA

Utilizing stiction, a common mode of failure in electromechanical systems, our work develops a method for the controlled fabrication of nanometer-thin gaps between electrodes. In this approach, through nanoscale force control, stiction promotes formation of nanogaps of controlled widths within the range as small as sub-15 nm. Our work demonstrates that through modifications of device design, the nanogaps can be optimized for applications in nanoelectromechanical and molecular devices.

11:15 **DIFFERENTIAL MICRO-PIRANI GAUGE FOR MONITORING MEMS WAFER-LEVEL PACKAGE** 89
Y.-C. Chen, W.-C. Lin, H.-S. Wang, C.-C. Fan, K.C.-H. Lin, B.C.S. Chou, and M.C.-M. Liu
Taiwan Semiconductor Manufacturing Company, TAIWAN

We proposed a multiple-sensor-solution, where two Pirani gauges were constructed under different pressures; one in sealed micro-cavity for measuring pressures and the other one in opened micro-cavity as a reference. The differential scheme compensates errors, allowing accurate pressure determinations, and it was successfully used in examining reliabilities and monitoring processes of wafer-level packages.

11:30 **MICROFABRICATED ELECTROSTATIC PLANAR LENS ARRAY AND EXTRACTORS FOR MULTI-FOCUSED ION BEAM SYSTEM USING IONIC LIQUID ION SOURCE EMITTER ARRAY** 93
R. Yoshida, M. Hara, H. Oguchi, and H. Kuwano
Tohoku University, JAPAN

To develop multi-focused ion beam system, we fabricated the electrostatic planar extractors and lenses, and integrated those with the field-emission ion source emitter array. Focusing ability of the integrated device was verified by confirming divergence angle reduction to ~70% of that without focusing effect. In addition, to find suitable ionic liquid for Si etching, we adopted various ionic liquid as an ion source and investigated etching characteristics by microscopy and mass spectrometry.

11:45 **NOVEL IONIC LIQUID - POLYMER COMPOSITE AND AN APPROACH FOR ITS PATTERNING BY CONVENTIONAL PHOTOLITHOGRAPHY** 97
N.A. Bakhtina[1], A. Voigt[2], N. MacKinnon[1], G. Ahrens[2], G. Gruetzner[2], and J.G. Korvink[1]
[1]University of Freiburg-IMTEK, GERMANY and [2]micro resist technology GmbH, GERMANY

* Award Nominee

We report a novel composite material based on an ionic liquid and a photoresist. In addition, an approach for the patterning of the composite material by conventional photolithography is introduced. The unique properties of the material are utilized for direct manufacturing of highly transparent, electrically conductive microcomponents.

12:00 **IEEE 2015 Andrew S. Grove Award Recipient**
Dr. Masayoshi Esashi, Ph.D.
Tohoku University, JAPAN

12:15 **Lunch & Exhibit Inspection**

13:15 **Poster/Oral Session II**
Poster presentations are listed by topic category with their assigned number starting on page 14.

15:15 **Break & Exhibit Inspection**

Session VII - Power & Energy I
Session Chairs:
H. Kim, *University of Utah, USA*
P. Woias, *Albert-Ludwig-University Freiburg, GERMANY*

15:30 TRIBOELECTRIFICATION BASED ACTIVE SENSOR FOR POLYMER DISTINGUISHING 102
B. Meng, X.L. Cheng, M.D. Han, H.T. Chen, F.Y. Zhu, and H.X. Zhang   * Award Nominee
Peking University, CHINA

We present a novel sensor to realize polymer distinguishing based on triboelectrification and electrostatic induction. Multiple cells of single friction layer and electrode are integrated on a flexible substrate. For different polymer groups, the friction layers can be selected according to the triboelectric serials. As an example, the distinguishing of PDMS, PE and PET has been well demonstrated by employing PI and PS friction layers, showing potential applications in robotics and industry use.

15:45 SKIN BASED FLEXIBLE TRIBOELECTRIC NANOGENERATORS
WITH MOTION SENSING CAPABILITY 106
L. Dhakar, F.E.H. Tay, and C. Lee   * Award Nominee
National University of Singapore, SINGAPORE

This paper presents a novel triboelectric nanogenerator (TENG) using outermost layer of human skin i.e. epidermis as an active triboelectric layer for device operation. The human skin also has an advantage of high tendency to lose electrons relative to PDMS, which leads to improved performance of triboelectric mechanism. TENG is also demonstrated as a wearable self-powered sensor to track human motion/activity.

16:00 FLEXIBLE TRIBOELECTRIC AND PIEZOELECTRIC COUPLING NANOGENERATOR
BASED ON ELECTROSPINNING P(VDF-TRFE) NANOWIRES 110
X. Wang, B. Yang, J. Liu, Q. He, H. Guo, C.S. Yang, and X. Chen
Shanghai Jiaotong University, CHINA

We fabricate and characterize a triboelectric and piezoelectric coupling nanogenerator based on MEMS technology which is more flexible and thickness controllable than conventional process fabricated. The electrospinning PVDF-TrFE nanofibers and MWCNTs doped PDMS films are function and friction layers. To characterize the sandwich-shaped nanogenerator's performance, its open-circuit output voltage and energy power density under two kinds of energy generation mechanism are tested and discussed.

16:15 HIGHLY RELIABLE MEMS RELAY WITH TWO-STEP SPRING SYSTEM
AND HEAT SINK INSULATOR FOR POWER APPLICATIONS 114
Y.-H. Yoon, Y.-H. Song, S.-D. Ko, C.-H. Han, G.-S. Yun, M.-H. Seo, and J.-B. Yoon
Korea Advanced Institute of Science and Technology (KAIST), SOUTH KOREA

This paper reports remarkably reliable MEMS relays having a unique two-step spring system and a heat sink insulator. The two-step spring system is designed to reduce Joule-heating by lowering contact resistance. The heat sink insulator is proposed for efficiently removing heat generated in the contact area. These two features are adopted in the MEMS relay for minimizing thermal damage in high current level, thus enhancing reliability significantly.

Session VIII - Power & Energy II
Session Chairs:
H. Kim, *University of Utah, USA*
P. Woias, *Albert-Ludwig-University Freiburg, GERMANY*

16:30 SOLIDIFIED IONIC LIQUID FOR HIGH-POWER VIBRATIONAL ENERGY HARVESTERS 118
S. Yamada[1], H. Mitsuya[2], S. Ono[3], K. Miwa[3], and H. Fujita[1]
[1]*University of Tokyo, JAPAN,* [2]*SAGINOMIYA SEISAKUSHO, Inc., JAPAN, and*
[3]*Central Research Institute of Electric Power Industry, JAPAN*

We propose a high power-output vibrational energy harvester based on ionic liquid. Ionic liquid enables very large capacitance (1.0-10 µF cm-2) on the electrode at bias voltage less than 1.9 V due to its extremely thin (~ 1 nm) electrical double layer. By mechanical squeezing and drawing the ionic liquid solidified with a polymer addictive between a pair of electrodes at 15 Hz, we stably obtained the current output of 22 µAp-p cm-2 at 1.5 V.

16:45 FERROFLUID LIQUID SPRING FOR VIBRATION ENERGY HARVESTING 122
Y. Wang, Q. Zhang, L. Zhao, and E.S. Kim
University of Southern California, Los Angeles, USA

We developed a ferrofluid liquid spring to suspend a magnet array for harvesting vibration energy. For the first time, the concept of ferrofluidbased suspension is demonstrated for low resonant frequency and high reliability. A ferrofluid liquid spring has reduced the resonant frequency of a microfabricated electromagnetic energy harvester to around 340 Hz. 36 nW is delivered into a load of 2.3 Ω from 7 g acceleration.

17:00 **AN ELECTROSTATIC ENERGY HARVESTER EXPLOITING VARIABLE-AREA WATER ELECTRODE BY RESPIRATION** 126

M.-H. Seo, D.-H. Choi, C.-H. Han, J.-Y. Yoo, and J.-B. Yoon

Korea Advanced Institute of Science and Technology (KAIST), SOUTH KOREA

* **Award Nominee**

This paper reports an electrostatic energy-harvester exploiting water-layer formed by respiration as a variable-area electrode. We discover that electrically conductive water-layer (~45 Mohms/sq under 5 V) is instantly formed on a silicon-dioxide surface by exhaled-breath. We adopt this layer as variable capacitive electrodes for electrostatic energy-harvesting. The capacitance change was evaluated using a theoretical modeling and finite-element-method (FEM) simulation, and theoretical power-generation was estimated (~2 μW/cm^2 at 1 V). We then fabricated the prototype device and verified the capacitance change experimentally. Finally, the prototype showed charging and discharging characteristics by respiration successfully for being used as an energy-harvester driven by human-breath solely.

17:15 **SILICON ANODE SUPPORTED BY CARBON SCAFFOLD FOR HIGH PERFORMANCE LITHIUM ION MICRO-BATTERY** 130

X. Li[1,2], X. Wang[1,2], and S. Li[1,2]

[1]*Tsinghua National Laboratory for Information Science and Technology, CHINA and* [2]*Tsinghua University, CHINA*

This paper reports a novel Si/void/C anode for lithium ion batteries, with high specific capacity and superior cyclability. SiNPs at anode make a significant contribution to specific capacity increase. Nano void spaces provide enough space for expansion and contraction of SiNPs to ensure a long cycle life. The porous carbon scaffold is obtained from a photoresist (SU-8) with SiNPs templates. As such, it's possible to implement direct prototyping of three dimensional (3D) micro-battery on chip.

17:30 **Adjourn for the Day**

Wednesday 21, January

08:25 Announcements

Invited Plenary Speaker III
Session Chairs:
W. van der Wijngaart, *KTH - Royal Institute of Technology, SWEDEN*
X. Wang, *Tsinghua University, CHINA*

08:30 STATUS AND FUTURE TRENDS OF THE MINIATURIZATION OF MASS SPECTROMETRY 134
Richard R.A. Syms, Ph.D.
Imperial College London, UK

Session IX - Acoustic Sensors
Session Chairs:
Y.-K. Kim, *Seoul National University, SOUTH KOREA*
T. Seki, *OMRON Corporation, JAPAN*

**09:15 SHORT-RANGE AND HIGH-RESOLUTION ULTRASOUND IMAGING
USING AN 8 MHZ ALUMINUM NITRIDE PMUT ARRAY** 140
Y. Lu[1], H.-Y. Tang[2], S. Fung[1], B.E. Boser[2], and D.A. Horsley[1]   * Award Nominee
[1]University of California, Davis, USA and [2]University of California, Berkeley, USA

We demonstrate short-range (~mm) and high-resolution (<100μm) imaging based on piezoelectric micromachined ultrasonic transducers (PMUTs) and a 1.8V interface ASIC. The PMUTs use piezoelectric Aluminum Nitride (AlN), which has the advantages of low-temperature (<400°C) deposition and compatibility with CMOS fabrication but has a relatively low piezoelectric constant (e31=-0.5C/m2), making detection of ultrasound signals from tiny (50μm) PMUTs a challenging task.

**09:30 MEASUREMENT OF SURFACE ACOUSTIC WAVES PROPAGATION
USING A PIEZORESISTIVE CANTILEVER ARRAY** 144
N. Minh-Dung, P. Quang-Khang, N. Thanh-Vinh, K. Matsumoto, and I. Shimoyama
University of Tokyo, JAPAN

We presents an approach for measuring the propagation of surface acoustic waves (SAW), using a piezo-resistive cantilever array. SAW was measured by cantilevers designed at the liquid-substrate interaction area by utilizing the structure of liquid-cantilever-air, which is highly sensitive and is able to measure acoustic wave at high frequency. Experiment results demonstrate that the measurable range was from 0.1 MHz up to 100 MHz.

**09:45 BROADBAND PIEZOELECTRIC MICROMACHINED ULTRASONIC
TRANSDUCERS BASED ON DUAL RESONANCE MODES** 146
Y. Lu[1], O. Rozen[1], H.-Y. Tang[2], G.L. Smith[3], S. Fung[1], B.E. Boser[2], R.G. Polcawich[3], and D.A. Horsley[1]
[1]University of California, Davis, USA, [2]University of California, Berkeley, USA, and [3]US Army Research Lab, USA

We demonstrate broadband PZT pMUTs that achieve a 97% fractional bandwidth by utilizing a thinner structure excited at two adjacent mechanical vibration modes. A front side XeF2 release process reduces fabrication cost and allows higher fill factor relative to pMUTs requiring through wafer DRIE. To reduce the frequency variations that result from non-uniform or inaccurately-controlled release etching, a 4μm thick metal layer defines the effective boundary of each pMUT.

**10:00 WHEN CAPACITIVE TRANSDUCTION MEETS THE THERMOMECHANICAL LIMIT:
TOWARDS FEMTO-NEWTON FORCE SENSORS AT VERY HIGH FREQUENCY** 150
S. Houmadi[1,2], B. Legrand[3,4], J.P. Salvetat[2], B. Walter[5], E. Mairiaux[5], J.P. Aimé[1], D. Ducatteau[5],
P. Merzeau[2], L. Buisson[2], J. Elezgaray[1], D. Théron[5], and M. Faucher[5]
*[1]CNRS, CBMN, FRANCE, [2]CNRS, CRPP, FRANCE, [3]CNRS, LAAS, FRANCE,
[4]University de Toulouse, FRANCE, and [5]IEMN, CNRS, FRANCE*

We show that the capacitive transduction associated with a microwave detection scheme achieves the measurement of the thermomechanical noise spectrum of high-frequency (>10 MHz) high-stiffness (>105 N/m) microresonators, reaching the outstanding displacement resolution of 1fm/√Hz. This paves the way for vibrating sensors with exquisite force resolution in the fN/√Hz range, enabling large-bandwidth measurements of mechanical interactions at small scale and rheology of fluids at high frequency.

10:15 Break & Exhibit Inspection

Session X - Medical Microdevices
Session Chairs:
A. Hierlemann, *ETH Zurich, SWITZERLAND*
W. Li, *Michigan State University, USA*

10:45 MEMS OXYGEN TRANSPORTER TO TREAT RETINAL ISCHEMIA 154
D. Kang[1], K. Murali[2], N. Scianmarello[1], J. Park[1], J.H.-C. Chang[1], Y. Liu[1],   * **Award Nominee**
K.-T. Chang[1], Y.-C. Tai[1], and M.S. Humayun[2]
[1]California Institute of Technology, USA and [2]University of Southern California, Los Angeles, USA

A paradigm shift of treating diabetic retinopathy is proposed in the sense of using MEMS devices to bring more oxygen to retina. A passive MEMS oxygen transporter was designed, built, and tested both in mechanical models and pig eyes to confirm its feasibility. The results predict that the proposed approach can even treat a complete retinal ischemia, although our current on-going live animal experiments must finish before going for human trials.

**11:00 A NEW MONOLITHICALLY INTEGRATED MULTI-FUNCTIONAL MEMS
NEURAL PROBE FOR OPTICAL STIMULATION AND DRUG DELIVERY** 158
Y. Son[1], H.J. Lee[1], J. Kim[1], C.J. Lee[1], E.-S. Yoon[1], T.G. Kim[2], and I.-J. Cho[1]
[1]Korea Institute of Science and Technology (KIST), SOUTH KOREA and [2]Korea University, SOUTH KOREA

We present a monolithically integrated multifunctional MEMS neural probe by integrating both an embedded microfluidic channel for drug delivery and a SU-8 waveguide for optical stimulation using controlled glass reflow process. Using our multifunctional neural probe, we conducted successful in vivo experiments and recorded neural spike signals from individual neurons with a good SNR. In this work, we present distinctive changes in neural signals induced by both optical and chemical stimulation.

**11:15 MINIATURIZED 3×3 OPTICAL FIBER ARRAY FOR OPTOGENETICS WITH INTEGRATED
460 NM LIGHT SOURCES AND FLEXIBLE ELECTRICAL INTERCONNECTION** 162
M. Schwaerzle, P. Elmlinger, O. Paul, and P. Ruther
University of Freiburg-IMTEK, GERMANY

We report on the design, fabrication, assembly, and optical and thermal characterization of a novel MEMS-based optical probe array for optogenetic research in neuroscience. Nine high-efficiency light emitting diodes (LED) are integrated as a 3×3 array in a micromachined silicon (Si) housing that ensures the mechanical stability and precise alignment of the nine 5-mm-long optical fibers. The overall housing volume of less than 2.2 mm^3 is compatible with chronic implantation.

**11:30 FLEXIBLE END-EFFECTOR INTEGRATED WITH SCANNING ACTUATOR AND OPTICAL
WAVEGUIDE FOR ENDOSCOPIC FLUORESCENCE IMAGING DIAGNOSIS** 166
Y. Muramatsu, T. Kobayashi, and S. Konishi
Ritsumeikan University, JAPAN

This paper presents a flexible end-effector integrated with a scanning actuator and optical waveguide for endoscopic fluorescence imaging diagnosis. Pneumatic balloon actuator (PBA) is used as the scanning actuator for the end-effector in consideration of its soft and safe features. SU-8 optical waveguides are integrated onto the PBA structure made of PDMS. Our developed device has successfully scanned, excited, and detected fluorescence beads distributed in a pseudo tissue.

**11:45 FULLY IMPLANTABLE AND RESORBABLE WIRELESS MEDICAL
DEVICES FOR POSTSURGICAL INFECTION ABATEMENT** 168
H. Tao[1], S.W. Hwang[2], B. Marelli[3], B. An[3], J.E. Moreau[3], M. Yang[3],
M.A. Brenckle[3], S. Kim[2], D. Kaplan[3], J.A. Rogers[2], and F.G. Omenetto[3]
*[1]Shanghai Institute of Microsystem and Information Technology, CAS, CHINA,
[2]University of Illinois, Urbana-Champaign, USA, and [3]Tufts University, USA*

We present a therapeutic application of a microfabricated implantable and resorbable medical device by demonstrating in vivo elimination of bacterial infection by wireless activation of the device after implantation. The device disappears upon its completion, requiring no retrieval.

12:00 MEMS 2016 Announcement

12:15 Lunch & Exhibit Inspection

13:15 Poster/Oral Session III
Poster presentations are listed by topic category with their assigned number starting on page 14.

15:15 Break & Exhibit Inspection

Session XI.a - Resonant Sensors

Session Chairs:
C. Nguyen, *University of California, Berkeley, USA*
L. Sorenson, *HRL Laboratories, LLC, USA*

15:30   TUNING THE FIRST INSTABILITY WINDOW OF A MEMS MEISSNER PARAMETRIC RESONATOR USING A LINEAR ELECTROSTATIC ANTI-SPRING 172
S. Shmulevich, I. Hotzen, and D. Elata
Technion - Israel Institute of Technology, ISRAEL

We demonstrate frequency tuning of the first instability window of a MEMS Meissner parametric resonator. We achieve parametric excitation by time-modulation of a negative electrostatic stiffness. In our device this negative electrostatic stiffness is not affected by motion. In contrast, all state of the art MEMS parametric resonators are either detrimentally affected by a nonlinear electrostatic stiffness, or are more sensitive to fabrication tolerances relative to our design.

15:45   A VISCOMETER BASED ON VIBRATION OF DROPLETS ON A PIEZORESISTIVE CANTILEVER ARRAY 176
N. Thanh-Vinh, K. Matsumoto, and I. Shimoyama
University of Tokyo, JAPAN

This paper reports a method to measure viscosity based on the resonant vibration of droplets on a piezoresistive cantilever array. We demonstrate that viscosity of small droplets (3 μL) can be estimated from the attenuation rate of the cantilever output during free-decay of the droplet vibration. Moreover, we show that the optimized location of a cantilever should be on the periphery of the contact area where the force change is largest.

16:00   HOLLOW MEMS MASS SENSORS FOR REAL-TIME PARTICLES WEIGHING AND SIZING FROM A FEW 10 NM TO THE μM SCALE 180
C. Hadji[1,2], C. Berthet[1,2], F. Baléras[1,2], M. Cochet[1,2], B. Icard[1,2], and V. Agache[1,2]
[1]University Grenoble, FRANCE and [2]CEA, LETI, MINATEC, FRANCE

We report hollow MEMS plate oscillators for mass sensing in liquid with an expected mass resolution of 3 femtograms. The performances reached by our sensors – 10,000-range Q factor and ppb-range frequency stability – make them amenable to individual particles metrology from a few 10 nanometers up to the micrometre diameter range. Our devices are operated in air inside a customized "plug and play" test platform and do not need to work in vacuum sealed package.

16:15   DEVELOPMENT OF MICROFLUIDIC RESONATORS VIA SILICON-ON-NOTHING TECHNIQUE 184
J. Kim and J. Lee
Sogang University, SOUTH KOREA

We report wafer-level batch fabrication of microfluidic resonators based on Silicon-on-Nothing (SON) structures resulting from high temperature argon annealing of silicon wafers with periodic cylindrical pits. Besides the process optimization of SON structures, elemental fabrication techniques such as planarization, device release, metallization, and packaging are developed. These techniques reduce fabrication cost and time significantly and enable switching of the resonator materials.

Session XI.b - Cell Handling

Session Chairs:
N. Tas, *University of Twente, THE NETHERLANDS*
W. Wang, *Peking University, CHINA*

15:30   A MICROFABRICATED, FLOW DRIVEN MILL FOR THE MECHANICAL LYSIS OF ALGAE 188
J. Millis, L. Connell, S.D. Collins, and R.L. Smith
University of Maine, USA

We report the micrfabrication, computer modeling and experimental testing of a novel, new means of mechanically lysing algae using a flow-driven grinding mill. The mill is demonstrated to have 96% efficiency in lysing the dinoflagellate genus Alexandrium, a neurotoxin producing algae, responsible for Red Tide and paralytic shellfish poisoning. Lysate DNA is demonstrated to be viable through successful PCR amplification using primers specific to Alexandrium.

15:45   CLUSTER SIZING OF CANCER CELLS BY RAIL-BASED SERIAL GAP FILTRATION IN STOPPED-FLOW, CONTINUOUS SEDIMENTATION MODE 192
M. Glynn, C. Nwankire, D. Kinahan, and J. Ducrée
Dublin City University, IRELAND

We have developed a centrifugal microfluidic strategy for the isolation and sizing analysis of multicellular clusters from a blood sample. The strategy is based on passing the sample over a size exclusion rail that gates entry of clusters to underlying bins, allowing estimation of clustering extent in the sample.

16:00   TUNABLE DIELECTROPHORESIS FOR SHEATHLESS 3D FOCUSING 196
Y.-C. Kung, D.L. Clemens, B.-Y. Lee, and P.-Y. Chiou
University of California, Los Angeles, USA

We report on a novel tunable insulator-based dielectrophoresis (TiDEP) for three-dimensional, sheathless, single-stream cell and bacteria focusing. For the first time, objects as small as sub-micron sized infectious bacteria are continuously focused in the center of a channel without sheath flows. Compared to prior DEP works, this new TiDEP can provide an extremely long DEP interaction distance to migrate small objects with weak DEP forces to a focused single stream at high speed flows.

16:15 CELL MANIPULATION METHOD BASED ON VIBRATIN-INDUCED LOCAL FLOW CONTROL IN OPEN CHIP ENVIRONMENT 200
T. Hayakawa, S. Sakuma, and F. Arai
Nagoya University, JAPAN

We present a novel cell manipulation method using vibration-induced local flow control in open-chip environment. By applying circular vibration to micropillars on a chip, whirling flow is induced around it. This phenomenon is theoretically analyzed considering an effect of convective flow. Based on it, we show two important applications; transport and 3D rotation of oocytes for 3D observation of oocyte. We succeeded in the transportation of oocytes with 25 µm/s and rotation with 184 degrees/s.

Session XII.a - Magnetic & Resonant Sensors
Session Chairs:
M. Rais-Zadeh, *University of Michigan, USA*
M. Sasaki, *Toyota Technological Institute, JAPAN*

16:30 ULTRA SENSITIVE LORENTZ FORCE MEMS MAGNETOMETER WITH PICO-TESLA LIMIT OF DETECTION 204
V. Kumar, M. Mahdavi, X. Guo, E. Mehdizadeh, and S. Pourkamali   * **Award Nominee**
University of Texas, Dallas, USA

This work presents ultra-high sensitivities for Lorentz Force resonant MEMS magnetometers enabled by internal thermal-piezoresistive vibration amplification. Up to 2400X sensitivity amplification has been demonstrated with the noise floor calculated to be as low as 18 pt/$\sqrt{Hz}$. This is by far the most sensitive MEMS Lorentz force magnetometer demonstrated to date.

16:45 HIGH FREQUENCY MICROWAVE ON-CHIP INDUCTORS USING INCREASED FERROMAGNETIC RESONANCE FREQUENCY OF MAGNETIC FILMS 208
K. Koh[1], D.S. Gardner[2], C. Yang[1], K.P. O'Brien[2], N. Tayebi[2], and L. Lin[1]
[1]*University of California, Berkeley, USA and* [2]*Intel Corporation, USA*

The fabrication and characterization of high frequency on-chip inductors using sputtered magnetic films with an improved frequency range is presented. Reducing the sputtering power in the deposition process was found to result in smoother film surfaces and stronger uniaxial magnetic anisotropy and increased the FMR of CoZrTaB from 1.48 GHz to 2.13 GHz. A magnetic-core, on-chip inductor was fabricated using the CoZrTaB films. Results have shown 150% higher inductance and a larger Q-factor up to 1.2 GHz as compared to an air-core inductor.

17:00 FUSION OF CANTILEVER AND DIAPHRAGM PRESSURE SENSORS ACCORDING TO FREQUENCY CHARACTERISTICS 212
R. Watanabe, N. Minh-Dung, H. Takahashi, T. Takahata, K. Matsumoto, and I. Shimoyama
University of Tokyo, JAPAN

This paper reports on an approach to measure sensitive barometric pressure by fusing a cantilever-based differential pressure sensor (DPS) and a commercial available diaphragm-based absolute pressure sensor (APS). At high frequency, the DPS can detect smaller absolute pressure change than the APS. At low frequency, the APS show absolute pressure of less drift than the DPS. By utilizing the DPS and APS at each advantageous frequency, we propose high sensitive measurement of barometric pressure.

17:15 DYNAMICALLY-BALANCED FOLDED-BEAM SUSPENSIONS 215
S. Shmulevich, I. Hotzen, and D. Elata
Technion - Israel Institute of Technology, ISRAEL

We present a design methodology and experimental evidence of a dynamically balanced folded-beam suspension. This suspension responds as a linear spring at the fundamental resonance, which is in sharp contrast to the response of standard folded-beam suspensions. The dynamic response of standard suspensions becomes strongly nonlinear for motions larger than the width of the flexure beams. The resonance response of the new dynamically-balanced suspension is linear over a wider range of motions.

Session XII.b - BioSensing
Session Chairs:
E. Iwase, *Waseda University, JAPAN*
R. Yokokawa, *Kyoto University, JAPAN*

16:30 MEMBRANE-BASED CHEMOMECHANICAL TRANSDUCER FOR THE DETECTION OF APTAMER-PROTEIN BINDING 219
J.-K Choi[1] and J. Lee[2]
[1]*Small Machines Incorporation, SOUTH KOREA and* [2]*Seoul National University, SOUTH KOREA*

We report a membrane-based chemomechanical transducer for the sensitive detection of surface molecular reaction through a highly reliable common mode rejection (CMR) technique. Chemomechanical transduction, originally based on the micro-cantilever, offers potential benefits: label-free assay, and real-time monitoring of molecular interaction via mechanical deformation. Here we show clear-cut detection of molecular binding using a membrane transducer fabricated with conventional MEMS technology.

16:45   **HIGHLY SENSITIVE SERS DIAGNOSIS FOR BACTERIA BY THREE DIMENSIONAL NANO-MUSHROOMS AND NANO-STARS-ARRAY SANDWICHED ON BACTERIAL AGGREGATION** 223

C.-W. Lee[1], J.-K. Wu[1], and F.-G. Tseng[1,2]

[1]National Tsing Hua University, TAIWAN and [2]Research Center for Applied Sciences, Academia Sinica, TAIWAN

This paper reports a highly sensitive SERS Diagnosis system by incorporating three dimensional Nano-Mushrooms and Nano-Stars-Array sandwiched on Bacterial Aggregation. Through the action of ACEOF and nano-mushroom/bacteria/nano-stars-array self-aggregation process, the signal can be much enhanced by 5 orders of magnitude in 5 minutes. Detection limit can approach 1 bacterium/ml from the analysis result.

17:00   **SIMULTANEOUS IMPEDANCE SPECTROSCOPY AND STIMULATION OF HUMAN IPS-DERIVED CARDIAC 3D SPHEROIDS IN HANGING-DROP NETWORKS** 226

S.C. Bürgel[1], Y. Schmid[1], I. Agarkova[2], D.A. Fluri[2], J.M. Kelm[2], A. Hierlemann[1], and O. Frey[1]

[1]ETH Zurich, SWITZERLAND and [2]InSphero AG, SWITZERLAND

We present a platform for in-situ electrical impedance spectroscopy (EIS) measurements and electrical stimulation of human iPS-derived cardiac 3D microtissue spheroids in hanging drop networks. Electrical stimulations and EIS measurements were performed in parallel through the same electrodes. Stimulation was performed with sine-wave signals. Our results reveal beating frequency modulation upon tuning the stimulation signal amplitude.

17:15   **DROPLET FLOW FOCUSING FOR MOLECULAR BINDING DETECTION** 230

S. Kim and J. Lee

Seoul National University, SOUTH KOREA

We report the detection of interfacial molecular binding via droplet generation in a flow focusing device. We introduce the detection of DNA hybridization based on the mode of droplet production caused by interfacial tension shift at the oil-water boundary. Our report includes a molecular protocol to functionalize the interface with single-strand (ss) DNA as receptor, and flow condition tuning for an unambiguous distinction of complementary binding.

17:30   **Adjourn for the Day**

19:00 -   **Conference Banquet**
22:00

Thursday 22, January

08:40   **Announcements**

Invited Plenary Speaker IV
Session Chairs:
X. Wang, *Tsinghua University, CHINA*
W. van der Wijngaart, *KTH - Royal Institute of Technology, SWEDEN*

08:45   **SEMICONDUCTOR IC PACKAGING, THE NEXT WAVE** 234
CP Hung, Ph.D.
Advanced Semiconductor Engineering Inc., TAIWAN

Session XIII - BioMEMS
Session Chairs:
T. Kawano, *Toyohashi University of Technology, JAPAN*
N. Roxhed, *KTH - Royal Institute of Technology, SWEDEN*

09:30   **ROTATIONAL CHAMBERS ON FLUIDIC CHANNELS FOR THE REPETITIVE
FORMATION OF OPTICALLY OBSERVABLE LIPID-BILAYER MEMBRANES** 235
F. Tomoike[1], T. Tonooka[1], T. Osaki[2], and S. Takeuchi[1]
[1]University of Tokyo, JAPAN and [2]Kanagawa Academy of Science and Technology, JAPAN

We develop a device adapted for repetitive formation of horizontal lipid bilayer membranes. This device enables simultaneous optical and electrophysiological measurements of the membranes. We integrated a rotational chamber on a fluidic channel via parylene micropores. The rotational motion enables us to form/reform the bilayer repeatedly. The formation of the bilayers was confirmed from the bilayer thickness and nanopore incorporation into the bilayer.

09:45   **MICROTUBULE SORTING WITHIN A GIVEN ELECTRIC
FIELD BY DESIGNING FLEXURAL RIGIDITY** 238
N. Isozaki[1], H. Shintaku[1], H. Kotera[1], E. Meyhöfer[2], and R. Yokokawa[1]
[1]Kyoto University, JAPAN and [2]University of Michigan, USA

We propose a method to control gliding directions of kinesin-propelled microtubules (MTs) corresponding to their flexural rigidity (EI). We prepared two kinds of MTs having different EI and their trajectories within an electric field were clearly separated. Therefore, this study demonstrated the EI-altered MTs can be workhorses to sort/concentrate various combinations of molecules with techniques of loading MTs with target molecules and capturing the sorted MTs.

10:00   **CARVING OF PROTEIN CRYSTAL BY HIGH-SPEED MICRO-BUBBLE
JET USING MICRO-FLUIDIC PLATFORM** 242
S. Takasawa[1], T. Syu[1], and Y. Yamanishi[1,2]
[1]Shibaura Institute of Technology, JAPAN and [2]Japan Science and Technology Agency (JST), JAPAN

This paper reports a novel processing method for protein crystal with electric-induced-bubble. This minimally invasive micro-processing method overcomes the difficulties of processing fragile material such as protein crystal under water. The combination of electrically-induced bubble knife and glass capillary provide effective carving of protein crystal by ablation crystal and suction of chips respectively. Also, we successfully made a new micro-fluidic platform for processing protein crystal.

10:15   **Break & Exhibit Inspection**

Session XIV - Tactile and Force Sensors
Session Chairs:
N. Miki, *Keio University, JAPAN*
E. Sarajlic, *SmartTip B.V., THE NETHERLANDS*

10:45 **A NOVEL CONFIGURATION OF TACTILE SENSOR TO ACQUIRE THE CORRELATION BETWEEN SURFACE ROUGHNESS AND FRICTIONAL FORCE** 245
R. Kozai, K. Terao, T. Suzuki, F. Shimokawa, and H. Takao
Kagawa University, JAPAN

We propose a novel configuration of MEMS tactile sensor that can interact with micro surface roughness at a high resolution are proposed and reported for quantification of the fingertip sense. Two-axis movements of the contactor-tip are independently detected by the two independent suspensions. In the evaluation experiments, correlation between the surface shape and the local frictional force were successfully obtained at the same time and at the same point for many samples for the first time.

11:00 **TUNNELING PIEZORESISTIVE TACTILE SENSING ARRAY FABRICATED BY A NOVEL FABRICATION PROCESS WITH MEMBRANE FILTERS** 249
C.-W. Ma, T.-H. Ling, and Y.-J. Yang
National Taiwan University, TAIWAN

In this work, a highly-sensitive tactile sensor array using the tunneling piezoresistive effect is presented. The sensing element, which is made of multi-wall carbon nanotubes and polydimethylsiloxane conductive polymer, was patterned with microdome structures by a novel fabrication process on a membrane filter substrate. The tunneling piezoresistive effects of the interlocked microdome structures with different MWCNT concentrations are demonstrated.

11:15 **A TACTILE SENSOR WITH THE REFERENCE PLANE FOR DETECTION ABILITIES OF FRICTIONAL FORCE AND HUMAN BODY HARDNESS AIMED TO MEDICAL APPLICATIONS** 253
Y. Maeda, K. Terao, T. Suzuki, F. Shimokawa, and H. Takao
Kagawa University, JAPAN

In this study, a highly sensitive tactile sensor with detection abilities of both human body hardness and frictional force is reported. Employing the new structure, hardness signal becomes less sensitive to the contact pressure, and hardness is stably measured even under an unstable contact force. Surface frictional force was successfully measured in real time. Also, shore A hardness was successfully measured in the range from 1HS to 54HS corresponding to the human organs.

11:30 **6-AXIS FORCE/TORQUE SENSOR FOR SPIKE PINS OF SPORTS SHOES** 257
H. Ishido, H. Takahashi, A. Nakai, T. Takahata, K. Matsumoto, and I. Shimoyama
University of Tokyo, JAPAN

This paper reports on the method of force measurement of spike pins for sports shoes. It is important to measure force acting on spike pins, because they are related to the increase of GRF (ground reaction force) in running. We fabricated a $2 \times 2 \times 0.3$ mm^3 size 6-axis force/torque sensor chip that can be embedded in spike pins. The sensor chip consists of 6 straight piezo-resistive beams. We calibrated the spike-pin-shaped sensor, and confirmed that 6-axis force/torque was able to be detected.

11:45 **Award Ceremony**

12:00 **IEEE MEMS 2015 Conference Adjourns**

Poster/Oral Presentations

M – Monday (13:00 - 15:00) **T** – Tuesday (13:15 - 15:15)

W – Wednesday (13:15 - 15:15)

Generic MEMS and Nanotechnologies
Generic MEMS and NEMS Manufacturing Techniques

M-001 3D PRINTED RF PASSIVE COMPONENTS BY LIQUID METAL FILLING 261

C. Yang[1], S.-Y. Wu[1,2], C. Glick[1], Y.S. Choi[3], W. Hsu[2], and L. Lin[1]

[1]University of California, Berkeley, USA, [2]National Chiao Tung University, TAIWAN, and
[3]Samsung Electronics Co., Ltd., SOUTH KOREA

We present a novel method to form three-dimensional (3D) micro-scale electrical devices by using 3D printing and a liquid-metal-filling technique. Various RF passive components including inductors, capacitors and resistors are fabricated and characterized as proof-of-concepts. This technique establishes an innovative way to form arbitrary 3D structures with highly efficient, labor-saving metallization process.

**T-002 A CONVENIENT METHOD TO FABRICATE MULTILAYER
INTERCONNECTIONS FOR MICRODEVICES** 265

J. Li, S. Chen, and C.-J. Kim

University of California, Los Angeles, USA

We report a new method to fabricate multilayer interconnection without wet or dry etching or deposition of insulating layers. Electrical connections are completed by merely depositing metal layers and anodizing them after lithographically defining a photoresist. Without the need to etch metal layers or deposit and pattern insulation layers, the overall process is simple, cheap, safe, and of low temperature.

**W-003 DEVELOPMENT OF AN ATMOSPHERIC PRESSURE AIR MICROPLASMA
JET FOR THE SELECTIVE ETCHING OF PARYLENE-C FILM** 268

H. Guo, J. Liu, Z. Wang, X.Z. Wang, X. Chen, B. Yang, and C. Yang

Shanghai Jiao Tong University, CHINA

We develop a novel and simple process device based on the atmospheric pressure air microplasma jet for the selective etching of parylene-C film. The main feature of this process device is that it can be easily integrated with roll-to-roll systems for large-scale manufacturing of flexible electronic devices due to its operating at ambient conditions. In order to realize the selective etching, a quartz glass microtube (100 μm I.D) is employed to fabricate the microplasma jet source.

**M-004 ELECTROPLATED STENCIL REINFORCED WITH ARCH STRUCTURES
FOR PRINTING FINE AND LONG CONDUCTIVE PASTE** 272

P.-H. Chen and C.-H. Lin

National Sun Yat-sen University, TAIWAN

This study presents an electroplated stencil reinforced with arch structures and a surrounding buffer reservoir for printing conductive paste of fine and long lines. The developed reinforced stencil successfully solves the problems came with the conventional stencil structure including limited printable line width and ease of facture. This work presents a novel process for fabricating a thin yet robust MEMS-based stencil by using two AZ4620 layers and one SU-8 layer as the electroplating molds.

**T-005 FABRICATION AND CHARACTERIZATION OF SINGLE CRYSTALLINE
4H-SIC MEMS DEVICES WITH N-P-N HOMOEPITAXIAL STRUCTURE** 276

F. Zhao and A.V. Lim

Washington State University, Vancouver, USA

We report single crystalline 4H-SiC MEMS with homoepitaxial n-p-n structure. Single crystalline fully exploits the superior material properties of SiC for operations in harsh environments. The n-p-n structure makes electrostatic actuation applicable which is essentially important for applications of resonators and actuators to sensor devices, and extends the capability of monolithic integration between SiC MEMS and electronic devices/circuits with n-p-n configurations such as BJTs and MOSFETs.

**W-006 INTEGRATION OF DISTRIBUTED GE ISLANDS ONTO SI WAFERS BY ADHESIVE
WAFER BONDING AND LOW-TEMPERATURE GE EXFOLIATION** 280

F. Forsberg[1], N. Roxhed[1], C. Colinge[2], G. Stemme[1], and F. Niklaus[1]

[1]KTH Royal Institute of Technology, SWEDEN and [2]California State University, USA

We present a novel and highly efficient wafer-level batch transfer process for populating silicon (Si) wafers with distributed islands of 1 micrometer-thick single-crystalline germanium (Ge). This is achieved by transferring Ge from a Si donor wafer containing thick Ge dies to a Si target wafer by adhesive wafer bonding and subsequent low-temperature Ge exfoliation.

M-007 MULTILAYER ETCH MASKS FOR 3-DIMENSIONAL FABRICATION OF ROBUST SILICON CARBIDE MICROSTRUCTURES 284

K.M. Dowling, A.J. Suria, A. Shankar, C.A. Chapin, and D.G. Senesky
Stanford University, USA

This paper demonstrates the fabrication of 3-D microstructures in 4H-silicon carbide (4H-SiC) substrates for the first time using a plasma etch and multilayer etch masks. This process was developed using a variety of thin film masks and demonstrated SiC etch rates as high as ~1 μm/min, a SiC:Ni etch selectivity as high as 60:1, and aspect ratio dependent etch characteristics. The microfabrication of complex SiC microstructures (mechanical gears, Lego®-like bricks, and poker chips) is presented.

T-008 NEMS BY MULTILAYER SIDEWALL TRANSFER LITHOGRAPHY 288

D. Liu, R.R.A. Syms, and M.M. Ahmad
Imperial College London, UK

We report an extension of a recently demonstrated technique to fabricate NEMS based on sidewall transfer lithography (STL). The process uses two STL steps to form intersecting nanoscale features such as suspension beams, breaking an important restriction of single-layer STL NEMS. The new process only requires optical lithography, making it suitable for low-cost mass parallel fabrication for complex NEMS on wafer scale. Current nanoscale features have a width of 100 nm and 50:1 aspect ratio.

W-009 NEW SCALABLE MICROFABRICATION METHOD FOR SURFACE ION TRAPS AND EXPERIMENTAL RESULTS WITH TRAPPED IONS 292

S. Hong[1], M. Lee[1], H. Cheon[1], J. Ahn[2], M. Kim[2], T. Kim[2], and D.D. Cho[1]
[1]Seoul National University, SOUTH KOREA and [2]SK Telecom, SOUTH KOREA

This paper presents a new microfabrication method for surface ion traps and experimental results with trapped ions. Using SiO_2 timed-etch method or copper sacrificial layer method, the ion trap chips with electrode overhang structures are fabricated. The ion trap chips are implemented in a 1×10^{-11} Torr vacuum environment for ion trapping experiments. Successful results in trapping strings of $^{171}Yb^+$ and $^{174}Yb^+$ ions as well as manipulating $^{171}Yb^+$ ions for qubit operation are demonstrated.

M-010 POST-RELEASE STRESS-ENGINEERING OF SURFACE-MICROMACHINED MEMS STRUCTURES USING EVAPORATED CHROMIUM AND IN-SITU FABRICATED RECONFIGURABLE SHADOW MASKS 296

R. Majumdar, V. Foroutan, and I. Paprotny
University of Illinois, Chicago, USA

We develop a novel post-release stress engineering process to provide an out-of-plane curvature to initially plane MEMS microstructures. The stressor layer(Chromium) is applied post-release and patterned appropriately with help of in-situ fabricated reconfigurable shadow masks. In addition to avoiding photolithography step, these shadow masks also provide variable coverage for underlying structures. A modified model is also shown which considers non-uniform Cr thickness on released structures.

T-011 SELF-ASSEMBLY OF MICROCOMPONENTS USING THE ENTROPIC EFFECT 300

U. Okabe, T. Okano, and H. Suzuki
Chuo University, JAPAN

We propose the use of the entropic effect (the depletion volume effect) for the self-assembly of microcomponents. When the solution contains macromolecule at relatively high concentration, microcomponents formed assembled structures. The bonding energy is not originated from the surface; it is generated by increasing the translational entropy of macromolecules in the solution. We expect that use of the depletion volume effect promotes the search of the global free-energy minima in the system by avoiding being trapped to the local minima in the self-assembly process.

W-012 THREE-DIMENSIONAL INTEGRATION OF SUSPENDED SINGLE-CRYSTALLINE SILICON MEMS ARRAYS WITH CMOS 304

Z. Song[1], Y. Du[1], M. Liu[1], S. Yang[1], D. Wu[1], and Z. Wang[1,2]
[1]Tsinghua University, CHINA and [2]Innovation Center for MicroNanoelectronics and Integrated System, CHINA

We present a generic three-dimensional (3-D) integration method to fabricate suspended single-crystalline silicon (SCS) MEMS arrays on CMOS. This method is applicable to a large variety of SCS MEMS including accelerometers, gyroscopes, micromirrors, RF MEMS switches, and resonators. Key challenges including fabrication process, mechanical reliability, and residue stress induced deflection have been addressed.

M-013 TRAJECTORY CONTROL OF MEMS FALLING OBJECT FABRICATED BY SU-8 MULTILAYER STRUCTURE 308

H. Yamane and S. Nagasawa
Shibaura Institute of Technology, JAPAN

We propose a trajectory control method for a MEMS falling object. The MEMS falling object is consisted of two units, an autorotation part and a non-rotation part. By using large falling objects, aerodynamics of the falling object was characterized. Then the MEMS falling object was designed considering with aerodynamics. The MEMS falling object was fabricated with a method of the SU-8 multi-layer structure. A MEMS autorotation part was fabricated and it rotated successfully in the wind-tunnel.

T-014 WAFER-SCALE INTEGRATION OF CARBON NANOTUBE TRANSISTORS
AS PROCESS MONITORS FOR SENSING APPLICATIONS 312
K. Chikkadi, W. Liu, C. Roman, M. Haluska, and C. Hierold
ETH Zurich, SWITZERLAND

We report on the fabrication of carbon nanotube transistors designed as process control monitors for applications such as gas and pressure sensing. We demonstrate the concept for an integration process used for gas sensor fabrication on a 100 mm wafer. Our analysis on 4463 (including 2702 semiconducting) devices allows the extraction of distributions of threshold voltage, minimum device resistance, hysteresis width, process yield and wafer uniformity data on a 100 mm wafer.

Manufacturing for Bio- and Medical MEMS and Microfluidics

W-015 FABRICATION OF A MONOLITHIC CARBON MOLD FOR PRODUCING
A MIXED-SCALE PDMS CHANNEL NETWORK USING A SINGLE MOLDING PROCESS 316
Y. Lee, Y. Lim, and H. Shin
Ulsan National Institute of Science & Technology (UNIST), SOUTH KOREA

We introduce a batch fabrication technique for a mixed-scale monolithic carbon mold producing a mixed-scale PDMS channel. A SU-8 structure fabricated by UV lithography was converted to a carbon structure trough the pyrolysis. The carbon mold dimension could be easily controlled from micrometer-scale to nanometer-scale when the pyrolysis accompanies enormous volume reduction. The mixed-scale PDMS channel network has a functionality for the nanochannel electroporation (NEP) without roof collapse.

M-016 FABRICATION OF PATTERNABLE NANOPILLARS FOR MICROFLUIDIC
SERS DEVICES BASED ON GAP-INDUCED UNEVEN ETCHING 320
Y. Wang[1,2], L.C. Tang[1,3], H.Y. Mao[1,2], C. Lei[1,3], W. Ou[1,2], J.J. Xiong[3], Y. Ou[1,2],
A.J. Ming[1,2], D. Li[4], and D.P. Chen[1,2]
*[1]Chinese Academy of Sciences, CHINA, [2]Jiangsu R&D Center for Internet of Things, CHINA,
[3]North University of China, CHINA, and [4]Stanford University, UNITED STATES*

We report a novel, simple and time-saving lithography-free approach for fabricating patternable nanopillars. The key technique of the approach is to introduce a gap by covering it with a cap, which contains through holes and the material on its lower surface has a similar etching rate with the substrate. By adjusting sizes and profiles of the perforations, nanopillars with desirable patterns can be obtained. Thus a new way for fabricating microfluidic SERS devices is further developed.

T-017 A LOW-COST AND LABEL-FREE ALPHA-FETOPROTEIN SENSOR
BASED ON SELF-ASSEMBLED GRAPHENE ON SHRINK POLYMER 324
S. Sando, B. Zhang, and T. Cui
University of Minnesota, USA

We develop a shrink-induced grapheme sensor for label-free biomolecule detection. While Enzyme-linked immunosorbent assay (ELISA) is the most popular method to detect specific proteins in the current medicine, label-free biosensors have attracted great attention due to simplicity and ease of use. The sensor described in this work demonstrates the ability to detect alpha-fetoproteins (AFP), one of the most important tumor markers associated with liver cancer and ovarian cancer.

W-018 HIGH-TOPOGRAPHY SURFACE FUNCTIONALIZATION BASED ON
PARYLENE-C PEEL-OFF FOR PATTERNED CELL GROWTH 328
F. Larramendy[1], D. Serien[2], S. Yoshida[2], L. Jalabert[2], S. Takeuchi[2], and O. Paul[1]
[1]University of Freiburg - IMTEK, GERMANY and [2]University of Tokyo, JAPAN

We develop a new technique for patterning functionalization layers on substrates with high topography. The method is based on a parylene-C template shaped by a structured, sacrificial photoresist layer and attached to the substrate where functionalization is not intended. We successfully demonstrate the technique with the guided growth of PC12 cells on honeycomb-shaped protein patterns on micropillars and microwells.

M-019 LASER TREATED GLASS PLATFORM WITH RAPID WICKING-DRIVEN
TRANSPORT AND PARTICLE SEPARATION CAPABILITIES 332
M. Ochoa, H. Jiang, R. Rahimi, and B. Ziaie
Purdue University, USA

Wicking and particle separation are two required capabilities for many microfluidics and lab-on-a-chip devices, but they often require multiple materials and structures (e.g., paper, polymer filters) which are difficult to integrate with established microfabrication techniques and materials. In this work, we combine both properties into a single glass platform with a straightforward and economical fabrication process. By laser machining soda lime glass with a specific power and laser speed, we create channels defined by an array of micro cracks (3–4 μm) which provide particle separation properties and simultaneously enable rapid liquid transport (up to 24.2 mm/s) as a result of capillary forces from the crevices and laser-induced surface hydrophilization.

T-020 LIPOSOME ARRANGEMENT CONNECTED WITH AVIDIN-BIOTIN
COMPLEX FOR CONSTRUCTING FUNCTIONAL SYNTHETIC TISSUE 336
H. Hamano[1], T. Osaki[2], and S. Takeuchi[1]
[1]University of Tokyo, JAPAN and [2]Kanagawa Academy of Science and Technology (KAST), JAPAN

This paper describes a method to arrange liposomes into organized structures with biochemical binding of avidin biotin complex, inspired by the biological anchoring junction. This approach enhances the stability of the liposome structure, which would provide an improved model for a liposome-based tissue in synthetic biology.

W-021 MAGNESIUM-EMBEDDED LIVE CELL FILTER FOR CTC ISOLATION 340
Y. Liu[1], J. Park[1], T. Xu[2], Y. Xu[2], J.H.-C Chang[1], D. Kang[1], X. Zhang[1], A. Goldkorn[2], and Y.C. Tai[1]
[1]California Institute of Technology, USA and [2]University of Southern California, USA

We develop a novel Magnesium-embedded cell filter for Circulating Tumor Cell (CTC) capture, release and isolation. The new and novel feature is the use of thin-film Mg to release the captured CTCs based on the fact that any Cl- containing culture medium can readily etch Mg away. The releasing and the isolation of each individual CTC are demonstrated. The top PA-C filter pieces break apart from the bottom after Mg completely dissolves, enabling captured CTCs to detach from the filter.

M-022 MICRO FLUIDIC CHAMBER WITH THIN SI WINDOWS FOR
OBSERVATION OF BIOLOGICAL SAMPLES IN VACUUM 344
H. Hayashi, M. Toda, and T. Ono
Tohoku University, JAPAN

A micro fluidic chamber with 178 nm-thick thin Si windows on a micro channel has been developed. Using this windows, the aquatic structure inside of the channel has been monitored by scanning electron microscopy. Secondary electrons from a sample in the channel are able to be detected in vacuum with an acceleration voltage of 15 kV. The micro fluidic chamber is possibly applied to cell imaging via the single crystal Si thin window in vacuum using magnetic resonance force microscopy.

T-023 RAPID, LOW COST FABRICATION OF CIRCULAR CROSS-SECTION
MICROCHANNELS BY THERMAL AIR MOLDING 348
T.-Q. Nguyen and W.-T. Park
Seoul National University of Science and Technology, SOUTH KOREA

This paper demonstrates a simple fabrication process of polydimethylsiloxane(PDMS) circular cross section microfluidic channels by using a PDMS master mold and thermal air molding. Based on this technique, circular cross section microchannel can be easily produced in a wide range of dimensions from 10μm to 500μm with simple bench top equipment. This technique can create perfect circular channels without any plasma activated bonding and alignment process. We can also apply this technique to fabricate micro concaves for spheroid culture, micro nozzles for droplet generation, and micro patch clamps for cell immobilization.

W-024 SKIN-EQUIVALENT INTEGRATED WITH PERFUSABLE CHANNELS ON CURVED SURFACE 351
N. Mori[1], Y. Morimoto[1,2], and S. Takeuchi[1,2]
[1]University of Tokyo, JAPAN and [2]Japan Science and Technology Agency (JST), JAPAN

We constructed a skin-equivalent on a curved surface. The skin-equivalent consists of not only the epidermis/dermis but also perfusable channels. We embedded an anchoring structure in our device to prevent horizontal contraction of the tissue. Owing to perfusion, we can culture epidermis at the air-liquid interface for cornification. Our method enables the skin-equivalent construction on a curved surface that is necessary for the construction of 3D skin surface such as biohybrid robots' skin.

M-025 TFT DISPLAY PANEL TECHNOLOGY AS A BASE FOR BIOLOGICAL CELLS
ELECTRICAL MANIPULATION - APPLICATION TO DIELECTROPHORESIS 354
A. Tixier-Mita, B.-D. Segard, Y.-J. Kim, Y. Matsunaga, H. Fujita, and H. Toshiyoshi
University of Tokyo, JAPAN

This paper reports for the first time the use of TFT (Thin Film Transistor) technology of display panels for biological cells electrical manipulation. This technology allows to have high density distributed transparent micro-electrodes, independently controllable, covering centimeter-size glass substrates. This technology is much superior to usual micro-technology used for Multielectrode Arrays (MEAs).The chosen application, to demonstrate the capability of such technology, is dielectrophoresis.

Materials for MEMS and NEMS

T-026 3D MORPHOLOGY RECONSTRUCTION OF HIGH ASPECT RATIO MEMS
STRUCTURE BY USING AUTOFLUORESCENCE OF PARYLENE C 358
L. Zhang, Y. Liu, F. Yang, W. Wang, D. Zhang, and Z. Li
Peking University, CHINA

We reported a MEMS fabrication compatible, damage free method for in-process 3D morphology reconstruction of high aspect ratio microstructure. As a novel morphology tracer, Parylene C thin film was conformally deposited onto the structure and annealed at high temperature under N2. By scanning with a confocal microscopy, 3D morphology of the microstructure was reconstructed from the autofluorescence information of Parylene C.

W-027 A NANOWIRE GAUGE FACTOR EXTRACTION METHOD FOR
MATERIAL COMPARISON AND IN-LINE MONITORING 361
I. Ouerghi, J. Philippe, C. Ladner, P. Scheiblin, L. Duraffourg, S. Hentz, and T. Ernst
CEA-LETI, FRANCE

We propose a new extraction method of gauge factor of nanowires for in-line monitoring of this parameter and piezoresistive material properties comparisons. Unlike conventional techniques which are destructive and suffer from reproducibility issues, this method allows a direct measurement of the GF locally at the nanoscale and at the wafer level. GFs have been reliably measured on a wide range of silicon-based NEMS resonators with different designs,crystalline structure and doping level.

M-028 A THREE-STEP MODEL OF BLACK SILICON FORMATION IN DEEP REACTIVE ION ETCHING PROCESS 365

F. Zhu[1], C. Wang[2], X. Zhang[1], X. Zhao[2], and H. Zhang[1]
[1]Peking University, CHINA and [2]Nankai University, CHINA

A three-step model used for modeling and simulation of black silicon formation in DRIE (Deep Reactive Ion Etching) process is presented. It combines quantum mechanics, sheath dynamics and diffusion theory together based on plasma environment. The simulation results show very good coincidence with experimental SEM images, proving the applicability of this theory and it's very promising to make black silicon formation in DRIE process to be controllable.

T-029 ANOMALOUS RESISTANCE CHANGE OF ULTRASTRAINED INDIVIDUAL MWCNT USING MEMS-BASED STRAIN ENGINEERING 369

K. Yamauchi, T. Kuno, K. Sugano, and Y. Isono
Kobe University, JAPAN

This research clarified the anomalous electric resistance change of ultrastrained multi-walled carbon nanotube (MWCNT), as well as its mechanical properties, using the in-situ SEM nanomanipulation system with Electrostatically Actuated NAnotensile Testing device (EANAT). Although the resistance change ratio was almost constant during the interlayer sliding of MWCNT, it showed a sharp raise at the end of the sliding in spite of the MWCNT not breaking mechanically.

W-030 FABRICATION OF TETRAPOD-SHAPED AL/NI MICROPARTICLES WITH TUNABLE SELF-PROPAGATING EXOTHERMIC FUNCTION 373

K. Inoue[1], T. Fujito[1], K. Fujita[2], Y. Kuroda[2], K. Takane[2], and T. Namazu[1]
[1]University of Hyogo, JAPAN and [2]Gauss Co., Ltd., JAPAN

This paper reports tetrapod-shaped Al/Ni microparticles fabricated by injection-molding and electroless-plating. We realize, for the first time, fabricating the 3D microtetrapods with self-propagating exothermic function that is able to tunable by changing Al powder's diameter and simultaneously by keeping porous Al tetrapod's void fraction constant. The maximum surface temperature and high-temperature duration during the reaction differ from sputtered Al/Ni multilayer film's exothermic performances. The tetrapod's exothermic performances can be freely controlled in response to the applications.

M-031 HIGH-PRODUCTIVE FABRICATION METHOD OF FLEXIBLE PIEZOELECTRIC SUBSTRATE 377

H. Hida, S. Yagami, Y. Sakurai, and I. Kanno
Kobe University, JAPAN

This paper reports a simple and high-productive fabrication method of flexible piezoelectric substrate using a new transfer technique. We experimentally clarified that Pb(Zr,Ti)O3 (PZT) thin films deposited on metal substrates can be transfer to PDMS substrates by using metal wet etching process. By characterizing crystal structures and electric properties, we confirmed that transferred PZT thin films have piezoelectric properties.

T-032 MECHANICAL CHARACTERIZATION OF THIN FILMS USING A MEMS DEVICE INSIDE SEM 381

C. Cao, B. Chen, T. Filleter, and Y. Sun
University of Toronto, CANADA

A MEMS device was developed for mechanical characterization of 2D ultra-thin films. The device utilizes electrothermal actuators to apply uniaxial tension. The robust design makes the device capable of withstanding both dry and wet transfer of 2D ultra-thin film materials onto the suspended structures on the device. Fracture stress of thin graphene oxide (GO) films was measured.

W-033 ROOM TEMPERATURE SYNTHESIS OF SILICON DIOXIDE THIN FILMS FOR MEMS AND SILICON SURFACE TEXTURING 385

A. Ashok and P. Pal
Indian Institute of Technology Hyderabad, INDIA

In this paper, SiO2 thin films deposited at room temperature using anodic oxidation method are explored for the fabrication of MEMS components and the surface texturing for solar cells applications. The anodic oxide is used as structural layer for the formation of freestanding structures (e.g. cantilever) of nanometer thickness on Si{100} wafers using anisotropic etchants. Further, the oxide film is employed as mask in KOH to texturize the silicon surface without using lithography.

M-034 SIZE EFFECT ON BRITTLE-DUCTILE TRANSITION TEMPERATURE OF SILICON BY MEANS OF TENSILE TESTING 389

A. Uesugi, Y. Hirai, T. Tsuchiya, and O. Tabata
Kyoto University, JAPAN

We report the size effect on BDTT of single crystal silicon using different width specimens (120-μm long, 5-μm thick and 4 or 9-μm wide). The BDTT was characterized by tensile testing in vacuum up to 600 °C with IR light heating. The fractured specimens showed slips on (111) planes at 500 °C and above, which indicated that the BDTT of micrometer-sized silicon decreased from millimeter-sized specimens. We found that the length along slip-propagation direction might dominate the BDTT.

T-035 STICTION FORCES AND REDUCTION BY DYNAMIC CONTACT IN ULTRA-CLEAN ENCAPSULATED MEMS DEVICES 393

D.B. Heinz[1], V.A. Hong[1], T.S. Kimbrell[1], J. Stehle[2], C.H. Ahn[1], E.J. Ng[1], Y. Yang[1], G. Yama[2], G.J. O'Brien[2], and T.W. Kenny[1]
[1]Stanford University, USA and [2]Robert Bosch RTC, USA

We demonstrate the consistent and manageable surface adhesion and stiction forces in epitaxially encapsulate MEMS devices. Data from over 2000 test structures of 80 design variations from three different fabrication runs were gathered. The measured adhesion forces (18- 25uN) are small enough for inertial sensors and are independent of contact geometry. In addition, we demonstrate anti-stiction bump stops with springs for a sliding contact and reduce the probability of stiction by over 50%.

W-036 STUDY OF THE HYBRID PARYLENE/PDMS MATERIAL 397
D. Kang[1], S. Matsuki[2], and Y.-C. Tai[1]
[1]California Institute of Technology, USA and [2]Northeastern University, USA

This paper reports the mechanical behaviors and barrier properties of hybrid PDMS-Parylene materials and presents a novel approach of implementing in-situ heating deposition to facilitate the diffusion and penetration of parylene coatings into PDMS, demonstrating enhanced pore sealing capability, for which a mathematical model was proposed and the average PDMS pore size was determined.

**M-037 SYNTHESIS OF CARBON NANOTUBES-NI COMPOSITE FOR
MICROMECHANICAL ELEMENTS APPLICATION** 401
Z. An, M. Toda, G. Yamamoto, T. Hashida, and T. Ono
Tohoku University, JAPAN

We present the fabrication and characterization of a silicon micromirror with carbon nanotubes -nickel composite beams. A novel electroplating method is developed for synthesis of the CNTs-Ni composite. The maximum variation of the resonant frequency of the fabricated micromirror during a long term stability test is about 0.3%, and its scanning angle is about 20o. It shows the potential ability of the CNTs-Ni composite for micromechanical elements application.

Packaging and Assembly

T-038 3D STRUCTURAL FORAMATION UTILIZING GLASS TRANSITION OF A PARYLENE FILM 405
T. Kan, A. Isozaki, H. Takahashi, K. Matsumoto, and I. Shimoyama
University of Tokyo, JAPAN

We reports on a fabrication of 3D micro structures supported by a Parylene thin film. The 3D structure formation procedure starts with an out-of-plane actuation of the Si micro structure coated with a Parylene film. At the same time, the environmental temperature was elevated above the glass transition temperature of the Parylene, and then cooled down below Tg. The rearrangement of the Parylene film happens above Tg, and the consolidated Parylene after the cooling down maintains the structure.

**W-039 A NOVEL FABRICATION AND WAFER LEVEL HERMETIC
SEALING METHOD FOR SOI-MEMS DEVICES USING SOI CAP WAFERS** 409
M.M. Torunbalci, S.E. Alper, and T. Akin
Middle East Technical University (METU), TURKEY

This paper presents a novel and inherently simple all-silicon (glass-free) fabrication and hermetic packaging method developed for SOI-MEMS devices, enabling lead transfer using vertical feedthroughs formed on an SOI cap wafer. The processes of the SOI cap wafer and the SOI-MEMS wafer require a total of five inherently simple mask steps, providing a combined process and packaging yield as high as 95%.

**M-040 BONDING MECHANISM IN THE VELCRO CONCEPT SI-SI
LOW TEMPERATURE DIRECT BONDING TECHNIQUE** 413
S. Keshavarzi[1,2], U. Mescheder[1], and H. Reinecke[2]
[1]Furtwangen University - IAF, GERMANY and [2]University of Freiburg - IMTEK, GERMANY

In this paper, we present the bonding mechanism of two Velcro-like (needle-like) surfaces for low temperature Si-Si direct bonding at ambient environment based on capillary force approach. The model considers both deformation and interaction mechanisms of the needles during bonding which makes it superior to other presented models.

**T-041 FET PROPERTIES OF SINGLE-WALLED CARBON NANOTUBES
INDIVIDUALLY ASSEMBLED UTILIZING SINGLE STRAND DNA** 417
K. Hokazono, Y. Hirai, T. Tsuchiya, and O. Tabata
Kyoto University, JAPAN

A new assembly process for isolated single-walled carbon nanotubes (SWCNTs) on MEMS structures and the electrical properties of SWCNT field effect transistor (FET) are reported. Mono-dispersed SWCNT solution is prepared by biotin modified single strand DNA's wrapping and the tubes are assembled onto gold electrodes using biotin-avidin bindings. The isolated SWCNT bridges over electrode gaps are successfully demonstrated. The Id-Vg curves of SWCNT show both conductor and semiconductor properties.

**W-042 IMPACT-INDUCED HARDENING PACKAGE FOR
TACTILE SENSORS USING DILATANT FLUID** 421
T. Takahata, K. Matsumoto, and I. Shimoyama
University of Tokyo, JAPAN

To realize high-sensitive and shock-resistant tactile sensor, the sensing element was surrounded by dilatant fluid, which is soft to a static force and hard to an impact, force. The applied static force was concentrated to the sensor, whereas the impact force was dispersed to the substrate. We have experimentally shown that the shock-resistant nature of the sensor with dilatant fluid package was 4 to 16 times as large as that of without the fluid.

M-043 MICRO DEVICES INTEGRATION WITH STRETCHABLE SPRING
EMBEDDED IN LONG PDMS-FIBER FOR FLEXIBLE ELECTRONICS 423
W.-L. Sung, C.-L. Cheng, and W. Fang
National Tsing Hua University, TAIWAN

This study presents a PDMS (Polydimethylsiloxane) fiber integrated with multi-devices scheme using stretchable electroplating copper spring. Each device was located on the node and embedded in PDMS-fiber. Thus, devices are mechanically connected by PDMS-fiber and electrically connected by inner stretchable spring. Advantages of this approach: (1) length magnification by stretchable spring; (2) thicker stretchable spring embedded in PDMS provides well mechanical/electrical characteristics; (3) node acts as a hub for devices implementation and integration; (4) partially stretched spring could reduce the resistance variation by external loads.

T-044 PREPARATION OF WAFER LEVEL GLASS-EMBEDDED HIGH-ASPECT-RATIO
PASSIVES USING A GLASS REFLOW PROCESS 427
M. Ma, J. Shang, and B. Luo
Southeast University, CHINA

We develop an innovative, uncomplicated and inexpensive method based on a glass reflow process to fabricate void-free glass-embedded passives for 3D MEMS packaging. The embedded structures include cylindrical, annular cylindrical and coaxial cylindrical conductive through-holes, plate and coaxial torus trench capacitors, square spiral, circular spiral and folding type trench inductors, and filters. These void-free structures have vertical and smooth sidewall, large signal pathways.

Micro- and Nanofluidics
Lab-on-Chip Medical Diagnostic Devices

W-045 "CELL PINBALL": WHAT IS THE PHYSICS? 431
R. Murakami[1], M. Kaneko[1], S. Sakuma[2], and F. Arai[2]
[1]Osaka University, JAPAN and [2]Nagoya University, JAPAN

During the deformability test of Red Blood Cell (RBC) by utilizing a micro fluidic channel, we found an interesting phenomenon where some RBCs behave just like elastic pinball. This phenomenon is called "Cell Pinball". Through visualization, we found that the RBC being in cell pinball mode rotates around the perpendicular axis to the flow line and its direction is one-to-one relationship with the moving direction. We also found that the rotating axis exists slightly behind the center of gravity

M-046 A FULLY MONOLITHIC MICROFLUIDIC DEVICE FOR COUNTING
BLOOD CELLS FROM RAW BLOOD 435
J. Nguyen[1], Y. Wei[1], Y. Zheng[1], C. Wang[2], and Y. Sun[1]
[1]University of Toronto, CANADA and [2]Mount Sinai Hospital, CANADA

We develop a monolithic microfluidic device capable of on-chip sample preparation for complete blood count from raw blood. For the first time, on-chip sample processing (e.g. dilution, lysis, and filtration) and downstream measurements were fully integrated to enable sample preparation and single cell analysis from raw blood on a single device. RBC and WBC concentration, WBC differential, mean corpuscular volume and cell distribution width are determined by electrical impedance measurements

T-047 A MICROWELL DEVICE FOR MEASUREMENT OF
MEMBRANE TRANSPORT OF ADHERENT CELLS 439
Y. Okada, M. Tsugane, and H. Suzuki
Chuo University, JAPAN

We developed the microwell device for measurement of membrane transport for single adherent cells. When cells were cultured on the microwells with ~10 μm opening, they spread to form the closed picoliter space. Thus, molecules exported from cells accumulate in such a space and be detected by imaging. We show that, by employing horizontal microwell design, materials exported from the cell membrane can be visualized. Efflux of the cancer drug transported by the multidrug resistance protein was tested.

W-048 ALGINATE HYDROGEL BASED 3-DIMENSIONAL CELL CULTURE AND
CHEMICAL SCREENING PLATFORM USING DIGITAL MICROFLUIDICS 443
S.M. George and H. Moon
University of Texas, Arlington, USA

We develop a method for creating arrays of individually addressable cell seeded calcium alginate hydrogels for 3D cell culture using electrowetting on dielectric (EWOD) digital microfluidics (DMF). Combined with EWOD DMF's multiplexing abilities, we demonstrate how a single integrated DMF device is capable of forming cell seeded alginate hydrogels, generating different concentrations of chemicals and delivering these to different gels to observe the effect of chemical concentrations on 3D tissue.

M-049 AN ELECTROCHEMICAL MICROFLUIDIC PAPER-BASED GLUCOSE
SENSOR INTEGRATING ZINC OXIDE NANOWIRES 447
X. Li, C. Zhao, and X. Liu
McGill University, CANADA

We develop an electrochemical microfluidic paper-based analytical device, featuring a working electrode decorated with semiconductor zinc oxide nanowires (ZnO NWs), for glucose detection in human serum. The integration of ZnO NWs into the paper device is realized via facile, low-cost hydrothermal synthesis of ZnO NWs on paper, and leads to superior analytical performance (high sensitivity and low limit of detection) and enhanced device stability (by removing light-sensitive electron mediators).

T-050 DRY REAGENT STORAGE IN DISSOLVABLE FILMS AND LIQUID TRIGGERED RELEASE FOR PROGRAMMED MULTI-STEP LAB-ON-CHIP-DIAGNOSTICS 451
G. Lenk, G. Stemme, and N. Roxhed
KTH Royal Institute of Technology, SWEDEN

A capillary driven lab on a chip system using dissolvable films for on-chip reagent storage, volume metering and timing of a multi-step sequence is demonstrated. Activation of the chip with a single liquid causes rehydration of four different reagents stored in dissolvable polymer layers and sequential release of the reagents to a common reaction zone. This capillary driven, single-liquid triggered multi-reagent sequence can potentially be used for multi-step PoC immunoassays.

W-051 ON-CHIP DETECTION OF WILD 3R, 4R AND MUTANT 4R TAU THROUGH KINESIN-MICROTUBULE BINDING 455
S.P. Subramaniyan[1], M.C. Tarhan[2], S.L. Karsten[3], H. Fujita[2], H. Shintaku[1], H. Kotera[1], and R. Yokokawa[1]
[1]Kyoto University, JAPAN, [2]University of Tokyo, JAPAN, and [3]NeuroInDx Inc., USA

We report demonstration of on-chip tau detection based on difference of landing rate and binding density of microtubules on kinesin surface. Tau detection device comprises of a MT reservoir, channel and collector region with overhung structures. We assayed MTs decorated with three tau types in the kinesin coated device. Since the increase of fluorescent intensity at collector regions reflected the type of tau decorated on MTs, thus by measuring FI we distinguish wild 3R, 4R and P301L mutant tau.

M-052 THE CAPILLARY NUMBER EFFECT ON THE CAPTURE EFFICIENCY OF CANCER CELLS ON COMPOSITE MICROFLUIDIC FILTRATION CHIPS 459
C. Zhao[1], R. Xu[1], K. Song[1], D. Liu[2], S. Ma[1], C. Tang[1], C. Liang[1], Y. Zohar[3], and Y.-K. Lee[1]
[1]Hong Kong University of Science and Technology (HKUST), HONG KONG,
[2]Guangzhou First Municipal People's Hospital, CHINA, and [3]University of Arizona, USA

We present a systematic study of the Capillary number effect on the capture efficiency of cancer cells on a composite microfluidic filtration chip. A phase diagram for the capture efficiency of microfiltration chips as a function of normalized cell diameter and Capillary number has been obtained, which will be useful for the designing the next generation of microfiltration devices for isolating circulating tumor cells.

Materials for Bio- and Medical MEMS and Microfluidics

T-053 3D CULTURE OF MOUSE IPSCS IN HYDROGEL CORE-SHELL MICROFIBERS 463
K. Ikeda[1,2,3], T. Okitsu[1,2], H. Onoe[1,2], and S. Takeuchi[1,2]
[1]University of Tokyo, JAPAN, [2]Japan Science and Technology Agency (JST), JAPAN, and
[3]University of Tsukuba, JAPAN

This paper reports the culturing and expansion of mouse induced pluripotent stem cells (iPSCs) in hydrogel core-shell microfibers; the core consists of iPSCs with or without extracellular matrix (ECM) proteins, and the shell is composed of calcium alginate. We revealed mouse iPSCs cultured in the micro-scale space with ECM proteins sustain their pluriotency efficiently. This 3D culture system may be useful tool to expand iPSCs for clinical use.

W-054 ALIGNMENT OF COLLAGEN NANOFIBERS IN 2D SUBSTRATES USING CYCLIC STRETCH 465
E.R. Nam[1,2], W.C. Lee[1,2], and S. Takeuchi[1,2]
[1]University of Tokyo, JAPAN and [2]Japan Science and Technology Agency (JST), JAPAN

In this work, collagen nanofibers are self-aligned in fully 2-dimensional substrates by applying cyclic stretch during the gelation process of collagen solution, and the fabricated collagen sheet induce the alignment of cells without any mechanical force within the cultivation period. We believe that our new aligned collagen sheet contributes to the regenerative medicine, which needs a scaffold that has biological structures and microenvironment.

M-055 BONDING-FRIENDLY PCPDMS: DEPOSITING PARYLENE C INTO PDMS MATRIX AT AN ELEVATED TEMPERATURE 467
Y. Liu[1], L. Zhang[1], W. Wang[1,2,3], and W. Wu[1,2,3]
[1]Peking University, CHINA, [2]National Key Laboratory of Science and Technology on Micro/Nano Fabrication, CHINA, and [3]Innovation Center for MicroNanoelectronics and Integrated System, CHINA

This paper reported a simple and effective process of bonding-friendly Parylene C-caulked PDMS (pcPDMS) for low-permeability required microfluidics. Parylene C was deposited into PDMS matrix at an elevated temperature (higher than 135°C) to caulk the permeable sites. The so-prepared pcPDMS can be directly bonded with oxygen plasma treatment just as pristine PDMS. SEM EDAX and Laser scanning confocal microscopy (LSCM) were introduced to characterize the Parylene C caulked status in the PDMS.

T-056 CHEMICALLY RESPONSIVE PROTEIN-PHOTORESIST HYBRID ACTUATOR 470
D. Serien[1] and S. Takeuchi[1,2]
[1]University of Tokyo, JAPAN and [2]Japan Science and Technology Agency (JST), JAPAN

We report the multiphoton fabrication of hybrid microstructures of photoresist and chemically responsive protein hydrogel for microactuation, such as a lever and a rotary stepper.

W-057 CORE-SHELL MICROPARTICLE SYNTHESIS IN DROPLET MICROFLUIDICS USING A SINGLE STEP POLYMERIZATION 472
X. Zhou, Y. Sun, A. Finne-Wistrand, W. van der Wijngaart, and T. Haraldsson
KTH Royal Institute of Technology, SWEDEN

We present, for the first time, a method for the synthesis of core-shell microparticles in a single polymerization step using two-phase droplet microfluidics. We verify the successful generation of core-shell microparticles using the novel synthesis approach.

M-058 VASCULAR NETWORK FORMATION FOR A LONG-TERM SPHEROID CULTURE BY CO-CULTURING ENDOTHELIAL CELLS AND FIBROBLASTS 476

T. Hayashi[1], H. Takigawa-Imamura[2], K. Nishiyama, H. Shintaku[1], H. Kotera[1], T. Miura[2], and R. Yokokawa[1]

[1]Kyoto University, JAPAN, [2]Kyushu University, JAPAN, and [3]Kumamoto University, JAPAN

We developed a PDMS microfluidic device to create a vascular network for a long-term spheroid culture, which consists of co-culture of human umbilical vein endothelial cells (HUVEC) and normal human lung fibroblasts (LF). Although network formation has been reported in several microfluidic devices, we successfully visualized that HUVEC networks formed by the co-culture with LFs reached a LF-based spheroid. Moreover, perfusability of the network was evaluated by injecting fluorescent microbeads.

T-059 HIGHLY CONTROLLABLE THREE-DIMENSIONAL SHEATH FLOW DEVICE FOR FABRICATION OF ARTIFICIAL CAPILLARY VESSELS 480

J. Ito, R. Sekine, D.H. Yoon, Y. Nakamura, H. Oku, H. Nansai, T. Chikasawa, T. Goto, T. Sekiguchi, N. Takeda, and S. Shoji

Waseda University, JAPAN

We developed a highly controllable three-dimensional (3D) sheath flow device for fabrication of artificial capillary vessels. In order to fabricate double-layer coaxial Core-Sheath microfibers applicable to long micro capillary vessels, three step sheath injection type 3D flow device which realizes wide core and sheath structure variations by simply flow rate control was applied, and vascular endothelial cells embedded to inside of the fabricated microfiber were cultured.

W-060 MODELING NON-WETTING PERFORMANCES OF SUPERLYOPHOBIC SURFACES BASED ON LOCAL CONTACT LINE 484

Z. Wang and T. Wu

Shenzhen Institutes of Advanced Technology, Chinese Academy of Sciences, CHINA

Superlyophobic surfaces (SLS) are promising as the novel universal platform for microfluidics due to the unique super-repellency and extremely low adhesion for almost all liquids. This paper proposed the pressure stability criteria and CAH estimation based on the local contact line analysis and were in excellent agreement with the experimental results. The achievements may shed new light for designing high-performance SLS for microfluidics and other MEMS fields.

Microfluidics and Nanofluidics

M-061 MICROFLUIDIC SELECTION OF APTAMERS USING COMBINED ELECTROKINETIC AND HYDRODYNAMIC MANIPULATION 488

T. Olsen, J. Zhu[1], J. Kim[1], R. Pei[2], M.N. Stojanovic[1], and Q. Lin[1]

[1]Columbia University, USA and [2]Chinese Academy of Sciences, CHINA

We present a microfluidic device that is capable of closed-loop, multi-round SELEX without manual intervention or use of off-chip instruments, as demonstrated by selection of DNA aptamers against the protein IgE with high affinity (KD = 12 nM) in a rapid manner (4 rounds in 10 hours).

T-062 A NOVEL DENSITY-BASED DIELECTROPHORETIC PARTICLE FOCUSING TECHNIQUE FOR DIGITAL MICROFLUIDICS 492

E. Samiei, H. Rezaei Nejad, and M. Hoorfar

University of British Columbia, CANADA

A new particle focusing technique based on negative dielectrophoretic (DEP) manipulation of non-buoyant particles is developed for digital microfluidic (DMF) platforms. This technique is compatible with conventional DMF electrode designs and does not require geometrical modification. Non buoyant particles can be concentrated on an electrode, followed by droplet splitting, resulting in two daughter droplets one with a high, and the other with a very low concentration of the particles.

W-063 ACTIVE CONTROL OF CHEERIOS EFFECT FOR DIELECTRIC FLUID 496

J. Yuan and S.K. Cho

University of Pittsburgh, USA

Using di-electrowetting we achieved on-demand control of Cheerios effect for dielectric(non-conductive) fluids. Additionally, our theory and experiment discovered that the tilting angle of wall is critical in this control. The present control would provide an efficient tool in many micro/nano particle manipulation/processes on phase interfaces.

M-064 BIOMECHANICAL CANAL SENSORS INSPIRED BY CANAL NEUROMASTS FOR ULTRA SENSITIVE FLOW SENSING 500

A.G.P. Kottapalli[1,2], M. Asadnia[1,2], J.M. Miao[1], and M.S. Triantafyllou[2,3]

[1]Nanyang Technological University, SINGAPORE,
[2]Singapore-MIT Alliance for Research and Technology, SINGAPORE, and
[3]Massachusetts Institute of Technology, USA

Fishes use their mechanosensory lateral-line system to detect minute disturbances underwater. The lateral-lines consist of superficial and canal neuromast (SN and CN) sensory sub-systems. In this paper, for the first time, we present the design, fabrication and experimental characterization of arrays of zero-powered and ultrasensitive MEMS piezoelectric haircell sensors encapsulated into biomimetic canals.

T-065 BAKING-POWDER DRIVEN CENTRIPETAL PUMPING CONTROLLED BY EVENT-TRIGGERING OF FUNCTIONAL LIQUIDS 504

D.J. Kinahan[1], R. Burger[2], A. Vembadi[1], N.A. Kilcawley[1], D. Lawlor[1], M.T. Glynn[1] and J. Ducrée[1]
[1]Dublin City University, IRELAND and [2]Technical University of Denmark (DTU), DENMARK

This paper reports radially inbound pumping by the event-triggered addition of water to on-board stored baking powder in combination with valving by an immiscible, high-specific weight liquid on a centrifugal microfluidic platform. This technology allows making efficient use of precious real estate near the center of rotation by enabling the placement of early sample preparation steps as well as reagent reservoirs at the spacious, high-field region on the perimeter.

W-066 CONTINUOUS-FLOW DIELECTROPHORETIC SORTING OF PARTICLES VIA 3D SILICON ELECTRODES FEATURING CASTELLATED SIDEWALLS 508

X. Xing and L. Yobas
Hong Kong University of Science and Technology (HKUST), HONG KONG

This paper for the first time describes continuous-flow dielectrophoretic (DEP) separation of particles using a simple microfluidic design incorporating 3D silicon electrodes featuring castellated sidewalls. The 3D electrodes generates non-uniform electric field along the channel depth which drive the particles into distinct layer. Meanwhile, continuous flow transport the separated particles into different outlets simultaneously.

M-067 ELECTRON BEAM SWITCHED TRAPPING AND RELEASE OF NANOPARTICLES ON NANOPORE ARRAY 512

T. Hoshino and K. Mabuchi
University of Tokyo, JAPAN

We demonstrated switching of trap and release of nanoparticles using an inverted-electron beam lithography (I-EBL). 240-nm nanobeads suspended in pure water were trapped and released on nanopore array. The nanopores in a silicon nitride membrane generated trapping flows for the beads by infinitesimal leakages toward directly connected high vacuum via the nanopores. The incident electron beam selectively induced release of the trapping beads into the solution by Coulomb force.

T-068 ENCODING AND MANIPULATING MICROCOMPONENT ON ELECTROMICROFLUIDIC PLATFORM 516

M.-Y. Chiang[1], S.-Y. Chen[1], and S.-K. Fan[2]   * Award Nominee
[1]National Chiao Tung University, TAIWAN and [2]National Taiwan University, TAIWAN

Microcomponent encoding and manipulation were performed on an electromicrofluidic platform using electrowetting and dielectrophoresis to drive particles and cells on micrometer scale for encoding and liquid or solid microcomponents on millimeter scale for larger structures assembly. 3D cell culture with reorganized fibroblasts in hydrogel microcomponents is demonstrated on the platform. The technology is applicable to heterogeneous structure formation and alternative 3D bioprinting.

W-069 IN VITRO DYNAMIC FERTILIZATION BY USING EWOD DEVICE 519

L.-Y. Chung[1], H.-H. Shen[1], Y.-H. Chung[1], C.-C. Chen[1], C.-H. Hsu[2], H.-Y. Huang[3], and D.-J. Yao[1]
[1]National Tsing Hua University, TAIWAN, [2]National Health Research Institutes, TAIWAN, and
[3]Chang Gung Memorial Hospital, TAIWAN

The result of EWOD chip culturing 2-cell to Blastocyst (B.C.) is shown on Table 1. The probability of 2-cell to 4-cell is 76.19%, 2-cell to 8-cell is 42.85%, and 2-cell to B.C. is 33.33%. This result is proved that the chip biocompatibility of EWOD has kept as a stable ratio. Figure 3 shows oocyte and sperm droplet mixed inside EWOD chip with the process from (A) to (D) with smooth situation prospectively. Table 2 is the comparison result of EWOD (Dynamic) and traditional IVF (Static) development.

M-070 LIPOPHILIC-MEMBRANE BASED ROUTING FOR CENTRIFUGAL AUTOMATION OF HETEROGENEOUS IMMUNOASSAYS 523

R. Mishra[1], R. Alam[2], D.J. Kinahan[1], K. Anderson[2], and J. Ducrée[1]
[1]Dublin City University, IRELAND and [2]Arizona State University, USA

We have devised strategic routing of flow from the reaction chamber in heterogeneous bead based ELISA to a distinct optical measurement chamber using lipophilic film valves which remain intact in aqueous solutions and selectively dissolve only when exposed to an ancillary, oleophilic solvent. We have integrated this routing feature on a "Lab-on-a-Disc" platform for the multi-step detection of anti-p53 antibodies from whole blood using event-triggered rotational flow control.

T-071 LOW RESISTANCE LIQUID MOTION FOR ENERGY HARVESTING 527

A. Goswami, S. Gowda, A. Tripathy, D. Roy, V. Bharadwaja, and P. Sen
Indian Institute of Science, INDIA

We demonstrate low resistance motion of liquid bulge on a well-defined path for energy harvesting application. A liquid bulge which arises due to an instability in a pre-wetted strip moves with very small hysteresis due to absence of a contact line. The pre-wetted strip confines the bulge and defines it motion path. Resistance to initiate motion of a bulge was studied experimentally and compared to other cases. An electrostatic energy harvesting device based on bulge motion is also demonstrated.

W-072 MICROFLUIDIC SWITCHING DEVICES SHOWING CONTROLLABLE HYSTERESIS 531

S. Ortiz and J.A. Lell
Comisión Nacional de Energía Atómica, ARGENTINA

We present a microfluidic MEMS device that is capable of switching among different states, thus behaving as an effective flip-flop. The devices consist of linear microfabricated deLaval nozzles with three exit channels, and we analyze the response as a function of input and output pressures, feed gas, and dimensions. In all cases, we have seen the appearance of a vortex, whose direction of swirl changes sign according to the input pressure, and showing hysteresis.

M-073 MICROFLUIDIC HANGING-DROP PLATFORM FOR PARALLEL CLOSED-LOOP MULTI-TISSUE EXPERIMENTS 535

S. Rismani Yazdi[1,2], A. Shadmani[1], A. Hierlemann[1], and O. Frey[1]
[1]ETH Zurich, SWITZERLAND and [2]Politecnico de Milano, ITALY

We present a new on-chip pumping approach for microfluidic hanging-drop networks that are used for experiments with 3D microtissue spheroids. Several independent hanging drop networks can be operated in parallel with only one single pneumatic actuation line. The pump concept enables closed-loop medium circulation between different organ models for body-on-a-chip applications and allows for multiple simultaneous assays in parallel.

T-074 ONE CORE-FIVE SHEATHS COAXIAL FLOW FORMATION USING MULTILAYER STACKED FLOW FOCUSING STRUCTURE 539

D.H. Yoon, J. Ito, N. Takeda, T. Sekiguchi, and S. Shoji
Waseda University, JAPAN

We proposed multilayer coaxial sheath flow formation by stacking multilayers of a single flow focusing structure. One core and five sheaths are simply formed with low diffusion between different core and sheaths. Only one point alignment for the sheath area is relatively free from misalignment, and the number of samples is infinitely expandable by increase in the number of stacking layers. The coaxial sheath flow is useful for biological fiber formation such as artificial blood vessel.

W-075 RECONFIGURABLE MICROFLUIDIC DILUTION GENERATION FOR QUANTITATIVE ASSAY 543

J. Fan[1], B. Li[2], and T. Pan[1]
[1]University of California, Davis, USA and [2]University of Science and Technology of China, CHINA

We report the first reconfigurable microfluidic dilution generator, producing discrete logarithmic dilution concentrations from a fixed sample volume of 10uL, without assistance of continuous fluid pumps or vacuum source. This portable chip serves as a facile tool for automatic generation of standard curves, indicating that it could be potentially employed for running generic quantitative assays for daily monitoring tasks in fields and biochemical laboratories.

M-076 SELF-MIXING BY ON-CHIP PREPARATION OF AQUEOUS TWO PHASE SYSTEMS AND ITS INFLUENCE ON EXTRACTION KINETICS 547

P.A.L. Wijethunga and H. Moon
University of Texas, Arlington, USA

This paper introduces an advantageous self-mixing phenomenon created on a digital microfluidic (DMF) device, and its influence on enhancing on-chip extraction kinetics, for the first time, highlighting a significant mixing capability in the absence of forced mixing. Such self mixing could contribute to achieve portable micro fluidics where only basic operations should be implemented and powered.

T-077 SINGLE-LAYER MICROFLUIDIC CURRENT SOURCE *VIA* OPTOFLUIDIC LITHOGRAPHY 551

C.C. Glick, S. Peng, M. Chung, K. Korner, M. Veale, C. Liu, J. Moore, A. Chu, A. Buckley, K. Iwai, R.D. Sochol, and L. Lin
University of California, Berkeley, USA

We develop and test a microfluidic current source which auto-regulates fluidic flow rate. We construct the device using optofluidic lithography in a single-layer PDMS channel.

W-078 THIN-FILM EDGE ELECTRODE LITHOGRAPHY ENABLING LOW-COST COLLECTIVE TRANSFER OF NANOPATTERNS 555

Y. Li[1], A. Goryu[1], K. Chen[2], H. Toshiyoshi[2], and H. Fujita[2]
[1]Toshiba Corporation, JAPAN and [2]University of Tokyo, JAPAN

This paper reports a new lithography method using thin-film edge electrodes (TEEs) to collectively transfer nanopatterns by generating oxide on the substrate surface via an electrochemical reaction (ECR). Nanometric thick TEEs are formed on the sidewall of insulating structures. ECR-based oxide patterns have the same width and shape as the TEEs because ECR is induced only between the conductor and the substrate. Oxide nanopatterns of 300nm and 70nm wide were collectively transferred on Si substrate in millimeter-scale area

M-079 ULTRAFINE PARTICLE COUNTER USING A MEMS-BASED PARTICLE PROCESSING CHIP 559

H.-L. Kim, J.S. Han, S.-M. Lee, H.B. Kwon, J. Hwang, and Y.-J. Kim
Yonsei University, SOUTH KOREA

We develop a microfluidic chip based ultrafine particle counter which is more compact and cost-effective than commercially available particle detection instruments. Unlike a conventional liquid-based microfluidic chip, the proposed particle processing chip handles a mixture of particles and air. We also develop a signal processing circuit which can process output signal from the microfluidic chip.

T-080 VERTICAL MEMBRANE MICROVALVES IN PDMS 563

J. Hansson, M. Hillmering, T. Haraldsson, and W. van der Wijngaart
KTH Royal Institute of Technology, SWEDEN

We present the design, realization and evaluation of the first leak-tight vertical membrane pneumatic microvalve. In comparison to horizontal membrane valves, our novel design features a 3D, instead of 2D, microchannel design, in which a vertical membrane actuates a vertical flow channel section. The valve closes under similar pneumatic control pressures to those for horizontal membrane microvalves but allows for a flow throughput per footprint area that is increased two orders of magnitude.

Bio and Medical MEMS
Biochemical Sensors

**W-081 3D HUMAN CARDIAC MUSCLE ON A CHIP: QUANTIFICATION OF
CONTRACTILE FORCE OF HUMAN IPS-DERIVED CARDIOMYOCYTES** 566
Y. Morimoto[1,2], S. Mori[1,2], and S. Takeuchi[1,2]
[1]University of Tokyo, JAPAN and [2]Japan Science and Technology Agency (JST), JAPAN

We propose a method for constructing fiber-type 3D tissue of human iPS-derived cardiomyocytes and quantifying its contractile force in response to the addition of drug. By culturing the cardiomyocytes in micropatterned hydrogel with anchors, we successfully obtained the fibers with aligned cardiomyocytes and fixed the fiber edges to the anchors. The contraction of aligned fibers in a single direction provides us to measure the contractile force reproducibly.

**M-082 A MICROFLUIDIC APTASENSOR INTEGRATING SPECIFIC ENRICHMENT WITH A
GRAPHENE NANOSENSOR FOR LABEL-FREE DETECTION OF SMALL BIOMOLECULES** 569
J. Yang[1], C. Wang[1], Y. Zhu[1], G. Liu[2], and Q. Lin[1]
[1]Columbia University, USA and [2]Nankai University, CHINA

We develop a microfluidic biosensor that combines aptamer-based specific enrichment and graphene conductance-based nanosensing on a single microchip, allowing label-free, specific, and quantitative detection of small biomolecules at low concentrations via aptamer-based competitive binding assay.

**T-083 A MICROSIZED MICROBIAL FUEL CELL BASED BIOSENSOR FOR FAST
AND SENSITIVE DETECTION OF TOXIC SUBSTANCES IN WATER** 573
H. Lee, W. Yang, X. Wei, A. Fraiwan, and S. Choi
State University of New York-Binghamton, USA

We demonstrate a microliter-sized (140 μL) microbial fuel cell (MFC)-based biosensor integrated with electrochemical sensing functionality and air-bubble trap, in which microorganisms act as the sensor for toxic substances in water. The small-scale MFC biosensor (i) reduces measurement time by increasing the probability of cell attachment and biofilm formation in the micro-sized chamber and (ii) enhances sensitivity by preventing air-bubbles on the sensing surface.

**W-084 A PAPER-BASED 48-WELL MICROBIAL FUEL CELL ARRAY FOR RAPID AND
HIGH-THROUGHPUT SCREENING OF ELECTROCHEMICALLY ACTIVE BACTERIA** 577
G. Choi[1], A. Fraiwan[1], D.J. Hassett[2], and S. Choi[1]
[1]State University of New York-Binghamton, USA and [2]University of Cincinnati College of Medicine, USA

We demonstrate the use of paper-based sensing platform for rapid and high-throughput characterization of microbial electricity-generating capabilities. A 48-well microbial fuel cell (MFC) array was fabricated on paper substrates, providing 48 high-throughput measurements and highly comparable performance characteristics in a reliable and reproducible manner. Within just 15 minutes, we successfully determined the electricity generation capacity of ten bacterial species with two controls.

**M-085 AN INTEGRATED MICROFLUIDIC SYSTEM WITH FIELD-EFFECT-TRANSISTOR-BASED
BIOSENSORS FOR AUTOMATIC HIGHLY-SENSITIVE
C-REACTIVE PROTEIN MEASUREMENT** 581
C.-H. Chu, W.-H. Chang, W.-J. Kao, C.-L. Lin, K.-W. Chang, Y.-L. Wang, and G.-B. Lee
National Tsing Hua University, TAIWAN

In this study, a new microfluidic device with a new methodology for measuring field-effect-transistor (FET)-based biosensors is presented. Not only can the proposed system work in a solution with physiological salt concentration but it also detects C-reactive protein with ultra-high sensitivity in an automatic fashion. This is the first time that a FET-based biosensor can effectively and automatically detect proteins in a physiological salt concentration without decreasing the sensitivity.

**T-086 BATCH-FABRICATED HYDROGEL/POLYMERIC-MAGNET
BILAYER FOR WIRELESS CHEMICAL SENSING** 585
J.H. Park, A. Kim, and B. Ziaie
Purdue University, USA

We introduce a fabrication and wireless chemical sensing scheme using a hydrogel/polymeric-magnet bilayer. Polymeric permanent magnets are batch fabricated/integrated on top of a hydrogel thin film. The swelling/shrinking of the hydrogel in response to chemical stimuli is remotely detected by a giant magneto resistance (GMR) sensor. The described device is the first integrated wireless hydrogel/polymeric-magnet transducer with potential applications in biomedical and environmental sensing areas

**W-087 CELL-LADEN HINGED MICROPLATES FOR MEASURING
THE CONTRACTILE FORCES OF CARDIOMYOCYTES** 589
H. Matsumoto[1], S. Yoshida[1], Y. Morimoto[1,2], N. Mori[1], D. Serien[1], and S. Takeuchi[1,2]
[1]University of Tokyo, JAPAN and [2]Japan Science and Technology Agency (JST), JAPAN

We report a method to measure contractile forces of cardiomyocytes at cellular level using microplates; pairs of microplates are connected by a flexible hinge at the center. Cardiomyocytes repeatedly contract and expand, and thereby fold the microplates at the flexible hinge. By measuring the change of the angle between folded microplates, we estimate contractile forces of cardiomyocytes. We believe that this method is a useful tool to study the dynamics of cardiomyocytes.

M-088 DROP TO MEASURE: A NOVEL NANOFLUIDIC CRYSTAL SENSING SCHEME WITH IMPROVED CHIP-TO-CHIP DATA CONSISTENCY FOR PORTABLE BIOCHEMICAL DETECTION 593
B. Wang[1], W. Zhao[1], R. Zhang[1], and W. Wang[1,2,3]
[1]*Peking University, CHINA,* [2]*National Key Laboratory of Science and Technology on Micro/Nano Fabrication, CHINA, and* [3]*Innovation Center for MicroNanoelectronics and Integrated System, CHINA*

We proposed a novel "drop to measure" nanofluidic crystal sensing scheme with improved chip-to-chip data consistency. Nanoparticles were self-assembled in a confined space guided by a well-designed surface chemistry treatment. The electrical readouts from different chips (n=5) varied within 4.8%. Biotin (using streptavidin-modified nanoparticles) and Pb^{2+} (using DNAzyme probed nanoparticles) were successfully detected by the present nanofluidic crystal sensor with a limit of detection of 1 nM.

T-089 ELECTRICAL DETECTION OF PESTICIDE VAPORS BY BIOLOGICAL NANOPORES WITH DNA APTAMERS 596
A. Nobukawa[1,2], T. Osaki[1,2], T. Tonooka[1], Y. Morimoto[1], and S. Takeuchi[1,2]
[1]*University of Tokyo, JAPAN and* [2]*Kanagawa Academy of Science and Technology (KAST), JAPAN*

We developed a vapor sensor using two robust biological molecules: A biological nanopore formed in a lipid bilayer and a DNA aptamer. The aptamer selectively binds to the target molecule, while the target molecule-aptamer complex clogs at the nanopore and blocks ionic current under electrical detection. A feasibility test was performed using a vapor phase sample, omethoate, demonstrating long-and-deep current blockades.

W-090 HIGH-PERFORMANCE AND LOW-COST LUNG CANCER SENSOR ARRAY BASED ON SELF-ASSEMBLED GRAPHENE 600
B. Zhang and T. Cui
University of Minnesota, USA

We develop a lung cancer sensor array (LCSA) based on layer-by-layer (LbL) self assembled grapheme, showing features including high performance and low cost in lung cancer biomarker detection due to graphene material properties in nature, self assembly technique, and multiple antigens detection within a single chip.

M-091 NOBEL DETECTION PLATFORM FOR ALZHEIMER'S AMYLOID-BETA USING MAGNETIC BEADS IN ELECTROCHEMICAL IMPEDANCE SPECTROSCOPY 604
K.-S. Shin, M.J. Kim, S.H. Lee, and J.Y. Kang
Korea Institute of Science and Technology (KIST), SOUTH KOREA

In this work, we proposed noble detection platform to detect Alzheimer's amyloid-beta (A-beta) using pre-treated magnetic beads in Electrochemical Impedance Spectroscopy (EIS), for the first time. Without any immobilization on the electrodes of the EIS device, it shows ability to detect a few pg/ml of amyloid-beta oligomers and compared to the result of a conventional ELISA, which allows to simplify the measurement procedure, recycle the device by only changing magnet beads.

T-092 ULTRASENSITIVE SURFACE-ENHANCED RAMAN SPECTROSCOPY USING DIRECTIONALLY ARRAYED GOLD NANOPARTICLE DIMERS 608
K. Sugano[1], D. Matsui[2], T. Tsuchiya[2], and O. Tabata[2]
[1]*Kobe University, JAPAN and* [2]*Kyoto University, JAPAN*

This paper reports an ultrasensitive nanostructure for surface-enhanced Raman spectroscopy (SERS). The gold nanoparticle dimer, which has been reported as the highest Raman enhancing structure, was directionally arrayed on a substrate for the first time, in order to match all dimers direction to polarization direction of the incident light. The strong enhancement can be achieved at all dimers. Optimizing the dimer arrangement, 10 pM limit of detection and 0.5 s rapid detection were achieved.

W-093 WATER-PROOF 'μ-DIVING SUIT' DRESSED ON RESONANT BIOCHEMICAL SENSOR FOR ONLINE DETECTION IN SOLUTION 612
H. Yu, Y. Chen, P. Xu, F. Yu, and X. Li
Shanghai Institute of Microsystem and Information Technology, Chinese Academy of Sciences, CHINA

This paper reports a new method to ensure resonant micro-sensor long-time resonating in solution for real-timebiochemical sensing/analysis. By design of a water-proof 'diving-suit' for the cantilever resonator and an antileakagenarrow 'slit' to free the cantilever vibration, only the sensing-region of the cantilever contacts to analytesolution, while the other parts remained in air for free resonance. The sensor experimentally realizes liquid-phase detection to ppb-level pesticide residue.

Medical Microsystems (Probes, Implantables, Minimally Invasive, Etc.)

M-094 A POLYCRYSTALLINE DIAMOND-BASED, HYBRID NEURAL INTERFACING PROBE FOR OPTOGENETICS 616
B. Fan[1], K.-Y. Kwon[1], R. Rechenberg[2], A. Khomenko[1], M. Haq[1], M.F. Becker[2], A.J. Weber[1], and W. Li[1]
[1]*Michigan State University, USA and* [2]*Fraunhofer USA-CCL, USA*

This paper reports a hybrid optoelectronic neural interfacing probe, combining microscale light emitting diode (μLED) and microelectrodes on a polycrystalline diamond (PCD) substrate for optogenetic stimulation and electrical recording of neural activity. PCD has superior thermal conductivity, which allows rapid dissipation of localized LED heat to a larger area to improve heat exchange with surrounding perfused tissues, and thus significantly reduce the risk of thermal damage to nerve tissue.

T-095   AN IMPLANTABLE TIME OF FLIGHT FLOW SENSOR 620
L. Yu, B.J. Kim, and E. Meng
University of Southern California, Los Angeles, USA

A micro time of flight (TOF) electrochemical impedance (EI) flow sensor was developed for characterization of in vivo flow dynamics. The transducer utilizes EI measurement between electrode pairs to monitor the passage of an electrolytically generated gas bubble within flowing solution. Biocompatible construction, low power consumption, and low profile thin film format make it ideally suited for chronic *in vivo* monitoring with immediate application in monitoring of hydrocephalus.

W-096   APPLICATION OF PERIODIC LOADS ON CELLS FROM MAGNETIC
MICROPILLAR ARRAYS IMPEDES CELLULAR MIGRATION 624
F. Khademolhosseini[1], C.-C. Liu[1,2], C.J. Lim[1,2], and M. Chiao[1]
[1]*University of British Columbia, CANADA and* [2]*Child and Family Research Institute, CANADA*

We conduct an experimental study on the application of active micropillar structures to control cell migration. In contrast to passive micropillar structures which cause no significant alterations in cell migration rates, active micropillar structures actuated at 1 Hz decrease cell migration rates by up to 80%. The magnetic micropillar structures presented can be actuated remotely, making them a viable candidate for the development of smart materials for tissue engineering applications in vivo.

M-097   CELL MOTILITY REGULATION ON STEPPED MICRO PILLAR
ARRAY DEVICE (SMPAD) WITH DISCRETE STIFFNESS GRADIENT 628
S. Lee, B. Saha, and J. Lee
Seoul National University, SOUTH KOREA

We report a micro pillar array device that provides discrete rigidity gradient to a cell with constant focal adhesion area. This goal is achieved through the use of "stepped" micro pillar array device (SMPAD) whose top area in contact with a cell is kept constant while the diameter of pillar bodies vary for variable mechanical stiffness. We show manipulating cell behavior using this simple, artificial platform that produces a pure physical stimulus.

T-098   FIBERED REFLECTIVE MICRO OBJECTIVES FOR MINIATURIZED
SCANNING CONFOCAL FLUORESCENCE MICROSCOPY 632
S. Xie[1], E. Shaffer[1], L. Jacot-Descombes[1], D. Joss[1], B. Rachet[1], D. Kosanic[2], and J. Brugger[1]
[1]*École Polytechnique Fédérale de Lausanne (EPFL), SWITZERLAND and*
[2]*SamanTree Technologies AG, SWITZERLAND*

We report on a design, fabrication and characterization of a novel micro-optical system for imaging based on a miniaturized reflective objective, which is fabricated by combing two additive micro-fabrication techniques, inkjet printing to create the spherical mirror shape and stencil lithography for local metal deposition. This novel fabrication process produces reflective micro-objectives of different optical properties tailored for targeted bio-imaging application.

W-099   FLEXIBLE SILICON-POLYMER NEURAL PROBE RIGIDIFIED BY DISSOLVABLE INSERTION
VEHICLE FOR HIGH-RESOLUTION NEURAL RECORDING WITH IMPROVED DURATION 636
F. Barz[1], P. Ruther[1], S. Takeuchi[2], and O. Paul[1]
[1]*University of Freiburg-IMTEK, GERMANY and* [2]*University of Tokyo, JAPAN*

We present a novel concept for flexible, intracortical neural probes delivered into the neural tissue by bio-dissolvable insertion vehicles. A completely implantable, silicon-based electrode array constitutes the probe tip. It is interfaced by a flexible ribbon cable that reduces stiffness and volume of the probe system. This is expected to increase the longevity of high resolution neural recording. The probes are encased in the insertion vehicles by means of a centrifuge-based molding process.

M-100   FLOW SPEED MEASUREMENT WITH DOPPLER EFFECT USING
ULTRASONIC RECEIVER FOR SMALL-SIZED SMART CATHETER 640
R. Matsui, Y. Takei, N. Minh-Dung, T. Takahata, K. Matsumoto, and I. Shimoyama
University of Tokyo, JAPAN

We propose a wide range frequency receiver which has "liquid / piezoresistive cantilever / air" multilayer structure. This structure can measure acoustic waves from Hz to MHz order frequency because the cantilever vibrates obeying the surface waves on liquid. Experimental results demonstrated that our device can measure flow speed in a cylindrical pipe ranging from 6 to 25 mm/s, which is equal to blood flow speed of an arteriole, with MHz order Doppler Effect within one percent error.

T-101   HUMAN ADIPOSE-DERIVED STEM CELL FIBER FOR BREAST RECONSTRUCTION 643
A.Y. Hsiao[1], T. Okitsu[1,2], and S. Takeuchi[1,2]
[1]*University of Tokyo, JAPAN and* [2]*Japan Science and Technology Agency (JST), JAPAN*

We describe the construction and differentiation of human adipose-derived stem cell (ADSC) fibers into the adipocyte lineage for breast reconstruction. Human ADSCs cultured as fiber-shaped constructs were induced for adipogenic differentiation. Accumulation of lipid droplets of significant size was observed in the cells, and viability assay showed that most of the cells were alive. These findings suggest the use of ADSC fibers as a promising approach for breast reconstruction.

W-102   MEASURING THE PROPAGATING TEETH VIBRATION OF HUMAN CHEWING 646
C. Suzuki, Y. Takei, T. Takahata, K. Matsumoto, and I. Shimoyama
University of Tokyo, JAPAN

We measured the propagation waves of teeth's vibrations when chewing food. Human senses texture of chewing food by teeth's vibrations. Therefore, measuring the teeth's vibrations will allow us to quantify the food texture which human actually senses. We made the acoustic sensor that is small enough to be attached to teeth. For sensor evaluation, we conducted the rice cracker chewing test with our sensor attached to the real scale 3D jaw model, and propagation waves of around 500Hz are observed.

M-103 MICRO-NEEDLE-BASED ELECTRO TACTILE DISPLAY TO PRESENT VARIOUS TACTILE SENSATION 649
N. Kitamura[1] and N. Miki[1,2]
[1]*Keio University, JAPAN and* [2]*Japan Science and Technology Agency (JST), JAPAN*

In prior work, micro-needle electrodes were developed, however they could only stimulate the tactile receptors that located very close to the surface and therefore, they could only present stinging sensation. We developed a newly electrotactile display which has a micro-needle electrode array and a counter flat electrode. The display can stimulate all the tactile receptors, which resulted in even lower voltage required to tactile stimulation and successful display of various tactile sensation.

T-104 MICROMACHINED ULTRASOUND TRANSDUCER ARRAY FOR CELL STIMULATION WITH HIGH SPATIAL RESOLUTION 651
K. Ko[1], J.-H. Lee[1], H.J. Lee[1], S.-J. Oh[1], Y.E. Chun[1], T.S. Kim[1], C.J. Lee[1], E.-S. Yoon[1], K.-S. Yun[2], and I.-J. Cho[1]
[1]*Korea Institute of Science and Technology (KIST), SOUTH KOREA and* [2]*Sogang University, SOUTH KOREA*

We present a piezoelectric micromachined ultrasonic transducer (pMUT) array for localized stimulating on cultured cells or brain slice with high spatial resolution for the first time. We observed an increase in the level of Ca2+ in more than 15 percent of TRPA1 expressing HEK293T cells under ultrasound irradiation, which confirms that TRPA1 channel in HEK293T cells is activated by ultrasound which produced mechanical stress on cells.

W-105 MINIMALLY INVASIVE NEEDLE-FREE BUBBLE INJECTOR FOR GENE THERAPY 655
K. Takahashi[1], S. Omi[1], and Y. Yamanishi[1,2]
[1]*Shibaura Institute of Technology, JAPAN and* [2]*Japan Science and Technology Agency (JST), JAPAN*

We have successfully developed minimally-invasive needle-free bubble injector designed for the usage in air. The novelty is that the minimally-invasiveness of injection whose resolution is less than 10 μm, and injection can be possible without any pain. The injector can be used for any kind of materials with various hardness, owing to the strong impact of cavitation phenomenon when the high-speed micro-bubbles are collapsed. The developed injector can be used for wide range of biomedical study.

M-106 OPTIMIZATION OF DRUG COCKTAIL ON AN INTEGRATED MICROFLUIDIC SYSTEM 658
W.-Y. Huang[1], K. Wang[2], and G.-B. Lee[1]
[1]*National Tsing Hua University, TAIWAN and* [2]*Academia Sinica, TAIWAN*

The present study demonstrates a new microfluidic platform capable of automatically dispensing a small amount of drugs to expedite screening of drug cocktails. It could significantly decrease manual bias and enhance the throughput of drug cocktail formulation on an automated and minituriazed microfluidic system.

T-107 MEMS ELECTROCHEMICAL PATENCY SENSOR FOR DETECTION OF HYDROCEPHALUS SHUNT OBSTRUCTION 662
B.J. Kim, W. Jin, L. Yu, and E. Meng
University of Southern California, USA

We present the first Parylene-based electrochemical (EC)-MEMS patency sensor module for direct and quantitative diagnosis of patency in hydrocephalus drainage shunts. The impact of electrode size, temperature, flow conditions, and H_2O_2 plasma sterilization on sensor functionality was evaluated and sensor operation in the presence of static and dynamic obstruction was demonstrated. This device will enable simple quantitative monitoring of shunt state and more importantly, a more accurate and timely diagnosis of shunt failure.

W-108 PDMS BALLOON PUMP WITH A MICROFLUIDIC REGULATOR FOR THE CONTINUOUS DRUG SUPPLY IN LOW FLOW RATE 666
Y. Mukouyama, Y. Morimoto, S. Habasaki, T. Okitsu, and S. Takeuchi
University of Tokyo, JAPAN

This paper describes small sized balloon pump for providing liquid in low flow rate without batteries. The balloon pump is composed of a balloon tank and a microfluidic regulator with a micro valve. The balloon tank can work as a driving source to pump liquid. By connecting the micro valve to the balloon tank, we achieved extremely low flow rate of the liquid. Therefore, our system will be applicable to implantable passive pumps for the continuous drug supply in low flow rate without batteries.

M-109 PULSE WAVE MEASUREMENT IN HUMAN USING PIEZORESISTIVE CANTILEVER ON LIQUID 670
T. Kaneko, N. Minh-Dung, P. Quang-Khang, Y. Takei, T. Takahata, K. Matsumoto, and I. Shimoyama
University of Tokyo, JAPAN

We propose a device that can measure pulse waves at various points on human body with high sensitivity. Pulse wave velocity was calculated from a synchronized measurement on two points. The device has a piezoresistive cantilever placed on silicone oil. Pressure waves from arteries can be well conveyed to the cantilever through human issues, for the human-skin-like acoustic impedance of the silicone oil. The SNR of the device was ~80 dB in 10–100 Hz, when excited ~1 μm of displacement.

T-110 THE MICRO SADDLE COIL WITH SWITCHABLE SENSITIVITY FOR MAGNETIC RESONANCE IMAGING 674
K. Murashige and T. Dohi
Chuo University, JAPAN

We fabricated a micro saddle coil with switchable sensitivity for MRI (magnetic resonance imaging). Since the coil is embedded in polydimethylsiloxane (PDMS) tube, the saddle-shaped coil deforms to planar shape by pushing. By placing the saddle-shaped coil in the luminal tissue, we can take large area MR images. By deforming the coil, the sensitive area is concentrated in one side and the sensitivity becomes higher. Therefore we can take both large area MR images and high sensitive MR images.

W-111 ULTRA-SENSITIVE AND STRETCHABLE STRAIN SENSOR BASED ON PIEZOELECTRIC POLYMERIC NANOFIBERS 678

M. Asadnia[1], A.G.P. Kottapalli[2], J.M. Miao[1], and M.S. Triantafyllou[3]
[1]*Nanyang Technological University, SINGAPORE,*
[2]*Singapore for MIT Alliance for Research and Technology (SMART), SINGAPORE, and*
[3]*Massachusetts Institute of Technology (MIT), USA*

There have been increasing demands for stretchable and high-sensitivity sensors for use in structure health monitoring, human motion capture, sport performance monitoring and rehabilitation. Here, we present a novel, highly stretchable, self-powered and ultra-sensitive strain sensor based on piezoelectric PVDF electrospun nanofiber. Complete studies on mechanical and piezoelectric characteristics of the single PVDF nanofiber are presented.

M-112 ULTRACOMPACT OPTOFLEX NEURAL PROBES FOR HIGH-RESOLUTION ELECTROPHYSIOLOGY AND OPTOGENETIC STIMULATION 682

M. Chamanzar[1], D.J. Denman[2], T.J. Blanche[2], and M.M. Maharbiz[1]
[1]*University of California, Berkeley, USA and [2]Allen Institute for Brain Science, USA*

Here we report on our recent development of high-density neural probes for high resolution, multiscale electrophysiology. Our 64-channel hybrid silicon-parylene probes provide at least three-fold better spatiotemporal resolution compared to the state of the art and minimize the tethering force on the brain tissue by two orders of magnitude. We demonstrate, for the first time, the design of ultracompact polymer optical waveguides that can be monolithically integrated with our neural probes.

T-113 VERTICALLY ALIGNED EXTRACELLULAR MICROPROBE ARRAYS/(111) INTEGRATED WITH (100)-SILICON MOSFET AMPLIFIERS 686

H. Makino, K. Asai, M. Tanaka, S. Yamagiwa, H. Sawahata, I. Akita, M. Ishida, and T. Kawano
Toyohashi University of Technology, JAPAN

We report a heterogeneous integration of vertically aligned extracellular microscale silicon (Si)-probe arrays/(111) with MOSFET amplifiers/(100), by IC processes and subsequent vapor–liquid–solid (VLS) growth of Si-probes. To improve the extracellular recording capability of the microprobe with a high impedance of > 1 Mohm at 1 kHz, here we integrated (100)-Si source follower buffer amplifiers by ~700 degree VLS growth compatible (100)-Si MOSFET technology.

W-114 WEARABLE FLEXIBLE MICRO ELECTRODE FOR ADULT ZEBRAFISH LONG TERM ECG MONITORING 690

X. Zhang[1], T. Beebe[2], Y. Liu[1], J. Park[1], T. Hsiai[2], and Y.-C. Tai[1]
[1]*California Institute of Technology, USA and [2]University of California, Los Angeles, USA*

All published adult zebrafish ECG recorded to this date have been done acutely with anesthetized fish. This work presents, for the first time a wearable flexible parylene (PA) micro-electrode that monitors the Adult Zebrafish ECG in longer term. We show here the design, fabrication and testing of the flexible electrode along with a micro-molded ultrasoft density adjusted silicone jacket, allowing ECG recording to be carried under water, in the fish's natural habitat with no need for anesthesia.

Nanobiotechnology

M-115 A METHOD FOR CONTROLLING MICROTUBULE VELOCITY USING LIGHT IRRADIANCE ON A PATTERNED GOLD SURFACE 694

T. Nakahara, H. Shintaku, H. Kotera, and R. Yokokawa
Kyoto University, JAPAN

We report a method to control the velocity of gliding microtubules by light irradiance and a gold pattern. The irradiance controlled a temperature in the assay condition by heat transfer from the gold pattern. The result showed that the velocity of microtubule increased approximately 1.8 folds from initial velocity at the irradiance of 13.5 W/cm2 in the gold pattern. This is first demonstration to perform the control and the switching velocity on the patterned gold surface.

T-116 CULTURING AND PROBING PHYSICAL BEHAVIOR OF INDIVIDUAL BREAST CANCER CELLS ON SiC MICRODISK RESONATORS 698

H. Jia, X. Wu, H. Tang, Z.-R. Lu, and P.X.-L. Feng   * Award Nominee
Case Western Reserve University, USA

This work describes the first exploration of directly culturing and measuring breast cancer cells, at single-cell level, by using silicon carbide (SiC) microdisk resonators. Enabled by the superior biocompatibility of SiC, individual breast cancer cells are observed to attach and spread on device surface within 3hrs of culturing. Multimode responses of SiC microdisk resonators (20-30μm in diameters) to single MDA-MB-231 cell loading are characterized by taking advantage of their robust high-frequency multimodality in water and biological solutions.

W-117 EARLY CHARACTERIZATION METHOD OF PLANT ROOT ADAPTABILITY TO SOIL ENVIRONMENTS 702

K. Ozoe[1], H. Hida[1], I. Kanno[1], T. Higashiyama[2,3], and M. Notaguchi[2,3]
[1]*Kobe University, JAPAN, [2]Japan Science and Technology Agency (JST), JAPAN, and [3]Nagoya University, JAPAN*

This paper reports a microfluidic platform for studying physical mechanisms of plant root at early growth stage. To measure driving force of root growth precisely and quantitatively, we developed a silicon microchannel device integrated with force displacement sensor which mimics a barrier in soil. By using developed microsystem, we successfully measured the driving forces of root growth in three kinds of plants including Arabidopsis thaliana known as model organism.

Physical Sensors
Fluidic Sensors (Flow, Pressure, Density, Viscosity, Etc.)

**M-118 A CANTILEVER WITH COMB STRUCTURE MODELED BY
A BRISTLED WING OF THRIPS FOR SLIGHT AIR LEAK** 706
H. Takahashi, A. Isozaki, K. Matsumoto, and I. Shimoyama
University of Tokyo, JAPAN

This paper reports a cantilever with comb structure, mimicking a bristled wing of thrips which acts as a continuous membrane wing because of the effects of low Reynolds numbers. The comb structures are formed at the edges of the cantilever and its surrounding. When differential pressure is applied to the cantilever, both comb structures act as airflow suppression through the gap. The leakage of the fabricated comb cantilever was smaller than the normal cantilever.

T-119 AIRFLOW SHEAR STRESS SENSOR USING SIDE-WALL DOPED PIEZORESISTIVE PLATE 710
R. Kazama, H. Takahashi, T. Takahata, K. Matsumoto, and I. Shimoyama
University of Tokyo, JAPAN

We report an airflow wall shear stress sensor consisting of a plate and surrounding membrane with narrow gap. The plate is supported by side-wall doped beams, which can detect the horizontal deformation of the plate due to airflow wall shear stress. The sensor structure does not disturb target airflow circumstance because of flat surface, and measures shear stress directly. Wind tunnel test shows our sensor was able to measure laminar airflow shear stress with the resolution under 1.0 Pa.

**W-120 MICRO TRIPLE-HOT-WIRE ANEMOMETER ON SMALL SIZED
GLASS TUBE FABRICATED IN 5DOF UV LITHOGRAPHY SYSTEM** 714
S. Liu[1], Z. Yang[2], Y. Zhang[3], F. Xue[1], S. Pan[1], J. Miao[1], and L.K. Norford[4]
*[1]Nanyang Technological University, SINGAPORE, [2]Shanghai Jiao Tong University, CHINA,
[3]National Institute of Advanced Industrial Science and Technology, JAPAN, and
[4]Massachusetts Institute of Technology, USA*

We develop novel designed and fabricated micro airflow sensors based on the hot-wire sensing principle, i.e. gas cooling of electrically heated resistance. With three micro Ti/Pt hot-wire components fabricated on a glass tube in five degrees of freedom (5DOF) UV lithography system with multi-layer alignment, the sensors on a cylindrical base have demonstrated high sensitivity, fast response time and ability to detect wind speed and direction.

**M-121 MULTI ROOF TILE-SHAPED VIBRATION MODES IN MEMS
CANTILEVER SENSORS FOR LIQUID MONITORING PURPOSES** 718
G. Pfusterschmied[1], M. Kucera[1,2], V. Ruiz-Díez[3], A. Bittner[1], J.L. Sánchez-Rojas[3], and U. Schmid[1]
*[1]Vienna University of Technology, AUSTRIA, [2]AC2T research GmbH, AUSTRIA, and
[3]Universidad de Castilla-La Mancha, SPAIN*

We realized piezoelectrically self-actuated self-sensing cantilever sensors for liquid monitoring purposes excited in higher roof tile-shaped modes. This advanced class of vibration modes supports very high Q-factors in liquid media and very high volume strain values which result in combination with an optimized electrode design in very high strain related conductance peaks. Therefore, precise fluid property measurements even for highly viscous liquids like D500 (~ 430 cP) are feasible.

**T-122 ON-CHIP PRESSURE SENSING BY VISUALIZING
PDMS DEFORMATION USING MICROBEADS** 722
C.-H.D. Tsai and M. Kaneko
Osaka University, JAPAN

A novel pressure sensing technique is proposed here for measuring local pressure inside a microfluidic device. By the proposed method, the local pressure can be directly "seen" without any wire foils but simply microbeads patterns. The experimental results show that microbeads pattern is stable and repeatable where the variation for the same given pressure is less than 1%. The correlation between the pressure obtained from the proposed method and a commercial pressure connected outside is 0.995.

Force and Displacement Sensors (Tactile, Force, Torque, Stress and Strain Sensor)

W-123 3-AXIS ALL ELASTOMER MEMS TACTILE SENSOR 726
A. Charalambides, J. Cheng, T. Li, and S. Bergbreiter
University of Maryland, College Park, USA

This paper reports the first 3-axis (normal and shear force) all-elastomer capacitive MEMS tactile sensor. Sensor area is 1.5 x 1.5 mm and uses vertical capacitive structures with 20 μm electrode gaps to achieve high shear force sensitivities of 8.8 fF/N, shear force resolutions of 50 mN, and shear range up to 700 mN; this aligns with the developed finite element prediction. Fabrication utilizes a simple elastomer molding process with reusable DRIE silicon molds for inexpensive manufacturing.

M-124 6-AXIS FORCE-TORQUE SENSOR CHIP COMPOSED OF 16 PIEZORESISTIVE BEAMS 730
A. Nakai[1], Y. Morishita[2], K. Matsumoto[1], and I. Shimoyama[1]
[1]University of Tokyo, JAPAN and [2]Touchence Inc., JAPAN

We propose the 6-axis force-torque sensor chip composed of 16 piezoresistive beams whose area, 2 mm square in size, is one-third of that of the state of the art, which will enhance the mounting density of sensor array and also reduce the cost in case of volume production. This paper will show the design, a part of the fabrication process in MNOIC, 8-inch MEMS foundry in Japan, especially ion doping method by oblique ion implantation, the calibration method and experimental results.

T-125 A 0.25mm³ ATOMIC FORCE MICROSCOPE ON-A-CHIP 732
N. Sarkar[1,2], D. Strathearn[1,2], G. Lee[1,2], M. Olfat[1,2], and R.R. Mansour[1,2]
[1]*University of Waterloo, CANADA and* [2]*ICSPI Corp, CANADA*

We report the highest resolution achieved with a single-chip atomic force microscope (sc-AFM). Images of a 20nm AFM calibration standard were obtained to show, for the first time, that a single-chip instrument may obtain a vertical resolution comparable to state-of-the-art instruments at a minuscule fraction of the size (1/1,000,000) and cost (1/1000). The reported performance represents a four-fold improvement in resolution when compared to previously reported sc-AFMs.

W-126 AN INSTRUMENTED TOOTH 736
F. Becker[1], C. Sander[1], F. Schmidt[2], B. Lapatki[2], and O. Paul[1]
[1]*University of Freiburg - IMTEK, GERMANY and* [2]*University of Ulm, GERMANY*

We developed a tool for orthodontic research and education, namely an instrumented tooth (IT) that allows measuring all six applied force and moment components. The core component is an 11.6-mm-high and 3.5-mm-diameter sensor module based on a CMOS stress sensor chip sandwiched between two metal pins. In the IT, the sensor module constitutes the root. The stiff and robust sensor module is capable of measuring orthodontically relevant forces up to 60 N and moments up to 10 Ncm in all directions.

M-127 ASYMMETRIC FAN-SHAPE-ELECTRODE FOR
HIGH-ANGLE-DETECTION-ACCURACY TACTILE SENSOR 740
S.-T. Chuang, T.-Y. Chen, Y.-C. Chung, R. Chen, and C.-Y. Lo
National Tsing Hua University, TAIWAN

This paper reports an up to 95.9% angle detection accuracy enhancement for capacitive tactile sensors, which entails asymmetric and intentionally shifted electrodes. The asymmetric electrodes containing one fan- and one square-shape in capacitors reduced unexpected and rotational-shift induced errors, by keeping the same overlap area of the electrodes. The minimal angle detection resolution was improved from 5.8 to 0.3-degree, making the tactile sensor practical and reliable in artificial skins.

T-128 CARBON NANOTUBES-ECOFLEX NANOCOMPOSITE FOR
STRAIN SENSING WITH ULTRA-HIGH STRETCHABILITY 744
M. Amjadi[1] and I. Park[2]
[1]*Electronics & Telecommunications Research Institute (ETRI), SOUTH KOREA and*
[2]*Korea Advanced Institute of Science & Technology (KAIST), SOUTH KOREA*

We developed ultra-stretchable, flexible and very soft conductors based on the carbon nanotubes (CNTs)-siliconrubber (Ecoflex®) nanocomposite thin films. Highly stretchable conductors were utilized as skin-mountable and wearable strain sensors. The resistance of the CNTs-Ecoflex nanocomposite thin film was fully recovered undercyclic loading/unloading for strains as large as 510%. Finally, motion detection of finger and wrist joints was conducted using CNTs-Ecoflex nanocomposite thin film.

W-129 DEVELOPMENT OF A MINIATURIZED LASER DOPPLER VELOCIMETER
FOR USE AS A SLIP SENSOR FOR ROBOT HAND CONTROL 748
N. Morita[1], H. Nogami[1], Y. Hayashida[1], E. Higurashi[2], T. Ito[3], and R. Sawada[1]
[1]*Kyushu University, JAPAN,* [2]*University of Tokyo, JAPAN, and* [3]*Kyushu Institute of Technology, JAPAN*

We have developed a miniaturized laser Doppler velocimeter (LDV), designed for use as a slip sensor in the control of a robot hand. This sensor is only 1/10,000th of the volume of commercial LDVs, which enables the sensor to be attached to a robot hand. Our LDV was able to detect scattering objects moving at velocities ranging from 10 μm/s to 20,000 μm/s. The output of this sensor is independent of the type of material measured, which included aluminum, cardboard, or rough-surface black plastic.

M-130 MICROCANTILEVER SYSTEM INCORPORATING INTERNAL RESONANCE
FOR MULTI-HARMONIC ATOMIC FORCE MICROSCOPY 752
C. Pettit[1], B. Jeong[2], H. Keum[2], J. Lee[1], J. Kim[1], S. Kim[2], D.M. McFarland[2], L.A. Bergman[2],
A.F. Vakakis[2], and H. Cho[1]
[1]*Texas Tech University, USA and* [2]*University of Illinois, Urbana-Champaign, USA*

We report a new design concept of a micromechanical cantilever system incorporating the internal resonance during dynamic mode AFM. The passive amplification of nth harmonic triggered through the mechanism of 1:n internal resonance enables AFM to utilize multiple harmonics. Detailed theoretical and experimental studies of the proposed design demonstrate that the multi-harmonic AFM is capable of simultaneous topography and compositional mapping with 10-fold enhanced sensitivity.

T-131 PRINTABLE FLEXIBLE TACTILE PRESSURE AND TEMPERATURE
SENSORS WITH HIGH SELECTIVITY AGAINST BENDING 756
K. Kanao, S. Harada, Y. Yamamoto, W. Honda, T. Arie, S. Akita, and K. Takei
Osaka Prefecture University, JAPAN

Conventional flexible tactile sensors detect the bending of a substrate in addition to a tactile pressure, and that is the one of bottlenecks to realize stable operation of flexible device such as an artificial electronic skin. To achieve high sensitivity, a cantilever type strain sensor and a temperature sensor in a flexible substrate are developed by using a fully printed method.

M-132 SOFT FLEXION SENSORS INTEGRATING STRETCHABLE METAL CONDUCTORS ON A SILICONE SUBSTRATE FOR SMART GLOVE APPLICATIONS 760
H.O. Michaud, J. Teixidor and S.P. Lacour
École Polytechnique Fédérale de Lausanne (EPFL), SWITZERLAND

We have designed and implemented a sensory skin that monitors in real time finger flexure (3 sensors per finger) of a user's hand. Compared to current technologies, the electronic skin is made entirely of stretchable materials integrating silicone rubber, low resistivity liquid metal interconnects and highly strain sensitive, microstructured thin metal films. We incorporated the skin on a textile glove and demonstrated its function as an interface for finger motion detection.

Gas and Chemical Sensors

W-133 A CIRCUMFERENTIALLY GROWN ZNO NANOWIRE FOREST ON A SUSPENDED CARBON NANOWIRE FOR A HIGHLY SENSITIVE GAS SENSOR 764
Y. Lim, Y. Lee, J. Lee, and H. Shin
Ulsan National Institute of Science and Technology (UNIST), SOUTH KOREA

We develop a suspended ZnO nanowire forest as a highly sensitive gas sensor. The nanowires were grown selectively on a suspended single glassy carbon nanowire using hydrothermal method so that the detrimental effects from the substrate inclusive of contamination, stagnant layer and limited mass transfer could be alleviated. The novel geometry of the radially grown ZnO nanowires resembling burs of a chestnut is expected to enhance the gas sensing capability because of enhanced mass transfer.

M-134 A CMOS CAPACITIVE VERTICAL-PARALLEL-PLATE-ARRAY HUMIDITY SENSOR WITH RF-AEROGEL FILL-IN FOR SENSITIVITY AND RESPONSE TIME IMPROVEMENT 767
V.P.J. Chung, C.-L. Cheng, M.-C. Yip, and W. Fang,
National Tsing Hua University, TAIWAN

This paper reports a high-sensitivity and high-speed capacitive humidity sensor with resorcinol-formaldehyde (RF) aerogel fill-in. A novel capacitive vertical parallel-plate (VPP) array topology was designed and implemented based on standard TSMC 0.18μm CMOS process and subsequent in-house post-processes.

T-135 AN INTEGRATED CHROMATOGRAPHY CHIP FOR RAPID GAS SEPARATION AND DETECTION 771
M. Akbar, H. Shakeel, and M. Agah
Virginia Tech, USA

This paper reports the first implementation of a highly sensitive micro helium discharge photoionization detector in a silicon-glass architecture and its monolithic integration with a separation column. The new detector requires a two-mask fabrication process, is universal, non-destructive, low power (<5mW), and insensitive to flow and temperature variations. It has yielded a minimum detection limit of ~10pg which is on par with the widely used destructive flame ionization detector.

W-136 FABRICATION OF SILVER NANOPARTICILES ON CYLINDRICAL SURFACE OF U-SHAPED FIBER ATR SENSOR BY MATERIAL REDUCTION 775
D. Li, C. Sun, S. Yu, H. Yu, and K. Xu
Tianjin University, CHINA

An implantable fiber ATR sensor enhanced by silver nanoparticles on circumferential surface was presented for continuous glucose monitoring. U-shaped structure was addressed to increase optical length for sensitivity enhancement. A novel method to fabricate silver nanoparticles on circumferential surface of fiber sensor based on chemical reduction of its silver halide material directly without any preliminary nanoparticles synthesis and the following covalent bond or self-assembly was proposed.

M-137 INTRINSIC ZnO NANOWIRES WITH NEW SENSING MECHANISM OF SULFURATION-DESULFURATION TWO-STEP REACTION FOR HIGH PERFORMANCE SENSING TO ppb-LEVEL H₂S GAS 779
P. Xu[1], H. Huang[2], D. Zheng[2], and X. Li[1]
[1]Chinese Academy of Sciences, CHINA and [2]Shanghai Institute of Technology, CHINA

The paper reports a novel H2S sensing-effect for intrinsic ZnO nanowires (NWs) chemiresistive sensor. Herein50nm-diameter ZnO-NWs are found and verified to be sulfurized by H2S to form ZnS that can be latterlydesulfurized back to ZnO by ambient oxygen, which is different from conventional ZnO sensing-mechanism where resistance of semiconductor ZnO is changed via electron depletion-layer variation by surface adsorbed ambientoxygen.The ZnO-NWs have realized detection to 50ppb H2S.

T-138 IONIC-GEL-COATED FABRIC AS FLEXIBLE HUMIDITY SENSOR 783
Y. Takei, K. Matsumoto, and I. Shimoyama
University of Tokyo, JAPAN

We fabricated flexible humidity sensor which responds 10 times faster than commercial CMOS humidity sensor. Our sensor is based on Ionic-Gel-coated Fabric (IG-Fabric). IG-Fabric has wide surface area and high gas permeability so that gases can be easily absorbed and detached. As a demonstration, we fabricate flu-mask-type IG-fabric humidity sensor and measured the relative humidity change caused by human breath.

W-139 DOG-BONE RESONATOR WITH HIGH-Q IN LIQUID FOR LOW-COST QUICK 'TEST-PAPER' DETECTION OF ANALYTE DROPLET 785

F. Yu, P. Xu, J. Wang, and X. Li

Shanghai Institute of Microsystem and Information Technology, Chinese Academy of Sciences, CHINA

Imitating on-site liquid-droplet analysis/assay with test-papers, a novel tri-beam structure dog-bone resonator is proposed for low-cost quick detection of trace-amount biochemical liquid sample. Effective depression to signal feed-through effect is realized by independent piezoresistive readout with the specifically designed central beam, thereby, the new tri-beam resonator exhibits high-Q resonance of extension-mode in liquid and liquid-droplet detection of ppb-level mercury ion in water.

Inertial Sensors (Gyros, Accelerometers, Resonators, Etc.)

M-140 A DYNAMICALLY MODE-MATCHED PIEZOELECTRICALLY TRANSDUCED HIGH-FREQUENCY FLEXURAL DISK GYROSCOPE 789

M. Hodjat-Shamami, A. Norouzpour-Shirazi, R. Tabrizian, and F. Ayazi

Georgia Institute of Technology, USA

This paper presents, for the first time, the design and implementation of a dynamically mode-matched high-frequency piezoelectric silicon disk gyroscope utilizing a unique pair of in-plane flexural gyroscopic modes. A linear bidirectional frequency tuning scheme compatible with piezo-only transduction is introduced to achieve dynamic mode-matching via electromechanical feedback. A fabricated AlN-on-Silicon solid disk gyroscope was frequency tuned by 500 ppm to yield a sensitivity of 410 pA/°/s.

T-141 A TEMPERATURE-STABLE MEMS OSCILLATOR ON AN OVENIZED MICRO-PLATFORM USING A PLL-BASED HEATER CONTROL SYSTEM 793

Z. Wu and M. Rais-Zadeh

University of Michigan, USA

In this work, an oxide-refill process is used to realize passive TCF compensation for silicon MEMS resonators as well as integrated thermal isolation structures. The technology enables fabrication of a low-power ovenized micro-platform on which multiple MEMS devices can be integrated. Intrinsic frequency drifts of two MEMS resonators are utilized for temperature sensing, and closed-loop oven control is realized by phase-locking two MEMS oscillators at a specific temperature.

W-142 ALN PIEZOELECTRIC ON SILICON MEMS RESONATOR WITH BOOSTED Q USING PLANAR PATTERNED PHONONIC CRYSTALS ON ANCHORS 797

H. Zhu and J.E.-Y. Lee

City University of Hong Kong, HONG KONG

We report an approach to suppress anchor loss in thin-film piezoelectric-on-silicon MEMS resonators by patterning 2D phononic crystals (PnCs) externally on the anchors. According to our measurements, adding the PnCs helps to double the unloaded quality factor (Q), while reducing the motional resistance by half. The results suggest significant reduction of acoustic leakage to the substrate by the PnCs.

M-143 BATCH-FABRICATED HIGH Q-FACTOR MICROCRYSTALLINE DIAMOND CYLINDRICAL RESONATOR 801

D. Saito[1], C. Yang[1], A. Heidari[2], H. Najar[2], L. Lin[1], and D.A. Horsley[2]

[1]University of California, Berkeley, USA and [2]University of California, Davis, USA

We report, for the first time, a 1.5mm batch-fabricated polycrystalline diamond Cylindrical Resonator (CR) for gyroscope applications. A quality factor (Q) of 313,100 is measured at the 23kHz 2theta elliptical wineglass modes, producing a ring-down time of 4.32seconds. Annealing CRs at 700°C in a nitrogen atmosphere improved Q from 75,000 to over 300,000 with an excellent frequency mismatch of 3Hz (130ppm) between the 2theta degenerate wineglass modes without applying any tuning voltage.

T-144 DESIGN, FABRICATION, AND CHARACTERIZATION OF A MICROMACHINED GLASS-BLOWN SPHERICAL RESONATOR WITH IN-SITU INTEGRATED SILICON ELECTRODES AND ALD TUNGSTEN INTERIOR COATING 805

J. Giner[1], J.M. Gray[2], J. Gertsch[2], V.M. Bright[2], and A.M. Shkel[1]

[1]University of California, Irvine, USA and [2]University of Colorado, Boulder, USA

The paper reports on design, fabrication, and characterization of micromachined spherical resonators with integrated high-aspect ratio silicon electrodes. The electrical connection of the spherical, non-conductive micromechanical resonator is possible thanks to Atomic Layer Deposition (ALD) of Tungsten inside the shell. Operating frequencies in the range of MHz have been measured providing one of the highest frequencies in spherical resonator shells up to date.

W-145 IN-SITU OVENIZATION OF LAMÉ-MODE SILICON RESONATORS FOR TEMPERATURE COMPENSATION 809

Y. Chen, E.J. Ng, Y. Yang, C.H. Ahn, I. Flader, and T.W. Kenny

Stanford University, USA

We demonstrates an inside-encapsulation ovenization method for the temperature compensation of Lamé-mode epi-sealed silicon resonators. With this method, the square Lamé-mode resonator itself acts both as a thermometer and a heater, which allows for simultaneous in situ sensing and control of the operating temperature. In this device, only the resonating element is heated, minimizing the time constant and the heating power.

M-146 ON-CHIP CHARACTERIZATION OF STRESS EFFECTS ON GYROSCOPE ZERO RATE OUTPUT AND SCALE FACTOR 813
E. Tatar, T. Mukherjee, and G.K. Fedder
Carnegie Mellon University, USA

This paper presents stress effects on a vacuum packaged MEMS gyroscope zero rate output (ZRO), scale factor (SF), and resonance frequencies by using on-chip environmental sensors measuring the temperature and stress separately, for the first time. Environmental sensors comprise released SOI-silicon resistors. Experimental results show that a system model can be established to compensate the gyroscope ZRO using the environmental sensor outputs.

Manufacturing Techniques for Physical Sensors

T-147 A FACILE FABRICATION TECHNIQUE FOR STRETCHABLE INTERCONNECTS AND TRANSDUCERS VIA LASER CARBONIZATION 817
R. Rahimi, M. Ochoa, W. Yu, and B. Ziaie
Purdue University, USA

We present a facile, low-cost approach to fabricate highly porous conductive carbon patterns on elastomeric substrates using laser carbonization of polyimide and subsequent transfer of locally pyrolyzed features onto a PDMS sheet. Using this technique, we fabricated stretchable interconnects and an array of piezoresistive tactile sensors. Characterizations of the stretchable patterns showed linear sensitivities of $8.912k\Omega/\epsilon\%$ and $518\Omega/N$ to strain and normal force, respectively.

W-148 A HIGH-Q ALL-FUSED SILICA SOLID-STEM WINEGLASS HEMISPHERICAL RESONATOR FORMED USING MICRO BLOW TORCHING AND WELDING 821
J.Y. Cho and K. Najafi
University of Michigan, USA

We report a new fabrication technology for making complete wineglass resonators through forming thin fused silica (FS) shell resonators integrated with arbitrarily sized FS solid stems through a simultaneous process of micro blow-torching and micro welding. The fabricated wineglass resonator operates at 22.6 kHz with long ring down time (35.9 s) and high quality factor (2.55 million). The ring down time and Q are the best reported values for micro FS devices.

M-149 GRAPHITE-ON-PAPER BASED TACTILE SENSORS USING PLASTIC LAMINATING TECHNIQUE 825
H.-P. Phan, D.V. Dao, T. Dinh, H. Brooke, A. Qamar, N.-T. Nguyen, and Y. Zhu
Griffith University, AUSTRALIA

We report for the first time a highly sensitive paper-based tactile sensor using laminated graphite drawn on paper.Due to a high gauge factor of 26.2, as well as its excellent humidity-resistance, plastic-laminated graphite-on-paper has a high potential for mechanical sensors. Additionally, the plastic lamination combined with the laser cutting technique proposed in this study will bring a step forward to the mass production of cleanroom-free fabrication and low-cost MEMS devices.

T-150 HIERARCHICAL WRINKLE STRUCTURING ON INSIDE WALL OF CLOSED MICRO CHANNEL 829
A. Takei and H. Fujita
University of Tokyo, JAPAN

This paper presents a surface structuring method for a closed micro channel. Because fine photolithography can only be made on a flat surface, it has been difficult to make complicated patterns on the inside walls of non-planar microfluidic channels. Here, we demonstrated that micro-scale hierarchical patterns can be formed on the inside walls of the micro channel simply by depositing a plastic thin film in the channel and stretching the micro channel.

W-151 MICROFABRICATION OF WIDE-MEASUREMENT-RANGE LOAD SENSOR USING QUARTZ CRYSTAL RESONATOR 833
Y. Murozaki, S. Sakuma, and F. Arai
Nagoya University, JAPAN

We present a wafer level fabrication process of the QCR load sensor that has three-layer structures; two Si-hold layers and a quartz layer. Using microfabrication and atomic diffusion bonding, the assembly process was simplified. The proposed sensor is easily integrated in outer package and design the measurement range. We succeeded in multi-biosignals (heartbeat, body motion) detection using fabricated QCR sensor and outer case.

M-152 SELF-CURVED DIAPHRAGMS BY STRESS ENGINEERING FOR HIGHLY RESPONSIVE PMUT 837
S. Akhbari[1], F. Sammoura[1,2], C. Yang[1], A. Heidari[3], D. Horsley[3], and L. Lin[1]
[1]University of California, Berkeley, USA, [2]Masdar Institute of Science and Technology, UAE, and [3]University of California, Davis, USA

A process to make self-curved diaphragms by engineering residual stress in thin films has been developed to construct highly responsive piezoelectric micromachined ultrasonic transducers (pMUT). This process enables high device fill-factor to achieve better than 95% wafer utilization with controlled formation of curved membranes.

T-153 THERMOCOUPLES ON TRENCH SIDEWALL IN CHANNEL FRONTING ON FLOWING MATERIAL 841
M. Shibata, T. Yamaguchi, S. Kumagai, and M. Sasaki
Toyota Technological Institute, JAPAN

Thermocouples on the trench sidewall fronting on the flowing material are fabricated by applying 3D photolithography. The first novelty is the fabrication technique to realize the device. And, the fabricated thermocouple is confirmed to work having the advantage sensing the flow temperature directly. The metals on the sidewall do not make the shadow allowing the observation inside the microchannel using the optical microscope.

W-154 VACUUM CAVITY FORMATION FOR HIGH THERMAL ISOLATION IN FLEXIBLE THERMAL SENSOR 845

P. Kim[1], S Shibata[1], and M. Shikida[2]
[1]Nagoya University, JAPAN and [2]Hiroshima City University, JAPAN

To realize MEMS sensors in the flexible fashion, we proposed to apply a Cu On Polyimide (COP) substrate as a starting material, and introduced a sacrificial etching for producing a cavity and an electrical feed through into the COP substrate. We also newly introduced a vacuum cavity realizing high thermal isolation in flexible thermal sensor, for the first time.

Materials for Physical Sensors

M-155 A STUDY OF ADHESION FORCES IN THICK EPITAXIAL POLYSILICON UNDER DYNAMIC IMPACT LOADING 849

S. Dellea[1], R. Ardito[1], B. De Masi[1], F. Rizzini[2], A. Tocchio[2], and G. Langfelder[1]
[1]Politecnico di Milano, ITALY and [2]ST Microelectronics, ITALY

We present a structure and a method for the characterization of impact and adhesion between MEMS moving and fixed parts: the focus is to monitor an inertial mass colliding with a stopper. From the measurements we evaluate the energy balance during impacts. The work analyzes the adhesion evolution after a number of collisions comparable to a 5-year operation. Results show growing and stabilizing adhesion forces of 170 nN. We also show the possibility to change and track the impact kinetic energy.

T-156 EXPLORING THE Q-FACTOR LIMIT OF TEMPERATURE COMPENSATED CMOS-MEMS RESONATORS 853

M.-H. Li, C.-S. Li, and S.-S. Li
National Tsing Hua University, TAIWAN

This work presents an in-depth study on the Q-factor of the passively temperature compensated CMOS-MEMS resonators through the collected material/experimental database and finite-element simulation. By adapting an anchor-loss-free double-ended tuning fork (DETF) resonator design, the intrinsic material loss is expected to be the major loss mechanism in CMOS-MEMS resonators that limits the maximum Q-factor below 3,400 at the frequency of interest (300 kHz–3 MHz).

W-157 NANOFIBER FORESTS AS A HUMIDITY-SENSITIVE MATERIAL 857

C. Lei[1,2], L. C. Tang[1,2], H.Y. Mao[1,3], Y. Wang[1,2], J.J. Xiong[2], W. Ou[1,3], Y. Ou[1,3],
A. J. Ming[1,3], D. Li[4], Q.L.Tan[2], W. B. Wang[1,3], D. P. Chen[1,3], and T. Liang[2]
*[1]Chinese Academy of Sciences, CHINA, [2]North University of China, CHINA,
[3]Jiangsu R&D Center for Internet of Things, CHINA, and [4]Stanford University, USA*

Nanofiber forests with high hydrophilicity are reported in this work. They are fabricated from polyimide(PI) by a plasma-stripping technique. In a relative humidity range of 50%-80%, nanofiber forest-based devices have a capacity ~50% larger than those of PI-based sensors. Besides, the absorption and desorption of moisture take less time. It is expected that the performance of such devices can be improved owing to the simple and fast processes for both nanofiber forests and the humidity sensors.

M-158 PIEZOELECTRIC PAPER FOR PHYSICAL SENSING APPLICATIONS 861

S.K. Mahadeva, K. Walus, and B. Stoeber
University of British Columbia, CANADA

We have developed robust and mechanically flexible piezoelectric paper and we have demonstrated its suitability for physical sensing at the example of a tactile sensor. This piezoelectric paper is mechanically strong and has the largest piezoelectric coefficient reported for paper to date (d_{33}=45.7□4.2 pC/N); this coefficient is comparable to that of commercially available piezoelectric polymers (polyvinylidene fluoride; PVDF d_{33}=30 pC/N).

Nanoscale Physical Sensors

T-159 A GRAPHENE ACCELEROMETER 865

A.M. Hurst[1,2], S. Lee[1], W. Cha[1], and J. Hone[1]
[1]Columbia University, USA and [2]Kulite® Semiconductor Products Inc., USA

This work presents an SU-8 clamped graphene nano-electro-mechanical-systems (GNEMS) accelerometer, with a SU-8 proof mass located at the center of the membrane. This GNEMS accelerometer is approximately three orders of magnitude smaller than state-of-the-art MEMS accelerometers with the graphene diameter of 3-5 μm and its proof mass diameter of 1-3 μm. The fabrication and experimental periodic calibration results show a repeatable response to a periodic input acceleration levels of ~40 gs.

W-160 A SOLID-GATED GRAPHENE FET SENSOR FOR PH MEASUREMENTS 869

Y. Zhu[1], C. Wang[1,2], N. Petrone[1], J. Yu[1], C. Nuckolls[1], J. Hone[1], and Q. Lin[1]
[1]Columbia University, USA and [2]Nankai University, CHINA

We develop and model a graphene field effect transistor (GFET) nanosensor that, with a back gate provided by a high-κ solid dielectric allows analyte detection in liquid media at low gate voltages (~1.5 V). On the basis of the experimental observations and quantitative analysis, we are able to propose that the charging of the electrical double layer capacitor, instead of the surface transfer doping, is the major mechanism responsible for the pH sensing.

**M-161 MULTI-MODAL GRAPHENE POLYMER INTERFACE CHARACTERIZATION
PLATFORM FOR VAPORIZABLE ELECTRONICS** 873
V. Gund, A. Ruyack, K. Camera, S. Ardanuc, C. Ober, and A. Lal
Cornell University, USA

We report a novel graphene-based micromechanical resonant platform with resistive and mass-dependent frequency-sensing for thermal-response measurements of thin-film analytes. Resistance-temperature variation of atomically thin graphene, which also serves as the resistive heater, due to surface-interactions with spun-on analytes, and mass-sensing with silicon-nitride as structural layer provides unique dual-signal electrical and mechanical signatures of analytes.

**T-162 TWO-DIMENSIONAL MoS₂ NANOMECHANICAL RESONATORS FREELY-SUSPENDED
ON MICROTRENCHES IN FLEXIBLE SUBSTRATE** 877
R. Yang[1], Z. Wang[1], P. Wang[1], R. Lujan[2], T.N. Ng[2], and P.X.-L. Feng[1]
[1]Case Western Reserve University, USA and [2]Palo Alto Research Center, USA

We demonstrate the first high-frequency ultrathin MoS2 nanomechanical resonators, freely-suspended on microtrenches (~13μm wide and 14μm deep) fabricated on flexible substrates, with bendability and stretchability. Through investigations of the device resonances via optical excitation and detection, we observe multimode resonances up to ~50MHz with the PDMS substrate under different bending and stretching conditions. This platform will enable studies of strain coupling effects in 2D crystals.

Other Physical Sensors

**W-163 A SENSOR FOR STIFFNESS CHANGE SENSING BASED ON THREE
WEAKLY COUPLED RESONATORS WITH ENHANCED SENSITIVITY** 881
C. Zhao[1], G.S. Wood[1], J. Xie[2], H. Chang[2], S.H. Pu[1,3], H.M.H. Chong[1], and M. Kraft[4]
*[1]University of Southampton, UK, [2]Northwestern Polytechnical University, CHINA, and
[3]University of Southampton Malaysia Camput, MALAYSIA, and [4]University of Liege, Montefiore Institute, BELGIUM*

A novel MEMS resonant sensing device consisting of three weakly coupled resonators that is ultra-sensitive to stiffness change was designed, fabricated and electrically tested. By measuring amplitude ratio change of two resonators caused by mode localization, due to a change of spring stiffness of one resonator, a 49 times improvement in sensitivity compared to a previously reported 2DoF resonator sensor, and 4 orders magnitude enhancement compared to a 1DoF resonator sensor has been achieved.

**M-164 CHIP-SCALE AEROSOL IMPACTOR WITH INTEGRATED RESONANT MASS BALANCES
FOR REAL TIME MONITORING OF AIRBORNE PARTICULATE CONCENTRATIONS** 885
M. Maldonado-Garcia[1], E. Mehdizadeh[1], V. Kumar[1], J.C. Wilson[2], and S. Pourkamali[1]
[1]University of Texas, Dallas, USA and [2]University of Denver, USA

This work presents chip-scale integration of a MEMS resonant mass balance along with an aerosol impactor on a single SOI. A three mask microfabrication process has been developed to produce the main components; mass balance, impactor nozzle, and impaction micro-chamber. In addition to extreme miniaturization of a conventionally bulky setup and allowing real-time particulate mass concentration data collection, this approach addresses misalignment issues between MEMS resonators and nozzle.

**T-165 HARBOR SEAL WHISKER INSPIRED FLOW SENSORS
TO REDUCE VORTEX-INDUCED VIBRATIONS** 889
A.G.P. Kottapalli[1,2], M. Asadnia[1,2], J.M. Miao[1], and M. Triantafyllou[2,3]
*[1]Nanyang Technological University, SINGAPORE, [2]Singapore-MIT Alliance for Research and Technology,
SINGAPORE, and [3]Massachusetts Institute of Technology, USA*

Harbor seals (Phoca vitulina) are able to track their prey underwater by detecting minute water movements using their whiskers. Through comparative experimental study conducted using two MEMS sensors -one possessing a circular cylindrical haircell and the other processing a haircell with whisker-like undulations, we validate the VIV reduction of in case of whisker geometry to be 50 times lower.

W-166 ISOTROPIC 3D SILICON HALL SENSOR 893
C. Sander, C. Leube, T. Aftab, P. Ruther, and O. Paul
University of Freiburg-IMTEK, GERMANY

This paper reports the first 3D Hall sensor with isotropic sensitivity for the three spatial components of the magnetic field. The silicon device has the shape of a hexagonal prism with symmetric sets of three contacts on its top and bottom surfaces. Sending currents obliquely across the device allows one to operate it as three mutually crossing, identical, and effectively orthogonal Hall sensors. We demonstrate a design achieving sensitivities of Sx=33.0 mV/VT, Sy=33.9 mV/VT and Sz=33.3 mV/VT.

**M-167 MASS-FABRICATION COMPATIBLE MECHANISM FOR
CONVERTING IN-PLANE TO OUT-OF-PLANE MOTION** 897
I. Hotzen, O. Ternyak, S. Shmulevich, and D. Elata
Technion - Israel Institute of Technology, ISRAEL

We present a mechanism that converts in-plane to out-of-plane motion, which is fully compatible with mass-fabrication technology. The motion conversion ratio of the mechanism is constant over a wide range of motion, and this ratio can be easily tuned by adding or subtracting modular elements into the design of an otherwise unchanged planform. The mechanism enables harnessing well behaved in-plane comb-drive actuators to achieve a well behaved out-of-plane motion.

T-168 MONOLITHIC INTEGRATION OF MICRO MAGNETIC PILLAR
ARRAY WITH ANISOTROPICMAGNETO-RESISTIVE (AMR) STRUCTURE
FOR OUT-OF-PLANE MAGNETIC FIELD DETECTION 901
W.-M. Lai, F.-M. Hsu, W.-L. Sung, R. Chen, and W. Fang
National Tsing Hua University, TAIWAN

A novel integrate micro magnetic pillar array with anisotropic magneto-resistive (AMR) structure for out-of-plane magnetic field detection has been proposed and demonstrated. Through the Nickel pillar to be as magnetic concentrator, the out-of-plane magnetic field can be detected by AMR sensor, and this study propose the micro magnetic pillar array to enhance the magnetic conversion efficiency.

W-169 MULTI-COLOR IMAGING WITH SILICON-ON-INSULATOR DIODE
UNCOOLED INFRARED FOCAL PLANE ARRAY USING THROUGH-HOLE
PLASMONIC METAMATERIAL ABSORBERS 905
D. Fujisawa[1], S. Ogawa[1], H. Hata[1], M. Uetsuki[1], K. Misaki[1], Y. Takagawa[2], and M. Kimata[2]
[1]Mitsubishi Electric Corporation, JAPAN and [2]Ritsumeikan University, JAPAN

We report a silicon-on-insulator (SOI) diode uncooled infrared focal plane array (IRFPA) with through-hole plasmonic metamaterial absorbers (TH-PLMAs) for multi-color imaging with a 320x240 array format. Through-holes formed on the PLMA can reduce the thermal mass while maintaining both the single-mode and high absorption due to the plasmonic metamaterial structures, which realizes fast response and high responsivity.

M-170 PASSIVE WIRELESS TEMPERATURE SENSING WITH
PIEZOELECTRIC MEMS RESONATORS 909
H. Fatemi, M.J. Modarres-Zadeh, and R. Abdolvand
University of Central Florida, USA

For the first time, a piezoelectric MEMS resonator is utilized for passive wireless temperature sensing with an accuracy of less than 0.1°C at 1m with a signal power of 500mW and 5dBi gain antennas. The high quality factor and low motional resistance of a 991MHz thin-film piezoelectric-on-silicon (TPoS) resonator are exploited to accurately determine the temperature from the change in the resonance frequency by taking the Fourier transform of the resonator's time-gated response.

W-171 SNR IMPROVEMENT IN AMPLITUDE MODULATED RESONANT MEMS
SENSORS VIA THERMAL-PIEZORESISTIVE INTERNAL AMPLIFICATION 913
M. Mahdavi, A. Ramezany, V. Kumar, and S. Pourkamali
University of Texas, Dallas, USA

We studied, the effect of thermal-piezoresistive internal amplification on signal to noise ratio (SNR) of amplitude modulated resonant MEMS sensors showing the possibility to significantly improve the detection limit. It has been shown that as the thermal-piezoresistive amplification sets in, noise rms value increases with a slower rate than the boost in quality factor (Q) and output signal level, therefore the SNR value increases.

Sonic and Ultrasonic MEMS Transducers (Microphones, PMUTs, Etc.)

T-172 A RESONANT PIEZOELECTRIC MICROPHONE ARRAY FOR
DETECTION OF ACOUSTIC SIGNATURES IN NOISY ENVIRONMENTS 917
A.A. Shkel, L. Baumgartel, and E.S. Kim
University of Southern California, USA

We report a MEMS acoustic resonator array with improved Automatic Speech Recognition (ASR) and signature detection characteristics in environments with high levels of acoustic interference. ASR experiments are performed, showing an increase of 62.7 percentage points in transcription accuracy with -15 dB Signal-to-Noise Ratio (SNR). The results of this study support the development of highly resonant acoustic sensors for a variety of pattern recognition applications.

W-173 AIR-COUPLED ALUMINUM NITRIDE PIEZOELECTRIC MICROMACHINED
ULTRASONIC TRANSDUCERS AT 0.3 MHZ TO 0.9 MHZ 921
O. Rozen[1], S.T. Block[2], S.E. Shelton[2], R.J. Przybyla[2], and D.A. Horsley[1]
[1]University of California, Davis, USA and [2]Chirp Microsystems, Inc., USA

For the first time, air-coupled piezoelectric micromachined ultrasonic transducers (PMUTs) operating at frequencies ranging from 300 to 900 kHz were designed, fabricated and characterized. We also increased the fractional bandwidth by about 35% by patterning the diaphragm center into a ring or structural ribs, resulting in a reduction of the PMUT's mass. Fabrication was conducted using wafer-level bonding of a MEMS PMUT wafer to a CMOS wafer using a conductive metal eutectic bond. This process allows for close integration of PMUT arrays and signal processing circuitry and is used here to study the effects of wafer-level packaging on acoustic performance.

M-174 SOUND FOCUSING IN LIQUID USING A VARIFOCAL ACOUSTIC MIRROR 925
R. Aoki, N. Thanh-Vinh, K. Noda, T. Takahata, K. Matsumoto, and I. Shimoyama
University of Tokyo, JAPAN

This paper reports a method to concentrate sound in liquid in a desired location using a PDMS-based varifocal acoustic mirror. We used PDMS–air boundary as a parabolic sound reflector to concentrate sound. By adjusting the curvature radius of the acoustic mirror, we could change the position where sound was concentrated. We confirmed that our method was able to make the output of the acoustic sensor ten times larger than that without focusing in water.

T-175 BIMORPH PMUT WITH DUAL ELECTRODES 928

S. Akhbari[1], F. Sammoura[1], C. Yang[1], M. Mahmoud[2], N. Aqab[2], and L. Lin[1]

[1]*University of California, Berkeley, USA and* [2]*Masdar Institute of Science and Technology, UAE*

We have successfully demonstrated "bimorph" piezoelectric micromachined ultrasonic transducers (pMUT) with unique advantages, dramatically improving the device capabilities in the process. The bimorph pMUT utilizes two active AlN layers in a CMOS-compatible process. This innovative design is the first bimorph pMUT with two active piezoelectric layers separated by a common electrode.

MEMS for Electromagnetics
DC and Low Frequency Magnetic and Electromechanical Components and Systems

**W-176 A LOW-NOISE SUB-500μW LORENTZ FORCE BASED INTEGRATED
MAGNETIC FIELD SENSING SYSTEM** 932

S. Brenna, P. Minotti, A. Bonfanti, G. Laghi, G. Langfelder, A. Longoni, and A.L. Lacaita

Politecnico di Milano, ITALIA

A complete 1-D magnetic field sensing system including a z-axis Lorentz force MEMS sensor and an integrated circuit (ASIC) is presented. Measurement results show an achievable sensor resolution of 220 nT·mA/√Hz with an achievable bandwidth 100 Hz. The ASIC low-noise performance does not impair the minimum detectable magnetic field, mainly limited by the MEMS thermo-mechanical noise. Dissipating only 400 μW, the circuit satisfies the consumer electronics low-power requirements.

M-177 CHIP-SCALE ELECTRODYNAMIC SYNTHETIC JET ACTUATORS 936

S.G. Sawant[1], E.A. Deem[2], D.J. Agentis[3], L.N. Cattafesta[2], and D.P. Arnold[1]

[1]*University of Florida, USA,* [2]*Florida State University, USA, and* [3]*Virginia Tech, USA*

We report the first chip-scale electrodynamic synthetic jet actuator that integrates both a coil and permanent magnet via micro-fabrication. This is achieved by integrating bonded NdFeB powder magnets into standard silicon micro-machining processes. The device has a volume of 7.5 mm x 7.5 mm x 1.1 mm and generates a fluidic jet with a peak velocity of 2.1 m/s while operating at 180 Hz with 20 mW input power. The actuator has applications in flow control and active cooling of electronic devices.

**T-178 FULLY-POLYMERIC NEM RELAY FOR FLEXIBLE, TRANSPARENT,
ULTRA-LOW POWER ELECTRONICS AND SENSORS** 940

Y. Pan, F. Yu, and J. Jeon

Rutgers, The State University of New Jersey, USA

A fully-polymeric NEM relay based on a conductive polymer, Poly(3,4-Ethylenedioxythiophene):Polystyrene-Sulfonate (PEDOT:PSS), and dielectric polymers is proposed for the first time to enable flexible, transparent, ultralow-power electronics and sensors, and the first functional prototype fabricated using a five-mask low-thermal-budget process is demonstrated. Exploiting the water-absorption behavior of PEDOT:PSS, the potential use of the relay as a biochemical sensor is also demonstrated.

**W-179 UHF PIEZOELECTRIC QUARTZ MEMS MAGNETOMETERS BASED ON
ACOUSTIC COUPLING OF FLEXURAL AND THICKNESS SHEAR MODES** 944

H.D. Nguyen, J.A. Erbland, L.D. Sorenson, R. Perahia, L.X. Huang, R.J. Joyce, Y. Yoon,
D.J. Kirby, T.J. Boden, R.B. McElwain, and D.T. Chang

HRL Laboratories, USA

We report for the first time piezoelectric Quartz MEMS magnetometers based on acoustic coupling between resonance modes. The magnetic sensors employ a novel transduction scheme to upconvert the desired near-DC magnetic field signal (using the fundamental flexural mode) onto frequency modulated (FM) sidebands of the primary quartz thickness shear (TS) oscillation. First-generation devices exhibit flexural and TS resonances at 2.77kHz and 583.31MHz, respectively, and a magnetic sensitivity of 63.6V/T.

Free Space Optical Components and Systems (Displays, Lenses, Detectors)

**M-180 A 45°-TILTED 2-AXIS SCANNING MICROMIRROR INTEGRATED
ON A SILICON OPTICAL BENCH FOR 3D ENDOSCOPIC OPTICAL IMAGING** 948

C. Duan[1], W. Wang[1], X. Zhang[1], J. Ding[2], Q. Chen[2], A. Pozzi[1], and H. Xie[1]

[1]*University of Florida, USA and* [2]*WiO Technology Limited, CHINA*

This paper presents a 2-axis electrothermal single-crystal-silicon (SCS) micromirror that is tilted 45° out of plane on a silicon optical bench (SiOB). The tilt of the mirror is achieved with the bending of a set of stressed bimorph beams and the stop provided by the silicon sidewall. To the best of our knowledge, this is the first demonstration of an integrated SiOB with a 2-axis SCS mirror tilted at a fixed angle without assembly.

**T-181 COMPACT NEAR-EYE DISPLAY SYSTEM USING A SUPERLENS-BASED
MICROLENS ARRAY MAGNIFIER** 952

H.S. Park[1], R. Hoskinson[2], H. Abdollahi[2], and B. Stoeber[1]

[1]*University of British Columbia, CANADA and* [2]*Recon Instruments Inc., CANADA*

We present a new approach to make a very compact near-eye display (NED) using only two layers of microlens arrays (MLA) working in conjunction as a magnifying lens (MLA magnifier). The purpose of the MLA magnifier is to generate a virtual image of a display, positioned within several centimeters from the eye, at optical infinity to minimize the optical disparity between the surrounding scenery and the image on the display.

**M-182   OPEN-STRUCTURE ELECTROWETTING DISPLAY
WITH CAPACITIVE SENSING FEEDBACK SYSTEM** 956
S. Choi and J. Lee
Seoul National University, SOUTH KOREA

We report an open-structure electrowetting-based reflective display with capacitive sensing feedback that enables an effective self-dosing of ink, high contrast, and the precise control of color level. We introduce an display that can achieve such improvements via an open structure design and a capacitive feedback system including the effective ink dosing process, off color area being ~ 8% of viewable area, and precision control of color area even under a large variation of interfacial tension.

W-183   TUNABLE METAMATERIAL LENS ARRAY VIA METADROPLETS 960
Q.H. Song[1,2,] W.M. Zhu[2], W. Zhang[2], P.C. Wu[2], Z.X. Shen[2], Z.C. Yang[4], Y.F. Jin[4],
Y.L. Hao[4], T. Bourouina[3], Y. Leprince-Wang[1], and A.Q. Liu[2]
[1]*UPEM, Université Paris-Est, FRANCE,* [2]*Nanyang Technological University, SINGAPORE,*
[3]*ESIEE, Université Paris-Est, FRANCE, and* [4]*Peking University, CHINA*

This paper reports a liquid based tunable metamaterial which is using droplets as unit cell structures. It can function as a tunable lens array, whose focus spot can be continuously tuned, for the first time, from defocusing to sub-wavelength focusing in THz region. This work also develops a new tuning method to reconfigure the shapes of the liquid micro-droplets, which is using air pressure to expand the height of the micro channel so that the height of the droplets will be enlarged.

Manufacturing for Electromagnetic Transducers

**M-184   FABRICATION OF PATTERNED MAGNETIC MICROSTRUCTURES
USING MAGNETICALLY ASSEMBLED NANOPARTICLES** 964
C. Velez, I. Torres-Díaz, L. Maldonado-Camargo, C. Rinaldi, and D.P. Arnold
University of Florida, USA

Modeling and experimental characterization of a fabrication method for forming magnetic microstructures with complex shapes using self-assembled iron oxide (Fe3O4) magnetic nanoparticles. This method can potentially be used in roll-to-roll production of magnetic structures either patterned onto substrates or lifted off to create free-floating micromagnetic actuators.

Other Electromagnetic MEMS

**T-185   ELECTRIC CONTACT STABILITY AND READOUT RESOLUTION
OF THE ANTIWEAR PROBE WITH A GROOVE AND OIL LUBRICATION
SLIDING SYSTEM FOR PROBE-BASED ARCHIVE MEMORIES** 968
Y. Tomizawa, K. Toya, A. Oonishi, Y. Li, J. Hirota, M. Yabuki, I. Kunishima, and H. Shinomiya
Toshiba Corporation, JAPAN

The authors have proposed the novel concept of a sliding system called "AGO" (antiwear probe with a groove and oil lubrication) for probe-based archive memories. The system has been proven to have the ability to endure a meter-scale probe slide, not only in terms of electric contact stability but also in regards to the readout resolution degradation of the recorded pattern. This demonstrates the possibility of bringing probe-based memories into actual products used in data archiving, which requires limited time data access.

W-186   NEMS VARIABLE OPTICAL ATTENUATOR (VOA) DRIVEN BY OPTICAL FORCE 972
B. Dong[1,2], H. Cai[2], Y.D. Gu[2], Z.C. Yang[3], Y.F. Jin[3], Y.L. Hao[3], D.L. Kwong[2], and A.Q. Liu[1]
[1]*Nanyang Technological University, SINGAPORE,* [2]*Agency for Science, Technology and Research (A*STAR),*
SINGAPORE, and [3]*Peking University, CHINA*

We develop a NEMS optomechanical VOA driven by optical gradient force. The VOA is realized via waveguide based optical directional coupler. The gap between the directional coupler is controlled via optical force driven actuator. The doubly clamped silicon beam actuator is controlled by tuning the wavelength of control light. The NEMS VOAs have merits such as small dimension, low power consumption and good capability for all optical integration as compared with conventional MEMS based fiber VOA.

Photonic Components and Systems

**M-187   A SUPER-REGENERATIVE OPTICAL RECEIVER BASED
ON AN OPTOMECHANICAL OSCILLATOR** 976
T. Beyazoglu, T.O. Rocheleau, A.J. Grine, K.E. Grutter, M.C. Wu, and C.T.-C. Nguyen
University of California, Berkeley, USA

We present a super-regenerative optical receiver that detects on-off key modulated light input via the radiation-pressure gain of a self-sustained electro-opto-mechanical oscillator (EOMO). With oscillation amplitude a function of the intensity of light coupled into the oscillator, this device now allows data to be directly demodulated using only silicon-compatible materials, i.e., without the expensive III-V compound semiconductor materials often used in conventional optical receivers.

T-188 INTEGRATED PIEZOELECTRICALLY DRIVEN ACOUSTO-OPTIC MODULATOR 980

S. Ghosh and G. Piazza
Carnegie Mellon University, USA

This paper presents a new type of acousto-optic modulator based on the conjunction of a piezoelectric contour mode resonator with a photonic whispering gallery mode resonator. The monolithic aluminum nitride device exhibits coupling of piezoelectrically-generated lateral vibrations into a traveling-wave photonic ring resonator in a fully-integrated platform with electrodes directly patterned on the resonator body. We demonstrate the optical sensing of an actuated mechanical mode at 654 MHz.

**W-189 SPECTRALLY SELECTIVE INFRARED DETECTOR BASED
ON AN ULTRA-THIN PIEZOELECTRIC RESONANT METAMATERIAL** 984

Y. Hui and M. Rinaldi
Northeastern University, USA

We report on the first demonstration of a spectrally selective uncooled MEMS resonant IR detector based on an ultra-thin piezoelectric resonant metamaterial. High quality factor of 1407 and electromechanical coupling coefficient of 1.9%, and spectrally selective absorption (~40%) of long wavelength infrared radiation (8.8 μm with FWHM of 1.88 μm) in an ultra-low volume device were achieved, resulting in a fast (~650 μs) and high resolution (NEP ~7 nW/rt-Hz at 200 Hz bandwidth) MEMS IR detector.

RF MEMS Components and Systems

**M-190 A CMOS-MEMS ARRAYED RGFET OSCILLATOR USING
A BAND-TO-BAND TUNNELING BIAS SCHEME** 988

C.-H. Chin, C.-S. Li, M.-H. Li, and S.-S. Li
National Tsing Hua University, TAIWAN

This work reports a CMOS-MEMS Resonant-Gate Field Effect Transistor (RGFET) oscillator comprising only one single transistor. A band-to-band tunneling (BTBT) charging technique is implemented for the first time. Furthermore, this charging phenomenon on the floating gate can be well preserved for more than one day. Finally, a CMOS-MEMS RGFET self-sustained oscillator with only one active transistor is demonstrated with a decent far-from-carrier phase noise of -122 dBc/Hz.

**T-191 ACTIVE REFLECTORS FOR HIGH PERFORMANCE
LITHIUM NIOBATE ON SILICON DIOXIDE RESONATORS** 992

L. Shi and G. Piazza
Carnegie Mellon University, USA

We design, demonstrate and optimize active reflectors for enhancing the electromechanical coupling (k_t^2) and suppressing spurious modes in Laterally Vibrating Resonators (LVRs) based on X-cut ion-sliced Lithium Niobate (LN) thin film on silicon dioxide (SiO_2). Optimized active reflectors that resort to 100% metal coverage of the λ/4 extensions at the two ends of the resonant plate enable: (i) a considerable improvement of k_t^2 (up to 13%) (ii) spurious mode suppression, robustness to processing (iii) misalignment and (iv) over/under-etching.

**W-192 APPLICATION OF STATISTICAL ELEMENT SELECTION TO 3D INTEGRATED
ALN MEMS FILTERS FOR PERFORMANCE CORRECTION AND YIELD ENHANCEMENT** 996

A. Patterson[1], E. Calayir[1], G.K. Fedder[1], G. Piazza[1], B.W. Soon[2], and N. Singh[2]
*[1]Carnegie Mellon University, USA and
[2]Agency for Science, Technology and Research (A*STAR), SINGAPORE*

By 3D integration of an array of 12 nominally identical AlN MEMS sub-filters with a CMOS switching matrix and application of statistical element selection to the same system, we have built a self-healing filter offering 495 unique filter responses and a tuning range of 500 kHz for both center frequency and bandwidth. This system enables correction of intrinsic, fabrication-induced variation in filter performance that would otherwise severely limit the manufacturing yield of standalone filters.

M-193 DAMPING IN 1 GHZ LATERALLY-VIBRATING COMPOSITE PIEZOELECTRIC RESONATORS 1000

J. Segovia-Fernandez and G. Piazza
Carnegie Mellon University, USA

This work experimentally proves the physics of damping in this class of MEMS resonators. We first confute a previously developed theory of interfacial dissipation that assumed a stress (or Young modulus) jump between different materials and then find that damping is instead related to either interfacial dissipation due to a velocity jump or thermoelastic dissipation (TED) in the electrodes.

**T-194 DUAL-CLOCK WITH SINGLE AND MONOLITHICAL 0-LEVEL
VACUUM PACKAGED MEMS-on-CMOS RESONATOR** 1004

A. Uranga[1], G. Sobreviela[1], N. Barniol[1], E. Marigó[2], C. Tay-Wee-Song[2], M. Shunmugam[2], A.A. Zainuddin[2],
A. Kumar-Kantimahanti[2], V. Madhaven[2], and M. Soundara-Pandian[2]
[1]Universitat Autònoma de Barcelona, SPAIN and [2]Silterra, MALAYSIA

This paper demonstrates the feasibility of a novel fabrication approach of MEMS resonators above standard CMOS circuitry and with zero-level vacuum package. As a proof of concept a monolithical CMOS-MEMS-closed loop oscillator showing dual-clock capabilities (11.9 MHz and 24.5 MHz) is presented. These two frequencies correspond to two different resonator modes, specifically the torsional and vertical out of plane, of a paddle shaped MEMS resonator.

W-195 EXPERIMENTAL INVESTIGATION ON MODE COUPLING OF BULK MODE SILICON MEMS RESONATORS 1008

Y. Yang[1], E. Ng[1], P. Polunin[2], Y. Chen[1], S. Strachan[2], V. Hong[1], C.H. Ahn[1], O. Shoshani[2], S. Shaw[2], M. Dykman[2], and T. Kenny[1]
[1]Stanford University, USA and [2]Michigan State University, USA

We present the effect of nonlinear elasticity on the coupling between different bulk modes of silicon MEMS resonators. From experimental data, the coupling has a strong dependence on the order and the shape of the coupled resonant modes, as well as the doping type/concentration, and crystal orientation, leading to a variety of complex and potential useful phenomena.

M-196 MEMS-BASED RF PROBES FOR ON-WAFER MICROWAVE CHARACTERIZATION OF MICRO/NANOELECTRONICS 1012

J. Marzouk, S. Arscott, A. El Fellahi, K. Haddadi, T. Lasri, L. Buchaillot, and G. Dambrine
University Lille 1, FRANCE

We demonstrate a radio frequency (RF) probe based on microelectromechanical systems (MEMS) design and processing technologies. The probe responds to the current needs of microelectronics requiring microwave characterization of nanoscale devices and systems having sub-micron pad sizes. The use of MEMS technologies enables the probe contact pad area dimensions to be reduced by a three orders of magnitude compared to existing commercial RF probes.

T-197 MICROMECHANICAL RING RESONATORS WITH A 2D PHONONIC CRYSTAL SUPPORT FOR MECHANICAL ROBUSTNESS AND PROVIDING MASK MISALIGNMENT TOLERANCE 1016

B. Figeys[1,2], B. Nauwelaers[2], H.A.C. Tilmans[1], and X. Rottenberg[1]
[1]imec, BELGIUM, [2]KU Leuven, BELGIUM

This paper reports on the design of ring-type electrostatically transduced bulk acoustic wave resonators designed for increased shock and vibration resistance. This was achieved through a 2D Phononic Crystal (PnC) support, simultaneously a mechanically strong and acoustically well-confined support. We manufactured SiGe-resonators at 137.8MHz with a Q-factor around 15k. Another feature to this design is the process tolerance of the Q-factor towards mask misalignment for the center support.

M-198 SILICON-MICROMACHINED SPACERS FOR UHF CAVITY RESONATORS 1020

D. Psychogiou, M.D. Sinani, and D. Peroulis
Purdue University, USA

This paper reports on a novel hybrid integration concept that enables the realization of high-quality (*Q*) factor, low-frequency cavity resonators with well-defined capacitive-loading and variable center frequency. It is based on a silicon-micromachined spacer that is mounted on top of a conventional CNC-machined metallic cavity to functionalize the resonator's capacitance. For the first time, it is demonstrated that low-frequency resonators with micrometer-scale gaps (10s of microns), relatively large *Q*-factor (459-505) and tunable response (18.5%) can be constructed without the need for post-fabrication tuning. To demonstrate these benefits, a resonator assembly was designed, built and experimentally tested at UHF band and for a frequency tuning range between 1424-1711 MHz.

W-199 SIMULTANEOUS MULTI-FREQUENCY SWITCHABLE OSCILLATOR AND FSK MODULATOR BASED ON A CAPACITIVE-GAP MEMS DISK ARRAY 1024

T.L. Naing, T.O. Rocheleau, and C.T.-C. Nguyen
University of California, Berkeley, USA

An array of capacitive-gap MEMS resonators with different frequencies combined with an ASIC amplifier, provides a first MEMS-based multi-frequency oscillator generating simultaneous oscillation outputs around 62MHz while employing only a single amplifier. Enabled via a softening non-linearity, amplitude is limited here for each MEMS resonator individually. Furthermore, electrical stiffness frequency tuning enables FSK modulation of the output waveform, offering a simple multichannel transmitter.

M-200 TAPERED PHONONIC CRYSTAL SAW RESONATOR IN GAN 1028

S. Wang, L.C. Popa, and D. Weinstein
Massachusetts Institute of Technology, USA

This paper presents a new Phononic Crystal (PnC) resonator design in which a tapered PnC is used to confine a 970 MHz SAW resonance in a GaN-on-Si platform. The use of a tapered PnC reflector in this work reduces the footprint of SAW resonators by >100X relative to the case of conventional metal grating reflectors while maintaining high Q. A 3.5X improvement in Q is experimentally demonstrated relative to uniform PnC reflectors of comparable dimensions.

THz MEMS Components and Systems

T-201 ENHANCED CONTROLLABILITY IN MEMS METAMATERIALS 1032

P. Pitchappa[1], C.P. Ho[1], Y. Qian[1], Y.-S. Lin[1], N. Singh[2], and C. Lee[1]
*[1]National University of Singapore, SINGAPORE and
[2]Agency for Science, Technology and Research (A*STAR), SINGAPORE*

We demonstrate a method to improve the controllabilty of the MEMS tunable metamaterials by individually actuating the alternate lines in the metamaterial array. This is the first step towards the realization of Programmable metamaterial, where each of the unit cell can be addressed independently.

PowerMEMS and Actuators
Actuator Components and Systems

W-202 **BI-DIRECTIONAL EXTENDED RANGE PARALLEL PLATE ELECTROSTATIC ACTUATOR BASED ON FEEDBACK LINEARIZATION** 1036

E.E. Moreira[1], F.S. Alves[1], R.A. Dias[2], M. Costa[2], H. Fonseca[2], J. Cabral[1], J. Gaspar[2], and L.A. Rocha[1]

[1]University of Minho, PORTUGAL and [2]International Iberian Nanotechnology Laboratory, PORTUGAL

A bi-directional extended range parallel-plate electrostatic actuator using feedback linearization control is presented in this paper. The actuator can have stable displacements up to 90% of the full-gap (limited by mechanical stoppers) on both directions, i.e, the device can move ±2um within a ±2.25um gap. The system has successfully tracked references until 1 kHz (limited by the dynamics of the device) and it presents a capacitor tuning rage of 17, using an actuation voltage from 0 to 10V.

M-203 **BIOCOMPATIBLE CIRCUIT-BREAKER CHIP FOR TEMPERATURE REGULATION OF ELECTROTHERMALLY DRIVEN SMART IMPLANTS** 1040

Y. Luo, M. Dahmardeh, and K. Takahata

University of British Columbia, CANADA

We present a thermoresponsive circuit breaker micromachined in a form of titanium-packaged chip for biomedical applications with a focus on electronic implants. This micro breaker has a temperature-sensitive cantilever actuator to serve as an absolute temperature limiter for the device of interest being protected from overheating, a critical safety feature for smart implants including those that are electrothermally active. Temperature regulation of a wireless heater powered by external RF field

T-204 **CONTROLLABLE 'SOMERSAULT' MAGNETIC SOFT ROBOTICS** 1044

T.S. Zhang, A. Kim, M. Ochoa, and B. Ziaie

Purdue University, USA

This paper reports controllable somersault magnetic soft robotics consisting of polymeric magnet embedded in a high friction silicone polymer. The soft structures are actuated by the rotation of a permanent magnet at fixed position and exhibit controllable linear movement in a flip-and-forward manner for extended distance on both horizontal and vertical non-magnetic surfaces. The control of direction of motion is also achieved.

W-205 **CYBORG BEETLE: THRUST CONTROL OF FREE FLYING BEETLE VIA A MINIATURE WIRELESS NEUROMUSCULAR STIMULATOR** 1048

T.T. Vo Doan, Y. Li, F. Cao, and H. Sato

Nanyang Technological University, SINGAPORE

We have developed a cyborg beetle, which is the hybrid of a miniature wireless communication system and a living beetle platform. We can remotely stimulate neuromuscular sites of the living beetle platform via the miniature system. In this study, we stimulated the subalar flight muscle, a major muscle directly inserted to the wing base of beetle, and demonstrated the thrust control of the cyborg beetle with graded response.

M-206 **CYLINDRICAL HALBACH MAGNET ARRAY FOR ELECTROMAGNETIC VIBRATION ENERGY HARVESTERS** 1051

I. Shahosseini and K. Najafi

University of Michigan, Ann Arbor (WIMS[2]), USA

This paper reports the design, optimization, and test results of a new magnetic structure for kinetic energy harvesters allowing seven-fold increase in power density compared to single-magnet configuration. For the first time, electromagnetic energy harvesters with "single cylindrical Halbach array" and "double-concentric Halbach array" magnetic structures are fabricated and tested.

T-207 **FLUID SEPARATED VOLUMETRIC FLOW CONVERTER (FSVFC) FOR HIGH SPEED AND PRECISE CELL POSITION CONTROL** 1055

T. Monzawa[1], S. Sakuma[2], F. Arai[2], and M. Kaneko[1]

[1]Osaka University, JAPAN and [2]Nagoya University, JAPAN

This paper proposes the on-chip Fluid Separated Volumetric Flow Converter (FSVFC) capable of high speed cell position control with high resolution, while the actuation fluid is physically separated from working fluid for biological considerations. By utilizing the newly developed on-chip comb shaped FSVFC, an online high speed vision sensor and a high speed PZT, we succeeded in controlling the position of a cell in microfluidic channel with the time constant of 12 ms and the resolution of 240 nm

W-208 **ON-CHIP ENUCLEATION USING AN UNTETHERED MICROROBOT INCORPORATED WITH AN ACOUSTICALLY OSCILLATING BUBBLE** 1059

I.S. Park, Y.R. Lee, S.J. Hong, K.Y. Lee, and S.K. Chung

Myongji University, SOUTH KOREA

This paper reports a novel on-chip enucleation method using an untethered microrobot incorporated with an acoustically excited microbubble, which will allow minimally invasive cell surgery for cloning techniques and biomedical applications. The proposed microrobot mainly consists of a compressible bubble for the manipulation of cells and twin permanent magnets for the manipulation of the microrobot in an aqueous medium.

M-209 OUT-OF-PLANE MICRO-FORCE FUNCTION GENERATOR WITH INHERENT SELF-FEEDBACK FOR MICRO-DEFORMATION MODIFYING 1063

X. Wang, D. Xiao, X. Wu, Z. Hou, Z. Chen, and H. He
National University of Defense Technology, CHINA

We propose a novel concept of out-of-plane micro-force function generator for micro-deformation modifying. The proposed generator is based on batch micro-fabricated polymer thermal actuators array and could actively modify micro-substrate warpage. This strategy constructively utilizes the inherent self-feedback for in-situ deformation control and has the potential for solving stress-induced problems of micro-fabricated devices.

Manufacturing for Actuators and PowerMEMS

T-210 A PAPER-LIKE MICRO-SUPERCAPACITOR WITH PATTERNED BUCKYPAPER ELECTRODES USING A NOVEL VACUUM FILTRATION TECHNIQUE 1067

C.-W. Ma, P.-C. Huang, and Y.-J. Yang   *Award Nominee*
National Taiwan University, TAIWAN

This study reports a paper-like micro-supercapacitor with in-plane interdigital buckypaper electrodes on a filter membrane substrate. A vacuum filtration method assisted by lithography techniques is proposed for patterning buckypaper. The proposed micro-SC features advantages including a flexible structure, simple fabrication, easy chip integration, and high specific capacitance. The specific capacitance measured by cyclic voltammetry was 107.27 mF/cm2 at a scan rate of 20 mV/sec.

W-211 A POTASSIUM ELECTRET ENERGY HARVESTER FOR 3D-STACK ASSEMBLY 1071

K. Misawa[1], T. Sugiyama[2], G. Hashiguchi[2], H. Fujita[1], and H. Toshiyoshi[1]
[1]University of Tokyo, JAPAN and [2]Shizuoka University, JAPAN

We report an electrostatic energy harvester based on the potassium ion (K+) electret that could be stacked up into a 3D structure to multiply the output power. Vertical comb electrodes are implemented in a silicon-on-insulator (SOI) wafer with a relatively heavy mass in the handle layer to lower the resonance. A single substrate formation exhibited a 0.34 μW output at 310 Hz for a load resistance of 1 MOhm.

M-212 FABRICATION OF A THREE DIMENSIONAL CANTILEVERED VIBRATIONAL ENERGY HARVESTER USING SILVER INK 1075

S.-J. Chen, Y.-Y. Feng, and S.-Y. Liu
National Central University, TAIWAN

This paper reports a 3D micro electromagnetic energy harvester. Compared to state of the art, multiple layers of conductive coils are dispensed on the micro-machined cantilever diaphragm by an injecting machine, which will increase the potential power density of the harvester.

T-213 WAFER-LEVEL FABRICATION OF A TRIBOELECTRIC ENERGY HARVESTER 1078

M. Han, B. Yu, Z. Su, B. Meng, X. Cheng, X.-S. Zhang, and H. Zhang
Peking University, CHINA

We present a wafer-level fabrication method for triboelectric energy harvester (TEH), which, for the first time, fabricates the TEH completely in MEMS process, without any manually assembly. Compared to state of the art, the proposed TEH is batch fabricated in CMOS-compatible process and the reduced size allows it to be integrated with other electronic devices (e.g., keyboards). This device can produce 235 mV peak voltage at the frequency of 30 Hz, under the 100 MΩ external resistance.

Materials for Actuators and PowerMEMS

W-214 DISPLACEMENT MAGNIFICATION OF GEL ACTUATOR USING pNIPAAm-SU8 HYBRID STRUCTURE 1082

T. Kogure, R. Okada, S. Maeda, and S. Nagasawa
Shibaura Institute of Technology, JAPAN

A fabrication method for a hybrid structure of the poly-N-Isopropylacrylamide (pNIPAAm) gel as a soft material and the SU-8 as a solid material is proposed. This pNIPAAm-SU8 hybrid structure is utilized various applications such as gel actuators. Our hybrid structure was not broken through repeating swelling-shrinking states for 10 times. The SU-8 solid structure magnified a minute displacement of the gel which is occurred by the phase transition.

M-215 LIQUID-TOLERANT ELECTRET USING SUPER-LYOPHOBIC PILLAR SURFACE 1086

Y.-C. Chen, K.-Y. Song, K. Morimoto, and Y. Suzuki
University of Tokyo, JAPAN

For the first time, electret that can be used in the liquid environment has been realized toward higher energy density of electret generators and actuators. By using super-lyophobic overhanging pillar surface with SiO2 layer, stable Cassie-Baxter state for low-surface-tension liquid is sustained even with high surface potential. The pillar surface is successfully charged with soft X-ray photoionization. Surface potential after liquid contact has been significantly improved with the pillars.

T-216 GRAPHENE ONE-SHOT MICRO-VALVE: TOWARDS VAPORIZABLE ELECTRONICS 1090

V. Gund, A. Ruyack, S. Ardanuc, and A. Lal
Cornell University, USA

We report a micro-scale arrayable single-trigger valve of graphene transferred on silicon-nitride for vaporizable electronics. Graphene serves as a nanoscale barrier to oxygen diffusion and as a resistive heater for pulsed-power thermomechanical cleaving to expose sealed alkali metals for heat generation to vaporize polymer electronics. Our valve demonstrates long-storage lifetime, durability under pressure and low-power triggering.

W-217 PIEZOELECTRIC MICRO ENERGY HARVESTERS EMPLOYING ADVANCED (MG,ZR)-CODOPED ALN THIN FILM 1094
L.V. Minh[1], M. Hara[1], H. Kuwano[1], T. Yokoyama[2], T. Nishihara[2], and M. Ueda[3]
[1]Tohoku University, JAPAN, [2]Taiyo Yuden Co., Ltd., JAPAN, and [3]Taiyo Yuden Mobile Technology Co., Ltd., JAPAN

We report the new doped-AlN thin film, (Mg,Zr)AlN, based micro energy harvester. By co-doping Mg and Zr into AlN crystal, (Mg,Zr)AlN shows giant piezoelectricity and preserves low permittivity. (Mg,Zr)AlN has higher figure of merit (FOM=$e_{31}^2(\varepsilon_0\varepsilon)$)) than conventional PZT. The 13 at%-(Mg,Zr)AlN had the experimental FOM of up to 16.7 GPa. The micromachining harvester provided the high normalized power density of 3.72 mW.g^{-2}.cm^{-3}. This achievement was 1.5-fold increase compared to state of the art.

M-218 PULSE POLING WITHIN 1 SECOND ENHANCE THE PIEZOELECTRIC PROPERTY OF PZT THIN FILMS 1098
T. Kobayashi[1], Y. Suzuki[1,2], N. Makimoto[1], H. Funakubo[3], and R. Maeda[1]
[1]National Institute of Advanced Industrial Science and Technology (AIST), JAPAN,
[2]Ibaraki University, JAPAN, and [3]Tokyo Institute of Technology, JAPAN

We present simple but fast poling technique to enhance the piezoelectric property of PZT thin films. Application of pulse voltage to the PZT thin films on MEMS microcantilevers has resulted in large piezoelectric constant (d31) as high as 105 pm/V. It took only 1 second for poling the PZT thin films.

Nanoscale Actuators and PowerMEMS

T-219 A BETAVOLTAIC MICROCELL BASED ON SEMICONDUCTING SINGLE-WALLED CARBON NANOTUBE ARRAYS/SI HETEROJUNCTIONS 1102
M.G. Li and J. Zhang
Peking University, CHINA

This paper reports a novel betavoltaic microcell based on semiconducting single-walled carbon nanotubes (s-SWCNTs). The aligned arrays of p-type s-SWCNTs were prepared onto n-type silicon forming the p-n heterojunction as the energy conversion. This heterojunction displays good rectification characteristics with I_0=1.5pA and n=1.83. Under 7.8mCi/cm^2 ^{63}Ni irradiation, the microcell achieves higher performance of V_{OC}=62mV, J_{SC}=3.8μA/cm^2, FF=33.4% and η=9.8% compared with our previous devices.

Other Actuators and PowerMEMS

W-220 A MICROGRIPPER WITH A RATCHET SELF-LOCKING MECHANISM 1106
Y.C. Hao, W.Z. Yuan, H.M. Zhang, and H.L. Chang
Northwestern Polytechnical University, CHINA

This paper reports a novel electrostatic actuated microgripper with a ratchet self-locking mechanism which enables the longtime gripping without continuously applying the external driving signal such as electrical, thermal or magnetic fields. This greatly reduces the influence and damage on the gripped micro objects induced by the external driving signals.

PowerMEMS Components and Systems

M-221 A HIGH-EFFICIENT BROADBAND ENERGY HARVESTER BASED ON NON-CONTACT COUPLING TECHNIQUE FOR AMBIENT VIBRATIONS 1110
X. Wu and D.W. Lee
Chonnam National University, SOUTH KOREA

In this work, a high-efficient piezoelectric energy harvester based on non-contact coupling technique is proposed and characterized, which allows it, for the first time, to take advantage of multi-cantilevers and frequency-up conversion technique to enhance the power generation efficiency for ambient excitation. The unique energy harvester can effectively scavenge environmental vibration energy with a wide bandwidth. Aiming for high space efficiency, folded cantilevers are designed.

T-222 BIDIRECTIONAL THERMOELECTRIC ENERGY GENERATOR BASED ON A PHASE-CHANGE LENS FOR CONCENTRATING SOLAR POWER 1114
M.S. Kim, M.K. Kim, H.R. Ahn, and Y.J. Kim
Yonsei University, SOUTH KOREA

This paper reports a bidirectional thermoelectric energy generator (TEG) with double type lenses for concentrating solar power. When solar power was applied to the TEG, solar energy is concentrated by PMMA lens firstly. The concentrated energy is absorbed as heat energy through phase-change of PCM. And then, the liquid PCM lens focuses energy on the TEG. After removing energy source, the latent heat in PCM is released. Therefore, the proposed TEG generates energy steadily.

W-223 BIOTEMPLATED HIERARCHICAL NICKEL OXIDE SUPERCAPACITOR ELECTRODES 1118
S. Chu, K. Gerasopoulos, and R. Ghodssi
University of Maryland, College Park, USA

We present hierarchical Ni/NiO supercapacitor electrodes utilizing Tobacco mosaic virus (TMV) as bio-nanotemplates. The hierarchical electrodes were fabricated by integrating high aspect ratio silicon micropillars with thermally oxidized nickel-coated TMVs. An ultra-high areal capacitance of 585.9mF/cm2 was achieved with hierarchical Ni/NiO electrodes, exceeding the capacitance of nanostructured only and planar Ni/NiO by a factor of 3.4 and 29.7, respectively.

M-224 DOUBLY RE-ENTRANT CAVITIES TO SUSTAIN BOILING NUCLEATION IN FC-72 1122
T. Liu and C.-J. Kim
University of California, Los Angeles, USA

We report a micro/nano-machined surface cavity on which boiling nucleation resumes after ceasing in refrigerant FC-72 for a short time. Having the lowest surface tension of all liquids, FC-72 completely wets any existing material including Teflon so that all existing cavities get flooded once nucleation stops and could not restart boiling without excessive heat. We experimentally confirm the half-century old idea of doubly re-entrant cavities as a boiling site, encouraging further development.

T-225 EXPERIMENTALLY VERIFIED MODEL OF ELECTROSTATIC ENERGY HARVESTER WITH INTERNAL IMPACTS 1125
B.D. Truong, C.P. Le, and E. Halvorsen
Buskerud and Vestfold University College, NORWAY

We present experimentally verified progress on modeling of MEMS electrostatic energy harvesters with internal impacts on transducing end-stops. The model includes nonlinearities of the electromechanical transduction, the squeezed-film damping and the impact force. The comparison between simulation and measurement shows that these effects are crucial and gives good agreement for phenomenological parameters. This is a significant step towards accurate modeling of this complex system.

W-226 MONOLITHIC 2-AXIS IN-PLANE PZT LATERAL BIMORPH ENERGY HARVESTER WITH DIFFERENTIAL OUTPUT 1129
S. Nadig, S. Ardanuç, and A. Lal
Cornell University, USA

We report a novel 2-axis (X-Y) piezoelectric energy harvester, whose sensitive axis in-plane is rotationally invariant, a result achieved by spiral cascading of lateral bimorphs. This is different than conventional piezoelectric energy harvesters that are sensitive only along one axis or can realize multi-axis sensitivity through integration or assembly of multiple devices at different orientations.

M-227 SOLID-STATE FLEXIBLE MICRO SUPERCAPACITORS BY DIRECT-WRITE POROUS NANOFIBERS 1133
C. Shen[1], G. Luo[1], A. Kozinda[1], M. Sanghadasa[2], and L. Lin[1]
[1]University of California, Berkeley, USA and [2]US Army, USA

We report solid-state flexible micro supercapacitors based on direct-write porous polymer nanofibers. Compared with state-of-art supercapacitors, key innovations include: 1) porous 3D nanostructure of conductive nanofibers via the near-field electrospinning process; 2) flexible solid-state micro electrodes with high energy density using the pseudocapacitive effect; (3) simple and versatile process compatible with different substrates and surfaces.

T-228 STRETCHABLE WIRELESS POWER TRANSFER WITH A LIQUID ALLOY COIL 1137
S.H. Jeong[1] and Z.G. Wu[1,2]
[1]Uppsala University, SWEDEN and [2]Huazong University of Science and Technology, CHINA

A stretchable wireless power transfer (WPT) device was fabricated with a liquid alloy coil, which was integrated with rigid electronic chips in elastomer packaging. Tape transfer masking with spray deposition was applied for patterning a long coil of the liquid alloy. The WPT efficiency reached 10% at 140 kHz and worked with 25% strain. Different sizes of liquid alloy coils and soft magnetic composite cores were tested for a higher efficiency system.

W-229 THREE-AXIS PIEZOELECTRIC VIBRATION ENERGY HARVESTER 1141
E.E. Aktakka and K. Najafi
University of Michigan, USA

This paper reports for the first time a piezoelectric harvester for scavenging vibrational energy in all three-dimensions. The device is formed of optimized PZT/Si unimorph crab-legs such that the first three resonance modes are linear in-plane and out-of-plane vibrational modes with closely spaced frequencies. Partitioned electrodes collect vibrational energy in the transverse piezoelectric mode, and have different phases in their outputs according to the axis of the applied vibration.

M-230 VERTICAL CAPACITIVE ENERGY HARVESTER POSITIVELY USING CONTACT BETWEEN PROOF MASS AND ELECTRET PLATE - STIFFNESS MATCHING BY SPRING SUPPORT OF PLATE AND STICTION PREVENTION BY STOPPER MECHANISM 1145
T. Takahashi[1], M. Suzuki[1], T. Nishida[2], Y. Yoshikawa[2], and S. Aoyagi[1]
[1]Kansai University, JAPAN and [2]ROHM Co. Ltd., JAPAN

In a vertical capacitive energy harvester, two methods to effectively use the contact between proof mass and electret plate are proposed; one is to match the stiffness between plate and mass, which is effective to increase their contact duration. For this purpose, instead of a gel shock-absorber, a soft spring for supporting plate is employed. Another is to surely detach mass from plate after their contact, opposing electrostatic attraction. The output power was improved by 5 times up to 50 μW.

A MICROFLUIDIC APTASENSOR INTEGRATING SPECIFIC ENRICHMENT WITH A GRAPHENE NANOSENSOR FOR LABEL-FREE DETECTION OF SMALL BIOMOLECULES

Jaeyoung Yang[1], Cheng Wang[1,2], Yibo Zhu[1], Guohua Liu[2] and Qiao Lin[1]
[1]Columbia University, New York, USA
[2]Nankai University, Tianjin, China

ABSTRACT

We here present a microfluidic aptasensor that integrates aptamer-based selective analyte enrichment, isocratic elution and conductance-based graphene nanosensing, achieving sensitive and label-free detection of small biomolecules. An aptamer specific to a target analyte is immobilized on microbeads for selective enrichment and isocratic elution of the analyte. A conductance-based graphene nanosensor using a competitive assay format achieves label-free detection, with a high sensitivity due to surface binding-induced changes in carrier concentration in the bulk of graphene. Experimental results show that our integrated device is capable of detecting arginine vasopressin (AVP), a small peptide, at clinically relevant low concentrations (1-500 pM).

INTRODUCTION

Graphene is a one-atom thick nanomaterial consisting of carbon atoms arranged as a two-dimensional (2D) hexagonal crystalline form [1]. With its charge carriers confined within an atomically thin layer, graphene exhibits high carrier mobility/capacity [2] and intrinsically low electrical noise [3]. Owing to these superior electrical properties in addition to its large surface-area-to-volume ratio and good biocompatibility, graphene has been used in biosensors that transduce biological reactions into measurable electrical signals [4]. Due to the fact that every carbon atom is located on the graphene surface with its 2D structure, the adsorption of charged molecules onto the surface induces changes in carrier concentration in the bulk of graphene, thus enabling the sensitive analyte detection based on conductance changes. For specific analyte detection in the conductance-based sensing scheme, graphene has been functionalized with molecular probes, such as antibodies [5] or aptamers [6, 7]. However, their application to detection of small biomolecules is still limited, as the binding of low charged small molecules does not directly induce detectable changes in graphene conductance, hence suffering low sensitivity.

This paper presents an integrated microfluidic aptasensor which combines aptamer-based specific enrichment and conductance-based graphene nanosensing, allowing detection of a low charged and small biomolecule with high sensitivity. An aptamer specific to a target analyte is immobilized on microbeads for solid phase-based selective enrichment and isocratic elution of the analyte. A field-effect transistor (FET) configuration is used to measure changes in graphene conductance, which are used to determine the analyte concentration. Microfluidic integration of enrichment and sensing on a single chip eliminates the need for off-chip sample handling, thus minimizing potential sample contamination or loss. For demonstration, arginine vasopressin (AVP), a small peptide and a clinically important biomarker for septic and hemorrhagic shocks, is chosen as a target analyte. Experimental results show that our integrated aptasensor is capable of detecting label-free AVP at clinically relevant low concentrations (1-500 pM).

PRINCIPLE AND DESIGN

The principle of AVP detection in our device is schematically shown in Fig. 1. AVP molecules in a sample ("sample AVP") are selectively captured by an aptamer immobilized on microbead surfaces at 37 °C (Fig. 1a), and are enriched by continuous sample infusion into the device. After the removal of nonspecifically adsorbed AVP molecules and impurities through buffer washing (Fig. 1b), the enriched AVP molecules are released at an elevated temperature (55 °C) and eluted with solution-borne aptamer molecules ("free aptamer"), causing isocratic elution of AVP with the free aptamer (Fig. 1c).

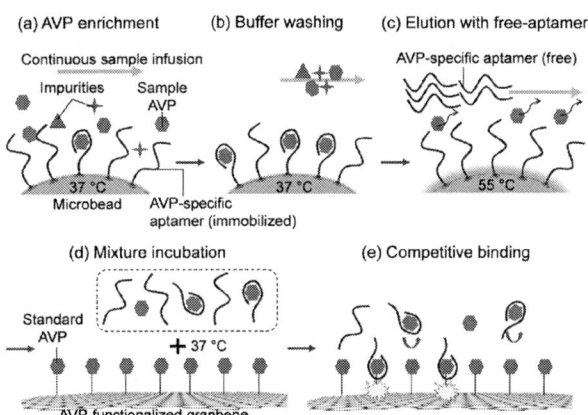

Figure 1: Principle of AVP detection. Sample AVP is enriched on microbead surfaces by aptamer binding (a). After the buffer washing (b), the temperature is raised to 55 °C, which disrupts aptamer-AVP complexes and release sample AVP into a free aptamer solution (c). The mixture of the free aptamer and released sample AVP is incubated with graphene functionalized with standard AVP (d), inducing the binding of the free aptamer to the standard AVP on graphene via competitive binding (e), thus changing the graphene conductance.

The eluate, a mixture of the free aptamer and released AVP originally captured from the sample ("sample AVP"), is then incubated with graphene pre-functionalized with reagent AVP ("standard AVP") at 37 °C (Fig. 1d). The graphene-bound standard AVP competes with the solution-phase sample AVP to bind to and capture some of the free aptamer molecules, causing an increase in the charge on the graphene (Fig. 1e). Hence, more sample AVP eluted in a mixture results in less aptamer molecules bound on the graphene surface, inducing smaller changes in graphene conductance.

As shown in Fig. 2, our integrated microdevice consists of an enrichment microchamber (height: 200 μm, volume: 2.5 μl), a graphene-based sensing microchamber (height: 1 mm, volume: ~5 μl) combined with a graphene nanosensor, three temperature control units (Cr/Au resistive microheater and temperature sensor), and a flow gate connected to a serpentine channel (width: 500 μm, height: 20 μm) for sample transfer to the sensing microchamber. A weir structure (height: 20 μm) in the enrichment microchamber retains microbeads (diameter: 50 to 80 μm) within the chamber during the selective enrichment process. Three sets of a resistive heater and a temperature sensor separately control temperatures inside two microchambers and the serpentine channel for forming or disrupting aptamer-AVP complexes.

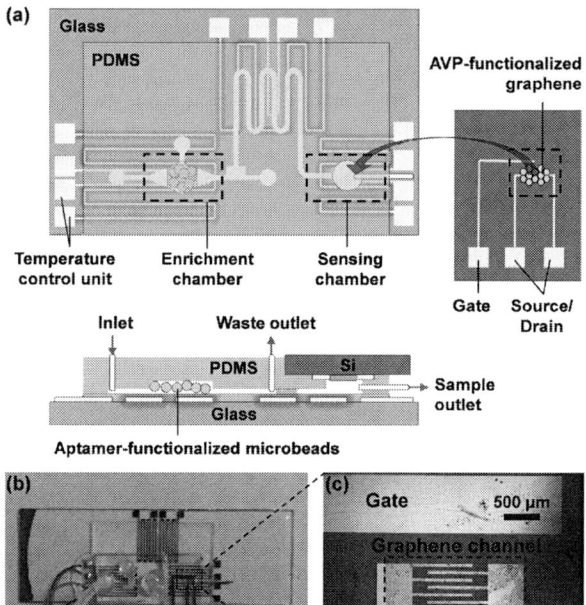

Figure 2: Schematics (a) and a photo (b) of the device. A micrograph of the graphene nanosensor (c).

A graphene nanosensor of a FET configuration is shown in Fig. 2c. The source and drain electrodes are connected by graphene which serves as a conducting channel. A voltage is applied to the gate electrode forming a gate voltage (V_{gs}) through a sample solution. In this FET configuration, the electrical conductance through graphene, which depends on the charge on the graphene surface, was measured from the drain current (I_{ds}) at a fixed drain voltage (V_{ds}).

EXPERIMENTAL

The microfluidic device was fabricated as follows. Graphene was synthesized via chemical vapor deposition

(CVD), and transferred onto gold electrodes that were fabricated on a silicon dioxide-coated silicon substrate. The graphene surface was then functionalized with AVP by incubating in 1-pyrenebutanoic acid succinimidyl ester (PASE linker, 2 mM) solution for 1 hr and AVP solution (1 μM) for 4.5 hrs, sequentially. A polydimethylsiloxane (PDMS) sheet defining microchambers and channels was fabricated via soft lithography, followed by punching through-holes which are used as inlet/outlet ports and an open chamber. The sheet was then bonded onto a glass substrate on which a resistive microheater and temperature sensor was patterned via photolithography. The graphene sensor chip was finally stacked and bonded on the PDMS layer, while the AVP-functionalized graphene region was included within the open sensing microchamber.

In experiments, the device was initially rinsed with buffer (10 μl/min), and the sensing microchamber was filled with a buffer solution to measure a reference signal ($I_{ds,ref}$). A sample solution (10 μl/min for 2 hrs) prepared in buffer with varying AVP concentrations was continuously introduced into the enrichment microchamber using a syringe pump (KD210P, KD Scientific Inc.) while the chamber was maintained at 37 °C using a temperature control unit (an integrated microheater and a temperature sensor controlled by LabVIEW) integrated on the bottom glass substrate. Following the enrichment, the chamber was thoroughly rinsed with a buffer solution (10 μl/ min) at 37 °C. After the waste outlet was closed, a plug of the free aptamer solution at 25 nM was flowed through (10 μl at 10 μl/ min) the enrichment microchamber at an elevated temperature, 55 °C. The mixture of eluted sample AVP and the free aptamer was then transferred to the sensing microchamber whose temperature is kept at 37 °C.

The measurement circuit used a DC power supply (E3631A, Agilent) to provide the drain voltage V_{ds}, a function generator (33220A, Agilent) to supply the gate voltage V_{gs}, and a digit multimeter (34410A, Agilent) to measure the drain current (I_{ds}). During the electrical measurement, I_{ds} values (at fixed V_{gs} and V_{ds}) were automatically collected once per second for a determined period through a PC-based LabVIEW.

RESULTS AND DISCUSSION

We first characterized the aptamer binding-induced change in graphene conductance using our graphene nanosensor. A free aptamer solution at different concentrations in the range of 0-1000 nM was incubated with the graphene nanosensor until I_{ds} reached a saturated level. The magnitude of the conductance difference, $\delta I_{ds} = |I_{ds} - I_{ds,ref}|$, was found to increase with the free aptamer concentration and saturate above 100 nM, indicating that the increase of surface charge arises from the affinity binding of highly-charged nucleic acid aptamer molecules to standard AVP on the graphene surface. Least square fitting of a monovalent binding model [8] to the experimental data yielded a dissociation constant (K_d) of 16.7 nM for aptamer-AVP binding, approximately 10 times higher than the value previously obtained in solution (1.7 nM [9]), which could be attributed to the restriction of configurational freedom of surface-immobilized AVP, which substantially reduces their binding kinetics.

978-1-4799-7956-1/15 $31.00 © 2015 IEEE   570

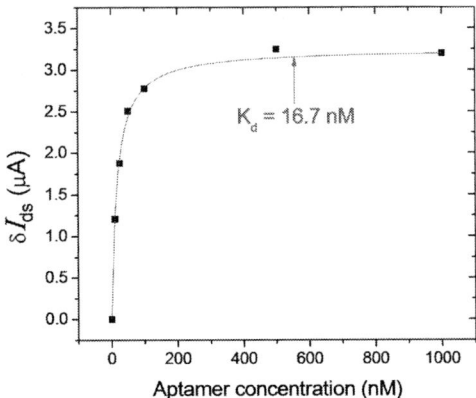

Figure 3: Characterization of the graphene nanosensor by measuring changes in graphene conductance with varying free aptamer concentrations (0-1000 nM).

The nanosensor was then tested with standard mixtures of the free aptamer at a fixed concentration (25 nM) and sample AVP at a varying concentration (c_{AVP}, 1-100 nM). Results showed that δI_{ds} decreased with increasing c_{AVP} in the range of 1-100 nM, which is attributed to that a higher AVP concentration in a standard mixture causes a higher aptamer occupancy, which impeded the aptamer binding to surface-immobilized standard AVP (Fig. 4). However, no appreciable variation in δI_{ds} with varying c_{AVP} was observed below 1 nM (data not shown), necessitating the enrichment of sample AVP prior to graphene nanosensing to achieve the AVP detection in the low picomolar range, required in clinical settings.

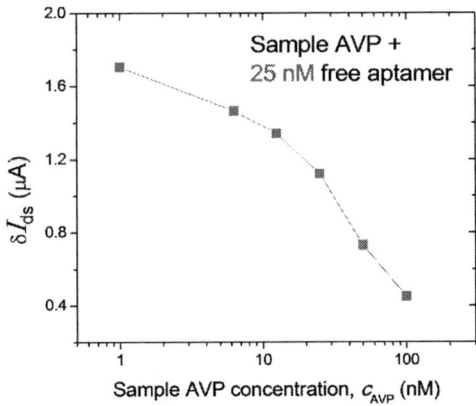

Figure 4: Measurements with standard mixtures of free aptamer at 25 nM and sample AVP at varying concentrations (1-100 nM).

We characterized the aptamer-based selective enrichment of AVP on microbead surfaces via fluorescence measurements. A TAMRA (as a fluorophore)-labeled AVP (TVP) solution at 100 pM was continuously infused into the enrichment chamber (10 μl/min) while the chamber temperature was maintained at 37 °C, and the fluorescence intensity of the bead surface was measured every 15 min. It was observed that mean fluorescence intensities increased over time and saturated after 1.5 hr (Fig. 5), suggesting that continuous sample infusion for more than 1.5 hr is needed to attain sufficient

analyte enrichment. The thermally activated release of captured AVP molecules was also tested by flushing beads with a free aptamer solution at 55 °C. The fluorescence intensity was quickly (< 1 min) dropped down to the baseline, indicating that enriched sample AVP molecules were released from aptamer at 55 °C which disrupts aptamer-AVP complexes [10]. This temperature-dependent behavior of aptamer-AVP binding enables isocratic elution of the analyte, allowing subsequent graphene nanosensing.

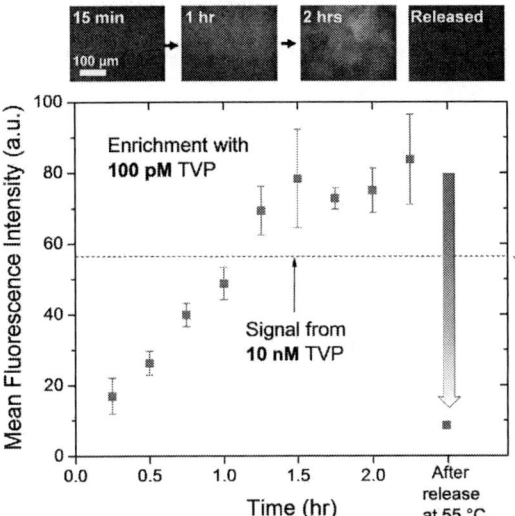

Figure 5: Fluorescence based time resolved measurements of aptamer-based specific enrichment during continuous introduction of a 100 pM TAMRA-labeled AVP (TVP) solution at 37 °C, followed by infusing a free aptamer solution at 55 °C for thermally-activated release of AVP.

To verify the specificity of aptamer-based enrichment, we tested integrated devices with and without aptamer on microbead surfaces for AVP detection. Devices with bare beads produced higher δI_{ds} values than those with aptamer-functionalized beads (Fig. 6) at 100 pM AVP. This indicates that the sample AVP was significantly enriched by aptamer-based specific recognition on the bead surface, thus resulting in the higher sample AVP concentration in a mixture.

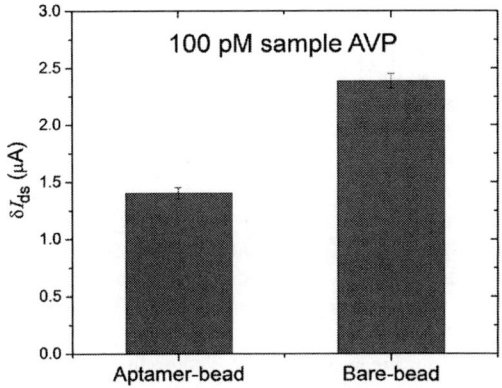

Figure 6: Bead-control experiments. Integrated devices with and without aptamers on microbeads were tested to verify aptamer-based specific enrichment.

We finally tested detection of AVP at physiologically relevant picomolar concentrations (1-500 pM) using the integrated device. To account for experimental variability caused by device-to-device variation of the graphene nanosensor, we normalized δI_{ds} with respect to $\delta I_{ds,max}$, the maximum conductance change of an individual graphene nanosensor. Experimental results showed that δI_{ds} decreased with increasing c_{AVP} (Fig. 7), implying that a higher c_{AVP} results in a higher aptamer occupancy in the mixture, thus less free aptamer molecules can bind to standard AVP on graphene. The results also confirmed the effectiveness of aptamer-based selective enrichment by demonstrating the picomolar detection that was difficult to attain without the enrichment process. It can thus be concluded that our approach is capable of quantitatively detecting AVP at picomolar levels, confirming the potential of our device as a quantitative AVP assay in clinical diagnostics. We speculate that the large error bars could arise from a variation of microbead packing density in the enrichment chamber between devices.

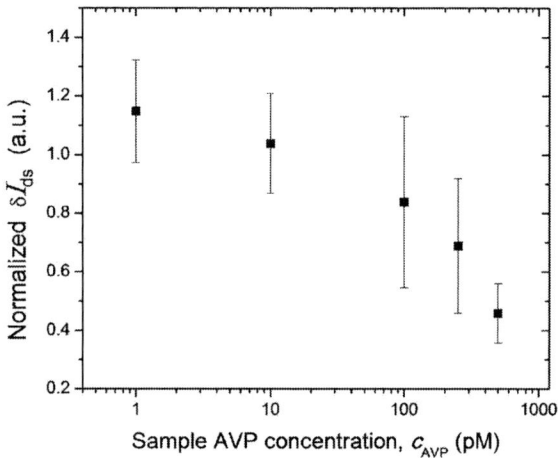

Figure 7: Normalized signals obtained from experiments with varying sample AVP concentrations (1-500 pM). Three independent experiments were conducted at each concentration.

CONCLUSION

We have developed a microfluidic aptasensor for conductance-based graphene nanosensing of AVP with high sensitivity. The device integrates aptamer-based specific enrichment and graphene nanosensing on a single chip, achieving label-free detection of low-abundance AVP. A bead-immobilized AVP-specific aptamer conducted the selective enrichment of AVP molecules from a dilute sample, thus improving the detection sensitivity of the graphene nanosensor. After the enrichment, the AVP concentration was measured by graphene nanosensing via competitive aptamer binding between enriched sample AVP in solution and standard AVP immobilized on graphene. Following their individual characterization, the processes of aptamer-based enrichment and graphene nanosensing were combined to enable quantitative detection of label-free AVP at physiologically relevant concentrations (1-500 pM),

showing that conductance changes on graphene were inversely related to the AVP concentration. These results suggest that our integrated device can be used to sensitively detect small biomolecule analytes in a label-free and quantitative manner.

ACKNOWLEDGEMENTS

We gratefully acknowledge the financial support from Raymond and Beverly Sackler program at the interfaces of biophysical and medical sciences at Columbia University.

REFERENCES

[1] K. S. Novoselov, A. K. Geim, S. V. Morozov, D. Jiang, Y. Zhang, S. V. Dubonos, I. V. Grigorieva, A. A. Firsov, "Electric Field Effect in Atomically Thin Carbon Films", *Science*, vol. 306, pp. 666-669, 2004.

[2] A. H. Castro Neto, F. Guinea, N. M. R. Peres, K. S. Novoselov, A. K. Geim, "The Electronic Properties of Graphene", *Rev. Mod. Phys.*, vol. 81, pp. 109-162, 2009.

[3] Y. Lin, P. Avouris, "Strong Suppression of Electrical Noise in Bilayer Graphene Nanodevices", *Nano Lett.*, vol. 8, pp. 2119-2125, 2008.

[4] Y. Shao, J. Wang, H. Wu, J. Liu, I. A. Aksay, Y. Lin, "Graphene Based Electrochemical Sensors and Biosensors: A Review", *Electroanal.*, vol. 22, pp. 1027-1036, 2010.

[5] S. Mao, G. Lu, K. Yu, Z. Bo, J. Chen, "Specific Protein Detection Using Thermally Reduced Graphene Oxide Sheet Decorated with Gold Nanoparticle-Antibody Conjugates", *Adv. Mater.*, vol. 22, pp. 3521-3526, 2010.

[6] Y. Ohno, K. Maehashi, K. Matsumoto, "Label-Free Biosensors Based on Aptamer-Modified Graphene Field-Effect Transistors", *J. Am. Chem. Soc.*, vol. 132, pp. 18012-18013, 2010.

[7] C. Wang, J. Kim, J. Zhu, R. Pei, G. Liu, J. Hone, M. Stojanovic, Q. Lin, "A Graphene Nanosensor for Detection of Small Molecules", *Proc. MEMS 2014*, pp. 1075-1078, 2014.

[8] D. A. Lauffenburger, J. J. Linderman, "Receptors: Models for Binding, Trafficking, and Signaling", *Oxford University Press*, 1996.

[9] W. G. Purschke, D. Eulberg, K. Buchner, S. Vonhoff, S. Klussmann, "An L-RNA-Based Aquaretic Agent that Inhibits Vasopressin In Vivo", *Proc. Natl. Acad. Sci.*, vol. 103, pp. 5173-5178, 2006.

[10] T. H. Nguyen, R. Pei, D. W. Landry, M. N. Stojanovic, Q. Lin, "Label-Free Microfluidic Characterization of Temperature-Dependent Biomolecular Interactions", *Biomicrofluidics*, vol. 5, pp. 034118, 2011.

CONTACT

*J. Yang, tel: +1-212-8544981; jy2344@columbia.edu

A MICROSIZED MICROBIAL FUEL CELL BASED BIOSENSOR FOR FAST AND SENSITIVE DETECTION OF TOXIC SUBSTANCES IN WATER

H. Lee, W. Yang, X. Wei, A. Fraiwan, and S. Choi[*]

Bioelectronics & Microsystems Lab., Department of Electrical & Computer Engineering,
State University of New York-Binghamton, USA

ABSTRACT

We report a microliter-sized (140 µL) microbial fuel cell (MFC)-based biosensor integrated with a three-electrode configuration and an air-bubble trap, in which microorganisms act as the sensor for toxic substances in water. The small-scale MFC biosensor produced favorable conditions for (i) reducing measurement time by increasing the probability of cell attachment and biofilm formation in the micro-sized chamber and (ii) enhancing sensitivity and reliability by providing a stable anodic potential and preventing air bubbles on the sensing surface. Using formaldehyde as a toxic component, the rapid current responses were obtained over a concentration range from 0.001% to 0.1% in a single chambered MFC biosensor with 0.2 V (versus Ag/AgCl reference electrode) applied on the anode.

INTRODUCTION

The National Academy of Engineering has identified "provide access to clean water" as one of fourteen "grand challenges" for engineering in the coming decades. The monitoring of water quality is important for providing safe and clean drinking water to the public [1]. Conventional techniques for monitoring the water quality are commonly based on physicochemical analyses, allowing accurate and sensitive detection of the various chemical compounds. However, conventional monitoring techniques are time-consuming, cumbersome and require a wide range of instrumentation. Moreover, these techniques cannot measure synergistic and antagonistic toxic effects that may be associated with biological and chemical mixtures [2]. Another approach for water quality monitoring is based on the use of living organisms such as fish, protozoans, algae, bivalves, and daphnids [3, 4]. This technique monitors a wide range of toxic components by detecting various changes in the behavior, survival, growth, or physiological conditions in those water organisms. However, it still suffers from evaluating the enormous amounts of information of continuously monitored organisms [1]. Therefore, simple, fast, sensitive and generic biosensors are needed for real-time applications.

Microbial fuel cells (MFCs) can potentially be used as a biosensor for the detection of water toxicity because a broad range of toxic components can inhibit bacterial metabolic activity (Fig. 1) [5, 6]. MFCs are bioelectrochemical systems that produce a current through bacterial metabolism [7, 8]. Consequently, the current generated from the MFC can be used as a measure of the water quality because a change in electrical current depends on several environmental factors including pH, dissolved oxygen, biochemical oxygen demand, and other organic compounds. In particular, the presence of toxic substances in water, such as formaldehyde, benzene, hexane toluene and heavy metals, significantly affect the

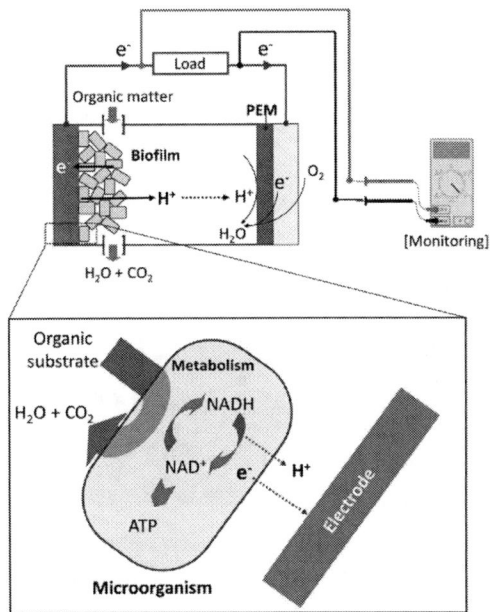

Figure 1: Schematic diagram of working principle of the MFC-based biosensor

bacterial metabolism and growth. Therefore, if the current drop is monitored from the MFC biosensor, an alarm is given and proper measures can be taken to protect our waterways from pollution [9]. MFCs have distinct advantages as a toxicity biosensor: (i) bacteria in MFCs will have fast response to changes in toxins, (ii) The self-sustainable nature of the MFCs will allow long-term online monitoring, (iii) MFCs will not need an external transducer and power source because the electrical output generated from the MFC can be a measure for toxic substances in water, (iv) mixed bacterial communities will have responses to a broad range of toxic components and (v) MFC device materials/fabrication/operation are cost-effective. Although the validation of the conceptual macro-sized MFCs as the biosensor was successful, efforts to miniaturize standard macro-sized MFCs for biosensor applications have been pursued by only a very small number of research groups [10,11]. Micro-sized MFCs offer inherently advantageous features such as large surface area-to-volume ratio, short electrode distance, and fast response time. However, micro-sized MFCs have been generally limited as a biosensor because of difficulties in stabilizing the baseline current with conventional two-electrode configuration and interferences with unwanted microbubbles in the chamber.

In this work, two novel technologies were integrated into the micro-sized MFC-based biosensor. First, the Ag/AgCl reference electrode was implemented into the device to avoid unwanted system perturbations and to provide more reliable and accurate information for a

978-1-4799-7956-1/15 $31.00 © 2015 IEEE

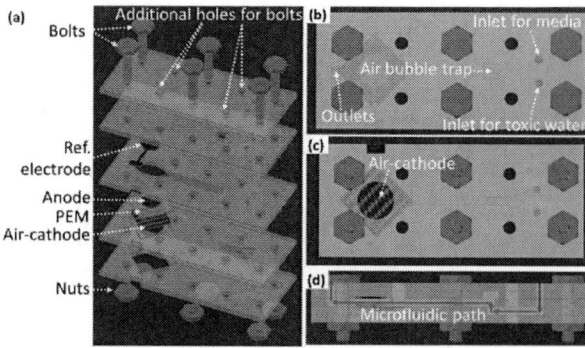

Figure 2: Schematic diagram of (a) the individual layers of the MFC biosensor, (b) top view, (c) bottom view, and (d) side view of the assembled device

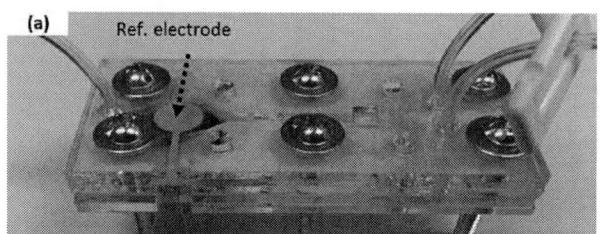

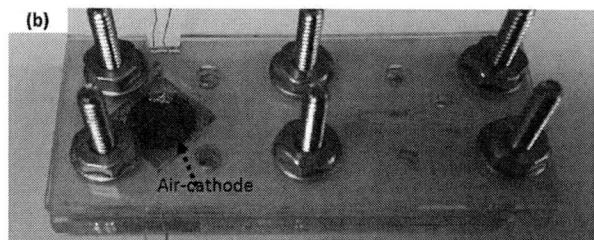

Figure 3: Photo-images of the assembled MFC biosensor. (a) Top and (b) bottom view

quantitative monitoring of the toxic substances in water [12,13]. Second, a microscale air bubble trap was integrated into the device. While macro-sized MFCs are not vulnerable to invading bubbles, micro-sized MFCs become sensitive to even tiny air bubbles [14]. This is because the size of the air bubble becomes comparable to that of the chambers and a small number of bacteria in the anode are more susceptible to bubbles. Moreover, they are difficult to be removed due to the slow flow rate of solutions. The proposed platform provides a dramatic increase in the detection speed and sensitivity.

MATERIALS AND METHODS
Device fabrication

Fig 2 & 3 shows schematic diagrams and photo images of the micro-sized MFC biosensor. Screen-printed Ag/AgCl ink covered by Nafion solution, carbon-cloth anode and air-cathode containing 0.5 mg/cm^2 of Pt catalyst formed a three-electrode configuration in a microfluidic MFC biosensor. The vertically-stacked anode and air-cathode were separated by a proton exchange membrane (Nafion 117). Unlike other conventional two-chambered devices, the MFC biosensor utilized the air-cathode to allow freely available oxygen to act as an

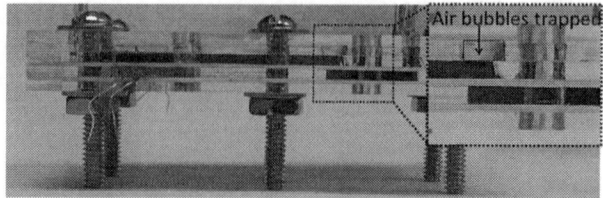

Figure 4: Photo-image of the device showing the microfluidic path and air bubbles trapped in a designed space

electron acceptor by the installation of the catalyst side of the air-cathode face toward the chamber while the opposite side was exposed to air [15]. Each Polymethlymethacrylate (PMMA) substrate, the main layer material, and silicone gaskets (0.02 inch) were precisely laser-machined to define the 140 µL chamber. The device had 3 holes for fluidic inlet/outlet and 10 holes for screws. All layers were manually stacked in sequence while carefully aligning the tubing holes for microfluidic channels. 3 tubes (CODAN, 0.35 mL volume) were plugged into the holes to form a Y-shape fluidic channel. The assembled device was sterilized with 70% ethanol and ultraviolet light for 24 hours.

Inoculum

We used wild-type *Shewanella oneidensis* as a model microorganism and cultured them in L-broth medium on a shaker at 35°C for 12 hours. The L-broth media contained 10.0 g triptone, 5.0 g yeast extract and 5.0 g NaCl per liter. To remove biomass, the cells were harvested by centrifugation and resuspended in a fresh L-broth medium.

RESULTS AND DISCUSSION
Air-bubble trap

Our previous report showed that intentionally induced bubble generation in the MFC anodic chamber reduced the current density by 33% and the MFC needed 4 days to recover its initial performance [14]. This is mainly because minute bubbles in micro-sized chambers can negatively affect bacterial growth and their subsequent electron transfer efficiency. Here, a power-free and passive air-bubble trap was designed to prevent microscale air bubbles from entering the MFC biosensor. In microfluidic channels, we provided a space for the upward floated bubbles and utilized the buoyancy of air bubbles in a liquid (Fig 4). Color ink was used to visualize the air bubbles being partially filled in the bubble trap. During the experiment, all unwanted bubbles were captured by the bubble trap, not interfering with sensor functionality.

Three-electrode MFC-based biosensor

Typically, small-scale MFCs consist of a two-electrode configuration; an anode and a cathode, which leads to an optimized solution for miniaturizing MFCs with modest power generation. However, three-electrode configuration having the reference electrode would provide more reliable, controllable, and accurate manipulation of the MFCs by providing optimized anodic potential. For example, by applying the anode potential at 0.1 V versus Ag/AgCl reference in the

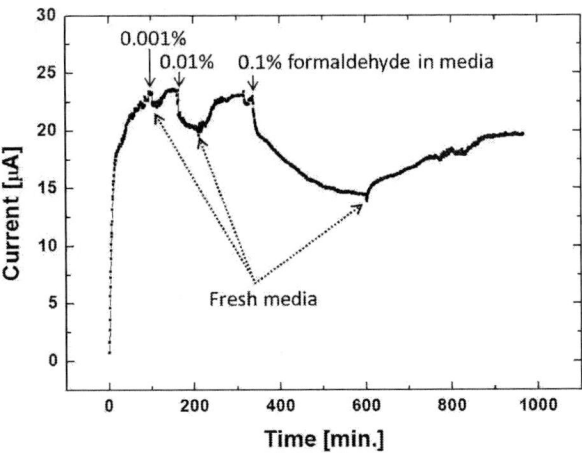

Figure 5: Current generation of the MFC biosensor. Real-time toxicity measurement response obtained with different concentration of the formaldehyde.

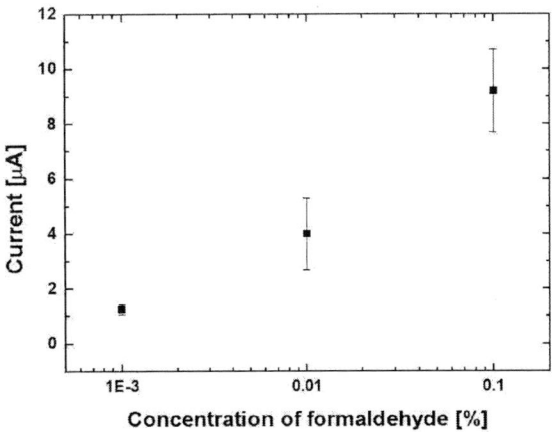

Figure 6: Calibration curve of the MFC biosensor for the formaldehyde.

Geobacter-based MFC, a c-type cytochrome with a formal potential of -0.08 V was identified to be the major electron transfer agent for current generation [16]. However, another c-type cytochrome with its potential centered at 0.48 V was monitored by applying the anode potential at 0.6 V, which resulted in a doubled current output in comparison to that of applying the electrode potential at 0.1 V [16]. Nevertheless, integrating the reference electrode into the microscale MFC platform poses a scientific challenge. The most effective and practical reference electrode is Ag/AgCl, and many research groups have attempted to reduce its footprint for miniaturized devices. Among many techniques, a planar type solid-state Ag/AgCl reference electrode has particularly attracted a great deal of interest during the last decade [17]. By integrating a planar Ag/AgCl reference electrode into the MFC biosensor, we avoided unwanted perturbations to the system and manipulated the device under controlled conditions. Moreover, the reference electrode can be used as an electrochemical sensor to quantify understanding of the bacterial redox potential, mass transport, and electrode/bacteria kinetics [18]. In this experiment, we poised 0.2 V versus Ag/AgCl reference by using 1-channel potentiostat (DY2300, Digi-Ivy) because wild-type *Shewanella oneidensis* expressed cytochromes OmcA/MtrC at the bacteria-anode interface at this potential [19, 20].

MFC operation and current generation

We first accumulated and acclimated bacteria on the anode of the MFC for 12 hours to ensure a mature biofilm under the closed circuit operation with a 1 kΩ resistor (data not shown) before the inoculum was switched to a fresh medium. The MFC biosensor was operated at 30 °C and the anolyte solution was continuously supplied using the syringe pump at a rate of 1 μL min⁻¹. After the biofilm was fully formed, the resistor was removed and the anodic potential (0.2V) was poised to the MFC biosensor to enable stable current generation with 10 μL min⁻¹ flow rate of the fresh media. The current was instantly produced and reached 23 μA.

Current responses to toxic substances

Formaldehyde was chosen as the toxic to be added to the MFC biosensor to allow observing the inhibition of bacterial metabolic activity. The formaldehyde has been utilized as a model toxin because it has a fast biological toxicity and stability with other chemicals in media [20]. We continuously injected the toxic sample through another pathway in a Y-shape microfluidics at a rate of 10 μL min⁻¹. The rapid current responses were obtained over a concentration range from 0.001% to 0.1% in the MFC biosensor with 0.2 V (versus Ag/AgCl reference electrode) applied on the anode (Fig. 5). The current decreased in proportion with the concentration of the toxic substance.

With fresh media, the anode bacterial activity was able to be recovered from the toxic samples with 0.001% and 0.01% formaldehyde in media. However, some of the bacteria were deactivated after 0.1% formaldehyde disturbance and could not be recovered even after a relatively long revival time. The reusability of MFCs as the biosensor is very critical for on-line and real-time monitoring of water quality. If the MFC cannot be recovered from toxic substances, the biosensors should be replaced, posing a problem for real-world applications. As shown in Fig. 5, this result suggests that our MFC biosensor can be reused under the toxic level of 0.01% formaldehyde in media. In Fig. 6, the current output after toxic solution injection is plotted against the relative concentration. In the logarithmic scale, a linear trend was observed in the range of 0.001% - 0.1% formaldehyde.

CONCLUSION

In this work, we developed a single-chambered micro-sized MFC biosensor. The biosensor included (i) three-electrode configuration for posing optimized anodic potential and (ii) air-bubble trap for preventing microscale air bubbles from entering the MFC biosensor. This MFC provided a novel analytical method to monitor water quality with controllable measurement and better sensitivity. Moreover, this conceptual micro-sized MFC biosensor showed how the micro-scale device can provide an instantaneous response when exposed to the presence of a toxic substance. Further studies should be done in order to (i) determine the responses to different toxic substances in water and (ii) make this technique suitable for practical applications.

ACKNOWLEDGEMENT

We thank Prof. Daniel Hassett for providing wild-type *Shewanella oneidensis* and stimulating discussions.

REFERENCES

[1] M. Bae, Y. Park, "Biological early warning system based on the response of aquatic organisms to disturbance: A review", *Science of the Total Environment*, vol. 466-467, pp.635-649, 2014.

[2] S. Yagur-Kroll, E. Schreuder, C.J. Ingham, R. Heideman, R. Rosen, S. Belkin, "A miniature porous aluminum oxide-based flow-cell for online water quality monitoring using bacterial sensor cells," *Biosensors and Bioelectronics*, vol. 64, pp.625-632, 2015.

[3] W.H. van der Schalie, T.R. Shedd, M.W. Widder, L.M. Brennan, "Response characteristics of an aquatic biomonitor used for rapid toxicity detection," *Journal of Applied Toxicology*, vol. 24, pp.387-394, 2004.

[4] A. Gerhardt, M.K. Ingram, I.J. Kang, S. Ulitzur, "In situ on-line toxicity biomonitoring in water: recent developments," *Environmental Toxicology and Chemistry*, vol. 25, pp.2263-2271, 2006.

[5] X.C. Abrevaya, N.J. Sacco, M.C. Bonetto, A. Hilding-Ohlsson, E. Corton, "Analytical applications of microbial fuel cells. Part II: Toxicity, microbial activity and quantification, single analyte detection and other uses, *Biosensors and Bioelectronics*, vol. 63, pp.591-601, 2015.

[6] N.E. Stein, H.M.V. Hamelers, G. Van Straten, K.J. Keesman, "On-line detection of toxic components using a microbial fuel cell-based biosensor," *Journal of Process Control*, vol. 22, pp.1755-1761, 2012.

[7] S. Mukherjee, S. Su, W. Panmanee, R.T. Irvin, D.J. Hassett, and S. Choi, "A microliter-scale microbial fuel cell array for bacterial electrogenic screening," *Sensors and Acuators: A. Physical*, vol. 201, pp.532-537, 2013.

[8] S. Choi, H.-S. Lee, Y. Yang, P. Parameswaran, C.I. Torres, B.E. Rittmann & J. Chae, "A μL-scale Micromachined Microbial Fuel Cell Having High Power Density," *Lab on a Chip*, vol.11, pp.1110-1117, 2011.

[9] N.E. Stein, H.M.V. Hamelers, C.N.J. Buisman, "Stabilizing the baseline current of a microbial fuel cell-based biosensor through overpotential control under non-toxic conditions," *Bioelectrochemistry*, vol. 78, pp.87-91, 2010.

[10] D. Davila, J.P. Esquivel, N. Sabate, J. Mas, "Silicon-based microfabricated microbial fuel cell toxicity sensor," *Biosensors and Bioelectronics*, vol. 26, pp.2426-2430, 2011.

[11] M. Lorenzo, A.R. Thomson, K. Schneider, P.J. Cameron, I. Ieropoulos, "A small-scale air-cathode microbial fuel cell for on-line monitoring of water quality," *Biosensors and Bioelectronics*, vol. 62, pp.182-188, 2014.

[12] F. Qian, D.E. Morse, "Miniaturizing microbial fuel cells," *Trends in Biotechnology*, vol. 29, pp.62-69, 2011.

[13] A. Fraiwan, C. Dai, N.K. Sidhu, A. Rastogi, S.Choi, "A Micro-sized Microbial Fuel Cell with Electrochemical Sensing Functionality," in *Proceedings of IEEE International Conference on Nano/Micro Engineering and Molecular Systems (NEMS)*, Hawaii, USA, Apr. 13 - 16, 2014, pp. 635-638.

[14] A. Fraiwan, S. Sundermier, D. Han, A. Steckl, D.J. Hassett, S. Choi, "Enhanced Performance of MEMS Microbial Fuel Cells using Electrospun microfibrous anode and optimizing operation," *Fuel Cells*, vol. 13, pp.336-341, 2013.

[15] H. Liu and B. E. Logan, "Electricity Generation Using an Air-Cathode Single Chamber Microbial Fuel Cell in the Presence and Absence of a Proton Exchange Membrane," *Environmental Science & Technology*, vol. 38, pp. 4040-4046, 2004.

[16] J.P. Busalmen, A. Esteve-Nunez, and J. M. Feliu, "Whole Cell Electrochemistry of Electricity-Producing Microorganisms Evidence an Adaptation for Optimal Exocellular Electron Transport," *Environmental Science & Technology*, vol. 42, pp. 2445-2450, 2008.

[17] M. W. Shinwari, D. Zhitomirsky, I. A. Deen, P. R. Selvaganapathy, M. J. Deen, D. Landheer, "Microfabricated reference electrodes and their biosensing applications," *Sensors*, vol. 10, pp. 1679-1715, 2010.

[18] C.I. Torres, A.K. Marcus, H. Lee, P. Parameswarn, R. Krajmalnik-Brown, B.E. Rittmann, "A kinetic perspective on extracellular electron transfer by anode-respiring bacteria," *FEMS Microbiology Reviews*, vol. 34, pp. 3-17, 2010

[19] L. Peng, S. You, J. Wang, "Electrode potential regulates cytochrome accumulation on Shewanella oneidensis cell surface and the consequence to bioelectrocatalytic current generation," *Biosensors and Bioelectronics*, vol. 25, pp. 2530-2533, 2010.

[20] X. Wnag, N. Gao, Q. Zhou, "Concentration response of toxicity sensor with Shewanella oneidensis MR-1 growing in bioelectrochemical systems," *Biosensors and Bioelectronics*, vol. 43, pp. 264-267, 2013.

CONTACT

*S. Choi, Assistant Professor, SUNY Binghamton, tel: +1-607-777-5913; sechoi@binghamton.edu

978-1-4799-7956-1/15 $31.00 © 2015 IEEE

A PAPER-BASED 48-WELL MICROBIAL FUEL CELL ARRAY FOR RAPID AND HIGH-THROUGHPUT SCREENING OF ELECTROCHEMICALLY ACTIVE BACTERIA

G. Choi[1], A. Fraiwan[1], D.J. Hassett[2] and S. Choi[1*]

[1]Department of Electrical and Computer Engineering, State University of New York-Binghamton, USA
[2]Department of Molecular Genetics, Biochemistry and Microbiology, University of Cincinnati College of Medicine, USA

ABSTRACT

We demonstrate the use of a paper-based sensing platform for rapid and high-throughput characterization of microbial electricity-generating capabilities. For the first time, a 48-well microbial fuel cell (MFC) array was fabricated on paper substrates, providing 48 high-throughput measurements and highly comparable performance characteristics in a reliable manner. Spatially distinct 48 wells of the sensor array were prepared by patterning 48 hydrophilic reservoirs in paper with hydrophobic wax boundaries. The paper-based platform exploited the ability of paper to quickly wick fluid and promote bacterial attachment to the gold anode pads, resulting in an instant current generation upon loading of bacterial inoculum and catholyte. This paper-based 48-well MFC array does not require external pumps/tubings and represents the most rapid and the highest throughput test platform for electrogenic bacterial screening.

INTRODUCTION

Microbial fuel cells (MFCs) represent an emerging technology for generating electricity from renewable biomass [1]. MFCs are gaining acceptance as a future alternative "green" energy technology and energy-efficient wastewater treatment technique [2]. MFCs are operated by living bacteria, fungi or algae which catalyse the degradation of a broad range of organic substrates under natural conditions [3]. Despite their potential, however, MFC technology has not yet been applied to any practical setting as their power generation is not sufficiently high compared to other fuel cell technologies [2]. Important strategies for enhancing MFC performance include genetically engineering microbes, optimizing microbial communities, and improving cultivation practices [3]. To date, however, a surprisingly small number of bacteria ($\sim$ 35) strains and their optimal growth conditions have been investigated for use in MFCs, revealing a crucial lack of fundamental knowledge as to which bacteria species or consortia may be best suited for generating power in MFCs [4,5]. This limitation originates from the fact that there are no screening techniques that characterize bacterial electricity generation capabilities in a high-throughput, rapid manner. Currently, no technology can even conceptually provide independent access to more than 24 spatially distinct microbial sensing units with dramatic reductions in measurement speed [6-8]. This is mainly because conventional MFC array is based on microfluidic techniques requiring complex MFC configurations with many tubings/channels and their operation with external pumps. Furthermore, MFC arrays require long start-up times for bacterial accumulation and acclimation as biofilms adhering to the anode [5]. These limitations have motivated us to develop a new conceptual MFC array, such that the parallelization and rapid power assessment can be significantly improved.

Recently, MFCs and MFC arrays were fabricated on paper [9-11]. This paper MFC showed the rapid electricity generation while conventional MFCs require long start-up time that is attributed to the accumulation and acclimation of bacteria on the anode. This is because the hydrophilic paper reservoir rapidly absorbs the anolyte and immediately promotes the attachment of a number of bacteria cells to the anode [9]. In addition, the paper-based microfluidics removes all the necessary tubings for inlets and outlets due to the ability of capillary action from paper structure. In this paper, this prominent feature of the paper-based devices for bacteria was used to develop a high-throughput 48-well MFC array for rapid screening of electrogenic bacteria. The 48-well MFC array was developed by patterning 48 hydrophilic reservoirs in paper with hydrophobic wax boundaries. Within just 20 minutes, we successfully determined the electricity generation capacity of two known bacterial electrogens and another metabolically more voracious organism with 8 isogenic mutants. This work is the potential realization of a practical tool for efficient high-throughput bacterial electrogenic behavior and fundamental MFC understanding, which may further improve power extraction in MFCs.

EXPERIMENTAL SETUP

Device Fabrication

Spatially-distinct 48 wells of the sensor array consisted of seven functional layers: anodic/cathodic PCB boards, anode/cathode layer (Au/Cr/Cu on polymethyl methacrylate (PMMA)), anodic/cathodic paper reservoir layers, and a paper-based proton exchange membrane (PEM) (Fig. 1 & Fig. 2). We used a commercial solid wax printer (Xerox® ColorQube 8570DN-37) to rapidly deposit wax on paper (Whatman #1 filter paper). The paper was then heated to re-melt the wax that penetrates the paper to generate complete hydrophobic barriers. The customized PCB board (80 x 85.68 mm) was designed to measure the electricity generation of each well in the array. Anodic/cathodic PCB boards had 1.63 mm-wide metal pads with 48 through-holes in the center to introduce anolyte and catholyte. The metal pads had direct contacts with the anode/cathode layer (Fig. 2 & Fig. 3). There were 24-pin holes on each side of the board for four 12-pin wire connectors (Fig. 3). PMMA substrates for anode and cathode layers were initially patterned by micromachining

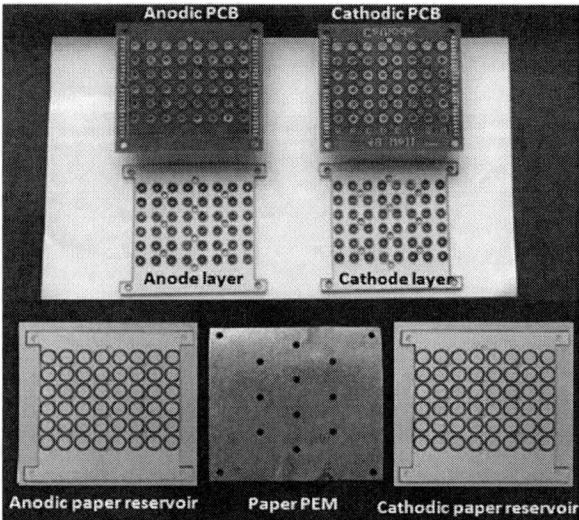

Figure 1: Photo-images of individual layers of the 48-well sensor array

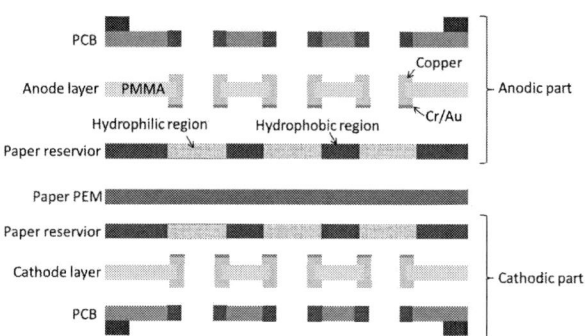

Figure 2: Schematic diagram of individual layers of the 48-well sensor array

Figure 3: Photo-image of the assembled sensor array. Voltage recordings from all 48 wells were conducted by connecting the load resistor to each well through customized LabView interface.

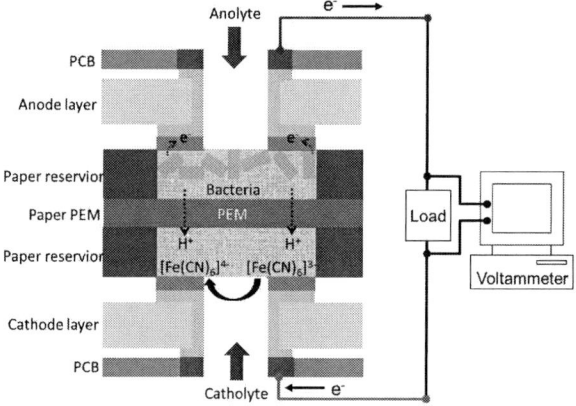

Figure 4: Schematic diagram of test set-up.

(Universal Laser System VLS 3.5). By utilizing magnetron sputtering, 500 nm-thick Cu was deposited on both the front and back sides of the anode/cathode layers to cover the side-walls of the 48 holes. Finally, 100 nm gold was deposited with 20 nm Cr as an adhesion layer over the back side of Cu layer by E-beam evaporation.

Inoculum and Catholyte

Ten microorganisms were prepared with two negative controls (water and media); *S. oneidensis*, *P. aeruginosa* and eight isogenic *pmpR* (encoding a negative regulator of the *Pseudomonas* quinolone signal (PQS) involved in the process of quorum sensing), *rpoS* (encoding the stationary phase sigma factor, RpoS, [12]), *lasR* (encoding LasR, the master regulator of quorum sensing, [13]), *pilT* (encoding PilT, that, when absent, allows for overproduction of type IV pili that are incapable of retraction and thus are always fully extended, thereby enhancing mediatorless electrogensis, [14]), *bdlA*, (encoding BdlA, protein involved in biofilm dispersion, [15]), *katA*, (encoding KatA, the major catalase, that functions to degrade H_2O_2 [16] and buffer anaerobically produced nitric oxide (NO) [17]), *phzS* (a strain that overproduces the mediator, pyorubrin) and *fliC pilA* (a strain that lacks the surface appendages flagellum and type IV pili [18]). The mutants

were generated using classical allelic replacement techniques with sucrose counter-selection as described by Hoang et al. [19]. All species were grown in L-broth medium (10.0 g tryptone, 5.0 g yeast extract and 5.0 g NaCl per liter). Each sample was loaded on four anodic chambers for generating error bars. The catholyte was 50 mM ferricyanide in a 100 mM phosphate bufferin which pH was adjusted at 7.5 ± 0.2 with 0.1 M NaOH.

Measurement Setup

We measured the potentials between the anodes and the cathodes with a data acquisition system (National instrument, USB-6212), and recorded the readings every 30 sec via a customized LabView interface (Fig. 4). External resistor (120 Ω) was connected between the anode and cathode to obtain the current flow through the resistor by Ohm's law (I = V/R). Anodic inoculum (30 μl) was injected by 8-channel pipette and quickly absorbed in hydrophilic region of the paper reservoir (Fig. 3).

RESULTS AND DISCUSSION
Open Circuit Voltages

Before connecting the MFC array with 120 Ω resistors, we first compared the open circuit voltages

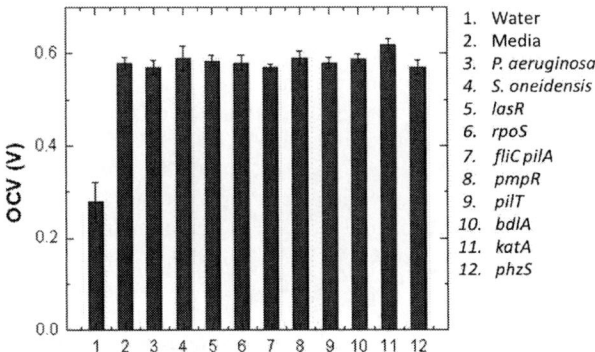

1. Water
2. Media
3. *P. aeruginosa*
4. *S. oneidensis*
5. *lasR*
6. *rpoS*
7. *fliC pilA*
8. *pmpR*
9. *pilT*
10. *bdlA*
11. *katA*
12. *phzS*

Figure 5: Open circuit voltages (OCVs) of ten bacterial species with two negative controls.

(OCVs) in each MFC unit in the array. No bacteria were in the anode reservoirs for these experiments, in which the cathode potentials were controlled by ferricyanide at ~300 mV. The OCVs of the 48 MFCs increased and reached a value of approximately 540 mV with less than 2.5% variation, which is far less than that of other MFC array (25%). After we confirmed that our device had such a low percent deviation, we began our experiment with actual bacteria samples, the OCVs were first recorded for 3 minutes (Fig. 5). The measured voltages varied between the MFC units, which clearly indicates performance variations according to the bacterial species. The OCVs ranged from 580 to 650 mV while the MFC unit with water as an anolyte showed very low OCV.

Current Measurement

After measuring the OCVs under no-load conditions for 3 minutes, the 120 Ω resistors were connected to enable current production and the voltage differences under the resistor were recorded. Each sample was loaded on four MFC units in the array and the experiment was repeated twice. The current with the external resistor gradually decreased for 20 minutes due to the depletion of the solution. The comparison of the current generations from each sample was made two times, the first after 1 minute of operation and the second at 20 minutes (Fig. 6).

The current generated from negative controls (#1 #2 in Fig. 6) showed distinct differences from bacterial samples, indicating that the current was generated by bacterial metabolisms. It should be noted that media-only sample without bacteria produced certain amount of current at 1 minute (Fig. 6, #2) because some chemical ions presented in the media contributed to the current generation. However, the value shortly reached zero due to their depletion.

At 1 minute, the *fliC pilA* and *phzS* mutants generated higher electric currents while *P. aeruginosa* and *pilT* showed lower ones. However, most samples have small differences between the various species used in terms of current generation, leading to difficulties in determining which bacterial species are superior in their current generation. This might be due to the instant current generation from the media, not from the bacteria.

At 20 minutes, Fig. 6 (b) clearly shows significant differences between MFC units because the current generated from the chemical ions in media-only sample

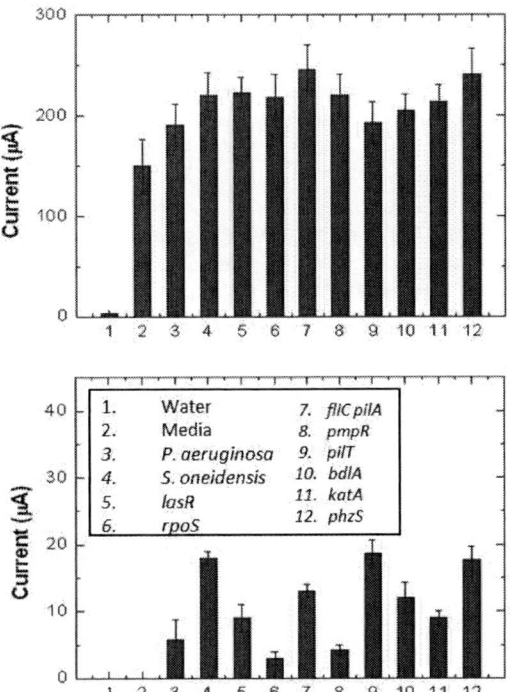

1. Water
2. Media
3. *P. aeruginosa*
4. *S. oneidensis*
5. *lasR*
6. *rpoS*
7. *fliC pilA*
8. *pmpR*
9. *pilT*
10. *bdlA*
11. *katA*
12. *phzS*

Figure 6: Currents calculated from voltage measurements in (a) 1 min. and (b) 20 min. At 1 min, the fliC pilA mutant has higher current generation followed by phzS. At 20 min, however, the pilT mutant showed the highest current generation.

was limited. The output data demonstrated the differences in the electricity generation capabilities of each microorganism, showing that our device proves useful for bacterial screening and characterization even after this relatively short operation time compared to previous MFC array techniques which require longer periods of time (several days to weeks) [5]. The *pilT* mutant of *P. aeruginosa* displayed the highest current generation than the other bacterial species. This finding is in good agreement with our previous report [5]. The *pilT* mutant might increase the current generation due to overproduction of electron-conductive pili that are incapable of retraction. Also, the *phzS* mutant produced much higher current than that of *P. aeruginosa*. The *phzS* mutant is known to generate of the merlot-colored redox cycling agent pyorubrin at very high levels [20, 21]. This work validates the utility of our MFC array by studying how strategic genetic modifications impact the electrochemical activity of bacteria.

In summary, within 20 minutes, the paper-based MFC sensor array was able to successfully characterize the electricity-generating capability of ten different microorganisms; *S. oneidensis*, *P. aeruginosa* PAO1 and eight isogenic *pmpR*, *rpoS*, *lasR*, *pilT*, *bdlA*, *katA*, *phzS* and *fliC pilA* mutants.

CONCLUSION

We developed a 48-well paper-based MFC array for characterization of microorganisms' electricity-generating capability. We could rapidly determine the two known bacterial and eight isogenic mutants' electricity-generating capability. The device was successively fabricated on paper without external pumps and tubings. The use of paper decreased the operating time considerably, and reduced total cost due to the operation with no expensive external equipment and only small amounts of samples required. This paper-based MFC array can be easily directed toward the development of much higher throughput array by simply patterning hydrophilic reservoirs in paper with hydrophobic wax boundaries.

ACKNOLOWDGEMENTS

We would like to express our sincere gratitude to the nano-fabrication lab and the Center for Autonomous Solar Power (CASP) at SUNY-Binghamton for providing the fabrication facilities. We also thank Amin Emrani for his kind contributions to this study.

REFERENCES

[1] B.E. Logan, J.M. Regan, "Electricity-producing bacterial communities in microbial fuel cells," *Trends in Microbiology*, vol. 14, pp.512-518, 2006.

[2] B.E. Rittmann, "Opportunities for renewable bioenergy using microorganisms," *Biotechnology and Bioengineering*, vol. 100, pp. 203-212, 2008.

[3] K.Rabaey, W. Verstraete, "Microbial fuel cells: novel biotechnology for energy generation," *Trends in Biotehcnology*, vol. 23, pp.291-298, 2005.

[4] H. Hou, L. Li, Y. Cho, P. de Figueiredo, A. Han, "Microfabcricated microbial fuel cell arrays reveal electrochemically active microbes," *PLOS ONE*, vol. 4, pp. e6570, 2009.

[5] S. Mukherjee, S. Su, W. Panmanee, R.T. Irvin, D.J. Hassett, & S. Choi, "A microliter-scale microbial fuel cell array for bacterial electrogenic screening," *Sensors and Actuators: A. Physical*, vol. 201, pp.532-537, 2013.

[6] H. Hou, L. Li, C.U. Ceylan, A. Haynes, J. Cope, H. H. Wilkinson, C. Erbay, P. de Figueiredo, A. Han, "A microfluidic microbial fuel cell array that supports long-term multiplexed analyses of electricigens," *Lab on a Chip*, vol. 12, pp.4151-4159, 2012.

[7] H. Hou, L. Li, P. de Figueiredo, A. Han, "Air-cathode microbial fuel cell array: A device for identifying and characterizing electrochemically active microbes," *Biosensors and Bioeletronics*, vol. 26, pp. 2680-2684, 2011.

[8] A. Fraiwan, D.J. Hassett, S. Choi, "Effects of light on the performance of electricity-producing bacteria in a miniaturized microbial fuel cell array," *Journal of Renewble and Sustainable Energy*, vol. 6, pp.063110, 2014.

[9] A. Fraiwan, S. Choi, "Bacteria-Powered Battery on Paper," *Physical Chemistry Chemical Physics*, vol. 16, pp. 26288-26293, 2014.

[10] A. Fraiwan, S. Mukherjee, S. Sundermeier, H.-S. Lee, S. Choi, "A paper-based Microbial Fuel Cell: Instant battery for disposable diagnostic devices," *Biosensors and Bioelectronics*, vol. 49, pp.410-414, 2013.

[11] A. Fraiwan, H. Lee, S. Choi, "A multi-Anode paper-based microbial fuel cell: A potential power source for disposable biosensors," *IEEE Sensors Journal*, vol. 14, pp.3385-3390, 2014.

[12] S.J. Suh, L. Silo-Suh, D.E. Woods, D.J. Hassett, S.E. West, D.E. Ohman, "Effect of rpoS mutation on the stress response and expression of virulence factors in Pseudomonas aeruginosa," *Journal of Bacteriology*, vol. 181, pp.3890-3897, 1999.

[13] L. Passador, J.M. Cook, M.J. Gambello, L. Rust, B.H. Iglewski, "Expression of Pseudomonas aeruginosa virulence genes requires cell-to-cell communication," *Science*, vol. 260, pp.1127-1130, 1993.

[14] C.B. Whitchurch, M. Hobbs, S.P. Livingston, V. Krishnapillai, J.S. Mattick, "Characterisation of a Pseudomonas aeruginosa twitching motility gene and evidence for a specialised protein export system widespread in eubacteria," *Gene*, vol. 101, pp. 33-44, 1991.

[15] R. Morgan, S. Kohn, S.H. Hwang, D.J. Hassett, K. Sauer, "BdlA, a chemotaxis regulator essential for biofilm dispersion in Pseudomonas aeruginosa," *Journal of Bacteriology*, vol. 188, pp. 7335-7343, 2006.

[16] J-F. Ma, U.A. Ochsner, M.G. Klotz, V.K. Nanayakkara, M.L. Howell, Z. Johnson, J. Posey, M.L. Vasil, J.J. Monaco, D.J. Hassett, "Bacterioferritin A modulates catalase A (KatA) activity and resistance to hydrogen peroxide in Pseudomonas aeruginosa," *Journal of Bacteriology*, vol. 181, pp. 3730-3742, 1999.

[17] S. Su, W. Panmanee, J.J. Wilson, H.K. Mahtani, Q. Li, B.D. Vanderwielen, T.M. Makris, M. Rogers, C. McDaniel, J.D. Lipscomb, R.T. Irvin, M.J. Schurr, J.R. Lancaster, R.A. Jr., Kovall, D.J. Hassett, "Catalase (KatA) Plays a Role in Protection against Anaerobic Nitric Oxide in Pseudomonas aeruginosa," *PLoS One* vol. 9, pp. e91813, 2014.

[18] M.A. Farinha, S.L. Ronald, A.M. Kropinski, W. Paranchych, "Localization of the virulence-associated genes pilA, pilR, rpoN, fliA, fliC, ent, and fbp on the physical map of Pseudomonas aeruginosa PAO1 by pulsed-field electrophoresis," *Infection and Immunity*, vol. 61, pp. 1571-1575, 1993.

[19] T.T. Hoang, A.J. Kutchma, A. Becher, H.P. Schweizer, "Integration-proficient plasmids for *Pseudomonas aeruginosa*: site-specific integration and use for engineering of reporter and expression strains," *Plasmid*, Vol. 43, pp. 59-72, 2000.

[20] S. Su S, E.A. Amba, R.E. Boissy, A. Greatens, W.R. Heineman, D.J. Hassett, "Cyclic Voltammetric, Fluorescence and Biological Analysis of Purified Pyorubrin (Aeruginosin A), the Secreted Red Pigment of Pseudomonas aeruginosa," *Microbiology*, vol. 159, pp. 1736-1747, 2013.

[21] D.V. Mavrodi, R.F. Bonsall, S.M. Delaney, M.J. Soule, G. Phillips, L.S. Thomashow, "Functional analysis of genes for biosynthesis of pyocyanin and phenazine-1-carboxamide from Pseudomonas aeruginosa PAO1," *Journal of Bacteriology*, vol. 183, pp. 6454-6465, 2001.

CONTACT

*S.Choi, tel: +1-607-777-5913; sechoi@binghamton.edu

AN INTEGRATED MICROFLUIDIC SYSTEM WITH FIELD-EFFECT-TRANSISTOR-BASED BIOSENSORS FOR AUTOMATIC HIGHLY-SENSITIVE C-REACTIVE PROTEIN MEASUREMENT

Chia-Ho Chu[1], Wen-Hsin Chang[2], Wei-Jer Kao[3], Chih-Lin Lin[2], Ko-Wei Chang[2], Yu-Lin Wang[1]* and Gwo-Bin Lee[1,2,3]*

[1]Institute of NanoEngineering and Microsystems, [2]Department of Power Mechanical Engineering, [3]Institute of Biomedical Engineering, National Tsing Hua University, Hsinchu, Taiwan
*Co-corresponding authors

ABSTRACT

Rapid and accurate diagnosis of C-reactive protein (CRP) is crucial for preventing cardiovascular diseases because it is a well-known biomarker for evaluating risks of cardiovascular diseases. Our previous work has shown that a microfluidic system equipped with a field-effect-transistor (FET)-based biosensor could detect CRP in 0.1X PBS and provided a limit of detection (LOD) of 26 pM CRP without gate bias. To improve the LOD, a new microfluidic device with a new methodology for measuring FET-based biosensors is presented in this study. Not only can the proposed system work in a solution with a physiological salt concentration but it also detects CRP with ultra-high sensitivity in an automatic fashion. This is the first time that a FET-based biosensor can effectively and automatically detect CRP in a physiological salt concentration without decreasing the sensitivity. The LOD of CRP using aptamer-immobilized AlGaN/GaN high-electron-mobility transistors (HEMTs) was experimentally found to be 1fM, demonstrating the superior performance of this new technique. It may be used as a point-of-care device for CRP detection in the near future.

INTRODUCTION

Cardiovascular diseases are responsible for 25-million deaths worldwide on a yearly basis. C-reactive protein (CRP) is a general biomarker for inflammation and infection, and has become a good indicator for evaluating risks of cardiovascular diseases. For faster detection of CRP, field-effect-transistor (FET)-based biosensors was employed in our previous study with a limit of detection (LOT) of 26 pM [1]. FET-based biosensors have been widely used for all kinds of biomolecule detection including proteins, enzymes, DNA etc. The materials of the FET devices may include Si [2], carbon nano-tubes [3, 4], or GaN [5, 6]. Due to the superior sensitivity and the smaller size, the FET-based sensor plays an important role in biosensing applications nowadays. However, no matter what type or what dimension of the FET-based sensor is, FET-based sensors currently all face an intrinsic problem, that is created by the nature, the screening effect and the Debye length [7]. Due to the severe screening effect in the physiological environment, the Debye length is extremely small (as small as 0.78 nm) [7]. Because most FET-based biosensors are the affinity type, which relies on the capture of the target molecules by the receptors immobilized on the gate region of the FET, the screening effect becomes a major inherent problem for achieving high sensitivity in a physiological environment. In this study, a new microfluidic system integrated with FET-based biosensors is therefore reported by using a new methodology, which allows the transistors to detect CRP in a physiological condition automatically with high sensitivity. The developed compact system has achieved a wide range of sensing concentrations and an extremely low LOD (1fM).

Materials and methods

Experimental procedures

The experimental procedure of the proposed integrated microfluidic system equipped with FET sensors is illustrated in Figure 1. First, the CRP-specific aptamer was immobilized on the gold surface of the gate region of the FET. Next, free aptamer was removed by activating a micro-pump with washing buffer. Then, samples with different concentrations of CRP were transported into the measuring chamber. After that, unbound CRP was washed out by activating the micro-pump. Finally, FET electric signals were measured.

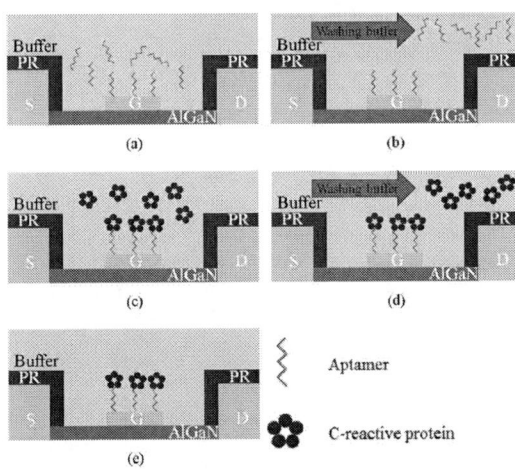

Figure 1: Experimental procedure for aptamer immobilization and CRP detection performed on the integrated microfluidic system. (a) Immobilization of aptamer; (b) Removal of free aptamer; (c) Addition of CRP; (d) Removal of unbound CRP; (e) Detection of FET signals. (S: source, G: gate, D: drain.)

Integrated microfluidic chip design

A schematic diagram and a photograph of the integrated microfluidic chip designed for automating the entire experimental procedure were presented in Figure 2. The chip was composed of two polydimethylsiloxane (PDMS) layers including a liquid channel layer (thick-film PDMS layer) and a pneumatic layer (thin-film PDMS layer), a double-sided tape used to bind the PDMS layers and a printed circuit board (PCB), an AlGaN/GaN HEMT-based FET sensor and a PCB layer.

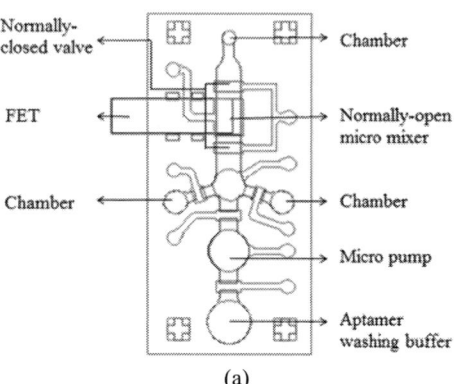

(a)

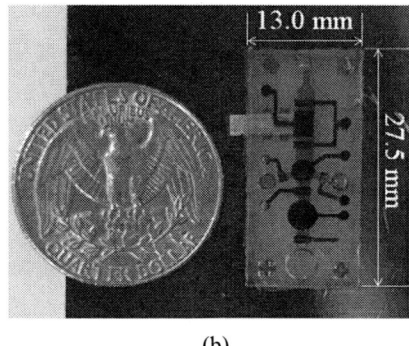

(b)

Figure 2: (a) A schematic diagram of the microfluidic chip. (b) A photograph of the integrated microfluidic chip. The dimensions of the integrated microfluidic chip were measured to be 13 mm × 27.5 mm × 4.5 mm. The blue color indicates the pneumatic layer and the red color indicates the liquid channel layer.

Packaging method for bonding hetero-materials

A new packaging method was developed in this study because FET, PCB and PDMS cannot be bound together by simple oxygen plasma treatment. The packaging procedure was illustrated in Figure 3. At first, the FET was covered by PDMS and placed on the surface of a PMMA sheet. At the same time, PCB was carved by using a computer-numerical-control (CNC) machining process to form a cavity for FET. Then, the double-side tape was pasted on the PCB and the cavity of the PCB was filled with PDMS.

Next, the PMMA plate with FET was aligned with the cavity of the PCB, which was placed on top of it afterwards. After baking for 1 hour for PDMS curing process, the double-side tape and the PMMA were removed and a flat surface of PDMS was reveled while bounding with other layers of PDMS to create the integrated microfluidic chip.

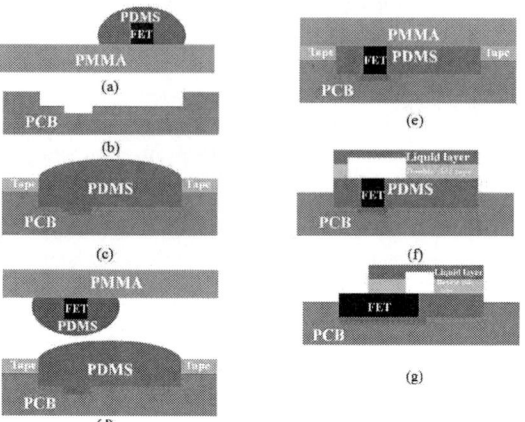

Figure 3: A new packaging method for bonding hetero-materials. (a) The FET covered by PDMS was placed on the surface of PMMA. (b) PCB was carved for mounting the FET. (c) The tape was pasted on top of the PCB and the cavity of the PCB was filled with PDMS. (d) The PMMA with FET was aligned with the cavity of PCB. (e) The whole chip was baked for 1 hour. (f) The cross section view of the chip which was parallel with the microfluidic channel. (g) The cross section view of the chip which was vertical with the microfluidic device.

AlGaN/GaN HEMT-based sensor design and signal measurement

Figures 4(a) and (b) show the schematics of an AlGaN/GaN high electron mobility transistor (HEMT) sensor and the top-view of the device, respectively. AlGaN and GaN thin films were grown on a sapphire substrate and the high concentration of the two-dimensional electron gas was formed between the AlGaN and GaN interface. Photoresist was then spin-coated to passivate the device, while only the sensing region was opened to allow the liquid solutions to flow across the surface. The CRP-specific aptamer was immobilized on the gate region followed by surface blocking with bovine serum albumin (BSA). The CRP solutions were then added and flew through the sensing region of the sensor, and then waited for 5 min for CRP binding. After CRP binding with aptamers, the sensor was washed with 1X PBS to remove the unbound CRP. 1X PBS was added on the sensor for electrical measurements. The sensor was measured by applying a dc bias of drain-source voltage (Vds) at 0.5V first, and then the gate voltage (Vg) was applied in a squared pulse of 0.5V for 50 micro-seconds on the gate electrode. The drain current (Id) of the transistor was recoded in real-time when the drain-source voltage and the gate voltage were applied. Different surface immobilizations of the sensor were compared with their drain currents and the total charges. The total charge was calculated by integrating the Id with the increasing time.

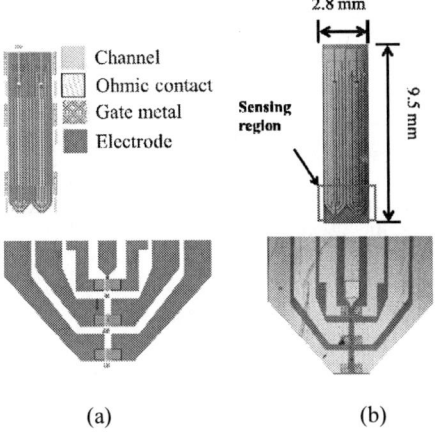

(a)   (b)

Figure 4: (a) The top-view of the FET and (b) the sensing region.

RESULTS AND DISCUSSION

The pumping rate of the micro-pump

The pumping rate of the micro-pump was first measured to optimize the operating condition of the proposed system. The pumping rate of the micro-pump on the integrated microfluidic system for liquid transportation was shown in Figure 5. The maximum pumping rate was measured to be 203 μl/min at an applied gauge pressure of -70 kPa and a driving frequency of 0.5 Hz.

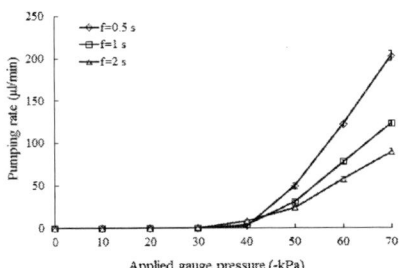

Figure 5: Relationship between the pumping rate and the applied gauge pressure on the micro-pump in the integrated microfluidic system. (f: driving frequency)

The drain current of the HEMT for different surface modifications

Figure 6 shows the drain current of the HEMT at Vds=0.5V versus time from 0~50 micro-seconds. The Vg was not applied in the beginning until 2 micro-seconds. A sharp peak of the drain current was observed after the Vg was applied. The drain current then quickly relaxed and gradually became steady. The drain current was measured for different

surface modification on the gate, including aptamer, BSA, and different CRP concentrations, ranging from 1 fM to 100 nM. It is obvious that before the gate voltage was applied, the drain current was almost the same for every surface modification, which was attributed to the severe charge screening effect in 1X PBS. When the gate voltage was applied, a typical capacitive charging process was observed and shown as the drain current change. Different surface modifications caused different relaxation of the drain currents, leading to the larger separation of the drain currents for different surface modifications.

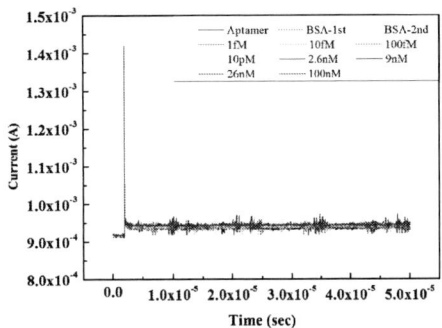

Figure 6: The current of FET sensors immobilized with CRP-specific aptamer, and followed by treating with BSA and different CRP concentrations from 0 s to 50 micro-seconds.

Signal amplification by counting total charges

Although the applied gate voltage generates a larger difference in drain currents for different CRP concentrations in the dynamic range during the capacitive charging process, to amplify the signals as largely as possible, we chose to count the total charges by integrating the drain current with time. The total charge for different surface modifications versus different time is shown as in Figure 7(a). It is clearly seen that the difference in total charges among different surface modification increases with time, as we expected. The total charge counted at 50 micro-seconds for each surface condition is shown in Figure 7(b). There is no too much difference in total charge for the aptamer-immobilized and the following BSA-blocked sensing. Significant change in total charge for 1fM CRP binding was observed. Higher CRP concentrations, including 10fM, 100fM, 10pM, 2.6nM, 9nM, 26nM, and 100nM were tested and caused more accumulated charges. The accumulated charges induced by CRP compared to aptamer/BSA are shown in Figure 8. The sensor detected a wide range of CRP concentration spanning 8 orders of magnitude. The LOD was then determined to be 1fM for CRP measurement. The high sensitivity and the low LOD of the sensor resulted from the charge accumulation and the less screening effect during the dynamic change of the drain current.

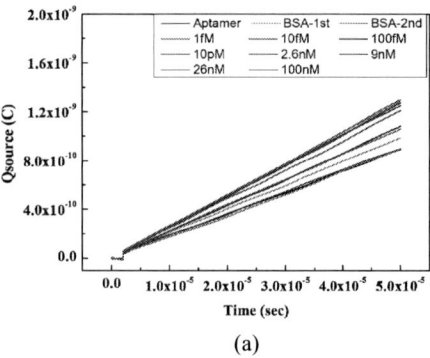

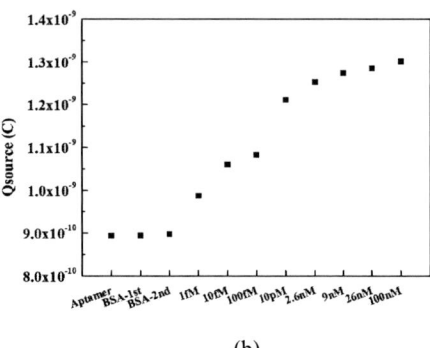

Fig 7: (a) The accumulated charges of transistors immobilized with CRP-specific aptamers, and followed by treating with BSA and different CRP concentrations from 0s to 50 micro-seconds. (b) The accumulated charges of different CRP concentrations at 50 micro-seconds.

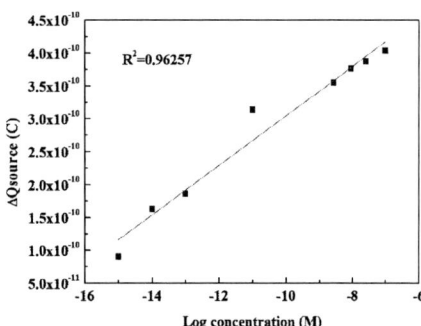

Fig 8: The accumulated charge change induced by different CRP concentrations.

CONCLUSION

In this study, a new microfluidic device using a new methodology for measuring field-effect-transistor (FET)-based biosensors has been presented. The compact system shows a wide-range concentration detection of CRP, which covers the entire range for the three categories of CRP

concentrations as cardiovascular disease markers, and the ultra-low detection limit (1fM) was observed. The developed system therefore has a great potential to assess the risks of cardiovascular diseases at the point of care in the future. The capability of the developed sensor for measuring an extremely low concentration of proteins may be extended to more biomedical applications.

ACKNOWLEDGEMENTS

The authors would like to thank the financial support from Ministry of Science and Technology in Taiwan and the technical support form National Nano Device Laboratories (NDL) in Hsinchu, Taiwan.

REFERENCES

[1] C. L. Lin, Y. W. Kang, K. W. Chang, W. H. Chang, Y. L. Wang, G. B. Lee, "An integrated microfluidic system using field-effect transistors (FET) for CRP detection" IEEE *MEMS'14 Conference*, San Francisco, January 26-30, 2014, pp. 10-15.

[2] O. Knopfmacher, A. Tarasov, Wangyang Fu, M. Wipf, B. Niesen, M. Calame, C. Schönenberger, "Nernst Limit in Dual-Gated Si-Nanowire FET Sensors", *Nano Lett.*, vol. 10, pp. 2268-2274, 2010.

[3] S. Subramanian, K. H. Aschenbach, J. P. Evangelista, M. B. Najjar, W. Song, R. D.Gomez, "Rapid, sensitive and label-free detection of Shiga-toxin producing Escherichia coli O157 using carbon nanotube biosensors", *Biosens. Bioelectron.*, vol. 32, pp. 69-75, 2012.

[4] K. Besteman, J. O. Lee, F. G. M. Wiertz, H. A. Heering, C. Dekker, "Enzyme-Coated Carbon Nanotubes as Single-Molecule Biosensors", *Nano. Lett.*, vol. 3, pp. 727-730, 2003.

[5] C.C. Huang, G. Y. Lee, J. L. Chyi, H. T., C.P. Hsu, Y. R. Hsu, C. H. Hsu, Y. F. Huang, Y.C. Sun, C.C. Chen, S.S Li, J. A. Yeh, D. J. Yao, F. Ren, Y. L. Wang," AlGaN/GaN high electron mobility transistors for protein–peptide binding affinity study", *Biosens. Bioelectron.*, vol. 41, pp. 717-722, 2014.

[6] Y. W. Kang, G. Y. Lee, J. I. Chyi, C.-P. Hsu, Y. R. Hsu, C. H. Hsu, Y. F. Huang, Y. C. Sun, C. C. Chen, S. C. Hung, F. Ren, J. A. Yeh and Y. L. Wang, Human immunodeficiency virus drug development assisted with AlGaN/GaN high electron mobility transistors and binding-site models. *Appl. Phys. Lett.*, vol. 102, pp. 173704-1-173704-5, 2013.

[7] K. I. Chen, B. R. Li, Y. T. Chen, "Silicon nanowire field-effect transistor-based biosensors for biomedical diagnosis and cellular recording investigation", *Nano. Today*, vol. 6, pp. 131-154, 2011.

CONTACT

*Dr. Gwo-Bin Lee, tel: +886-3-5715131 ext.33765; gwobin@pme.nthu.edu.tw
*Dr. Yu-Ling Wang,
tel:+886-3-5162405; ylwang@mx.nthu.edu.tw

BATCH-FABRICATED HYDROGEL/POLYMERIC-MAGNET BILAYER FOR WIRELESS CHEMICAL SENSING

J. H. Park[1,2], A. Kim[1,2], and B. Ziaie[1,2]

[1] Birck Nanotechnology Center, Purdue University, West Lafayette, IN 47907, USA
[2] School of Electrical and Computer Engineering, Purdue University, West Lafayette, IN 47907, USA

ABSTRACT

This paper introduces a fabrication and wireless chemical sensing scheme using a hydrogel/polymeric-magnet bilayer. The swelling/deswelling of the hydrogel in response to chemical stimuli (e.g. pH, glucose etc.) results in vertical movement of the polymeric permanent magnet located on top of the hydrogel film. The hydrogel volume change is wirelessly detected by a giant magnetoresistance (GMR) sensor. A pH-sensitive hydrogel magnet bilayer is fabricated and tested as proof of the concept. The sensor shows a sensitivity of 0.01 Oe/pH, a resolution of 0.05 pH unit between pH 4 and pH 6, and a response time of 30 minutes. The described device is the first integrated wireless hydrogel/polymeric-magnet transducer with potential applications in biomedical and environmental sensing areas.

INTRODUCTION

Environmentally-sensitive hydrogels offer a unique chemo-mechanical transduction mechanism based on reversible swelling/deswelling of the polymeric network in response to various physical (temperature) and chemical (pH, glucose, antibody, etc.) stimuli [1]–[5]. They are in particular attractive for implantable applications since they do not require an on-board power source and/or electronic circuitry. One such reported device is a MEMS LC resonator whose capacitor is coupled to a pH/glucose-sensitive hydrogel [6]. The sensor, however, requires complicated fabrication techniques in order to provide a hermetic seal for the capacitive sensor cavity. In addition, achieving a tight mechanical coupling between the hydrogel and the movable plate of the capacitor is not trivial. Furthermore, passive LC sensing methods require sophisticated custom-made readout electronics. We previously reported on a simpler scheme using a thin ferrogel (magnetically functionalized hydrogel) film bonded on top of a planar inductor [5]. Swelling-deswelling of the gel resulted in a change of the inductance that was remotely monitored using the standard LC sensing scheme (phase-dip method). Although solving most of the fabrication problems associated with MEMS-based LC device, the sensor still required the same sophisticated readout apparatus. In this paper, we present a potential solution to all of the above-mentioned problems by placing a light polymeric permanent magnet on top a hydrogel film. The swelling/deswelling of the hydrogel in response to chemical stimuli (e.g. pH, glucose etc.) results in vertical movement of the polymeric permanent magnet which can be detected by a giant magnetoresistance (GMR) sensor. GMR sensors are uniquely suited for this application, since they can detect small changes in a magnetic field [7]. They are also low cost and can be easily integrated into wearable devices.

Figure 1 shows a schematic of an envisioned wireless chemo-mechanical sensing system. The hydrogel/polymeric-magnet bilayer sensor is implanted in the subcutaneous tissue. The volumetric response of the hydrogel is externally monitored by the GMR sensor housed in a wrist watch.

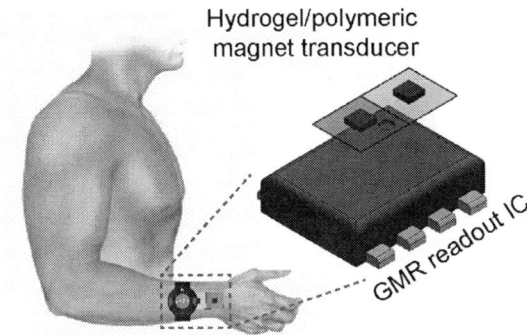

Figure 1: Schematic of a wireless chemical sensing system based on chemical modulation of magnetic field.

SENSOR STRUCTURE AND OPERATION

Figure 2 shows a cross sectional schematic of the sensor and its operation. The sensor consists of two polymeric permanent magnets, one directly bonded to the substrate and acting as a reference while the second one is located on top of a hydrogel film and can move up and down when hydrogel is responding to the chemical stimuli. The differential magnetic field intensity, indicative of the amount of swelling, between the sensor and the reference is measured by external GMR sensors located in close proximity. In our experiments, we used a pH sensitive hydrogel (poly (methacrylic acid-co-acrylamide) (mAA-co-AAm)) that swells with increasing pH levels.

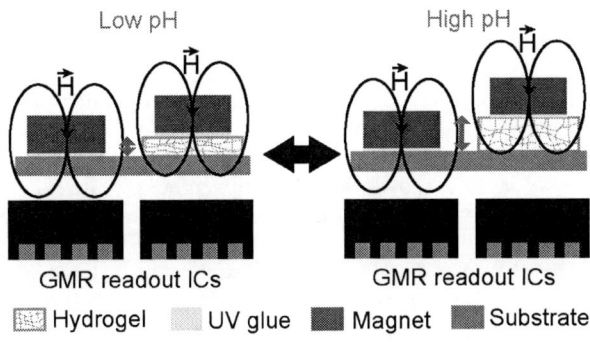

Figure 2: Cross section of the sensor showing the magnet pair height modulation in response to pH.

978-1-4799-7956-1/15 $31.00 © 2015 IEEE

EXPERIMENT RESULTS

For the proof of concept, we used poly (mAA-co-AAm) hydrogel which exhibits pH sensitivity. The hydrogel swelling is due to the electrostatic repulsion force between polymer chains resulting from ionization of COOH groups of poly (mAA) in alkaline pH levels. In contrast, the ionized COOH groups recombine with H^+ ions in lower pH levels resulting in the gel shrinkage.

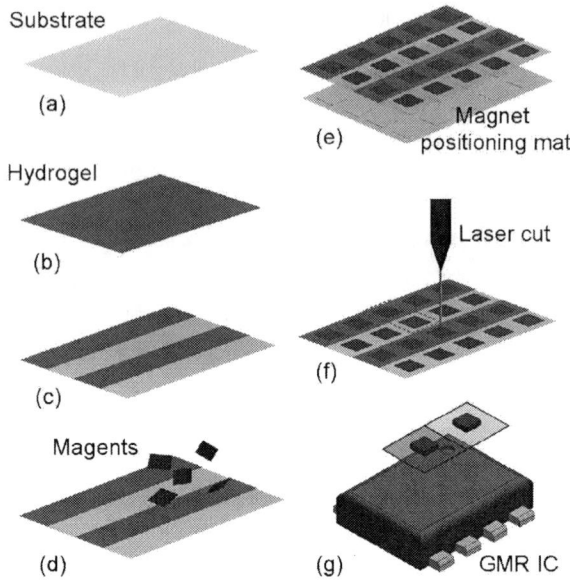

Figure 3: Fabrication process: (a) start with the GelBond substrate, (b) cast hydrogel layer, (c) laser-machine reference magnet area (d) disperse 1mm² magnets on top of the hydrogel covered with UV epoxy (e) place a magnet positioning mat and cure the epoxy, (f) separate the dies, (g) final device and the GMR readout IC.

Prior to the device fabrication, two pregel solutions were prepared to create the pH-sensitive poly (mAA-co-AAm) hydrogel. The pregel solution A was produced by mixing 20.16 µL of methacrylic acid (mAA, Sigma Aldrich), 66.9 mg of acrylamide (AAm, Sigma Aldrich), 20 µL of N,N,N',N'-tetramethylethylenediamine (TEMED, Sigma Aldrich), 3.27 mg of N,N'–Methylene Bisacrylamide (BiS, Polysciences Inc.) in 0.24 ml of deionized (DI) water. The pregel solution B was prepared by dissolving ammonium persulfate (APS, Polyscience Inc.) in DI water (80 mg/ml). The pregel solutions A and B were vortex-mixed in a volume ratio of 5.9:1 to form the hydrogel.

Figure 3 shows the fabrication process. First, a thin poly (mAA-co-AAm) hydrogel film (~50 µm) was cast on a GelBond® substrate using squeeze-film method (Fig. 3(a)-(b)) [8]. The film was allowed to dry overnight at room temperature. Next, the hydrogel film was laser-machined using a laser engraving system (Universal Laser System, Inc. PLS 6.75) to create the space for the reference magnet (Fig. 3(c)). A thin layer of UV-curable epoxy was then spin-coated on the hydrogel. Afterwards, 1 mm² pieces of laser-cut polymeric permanent magnets were dispersed while a magnetic positioning mat was placed under the sheet to attract and align the magnets on the desired and designed locations (Fig. 3(d)-(e)). The UV epoxy was completely cured under the UV floodlight for 10 minutes. Finally, the hydrogel/polymeric magnetic bilayers were cut into 4 mm × 6 mm dies that contains the sensor and the reference pair (Fig. 3(g)).

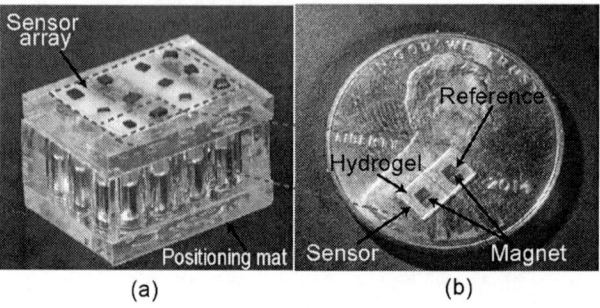

Figure 4: (a) Photograph of the batch-fabrication showing a sensor array and a magnet positioning mat (b) Photograph of a final device showing both the sensing and reference magnets.

Figure 4 shows the optical photograph of a batch-fabricated sensor array with a magnet positioning mat, Figure 4(a), and a final prototype, Figure 4(b). The positioning mat was prepared by encapsulating an array of neodymium magnets in a laser-machined acrylic container. The final hydrogel/polymeric magnetic bilayer composing of a sensor (a polymeric magnet on the hydrogel (left)) and a reference (a polymeric magnet on the substrate (right)) is displayed in Figure 4(b).

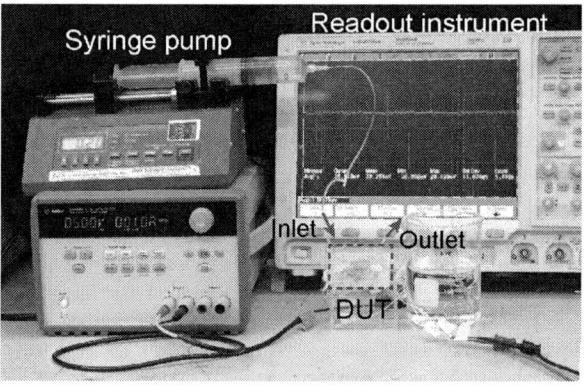

Figure 5: Experimental setup with the flow system.

Following the fabrication, the hydrogel/polymeric magnetic bilayers were placed in a laser-machined acrylic fluidic chamber. The chamber was firmly attached to a commercial GMR-IC circuitry (AAH002-02E, NVE Corp.). The distance between sensor/reference to GMR-IC was 1.6 mm. Prior to sensor response characterization, DI water was flown through the fluidic chamber to withdraw the unreacted monomers. Afterwards, two different pH solutions (pH 4 and pH 6) were alternatively flown into the chamber by a syringe pump at a constant flow rate (0.3 ml/min). The differential output voltages of the GMR-IC circuitry were measured by an oscilloscope

(InfiniiVision DSO7034B, Agilent Technologies) while the pH solutions inside of the chamber were switch from 4 to 6. The measured differential voltages were converted into the magnetic field intensities. Figure 5 shows a photograph of the experimental test setup with the flow system.

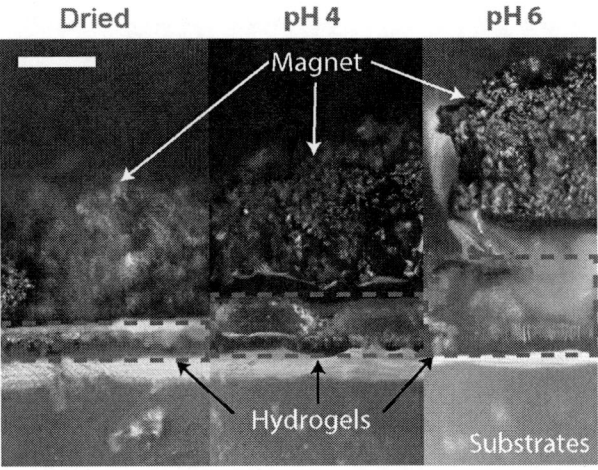

Figure 6: Cross sectional optical photographs showing the hydrogel/magnet bilayer in dry state, pH 4, and pH 6. (scale bar is 100 μm)

The magnetic field modulation by hydrogel swelling and deswelling was confirmed from microscopic evaluations of the device cross-section at different pH levels, Figure 6. The lightweight polymeric magnet does not impede the swelling and deswelling of the hydrogel. The hydrogel film exhibited swelling ratios of 1.4 and 2 at pH 4 and pH 6 (as compared to the dried hydrogel film). The swelling capacity of the hydrogel film might be to some extent compromised as the result of the mass of the magnet and UV epoxy.

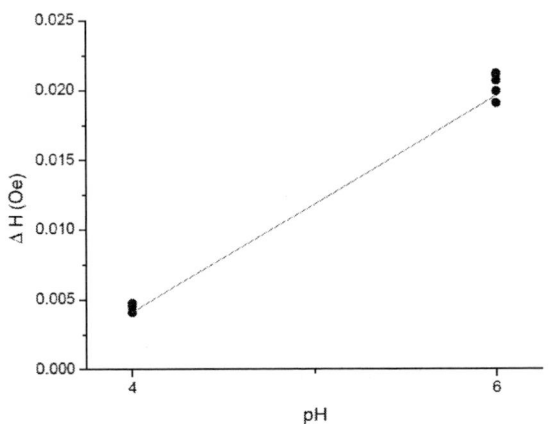

Figure 7: Calibration curve between pH 4 and pH 6 where this particular hydrogel has it maximum sensitivity and linearity.

Figure 7 shows the calibration curve of the device for pH values between 4 and 6. The differential magnetic field strength varied from 4.1 mOe to 19.6 mOe. The 15.5 mOe of differential magnetic field strength represents approximately 30 μm of the hydrogel film vertical movement, Figure 6. This particular hydrogel composition has its maximal sensitivity and linearity between pH 4 and 6 [3]. However, this range can be shifted to physiological pH levels by changing the ionizable monomer [9]. In this range, the sensor exhibited a sensitivity of 0.01 Oe/pH and a resolution of 0.05 pH unit.

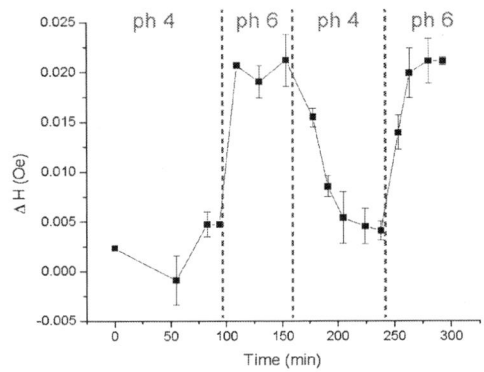

Figure 8: GMR sensor output for pH steps between pH4 and pH6.

The temporal changes in the differential magnetic field strength of the hydrogel/polymeric magnetic bilayer sensor in response to different pH values are shown in Figure 8. As the pH of the solution inside of the chamber dropped; the shrunken hydrogel film resulted in the reduced differential magnetic field strength. On the other hand, the differential magnetic field strength increased when the hydrogel swelled at pH of 6. It took about 30 minutes for the differential magnetic field strength to be stable after the pH solution was entirely replaced (~ 4 *min*).

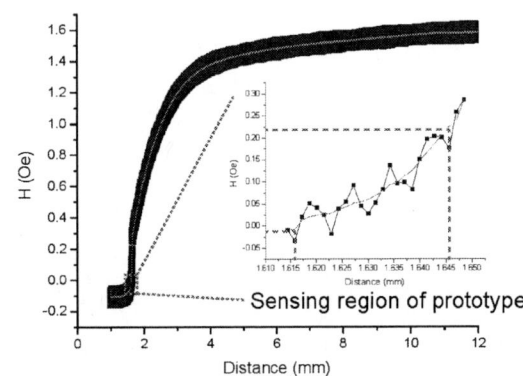

Figure 9: GMR output as a function of its separation from the polymeric magnets.

Finally, we should mention that the sensitivity is also a function of separation between the sensor and the GMR.

Figure 9 shows the GMR readout as a function of distance to the magnets. As can be seen, the output is a linear function of distance from 1.6 mm to 3 mm. At separations < 1.6 mm and > 3 mm the GMR exhibits a lower sensitivity (the reported results for sensitivity were obtained at 1.6 mm separation).

The results of GMR measurements confirmed the sensing principle; however, further improvements are required in order for the sensor be of practical value. The sluggish response time of the sensor can be improved by perforating the polymeric magnet in order to allow the solution to reach and penetrate the hydrogel more easily. Moreover, as mentioned previously, the maximum sensitivity range must be shifted to relevant physiological pH levels, thus allowing the sensor to monitor acidosis and alkalosis. Although the sensing method was demonstrated using a pH-sensitive hydrogel, a similar structure and approach can be employed to measure other chemical species such as glucose.

CONCLUSIONS

In this paper, we introduced a wireless chemical sensing method utilizing a hydrogel/polymeric-magnet bilayer. This method is based on differential magnetic field strength measurement between the sensing element (a hydrogel with polymeric magnet on top) and the reference magnet. The bilayer sensor can be fabricated in a batch scale manner using laser engraving and a magnet positioning mat. The sensing scheme was characterized by monitoring the change of the differential magnetic field strength at different pH levels.

ACKNOWLEDGEMENTS

The authors would like to thank the staff of Birck Nanotechnology Center at Purdue University for their assistance in fabrication.

REFERENCES

[1] D. Beebe, J. Moore, J. Bauer, Q. Yu, R. Liu, C. Devadoss, and B. Jo, "Functional hydrogel structures for autonomous flow control inside microfluidic channels," *Nature*, vol. 404, pp. 588–590, 2000.

[2] M. Lei, A. Baldi, E. Nuxoll, R. A. Siegel, and B. Ziaie, "A hydrogel-based implantable micromachined transponder for wireless glucose measurement.," *Diabetes Technol. Ther.*, vol. 8, pp. 112–122, 2006.

[3] C. Chang, Z. Ding, V. N. L. R. Patchigolla, B. Ziaie, and C. A. Savran, "Reflective Diffraction Gratings From Hydrogels as Biochemical Sensors," *IEEE Sens. J.*, vol. 12, no. 7, pp. 2374–2379, 2012.

[4] X. Huang, S. Li, J. S. Schultz, Q. Wang, and Q. Lin, "A MEMS affinity glucose sensor using a biocompatible glucose-responsive polymer.," *Sens. Actuators. B. Chem.*, vol. 140, pp. 603–609, 2009.

[5] S. H. Song, J. H. Park, G. Chitnis, R. A. Siegel, and B. Ziaie, "A wireless chemical sensor featuring iron oxide nanoparticle-embedded hydrogels,"

Sensors Actuators B Chem., vol. 193, pp. 925–930, 2014.

[6] M. Lei, A. Baldi, E. Nuxoll, R. a Siegel, and B. Ziaie, "Hydrogel-based microsensors for wireless chemical monitoring.," *Biomed. Microdevices*, vol. 11, pp. 529–538, 2009.

[7] C. Reig, M.-D. Cubells-Beltran, and D. R. Muñoz, "Magnetic Field Sensors Based on Giant Magnetoresistance (GMR) Technology: Applications in Electrical Current Sensing.," *Sensors (Basel).*, vol. 9, pp. 7919–7942, 2009.

[8] Z. Ding, A. Salim, and B. Ziaie, "Squeeze-film hydrogel deposition and dry micropatterning," *Anal. Chem.*, vol. 82, pp. 3377–3382, 2010.

[9] K. Varaprasad, N. N. Reddy, N. M. Kumar, K. Vimala, S. Ravindra, and K. M. Raju, "Poly(acrylamide-chitosan) Hydrogels: Interaction with Surfactants," *Int. J. Polym. Mater.*, vol. 59, pp. 981–993, 2010.

CONTACT

*J.H. Park, tel: +1-765-4095166; park1@purdue.edu

CELL-LADEN HINGED MICROPLATES FOR MEASURING THE CONTRACTILE FORCES OF CARDIOMYOCYTES

Hiroaki Matsumoto[1], Shotaro Yoshida[1], Yuya Morimoto[1,2],
Nobuhito Mori[1], Daniela Serien[1], and Shoji Takeuchi[1,2]
[1]Institute of Industrial Science, The University of Tokyo, JAPAN
[2]Takeuchi Biohybrid Innovation Project, ERATO, JST, JAPAN

ABSTRACT

We report a method to measure contractile forces of cardiomyocytes at cellular level using microplates; pairs of single-cell sized plates are connected by a flexible hinge at the center. Cardiomyocytes repeatedly contract and relax, and thereby fold and stretch the microplates at the flexible hinge. By measuring the change of the bending angle in folded microplates, we can estimate contractile forces of cardiomyocytes. We believe that this method is a useful tool to study the biomechanics of cardiomyocytes.

INTRODUCTION

Cardiomyocytes are involuntary striated muscle cells of hearts and generate contractions synchronized with each other. One of the main reasons of heart failure is a reduction of the contractility to pump blood to the body. The reduction of the contractility arises from a difference between contractile forces of normal cardiomyocytes and those of diseased cardiomyocytes. Even in the early stages of the heart failure, the contractility of the cardiomyocytes is decreased [1]. Thus, in order to study heart diseases, the quantitative analysis of the biomechanics of the cardiomyocytes is required. The analysis of the cardiomyocytes enables us to extend our knowledge of cardiac function and advances the therapeutic interventions to treat heart failure. For the analysis, it is important to study cardiomyocytes at cellular level since cellular level analysis without effects from the surrounding cells or tissues allows us to reveal how the cardiomyocytes individually contribute to cardiac function [2]. There are several studies to measure the biomechanics of the cardiomyocytes [3-6]. In particular, micropillars can measure contractile force of cardiomyocytes at cellular level quantitatively. Micropillars help measuring the force by observing the deformation of the pillars generated from the contractions of the cardiomyocytes [6]. However, it is difficult to measure contractile force of cardiomyocytes with micropillars because the deformation of micropillars is small. An improved tool of estimating contractile force of cardiomyocytes by measuring a more accessible change is required.

In this study, we propose a method to measure the contractile force of the cardiomyocytes from the deformation of a microplate. Pairs of single-cell sized plates are connected by a flexible hinge at the center [7]. Since the cardiomyocytes can bend the microplates at the hinge, the microplates are folded and stretched at the hinge as the cardiomyocytes repeatedly contract and relax (Figure 1). Since the motion of the hinged microplates is not a linear motion but a bending motion around the hinge, we observe large movement of the microplates due to the contractions of the cardiomyocytes. Thus, by measuring

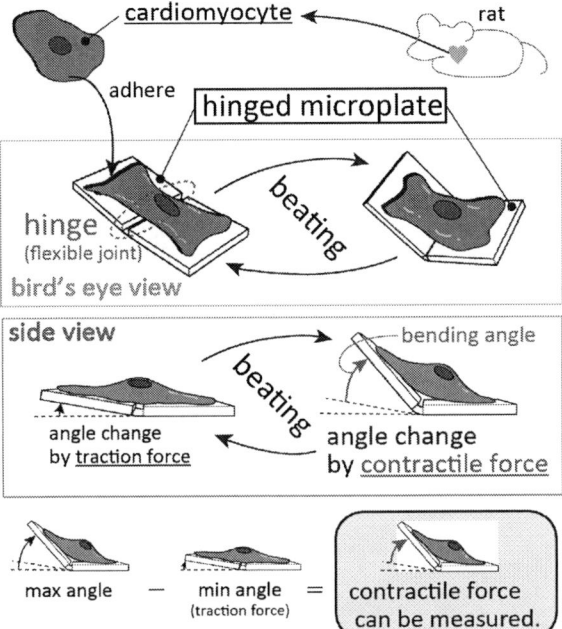

Figure 1: Conceptual illustration of this study. Cardiomyocytes derived from rats adhered onto a microplate with a flexible joint, hinge. On the hinged microplate, the cardiomyocytes were beating. By measuring the change of the bending angle of the microplates, the contractile force of cardiomyocytes were measured.

the change of the bending angle in folded microplates, we can easily estimate contractile forces of cardiomyocytes.

MATERIALS AND METHODS
Fabrication of hinged microplates

We first fabricated the hinged microplates by using photolithography techniques [7]. Figure 2 shows the fabrication process of the microplates. Firstly, we spin-coated gelatin solution (0.5% (w/v), Sigma-Aldrich, USA) on a glass substrate as a sacrificial layer, and deposited parylene C (Specialty Coating Systems, USA) using a parylene deposition machine (LABCOTER PDS2010, Specialty Coating Systems, USA) on the gelatin layer (Figure 2(a)). After patterning a SU-8 (Micro Chem, USA) layer and depositing another lawyer of parylene (Figure 2(b)), we masked microplates with aluminum to remove gelatin and parylene outside of the aluminum mask by O_2 plasma (RIE-10NR, SAMCO, Japan) (Figure 2(c, d)). Next, we spin-coated 2-methacryloyloxyethyl phosphorylcholine (MPC) polymer [8] for blocking protein coating and cellular adhesion to the surface of the parylene

978-1-4799-7956-1/15 $31.00 © 2015 IEEE

and removed the aluminum mask (Figure 2(e, f)). By coating collagen solution (3% (w/v), Nitta Gelatin Inc., Japan) on the surface of the microplates (Figure 2(g)), the cardiomyocytes could adhere on them (Figure 2(h)).

(a) Coat gelatin and deposit parylene

(b) Pattern SU-8 and deposit parylene

(c) Pattern Aluminum as a mask

(d) Etch parylene and gelatin

(e) Coat MPC polymer

(f) Remove Aluminum mask (and lift-off MPC polymer)

(g) Coat collagen

(h) Seed cardiomyocytes on hinged microplate

Figure 2: Fabrication process flow for the hinged microplates with cardiomyocytes. Microplates are fabricated by using photolithography techniques. MPC polymer coated in the figure (e) prevents collagen, cardiomyocytes and other cells from adhering on glass substrate.

Cell culture

We used primary rat cardiomyocytes (CMC02, Primary Cell, Japan). Cardiomyocytes were maintained in Dulbecco's modified eagle medium: nutrient mixture F-12 (DMEM/F-12 medium, Primary Cell, Japan) supplemented with 10% fetal bovine serum, 10 units/mL penicillin, and 10 μg/mL streptomycin at 37°C under a humidified atmosphere of 5% CO_2. We resuspended cardiomyocytes in DMEM/F-12 medium, and seeded them on the microplates. The cell concentration for seeding was 7.6×10^4 cells/mL.

Measurement of the change of the bending angle

After the cardiomyocytes began to beat, we detached the microplates from the glass substrate completely using a micromanipulator. Peeled from the glass substrate, the microplates were folded and stretched repeatedly by the beating of the cardiomyocytes. We observed and recorded this motion of the hinged microplates with a microscope (IX81, Olympus, Japan), and measured the change of the bending angle in folded microplates using an image analysis software (ImageJ, U. S. National Institutes of Health, USA).

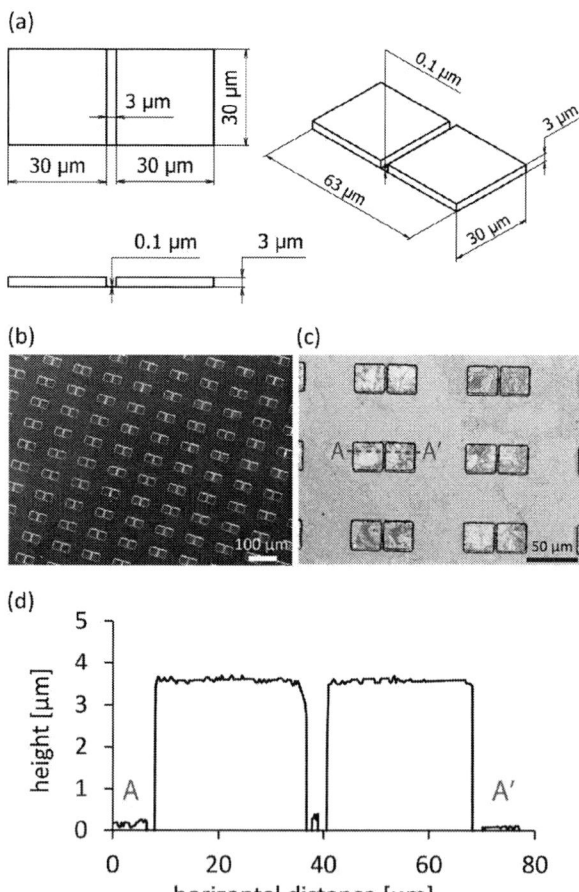

Figure 3: (a) Designed dimensions of the hinged microplates. (b,c) Bright field images of the fabricated microplates. (d) Cross section of a hinged microplate measured by the 3D laser scanning microscope (VK-X200, Keyence, Japan).

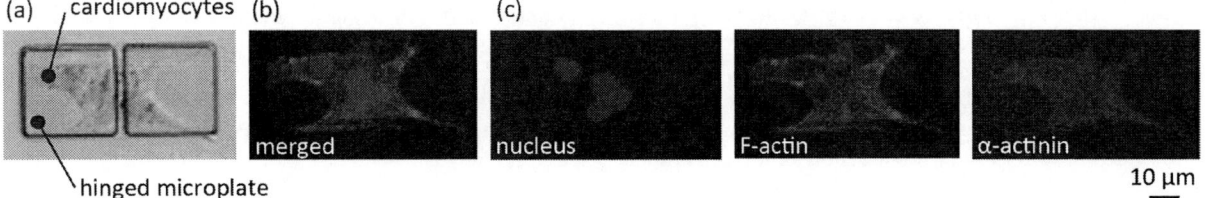

Figure 4: Adhered cardiomyocytes on a hinged microplate. (a) Phase contrast image. (b, c) Immunocytochemistry of cardiomyocytes with nucleus, F-actin and α-actinin.

RESULTS

Fabrication of hinged microplates

Figure 3(a) shows the design of microplates and Figure 3(b, c) shows fabricated microplates that were consequently used for culturing cardiomyocytes. Length and width of microplates were identical to the initial design. Measured height of microplates was similar to designed height (Figure 3(d)). The conformation of the dimensions indicates that our fabrication method is appropriate for preparing microplates.

Culture of cardiomyocytes on hinged microplates

To investigate whether the cells on the microplate were cardiomyocytes, we conducted immunostaining of nucleus, F-actin and α-actinin (Figure 4). Expression of α-actinin indicated that the cells on the microplate were cardiomyocytes rather than cardiac fibroblasts.

Beating cardiomyocytes on hinged microplates

We recorded a movie that a few cardiomyocytes were beating on the hinged microplate (Figure 5(a)). The change of the bending angle over time could be measured clearly from the analysis of the recorded movie (Figure 5(b, c)). We could observe the large deformation, 20 – 50 degrees, which corresponds to 11 – 23 µm displacement. Since the displacement is approximately 10 times larger than the previous work (1 – 2 µm) [6], the microplates enable estimation of the contractile force easier.

Estimation of contractile force

In order to estimate the contractile force of the cardiomyocytes from the change of the bending angle in folded microplates, we assumed a model described in Figure 6(a). According to this model, the relationship between the force induced by the cardiomyocytes at the center of the microplate F and the bending angle of the microplate θ was represented by the equation (1) shown below,

$$F = \frac{EI}{l} \cdot \frac{\theta}{d\sin\frac{\theta}{2} + t\cos\frac{\theta}{2}} \quad (1)$$

where E is the Young's modulus of the hinge, I is the second moment of the area of the hinge, l is the length of the hinge, d is the distance between the center of the hinge and the center of the microplate, and t is the thickness of the microplate. Figure 6(b) illustrates the curve of the estimated force by measured bending angle. For derivation of the contractile force, we subtracted the force estimated by the minimum bending angle from the force estimated by the maximum bending angle. In addition, we also

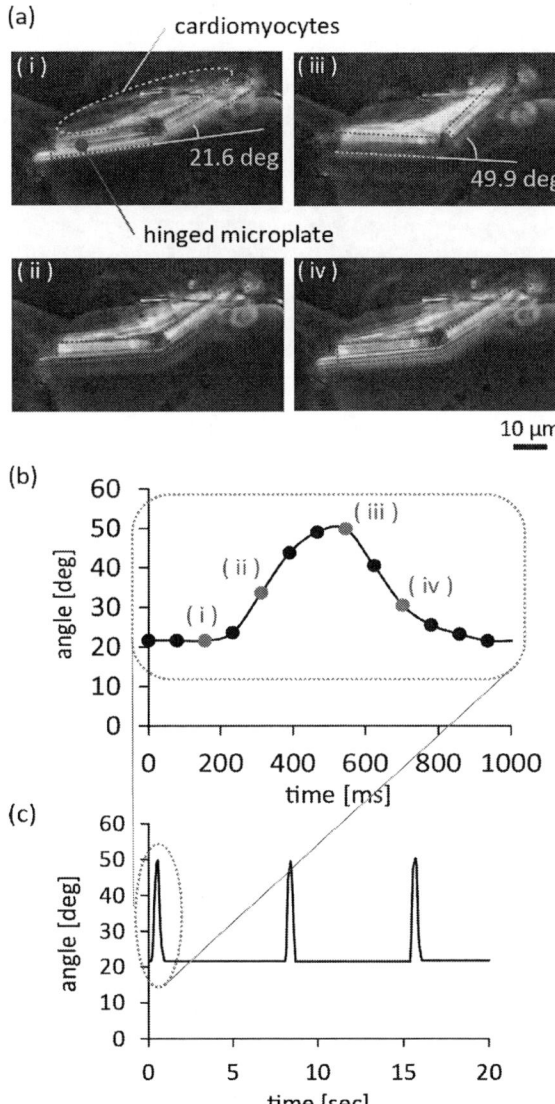

Figure 5: (a) Time lapse pictures of beating cardiomyocytes on a hinged microplate. The pictures (i) – (iv) are recorded at each time (i) – (iv) in the graph (b). (b,c) The change of the bending angle over time in folded microplates. The graph (b) indicates the first 1000 ms of the graph (c).

estimated the corresponding work of the cardiomyocytes by the strain energy stored in the hinge of the microplate based on the model. Table 1 represented the estimated force and work generated by a few of cardiomyocytes on a hinged microplate comparable in our experimental condition (Figure 5). The contractile force of the cardiomyocytes was estimated to be approximately 64 nN. This value is in the range of 0 – 150 nN. This range was estimated by previous work [6] for the contractile force of cardiomyocytes at cellular level. Due to this agreement with pervious work, it appears our model and method are appropriate for determining contractile forces at cellular level.

CONCLUSION

We fabricated hinged microplates for measuring the contractile forces of cardiomyocytes at cellular level. We confirmed that cardiomyocytes were able to adhere and beat on the surface of the microplates. By measuring the change of the bending angle in folded microplates, we estimated the contractile force of the cardiomyocytes. We obtained clear experimental data for the estimation of contractile force of cardiomyocytes owing to the large deformation of the microplates. We believe that hinged microplates are a useful tool to study the dynamics of cardiomyocytes.

ACKNOWLEDGEMENTS

The authors acknowledge Prof. Kazuhiko Ishihara at the University of Tokyo appreciatively for providing the MPC polymer. This work was supported by Takeuchi Biohybrid Innovation Project, ERATO, Japan Science and Technology Agency (JST).

REFERENCES

[1] R. V. Yelamarty, R. L. Moore, F. T. Yu, M. Elensky, A. M. Semanchick, and J. Y. Cheung, "Relaxation abnormalities in single cardiac myocytes from renovascular hypertensive rats", *American Journal of Physiology*, vol. 262, pp. C980-C990, 1992.

[2] G. Iribe, M. Helmes, and P. Kohl, "Force-length relations in isolated intact cardiomyocytes subjected to dynamic changes in mechanical load", *American Journal of Physiology - Heart and Circulatory Physiology*, vol. 292, pp. H1487-H1497, 2007.

[3] M. G. Garcia-Webb, A. J. Taberner, N. C. Hogan, and I. W. Hunter, "A modular instrument for exploring the mechanics of cardiac myocytes", *American Journal of Physiology - Heart and Circulatory Physiology*, vol. 293, pp. 866-874, 2007.

[4] G. Lin, R. E. Palmer, K. S. Pister, and K. P. Roos, "Miniature Heart Cell Force Transducer System Implemented in MEMS Technology", *IEEE Transactions on Biomedical Rngineering*, vol. 48, pp. 996-1006, 2001..

[5] S. Yin, X. Zhang, C. Zhan, J. Wu, J. Xu, and J. Cheung, "Measuring Single Cardiac Myocyte Contractile Force via Moving a Magnetic Bead", *Biophysical Journal*, vol. 88, pp. 1489-1495, 2005.

[6] A. Kajzar, C. M. Cesa, N. Kirchgeβner, B. Hoffmann, and R. Merkel, "Toward Physiological Conditions for Cell Analyses: Forces of Heart Muscle Cells Suspended Between Elastic Micropillars", *Biophysical Journal*, vol. 94, pp. 1854-1866, 2008.

[7] K. Kuribayashi-Shigetomi, H. Onoe, and S. Takeuchi, "Cell Origami: Self-Folding of Three-Dimensional Cell-Laden Microstructures Driven by Cell Traction Force", *PLoS ONE*, vol. 7, no. 12, e51085, 2012.

[8] K. Ishihara, Y. Iwasaki, S. Ebihara, Y. Shindo, and N. Nakabayashi, "Photoinduced graft polymerization of 2-methacryloyloxyethyl phosphorylcholine on polyethylene membrane surface for obtaining blood cell adhesion resistance", *Colloids and Surfaces B: Biointerfaces*, vol. 18, pp. 325-335, 2000.

(a)

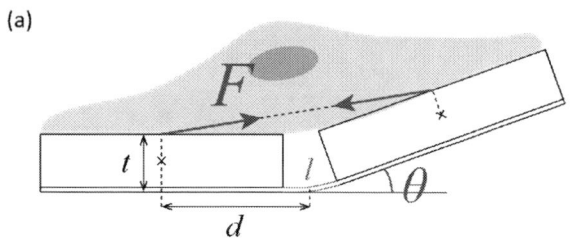

(b)

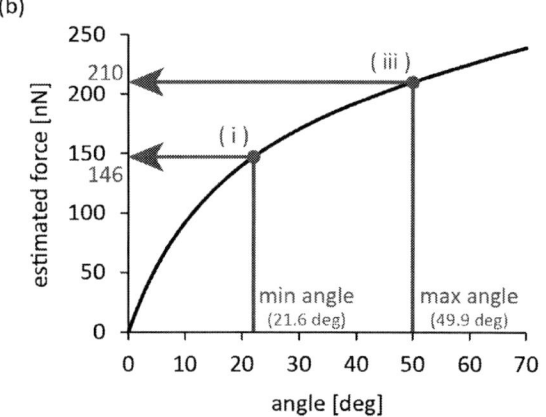

Figure 6: (a) A model of folded microplate due to contractile force. (b) A graph of the relationship between force and bending angle of a microplate. Measured minimal and maximal angles were used for the estimation of the corresponding estimated force.

Table 1: Estimated force and work generated by a few cardiomyocytes on hinged microplates.

measured/estimated contents		value
measured angle	max	49.9 deg
	min	21.6 deg
estimated force	max	210 nN
	min	146 nN
estimated contractile force	max – min	64 nN
estimated strain energy	max	0.885 nJ
	min	0.166 nJ
estimated work by contractile force	max – min	0.719 nJ

CONTACT

*Hiroaki Matsumoto,
tel: +81-3-5452-6650; fax: +81-3-5452-6649;
E-mail: masyu@iis.u-tokyo.ac.jp

DROP TO MEASURE: A NOVEL NANOFLUIDIC CRYSTAL SENSING SCHEME WITH IMPROVED CHIP-TO-CHIP DATA CONSISTENCY FOR PORTABLE BIOCHEMICAL DETECTION

Baojun Wang[1], Wenda Zhao[1], Rui Zhang[1] and Wei Wang[1, 2, 3]*

[1]Institute of Microelectronics, Peking University, 100871, China;
[2]National Key Laboratory of Science and Technology on Micro/Nano Fabrication, 100871, China;
[3]Innovation Center for Micro-Nano-electronics and Integrated System, Beijing, 100871, China

ABSTRACT

We propose a novel "drop to measure" nanofluidic crystal sensing scheme with improved chip-to-chip data consistency here. Nanoparticles were self-assembled in a confined space guided by a well-designed surface chemistry treatment. The electrical readouts from different chips (n=5) varied within 4.8% (compared to 7.5% for 3 chips in the previous work). Biotin (using streptavidin-modified nanoparticles) and Pb^{2+} (using DNAzyme probed nanoparticles) were successfully detected by the present "drop to measure" nanofluidic crystal sensor with a limit of detection of 1 nM.

INTRODUCTION

Sensing based on nanoscale electrokinetics received increasing attentions in the past ten years [1, 2]. The interstices of the nanoparticle crystal can form a nanochannel network. When the electrical double layers are overlapped in the nanochannel inside the nanofluidic crystal, the ionic conductance across the device will be dominated by the surface charge density of the nanoparticles, which can then be used to fulfill a biochemical sensing. Compared to the devices prepared by traditional nanofabrication techniques, nanofluidic crystal holds advantages of easy surface modification, fast and low cost prototyping [3]. However, large chip-to-chip data variations were found in nanofluidic crystal sensor because of the low geometrical reproducibility of self-assembly nanoparticles crystal in a microstructure. A complicated self-calibration is thereby required for every sensing operation, which considerably restricts its applications in portable biochemical detection. Our previous work [4] designed a new chip for controllable nanoparticle assembly and achieved a chip-to-chip variation of 7.5% (n=3), but the sample loading process need accurate location and the electrodes structure is not optimal. This paper further improved the chip performance by elaborating the nanoparticle self-assembling space, the measuring electrodes along with the well-designed surface chemistry treatment. The chip-to-chip data consistency was improved to 4.8% (n=5). Also, utilizing the coffee ring, the circle sample-loaded structure economizes the nanoparticle suspension and makes the nanoparticles self-assembly process more easily by simply dropping it onto the chip without precise location. This provides more possibility for portable outdoor detection.

DESIGN AND FABRICATION

The present chip is schematically illustrated in Figure 1a. Four testing cells were located on the edge of a circular sample-loading area. The fabrication is briefly shown in Figure 1b-e. In short, 150/3000 Å Cr/Au electrode pairs were fabricated onto a glass substrate and Parylene C confined spaces (50 μm×120 μm×3 μm for each) were achieved by using Al as sacrificial layer. Finally, the chip was treated by SF_6 for 150 s to obtain a distinct hydrophobicity difference between the Parylene C and the exposed glass surface.

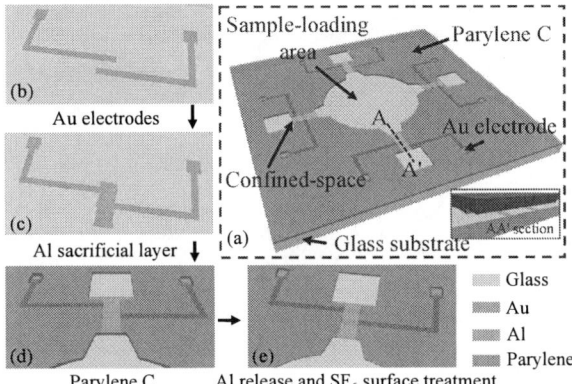

Figure 1: The structure and fabrication process of the chip. (a)The schematic structure of the whole chip. The right bottom inserts shows the detailed AA' section. (b-e) The fabrication process of one cell in the chip.

Figure 2 (a) shows the SEM photo of the testing structure. A microstructure was formed to define the nanoparticles self-assembling space. Figure 2 (b-c) shows the confined space before and after the nanoparticle packed, with a close-up detail of the nanofluidic crystal in Figure 2 (d). The elaborated nanoparticle self-assembling space guarantee a high geometrical reproducibility of self-assembly nanoparticles crystal between different chips.

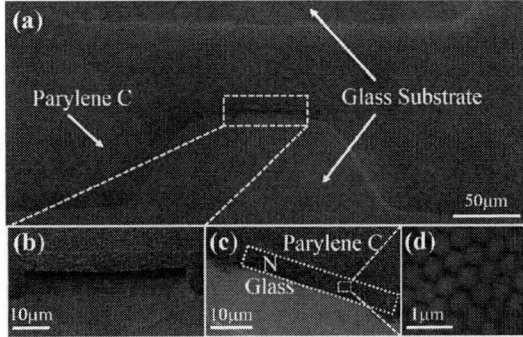

Figure 2: (a) SEM photo of the testing structure. (b) and (c) SEM photos of the confined space before and after the nanoparticle packed. (d) A close-up detail of the nanofluidic crystal in the Parylene C confined space.

Assuming that inside and outside the microstructure were filled with water, the electrical signal distribution was simulated based on the COMSOL™ and the distribution of the current density inside the Parylene C confined-space was indicated in Figure 3. The results demonstrate that the ionic conductance measured from the chip was contributed mainly by the sample between the two electrodes in the confined space, which also guaranteed an improved chip-to-chip data consistency.

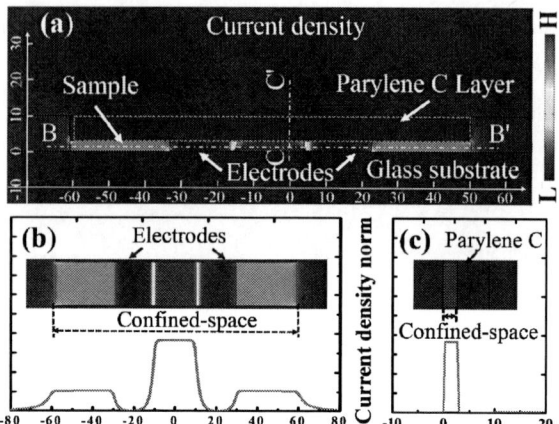

Figure 3: (a) COMSOL™ simulation of the current density inside the Parylene C confined-space. The distribution of the current density corresponding to the BB' section (b) and CC' section (c) of (a) respectively.

EXPERIMENT

Device Preparation

A droplet of nanoparticle suspension was dropped onto the circular sample-loading area to form nanofluidic crystal, which is more portable and without precise location. The chips were then kept in refrigerator overnight at 4 ℃ before the following experiments. As shown in Figure 4 (a), an 8×9 or larger array can be operated at the same time. Rhodamine was used as auxiliary dye to demonstrate the process. Figure 4 (b) shows the photos before and after the nanoparticles self-assembled. The coffee ring in the circle sample-loaded structure economizes the nanoparticle suspension and makes the nanoparticles self-assembly process more easily.

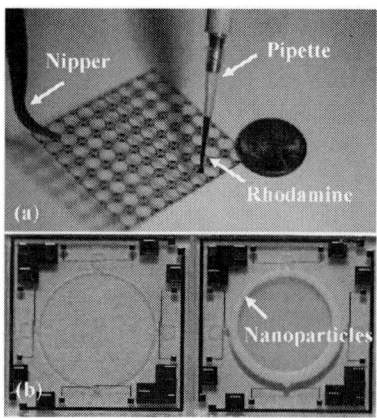

Figure 4: (a) The view of the sample dropping process. (b) Photos before and after the nanoparticles self-assembled.

To make sure the nanoparticles flow into the confined space, the chip have to be treated by SF_6 for 150 s to obtain a distinct hydrophobicity difference between the Parylene C and the exposed glass surface. Figure 5 shows the microphotograph of different results of the nanaoparticles self-assembly before and after SF_6-treatment. The contact angles in the inserts shows the hydrophobic property of parylene C is greatly improved to prevent the sample flow over the circular sample-loading area.

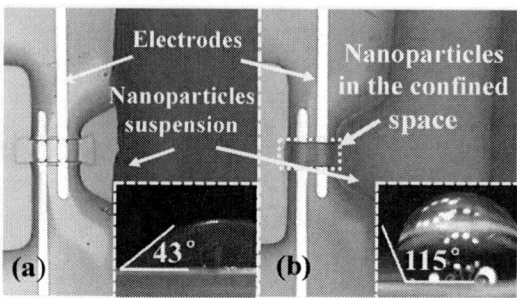

Figure 5: The different results of the self-assembly between the chips (a) before and after SF_6-treatment. The right bottom inserts show the contact angle on the Parylene C surfaces respectively.

Characterization and Results

In all experiments, DI water was used as the sensing buffer because of its low ionic concentration, which can guarantee an overlapped electrical double layer. The conductance of the chip was measured by a probe station along with HP 4156B. Impedance signals from different chips with/without nanoparticles packed were measured firstly and typical results were shown in Figure 6. A high chip-to-chip data consistency of 4.8% (σ/A: Standard Deviation/ Average) was achieved in five devices for each measurement.

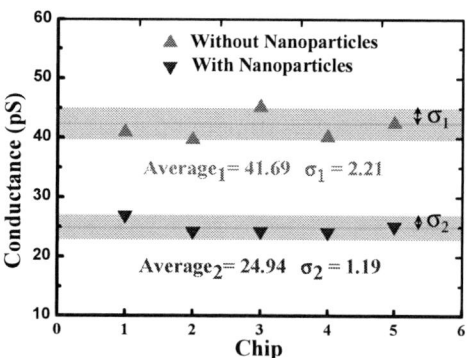

Figure 6: Typical conductance measurements of different chips. 540 nm streptavidin-modified nanoparticles constructed the nanofluidic crystal devices. DI water was used as the testing buffer.

The sensing performance of the present "drop to measure" nanofluidic crystal sensors was also detected by analyzing the conductance variation caused by biotin (using loading in the streptavidin-modified nanoparticles) and Pb^{2+} (using catalyzing DNAzyme probed nanoparticles) respectively. For the biotin sensing, the biotin whose concentrations varied from 0.1nM to 1 mM in DI water were loaded onto the self-assembled nanofludic crystal and reacted about 10 min. The conductance variation caused by the loaded biotin on the streptavidin modified nanofluidic crystal at different ionic concentrations is shown in Figure 7. Each data was obtained from 3 chips and no self-calibration for every sensing operation was required. It indicates that the sensor has a limit of detection of 1 nM for biotin and has a high chip-to-chip data consistency.

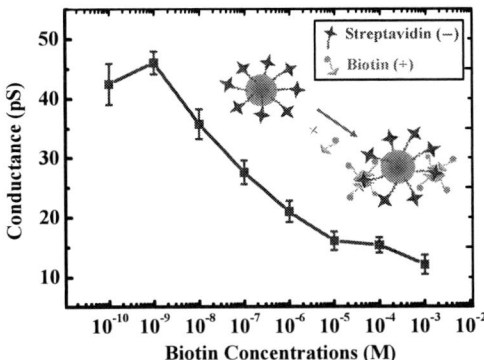

Figure 7: Biotin sensing performance. (Each point stands for three independent chips.) A limit of detection of 1 nM was obtained.

Recently, DNA probes have shown its unique advantages in lead detection, such as slow detection limit and selectivity [5]. Sequences of the DNA substrate and the DNAzyme were showed in Figure 8. If Pb^{2+} existed, it could cut the DNA chain off from the nanoparticle surface, which led to surface charge decrement and could be traced by present nanofluidic crystal sensor.

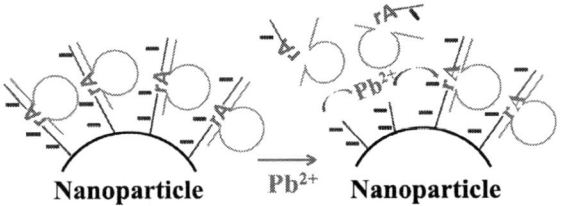

Figure 8: Schematic of the present nanofluidics crystal based lead sensing.

Figure 9 showed the Pb^{2+} concentration sensing performance. Mg^{2+} detection was used as control experiment. The experimental results showed a limit of detection of 1 nM was achieved with an excellent ion selectivity. Each point stands for one independent chip. Further study of the lead detection in this chip will be continued.

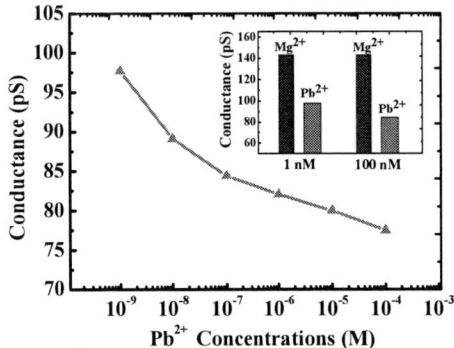

Figure 9: Pb^{2+} sensing performance.

CONCLUSION

In short, having a high chip-to-chip data consistency and a limit of detection of 1 nM, the present "drop to measure" nanofluidic crystal sensing scheme is ready for portable biochemical detection.

ACKNOWLEDGEMENTS

This work was financially supported by the National Natural Science Foundation of China (Grant No. 81471750 and 91023045), the Major State Basic Research Development Program (973 Program) (Grant Nos. 2009CB320300 and 2011CB309502), and the 985-III program (clinical applications) in Peking University.

REFERENCES

[1] R.Karnik et al., "Effects of Biological Reactions and Modifications on Conductance of Nanofluidic Channels", Nano Lett., 5 (9), pp. 1638–1642, 2005.

[2] Ruoshan Wei, Ulrich Rant et al., "Stochastic sensing of proteins with receptor-modified solidstate nanopores", Nature nanotechnology, 7 (4), pp. 257-263, 2012.

[3] Jianming Sang, Wei Wang, et al., "Protein sensing by nanofluidic crystal and its signal enhancement", Biomicrofluidics, 7 (2), 024112, 2013.

[4] Baojun Wang, Wei Wang et al., "Nanofluidic crystal in a Parylene C confined space for high consistent biosensing", in Digest Tech. Papers MicroTAS 2013 Conference, pp. 874-876, 2013.

[5] Tian Lan et al., "A highly selective lead sensor based on a classic lead DNAzyme", Chem.Commun., 46, pp. 3896-3898, 2010.

CONTACT

*Wei Wang, Tel: +00-010-62769183;
w.wang@pku.edu.cn.

ELECTRICAL DETECTION OF PESTICIDE VAPORS BY BIOLOGICAL NANOPORES WITH DNA APTAMERS

Aiko Nobukawa[1, 2], Toshihisa Osaki[1, 2], Taishi Tonooka[1], Yuya Morimoto[1], and Shoji Takeuchi[1, 2]
[1]Institute of Industrial Science (IIS), The University of Tokyo, JAPAN
[2]Kanagawa Academy of Science and Technology (KAST), JAPAN

ABSTRACT

This paper describes a vapor detecting system that applies two robust biological elements: A biological nanopore formed in a lipid bilayer and a DNA aptamer. The principle of the sensor is as follows: 1) DNA aptamer selectively captures the target molecules, 2) builds up a molecular complex larger than the nanopore size, and 3) inhibits the ionic current through the nanopore by blocking. We integrated these biological molecules into a previously developed device. A feasibility test was performed using a vapor phase sample, an organophosphorus pesticide, and represented the results demonstrating long-and-deep current blockades with the presence of the omethoate.

INTRODUCTION

Vapor Molecule Detection

Vapor molecule detection methods have long been relying on chromatography coupled with mass spectroscopy. Despite its high sensitivity and capability of multiple vapor detection, the chromatography has several disadvantages, such as extensive sample preparation protocols, technical expertise necessary for operation, and high cost. Therefore, many alternative devices, such as semiconductor-based sensors [1], and biosensors [2], have been developed for rapid and/or sensitive detection of vapor molecules. However, such systems need to identify the captured target species separately.

Residual Pesticides

Various pesticides are widely used in agriculture to maintain the conditions of crops and increase their yields. On the one hand, there is a global concern over the problem of pesticide residues to our environment (geo-pollution) as well as food contamination. For these reasons, it is essential to develop a rapid, sensitive, and easy-to-use detection system of pesticides.

Molecular Detection Using a Biological Nanopore and a DNA Aptamer

One of the most famous biological nanopores is the α-hemolysin (αHL) nanopore, consisting of seven αHL protein monomers, which autonomously assemble to the nanopore form in a lipid bilayer. The inner diameter of the nanopore is around 1.5 nm, which is known to pass through a single-stranded DNA (1.0 nm in diameter) but not a double-stranded DNA [3]. These size discrimination characteristics allows the nanopore to be used as a sensor element to detect and analyze DNA, RNA, and proteins.

Aptamers are one or a few strings of single-stranded DNA or RNA that bind to a target molecule, such as small organic molecules and proteins, with high affinity and specificity.

When the aptamer binds to the target molecule, the size may become larger than the inner diameter of the nanopore. Therefore, the nanopore formed within a lipid bilayer platform coupled with a DNA aptamer will further extend the possibilities to detect various small molecules other than DNA, RNA, and proteins. Our previous work chose the "cocaine" molecule as the target and successfully proved the concept of the rapid detection [4].

In this work, we aim to develop a vapor-sensing device taking advantage of our previous study using the biological nanopore in a lipid bilayer and a DNA aptamer designed to bind with the target molecule, a pesticide vapor, omethoate. Figure 1 shows the concept of the residual vapor pesticide detection. 1) Pesticide vapor adsorbed on crops are collected by a syringe. 2) The collected vapor sample is dissolved in a buffer and mixed. 3) The sample is injected into the vapor detection device. 4) When the pesticide-DNA aptamer complex clogs to the nanopore, the ionic current switches ON to OFF (Figure.2). Here, we design the DNA aptamer for the omethoate molecule, examine whether the aptamer forms a complex with the "liquid state" omethoate, and investigate the characteristics

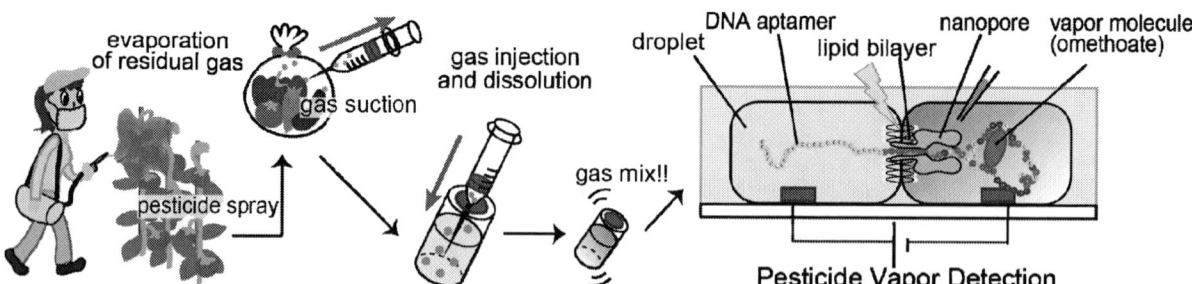

Figure 1: The conceptual diagram of the vapor omethoate sensor. The detection principle is as follows: 1) Most vegetables are sprayed pesticides for growth. 2) Residual vapor, which remains in vegetables, evaporates. 3) Residual vapor is collected with a syringe. 4) Collected vapor is dissolved into a buffer solution. 5) The sample solution is injected into the pesticide-vapor detection device. 6) When the target molecule-DNA aptamer complex clogs the nanopore, the ionic current switches from ON to OFF.

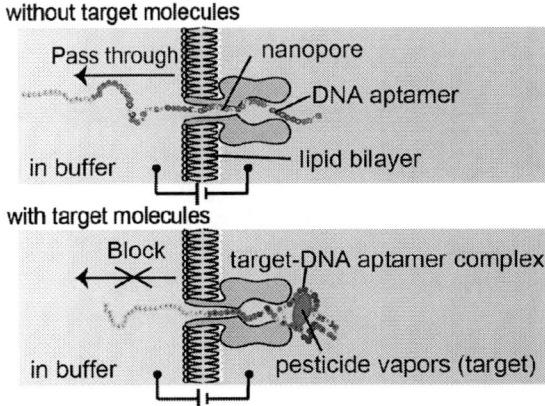

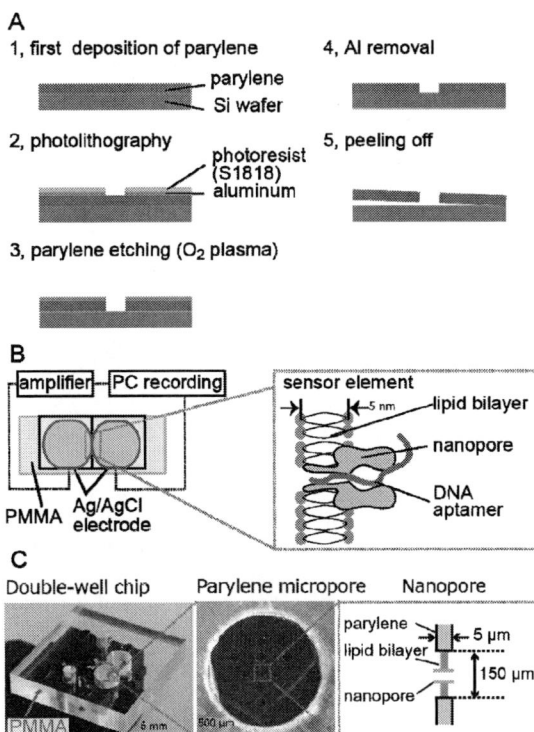

Figure 2: Schematics of the detection mechanism using the nanopore and the target-DNA aptamer complex.

of the nanopore signals with the presence/absence of the omethoate molecules. Finally, "vapor phase" omethoate is applied to the device and performed a feasibility test.

EXPERIMENTAL

Chemicals and Materials

The DNA aptamer (SS4-54) binding to an omethoate pesticide was previously reported by L. Wang *et al.* [5]. The sequence is 5'-AAG CTT TTT TGA CTG ACT GCA GCG ATT CTT GAT CGC CAC GGT CTG GAA AAA GAG -3'. In this study, we added cytosine (27-mer) at the 3' end of the SS4-54. According to the previous work [4], the aptamer, a single-stranded DNA, specifically binds to four organophosphorus pesticides, i.e. phorate, profenofos, isocarbophos and omethoate. The affinity for the omethoate (K_d) is 2.0 μM and the underlined nucleotides-sequences are considered as the binding sites, but the tertiary structure has not yet been determined. The HPLC-grade DNA oligonucleotide was purchased from Sigma-Aldrich (Tokyo, Japan), and stored at -20°C.

The omethoate was obtained from Wako Pure Chemical Industries (Osaka, Japan), and stored at -20°C before use. A diluted omethoate solution was dissolved in the buffer solution described below and stored at 4°C. Omethoate is famous as an organophosphorus pesticide and its physical properties exhibit colorless/pale-yellowish clear liquid state at room temperature but it is easy to volatilize.

A buffered electrolyte solution of 1.0 M KCl with 10 mM phosphate buffer (adjusted at pH7.2) was prepared from 18 MΩ·cm Milli-Q water (Merck Millipore, Tokyo, Japan). All chemicals for the buffer solution were obtained from Wako Pure Chemical Industries.

The nanopore was formed using the transmembrane toxin α-hemolysin (αHL) from *Stapylococcus aureus*. The αHL was obtained from Sigma-Aldrich (MO, USA).

The lipid bilayer membrane was prepared using 1, 2-diphytanoyl-*sn*-glycero-3-phosphocholine (DPhPC, Avanti Polar Lipids, Inc., AL, USA) dissolved in *n*-decane (Sigma-Aldrich), and stored at 4°C.

Fabrication of Device

The fabrication process for the poly (chloro-*p*-xylylene) (parylene C) micropore is shown in

Figure 3: (A) Fabrication process of the parylene C film with micropores. (B) The device overview and the nanopore coupled with a DNA aptamer as the sensor element. (C) The images of the double-well chamber and parylene micropores, and a schematic of the micropore.

Figure 3A. First, a 5 μm-thick film of parylene C was deposited on a single-crystalline silicon substrate by a chemical vapor deposition (PDS2010, Specialty Coating Systems, Inc., IN, USA). A thin aluminum layer was then deposited on the parylene C film, and a positive photoresist S1818 (Shipley, MA, USA) was sequentially spin-coated. The S1818 and aluminum layers were patterned using a standard photolithographic process. Using the aluminum layer as a mask, the parylene C film was patterned by oxygen plasma (FA-1, Samco Inc., Kyoto, Japan). After the aluminum mask was removed, the parylene C film with the patterned micropores was peeled off from the silicon substrate using tweezers.

The patterned parylene C film and a poly (methyl methacrylate) (PMMA) plate (Mitsubishi Rayon, Tokyo, Japan) was used as the device materials. The device was assembled as follows (see Figure 3): First, the patterned parylene C film (consisting of five pores with 150 μm in diameter) was sandwiched between 200 μm-thick PMMA films with adhesion bond (Aron Alpha, Toagosei Co., Tokyo, Japan), used as a separator below. Next, two circular wells (4 mm in diameter, 3 mm in depth) were micromachined using a mini miller (MM100, Modia Systems, Saitama, Japan). At the bottom of each well, a through hole was drilled to embed the Ag/AgCl electrode. The separator was then inserted between the two circular wells and fixed with the adhesion bond. Finally, the electrodes were fixed to a BNC (Bayonet Neill-Concelman) connector to connect to a patch clamp

978-1-4799-7956-1/15 $31.00 © 2015 IEEE

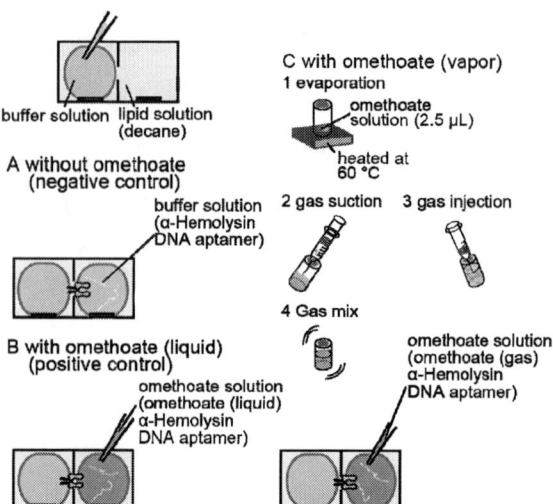

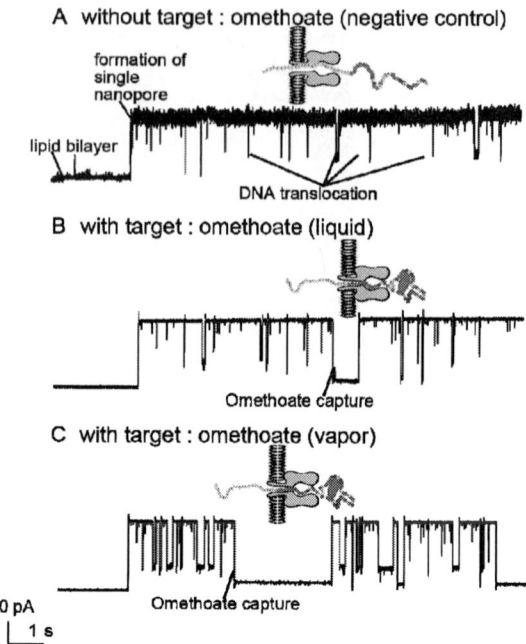

Figure 4: Protocol of the pesticide vapor (omethoate) detection. First, a lipid solution was injected in both chambers, and a buffer solution was put in the left chamber. The sample buffer, which is placed at the right chamber, is prepared as follows. Condition A: αHL and DNA aptamer in a buffer solution. Condition B: "liquid" omethoate, αHL, and DNA aptamers in a buffer solution. Condition C: "vapor" omethoate, αHL, and DNA aptamers in a buffer solution. The vapor omethoate was prepared as described.

amplifier.

Bilayer Formation and Electrical Measurements

The contact method with aqueous droplets is the way to form a planar lipid bilayer membrane, firstly reported by K. Funakoshi *et al.* [6]. The principle of the bilayer formation with this method is based on the monolayer self-assembly of lipids at the interface between water and hydrophobic oil. When a pair of the interfaces approaches and contacts each other, the pair of monolayers forms a bilayer with hydrophobic acyl chains facing each other. The double-well chamber in Figure 3 is designed to settle the aqueous droplets at the suitable positions. The Ag/AgCl electrodes are placed at the bottom of the chambers to contact with each droplet, and applied to measure the membrane conductance using a patch clamp amplifier (CEZ-2400, Nihon Kohden Co., Tokyo, Japan).

Nanopore formation at the planar lipid bilayer is performed on the device as follows: First, we inject a DPhPC lipid solution to the two wells (5 μL each), and sequentially added two aqueous droplets in each well (20 μL of the buffered electrolyte solution). Gradually the two droplets come together and form a bilayer. The final concentration of 30 μM αHL is mixed in one of the droplets, in which soluble αHL monomers insert into the lipid bilayer and form the nanopores. The ionic current across the membrane is observed using voltage clamp at 100 mV. Typically, 100 pA steps, which correspond to 1 nS of conductance in the condition of 1.0 M KCl buffer solution, appear a few minutes after the droplet injections. This step size is consistent with the previously reported nanopore conductance assembled into a lipid bilayer [7].

Figure 5: Typical current signal vs. time traces with the three conditions. Condition A: Negative control without the target omethoate molecules. The fast blocking signals observed are due to translocation of DNA through the nanopore. Condition B: Positive control with the "liquid" omethoate molecules. The long-and-deep blocking signals were observed because of the clogging of the omethoate-DNA aptamer complex. Condition C: The feasibility test result using the "vapor" omethoate molecules. The long-and-deep blockades similar to the condition B were observed.

Pesticide Injection

A volatile pesticide, omethoate, was chosen as the target molecule to examine whether monitoring of the ionic current through the αHL nanopore accurately discriminates the presence and the absence of the target molecule-DNA aptamer complex.

The protocol of the target detection is as described in Figure 4. Here we define the electrode direction as the cathode at the 'right' chamber and the anode at the 'left' chamber. Since DNA is negatively charged, DNA aptamer was dissolved in the right droplet to capture at the nanopore. For vapor molecule detection, three conditions were examined. As mentioned above, the lipid solution is firstly injected in the both chambers, and the buffer solution is put in the left chamber. Then the sample buffer is put in the right chamber with setting the following three conditions: A) Negative control condition without the target omethoate (αHL and DNA aptamers only); B) Positive control condition with "liquid" omethoate dissolved in the sample buffer (αHL, DNA aptamers, and 100 μM liquid omethoate); C) Feasibility test using "vapor" omethoate. In this case, first, the liquid omethoate (2.5 μL) was put in a closed vial and heated at 60°C, resulting in evaporation of the entire liquid omethoate. Then, the vapor omethoate in the vial was taken by a gas-tight syringe and moved to another vial filled Milli-Q water. The vial is vigorously shaken to mix the vapor omethoate and water and kept for a couple of hours to dissolve the vapor omethoate into the water phase.

RESULTS & DISCUSSION
Electrical Signals without Omethoate
Condition A was performed as the negative control experiment. As shown in Figure 5A, a 100 pA stepwise current increase was observed, and then spike-like blockades occurred under the flow of a constant current. The baseline current signal is associated with the formation of a lipid bilayer, while incorporation of αHL in the lipid bilayer and formation of the nanopore caused the 100 pA step. When the DNA aptamers interact with the nanopore, the spike-like blockades occur; the blockade signals can be divided to two types: 1) Fast blockades (<500 ms) that reduced the current level to 10 pA (90% blockades). 2) Relatively short blockades (<500 ms) that reduced the current level to 55-65 pA (around 50% blockades). The fast blockades probably indicate that DNA aptamers pass through the nanopore. The relatively short blockades may be caused by collision or transient partial entry of DNA into αHL vestibule [8]. These spike-like signals ensure that the DNA aptamer does not form any aggregated structure by itself, and the validity of our aptamer design.

Electrical Signals with Liquid Omethoate
Condition B was performed as the positive control experiment. It is considered that the DNA aptamer forms an omethoate-DNA aptamer complex because of the presence of the target omethoate. Since the complex becomes larger than the inner diameter of the nanopore in the lipid bilayer, the complex is expected to stack at the nanopore and the ionic current will be blocked. This feature due to the omethoate-DNA aptamer complex was observed as the long and deep current blockades (Figure 5B). The electrical signals with omethoate showed two types of blockades: 1) Long blockades (>500 ms) that reduced the current level to 10 pA (90% blockades) and 2) short to long blockades that reduced the current level to 55-65 pA (around 50% blockades). The long blockades probably indicate the stacking of the omethoate-DNA aptamers complex, while the other blockades may be caused by the collision/transient partial entry of the complex/single-stranded DNA.

Electrical Signals with Vapor Omethoate
Condition C used the vapor omethoate sample that was once volatilized. In this case, we observed similar electrical signals, the long and deep blockades, to the liquid omethoate (Figure 5C). This result indicates that our system using the nanopore and DNA aptamer has a potential for detecting omethoate vapor.

Threshold of Pesticide Detection
For vapor omethoate sensing, the threshold parameters have to be determined. When the absence of the omethoate molecules, the 90% and 50% blockades (both of them were <500 ms) were observed. On the other hand, when the presence of the omethoate molecules, the 90% blockades (>500ms) and 50% blockades were observed. Based on these features, the threshold parameters, i.e. the blocking level and the duration, can be determined as the 90% blockades with the duration of 500 ms or longer.

CONCLUSIONS
We detected the vapor omethoate molecules, an organophosphorus pesticide, using the robust biological elements: a biological nanopore and a DNA aptamer. The vapor sample clearly showed long-and-deep ionic current blockades of the nanopore by the omethoate-DNA aptamer complex. For sensing the target molecule, the threshold parameters may be determined as over 90% current blockades with 500 ms periods or longer.

ACKNOWLEDGEMENTS
The authors deeply acknowledge the technical supports provided by Ms. U. Nose, M. Uchida, A. Nozaki, and Y. Kagamihara (KAST), and the valuable discussion with Dr. K. Kamiya and Mr. K. Inoue (KAST). This work was partly supported by the Regional Innovation Strategy Support Program of MEXT, Japan.

REFERENCES
[1] G. Sakai, N. Matsunaga, K.Shimanoe, N. Yamazoe. *"Theory of gas-diffusion controlled sensitivity for thin film semiconductor gas sensor"*, Sensors and Actuators B, vol. 80, pp. 125-131, 2001.

[2] K. Sato and S.Takeuchi, *"Chemical Vapor Detection Using a Reconstituted Insect Olfactory Receptor Complex"*, Angew. Chem. Int. Ed., vol. 53, pp. 11798-11802, 2014.

[3] Kasianowicz, J. J., E. Brandin, D. Branton, and D. W. Deamer., *"Characterization of individual polynucleotide molecules using a membrane channel"*, Proc. Natl. Acad. Sci. USA., vol. 93, pp. 13770-13773, 1996.

[4] R. Kawano, T. Osaki, H. Sasaki, M. Takinoue, S. Yoshizawa, and S. Takeuchi., *"Rapid Detection of a Cocaine-Binding Aptamer Using Biological Nanopores on a Chip"*, J. Am. Chem., vol. 133, pp. 8474-8477, 2011.

[5] L. Wang, X.Liu, Q. Zhang, C. Zhang, Y. Liu, K. Tu and J. Tu, *"Selection of DNA aptamers that bind to four organophosphorus pesticides"*, Biotechnol Lett., vol. 34, pp. 869-974, 2012.

[6] K. Funakoshi, H.Suzuki, and S.Takeuchi, *"Lipid Bilayer Formation by Contacting Monolayers in a Microfluidic Device for Membrane Protein"*, Anal. Chem., vol. 78, pp. 8169-8174, 2006.

[7] D. Deamer, D. Branton, *"Characterization of Nucleic Acids by Nanopore Analysis"*, Acc.Chem. Rev., vol. 35, pp. 817-825, 2002.

[8] T. Z. Butler, J. H. Gundlach, and M.Troll, *"Ionic Current Blockades from DNA and RNA Molecules in the α-Hemolysin Nanopore"*, Biophysical Jounal, vol. 93, pp. 3229-3240, 2007.

CONTACT
*Aiko Nobukawa, tel +81-3-5452-6650; E-mail: nobukawa@iis-u.tokyo.ac.jp

HIGH-PERFORMANCE AND LOW-COST LUNG CANCER SENSOR ARRAY BASED ON SELF-ASSEMBLED GRAPHENE

B. Zhang and T. Cui
University of Minnesota, Minneapolis, USA

ABSTRACT

The lung cancer sensor array (LCSA) based on layer-by-layer (LbL) self-assembled graphene presented in this paper is capable of detecting different lung cancer biomarkers selectively with advantages of high sensing performance and low cost due to graphene inherent properties and self assembly technique. According to the resistance change of graphene, the detection limit of LCSA is down to 0.1 pg/mL.

INTRODUCTION

In personalized medicine, the goal is to segregate groups of patients by defining a diagnosis, the risk for death or recurrence of disease (prognosis), or the proper therapy for the appropriate patient to maximize treatment response (prediction). Discovery and validation of biomarkers is central to this goal [1]. Therefore, testing and monitoring concentration of cancer biomarkers in human body play a very significant role for an early diagnose and control of relevant diseases. Among various cancers, lung cancer is always the top killer to patients. Many researchers have proposed to use three typical cancer markers, such as VEGF, ENO1, and ANXA2, to early diagnose and control lung cancer.

As observed in tumor malignancy in general, VEGF may play a critical role in tumor growth and metastasis of lung cancer. It has been shown that VEGF is significantly associated with increased micro vessel density in lung cancer. VEGF is also expressed in a majority of non-small cell lung cancer (NSCLC) tumors. In addition, VEGF is expressed at higher levels as lung cancer progresses. VEGF may be correlated with carcinogens in lung cancer. What's more, VEGF expression has been associated with decreased survival in lung cancer [2]. Therefore, VEGF concentration detection is very meaningful for lung cancer diagnosis.

ENO1 is a glycolytic enzyme that converts 2-phosphoglycerate into phosphoenolpyruvate in glycolysis and a multifunctional protein that play a crucial role in a variety of biological and pathophysiological processes. ENO1 may act as a stress protein that promotes hypoxic tolerance in tumor cells by increasing anaerobic metabolism. ENO1 may also function as a plasminogen receptor on the surface of a variety of tumor, hematopoetic, epithelial and endothelial cells. Increased cell-surface expression of ENO1 promotes cell transformation and invasion in non-small cell lung cancer, brain cancer and neck cancer. Therefore, it is very important to identify the concentration of ENO1 to diagnose lung cancer [3].

ANXA2, a 36 kDa protein, is expressed in lung tumor cells, endothelial cells, macrophages, and mononuclear cells. ANXA2 contains three distinct functional regions: the N-terminal region, the C-terminal region, and the core region. In ANXA2, the core domain possesses two annexin-type calcium-binding sites. ANXA2 is activated in a calcium-dependent manner and undergoes a conformational change that exposes a hydrophobic amino acid to form a heterotetramer [4]. In general, higher ANXA2 concentration will represent higher possibility of lung cancer.

This lung cancer sensor array (LCSA) based on layer-by-layer (LbL) self assembled graphene presented in this paper shows features including high performance and low cost in lung cancer biomarker detection due to graphene material properties in nature, self assembly technique, and multiple antigens detection on a single chip.

To overcome the hurdles of these previous ion sensitive sensing methods, the layer-by-layer (LbL) self-assembled graphene is introduced to the lung cancer biomarker sensing applications in this work. Due to its unique structural, electrical, chemical, and mechanical properties [5], graphene has attracted more and more attention these days. With rational chemical and/ or physical modification, graphene is capable of detecting many types of molecules and ions [6]. The LCSA based on LbL self assembled graphene provides a promising way to selectively detect different ions simultaneously with a very high detection resolution due to its high chemical sensitivity and low electrical noise [7]. The LCSA based on LbL self assembled graphene is designed to selectively detect multiple lung cancer biomarkers in a testing solution. Moreover, with the LbL self-assembly technique, LCSA has a high performance at very low cost.

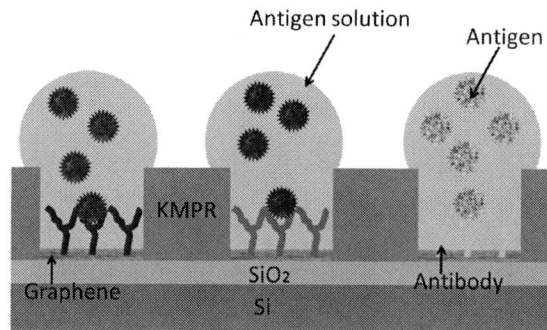

Figure 1: Schematic of LCSA sensor array. When a testing solution contains one type of lung cancer biomarkers, only the sensing channel with a matched antibody triggers a signal.

DESIGN AND FABRICATION

As shown in Fig. 1, different polyions and graphene

nanoplatelets are deposited into the 20 μm gaps between the sensor electrode arrays by self assembly technique. A layer of KMPR confining the sensing region of LCSA acts as a confinement layer. Different lung cancer biomarker antibodies coated on the sensing region only capture matched antigens to trigger the biochemical reactions. Due to the absorption of charged antigens, the resistance of graphene will change with antigen concentrations [8].

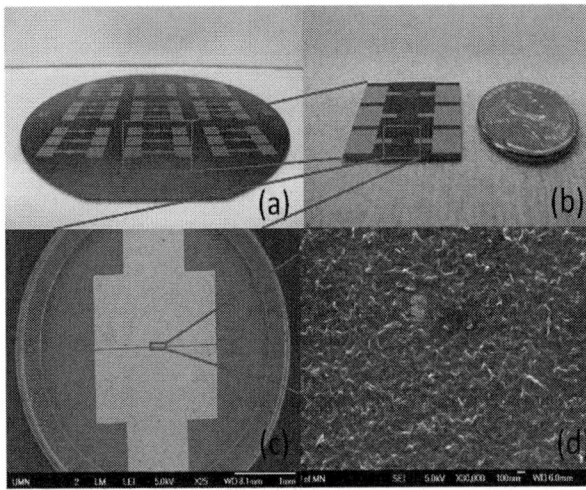

Figure 2: (a) image of LCSA wafer; (b) image of single chip of LCSA; (c) SEM image of sensing well confined by KMPR; (d) LbL self-assembled graphene layer, showing the porous surface profile of graphene composites, the average size of graphene sheets is about 100 nm by 100 nm.

Microfabrication was utilized to pattern the electrodes of LCSA, and graphene layers were deposited by self assembly technique. Chromium/ gold layers 50/200 nm thick were firstly deposited on a clean silicon/ silicon dioxide wafer with an AJA sputter system. Subsequently, sensor electrodes were patterned by photolithograph. Another lithographic step was used to fabricate a window area on which the graphene film was self assembled, while protecting the testing pads from the adsorption of graphene solutions. The polyelectrolytes used in this study were poly(diallyldiamine chloride) (PDDA) and poly(styrene sulfonate) (PSS). The concentrations of aqueous PDDA and PSS were 1.5 and 0.3 wt% respectively, with an addition of 0.5 M sodium chloride to enhance the surface properties. The concentration of graphene suspension solution was 0.25 mg/ml. Next, the substrate was immersed into the charged suspensions with a sequence of the immersion [PDDA (10 min) + PSS (10 min)]₂ + [PDDA (10 min) + graphene suspension (20 min)]₅. Afterwards the substrate was immersed into acetone for 5 minutes to lift off the photoresist mask, and the graphene sensors were inspected by scanning electron microscopy (SEM) (Fig. 2d). As shown in Fig.2, due to the hydrophobic property of KMPR with a contact angle of 97° and

hydrophilic property of grapheme with a contact angle of 64°, the applied solution can be easily confined in the sensing area. Different antibodies can be immobilized on the single chip due to this phenomenon.

The more important factor is that the different surface wetting ability can confine the solution area. In this way, the different antibody solutions can be deposited on the same chips without any mixture. This will make the LCSA able to detect various biomarkers at the same time on one single chip. As shown in Fig. 3, different antigen solutions were confined in the sensing well by surface wetting ability

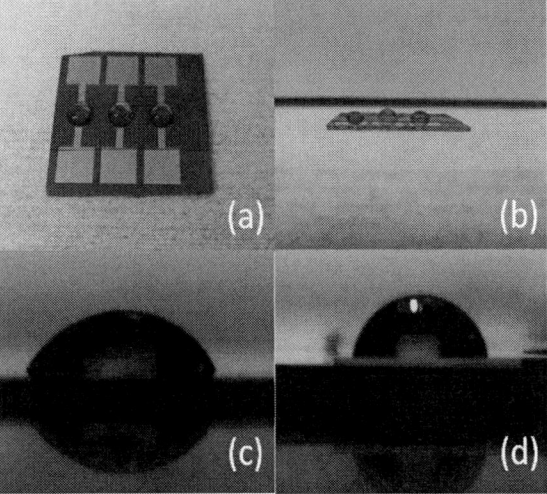

Figure 3: (a) and (b) different antigen solutions were confined in the sensing well by surface wetting ability; (c) contact angle of LbL self assembled graphene is 64° ; (d) contact angle of KMPR is 97° .

RESULTS AND DISCUSSIONS

For the characterization of LCSA, different concentrations of various lung cancer biomarkers were introduced to the sensing regions, and the resistance shift of LbL self assembled graphene was monitored by an Agilent data logger.

After the fabrication of LCSA, the biosensor was immunized by immobilization of antibody and antigen on the surface. 0.1% PLL aqueous solution (Sigma–Aldrich Inc. without further treatments) was first applied on the three wells for 1 hour. Next, the LCSA was incubated for overnight at 4°C in ANXA2 (Santa Cruz Biotechnology Inc.), ENO1 (Santa Cruz Biotechnology Inc.), VEGF (Sigma–Aldrich Inc.) capture antibody solutions respectively at a concentration of 10 μg/ml. The sensor was immersed in a PBS solution (Dulbecco's phosphate buffered saline, Invitrogen Inc.) for 10 minutes to rinse the biosensors. Next, the sensor was incubated in 3% BSA blocking solution (Santa Cruz Biotechnology Inc.) at room temperature for 5 hours to block nonspecific binding sites. After repeating the rinsing step, the label free sensor was ready for testing.

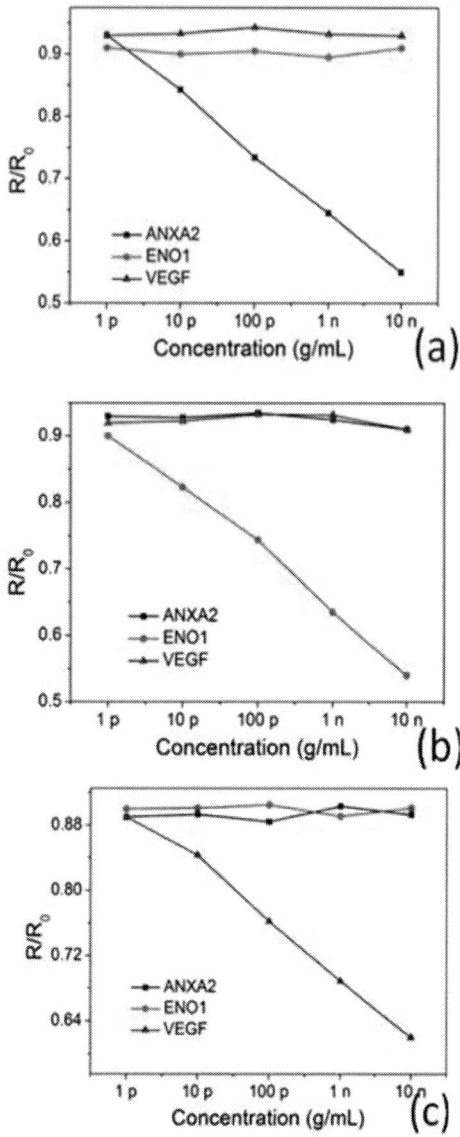

To investigate the selectivity of the LCSA, the testing solution containing one type of lung cancer biomarkers was applied to the LCSA, and the resistance change of channels coated with different antibodies were recorded.

The signal shifts of graphene devices were monitored using Agilent Data Logger (34970A, Agilent Inc.). The biosensors were measured in a DC mode, where bias voltage of 30 mV were applied by analog outputs of a Wavetek 164 (Tucker Electronics). Data was acquired at a frequency of 10 kHz by TDS 2024B (Tektronix Inc).

As shown in Fig. 4, the LbL self-assembled graphene resistance decreases when capturing these matched antigens. Different sensing antibody channels had the corresponding signals representing the antigen concentrations in the testing solution. To get an obvious readout, a normalized resistance was deduced. Resistance for a solution of PBS was used as an initial conductance, R_0, and other conductance tested under different concentrations R was divided by R_0. Normalized resistance was represented as R / R_0.

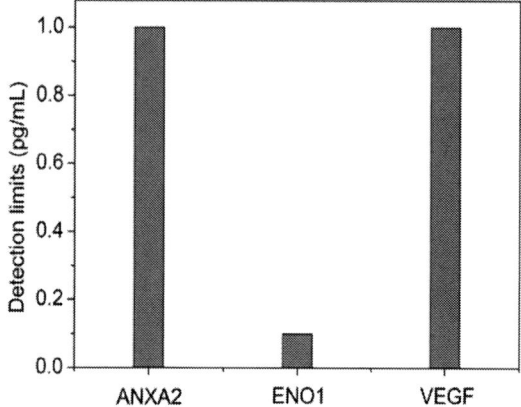

Figure 5: Detection limits of different antigens were investigated. LCSA can achieve 0.1~1 pg/mL, ultralow detection limits.

It is evident that a change in density and/ or mobility of charge carriers must be responsive when charged antigens are absorbed by graphene, reflected on the shift of resistance of graphene. The equation $\sigma = nqv$ can demonstrate the relationship, where σ is conductance, n is carrier density, q is charge per carrier, and v is the carrier mobility. The charged antigens captured by LbL self assembled graphene increase the carrier density of graphene, in the chemical equivalent of the electric-field effect. Moreover, they are serving to partially neutralize the Coulomb scatters induced by the substrate, allowing a rapid increase in electrical mobility. According to the resistance calculation equation above, both of the mechanisms increase the conductance with the absorption of hydrogen from the bio-catalyzed reaction. The conductance of the graphene shifts with the concentration changes of different ion solutions.

Figure 4: Testing solution containing only (a) ANXA2, (b) ENO1, (c) VEGF was introduced to the LCSA. The antigens, matched the corresponding antibodies, trigger the corresponding signals, representing the antigen concentrations in the testing solution. To get an obvious readout, a normalized resistance was deduced. Resistance for a solution concentration of 1 pg/mL was used as an initial conductance, R_0, and other resistance tested under different concentrations R was divided by R_0. Normalized resistance was represented as R / R_0.

As shown in Fig. 5, the resolution of LbL self-assembled graphene based LCSA was characterized, and compared with the results of different lung cancer biomarkers under the same conditions of design, manufacture, and measurement. When changing ions with different concentrations, the graphene conductance was recorded by an Agilent data logger. The resolution for different ions was determined by the lowest concentration detected by the LCSA. It is observed that the graphene LCSA has a good detection limit.

The detection limit of a sensor that can be processed is ultimately determined by its signal-to-noise ratio. If the input signal and sensor structure are constant, the output signal is supposed to be more stable and sensitive to the lower 1/f noise. Thus the noise spectra of suspended and unsuspended graphene devices were further investigated under the same experimental conditions, and concentrated on low-frequency (below 1 kHz) noise reported to be dominant in limiting performance of nanodevices. The current power spectra can be expressed as $S_I = AI^2 f^\beta$, where S_I is the noise power density, I is current, f is the frequency, A is defined as the 1/f noise amplitude, and β is the frequency exponent with a value close to -1. It was reported that trapped charges at the interface and in the substrate degrade transport characteristics of a single-layer graphene, which exhibits the effect in our experiment results. In addition, due to the self assembly technique and polymers, the surface profile of the graphene layers shows great porous topography, more suitable to decorate capture proteins, providing the greatest sensing surface area per unit volume.

The high detection limit of graphene was attributed to its inherent properties. Due to the crystal lattice and two-dimensional nature, graphene tends to screen charge fluctuations more than one-dimensional materials such as carbon nanotubes. The low 1/f noise of graphene results in the better performance in sensing resolution. Moreover, two dimensional structure gives rise to more exposure to ions, providing the greatest sensing area per unit volume for graphene. These two factors operated together to promote the good detection limits of graphene sensors.

CONCLUSIONS

Lung cancer is the most common cause of cancer-related death worldwide. In 2013, approximately 1.6 million new lung cancer cases and 1.4 million lung cancer deaths have occurred all over the world [9]. The mortality rate has not decreased during the last decade [9], because the lack of clinical symptoms in early-stage lung cancer leads to diagnosis at a late stage. Low-cost early diagnosis methods are very important to fight with lung cancer. The earlier cancer can be detected, the better chance of a cure will be. Effective, accurate methods of cancer detection and clinical diagnosis are urgently needed. The use of biosensors in cancer detection and monitoring holds vast potential. Biosensors can be designed to detect emerging cancer biomarkers and to determine drug effectiveness at various target sites. Biosensor technology has the potential to provide fast and accurate detection, reliable imaging of cancer cells,

and monitoring of angiogenesis and cancer metastasis, and the ability to determine the effectiveness of anticancer chemotherapy agents.

A low-cost and high-performance LCSA was investigated in this paper. It was characterized by applying testing solutions containing certain lung cancer biomarker to the sensor array. The LbL self-assembled graphene resistance decreased when capturing the charged antigen. Different lung cancer biomarker antibody channels output the corresponding signals representing the antigen concentrations in the testing solution. Three different lung cancer biomarkers including VEGF, ENO1 and ANXA2 were successfully detected by LCSA. The selectivity of the LCSA was verified, providing a potential protocol for lung cancer diagnosis. In addition, the detection limits of the LCSA based on graphene is 0.1~1 pg/mL. The results presented herein suggest a new route to an inexpensive and high-performance LCSA for medical sensing applications.

ACKNOWLEDGEMENT

The authors acknowledge the assistance of fabrication and characterization from the Minnesota Nano Center and the Characterization Facility at the University of Minnesota. The authors also acknowledge the supports from Minnesota Partnership foundation.

REFERENCES

[1] H. Pass, D. Beer, S. Joseph, P. Massion, *Thoracic Surgery Clinics*, 23(2013), pp. 211-224.

[2] R. Rosa, F. Monteleone, N. Zambrano, and R. Bianco, *Current Medicinal Chemistry*, 21 (2014), pp. 1595-1606.

[3] J. Ho, H. Chang, N. Shih, L. Wu, Y. Chang, C. Chen, C. Chou, *Analytical Chemistry*, 82 (2010), pp. 5944-5950.

[4] C. Wang and C. Lin, *Disease Markers*, 11 (2014), pp. 1-7.

[5] K. S. Novoselov, A. K. Geim, S. V. Morozov, D. Jiang, M. I. Katsnelson, I. V. Grigorieva, S. V. Dubonos, A. A. Firsov, *Nature*, 438 (2005) pp. 197-200.

[6] J. D. Fowler, M. J. Allen, V. C. Tung, B. H. Weiller, *ACS Nano.*, vol. 3, pp. 301-306, 2009.

[7] K. Ratinac, W. Yang , S. Ringer, F.Braet, *Environ. Sci. Technol.* vol. 44, pp. 1167-1176, 2010.

[8] F. Schedin, A. K. Geim, S. V. Morozov, E. W. Hill, P. Blake, M. I. Katsnelson, K. S. Novoselov, *Nat. Mater.* vol. 6, pp. 652– 655, 2007.

[9] Y. Zhang, D. Yang, L. Weng, L.Wang, *Int. J. Mol. Sci.*, vol. 14, pp. 15479-15509, 2013.

CONTACT

* T. Cui, tel: +1- 612-626-1636; tcui@me.umn.edu

NOVEL DETECTION PLATFORM FOR ALZHEIMER'S AMYLOID-BETA USING MAGNETIC BEADS IN ELECTROCHEMICAL IMPEDANCE SPECTROSCOPY

Kyeong-Sik Shin, Moo Jong Kim, Soo Hyun Lee and Ji Yoon Kang

Center for Biomicrosystems, Korea Institute of Science Technology, Seoul, KOREA

ABSTRACT

In this paper, we proposed novel detection platform to detect Alzheimer's amyloid-beta (Aβ) using pre-treated magnetic beads in Electrochemical Impedance Spectroscopy (EIS), for the first time. Without any immobilization on the electrodes of the EIS device, it shows ability to detect a few pg/ml of amyloid-beta oligomers and compared to the result of a conventional ELISA, which allows to simplify the measurement procedure, recycle the device by only changing magnet beads.

INTRODUCTION

Alzheimer's disease (AD) is one of the neurodegenerative diseases, characterized by severe cognitive degradation and memory impairment [1]. When clinical symptoms of dementia were occurred, it implies that significant and irreversible brain damage has occurred. Unfortunately, there are no way to cure this disease. So, it is very important to diagnose it in early stage to do therapeutic treatment and preventative therapies [2, 3]. Amyloid-beta (Aβ) is a major causative protein, which is one of the pathological hallmarks of AD [4]. It has been reported that Aβ oligomers are toxic to neuronal cells and major pathogenic species [5-7]. Therefore, it is promising way to detect Aβ oligomers for early detection of AD. Aβ has been usually measured using enzyme-linked immune sorbent assay (ELISA) but its sensitivity is not high enough to detect Aβ at levels below 10 pg/ml [8, 9]. In addition, several label-free biosensors, such as STM-based sensor [10], reduced grapheme oxide-based biosensor [11] and field effect transistor (FET)-based sensors [1, 12], were reported to measure Aβ in PBS as low as 1 pg/ml. Another more popular way to detect Aβ is using electrochemical impedance spectroscopy (EIS) since it is low cost method with high sensitivity, reliability and convenience in constructing biosensors [13, 14]. Therefore, it has been extensively exploited to use EIS sensor to detect Aβ oligomers or Aβ monomers with intermediate linker such as oligopeptide or biotin/NeutrAvidin bridge [13, 14]. In general, all devices for EIS measurement required to modify the surfaces of devices to bind some antibody on electrodes or gap between electrodes, and measure a change of the charge transfer resistance between surface modified electrodes and Aβ. Those limit of detection are around a few pg/ml. But, this surface modification has disadvantages in recycling and calibration of the EIS devices. Therefore, we tried to carry out immobilization of antibody on magnetic beads rather than on electrodes of EIS device because we notice that the effective permittivity/or surface charge of beads is changed with binding Aβ oligomers on beads. So, we proposed a new novel detection platform including microarray EIS device,

fluidic channel and magnetic bar, and will show it can effectively detect Aβ oligomer.

MATERIALS AND METHODS

Preparation of Magnetic beads with amyloid-beta

For the detection of Aβ oligomer using magnetic beads, we basically followed the protocol of Kim et al. [15]. Briefly, it will be explained in this section. For an assay, two different monoclonal antibodies of human Aβ were used. One of the antibodies was conjugated with the magnetic beads as a capture bead for Aβ, and the other was conjugated with the horseradish peroxidase (HRP) to detect Aβ and generate a fluorescence signal by HRP substrate for reference experimental data with ELISA. The magnetic beads (Dynabeads M-280, 2.8 μm) were conjugated with capture antibodies in 0.1 M phosphate buffer by incubation for 24 h at 37 °C. After conjugation of the antibodies, the beads were washed with phosphate buffer solution (PBS). The magnetic beads were diluted at a concentration of 0.56 mg/ml, and the detection antibodies (1 mg/mL) were added to the mixture at a concentration of 3 ng/mL. For stable and reproducible assay of Aβ oligomer, we used synthesized oligomer mimicking standard protein (OMSP), a stable synthesized molecule where several Aβ monomers are attached to an albumin preventing the aggregation of Aβ (Gachon Univ, Korea). OMSP was diluted in the buffer to proper concentrations (0– 370 pg/mL) just before the experiments. A mixture (7 mL with about 3.94 mg beads) was added to the incubation for 45 min.

Methods

New novel detection platform has multiple microarray EIS devices on a slide glass, a microfluidic channel and a magnetic bar beneath a slide glass. A microarray EIS device consisted of an array of individual hole (SU-8 patterned) on each electrode pair (working and counter electrodes, gap ~ 2 μm). The individual hole size was less than 7 μm, and a magnetic bead size was around 2.8 μm. Therefore, each hole can capture 2 or 3 pre-treated beads which was prepared by incubation with capture beads, horseradish peroxidase (HRP), and Aβ oligomer (OMSP) as described in previous section. Experimental procedures using novel detection platform and ELISA were shown in Fig. 1. First, we will explain the experimental procedure using ELISA. After preparation of pre-treated beads with various concentration of Aβ oligomer, they were washed in 96 wells to remove nonspecific binding by TBST (0.01 % Tween-20), and the chemifluorescent HRP substrate with excitation/ emission of 570/585 nm was reacted with the HRP enzyme. The fluorescence detection was monitored by an inverted fluorescence. The other procedure using novel detection platform can be summarized as follows.

978-1-4799-7956-1/15 $31.00 © 2015 IEEE

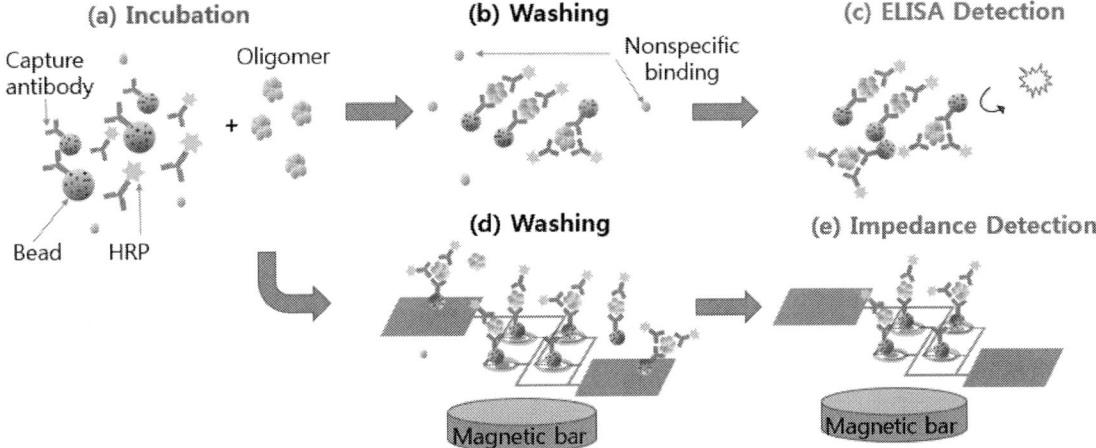

Figure 1. Schematics of experimental procedures using ELISA and novel detection platform (a) magnetic beads with antibody, secondary antibody with HRP and Aβ oligomer (OMSP) were incubated for 45 min (b) washing step to remove nonspecific bindings for ELISA (c) Injection of HRP substrate and ELISA detection (d) washing step to remove nonspecific binding and the beads on out of holes (d) electrochemical impedance detection

After preparation of pre-treated magnetic beads, they were injected into the fluidic channel and trapped at the array of individual hole by magnetic bar. Then, washing step was carried out to remove non-specific binding and beads located on SU-8 surface by injection of buffer solution. After washing step, the impedance of a microarray EIS device was measured to detect a concentration of Aβ oligomers.

RESULTS AND DISCUSSION

Whole novel detection platform with multiple microarray EIS devices, microfluidic channel and magnetic bar, was shown in Fig 2(a). Each microarray EIS device has 10 × 10 holes with electrode pairs made of Pt/Ti

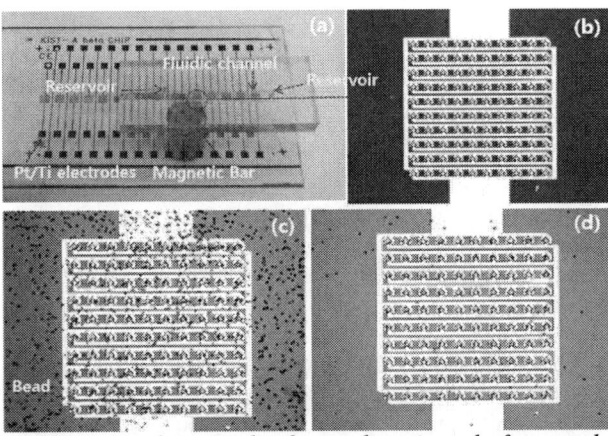

Figure 2. A photograph of new detection platform and images of pre-treated beads in holes by magnetic bar (a) a microarray EIS device on a slide glass with fluidic channel and magnetic bar (b) microscope image of a microarray EIS device (c) after injection of pre-treated beads in to the device thorough fluidic channel (d) After washing step, only a few beads were remained on SU-8 surface.

(counter and working electrodes), and it was presented in Fig. 2(b). After injection of pre-treated beads in to microfluidic channel, pre-treated beads were captured by magnetic bar as shown in Fig. 2(c). The distribution of beads was fairly uniform on a microarray. The number of beads in each hole will be increased with injection time and lower flowing speed. Then, washing buffer was introduced to remove nonspecific binding and beads on out of holes with slightly higher flowing speed than that of injection of beads. In this case, the flowing speeds of 2 μl/min and 10 μl/min were used for bead injection and washing steps, respectively. Fig. 2(d) show the image of a microarray EIS device after washing step. Still we can observe that some of the beads were remained on SU-8 passivated surface. But it does not affect impedance measurement. To verify the distribution of beads in each hole of a microarray, the number of beads was counted as shown in Fig. 3. Although the number of beads was slightly changed in different experiments (N=3), but its distribution was almost same shape and it was hard to find holes with more than 4 or 5 beads in several experiments. In most case, each hole was

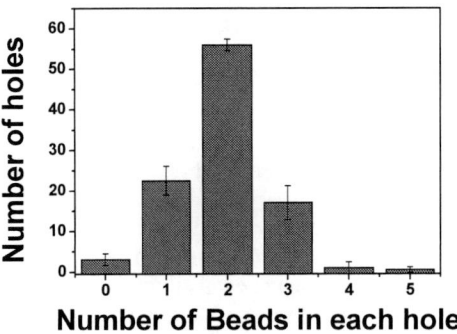

Figure 3 Number of beads in each hole of micro EIS array after washing procedure (N=3)

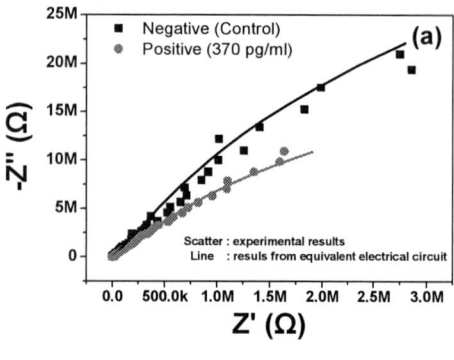

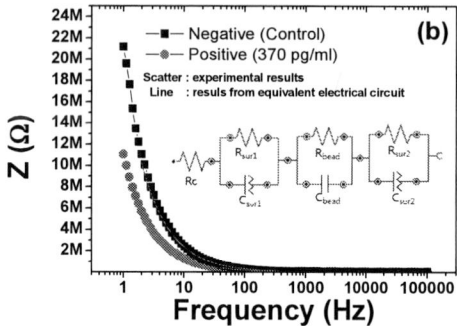

Figure 4. Characterizations of a new detection platform. (a) Nyquist plots for control sample and positive sample (370 pg/ml of Aβ oligomer) (b) impedance as a function of frequency with control sample and positive sample (inset : equivalent electrical circuit)

occupied by 1-3 beads, and the number of holes with two beads were dominant in all experiments.

Prior to detect varying concentration of Aβ oligomer, a microarray EIS device was tested with a control sample (with only capture-antibody on beads) and a positive sample (370 pg/ml of Aβ oligomer). The results were clearly distinguishable between two samples in both the Nyquist plot and the impedance plot as shown in Fig. 4. As expected, the imaginary impedance and resistance are significantly changed between two different samples. It could be understood by surface charges of Aβ. In general, Aβ has negative charges in PBS buffer (pH ~7.4). Therefore, it implied that the impedance will be modulated by the increment of surface charges caused by absorption of Aβ on beads. Therefore, the equivalent electrical circuit will be composed of capacitors and resistors as shown in Fig. 4(b). In an equivalent electrical circuit, R_c, R_{bead}, C_{bead}, R_{sur1}, R_{sur2}, C_{sur1}, C_{sur2} represent a contact resistance between electrodes and beads, a resistance of beads, a capacitance of beads, a varying resistance at surface of beads closed to counter/working electrode, a varying resistance at surface of beads closed to working/counter electrode, a varying capacitance at surface of beads closed to counter/working electrode and a varying capacitance at surface of beads closed to working/counter electrode, respectively. From this equivalent circuit, the impedance is mainly determined by changing of R_{sur1}, R_{sur2}, C_{sur1}, C_{sur2}. In Fig. 4, the scatters represent experimental results, and lines show the calculated results from an equivalent circuit.

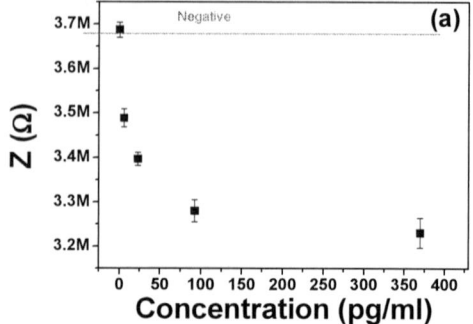

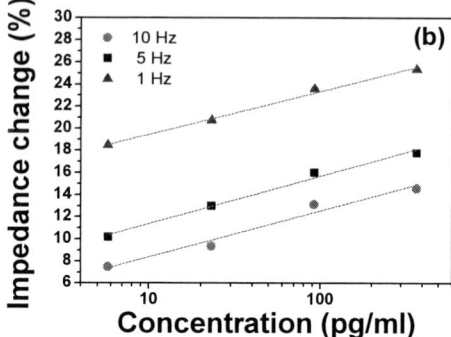

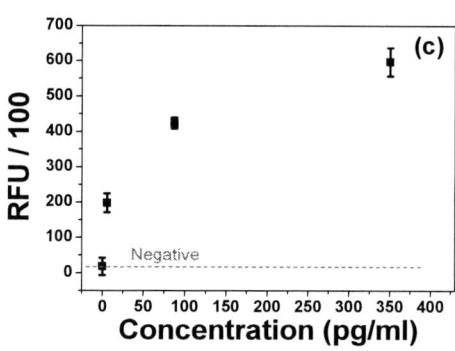

Figure 5. The measurement results from new detection platform and ELISA with varying concentrations (a) impedance as a function of concentration at 5 Hz (b) EIS impedance change as a function of concentration at different frequency (c) ELISA result with varying concentration

Those two results are well matched between two different samples. Next, we compared an ELISA and a novel detection platform with various concentrations of Aβ, and these results are shown in Fig. 5. The results from EIS measurement show that impedance change is more increased in lower frequency. It means that it is desirable for new detection platform to use lower frequency to detect Aβ with magnetic beads if background noise can be acceptable. When these EIS results are compared with those of ELISA, it can be confirmed that novel platform can detect Aβ without any immobilization on electrodes by coincidence between EIS and ELISA results. Also, it is remarkable that novel platform has still margin to push the LOD to a few hundred fg/ml range after optimization of devices such as increment of array for averaging effect, and minimization of hole size for improving uniformity of number of beads in holes. We show the possibility that novel detection platform can detect less than or comparable to pg/ml of Aβ oligomer after capturing pre-treated beads in the array of holes by magnetic bar, without any immobilization on electrodes

CONCLUSIONS

In this paper, we proposed a novel detection platform to detect Alzheimer's amyloid-beta (Aβ) using pre-treated magnetic beads by simple measurement of impedance change. Proposed detection platform consisted of multiple microarray EIS device, a microfluidic channel and a magnetic bar. Pre-treated magnetic beads were injected into fluidic channel, and those injected magnetic beads were positioned on microarray EIS device by magnetic bar. Then, impedance was measured to detect Aβ. Without any immobilization on the electrodes of the EIS device, it shows ability to detect a few pg/ml of Aβ and compared to the result of a conventional ELISA, which allows to simplify the measurement procedure, recycle EIS device by only changing magnet beads.

.

REFERENCES

CONTACT

*J. Y. Kang; jykang@kist.re.kr

[1]   S. Hideshima, M. Kobayashi, T. Wada, S. Kuroiwa, T. Nakanishi, N. Sawamura, et al., "A label-free electrical assay of fibrous amyloid [small beta] based on semiconductor biosensing," Chemical Communications, vol. 50, pp. 3476-3479, 2014.

[2]   S. Seshadri, A. Beiser, R. Au, P. A. Wolf, D. A. Evans, R. S. Wilson, et al., "Operationalizing diagnostic criteria for Alzheimer's disease and other age-related cognitive impairment—Part 2," Alzheimer's & Dementia, vol. 7, pp. 35-52, 1// 2011.

[3]   M. Ewers, R. A. Sperling, W. E. Klunk, M. W. Weiner, and H. Hampel, "Neuroimaging markers for the prediction and early diagnosis of Alzheimer's disease dementia," Trends in Neurosciences, vol. 34, pp. 430-442, 8// 2011.

[4]   H. Fukumoto, T. Tokuda, T. Kasai, N. Ishigami, H. Hidaka, M. Kondo, et al., "High-molecular-weight β-amyloid oligomers are elevated in cerebrospinal fluid of Alzheimer patients," The FASEB Journal, vol. 24, pp. 2716-2726, August 1, 2010 2010.

[5]   P. N. Lacor, M. C. Buniel, L. Chang, S. J. Fernandez, Y. Gong, K. L. Viola, et al., "Synaptic Targeting by Alzheimer's-Related Amyloid β Oligomers," The Journal of Neuroscience, vol. 24, pp. 10191-10200, November 10, 2004 2004.

[6]   J. P. Cleary, D. M. Walsh, J. J. Hofmeister, G. M. Shankar, M. A. Kuskowski, D. J. Selkoe, et al.,

"Natural oligomers of the amyloid-[beta] protein specifically disrupt cognitive function," Nat Neurosci, vol. 8, pp. 79-84, 01//print 2005.

[7]   G. M. Shankar, B. L. Bloodgood, M. Townsend, D. M. Walsh, D. J. Selkoe, and B. L. Sabatini, "Natural Oligomers of the Alzheimer Amyloid-β Protein Induce Reversible Synapse Loss by Modulating an NMDA-Type Glutamate Receptor-Dependent Signaling Pathway," The Journal of Neuroscience, vol. 27, pp. 2866-2875, March 14, 2007 2007.

[8]   A. M. Fagan, D. Head, A. R. Shah, D. Marcus, M. Mintun, J. C. Morris, et al., "Decreased cerebrospinal fluid Aβ42 correlates with brain atrophy in cognitively normal elderly," Annals of Neurology, vol. 65, pp. 176-183, 2009.

[9]   N. Schupf, M. X. Tang, H. Fukuyama, J. Manly, H. Andrews, P. Mehta, et al., "Peripheral Aβ subspecies as risk biomarkers of Alzheimer's disease," Proceedings of the National Academy of Sciences, vol. 105, pp. 14052-14057, September 16, 2008 2008.

[10]   D.-Y. Kang, J.-H. Lee, B.-K. Oh, and J.-W. Choi, "Ultra-sensitive immunosensor for β-amyloid (1–42) using scanning tunneling microscopy-based electrical detection," Biosensors and Bioelectronics, vol. 24, pp. 1431-1436, 1/1/ 2009.

[11]   T. Kurkina, S. Sundaram, R. S. Sundaram, F. Re, M. Masserini, K. Kern, et al., "Self-Assembled Electrical Biodetector Based on Reduced Graphene Oxide," ACS Nano, vol. 6, pp. 5514-5520, 2012/06/26 2012.

[12]   J. Oh, G. Yoo, Y. W. Chang, H. J. Kim, J. Jose, E. Kim, et al., "A carbon nanotube metal semiconductor field effect transistor-based biosensor for detection of amyloid-beta in human serum," Biosensors and Bioelectronics, vol. 50, pp. 345-350, 12/15/ 2013.

[13]   A. J. Veloso, A. M. Chow, H. V. S. Ganesh, N. Li, D. Dhar, D. C. H. Wu, et al., "Electrochemical Immunosensors for Effective Evaluation of Amyloid-Beta Modulators on Oligomeric and Fibrillar Aggregation Processes," Analytical Chemistry, vol. 86, pp. 4901-4909, 2014/05/20 2014.

[14]   J. V. Rushworth, A. Ahmed, H. H. Griffiths, N. M. Pollock, N. M. Hooper, and P. A. Millner, "A label-free electrical impedimetric biosensor for the specific detection of Alzheimer's amyloid-beta oligomers," Biosensors and Bioelectronics, vol. 56, pp. 83-90, 6/15/ 2014.

[15]   J. A. Kim, M. Kim, S. M. Kang, K. T. Lim, T. S. Kim, and J. Y. Kang, "Magnetic bead droplet immunoassay of oligomer amyloid β for the diagnosis of Alzheimer's disease using micro-pillars to enhance the stability of the oil–water interface," Biosensors and Bioelectronics, In Press.

978-1-4799-7956-1/15 $31.00 © 2015 IEEE

ULTRASENSITIVE SURFACE-ENHANCED RAMAN SPECTROSCOPY USING DIRECTIONALLY ARRAYED GOLD NANOPARTICLE DIMERS

Koji Sugano[1], Daimon Matsui[2], Toshiyuki Tsuchiya[2] and Osamu Tabata[2]
[1]Department of Mechanical Engineering, Kobe University, Kobe, JAPAN
[2] Department of Micro Engineering, Kyoto University, Kyoto, JAPAN

ABSTRACT

This paper reports an ultrasensitive nanostructure for surface-enhanced Raman spectroscopy (SERS). The gold nanoparticle dimer, which has been reported as the highest Raman enhancing structure, was directionally arrayed on a substrate for the first time. The highest enhancement can be achieved when a particle connection direction of a dimer is matched to polarization direction of incident light. Therefore the huge enhancement can be achieved at all dimers in total. Optimizing the dimer arrangement, 10 pM limit of detection and 0.5 s rapid detection were achieved.

INTRODUCTION

Molecular trace analysis has been widely used for various fields such as environmental measurements, security, biology, and medicine. In such applications, highly sensitive, rapid, and in-situ detection of molecules has been expected. In order to satisfy these demands, we focus on Surface-enhanced Raman spectroscopy (SERS) in this study. Raman spectroscopy is an extremely useful analytical tool since Raman spectra provide molecular structural information, enabling label-free identification of molecules. SERS has been expected for in-situ analysis since a commercialized handheld-size Raman spectrometer is available.

Although Raman scattering is significant weak, it can be strongly enhanced by bringing the molecules into contact with metal nanostructures [1]. The extraordinary enhancement can be obtained from nanogap between metal nanostructures, less than 1 nm, so called hot spot [2-4]. Electron-beam (EB) lithography-based process cannot fabricate such narrow nanogaps [5-7]. Therefore, various nanogaps using self-organization process have been reported so far, which are constructed with nanoparticles, nanoporous and so on. A nanoparticle dimer have been frequently used as the nanostructure which achieves the highest enhancement at particle–particle contact when the connection direction of the particles is matched to the polarization direction of the incident light [1,2]. Single-molecule-level analysis can be achieved using this geometry. However, the particles form random geometry in random directions on a substrate [8-10]. We therefore need to find an optimal geometry and then analyze target molecules by locating a laser spot on the structure and coupling with the polarization direction. This procedure is inefficient way. Furthermore sensitivity is limited because of only a few matched structure in the laser spot.

In this study, we developed and evaluated an advanced SERS substrate on which particle dimers are regularly and directionally arrayed in order to match all dimers axes to a polarization direction as shown in Fig. 1. It is expected that the strong enhancement can be achieved at all dimers. In this manuscript, we report on fabrication and evaluation of the proposed SERS substrate. We evaluated the fabricated

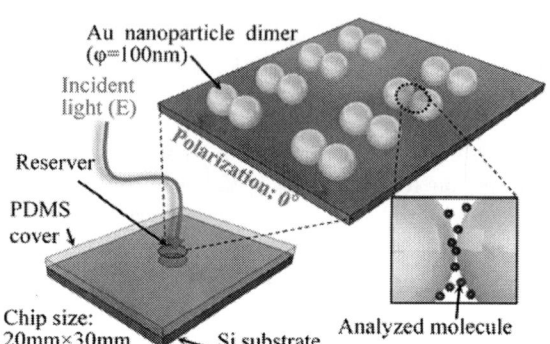

Figure 1: Schematic image of an array of directionally arranged nanoparticle dimer.

SERS substrate on the dependency of Raman intensity on the polarization angle, and optimized the dimer arrangement. Finally limits of molecular concentration and measurement time of SERS detection were evaluated.

EXPERIMENTAL

Fabrication of Nanoparticle Dimer Array

The proposed structure was fabricated according to the process shown in Fig. 2. Gold nanoparticles were arranged onto nanotrenches, using nanotrench-guided self-assembly [11, 12]. Colloidal gold nanoparticles of mean diameter 100 nm dispersed in water were purchased from BBI Solutions, UK. A gold solution of concentration 0.0004 wt% was prepared and introduced into two substrates, the Si substrate with the fabricated nanotrenches and a glass substrate. Drying the aqueous dispersion between the substrates causes the water surface line to move backward, and the particles are concentrated near the meniscus edge. The interfacial forces drag and press the particles onto the Si substrate. When the meniscus passes over the nanotrenches, the particles are trapped on them, and then water bridges form between the trapped particles. During removal of the remaining water between the particles, the particles attract each other and form particle–particle

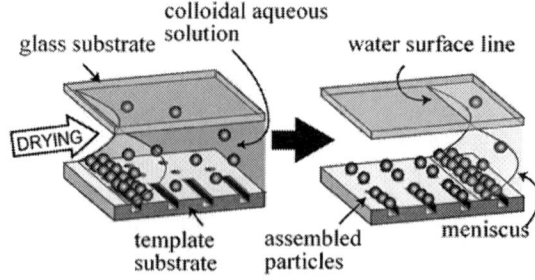

Figure 2: Experimental method for nanoparticle dimer arrangement using nanotrench-guided self-assembly.

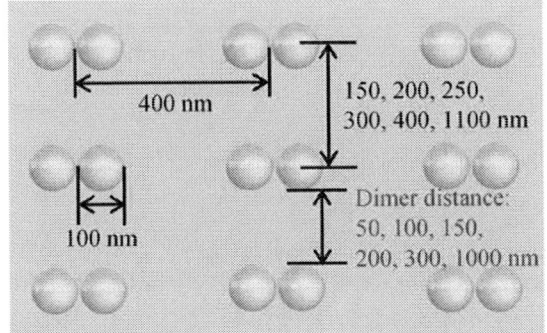

Figure 3: Schematic image of experimental and analytical dimer arrangement of 100 nm diameter nanoparticles.

contacts, which act as hot spots.

The nanotrenches are fabricated on a Si substrate by EB lithography and subsequent Si dry etching. The nanotrench array was designed in order to fabricate the particle dimer array as shown in Fig. 3. A pitch of dimers is set to 400 nm in longitudinal direction of dimers. A pitch of dimers in transverse direction is set to 150, 200, 250, 300, 400, and 1100 nm. It indicates dimer surface distances of 50, 100, 150, 200, 300, and 1000 nm as a parameter. The measured length, width, and depth of nanotrenches are 265, 89, and 38 nm in average, respectively. The nanotrenches are arranged in 5 μm × 5 μm region.

Analytical Method

Raman enhancement factor was calculated based on FDTD (Finite Differential Time Domain) simulation. The analytical structure with the particle arrangement as shown in Fig. 3 was used. The nanogap between particles in a dimer was set to 1 nm.

In this analysis, the polarization angle of the incident light and the dimer distance were used as a parameter. The polarization angle was ranged from 0° to 90° with the dimer distance of 200 nm. Then the dimer distance was varied from 0 to 1000 nm with the polarization angle of 0°.

As an analytical result, the maximum electromagnetic enhancement $|E|^2$ at the hotspot was obtained. Then Raman enhancement was calculated as $|E|^4$ [13]. In the case of the dimer distance evaluation, the normalized Raman enhancement factor was calculated multiplying a particle dimer density by Raman enhancement $|E|^4$.

Raman Spectroscopy Experiments

Raman spectra were obtained using a micro-Raman spectrometer equipped with a 632.8 nm wavelength laser with a 2 μm beam spot. Dicarboxyacetone molecule attached uniformly to the chemically synthesized colloidal particles [14] was used for the nanostructure evaluation and optimization taking the polarization angle (Fig. 1) and the dimer distance as a parameter (Fig. 3).

4,4'-Bipyridine molecule was used for evaluating limits of molecular concentration and measurement time of detection. The laser spot was located at the center of the array region 5 μm × 5 μm. Before the SERS experiments for 4,4'-bipyridine molecule, ultraviolet (UV)/O₃ treatment was performed for 90 min at 80 °C to remove the dicarboxyacetone molecules attached to the particles. We

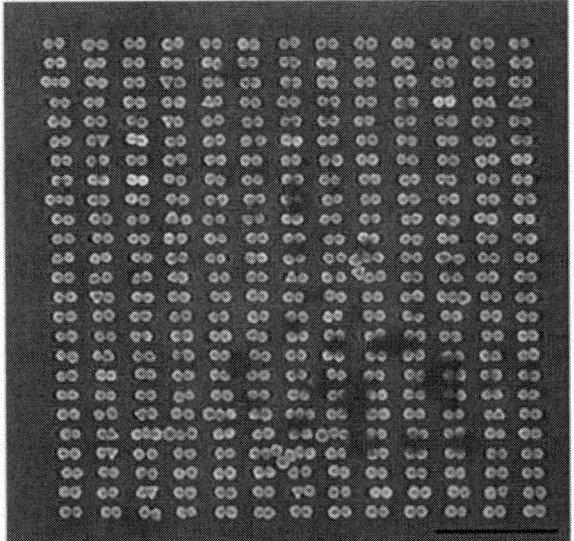

Figure 4: SEM image of dimer array of 100 nm particles with dimer distance of 100 nm. Arrangement area is 5 μm×5 μm. Scale bar indicates 1 μm.

confirmed that this treatment was able to remove the dicarboxyacetone molecules.

RESULTS AND DISCUSSION

Fabrication of Nanoparticle Dimer Array

Figure 4 shows the SEM image of the arrayed dimers with the dimer distance of 100 nm. We confirmed that dimers were arrayed with high yield in one direction and two particles in a dimer connected each other acting as a hotspot. The particle–particle contacts was formed by the water bridge.

Experimental and Analytical Results of Raman Spectroscopy

Figure 5 shows Raman spectra of the dicarboxyacetone molecule as a function of the polarization angle. Some peaks derived from the molecule were clearly observed in the spectra. Figure 6 shows Raman intensities of peaks at around 1580 cm⁻¹ depending

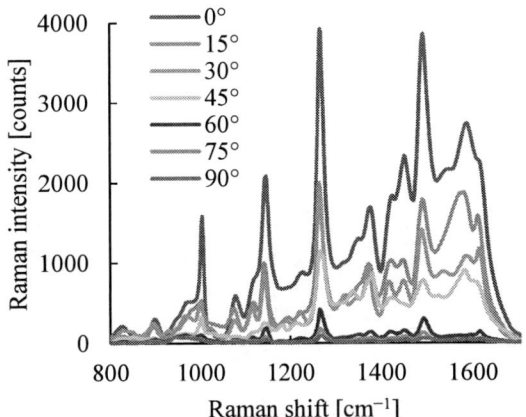

Figure 5: Raman spectra of dicarboxyacetone molecule attached on gold nanoparticles depending on the polarization angle.

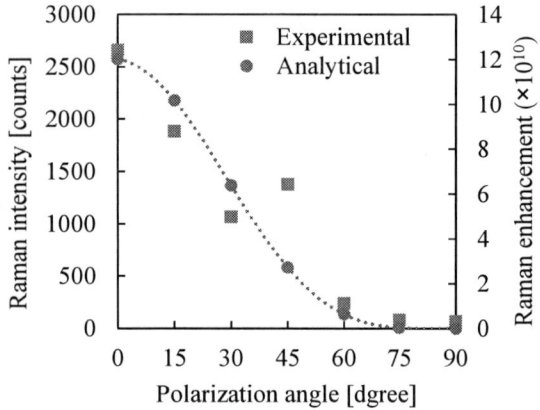

Figure 6: Experimental Raman intensity and simulated Raman enhancement factor $|E|^4$ as a function of the polarization angle.

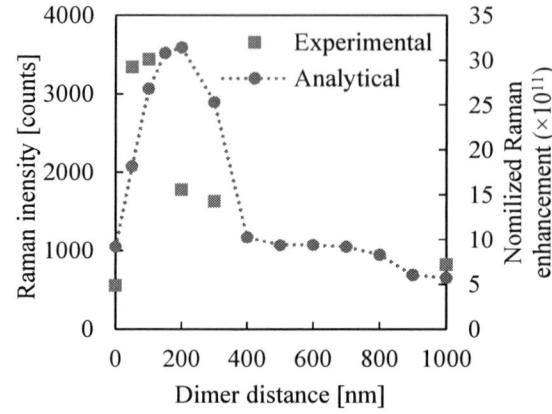

Figure 7: Experimental Raman intensity and simulated normalized Raman enhancement factor as a function of the dimer distance.

on the polarization angle with the integration time of 2 s. The analytical results of Raman enhancement factor were also shown. In both experimental and analytical results, the Raman intensities decreased with increasing the polarization angle. The experimental and analytical results show same tendencies for the polarization angle. An error at 45° is thought to be due to molecular resonance Raman scattering depending on a molecular orientation, which was not considered in the analysis. Also a chemical enhancement by charge transfer was not considered [15]. However, the results indicate that the most of Raman enhancement is determined by the electromagnetic enhancement. In a case of random geometry of particles, the angle dependency should not be shown. Therefore it was confirmed that the directionally arrayed dimers enable us to use huge electromagnetic enhancement effectively.

Figure 7 shows peak Raman intensities at around 1580 cm^{-1} and analytical results of the normalized Raman enhancement factor depending on the dimer distance. The experimental result of fully-packed particle aggregation structure was used as the dimer distance of 0 nm. The

experimental intensity increased from 0 nm and decreased from 100 nm with the dimer distance. The dimer distance of 100 nm showed the maximum intensity. In the analytical result, the distance of 200 nm showed the maximum enhancement. However, the experimental and analytical results showed the same tendency.

This tendency is due to an electromagnetic interaction between dimers. A strong electromagnetic enhancement occurs at the particle–particle contacts when a polarization angle is parallel to the particle connection direction. This is because of electromagnetic resonance between two particles. In the case of the perpendicular direction, enhancement is small since this geometry diminishes the enhancement each other. Considering the particle dimer, the interaction between adjacent dimers depends on the distance. With increasing the dimer distance, the diminishing effect decreased. As the density of the particle dimers decreased, the total enhancement decreases. Therefore the optimal dimer distance appears. Controlling the dimer distance, we obtained high Raman enhancement.

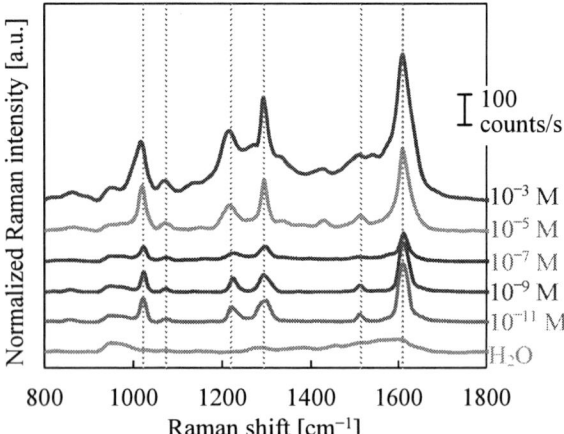

Figure 8: Raman spectra of 4,4'-bipyridine depending on molecule concentration. Red dotted lines indicate 4,4'-bipyridine derived Raman peaks. Vertical axis indicates Raman intensity per measurement time.

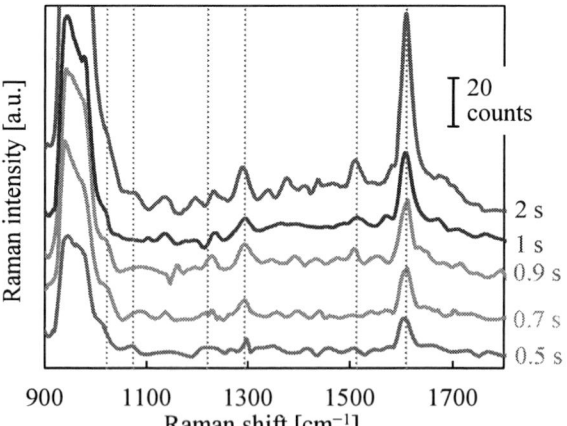

Figure 9: Raman spectra of 10^{-11} M 4,4'-bipyridine depending on measurement time. Red dotted lines indicate 4,4'-bipyridine derived Raman peaks.

Experimental Results of Limit of Detection

The dimer array with the distance of 100 nm was used for evaluating concentration and measurement time limits of detection.

Figure 8 shows the Raman spectra depending on the concentration from 10^{-3} to 10^{-11} M. The y-axis indicates the Raman intensity per measurement time. The red dotted lines indicate 4,4'-bipyridine derived peaks. The Raman spectrum from water without the analyte shows broad peaks at around 950 cm^{-1} and 1600 cm^{-1} derived from Si and water, respectively. The dicarboxyacetone molecule derived peaks were not observed. This indicates that the dicarboxyacetone molecules were adequately removed by UV/O$_3$ treatment. According to the figure 8, the peaks derived from 4,4'-bipyridine molecule were clearly observed from all cases of concentrations including 10^{-11} M solution using the optimized structure. This concentration corresponds to one molecule per volume of a cube with a side 5 μm.

Figure 9 shows the Raman spectra of 10^{-11} M solution as a function of the measurement time. The detection for the 10^{-11} M solution was possible at the measurement time of 0.5 s. We confirmed that the developed SERS substrate enables us ultra-sensitive and ultra-rapid SERS analysis.

CONCLUSIONS

In this study, we fabricated the directionally arrayed gold nanoparticle dimers in order to achieve higher Raman enhancement, and evaluated the structure experimentally and analytically.

The dimer array was fabricated on the Si substrate by the nanotrench-guided self-assembly of 100 nm diameter gold nanoparticles. We confirmed the high yield arrangement of the particle dimers with hot spots in one direction.

The fabricated structure showed the polarization angle dependency of Raman intensity in Raman spectroscopy experiments. We confirmed that the directionally arrayed dimers enable us to use huge electromagnetic enhancement effectively. Then the dimer distance was optimized. We found the optimized dimer distance of 100 nm. This can be explained by electromagnetic interaction between nearby dimers. Finally, ultra-sensitive and ultra-rapid SERS detection with 10^{-11} M and 0.5 s limit of detection were achieved using the optimized SERS structure.

REFERENCES

[1] E. J. Blackie, E. C. Le Ru and P. G. Etchegoin, "Single-Molecule Surface-Enhanced Raman Spectroscopy of Nonresonant Molecules", *J. Am. Chem. Soc*, vol.131, pp. 14466–14472, 2009.

[2] K. Yoshida, T. Itoh, H. Tamaru, V. Biju, M. Ishikawa and Y. Ozaki, "Quantitative evaluation of electromagnetic enhancement in surface-enhanced resonance Raman scattering from plasmonic properties and morphologies of individual Ag nanostructures", *Phys. Rev. B*, vol.81, pp.115406.1–9, 2010.

[3] A. Dhawan, S. J. Norton, M. D. Gerhold and T. Vo-Dinh: "Comparison of FDTD numerical computations and analytical multipole expansion method for

plasmonics-active nanosphere dimers", *Opt. Express*, vol.17, pp.9688–9703, 2009.

[4] D. P. Fromm, A. Sundaramurthy, P. J. Schuck, G. Kino and W. E. Moerner: "Gap-Dependent Optical Coupling of Single "Bowtie" Nanoantennas Resonant in the Visible", *Nano Lett.*, vol.4, pp.957–961, 2004.

[5] E. C. Le Ru and P. G. Etchegoin: "Quantifying SERS Enhancements", *MRS Bull.*, vol.38, pp.631–640, 2013.

[6] P. J. Schuck1, D. P. Fromm, A. Sundaramurthy, G. S. Kino and W. E. Moerner: "Improving the Mismatch between Light and Nanoscale Objects with Gold Bowtie Nanoantennas", *Phys. Rev. Lett.*, vol.94, pp.017402.1–4, 2005.

[7] Q. Sun, K. Ueno, H. Yu, A. Kubo, Y. Matsuo and H. Misawa, "Direct imaging of the near field and dynamics of surface plasmon resonance on gold nanostructures using photoemission electron microscopy", *Light Sci. Appl.*, vol.2, e118, 2013.

[8] L. Lu, G. Sun, H. Zhang, H. Wang, S. Xi, J. Hu, Z. Tian and R. Chen: "Fabrication of core-shell Au-Pt nanoparticle film and its potential application as catalysis and SERS substrate", *J. Mater. Chem*, vol.14, pp.1005–1009, 2004.

[9] R. G. Freeman, K. C. Grabar, K. J. Allison, R. M. Bright, J. A. Davis, A. P. Guthrie, M. B. Hommer, M. A. Jackson, P. C. Smith, D. G. Walter and M. J. Natan, "Self-Assembled Metal Colloid Monolayers: An Approach to SERS Substrates", *Science*, vol.267, pp.1629–1632, 1995.

[10] K. C. Grabar, R. G. Freeman, M. B. Hommer and M. J. Natan: "Preparation and Characterization Monolayers", Anal. Chem, vol.67, pp.735–743, 1995.

[11] T. Ozaki, K. Sugano, T. Tsuchiya and O. Tabata: "Versatile Method of Submicroparticle Pattern Formation Using Self-Assembly and Two-Step Transfer", *J. Microelectromechanical Syst.*, vol.16, pp.746–752, 2007.

[12] K. Sugano, T. Ozaki, T. Tsuchiya and O. Tabata: "Fabrication of gold nanoparticle pattern using combination of self-assembly and 2-step transfer", *Sensor. Mater.*, vol.23, pp.263–275, 2011.

[13] J. M. McMahon, A.-I. Henry, K. L. Wustholz, M. J. Natan, R. G. Freeman, R. P. van Duyne and G. C. Schatz: "Gold nanoparticle dimer plasmonics: finite element method calculations of the electromagnetic enhancement to surface-enhanced Raman spectroscopy", *Anal. Bioanal. Chem.*, vol.394, pp.1819–1825, 2009.

[14] S. Kumar, K. S. Gandhi and R. Kumar: "Modeling of Formation of Gold Nanoparticles by Citrate Method", *Ind. Eng. Chem. Res*, vol.46, pp.3128–3136, 2007.

[15] A. D. McFarland, M. A. Young, J. A. Dieringer and R. P. van Duyne, "Wavelength-Scanned Surface-Enhanced Raman Excitation Spectroscopy", J. Phys. Chem. B, vol.109, pp.11279–11285, 2005.

CONTACT

*K. Sugano, tel: +81-78-803-7214;
sugano@mech.kobe-u.ac.jp

WATER-PROOF 'μ-DIVING SUIT' DRESSED ON RESONANT BIOCHEMICAL SENSOR FOR ONLINE DETECTION IN SOLUTION

Haitao Yu, Ying Chen, Pengcheng Xu, Feng Yu, and Xinxin Li

State Key Lab of Transducer Technology, Shanghai Institute of Microsystem and Information Technology, Chinese Academy of Sciences, Shanghai 200050, CHINA

ABSTRACT

This paper reports a new method to ensure resonant micro-sensor long-time resonating in solution for real-time biochemical sensing/analysis. By designing a water-proof 'diving-suit' for the cantilever resonator and an anti-leakage narrow 'slit' to free the cantilever vibration, only the sensing-region of the cantilever contacts to analyte solution, while the other parts are remained in air to keep free-resonance. With this 'μ-diving suit' design, the resonant cantilever with electrothermal driven and piezoresistive frequency readout integrated can achieve a Q factor of 17 in water. With specific sensing-material loaded at the sensing-region, the cantilever sensor experimentally realizes liquid-phase real-time detection to ppb-level organophosphorous pesticide of acephate.

INTRODUCTION

Resonant micro-cantilever bio/chemical sensor, which is emerged in the mid-1990s [1], is today a well-known technique for sensitive, cheap and portable analysis systems [2-4]. The working principle lies in that the resonant frequency of the cantilever depends on the mass of the cantilever and the resonant frequency drops as the mass increases due to analyte absorption. Thus, it is possible to make indirect mass change estimations by following the resonant frequency change of the cantilever. Hence, it is not difficult to deduce that resonant microcantilevers can be easily used in air for gas detection but are hardly operated in solution for real-time biochemical detection, due to the significantly decreased Q-factor caused by severe viscous drag of liquid [5, 6]. To fulfill the demand of bio/chemical detection in liquid, many kinds of efforts were made to reduce the liquid effect. Some examples are shown below: (1) using higher vibrating modes [7]; (2) vibrating at air-liquid interface [8]; (3) integrating micro-channel in cantilever [9]. Unfortunately, each of the three strategies has own drawbacks: (1) higher-mode is hard to excite; (2) with one cantilever-surface entirely contacting the liquid, resonance is still difficult; (3) big-size bio-substance like cell cannot access into the micro-channel. Besides, (1) and (2) are both with the whole senor surface suffering from nonspecific adsorption.

Inspired by Ref.[10] that reported an encased AFM probe for tapping-mode scanning in liquid, herein we propose a new solution that is able to achieve resonant micro-sensors for liquid-phase real-time detection. By designing a water-proof 'diving-suit' for the cantilever resonator and an anti-leakage narrow 'slit' to free the cantilever vibration, only the sensing-region of the cantilever contacts to analyte solution, while the other parts are remained in air to keep free-resonance. The design can greatly decrease the influence of the liquid media damping, thereby increase the Q factor of the resonance. More importantly, this technique features unique advantages of: (1) protecting non-sensing region of sensor from nonspecific adsorption for reliable detection; (2) capability of being spread to various liquid-sensing devices and suitable for various sensor-geometries with the solution-contact sensing-region independently designed; (3) the wafer-level batch fabrication of the sensor is low-cost and the sensor can be embedded in various micro-fluidic lab-chip system for real-time biochemical analysis.

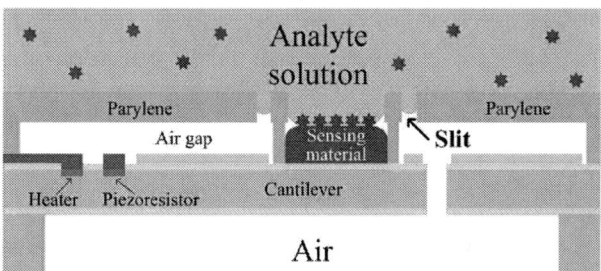

Figure 1: Cross-section schematic of the resonant sensor structure for biochemical analyte detection in liquid.

SENSOR DESIGN AND FABRICATION

The design of the sensor is schematically shown in Figure 1. After electro-thermal resonance-exciting and piezoresistive self-sensing elements integrated on a micro-cantilever, a hydrophobic parylene film is built above the resonator. Gapped with a released distance for free resonance in air, the front-side of the resonator is protected by the water-proof film from liquid, with an exception that the small area above a parylene-made 'pool' built at the cantilever-end, which is opened to liquid for specific biochemical binding. The narrow 'slit' constructed between the walls of the 'pool' and the parylene opening edge can prevent liquid leakage. This design features following advantages: (1) the geometries of the 'pool' for accommodating sensing material can be flexibly designed and the cover-cantilever gap-distance can be independently constructed for high Q-factor; (2) the integrated resonance-excitation micro-heater keeps no-touch with solution, thereby avoiding exciting-force attenuation cause by high heat-conduction of liquid; (3) experiment has proved more than 72hrs water-proof function of the sensor, and absorbent material can be put in the cantilever backside to further absorb water vapor, thereby prolonging this period.

The sensor fabrication steps are shown in Figure 2 and described as follows. Firstly, a micro-cantilever with electro-thermal resonance-exciting and piezo-resistive self-sensing elements integrated is fabricated with SOI water (detailed in our previous literature [11], size=200μm×100μm×4μm). Then thick photoresist is sprayed and patterned as sacrificial-layer. It should be

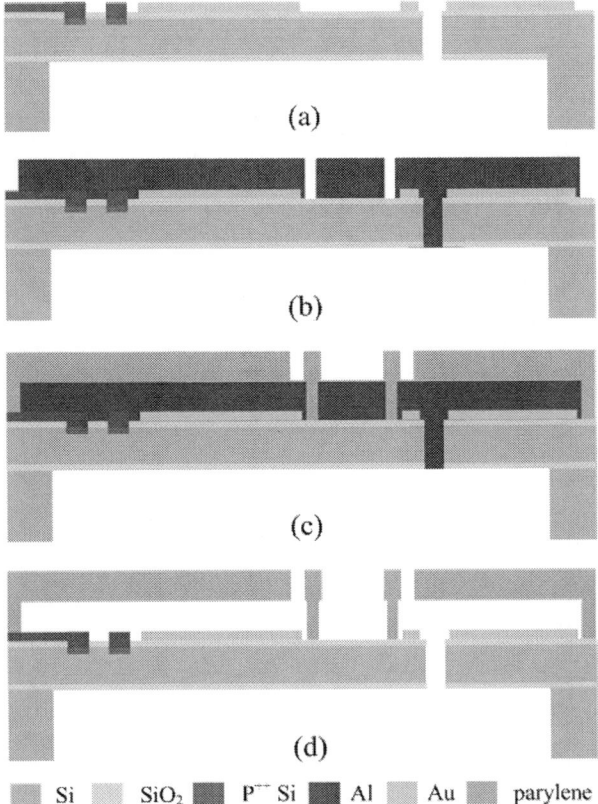

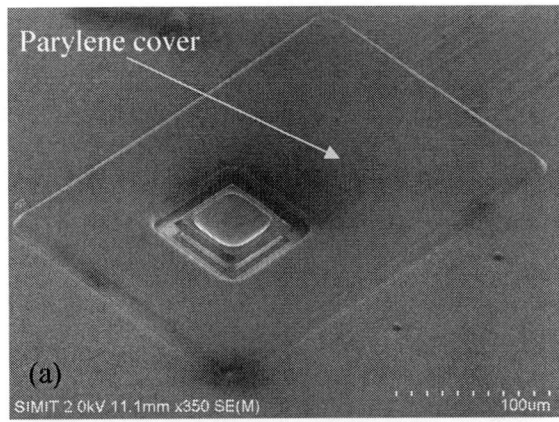

| Si | SiO$_2$ | P^{--} Si | Al | Au | parylene |

Figure 2: Illustration of the processes to fabricate the water-proof resonant biochemical sensor. (a) Formation of the integrated resonant cantilever. (b) Thick photoresist sprayed and patterned as sacrificial layer. (c) Parylene thin-film deposition and patterned with oxygen plasma RIE. (d) Structure release by removal of the photoresist.

pointed out that the thickness of the photoresist determines the size of the gap between the parylene layer and the cantilever, which will further limit the Q factor of the resonance. Thirdly, a parylene water-proof cover-layer is formed and patterned. Then, the resonator is freed by removing the sacrificial-layer. Finally, acephate sensing material of fluorinated-phenol modified hyper-branched polymer is prepared and loaded into the 'pool' at the cantilever using a commercial micro-manipulator (Eppendorf, model PatchMan NP2). The SEM images of the fabricated cantilever sensor are shown in Figure 3(a) and (b), and the acephate pesticide specific sensing material of fluorinated phenol modified hyper-branch polymer is schematically shown in Figure 3(c). To give a complete view of the cantilever beneath the 'μ-diving suit', the parylene film of a cantilever is striped off by oxygen plasma etching, and the SEM image of the buried cantilever is shown in Figure 4(a). With the 'μ-diving suit' of the cantilever partly cut by FIB, the parylene cover and the beneath cantilever are shown in Figure 4(b), and the close-up view in Figure 4(c) shows that the gap distance between parylene and cantilever is about 7.32μm.

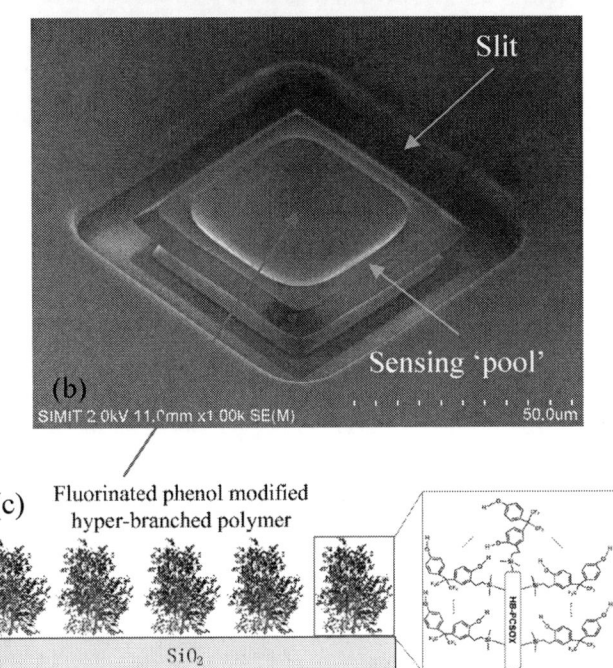

Figure 3: (a and b) SEM images of the fabricated water-proof sensor with the sensing 'pool' area loaded with acephate pesticide specific sensing material of fluorinated phenol modified hyper-branch polymer, whose schematic is in (c).

EXPERIMENTS AND RESULTS

The open-loop amplitude-frequency characteristics of the sensor is tested and recorded. Figure 5 shows the

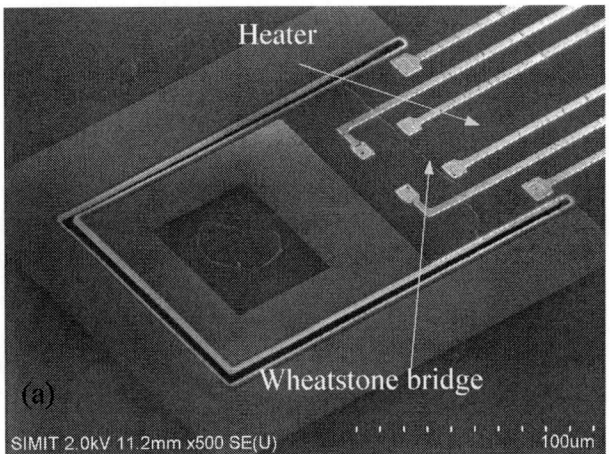

978-1-4799-7956-1/15 $31.00 © 2015 IEEE

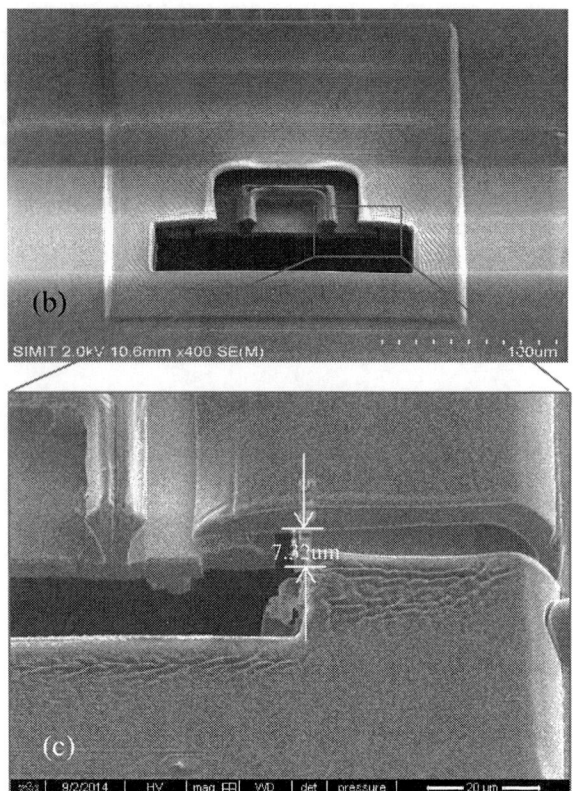

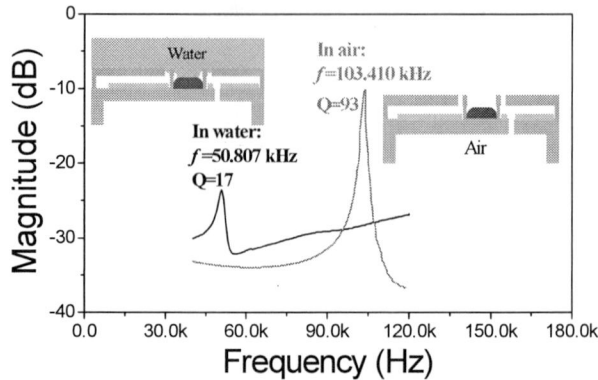

Figure 5: Tested amplitude-frequency characteristics of the sensor in air (red curve) and water (black curve).

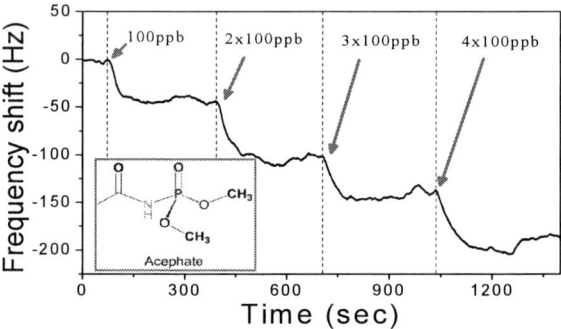

Figure 6: In aqueous solution, sensor response to trace-level pesticide of acephate, whose concentration is stepwise increased by the increment of 100ppb.

Figure 4: SEM images of the fabricated sensor structure. (a) The buried cantilever is exposed after the parylene cover film is peeled off. (b) With the 'pool' cut by FIB, the parylene cover and the beneath cantilever are shown. (c) Close-up view of the parylene-cantilever gap distance.

tested results in air and water, respectively. The resonant frequency and Q factor of the cantilever resonating in air is 103.410 kHz and 93, respectively. When the same cantilever is immersed into water, the resonant frequency and Q factor decreased to 50.807 kHz and 17. Considering that when the bare cantilever was ever directly immersed in water (i.e. without the parylene 'diving-suit'), resonance could not be excited, so herein achieved Q=17 in water makes significant breakthrough for such big-size cantilevers.

Equipped with a close-loop interface circuit, the micro-cantilever sensor is used to detect acephate solution in real-time. Firstly, the sensor is immerged in 1mL of water, and then 1μL of acephate solution is injected into the water and repeated for 3 times. After each time of injection, the concentration of the obtain acephate solution is stepwise increased by the increment of 100 ppb. The experimental results are shown in Figure 6. With the concentration increasing stepwise in 100 ppb increments, the sensor outputs serial frequency-shift signals. The sensitivity of the sensor is obtained as about 50 Hz per 100 ppb, and there is no obvious attenuation in sensitivity during the multi-concentration continuing detection. The response time of the sensor is less than 3 minutes, and the limit of detection is estimated as about dozens of ppb.

CONCLUTION

We have developed a new method to ensure resonant micro-sensor long-time resonating in solution for real-time biochemical sensing/analysis. With a 'μ-diving suit' design, a hundreds-micron-sized resonant cantilever with electrothermal driven and piezoresistive frequency readout integrated can vibrate in water and the Q factor can achieve as high as 17. After acephate pesticide specific sensing material of fluorinated phenol modified hyper-branch polymer loaded onto the sensing region of the cantilever, sensing experiments are carried out, with the results verifying that the sensor exhibits real-time and rapid detection to 100 ppb acephate solution. Such a novel sensing technique is promising in various bio/chemical liquid detection applications.

ACKNOWLEDGEMENTS

This research is supported by NSF of China (91323304, 91023046, 61161120322, 61401446) and the Chinese 973 Project (2011CB309503).

REFERENCES

[1] T. Thundat, E. A. Wachter, S. L. Sharp, R. J. Warmack, "Detection of Mercury Vapor Using Resonating Microcantilevers", *Appl. Phys. Lett.*, vol. 66, pp. 1695-1697, 1995.

[2] H. T. Yu, P. C. Xu, X. Y. Xia, D.-W. Lee, X. X. Li, "Micro-/nanocombined Gas Sensors with

Functionalized Mesoporous Thin Film Self-assembled in Batches onto Resonant Cantilevers", *IEEE Trans. Ind. Electron.*, vol. 59, pp. 4881-4887, 2012.

[3] H. T. Yu, P. C. Xu, D.-W. Lee, X. X. Li, "Porous-layered Stack of Functionalized AuNP-rGO (Gold Nanoparticles–Reduced Graphene Oxide) Nanosheets as a Sensing Material for the Micro-gravimetric Detection of Chemical Vapor", *J. Mater. Chem. A*, vol. 1, pp. 4444-4450, 2013.

[4] P. C. Xu, H. T. Yu, X. X. Li, "Functionalized Mesoporous Silica for Microgravimetric Sensing of Trace Chemical Vapors", *Anal. Chem.*, vol. 83, pp. 3448–3454, 2011,

[5] C. Vancura, I. Dufour, S. M. Heinrich, F. Josse, A. Hierlemann, "Analysis of Resonating Microcantilevers Operating in a Viscous Liquid Environment", *Sensor. Actuat. A-Phys.*, vol. 141 pp. 43–51, 2008.

[6] Y. H. Tao, X. X. Li, T. G. Xu, H. T. Yu, P. C. Xu, B. Xiong, C. Z. Wei, "Resonant Cantilever Sensors Operated in a High-Q In-plane Mode for Real-time Bio/chemical Detection in Liquids", *Sensor. Actuat. B-Chem.*, vol. 157, pp. 606-614, 2011.

[7] S. Truax, K. Demirci, L. Beardslee, Y. Luzinova, A. Hierlemann, B. Mizaikoff O. Brand, "Mass-sensitive Detection of Gas-phase Volatile Organics Using Disk Microresonators", *Anal. Chem.*, vol. 83, pp. 3305-3311, 2011.

[8] J. Park, S. Nishida, P. Lambert, H. Kawakatsu, H. Fujita, "High-resolution Cantilever Biosensor Resonating at Air–liquid Iin a Microchannel", *Lab Chip*, vol. 11, pp. 4187-4193, 2011.

[9] T. Burg, M. Godin1, S. Knudsen, W. Shen, G. Carlson, J. Foster, K. Babcock, S. Manalis, "Weighing of Biomolecules, Single Cells and Single Nanoparticles in Fluid", *Nature*, vol. 446, pp. 1066-1069, 2007.

[10] D. Ziegler, A. Klaassen, D. Bahri, D. Chmielewski, A. Nievergelt, F. Mugele, J. Sader, P. Ashby, "Encased Cantilevers for Low-noise Force and Mass Sensing in Liquids", in *Digest Tech. Papers IEEE MEMS 2014 Conference*, San Francisco, January 26-30, 2014, pp. 128-131.

[11] H. T. Yu, X. X. Li, X. H. Gan, Y. J. Liu, X. Liu, P. C. Xu, J. G. Li, M. Liu, "Resonant-cantilever Bio/chemical Sensors with an Integrated Heater for Both Resonance Exciting Optimization and Sensing Repeatability Enhancement", *J. Micromech. Microeng.*, vol. 19, 2009, p. 045023.

CONTACT

*X.X. Li, tel: +86-21-62131794; xxli@mail.sim.ac.cn

A POLYCRYSTALLINE DIAMOND-BASED, HYBRID NEURAL INTERFACING PROBE FOR OPTOGENETICS

Bin Fan[1], Ki-Yong Kwon[1], Robert Rechenberg[2], Anton Khomenko[1], Mahmoodul Haq[1],
Michael F. Becker[2], Arthur, J. Weber[1] and Wen Li[1]
[1]Michigan State University, MI, USA
[2] Fraunhofer USA-CCL, MI, USA

ABSTRACT

This paper reports a hybrid optoelectronic neural interfacing probe, combining micro-scale light emitting diode (μLED) and microelectrodes on a polycrystalline diamond (PCD) substrate for optogenetic stimulation and electrical recording of neural activity. PCD has superior thermal conductivity (up to 1800 $Wm^{-1}K^{-1}$) [1], which allows rapid dissipation of localized LED heat to a larger area to improve heat exchange with surrounding perfused tissues, and thus significantly reduce the risk of thermal damage to nerve tissue. During repetitive stimulation with 100ms and 1Hz pulses, the maximum rise in surface temperature of the PCD probe is less than 1 °C, which is ~90% lower than that of a polymer-based probe. A PCD based probe with two stimulating sites and four recording sites was fabricated. The capacity of the probe for neural stimulation and recording has also been demonstrated *in vivo* by successfully observing light evoked action potentials.

INTRODUCTION

Optogenetics has become a hotspot in the field of neuroscience because it provides the ability to express specific opsins in select neurons, and then use light to modulate their electrophysiological responses [2]. There are several types of light sources for optogenetics, such as incandescent sources [3], laser [4] and micro-LEDs (μLEDs) [5]. Micro-LEDs, in particular, show promise with respect to device miniaturization, simplicity, and low cost of system implementation. Low power μLEDs also have the potential to be integrated with wireless telemetries in order to achieve truly un-tethered systems for studies involving freely behaving animals. However, tissue heating during the operation of μLEDs remains a major challenge. In addition, thermal effects may bias the outcomes of optogenetic experiments, especially when μLEDs are used near tissues receiving continuous light as opsin-negative controls [3].

To address these challenges, we propose a hybrid optoelectronic probe, which utilizes a PCD heat spreader to minimize focal temperature increases during optical stimulation, as shown in Figure 1. Compared to SU-8 that has a thermal conductivity of 0.3 $Wm^{-1}K^{-1}$ [6], PCD has a thermal conductivity up to 1800 $Wm^{-1}K^{-1}$. Therefore, the PCD heat spreader of the neural probe can dissipate electrically-induced heat rapidly and uniformly to reduce localized hot spots, thereby minimizing tissue damage during optical stimulation.

To validate the efficacy of the PCD heat spreader, the heat distribution and temperature variance of single-shank probes made of SU-8 and PCD were investigated in air using a high-resolution infrared camera. Then, a PCD based probe with two shanks was fabricated, with each shank containing one stimulating site and two recording sites. Finally, the functionality of the PCD probe was demonstrated by recording light-induced action potential from the primary visual cortex (V1) of a channelrhodopsin-2 (ChR2) transfected rat.

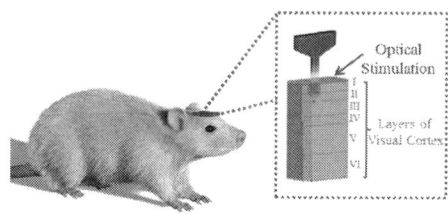

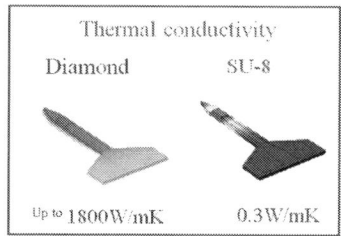

Figure 1: Concept diagram of the proposed neural probe with a PCD heat spreader.

THERMAL PROPERTY

To demonstrate the high thermal conductivity of PCD, a SU-8 probe without a top coating was first fabricated using the method reported in [7], and compared with a PCD based probe with the same dimensions fabricated according to the protocol described later in this paper. A μLED (Samsung, Inc) was mounted onto the probe tip and driven by 1 Hz, 100 ms pulses using different input voltages. Thermal images were taken using a high-resolution infrared camera (Delta Therm HS1570 and DT v2.19 software) and processed using MATLAB® (R2011a, The MathWorks). As shown in Figure 2 (a)-(b), with a 3.4 V input voltage, the SU-8 probe accumulated heat at the tip due to the poor thermal conductivity of SU-8, while the PCD probe dissipated heat throughout the entire shank in less than 0.5 sec without creating a localized hot spot on the probe. Figure 3 (c) shows the cooling curves of the probes after activating the μLED for 60 sec with different input voltages. Figure 3 (d) and (e) show the instantaneous changes in the tip temperatures and steady state temperature variations (after 1 min activation) of the probes for six On-Off cycles, respectively. The maximum temperatures of the SU-8 probe with input voltages of 3.0 V, 3.2 V and 3.4 V increased from 22 °C to 24.5 °C, 26.5 °C and 27 °C during the first duty cycle (100 mS), continued to increase to 25.5 °C, 29 °C and 31 °C within the first 7 sec, and then stabilized at 26 °C, 30 °C and 34 °C, respectively. On the contrary, the temperature rises of the PCD probe with input voltages of 3.0V, 3.2V and 3.4V were within 1 °C of the baseline temperature. These results demonstrate that application of the PCD probe will not only reduce the risk of thermally-induced tissue damage, but also improve the accuracy of optogenetic experiments by minimizing biological interferences due to thermal effects.

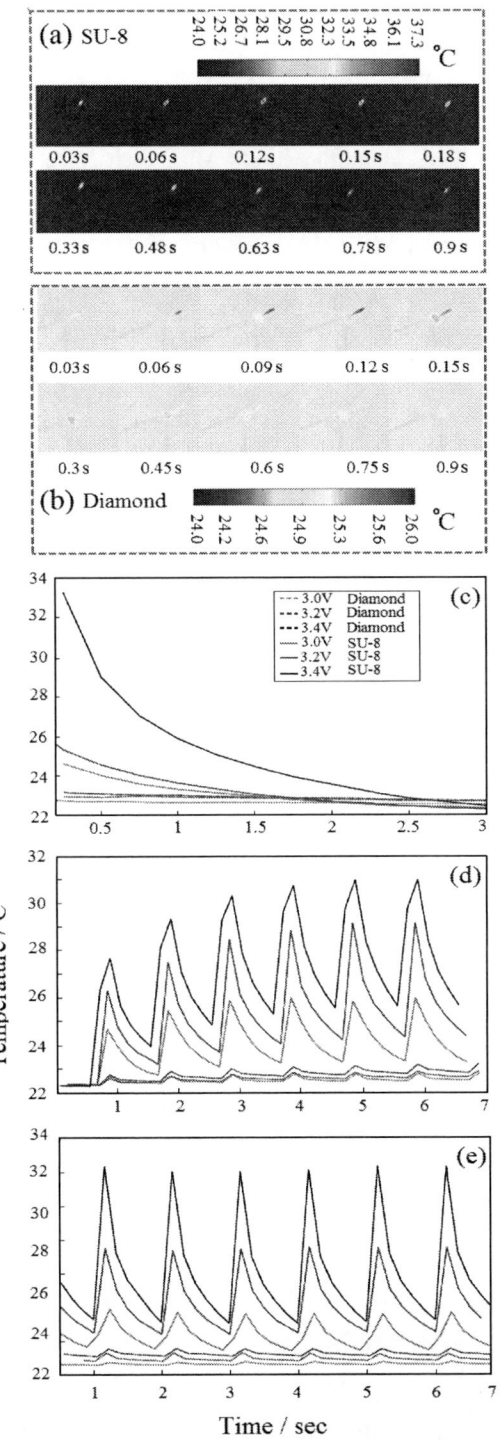

Figure 2: (a)-(b) Heat distribution of the SU-8 and PCD probes. (c) Cooling curves of the probes after activating μLED for 60sec with different inputs. (d) Instantaneous change in the tip temperature. (e) Steady-state temperature variations of the probes.

FABRICATION PROCESS

The fabrication process is shown in Figure 3. Specifically, (a) PCD was grown on a molybdenum substrate using a 2.45GHz microwave plasma assisted chemical vapor deposition (MW-PACVD) reactor with 2-3kW microwave power in a methane and hydrogen mixture atmosphere (4% CH_4, 160-240Torr) and released by thermal stress during cooling from growth temperature to room temperature. (b) The diamond substrate was cleaned by sonication in isopropanol (IPA) and deionized (DI) water for 30 min each and then in nitric acid at 80 °C for 30 min. (c) A 0.5 μm layer of Cu and 3 nm layer of Ti were deposited using a thermal evaporator (Auto 306, Edward, Inc). Ti was used as an adhesion layer to improve the bonding strength between Cu and PCD. A photoresist mask was patterned using a mask aligner (ABM, Inc) for metal patterning. Then the Cu/Ti layer was wet etched using ferric chloride (to remove Cu) and buffered oxide etchant (to remove Ti) to form microelectrodes, contact pads, and interconnect wires. (d) Photoresist (S1813, Microchem) was spun on and selectively patterned to expose the μLED pad areas. Oxygen plasma (PX-250 plasma system, Nordson March, Inc) at power of 100W and pressure of 0.5Torr was used to remove photoresist residue on the pad areas for 5min. (e) Low melting point (LMP) solder (62 °C, 144 ALLOY Field's Metal) was applied in an acid bath [8] and the μLEDs were self-assembled onto the contact pads wetted with LMP solder. (f) Then the substrate was rinsed with acetone, IPA and DI water to remove the photoresist layer. (g) A ~5 μm Parylene C layer was deposited on top of the probe as an encapsulation using a CVD evaporator (PDS 2010, Specialty Coating System, Inc). (h) Parylene C was then patterned using oxygen plasma (RIE-1701, Nordson March, Inc) and a photoresist mask at power of 300W and pressure of 0.25Torr, in order to open the recording sites and contact pads for electrical interconnects. (i) Finally, the probe was shaped using a Nd:YAG laser with power of 8-14W (UltraShape 5xs, Bettonville Inc).

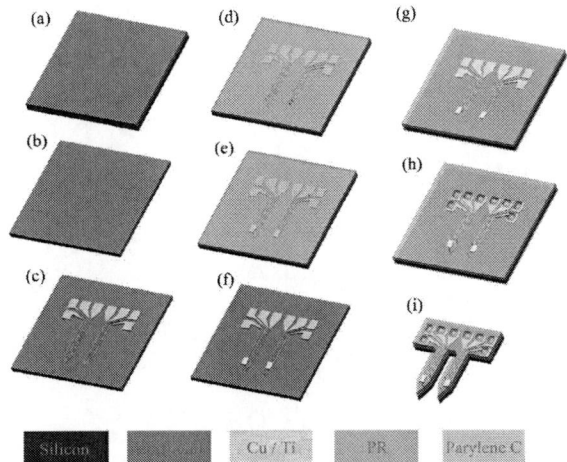

Figure 3: Fabrication process for making the proposed PCD probe.

Figure 4 (a) shows a micro-fabricated, double-shank probe with one LED and two microelectrodes on each shank. To connect the probe to external powering and recording electronics, the pad areas were covered with the LMP solder, and the thin wires were assembled to the probe, as shown in Figure 4 (b). Epoxy was used to strength the bonding between the pads and wires. The electrochemical impedances of the electrodes at 1KHz (Channels 1-4) were 89.0, 20.6, 38.5, and 54.8 Kohm, respectively. Channel-1 had higher impedance due to incomplete Parylene removal, and therefore was not used in further signal analysis. The dimensions of the PCD probe and µLED are listed in Table 1. The light intensity of the µLED chip was measured using a digital power meter (Model 815 Series, Newport, Inc) and read through the RHA 2000 evaluation board, as shown in Figure 4 (c).

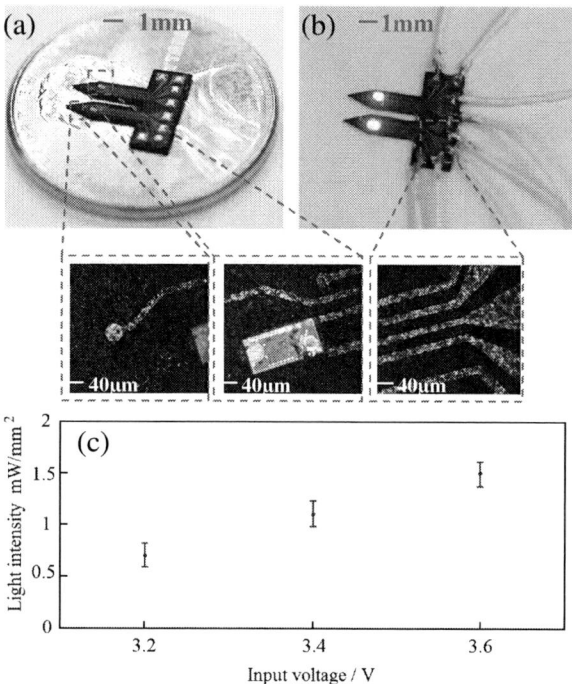

Figure 4: (a) A fabricated prototype with close-up views on different segments of the probe. (b) The assembled probe showing that the LEDs were powered up. (c)Light intensity of µLED driven by different input voltages.

Table 1 Dimensions of the PCD probe and µLED
(L: length, W: width, H: height)

µLED (L×W×H)	0.55mm×0.29mm×0.1mm
Shank (L×W)	5mm×1mm
Total dimension (L×W×H)	8mm×6.8mm×0.25mm

IN-VIVO SIGNAL RECORDING

In vivo acute experiments were conducted in V1 of a rat to demonstrate the functionality of the as-fabricated PCD probe. All procedures were approved by the Institutional Animal Care and Use Committee (IACUC) at Michigan State University. Prior to the *in vivo* testing, the rat was transfected with channelrhodopsin-2 (CHR2) to enable the functionality of light sensitivity upon blue light illumination. For viral transduction, the rat was anesthetized with ketamine and xylazine, and then placed in a stereotaxic apparatus. A rostral - caudal incision was made in the skin to expose the skull. Four holes were made medial-laterally with two over each hemisphere of V1. Each site was injected with 1.0 µL ($10×10^{11}$~$10×10^{12}$ vector genome (vg/ml) of the virus solution (AAV-hSyn-hCHR2(H134R)-mCherry, UNC Vector Core), with an injection rate of 0.1 µL/min, using a micro syringe (5 µL, Model 75 RN SYR and 100 µL, Neuros Adapter Kit, Hamilton, Inc). After each injection, the syringe was maintained in place for an additional 5 minutes to allow the viral vector to diffuse from around the injection site. Once the injections were completed, the cortical openings were plugged with bonewax, covered with Gelfoam, and the overlying skin was sutured. Then the rat was given 5 µL of sterile saline (0.9% NaCl solution) subcutaneously to prevent dehydration during recovery, as well as an injection of buprenorphine for pain relief.

The functionality of simultaneous optical stimulation and recording was tested 3-4 weeks post-surgery to allow for expression of the ChR2 gene in the targeted cortical neurons. Following the same procedure of the viral vector injection, the rat was anesthetized and an opening covering the two injection sites was made for inserting PCD probe. The PCD probe was inserted into V1 of the left hemisphere of the rat with right hemisphere serving as a vector-injected control, as shown in Figure 5 (a). In this case, the µLED on the left side of the probe was driven by repetitive pulses with a frequency of 1 Hz, pulse width of 10 ms, and amplitude varying from 3.2 V to 3.6 V, as shown in Figure 5 (b). Three functioning electrodes were used to record neural activity, through a RHD 2132 32-channels headstage and RHD 2000 USB Evaluation System. The µLED on the right side probe was not tested in this study due to malfunction of the probe before device insertion.

Figure 6 (a)-(c) show the neural signals recorded from Channel 2 with different applied voltages. Light-evoked action potentials were observed when the optical stimuli were switched from On to Off with high input voltage of 3.6 V, whereas no action potential was observed with 3.2 V and 3.4 V inputs. The light intensities with 3.2 V, 3.4 V and 3.6 V input voltage were 0.6 mW/mm², 1 mW/mm² and 1.5 mW/mm², respectively. The minimum light intensity to evoke a ChR2 transfected ion channel has been reported to be around 1 mW/mm² [9]. Considering the coupling efficiency between a µLED and the brain tissue, the input voltage of 3.2 V and 3.4 V of the µLED were not sufficient to evoke any action potential.

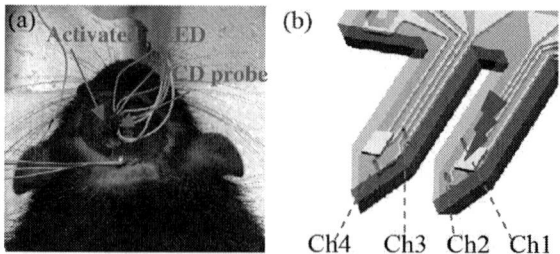

Figure 5: (a) In-vivo testing setup. (b) Schematic design of the probe.

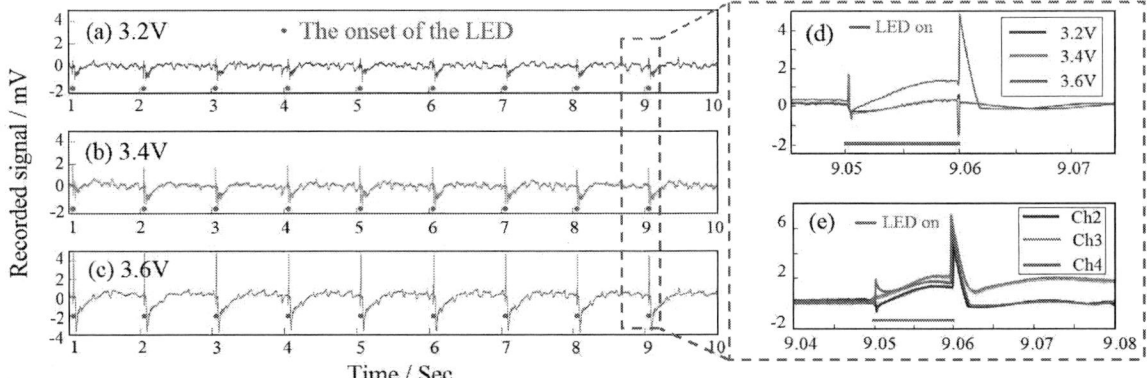

Figure 6: (a)-(c) Neural activity recorded from Channel 2 with µLED input voltage of 3.2 V, 3.4 V and 3.6 V, respectively. (d) A close-up view of signals from Channel 2 with different µLED input voltages. (e) Action potentials recorded from different channels at input voltage of 3.6 V.

CONCLUSION

This paper reports the design, fabrication and properties of a PCD-based, hybrid optoelectronic neural interfacing probe, which is capable of optical stimulation and on-chip recording of neural activity. Experimental results show the PCD probes have superior thermal dissipation performance, which not only allows localized LED heat exchange with surrounding tissues, but also minimizing abnormal neural activity due to thermal effects during optogenetic neuromodulation. *In vivo* testing was performed to demonstrate the functionality of optical stimulating and electrical recording of action potentials of ChR2 transfected neurons. Action potentials were observed with 3.6 V input, 10 ms duration, 1 Hz repetitive pulses. No action potentials were observed with 3.2 V and 3.4 V input. The light intensity testing has proven that only 3.6 V input has exceeded the minimum requirement of opening opsin ion channels, which is 1 mW/mm^2. The mechanical properties of the proposed neural probe will be tested in a future study. After the *in vivo* experiment, the animal was euthanized with an overdose of pentobarbital sodium and the brain tissue perfused and prepared for histological analysis. The gene transfection, actual penetration depth and any tissue injury produced by insertion of the PCD probe will be examined in the future.

ACKNOWLEDGEMENT

This work was supported by the National Science Foundation under the Award Numbers CBET-1264772 and ECCS-1407880. The authors would like to thank Dr. Baokang Bi for the help on micro-fabrication and Mr. Yue Guo for the help on drawing the process flow.

REFFERNCES

[1] Advanced Diamond Technology, "The CVD diamond booklet." [Online]. Available: http://www.diamond-materials.com/downloads/cvd_diamond_booklet.pdf.

[2] F. Zhang, V. Gradinaru, A. R. Adamantidis, R. Durand, R. D. Airan, L. de Lecea, and K. Deisseroth, "Optogenetic interrogation of neural circuits: technology for probing mammalian brain structures," *Nat. Protoc.*, vol. 5, no. 3, pp. 439–456, Mar. 2010.

[3] O. Yizhar, L. E. Fenno, T. J. Davidson, M. Mogri, and K. Deisseroth, "Optogenetics in Neural Systems," *Neuron*, vol. 71, no. 1, pp. 9–34, Jul. 2011.

[4] A. M. Aravanis, L.-P. Wang, F. Zhang, L. A. Meltzer, M. Z. Mogri, M. B. Schneider, and K. Deisseroth, "An optical neural interface: in vivo control of rodent motor cortex with integrated fiberoptic and optogenetic technology," *J. Neural Eng.*, vol. 4, no. 3, pp. S143–156, Sep. 2007.

[5] K. Kwon and W. Li, "Integrated multi-LED array with three-dimensional polymer waveguide for optogenetics," in *2013 IEEE 26th International Conference on Micro Electro Mechanical Systems (MEMS)*, 2013, pp. 1017–1020.

[6] "SU-8 2000 MSDS." [Online]. Available: http://www.microchem.com/pdf/SU-82000DataSheet2025thru2075Ver4.pdf.

[7] B. Fan, K. Y. Kwon, A. J. Weber, and W. Li, "An implantable, miniaturized SU-8 optical probe for optogenetics-based deep brain stimulation," in *2014 36th Annual International Conference of the IEEE Engineering in Medicine and Biology Society (EMBC)*, 2014, pp. 450–453.

[8] K. Y. Kwon, B. Sirowatka, W. Li, and A. Weber, "Opto-µ ECoG array: Transparent µECoG electrode array and integrated LEDs for optogenetics," in *2012 IEEE Biomedical Circuits and Systems Conference (BioCAS)*, 2012, pp. 164–167.

[9] K. Y. Kwon, B. Sirowatka, A. Weber, and W. Li, "Opto- µECoG array: a hybrid neural interface with transparent µECoG electrode array and integrated LEDs for optogenetics," *IEEE Trans. Biomed. Circuits Syst.*, vol. 7, no. 5, pp. 593–600, Oct. 2013.

AN IMPLANTABLE TIME OF FLIGHT FLOW SENSOR

Lawrence Yu, Brian J. Kim, and Ellis Meng

Department of Biomedical Engineering, University of Southern California, Los Angeles, CA, USA

ABSTRACT

A micro time of flight (TOF) electrochemical impedance (EI) flow sensor μEIFS suitable for implantation and integration with a catheter was developed. The transducer utilizes two pairs of electrodes to monitor using EI measurement the passage of a gas bubble generated upstream. High precision measurement of bubble TOF (SD < 6% of mean) was achieved over the velocity range 0.83-83 μm/s. The volumetric flow rate was inversely proportional (linearized, $r^2 = 0.99$) to time of flight. Biocompatible construction (only Parylene C and platinum), low power consumption, and low profile thin film format make the μEIFS ideally suited for chronic *in vivo* monitoring with immediate application in tracking flow within hydrocephalus shunts.

INTRODUCTION

The commonly used thermal TOF flow sensor is limited by the dissipation of its heat tracer signal [1] which in turn reduces sensitivity, resolution, and flow measurement range. To realize high resolution *in vivo* TOF flow measurement, the tracer should accurately track flow with minimal degradation such as thermal dissipation or dissolution. Previously, dissolved oxygen was investigated as a flow tracer [2], but its low sensitivity limited resolution and restricted use to a small range of flow rates (0.83–12.45 μm/s).

Here, electrolytically generated bubbles were selected as the tracer because they have been shown to remain in solution for sufficiently long durations of time (> 15 min) [3, 4] and are advantageous for measuring low flow rates with long time of flight. In addition, microbubble size can be measured in real time with high sensitivity (> 20 SNR) using an electrochemical impedance sensing technique [5]. Therefore, the μEIFS employs electrolytically generated bubbles as tracers for the first time in TOF flow measurement.

DESIGN

The μEIFS employs three pairs of electrodes positioned in parallel to the direction of fluid flow (fig 1). One pair of electrodes was designated for electrolytic bubble generation and the second and third pairs were used solely for measurement. This arrangement improves temporal resolution compared to having electrodes with shared EI sensing and electrolysis functions. The electrode arrangement is symmetric along the direction of flow and thus the sensing and electrolysis roles can be reassigned in the case of reverse or pulsating flow.

To perform a flow measurement, a bubble is first electrolytically generated with the upstream pair of electrodes (fig. 1a). At a sufficiently large size, the electrolysis current is terminated, the bubble detaches (fig. 1b), and the bubble is carried along with the flow through the sensing electrode pairs.

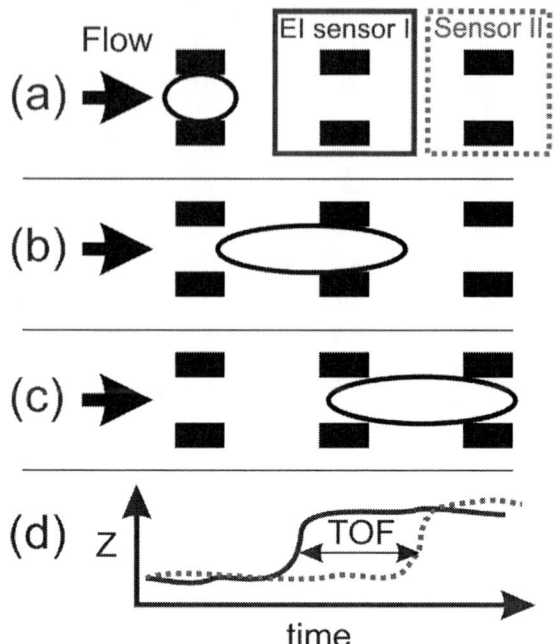

Figure 1: Schematic of sequence for time of flight (TOF) measurement. (a) Bubble is electrolytically generated at the leftmost electrode pair. (b) The electrochemical impedance magnitude increases as bubble passes through sensor I. (c) After bubble passes through sensor II, difference in time of electrochemical impedance rising edge is used to calculate TOF and flow rate. (d) Representative temporal EI response, illustrating the time of flight measurement.

Bubble sensing using electrochemical impedance methods has been studied and reported previously [3, 6, 7]. Briefly, a small AC signal is passed between two platinum electrodes residing within an electrolyte solution, and this system can be modeled with the Randles circuit in which the parallel combination of electrode charge transfer resistance and double layer capacitance is in series with the solution resistance R_s [8]. At a suitably high frequency, the measured impedance approximates the solution resistance R_s. As the bubble travels downstream, it disrupts the path of ionic current between each measurement electrode pair and this manifests as increased electrochemical impedance. The TOF measurement is observed as the difference in time between the increases of electrochemical impedance measured from the rising edge of the impedance signals.

Given the predefined flow channel and electrode geometry, flow velocity is derived from the difference in the time of the onset of measured impedance change at these two electrode pairs, which corresponds to the passage of the leading edge of the bubble and is expressed by the following formula:

978-1-4799-7956-1/15 $31.00 © 2015 IEEE   620   MEMS 2015, Estoril, PORTUGAL, 18 - 22 January, 2015

$$flow\ velocity = \frac{d}{t_{TOF}} \qquad (1)$$

where d is the distance between the electrochemical impedance sensors and t_{TOF} is the measured time of flight. Volumetric flow rate may then be determined given the cross sectional area of the flow channel. Two electrode spacing distances, 500 and 2500 µm, were fabricated to enable flow measurement in the range of 1-100 µm/s, which includes the clinically relevant flow velocities found in patients with hydrocephalus.

METHODS
Fabrication

The µEIFS consists of a Parylene-metal-Parylene sandwich created using standard Parylene microfabrication processes. To start, thin film platinum (200 nm thick, 20000 µm^2) electrodes were defined on a Parylene C (10 µm) coated silicon support wafer. An additional layer of Parylene (10 µm) was deposited for insulation, and the electrodes were subsequently exposed via oxygen plasma reactive ion etching (RIE) process. The die was etched out with a switched chemistry deep RIE [9] and then released in an acetone bath. A fabricated sensor is shown in fig. 2 along with other EI sensors under development. The generation and measurement electrodes are indicated.

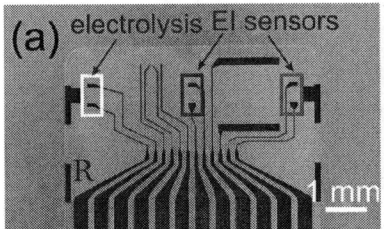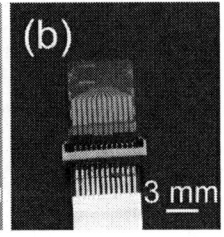

Figure 2: (a) Micrograph of sensor as fabricated and released from wafer and (b) attached to flat flexible cable for testing via zero insertion force connector. Additional electrodes on die were utilized for other electrochemical impedance measurements.

Experimental Setup

The µEIFS was designed for integration within the lumen of a commercially available luer lock connector (4 mm ID, fig. 3b). A thin slit was milled into the side of the connector, the sensor was inserted into the flow path through the slit, and subsequently the sensor cable-slit interface was sealed with biocompatible epoxy (EPO-TEK 353ND). This configuration was selected to allow subsequent testing of sensors with an extraventricular drain (EVD) used in acute clinical settings for managing elevated intracranial pressure and draining cerebrospinal fluid (CSF). For benchtop testing, the integrated sensor module was positioned such that the flow was parallel to ground level and was attached to a peristaltic pump (Watson-Marlow 400DM3) for flow calibration.

To mimic biological fluids, phosphate buffer solution (1X PBS) was used for sensor characterization. Electrochemical impedance spectroscopy yielded 10 kHz as the optimum frequency (minimum system phase) for EI

measurement (fig 4). Multiplexed EI measurement (1 Vp-p, < 1 nW) was conducted with a custom PCB attached to a precision LCR meter (Agilent E4980A). Data acquisition was carried out in a LabVIEW environment to realize high temporal resolution (50 ms sampling, limited by software) for TOF measurement.

Additonal flow channel geometries and electrode arrangements were also tested (fig 3). The sensor was also directly inserted into a silicone catheter (1 mm ID) and clamped into a custom acrylic jig having a rectangular cross section channel (fig 3d). Spacing between EI sensing electrode pairs was assessed with two different electrode layouts (500 and 2500 µm separation).

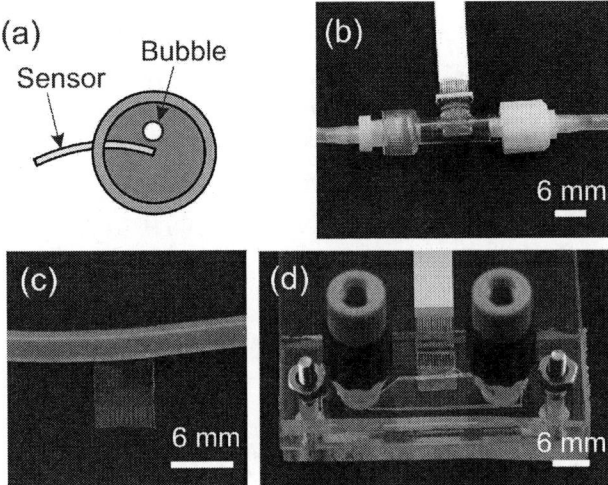

Figure 3: (a) Cross sectional view of sensor integrated into lumen of catheter, illustrating relative position of bubble within fluid. (b) Integration of sensor with luer lock interconnect. (c) Sensor directly inserted into silicone catheter. (d) Custom acrylic jig for flow calibration.

A bubble was electrolytically generated (50-70 µA, 30-45 s) within flowing PBS (10-1000 µL/min) at the upstream electrode pair and EI measurements were downstream at the two sensing electrode pairs. The TOF was calculated by noting the difference in time values when the measured impedance rose 10% above the baseline reading at each sensor.

RESULTS

High precision measurement of bubble TOF (SD < 6% of mean) was achieved with a flow velocity range of 8.3-830 x 10^{-7} m/s. Results were achieved using the luer lock integrated sensor module (additional testing underway). The volumetric flow rate was inversely proportional (linearized, r^2 = 0.99) with the bubble time of flight (fig. 7), confirming the relationship described in equation 1.

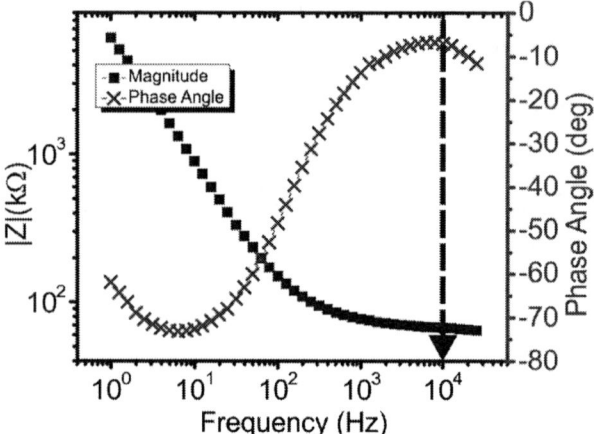

Figure 4: EI spectrotroscopy magnitude and phase plots of 1X PBS. To measure the solution resistance with minimum capacitive effects, measurement frequency was selected where phase is nearest 0 degrees.

Electrode spacing was selected to be 2500 µm because bubble TOF was indiscernible at most flow conditions (above 25 µm/s flow velocity) with electrode spacing of 500 µm. This is largely attributed to the need to generate large bubbles relative to the diameter of the flow channel to encourage detachment from the hydrophobic Parylene surface. By further improving bubble generation via electrolysis, smaller bubbles may be used to thereby decrease the spacing requirement. With the 2500 µm spaced electrodes, flow measurements were taken at the rate of < 0.1 Hz (fig. 6).

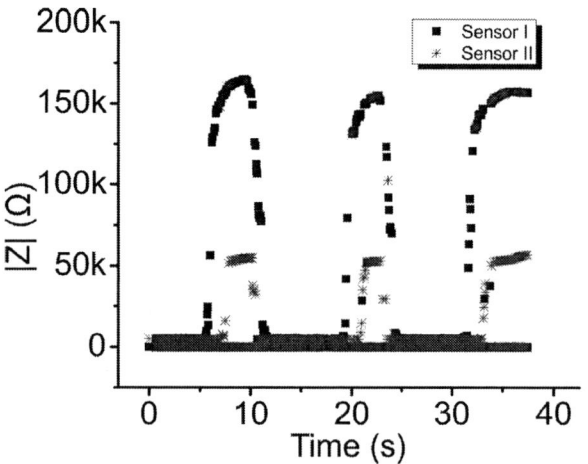

Figure 5: Impedance response of multiple bubbles passing through luer lock interconnect with flow sensor at 300 µL/min.

The discrepancy between the measured impedance of the two EI sensors (fig. 6) may be a result of process variation or variation in the position of the bubble from the first to the second electrode pair. An alternate electrode arrangement along the catheter walls is under development to reduce the position dependency of the bubble relative to the sensing electrodes.

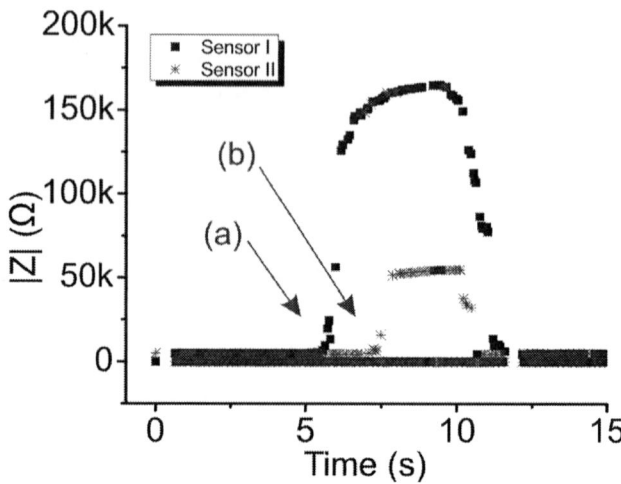

Figure 6: Passage of bubble is sensed as an increase in measured EI at sensor I (a), and eventually sensor II (b). Detection of bubble at each sensor initiates a rise in measured electrochemical impedance, and is used to derive time of flight.

Detachment from the Parylene surface of the µEIFS did not occur until the bubble was of a sufficiently large size. Smaller tracer bubbles require less power and time to create via electrolysis, and would be greatly beneficial to improving sensor performance. Modification of the surface chemistry [10] is a technique that has been explored and is currently under assessment. Variation of the electrode arrangement is also being considered to encourage detachment.

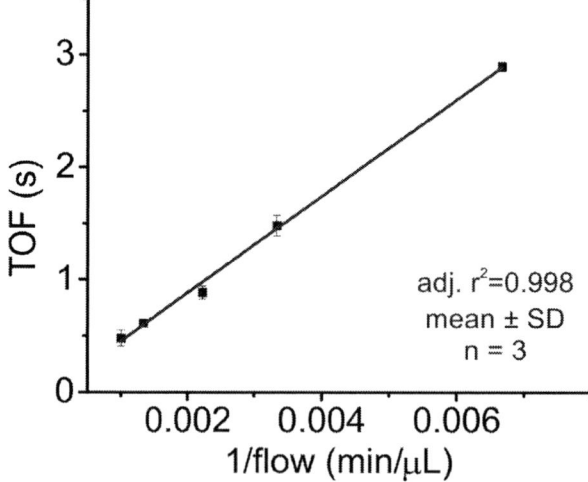

Figure 7: Relationship between flow rate and bubble time of flight is inversely proportional. Adjusted r² includes 10 µL/min data (data not shown). Data taken from device integrated into 4 mm ID luer lock interconnect.

CONCLUSIONS

We have designed, fabricated, and tested a Parylene based liquid flow sensor utilizing an electrochemical impedance transduction technique. Microbubbles were employed as tracers to enable high dynamic range, high sensitivity monitoring of flow. Additional sensor characterization is underway (varying electrolyte composition such as use of artificial and human CSF, sensor orientation and placement, electrode layout). Sensors will be attached to the external ventricular drain in clinical settings to achieve first-in-human demonstration.

ACKNOWLEDGEMENTS

This work was funded by the NSF under award number EFRI-1332394. The authors would like to thank the members of the USC Biomedical Microsystems Laboratory for their assistance.

REFERENCES

[1]   J. R. Madsen, G. S. Abazi, L. Fleming, M. Proctor, R. Grondin, S. Magge, *et al.*, "Evaluation of the shuntcheck noninvasive thermal technique for shunt flow detection in hydrocephalic patients," *Neurosurgery,* vol. 68, pp. 198-205, 2011.

[2]   J. Wu and W. Sansen, "Electrochemical time of flight flow sensor," *Sensors and Actuators A: Physical,* vol. 97, pp. 68-74, 2002.

[3]   C. A. Gutierrez and E. Meng, "A subnanowatt microbubble pressure sensor based on electrochemical impedance transduction in a flexible all-parylene package," in *24th IEEE International Conference on Micro Electro Mechanical Systems, MEMS 2011, January 23, 2011 - January 27, 2011,* Cancun, Mexico, 2011, pp. 549-552.

[4]   P. S. Epstein and M. S. Plesset, "On the Stability of Gas Bubbles in Liquid-Gas Solutions," *Journal of Chemical Physics,* vol. 18, pp. 1505-1509, 1950.

[5]   C. A. Gutierrez, C. McCarty, B. Kim, M. Pahwa, and E. Meng, "An Implantable All-Parylene Liquid-Impedance Based MEMS Force Sensor," in *MEMS 2010: 23rd IEEE International Conference on Micro Electro Mechanical Systems, Technical Digest,* ed New York: IEEE, 2010, pp. 600-603.

[6]   D. A. Ateya, A. A. Shah, S. Z. Hua, and F. Sachs, "Bubble based microfluidic sensors," in *2004 ASME International Mechanical Engineering Congress and Exposition, IMECE 2004, November 13, 2004 - November 19, 2004,* Anaheim, CA, United states, 2004, pp. 437-441.

[7]   L. Yu and E. Meng, "A microbubble pressure transducer with bubble nucleation core," in *Micro Electro Mechanical Systems (MEMS), 2014 IEEE 27th International Conference on,* 2014, pp. 104-107.

[8]   J. E. B. Randles, "Kinetics of Rapid Electrode Reactions," *Discussions of the Faraday Society,* vol. 1, pp. 11-19, 1947.

[9]   E. Meng and Y.-C. Tai, "Parylene etching techniques for microfluidics and bioMEMS," in *Micro Electro Mechanical Systems, 2005. MEMS 2005. 18th IEEE International Conference on,* 2005, pp. 568-571.

[10]   A. Volanschi, W. Olthuis, and P. Bergveld, "Gas bubbles electrolytically generated at microcavity electrodes (MCE) used for the measurement of the dynamic surface tension in liquids," in *Proceedings of the International Solid-State Sensors and Actuators Conference - TRANSDUCERS '95, 25-29 June 1995,* Stockholm, Sweden, 1995, pp. 385-8.

CONTACT

*E. Meng, tel: +1-213-7406952; ellis.meng@usc.edu

APPLICATION OF PERIODIC LOADS ON CELLS FROM MAGNETIC MICROPILLAR ARRAYS IMPEDES CELLULAR MIGRATION

Farzad Khademolhosseini[1], Chi-Chao Liu[1,2], Chinten J. Lim[1,2], and Mu Chiao[1]
[1]The University of British Columbia, Vancouver, Canada
[2]Child and Family Research Institute, Vancouver, Canada

ABSTRACT

This paper presents a study on the application of active micropillar surfaces to control cell migration. We present experimental results on the migration of confluent sheets of cells subject to periodic mechanical forces from actuated magnetic polymer micropillars. We show that in contrast to passive micropillar surfaces which cause no significant alterations in cell migration rates, active micropillar surfaces actuated at a frequency of 1 Hz can decrease cell migration rates by 80%. The magnetic micropillar structures presented can be actuated remotely with small magnetic fields making them a viable candidate for the development of smart materials for in vivo tissue engineering applications.

INTRODUCTION

Passive polymer micropillar structures have been widely used as sensors for cell mechanotransduction studies, to quantify the traction forces generated by cells during cellular locomotion [1]-[3]. More recently, active micropillar structures that can be actuated with magnetic fields have been used to apply loads on individual cells and study the propagation of force through the cytoskeleton [4, 5]. This work builds on previous studies by looking at the application of active micropillar structures in controlling the migration behavior of large clusters/sheets of confluent cells, to assess their usefulness in tissue engineering applications. Using magnetic polymer micropillar chips fabricated with a dry nanoparticle embedding technique [6], we are able to remotely actuate magnetic micropillars over large areas and apply cyclic mechanical loads on sheets of confluent cells in order to study their collective migration behavior. We present experimental results of cell migration in wound healing assays, controlling for variations in micropillar topology and the relative directions of micropillar actuation vs. cell migration.

DESIGN

When looking at the migration of cells among magnetic micropillar surfaces, several factors can affect cell migration behavior and cell migration rates. The first factor is the topology of the micropillar surfaces, i.e., the surface density and relative locations of the magnetic micropillars. The second factor is the application of forces on cells, i.e., actuated vs. non-actuated micropillars. The third factor is the direction of force application with respect to the direction of cell migration. To independently assess the net effect of each of these three factors on cell migration rates, multiple levels of control are needed. To control for the first two factors, i.e., surface topology and force application, we designed the cell-migration chips to have nine different regions, as demonstrated in Figure 1. Regions 1 to 4 have non-

magnetic pillars, with pillars located either in a square or diamond pattern, and spaced either 110 or 150 μm apart. Regions 6-9 have the exact same topology as regions 1-4, but are comprised of magnetic pillars that allow application of forces on cells. Region 5 is flat and does not contain any pillars.

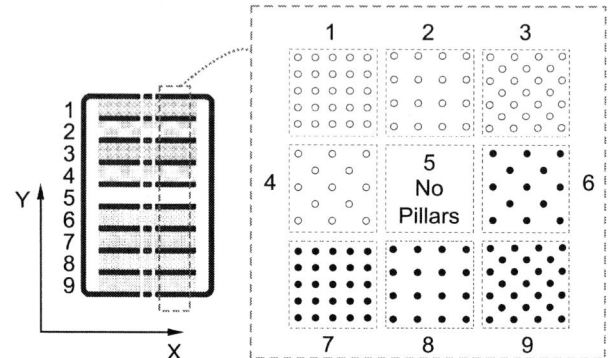

Figure 1: Schematic of cell-migration chip, designed with nine different regions; four regions/patterns with magnetic pillars, four regions/patterns with non-magnetic pillars and one region with no pillars (flat PDMS).

Finally, to control for the relative direction of force application (pillar bending) vs. direction of cell migration, chips were designed with walls to separate the cells in each of the regions and confine the net cell migration to the X-direction, whereas pillar bending and force application could be applied in the X or Y directions.

FABRICATION

To fabricate the cell-migration chips containing both magnetic and non-magnetic micropillars, we used a variation of the dry nanoparticle embedding technique previously demonstrated by Khademolhosseini et al. [6]. The dry nanoparticle embedding technique allows for the patterning of magnetic and non-magnetic polymer micropillars on one chip in one polymer casting step, and is ideally suited for the fabrication of our multi-region cell-migration chips.

To fabricate the molds required for the dry nanoparticle embedding technique, first, an SU8 master of the micropillar chip was developed using conventional photolithography, followed by casting and curing of a Polydimethylsiloxane (PDMS) layer (Sylgard 184, Dow Corning Corp.) on the SU8 master to obtain a PDMS negative. This PDMS negative was then treated with oxygen plasma and silanized with Hexamethyldisilazane (HMDS) overnight, and a second PDMS casting and curing was performed to obtain a PDMS replica of the SU8 master. Next a polyurethane (PU) mixture (Smoothcast 310, Smooth-On Corp.) was cast on the

PDMS replica, cured and de-molded to obtain the reusable mold for fabrication of the cell-migration chips.

Next, employing the dry nanoparticle embedding technique, masking tape was used to cover parts of the PU mold corresponding to regions 1 to 4 on the migration chip, to prevent embedding of magnetic particles in those regions. Carbonyl Iron (FeC) magnetic particles were then applied to the surface of the mold, and a permanent magnet was used to pull the particles into the cavities of the unmasked regions (regions 6 to 9). Excess particles were then removed by wiping off with a cotton applicator or kim-wipes. Next, the masking tapes were removed and a PDMS polymer solution was cast on the PU mold and allowed to cure at room temperature overnight. Once cured, the PDMS was de-molded from the PU mold. This final PDMS layer is the cell-migration chip containing regions with magnetic and non-magnetic pillars. Figure 2 shows a schematic of the fabrication steps.

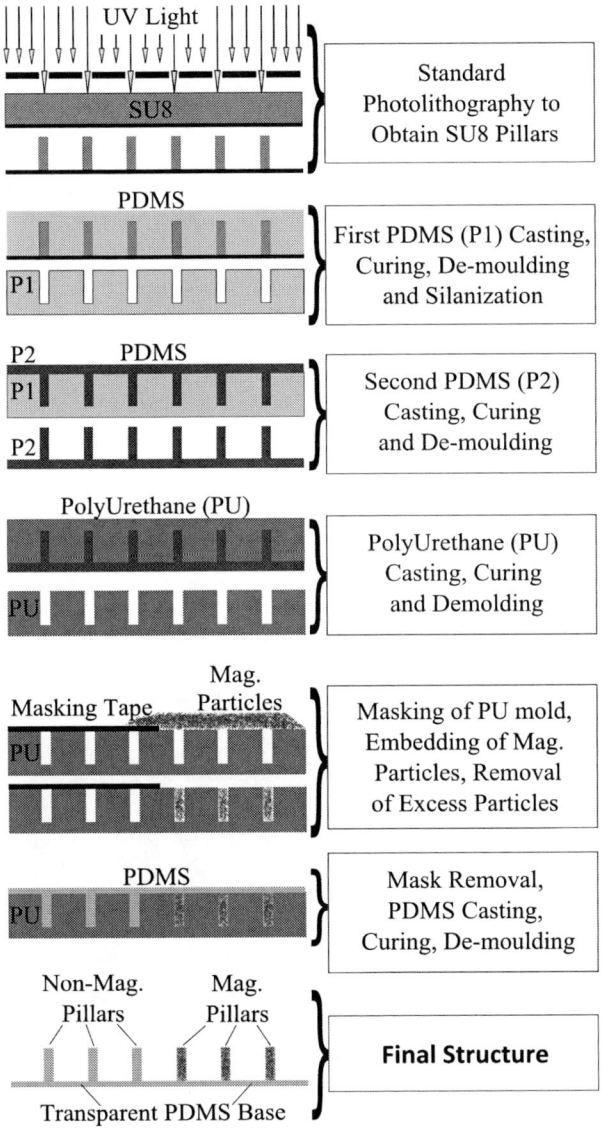

Figure 2: Schematic of fabrication steps for cell-migration chip. The dry nanoparticle embedding technique is demonstrated.

Figure 3 shows an image of the cell-migration chip fabricated using the presented fabrication method. Magnified insets of the pillars of region 7 before and after magnetic actuation are also shown.

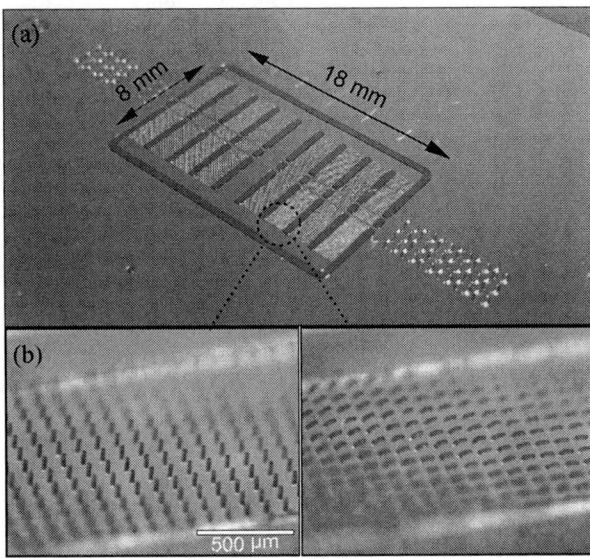

Figure 3: (a) Image of a multi-region cell-migration chip having magnetic and non-magnetic pillars and a combination of various micropillar topologies. Chip was fabricated using standard photolithography followed by replica molding and the dry nanoparticle embedding technique, (b) Magnified images showing the magnetic pillars of region 7, before actuation (left) and after actuation in X-direction using external magnetic field (right).

CELL MIGRATION EXPERIMENTS
Cell Culture and Chip Preparation

Human Umbilical Vein Endothelial Cells (HUVECs) were chosen as the cell line for our migration experiments. Endothelial cells have been shown to be very responsive to substrate strains and are therefore a good candidate for cell strain experiments assessing the effect of periodic loads from micropillar arrays on cell migration behavior. HUVECs (EGM-2, cryo amp, code cc-2517A) and culture media (EGM-2 BulletKit, code cc-3162) were obtained from Lonza. The cells were expanded and passaged in culture dishes in the EGM-2 culture media. Cells from passages 3 to 8 were used for conducting the experiments.

To facilitate cell attachment to the hydrophobic PDMS micropillar arrays, the arrays were treated with Oxygen plasma for 120 seconds and immersed in a 22 µg/ml human Fibronectin/PBS solution for 24 hours. The arrays were rinsed twice with PBS prior to cell culture.

Wound Healing Assay

To simulate a scratch wound, HUVECs were seeded at sub-confluent densities on the outer sides of two vertically placed, parallel, micro cover glasses spaced 0.9 mm apart and incubated in EGM-2 under 5% CO_2 for 12 hours until confluent sheets of cells were observed. The micro cover glasses were then removed, and the cells were allowed to migrate onto the area of the simulated scratch wound. An external magnetic field was used to actuate the magnetic pillars in either the X or Y direction at a cyclic frequency of 1 Hz during the experiment. At different time intervals during the experiment, the scratch wound corresponding to regions 1 to 9 on the micropillar chip was imaged using an Olympus IX81 inverted microscope equipped with a 4x/0.13 NA objective and a CoolSnap HQ2 ccd camera. Post-acquisition image analysis was performed in ImageJ to track the cell front as cells migrated into the wound area.

Long Term Study

To see the long term effect of cyclic loading from the micropillar arrays on HUVEC migration rates, HUVECs were cultured to confluency on the micropillar arrays on one side of a micro cover glass barrier. The micro cover glass was then removed and the migration of HUVECs among the micropillars was studied over a 12 day period.

RESULTS

Figure 4 shows a sample image of a simulated scratch wound assay on a micropillar array. Movement of the cells into the wound area after 24 hours can be clearly observed. Figure 5 shows the average migration rates of a 24 hour wound healing assay for three different migration chips, one where the magnetic pillars were actuated in X-direction, one where the magnetic pillars were actuated in Y-direction and a control chip with no actuation. In all chips cell migration was confined to and measured in the X-direction.

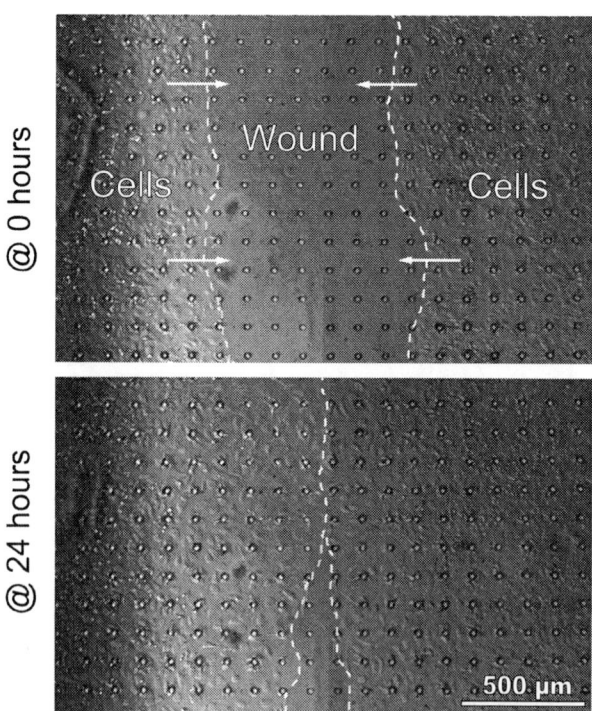

Figure 4: Wound healing assay conducted on a PDMS micropillar chip. The migration rate of the sheet of cells into the wound area and the wound closure rate were studied over a 24 hour period using live imaging.

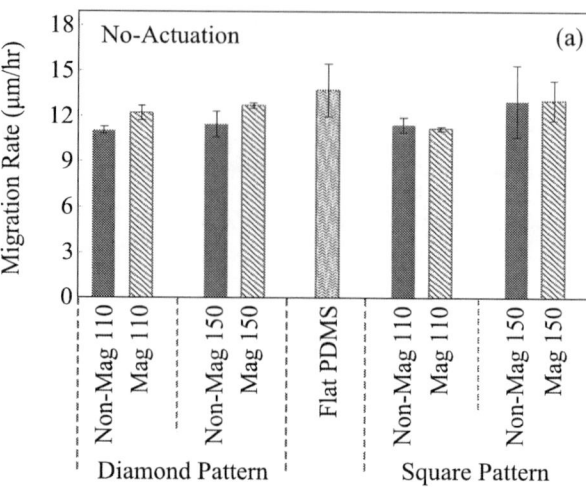

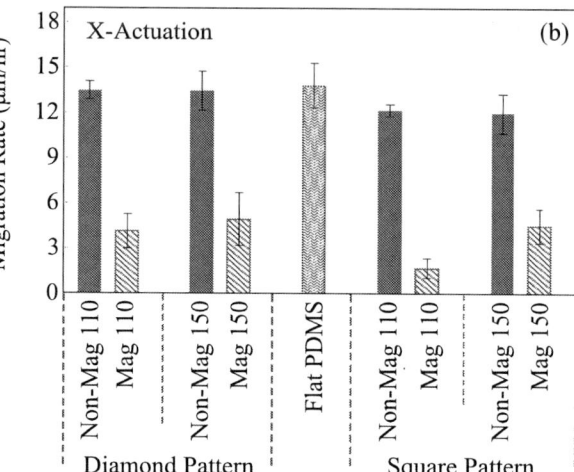

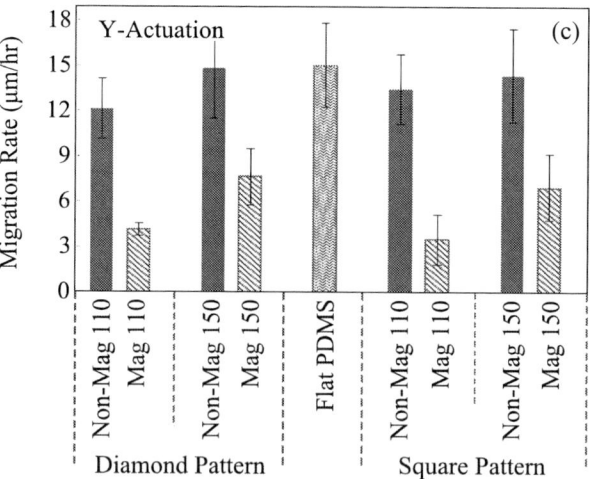

Figure 5: Effects of micropillar actuation on the migration rate of cells; (a) No magnetic field applied (no micropillar actuation), (b) Magnetic pillars were actuated in X-direction at 1Hz, (c) Magnetic pillars were actuated in Y-direction at 1Hz. In (b) and (c) all 9 regions were subjected to the same magnetic field. As observed, application of periodic forces from magnetic pillars on the cells impeded cell migration.

Based on results observed from the control chip, in the absence of an external magnetic field (Figure 5(a)), i.e., when no periodic loads were applied on the cells from the magnetic pillars, HUVECs growing among magnetic and non-magnetic pillars showed similar migration rates. Furthermore, the migration rate of HUVECs among the micropillars was smaller than the migration rate of HUVECs on flat PDMS (with no pillars) and consistent with previously reported values for endothelial cell migration rates [7]. Variations in surface topology, i.e., surface density of micropillars and micropillar patterns, caused minimal changes in cell migration rates. The existence of micropillars reduced cell migration rates for all the different topologies studied, and topologies with a higher density of micropillars showed higher reductions in cell migration rates. The maximum reduction in cell migration rates due to surface topology was approximately 17% for micropillars with center to center spacing of 110 microns arranged in a square pattern.

In the presence of an external magnetic field, application of periodic loads on cells from magnetic pillars caused approximately a 50% to 85% reduction in cell migration rates, compared to regions with non-actuated non-magnetic pillars (Figure 5(b) and 5(c)). Since both magnetic and non-magnetic pillars were subjected to the same external magnetic field, this reduction in cell migration rates among the magnetic pillars can be attributed to the periodic loads applied on the cells. Among the different patterns and directions of force application studied, magnetic pillars with center to center spacing of 110 μm, arranged in a square pattern and actuated in X-direction caused the largest reduction in cell migration rates (Figure 5(b)).

Interestingly, all regions with magnetic pillars showed lower cell migration rates when pillars were actuated in X-direction compared to when pillars were actuated in Y-direction, i.e., X-actuation was more effective in impeding cell migration in the X-direction. This behavior is consistent with previous reports from membrane stretching experiments where endothelial cells showed a tendency to migrate transverse to the direction of strain/force application [8].

Results from long-term experiments conducted over a twelve-day timespan were in good agreement with the results obtained from the one-day wound healing assays. Table 1 presents the 1-day and 12-day non-dimensional migration rates, i.e., the ratio of migration rate among magnetic pillars to the migration rate among non-magnetic pillars, for X-actuated and non-actuated magnetic pillars of square pattern and 110 μm spacing. At 12-days X-actuated magnetic pillars showed more than a 60% reduction in the migration rate of HUVECs (based on the measured non-dimensional migration rate of 0.38).

Table 1: 1-day vs. 12-day non-dimensional migration rates for X-actuated and non-actuated magnetic pillars of square pattern and 110 μm spacing.

Non-Dimensional Migration Rates	1-Day	12-Day
Control (No actuation)	1.10±0.06	1.25±0.03
X-actuation @ 1 Hz	0.14±0.05	0.38± 0.09

CONCLUSIONS

Application of 1 Hz cyclic loads on cells from magnetic micropillar arrays reduces cell migration rates and impedes cell migration. The highest reduction in cell migration rates occurs when force application is in the direction of cell migration. Since actuation of magnetic micropillars can be achieved remotely with the use of an external magnetic field, magnetic micropillar arrays present great possibilities for future development of smart materials that can be integrated with implants for in vivo tissue engineering applications.

ACKNOWLEDGEMENTS

Fabrication for this work was partly funded by CMC Microsystems. F. Khademolhosseini was partly funded by the NSERC Vanier Canada Graduate Scholarship and the Izaak Killam Doctoral Fellowship.

REFERENCES

[1] J. L. Tan, J. Tien, D. M. Pirone, D. S. Gray, K. Bhadriraju and C. S. Chen, "Cells lying on a bed of microneedles: an approach to isolate mechanical force," *Proc. Natl. Acad. Sci. U. S. A.,* vol. 100, no. 4, pp. 1484-1489, 2003.

[2] F. Zhang, S. Anderson, X. Zheng, E. Roberts, Y. Qiu, R. Liao and X. Zhang, "Cell force mapping using a double-sided micropillar array based on the moiré fringe method," *Appl. Phys. Lett.,* vol. 105, pp. 033702, 2014.

[3] M. Ghibaudo, J. Di Meglio, P. Hersen and B. Ladoux, "Mechanics of cell spreading within 3D-micropatterned environments," *Lab Chip,* vol. 11, pp. 805-812, 2011.

[4] N. J. Sniadecki, A. Anguelouch, M. T. Yang, C. M. Lamb, Z. Liu, S. B. Kirschner, Y. Liu, D. H. Reich and C. S. Chen, "Magnetic microposts as an approach to apply forces to living cells," *Proc. Natl. Acad. Sci.,* vol. 104, no. 37, pp. 14553-14558, 2007.

[5] J. le Digabel, N. Biais, J. Fresnais, J. F. Berret, P. Hersen and B. Ladoux, "Magnetic micropillars as a tool to govern substrate deformations," *Lab Chip,* vol. 11, pp. 2630-2636, 2011.

[6] F. Khademolhosseini and M. Chiao, "Fabrication and Patterning of Magnetic Polymer Micropillar Structures Using a Dry-Nanoparticle Embedding Technique," *J. Microelectromech. Syst.,* vol. 22, pp. 131-139, 2013.

[7] P. Vitorino and T. Meyer, "Modular control of endothelial sheet migration," *Genes Dev.,* vol. 22, pp. 3268-3281, 2008.

[8] Y. C. Yung, J. Chae, M. J. Buehler, C. P. Hunter and D. J. Mooney, "Cyclic tensile strain triggers a sequence of autocrine and paracrine signaling to regulate angiogenic sprouting in human vascular cells," *Proc. Natl. Acad. Sci.,* vol. 106, no. 36, pp. 15279-15284, 2009.

CONTACT

F. Khademolhosseini, tel: +1-604-822-4151; khadem@mail.ubc.ca

978-1-4799-7956-1/15 $31.00 © 2015 IEEE

CELL MOTILITY REGULATION ON STEPPED MICRO PILLAR ARRAY DEVICE (SMPAD) WITH DISCRETE STIFFNESS GRADIENT

Sujin Lee[1], Biswajit Saha[1,2], and Junghoon Lee[1]*

[1]School of Mechanical & Aerospace Engineering, Seoul National University, Seoul, South Korea
[2]BK21 PLUS Transformative Training Program for Creative Mechanical & Aerospace Engineers, Seoul National University, Seoul, South Korea

ABSTRACT

Here we have shown a new variation of the microfabricated pillar array detector (MPAD) that can decouple the stiffness gradient from the focal adhesion area of a cell. This goal is achieved through the use of "stepped" micro pillar array device (SMPAD) whose top contact area with a cell is kept constant while the diameter of pillar bodies vary for variable mechanical stiffness. We have observed manipulating cell behavior using this simple, artificial platform that produces a pure physical stimulus. This report includes a new discovery of gradient-dependent cell motility enhancement as well as the "classical" demonstration of durotaxis on the SMPAD.

INTRODUCTION

Cell migration guided by stiffness gradient

The research outcomes of directional cell migration have direct implication on understanding many physiological processes such as morphogenesis [2], immune response, and wound healing [3]. It is well known that cell movements can be guided by gradients of various chemical [4,5], mechanical and topological signal [6]. These findings are critical in many biomedical applications including implantable medical devices, biomaterials, and tissue engineering [7].

Cell movement involves a number of related events, such as the protrusion of pseudopodia, the formation of new adhesions, the development of traction, and the release of old adhesions [8]. To achieve appropriate physiological outcomes, cell movement must maintain a distinct direction and speed in response to environment stimuli. Cell migration, which is controlled by the gradients of dissolved or substrate attached chemicals (chemotaxis and haptotaxis, respectively) has been investigated for many years. In addition, cells are known to orient and migrate in response to gradients of light intensity, electrostatic potential, and gravitational potential [9,10].

Cell migration guided by substrate rigidity is known as durotaxis. Durotaxis have been studied using soft and stiff 2D substrates with several materials *in vitro* [1,3,6,7]. These prior works utilized numerous methods to create polymer based substrates with varying rigidity. However, it was, challenging to maintain an accurate control over the arrangement of substrate stiffness such as the amount and direction of gradient.

Micro pillar array

Micro-scale polymeric pillar arrays were normally used as cell force detectors [11,12]. This polymeric micro pillar array has been recently used to study cell movement and shape with various methods including the gradient of rigidity provided by varying pillar diameters [13]. In this case, the adhesion area of extra cellular matrix (ECM) also varied on different pillars because of the change in diameters. The variation of the adhesion area is one of the essential cues associated with the formation of focal adhesion that may affect cell migration and proliferation [1,14], thus needs to be decoupled from the effect of pure rigidity gradient. One extreme example would be the different modes of growth and movement of cells on patterned ECM vs. a flat substrate with a simple ECM coating [15].

CONCEPT

We present a stepped micro pillar array device (SMPAD) that offers the rigidity gradient in a discrete fashion without varying the top adhesion area. Figure 1 is the 3-D modeling view of a stepped micropillar array device. The inset of Figure 1 shows deformation of the pillars due to the attachment of focal adhesion and the force applied by the cell.

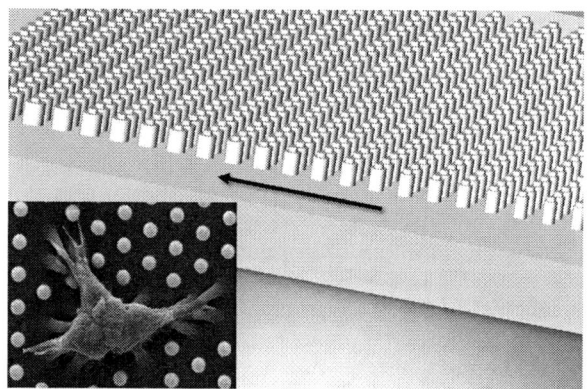

Figure 1: 3-D view of stepped micro pillar array device. The black arrow represents the direction of increased stiffness gradient.

The bottom layer of micro pillars have varying diameters (3 μm ~ 7 μm) while the diameter of the top layer is maintained identical (3 μm) by the double layers structure. Spring constant of the pillar is calculated by equation (1) and (2)

$$k = \frac{3\pi E}{4 L^3} r^4 \qquad (1)$$

$$\frac{1}{k_{pi}} = \frac{1}{k_a} + \frac{1}{k_b} \qquad (2)$$

,where k_{pi} is the spring constant of a single pillar, k_a and k_b are the spring constants of top and bottom portions respectively. The variations in bottom layer diameters from 3 μm to 7 μm with a total height of 7 μm correspond to the stiffness variations from 0.067 to 0.76 μN/μm.

FABRICATION

Figure 2 shows the fabrication process to implement this design via "stepped" molding. Fibronectin was "stamped" on the top area as an extracellular matrix that facilitates biological contacts with cells. The fabrication of the SMPAD begins with a double-step patterning of SU-8 photoresist (MicroChem, USA). After developing and hard baking the first patterned SU-8 layer with the thickness of 2 μm, additional layer was coated and patterned again as a second mold layer with the thickness of 5 μm. Degassed polydimethylsiloxane PDMS (Dow Corning, USA) was poured and cured at 60 °C overnight. The PDMS substrate was peeled off from the mold to complete the device. Oxygen plasma was used to render the surface hydrophilic for enhancing the adhesion of the ECM (Sigma Aldrich, USA). The ECM was placed on the top pillar surface by stamping on a flat PDMS substrate with ECM coating [11]. It should be emphasized that the size and arrangement of the stamped ECM (FITC labeled), thus the adhesion area and the pattern, was kept uniform while the size of the pillar body varies (Figure 2(f)). The remaining area of the pillar and the bottom substrate were blocked by 2% Pluronic F127 (BASF, USA) to prevent the interference of cell behavior due to unnecessary attachment. Some non-specific adhesion of cells occurred on the pluronic treated surface. The efficiency of the non-specific adhesion, however, was negligible especially when a single cell was used in the experiment.

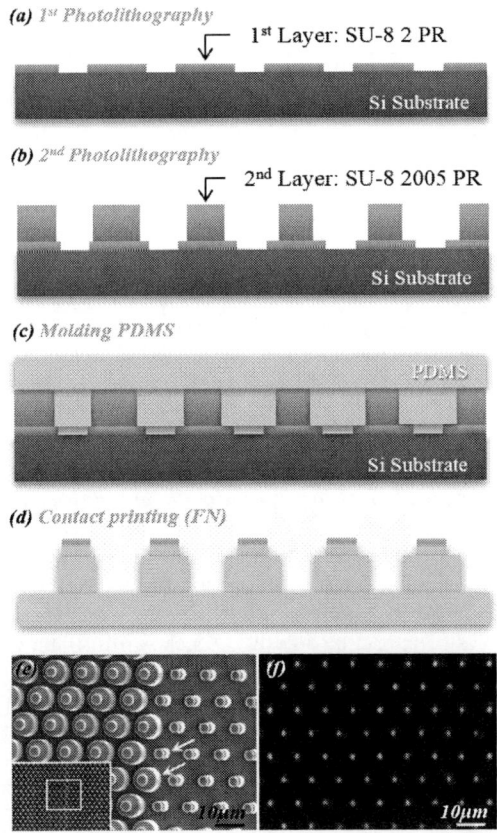

Figure 2: *Process flow of SMPAD substrate fabrication (a-d). The result of process is shown in SEM image (e). Fluorescence-labeled FN image of pillar top shows constant contact area (f).*

EXPERIMENT AND RESULTS

NIH3T3 cells were used for migration experiments. NIH3T3 cell is a mouse fibroblast with high motility, frequently used for such experiments [16,17]. A live imaging instrument (Nikon, Japan) was used to monitor the trajectory of migration. In order to minimize cell-to-cell communication, small number of cells (1×10^3 cells/mL) was plated sparsely on the SMPAD contained in a culture dish. Then we focused on the migration of an individual cell on the array.

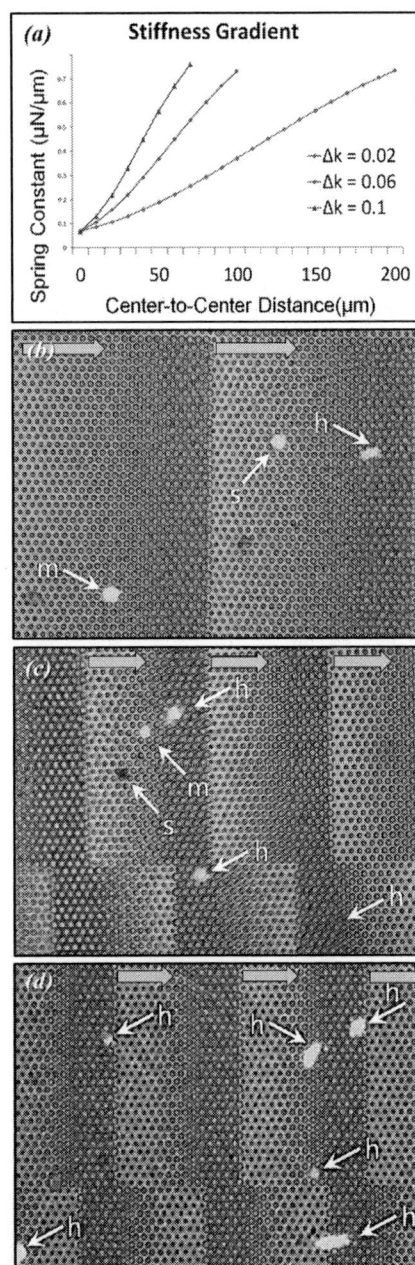

Figure 3: *Values for stiffness gradient of three type of platforms (a), Cell test results on platforms with Δk = 0.02 (b), 0.06 (c), and 0.1 (d) μN/μm. Cells randomly distributed initially, ended up with the vicinity of indicated locations after 18 hours of migration. White arrows indicate the stiffness of the final locations. (h: hard, m: medium, s: soft area).*

In a control experiment a stepped micro pillar array without variation in diameter was tested. It was found that the cells migrated randomly in this case. Figure 3, in contrast, indicates that the cell on the SMPAD substrate migrates toward the direction of stiffer pillars. Figures 3 (b), (c), and (d) show the locations of the enhanced green fluorescence protein (eGFP) expressed NIH3T3 cells 60 hours after launching the experiment. When the rigidity gradient was 0.02 µN/µm, the cells wandered almost randomly without a specific orientation. The cells started showing directional migration toward higher stiffness region when the gradient became 0.06 µN/µm. This stiffness guided migration was highly repeatable and consistent on the pillar array with the stiffness gradient of 0.1 µN/µm. The final locations of cells in Figures 3 (b), (c), and (d) manifests these behavior. It was hard to design the gradient higher than 0.1 µN/µm because the inter-pillar distance became too small to be fabricated.

This stiffness-guided migration was more consistent on the pillar array with the stiffness gradient of 0.1 µN/µm. Figure 4 shows that the motility of cells depending on the stiffness gradient. After 6 hours, cells moved much faster on the pillar array with a stiffness gradient than on controls (Figure 4 (a), n=15). This motility enhancement became more pronounced on the larger stiffness gradient as illustrated by Figure 4 (b). This is an important finding that verifies some physiological behaviors in-vivo [6], possibly leading to the design of effective scaffolds.

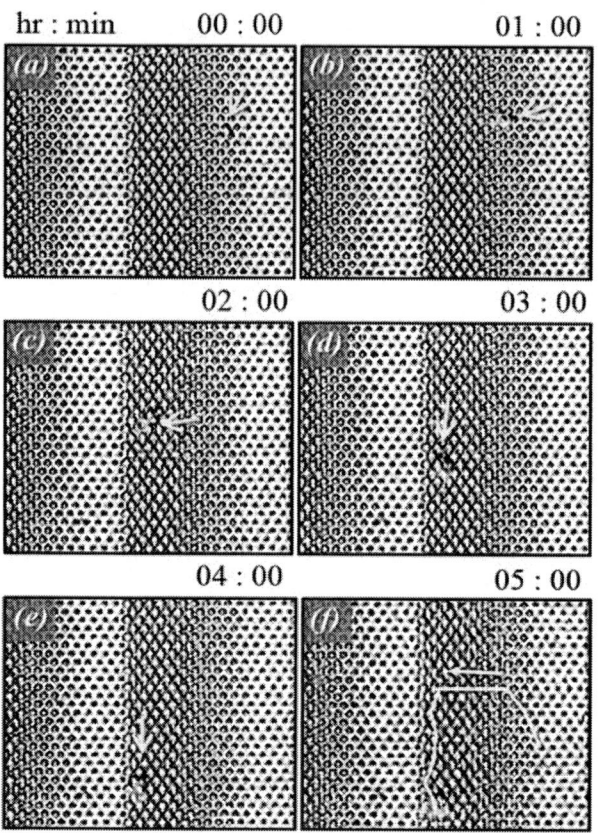

Figure 5: The cell initially moves to the direction of stiffer pillars (Δk = 0.1) (a). When the cell reaches the edge of the stiff region (c), it changes the direction of motion, moving along the boundary of the stiff region (d)-(f). The trajectory of the cell is marked by the yellow arrow (f). The cell migrates for about 180 µm within 5 hours in this case.

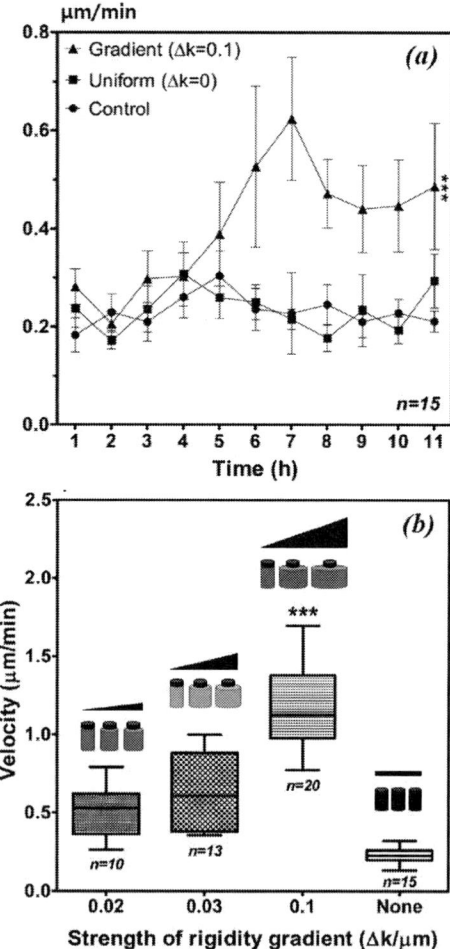

Figure 4: Motility of cells depending on the stiffness gradient. Cells moved faster on the pillar array with stiffness gradient after 6 hours of observation (n=15) (a). Average velocity of each cell is traced for 24 hours (n~20) and error bar represents V_{max} and V_{min} of each platform (b). It is clearly shown that the strength of stiffness gradient highly affects cell motility.

Due to design constraint several rows of pillars with the same diameter were arranged at the end of the rigidity gradient. The behavior of the cell in this constant rigidity region is of particular interest. As shown in Figure 5, the cell in the constant stiffness area may change the direction. However, when the cell approached the edge of the high stiffness area, its protrusion of pseudopodia started touching the soft area as a way of probing, and the cell instantly turned back and migrated parallel to the boundary, staying in the original area. We believe this interesting behavior is attributed to the step rigidity gradient along the boundary.

CONCLUSION

In conclusion, cell migration was demonstrated along the discrete rigidity gradient created by pillars with varying diameters. By maintaining the top area of the pillars constant we could eliminate the concern for the size effect of adhesion area. Double step fabrication technique of the double-layered pillar enabled this special configuration. The SMPAD introduced can be further used to study the cell migration for many other biological applications such as artificial tissue engineering, plastic surgery, and studying the migrational behavior of cancerous cells.

ACKNOWLEDGEMENTS

This work was supported by the ICT & Future Planning as Global Frontier Project (CISS-2012M3A6A 6054193). And the fabrication was performed at the Interuniversity Semiconductor Research Center (ISRC) in Seoul National University.

REFERENCES

[1] M. Allioux-Guérin, D. Icard-Arcizet, C. Durieux, S. Hénon, F. Gallet, J. C. Mevel, M. J. Masse, M. Tramier, M. Coppey-Moisan, "Spatiotemporal Analysis of Cell Response to a Rigidity Gradient: A Quantitative Study Using Multiple Optical Tweezers", *Biophys. J.*, vol. 96(1), pp. 238-247, 2009.

[2] R. L. Juliano and S. Haskill, "Signal transduction from the extracellular matrix", *J. Cell Biol.*, vol. 120, pp. 577-585, 1993.

[3] P. Martin, "Wound healing - aiming for perfect skin regeneration", *Science*, vol. 276, pp. 75-81, 1997.

[4] R. B. Dickinson, S. Guido, and R. T. Tranquillo, "Biased cell migration of fibroblasts exhibiting contact guidance in oriented collagen gels", *Ann Biomed Eng.*, vol. 22(4), pp. 342-56, 1994.

[5] N. L. Jeon, S. K. Dertinger, D. T. Chui, I. S. Choi, A. D. Stroock, and G. M. Whitesides, "Generation of solution and surface gradients using microfluidic systems" *Langmuir*, vol. 16, pp. 8311-16, 2000.

[6] P. Clark, P. Connolly, A. S. G. Curtis, J. A. T. Dow, and C. D. W. Wilkinson, "Cell guidance by ultrafine topography in vitro", *J Cell Sci.*, vol. 99, pp. 73-77, 1991.

[7] C. M. Lo, H. B. Wang, M. Dembo, and Y. l. Wang, "Cell Movement Is Guided by the Rigidity of the Substrate", *J. Biophy.*, vol. 79, pp. 144-152, 2000.

[8] D. A. Lauffenburger, and A. F. Horwitz, "Cell migration: a physically integrated molecular process", *Cell*, vol. 84(3), pp. 359-369, 1996.

[9] J. Saranak, and K. W. Foster, "Rhodopsin guides fungal phototaxis", *Nature*, vol. 387, pp. 465-466, 1997.

[10] C. A. Erickson, and R. Nuccitelli, "Embryonic cell motility can be guided by weak electric fields", *J. Cell Biol.*, vol. 95, pp. 314a, 1982.

[11] N. Q. Balaban, "Force and adhesion assembly: a close relationship between studies using elastic micropatterned substrates", *Nat. Cell Biol.*, vol. 3, pp. 466-473, 2001.

[12] J. L. Tan, J. Tien, D. M. Pirone, D. S. Gray, K. Bhadriraju, and C. S. Chen, "Cells lying on a bed of microneedles: An approach to isolate mechanical force", *Proceedings of the National Academy of Science*, vol. 100 (4), pp. 1484-1489, 2003.

[13] R. Sochol, and L. Lin, "Microscale Control of Micropost Stiffness to Induce Cellular Durotaxis", *Twelfth International Conference on Miniaturized Systems for Chemistry and Life Sciences*, pp. 1335-1337, 2008.

[14] M. A. Wozniak, K. Modzelewska, L. Kwong, P. J. Keely, "Focal adhesion regulation of cell behavior", *Biochim Biophys Acta.*, vol. 1692(2-3), pp. 103-119, 2004.

[15] D. A. Rubenstein and M. D. Frame, "Micro-stamped ECM proteins enhance endothelial cell adhesion and directed growth", *The FASEB Journal*, vol. 21, pp. 897.1, 2007.

[16] T. J. Bos, P. Margiotta, L. Bush, and W. Wasilenko, "Enhanced cell motility and invasion of chicken embryo fibroblasts in response to JUN over-expression", *Int. J. Cancer*, vol. 81, pp. 404-410, 1999.

CONTACT

*J. Lee, tel: +82-2-880-9104; jleenano@snu.ac.kr

FIBERED REFLECTIVE MICRO OBJECTIVES FOR MINIATURIZED SCANNING CONFOCAL FLUORESCENCE MICROSCOPY

Shenqi Xie[1,], Etienne Shaffer[1,2,*], Loïc Jacot-Descombes[1], Diego Joss[1], Bastien Rachet[1,2], Davor Kosanic[2], and Jürgen Brugger[1,*]*

[1]Microsystems Laboratory, EPFL, 1015 Lausanne, SWITZERLAND
[2]SamanTree Technologies AG, Zurich, SWITZERLAND

ABSTRACT

This paper reports on the design, fabrication and characterization of a novel micro-optical system for imaging that is based on a miniaturized reflective objective. The fabrication process innovatively combines two additive micro-fabrication techniques: inkjet printing, to create the smooth spherical mirror shape, and stencil lithography, for local metal deposition. The process is highly flexible and thus allows the fabrication of reflective micro-objectives of different optical properties, which can be tailored for targeted applications. The fabricated micro-objectives presented here are designed for endoscopic microscopy and exhibit a spot size of 1 micrometer, which represents a two-fold improvement over commercial spherical microlenses of comparable size.

INTRODUCTION

Confocal fluorescence microscopy is an optical imaging technique for obtaining depth-selective information in thick biological tissue samples [1]. Recent development in micro-optical components has enabled the miniaturization of such systems and the emergence of new fields of application like endoscopic microscopy for in-*vivo* diagnosis inside of a patient, a growing trend in the medical field [2]. However, when miniaturizing optical system, refractive and diffractive micro-optical components face limitations for demanding optical performances in high-end application of fluorescence microscopy [3]. Reflective optics on the other hand can overcome these limitations, yet their miniaturization introduces several manufacturing challenges. It is in particular extremely difficult to fabricate rounded metallic mirror surfaces by MEMS processes that have well-controlled overall shape and radius of curvature (ROC) and that are also smooth for optical reflection.

In this paper, we present novel fabrication steps that allow us to overcome these challenges, by combining two additive micro-fabrication methods, inkjet printing (IJP) and stencil lithography (SL). Drop-on-demand IJP is a very promising technique to fabricate polymeric microlenses [4, 5]. With its ability to produce well-controlled spherical cap shapes [6], IJP offers a high flexibility and thus stands out for rapid and cost-efficient prototyping. SL is another versatile nonconventional micro patterning technique, with the unique advantage that it allows for direct and local metal deposition on high topographical substrates and 3D structures [7]. By combining IJP and SL, and benefiting from their respective advantages, we fabricated novel micro-optical components: reflective micro objectives (RMOs). Furthermore, the flexibility inherent to these techniques makes possible the fabrication of RMOs of different optical designs and properties.

DESCRIPTION OF NEW DEVICE

RMOs consist of two spherical mirrors with coincident optical axes (Figure 1), in a design similar to Schwarzschild objectives [8]. In the proposed point-scanning confocal imaging configuration, the laser light emerging from the core of an optical fiber enters the RMO by the aperture located in the center of the large spherical mirror. Light is reflected off both mirrors and focused on the sample. An image is formed by scanning the focus with respect to the sample with an appropriate scanning system, while collecting light emitted from the sample through the same optical path. In this configuration, the fiber core acts as a confocal pinhole.

When miniaturizing optical elements, spherical shapes are easier to produce, since they more readily occur in nature as the result of physical phenomena like surface tension to name only one. When considering optical elements of planar and spherical shapes only, RMOs offer several advantages over microlenses. For one, the two reflective surfaces provide more degrees of freedom to design (aberration-) corrected optical systems, a requirement for achieving the high resolution and sensitivity needed for fluorescence microscopy. Also, the folded light path makes RMO a more compact and effective solution for use in conjugation with the optical fibers needed in miniaturized distal devices. Finally, as they are based on reflection rather than refraction, RMOs are more suitable for achromatic systems. These advantages, and their small size, make RMO very promising for endoscopic microscopy.

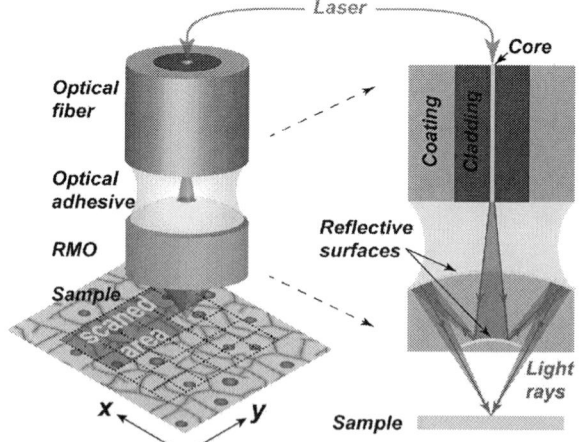

Figure 1: 3D illustration (left) of the imaging principle using a single fibered RMO and close-up cross-sectional view (right) of the working principle. The RMO consists of two spherical domes with metallic coating as reflective surfaces.

DEVICE FABRICATION
Micro-fabrication of RMO

The RMO is created by combining conventional lithography with IJP and SL to sequentially fabricate the two spherical micro-mirrors, as shown in the schematic process flow in Figure 2. The fabrication starts with the coating of 300 nm dextran acting as a water-soluble sacrificial layer [9] for later releasing the RMO from its substrate. A 5 μm thick inner SU-8 platform with 100 μm in diameter is then structured by conventional lithography (Figure 2a) and spatially confines the SU-8 printed by IJP within its edges [10] (Figure 2b). The surface tension of the liquid SU-8 creates a smooth surface, which ensures a perfect spherical cap shape. The local metallization on the dome is achieved by aligning a stencil aperture with the platform during Al deposition in an E-beam evaporator (Figure 2c). A metal film thickness of 200 nm has been selected in order to ensure an excellent reflectivity. The fabrication of the second large spherical mirror (with a platform of 200 μm in diameter and 175 μm in height) follows the same sequence as the small one, except that for the reflective coating on the large dome, two complementary stencils are successively used to create the reflective metal film (400 nm Al) with the isolated aperture in the center of the dome, as illustrated in Figure 2d, 2e and 2f. Figure 3 shows an artificially colored SEM image of a RMO in an array after fabrication.

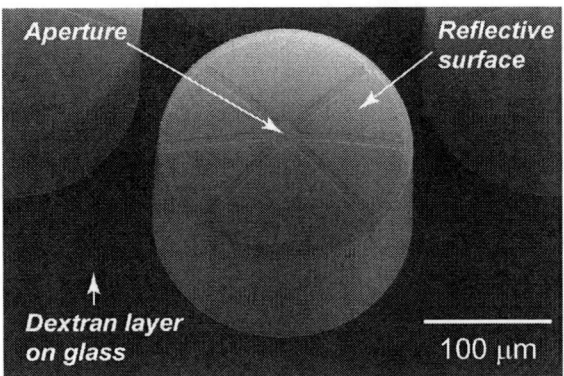

Figure 3: Colorized SEM image of a RMO in an array after fabrication. Aluminum reflective coating is colorized in yellow and dextran coating is colorized in blue.

IJP for the spherical dome

The IJP system (MicroFab Technologies Inc., USA) allows aligning between the nozzle and platforms. First, an offset correction in XY and θ is performed to calibrate the alignment, which can be measured and visualized by a top view camera. The epoxy-based polymeric ink (InkEpo, Micro Resist Technology GmbH, Germany) is then locally deposited onto each platform, with a lateral precision of 5 μm. Once the platforms are covered with the precisely controlled pre-calculated volume, a soft bake at 60 °C for 30 min is applied to evaporate the solvents, following by a flood exposure to UV-light and post-exposure bake (PEB) at 90 °C for 1 hour. The printed epoxy-based polymer is then completely polymerized and ready for the following process steps. In order to adapt the printing process to the platform sizes and printed volume, the IJP nozzles with diameter of either 30 μm or 50 μm are selected.

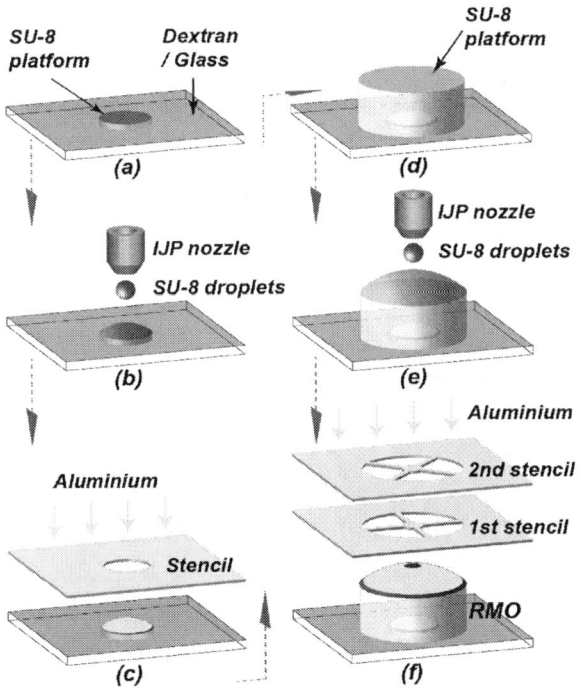

Figure 2: Schematic process flow. (a) SU-8 lithography to create the first inner platform. (b) IJP to form the spherical cap shape on the inner platform. (c) Aligned stencil to locally metallize the small reflector. (d) SU-8 lithography to create the second outer platform. (e) IJP to form the special cap shape on the outer platform. (f) Successive metallization with two complimentary stencils aligned with RMO creates the reflective surface for the outer mirror.

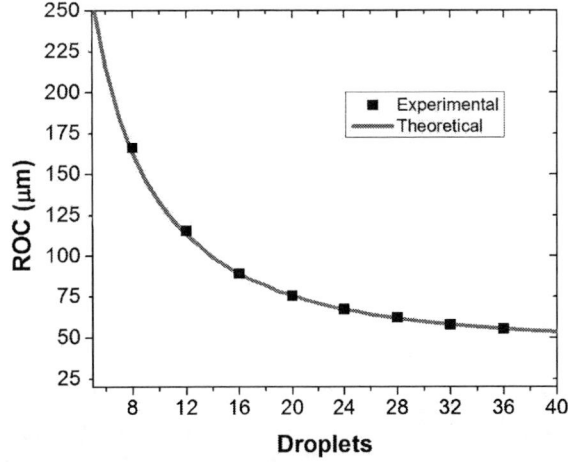

Figure 4: A representative example of the relationship between ROC of printed microlenses and the number of droplets in IJP. An IJP nozzle with an orifice of 30 μm in diameter was used. The diameter of the confinement platforms is 100 μm.

Precision and accuracy on the fabrication of the spherical caps of the reflective surfaces is essential for the fabricated RMO to achieve the designed optical properties. Figure 4 plots an example of the radius of curvature (ROC) of the printed spherical dome versus the number of droplets in IJP. The experimental data agrees very well with the theoretical expectation, which indicates the capability of controlling ROC with droplet counts. Furthermore, fine-tuning the drop generation parameters in IJP, printed volume can be finely tuned, granting access ROC values located within the discrete steps between two drop counts in Figure 4. The wide range of spherical shapes accessible by IJP of optical epoxy-based ink on confinement platforms [6] and the fine control we have reached allows us fabricating RMO of different optical designs and properties, tailored to desired optical functionality and application.

SL for the reflective surface

A unique advantage of SL is to locally create the reflective surfaces on top of high topography structure, such as the RMO spherical domes. The stencil is made of low stress LPCVD SiN with 500 nm in thickness [11]. Having an alignment accuracy of down to 2 μm, the stencil is able to precisely align with the optical axis of the spherical dome, hence guaranteeing a good optical performance in comparison with the optical design.

The SiN stencil is robust, yet the membrane could be damaged due to the deformation induced from the contact with the spherical dome. To extend the life-time of the stencils, spacers were fabricated on the substrate to maintain a small distance (around 50 μm) between the stencils and the spherical dome. The Al deposited (Leybold Optics LAB600H) on the stencil membrane can be removed by selective wet etching after each deposition [12], thus allowing us to reuse the stencils in a cost efficient way.

Fibering RMO

The RMO is assembled to an optical fiber after its fabrication. In this crucial assembly step, the core of the optical fiber has to be precisely aligned with the aperture on top of the RMO and then securely connected in place, using a UV-curable optical adhesive for instance. After dissolving the dextran layer in water, the RMO is finally released from the substrate and remains assembled to the optical fiber, as shown in Figure 5.

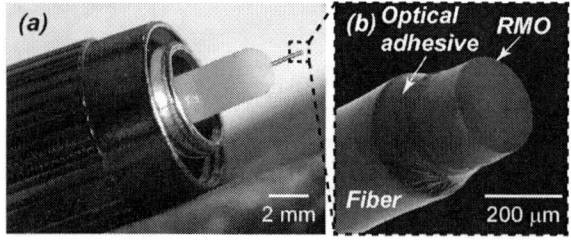

Figure 5: (a) RMO attached to an optical fiber emerging from a FC/PC connector and (b) close-up colorized SEM image. Optical adhesive is colorized in blue.

Diffusion of water into SU-8 microstructure is a known issue [13] and could become problematic for the metal/SU-8 interface when it is immersed in water for few hours [14]. Keeping immersion time sufficiently short is critical to preserve the reflective surfaces. In our case, thanks to the small bottom contact area of the RMO, releasing the fibered RMO in water takes less than five minutes. Hence the metal layers remain intact after releasing.

RESULTS
Characterization of RMO

The RMO footprint presented here is very small, which is only 200 μm some 50 μm less than the coating diameter of a typical single mode fiber. Additionally, its small height (<200 μm) makes it an extremely compact optical system for focusing light emerging from an optical fiber.

Moreover, we have characterized the spot size of fibered RMO using a commercial digital microscope equipped with a 50X 0.80NA objective designed for operation in air and not corrected for glass coverslips. Accordingly, spot size measurement was performed in air environment, even though the RMO was designed to work in an immersion medium of higher refractive index. For these measurements, a broadband white light source was coupled in the single-mode fiber (NuFern, 780-HP) to which the RMO had been connected. We have measured the spot full width at half maximum (FWHM) to be 1 micrometer (Figure 6), more than two times better than off-the-shelf commercial spherical microlenses of comparable footprint.

Together, compactness and reasonable resolution are highly desirable for endoscopic microscopy. Of course, other RMO designs with smaller focus or longer working distance can also be fabricated with the proposed process.

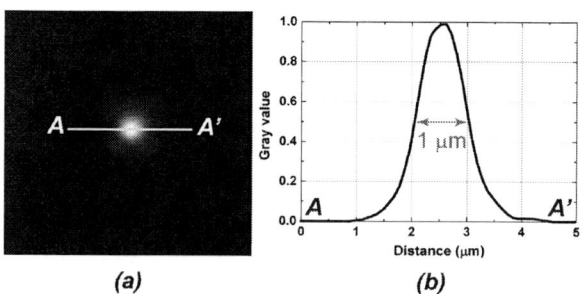

Figure 6: (a) Focus spot by RMO and (b) cross-sectional view to show the spot size. The spot size, defined by its full-width at half maximum (FWHM), is 1 μm.

Imaging using RMO

Finally, proof-of-principle images of fluorescent samples with a fibered RMO are presented. The wavelength of the incident laser used is 488 nm (JDSU Diode-pumped Solid State Laser). Figure 7a illustrates a high resolution imaging of monolayer fluorescent polystyrene microspheres with 1.34 μm in diameter (Fluoresbrite®, Polysciences Inc.), using a scanning pitch of 100 nm per pixel. Using the same RMO, figure 7b shows lithographically defined test patterns in SU-8 doped with fluorescein sodium salt (Fluka, SIGMA-ALDRICH), this time with a larger field of view and a larger scanning pitch of 5 μm per pixel. The image allows visualizing very fine details of the test patterns in different pitches, especially the failed patterns with smallest feature size on the left column. The imaging acquisition tests demonstrate the flexibility of the system for targeting either high resolution or large field of view. Our current effort focuses on the implementation of the new RMO systems for the imaging of biological samples.

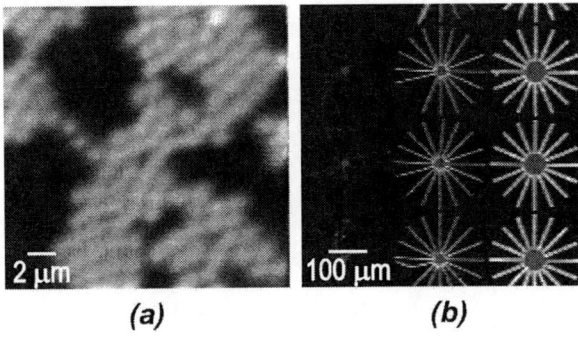

Figure 7: Fluorescent images acquired by a fibered RMO with different laser wavelength. (a) Fluorescent microspheres (Scanning pitch: 100 nm per pixel; Field of view: 20 μm x 20 μm) and (b) Lithographied patterns in - fluorescent photoresist (Scanning pitch: 5 μm per pixel; Field of view: 750 μm x 750 μm).

CONCLUSION

We have demonstrated in this paper a novel micro-optical system based on a reflective micro objective (RMO) for miniaturized scanning confocal fluorescence microscopy which targets bio-imaging application. The fabricated micro-objectives show spot size of 1 micrometer, a 2x improvement over commercial spherical microlenses of comparable size. Fluorescent images were successfully acquired on testing structures with different scanning pitches.

The novel fabrication process for RMO, combing two additive micro-fabrication techniques, inkjet printing (IJP) and stencil lithography (SL), opens new possibilities to fabricate other reflective micro-optical components for optical MEMS application.

ACKNOWLEDGEMENT

The authors would like to acknowledge the financial support from the Swiss Commission for Technology and Innovation (CTI) under grant No.15434.2.

REFERENCES

[1] J. G. White, W. B. Amos, M. Fordham, "An evaluation of confocal versus conventional imaging of biological structures by fluorescence light microscopy", *The Journal of cell biology*, Volume 105, 1987 41-48.

[2] L. Thiberville, S. Moreno-Swirc, T. Vercauteren, E. Peltier, C. Cave, G. B. Heckly, "In Vivo Imaging of the Bronchial Wall Microstructure Using Fibered Confocal Fluorescence Microscopy", *American journal of respiratory and critical care medicine*, Vol 175, 2007.

[3] B. A. Flusberg, E. D. Cocker, W. Piyawattanametha, J. C. Jung, E. L. M. Cheung, M. J. Schnitzer, "Fiber-optic fluorescence imaging", *Nature Methods*, Vol 2, 941 – 950 (2005).

[4] A. Voigt, U. Ostrzinski, K. Pfeiffer, J.Y. Kim, V. Fakhfouri, J. Brugger, G. Gruetzner, "New inks for the direct drop-on-demand fabrication of polymer lenses", *Microelectron. Eng.* 88 (2011) 2174–2179.

[5] V. J. Cadarso, J. Perera-Núñez, L. Jacot-Descombes, K. Pfeiffer, U. Ostrzinski, A. Voigt, A. Llobera, G. Grützer, and J. Brugger, "Microlenses with defined contour shapes", *Optics Express*, Vol. 19, Issue 19, pp. 18665-18670 (2011).

[6] L. Jacot-Descombes, M. R Gullo, V. J Cadarso and J. Brugger, "Fabrication of epoxy spherical microstructures by controlled drop-on-demand inkjet printing", *J. Micromech. Microeng.*, 22 (2012) 074012.

[7] G. Villanueva, O. Vazquez-Mena, C. Martin-Olmos, V. Savu, K. Sidler, J. Montserrat, P. Langlet, C. Hibert, P. Vettiger, J. Bausells and J. Brugger, "All-stencil transistor fabrication on 3D silicon substrates", *J. Micromech. Microeng.*, 22 (2012) 095022.

[8] Igor A. Artyukov, "Schwarzschild objective and similar two-mirror systems", *Proc. of SPIE* Vol. 8678 86780A-1.

[9] V. Linder, B. D. Gates, D. Ryan, B. A. Parviz, and G. M. Whitesides, "Water-Soluble Sacrificial Layers for Surface Micromachining", *Small*, 1 (2005) 730-736.

[10] D. Quéré, "Non-sticking drops", *Reports on Progress in Physics*, Volume 68, Issue 11, pp. 2495-2532 (2005).

[11] M. A. F. van den Boogaart, G. M. Kim, R. Pellens, J.-P. van den Heuvel and J. Brugger, "Deep-ultraviolet–microelectromechanical systems stencils for high-throughput resistless patterning of mesoscopic structures", *J. Vac. Sci. Technol. B* 22, 3174 (2004).

[12] O. Vázquez-Mena, G. Villanueva, M.A.F. van den Boogaart, V. Savu and J. Brugger, "Reusability of nanostencils for the patterning of Aluminum nanostructures by selective wet etching", *Microelectronic Engineering* 85 (2008) 1237–1240.

[13] C. Liu, Y. Liu, M. Sokuler, D. Fell, S. Keller, A. Boisen, H. Butt, G. K. Auernhammer and E. Bonaccurso, "Diffusion of water into SU-8 microcantilevers", *Phys. Chem. Chem. Phys.*, 2010, 12, 10577–10583.

[14] G. C. Hill, R. Melamud, F. E. Declercq, A. A. Davenport, I. H. Chan, P. G. Hartwell, B. L. Pruitt, "SU-8 MEMS Fabry-Perot pressure sensor", *Sensors and Actuators A*, 138 (2007) 52–62.

CONTACT

*S. Xie, shenqi.xie@epfl.ch
*E. Shaffer, etienne@samantree.com
*J. Brugger, juergen.brugger@epfl.ch

FLEXIBLE SILICON-POLYMER NEURAL PROBE RIGIDIFIED BY DISSOLVABLE INSERTION VEHICLE FOR HIGH-RESOLUTION NEURAL RECORDING WITH IMPROVED DURATION

Falk Barz[1,2,*], Patrick Ruther[1], Shoji Takeuchi[2] and Oliver Paul[1]

[1]Department of Microsystems Engineering – IMTEK, University of Freiburg, Freiburg, GERMANY
[2]Institute of Industrial Science, The University of Tokyo, Tokyo, JAPAN

ABSTRACT

This study reports on a novel concept for silicon (Si)-based intracortical neural probes with improved mechanical flexibility and reduced dimensions. Microelectrode arrays with cross-sections as small as 50×120 µm2 and polyimide ribbon cables of similar width but a thickness of only 11 µm are assembled into slender, flexible recording systems. As their interfacing section is reduced to the width of the probe shank, the Si-based electrode array can be completely implanted into brain tissue while the cable creates the interface to the external instrumentation. This hybrid probe concept allows the reduction of the implanted volume and probe stiffness by factors of ca. 5 and 1400, respectively, compared to conventional Si probes. As a consequence, the stability of neural recording is expected to be increased. For probe implantation, the probe is stiffened by a bio-dissolvable polymer mold around the probe using the centrifuge-based molding of polyethylene glycol.

INTRODUCTION

The longevity of the electrode/tissue interfaces of neural probes is a highly relevant topic in the development of high-resolution brain machine interfaces. Chronically implantable devices may serve for the treatment of neurological disorders, the restoration of sensory or motor functions, and as surrogates for pharmaceutical therapies. The stability of electrical extracellular recording is limited by the lifetime of the neural probe itself and the foreign body response of the brain tissue. Neuronal cell death and glia formation around the recording sites increase the distance to active neurons and thus reduce the achievable signal-to-noise ratio. Recording stability strongly depends on the mechanical properties and geometry of the probe. Cortical implants of smaller volume floating in the tissue cause less mechanical trauma and elicit notably less pronounced host reaction than larger probes fixed to the skull [1]. Furthermore, floating probes are also less prone to electrode shifts relative to the neural tissue affecting the recording quality.

Silicon (Si)-based neural probes fabricated by means of microsystems technologies are established tools in neuroscience [2]. At comparable or even smaller system dimensions, they offer larger numbers of recording channels than wire-based implants. The mismatch in mechanical properties between the Si substrate and the brain tissue is however disadvantageous.

Recently, the use of flexible substrate materials like polyimide (PI) and parylene instead of Si was demonstrated to improve the mechanical properties of the probes. The implantation of flexible probes is facilitated either by auxiliary tools like stiff shuttles [3,4] or by bio-dissolvable polymer-based coatings serving for temporary probe stiffening [5,6].

CMOS circuitry integrated on Si-based neural probes has also been shown to improve signal quality and stability by reconfiguring the location of recording site in-situ without taking physical influence on the location of the probe [7].

A concept for neural probes with improved mechanical and geometrical properties and compatible with CMOS integration was recently proposed [8]. The monolithic Si probe with flexible joints connecting slender silicon shafts to a broad probe base was rigidified for implantation by sucrose. Still, in this concept a rigid shaft projects into the tissue and through the brain meninges.

Our approach aims at a further reduction of stiffness and mechanical mismatch without sacrificing CMOS integration. Further, we aim for CMOS integration by hybrid assembly of CMOS components with the extended polymer ribbon cables necessary for interfacing chronic implants in deep brain areas.

PROBE CONCEPT

In contrast to existing probes {Fig. 1(a)}, our novel concept offers a pronounced reduction of the interface area between electrode array and the highly flexible ribbon cable. The slender shape allows the complete implantation of the electrode array into the brain tissue, as illustrated in Fig. 1(b). Replacing large sections of the Si probe shaft with a thickness of 50 µm by the 11-µm-thin PI cable results in a reduction of the implant volume by a factor of

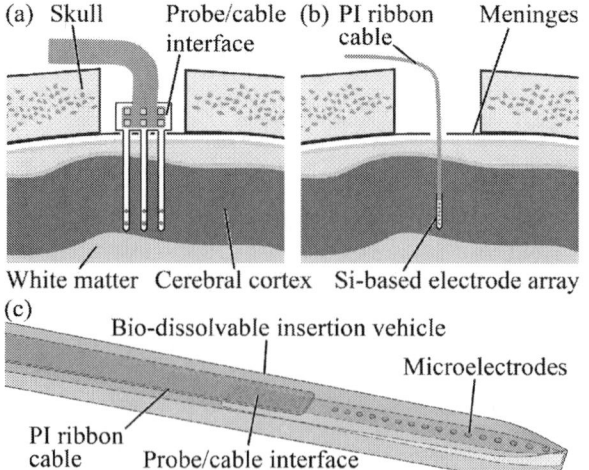

Figure 1: Schematic of (a) conventional and (b) slender, flexible hybrid probe in-situ, and (c) assembled hybrid probe with insertion vehicle.

978-1-4799-7956-1/15 $31.00 © 2015 IEEE   636   MEMS 2015, Estoril, PORTUGAL, 18 - 22 January, 2015

around 5. Switching from Si as a substrate material to the thin PI layer also decreases the bending stiffness defined as $Ebh^3/12$. Here, E, b and h denote the elastic modulus of the substrate and the probe width and thickness, respectively. Using representative elastic moduli for PI and silicon of 9 and 130 GPa, respectively [9,10] and the above-mentioned probe dimensions, the probe stiffness is reduced by a factor of about 1400. This will allow the Si-based electrode array to float deep in the brain while the flexible cable exerts minimal forces onto it.

However, the implantation of such a highly flexible probe is possible only by temporarily increasing its stiffness. In this study, we focused on the idea of a bio-dissolvable polyethylene glycol (PEG) cast embedding the probe and forming a needle-like insertion vehicle {Fig. 1(c)}. The rapid dissolution of the PEG coating in the neural tissue re-establishes the initial flexibility of the probe soon after implantation. The applied prototyping scheme relies on cost-effective transparency masks enabling the fast and flexible modification of the vertical and lateral mold dimensions for the PEG cast.

FABRICATION

Fabrication of probe components and probe assembly

Electrode arrays as shown in Fig. 2(a) are fabricated using an established MEMS process relying on deep reactive ion etching (DRIE) and wafer grinding [11]. As a result, 50-μm-thick electrode arrays are released from a silicon substrate. Arrays with minimal width and length of 120 μm and 1.3 mm, respectively, carrying 16 or 32 electrodes were realized. The PI ribbon cables (thickness ca. 11 μm, {Fig. 2(b)}) comprise a two metal layers sandwiched between three polyimide layers (U-Varnish S, UBE Industries Ltd., Tokyo, Japan). A similar width of the cables and their Si counterparts was achieved.

Test structures with comparable mechanical properties were used for process development of the insertion vehicle. Their metallization is patterned using transparency masks of lower resolution enabling the integration of only up to three interconnection lines terminating in platinum (Pt) electrodes with a diameter of 25 μm. Figure 2(c) compares a conventional 16-ch probe with such a test structure.

The probes are assembled using a flip-chip (FC) bonding process. A commercial FC bonder (Fineplacer 96λ,

Finetech, Berlin, Germany) with interposer structures fixed to the vacuum chucks of the ultrasonic (US) tool and bonder hotplate serves to assemble these delicate micro-components (Fig. 3). The interposers were fabricated from silicon substrates using two-sided DRIE. Bonding is performed at 100 °C at a force of 4 N or 8 N for the test structure with 16 or 32 bond pads, respectively, and a US power of 400 mW {Fig. 3(b)}. After FC bonding, the bond sites are mechanically reinforced by underfilling the interface with a low-viscosity, medical grade epoxy (EPO-TEK 301, Epoxy Technology Inc., Billerica, MA, USA).

Mold for insertion vehicle patterning

Mold fabrication is summarized in Fig. 4. It starts with the patterning of a 75-μm-thick dry resist (Ordyl Alpha 375, Elga Europe s.r.l, Milan, Italy) which is subsequently combined with 3D printed high-aspect-ratio (HAR) ABS structures adhesively bonded to the Si substrate. The resulting inverse mold shown in Fig. 6 is then transferred into silicone rubber (Elastosil M 4600, Wacker Chemie AG, Munich, Germany) using a casting process. The dry resist is laminated at 100°C onto 4" silicon substrates pre-treated with HMDS {Fig. 4(a)}. In case layers thicker than 75 μm are required, the lamination step is repeated up to four times resulting in a total thickness of ca. 220 μm. The exposure of the resists film is performed at 220 mJ/cm² per layer using transparency masks {Fig. 4(b)}. This is followed by the development at 33°C in an aqueous 0.9 wt% sodium carbonate solution (EMSURE ISO, Merck Chemicals GmbH, Schwalbach, Germany) {Fig. 4(c)}. The film development completed by a thorough rinsing in DI water. The inverse mold is finished by adhesively bonding the HAR structures onto the resist layers and the wafer using a high-viscosity epoxy (EPO-TEK 353 NDT, Epoxy Technology Inc., Billerica, MA, USA) {Fig. 4(d)}. The HAR structures define vias and reservoirs in the silicone rubber and an outer frame for the mold fabrication (cf. Figs. 4 and 5).

Silicone rubber is dispensed onto the inverse mold, degassed, and cured for 45 min at 70 °C. Subsequently, a second, thicker layer is added only to the center of the mold to create a larger central reservoir {Fig. 4 (e)}; it is cured for 1 hour at 70°C. The two-stepped casting allows a thin and highly transparent section of the mold to be created where

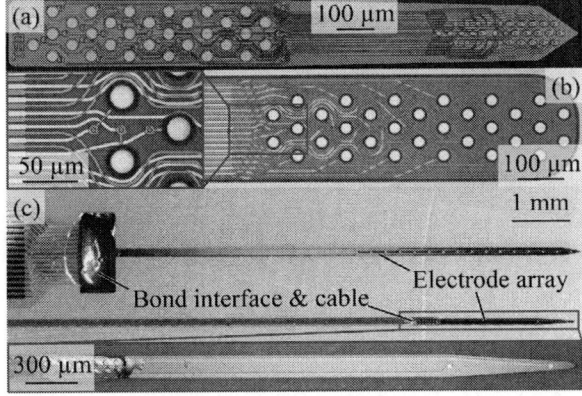

Figure 2: Micrographs of (a) electrode array, (b) PI cable and (c) comparison between conventional and slender probe.

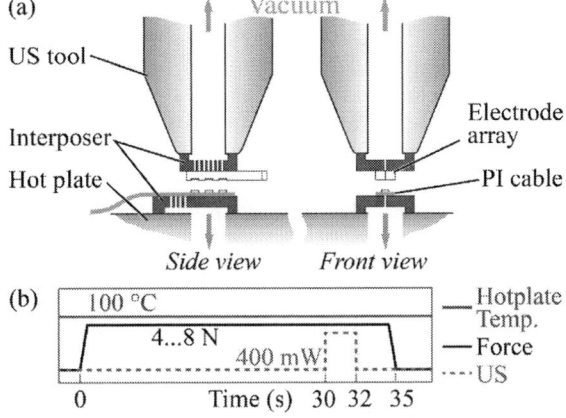

Figure 3: Schematic of (a) probe bonding using vacuum interposers for FC bonding and (b) process diagram.

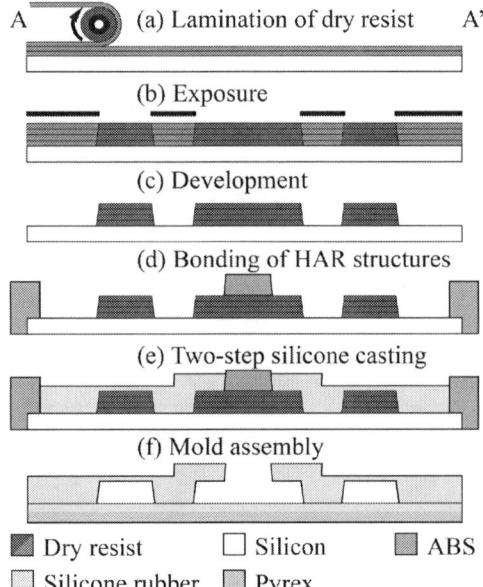

(a) Lamination of dry resist

(b) Exposure

(c) Development

(d) Bonding of HAR structures

(e) Two-step silicone casting

(f) Mold assembly

■ Dry resist   □ Silicon   ▨ ABS
□ Silicone rubber   ▨ Pyrex

Figure 4: Schematic process flow of mold fabrication; cross-sectional view along line A-A' in Fig.5.

the channels and cavities are located. This allows one later to optically examine the probes and the stiffening polymer in the assembled mold (Fig. 6). The silicon rubber is finally peeled off the inverse mold. To finish the mold, the open trenches in the silicone rubber are closed with a capping wafer thereby becoming buried channels and the mold cavities {Fig. 4(f)}.

Stiffening of highly-flexible neural probe

For probe stiffening, the hybrid probes are positioned in the respective trenches of the silicone mold prior to capping. Then, a Pyrex wafer coated with a layer of silicone is positioned on the mold to close the trenches. To achieve a temporary fixation between the cap wafer and the patterned rubber disc, the silicone layer is wetted by DI water followed by exposing the assembly to vacuum to remove bubbles trapped at the interface of the two components. After drying at ambient pressure, the bond between the two rubber components is strong enough for the PEG molding.

PEG-1500 (Merck Chemicals GmbH, Schwalbach, Germany) was used for the probe stiffening experiments. It is heated to 75 °C and dispensed with a syringe into the mold heated to 75 °C as well. Upon centrifuging on a spin coater at 2000 rpm the molten PEG spreads into the microchannels leading to the probe cavities as indicated in Fig. 6 These channels are designed in such a way that excessive material is discarded sideways into neighboring cavities. After approximately 3 min of centrifuging the PEG is solidified since the mold cools down to room temperature during spinning.

The quality of the molded structures was optically controlled in the thinner mold areas using a microscope. Whenever bubbles formed in the cavities, the mold was reheated and the spinning continued until the trapped air was displaced by the PEG and removed through the vias. Finally, the mold was carefully disassembled by peeling off the top part containing the channels. The stiffened probes can then be removed from the channels using tweezers.

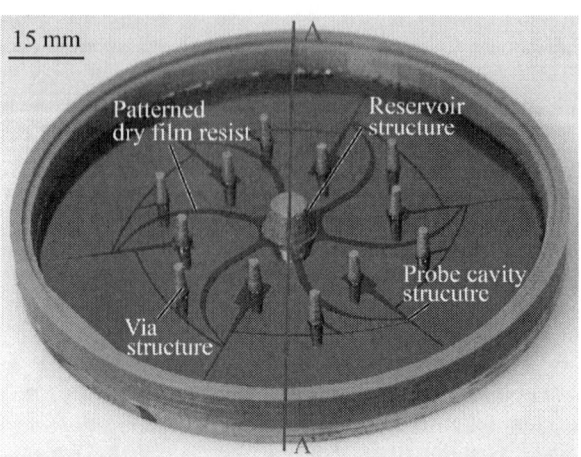

Figure 5: Photograph of inverse mold used for silicone rubber molding.

A representative result of the molding process is shown in Fig. 7(a) indicating the precise replication of the tapered insertion vehicle with the Si probe in the center. The average thickness of the coating was determined to be 215 µm with a slight taper towards the distal end of the probe.

INSERTION EXPERIMENT

The insertion capability of the stiffened hybrid silicon-polymer neural probes was demonstrated by implantation into an agar-based brain model {Fig. 7(b}. The brain phantom is intended to model only the cortical fraction of the brain. It was prepared by mixing agar with DI water at a concentration of 0.6 wt% [12]. A transparent plastic container was filled with agar solution to allow optical inspection of the neural probes after implantation. The probes were attached to a custom-made insertion tool with vacuum chuck and descended into the brain model using a microdrive to a depth of almost 6 mm. After dissolution of the PEG, the cable recovers its initial flexibility and bends to the surface of the brain model being mechanically decoupled from the electrode tip {Fig. 7 (c,d)}.

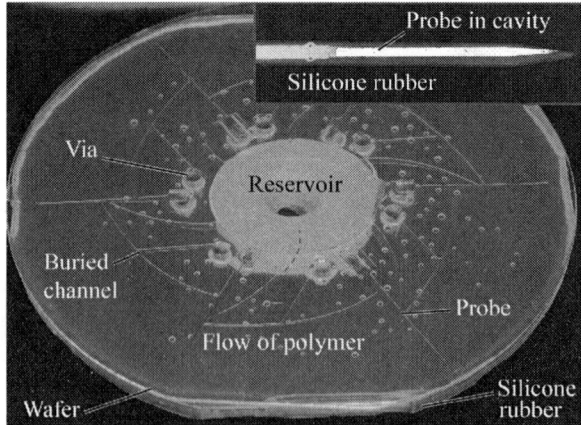

Figure 6: Photograph of mold, inset showing micrograph of probe cavity; during spinning the stiffening polymer flows from the reservoir into the probe cavities, while excess material is discarded sideways.

978-1-4799-7956-1/15 $31.00 © 2015 IEEE

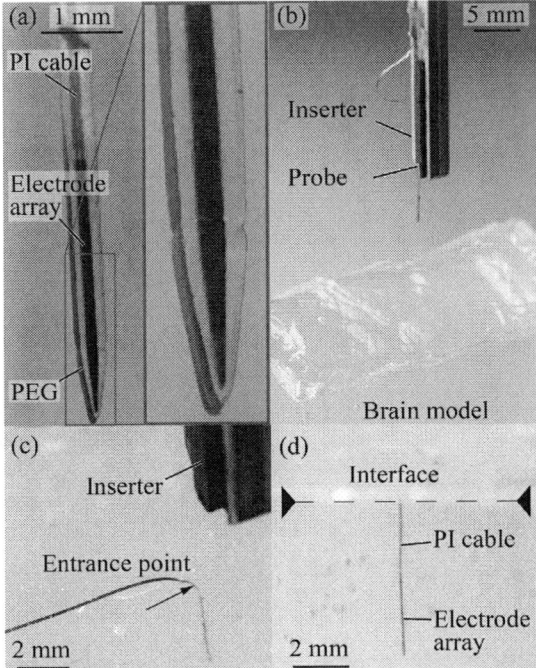

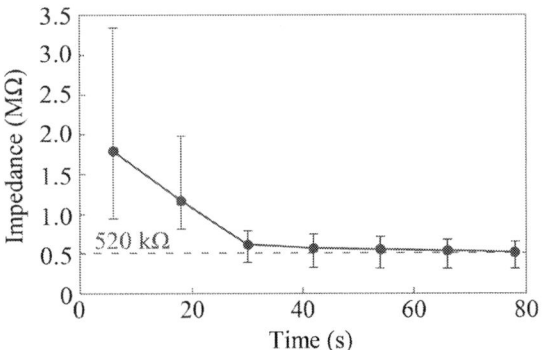

Figure 8: Electrode impedance (mean, upper and lower limits) at 1 kHz of four PEG coated Pt electrodes over time in the brain model.

EU 7th Framework Program and the Cluster of Excellence BrainLinks-BrainTools funded by the German Research Foundation (DFG, grant no. EXC 1086).

REFERENCES

[1] J. Thelin et al., "Implant size and fixation mode strongly influence tissue reactions in the CNS.," *PLoS One*, vol. 6, p. e16267, 2011.

[2] P. Ruther et al., "Recent progress in neural probes using silicon MEMS technology," *IEEJ Trans. Electr. Electron. Eng.*, vol. 5, pp. 505–15, 2010.

[3] J.T.W. Kuo et al., "Novel flexible Parylene neural probe with 3D sheath structure for enhancing tissue integration.," *Lab Chip*, vol. 13, pp. 554–61, 2013.

[4] S. Felix et al., "Removable silicon insertion stiffeners for neural probes using polyethylene glycol as a biodissolvable adhesive.,", in *Proc. IEEE EMBC 2012*, pp. 871–4.

[5] S. Takeuchi et al., "Parylene flexible neural probes integrated with microfluidic channels.," *Lab Chip*, vol. 5, pp. 519–23, 2005.

[6] P.J. Gilgunn et al., An ultra-compliant, scalable neural probe with molded biodissolvable delivery vehicle," in *Proc. IEEE MEMS 2012*, pp. 56–9.

[7] K. Seidl et al., "CMOS-Based High-Density Silicon Microprobe Array for Electronic Depth Control in Neural Recording," in *Proc. IEEE MEMS 2009*, pp. 232–5.

[8] M. Jeon et al., "Partially flexible MEMS neural probe composed of polyimide and sucrose gel for reducing brain damage during and after implantation," *J. Micromech. Microeng.*, vol. 24, p. 025010, 2014.

[9] B. Rubehn et al., "In vitro evaluation of the long-term stability of polyimide as a material for neural implants.," *Biomaterials*, vol. 31, pp. 3449–58, 2010.

[10] S. Sze and K. Ng, Physics of Semiconductor Devices, 3rd ed. Hoboken: John Wiley & Sons, 2007.

[11] S. Herwik et al., "Ultrathin Silicon Chips of Arbitrary Shape by Etching Before Grinding," *J. Microelectromech. Sys.*, vol. 20, pp. 791–793, 2011.

[13] N. H. Hosseini et al., "Comparative study on the insertion behavior of cerebral microprobes.," in *Proc. IEEE EMBC 2007*, pp. 4711–4.

Figure 7: (a) Micrograph of stiffened neural probe; photographs of (b) insertion setup and inserted probe shown (c) from above the interface of air and brain model and (d) in cross-section.

To estimate the duration until recording experiments can be performed, a time-dependent impedance spectroscopy of the encapsulated Pt microelectrodes was done in an electrically conducting agar-based gel. Instead of DI water, a 1 M Ringer's solution was used to prepare a gel providing an appropriate conductivity. Impedance spectra between 500 Hz and 2 kHz were recorded at time intervals of 12 s starting 6 s after immersing the probes in the gel. Figure 8 shows average, minimum and maximum of the recorded absolute impedance values of four electrodes at 1 kHz. As indicated in Fig. 8, the impedances stabilize after roughly 1 min to an average of 520 kΩ indicating that the probe can be used almost immediately after implantation.

CONCLUSION

This paper presented a novel hybrid silicon-polymer neural probe stiffened by a bio-dissolvable PEG coating. The fully-implantable Si-based probe tip allows the integration of CMOS-based high-density electrode arrays while the highly flexible PI cable enables the floating operation of the Si probe in the brain tissue. The efficacy of the bio-dissolvable insertion vehicle was successfully demonstrated by implanting the probe into a brain model to a depth of around 6 mm. It was further shown that the probe can be used for neural recording within 1 min following implantation. Our molding scheme for the insertion vehicle is fast and allows rework with minimal effort. Lateral as well as vertical dimensions of the mold cavities can simply be adjusted. The resulting microstructures closely replicate the shape of mold.

ACKNOWLEDGEMENTS

The authors gratefully acknowledge financial support by the project EUJO-LIMMS (no. 295089) funded by the

CONTACT

*F. Barz, tel: +49-761-203-7202; falk.barz@imtek.de

FLOW SPEED MEASUREMENT WITH DOPPLER EFFECT USING ULTRASONIC RECEIVER FOR SMALL-SIZED SMART CATHETER

Ryo Matsui, Yusuke Takei, Nguyen Minh-Dung, Tomoyuki Takahata,
Kiyoshi Matsumoto and Isao Shimoyama
The University of Tokyo, Tokyo, Japan

ABSTRACT

We measured flow speed in a cylindrical pipe with Doppler effect using an ultrasonic receiver. We proposed a wide frequency range receiver to measure both blood flow speed and blood pressure at the same time in a blood vessel. Our sensor has a piezoresistive cantilever which is put on the boundary between air and liquid in a device. According to this structure, proposed receiver can measure acoustic waves from Hz to MHz order frequency, so that we can obtain blood flow speed (MHz order), and blood pressure change (Hz order) with the same device. We measured flow speed ranging from 6 to 25 mm/s, which was equal to blood flow speed of an arteriole, in a cylindrical pipe, of MHz order Doppler effect.

INTRODUCTION

There are lots of medical indexes to monitor health conditions. Especially, blood flow speed and blood pressure are important indexes. A previous research has shown that blood flow speed of subarachnoid hemorrhage patients is significantly different from that of healthy people [1]. Blood pressure is also important to maintain health, and people who have high-normal blood pressure have more risk of cardiovascular disease than those who have optimal blood pressure [2].

Simultaneous measurement of blood flow speed and blood pressure with the same device should be a smart solution for efficient medical diagnosis because of shortening diagnosis time. Moreover, a miniaturized device enables health care provider to diagnose low invasively and to utilize in a sub millimeter thin blood vessel.

Conventional researches realized miniaturized devices for measurement of blood flow speed or blood pressure. Some of them for blood speed measurement consist of fiber optic techniques or ultrasonic Doppler effect [3] [4]. A previous research realized the development of miniaturized pressure transducer for intravascular measurement [5].

However, previous devices were not able to measure both blood flow speed and blood pressure because each frequency to detect is different.

The objective of this research is to measure flow speed with a small-size MEMS receiver which can detect wide frequency range for simultaneous measurement of blood flow speed and blood pressure.

PRINCIPLE AND FABRICATION

Measurement of flow speed utilizes Doppler effect, and a receiver detected ultrasonic waves of MHz order emitted by an oscillator (Figure 1(a)). Blood pressure is also measured by detecting hydraulic pressure change of Hz order caused by pulsation of the heart (Figure 1(b)). A previous research shows that an acoustic receiver with a

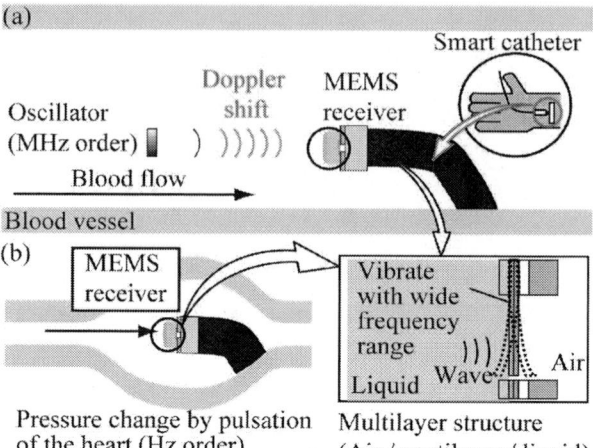

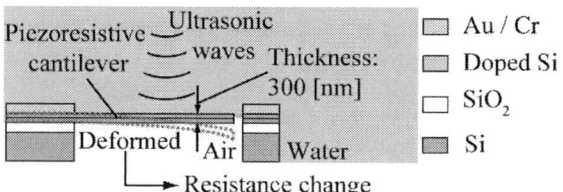

Figure 1: Concept of smart catheter. (a) The catheter measures blood flow speed with Doppler effect by detecting MHz order ultrasonic waves from an oscillator. (b) The catheter measures blood pressure.

Figure 2: Structure of the ultrasonic receiver.

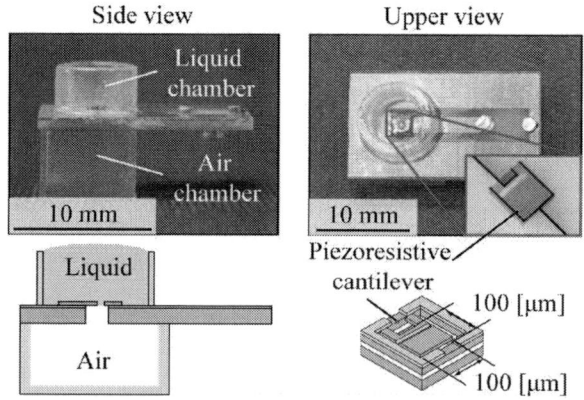

Figure 3: Outline of the fabricated ultrasonic receiver.

piezoresistive cantilever on liquid was able to detect Hz order acoustic waves [6]. The structure of the receiver is shown in Figure 2. This receiver consists of a piezoresistive cantilever and two chambers, and it has air/cantilever/liquid structure [7]. This structure makes it possible to detect ultrasonic waves with wide frequency range because the vibration of a cantilever obeys the surface waves on liquid, not its resonant frequency.

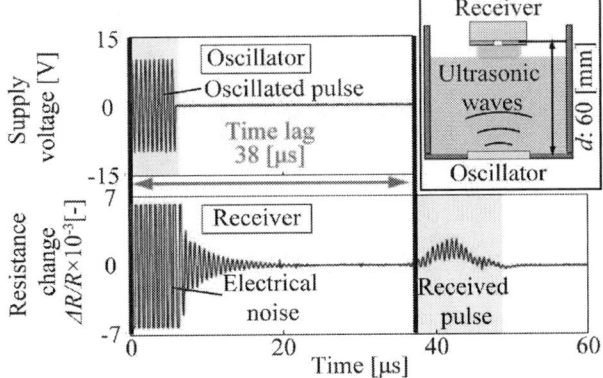

Figure 4: Experiment of flow speed measurement.

Table 1: Experimental conditions of ultrasonic reception experiment.

Frequency	1.6 MHz
Pulse wave	10 Cycles
Supply voltage	10 V
Distance (Oscillator-Receiver)	60 mm

Outline of the fabricated ultrasonic receiver is shown in Figure 3. We fabricated the piezoresistive cantilever from 0.3 μm/0.4 μm/300 μm SOI wafer, and the size of the cantilever was 100 μm×100 μm. Thickness of the cantilever is 300 nm. Previous research shows that this structure detects acoustic wave with high sensitivity from resistance change caused by its deformation [7]. After the fabrication of the cantilever, it was assembled with the air chamber and the liquid chamber. Finally, water was poured into the liquid chamber for acoustic impedance matching and efficient transmission of ultrasonic waves.

EXPERIMENTS AND RESULTS
Reception of MHz order acoustic waves

We conducted two experiments with the proposed sensor. First experiment was the detection of MHz order ultrasonic wave, which was used in medical diagnosis field. The proposed sensor was put on the water surface of the plastic vessel with an acoustic oscillator on the bottom (Figure 4). The oscillator emitted 1.6 MHz, 10 cycle burst waves. After 38 μs from the ultrasonic emission, the sensor responded along the arrival of ultrasonic waves. The distance between the oscillator and the receiver was 60 mm, and sonic speed in water is 1,500 m/s (Table 1), so theoretical time lag was 40 μs. The difference of both time is within 5 percent. The result demonstrates that the proposed sensor could receive ultrasonic waves.

Flow speed measurement

We conducted flow speed measurement with Doppler effect. We fabricated a cylindrical flow channel of 4 mm in diameter. The oscillator and the proposed sensor were installed at both ends respectively (Figure 5). Water flowed in the flow channel with a pump (ATS Lab, Model700), and the oscillator emitted 1.6 MHz continuous acoustic waves (Table 2, Figure 6). While acoustic waves passed the channel, Doppler Effect was caused, and the

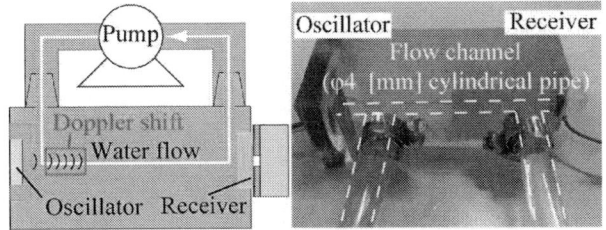

Figure 5: Experimental setup of flow speed measurement experiment.

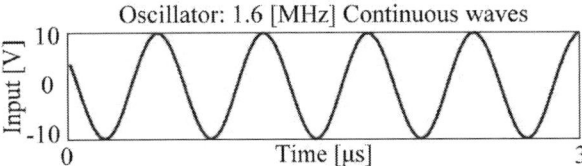

Figure 6: Supplied voltage to the ultrasonic oscillator

Table 2: Experimental conditions of flow speed measurement experiment.

Frequency	1.6 MHz
Supply Voltage	10 V
Flow Speed	6 - 25 mm/s

acoustic waves changed their frequency because of the water flow. The shift frequency f_d is proportional to water flow speed v, and f_d is written as Eqn. 1 with sonic speed c and frequency of oscillator f_0. We experimented flow speed measurement by changing flow speed ranging from 6 to 25 mm/s.

$$f_d = \frac{v}{c} f_0 \qquad (1)$$

When acoustic waves reached the receiver, piezoresistive cantilever deforms. A Half bridge circuit and a 200 times amplifier detects its deformation as a change of electrical resistivity. After the detection of acoustic waves with the receiver, Doppler shift frequency was extracted with a mixing circuit and FFT analysis. Figure 7 shows two result examples of flow speed measurement. When the flow speed was 8 mm/s or 10 mm/s, each analyzed FFT peak frequency was 13 Hz or 14 Hz. From the peak frequency from FFT analysis, the measured Doppler frequency was proportional to flow speed (Figure 8). The constant of proportionality between Doppler frequency and flow speed obtained from measured Doppler frequency was almost the same as the theoretical constant of proportionality. The error was within 1 percent. However, calculated flow speed was faster than setting speed. We suppose that the result was caused by pulsation of the pump or the bent shape of the flow channel.

CONCLUSION

We proposed an approach for measuring flow speed using Doppler Effect in liquid. Our proposed sensor is suitable for detecting ultrasonic waves because of the air/piezoresistive cantilever/liquid structure.

Experimental result shows that the proposed sensor

978-1-4799-7956-1/15 $31.00 © 2015 IEEE

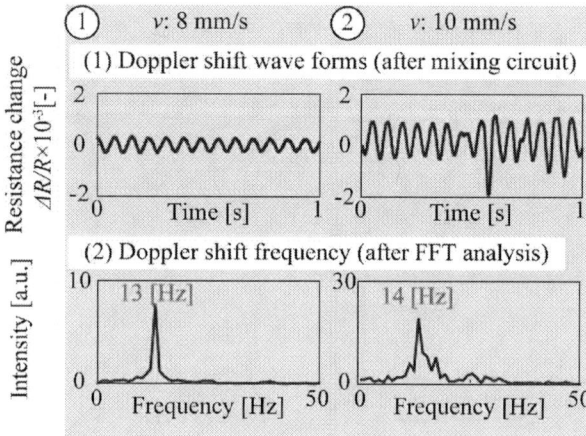

Figure 7: Detected Doppler Shift wave forms and FFT analysis result

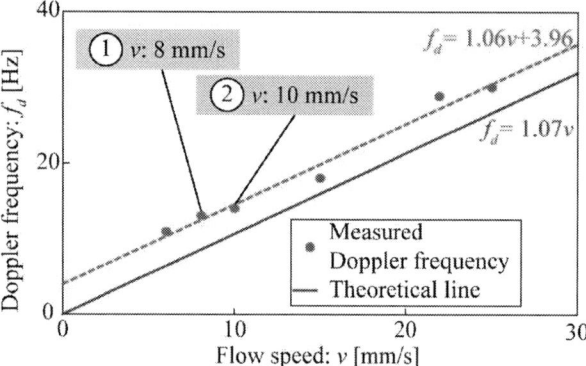

Figure 8: Relationship of measured Doppler Frequency and actual flow speed.

was able to detect 1.6 MHz ultrasonic waves and calculate flow speed by detecting Doppler frequency.

From those results, we suppose that our MEMS ultrasonic receiver makes it possible to measure blood flow speed and blood pressure in blood vessel at the same time, and to develop a small-sized smart catheter.

ACKNOWLEDGEMENT

The photolithography masks were made using the University of Tokyo VLSI Design and Education Center (VDEC)'s 8 inch EB writer F5112 + VD01 donated by ADVANTEST Corporation.

REFERENCES

[1] Venkatesh Bala MD, Shen Quiaomei and Lipman Jeffrey, "Continuous measurement of cerebral blood flow velocity using transcranial Doppler reveals significant moment-to-moment variability of data in healthy volunteers and in patients with subarachnoid hemorrhage," *Critical Care Medicine*, Vol. 30, No. 3, pp.563-569, 2002.

[2] Ramachandran S. Vasan, M.D., Martin G. Larson, Sc.D., Eric P. Leip, M.S., Jane C. Evans, Ph.D., Christopher J. O'Donnell, M.D., M.P.H., William B. Kannel, M.D., M.P.H. and Daniel Levy, M.D., "Impact of high-normal blood pressure on the risk of cardiovascular disease," *The New England Journal of Medicine*, Vol. 345, No. 18, pp. 1291-1297, 2001.

[3] Toyoichi Tanaka and George B. Benedek, "Measurement of the Velocity of Blood Flow (in vivo) Using a Fiber Optic catheter and Optical Mixing Spectroscopy," *Applied Optics*, Vol. 14, No. 1, pp. 189-196, 1975.

[4] C. J. Hartley and J. S. Cole, "An ultrasonic pulsed Doppler system for measuring blood flow in small vessels," *Journal of Applied Physiology*, Vol. 37, No. 4, pp. 626-629, 1974.

[5] Lindstrom Lars H., "Miniaturized Pressure Transducer Intended for Intravascular Use," *IEEE Transactions on Biomedical Engineering*, Vol. BME-17, No. 3, pp. 207-219, 1970.

[6] Tomonori Kaneko, Nguyen Minh-Dung, Ryo Aoki, Tomoyuki Takahata, Kiyoshi Matsumoto and Isao Shimoyama, "Measurement of Mechanomyogram," *IEEE Proceedings of MEMS2014*, pp. 845-848, 2014.

[7] Nguyen Minh-Dung, Phan Hoang-Phuong, Kiyoshi Matsumoto and Isao Shimoyama, "A hydrophone using liquid to bridge the gap of a piezo-resistive cantilever," *Proceedings of Transducers2013*, pp. 70-73, 2013.

CONTACT

*Ryo Matsui, Mechano-Informatics Department, Graduate School of Information Science and Technology, The University of Tokyo, 7-3-1 Hongo, Bunkyo-ku, Tokyo, 113-8656, Japan.
E-mail: r_matsui@leopard.t.u-tokyo.ac.jp
Tel: +81-3-5841-6318, Fax: +81-3-5841-6341

HUMAN ADIPOSE-DERIVED STEM CELL FIBER FOR BREAST RECONSTRUCTION

Amy Y. Hsiao[1], Teru Okitsu[1,2], and Shoji Takeuchi[1,2]
[1]Institute of Industrial Science, The University of Tokyo, JAPAN
[2]ERATO Takeuchi Biohybrid Innovation Project, Japan Science and Technology Agency, JAPAN

ABSTRACT

This paper describes the construction and differentiation of human adipose-derived stem cell (ADSC) fibers into the adipocyte lineage for breast reconstruction. Human ADSCs were encapsulated into the core region of hydrogel core-shell fibers by the cell fiber technology [1], induced for adipogenic differentiation, and maintained for over 2 months. After adipogenic induction, accumulation of lipid droplets of significant size was observed in the cells, and viability assay showed that most of the cells were alive. These findings suggest the use of ADSC fibers as a promising approach for breast reconstruction.

INTRODUCTION

Autologous fat grafting is a promising treatment for breast reconstruction [2,3]. However, conventional methods suffer from problems such as necrosis of the fat cells leading to low rate of graft survival and difficulties in controlling the shape of the grafted fat [2,3]. To improve graft survival rate, autologous ADSCs have been used in conjunction with lipoinjection [2]. ADSCs have the potential to differentiate into various lineages, including chondrogenic, osteogenic, and adipogenic lineages [2]. Recent findings in the field show the promise of ADSCs in becoming valuable tools in a wide range of cell-based therapies [2]. With their multipotency, ADSCs have been shown to enhance angiogenesis and thus improving the survival rate of grafts [2].

Here, our approach is to culture ADSCs as 3D fiber shaped constructs using the cell fiber technology [1], induce the cells for adipogenic differentiation, and implant the differentiated cell fiber for breast reconstruction. Instead of isolating adipocytes in huge mass and subjecting them to higher risks of necrosis, obtaining adipocytes from adipogenic differentiation of ADSCs in a 3D microenvironment with sufficient oxygen and nutrients are expected to minimize necrosis. The ADSCs that remain undifferentiated in the cell fibers may further contribute to the enhancement of angiogenesis. Finally, the already 3D and handelable long cell fiber constructs can be entangled into arbitrary shapes while leaving sufficient interstitial space for oxygen and nutrient diffusion.

MATERIALS AND METHODS

Materials

Human ADSCs and ADSC growth medium were obtained from Lonza. Phosphate buffered saline (PBS), insulin, 3-isobutyl-1-methylxanthine (IBMX), dexamethasone, fibrinogen powder, and thrombin were purchased from Sigma-Aldrich Japan. Type I native collagen was purchased from Koken Co. Ltd. Accutase cell dissociation reagent, BODIPY 493/503, and LIVE/DEAD Viability Kit were obtained from Life Technologies. Calcium chloride ($CaCl_2$) was purchased from Kanto Chemicals. Sodium chloride (NaCl), sodium alginate (Na-alginate), and rosiglitazone were purchased from Wako Pure Chemical Industries. Sucrose was obtained from Nacalai Tesque. Aprotinin was obtained from Roche Applied Science.

Cell Culture

Human ADSCs were routinely cultured in ADSC growth medium and passaged at 70% confluence until P5. Only passages before P5 were used for cell fiber experiments. ADSC fibers were maintained in control or differentiation induction and maintenance medium in petri dishes. Control medium is composed of ADSC growth medium supplemented with 20 µg/ml of aprotinin. Adipogenic differentiation induction medium [4] is composed of ADSC growth medium + 10 µg/ml insulin + 0.5 mM IBMX + 0.25 µM dexamethasone + 1 µM rosiglitazone supplemented with 20 µg/ml of aprotinin, whereas adipogenic differentiation maintenance medium is composed of only ADSC growth medium + 1 µg/ml insulin supplemented with 20 µg/ml of aprotinin. All cultures were maintained in a humidified incubator at 37°C with 5% CO_2.

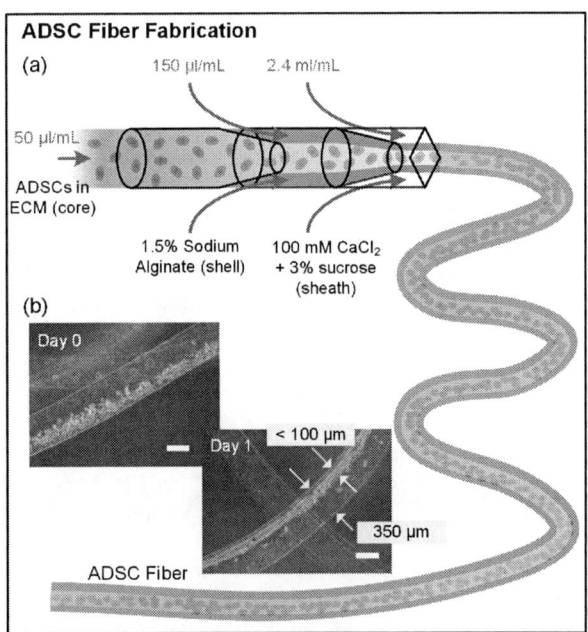

Figure 1: (a) Schematic illustration of the microfluidic double co-axial device used to fabricate ADSC fiber. (b) Images of ADSC fiber 0 and 1 day after fabrication. Scale bar = 200 µm.

Device Fabrication

The double co-axial microfluidic device for fiber fabrication was made from two pulled glass capillary tubes with 300 μm tips aligned serially and joined by custom-made three-way connectors as previously described [1] (Figure 1a).

Fiber Fabrication

To prepare the core solution, human ADSCs were suspended in a 1:1 mixture of fibrinogen (33.3 mg/ml) and 0.4% collagen at 1 x 10^7 cells/ml. The shell solution is 1.5 wt% sodium alginate solution, and the sheath solution is 100 mM CaCl₂ with 3% sucrose. To fabricate fibers, the shell solution is first introduced into the device at 150 μl/min by a syringe pump, followed by the sheath solution at 3.6 ml/min. After checking that the generated stream is stable, the core solution is finally introduced at 50 μl/min. During this process, ADSCs were encapsulated inside calcium alginate shell fibers. Immediately following fabrication, to convert fibrinogen into fibrin and to allow collagen to solidify, the fibers were incubated in thrombin (4.2 units/ml) for 15 min inside 37°C humidified incubator. The fibers were then cultured in petri dishes filled with control medium.

Adipogenic Differentiation Induction

One day after fabrication of the fibers (Day 1), ADSC fibers were separated into two groups: control and differentiation groups. In the control group, the medium was replaced with fresh control medium. In the differentiation group, the medium was changed to differentiation induction medium. On Day 3, medium was replaced in both groups with their respective medium. On Day 5, medium was replaced again: the medium was simply replaced with fresh control medium for the control group, but was changed to differentiation maintenance medium for the differentiation group. For the rest of the culture period, media exchanges were performed every 3 or 4 days (control: control medium, differentiation: differentiation maintenance medium).

Live/Dead Viability Assay

On Day 65 of culture, live/dead viability assay was performed on cell fibers in both control and differentiation groups. The cell fiber were incubated in live/dead viability assay working solution consisting of PBS (+) + 1 μM calcein AM + 1 μM ethidium homodimer-1 for 30 min at 37 °C.

RESULTS AND DISCUSSION

Cell Fiber Formation

Using the double co-axial microfluidic device (Figure 1a), human ADSCs were successfully encapsulated into the core region of hydrogel core-shell fibers (Figure 1b). The outer shell is composed of calcium alginate, and the core region contains ADSCs in a mixture of collagen and fibrin. The diameter of the entire fiber (including the alginate shell) is approximately 350 μm, and the core region is less than 100 μm. One day after ADSCs were encapsulated into hydrogel core-shell fibers, ADSCs aggregated into fiber-shaped constructs with the cells

stretched and aligned uniformly in the longitudinal direction (Figure 1b). Over time, as the cell traction force exhibited by the ADSCs in the cell fibers slowly increased, the entire hydrogel core-shell fiber started to curve and eventually self-assembled into spring-shaped structures (Figure 2a) that is maintained throughout the entire culture period.

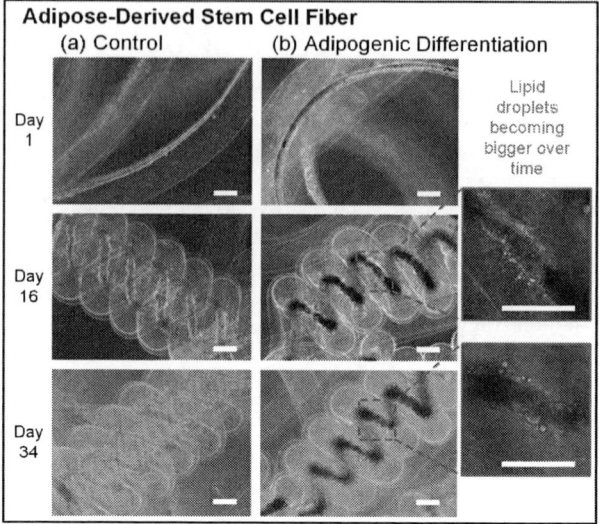

Figure 2: Images of ADSC fibers cultured in control media (a) and induced for adipogenic differentiation (b) over time. Scale bar = 200 μm.

Adipogenic Differentiation

Upon successful cell fiber formation, ADSC fibers were subsequently induced to differentiate to the adipocyte lineage in differentiation induction medium for 4 days and maintained in maintenance medium for over 2 months. Similar to the control fibers, the ADSC fibers induced for adipogenic differentiation also self-assembled into spring-shaped structures over time. Furthermore, Intracellular accumulation of lipid droplets was clearly visible in the cell fibers induced for adipogenic differentiation by Day 16, but not in those cultured in control media (Figure 2). Over time, the lipid droplets became bigger, and by Day 34, lipid droplets reached average diameters of approximately 25 μm and remained at such size for the rest of the culture period. The fibers further naturally entangled into compact shapes through their self-assembly into coiled spring structures. Such coiled spring structures might be able to provide better control of fat grafts.

After 65 days of culture, both the cell fibers cultured in control media and induced for adipogenic differentiation showed high cellular viability as confirmed by live/dead viability assay. The majority of the cells in both groups were alive as identified with green fluorescence, and only a few individual dead cells were observed with red fluorescence.

CONCLUSION

Altogether, these results indicate that human ADSCs cultured as 3D fiber constructs successfully differentiated toward the adipocyte lineage (as characterized by intracellular accumulation of lipid droplets) and were maintained for over 2 months with high viability. Given the fibers' high handelability, their ease of being assembled into compact shapes, and our new findings, the human ADSC fibers possess promising potential in reconstructing breasts for breast cancer patients after surgery.

ACKNOWLEDGEMENTS

A.Y.H. is supported by Japan Society for the Promotion of Science (JSPS) Postdoctoral Fellowship for Foreign Researchers.

REFERENCES

[1] H. Onoe, T. Okitsu, A. Itou, M. Kato-Negishi, R. Gojo, D. Kiriya, K. Sato, S. Miura, S. Iwanaga, K. Kuribayashi-Shigetomi, Y. Matsunaga, Y. Shimoyama, S. Takeuchi, "Metre-long Cell-laden Microfibres Exhibit Tissue Morphologies and Functions, *Nat. Mater.*, vol. 12, pp. 584-590, 2013.

[2] K. Yoshimura, K. Sato, N. Aoi, M. Kurita, T. Hirohi, K. Harii, Cell-assisted Lipotransfer for Cosmetic Breast Augmentation: Supportive Use of Adipose-derived Stem/Stromal Cells, *Aesth. Plast. Surg.*, vol. 32, pp. 48-55, 2008.

[3] H. Uda, Y. Sugawara, S. Sarukawa, A. Sunaga, Brava and Autologous Fat Grafting for Breast Reconstruction after Cancer Surgery, *Plast. Reconstr. Surg.*, vol. 133, pp. 203-213, 2014.

[4] C. Pantoja, J.T. Huff, K.R. Yamamoto, Glucocorticoid Signaling Defines a Novel Commitment State during Adipogenesis, *In Vitro, Mol. Biol. Cell*, vol. 19, pp. 4032-4041, 2008.

CONTACT

*A.Y. Hsiao, tel: +80-3-5452-6650; ahsiao@iis.u-tokyo.ac.jp

MEASURING THE PROPAGATING TEETH VIBRATION
OF HUMAN CHEWING

Chie Suzuki, Yusuke Takei, Tomoyuki Takahata, Kiyoshi Matsumoto, and Isao Shimoyama
The University of Tokyo, Tokyo, JAPAN

ABSTRACT

We measured the propagation waves of teeth's vibrations when chewing food. The objective of this measurement is to find out the difference of perceived texture resulted from the difference of teeth alignment. Human senses teeth's vibrations caused by chewing food with periodontal membranes and regards it as texture. Therefore, measuring the teeth's vibrations will allow us to quantify the food texture which human actually senses. We fabricated an acoustic sensor that is small enough to be attached to teeth. For sensor evaluation, we conducted the rice cracker chewing test with our sensor attached to a real scale 3D jaw model, and propagation waves of around 500 Hz are observed.

INTRODUCTION

The food texture is one of the main factors that affect the taste of food. It is thought that human perceives the texture by vibrations of teeth and sound of food breaking when human is chewing [1]. At the root of the teeth, there are periodontal membranes which sense force and vibrations acting on teeth [2]. Therefore, measuring the teeth's vibrations directly allows us to know the signals which periodontal membranes detect and the texture which human perceives.

There are many studies to quantify the food texture. Most of them are based on measurement of the vibrations of a probe or the sound of food breaking when the probe is being inserted into the food [3-5]. Using this method, the quality of food can be evaluated. On the other hand, real teeth are lined and contact each other, and food is chewed by more than one tooth at the same time. Therefore, the vibrations by breaking food are transferred through other teeth. It means that the probe's vibrations are different from the vibrations, which periodontal membranes detect. In addition, the arrangements of the teeth also affect the propagations of vibrations. Therefore, to investigate the food texture sensed by human, it is also important to measure the vibrations of human's real teeth.

In this paper, we measured transferred vibrations of a tooth from other teeth when human is chewing food to understand interaction of teeth's vibrations (**Fig. 1**). Especially, we focused on the relationships between the texture sensation and the alignment of teeth. Measurement is carried out using a MEMS acoustic sensor that is small and safe enough to put it into mouth. Our final goal is measuring the real texture which human perceives by attaching sensors to real teeth of human. As the first step to reach the final goal, the experiments were carried out using the real scale 3D printer jaw model, which has the same size that of human's jaw.

SENSOR DESIGN AND FABRICATION

Fig. 2 shows the conceptual sketch of our fabricated acoustic sensor. Our acoustic sensor is based on a

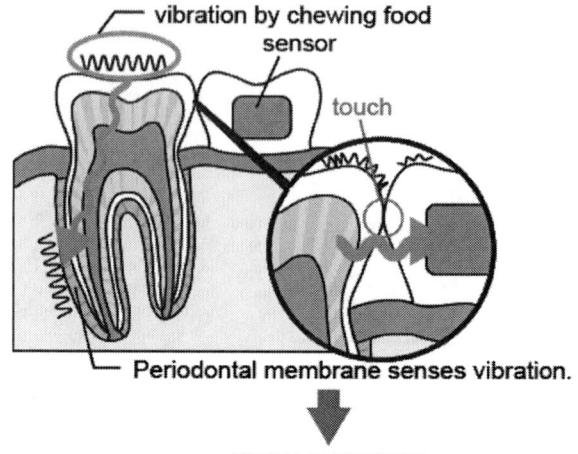

Figure 1: Food texture is perceived when periodontal membrane gets the vibration of teeth. The vibration is transferred to near teeth.

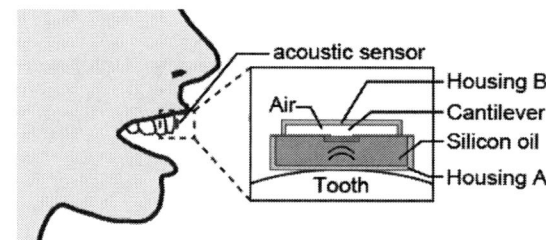

Figure 2: Mechanism and structure of our acoustic sensor. The vibrations from the teeth transferred through silicon oil to cantilever.

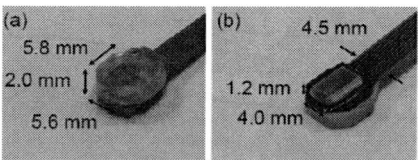

Figure 3: The size of our acoustic sensor. (a) the upper side (b) the under side

differential pressure sensor using a piezo-resistive cantilever [6-7]. The sensor chip was attached to the gold wiring patterned on flexible polyimide substrate (thickness 80 μm) by epoxy glue (Loctite 4304, Loctite). After that, we connected the wiring between the sensor chip and the polyimide substrate by aluminum wire bonding. The wire and the edges of the sensor chip were covered with medical light cure instant adhesive (Loctite 4305, Loctite). The sensor chip is covered with housings A and B (thickness 0.3mm) as shown in **Fig 2**. The housings are fabricated by 3D printer (Stratasys Objet Eden260V) and the material is FullCure 720 (Young's modulus 2000-3000 MPa). The housing A was directly

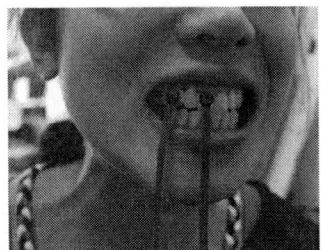

Figure 4: The sensor is small enough to be attached to human's real teeth.

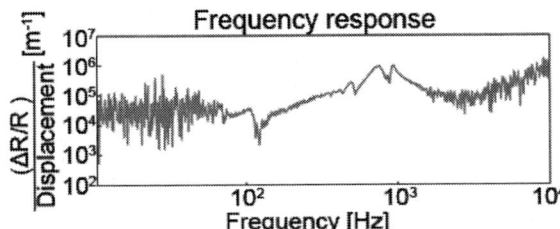

Figure 5: Frequency response of our sensor. It is the result in consideration in a characteristic of exciter.

attached to a tooth using medical glue. Inside the housing A is filled with silicone oil (HIVAC F-4, Shin-Etsu Silicone). When a tooth vibrates, its vibration propagates to the cantilever through the oil.

The size of the fabricated sensor is 5.8 mm × 5.6 mm × 3.2 mm (**Fig. 3**). It is small enough to be attached on human's front tooth whose surface is about 8 mm × 9 mm as shown in **Fig. 4.**

To put the sensor in a mouth, the sensor is needed to be coated with biocompatible materials. We coated the sensor with parylene and the gap is filled with instant adhesive (Loctite 4305, Loctite) to prevent silicone oil from leaking.

As the sensor evaluation, we measured the frequency response of the sensor using an exciter. The result shows that our sensor has good response from 100 Hz to 1 kHz (**Fig. 5**). We thought this response range was enough because the vibration receptor of human sense only low frequency waves under around 500 Hz [8].

EXPERIMENTS AND RESULTS

We conducted the experiments to investigate the vibration propagation through teeth. In this experiments, we used a real scale 3D jaw model made by a 3D printer (**Fig. 6**). The model is superior to human real teeth in repeatability and ease of preparing variety of experiment conditions .The material is FullCure 720. We made parts of teeth and gums separately and bonded those parts together. We attached two sensors to model's teeth. **Fig. 7(a) ~ (d)** shows the four cases of experiment conditions. By using those sensors, we measured the teeth vibration while the 3D model was being inserted into rice crackers. We selected rice crackers as a test piece because their crunchy textures were thought to be detected easily. The model of upper jaw was attached to the Z-axis stage that moved at the constant speed of 5 mm/s (**Fig. 8**). The rice cracker was put on the board under the upper jaw model. When the model was being inserted into rice crackers, the tooth with sensor 1 and the teeth to the left side of the

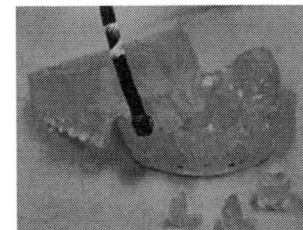

Figure 6: A picture of 3D jaw's model. The part of gums and teeth are separated.

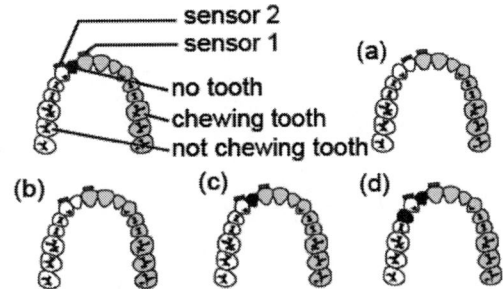

Figure 7: The experimental conditions. (a)The tooth with sensor 1 and the tooth with sensor 2 lie next to each other. (b) One tooth lies between the tooth with sensor 1 and the tooth with sensor 2. (c)There is no tooth between the tooth with sensor 1 and the tooth with sensor 2. (d) There is no tooth at the sides of the tooth with sensor 2.

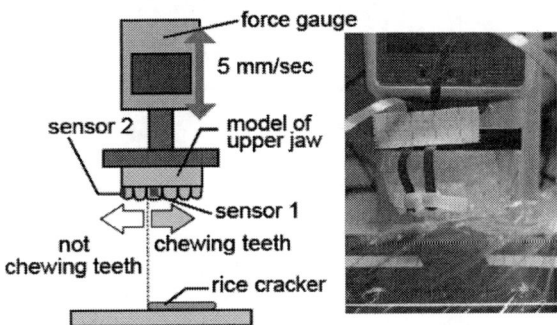

Figure 8: Experimental setup.

tooth were inserted into the rice cracker while other teeth including the tooth with sensor 2 were not inserted (**Fig. 7**).

The data of output of voltage value was obtained using out sensor during the model moving. At the same time, we measured the pushing force to know when the insertion of the model into rice cracker started and when the insertion finished using force gauge (IMADA ZP-500N). We conducted FFT (Fast Fourier Transform) analysis about the data while the model was inserted into rice cracker (**Fig. 9**). **Fig. 10** shows the result of FFT analysis. There are 2 peaks over the range of 200 Hz and 500 Hz in both sensors. When the sensor 2 was attached to the nearest teeth from the front tooth, it measured the strongest wave of 500 Hz. The result indicates that the vibrations of 500 Hz were transferred through the teeth. Therefore, the vibration was weaker when the tooth with our sensor did not contact with vibrating tooth. On the other hand, the vibrations of 200 Hz were transferred to

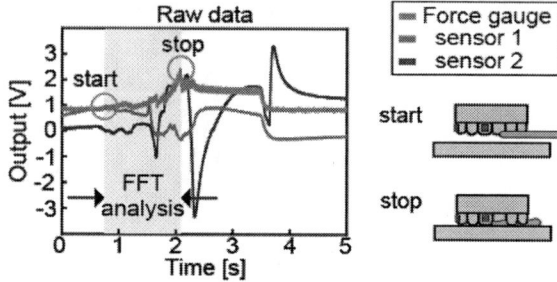

Figure 9: Output of voltage data used in FFT.

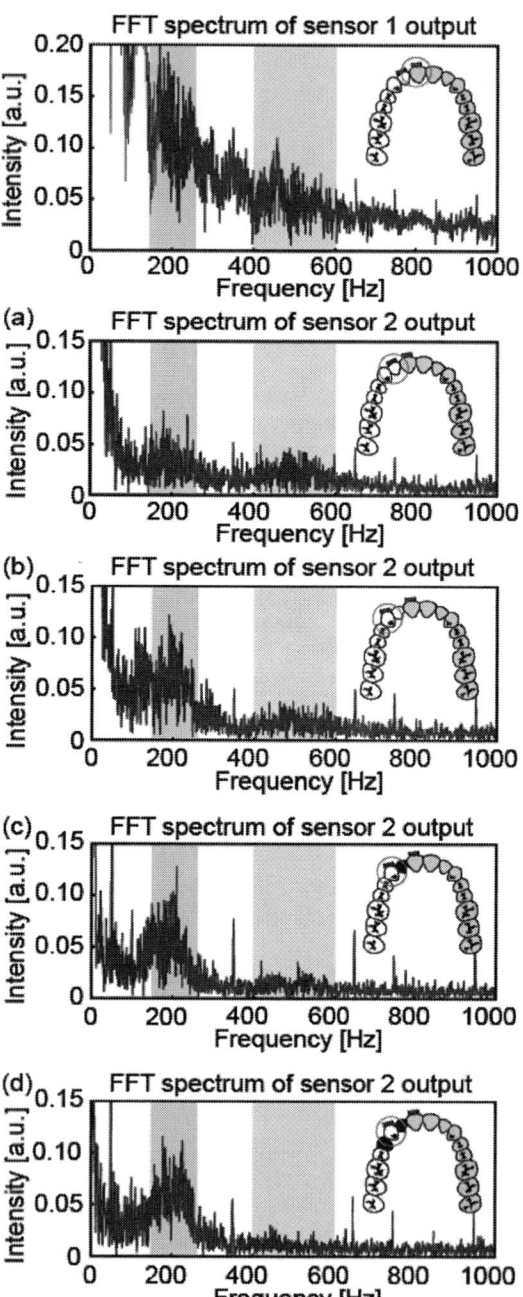

Figure 10: The spectrum of the vibration showing the vibrations around 500Hz were transferred through teeth.

other teeth regardless of contact of teeth. It is thought that this vibrations was transmitted through the part of gums.

CONCLUSION

We fabricated the acoustic sensor and measured the propagation of teeth vibrations during human chewing. When the jaw model was being inserted into the rice cracker, we observed strong signal around 200 Hz and 500 Hz. The spectrums around 500 Hz were weak when the tooth with sensor did not contact with other teeth. The results indicate the possibility that the arrangement of teeth and the gap between the teeth affected the transmission of vibration related to the texture perception.

ACKNOWLEDGEMENTS

This research is supported by Grant-in-Aid for Young Scientists (B) (24700589), the Ministry of Education, Culture, Sports, Science and Technology (MEXT), Japan. The photolithography masks were made using the University of Tokyo VLSI Design and Education Center (VDEC)'s 8 inch EB writer F5112 + VD01 donated by ADVANTEST Corporation.

REFERENCES

[1] Christensen. C. M., Vickers. Z. M., "Relationships of chewing sounds to judgments of food," *Journal of Food Science,* vol. 46, no. 2, pp. 574-578, 1981.

[2] JB Hutchings, et al., "THE PERCEPTION OF FOOD TEXTURE-THE PHILOSOPHY OF THE BREAKDOWN PATH." *Journal of Texture Studies,* vol. 19, no. 2, pp. 103-115, 1988.

[3] Maruyama Thais Terumi, et al., "Time-frequency analysis of acoustic noise produced by breaking of crisp biscuits," *Journal of Food Engineering,* vol. 86, no. 1, pp. 100-104, 2008.

[4] Zdunek Artur, et al., "New contact acoustic emission detector for texture evaluation of apples," *Journal of Food Engineering,* vol. 99, no. 1, pp. 83-91, 2010.

[5] S. Iwatani, et al., "Evaluation of grape flesh texture by an acoustic vibration method," *Postharvest Biology and Technology,* vol. 62, no. 3, pp. 305-309, 2011.

[6] K. Tomonori, et al., "Measurement of mechanomyogram," *Micro Electoro Mechanical Systems(MEMS), 2014 IEEE 27th International Conference on.* IEEE, 2014, pp. 845-848.

[7] H. Takahashi, et al., "Differential pressure sensor using a piezoresistive cantilever." *Journal of Micromechanics and Microengineering,* vol. 22.5, no. 5, article no. 055015, 2012.

[8] Bolanowski Jr, et al., "Four channels mediate the mechanical aspects of touch," *The Journal of the Acoustical Society of America,* vol. 84, no. 5, pp. 1680-1694, 1988.

CONTACT

*Chie Suzuki, Mechano-Informatics Department, Graduate School of Information Science and Technology, The University of Tokyo, 7-3-1 Hongo, Bunkyo-ku, Tokyo, 113-8656, Japan.
E-mail: c_suzuki@leopard.t.u-tokyo.ac.jp
Tel: +81-3-5841-6318, Fax: +81-3-5841-6341

MICRO-NEEDLE-BASED ELECTRO TACTILE DISPLAY TO PRESENT VARIOUS TACTILE SENSATION

Norihide Kitamura[1] and Norihisa Miki[1,2]
[1]Keio University, Kanagawa, Japan
[2]JST PRESTO, Japan

ABSTRACT

In prior work, micro-needle electrodes were developed, however they could only stimulate the tactile receptors that located very close to the surface and therefore, they could only present stinging sensation. We developed a newly electrotactile display which has a micro-needle electrode array and a counter flat electrode. The display can stimulate all the tactile receptors, which resulted in even lower voltage required to tactile stimulation and successful display of various tactile sensation.

INTRODUCTION

Tactile displays are the device which can show various tactile sensation through artificial stimulation. The tactile sensation can be provided by many ways such as mechanical vibration [1], ultrasonic vibration [2], electrostatic force [3], Ultra Sound [4] and electric stimulation [5]. Electric stimulation type of display is a device for displaying tactile sensations using electric stimulation and called electrotactile display [5]. This type of the device was superior to other types of tactile displays in miniaturization, device flexibility, and light weight because they need only electrodes to display tactile sensation and thus, have good affinity with MEMS technologies. However, on the other hand, electrotactile displays still requires high voltages as large as 100 to 300 V because the electric impedance of the stratum corneum which located at the surface of the finger skin is very high. Concerning this problem, "Micro-needle electrode arrays for electrotactile displays" were reported at MEMS2014 conference [6]. The electrotactile display with micro-needle electrode array drastically lowered the necessary voltage to provide tactile sensation to the subject (as low as 5% of prior work with flat electrodes) and repeatability of displayed sensation was improved. However, since the needle length of 600 μm was set not to reach the pain points, the devices could stimulate only the tactile receptors locating at the shallow region (merkel's discs and meissner corpuscles) and therefore, presented tactile sensation was limited to a stinging one. Solving this problem, we designed the electrotactile display such that the flat electrode that worked as the ground to be positioned onto the backside of the finger as shown in Fig. 1. The needle electrodes that inserted into the fingertip were connected to a pulse generator through controlling circuit. The electric current flew inside the finger and stimulated all the tactile receptors that included pacini corpuscles locating in the deep area of the finger. Since the display can exploit all the tactile receptors, it is considered to be able to display tactile sensation even more efficiently (at lower voltage) and in various ways (including surface textures). 5 lined-up needle electrodes made of titanium were manufactured by electrochemical etching.

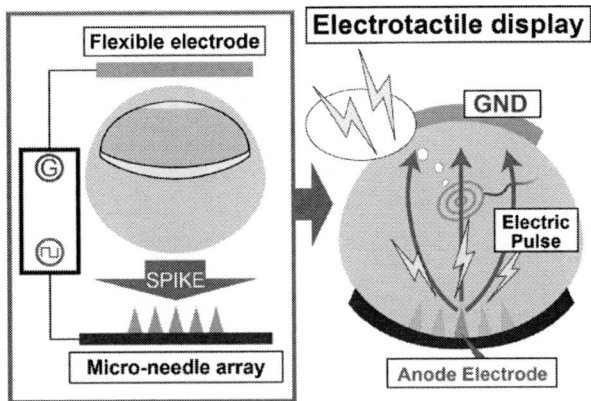

Figure 1: Schimatic images of electrotactile display with micro-needle electrode with nail-side electrode. The current can stimulate the receptor locates deep part.

FABRICATION

Micro-needle Array

The micro-needle array was fabricated by two steps: arraying wires and processing wires to the needle. The wire array consists of Titanium wires and Polydimethylsiloxane (PDMS). The wires which were connected to the flat cable were arrayed with the holder that has 5 holes and fixed with PDMS. After that, we peeled off the array from the holder. Then, the titanium wire array connected to the flat cable was processed by electrochemical etching which was used as electro polishing [7]. When the array was processed electrochemical etching, the initial length, the etching distance and etching time is the important factor of the final shape of the needles. Therefore, we controlled the factors by using the holder, spacer which depth and thickness were adjusted to the needed length. With these method, we could successfully fabricated the 5-micro-needle

Experimental System

Displaying tactile sensation was achieved by applying the voltage with two factors: electric pulse shape (frequency and peak voltage) and the speed of the stimulation flow on the finger. The stimulation flow was generated by changing the activated electrode in order. The system of the tactile display was shown in Fig. 2. The applied frequency was positive square pulse and was provided by a pulse generator. The generated electric pulse flew to the needles but the photo MOS relays switched the destination of the pulse. Then, the pulse flew to the GND flat electrode positioned on the backside of the finger through the tactile receptors and stimulated them. In this system, photo MOS relays were controlled by the micro-controller (Aruduino MEGA) and each photo MOS relay switched the pulse flow by the programed switching time.

978-1-4799-7956-1/15 $31.00 © 2015 IEEE   649   MEMS 2015, Estoril, PORTUGAL, 18 - 22 January, 2015

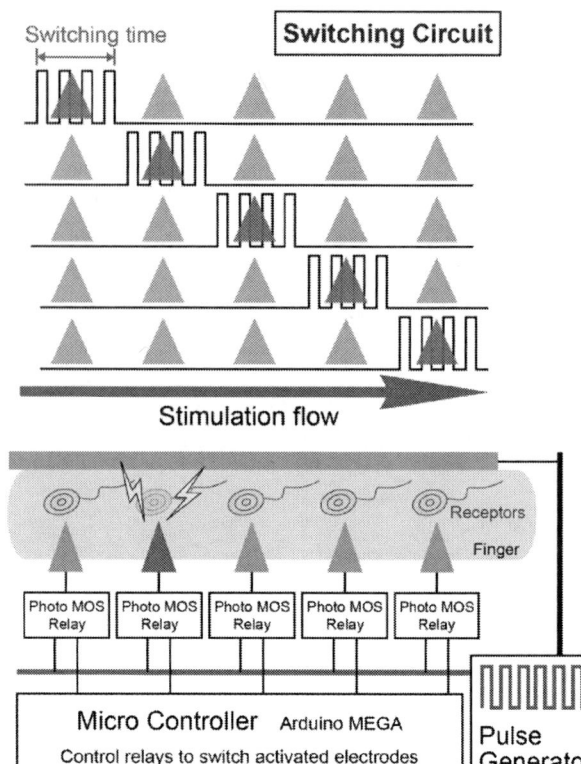

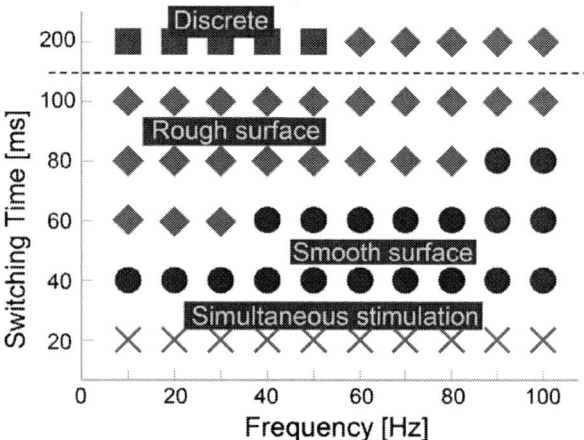

Figure 2: Schimatic images of the System of electrotactile display. The positive square pulses were applied from pulse generator and activated needle electrodes were switched by photo MOS relays which controlled with micro controller (Arduino MEGA).

Figure 3: The results of the relations: Switching Time vs. Frequency. Applied voltage in the experiment was 3.0V.

EXPERIMENTAL

First of all, we measured the thresholds voltage to display tactile sensation. The newly proposed electrotactile display could present the tactile sensation at even lower voltage than the previous work. Given the required low voltage below 5 V, the device can potentially be designed wearable and it strengthens advantages of the needle-type electrotactile device. In the next experiments, we applied

voltage to the needle electrodes in sequence with a certain time lag (switching time) as shown in Fig.2. We investigated the presented tactile sensation with respect to the switching time and pulse frequency. Electric pulse was positive square pulse and the voltage was fixed to 3.0V. The experimental results were shown in Fig. 3. When the switching time was below 20 ms, the subject perceived only pulse-like stimulation. However, when the switching time was between 20 ms and 100 ms, the subject could recognize continuous surfaces whose roughness depended on the switching time. The sensed roughness increased with the switching time. When the switching time was above 100 ms, the subject felt discrete electrical stimulation. From these Effects of the pulse frequency were not dominant.

CONCLUSION

The electrotactile display which has micro-needle array and backside flat electrode was presented. The newly type electrotactile display could present tactile sensation at extremely low voltage and successfully display the surface textures.

ACKNOWLEDGEMENTS

This work was supported in part by JST PRESTO (Information Environment and Humans)

REFERENCES

[1] J. Watanabe, H. Ishikawa, X. Arouette, Y. Matsumoto, N. Miki, "Demonstration of vibrational braille code display using large displacement micro-electro-mechanical systems actuator", *Japanese Journal of Applied Physics*, Vol.51, 2012

[2] T. Watanabe, S. Fukui, "A method for controlling tactile sensation of surface roughness using ultrasonic vibration", *IEEE International Conference on Robotics and Autommation*, pp. 1134-1139, 1995

[3] T. Ishii, N. Hida, A. Yamamoto, T. Higuchi, "Electrostatic Tactile Display Using Thin Film Slider", *6th International Conference on Motion and Vibration Control*, pp.547-552, 2002

[4] T. Iwamoto, M. Tatezono, H. Shinoda, "Non-Contact Method for Producing Tactile Sensation Using Airborne Ultrasound", *Eurohaptics 2008*, LNCS 5024, pp. 504-513, 2008

[5] H. Kajimoto, "Electrotactile display with real-time impedance feedback using pulse width modulation", *IEEE Transactions on Haptics*, pp.184-188, 2012

[6] N. Kitamura, J. Chim, N. Miki, "Effect of needle shape on performance of needle –type electro tactile display", *27th IEEE International Conference on Micro Electro Mechanical Systems*, pp.1183-1184, 2014

[7] T. Deguchi, "Electrolytic Etching Machining by Ethylene Glycol Solutions", *The journal of the Surface Finishing Society of Japan*, Vol.61, No.4, pp.305-306, 2010

CONTACT

*N. Kitamura, tel: +81-45-563-1141;
norihide.kitamura@z2.keio.jp

MICROMACHINED ULTRASOUND TRANSDUCER ARRAY FOR CELL STIMULATION WITH HIGH SPATIAL RESOLUTION

Kyungmin Ko[1,2], Jin-Hyung Lee[1], Hyunjoo Jenny Lee[1], Soo-Jin Oh[1], Ye Eun Chun[1], Tae Song Kim[1], C. Justin Lee[1], Eui-Sung Yoon[1], Kwang-Seok Yun[2], and Il-Joo Cho[1]

[1]KIST (Korea Institute of Science and Technology), Seoul, South Korea
[2]Sogang University, Seoul, South Korea

ABSTRACT

In this work, we present a piezoelectric micromachined ultrasonic transducer (pMUT) array for local stimulation on cultured cells with a high spatial resolution for the first time. We used a bulk piezoelectric film that has a higher piezoelectric coefficient (d_{31}) than that of the thin film materials to achieve high acoustic power within a small membrane. The 500-μm-wide and 55-μm-thick transducer membrane exhibits a resonant frequency of 780 kHz, which is within the effective frequency range for stimulating cells. We also demonstrate successful *in vitro* experiments by stimulating cells using the fabricated pMUT array and verifying the stimulation using fluorescence calcium imaging. Using fluorescence imaging, we observed an increase in Ca^{2+} level of TRPA1 expressing HEK293T cells under ultrasound sonication, which confirms that TRPA1 channel in HEK293T cells was activated by ultrasound.

INTRODUCTION

Extensive studies on the biological effects of brain stimulation using various methods have advanced our basic understanding of biological mechanisms and expanded brain stimulation into clinical therapy or diagnosis. Various techniques such as deep brain stimulation (DBS) [1], electrical stimulation [2], transcranial magnetic stimulation (TMS) [3], and optical stimulations [4] are widely used for clinical or research purposes. However, there are significant drawbacks of previous techniques such as poor spatial resolution [2,3] and invasiveness [1,4]. Therefore, ultrasound has recently emerged as a new stimulation modality in neuroscience because of its non-invasiveness and capability for deep-brain stimulation with higher spatial resolution [5].

However, the mechanism behind the biological effects of ultrasound is still unknown. Thus, recent research efforts have been focused on *in vitro* ultrasound stimulation on hippocampal slices or intact brain circuits to investigate the mechanism [6]. Most of these previous works used commercially available bulky single transducers [6-10] and thus suffered from a few drawbacks such as poor spatial resolution (larger than a few mm²) [7,8], lack of capability to deliver simultaneous ultrasound stimulation on multiple areas, and difficulty in maintaining stable stimulating conditions (*e.g.* power and gap) [9,10]. The development of micro-electro mechanical system (MEMS) technology has opened up new possibilities for improved approaches with precise spatial and temporal investigation of cellular mechanotransduction processes at molecular and single-cell levels [11,12]. MEMS ultrasound transducer array enables not only localized stimulation with small transducers but also stable stimulation by seeding cells on top of the transducers.

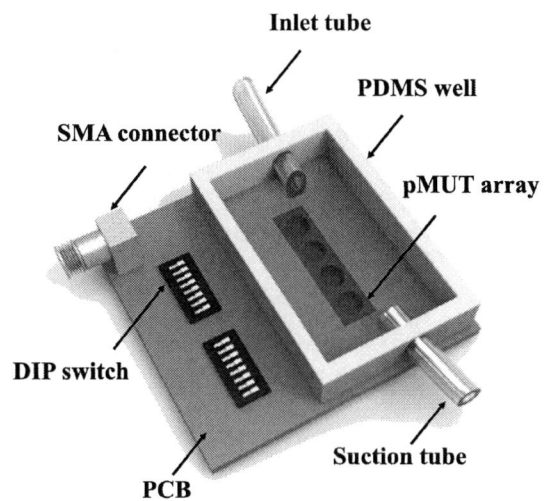

Figure 1: Conceptual diagram of the proposed device platform integrated with a pMUT array.

In this work, we present a pMUT-based ultrasound stimulation platform which is developed with a goal to investigate how the ultrasound is affected to cells *in vitro* (Fig. 1). The stimulation platform is integrated with a cell culture chamber for local stimulation of cultured cells in stable stimulation conditions. The 500-μm-diameter single transducer enables stimulation with high spatial resolution and provides direct ultrasonic stimulation by coating collagen layers on top of the transducer array where cells are seeded. Also, simultaneous stimulations on distinct areas of seeded cells are possible by selectively controlling the transducer array.

FABRICATION PROCESS

We fabricated the proposed pMUT array for cell stimulation using a bulk piezoelectric (PZT) film coated with CuNi electrodes on both sides (PIC 151, PI, Germany). The fabrication process starts with a 4-inch silicon-on-insulator (SOI) wafer and the wafer is bonded with a 1-mm-thick bulk PZT film. As a bonding layer between SOI wafer and bulk PZT film, CYTOP (CTL-809; AGC, Japan) is spin-coated on both sides of the SOI wafer and the PZT film. Then, they are bonded together in a wafer-level bonder at the temperature of 160°C with an applied pressure of 3.5 kg·f/cm². Next, the bonded PZT film is thinned down to 40 μm by chemical mechanical polishing (CMP) process. Then, 300-nm-thick platinum and 20-nm-thick chrome layers are deposited using a sputter and patterned by lift-off process to form top electrodes. To pattern the bulk PZT layer, we use a combination of 150-nm-thick chrome layer and a 10-μm-thick AZ 9260 photoresist layer as an etch mask. The PZT layer is etched in a mixture of H_2O:HCl:HF with a ratio of 250:10:1 until the bottom electrode is exposed. After the bottom electrodes are exposed by the PZT etching process, a 300-nm-thick gold and 20-nm-thick chrome adhesion layers are deposited and patterned using lift-off process to provide electrical connections between bottom electrodes and pads. Next, a 100-nm-thick aluminum layer is deposited and patterned on the backside of the SOI wafer as an etch mask for DRIE process. Finally, the membrane of the transducer is released by DRIE process by using the 0.7-μm-thick buried oxide layer as an etch stop layer (Fig. 2 f).

Figure 3 shows SEM images of the successfully fabricated 16-pMUT array. The top electrodes are isolated while the bottom electrodes are connected through the silicon substrate to allow for individual control of each element. The fabricated pMUT array was mounted on a printed circuit board (PCB) and pads on the pMUT array were connected to the pads on the PCB using wire bonding.

EXPERIMENTAL RESULTS
Device characterization

A customized experimental setup was developed for

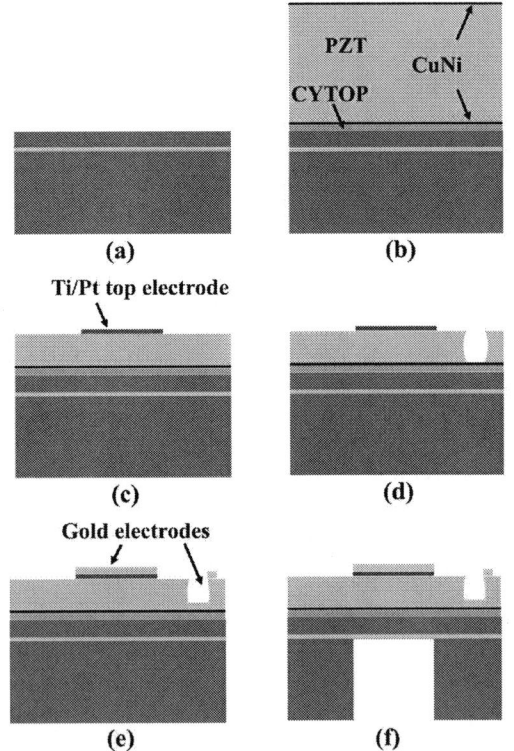

Figure 2: Fabrication process of the proposed pMUT array: (a) Starting SOI wafer with a 15-μm-thick top silicon, (b) bonding of PZT thick film on top of the SOI wafer using CYTOP, (c) top electrode patterning (Ti/Pt) after CMP of PZT thick film, (d) PZT etching for electrical connection of bottom electrode, (e) Au deposition and patterning for electrical connection, and (f) Release from the backside using DRIE.

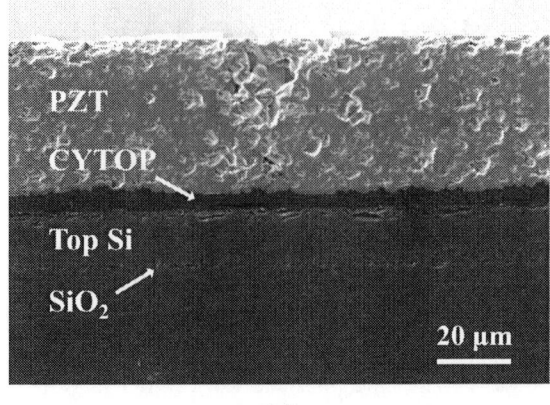

(a)

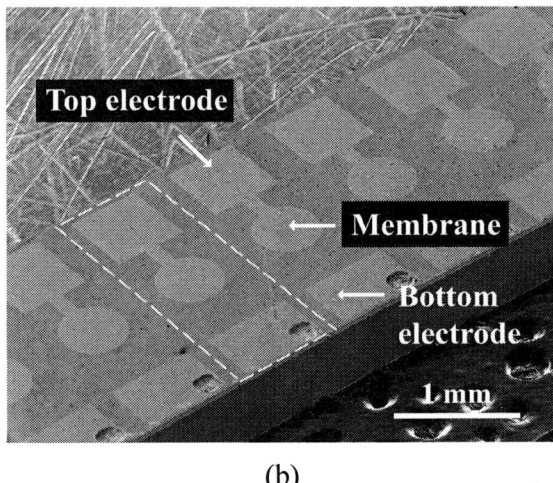

(b)

Figure 3: SEM pictures of the fabricated one-dimensional pMUT array containing 16 elements showing (a) a cross-section of the pMUT, and (b) an angled view of the front-side of the pMUT.

ultrasound stimulation on cells *in vitro* (Fig. 4). Since Type I collagen is a favorable for cell adhesion [13], we prepared 0.5% Type I collagen sheet with a thickness of 300 μm and attached the collagen sheet on top of the pMUT array. To maintain stable environments during the experiments, we integrated a biocompatible polydimethylsiloxane (PDMS) well with inlet/outlet tubes to flow medium for convenient cell culturing protocol and minimal contamination (Fig. 4). The medium that was used in the experiment is a water-like solution (150 NaCl, 10 Hepes, 3 KCL, 2 $CaCl_2$, 2 $MgCl_2$, 5.5 glucose in mM) and the medium has acoustic impedance close to 1.5 MRayls. The ultrasound wave generated from the transducer array was propagated through a collagen sheet to the HEK293T cells. The acoustic impedance of collagen is also approximately 1.5 MRayls, which prevents any reflection of ultrasound.

We used our pMUT device with the first fundamental frequency of 780 kHz, which is within the effective frequency range for stimulating cells (Fig. 5 a). Thus, we constructed waveforms for generating ultrasound with a frequency of 780 kHz to achieve maximum output sound

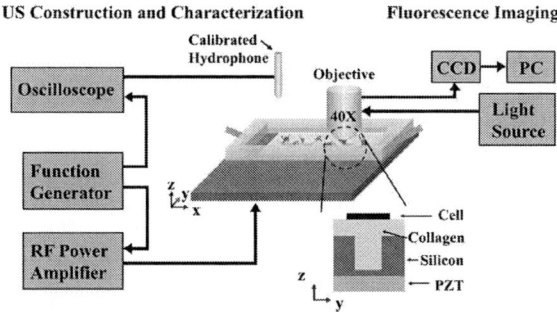

Figure 4: Experimental setup for in vitro ultrasound cell stimulation system that generates waveforms for ultrasound stimulation and monitors fluorescence intensity.

pressure. We modulated the ultrasound carrier signal in kilohertz range to control the duty cycle of the ultrasound stimulation without generating macroscopic heating [14]. Single ultrasound pulses contained 100 acoustic cycles per pulse for pulse duration (PD) lasting 0.13 ms and the pulses were repeated at pulse repetition frequencies (PRF) of 2 kHz (Fig. 5 b). The peak sound pressure of the pulsed ultrasound waveforms was 0.024 MPa (Fig. 5 c). From the measured peak sound pressure, pulse intensity integrals (PII) is calculated to be 0.004mJ/cm^2. While spatial-peak pulse-average intensities (I$_{SPPA}$) is 0.032W/cm^2, spatial-peak temporal-average intensities (I$_{SPTA}$) of our system is calculated to be 8mW/cm^2.

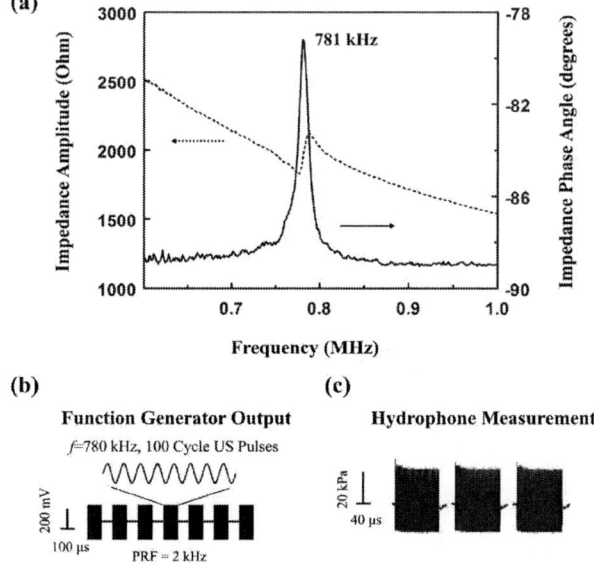

Figure 5: Characterization of the fabricated pMUT array: (a) plot of input impedance measurement, (b) diagram of output voltage of function generator to drive the ultrasound transducer, and (c) plot of the measured sound pressure of a single pMUT.

In vitro experiment

To verify the ultrasound stimulation on cells using the fabricated pMUT array, we monitored fluorescence calcium imaging. The transient receptor potential ankyrin 1 (TRPA1) channels are known as the mechanical receptors that mediate intracellular Ca^{2+} transient in astrocyte-neuron co-culture environment [15]. Therefore, we obtained time-resolved measurements of intracellular calcium transient in TRPA1 expressing HEK293T cells under ultrasound sonication (Fig. 4). The intensity changes of Fura-2 fluorescence in cells during ultrasound stimulation (Fig. 6 (yellow arrow)) indicates that Fura-2 fluorescence increased during the ultrasound stimulation then decreased back to the initial calcium level when the stimulation was turned off. The corresponding fluoroscence intensities are graphically shown for 340 and 380 nm excitation during ultrasound stimuli from 150 to 350 s. When Ca^{2+} increases in the intracellular cells, Fura-2 fluorescence intensity increases with 340 nm excitation while Fura-2 fluorescence intensity decreases with 380 nm excitation. Therefore, when observing the ratio (340/380 excitation) of the fluorosence intensties, we observed an increase in the intracellular calcium level during the ultrasound stimuli with the peak intensity ratio at 270 s. Using fluorescence imaging, we observed an increase in Ca^{2+} level of HEK293T cells under ultrasound sonication, which confirms that TRPA1 channels in HEK293T cells are activated by ultrasound.

After the experiment, we applied TFLLR to the cells to confirm the activity of the cells and we observed an increase in intracellular Ca^{2+} level (Fig. 7 b).

CONCLUSIONS

In this work, we present a pMUT array, its experimental setup and its use for the first time in local stimulation *in vitro* with a high spatial resolution. We observed an increase in Ca^{2+} level of TRPA1 expressing HEK293T cells under ultrasound sonication, which confirms that TRPA1 channel in HEK293T cells is activated by ultrasound (*i.e.* through mechanical stress on cells). This new device for ultrasound stimulation is a promising platform to study the effects of ultrasound on cells and brain slices *in vitro* and to investigate the effects of ultrasound stimulation. Selectively stimulating different

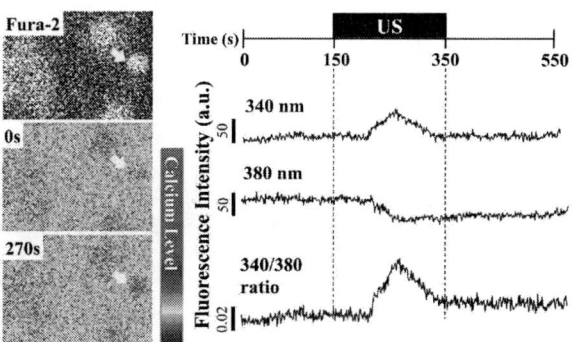

Figure 6: Fura-2 florescence optical image of HEK293T cells and intracellular calcium increase during the ultrasound sonication (US).

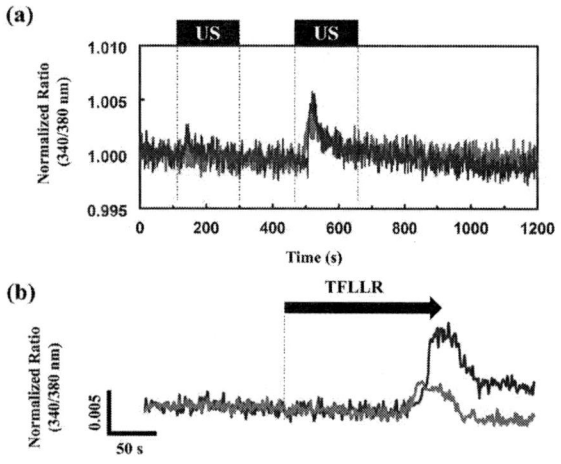

Figure 7: Transient plots of calcium imaging: (a) intracellular calcium increase during the multiple ultrasound sonication (US) and (b) TFLLR induced intracellular calcium increase. (Each red and blue line represents two distinct HEK293T cells.)

regions of cells with high spatial resolution is part of our future work.

REFERENCES

[1] J. Perlmutter, J. Mink, "Deep Brain Stimulation", *Annu. Rev. Neurosci.*, vol. 29, pp. 229-257, 2006.

[2] M. Histed, A. Ni, J. Maunsell, "Insights into Cortical Mechanisms of Behavior from Microstimulation Experiments", *Progress in neurobiology*, vol. 103, pp. 115-130, 2013.

[3] T. Wagner, A. Valero-Cabre, and A. Pascual-Leone, "Noninvasive Human Brain Stimulation", *Annual Review of Biomedical Engineering*. vol. 9, pp. 527-565, 2007.

[4] P. Kalanithi and J. Henderson, "Optogenetic Neuromodulation," *Emerging Horizons in Neuromodulation: New Frontiers in Brain and Spine Stimulation*, vol. 107, pp. 185-205, 2012.

[5] T. Wagner, A. Valero-Cabre, "Noninvasive Human Brain Stimulation", *Annu. Rev. Biomed*, vol. 9, pp. 527-565, 2007.

[6] W. Tyler, Y. Tufail, M. Finsterwald, M. Tauchmann, "Remote Excitation of Neuronal Circuits using Low-intensity, Low-frequency Ultrasound", *PLoS One 3*, vol. 10, e3511, 2008.

[7] Y. Hu, W. Zhong, J. Wan, A. Yu, "Ultrasound can Modulate Neuronal Development: Impact on Neurite Growth and Cell Body Morphology", *Ultrasound in medicine & biology*, vol. 5, pp. 915-925, 2013.

[8] R. King, J. Brown, W. Newsome, K. Pauly, "Effective Parameters for Ultrasound-Induced *In Vivo* Neurostimulation", *Ultrasound in medicine & biology*, vol. 2, pp. 312-331, 2013.

[9] A. Schuster, T. Schwab, M. Bischof, M. Klotz, "Cell Specific Ultrasound Effects are Dose and Frequency Dependent", *Annals of Anatomy-Anatomischer Anzeiger*, vol. 195, pp. 57-67, 2013.

[10] A. Katiyar, R. Dunca, K. Sarkar, "Ultrasound Stimulation Increases Proliferation of MC3T3-E1 Preosteoblast-like Cells", *Journal of Therapeutic Ultrasound*, vol. 2, pp. 1-10, 2014.

[11] F. Kurth, K. Eyer, A. Franco-Obregon and P. S. Dittrich, "A New Mechanobiological Era: Microfluidic Pathways to Apply and Sense Forces at the Cellular Level", *Current Opinion in Chemical Biology*, vol. 16, 2012, pp. 400-408.

[12] T. Shibata, G. Umegaki, Y. Ishihara, M. Nagai, "Fabrication and Characterization of Cell Culture Microdevice for Nanomechanical Stimulation of Living Cells", in *The IRAGO CONFERENCE 2013*, Aichi, Oct. 24-25, vol. 1585, AIP Publishing, 2014, pp. 108-116.

[13] V. L. Cross, Y. Zheng, N. W. Choi, S. S. Verbridge, B. A. Sutermaster, L. J. Bonassar, *et al.*, "Dense Type I Collagen Matrices that Support Cellular Remodeling and Microfabrication for Studies of Tumor Angiogenesis and Vasculogenesis *In Vitro*," *Biomaterials*, vol. 31, pp. 8596-8607, Nov 2010.

[14] M. D. Menz, O. Oralkan, P. T. Khuri-Yakub, and S. A. Baccus, "Precise Neural Stimulation in the Retina Using Focused Ultrasound," *Journal of Neuroscience*, vol. 33, pp. 4550, 2013.

[15] D. Corey, J. García-Añoveros, J. Holt, K. Kwan, "TRPA1 is a Candidate for the Mechanosensitive Transduction Channel of Vertebrate Hair Cells", *Nature*, vol. 432, pp. 723-730, 2004.

MINIMALLY INVASIVE NEEDLE-FREE BUBBLE INJECTOR FOR GENE THERAPY

Kazuki Takahashi[1], Shun Omi[1] and Yoko Yamanishi[1, 2]
[1]Shibaura Institute of Technology, JAPAN
[2]Japan Science and Technology Agency (JST) PRESTO, Japan

ABSTRACT

We have successfully developed minimally-invasive needle-free bubble-injector designed for usage in air. The novelty is that minimally-invasiveness of injection whose resolution is less than 10 µm, and hence cellular-scale and painless injection can be obtained. The bubble-injector can be employed to any kind of materials with wide range of hardness using strong impact of cavitation. The fine adjustment of invasiveness of injection can be controlled by the number of applied electric pulses. The developed injector can be used for wide range of biomedical study, especially in gene therapy.

BACKGROUND

Needle-free injection systems were initially described by Marshall Lockhart in 1936 in his patent jet injection. Then in the early 1940's Higson and others developed high pressure "guns" using a fine jet of liquid to pierce the skin and deposit the drug in underlying tissue [1, 2]. The technology for needle-free injection keeps developing since then, however the minimally invasiveness with higher resolution as well as sufficient penetration ability is still required for painless injection. The authors has developed electrically-induced bubble knife for cell surgery and gene injection for animal and plant cell recently. However, it can be used only under electrolysis solution such as medium environment [3]. For the present study, the design of bubble knife is revised to fit to use under air environment by using a "shielding chamber function" which can adjust the contact pressure to cell.

CONCEPT

Figure 1 shows the concept of the needle-free bubble injector. The size of the edge of the inner electrode was about 30 µm and diameter of orifice was about 5 µm. The orifice has a thick structure to gain the robustness of bubble injector as well as preventing from penetration of object. Reagent is filled in the space between outer glass shell and inner glass electrode. Also, damper structure is designed to improve positioning performance and to prevent from break by oscillations which tend to be occurred in-vivo injection such as heart beat. Finally, constriction structure is designed to keep coaxially of inner glass electrode. By combination of using the revised design enables to have "shielding chamber function" at the tip of the bubble knife no matter what the ambient environment is air or any other liquid. Hence the reagent can be filled in only inside of the shielding chamber without any leakage to create same environment as the conventional bubble knife injection within the local area. The contact pressure of the chamber can be adjusted by the hardness of the targeting object.

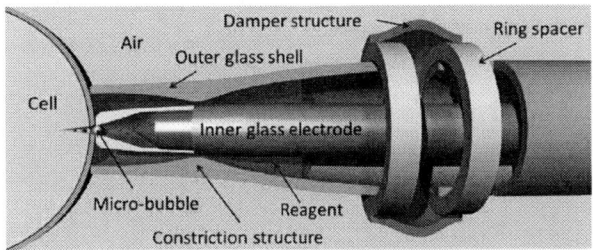

Figure 1: Schematics of minimally-invasive needle-free bubble-injector with damper control.

The mechanism of injection by using bubble cavitation is described as follows [3]. When the voltage is applied to the inner glass electrode, a bubble is generated in the small closed-space mounted at the tip of the inner glass electrode by electrolysis or by heat, and the bubbles are accumulated and glow to fit to the inner glass diameter in the space. Continuous applied electric pluses provide bubble ejection eventually. Then the generated bubble is collapsed suddenly as soon as after the ejection. Immediately after that, micro-jet caused by collapse of the bubble which is shown in Figure 2, and this needle-like sharp edge of jet perforates the cell. At the same time, reagent is transported into cell by adsorption force of the air liquid interface of bubble.

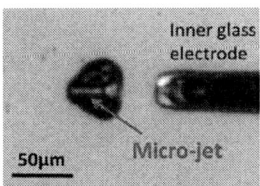

Figure 2: Micro-jet generated from needle-free bubble-injector (1,000,000fps taken by Shimadzu HPV-X)

FABRICATION

Figure 3 shows the process flow to fabricate the needle-free bubble injector. Figure 3(a) shows how to fabricate the robust thick structure at the edge. First of all, the glass tube with inserted copper wire is pulled by using glass puller (P-1000IVF, Sutter Co. Ltd.). Then the tip of the glass tube is rounded by using micro-forge (MF-900, Narishige Co. Ltd.) to be robust thick structure to prevent from break by oscillations which tend to be occurred in-vivo injection such as heart beat.

Figure 3(b) shows the fabrication process to have accurate coaxial structure of outer glass shell. The constriction region was created to align the electrode position at the center. The constriction part was fabricated by glass puller and the inner diameter of constriction part of outer glass shell is just fit to the inner glass electrode to support accurate positioning.

Figure 3(c) shows the how to fabricate the damper structure in order to adjust the contact pressure on the surface. Thin PDMS sheet was employed to wrap the gap between the two ring spacer mounted at the outer glass shell and reagent supply tube. By controlling the diameter of the ring or the thickness of PDMS sheet, the contact pressure to the target material can be adjusted. For example, if the material to be injected is hard material, the diameter of the ring should be large.

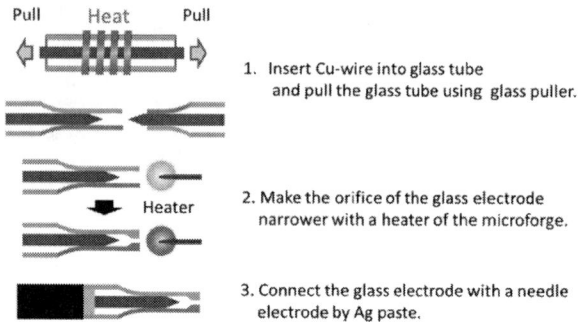

Figure 3(a): Fabrication process to produce inner glass electrode with robust thick structure at the edge.

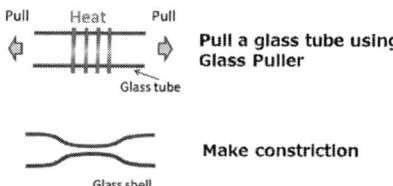

Figure 3(b): Fabrication process to have accurate coaxial structure of injector using glass puller.

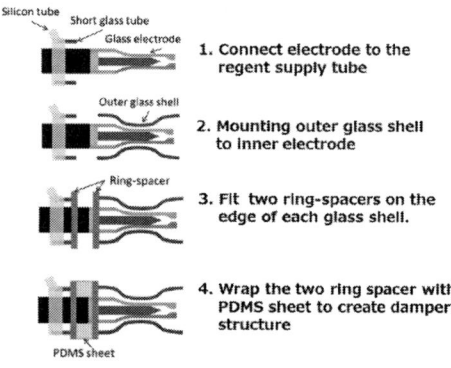

Figure 3(c): Fabrication process to have damper structure to adjust the contact pressure on the surface.

Figure 3(a)-(c): Fabrication process to produce needle-free bubble-injector.

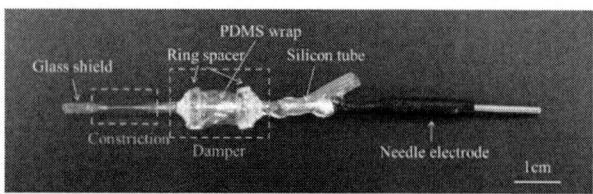

Figure 4: Fabricated needle-free bubble-injector .

Figure 4 shows a photo of fabricated needle-free bubble-injector. Outer glass tube is connected to the reagent supplying silicon tube and which connect to the peristaltic pump (RP-TX, Aquatech Co. Ltd.). Also, the inner glass electrode is connected to the needle electrode which is connected to the power supply (Hyfrecator2000). A glass shield is optionally mounted to inject harder material to protect the tip of probe.

EXPERIMENT

The electrical circuit used for the current study is identical to the previous work [3]. Figure 5 shows the electrical circuit of the bubble injector which is modification of the normal electric knife. Non-inductive resistance which was installed to the circuit to adapt to cellular-scale ablation was 10.82 kΩ. Discharging time and timing was controlled by Digital input output board which time resolution was 1ms and Electrically-induced micro-bubble whose generation speeds is typically 30.8 kHz.

First of all, we have evaluated the robustness of the injector. It is important to note that we have revised the design of the tip of the probe to the thick structure. Figure 6 shows the comparison between the conventional design and revised design. For the conventional orifice shape which is shown as upper photos, bubble-injector was broken at the input voltage of 9 W. On the other hand the revised bubble-injector which has narrow orifice channel and thick structure at the edge can be operated perfectly until the input voltage is reached to 19 W.

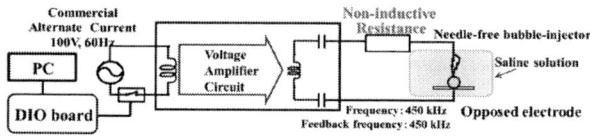

Figure 5 :Electrical circuit to eject bubbles.

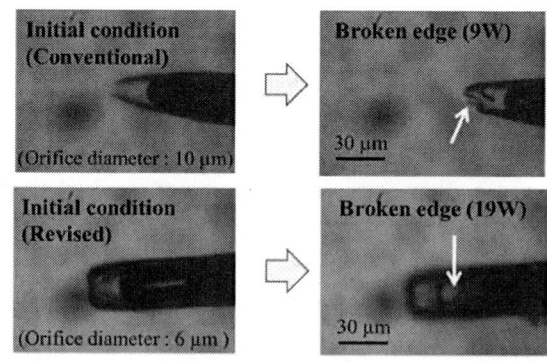

Figure 6 : Evaluation of robustness of thick structure at the edge.

Secondly, we have evaluated the perforation diameter of bovine oocyte which is shown in Figure 7. It is important to note that the cross-section of the perforated hole has minimal damage. Next Figure 8 shows the perforation diameter as a function of input power.

978-1-4799-7956-1/15 $31.00 © 2015 IEEE

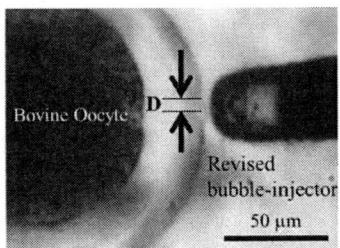

Figure 7: Evaluation of perforation diameter of bovine oocyte.

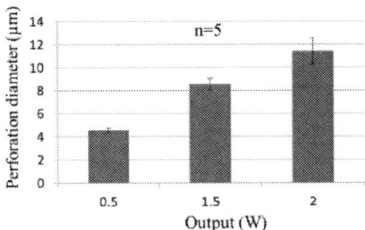

Figure 8: Evaluation of perforation diameter as a function of input power of power supply.

It was observed that the increase of the perforation diameter with increase of input power of power supply. The range of perforated diameter was about 5 μm to 11μm for the input voltage of 0.5 W-2 W as shown in Figure 8.

Thirdly, the contact force of needle-free bubble injector against target object was evaluated. Figure 9 shows the evaluation of the contact force using the damper structure. For the testing of evaluation, a precise pressure sensor is employed and the spring coefficient was evaluated. According to the graph, the damper is movable up to mN order. As authors have previous described, the contact pressure can be adjusted by controlling the diameter of the ring or the softness of the PDMS sheet to adapt to the wide range of object.

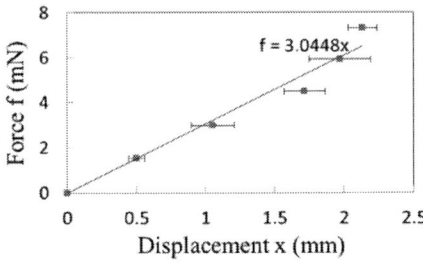

Figure 9: Evaluation of displacement of the damper as a function of applied force

Finally, the operation of needle-free injector is carried out by using chicken meat. Figure 10 shows the demonstration of the fluorescent bead(ϕ 2.1 μm) injection. First of all the outer glass shield contact to the surface of the meat, then the regent is filled in the closed space. After the reagent is filled in the closed space (Figure 10(a)), the high-speed bubble is ejected. Then, Figure 10 (b) and Figure 10(c) show a hole whose size is less than 10μm. It

was confirmed that the damper structure worked well. It is important to note that air-liquid interface was maintained at the edge of the probe so that preventing leakage of reagent. Figure 11 shows a photo when the florescent bead was injected successfully. The bright field photo showed a hole was produced in the chicken meat and dark field photo shows the injected fluorescent beads.

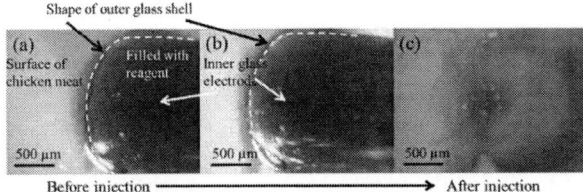

Figure 10: Demonstration of reagent injection to the chicken meat under air environment.

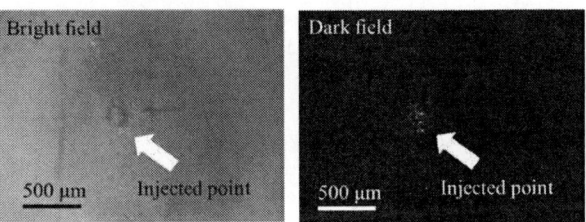

Figure 11: Confirmation of injected fluorescent bead in the chicken meat.

CONCLUSION

The paper shows the fabrication and operation of needle-free bubble-injector for the first time. The perforation diameter of bovine oocyte was about in the range of 5 μm to 11 μm. We have successfully injected fluorescent bead to chicken meat under air environment by using needle-free bubble-injector. This novel injection method can be applied to wide range of study such as gene therapy and direct gene injection to the hard plant materials, and also it is contribute to minimally-invasive injection with less consumption of reagent for wide biological target and biomedical researches.

ACKNOWLEDGEMENTS

This work was supported in part by JST PRESTO program the Ministry of Education, Culture, Sports, Science and Technology (25289059, 25630091 and 25108504). The authors thank Prof. Motoji Kaya, Tokyo Univ. for valuable advices and Shimadzu co. ltd. Japan for high-speed camera demonstration.

REFERENCES

[1] C.C. L. Chase., et al., Journal of Swine Health and Production, 16(5), .pp.254-261, (2008).
[2] T. R. Kale, M. Momin, Innovations in pharmacy, 5(1), 148, (2014).
[3] H. Kuriki, *et al., Proc. MEMS 2013,* pp. 209-211.

CONTACT

*Y.Yamanishi, tel: +81-3-5859-8013;
 yoko@shibaura-it.ac.jp

978-1-4799-7956-1/15 $31.00 © 2015 IEEE

OPTIMIZATION OF DRUG COCKTAIL ON AN INTEGRATED MICROFLUIDIC SYSTEM

Wen-Yen Huang[1], Kuan Wang[4] and Gwo-Bin Lee,[1,2,3]*

[1]Department of Power Mechanical Engineering, [2]Institute of NanoEngineering and Microsystems,
[3]Institute of Biomedical Engineering National Tsing Hua University, Hsinchu, Taiwan
[4]Institute of Biological Chemistry, Academia Sinica, Taipei, Taiwan

ABSTRACT

Optimization of drug cocktails has been an important issue for a number of therapies of complicated diseases; however, it is an extremely task using traditional methods. Here we demonstrate a new microfluidic platform capable of automatically dispensing, mixing a small amount of drug combinations, and adding them to the cell cultures such that a precise and fine-tuned drug cocktails may be formed for the subsequent cell-level testing. Our proposed system could decrease manual bias and enhance the throughput of drug cocktail formulation on an automated and minituriazed microfluidic system.

INTRODUCTION

Combination of different therapeutic agents has been extensively used to improve single-drug efficacy and lower side effects in clinical applications. These drug formulations, so called drug cocktails, are becoming the standards of treatment for a number of complicated diseases, such as cancers, infectious diseases, neurodegenerative diseases, and metabolic syndromes. Optimization of drug cocktail formulation by using traditional trial-and-error methods is a relatively labor-intensive and time-consuming task due to the considerable amounts of combinations. For instance, finding the optimal conditions for six different drugs at 10 different concentrations requires 1 million potential tests. Furthermore, a complete search for the most optimal combination in *in-vivo* or clinical tests is not practical even for combinations of only 2 to 3 drugs. Recently, a feedback system control (FSC) scheme [1] was reported which could rapidly identify the best combination of drug dosage in fewer tests and bypass the need to test all the potential test trials [2, 3]. The approach can save up to 3-5 orders of magnitude in experimental expenditure. This FSC platform has been successfully applied to the identification of drug combinations for various biological applications including inhibition of vesicular stomatitis virus, activation of Kaposi's Sarcoma-associated herpesvirus, and defining a pathogen-free culture system for the long-term maintenance of human embryonic stem cell [4]. In addition, several studies using the FSC scheme have succeeded to identify synergistic antiviral drug combinations or useful drug ingredients in herbal medicine [5, 6]. Briefly, this FSC scheme, consisting of an iterative loop of three operations, includes the formulation of therapeutic agents, experimental readouts, and a search algorithm linking the cellular readouts and the cocktail combinations to generate new combinations for the subsequent iteration of next experimental tests.

Although FSC avoids tremendous efforts to test all potential trials, a few iterative cycles of close-loop optimization are required, indicating that tedious drug combination processes using a small amount of drugs is still inevitable. Therefore, a new system to combine the FSC scheme and automatic drug dispensing is greatly needed. It is envisioned that the development of an integrated microfluidic and biomolecular sensing systems to automatically perform drug cocktail optimization could tackle this issue. Recently, microfluidic systems have become available to perform several crucial operations on a single, integrated system including sample pretreatment, transporting, mixing, reactions, and separation and detection [7, 8]. The development of an integrated microfluidic system dedicated for FSC-based drug cocktail optimization may offer several advantages including a significant decrease in sample and reagent consumption (which could be relatively expensive), high throughput, low power consumption and low costs. This work therefore reports our recent process on the development of the integrated microfluidic system for fast formation of drug cocktail formulation.

MATERIALS AND METHODS
Experimental Procedure

In this work, we designed and fabricated a micro-dispenser and a micro-mixer for automatically dispensing different drug combinations to cell cultures for the subsequent FSC testing. Figure 1 shows a schematic illustration of the integrated drug dispensing and mixing microdevices for one FSC ietration. Briefly, six different drug stock solutions were first pre-loaded in the drug storage chambers of the micro-dispenser. Then, the desired volumes of drugs were injected into the central chamber automatically by activating the pneumatically-driven micro-dispenser. The unmixed drugs were subsequently loaded to the single chamber of the micro-mixer through the funnel-like structure with gravity. Each drug combination was then dispensed to one chamber of the micro-mixer until all 16 chambers were loaded with specified combinations of drugs (Fig. 1(a)). Next, all the drug combinations were automatically mixed by activating the pneumatically-driven micro-mixer simultaneously (Fig. 1(b)). The resulting drug mixtures were then injected from the micro-mixer to the cell cultures on a 96-well microliter plate by using an air shower scheme (Fig. 1(c)). Finally, the cells treated with different combinations of drugs were cultured and observed with a time-lapse imaging system for cell viability or other markers for 3 to 7 days. The data of cell viability assays were finally calculated by the FSC algorithm to determine the amount for each drug tested in the next round of drug dispensing (Fig. 1(d)).

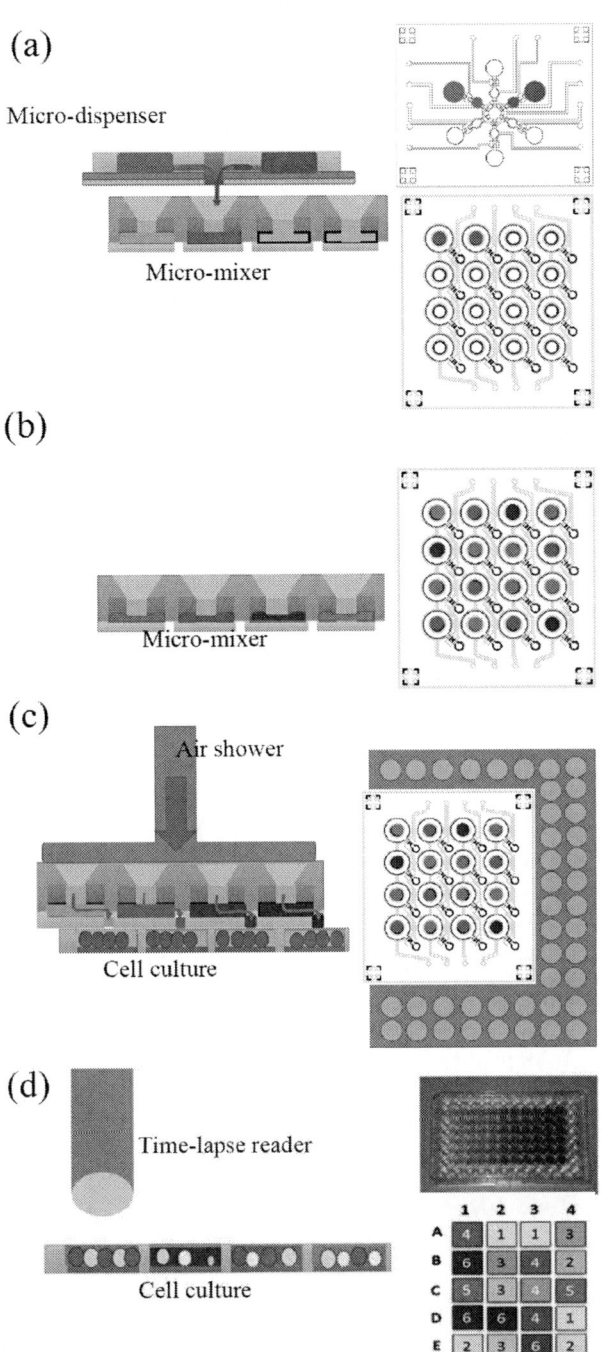

Figure 1: Schematic illustration of a six-drug micro-dispenser and a sixteen-combination micro-mixer for one FSC iteration. (a) drug dispensing; (b) drug mixing; (c) drug injection and (d) cell analysis.

Chip Design and Fabrication

The integrated microfluidic chips are shown in Figure 2. Figures 2(a) and 2(b) are the photographs of the micro-dispenser and the micro-mixer, respectively. The chip was composed of five layers as indicated by an exploded view of the microfluidic chip, as shown in Figures 2(c) and 2(d). Briefly, both of them were composed of two polydimethylsiloxane (PDMS) layers including an air control layer (thick-film PDMS layer) and a liquid channel layer (thin-film PDMS layer), and a glass substrate. The dimensions of the micro-dispenser chip were measured to be 38.6 mm x 46.0 mm, while the dimensions of the micro-mixer chip were 52.6 mm x 46.0 mm. Note that red dye indicates the chambers in the liquid channel layer and blue dye indicates the chambers in the air control layer (Figs. 2(a) and (b)). In addition, the backside of micro-dispenser was covered with a polyethylene film, which is a hydrophobic material, on the glass substrate to prevent the spreading and accumulation of liquids on the glass substrate. In order to avoid the loss of liquids while injecting them from the micro-dispenser to the micro-mixer, a PDMS-made funnel-like structure was added and bound on the top of the micro-mixer (Fig. 2 (e)).

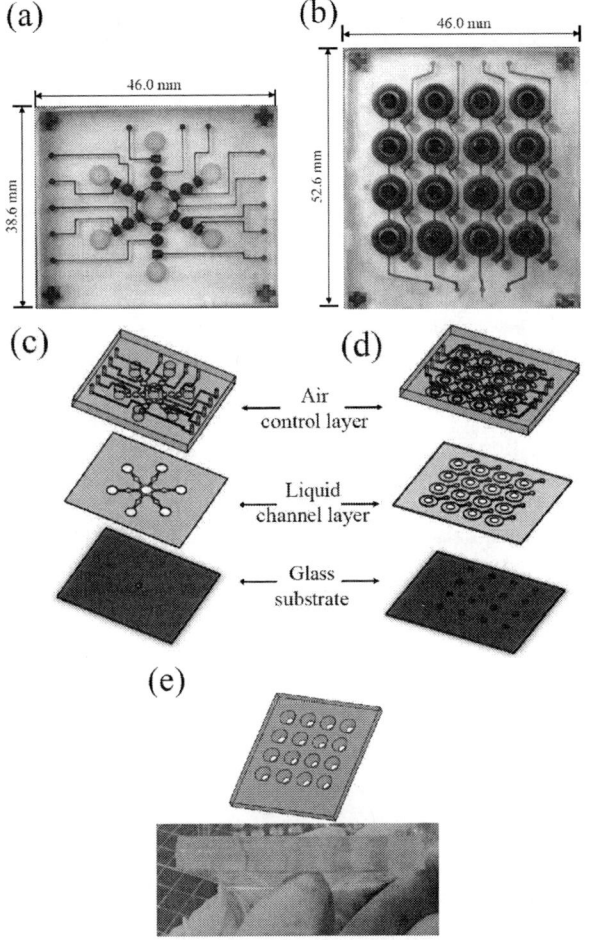

Figure 2: Photographs of (a) the drug micro-dispenser and (b) the drug micro-mixer chips. Red indicates the liquid chamber and blue indicates the air chamber; (c) and (d) are exploded views of the three-layer chips. An additional funnel-like structure is shown in (e).

Pumping Volume and Mixing Index Measurement

The pumping volume was measured as previously described [9]. Briefly, the volume of the different liquid solvents was measured after twenty pneumatically-driven cycles of the micro-pump in the micro-dispenser, and the average value of the pumping volume were calculated accordingly. To measure the mixing efficiency of the micro-mixer, the mixing index was measured as previously described [10]. Briefly, 1 μl of red ink was added into a chamber of the micro-mixer containing 30 μl of water. A high-speed digital camera was used to acquire the optical images while the pneumatically-driven mixing was launched, and these images were subsequently analyzed for the color distribution by using imaging process software.

DNA Plasmid Tests and Gel Electrophoresis

In order to evaluate the accuracy of dispensing by using our microdevices, DNA plasmids were used as the solutes, which were mimicked as drugs, for the pilot experiments. DNA plasmids are relatively stable in water at room temperature and could be easily measured in standard molecular biology laboratories. Most importantly, two different sizes of DNA plasmids do not interact crossly into another size or form under a normal condition. According, we prepare two DNA plasmids (pEGFP-N1 and pX330), which are 4.7 and 8.5 kilo-bases, respectively. DNA plasmids were linearized with *Eco*RI restriction enzyme digestion and then were subjected to the micro-dispenser. After dispensing, the DNA mixture were subsequently performed on slab-gel electrophoresis and analyzed with an imaging densitometer for semi-quantification of the two DNA plasmid amounts.

RESULTS AND DISCUSSION

Most of chemical drugs are dissolved in liquid solvents. To evaluate the capacities of the micro-dispenser in different solvents, pumping volumes of the micro-dispenser were measured. Figure 3 shows the results of liquid injection by the micro-dispenser. Three kinds of liquid solvents (ethanol, double distilled water (ddH$_2$O), and dimethyl sulfoxide) were tested, showing no significant differences of transported volumes among the solvents, which are commonly used for drugs. Additionally, the transported volume was observed to increase along with the applied gauge pressure, demonstrating that the desired volume of injected liquids could be precisely fine-tuned.

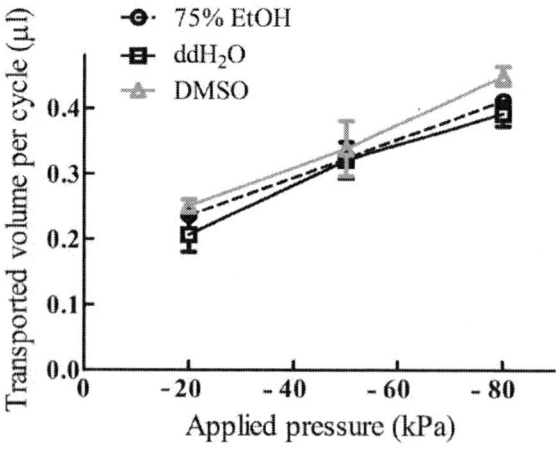

Figure 3: Tests of pneumatically-driven drug micro-dispenser. The injected volume of seventy-five percents of ethanol (EtOH), ddH$_2$O, and dimethyl sulfoxide (DMSO) were measured under different applied gauge pressures

Furthermore, two different kinds of DNA plasmids with different lengths were then injected for different combinations in order to verify if the micro-dispenser is capable of dispensing specified volumes of two different solutions precisely. Note that these two DNA plasmids were dispensed with indicated ratios on bench (manual) or on chips for comparison. The experimental results (Fig. 4) indicate that the micro-dispenser has better performance than manual operation. In addition to the DNA plasmids for pilot experiments, other substances, for example, dyes or chemical drugs are also tested. The experiments are undergoing.

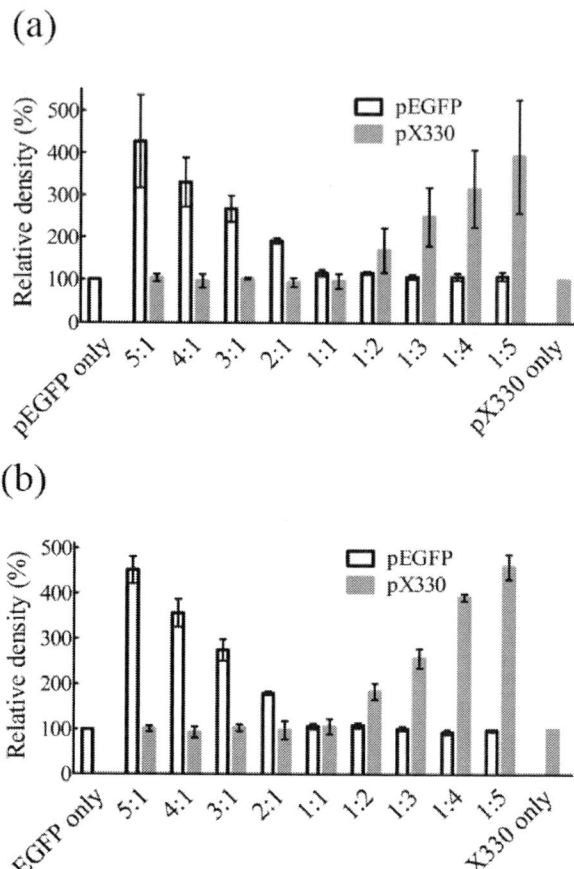

Figure 4: Dispensing test of the micro-dispenser using two DNA plasmids (pEGFP and pX330), which were dispensed with indicated ratios (a) on bench (manual) (a) or (b) on chips. The DNA mixture were subsequently performed on slab gel electrophoresis and analyzed with an imaging densitometer for semi-quantification of the two DNA plasmid amounts.

After the precise injection of drugs, the micro-mixer was activated to efficiently mix the drug cocktails. Note that the precipitation of the drugs could be alleviated by this approach as well. As shown in Figure 5, the mixing index could achieve 96% within 10 sec under a driving frequency of 5 Hz.

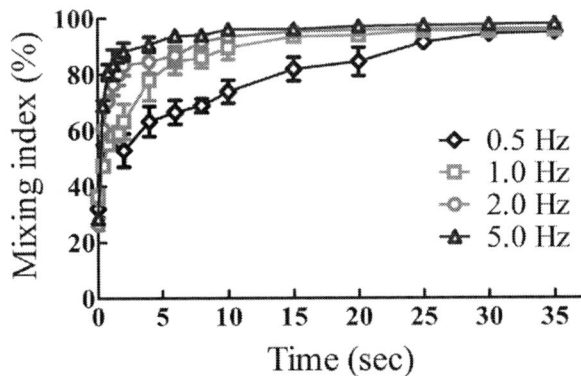

Figure 5: The mixing index of drug micro-mixer at different driving frequencies. Ninety-six percents of mixing index was achieved in 10 sec at a frequncy of 5.0 Hz under a gauge pressure of 70 kPa.

CONCLUSION

In the present study, an integrated system that combines the advantages of microfluidic devices and FSC iteration scheme was established to optimize drug cocktails automatically with both less manual bias and few redundant trial-and-error efforts. The preliminary data showed that the developed device is capable of dispensing drug cocktail formulation with reasonable precision. Furthermore, the system demonstrates a great potential to expedite the screening of therapeutic drug cocktails.

ACKNOWLEDGEMENTS

The authors would like to thank the financial support from Ministry of Science and Technology in Taiwan. Partial financial support from the "Towards A World-class University Project" and Nanomedicine Program, Academia Sinica are also greatly appreciated.

REFERENCES

[1] P. K. Wong, F. Yu, A. Shahangian, G. Cheng, R. Sun, C. M. Ho, "Closed-loop Control of Cellular Functions Using Combinatory Drugs Guided by a Stochastic Search Algorithm", *Proc. Natl. Acad. Sci. U.S.A.*, vol. 105, pp. 5105-5110, 2008.

[2] F. Wei, B. Bai, C.M. Ho, "Rapidly Optimizing an Aptamer Based BoNT Sensor by Feedback System Control (FSC) scheme", *Biosens. Bioelectron.*, vol. 30, pp. 174-179, 2011.

[3] C. P. Sun, T. Usui, F. Yu, I. Al-Shyoukh, J. Shamma, R. Sun, C. M. Ho, "Integrative Systems Control Approach for Reactivating Kaposi's Sarcoma-associated Herpesvirus (KSHV) with Combinatory Drugs", *Integrative Biol. (Camp)*, vol. 1, pp. 123-130, 2009.

[4] H. Tsutsui, B. Valamehr, A. Hindoyan, R. Qiao, X. T. Ding, S. L. Guo, O. N. Witte, X. Liu, C. M. Ho, H. Wu, "An Optimized Small Molecule Inhibitor Cocktail Supports Long-term Maintenance of Human Embryonic Stem Cells", *Nat Commun*, vol. 2, pp. 167, 2011.

[5] X. Ding, D. J. Sanchez, A. Shahangian, I. Al-Shyoukh, G. Cheng, C. M. Ho, "Cascade Search

for HSV-1 Combinatorial Drugs with High Antiviral Efficacy and Low Toxicity", *Int. J. Nanomed.*, vol. 7, pp. 2281-2292, 2012.

[6] H. Yu, W. L. Zhang, X. Ding, K. Y. Z. Zheng, C. M. Ho, K. W. K. Tsim, Y. K. Lee, "Optimizing Combinations of Flavonoids Deriving from Astragali Radix in Activating the Regulatory Element of Erythropoietin by a Feedback System Control Scheme", *Evid. Based. Complement. Alternat. Med.*, Article ID 541436, 2013.

[7] P. A. Auroux, D. Iossifidis, D. R. Reyes, A. Manz, "Micro Total Analysis Systems. 2. Analytical Standard Operations and Applications", *Anal. Chem.*, vol. 74, pp. 2637-2652, 2002.

[8] G. B. Wisdom, A. Wochner, M. Menger, D. Orgel, B. Cechet, M. Rimmele, V. A. Erdmann, "Enzyme-immunoassay", *Clin. Chem.*, vol. 22, pp. 1243-1255, 1976.

[9] S. Y. Yang, J. L. Lin, G. B. Lee, "A Vortex-type Micromixer Utilizing Pneumatically Driven Membranes", *J. Micromech. Microeng.*, vol. 19, pp. 035020, 2009.

[10] C. H. Weng, K. Y. Lien, S. Y. Yang, G. B. Lee, "A Suction-type, Pneumatic Microfluidic Device for Liquid Transport and Mixing," *Microfluidics and nanofluidics*, vol. 10, pp. 301-310, 2011.

CONTACT

*Dr. Gwo-Bin Lee, tel: +886-3-5715131 Ext.33765; gwobin@pme.nthu.edu.tw

MEMS ELECTROCHEMICAL PATENCY SENSOR FOR DETECTION OF HYDROCEPHALUS SHUNT OBSTRUCTION

Brian J. Kim, Willa Jin, Lawrence Yu, and Ellis Meng
Department of Biomedical Engineering, University of Southern California, Los Angeles, CA, USA

ABSTRACT

Currently there are no practical methods to definitively diagnose the obstruction failure of hydrocephalus shunts without surgical intervention, resulting in suboptimal treatment and needless patient suffering. We present the first MEMS electrochemical patency sensor for *direct and quantitative* tracking of shunt patency and obstruction. The impact of electrode size, temperature, flow, and hydrogen peroxide (H_2O_2) plasma sterilization on sensor function was evaluated and sensor performance in the presence of static and dynamic obstruction was demonstrated. Electrode size was found to have a minimal effect on sensor performance and increased temperature and flow resulted in a slight decrease in the baseline impedance (~8.5%) due to an increase in ionic mobility. H_2O_2 plasma sterilization also had no effect on sensor performance. This low power and simple format sensor was developed with the intention of future integration into shunts for wireless monitoring of shunt state and more importantly, a more accurate and timely diagnosis of shunt failure.

INTRODUCTION

Hydrocephalus is a chronic, incurable condition characterized by the accumulation of excess cerebrospinal fluid (CSF) within the ventricles of the brain, resulting in elevated intracranial pressure. Currently, hydrocephalus is treated by implanting a shunt, a multi-holed silicone catheter and valve system, into the ventricles to drain the excess CSF. Though effective, shunts fail at an alarming rate of 40% within the first year [1]. There are many causes of shunt failure, including mechanical issues [2] and infection [3], but the most common is blood and tissue obstruction of the drainage ports [2].

Though these failure rates are high, there are currently no reliable and convenient methods to detect shunt obstruction failure. Obstruction can present itself in only vague symptoms, such as headaches and nausea, which can be confused with other medical conditions. There is a need for a non-invasive method to periodically assess shunt patency so that treatment can be improved through the timely and accurate diagnoses of shunt failure.

We developed a Parylene C-based electrochemical patency sensor capable of providing periodic measurements of shunt obstruction. This sensor utilizes a very simple transduction scheme to assess the degree of shunt patency by measuring changes in electrochemical impedance in the conductive path through the drainage ports in the shunt catheter. As the sensor is constructed on a flexible, thin film Parylene C substrate, it is easily integrated with current shunt systems either as a modular add-on or by direct integration into the shunt.

DESIGN AND OPERATION

The sensor consists of two electrodes, one positioned on each internal and external surface of the catheter, such that the catheter ports establish an ionic conductive path between them. When measuring the electrochemical impedance at a sufficiently high frequency (f_m) to isolate the solution resistance, any disturbances in the volumetric conduction path between the two electrodes (i.e. port blockage) will register as impedance changes. Thus, upon partial or complete blockage of the catheter ports by tissue, the fluidic conduction pathway between the internal and external electrodes is perturbed and an increase in the electrochemical impedance is measured, similar to the mechanism of the Coulter counter (Figure 1).

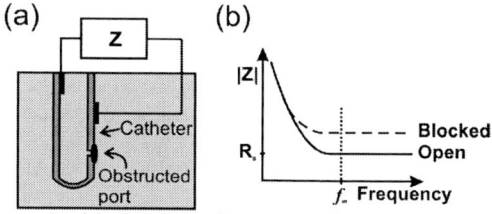

Figure 1: (a) Conceptual cartoon of impedance sensing mechanism of patency sensor. Two electrodes on the internal and external surfaces of the catheter are fluidically connected via the drainage ports. Obstruction of these ports impedes the ionic conduction path between the electrodes, and (b) the electrochemical impedance between the electrodes increases for measurements above a certain frequency (f_m).

The patency sensor was constructed using standard surface micromachining processes for Parylene MEMS devices. Platinum electrodes (2000 Å) were deposited and patterned using the liftoff method on a Parylene C substrate (15 μm) supported by a silicon carrier wafer. Platinum was chosen as it has shown outstanding inert performance for *in vivo* and electrochemical sensing applications [4]. The electrodes were insulated using Parylene C (2 μm) and electrode sites were exposed via oxygen plasma etching. In the current design, four electrodes having different surface areas were fabricated on a single device to evaluate effects of electrode size on sensor performance (Figure 2a). Devices were released and electrically packaged using a zero insertion force (ZIF) connector and flat flexible cable (FFC) used for Parylene devices [5] before further assembly into the module (Figure 2b).

MEMS 2015, Estoril, PORTUGAL, 18 - 22 January, 2015

978-1-4799-7956-1/15 $31.00 © 2015 IEEE

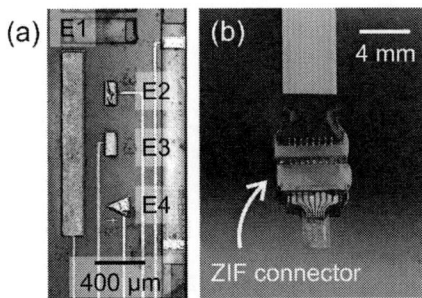

Figure 2: (a) Optical micrograph of fabricated device with four electrodes (E1-E4) of varying electrode surface areas for design optimization. (b) Image of packaged device using a ZIF connector and FFC for assembly into module.

EXPERIMENTAL METHODS

The sensor platform was assembled into different luer-lock modules designed for early acute validation studies in human using an external ventricular drain system: a cap and inline module (Figure 3). To form the cap module, the Parylene device was first affixed within a slit of a rubber stopper on top of the cap housing, which was then filled with artificial CSF (aCSF); a 3-way valve system allowed for attachment of the cap module to the rest of the testing system. The inline module was formed by affixing the sensor platform within a milled slit in a luer-lock adaptor. The inline design removed the need to prefill the module and also allowed for simpler integration with the drainage system, with necessitating an extra 3-way valve component. In both modules, the FFC was used as the connection scheme to the impedance measurement system.

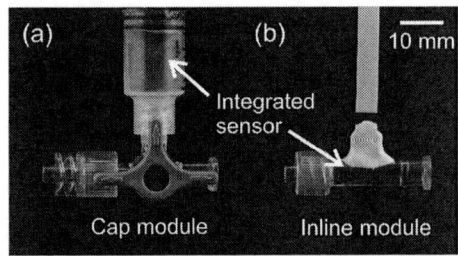

Figure 3: Assembled (a) cap and (b) inline modules for simple integration of the Parylene-based patency sensor with external ventricular drains within the clinic using standard luer-lock connections.

In benchtop tests, blockages of the catheters were simulated by mock catheters (silicone) of varying numbers of holes, with 16 holes simulating 100% open (a 4-holed catheter would be classified as 75% blockage, 8 holes as 50%, etc.) (Figure 4a). To construct the mock catheters, one end of a silicone tube (1.5 mm ID) was plugged and holes were manually punched using a 15 gauge coring needle to create 1 mm diameter holes (Figure 4b), similar to those used in a proximal catheter of a hydrocephalus shunt.

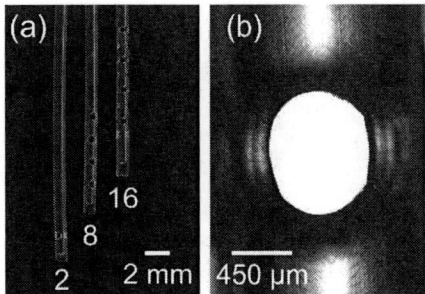

Figure 4: (a) Mock silicone catheters with varying number of holes. (b) Magnified image of manually punched hole of 1 mm diameter.

The catheter was then placed within a beaker of aCSF and connected to the sensor module. The assembly was filled via a syringe or peristaltic pump (for static and flow conditions, respectively) prior to testing (Figure 5). A platinum wire electrode was placed within the beaker to close the circuit and complete the sensing setup. Impedance measurements were acquired using a Gamry R600 potentiostat for experiments requiring frequency ranges or an Agilent e4980 LCR meter for a measurement at a single frequency.

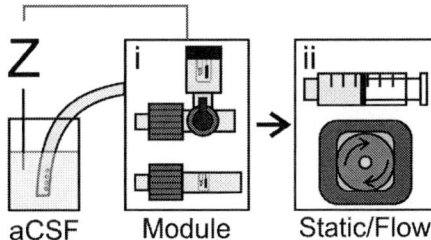

Figure 5: Cartoon schematic of experimental setup. Experiments were conducted using either the (i) cap or inline module with a (ii) syringe or peristaltic pump for static or flow conditions, respectively. Impedance was measured between the sensor electrode and platinum ground electrode in the beaker.

First, the f_m that isolates the solution resistance within the sensor's impedance response was determined so as to have the optimal sensing performance for patency. Then sensor sensitivity measurements were performed out to assess the relationship between measured electrochemical impedance and number of holes (i.e. percent shunt blockage). These experiments were conducted using all four electrode sizes to evaluate electrode size effects on sensor performance.

As these devices are to be used within the body, testing at body temperature (37°C) and in the presence of clinically relevant catheter drainage flow rates (0.3 ml/min [6]) was conducted. The effect of temperature on sensor performance was evaluated by using a water bath to maintain the beaker of aCSF at a constant 37°C; flow was generated in the system

978-1-4799-7956-1/15 $31.00 © 2015 IEEE   663

using a peristaltic pump. Also, for use of this device in clinical applications, it is vital that the sensor modules are sterilized to eliminate any source of infection. Considering this, the functionality of the devices following hydrogen plasma (H_2O_2) sterilization, a commonly used sterilization technique used by hospitals, performed with a Sterrad 100 NX system was also assessed.

RESULTS AND DISCUSSION

Sensor Characterization

Initial experiments were conducted at f_m between 0.1 Hz – 1 MHz and the measured impedance over certain frequency ranges ($>f_m$, corresponding to where the solution resistance dominates the impedance response) correlated well with catheter blockage over all electrode sizes and types (data not shown). An optimal measured f_m was determined for each sensor size (Table 1).

Table 1: Obtained optimal impedance measurement frequencies and sensitivities for electrodes of the Parylene-based EC-MEMS patency sensor.

Electrode design	Surface area (μm^2)	Measurement frequency (kHz)	Sensitivity (% Z / %Blockage)
E1	300,000	10	0.183
E2	20,000	30	0.157
E3	20,000	30	0.168
E4	17,320	30	0.161

By analyzing the data for each electrode at its corresponding optimal measurement frequency, calibration curves were formed and results indicated that the impedance varied inversely with the percent blockage of the catheter (Figure 6). The inverse relationship follows from the equation for solution resistance (R_s), where ρ is the conductivity of the solution, l, the distance between the electrodes, and A, the cross-sectional area between the electrodes:

$$R_s = \frac{\rho l}{A} \qquad (1)$$

The results suggest that hole blockages alter the area term of equation 1. Though all electrode sizes were similar in performance, the sensitivity of the largest electrode (E1) was slightly higher than the others (0.183 % Δ impedance/% blockage), and thus was chosen for further sensor development.

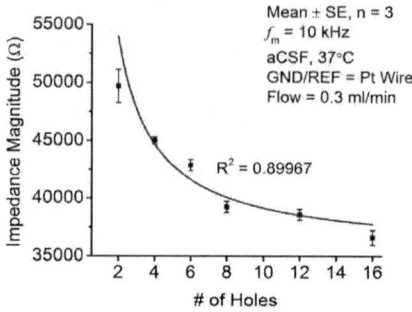

Figure 6: Representative calibration curve for E1 sensor electrode indicated that the response of the sensor was inversely proportional ($R^2 = 0.89967$) to the number of holes.

Temperature and Flow Characterization

Both an increase in fluidic temperature from room temperature (20°C) to body temperature (37°C) and the addition of flow in the system decreased the baseline impedance measured (~8.5% for both cases), but still retained a similar inverse relationship and sensitivity (Figure 7). The decrease in baseline impedance is due to an increase in ionic mobility that occurs at elevated temperatures and increased fluidic flow as observed within literature [7, 8].

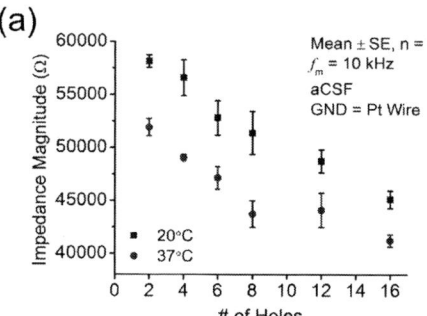

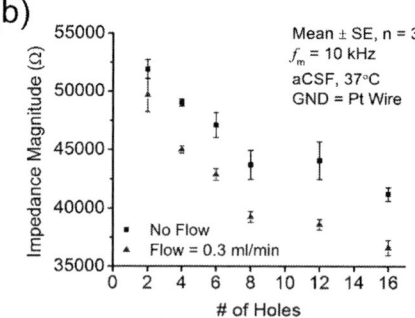

Figure 7: Sensor performance at (a) elevated temperature and (b) with flow conditions demonstrated a decrease in the baseline impedance due to increased ion mobility, but the functionality of the sensor was maintained.

Sterilization Characterization

H_2O_2 sterilization of inline module packaged devices did not alter sensor performance (Figure 8). Electrode characterization using electrochemical impedance spectroscopy and cyclic voltammetry also indicated no changes in electrode surface area or properties following sterilization (data not shown). Further studies are underway to assess the efficacy of the method in sterilizing the modules.

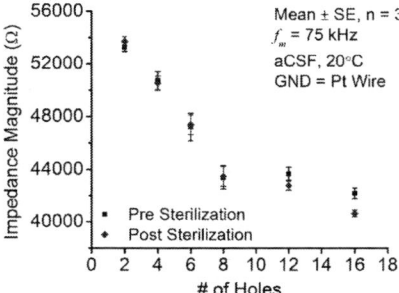

Figure 8: Plots of sensitivity pre and post sterilization, indicating no change to devices following hydrogen peroxide sterilization.

Transient Patency Measurement

The ability of the sensor to measure dynamic blockages in the benchtop system was also assessed. A 16-holed catheter was used, and transient blockages were simulated by sheathing/unsheathing the drainage ports using larger diameter tubing. Results indicated that the sensor was capable of repeatedly measuring blockage events over time (Figure 9).

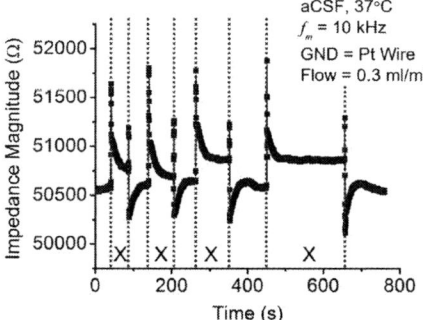

Figure 9: Transient blockage experiments of sheathing/unsheathing a 16-holed catheter to simulate dynamic blockage illustrated the real-time measurement capabilities of the device. Obstruction events are labeled with an X.

CONCLUSION

We designed, fabricated, packaged, and characterized a MEMS electrochemical patency sensor to detect obstruction of hydrocephalus shunts. A cap and inline module for integration of the sensor with current catheter systems was developed for testing, and percent shunt blockage was found to correlate inversely with electrochemical impedance measured at an optimal measurement frequency varying with electrode size. The integration of patency sensors into shunts will enable quantitative monitoring of shunt performance and more importantly, provide accurate and timely diagnosis of impending failure to improve treatment for hydrocephalus patients.

ACKNOWLEDGEMENTS

The authors thank Dr. Eisha Christian of Children's Hospital Los Angeles for clinical consultation during device design, Dr. Donghai Zhu of the Keck Photonics Laboratory for help with fabrication, and members of the Biomedical Microsystems Laboratory of USC for their assistance. An OAI model 30 light source was used for processing of Parylene devices.

This work was funded in part by the NSF under award number EFRI-1332394 and the University of Southern California Provost Ph.D. Fellowship (BK).

REFERENCES

[1] J. Drake, *et al.*, "CSF shunts 50 years on–past, present and future," *Child's Nervous System*, vol. 16, pp. 800-804, 2000.

[2] S. R. Browd, *et al.*, "Failure of cerebrospinal fluid shunts: part I: obstruction and mechanical failure," *Pediatric neurology*, vol. 34, pp. 83-92, 2006.

[3] A. V. Kulkarni, *et al.*, "Cerebrospinal fluid shunt infection: a prospective study of risk factors," *Journal of Neurosurgery*, vol. 94, pp. 195-201, 2001.

[4] Y. Nam, "Material considerations for in vitro neural interface technology," *MRS bulletin*, vol. 37, pp. 566-572, 2012.

[5] C. A. Gutierrez, *et al.*, "Epoxy-less packaging methods for electrical contact to parylene-based flat flexible cables," in *Solid-State Sensors, Actuators and Microsystems Conference (TRANSDUCERS), 2011 16th International*, 2011, pp. 2299-2302.

[6] M. E. Wagshul, *et al.*, "Amplitude and phase of cerebrospinal fluid pulsations: experimental studies and review of the literature," *Journal of Neurosurgery*, vol. 104, pp. 810-819, 2006.

[7] H. E. Ayliffe and R. Rabbitt, "An electric impedance based microelectromechanical system flow sensor for ionic solutions," *Measurement science and Technology*, vol. 14, p. 1321, 2003.

[8] C. A. Gutierrez, "Development of Flexible Polymer-Based MEMS Technologies for Integrated Mechanical Sensing in Neuroprosthetic Systems," Ph.D. Dissertation, Biomedical Engineering, University of Southern California, Los Angeles, 2011.

CONTACT

*E. Meng, tel: +1-213-8213949; ellis.meng@usc.edu

PDMS BALLOON PUMP WITH A MICROFLUIDIC REGULATOR FOR THE CONTINUOUS DRUG SUPPLY IN LOW FLOW RATE

Yumi Mukouyama, Yuya Morimoto, Shohei Habasaki, Teru Okitsu, and Shoji Takeuchi

Institute of Industrial Science, The University of Tokyo, JAPAN

ABSTRACT

This paper describes a small sized balloon pump for providing liquid in low flow rate without a driving source. The balloon pump is composed of a balloon tank and a microfluidic valve. The balloon tank can work as a reservoir to store liquid and an actuator to pump liquid. By connecting the microfluidic valve to the balloon tank, we achieved extremely low flow rates of the liquid. Therefore, our balloon pump will be applicable to implantable pumps for the continuous drug supply in low flow rates without batteries.

INTRODUCTION

Drug supply pumps with batteries are used for the medical treatment of diabetes to supply insulin [1, 2], relieve pain. These pumps generally are composed of an actuator, a reservoir, a regulator and a driving source for the actuator and the regulator. Although these pumps can control accurate flow rates, it is problematic that large sizes of actuators and driving sources prevent miniaturization of these pumps. To miniaturize actuators and driving sources, osmotic pressure pump [3] and magnetic-driven pump [4] have been proposed because a semi-permeable membrane and a magnet require less space than a battery and a motor. However, each pump has limitations of miniaturization because the volume of semi-permeable membrane and magnet are not be cut to zero.

Here, we proposed a balloon pump composed of a balloon tank with a microfluidic regulator to achieve pneumatically-driven pumping. The balloon tank works as reservoir and actuator because the expansion of the tank allows liquid storage and the shrinkage of the tank allow liquid discharge. Therefore, integration of reservoir and actuator as balloon tank can miniaturize the size of the balloon pump. The balloon pump was made of a thin poly(dimethylsiloxane) (PDMS) film and a thick PDMS substrate with a microchannel. To control flow rate, we placed a microfluidic valve in the microchannel. Although many techniques for conventional fabrication methods of the regulator in microchannels were proposed [5]-[9], it was difficult for them to be integrated with actuators because of their shape and fabrication methods. In this conference paper, we solved the problem by exposing a photoreactive resin filled in the microchannel to light for the fabrication of microfluidic valve in a designed shape and at an arbitrary space. Our balloon pump composed of the balloon tank and the microfluidic valve achieved continuous liquid supply in low flow rates by uniting the actuator, the reservoir and the driving source. The balloon pump might be applicable to small-sized supply pumps in low flow rates.

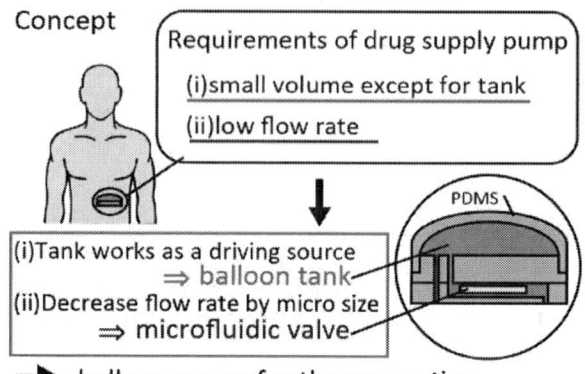

Concept

balloon pump for the generation of low flow rates

Figure 1: Concept of our work
There are two requirements of drug supply pump. To miniaturize the pump of all components, it is necessary to decrease volume except for its tank. Therefore, integration of a reservoir and an actuator in the balloon tank can miniaturize the size of the balloon pump. To control the flow rates from the balloon tank, we placed a microfluidic valve in a microchannel. Our balloon pump composed of the balloon tank and microfluidic valve achieved continuous liquid supply in low flow rates.

FABRICATION METHOD

Fabrication of the balloon tank

The balloon pump was composed of the balloon tank and the microfluidic valve. The balloon tank consists of three PDMS layers; balloon layer, intermediate layer and regulator layer shown in Figure 2(a). Three PDMS layers were shaped by molding. Molds for the balloon layer and the intermediate layer were made from photoreactive acrylate resin ("R11, 25-50 µm layers") using a stereolithography modeling machine (Perfactory, Envision Tec, Germany). We designed the molds using a 3D modeling software (Rhinoceros, Applicraft). After fabrication of the molds by stereolithography, we exposed them to ultra violet (UV) light for complete curing. Then we coated them with a 2 µm thick parylene layer to avoid adhesion by chemical vapor deposition.

A mold of the regulator layer was made by standard softlithography techniques with a SU-8 mold (SU-8 Series; Microchem Corp., USA). SU-8 was spin-coated on a silicon wafer to form a thin layer at 1300 rpm for 55 sec, and pre-baked to remove unwanted moisture or organic solvents. After designing a photo-mask with a microchannel by using a mask exposure machine (NanoSystem Solutions, Inc), the SU-8 layer was exposed to UV light through the photo-mask (Clean Surface Technology Co., Ltd), using a mask aligner machine (Union Optical Co, Ltd). After the exposure process, the SU-8 layer on the silicon wafer was post-baked. Finally, by dipping the SU-8 layer in SU-8 developer and rinsing it in isopropanol, we obtained the mold for the regulator layer.

A mixture of 10:1 (w/w) of PDMS and curing agent (Sylgard 184, Toray) were filled into the molds, degassed in a vacuum chamber and baked for curing. Cured PDMS layers were peeled off from the molds without any organic solvents. To bond three layers, the intermediate layer was coated with pre-cured PDMS (Figure 2a) and heated for 55 min at 60°C to fabricate semi-cured PDMS layer. Finally, we bonded all layers by heating them again for 90 min at 75°C shown in Figure 2b.

Fabrication of the microfluidic valve

We constructed the microfluidic valve in the microchannel at the regulator layer. First, we prepared polyethylene glycol diacrylate (PEGDA) with 1%(w/v) of phenylbis (2,4,6-trimethylbenzoyl)phosphine oxide) as photoreactive resin. The microchannel was filled with the photoreactive resin (Figure 2c) and was exposed to UV light in shape of the valve. We used a microscope (IX71, Olympus) with a digital micromirror device (DMD) to expose the regulator layer to light with the desired shape. After the exposure process, we washed the microchannel with ethanol and water (Figure 2d) to wash non-cured resin out of the microchannel. We could build the microfluidic valve with the designed shape in the microchannel (Figure 2e). The microfluidic valve did not stick to the wall of the microchannel because an oxygen layer on the PDMS surface works as PEGDA polymerization inhibitor [10]. A side of the valve was larger than that of the exposure pattern in flow direction because the photoreactive resin moved in the microchannel during the exposure process (Figure 2f). From the difference of length between the valve and the exposure pattern, we obtained the valve in the desired shape by adjusting exposure patterns.

RESULTS AND DISCUSSION

Characteristics of the balloon tank

We checked the property of a balloon tank by bonding the balloon layer with a thin PDMS film and the intermediate layer. We made balloon layers of five different thicknesses; 0.25 mm, 0.30 mm, 0.35 mm, 0.40 mm, and 0.45 mm. Figure 3a shows pictures of a balloon tank with 0.25 mm-thick balloon layer storing 0 or 4 mL liquid. We confirmed that the balloon tank can store liquid between both layers by the expansion of the thin PDMS film without leakage. Figure 3b shows the relation between changes of flow rates over time and the thickness of the thin PDMS films. The flow rates of discharged liquid from the tank were large and instable. That is why, we added a regulator layer.

Characteristics of the microfluidic valve

We checked the characteristics of the microfluidic valve inside the microchannel. The valve was free-floating in the microchannel and moved according to flow directions (Figure 4a). When the valve was pressed to the microchannel wall by the flow, the valve decreased the flow rate even at high input pressure (Figure 4b) because the flow path became narrow. In addition, when we fabricated a larger valve inside a microchannel, we

Fabrication

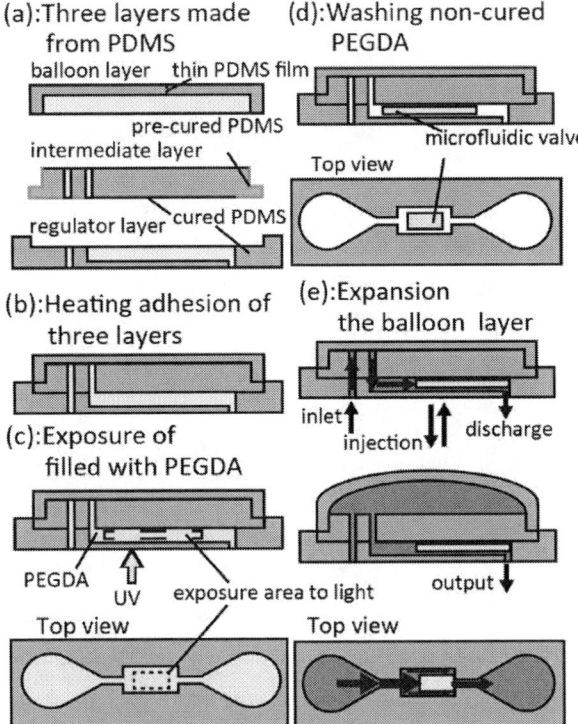

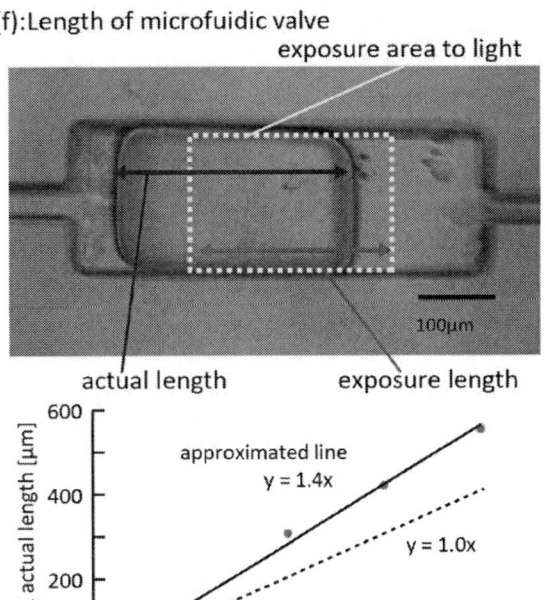

Figure 2: Fabrication methods.
(a) Balloon layer, intermediate layer and regulator layer were made from PDMS. (b) They were bonded by pre-cured PDMS. (c) Microchannel with valve was fabricated by optofluidic lithography. (d) We washed the micro- channel with ethanol and water. (e) Expansion of the balloon tank allows liquid storage and shrinkage of the balloon tank allows liquid discharge. (f) A size of the valve was larger than that of the exposure pattern.

Balloon tank

(a):Pictures of the balloon tank

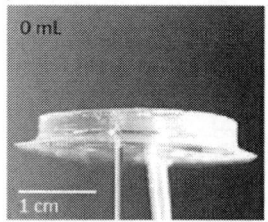 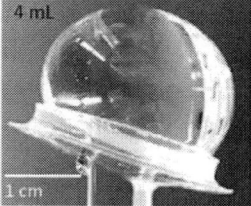

(b):Relation between the thickness of the thin PDMS film and characteristics of output pressure

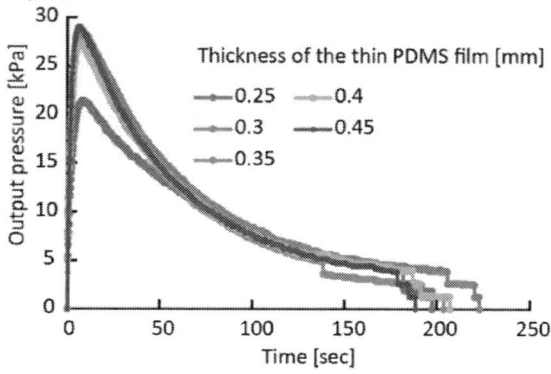

(c):Relation between the thickness of the thin PDMS film and characteristics of remaining of liquid

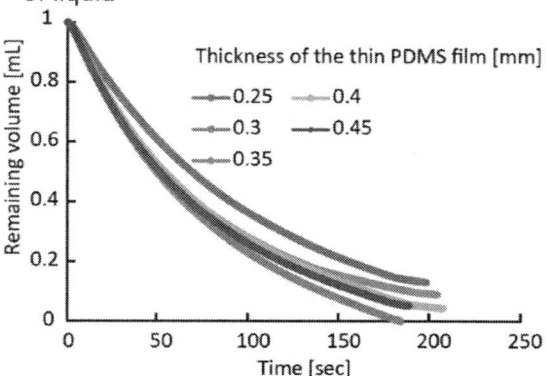

Figure 3: Characteristics of the balloon tank. (a) Pictures of the balloon tank with 0 or 4 mL liquid. (b) Relation between the thickness of the thin PDMS film and characteristics of output pressure. (c) Relation between the thickness of the thin PDMS film and characteristics of remaining volume.

decreased flow rates. These results indicate that the microfluidic valve can work as a regulator in microfluidic channel and regulation are controlled by the dimension of the valve.

Characteristics of the balloon pump

We made the balloon pump integrated with the balloon layer, the intermediate layer and the regulator layer. We compared stream flows of a balloon pump with the regulator layer (Figure 5a) and a balloon pump without the regulator layer (Figure 5b). The tank without the regulator

Microfluidic valve in microchannel

(a):Pictures of the microfluidic valve

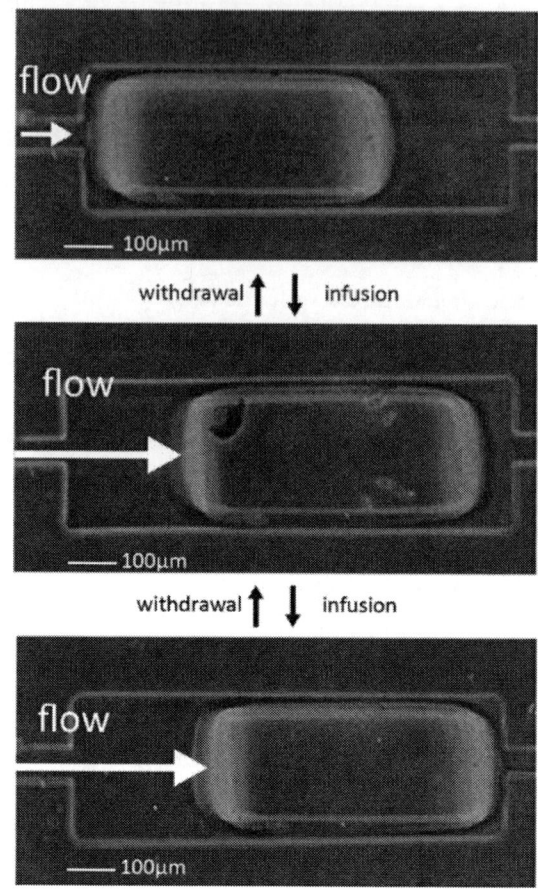

(b):Relation between flow rate and input pressure

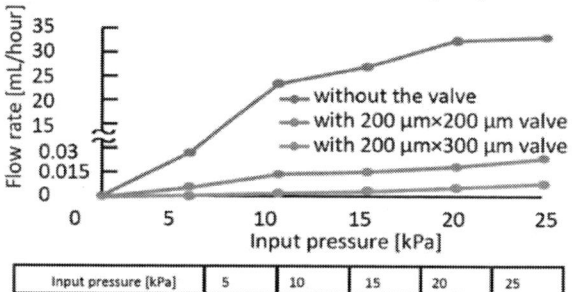

Input pressure [kPa]	5	10	15	20	25
without the valve [mL/hour]	9	23.4	27	32.4	33.12
with 200 µm×200 µm valve	0.006	0.0144	0.016	0.0192	0.024
with 200 µm×300 µm valve	0.0009	0.0024	0.0036	0.006	0.0084

Figure 4: Characteristics of microfluidic valves in microchannel as regulators. (a) Pictures of micro fluidic valves. (b) Relation of flow rates with input pressure. The flow rate dropped by existence of the valve.

layer discharged liquid at large flow rates as jet stream. On the other hand, the tank with the regulator layer discharged liquid at low flow rate as droplets of liquid. From these results, we confirmed that a regulator layer is necessary to decrease flow rates of the balloon pump discharge.

CONCLUSION

We confirmed that the balloon tank stores and

Integration of three layers

(a): Pictures of a balloon pump **with** a regulator layer

(b): Pictures of a balloon pump **without** a regulator layer

(i) Top views

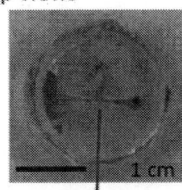 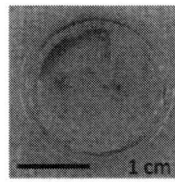

microchannel in a regulator layer

(ii) Side views after infusion

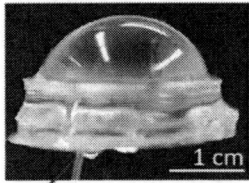 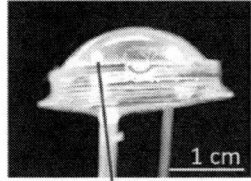

inlet   balloon tank

(iii) Side views during liquid discharge

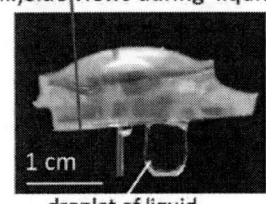 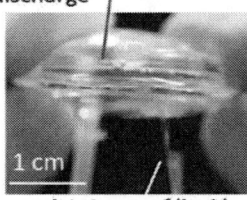

droplet of liquid   jet stream of liquid

Figure 5: Pictures of (a) the balloon pump with a regulator layer and (b) the balloon pump without a regulator layer. (i) Top views. (ii) Side views after injection of liquid. (iii) Side views during discharging liquid. Due to the existence of the regulator layer, stream flow of the balloon pump with the regulator layer is slower than stream flow of the balloon pump without the regulator layer.

discharges liquid without leakage. Moreover, we confirmed that a microfluidic valve functions as a regulator generating low flow rate. When we integrated the balloon layer, the intermediate layer and the regulator layer, the regulator layer decreased flow rates from the balloon pump. We believe that the balloon pump might be a useful implantable pump to supply drug via continuous infusion.

REFERENCES

[1] C. D. Saudek, J. L. Selam, H. A. Pitt, K. Waxman, M. Rubio, N. Jeandidier, D. Turner, R. E. Fischell and M. A. Charles, "A preliminary trial of the programmable implantable medication system for insulin delivery" *The New England Journal of Medicine*, vol. 321, pp. 574-579, 1989

[2] J. L. Selam, P. Micossi, F. L. Dunn and D. M. Nathan, "Clinical trial of programmable implantable insulin pump for Type I diabetes" *Diabetes Care*, vol. 15, No. 7, pp877-885, 1992

[3] F. Theeuwes, "Elementary osmotic pump", *Jornal of Pharmaceuticcal Sciences,* vol. 64, No. 12, pp. 1987-1991, 1975.

[4] J. Casals-Terre, M. Duch, J. A. Plaza, J. Esteve, R. `erex-Castillejos, E. Valles and E. Gomez, "Design, fabrication and characterization of an externally actuated ON/Off microvalve", *Sensors and actuators A,* vol. 147, pp. 600-606, 2008.

[5] P. Cousseau, R. Hirschi, B. Frehener, S. Gamper and D. Maillefer, "Improved micro-flow regulator for drug delivery systems". *Proc. of the14th MEMS,* pp. 527-530, 2001.

[6] B. Yang, J. W. Levis, and Q. Lin, "A PDMS-based constant-flow rate microfluidic control device" *Proc. of the17th MEMS,* pp. 379-382, 2004.

[7] M. Mescher, C. Dub, and M. Varghese, "Surface mount microfluidic flow regulator on a polymer substrate", *Proc. of the 7th Micro TAS 2003,* pp. 947-950

[8] A. Unger, H. Chou, T. Thorsen, A. Scherer and S. Quake, "Monolithic microfabricated valves and pumps by multilayer soft lithography" *Science,* vol. 288, pp. 113-117, 2000.

[9] S. E. Chung, W. Park, H. Park, K. Yu, N. Park, S. Kwon, "Optfluidic maskless lithography system for real-time synthesis of photopolymerized microstructures in microfluidic channels", *Appl. Phys. Lett,*, vol. 91, pp. 041106, 2007.

[10] D. Dendukuri, D. C. Pregibon, J. Collins, T. A. Hatton, P. S. Doyle, "Continuous-flow lithography for high-throughput microparticle synthesis", *Nat. Mater.,* vol. 5, pp. 365-369, 2006.

CONTACT

*Y. Mukouyama, tel: +81-3-5452-6650;
mukouya@iis.u-tokyo.ac.jp

PULSE WAVE MEASUREMENT IN HUMAN USING PIEZORESISTIVE CANTILEVER ON LIQUID

T. Kaneko, N. Minh-Dung, P. Quang-Khang, Y. Takei,
T. Takahata, K. Matsumoto, and I. Shimoyama
The University of Tokyo, Tokyo, Japan

ABSTRACT

We propose a device that can measure pulse waves at various points on human body with high sensitivity. Pulse wave velocity was calculated from a synchronized pulse wave measurement on two points. The device had a piezoresistive cantilever placed on silicone oil. The cantilever with oil was embedded in polydimethylsiloxane (PDMS). Pressure waves from arteries can be well conveyed to the cantilever, for the human-skin-like acoustic impedance of the silicone oil and PDMS. The signal to noise ratio of the device was ~80 dB in 10–100 Hz, when excited ~1 µm in displacement.

INTRODUCTION

Cardiovascular diseases are the top cause of death all over the world [1]. Atherosclerosis, i.e., the stiffening of arteries in human body, is deeply related to these diseases [2]. For human health monitoring, it has been significantly important to know the arterial condition continuously [3].

Recently, various health information is obtained from pulse wave measurement [2], [3]. For instance, by considering the reflect principle of pulse wave, one's pulse waveform can tell the stiffness of his/her arteries [4]. Moreover, pulse wave velocity (PWV) is one of the useful methods for the arteriosclerosis assessment [2] since PWV tends to be faster when the arteries of the patient get stiff. The pulse wave velocity is defined as following equation,

$$PWV = \frac{d}{\Delta t} \qquad (1)$$

where d is the distance between two measurement points and Δt is the time difference of pulse waves that reach the measurement points.

Human pulse wave has been measured by various sensors such as lead zirconium titanate (PZT) sensors [5], optical sensors [6], or ultrasonic probes [7]. In previous researches, there were problems to overcome such as the size of detecting system or the sensitivity of the sensor. To achieve an instant and precise measurement of pulse wave, a small-size and highly sensitive vibration-detectable transducer is necessary.

In this paper, we propose a device that can measure pulse waves in human with high sensitivity. Schematic overview of this study is shown in Figure 1. When a heart ejects blood into arteries, the radii of arteries change due to the pressure changes. These radial changes create pulse wave, which propagates through human tissue, polydimethylsiloxane (PDMS) and silicone oil to a piezoresistive cantilever. A piezoresistive cantilever placed on liquid has been reported to be able to detect pressure waves with high sensitivity due to the surface tension of the liquid [8].

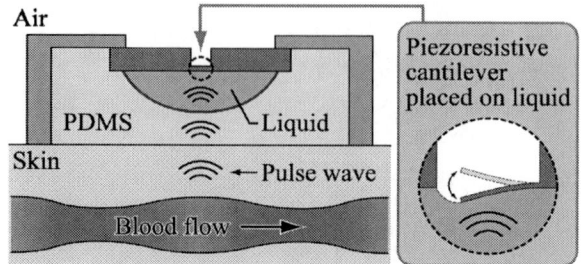

Figure 1: Schematic overview of pulse wave measurement on human body. A piezoresistive cantilever placed on liquid can detect pulse waves with high sensitivity.

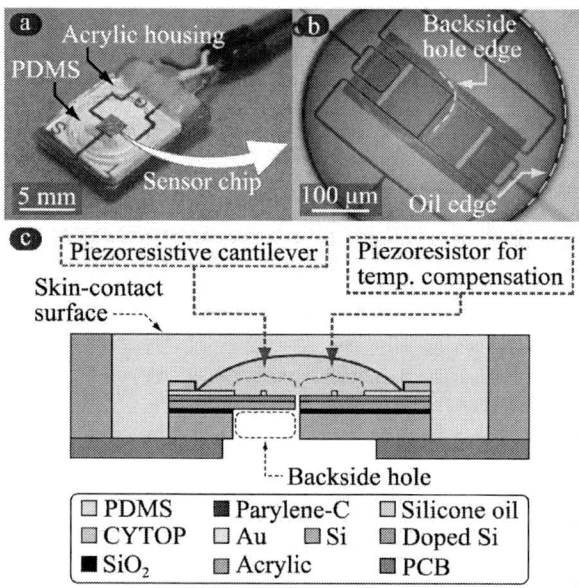

Figure 2: Images of (a) whole device and (b) pressure sensor chip. The chip consisted of a piezoresistive cantilever for vibration measurement and a piezoresistor for temperature compensation. (c) Cross-sectional diagram of the device. Hydrophobic CYTOP layer holds the silicone oil on the piezoresistors.

We fabricated the device and conducted the validation of temperature compensation, then measured the pulse waves on various points in human.

DESIGN AND FABRICATION

The photograph of whole device is in Figure 2(a). The device size was 11 mm×14 mm×3 mm. An acrylic housing was attached to the printed circuit board (PCB) substrate using ultraviolet (UV) curable resin. PDMS was poured into the housing and cured.

Since PDMS is bio-compatible material, it creates a safe contact surface to the skin. An acrylic housing would restrict lateral deformation of PDMS when the device was

Table 1: Acoustic impedances of materials.

Media	Acoustic impedance (kg/m^{2}s)	Reference
Skin	1.99×10^6	[10]
PDMS	1.05×10^6	[11]
Silicone oil (HIVAC F-4)	1.07×10^6	[12], [13]
Air	4.07×10^2	

pushed onto the skin. To reduce the electrical noise when the device was attached to the skin, the cables were soldered onto PCB and immersed in epoxy resin.

Figure 2(b) shows a photograph of pressure sensor chip. The chip size was 2 mm×2 mm×0.3 mm. We fabricated the sensor chip using a 0.3 μm/0.4 μm/ 300 μm-thick SOI wafer, following the previously reported procedure [9].

The pressure sensor chip had a piezoresistive cantilever for vibration measurement and a piezoresistor for temperature compensation (below here, we call them "piezoresistors"). The gap between the cantilever and the surrounding wall was designed to be 1.5 μm, which prevented the liquid from leaking.

The cross-sectional diagram of the device is in Figure 2(c). The hydrophobic layer of CYTOP (CTL-809M, Asahi Glass), which holds liquid in the specific area, was formed around the piezoresistors. The radius of the non-hydrophobic area was 500 μm. Silicone oil (HIVAC F-4, Shin-Etsu Chemical) was put on the non-hydrophobic area and kept by coating it with 1-μm-thick parylene film.

In order to convey pulse wave from the skin to the cantilever without much attenuation, PDMS was chosen as the propagation medium. The acoustic impedances of materials are shown in Table 1. The acoustic impedance of air is different from that of the skin, pressure wave from the arteries would reflect at the boundary surface of the skin and air. However, PMDS and silicone oil have human-skin-like acoustic impedance, pulse wave can be well conveyed to the cantilever.

FREQUENCY CHARACTERISTICS

We obtained the frequency response of the device with following procedure.

At first, we measured the displacement of the exciter (4810, Brüel & Kjær) using microscopic vibrometer (MSA-500, Polytec) in the frequency range from 10 Hz to 10 kHz. Figure 3(a) is the schematic setup of displacement measurement and Figure 3(c) is the result. The displacement of the exciter was descending from ~100 Hz. This result matched the instrumental datasheet of the exciter.

Figure 3(b) shows the schematic setup of frequency characteristics measurement of the device. The PDMS surface of the device was attached to the exciter and fixed by a jig. We conducted sweep excitation from 10 Hz to 10 kHz and measured cantilever-response using a network analyzer (4395A, Agilent). The results shown in Figure 3(d) indicates that the cantilever has ~80 dB signal to noise ratio (SNR) at 10–100 Hz when excited ~1 μm in displacement. The response was descending from ~100 Hz

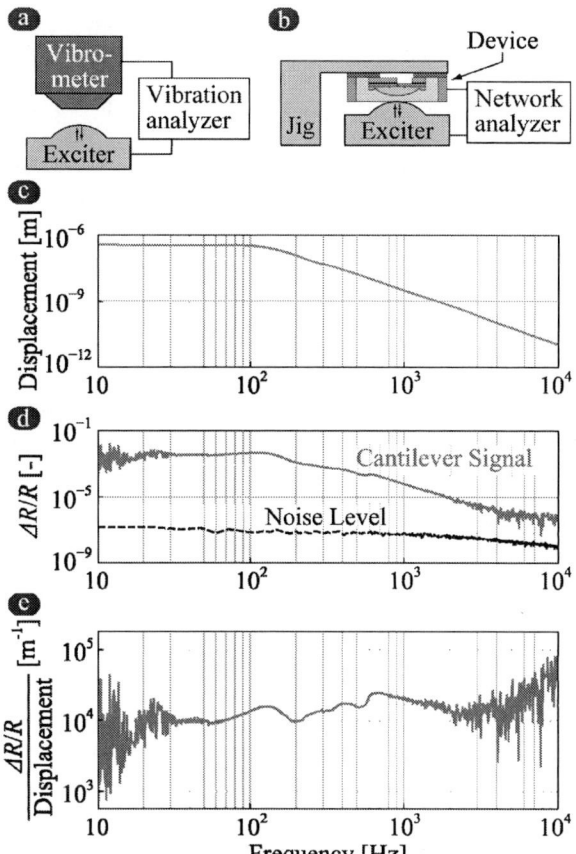

Figure 3: Measurement setup of (a) displacement of the exciter and (b) frequency characteristics of the device. (c) Displacement of the exciter versus frequency. (d) Response of the device versus frequency. (e) Response of the device normalized by the excitation displacement.

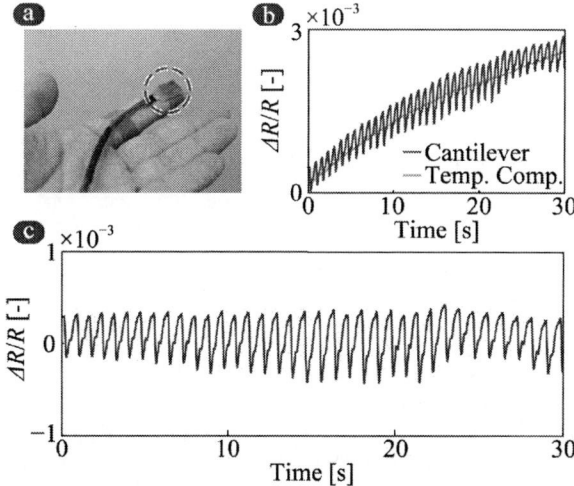

Figure 4: (a) The device and cable were attached to the subject's left-index finger by adhesive tape. (b) Response of the piezoresistive cantilever and the piezoresistor for temperature compensation. (c) Waveform gained by subtracting the response of temperature compensation from that of the cantilever.

Considering the result of the displacement of the exciter, i.e., from dividing the cantilever response by the displacement of the exciter, we confirmed that the cantilever has a flat frequency response in 10 Hz to 10 kHz range (Figure 3(e)).

EXPERIMENTS AND RESULTS

Temperature compensation

We recorded the resistance changes of the vibration-detecting cantilever and the temperature-compensation piezoresistor simultaneously. The device and its cable were attached to the left-index finger and fixed by adhesive tape (Figure 4(a)). The subject was a 24 years old healthy male, whose height and weight were 170 cm and 63 kg, respectively.

In Figure 4(b), both of the signals of the cantilever and the piezoresistor were gradually ascending in the same proportion. There was vibrating response in the cantilever signal, while no signals were in the piezoresistor signal. This results indicates that the piezoresistor would measure only the temperature change of the skin.

We conducted the temperature compensation by subtracting the piezoresistor output from the cantilever output. Obtained pulse wave is shown in Figure 4(c). The frequency of the wave was 1.2 Hz, thus the period of the pulse wave was 0.83 s. Fluctuation of amplitude in the signal probably stemmed from body movements or aspiration of the subject.

Non-synchronized pulse wave measurement

We conducted pulse wave measurements on the points at angular, carotid, apical and finger arteries. The pulse waves obtained after temperature compensation are shown in Figure 5. From the results, we can say that pulse wave measurement with high sensitivity was achieved compared with the previous research [5]. Note that each signal is not synchronized with other signals.

Synchronized pulse waves measurement

Figure 6 shows the result of a synchronized pulse waves measurements on angular and apical arteries. The distance between the two points was 40 cm, and the time difference of the two peaks in each pulse wave was 42 ms. From these data, the PWV between the heart and the forehead was calculated as 9.5 m/s.

CONCLUSION

We fabricated a device that can measure human pulse waves with high sensitivity. The device was composed of a piezoresistive cantilever on liquid and a piezoresistor for temperature compensation.

Frequency characteristics of the device was flat within the range of 10 Hz to 10 kHz and SNR was ~80 dB at 10–100 Hz when excited ~1 μm in displacement.

We confirmed the validity of the temperature compensation and conducted pulse wave measurements on angular, carotid, apical and finger arteries. From a synchronized pulse waves measurements on angular and apical arteries, PWV was calculated as 9.5 m/s.

The experimental results indicates the proposed device is desirable for human health monitoring, in respect

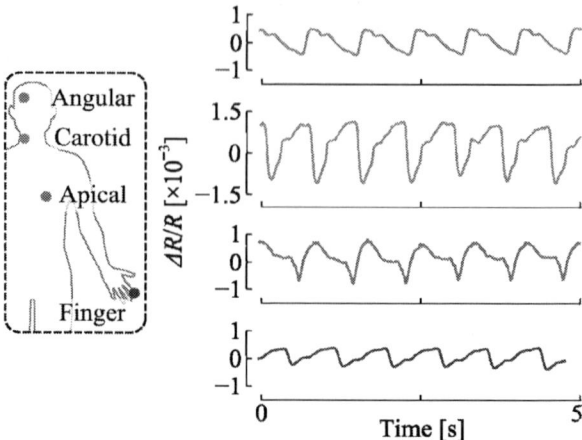

Figure 5: Non-synchronized pulse wave measurements on angular, carotid, apical and finger arteries.

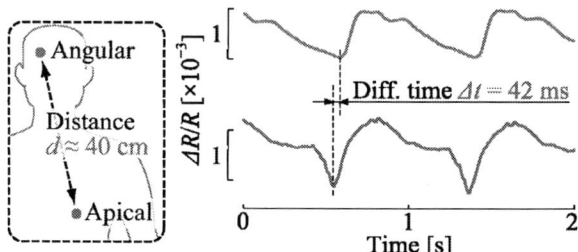

Figure 6: Synchronized pulse wave measurement on angular and apical arteries. The distance between two measurement points was 40 cm and the differential time of two waveform was 42 ms, thus pulse wave velocity (PWV) was calculated as 9.5 m/s.

of the size and the sensitivity.

ACKNOWLEDGEMENTS

The photolithography masks were made using the University of Tokyo VLSI Design and Education Center (VDEC)'s 8 inch EB writer F5112 + VD01 donated by ADVANTEST Corporation.

REFERENCES

[1] World Health Organization. (2013, March) Cardiovascular diseases (CVDs) [Online]. Available: http://www.who.int/mediacentre/factsheets/fs317/en/

[2] R. Asmar, A. Benetos, J. Topouchian, P. Laurent, B. Pannier, A. M. Brisac, R. Target, B. I. Levy, "Assessment of Arterial Distensibility by Automatic Pulse Wave Velocity Measurement – Validation and Clinical Application Studies," *Hypertension*, vol. 26, no. 3, pp. 485–490, 1995.

[3] Y. P. Hsu, D. J. Young, "Skin-Coupled Personal Wearable Ambulatory Pulse Wave Velocity Monitoring System Using Microelectromechanical Sensors," *Sensors*, vol. 14, no. 10, pp. 3490–3497, 2014.

[4] R. Kelly, C. Hayward, A. Avolio, M. O'Rourke, "Noninvasive Determination of Age-Related Changes in the Human Arterial Pulse," *Circulation*, vol. 80, no. 6, pp. 1652–1659, 1989.

[5] H. J. Tseng, W. C. Tian, W. J. Wu, "Flexible PZT Thin Film Tactile Sensor for Biomedical Monitoring," *Sensors*, vol. 13, no. 5, pp. 5478–5492, 2013.

[6] P. A. Kyriacou, K. Shafqat, S. K. Pal, "Pilot investigation of photoplethysmographic signals and blood oxygen saturation values during blood pressure cuff-induced hypoperfusion," *Measurement*, vol. 42, pp. 1001–1005, 2009.

[7] M. Sato, H. Hasegawa, H. Kanai, "Correction of change in propagation time delay of pulse wave during flow-mediated dilation in ultrasonic measurement of arterial wall viscoelasticity," *Japanese Journal of Applied Physics*, vol. 53, no. 7, 07KF03, 2014.

[8] N. Minh-Dung, P. Hoang-Phuong, K. Matsumoto, I. Shimoyama, "A Sensitive Liquid-Cantilever Diaphragm for Pressure Sensor," *The 26th IEEE International Conference on Micro Electro Mechanical Systems*, Taipei, January 20–24, 2013, pp. 617–620.

[9] P. Quang-Khang, N. Minh-Dung, N. Binh-Khiem, P. Hoang-Phuong, K. Matsumoto, I. Shimoyama, "Multi-axis force sensor with dynamic range up to ultrasonic," *The 27th IEEE International Conference on Micro Electro Mechanical Systems*, San Francisco, January 26–30, 2014, pp. 769–772.

[10] H. Azhari, "Typical Acoustic Properties of Tissues," in *Basics of Biomedical Ultrasound for Engineers*, New York: Wiley, 2010, pp. 313–314.

[11] I. Leibacher, S. Schatzer, J. Dual, "Impedance matched channel walls in acoustofluidic systems," *Lab on a Chip*, vol. 14, no. 3, pp. 463–470, 2014.

[12] Shin-Etsu Chemical. (2014, April) Oil catalog [Online]. Available: http://www.silicone.jp/j/catalog/pdf/fluid_j.pdf

[13] Shin-Etsu Chemical. (2014, May) Technical document [Online]. Available: http://www.silicone.jp/j/catalog/pdf/kf96_j.pdf

CONTACT

*T. Kaneko, Mechano-Informatics Department, Graduate School of Information Science and Technology, The University of Tokyo, 7-3-1 Hongo, Bunkyo-ku, Tokyo, Japan.

E-mail: kaneko@leopard.t.u-tokyo.ac.jp

Tel: +81-3-5841-6318, Fax: +81-3-5841-6341.

THE MICRO SADDLE COIL WITH SWITCHABLE SENSITIVITY FOR MAGNETIC RESONANCE IMAGING

Kosuke Murashige[1], and Tetsuji Dohi[1]

[1]Faculty of Science and Engineering, Chuo University, Tokyo, Japan

ABSTRACT

This paper reports a micro saddle coil with switchable sensitivity for Magnetic Resonance Imaging (MRI). The coil comprises a polydimethyilsiloxane (PDMS) tube and a flexible substrate with a coil pattern. The coil deforms from a saddle-shaped mode to a planar-shaped mode. In MR images acquired in saddle-shaped mode, a large sensitive area existed around the coil. Although the area considerably reduced in the planar-shaped mode, clear MR images were obtained. The signal-to-noise ratios (SNR) of the saddle-shaped and planar-shaped modes were 194.9 and 505.9, respectively, at $2.0 \times 2.0 \times 2.0$ mm^3 voxel size and 11.7 and 37.4, respectively, at $0.5 \times 0.5 \times 1.0$ mm^3. Thus, the micro saddle coil enabled highly sensitive and large-area imaging.

INTRODUCTION

Medical MRI is a promising method to visualize the inside of a human body from NMR (Nuclear Magnetic Resonance) signal. Unfortunately, medical MRI cannot acquire the MR images of features such as small tumors, because of the low SNR. The resolution of MR images can be improved by installing a micro coil as the antenna of the NMR signal receiver. By selecting the size of the micro coil to suit the imaging sample, or by increasing the number of turns of the micro coil, we can acquire sufficiently clear NMR signals to observe small tumors.

Many studies on micro coils for MRI to take MR images of small tumors were reported. Since a micro coil has a small sensitive area, the low noise NMR signal can be measured. Therefore, high resolution MR images can be observed [1]. The saddle coil for MRI was reported which can take MR images of luminal tissue [2]. Although the saddle coil has the large sensitive area around the coil, the sensitivity of the saddle coil was still low. In another study, the micro planar coil was reported [3]. Since the planar coil has many wiring per unit length, the sensitivity of the coil was very high. However, it was difficult to take both large area MR images of luminal tissue and high sensitive MR images of small tumors.

Underlying this difficulty is the different design of the saddle and planar coils. In this study, we propose a micro saddle coil with switchable sensitivity for MRI. The micro saddle coil adopts two shape modes; saddle-shaped and planer-shaped. The coil shape is switched by deforming the coil. The altered shape of the micro saddle coil changes the sensitivity and sensitive area of the coil. In this way, the proposed micro saddle coil can acquire both large-area as well as highly sensitive MR images.

CONCEPT AND FABRICATION

Concept of the micro saddle coil

Figure 1 shows our concept of the micro saddle coil. The coil consists of a PDMS tube embedded with a flexible substrate with a coil pattern. The micro coil was designed

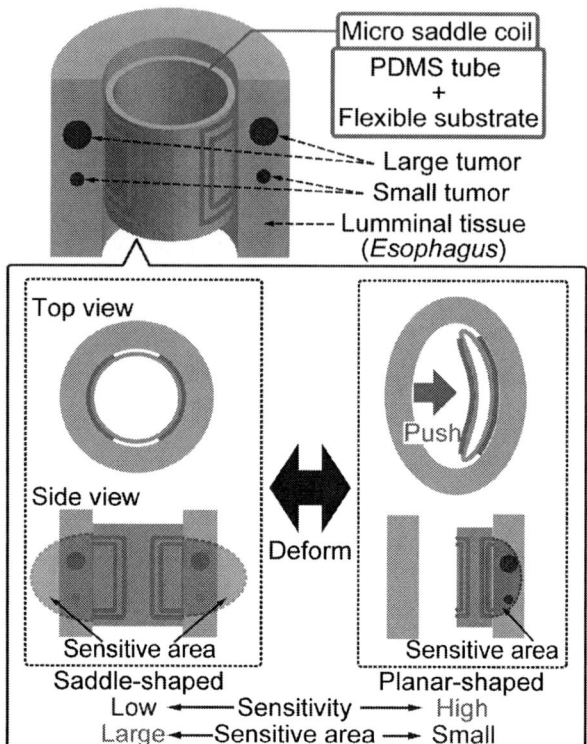

Figure 1: Concept of the micro saddle coil. The saddle coil consists of a PDMS tube and a flexible substrate. The coil pattern is deformed into the planar shape by pushing. In saddle-shaped mode, the coil has a large sensitive area, and we can observe many tumors. In planar-shaped mode the sensitive area is concentrated at one side, and the coil becomes sufficiently sensitive to observe small tumors.

for observing luminal tissue such as the *esophagus*. The micro saddle coil has a flexible architecture, being deformable into a saddle-shaped mode and a planar-shaped mode. The saddle-shaped mode has a large sensitive area, enabling large MR images of luminal tissue covering many tumors. The coil is deformed into the planar-shaped mode by pushing. Since the coil is flattened in this mode, the number of wire turns per unit length is increased, and the sensitive area becomes concentrated at one side. This configuration sufficiently increases the sensitivity of the coil to observe small tumors.

Thus, the micro saddle coil achieves large area MR images and highly sensitive MR images by deforming its shape to switch the coil sensitivity.

Fabrication process

Figure 2 shows the fabrication process of the micro saddle coil. The flexible substrate is layered Cu/polyimide/Cu. The thickness of the Cu layers and poly-imide layers are 12 μm and 25 μm. The coil wiring is patterned by etching the top and bottom Cu layers. A

978-1-4799-7956-1/15 $31.00 © 2015 IEEE

1. Use flexibre substrate (Cu/Poly-imide/Cu) and pattern Cu layers to coil wiring.

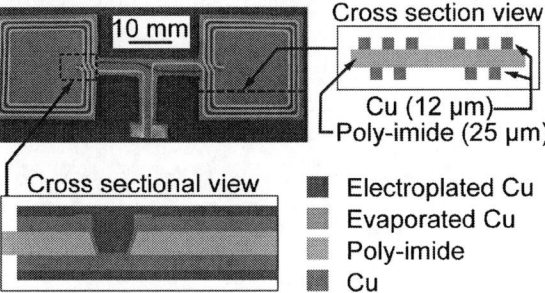

2. Use plastic column and acrylic pipe to make PDMS tube.

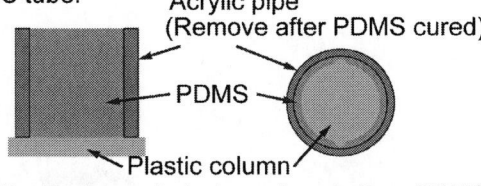

3. Fold the flexible substrate and paste it on PDMS tube.

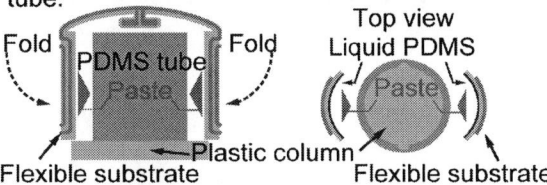

4. Paste thin PDMS films on the coil.

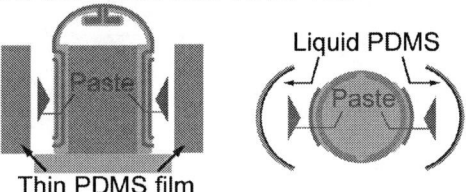

5. Remove the plastic column from the coil.

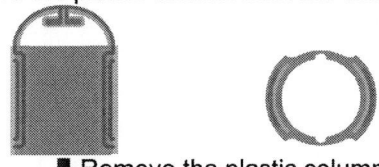

Figure 2: Fabrication process of the micro saddle coil.

through hole is inserted in the top Cu and polyimide layers, and the top and bottom Cu layers are connected by evaporating the Cu and electroplating another Cu layer (Step 1 of Figure 2). The thickness of the Cu layers are 50 μm after electroplating. The excess polyimide layers are cut away, and liquid PDMS is applied as a paste to the back of the flexible substrate. Next, the flexible substrate is pasted onto the PDMS tube, and the PDMS tube is affixed to the plastic column. The coil is covered by a thin PDMS film to prevent it from peeling off. Once the PDMS is cured, the plastic column is removed.

Figure 3 shows the fabricated micro saddle coil. The coil deforms from saddle-shaped mode to planar-shaped mode. Since the PDMS tube is constricted, the coil deformations are reproducible. The diameter and length of the coil is 20 mm and 30 mm, respectively, and 10 turns of wiring are used.

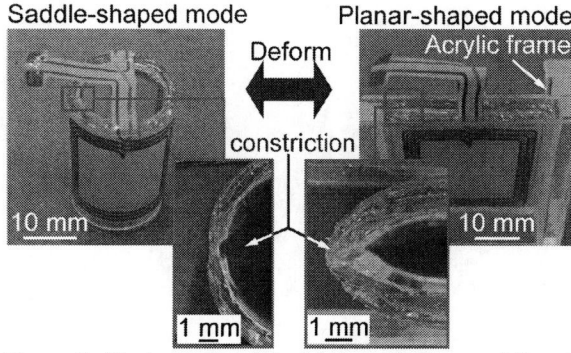

Figure 3: Photographs of the fabricated micro saddle coil. The coil deforms from saddle-shaped mode to planar-shaped mode. Since the PDMS tube is constricted, the coil deformations are reproducible.

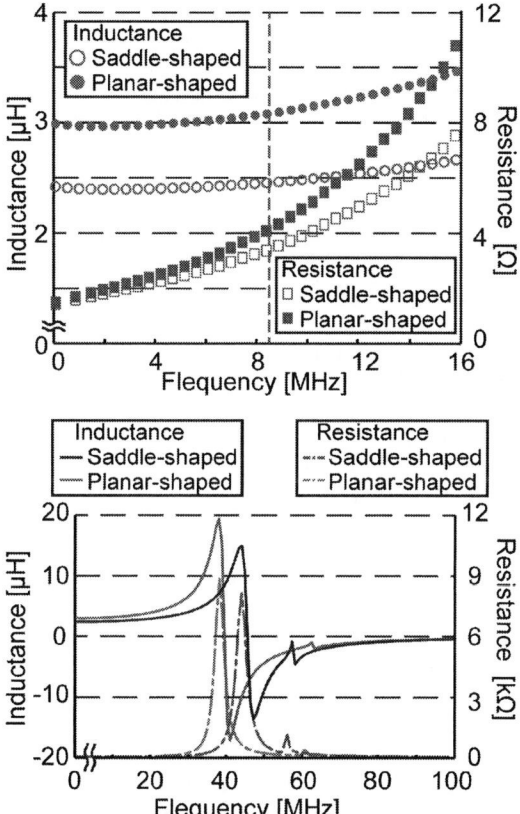

Figure 4: Characteristics of the micro saddle coil. The resistance and inductance of the coils is plotted from 1 to 16 MHz (upper panel) and from 1 to 100 MHz, (lower panel).

Table 1: Electrical characteristics of the coil at 8.5 MHz.

Coil shape	Saddle-shaped	Planer-shaped
Inductance [μH]	2.45	3.07
Resistance [Ω]	3.31	3.92
Q-factor [-]	39.9	42.9
Self-resonant frequency[MHz]	44.8	39.5

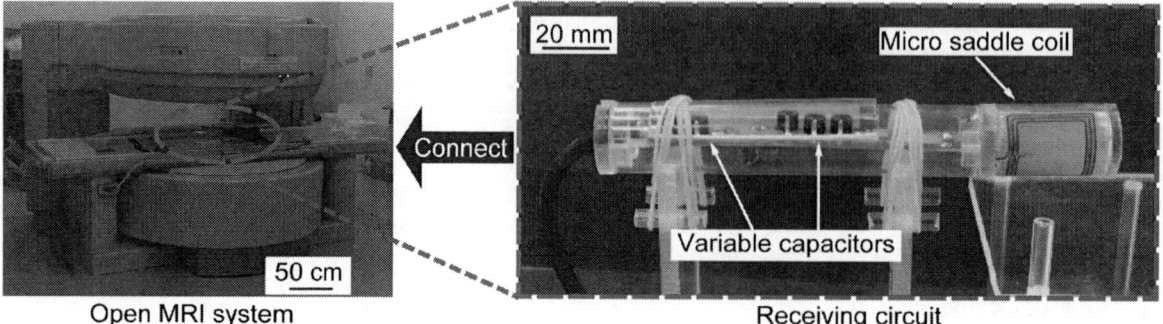

Figure 5: Experimental setup for measuring the MR image. The micro saddle coil is attached to the receiving circuit. The receiving circuit is connected to MRI system.

EXPERIMENT

Electrical characteristics of the coil

We measured the electrical characteristics of the micro saddle coil in both saddle-shaped and planar-shaped modes. The coil was exposed to a static magnetic field of 0.2 T, giving an MRI frequency of 8.5 MHz. The electrical characteristics of the micro saddle coil are presented in Figure 4 and Table 1. At 8.5 MHz, the inductance of the saddle-shaped and planar-shaped mode was 2.45 μH and 3.07 μH, respectively, and their respective resistance was 3.31 Ω and 3.92 Ω. The Q-factor was 39.9 and 42.9, respectively. All three of the resistance, inductance, and Q-factor were increased. Since the Q-factor was increased, deformation into planar form increased the sensitivity of the micro saddle coil. The self-resonant frequency was higher in saddle-shaped mode than in the planar-shaped mode (44.8 MHz vs. 39.5 MHz). It can be considered that coil deformation reduced the gaps between the coil wirings and increased the parasitic capacitance of the coil.

Taking MR images

To evaluate the sensitivity and the range of the sensitive area, we took MR images by using the micro saddle coil. Figure 5 shows the experimental setup for measuring the MR images. The coil was attached to a receiving circuit comprising the micro saddle coil and two variable capacitors for tuning and impedance matching. The receiving circuit was connected to an open-type medical MRI system. All MR images were acquired under the following conditions: spin echo (SE) sequence with a repetition time (TR) and echo time (TE) of 250 ms and 22 ms, respectively, and a repeat count of 16 times per unit time. The voxel sizes of the images were $0.5 \times 0.5 \times 1.0$ mm^3 and $2.0 \times 2.0 \times 2.0$ mm^3.

With the coil in saddle-shaped and planar-shaped modes, we acquired MR images of gelatin containing an inserted grid. Figure 6 is a schematic of the gelatin placement. In the saddle-shaped mode, the micro saddle coil was inserted into the gelatin. In planar-shaped mode, it was pushed against the gelatin. The acquired images were compared against images of gelatin placed at the center of a medical head coil.

The measured MR images are shown in Figure 7. The sensitive area of the medical coil was large, but the MR image was excessively noisy at the small voxel size ($0.5 \times 0.5 \times 1.0$ mm^3). The saddle-shaped mode admits a large

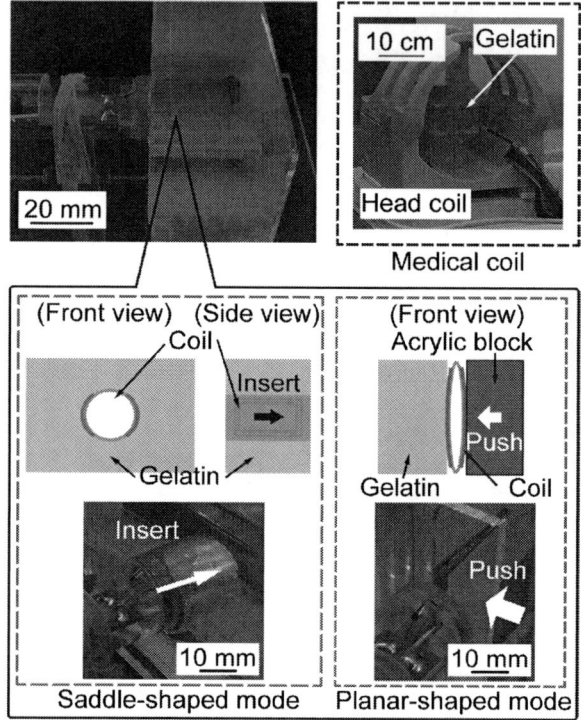

Figure 6: Schematic of the gelatin placement. The micro saddle coil was inserted into the gelatin in saddle-shaped mode, and pushed against the gelatin in planar-shaped mode. The medical head coil is evaluated for comparison.

sensitive area around the coil. On the other hand, the sensitive area of the planar-shaped mode was small, but clear MR images were taken.

Table 2 presents the SNRs of the medical and micro saddle coils. The SNR of the medical coil was 41.7 and 3.5 at voxel sizes of $2.0 \times 2.0 \times 2.0$ mm^3 and $0.5 \times 0.5 \times 1.0$ mm^3, respectively. In contrast, the SNR of the saddle-shaped and planar-shaped modes were 194.9 and 505.9 respectively at $2.0 \times 2.0 \times 2.0$ mm^3, and 11.7 and 37.4 respectively at $0.5 \times 0.5 \times 1.0$ mm^3. The SNR of the saddle-shaped and planar-shaped modes were improved by 259 % at a voxel size of $2.0 \times 2.0 \times 2.0$ mm^3, and by 319 % at $0.5 \times 0.5 \times 1.0$ mm^3. This nearly threefold improvement in the SNR at both voxel sizes confirms the high sensitivity of our coil, achieved by deforming and concentrating the sensitive area.

978-1-4799-7956-1/15 $31.00 © 2015 IEEE

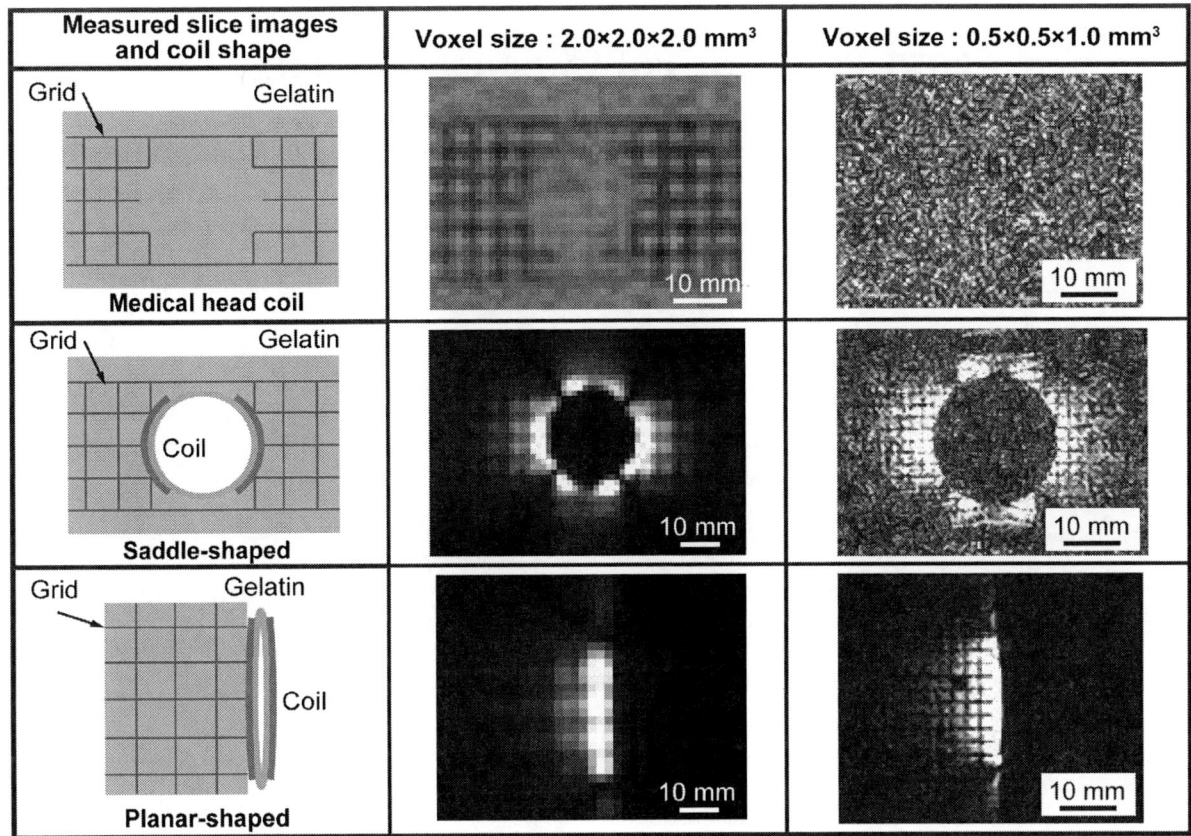

Figure 7: The MR images of medical coil, saddle-shaped coil, and planar-shaped mode coils with the voxel size of 0.5×0.5×1.0 mm³ and 2.0×2.0×2.0 mm³.

Table 2: The SNRs of medical coil and micro saddle coil.

Coil shape	Voxel size 2.0×2.0×2.0 [mm³]	Voxel size 0.5×0.5×1.0 [mm³]
Medical coil	SNR:41.7	SNR:3.5
Saddle-shaped mode	SNR:194.9	SNR:11.7
Planar-shaped mode	SNR:505.9	SNR:37.4
Increase rate	259 %	319 %

CONCLUSION

In conclusion, we fabricated a micro saddle coil with switchable sensitivity for MRI. Since the coil is embedded in the PDMS tube, the micro saddle coil can deform from saddle-shaped mode to planar-shaped mode by pushing. The diameter and length of the coil is 20 mm and 30 mm, respectively, and 10 turns of wiring were used. The inductances and resistances of the saddle-shaped and planar-shaped modes were quite similar. Since deformation increases the Q-factor of the coil from 39.9 to 42.9, it intrinsically alters the coil sensitivity. The performance of the micro saddle coil was evaluated from MR images acquired by the coil. In saddle-shaped mode, a large sensitive area exists around the coil. On the other hand, the sensitive area is much reduced in planar-shaped mode, but this mode ensures clear MR images. Our coil concentrates the sensitive area to achieve high sensitivity. This is evidenced by the nearly threefold increase in the SNR of our proposed coil (relative to the standard medical coil) at both voxel sizes. Thus, by switching the coil sensitivity, we acquired MR images over a large area and over a small area with greater sensitivity.

ACKNOWLEDGEMENTS

This work was supported by JSPS KAKENHI Grant Number 26709015.

REFERENCES

[1] C. Massin et al., "Planar microcoil-based microfluidic NMR probes", *Journal of Magnetic Resonance*, vol. 164, pp. 242-245, 2003.

[2] S. Goto, T. Matsuoka, K. Kuroda, M. Esashi, Y. Haga, "Development of High-Resolution Intraluminal and Intravascular MRI Probe Using Microfabrication on Cylindrical Substrates", in *Proceedings IEEE 20th MEMS Conference*, Kobe, January 21-25, 2007, pp. 329-332.

[3] H. Takahishi, T. Dohi, K. Matsumoto, I. Shimoyama, "A micro planar coil for local high resolution magnetic resonance imaging", in *Proceedings IEEE 20th MEMS Conference*, Kobe, January 21-25, 2007, pp. 549-552.

CONTACT

*K. Murashige, tel: +81-3-3817-1832
murashige@msl.mech.chuo-u.ac.jp
T.Dohi, dohi@mech.chuo-u.ac.jp

ULTRA-SENSITIVE AND STRETCHABLE STRAIN SENSOR BASED ON PIEZOELECTRIC POLYMERIC NANOFIBERS

M. Asadnia[1,2,], A.G.P. Kottapalli[1,2], J.M. Miao[1] and M.S. Triantafyllou[2,3]*

[1]School of Mechanical and Aerospace Engineering, Nanyang Technological University, SINGAPORE

[2]Center for Environmental Sensing and Modeling (CENSAM), Singapore-MIT Alliance for Research and Technology (SMART), SINGAPORE

[3]School of Mechanical Engineering, Massachusetts Institute of Technology, USA

ABSTRACT

There have been increasing demands for stretchable and highly sensitive sensors for use in structural health monitoring, human motion capturing, sport performance monitoring and rehabilitation. Such high performance myoelectric sensors are also remarkably important in developing artificial limbs where a combination of robotic actuators and sensory systems are required to provide lifelike, affordable, functional and easy to use devices that can interact with the human body. Here, we present a highly stretchable, self-powered and ultra-sensitive strain sensor based on piezoelectric PVDF nanofibers. We demonstrate the performance of the proposed device in response to an oscillatory load at very low frequency (0.5Hz). We also examined the applicability of our strain sensor on human's motion recognition by fabricating a glove with two sensors mounted on the middle and index fingers.

INTRODUCTION

There is an immense need for building myoelectric limbs that could help amputees to regain independence in their everyday lives. Myoelectric sensors are required to provide signals for control of artificial limbs. In the recent years, there has been a considerable research progress towards developing stretchable strain sensors using materials such as carbon nanomaterial or silver nanowires [1]. Although devices employing these materials show superior performance, lack of stretchability and high power consumption pose a major disadvantage for use in real-time applications. The proposed PVDF nanofiber strain sensor demonstrates a higher sensitivity and excellent stretchability while not requiring power supply to operate.

In the past, piezoresistive and piezoelectric materials such as PZT and PVDF in the form of bulk and thin films have been extensively used for developing various MEMS sensors [1-4]. During the last decade, (PVDF) nanofiber has gained remarkable interest in many applications such as energy harvesting [5], tissue engineering [6] and sensors [7]. PVDF exhibits impressive, mechanical and electrical characteristics such as piezoelectricity (highest among the synthetic polymers), nonlinear optical properties and flexibility [8]. In the past, various methods were developed to fabricate PVDF nanofibers such as conventional far field electrospinning (CFFES) [9], modified far field electrospinning (FFES) [10] and near field electrospinning (NFES) [11, 12]. The main difference between these methods is the distance between needle and collector, which is higher for FFES (around 100mm) as compared to that of the NFES (1mm). Having a small emitting distance for fibers in NFES method allows us to provide well aligned fibers in desired forms [7]. Figure 1 shows a schematic view of far field electrospinning process. After the jet of fibers is dispensed from the needle to the collector, they bend in a complex shape and form a chaotic path (see figure 2). In order to determine the molecular and crystalline structure of the PVDF after electrospinning process, various characterization methods such as X-ray Diffraction (XRD), Fourier transform infrared spectroscopy (FTIR), Raman spectra can be used which are explained in [8].

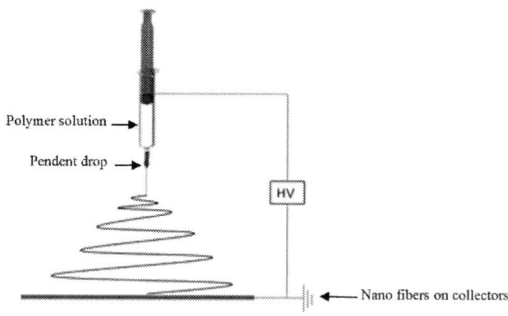

Figure 1. Schematic view of the far field electrospinning process

PVDF ELECTROSPUN NANOFIBER

PVDF is a semicrystalline polymer with a structure consisting of linear chains with sequence hydrogen and fluoride along with carbon backbone with a simple chemical formula (CH_2–CF_2). The chemical structure of PVDF falls between structure of Polytetrafluoroethylene (PTFE) which is (CF_2-CF_2) and Ethylene (CH_2-CH_2). While having close structure to Ethylene provides a great flexibility for PVDF, the crystalline similarity with PTFE gives stereochemical constraint to PVDF [13]. Due to this structural characteristic, PVDF forms in different crystal structures depending on sample preparation conditions. In nature, PVDF appears in different phases which are known as α, β, γ and δ. Each of these phases is transferable to the others under certain external conditions. In general, α-phase is the most available phase in nature which typically obtained when the PVDF is cooled and solidified from melt. While the α-phase is known as a non-polar structure which does not show piezoelectricity, β-phase is understood as the only PVDF ferroelectric crystalline structure (polar) with strong piezoelectric effect [5]. In general, high mechanical (approximately 50%) and electrical stretches (to align the dipoles) are required to predominantly convert the PVDF from α-phase to that of with β-phase [13]. In the past various methods have been proposed to increase the ratio of the β-phase in the materials. For instance, annealing the sample at high pressure and high temperature or adding strongly polar hexamethylphosphorictriamide (HMPTA) in the solution [14]. It is also reported that adding carbon nanotubes in the PVDF solution can increase the Young's modulus of the material and enhance the growth of the β-phase structure and provide PVDF composites fiber with improved piezoelectric properties [11].

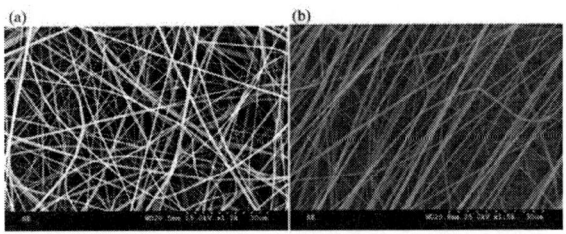

Figure 2. Electrospun nanofiber collected by (a) stationary collector which forms a chaotic fibers (b) rotatory collector (500 rpm) which forms aligned nanofibers.

Developing PVDF with β-phase requires careful optimization of electrospinning process and solution preparation. In order to observe the material phase after each step of optimization, we performed XRD. Figure 3 shows the XRD patterns of the PVDF nanofibers which are observed with a Siemens D5000 X-ray diffractometer with Cu Kα radiation ($\lambda = 1.54 \overset{\circ}{A}$). The tests are conducted in reflection mode at an ambient temperature with two theta (degree) varying between 10° and 50°. Figure 3 reveals a strong peak at 20.2° for the nanofibers which shows the β-phase is the dominate structure in

the material. α-phase absorption bands, such as, 18.3° and 41.1° are not evident in the XRD pattern which indicates the existence of a very small portion of α-phase [15].

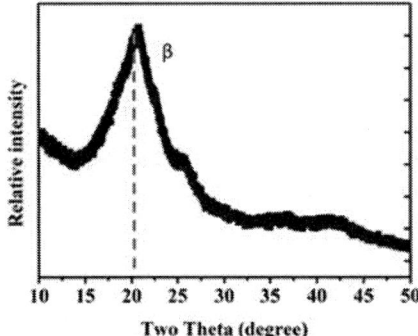

Figure 3. XRD patterns recorded for PVDF nanofibers of diameter of 800nm.

SENSOR DEVEOPMENT

The fabrication process flow of the sensor is depicted in figure 4. Initially, aligned PVDF nanofibers are collected on an aluminium foil substrate. Later on, the fibers are carefully transferred to a 25μm thick flexible LCP layer of dimensions 10mm in width, and 20mm in length. Gold electrodes of 2mm in width and 10mm in length are fixed on two ends of the sensors and fibers are protected using carbon tape.

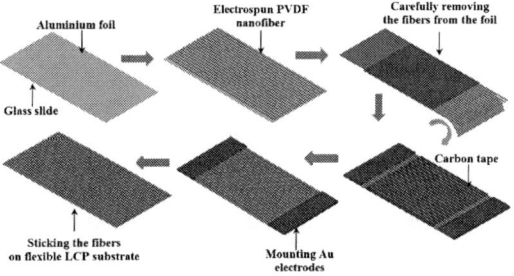

Figure 4. Fabrication process of the ultra-sensitive, stretchable strain sensor based on PVDF electrospun nanofibers.

DYNAMIC PRESSURE DETECTION USING DIPOLE STIMULUS

In order to evaluate the performance of the proposed sensor under dynamic pressure a vibrating sphere (dipole) is used to generate oscillatory pressure and the sensor output is acquired at various frequencies of vibration. The details of the vibrating sphere oscillator system are described in [16]. The dipole is positioned at the distance of 2mm above the sensor. The amplitude of vibration is kept constant (250mVrms) while the frequency is varied in steps from 0.5Hz to 5 Hz. Figure 5 shows the schematic of the experimental set-up. To ensure the repeatability of the results, the experiment is repeated on four different sensors. Figure 5 shows the sensor output as a function of time for various frequencies. In order to ensure that the voltage generated is actually from the PVDF nanofibers, a dummy sensor with the same electrode set-up but with no nanofibers is tested under the same experimental conditions as that of the nanofiber sensor and the output of the device without sensing element (marked as paper in figure 5) is recorded. The experimental results conducted using the dipole stimulus is shown in figure 5.

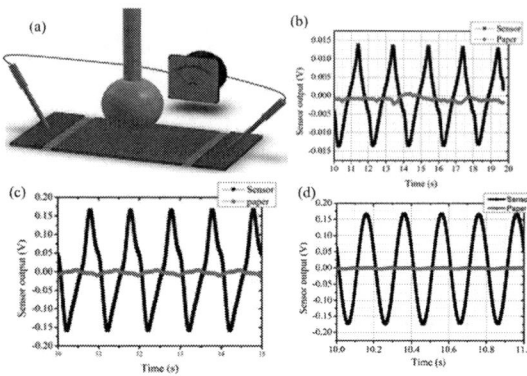

Figure 5. Testing the sensor under low frequency dynamic loads. (a) schematic diagram of the experimental setup (b) experimental result when the dipole is excited at frequency of 0.5Hz and amplitude of 250mVrms.

APPLICATIONS IN ARTIFICIAL LIMBS

There is immense need for producing myoelectric limbs that could help amputees to regain independence in their everyday lives. Myoelectric sensors are required to provide signals for control of artificial limbs. In order to investigate the performance of the proposed sensor in detecting the movement of human limb joints we developed a smart glove by mounting the sensors on the fingers of the glove (figure 6). This glove is integrated with a data acquisition system that can transfer the sensor output to a computer. Two sensors are mounted on index and middle fingers of the glove. Figure 7 shows the sensor output in response to the bending of the index and middle fingers. For example, in the first case, the middle finger is bent while the index finger is kept straight. The sensor on the index finger generated a clear voltage peak (plotted by red) due to the displacement of this finger while the sensor output on the index finger (plotted in black) remained unchanged. Increased bending of the fingers lead to an increase in mechanical stress induced on the nanofibers which led to a higher sensor output. Similar explanation can be applied for the other cases when the middle finger remained straight, both fingers stayed unchanged and for both fingers bend together (see figure 6). The sensor exhibited an excellent stability, response speed, and repeatability.

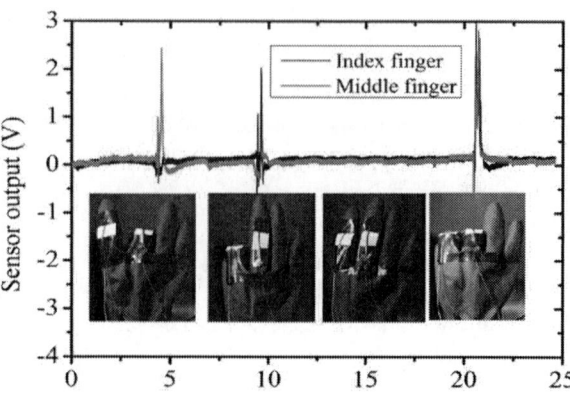

Figure 6. Human motion detection by using the proposed nanofiber sensor. Each case shows the response of the strain sensor to the bending of corresponding finger.

CONCLUSION

In this work, we developed piezoelectric PVDF nanofibers by using electrospinning process. We developed a flexible strain sensor with high sensitivity, stretchability with simple and low cost of fabrication process by using PVDF nanofiber as the sensing element. Performance of the proposed sensor under low frequency dynamic load and in artificial limbs presented. The sensory feedback provided by these devices can open-up new avenues in rehabilitation engineering and brings a sense of touch to myoelectric limbs.

ACKNOWLEDGEMENT

This research was funded by Singapore National research foundation (NRF) through the Singapore-MIT Alliance for Research and Technology (SMART) Centre, Centre for Environmental Sensing and Modeling (CENSAM) IRG

REFERENCES

[1] M. Asadnia, A. G. P. Kottapalli, J. M. Miao, A. B. Randles, A. Sabbagh, P. Kropelnicki, *et al.*, "High temperature characterization of PZT(0.52/0.48) thin-film pressure sensors," *J. Micromech. and Microeng,* vol. 24, Jan 2014.

[2] M. Asadnia, A. G. P. Kottapalli, Z. Shen, J. Miao, and M. Triantafyllou, "Flexible and Surface-Mountable Piezoelectric Sensor Arrays for Underwater Sensing in Marine Vehicles," *Ieee Sensors J,* vol. 13, pp. 3918-3925, 2013.

[3] M. Asadnia, A. G. P. Kottapalli, Z. Shen, J. M. Miao, G. Barbastathis, M. S. Triantafyllou, *et al.*, "Flexible, zero powered, piezoelectric MEMS pressure sensor arrays for fish-like passive underwater sensing in marine vehicles," in *26th IEEE MEMS' 13, Conference*, Taipei, January 20-24, pp. 126-129, 2013.

[4] J. Dusek, A. G. P. Kottapalli, M. E. Woo, M. Asadnia, J. Miao, J. H. Lang, *et al.*, "Development and testing of bio-inspired microelectromechanical pressure sensor arrays for increased situational awareness for marine vehicles," *Smart. Mater. Struc,* vol. 22, 2013.

[5] C. Chang, V. H. Tran, J. Wang, Y.-K. Fuh, and L. Lin, "Direct-Write Piezoelectric Polymeric Nanogenerator with High Energy Conversion Efficiency," *Nano Letters,* vol. 10, pp. 726-731, 2010.

[6] Y.-F. Goh, I. Shakir, and R. Hussain, "Electrospun fibers for tissue engineering, drug delivery, and wound dressing," *J. Materials Science,* vol. 48, pp. 3027-3054, 2013.

[7] D. Mandal, S. Yoon, and K. J. Kim, "Origin of Piezoelectricity in an Electrospun Poly(vinylidene fluoride-trifluoroethylene) Nanofiber Web-Based Nanogenerator and Nano-Pressure Sensor," *Macromolecular Rapid Communications,* vol. 32, pp. 831-837, 2011.

[8] J. Chang, M. Domnner, C. Chang, and L. Lin, "Piezoelectric nanofibers for energy scavenging applications," *Nano Energy,* vol. 1, pp. 356-371, 2012.

[9] D. H. Reneker, A. L. Yarin, H. Fong, and S. Koombhongse, "Bending instability of electrically charged liquid jets of polymer solutions in electrospinning," *J. Applied Physics,* vol. 87, pp. 4531-4547, 2000.

[10] E. Hollenstein, M. Davis, D. Damjanovic, and N. Setter, "Piezoelectric properties of Li- and Ta-modified (K0.5Na0.5)NbO3 ceramics," *Applied Physics Letters,* vol. 87, Oct 31 2005.

[11] Z. H. Liu, C. T. Pan, L. W. Lin, and H. W. Lai, "Piezoelectric properties of PVDF/MWCNT nanofiber using near-field electrospinning," *Sensors and Actuators A-Physical,* vol. 193, pp. 13-24, Apr 15 2013.

[12] D. H. Sun, C. Chang, S. Li, and L. W. Lin, "Near-field electrospinning," *Nano Letters,* vol. 6, pp. 839-842, Apr 2006.

[13] H. S. Nalwa, "Ferroelectric Polymers: Chemistry: Physics, and Applications," *CRC Press,* vol. 20 Jun 1995.

[14] C. Ribeiro, V. Sencadas, J. L. Gomez Ribelles, and S. Lanceros-Mendez, "Influence of Processing Conditions on Polymorphism and Nanofiber Morphology of Electroactive Poly(vinylidene fluoride) Electrospun Membranes," *Soft Materials,* vol. 8, pp. 274-287, 2010.

[15] A. Baji, Y.-W. Mai, Q. Li, and Y. Liu, "Electrospinning induced ferroelectricity in poly(vinylidene fluoride) fibers," *Nanoscale,* vol. 3, pp. 3068-3071, 2011.

[16] A. G. P. Kottapalli, M. Asadnia, J. M. Miao, G. Barbastathis, and M. S. Triantafyllou, "A flexible liquid crystal polymer MEMS pressure sensor array for fish-like underwater sensing," *Smart. Mater. Struct,* vol. 21, Nov 2012.

CONTACT

*M. Asadnia, Tel: (+65) 6790 4264, School of Mechanical & Aerospace Engineering, Nanyang Technological University, mohs0002@e.ntu.edu.sg

ULTRACOMPACT OPTOFLEX NEURAL PROBES FOR HIGH-RESOLUTION ELECTROPHYSIOLOGY AND OPTOGENETIC STIMULATION

Maysamreza Chamanzar[1]*, Daniel J. Denman[2], Timothy J. Blanche[2], and Michel M. Maharbiz[1]
[1]EECS Department, University of California Berkeley, Berkeley, USA
[2]Allen Institute for Brain Science, Seattle, USA

ABSTRACT

We report on the development of high-density neural probes for distributed neuronal recording and stimulation. Our hybrid silicon-parylene probes provide high spatial resolution and incorporate a monolithically integrated flexible cable to address the challenge of stable recordings in chronic neural implants. We address a long-standing but often overlooked issue in parylene processing to realize reliable multilayer interconnects. We also discuss the design of ultracompact parylene optical waveguides for localized optogenetic stimulation of neurons. We demonstrate in-vivo electrophysiology recordings in mice.

INTRODUCTION

Understanding the neural basis of brain function remains an elusive goal of systems neuroscience [1]. Existing neural probe technologies: lack the electrode array density necessary for reliably isolating action potentials (APs) from single neurons while also providing millimetric scale coverage of brain nuclei; are large and stiff which causes damage to the neuropil and vasculature; are prohibitively expensive, and are not scalable in a way that enables recordings from a large population of neurons in multiple adjacent brain areas. High-density silicon probes have been fabricated using electron beam lithography (EBL) [2], however EBL is a serial and low-throughput method that cannot be easily scaled up to mass-produce high-yield probes. Additionally, the rigid connection of probes to the recording circuitry is a contributing factor in the chronic inflammatory response that results in the formation of glial scarring and degradation of recording quality over time. To overcome this problem, polymer-based probes have been recently proposed [3]. The low stiffness of such probes makes implantation difficult and the density of recording channels is low because of the limited resolution of lithography on polymer substrates. To address these challenges, we have designed a process based on high-throughput deep UV (DUV) lithography to realize ultra compact implantable neural probes in a hybrid parylene-silicon platform [4]. Our neural probes thus inherit the benefits of both silicon-based and all-polymer neural implants.

Optogenetics is a powerful tool for perturbation of neural circuits [5] that would be even more powerful if light could be delivered locally to small subsets of neurons instead of illuminating whole brain regions using disruptive methods such as optical fibers. We performed quantitative modeling to show the possibility of realizing ultra compact all-parylene waveguides that can be monolithically integrated on the probe. We characterized the electrical properties of our probes and demonstrated their use in acute electrophysiology recordings in mice.

A schematic of our neural probe is shown in Figure 1. The probe shank is made of silicon with an insulating layer of parylene C and incorporates 64 recording sites. The shank is 20 μm thick, and 30-95 μm wide depending on the recording site configuration (which is usually the case for other probes where the shank width is determined by the size and number of interconnects). The 64 recording sites are routed through a high-density array of interconnects over the silicon shank, which are then connected monolithically to the corresponding array of interconnects embedded in a thin (8 μm) parylene C ribbon cable. This compliant cable conducts the recorded signals through the skull to the head-mounted recording electronics. The flexible nature of this cable allows the implanted probes to 'float in the brain' and move coherently with the brain micromotion, thus minimizing the tethering force on the brain tissue. The end of the parylene cable is also monolithically connected to a stiff silicon backend, where the traces expand and end in an array of bond pads on silicon to allow integration with the recording circuitry possible using standard packaging methods such as wire- or flip-chip bonding.

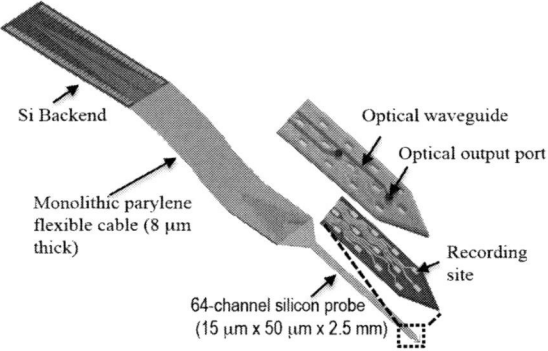

Figure 1: Schematic of the hybrid silicon-parylene probe

FABRICATION PROCESS

To realize high-yield and compact neural probes, we designed a fabrication process in which the silicon layer and the parylene layers are realized in different steps on the same wafer. We optimized the process so that more than 500 probes can be made on a 6-inch wafer. A simplified version of the fabrication process highlighting the important steps is shown in Figure 2. First, we grow thermal oxide at 1000C in a Tystar furnace to a thickness of 1 μm on a SOI wafer with a device layer of 15 μm that defines the thickness of the probe shank and the backend. Then, using a DUV ASML5500/300 stepper, we define the recording sites and the high resolution interconnects. A stack of Ti/Au/Pt with thicknesses of

978-1-4799-7956-1/15 $31.00 © 2015 IEEE   682   MEMS 2015, Estoril, PORTUGAL, 18 - 22 January, 2015

100A/1400A/100A is deposited using a CHA evaporation system. Ti provides good adhesion to the oxide layer and Pt is used to ensure a strong adhesion to parylene. Following a lift off process, the recording sites and interconnects are realized. We optimized the process to realize ~250 nm space and trace interconnects along the 2 mm length of the probe shank with a near perfect yield (Figure 3). In the next step, parylene is conformally deposited to a thickness of 4 μm in a SCS Labcoater 2010 machine. A-174 silane in vapor phase is used to enhance the adhesion of parylene to the underlying layer. Next, vias are etched in parylene using an oxygen plasma process to connect the first layer interconnects to the cable tracks, subsequently defined using another lithography and lift off process. This way the interconnects on SiO$_2$ layer are routed on top of the parylene layer through etched vias.

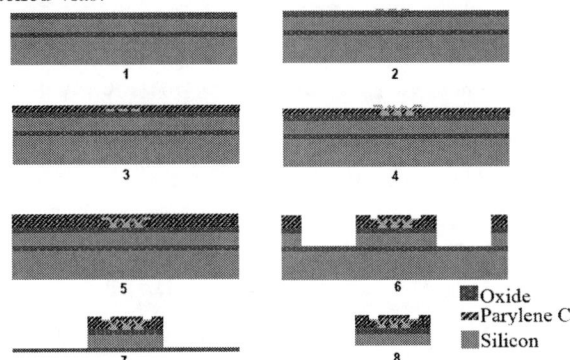

Figure 2: The fabrication process starts with patterning the first layer interconnects on an SOI wafer, followed by parylene C deposition and a second layer of interconnects on parylene C. Then the recording sites are etched and exposed; the probe outline is defined and etched; and finally the probes are released.

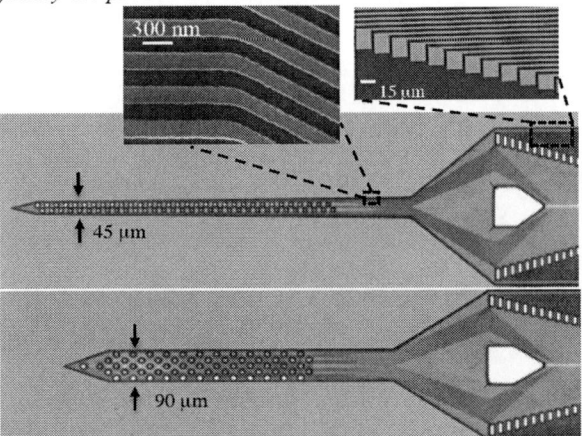

Figure 3: Optical micrographs of two different probe shanks. Left inset shows a scanning electron micrograph of interconnects on the shank. Right inset shows 4 μm tacks at the silicon-parylene cable interface.

Another layer of parylene is deposited to a thickness of 4 μm to completely embed interconnects in parylene thus forming the cable. The resulting cable is only 8 μm thick and extremely compliant (with the cantilever stiffness of k_{cant}~10^{-5} N/m). In the next step, the outlines of the devices are etched in parylene, oxide, and silicon layers. The wafer is mounted on a carrier wafer to release the devices from the backside in parallel. This process is high throughput. Based on probe station tests on a few different wafers, more than 90% of the devices on the wafer were functional, with some devices close to the rim partially damaged mostly due to etching non-uniformities.

PARYLENE INTERCONNECT PROCESS CONSIDERATIONS

Parylene C has been widely used as a highly inert and biocompatible material in a range of implantable devices. To realize functional devices, parylene needs to be patterned, etched, and integrated with other materials. Despite the favorable properties of parylene films, processed parylene-based devices suffer from stability and longevity issues, especially when implanted in the body. We have identified two main causes that potentially contribute to the failure of parylene devices. One problem is the residue that remains on vias after plasma etching of parylene. This microstructured residue (shown in Figure 4a) is extremely difficult to remove, even after extensive overetching using oxygen plasma. The poor electrical connection and a high chance of delamination result in a very unstable layer-to-layer connection. To solve this issue, we have designed a process based on an aluminum sacrificial etch stop layer to lift off the remnant residue. The second problem is the very poor quality worm-holed sidewalls when etching parylene using standard oxygen plasma recipes (Figure 4a). Over-etching damages sidewalls by aggressively undercutting while leaving behind worm-holed walls, which are hard to see without SEM imaging. These porous etched sidewalls allow for permeation of liquid, proteins and other molecules and therefore failure and delamination of the patterned and etched parylene layers. We have optimized a custom DRIE recipe consisting of alternating O$_2$ plasma (600W, 60 mTorr)/ fluoropolymer deposition steps (6 sec / 7 sec) to achieve smooth and seamless sidewalls (Fig. 4b).

Figure 4: (a) A via etched in Parylene C using anisotropic O$_2$ plasma etching (200W, 80 mTorr), where rough sidewalls and residues at the bottom are evident. (b) the same structure etched using our DRIE recipe with alternate C$_4$F$_8$ deposition and O$_2$ plasma etching and an Al sacrificial etch stop layer. The sidewalls are smooth and no residue is left at the bottom.

We verified the reliability of the vertical metal interconnects through different layers of parylene C after these process innovations. Probe station tests on three different wafers revealed that the yield of the interconnections at the junction of the cable and the probe shank were increased from 30% to more than 96% after we optimized the parylene etching process. Moreover, we tested the mechanical stability of the released probes in Phosphate Buffered Saline (PBS 1X) for 8-hour periods over 30 days. No degradation of impedance was observed. In addition, we measured the impedance spectrum before and after several implantations, where again the electrical impedances of the channels remained stable. These observations further corroborate the importance of having high-quality etched layers in parylene C to enhance the mechanical stability and prevent failures in vivo.

FLEXIBLE OPTICAL WAVEGUIDES FOR LOCAL OPTOGENETIC STIMULATION

Despite recent advances in neural modulation techniques, including a rapidly expanding optogenetic toolset [5, 6], still we lack a robust, minimally-invasive optogenetic stimulation platform [6]. The ability to independently deliver light to multiple, highly localized (~50 μm^3) subsets of neurons that are simultaneously recorded on the electrode array would drastically improve the sensitivity and interpretability of in vivo optogenetic experiments. Illuminating a large volume of neuropil using light sources on the surface of the brain does not provide the requisite spatial resolution and since the intensity falls off rapidly, only a small fraction of target neurons in the vicinity of the light source (~200 μm) will be excited. Increasing the light source power, on the other hand, results in the generation of excessive heat in the brain and the potential for phototoxicity [7]. Given a scattering coefficient of 11 mm^{-1} in the mouse brain and the minimum threshold intensity of 1 mW/mm^2 for a channelrhodopsin to evoke action potentials [8], an input power of 2.25 mW is required in a fiber optic (200 μm, NA=0.37) to excite a neuron at a depth of 2 mm into the cortex resulting in a (very high) intensity of 71.6 mW/mm^2 at the output aperture of the fiber, sufficient to cause damage to the brain tissue. This trade-off between the range of stimulation and the required optical power results in an inherently low spatial resolution. To reach deeper brain regions, bare fibers (typically 100~200 μm in diameter) are inserted, causing a large tissue displacement and introducing a tethering force on the brain. Recently, photonic waveguide devices have been used to deliver light locally deep into the brain [7, 10]. These implementations are based on semiconductor and dielectric materials such as SiO_2 and Oxynitride [7, 10]. Such waveguides, although compact, are stiff and cannot be integrated with a neural implant such as the one we have introduced in this paper with both stiff and flexible parts. Material biocompatibility and mechanical stability is always a point of concern, especially for chronic implants.

Here we introduce a novel integrated photonic device platform to realize ultra-compact all-parylene photonic waveguides. Parylene waveguides are flexible, and can be monolithically integrated with the insulating parylene C layer to deliver light from outside the brain to specific locations along the probe shank. The core of the waveguide is made of parylene N, which has a higher refractive index than the parylene C cladding layer. The parylene C cladding serves as an intermediate layer between the electrical layer of the probe and the optical layer (Figure 1). Ellipsometry of the parylene films

revealed a refractive index of 1.66 for parylene C and 1.786 for parylene N at the wavelength of $\lambda = 480$ nm, which is the excitation wavelength for channelrhodopsins. The fabrication process consists of implementing trenches in parylene C and depositing parylene N to form ridge waveguides. An etchback step is used to remove the rib layer that forms in between the waveguides as a result of the conformal deposition of parylene N. Finally, the waveguide array is encapsulated in a cladding layer of Parylene C. An array of fabricated optical waveguides on a stand-alone parylene substrate is shown in Figure 5a. Finite Element simulations show that a very compact waveguide with a cross section of 3 μm x 3 μm can support a confined guided mode at the wavelength of $\lambda = 480$ nm (Figure 5b). This novel integrated photonic material consisting of only parylene layers is seamless, biocompatible, and compliant. We believe such a parylene-in-parylene (PiP) integrated photonic device platform can find interesting applications for a new class of implantable integrated photonic devices, beyond neural probes, where different functional devices such as waveguides, resonators, and filters can be easily implemented.

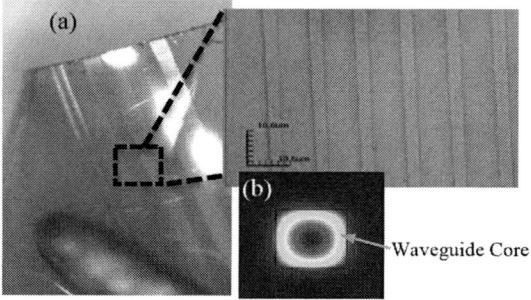

Figure 5: (a) optical micrograph of an array of optical waveguides on a parylene film. (b) The field profile of the confined optical mode of a 3 μm x 3 μm waveguide.

ELECTRICAL CHARACTERIZATION

After the probes are released from the wafer, we package them with a custom-designed adaptor PCB that connects the probe to the headstage recording circuitry (e-cube, White Matter LLC). We designed a die-attach robot capable of finding and assembling released probes from a tray. A fine pitch wire bonder was used to wirebond the probe to the PCB. A packaged probe is shown in Figure 6.

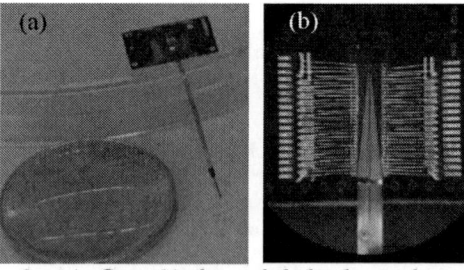

Figure 6: (a) Our 64-channel hybrid parylene-silicon probe, consisting of a high-density silicon shank (50 μm ×15 μm ×2 mm) and a flexible parylene cable (8 μm × 500 μm ×20 mm) assembled to the headstage adaptor PCB. (b) Backend wirebonded with the adaptor PCB.

978-1-4799-7956-1/15 $31.00 © 2015 IEEE

We characterized the electrical properties of the probes by measuring the recording site impedances in Phosphate Buffered Solution (PBS 1X) using a nanoZ impedance spectrum analyzer (White Matter LLC). To improve the signal-to-noise-ratio (SNR), conductive polymer Poly(3,4-ethylenedioxythiophene) Polystyrene sulfonate (PEDOT:PSS) was electroplated on the recording sites. The impedance spectra of a typical recording site is plotted in Figure 7, before and after electroplating with a 0.2 M concentration of PEDOT:PSS and a total charge deposition of 6 µC. The impedance was reduced by an order of magnitude. Since the recording sites are recessed in parylene, the parylene sidewalls prevent enlargement of the sites, which prevents possible shorting with other sites during plating. PEDOT-filled recording sites have a lower noise floor without increase in the electrode cross-sectional area, which effectively increases the SNR of extracellular neural recordings.

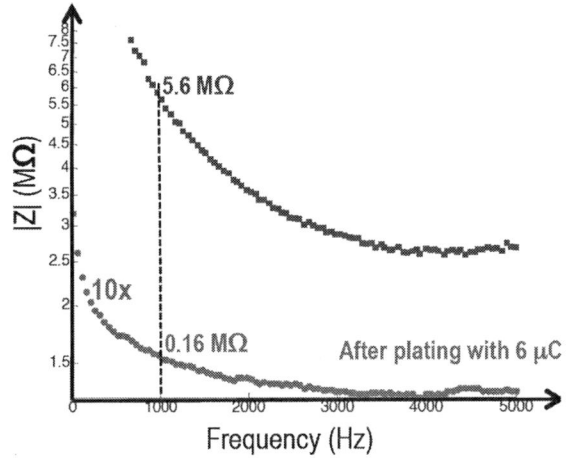

Figure 7: The impedance spectra for a ~113 µm² recording site before (blue) and after (red) electroplating in PEDOT:PSS. The spectrum after electroplating is plotted with 10x amplification for legibility.

Another advantage of recessed recording sites is the stability of the electroplated PEDOT layer in the brain. As can be seen in Figure 8, the deposited PEDOT polymer layer is protected by the parylene sidewalls and does not easily come off during implantation. We performed repeated insertion tests that showed that the impedances of the sites are stable with multiple insertions in the brain.

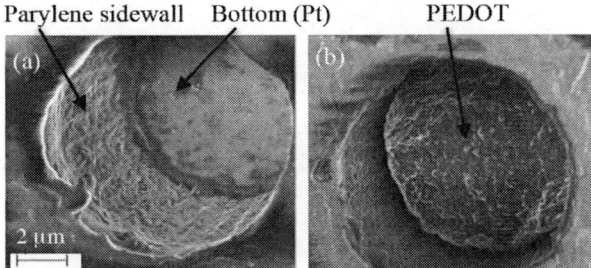

Figure 8: (a) The native Pt recording site at the bottom of a parylene hole. (b) Electroplated PEDOT conductive polymer on the recording site.

IN-VIVO TESTS

We implanted our neural probes in the visual cortex of wildtype (C57) mice using a robotic stereotaxic micromanipulator (Neurostar, Germany) equipped with piezo microtweezers (Figure 9a). Acute in-vivo recordings with excellent SNR were demonstrated (Figure 9b).

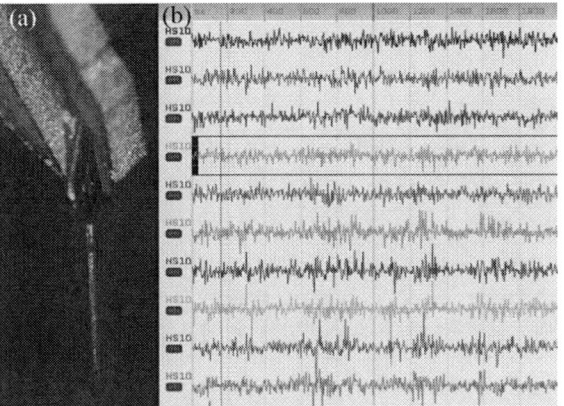

Figure 9: (a) a probe held by piezo microgrippers above the cortex prior to implantation. (b) acute extracellular recordings of ten representative channels, high-pass filtered (300-6kHz) to show neural action potentials.

ACKNOWLEDGEMENTS

This project was supported by the National Science Foundation, IDBR grant # 1152658.

REFERENCES

[1] G. Buzsaki, Nature Neuroscience vol. 7, p. 446, 2004.
[2] J. Du, T. J. Blanche, R. R. Harrison, H. A. Lester, and S. C. Masmanidis, PLoS One, vol. 6, p. e26204, 2011.
[3] J. P. Seymour, D. R. Kipke, Engineering in Medicine and Biology Society, 2006. EMBS '06, p. 4606, 2006.
[4] C. Pang; J.G. Cham, Z. Nenadic, S. Musallam, Y. Tai; J. W. Burdick, R. A. Andersen, Engineering in Medicine and Biology Society, 2005. EMBS '05, p. 7114, 2005.
[5] L. Fenno, O. Yizhar, K. Deisseroth, Annual review of neuroscience, vol. 34, p. 389, 2011.
[6] E. S. Boyden, F1000 Biology reports 2011.
[7] A. N. Zorzos, J. Scholvin, E. S. Boyden, Optics letters, vol. 37, p. 4841, 2012.
[8] O. Yizhar, L. E. Fenno, T. J. Davidson, M. Mogri, K. Deisseroth, Neuron, vol. 71, p. 9, 2011.
[9] A. M. Aravanis, L. P. Wang, F. Zhang, L. Meltzer, M. Mogri, M. B. Schneider, K. Deisseroth, Journal of neural engineering, vol. 4, p. S143, 2007.
[10] T. V. Abaya, S. Blair, P. Tathireddy, L. Rieth, F. A. Solzbacher, Biomedical optics express, vol. 3, p. 3087, 2012.

CONTACT

*M. Chamanzar, chamanzar@berkeley.edu

VERTICALLY ALIGNED EXTRACELLULAR MICROPROBE ARRAYS/(111) INTEGRATED WITH (100)-SILICON MOSFET AMPLIFIERS

Hiroki Makino[1], Kohei Asai[1], Masahiro Tanaka[1], Shota Yamagiwa[1], Hirohito Sawahata[1], Ippei Akita[1], Makoto Ishida[1, 2] and Takeshi Kawano[1]

[1]Department of Electrical and Electronic Information Engineering, Toyohashi University of Technology, Japan, [2]Electronics-Inspired Interdisciplinary Research Institute (EIIRIS), Toyohashi University of Technology, Japan.

ABSTRACT

We report a heterogeneous integration of vertically aligned extracellular micro-scale silicon (Si)-probe arrays/(111) with MOSFET amplifiers/(100), by IC processes and subsequent vapor-liquid-solid (VLS) growth of Si-probes. To improve the extracellular recording capability of the microprobe with a high impedance of > 1 MΩ at 1 kHz, here we integrated (100)-Si source follower buffer amplifiers by ~700°C VLS growth compatible (100)-Si MOSFET technology. Without on-chip source follower, output/input signal ratio of the microprobe in saline was 0.59, which was improved to 0.72 by the on-chip source follower configuration, while the signal-to-noise ratio (SNR) was improved to 12.5 dB in the frequency of extracellular recording. These results indicate that the integration of the source follower buffer amplifiers becomes a powerful way to enhance the performance of high impedance microprobe electrodes in neural recordings.

INTRODUCTION

An array of microelectrodes, which enables simultaneous electrical recordings of large numbers of neurons/cells in a tissue, has been made significant contributions to fundamental neuroscience and medical applications. A next challenge of the microelectrode technology is the realization of low invasive electrode array, offering the stable recordings and chronic applications, including neuronprosthetics and epileptic focus localization[1, 2].

A way to realize the low invasive electrode technology is minimization of the diameter or cross-sectional area of the needle-electrode. However, the electrical metal/electrolyte interfacial impedance of the electrode increases with decreasing the area of the recording-site, resulting in neuronal signals with a high SNR impossible to record. To address the issue of the high impedance electrode, integrating microelectronics amplifiers is a possible way. Such on-chip electronics configuration also offers integrations of other circuitry (e.g., filter, selector, and RF circuit), enhancing the performance of the electrode device.

We have proposed low invasive, high-spatial-resolution, <5 μm diameter Si probe electrodes/(111)-Si fabricated by vapor-liquid-solid (VLS) growth of Si at ~700°C[3], and the recording capability has been demonstrated using the retina of a carp[4, 5] and the cortex of a rat[6]. However, as aforementioned, amplitude attenuations of the recorded neuronal signals (20–60%) have been observed, due to the high impedance characteristics of the VLS-Si microprobe (e.g., >1 MΩ at 1

kHz for Pt-tipped VLS-Si microprobe) and the parasitic impedances embedded in the measurement system including cables (1–10 MΩ at 1 kHz)(Figs. 1a and 1b)[5, 6]. To improve the recording capability of the VLS-Si microprobe, here we designed and fabricated the VLS-Si probe arrays/(111)-Si integrated with (100)-Si buffer source follower amplifiers by ~700°C VLS growth compatible (100)-Si MOSFET process (Figs. 1c and 1d).

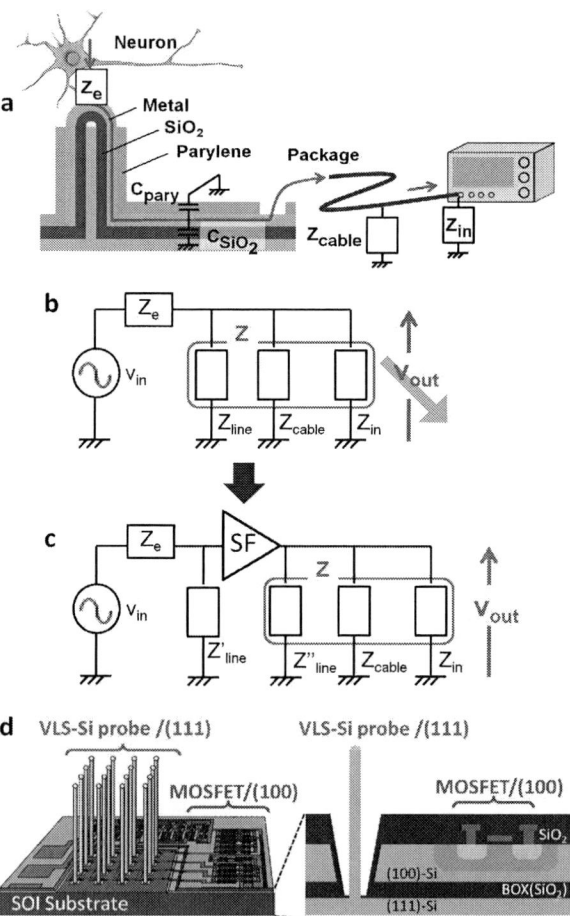

Figure 1: Extracellular recording via a VLS-Si microprobe. (a) Schematic of the probe with impedance, Z_e. (b, c) Circuit diagrams for device parasitic impedances of device interconnection, Z'_{line} and Z''_{line}, recording cable from the chip-bonding pad to the buffer amplifier, Z_{cable}, and input impedance of the buffer amplifier, Z_{in}(b) and the device with source follower (SF) configuration(c). (d) Schematic and cross-sectional image of a heterogeneously integrated MOSFET/(100) and VLS-Si microprobe array/(111) using a hybrid (100)-top-Si/BOX/ (111)-handle-Si system SOI substrate.

978-1-4799-7956-1/15 $31.00 © 2015 IEEE

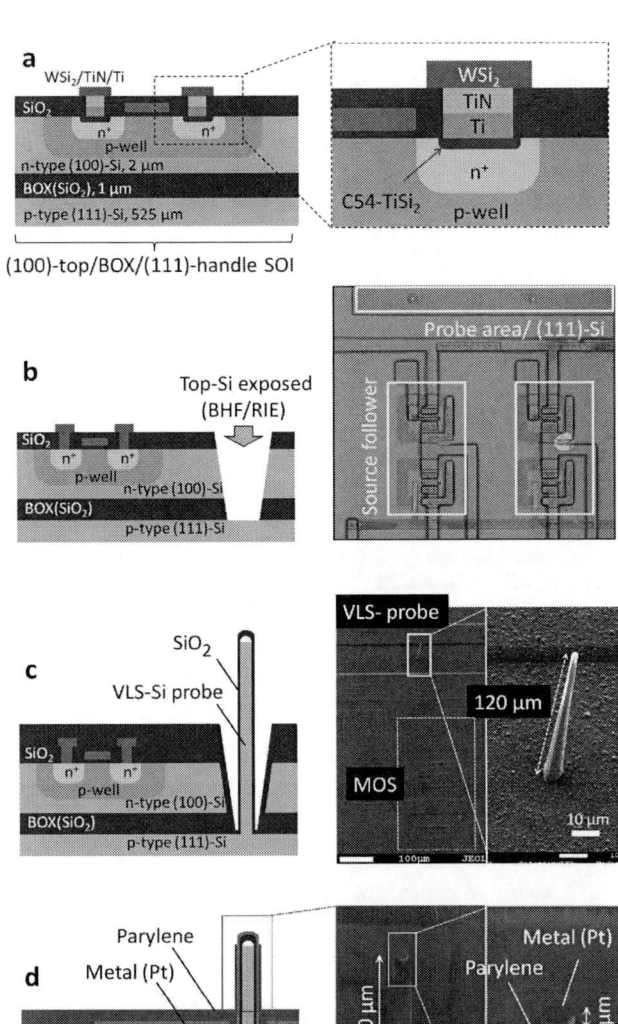

Figure 2: Process flow: (a) MOFET/(100)-top-Si with WSi/TiN/Ti metal and C54-TiSi₂ barrier layer system, (b) exposing the top-layer, (c) VLS growth of Si probe/(111)-handle, and (d) probe metallization, parylene coating, and tip exposing. The length and diameter of the probe are 120 μm and 5 μm, respectively.

FABRICATION PROCESS

The fabrication of vertically aligned VLS-Si microprobes requires a (111)-Si-surface oriented substrate. However, characteristics of the MOSFET fabricated on (111) are inferior to those on (100), due to properties of the MOS system such as the lower electron mobility, higher values of interface trapped charges, and fixed oxide charges compared to (100)-Si-surface-orientated substrate[3, 4]. To overcome the technological issue, we have proposed the VLS-Si probes/(111) integrated with MOSFETs/(100) by utilizing a (100)-top-Si/buried oxide (BOX)/(111)-handle-Si wafer system[7].

The Si microprobe array/(111) can be assembled by ~ 700°C VLS growth after the MOS/(100) process. The multiple layered metal system of WSi/TiN/Ti with a high

melting point is a candidate material as the MOSFET interconnections prior to the VLS growth[8]. However, we have confirmed the increased MOSFET off-current between the source and the drain, due to that the bottom layer of Ti reaches to the Si-source/drain after annealing at the VLS growth temperate (~700°C). The TiSi₂ near the *pn* junction of the source/drain resulted in the significantly increased off-currents (*pn* junction depth designed at 0.5 μm in our in-house NMOS technology). Note that TiSi₂ has two stable states of C49 at 500 – 750°C and C54 at 800 – 850°C[9, 10], and the TiSi₂ annealed at the VLS growth temperature is assumed to be C49-TiSi₂. To prevent the *pn* junction shorts by the TiSi₂, here we formed a barrier layer of C54-TiSi₂ prior to the VLS growth.

Figure 2 shows the process flow for the integration of Si probe arrays/(111) and MOSFET source follower arrays/(100). To fabricate the device, a hybrid 4-in SOI substrate consisting of 2-μm-thick (100)-top-Si/1-μm-thick BOX/525-μm-thick (111)-handle-Si wafer system was used. The on-chip source followers were fabricated on the (100)-top, based on in-house 5 μm NMOS technology[3, 7]. As the device interconnections, a high melting point WSi/TiN/Ti metal system with a barrier layer of C54-TiSi₂ [9, 10] is used for the ~700°C VLS-growth (Fig. 2a). After opening the contacts of the source/drain, a 20-nm-thick Ti was formed by sputtering, followed by the formation of C49-TiSi₂ by annealing at 620°C for 1 min with rapid thermal annealing (RTA). After removing the unreached Ti by ammonia-peroxide mixture (APM), the C54-TiSi₂ was formed by an additional annealing at 850°C for 30 sec with RTA. Finally, the MOSFETs with the C54-TiSi₂ contact barrier were metalized with WSi₂/TiN/Ti.

The array of 120-μm-height, 5-μm-diameter Si probes was fabricated at the exposed (111)-handle wafer by the VLS growth (Figs. 2b and 2c)[11]. For the electrical functionality of the Si-probe, each probe was encapsulated with 1-μm-thick SiO₂ insulator, followed by the metallization with 200-nm Pt and the device encapsulation with biocompatible 1-μm-thick parylene-C. Finally, the probe-tips with the exposed height of < 4 μm were opened by O₂-plasma for extracellular recordings (Fig. 2d).

RESULTS AND DISCUSSION

We integrated VLS-Si microprobes/(111) with on-chip source follower buffer amplifiers/(100) by (100)-Si NMOS technology. Figure 3a is the drain current I_{DS} – gate voltage V_{GS} corves of the NMOSFET after the processes completion, showing the threshold voltage of 1.15 V with a subthreshold swing of 140 mV/decade and the leakage current of 2.2 nA (drain voltage = 5 V). After the probe metallization process (Fig. 2d, before the parylene deposition), hydrogen (H₂) annealing at ambient conditions containing 4% H₂ in nitrogen was carried out at 400°C for 60 min, in order to terminate the dangling bonds and to decrease the interface states of the MOS system due to plasma processes[12] and thermal annealing in an vacuum environment during the VLS growth (Fig. 2c)[4, 7]. Compared to the designed threshold voltage of 0.8 V, the fabricated NMOS exhibits the increased value of 1.15 V, which can be improved by increasing the duration of the H_2 annealing (> 60 min). Figure 3b shows the drain current I_{DS} – drain voltage V_{DS} curves of the NMOSFET shown in Fig. 3a. The length and width of the measured NMOSFET are 10 μm and 50 μm, respectively.

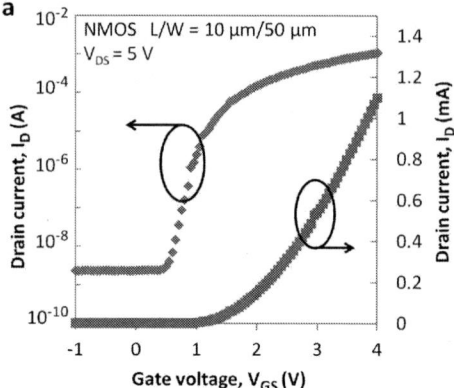

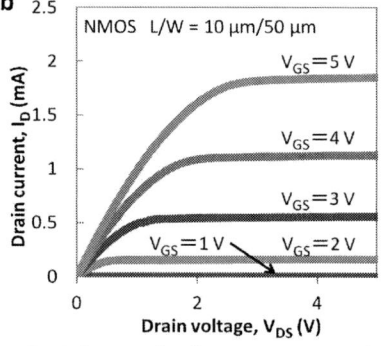

Figure 3: Electrical characteristics of an integrated NMOSFET/(100)-top. (a) Drain current I_{DS} – gate voltage V_{GS} characteristic and (b) Drain current I_{DS} – drain voltage V_{DS} characteristic of the NMOSFET/(100) after the device process completion. The gate length and width of the MOSFET are 10 μm and 50 μm, respectively.

The output/input (O/I) signal amplitude ratio of the probe device in saline was analyzed by applying test signals to the device in a saline solution bath (Fig. 4). Herein, 1 kHz, 200 μV$_{p-p}$ input sinusoidal wave signals as extracellular signals were applied to the solution bath via a counter electrode[5, 6, 8]. The fabricated microprobe with the on-chip source follower exhibits the O/I ratio of 0.72, while the other probe without the source follower shows the lower ratio of 0.59 due to the aforementioned parasitic impedances-induced signal attenuation[5, 6].

Figure 5a shows the power spectrum of the signals: input signal and recorded signals with and without source follower. Figure 5c shows SNRs obtained with root-mean-square (RMS) voltages of output signals (Fig. 5b). The signals recorded with source follower exhibits the SNR of 12.5 dB in the actual frequency of extracellular recordings of 500 Hz – 3 kHz, while the SNR of 6.81 dB is obtained without source follower configuration.

The power spectrum obtained without source follower shows the increased noise at 500 Hz – 3 kHz (green line in Fig. 5a). Because the input signals were applied to the

whole saline bath in the used measurement scheme, the noise is probably due to the input-signal dependent noise. On the other hand, with the source follower configuration, the noise waveform becomes relatively input signal-independent, obeying a flicker noise (1/f noise)(blue line in Fig. 5a). Although further discussion will be required to clarify the noise source, these results indicate that the integrating source follower improves the SNR in signal recordings.

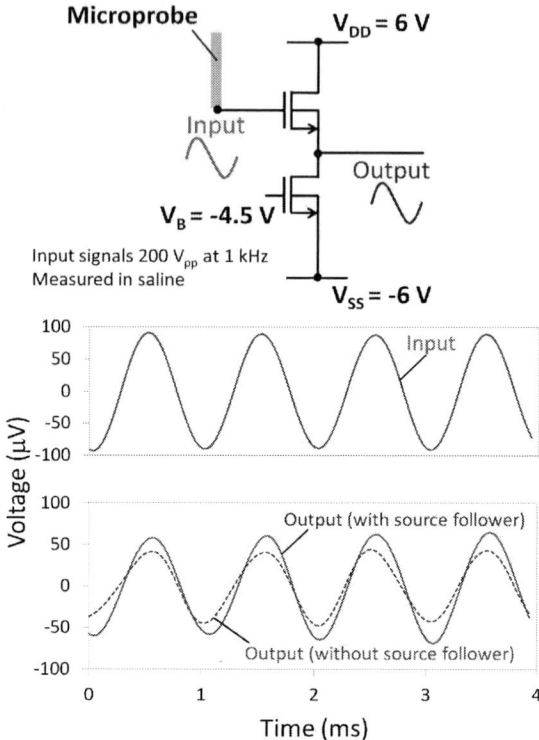

Figure 4: Test signal recording in saline using 1 kHz. 200 μV$_{p-p}$ input sinusoidal wave signals (red). Blue and green curves are with and without source follower, respectively.

CONCLUSIONS

To improve the recording capability of the microprobe with a high impedance characteristics, we demonstrated the integration of the VLS-Si probe arrays/(111)-Si with (100)-Si buffer source follower amplifiers by ~700°C VLS growth compatible (100)-Si MOSFET process. The metal system of WSi/TiN/Ti as the device interconnection results in the electrical *pn* junction shorts, which was solved by forming a buffer layer of C54-TiSi$_2$. Although the electrical characteristics of the fabricated on-chip NMOSFETs were inferior to the designed values, increasing the duration of the H_2 anneal will improve these characteristics. We also demonstrated the test signal recording in saline using input sinusoidal wave signals as extracellular signal, confirming that SNRs of the output signals via VLS-probe can be improved with on-chip source follower configuration. These results indicates that the integration of (100)-Si source follower arrays as the on-chip buffer amplifiers becomes a powerful way to enhance the performance of such high impedance microelectrodes in the neural recording.

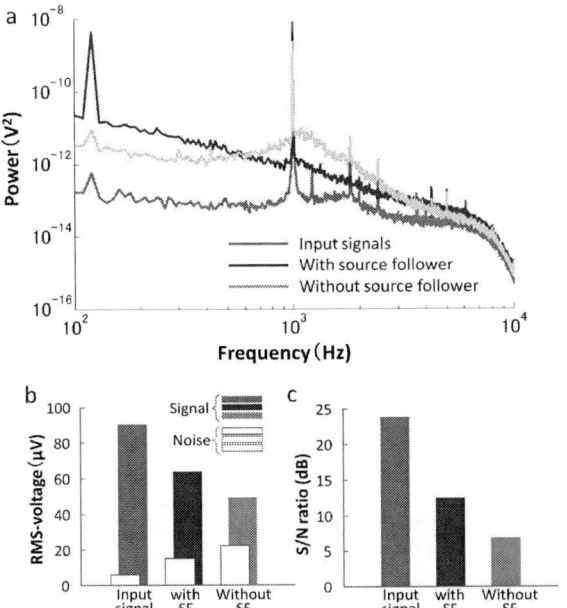

Figure 5: Power spectrum for input signal, with and without source follower measured in saline. (a) Power spectrums showing the input signal (red line), output signals of with source follower (blue line), and without source follower (green line). (b, c) RMS voltages and SNRs of input signal, output signals with and without source follower taken from the spectrums shown in (a).

The proposed on-chip source follower is also applicable to numerous high impedance electrodes, including nano-scale electrode[13] which penetrates a neuron and record the intracellular signals such as postsynaptic potentials (EPSP and IPSP). Further requirement of the source follower integrated electrode for implantable device application is the flexible substrate, which will be realized by assembling Si-MOSFETs on a flexible material film[14].

ACKNOWLEDGEMENTS

This work is supported by a Grant-in-Aid for Scientific Research (S, A), Young Scientists (A, B), the Strategic Research Program for Brain Sciences (SRPBS) from MEXT, and the PRESTO Program from JST.

REFERENCES

[1] L. R. Hochberg, D. Bacher, B. Jarosiewicz, N. Y. Masse, J. D. Simeral, J. Vogel, S. Haddadin, J. Liu, S. S. Cash, P. van der Smagt, and J. P. Donoghue, "Reach and Grasp by People with Tetraplegia using a Neurally Controlled Robotic arm," Nature, Vol. 485, No. 7398, pp. 372–375, 2012.

[2] W. C. Stacey and B. Litt, "Technology Insight: Neuroengineering and Epilepsy—Designing devices for seizure control", Nature Clinical Practice Neurology, Vol. 4, pp. 190–201, 2008.

[3] T. Kawano, Y. Kato, R. Tani, H. Takao, K. Sawada and M. Ishida, "Selective Vapor-Liquid-Solid Epitaxial Growth of Micro-Si Probe Electrode Arrays with On-Chip MOSFETs on Si (111) Substrates."

IEEE Transactions on Electron Devices, Vol. 51, pp. 415–420, 2004.

[4] T. Kawano, T. Harimoto, A. Ishihara, K. Takei, T. Kawashima, S. Usui and M. Ishida, "Electrical Interfacing between Neurons and Electronics via Vertically-integrated Sub-4 Micron-diameter Silicon Probe Arrays Fabricated by Vapor-liquid-solid Growth," Biosensors and Bioelectronics, Vol. 25, No. 7, pp. 1809–1815, 2010.

[5] T. Harimoto, K. Takei, T. Kawano, A. Ishihara, T. Kawashima, H. Kaneko, M. Ishida and S. Usui, "Enlarged Gold-tipped Silicon Microprobe Arrays and Signal Compensation for Multi-site Electroretinogram Recordings in the Isolated Carp Retina." Biosensors and Bioelectronics, Vol. 26, pp. 2368–2375, 2011.

[6] A. Fujishiro, H. Kaneko, T. Kawashima, M. Ishida and T. Kawano, "*In-vivo* Neuronal Action Potential Recordings via Three-dimensional Microscale Needle-electrode Arrays," Scientific Reports, Vol. 4, No. 4868, 2014.

[7] A. Okugawa, K. Mayumi, A. Ikedo, M. Ishida and T. Kawano, "Heterogeneously Integrated Vapor-liquid-solid Grown Silicon probes/(111) and Silicon MOSFETs/(100)," IEEE Electron Device Letters, Vol. 32, No. 5, pp. 683–685, May 2011.

[8] K. Takei, T. Kawashima, T. Kawano, H. Takao, K. Sawada and M. Ishida, "Integration of Out-of-plane Silicon Dioxide Microtubes, Silicon Microprobes, and On-chip NMOSFETs by Selective Vapor-Liquid-Solid Growth," Journal of Micromechanics and Microengineering, Vol. 18, No. 3, 035033, 2008.

[9] R. W. Mann and L. A. Clevenger, "The C49 to C54 Phase Transformation in TiSi2 Thin Films." Journal of The Electrochemical Society, Vol. 141, pp. 1347-1350, 1994.

[10] C. Cabral, Jr., L. A. Clevenger, J. M. E. Harper, F. M. d'Heurle, R. A. Roy, C. Lavoie, K. L. Saenger, G. L. Miles, R. W. Mann, and J. S. Nakos, "Low Temperature Formation of C54-TiSi2 Using Titanium Alloys." Applied Physics Letters, Vol. 71, pp. 3531–3533, 1997.

[11] A. Ikedo, T. Kawashima, T. Kawano and M. Ishida, "Vertically Aligned Silicon Microwire Arrays of Various Lengths by Repeated Selective Vapor-Liquid-Solid Growth of n-type Silicon/n-type Silicon," Applied Physics Letters, Vol. 95, 033502, 2009.

[12] D. Park and C. Hu, "Plasma Charging Damage on Ultrathin Gate Oxides," IEEE Electron Device Letters, Vol. 19, No. 1, pp. 1–3, 1998.

[13] Y. Kubota, H. Oi, H. Sawahata, A. Goryu, Y. Ando, R. Numano, M. Ishida and T. Kawano "A Vertically Integrated Nanoscale Tipped Microprobe Intracellular Electrode Array", IEEE Micro Electro Mechanical Systems (IEEE-MEMS) Conference 2014, San Francisco, USA, January 2014.

[14] A. Fujishiro, S. Takahashi, K. Sawada, M. Ishida and T. Kawano, "Flexible Neural Electrode Arrays with Switch-matrix based on a Planar Silicon Process," IEEE Electron Device Letters, Vol. 35, No. 2, pp. 253–255, February 2014.

978-1-4799-7956-1/15 $31.00 © 2015 IEEE

WEARABLE FLEXIBLE MICRO ELECTRODE FOR ADULT ZEBRAFISH LONG TERM ECGMONITORING

Xiaoxiao Zhang[1], Tyler Beebe[2], Yang Liu[1], Jungwook Park[1], Tzung Hsiai[2], and Yu-Chong Tai[1]
[1]California Institute of Technology, Micro Machining Laboratory, Pasadena, CA, USA
[2]University of California Los Angeles, School of Medicine, Los Angeles, CA, USA

ABSTRACT

During the last decade, close resemblance between the zebrafish heart and human heart physiology has been discovered [1] andzebrafish (Danio rerio) has become an emerging animal model for studying side effects developmental drugs may impose on the heart [2][3]. More interestingly, contrary to human heart the zebrafish heart has a remarkable ability to "regenerate" after severe injury [4], making it also a popular model for studies of regenerative medicine. On the other hand, Electrocardiogram (ECG) is a widely used tool to monitor the physiological changes of the zebrafish heart. However, due to its in-water habitat, a long term ECG monitoring solution, although very much needed, was not present. All published adult zebrafish ECG recorded to this date have been done acutely with anesthetized fish. This work presents, for the first time a wearable flexible parylene (PA) micro-electrode that monitors the Adult Zebrafish ECG long term. We show here the design, fabrication and testing of the flexible electrode along with a micro-molded ultra-soft, density adjusted silicone jacket, allowing ECG recording to be carried under water, in the fish's natural habitat with no need for anesthesia.

INTRODUCTION

Our previous work has demonstrated multiple-lead zebrafish ECG recording with MEMS MEAs [5]; however, the work was based on the acute scheme on anesthetized animals. It has been widely known that anesthesia drug itself affects both the heart rate and the morphology of the ECG, but since the quantified effects are unknown and varies heavilybetweenindividuals, the distortion to heart rate and the ECG waveform morphology cannot be systematically subtracted. The proposed wearable system builds on the fabrication process published in [5][6][7][8][9], however with a significant difference in electrode design, and an additional ultra-soft silicone mounting jacket. This work also includes a set of non-linear signal processing techniques to further filter out breathing and muscle EMG artifacts from the recorded signals.

DEVICE DESIGN AND FABRICATION

The Device design and fabrication process are shown in detail in Fig. 1. Two Au electrodes (chest electrode and reference electrode) are 3mm apart. Each electrode has a micro meshed structure of 25μm x 25μm openings; this is to reduce the stress on the Au thin film during stretching and to lengthen the electrode's lifetime. The ultra-soft silicone jacket utilizes a zip-tie structure with multiple teeth to adjust to different sizes of fish; the extended zip tie length will be

removed with micro surgical scissors once threading is complete. A back padding is also in place to reduce the stress on the animal's back where stress is concentrated during forward swimming motion. The chest electrode will be threaded through the jacket's center "cross" opening and placed at the tip of the jacket. The electrode is then fixed onto the jacket with an additional application of uncured silicone around the anchor holes on the electrodes.

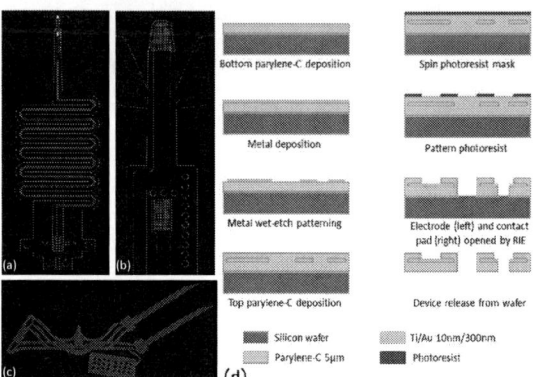

Figure 1: (a) Parylene electrode and cable design. The blue pattern is Parylene outline, yellow pattern is Au trace. Two Au electrodes (chest electrode and reference electrode) are 3mm apart. Each electrode has a micro meshed structure of 25μm x 25μm openings, this is to reduce the stress on the Au thin film during stretching and lengthen the electrode's lifetime. The "wings" next to the chest electrode are to help with threading into the silicone jacket and are torn off after the integration with jacket is done. The PA holes around the reference electrode are used to anchor the MEMS electrode/cable onto the jacket. Uncured silicone will be applied at these holes to adhere the electrode onto the jacket. (b) Zoomed in view of chest electrode and reference electrode. (c) Ultra-soft silicone jacket design; the jacket utilizes a zip tie structure with multiple teeth to adjust to different sizes of fish; the extended zip tie length will be cut after threading. A back padding is in place to reduce the stress onto the fish skin due to Yong's modulus mismatch of the skin and the silicone jacket. The chest electrode will be threated through the center "cross" opening and be placed at the tip of the jacket. The electrode is anchored at the PA opening holes as explained in (a). (d) Fabrication process of the PA-Au/Ti-PA electrode.

The ECG electrode has a PA-C/Ti/Au/PA-C stacked structure (Fig. 1d) and has a fabrication process similar to that in [5]; the complete device is shown in Fig. 2. The micro electrode cable is fixed to a flat FFC cable which will be used as a guide to plug into a zero insertion force FFC connector. Fig. 3 shows the fish wearing the device in water.

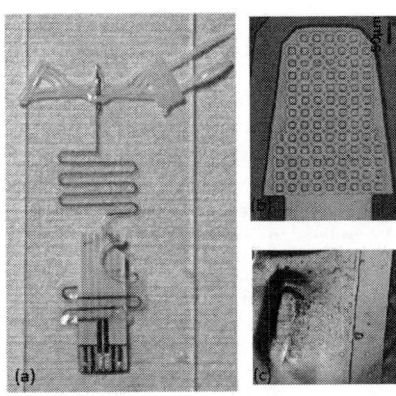

Figure 2: (a) complete device with PA electrode and cable, Silicone jacket, and flexible flat connector (b) zoomed in view of chest electrode with micro mesh structure; each micro mesh opening is 25µm x25µm (c) Au electroplating thickened connection pads, the denture mark shows the thickened contact pad after connecting with the zero insertion force connector 3 times, the thickened gold pad is reusable and is less time consuming to prepare for batch production.

EXPERIMENT

The recording setup is shown in Fig.4. Differential signals are recorded between the chest electrode and the reference electrode on the fish body. The fish is placed in a transparent tube (to reduce EMG and motion artifacts from excessive muscle movement) and in regular tank water. ECGs over 26 days post ~20% ventricle amputation are obtained, more specifically, four 24-hour continuous ECG data sets, control (pre-amputation), at day 10, day 18 and day 26 post amputation are recorded. Day 10 heart rates are compared with the control result over 24 hours, while all three post amputation ECG morphologies are compared with the control.

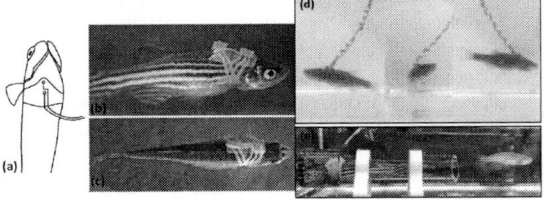

Figure 3: (a) electrode placement on Adult zebrafish chest. (b)(c) Zebrafish wearing the silicone jacket. The silicone is dry film micro casted with medical grade PDMS elastomer (MDX4210, Dow Corning, USA) with a mixing ratio of 20:1 to reduce young's modulus, and mixed with hollow micro

glass beads (3M-S38 Microspheres, 3M, USA) to obtain a density of 1g/ml for minimal weight burden. (d) Free-swimming fish wearing the recording system. (e) fish in confinement tube during recording.

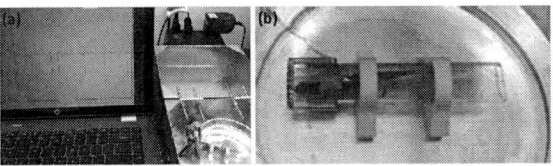

Figure 4: (a) The recording setup. In shielding cage: zebrafish under testing (in transparent plastic tube), amplifying and multiplexing circuitry; digitized signal then goes out of the cage and into the signal routing board and then into the PC where the control software is installed.(b) zoomed in view of fish in the tube during recording.

SIGNAL PROCESSING

Due to the inherent gill motion artifact, the raw data requires further processing to subtract the large breathing baseline for the morphology study. Rawdata is firstrun with a peak detection algorithm [10][11] to detect the R-peaks in order to index the data into single ECG wave segments. All the segments (roughly 500 ECG waves) are then aligned at their R-peaks and averaged to eliminate the non-correlated breathing baselines. The averaged waveform is then baseline subtracted with a cubic spline fitted curve to obtain the ECG wave morphology (Fig.5).

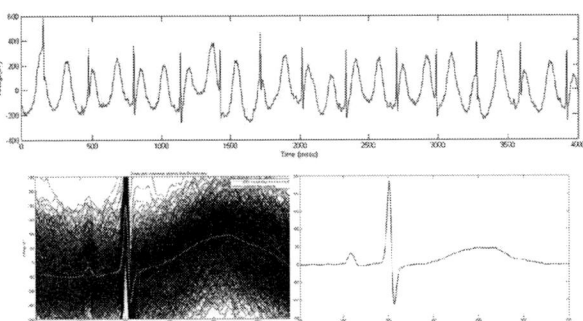

Figure 5: Up: raw data of a typical recorded waveform, ECG contaminated with breathing artifacts and low frequency baseline wander. R waves (peaks) are detected with a peak detection algorithm and all individual ECG's are aligned at the peak and summed together to remove the un-correlated breathing artifact and baselines. Down: left: a typical summed ECG waveform over 5 minutes, the red is the summed ECG waveform and the black lines are each individual ECG segment. The residue baseline is then fitted with a cubic spline curve (blue) and subtracted from the summed ECG. Right: cubic spline fitting subtracted ECG waveform.

RESULT

978-1-4799-7956-1/15 $31.00 © 2015 IEEE   691

In the heart rate study, we compare the 24 hour heart rate data for the control and 10 day post amputation (Fig. 6). Each heart rate data point is averaged with 5-minute raw ECG data; the corresponding heart rate variability data point is the standard deviation of the R-peak intervals within that 5 minutes.

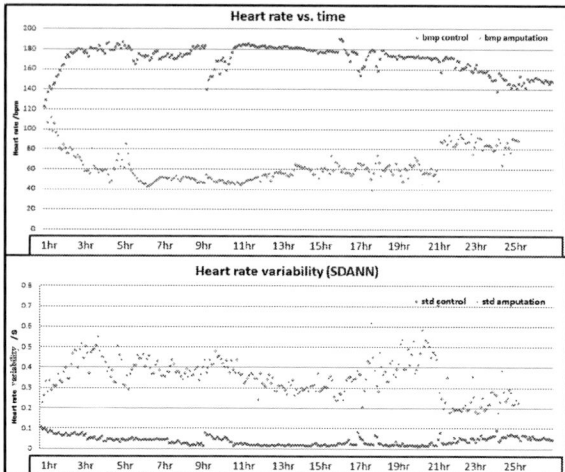

Figure 6: Up: Heart rate over 24 hours of control and 10 day post amputation. Down: Heart rate variability over 24 hours of control and 10 day post amputation.

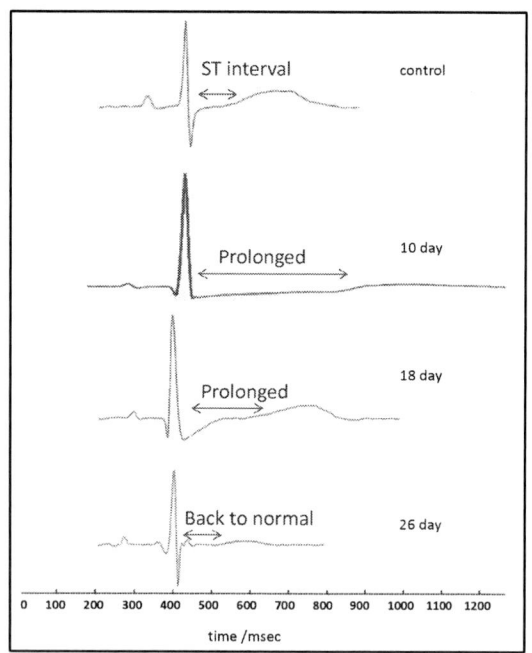

Figure 7: ST segment depression (sign of ventricle injury). We see on 26th day, the ST interval has shortened back to normal.

In Fig. 6 we see both the heart rate of the fish in the control and post amputation states, and the heart rate variability (level of arrhythmia) at the two different states. At 10 day post-amputation, the heart rate is significantly lower, with higher heart rate variability. Both are indications that the heart is not in a normal state and recovery is not yet complete. Furthermore, we see clearly the anesthesia drug's effect on the heart rate over the first 2 to 3 hours after bringing the animal out of anesthesia, which confirms the concern that acute recording does not reveal accurate information of the heart's true physiology. In the morphology study we see clearly the heart's regeneration process over a month-long period from less depressed ST segment over time. On the 26th day, the ST interval has recovered back to normal, which is in agreement with results from other regeneration studies [4].

CONCLUSION

We presented, for the first time, a wearable flexible parylene (PA) micro-electrode that monitors the Adult Zebrafish ECG long term. We demonstrated the design, fabrication and testing of the flexible micro electrode along with a micro-molded ultra-soft, density adjusted silicone jacket, allowing ECG recording to be carried out under water, in the fish's natural habitat with no need for anesthesia. To the authors' knowledge, this is the first time non-anesthetized zebrafish ECG data has been continuously recorded over 24-hour period. The heart rate and heart rate variability data obtained from raw data confirms that anesthesia indeed altered the heart rate and the level of arrhythmia. The wearable flexible electrode shows potential for an effective tool for long-term adult zebrafish drug screening and other heart studies.

ACKNOWLEDGEMENT

This work is partially funded by the NIH, grant number R01 HL111437.The author would like to thank all Caltech MEMS group members for their fruitful discussions and suggestions.

REFERENCES

[1] P. Nemtsas, E. Wettwer, T. Christ, G. Weidinger, and U. Ravens. "Adult Zebrafish Heart as a Model for Human Heart? An Electrophysiological Study." In *J. Molecular and Cellular Cardiology*, 48(1): 161-71, 2010.

[2] I.U. Leong, J.R. Skinner, A.N. Shelling, and D.R. Love, "Zebrafish as a model for Long QT syndrome: the evidence, and the means of manipulating zebrafish gene expression", in *Acta Physiol.*, Oxf. 199, pp. 257–276, 2010.

[3] DJ. Milan, IL. Jones, PT. Ellinor, CA. MacRae, "In vivo recording of adult zebrafish electrocardiogram and assessment of drug-induced QT prolongation", in *Am. J. Physiol. Heart Circ. Physiol.* 291(1):269-273, 2006.

[4] K. Poss, L. Wilson, and M. Keating. "Heart Regeneration in Zebrafish", in *Science.* 298(5601): 2188-190, 2002.

[5] X. Zhang, J. Tai, J. Park and YC. Tai, "Flexible MEA for adult zebrafish ECG recording covering both ventricle and atrium", *in IEEE MEMS Conference,2014.*

[6] JH. Chang, B. Lu, and YC. Tai, "Adhesion-enhancing surface treatments for parylene deposition", *in IEEE TRANSDUCERS Conference,2011.*

[7] Han-Chieh Chang, "Wireless parylene-based retinal implant" (*PhD diss., California Institute of Technology*, 2014).

[8] YC. Tai, and HC. Chang, "High lead count implant device and method of making the same" (*US Patent App.* 13/830,272, 2013)

[9] JH. Chang, Y. Liu, D. Kang, M. Monge, Y. Zhao, CC. Yu, A. Emami, J. Weiland, M. Human and YC. Tai, "Packaging study for a 512-channe intraocular epiretinal implant", *in IEEE MEMS Conference,2013*

[10] J. Pan, W. Tompkins, "A real-time QRS detection algorithm", in *IEEE Biomedical Engineering*, vol. BME-32 NO. 3. 1985.

[11] P. Hamilton, W. Tompkins. "Quantitative Investigation of QRS Detection Rules Using the MIT/BIH Arrythmia Database", *IEEE Biomedical Engineering*, vol. BME-33, NO. 12.1986.

CONTACT

*Xiaoxiao Zhang, +1-626-395-2254,xzzhang@caltech.edu

A METHOD FOR CONTROLLING MICROTUBULE VELOCITY USING LIGHT IRRADIANCE ON A PATTERNED GOLD SURFACE

Tasuku Nakahara, Hirofumi Shintaku, Hidetoshi Kotera and Ryuji Yokokawa
Kyoto University, Kyoto, JAPAN

ABSTRACT

This paper describes a method for controlling the velocity of gliding microtubules by light irradiance on a gold-patterned surface. The light was used to control the temperature of the buffer solution by heat transfer from the thin gold film. Microtubule velocity increased approximately 1.8-fold from the initial velocity at the 13.5 W/cm^2 on the gold-coated area. The proposed method also allows cyclical control of gliding velocity between 0.53 µm/s and 1.01 µm/s by switching irradiance between 0.7 W/cm^2 and 13.7 W/cm^2. This is the first demonstration of the control and switching of velocity on a gold-patterned surface.

INTRODUCTION

The motor protein kinesin moves along cytoskeleton microtubules via hydrolysis of adenosine triphosphate (ATP) *in vivo*. The motor protein plays important roles in the transport of intracellular cargoes [1,2]. Since kinesin acts as a nanoscale motor, it may have applications as a nanotransporter. *In vitro* gliding assays are widely used in various molecular applications [3] wherein the microtubule glides on a kinesin-coated substrate; however, the system does not enable the control direction and velocity as seen *in vivo*. Many methods have been proposed to address these limitations.

In the gliding assay, the minus end of a microtubule becomes the leading tip when gliding on a kinesin-coated surface. Since the tip is exposed to thermal fluctuation, the gliding direction is random. Therefore, several methods have been developed to control gliding direction. For example, Hutchins *et al.* reported a method that uses magnetic forces [4] that lead the microtubule tip in the desired direction with labeled magnetic nanoparticles. Applying a fluidic shear force to gliding microtubules can also be used to manipulate gliding direction because the leading tip of the microtubule is oriented according to the direction of flow. Kim *et al.* reported that nearly all microtubules orient in parallel to the flow field when a fluidic shear force is applied [5]. Microtubules can also be controlled by rectifier structures [6]. Thus, gliding direction can be controlled in reconstructed systems when facilitated by external factors.

Another essential technique is velocity control. Proposed methods for controlling velocity include modulations of ATP concentration, temperature, and buffer solution [7,8]. Electric fields and fluid shear force have also been used in engineering approaches to velocity control. Dujovne *et al.* controlled gliding microtubule velocity by altering an electric field [9]. Chemical controls include thermoresponsive polymers, photoisomerizable monolayers, and ion-reactive proteins [10,11]. Recently, Kumar *et al.* reported complete on-off switching of motors by UV irradiation of caged peptides [12]. Although the reported methods enable control of microtubule gliding

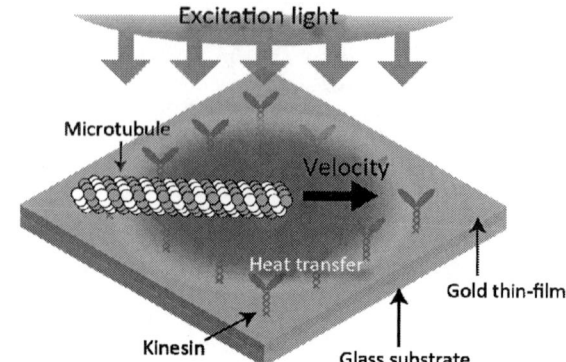

Figure 1: Schematic illustration of our proposed method

velocity, the controlled area is limited to a millimeter-range flow cell. A method for controlling gliding velocity at high spatial resolution is in high demand.

We propose a method for controlling microtubule velocity using light irradiance on a patterned gold surface, illustrated in Figure 1. The proposed method enables simultaneous velocity control and observation of microtubules in selectively irradiated areas. The temperature of the focus area changed due to heat transfer from the thin gold film. We successfully used this method to control microtubule velocity.

EXPERIMENTAL METHOD

Device fabrication

To evaluate the change in temperature, we fabricated a thermal sensor on a glass substrate. Figure 2 illustrates the process. First, the substrate (76 mm × 26 mm, No. 1 thickness, Matsunami) was cleaned with piranha solution (H_2SO_4:H_2O_2 = 3:1) at 80°C for 20 min. The cleaned substrate was coated with chromium (20 nm) and gold (100 nm) by thermal deposition (VPC-260F, ULVAC). Since gold is chemically inert, biocompatible, and has an appropriate temperature coefficient (~0.0034 K^{-1}), we selected gold as the sensing and heater material. A photoresist coating (S1813, Rohm and Haas Electronic Materials) was added to the substrate using a spin-coater at 4000 rpm for 30 s. The substrate was then exposed to UV light at 40 mJ/cm^2 over a photo-mask pattern of 4-wire resistive temperature detectors (RTDs). The exposed parts were developed for 1.5 min, and the bare gold and chromium layers were etched with three-fold diluted aqua regia (HCl:HNO_3 = 3:1) and chromium etchant (S-24, Sasaki Chemical) for 10 s each. Finally, the residual photoresist material was removed with piranha solution. For the *in vitro* assay, we prepared a flow cell from double-sided tape (50 µm thickness, 400P50, KGK, Saitama, Japan) and a coverslip (18 mm × 18 mm, No. 1

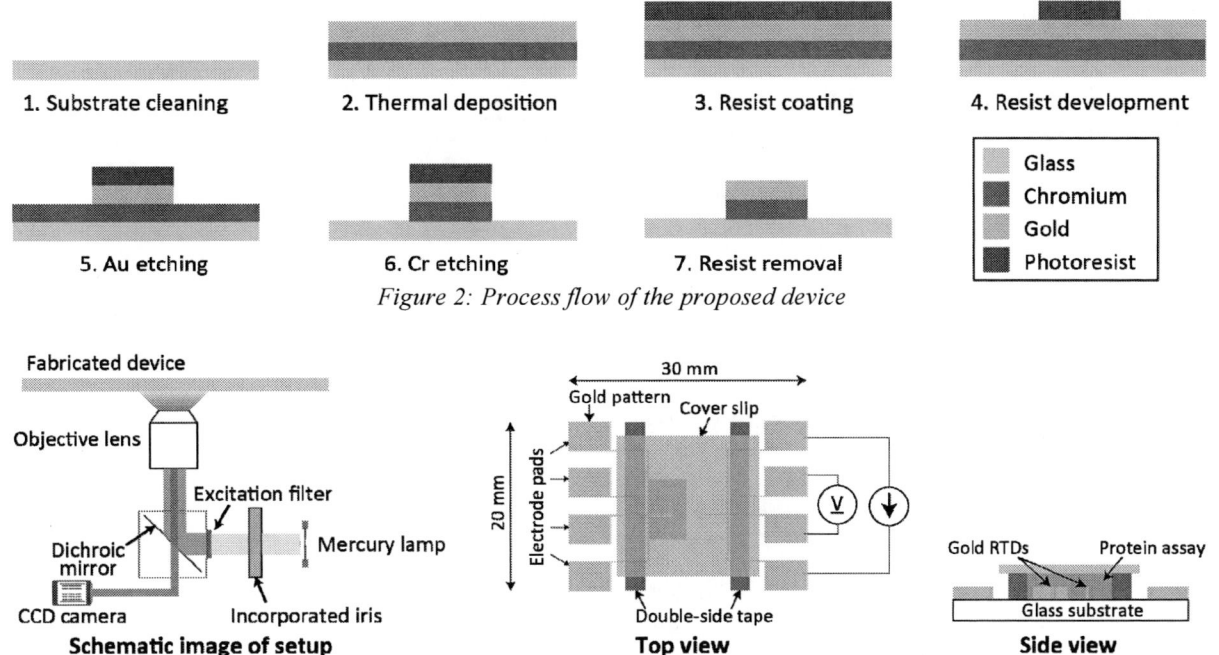

Figure 2: Process flow of the proposed device

Figure 3: Schematic illustration of experimental setup

thickness, Matsunami, Japan) on the fabricated device.

Device characterization

To evaluate changes in temperature, we adopted four-wire RTDs for high accuracy and reliability. One pair of wires connected to a sensor to supply a constant current of 1.0 mA using a source meter unit (2401, Keithley, Cleveland, OH). The other pair was used to measure voltage on an electrometer (6517B, Keithley). To determine the relationship between resistance and temperature, a first calibration was performed in a water bath (SM-05R, Taitec Corp., Saitama, Japan). The range of calibration was set between 20°C and 50°C with 5°C intervals. A second calibration was performed to relate temperature and irradiance. The gold-coated and bare glass areas were excited with various fluorescence intensities, which were altered using an incorporated six-step iris (3%, 6%, 12%, 25%, 50%, 100%; U-HGLGPS, Olympus, Tokyo, Japan). The irradiance was calibrated using a power meter (PM-247A, Neoark, Tokyo, Japan).

Protein preparation

Tubulin was purified from porcine brains through two cycles of polymerization and depolymerization, followed by phosphocellulose chromatography [13]. Fluorescence-labeled tubulin was prepared by labeling tetramethyl rhodamine (TMR; C1171, Invitrogen, Carlsbad, CA) for tubulin. TMR-labeled microtubules were prepared by polymerizing unlabeled tubulin and TMR-labeled tubulin (5:1) for 30 min at 37°C. The kinesin motor was purified as described [14].

In vitro motility assay

The gliding assay was constructed in a flow cell. All motility assays were performed in BRB80 buffer solution (80 mM PIPES, 1 mM EGTA, 1 mM $MgCl_2$, pH 6.8). A 50 µg/mL kinesin solution containing 0.1 mg/mL casein was prepared and introduced into the flow cell. After a 5 min incubation, 0.01 mg/mL TMR-labeled microtubule solution was introduced and incubated for an additional 5 min. The solution was exchanged with 1 mM ATP solution (0.03 mg/mL catalase, 0.2 mg/mL glucose oxidase, 20 mM DTT, 25 mM glucose, and 1% 2-mercaptoethanol). Then, we began to measure the velocity and monitor changes in temperature.

Figure 3 shows the experimental setup. We used an inverted microscope (IX73 Olympus, Japan) with a 100× oil-immersion objective lens (UPlanSApo 100×, Olympus, Japan). Excitation light (wavelength: 523 nm) was used to excite the labeled microtubules and heat the gold-patterned surface. The area of irradiation was 0.265 mm, which was determined by the magnification of the objective lens and the field diaphragm diameter of the microscope eyepiece. The irradiance was controlled by the integrated six-step iris. To observe the microtubules, we used a complementary metal-oxide semiconductor (CMOS) camera (ORCA-Flash4.0 V2, Hamamatsu, Shizuoka, Japan) and a charge-coupled device (CCD) on IX73 and IX71 microscopes, respectively. The images were obtained using image-processing software (HCImage, Hamamatsu) at 2 fps.

Repetitive accelerations and decelerations were performed by switching irradiance between a neutral density (ND) filter (transmittance of 5%, Sigma Koki, Saitama, Japan) and blank (transmittance of 100%), which correspond to 0.7 W·cm^{-2} and 13.7 W·cm^{-2}, respectively. The filter was mounted in a filter wheel (99A041, Ludl, Hawthorne, NY, USA) integrated into the inverted microscope (IX71). The time of exposure was 20 s for each filter. Filter changes and temperature monitoring were synchronized using a custom-written LabVIEW routine (National Instruments, Austin, TX). Images and videos were processed with ImageJ software (NIH, Bethesda, MD), MATLAB (MathWorks, Natick, MA), and FIESTA

978-1-4799-7956-1/15 $31.00 © 2015 IEEE   695

to measure velocity by tracking the ends of the gliding microtubules. Tracking time was 20 s for each experiment.

RESULTS AND DISCUSSION

RTDs characterization

Figure 4 shows the fabricated RTDs. The fabricated sensor size was 30 μm × 35 μm and the line pattern width was approximately 4 μm. Since the sensor size was smaller than the irradiated area, this sensor enabled us to evaluate the localized changes in temperature.

The calibration results showed the linear relationship between resistance and temperature (Figure 5). As expected, there was a difference in the initial resistance between gold and glass because the initial resistance is affected by differences in pattern width. Since we were able to obtain resistances that were proportional to the temperature from 20°C to 50°C, the fabricated sensor could be used for measurement of thermal effects by irradiance. For the second calibration, we measured the resistance of RTDs that could be translated to temperature using the results of the first calibration. Figure 6 shows the changes in temperature with irradiance. The temperature in the gold-patterned area increased 10.4 ± 1.2°C with irradiance from 0 to 13.5 W/cm², while the temperature in the glass area increased only 2.1 ± 1.0°C at the same irradiance range. It is possible that the absorbed energy of the gold, which was converted to heat, was larger than that of glass. From these calibration results, we assumed that irradiation enabled us to control microtubule velocity, which has been shown to depend on temperature [7].

Measurement of gliding velocity

Figure 7 shows sequential observations of the gold-coated and bare glass areas. Microtubule velocity was faster in the gold-coated area than in the bare glass area, and we observed a difference in mobility after 20 s. The velocity increased with increasing irradiance by approximately 1.8-fold at 13.5 W/cm² in the gold area (Figure 8). However, the increase in velocity in the bare glass area was only 1.1-fold at the same irradiance range. Tukey-Kramer analysis with a 5% significance level revealed a significant difference between the initial value and normalized gliding velocity on the gold-coated area at 13.5 W/cm².

Repetitive velocity control was achieved by switching irradiance between 0.7 W·cm⁻² and 13.7 W·cm⁻² (green-colored area) (Figure 9). The gliding velocities

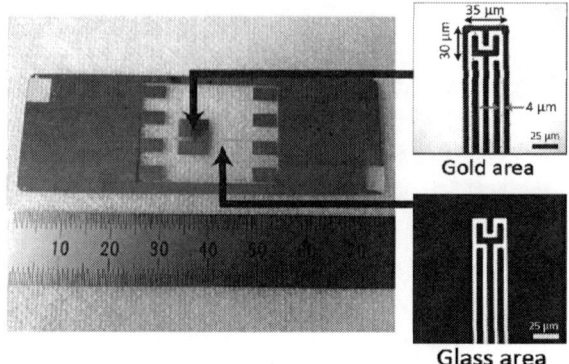

Figure 4: The device

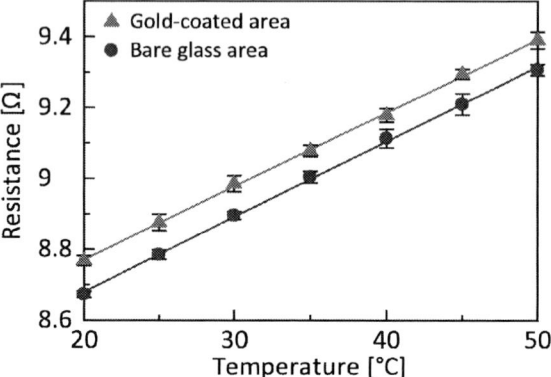

Figure 5: Calibration in water bath

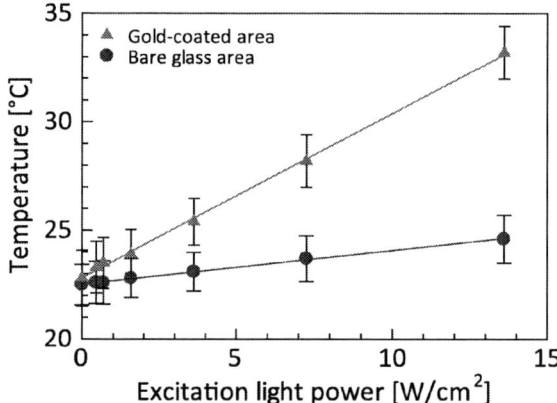

Figure 6: Relationship between irradiation and temperature

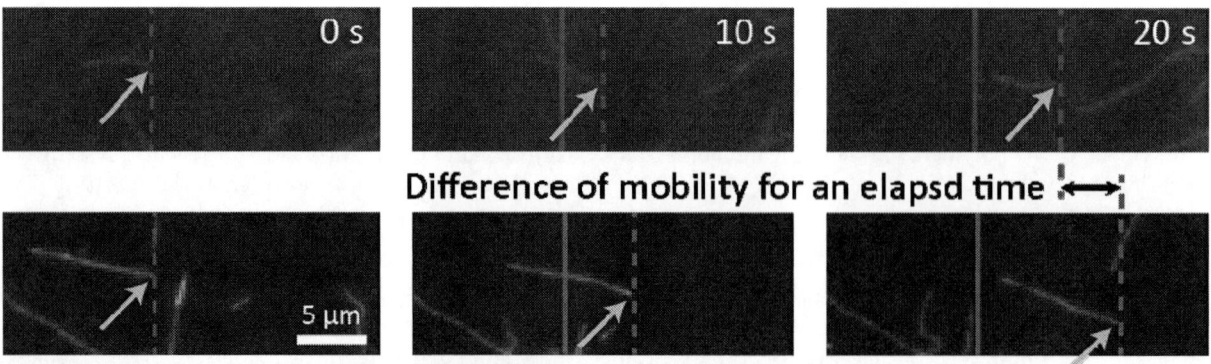

Figure 7: Sequential images-mobility assay

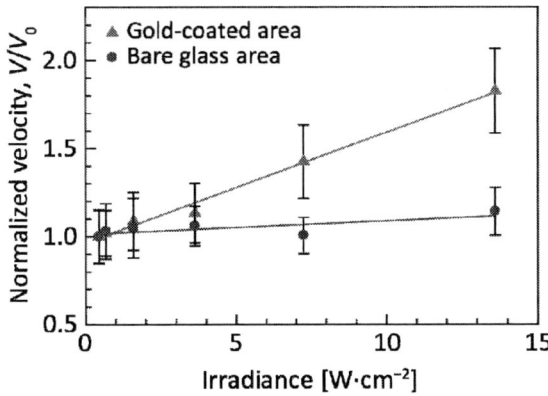

Figure 8: Irradiance and gliding velocity

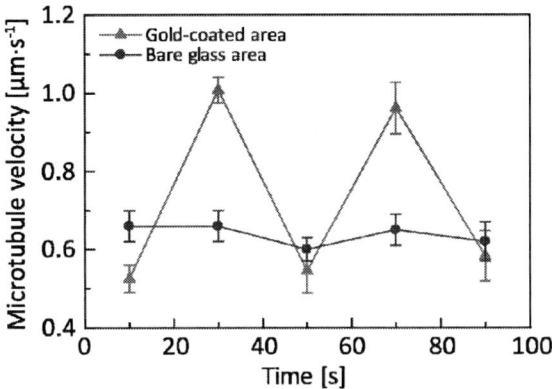

Figure 9: Switching property

switched between 0.53 ± 0.03 $\mu m \cdot s^{-1}$ and 1.01 ± 0.03 $\mu m \cdot s^{-1}$. Sequential switching was demonstrated without loss of kinesin activity. In contrast, switching did not occur in the bare glass area because the gliding velocities remained nearly the same with and without irradiation.

CONCLUSIONS

In this study, we developed a method for controlling microtubule velocity using light irradiance on a patterned gold surface. To evaluate the change in temperature, we fabricated a thermal sensor by patterning the gold layer on the substrate. The fabricated sensors showed linear resistance properties for the change in irradiance. The maximum change in temperature was approximately $10.4 \pm 1.2°C$ at 0 to 13.5 W/cm^2. Velocity measurements showed that the gliding velocity changed in the localized target area, achieving a maximum change of 1.8-fold. We also demonstrated repetitive velocity switching on the gold-coated area. We believe our proposed system will be useful for various molecular applications.

ACKNOWLEDGMENTS

This work was partly supported by JSPS KAKENHI Grants (Number 25709018 to R. Y. and Number 25·1305 to T. N.).

REFERENCES

[1] R. D. Vale, "The molecular motor toolbox for intracellular transport", *Cell*, Vol. 112, pp. 467-480,
2003.

[2] K. J. Verhey and J. W. Hammond, "Traffic control: regulation of kinesin motors", *Nature Reviews Molecular Cell Biology*, Vol. 10, pp. 765-777, 2009.

[3] J. Howard, A. J. Hudspeth, and R. D. Vale, "Movement of microtubules by single kinesin molecules", *Nature*, Vol. 342, pp. 154-158, 1989.

[4] B. M. Hutchins, M. Platt, W. O. Hancock and M. E. Williams, "Directing transport of $CoFe_2O_4$-functionalized microtubules with magnetic fields", *Small*, Vol. 3, pp. 126-31, 2007.

[5] T. Kim, M. T. Kao, E. Meyhöfer, and E. Hasselbrink, "Biomolecular motor-driven microtubule translocation in the presence of shear flow: analysis of redirection behaviours" *Nanotechnology*, Vol. 18, 025101 (9 pp), 2007.

[6] C. T. Lin, M. T. Kao, K. Kurabayashi, and E. Meyhöfer, "Efficient designs for powering microscale devices with nanoscale biomolecular motors", *Small*, Vol. 2, No. 2, pp. 281-287, 2006.

[7] K. J. Böhm, R. Stracke, M. Baum, M. Zieren, and E. Unger, "Effect of temperature on kinesin-driven microtubule gliding and kinesin ATPase activity", *FEBS Letters*, Vol. 466, pp. 59-62, 2000.

[8] K. J. Böhm, R. Stracke, and E. Unger, "Speeding up kinesin-driven microtubule gliding in vitro by variation of cofactor composition and physicochemical parameters", *Cell Biology International*, Vol. 24, No. 6, pp. 335-341, 2000

[9] I. Dujovne, M. Van Den Heuvel, Y. Shen, M. De Graaff, and C. Dekker, "Velocity modulation of microtubules in electric fields", *Nano Letters*, Vol. 8, No. 12, pp. 4217-4220, 2008.

[10] L. Ionov, M. Stamm, and S. Diez, "Reversible switching of microtubule motility using thermoresponsive polymer surfaces", *Nano Letters*, Vol. 6, No. 9, pp. 1982-1987, 2006.

[11] M. K. Rahim, T. Fukaminato, T. Kamei and N. Tamaoki, "Dynamic photocontrol of the gliding motility of a microtubule driven by kinesin on a photoisomerizable monolayer surface", *Langmuir*, Vol. 27, pp. 10347-10350, 2011.

[12] K. R. Sunil Kumar, T. Kamei, T. Fukaminato, and N. Tamaoki, "Complete ON/OFF Photoswitching of the Motility of a Nanobiomolecular Machine", *ACS Nano*, Vol. 8 No. 5, pp. 4157-4165, 2014.

[13] R. C. Williams, and J. C. Lee, "Preparation of tubulin from brain", *Methods in Enzymology*, Vol. 85, pp. 376-385, 1982.

[14] R. Yokokawa, M. C. Tarhan, T. Kon, and H. Fujita, "Simultaneous and bidirectional transport of kinesin-coated microspheres and dynein-coated microspheres on polarity-oriented microtubules", *Biotechnology and Bioengineering*, Vol. 101, No. 1, pp. 1-8, 2008.

CONTACT

T. Nakahara, Tel: +81 75 383 3687,
Fax: +81 75 383 3681;
nakahara.tasuku.28x@st.kyoto-u.ac.jp

CULTURING AND PROBING PHYSICAL BEHAVIOR OF INDIVIDUAL BREAST CANCER CELLS ON SiC MICRODISK RESONATORS

Hao Jia[1†], Xiaohui Wu[2], Hao Tang[1], Z.-R. Lu[2] and Philip X.-L. Feng[1†]

[1]Electrical Engineering, [2]Biomedical Engineering, Case School of Engineering,
Case Western Reserve University, Cleveland, OH 44106, USA
[†]Email: hao.jia2@case.edu; philip.feng@case.edu

ABSTRACT

This digest paper reports the first experimental exploration of directly culturing and measuring breast cancer cells at single-cell level, on the surfaces of silicon carbide (SiC) microdisk resonators. Enabled by the superior biocompatibility of SiC, individual breast cancer cells are observed to attach and spread on surfaces of SiC devices within only 3 hours of culturing. Multimode resonances at very high frequencies (up to ~100MHz) of SiC microdisks (with diameter d~20–30μm), and their responses to single attached MDA-MB-231 cells are characterized by taking advantage of the robust presence of multiple resonance modes in biological solutions. Such devices provide a useful biosensing platform for probing physical properties and behaviors of breast cancer cells *in vitro*, at single-cell level and in real time.

INTRODUCTION

Probing basic mechanical properties and physical behavior of breast cancer cells at single-cell level may lead to new understandings of cancer metastatic mechanisms and motivate therapeutical strategies to prevent, relieve and cure the disease. Among all the emerging techniques, micro/nanoelectromechanical systems (MEMS/NEMS) offer very small device dimensions and very high sensitivities (down to single-cell and single-molecule levels), thus providing new opportunities to measure forces, movements, and mass changes during cellular and subcellular processes *in vitro* [1-4]. A number of interesting approaches on studying mechanical properties of single cells have been explored. In static mode, micro-post arrays enable the spatial mapping of single-cell traction forces [2]. In dynamic mode, trampoline resonators can monitor mass changes during cell growth and division [3]; hollow cantilevers with embedded nanochannels can quantify cell growth by measuring the buoyant mass [4]. Toward developing very high frequency (VHF) resonators that are easy to fabricate yet high performance in biofluids, very limited types of devices have been explored, because the cell-living environment imposes not only significant mass loading (much lower frequency and smaller signal) and viscous damping (much lower quality (Q) factor), but also great challenges on choosing material (chemically inert and biocompatible), designing robust resonating structure, and finding actuation/detection scheme that is compatible with conductive biosolutions while reasonably minimizing structure complexity.

In this work, we take a novel initial step to culture and probe physical behavior of breast cancer cells at single-cell level, directly on surfaces of SiC microdisk resonators. Compared with previous device prototypes in single-cell studies, we offer the following features. (i) From material perspective, superior optical, thermal, mechanical properties [5-7] and more interestingly, biocompatibility [8-9]. Individual breast cancer cells are observed to fix and spread on the device surface within 3 hours of culturing without surface functionalization or coating. (ii) From structural perspective, 2D microdisk structures with large cell-capturing areas that are totally immersed and optically-operated in biosolutions. Such simple structures can be easily fabricated using a "resistless" process. Cancer cells can be directly cultured on such planar surfaces and their properties (*e.g.,* mass) and behavior (*e.g.,* migration) can be further studied *in-vitro*, without causing undesired extracellular stimulation. (iii) From device performance in biofluids, multimode responses at high and very high frequency bands (HF/VHF, up to ~100MHz) with ~6 flexural modes and Q factors of ~15 are measured in different biological solutions. Multimode sensing provides not only information from more degrees of freedom, but also higher Q factor and mass responsivity from higher modes without suffering from f-Q tradeoff (Q drops as f increases) due to single-mode sensing. Single breast cancer cell mass loading can be successfully characterized by taking advantage of such robust multimode responses. Hence, SiC microdisk resonators may provide a new and attractive biosensing platform for probing cell mechanical properties and physical behavior *in vitro*, at single-cell level and in a real-time fashion.

DEVICE DESIGN AND FABRICATION

SiC microdisks with diameters of $d{\approx}20$–30μm are deliberately designed and fabricated for capturing individual breast cancer cells (MDA-MB-231, ~10μm in diameter on average when suspended). Figure 1 illustrates an array of the fabricated devices and a zoom-in view of a typical device with $d{\approx}20$μm.

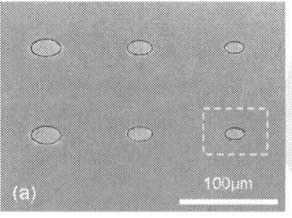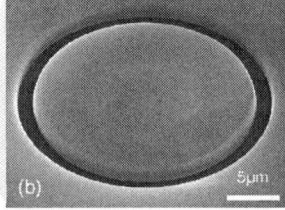

Fig. 1: SEM images of (a) a small array of SiC microdisk resonators with $d{\approx}30$μm, 25μm and 20 μm (aerial view), and (b) a typical $d{\approx}20$μm SiC microdisk resonator (zoom-in view).

The SiC microdisks are nanomachined in a SiC-on-SiO$_2$ platform [10-11], as shown in Fig. 2. A

500nm-thick SiO$_2$ is grown on the Si substrate as a sacrificial layer. A 1μm-thick SiC film is then deposited atop SiO$_2$ by low-pressure chemical vapor deposition (LPCVD) using dichlorosilane (SiH$_2$Cl$_2$) and acetylene (C$_2$H$_2$) dual precursors at a temperature of $T\approx900°C$ and pressure of $p\sim0.5$Torr [10]. SiC microdisks are patterned by high-throughput focused ion beam (FIB). The FIB time is controlled so that the ion beam mills through SiC and SiO$_2$ layers until Si surface is exposed. We use buffered oxide etching (BOE) to etch the underneath SiO$_2$ layer to release the SiC microdisks. Controlling the etching time tunes the size of the anchoring SiO$_2$ pedestal beneath the center. Such fabrication process is totally "resistless", and we can easily engineer the multiple modes of SiC microdisk resonators by controlling the scaling factor $\eta=b/a$, where b is the diameter of the SiO$_2$ pedestal and a is the diameter of the SiC microdisk [10].

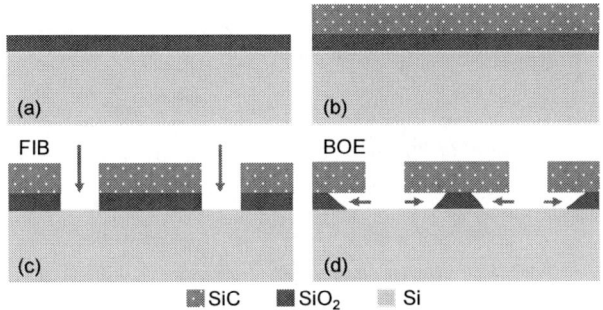

SiC ■ SiO$_2$ ■ Si

Fig. 2: Schematic illustration of the fabrication process of SiC microdisk resonators. (a) 500nm-thick sacrificial SiO$_2$ layer is initially grown on the Si substrate; (b) 1μm-thick SiC layer is deposited on top of SiO$_2$ using LPCVD; (c) diameter of a SiC microdisk is patterned by FIB; (d) SiO$_2$ pedestal is formed by BOE, making a suspended structure.

MEASUREMENT TECHNIQUES

Optical actuation/detection techniques have been effectively applied to excite and measure the resonances of MEMS/NEMS resonators in both air and viscous fluids [12]. Compared to electrostatic, piezoelectric, and electromagnetic signal transduction schemes, optical techniques are widely compatible with various biofluids without issues of electrical leakage and electrochemical reactions imposed by the conductive nature of biosolutions. Optical techniques are particularly attractive with SiC because of the large band gap (E_g>2.3eV) and high laser power handing ability enabled by its very high thermal conductivity ($k\sim360$-490 W/[m·K]).

Figure 3 illustrates the experimental system of optically operating SiC microdisk resonators in biosolutions for characterizing the mass loading effects of attached single breast cancer cell. The flexural vibrational modes are excited by an amplitude-modulated 405nm blue laser (modulated signal supplied by a network analyzer) and interferometrically measured by a 633nm He-Ne red laser focused on the device surface (reflected light intensity is simultaneously tuned as the gap between SiC and Si layers periodically changes during flexural vibrations). By sweeping the driving frequency, we record the output signal using the same network analyzer, from which we can further extract the resonance frequencies and Qs of multiple modes based on the damped simple harmonic resonator model.

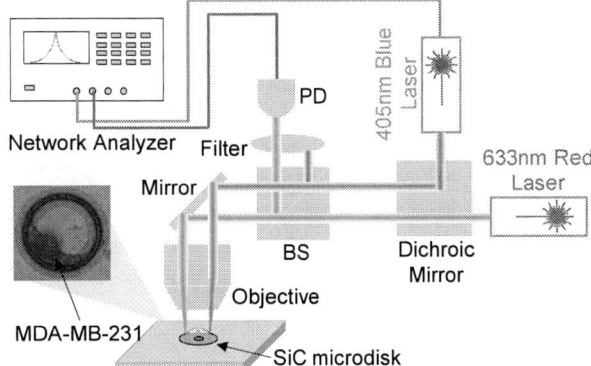

Fig. 3: Schematic illustration of experimental system for optically exciting and detecting SiC microdisk resonances in biosolutions for quantifying single breast cancer cell loading effects. Inset shows optical image of a single breast cancer cell cultured on device surface.

CELL CULTURE AND IMAGING

Breast cancer (MDA-MB-231) cells are initially cultured in a petri dish filled with culture solution (RPMI medium 1640+10% FBS+1% Penstrep), and kept in an incubator (with 37°C and 5% CO$_2$) for 2 days. During the experiment, breast cancer cells are first detached from the petri dish and become individually suspended through (trypsin-induced) de-adhesion, centrifuge, and filtration. We then dilute the culture solution and control the cell concentration at ~25,000–50,000 cells/mL so as to realize single-cell attachment on certain devices. The SiC chip with patterned microdisk resonators is put into a new petri dish and then immersed by the culture solution with suspended cells. The petri dish is finally cultured in the incubator again for 3 hours to allow individual breast cancer cells fully attach and fix onto the device surface. No surface functionalization or coating is used during the fixation process.

In order to take scanning electron microscopy (SEM) images of single breast cancer cell morphology on the device, the chip is immediately taken out from the culture solution and immersed in 4% paraformaldehyde (PFA) solution (cell are dead), and then immersed in 50%, 60%, 70%, 80%, 95%, 100% ethyl alcohol solutions (2 times for each solution, and 5min for each time) for dehydration. Subsequently, the whole chip is kept in vacuum overnight, and the "dry cell" will be well-preserved on the device with exactly the same morphology as when it was alive. A very thin layer of Pt (5nm) is locally deposited on the cells of interest before SEM imaging since these biological cells are non-conductive.

EXPERIMENTAL RESULTS

We first characterize the SiC microdisk resonators by directly operating them in air using laser excitation/detection techniques. Figure 4 shows the multimode resonance characteristics of a $d\approx30$μm SiC microdisk. We observe up to 17 flexural modes within HF/VHF (~10–200MHz) bands, and Qs up to ~2200 in air.

978-1-4799-7956-1/15 $31.00 © 2015 IEEE   699

The Q tends to increase with f, indicating that higher modes can offer higher Qs than the fundamental mode. This defies the conventional f-Q tradeoff, and facilitates robust multimode responses in water and biofluids.

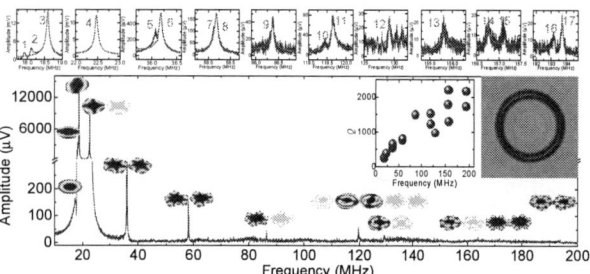

Fig. 4: *Multimode resonance characteristics of a typical $d{\approx}30\mu m$ SiC microdisk resonator operating in air. f and Q values for each flexural mode are extracted by fitting to a damped simple harmonic resonator model. Mode shapes are simulated by COMSOL. Insets are corresponding optical image, and f-vs-Q plot of this device.*

We directly culture MDA-MB-231 cells on SiC chip patterned with microdisk resonators. Individual breast cancer cells randomly attach to the SiC surface *without any assistance of functionalization or coating*. SEM images in Fig. 5 show morphology details of breast cancer cells on the SiC microdisks and the remaining SiC substrate. Enabled by the superior biocompatibility of SiC, MDA-MB-231 cells are observed to fix and spread on the SiC surface (Fig. 5a, 5b, 5c, & 5d) after merely 3 hours of culturing. More interestingly, we observe, for the first time, that individual breast cancer cells are extending on SiC microdisk resonators with an average coverage of 1100 μm^2 after only 3 hours of culturing (previous studies only showed cell proliferation on SiC substrate after 24 hours or days of culturing [8]), and morphological structures such as *lamellipodia* and *filopodia* are visible (Fig. 5e & 5f), which indicate strong interactions of individual breast cancer cells with SiC devices.

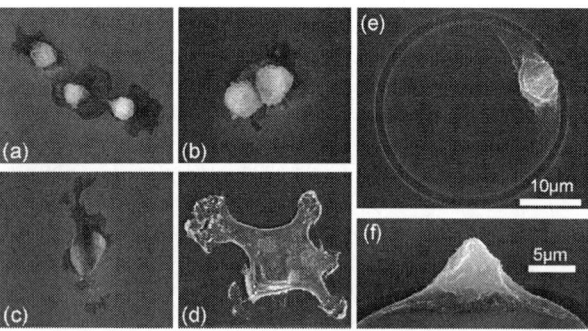

Fig. 5: *MDA-MB-231 cells are observed to (a), (b) fix and (c), (d) spread on SiC substrate. (e) top-view and (f) zoom-in view of single MDA-MB-231 cell extending on a SiC microdisk surface. Morphological details, such as lamellipodia and filopodia, are visible (after only 3 hours of culturing).*

The pretreatments for cell SEM imaging also allow us to quantify single breast cancer cell "dry mass" loading in air by using SiC microdisk resonators. Figure 6 shows an example of preserving single MDA-MB-231 "dry" cell on a $d{\approx}30\mu m$ microdisk resonator. Through recording the resonance frequencies before and after dry mass loading, we clearly observe frequency shifts of 3.79, 2.84, 4 and 3.47MHz, respectively, from the first 4 flexural modes in the range of ~60–90 MHz (in VHF band). The frequencies of the first 4 modes are much higher compared with those in Fig. 4 because of the much larger pedestal diameter due to shorter etching time.

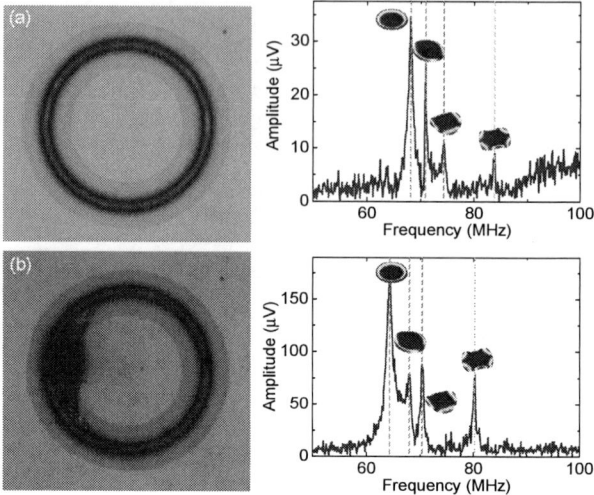

Fig. 6: *Single breast cancer cell "dry mass" loading effect measured by a $d{\approx}30\mu m$ SiC microdisk resonator in air. Optical images and multimode resonances are compared (a) before and (b) after single breast cancer cell fixation.*

Before characterizing single breast cancer mass loading effects in biosolutions, we operate the SiC microdisk resonators in water and biosolutions including phosphate buffered saline (PBS), culture solution and 4% PFA to prove the compatibility of such device platform and operation scheme in different liquid environment.

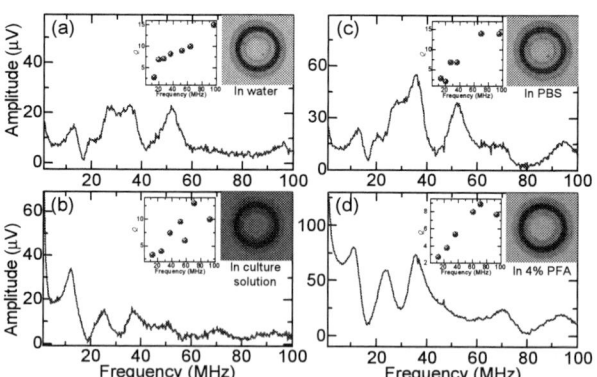

Fig. 7: *Multimode resonance characteristics of a $d{\approx}30\mu m$ SiC microdisk resonator operating in (a) water, (b) breast cancer cell culture solution, (c) PBS and (d) 4% PFA. Insets are corresponding optical images and f-vs-Q plots.*

To distinguish the resonance modes of such device in liquid, we first place the red detecting laser on the device surface close to the edge (where the largest displacements of most flexural modes occur) and measure the multimode resonance characteristics in liquid with blue laser driving (as shown in Fig. 7). We then block the driving signal, and all peaks immediately disappear. By carefully performing such control tests, we have observed and discerned 6 resonance modes in the range of 1–100MHz, with Qs up to 15 from a $d\approx30\mu m$ microdisk resonator. Similar to Fig. 4, Q tends to increase with f, indicating the advantage of multimode sensing in liquid by taking advantage of information from higher modes with higher Qs.

Figure 8 shows an example of using a $d\approx25\mu m$ device to quantify single breast cancer cell mass loading in the culture solution (while cell is alive) and then in 4% PFA (cell subsequently expires), from which we have measured 2 mode-dependent frequency shifts of ~1.6MHz and ~2.1MHz in the culture solution and ~1.8MHz and ~1.85MHz in 4% PFA, respectively.

In all the above, such obvious and robust multimode responses to single breast cancer cell loading indicates high mass sensitivity of such devices as potential biosensing platform for single-cell biophysical studies in biofluids.

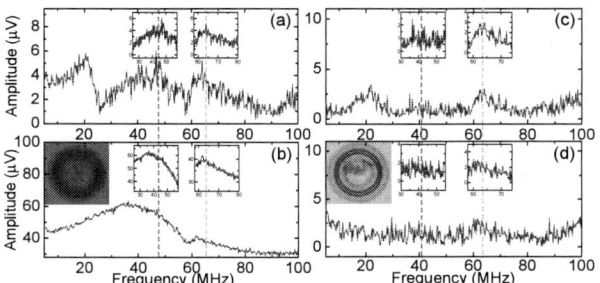

Fig. 8: Single breast cancer cell mass loading effect measured by a $d\approx25\mu m$ SiC microdisk resonator. Multimode resonance characteristics are measured in culture solution (a) before and (b) after single cell is fixed on the device surface and (c) before and (d) after single cell is fixed in 4% PFA. Insets are corresponding optical images and zoom-in view of fitted resonance modes.

CONCLUSIONS

We describes the first exploration of directly culturing and measuring breast cancer cells, at single-cell level, by using silicon carbide (SiC) microdisk resonators. Individual breast cancer (MDA-MB-231) cells are observed to fix and spread on SiC microdisks (with an average coverage of 1100 μm^2) after only a 3hr culturing, and clear morphological structures such as *lamellipodia* and *filopodia* indicate superior biocompatibility of SiC and strong interactions of breast cancer cells with device surfaces. We then experimentally demonstrate the operation of SiC microdisk resonators in air and different biosolutions, from which we have observed robust multiple modes at HF/VHF. We further probe single breast cancer cell mass loading using SiC microdisk resonators, and we observe obvious and robust multimode responses, indicating high mass responsivity of such resonating devices. Such SiC microdisk resonators may provide a new and attractive biosensing platform for studying mechanical properties and physical behavior (*e.g.*, growth, division, and migration) of breast cancer cells *in vitro*, at single-cell level and in real-time fashion.

ACKNOWLEDGEMENTS

We thank the financial support and resources from National Science Foundation (ECCS-1408494), Case School of Engineering, and Swagelok Center for Surface Analysis of Materials (SCSAM). We thank J. Lee, Z. Wang, R. Yang, and T. He for technical support in the labs.

REFERENCES

[1] J.L. Arlett, E.B. Myers, M.L. Roukes, "Comparative Advantages of Mechanical Biosensors", *Nature Nanotech.*, vol. 6, pp. 203-215, 2011.

[2] J.L. Tan, J. Tien, D.M. Pirone, D.S. Gray, K. Bhadriraju, C.S. Chen, "Cells Lying on a Bed of Microneedles: An Approach to Isolate Mechanical Force", *Proc. Natl. Acad. Sci. U.S.A.*, vol. 100, pp. 1484-1489, 2003.

[3] K. Park, L.J. Millet, N. Kim, H. Li, X. Jin, G. Popescu, N.R. Aluru, K.J. Hsia, R. Bashir, "Measurement of Adherent Cell Mass and Growth", *Proc. Natl. Acad. Sci. U.S.A.*, vol. 107, pp. 20691-20696, 2010.

[4] M. Godin, F.F. Delgado, S. Son, W.H. Grover, A.K. Bryan, A. Tzur, P. Jorgensen, K. Payer, A.D. Grossman, M.W. Kirschner, S.R. Manalis, "Using Buoyant Mass to Measure the Growth of Single Cells", *Nature Methods*, vol. 7, pp. 387-390, 2010.

[5] H. Zamani, J. Lee, S. Rajgopal, C.A. Zorman, M. Mehregany, P.X.-L. Feng, "Radio-Frequency Multimode Micromechanical Disk Resonators in 500nm Thin Silicon Carbide (SiC)", in *Tech. Digest, the 15th Solid-State Sensors, Actuators & Microsystems Workshop (Hilton Head 2012)*, pp. 363-366, Hilton Head Island, SC, June 3-7, 2012.

[6] X. Lu, J.Y. Lee, P.X.-L. Feng, Q. Lin, "Silicon Carbide Microdisk Resonator", *Opt. Lett.*, vol. 38, pp. 1304-1306, 2013.

[7] X. Lu, J.Y. Lee, P.X.-L. Feng, Q. Lin, "High Q Silicon Carbide Microdisk Resonator", *Appl. Phys. Lett.*, vol. 104, pp. 181103, 2014.

[8] C. Coletti, M.J. Jaroszeski, A.M. Hoff, S.E. Saddow, "Culture of Mammalian Cells on Single Crystal SiC Substrates", in *MRS Proc.*, 0950-D04-22, Boston, November 27-December 1, 2006.

[9] C. Coletti, M.J. Jaroszeski, A. Pallaoro, A.M. Hoff, S. Iannotta, S.E. Saddow, "Biocompatibility and Wettability of Crystalline SiC and Si Surfaces", in *Proc. IEEE. Eng. Med. Biol. Soc.*, pp. 5850-5853, Lyon, August 23-26, 2007.

[10] Z. Wang, J. Lee, P.X.-L. Feng, "Spatial Mapping of Multimode Brownian Motions in High Frequency Silicon Carbide (SiC) Microdisk Resonators", *Nature Communications*, vol. 5, art. no. 5158, 2014.

[11] H. Jia, J. Lee, Z. Wang, P.X.-L. Feng, "High-Frequency SiC Microdisk Resonators Operating in Water with Responses to H_2O_2 and NH_4OH", in *Proc. IEEE Int. Freq. Contr. Symp. (IFCS 2014)*, pp. 24-27, Taipei, May 19-22, 2014.

[12] S.S. Verbridge, L.M. Bellan, J.M. Parpia, H.G. Craighead, "Optically Driven Resonance of Nanoscale Flexural Oscillators in Liquid", *Nano Lett.*, vol. 6, pp. 2109-2114, 2006.

EARLY CHARACTERIZATION METHOD
OF PLANT ROOT ADAPTABILITY TO SOIL ENVIRONMENTS

Katsuya Ozoe[1], Hirotaka Hida[1], Isaku Kanno[1], Tetsuya Higashiyama[2,3,4], and Michitaka Notaguchi[2,3]*

[1]Department of Mechanical Engineering, Kobe University, Kobe, JAPAN
[2]JST, ERATO, Higashiyama Live-Holonics Project, Nagoya University, Nagoya, JAPAN
[3] Graduate School of Science, Nagoya University, Nagoya, JAPAN
[4] Institute of Transformative Bio-Molecules (WPI-ITbM), Nagoya University, Nagoya, JAPAN

ABSTRACT

This paper reports an on-chip analytical method for studying physical mechanisms of plant root growth in soil environments. To quantitatively evaluate physical interaction between root and soil, we developed a silicon-based microchannel device integrated with force displacement sensor which mimics a barrier in soil. By using developed microsystem, we successfully characterized the driving forces of root growth in three kinds of plants including *Arabidopsis thaliana*, which is known as a model organism. This analytical method allows us to efficiently characterize potential adaptability of root system to soil environments at early growth stage. The gaining knowledge might contribute for breed improvement, in terms of increasing crop productivity and plant biomass.

INTRODUCTION

A root is one of the most important organs for plants to take up water and nutrients from the soil, in addition, to support body of plant physically. During a lifetime of plant, the roots are affected by various environmental conditions in the soil such as humidity, chemicals, soil bacteria and physical interaction. For characterizing root growth under varied environments, agarose plates have been used to grow plants [1, 2]. In recent years, microfluidic platforms [3] have been applied to study on root growth mechanisms because of their advantages. For example, local environments can be controlled with high space and time resolution, and furthermore, responses of

root to environment changes can be easily observed at a cellular level with optical microscope. Recently, several microfluidic platforms for characterizing root growth have been reported; multiple phenotype assay using high-throughput seed sorting mechanism [4], chemical stimulation assay [5, 6] and pathogenic interaction assay [7]. However, there is still challenge using microfluidic platform. Almost all of these analytical methods were specialized for chemical and biochemical characterization of root growth at present. In fact, however, physical interaction between roots and soil conditions should be considered for studying mechanisms of root growth deeply because when roots face physical barriers such as stones and rotten leaves in soil, they may show their own strategy to avoid or penetrate them by driving force with growth. Here, we developed a novel microsystem to evaluate responses of root to physical barrier and the driving force of root growth quantitatively.

METHOD AND EXPERIMENTS

Microchannel device for characterization root growth

Figure 1 shows a schematic view of the developed microsystem for characterizing root growth. The microsystem mainly consists of two components; silicon tapered microchannel and PDMS (Poly-dimethylsiloxane-based) top/bottom covers. A force displacement sensor, which has V-shaped trap root tip and mimics a physical barrier in soil, is integrated into the microchannel. We formed micro gaps on both PDMS top/bottom covers for allowing the force sensor to deform by generating force with root growth.

During growth process of plant on microsystem, first, top/bottom PDMS covers are attached on the silicon microchannel, which is filled with gel-medium containing nutrients for plant growth. Then, a seed is placed in a seed pocket and root grows along tapered microchannel while shoot and leaves can freely grow in opened space. Finally, we evaluate physical interaction between the root and the force sensor by observing through transparent PDMS covers with optical micro scope.

Measurement principle

A force detection mechanism on developed microsystem is as illustrated in Fig. 2. Firstly, the root grows in microchannel with a constant speed before reaching the sensor. After the root reaches the force displacement sensor, growth speed of root is decreased due to physical containment by V-shaped trap. Thus, we can clearly define a changing point of growth speed as start time of contact between root and sensor. During root contacts to sensor, sensor deformation enhanced by

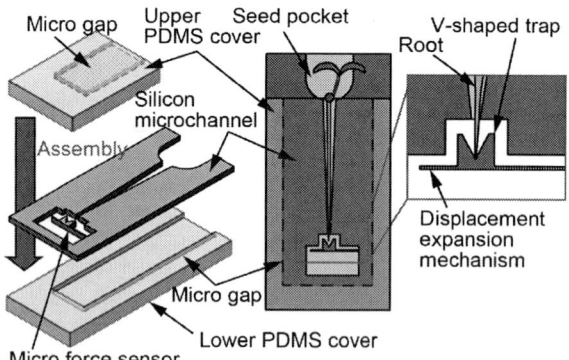

Figure 1: Illustration of microsystem for characterizing root at early stage. Assembly of microsystem (left) and magnified view of force displacement sensor (right). Silicon microchannel and micro gap is 300- and 60-μm-high respectively.

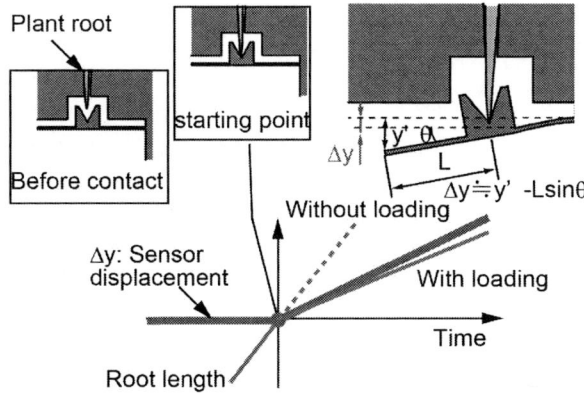

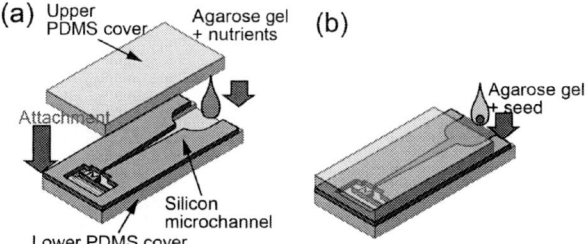

Figure 3: Schematics of protocol for plant growth on developed microsystem. (a)Agarose gel containing nutrients was loaded into microchannel, and (b) Seeding on microsystem with micropipette.

Figure 2: Principle for measurement of driving force of root growth. Sensor displacement is detected with displacement expansion mechanism (Upper right).

displacement expansion mechanisms as shown in Fig. 2. Finally, the driving force is measured by sensor displacement and its mechanical stiffness.

Design and fabrication

We designed the microchannel according to typical size of root of plant *Arabidopsis thaliana*, which is a model plant, at early growth stage. The microchannel was 300 µm in height, 6 mm in length and width of upper/lower tapered were 500 and 200 µm, respectively. We also designed force displacement sensors having varied stiffness ranging 1 to 200 N/m based on finite element method (FEM) simulation results. The silicon microchannel device was fabricated by using typical photolithography and deep reactive ion etching processes, and PDMS (Sylgard 184, Dow Corning, USA) covers were manufactured by soft lithography process with 60-µm-high mold, which formed micro gaps on PDMS covers.

Protocol for plant growth on device

For demonstrating characteristics of root growth on developed microsystem, we used three kinds of plants: *Arabidopsis thaliana*, one of the representative model organisms, *Capsella rubella*, closely related specie of *Arabidopsis thaliana* and *Nicotiana benthamiana*, namely tobacco, as an example of agricultural crops. Before sowing, all seeds were sterilized in mixed aqueous solution contains 5 % (V/V) sodium hypochlorite and 0.02 (V/V) Triton X-100 (Polyoxyethylene(10) octylphenyl ether) and then washed with DI water.

We have established a protocol for directly growing plant on the microsystem as shown in Fig. 3. First, silicon microchannel device was attached onto PDMS bottom cover. Then 1 % (W/V) agarose containing half-strength Murashige and Skoog (MS) medium which consists of nutrients for plant growth was loaded into microchannel by attaching PDMS top cover (Fig.3 (a)). After that, 0.1 % (W/V) agarose gel including a seed was injected into seed pocket (Fig. 3 (b)). The placed seeds were also treated with dormancy breaking process to control germination period in following procedures. The seeds were firstly stored in a light place at room temperature (RT, 25±3°C) for one day,

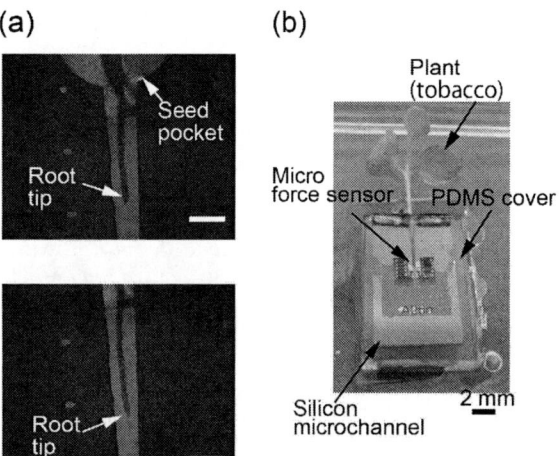

Figure 4: Observation results of plants on microsystem. (a) Root of Arabidopsis thaliana two (upper) and three days (lower) after dormancy breaking process (Bar=500 µm), and (b) Photograph of grown plant Nicotiana benthamiana on device two weeks after dormancy breaking process.

then, in dark place at 5 °C for two or three days. Finally, the seeds were vertically placed in light place at RT.

We observed developing root growth on microsystem with optical microscope and evaluated the physical interaction between roots and the sensors.

RESULTS AND DISCUSSION

Figure 4(a) and (b) show typical observation results of grown roots of *Arabidopsis thaliana,* and a whole plant *Nicotiana benthamiana* on developed microsystem respectively. We experimentally confirmed that all three plants *Arabidopsis thaliana, Capsella rubella* and *Nicotiana benthamiana* can grow on the microchannel device.

We measured root width at root tip region a few days after rooting and measurement result was shown in Fig. 5. By using the device, we easily observed roots at a cellular level with optical microscope due to transparent property of PDMS covers.

Figure 6 shows typical measurement results of driving force with root growth by using developed devices. As shown in Fig. 6 (A), the sensor displacement was nearly equal to an elongation of root while the root was captured

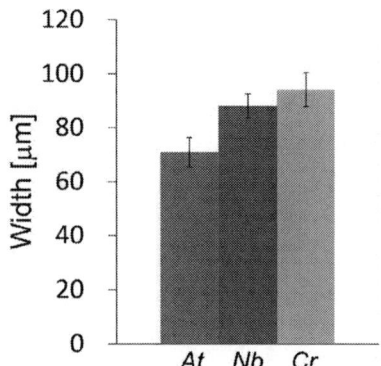

Figure 5: Root width of three kinds of plants on microsystem (N>4). At, Nb and Cr respectively mean *Arabidopsis thaliana*, *Nicotiana benthamiana* and *Capsella rubella*.

in V-shaped trap. However, some of plant roots passed between the sensor and micro gap in PDMS cover after deforming the sensor as shown in Fig. 6 (b). This result related to mechanisms of driving force with root growth is discussed as below.

We consider that driving force with root growth is generated by cell division process at root tip region only. Moreover, after division process, specialized cells were physically fixed in their own positions. In Fig. 6 (b), the root tip region passed the sensor from 24 to 48 hours after the root cap was captured at V-shaped. As a result, the sensor was fixed by specialized cells. On the other hand, some of plant roots, particularly *Arabidopsis thaliana*, detected driving forces was much smaller than that of other roots (Fig. 6 (c)). From these results, we defined the calculated force from total displacement of sensor during the root trapped and stiffness of the sensor as driving force with root growth.

Finally, we characterized the driving force of three kinds of plants by using developed microchannel device. Figure 7 (a) shows relationship between measured driving forces of root and width root tip. We clarified that the driving force of roots was increased by increasing width of root tip in all of three kinds of plant. From this result, we considered that the driving force with root growth was depended on sectional area of root. However, the calculated driving forces of *Arabidopsis thaliana* were typically smaller than those of *Nicotiana benthamiana* and *Capsella rubella* while the root width was almost same among three plants as shown in Fig. 7 (b). These results suggest that plants may have their own physical characteristics, for example driving force with growth and mechanical stiffness of cells.

CONCLUSIONS

We have demonstrated a new method for characterizing plant root growth by using microsystem and successfully characterized the driving forces of root growth in three kinds of plants including *Arabidopsis thaliana*. This analytical method allows us to efficiently characterize potential adaptability of root system to soil environments and obtained knowledge might contribute

for breed improvement and increasing crop productivity

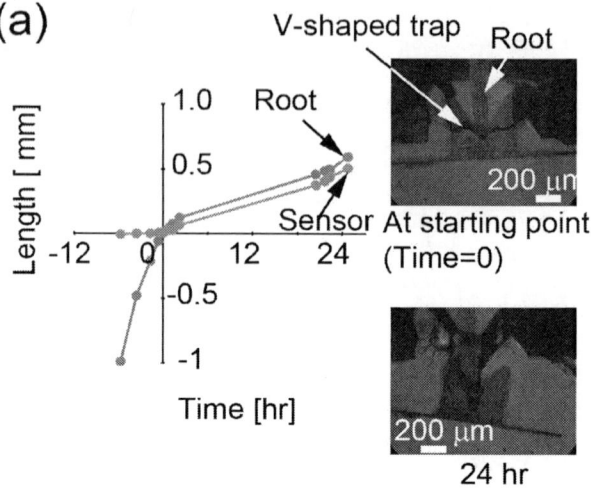

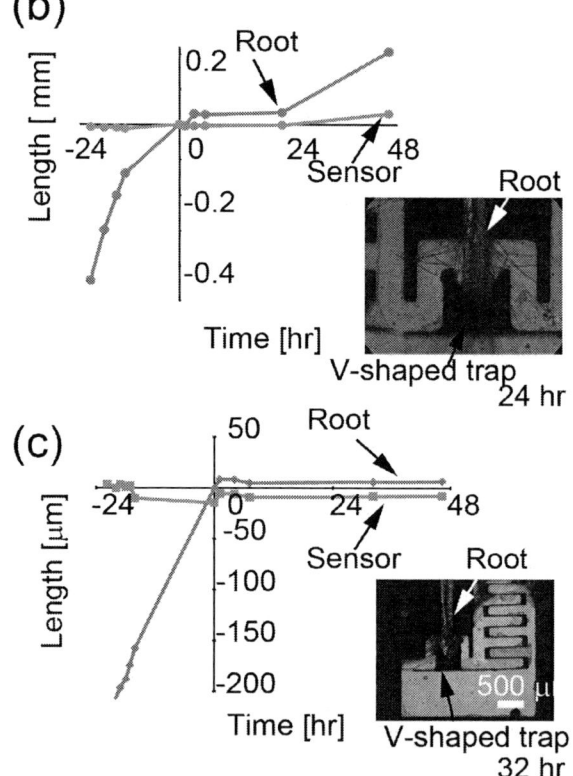

Figure 6: Time course of plant root length and sensor displacement. (a) Capsella rubella (Sensor k=5 N/m), (b) Nicotiana benthamiana (k=3.5 N/m), and (c) Arabidopsis thaliana (k=3.5 N/m).

and plant biomass in future.

ACKNOWLEDGEMENTS

This work was partly supported by JSPS KAKENHI, Grant Number 25790035.

REFERENCES

[1] X. Fu and N. P. Harberd, "Auxin promotes

978-1-4799-7956-1/15 $31.00 © 2015 IEEE

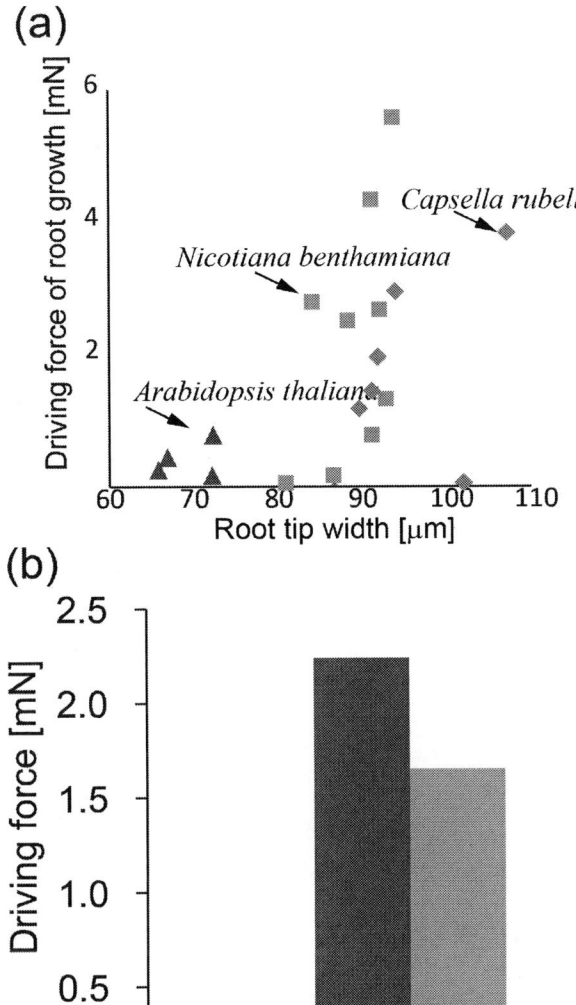

Figure 7: *Measurement of driving force of root with growth. (a) Relationship between the root tip width and driving force, and (b) Averaged value of driving forces in all three plants (N>4).*

Arabidopsis root growth by modulating gibberellin response.," *Nature*, vol. 421, no. 6924, pp. 740–743, 2003.

[2] C. Buer, J. Masle, and G. Wasteneys, "Growth conditions modulate root-wave phenotypes in *Arabidopsis*," *Plant Cell Physiol.*, vol. 41, no. 10, pp. 1164–1170, 2000.

[3] G. M. Whitesides, "The origins and the future of microfluidics.," *Nature*, vol. 442, no. 7101, pp. 368–73, 2006.

[4] H. Jiang, Z. Xu, M. Aluru, and L. Dong, "A microfluidic whole-plant phenotyping device", in *Digest Tech, Paper Transducers '13 Conference*, Barcelona, June 16-20, 2013, pp. 1539–1542.

[5] M. Meier, E. Lucchetta, and R. Ismagilov, "Chemical stimulation of the Arabidopsis thaliana root using multi-laminar flow on a microfluidic chip," *Lab Chip*, vol. 10, no. 16, pp. 2147–2153, 2010.

[6] G. Grossmann, W.-J. Guo, D. W. Ehrhardt, W. B. Frommer, R. V Sit, S. R. Quake, and M. Meier, "The RootChip: an integrated microfluidic chip for plant science.," *Plant Cell*, vol. 23, no. 12, pp. 4234–4240, 2011.

[7] A. Parashar and S. Pandey, "Plant-in-chip: Microfluidic system for studying root growth and pathogenic interactions in Arabidopsis," *Appl. Phys. Lett.*, vol. 98, no. 26, p. 263703 (3p), 2011.

CONTACT
*H. Hida, tel: +81-78-803-6058; hida@mech.kobe-u.ac.jp

A CANTILEVER WITH COMB STRUCTURE MODELED BY A BRISTLED WING OF THRIPS FOR SLIGHT AIR LEAK

H. Takahashi, A. Isozaki, K. Matsumoto and I. Shimoyama
The University of Tokyo, Tokyo, JAPAN

ABSTRACT

This paper reports a cantilever with comb structure, inspired by a bristled wing of thrips which acts as a continuous membrane wing because of the effects of low Reynolds numbers. The comb structures are formed at the edges of the cantilever and its surrounding frame. When differential pressure is applied to the cantilever, both comb structures act as airflow suppression through the gap. The leakage of the fabricated comb cantilever was smaller than the normal cantilever due to the overlapping area of the combs, even the gap area was twice larger. The proposed structure will be utilized as a high sensitive barometric pressure change detector.

INTRODUCTION

A differential pressure sensor, which is composed of a cantilever with narrow gap [1-4], has been utilized as a barometric pressure change detector by combining with an unsealed chamber [5-6].

The cantilever type differential pressure sensor has high sensitivity because of its deformable structure. Then, the detectable range of the pressure change is small compared with a conventional diaphragm type sensor.

As the other characteristics of the cantilever type sensor, there is small air leak from the cantilever gap. The air leak does not affect the sensitivity to differential pressure when the cantilever is used alone [3-4]. However, the air leak would decrease the sensitivity to barometric pressure change when the cantilever is used with a chamber because the chamber's pressure varies with the air leak [5]. Even if the gap is fabricated to narrower, the vertical deformation due to large differential pressure results in large air leak area, especially at the cantilever tip.

In nature, it is known that a thrips (*Thysanoptera*), which is a small insect, uses a micro-size-bristled wing for flapping flight. The wing length is approximately 1 mm. The length and width of the bristled structures are approximately 300 μm and 2 μm, respectively. A thrips flaps its wings at several hundred Hertz, and the Reynolds number around its wings is less than 100. A number of studies have been performed to estimate the aerodynamics acting on bristled wings of a thrips [7-9]. Previous studies have shown that a bristled wing can produce a force similar to that of a continuous membrane wing because of the effects of viscosity. The air next to the bristles is dragged with the bristle due to boundary layer effects at low Reynolds numbers. We have also demonstrated that a MEMS cantilever with comb structure generated aerodynamic force the same as that of a continuous one plate when airflow is applied to the surface [10].

In this paper, we propose a cantilever with comb structure modeled by a bristled wing of a thrips for a differential pressure sensor. Both cantilever and its surrounding frame have comb structure by etching with zigzag line as shown in Figure 1(a). The double comb

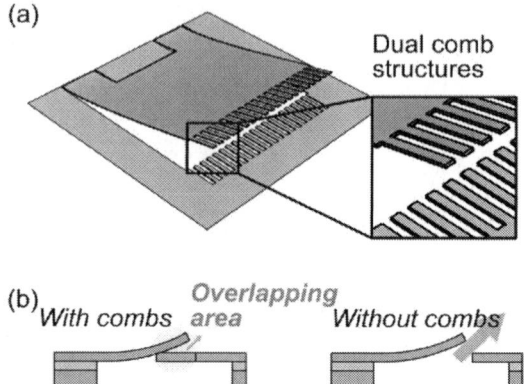

Figure 1 : (a) Concept sketch of a cantilever with comb structure. (b) Airflows which pass through the gaps with / without combs.

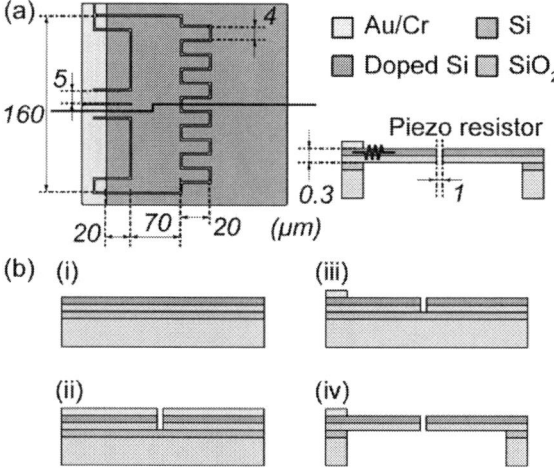

Figure 2 : (a) Design of the cantilever with comb structure. The cantilever acts as a differential pressure sensor with piezo resistive element. (b) Fabrication process.

structures act as overlapping continuous plates when differential pressure is applied. Then, the imaginary plates lead more sufficient airflow suppression than a normal cantilever gap as shown in Figure 1(b). Thus, the proposed structure decreases air leak from the gap, which will result in a high sensitive barometric pressure change detector.

DESIGN AND FABRICATION

The proposed cantilever is designed to be 100 μm × 160 μm × 0.3 μm in size as shown in Figure 2(a). The comb part is formed on the cantilever tip. The piezoresistor is formed on the cantilever surface. We measure the resistance change between the inside two-leg component to detect differential pressure. The comb width is designed to be 4 μm, which is corresponding to that of a bristle of a thrips wing [7]. The length and number of the comb are 20

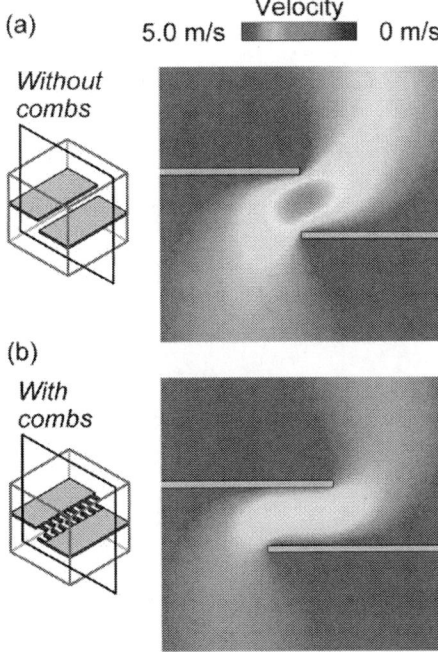

Velocity
5.0 m/s ▦ 0 m/s

(a) Without combs

(b) With combs

Figure 3 : Simulation of airflow velocity when differential pressure is applied to vertically separated two plates (a) without and (b) with comb structure.

μm and 17, respectively. The initial gap between the cantilever and surround, which is corresponding to the etching widths, is designed to be 1 μm. A comb-less normal type cantilever is also designed as a reference. The surface area of the normal cantilever is same as the comb cantilever. The initial gap areas of the comb and normal cantilevers are approximately 1200 μm² and 600 μm², respectively.

We simulated airflow velocity between vertically separated two plates with/without comb structure using finite element modeling (FEM) software (COMSOL Multiphysics, COMSOL). The simulation was performed as semi-infinite plates. The sizes of the comb are the same as the designed values. The thickness of the plate is designed to be 1 μm instead of 0.3 μm. The vertical distance and differential pressure are 20 μm and 50 Pa, respectively. This relation is similar to the calculated relation of the deformation of the tip and the applied differential pressure of the designed cantilever.

The cross sectional views of the airflow velocity around the two plates are shown in Figure 3. The result suggests that the airflow velocity with the comb plates is smaller than that of the normal plates because the comb structure acts as a continuous plate. It is observed that the volume of the air leak of the comb plates is also smaller than that of the normal plates.

The fabrication process of the cantilever is described in previous work in detail [3]. The process flow is shown in Figure 2(b). The device was fabricated on a p-type SOI (Silicon on Insulator) wafer.

The fabricated sensor chip is shown in Figure 4 (a). The chip size was 1.5 mm × 1.5 mm × 0.3 mm. The cantilever was formed on the center of the chip. Photographs of the comb and normal cantilevers are shown

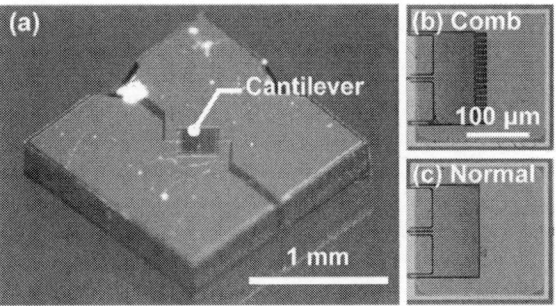

Figure 4 : Photographs of (a) the fabricated device chip,(b) the comb cantilever, and (c) the reference normal cantilever.

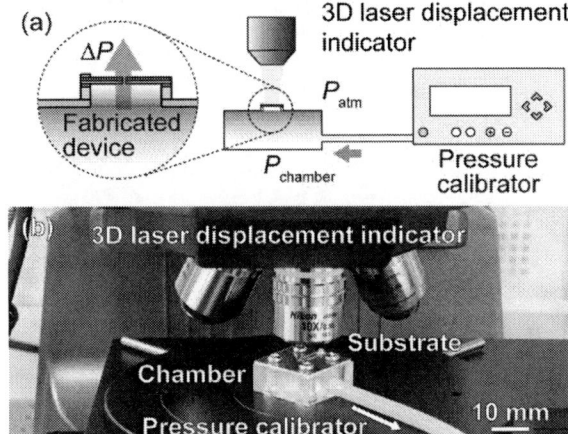

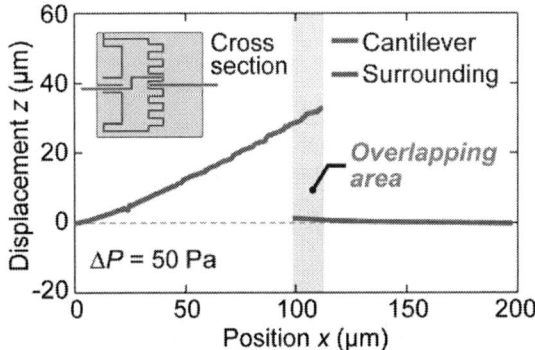

Figure 5 : (a) Concept sketch and (b) photograph of the experimental setups to measure the deformation of the cantilever.

Figure 6 : Displacement of the cantilever beam.

in Figure 4 (b) and (c), respectively. The initial resistances of both cantilevers were approximately 1 kΩ.

EXPERIMENT AND RESULT

The experimental setup to measure the deformation of the cantilever with differential pressure is shown in Figure 5. The cantilever chip was attached on a substrate with an air hole. Then, the substrate was mounted on a chamber box, which had a connector for pressure supply. We can apply positive constant differential pressure based on barometric pressure to the chamber from a pressure calibrator (KAL200, Halstrup-Walcher GmbH). Then, the

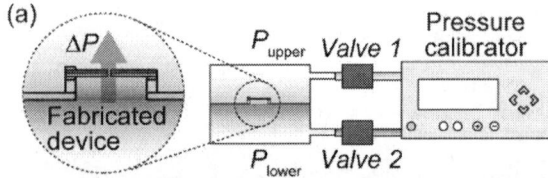

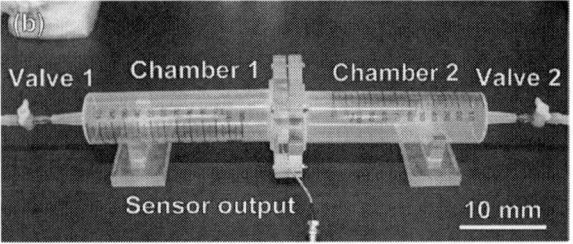

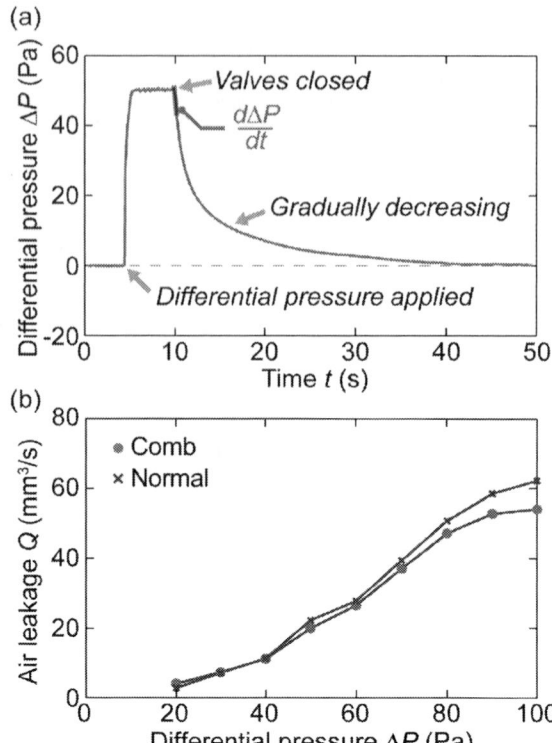

Figure 7 : (a) Concept sketch and (b) photograph of the experimental setups to measure the fractional resistance change when differential pressure is applied.

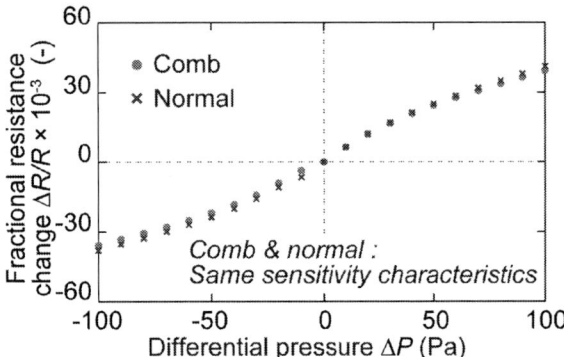

Figure 8 : Relationship between the differential pressure and the fractional resistance change.

chamber was fixed under a 3D laser displacement indicator (VK-8700, Keyence).

The displacement of the comb cantilever, when differential pressure of 50 Pa was applied, is shown in Figure 6. The overlapping area was observed in the comb part. The vertical distance between the cantilever tip and surrounding frame was approximately 30 μm.

The experimental setup to measure the relationship between the applied differential pressure and the fractional resistance change is shown in Figure 7. The substrate with the chip was mounted to a jig that separated two syringes (each 150 ml in volume). The sensor output was connected to Wheatstone-bridge and instrumentation amplifier circuits (AD623, Analog Devices). We applied step responses of constant differential pressure from −100 to 100 Pa at 10 Pa intervals in this experiment.

The experimental results of both comb and normal cantilevers are shown Figure 8. It was observed that the comb and normal cantilevers had similar relationship. The sensitivities were approximately equal to the previous sensors [3,5]. This result indicated that the comb cantilever responded to differential pressure same as the normal cantilever because of the same surface area.

Finally, the air leak through the cantilever gap was evaluated. To obtain the value of the air leakage, we measured the differential pressure changing between two chambers after closing the pressure supply valves. The

Figure 9 : (a) Graph of the differential pressure change caused by the air leak. (b) Relationship between the differential pressure and air leakage.

differential pressure was able to be monitored by the calibrated cantilever itself.

Figure 9(a) shows the differential pressure changing of the comb cantilever when the initial constant differential pressure was 50 Pa. After closing the valves, the differential pressure decreased gradually, and became approximately zero approximately 40 second later. The derivative value of the differential pressure right after closing valves, $d\Delta P/dt$, was obtained as shown in the graph, which was 26.6 Pa/s. The air leakage Q was calculated using the following equation:

$$Q = \frac{V_0}{2P_{\mathrm{atm}}} \times \frac{d\Delta P}{dt} \qquad (1)$$

where V_0 and P_{atm} are the volume of the chambers and barometric pressure, respectively. Thus, the air leakage Q with differential pressure of 50 Pa was calculated as 20.0 mm³/s.

The air leakages from both comb and normal cantilevers were calculated, when the initial differential pressure was from 20 Pa to 100 Pa at 10 Pa intervals. We evaluated the air leakage of each initial differential pressure as above.

Figure 9(b) shows the relationship between the calculated air leakage and the initial differential pressure. When the initial differential pressure was 20 Pa, the air leakage was approximately 5 mm³/s. The air leakage increased with increasing initial differential pressure. When the initial differential pressure was 100 Pa, the air leakage became approximately 50 mm³/s, which was 10 times larger than that of 20 Pa.

It was observed that the leakage of the comb cantilever was larger than that of the normal cantilever in small differential pressure (< 40 Pa). This result suggested that the difference of the initial gap areas was a dominant factor to the air leakage because the vertical deformation was small. On the other hand, the leakage of the comb cantilever became smaller in large differential pressure (> 40 Pa), which was thought to be the effect of the overlapping comb structure because of the large deformation.

The ratio of the leakage decrement in this design was up to 10%. The optimization of the comb and cantilever structure will lead more sufficient overlapping effect and reduce air leak, which enhances the sensitivity of barometric pressure change detection.

CONCLUSION

In conclusion, the comb type cantilever was designed and fabricated as a differential pressure sensor element. The overlapping of the comb structure was designed to decrease the air leak from the gap between the cantilever and surrounding frame. The air leakage of the fabricated comb cantilever was smaller than that of the normal cantilever when the differential pressure was over 40 Pa, whereas the sensitivity to constant differential pressure was equal to each other. The proposed comb cantilever will be utilized to the enhancement of the sensitivity of barometric pressure change detection using a cantilever type sensor.

ACKNOWLEDGEMENTS

A part of this work was supported by New Energy and Industrial Technology Development Organization (NEDO). The photolithography masks were made using the University of Tokyo VLSI Design and Education Center (VDEC)'s 8 inch EB writer F5112 + VD01 donated by ADVANTEST Corporation.

REFERENCES

[1] J. Kauppinen, K. Wilcken, I. Kauppinen, and V. Koskinen, "High sensitivity in gas analysis with photoacoustic detection," *Microchem. J.*, vol. 76, pp. 151-159, 2004.

[2] P. Sievila, V. P. Rytkonen, O. Hahtela, N. Chekurov, J. Kauppinen, and I. Tittonen, "Fabrication and characterization of an ultrasensitive acousto-optical cantilever," *J. Micromech. Microeng.*, vol. 17, pp. 852-859, 2007.

[3] H. Takahashi, N. M. Dung, K. Matsumoto, and I. Shimoyama, "Differential pressure sensor using a piezoresistive cantilever," *J. Micromech. Microeng.*, vol. 22, p. 055015, 2012.

[4] Y. Tomimatsu, H. Takahashi, T. Kobayashi, K. Matsumoto, I. Shimoyama, T. Itoh, *et al.*, "A piezoelectric cantilever-type differential pressure sensor for a low standby power trigger switch," *J. Micromech. Microeng.*, vol. 23, p. 125023, 2013.

[5] N. Minh-Dung, H. Takahashi, T. Uchiyama, K. Matsumoto, and I. Shimoyama, "A barometric pressure sensor based on the air-gap scale effect in a cantilever," *Appl. Phys. Lett.*, vol. 103, pp. 143505-4, 2013.

[6] Y. Kaiho, H. Takahashi, Y. Tomimatsu, T. Kobayashi, K. Matsumoto, I. Shimoyama, *et al.*, "An AlN cantilever for a wake-up switch triggered by air pressure change," *JPCS*, vol. 476, p. 012122, 2013.

[7] C. P. Ellington, "Wing Mechanics and Take-Off Preparation of Thrips (*Thysanoptera*)," *J. Exp. Biol.*, vol. 85, pp. 129-136, 1980.

[8] S. Sunada, H. Takashima, T. Hattori, K. Yasuda, and K. Kawachi, "Fluid-dynamic characteristics of a bristled wing," *J. Exp. Biol.*, vol. 205, pp. 2737-2744, 2002.

[9] E. Barta and D. Weihs, "Creeping flow around a finite row of slender bodies in close proximity," *J. Fluid Mech.*, vol. 551, pp. 1-17, 2006.

[10] K. Sato, N. Minh-Dung, H. Takahashi, K. Matsumoto, and I. Shimoyama, "Effectiveness of bristled wing of thrips," in *Digest Tech. Papers MEMS2013 conference*, Taipei, Taiwan, Jan. 20-24, 2013, pp. 21-24.

CONTACT

*H. Takahashi, Tel: +81-3-5841-0461;
takahashi@leopard.t.u-tokyo.ac.jp

AIRFLOW SHEAR STRESS SENSOR
USING SIDE-WALL DOPED PIEZORESISTIVE PLATE
R. Kazama, H. Takahashi, T. Takahata, K. Matsumoto and I. Shimoyama
The University of Tokyo, Japan

ABSTRACT

This paper reports on a shear stress sensor for airflow. The sensor consists of a plate, side-wall doped beams and surrounding membrane with narrow gaps. The sensor plate was 300 μm×350 μm×3 μm in size. The plate was supported by the side-wall doped beams, which can detect their horizontal deformation. The sensor structure was designed not to disturb target airflow circumstance owing to its flat surface. The fabricated sensor was able to measure shear stress of laminar airflow with the resolution under 1.0 Pa.

INTRODUCTION

In the fields of development of micro air vehicles (MAV) or research of insect flight, the evaluation of drag force on wings is an important key factor, because the drag force is related to a condition of flight. During flight, the drag force is affected by flow velocity, angle of attack, skin friction, and so on. Wall shear stress at the leading edge of a wing is one of the essential parameters to determine the drag force originating skin friction [1]. Previous reserches indicated that the shear stress less than several pascals is generated during the insect or MAV flight [2,3].

As a method to measure shear stress, there were many approaches using MEMS technology. For instanse, previous researches have developed a micro flow sensor converting temperature distribution around heater [4], or converting pressure drag working on small pillars [5] to shear stress. However, the wall shear stress varies by the slight disturbance of airflow. Therefore, these sensing elements themselves have possibility to disturb airflow circumstance.

Figure 1 shows the concept of our proposed shear stress sensor without disturbing target airflow. There were flat surface and small gap between the sensing element and surrounding membrane. This sensor detected horizontal stress on the surface of the sensor plate. As stress detectors, side-wall doped piezoresistive beams were used. Measurement without disturbing airflow was possible by embedding the proposed sensor in the target object such that both surfaces were in the same plane.

In this paper, we designed and fabricated an airflow shear stress sensor aimed at 1 Pa order measurement. By wind-tunnel experiment, we measured the wall shear stress of laminar airflow whose velocity was several meter per second. The value of skin friction drag of each wind velocity was calculated from the measured shear stress.

DESIGN AND PRINCIPLE

The design of the sensor was shown in Figure 2(a). The proposed sensor was designed to measure shear stress without disturbing airflow circumstance. For this specification, the sensor was composed of a plate, supporting beams and surrounding membrane whose surfaces were flat to each other. The sensor plate had a role

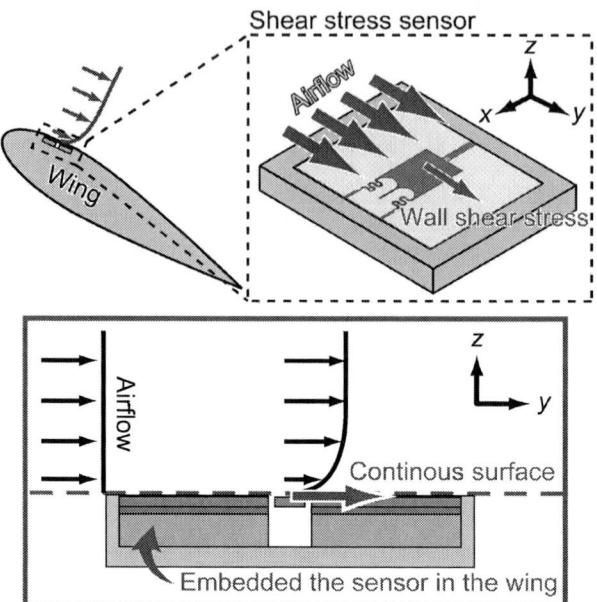

Figure 1: Concept of the measurement shear stress of airflow. The proposed sensor is able to measure the shear stress without disturbing airflow.

of receiving force of airflow. The size of the plate was designed to be 300 μm×350 μm. Three beams sustained this sensor plate. Each beam was connected to two pads and ground, respectively. The width of the beams was 5 μm. A center beam was an electric wire connected to the ground. The piezoresistive layer was doped to the side-wall of the other two beams. The gap between the plate and surrounding membrane was 5 μm so as not to disturb airflow. When airflow shear stress was applied to the plate, the plate deformed horizontally. The deformation was detected by two symmetrical side-wall doped beams. As shown in Figure 2(b), the responses were in the opposite sign to each other. One was compressive, and the other was tensile. The fractional resistance change due to vertical deformation was in the same sign. Also, other factors like light and temperature transition caused the similar resistance change. So these signal could be canceled because of the symmetrical design.

The proposed sensor was fabricated following procedures shown in Figure 3. The details of the fabrication process were reported in the previous work [6]. In fabricated process, a P-type silicon-on-insulator (SOI) wafer was used. The thickness of the SOI wafer layers were 3 μm, 2 μm and 200 μm, respectively. First step was etching two hole for side-wall doping on the device Si layer as shown in Figure 3(a). Second step was doping a piezoresistive layer on the side-wall of the holes and surface of the device Si layer with rapid thermal diffusion as shown in Figure 3(b). Third step was deposition of an

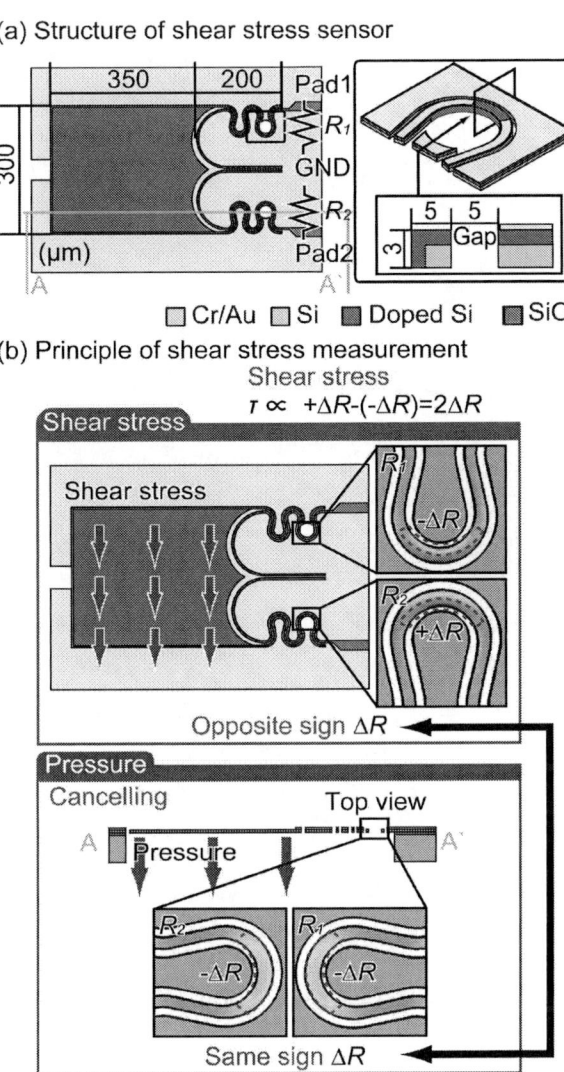

Figure 2: (a) Design of the shear stress sensor. (b) Principle of shear stress detection.

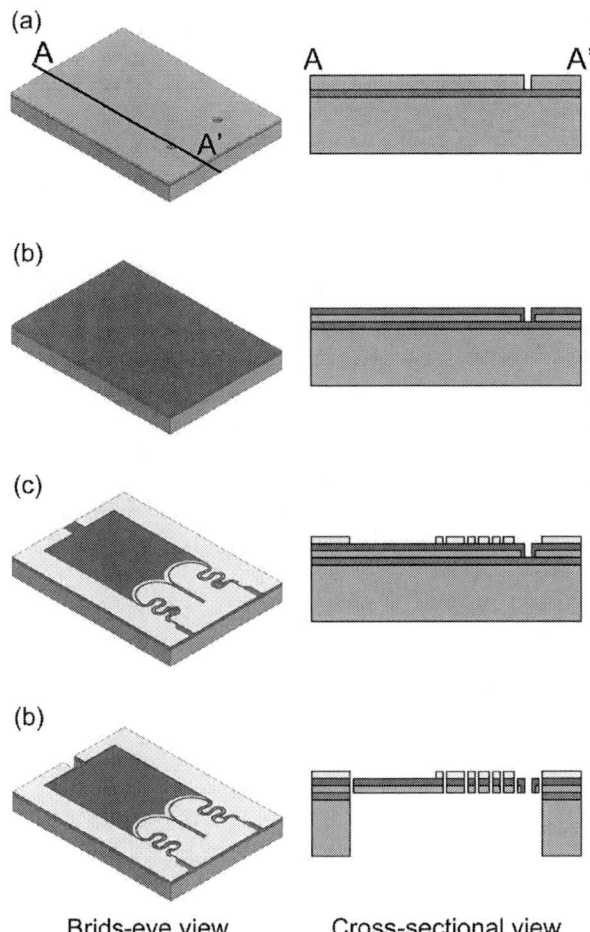

Figure 3: The fabrication process of the sensor chip (a) Etching of the device Si layer. (b) Doping piezoresistive layer with thermal diffusion. (c) Depositing Au/Cr layer for electric wire. (d) Etching doped Si layer and removing handle Si and SiO2 layers.

Au/Cr layers as shown in Figure 3(c). The thickness of the Au/Cr layers were 10 nm and 30 nm. This layer was used as electric wire. The electric wire shape was formed by lift off process. Forth step was etching on the device Si layer. The shape of sensing element was formed in this step. Final step was etching the handle Si layer and releasing SiO₂ layer as shown in Figure 3(d).

EXPERIMENT AND RESULT

The fabricated sensor chip is shown in Figure 4. The size of sensor chip was 2 mm×2 mm×0.2 mm. The resistance values of both beams were approximately 2 kΩ. As the sensor calibration, we measured the response to concentrated force to the side of the plate (Figure 5(a)). A glass needle whose diameter was smaller than the sensor gap was used to apply concentrated force. The range of the applied force was 6 μN. The point where the glass needle touched the sensor was the center of the side-wall of the plate. The grass needle was attached to a force gauge (Kyowa Electronic Instrument, LVS-5GL).

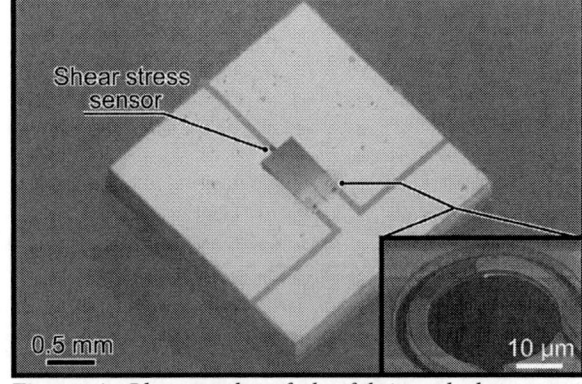

Figure 4: Photographs of the fabricated shear stress sensor.

The measurement system is as shown in Figure 5(b). A lock-in amplifier (Zurich Instruments, HF2LI-H) was used for the measurement to decrease noise level. Both piezoresistive beams connected to a bridge circuit. The applied voltage to the bridge circuit was alternating current whose frequency was 1 kHz. We assumed the

(a) Caribration setup

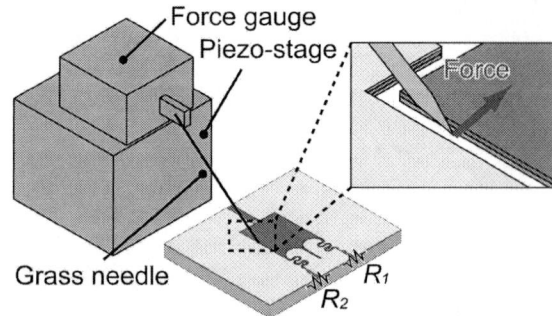

(b) Measurement system

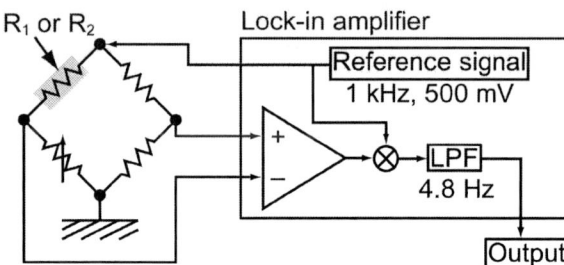

(c) Response to the applied force

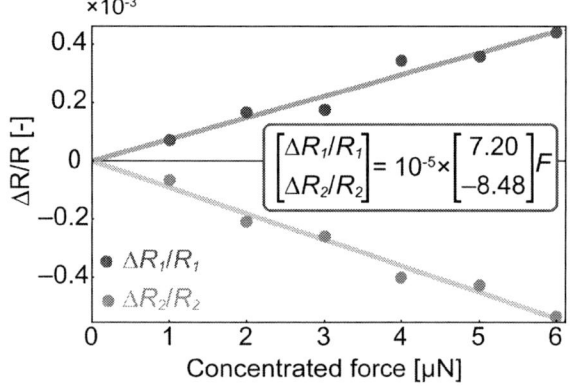

Figure 5 (a) Calibration setup (b) relationship between the applied concentrated force and resistance changes of the sensor

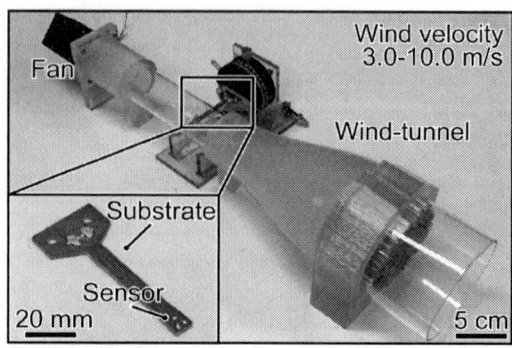

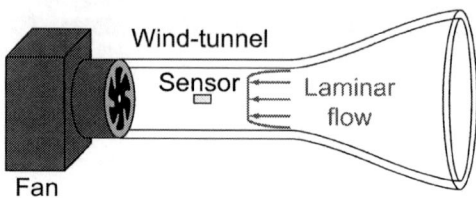

Figure 6: Photographs and concept of experimental setup.

(a) Wall shear stress measurement

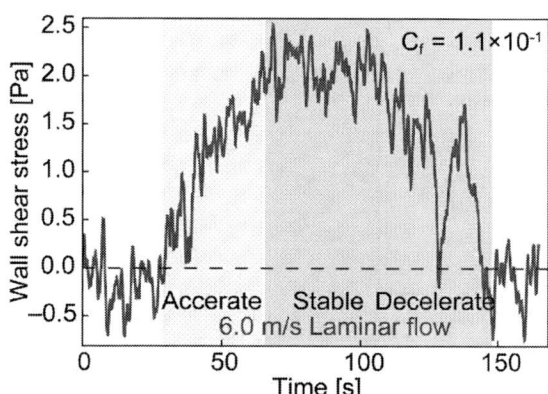

(b) Relationship between wind velocity and wall shear stress

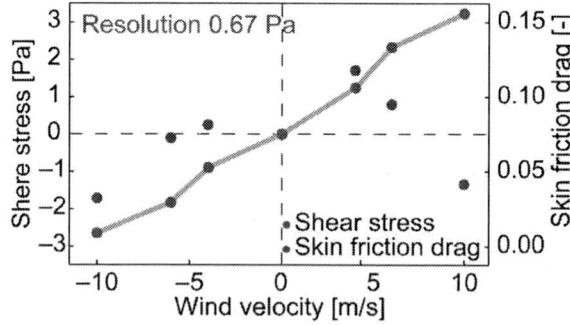

Figure 7: (a) Wall shear stress measurement under the condition of 6.0 m/s airflow velocity. (b) Relationship between airflow velocity and wall shear stress.

measurement signal under 1 Pa. Therefore, the frequency of the applied voltage was high enough compared to that of the measurement signal.

A symmetrical response was observed in two beams as shown in figure 5(c). The stress sensitivity of the two beams corresponding to the fractional resistance values were 7.20×10^{-5} μN^{-1} and -8.48×10^{-5} μN^{-1}, respectively. The applied force was converted to the shear stress to divide the force by the surface area of the sensor plate. In this time, we assumed that the applied force was distributed uniformly to the surface of the sensor plate. Using this converted shear stress, we calculated the calibration matrix between the fractional resistance change and the shear stress defined as τ. The calibration matrix is as follows;

$$\tau = \begin{pmatrix} -0.41 & 0.39 \end{pmatrix} \times 10^6 \begin{pmatrix} \Delta R_1/R_1 \\ \Delta R_2/R_2 \end{pmatrix}. \quad (1)$$

The unit of shear stress is Pa.

Figure 6 shows the setup of wind-tunnel experiment to measure wall shear stress of laminar airflow. A wind

tunnel was designed to intake type. A fan was attached to the exhaust duct. Inlet and exit diameters of the wind tunnel nozzle were 10 cm and 4 cm, respectively. The nozzle was fabricated using a 3D printer (Stratasys Ltd., EN260V). A bundle of straws in the nozzle was a rectifier. The sensor, which was fixed to the tip of a substrate, was set in the center of the test section made of acrylic pipe. Using Dotite, the sensor chip was wired to the substrate. The substrate was 1.5-mm-thick. The distance between the sensor plate and both edges of the substrate was 2 mm. The sensor surface and airflow direction were parallel. The measurement system was the same one using at the calibration as shown in Figure 3(b).

The wall shear stress measurement was done as shown in the following method. We started to measure the sensor response with windless condition. After 30 seconds, the fan of the wind tunnel was operated for 80 seconds. After that, the fan was switched off. We measured wall shear stress under the condition of three airflow velocity ($\pm$ 4.0, 6.0, 10.0 m/s).

The experimental result when airflow velocity was 6.0 m/s (Reynolds number: 3.4×10^2) is shown in Figure 7(a). When the fan was switched on, the sensor output gradually increased. After that, the sensor output became stable. When the fan was switched off, the sensor output gradually returned to the default. The wall shear stress was evaluated to be 2.3 Pa from fractional resistance value when the sensor output was stable. The skin friction drag C_f was calculated to be 1.1×10^{-1} from this shear stress. Then, the shear stress resolution was defined 0.67 Pa from the noise level.

Figure 6(b) shows the relationship between the applied airflow velocity with the measured wall shear stress and skin friction drag. There was linearity in the measured wall shear stress and airflow velocity. The skin friction drag increased according to the absolute value of airflow velocity decrement, which was corresponding to Reynolds number decrement. These result agreed with the previous study [7].

These results indicated the fabricated sensor was able to measure airflow wall shear stress under 1.0 Pa resolution. Thus, the shear stress sensor was thought to be useful to measure the wind whose velocity was several meter per second.

CONCLUSION

In conclusion, we proposed the high resolution shear stress sensor for airflow. The sensor was designed to detect wall shear stress without disturbing airflow by their flat surface and small gap. The sensor plate was 300 μm×350 μm×3 μm in size. In the wind-tunnel experiment, we confirmed that the proposed sensor was able to detect wall shear stress of airflow whose Reynolds number was 10^2 order.

ACKNOWLEDGEMENTS

The photolithography masks were made using the University of Tokyo VLSI Design and Education Center (VDEC)'s 8 inch EB writer F5112 + VD01 donated by ADVANTEST Corporation. This work was supported by JSPS KAKENHI Grant Number 25000010.

REFERENCES

[1] U. Ehrenstein, M. Marquillie, C. Eloy, "Skin friction on a flapping plate in uniform flow," *Phil. Trans. R. Soc. A*, vol. 372, no. 2020, 2014.

[2] E. Swanton, B. Vanier, K. Mohseni, "Flow visualization and wall shear stress of a flapping model hummingbird wing," *Experiments in Fluids*, vol. 49, pp. 657-671, 2010.

[3] J. Tang, D. Viieru, W. Shyy "Effects of Reynolds Number and Flapping Kinematics on Hovering Aerodynamics," *AIAA Journal*, vol. 46, pp. 967-976, 2008.

[4] L. Löfdahl, E. Kälvesten, T. Hadzianagnostakis, G. Stemme, "An Integrated Silicon based Wall Pressure-Shear Stress Sensor for Measurements in Turbulent Flows," *Fluid Mechanics and its Applications*, vol. 36, pp. 465-469, 1996.

[5] E. Gnanamanickam, "Unsteady distributed wall shear stress measurements in fluid flows," *A Dissertation of Purdue University*, 2010.

[6] H. Takahashi, A. Nakai, N. Thanh-Vinh, K. Matsumoto, I. Shimoyama, "A triaxial tactile sensor without crosstalk using pairs of piezoresistive beams with sidewall doping," *Sensors and Actuators A: Physical*, vol. 199, pp. 43-48, 2013.

[7] S. Hoerner, "SKIN-FRICTION DRAG," in *FLUID-DYNAMIC DRAG*, 1965.

CONTACT

*Ryohei Kazama, Mechano-Informatics Department, Graduate School of Information Science and Technology, The University of Tokyo, 7-3-1 Hongo, Bunkyo-ku, Tokyo, 113-8656, Japan.
 Tel: +81-3-5841-6318, Fax: +81-3-5841-6341
 E-mail:kazama@leopard.t.u-tokyo.ac.jp

MICRO TRIPLE-HOT-WIRE ANEMOMETER ON SMALL SIZED GLASS TUBE FABRICATED IN 5DOF UV LITHOGRAPHY SYSTEM

Shuwei Liu[1], Zhuoqing Yang[2], Yi Zhang[3], Fei Xue[1], Shanshan Pan[1], Jianmin Miao[1], and Leslie K. Norford[4]

[1]Nangyang Technological University, Singapore
[2]Shanghai Jiao Tong University, China
[3]National Institute of Advanced Industrial Science and Technology, Japan
[4] Massachusetts Institute of Technology, USA

ABSTRACT

This paper reports novel designed and fabricated micro airflow sensors based on the hot-wire sensing principle, i.e. gas cooling of electrically heated resistance. With three micro Ti/Pt hot-wire components with width of 50 µm and thickness of 130 nm fabricated on a glass tube (outer diameter of 2.8mm and inner diameter of 2.1mm) in five degrees of freedom (5DOF) UV lithography system with multi-layer alignment, the sensors on a cylindrical base have demonstrated high sensitivity, fast response time and ability to detect wind speed and direction. Environmental characterization of sensor at different levels of temperature, humidity and solar radiation has been carried out.

INTRODUCTION

MEMS sensors can reduce power consumption and have lower cost owing to the nature of batch fabrication and small size. Micro airflow sensors based on the thermal principle are commonly implemented because of their fast response and robust and simple structure.

In order to detect airflow direction, the micro airflow sensors must have more than one sensing component to establish a flow-dependent thermal gradient. An out-of-plane airflow sensor structure has manifested greater sensitivity than an in-plane sensor structure by elevating the thermal element away from the bottom of the fluid boundary layer and therefore the thermal element is exposed to greater flow speed [1]. We obtained high sensitivity for airflow speed and direction detection is achieved from a cylindrical sensor structure with three manually assembled MEMS-based hot-wire resistors arranged 120 degrees apart on the structure [2]. To improve the assembly effectiveness, we fabricated three self-assembled micro hot-wire resistors on a borosilicate glass bubble that provides good thermal isolation [3]. To elevate the self-assembly resistors to a higher position on an axisymmetric circular structure, this paper will present a novel micro airflow sensor with self-assembled micro hot-wire resistors arranged 120 degrees apart on a cylindrical base fabricated in a 5DOF programmable UV lithography system with multi-layer alignment, as shown in Figure 1(a).

FABRICATION

A programmable UV lithography equipment with five degrees of freedom (5DOF) for cylindrical substrates has been developed [4]. The equipment mainly consists of four parts: a uniform illumination system, a reduced projection lithography system, a synchronized motion stage system,

and a CCD multilayer alignment system as shown in Figure 1(b).

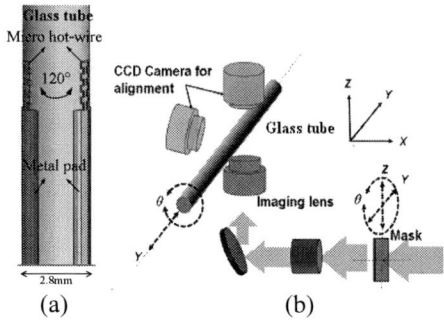

(a)   (b)

Figure 1: (a) Glass tube with micro hot-wire resistors; (b) Sketch of five degrees of freedom (5DOF) UV lithography system for cylindrical base with multi-layer alignment)

Incident light from a UV-light source is focused onto a photoresist coated cylindrical substrate with a reduced mask image by passing it through the optical elements of lenses, mask, and mirror. The incident UV-light from an Hg–Xe light source (LC8, Hamamatsu Photonics) has a broad wavelength range of 250–600 nm. This broad wavelength of incident light results in chromatic aberration in the focal plane. In order to remove this aberration, a band-pass interference filter with a central wavelength of 436 ± 10 nm was inserted between the light source and the shutter. Furthermore, to obtain a high resolution, a reduced projection lithography system consisting of an imaging lens and a biconvex lens, which is commonly known as an "infinity corrected optical system (ICOS)," was used to precisely focus and transport the mask patterns. The ICOS consists simply of an objective lens used as an imaging lens, and a biconvex lens. Therefore, the overall magnification of the lithography equipment is directly determined by the magnifying power of the objective lens. Considering the required size and complexity of the microstructures to be developed, the optical magnification of the ICOS was selected as 0.5, by applying an objective lens with a magnifying power of two times as an imaging lens. Finally, the obtained imaging focal depth of the projection patterning is $\pm45\mu$m. To realize the coaxial rotation of the photoresist coated cylindrical substrate (a glass tube in this work), two chucks installed in the motorized rotation stages are used to fix the substrate. At the same time, the laser path from a He-Ne laser was used as a reference to align all these elements, tune the optical axes and make sure their coaxial rotations. In order to realize the secondary or multilayer alignment during the exposure fabrication, two CCD cameras are used to observe simultaneously the projected mask's patterns and those ever fabricated on the capillary. This operation can be realized by real-timing monitoring software.

The mask and the glass tube substrate can be simultaneously driven along the 5 DOF. All motorized XYZ-θ stages and exposure time were automatically driven and precisely controlled, respectively. In the each axial direction, the movement accuracy was set to 0.1μm, and the rotation accuracy in the θ-direction was set to 0.01°, and these values were defined as the repeat positioning accuracy of the automatic stages used. The expected micropatterns on the photoresist-coated glass tube can be drawn by executing the programmable sentences in PC lithography controlling window. Finally, the photoresist micro hot-wire patterns will be obtained on the glass tube surface after the development. This equipment can be used not only for glass tube exposure, but also for those similar cylindrical substrates, such as optical fibers, metal wires and fabric wires, etc.

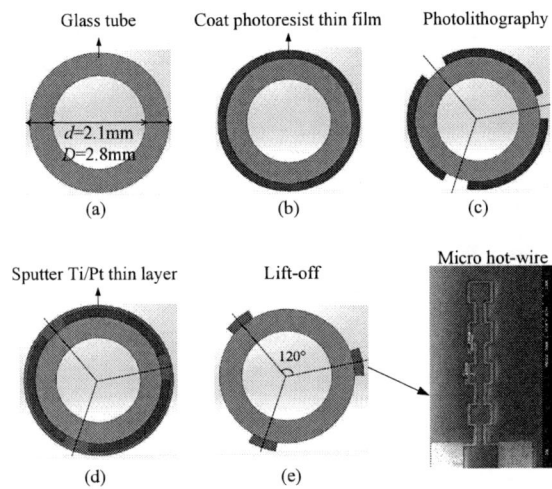

Figure 2: Fabrication process of three micro hot-wire resistors arranged 120 ° apart on a glass tube with diameter of 2.8mm (cross-section view)

Figure 2 shows the fabrication steps:

(a) Glass tube (2.8mm diameter) substrate ready and cleaning the surface of the tube using plasma;

(b) Spray coating photoresist film by bove-mentioned method;

(c) Photolithography and patterning of the tube with coated resist film by developed lithography method;

(d) Magnetron sputtering Ti/Pt film (10nm/130nm) onto the patterned glass tube surface after development;

(e) Resist film was removed by lift-off and obtaining the three micro hot-wire resistors arranged 120 ° apart on a glass tube. Finally, the micro triple-hot-wire airflow sensor on the glass tube has been successfully fabricated, as shown in Figure 2.

MEASURMENT

Response time

Response time measures how fast the sensor responds to the change of wind speed. However, in practical case, it is difficult to generate a speed pulse or step flow profile sharp transient, electrical method will be adopted to simulate mechanical perturbation and measure the response time of sensor output. Each resistor R_s of hot-wire is connected to a constant temperature anemometry (CTA) circuit channel based on a Wheatstone bridge, as shown in Figure 3(a). Constant temperature means that the circuit actively keeps the resistance of each resistor constant by a closed-loop feedback. The input impedance R_{in} functions as the perturbing factor. During the measurement, the sensor is exposed to wind with certain speed U. The output voltage from one channel is Vo. At the onset of switching on R_{in}, the feedback circuit will try to balance the bridge by increasing the voltage output (and hence the heating current), this is similar to a sudden increase in wind speed. Output voltage will increase to V_{m1} and then stabilize at $V'o$. Then based on V_O-U_0 curve in Figure 4, we would be able to obtain the equivalent mechanical perturbation, in other words, the value of increased wind speed. The response time for increasing wind speed is denoted as τ_{inc}, defined here as the 90% value of the time taken for the change of the output voltages from V_{m1} to V'_o. Switching off R_{in} is similar to a sudden decrease in wind speed. Similarly, the response time for decreasing wind speed is denoted as τ_{inc}, defined as the 90% value of the time taken for the change of the output voltages from V_{m2} to V_o. By varying R_{in}, different level of perturbation is introduced. Table 1 lists the response time at varied R_{in}, when the sensor is exposed to wind speed 2.5m/s. The higher level of perturbation, the faster the response is. In addition, the sensor respond faster to the increase of wind speed than to the decrease of wind speed.

Constant temperature anemometry (CTA) circuit

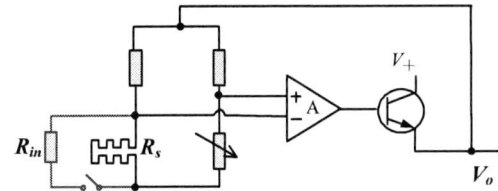

R_s: resistance of hot-wire

R_{in}: input resistance to create perturbation simulating speed change

(a)

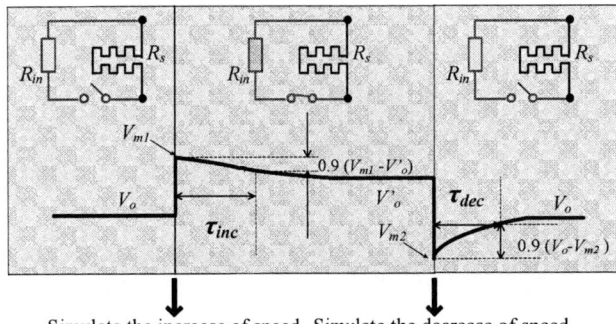

Simulate the increase of speed   Simulate the decrease of speed

(b)

Figure 3: (a) Constant temperature anemometry circuit with input resistance to simulate the mechanical perturbation; (b)The mechanism of measurement of response time using electrical method

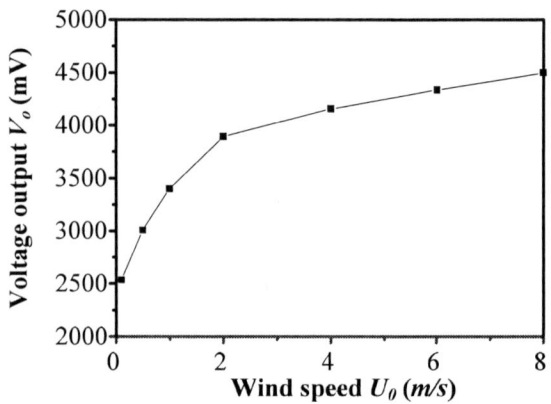

Figure 4: Voltage output V_o with respect to wind speed U_0

Table 1. Measured response time when the sensor is exposed to wind speed U_0 of 2.5m/s

Input resistance R_{in}(kΩ)	Output voltage with R_{in} V'_o (V)	change of speed Δv (m/s)	τ_{inc} (ms)	τ_{dec} (ms)
218.3	3.8	0.5	600	515
100	4.1	3.3	119	108
47	4.35	4.5	65	62

Wind speed and direction

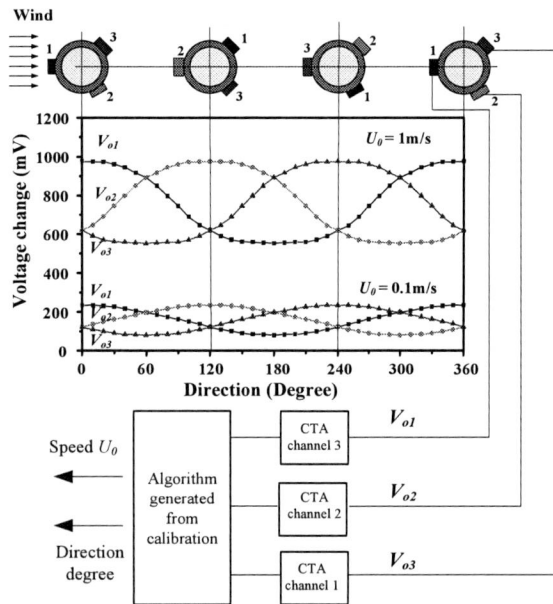

Figure 5: Output voltages (V_{o1}, V_{o2}, V_{o3}) from three CTA channels with respect to direction when wind speed U_0 varies

The flow direction measurement is based on the relative output difference of the three sensing elements in response to temperature variation induced by airflow. The principle of detection of airflow speed and direction is illustrated in Figure 5. A detailed description of the principle can be found in [2]. Each resistor is connected to a constant temperature anemometry (CTA) circuit channel.

The voltage outputs, V_{o1}, V_{o2} and V_{o3}, from channels are processed by an algorithm, and then the corresponding airflow speed and direction are derived. Figure 5 shows the voltage outputs from three channels at wind speed 0.1m/s and 1m/s. To generate the algorithm, sensors need to be calibrated by conducting measurement in wind tunnel and voltage outputs are measured at different speeds and directions. During the calibration, the sensor chip is placed on the top of a pillar, which is mounted on a rotary station for flow direction characterization. The sensor is positioned in a wind tunnel such that the airflow is parallel to the sensor surface. The position at which hot-wire resistor 1 is facing the airflow direction is marked as 0°. Flow direction characteristics are analyzed as the supporting pillar is rotated in the wind tunnel.

Environmental characterization

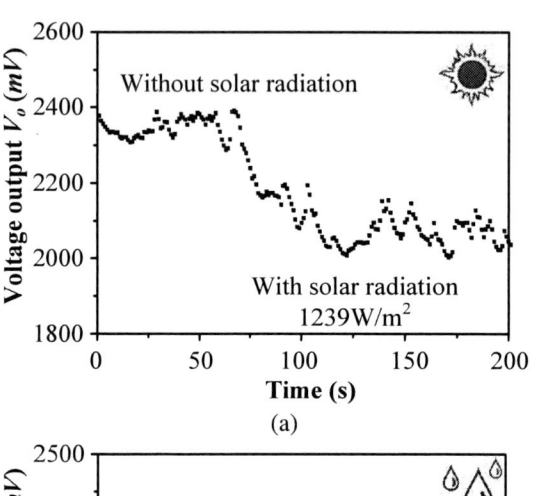

(a)

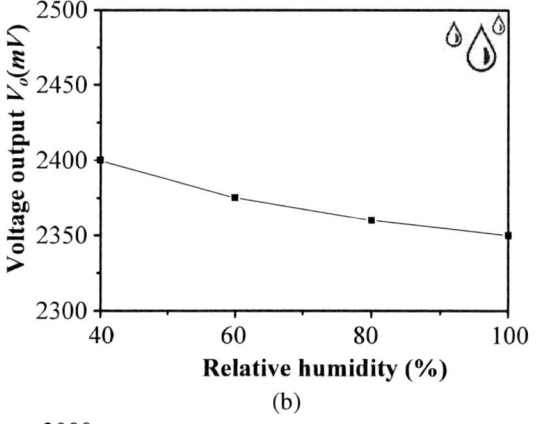

(b)

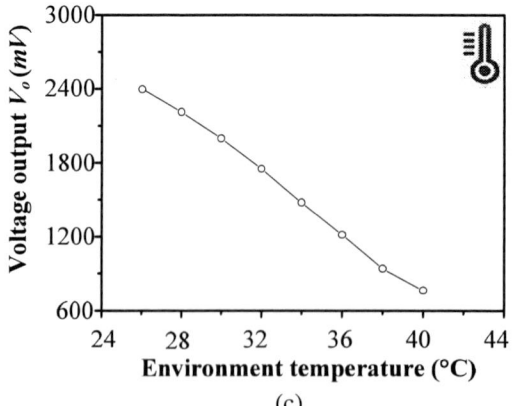

(c)

Figure 6: The influence of environmental factors on sensor outputs: (a) sola irradiation; (b) humidity; (c) temperature

Environmental characterization of sensor at different levels of solar irradiation, humidity, and temperature has been carried out in Figure 6(a), Figure 6(b), and Figure 6(c), respectively. Initial voltage is set at 2400mV. The errors introduced by the environmental factors are listed in Table 2. The influence of humidity is negligible. Strong solar irradiation of 1239W/m^2 will introduce error of 8%. Therefore airflow sensor should be installed in place in the shade. Temperature fluctuation has a significant influence on the sensor output. Change of 1°C will introduce error of 5%. Therefore, temperature compensation is needed for accurate measurement.

Table 2: Error introduced by environmental factors

Environment	error
Solar irradiation	8%@1239W/m^2
Humidity	0.03%/RH%
Temperature	5%/°C

Wireless sensor system for field test

Three Pt thin-film temperature sensors are incorporated into the sensing system for temperature compensation, as shown in Figure 7. The overall wireless system consists of a MEMS-based airflow sensor network and an android phone. The airflow sensor network uses star topology, in which the data of airflow sensors can reach the android phone within its one-hop distance via Bluetooth transimission. In other words, no multi-hop routing protocols are utilized in the sensor network. The airflow sensor reports airflow direction and velocity periodically. The Android phone collects the information from the airflow sensor network passively.

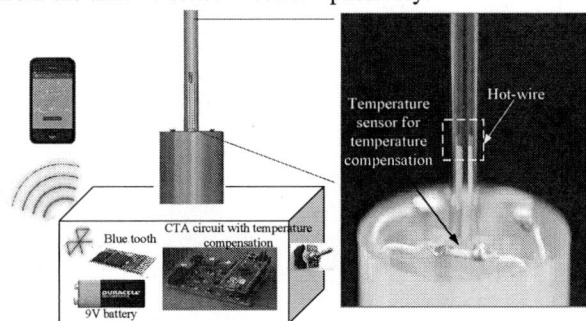

Figure 7: Wireless sensor system integrated with temperature compensation function

After the tests, wind speed and direction were calculated using the analytical procedure for every second in the time frame. As shown in Figure 8, the measurement results by the MEMS based anemometer were compared with those recorded by the commercial Vaisala ultrasonic anemometer. From the figures we can see that the results of the MEMS based anemometer and the Vaisala anemometer agree well with each other. The achieved mean error of direction is 13.5° and the mean error of speed is 0.02m/s.

The mean error, especially of the direction, is caused by slower response of hot-wire anemometer than ultrasonic anemometer, leading to some sudden changes displayed in the measurement results of the ultrasonic anemometer are not captured by micro triple-hot-wire anemometer

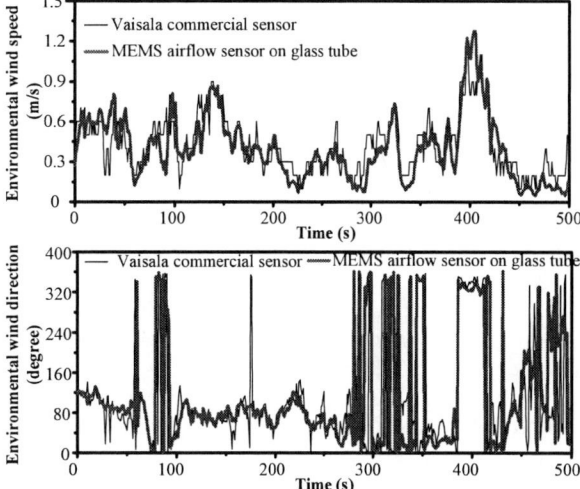

Figure 8: The comparison of wind speed and direction between Vaisala ultrasonic anemometer and MEMS sensor on glass tube

ACKOWLEDGEMENT

The research was funded by the Singapore National Research Foundation (NRF) through the Singapore-MIT Alliance for Research and Technology (SMART) Center for Environmental Sensing and Modeling (CENSAM).

REFERENCES

[1] J. Chen and C. Liu, "Development and characterization of surface micromachined, out-of-plane hot-wire anemometer," *J. Microelectromech. Syst.*, vol. 12, pp. 979-988, December 2003

[2] H.B, Liu, N. Lin, S.S. Pan, J.M. Miao, and L.K. Norford, " High sensitivity, miniature, full 2-D anemometer based on MEMS hot-film sensors", *IEEE Sensors J.* vol.13, pp.1914-1920, 2013

[3] S.W. Liu, S.S. Pan, F. Xue, L. Nay, J.M. Miao and L. K. Norford, "Optimization of hot-wire airflow snsors on an out-of-plane galss bubble for 2-D detection," *J. Microelectromech. Syst.*, October 2014

[4] Z. Yang, Y. Zhang, T. Itoh, R. Maeda, "Flexible Implantable Micro Temperature Sensor Fabricated on Polymer Capillary by Programmable UV Lithography with Multi-layer Alignment for Biomedical Applications", *J. Microelectromech. Syst.*, vol. 23, pp.21-29, 2013

CONTACT

* J.M. Miao, Tel: 65-6790 6038; mjmmiao@ntu.edu.sg

MULTI ROOF TILE-SHAPED VIBRATION MODES IN MEMS CANTILEVER SENSORS FOR LIQUID MONITORING PURPOSES

Georg Pfusterschmied[1], Martin Kucera[1,2], Víctor Ruiz-Díez[3], Achim Bittner[1], José Luis Sánchez-Rojas[3], Ulrich Schmid[1]

[1]Institute of Sensor and Actuator Systems, Vienna University of Technology, AUSTRIA
[2]Austrian Center of Competence for Tribology, AC2T research GmbH, AUSTRIA
[3]Group of Microsystems, Actuators and Sensors, E.T.S.I. Industriales, Universidad de Castilla-La Mancha Ciudad Real, SPAIN

ABSTRACT

We realized piezoelectrically self-actuated self-sensing cantilever sensors for liquid monitoring purposes excited in higher orders of the roof tile-shaped mode. This advanced class of vibration mode supports very high Q-factors in liquid media and high volume strain values which result in combination with an optimized electrode design in enhanced strain related conductance peaks. Therefore, precise fluid property measurements even for highly viscous liquids like D500 (~ 430 cP) are feasible.

INTRODUCTION

Since many years, the demand of micromachined cantilever-based sensors, being capable to measure physical properties such as density and viscosity [1-3], is continuously increasing. The fabrication process used in this work is based on standard silicon technology, allowing smaller package size, lower costs due to large-scale integration and low power consumption. This makes it well suited for providing sensors not only for laboratory equipment but even for complete new types of mobile low-power systems. Latest research has introduced a special class of excitation mode exceeding the overall performance of commonly used out-of-plane vibration modes, showing the highest Q-factor of a cantilever-based MEMS resonator in liquid media up to now [4]. The cantilever sensors realized in Ref. [4-5] use two electrode pairs, covering half of the sensor surface, allowing in-parallel and anti-parallel actuation exciting apart from standard in-plane [6] and out-of-plane [7] modes either odd (e.g. 1st) or even (e.g. 2nd) roof-tile shaped modes [4]. This new class of vibration modes (see Figure 1) can be described as a transversal out-of-plane vibration mode with a free-free boundary condition along the length of a single-sided clamped beam. Considering Leissa's nomenclature [8] by counting the number of nodal lines in x- and y-direction, the multi roof tile-shaped modes are named in the following text 1X-mode. The 12-mode (Figure 1 (a)) e.g. has two longitudinal nodal lines, whereas the 13-mode (Figure 1 (b)) has three nodal lines, the 14-mode (Figure 1 (c)) already four nodal lines and so on. When exciting higher order modes, due to the increase in nodal lines, the number of areas with different curvature and thus, different sign of volumen strain, grows. Consequently, a standard electrode design covering the complete cantilever surface used in Ref. [4] acts as a filter for higher modes [9]. This belongs to the partial cancellation of surface charges caused by electrodes covering areas with opposite volume strain.

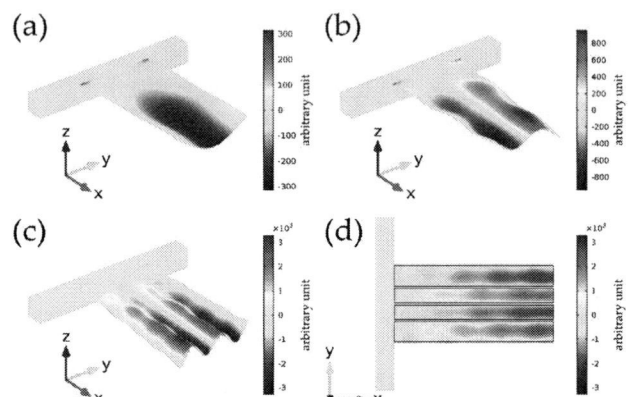

Figure 1: Visualization of the (a) fundamental (1st) roof tile-shaped modal shape (12-mode), (b) the 2nd order of the roof tile-shaped modal shape (13-mode) and (c) the 4th order of the roof tile-shaped modal shape (15-mode). (d) Top view on the 15-mode with tailored electrode design. The colored areas on the cantilever surface represent the local volume strain distribution.

In this work we present an optimized electrode design for micromachined self-sensing and self-actuated aluminum-nitride (AlN) cantilevers excited in higher orders of the 1X-mode. The electrode design allows an anti-parallel actuation (+-+-), which increases the deflection and prevents charge compensation resulting in very high strain related conductance peaks. A complete set of experimental results in combination with finite element method (FEM) simulations are presented, which provides design guidelines for adapting this method to other types of resonator-based sensors.

DEVICE FABRICATION

The micromachined cantilevers (see Figure 2) used in this work are fabricated on 4 inch SOI wafers and are based on the fabrication procedure reported in Ref. [10]. The cantilevers with a length of $L = 2524$ µm and a width of $W = 1274$ µm are originally designed for an in-plane study [2] to achieve the same in-plane resonance frequency as a 50×500 µm² cantilever, but with a different scaled aluminum-nitride area. More information on this electrode design is given in Kucera *et al.* [5]. The electrode design shown within this article is optimized for the 15-mode and results in four anti-parallel (+-+-) connected electrode stripes to ensure a collection of all charges without cancellation. After the finished fabrication process the die is packaged in a 24-DIP (Dual in-line package) and bonded via gold wires (see Figure 2).

Figure 2: Close up view of the in-house fabricated silicon die (6 x 6 mm²), containing a released cantilever and a non-released counterpart for parasitic effect compensation purposes (not used in this work) with the top area dimensions of 2524 x 1274 µm². This cantilever uses an optimized electrode patterning, considering the volume strain of the modal shape presented in Figure 1 (c) and (d).

DEVICE CHARACTERIZATION

Figure 1 presents the results of finite element method (FEM) eigenmode analyzes for higher orders of the roof tile-shaped mode, starting with the 12-mode (a), the 13-mode (b) and the 15-mode (c). The color scale represents the volume strain and thus, the piezoelectrically generated surface charge distribution. This figure illustrates the increase of longitudinal nodal lines when increasing the order of the mode.

The results from the FEM eigenmode analyzes are used for optimizing the electrode pattering. This is indicated for the 15-mode by the top view shown in Figure 1(d). Figure 2 depicts a typical die layout after packaging and wire-bonding including a released cantilever and a non-released counterpart, both with the optimized electrode stripes for 15-mode. The non-released counterpart is not used in this study.

Figure 3 compares the cantilever deflection and electrical behavior in the "non-optimized" anti-parallel configuration (++--) and the "optimized" anti-parallel configuration (+-+-). It can be seen that a tailored electrode design, which considers the modal shape of the 15-mode, achieves a 10 times higher deflection value and a 100 times higher conductance peak. In contrast, the 15-mode without "non-optimized" anti-parallel connection (++--) suppresses this vibration mode almost completely.

This follows from the charge cancellation due to almost equally strained areas of opposite sign, resulting in the presence of contrary charge polarities (see Figure 1 (c) and (d)). Beside the 15-mode at 990 kHz, the 25- and the 35-mode can be detected at 1020 kHz and 1070 kHz, respectively. From these results, it is concluded that even for the 25- and 35-modes, the optimized electrode actuation shows superior performance when compared to the non-optimized.

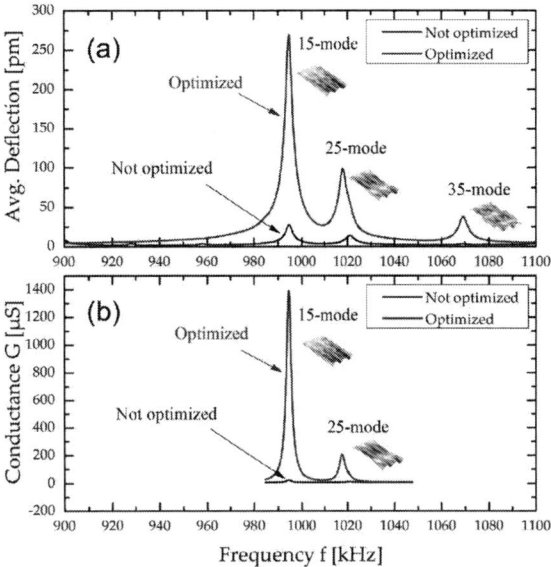

Figure 3: Optical (a) and electrical (b) characterization of the 4th order (15-mode) in air, exciting the proposed cantilever device in optimized (+-+-) and non-optimized (++--) configuration. The deflection spectrum is measured with a Polytec laser Doppler vibrometer MSV-400 whereas the electrical characterization is performed with an Agilent impedance analyzer 4294A. The side peaks show higher harmonics of the new vibration mode, which are the 25-mode and the 35-mode, respectively.

Figure 4 presents the quality factor as a function of the inverse square root of the viscosity-density product for several liquids (Isopropanol, N10, N100, D500) with dynamic viscosities up to ~ 430 cP (see Table 1).

Table 1: Density and dynamic viscosity values for the used liquids, determined with Stabinger viscometer SVM3000 and with the Ubbelohde-Walther equation at 27°C.

Name	Dynamic viscosity [cP]	Density [g/ml]
Isopropanol	1.98	0.780
N10	13.50	0.846
N100	181.62	0.862
D500	432.96	0.867

For the following characterization, two different devices were used, one optimized for the 12/13-mode (2 electrodes) and one optimized for the 15-mode (4 electrodes). The results in Figure 4 show the increase in quality factor when exciting higher orders of the 1X-

978-1-4799-7956-1/15 $31.00 © 2015 IEEE   719

mode. This enhancement in quality factor is related to the increase in resonance frequency for higher modes, leading to a very high quality factor of 120 in isopropanol when excited in the 15-mode. Compared to literature, the slope in the quality factor for the 15-mode exceeds also the overall performance of standard out-of-plane actuated resonators [4, 11] promising an enhanced sensor sensitivity.

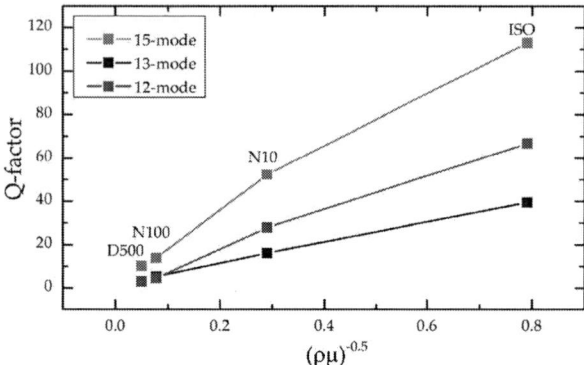

Figure 4: Electrical characterization of two different cantilevers in different liquids (isopropanol and viscosity standards N10, N100 and D500). The 12- and 13-mode are excited with only 2 electrode stripes connected in-parallel (++) as in Ref. [4] and in anti-parallel connection (+-) as in Ref. [5], respectively. The 15-mode is excited with optimized actuation (+-+-).

CONCLUSION

This paper investigated for the first time higher orders of the roof tile-shaped out-of-plane vibration mode (1X-modes) in piezoelectrically actuated self-sensing MEMS cantilever sensors for liquid monitoring purposes. This advanced class of vibration modes supports very high Q-factors in liquid media and very high volume strain values, which results, in combination with an optimized electrode design, in very high strain related conductance peaks as demonstrated for the 4th order mode (15-mode). These features predestinated this superior class of vibration modes for a large variety of challenging resonator-based sensing applications in liquid media, exceeding the overall performance of commonly used out-of-plane vibration modes. The generation of the conductance peak was discussed, showing a great potential for increasing by introducing electrode stripes which are optimized for the specific modal shape. Finite element method (FEM) eigenmode analyses for higher orders of the roof tiled-shaped mode were presented, starting with the 12-mode, the 13-mode and the 15-mode. The visualized partial cancellation of surface charges was presented for the 15-mode by a XY-plot with indicated electrode pairs. A comparison of the cantilever deflection and the electrical behavior in the "non-optimized" anti-parallel configuration (++--) and in the "optimized" anti-parallel configuration (+-+-) was shown. It could be revealed that a tailored electrode design, which considers the modal shape of the 15-mode, achieves a 10 times higher deflection value and a 100 times higher conductance peak. In contrast, the 15-mode with "non-optimized" anti-parallel connection (++--) suppressed this

vibration mode, due to charge compensation, almost completely. The quality factor was evaluated for the 12-, 13- and 15-mode in several liquids (Isopropanol, N10, N100 and D500) showing superior performance of the 15-mode (quality factor about 120 in isopropanol) followed by the 13-mode and the 12-mode. It was even possible to excite the 15-mode in a liquid of 430 cP (D500) with a quality factor greater than 10. The improved performance of the resonator, shown in this letter, predestines the use of the 15-mode for challenging chemical bio-sensing, but also for atomic force microscopy (AFM) in liquid media.

ACKNOWLEDGEMENTS

This work has been supported by the Austrian Research Promotion Agency within the COMET-K2 Project XTribology (Project-No. 824187). The financial support given by the Spanish Ministerio de Economia y Competitividad: Project Ref. DPI2009-31203, FPI Grant (Ref. BES-2010-030770) awarded to Tomas Manzaneque and FPU Grant (Ref. AP2010-6059) awarded to Victor Ruiz is gratefully acknowledged.

REFERENCES

[1] C. Riesch, E.K. Reichel, A. Jachimowicz, J. Schalko, P. Hudek, B. Jakoby, F. Keplinger, "A suspended plate viscosity sensor featuring in-plane vibration and piezoresistive readout", in *J. Micromech. Microeng.*, vol. 19, 2009.

[2] M. Kucera, "Performance of cantilever-based piezoelectric MEMS resonators in liquid environment", PhD-Thesis, Vienna University of Technology, 2014

[3] C. Riesch, E.K. Reichel, F. Keplinger, B. Jakoby, "Characterizing vibrating cantilevers for liquid viscosity and density sensing", in *J. Sens. Volume 2008*.

[4] M. Kucera, E. Wistrela, G. Pfusterschmied, V. Ruiz-Díez, T. Manzaneque, J. L. Sánchez-Rojas, J. Schalko, A. Bittner, and U. Schmid, "Characterization of a roof tile-shaped out-of-plane vibrational mode in aluminum nitride-actuated self-sensing micro-resonators for liquid monitoring purposes", in *Appl. Phys. Lett.*, vol. 104, 2014.

[5] M. Kucera, E. Wistrela, G. Pfusterschmied, V. Ruiz-Díez, T. Manzaneque, J. Hernando-Garcia, J.L. Sánchez-Rojas, A. Jachimowicz, J. Schalko, A. Bittner and U. Schmid., "Design-dependent performance of self-actuated and self-sensing piezoelectric-AlN cantilevers in liquid media oscillating in the fundamental in-plane bending mode.", *Sens. Actuator B*, vol. 200, pp. 235-244, 2014.

[6] L. A. Beardslee, A. M. Addous, S. Heinrich, F. Josse, I. Dufour, and O. Brand, "Thermal Excitation and Piezoresistive Detection of Cantilever In-Plane Resonance Modes for Sensing Applications", in *J. Microelectromech. Syst.*, vol. 19, pp. 1015-1017, 2010.

[7] T. Manzaneque, J. Hernando, L. Rodriguez-Aragon, A. Ababneh, H. Seidel, U. Schmid, and J. L. Sanchez-Rojas, "Analysis of the quality factor of

978-1-4799-7956-1/15 $31.00 © 2015 IEEE

AlN-actuated micro-resonators in air and liquid", in *Microsyst. Technol.*, vol. 16, pp. 837-845, 2010

[8] A. W. Leissa, "Vibration of Plates", in *Scientific and Technical Information Division, National Aeronautics and Space Administration*, NASA SP-160, 1969.

[9] J. L. Sanchez-Rojas, J. Hernando, A. Donoso, J. C. Bellido, T. Manzaneque, A. Ababneh, H. Seidel, and U. Schmid, "Modal optimization and filtering in piezoelectric microplate resonators.", in *J. Micromech Microeng.*, vol. 20, 2010

[10] M. Kucera, F. Hofbauer, E. Wistrela, T. Manzaneque, V. Ruiz-Díez, J.L. Sánchez-Rojas, A. Bittner, U. Schmid. "Lock-in amplifier powered analogue Q control circuit for self-actuated self-sensing piezoelectric MEMS resonators.", in *Microsyst. Technol.*, vol. 20, pp. 615-625, 2014.

[11] T. Manzaneque, V. Ruiz, J. Hernando-García, A. Ababneh, H. Seidel and J. L. Sánchez-Rojas, "Characterization and simulation of the first extensional mode of rectangular micro-plates in liquid media", in *Appl. Phys. Lett.*, vol. 101, 2010.

CONTACT

*G. Pfusterschmied, tel: +43-58801-36649; georg.pfusterschmied@tuwien.ac.at

ON-CHIP PRESSURE SENSING BY VISUALIZING PDMS DEFORMATION USING MICROBEADS

Chia-Hung Dylan Tsai* and Makoto Kaneko

Department of Mechanical Engineering, Osaka University, Suita, JAPAN

ABSTRACT

A novel pressure sensing technique based on visualizing Polydimethylsiloxane (PDMS) deformation using microbeads is proposed here for measuring local pressure inside a microfluidic device. By the proposed method, the pressure can be directly "seen" without attaching any wire foils, such as a strain gauge, nor complex fabrication process, such as multilayer design or surface grating. Experimental results are shown and analyzed based on brightness value from captured images of microbeads pattern. The developed sensor is firstly calibrated by a commercial pressure sensor with feedback controlled syringe pump connected externally. According to the experimental results, the proposed sensing method is stable and repeatable in the steady state under dynamic pressure change, and the variation for the same given pressure from time to time is less than 1%. The correlation, R, between the pressure obtained from the proposed method and the reference pressure connected outside is up to 0.9953.

INTRODUCTION

On-chip pressure measurement for local pressure inside a microchannel is in great demand for microfluidic applications because even a small change of pressure inside a microchannel could significantly affect the accuracy of cell manipulation or the evaluation of cell properties. For example, the transit time of a cell passing through a constriction channel under constant pressure difference has been used as an index of cell deformability.[1][2][3] The evaluation results become inadequate if the pressure is changed and no longer a constant. A common instance of local pressure change happens while a cell is squeezing into a constriction channel. Since the channel is partially blocked by the cell and the flow resistance changes, it is expected to have a sudden change of pressure which could significantly affect cell motion as well as evaluation result. A reasonable solution to this case is to neglect the entering phase and only focusing on the equilibrium phase of cell motion inside the constriction for stiffness-based deformability.[1] However, the solution still cannot help obtain viscosity-based deformability. Thus, information of local pressure change is necessary, and that is the main motivation of the proposed local pressure sensing technique.

Figure 1 illustrates the overview of the proposed idea and setup. Figure 1(a) shows the overview of the validation system where one end of the Y-shape connector is connected to a controlled pressure source, and the other two ends are connected to the inlets of the proposed sensing channel. Figure 1(b) illustrates a pattern formed by the microbeads inside the sensing channel. The volume inside the channel changes due to the deformation, and the resulting fluid flow changes the

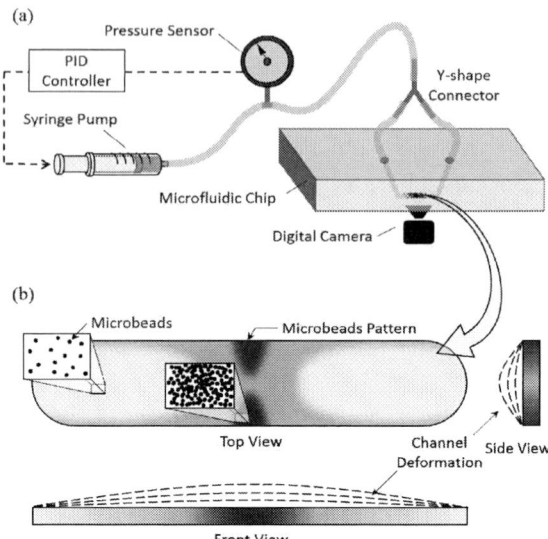

Figure 1: *The key idea of the local pressure sensing. (a) A pressure sensing channel is fabricated inside a microfluidic chip with calibration by a feedback controlled pressure system connected externally. (b) The pattern of microbeads in the channel changes with respect to applied pressure according to the amount of PDMS deformation.*

pattern correspondingly. According to the microbeads pattern, the pressure inside the channel can be estimated.

Simulations and experiments have be conducted according to the proposed idea. In simulation, finite element method (FEM) is applied for realizing the PDMS deformation under a given pressure with PDMS properties. In experiments, the brightness value of the center column of the pattern is firstly calibrated with fixed values of pressure. Next, a periodic pressure change between 100 [kPa] to 200 [kPa] is applied from the external pressure source, and the dynamic response of the proposed sensing method is tested and discussed.

RELATED WORKS

Different devices for measuring the local pressure inside a microfluidic device have been developed. For example, one of the common approaches is to measure local pressure by the pattern fringes formed from the grating on the surface of microfluidic device while the device is deformed due to different pressure.[4][5] In order to generate optical diffraction for the patterns, an additional light source is needed in this approach. Another approach is to estimate the pressure using trapped air compression.[6] The problem with this method is that it is not applicable for the microfluidic device made of permeable substrates, such as PDMS. Another group coated piezoelectric material at the specific location of

978-1-4799-7956-1/15 $31.00 © 2015 IEEE

TABLE I *SIMULATION PARAMETERS*

PROPERTY	VALUE
Young's Modulus	$7.5 \times 10^5 [Pa]$
Poisson Ratio	0.5
PDMS Density	$9.2 \times 10^2 [kg/m^3]$
Applied Pressure	$1.0 \times 10^5 [Pa]$

measuring the local pressure by electrical signals. However, these methods usually requires additional instruments, such as additional light source, mask alignment for multi-layer microfluidic design, or is too complex to be directly implemented into exist on-chip applications.

THE IDEA AND SIMULATION

The key idea of the proposed method is to visualize the deformation of the microfluidic device with microbeads patterning due to the local pressure inside the device. While the fluid pressure is increased from the external system, the microchannel made of PDMS is deformed, and the microbeads are moved with flow toward the symmetric center of the channel geometry. As a result, the density of the microbeads at the center of the channel increases and the brightness value decreases. On the other hand, the density of the microbeads around the inlets of fluid decreases due to the additional fluid flow in. Different patterns of microbeads are expected to be formed by different amount of applied pressure, and thus, the patterns can be used for measuring pressure inside microfluidic channel. The advantages of the proposed method are simplicity, electricity-free and requiring no additional instrument but only microscope vision which already implemented in most of microfluidic application. All it needs is a microchannel filled up with microbeads, and both open ends connect to the point of interest.

Figure 2 shows the simulation of deformation for a microchannel fabricated inside a PDMS microfluidic device. Figure 2(a) shows the geometry of the PDMS device in 3D space. The device is a 30 [mm] by 20 [mm] block with a microchannel located at the center of the bottom surface. The length, width and height of the microfluidic channel are 5, 1 and 0.01 [mm], respectively. For simulating the actual microchannel environment, two cylindrical holes of inlets are penetrating from the top to the bottom of the device as shown. The PDMS properties used for the simulation are tabulated in Table I.[6] Figure 2(b) shows the top view and side view of the channel deformation while applying the a given pressure of 200 [kPa]. The cross-sections in Figure 2(b) are indicated in Figure 2(a) as S_1 and S_2, respectively. The color bar on the left specifies the amount of deformation under a given pressure. Figure 2(c) shows the numerical results of the deformation along the central line of the channel ceiling, the line $\overline{AA'}$ in Figure 2(b). According to the results, the maximum deformation on the ceiling of the channel is almost 0.1 [mm], which is about ten times to its original height, 0.01 [mm]. Such great deformation provides decent sensitivity for pressure sensing.

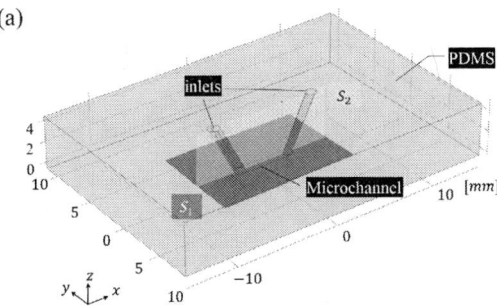

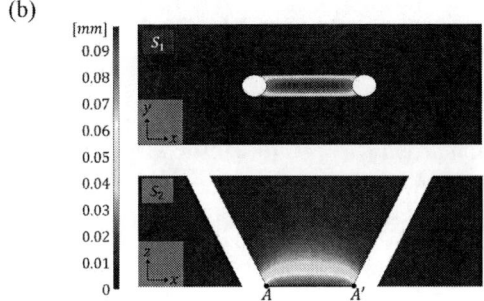

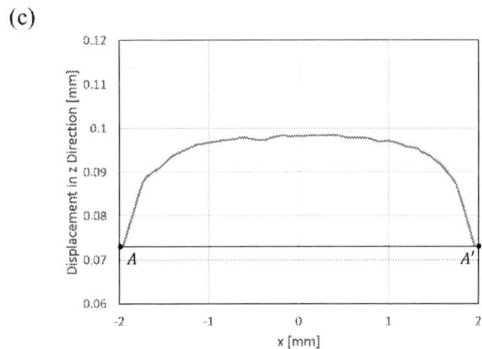

Figure 2: *Simulation of PDMS channel deformation under static pressure in COMSOL. (a) The model for simulation. (b) The displacement in z-direction on the surfaces S_1 and S_2. (c) The displacement on the ceiling of the microchannel.*

EXPERIMENTS

Experimental System

Figure 3 show the experimental system, which includes a microfluidic chip made of PDMS, a syringe pump (SGE Corp.), a pressure sensor (FP101A, COPAL ELECTRONICS Inc.), a microscope (IX71, OLYMPUS Corp.) and a camera (IDP, PHOTRON Inc.). The geometries of the microfluidic channel inside the chip is the same as the one used in the simulation whose length, width and height are 5, 1 and 0.01 [mm], respectively. The pressure sensor and syringe pump are read and controlled by the computer to perform as a controllable pressure source. A Y-shape connector is connecting the pressure source and two inlets of the microchannel, so the applied pressure are simultaneously propagated into the channel from the two inlets. The patterns and microbeads

978-1-4799-7956-1/15 $31.00 © 2015 IEEE

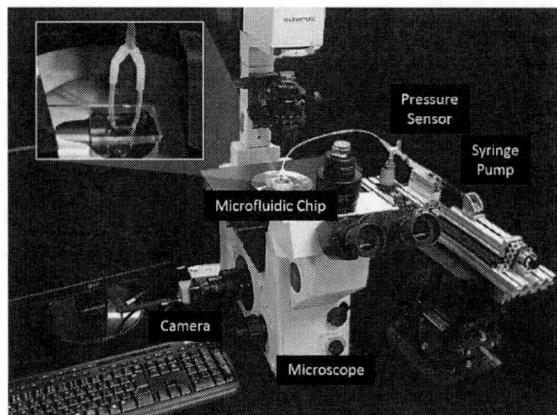

Figure 3: *Experimental setup, which includes a microfluidic chip, a pressure sensor, a syringe pump, a microscope and a camera.*

motion are captured by the video camera through the microscope.

Experimental Procedure and Results

Figure 4(a) shows both the target pressure and the measured pressure, and they are in red dashed line and blue curves, respectively. The pressure is specified to change from 100 to 200[kPa]. According to the pressure reading from the external pressure sensor, we can find that the actual pressure reaches to 200 [kPa] from 100 [kPa] in about 40 [ms], and settles at 200 [kPa] in about 100 [ms]. Figure 4(b)-(e) shows the corresponding patterns of microbeads at 0, 33, 67 and 100 [ms], respectively. The diameter of each microbeads is 1[μm]. As mentioned previously, the pressure are applied to the channel from two sides of channel, so that the microbeads are pushed to the center, and form a darker region with high density of microbeads. It is noted that the clearest change of the patterns is around the area around the symmetric center of the channel. Therefore, we focus on the middle column for pressure estimation and calibration.

Figure 5(a) shows the mean value of brightness over each column from two captured images under given pressure of 100 and 200 [kPa]. The mean brightness value is denoted by h. The h value at the middle column, where $x=0$ is selected for pressure sensing, and is linearly calibrated with two reference pressure values, 100 and 200 [kPa], while the patterns are settled. The linear equation for the calibration is

$$P = C_0 h + C_1$$

where P, h, C_0 and C_1 are the calibrated pressure, mean brightness value over a column, and two coefficients, respectively. In this case, we obtained

$$C_0 = -1.22 \quad and \quad C_1 = 325.86$$

Figure 5(b) shows the comparison between the calibrated pressure value, from the proposed method, and the reference pressure, from the commercial pressure sensor outside the PDMS chip. Even under the dynamic

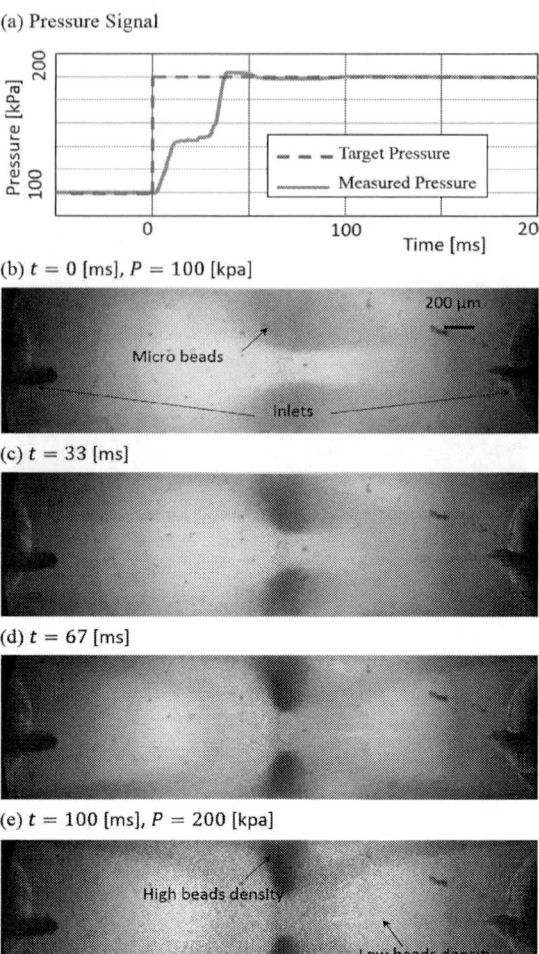

Figure 4: *Experimental results. (a) The target for the controller and the measured pressure from the pressure sensor. (b)-(e) The microbeads inside the channel forms different patterns while pressure is changed from 100 [kPa] to 200[kPa].*

change of pressure, two measures match well with each other with about 0.2 [sec] lag during the transition from 200 to 100 [kPa]. The variation of the proposed sensing method in the steady state is less than 1%, and that indicates the proposed method is stable with low noise ratio. On the other hand, the responses in the transient state show similar tendency between the reference pressure and calculated pressure, that is, they both show underdamped-like response after pressure rising from 100 [kPa] to 200[kPa] and overdamped-like response after pressure falling from 200[kPa] to 100[kPa] as circled in Figure 5(b).

A special note would like to be added here that the reason of 0.2 [sec] lag is not necessary coming from the nature of the proposed sensing method but other possibilities. One possibility is that the commercial pressure sensor which measures the reference pressure measured, is located a few centimeters away from the PDMS chip. The distance between the pressure sensor and the PDMS chip could also be the reason causing such a delay due to deformation of connecting tubes. Since the reference pressure is not necessary the actual pressure inside the microchannel, the current setup is limited for further evaluating the dynamic response of the proposed sensing method.

Although the current experimental setup has limitation on evaluating dynamic response of the proposed sensing method as pointed out in previous paragraph, it is still capable to evaluate the static response of the method. Figure 5(c) shows the correlation between calculated and reference pressure under static pressure from 100[kPa] to 220[kPa] with the increment of 20[kPa]. Linear regression is applied to the data, and the fit function is shown in Figure 5(c) with $R^2 = 0.9906$. The correlation between the reference pressure and calculated pressure under given static pressures is 0.9953.

CONCLUSION

A new pressure sensing method for local pressure inside a microfluidic device is proposed and developed in this work. Three concluding remarks are listed as follows:

1. The fluid pressure on a chip can be directly seen from the microbeads pattern without any additional devices.
2. In this work, the mean brightness value of the central column of the pattern is used for pressure sensing. The proposed method is shown stable under a given static pressure with variation less than 1%.
3. The propose method can catch the tendencies of dynamic responses during the transient phase while the correlation under static responses is 0.9953.

ACKNOWLEDGEMENTS

This work is supported by Grant#23106003 and Grant#26820086 of The Ministry of Education, Culture, Sports, Science and Technology (MEXT) of Japan.

REFERENCES

[1] C. D. Tsai, S. Sakuma, F. Arai and M. Kaneko, "A New Dimensionless Index for Evaluating Cell Stiffness-Based Deformability in Microchannel", IEEE Transactions on Biomedical Engineering, vol.61, no.4, pp1187-1195, 2014.
[2] A. Adamo, A. Sharei, L. Adamo, B. Lee, S. Mao and K. F. Jensen, "Microfluidics-Based Assessment of Cell Deformability", Analytical Chemistry, vol.84, no.15, pp6438-6443, 2012.
[3] Y. Zheng, E. Shojaei-Baghini, A. Azad, C. Wang and Y. Sun, "High-throughput biophysical measurement of human red blood cells", Lab on a chip, vol.12, no.14, pp2560-2567,2012.
[4] K. Hosokawa, K. Hanada and R. Maeda, "A polydimethylsiloxane (PDMS) deformable diffraction grating for monitoring of local pressure in microfluidic devices", Journal of Micromechanics and Microengineering, vol.12, pp1-6, 2002.
[5] S. Foland, K. Liu, D. MacFarlane and J. Lee, "High-sensitivity microfluidic pressure sensor using a

(a) Mean brightness value over each column

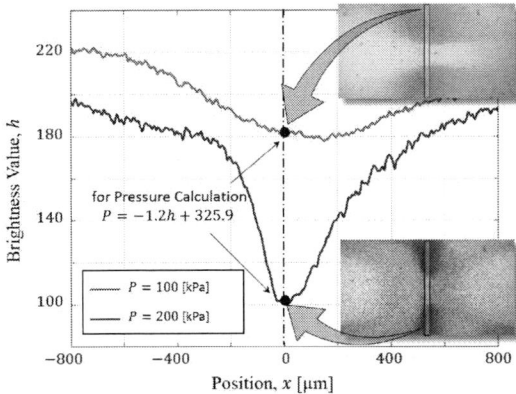

(b) Dynamic performance

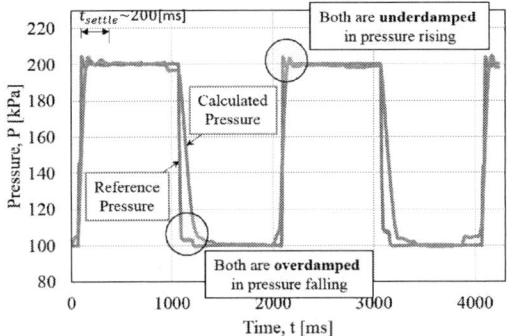

(c) Static response

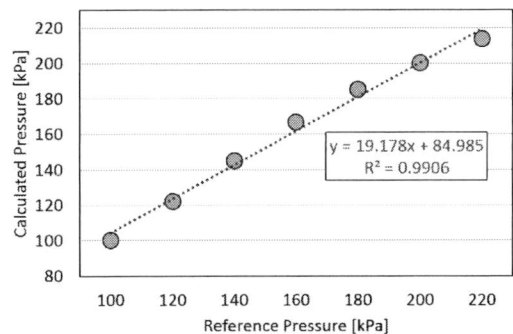

Figure 5: *Analysis and comparison to the external reference pressure from a commercial sensor. (a) The mean value of the brightness, h, is used for pressure calculation. (b) The dynamic response of the calculated and reference pressure. (c) The correlation between the calculated and reference pressure under static pressure.*

membrane-embedded resonant optical grating", IEEE SENSORS 2011, pp101-104.
[6] D. Armani, C. Liu and N. Aluru, "Re-configurable fluid circuits by PDMS elastomer micromachining", IEEE MEMS 1999, pp222-227.

CONTACT

*Chia-Hung Dylan Tsai, tel: +81-6-6879-7333; tsai@hh.mech.eng.osaka-u.ac.jp

3-AXIS ALL ELASTOMER MEMS TACTILE SENSOR

Alexi Charalambides[1], Jian Cheng, Teng Li, and Sarah Bergbreiter[1]*

Department of Mechanical Engineering

[1]Institute for Systems Research

University of Maryland, College Park, MD 20742, USA

ABSTRACT

This paper reports the first 3-axis (normal and shear force) all-elastomer capacitive MEMS tactile sensor. A multiphysics finite element model was developed and was used to tailor sensor geometry for high shear force sensitivity. Sensor area was 1.5 x 1.5 mm and used vertical capacitive structures with 20 μm electrode gaps to achieve high shear force sensitivities of 8.8 fF/N, shear force resolutions of 50 mN, and shear range of more than 2000 mN, with a normal force sensitivity of 0.9 fF/N. Fabrication utilized a simple elastomer molding process with reusable DRIE silicon molds for inexpensive manufacturing.

INTRODUCTION

For over 30 years, scientists and engineers have attempted to solve the tactile sensing challenges related to robotic grippers [1], [2], [3]. Inevitably, these tactile sensors struggle with a wide array of issues, such as insufficient force range and/or resolution, high structural stiffness ill-suited for anthropomorphic appendages, and complex fabrication [4]. In addition, systems level challenges such as wiring and local processing remain scarce in the literature [4]. The ultimate goal of this work is to create a "robotic skin" with desirable characteristics, and to address the systems level challenges. This paper focuses on the challenge of creating high area density elastomer-based sensors that can sense both normal and shear forces in a mechanically robust way.

Recent work has sought to incorporate shear force sensing in addition to normal force due to its importance in slip detection. In addition, sensors have been made using microfabrication techniques to achieve high area density. Previous work by Hwang demonstrated a resistive-based 3-axis sensor on a polyimide substrate to measure reaction forces, but sacrificed sensitivity for force range [5]. Pyo also used a piezoresistive design with a range better suited for robot fingers but in a much larger form factor [6]. Lee used air-gap capacitors embedded in PDMS for 3-axis force sensing, but with very low ranges (10 mN) and relatively complex fabrication [7]. All of these sensors used a laminated out-of-plane 'bump' structure to transfer shear forces to the sensing elements.

The goal of this paper is to create a flexible 3-axis tactile sensor with high spatial density and force range for robotic fingertip sensing. It improves the state-of-the-art by using nonplanar electrode geometries for capacitive sensing to improve force range and sensitivity, while enabling 3-axis force transduction without the need of an out-of-plane 'bump' structure.

SENSOR DESIGN

Architecture

Capacitive tactile sensors have typically utilized

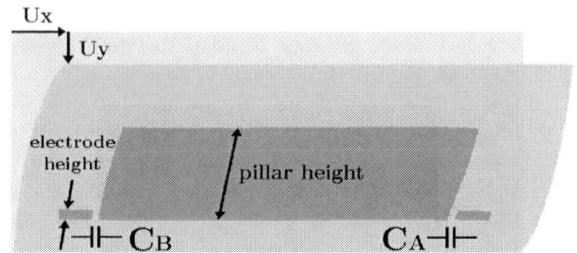

Fig. 1: Proposed sensor architecture. Displacements U_x and U_y induce unique changes in capacitance of C_A and C_B in order to detect normal and shear deformation modes.

parallel plate style electrodes, oriented orthogonal to the direction of applied force, to transduce normal force or pressure; as a normal displacement is applied to the sensor, the distance between the two plates decreases and therefore increases capacitance [7]. Naturally, this technique requires multi-layer assembly which can greatly increase fabrication complexity. Electrode materials are typically metal and can limit the overall sensor compliance and render it incompatible with magnetic resonance imaging (MRI). In addition, sensor architecture is further complicated by the need for an out-of-plane structure to induce asymmetric deformation when the sensor is subject to shear forces.

In order to circumvent these complications, the proposed tactile sensor utilizes conductive elastomeric features of varying heights, "pillars" and "electrodes", which enable unique shear and normal deformation modes detectable using capacitive sensing. Under shear loading, the pillar deforms towards one electrode and away from the other, Fig. 1, while under normal loading, the sensor flattens and the electrode gaps uniformly increase. Thus, the "capacitance differential" between the two electrode pairs, C_A and C_B, indicates the type of deformation occurring. These pairs of electrodes are placed in both shear directions, in order to achieve 3-axis force sensing (2 shear, 1 normal).

Multiphysics Finite Element Modeling

To guide design, 2D nonlinear large deformation finite element simulations were conducted using ANSYS to study the effect of sensor geometry on capacitance. An uncoupled multiphysics simulation was developed such that: first, the geometry was mechanically deformed (element type PLANE182), and secondly, the electric potential field was solved to find capacitance in the new configuration (element type PLANE121). Capacitance was solved using Eq. 1,

$$C = \frac{\int_\Omega \varepsilon |\overline{E}|^2 d\Omega}{V^2}$$

[1]

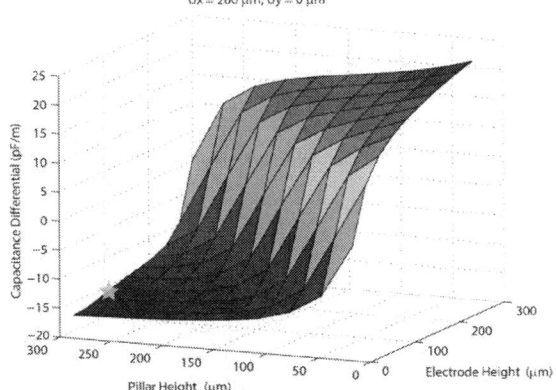

Fig. 2: Parametric study of the effects of sensor geometry on the capacitance differential when subject to shear displacement. Gold star is the geometry selected.

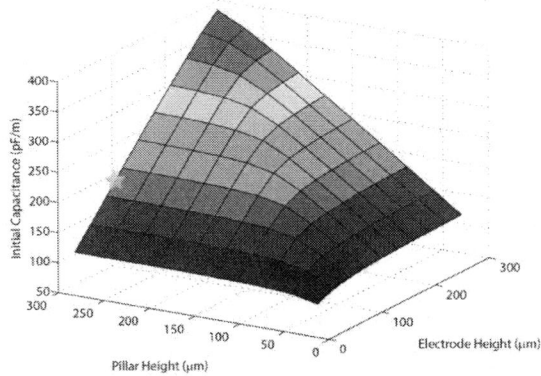

Fig. 3: Parametric study of the effects of sensor geometry on initial (undeformed) capacitance. Gold star is the geometry selected.

where ε is the dielectric constant, $\bar{E}$ is the electric field intensity, Ω is the volume of the dielectric, and V is voltage. Experimentally obtained Neo-Hookean hyperelastic constitutive models were used for a 10 wt.% carbon/polydimethylsiloxane (C/PDMS) composite and plain polydimethylsiloxane (PDMS), with shear moduli of 1.01 MPa and 0.54 MPa, respectively. For the electrical simulations, a dielectric constant of 2.5 was assumed for PDMS.

Parametric studies were conducted to find favorable electrode geometries that provided large capacitance differential under shear, Fig. 2. In previous work, high normal force sensitivity was related to small electrode gaps which translate to high initial capacitance [8]. Therefore, initial capacitance was also parametrically studied to improve normal force sensitivity, Fig. 3. Note that due to being a 2D simulation, the z-axis of both figures is in pF/m, and the actual capacitance can be found by multiplying by the out-of-plane electrode width, which in the presented design was 1 mm.

For shear sensing, two extreme regions were found: a tall pillar and short electrode (blue region), and vice versa (red region). In addition, when the pillar height equals the electrode height, a capacitance differential of zero is found (green region). Although the magnitude of the red region is larger than the blue, a geometry within the blue region was favorable due to fabrication reasons. By contrast, for normal sensing it was found that a large pillar height and electrode height lead to the highest initial capacitance. Due to the competing phenomena, a geometry which maintains both high shear force sensitivity and sufficient initial capacitance for normal force sensitivity was selected, as represented by the gold star in Fig. 2 and 3; electrode height was 100 μm, and pillar height was 300 μm.

MICROFABRICATION

Sensors were made from a reusable multi-depth silicon mold, which was fabricated using a two mask microfabrication process, Fig. 4. First, a silicon wafer was deposited with silicon dioxide, patterned, and etched with the first mask, which contained the pillar geometry. Then, a second pattern and partial oxide etching was done using the second mask, which contained the electrode and electrical lead geometries. Next, an oxygen plasma was used to clean the surface of photoresist, followed by a deep reactive ion etch (DRIE) to create the mold, which was finally coated with a DuPont amorphous fluoroplastic solution (Grade 400S2) as an antistick agent. This completed the reusable silicon mold. After each use, additional fluoroplastic could be applied as needed.

10 wt.% C/PDMS was prepared by mixing 1 g of carbon black powder (39724, Alfa Aesar), 8.18 g of PDMS base and 0.818 of PDMS curing agent (Sylgard 184, Dow Corning), and 22.5 g of hexane for 30 min. It was then poured on the mold, vacuumed for 2 min at 1 Torr, and planarized by hand using an industrial screen printing squeegee (Ryonet). After curing, a layer of PDMS was poured on the mold, vacuumed for 15 min at 1 Torr, and cured on a hot plate for 15 min at 120 C. Then, the elastomeric sensor was peeled from the wafer as one whole piece. Lastly, the sensing area was encapsulated in another layer of PDMS using the same process, and resulted in a total thickness of 1.06 mm. In the future, PDMS thickness can be decreased by spin coating instead of gravity leveling.

EXPERIMENTAL RESULTS

Test Setup

Testing was conducted by applying a displacement to the sensor and reading the resultant capacitances of each electrode pair as well as the reaction forces. Micron-scale displacements were applied using a Thorlabs PT3-Z8 3-axis stage equipped with a rapid prototyped delrin probe, with a square probe tip area of 3 x 3 mm. Capacitance was measured using an AD7745/46 evaluation board with a resolution of 0.1 fF at a sampling rate of 16 Hz. Lastly, forces were acquired using an ATI Nano17 transducer, and the assembled test setup can be seen in Fig. 5.

Results

Shear force sensitivity and resolution were determined by conducting incremental shear displacement tests; in this case, increments of 10 μm shear displacements were applied up to 100 μm. A linear capacitive response was observed up to 700 mN shear

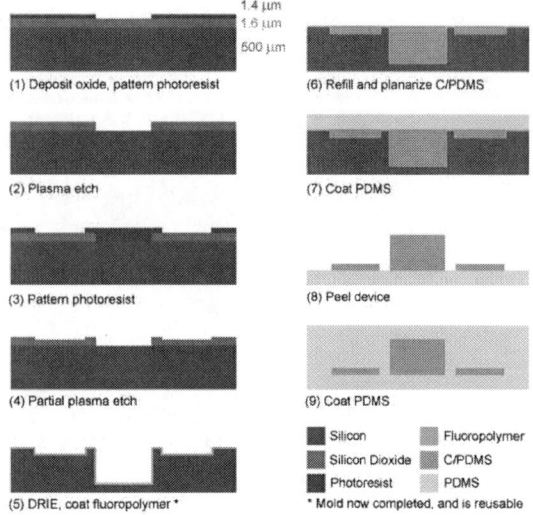

Fig. 4: Microfabrication flow chart detailing the creation of the silicon mold and subsequent polymer refilling.

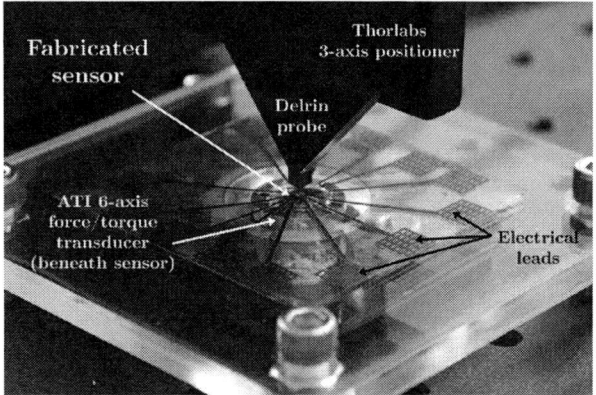

Fig. 5: Test setup of the Thorlabs and ATI equipment. The AD7745/46 capacitive board, not shown, has pen-style probes which may be pressed against the fabricated sensor's electrical leads to acquire data.

force, Fig. 6, and was found to agree well with the predicted (nonlinear) ANSYS behavior for Electrode A. As intended, Electrode A increased in capacitance, Electrode B decreased, while Electrode C remained relatively constant. A shear force sensitivity and resolution of 8.8 fF/N and 50 mN were calculated, respectively. In addition, highly repeatable behavior was found over four successive tests up to 2000 mN of shear force (300 μm shear displacement), Fig. 7.

Next, a normal displacement was applied to the sensor and shear testing was conducted again, Fig. 8. A net decrease in capacitance is apparent, consistent with the intended design and previous work [8]. Fig. 8 also demonstrates the clarity of the signal (ie: low noise) in the time domain. Normal force only tests were also conducted, Fig. 9, and a net decrease in each electrode pair was observed up to 10 N with an average sensitivity of approximately 0.9 fF/N. Lastly, cyclic testing was conducted for a 100 μm shear displacement, Fig. 10. At 100 cycles over 160 s, no hysteresis was observed, and the signal remained robust enough to even discern overshoot from the Thorlabs stage controller.

Limitations

One limitation was the needed normal force to enable shear force to be detected; about 1.8 N of normal force was applied during shear testing in order to ensure sufficient contact between the delrin probe and fabricated sensor. Also, due to tight space restrictions, each pillar had two electrodes min one shear direction, but only one in the other. This was due to the low conductivity of C/PDMS which needed significantly wider electrical leads, which limited the space available. Measured resistance from the location of the sensor to the end of the electrical leads was on average 850 k over a length of 3 cm and cross-section of approximately 1 mm x 100 μm (the electrical leads had a tappered crosssection). A high conductivity composite polymer, such as silver-polydimethylsiloxane (Ag/PDMS) [9], may potentially solve this issue; it was also utilized in previous work [8]. Another challenge was interfacing reliably with the conductive polymer surface. Traditional solder and copper wire does not bond to the polymer surface, so pen-style probes were used. Integrating processing with the elastomer skin will require a new connection approach.

CONCLUSIONS AND FUTURE WORK

This work demonstrated an all-elastomer, high area density tactile sensor capable of sensing in both normal and shear while alleviating many complications in previous tactile sensors. A finite element model was used to demonstrate the effects of sensor geometry on shear and normal force sensitivities. In addition, a simple microfabrication technique was used to rapidly and inexpensively create sensors, which achieved high shear and normal force ranges of 2 N and 10 N, respectively. Shear and normal force sensitivities of 8.8 fF/N and 0.9 fF/N were found, with shear force resolutions as low as 50 mN. This work serves as the foundation to address systems level challenges in future work.

ACKNOWLEDGEMENTS

This work was supported by NASA under Award #NNX12AM02G.

REFERENCES

[1] S. C. Jacobsen, I. McGammon, K. B. Biggers, and R. P. Phillips, "Design of tactile sensing systems for dextrous manipulators," Control Systems Magazine, IEEE, vol. 8, no. 1, pp. 3–13, 1988.

[2] M. H. Lee and H. R. Nicholls, "Review article tactile sensing for mechatronics a state of the art survey," Mechatronics, vol. 9, no. 1, pp. 1–31, 1999.

[3] H. Yousef, M. Boukallel, and K. Althoefer, "Tactile sensing for dexterous in-hand manipulation in roboticsa review," Sensors and Actuators A: physical, vol. 167, no. 2, pp. 171–187, 2011.

[4] R. S. Dahiya, G. Metta, M. Valle, and G. Sandini, "Tactile sensing from humans to humanoids," Robotics, IEEE Transactions on, vol. 26, no. 1, pp. 1–20, 2010.

[5] E.-S. Hwang, J.-h. Seo, and Y.-J. Kim, "A polymer-based flexible tactile sensor for both normal and

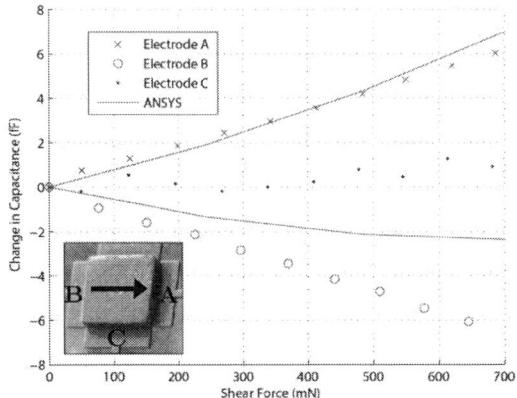

Fig. 6: High resolution shear testing compared to the finite element model. Each data point was gathered after a shear displacement of 10 μm was applied. [Inset] Macro of an individual pillar and surrounding electrodes taken just after Step (8) from Fig. 4

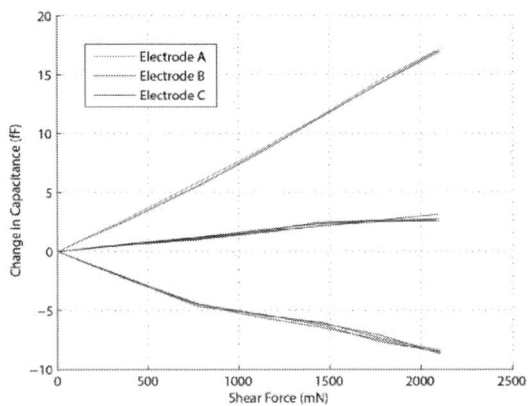

Fig. 7: High shear force testing, displaced up to 300 μm. 4 trials of a single sensor are shown.

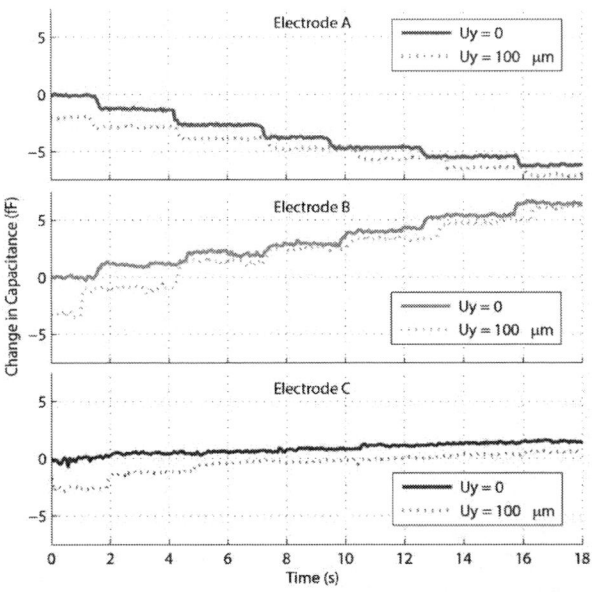

Fig. 8: Shear performance of each electrode of a single sensor with and without an applied normal displacement of 100 μm. Each step is an applied shear displacement of 20 μm.

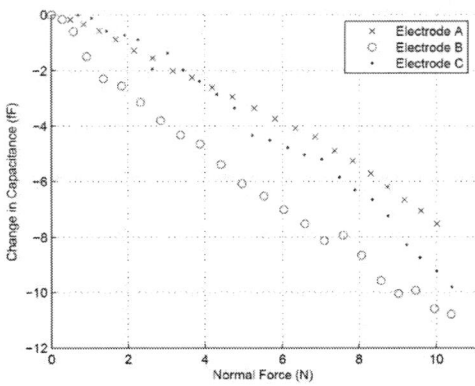

Fig. 9: Normal force tests of the sensor without any shear force applied.

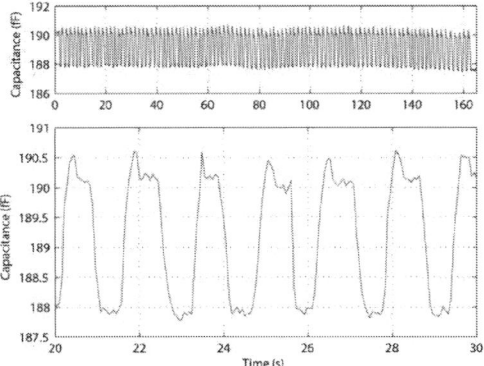

Fig. 10: Cyclic testing of a single pillar, Electrode A. (Top) 100 cycles of 100 μm shear displacement. (Bot) Detailed view of the signal revealing overshoot from the Thorlabs controller.

shear load detections and its application for robotics," Microelectromechanical Systems, Journal of, vol. 16, no. 3, pp. 556–563, 2007.

[6] S. Pyo, J.-I. Lee, M.-O. Kim, T. Chung, Y. Oh, S.-C. Lim, J. Park, and J. Kim, "Development of a flexible three-axis tactile sensor based on screen-printed carbon nanotube-polymer composite," Journal of Micromechanics and Microengineering, vol. 24, no. 7, p. 075012, 2014.

[7] H.-K. Lee, J. Chung, S.-I. Chang, and E. Yoon, "Normal and shear force measurement using a flexible polymer tactile sensor with embedded multiple capacitors," Microelectromechanical Systems, Journal of, vol. 17, no. 4, pp. 934–942, 2008.

[8] A. Charalambides and S. Bergbreiter, "All-elastomer in-plane MEMS capacitive tactile sensor for normal force detection," in Sensors, 2013 IEEE. IEEE, 2013, pp. 1–4.

[9] X. Niu, S. Peng, L. Liu, W. Wen, and P. Sheng, "Characterizing and patterning of pdms-based conducting composites," Advanced Materials, vol. 19, no. 18, pp. 2682–2686, 2007.

CONTACT

*Sarah Bergbreiter, sarahb@umd.edu

6-AXIS FORCE-TORQUE SENSOR CHIP
COMPOSED OF 16 PIEZORESISTIVE BEAMS

Akihito Nakai[1], Yasuhiko Morishita[2], Kiyoshi Matsumoto[1], and Isao Shimoyama[1]
[1]The University of Tokyo, Tokyo, JAPAN
[2]Touchence Inc., Tokyo, JAPAN

ABSTRACT

This paper reports the design, a part of the fabrication process, especially ion doping method by oblique ion implantation, and experimental results of a 6-axis force-torque sensor chip composed of 16 piezoresistive beams. The sensor chip's area, 2mm square in size, is one-third of that of the minimum 6-axis sensor chip ever reported. It will enhance the mounting density of sensor array and also reduce the cost in case of volume production. These sensor chips were fabricated by MNOIC, 8-inch MEMS foundry in Japan, and their characteristic variations are enough small to make practical use of them.

INTRODUCTION

Low-cost 6-axis force-torque sensors are needed in the fields of robotics, human interface and healthcare applications. Large-size, which means non-MEMS, 6-axis force-torque sensors with more than 17 mm in diameter and 14 mm in height using strain gauges or capacitance sensors are commercially available. However, even the ATI Nano17, the smallest one in this category, is unsuitable for the preceding fields because of its size, especially its height. 6-axis force-torque sensors with contact probe or cross-shaped beam fabricated by MEMS technologies were reported [1-3]. These sensors are suitable for the cell manipulation or material characterization, but unsuitable for the preceding applications, because the area where a load can be applied is limited to the tip of the probe or the center of the cross-shaped beam. 6-axis force-torque sensor chips which were several millimeters in size were also reported [4, 5]. One was achieved by using 32 FET based piezoresistive sensors and 4 mm x 3 mm in size. The other was achieved by using capacitance sensors and 9 mm square in size.

In this paper, we propose a 6-axis force-torque sensor chip which will be embedded in the designed resin with proper size and hardness such as silicone rubber or epoxy resin for each practical purpose. The sensor chip's area is 2mm square in size so that it's enough small to embedded in the resin and we can reduce the cost in case of volume production.

DESIGN AND FABRICATION

Two pairs of piezoresistive beams with surface doping and side-wall doping which function as a pressure sensor and a shear stress sensor, respectively [6], were placed point-symmetrically on every side of the sensor chip, and the center of symmetry is equal to that of the sensor chip (Figure 1). By configuring the half bridge using each pair of beams, temperature compensation is achieved by itself. Therefore, we can simply derive six representative values V_{Fx}, V_{Fy}, V_{Fz}, V_{Mx}, V_{My}, V_{Mz} from measured voltage

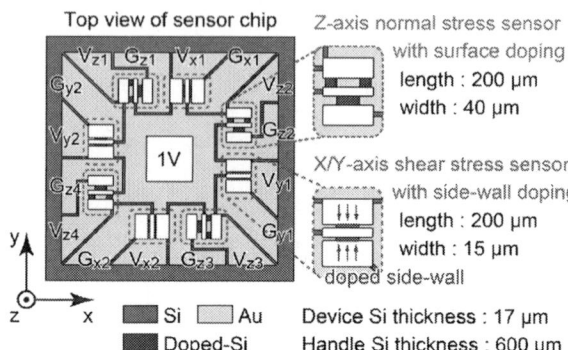

Figure 1: Design of the 6-axis force-torque sensor chip.

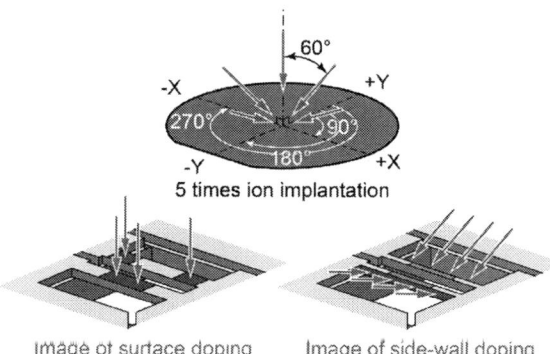

Figure 2: Oblique ion implantation was carried out 4 times for side-wall doping, and ordinary one was 1 time for surface doping.

changes ΔV_{x1}, ΔV_{x2}, ΔV_{y1}, ΔV_{y2}, ΔV_{z1}, ΔV_{z2}, ΔV_{z3}, ΔV_{z4} as follows.

$$V_{Fx} = (\Delta V_{x1} - \Delta V_{x2})/2 \tag{1}$$

$$V_{Fy} = (\Delta V_{y2} - \Delta V_{y1})/2 \tag{2}$$

$$V_{Fz} = (\Delta V_{z1} + \Delta V_{z2} + \Delta V_{z3} + \Delta V_{z4})/4 \tag{3}$$

$$V_{Mx} = (\Delta V_{z2} - \Delta V_{z4})/2 \tag{4}$$

$$V_{My} = (\Delta V_{z1} - \Delta V_{z3})/2 \tag{5}$$

$$V_{Mz} = (\Delta V_{x1} + \Delta V_{x2} + \Delta V_{y1} + \Delta V_{y2})/4 \tag{6}$$

The sensor chips were fabricated by MNOIC (MicroNano Open Innovation Center), 8-inch MEMS foundry in Japan, to bring the trial production into view. In ion implantation process for side-wall doping, a wafer was set so that the incidence angle of the ion beam was inclined at 60 degrees relative to a vertical axis to the wafer surface, and rotated by 90 degrees around the vertical axis to the wafer 4 times so that the ion beam can be implanted equally into every orthogonal side-wall (Figure 2).

Photos of the fabricated 8-inch wafer and a sensor chip are shown in Figure 3(a) and (b). The yield of the sensor chip is about 5,000.

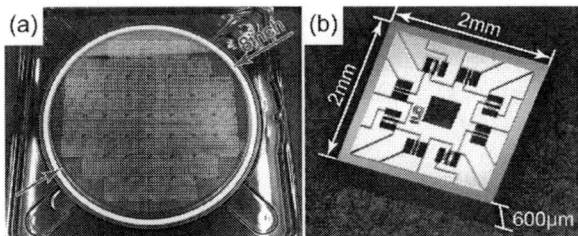

Figure 3: Photos of (a) the fabricated 8-inch wafer and (b) a 6-axis force-torque sensor chip.

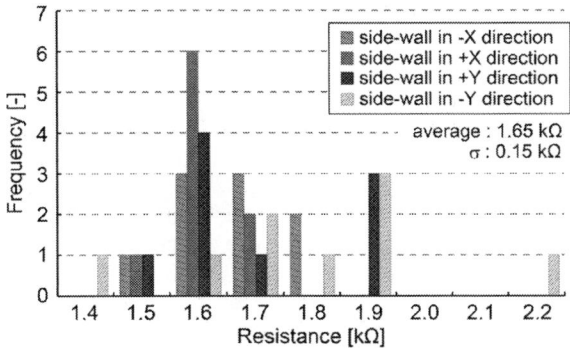

Figure 4: Histogram of the side-wall doped beams resistance in 4 directions at 9 points in the 8-inch wafer.

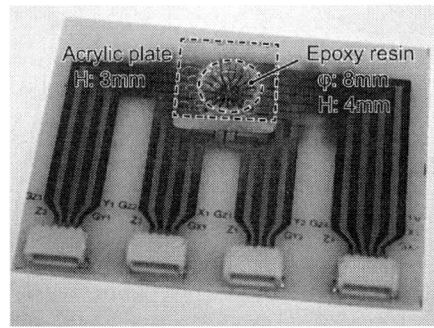

Figure 5: A photo of fabricated 6-axis force-torque sensor.

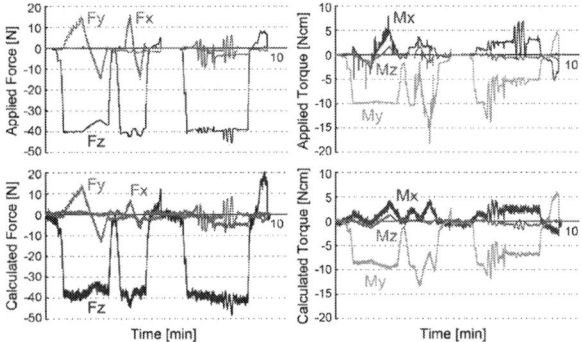

Figure 6: Comparison of applied forces and torques measured by the commercial 6-axis force-torque sensor with calculated ones by fabricated sensor.

EXPERIMENTS AND RESULTS

Resistance of side-wall doped beams in 4 directions at 9 points in the 8-inch wafer was measured and analyzed. The histogram of measured data is shown in Figure 4. The average and standard deviation of resistance are 1.65 kΩ and 0.15 kΩ, respectively.

The 6-axis force-torque sensor made by embedding the sensor chip into epoxy resin is shown in Figure 5. It was calibrated using a commercial 6-axis force-torque sensor, motorized 3-axis linear stages and $\alpha\beta$-goniometer stages, and a calibration matrix was determined by a least squares fitting. The comparison of applied forces and torques measured by the commercial 6-axis force-torque sensor with calculated ones by fabricated sensor is shown in Figure 6.

CONCLUSIONS

We proposed a 6-axis force-torque sensor chip composed of 16 piezoresistive beams. The sensor chips were fabricated by MNOIC, 8-inch MEMS foundry in Japan, and oblique ion implantation method was adopted for side-wall doping of shear stress sensors beams. The characteristic variations of sensor chips are enough small to make practical use of them. Fabricated 6-axis force-torque sensor embedded in epoxy resin was calibrated and compared with the commercial 6-axis force-torque sensor.

ACKNOWLEDGEMENTS

This work was supported by JSPS KAKENHI Grant Number 25000010, Grant-in-Aid for Specially Promoted Research.

REFERENCES

[1] P. Estevez, J.M. Bank, M. Porta, J. Wei, P.M. Sarro, M. Tichem, and U. Staufer, "6 DOF force and torque sensor for micro-manipulation applications," *Sensor. Actuat. A-Phys.*, vol. 186, pp. 86-93, 2011.

[2] F. Beyeler, S. Muntwyler, and B.J. Nelson, "A Six-Axis MEMS Force-Torque Sensor With Micro-Newton and Nano-Newtonmeter Resolution," *J. Microelectromech. S.*, vol. 18, no. 2, pp. 433-441, 2009.

[3] D.V. Dao, T. Toriyama, J. Wells, and S. Sugiyama, "Silicon Piezoresistive Six-Degree of Freedom Micro Force-Moment Sensor," *Sensor. Mater.*, vol. 15, pp. 113-135, 2002.

[4] J. Handwerker, P. Gieschke, M. Baumann, and O. Paul, "CMOS Integrated Silicon/Glass-bonded 3D Force/ Torque Sensor," *Proc. MEMS2012*, pp. 136-139.

[5] R.A. Brookhuis, H. Droogendijk, M.J.de Boer, R.G.P. Sanders, T.S.J. Lammerink, R.J. Wiegerink, and G.J.M. Krijnen, "Six-axis force-torque sensor with a large range for biomechanical applications," *J. Micromech. Microeng.*, vol. 24, no. 3, pp. 1-10, 2014.

[6] H. Takahashi, N. T. Vinh, A. Nakai, K. Matsumoto, and I. Shimoyama, "A triaxial tactile sensor without crosstalk using pairs of piezoresistive beams with sidewall doping," *Sensor. Actuat. A-Phys.*, vol. 199, pp. 43-48, 2013.

CONTACT

*A. Nakai, e-mail: nakai@leopard.t.u-tokyo.ac.jp

A 0.25mm³ ATOMIC FORCE MICROSCOPE ON-A-CHIP

Neil Sarkar[1,2], Duncan Strathearn[1,2], Geoffrey Lee[1,2], Mahdi Olfat[1,2], Raafat R. Mansour[1,2]
[1]University of Waterloo, Waterloo, CANADA
[2]ICSPI Corp., Waterloo, CANADA

ABSTRACT

This paper reports the highest resolution achieved with a single-chip Atomic Force Microscope (sc-AFM). Images of a 20nm AFM calibration standard were obtained to show, for the first time, that single-chip instruments may obtain a vertical resolution comparable to state-of-the-art instruments at a minuscule fraction of the size (volume=1/1,000,000) and cost (1/1000). A maskless, 2-step release process is performed on CMOS chips in order to obtain devices that can image a sample without the need for any off-chip scanning or sensing components. We report a four-fold improvement in resolution when compared to previously reported sc-AFMs, enabling metrology for nanoscale manufacturing using MEMS AFM technology.

INTRODUCTION

Background

The concept of AFM was invented by Binnig [1] in 1986. The first images were acquired with the instrument and published by Binnig, Quate and Gerber [2] soon thereafter. Although the inventors anticipated that the instrument could achieve atomic resolution, five years elapsed before Giessbil et. al. [3] demonstrated it using a dynamic mode of operation. In dynamic AFM, the force measurement is enhanced by a gain of roughly Q (the quality factor of a resonator) that is achieved in the mechanical domain by placing a cantilever in resonance. Today, AFMs are regarded as workhorse instruments of nanoscience, and are among the highest resolution microscopes available. Amplitude Modulation AFMs (AM-AFMs) have enjoyed the most widespread use out of the entire family of scanning probe microscopes (SPMs) because of their versatility and ease-of-use.

Conventional AFMs comprise macroscale components in a bulky construction that exhibits poor vibration immunity and significant thermal drift. In addition, state-of-the-art AFM scanners employ piezoelectric materials that suffer from creep and hysteresis, and they rely on laser-based position sensing with lengthy free-space optical paths. Despite these shortcomings, the price of modern AFMs may exceed USD $100,000. The benefits of scaling such instruments downward in size have long been known. For example, an array of 1 Degree-Of-Freedom (DOF) AFM cantilevers with integrated electronics was presented in [4], and a 2 DOF AFM scanner was reported in [5]. An important distinguishing feature of sc-AFMs when compared to prior work is that the present devices integrate all of the necessary electromechanical components that are required to obtain an AFM image onto a single CMOS chip. In contrast, prior AFM chips required a conventional AFM to perform scanning or force-sensing functions, limiting their utility as standalone tools. Although some examples of sc-AFMs exist in the literature [6],[7], the best resolution reported by the authors was 90nm, which does not satisfy modern metrology requirements met by commercial instruments.

Device Geometry and Operation

The small volume of sc-AFMs (0.25mm³) shifts the paradigm of putting one's sample "inside" the AFM, and introduces the notion of placing one's AFM onto a sample (Figure 1). Several benefits arise from scaling the instrument's dimensions downwards. The short mechanical path between the tip and sample improves immunity to thermal drift. The devices are robust to ambient vibrations because their mass, which scales volumetrically, is significantly reduced. Microfabricated electrothermal actuators enjoy a small footprint, low thermal mass, high stiffness, and CMOS-compatible drive voltages, all of which combine into a system that is well-suited to video-rate imaging. To further alleviate the bottleneck on imaging bandwidth that is imposed in conventional instruments, these devices may be manufactured as arrays at the wafer scale to achieve parallel operation. If the CMOS process is augmented with through-silicon vias, the sc-AFM devices may be placed directly in contact to a sample surface (Figure 1).

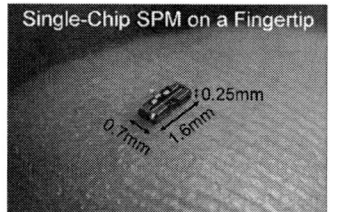 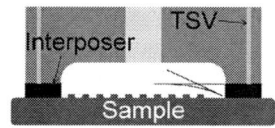

Figure 1: Left: Single chip SPM die after singulation, Right:With the use of TSVs, the SPM may be placed directly on a sample under study.

The device structure of the sc-AFM consists of lateral scanners, a vertical actuator to track topology, a thermal-piezoresistive resonator to detect tip-sample interaction forces, and thermal management features (Figure 2).

The instrument may be operated in various modes including AM-AFM, FM-AFM, and forced oscillation with higher-harmonic detection[8]. A conductive path to the tip is included to support electrical measurement modes such as Kelvin Probe Force Microscopy (KPFM) and conductive AFM. The results presented here were obtained in the AM-AFM mode, at force settings that correspond to non-contact and intermittent contact.

In AM-AFM, a resonant cantilever is driven near its resonant frequency with a constant-amplitude drive signal. In the presence of a tip-sample force gradient, the natural frequency of the resonator shifts and alters the measured (mechanical) amplitude of oscillation. A closed-loop controller varies the tip-sample distance in order to hold the amplitude constant. The cantilever may be excited into resonance by directly driving the piezoresistive beams (self-heating) with a small AC signal superposed on a large DC signal. In this scenario (self-actuated resonant cantilever), the controller output may drive the z-actuator to track topology. Alternatively, the z-actuator may externally shake the proof mass into resonance, with a DC input from the controller (to track topology) and an AC input to drive resonance. In this scenario (externally actuated resonant cantilever), a constant DC bias is applied to the piezoresistors to measure oscillation amplitude. The images reported here correspond to the latter scenario, in which vertical bimorphs shake the cantilever into resonance. The DC bias may be used to tune the Q-Factor of the resonator, because the dynamic variation in resistance provides a mechanism to pump power back into the cantilever to compensate damping [7].

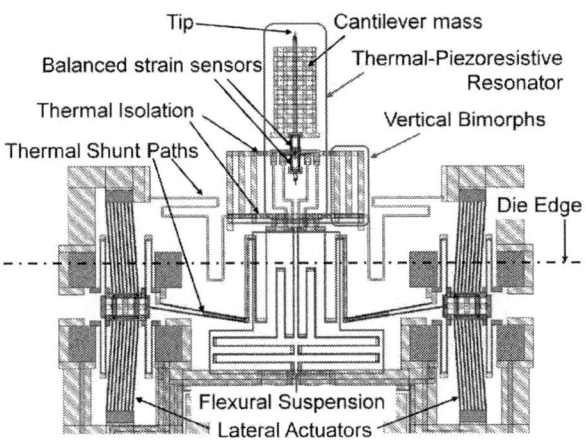

Figure 2: Layout capture reveals the device geometry of an sc-AFM.

Lateral electrothermal actuators make use of beams that are arranged in a chevron pattern, as described in [9]. Flexures provide geometric advantage to provide the desired scan range. A pair of lateral actuators scans the tip in polar coordinates (r, θ): a common mode voltage displaces the tip in the 'r' direction, while differential voltage deflects the tip in the 'θ' direction.

The actuators undergo temperature excursions from Joule heating ($\Delta T \propto V^2/R = Pwr$), affecting deleterious thermal cross-talk to the piezoresistors through the Temperature Coefficient of Resistivity (TCR). It is therefore desirable to hold piezoresistor temperature constant while scanning the sample ("isothermal scanning"). A nonlinear inversion compensates the V^2 dependence of Joule heating; however, this does not capture the TCR effect, which introduces a second nonlinearity. A calibration procedure captures the TCR nonlinearity to further suppress thermal coupling, yielding the appropriate voltage drive waveforms (Figure 3). Each line in the scan consists of a scaled pair of voltage waveforms like those in the Figure to hold the piezoresistors at a constant temperature. A plot of the tip

trajectory over a typical scan is shown below (Figure 3).

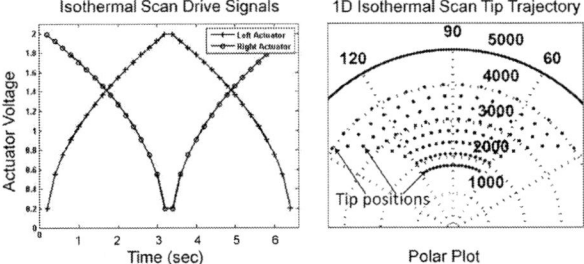

Figure 3: Left, isothermal scan waveform obtained after calibration. Right: Polar plot of tip positions over the course of several 1D isothermal line-scans.

Note that the scan range of the bottom-most linescan is lower than that of the top-most linescan, implying that the lateral resolution varies as the scan progresses, with a maximum resolution attained at the lowest temperature.

Microfabrication and Assembly

CMOS devices are received from the foundry, typically as 5x5mm dies fabricated through multi-project wafer services. A laser scribing process is first performed to etch trenches underneath the cantilevers. Following the scribe step, a maskless, 2-step CMOS-MEMS release is performed [10]. The intermetal dielectric may first be anisotropically etched using a combination of CF_4 and H_2 to define the structure of the device. The Silicon substrate may then be etched isotropically with SF_6 as a release step. Following these RIE steps, the chip is singulated to obtain cantilevers that are suspended over the die edge. This is an important feature, as it ensures line of sight to the tip-sample region while the instrument is scanning.

Following the release and singulation steps, the dies are attached and wirebonded onto a carrier PCB (Figure 4). The carrier PCB is then mounted onto a coarse approach mechanism to engage the tip and sample, which is mounted on a standard SEM stub. Alignment holes allow the user to rapidly exchange AFM chips within the system.

Figure 4: sc-AFM die-bonded and wirebonded onto a carrier PCB for facile, rapid exchange

Two examples of inexpensive coarse approach mechanisms are shown below (Figure 5). A PCB motor is used to slide the sample holder in 20nm increments until

978-1-4799-7956-1/15 $31.00 © 2015 IEEE   733

contact is made. Alternatively, a 3D-printed AFM holder may be used to convert a manual micropositioner into an AFM with a footprint-compatible probe replacement. The coarse approach mechanism design constraints include short tip-sample mechanical paths, ease of use, and rapid sample approach. In air, the sc-AFM's piezoresistive output may be used to detect the presence of the sample long before contact (100μm range), due to thermal interactions between the self-heated resistors and the sample, thereby enabling a rapid automated approach. Note that cumbersome laser alignment steps are obviated by piezoresistive detection, which also enables the use of sc-AFMs in confined spaces (e.g. SEMs).

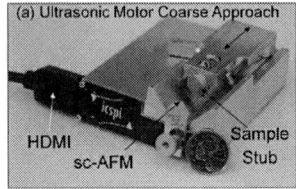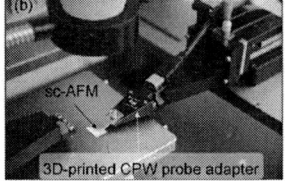

Figure 5: Left, an automated coarse approach mechanism with an integrated PCB motor. Right, a 3D printed adaptor that may be attached to a manual micropositioner as a foot-print compatible AFM replacement for a conventional probe.

Measurement Results

The tip-sample interaction force experienced by the AFM is a nonlinear function of tip-sample separation. A standard procedure to elucidate the operating parameters of the instrument is to obtain approach curves, as shown in Figure 6. Several features are visible on the curves. Unstable regions correspond to the portion of the force curve in which the force gradient is close to the cantilever spring constant. In these regions, there are 2 equilibrium positions that may be occupied by the tip. The phase of the resonator is a clear indicator of whether the tip is experiencing a net attractive or repulsive force gradient. Also visible are the pull-in and snap-out features that indicate that that tip experiences adhesive, stiction-like forces when in contact.

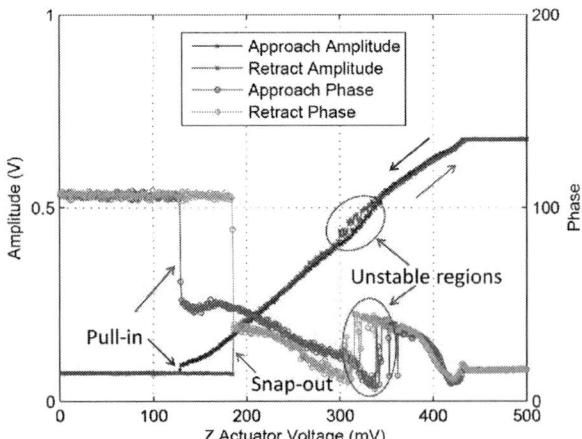

Figure 6: Dynamic AM-AFM approach curves

It is important to note that these curves were obtained under ambient conditions, with no vibration or acoustic

isolation. The alignment between the approach and retract curves indicates that creep and hysteresis do not manifest in the CMOS-MEMS actuators, suggesting that these instruments may be well-suited to long-term force spectroscopy measurements.

We imaged a 20nm calibration standard with the sc-AFM. Ridge artifacts appeared in images that were intentionally calibrated over a shorter range than the scan range; they did not appear upon calibration over the full range (Figure 7). The curvature in the image arises from the circular tip trajectory. A polynomial subtraction using an open-source AFM image processing tool (Gwyddion) removes the curvature. A line profile extracted from the flattened image shows that the instrument achieves ~3nm pk-pk position noise. Importantly, these images were obtained without any vibration or acoustic isolation, and with inexpensive electronics (<$300USD).

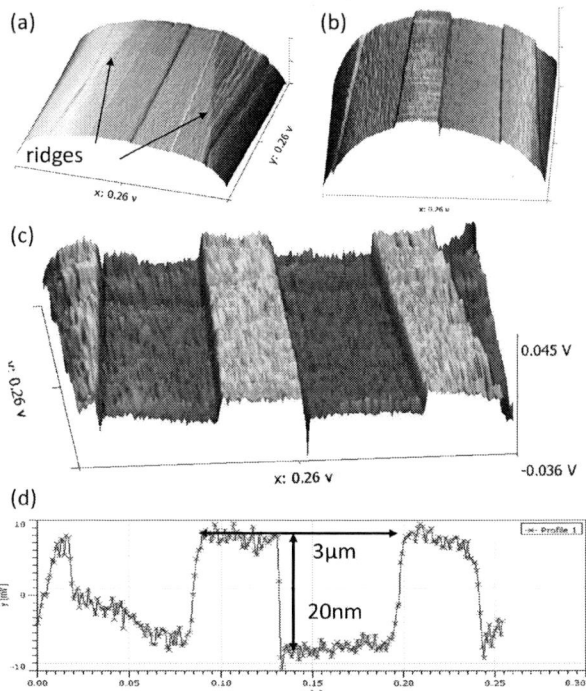

Figure 7: a) Raw image of a 20nm calibration grating obtained with partial calibration, b) raw image of same grating with calibration performed over entire scan regions, c) image post-processed to remove curvature, d) line profile extracted from the image in c).

The images shown above represent a rectangular sliver of the calibration sample that is ~9μm x 1μm. These images do not suffer noticeably from the distortion from the polar coordinate scan tip trajectory of the instrument. A larger scan area is shown below (Figure 8). Imaging was performed on a 2-D calibration standard that consists of 50nm-deep pits on a 2μm pitch.

The distortion is clearly visible, as the image spans a large range (~25μm) at the top, and a smaller range at the bottom (~4μm). This artifact may be removed in software with a coordinate transformation; however, the resolution of the image varies with position, rendering the post-processing cumbersome.

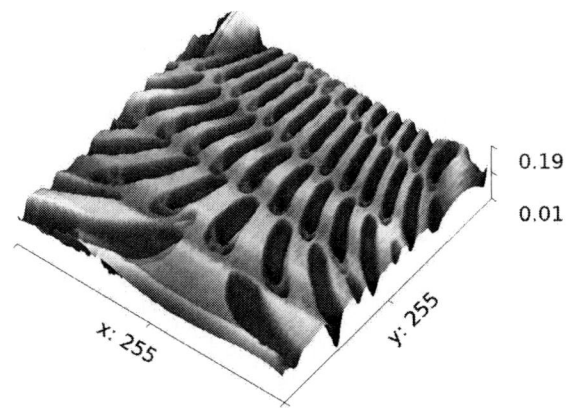

Figure 8: sc-AFM image of an AFM calibration standard with 50nm deep pits on a 2 micron pitch. Scan lines become progressively narrower as the scan progresses downwards in the image.

To mitigate this distortion issue, a new scanner geometry was designed in order to provide 2DOF isothermal scanning over a square region of the sample. Characterization of this new scanner design is underway, with early results (obtained on the same calibration sample) shown below (Figure 9). No post-processing was required to produce this image.

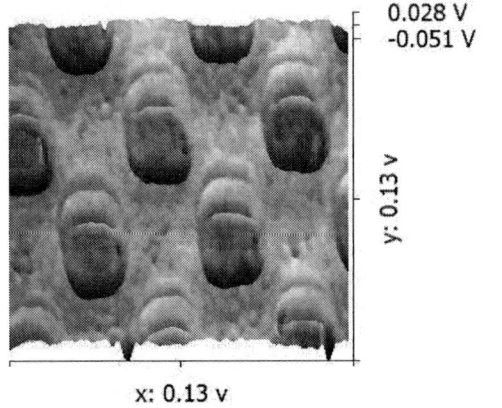

Figure 9: Image of same calibration grating obtained with 2D isothermal scanner design.

Conclusion

AM-AFM is by far the most popular form of SPM because of its versatility and robustness. This work demonstrates, for the first time, that a complete AFM instrument that supports AM operation can be integrated onto a single chip, with performance that is commensurate with state-of-the-art tools. The high cost of conventional instruments has been a longstanding barrier to the widespread adoption of SPMs in metrology and education.

These results imply that sc-AFMs may replace conventional AFMs in metrology for advanced manufacturing at the nanometer scale. The integration of arrays of CMOS-MEMS AFMs into roll-to-roll and additive manufacturing systems is an enabling technology that stands to improve the precision and throughput of nanomanufacturing.

ACKNOWLEDGEMENTS

The authors would like to acknowledge the support of TowerJAZZ and the Canadian Microelectronics Corporation. This work was supported by DARPA, the Texas ETF, and the Ontario Research Fund (ORF-RE).

REFERENCES

[1]   G. K. Binnig, "Atomic force microscope and method for imaging surfaces with atomic resolution, US Patent 4724318." Google Patents, 1988.

[2]   G. Binnig, C. Quate, and C. Gerber, "Atomic force microscope," *Phys. Rev. Lett.*, 1986.

[3]   F. Giessibl and H. Bielefeldt, "Imaging of atomic orbitals with the Atomic Force Microscope-experiments and simulations," *arXiv Prepr. cond-mat/ ...*, no. 111, pp. 1–21, 2001.

[4]   S. Hafizovic, D. Barrettino, T. Volden, J. Sedivy, K.-U. Kirstein, O. Brand, and A. Hierlemann, "Single-chip mechatronic microsystem for surface imaging and force response studies.," *Proc. Natl. Acad. Sci. U. S. A.*, vol. 101, no. 49, pp. 17011–5, Dec. 2004.

[5]   A. G. Fowler, A. N. Laskovski, A. C. Hammond, and S. O. R. Moheimani, "A 2-DOF Electrostatically Actuated MEMS Nanopositioner for On-Chip AFM," *J. Microelectromechanical Syst.*, vol. 21, no. 4, pp. 771–773, Aug. 2012.

[6]   N. Sarkar, G. Lee, and R. Mansour, "CMOS-MEMS dynamic FM atomic force microscope," in *Solid-State Sensors, Actuators and Microsystems (TRANSDUCERS & EUROSENSORS XXVII), 2013 Transducers & Eurosensors XXVII: The 17th International Conference on. IEEE*, 2013.

[7]   N. Sarkar and R. R. Mansour, "Single-chip atomic force microscope with integrated Q-enhancement and isothermal scanning," in *2014 IEEE 27th International Conference on Micro Electro Mechanical Systems (MEMS)*, 2014, pp. 789–792.

[8]   N. Sarkar, R. Mansour, and K. Trainor, "Forced oscillation and higher harmonic detection in an integrated cmos-mems scanning probe microscope," *Hilt. Head*, 2012.

[9]   Y. B. Gianchandani, "Bent-beam electrothermal actuators-Part I: Single beam and cascaded devices," *J. Microelectromechanical Syst.*, vol. 10, no. 2, pp. 247–254, Jun. 2001.

[10]   G. K. Fedder, S. Santhanam, M. L. Reed, S. C. Eagle, D. F. Guillou, M. S.-C. Lu, and L. R. Carley, "Laminated high-aspect-ratio microstructures in a conventional CMOS process," in *Proceedings of Ninth International Workshop on Micro Electromechanical Systems*, pp. 13–18.

CONTACT

*N. Sarkar, tel: +1-519-888-4567; nsarkar@uwaterloo.ca

AN INSTRUMENTED TOOTH

Felix Becker[1], Christian Sander[1], Falko Schmidt[2], Bernd Lapatki[2], and Oliver Paul[1]

[1]Department of Microsystems Engineering (IMTEK), University of Freiburg, GERMANY
[2]Department of Orthodontics, University of Ulm, GERMANY

ABSTRACT

This paper presents a tool for orthodontic research and education, namely an instrumented tooth enabling all six force and moment components exerted on the tooth to be measured. The mechanical load values are inferred from mechanical stress values measured at the surface of a CMOS stress sensor system sandwiched between two pins dimensioned as a tooth root. The resulting sensor module measures 11.6 mm in height and 3.5 mm in diameter. It allows the measurement of forces up to 60 N and moments up to 10 Ncm and thus covers the range of forces and moments relevant for orthodontics.

INTRODUCTION

The real-time measurement of forces and moments acting on teeth is of great interest in the scope of orthodontic treatments [1]. An appropriate force and moment (FM) transducer provides researchers with a tool to elaborate novel treatment strategies and validate or optimize already established methods. Furthermore a real-time feedback of the applied loads, for example on the teeth of a phantom head, provides interesting perspectives for the education of prospective orthodontists.

The orthodontic simulator (OSIM) [2] based on the 17-mm-wide and 14.5-mm-high FM transducer Nano17 from ATI, the smallest commercial sensor for six-dimensional (6D) FM measurements, addresses this demand. However OSIM is bulky, expensive, and unrealistically resilient. An approach to measure 6D loads applied by fixed orthodontic appliances, is to integrate the FM transducer unit directly into the so-called smart bracket [3]. Integrating the transducer into fixed orthodontic appliances precludes wired data and energy transfer. Even well-established near field communication (NFC) techniques cannot be used because of the modest dimensions of standard brackets. Orthodontic simulators address the field of orthodontic research and education and, unlike smart brackets, are free of aesthetic and biocompatibility-related considerations. Energy and data transfer by wire is therefore a viable option.

In this contribution we present a wired instrumented tooth (IT), where the FM measurement is done at the location of the root of a commercial artificial tooth. The single IT and the assembly of several ITs into a so-called instrumented dentition enables the in-vitro simulation of a wide range of orthodontic situations. Compared to NFC the parallel communication with all ITs is easy to establish and there are fewer restrictions on how the load is applied.

Figure 1 shows an IT with a crown of a lower canine. The artificial crown is taken from a commercial artificial tooth. It is placed on the IT sensor module consisting of two metal pins sandwiching a mechanical stress sensing CMOS chip. After proper calibration the continuously transmitted signals of the 32 sensors distributed over the surface of the CMOS chip enable the load components of arbitrary load states applied to the crown to be calculated.

STRESS SENSOR MODULE

Figure 2 shows an assembled IT sensor module. The overall length and diameter of the device without polyimide (PI) cable are 11.6 mm and 3.5 mm, respectively. The stress sensing CMOS chip is adhesively bonded to the

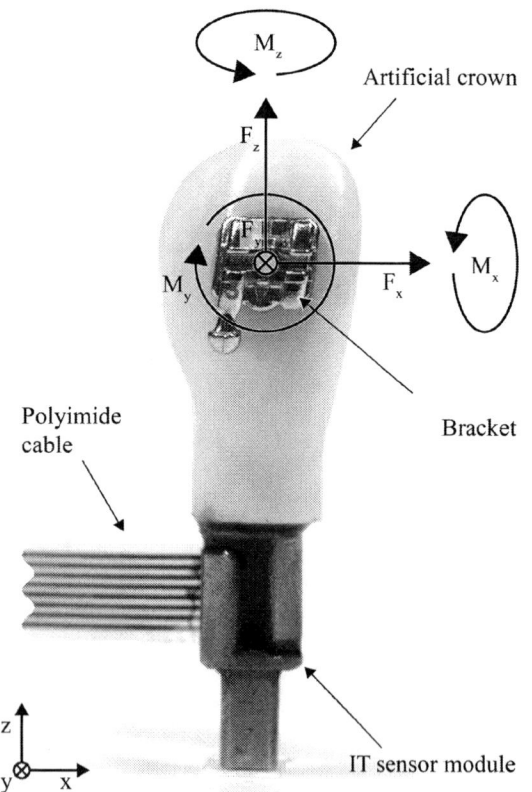

Figure 1: IT with its sensor module and an artificial lower canine crown equipped with a bracket.

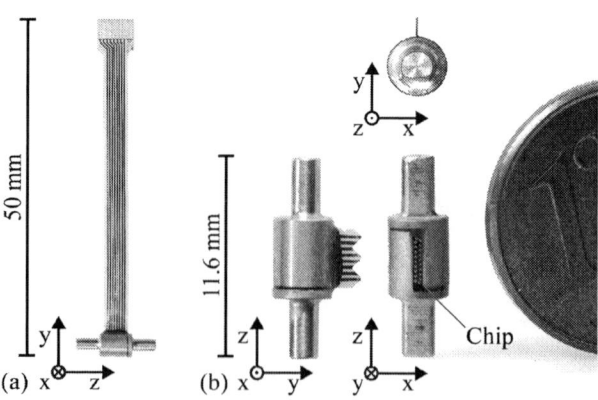

Figure 2: Overall (a) and detailed (b) views of the IT sensor module.

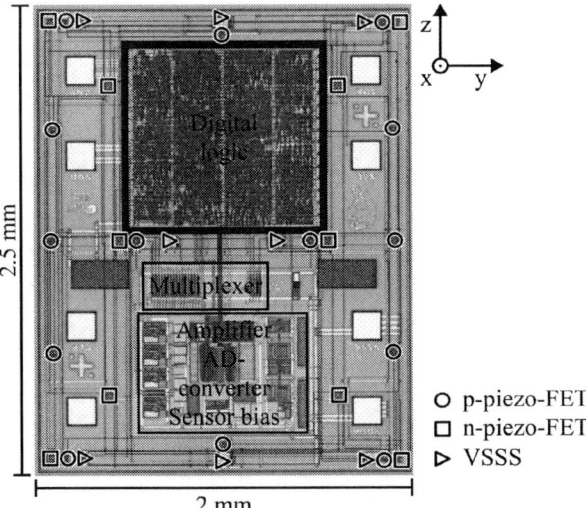

Figure 3: The utilized CMOS sensor chip with the positions of the 14 p-piezo-FET sensors, the 10 n-piezo-FET sensors and the 8 VSSS.

metal pins. Its position is indicated by the red dashed rectangle in figure 2 (b).

The conducting paths of the PI cable consist of a 250-nm-thick platinum metallization with a 50-nm-thick titanium tungsten layer on top. For protection the conducting paths are sandwiched by two 5-µm-thick PI layers. At the location of the contact the upper PI layer is opened and gold contacts are electroplated. The cables are processed on standard four-inch handle wafers; thus the maximum cable length is limited to about 80 mm. Thanks to its high flexibility the 10-µm-thick PI cable transmits only a negligible mechanical load from the environment to the sensor module.

CMOS chip

Figure 3 shows the mechanical stress mapping CMOS sensor chip which has been published elsewhere [4]. Into its surface 14 piezoresistive *p*-channel field effect transistor (FET) devices, 10 piezoresistive *n*-channel FET devices [5], and 8 vertical shear stress sensors [6] have been integrated in combination with drive, signal conditioning, and communication circuitry. The piezoresistive devices are abbreviated as p-piezo-FET, n-piezo-FET and VSSS, respectively. The sensors map the in-plane normal, in-plane shear, and out-of-plane shear stress components, respectively. The sensors are placed at positions exhibiting high induced mechanical stress when loads are applied. As a result, assuming the linear elastic response of the mechanical structure, when a load vector with components $l_1, \ldots, l_6$ corresponding to the three components F_x, F_y and F_z of the force and the three components M_x, M_y, M_z of the moment is applied, an output of 32 sensor signals $s_1, \ldots, s_{32}$ related to $l_1, \ldots, l_6$ via

$$\begin{pmatrix} s_1 \\ \vdots \\ s_{32} \end{pmatrix} = \begin{pmatrix} T_{1,1} & \cdots & T_{1,6} \\ \vdots & \ddots & \vdots \\ T_{32,1} & \cdots & T_{32,6} \end{pmatrix} \begin{pmatrix} l_1 \\ \vdots \\ l_6 \end{pmatrix} \qquad (1)$$

is obtained. The values $T_{i,j}$ denote the components of the

transfer or sensitivity matrix of the system.

The read-out of the sensors is done sequentially using a multiplexer. The signals proportional to the applied load are amplified in the analog domain, digitized on-chip and transmitted using an I²C protocol, whereby the amplification factor is set externally.

Assembly

The assembly of the IT sensor module is schematically shown in figure 4. The CMOS chip and the cable wafer are cleaned in an ultrasonic bath. Gold stud bumps are placed on the contact pads of the chip by thermosonic wire bonding. The cable is pretreated on the handle wafer with the NanoFlame Set NF05 (Polytec) and then peeled off the wafer. Using a flip-chip bonder the cable is then thermosonically bonded to the chip {figure 4 (a)}. A droplet of epoxy adhesive U300 (Epotek) is placed on an opening in the cable. Due to the decreased viscosity at the elevated temperature the capillary forces draw the resin into the gap between cable and chip {figure 4 (b)}, thus filling it uniformly.

After a curing step the chip/cable assembly is adhesively bonded to the metal pins {figure 4 (c)} with the epoxy resin 353-ND (Epotek). This is done in a custom-made assembly setup assuring proper alignment of the components. An additional curing completes the assembly {figure 4 (d)}. All curing steps are done with a temperature profile tailored for minimal thermally induced stress.

CHARACTERIZATION
Calibration

A typical measurement realizes the reverse of the process described by equation (1): In fact a load vector is inferred from a sensor signal vector via

$$\begin{pmatrix} l_1 \\ \vdots \\ l_6 \end{pmatrix} = \begin{pmatrix} C_{1,1} & \cdots & C_{1,32} \\ \vdots & \ddots & \vdots \\ C_{6,1} & \cdots & C_{6,32} \end{pmatrix} \begin{pmatrix} s_1 \\ \vdots \\ s_{32} \end{pmatrix}. \qquad (2)$$

Here the elements $C_{i,j}$ define the compliance or calibration matrix C. For the experimental determination of C we proceed as follows. A number of $n_L \geq 6$ loads quantified by a reference sensor and fully spanning the 6D load space is applied to the IT. Simultaneously the corresponding sensor

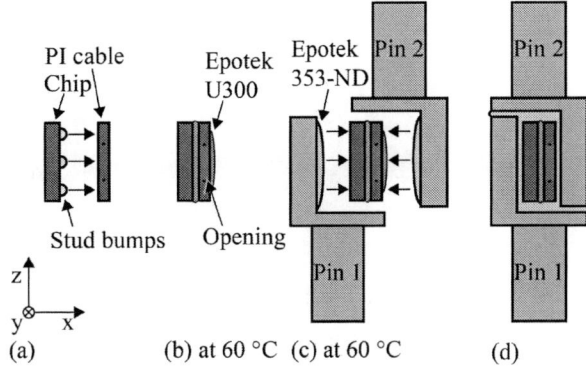

Figure 4: Assembly of the IT sensor module: (a) flip-chip-bonding, (b) underfilling, (c) pin assembly, (d) curing.

978-1-4799-7956-1/15 $31.00 © 2015 IEEE

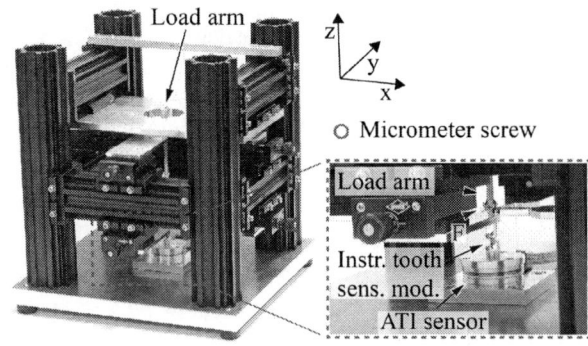

Figure 5: Experimental setup serving for the calibration and characterization of the IT sensor module. Different load cases are applied via a load arm and measured by a reference sensor from ATI.

signals are recorded. In substance C is determined by linear regression between the load and sensor vector data. In practice the resulting regression matrix is computed as

$$C = \begin{pmatrix} l_{1,1} & \cdots & l_{1,n_L} \\ \vdots & \ddots & \vdots \\ l_{6,1} & \cdots & l_{6,n_L} \end{pmatrix} \begin{pmatrix} s_{1,1} & \cdots & s_{1,n_L} \\ \vdots & \ddots & \vdots \\ s_{32,1} & \cdots & s_{32,n_L} \end{pmatrix}^+ , \quad (3)$$

where the first matrix lists the applied calibration load vectors as columns, while the second matrix does so with the corresponding measured sensor signals. The second matrix with the "+" symbol attached in superscript denotes the Moore-Penrose pseudoinverse [7] of the sensor signal matrix.

Figure 5 shows the experimental setup used for the calibration procedure. As shown in the close-up view, the IT sensor module is fixed between the load arm and the commercial FM transducer Nano43 SI-72-1 from ATI, USA, serving as a reference sensor. Forces and moments in the horizontal plane are applied by loading the load arm from all four horizontal directions at both ends by means of eight precision linear stages acting on ball bearings mounted on

the load arm. The ball bearings enable the application of loads in the z direction.

During the calibration procedure and subsequent measurements a history of load cases with different force and moment combinations is applied to the sensor module while the corresponding sensor signals are recorded.

Measurement

To assess the quality of the presented IT sensor module the module is exposed to forces and moments as described above. At the same time the signal of the reference sensor is recorded for subsequent comparison with the load values inferred from the sensor signals using equation (2).

Figure 6 shows a representative measurement series generated from over 8000 applied load cases. For clarity only every 10th data point is displayed. The colored square symbols represent the load values determined by the calibrated IT sensor module, whereas the solid black lines represent the corresponding load components measured using the reference transducer. Quite obviously the load values measured with the IT sensor module are in good agreement with the reference data. Also worth mentioning is the negligible cross-sensitivity between the various components of force and moment.

The small size of the cross-sensitivity is evidenced better in figure 7, where the load values extracted using the IT sensor module are plotted against the reference signal. In each case the data are fitted by straight lines, with small standard deviations in four out of six cases and a slope close to one for all cases. Reasons for the larger variances of the F_z and M_y data are that the sensor module is less sensitive to these load components and that F_z and M_y show a higher mutual cross-sensitivity than all other load component combinations. Both reasons are due to the mechanical design of the present IT.

Mechanical robustness

To estimate the mechanical robustness of the IT sensor module, 24 modules were exposed to increasing loads. Loads were applied to each module in the sequence $-F_x$,

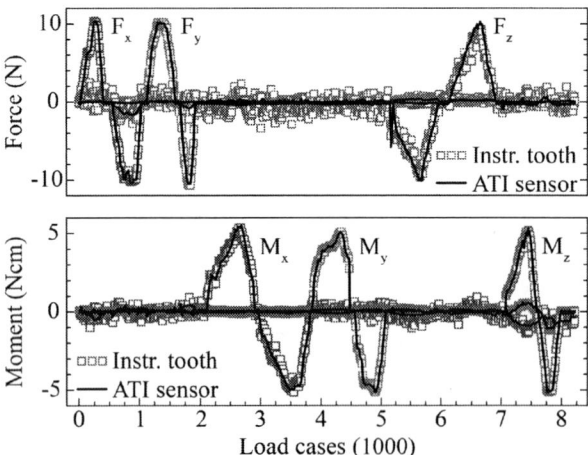

Figure 6: Series of over 8000 load cases measured with the IT sensor module (square symbols) and the data of the ATI reference sensor (solid black lines); only every 10th data point is displayed.

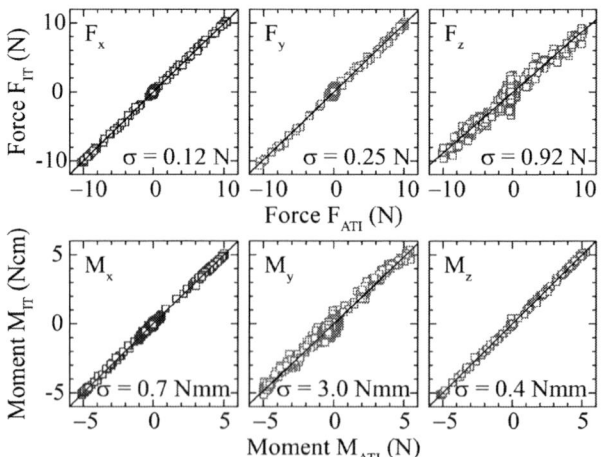

Figure 7: Load components determined using the IT sensor module, F_{IT} and M_{IT}, vs. the corresponding load components measured using the reference transducer, F_{ATI} and M_{ATI}.

978-1-4799-7956-1/15 $31.00 © 2015 IEEE

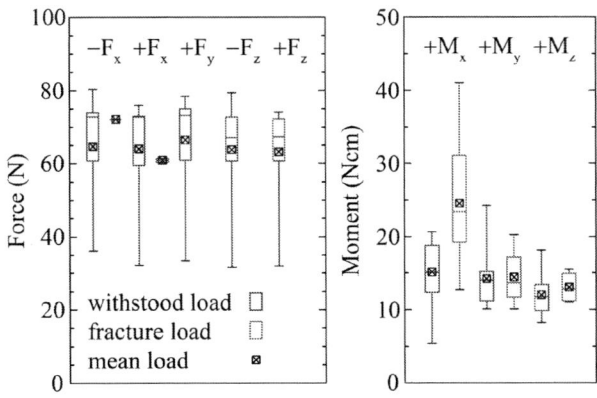

Figure 8: Distributions of the maximum withstood loads (blue) and of the loads applied when the module broke (red). The horizontal boundaries of the boxes and the lines within them represent the first, second, and third quartiles; the whiskers indicate the extremal values.

$+F_x$, $+F_y$, $-F_z$, $+F_z$, $+M_x$, $+M_y$, $+M_z$, with increasing magnitude of each load component until fracture terminated the test for the module. This procedure was chosen in order to extract as much statistically significant information as possible from the modest group of test samples.

The results are summarized in the box-and-whisker plots of figure 8. While the ranges in blue show the distributions of the maximum load values the modules withstood, those in red show the distribution of loads at which the modules broke. Whiskers mark the extremal values of the distributions while the horizontal boundaries of each box and the line within it mark the first, second, and third quartiles, respectively.

The small spread of the F_x loads is explained by the fact that only a single module broke while loaded with $-F_x$ and two modules broke while loaded with $+F_x$. In conclusion the tested modules withstood forces up to 60 N and moments up to 10 Ncm. These values are larger than the typical loads in the context of orthodontic therapy.

SUMMARY

We reported a miniaturized FM transducer of interest for orthodontic research and education. In addition to its high accuracy with respect to all six load components, its mechanical robustness up to 60 N for forces and up to 10 Ncm for moments was demonstrated. Since there is no commercial FM transducer with comparably small dimensions, the presented module may be interesting for other applications as well. We expect the mechanical robustness to be further increased by an optimized design of the pins and the bonding areas and by the use of alternative epoxy resins.

ACKNOWLEDGEMENTS

Funding of this project by the Deutsche Forschungsgemeinschaft (DFG, German Research Foundation, research grants PA 792/5-3, LA 2418/1-3 and MA 2193/7-3) is gratefully acknowledged.

REFERENCES

[1] S.J. Lindauer, "The Basics of Orthodontic Mechanics", *Semin. Orthod.*, vol. 7, no. 1, pp. 2-15, 2001.

[2] H.M. Badawi et al., "Three-Dimensional Orthodontic Force Measurements", *AJO-DO*, vol. 136, no. 4, pp. 518-528, 2009.

[3] B.G. Lapatki, O. Paul, "Smart Brackets for 3D-Force-Moment Measurements in Orthodontic Research and Therapy – Development Status and Prospects", *J. Orofac. Orthop.*, vol. 68, pp. 377-396, 2007.

[4] P. Gieschke, O. Paul, "CMOS-Integrated Sensor Chip for In-plane and Out-of-plane Shear Stress", *Procedia Engineering*, vol. 5, pp. 1364-1367, 2010.

[5] P. Gieschke et al., "CMOS-Integrated Stress Mapping Chips with 32 n-type or p-type Piezoresistive Field Effect Transistors", in *Tech. Dig. IEEE MEMS*, pp. 769-772, 2009.

[6] B.Lemke et al., "Piezoresistive CMOS-Compatible Sensor for Out-of-Plane Shear Stress", in *Sens. Actuators A*, vol. 189, pp. 488-495, 2013.

[7] A. Albert, *Regression and the Moore-Penrose Pseudoinverse*, ACADEMIC PRESS New York and London, 1972.

CONTACT

F. Becker, tel: +49-761-203-7195
felix.becker@imtek.de

978-1-4799-7956-1/15 $31.00 © 2015 IEEE

ASYMMETRIC FAN-SHAPE-ELECTRODE FOR HIGH-ANGLE-DETECTION-ACCURACY TACTILE SENSOR

Shi-Te Chuang[1], Tsun-Yi Chen[2], Yi-Cheng Chung[1], Rongshun Chen[1, 2], and Cheng-Yao Lo[1, 2, *]

[1]Department of Power Mechanical Engineering,
[2]Institute of NanoEngineering and MicroSystems,
National Tsing Hua University, Hsin Chu, TAIWAN (R. O. C.)

ABSTRACT

This paper reports an up to 95.9% angle detection accuracy enhancement for capacitive tactile sensors, which entails asymmetric and intentionally shifted electrodes. The rotational shift between two electrodes of a capacitor that greatly contributed to angle detection errors was theoretically analyzed and examined by experiments with simulations. The asymmetric electrodes containing one fan- and one square-shape in capacitors reduced unexpected and rotational-shift induced errors, by keeping the same overlap area of the electrodes. The minimal angle detection resolution was improved from 5.8° to 0.3°, making the tactile sensor practical and reliable in artificial skins.

INTRODUCTION

Capacitive tactile sensors have been widely studied for their potential application on artificial skins. Nevertheless, attentions were only paid on sensitivity and response time on the xy plane with limited angle detection capability in only quadrants. With special arrangements, capacitive tactile sensors take advantage of capacitance changes between capacitors to indicate shear force (F) direction [1]. Many research results imply that shear force detection is only eligible in 0°, 90°, 180°, and 270° on the xy plane when two adjacent capacitors change their capacitances with shear forces. With a four-capacitor array in 2×2 format, those four angles can be distinguished [2-5].

However, for artificial skin applications, precise angle detection is necessary to simulate the touch sense of human beings. Current capacitive tactile sensor proposals support only limited angle resolution, which in turn limits their applications.

Our previous works with intentionally xy-shifted square-electrodes (Figure 1(a)) [6] provided an arbitrary angle detection solution. The electrodes of capacitors were intentionally designed with shifts with each other. As a result, precise angle detection became possible because of the introduction of the area (A) change from the capacitance (C) viewpoint: $C = \varepsilon A/d$, where ε and d represent the dielectric constant and the distance between capacitor's two electrode, respectively. Nevertheless, the angle detection accuracy (ADA) is greatly influenced by the intentional area shift, which is limited by the process controllability. From experimental results, the angle detection error (ADE) could reach 80° when the process as well as the operation generates combinational shifts between the electrodes [6].

Because the process tolerance would never be zero and the algorithm cannot be efficiently customized for individual sensors of a large array, a structural novelty that enhances the ADA is important and necessary.

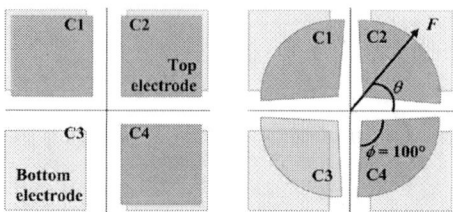

Figure 1: The fundamental symmetric square-electrode (a) and the proposed asymmetric fan-electrode (b). The shear force (F) is applied with an angle (θ) on the xy-plane and a single fan-electrode is with opening angle (φ).

DESING

Different overlap areas of electrodes in a capacitor contribute to different capacitances and alter the results. Although d also contributes to C, it is negligible when the dimensional difference in the xy-direction and in the z-direction is large. The effort for enhancing the ADA consequently fell on the minimization of the overlap area (A) variation during process and operation.

Also from the viewpoint of 360° angle detection requirement, a circular electrode helps on reducing electrode area variation. As a result, a capacitor with asymmetric electrodes on the top and at the bottom was designed. The two electrodes of the capacitor appear with one square- and one fan-shape. Four capacitors finally consist a unit tactile sensor (Figure 1(b)), which is capable for shear force angle (θ) detection in 360°.

The improved ADA could be understood by rotating the upper electrode clockwise or counterclockwise. In the conventional design shown in Figure 1(a), slight process or operation rotation of the upper electrode shows misleading capacitance change because with different shear force angles the overlap area sizes of capacitors will be different. This misleading judgment could be alleviated when the fan-electrode is introduced. Slight process or operation rotation of the upper electrode will not alter the capacitance because the overlap areas of the capacitors are the same. Larger opening angle (ϕ) assures wider process and operation rotation tolerance but it is also limited by its dimensions in the x- and y-direction as well as the spacings between capacitors. In this study, ϕ of the fan-electrode was set to 100°.

PRINCIPLE

During operation, the capacitance of C2 (C4) and C1 (C3) increased and reduced, respectively, when a shear force is applied with $\theta = 0°$ in Figure 1(b). The corresponding increment and decrease of the capacitances are the result of torsion based on spacers supporting at four corners of each individual capacitor as shown in Figure 2.

978-1-4799-7956-1/15 $31.00 © 2015 IEEE   740   MEMS 2015, Estoril, PORTUGAL, 18 - 22 January, 2015

Two adjacent capacitors share two spacers in between.

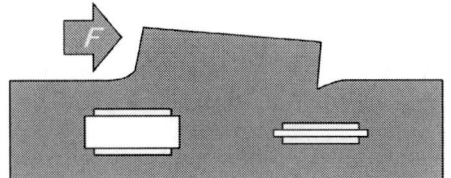

Figure 2: The shear force applied on the tactile sensor, generating a torsion based on spacers supporting around capacitors. The dimensions are not in scale.

In the case of shear force, only distance between two electrode changes are considered and the overlap area changes are neglected because of large dimensional different in the *xy*- and in the *z*-direction.

With $\theta = 0°$, C2 and C4 increase simultaneously while C1 and C3 decrease simultaneously. The increment of C2 and C4 should be identical, and the decrease of C1 and C3 should also be identical. Similar situations take place for $\theta = 90°$, 180°, and 270°, but their corresponding capacitors with decreasing and increasing capacitances are different. However, when θ is different from those four specific angles, the increment of C2 and C4 as well as the decrease of C1 and C3 is not identical. This non-uniform capacitance changes can be further understood by defining θ between 0°-45° and between 45°-90°. The capacitance behaviors of $\theta = 0°$-45° is also a replica of that of $\theta = 45°$-90°. As a result, when separating the 360° with eight 45° regions, the shear force angle detection becomes possible with the help of judgment algorithm $\theta = \tan^{-1}(\Delta C1+\Delta C3)/(\Delta C3+\Delta C4)$ [6].

PROCESS

Electrodes in the capacitors were prepared with 20 nm gold (Au) by conventional photolithography steps as shown in Figure 3. Polyethylene terephthalate (PET) of 188 μm was chosen as the substrate to support possible applications in displays.

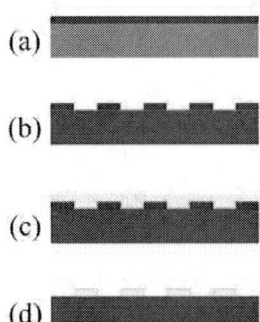

Figure 3: The photolithography steps for the electrodes. (a) Photo resist coating, (b) pattern definition, (c) Au metal evaporation, and (d) PR lift-off. The dimensions are not in scale.

In order to have elastic capacitor for capacitive tactile sensing, first polydimethylsiloxane (PDMS) was prepared with 20 μm thickness and was patterned by silicon (Si) mold. The Si mold depth which represents the spacer height shown in Figure 4. The structured first PDMS was

detached from Si mold and was laminated to the electrode previously prepared in Figure 3. Similar second PDMS processes were repeated with different structure as bumps on top of the capacitors and the second PDMS was also laminated to another electrode previously prepared in Figure 3.

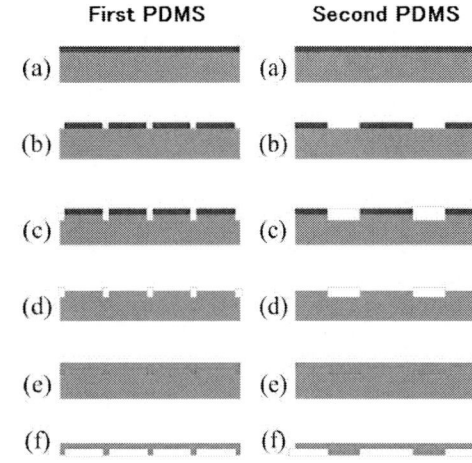

Figure 4: The photolithography steps for the PDMSs. (a) Photo resist coating on silicon mold, (b) pattern definition, (c) silicon mold dry etch, (d) photo resist removal, (e) PDMS coating, and (e) PDMS detachment.

Finally, the first and the second PDMSs were laminated with care to have intentional electrode shifts mentioned in Figure 1(b) as shown in Figure 5.

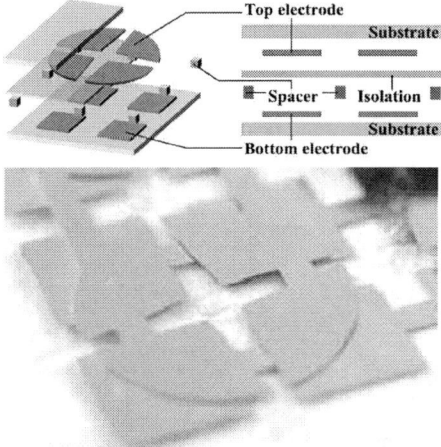

Figure 5: (a) The schematic plot for the proposed capacitive tactile sensor and (b) the tactile sensor after fabrication. In (b), asymmetric electrodes are clearly visible and a transparent bump indicates the center of the four capacitors. The top fan-electrode in (b) was intentionally rotated counterclockwise for ADE studies. The dimensions are not in scale.

SIMULATIONS

Extensive simulations were performed to understand the ADE with different shear force angles as well as different shear forces. The simulation results suggest negligible rotational-shift (δ-shift) ADE if the δ-shift is less than 1.5°. Note that clockwise and counterclockwise rotations represent negative and positive ADEs, respectively, and only clockwise δ-shift ADEs were used in this study.

Simulation results (Figure 6(a)) suggest that larger applied shear force induces slightly larger ADE. This was the influence of the distance between two electrodes change, which was neglected in the model but was magnified under larger shear force. Although the ADE differences between different shear forces were negligible, the absolute ADE values easily reached more than 10°. This was not acceptable because 10° is perceptible by human being. Further simulation results (Figure 6(b)) also suggest that the shear force angle together with the δ-shift issue contributes to unacceptable ADEs. The ADE becomes smaller when θ approaches 45° and it becomes larger when θ approaches 0°.

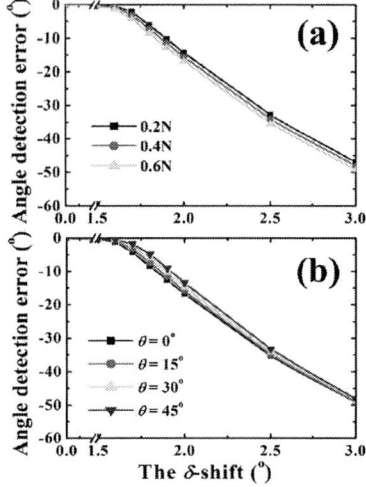

Figure 6: The δ-shift induced A variations resulted in ADEs with (a) various applied forces and (b) various angles. Smaller ADE is irrelevant to applied shear force and its angle, and is effective on all conditions. Both x-axes are δ-shift.

RESULTS AND DISCUSSION

Measurement was performed with the help of force gauge, which provided shear forces to the bump of the tactile sensor. The tactile sensor was connected to a multiplexer (Analog Devices, ADG1208). The multiplexer was controlled by a microprocessor (Microchip, PIC24FJ128GB106), which is connected through the RS232 interface to a computer. The multiplexer selected a specific capacitor from the tactile sensor array and passed the 16-bit digital signal to the microprocessor (Figure 7) and the capacitance value was collected. Because the proposed tactile sensor has four capacitors, a sequential scanning is required. Comparing the initial capacitance with the capacitance under force of the same capacitor, the capacitance change was obtained.

The capacitance was then calculated by algorithm for shear force angle. The shear force was applied with $\theta = 0°$, 15°, 30°, and 45° to examine the ADA. The measured results show identical behavior as the simulations suggested: different C1 to C4 capacitance changes. With

these fan-electrode relative capacitance changes, the algorithm judged the shear force angles. Compared with those measured with square-electrode (Figure 8), a consistent capacitance tactile sensing mechanism was proved (Figure 9). Although the absolute capacitance change values were different in square- and in fan-electrode, the algorithm only verified the relative capacitance changes and the judgment integrity was not influenced by their absolute values.

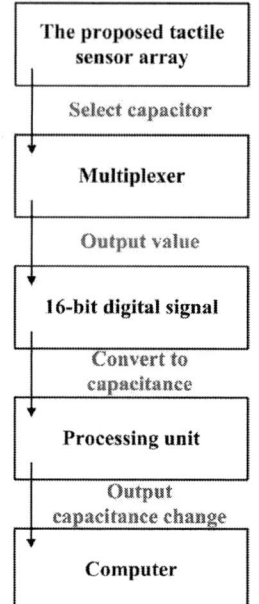

Figure 7: The structure of the sensing module of the proposed tactile sensor.

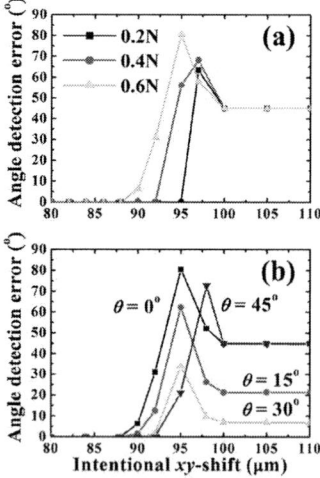

Figure 8: ADEs of 7°–80° appeared in square-electrode with F of 0.2–0.6N (a) and its angle (θ) of 0°–45° (b). Both x-axes are xy-shift.

The average ADE with shear forces (0 N to 1 N) at demonstrative angles (θs) of 0°, 15°, 30°, and 45° were 1.9°, 1.0°, 1.0°, and 0.4°, respectively; while that of the square-electrode were 3.3°, 3.3°, 2.7°, and 0.8°. The experiment results demonstrated enhanced ADA with the fan-electrode, which improved the ADE from 80° (Figure

7) to less than 3.3° (95.9% improvement).

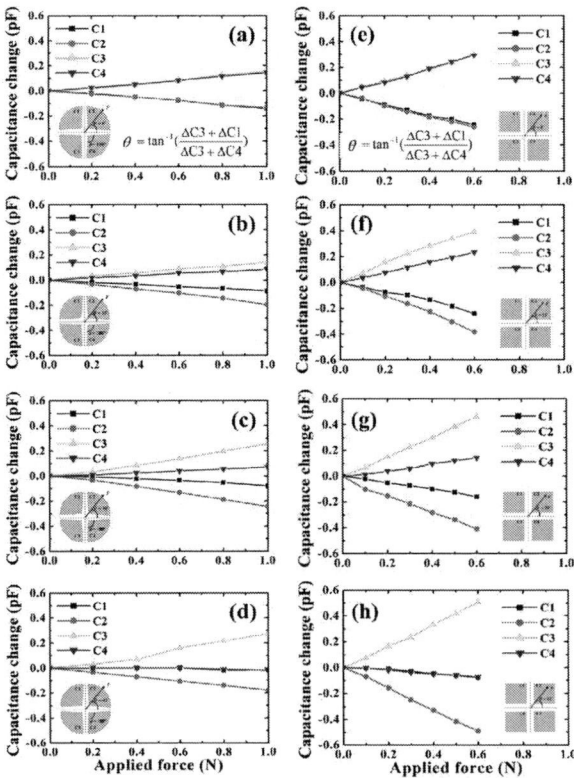

Figure 9: Experiment results for fan- (a-d) and square- (e-h) electrodes of $\theta = 0°$ ((a) and (e)), $\theta = 15°$ ((b) and (f)), $\theta = 30°$ ((c) and (g)), and $\theta = 45°$ ((d) and (h)). All x-axes are the applied force (F). With identical measurement setup and algorithm, the fan-electrode represented smaller capacitance change (a-d), which resulted in smaller ADEs or higher ADAs.

Furthermore, the fan-electrode was manufactured with rotational shift from 3° to 6.5° intentionally, to verify the correctness of theoretical analysis. The average ADE ($\theta = 0°$ to 45°) with 3°, 6°, and 6.5° δ-shifts were 1.5°, 1.1°, and 2.3°, respectively. These results demonstrated effective ADA enhancement and the results were identical to the theoretical analysis, which suggested a substantial ADE with the δ-shift larger than 6.4° (Figure 10).

SUMMARY

In summary, this work reports an asymmetric fan-electrode for capacitive tactile sensors to enhance the ADA. Theoretical analysis, simulation, and experiment were performed, proving it effective. Intentional δ-shift was applied to explain the correctness of the theoretical analysis, and the fan-electrode makes the capacitive tactile sensors practical and reliable in angle detection.

ACKNOWLEDGMENT

This work was partially supported by the Ministry of Science and Technology (MOST, formerly the National Science Council, NSC) with project IDs: NSC 99 – 2221 – E – 007 – 041 - MY3, NSC 101 – 2221 – E – 007 – 054 -

MY2, and MOST 103 – 2221 – E – 007 – 080. This work was also partially supported byNational Tsing Hua University's "Toward a World-Class University Project." The authors appreciate the use of facilities at the Center for Nanotechnology, Material Science, and Microsystems (CNMM) of National Tsing Hua University, which is partly supported by the Ministry of Science and Technology.

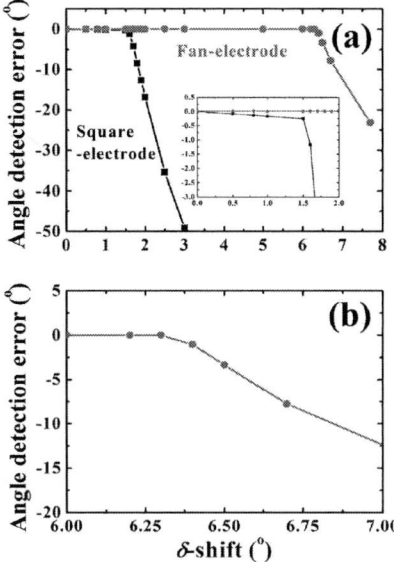

Figure 10: (a) Fan-electrode showed smaller ADE compared with square-electrode with detail in inset and in (b). Both x-axes are δ-shift.

REFERENCE

[1] Yung-Chen Wang, Tsun-Yi Chen, Rongshun Chen, and Cheng-Yao Lo, "Mutual Capacitive Flexible Tactile Sensor for 3-D Image Control", *J. Microelectromech. Syst.*, vol. 22, pp. 804-814, 2013.

[2] Tsun-Yi Chen, Yung-Chen Wang, Cheng-Yao Lo, and Rongshun Chen, "Friction-Assisted Pulling Force Detection Mechanism for Tactile Sensors", *J. Microelectromech. Syst.*, vol. 23, pp. 471-481, 2014.

[3] T. –Y. Chen, *Proc. Transducers* 2013, pp. 1008-1011.

[4] H. K. Lee, J. Chung, S. I. Chang, and E. Yoon, "Normal and Shear Force Measurement Using a Flexible Polymer Tactile Sensor with Embedded Multiple Capacitors", *J. Microelectromech. Syst.*, vol. 17, pp. 934-942, 2008.

[5] M. Liu, J. Xiong, and T. Cui, "Real-Time Measurement of the Three-Axis Contact Force Distribution Using a Flexible Capacitive Polymer Tactile Sensor", *J. Micromech. Microeng.*, vol. 21, 035010, 2011.

[6] Unpublished work.

CONTACT

*C. –Y. Lo, tel: +886-3-5162404; chengyao@mx.nthu.edu.tw

CARBON NANOTUBES-ECOFLEX NANOCOMPOSITE FOR STRAIN SENSING WITH ULTRA-HIGH STRETCHABILITY

Morteza Amjadi[1, 2] and Inkyu Park[2]

[1]Future Research Creative Laboratory, Electronics & Telecommunications Research Institute (ETRI), South Korea

[2]Department of Mechanical Engineering, Korea Advanced Institute of Science & Technology (KAIST), South Korea

ABSTRACT

We developed highly stretchable, flexible and very soft conductors based on the carbon nanotubes (CNTs)-silicone rubber (Ecoflex®) nanocomposite thin films. The resistance of the CNTs-Ecoflex nanocomposite thin film was recovered to its original value under cyclic loading/unloading for strains as large as 510%. Failure strain of the CNTs-Ecoflex nanocomposite was measured to be about ~ 1380% showing its ultra-high stretchability and robustness. As an application of our highly stretchable conductors, we utilized them as skin-mountable and wearable strain sensors for human motion detection. The strain sensors possess high linearity and low hysteresis performance. We observed overshoot behavior of the strain sensors with maximum normalized overshooting peaks 15%. Finally, motion detection of the finger and wrist joints was conducted by using CNTs-Ecoflex nanocomposite thin film strain sensors.

INTRODUCTION

Resistive type strain sensors respond to the mechanical deformation by the change in electrical resistance. Resistive type flexible strain sensors have been reported by coupling of polymers and nanomaterials such as silver nanoparticles (AgNPs) [1], silver nanowires (AgNWs) [2, 3], carbon black particles (CBs) [4, 5], carbon nanotubes (CNTs) [6, 7], and graphene [8, 9]. Moreover, polymeric materials and nanomaterials have been utilized as flexible substrate and strain sensing elements, respectively. For instance, highly stretchable strain sensors have been reported by using AgNWs/PDMS nanocomposite thin films [2, 3], CNTs/PDMS composites [6], CBs/ thermoplastic elastomer (TPE) nanocomposites [4], CBs/silicone rubber (Ecoflex®) composites [10], and graphite/natural rubber composite thin films [7]. However, nonlinear response with high hysteresis is one of the main drawbacks of the reported resistive type sensors [2-4, 6, 7, 10]. On the other hand, highly stretchable ($\varepsilon > 100\%$) and high performance strain sensors are needed for the wearable and skin-mountable applications such as rehabilitation and personal health monitoring [11-13], sport performance monitoring [14, 15], and entertainment fields (e.g. motion capture for games and animation) [2, 16]. The change from "homogeneous" thin film to the "inhomogeneous" thin film is found to be the reason of high nonlinearity. For example, in the case of the AgNW network/PDMS nanocomposite based strain sensors, we found that homogeneous AgNW network changes to the inhomogeneous network under high strains inducing bottleneck locations in the percolation network [2]. These bottlenecks critically limit the electrical conductivity through percolation network causing a high nonlinear resistance-strain dependency of the strain sensors. Highly non-uniform crack propagation in the thin film is the reason of highly nonlinearity of the graphene woven fabrics (GWFs)/PDMS composite based strain sensors [8]. In addition, high hysteresis behavior of flexible strain sensors could be explained by two mechanisms: (i) intrinsic hysteresis behavior of polymeric materials itself due to their viscoelastic properties, and (ii) interaction between nanomaterials and polymer substrates [2, 6]. Specially, in the case of nanocomposite films, friction force between nanomaterials and polymers plays an important rule on the hysteresis performance. Moreover, under stretching, sliding of nanomaterials occurs within polymer medium. Upon releasing cycle, percolation network of nanomaterials cannot re-establish itself suddenly due to the friction force causing high hysteresis behavior in the strain sensors [2]. Therefore, achieving high performance resistive type strain sensors is still challenging.

Herein, we report highly flexible and stretchable thin films based on the nanocomposite of CNT percolation network and Ecoflex. The CNTs-Ecoflex nanocomposite thin films have been utilized as wearable and skin-mountable strain sensors for human motion detection. The characteristics of strain sensors such as linearity, hysteresis performance, stretchability, and overshoot behavior have been investigated. Strain sensors possess high stretchability ($\varepsilon \sim 510\%$) with very good resistance recovery under cyclic loading/unloading. Strain-resistance dependent behavior of the strain sensors is highly linear with low hysteresis. Strain sensors show overshooting upon stretching with maximum overshoot peak value of 15%. As applicability of our strain sensors for the human motion detection, we conducted finger and wrist joint motion detections. The resistance changes caused by the bending and relaxation of strain sensors on the finger and wrist were measured in real-time with high repeatability, stability and fast response.

MATERIALS AND FABRICATION

Multi-walled carbon nanotubes (MWCNTs) with an average length and diameter of 5~20 μm and 16 ± 3.6 nm were purchased from Hyosung Co., South Korea. 0.05 %wt. of non-functionalized CNTs was added into isopropyl alcohol (IPA) and sonicated for an hour. The CNTs-IPA solution was further stirred for another hour to release agglomerated CNTs and perfectly suspend all CNTs into the IPA medium. The uniform suspension of CNTs in IPA was stored for further experiments.

Figure 1 shows fabrication process of the highly

978-1-4799-7956-1/15 $31.00 © 2015 IEEE

stretchable strain sensors based on the CNTs-Ecoflex nanocomposite thin films. CNT solution was first drop-cast on the patterned polyimide (PI) substrate (i.e. rectangular shape: 3×40 mm) and the solution was dried under light heating. Light heating (Lamp light: OSRAM DR 51 50W 12V with Luminous intensity of 1450 cd) was utilized to provide a uniform evaporation of IPA and homogeneous deposition of the CNT thin film [2, 3]. After formation of the CNT thin film, it was annealed under 150 °C for 30 min to remove residual organics. Filtration method was used to transfer the CNT thin film to the polymer medium [2, 17]. To avoid soaking of the liquid Ecoflex between connected CNTs, CNT thin film was pressed by a PI stamp to make it more dens and uniform [17]. Next, liquid Ecoflex with thickness of 0.5 mm was cast on the CNT thin film. Liquid Ecoflex penetrated into the porous network of the CNT thin film due to its very low viscosity. After curing the liquid Ecoflex, a robust nanocomposite of CNTs and Ecoflex was fabricated. All CNT thin film were successfully transferred to the surface of Ecoflex simply by peeling-off the Ecoflex layer from PI substrate due to the strong interfacial adhesion between CNTs and Ecoflex. Then, copper wires were attached to the two ends of the thin film by silver paste and another layer of Ecoflex with the same thickness (~ 0.5 mm) was cured on the top of the CNT thin film to form sandwich structured samples (i.e. Ecoflex layer/CNTs-Ecoflex nanocomposite thin film/Ecoflex layer).

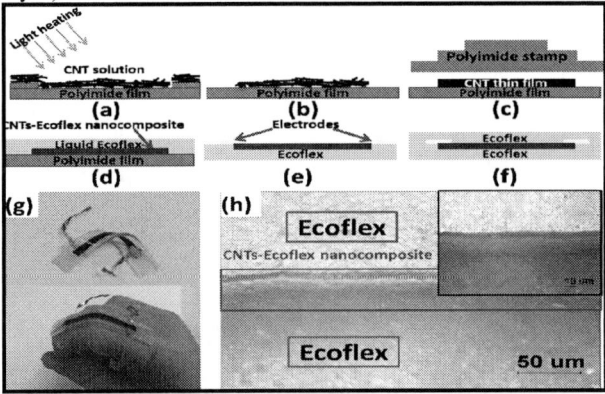

Figure 1: a-f) Fabrication processes of the CNTs-Ecoflex nanocomposite thin film. a) Deposition of CNTs on the patterned PI film under light heating. b) Annealing the thin film at 150 °C for 30 min. c) Pressing the thin film by a PI stamp to make it more dense and uniform. d) Casting the liquid Ecoflex on the CNT thin film and curing it at 70 °C for 2 hours. e) Peeling-off and flipping the Ecoflex layer and attaching the electrodes by silver paste. f) Casting another layer of Ecoflex and curing it at 70 °C for 2 hours. g) Photographs of the fabricated samples under twisting and bending. h) Cross-sectional optical images of the sandwich structured sample; nanocomposite layer is embedded between two layers of Ecoflex.

RESULTS AND DISCUSSION

Figure 1g shows the fabricated strain sensors under twisting and bending with high flexibility and softness. They could easily be attached onto the clothing or directly mounted on the skin. Figure 1h depicts cross-sectional optical image of the sandwich structured sample. As the

figure shows, thin layer of the CNTs-Ecoflex nanocomposite is sandwiched between two layers of Ecoflex.

Electromechanical tests were conducted by attaching the strain sensors to a motorized moving stage (Future Science Motion Controller, FS100801A1P1). Cyclic loading/unloading was applied to the strain sensors while their current was measured by a potentiometer (CH Instruments, Electrochemical Workstation, CHI901D). Figure 2 shows the typical curve for the relative change of resistance versus applied strain. Different strain levels (i.e. 30, 60, and 90%) were applied to the strain sensor and response of the strain sensors under both stretching and releasing cycles was measured. As figure illustrates the resistance of sample was recovered very well after releasing it from strain with a good linear response. Moreover, there is a small hysteresis in the response of the strain sensor mainly caused by viscoelastic behavior of Ecoflex.

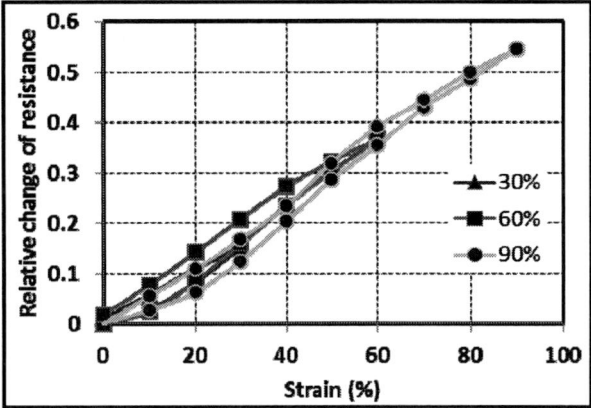

Figure 2: Resistance-strain dependent and hysteresis curve for a CNTs-Ecoflex nanocomposite based strain sensor.

The sensitivity or gauge factor (GF)-slope for curve of the relative change of resistance against applied strain- was calculated to be around ~ 0.61. The main mechanism of resistance-strain dependent behavior of the CNTs/polymers nanocomposite was found to be phenomenon known as "tunneling effect" [18, 19]. Moreover, electrons can tunnel through polymer matrix when the distance between two adjacent CNTs is very small (e.g. 1.8 nm) [18]. When strain sensors are stretched out, re-positioning and re-orientation of CNTs within Ecoflex matrix occur (See inset of Figure 4). Change in the percolation network of CNTs upon stretching increases the tunneling distance between some neighboring CNTs increasing tunneling resistance and consequently increasing the electrical resistance in the whole thin film. Under releasing cycles, recovery and re-establishment of the CNT percolation network leads to recovery of the resistance.

Figure 3 illustrates performance of a strain sensor to the cyclic loading/unloading from 0% to 510%. As shown in figure, there is an excellent overlap between loading profile and response of the strain sensors indicating high stretchability of the CNTs-Ecoflex nanocomposite based strain sensors. The resistance of strain sensors increases linearly with the applied strain and fully recovers to its initial value after releasing with a small hysteresis.

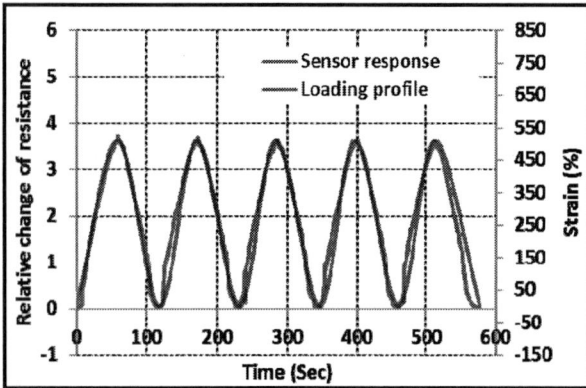

Figure 3: Response of a CNTs-Ecoflex nanocomposite strain sensor to 510% of stretch and release cycles; there is an excellent agreement between sensor's response and loading profile.

To measure the failure strain for a strain sensor, strain was continuously applied to the strain sensor while its resistance changes were monitored. As shown in Figure 4, strain sensor responded to the applied strain with a linear manner and strain sensor lose its electrical conductivity for a strain as large as 1380% indicating ultra-robustness of the CNTs-Ecoflex nanocomposite based strain sensors.

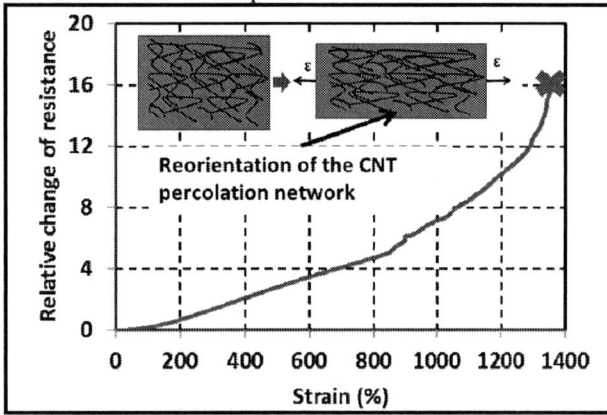

Figure 4: Failure strain measurement for a CNTs-Ecoflex nanocomposite based strain sensor; inset, re-positioning and re-orientation of CNTs in the CNT percolation network upon stretching.

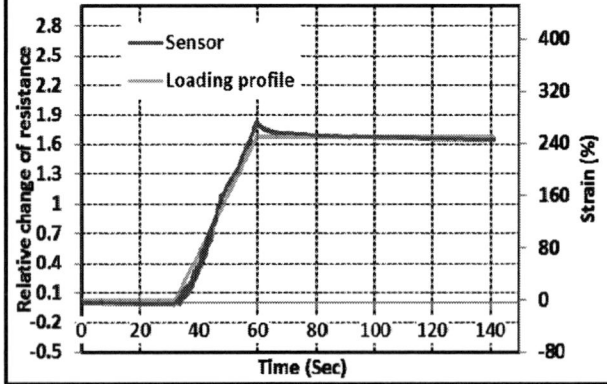

Figure 5: Overshoot behavior of the strain sensor; a small overshoot peak due to the viscoelastic behavior of Ecoflex substrate.

Overshooting was observed in the response of the strain sensors due to the acceleration and viscoelastic

properties of the polymer matrix [6, 10]. A strain sensor was subjected to a continuous strain from 0% to 270% and held at strain of 270% while its resistance changes were measured. As indicated in Figure 5, the strain sensor shows overshooting behavior with normalized peak values about 15%.

Since CNTs-Ecoflex nanocomposite based strain sensors are very flexible and stretchable, they could be utilized as wearable and skin-mountable electronic devices. To demonstrate the applicability of our strain sensors, we conducted human motion detection by mounting of strain sensors on the human body. As depicted in the insets of Figure 6, strain sensors are mounted on the finger and wrist for the joint motion detection. Strain sensors are well-attached to the body without any delamination or sliding due to the softness of the Ecoflex matrix. Figure 6 shows the responses of the strain sensors to the bending of finger and wrist. More bending accommodates more strain on the strain sensors and consequently increases the resistance of the strain sensors. Strain sensors responded to the cyclic bending/relaxation with very good repeatability, stability, and fast response.

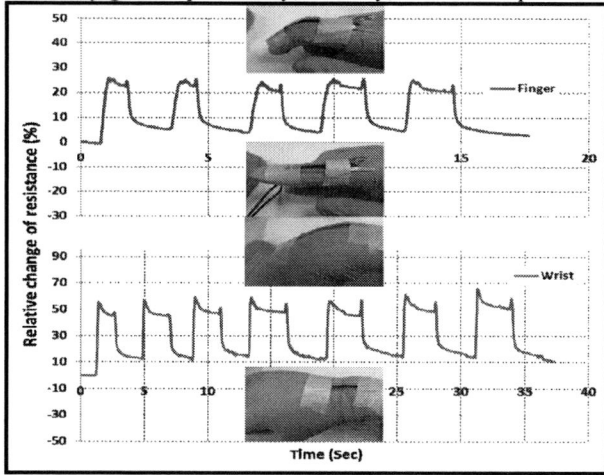

Figure 6: Finger and wrist joints' motion detection using ultra-stretchable and wearable strain sensors; insets of figure, strain sensors mounted on the finger and wrist.

CONCLUSIONS

In this paper, highly stretchable strain sensors based on the CNTs-Ecoflex nanocomposite thin films were fabricated. Stretchability and GFs of the sensors are 510% and 0.61, respectively. The strain sensors exhibit a good linearity with small hysteresis behavior. Maximum normalized overshoot peak was 15% when strain sensors were stretched. As a wearable application, strain sensors were mounted on the finger and wrist for the joint bending measurements. Strain sensors possess high repeatability, stability and fast response under cyclic loading/unloading.

ACKNOWLEDGEMENTS

This research was supported by the Fundamental Research Program (PNK3771) of the Korean Institute of Materials Science (KIMS), Research Program (KM3330) of Korea Institute of Machinery & Materials (KIMM), and the Nano•Material Technology Development Program through the National Research Foundation of Korea (NRF) funded by the Ministry of Science, ICT & Future Planning (No. 2013043661).

REFERENCE

[1] J. Lee, S. Kim, J. Lee, D. Yang, B. C. Park, S. Ryu, *et al.*, "A stretchable strain sensor based on a metal nanoparticle thin film for human motion detection," *Nanoscale,* 2014.

[2] M. Amjadi, A. Pichitpajongkit, S. Lee, S. Ryu, and I. Park, "Highly stretchable and sensitive strain sensor based on silver nanowire-elastomer nanocomposite," *ACS Nano,* vol. 8, pp. 5154-63, 2014.

[3] M. Amjadi, A. Pichitpajongkit, S. Ryu, and I. Park, "Piezoresistivity of AG NWS-PDMS nanocomposite," in *Micro Electro Mechanical Systems (MEMS), 2014 IEEE 27th International Conference on,* 2014, pp. 785-788.

[4] C. Mattmann, F. Clemens, and G. Tröster, "Sensor for measuring strain in textile," *Sensors,* vol. 8, pp. 3719-3732, 2008.

[5] N. Lu, C. Lu, S. Yang, and J. Rogers, "Highly Sensitive Skin-Mountable Strain Gauges Based Entirely on Elastomers," *Advanced Functional Materials,* vol. 22, pp. 4044-4050, 2012.

[6] T. Yamada, Y. Hayamizu, Y. Yamamoto, Y. Yomogida, A. Izadi-Najafabadi, D. N. Futaba, *et al.*, "A stretchable carbon nanotube strain sensor for human-motion detection," *Nature nanotechnology,* vol. 6, pp. 296-301, 2011.

[7] S. Tadakaluru, W. Thongsuwan, and P. Singjai, "Stretchable and Flexible High-Strain Sensors Made Using Carbon Nanotubes and Graphite Films on Natural Rubber," *Sensors,* vol. 14, pp. 868-876, 2014.

[8] X. Li, R. Zhang, W. Yu, K. Wang, J. Wei, D. Wu, *et al.*, "Stretchable and highly sensitive graphene-on-polymer strain sensors," *Scientific reports,* vol. 2, 2012.

[9] C. Yan, J. Wang, W. Kang, M. Cui, X. Wang, C. Y. Foo, *et al.*, "Highly Stretchable Piezoresistive Graphene–Nanocellulose Nanopaper for Strain Sensors," *Advanced Materials,* 2013.

[10] J. T. Muth, D. M. Vogt, R. L. Truby, Y. Mengüç, D. B. Kolesky, R. J. Wood, *et al.*, "Embedded 3D Printing of Strain Sensors within Highly Stretchable Elastomers," *Advanced Materials,* 2014.

[11] C.-X. Liu and J.-W. Choi, "An embedded PDMS nanocomposite strain sensor toward biomedical applications," in *Engineering in Medicine and Biology Society, 2009. EMBC 2009. Annual International Conference of the IEEE,* 2009, pp. 6391-6394.

[12] T. Giorgino, P. Tormene, F. Lorussi, D. De Rossi, and S. Quaglini, "Sensor evaluation for wearable strain gauges in neurological rehabilitation," *Neural Systems and Rehabilitation Engineering, IEEE Transactions on,* vol. 17, pp. 409-415, 2009.

[13] F. Lorussi, E. P. Scilingo, M. Tesconi, A. Tognetti, and D. De Rossi, "Strain sensing fabric for hand posture and gesture monitoring," *Information Technology in Biomedicine, IEEE Transactions on,* vol. 9, pp. 372-381, 2005.

[14] R. Helmer, D. Farrow, K. Ball, E. Phillips, A. Farouil, and I. Blanchonette, "A pilot evaluation of an electronic textile for lower limb monitoring and interactive biofeedback," *Procedia Engineering,* vol. 13, pp. 513-518, 2011.

[15] C.-X. Liu and J.-W. Choi, "Patterning conductive PDMS nanocomposite in an elastomer using microcontact printing," *Journal of Micromechanics and Microengineering,* vol. 19, p. 085019, 2009.

[16] S. S. Rautaray and A. Agrawal, "Interaction with virtual game through hand gesture recognition," in *Multimedia, Signal Processing and Communication Technologies (IMPACT), 2011 International Conference on,* 2011, pp. 244-247.

[17] H. Eom, J. Lee, A. Pichitpajongkit, M. Amjadi, J. H. Jeong, E. Lee, *et al.*, "Ag@ Ni Core–Shell Nanowire Network for Robust Transparent Electrodes Against Oxidation and Sulfurization," *Small,* 2014.

[18] C. Li, E. T. Thostenson, and T.-W. Chou, "Dominant role of tunneling resistance in the electrical conductivity of carbon nanotube–based composites," *Applied Physics Letters,* vol. 91, p. 223114, 2007.

[19] N. Hu, Y. Karube, C. Yan, Z. Masuda, and H. Fukunaga, "Tunneling effect in a polymer/carbon nanotube nanocomposite strain sensor," *Acta Materialia,* vol. 56, pp. 2929-2936, 2008.

CONTACT

*I. Park, tel: +82-42-350-3240; inkyu@kaist.ac.kr

DEVELOPMENT OF A MINIATURIZED LASER DOPPLER VELOCIMETER FOR USE AS A SLIP SENSOR FOR ROBOT HAND CONTROL

N. Morita[1], H. Nogami[2], Y. Hayashida[2], E. Higurashi[3], T. Ito[4], and R. Sawada[1, 2]

[1]Graduate School of Systems Life Sciences, Kyushu University, Fukuoka, JAPAN

[2] Department of Mechanical Engineering, Kyushu University, Fukuoka, JAPAN

[3]Department of Precision Engineering, University of Tokyo, Tokyo, JAPAN

[4]Faculty of Computer Science and Systems Engineering, Kyushu Institute of Technology, Fukuoka, JAPAN.

ABSTRACT

We developed a miniaturized laser Doppler velocimeter (LDV) for use as a slip sensor in the control of a robot hand. This sensor is only 1/10,000th of the volume of commercial LDVs, which enables the sensor to be attached to a robot hand. Our LDV is able to detect objects moving at velocities ranging from 10 μm/s to 20,000 μm/s. The output of this sensor is independent of the type of material measured; aluminum, cardboard, and rough-surfaced black plastic were all tested.

INTRODUCTION

In the near future, many countries in the world will face ageing populations. As a result, many problems will emerge, such as an increase in the number of patients, and a reduced workforce. Thus, we need robots that can assist us in the same manner as a human, i.e., over a wide range of tasks and environments. A robotic hand that is capable of gripping various objects with an appropriate force and without slipping, similar to a human hand, is thus a necessity. To realize a practical robotic hand, a slip detection sensor that can detect slipping of various materials, as may be found in hospitals, offices, factories, or in the home, is necessary for effective control, and it must be small enough to be attached to the fingers of this hand (Fig. 1).

However, there is as yet no slip detection method that is generally applicable and versatile. For example, two types of small slip detection sensors have been developed: the piezoelectric polyvinylidene fluoride (PVDF) film type [1] and the pressure-conductive rubber type [2]. These sensors can determine if an object is slipping or not by measuring rapid changes in shear force. However, the magnitude of the shear force depends on the surface material of the object, and thus it is difficult to set a threshold value of the shear force as a slip discriminant.

Fortunately, another slip detection method is available for various types of surface material. The laser Doppler velocimeter (LDV) has been widely used to measure the velocities of objects made of various materials, such as reflective metal, matte paper, and liquid, in fields ranging from industrial production to medicine [3]. An LDV has the advantage of enabling a slip detection sensor to measure the velocity of various objects that scatter light. Thus, we propose using laser Doppler velocimeter as a slip detection method by measuring the motion of objects. However, the sizes of commercially available LDVs range from several square centimeters to tens of square centimeters, and are not small enough to be embedded in robotic fingertips.

In our previous research, we developed an integrated micro-optical encoder a few millimeters in size, which consisted of optical elements such as a laser-diode or a photo-diode, mirrors, and lenses [4-6]. Since the components of a micro-optical encoder are similar to those of an LDV, we can apply our miniaturization technique for optical sensors towards miniaturizing LDVs.

In this research, we developed an extremely compact LDV, called a miniaturized LDV, using the fabrication process for our micro-optical encoder. It can measure surface velocity of a moving object, and is small enough to be embedded in the fingertips of a robotic hand.

MINIATURIZED LDV

Design of the miniaturized LDV

Figure 2 shows a picture of the miniaturized LDV. The sensor is 2.8 × 2.8 mm and 1.0 mm thick. It is only 1/10,000th of the volume of commercial LDVs. It consists of two layers: 1) a micro-machined Si optical bench incorporating a bonded laser diode (LD), Au micromirrors, and electrodes (including through-hole electrodes), as shown in Fig. 3(a); and 2) a glass substrate incorporating a bonded photo-diode (PD), electrical interconnection, and refractive microlenses, as shown in Fig. 3(b). The Si substrate has a cavity that was fabricated using an anisotropic silicon etching technique. The micromirrors and electrodes were formed by depositing gold onto the silicon (111) facet. Laser beams are collimated within a radius of 0.4 mm by refractive lenses, which were fabricated using gray-scale lithography.

Principles of the LDV

Figure 4 shows a schematic of the A–A′ section shown in Fig. 2. Two laser beams are simultaneously emitted from the LD, reflected by the Au mirrors, collimated by the refractive microlenses, and hit the moving object. The two laser beams, which come from the left and right sides, are scattered and their frequencies are Doppler shifted by Δf_L and Δf_R, respectively:

$$\Delta f_L = -\frac{V}{\lambda} cos\theta \text{ and} \tag{1}$$

$$\Delta f_R = \frac{V}{\lambda} cos\theta, \tag{2}$$

where V is the velocity of the moving object, θ (= 66°) is the angle between the direction of the laser beam and the direction of the object's motion, and λ (= 1.31 μm) is the

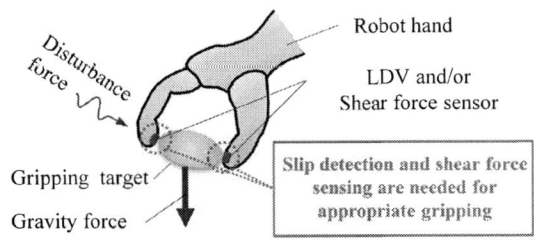

Figure 1: The application of the miniaturized LDV.

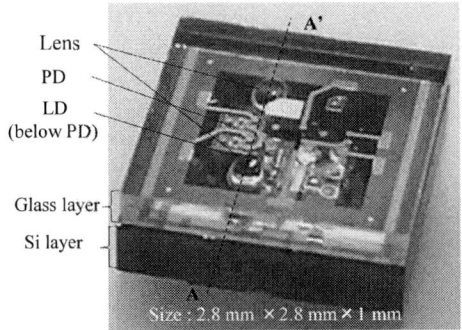

Figure 2: Picture of the miniaturized LDV.

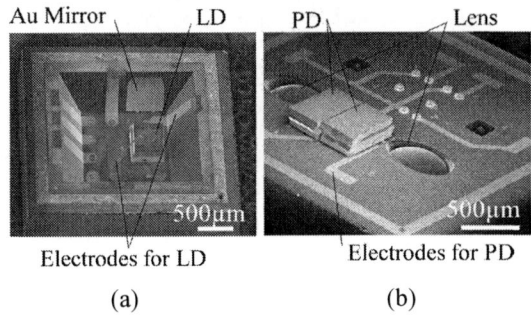

(a)　　　　　　　　(b)

Figure 3: SEM images of the components of the miniaturized LDV: (a) the glass layer and (b) the Si layer.

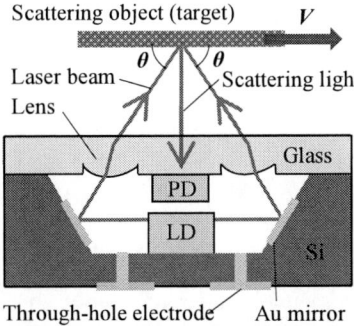

Figure 4: Schematic of the miniaturized LDV (along the A–A' cross-section shown in Fig. 1).

laser's wavelength. After the beams are scattered, they interfere with each other because the two beams do not have the same frequency as when they hit the moving object. This interference causes the emergence of a beat frequency, f_d, which is equal to the difference in frequencies:

$$f_d = (f + \Delta f_R) - (f + \Delta f_L) = 2\frac{V}{\lambda}\cos\theta \qquad (3)$$

In Eq. (3) [3], f_d is linearly proportional to V, and f is the initial laser beam frequency before it impacts the moving object. Although the frequency of a laser cannot be detected directly as it is extremely high (230 THz for a wavelength of 1.31 μm) compared to the PD's response speed, the PD can detect f_d because the beat frequency is significantly lower than the laser frequency f.

EXPERIMENT

Figure 5 shows a schematic of the experimental setup for measuring the characteristics of the miniaturized LDV. The miniaturized LDV was placed opposite a scattering object, which was attached to a voice coil motor stage. Using the voice coil motor, the scattering objects were moved horizontally past the sensor surface to observe the difference in the signal. As we are simulating a robotic hand detecting the initial slip of a grasped object, the object was moved at velocities ranging from 10 μm/s to 20,000 μm/s. We used three types of scattering boards, made of aluminum, cardboard, and rough-surfaced black plastic, as shown in Fig. 6. An LD current was supplied by an LD driver. The output voltage from the PD was amplified and then transformed into a power spectrum via Fast Fourier Transform (FFT) to obtain the beat frequency, which is the frequency at the peak power.

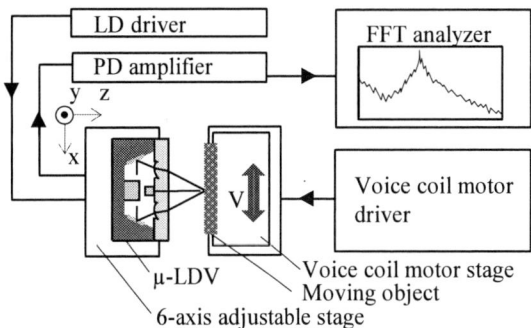

Figure 5: Experimental setup for measuring the velocimeter characteristics.

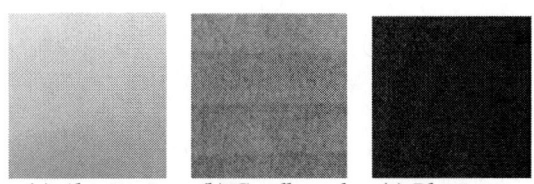

(a) Aluminum　(b) Cardboard　(c) Plastic

Figure 6: Images of the scattering boards used as moving objects.

RESULTS

Figures 7 and 8 show example results for the power spectra obtained from black plastic at low and high velocities, respectively. The velocity, frequency, and power at each peak are shown for each graph. The peaks are proportional to the velocity and are almost the same as

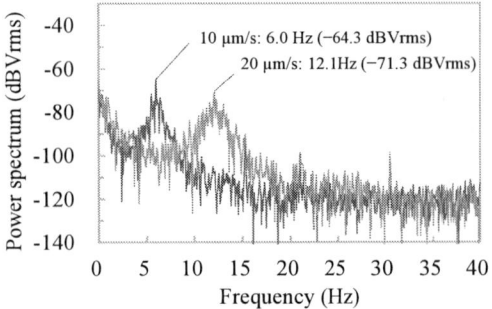

Figure 7: Power spectra from black plastic at low velocities.

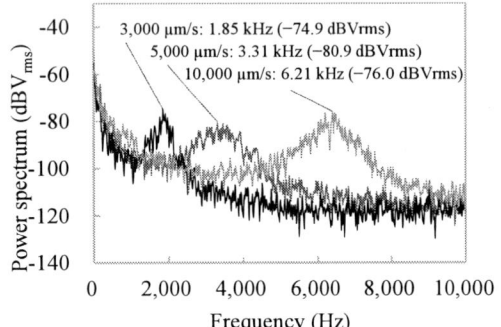

Figure 8: Power spectra from black plastic at high velocities.

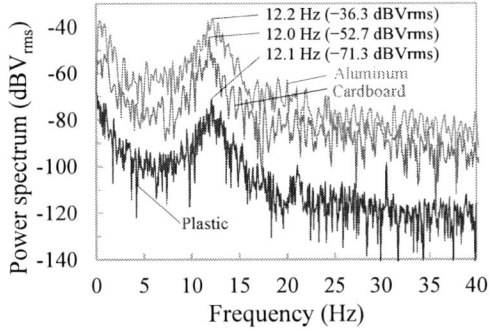

Figure 9: Power spectra from three moving object material types at a velocity of 20 μm/s.

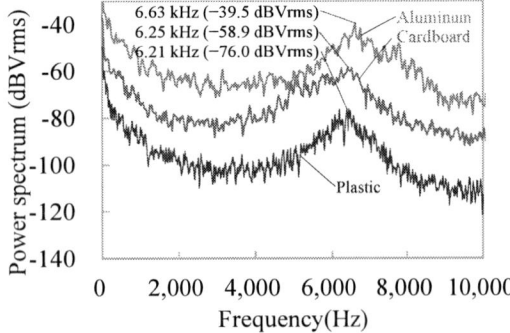

Figure 10: Power spectra from three moving object material types at a velocity of 10,000 μm/s.

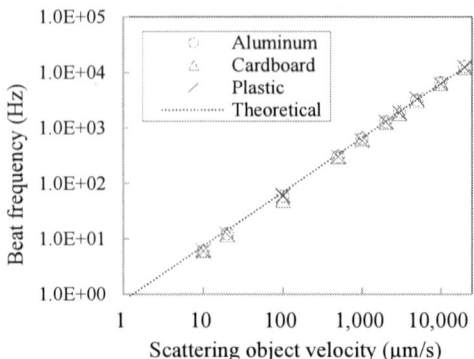

Figure 11: Beat frequencies for all measured velocities and materials.

the theoretical beat frequencies determined by Eq. (3). There is a difference of more than 20 dB between the peak and the noise of the spectrum, except below 4 Hz for low velocity measurements and below 1 kHz for high velocity measurements. We obtained a similar pattern of power spectrum peaks for other velocities and objects.

Figures 9 and 10 show example results for the power spectra obtained from the three scattering board types (aluminum, cardboard, and black plastic) at low and high velocities, respectively. The frequency and power at each peak are shown for each graph. Although the power at each peak depends on the type of material and surface, the frequencies at each peak for the same velocity are nearly identical.

The beat frequencies for all measured velocities for each material and the theoretical beat frequencies expressed by Eq. (3) are plotted in Fig. 11. Good linearity between the object velocity and the measured beat frequency is observed for all three material types. The measured results are very close to the theoretical beat frequencies over a wide range of velocities. According to this, by reading the frequency at the peak of the power spectrum, the miniaturized LDV can measure the velocity of various objects, without requiring any information about the type of material and surface conditions.

CONCLUSION

We developed a miniaturized laser Doppler velocimeter (LDV), which is 2.8 × 2.8 mm and 1.0 mm thick. It is only 1/10,000th of the volume of a commercial LDV. In this way, owing to its small size and light weight, our miniaturized LDV can be embedded in robotic fingertips. By reading the beat frequency, without requiring additional surface information, the sensor can detect velocities ranging from 10 μm/s to 20,000 μm/s for objects made of various materials, such as aluminum, cardboard, or black plastic. We expect that the miniaturized LDV can be used in the near future for the control of robotic hands in gripping objects of various materials with an appropriate gripping force without slipping.

REFERENCES

[1] S. Shirafuji, K. Hosoda, "Detection and prevention of slip using sensors with different properties embedded in elastic artificial skin on the basis of previous experience", *Robotics and Autonomous Syst.,* vlo. 62, pp.46-52, 2014.

[2] S. Teshigawara, et al, "Highly Sensitive Sensor for Detection of Initial Slip and Its Application in a Multi-fingered Robot Hand", *IEEE Intern. Conf. on Robotics and Automation,* pp.1097-1102, 2011.

[3] J.B. Abbiss, T.W. Chubb, and E.R. Pike, "Laser Dppler anemometry", Optics and Laser Technol., 6.6 (1976), pp. 249–261.

[4] E. Higurashi et al., "Optical Microsensors Integration Technologieds for Biomedical Applications", *IEICE Trans. Electron.,* 92.2 (2009), pp. 231–238.

[5] R. Sawada, Eiji Higurashi, and Yoshito Jin "Hybrid Microlaser Encoder", *Jurnal of lightwave technology,* Vol. 21, No. 3, pp815-820, 2003.

[6] T. Takeshita, T, Iwasaki, E. Higurashi, and R. Sawada, "Application of Nanoimprint Technology to Diffraction Grating Scale for Microrotary Encoder", *Sensor and Materials,* Vol. 25, pp609-618, 2013.

MICROCANTILEVER SYSTEM INCORPORATING INTERNAL RESONANCE FOR MULTI-HARMONIC ATOMIC FORCE MICROSCOPY

Chris Pettit[1], Bongwon Jeong[2], Hohyun Keum[2], Joohyung Lee[1], Jungkyu Kim[1], Seok Kim[2]
Donald Michael McFarland[3], Lawreence A. Bergman[3], Alexander F. Vakakis[2], Hanna Cho[1]

[1]Department of Mechanical Engineering, Texas Tech University, TX, USA
[2]Department of Mechanical Science and Engineering, University of Illinois at Urbana-Champaign, IL, USA
[3]Department of Aerospace Engineering, University of Illinois at Urbana-Champaign, IL, USA

ABSTRACT

We report a new design concept of micromechanical cantilever system incorporating the 1:3 internal resonance during dynamic mode operation of atomic force microscopy (AFM). The passive amplification of third harmonic triggered through the mechanism of 1:3 internal resonance enables AFM to utilize multiple harmonics in an air environment. Detailed theoretical and experimental studies of the proposed design demonstrate that the multi-harmonic AFM (MH-AFM) is capable of simultaneous topography imaging and compositional mapping with more than 10-fold enhanced sensitivity.

INTRODUCTION

AFM is widely used to image the topography and study material properties with nanoscale resolution. In dynamic AFM mode, the microcantilever is excited at a frequency, usually near its fundamental resonant frequency, and the response of the cantilever is measured as the tip interacts with the sample surface. Multi-frequency AFM, exploiting more than one frequency in cantilever dynamics, is one of the leading trends to improve the state-of-art AFM [1]. Such a trend is motivated by the fact that the higher harmonic terms, generated due to the nonlinear tip-sample interaction in dynamic AFM operation, are expected to have more information about the sample due to their higher temporal resolution. However, a special experimental scheme is required to get reliable results because of their low signal-to-noise ratio. Intrinsic enhancement of higher harmonics can sometimes be utilized in a liquid environment [2]. In air, however, multi-frequency operation is achieved not in a passive way but in an active way by exciting multiple-modes of cantilever [3]. Here we realize MH-AFM measurement in air by introducing a new design of microcantilever system that enables passive amplification of a higher harmonic through the mechanism of internal resonance.

NEW CANTILEVER SYSTEM

Figure 1a shows a scanning electron microscopy (SEM) image of the modified AFM cantilever system. The new cantilever system consists of an inner paddle attached in the form of a silicon membrane. The inner paddle sits in the cavity of the modified commercial AFM cantilever. The outer and inner cantilevers act like two coupled linear oscillators. The first two bending modes correspond to in-phase and out-of-phase oscillations between the base microcantilever and the inner paddle. The dimensions of the inner paddle are carefully selected

to achieve a 1:3 internal resonance between the two leading bending modes of the combined cantilever system. The higher harmonic frequency response can be studied by the activation of the higher harmonics due to nonlinear energy transfer triggered by the inherent nonlinear cantilever tip-surface interactions.

Figure 1b shows the lumped-parameter, reduced order model of the cantilever system in Fig. 1a. Due to the large discrepancy in the dimensions between the inner and outer cantilever, and assuming each oscillates in their respective fundamental modes, each cantilever can be considered as a discrete mass-spring-damper system linearly coupled with each other. The equations of motion can be described as the following:

$$m_1 \ddot{x}_1 + c_1 \dot{x}_1 + k_1 x_1 + c_2(\dot{x}_1 - \dot{x}_2) + k_2(x_1 - x_2)$$
$$= k_1 y_0 \cos(\omega_d t) + c_1 y_0 \omega \cos(\omega_d t) + F_{ts}(x_1) \quad (1)$$

$$m_2 \ddot{x}_2 + c_2(\dot{x}_2 - \dot{x}_1) + k_2(x_2 - x_1) = 0 \quad (2)$$

(a)

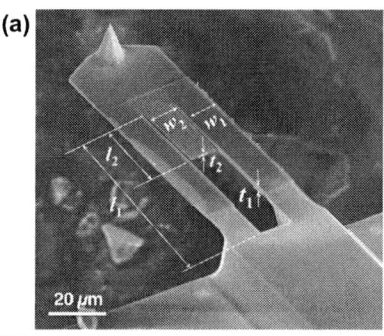

(b)

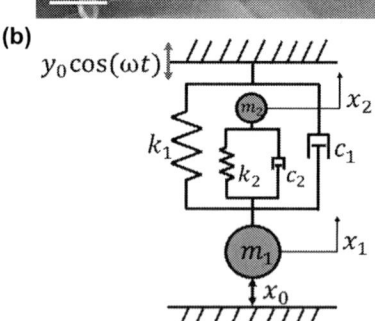

Figure 1: Microcantilever system with an inner paddle designed to incorporate 1:3 internal resonance. (a) SEM image, (b) simplified lumped-parameter model; m, k, c, and x denote the effective mass, stiffness, damping coefficient, and displacement, where the subscripts 1 and 2 denote the fundamental bending mode of the base microcantilever and inner paddle, respectively; x_0 is tip-sample clearance.

(a)

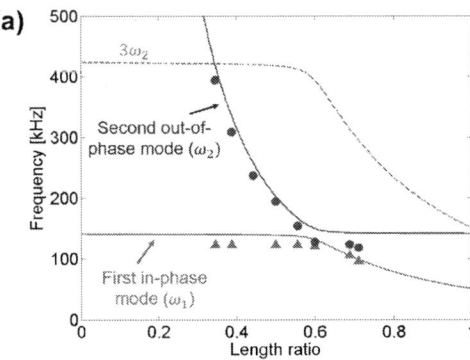

(b)

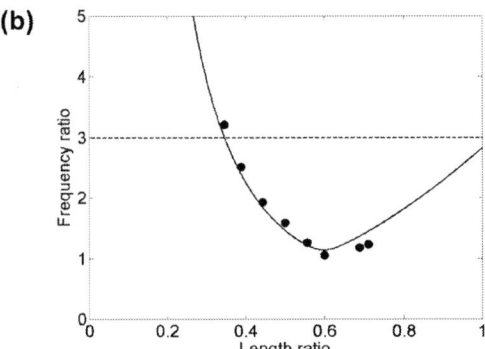

Figure 2: Linearized eigenfrequencies predicted by linearized model by varying length ratio of inner paddle and external cantilever. (a) the red and blue lines show the first (in phase) and second (out of phase) modes of the system. The dashed-red line is simply the first eigenfrequency multiplied by 3. The intersection of the dashed-red and blue line show the desired 1:3 internal resonance. The dots show experimental data results. (b) The solid black line shows the eigenfrequency ratio with respect to length ratio of the inner paddle to outer cantilever, compared with experimental results as dots.

where m, c, k, represent the effective modal mass, damping coefficient, and spring constant, respectively. The subscripts 1 and 2 represent inner and outer cantilever, respectively. During dynamic AFM operation, the cantilever is excited at the base expressed as $y_0 cos(\omega t)$. The term F_{ts} is the nonlinear cantilever tip-sample interaction forces applied to mass m_1. The tip sample interaction was mathematically modeled by the Derjaguin-Müller-Toporov (DMT) contact model [5,6]. The tip sample force is a function of tip sample separation described by the following equations:

$$
F_{ts}(x_1) = \begin{cases} \dfrac{-Hr}{6(x_0 + x_1)^2} & (x_0 + x_1 > a_0) \\[2ex] \dfrac{-Hr}{6a_0^2} + \dfrac{4}{3}E^*\sqrt{r}\left[a_0 - (x_0 + x_1)\right]^{3/2} & (x_0 + x_1 \le a_0) \end{cases} \quad (3)
$$

where H, r, E^*, a_0, and $(x_0 + x_1)$ are the Hamaker constant, tip radius, effective Young's modulus, molecular distance, and tip-sample separation, respectively.

Modal analysis was performed in order to find the design parameters of the inner paddle fabrication in order

to achieve the desired 1:3 internal resonance [7,8]. In order to relate the mass and spring constant of the inner cantilever to the external cantilever, the leading eigenfrequencies and normalized parameters are expressed as:

$$
\omega_{1,2} = \omega_L \left[\frac{\left(\alpha^2\left(1 + \alpha\beta\gamma + \mu\right)\beta + ^4\right)}{2\beta^4(1+\mu)} \pm \frac{\sqrt{\left(\alpha^2\left(1 + \alpha\beta\gamma + \mu\right) + \beta^4\right)^2 - 4\alpha^2\beta^4\mu(1)}}{2\beta^4(1+\mu)} \right] \quad (4)
$$

$$
\alpha = t_2/t_1 \quad \beta = l_2/l_1 \quad \gamma = w_2/2w_1 \quad \mu = M/m_1 \quad (5)
$$

Since the ratios of thickness and width are known, the length scale ratio to achieve the 1:3 internal resonance must be determined. Figure 2a shows the linearized eigenfrequencies as a function of varying the length ratio of the inner paddle and outer cantilever. The red and blue lines show the in-phase and out-of-phase modes, respectively, calculated based on the analytical estimation of Eqs. (4-5). The dashed-red line is the first mode multiplied by three, the desired ratio between bending modes. The internal resonance would only be possible to achieve where the blue and dashed-red lines intersect, giving the proper length ratio for the fabrication process. Figure 2b shows the normalized eigenfrequency ratio as a function of varying the length ratio. The experimentally measured eigenfrequencies are denoted by red (in-phase), blue (out-of-phase), and black (ratio) markers. The good agreement between the analytical and experimental results verifies the above model to be valid to determine the geometric parameters of the designed system.

The fabrication of the cantilever system with the inner paddle was fabricated by modifying a commercially available AFM cantilever. The base cantilever was selected as a MikroMasch NSC-14 AFM cantilever. The basic idea behind the fabrication involves transfer printing, annealing, and cutting with a focused ion beam. First, the ion-milling process was performed in a focused ion-beam (FIB) to carve out the middle part of the cantilever. A silicon on insulator (SOI) wafer was used as a 'donor' where thin layers are cut to achieve the Si membrane. The thin, 300 nm-sheet of silicon membrane, also called 'inks,' was placed on the back side of the base cantilever. The adhesion was done by Si-SiO$_2$-Si bonding and then annealed in a furnace at 1000°C. After the membrane fully bonded to the cantilever, the inner paddle was then carved out by the FIB. Then the inner paddle length was gradually adjusted by trimming the free end down, adjusting the eigenfrequencies to have an internal resonance ratio of 1:3.

RESULTS

Numerical simulations were performed to estimate the response of the amplitude and phase at the first and third harmonics while being driven at the fundamental in-phase mode frequency. In Fig. 3a the amplitude at first harmonic (1Ω) and third harmonic (3Ω) of the driving frequency are shown as a function of average tip-sample distance. There is no tip-sample interaction when the average tip-sample distance is greater than 200 nm. At

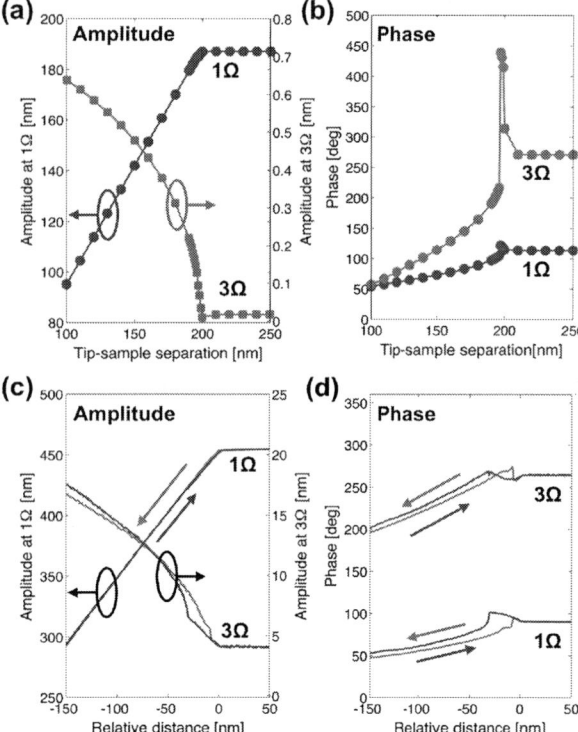

Figure 3: (a) Numerical simulation results based on the model, imposing the Derjaguin-Müller-Toporov (DMT) model in the tip-sample interaction where the effective Young's modulus is 10 GPa. (b) Experimentally obtained results on a silicon specimen for a cantilever system having 1:2.999 ratio between its first two mode frequencies; red and blue lines indicate the measurement as the tip approached to, and retreated from the surface of the sample, respectively.

this point it is obvious that the amplitude at 1Ω is not being affected and there is no 3Ω amplitude response. The higher harmonic amplitude is not observed because of the absence of tip-sample interaction forces. Once the cantilever tip has engaged to the sample, when the average tip-sample distance is less than 200 nm, it is obvious that the 3Ω higher harmonic amplitude has been enhanced. This shows that the 1:3 internal resonance has been triggered with tip-sample interaction. As the tip-sample distance is lowered, the amplitude at 1Ω is gradually decreased and the amplitude at 3Ω is increased because the nonlinearity in the system becomes stronger. This further proves that the higher harmonic is sensitive to tip-sample interaction forces. In the phase plots of Fig. 3b, the shift in phase also occurs as the nonlinear tip-sample interaction affects increasingly the dynamics of the AFM system. Comparing the phase shift of the first and third harmonic terms, the third harmonic phase is about three times more sensitive to the change of tip-sample interaction. The sudden jump in both the first and third harmonic phase responses represents the transition of the dominant tip-sample interaction from the attractive to the repulsive regimes.

Figures 3c-d also show the experimental responses of the first and third harmonic frequencies, represented as Ω and 3Ω respectively, with respect to the tip-sample

distance. The cantilever selected has 1:2.95 eigenfrequency ratio, with the first bending mode at 120.0 kHz and the second bending mode at 353.8 kHz. The measurement was performed on a hard silicon sample. It is noted that the third harmonic was not excited, and the response is only observable due to the nonlinearities of the tip-sample interaction. The amplitude is plotted as a function of average cantilever tip-sample distance, with red and blue lines indicating the approaching and retreating measurement, respectively. As predicted by the numerical simulation, the amplitude at 1Ω decreases and the amplitude at 3Ω increases with decreasing tip-sample distance. This shows that higher harmonic response has been amplified from the repulsive forces of the tip-sample interaction, due to a nonlinear energy transfer between flexural bending modes. The low signal to noise ratio, which is easily perceived, is due largely in part to the smaller, inner paddle. Another important factor is that the phase signal sensitivity at 3Ω is much more sensitive than that of the fundamental frequency.

In order to evaluate the effectiveness of the proposed cantilever with 1:3 internal resonance, MH-AFM imaging was performed on a thin PDMS film embedded with 200 nm polystyrene (PS) nanoparticles. The sample was prepared by randomly distributing PS nanoparticles into a polydimethylsiloxane (PDMS) substrate mixed with hexane (at ratios of curing agent/base/hexane of 1:10:1000), based on drop casting method. The PDMS films with nanoparticle-coated glass slides were cured on a hot plate at 150° C for ten minutes. The fabricated microcantilever, with 1:2.95 eigenfrequency ratio (i.e., the first bending mode at 120.0 kHz and the second bending mode at 353.8 kHz), was mounted on a conventional AFM system (model MFP-3D AFM, Asylum Research).

The images were acquired at the first and third harmonics while exciting the cantilever near the fundamental resonance. In Fig. 4, a PS nanoparticle is clearly visible in the height image completed by the feedback of the first harmonic signal. In dynamic AFM, the amplitude of the driving frequency is to be held constant by the feedback system. However, sudden changes in height or physical properties of the sample result in changes of amplitude. Therefore amplitude at the fundamental frequency can be thought of as the error of the feedback controller. The phase image is acquired simultaneously, plotting the phase shift of the oscillation. While the amplitude mainly depends on the topography and composition of the sample surface, the phase depends on the energy transferred from the tip to the sample surface.

The image taken by the third harmonic amplitude gives enhanced results by providing a cleaner shape of amplitude variation around the particle. More notably, the third harmonic phase map exhibits more than 10-fold better sensitivity than the first harmonic phase map. Besides, the bump observed in the first harmonic phase map at the particle boundary where an abrupt first harmonic amplitude change exists disappears in the third harmonic phase map. This means the higher phase harmonic signal is ideal to study the material composition in terms of better sensitivity and less crosstalk with height variations.

978-1-4799-7956-1/15 $31.00 © 2015 IEEE

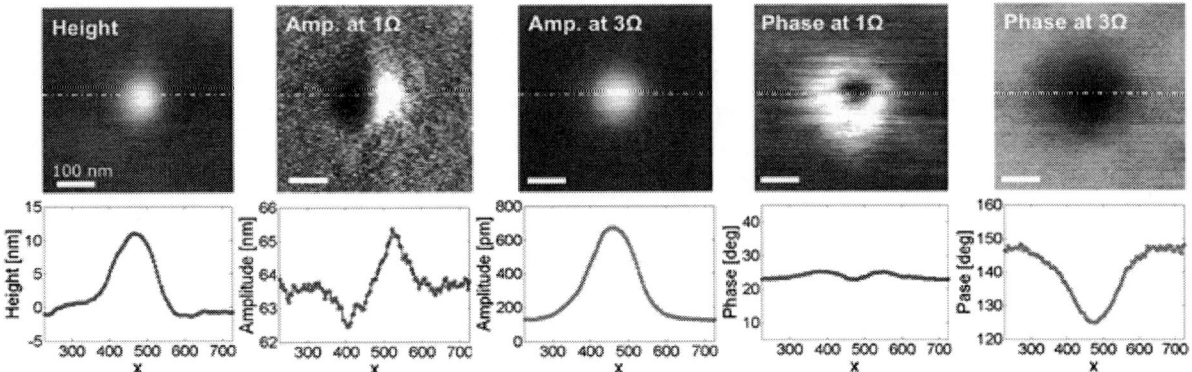

Figure 4: AFM images of 200 nm polystyrene nanoparticles embedded in a thin PDMS film using new microcantilever system with proposed 1:3 internal resonance. The leading bending modes are at 120.0 kHz and 353.8 kHz, thus the eigenfrequency ratio being 1:2.95. The bottom row shows the curve of a single scan line, marked with a line on the top row, showing the magnitude change of amplitude and phase at the first and third harmonics.

CONCLUSION

The modified AFM cantilever system significantly amplifies the response of higher harmonics in dynamic AFM. The modification is a physical alteration of a typical AFM microcantilever as an addition of an internal paddle used to incorporate internal resonance between the first two bending modes. With the results shown, this cantilever was designed to have an intrinsic 1:3 internal resonance, significantly amplifying the third harmonic response. We note that extension of this concept is possible either by implementing higher-order internal resonances (*e.g.*, 1:5, 1:7,...) or by introducing multiple internal paddles designed for simultaneous excitation of multiple internal resonances.

The reduced order model and mode analysis proved to be a good tool in the prediction of the cantilever dimensions to achieve the targeted internal resonance. The theoretical simulations were validated by experimental results, together confirming the nonlinear transfer of energy resulting in the amplification of higher modes. The experimental observations were performed with an inhomogeneous polymer sample, with the higher harmonics response having a 10-fold enhanced sensitivity.

Further theoretical studies are needed to understand how the sample composition is related to the amplitude and phase information, specifically at the higher harmonics.

ACKNOWLEDGEMENTS

This work was financially supported in part by National Science Foundation Grant CMMI-100615and by Texas Tech University through new investigator start-up funding, and was carried out in part in the Frederick Seitz Materials Research Laboratory Central Research Facilities, University of Illinois.

REFERENCES

[1] R. Garcia and E. T. Herruzo, "The Emergence of Multifrequency Force Microscopy", *Nature Nanotech.*, vol. 7, pp. 217-226, 2012.

[2] A. Raman, S. Trigueros, A. Cartagena, A.P.Z. Stevenson, M. Susilo, E. Nauman, and S. Antoranz Contera, "Mapping Nanomechanical Properties of Live Cells Using Nulti-harmonic Atomic Force Microscopy", *Nature Nanotech.*, vol. 6, pp.809-814, 2011.

[3] D. Ebeling, B. Eslami, and S. D. J. Solares, "Visualizing the Subsurface of Soft Matter: Simultaneous Topographical Imaging, Depth Modulation, and Compositional Mapping with Triple Frequency Atomic Force Microscopy", *ACS Nano*, vol. 7, pp. 10387-10396, 2013.

[4] H. Keum, A. Carlson, H. Ning, A. Mihi, J. Eisenhaure, P.V. Braun, J.A. Rogers, and S. Kim, "Silicon Micro-Masonry Using Elastomeric Stamps for Three-Dimensional Microfabrication", *J. Micromech. Microeng.*, vol. 22, pp. 055018, 2012.

[5] M. Baumann, R.W. Stark, "Dual Frequency Atomic Force Microscopy on Charged Surfaces", *Ultramicroscopy*, vol. 110, pp. 578-581, 2010.

[6] C. Dietz, M. Zerson, C. Riesch, A.M. Gigler, R. W. Stark, N. Rehse, R. Margerle, "Nanotomography with Enhanced Resolution Using Bimodal Atomic Force Microscopy", *Appl. Phys. Lett.*, vol. 92, pp. 143107, 2008.

[7] A. F. Vakakis, O. Gendelman, L. A. Bergman, D. M. McFarland, G. Kerschen, Y. S. Lee, *Passive nonlinear targeted energy transfer in mechanical and structural systems*, Springer Verlag, 2008.

[8] A. H. Nayfeh, D. T. Mook, *Nonlinear oscillations*, Wiley, 1995.

CONTACT

*H. Cho, tel: +1-806-834-7663; hanna.cho@ttu.edu

PRINTABLE FLEXIBLE TACTILE PRESSURE AND TEMPERATURE SENSORS WITH HIGH SELECTIVITY AGAINST BENDING

Kenichiro Kanao, Shingo Harada, Yuki Yamamoto, Wataru Honda, Takayuki Arie, Seiji Akita, and Kuniharu Takei

Department of Physics and Electronics, Osaka Prefecture University, Osaka, JAPAN

ABSTRACT

Flexible electronics are of great interest in a future electric device such as artificial electronic devices. Especially, artificial electronic skin (e-skin) is widely studied by developing a tactile pressure sensor on a flexible substrate. However, conventional flexible tactile sensors also detect the bending of substrate without applying a tactile pressure, and that is the one of bottlenecks to realize stable operation of a flexible device. This study demonstrates the high selectivity of tactile pressure and temperature sensors against bending based on strain engineering. To achieve high selectivity, a cantilever type strain sensor in a flexible substrate is developed. In addition, the temperature sensor is also mechanically stable. It should be worth to note that these sensors are fabricated by a fully printing method using a screen printer. This finding and demonstration eventually allow us to apply the flexible devices on versatile surfaces with accurate sensing.

INTRODUCTION

Flexible electronics have attracted much attention for future versatile applications. In particular, strain sensor to detect tactile pressure has been studied widely for the application of e-skin [1-6]. However, there are still several challenges to overcome to move forward to applying the flexible tactile pressure sensor to practical e-skin (*i.e.* Robotics, etc.) such as (1) high selectivity between a tactile pressure and a bending of a substrate, (2) an economical fabrication method on a macro-scale, flexible substrate, (3) multi-functionality to imitate a human skin, and (4) system integration.

In this report, we especially focus on (1) high selectivity between tactile pressure and bending of a substrate. Without achieving the high selectivity of them, this problem prevents the precise monitoring from a tactile pressure sensor. However, due to difficulty to realize high selectivity between them, the device has yet to be

demonstrated. To address this issue, this paper proposes and demonstrates a strain engineered flexible tactile pressure sensor on a cantilever structure as shown in Figure 1. Furthermore, for the application of e-skin, a temperature sensor is also integrated as the first approach to realize multi-functionality of a flexible device. In addition, the tactile pressure sensor, temperature sensor, and metal interconnection are completely formed by using printing methods for macro-scale and low cost devices as the first proof of concept.

METHOD

Figure 2 exhibits the device structure of the fully printed resistive strain and temperature sensors. First, silver (Ag) interconnection was screen printed on both top and bottom surface of a polyethylene terephthalate (PET) substrate for strain and temperature sensors, respectively. The Ag film was then cured at 130 °C for more than 30 mins. Next, strain sensor was printed with alignment of the top surface of Ag interconnection, and cured at 70 °C for more than 1 hour. For strain sensor, mixture ink of a carbon nanotube (CNT) ink (SWeNT, USA) and Ag nanoparticle (AgNP) ink (Paru, Korea) with a weight ratio of 5:3 was used to realize high sensitivity based on our previous study [7]. Subsequently, a temperature sensor using a mixture of CNT ink and (3,4-ethylenedioxythiophene)-poly(styrenesulfonate) (PEDOT:PSS, 1.3 wt% in water) solution (Sigma-Aldrich, USA) [8] was printed on a backside of the PET substrate with the alignment of Ag interconnection. To electrically connect Ag interconnections on top and bottom surfaces and form a cantilever structure, a laser cutter tool was used to cut the PET substrate as shown in Figure 2. Ag ink was dipped in the vias to connect Ag interconnection between top and bottom surfaces. When applying a tactile pressure,

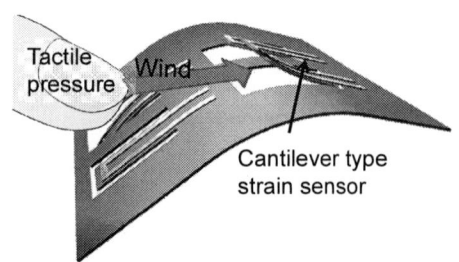

Figure 1: Schematic of a proposed cantilever-type flexible tactile pressure sensor.

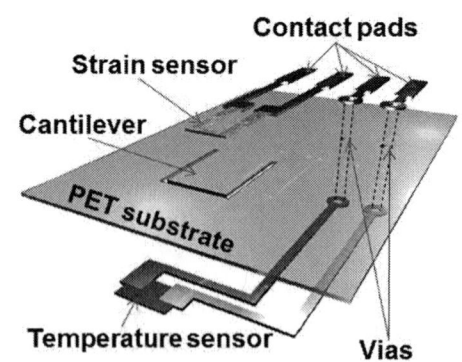

Figure 2: Schematic of the proposed device structure integrated with strain and temperature sensors.

978-1-4799-7956-1/15 $31.00 © 2015 IEEE   756   MEMS 2015, Estoril, PORTUGAL, 18 - 22 January, 2015

the cantilever with strain sensor is bent as shown in Figure 1. Due to the strain in the sensor by bending the cantilever structure, the electrical resistance is changed as a function of applied pressure.

Although the strain sensor has been optimized to achieve high sensitivity [7], the temperature sensor using the inks has not yet been optimized. To realize high sensitivity of temperature, sensitivity dependence of the ink composition ratio is studied by changing weight ratio of PEDOT:PSS and CNT inks. The curing temperature was used at 70 °C for all composition ratios.

RESULTS
Finite element method simulation

First, a finite element method (FEM) simulation was conducted. Figure 3 shows that almost zero strain distribution in a cantilever structure, where a strain sensor is integrated, is observed. This concludes that the sensor is independent of the bending of a substrate by applying the cantilever structure.

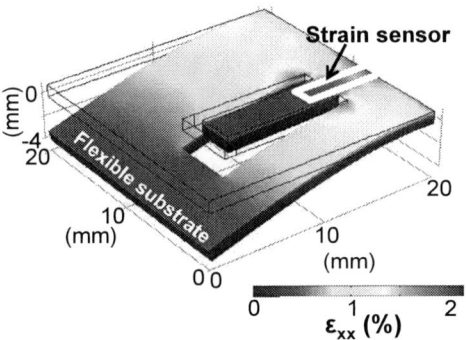

Figure 3: FEM simulation result by bending a flexible substrate, showing that the cantilever with strain sensor is independent of the bending of substrate.

Device characteristics

Figure 4 exhibits the final device structure after four printing processes and one laser cutter process. As shown

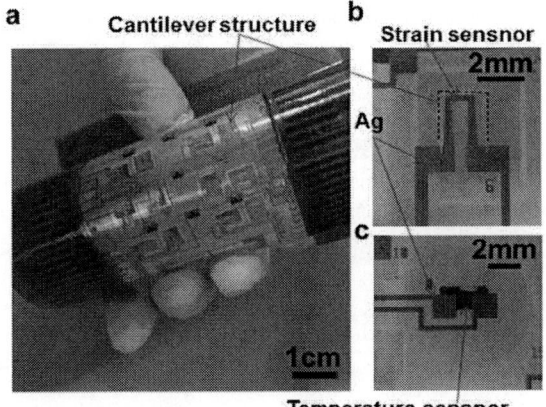

Figure 4: (a) Photo of a fully printed cantilever-type tactile and temperature sensor. Zoom-up image photos of (b) strain sensor and (c) temperature sensor.

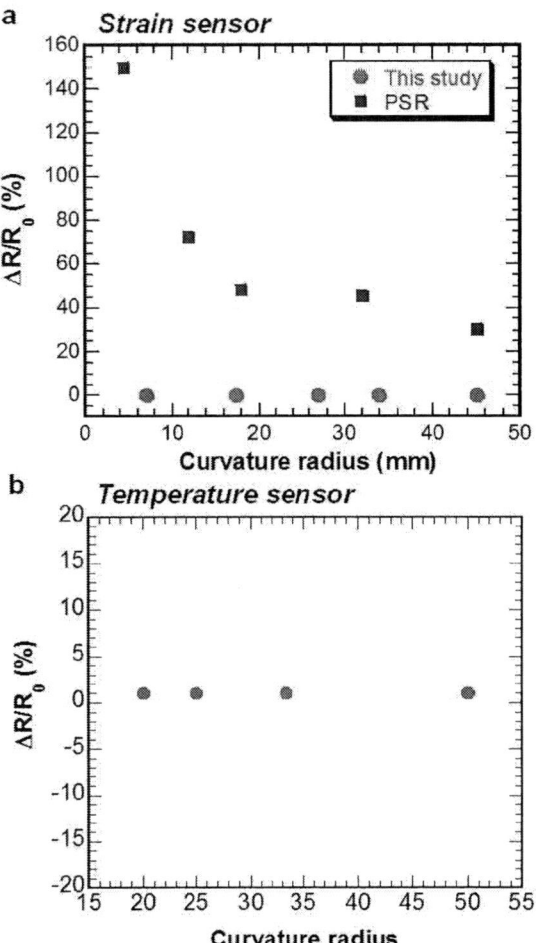

Figure 5: (a) Normalized resistance change of tactile pressure sensor as a function of curvature radius of the substrate. As the control, results of conventional PSR are also shown. (b) Normalized resistance change of temperature sensor at different curvature radius.

in Figure 4a, the device can be readily fabricated and integrated on a macroscale (8 cm × 8 cm) flexible substrate (PET). It is confirmed that all layers (i.e. strain sensor, temperature sensor, and Ag interconnection) are not delaminated or cracked under bending of the substrate as shown in Figure 4.

Mechanical stability of each strain and temperature sensor was characterized. Figure 5a shows the normalized resistance change ($\Delta R(=R\text{-}R_0)/R_0$) as a function of curvature radius, where R_0 and R are the electrical resistance of strain and temperature sensors at a flat and a bent state of a substrate. Due to the cantilever structure by considering the strain engineer as simulated in Figure 3, strain sensor does not show any electrical resistance change by bending the substrate up to 7 mm curvature radius (Figure 5a). As the control to show the advantage of this cantilever type sensor, the resistance change of a conventional pressure sensitive rubber (PSR) used for e-skin application [1,2] is also shown in Figure 5a. This indicates that PSR has a strong dependence on the substrate bent. Subsequently, the mechanical stability of a

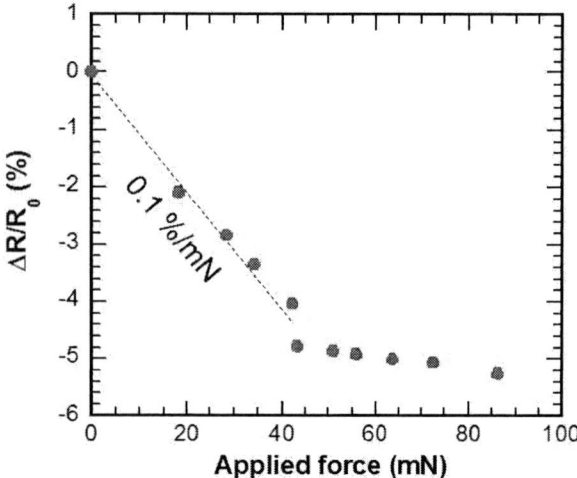

Figure 6: Normalized resistance change of the tactile sensor by applying a tactile force on top of cantilever structure.

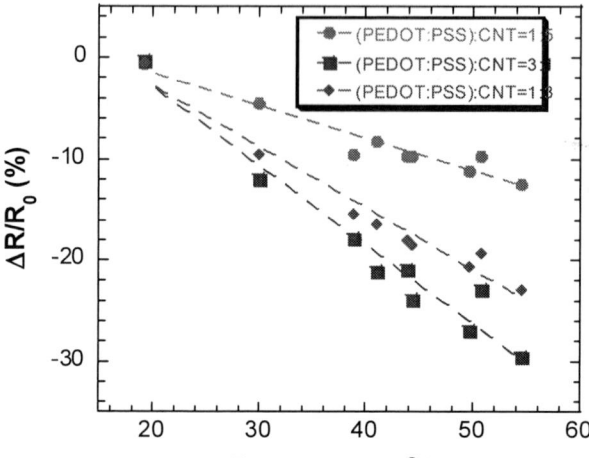

Figure 7: Normalized resistance change of the temperature sensor at ~19 °C – 52 °C with different composition ratio of PEDOT:PSS and CNT.

temperature sensor is also confirmed that the resistance does not change by bending the substrate (Figure 5b). Based on Figure 5, the strain sensor and temperature sensor can be realized with high selectivity of mechanical bending of substrate that is one of the biggest problems for e-skin applications.

By applying the cantilever structure, the sensor can detect each stimulus of tactile force and temperature precisely. The next step is to confirm the sensitivity of each sensor. Figure 6 shows the normalized resistance change $(\Delta R(=R-R_0)/R_0)$ as a function of applied tactile force onto the cantilever structure, where R and R_0 are the electrical resistance of a tactile sensor with and without an applied force. The tactile sensor has a linear dependence between the applied force and a resistance change up to around 40 mN. The sensitivity extracted from linear fitting of the data points is ~0.1 %/mN. That sensitivity is high enough to detect a human touch as an e-skin. However, this sensitivity is not the state-of-art because of most likely a geometric problem such as the length of the cantilever and thickness of PET substrate [7]. To improve the force sensitivity, structural engineering will be further required to apply higher strain at the same applied force in the future. When the force is applied more than 40 mN, the normalized resistance change is saturated, this is due to out of sensing range of the strain sensor reported previously [7].

Next, the temperature sensor is optimized and characterized by changing the composition ratio of PEDOT:PSS and CNT inks. Mixture ink was varied with 1:5, 1:3, and 3:1 wt% of PEDOT:PSS and CNT. Figure 7 depicts the sensitivity difference of each temperature sensor. The sensitivities extracted from linear fitting of the data points are ~0.32 %/°C, ~0.6 %/°C, and ~0.78 %/°C for 1:5, 1:3, and 3:1 wt% of the inks, respectively. The reason of the sensitivity difference is probably due to interface difference. Since this temperature sensor is based on an electron hopping at the interface between PEDOT:PSS and CNT [8], there should be the optimal composition ratio to achieve high sensitivity in terms of temperature. Figure 7 suggests that the optimal interface and condition as a temperature sensor is 3:1 wt% of the inks. This sensitivity is relatively high compared to other metal-based flexible temperature sensor (eg. 0.53 %/°C for Pt temperature sensor) [9].

CONCLUSION

This study demonstrates high selective strain sensor against bending of the flexible substrate. Based on the strain engineering, a cantilever type structure is optimal to prevent the effect of substrate bending confirmed by FEM simulation and experimental results. Furthermore, the sensors are fabricated by using printing methods with different nanomaterial inks. This allows us to realize economical macroscale flexible devices that are required for flexible and wearable electronics in the near future. Finally, we also show the possibility of strain and temperature sensor integration using printing method. Although some requirements for practical flexible electronics are still remained to address, these developments in this study should open a next class of flexible electronic to move forward to realizing practical devices.

ACKNOWLEDGEMENTS

This work was partially supported by JSPS KAKENHI Grant (#26630164 & #26709026), the Mazda Foundation, and Tateishi Science and Technology Foundation.

REFERENCES

[1] T. Someya, T. Sekitani, S. Iba, Y. Kato, H. Kawaguchi, T. Sakurai, "A Large-Area, Flexible Pressure Sensor Matrix with Organic Field-Effect Transistors for Artificial Skin Applications", Proc. Natl. Acad. Sci. USA, vol. 101, pp. 9966-9970, 2004.

[2] K. Takei, T. Takahashi, J. C. Ho, H. Ko, A. G. Gillies, P. W. Leu, R. S. Fearing, A. Javey, "Nanowire Active

Matrix Circuitry for Low-Voltage Macro-Scale Artifical Skin", *Nat. Mater.*, vol. 9, pp. 821-826, 2010.

[3] G. Schwartz, B. C.-K. Tee, J. Mei, A. L. Appleton, D. H. Kim, H. Wang, Z. Bao, *Nat. Commun.*, vol. 4, p. 1859, 2013.

[4] D.-H. Kim *et al.*, "Epidermal Electronics", *Science*, vol. 333, pp. 838-843, 2011.

[5] W. Wu, X. Wen, Z. L. Wang, "Taxel-Addressable Matrix of Vertical-Nanowire Piezotronic Transistors for Active and Adaptive Tactile Imaging", *Science*, vol. 340, pp. 952-957, 2013.

[6] J. Park, Y. Lee, J. Hong, M. Ha, Y.-D. Jung, H. Lim, S. Y. Kim, H. Ko, "Giant Tunneling Piezoresistance of Composite Elastomers with Interlocked Microdome Arrays for Ultrasensitive and Multimodal Electronic Skins", *ACS Nano*, vol. 8, pp. 4689-4697, 2014.

[7] S. Harada, W. Honda, T. Arie, S. Akita, K. Takei, "Fully Printed, Highly Sensitive Multi-Functional Artificial Electronic Whisker Array Integrated with Strain and Temperature Sensors", *ACS Nano*, vol. 8, pp. 3921-3927, 2014.

[8] W. Honda, S. Harada, T. Arie, S. Akita, K. Takei, "Wearable Human-Interactive Health-Monitoring Wireless Devices Fabricated by Macroscale Printing Techniques", *Adv. Funct. Mater.*, vol. 24, pp. 3299-3304, 2014.

[8] D.-H. Kim *et al.*, "Materials for Multifunctional Ballon Catheters with Capabilities in Cardiac Electrophysiological Mapping and Ablation Therapy", *Nat. Mater.*, vol. 10, pp. 316-323, 2011.

CONTACT

*K. Takei, tel: +81-72-254-9497;
takei@pe.osakafu-u.ac.jp

SOFT FLEXION SENSORS INTEGRATING STRECHABLE METAL CONDUCTORS ON A SILICONE SUBSTRATE FOR SMART GLOVE APPLICATIONS

Hadrien O. Michaud, Joan Teixidor, and Stéphanie P. Lacour[*]

Laboratory for Soft Bioelectronic Interfaces (LSBI), École Polytechnique Fédérale de Lausanne (EPFL), Lausanne, Switzerland

ABSTRACT

We design and implement a sensory skin that monitors in real time finger flexure (three sensors per finger) of a user's hand. Compared to current technologies, the electronic skin is made entirely of stretchable materials integrating silicone rubber, low resistivity liquid metal interconnects and high strain sensitivity, microstructured thin metal films. Microfabrication of the sensors combines traditional thin film process and additive manufacturing techniques. We incorporate the skin on a textile glove and demonstrate its function as an interface for finger motion and posture detection using a robotic test platform.

INTRODUCTION

Data gloves are sensor systems capable of encoding and reconstructing the posture and the movement of the human hand. Such devices were initially developed for animation and machine interaction purposes but they are now gaining attention for health monitoring and rehabilitation applications [1]. Integrating multiple sensors based on thin and soft materials, such as textile, in order to monitor joints position with minimal motion impediment is a desirable feature that has been recently investigated [2].

Electronic skins incorporating multiple functionalities on ultra-compliant substrates are a promising research area for smart gloves. Recent advances demonstrated sensing of multiple stimuli - including movement, pressure or temperature - via thin and soft systems that can be worn as a second skin by human and robots [3].

As far as movement or strain sensing is concerned, strain gauges appear as a relevant solution since they can be easily interfaced with a wide variety of off-the-shelf micro-electronic components and do not require multilayer architecture or shielding against electromagnetic noise sources. In addition, they are insensitive to other stimuli such as normal pressure and can be temperature compensated.

Soft piezoresisitive composites based on quantum tunneling and percolation between conductive micro-nanoparticles trapped in a polymer matrix have been extensively reported as potential strain sensing elements for artificial skin with application to human motion detection [4]. However, these material assemblies are prone to creep. It results in a large viscoelastic response to tensile and compressive strains, hindering the dynamic response of the sensors due to overshoot after unloading and large relaxation time [5,6].

Liquid metal microstructures are another popular solution for soft strain sensors. Hammond *et al.* demonstrated microchannels filled with a liquid metal alloy as strain gauges embedded in a glove-like system for monitoring human finger flexion [7]. However, the need to use rigid, strain insensitive wires as interconnects in order to detect locally the change in resistance of the liquid metal gauges may limit this approach. Moreover, the interface between soft sensors and hard wires is reported as a cause of failure for the system due to leaks and delamination.

In this work, we propose a combination of gold thin films deposited on a soft rubber substrate [8] as strain sensing elements with directly plotted gallium and indium eutectic alloy (EGaIn) micro-wires as interconnects [9]. We assembled three flexion sensors into a thin (<0.5 mm) and soft silicone membrane. The system was mounted on a humanoid hand and we monitored its static and dynamic performance. The iCub robot was chosen as the test platform since its hand mimics the kinematic of the human fingers and embeds its own commercial rotation sensors [10].

CONCEPT AND DESIGN
Sensing mechanism and material

In our implementation, the sensors were integrated in an elastomer membrane placed in tight contact with the dorsal side of the finger. Let's consider an articulated finger with three cylindrical joints (metacarpophalangeal

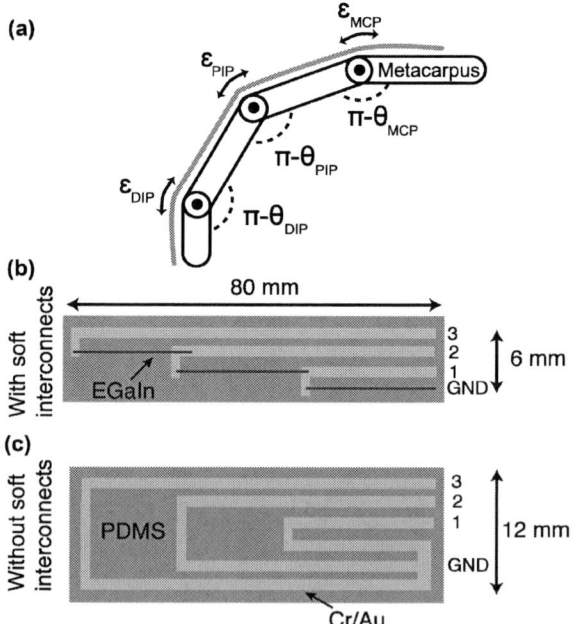

Figure 1: Design and principle of the soft flexion sensing strip with soft interconnects for humanoid fingers.

MCP, proximal interphalangeal PIP, and distal interphalangeal DIP) of radius r and rigid phalanxes. When a joint of the finger rotates by an angle θ, the length of the dorsal part the finger increases by θr. It results in a tensile strain ε across the surface of the soft membrane located on top of the joint (Figure 1a). In order to transduce this mechanical stimulus, we chose to form strain gauges from patterns of stretchable gold thin films [8]. Such films can withstand large deformations imposed by the finger's motion. Strain accommodation mechanism results from microscopic rearrangement of interconnected gold ligaments. The resulting resistance change (hundreds to thousands of Ohms depending on initial resistance R_0) is essentially linear for tensile strain regimes up to 15%, with gauge factors of the order of 2 to 5.

Design of sensors and soft interconnects

Soft liquid metal micro-wires embedded in silicone have small resistance at rest (less than 1 Ω for centimeter-long wires) and their resistance increases by less than 150% for strains up to 50% [11]. These characteristics were ideal for building a stretchable interconnects network. Three gold-film strain gauges with graded lengths aligned with the middle of the phalanxes of the robotic finger were interconnected as shown Figure 1b. The staggered EGaIn interconnect design was laid down because of a limitation from our printing set-up (which restricted to 35mm the maximum length of a printed wire). Figure 1c presents an equivalent design using only stretchable thin gold films. Although such pattern could have been produced in a single evaporation step, it would have had a larger footprint (discarding it for mounting on the iCub finger) and the implementation of the corresponding shadow mask would have been challenging (thin polyimide features). Furthermore, the liquid metal interconnects provided a fully elastic wire-sensor interface thereby avoiding interfacing mechanically soft sensors with hard wiring and preventing mechanical stress concentration at the soft-hard junctions and failure of the sensory skin.

Equivalent electrical circuit

Figure 2 presents the equivalent electrical circuit of the soft sensing system. For each gauge j, $R_j >> \Sigma R_{int,k}$ at rest. When the finger flexed, we had $\Delta R_j >> \Sigma \Delta R_{int,k}$, so the sensors could be addressed independently through the interconnection network.

FABRICATION METHODS

The fabrication process flow for the soft metal sensing strips is detailed Figure 3. Polydimethylsiloxane (PDMS, Sylgard 184, Dow Corning) prepolymer and curing agent were mixed in a 10:1 ratio and spun at 500 RPM on a 100 mm silicon wafer and cured at 80°C for

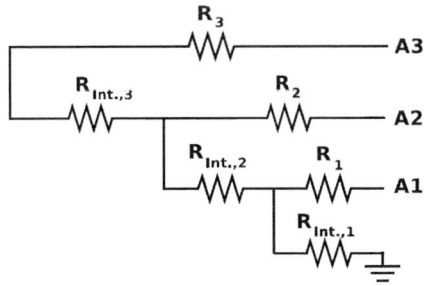

Figure 2: Equivalent electrical circuit of the soft integrated flexion sensing strip.

two hours. 5/25 nm Cr/Au thin films were thermally evaporated through a polyimide shadow mask to define the strain sensing patterns. Next, a syringe was loaded with the EGaIn alloy (Sigma Aldrich) and attached to an X-Y-Z stage (GIX microplotter II, Sonoplot) via a custom holder. The syringe's tip inner diameter was 360 μm. A controlled pressure (approx. 0.02 bar) was applied inside the syringe to force the EGaIn to reach the tip's end without forming a droplet. The tip was lowered about 70 μm above the PDMS layer in order to form a liquid metal meniscus between the tip and the substrate. The stage then followed predefined 2D patterns in the XY plane while pressure was maintained to plot the soft liquid metal interconnects.

Next, solid wires were mounted at the proximal extremity of the strip, connected to soft interconnects or strain gauges using small liquid metal droplets, and secured using silicone sealant (Dow Corning 734).

Finally, a PDMS encapsulation layer was spin-coated at 250 RPM. The final thickness of the strips was below 500 μm. Figure 4 provides detailed pictures of the soft materials assembly. EGaIn microwires had a semi-circular cross-section and their typical diameter was 250 μm (Figure 4a).

We observed the micro-cracked structure in the gold film on PDMS that enabled accommodation of large mechanical deformation and finite electrical resistance (Figure 4b). Gallium and gold formed an electrically conductive alloy that was visible within a few hundred microns around the interconnection area (Figure 4c) [12].

Overall, the skin fabrication process only required four steps. Designs may be easily customized since interconnects were patterned using an additive, mask-less and mold-less process.

RESULTS AND DISCUSION

Single sensor characterization on the iCub hand

After microfabrication, the strip was attached to a textile glove using EcoFlex (Smooth-On) as a curable adhesive. When the robot was wearing the sensorized glove, we observed it could rotate normally all finger's joints and open and close completely its hand (Figure 5).

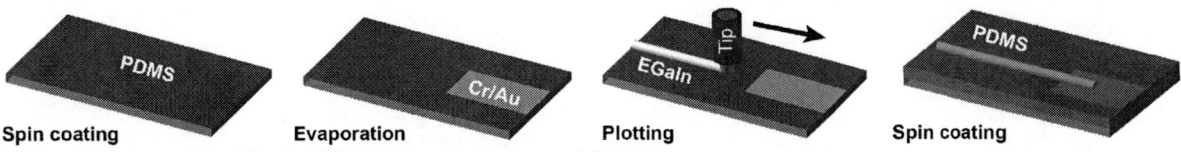

Figure 3: Process flow for the fabrication of a soft metal flexion sensor.

978-1-4799-7956-1/15 $31.00 © 2015 IEEE   761

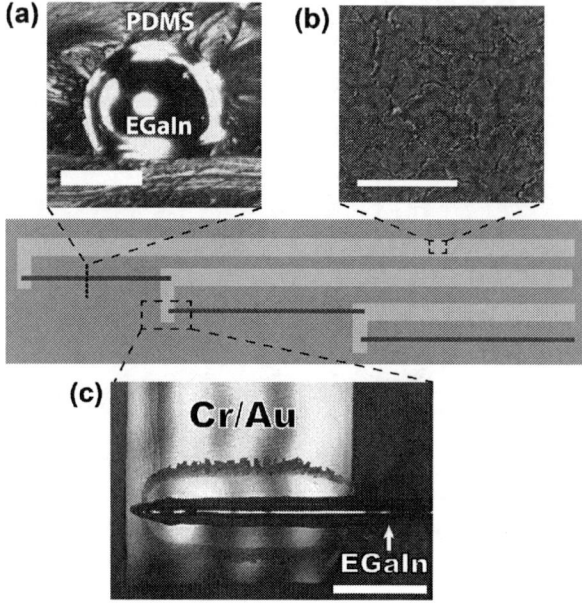

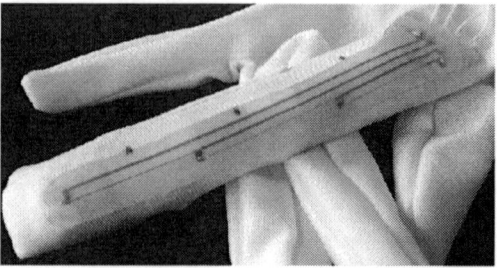

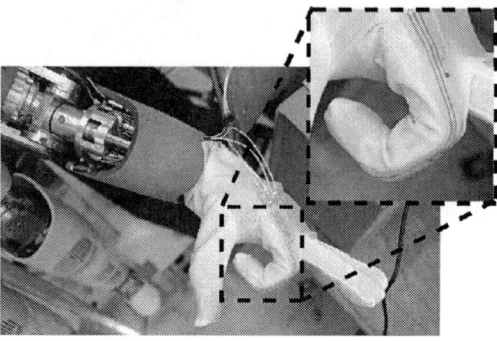

Figure 4: Materials constituting the soft metal flexion-sensing strip. a) Cross section of an EGaIn microwire (scale bar: 100 µm). b) SEM image of the stretchable 5/25 nm Cr/Au thin film (scale bar: 1 µm). c)Top view of the interconnection zone between the soft micro-wire and thin film (scale bar: 500 µm).

Figure 5: Soft flexion sensing trip after mounting on a textile glove and worn on the iCub robot with its finger fully flexed.

Each sensor output was read in real time using three voltage dividers connected to the analog inputs of an Arduino Micro board. The board was addressed from a PC through the serial communication bus. We assessed the repeatability and linearity of the sensors by recording the relative increase in resistance of sensor 1 when the MCP joint was incrementally closed multiple times (Figure 6). Small standard deviations and high R^2 value indicated good performance of the sensors.

Computation of the joint's angles

We made the following assumptions in order to compute the three joints angle from the output of the sensors:

- The rotation of each joint resulted in a linear increase of the electrical resistance of each strain gauge.
- This increase in resistance was independent from the position or rotation of the other joints.

Hence, the relative increase in resistance of each sensor was a linear combination of the three joints' angles, with fixed coefficients. We defined the vectors $\theta=[\theta_{MCP};\theta_{PIP};\theta_{DIP}]$ and $R=[\Delta R_1/R_{1,0};\Delta R_2/R_{2,0};\Delta R_3/R_{3,0}]$. We had:

$$R = J\theta \qquad (1)$$

where $J = K^{-1}$ was a 3x3 matrix.

The calibration matrix K ($\theta=KR$) was determined with the following calibration scheme. The outputs of the sensors were recorded for six different known positions of the finger and arranged in an 18x9 matrix R_{cal}. The angles of the three joints for the six known positions were arranged in an eighteen elements vector θ_{cal}. The nine

elements vector containing the coefficients of matrix K, C_K, was computed using the least square method:

$$C_K = (R_{cal}^T R_{cal})^{-1} R_{cal}^T \theta_{cal} \qquad (2)$$

This calibration routine was implemented in Matlab (MathWorks).

Real-time acquisition of the hand posture

After determination of the calibration matrix K, a Matlab script converted the sensors' outputs into joint angles and displayed the finger posture. Figure 8 presents the computed angles and reconstructed finger posture for repeated complete closings of the robot's hand. The MCP joint was first rotated. Then, PIP and DIP joints were completely closed at the same time. The joints were actuated in the inverse order for opening. Data show the

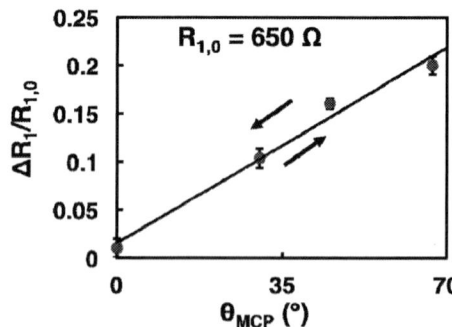

Figure 6: Response of sensor 1 to flexion of the MCP joint. 12 flexion cycles from 0 to 67° are represented. Error bars represent 99% confidence interval. Solid line represents linear interpolation (R^2=0.98).

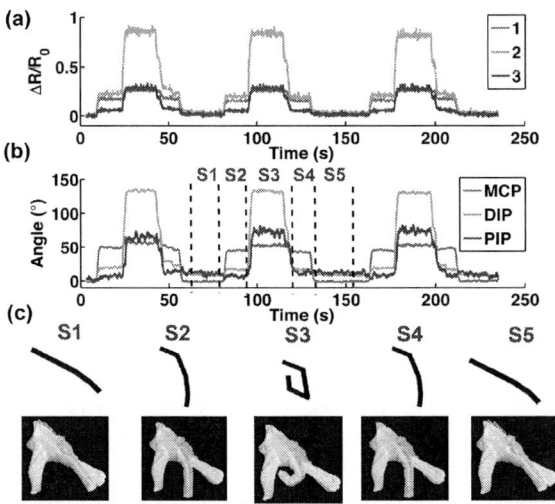

Figure 7: Relative increase in resistance of the sensors and computed joints angles as a function of time when the iCub hand is completely closed three times. Comparison between reconstructed finger profile and actual hand profile.

sensors were stable, quickly adapted to finger motion with good repeatability, and did not overshoot (Figure 7). We also observed a systematic overestimation of θ_{PIP} and underestimation of θ_{DIP} when the finger was completely closed. This might come from non-linearity or global strain (not only in the vicinity of the joints) occurring in the skin during finger bending. These phenomena were not taken into account in the calibration scheme. With longer interconnects, the accuracy of the sensors could have been improved by patterning a single strain sensing area per joint, thus mechanically reducing crosstalk between sensors. For example, as sensor 1 was covering the MCP joint only, it was mainly sensitive to the closing of the MCP joint and not the rotation of the other two joints (Figure 7a). However, the calibration scheme was general and could be used for compensating crosstalk in alternative sensor's layout.

CONCLUSION

We reported the fabrication and characterization of stretchable, skin-like flexion sensors that encoded the relative rotation of finger's joints. Ad-hoc choice of materials and fabrication methods resulted in a thin and fully stretchable system with soft integrated interconnects and strain gauges. We mounted the skin system on the iCub humanoid hand to assess the sensors' linearity, repeatability and dynamic behavior. We also proposed a calibration algorithm that overcame design limitations and enabled reconstruction of the finger posture in real time. Modification of the sensor's design could enhance accuracy of the sensors. Long term response and fatigue resistance of the proposed sensing skin still need to be investigated and quantified.

Future applications could involve development of a data glove for the human hand and coupling with other soft transducers such as tactile sensors.

ACKNOWLEDGEMENTS

This work was sponsored by the Nanotera.ch initiative within the WiseSkin project. We warmly acknowledge Nicolas Sommer and Prof. Aude Billard from LASA (EPFL) for access and help with the iCub robot.

REFERENCES

[1] L. Dipietro, A. Sabatini, and P. Dario, "A survey of glove-based systems and their applications," *IEEE Trans. Syst., Man, Cybern. C*, Appl. Rev., vol. 38, no. 4, pp. 461–482, 2008.

[2] G. D. Mura, F. Lorussi, A. Tognetti, G. Anania, N. Carbonaro, M. Pacelli, R. Paradiso, and D. De Rossi, "Piezoresistive Goniometer Network for Sensing Gloves", in *IFMBE Proc. 41*, 2014, pp. 1547–1550.

[3] M. L. Hammock, A. Chortos, B. C.-K. Tee, J. B.-H. Tok, and Z. Bao, "The evolution of electronic skin (e-skin): a brief history, design considerations, and recent progress.," *Adv. Mater.*, vol. 25, no. 42, pp. 5997–6038, 2013.

[4] M. Park, J. Park, and U. Jeong, "Design of conductive composite elastomers for stretchable electronics", *Nano Today*, vol. 9, no. 2, pp. 244–260, 2014.

[5] Q. Zheng, J. F. Zhou, and Y. H. Song, "Time-dependent uniaxial piezoresistive behavior of high-density polyethylene/short carbon fiber conductive composites", *J. Mater. Res.*, vol. 19, no. 9, pp. 2625–2634, Mar. 2011.

[6] J. T. Muth, D. M. Vogt, R. L. Truby, Y. Mengüç, D. B. Kolesky, R. J. Wood, and J. a. Lewis, "Embedded 3D Printing of Strain Sensors within Highly Stretchable Elastomers," *Adv. Mater.*, vol. 26, pp. 6307–6312, 2014.

[7] F. L. Hammond, Y. Mengüç, and R. J. Wood, "Toward a Modular Soft Sensor - Embedded Glove for Human Hand Motion and Tactile Pressure Measurement," in *IROS 2014*, 2014, pp. 4000-4007

[8] S. P. Lacour, S. Wagner, Z. Huang, and Z. Suo, "Stretchable gold conductors on elastomeric substrates", *Appl. Phys. Lett.*, vol. 82, no. 15, p. 2404, 2003.

[9] J. W. Boley, E. L. White, G. T.-C. Chiu, and R. K. Kramer, "Direct Writing of Gallium-Indium Alloy for Stretchable Electronics," *Adv. Funct. Mater.*, pp. 3501–3507, Feb. 2014.

[10] A. Schmitz, U. Pattacini, F. Nori, L. Natale, G. Metta, and G. Sandini, "Design, realization and sensorization of a dextrous hand: the iCub design choices", in *Humanoids 2010*, 2010, pp. 186–191

[11] H.O. Michaud, J. Teixidor, and S.P. Lacour, "Soft metal constructs for large strain sensor membrane", under revision, 2014.

[12] H.J. Kim, "Stretchable Interconnects Using Room Temperature Liquid Alloy on Elastomeric Substrate", *Ph.D. thesis Purdue University*, USA, 2007.

CONTACT

*S.P. Lacour, stephanie.lacour@epfl.ch

A CIRCUMFERENTIALLY GROWN ZNO NANOWIRE FOREST ON A SUSPENDED CARBON NANOWIRE FOR A HIGHLY SENSITIVE GAS SENSOR

Yeongjin Lim[1], Yunjeong Lee[1], Jongmin Lee[1] and Heungjoo Shin[1]
[1]Ulsan National Institute of Science and Technology (UNIST), KOREA

ABSTRACT

We present a suspended ZnO nanowire forest as a highly sensitive gas sensor. ZnO nanowires were grown selectively on a suspended single glassy carbon nanowire using hydrothermal method so that the detrimental effects from the substrate inclusive of contamination, stagnant layer and limited mass transfer could be alleviated. The novel geometry of the radially grown ZnO nanowires resembling burs of a chestnut is expected to enhance the gas sensing capability because of enhanced mass transfer.

INTRODUCTION

The development of highly sensitive detection tools of biological and chemical species in atmosphere is critical in the fields of environmental monitoring, medical diagnosis and a variety of industries [1-3]. To enhance the limit of detection of gas sensing platforms, nanowires have been extensively researched as detecting materials [3-4].

Nanowire-based gas sensors have shown special characteristics owing to their high surface to volume ratio and the quantum confinement effect [1-6]. The sensitivity of nanowire-based gas sensor can be improved by enhancing mass transport, increasing temperature and loading catalysts [5-8]. Even with catalyst loading, nanowire-based gas sensors built in contact with the substrate did not achieve theoretical sensing limit because of the insufficient mass transport.

To improve mass transport in nanowire-based gas sensor, various nanowire morphologies [5, 6] and suspended structure [9] were introduced. Nanowires with core-shell structure, heterostructure, nanotapes and hierarchical nanowire have been investigated to increase the surface to volume ratio and improve mass transport [1]. Suspended-nanowire-based hydrogen gas sensor was fabricated using focused ion beam (FIB) [9]. By scanning focused ion-beam between two electrodes as the beam point was guided in the air, the tungsten nanowire could bridge the two electrodes. Although this FIB nanofabrication enabled successful implementation of suspended nanowire, the expensive and complex fabrication steps are drawbacks, and the relatively low structural rigidity limits further functionalization on to the suspended nanowire.

Metal oxide nanowires such as ZnO, In_2O_3, CuO and SnO_2 are promising candidates for gas detection materials because of their high aspect ratio, reasonable detection limit and a wide range of detectable gases such as NO_x, CO, CO_2, H_2, NH_3 [5-8, 14]. Some researchers have developed hierarchical ZnO nanowires that showed better sensing capability compared to one-dimensional nanowires [5, 6]. However, the effect of the hierarchical architecture of the nanowires can be maximized when the structure is positioned away from the substrate and the batch fabrication technology for these complex architecture is not well-developed yet.

We introduced a novel fabrication method building a monolithic suspended single carbon nanowire using carbon-MEMS consisting of UV-lithography and the polymer pyrolysis [10-12]. Although the UV-lithography process defines a micro-sized photoresist structures , the micro-sized polymer structures is converted to a nano-sized glassy carbon structure owing to high volume reduction in pyrolysis. In addition, longitudinal tensional stress is generated along the suspended carbon nanowire because of the volume reduction of carbon post supporting the suspended nanowire [13]. Because of the longitudinal tensional stress, relatively high Young's modulus, and bent ends of the suspended carbon nanowire, the nanowire can be selectively coated with functional materials using simple photolithography and deposition processes.

In this paper, we developed a circumferentially grown ZnO nanowire forest on a suspended carbon nanowire for a gas sensor. Furthermore, the metal oxide nanowires are grown relatively at low temperature using the hydrothermal method that is compatible with CMOS process. This paper shows a complete integration of a ZnO nanowire forest on a suspended carbon nanowire using only batch microfabrication processes ensuring high gas sensing capability and good manufacturability.

EXPERIMENTAL

Fabrication

Before fabricating carbon structures, trenches were patterned on a 6-inch Si substrate using an isotropic DRIE (deep reactive-ion etching) process under a 1-μm-thick thermally grown a SiO_2 mask (step (1)-(4) in Figure 1).

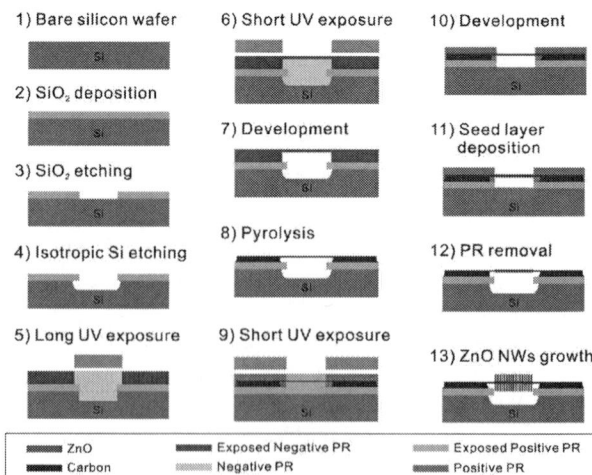

Figure 1: Fabrication steps for ZnO nanowires grown on a suspended single carbon nanowire.

978-1-4799-7956-1/15 $31.00 © 2015 IEEE

After coating a single SU-8 photoresist layer on the Si substrate, a monolithic suspended photoresist micro-sized wire was fabricated using two-step UV exposure processes and a single development process.

Via pyrolysis, a suspended polymer micro-sized wire was shrunk dramatically and thus converted into a nano-sized carbon wire of which size ranged from 120 to 250 nm in diameter. Because of the anisotropic volume reduction of carbon posts supporting the suspended carbon nanowire, the suspended carbon nanowire was elongated and bent downward so that a suspended carbon nanowire became strong enough to overcome stiction problems which could be found in wet processes such as photolithography and hydrothermal growth that were used for the integration of ZnO nanowires into the carbon nanowire.

Then, a 5-nm-thick ZnO seed layer was selectively deposited on the suspended single carbon nanowire using photolithography and sputtering. In the photolithography process, the (positive) photoresist was exposed with lower does than what was for the full exposure across the total photoresist thickness so that the ZnO seed layer remained selectively along the carbon nanowire after photoresist stripping (step (9-12) in Figure 1). Finally, ZnO nanowires were grown on the selective-coated ZnO layer in a growth solution (10 mM zinc nitrate hexahydrate ($Zn(NO_3)_2 \cdot 6H_2O$) and 10 mM hexamethylene tetramine (($CH_2)_6N_4$ in deionized water).

Gas sensing measurement

Gas sensing capability of a suspended ZnO nanowire forest was evaluated using a two-probe measurement. Because the size of the carbon posts supporting the suspended nanowire is much larger than that of the nanowire, the effect of the contact resistance and spreading resistance on the resistance can be neglected. First, the gas sensing chamber was purged with dry air gas for the 40 minutes. After the purging period, NO_x gas in various concentrations was injected and the electrical resistance change along the suspended nanowire integrated with the ZnO nanowire forest was measured at 200 °C. Between the NO_x gas injection periods, the dry air gas was purged in the chamber to make the nanowire resistance return to its original value.

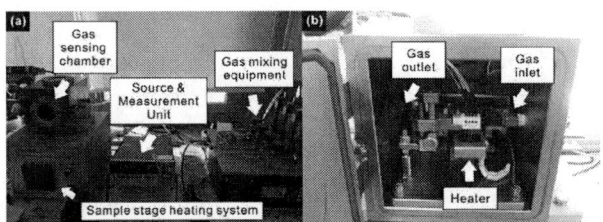

Figure 2: (a) Overview of the gas sensing measuring system consisting of a gas sensing chamber, an electrical measurement equipment and a gas mixing system; (b) the detailed view of the two-probe-based electrical resistance measurement system.

RESULTS
Crystallinity and morphology of ZnO nanowires

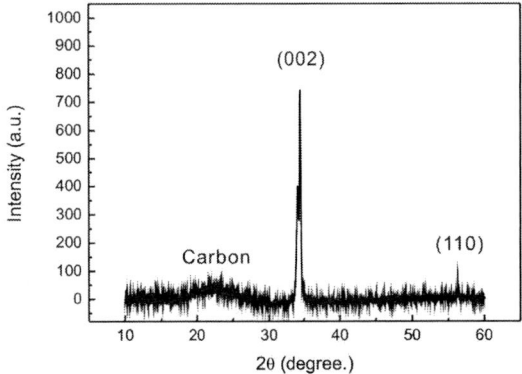

Figure 3: XRD pattern of ZnO nanowires grown on a pyrolyzed carbon pad using hydrothermal method.

The crystallinity of ZnO nanowires grown on a suspended glassy carbon nanowire was analyzed using an XRD (X-ray diffraction) experiment on ZnO nanowires grown on a pyrolyzed carbon pad that were prepared using the same methods as the integrated ZnO-C nanowires. The ZnO nanowires showed distinct crystal faces of (002) and (110), and the diffraction patterns of XRD could be readily indexed to a hexagonal structure that could be also found in SEM images as shown in Figure 4.

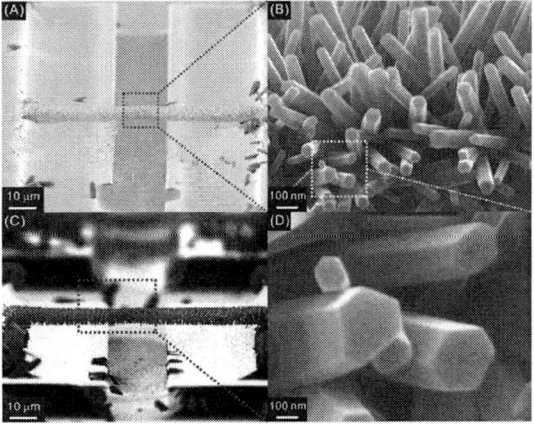

Figure 4: SEM images of a ZnO nanowire forest grown uniformly along a suspended carbon wire; (A, B, D) top-view, (C) side-view (ZnO nanowire width = ~ 80 nm, length = ~1.2 μm).

ZnO nanowires were successfully integrated on a suspended single carbon nanowire as shown in Figure 4. The diameter and length of the ZnO nanowires ranged from 30 ~ 80 nm and 0.5 ~ 2 μm respectively depending on the growth time. ZnO nanowires were grown radially on a suspended single carbon nanowire so that the gap among ZnO nanowires increased along the radial direction. These radially-directed nanowires ensure efficient mass transport of gas analyte to the ZnO sensing sites. As explained in the experimental section, the ZnO seed layer was selectively coated only along the suspended carbon nanowire as shown in Figure 4. The Si trench below the suspended nanowire prevented the formation of ZnO residue after ZnO growth, and the large spacing between the suspended nanowire and the Si trench ensured partial exposure around the suspended nanowire in the photolithography process for the selective ZnO seed layer coating

NOx gas sensing

Figure 5 shows the results of the gas response measurement at various NO_x concentrations from 1000 ppm to 1 ppm. Gas sensing sensitivity ($S_g = \Delta R/R$) of the suspended nanowire integrated with ZnO nanowire forest is changed in proportional to the concentration of NO_x gas mixed in dry air at 200 °C. Owing to the enhanced mass transport caused by the architecture of suspended radially grown ZnO nanowires, the gas sensor showed the short gas reaction time and recovery time.

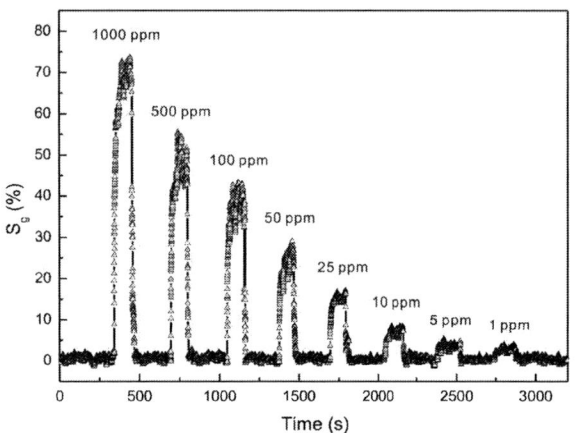

Figure 5: Gas sensing sensitivity ($S_g = \Delta R/R$) of the ZnO-nanowire-forest based gas sensor at various NO_x gas concentrations mixed in dry air at 200 °C.

CONCLUSION

In this research, ZnO nanowires were successfully integrated onto a suspended carbon nanowire using batch microfabrication processes such as carbon-MEMS and hydrothermal method. By modulating exposure time in the photolithography and integrating Si trenches, the ZnO nanowire forest could be selectively grown only along the suspended nanowire. The ZnO nanowires grown on a glassy carbon substrate showed good crystal growth in the direction of the c-plane ensuring good gas sensing capability. The batch fabricated ZnO-nanowire-based gas sensor showed good gas sensing capability for NOx.

ACKNOWLEDGEMENTS

This research was supported by the Presidential Research Grant Fund (1.140082.01) of UNIST (Ulsan National Institute of Science and Technology). We are grateful for technical assistance from the staff members at UCRF (UNIST Central Research Facilities) in UNIST.

REFERENCES

[1] F. Patolsky, C. M. Lieber, "Nanowire nanosensors",
Mater. Today, vol. 8, pp.20-28, 2005.

[2] Y. Chi, Q. Wei, C. M. Lieber, "Nanowire Nanosensors for Highly Sensitive and Selective Detection of Biological and Chemical Species", Science, vol. 17, pp-1289-1292, 2001.

[3] C. M. Lieber, Z. L. Wang, "Functional Nanowires", vol. 32, pp. 99-108, 2007.

[4] X. Chen, C. K. Y. Wong, C. A. Yuan, G. Zhang, "Nanowire-based gas sensors", Sens. Actuat. B-Chem., vol. 177, pp. 178-195, 2013.

[5] S. Baruah, C. Thanachayanont, J. Dutta, "Growth of ZnO nanowires on nonwoven polyethylene fibers", Sci. Technol. Adv. Mater., vol. 9 pp.025009-8, 2008.

[6] M. R. Alenezi, S. J. Henley, N. G. Emerson, S. R. P. Silva, "From 1D and 2D ZnO nanostructures to 3D hierarchical structures with enhanced gas sensing properties", Nanoscale, vol. 6, pp.235-247, 2014.

[7] H. Kim, J. Lee, "Highly sensitive and selective gas sensor using p-type oxide semiconductors: Overview", Sens. Actuat. B-Chem., vol. 192, pp.607-627, 2014.

[8] N. Ramgir, N. Datta, M. Kaur, S. Kailasaganapathi, A. K. Debnath, D. K. Aswal, S. G. Gupta, "Metal oxide nanowires for chemiresistive gas sensor: Issues, challenges and prospects", Colloid Surf. A-Physicochem. Eng. Asp., vol. 439, pp.101-116, 2013.

[9] J. Choi, J. Kim, "Highly sensitive hydrogen sensor based on suspended, functionalized single tungsten nanowire bridge", Sens. Actuat. B-Chem., vol. 136, pp.92-98, 2009.

[10] J. –I. Heo, Y. Lim, H. Shin, "The effect of channel height and electrode aspect ratio on redox cycling at carbon interdigitated array nanoelectrodes confined in a microchannel", Analyst, vol. 138, pp.6404-6411, 2011.

[11] Y. Lim, J. –I. Heo, M. J. Madou, H. Shin, "Monolithic carbon structures including suspended single nanowires and nanomeshes", Nanoscale Res. Lett., vol.8, pp.492-9, 2013.

[12] Y. Lim, J. Heo, H. Shin, "Fabrication and application of a stacked carbon electrode set including a suspended mesh made of nanowires and a substrate-bound planar electrode toward for an electrochemical/biosensor platform", Sens. Actuat. B-Chem., vol. 192, pp.796-803, 2014.

[13] M. F. L. De Volder, R. Vansweevelt, P. Wagnerm D. Reynaerts, C. V. Hoof, A. J. Hart, "Hierarchical Carbon Nanowire Microarchitectures Made by Plasma-Assisted Pyrolysis of Photoresist", ACS Nano, vol. 5, pp.6593-6600, 2011.

[14] A. Z. Sadek, S. Choopun, W. Wlodarski, S. J. Ipppolito, K. Kalantar-zadeh, "Characterization of ZnO Nanobelt-Based Gas Sensor for H2, No2, and Hydrocarbon Sensing", IEEE Sens. J., vol. 7, pp.919-924.

CONTACT

*H. Shin, tel: +82-52-217-2315; hjshin@unist.ac.kr

A CMOS CAPACITIVE VERTICAL-PARALLEL-PLATE-ARRAY HUMIDITY SENSOR WITH RF-AEROGEL FILL-IN FOR SENSITIVITY AND RESPONSE TIME IMPROVEMENT

Vincent P.J. Chung, Chao-Lin Cheng, Ming-Chuen Yip and Weileun Fang

Power Mechanical Engineering Department, National Tsing Hua University, Hsinchu, Taiwan

ABSTRACT

This paper reports a high-sensitivity and high-speed capacitive humidity sensor. A novel capacitive vertical parallel-plate (VPP) array had been monolithically integrated with an on-chip ring oscillator (RO) using the standard TSMC 0.18μm CMOS process and subsequent in-house post-processes. Resorcinol-formaldehyde (RF) aerogel, a new kind of moisture-sensitive polymer, was deposited and defined in columns in the VPP array. Measurements show that the typical fabricated device has a sensitivity of 0.566% capacitance change per percent-relative-humidity (%RH) and a response time of 6s. The monolithic integration of the humidity sensor with RO is also demonstrated to provide a sensitivity of 3.8 kHz/%RH. Further improvement in sensitivity is required by varying the operating frequency of RO.

Keywords – CMOS MEMS; Capacitive sensor; Vertical-parallel-plate array; RF-aerogel; Humidity sensor

INTRODUCTION

Presently, humidity sensors find various applications in mobile and wearable devices. State-of-the-art sensors are mainly polymer-coated micro-capacitors integrated with on-chip circuitry through industrial CMOS technology. A number of capacitive topologies have been reported and some were fabricated using the standard CMOS platform for the integration of circuitry. The structural design accordingly can be categorized into interdigitated electrodes (IDE) [1, 2] and vertical parallel plate (VPP) [3, 4]. The VPP design inherently provides better sensitivity over interdigitated one [3], nonetheless, at the expense of response time (RT). In most cases, a tradeoff between sensitivity and response time arises. Therefore, it remains a challenge for a sensor to possess high sensitivity and high speed at the same time. In [5], an amalgam of both VPP and interdigitated topologies was reported with the sensitivity up to 0.37%/%RH and the response time as low as 1s. Herein, a commonly used VPP topology was adopted to provide high sensitivity. Unlike previous studies, however, hundreds of VPP elements with dozens of microns in width were employed in a formation of a unitary structure – an array. Compared with a single VPP structure in the same footprint, this technique allowed water vapor to diffuse through a much smaller periphery of each VPP element and thereby significantly reduced the diffusion path of water vapors.

As for the sensing material, resorcinol-formaldehyde (RF) aerogels, shown great potential in humidity sensing and the capability to integrate with CMOS platforms [4], was adopted and defined in between the VPP electrodes. This new kind of porous polymer presented numerous benefits in moisture sensing as follow: (1) Hydrophilic in

nature (with hydroxyl sites on the backbones). (2) Low curing temperature (can be cured at room temperature). (3) High porosity facilitates the diffusion process of water vapor into/out of the substance. (4) High surface area and low dielectric constant could provide high capacitance changes upon water absorption/desorption. In this study, RF-aerogels were prepared using the sol-gel method [6, 7]. Along the preparation process, supercritical carbon dioxide (CO_2) drying was applied to maintain the volume and porous structures of the gel network. The prepared RF-aerogels exhibited high porosity (>80%), high surface areas (400~900 m^2g^{-1}), and low densities (0.03~0.079 gcm^{-3}).

In general, the proposed design aims for the enhancement of sensitivity and response time of humidity sensors. Where, VPP array and RF-aerogels together provide the merits of high speed and high sensitivity for the present humidity sensor.

DESIGN CONCEPT

Fig.1a presents a perspective-view schematic of the proposed humidity sensor. The integrated VPP array is implemented through the TSMC 0.18μm 1-polysilicon and 6-metal (1P6M) CMOS process and in-house post-processes. Fig.1b further shows the detailed cross-section view of the VPP array along A-A'. Note that

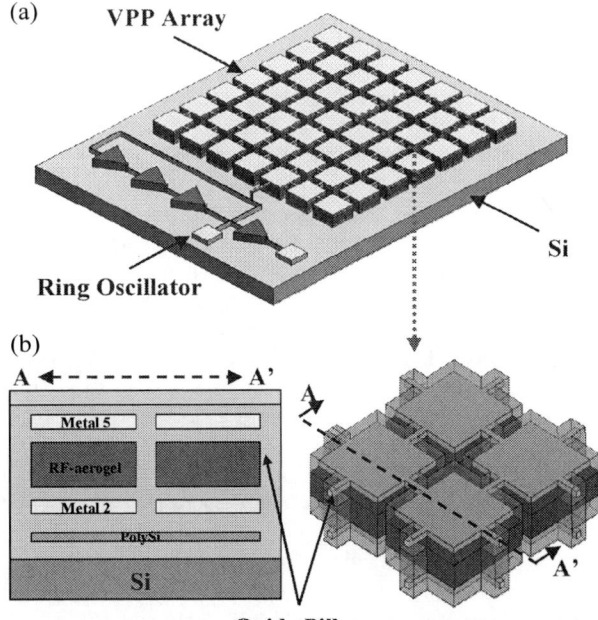

Figure 1: Schematic drawings of the proposed design: (a) perspective-view of a CMOS-MEMS capacitive humidity sensor with a VPP array and an on-chip interface circuitry and (b) cross-section view of the VPP array along A-A'.

978-1-4799-7956-1/15 $31.00 © 2015 IEEE 767 MEMS 2015, Estoril, PORTUGAL, 18 - 22 January, 2015

each VPP element is supported by four oxide pillars with metal and RF-aerogel layers served respectively as the parallel-plate sensing electrodes and dielectrics. A dispensing target (not showing in the schematics) was designed as the inlet hole for aerogel solution. Thereby, RF-aerogels can fill directly and fully into the gap between each VPP element of the array and to achieve a maskless post-processing means. Periodic slots are deployed as release holes and vent holes for the etching process and access paths for water vapor to the aerogels, respectively. The excess RF-aerogels in the slots are etched away by an anisotropic plasma etching, leaving each VPP element as a standalone structure. In other words, RF-aerogels are defined within the geometry of VPP elements. The resulting column structure is beneficial to the sensor's response since the diffusion path of water vapor is drastically shortened to the level of a few microns.

Upon water absorption and desorption, the effective dielectric constant of aerogels varies and thus induces changes in capacitance under different humidity levels. The fill factor of a VPP is defined as the electrode area divided by the device area. Essentially, high fill factor designs provide better area efficiency than low fill-factor ones. In our previous work [4], a high fill-factor (up to 98%) VPP structure was utilized in pursuit of high area efficiency and sensitivity but at the expense of response time. The current work presents a novel structural topology, viz. VPP array, with a significant improvement in response time while keeping the fill factor up to 80%.

FABRICATION PROCESS

Based on the concept in Fig.1, this study designed VPP-array humidity sensors with different sizes and numbers of VPP elements. Fig.2 shows the process steps of the sensor. As shown in Fig.2a, the fabrication begins with a standard CMOS chip prepared by TSMC. In Fig.2b, metal wet-etching was performed to realize vertical sensing gaps and reactive-ion etch (RIE) was then performed for pad opening. Along the metal wet etch, the VPP structures were released by deploying two via-connected metal layers (M3 and M4) in the CMOS stack as sacrificial layers to create a 1.91μm vertical gap that can be filled subsequently with RF-aerogels. The electrodes of the VPP capacitor thus formed are comprised of M2 and M5 layers with dielectric coverage. The dielectric layers serve as the protective coverage of the electrodes during the wet etching and after the device have been fabricated, as well. Afterwards, RF-aerogel solution was dispensed into the dispensing target and then filled into the gap between VPP structures as shown in Fig.2c. After condensation in room-temperature, RF-aerogels were dried with supercritical CO_2 fluid to obtain open-celled highly porous nanostructures. In order to pattern RF-aerogels into columns in between micro-capacitors, an anisotropic RIE recipe (CF4/O2) with an etch rate of 5 μm/min was developed and applied. Finally, the RF-aerogels inside the etch/vent holes were etched away, forming a VPP array in Fig.2d.

Micrograph in Fig.3a shows a fabricated chip with four different VPP-array sensors and a dispensing target as the inlet entrance for RF-aerogel solution. Fig.3b further shows the typical one (an array of 400 VPP capacitors, and the planar size of each VPP element is 12.3μm×12.3μm) for the following tests. Micrographs in Fig.3d-e display the FIB sectioning before and after the anisotropic RIE process, respectively. The aerogels in the slots were etched away and the remains formed as columns in between each VPP.

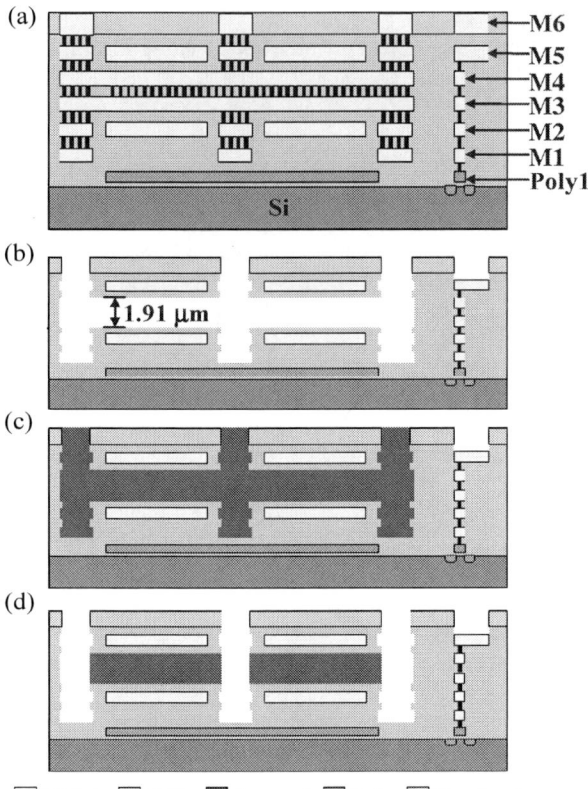

☐Metal ☐SiO$_2$ ■Aerogel ■Poly ☐Passivation

Figure 2: Fabrication process of the proposed sensor.

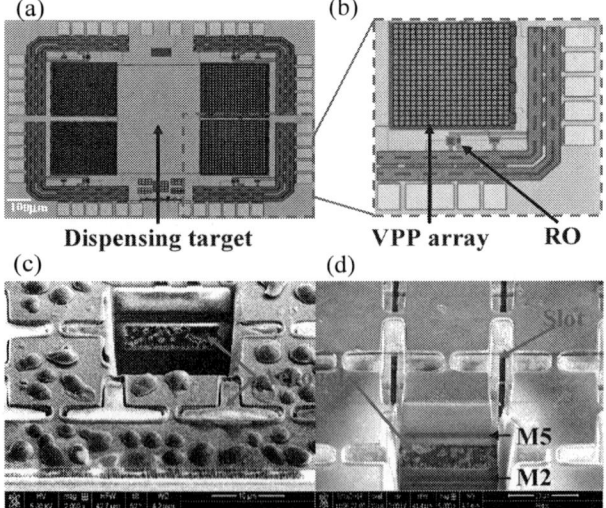

Figure 3: Fabrication micrographs: (a) Top views of the chip before RF solution was dispensed, (b) close-up view of the fabricated VPP array (400 VPPs in aggregate with each size of 12.3μm×12.3μm) and (c-d) cross-section view before and after the anisotropic RIE etching process.

MEASUREMENT RESULTS

Fig.4 shows the surface profile of the VPP array characterized by a 3-D optical interferometer. The maximum deflection along the AA' cross section of the fabricated sensor is less than 0.3 μm.

Experimental setup in Fig.5 was established to characterize capacitance and frequency changes of fabricated devices. The device under test (DUT) was bonded to a printed circuit board for sensor characterization. The DUT mounted on a side-brazed ceramic DIP was first wire-connected to a LCR meter (Agilent E4980A) for capacitance measurements of the VPP array. Later, the on-chip integrated RO was used to convert the capacitance of the VPP array into a oscillation frequency output. The capacitance and frequency values at different humidity set-points were both acquired through a computer. In order to monitor the precise RH inside the chamber, a Sensirion SHT21 humidity sensor was used as a reference. The characterizations of the proposed sensor are presented as follow.

SENSITIVITY

The capacitance changes of the VPP array was tested as shown in Fig.6. Measurements have been taken at 30°C and measured frequencies ranging from 250Hz to 1MHz.

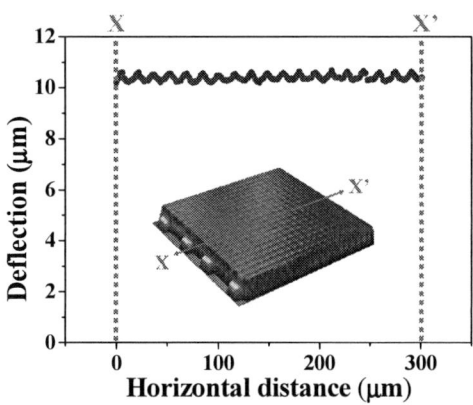

Figure 4: Optical interferometer surface measurement of the typical VPP array in Fig3b.

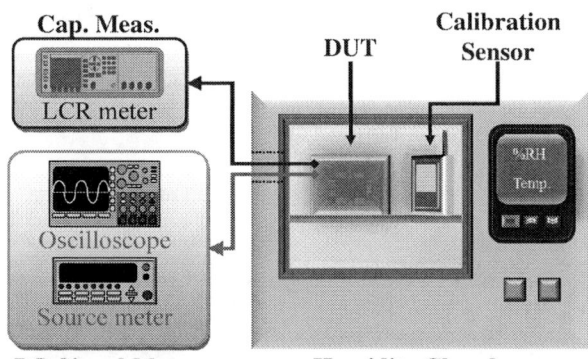

Figure 5: Schematic diagram of the measurement setup for VPP capacitor array and RO readout of the sensor. The device under test (DUT) was bonded to a side-brazed ceramic DIP (dual in-line package), and placed inside a humidity chamber for sensor characterization.

The sensitivity of the sensor abates as the measured frequency rises. The dependence on the measured frequency can be ascribed to the finite migration velocity of ions dissolved in water layers. In other words, the maximal polarization of RF-aerogels could not be developed if the electric field of a capacitor changes too fast. Therefore, at high measured frequency levels, a slightly changes in effective dielectric constant and capacitance was observed. Also, the measurements exhibited non-linear characteristic curves and escalation of capacitance at high humidity levels (above 80%RH). Typically, these issues could be ameliorated by heating up the sensor [5]. At 250Hz, in the linear region below 80%RH, a nominal sensitivity of 0.566%/%RH was given.

RESPONSE TIME

In Fig.7, the response time of the sensor was determined through a step change of 60%RH (from 20%RH to 80%RH) at 30°C. The time constant to reach 90% final capacitance value is 6s. In assessment, RF-aerogels diffusion constant of 5×10^{-11} m^2/s derived from our previous work [4] was substituted into COMSOL diffusion model. According to the simulation displayed in Fig. 7, the response time of presented sensor can be further improved within 1s.

Table1 summarizes the performances of presented and other humidity sensors for comparison.

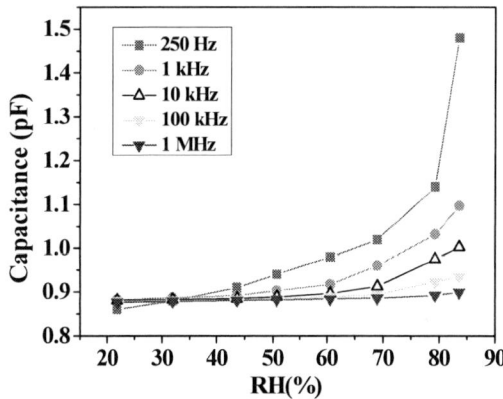

Figure 6: Characteristic curves of the proposed sensor at different frequency.

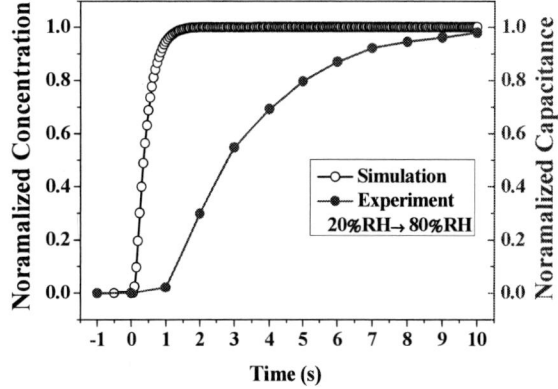

Figure 7: Optical interferometer surface measurement of

978-1-4799-7956-1/15 $31.00 © 2015 IEEE

Table 1: Comparison of the presented and existing humidity sensors

		Chung [4]	Lazarus [3]	Kang [5]	This study
S (%/%RH)		0.571	0.24	0.37	0.566
RT (s)	Sim	-	4 (%3)	-	1 (%60)
	Exp	19(%60)	-	1 (%57)	6 (%60)

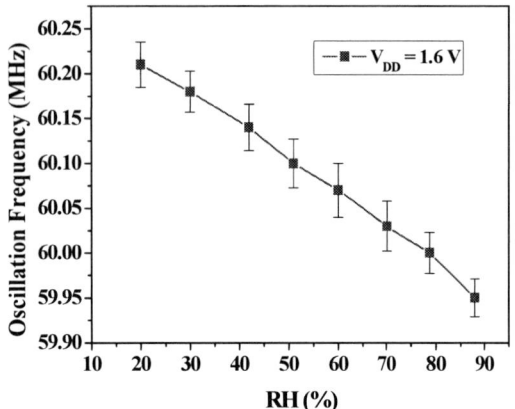

Figure 8: Oscillation frequency of the integrated sensor at different humidity levels.

INTEGRATED RING OSCILLATOR

In Fig.8, measurements of the integrated RO circuit have also been performed at 30°C to investigate the frequency shift at different RH levels. At high operating frequency level up to 60 MHz, the sensor exhibited a sensitivity of 3.8 kHz/%RH. Feasibility of the monolithic integration of humidity sensor and ring oscillator is demonstrated. Since high oscillation frequency in our current design is not favorable for the development of polarization. Further improvement in sensitivity is required by lowering the operating frequency of the integrated sensor.

CONCLUSION

In this paper, a capacitive type CMOS-MEMS humidity sensor comprised of a VPP array filled with column-defined RF-aerogels has been proposed. The humidity sensor was successfully implemented using the TSMC 0.18μm 1P6M CMOS standard process along with the in-house post-processing of the CMOS chip. RF-aerogels were filled directly into the sensing gap of VPP array and defined into columns through development

of a CF4/O2 RIE etching recipe. The sensor was measured to have sensitivity in capacitance change of 0.566% per percent RH and a response time of 6s. Additionally, an on-chip RO circuit was employed to convert capacitance into frequency readout. The integrated sensor presented sensitivity of 3.8 kHz/%RH in frequency shift.

ACKNOWLEDGEMENT

The authors would like to thank the TSMC and National Chip Implementation Center (CIC), Taiwan, for manufacturing of the CMOS chip and the Central Regional MEMS Research Center of National Science Council, and the National Nano Device Laboratory of NSC for providing the fabrication facilities. This research was sponsored in part by the Ministry of Science and Technology under contracts of NSC 102-2221-E-007-027 and NSC 103-2811-E-007-008-MY3..

REFERENCES

[1] T. Boltshauser, C. Azeredo Leme and H. Baltes, "High sensitivity CMOS humidity sensors with on-chip absolute capacitance measurement system," *Sens. Actuators B,* vol. **15**, pp. 75-80, 1993.

[2] C.-L. Dai, "A capacitive humidity sensor integrated with micro heater and ring oscillator circuit fabricated by CMOS–MEMS technique," *Sens. Actuators B,* vol. **122**, pp. 375-380, 2007.

[3] N. Lazarus and G. K. Fedder, "Designing a robust high-speed CMOS-MEMS capacitive humidity sensor," *J.Micromech. Microeng.,* vol. **22**, p. 085021, 2012.

[4] V. P. J. Chung, J. K.-C. Liang, C.-L. Cheng, M.-C. Yip and W. Fang, "Development of a CMOS-MEMS RF-aerogel-based Capacitive Humidity Sensor," *in Proc. IEEE Sensors,* pp. 190-193, 2014.

[5] U. Kang and K. D. Wise, "A high-speed capacitive humidity sensor with on-chip thermal reset," *IEEE Trasn. Electron Devices,* vol. **47**, pp. 702-710, 2000.

[6] R. W. Pekala, "Organic aerogels from the polycondensation of resorcinol with formaldehyde," *J. Mater. Sci.,* vol. **24**, pp. 3221-3227, 1989.

[7] L. M. Hair, R. W. Pekala, R. E. Stone, C. Chen and S. R. Buckley, "Low- density resorcinol–formaldehyde aerogels for direct- drive laser inertial confinement fusion targets," *J. of Vacuum Sci. & Tech. A,* vol. **6**, pp. 2559-2563, 1988.

CONTACT

* W. Fang, Tel: +886-3-5742923; fang@pme.nthu.edu.tw

AN INTEGRATED CHROMATOGRAPHYCHIP FOR RAPID GAS SEPARATION AND DETECTION

Muhammad Akbar, Hamza Shakeel, and Masoud Agah
VT MEMS Laboratory, Virginia Tech, Blacksburg, USA

ABSTRACT

This paper reports the first implementation of a highly sensitive micro discharge photoionization detector (μDPID) in a silicon-glass architecture and its monolithic integration with a high performance micro separation column. The new detector requires a two-mask fabrication process, is universal, non-destructive, low power (<2.5mW), and insensitive to flow and temperature variations. It has yielded a minimum detection limit of ~10pg which is on par with the widely used destructive flame ionization detector. The integrated chip comprises a newly developed semi-packed column with atomic layer deposited (ALD) stationary phases and is capable of multi-analyte gas mixture separation and their trace level identification while allowing the use of temperature programming for enhanced separation performance and speed.

INTRODUCTION

Gas chromatography (GC) is one of the premium analytical techniques primarily utilized for the analysis of volatile organic compounds (VOCs). It has numerous applications such as environmental monitoring, biomedical diagnostics, food processing, space exploration and homeland security. Through advancement in micro fabrication technology, numerous research efforts have been focused on the development of an easily field-deployable micro gas chromatography (μGC) system. Additional desired features of μGC include reduced weight and size, less consumable usage, minimal production cost, and novel designs to improve the performance.

A typical μGC system includes a microfabricated preconcentrator (μPC), a separation column (μSC), and a detector. The μPC serves to amplify the concentration of VOCs present at a diluted concentration (usually in parts-per-billion range) in a sample. This is accomplished by passing the sample through the device and capturing the VOCs on the adsorbent surface over a period of time. After sufficient amount of VOCs collection, the μPC is rapidly desorbed to inject the sample as a narrow plug increasing the concentration of VOCs. The VOCs then enter the μSC which is a microfluidic channel coated with a solid or liquid stationary phase. The VOCs are carried through μSC by an inert carrier gas known as the mobile phase. The sample partitions between the mobile and stationary phases depending upon their relative solubility and vapor pressure in the stationary phase. This results in their separation into the individual components. The detector finally detects the separated compounds generating an electrical signal known as the chromatogram. The compounds are identified by their respective retention times (time to pass through the separation column).

Since the conceptual demonstration of first the μGC

nearly 35 years ago [1], less attention has been directed towards their integration on a single chip referred to as monolithic integration [2]. Most of the research efforts have been directed towards the development and improving the performance of individual μGC components [3-10]. These components are then assembled manually using fluidic interconnections to realize a complete μGC unit [11-14]. The assembling of these components is a labor intensive and delicate process. The existence of leaks, fluidic blockages and dead volumes can further deteriorate the performance of the overall system. Additionally, the presence of cold spots between the μSC and the detector interconnection can lead to sample condensation or extensive band-broadening of VOCs degrading the efficiency of the μSC. One way to alleviate this issue requires assembling heating wires around the capillary tubes to elevate the temperature of the transfer lines which increases the overall power consumption. Thus, there is a critical need to monolithically integrate the μSC and the detector. However, such integration demands for thoughtful selection of the detector which should be easy to integrate with the μSC to lower the fabrication cost. Additionally, it

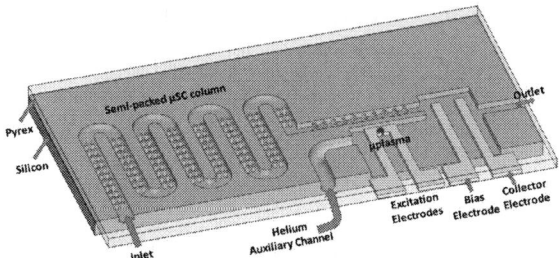

Figure 1: Schematic showing the conceptual image of the monolithic integrated chip

should be able to withstand the temperature and flow rate fluctuations necessary for the temperature programming of μSC.

Monolithic integration of the μSC and the detector has been implemented for Fabry-Perot optical sensors and micro thermal conductivity detectors (μTCD) to alleviate the loss in separation efficiency due to the presence of cold transfer lines and to reduce cost [15, 16]. While the former is sensitive enough to achieve picogram (pg) level of detection, it is not energy efficient and requires sophisticated accessories such as a laser, spectrometer or photodetectors, and beam splitter or collimator. The latter, however, is simple to fabricate and operate but lacks high sensitivity (nanograms). Using a specially designed surface micromachining process, detection limits down to 100pg have been achieved for μTCDs [17]. Nevertheless, both of the aforementioned detectors are prone to flow and/or temperature variations which limits the μSC operation only to isothermal programming conditions upon integration.

The new chip shown schematically in Figure 1 addresses these deficiencies. Herein, we have integrated high performance semi-packed μSC and μDPID in a single 1.5cm× 3cm microchip. The discharge is created in helium by applying 550V to the excitation electrode pair while the alumina is used as a stationary phase in μSC. The high energy component of the discharge including the photons (energy>10eV) and metastable He atoms (energy>19.8eV) are responsible for the ionization of VOCs emerging from the outlet of the μSC. The μSC outlet bypasses the micro-plasma to enable non-destructive analysis of VOCs. The ionized species induce a current which is measured by the picoammeter connected to the collector electrode. The chip is capable of providing picogram level minimum detection limit and rapid analysis (<1.5 min) facilitated through temperature programmed run for seven compounds spanning over a boiling points ranging from 98˚C-151˚C.

CHIP DEVELOPMENT
Fabrication

The integrated chip is realized through standard micro electro mechanical systems (MEMS) fabrication processes including lithography, deep reactive ion etching (DRIE), ALD coating, and electron-beam evaporation. The fabrication is performed on a 4 inch, 500μm thick single-side polished silicon and double-side polished Borofloat wafers and is accomplished using two masks. The processing starts with the standard RCA cleaning of silicon wafer followed by wafer priming. AZ9260 photoresist is spin-coated at 2000 rpm to achieve ~8 μm thick photoresist layer (Figure 2a). The wafer is then patterned using mask aligner followed by development in AZ400K developer. The wafer is hard-baked for 3 min at 110 ˚C. Afterwards, anisotropic etching of silicon is performed using DRIE which results in the creation of

Figure 3: SEM images of the semi-packed μSC column showing (a) top view of the channel with embedded 20μm circular micropillars (b) cross-sectional view showing high aspect ratio pillars (c) Optical image of the packaged chip with the close- up of micro plasma across 20μm gap in the inset.

190 μm-wide and 240 μm-deep channels with 20 μm in diameter embedded circular micropillars and a cavity for electrode transfer (Figure 2b). The photoresist is then stripped off using acetone. For coating the columns, an atomic layer deposition (ALD) technique is utilized to deposit thin layers of alumina on the high aspect ratio (HAR) pillars and other surfaces of μSC (Figure 2c). Approximately 10 nm of alumina is deposited at 250 ˚C using trimethylaluminum (TMA) and water as precursors in the process.

$$(A) -Si-OHAl(CH_3)_3 \rightarrow -Si-O-Al(CH_3)_2 + CH_4$$
$$(B) -Si-Al(CH_3)_2 + 2H_2O \rightarrow -Si-O-AlO*(OH) + 2CH_4$$

The asterisk in reaction (B) represents oxygen shared between adjacent aluminum atoms on the surface. ALD

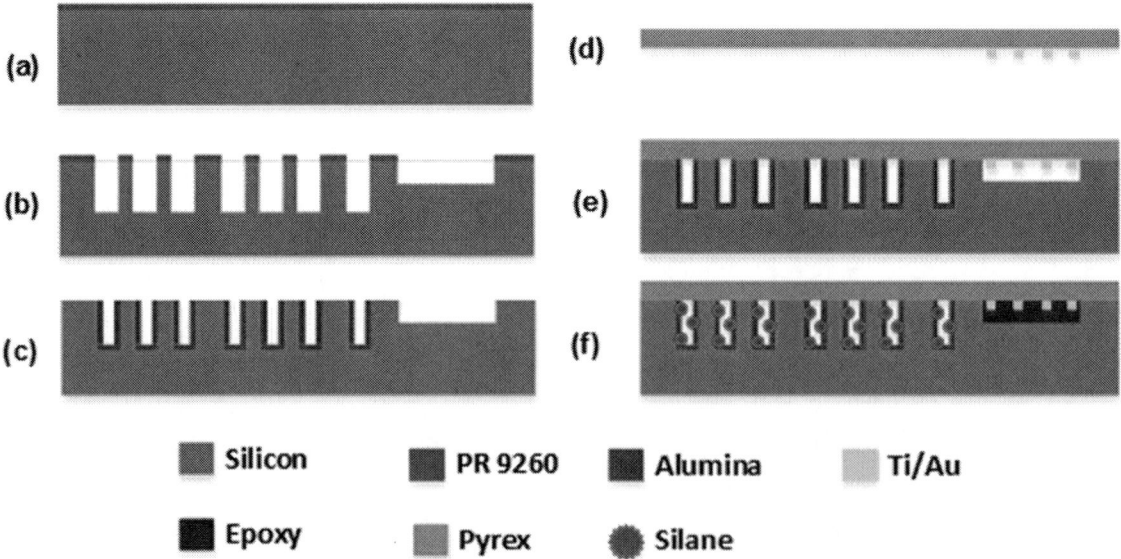

Figure 2: Process flow for the fabricated of integrated chip (a) photolithography (b) deep reactive ion etching (c) aluminum oxide coating using ALD (d) Ti/Au evaporation (e) anodic bonding and (f) functionalization with silane.

provides the best available control over the layer thickness and uniformity irrespective of the topographical changes in the channels. The second mask is utilized to fabricate the detector electrodes on the Borofloat wafer with similar lithography conditions as described previously. After depositing a 700 nm/40 nm thick Ti/Au metal stack using electron-beam evaporator, acetone is used to facilitate a lift-off process for patterning the electrodes (Figure 2d). Both silicon and Borofloat wafers are then diced to release individual devices. Anodic bonding is performed next after aligning the electrodes inside the silicon etched cavity (Figure 2e). Anodic bonding is carried out at 370 °C and 1000 V for 45 min. After bonding, the cavity is sealed with the epoxy and the electrical wires are soldered to the bond pads. Finally, functionalization with 10 mM chloro-dimethyloctadecylsilane (CDOS) solution in toluene is performed for 24 h at room temperature (Figure 2f). The column temperature conditioning is then performed in the GC oven for 1 h (35 °C ramped at 2 °C/min to 150 °C) at a constant inlet pressure of 10 psi. Figure 3 shows the scanning electron microscopy (SEM) and optical images of the fabricated chip.

RESULTS AND DISCUSSION
Testing Setup

The chip performance was evaluated experimentally by connecting the inlet and auxiliary channel of the chip to the injection ports of a GC 5890. Ultra high purity helium was used as the carrier and auxiliary gas for the plasma generation. The pressure was set to 10 psi at both ports. Plasma was generated across 20 μm gap by applying a 550 V DC voltage to the excitation electrode establishing the baseline current which was recorded on a picoammeter. The output of picoammeter was fed to the Keithley 2700 multimeter. A LABVIEW program recorded the voltage measurements from the rear-terminal output of the multimeter. A 5V DC voltage was also applied to the bias electrode. The outlet of the chip was connected to a flame ionization detector (FID) for cross examination.

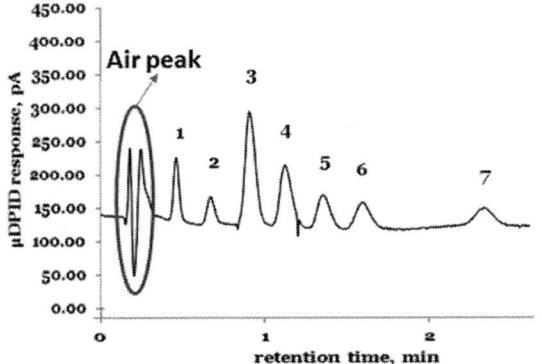

Figure 4: Separation and detection of seven compounds by the integrated chip at isothermal temperature of 40°C. Compounds identification 1) n-heptane 2) toluene 3) tetrachloroethylnene 4) chlorobenzene 5) ethylbenzene 6) p-xylene and 7) n-nonane.

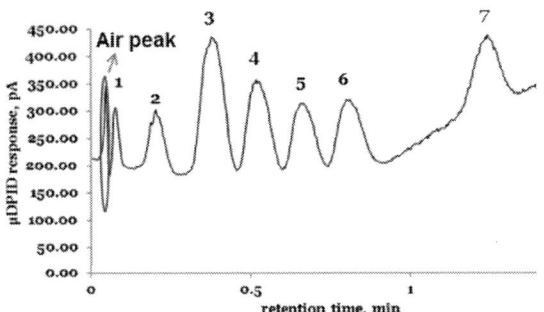

Figure 5: Chromatogram showing the separation of seven compounds in 1.4 min run through temperature programming (40°C-30°C/min-65°C).

Chromatographic Results

The performance of μSC is characterized by the parameter known as the number of theoretical plates (N) for a particular compound. N is calculated from the chromatogram by the following relationship:

$$N = 5.54 * (\frac{t_r}{w_{1/2}})^2 \qquad (1)$$

where t_r is the retention time and $w_{1/2}$ is the peak width at half height. ALD-treated/silane-functionalized μSC used in this paper are similar to our previously developed columns which have yielded 4200 plates per meter at 50 °C at the inlet pressure of 7.5 psi [7].

The performance of the integrated chip was evaluated by injecting a 5 μl headspace volume of a mixture of seven compounds into the injector of GC system. The split ratio was maintained at 100:1. The chip temperature was maintained at an isothermal temperature of 40 °C utilizing conventional GC oven. The base line was recorded to be around 130 pA signal level. Figure 4 shows the successful separation and detection of the VOCs in the mixture by the integrated chip within 2.5 min. The detection of air peak illustrates the universality of the μDPID. The tested compounds were eluted in order of increasing boiling points with the most volatile compound eluting first. Similar response was observed on the FID except for the air peak which could not be detected due to the selectivity of the detector to only organic compounds.

Temperature programming is considered as an effective method for optimizing the analysis. In this process the column temperature is increased which provides advantage of rapid analysis of complex samples. Nevertheless, it requires a detector which can tolerate the temperature variations and maintain a stability baseline when subjected to temperature programmed runs. For that purpose, the stability of the detector towards the variation in column temperature was studied. The chip temperature was increased from 40 °C to 65 °C at the rate of 30 °C/min. As evident from Figure 5 the analysis time was reduced to less than 1.5 min by employing temperature programming technique. After 1 min run time a drift in the baseline signal was observed; however, the noise level noted was much less (<5 pA) compared to the signal level. Thus, the detection of n-nonane was not hampered by the fluctuations in the baseline. It is worth-mentioning that a relatively comparable signal in terms of signal to noise ratio was observed on the FID which makes our detector performance (~10 pg of minimum detection limit) comparable to the conventional detectors.

CONCLUSION

A highly sensitive detector referred to as μDPID suitable for μGC applications was presented in this paper. The detector was fabricated using silicon and glass substrates. The single-chip integration of this detector with a highly efficient semi-packed column using simple two-mask fabrication process was also demonstrated. The separation and detection of seven VOCs by the integrated chip was validated both under isothermal and temperature programmed runs. The insensitivity of μDPID to temperature variation would provide rapid chromatographic screening of the complex sample.

ACKNOWLEDGEMENTS

The authors would like to thank Mr. Donald Leber with the MicrOn Cleanroom at Virginia Tech. This work was supported by National Institute for Occupational Safety and Health (NIOSH) and National Science Foundation (NSF) under Award No. 1R21OH010330 and ECCS-1002279 respectively.

REFERENCES

[1] S. C. Terry, J. H. Jerman, and J. B. Angell, "A Gas Chromatographic Air Analyzer Fabricated on a Silicon Wafer", *IEEE Transactions on Electron Devices*, vol. 26, pp. 1880-1886, 1979.

[2] B. C. Kaanta, H. Chen, and X. Zhang, "A Monolithically Fabricated Gas Chromatography Separation Column with an Integrated High Sensitivity Thermal Conductivity Detector", *Journal of Micromechanics and Microengineering*, vol. 20, pp. 055016, 2010.

[3] R. Haudebourg, J. Vial, D. Thiebaut, K. Danaie, J. Breviere, P. Sassiat, I. Azzouz, and B. Bourlon, "Temerature-Programmed Sputtered Micromachined Gas Chromatography Columns: An Approach to Fast Separations in Oilfield Applications", *Analytical Chemistry*, vol. 85, pp. 114-20, 2012.

[4] M. Akbar, D. Wang, R. Goodman, A. Hoover, G. Rice, J. R. Heflin, and M. Agah "Improved Performance of Micro-Fabricated Preconcentrators Using Silica Nanoparticles as a Surface Template", *Journal of Chromatography A*, vol. 1322, pp. 1-7, 2013.

[5] M. Akbar, and M. Agah "A Microfabricated Propofol Trap for Breath-Based Anesthesia Depth Monitoring", *Journal of Microelectromechanical Systems*, vol. 22, pp. 443-451, 2013.

[6] H. Shakeel, D. Wang, R. Heflin, and M. Agah "Width-Modulated Microfluidic Columns for Gas Separations", *IEEE Sensors Journal*, vol. 14, pp. 3352-3357, 2014.

[7] H. Shakeel, G. Rice, and M. Agah "Semipacked Columns with Atomic Layer-Deposited Alumina as a Stationary Phase", *Sensors and Actuators B: Chemical*, vol. 203, pp. 641-46, 2014.

[8] D. Wang, H. Shakeel, J. Lovette, G. Rice, J. R. Heflin, and M. Agah "Highly Stable Surface Functionalization of Microgas Chromatography Columns Using Layer-by-Layer Self-Assembly of Silica Nanoparticles'", *Analytical Chemistry*, vol. 85, pp. 8135-41, 2013.

[9] T. Sukaew, and E. T. Zellers "Evaluating the Dynamic Retention Capacities of Microfabricated Vapor Preconcentrators as a Function of Flow Rate", *Sensors and Actuators B: Chemical*, vol. 183, pp. 163-71, 2013.

[10] M. Agah, and K. D. Wise "Low-Mass PECVD Oxynitride Gas Chromatographic Columns", *Journal of Microelectromechanical Systems*, vol. 16, pp. 853-60, 2007.

[11] Y. Mohsen, H. Lahlou, J.-B. Sanchez, F. Berger, I. Bezverkhyy, G. Weber, and J.-P. Bellat, "Development of a Micro-Analytical Prototype for Selective Trace Detection of Orthonitrotoluene", *Microchemical Journal*, vol. 114, pp. 48-52, 2014.

[12] S. Zampolli, I. Elmi, F. Mancarella, P. Betti, E. Dalcanale, G. C. Cardinali, and M. Severi, "Real-Time Monitoring of Sub-ppb Concentrations of Aromatic Volatiles with a Mems-Enabled Miniaturized Gas-Chromatograph", *Sensors and Actuators B: Chemical*, vol. 141, pp. 322-28, 2009.

[13] W. R. Collin, G. Serrano, L. K. Wright, H. Chang, N. Nuñovero, and E. T. Zellers "Microfabricated Gas Chromatograph for Rapid, Trace-Level Determinations of Gas-Phase Explosive Marker Compounds", *Analytical Chemistry*, vol. 86, pp. 655-63, 2013.

[14] M. Akbar, S. Narayanan, M. Restaino, and M. Agah "A Purge and Trap Integrated MicroGC Platform for Chemical Identification in Aqueous Samples", *Analyst*, vol. 139, pp. 3384-92, 2014.

[15] S. Narayanan, B. Alfeeli, and M. Agah "Two-Port Static Coated Micro Gas Chromatography Column With an Embedded Thermal Conductivity Detector", *IEEE Sensors Journal*, vol. 12, pp. 1893-1900, 2012.

[16] K. Reddy, Y. Guo, J. Liu, W. Lee, M. K. Oo, and X. Fan "Rapid, Sensitive, and Multiplexed on-Chip Optical Sensors for Micro-Gas Chromatography", *Lab on a Chip*, vol. 12, pp. 901-05, 2012.

[17] S. Narayanan, and M. Agah "Fabrication and Characterization of a Suspended TCD Integrated with a Gas Separation Column", *Journal of Microelectromechanical Systems*, vol. 22, pp. 1166-73, 2013.

CONTACT

*M. Agah, tel: +1-540-2312653; agah@vt.edu

FABRICATION OF SILVER NANOPARTICILES ON CYLINDRICAL SURFACE OF U-SHAPED FIBER ATR SENSOR BY MATERIAL REDUCTION

Dachao Li[1], Changyue Sun[2], Songlin Yu[1], Haixia Yu[2], and Kexin Xu[1]

[1] State Key Laboratory of Precision Measuring Technology and Instruments, Tianjin, CHINA
[2] Tianjin Key Laboratory of Biomedical Detecting Techniques and Instruments, Tianjin, CHINA

ABSTRACT

An implantable U-shaped fiber ATR (attenuated total reflection) sensor enhanced by silver nanoparticles (AgNPs) on cylindrical surface was presented for continuous glucose monitoring. U-shaped structure was addressed to increase effective optical length at limited implantable space to enhance the sensitivity of fiber ATR sensor. A novel method to fabricate silver nanoparticles on cylindrical surface of U-shaped fiber ATR sensor based on chemical reduction of its silver halide material directly without any preliminary nanoparticles synthesis and the following covalent bond or self-assembly was proposed. The silver nanoparticles will further enhance infrared absorption signal of fiber ATR sensor. The experiment result shows that the sensitivity of enhanced fiber ATR sensor enhanced by the silver nanoparticles is about three times than the normal one.

INTRODUCTION

Diabetes mellitus is a serious human disease, and it is important to monitor blood glucose levels continuously to provide guidance for diagnosis and therapy. To date, the implantable enzyme electrode sensing technique is the only method that has been used in clinical settings for continuous glucose determination by measuring the electric current generated by the glucose oxidase (GOD) enzyme reactions in subcutaneous tissue [1]. However, the significant drift caused by bioelectricity and the effect of electrochemical reactions under hypoxia reduce the accuracy of glucose measurements. In addition, the local glucose level close to the enzyme electrode is irreversibly depleted by the GOD enzyme reaction, resulting in a measured glucose value that is lower than the true concentration, especially in the case of hypoglycemia.

The fiber-based technique provides an excellent approach to fabricate small ATR sensors, which makes it possible to implant the sensors in subcutaneous tissue for continuous glucose monitoring [2]. Compared with enzyme electrode sensors, the fiber ATR sensor is the only part implanted under the skin, which means only the optical signal passes through the subcutaneous tissue, so the glucose measurement by fiber ATR sensor is not affected by bioelectricity in the body. In addition five emission wavelengths 1081, 1076, 1051, 1041 and 1037 cm^{-1} around the glucose "finger print" peaks (1080 and 1035 cm^{-1}) were selected as the working wavelengths for glucose specific measurement instead of GOD enzyme [3], which means the absorbance of glucose at five absorption peak could replace the GOD enzyme and it does not consume the glucose molecule during measurement. These characteristics make the fiber ATR sensor more suitable and promising for implantation in tissue for continuous glucose monitoring. However, the miniaturization of fiber ATR sensor could bring about low sensitivity and resolution, so how to improve the sensitivity and measurement resolution of the fiber ATR sensor is the key point for continuous glucose monitoring in clinic.

In this paper, bent structure was designed to increase the sensing optical length to improve the sensitivity in a limited space for implanted continuous glucose monitoring [4]. Silver nanoparticles were used to modify the cylindrical surface of the fiber ATR sensor and thereby enhance the infrared absorption of glucose molecules. The fabrication of silver nanoparticles was based on chemical reduction of its silver halide material directly without any preliminary nanoparticles synthesis and the following covalent bond or self-assembly, which is more environmental and easier to operate than physical vapor deposition (PVD), chemical vapor deposition (CVD) or other electroless deposition approaches [5-6].

DESIGN OF FIBER ATR SENSOR

Structure of Implantable Fiber ATR Sensor

Fig. 1 shows the silver-nanoparticle-enhanced fiber ATR sensor which can be implanted into subcutaneous tissue for continuous glucose determination. A biocompatible semipermeable membrane with a selectable molecular weight cut-off was used as a protective cover to separate the implanted sensor from the tissue, filter out large biological molecules within the ISF and allow glucose molecules to pass through. The sensor was fabricated in three steps: (1) the structure of the bent fiber ATR sensor; (2) the fabrication of silver nanoparticles; (3) the biocompatible encapsulation of the sensor.

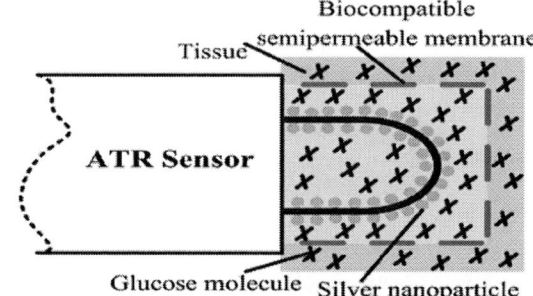

Figure 1: Schematic diagram of implantable U-shaped fiber ATR sensor.

Enhancement of Fiber ATR Sensor by Bent Structure

The numerous excellent properties of polycrystalline infrared (PIR) fibers, such as broad transmission band (4–18 μm), non-hygroscopic behavior, no brittleness and no toxicity, have enabled their widespread use in bio-sample measurement. Multimode silver halide PIR fibers (A.R.T. Photonics GmbH, Berlin, Germany) were

978-1-4799-7956-1/15 $31.00 © 2015 IEEE

used in this study. According to the Lambert-Beer law, increasing the sensing optical length of a fiber ATR sensor is conducive to improving its sensitivity, where the sensing optical length is proportional to each penetration depth dp and total number of internal reflections N_m. For a conventional straight fiber ATR sensor, the penetration depth d_p and total number of reflections N_m can be determined as follows:

$$d_p = \frac{\lambda_0}{2\pi n_1 \left[\sin^2 \theta - \left(n_2 / n_{core}\right)^2\right]^{1/2}} \quad (1)$$

$$N_m = \frac{L}{2\rho} \cot \theta \quad (2)$$

where ρ is the core radius; L is the sensing length (unclad region length) and θ is the angle of incidence; and n_2 is the refractive index of the measurement sample. Because $n_2 \ll n_{core}$ and L, ρ and the numerical aperture of the fibers are restricted, the sensitivity and resolution of straight fiber ATR sensors are low. A U-shaped sensor could be used to increase the number of reflections and the penetration depth at each reflection, increasing the effective sensing optical length. Thus, a U-shaped fiber sensor with a bending radius of 2.5 mm, as shown in Fig. 2, was used in this study. In addition, silver nanoparticles were used to further enhance the sensitivity and resolution of the fiber ATR sensor.

Enhancement of Fiber ATR Sensor by AgNPs

The infrared absorption of glucose molecules would be more intense when they were adsorbed on AgNPs than expected from conventional measurements without AgNPs [7]. The intense enhancement were mainly contributed by the electromagnetic and chemical mechanisms .The infrared absorption (A) may be written [7] as (3):

$$A \propto |\partial\mu/\partial Q \cdot E|^2 = |\partial\mu/\partial Q|^2 |E|^2 \cos^2\theta \quad (3)$$

Where $\partial\mu/\partial Q$ is the derivative of the dipole moment with respect to a normal coordinate Q, E is the electric field that excites the silver molecule, and θ is the angle between $\partial\mu/\partial Q$ and E. The electromagnetic mechanism assumes an increase of the local electric field (E) at the surface of AgNPs [8] .On the other hand, the chemical mechanism assumes an increase of $|\partial\mu/\partial Q|^2$ (i.e. the absorption coefficient) due to chemical interactions between the absorbed glucose molecules and the surface of nanoparticles [9].Therefore, the enhancement of infrared absorption signal is the collaborative result of both the electromagnetic and chemical mechanisms.

FABRICATION OF AgNPs ON SURFACE OF FIBER ATR SENSOR

The PIR fiber was composed of $AgCl_xBr_{1-x}$ [10], and material reduction method was used to prepare silver nanoparticles on the cylindrical surface of the fiber ATR sensor as was shown in Fig.2, and the absorption spectrum of the glucose molecules near the silver nanoparticles grown on the surface of fiber ATR sensor was enhanced.

Glucose was employed as the reducing agent, and the chemical reduction reaction between ions in the fiber material can be expressed as follows [11]:

$$Ag^+Cl^- + Ag^+Br^- + CH_2OH\left(CHOH\right)_4 CHO + NaOH$$
$$\rightarrow Ag + CH_2OH\left(CHOH\right)_4 COOH + Na^+Cl^- + Na^+Br^-$$

(4)

Figure 2: The AgNPs on cylindrical surface of fiber ATR sensor by material reduction

The dynamic equilibrium of glucose in water is shown in Fig. 3. The concentration of glucose with aldehyde groups (-CHO) is very low. However, the silver ions can be reduced by -CHO at normal temperature in alkaline solution. A sodium hydroxide (NaOH) reagent was added to the reaction solution to adjust the PH value to 12 which offers a good alkaline environment. The added alkaline reagent caused the reduction reaction to take place at room temperature and enhanced the reduction ability of -CHO. The temperature was controlled by a water bath to stabilize the reaction conditions. The optimal conditions were observed to be the addition of 15 mM NaOH to 0.15 and 0.25 mM glucose solutions; the reaction times for these solutions were 100 and 80 min, respectively.

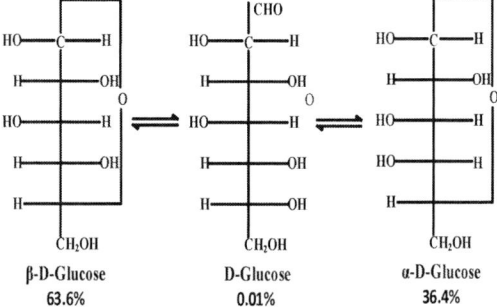

Figure 3: Dynamic equilibrium of glucose in water

This method to grow AgNPs on cylindrical surface of U-shaped fiber ATR sensor which was based on chemical reduction of its silver halide material directly without any preliminary nanoparticles synthesis and the following covalent bond or self-assembly. This method offers an option to fabricate nanoparticles on cylindrical surface of the fiber sensor based on silver halide materials.

EVALUATION OF AgNPs ON SURFACE OF FIBER ATR SENSOR

Only the sensing part (bent region) of the U-shaped fiber ATR sensor was placed in a beaker containing the specified concentration of reducing agents and sodium hydroxide. After the desired reaction time arrived, the sensor was taken out and immersed in deionized water for

978-1-4799-7956-1/15 $31.00 © 2015 IEEE

1~2 min to stop the reaction, and the reaction product covering the fiber was removed to facilitate the reaction to proceed. The reaction time was optimized and given as following: (1) when the reducing agent concentration is of 0.25 mM, the total reaction time is of 80 min. After reaction begins 30 min, the sensor was taken out and dipped into deionized water for 1 min washing, then immersed in the reaction solution again. The next washing time is of 55 min after reaction begins. (2) When the reducing agent concentration is of 0.15 mM, the total reaction time is of 100 min and the washing time is 40, 60 and 80 min after reaction begins. All the experimental procedure was processed in dark room to avoid the degradation of silver halide fiber.

The size and distribution of silver nanoparticles growing on the cylindrical surface of fiber ATR sensor were preliminarily evaluated using scanning electron microscope (Nanosem 430, FEI Co., Oregon, USA). The experimental results were shown in Fig. 4 (a) and 4 (b). As shown in Fig. 4 (a), the reducing agent concentration of 0.15 mM leads to sparse and clustered silver nanoparticles. It maybe caused by low reduction rate of silver atoms due to low concentration of a-CHO. The general sizes of these nanoparticles were between 85~140 nm. According to the previous studies, the stacked nanoparticles result in SEIRA (surface-enhanced infrared absorption) [11].

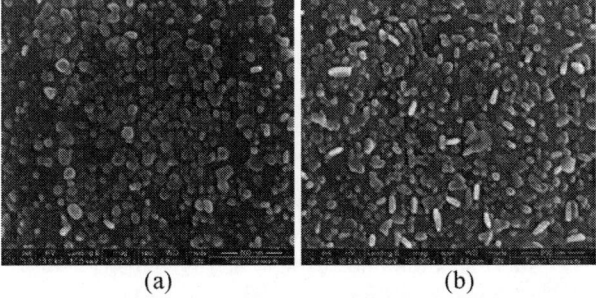

(a)   (b)

Figure 4: SEM images of nanoparticles on cylindrical surface of silver halide fiber. (a) Reducing agent: 0.25 mM; NaOH: 15 mM. The total reaction time is of 80 min. (b) Reducing agent: 0.15 mM; NaOH: 15 mM. The total reaction time is of 100 min.

(a)   (b)

Figure 5: Nanoparticles reduced by 0.25 mM reducing agent and 15 mM NaOH in 25 ℃ water bath. Images (a) and (b) show nanoparticles growing on different regions of the silver halide fiber. The total reaction time is of 80 min.

Water bath (25 ℃) was employed to control the temperature further to strengthen the repeatability of material reduction method. The following study will focus on this experimental condition. Nanoparticles distributed in different regions on surface of silver halide fiber were shown in Fig. 5 (a) and (b). The most nanoparticles are near

spheres and the size are between 60~100 nm, as shown in Fig. 5 (a). A fewer nanoparticles are near ellipsoid, as shown in Fig. 5 (b). The long axis are between 85~120 nm, and the short axis are between 35~50 nm. Therefore, the ratio of most particles are between 2~2.5.

CHARACTERIZATION OF BENT FIBER ATR SENSOR ENHANCED BY AgNPs

Experimental System

As shown in Fig. 6, a dual-path laser-measurement set-up was established for glucose determination using a tunable CO_2 laser [12]. The CO_2 laser used in this study （Merit-G, Access Laser Co., Washington, USA）operates in TEM00 mode, and the polarization state is perpendicular to the horizontal plane, with a waist diameter of < 2 mm and divergence angle of < 5 mrad. The maximum output power is approximately 800 mW.

To achieve line tuning, a linear step motor and a grating fixed on the substrate of the grating turret by a piezoelectric ceramic actuator were employed. An infrared attenuator (Model 401, Lasnix, Berg, Germany) was used to attenuate the high laser output power (maximum power of 800 mW) to a reasonable level. The attenuated laser beam was divided into dual paths by a zinc selenide (ZnSe) beam splitter (BS), one for reference and the other for sample measurement. The dual-path incidence laser beams were coupled into reference and sample detectors (Det. R and Det. S) by a ZnSe lens. The two infrared detectors (LME-353, InfraTec GmbH, Dresden, Germany) were matched with two lock-in amplifiers of the same model (SR830, Stanford Research Systems, Inc., California, USA), the synchronous reference frequency (750 HZ) of which was provided by the RF circuit of the laser system. The lock-in signals of the dual paths were then recorded by a data-acquisition card synchronously. The "sandwich" measurement method was adopted to overcome the fluctuation of light intensity caused by CO_2 laser and the instrument drift.

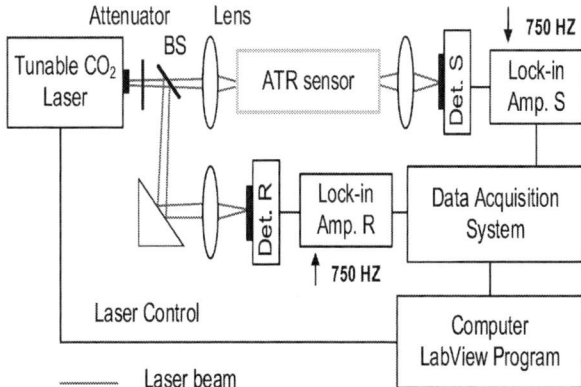

Figure 6: The sketch of the dual-path laser-measurement set-up for glucose detection.

Performance of Bent Fiber ATR Sensor with AgNPs

A conventional fiber ATR sensor and a nanoparticle-enhanced fiber ATR sensor were used for glucose measurement. As shown in Fig. 7 and 8, the absorbance correlates well with the glucose concentration at 1081 and 1037 cm^{-1}, and the linearity of the two types of fiber ATR sensors is such that $R^2 \geq 0.98$. The correlations between absorbance and glucose concentration in the low range at 1081 and 1037 cm^{-1} are presented in the inserts of Fig. 7 and 8. The slope of the linear fit of the glucose absorbance versus the glucose concentration was employed to calculate the enhancement factor of the nanoparticles. The enhancement factors at 1081 and 1037 cm^{-1} were calculated to be $R_{1081}=2.97$ and $R_{1037}=3.18$, respectively. For the conventional U-shaped fiber ATR sensor, the measurement resolution was approximately 45 mg/dL, whereas the measurement resolution of the silver-nanoparticle-enhanced U-shaped fiber ATR sensor was approximately 15 mg/dL.

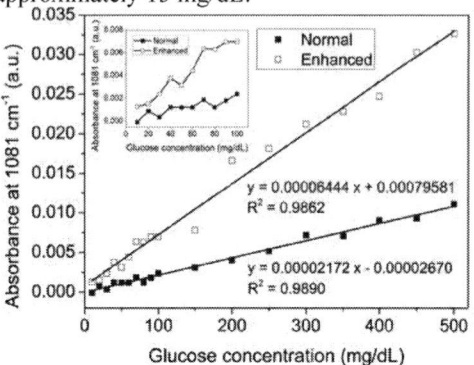

Figure 7: Plot of absorbance vs. glucose concentration measured by normal and enhanced sensors at 1081 cm^{-1}.

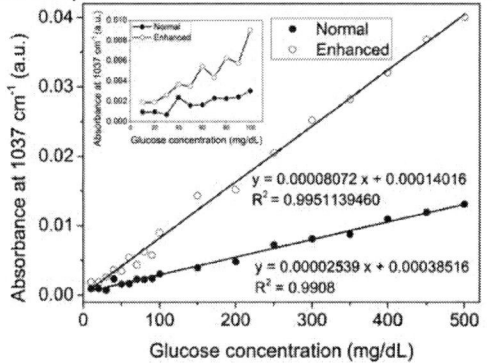

Figure 8: Plot of absorbance vs. glucose concentration measured by normal and enhanced sensors at 1037 cm^{-1}.

CONCLUSIONS

In this paper, an implanted fiber ATR sensor was reported for continuous glucose monitoring. Bent structure was designed to increase the sensing optical length to improve the glucose sensitivity. A method of fabricating nanoparticles on the cylindrical surface of fiber ATR sensor chemical reduction of its silver halide material directly without any preliminary nanoparticles synthesis and the following covalent bond or self-assembly was also proposed to increase the sensitivity for glucose. The experiment results indicate that the measurement resolution of the silver nanoparticles enhanced fiber ATR sensor is about three times than the normal one. The future work will focus on the sensor biocompatible encapsulation using biological materials and animal experiment.

ACKNOWLEDGEMENTS

This work was supported by the National Natural Science Foundation of China (No. 61176107, No. 51350110233, No. 11204210, No. 61428402 and No.61201039), the Key Projects in the Science & Technology Pillar Program of Tianjin (No. 11ZCKFSY01500), and the National Key Projects in Non-profit Industry (No. GYHY200906037), and the National High Technology Research and Development Program of China (No. 2012AA022602).

REFERENCES

[1] N. Oliver, C. Toumazou, A. Cass, D. Johnston, "Glucose sensors: a review of current and emerging technology," *J. Diabetic. Med.*, vol. 26, pp. 197-210, 2009.

[2] A. Lambrecht, T. Beyer, K. Hebestreit, R. Mischler,W. Petrich, "Continuous glucose monitoring by means of fiber-based, mid-infrared laserspectroscopy,"*J.Appl.Spectrosc.*,vol.60,pp.729-736,2006.

[3] S. Yu, D. Li, H. Chong, C. Sun,K. Xu, "Continuous glucose determination using fiber-based tunable mid-infrared laser spectroscopy, " *J. Opt. Laser. Eng.*, vol.55, pp.78-83, 2014.

[4] Y. Raichlin, A. Katzir, "Fiber-optic evanescent wave spectroscopy in the middle infrared," *J. Appl. Spectrosc.* , vol. 62, 55A, 2008.

[5] S. Sanchez-Cortes, C. Domingo, J. Garcia-Ramos, J. Aznarez, "Surface-enhanced vibrational study (SEIR and SERS) of dithiocarbamate pesticides on gold films", *Langmuir*, vol. 17, pp.1157-1162, 2001.

[6] Á. López-Lorente, M. Sieger, M. Valcarcel. "Infrared attenuated total reflection spectroscopy for the characterization of gold nanoparticles in solution." *J. Anal. Chem.*, vol. 86, pp.783-789, 2013.

[7] M. Osawa, "Surface-enhanced infrared absorption [M]//Near-Field Optics and Surface Plasmon Polaritons." in *Springer*，Berlin Heidelberg, 2001，pp. 163-187.

[8] M. Osawa, "Dynamic Processes in Electrochemical Reactions Studied by Surface-Enhanced Infrared Absorption Spectroscopy (SEIRAS)." *J.B. Chem. Soc. Jpn.*, vol.70, pp.2861-2880, 1997.

[9] G. Merklin, P. Griffiths. "Influence of chemical interactions on the surface-enhanced infrared absorption spectrometry of nitrophenols on copper and silver films." *J.Langmuir*, vol.13, pp.6159-6163, 1997.

[10] V. Artjushenko, P. Baskov, G. Kuz'micheva, M. Musina, V. Sakharov,T. Sakharova, "Structure and properties of AgCL$_{1-x}$Br$_x$ (x= 0.5–0.8) optical fibers, " *J. Inorg. Mater+.*, vol.41, pp.178-181, 2005.

[11] G. Rao, J.Yang, "Preparation of high-capacity substrates from polycrystalline silver chloride for the selective detection of tyrosine by surface-enhanced infrared absorption (SEIRA) measurements," J.Anal.Bioanal.Chem. , vol. 401, pp.2935-2943, 2011.

[12] Y. Ma, D. Liang, "Tunable and frequency-stabilized CO$_2$ waveguide laser,"*J.Opt.Eng.* , vol. 41, pp.3319-3323, 2002.

CONTACT

*D.C. Li, tel: +86-22-27403916; dchli@tju.edu.cn

978-1-4799-7956-1/15 $31.00 © 2015 IEEE

INTRINSIC ZnO NANOWIRES WITH NEW SENSING MECHANISM OF SULFURATION-DESULFURATION TWO-STEP REACTION FOR HIGH PERFORMANCE SENSING TO ppb-LEVEL H₂S GAS

Pengcheng Xu[1], Haiyun Huang[2], Dan Zheng[2], and Xinxin Li[1]

[1]State Key Lab of Transducer Technology, Shanghai Institute of Microsystem and Information Technology, Chinese Academy of Sciences, Shanghai 200050, CHINA
[2]School of Chemical and Environmental Engineering,
Shanghai Institute of Technology, Shanghai 201418, CHINA

ABSTRACT

The paper reports a novel H_2S sensing-effect for intrinsic ZnO nanowires (NWs) chemiresistive sensor. Herein 50nm diameter ZnO-NWs are found and verified to be sulfurized by H_2S to form ZnS that can be latterly desulfurized back to ZnO by ambient oxygen, which is different from conventional ZnO sensing-mechanism where resistance of semiconductor ZnO is changed via electron depletion-layer variation by surface adsorbed ambient oxygen. The ZnO-NWs have realized detection to 50ppb H_2S.

INTRODUCTION

Metal oxide semiconductors, such as ZnO, In_2O_3 and SnO_2, have been widely applied as gas sensing materials for detection of toxic and inflammable gases in ambient air, due to their various nanostructures, low cost, and high compatibility with MEMS fabrication process [1]. Among the investigated metal oxide sensing materials, ZnO has been extensively researched because of its advantageous properties like high mobility of conductive electron and good chemical/thermal stability. Moreover, it can be constructed into lots of high specific surface area nanostructures that include nanowires, nano-brushes and nanotubes [2]. By now, many kinds of the gases like C_2H_5OH, H_2, NH_3, and CO have been detected by ZnO-based sensors. Compared with these gases, hydrogen sulfide (H_2S) is comparably poisonous to CO and most flammables. If its concentration reaches *ppm* level, the exposed gas could cause desensitized human nose, declined smell ability or even impaired nervous system, thereby becoming an important concern of environmental pollution. On the other hand, H_2S recently attracts intensive interest in biomedical field, where it is studied as an endogenous gaseous mediator and potential target for pharmacological manipulation [3]. H_2S in animals and humans has been involved in diverse physiological and pathophysiological processes, such as learning and memory, neurodegeneration, regulation of inflammation and blood pressure and metabolism. H_2S in human adult has been reported to be at *ppm* level. With the concentration change as a biomarker of lesion stage, direct H_2S detection with sub-ppm resolution is highly in demand. However, the lack of effective detecting and analytical tools to trace-level H_2S has hindered the biomedical researches correlative to the endogenous-gas. Therefore, it is extremely crucial to develop gas sensing technologies to detect and monitor H_2S at ultra-low concentration.

Inspired by the reported studies for syngas, where the ZnO particles with decreased nanoscale diameter show greatly lowered vulcanization temperature, we conceive to explore nanoscale size-effect induced new sensing effect of intrinsic ZnO material for achieving higher H_2S sensing performance at lowered working temperature. For this purpose, dense array of ~50nm diameter intrinsic ZnO nanowires (NWs) is regioselectively grown and inter-linked at the gap between two comb-electrodes to form chemiresistive sensing structure. In environmental air, the sensor experimentally exhibits ultra-high sensitivity, repeatability and selectivity to ppb level H_2S. Based on the experimentally observed temporary generation of ZnS as intermediate product, a new sulfuration-desulfuration sensing effect for H_2S is proposed and verified by experiment and analysis. Originated from size-effect of intrinsic ZnO NWs, the two-step reaction sensing mechanism directly induces the drastically improved sensing performance at low temperature of 150°C.

EXPERIMENTS

Fabrication of Chemiresistive Sensing Chips

Firstly, Pt/Ti double layer of 100nm/10nm in thickness is sputtered on Pyrex 7740 glass wafer and patterned with photolithographic steps and physical ion-beam etch. As is shown in Figure 1, the Pt/Ti heating resistor is laid at the double sides of the comb-fingers shaped sensing electrode-pair, thereby providing elevated working temperature for the nanowire-array chemical gas sensor. Thereafter, Au/Cr composite layer of 60nm/30nm in thickness is deposited by electron-beam evaporation and patterned into comb-electrodes through lift-off process. The narrow gap areas between the cross-finger electrodes are electrically bridged into a chemiresistor with the *in situ* synthesized intrinsic ZnO NWs that form inter-linked dense array.

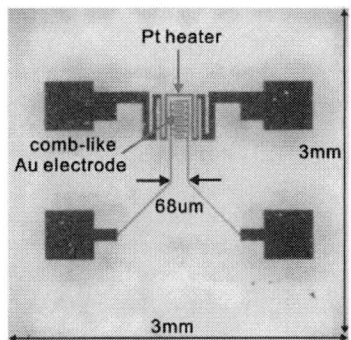

Figure 1: Micrograph showing the layout of the micro-sensor chip before growth of the ZnO NW array.

In situ synthesis of ZnO NW array

The ZnO NWs are bottom-up synthesized from aqueous solution and directly grown at the sensing area of the sensor by using a modified hydrothermal method. Firstly, the senor chips (see Figure 1) are cleaned by ultrasonic rinsing in acetone, ethanol and deionized water, with each for 5min. Then, the chips are blowing dried and plasma pre-treated (with Harrick PDC-002 equipment) for 5 min to further eliminate surface contamination. After 5mM zinc acetate in ethanol is drop-wise coated at the comb-electrode region of the chip, the chip is dried in air for 10s, rinsed with ethanol and, then, blowing dried with pure nitrogen stream. This coating step needs to be repeated for 3~5 times. Covered with the seed film of zinc acetate crystallites, the seeded chip is heated to 350°C in air for 20 min to yield a nano-layer of ZnO seeds. Then ZnO NWs are *in situ* grown hydrothermally by face-down suspending the chip at the surface of an aqueous solution that contains zinc nitrate (25mM), hexamethylenetetramine (25mM) and polyethylenimine (PEI, 6mM). The bottom-up NWs growing process is implemented at 90°C for 2h, during which the solution should be tightly sealed in a bottle. Finally, the chip is taken out from the solution, washed with deionized water and dried for sensing experiment.

Characterization

The morphology and nanostructure of ZnO NWs are observed with scanning electron microscope (SEM, Hitachi S4800) and transmission electron microscope (TEM, JEM-2010F). The crystal structures of ZnO NWs are obtained by using X-ray diffraction (XRD, Bruker D8 DISCOVER). The gaseous products in the sensing experiment are analyzed with a gas chromatograph-mass spectrometer (GC-MS, Agilent 7890A-5975C). Gas sensing experiment is carried out in a lab-made testing chamber. A multi-meter (Agilent-34401A) is used to record the sensor resistance. During experiment, the sensor is sequentially exposed to the H_2S gas with varied concentrations of 50ppb, 100ppb and 1ppm~2ppm as well. The sensing response is defined as the relative resistance change of $\Delta R/R_0 = (R_0 - R_{gas})/R_0$.

RESULTS & DISCUSSION

As is shown in Figure 1, the Pt resistor serves as a micro-heater to provide desired working temperature for the sensing material by applying DC voltage. The relationship of heating voltage versus temperature has been calibrated by putting the sensor chip into a temperature programmable oven to detect the resistance value. Experimental results show that, the sensor is sensitive to H_2S at about 150 °C, where 10V heating voltage is applied. By using finite-element simulation of ANSYS, the DC 10V heating induced uniform 150 °C at the sensing area is confirmed.

The SEM images in Figure 2 show the ZnO NW-array grown in the sensor chip. It can be observed that the densely arrayed ZnO NWs are fully filled into the gap space between the two interlacing electrode-combs. More importantly, the dense ZnO NWs are regioselectively grown at the gap areas where the substrate is ≡Si-OH covered glass. In contrast, extremely rare NWs are grown on top of the Au electrodes. In the SEM images, the NWs are generally a couple of microns in length and 50~60nm in diameter. Stable resistance value of the sensing chemi-resistor is measured that indicates effective carrier conductivity of the ZnO NW array. Attributed to the densely and direction-orderly grown NWs, physical cross-linking of adjacent NWs builds the electric current-flow routes.

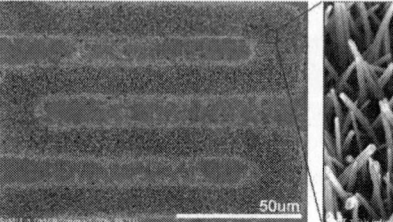

Figure 2: SEM image of the ZnO NW array region-selectively grown on the sensor chip.

The dense and highly inter-crossed ZnO NW array forms a resistor at the gap area between the two comb-finger electrodes. With the resistor heated to 150 °C, the typical resistance is in the range of several tens of $k\Omega$ when the device is stored in N_2. As an environmental monitoring sensor, the H_2S sensing chemi-resistor needs to work in natural ambient air. In air environment, the resistance arises and stabilizes at $M\Omega$ level that will be discussed in details in following subsection. Figure 3a shows the H_2S sensing experiment results of the sensor operated at 150°C. The standard H_2S gas is diluted to desired concentration by volume mixing with air. When H_2S introduced, the resistance of the ZnO NW array decreases rapidly and, after H_2S switched off, increases back to the base-line soon. For the very low concentration H_2S of 50ppb, the response signal is still as high as about 2% and much higher than noise-floor. Since H_2S with even lower concentration cannot be precisely prepared in our lab, herein the limit of detection (LOD) of a couple of tens ppb can be reasonably estimated. Without any catalyst used, the intrinsic ZnO NW array has exhibited surprisingly high sensitivity and amazingly fine LOD under so low working temperature of 150°C. The results indicate high applicability of the sensing material. Selectivity of the sensor is also examined by comparing the H_2S sensing response with the interfering signals of eight kind of flammable gases of methanol, ethanol, acetone, hexane, ethyl acetate, dichloromethane, benzene and toluene (the concentrations of all the gases are 20ppm). With the results shown in Figure 3b, satisfactory selectivity of the H_2S sensor can be apparently observed.

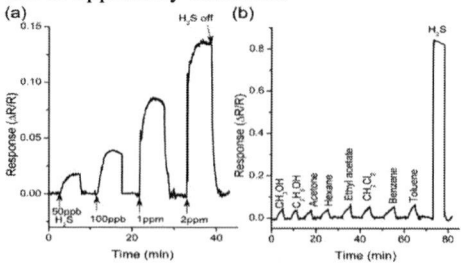

Figure 3: (a) Sensing experimental results of the ZnO NW-array sensor to H_2S of 50ppb-2ppm concentrations. (b) Testing results for assessment of selectivity.

If our ZnO nanomaterial still follows the conventional H_2S sensing mechanism that was explained in many published literatures, *i.e.*, the decrease of ZnO resistance is caused by reaction between H_2S and adsorbed oxygen that weakens the electron depletion-layer, there would be no reason for our sensor to exhibit so good performance. Not to mention our pure ZnO material (without catalyst) and the low operation temperature of only 150°C. Viewed from another angle, H_2S and the eight kinds of interfering gases are all reductive gases. If they all feature the same conventional sensing effect of oxidization weakened depletion-layer and enhanced electron conductivity, similar shape of response-curve should be obtained for all the gases. However, Figure 3b shows great difference between the response value to H_2S and those to other reductive gases. Therefore, we generate interest to explore the sensing effect of this ZnO NW-array material.

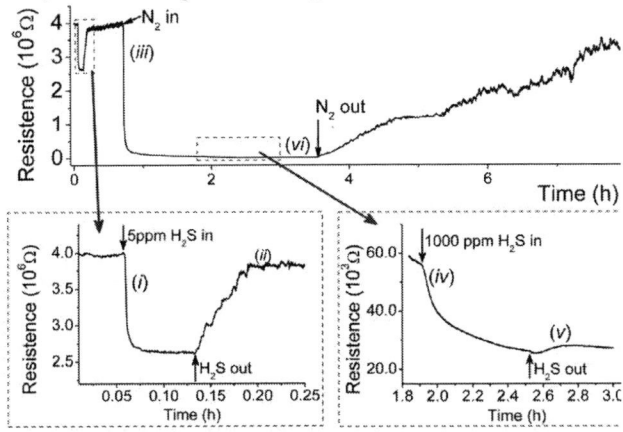

Figure 4: HR-TEM images for the ZnO NW before exposed to H_2S in (a) and after exposed to H_2S in (b).

We have conducted an experiment to identify the products generated after sensing to H_2S in ambient air, resulting in complete consistency with the known theory described above. The TEM images in Figure 4a shows the structure of a ZnO NW that has been long-time stored in air. Referred to previous work, the ZnO NW shows high-quality single-crystal structure with smooth surface and (001)-oriented growing direction. However, when another ZnO NW is imaged with TEM, which has been stored in 1000ppm H_2S, Figure 4b shows a lot of surface stained crystalline spots. With the stained areas close-up viewed, an extra structure is found being newly generated on top of the ZnO surface. According to the JCPDS crystal data, the surface spots are recognized as ZnS.

In order to further discover the ZnS generation procedure, we specifically design an experiment to examine the ZnO NW-array chemiresistive sensor. The resistance variation of the sensor in terms of sequential change of gaseous environment exposed to the sensor is real-time recorded and plotted in Figure 5. The examined sensor is located in a hermetic chamber that can be fulfilled by different gases via pipe linking. The sensor is heated stably to 150°C. At each state of the gas changing process (see Figure 5), the corresponding XRD pattern is characterized and plotted in Figure 6. It needs to be noted that the XRD results are obtained from the ZnO NW-array grown on larger sized glass substrate instead of the micro-sensor to obtain adequate sample for characterization. The sensor is firstly exposed to natural air and its resistance is stabilized at about 4MΩ, as is shown in Figure 5. When 5ppm H_2S gas is injected at the moment (i), the resistance steeply drops to 2.6MΩ. Then, the resistance rapidly recovers to the original value after the circumstance is replaced by fresh air again at the moment (ii). From the moment (iii) on, pure nitrogen gas is introduced to gradually replace the oxygen contained air. During the period of time the resistance drastically decreases for longer than 1h and the finally stable value is below 60kΩ. Very high concentration H_2S (1000ppm) is injected at the moment (iv) and the space is fulfilled again by pure nitrogen at the moment (v). During the period of H_2S introduced in N_2, the resistance undergoes a very slight decrease. After natural air begins to replace the N_2 environment from the moment (vi), more than 4h are needed to allow the resistance recovered to the initial value, *i.e.* before the moment (i). With each state associated to the corresponding XRD pattern in Figure 6, interesting phenomena are found. When the ZnO NW array is in the oxygen contained air, the new material of ZnS appears after the moment (i) but disappears after (ii). The appearance of ZnS is confirmed by the new peak of XRD at the 2θ angle of 29°. When the ZnO is in pure N_2 environment however, ZnS begins to appear at the moment (iv), and keeps remained at the moment (v) until the moment (vi) when natural air is introduced again. In other words, the generation of ZnS not only relies on the input of H_2S gas, but also is subject to the inexistence of oxygen. After removal of H_2S, the ZnS will remain in N_2 environment but disappear in O_2 contained air atmosphere.

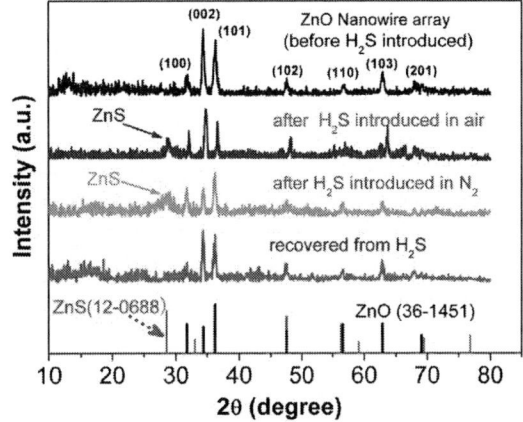

Figure 5: The resistance changing curve of ZnO in pure air, H_2S of 5ppm at air atmosphere, pure N_2 and H_2S of 1000ppm at N_2 atmosphere, respectively.

Figure 6: XRD patterns of ZnO nanowire arrays.

978-1-4799-7956-1/15 $31.00 © 2015 IEEE

All the experiments and characterization above clearly depict the brand new H_2S sensing effect as follows. Produced at the surface of ZnO NWs by sulfuration reaction with the targeted H_2S, the metastable ZnS intermediate will resist against the electron transfer from ZnO to the surface adsorbed O_2 molecules. The thinned electron depletion layer causes steep decrease of ZnO resistance, thereby forming the rapid and significant sensing response to H_2S. When H_2S is moved off, however, the ZnS intermediate can be instantaneously desulfurized back to ZnO by ambient O_2 molecules and, thus, SO_2 is generated (proved by GC-MS analysis). The new H_2S sensing effect of sulfuration-desulfuration reversible reaction would not have been found easily without the above-described complex experiment and the analysis. It is worthy notifying that, when H_2S is exposed to the ZnO NWs in N_2 atmosphere [*i.e.*, between the moments (iv) and (v) in Figure 5], the slightly decreased resistance is possibly caused by the resistance shunting effect of the surface generated ZnS to ZnO.

Since the reaction is reversible and the sulfuration and desulfuration processes occur simultaneously, we cannot observe the stable intermediate product of ZnS in air environment. However, we have validated that the sulfuration reaction indeed exists. The sulfuration reaction will do cause influence to the capability of oxygen extracting electron from ZnO. Viewed from another point, the semiconductor ZnO nanowire plays a role that is analogous to catalyst. Herein ZnO is ever involved in the H_2S sensing process but finally quit out.

H_2S is really an exception among various kinds of reductive gases, since other kinds of reductive gases have no luck to earn profit from this two-step reaction sensing-mechanism. Still abiding by the conventional sensing mechanism, the ZnO NW sensor has experimentally exhibited much slower and weaker response to other reductive gases like methanol, ethanol, acetone, hexane, benzene and toluene, which can be seen from Figure 3b.

The reported experiment gave an evidence for nanosize-effect induced ZnO sulfuration at lower temperature than usual. Along with size shrinkage of a nano-material, its surface becomes more and more active. Herein we still attribute the new H_2S sensing mechanism to surface size-effect. To verify this point, we use commercially available ZnO powder to test H_2S gas in air. Made of the coated ZnO micro-particles (particle size ranged in 0.15~0.8μm), the chemiresistive sensor exhibits no obvious sensing signal at 150°C. In our ZnO-NW senor, the diameter of the NWs is as small as 50~60nm. It may be just the small nano-dimension that enables the ZnO-NWs more reactive capability than normal micro-scale ZnO powder.

So far, the brand new H_2S sensing mechanism of the ZnO NWs sensor is raised, as is schematically sketched in Figure 7. The nano-diameter ZnO NW-array features high reactive capability at 150°C to induce the reaction of ZnO sulfurized by H_2S (during sensing process) and desulfuration reaction by ambient oxygen (during recovery). The sulfuration generated ZnS leads to quick and drastic decrease of the resistance, as the generated surface ZnS intermediate act as shield layer to effectively depress the capture of free electron from ZnO to the adsorbed oxygen species. When H_2S is removed, the temporarily generated ZnS is desulfurized back to ZnO, and thereby the resistance returns rapidly.

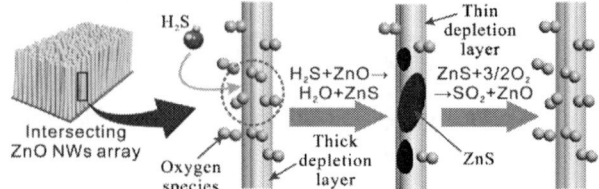

Figure 7: Sulfuration-desulfuration two-step reaction mechanism for nanosize ZnO.

CONCLUSION

We reported a new sensing effect on H_2S gas, which is based on a sulfuration-desulfutation two-step reaction. At the relatively low temperature of 150°C, experimental sensing results show unusually fine H_2S resolution of tens of *ppb*, as well as, ultra-high and repeatable sensitivity to H_2S. By conducting a series of specially designed experiments, characterizations and analysis, the greater than ever sensing performance is attributed to a new sensing effect for semiconductor ZnO chemiresistive sensor, which is nanosize-effect induced sulfuration-desulfuration two-step reaction.

ACKNOWLEDGEMENTS

This research is supported by NSF of China (91323304, 91023046, 61161120322, 61401446) and the Chinese 973 Project (2011CB309503).

REFERENCES

[1] M. E. Franke, T. J. Koplin and U. Simon, "Metal and Metal Oxide Nanoparticles in Chemiresistors: Does the Nanoscale Matter?", *Small*, vol. 2, pp. 36-50, 2006.

[2] Y. Zhang, J. Xu, Q. Xiang, H. Li, Q. Pan and P. Xu, "Brush-Like Hierarchical ZnO Nanostructures: Synthesis, Photoluminescence and Gas Sensor Properties", *J. Phys. Chem. C*, vol. 113, pp. 3430-3435, 2009.

[3] S. Choi, B. Jang, S. Lee, B. Min, A. Rothschild, I. Kim, "Selective Detection of Acetone and Hydrogen Sulfide for the Diagnosis of Diabetes and Halitosis Using SnO_2 Nanofibers Functionalized with Reduced Graphene Oxide Nanosheets", *ACS Appl. Mater. Interfaces*, vol. 6, pp. 2588−2597, 2014.

CONTACT

*X.X. Li, tel: +86-21-62131794; xxli@mail.sim.ac.cn
*D. Zheng, tel: +86-21-69877214; zhengdan@sit.edu.cn

IONIC-GEL-COATED FABRIC AS FLEXIBLE HUMIDITY SENSOR

Yusuke Takei, Kiyoshi Matsumoto, and Isao Shimoyama
The University of Tokyo, Tokyo, Japan

ABSTRACT

We fabricated flexible humidity sensor which responds 10 times faster than commercial CMOS humidity sensor. Our sensor is based on Ionic-Gel-coated Fabric (IG-Fabric). We use $EMIMBF_4$ as ionic liquid and mixed with PVDF to fabricate ionic gel. The $EMIMBF_4$ has a characteristic that it absorbs H_2O and changes its impedance. IG-Fabric has wide surface area and high gas permeability so that gases can be easily absorbed and detached. This sensor has many applications such as flu-mask-type human breath sensor, wearable humidity sensor.

INTRODUCTION

Humidity is an important indicator to know human health and living environment. For example, respiratory organs are desirable to keep in appropriate humidity (40 to 70 %RH (relative humidity)), because virus may increase in low humidity environment. Many types of humidity sensors are developed, such as capacitive- / thermal conductive- / resistive-humidity sensor [1][2]. When we think of monitoring human breath humidity or skin humidity (sweat) without disturbing their movement, wearable humidity sensors are strongly required.

Recently ionic liquid was focused in gas sensor field, because it absorbs many kinds of gases including H_2O vapor [3][4][5]. So we decided to fabricate ionic liquid based humidity sensor. In addition, we applied gelation technic to ionic liquid and use them as a coating material. To fabricate wearable humidity sensor, we coated fabric with ionic gel and use them as sensing element (Figure 1)

FABRICATION OF IG-FABRIC

We use non-woven fabric "Bemcot TR-7F" (Asahi Kasei corp., Japan) as IG-Fabric basement. Fabrication steps of IG-Fabric are shown in Figure 2. First, we prepare the mixture of ionic liquid and Poly Vinylidene DiFluoride (PVDF) and Dimethylacetamide (DMAc) as ionic gel solution. Mixture rate of the liquids are Ionic Liquid ($EMIMBF_4$) : PVDF : DMAc = 1 : 1 : 20, in volume. Then we dip the fabric into the ionic gel solution for 5 minutes in room temperature 25 °C. After picking up the fabric from the solution, we baked the fabric on the 80 °C hotplate for 30 minutes.

EXPERIMENT

To evaluate IG-Fabric humidity sensor, we conducted three experiments as listed below.

(a) IG-Fabric impedance change rate to humidity
(b) IG-Fabric impedance change rate to temperature
(c) IG-Fabric reaction speed

In this research, the impedance was measured at 1 kHz. Figure 3 (a) is showing the relation between relative humidity and IG-fabric impedance change at 30 °C. There is linearity in humidity and impedance change. Figure 3 (b) is showing the relation between temperature and IG-fabric

impedance change at 40 %RH. When we think of the application for our IG-fabric humidity sensor, operating environment is around human body temperature (30~40 °C), and humidity (30~80 %RH). Focus on Figure 3 (b), the impedance change of 30 °C to 40 °C is 0.01, and on the contrary, in Figure 3 (a) the impedance change of 40 %RH to 80 %RH is 0.1. These results indicate that impedance change caused by humidity is 10 times larger than that of temperature in our application (measuring human body temperature, breath, sweat). So we think impedance change caused by temperature can be compensated by the measured data. Figure 4 is showing raw data when we sprayed the N_2 gas (0 %RH) to the IG-fabric held tightly in the air (40 %RH). From the measurement data, the response speed τ_{63} estimated as around 0.7 seconds. This response speed is 10 times faster than commercial CMOS humidity sensor [6]. This is because IG-fabric has a large surface area and high gas permeability, compare to silicon wafers. We also evaluated the flexibility of our humidity

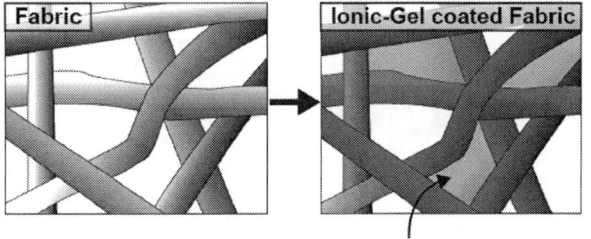

Figure 1: Concept sketch of proposed Ionic-Gel coasted fabric. Surface of the fibers are coated with ionic gel and gaps between the fibers are also filled in with ionic gel.

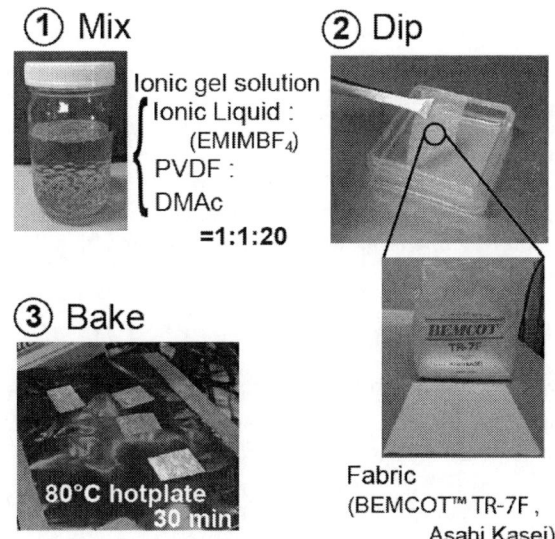

Figure 2: IG-fabric fabrication. Ionic gel solution is consisting of $EMIMBF_4$, PVDF, and DMAc. We use non-woven fabric as sensor basement.

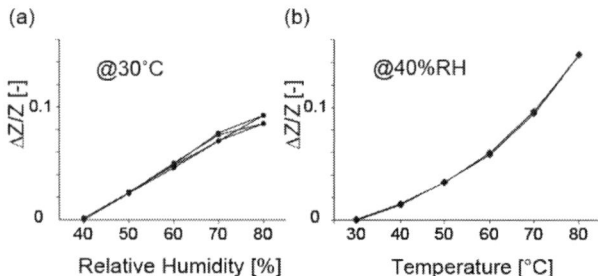

Figure 3: (a) IG-Fabric impedance change to relative humidity. (b) IG-Fabric impedance change to temperature.

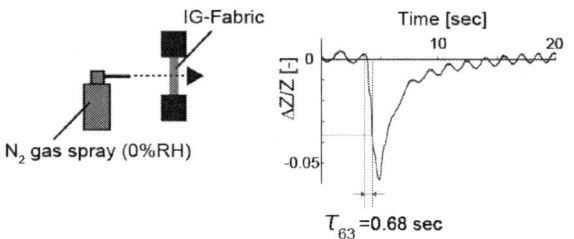

Figure 4: IG-Fabric response speed. τ_{63} is around 0.7 seconds.

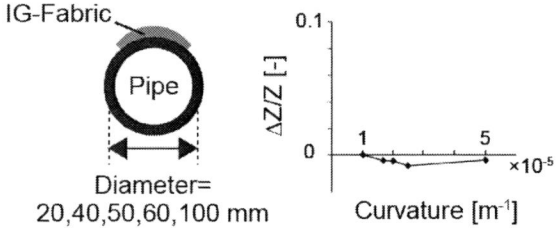

Figure 5: Flexibility of IG-Fabric. Impedance is stable against bending.

sensor as shown in Figure 5. When we set the IG-fabric on the column which diameter is 20, 40, 50, 60, 100 mm, impedance change is almost same. This result shows high flexibility of our sensor. As a demonstration, we fabricate flu-mask-type IG-fabric humidity sensor (Figure 6). Figure 7 is the relative humidity change measurement of human breath. Red line is showing the expiration and blue line is showing inhalation of the examinee.

CONCLUSION

In conclusion, flexible humidity sensor based on Ionic-gel-coated fabric was fabricated. The IG-Fabric impedance showed linearity between 40%RH to 80%RH. The impedance change rate is around 2.5×10^{-3} [/%RH]. The proposed IG-Fabric humidity sensor will be utilized to the flu-mask-type human breath sensor, wearable humidity sensor.

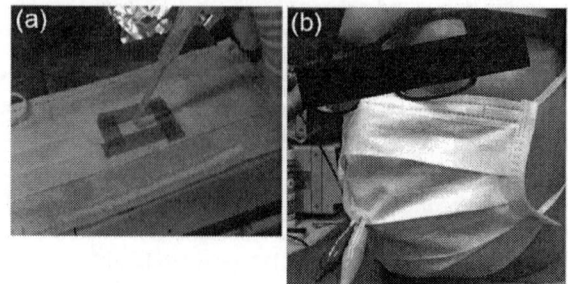

Figure 6: Demonstration of IG-Fabric humidity sensor. We fabricated flu-mask type humidity sensor (a), (b).

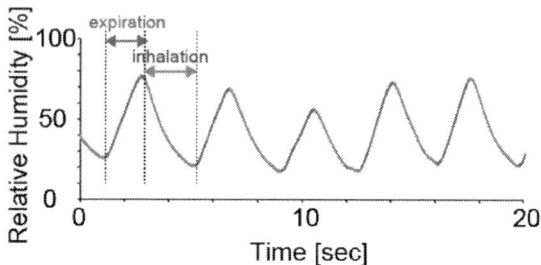

Figure 7: Detection of human breath (inhalation and expiration).

ACKNOWLEDGEMENTS

A part of this work was supported by New Energy and Industrial Technology Development Organization (NEDO).

REFERENCES

[1] H. Farahani, R. Wagiran, and M. N. Hamidon, "Humidity Sensors Principle, Mechanism, and Fabrication Technologies: A Comprehensive Review," *Sensors*, vol. 14, pp. 7881-7939, 2014.

[2] Z. M. Rittersma, "Recent achievements in miniaturized humidity sensors — a review of transduction techniques," *Sensors and Actuators A*, vol. 96, pp. 196-210, 2002.

[3] D. S. Silvester, "Recent advances in the use of ionic liquids for electrochemical sensing," *Analyst*, vol. 136, pp. 4871-4882, 2011.

[4] T. Welton, "Room-temperature ionic liquids. Solvents for synthesis and catalysis," *Chemical Reviews*, vol. 99, pp. 2071-2084, 1999.

[5] A. Finotello, J. E. Bara, D. Camper, R. D. Noble, "Room-temperature ionic liquids: temperature dependence of gas solubility selectivity," *Industrial and Engineering Chemistry Research*, vol. 47, pp. 3453-3459, 2007.

[6] Sensirion SHT series, http://www.sensirion.com

CONTACT

*Yusuke Takei, tel: +81-3-5841-0461; takei@leopard.t.u-tokyo.ac.jp

DOG-BONE RESONATOR WITH HIGH-Q IN LIQUID FOR LOW-COST QUICK 'TEST-PAPER' DETECTION OF ANALYTE DROPLET

Feng. Yu[1,2], Pengcheng. Xu[1], Jiachou. Wang[1], and Xinxin. Li[1,2]

[1] Key Lab of Transducer Technology, Shanghai Institute of Microsystem and Information
Technology, Chinese Academy of Sciences, Shanghai 200050, China
[2] University of Chinese Academy of Sciences, Beijing100049, China

ABSTRACT

Imitating on-site liquid-droplet analysis/assay with test-papers, a novel dog-bone resonator of tri-beam structure is proposed and developed for low-cost quick detection of trace amount biochemical liquid sample. With the specifically designed central beam, signal feed-through effect is effectively depressed by independent piezoresistive readout. Thereby, the new tri-beam resonator exhibits high-Q (=256) resonance factor of length-extension mode in liquid-droplet. Besides, non-SOI wafer and low-cost single-wafer single-side fabrication are developed to replace the previously used expensive SOI process for the cheap 'test-paper' disposable applications. With the low-cost fabricated MEMS 'test-paper', liquid-droplet detection function is experimentally demonstrated by on-site reorganization of $500ppb$ Hg^{2+} ion in water solution.

INTRODUCTION

Resonant sensors mainly consist of a specific sensing layer and a resonant transduction element to convert the adsorbed mass into frequency shift signal [1]. Most traditional resonant sensors, like cantilevers, are operated in flexural bending or other out-of-plane resonance modes. Such out-of-plane resonance mode generally suffers strong damping force in liquid media and high loss of vibration energy. The strong liquid drag force induced technical shortcomings and hindered in-liquid application of the resonant micro-sensors. In-plane resonance mode can obviously decrease the liquid drag force by shearing with the surrounding media rather than compressing the media molecules, which is a promising candidate for on-the-spot bio/chemical monitoring and recognition [2].

Operated in length-extensional bulk resonance-mode, the recently developed 'dog-bone' resonating structure has been mainly proposed as telecommunication devices for timing and frequency reference due to its high quality-factor (Q-factor) in air [3]. Individual air-borne particles are also detected by utilizing this dog-bone resonator [4]. With sensing materials coated, trace-level chemical gas sensors have been developed for detection in air environments [5].

As use the same beams both for resonance-excitation and frequency-readout, this kind one-port detection features ultra-high feedthrough [6]. The resonance peak still can be detected for the high Q-factor in air, but signal feed-through is too strong to submerge the resonance-peak with sharp decrease of Q-factor in liquid.On the other hand, not only the developed 'dog-bone' resonators but also traditional cantilever-resonators are used by expensive SOI-wafer-based fabrication, thereby hindering low-cost MEMS applications.

In this paper, a new tri-beam dog-bone resonating structure is designed and proposed to depress the feed-through, therefore, to enhance high-Q resonating in liquid environment As the concept schematically shown in Fig.1, the new low-cost fabricated tri-beam resonator is utilized for bio/chemical detection in liquid-droplets, just like pH-test-paper.

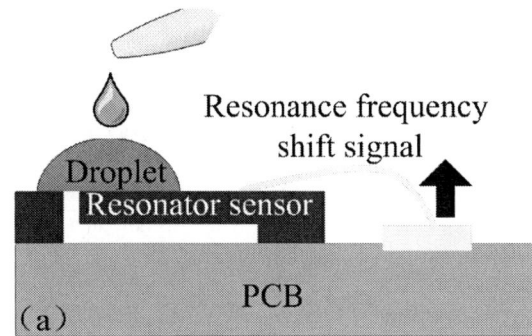

Figure 1: Concept of the resonant micro-sensor as 'test paper''.

TRI-BEAM RESONATOR DESIGN

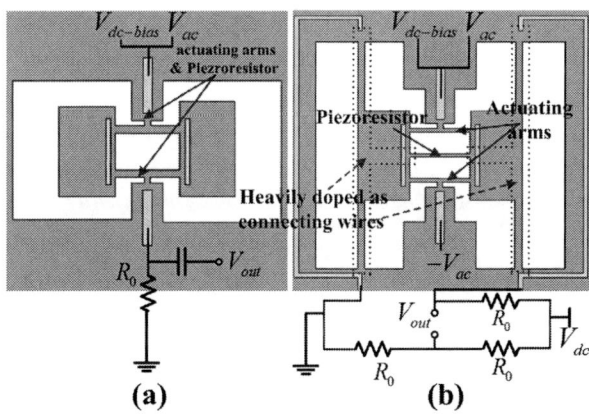

Figure 2: (a) Top-view of traditional dual-beam dog-bone resonator. (b) Novel tri-beam resonator with independent thermal-electric actuation and piezoresistive readout for feed-through depression.

Operated in length-extensional resonance mode, the developed conventional dual-beam dog-bone resonator with one-port signal detection scheme is schematically shown in Fig. 2(a). Supplied with a DC-biased AC current into the actuating beams, the two dog-bone arms are electro-thermally excited into length-extensional mode. The generated alternating tensile and compressive stress of the arms results in synchronous change in the electrical resistance by piezoresistive effect. For the DC component of the actuating current that flows through the arms, the piezoresistive-effect induced alternating change of the resistance will produce a feed-through voltage noise in the output signal.

Dislike the conventional resonator where the two beams are used both for thermo-electric actuation and piezoresistive signal-pickup, the trim-beam resonator employs an additional central piezoresistive beam for independent Wheatstone-bridge signal readout (see Figs. 2b). The four slim beams linked to the sensing-plates are designed for signal lines.

The two beams of conventional dog-bone resonator play as both the actuator and the piezoresistive detector, and this kind one-port detection circuit is plotted in Figs. 3a. Apparently, the actuating signal is mixed into the output signal path, forming a high signal feed-through. Resonance peak will be weaken by the severe feed-through significantly, and needs to use post Matlab code to calculate Q-factor [7]. Thanks to the ultra high Q-factor resonance mode, the resonance characteristics of dual-beam resonator still can be directly detected by using a network analyzer. Unfortunately, the Q-factor will drastically reduce when the operating environment is changed to liquid. Therefore, the signal feed-through induced noise will submerge the resonating peak signal. This is the reason why this conventional dual-beam dog-bone resonator failed to resonant in liquid environment.

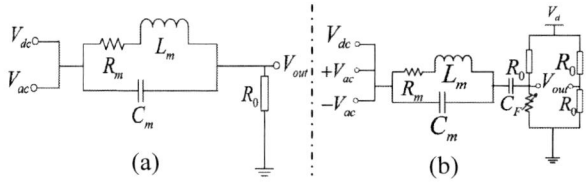

(a)   (b)

Figure 3: (a) One-port detection equivalent circuit of the traditional dual-beam resonator. (b) Wheatstone-bridge detection equivalent circuit of the new triple-beam resonator, which effectively depresses the feed-through effect.

Different with the one-port detection, the novel tri-beam resonator can eliminate (or significantly depress) the feedthrough by its Wheatstone-bridge detection circuit, as is shown in Figs. 3b. Apparently, the out-put signal is separated from the actuating signal by substrate capacitance of C_F. In addition, the designed positive/negative symmetric layout of the AC actuating currents in the new device helps to further eliminate the feed-through by cancelling out the influence from substrate capacitive coupling.

LOW-COST FABRICATION

IC-foundry compatible, high-yield and low-cost MIS (micro-openings inter-etch & sealing) micro-machining technology is recently developed and used to process pressure sensors in a single-side polished (111)-silicon wafer [8], [9]. In this paper, an up-graded MIS process

scheme is developed for the design of micro-resonator. In this case, resonator sensor can be used in a disposable way. The detailed fabrication steps are shown in Fig. 4 and described as follows:

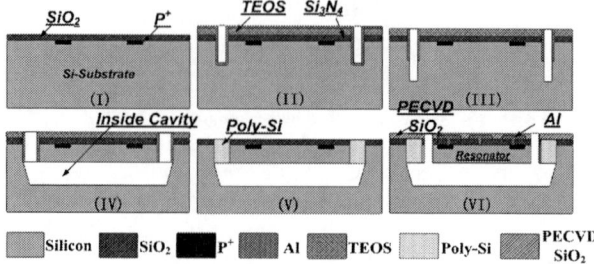

Figure 4: Non-SOI low-cost fabrication processes for design of resonator

(I) After thermal oxidization, the first photolithography is performed from the wafer front-side to define the piezoresistors. Then buffered wet HF is used to etch the SiO$_2$ on top of the piezoresistors location, with the photoresist as an etching mask. The piezoresistors are formed by boron ion-implantation and the following drive-in process.

(II) After low-pressure chemical vapor deposition (LPCVD) for growing TEOS (tetraethoxysilane) layer, the second photolithography is employed to pattern the cavity releasing micro-hole array. The deep reactive ion etching (deep-RIE) etched micro-holes, and the depth defines the thickness of the resonator structure. Then TEOS film is again deposited with LPCVD to protect the deep-trench sidewall surface of the micro-holes from the following anisotropic wet etch.

(III) RIE is used to selectively etch off the SiO$_2$ at the trench bottom to expose bare silicon there, while the passivation layer at the vertical sidewalls is still retained. Then, silicon deep-RIE is processed again to deepen the holes, which will be laterally under-etched by anisotropic wet etch.

(IV) The wafer is etched in TMAH to complete the inter-hole cavity-release by lateral under-etch. Without protection of the passivation layer, the bottom segment of the micro-hole trench-sidewalls will be under-etched along lateral direction, while the top segment of the sidewalls remains not etched due to the protection of the passivation layer.

(V) LPCVD poly-silicon is deposited to refill the micro-holes. The excessive poly-silicon at the front surface of the wafer is stripped by using RIE.

(VI) Al thin-film is sputtered, patterned, wet etched and sintered to form the interconnection wires. A high-quality plasma enhanced chemical vapor deposition (PECVD) silicon-dioxide thin-film is deposited on top of the structure and patterned by RIE for passivation. Finally, silicon deep-RIE is employed to etch the silicon diaphragm to form the resonator structure.

The finally fabricated tri-beam dog-bone resonator is shown in Fig. 5. The non-SOI chip size is as small as

0.5mm×0.5mm with the use of this new single-side fabrication. Therefore, the volume-fabrication cost in IC-foundry can be as low as 1 US-cent/sensor. Obviously decreased the volume cost compared with the traditional SOI-based fabrication.

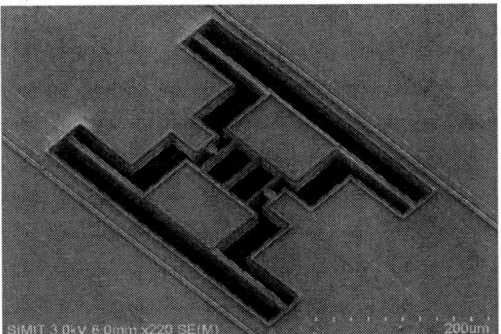

Figure 5: SEM image of the new tri-beam resonator.

RESONANCE CHARACTERISTICS

After the sensor frame is coated with insulating silica glue to protect the metal signal lines, the devices are dripped with deionized water droplets for the resonance characterization detection in liquid. The testing results of S21-parameters are shown in Fig. 6. As the reason mentioned before, the dual-beam device exhibits no obvious resonance-peak for its high feedthrough. In contrast, the new tri-beam dog-bone resonator still exhibits 1dB peak-signal and high-Q (=256). Thus, based this new designed tri-beam structure, high performance resonant sensor in liquid environment for on-the-site bio/chemical detection can be achieved.

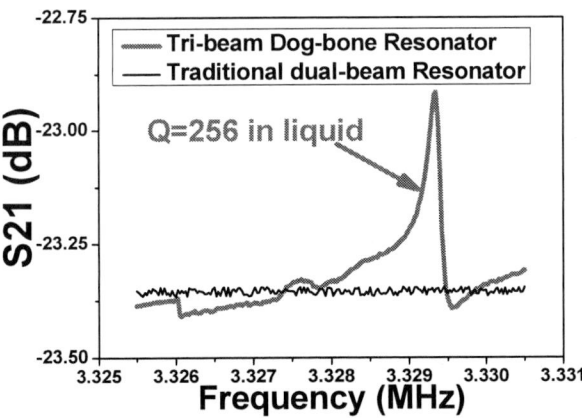

Figure 6: Measured S21 parameters of the two types of resonators submerged in liquid droplets. The Q-factor of the tri-beam resonator is as high as 256 in liquid, while the resonant peak of the traditional dual-beam resonator cannot be observed due to the high signal feed-through induced noise.

WATER-DROPLET DETECTION

With the non-SOI single-sided fabrication technique, the low-cost MEMS-sensors can be used for 'test-paper' like detection in disposable way. Similar to the disposable pH-value test paper, the low-cost and volume fabricated sensor can be used to on-the-site by dropping a liquid droplet onto it.

Detection of trace-level Hg^{2+} ions is meaningful for monitoring heavy metal ion pollution to water resource. Herein, mesoporous-silica is utilized as sensing materials, since the nano-material features as high specific surface area as $1500m^2/g$ [10]. Thiol (−SH) functionalized mesoporous silica that has been proved for specific adsorption of mercury ions [11]. After modifying mesoporous with −SH sensing-group to specifically capture Hg^{2+} ions, this functionalized mesoporous is loaded onto the mass-region by miro-manipulator. The fabricated resonator sensor is shown in Fig.7.

Figure 7: SEM image of the tri-beam resonator, loaded with the mesoporous-silica sensing material at the sensing plates.

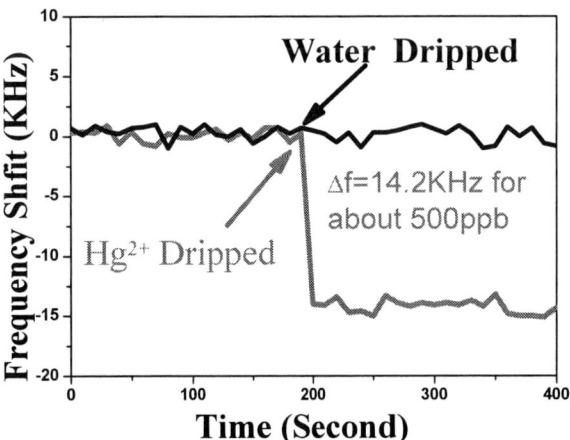

Figure 8: Real-time liquid-droplet detection data for 500ppb $Hg2+$ ion. A drop of water is firstly dripped on the device. After the frequency stabilized, a drop of 1000ppb aqueous $Hg(NO3)_2$ (with identical drop volume) is dripped on the top to form 500ppb aqueous Hg^{2+} ion. 14.2KHz frequency decrease is measured within 15sec.

A drop of deionized water is firstly dripped on the resonant chip for frequency stabilization in solution. Then, a drop of aqueous $Hg(NO3)_2$ with 1000ppb concentration is dripped at the same area. The equivalent concentration of Hg^{2+} ion is about 500ppb as the drop volume is kept constant. Fig.8 shows the recorded frequency-shift of -14.2kHz within 15sec. A control experiment is also performed and compared, where the second drop is deionized water instead of the Hg^{2+} solution. As the experimental results clearly show, it is just the specifically adsorbed mass of Hg^{2+} ions that cause the quick frequency decrease. Compared to the noise determined base line of the recorded signal, a noise-limited resolution of finer than 500 *ppb* can be estimated.

CONCLUSION

A novel tri-beam 'dog-bone' micro-resonator scheme with Wheatstone-bridge detection is proposed and developed in this paper. Compared with the conventional dual-beam resonator, the new tri-beam structure depressed signal feed-through and improved resonance performance (including Q=256) in solution for bio/chemical analyte detection. Low-cost non-SOI single wafer single-side process is developed to replace the previously used expensive SOI fabrication for realizing the cheap 'test-paper' disposable applications. Droplet detection function is experimentally demonstrated by on-the-site reorganization of 500*ppb* Hg^{2+} ion in water solution.

ACKNOWLEDGEMENTS

The research is supported by Chinese 973 Program (2011CB309503), NSF of China (91023046, 61161120322, 51205388, and 61021064).

REFERENCES

[1] A. Boisen, S. Dohn, S. S. Keller, S. Schmid, and M. Tenje, "Cantilever-like micromechanical sensors," Rep. Prog. Phys. vol. 74, 036101, 2011.

[2] Y. H. Tao, X. X. Li, T. G. Xu, H. T. Yu, P. C. Xu , and B. Xiong, "Resonant cantilever sensors operated in a high-Q in-plane mode for real-time bio/chemical detection in liquids," Sensors Actuators B vol. 157, no. 2, pp. 606-614, 2011.

[3] J. T. M. van Beek, P. G. Steeneken, B. Giesbers, "A 10MHz piezoresistive MEMS resonator with high Q," in proc.IEEE Int. Freq. Control symp. Miami, Jun. 2006, pp. 475-480.

[4] A. Hajjam, J. C. Wilson, and S. Pourkamali, "Individual air-borne particle mass measurement using high-frequency micromechanical resonators," J. Sensor, vol. 11, no.11, pp, 2883-2890, 2011.

[5] F. Yu, H. T. Yu, P. C. Xu, T. G. Xu, C. Yang, and X. X. Li, "A high-Q bulk-resonator equipped with ultra-sensitive mesoporous nanomaterial for detection of trace-level chemicals," in Proc. Transducers'13, Barcelona, Jun. 2013, pp. 842-845.

[6] Y. Xu, and J. E. Y. Lee, "Single-device and on-chip feedthrough cancellation for hybrid MEMS resonators," IEEE Trans. Ind. Electron. Vol. 59, no. 12, pp. 4930-4937, 2012.

[7] J. E. Y, Lee, and Y. Xu, "Direct inference of parameters for piezoresistive micromechanical resonators embedded in feedthrough," Sens. Actuators, A vol. 186, no. 19, pp. 257-263, 2012.

[8] J. C. Wang and X. X. Li, "Single-Side Fabricated Pressure Sensors for IC-Foundry-Compatible, High-Yield, and Low-Cost Volume Production," IEEE Elect. Dev. Lett. vol. 32, no. 7, pp. 979-981, 2011.

[9] J. C. Wang, X. Y. Xia, and X. X. Li, "Monolithic Integration of Pressure Plus Acceleration Composite TPMS Sensors With a Single-Sided Micromachining Technology," J.Microelectromech.Syst. vol. 21, no. 2. pp. 284-293, 2012.

[10] P. C. Xu, H. T. Yu, and X. X. Li, "Functionalized mesoporous silica for microgravimetric sensing of trace chemical vapors," Anal. Chem. vol. 83, no. 9, pp. 3448-3454, 2011.

[11] C. Thompson, J. Hu, S. N. Kaganove, S. E. Keinath, D. L. Keeley, and P. R. Dvornic, "Hydrogen-bond acidic hyperbranched polymers for surface acoustic wave (SAW) sensors," Chem. Mater. vol. 16, no. 25, pp. 5357-5364, 2004.

CONTACT

*Xinxin. Li, tel: +86-21-62131794;
Email: xxli@mail.sim.ac.cn

A DYNAMICALLY MODE-MATCHED PIEZOELECTRICALLY TRANSDUCED HIGH-FREQUENCY FLEXURAL DISK GYROSCOPE

Mojtaba Hodjat-Shamami, Arashk Norouzpour-Shirazi, Roozbeh Tabrizian, and Farrokh Ayazi
Georgia Institute of Technology, Atlanta, GA, USA

ABSTRACT

This paper presents, for the first time, the design, implementation, and characterization of a dynamically mode-matched high-frequency piezoelectrically transduced silicon disk gyroscope utilizing a unique pair of degenerate orthogonal in-plane flexural gyroscopic modes. To eliminate the need for submicron capacitive gaps and large DC polarization voltages that are needed for electrostatic tuning, a linear bidirectional frequency tuning scheme, compatible with all-piezoelectric transduction, is introduced to achieve dynamic mode-matching via active electromechanical feedback. A fabricated 4.34 MHz AlN-on-Silicon 1-mm-diameter solid disk gyroscope supported by a network of peripheral beams was frequency-tuned by 500 ppm (~2.2 kHz) and achieved a high measured sensitivity of 410 pA/°/s with an effective in-air quality factor of ~4000, corresponding to a large open-loop operation bandwidth of ~550 Hz under mode-matched conditions.

INTRODUCTION

In recent years, MEMS gyroscopes have been adopted rapidly in a variety of consumer applications due to significant reduction in their size, cost and power consumption. Conventional vibratory rate gyroscopes use a pair of low-frequency rigid-body resonance modes in a microstructure for rotation rate detection, by actuating the primary resonance mode of the device and detecting the rate-proportional Coriolis displacement signal along the secondary resonance mode [1,2]. Although these vibratory gyroscopes provide the degree of functionality required by some consumer applications, they fail to offer the performance level demanded by many high-end applications, such as short-range inertial navigation, while maintaining a micro-scale physical size.

By taking advantage of the stiff bulk resonance modes of the device structure, high-frequency resonant bulk acoustic wave (BAW) gyroscopes [3,4] have overcome many limitations of low-frequency gyroscopes, such as vibration sensitivity, susceptibility to mechanical shock, and inadequate bandwidth and dynamic range under mode-matched conditions. However, the mechanical rate sensitivity of gyroscopes decreases at higher resonance frequencies due to the smaller vibration amplitude and the distribution of mass and stiffness of bulk resonance modes over the volume of the device.

High-sensitivity capacitive BAW gyroscopes utilize submicron air gaps and large DC voltages to provide efficient transduction at high frequencies. They may also require vacuum encapsulation to avoid squeeze-film damping, which in turn necessitates special design considerations for co-integration of capacitive gyroscopes with static accelerometers, where low-pressure requirements for gyroscope packaging conflicts with the desired over-damped performance of the accelerometer

needed for fast settling time and small overshoot [5].

The quest for implementation of BAW gyroscopes that can provide efficient in-air transduction to minimize packaging complexity of multi-degree-of-freedom sensors, without the need for narrow gaps and large DC polarization voltages to further reduce fabrication cost and high voltage requirements of the sensor, has led to the implementation of piezoelectrically transduced high-frequency resonant gyroscopes [6]. Inherent linearity and high efficiency of the piezoelectric transduction combined with superior power handling of thick single-crystal silicon acoustic platform facilitate actuation of the piezoe-on-silicon gyroscopes with adequate vibration amplitudes, paving the way towards significant enhancement of rotation rate sensitivity and total signal-to-noise ratio.

Although the piezoelectric thin film provides effective transduction and thus large drive amplitude, an efficient frequency tuning mechanism is needed to enable mode matching of all-piezoelectric high-frequency resonant gyroscope in the presence of process non-idealities. This work introduces a multi-port AlN-on-Si BAW disk gyroscope, utilizing a novel gyroscopic mode pair, with mode matching capability, enabled by a dynamic frequency tuning technique based on electrical feedback of the drive-mode displacement signal [7].

PIEZOELECTRIC GYROSCOPE DESIGN

High-frequency capacitive gyroscopes typically use the elliptical bulk resonance modes of a disk microstructure to enable Coriolis energy transfer between the two degenerate modes. However, the distortional stress-field pattern of such mode shapes prevents their efficient transduction by surface piezoelectric thin films. Identifying distinct resonance modes that not only allow for piezoelectric transduction but also demonstrate gyroscopic coupling is key to the implementation of high-frequency piezoelectric gyroscopes.

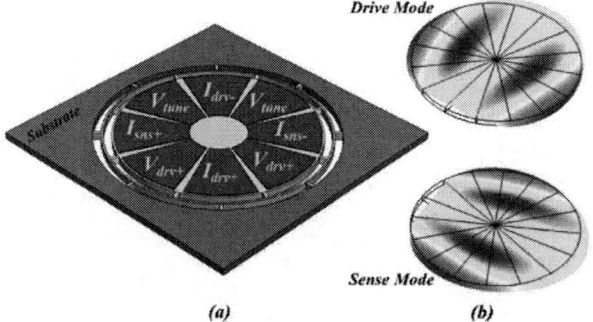

Figure 1: (a) Schematic view of the AlN-on-Si disk gyroscope anchored to the substrate using 16 circularly symmetric T-supports to ensure no structural mode split between the drive and sense modes (b) A novel in-plane flexural resonance mode pair of the disk structure provides energy transfer path for the Coriolis signal.

The degenerate in-plane flexural mode pair of the disk structure (Fig. 1), introduced here for gyroscopic application, exhibits both of these properties and is utilized in this work to realize a Coriolis BAW gyroscope. The degenerate in-plane flexural modes show 10x larger electromechanical coupling coefficient compared with conventional elliptical modes of the disk, and thus significant improvement in motional resistance, rate sensitivity, and thermomechanical noise performance.

The schematic view of the AlN-on-Si disk gyroscope, along with the orthogonal degenerate drive and sense flexural mode shapes are depicted in Fig. 1. The device is actuated, sensed, and tuned using 8 identical top electrodes and anchored to the substrate by a network of 16 circularly symmetric peripheral T-supports to ensure no structural mode split between the drive and sense modes and also facilitate integration of multiple isolated transduction ports required for device operation.

Fig. 2 shows the SEM images of the fabricated 1-mm-diameter piezoelectrically transduced disk gyroscope comprised of a 1.3-μm thin film of AlN sandwiched between Molybdenum (Mo) electrode layers and stacked upon a 35-μm-thick (100) plate of single-crystal silicon. The bottom Mo electrode is used as the common terminal and the electrically-isolated top Mo electrodes are used for actuation, sensing and tuning of the gyroscope. The device was fabricated using a 4-mask AlN-on-Si process similar to the one described in [8].

The gyroscopic coupling between the two degenerate in-plane flexural modes of a 1-mm-diameter disk structure having a resonance frequency of 4.34 MHz was verified by finite element analysis using COMSOL Multiphysics. A quality factor (Q) of 4000, obtained from experimental results of the fabricated gyroscope, operating in air, was assumed in the simulations. Special drive and sense electrode configuration and placement have been used to guarantee orthogonal transduction of the drive and sense modes. As a result, application of the drive excitation to the drive electrode pair does not cause any undesired electromechanical coupling to the sense mode, because of incompatibility with piezoelectric stress-field pattern. The simulated sense-mode zero-rate output current level is more than 60 dB lower than the drive mode current.

The device was actuated by applying a 1-V_p signal to induce a maximum drive-mode kinetic energy of 5.5 nJ. The Coriolis-induced output current is sensed differentially at the location of zero-stress drive signal to further improve modal decoupling. The simulated rotation rate sensitivity of the piezoelectric gyroscope was extracted to be 458 pA/°/s, as shown in Fig. 3, demonstrating a wide linear input range. The gyroscopic modal coupling factor was calculated to be 0.37 for the in-plane flexural modes of the disk structure. Although this is smaller than the gyroscopic coupling factor for the secondary elliptical mode pair (~0.6), the significant improvement in the electromechanical transduction of the flexural mode over the elliptical mode notably enhances the performance of the gyroscope.

The mechanical noise equivalent rotation rate (MNEΩ) of a resonant gyroscope can be expressed as a function of the kinetic energy of the drive mode

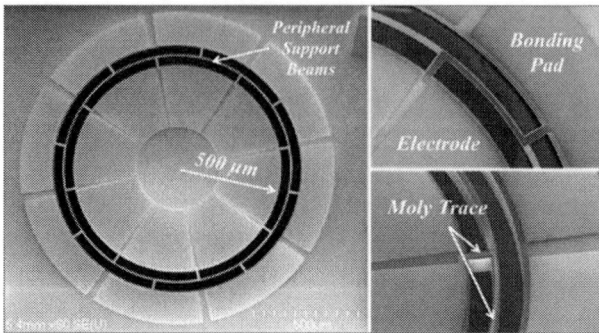

Figure 2: SEM view of the AlN-on-Si disk gyroscope fabricated using a simple 4-mask process described in [8] and the close-up capture of the electrodes and support structure.

$$MNE\Omega = \frac{1}{\lambda}\sqrt{\frac{k_B T \omega_o}{2 E_k Q_s}}\ (rad/s)/\sqrt{Hz}, \qquad (1)$$

where λ, ω_o, E_k, and Q_s are, gyroscopic coupling factor, resonance frequency, drive-mode kinetic energy and sense-mode quality factor, respectively. Although the current design has a Brownian noise of 27 °/hr/√Hz, the noise performance can be significantly improved through optimization of the device size, quality factor, and applied drive voltage.

DYNAMIC MODE MATCHING

Capacitive resonant gyroscopes, take advantage of the Q-amplification of the Coriolis signal by matching the drive and sense resonance frequencies through the electrostatic spring softening effect. However, generating the DC tuning voltages puts a large burden on the design of the interface circuitry and also accounts for a significant portion of the total power consumption. Moreover, implementation of submicron capacitive gaps, needed for adequate frequency tuning of BAW gyroscopes, adds to the overall complexity of the fabrication process. Passive tuning of lateral piezoelectric resonators utilizing the piezoelectric stiffening effect through variation of the termination load has been demonstrated [9]. However, the efficiency of this technique diminishes dramatically for resonators with a

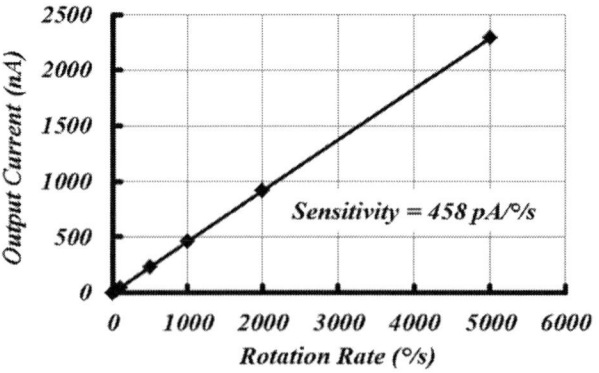

Figure 3: Simulated output current amplitude of the gyro due to applied z-axis rotation rate showing a coupling factor of 0.37 and sensitivity of 458 pA/°/s for the drive voltage of 1 V_p.

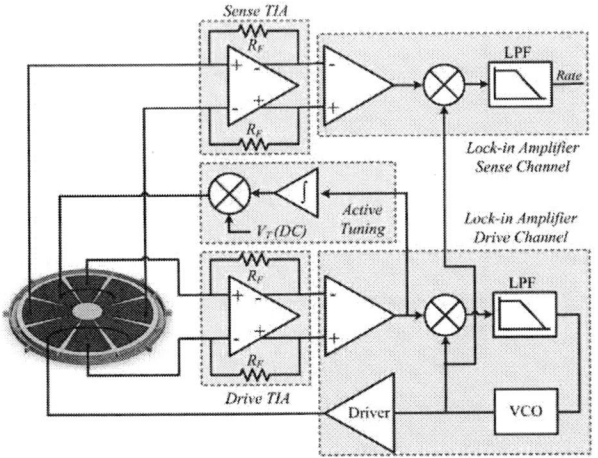

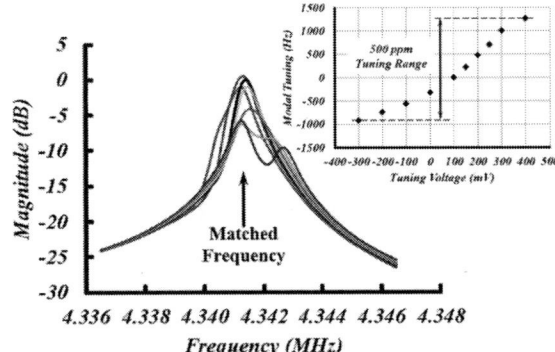

Figure 4: Architecture of the interface circuitry. The common-mode excitation of the drive mode prevents actuation of the sense mode and differential readout of the drive signal suppresses symmetric spurious modes. Modal tuning is achieved via active displacement feedback.

device layer that is much thicker than the piezoelectric active layer.

To circumvent all of the challenges mentioned above and to take full advantage of an all-piezoelectric implementation, a dynamic mode-matching technique has been developed that contrary to electrostatic tuning scheme can provide bidirectional linear tuning capability, thus further simplifying mode matching of the device. The dynamic mode-matching is accomplished via electromechanical feedback of the drive-mode displacement signal to the drive-tuning electrodes as described in [7]. The drive-mode displacement signal is replicated in the electrical domain, by analog integration of the velocity-proportional output current of the drive mode. The displacement signal is then scaled by a tuning voltage using an analog multiplier and is fed back to the common-mode tuning electrodes to dynamically modify the drive-mode effective stiffness, thereby provide linear, bidirectional tuning to the drive-mode resonance frequency, without affecting the sense-mode dynamics.

Considering an equivalent spring-mass-dashpot model for the drive mode resonator, the effect of the electromechanical active tuning can be explained by inclusion of the feedback force in the equation of motion,

$$M\frac{d^2x}{dt^2} + D\frac{dx}{dt} + Kx = F_{drive} + F_{tune}, \tag{2}$$

where M, D, and K are the equivalent mass, damping and stiffness parameters of the drive mode, x is the drive mode displacement, F_{drive} is the actuation force, and F_{tune} is the tuning force fed back to the drive mode. Since F_{tune} is generated by integration and scaling of the output current of the drive mode, which is proportional to the drive-mode velocity, this force is proportional to the drive-mode displacement, i.e.,

$$F_{tune} = V_T.R_F.\eta^2.x, \tag{3}$$

where V_T, R_F, and η are the DC scaling voltage, transimpedance gain, and transduction coefficient,

Figure 5: Bidirectional tuning response of the gyro. The effective quality factor increases as the drive peak approaches the sense resonance frequency. The inset shows a 500 ppm tuning range for the drive mode.

respectively. Therefore, the equation of motion can be rewritten to show the dynamic softening or stiffening of the equivalent spring constant, or in other words, bidirectional tuning of the drive mode, based on the polarity of the scaling parameter.

$$M\frac{d^2x}{dt^2} + D\frac{dx}{dt} + (K - V_TR_F\eta^2)x = F_{drive}. \tag{4}$$

The interface architecture for the gyroscope is shown in Fig. 4. Discrete differential transimpedance amplifiers (TIAs) are utilized for current pick-off from both the drive and sense modes. An HF2LI lock-in amplifier is used to process the output of the sense TIAs for detection of the angular rotation rate, and the output of the drive TIAs to implement an oscillator loop to actuate the drive mode at its resonance frequency. Common-mode application of the drive signal and differential readout of the drive- and sense-mode output currents, as labeled in Fig. 1, allow for independent actuation of the drive mode, without the excitation of the sense mode, as well as suppression of symmetric spurious modes. The actuation signal is used in a coherent AM demodulation architecture to extract rotation rate information from the Coriolis component of the sense-mode output.

EXPERIMENTAL RESULTS

The bidirectional tuning capability of the drive mode was successfully implemented on a PCB prototype using discrete components. The drive-mode velocity signal taken from the lock-in amplifier is integrated and scaled on the board and fed back to the device to tune the drive-mode resonance frequency by a small DC tuning voltage.

Fig. 5 shows the frequency response of the gyroscope from the drive input to the sense output for different applied tuning voltages, displaying a sweep of the drive peak across the sense-mode resonance frequency. The electromechanical feedback tuning technique demonstrates resonance frequency tuning range of 2200 Hz (~500 ppm), to compensate for the modal frequency split caused by process non-idealities and crystallographic misalignments. As can be seen in the frequency response in Fig. 5, the effective quality factor of the device increases as the drive peak approaches the sense

978-1-4799-7956-1/15 $31.00 © 2015 IEEE   791

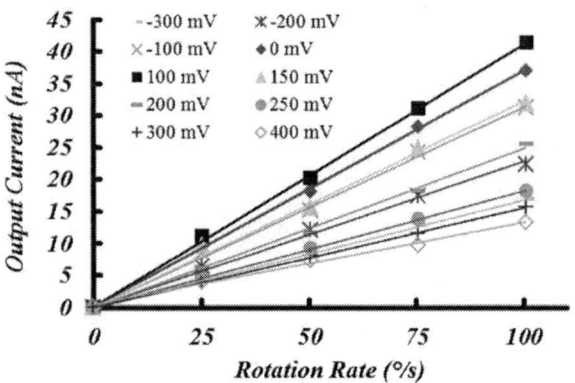

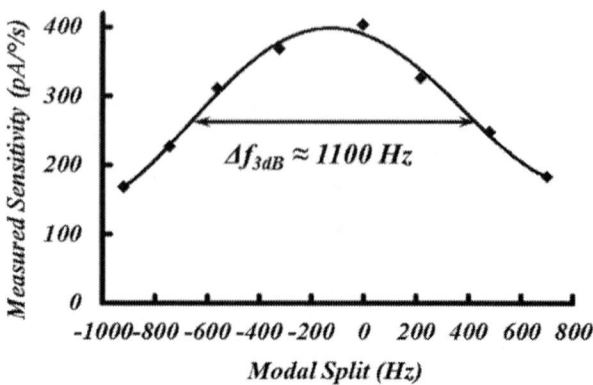

Figure 6: Measured rotation rate response of the gyro at different tuning voltages showing a maximum sensitivity of 410 pA/°/s under mode-matched conditions.

Figure 7: Measured sensitivity of the gyro as a function of modal split shows a 3-dB bandwidth of ~1100 Hz, corresponding to the quality factor of the device.

resonance frequency, improving the sensitivity of the gyroscope. The inset shows the dynamic tuning curve of the drive mode resonance frequency for tuning voltages of -300 mV to +400 mV.

Fig. 6 shows the rotation rate response of the gyroscope for different applied tuning voltages demonstrating a maximum measured sensitivity of 410 pA/°/s under mode-matched conditions, for an applied drive signal of 1 V_p, which shows very close agreement with simulation results (Fig. 3).

The rotation rate sensitivity of the gyroscope is plotted in Fig. 7, against the frequency split controlled by the applied tuning voltage, signifying a 3-dB bandwidth of ~1100 Hz which corresponds to the sense-mode quality factor, confirming that as the drive-mode frequency is tuned dynamically, the Coriolis-induced displacement and so the angular rate sensitivity, varies according to the frequency response of the sense mode.

CONCLUSIONS

This work introduced a dynamically mode-matched high-frequency piezoelectrically transduced silicon disk gyroscope utilizing a novel degenerate flexural gyroscopic mode pair to enable efficient transduction and tuning of the device, compatible with an all-piezoelectric platform. A fabricated 4.34 MHz AlN-on-Silicon 1-mm-diameter solid disk gyroscope demonstrated, for the first time, the mode matched operation of a thin-film piezoelectric gyroscope with frequency tuning capability of 500 ppm and achieved a high measured sensitivity of 410 pA/°/s. An effective quality factor of ~4000 was measured in air corresponding to a large open-loop bandwidth of ~550 Hz. The efficient thin-film piezoelectric transduction facilitates large rotation rate sensitivity of the gyroscope and paves the way for implementation of low-power multi-degree-of-freedom sensors at lower fabrication cost and smaller physical size.

ACKNOWLEDGEMENT

This work was supported in part by the DARPA Microsystems Technology Office, Single-Chip Timing and Inertial Measurement Unit (TIMU) program through SSC pacific contract #N66001-11-C-4176. The authors

would like to thank the OEM Group for AlN deposition and the staff at Georgia Tech Institute for Electronics and Nanotechnology for fabrication support.

REFERENCES

[1] J. Seeger et al., "Development of high-performance, high-volume consumer MEMS gyroscopes," in *Proc. Solid-State Sensors, Actuators, and Microsystems Workshop*, pp. 61-64, June 2010.

[2] L. Prandi et al., "A low-power 3-axis digital-output MEMS gyroscope with single drive and multiplexed angular rate readout," in *Tech. Digest Solid-State Circuits Conference (ISSCC)*, pp.104, 106, Feb. 2011

[3] H. Johari; F. Ayazi, "Capacitive Bulk Acoustic Wave Silicon Disk Gyroscopes," *Electron Devices Meeting (IEDM)*, pp.1, 4, Dec. 2006

[4] F. Ayazi, "Multi-DOF inertial MEMS: From gaming to dead reckoning," in *Tech. Digest Solid-State Sensors, Actuators and Microsystems Conference (TRANSDUCERS)*, pp.2805, 2808, June 2011

[5] Y. Jeong et al., "Wafer-level vacuum-packaged triaxial accelerometer with nano airgaps," in *Proc. Micro Electro Mechanical Systems (MEMS)*, pp.33, 36, Jan. 2013

[6] R. Tabrizian et al., "High-Frequency AlN-on-Silicon Resonant Square Gyroscopes," *J. Microelectromech. Systems*, vol.22, no.5, pp.1007, 1009, Oct. 2013

[7] A. Norouzpour-Shirazi et al., "Dynamic Tuning of MEMS Resonators via Electromechanical Feedback," *Ultrasonics, Ferroelectrics, and Frequency Control, IEEE Transactions on*, to be published.

[8] W. Pan et al., "Thin-film piezoelectric-on-substrate resonators with Q enhancement and TCF reduction," in *Proc. Micro Electro Mechanical Systems (MEMS)*, pp.727, 730, Jan. 2010

[9] M. Shahmohammadi et al., "Passive tuning in lateral-mode thin-film piezoelectric oscillators," in *Proc. Frequency Control and the European Frequency and Time Forum (FCS)*, pp.1,5, May 2011

CONTACT

*M. Hodjat-Shamami, shamami@gatech.edu

A TEMPERATURE-STABLE MEMS OSCILLATOR ON AN OVENIZED MICRO-PLATFORM USING A PLL-BASED HEATER CONTROL SYSTEM

Zhengzheng Wu and Mina Rais-Zadeh*
Department of Electrical Engineering and Computer Science
University of Michigan, Ann Arbor, MI 48109, USA

ABSTRACT

In this work, an oxide-refill process is used to null the first-order temperature coefficient of frequency (TCF) of silicon MEMS resonators and to achieve high thermal resistance isolation structures. The technology enables fabrication of a low-power ovenized micro-platform on which multiple MEMS devices can be integrated. The intrinsic frequency-temperature characteristic of two resonators is utilized for temperature sensing, and closed-loop oven control is realized by phase-locking two MEMS oscillators at a specific temperature. PLL-based control circuitry is implemented in 0.18 μm CMOS to interface with the MEMS resonators. The ovenized MEMS oscillator exhibits an overall frequency drift of ± 5.5 ppm over -40 °C to 70 °C. The MEMS oscillator exhibits near zero phase noise degradation in closed-loop operation.

INTRODUCTION

Because of the intrinsic temperature coefficient of elasticity (TCE) of silicon, an uncompensated moderately doped silicon micromechanical resonator shows a linear TCF of ~-30 ppm/°C [1], which typically dominates the environmentally induced frequency drift of a silicon MEMS oscillator. This relatively large TCF needs to be compensated when making silicon-based timing units or resonant-type physical sensors. One solution is to actively change the frequency of the resonator to compensate for environmental changes. For tight control of the frequency, this method requires accurate monitoring of the resonator temperature. Techniques utilizing a resistive temperature detector on chip are inaccurate in estimating the true local temperature of the resonator [2]. Among other sensing methods, frequency-based temperature sensing has been used in quartz crystal references with excellent accuracy [3]. Such a frequency detection method has also been adopted in miniature MEMS timing units by engineering the TCF values of two resonators with a common thermal isolation structure and heating only the two resonators to achieve stable operation [4]. Using self-sensing method based on frequency drift of the MEMS resonator, excellent stability has been demonstrated for ovenized MEMS oscillators. However, when employing a frequency-based sensing method on two thermally isolated resonators, the need for having controlled TCF and high thermal isolation to the platform sets additional constraints on the resonator design, making it challenging to simultaneously achieve high-Q, low resistance, and high thermal isolation.

In this work, we study a different approach in implementing ovenized silicon MEMS oscillators. A general-purpose thermally isolated micro-platform is implemented, inside which multiple micro-devices can be integrated. The platform does not add any constraint to the design of individual devices on the platform. The active compensation concept is sketched in Figure 1. Two MEMS resonators in a thermally isolated platform are designed to have different frequency drift characteristics as the temperature of the platform changes. The TCF values of the resonator dominate the frequency-temperature characteristic of the oscillators. To null the TCF of the output frequency, the feedback control signal is extracted from the phase offset between the two MEMS oscillators in a phase-locked loop (PLL). In the PLL circuitry, the oscillators are locked into a stable operating point, where the MEMS resonators are heated to a desired oven-set temperature. This way, the oscillators are stabilized regardless of external temperature variations. The control loop also stabilizes the temperature of the whole micro-platform, which favors integration of other devices (such as gyroscopes and accelerometers) for temperature-stable operation.

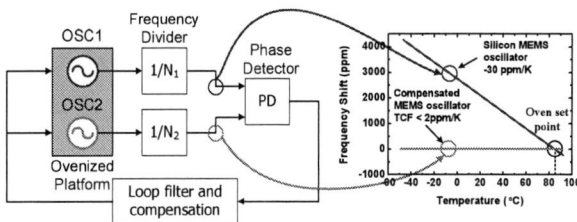

Figure 1: Principle of using two MEMS resonators on an ovenized platform to realize oscillators, temperature sensing, and closed-loop oven-control.

MEMS PROCESS FOR THERMAL ISOLATION AND TCF-COMPENSATION

The frequency-based temperature sensing method requires two MEMS resonators with different TCF values. The TCF of silicon resonators can be engineered by including silicon dioxide-refilled trenches in areas of high strain energy density [5]. This technique is adopted to fabricate a passive TCF-compensated resonator having zero first-order TCF simultaneously with an uncompensated resonator. The oxide-refilled trenches or islands benefiting from low thermal conductivity of silicon dioxide (1.3 W·m⁻¹K⁻¹) are also ideal for fabricating thermal isolation structures. Embedding oxide islands in silicon structures results in substantial reduction of the overall thermal conductivity (thermal conductivity of silicon is 131 W·m⁻¹K⁻¹) and thus the power consumption for ovenization.

Oxide islands are formed by silicon deep reactive ion etch and a subsequent oxide-refill process (Figure 2). The process adopts the steps in fabricating TCF-compensated piezoelectric-on-silicon MEMS resonators [5]. Silicon-on-insulator (SOI) wafers with a 20 μm thick high-resistivity

(>1000 Ω.cm) device layer are used. A scanning electron microscope (SEM) image of a fabricated micro-platform is shown in Figure 3. The platform includes two MEMS resonators as a proof-of-concept; more devices can be integrated on the same temperature-stable platform. The active area incorporating a built-in metal heater and two resonators is supported by four thermal isolation legs. A cross-sectional view of the fabricated oxide-refilled structures is shown in Figure 3.

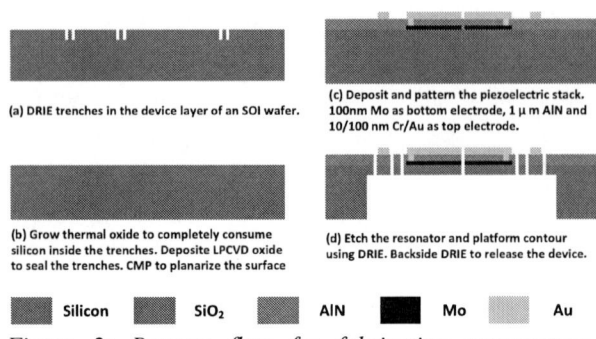

Figure 2: Process flow for fabricating temperature-compensated resonators on a thermal isolation platform.

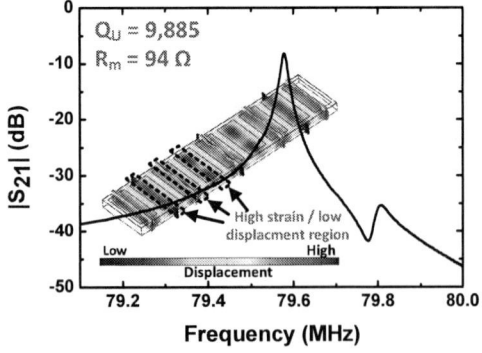

Figure 3: Left: An SEM image of a MEMS platform, which includes two resonators; Right: A cross-sectional SEM of an oxide island formed using the process.

HIGH PERFORMANCE RESONATORS ON A MICRO-PLATFORM

Both an uncompensated and a TCF-compensated resonator are fabricated on the platform to realize the PLL-based oven control. For the uncompensated AlN-on-Si resonator, a 9th-order length extensional mode bulk acoustic resonator (LBAR) has been designed (Figure 4).

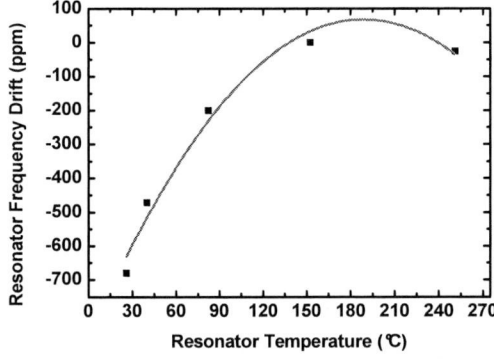

Figure 4: Measured response of an uncompensated 9th-order length extensional mode resonator (LBAR).

Nine electrodes are patterned to cover the high stress regions for proper piezoelectric coupling. Six tethers are located on the nodal points to minimize anchor loss and

provide robust support. The measured frequency response of the LBAR is shown in Figure 4. The unload-Q (Q_U) of this vibration mode is extracted as 9,885 (f×Q ~ 8×10^{11}). The resonator exhibits a low insertion loss of 8.6 dB, indicating a low motional impedance of 94 Ω.

For the TCF-compensated resonator, a coupled-ring resonator design is adopted from [5]. Oxide islands are embedded inside two vibrating rings. The measured response of the coupled-ring resonator is plotted in Figure 5, showing the resonance mode near 19.2 MHz. This coupled-ring resonator exhibits a Q_U of 7,354 (f×Q ~ 1.4×10^{11}) and a motional impedance of 443 Ω. As shown in Figure 6, the extracted TCF of the resonator is +5 ppm/K near the desired oven-set temperature (~90 °C). The turn-over point (maxima of the parabolic curve) is near 190 °C. The over-compensation can be reduced by improving the design and the fabrication control.

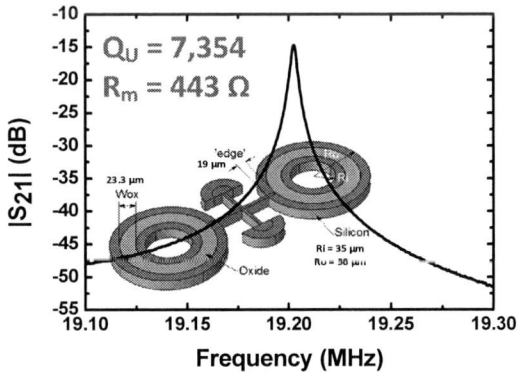

Figure 5: Schematic of a 19.2 MHz TCF-compensated coupled-ring resonator.

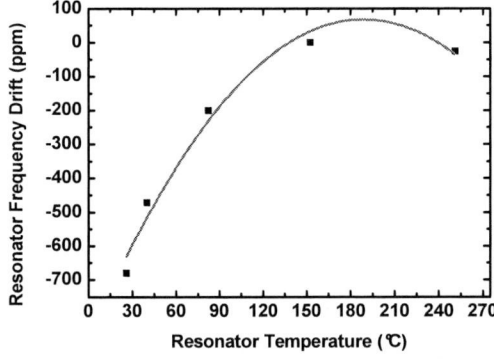

Figure 6: Measured frequency shift of the 19.2 MHz coupled-ring resonator with temperature (fitted to a second-order polynomial curve).

THERMAL PROPERTIES OF THE MICRO-PLATFORM

The oxide-refill process is also used to implement high thermal resistance support structures for the platform. The top view of a thermal isolation leg attached to a square-shape platform is sketched in Figure 7. Oxide islands are embedded in the supporting legs of the platform, as highlighted. Using oxide islands, the legs can be designed with high thermal resistance while having sufficient stiffness. Also, wide supporting legs allow routings of multiple low-resistance electrical connections using a thin-film metal layer, which is favorable for integrating multiple devices on the platform. The effective thermal resistance from the platform center to the external thermal boundary is measured to be ~13.4 K/mW at ~ 1 mTorr ambient pressure ☐ an improvement of 22× compared to an all-silicon design (Figure 7) ☐ allowing for low-power ovenization (in mW range) in a typical MEMS package.

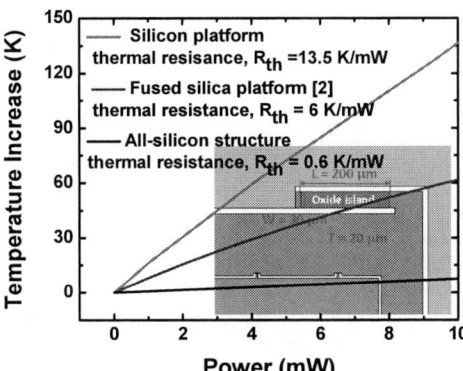

Figure 7: Temperature increase of the platform active area (red line) vs. heater power extracted from measurement. Results are compared to a fused silica platform [2] (blue line) and a silicon platform without oxide islands (black line). The inset shows dimensions of the thermal isolation leg.

The thermal property can be modeled using an RC equivalent circuit as shown in Figure 8. In the equivalent circuit model, the thermal resistance introduced by the isolation legs of the platform is represented by a resistor, $R_{th,leg}$. Also, $R_{th,RES1}$ and $R_{th,RES2}$ model the thermal resistances of the coupled-ring resonator and LBAR, respectively. Heat capacity (thermal mass) of a solid structure is modeled using capacitors. In Figure 8, capacitor C_{pl}, models the heat capacity of the large platform and two capacitors, C_{RES1} and C_{RES1}, model the heat capacity of the resonators on the platform.

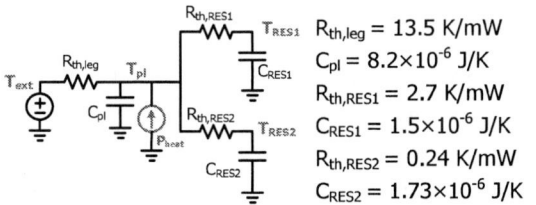

$R_{th,leg}$ = 13.5 K/mW
C_{pl} = 8.2×10⁻⁶ J/K
$R_{th,RES1}$ = 2.7 K/mW
C_{RES1} = 1.5×10⁻⁶ J/K
$R_{th,RES2}$ = 0.24 K/mW
C_{RES2} = 1.73×10⁻⁶ J/K

Figure 8: A simplified thermal model for the micro-platform and two resonators.

PLL-BASED OVEN CONTROL SYSTEM

The design of the PLL-based oven-control system can be studied using a linear control model sketched in Figure 9. In this model, temperature-induced frequency drift of two oscillators are modeled with coefficients, TCF_1 and TCF_2, respectively. Values of TCF_1 and TCF_2 transduce temperature in the thermal domain to frequency domain signals. Using programmable frequency dividers, the divided-down frequencies of the two oscillators are configured to be matched at the desired oven-set temperature. A phase-frequency detector (PFD) detects the phase or frequency difference, and the average voltage output from the PFD indicates phase offset in lock. The phase offset information translates into a control signal to heat the platform to the oven-set temperature. The heater driver is designed using a square-root generator to linearize the transfer characteristic from the control voltage (V_{CTRL}) to the heater power (P_{heat}). Using such a design, the PLL can be treated as a linear control system, and the loop gain is near constant regardless of the operating point [2]. The analog square-root generator circuit is designed using CMOS translinear circuits [6].

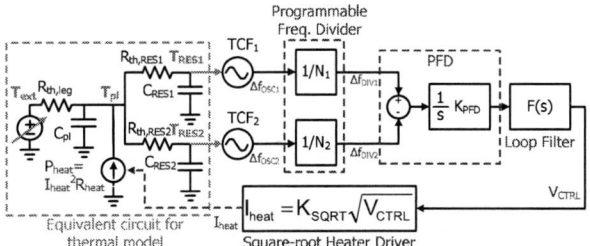

Figure 9: Linear model for the PLL using two MEMS oscillators for temperature sensing and oven-control.

The PLL is designed to have two integrators in the loop (type-II PLL), including one integrator from the PFD and another integrator in the loop filter. Such a design ensures a high DC loop gain to eliminate static errors. However, the feedback loop is potentially unstable with two integrators and a thermal pole (f_{pl} = $1/2\pi R_{th,leg}·C_{pl}$) at low frequencies. In this work, a special loop filter design with two compensation zeroes is employed. Two zeroes are used to cancel out the low frequency thermal pole (f_{pl}) and create a phase margin at the unity-gain frequency (Figure 10). Therefore, by strategically placing the frequency of zeros, the loop dynamics can be defined to be independent of the thermal pole frequency. In this prototype design, the control system has a unity-gain bandwith of more than 10× the thermal pole frequency (f_{pl}), which indicates the control loop can respond fast to correct any dynamic error.

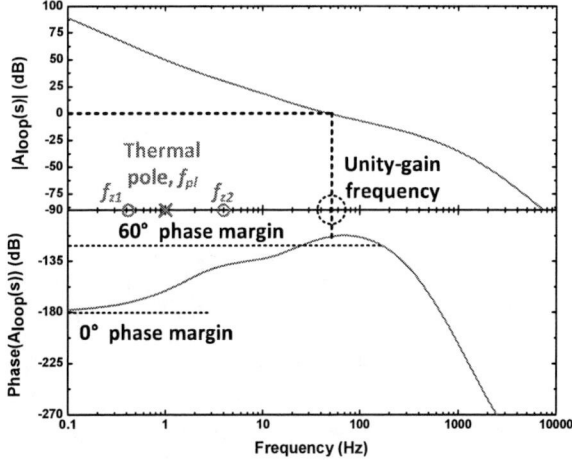

Figure 10: Loop gain of the temperature controller with two compensation zeros on a Bode plot; the system has two integrator, a low-frequency thermal pole (f_{pl}), and two compensation zeroes (f_{z1} and f_{z2}) from the loop filter.

978-1-4799-7956-1/15 $31.00 © 2015 IEEE

MEASUREMENT RESULTS

The CMOS PLL circuitry for the oven-control system is implemented using TSMC 180 nm CMOS technology. In measuring the prototype system, the CMOS chip and the MEMS chip are mounted separately in ceramic packages, and the packages are assembled on a PCB. The frequency stability of MEMS oscillators under external temperature change is measured. During the measurements, the PCB containing both the MEMS chip and the CMOS chip is placed in a vacuum chamber with a pressure level of less than 10 mTorr. The chamber temperature is swept from -40 °C to 70 °C while the output frequency of the MEMS oscillators in the PLL is monitored using a frequency counter (Agilent 53181A). The chamber temperature ramp is a relatively slow process and the PLL-based oven control system loop has a sufficiently large bandwidth. Therefore, the two oscillators used for active compensation are locked during the chamber temperature ramp. As a result, the frequency counter records identical frequency outputs from the oscillators. The frequency stability of the MEMS oscillators on the micro-platform is measured when the PLL-based oven-control system is active. The frequency drift of the MEMS oscillators is recorded and plotted in Figure 11. The oscillator has an overall frequency drift of ± 5.5 ppm over the chamber temperature of -40 to 70 °C. Recalling that an uncompensated silicon MEMS oscillator has a TCF of -30 ppm/°C, the oven-control regulates the temperature fluctuation of the uncompensated resonator on the platform to within 367 m°C.

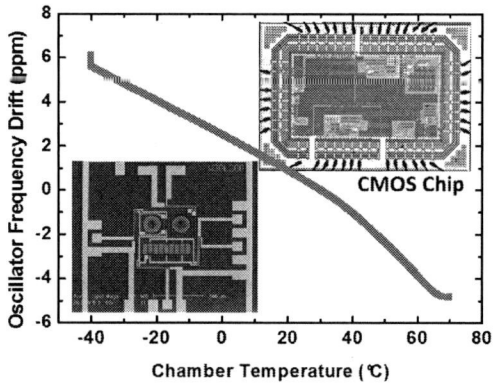

Figure 11: Measured output frequency drift versus temperature (inset shows microscopic photographs of the MEMS chip and the CMOS chip for the PLL-based control system).

The phase noise when the oscillator is in the PLL for ovenized operation is measured and compared to the phase noise when the MEMS oscillator is free running in Figure 12. It can be seen that the PLL-based active compensation does not degrade the far-from-carrier noise, while the measured close-in-carrier phase noise is even improved. In the phase noise measurement using an Agilent E5500 system, a frequency scan of the close-in-carrier region (offset frequency in the range of 1-100 Hz from carrier) takes approximately 5-10 s to complete. In such measurements, slow frequency fluctuations due to temperature variations cannot be distinguished from 1/f

noise in the measured phase noise plot. In fact, slow temperature variations tend to dominate the fluctuations in the close-in-carrier region. When the MEMS oscillator is are placed in the PLL, the active compensation loop effectively regulates temperature variations of the MEMS resonator. Therefore, we observe improvement in the phase noise when MEMS oscillators are in the PLL-based active compensation system.

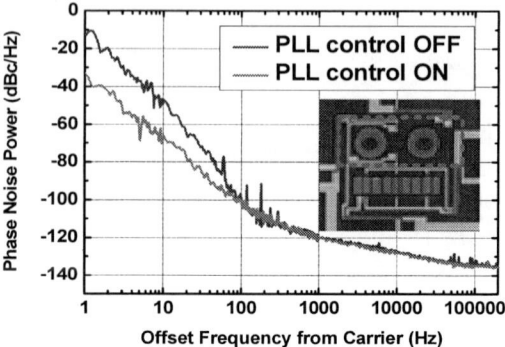

Figure 12: Measured phase noise performance of the 19.2 MHz MEMS oscillator using a TCF-compensated coupled-ring resonator on the platform. The phase noise performance with PLL-based compensation is compared to the same free-running oscillator performance.

ACKNOWLEDGEMENTS

The authors would like to acknowledge the staff at Michigan Lurie Nanofabrication Facilities (LNF) for their support and Dr. Jong Kwan Woo for helpful discussions on the CMOS IC implementation. This work is supported by DARPA and NASA.

REFERENCES

[1] R. Melamud *et al.* "Temperature-insensitive composite micromechanical resonators," *J. Micro-electromech. Syst.* vol. 18, no. 6, pp.1409-1419, Dec. 2009.

[2] Z. Wu *et al.*, "Low-power ovenization of fused silica resonators for temperature-stable oscillators," *Proc. IFCS 2014,* Taipei, Taiwan, May 2014, pp. 1-5.

[3] E. Jackson, "The microcomputer compensated crystal oscillator-practical application of dual-harmonic mode quartz thermometry," *Proc. IFCS, Aug. 2004,* pp. 401-405.

[4] J. C. Salvia *et al.*, "Real-time temperature compensation of MEMS oscillators using an integrated micro-oven and a phase-locked loop," *J. Microelectromech. Syst.,* pp. 192-201, 2010.

[5] V. Thaker *et al.*, "A temperature-stable clock using multiple temperature-compensated micro-resonators," *Proc. IFCS 2014,* Taipei, Taiwan, May 2014, pp. 1-4.

[6] D. Barrettino *et al.* "CMOS-based monolithic controllers for smart sensors comprising micromembranes and microcantilevers," *IEEE Trans. Circuits Syst. I, Reg. Papers,* vol. 54, no. 1, pp. 141-152, Jan. 2007.

CONTACT

*M. Rais-Zadeh, email: minar@umich.edu

ALN PIEZOELECTRIC ON SILICON MEMS RESONATOR WITH BOOSTED Q USING PLANAR PATTERNED PHONONIC CRYSTALS ON ANCHORS

Haoshen Zhu[1] and Joshua E.-Y. Lee[1,2]

[1] Department of Electronic Engineering, City University of Hong Kong, HONG KONG
[2] State Key Laboratory of Millimeter Waves, City University of Hong Kong, HONG KONG

ABSTRACT

We report an approach to suppress anchor loss in thin-film piezoelectric-on-silicon (TPoS) micromechanical (MEMS) resonators by patterning 2D phononic crystals (PnCs) externally on the anchors. The PnCs serve as a frequency-selective reflector for outgoing acoustic waves through the tethers of the TPoS resonator. According to our experimental results, combining the PnCs with the conventional TPoS resonator significantly enhances the quality factor (Q) and correspondingly lowers the insertion loss (IL). The measured improvement is reproducible over multiple samples and consistent with the simulations by tuning the PnC bandgaps, suggesting significant reduction of acoustic leakage to the substrate by adopting the PnCs.

INTRODUCTION

TPoS MEMS resonators [1] have become highly attractive for timing applications owing to their excellent electromechanical coupling and power handling, as well as low intrinsic mechanical loss. As such, real-time clocks [2], high-frequency oscillators [3], filters [4], gyroscopes [5], pressure and temperature [6, 7] sensors based on TPoS MEMS resonators have been demonstrated recently. What is common to all the above applications is the need for high Q. However, the Q's of TPoS MEMS resonators still lag behind those of capacitive pure silicon resonators. In the case of MEMS oscillators, according to the Leeson's model, Q determines the close-to-carrier phase noise performance. Therefore improving Q enables TPoS MEMS oscillators to better meet more stringent phase noise requirements. On this note, bulk-mode TPoS resonators [1] are particularly of interest owing to their higher fQ product and better power handling capability than flexural-modes.

Q is an indicator of energy loss in a resonant system. Various loss mechanisms have been identified for MEMS resonators. Thus the total mechanical (unloaded) Q can be expressed as:

$$\frac{1}{Q_{tot}} = \frac{1}{Q_{anc}} + \frac{1}{Q_{inter}} + \frac{1}{Q_{AKE}} + \frac{1}{Q_{TED}} + \frac{1}{Q_{air}} + \frac{1}{Q_{other}}. \quad (1)$$

It has been postulated that for silicon, phonon interaction induced Akhiezer loss sets the fundamental limit on the fQ product. However, Akhiezer loss is not the major limiting factor for TPoS resonators since their typically measured fQ is still much lower than the theoretical material limit. Thermoelastic damping (TED) dominates in flexural-mode resonators but not bulk-modes. Air damping (Q_{air}) is also trivial given the little difference in Q between measuring a TPoS bulk-mode resonator in air and in vacuum. For TPoS bulk-mode resonators, the anchor loss (Q_{anc}) is believed to be one of the major loss mechanisms. Another competing loss factor could be interface loss (Q_{inter}). Anchor loss is attributed to acoustic waves leaking out from the resonator

to the support plane and substrate. While interface loss is associated more with the fabrication process, anchor loss can be reduced by careful design of mechanical supports.

Reported approaches to suppress anchor loss include adding acoustic reflectors on the anchor [9] and also using thicker silicon device layers [10]. More recently, 1D PnC rings were adopted as tethers [11-12]. Theoretically, the acoustic energy can be perfectly confined within the resonator body, once the resonant frequency sits in the bandgap. But the effectiveness of this method relies on the fabrication of long and narrow 1D PnC strings, making the device vulnerable to mechanical shock. In this work, we propose a novel approach by deploying 2D PnC arrays on the anchors to block acoustic leakage and hence boost Q.

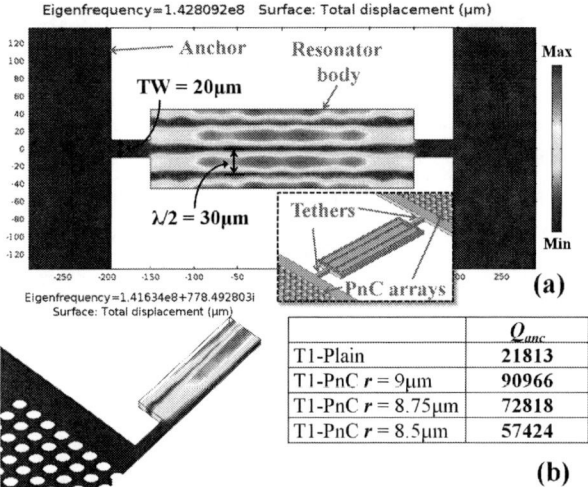

Figure 1: (a) 2D FE simulation of the 3rd order vibration mode shape and associated eigenfrequency of the TPoS resonator (inset: 3D view of the proposed device); (b) The simulated Q_{anc} via a 3D PML-based simulation.

	Q_{anc}
T1-Plain	21813
T1-PnC $r = 9\mu m$	90966
T1-PnC $r = 8.75\mu m$	72818
T1-PnC $r = 8.5\mu m$	57424

DEVICE DESIGN

Although anchor loss can be substantially reduced in lateral shear-wave (e.g. Lamé/wine-glass mode) resonators, the resulting piezoelectric coupling efficiency is extremely weak due to the strain profiles of lateral shear-wave modes. Therefore, longitudinal-wave modes are normally adopted for TPoS bulk-mode resonators. Our AlN TPoS resonator was designed to resonate in the 3rd-order width-extensional mode, as shown in Fig. 1a. As with other TPoS resonators reported previously, a pair of straight beam tethers support the resonator and provides electrical feedthrough. Due to the Poisson's effect, the standing longitudinal wave in the width direction will couple into the length direction and propagate to the anchor through the tethers, resulting in anchor loss. The device was aligned to the <110> direction in the silicon (100) plane, preferred for its low Poisson's

978-1-4799-7956-1/15 $31.00 © 2015 IEEE   797   MEMS 2015, Estoril, PORTUGAL, 18 - 22 January, 2015

ratio. The resonant frequency (f_0) can be determined by:

$$f_0 = \frac{1}{\lambda}\sqrt{\frac{E}{\rho}}, \qquad (2)$$

where λ is the longitudinal wave length, E is the Young's modulus in <110> axes and ρ is the density of SCS. The f_0 calculated from (2) is ~142.2MHz in our case ($\lambda = 60\mu m$). Normally, a narrower tether width (TW) should help to reduce loss through the anchors. To test the effect of PnCs on suppressing anchor loss in relation to TW, two versions of the same resonator but different TWs (T1: 20μm; T2: 40μm) were adopted in our designs.

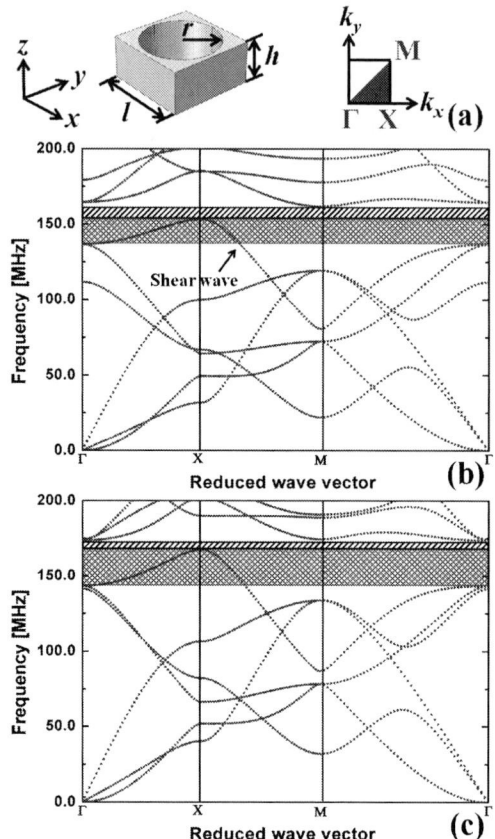

Figure 2: (a) Schematic of a unit cell in the PnC structure and its irreducible Brillouin zone in the **k**-space, where **k**(k_x, k_y) is the wave vector in 2D. FE-simulated band structure of the 2D PnC lattice with unit cell dimension: $l = 20 \ \mu m$, $h = 10 \ \mu m$, (b) $r = 9 \ \mu m$; (c) $r = 8.5 \ \mu m$.

Complete acoustic bandgaps have been predicted [13] and experimentally verified [14] for 2D silicon PnCs based on the square unit cell (see Fig. 2a). Due to the anisotropy of silicon, the band structure for silicon PnCs will depend on the crystal orientation. In our design, the PnC extends in x and y directions along the <110> axes within the (100) plane. By applying Floquet periodicity, we simulated the bandgap structure in the reduced **k**-space by finite-element (FE) analysis as shown in Fig. 2b and 2c. Fig. 2b shows that the PnC lattice creates a complete acoustic bandgap from 153-162MHz (black shaded bar). Since only shear waves can propagate in the partial bandgap (red shaded bar), the 2D PnC forms a stop band for leaking longitudinal waves from 137-162MHz. The acoustic bandgap has been

found to be sensitive to the unit cell dimensions [13]. In our simulation, shrinking r by 0.25 μm and 0.5μm while keeping l and h constant shifts the lower bound of the stop band to around 140MHz and 143MHz, respectively. As shown in Fig. 2c, when $r = 8.5\mu m$, the stop band is slightly above the designed f_0. In this sense, we can tune and verify the effectiveness of a given PnC as an acoustic reflector. Fig. 1b lists the FE simulated Q_{anc} for different PnC cell dimensions, computed using a perfectly matched layer (PML). The results show that the PnC-on-anchor approach we here propose greatly reduces anchor loss and the characteristics of the PnC can be adjusted by varying r.

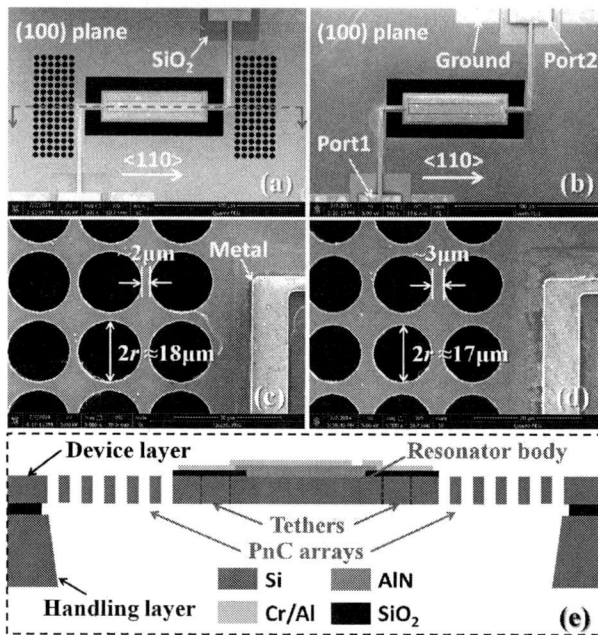

Figure 3: Scanning electron micrographs (SEMs) of the fabricated TPoS resonators (a) with and (b) without PnC on anchors; fabricated PnCs with designed (c) $r = 9$ and (d) $r = 8.5\mu m$, respectively; (e) transverse view schematic taken along the red dashed line in (a).

FABRICATION & CHARACTERIZATION

The devices were fabricated in a 5-mask foundry AlN-on-SOI MEMS process. First, 200nm thermal oxide was grown and patterned to provide electrical isolation. Then 500nm of piezoelectric layer AlN was sputtered and patterned. Next an e-beam evaporated metal stack of 20nm Cr 1000nm Al was deposited and patterned by lift-off. The 10μm Si device layer was then etched by deep reactive-ion etching (DRIE) to define the resonator and PnC structures. Finally, the Si handle layer was etched through from the backside by DRIE. The device was released after wet HF etching of the exposed buried oxide layer. Fig. 3 shows the fabricated devices with and without PnCs, close-up views of PnC versions with different r. The silicon device layer is highly phosphorus-doped on the surface and can be used as the bottom electrode. The transverse view schematic (Fig. 3e) of the device with PnCs illustrates that the top metal electrode is uniquely insulated from the silicon device layer at the tethers by the oxide layer instead of AlN, which is commonly the case in previous TPoS resonators [1-7]. Replacing the AlN with the oxide layer avoids unwanted excitation at the anchors, which could in turn help further reduce anchor loss.

Electrical characterization of all the fabricated devices was conducted at room temperature in a vacuum probe station equipped with GSG probes. The input/output ports were connected to a network analyzer to measure the S21 electrical transmission. The typical two-port setup was applied for device characterization after a standard SLOT calibration.

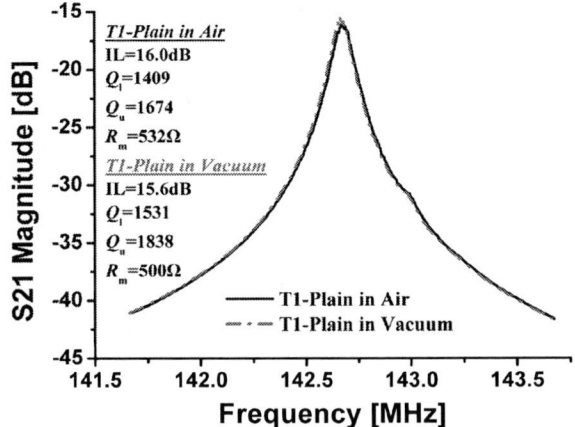

Figure 4: Measured S21 transmission of a TPoS resonator without PnCs on the anchor in air and in vacuum.

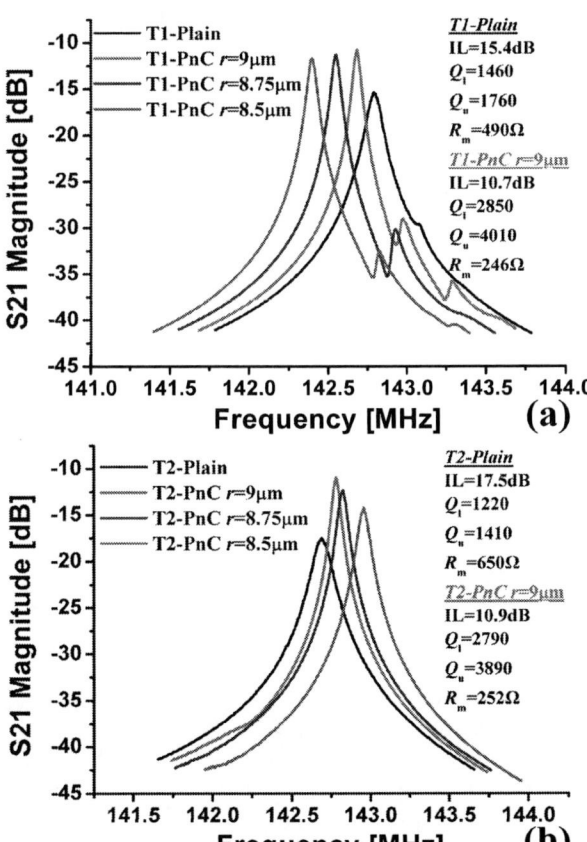

Figure 5: Typical measurement of the S21 transmission of TPoS resonators with PnCs of different hole-radii (r) on the anchors and also without (referred to as 'plain') tested with two different tether widths: (a) 20μm tethers (T1); (b) 40μm tethers (T2).

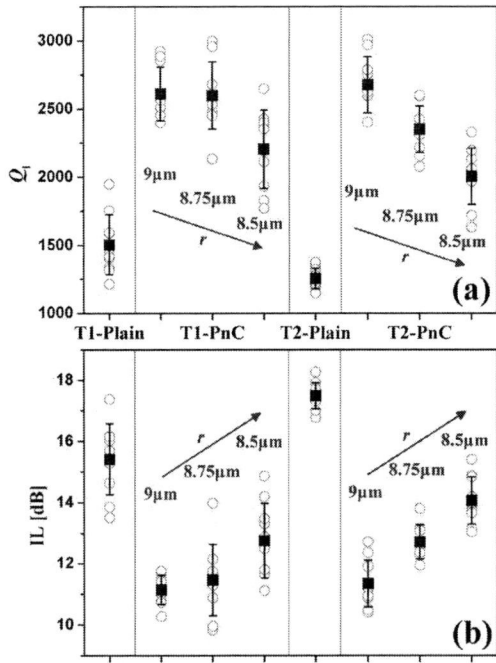

Figure 6: Measured (a) loaded Q (Q_l) and (b) insertion loss (IL) of TPoS resonators with and without PnCs on anchors. Red circle: measured value for each respective device. Black square and error bar: mean and standard deviation over 10 die samples measured.

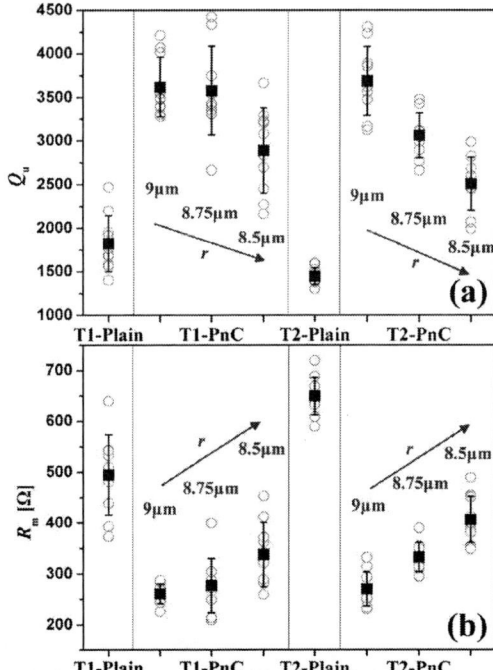

Figure 7: Extracted corresponding (a) unloaded Q (Q_u) and (b) motional resistance (R_m) of TPoS resonators with and without PnCs on anchors. Red circle: extracted value for each respective device. Black square and error bar: mean and standard deviation over 10 die samples.

From Fig. 4, given that measuring in air and vacuum shows little difference on Q, the effect of air damping on Q appears to be negligible for these TPoS resonators. We categorized the devices into 8 design variants in terms of

TW, PnCs of different *r* and without PnCs. A total of 80 devices (10 die samples with the 8 design variants on each die sample) were all electrically characterized in air. The typical two-port measurement results of the electrical transmission for resonators with and without (referred to as "plain") PnCs are compared in Fig. 5. The measured f_0s all closely agree with both the analytical and also numerical predictions. Notably, for either of the tether widths, the PnC-enhanced devices exhibited higher loaded Q (Q_l) and lower IL. It can also be seen that the measured Q_l drops and IL increases as a result of reducing *r* from 9 to 8.5µm. This is consistent with the theory which predicts that f_0 lies outside the bandgap in the case of *r*=8.5µm. Our PML FE simulations of Q_{anc} (Fig. 1b) also show that Q improves with the PnCs, though the values do not exactly match the measured Q. A possible explanation for the discrepancy is the existence of interfacial damping and its determination of the measured Q, for which further investigation would be needed to confirm. As depicted in Fig. 6, this observed improvement in both Q_l and IL is repeatable over multiple measurements on the 10 dies characterized. To exclude the loading effect from the 50Ω termination resistance (R_0) of the network analyzer, we extracted the unloaded Q (Q_u) and motional resistance (R_m) as follows:

$$Q_u = \frac{Q_l}{1 - 10^{-\text{IL}/20}}, \qquad (3)$$

$$R_m = 2R_0 \left(10^{\text{IL}/20} - 1\right). \qquad (4)$$

Here, Q_u reflects the total mechanical Q of the devices, referring to (1). Fig. 7 collects all the Q_u and R_m from the measurements of the 80 devices. The lower Q_u found in T2-Plain over T1-Plain devices indicates a higher anchor loss due to the wider tether. But the well-designed PnC (*r* = 9µm) on the anchor blocks acoustic energy propagating out from the tethers and in turn boosts Q_u to the same level as a T1-PnC device. Compared to the state-of-the-art TPoS resonators with 1D PnC tethers [11], our PnC-on-anchor devices demonstrate comparable performance in terms of *fQ* product and R_m.

CONCLUSION

In this work, we have designed and fabricated silicon 2D PnCs on the anchors of 142MHz TPoS resonators to improve their Qs. Using the same resonator dimensions, variations in the tether width and PnC unit cell details were applied to test the effect of the PnCs. The measurement results consistently show that a well-engineered PnC can effectively improve the Q_u from 1822 to 3620 (mean), and correspondingly reduces R_m by half from 495 to 261Ω (mean). Compared with a previous approach that uses 1D PnC tethers, our method here delivers the same high Q while potentially improving the robustness against mechanical shock that narrow PnC tethers must withstand.

ACKNOWLEDGEMENTS

This work was supported by grants from the Research Grants Council of Hong Kong (project numbers CityU 124312 and CityU 116113).

REFERENCES

[1] G. K. Ho, et al., "Piezoelectric-on-silicon lateral bulk acoustic wave micromechanical resonators", *J. Microelectromech. Syst.*, vol. 17, no. 2, pp. 512–520, Apr. 2008.

[2] D. E. Serrano, et al., "Electrostatically tunable piezo-electric-on-silicon micromechanical resonator for real time clock", *IEEE Trans. Ultrason., Ferroelect., Freq. Contr.*, vol. 59, no. 3, pp. 358-365, Mar. 2012.

[3] R. Abdolvand, H. M. Lavasani, G.K. Ho, and F. Ayazi, "Thin-film piezoelectric-on-silicon resonators for high-frequency reference oscillator applications", *IEEE Trans. Ultrason., Ferroelectr., Freq. Control*, vol. 55, no. 12, pp. 2596-2606, Dec. 2008.

[4] R. Abdolvand and F. Ayazi, "Monolithic thin film piezoelectric-on-substrate filters", in *Proc. IEEE Int. Microwave Symp.*, Honolulu, Hawaii, Jun. 2007, pp. 509-512.

[5] R. Tabrizian, et al., "High-frequency AlN-on-silicon resonant square gyroscopes", *J. Microelectromech. Syst.*, vol. 22, no. 5, pp. 1007-1009, Oct. 2013.

[6] R. Tabrizian and F. Ayazi, "Dual-mode vertical membrane resonant pressure sensor", *Proc. IEEE MEMS Conf.*, San Francisco, CA, Jan. 2014, pp. 120-123.

[7] J. L. Fu, R. Tabrizian, and F. Ayazi, "Dual-mode AlN-on-silicon micromechanical resonators for temperature sensing", *IEEE Trans. Electron Devices*, vol. 61, no. 2, pp. 591-597, Feb. 2014.

[8] J. T. M. van Beek, and R. Puers, "A review of MEMS oscillators for frequency reference and timing applications", *J. Micromech. Microeng.*, vol. 22, no. 1, pp. 013001-1–013001-35, 2012.

[9] B. P. Harrington, and R. Abdolvand, "Q enhancement through minimization of acoustic energy radiation in micromachined lateral-mode resonators", in *Digest Tech. Papers Transducers'09 Conf.*, Denver, CO, Jun. 2009, pp. 700-703.

[10] W. Pan and F. Ayazi, "Thin-film piezoelectric-on-substrate resonators with Q enhancement and TCF reduction", in *Proc. IEEE MEMS Conf.*, Hong Kong, Jan. 2010, pp. 104-107.

[11] L. Sorenson, et al. "One-dimensional linear acoustic bandgap structures for performance enhancement of AlN-on-Si micromechanical resonators" , in *Digest Tech. Papers Transducers'11 Conf.*, Beijing, China, Jun. 2011, pp. 918-921.

[12] C.-M. Lin, et al., "Anchor loss reduction in AlN Lamb wave resonators using phononic crystal strip tethers", in *Proc. IEEE Int. Freq. Contr. Symp.*, Taipei, Taiwan, May 2014, pp. 371-375.

[13] S. Mohammadi, et al., "Complete phononic bandgaps and bandgap maps in two-dimensional silicon phononic crystal plates", *Electron. Lett.*, vol. 43, no. 16, pp. 898-899, Aug. 2007.

[14] C.-Y. Huang, et al., "A two-port ZnO/silicon Lamb wave resonator using phononic crystals", *Appl. Phys. Lett.*, vol. 97, no. 3, pp. 031913-1–031913-3, Jul. 2010.

CONTACT

*Joshua E.-Y. Lee, telephone: +852-3442-9897; joshua.lee@cityu.edu.hk

BATCH-FABRICATED HIGH Q-FACTOR MICROCRYSTALLINE DIAMOND CYLINDRICAL RESONATOR

Daisuke Saito[1], Chen Yang[1], Amir Heidari[2], Hadi Najar[2], Liwei Lin[1], and David A. Horsley[2]
[1]University of California, Berkeley, USA
[2]University of California, Davis, USA

ABSTRACT

This paper reports a 1.5 mm batch-fabricated polycrystalline diamond Cylindrical Resonator (CR) for gyroscope applications. The device is fabricated in a cylindrical shape using silicon on insulator (SOI) wafers and deep reactive ion etching (DRIE), which allows flexibility of choosing different geometries and materials for the resonator structure. A quality factor (Q) of 313,100 is measured at the 23 kHz 2 theta elliptical wineglass modes, producing a ring-down time of 4.32 seconds. Annealing CRs at 700 °C in a nitrogen atmosphere improved Q from 75,000 to over 300,000. The highly symmetric fabrication results in CRs with an excellent frequency mismatch of 3 Hz (130 ppm) between the 2 theta degenerate wineglass modes without applying any tuning voltage.

INTRODUCTION

Micro vibratory gyroscopes (VG) have been used for consumer products because they offer a combination of small size, low power consumption and low cost but they have not yet achieved navigation-grade performance. The commercial hemispherical resonator gyroscope (HRG) has high precision and high reliability and is the only VG to achieve navigation-grade performance [1]. The HRG's highly axisymmetric, center-anchored hemispherical shell minimizes unwanted coupling of the HRG's vibration to the base substrate and, by using high-purity fused-quartz material, results in high Q. The HRG is one of the few VG that can be used as a rate-integrating gyroscope (RIG) which can measure rotation angle directly rather than the angular rotation rate which is commercialized as rate gyroscopes (RG) to date.

The characteristics of HRGs have inspired efforts to make micro-scale, low-cost, mass manufactured, navigation-grade gyroscopes with large dynamic range. To enable RIG-mode operation, the resonator's two Coriolis-coupled degenerate vibration modes must be closely matched in resonant frequency since any frequency mismatch (Δf) is indistinguishable from rotation rate and results in bias error. For example, a 1 Hz frequency mismatch between these modes produces a bias error of 360 deg/s, much larger than the bias error of a typical commercial MEMS rate gyroscope today.

Another important parameter for gyroscopes is the short-term noise, known as Angle Random Walk (ARW). ARW, in $(°/s)/\sqrt{Hz}$, is given by

$$ARW = \frac{1}{A_\theta} \sqrt{\frac{k_B T}{2(^1/_2\, k_m x^2)(^Q/_{\pi f_n})}} \frac{180}{\pi} \quad (1)$$

where A_θ is the angular gain, k_B is the Boltzmann constant, T is the temperature in K, k_m is the spring constant of a resonator, x is oscillation amplitude of the driven axis, and f_n is the resonant frequency (assuming mode-matched operation). In the square root of equation (1), the numerator is the thermal energy and the denominator is the product of the mechanical vibration energy and the decay time constant ($\tau = Q/\pi f_n$).

Therefore, according to equation (1), in order to achieve low ARW, A_θ, x, and Q must be increased and f_n must be decreased. In micro-scale gyroscopes, x is typically limited to a few microns and the lower limit for f_n is ~10 kHz, below which vibration sensitivity becomes problematic. A_θ is a constant determined by the resonator geometry, while Q is related to the resonator geometry and the material [2, 3].

Several groups have recently investigated micro-scale HRG's to achieve small Δf and high A_θ and Q. Devices made via glass molding [4-6] and polycrystalline diamond [7, 8] have excellent mechanical properties, but they used complex fabrication processes such as manual assembly of the wineglass onto a second substrate, laborious wet-etching of glass substrates, or Micro-Electro Discharge Machining (µEDM) and wet etching of silicon substrates. A cylindrical resonator (CR) made from thin-film SiO_2 [9] solves many of these problems but dielectric resonators require complicated fabrication to add capacitive drive/sense electrodes.

Here, we demonstrate a high Q CR fabricated from microcrystalline diamond (MCD), a high-stiffness, low-loss material [2, 3]. Relative to hemispherical resonators [4-8], the CR presented here has greater depth-to-diameter aspect-ratio, resulting in higher TED-limited quality factor (Q_{TED}) and higher A_θ. Moreover, the CR is fabricated via wafer-scale DRIE, ensuring suitability for mass manufacturing and producing more symmetric resonators with low Δf.

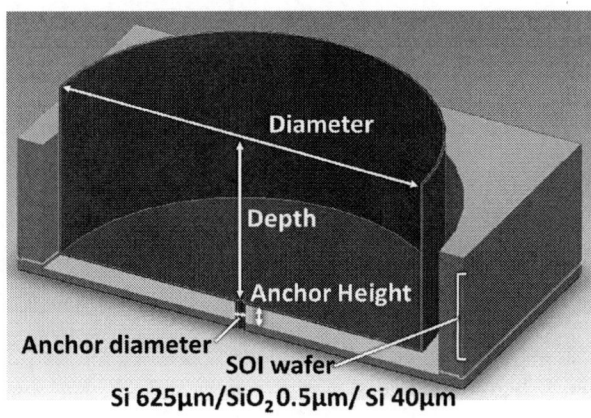

Figure 1: Schematic of cylindrical resonator (CR).

DEVICE DESIGN

The device consists of a 1.5 mm diameter, 625 µm-deep cylindrical shell resonator formed from 2.3 µm-thick hot filament chemical vapor deposition (HFCVD) polycrystalline diamond supported by a 15 µm diameter, 1.4 µm-high anchor at the center (Figure 1).

While micro-HRG's must be fabricated using unconventional isotropic etching processes, the CR has precise geometry because the resonator geometry is defined using DRIE of SOI wafers, both well-established technologies used in mass manufacturing of inertial sensors. The geometry of the CR is easy to change by making different diameters on the photomask and different thicknesses of SOI wafer. The CR can also use different structural materials in place of polycrystalline diamond. Because of this flexibility, the CR can achieve higher depth-to-diameter aspect-ratio, resulting in greater scale-factor due to increased angular gain (A_θ). The other advantage of the CR is that the direction of vibration is parallel to the base of the cylinder, leading to lower anchor loss and higher Q.

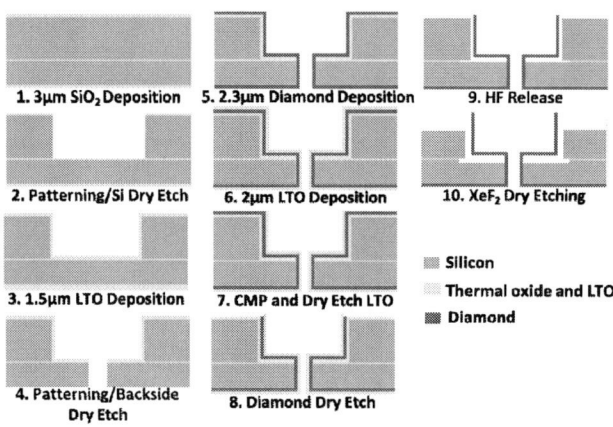

Figure 2: Fabrication process flow.

FABRICATION PROCESS

Novel process steps are employed to make symmetric and smooth CRs (Figure 2). Starting substrates are SOI wafers consisting of 625 µm handle layer, 0.5 µm SiO$_2$ layer and 40 µm device layer. First, cylindrical molds are etched into the low resistivity (1-20 Ω-cm) Si device layer using DRIE with a 3 µm thermal oxide as hard mask. Then, a 1.5 µm thick Low Temperature Oxide (LTO) sacrificial layer is conformally deposited inside the silicon molds via chemical vapor deposition (CVD). Then, front-to-backside lithography is performed and 15 µm diameter center anchors are etched in the device layer of the SOI wafer. The SiO$_2$ surface is seeded using an ultrasonic seeding suspension containing 5-50 nm diameter nanocrystalline diamond powder dispersed in solvent. Using a methane to hydrogen (CH$_4$/H$_2$) concentration of 1 % and a relative tetramethyl boron/CH$_4$ concentration of 3.3 %, the 2.3 µm thick boron-doped diamond structural layer is deposited via HFCVD (sp^3 Diamond Technologies). At this step, the diamond is conformally deposited within the device layer anchor molds, ensuring mechanical and electrical contact between the diamond resonator body and the Si device layer. Another 1.8 µm LTO layer is conformally deposited

to mask the topside of the diamond. This LTO layer covers the rough microcrystalline diamond surface. Chemical mechanical polishing (CMP) removes the oxide mask at a rate of ~300 nm/min from the wafer surface followed by a tetrafluoromethane (CF$_4$) plasma etch in an inductively coupled plasma (ICP) etcher (SPTS Inc., APS etcher) but retains the mask within the mold. The exposed diamond on the wafer surface is etched in an O$_2$/CF$_4$ plasma at a flow ratio of 50:1 in an ICP etcher at a process pressure of 30 mTorr, leaving diamond only within the molds. Finally, the diamond shells are released in 49% hydrofluoric acid (HF) to remove the sacrificial SiO$_2$, after which the silicon molds are partially removed to facilitate non-contact optical characterization of the resonator. In order to etch back the silicon mold from the top of the die, xenon difluoride (XeF$_2$) is used in these experiments. SEM images in Figure 3 (a) show a 1.5 mm CR and close-up of the diamond film of the resonator wall and Figure 3 (b) shows the optical images of an array of released diamond CRs attached to the silicon substrate. The DRIE process was optimized to provide smooth, radially-symmetric etching of the silicon mold in order to achieve close frequency matching between the two degenerate vibration modes. Figure 3 (c) shows that the roughness of the silicon mold is copied to the outer diamond surface of the resonator.

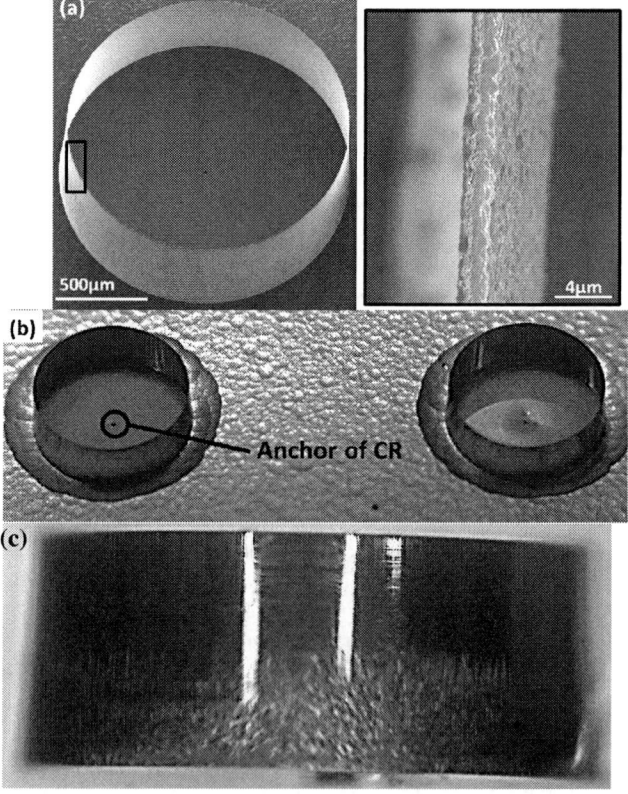

Figure 3: (a) SEM of the 1.5 mm CR and close-up of the 2.3 µm thick resonator wall; (b) Optical images of two adjacent CRs following XeF$_2$ etch of surrounding silicon; (c) Optical image of the CR's outside wall showing roughness resulting from the DRIE etch of the mold wafer.

EXPERIMENTAL RESULTS

The resonance frequency and Q of the first six wineglass vibration modes was measured using a laser doppler vibrometer (LDV, Polytec Inc.) in vacuum. The diamond shell was mounted vertically onto a shear mode piezoelectric actuator (Noliac A/S) with a bandwidth of 1.7 MHz to excite the shell. The single point LDV laser spot was focused through a 5x microscope objective onto the rim of the shell to measure the radial displacement of the vibrating shell. The whole setup except the LDV laser was mounted in a vacuum chamber to allow measurements at low pressures.

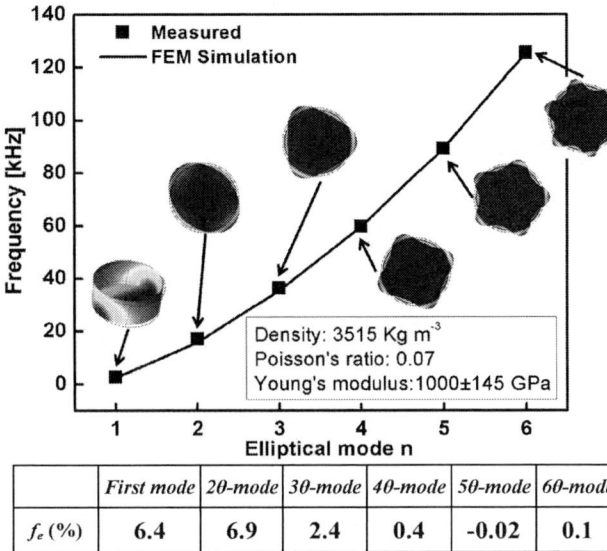

	First mode	2θ-mode	3θ-mode	4θ-mode	5θ-mode	6θ-mode
f_e (%)	6.4	6.9	2.4	0.4	-0.02	0.1

Figure 4: Measured frequencies of the first 6 elliptical modes compared with FEM model (figure) and the error between experimental results and FEM results ($f_e = (f_{FEM} - f_{exp}) / f_{FEM}$) (table).

The measured and FEM-simulated frequencies of the first six modes are shown in Figure 4. The error between the experimental value and the value from FEM simulation are listed in the table below the graph. The maximum error is 6.4%, which shows that measurements agree well with FEM simulation results. The Young's modulus of the diamond is estimated to be 1000±150 GPa. This value is reasonable compared to that of single crystalline diamond. The imprecise estimate for Young's Modulus results from the lack of precision in measuring the thickness of polycrystalline diamond film: 7.5% uncertainty in the film thickness measurement translates to 15% uncertainty in the estimate for Young's Modulus. FEM simulations show that the residual stress of the polycrystalline diamond causes a small shift (less than 1%) in the resonance frequency. While the as-deposited diamond film has -250 MPa compressive residual stress, the released resonator is free to expand and release the stress, resulting in a nearly stress-free resonator.

The frequency response of a CR in vacuum at 16 μTorr is shown in Figure 5. Resonance peaks at $f_1 = 23.071$ kHz and $f_2 = 23.074$ kHz match the FEM simulation result for the 2θ-mode frequency. The frequency mismatch between these two degenerate resonance modes was $\Delta f = 3$ Hz (130 ppm) without applying any tuning voltage. This frequency mismatch is smaller than the value reported for other micro-scale HRGs without using any trimming [4, 6, 7]. To extract Q, a Lorentzian was fit to the response, resulting in $Q = 313,100$ at f_1 and 309,000 at f_2.

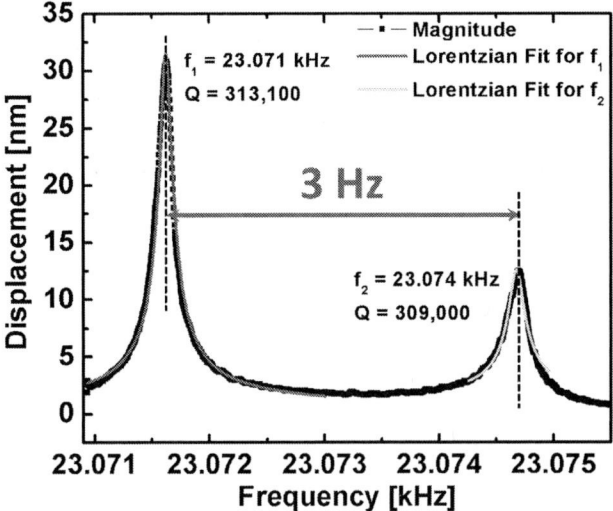

Figure 5: Measured CR frequency response showing 2 modes with 3 Hz frequency split (f) and high Q-factor ($f_1 = 23.071$ kHz, $Q_1 = 313,100$; $f_2 = 23.074$ kHz, $Q_2 = 309,000$ @ 16 μTorr).

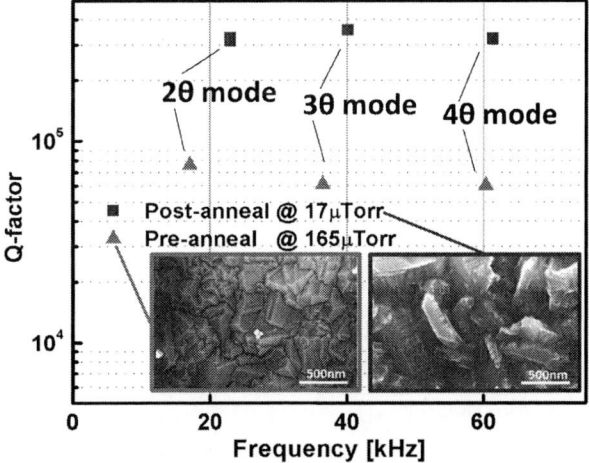

Figure 6: Pre- and post-anneal Q-factor of 2θ, 3θ and 4θ wineglass modes and SEM of the CR's diamond surface.

Measurements of 35 resonators showed an initial as-fabricated Q of up to 75,000 in the first three elliptical modes. Annealing at 700 °C for 2 hours in a nitrogen atmosphere increased Q by a factor of 4x to 6x to a maximum value of 355,100 (Figure 6). The increase in Q was almost the same in all three elliptical modes. We believe that the two intrinsic damping sources of the material, thermoelastic damping and surface loss, are the dominant sources of dissipation [2, 3]. The annealing results in an observable change in the diamond surface at the inner bottom surface of the CR (Figure 6). We attribute the increased Q to reduced surface loss arising from the removal of a high-loss layer of surface diamond.

Figure 7 shows a ring-down test of the 2θ vibration mode in vacuum at 16 μTorr, yielding τ of 4.32 sec, in good agreement with the Q observed in Figure 5 and Figure 6. Compared with prior work, Table 1, this work demonstrates the highest Q and lowest asymmetry (Δf) in a millimeter-scale 3D resonator to date.

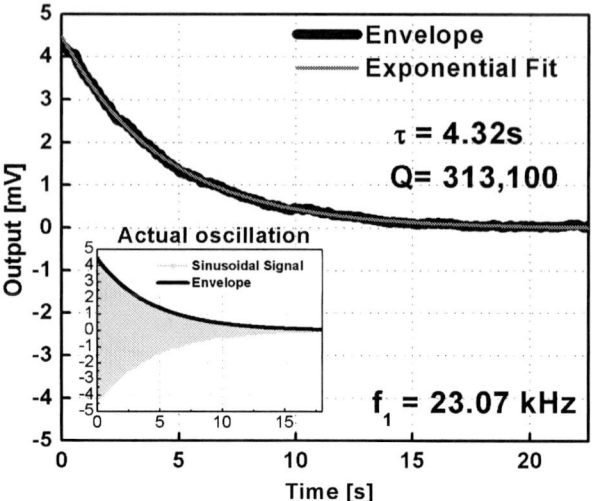

Figure 7: Ringdown test showing the decay of the oscillation amplitude following excitation at 2θ wineglass mode f_1 = 23.07 kHz (τ = 4.32 sec, Q = 313,100 @ 16 μTorr). The inset graph shows the sinusoidal signal based on the ring-down envelope.

Table 1. Comparison between this work and previous work

	This work	Draper[7]	Michigan[4]	Irvine[6]
Q-factor	**313,100**	143,000	1,200,000	1,140,000
Q/Volume (m⁻³)	**$2.8*10^7$**	$2.4*10^7$	$3.6*10^5$	$3.8*10^4$
$f_{no-trim}$ (Hz)	**3**	7	6.7	14 (0.16)
(s)	**4.3**	2.4	43.6	3.2
Fabrication	**Batch**	Semi-Batch	Hand made	Batch
Design-flexibility (Aspect-ratio d/D >0.5)	**High (DRIE)**	Low	High to Middle	High to Middle

CONCLUSIONS

The CR has high angular gain and the potential for high Q arising from low anchor loss. The fabrication method demonstrated here, using SOI wafers and DRIE, results in high symmetry and precise geometry, demonstrated by small mismatch between the vibration modes. A Q of 313,100 is measured at the 23 kHz 2θ vibration mode, producing a ring-down time of 4.32 seconds with an excellent frequency mismatch of 3 Hz (130 ppm) between the degenerate wineglass modes without applying any tuning voltage. Annealing at 700 °C for 2 hours in a nitrogen atmosphere increased Q by a factor of 4x to 6x to a maximum value of 355,100. Observable change in the diamond surface suggests that

the increase in Q is due to the removal of a lossy surface layer of diamond.

ACKNOWLEDGEMENTS

Devices were fabricated in the UC Berkeley Marvell Nanofabrication Laboratory and the authors would like to thank the Laboratory staff and users for discussion regarding fabrication.

REFERENCES

[1] D.D. Lynch, "Hemispherical resonator gyro", *IEEE Trans. Aerospace Electronic Syst.*, AES 20, No. 4 pp. 414-444, 1984.

[2] H. Najar, M.-L. Chan, H.-A. Yang, L. Lin, D.G. Cahill, and D.A. Horsley, "High quality factor nanocrystalline diamond micromechanical resonators limited by thermoelastic damping", *APL*, vol. 104, 151903, 2014.

[3] H. Najar, A. Thron, C. Yang, S. Fung, K. van Benthem, L. Lin, and D.A. Horsley, "Increased thermal conductivity polycrystalline diamond for low-dissipation micromechanical resonators", *MEMS 2014*, pp. 628-631.

[4] J. Cho, T. Nagourney, A. Darvishian, B. Shiari, J.-K. Woo, and K. Najafi, "Fused silica micro birdbath shell resonators with 1.2 million Q and 43 second decay time constant", *Solid-State Sensors, Actuators and Microsystems Workshop*, pp. 103-104, 2014.

[5] J. Cho, J.K. Woo, J. Yan, R.L. Peterson, and K. Najafi, "Fused-Silica Micro Birdbath Resonator Gyroscope (μ-BRG)", *JMEMS*, vol. 23, no. 1, pp. 66-77, 2014.

[6] D. Senkal, M.J. Ahamed, S. Askari, and A.M. Shkel, "1 million Q-factor demonstrated on micro-glassblown fused silica wineglass resonators with out-of-plane electrostatic transduction", *Solid-State Sensors, Actuators and Microsystems Workshop*, pp. 68-71, 2014.

[7] J. Bernstein, M. Bancu, E. Cook, T. Henry, P. Kwok, T. Nyinjee, G. Perlin, B. Teynor, and M. Weinberg, "Diamond hemispherical resonator fabrication by isotropic glass etch", *Solid-State Sensors, Actuators and Microsystems Workshop*, pp. 273-276, 2014.

[8] A. Heidari, M.-L. Chan, H.-A. Yang, G. Jaramillo, P. T.-T., P. Fonda, H. Najar, K. Yamazaki, L. Lin and D.A. Horsley, "Hemispherical wineglass resonators fabricated from the microcrystalline diamond", *J. Micromech. Microeng.*, vol. 23, no. 12, 125016, 2013.

[9] R. Perahia, H.D. Nguyen, L.X. Huang, T.J. Boden, J.J. Lake, D.J. Kirby, R.J. Joyce, L.D. Sorenson, D.T. Chang, "Novel touch-free drive, sense, and tuning mechanism for all dielectric micro-shell gyroscope", *Solid-State Sensors, Actuators and Microsystems Workshop*, pp. 383-386, 2014.

CONTACT

*D.A. Horsley, tel: +1-530-341-3236;
dahorsley@ucdavis.edu

978-1-4799-7956-1/15 $31.00 © 2015 IEEE

DESIGN, FABRICATION, AND CHARACTERIZATION OF A MICROMACHINED GLASS-BLOWN SPHERICAL RESONATOR WITH IN-SITU INTEGRATED SILICON ELECTRODES AND ALD TUNGSTEN INTERIOR COATING

Joan Giner[1], Jason M. Gray[2], Jonas Gertsch[2], Victor M. Bright[2], Andrei M. Shkel[1]

[1]University of California, Irvine, CA, USA
[2]University of Colorado, Boulder, CO, USA

ABSTRACT

We propose a new approach for integration of electrodes as a part of the micro glass-blown spherical resonator fabrication process as well as an ALD metallization of the inner side of the spherical shell enabling electrostatic conduction. We use a 500µm thick silicon electrode whose electrostatic gap width, defined during the glass blowing process, is not limited by lithographic effect and enables sub-micron gap possibilities. In addition, we introduce a metallization technique based on ALD of tungsten on the inner side of the shell. 35:1 aspect ratio electrodes demonstrated to excite a 500µm radius spherical resonator with an operating frequency of 1.66MHz and a Q factor above 2700 in vacuum (0.4mT)

INTRODUCTION

Micro-Electro-Mechanical Systems (MEMS) resonators are desirable for a wide variety of applications, including signal processing, timing, frequency control, and inertial sensing [1]. Several silicon MEMS resonators are currently on the market [2], and the vast majority of MEMS resonators are fabricated using silicon as a structural material, photolithography and DRIE techniques for defining the features. The dimensional resolution is due to fabrication tolerances introduced by etching, such as DRIE-induced scalloping, surface roughness, and the intrinsic limits of the aspect ratio of features. As a result, the fabrication of highly symmetric, frequency-matched devices with high quality factors becomes extremely challenging. These factors motivate the investigation of alternative fabrication approaches that allow the development of 3D MEMS resonator architectures with increased symmetry, reduced roughness, and increased aspect ratios—all accomplished simultaneously.

Recently, there has been significant interest in the development of 3D MEMS spherical and hemispherical resonators for use in timing and inertial sensing applications. Photolithography and DRIE based approaches have been used to fabricate silicon oxide hemispherical resonators using silicon molds, resulting in quality factors (Q) as high as 20,000 at a resonant frequency of 22kHz [3]. The same fabrication approach was demonstrated to build polysilicon resonators for inertial application with Q's around 8000 at a resonant frequency of 416kHz [4]. Plastic deformation of metallic glasses to achieve spherical structures has been explored by using a blow-molding technique [5]. The use of low

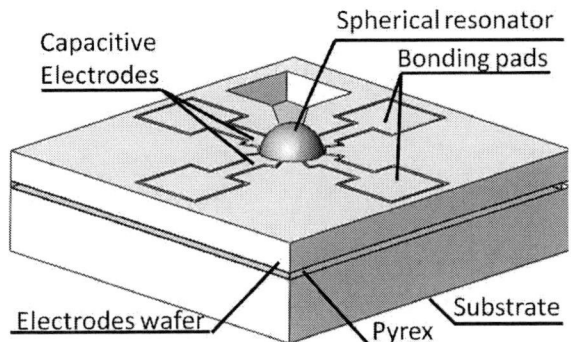

Figure 1 3D diagram of the spherical resonator with integrated silicon electrodes.

Thermo-Elastic Damping (TED) materials, such as Fused Silica (FS), as structural material has also being explored; Q factors above 100,000 at frequencies around 10kHz have been demonstrated [6]. Processing of these materials requires temperatures in excess of 1,600°C. A highly parallel batch fabrication glass-blowing process developed at the UC Irvine Microsystems Laboratory allows shaping of fused silica glass and Pyrex to create highly symmetric structures with resonant frequencies in the range of kHz [7] achieving outstanding Q factors on the level of 1M and axial symmetries with Δf less than 1Hz.

MEMS resonators for stable time applications require higher frequencies in order to reject the effect of the external acceleration. We fabricated quasi-spherical structures using micro glass-blowing techniques [8] in our implemented 3D spherical resonators demonstrated low order resonant frequencies about 1MHz for resonators [9] with improved thermal stability[10]

Instrumentation of spherical vibratory elements requires fabrication of electrostatic electrodes for actuation and sensing. For instance, assembled out-of-plane fold electrodes are used to actuate a 1.2M Q factor wine-glass shell [11]. In-plane assembled silicon electrodes are also used to excite SiO_2 hemispherical shells at 113KHz [12], however assembly approaches require extra fabrication steps that increase process complexity. Others developed integrated alternatives by sputtering and etching metal electrodes around the shell [3] or doping some areas of the silicon handle wafer [4]. With these techniques, electrodes as thick as several microns can be fabricated, but they do not take advantage of the large transduction area that the use of 3D spherical and hemispherical shells provide.

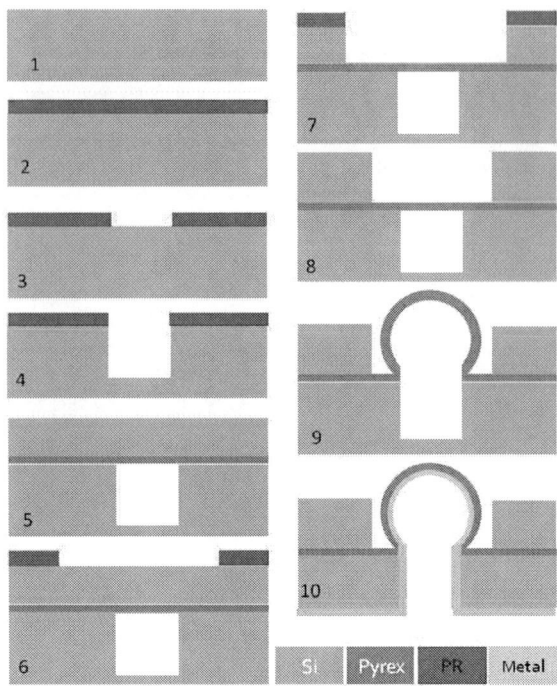

Figure 2 Fabrication flow of the spherical resonator with integrated electrode.

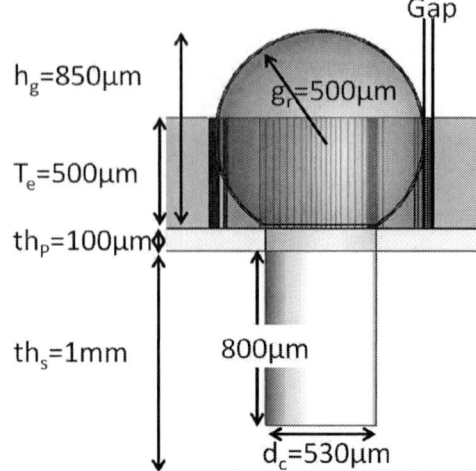

Figure 3 Cross-section diagram of the spherical resonator with the electrostatic gap between the shell and the electrode with the design parameters.

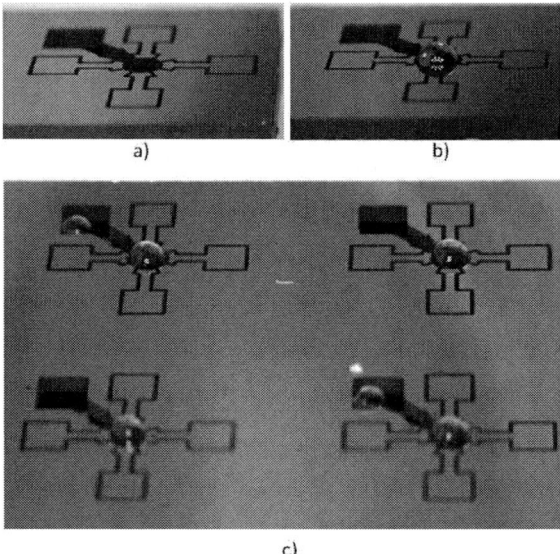

Figure 4 a) shell before glassblowing, b) shell after glass-blowing. c) Wafer level process.

In this paper we present the design, fabrication and test of high aspect ratio integrated silicon electrodes for quasi-spherical glass-blown Pyrex resonators. The silicon electrodes are designed according to the glass blowing dynamics to enable narrow-gap, high aspect ratio electrodes. Electrical connection to the shell is provided by Atomic Layer Deposition (ALD) of tungsten inside the shell, providing a homogeneous thin metal layer to preserve the geometric advantages of the spherical resonator.

DEVICE

The device (Figure 1) consists of a g_r=500μm radius micro glass-blown Pyrex spherical resonator with four electrostatic electrodes with thickness T_e=500, two for differential excitation and two for differential detection.

The distance from the sidewall of the electrode to the center of the shells is 520μm and is defined by applying the equation of the dynamics of the glassblowing process that defines the radius of the shell as a function of the air cavity size and environmental conditions such as temperature and pressure. After the glassblowing process the inner part of the shell is coated with ALD metal. To generate electrostatic force and to drive vibration, the resonator is polarized with a DC voltage while an AC voltage is applied to the excitation electrodes. When the frequency of the AC signal corresponds to the natural frequency of the shell, the electrostatic force induce mechanical vibration in the resonator. The shells is coated in the inner side by ALD. The atomic layer deposition coating from the back side proves to be the effective method to electrically separate the electrodes from the shell and to allow for electrostatic actuation. The transduction takes advantage of the large area of the spherical resonator significantly increasing the aspect ratio compared with conventional 2D approaches.

FABRICATION

Spherical Resonator with integrated electrodes

The fabrication of the spherical resonator with integrated electrodes (Figure 2) starts with the etching of circular cavities in a 1mm thick bare prime silicon wafer using a 24μm AZ4620 positive photo resist mask. The cavities, with 265nm radius and a depth of 800μm, contain the air at atmospheric pressure that facilitates the glass-blowing of the shell; therefore its volume will define the size and thus the resonant frequency of the spherical resonator. Figure 3 shows the schematic of the cross-section of a spherical shell with integrated electrodes showing the design parameters used to

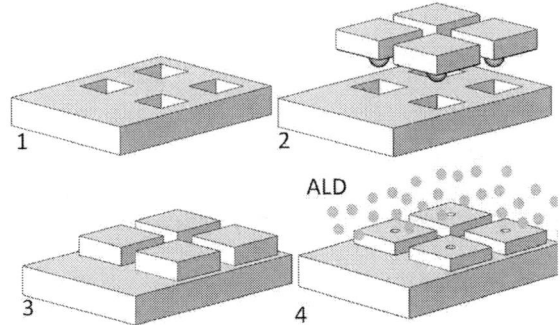

Figure 5 Diagram of the fabricated device holder used for backside etching and ALD deposition.

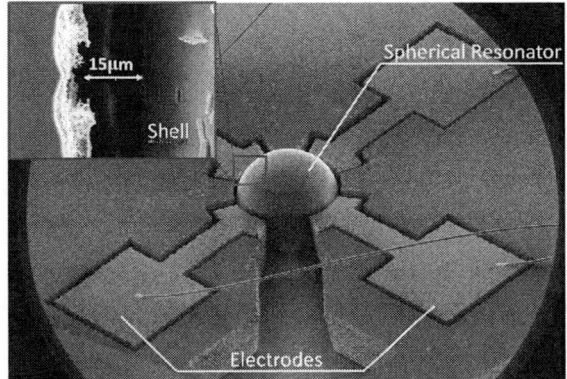

Figure 6 SEM micrograph of the 500um radius spherical shells with integrated electrodes.

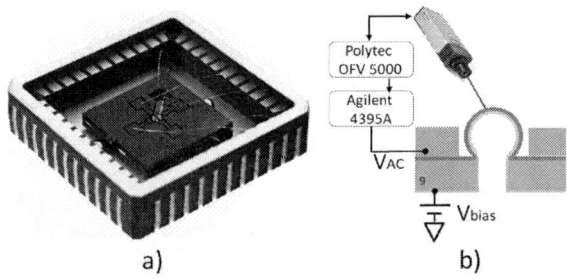

Figure 7 a) Packaged spherical resonator and characterization set-up schematic for electrostatic actuation and optical detection(b).

fabricated the spherical resonator. Deep Reactive Ion Etching (DRIE) is used to open the air pocket on the silicon wafer. After striping the photoresist, the drilled handle wafer, a 100 μm Pyrex wafer, and a double-sided polished 500 μm silicon wafer are cleaned using an RCA-1 solution to remove any contaminants. Wafers are stacked together under weight to assure intimate contact of the interfaces. Then, AC anodic bonding with a 0.1 Hz frequency and 800 V_{pp} is used to bond the three wafers. The electrode mask is patterned on the 500 μm silicon wafer. The glass-blowing process is carried out inside the annealing furnace in a nitrogen environment at 880 °C. Figure 4 shows a single device before (a) and after the glassblowing process (b). After glassblowing, 500μm radius spherical shells with 40μm shell thickness around the equator were obtained. Figure 4 (c) shows the wafer

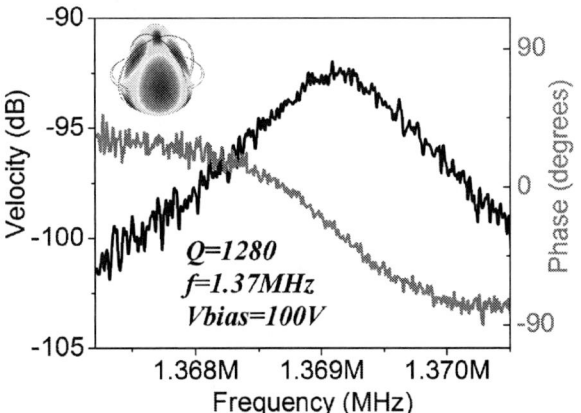

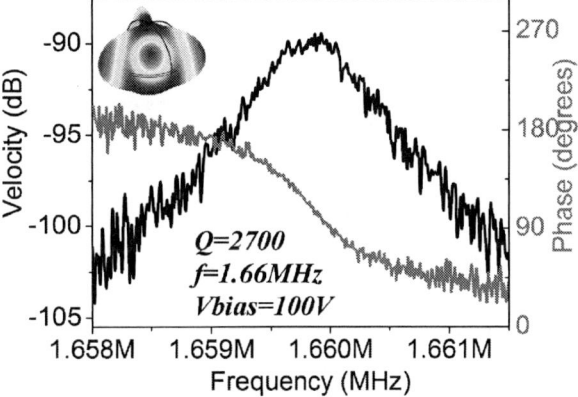

Figure 8 Characterization results of the electrostatic actuated spherical resonator. N=2 and N=3 resonant modes are shown.

level finalized fabrication process of spherical shells with integrated silicon electrodes. Note that the shells are not coated in the picture.

ALD Tungsten Deposition

After fabrication of the spherical resonator with integrated electrodes, metallization of the interior of the spherical shells has to be done in order to electrically connect the non-conductive glass shells. We do this in two steps. The first step consists of facilitating access to the inner side of the shell by back-side etching the silicon substrate. We fabricated a device holder (Figure 5). Squared cavities were dry etched in silicon (1), then shells are inserted in the cavities in order to expose the silicon substrate and to protect the shells during the dry etching step (2-3) Backside of the resonators is plasma etched until the air cavity is fully open. The same silicon holder is used to protect the top side of the shells from the ALD deposition sealing them using Van der Walls force. Atomic Layer deposition took place in a viscous flow ALD reactor at 130 °C with N_2 as the purge and carrier gas. A 3.6 nm aluminum oxide (Al_2O_3) ALD base layer, using trimethylaluminum and water precursors, was used to promote tungsten (W) nucleation. This was followed by a 20 nm deposition of ALD W using disilane (Si_2H_6) and tungsten hexafluoride (WF_6) chemistries. The metal film was capped with an additional 1.2 nm of Al_2O_3 to prevent

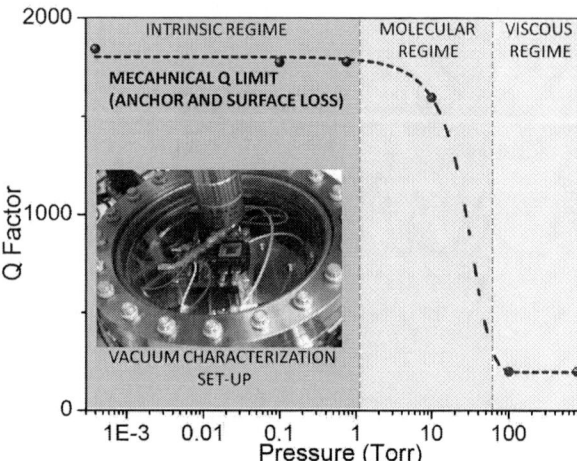

Figure 9. Measurement of the Q factor of the Pyrex spherical resonator as function of vacuum.

metal oxidation.

CHARACTERIZATION

Figure 6 shows a SEM micrograph of the coated spherical resonator with 15µm electrostatic gaps (inset in Figure 6). Resonators in Figure 7 a) were evaluated in a custom built vacuum chamber at 0.4mT with an optical port. For experimental characterization the input signal is generated with a network analyzer (Agilent 4395A) and divided in a balanced (0° phase shift) and unbalanced (180° phase shift). The frequency response was measured using a Polytec OFV5000 single-point Laser Doppler Vibrometer. Figure 8 shows the frequency response of the N=2 mode, and N=3 mode resonator frequency at 1.37 MHz and 1.66 MHz, respectively. Results on characterization of the shells in vacuum (Figure 9) show a maximum on the quality factor at the intrinsic regime (<1 Torr) and suggest that the Q factor is limited by material impurities and anchor losses, given that modeled thermoelastic dissipation Q factors (Q_{TED}) of Pyrex are around 10^8.

CONCLUSION

We have demonstrated the fabrication of operational integrated silicon transductors along with glass-blown spherical resonator fabrication. ALD of tungsten on the inner side of the spherical shells is used to electrically connect the non-conductive shell and to isolate it from the electrodes. The combination of the silicon electrodes and ALD enable 3D MEMS resonator transduction, taking advantage of the large transduction areas that such resonator. 35:1 aspect ratio electrodes have been fabricated to excite spherical resonators. Operational frequencies in the High Frequency range with Q factor around 2700 have been demonstrated.

ACKNOWLEDGEMENTS

This work was supported by the DARPA Microsystems Technology Office under contract N66001-10-1-4074. The authors would like to thank the clean room staff at University of California, Irvine and University of California Los Angeles, for their assistance in the fabrication.

REFERENCES

[1] C. T. C. Nguyen, "MEMS technology for timing and frequency control." *IEEE Transactions on Ultrasonics, Ferroelectrics, and Frequency Control*, vol. 54, pp. 251–270, 2007.

[2] S. Tabatabaei and A. Partridge, "Silicon MEMS Oscillators for High-Speed Digital Systems," *IEEE Micro*, vol. 30, no. 2, pp. 80–89, 2010.

[3] P. Pai, F. K. Chowdhury, H. Pourzand, and M. Tabib-Azar, "Fabrication and testing of hemispherical MEMS wineglass resonators," *IEEE 26th Int. Conf. on Micro Electro Mechanical Systems (MEMS), 2013* pp. 677–680.

[4] L. D. Sorenson, X. Gao, and F. Ayazi, "3-D micromachined hemispherical shell resonators with integrated capacitive transducers," *IEEE 25th Int. Conf. on Micro Electro Mechanical Systems (MEMS), 2012*, pp. 168–171.

[5] B. Sarac, G. Kumar, T. Hodges, S. Ding, A. Desai, and J. Schroers, "Three-Dimensional Shell Fabrication Using Blow Molding of Bulk Metallic Glass," *IEEE/ASME Journal of Microelectromechanical Systems*, vol. 20, pp. 28–36, 2011.

[6] J. Cho, J. Yan, J. A. Gregory, H. Eberhart, R. L. Peterson, and K. Najafi,"High-Q fused silica birdbath and hemispherical 3-D resonators made by blow torch molding," *IEEE 26th Int. Conf. on Micro Electro Mechanical Systems (MEMS)*, 2013, pp.177–180.

[7] D. Senkal, J. M. Ahamed, A. A. Trusov, and A. M. Shkel, "Demonstration of Sub-1Hz Structural Symmetry in Micro-Glassblown Wineglass Resonators with Integrated Electrodes," *IEEE Transducers Conference*, 2013.

[8] E. J. Eklund and A. M. Shkel, "Glass Blowing on a Wafer Level," *IEEE/ASME Journal of Microelectromechanical Systems* vol. 16, no. 2, pp. 232–239, 2007.

[9] I. P. Prikhodko, S. A. Zotov, A. A. Trusov, and A. M. Shkel, "Microscale Glass-Blown Three-Dimensional Spherical Shell Resonators," *IEEE/ASME Journal of Microelectromechanical Systems* vol. 20, no. 3, pp. 691–701, 2011.

[10] J. *Giner*, L. Valdevit, A. M. Shkel, "*Glass-Blown Pyrex resonator with* Ti Coating for *Reduction of* TCF", IEEE *Int. Symp. of Inertial Sensors and Systems* 2014,

[11] D. Senkal, M. J. Ahamed, S. Askari, and A. M. Shkel, "1MillionQ-Factor Demonstrated On Micro-Glassblown Fused Silica Wineglass Resonators With Out-Of-Plane Electrostatic Transduction," *Solid-State Sensors, Actuators and Microsystems Workshop*, Hilton Head,, 2014, pp. 68–71.

[12] P. Shao, L. D. Sorenson, X. Gao, and F. Ayazi, "Wineglass-on-a-chip," *Solid-State Sensors, Actuators, and Microsystems Workshop*, Hilton Head, 2012, pp. 275–278.

CONTACT

*J.Giner, tel: +1-562-277-4787; jginerde@uci.edu

IN-SITU OVENIZATION OF LAMÉ-MODE SILICON RESONATORS FOR TEMPERATURE COMPENSATION

Yunhan Chen[*], *Eldwin J. Ng, Yushi Yang, Chae Hyuck Ahn, Ian Flader, and Thomas W. Kenny*
Stanford University, Stanford, California, USA

ABSTRACT

This paper reports an inside-encapsulation ovenization method for the temperature compensation of Lamé-mode epi-sealed silicon resonators. With this method, the square Lamé-mode resonator itself acts both as a thermometer and a heater, which allows for simultaneous *in situ* sensing and control of the operating temperature. In this device, only the resonating element is heated, reducing the power consumption and the thermal time constant, relative to approaches which control the temperature of an entire MEMS chip or system. Preliminary results of real-time frequency compensation achieve a frequency stability of ~5ppm over -40~+80°C without the need for sophisticated control schemes.

INTRODUCTION

MEMS resonators have gained great attention as an alternative to quartz crystal resonators for timekeeping and communication applications due to their benefits such as cost, size and power consumption reduction, IC compatibility, and better aging performance [1, 2]. Temperature sensitivity is, however, one of the shortcomings that has often precluded MEMS resonators from being considered in high-precision applications such as navigation and radar. Single crystal silicon (SCS) MEMS resonators typically have large temperature coefficients of frequency (TCf) of about -30ppm/°C, and thus require compensation to achieve temperature stability that is comparable to quartz crystal oscillators for high-precision applications.

Various compensation schemes have been reported in the past, which generally can be categorized into passive and active methods. Passive compensation methods include the use of stress [3], variable gaps [4], composite materials [5], or doping [6] to compensate temperature effects. While passive compensation can effectively reduce the frequency deviation to the ppm range without consuming extra power, active temperature control methods have the potential to achieve a better frequency stability [7-12]. Active compensation methods usually consist of two steps: sensing and tuning. The device temperature is firstly sensed by a "thermometer" of some form. Previous works have demonstrated the use of quality factor, frequency difference between two resonators, or even a separate resistive device as a thermometer [9, 11, 12]. According to the measured device temperature, tuning/compensation methods, which include electrostatic tuning [7, 8], variable frequency multiplication [9], etc., are then applied to cancel the temperature effects on resonance frequency.

Ovenization is one of the active compensation methods that incorporates micro-ovens which utilize Joule heating to elevate and maintain the resonators at a desired temperature, and, as a result, stabilizes the resonant frequency. Ovenized MEMS resonators have been

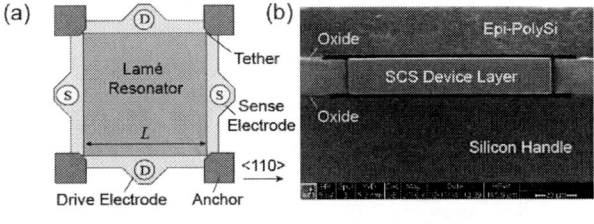

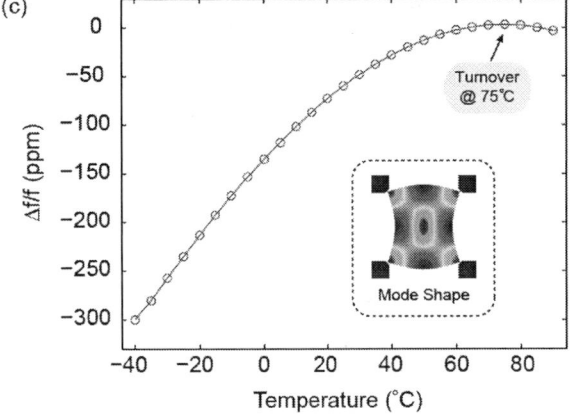

Figure 1: (a) Design of the Lamé-mode resonator; (b) A cross section SEM image of an epi-sealed MEMS device; (c) Frequency shift as a function of temperature for unheated devices.

demonstrated in [10-12]. In fact, the same concept has also been used for decades in quartz crystals as Oven Controlled Quartz Crystal Oscillators (OCXO) which are the highest precision mechanical-resonance-based oscillator available.

In this paper, we demonstrate a new method of ovenizing Lamé-mode silicon resonators where the 4-point measurement of the *in situ* resistance of the square resonator serves as the temperature measurement. The same current that is used for resistance measurement heats the resonator directly and is controlled to maintain an elevated target operating temperature.

DESIGN AND FABRICATION

The Lamé-mode resonator is designed to have a 400µm square mass aligned to the <110> crystal direction, and anchored at the four nodal points by tethers. Four electrodes are placed at the four sides, two of which are used to drive the resonators and the other two are used for sensing output current signals. The entire resonator is encapsulated in an ultra-clean cavity sealed with epitaxial polysilicon (Figure 1b). The *epi-seal* encapsulation process was proposed by researchers at the Robert Bosch Research and Technology Center in Palo Alto and then demonstrated in a close collaboration with Stanford

Table 1: Device parameters.

Geometry	L (μm)	400
	h (μm)	17
	Tether L (μm)	7.6
	Tether W (μm)	2.4
	Orientation	<110>
	Etch Hole	Without
Doping	Dopant	P/Boron
	Concentration (cm⁻³)	1.7e20
Measurement	f (MHz)	10.044885
	Q	1.7M

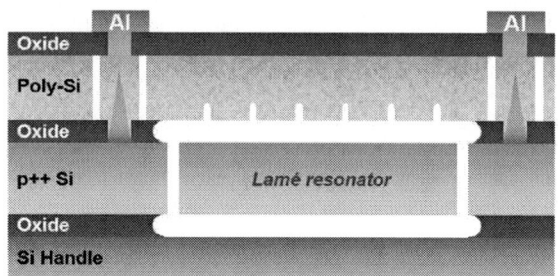

Figure 2: Final cross section view of the encapsulated Lamé-mode resonator.

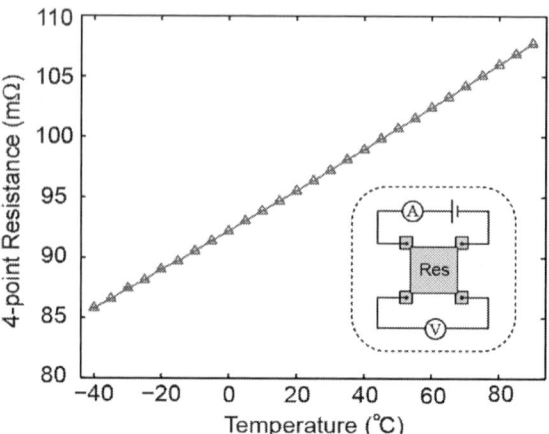

Figure 3: 4-point measured resistance as a function of temperature.

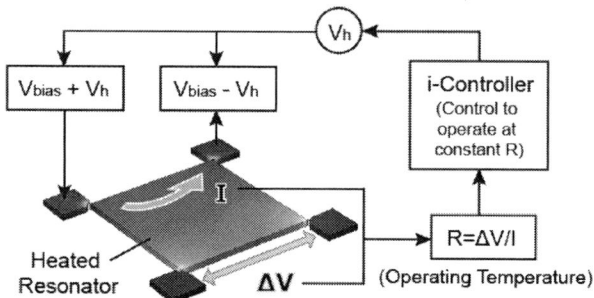

Figure 4: Schematic of the temperature compensation system. Colors are indicative of the temperature, blue: cold, red: hot. 4-point resistance measurement is used as an indicator of the device temperature.

University. This collaboration is continuing to develop improvements and extensions to this process for many applications, while the baseline process has been brought into commercial production by SiTime Inc., recently acquired by Megachips Inc. The design and parameters of the Lamé-mode resonator are illustrated and listed in Figure 1a and Table 1 respectively. The devices used in this work are passively compensated by using highly doped p-type single crystal silicon. In Figure 1c, the TCf curve of a highly doped p-type Lamé resonator with <110> orientation shows a frequency deviation of 300ppm across the -40 to +85°C temperature range with a turnover point at ~+75°C. Operating near the turnover point gives better control stability as the resonator is insensitive to control errors.

The devices are fabricated using an etch hole free *epi-seal* fabrication process reported in [13], which provides exceptional f*Q for high performance. Figure 2 illustrates the final cross section view of the device. The resonator is sealed in high temperature by a layer of epitaxial polysilicon cap immediately after vapor HF releasing the device from the surrounding oxide. Electrical vias are also defined for the electrodes. A cavity pressure of <1Pa is achieved by diffusing out the residual hydrogen. This process provides an ultra-clean and low pressure environment, which allows the resonators to achieve long-term stability.

METHOD

Here we propose an *in situ* ovenization method which utilizes the 4-point measured resistance as an indication of device operating temperature. The Lamé resonator

conveniently is one of the typical structures used for van de Pauw electrical resistivity measurements. As shown in Figure 3, we passed a small current between two adjacent anchors of the resonator and measured the voltage difference between the other two anchors. The 4-point measured resistance can thus be calculated by Ohm's law $R = \Delta V/I$. The resistance, as shown, has a linear correlation with temperature, which serves as a thermometer for measuring the temperature of the device itself.

Figure 4 schematically shows the proposed *in situ* ovenization method. Bidirectional offset voltages ($\pm V_h$) are imposed on the bias voltage (V_{bias}) and connected to two of the anchor electrodes respectively, resulting a current (I) flowing through and heating up the resonator, as well as causing a voltage difference (ΔV) between the other two anchor electrodes. The resistance R can thus be measured as $\Delta V/I$. A feedback controller is used to control the value of V_h to maintain a constant resistance value R in order to operate at the target operating temperature. In this work, a simple integral controller (i-Controller) was used. The error between target and measured resistance values was summed over time. The updated V_h can thus be calculated as the summed error multiplied by a constant integral gain.

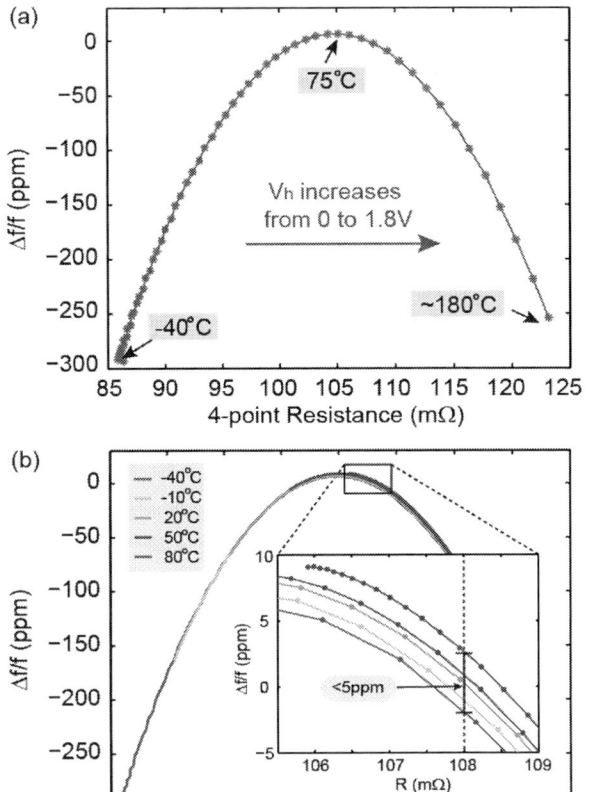

Figure 5: (a) Frequency deviation and resistance value change as offset voltage V_h increases. The ambient temperature is controlled to be -40°C; (b) Measured resistance vs frequency deviation at different ambient temperatures and heating currents;

RESULTS

Heating

The heating properties are firstly investigated. We placed the device into an oven with temperature controlled at -40°C. By increasing the offset voltage V_h, we can increase the current, and thus raise the device temperature, which results in changing the resonant frequency. The resonant frequency is extracted through the peak frequency of an open-loop frequency sweep using a network analyzer. The actual device temperature then can be evaluated by comparing with the TCf curve (Figure 1c).

As shown in Figure 5a, the resonant frequency changes in the same fashion with the TCf curve when V_h increases, indicating the device is indeed being heated. Particularly, when V_h=1.8V, i.e. the voltage difference between two anchors are 3.6V, the resonator is heated up to ~180°C, much higher than the turnover point temperature, which shows a sufficient heating capability even for the case where the ambient temperature is as low as -40°C.

The same characterization has been conducted in different ambient temperatures as shown in Figure 5b. The curves follow the TCf curve but start from the frequency corresponding to the particular ambient temperature. This agrees with the intuitive understanding that less current is

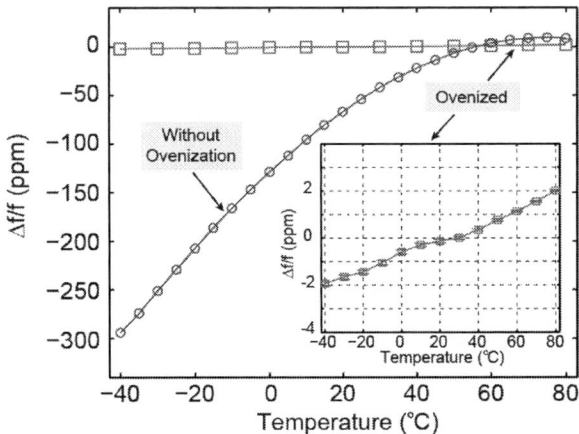

Figure 6: Measurements of steady-state frequency deviation of the ovenized device when operated at target resistance R = 108mΩ, compared against the device without ovenization.

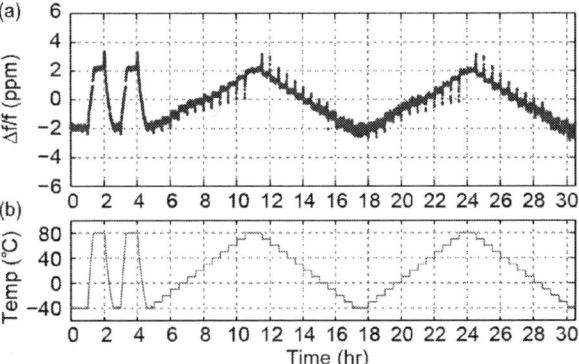

Figure 7: (a) Real-time frequency measurement of the ovenized Lamé-mode resonator subject to 5 °C/min ambient temperature ramps; (b) Real-time measurement of the ambient temperature.

required to heat the resonator to the same temperature when the ambient temperature is higher. Ideally, the curves for different ambient temperatures should all collapse into one identical curve. It, however, has been observed that for different ambient temperatures, the curves slightly shift. This shift of the R-f curve is suspected to be due to stress effects and inefficient thermal isolation. As the result, a residual frequency deviation of <5ppm is observed as shown in the zoomed-in figure, when operating at constant R = 108mΩ (~90°C). Hence the same amount of frequency deviation is expected when the resonator is actively compensated under a closed-loop control.

Stability with Closed-loop Control

Now we operate the Lamé resonator with the proposed closed-loop control. As we want to operate the device over an ambient temperature range of -40~+80°C, an operating device temperature higher than 80°C is required. Moreover, operating near the turnover point gives better control stability due to smaller second order temperature coefficient. Hence the target resistance value R = 108mΩ was chosen.

Figure 6 compares the steady-state T-f curve for resonators with and without ovenization. The frequency stability is improved by more than 100X. The inset figure shows the mean and standard deviation of the time-averaged performance. A residual frequency deviation of ~4ppm exists which agrees with our expectation discussed before.

Furthermore, a real-time frequency measurement of the compensated Lamé resonator subject to 5°C/min ambient temperature ramps is shown in Figure 7. A frequency stability of ~5ppm is achieved over the temperature range of interest, and no frequency hysteresis is observed for both ramping up and down the ambient temperature. It is noticed that relatively large frequency transients from expected values happened when the ambient temperature ramped up or down. This is due to the slow response time of the controller. Better temperature stability is expected with improvements on control schemes. Another way to improve the performance is to design the resonator with a different doping level to have a turnover point on the TCf curve that is higher than the target temperature range (in this case, 80°C). Operating at the turnover point minimizes the frequency change due to control errors.

The overall Joule heating power consumption is estimated to be around 13~140 mW for an ambient temperature of 80~-40°C. Furthermore, a thermal time constant of 11ms is measured.

CONCLUSION

Demonstrated in this work is a new method of ovenizing Lamé-mode silicon resonators where *in situ* temperature sensing and heating with a closed control loop actively compensate the temperature effects on resonance frequency. Preliminary results of real-time frequency compensation achieve a frequency stability of ~5ppm over -40~+80°C and over a time range of 30hr. This work shows the potential of MEMS resonator ovenization for high-precision applications. Future work will focus on achieving better frequency stability, circuit integration and power consumption.

ACKNOWLEDGEMENTS

This work was supported in part by the Defense Advanced Research Projects Agency Precision Navigation and Timing Program managed by Dr. A. Shkel and Dr. R. Lutwak under Contract N66001-12-1-4260, and in part by the National Science Foundation through the National Nanotechnology Infrastructure Network under Grant ECS-9731293. We appreciate the benefit of regular conversations about high-stability time references with Dr. John Vig. Y. Chen is supported by a Stanford Graduate Fellowship, and C. H. Ahn was supported by a Kwanjeong Foundation Scholarship.

REFERENCES

[1] C. T.-C. Nguyen, "MEMS technology for timing and frequency control", *IEEE UFFC Transactions*, Vol.54, No.2, pp. 251-270, 2007.

[2] www.sitime.com

[3] W.-T. Hsu, C. T.-C. Nguyen, "Geometric stress compensation for enhanced thermal stability in micromechanical resonators", in *Proc. IEEE Int. Ultrason. Symp.*, Sendai, Japan, Oct. 5–8, 1998, pp. 945–948.

[4] W.-T. Hsu, C. T.-C. Nguyen, "Stiffness compensated temperature insensitive micromechanical resonators," in *Proc. 15th IEEE MEMS*, Las Vegas, NV, Jan. 20–24, 2002, pp. 731–734.

[5] R. Melamud, S. A. Chandorkar, B. Kim, H. K. Lee, J. C. Salvia, G. Bahl, M. A. Hopcroft, and T. W. Kenny, "Temperature-Insensitive Composite Micromechanical Resonators", *J. Microelectromech. Syst.*, Vol. 18, No. 6, pp. 1409-1419, 2009.

[6] E. J. Ng, V. A. Hong, Y. Yang, C. H. Ahn, C. L.M. Everhart, and T. W. Kenny, "Temperature Dependence of the Elastic Constants of Doped Silicon", *J. Microelectromech. Syst.*, 2014.

[7] G. K. Ho, K. Sundaresan, S. Pourkamali, and F. Ayazi, "Temperature compensated IBAR reference oscillators," in *Proc. 19th IEEE MEMS*, Istanbul, Turkey, Jan. 22–26, 2006, pp. 910–913.

[8] H. K. Lee, M. A. Hopcroft, R. Melamud, B. Kim, J. Salvia, S. Chandorkar and T. W. Kenny, "Electrostatic-tuning of hermetically encapsulated composite resonator", in *Proc. Hilton Head Workshop*, Hilton Head, SC, Jun. 8-12, 2008, pp. 48-51.

[9] R. Melamud, P. M. Hagelin, C. M. Arft, C. Grosjean, N. Arumugam, P. Gupta, G. Hill, M. Lutz, A. Partridge, F. Assaderaghi, "MEMS Enables Oscillators with sub-ppm Frequency Stability and sub-ps Jitter" in *Proc. Hilton Head Workshop*, Hilton Head, SC, Jun. 3-7, 2012, pp. 66-69.

[10] C. T.-C. Nguyen, R. T. Howe, "Microresonator frequency control and stabilization using an integrated micro oven", in *Digest Tech. Papers Transducers'93 Conference*, Yokohama, Japan, Jun. 7-10, 1993, pp. 1040-1043.

[11] M. A. Hopcroft, H. K. Lee, B. Kim, R. Melamud, S. Chandorkar, M. Agarwal, C. M. Jha, J. Salvia, G. Bahl, H. Mehta, and T. W. Kenny, "A High-Stability MEMS Frequency Reference", in *Digest Tech. Papers Transducers'07 Conference*, Lyon, France, Jun. 10-14, 2007, pp. 1307-1309.

[12] J. C. Salvia, R. Melamud, S. A. Chandorkar, S. F. Lord, and T. W. Kenny, "Real-Time Temperature Compensation of MEMS Oscillators Using an Integrated Micro-Oven and a Phase-Locked Loop", *J. Microelectromech. Syst.*, Vol. 19, No. 1, pp. 192-201, 2010.

[13] E. J. Ng, Y. Yang, Y. Chen, and T. W. Kenny, "An etch hole-free process for temperature-compensated, high Q, encapsulated resonators" in *Proc. Hilton Head Workshop*, Hilton Head, SC, Jun. 8-12, 2014, pp. 99-100.

CONTACT

*Y. Chen, email: yunhanc@stanford.edu

ON-CHIP CHARACTERIZATION OF STRESS EFFECTS ON GYROSCOPE ZERO RATE OUTPUT AND SCALE FACTOR

Erdinc Tatar[1], Tamal Mukherjee[1], and Gary K. Fedder[1, 2]
[1]Department of Electrical and Computer Engineering
[2]The Robotics Institute, Carnegie Mellon University, Pittsburgh, USA

ABSTRACT

Stress effects on performance are quantified for a vacuum packaged silicon-on-insulator (SOI) MEMS gyroscope, including zero rate output (ZRO), scale factor (SF), and resonance frequencies. On-chip environmental sensors comprising released SOI-silicon resistors in a bridge measure the temperature and stress separately. Experimental results from a four-point bending test-bed lead to a system model to compensate the gyroscope ZRO using the environmental sensor outputs. The ZRO shift varied linearly with stress, with a measured maximum of 3.5 °/s for 533 kPa applied external stress.

INTRODUCTION

Achieving long term MEMS gyroscope stability and sub-degree per hour bias instability requires compensation of the gyroscope output with environmental changes such as temperature [1]. Slowly varying environmental conditions (e.g., temperature and stress) limit the gyroscope stability over long run times. Gyroscope bias compensation with temperature [1, 2], and quadrature signal [2] have been reported to improve the long term MEMS gyroscope stability. Measuring the temperature on-chip provides a better assessment of the actual gyroscope temperature compared to off-chip measurements [2]. However as the temperature changes, inevitable thermal coefficient of expansion (TCE) mismatches between the MEMS die and surroundings exert stress as large as 50MPa on the die through the device spring anchors [3]. The common approach for gyroscope compensation considers temperature as the only environmental variable [1-2]. This work focuses on characterizing the gyroscope bias by measuring the on-chip stress and temperature separately. Knowing the stress and temperature independently may lead to a more effective bias compensation.

Our previous works have demonstrated temperature driven stress effects on gyroscope resonance frequencies [4], and pure stress effects on gyroscope resonance frequencies with three- and four-point bending test-beds [5]. References [4] and [5] demonstrate that MPa level stress affects the gyroscope resonance frequencies at levels comparable to temperature effects. Gyroscope resonance frequencies shift as a result of anchor displacements changing the spring constant.

A system level simulation technique was developed relating stress to ZRO and SF in a circuit simulation environment [6]. Stress leads to drive and sense comb mismatches through non-equal anchor displacements of the stator and rotor that generate a ZRO signal indistinguishable from the Coriolis signal. This work uses a four-point bending test-bed similar to [5] but adds on-chip stress and temperature sensors to the SOI-MEMS gyroscope and experimentally validates the simulation trends in [6]. The gyroscope is vacuum packaged by using an in-house developed system and tested in the closed loop drive and open-loop sense configuration.

ON-CHIP STRESS SENSORS

Figure 1.a shows the SEM image of the three-fold-symmetric SOI-MEMS gyroscope along with the location of the stress sensors highlighted. The electrostatic combs are symmetric so either of the two modes can be used as drive or sense with the frequency tuning capability. The stress sensors are located at the four sides of the gyroscope. Each stress sensor comprises four released SOI-silicon resistors connected in a Wheatstone bridge configuration to cancel out the temperature variations as shown in Figure 1.d. Since the SOI device layer is highly doped, the resistors are meandered to have a high resistance in the constrained space. Figure 1.b presents the stress and temperature sensitive silicon resistors that are series connected fixed-fixed beams. The stress insensitive, temperature sensitive resistors are released folded silicon beams as in Figure 1.c. Alignment of the anchors determines the stress sensitivity. Each SOI-silicon resistor value is $1.7 \, k\Omega \pm 2\%$. The stress measurements were performed by applying a 5 V DC voltage across the bridge and reading the bridge output with a multimeter. Stress sensors at each side measure their respective axial component of the in-plane stress.

Figure 2 shows the oven test results of the resistance change vs. temperature for the stress sensitive and insensitive resistors. They both exhibit a positive temperature coefficient of resistance (TCR) of +0.0016 verifying the ability to cancel temperature with the bridge. Figure 3 shows the resistance changes for the stress sensitive and insensitive resistors when stress is applied to the die. The stress insensitive resistors do not exhibit any change proving their operation as pure temperature sensors.

Vacuum packaging of the gyroscope is critical in terms of testing and suppressing the Brownian noise. The gyroscope in this study is in-house vacuum packaged by using 40 pin off-the-shelf ceramic dual inline packages (DIP) and metal lids. Low temperature indium solder preforms are used to bond the lid and DIP. A titanium based getter, similar to [7], is deposited on the metal lid for long term stable vacuum. The vacuum packaged gyroscopes have quality factors (Q) varying between 4500 and 8000, and the variation is believed to be a result of the outgassing period before the packaging. Vacuum packaging has been successfully repeated on 5 gyroscope dies. The estimated pressure is in the rage of 150-300 mTorr with no leaks detected for more than one month.

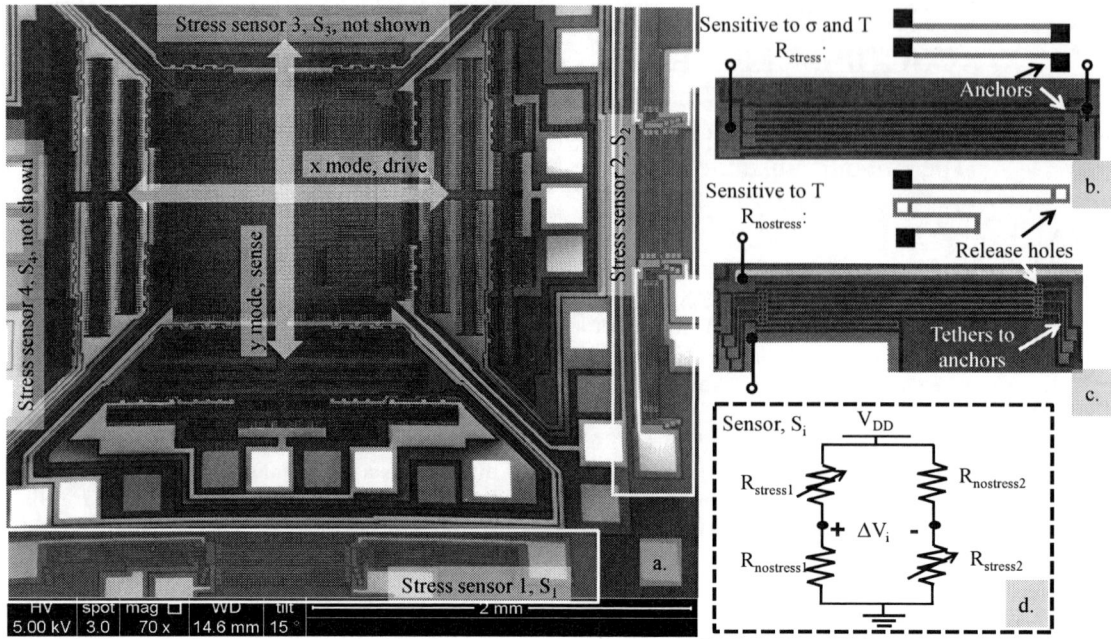

Figure 1: (a) Three-fold symmetric SOI-MEMS gyroscope highlighting the location of the stress sensor elements. (b) Stress and temperature sensitive fixed-fixed SOI-silicon beam resistor. (c) Stress-insensitive folded SOI-silicon beam resistor. (d) Wheatstone bridge.

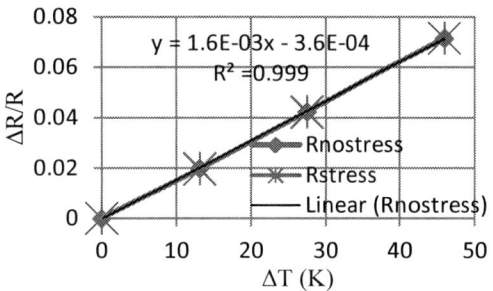

Figure 2: Resistance change vs. temperature test results in oven test, both resistors react similarly to the temperature verifying the Wheatstone bridge operation.

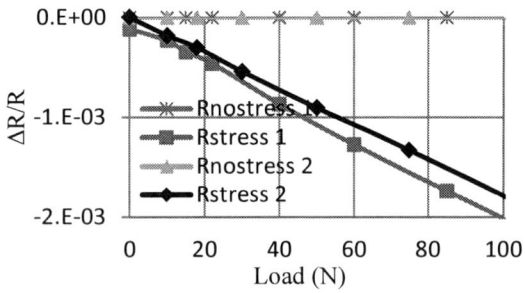

Figure 3: Resistance change with the stress applied to the die. Stress insensitive resistors do not react.

EXPERIMENTAL RESULTS

Although the two modes were identically designed in the layout, systematic frequency mismatches of 150-200 Hz have been observed in our previous designs [4]. Two possible reasons for the mismatch are a 30 nm raster-scanned mask linewidth mismatch or the 5° misaligned <111> silicon device layer. We have learned that although not specified, most commercial p-type <111> silicon wafers are cut 5° off axis. The frequency mismatch has been decreased by intentionally offsetting the beam width of springs of the corresponding mode as part of the design. The resonance frequencies for the gyroscope generating the results below are 8.6 kHz with 4 Hz mismatch without frequency tuning. Frequency mismatches as low as 2 Hz have been observed on the other dies.

The gyroscope is tested at a tuned 18 Hz mismatch under zero stress with $f_{drive} < f_{sense}$. A Zurich Instruments HF2LI lock-in amplifier was used to control the gyroscope. Transimpedance amplifiers on the PCB convert the gyroscope output current to voltage and the voltage outputs are fed to the lock-in amplifier for demodulation and control loops. The gyroscope is operated at closed loop drive with a PLL locking on the resonance frequency and amplitude stabilization loop and open loop sense mode.

Figure 4 shows the four-point bending stress test-bed used during the experiments along with the photograph of the vacuum packaged gyroscope. The gyroscope PCB is sandwiched in-between the steel cylinders that are held by the aluminum fixtures. Adding weights on top of the setup creates compressive bending stress on the gyro changing its characteristics. The goal of this test is finding the coefficients between the gyroscope ZRO and stress sensor outputs.

The weights ("load" in Figure 4) on the gyroscope are incremented with 5 lbs steps up to 15 lbs during the tests. At each step, gyroscope ZRO, stress on both axes and temperature (with a temperature sensing resistor) are recorded for 6 minutes, followed by a SF test on the rate table at ±5°/s and ±10°/s to compute the rate referred ZRO. Stress sensors 1 and 2 (S_1 and S_2 in Figure 1) are used capture the strain in the sense and drive axes, respectively. The stress test-bed was set up on the rate table for continuous testing. The primary drift source in the experiments is believed to be the external stress since the measured on-chip temperature variation was less than 0.02°C.

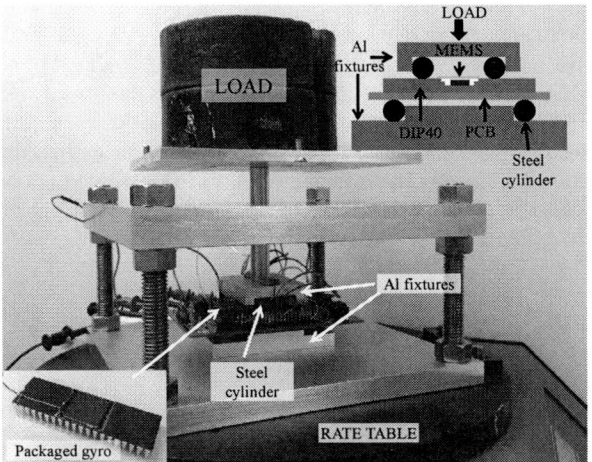

Figure 4: Four-point bending stress test-bed used to test the vacuum packaged gyroscope on the rate table. Adding weights on the gyroscope creates compressive bending stress.

Figure 5 shows the orientation of the die and transformation matrix that relates the measured stress and gyroscope parameters at constant temperature. This matrix is a step towards full model-based stress compensation. Each performance parameter has a dominant stress coefficient and a minor stress coefficient that is at least five times less sensitive. The dominant stress term for each gyroscope parameter is highlighted. The gyroscope is mounted 45° with respect to the package to result in nominally equal package-induced stress on both modes for better initial resonant frequency matching.

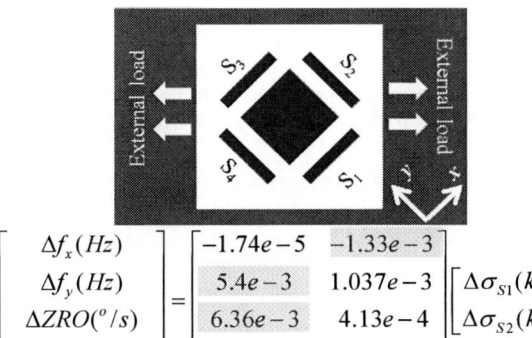

$$\begin{bmatrix} \Delta f_x(Hz) \\ \Delta f_y(Hz) \\ \Delta ZRO(°/s) \\ \Delta SF(\mu V/°/s) \end{bmatrix} = \begin{bmatrix} -1.74e-5 & -1.33e-3 \\ 5.4e-3 & 1.037e-3 \\ 6.36e-3 & 4.13e-4 \\ -2.74e-2 & -3.87e-3 \end{bmatrix} \begin{bmatrix} \Delta\sigma_{S1}(kPa) \\ \Delta\sigma_{S2}(kPa) \end{bmatrix}$$

Figure 5: Orientation of the die and the transformation matrix between stress and gyroscope parameters, each term is dominated by a single stress component.

Figures 6 and 7 show the resonance frequency vs. dominant stress for drive and sense modes, respectively. The stress sensor outputs provide an approximately linear estimation of the resonance frequency shifts from the baseline. The stress is negative since the four-point bending test exerts compressive stress on the die. Drive and sense frequencies primarily shift with stress along their respective orthogonal axis since their flexural spring

constant is primarily affected by longitudinal stress [4]. Both mode frequencies have a repeatable relationship with compressive stress; however the sign of drive frequency response is counterintuitive. A decrease in the resonance frequency is expected with compressive stress. Although the 45° mounting equalizes the stress, we believe the drive mode is pre-stressed due to the possible deviations from the 45° angle and variation in the die bonding. The relatively low stress (< 1MPa) applied during the tests is not sufficient for the drive frequency to start decreasing.

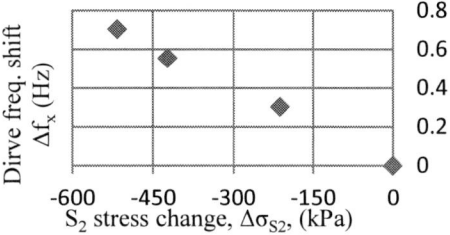

Figure 6: Linear drive mode resonance frequency shift vs. drive axis stress(S_2).

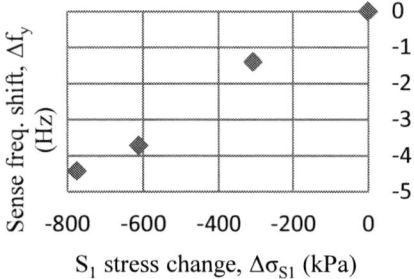

Figure 7: Linear sense mode resonance frequency shift vs. sense axis stress (S_1).

Figure 8 presents the SF test results as a function of the dominant sense mode stress (S_1) showing the increase with increasing stress. Since $f_{drive} < f_{sense}$ and the sense resonance frequency decrease is larger than drive frequency increase, reduced mismatch leads to increasing SF in open-loop operation. SF relates mainly to S_1 because the sense frequency shift is considerably larger than the drive frequency shift.

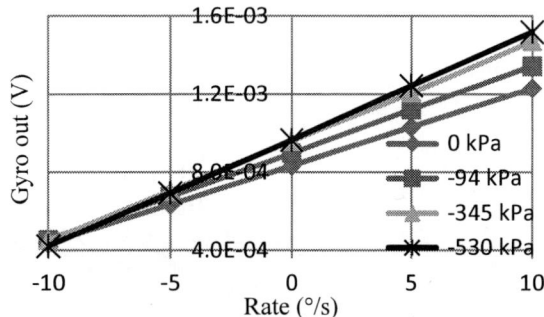

Figure 8: Scale factor test results for different stress levels on sense mode stress sensor (S_1).

Figure 9 shows the rate-referred ZRO change vs. the sense axis stress. The ZRO shift is believed to be originating from drive and sense comb gap mismatches. A drive comb gap mismatch generates a Coriolis in-phase force on the sense mode, and leaks to the rate output because of the finite drive axis spring constant of the sense axis springs [6]. A sense comb mismatch on the other hand, increases the SF since an imbalance in the combs results in higher capacitive sensitivity. As the sense axis stress is larger than the drive axis stress in the tested gyroscope, sense comb mismatches increase the SF faster than ZRO which is dominated by the drive comb mismatches leading to a decrease in rate referred ZRO. Although the impressed stress is small (< 1MPa), Figure 9 shows that external stress at 100 kPa levels leads to °/s bias shifts.

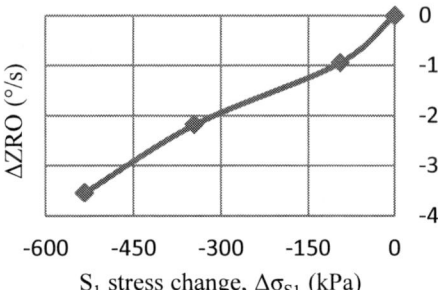

Figure 9: Zero rate output vs. sense axis stress.

After the matrix coefficients are obtained the logical next step is to explore output compensation. However, the stress sensors on the tested die failed due to electrostatic pull down to the substrate and subsequent short circuit. The 600 μm-long 5 μm wide 15 μm-thick stress sensor beams are suspended over the 2 μm SOI gap. Therefore, special attention must be paid to prevent electrostatic pull-down from the high DC potential (V_{PM}, 20V-40V DC) on the SOI substrate that is biased at the same potential as the proof mass. The problem is solved by biasing one side of the bridge to V_{PM} and the other side to $V_{PM} - 5V$, and powering the $V_{PM} - 5V$ source after V_{PM}.

Compensating the gyroscope ZRO with stress and temperature is an ongoing work. The uncompensated Allan deviation graph in Figure 10 shows the performance of a second SOI-MEMS gyroscope with a bias instability of 6.1°/hr and angle random walk (ARW) of 46.6°/hr/√Hz. The gyroscope was operated at 50 Hz mismatch with closed-loop sense and quadrature cancellation in this test. The stress and temperature compensation is expected to remove a significant portion of the long term rate random walk.

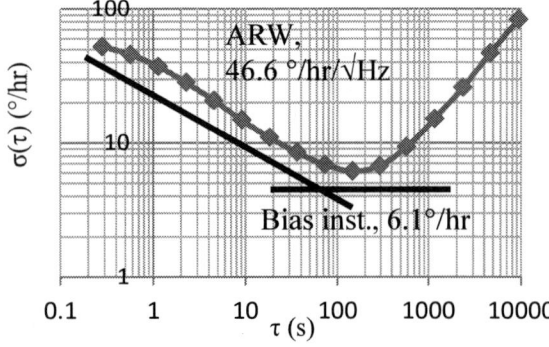

Figure 10: Uncompensated Allan deviation test results of a second gyroscope sample.

CONCLUSIONS

The importance of external stress on the MEMS gyroscope ZRO and SF has been shown using on-chip sensors that measure the stress and temperature separately and validated through a four-point bending test-bed. Drive and sense comb gap mismatches arising from stress-induced anchor displacements are believed to be the source of the ZRO shifts. The changes in the stress and ZRO were not captured by the temperature sensor indicating the importance of on-chip stress sensors on the die. This study shows the possible benefits of having on-chip stress sensors in addition to temperature sensors on the MEMS die for gyroscope bias compensation.

ACKNOWLEDGEMENTS

This work was supported by Defense Advanced Research Projects Agency (DARPA) under agreement number FA8650-08-1-7824. The work was also partially supported by the National Science Foundation Grant #CNS 0941497 as part of the Cyber-enabled Discovery and Innovation (CDI) program.

REFERENCES

[1] A. D. Challoner, H. H. Ge, and J. Y. Liu, "Boing disc resonator gyroscope", *Proc. PLANS 2014*, Monterey, May 5-8, 2014, pp. 504-514.

[2] S. A. Zotov, B. R. Simon, G. Sharma, A. A. Trusov, and A. M. Shkel, "Utilization of mechanical quadrature in silicon MEMS vibratory gyroscope to increase and expand the long term in-run bias stability", *Proc. ISISS 2014*, Laguna Beach, February 25-26, 2014, pp. 145-148.

[3] S. S. Walwadkar, and J. Cho, "Evaluation of die stress in MEMS packaging: Experimental and theoretical approaches", *IEEE Trans. on Components and Packaging Tech.*, vol. 29, no. 4, pp. 735-742, December 2006.

[4] E. Tatar, C. Guo, T. Mukherjee, and G. K. Fedder, "Interaction effects of temperature and stress on matched-mode gyroscope frequencies", *Proc. TRANSDUCERS 2013*, Barcelona, June 16-20, 2013, pp. 2527-2530.

[5] E. Tatar, T. Mukherjee, and G. K. Fedder, "Effects of stress on matched-mode gyroscope frequencies", *Proc. ISISS 2014*, Laguna Beach, February 25-26, 2014, pp. 1-4.

[6] E. Tatar, T. Mukherjee, and G. K. Fedder, "Simulation of stress effects on mode-matched MEMS gyroscope bias and scale factor", *Proc. PLANS 2014*, Monterey, May 5-8, 2014, pp. 16-20.

[7] V. Chidambaram, X. Ling, and C. Bangtao, "Titanium-based getter solution for wafer-level MEMS vacuum packaging", *J. Electronic Materials*, vol. 42, no.3, pp. 485-491, 2013.

CONTACT

E. Tatar, tel: +1-412-268-6606; etatar@andrew.cmu.edu

A FACILE FABRICATION TECHNIQUE FOR STRETCHABLE INTERCONNECTS AND TRANSDUCERS VIA LASER CARBONIZATION

Rahim Rahimi, Manuel Ochoa, Wuyang Yu, and Babak Ziaie
Purdue University, West Lafayette, IN, USA

ABSTRACT

This paper presents a facile, low-cost approach for fabricating highly-porous conductive carbon patterns on elastomeric substrates using laser carbonization of polyimide and its subsequent transfer onto a PDMS sheet. Using this technique, we fabricated stretchable interconnects and an array of piezoresistive tactile sensors. Raman spectra of pyrolyzed carbon regions revealed that under optimal laser settings, one can obtain highly porous carbon particles, nanotubes, and graphite sheets with sheet resistances as low as 50 Ω/ . Characterizations of the stretchable patterns showed linear sensitivities of 8.912 kΩ/ε% and 518Ω/N in response to strain and normal force, respectively.

INTRODUCTION

In recent years, flexible and stretchable electronics have attracted considerable attention due to their promise of creating more bio-friendly and biomimetic smart microsystems. Their advantage lies in their strong yet flexible and stretchable structure, which enables their seamless integration onto garments, biological tissue and robotic platforms. The development of these flexible/stretchable electronics is usually approached by 1) the creation of new stretchable materials [1], and/or 2) the design of novel stretchable structural constructs using established materials (e.g., buckled or meander conductors on elastomeric substrates). Owing to the difficulty in developing new organic materials with acceptable stretchability and adequate electrical performance, most previous attempts in the material design area have been focused on engineering new composites by mixing traditional elastomers such as PDMS, Ecoflex, and polyurethane, with conductive micro and nanoparticles. Among the various conductive particles available, carbon-based nanostructures (e.g., CNT, graphene) are ideal candidates for use in conductive composites; this is due to their unique properties such as high carrier mobility (conductivity), thermal stability, chemical inertness, and large surface area. These advantages have enabled the use of carbon-based nano-composites for fabricating stretchable batteries [2] [3] [4], supercapacitors [5] [6] [7], strain sensors [8] [9], tactile sensors [10] [11].

The second approach used for fabricating stretchable electronics relies on structural design utilizing traditional materials. For example, in 1998, Bowden et al. reported on the first stretchable buckled electrode fabricated by direct deposition of a thin gold film on the surface of a pre-heated PDMS substrate [12]. This resulted in buckled structures once the strain was removed, hence, imparting some stretchability to the structures. Another type of stretchable structure is the wavy/meander metallic patterns fabricated on top of elastomeric substrates [13]. Using this method, various stretchable devices have been reported; these include fingertip electronics [14], lithium-ion batteries [15], epidermal electronics [16]. In addition to these methods, various research groups (including our own) have investigated the use of microchannels filled with a eutectic liquid metal alloy (e.g., Galinstan, melting point of -19 °C) as stretchable interconnects [17], reconfigurable antennas [18], and strain sensors [19].

Despite the broad array of fabrication techniques for creating flexible and stretchable devices, the resulting systems typically suffer from one or more of the following shortcomings: 1) high costs associated with materials such as CNT and silver nanoparticles or required access to a clean room facility; 2) structural weakness; 3) limited stretchability; 4) environmental sensitivity/instability; and 6) inadequate reliability (liquid alloy leakage or structural failure). As an alternative approach that addresses these issues, we present a simple and low-cost technique to create stretchable interconnections and transducers using elastomer-embedded, laser-carbonized nanomaterials. At the core of this technology lie high-performance, carbon-based nanomaterials which are formed on a laser-pyrolyzed polymer and are subsequently transferred to an elastomeric substrate. Such platforms can be fabricated without the need of a clean room environment and are adaptable for high throughput and/or rapid-prototyping processes.

DESIGN

The stretchable carbon trace platform consists of PDMS embedded with patterns of inter-twined carbon micro/nano structures which form electrically conductive traces. The carbon material is created by directly laser-pyrolizing polyimide tape, resulting in highly porous traces. The carbon particles are encapsulated in PDMS and released from their original substrate, yielding arrays of stretchable, conductive traces which can be used as either conductors or piezoresistive elements.

FABRICATION

Based on the design concept described above, the fabrication process is illustrates in Figure 1. First, a polyimide tape (Polyimide tape, thickness: 60μm, VWR®) is attached to a commercial PET sheet for handling rigidity during the process. Next, using the optimal settings (power: 7 W and speed: 0.9 mm/ms) of a CO_2 laser engraver (PLS6MW, Universal Laser Systems®) the conductive carbon patterns are directly pyrolyzed on the surface of the polyimide tape. The laser pyrolization results in patterns of highly-porous microstructures on the surface of the polyimide. The traces are subsequently immersed in n-heptane for 20s; this improves the adhesion and increases the diffusion of the elastomeric materials into the carbon network. Next, the traces are dried at 70°C for 10 min. For transferring to an elastomer,

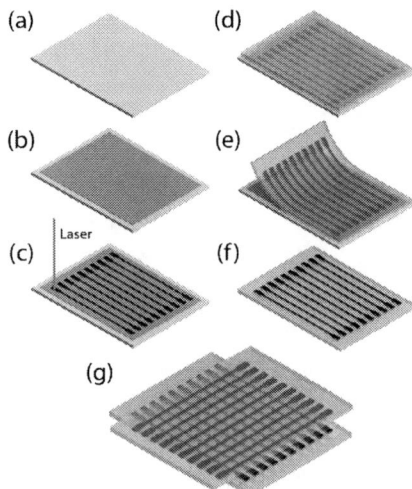

Figure 1: Fabrication process for direct laser pyrolization of polyimide. (a-b) attach polyimide tape to PET sheet; (c) laser pattern carbon traces; (d-f) transfer carbon traces to PDMS; (g) bond two layers by Ecoflex® with patterns orthogonal .

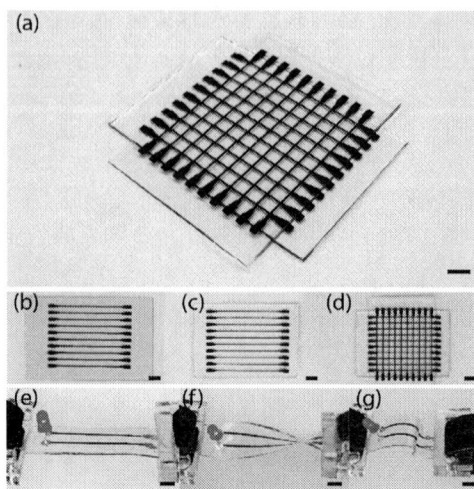

Figure 2: (a) photograph of final fabricated tactile sensor array, (b) carbon pattern on PI tape; (c) carbon patterns transferred to PDMS; (d) two layers orthogonally bonded; (e-g) a lit LED connected to stretchable carbon interconnection. Scale bars: 5mm.

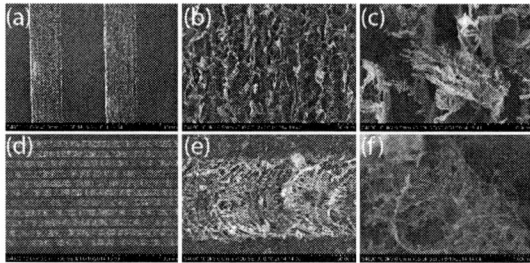

Figure 3: SEM of highly porous conductive carbon patterns. (a-c) images of 1 mm wide patterns. (d-f) images of smallest features with 90 µm width.

PDMS pre-polymer is prepared (Sylgard 184A and 184B, Dow Corning, 10:1 ratio) and degassed in a vacuum chamber. The pre-polymer is then poured over of the carbon traces, creating a layer up to 500 µm thick. The PDMS is then in a vacuum for 20 minutes and cured at 70°C for 2 hours. Due to the high porosity of the carbon traces, the PDMS is able to diffuse into them for proper integration of carbon micro/nano particles in PDMS. After cooling, the PDMS is peeled off from the polyimide, with the conductive carbon patterns embedded in the PDMS substrate. The particles form a dense conductive network of carbon nano/micro particles filled with PDMS elastomer between the gaps. To fabricate a tactile sensor array, two sets of conductive patterns are made by repeating the aforementioned process and bonded orthogonally using oxygen plasma treatment and Ecoflex® as an interlayer adhesive.

CHARACTERIZATION AND RESULTS
Qualitative characterization

As a proof of concept application of our technique, we created an array of strain/pressure sensors. Figure 2(a) shows an optical image of the final fabricated tactile sensor array with the intermediate fabrication steps. Figure 2(b) is the laser patterned porous carbon traces on the polyimide tape before transferring to PDMS. Figure 2(c) shows the carbon traces after being transfer into the PDMS elastomer. Figure 2(d) shows the alignment and bonding of two arrays of sensing elements. The final tactile sensor is an array of 10 ×10 force-sensing elements with an effective area of 35 × 35 mm². The width of each element is 1 mm with a spacing of 2.8 mm between each individual element. The size of the effective area and the width of the sensing elements can be easily modified by changing the laser carbonized patterns on the polyimide tape. The conductivity of the carbon patterns is qualitatively demonstrated in Figures 2(e-g), wherein the carbon traces are used as connections for lighting an LED.

The traces are sufficiently conductive to maintain the LED in its on state even upon exposure to twisting bending, Figures 2(f,g).

The morphology of the carbon patterns was investigated by scanning electron microscopy (SEM). The SEM images in Figure 3(a-c) reveal a high degree of alignment and porosity as well as uniformity among the conductive carbon traces. Figure 3(d-e) shows SEM images of the smallest features (width × pitch = 90 m × 120 µm) achievable with our laser system, which is limited to the beam size of the laser. Figure 3(f) shows a high-magnification top view of the carbon nano particles intertwining for increased conductivity.

Quantitative characterization

The presence of specific carbon nanostructures on the carbon traces was confirmed by Raman spectroscopy. The Raman spectra was measured before and after laser carbonization using an excitation laser source at 532 nm in the range of 1000–3000 cm⁻¹. The red and black lines in Figure 4 show the Raman spectra of the polyimide tape before and after laser carbonization respectively. The Raman characterization clearly shows the presents of

978-1-4799-7956-1/15 $31.00 © 2015 IEEE   818

three fundamental bands D, G and 2D after laser treatment, which are attributed to the presence of carbon nanotubes and graphite in the material. The peak intensity of I_D/I_G and I_{2D}/I_G imply that the material is composed of few layers of graphene with low defects.

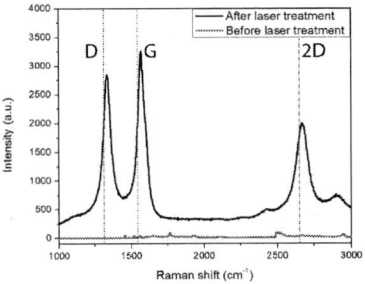

Figure 4: Raman spectra before and after laser treatment.

Figure 5 presents the sheet conductance of the carbon traces as a function of the processing power and speed of the laser system before and after transferring to PDMS. The sheet conductance (or sheet resistance) was measured by using the four probe method. The plot shows the strong dependence of the conductivity of the carbon traces on both power and speed of the laser system. The laser must produce a sufficient amount of energy to initiate the pyrolization of the polymer. Note that with higher energies (high power and low speeds setting of the laser) the polymer will burn into white ashes that have very low conductivity. The maximum point on each plot in Figure 5 corresponds to the optimal combination of power and speed of the laser for producing highly porous and conductive traces. The average optimal laser setting result in sheet conductivity of up to 0.02 Mhos/□ (R_s~50 Ω/□). Figure 5(b) shows a small decrease in conductivity of the carbon traces after transfer to PDMS, which is attributed to the incomplete transfer of all the carbon nano/micro particles.

Figure 6 illustrates the test setup used for measuring the sensitivity of the elastic carbon traces to strain and normal forces. The output contact ends of the sensing elements were electrically connected using silver paste. The resistance measurements of the sensing elements were conducted at room temperature using an Agilent 34401A digital multimeter with a data acquisition setup. Micromanipulators were used for applying strain and normal force. Figure 6(a) shows the test used for monitoring the resistance (R) variation with applied strain (ΔL/L). As shown in figure 6(b) the normal force applied to the tactile senor was measure with an electronic scale,

as the micro manipulator moves down the normal force applied to the sensor increases.

Figure 7a shows the measured resistance change in response to applied strain and normal force. For strain the range 0–28 %, the plot shows a positive change in the resistance of the sensing element with a linear slope of 8.912kΩ/ε% and r^2=0.9859. The resistance shows limit hysteresis in repeating cycles.

Figure 7(b) shows the measured resistance vs. applied normal force. The resistance change is characterized by two linear regions of 0 N to 1.6 N and 1.6 N to 2.4 N. The first region (0–1.6 N) shows a linear sensitivity of 518 Ω/N with r^2 = 0.994 and the second region (1.6–2.4 N) with a sharp increase shows a linear sensitivity of 1530 Ω/N and r^2=0.9892. The results by the presented method show comparable sensitivity to previously reported tactile sensors utilizing conductive composites with CNT and graphene particles.

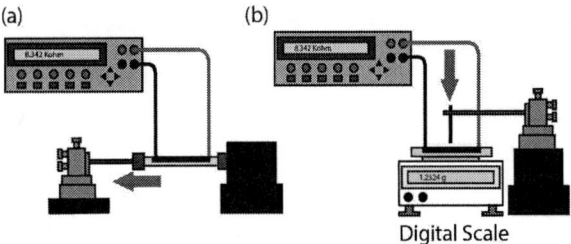

Figure 6: Experimental setup for measuring resistance change of the sensing element to (a) strain and (b) normal force.

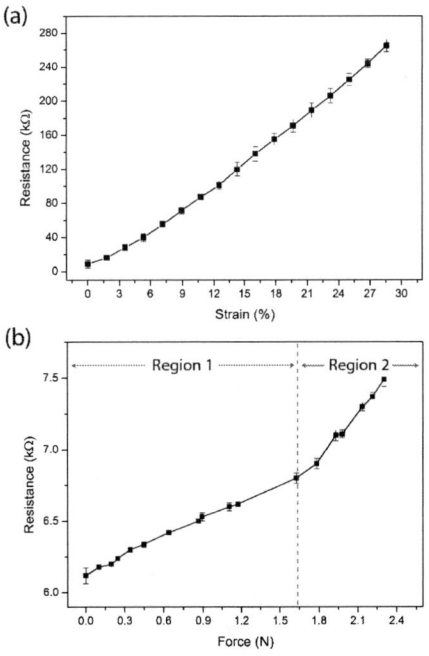

Figure 7: Measured sensitivity of the sensor to strain and applied normal force. (a) Plot of resistance versus strain with a linear slope of 8.912kΩ/ε% and r^2=0.9859; (b) plot of resistance versus normal force for regions 1 and 2 with linear slope of 518Ω/N, r^2=0.994and 1530Ω/N, r^2=0.9892 respectively.

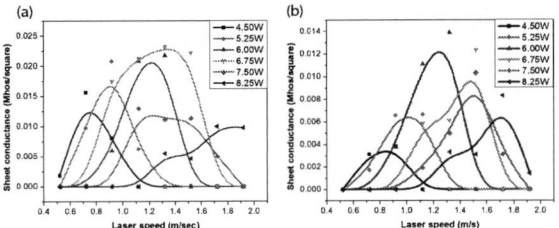

Figure 5: Sheet conductivity of carbon traces (a) before transfer to PDMS; (b) after transfer to PDMS.

978-1-4799-7956-1/15 $31.00 © 2015 IEEE

CONCLUSION

In this work, we presented a facile, low-cost approach for fabricating stretchable electrical interconnects and transducers. The fabrication is based on laser pyrolization of polyimide into highly porous conductive carbon traces and their subsequent transfer into a PDMS sheet. The transferred patterns form conductive PDMS-carbon composites with low sheet resistance (as low as $50\Omega/\square$), suitable for flexible/stretchable and wearable electronics. The characterization of the stretchable traces shows linear sensitivities to strains of up to 28% ($8.912\,k\Omega/\varepsilon\%$, $r^2=0.9859$) and normal forces of up to ($518\Omega/N$, $r^2=0.994$).

ACKNOWLEDGEMENTS

The authors thank the staff of the Birck Nanotechnology Center for their support. Funding for this project was provided by the National Science Foundation under grant EFRI-BioFlex #1240443.

REFERENCES

[1] T. Sekitani and T. Someya, "Stretchable, large-area organic electronics.," *Adv. Mater.*, vol. 22, no. 20, pp. 2228–46, May 2010.

[2] H. Lee, J.-K. Yoo, J.-H. Park, J. H. Kim, K. Kang, and Y. S. Jung, "A Stretchable Polymer-Carbon Nanotube Composite Electrode for Flexible Lithium-Ion Batteries: Porosity Engineering by Controlled Phase Separation," *Adv. Energy Mater.*, vol. 2, no. 8, pp. 976–982, Aug. 2012.

[3] M. Kaltenbrunner, G. Kettlgruber, C. Siket, R. Schwödiauer, and S. Bauer, "Arrays of ultracompliant electrochemical dry gel cells for stretchable electronics.," *Adv. Mater.*, vol. 22, no. 18, pp. 2065–7, May 2010.

[4] G. Kettlgruber, M. Kaltenbrunner, C. M. Siket, R. Moser, I. M. Graz, R. Schwödiauer, and S. Bauer, "Intrinsically stretchable and rechargeable batteries for self-powered stretchable electronics," *J. Mater. Chem. A*, vol. 1, no. 18, p. 5505, 2013.

[5] C. Yu, C. Masarapu, J. Rong, B. Wei, and H. Jiang, "Stretchable supercapacitors based on buckled single-walled carbon-nanotube macrofilms.," *Adv. Mater.*, vol. 21, no. 47, pp. 4793–7, Dec. 2009.

[6] J. Zang, C. Cao, Y. Feng, J. Liu, and X. Zhao, "Stretchable and high-performance supercapacitors with crumpled graphene papers.," *Sci. Rep.*, vol. 4, p. 6492, Jan. 2014.

[7] X. Li, T. Gu, and B. Wei, "Dynamic and galvanic stability of stretchable supercapacitors.," *Nano Lett.*, vol. 12, no. 12, pp. 6366–71, Dec. 2012.

[8] C. Yan, J. Wang, W. Kang, M. Cui, X. Wang, C. Y. Foo, K. J. Chee, and P. S. Lee, "Highly stretchable piezoresistive graphene-nanocellulose nanopaper for strain sensors.," *Adv. Mater.*, vol. 26, no. 13, pp. 2022–7, Apr. 2014.

[9] J.-H. Kong, N.-S. Jang, S.-H. Kim, and J.-M. Kim, "Simple and rapid micropatterning of conductive carbon composites and its application to elastic strain sensors," *Carbon N. Y.*, vol. 77, pp. 199–207, Oct. 2014.

[10] D. J. Lipomi, M. Vosgueritchian, B. C.-K. Tee, S. L. Hellstrom, J. a Lee, C. H. Fox, and Z. Bao, "Skin-like pressure and strain sensors based on transparent elastic films of carbon nanotubes.," *Nat. Nanotechnol.*, vol. 6, no. 12, pp. 788–92, Dec. 2011.

[11] Y.-T. Lai, Y.-M. Chen, T. Liu, and Y.-J. Yang, "A tactile sensing array with tunable sensing ranges using liquid crystal and carbon nanotubes composites," *Sensors Actuators A Phys.*, vol. 177, pp. 48–53, Apr. 2012.

[12] N. Bowden, S. Brittain, A. G. Evans, & J. W. H., and G. M. Whitesides, "Spontaneous formation of ordered structures in thin films of metals supported on an elastomeric polymer," Nature, vol. 393, no. May, pp. 146–149, 1998.

[13] M. Gonzalez, F. Axisa, M. Vanden Bulcke, D. Brosteaux, B. Vandevelde, and J. Vanfleteren, "Design of metal interconnects for stretchable electronic circuits," *Microelectron. Reliab.*, vol. 48, no. 6, pp. 825–832, Jun. 2008.

[14] M. Ying, A. P. Bonifas, N. Lu, Y. Su, R. Li, H. Cheng, A. Ameen, Y. Huang, and J. A. Rogers, "Silicon nanomembranes for fingertip electronics.," *Nanotechnology*, vol. 23, no. 34, p. 344004, Aug. 2012.

[15] S. Xu, Y. Zhang, J. Cho, J. Lee, X. Huang, L. Jia, J. a Fan, Y. Su, J. Su, H. Zhang, H. Cheng, B. Lu, C. Yu, C. Chuang, T.-I. Kim, T. Song, K. Shigeta, S. Kang, C. Dagdeviren, I. Petrov, P. V Braun, Y. Huang, U. Paik, and J. a Rogers, "Stretchable batteries with self-similar serpentine interconnects and integrated wireless recharging systems.," *Nat. Commun.*, vol. 4, p. 1543, Jan. 2013.

[16] D.-H. Kim, N. Lu, R. Ma, Y.-S. Kim, R.-H. Kim, S. Wang, J. Wu, S. M. Won, H. Tao, A. Islam, K. J. Yu, T. Kim, R. Chowdhury, M. Ying, L. Xu, M. Li, H.-J. Chung, H. Keum, M. McCormick, P. Liu, Y.-W. Zhang, F. G. Omenetto, Y. Huang, T. Coleman, and J. a Rogers, "Epidermal electronics.," *Science*, vol. 333, no. 6044, pp. 838–43, Aug. 2011.

[17] K. Hyun-joong, T. Maleki, P. Wei, B. Ziaie, and S. Member, "A biaxial stretchable interconnect with liquid-alloy- covered joints on elastomeric substrate A Biaxial Stretchable Interconnect With Liquid-Alloy-Covered Joints on Elastomeric Substrate," 2009.

[18] J.-H. So, J. Thelen, A. Qusba, G. J. Hayes, G. Lazzi, and M. D. Dickey, "Reversibly Deformable and Mechanically Tunable Fluidic Antennas," *Adv. Funct. Mater.*, vol. 19, no. 22, pp. 3632–3637, Nov. 2009.

[19] S. Cheng and Z. Wu, "A Microfluidic, Reversibly Stretchable, Large-Area Wireless Strain Sensor," *Adv. Funct. Mater.*, vol. 21, no. 12, pp. 2282–2290, Jun. 2011.

CONTACT

*B. Ziaie: bziaie@purdue.edu

A HIGH-Q ALL-FUSED SILICA SOLID-STEM WINEGLASS HEMISPHERICAL RESONATOR FORMED USING MICRO BLOW TORCHING AND WELDING

Jae Yoong Cho and Khalil Najafi

Center for Wireless Integrated MicroSensing and Systems (WIMS²)

University of Michigan, USA

ABSTRACT

We report a new fabrication technology for making fused silica (FS) wineglass resonators with arbitrarily sized FS solid stems through a simultaneous process of micro blow-torching and microwelding. The process allows the welding of multiple FS structures at controlled locations during blow torching. We demonstrate a new micro FS wineglass resonator with high quality factor (Q) and long ring down time (τ). The resonator is formed by blow torching and flowing a thin FS substrate using vacuum to form the resonator shell, and by welding the shell to a solid post at a controlled location. The flowing of the shell and welding to the rod is performed in one step and in a single mold. This solid-stem resonator offers low anchor loss due to the large stem length/stiffness, and small shell rim thickness. The device has a shell radius/height of ~2.8 mm, and a stem radius of 0.5 mm. At <10 μTorr vacuum, the n=2 wineglass mode, located at a frequency (f) of 22.6 kHz, has τ = 35.9 s and Q = 2.55 million.

INTRODUCTION

Three-dimensional (3D) axisymmetric shell resonators are attractive due to their stiffness and damping symmetry. FS is an excellent material for 3D micro shell resonators, because 1) it has very low thermoelastic damping (TED) [1] due to its very low linear thermal expansion coefficient (α_{FS} = 0.52 × 10^{-6} K^{-1}) and low thermal conductivity (k_{FS} = 1.38 Wm^{-1}K^{-1}), and 2) it has very low surface energy loss due to its very smooth surface (<1 nm) when reflow molded. We have recently reported a FS micro birdbath (BB) resonator with long τ = 43 s, Q = 1.2 million, frequency symmetry (Δf = 6 Hz), and a high angular gain (A_g (calculated) ~ 0.25) [2]. To improve performance further, the Q of the resonator has to be increased. Several reasons have been identified as the source of lower Q in these small shell resonators [2]. Among these is energy loss through the anchor.

The dependence of Q on anchor design of 3D micro axisymmetric resonators has been studied by several researchers. The dependence of Q on the anchor stem for a micro hemispherical resonator [3] and on stem length for a silicon oxide micro hemispherical resonator [4] have been studied numerically. The dependence of Q on stem length for a metallic-glass micro hemispherical resonator has been experimentally studied [5]. An analytic expression for reduction in Q due to coupling between the wineglass mode and the bending mode of the stem has been derived in [6]. These studies suggest that stiff, long, and small anchors like that of the hemispherical resonant gyro (HRG) could reduce anchor loss.

Other 3D micro shell geometries have been proposed (Table 1). Most of these resonator geometries have a small device height, short anchor length, small shell thickness near the anchor, or limited control over stem dimensions, which could increase anchor loss, reduce resistance to shock, and reduce angular gain, which is a very important parameter for high-performance gyroscopes. In this work, we present a new FS micro wineglass resonator with a long, solid stem, which is fabricated using microwelding of a FS rod and a FS shell during blowtorch molding. We present the fabrication process, resonance characteristics, and experimental results of the FS micro wineglass resonator.

FABRICATION PROCESS

The fabrication process is described in Figures 1a through 1e. A graphite mold is fabricated from a graphite wafer using a precision milling process and patterned into the shape of a dome, with a through hole (diameter: 1-1.5 mm) formed in the center of the dome. A FS rod is inserted into the through hole (Figure 1a). A FS substrate (100-200 μm) is then clamped on the top of the mold using vacuum. The shell is controllably heated and reshaped by blowtorching the FS substrate above its softening point (1585 °C) with a pressure difference of ~600 Torr across it for ~20s. The FS substrate flows down and where it touches the rod, it is fused to it

Table 1: Comparison of anchor geometries for micro axisymmetric resonators.

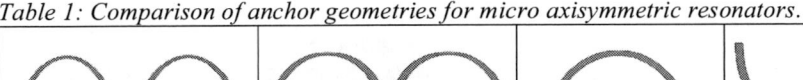

A. Hollow anchor for birdbath shell [2, 7].	B. Solid anchor for birdbath shell [7].	C. Boundary anchor for hemispherical shell [8].	D. Short solid anchor for hemispherical shell [4].	E. This work: long solid anchor for hemispherical shell [5, 9].
Challenges: 1) potentially large anchor loss, 2) low shock resistivity.	Challenges: 1) small device height, 2) low shock resistivity.	Challenges: 1) potentially large anchor loss.	Challenges: potentially large anchor loss.	Advantages: 1) potentially low anchor loss, 2) high angular gain, 3) good shock resistivity

permanently (Figure 1b). The rod and shell formed by the shape of the mold are thus self-aligned. The combined structure is then detached from the mold (Figure 1c). The shell is then mounted face-up on the bottom surface of a Si releasing jig and is embedded in a polymer (Figure 1d). The shell with its solid stem is released using lapping and chemical mechanical polishing (CMP) processes (Figure 1e).

This process has several attractive features: First, FS microstructures of arbitrary geometries can be fused to create more complex 3D micro sensors and actuators. Second, strong FS fusion bonding between the stem and the shell has very low energy loss at the bonding interface, which is critical for achieving high Q. Third, shells and stems made with different reflowable materials can be also welded. Fourth, sub-nanometer surface roughness can be obtained, which is critical for reducing surface energy loss.

The important resonator design parameters are: mold radius and height, FS rod diameter, and thickness of the FS substrate. The resonator radius (R) is determined by the radius of the dome formed in the mold. The mold used in this study has a radius of 2.5 mm and a height of 3.75 mm. The anchor diameter is determined by the

diameter of the FS rod. The thickness of the shell (t) is determined by the original thickness of the substrate.

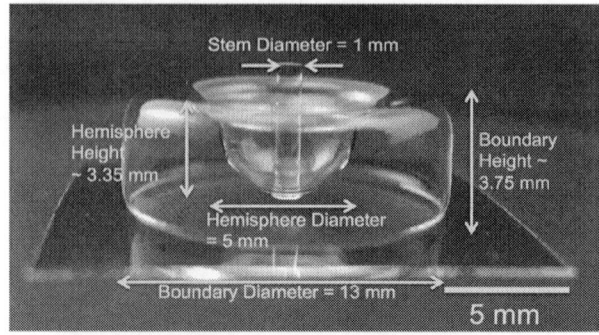

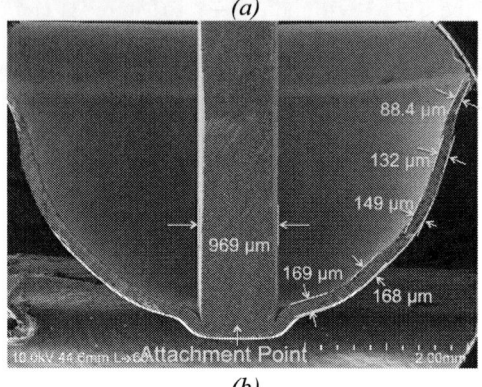

(b)

Figure 2: (a) Photograph of unreleased wineglass shell. (b) Cross-sectional scanning electron microscopy image of unreleased wineglass shell.

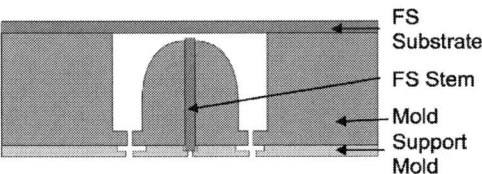

(a) Insert fused silica (FS) rod through the through hole at the center of the mold. Clamp a FS substrate (100-200 µm).

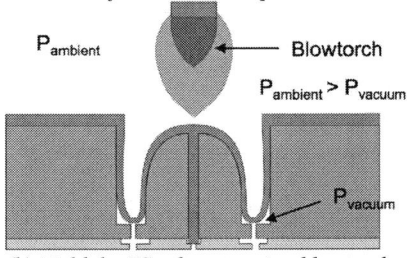

(b) Mold the FS substrate using blowtorch at controlled pressure, temperature, and duration.

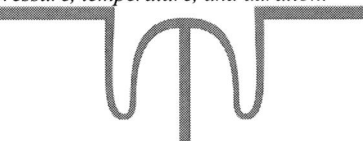

(c) Detach the shell from the mold.

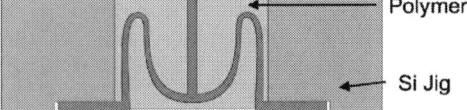

(d) Attach the shell on a Si jig and embed it in a polymer. Release the resonator using lapping and chemical mechanical polishing processes.

(e) Dissolve polymer.
Figure 1: Fabrication process flow.

Figure 3: Photograph of a wineglass resonator with an apple seed.

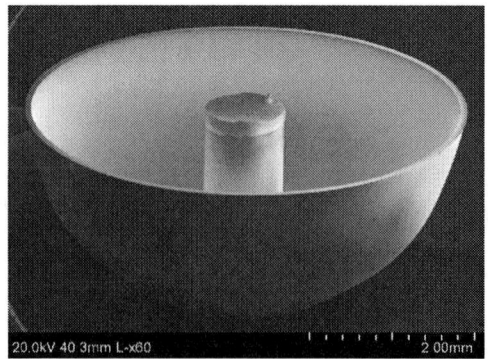

Figure 4: Scanning electron microscopy image of the fabricated wineglass resonator.

978-1-4799-7956-1/15 $31.00 © 2015 IEEE

Figure 2a shows the photograph of a reflown and molded shell. The hemisphere has a diameter of 5 mm, a height of 3.35 mm, and a stem diameter of 1 mm. The boundary has a diameter of 13 mm and a height of 3.75 mm. Figure 2b shows the cross-sectional scanning electron microscope (SEM) image of an unreleased shell. The homogeneous and seamless welded interface between the shell and the stem is clearly visible.

Figure 3 shows a micro hemispherical resonator. The resonator has an outer radius (R) of ~2.8 mm, an anchor radius (AR) of 0.5 mm, a height (H) of ~2.8 mm, and a rim thickness (t_{rim}) of ~90 µm. This resonator resembles the HRG. The SEM image the polished and smooth wineglass resonator rim is shown in Figure 4.

As shown in Figure 2b, the shell thickness increases from the top (t_{top} = 88.4 µm) to the bottom (t_{bottom} = 169 µm). This thickness profile is the opposite of that of the BB resonator [2]. This gradual increase in the shell thickness from the rim to the anchor will help improve the Q and reduce shock and vibration sensitivity due to the high mechanical stiffness near the anchor and low mechanical stiffness near the rim.

FINITE ELEMENT METHOD SIMULATION

A finite element method (FEM) model of the wineglass resonator is built based on the cross-sectional SEM image of a fully released resonator. The resonator has R = 2.76 mm, AR = 0.5 mm, H = 2.8 mm, t_{rim} = 90 µm, and t_{bottom} = 148 µm. The n=2 wineglass mode is found at 19.85 kHz. Below the wineglass mode, the rotating mode ($f_{rotating}$) is found at 5.3 kHz, the tilting modes ($f_{tilting}$) are found at 6.2 kHz, and the anchor bending modes ($f_{bending}$) are found at 11.7 kHz. These are called parasitic modes, because they couple to external

Table 2: Dimensions, resonance frequencies, and physical parameters for micro hemispherical resonator.

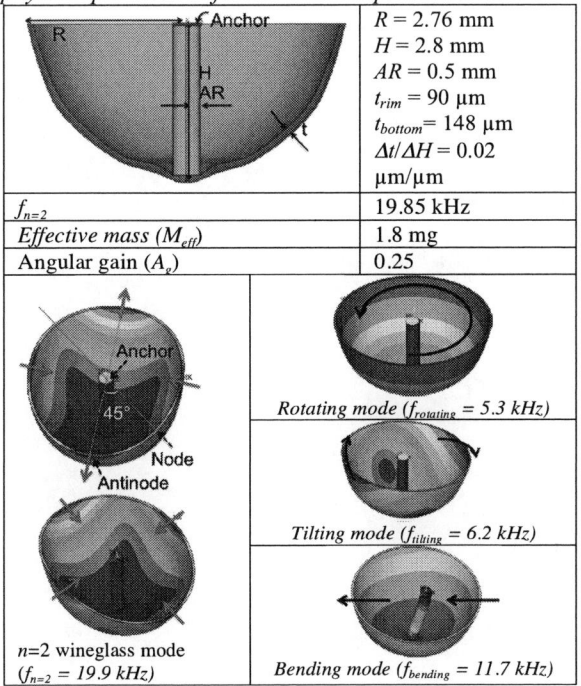

(diagram)	R = 2.76 mm H = 2.8 mm AR = 0.5 mm t_{rim} = 90 µm t_{bottom} = 148 µm $\Delta t/\Delta H$ = 0.02 µm/µm
$f_{n=2}$	19.85 kHz
Effective mass (M_{eff})	1.8 mg
Angular gain (A_g)	0.25

n=2 wineglass mode ($f_{n=2}$ = 19.9 kHz)

Rotating mode ($f_{rotating}$ = 5.3 kHz)

Tilting mode ($f_{tilting}$ = 6.2 kHz)

Bending mode ($f_{bending}$ = 11.7 kHz)

shock or vibration and can create additional error for the gyroscope. These frequencies can be increased by changing the size of the anchor.

The effective mass (M_{eff}) and A_g are key parameters affecting the gyro performance. M_{eff} is the mass of the gyro when it is modeled as a lumped-mass gyroscope, that is, when the entire proof mass is modeled to have the same vibration amplitude. M_{eff} increases with t. A_g is the scale factor of a rate-integrating gyroscope. A_g increases with the height-to-radius (H/R) ratio, because the shell moves increasingly perpendicularly to the yaw axis due to increased vertical stiffness. M_{eff} and A_g are numerically calculated using ANSYS [10]. Due to the large t (> 90 µm) and H/R (~1), this resonator has large M_{eff} (= 1.8 mg) and A_g (= 0.25). Table 2 summarizes the modal characteristics and physical parameters of the hemispherical resonator.

TESTING

The resonator is mounted face down on a Si substrate using glass frit. The Si anchor has ~300 µm step height between the pedestal and the boundary. The resonator is mounted on top of a piezoelectric actuator. The piezoelectric actuator is mounted on the sidewall of a metal block in a vacuum chamber. The vacuum chamber is pumped down to <10 µTorr pressure. The resonator is

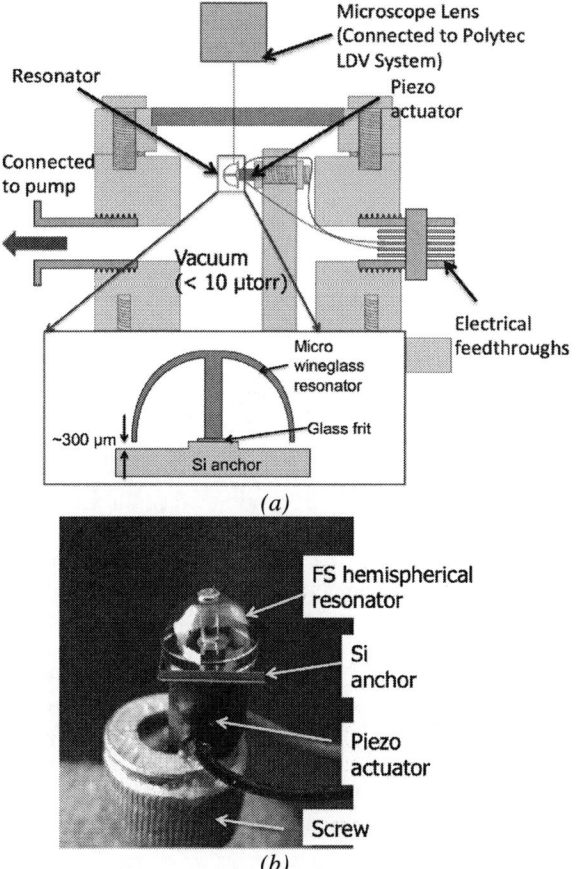

Figure 5: (a) Vacuum Laser Doppler Velocimetry (LDV) setup diagram. (b) Photograph of micro hemispherical resonator mounted on piezoelectric actuator.

978-1-4799-7956-1/15 $31.00 © 2015 IEEE

driven at the $f_{n=2}$ using the piezoelectric actuator to sustain a certain vibration amplitude. The driving voltage is then turned off, and the resonator amplitude decays exponentially. The ring-down time, τ, is determined as the time it takes for the vibration amplitude to reach 1/e ($\sim$ 0.3678) of the original amplitude. Figure 6 shows the ring down time of one of the wineglass modes found at 22.6328 kHz. The τ and Q ($Q = \pi f \tau$) are found to be 35.9 s and 2.55 million, respectively. The other $n=2$ mode is located at 22.496 kHz with $\tau \sim$ 5s ($Q = 0.35$ million). The reason for the large difference in τ between these two nodes has not been identified yet and is currently being investigated.

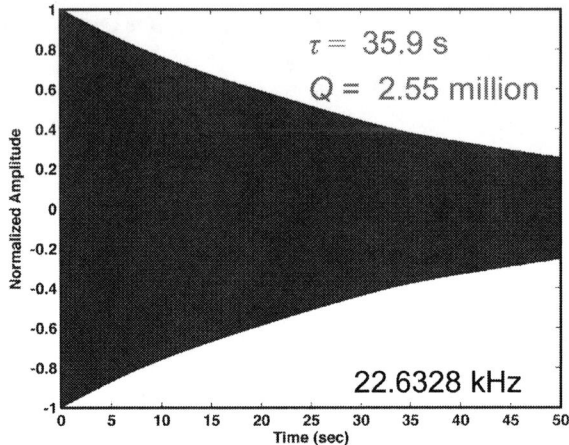

Figure 6: Ring down plot of the wineglass resonator measured using Laser Doppler Vibrometer (LDV) at <10 μTorr vacuum. $f_{n=2}$ = 22.633 kHz, τ =35.9 s, and Q = 2.55 million.

SUMMARY

We demonstrated a new fabrication technology for making FS wineglass resonators with arbitrarily sized FS solid stems through a simultaneous process of micro blow-torching and microwelding. The process allows the welding of multiple FS structures at controlled locations during blow torching. The resonator is formed by blow torching and flowing a thin FS substrate using vacuum to form the resonator shell, and by welding the shell to a solid post at a controlled location. The flowing of the shell and welding to the rod is performed in one step and in a single mold, so the shell and rod are self aligned. This solid-stem resonator offers low anchor loss due to the large stem length/stiffness, and small shell rim thickness. The device has a shell radius/height of ~2.8 mm, a stem radius of 0.5 mm, and a rim thickness of 90 μm, and a shell thickness 148 μm near the anchor. The resonator has large effective mass (= 1.8 mg) and high angular gain (= 0.25). At <10 μTorr vacuum, the $n=2$ wineglass mode, located at a frequency of 22.6 kHz, has τ = 35.9 s and Q = 2.55 million.

ACKNOWLEDGEMENTS

This work is supported by DARPA MRIG award #W31P4Q-11-1-0002. The authors thank Mr. Ali Darvishian, Dr. Behrouz Shiari, Mr. Robert Gordenker, and Mr. Tal Nagourney for the testing support. Portions of this work were done in the Lurie Nanofabrication Facility (LNF), a site of the National Nanotechnology Infrastructure Network (NNIN).

REFERENCES

[1] R. Lifshitz and M. L. Roukes, "Thermoelastic damping in micro and nanomechanical systems," *Phys. Rev. B*, vol. 61, n. 8, 2000, pp. 5600-5609.

[2] J. Cho *et al.*, "Fused silica micro birdbath shell resonators with 1.2 million Q and 43 second decay time constant," in *Proc. Hilton Head 2014*, pp. 103-104.

[3] D. Senkal *et al.*, "Titania silicate/fused quartz glass blowing for 3-D fabrication of low internal loss wineglass micro-structures," in *Proc. Hilton Head 2012*, pp. 267-270.

[4] L. Sorenson and F. Ayazi, "Effect of structural anisotropy on anchor loss mismatch and predicted case drift in future micro-hemispherical resonator gyros," in *Proc. IEEE/ION PLANS 2014*, pp. 493-498.

[5] Kanik *et al.*, "Metallic glass hemispherical shell resonators," *JMEMS* [Online]. Available: http://ieeexplore.ieee.org/xpls/abs_all.jsp?arnumber=6937058

[6] Y. K. Zhbanov and V. P. Zhuravlev, "Effect of movability of resonator center on the operation of a hemispherical resonator gyro," *Mechanics of Solids*, 2007, vol. 42, no. 6, pp. 851-859.

[7] D. Senkal *et al.*, "Design and modeling of micro-glassblown inverted wineglass structures," in *Proc. ISISS 2014*, pp. 13-16.

[8] A. Vafanejad and E. S. Kim, "Sub-degree angle detection using dome –shaped diaphragm resonator with wine-glass mode vibration," in *Proc. Hilton Head* 2014, pp. 391-394.

[9] E. J. Loper, Jr. and D. D. Lynch, "Vibratory rotation sensor," U.S. Patent 4 901 508.

[10] J. Cho, "High-performance micromachined vibratory rate- and rate-integrating gyroscopes," Ph.D. dissertation, Dept. EECS, Univ. Michigan, Ann Arbor, MI, 2012.

CONTACT

*K. Najafi, najafi@umich.edu.
 J. Cho, jycho@umich.edu.

GRAPHITE-ON-PAPER BASED TACTILE SENSORS USING PLASTIC LAMINATING TECHNIQUE

Hoang-Phuong Phan[1], Dzung Viet Dao[1,2], Toan Dinh[1], Harrison Brooke[2], Afzaal Qamar[1], Nam-Trung Nguyen[1], and Yong Zhu[1,2]

[1]Queensland Micro- and Nanotechnology Centre, Griffith University, QLD, Australia
[2]School of Engineering, Griffith University, QLD, Australia

ABSTRACT

We report for the first time a highly sensitive paper-based tactile sensor using laminated graphite drawn on paper. Thanks to the high gauge factor of 26.2, as well as its excellent robustness and humidity-resistance, plastic-laminated graphite-on-paper has a high potential for mechanical sensors. Additionally, the plastic lamination combined with the laser cutting technique proposed in this study would enable the mass production of cleanroom-free fabrication and low-cost MEMS devices.

INTRODUCTION

In the last two decades, humanoid robots have attracted a great attention from the research and development community. Humanoid robots assist humans in daily life and replace them in hazardous working environments. Along with the development of actuators which are responsible for robotic motion, various sensor systems also have been investigated to improve the performance of humanoid robots [1][2]. Among these sensing systems, Micro Electro Mechanical Systems (MEMS) tactile sensors have been widely utilized to sense the grasping/impacting forces between robots' fingers and the grasped objects. The information obtained from tactile sensors is significant for feedback controls, pattern recognition, and dexterous handling and gripping objects [3][4].

Several sensing mechanisms have been deployed in tactile sensors such as capacitive, piezoresistive, piezoelectric, and optical sensors. Among these techniques, the piezoresistive effect in semiconductors such as silicon (Si), germanium (Ge), and silicon carbide (SiC) is one of the most common methods for mechanical sensing, due to its high sensitivity and simple structures [1-10]. Silicon piezoresistive cantilevers embedded inside elastomers such as poly dimethylsiloxane (PDMS) and poly methyl methacrylate (PMMA) have been applied for pressure sensing on robots' finger tips [5][6]. The large gauge factor and well-established fabrication technology of Si make it one of the most common materials for robotic sensing applications. However, the fabrication of Si based MEMS transducers requires expensive equipment and complicated processes such as micro-machining using cleanroom facilities. This drawback is indeed one of the crucial reasons making the cost of Si based sensors relatively high [11].

Recent studies have been aiming at developing MEMS sensors using low cost materials to solve the above bottle neck [11][12]. Graphite-on-paper (GoP) has emerged as a promising candidate for mechanical transducers [11-16]. The main advantages of GoP are its low cost, world wide availability, and simple fabrication process. The piezoresistive effect of GoP has been intensively investigated recently. Liu *et al.* reported a GoP force sensor with a sensitivity of 120 µN, using a screen printing technique [11]. Employing pencil drawing on paper, Kang reported large, tunable gauge factors of GoP varying from 15 to 50, indicating the capability of GoP for highly sensitive mechanical sensors [12]. Most of the previous studies only focused on the characterization of the piezoresistive effect in GoP. Only a small number of studies has demonstrated the feasibility of using the piezoresistive effect in GoP for sensing applications. One significant reason for this limitation in the practical applications of GoP is due to its direct exposure to environments which can be easily affected by ambient conditions such as humidity.

In this work, we propose for the first time, a novel platform for tactile sensors using plastic-laminated graphite-on-paper (LGoP). Plastic lamination and laser machining have been used for making low-cost microfluidic devices [13]. Plastic lamination not only improves the robustness of paper but also enhances its humidity-resistance, making it possible to embed LGoP inside elastomers for tactile sensing purposes. The platform proposed in this work demonstrates that LGoP has a high potential for simple, low cost mechanical sensing applications.

SENSOR CONCEPT AND FABRICATION

Concept of tactile sensor and its working principle

Figure 1 (a) shows the concept of the LGoP tactile sensor. The sensor consists of three parts assembled using epoxy. The pressure-transfer layer is made of PDMS with a thickness of 2.5 mm. The piezoresistive effect of the GoP was utilized for sensing the applied pressure. The bottom acrylic layer was used as the frame for the sensor. Figure 1 (b) describes the working principle of tactile sensing using LGoP. When a pressure/force is applied to the top surface of the PDMS layer, it is transferred to the sensing layer through the deformation of PDMS, bending the LGoP cantilever underneath. Due to the deflection of the LGoP cantilever, a strain is induced on the graphite layer, changing the distance between the graphite grains. The electrical current flowing inside graphite layer is the tunneling current between the graphite grains. Accordingly, the relationship between the resistance of GoP and tunneling distance is [12][18]:

$$R(l) \propto e^{\beta l} \qquad (1)$$

where β is a function of the potential barrier height, and l is the distance between the graphite grains. Under strain, the tunneling distance is changed, leading to the change of tunneling resistance [18]:

(a) Concept of tactile sensor

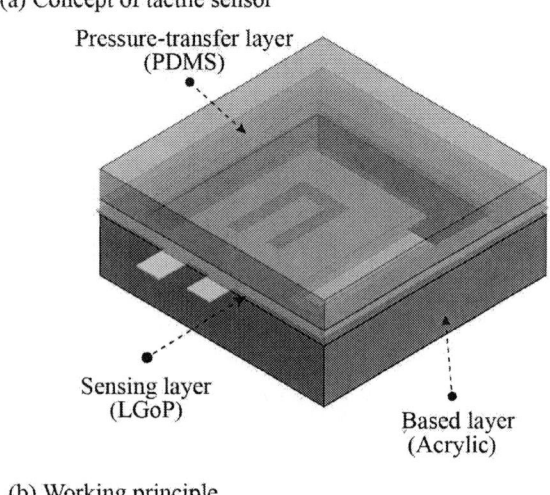

(b) Working principle

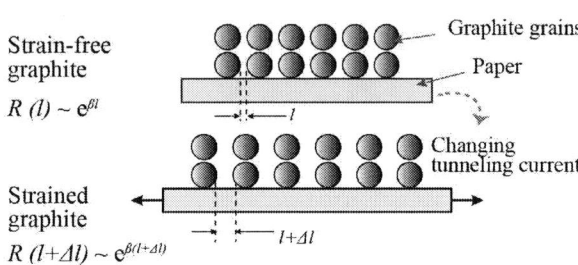

Figure 1: The concept of the tactile sensor using LGoP and its working principle.

$$R(l+\Delta l) \propto e^{\beta(l+\Delta l)} \qquad (2)$$

where Δl is the displacement between the graphite grains. As a consequence, the applied pressure can be obtained by measuring the resistance change of the GoP.

Fabrication of LGoP based tactile sensors

Figure 2 shows the fabrication process of the tactile sensor. The graphite layer was fabricated by drawing with a 2B pencil (Faber-Castel™) on commercial A4 paper. Graphite resistors were formed in a U-shape structure, with a length of 8 mm, a width of 2 mm, and their resistance was in the range of 100~200 kΩ. Aluminum tape (3M™) and highly conductive silver paste (186-3616, RS Components) were employed to make the electrodes of the graphite resistors (Fig. 2, step 1). Subsequently, a layer of paper was placed on the top surface of the graphite layer to prevent adhesion between the graphite and the laminated plastic film. Another layer of paper was also attached under the GoP layer to increase the distance from the graphite to the neutral axis of the composite beam (Fig. 2, step 2). These three paper layers were then laminated within two plastic films (Lowell Presentation System™) (Fig. 2, step 3). The thickness of each paper layer was 100 μm, while the thickness of the laminated layers of paper was 430 μm. A free standing LGoP cantilever with a dimension of 6 mm × 10 mm × 0.43 mm was formed by laser cutting (Speedy 300™,

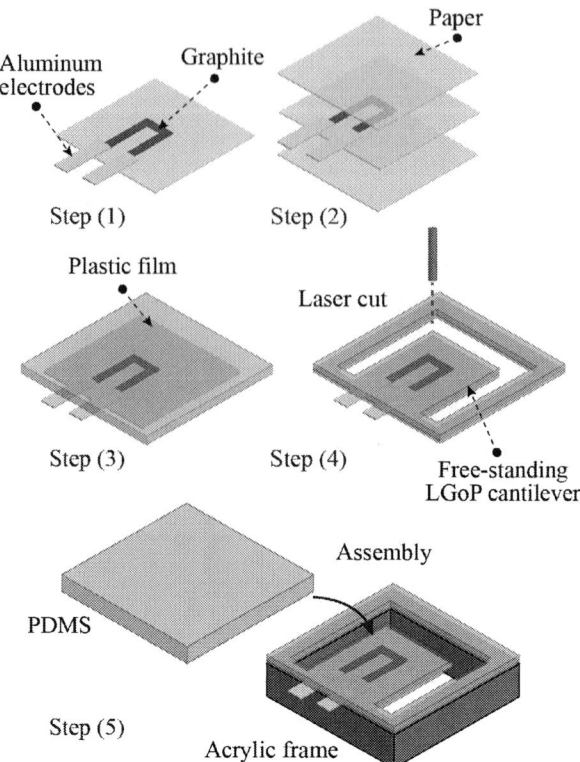

Figure 2: The fabrication process of the tactile sensor using LGoP

Trotecs) to improve the flexibility of the sensing layer (Fig. 2, step 4). In the next step, a PDMS layer with a thickness of 2.5 mm was coated using an acrylic mould. An acrylic frame was also made using a laser cutting technique. These three components of the LGoP, the PDMS top layer and the acrylic frame were then assembled using epoxy, forming the paper based tactile sensor (Fig. 2, step 5).

EXPERIMENTAL AND RESULTS
Characterization of LGoP and its piezoresistive effect

The mechanical properties of LGoP and the piezoresistive effect of graphite were characterized by using the bending beam method [7], in which an end of

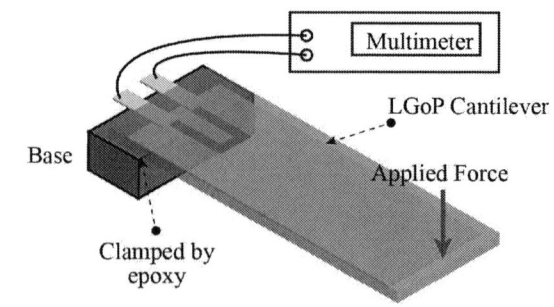

Figure 3: Experimental setup for characterizing the mechanical properties of LGoP and the piezoresistive effect of graphite.

LGoP cantilever was clamped, while the other end was deflected, Fig. 3. A LGoP-cantilever with dimensions of 10 mm × 40 mm × 0.43 mm was used in this experiment. The deflection of the tip of the cantilever was controlled using a two axis manipulator, while the applied force was measured by a high resolution balance (Ohaus™, Pioneer, PA4102). The resistance change of the graphite resistor was monitored using an Agilent™ 344110A multimeter. Figure 4 shows a good linear relationship between the applied force and the deflection of the LGoP. Accordingly, the stiffness (S) of the laminated paper was calculated using the following equation:

$$S = F/d \qquad (3)$$

where F and d are the applied force and the deflection of the LGoP cantilever, respectively. Consequently, the stiffness of the laminated paper was found to be 12 Nm^{-1}. Young's modulus (E) of the LGoP is:

$$E = -4Fl^3/(wh^3d) \qquad (4)$$

where l, w, h are the length, width and thickness of the LGoP, respectively. Using Eq. (4), Young's modulus of the laminated paper was found to be 3.5 GPa.

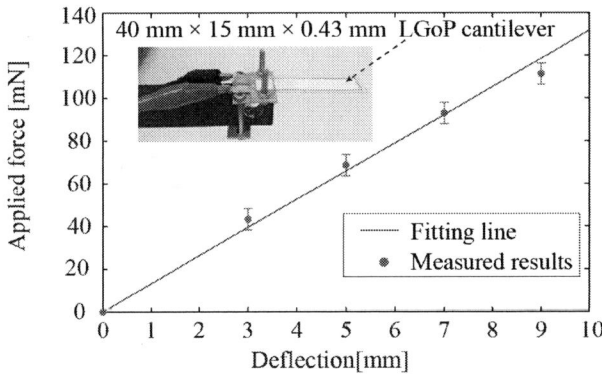

Figure 4: The relationship between the applied forces and the deflection of LGoP.

The strain induced on the graphite layer (ε) is:

$$\varepsilon = 12Flt/(Ewh^3) \qquad (5)$$

where t is the distance from the graphite layer to the neutral axis of the LGoP cantilever. Figure 5 presents the relationship between the applied strains and the resistance change of the 2B graphite resistors, which shows a relatively good linearity.
The gauge factor can be calculated as:

$$GF = (\Delta R/R)/\varepsilon \qquad (6)$$

where R, ΔR and ε are the graphite resistance, the resistance change, and the applied strain, respectively. As a result, the gauge factor of 2B graphite was calculated to be 26.2, which is consistent with the results reported previously by Kang [12]. This high gauge factor of

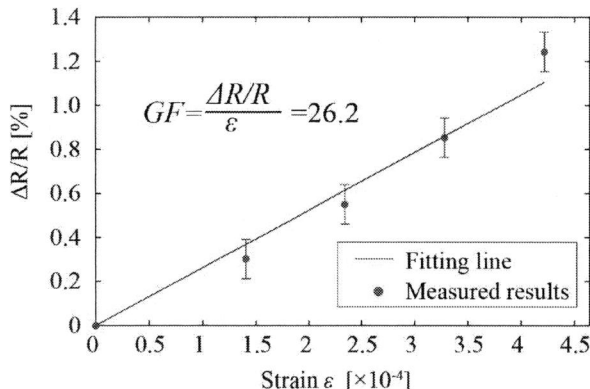

Figure 5: The relationship between relative resistance change of 2B graphite and the applied strains.

graphite indicates that GoP is a good candidate for low cost and highly sensitive transducers.

Demonstration of LGoP tactile sensors

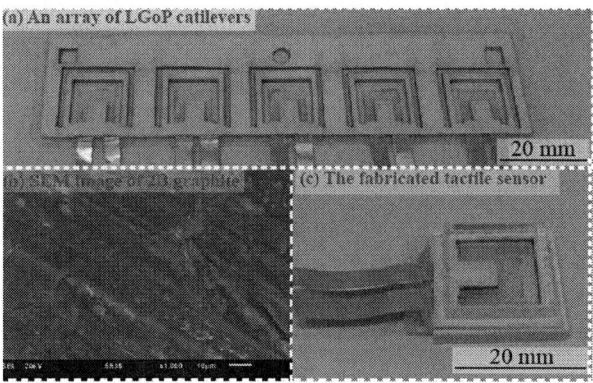

Figure 6: Photographs of fabricated tactile sensors.

Utilizing the piezoresistive effect of graphite and the lamination technique, LGoP tactile sensors have been developed with dimensions of 20 mm × 20 mm × 5 mm. Figure 6 shows photographs of the fabricated LGoP cantilevers and tactile sensors.

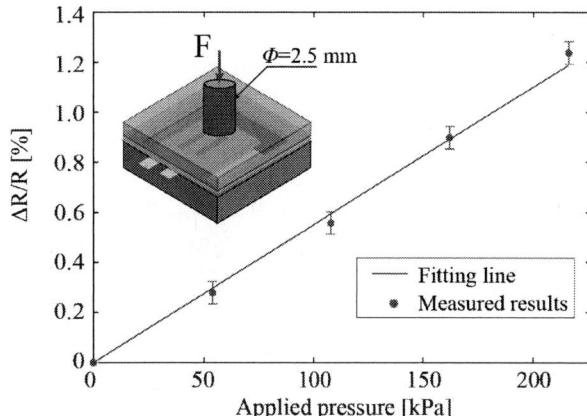

Figure 7: Response of the tactile sensors to the applied pressure.

By using the laser cutting technique, an array of LGoP can be easily fabricated, demonstrating the feasibility of mass production using the proposed method. The performance of the tactile sensor was characterized by applying pressure on the top surface of PDMS layer. Figure 7 shows the response of the developed tactile sensor to applied pressure varying from 0 to 200 kPa. As a result, the relative resistance change of LGoP was measured at 0.6×10^{-4} kPa^{-1}. From the signal to noise ratio, the resolution of the tactile sensor was 18 kPa.

CONCLUSION

In this paper, we reported on highly sensitive and low cost tactile sensors using plastic laminated graphite on paper. The laminated GoP not only enhances the robustness of paper but also improves its humidity-resistance, making it possible to embed LGoP inside elastomers for tactile sensing. Tactile sensors using LGoP sandwiched between a PDMS layer and an acrylic frame were fabricated and characterized, demonstrating the feasibility of using LGoP as a new platform for pressure sensing. With a high gauge factor graphite, a simple fabrication process, and a well-established laminating/laser cutting technique, paper based tactile sensors are good candidates for robotic applications in the future.

ACKNOWLEDGEMENTS

This work was performed in part at the Queensland node of the Australian National Fabrication Facility; a company established under the National Collaborative Research Infrastructure Strategy to provide nano and micro-fabrication facilities for Australia's researchers. This work has been partially supported by Griffith University's New Researcher Grants.

REFERENCES

[1] H. Yousef, M. Boukallel, and K. Althoefer, "Tactile sensing for dexterous in-hand manipulation in robotics—A review", *Sensors Actuators A: Physical.*, vol. 167, pp. 171-187, 2011.

[2] R. S. Dahiya, G. Metta, M. Valle, and G. Sandini, "Tactile Sensing—From Humans to Humanoids, *IEEE Trans. Robotics*, vol. 26, no. 1, pp. 1-20, 2010.

[3] V. Ho, D. V. Dao, S. Sugiyama, and S. Hirai, "Development and Analysis of a Sliding Tactile Soft Fingertip Embedded with a Micro Force/Moment Sensor", *IEEE Trans. Robotics*, vol. 27, no. 3, pp. 411-424, 2011.

[4] D. V. Dao, Q. Wang, and S. Sugiyama, "Fabrication and Characterization of 3-DOF Soft-Contact Tactile Sensor Utilizing 3-DOF Micro Force Moment Sensor", *IEEJ Trans. Sensors and Micromachines*, vol. 127, no. 3, pp. 177-181, 2007.

[5] K. Noda, K. Hoshino, K. Matsumoto, and I. Shimoyama, "A shear stress sensor for tactile sensing with the piezoresistive cantilever standing in elastic material", *Sensors Actuators A: Physical.*, vol. 127, pp. 295-301, 2006.

[6] M. Sohgawa, T. Mima, H. Onishi, T. Kanashima, M. Okuyama, K. Yamashita, M. Noda, M. Higuchi, and H. Noma, "Tactile sensor array sensor with inclined chromium/silicon piezoresistive catilevers embedded in elastomer", in *Proc. Int. Conf. Solid-State Sensors, Actuators and Microsystem (TRANDUCERS)*, 2009, pp. 284-287, 2009.

[7] H. P. Phan, P. Tanner, D. V. Dao, L. Wang, N. T. Nguyen, Y. Zhu, and S. Dimitrijev, "Piezoresistive Effect of p-Type Single Crystalline 3C-SiC Thin Film", *IEEE Electron Device Lett.*, vol. 35, no. 3, pp. 399-401, 2014.

[8] H. P. Phan, D. V. Dao, P. Tanner, L. Wang, N. T. Nguyen, Y. Zhu, and S. Dimitrijev, "Fundamental piezoresistive coefficients of p-type single crystalline 3C-SiC", *Appl. Phys. Lett.*, vol. 104, no. 11, pp. 111905, 2014.

[9] H. P. Phan, D. V. Dao, P. Tanner, J. S. Han, N. T. Nguyen, Y. Zhu, S. Dimitrijev, G. Walker, L. Wang, and Y. Zhu, "Thickness dependence of the piezoresistive effect in p-type single crystalline 3C-SiC nano thin films", *J. Matter. Chem. C*, vol. 2, no. 35, pp. 7176-7179, 2014.

[10] N. Minh-Dung, H. P. Phan, K. Matsumoto, and I. Shimoyama, "A hydrophone using liquid to bridge the gap of a piezoresistive cantilever", in *17th Int. Conf. Solid State Sensors, Actuators and Microsystem (TRANSDUCER)*, Barcelona, Spain, 2013, pp. 70-73.

[11] X. Liu, M. Mwangi, X. Li, M. O'Brien, and G. M. Whitesides, "Paper-based piezoresistive MEMS sensors", *Lab. Chip.*, vol. 11, pp. 2189, 2011.

[12] T. K. Kang, "Tunable piezoresistive sensors based on pencil-on-paper", *Appl. Phys. Lett.*, vol. 104, pp. 073117, 2014.

[13] N. T. Nguyen, and X. Y. Huang "Mixxing in microchannels based on hydrodynamic focusing and time-interleaved segmentation: Modelling and experiment", *Lab. Chip.*, vol. 5, pp. 1320-1326, 2005.

[14] C. W. Lin, Z. Zhao, J. Kim, and J. Huang, "Pencil drawn strain gauges and chemiresistors on paper", *Scientific Reports*, vol. 4, pp. 3812, 2014.

[15] T. Akter, J. Joseph, and W. S. Kim, "Fabrication of sensitivity tunable flexible force sensor via spray coating of graphite ink", *IEEE. Electron Device Lett.*, vol. 33, no. 6, pp. 902-904, 2012.

[16] T. L. Ren, H. Tian, D. Xie, and Y. Yang, "Flexible graphite-on-paper piezoresistive sensors", *J. Sensors.*, vol. 12, pp. 6685-6694, 2012.

[17] A. Bessonov, M. Kirikova, S. Haque, I. Gartseev, and M. J. A. Bailey, "Highly reproducible printable graphite strain gauge for flexible devices", *Sensor Actuator: A Physical.*, vol. 206, pp. 75-80, 2014.

[18] J. Zhao, C. He, R. Yang, Z. Shi, M. Cheng, W. Yang, G. Xie, D. Wang, D. Shi, and G. Zhang, "Ultra-sensitive strain sensors based on piezoresistive nanographene films", *Appl. Phys. Lett.*, vol. 101, pp. 063112, 2012.

CONTACT

* H. P. Phan, tel: +61 45 2423886; Queensland Micro and Nanotechnology Centre, Griffith University, 170 Kessels Road, Nathan, QLD 4111, Australia.
Email: hoangphuong.phan@griffithuni.edu.au

HIERARCHICAL WRINKLE STRUCTURING
ON INSIDE WALL OF CLOSED MICRO CHANNEL
A. Takei and H. Fujita
Center for International Research on Micronano Mechatronics, Institute of Industrial Science,
The University of Tokyo
4-6-1 Komaba, Meguro-ku, Tokyo, JAPAN
e-mail: {atakei,hfujita}@iis.u-tokyo.ac.jp

ABSTRACT

This paper presents a surface structuring method for a closed micro channel. Because fine photolithography can only be made on a flat surface, it has been difficult to make complicated patterns on the inside walls of non-planar microfluidic channels. Here, we demonstrated that micro-scale hierarchical patterns can be formed on the inside walls of the micro channel simply by depositing a plastic thin film in the channel and stretching the micro channel.

1. INTRODUCTION

In microfluidics, micro channels are fabricated using molds made of silicon or UV resin. In both case, the molds are made by using lithography, and the surface of the micro channels are generally flat. Stroock et al. reported that if a micro channel has micro structures on the surface, the mixing of fluids flowing in the channel is enhanced [1]. However, complicated processes such as multiple UV exposures with precise alignments are required to fabricate micro-structured molds for the structured micro channels [2]. In addition to the difficulty of the fabrication, complicated micro patterns such as three-dimensional hierarchical structures are difficult to be formed on the surface of the molds only by using photolithography based micromachining.

On the other hand, wrinkling phenomenon has been widely studied for non-photolithography based micro patterning. When the bi-layer system composed of a thin stiff film and an elastomeric substrate is deformed, wrinkle patterns appear on the surface. With compressive strain exerted to the bi-layer system, the elastomeric substrate shortens its length. Contrary, the thin stiff layer tends to keep its length by buckling, since the buckling is energetically more favorable than the simple compressing for the thin film. Consequently, the mismatch of the lengths between the film and the substrate is induced, and the wrinkle is formed on the surface to release the mismatch. The wavelength of the wrinkle is determined by the geometry and the mechanical properties of the bi-layer system (i.e. the thickness and the stiffness of the film and the substrate). Especially, if the thickness of the thin film is the order of micro-meter, the representative scale of the wrinkle becomes also the order of micro-meter. By controlling the direction of the strain applied to the bi-layer system, wavy one-dimensional wrinkles [3], random two-dimensional wrinkles [4] and also hierarchical wrinkle structures [5] can be obtained.

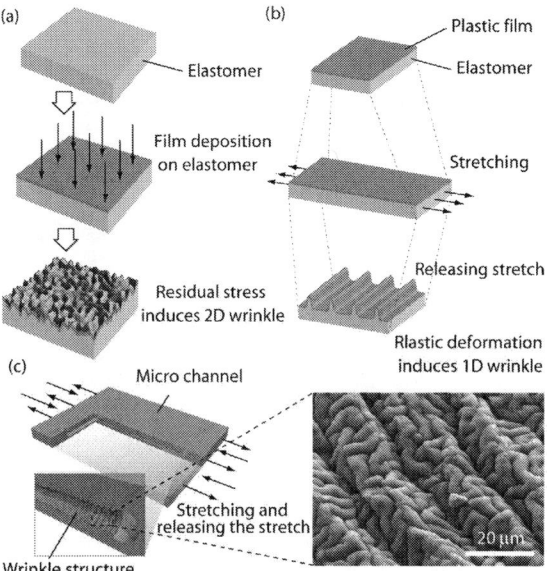

Figure 1 *Wrinkle structuring method.*
(a) Two-dimensional wrinkle structuring using residual stress: The residual stress induces the length difference between the film and the substrate, and the two-dimensional wrinkle is formed on the surface. (b) One-dimensional wrinkle structuring using plastic deformation: plastic film/elastomer substrate bi-layer system is stretched one-dimensionally, and the stretch is completely released. Due to the mismatch of the length, one-dimensional wrinkle is formed. (c) Hierarchical wrinkle structuring on inside wall: Plastic film is coated on the inside wall of the microchannel, and both one-dimensional and two-dimensional structuring method are applied.

By simply applying the strain, the micro pattern can be formed on the surface. Thus the wrinkle phenomenon is one of the promising methods of micro structuring.

In this paper, we propose a surface structuring method for inside walls on a closed micro channel. Hierarchical wrinkle pattern was formed on the surface of the micro channel by depositing Parylene (commercial polymer material, Special coating systems) thin film on it. The hierarchical wrinkle pattern was formed by the combination of two-dimensional and one-dimensional wrinkle formation induced by residual stress and plastic deformation, respectively. The deposition of Parylene on a

978-1-4799-7956-1/15 $31.00 © 2015 IEEE   829   MEMS 2015, Estoril, PORTUGAL, 18 - 22 January, 2015

heterogeneous material causes the residual stress in the Paryelene film [6]. We found that the two-dimensional wrinkle structure is formed on the surface of a bi-layer system composed of a Parylene film and PDMS substrate. If the Parylene film is deposited on the elastomeric substrate, the residual stress of the film compresses the substrate bi-axially, and the mismatch of the length between the film and the substrate induces the two-dimensional wrinkle formation as illustrated in Figure 1a. One-dimensional wrinkle pattern can be added to the two-dimensional wrinkle pattern by simply stretching the structure one-dimensionally. With excess stretch beyond the yield stress, Parylene is plastically deformed. At the release of the one-dimensional stretch, the elastomeric substrate recovers its initial length, while the film remains elongated due to the plastic deformation. Consequently, the mismatch of the length induces the one-dimensional wrinkle pattern (Figure 1b). If the two-dimensional and one-dimensional wrinkle patterns are formed sequentially on the same surface, the hierarchical wrinkle structure can be formed. The Parylene film can be deposited on the inside walls of the micro channel, and the hierarchical wrinkle structure can be formed on it by using our method as presented in Figure 1c. The paper present the typical morphology of the wrinkle patterns induced with our method. The relationship between the representative scale of the micro pattern and the geometry and the mechanical properties of the bi-layer system was quantified experimentally. In the end of the paper, we present the hierarchical pattern formed on the surface of the closed micro channel. The functionality of the structured micro channel is also presented in the paper. Two liquids flowing in the structured channel are mixed quickly. The proposed method opens the way for making structured walls in micro channels.

2. EXPERIMENT AND RESULT

Firstly, two-dimensional wrinkle formation was verified using a Parylene/PDMS bi-layer system. For simplification, a Parylene film was coated on a flat PDMS substrate. In this case, the base/catalyst ratio of the used PDMS was 20:1, and its shear modulus was 83 KPa. Before the deposition, the PDMS was previously spin-coated on a glass wafer and cured at 75degree C for 2hours to form 200μm thick substrate. Then 2μm thick Parylene film was deposited on the surface of the substrate. The Parylene deposition on a heterogeneous material surface causes two-dimensional residual stress in the film. The residual stress induces the length difference between the film and the substrate, and the two-dimensional wrinkle structure is formed. The surface of the sample is presented in Figure 2a and b. The morphology of the sample surface was analogous to a two-dimensional wrinkle structure on bi-layer systems reported in a previous work [7]. The representative width and the peak-to-peak height of the patterns were 10μm and 2μm, respectively. By simply depositing the Parylene film on the PDMS substrate, the two-dimensional

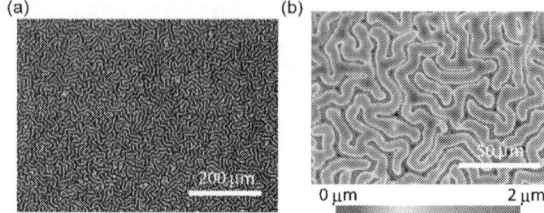

Figure 2 *Typical morphology of two-dimensional wrinkle induced by residual stress. 2μm thick Parylene film was deposited on 20:1 PDMS. (a) photograph and (b) profile. The width of the wrinkle pattern is the order of 10μm, and the peak-to-peak height is 2μm.*

structural surface can be obtained in micro-meter scale. The representative scale of the pattern was 10μm, and the substrate thick can be considered to be infinitely deep.

Secondly, an one-dimensional wrinkle structure was formed also by using a Parylene/PDMS bi-layer system. In a previous work [3], one-dimensional wrinkle structures were formed by depositing a stiff film on a stretched elastomeric substrate. Once the stretch was released, the length difference was induced between the film and the substrate, and the one-dimensional wrinkle pattern was formed on the surface. In this case, the elastomeric substrates were stretched with a mechanical stretcher during the film deposition. However, the space of the chamber of a Parylene deposition setup is limited, and the mechanical stretcher is not suitable to be entered with the bi-layer system. To overcome the spatial limitation during the deposition, we induced the length difference by using plastic deformation of the Parylene film. The Parylene film was deposited on a non-stretched PDMS substrate, and the bi-layer system was stretched one-dimensionally. By stretching the bi-layer system beyond the yield stress of the film, its length is not recovered to the initial length at the release of the stretch, while the elastomeric substrate recovers its initial length. This idea was verified experimentally. In the experiment, flat bi-layer system was used for the simplicity. By using the stiff PDMS (base/catalyst ratio=5:1 and 10:1), the emergence of the two-dimensional wrinkle structure caused by the residual stress was avoided. Figure 3a-c presents the morphology of the one-dimensional wrinkle structure with respect to compressive strain ε. By releasing the stretch completely, the wavy one-dimensional wrinkle was formed on the surface. The width of the wrinkle was quantified experimentally as a function of the thickness h_f, the stiffness of the film μ_F and the substrate μ_S, and the stretch λ, since no previous work has used plastic deformation to form wrinkle pattern. In the experiment, the thickness of the Parylene film was ranged from 0.9μm to 7.3μm, and the stretch λ was range from 1.4 to 2.4. The shear modulus of the used PDMS (5:1 and 10:1) were 460KPa and 270KPa, respectively. The experimental result was plotted in Figure 3d. Through the experiment, we found that the wrinkle width was given by

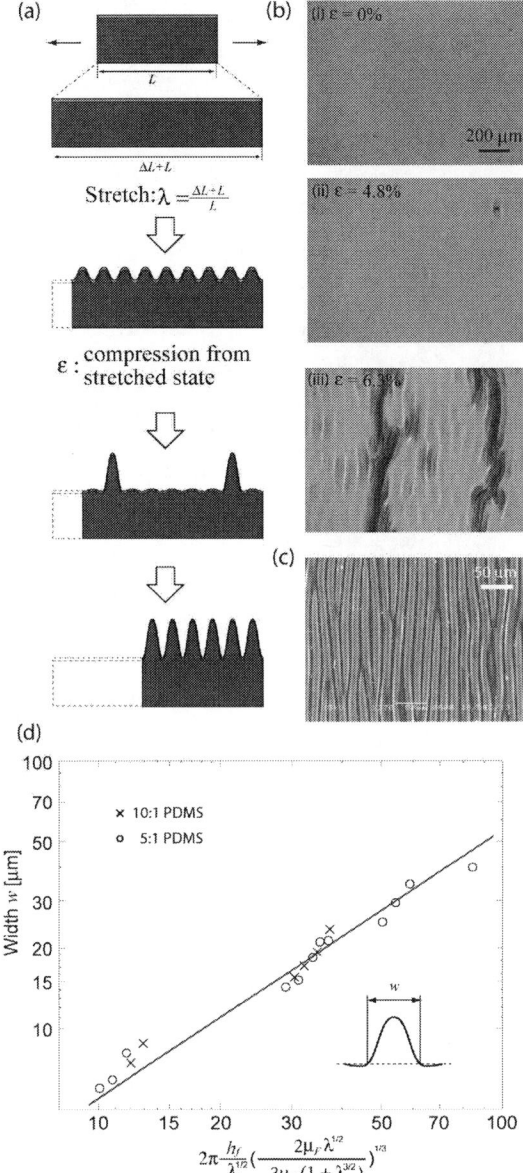

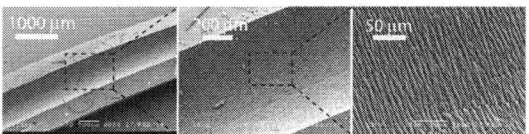

Figure 4 *Wrinkle structuring on inside wall of elastomeric tube. Parylene film was coated inside wall, and the tube was stretched. On the surface of the inside wall, the periodical pattern was formed perpendicular to the long axis of the tube.*

$$w = 2\pi \frac{h_f}{\lambda^{1/2}} \left(\frac{2\mu_F \lambda^{1/2}}{3\mu_S (1+\lambda^{3/2})} \right)^{1/3} \qquad (1)$$

The one-dimensional wrinkle whose width ranged from 4μm to 70μm was obtained. We set the shear modulus of the Parylene film beyond its yield stress μ_F to be 100 MPa. For the simplicity, the wrinkle width was measured when some peaks of the wrinkles increased as demonstrated in Figure 3b.iii. The width of the wrinkle was not changed significantly, as the compressive strain increased.

Thirdly, we verified our method is applicable to non-planer surface. Parylene film can be coated to the surface of non-planer structures such as inside walls of a tube or a microfluidic channel. 1μm thick Parylene film was coated on the surface of an elastomeric tube whose inner diameter was 1mm. The tube was stretched over $\lambda=2$, and the stretch was completely released. The surface of the tube is presented in Figure 4. Note that the tube was cut in half after the wrinkle formation for observation. One-dimensional wrinkle of 10μm period was formed on the inner surface of the tube. The surface structuring method using plastic deformation can be applied to non-planer structures.

Finally, we achieved the micro channel which has the hierarchical wrinkle patterns. The width and the height of the micro channel used in the experiment were 400μm and 100μm, respectively. The micro channel was fabricated by bonding an open PDMS channel and flat PDMS plate. The bonding was carried out by applying oxygen plasma to the surfaces of the two components and by contacting the surfaces. The base/catalyst ratio of the used PDMS was 20:1. The Parylene film was coated on the surface of the micro channel. During the deposition process, monomer of the Parylene goes through the micro channel from the inlet, and forms the polymer film of Parylene on the surface. ~2μm thick Parylene film was formed on the surface of the micro channel. Due to the residual stress of the Parylene film, the two-dimensional wrinkle structure was formed after the film deposition as shown in Figure 5a. After the formation of the two-dimensional wrinkle structure, the micro channel was stretched $\lambda=1.5$ parallel to the channel, then the stretch was completely released. The one-dimensional wrinkle pattern of 20μm period was formed perpendicular to the stretch direction. As a result of the combination of the one-dimensional wrinkle and the two dimensional structure, the hierarchical wrinkle pattern was formed on

Figure 3 *Formation of one-dimensional wrinkle structure. (a)Morphology of the wrinkle induced by plastic deformation: The compressive strain is applied to the bi-layer system by releasing the stretch. With small compressive strain (ε~5%), the aspect ratio between the wrinkle height and the wavelength is the order of 0.1. As the compressive strain increases (ε~6%), some peaks of the wrinkles increases their heights. When the stretch is released completely, the wavy wrinkle pattern is formed again. (b)Surface of the bi-layer system with (i)ε=0%, (ii)4.8%, (iii)6.3%. (c) SEM photograph of the bi-layer system when the stretch is completely released. (d) The width of the wrinkle pattern as a function of the thickness of the film, the stiffness of the film and the substrate, and the stretch.*

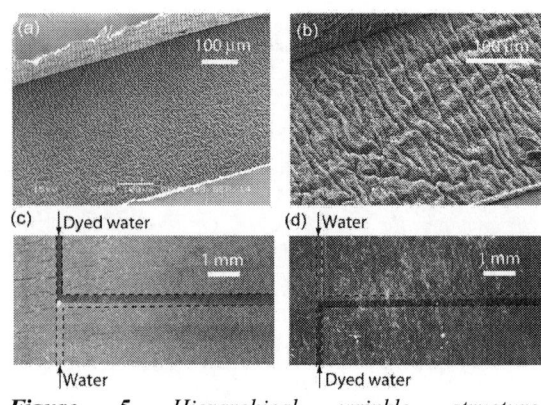

Figure 5 *Hierarchical wrinkle structured microchannel: SEM photograph of the micro channel (a) before and (b) after stretching. Before the stretch, two-dimensional wrinkle covered the entire bottom surface. After the stretch, the one-dimensional wrinkle structure of the wavelength of 20μm was added. Flow of water and dyed water in (c) the structured channel and (d) the non-structured channel. The mixing of the fluids in the structure channel was enhanced. In the non-structured channel, the mixing was not completed at the left edge of the image.*

the inner surface of the micro channel as shown in Figure 5b. Stroock et al. reported that surface structured microchannels enhance the mixing of heterogeneous liquids flowing in it [1]. To verify the mixing enhancement of our structured micro-channel, flow of DI water and dyed water in the structured micro channel was observed (Figure 6c). For the comparison, flow of the DI water and dyed water in a non-structured micro channel was also observed (Figure 6d). The fluids flowing in the structured channel was mixed quickly. The flow rate was ~1ml/min, and the Reynold's number was ~100.

5. CONCLUSION

In this paper, we proposed the surface structuring method for inside walls of closed channels. Using the wrinkle formation induced by the residual stress and the plastic deformation, the hierarchical structure was formed on the inside walls. The formation of the one-dimensional wrinkle was experimentally quantified by varying the thickness of the film, the stiffness of the substrate and the stretch. The mixing enhancement of the structured micro channel was experimentally confirmed. In literature, even though wrinkle patterns are formed on a flat structure, the surface can be used for a wide variety of functional surfaces such as a hydrophobic surface [8], a smart adhesion surface [9] and a template for cell cultivation [10]. Therefore, not only for mixing enhancement, the proposed method opens the way for making microchannels which have three-dimensional functional surfaces.

ACKNOWLEDGEMENT

We made the photo masks, using the University of Tokyo VLSI Design and Education Center (VDEC)'s 8-inch EB writer F5112+VD01 donated by ADVANTEST corporation.

REFERENCES

[1] A. D. Stroock *et al.*, "Chaotic Mixer for Microchannels," Science, 295, pp. 647-651, 2002.

[2] H. Sato *et al.*, "Improved Inclined Multi-lithography Using Water as Exposure Medium and Its 3D Mixing Microchannel Application," Sensors and Actuators A, 128, pp. 183-190, 2006.

[3] C. M. Stafford *et al.*, "A Buckling-based Metrology for Measuring the Elastic Moduli of Polymeric Thin Films," Nature Materials,. 3, pp. 545-550, 2004.

[4] N. Bowden *et al.*, "Spontaneous Formation of Ordered Structures in Thin Films of Metals Supported on an Elastomeric Polymer," Nature, 393, No. 1, pp. 146-149, 1998.

[5] C. Cao *et al.*, "Harnessing Localized Ridges for High-Aspect-Ratio Hierarchical Patterns with Dynamic Tunability and Multifunctionality," Advanced Materials, 26, pp 1763-1770, 2014.

[6] S. Dabral *et al.*, "Stress in Thermally Annealed Parylene Films," Journal of Electronic Materials, 21, pp. 989-994, 1992.

[7] P. C. Lin *et al.*, "Spontaneous Formation of One-dimensional Ripples in Transit to Highly Ordered Two-dimensional Herringbone Structures Through Sequential and Unequal Biaxial Mechanical Stretching," Applied Physics Letters, 90, 241903, 2007.

[8] J. Y. Chung *et al.*, "Anisotropic Wetting on Tunable Micro-Wrinkled Surfaces," Soft Matter, 3, pp. 1163-1169, 2007

[9] E. P. Chan *et al.*, "Surface Wrinkles for Smart Adhesion," Advanced Materials, 20, pp. 711-716, 2008

[10] M. Guvendiren *et al.*, "Stem Cell Response to Spatially and Temporally Displayed and Reversible Surface Topography," Advanced Healthcare Materials, 2, pp. 155-164, 2013

MICROFABRICATION OF WIDE-MEASUREMENT-RANGE LOAD SENSOR USING QUARTZ CRYSTAL RESONATOR

Yuichi Murozaki, Shinya Sakuma, and Fumihito Arai
Dept. of Micro-Nano Systems Eng., Nagoya University, Nagoya, JAPAN

ABSTRACT

We successfully established a wafer level fabrication process of the quartz crystal resonator (QCR) load sensor using atomic diffusion bonding. The proposed sensor has three-layer structures; two Si-hold layers and a quartz layer. Using microfabrication and atomic diffusion bonding, the assembly process was simplified. The fabrication process enables further miniaturization of the QCR sensor due to the simplified assembling method. The fabricated sensor is easily integrated in the outer package and can be designed the measurement range. Finally, we succeeded in multi-biosignals (heartbeat, body motion) detection using fabricated QCR sensor and the outer case.

INTRODUCTION

Load sensors using an AT-cut QCR have superior characteristics such as high accuracy, strength under compressive stress, long-term stability compared with commonly-used force sensors like strain gauge. QCRs have been researched as a force sensor employing retention mechanism that firmly support the QCR[1-5]. However, fabrication of the retention mechanism and the assembly process are complicated and miniaturization of the sensor is quite difficult in conventional sensors. Moreover the QCR is easily broken by bending. Under these circumstances, we previously proposed a novel design and fabrication method for a QCR load sensor[5]. The fabricated sensor was drastically miniaturized to 24.6 mm³ and had wide-measurement-range (5×10^5) (Figure 1). The proposed sensor has three-layer structures; two Si-hold layers and a quartz layer as shown in Figure 2. However, bonding process of the quartz layer and silicon layers is difficult, because quartz crystal and silicon have a large difference in coefficient of thermal expansion. In the past, adhesive was used to assemble the three-layers. In this case,

each sensors is bonded one by one. Therefore, this process makes the fabrication process complex and low throughput. Furthermore, adhesive attached on the surface of the QCR effects the property of the QCR. The fabricated sensors vary widely in the quality. Low quality of the QCR causes decreasing stability of the sensor output. In this work, by applying the wafer level bonding replacing the adhesive, we established the wafer level fabrication process of the QCR load sensor that enables mass production and expansion of the measurement-range. Also, we evaluated the fundamental characteristics and demonstrated the effectiveness of a fabricated QCR load sensor.

FABRICATION OF THE SENSOR
Method

As mentioned in introduction, the proposed QCR load sensor has three-layer structures. The quartz layer has gold electrodes deposited by sputtering. The slits of the quartz layer and the Si-hold layers were formed using sand blasting and DRIE, respectively. The fabrication process of the all layers is based on photolithography. It enables mass manufacturing of the sensor. Here, we employed atomic diffusion bonding of Au-Au film for the bonding process[6]. It enables to bond wafer under room temperature, low pressure and atmosphere condition. Therefore, we can avoid the thermal effect in bonding process. Which is caused by the big difference in coefficient between the quartz layer and the silicon layers, when we apply the commonly used

Process

The bonding process of the sensor fabrication is shown in Figure 3. The quartz layer and Si layers have arrayed patterns of sensors and holes for the alignment. The bonding condition is affected by cleanness and quality

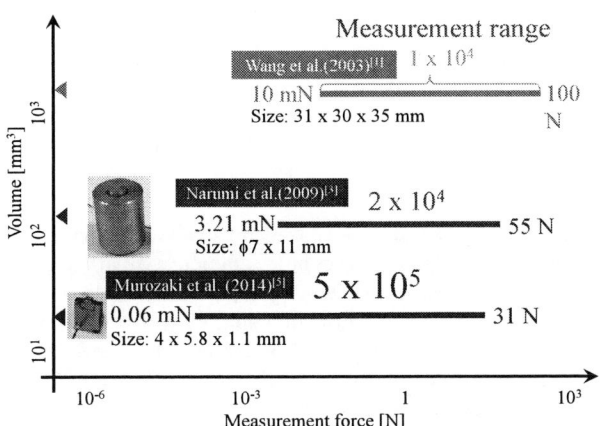

Figure 1: Comparison of the force sensor using QCR[1],[3],[5].

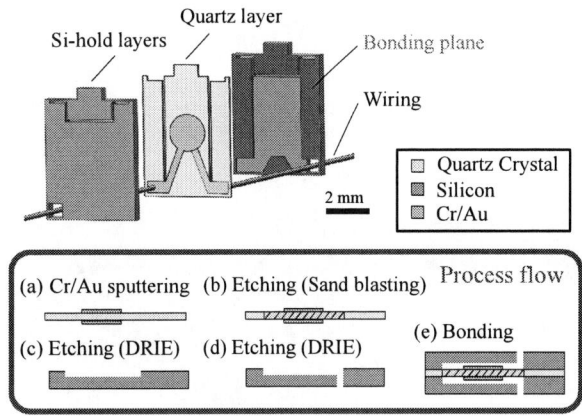

Figure 2: Schematic and process flow of QCR sensor.

978-1-4799-7956-1/15 $31.00 © 2015 IEEE   833   MEMS 2015, Estoril, PORTUGAL, 18 - 22 January, 2015

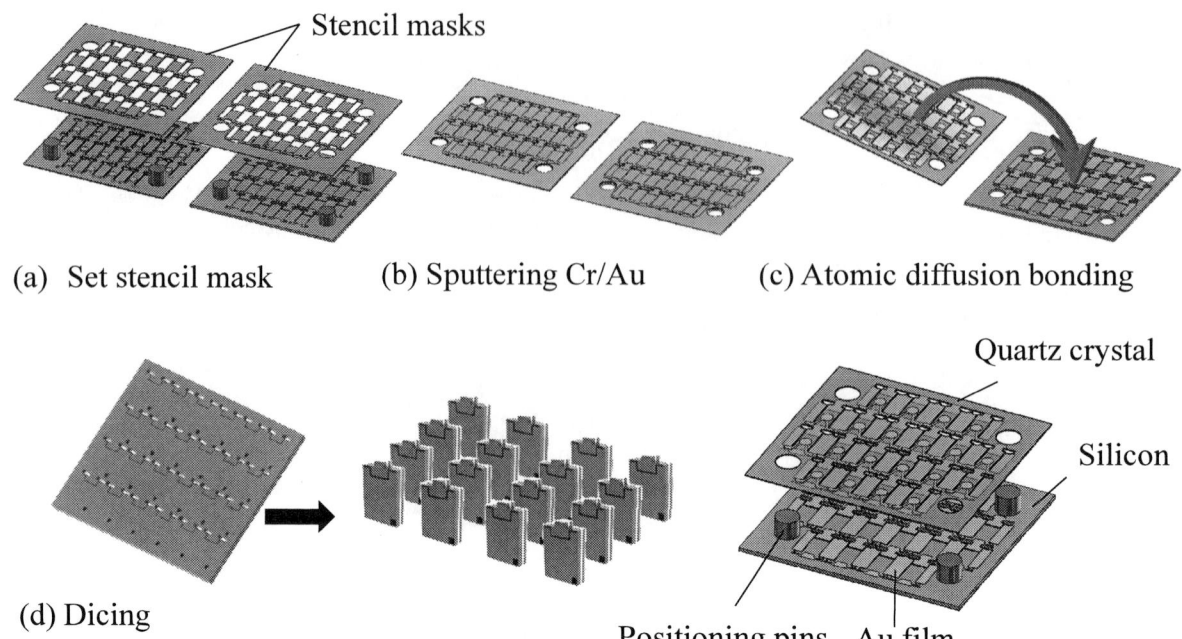

(a) Set stencil mask   (b) Sputtering Cr/Au   (c) Atomic diffusion bonding

(d) Dicing

Figure 3: Bonding process of the sensor fabrication.

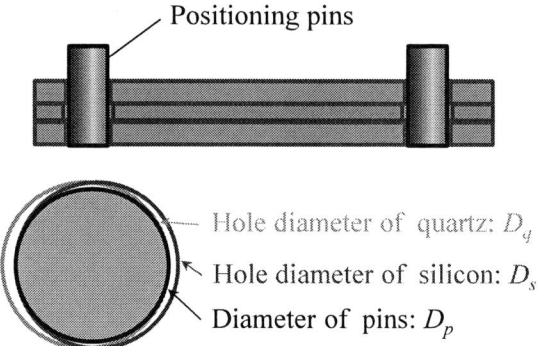

Figure 4: Accuracy of the alignment using positioning pions.

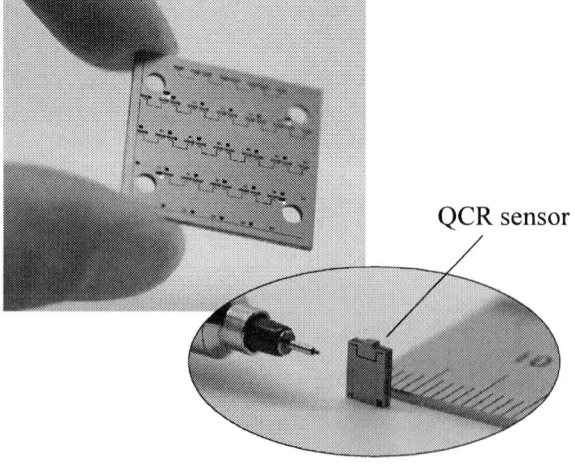

Figure 5: Pictures of fabricated QCR sensor.

of the metal film. Therefore, the etching process or lift off process is not suitable for patterning of the Au film in the atomic diffusion bonding process. Instead of the photoresist patterning, we employed a stencil mask to pattern the Au film. Au film is formed only the holed parts of stencil mask. The details of fabrication process are following. (a) First, the stencil mask was set on the quartz layer and Si layer using the positioning pins. (b) Cr film and Au film were deposited on the surface of the quartz layer and silicon layers thorough the lodes of stencil mask by sputtering. (c) After stencil masks were removed, two layers were placed together and pressed. Positioning pins were also used for alignment of the assembly. The positioning pins allow simplifying the alignment. Contamination of the metal file is minimized as a result of reducing assembling operation time. (d) Finally, the arrayed sensors were cut into each sensors using dicing saw.

In this fabrication process, the accuracy of the alignment is depending on the diameters of the pins and holes. Error range of the alignment (E_a) is given by following:

$$E_a < \{(D_s\text{-}D_p)\text{+}(D_q\text{-}D_p)\}/2 \qquad (1)$$

where, D_p, D_q, and D_s show the diameter of pins, hole diameter of quartz, and hole diameter of silicon. It is required to design the holes pattern considering accuracy of the sensor. For example, the quartz layer was formed by sand blasting. Using sand blasting, tapered bore was typically formed. In this time, D_p, D_q, and D_s were 3.15 mm, 3.29 mm, and 3.20 mm, respectively. Therefore, accuracy of the alignment was less than 95 μm calculated by equation (1). The assembled sensors and a single QCR load sensor are shown in Figure 4. We succeeded in assembly of the arrayed sensors using atomic diffusion bonding. The fabrication process is applicable to wafer level fabrication process.

EVALUATION OF THE SENSOR
Load Characteristics

We calibrated the loading characteristic of the fabricated QCR load sensor by the loading test. In the experiment, the sensor was put on an oscillation circuit and connected. The sensor output was measured using a frequency counter (53230A, Hewlett Packard). The QCR load sensor was fixed on the Z-stage, and electronic balance (UW4200A, SHIMADZU) was set under the sensor as reference signal of applied load. The results are shown in Figure 6. Load was applied in 1 N increments from 0 N to 20 N, and then unloaded in 1 N decrements from 20 N to 0 N by Z-stage moving up and down, respectively. The relationship between the applied load and the frequency shift was approximated by

$$y = 1524\,x + 16287243 \tag{2}$$

where, x and y are applied load [N] and the oscillating frequency [Hz], respectively. The correlation coefficient between the experimental results and the linear approximation is $R^2 = 0.9996$. The experimental results showed that the nonlinearity and hysteresis were 1.46 % of the full scale (F.S.) and 0.37 % F.S., respectively.

Stability of Sensor Output

Time stability of the sensor output was measured. The measurement was conducted after the sensor output reached steady state. Figure 7 shows fluctuation of the sensor output for 3 minute. Here, the error shows relative error between average value of the measurement frequency and measured frequency. The frequency fluctuations of the fabricated sensor were 1.9 ppb for 3 minutes. Compared to the sensor fabricated by adhesive (5.5 ppb[5]), the fabricated sensor had 3 times superior stability with regard to the sensor output. From these results, we can confirm that the stability of the sensor output is improved by omitted effect of adhesive, and the proposed sensor has the high stability feature. The fluctuation of the external load estimated from the fluctuation of the sensor output was 0.02 mN for 3 min.

Application to Biosignlas Detection

We conducted experiments of biosignal detection on a seat. In the experiment, we fixed the fabricated sensor on the oscillation circuit, which reduces sensor output error caused by parasitic capacitance of wiring connection between the sensor and the circuit, and packaged by an outer case to shift the measurement range upward and to enhance the durability against unexpected force. The packaged sensor was placed in a plate. Semilunar bar is fixed on the sensor by bolts. Applying load to the bar can be measured. The base with sensor is set on chair and covered by cushion for comfortableness as shown in Figure 8.

Figure 9 shows the typical measurement result on a subject (24 years old, male, 182 cm tall). During the subject sat on the chair, the sensor output periodically

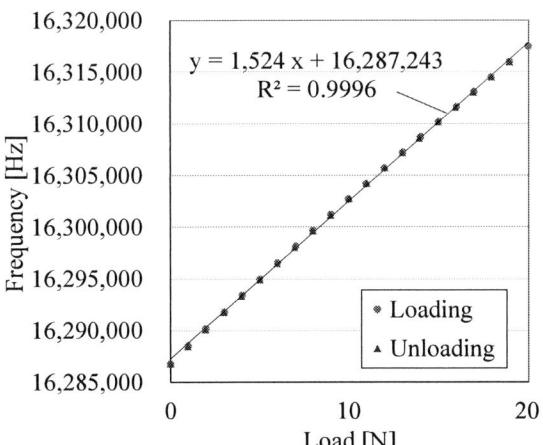

Figure 6: Result of loading test.

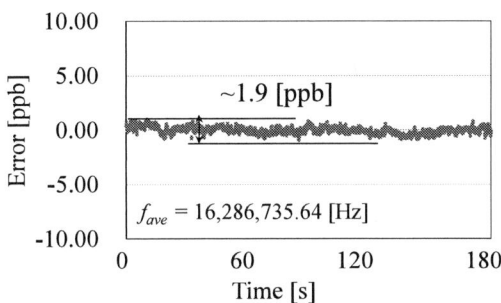

Figure 7: Stability of the sensor output.

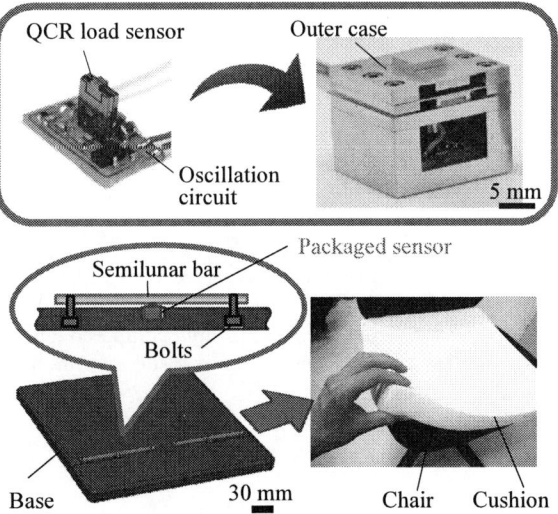

Figure 8: Measurement setup of biosignals detection.

changed approximately 5 s cycle. This periodic could be caused by breathing. Also, small and faster periodic changes were observed when the subject sat on the chair. This periodic changes were not observed while the subject stood up. The measured frequency has approximately 64 peaks per minute. This periodic change could be caused by heartbeat. The magnitude of the pulse signals is less than

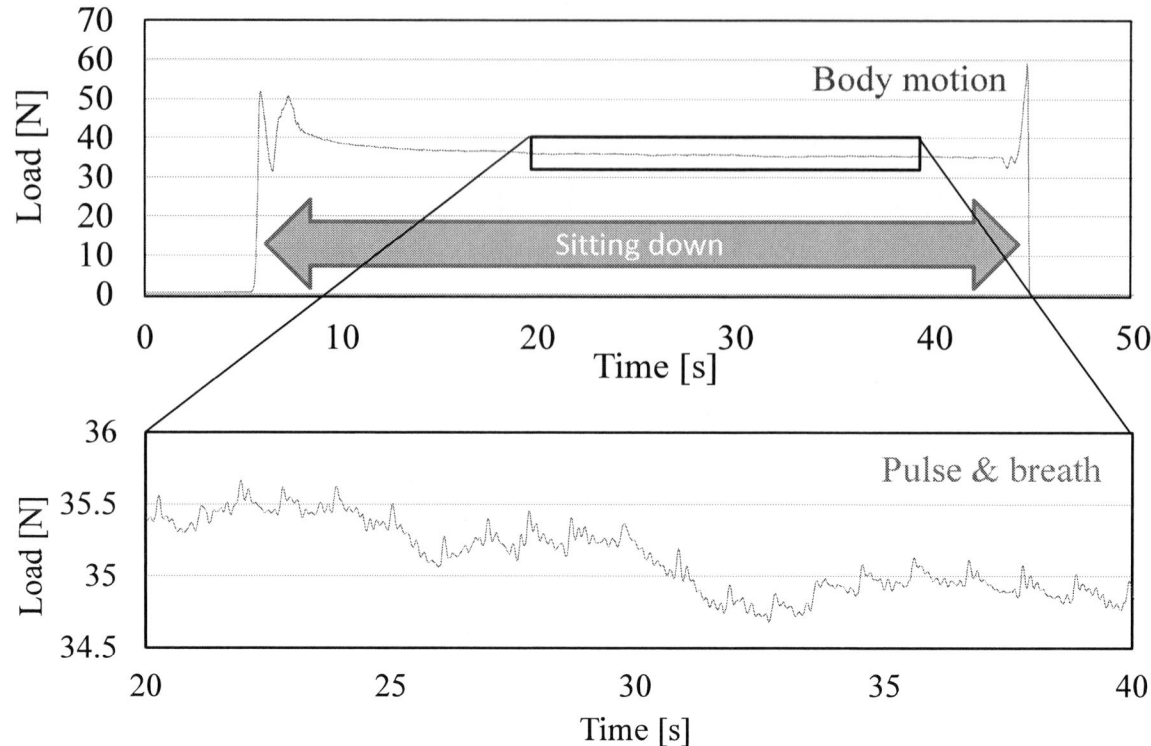

Figure 9: Result of biosignals detection in chair.

100 mN. The load was increasing with the motion of standing up, and approximately 60 N was applied to the sensor. It seems that body motion of the standing up can be measured. From these results, we concluded that we succeeded in multi-biosignals detection using fabricated QCR load sensor with the outer case.

CONCLUTION

We established the fabrication process of QCR load sensor that enables to fabricate wafer level arrayed sensors. Arrayed sensors were successfully assembled by employing atomic diffusion bonding of Au-Au film for the bonding process. The fabricated sensor had superior stability with regard to the sensor output compared with the sensor fabricated by adhesive. The measurement range of the sensor was expanded to 1.6×10^6 (0.02 mN to 31 N). Furthermore, we succeeded in multi-biosignals detection using fabricated QCR sensor with outer case.

ACKNOWLEDGEMENTS

This work was supported by Center of Innovation Program.

REFERENCES

[1] Z. Wang, H. Zhu, Y. Dong, and G. Feng, "A thickness-shear quartz resonator force sensor with dual-mode temperature compensation," *IEEE Sensors Journal*, vol. 3, pp. 490–496, 2003.

[2] A. Asakura, T. Fukuda, and F. Arai, "Design, fabrication and characterization of compact force sensor using AT-cut quartz crystal resonators," *J.* *Robot Mechatron,* vol. 21, No. 2, pp. 260-266, 2009

[3] K. Narumi, A. Asakura, T. Fukuda, and F. Arai, "Compact force sensor using AT-cut quartz crystal resonator supported by novel retention mechanism," *J. Rob. Mechatron.*, vol. 21, No. 2, pp. 260–266, 2009

[4] Y. Murozaki, and F. Arai, "Wide-range load sensor using quartz crystal resonator for biological signal detection," *in IEEE Inter. Conference on ICRA*, Hong Kong, May 31-June 7, 2014, pp. 4405-4410.

[5] Y. Murozaki, K. Nogawa and F. Arai, "Miniaturized load sensor using quartz crystal resonator constructed through microfabrication and bonding," *Robomech J.*, vol. 1, 2014.

[6] T. Shimatsu and M. Uomoto, "Atomic diffusion bonding of wafers with thin nanocrystalline metal films," *J. Vacuum Science & Technology*, vol. 28, pp. 706-714, 2010.

CONTACT

*Y. Murozaki, tel: +81-52-7895220;
murozaki@biorobotics.mech.nagoya-u.ac.jp

SELF-CURVED DIAPHRAGMS BY STRESS ENGINEERING FOR HIGHLY RESPONSIVE PMUT

Sina Akhbari[1], Firas Sammoura[1,2], Chen Yang[1], Amir Heidari[3], David Horsley[3], and Liwei Lin[1]

[1]Department of Mechanical Engineering, University of California at Berkeley, Berkeley, USA
[2]Department of Electrical Engineering and Computer Science, Masdar Institute of Science and Technology, Abu Dhabi, UAE
[3]Department of Mechanical and Aerospace Engineering, University of California at Davis, Davis, USA

ABSTRACT

A process to make self-curved diaphragms by engineering residual stress in thin films has been developed to construct highly responsive piezoelectric micromachined ultrasonic transducers (pMUT). This process enables high device fill-factor for better than 95% area utilization with controlled formation of curved membranes. The placement of a 0.65 µm-thick, low stress silicon nitride (SiN) film with 650 MPa of tensile residual stress and a low temperature oxide (LTO) film with 180 MPa of compressive stress sitting on top of a 4 µm-thick silicon film has resulted in the desirable self-curved diaphragms. A curved pMUT with 200 µm in nominal radius, 2 µm-thick aluminum nitride (AlN) piezoelectric layer, and 50% SiN coverage has resulted in a 2.7 µm deflection at the center and resonance at 647 kHz. Low frequency and resonant deformation responses of 0.58 nm/V and 40nm/V at the center of the diaphragm have been measured, respectively. This process enables foundry-compatible CMOS process and potentially large fill-factor for pMUT applications.

INTRODUCTION

Recent advancements in the pMUT technologies have attracted great attentions in potential applications in consumer electronics, such as gesture recognition, range finding, and medical imaging [1-4]. An ultrasound system consisting of a large array of pMUT elements [5] can carry out acoustic beam forming and focusing [6-7] with the assistance of microelectronics. From a system design perspective, there are two key criteria for the effective performance of ultrasonic systems: (1) optimizing the individual pMUT element for effective electromechanical coupling, and (2) designing the whole array system with high fill factor for efficient area utilization and enhanced output acoustic pressure generation [8].

Theoretical models for flat-shape pMUTs have been well reported. For example, the deflection equation of a pMUT with the circular plate and circular/ring electrode design has been explicitly derived using the approach of Green's function [9]. The largest center plate deflection per unit input voltage was achieved when the electrode radius coverage was 60% of the whole plate radius using a central electrode [10]. Previously, the electromechanical coupling efficiency of pMUTs was investigated for the design of multiple electrode structures [11]. It is found that 100% higher acoustic output per unit input voltage could be obtained for a two-port electrode design as compared to a conventional single-port pMUT [12]. Furthermore, a bimorph pMUT design with two active AlN layers can achieve 400% higher acoustic outputs as compared to a unimorph pMUT with similar sizes [13].

Our group has developed the "curved pMUT" devices both analytically and experimentally to realize improved electromechanical coupling as compared to flat-pMUTs [14-16]. Previously, a HNA wet etching step was used to construct the curved surface before the aluminum nitride piezoelectric thin film was deposited. Since only part of the curved surface is used as the structural diaphragm, some of the wafer areas are not utilized and the device fill-factor is low. This work presents the concept of self-curved diaphragms by stress engineering instead of the wet etching process with three features: (1) fabrication of self-curved diaphragms without the wet etching process; (2) controllable designs of diaphragm curvature by the combination of residual stresses in thin films and their sizes; and (3) high fill-factor to construct self-curved pMUT arrays.

CONCEPT

Figure 1 shows the 3D schematic diagram of the stress engineered, self-curved pMUT. The curved structure is realized by a piezoelectric AlN layer sandwiched between a bottom and a top metal electrode on top of a silicon diaphragm with a self-generated curvature due to residual stresses in the films. Specifically, the silicon nitride and silicon oxide layers with known tensile and compressive stress, respectively, are introduced on top of the device layer on a SOI wafer to induce the targeted concave-shape structure. The final curvature of the diaphragm is caused by the balance of stresses in various thin films and can be adjusted by the size and properties of the thin films. In the prototype demonstrations, suspended diaphragms can bend downward as illustrated without unutilized portions as those fabricated previously by the wet etching process. As such, high fill factor can be achieved.

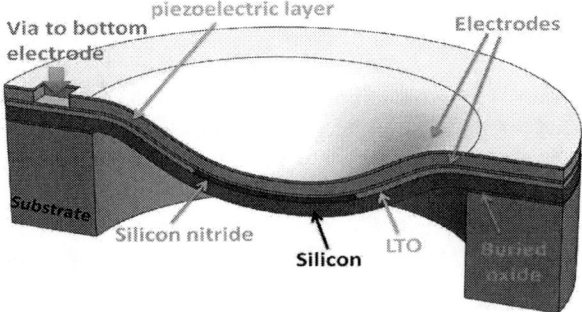

Figure 1: 3D cross-sectional view of a stress engineered curved pMUT fabricated in a CMOS-compatible process.

978-1-4799-7956-1/15 $31.00 © 2015 IEEE

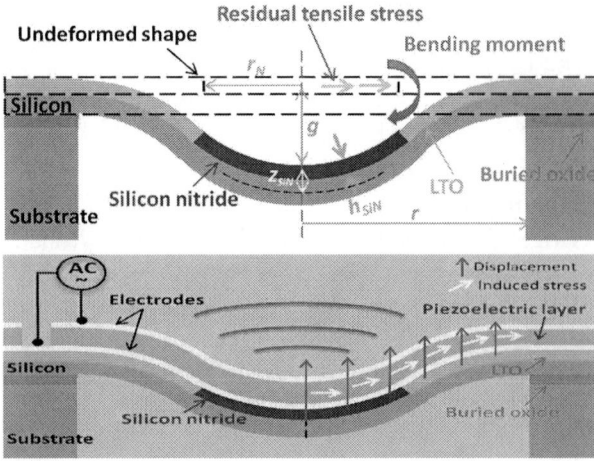

Figure 2: (a, top) Cross-sectional view of the concave-shape diaphragm by the stress engineering process due to the SiN and LTO films with tensile and compressive residual stresses, respectively; (b, bottom) After adding the bottom and top electrodes and the AlN layer to complete the stress engineered curved pMUT fabrication.

The cross-sectional diagram in Figure 2a details the stress engineering design. The combination of tensile stressed silicon nitride layer (partially covering the central region of the circular diaphragm) and the compressive stressed LTO (covering the rest of the diaphragm) results in the concave-shape structure. Analytically, if a flat, stress-free, clamped diaphragm is deflected downward, radial tensile stress is formed at the outer portion and radial compressive stress is established at the inner portion of the top surface of the diaphragm. The stress neutral line (zero stress) or the inflection circle is located at ~0.65r position, where r is the radius of the diaphragm. Therefore, the stress-free concave-shape diaphragm as illustrated in Fig. 2a can be achieved by placing a thin film with tensile residual stress at the inner portion and compressive stress in the outer portion of a flat diaphragm. Once the residual stresses are released, a curved downward diaphragm can be self-constructed.

The curvature of the self-curved diaphragm can be designed to achieve a desirable center deflection, g, by tuning the silicon nitride and oxide residual stresses σ_{SiN} & σ_{Ox}, the silicon nitride and oxide thickness h_{SiN} & h_{Ox}, their distances from neutral axes Z_{SiN} & Z_{Ox}, Poisson's ratios v_{SiN} & v_{Ox}, and the coverage radius r_N. The radial force per unit length due to the residual stress in the nitride film is simplified as $\sigma_{SiN}h_{SiN}$, and the moment per unit length generated by the nitride layer about the neutral axis of the diaphragm stack is $\sigma_{SiN}h_{SiN}Z_{SiN}$. The residual stresses in the thin films can generate high moments to bend the released diaphragm after the backside silicon is etched away [17]. The deformation profile, $W_s(r)$, is:

$$W_s(r) = \frac{\pi\sigma_{SiN}h_{SiN}Z_{SiN}}{rD(1-\upsilon_{SiN})}\sum_k \frac{O_k(r_N)}{\Lambda_k\Gamma_k}\Psi_k(r)$$
$$-\frac{\pi\sigma_{Ox}h_{Ox}Z_{Ox}}{rD(1-\upsilon_{Ox})}\sum_k \frac{O_k(r_N)}{\Lambda_k\Gamma_k}\Psi_k(r) \qquad (1)$$

where r and D are the diaphragm nominal radius and flexural rigidity, respectively and O_k, Ψ_k, Λ_k, and Γ_k are functions defined in [17].

By adding the bottom and top electrodes and the AlN layer to complete the fabrication process after Fig. 2a, the stress engineered curved pMUT can operate as shown in Fig. 2b in the transmission mode under an AC voltage. The induced stress in the piezoelectric layer due to the d_{31} effect stretches and compresses the diaphragm, such that it resonates in the flexural mode to emit acoustic waves. The induced stress due to d_{31} has a vertical component in the desired vertical motion to enhance electromechanical coupling of the device.

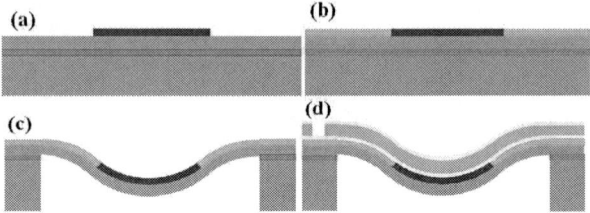

Figure 3: Process flow for the stress-engineered curved pMUT: (a) silicon nitride deposition and patterning; (b) LTO deposition and CMP; (c) backside DRIE to form the concave-shape diaphragm; (d) Mo/AlN/Mo sputtering and via opening to the bottom electrode.

FABRICATION

Process Flow

Figure 3 shows the process flow of the stress engineered self-curved pMUT. The process starts with the deposition and pattering of a 650 nm-thick silicon nitride layer with naturally inherent tensile residual stress (650MPa in our lab) on a SOI wafer with a 4 µm-thick device layer and a 1 µm-thick BOX layer (Fig. 3a). The next step is LTO deposition followed by chemical mechanical polishing (CMP) (Fig. 3b). There are two purposes for LTO deposition and CMP: (1) to smoothen out the surfaces for the future Mo/AlN/Mo sputtering on the diaphragm area, and (2) to further help the curvature formation by using the LTO residual compressive stress, which in our case is 180 MPa (compressive). Backside deep reactive ion etching (DRIE) is then used to release the self-curved diaphragm (Fig. 3c). After the BOX layer under the diaphragm is removed, the diaphragm bends in a concave form due to the residual stresses of the nitride and oxide thin films before the depositions of electrode layers and AlN piezoelectric layer using active sputtering of Mo/AlN/Mo as bottom electrode, piezoelectric layer, and top electrode with 150nm, 2µm, and 150nm in thickness, respectively. The via to the bottom electrode is opened using a combination of dry and wet AlN etching steps by chlorine based plasma and the MF-319 developer, respectively (Fig. 3d) in order to reduce the damages to the Mo bottom electrode layer. Top Mo was patterned beforehand using SF_6 plasma etching.

Fabrication Results

Figure 4 shows confocal laser scanned images captured using Olympus LEXT OLS4000 3D Confocal Laser Microscope of a fabricated, self-curved pMUT with

400μm in diameter, 50% nitride coverage, and measured center deflection of 2.7μm.

Figures 5a & 5b are SEM micrographs of two self-curved pMUTs portraying the clamped and curved diaphragm. Figure 5c shows cross-sectional view of the diaphragm stack composed of - from bottom to top, the buried oxide, silicon device, silicon nitride, and LTO layers as well as the Mo bottom electrode, AlN layer, and the top Mo electrode, respectively. Figure 5d is a close-up view on the AlN illustrating good crystal orientation.

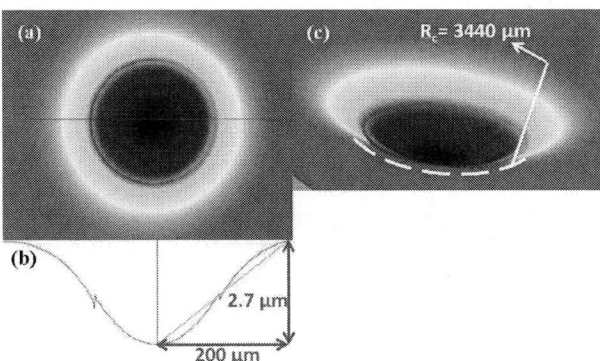

Figure 4: *Confocal laser scanned image of a fabricated curved pMUT (a) top view; (b) measured curvature profile; (c) 3D tilted view and the radius of curvature.*

RESULTS AND DISCUSSIONS

The center diaphragm deflection, g, versus the silicon nitride radial coverage percentage, r_N, is shown in Figure 6 for a diaphragm with a nominal radius of 200 μm and silicon thickness of 4 μm. The 650 nm–thick SiN has a tensile residual stress of 650 MPa and the LTO has a compressive residual stress of 180 MPa. Results show good consistency between theory (coded in Matlab™), simulation (COMSOL), and experimental data. It is observed that the higher nitride coverage results in higher center deflection for the range of nitride coverages between 40%-55%. Since the curvature of the diaphragm can affect both the resonant frequency and the excited deformation of the devices, the SiN radial coverage ratio can be used in the design process to optimize the device performances. If the coverage percentage increases to be above the inflection circle (roughly 65%-70% of the radius of the diaphragm), the center deflection will start to reduce as compressive regions of the diaphragm can start to reduce the bending moment. The optimal design values can be analyzed or simulated with known properties and parameters of the thin films.

The dynamic responses of a fabricated curved pMUT without (blue) and with (red) the bottom silicon layer are measured using Laser Doppler Vibrometer (LDV) and presented in Figure 7. Resonant frequency reduces from 646.7 to 520 kHz while low frequency displacement remains at 0.58 nm/V after the removal of the silicon layer. It is expected from Finite Element Modeling (FEM) that the released diaphragm without silicon would have lower resonant frequency of 381 kHz and higher low-frequency displacement of 8.5 nm/V as compared to the measured values. The discrepancy between the theoretical and experimental data is attributed to the

excessive residual stress in the as-deposited AlN layer (tensile 170 MPa).

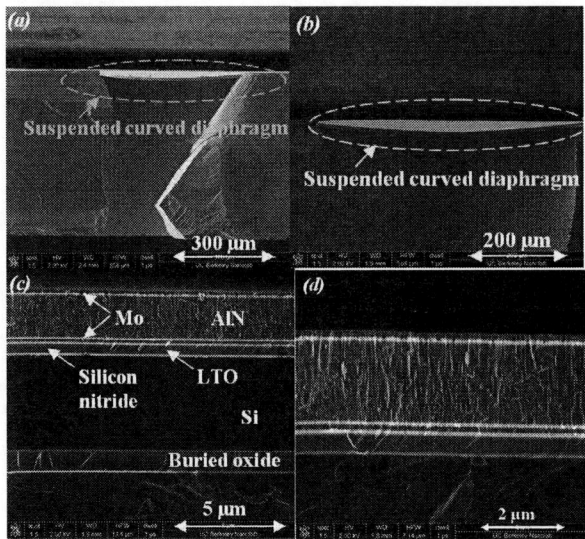

Figure 5: (a, b) Tilted and front view SEM micrographs of two self-curved pMUTs after the devices are cleaved; (c) a released diaphragm showing the stack of the pMUT layers, and (d) close-up view showing good crystal alignment of AlN on the curved diaphragm.

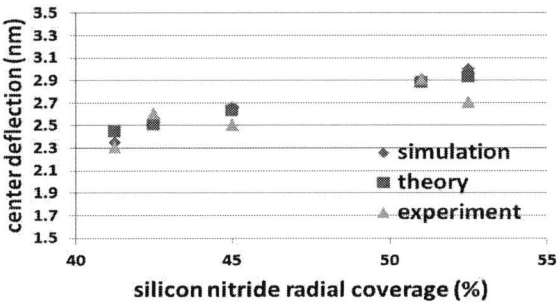

Figure 6: Center deflection versus nitride radial coverage (%) for devices with 200μm in nominal radii using a 650nm-thick nitride layer. Results show good consistency among simulation, theory, and experimental data.

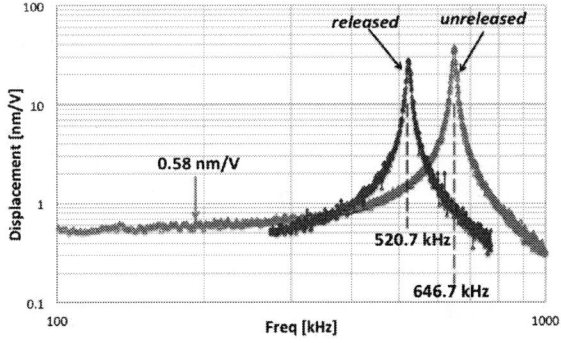

Figure 7: Measured dynamic responses of stress-engineered curved pMUTs without (blue) and with (red) the bottom silicon layer. The pMUTs have 200μm in nominal radius and 2.7μm center deflection before release. The AlN, Si, and BOX layer thicknesses are 2μm, 4μm, and 1μm, respectively.

Figure 8 shows the effects of residual stress in AlN

on the dynamic responses of stress engineered curved pMUT devices. As the residual stress in the AlN increases, the low-frequency displacement per unit input voltage drops and the resonant frequency increases. It is expected that the device performance would match with the simulated values when the stress in the sputtered AlN is controlled to be within 30 MPa.

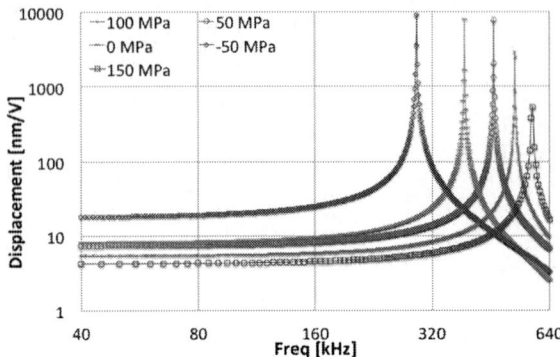

Figure 8: Simulated dynamic responses of a stressed engineered curved pMUT with 200 μm in nominal radius and 2.34 μm center diaphragm displacement for -50, 0, 50, 100, and 150 MPa residual stress in the AlN layer.

ACKNOWLEDGEMENTS

The authors would like to thank Jeffery Clarkson, Ryan Rivers, and Jay Morford from the UC Berkeley Marvell nanofabrication laboratory and Mr. Hadi Najar from the Mechanical and Aerospace Engineering Department of UC Davis for valuable helps. This project is supported in part by the Masdar Institute of Science and Technology, Abu Dhabi, UAE.

REFERENCES

[1] B. E. Boser *et al.*, "Ultrasonic transducers for navigation," *Proceedings of Meetings on Acoustics*, vol. 19, pp. 030087, Montreal, Canada, 2-7 June 2013.

[2] R. Przybyla *et al.*, "3D ultrasonic gesture recognition," in Proceeding of 2014 IEEE International Solid-State Circuits Conference Digest of Technical Papers (ISSCC), pp. 210-211, September 2, 2014.

[3] D. E. Dausch *et al.*, "Piezoelectric Micromachined Ultrasound Transducer (pMUT) arrays for 3D Imaging Probes," *in Proc. IEEE Int. Ultrason. Symp. (IUS)*, pp. 934-937, Vancouver, BC, Canada, October 2-6, 2006.

[4] R. Przybyla et al., "An ultrasonic rangefinder based on an aln piezoelectric micromachined ultrasound transducer", *in Proc. IEEE Sensors 2010*, pp. 2417-2421, Nov. 2010.

[5] S. Shelton *et al.*, "CMOS-compatible AlN piezoelectric micromachined ultrasonic transducers," *in Proc. IEEE Int. Ultrason. Symp. (IUS)*, pp. 402-405, Rome, Italy, September 20-23, 2009.

[6] D. E. Dausch *et al.*, "Improved pulse-echo imaging performance for flexural-mode pMUT arrays," *in Proc. IEEE Int. Ultrason. Symp. (IUS)*, pp. 451-454, San Diego, CA, USA, October 11-14, 2010.

[7] S. H. Wong *et al.*, "Advantages of capacitive micromachined ultrasonics transducers (CMUTs) for high intensity focused ultrasound (HIFU)," *in Proc. IEEE Ultrason. Symp.*, pp. 1313-1316, New York, USA, 2007.

[8] Y. Lu *et al.*, " A High fill-factor annular array of high frequency piezoelectric micromachined ultrasonic transducers", *Journal of Microelectromechanical Systems, vol. PP, issue 99, pp. 1-10, September 29, 2014.*

[9] K. Smyth *et al.*, "Analytic solution for N-electrode actuated piezoelectric disk with application to piezoelectric micromachnined ultrasonic transducers," *IEEE Trans. Ultrason. Ferroelectr. Freq. Control*, vol. 60, no. 8, pp. 1756–1767, Aug. 2013.

[10] F. Sammoura, K. Smyth, and S. G. Kim, "Optimizing the electrode size of circular bimorph plates with different boundary conditions for maximum deflection of piezoelectric micromachined ultrasonic transducers," *Ultrasonics*, vol. 53, pp. 328–334, Feb. 2013.

[11] F. Sammoura, S. Akhbari, N. Aqab, M. Mahmoud, and L. Lin, "Multiple electrode piezoelectric micromachined ultrasonic transducers," *in Proc. 2014 IEEE Int. Ultrason. Symp. (IUS)*, pp. 305-308, Chicago, IL, USA, September 3-6, 2014.

[12] F. Sammoura *et al.*, "A two-port piezoelectric micromachined ultrasonic transducer," *in Proc. 2014 Joint IEEE ISAF/IWATMD/PFM*, pp. 1-4, State College, PA, May 12-16, 2014.

[13] S. Akhbari *et al.*, "Bimorph pMUT with dual electrodes," in *the Proceedings of 28th IEEE Micro Electro Mechanical Systems Conference*, Estoril, Portugal, January 18-22, 2015, accepted.

[14] F. Sammoura, S. Akhbari, and L. Lin, "An analytical solution for unimorph curved piezoelectric micromachined ultrasonic transducers with spherical shape diaphragms," *IEEE Transactions on Ultrasonics, Ferroelectrics, and Frequency Control*, vol. 61, no. 9, pp. 1533-1544, September 2014.

[15] S. Akhbari, F. Sammoura, and L. Lin, "An Analytical Circuit Model for Piezoelectric Micromachined Ultrasonic Transducers with Spherical-shape diaphragms" *in Proc. IEEE Int. Ultrason. Symp. (IUS)*, pp. 301-305, Chicago, IL, USA, September 3-6, 2014.

[16] S. Akhbari *et al.*, "Highly responsive curved aluminum nitride pMUT," in *the Proceedings of 27th IEEE Micro Electro Mechanical Systems Conference*, pp. 124-127, San Francisco, CA, January 26-30, 2014.

[17] F. Sammoura *et al.*, "An analytical analysis of the sensitivity of circular piezoelectric micromachined ultrasonic transducers to residual stress", *IEEE Int. Ultrason. Symp. (IUS)*, pp. 580-583, Dresden, Saxony, Germany, October 07-10, 2012.

CONTACT

*S. Akhbari, tel: +1-510-9267150;
sina.akhbari@berkeley.edu

THERMOCOUPLES ON TRENCH SIDEWALL IN CHANNEL FRONTING ON FLOWING MATERIAL

Masahiro Shibata[1], Takahiro Yamaguchi[1], Shinya Kumagai[1], and Minoru Sasaki[1]
[1]Toyota Technological Institute, Nagoya, JAPAN

ABSTRACT

In a microfluidic channel, thermocouples were fabricated on the sidewall of microchannel using the three-dimensional photolithography. The thermocouples on the side wall can directly sense the microfluid in the channel, and the accurate temperature measurement can be achieved. Moreover, the thermocouple metals on the sidewall do not make the shadow allowing the observation using the optical microscopy.

INTRODUCTION

Microfluidic devices have been developed to conduct technical operations on a chip such as precise positioning of biological materials and controlling chemical and physical conditions for biological experiments [1,2]. The microfluidic devices can remove bottlenecks in handling liquid samples (ex. biological materials, chemically reactive solutions) compared with the conventional methods that are manual, laborious, and not well controlled [3,4]. Activities of the above liquid samples strongly depend on the temperature. Controlling temperature is one of the important issues in the microfluidics experiments. Careful treatments are required. The typical application is the polymerase chain reaction for amplification of DNA molecules. In the other experiments, the fluorescence signal for detecting the specific molecules was affected by the changes of the environment (e.g., temperature, pH, solvent polarity) [6]. Thus, there are strong demands for measuring and controlling the temperature of the sample in the microchannel.

So far, temperatures in the microchannel have been measured by microfabricated sensors such as resistive sensor and thermocouple. A thermocouple is one of the popular sensors used in the temperature measurement. The above microfabricated sensors were frequently integrated with micro-heater to maintain the temperature of microfluid.

The microsensors should be placed in the vicinity of the microchannel. Due to the structural restrictions. the microsensors were placed on top or bottom of microchannel. However, such arrangement will disturb the optical observation of the flowing material. The optical observation inside the channel will be hardly conducted. In this study, the thermocouple was fabricated on the sidewall of microchannel by three-dimensional (3D) photolithography, and characterized.

DESGIN

In the present study, a microchannel device has two thermocouples as shown in Fig. 1. One is located in the vicinity of microchannel. This thermocouple on the top surface is fabricated by the conventional planer photolithography. Temperature of microfluid is sensed through the thermal conduction of a substrate material. The other thermocouple is located on the sidewall of microchannel. This thermocouple on the sidewall is fabricated by 3D photolithography. The thermocouple on the sidewall can sense the microfluid directly. Accurate temperature monitoring can be achieved.

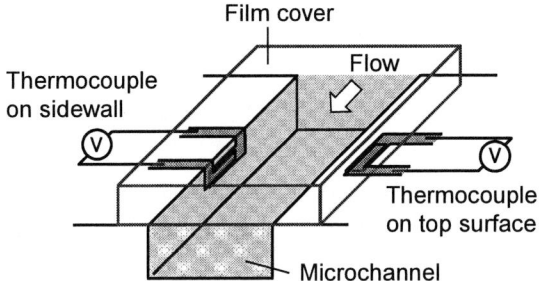

Figure 1: Structure of microchannel device with thermocouples (Left: thermocouple on sidewall, Right: thermocouple on top surface).

Figure 2 shows the layout of the microchannels with thermocouples. The length, width, and depth of microchannel are 16mm, 200μm, and 100μm, respectively. Both ends of microchannels have the chambers of triangular shape (2.6x1.5mm²). Cr and Al films were used for thermocouple materials. Both materials are frequently used in microfabrication. Line widths of Cr and Al films are 25 μm. Gap distance between Cr and Al lines are 20 μm. The thermocouple on the sidewall is used to measure the temperature of microfluid directly, while the thermocouple on the top surface is used as the reference. The thermocouples on the top surface and the sidewall face each other, and the microchannel is located in between. The thermocouples are arrayed along the flow of microfluid. Five groups of thermocouples are arranged along the microchannel. Pad electrodes (500x500μm²) are placed 1.28mm apart from the microchannel.

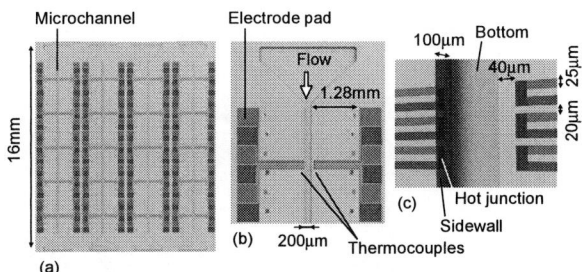

Figure 2: Device layout. (a) Whole image of device chip. (b) Image of a group of thermocouples. (c) Magnified image of a group of thermocouples. The hot junctions are on the sidewall or the top surface.

978-1-4799-7956-1/15 $31.00 © 2015 IEEE

PHOTOLITHOGRAPHY FOR THREE-DIMENTIONAL STRUCTURE

Photoresist Spray-Coating

Photoresist film is required for the patterning. However, the conventional spin-coating of the photoresist can not prepare the uniform film on the sample having 3D structure. The photoresist easily accumulates in the trenches migrating from the convex corners. The resultant film thickness is too different depending on the sample shape to transfer the pattern using the conventional exposure method. The uniformity of the photoresist thickness is required as much as possible.

We used the spray-coating for the uniform photoresist deposition. Briefly, photoresist liquid was mixed with N_2 gas to form the spray. Under the spray of photoresist, a sample was scanned and coated with the photoresist film. Details are described elsewhere [6].

Angled Exposure

In the fabrication of thermocouples on the sidewall, patterning the photoresist on the microchannel sidewall is important. Angled exposure technique was applied for the patterning [7]. By tilting the sample, the sidewall can be irradiated by UV light. However, we have to pay attentions to the lights reflected at the bottom or sidewall surfaces of microchannel. Incident light of UV exposed sidewall surface and then reflected light exposed the bottom surface of the microchannel, causing the unwanted exposure. The light reflection is one of the critical issues for 3D photolithography.

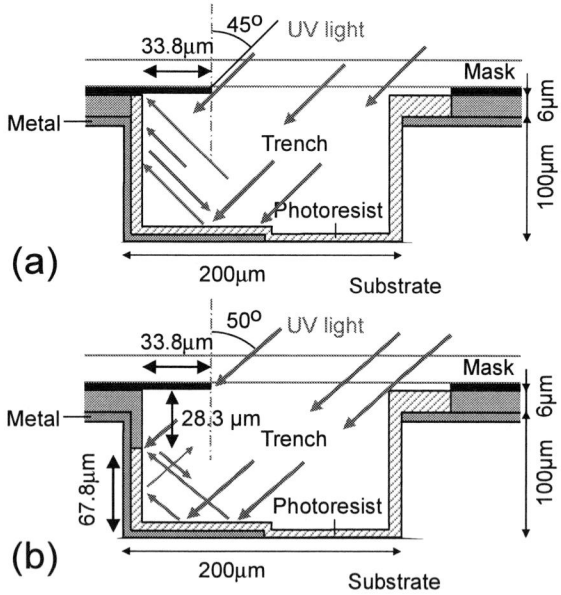

Figure 3: Geometry of angled exposure. (a) Irradiation angle of 45°. All area of sidewall is exposed by the reflected UV light. (b) Irradiation angle of 50°. A part of the top sidewall area is not exposed by the reflected UV light.

Thickness of the deposited photoresist tends to be thinner inside the microchannel. In our condition used, appropriate exposure energy was $300mJ/cm^2$ for the top surface and $150mJ/cm^2$ for the bottom. The exposure dose should be tuned for opening the pattern in the thicker photoresist. However, this exposure condition is over-dose for the thinner photoresist film deposited on the sidewall.

Considering the geometry of microchannel, the area of the sidewall where is to be patterned is determined. Assume that the aspect ratio of the trench is 0.5 (width: 200μm, depth: 100μm), thickness of photoresist film on the top surface is 6μm. Figure 3(a) shows the case when the incident angle of UV light is 45°, all area of trench sidewall is exposed by the reflected UV light. Figure 3(b) shows the case when the incident angle is 50°, the top part of the sidewall is not exposed by the reflected UV light. The UV light reflected at the bottom surface can reach the height of 67.8μm from the bottom surface. In other words, the area of 28.3μm from the top photoresist film surface is not exposed by the reflected UV light. Therefore, appropriately designing the metal patterns of photomask for shadowing the area of 28.3μm can prevent there from being exposed by the UV light.

FABRICATION

Figure 4 shows the fabrication sequence. Si chip of 24mmx24mm was used as the starting substrate. (1) The microchannels are patterned by conventional planer photolithography. Deep reactive ion etching is conducted to make 100μm-deep trench. (2) The etched Si substrate is thermally oxidized for making isolation layer between Si substrate and metal layers of thermocouple. (3) Tilting the oxidized Si substrate to 45° to the Cr metal source, Cr film is deposited (thickness: 570nm). The thermocouple on the sidewall is designed to be located on the one side of the trench. (4) To form the photoresist film on the substrate, the spray-coating is performed (Ushio USC-2000ST, modified by VIC international). Positive photoresist (Tokyo Ohka Kogyo, TMMR P-W1000 (180cp) is diluted to 10% by adding thinner of OK73 (Tokyo Ohka Kogyo, mixture of propyleneglycol monomethyl ether and propyleneglycol monomethyl ether acetate). The diluted photoresist is used. The spray-coating is performed using the condition as described before [7]. Generally, the photoresist film deposited on the corner edge of the trench tends to be thin due to the surface tension of the photoresist. Thus, during the spray-coating, the sample is heated at 110 ^{o}C to accelerate the drying of photoresist, which prevents the thinning of the photoresist. Angled exposure is performed to make first pattern of photoresist. (5) Using the patterned photoresist film, Cr film is wet-etched. (6) Similarly to Cr film deposition, Al film of 460nm is deposited. (7) Again, spray-coating, angled exposure and Al wet-etching are performed. After the Al and Cr stacking process, the samples are sintered to form the stable hot junctions of thermocouples. Figure 5 shows the fabricated device. (8) At the last step, the trench with thermocouples is covered by poly-di-methyl-siloxane (PDMS) film to form microchannel structure as shown in Fig. 6.

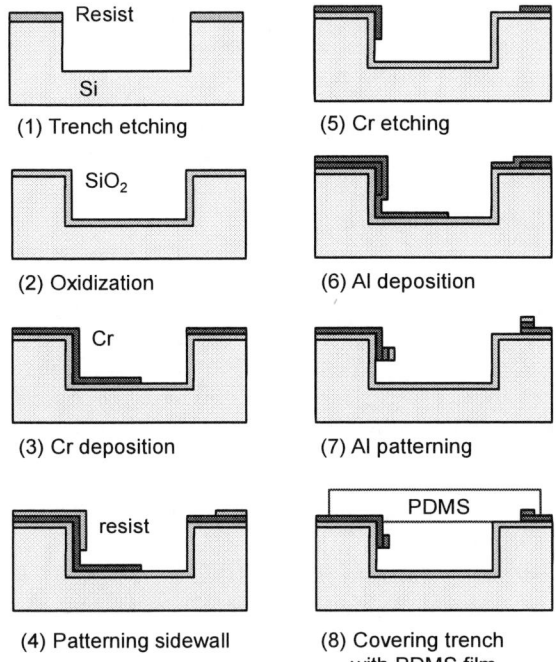

Figure 4: Fabrication sequence.

(1) Trench etching
(2) Oxidization
(3) Cr deposition
(4) Patterning sidewall
(5) Cr etching
(6) Al deposition
(7) Al patterning
(8) Covering trench with PDMS film

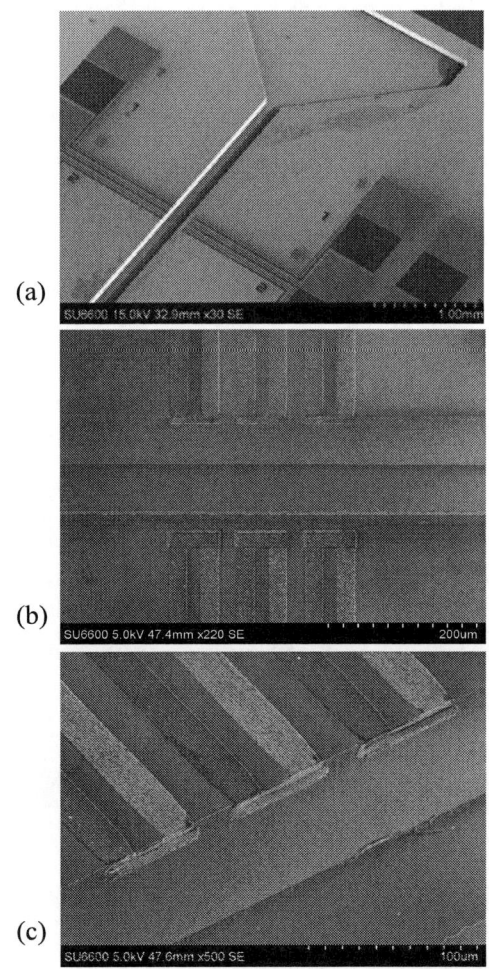

Figure 5: (a) A microchannel and the thermocouples. (b) Paired thermocouples placing on the opposite sides of the microchannel. (c) Thermocouples on the sidewall.

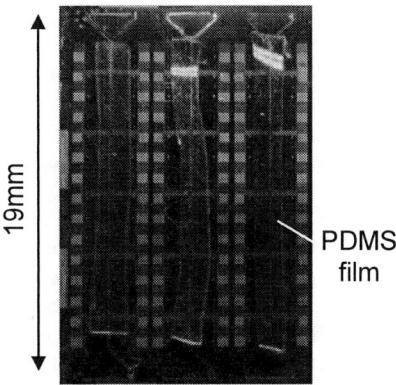

Figure 6: Whole image of fabricated microchannel device with thermocouples covered by PDMS film.

EXPERIMENTS

Thermocouples were analyzed by two experiments. Figure 7(a) shows the first experimental setup. The microchannel device was set on a hot plate and the air at the room temperature was supplied to the channel. The syringe pump was used to flow the air into the microchannel (flow rate: 30ml/min). Figure 7(b) shows the second experimental setup. The microchannel device was set at room temperature, and heated air was supplied into the microchannel. The output of the open voltage of the thermocouple is measured with nano-voltmeter (Keithley, 2182A). The thermocouples compared were placed on the sidewall and on the top surface facing each other against the microchannel.

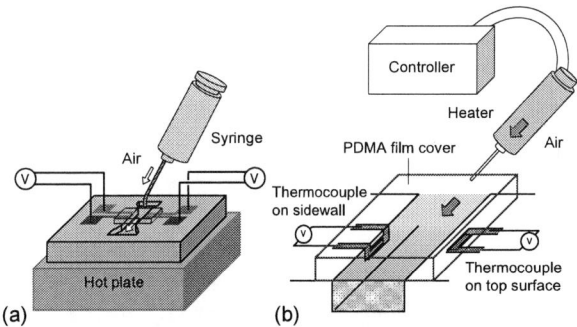

Figure 7: Experimental setup. (a) Setup for the first experiment. Air is supplied to the thermocouple devices heated by a hot plate. Static characteristics of a thermocouple are analyzed. (b) Setup for the second experiment. Dynamic response (time response) of a thermocouple is analyzed.

RESULTS AND DISCUSSIONS

In first experiments, open voltage was measured with increasing the hot plate temperature. Figure 8 shows the result. The open voltage increased linearly to the hot-plate temperature. The increasing rate was evaluated to 16μV/K. This value shows a good agreement with the reported Seebeck constant of Cr-Al junction [8,9]. Under the air flow, the open voltage decreased. The difference in the open voltage was 70μV, which corresponded to the temperature difference of 4°C. Although Si material had good thermal conductivity and the thermocouple on the top surface was placed only 40μm apart from the microchannel, the temperature difference indeed existed. The thermocouples should be placed inside the microchannel to measure accurate temperature.

In the second experiment, the time response of the thermocouple was measured. The open and filled circles in Fig. 9 correspond to the open voltage obtained from the thermocouples from the vertical sidewall and the top surface beneath the PDMS film cover, respectively. Generally, reflecting the room-temperature-measurement, the magnitude of open voltage was smaller compared with those in the first experiment. At 0 sec, the hot air was supplied. The open voltage from the sidewall increased. This indicated that the thermocouple on the sidewall detected the hot air quickly.

The open voltage showed the signal drift. The magnitude of the drift was 0.12μV/10sec. Considering the Seebeck coefficient, the drift correspond to 0.006K/10sec. This would be the limit of accuracy under the experimental setup used. With improving the stability of cold junction, resolution of temperature measurement could be improved.

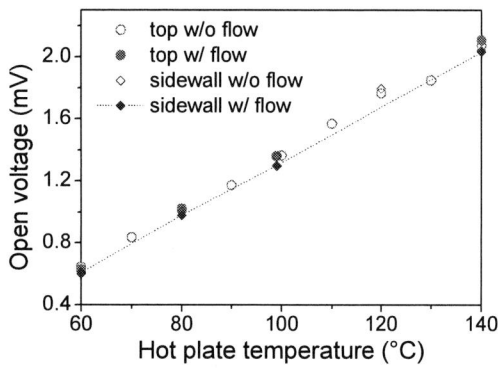

Figure 8: Open voltage of thermocouple as a function of hot plate temperature.

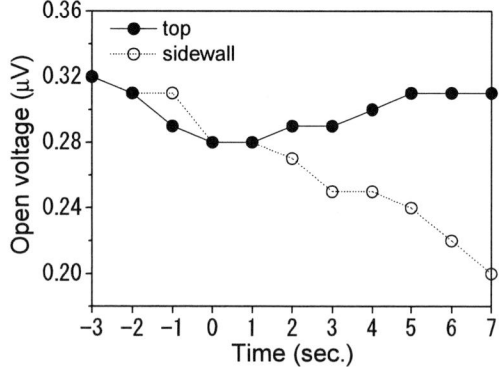

Figure 9: Time response of thermocouples. Open and filled circles correspond to thermocouples on the sidewall and on the top surface, respectively.

CONCLUSIONS

Thermocouples were fabricated on the sidewall of microchannels by 3D photolithography. The 3D structure

of thermocouple had the advantage that the device surface can directly sense the flow inside the microchannel. By designing the thermocouple at the upper position on the sidewall, the process condition can be relatively easy. The obtained signal confirms the advantage of the direct measurement of the flow inside the microchannel.

Acknowledgements

This study was supported by the MEXT program for forming strategic research infrastructure (S1101028), "Nanotechnology Platform Japan", and funding of Toyota Physical and Chemical Research Institute. Professor Jun Ohta (Nara Institute of Science and Technology, Japan) kindly allowed us to summarize this research with our previous student currently studying in his laboratory.

REFERENCES

[1] E. K. Sackmann, A. L. Fulton, and D. J. Beebe, "The present and future role of microfluidics in biomedical research", *Nature*, vol. 507, 2014, pp. 181-189.

[2] H. Tanaka, P. Fiorini, B. Jones, S. Peters, R. S. Wiederkehr, B. Majeed, H. Yaku, M. Hiraoka, T. Matsuno, and I. Yamashita, "Electrochemical sensor with dry reagents implemented in lab-on-chip for single nucleotide polymorphism detection", *Jpn. J. Appl. Phys.*, vol. 53, 2014, 05FS03.

[3] A. Hattori, T. Kaneko, and K. Yasuda, "Improvement of Particle Alignment Control and Precise Image Acquisition for On-Chip High-Speed Imaging Cell Sorter", *Jpn. J. Appl. Phys.*, vol. 50, 2011, 06GL06.

[4] W.-H. Tan, and S. Takeuchi, "A trap-and-release integrated microfluidic system for dynamic microarray applications", *PNAS*, 104 (2007) 1146–1151

[5] R. Zenobi, "Single-Cell Metabolomics: Analytical and Biological Perspectives", *Science*, vol. 342, 2013, 1243259.

[6] S. Kumagai, H. Tajima, M. Sasaki, "Flow Analysis of Photoresist Spray Coating towards Improving Coverage on Three-Dimensional Structures", *Jpn. J. Appl. Phys.* vol. 50, 2011, 106501.

[7] S. Kumagai, T. Yamamoto, H. Kubo, and M. Sasaki, "3-D WIRING ACROSS VERTICAL SIDEWALLS OF SI PHOTO CELLS FOR SERIES CONNECTION AND HIGH VOLTAGE GENERATION", Proc. 25th *International Conference on Micro Electro Mechanical Systems*, 2012, pp. 60-63.

[8] A. W. Van Herwaarden and P. M. Sarro, "Thermal sensors based on the seebeck effect", *Sens. Actuators*, vol. 10, 1986, pp. 321-346.

[9] T. Ono, C.-C Fan, and M. Esashi, "Micro instrumentation for characterizing thermoelectric properties of nanomaterials", *J. Micromech. Microeng.*, vol. 15, 2005, pp. 1–5.

CONTACT

*M..Sasaki, tel: +81-52-809-1840;
mnr-sasaki@toyota-ti.ac.jp

VACUUM CAVITY FORMATION FOR HIGH THERMAL ISOLATION IN FLEXIBLE THERMAL SENSOR

Pilyoung Kim[1], Shunji Shibata[2], and Mitsuhiro Shikida[3]

[1]Dept. of Mechanical Science Engineering, Nagoya University, Aichi, JAPAN
[2]Dept. of Micro-Nano Systems Engineering, Nagoya University, Aichi, JAPAN
[3]Dept. of Frontier Science, Hiroshima City University, Hiroshima, JAPAN

ABSTRACT

For producing a variety of flexible MEMS sensors, we previously developed a process that uses a Cu On Polyimide (COP) substrate as a starting material and sacrificial Cu etching to produce a cavity and electrical feed-through structures on the substrate [1]. In the current study, we introduced a vacuum cavity realizing high thermal isolation in a flexible thermal sensor, for the first time. Parylene thickness as a function of the amount of dimer usage was studied, and a preliminary experiment exploring the fabrication of the vacuum cavity structure was performed. These results, demonstrate that we successfully produced a vacuum cavity in the thermal sensor. To determine the effect of the thermal isolation by vacuum cavity to thermal sensor, we applied it to a flow sensor, and found that we can reduce the response time to one-half by introducing the vacuum cavity in thermal sensor.

INTRODUCTION

Thermal convective phenomena have been widely applied as sensing mechanisms in the flow, shear stress, and acceleration sensors used in Micro Electro Mechanical Systems (MEMS), because the thermal capacity of the sensors is reduced dramatically by their smaller body size. Three different thermal principles, i.e., thermal anemometry, calorimetric sensing, and time-of-flight sensing, have been used in miniaturized thermal sensors [2-7]. For example, a thermal sensor formed on a thin diaphragm was used to detect the shear force of a fluid acting on a plane surface [2]. These sensors have also been applied as miniaturized accelerometers [3-5]. A single heater and with two temperature sensors placed on the both sides of it was formed on a thin diaphragm, and the applied acceleration was detected by the change in the air temperature distribution pattern over the two sensors. Thermal type MEMS sensors have excellent space and time resolutions, and are also advantageous in that they can detect physical amounts (flow, shear stress, and acceleration) without the need for a moving element.

A Si substrate is generally used to produce thermal MEMS sensors, because it can easily fabricate the diaphragm as the thermal isolation structure. However, Si-based thermal MEMS sensors are difficult to mount on bendable surfaces, since Si material is so brittle. To overcome this problem, polymer materials, including polyimide, parylene, and silicone resin, have been introduced as substrate materials in MEMS for producing flexible devices. Polyimide is particularly attractive because it has excellent heat and chemical resistances and a high Young's modulus; it has thus been widely applied as the substrate for producing flexible MEMS sensors. Parylene has a unique biocompatibility aspect, so it is widely used in medical applications.

We previously proposed using a Cu On Polyimide (COP) substrate as a starting material as a way of realizing various flexible MEMS sensors [1]. In this work, we newly introduced a vacuum cavity formation under the thermal sensor to realize high thermal isolation for improving of the response time.

FLEXIBLE THERMAL SENSOR BASED ON COP SUBSTRATE

A flexible thermal sensor based on a COP substrate is shown in Figure 1. The substrate is composed of a base polyimide film and a Cu layer. A thin polyimide layer is also formed on the substrate, and the MEMS sensors are produced on its surface. A cavity and an electrical feed-through are produced by selectively etching part of the Cu layer in the same process (sacrificial etching). The advantages of a MEMS sensor based on a COP substrate are as follows:

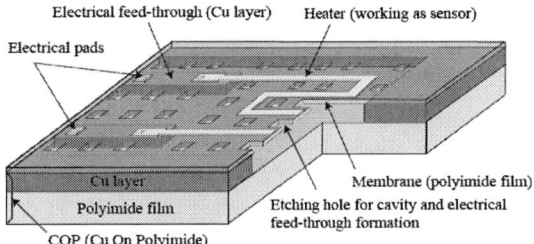

Figure 1: Flexible thermal sensor based on Cu On Polyimide (COP) substrate.

(1) It can produce various flexible thermal MEMS sensors, because the cavity works as a thermal isolation in thermally operated flow and acceleration sensors.
(2) It can easily fabricate electrical feed-throughs having low electrical resistance, because part of the Cu layer works as the electrical wiring after the sacrificial etching.

Even if the cavity under the membrane as for the thermal isolation, a part of the heat on the membrane propagates to the air in the cavity as shown in Figure 2(a), and thus it disturbed the response time shortening in the previous thermal MEMS sensor. To overcome this problem, we newly introduced a vacuum cavity formation (Fig. 2(b)) to realize high thermal isolation in flexible thermal MEMS sensor in this paper.

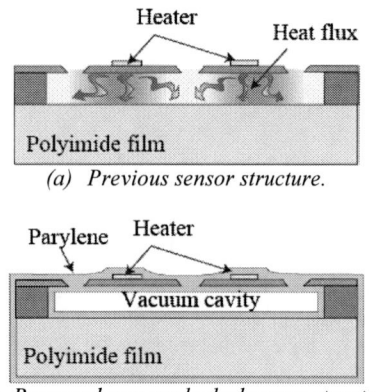

(a) Previous sensor structure.

(b) Proposed vacuum locked sensor structure.

Figure 2: Previous and proposed sensors structures.

VACUUM CAVITY FORMATION IN THERMAL FLEXIBLE SENSOR

The fabrication process of the flexible thermal MEMS sensor based on a COP substrate with a vacuum cavity is shown in Figure 3. The COP substrate (Ube Exsymo Co., Ltd.) was used as the starting material. The thickness of the copper and the polyimide was 18 μm and 50 μm, respectively. First, a photosensitive polyimide solution (Photoneece, Toray Industries, Inc.) was coated to a thickness of 3 μm and patterned to define a structure in the Cu layer. This photosensitive polyimide works as an etching mask for forming the cavity and feed-throughs in the Cu layer in the final sacrificial Cu etching, and it becomes a membrane structure. Part of the pattern is used to form a connection area between the deposited metal film and the electrical feed-through in the Cu layer. A negative photoresist (ZPN1150-90, Zeon Corporation), specially developed for the lift-off process, was placed on the film surface and patterned with UV light to define the shapes of the Au/Cr film that would function as the heater and temperature sensors. The Au/Cr film was deposited by sputtering and patterned by selectively removing the photoresist (lift-off process). The thickness of the Au and Cr was 250 nm and 10 nm, respectively. Finally, the cavity and electrical feed-throughs were formed by selectively etching part of the Cu layer. An iron (II) chloride solution (Sunhayato Corp.) was used as the etchant and the etching temperature was set to 40°C. The etching resulted in metal patterns working as sensors being formed on the polyimide membrane and electrical feed-throughs being formed on the COP substrate. Finally, etching holes formed on the polyimide membrane were completely sealed by depositing parylene C film in a vacuum condition to produce a vacuum cavity under the membrane.

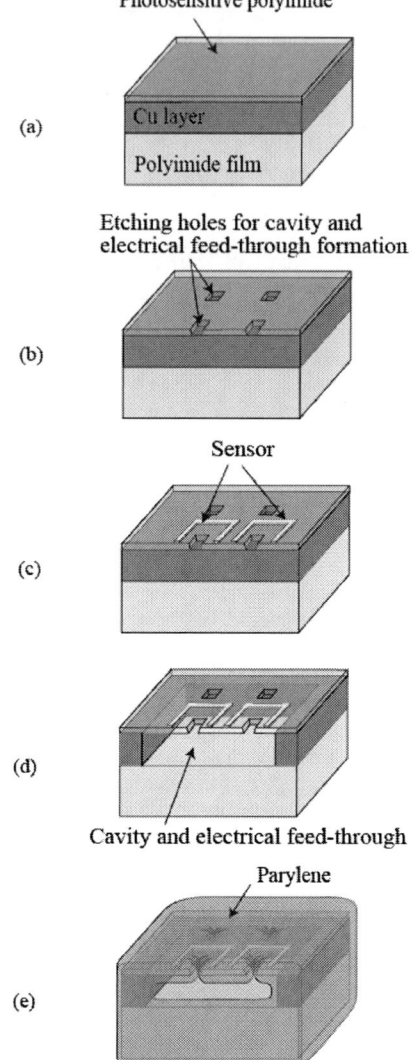

Figure 3: Fabrication process of locked flexible thermal sensor.

EXPERIMENTS

Parylene film deposition is used as an encapsulation of the etching holes in the vacuum cavity formation. We therefore investigated the parylene thickness as a function of the amount of parylene dimer usage to estimate the sealing thickness by the deposition. Parylene thickness is in proportion to the amount of dimer usage, as shown in Figure 4. The effect of the etching hole sealing differing amounts of the parylene deposition was experimentally studied and confirmed by scanning electron microscopy, shown in Figure 5.

978-1-4799-7956-1/15 $31.00 © 2015 IEEE

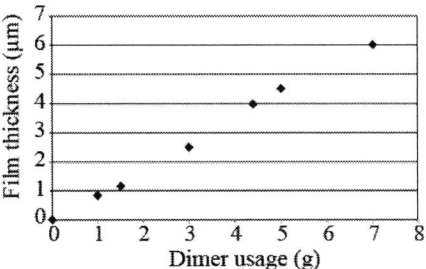

Figure 4: Parylene thickness as a function of amount of dimer usage.

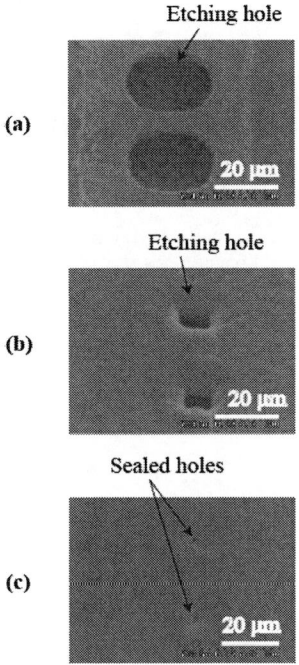

Figure 5: Etching hole sealing by deposition (dimer usage: (a) 0 g, (b) 7 g, (c) 12g).

Based on these results, we successfully produced a vacuum cavity under the membrane by sealing the etching hole (Figure 6). Since the parylene coating is a gaseous process, every surface is uniformly coated by the parylene film. As a result, the interior of the cavity is also covered with parylene film due to the parylene molecules passing through the etching holes. This means that the cavity space decreases with the increase of the parylene film thickness. Thus, the size of the etching hole should be decided in order to minimize the reduction of the vacuum cavity space by the parylene film deposition.

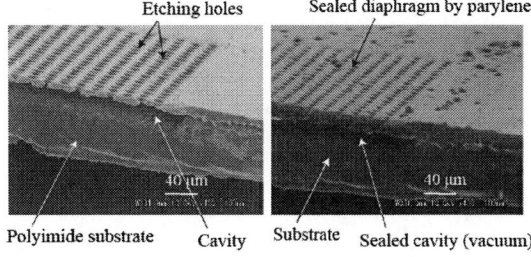

Figure 6: SEM side views of COP substrates: (a) before and (b) after parylene coating.

To demonstrate the effect of the thermal isolation by vacuum cavity formation to the thermal MEMS sensor, we applied it to a thermal flow sensor. The response waveforms of the thermal sensor were investigated to apply a flow rate raging from 0 to 300 cc/min. The response waveforms without and with the parylene film coating are shown in Figures 7(a) and 7(b), respectively. The sensor output rose sharply and became a constant within a short time in the case of the vacuum cavity condition by the parylene film deposition.

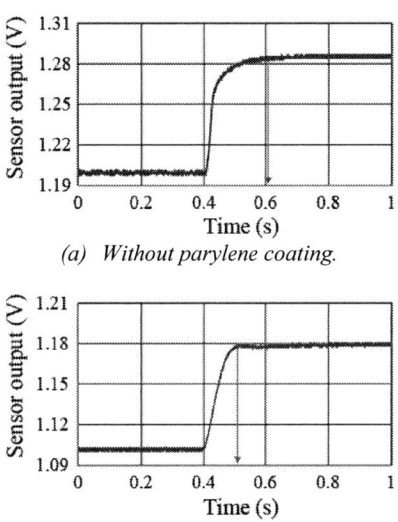

(a) Without parylene coating.

(b) With parylene coating (vacuum cavity).

Figure 7: Response wave form of thermal sensor used for flow sensing.

The response time was calculated from the elapsed time between the beginning and the end of the obtained waveform transition (Table 1). Since the heater was passively cooled by surrounding air, the response time at the cooling mode was a little bit longer than that at the heating mode. These results, demonstrate that the response time was reduced to one-half by introducing the vacuum cavity under the thermal sensor.

978-1-4799-7956-1/15 $31.00 © 2015 IEEE

Table 1: Summary of response time.

		Without parylene (no vacuum)	With parylene (vacuum cavity)
Response time (ms)	Rising	200	100
	Falling	220	130

CONCLUSION

We used a Copper On Polyimide (COP) substrate as a start material for producing flexible MEMS sensors, and introduced the a parylene coating process to fabricate a vacuum cavity to realize high thermal isolation. Parylene thickness as a function of the amount of dimer usage was investigated to estimate the sealing thickness by the parylene deposition, and a vacuum cavity under the membrane successfully produced by sealing the etching hole. To demonstrate the effect of the thermal isolation by vacuum cavity to thermal sensor, we applied it to a flow sensor, and confirmed that we can reduce the response time to one-half by parylene coating.

ACKNOWLEDGEMENTS

This research was supported by The Canon Foundation and a Grant-in-Aid for Scientific Research (B) No. 26286034 from the Ministry of Education, Culture, Sports, Science and Technology (MEXT), Japan

REFERENCES

[1] Y. Niimi, S. Shibata, M. Shikida, "Polymer micromachining based on Cu On Polyimide substrate and its application to flexible MEMS sensor," IEEE MEMS 2014, pp. 528-531.

[2] J. B. Huang, Steve Tung, C. M. Ho, C. Liu, and Y. C. Tai "Improved Micro Thermal Shear-Stress Sensor," IEEE Transactions on Instrumentation and Measurement, vol. 45, no. 2, 1996.

[3] U. A. Dauderstädt, P.H.S. de Vries, R. Hiratsuka, and P. M. Sarro, "Silicon accelerometer based on thermopiles," Sensors and Actuators A, vol. 46, pp. 201-204, 1995.

[4] F. Mailly, A. Giani, A. Martinez, R. Bonnot, P. Temple-Boyer, and A. Boyer, "Micromachined thermal accelerometer," Sensors and Actuators A, vol. 103, pp. 359-363, 2003.

[5] J. Courteaud, N. Crespy, P. Combette, B. Sorli, and A. Giani, "Studies and optimization of the frequency response of a micromachined thermal accelerometer," Sensors and Actuators A, vol. 147, pp. 75-85, 2008.

[6] Y. Gianchandani, O. Tabata, and H. Zappe, "Comprehensive MEMS," 2, Flow sensor, pp. 209-272, Elsevier, 2008.

[7] M. Elwenspoek and R. Wiegerink, "Mechanical microsensors," Springer, 2001.

CONTACT

P. Kim, tel: +1-52-789-5224;
kim.pil-young@g.mbox.nagoya-u.ac.jp

A STUDY OF ADHESION FORCES IN THICK EPITAXIAL POLYSILICON UNDER DYNAMIC IMPACT LOADING

Stefano Dellea[1], Raffaele Ardito[1], Biagio De Masi[1], Francesco Rizzini[2], Alessandro Tocchio[2], and Giacomo Langfelder[1]

[1]Politecnico di Milano, Italy
[2]ST Microelectronics, Cornaredo (MI), Italy

ABSTRACT

The work presents a structure and a method for the in-line characterization of impacts and adhesion phenomena between MEMS moving and fixed parts: the focus is on the monitoring of an inertial proof mass motion when colliding with a mechanical stopper. Through such measurements, one can evaluate the energy balance during impact events. The work analyzes the adhesion force evolution after a number of impact cycles comparable or larger than shocks in a 5-year operation. Results obtained on two different specimens show growing and then stabilizing adhesion forces of on average 170 nN, under impact cycles with about 500 fJ energy loss. No marked dependence on the specimen area is obtained. The possibility to change and track the impact kinetic energy is also demonstrated.

INTRODUCTION

Surface interactions between fixed and movable parts in MEMS devices were investigated in past works with the main goal of studying adhesion forces [1]. Adhesion indeed may represent either an *in-process* or an *in-operation* reliability issue. The major causes of in-process adhesion are represented by a combination of process generated stresses and capillary forces, and can be solved using suitably tailored process (e.g dry/vapor etching). In-operation adhesion may occur due to unwanted contact between a fixed and a suspended part. In case of inertial sensors, this may occur after mechanical shocks. Even if the devices operate in high or moderate vacuum (fractions of mbar to few hundred mbar), so that capillary effects are negligible, other sources of adhesion force are represented by proximity forces (Van der Waals, electrostatic...), hydrogen bonding or nano-asperities welding [1-4].

Investigating adhesion forces through optimized test structures [5] has a twofold relevance in that (i) it can help the scientific comprehension of the occurring phenomena, determining process and design guidelines for increased immunity to shocks, and (ii) from a practical point of view, it allows statistical in-line monitoring on every wafer during industrial device fabrication.

The methodologies used in previous works typically relied on quasi-stationary loading cycles: in such experiments each device was brought to contact to a fixed specimen through forward increasing forces, which were then decreased to capture the backward curve [2]. The difference between the contact position and the release position is an indication of the adhesion force. Resolution in the order of 60 nN were demonstrated [1]. Electrostatic forces can be applied through voltages and the position can be measured through a corresponding capacitance variation, so that a C-V curve is obtained. This quasi-stationary contact somewhat resembles the operation of mechanical switches.

However, for inertial devices, in-operation typical surface contact occurs through dynamic impacts. Such events may last more than the device time constant [6], bringing a slightly underdamped accelerometer (with e.g. a typical proof mass in the order of a few 10^{-9} kg and a typical stiffness in the order of a few N/m) to contact with a mechanical stopper, designed at a distance in the order of 1 μm. For the given orders of magnitude, a 100 g (gravity units) shock acceleration can result in a contact time of a few tens of μs, a contact velocity of a few cm/s and thus a kinetic energy at impact of about 500-2000 fJ.

This work presents a novel structure and a measurement method to study collisions and adhesion phenomena between a suspended mass characterized by typical parameters of consumer MEMS accelerometers, and fixed mechanical stoppers. The proof mass position, its impact velocity (and kinetic energy), and the energy loss during collisions can be continuously monitored. Their effects on surface evolution and adhesion forces can be as well monitored using interspersed C-V curves between the collision events. In addition, the achieved resolution in the adhesion force measurements (lower than 6 nN) improves by an order of magnitude with respect to the referenced works [1, 3].

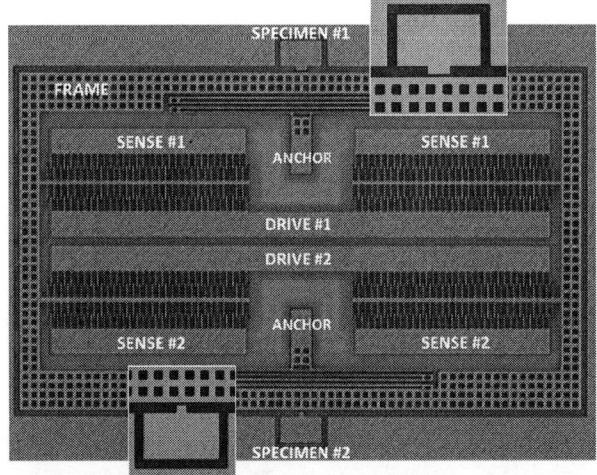

Figure 1: image of the test structure used in this work. The stoppers, with a nominal contact area of 168 μm² and 336 μm², are designed 1.5 μm away from each side.

DEVICE CONCEPT AND OPERATION

Fig. 1 is a top-view of the 24-μm-thick surface micromachined test structure, fabricated using the ThELMA (thick epitaxial layer for micro-actuators and accelerometers) process by ST Microelectronics [7]. A suspended rectangular frame faces four sets of comb-finger capacitors. One differential pair is used for in-plane push-pull actuation, so that the applied varying voltage causes a linear displacement of the proof mass. The other differential pair is used for capacitive sensing. At a distance of 1.5 μm on both the device sides, stoppers with different contact area are designed. The largest one has an area of 336 μm^2, the smallest one has half this area. The stoppers are electrically shorted to the proof mass.

With a mass in the order of $3 \cdot 10^{-9}$ kg, a quality factor of about 2.5, and a resonance frequency of about 3.5 kHz, the nominal device design mimics the parameters of consumer accelerometers mentioned in the Introduction.

For the tests, the push-pull drive electrodes need to be operated at $V_B \pm V_A$, where V_B is the polarization voltage and V_A is the differential driving voltage.

The capacitive readout is based on a high-frequency test signal (in the order of 1 MHz) applied around the ground potential to the seismic mass, and on a differential readout of the currents flowing through the sense electrodes, which are linear with the motion.

Both the test signal, the drive signals and the sense readout are obtained through the five differential I/O channels of an *ITmems* MCP-G characterization platform [7]. Such a tool enables both dynamic response real-time monitoring, and quasi-stationary C-V motion measurement, to estimate both the impact energy and its resulting effects on adhesion. For this work, an automatic routine specifically developed for adhesion force measurements was exploited.

The devices are packaged at a pressure of a few hundred mbar. An initial on-wafer electromechanical characterization on all the samples to be tested reveals quite symmetric rest capacitances of on average 126 fF, an mean resonance frequency of 3.6 kHz and an average quality factor of 2.6, all in line with the theoretical and design predictions.

Before beginning the impact cycles and adhesion measurement campaign, the native adhesion is evaluated.

Fig. 2 refers to C-V curves for both the specimens of a pristine sample, with outward curves in bold blue and return curves in light red. Continuous lines correspond to the large-area sample, dashed lines to the small-area one.

At a sensing resolution of 50 aF$_{rms}$, an almost complete overlap between outward and return curves is visible, with indistinguishable contact and release voltages. This indicates adhesion forces lower than 6 nN for both the specimen areas. None of the 30 tested samples revealed a distinguishable native adhesion force.

EXPERIMENTAL RESULTS

In a second phase of the measurement campaign, square waves at voltages larger than the found contact value are applied, while the system tracks the structure displacement. Fig. 3 is a sample experimental motion capture, clearly showing the mass impact and its consecutive bounces on the stopper. The dashed blue segment represents the theoretical impact point at a distance of 1.5 μm. The shown derivatives of the measured curve at the discontinuity instant (indicated as t_0 in the figure) represent the velocity before and after the collision, giving a description of the impact kinetic energy (650 fJ in this example) and of the energy loss (about 514 fJ). The secondary bounces are assumed as negligible in determining the surface evolution with respect to the primary ones. The square wave is applied at a frequency of 500 Hz, so that the system has the time to damp before the beginning of the following impact event.

After selected numbers of impacts, the square wave is stopped and the system automatically measures a high-resolution C-V curve in the region of the contact voltage. Then, the square wave begins again until the next C-V curve measurement. The procedure is repeated up to $\sim 7 \cdot 10^5$ impact cycles. Measurements obtained through this procedure show an increase and then a stabilization of the adhesion force.

A sample results is shown in Fig. 4 for one large specimen.

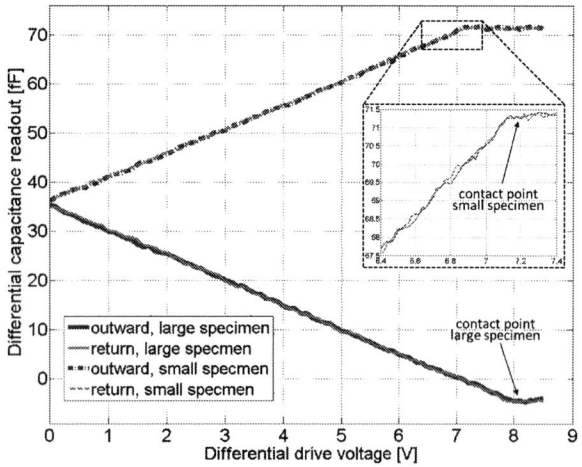

Figure 2: contact measurement on the two specimens of a pristine sample. Taking into account the used instrument resolution (< 50 aF$_{rms}$), the native adhesion force can be calculated to be lower than 6 nN for both the areas.

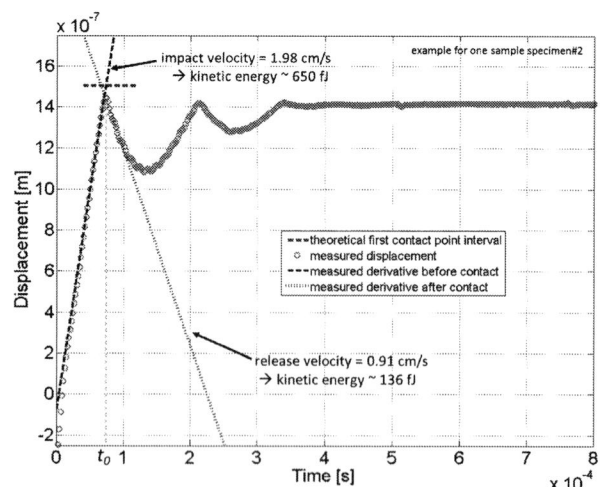

Figure 3: measured suspended mass motion during the impact on the stopper: at the collision time t_0, the velocity changes its sign. From the shown derivatives, it is possible to calculate the impact energy loss.

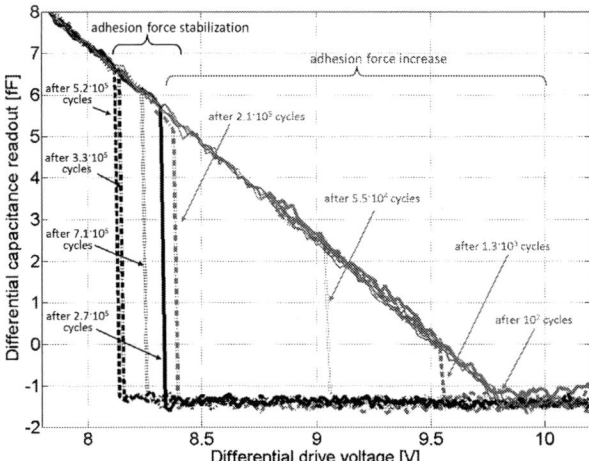

Figure 4: C-V curves after different numbers of impact cycles at the same collision energy on a large-area specimen. The adhesion force shows a phase of initial increase, followed by a stabilization.

Eight different C-V curves are reported, corresponding to different numbers of impacts between 10^2 and $7.1 \cdot 10^5$. For a comparison, an inertial device subject to 5-50 impacts per day would suffer 10^4-10^5 events in a 5-year operation. The outward curves are shown in light green. The more interesting return curves are shown in bold gray and black.

Fig. 5 collects the measured adhesion force evolution on all the 30 specimens, as a function of the number of impact cycles. Two sample results for a large and a small specimen are reported as bold curves. The results are quite similar for the two kinds of specimens, with the most frequent adhesion force value being around 60 nN in both cases. The results thus suggest no significant dependence on the contact area. Further, the statistical spread at the end of the test is relatively large: the standard deviation of the found adhesion force is indeed comparable to its mean value.

The behavior described by Fig. 4 and Fig. 5 is markedly different from what observed with similar specimens and process but with stationary cycles [2], and may be related not only to asperities plastic deformation, but also to damage accumulation and wear of surfaces at contact. Possible interpretations of the obtained results are indeed the following:

- *plastic deformation of asperities*: at the first cycle, the dominant asperity [9] is the only one hit by the proof mass and is deformed; in the next cycles, the secondary asperities are hit as well, and begin to be plastically deformed. The effective contact area stabilizes during cycles to a value for which the applied pressure (force per unit area) is no longer capable of further deformation. We expect therefore the adhesion force at regime to be independent on the nominal specimen area. We also expect a relatively small statistical spread of the regime adhesion force;

- *fatigue wear of surfaces*: this explanation assumes that asperities are damaged through micro cracks nucleation and propagation. This implies, on average, an increase of the contact area during cycles (dominant asperities are subject to wear in this case), but with the statistical behavior typical of fatigue experiments in micro and nano scale Silicon and polysilicon [10]. This would justify the observed spread.

Comparative analysis of pristine and tested specimens using scanning electron microscopy may help in understanding the ongoing phenomena. It however requires first the device dicing and sorting from the wafer, and then a careful disassembling of the package. It is planned as future work.

Table 1 summarizes the obtained results for the electromechanical characterization and for the adhesion in the two kinds of specimens. The statistics, so far limited to 30 samples, will be improved with further measurements.

Table 1: adhesion force results after $7 \cdot 10^5$ cycles.

Measured parameter	Value	Unit
Resonance frequency	3.6	kHz
Quality factor	2.6	-
Elastic stiffness	2	N/m
Overall device area	0.253	$(mm)^2$
Process thickness	24	μm
Large specimen		
Contact area	336	$(\mu m)^2$
Native adhesion force	< 6	nN
Mean adhesion force (@ $7 \cdot 10^5$ cycles)	192	nN
Most frequent bin (@ $7 \cdot 10^5$ cycles)	40-60	nN
Standard deviation (@ $7 \cdot 10^5$ cycles)	150	nN
Small specimen		
Contact area	168	$(\mu m)^2$
Native adhesion force	< 6	nN
Mean adhesion force (@ $7 \cdot 10^5$ cycles)	148	nN
Most frequent bin (@ $7 \cdot 10^5$ cycles)	60-80	nN
Standard deviation (@ $7 \cdot 10^5$ cycles)	90	nN

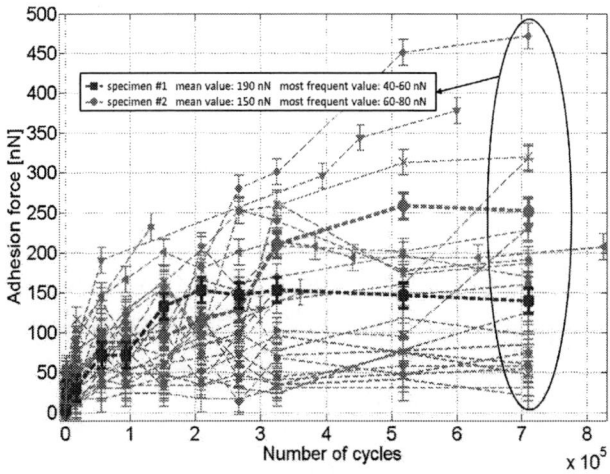

Figure 5: adhesion forces vs the number of impact cycles for both kinds of specimens. For sake of clarity, two typical behaviors are shown in bold. No marked dependence on the contact area is found.

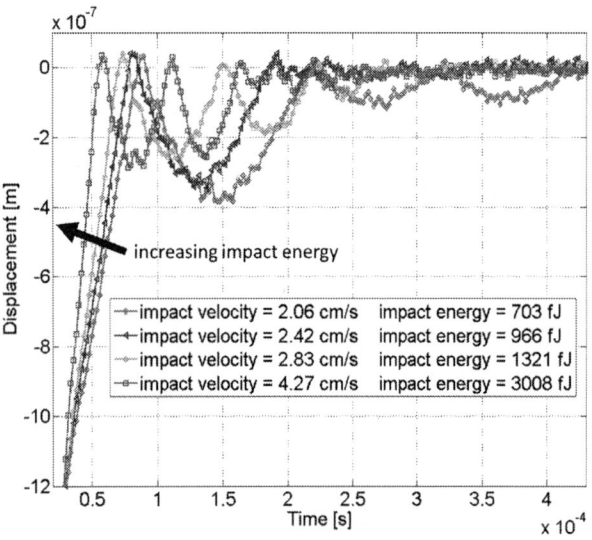

Figure 6: experimentally measured impact dynamics at different collision velocities (kinetic energies) of the suspended mass.

CONCLUSIONS

The work investigated the evolution of adhesion forces between a MEMS proof mass and two stoppers of different area, under repeated impact cycles with collision energies comparable to mechanical shocks at 100-1000 *g*. Results, obtained with a system resolution of 6 nN, demonstrate increasing and then stabilizing adhesion. Besides, the new measurement technique gives access to a large amount of information, which can be exploited, together with accurate finite element simulation [11], to have a deeper insight on the phenomena responsible for the surface adhesion.

As an anticipation of future experiments, measurements similar to those presented in this work were repeated at different, monitored, impact energies. Fig. 6 is a preliminary example of how the system enables changing the impact speed, to accelerate the in-line tests and to deepen these analyses as a function of the collision energy. By changing the square wave amplitude, different impact velocities in the range of 2 cm/s to 4 cm/s are obtained, corresponding to kinetic energies at the collision time of 700 fJ to 3 pJ.

Future work also includes measurement of more samples belonging to different wafers before and after dicing, comparative analysis of dynamic versus quasi-stationary cycling, and comparative microscopy inspection of native and stressed surfaces.

The presented procedure can be used also to optimize the stoppers geometry and tolerance to shocks and adhesion phenomena.

ACKNOWLEDGEMENTS

The author wish to thank Dr. N. Aresi and *ITmems* s.r.l. for helping with measurement routine preparation. Part of the experiments was done using the MEMS&3D lab facilities of Politecnico di Milano.

REFERENCES

[1] M. Shavezipur, W. Gou, M. Fisch, C. Carraro, R. Maboudian, "Inline Measurement of Adhesion Force Using Electrostatic Actuation and Capacitive Readout", *Journ. of Microelectromech. Syst.*, 21 (4) 2012, pp. 768-770.

[2] T. Friedrich, C. Raudzis, R. Müller-Fiedler, "Experimental Study of In-Plane and Out-of-Plane Adhesions in Microelectromechanical Systems", *Journ. of Microelectromech. Syst.*, 18 (6), 2009, pp. 1326-1334.

[3] M. Shavezipur, W. Gou, C. Carraro, Y. Tian, R. Maboudian, I. Gelmi, M. Azpeitia, "In-Line Adhesion Monitoring and the Effects of Process Variations on Adhesion in MEMS", in *Proc. Transducers 2013,* Barcelona, Spain, pp. 1464-1467.

[4] D. H. Alsem, H. Xiang, R. O. Ritchie, K. Komvopoulos, "Sidewall Adhesion and Sliding Contact Behavior of Polycrystalline Silicon Microdevices Operated in High Vacuum", *Journ. of Microelectromech. Syst.*, 21 (2), 2012, pp. 359-369.

[5] J. De Coster, F. Ling, A. Witvrouw, I. De Wolf, "Dedicated Test Structure for The Measurement of Adhesion Forces Between Contacting Surfaces in MEMS Devices", in *Proc. Transducers 2013,* Barcelona, Spain, pp. 2692-2695.

[6] S. Sundaram, M. Tormen, B. Timotijevic, R. Lockhart, T. Overstolz, R. P. Stanley, H. R. Shea1, "Vibration and shock reliability of MEMS: modeling and experimental validation", *J. Micromech. Microeng.* 21 (2011) 045022 (13pp).

[7] G. Langfelder, S. Dellea, F. Zaraga, D. Cucchi, and M. A. Urquia, "The dependence of fatigue in microelectromechanical systems on the environment and the industrial packaging", *IEEE Trans. on Ind. Electronics*, 59 (12), pp. 4938–4948, Dec. 2012.

[8] ITmems s.r.l., MEMS Characterization Platform, MCP-G, *Product Datasheet*, [Online]. Available: http://www.itmems.it, accessed Nov. 2014.

[9] R. Ardito, A. Corigliano, A. Frangi, "Modelling of Spontaneous Adhesion Phenomena in Micro-Electro-Mechanical Systems", *European Journal of Mechanics A/Solids*, 39, 2013, pp. 144-152.

[10] S. Dellea, G. Langfelder, A. F. Longoni, "Fatigue in Nanometric Single-Crystal Silicon Layers and Beams", *Journ. of Microelectromech. Syst.*, accepted for publication, DOI 10.1109/JMEMS.2014.2352792.

[11] R. Ardito, L. Baldasarre, A. Corigliano, B. De Masi, A. Frangi, "Experimental Evaluation and Numerical Modeling of Adhesion Phenomena in Polysilicon MEMS", *Meccanica*, 48, 2013, pp. 1835-1844.

CONTACT

*S. Dellea, tel: +39-02-23993744; mail: stefano.dellea@polimi.it

EXPLORING THE Q-FACTOR LIMIT OF TEMPERATURE COMPENSATED CMOS-MEMS RESONATORS

Ming-Huang Li[1], Cheng-Syun Li[1], and Sheng-Shian Li[1,2]
[1]Inst. of NanoEngineering and MicroSystems, [2]Dept. of Power Mechanical Engineering
National Tsing Hua University, Hsinchu, TAIWAN

ABSTRACT

This work presents a study on the quality factor (Q-factor) of the passively temperature compensated CMOS-MEMS resonators through the collected material/experimental database and finite-element (FEM) simulation. By adapting an anchor-loss-free double-ended tuning fork (DETF) resonator design, the intrinsic material loss is expected to be the major loss mechanism in CMOS-MEMS resonators that limits the maximum Q-factor below 3,400 at the frequency of interest (300 kHz – 3 MHz). The highest Q-factor of 3,029 is measured in a 1.17-MHz DETF resonator, which is in good agreement with the theoretical prediction. Moreover, the increase of Q-factor at high temperature is also observed, which makes the proposed resonator very attractive for oven-controlled CMOS-MEMS oscillator applications.

INTRODUCTION

The quality factor (Q-factor) is one of the most crucial parameter of a micro-electro-mechanical systems (MEMS) resonator for low-power, low-noise signal processing and sensor applications. Since silicon is the most common material in both MEMS and integrated circuits, the Q-factor for silicon-based resonators is comprehensively studied in literature [1][2]. It is well-known that silicon MEMS resonators promise high f–Q product ($>10^{13}$) [3] due to its single- or poly-crystalline properties, which offer a great potential for many critical applications. However, the integration of high-Q silicon resonators with interface circuits is still very challenging, that hinders the overall system size reduction. Although a newly developed MEMS-first platform [4] provides an approach to integrate silicon MEMS with circuits, the use of SOI-CMOS might not be very cost-effective.

On the other hand, adapting back-end of line (BEOL) material staking in standard bulk CMOS processes as mechanical structure is an alternative way for CMOS-MEMS co-fabrication. In CMOS process, the BEOL materials can be roughly divided into two groups, i.e., metal and oxide. It has been shown that the oxide layer (SiO$_2$) in CMOS is a very high-Q material that could offer a high f–Q product of 5.4×10^{11} and also provide a positive temperature coefficient of frequency (TC_f) for resonators [5]. However, the metals, mostly formed by AlCu and tungsten (W), present significantly lower Q-factors and provide a negative TC_f for resonators [6]. Since a high-performance resonator is not only judged by its Q-factor, the temperature stability is also crucial to be fitted into the timing application specs. The intrinsic TC_f compensation can be attained by mixing these two material groups (i.e., metal and oxide) with proper weighting ratios. As a result, most of CMOS-MEMS resonators have an average Q-factor of ~2,000 and TC_f of ~60 ppm/°C, as shown in Fig. 1. As also shown in Fig. 1, the recently

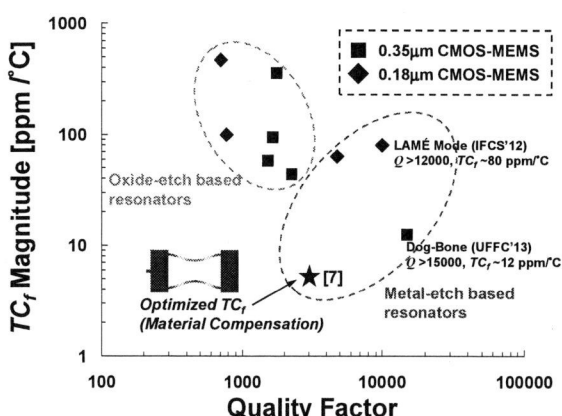

Figure. 1: Summary of the temperature coefficients of frequency (TC_f) and Q-factors of recently-developed CMOS-MEMS resonators [8].

developed CMOS-MEMS ovenized resonator [7] provides $Q > 3,000$ and decent $TC_f < 5.1$ppm/°C, suited for low-power oscillator applications.

Although high performance CMOS-MEMS resonators have already been demonstrated, the loss mechanism of the CMOS-MEMS resonators is still vague, therefore hindering the theoretical Q-factor prediction as well as the future CAD (computer-aided design) in general MEMS. To perform precise Q-factor evaluation at early design stages is of great importance. Therefore, the energy loss mechanism is studied in this work by extracting the "intrinsic material Q-factor" throughout a number of CMOS-MEMS resonators we developed in past few years [8]. Since interface loss [9] is very difficult to be extracted in multi-material resonators, an empirical model is used to evaluate the upper-bond of the Q-factor in this study.

CMOS-MEMS RESONATOR

In this work, a temperature-compensated double-ended tuning fork (DETF) CMOS-MEMS resonator is used to perform the Q prediction as a case study. The cross-sectional view of the resonator is depicted in Fig. 2 where the width of AlCu, W, and poly-Si layers are 0.9 µm, 0.5 µm, and 1.65 µm, respectively. The width and length of the resonator is around 5 µm and 136 µm, respectively. The frequency of the resonator is given by

$$f_o \approx \frac{(4.73)^2}{2\pi} \sqrt{\frac{\sum E_i I_i}{\sum \rho_i A_i}} \frac{1}{L_r^2}. \qquad (1)$$

With the material properties in Fig. 2, the simulated resonance frequency is 1.3 MHz, which is slight higher than measured resonance frequency of 1.2 MHz. The discrepancy between simulation and measurement comes from the uncertainty of material properties. Based on (1),

	E (GPa)	ρ (kg/m³)	TC_E (ppm/°C)
Metal (AlCu)	70	2700	-620
VIA (W)	411	19500	-6
Poly-Si	160	2300	-30
IMD (SiO₂)	70	2200	+180

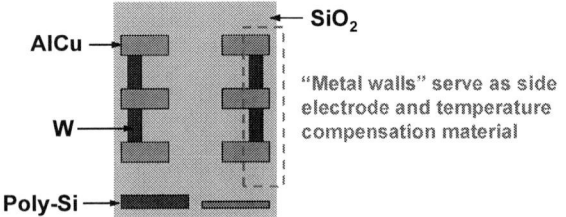

Figure. 2: Material properties and cross-sectional view of a temperature-compensated CMOS-MEMS resonator.

the temperature coefficient of frequency (TC_f) can be further expressed by

$$TC_f = \frac{1}{f(T_o)} \left(\frac{\partial f(T)}{\partial T} \right)_{T_o} \quad (2)$$

where $f(T)$ is the frequency equation and T_o is the reference temperature. In this resonator design, the "metal walls" are placed at the very edge of the resonator to serve as electrodes for electrostatic transduction as well as temperature compensation layers. Based on the material properties in Fig. 2, the simulated TC_f is +9.3 ppm/°C.

Q OF CMOS-MEMS RESONATORS

As reported in [7], the typical Q-factor of temperature compensated CMOS-MEMS resonator ranges from 1,500 to 3,000. The limitation on the Q-factor will be discussed in this section.

The resonator Q-factor can be calculated based on individual Q's contributed by each energy loss mechanism. In this work, we only consider the most significant loss mechanisms as indicated in (3). Therefore the total Q-factor can be approximated by

$$\frac{1}{Q_{Total}} \approx \frac{1}{Q_{Anchor}} + \frac{1}{Q_{TED}} + \frac{1}{Q_{MAT}} \quad (3)$$

where Q_{Anchor}, Q_{TED}, and Q_{MAT} are the Q-factors contributed by anchor loss, thermoelastic damping, and intrinsic material loss, respectively. Air damping is neglected here since the resonator is operated in vacuum.

Anchor (Acoustic) Loss

The acoustic loss is one of the most significant Q-limiting factors for MEMS resonators. Unlike intrinsic loss mechanism such as interface loss [9], the anchor loss of a MEMS resonator can be minimized by appropriate designs. For in-plane vibrating resonators, balanced dual-resonator design turns out to be an effective solution for high Q_{Anchor}. A DETF resonator, which is composed of two clamped-clamped beams, can attain very high Q by enhancing the acoustic energy confinement between resonant beams. To evaluate the Q_{Anchor} for CMOS-MEMS resonators, a finite-element simulation approach is adopted, as shown in Fig. 3. The perfect matched layer (PML) is placed at the ends of the supports to describe the effect of

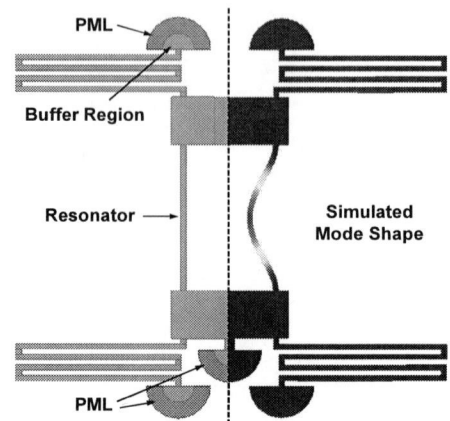

Figure. 3: Model of a DETF resonator for anchor loss simulation in COMSOL Multiphysics.

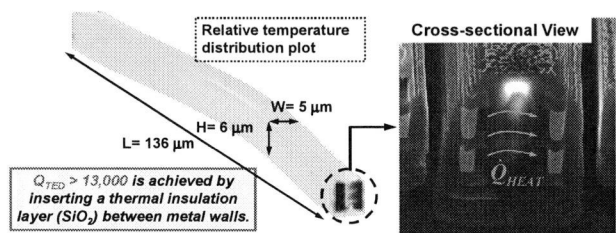

Figure. 4: Simulated thermoelastic dissipation of a temperature compensated CMOS-MEMS resonator.

elastic acoustic energy radiation to the substrate. As an extension of supporting beams, an interface buffer region is inserted in between PML and structural layers. In this design, the simulated Q_{Anchor} is well above 100,000; therefore, it would not be the dominant loss mechanism.

Thermoelastic Damping (TED)

Thermoelastic damping (TED) is described by the coupling between mechanical stress and temperature, which exists in every material with certain thermal expansion coefficients. For a simple and homogeneous beam, the TED effect can be described well by Zener's equation [10]. However, the theoretical prediction of Q_{TED} for composite material resonator is quiet difficult.

To address this issue, the FEM simulation is again used for Q_{TED} prediction. The simulated $Q_{TED} > 13,000$ is demonstrated, as shown in Fig. 4. Note that this Q_{TED} is comparable to that of silicon resonators operated at similar frequency [10]. Since SiO₂ is placed at the central region of this resonator, the low thermal conductivity of the oxide prevents the thermal flows from one side to the other side. This is similar to the hollow or slotted resonator that isolates the thermal flow by slots on the resonator body [10][11]. Since the Q_{TED} is in 10^4 level, it only has minor effects on the Q_{Total}.

Material and Interface Loss

Since Q_{Anchor} and Q_{TED} can be modeled by FEM simulation, Q_{MAT} is the only parameter that needs to be extracted from measurement data. For a multi-material CMOS-MEMS resonator, the Q_{MAT} can be expressed as [12]

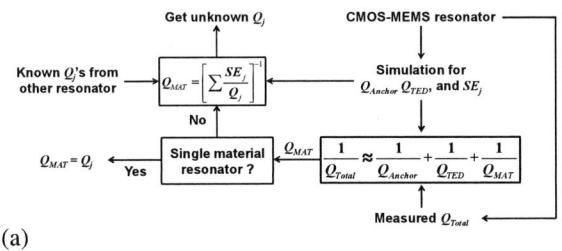

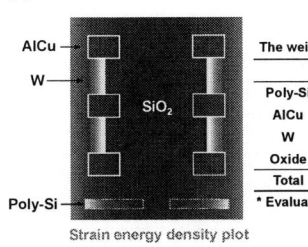

The weighting of strain energy on each material:		
	Strain Energy, *SE* (%)	*Qj*
Poly-Si	5%	> 100k
AlCu	8.6%	3000*
W	36.4%	2000*
Oxide	50%	>100k
Total	100%	~4600
* Evaluated from experiment data [5][6][8]		

Strain energy density plot

(b)

Figure. 5: (a) Flow chart for extracting the Q_j's for each material. (b) Simulated strain energy density plot of the CMOS-MEMS resonator for material Q-factor evaluation. The Q_j of each material is extracted from previous literature [5][6][8].

$$\frac{1}{Q_{MAT}} = \frac{SE_{AlCu}}{Q_{AlCu}} + \frac{SE_{SiO2}}{Q_{SiO2}} + \frac{SE_W}{Q_W} + \frac{SE_{Poly-Si}}{Q_{Poly-Si}} \quad (4)$$

where SE_j is defined as the fractional strain energy of the *j*-th structural material, which is defined as

$$SE_j = \left(\frac{\text{Strain Energy of j-th Mat.}}{\text{Total Strain Energy}} \right) \quad (5)$$

and Q_j is the equivalent Q-factor of the *j*-th structural material. By using (4), the material Q-factors can be extracted from the previously-developed CMOS-MEMS resonators [5][6][8]. Please note that the interface loss caused by the elastic modulus mismatch between materials (ΔE) [9] is merged with the extracted Q-factor of the *j*-th material layer in this empirical model since the interface loss cannot be solely extracted by experimental results.

The flow chart for Q_j extraction is shown in Fig. 5(a). The extraction of the Q_j starts with a single material resonator, such as mere-metal resonator in [6] and mere-oxide resonator in [5] for Q_{AlCu} and Q_{SiO2} evaluation. Note that Q_{SiO2} is also suggested to be in 10^5 level in [12]. After Q_{AlCu} and Q_{SiO2} are obtained, the Q_W is extracted from composite resonators, and Q_W of 2,000 is fitted to match the measurement results from different types of resonators. The $Q_{Poly-Si}$ is assumed to be the same level with Q_{SiO2}. The extraction result is depicted in Fig. 5(b).

Obviously, the Q_W is the dominant factor for the Q_{MAT} in this temperature compensated resonator since tungsten (W) occupies 36% of total strain energy. With the above equation, $Q_{MAT} = 4,600$ is expected for the DETF resonator.

EXPERIMENTAL RESULTS

Fig. 6(a) shows the typical resonant spectrum of the DETF resonator with on-chip readout electronics, where the Q-factor is around 2,000. Fig. 6(b) demonstrates the temperature stability of the resonator using its embedded micro-heater [7]. As the heating power increases, the frequency upper-shifts with an equivalent TC_f of

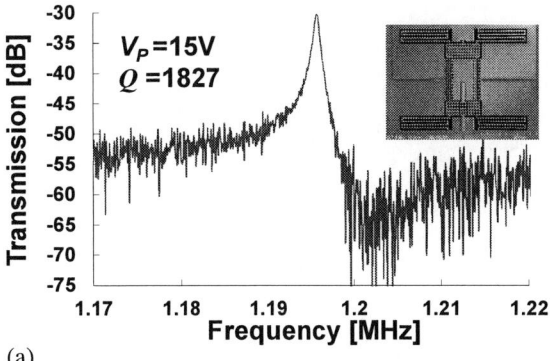

(a)

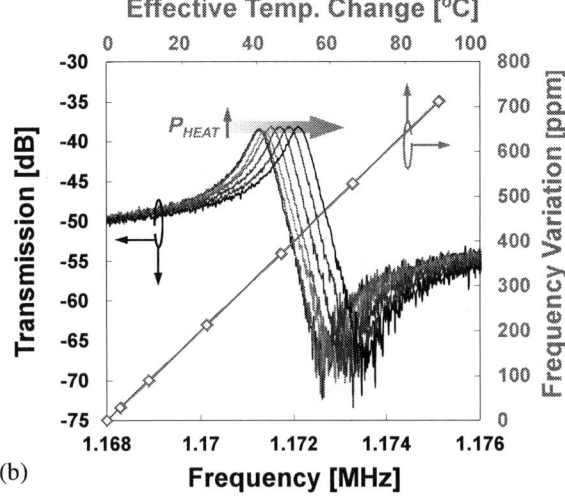

(b)

Figure. 6: (a) Measured transmission spectra of the DETF resonator circuit at $V_P=15V$ in vacuum environment. (b) Measurement of frequency stability using the embedded micro-heater.

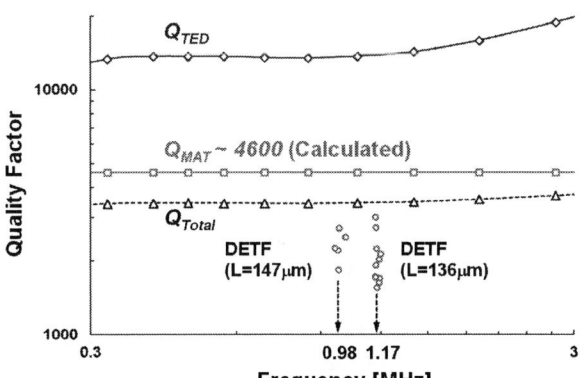

Figure. 7: Q-factor evaluation using simulated Q-factors. The circle legends indicate the measurement results. The highest Q-factor measured is 3,029 (beam length L=136μm).

+8 ppm/°C. In this work, we have measured 15 DETF resonators with different beam lengths. Fig. 7 shows the measurement results of the resonators with a highest Q of 3,029, which is very close to the theoretical prediction (Q_{Total}~3,400 at 1.17 MHz). Note that the simulated Q_{Anchor} is greater than 100,000; therefore, it does not appear in this figure. Since a composite material resonator is designed, simulated Q_{TED} possesses a flat region across 0.3 to 3 MHz.

978-1-4799-7956-1/15 $31.00 © 2015 IEEE

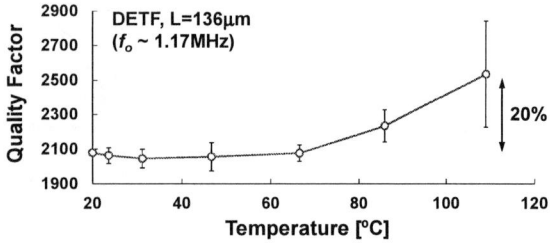

(a)

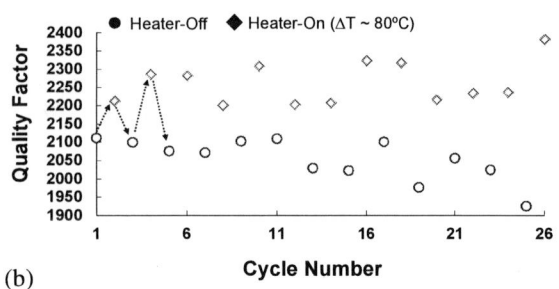

(b)

Figure. 8: (a) Measurement of temperature dependency of the Q-factors (averaged from 4 heating/cooling cycles). (b) Q-factor measurement under rapid temperature cycling tests using an integrated micro-heater.

Table 1: Summary of the Q-factor evaluation

Loss Mechanism	Quality factor	Comment
Anchor Loss	> 100,000	FEM simulation (*PML Method)
Thermoelastic Damping	> 13,000	FEM simulation
**Material Loss	~4,600	Evaluated from experimental data
Predicted Q_{Tot}	~3,400	Defines the "upper limit" of the measured Q-factor
Measurement Data (DETF, TC_f < 10 ppm °C)	1,557 (Min) 2,041 (Avg) 3,029 (Max)	The Q-factor shows a slight temperature dependency

* *PML: Perfect Matched Layer.*
** *Interface loss is incorporated into material loss in this study.*

Moreover, the temperature dependency of the Q-factor is also investigated, which shows a slightly increasing tendency (cf. Fig. 8(a)). Multiple rapid temperature cycling tests using the embedded micro-heater is also performed to confirm the repeatability of this phenomenon (cf. Fig. 8(b)). This effect is similar to that reported in [13]. This is possibly related to the temperature coefficient of quality factor (TC_Q) of the SiO$_2$.

Finally, Table 1 summarizes the findings of this work, which would provide new insights for future CMOS-MEMS resonator designs.

CONCLUSIONS

In this work, a study on the Q-factor for CMOS-MEMS resonators is presented. An empirical approach is conducted to estimate the contributions from various energy loss mechanisms. With the collected resonator database, the intrinsic material and interface loss is expected to be the dominant factor for the Q_{Total}. In summary, this work provides valuable information for "engineering" the Q-factor of CMOS-MEMS resonators, which will greatly benefit the computer-aided MEMS design in the future.

ACKNOWLEDGEMENTS

This research was sponsored by the MOST of Taiwan (MOST-103-2221-E-007-113-MY3). The chip fabrication is supported by CIC and TSMC, Taiwan. The authors are grateful to the Center for Nanotechnology, Materials Science and Microsystems of National Tsing Hua University for the use of fabrication and measurement facilities.

REFERENCES

[1] S. A. Chandorkar, *et al.*, "Limits of quality factor in bulk-mode micromechanical resonators," *Proc. MEMS'08*, pp. 74-77, Jan 2008.

[2] R. Tabrizian, *et al.*, "Effect of phonon interactions on limiting the *f-Q* product of micromechanical resonators," *Tech. Dig. Transducers'09*, pp. 2131-2134, June 2009.

[3] D. Weinstein and S. A. Bhave, "Internal dielectric transduction in bulk-mode resonators," *J. Microelectromech. Syst.*, vol. 18, no. 6, pp. 1401-1408, Dec. 2009.

[4] J. Philippe, *et al.*, "Fully monolithic and ultra-compact NEMS-CMOS self-oscillator based-on single-crystal silicon resonators and low-cost CMOS circuitry," *Proc. MEMS'14*, pp. 1071-1074, Jan 2014.

[5] W.-C. Chen, *et al.*, "VHF CMOS-MEMS oxide resonators with $Q > 10,000$," *2012 IEEE Int. Frequency Control Symposium*, pp. 1-4, May 2012.

[6] W.-C. Chen, W. Fang, and S.-S. Li, "A generalized CMOS-MEMS platform for micromechanical resonators monolithically integrated with circuits," *J. Micromech. Microeng.*, vol. 21, no. 6, pp. 065012, May 2011.

[7] M.-H. Li, *et al.*, "A monolithic CMOS-MEMS oscillator based on an ultra-low-power ovenized micromechanical resonator," *J. Microelectromech. Syst.*, 2014, *In Press*.

[8] S.-S. Li, "CMOS-MEMS resonators and their applications," *Proc. UFFC'13*, pp. 915-921, July 2013.

[9] L. G. Villanueva, *et al.*, "Interface losses in multimaterial resonators," *Proc. MEMS'14*, pp. 632-635, Jan. 2014.

[10] A. Duwel, *et al.*, "Engineering MEMS resonators with low thermoelastic damping," *J. Microelectromech. Syst.*, vol. 15, no. 6, pp. 1437-1445, Dec. 2006.

[11] S. Prabhakar and S. Vengallatore, "Thermoelastic damping in hollow and slotted microresonators," *J. Microelectromech. Syst.*, vol. 18, no. 3, pp. 725-735, June 2009.

[12] C.-C. Lo, "CMOS-MEMS resonators for mixer-filter applications," *Ph.D. Dissertation*, Dept. Electr. Comput. Eng., Carnegie Mellon Univ., 2008.

[13] R. Tabrizian, *et al.*, "Temperature-stable silicon oxide (SilOx) micromechanical resonators," *IEEE Trans. Electron Dev.*, vol. 60, no. 8, pp. 2656-2663, Aug. 2013.

CONTACT

*S.-S. Li, ssli@mx.nthu.edu.tw

NANOFIBER FORESTS AS A HUMIDITY-SENSITIVE MATERIAL

C. Lei[1,2], L. C. Tang[1,2], H.Y. Mao*[1,3], Y. Wang[1,2], J.J. Xiong[2], W. Ou[1,3], Y. Ou[1,3], A. J. Ming[1,3], D. Li[4]
Q.L.Tan[2], W. B. Wang[1,3], D. P. Chen[1,3] and T. Liang[2]

[1]Key Laboratory of Microelectronics Devices & Integrated Technology, Institute of Microelectronics, Chinese Academy of Sciences, Beijing 100029, P. R. China
[2]National Key Laboratory for Electronic Measurement Technology, North University of China, Taiyuan 030051 P. R. China
[3]Smart Sensor Engineering Center, Jiangsu R&D Center for Internet of Things, Wuxi 214135, P. R. China
[4]Department of Electrical Engineering, Stanford University, California, 94305, United States

ABSTRACT

Nanofiber forests with high hydrophilicity and humidity-sensitivity are reported in this work. They are fabricated from polyimide (PI) by a plasma-stripping technique, and are utilized as a humidity-sensitive material in humidity sensors. In a relative humidity range of 50%-80%, nanofiber forest-based devices have capacitance ~50% larger than that of PI-based sensors. Meanwhile, as the nanofibers construct a non-closed layer, the internal nanostructures are fully exposed to external environment, thus the absorption and desorption of moisture consume only a short time. Since the fabrication processes for both nanofiber forests and the devices are simple and fast, thus such a method may provide possibility for development of novel humidity sensors with high sensitivity and short response time.

INTRODUCTION

Humidity sensors plays an important part in various fields, including manufacturing process control for precision electronic components, meteorological watch, agricultural cultivation, as well as material and equipment storage [1, 2]. In these fields, an accurate and rapid measurement of relative humidity is of great importance as it directly affects functions of materials and equipments, qualities of process controlling and final products, and meanwhile, it has an effect on analysis results of meteorological data. Therefore, the demand for humidity sensors is increasing all the time [3]. However, applications of conventional humidity sensors are limited by their large volumes, low sensitivity, slow response and high cost [4]. Besides, they can hardly meet the trends of being intelligent and able to be integrated into systems. With the development of Micro-Electro-Mechanical systems (MEMS) in recent decades, miniaturized humidity sensors based on different principles, including capacitive, piezoresistive, resonating, mechanical and thermo-elemental ones, have been developed [5]. Among all these MEMS humidity sensors, capacitance-based devices are favored ascribed to a better linearity in a wide range of relative humidity and a shorter response time [6]. For capacitive humidity sensors, comb capacitors can magnify the capacitance by the number of the comb pairs, thus they have a relatively high sensitivity and become the mostly employed structure.

To achieve even better performance thus to meet various applications, selecting materials with high humidity-sensitivity has become a research focus for years [7]. For capacitive humidity sensors, there are three main types of sensing materials, including porous silicon, porous ceramics and organics [7, 8]. Porous silicon and porous ceramics like anodic aluminium oxide have been proven to exhibit highly humidity-sensitive properties, but fabrication processes related to these materials are relatively complex. Due to its ability to improve sensitivity, output range, and stability of the humidity devices, PI has been considered one of the most potential wet-sensing organic materials [9]. Even so, long periods of time for high-temperature curing and imidization are needed before PI can be brought into full play [10], thus other materials used in the device should be carefully chosen to suit the high temperature. Besides, PI and the other aforementioned materials are membrane-like materials, containing nanoscale internal pores, thus moisture absorption and desorption takes a relatively long time. Though progresses have been made to employ PI as a wet-sensing material in humidity sensors, and accordingly, performance of the sensors are being improved, still, there is a need to find a novel material for humidity sensors with better performance.

In this work, nanofiber forests with high humidity-sensitivity are fabricated using a plasma-stripping technique, as the process avoids high-temperature curing and imidization, thus the nanofiber forests can be easily integrated into humidity sensors. As has been tested, the nanofiber forest-based devices achieve capacitance ~50% larger than that of PI-based sensors in a relative humidity range of 50%-80%. Besides, the processes for moisture absorption and desorption are less time consuming.

FBRICATION

In conventional micromachining processes, plasma-stripping is a commonly utilized technique to remove unwanted polymers, such as different types of photoresist. As taken for granted, after long periods of bombardment, all the polymers would be removed, and the substrate surfaces should be very clean, however, observations of nanofibers have been reported [11-13]. Based on this phenomenon, an approach for fabricating patternable nanofiber forests is proposed, the process is schematically depicted in Fig. 1. To begin with the process, polymer is spin-coated and photopatterned, as shown in Fig. 1(a). Subsequently, plasma bombardment is performed on the polymer patterns for a few minutes (Fig. 1(b)). In this way, nanofiber forests are obtained in regions of the polymer patterns (Fig. 1(c)).

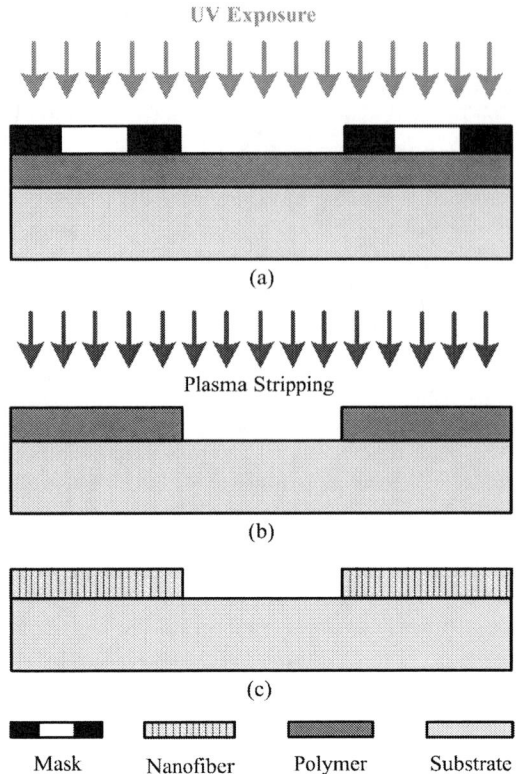

Figure 1: Process for fabricating nanofiber forests from polymers by plasma-stripping technique.

The generation of nanofibers in the plasma-stripping process can be ascribed to the combined actions of plasma ashing and plasma polymerization [14]. In step of the plasma ashing, monomer molecules are generated and subsequently energized and dissociated into neutral particles and reactant fragments, which are further repolymerized in the following plasma stripping process. As stripping continues, highly branched and cross-linked networks are gradually generated and consequently appear in morphologies of nanofibers.

It has been proved in our experiments that this method is applicable to different plasmas and polymers. Figure 2 shows scanning electron microscopy (SEM) images of the nanofiber forests acquired from different polymers by using different plasmas. Nanofiber in morphologies of bunches realized by oxygen-plasma-stripping-of-SPR photoresist is showed in Fig. 2(a). The heights of the fiberous bunches are 3.5 μm, while the thicknesses of original photoresist patterns were about 4.5 μm. By using oxygen-plasma-stripping technique, the nanofiber bushes shown in Fig. 2 (b) were generated from SU-8 patterns with an original thickness of about 8 μm. Figure 2(c) displays fibers from oxygen-plasma-stripping-of-PI. In this figure, the fibers present as individual nanowires with ~3.3 μm in length, and the thickness of original PI layer was about 4.5 μm. The fibers (2.6 μm in length) fabricated from argon-plasma stripped PI are shown in Fig. 2 (d), while the original PI patterns were about 3 μm in thickness. In the experiment of oxygen plasma stripping process, the RF power was 250 W, and the flow rate of O_2 was ~30 sccm. For argon bombardment, the RF power was 267 W

and flow rate of Ar was ~45 sccm.

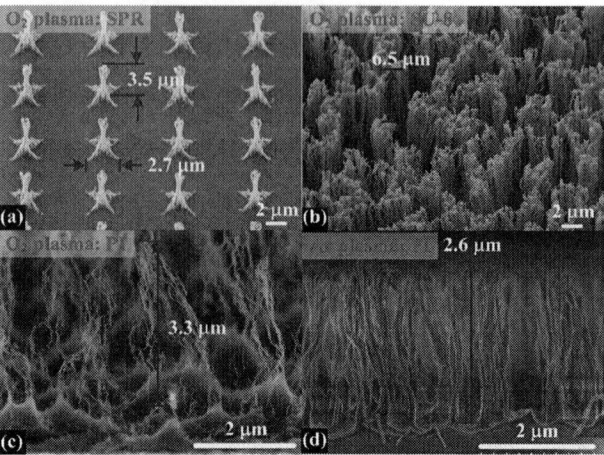

Figure 2: SEM images of the nanofiber forests. The nanofiber forests are obtained by (a) oxygen plasma stripping of SPR-220 photoresist; (b) oxygen plasma stripping of SU-8 photoresist; (c) oxygen plasma stripping of PI; (d) argon plasma stripping of PI.

CHARACTERIZATION

Due to the large capillary force caused by the large number of nano-crevices existing in the nanofibers, the nanofiber forests exhibit excellent hydrophilicity. This characteristic was testified by liquid flowing along microchannel-shaped nanofibers. A microchannel with a reservoir region and several branches was prepared. All the regions in the microchannel were fully composed of nanofibers. In the experiment, a droplet of DI water was dropped in the reservoir region, and because of the superhydrophilic features of the nanofibers, the fluid spread quickly by reaching the nanofibers at the droplet boundaries, seemed like flowing to the branches along the channels. A few seconds later, as a consequence, the channels a few millimeters away from the reservoir region were wetted, and obvious boundaries of dry and wet areas could be seen in the microchannel. Experimental details have been described elsewhere [13]. Figure 3 demonstrates the flowing results.

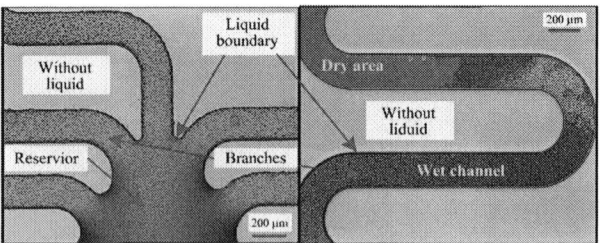

Figure 3: Liquids flowing along the microchannel consisting of nanofiber forests, demonstrating high hydrophilicity of the nanofibers.

APPLICATION

As the nanofiber forests are superhydrophilic and their fabrication process is compatible with microfabrication, therefore, it is possible to integrate the nanofiber forests into humidity sensors to improve performance. Herein, a comb-based humidity sensor using nanofiber forests as sensing material is designed, the structure and fabrication process of the device are illustrated in Fig. 4. In the structure, micro-combs are used as electrodes to form a capacitor, while an S-shaped resistor strip functions as a micro heater to release moisture before each measurement.

To fabricate the device, the surface of a Si substrate is covered with a SiO₂ film, which functions as an insulating layer (Fig. 4(a)). In our experiment, a SiO₂ film with thickness of 4000 Å was obtained by oxidation. After that, a metal layer is deposited and photo-patterned on the SiO₂ film, and the Al structures are utilized as the micro-combs and the resistor strip, as illustrated in Fig. 4(b). In this work, a 2 µm Al layer was sputtered and patterned into comb-pairs, a resistor strip, as well as electrodes. Subsequently, PI is spin-coated and patterned within the regions between each comb-pair (Fig. 4(c)). Herein, a PI film with thickness of ~3 µm was coated, then the film was suffered to steps of UV exposure and development, in this way, PI was patterned. As is exhibited in Fig. 4(c), the PI pattern is distributed along the resistor strip, and takes full use of the spaces between each comb-pair. Finally, plasma-stripping is adopted to bombard the PI pattern thus to generate nanofiber forests, shown in Fig. 4(d). To avoid oxidization of Al, Ar plasma was employed to bombard the PI layer in our experiment, using the parameters mentioned above. By doing this, nanofibers with heights of ~2.5µm were successfully integrated into the humidity sensor. Figure 5 demonstrates the SEM images of a nanofiber forest-based humidity sensor with 9 pairs of combs.

In order to characterize properties of the nanofiber forest-based humidity sensor, another sensor based on imidized PI with the same structure was also prepared. Then the two types of humidity sensors were placed together in a humidity and temperature controlling chamber. At the same time, the measurement system was equipped with a semiconductor parameter analyzer (Agilent technology, B1500A) and a digital power. The digital power was connected with the electrodes of the resistor strip for heating the nanofibers to release moisture, while the B1500A was used to detect the output capacitance signals. In the measurement, the humidity inside the chamber was varied from 50-80 RH%, while the temperature was kept at a constant value of 60 °C. The frequency of the measurement system was set to be 1 MHz, and meanwhile, a voltage of 30 mV was employed as bias voltage in B1500A. At the beginning of the test, the humidity of the chamber was set at 20RH% and both humidity sensors were desorbed of moisture by supplying the electrodes with 0.1 Volt as the power to ensure accuracy of the following tests. By adjusting the relative humidity of the chamber, the capacitance of both sensors was detected within the range from 50 RH% to 80 RH %, the results are shown in Fig. 6. As shown in this figure, the nanofiber forest-based device has capacitance ~50% larger than that of the PI-based sensor, and its sensitivity reaches 1.43 pF/10RH%, while that for the PI-based one is 0.96 pF/10RH%. The enhancement can be ascribed to the larger surface-volume ratio and higher hydrophilicity of the nanofiber forests when compared with imidized PI. Both of these reasons make the device absorb more moisture and

thus lead to a larger change in dielectric constant. Besides, a shorter response time of the nanofiber forest-based device was also observed, and this could be attributed to the highly hydrophilic internal fiberous structures and the non-closed morphologies as well.

(a) SiO₂ layer preparation on a Si substrate

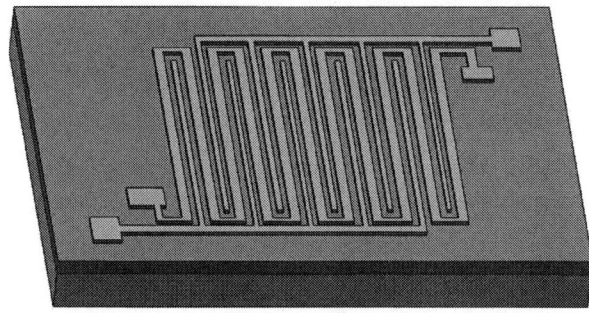

(b) Al sputtering and patterning to form micro-heater and combs

(c) PI layer spin-coating and photopatterning

(d) Plasma-stripping of PI to form nanofiber forests

Figure 4: Schematic structure and fabrication process of a nanofiber forest-based humidity sensor.

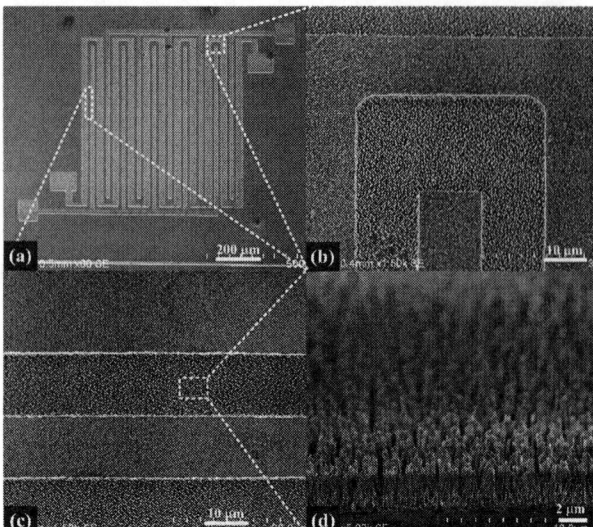

Figure 5: SEM images of a nanofiber forest-based humidity sensor. The nanofibers are distributed in gaps between each comb pair, the heights of the nanofibers are ~2.5 μm, which take full use of the space between comb pairs. The combs are 2 μm in height.

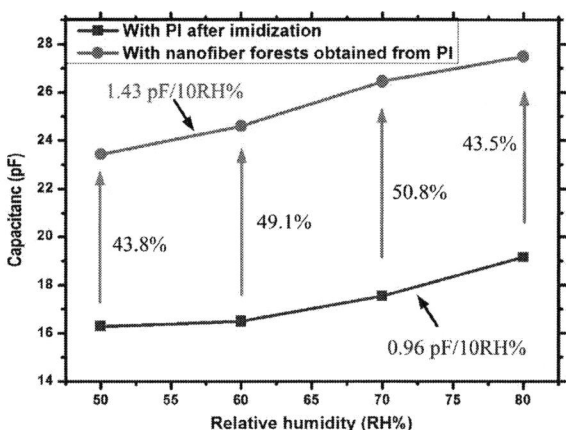

Figure 6: Humidity-sensitivity of the nanofiber forest-based humidity sensor and a PI-based sensor, at 60 °C, 30 mV and 1 MHz. Nanofiber forests show higher humidity-sensitivity.

CONCLUTION

By using a plasma-stripping technique, nanofiber forests with superhydrophilicity are fabricated and utilized as a humidity-sensitive material in humidity sensors. Compared with conventional PI-based sensors, the nanofiber forest-based devices have a much higher humidity-sensitivity and a shorter response time. It is expected that the performance of such devices can be further improved by optimizing structures of the nanofiber forest-based sensors, so that they will be capable of meeting more applications.

ACKNOWLEDGEMENTS

This work was supported in part by Jiangsu Natural Science Foundation (Grant No. BK20131098), the National Natural Science Foundation of China (Grant No.

61401458, 61335008 & 61176114). The authors thank Prof. W.G. Wu and Mr. Y. F. Mao from Institute of Microelectronics, Peking University for their help with the experiment and preparation of SEM images.

REFERENCES

[1] A. Tsigara *et al.*, "Hybrid polymer/cobalt chloride humidity sensors based on optical diffraction," *Sensor. Actuat.* B, vol. 120, pp. 481-486, 2007.

[2] F. W. Zeng, X. X. Liu, D. Diamond and K. T. Lau, "Humidity sesors based on polyaniline nanofibres," *Sensor. Actuat.* B, vol. 143, pp. 530-534, 2010.

[3] E. Traversa, "Ceramic sensors for humidity detection: the state-of-the-art and future developments," *Sensor. Actuat.* B, vol. 23, pp. 135-156, 1995.

[4] Y. Y. Qiu, C. A. Leme, L. R. Alcacer and J. E. Franca, "A CMOS humidity sensor with on-chip calibration," *Sensor. Actuat.* A, vol. 92, pp. 80-87, 2001.

[5] C. L. Dai, "A capacitive humidity sensor integrated with micro heater and ring oscillator circuit fabricated by CMOS-MEMS technique," *Sensor. Actuat.* B, vol. 122, pp. 375-380, 2007.

[6] G. J. W. Visscher and J. G. Kornet, "Long-term tests of capacitive humidity sensors," *Meas. Sci. Technol.*, vol. 5, pp. 1294-1302, 1994.

[7] J. Yeow and J. She, "Carbon nanotube-enhanced capillary condensation for a capacitive humidity sensor," *Nanotechnology*, vol. 17, pp. 5441-5448, 2006.

[8] A. Boukezzata *et al.*, "Investigation properties of Au-porous α-Si$_{0.70}$C$_{0.30}$ as humidity sensor," *Sensor. Actuat.* B, vol. 176, pp. 1183-1190, 2013.

[9] J. R. Cha and M. S. Gong, "Polyelectrolyte humid membranes anchored to the gold surface on flexible Polyimide substrate and their water durability," *Sensor. Actuat.* B, vol. 160, pp. 1082-1090, 2011.

[10] M Dokmeci and K. Najafi, "A high-sensitivity Polyimide capacitive relative humidity sensor for monitoring anodically bonded hermetic micropackages," *J. Microelectromech. Syst.*, vol. 10, pp. 197-204, 2011.

[11] H. Y. Mao *et al.*, "Microfluidic Surface-enhanced Raman Scattering sensors based on nanopillar forests realized by an oxygen-plasma-stripping-of-photoresist technique," *Small*, vol. 10, pp.127-134, 2014.

[12] H. Y. Mao, D.Wu, W. G. Wu, J. Xu and Y. L. Hao, "The fabrication of diversiform nanostructure forests based on residue nanomasks synthesized by oxygen plasma removal of photoresist," *Nanotechnology*, vol. 20, pp. 445304, 2009.

[13] H. Mao, W. Wu, Q. Liu, Y. Zhang, and Y. Li, "Nanofiber-based surface microfluidic structures for cell and nanoparticle patterning," *microTAS 2010*, pp. 500-502, Oct. 2010.

[14] H. Y. Mao, et al., "Nanofiber forests with high infrared absorptance," *IEEE MEMS 2014*, pp. 644-647, Jan. 2014.

CONTACT

*H.Y. Mao, tel: 86-10-82995934; maohaiyang@ime.ac.cn

PIEZOELECTRIC PAPER FOR PHYSICAL SENSING APPLICATIONS

Suresha K. Mahadeva, Konrad Walus, and Boris Stoeber

The University of British Columbia, Vancouver, BC V6T 1Z4, CANADA

ABSTRACT

We have developed robust and mechanically flexible piezoelectric paper. The fabrication process involves functionalization of barium titanate ($BaTiO_3$) nanostructures onto wood fibers, followed by activation in a suspension of the commercially available paper-strength-enhancing additive, carboxymethyl cellulose (CMC), which improves fiber-fiber bonding. This leads to piezoelectric paper with both high tensile strength and flexibility. We have investigated the effect of CMC concentration (2-6 wt%) on the tensile properties of the paper and found the highest tensile strength at 6wt% CMC. This piezoelectric paper has the largest piezoelectric coefficient reported for paper to date ($d_{33} = 37 - 45.7 \pm 4.2$ pC/N) and is comparable to that of commercially available piezoelectric polymers such as polyvinylidene fluoride with $d_{33} = 30$ pC/N. In addition, we have demonstrated the application of this paper as a tactile sensor.

INTRODUCTION

Significant research efforts have recently been dedicated to the development of low-cost paper-based sensing devices with broad applications in MEMS including: microfluidics [1]; strain sensing [2]; energy storage [3]; and volatile organic compound (VOC) detection [4]. It is expected that in these applications, paper would also have significant benefit of being disposable and biodegradable. Recently, a significant focus has been on developing flexible piezoelectric paper by adopting hydrothermal synthesis of zinc oxide nanostructures on paper substrate at low temperature and demonstrating their application in electronic devices [5], strain sensing [6], and energy harvesting [7]. More recently, piezoelectric paper has been fabricated by functionalizing wood fibers with $BaTiO_3$ nanoparticles prior to paper making [8, 9]. The fabrication process for this piezoelectric paper is cost-effective and compatible with the conventional paper-making process. However, this piezoelectric paper possesses a low piezoelectric coefficient ($d_{33} = 4.8 \pm 0.4$ pC/N) and suffers from poor mechanical properties including a 78% reduction in tensile strength compared to non-functionalized paper due to the reduced adhesion between the functionalized wood fibers. Here we report mechanically strong piezoelectric paper that possesses the largest piezoelectric coefficient reported for paper to date. Our piezoelectric paper fabrication process is similar to the process reported in [8, 9], and is modified to enhance adhesion between the functionalized wood fibers with paper-strength-enhancing additives. In addition, we demonstrated the material as the basis for a simple sensor. A tactile or touch sensor is a device that provides a response to mechanical interactions with its surrounding environment. Tactile sensors play a vital role in the field of electronics, robotics, security systems, industrial automation, and medicine [10], and

over the years their use in these areas increased tremendously. Over the past few years, tactile sensing technology witnessed multifold research efforts in the development of new material and transduction mechanisms including resistive, capacitive, optical, piezoresistive, magnetic, and piezoelectric [11].

MATERIALS AND METHODS

Poly(diallyldimethylammonium chloride) (PDDA, 20 wt% in water, MW 100,000 - 200,000) and poly(sodium 4-styrenesulfonate) (PSS, MW 70,000) were purchased from Sigma Aldrich, Sodium chloride (ACS certified) and $BaTiO_3$ nanoparticles (99.9% purity, tetragonal structure with particles size of 300 nm) were purchased from Fisher and US Research Nanomaterials Inc., respectively. Carboxymethyl cellulose (CMC-1161) was acquired from AkzoNobel Pulp and Performance Canada Inc. All the chemicals were used as received. Wood fibers from bleached softwood pulp (Canfor Prince George Mills) were obtained after disintegration and washing with distilled water.

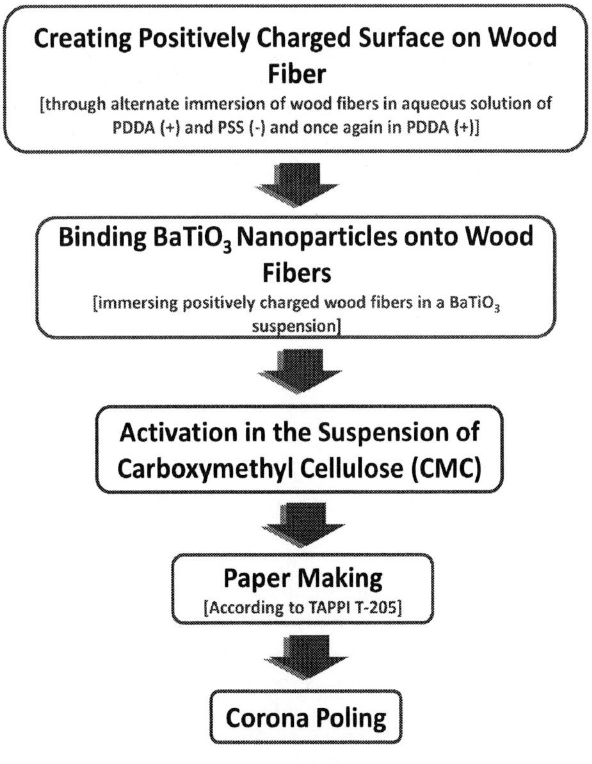

Creating Positively Charged Surface on Wood Fiber
[through alternate immersion of wood fibers in aqueous solution of PDDA (+) and PSS (-) and once again in PDDA (+)]

Binding BaTiO₃ Nanoparticles onto Wood Fibers
[immersing positively charged wood fibers in a BaTiO₃ suspension]

Activation in the Suspension of Carboxymethyl Cellulose (CMC)

Paper Making
[According to TAPPI T-205]

Corona Poling

Figure 1: Flow-chart showing piezoelectric paper fabrication process.

Piezoelectric Paper Fabrication

Our piezoelectric paper fabrication process builds on the process reported in [8, 9]. This fabrication process has been modified to improve the mechanical properties of the piezoelectric paper. Figure 1 shows the complete piezoelectric paper fabrication process including the final poling step, required to align the polarization of the barium titanate nanoparticles. In brief, wood fibers were first immersed in the aqueous solution of PDDA (+) and then in PSS (-) and once again in PDDA (+), to create a positively charged surface on the wood fiber. This is followed by immersion of the wood fibers into a $BaTiO_3$ suspension, leading to the functionalization of the wood fibers with nanostructured $BaTiO_3$. The next step of our approach involves overnight activation of the functionalized fibers in a suspension of commercially available paper-strength-enhancing additives. This process ensures the uniform coating of the CMC over the functionalized wood fibers (Figure 2) and results in improved fiber-fiber bonding. Paper hand sheets (approximate diameter: 16 cm) were made according to the TAPPI method T-205 [12]. Finally, the resulting paper is subjected to corona poling at $120\,°C$ for 4 hours with a needle voltage of 17 kV and a grid voltage of 5 kV to render it piezoelectric. A detailed description of the corona poling setup and procedure can be found elsewhere in the literature [13].

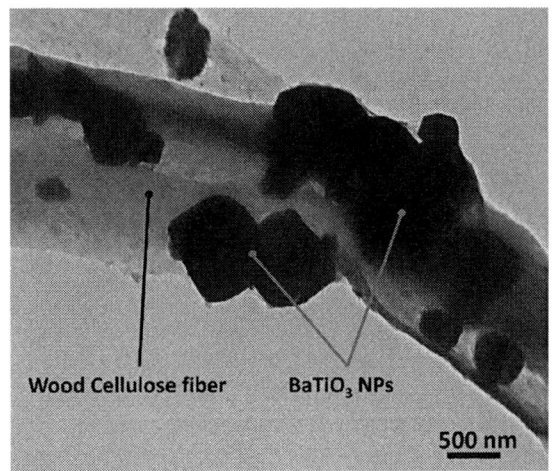

Figure 2: TEM image of functionalized wood cellulose fiber after activation in a CMC suspension.

Piezoelectric Paper Characterization

The tensile properties of the piezoelectric paper were evaluated by measuring the stress and strain. For that purpose, the piezoelectric paper was cut into strips with dimensions of 50 mm × 5 mm. These strips were then subjected to tensile testing using a BOSE ElectroForce® 3100 uniaxial tensiometer. The two ends of a test strip were fixed between the upper and lower grips of the instrument, with a gauge length of 20 mm. The test was performed at room temperature and a relative humidity of approximately 40% with a displacement rate of 0.01 mm/s.

A FEI Technai G2 transmission electron microscope (TEM) was employed to visualize the CMC coating over the functionalized wood fibers. The piezoelectric coefficient of this piezoelectric paper and its sensitivity in tactile sensing was measured using a charge meter (Kistler Charge meter 5015).

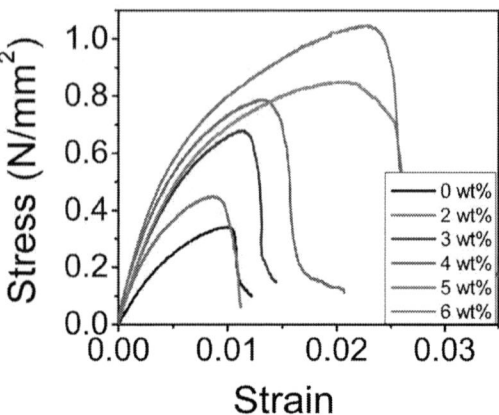

Figure 3: Stress-strain curves of piezoelectric paper showing the effect of CMC addition on tensile properties.

RESULTS AND DISCUSSIONS

The effect of CMC on the tensile properties of piezoelectric paper was evaluated by subjecting it to tensile load using a tensiometer. We investigated six different piezoelectric paper samples prepared by activation in a CMC suspension with increasing concentration (0, 2, 3, 4, 5, and 6 wt%). The stress-strain curves of the piezoelectric paper as a function of CMC concentration are shown in Figure 3. The tensile strength of the piezoelectric paper is shown to improve significantly with increased CMC concentration.

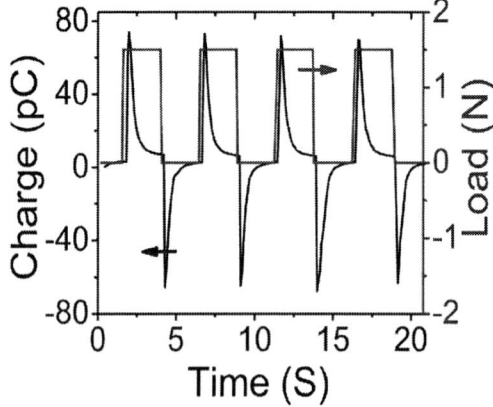

Figure 4: Typical piezoelectric response of piezoelectric paper subjected to a compressive load of 1.5 N.

This is in agreement with published reports. Ma *et al.* [14] reported a significant improvement in the mechanical properties of starch films due to added CMC, while Ankerfors *et al.* [15] report a significant improvement in the mechanical properties of paper with the addition of CMC. The breaking strength of pure piezoelectric paper (0 wt% CMC) of $0.34\ N/mm^2$ is improved to $0.45\ N/mm^2$ upon activation in the 2 wt% CMC suspension, and it tends to increase with increasing CMC concentration and reaches a maximum of $1.05\ N/mm^2$ at 6 wt% CMC (as shown in Figure 3). This may be attributed to the interfacial interaction between matrix and filler [16] that arises due to chemical similarities between wood fibers (cellulose) and CMC. Interestingly, CMC improved the paper strength without reducing flexibility; such a combined effect cannot be achieved with other paper strength additives such as clay, and synthetic resin at relatively large concentration [17]. We limited the CMC concentration to 6 wt%, as the handsheet formed on a polyester mesh screen of a handsheet maker is transformed to a jelly mass at higher CMC concentration (7 wt% and above), that is difficult to couch with blotters and lift off the screen.

We also determined the piezoelectric coefficient (d_{33}) of the CMC added piezoelectric paper by subjecting it to compressive load of 0.5 N to 3 N to 0.5 N in steps of 0.5 N and measuring the corresponding charge induced using a charge meter (Kistler 5015A). The typical piezoelectric response of the paper activated with 6 wt% CMC subjected to load of 1.5 N is shown in Figure 4. The paper produced piezoelectric charge of 72 pC when loaded at 1.5 N and exhibited repeatable piezoelectric response over time. We measured the d_{33} piezoelectric paper samples with different concentration CMC (2, 3, 4, 5, and 6 wt%) and is estimated to be 37 - 45.7 ± 4.2 pC/N, which is approximately 8-times larger than that of pure piezoelectric paper (0 wt% CMC). The large piezoelectric coefficient of CMC added piezoelectric paper is most likely due to a reduced absorption of the applied stress by the wood fiber network and a more effective transfer of strain to the nanoparticles in the piezoelectric paper as a result of the improved fiber-fiber bonding. Also, this piezoelectric coefficient is the largest d_{33} value reported for paper to date and is comparable to the piezoelectric coefficient of commercially available piezoelectric polymers such as polyvinylidene fluoride (PVDF) with d_{33} = 30 pC/N [18].

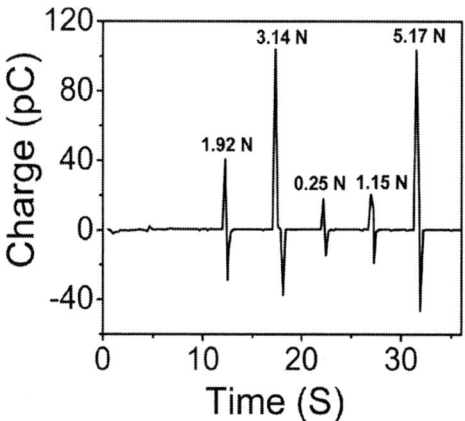

Figure 6: Response of piezoelectric paper to pressure exerted from a finger.

Figure 6 shows the piezoelectric response of the piezoelectric paper subjected to five touch events as well as the corresponding force exerted on it as measured by the reference sensor. Note that the magnitude of the charge induced by the paper is highly dependent on the dynamics of the force exerted in each touch event. For example, for the first event the charge induced by the paper is 41 pC (force exerted is 1.92 N), while it produced a low charge at low tactile pressure (third event, 18 pC at 0.25 N). As shown in Figure 6 the piezoelectric paper showed a similar signal amplitude for the second and the fifth event, as well as for the third and the fourth event. This indicates that, while each tactile event leads to a sensor signal, the signal amplitude depends on how fast the force is applied to the sensor.

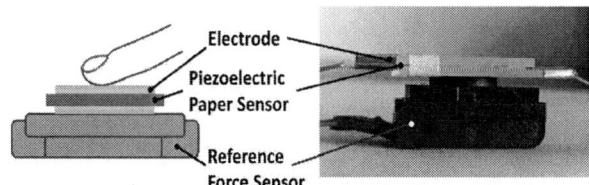

Figure 5: Schematic of the experimental setup for applying piezoelectric paper to tactile sensing.

To demonstrate the physical sensing capability of our piezoelectric paper, it is sandwiched between glass substrates with electrodes and placed on a reference force sensor (Force Sensor FS01 from Honeywell). Silver electrodes were deposited onto clean glass substrates using a CircuitWriter™ precision conductive ink dispenser from CAIG Laboratories Inc., Poway, CA, United States. The force exerted onto the piezoelectric paper due to the pressure from a finger was measured using the reference force sensor, while the response of the piezoelectric paper to that pressure was measured using the charge meter (Kistler 5015A). The schematic experimental setup for physical sensing of our piezoelectric paper is shown in Figure 5. The active area between the electrodes had a surface area of 400 mm^2.

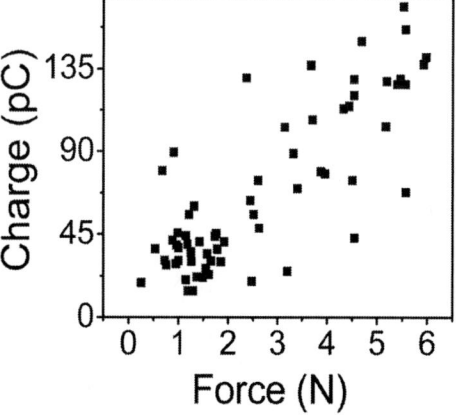

Figure 7: Piezoelectric paper used as a tactile sensor; while each tactile event developed a sensor signal, the signal amplitude depends on how fast the force has been applied to the sensor.

The rapid decay of the charge signal in Figure 6 is indicative of charge leakage. The correlation between the induced piezoelectric charge and the magnitude of the tactile force on the sensors is high with $R^2 = 0.893$ as shown in Figure 7. The scatter in the data in Figure 7 is due to force application at different rates.

SUMMARY AND CONCLUSIONS

We have developed robust and mechanically flexible piezoelectric paper. Our approach involves functionalization of the wood fibers with nanostructured barium titanate (BaTiO$_3$) followed with activation in a suspension of CMC. This activation step leads to a dramatic increase in strength and piezoelectric properties of this material.

The tensile properties such as yield strength, breaking strength, and breaking strain tend to increase with an increase of CMC concentration. We investigated six different piezoelectric paper types with different CMC suspension concentration (0, 2, 3, 4, 5, and 6 wt %); among these samples paper with 6 wt% CMC exhibited the highest tensile strength and the largest piezoelectric coefficient (d_{33} = 37 - 45.7 ± 4.2 pC/N). We demonstrated its suitability for physical sensing at the example of a tactile sensor.

Our study suggests that the piezoelectric paper may be a promising low-cost and environment-friendly substrate for building various physical sensors, and similar to other piezoelectric materials, its advantages are most significant when capturing dynamic events.

ACKNOWLEDGEMENTS

This work was supported by BCFIRST Natural Resources and Applied Sciences (NRAS) endowment through the Research Team Program. The authors thank Canfor Pulp Limited Partnership (CPLP), Prince George Mills, BC, Canada for providing the wood cellulose fiber used in this study. BS acknowledges the support from the Canada Research Chair Program.

REFERENCES

[1] K. M. Schilling, A. L. Lepore, J. A. Kurian, A. W. Martinez. "Fully Enclosed Microfluidic Paper-based Analytical Devices", *Anal. Chem.*, vol. 84, pp. 1579-1585, 2012.

[2] E. Khajeh, W. Lou, B. Stoeber. "Paper based Strain Sensing Material", *Proc. MEMS 2013*, pp. 473- 476.

[3] G. Zheng, L. Hu, H. Wu, X. Xie, Y. Cui. "Paper Supercapacitors by a Solvent-Free Drawing Method", *Energy Environ. Sci.,* vol. 4, pp. 3368- 3373, 2011.

[4] K. A. Mirica, J. G. Weiss, J. M. Schnorr, B. Esser, T. M. Swager. "Mechanical Drawing of Gas Sensors on Paper", *Angew. Chem. Int.*, vol. 51, pp. 10740-10745, 2012.

[5] A. Manekkathodi, M. Y. Lu, C. W. Wang, L. J. Chen. "Direct Growth of Aligned Zinc Oxide Nanorods on Paper Substrate for Low-cost Flexible Electronics", *Adv. Mater.,* vol. 22, pp. 4059- 4063, 2010.

[6] H. Gullapalli, V. S. M. Vemuru, A. Kumar, A. Botello-Mendez, R. Vajtai, M. Terrones, S. Nagarajaiah, P. M. Ajayan. "Flexible Piezoelectric ZnO-Paper Nanocomposite Strain Sensor", *Small,* vol. 6, pp. 1641- 1646, 2010.

[7] K. H. Kim, K. Y. Lee, J. S. Seo, B. Kumar, S. W. Kim. "Paper-based Piezoelectric Nanogenerator with High Thermal Stability", *Small*, vol. 7, no. 18, pp. 2577- 2580, 2011.

[8] S. K. Mahadeva, K. Walus, B. Stoeber. "Fabrication and Testing of Piezoelectric Hybrid Paper for MEMS Applications", *Proc. MEMS 2014*, pp. 620- 623.

[9] S. K. Mahadeva, K. Walus, B. Stoeber. "Piezoelectric Paper Fabricated via Nanostructured Barium Titanate Functionalization of Wood Cellulose Fibers", *ACS Appl. Mater. Interfaces,* vol. 6, no. 10, pp. 7547-7553, 2014.

[10] S. Omata, Y. Terunuma. "New Tactile Sensor Like the Human Hand and its Applications", *Sens. Act. A.,* vol. 35, pp. 9- 15, 1992.

[11] R. S. Dahiya, G. Metta, M. Valle, A. Adami, L. Lorenzelli. "Piezoelectric Oxide Semiconductor Field Effect Transistor Touch Sensing Device", *Appl. Phys. Lett.*, vol. 95, pp. 034105, 2009.

[12] M. Agarwal, Y. Lvov, K. Varahramyan, "Conductive Wood Microfibers for Smart Paper through Layer-by-Layer Nanocoating", *Nanotechnology.*, Vol. 17, pp. 5319- 5325, 2006.

[13] S. K. Mahadeva, J. Berring, K. Walus, B. Stoeber, "Effect of Poling Time and Grid Voltage on Phase Transition and Piezoelectricity of Poly(vinylidene fluoride) Thin Films using Corona Poling", *J. Phys. D:Appl. Phys.*, Vol. 46, pp. 285305, 2013.

[14] X. Ma, P. R. Chang, J. Yu. "Properties of Biodegradable Thermoplastic Pea Starch/Carboxymethyl Cellulose and Pea Starch/Microcrystalline Cellulose Composites", *Carbohyd. Polym.,* vol. 72, pp. 369- 375, 2008.

[15] M. Ankerfors, E. Dukers, T. Lindström. "Topo-Chemical Modification of Fibers by Grafting of Carboxymethyl Cellulose in Pilot scale", *Nordic Pulp Paper Res. J.,* vol. 28, no. 1, pp. 6- 14, 2013.

[16] B. Ghanbarzadeh, H Almasi, A. A. Entezami. "Physical Properties of Edible Modified Starch/Carboxymethyl Cellulose Films", *Innov. Food Sci. Emerg. Technol.,* vol. 11, pp. 697- 702, 2010.

[17] S. Ondaral, O. C. Kurtulus, M. Usta. "Effect of Fiber Modification with Carboxymethyl cellulose on the Efficiency of a Microparticle Flocculation System", *Chem. Papers,* vol. 65, no. 1, pp 16-22, 2011.

[18] http://www.piezotech.fr/image/documents/22-31-32-33-piezotech-piezoelectric-films-leaflet.pdf

CONTACT

*Suresha K. Mahadeva, tel: +1-604-827-4593; sure1977@mail.ubc.ca

A GRAPHENE ACCELEROMETER

Adam M. Hurst[1,3,+], Sunwoo Lee[2,+], Wujoon Cha[1], and James Hone[1]

[1]Department of Mechanical Engineering, Columbia University, New York, New York, USA
[2]Department of Electrical Engineering, Columbia University, New York, New York, USA
[3]Kulite® Semiconductor Products, Inc. Leonia, New Jersey, USA

ABSTRACT

This work presents an SU-8 clamped graphene nano-electro-mechanical-system (GNEMS) accelerometer. A suspended graphene membrane is circularly clamped by SU-8, with an additional proof mass made of either SU-8 or gold, located at the center of the membrane. This GNEMS accelerometer is approximately three orders of magnitude smaller than state of the art micro-electromechanical (MEMS) accelerometers with the diameter of the suspended graphene membrane being 3-10 µm and the proof mass diameter being 1-5 µm. Here, we present the fabrication, simulation, and experimental aperiodic calibration results of the GNEMS accelerometer, demonstrating a repeatable response to an input acceleration levels of ~1000-3000 g.

INTRODUCTION

Micro-electromechanical (MEMS) accelerometers are one of the most commonly employed sensors, which play increasingly critical roles in devices such as smart phones, computers and wearable electronics as well as in the control and health monitoring systems of automobiles, jet engines and power plants [1, 2]. MEMS accelerometers sense acceleration through deflection of a micro-machined suspended structure with a proof mass. Deflection is detected via a change in resistance of embedded piezoresistors, a change in capacitance or through a piezoelectric material. Typical commercial capacitive sensors have dimensions on the order of 500-3500 µm [2, 3].

While larger devices are more sensitive and suitable for low-g applications, devices with smaller dimensions, higher resonant frequencies and lower quality factors will provide faster rise times with less unwanted overshoot and shorter settling times, a specific requirement for high-g, shock testing applications [4]. However, such performance characteristics have been hard to achieve with current technologies because further scaling of MEMS devices typically leads to the reduced sensing area, hence driving down the sensitivity. In addition, fabrication also becomes more difficult with scaling, as bulk micromachining or etching, can be challenging to control with nanometer precision [5].

Graphene exhibits several parameters that make it an attractive structural and sensing material for a NEMS, high-g accelerometer. Graphene is the strongest known material with a Young's modulus of 1 TPa [6], 5-7 times that of Si. From a fabrication perspective, it is a single atomic layer with a constant thickness, which could be grown in a scalable chemical vapor deposition (CVD)

process [6]. These physical and mechanical properties enable substantial device scaling. In addition, graphene is a zero bandgap semiconductor, exhibiting a change in conductivity when exposed to an electric field which can be employed as a mechanism to amplify sensed displacements.

FABRICATION

The graphene accelerometer presented in this work consists of a suspended graphene membrane that is 3-20 µm in diameter. A SU-8 photoresist structure clamps the membrane in place and adds pretension, making the membrane highly taut. A local gate made of refractory metal – platinum in our case – is first patterned on an insulating substrate, before cladded by plasma enhanced chemical vapor deposited (PECVD) silicon oxide. The PECVD oxide surface is then chemical-mechanical polished (CMP) to reduce the surface roughness to < 0.5 nm rms, to guarantee adequate van der Waals adhesion between the CVD graphene and the substrate. The transferred graphene is patterned using electron-beam (e-beam) lithography followed by oxygen plasma etch. Electrical connection to the graphene is established by e-beam lithography followed by e-beam metal deposition of Ti/Pd/Au stack (1/20/50 nm). Finally, SU-8 clamping layer is e-beam patterned to enhance the device yield and pretension the graphene membrane [7]. A SU-8 or metal proof mass is fabricated at the center of the graphene membrane, either during the SU-8 patterning process or the source-drain metal deposition. This fabrication process flow enables the accelerometer proof mass to be incorporated into the device without additional processing steps [7].

Numerous device mass sizes were explored within this work ranging in diameter of 1-10 µm and thickness of ~1-1.5 µm, primarily made of SU-8. Figure 1a shows a schematic of the device architecture. Figure 1b provides false-colored scanning electron microscope (SEM) images of a fabricated graphene accelerometer, clearly illustrating the achieved device structure. It should be noted that this device design enables easy adjustment of accelerometer range by varying the diameter of the graphene membrane and the mass diameter and thickness. For example, increasing the diameter of the graphene membrane and mass size can lower the full-scale acceleration of a graphene accelerometer.

THEORY

We predict the mechanical and electrical response of graphene NEMS accelerometers using analytical models and experimental results. Within the usable range of this

---

+These authors contributed equally to this work.

accelerometer (~100 kg+), the mechanical response will be governed by the built-in pretension of the suspended graphene membrane, which directly impacts the spring constant, k, of the membrane [8]. The spring constant of the clamped, graphene membrane is estimated from the observed pull-in voltage of a graphene accelerometer applying Equation 1 [9].

$$k = V_{Pull-In}^2 \cdot \frac{27\varepsilon A}{8z_0^3} \quad (1)$$

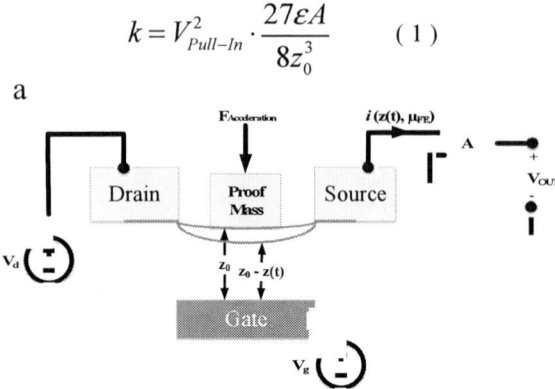

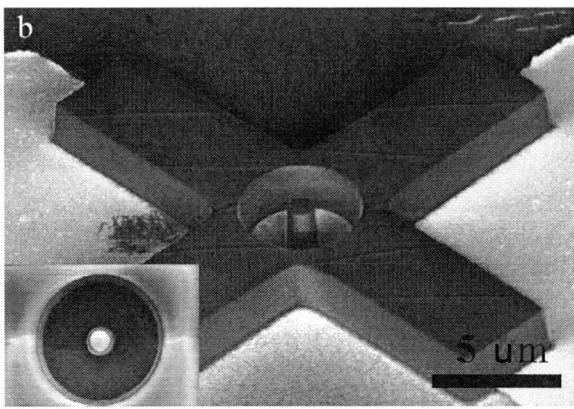

Figure 1: (a) Graphene accelerometer device schematic. (b) False colored SEM image of fabricated graphene accelerometer.

One device exhibited an experimental pull-in voltage of ~ 4.7 V. Applying this value yields an anticipated spring constant of ~2.5 N/m. This value is higher than anticipated, but is reasonable considering residual SU-8 resist on the surface of the suspended graphene will stiffen the membrane. With the experimentally derived k, the deflection of the graphene membrane is predicted using the input acceleration and mass of the SU-8 at the center of the graphene membrane. The typical aperiodic impulse acceleration drop test can be modeled as a half-

sine wave of frequency, ω, and magnitude, γ. Using this input acceleration waveform, the deflection of the graphene membrane is modeled using the equation below:

$$z(t) = \frac{-m\gamma}{k} e^{j\omega t} \quad (2)$$

Figure 2 shows the anticipated membrane deflection for a graphene accelerometer with a diameter of 3 μm, an SU-8 mass of 1.5 μm in diameter and 1.1μm in thickness to a 100,000 g input acceleration. Table 1 provides several accelerometer geometries fabricated and their predicted deflections at a full-scale range of 100kg.

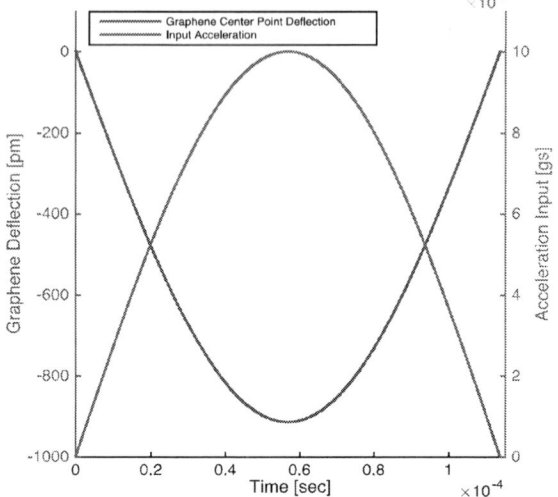

Figure 2: Model membrane deflection with acceleration.

With the expected mechanical deflection, we predict the electrical output of the device based upon graphene's trans-conductive sensing mechanism. As a zero band gap semiconductor, graphene's conductance, G, changes with respect to the applied electric field. Figure 3 provides experimental results of the trans-conductive response, dG/dV_g, of a graphene accelerometer at rest with a gate to membrane spacing, z_0, of 30-120 nm. Typical experimental values for graphene mobility and dG/dV_g are 1500-4000 cm^2/Vs and -5 to -35μS/V, respectively.

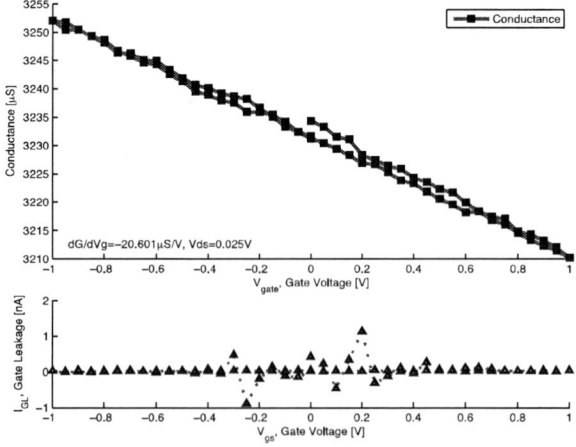

Figure 3: Electrical transport measurements.

The current modulation within the device is then predicted based upon the change in conductance, G, of the graphene membrane as it moves in relation to the gate electrode. In addition to the trans-conductive sensing

Table 1: Accelerometer Geometry and Deflection

Input Acceleration, [g]	Diameter, [μm]	Mass, [pg]	k, [N/m]	$z(t)$, [nm]	f_0, [MHz]
100k	3	1	.1	23	1
100k	3	1	1	2.3	3.3
100k	3	1	2.5	0.91	5.2

mechanism, the commonly employed capacitive sensing mechanism will also modulate the current within the membrane, albeit a substantially smaller change in current. By combining the trans-conductive and capacitive terms, we can model the modulation of current as a function of time and membrane deflection $z(t)$, as shown in the equation below [10].

$$I_d(t) = I_{d,Cap} + I_{d,Trans} = -C(t) \cdot j\omega \cdot V_g\left(\frac{z(t)}{z_0 + z(t)}\right) + V_d \cdot \left(\frac{z_0}{z_0 + z(t)}\right) V_g \frac{dG}{dV_g}$$

(3)

The first term, $I_{d,Cap}$, in Equation 3 is the anticipated current modulation due to the change in capacitance. The second term is the change in current based upon the trans-conductive sensing mechanism, $I_{d,Trans}$. With the small device dimensions and small predicted deflections even at high acceleration levels (~10^5 g), we anticipate the trans-conductive sensing mechanism to be six orders of magnitude larger than the capacitive term [10].

EXPERIMENTAL RESULTS

Following fabrication graphene accelerometers were screened for high field-effect mobility and absence of gate leakage. If gate leakage is observed, the membrane likely collapsed rendering the device useless. Functional chips were then mounted in a stainless steel package with EPO-TEK H70E epoxy and wired bonded to tabs mounted in the stainless steel package.

Devices were then experimentally characterized using a drop table, as the schematic in Figure 4 illustrates. A metal block is dropped from 1 m and then impacts the base of the drop table. The block then rebounds generating a significant increase and then decrease in acceleration as illustrated in Figure 4. The acceleration generated by the calibration setup is simultaneously measured by both the graphene accelerometer and a commercial MEMS piezoresistive accelerometer, Endevco 7270A-20k.

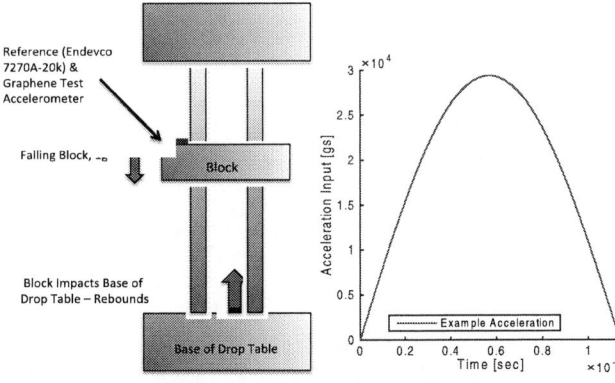

Figure 4: Accelerometer calibration setup.

Graphene accelerometer performance is evaluated based upon the device's ability to track the reference accelerometer specifically as the acceleration increases. Beyond the peak acceleration point, the reference accelerometer begins to resonate from the drop event, at which point the reference is considered unreliable. Figure 5, below, provides drop table experimental results for a graphene accelerometer with a membrane diameter of 3 μm and mass of 1 pg. As desired, the graphene accelerometer tracks the reference sensor as the acceleration increases. The graphene accelerometer additionally follows the theoretical parabolic trend with decreasing acceleration. On the other hand, the reference accelerometer is resonating as the acceleration decreases, which leads to measurement error. This superior frequency response of the graphene accelerometer is anticipated as device should respond as a over-damped (low quality factor [11]), second order system with a natural resonance of 1-10MHz, which is substantially higher than that of commercially available accelerometers.

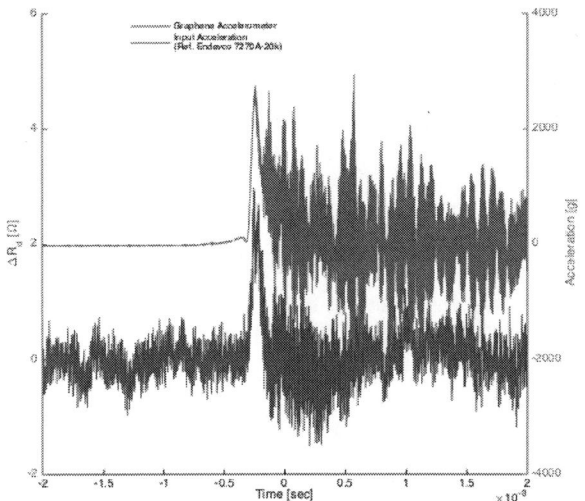

Figure 5: Experimental drop table response of graphene accelerometer.

These experimental results are magnified to the drop event and compared with theoretical predictions in Figure 6. The top plot within Figure 6 shows the reference accelerometers response along with the theoretical input acceleration. Beyond the peak acceleration point, the reference accelerometer experiences excitation of its resonant frequency of ~250kHz, and as an under-damped, second-order system the ringing distorts the acceleration measurement.

The bottom plot in Figure 6 examines the response of the graphene accelerometer to the same input acceleration. While the graphene accelerometer does track the input acceleration, a scaling factor of ~20 was added to the deflection term within the model in order to produce the observed change in resistance of the graphene device with acceleration. This scaling term is likely related to an error within the model or within the estimation of the spring constant of the device. The spring constant varies between devices due to fabrication variability and defects within the graphene, such as tearing.

While accelerometer frequency response performance improves with device scaling, a greater sensitivity to Brownian noise is expected. Brownian noise or thermally driven agitation of gas molecules surrounding the suspended structure causes the suspended structure to move defining the smallest measurable acceleration [5, 9]. There is a design tradeoff when considering the improvements in frequency response with device scaling and the undesirable increase in Brownian motion. For high-g accelerometers, specifically, it is highly desirable to increase the frequency response and reduce the quality factor of the device to limit the overshoot and settling time of the sensor at the cost of increased Brownian noise.

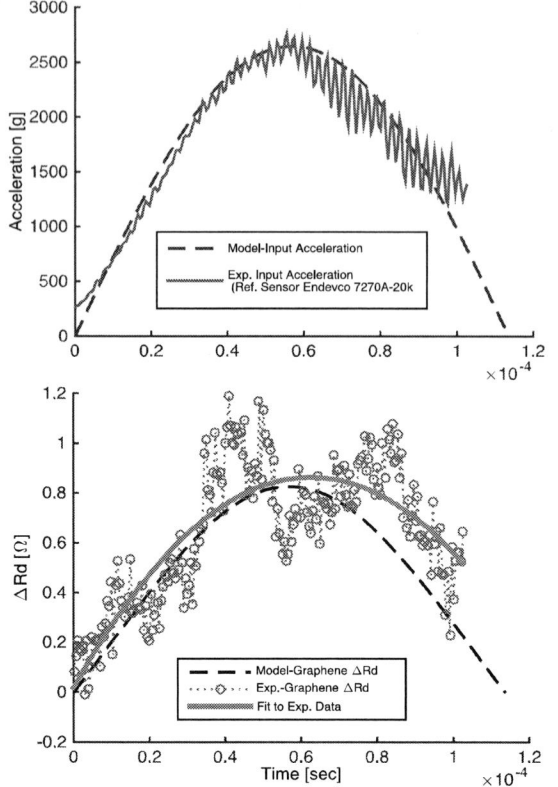

Figure 6: Theoretical and experimental data. Top: reference. Bottom: graphene accelerometer.

Since the capacitive induced output current modulation diminishes with the device scaling, we utilize graphene's charge-dependent conductance as a high-sensitivity transduction mechanism [12, 13], which acts as an amplification mechanism to the capacitive current modulation. This detection scheme is a unique mechanism to counteract the decreased sensing area, and is well suited for graphene, known for exceptionally high carrier mobility.

CONCLUSION

We present a high-g graphene-based accelerometer that uses its trans-conductive nature to break the trade-off between enhanced stability and reduced sensor area due to device scaling. While the full-scale range increases with shrinking device geometries, the trans-conductance amplified current is predicted to be six orders of magnitude larger than capacitive sensing mechanisms. This quantitative difference along with the improved frequency response demonstrates the benefits of GNEMS over conventional accelerometer technologies in high-g applications.

ACKNOWLEDGEMENTS

We would like to thank Dr. John P. Hilton and Dr. Arend van der Zande for insightful discussions. We would also like to thank Mrs. Nora Kurtz and Kulite Semiconductor Products, Inc. for providing financial support for this research effort.

REFERENCES

[1] A. Albarbar, A. Badri, J. K. Sinha, and A. Starr, "Performance evaluation of MEMS accelerometers," *Measurement,* vol. 42, pp. 790-795, Jul 01 2009.

[2] S. D. Senturia, *Microsystem design.* Boston: Kluwer Academic Publishers, 2001.

[3] S. Kal, S. Das, D. K. Maurya, K. Biswas, and A. R. Sankar, "CMOS compatible bulk micromachined silicon piezoresistive accelerometer with low off-axis sensitivity," *Microelectronics,* 2006.

[4] A. R. Szary and D. R. Firth, "Signal Conditioning Perspectives on Pyroshock Measurement Systems," *Sound and Vibration,* vol. 47, pp. 7-13, 2013.

[5] M. A. Cullinan, R. M. Panas, C. M. DiBiasio, and M. L. Culpepper, "Scaling electromechanical sensors down to the nanoscale," *Sensors and Actuators A: Physical,* vol. 187, pp. 162-173, Nov 01 2012.

[6] G. H. Lee, R. C. Cooper, S. J. An, S. Lee, A. van der Zande, N. Petrone, *et al.,* "High-Strength Chemical-Vapor-Deposited Graphene and Grain Boundaries," *Science,* vol. 340, pp. 1073-1076, Jun 30 2013.

[7] S. Lee, C. Chen, V. V. Deshpande, G.-H. Lee, I. Lee, M. Lekas, *et al.,* "Electrically integrated SU-8 clamped graphene drum resonators for strain engineering," *Applied Physics Letters,* vol. 102, p. 153101, 2013.

[8] M. K. Small and W. D. Nix, "Analysis of the Accuracy of the Bulge Test in Determining the Mechanical-Properties of Thin-Films," *Journal of Materials Research,* vol. 7, pp. 1553-1563, Jun 1992.

[9] V. Kaajakari, *Practical MEMS.* Las Vegas, Nev.: Small Gear Pub., 2009.

[10] Y. Xu, C. Chen, V. V. Deshpande, F. A. DiRenno, A. Gondarenko, D. B. Heinz, *et al.,* "Radio frequency electrical transduction of graphene mechanical resonators," *Applied Physics Letters,* vol. 97, p. 243111, 2010.

[11] C. Chen, S. Lee, V. V. Deshpande, G.-H. L. a. t. o. i. h. o. o. 0000-0002-3028-867X, M. Lekas, K. Shepard, *et al.,* "Graphene mechanical oscillators with tunable frequency," pp. 1-5, Nov 21 2013.

[12] A. M. Hurst, S. Lee, and N. Petrone, "A transconductive graphene pressure sensor," *TRANSDUCERS,* 2013.

[13] C. Chen and J. Hone, "Graphene Nanoelectromechanical Systems," *Proceedings of the IEEE,* vol. 101, pp. 1766-1779.

CONTACT

*A. Hurst, amh2003@columbia.edu

A SOLID-GATED GRAPHENE FET SENSOR FOR PH MEASUREMENTS

Yibo Zhu[1], Cheng Wang[1, 3], Nicholas Petrone[1], Jaeeun Yu[2],
Colin Nuckolls[2], James Hone[1], and Qiao Lin[1]
[1]Department of Mechanical Engineering and [2]Department of Chemistry,
Columbia University, New York, NY 10027, United States
[3]Department of Microelectronic Engineering, Nankai University, Tianjin 300071, China

ABSTRACT

This paper presents a graphene field effect transistor (GFET) nanosensor that, with a solid gate provided by a high-κ dielectric, allows analyte detection in liquid media at low gate voltages. The gate is embedded within the sensor and thus is isolated from a sample solution, offering a high level of integration and miniaturization and eliminating errors caused by the liquid disturbance, desirable for both in vitro and *in vivo* applications. We demonstrate that the GFET nanosensor can be used to measure pH changes in a range of 5.3-9.3. Based on the experimental observations and quantitative analysis, the charging of an electrical double layer capacitor is found to be the major mechanism of pH sensing.

INTRODUCTION
Background

Graphene is a two-dimensional nanomaterial consisting of a single layer of carbon atoms arranged in hexagonal crystalline form, and its unique properties have been exploited in various biosensors, such as electric [1], optical [2] and electrochemical [3] based sensors. In particular, graphene has been used to form a conducting channel in a field effect transistor (FET), allowing highly sensitive electric detection of analytes. Such graphene FET (GFET) sensors, when operating in liquid media, are generally constructed in a solution-gated or solid-gated configuration. In a solution-gated GFET sensor [1], an Ag/AgCl wire is inserted into the electrolyte solution in contact with graphene to serve as the gate electrode, while the electric double layer (EDL) [4] formed at the solution-graphene interface provides the gate dielectric. Using various compositions of an electrolyte solution or graphene-immobilized functional groups, such sensors have demonstrated the detection of physicochemical parameters such as pH [1] and biochemical analytes such as DNA [5]. Theses solution-gated sensors typically require an external electrode inserted into the electrolyte solution, which hinders the integration and miniaturization of the device [4]. In addition, the gate capacitance, or the capacitance across the EDL dielectric layer is susceptible to disturbances to liquid media, which can result in fluctuations in electrical measurements of properties of graphene including the conductance and the location of the Dirac point. In contrast, in a typical solid-gated GFET [6], the gate capacitance is provided by a SiO$_2$ dielectric layer sandwiched between graphene and the underlying silicon substrate serving as the gate electrode. By eliminating the need for the external wire insertion into the electrolyte solution, solid-gated sensors are amenable to integration and miniaturization. However, due to the intrinsically low capacitance of the SiO$_2$ layer, usually the solid-gated GFET sensors require undesirably high gate voltages (40~50 V) [7], consequently impeding their application to biosensing in liquid media.

In this paper, we present a GFET nanosensor that uses a thin layer of HfO$_2$ with a high dielectric constant (κ) as a gate dielectric layer, embedded between a gate electrode and graphene which serves as a conducting channel (Fig. 1). Embedding the gate within the sensor enables a high level of integration, particularly desirable for *in vivo* measurements. The use of the high κ dielectric material (HfO$_2$) provides two orders of magnitude higher specific capacitance than conventional silicon dioxide gated sensors, therefore rendering high transconductance and allowing the device to operate at low gate voltages. In addition, the gate dielectric is isolated from a sample solution, thus eliminating errors caused by the liquid disturbance. Furthermore, the sensor is amenable to time- and cost-effective microfabrication using photolithography without the need for manual assembly of discrete components (e.g., electrodes) with graphene, thereby simplifying the fabrication process. We demonstrate the pH sensing using our GFET nanosensor. Experimental results show that the device is capable of measuring pH in a range of 5.3 to 9.3 with a sensitivity of ~57.5 mV/pH, which is close to the Nernstian limit, at a gate voltage less than 1.5 V approximately a factor of 30 lower than that used in SiO$_2$ solid-gated GFET sensors. We also elucidate the mechanism of pH-dependent changes in graphene conductivity by analyzing experimental data using a theoretical model.

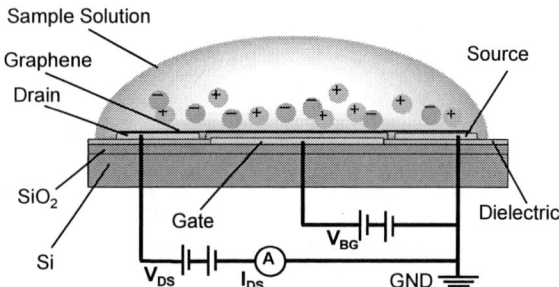

Figure 1: Schematic of the nanosensor. A HfO$_2$ layer between graphene and the gate electrode serves as a dielectric layer; graphene on top of the sensor serves as the conducting channel.

Principle

The device is configured as a FET, in which a graphene sheet connects the source and drain electrodes on an HfO$_2$ dielectric layer, under which the solid gate electrode is embedded (Fig. 1). When an electrolyte solution is introduced onto the graphene, the electrostatic field in the solution near the surface will be dependent on the ions concentration in the solution. The variations in the electrostatic potential will induce changes in the carrier density in the bulk of the graphene, and therefore in the conductivity of graphene. As a result, the conductance of the graphene can be measured to determine the concentrations of ions of a solution.

In a FET, the gate capacitance can be estimated by $C_{ox} = \kappa \varepsilon_0 / t_{ox}$, where ε_0 is the dielectric constant, and t_{ox} is the thickness of the dielectric layer. Therefore, by using a high κ material in our device (HfO$_2$ with $\kappa = 20\sim25$, compared to $\kappa = 3.9$ for SiO$_2$), a high gate capacitance is achieved which allows the device to operate at low gate voltages with improved transconductance. In a SiO$_2$ gated GFET sensor, the thickness of SiO$_2$ is typically 285-nm while the dielectric constant is approximately 3.9 at low frequencies, yielding a specific gate capacitance of 0.012 μF/cm^2. In our solid gated GFET sensor with a 20-nm HfO$_2$ as the gate dielectric, the specific gate capacitance is 0.88~1.1 μF/cm^2, which is two orders of magnitude higher than that of that in the typical SiO$_2$ gated GFET sensor, which results in higher transconductance and allows device operation at low gate voltages (< 1.5 V).

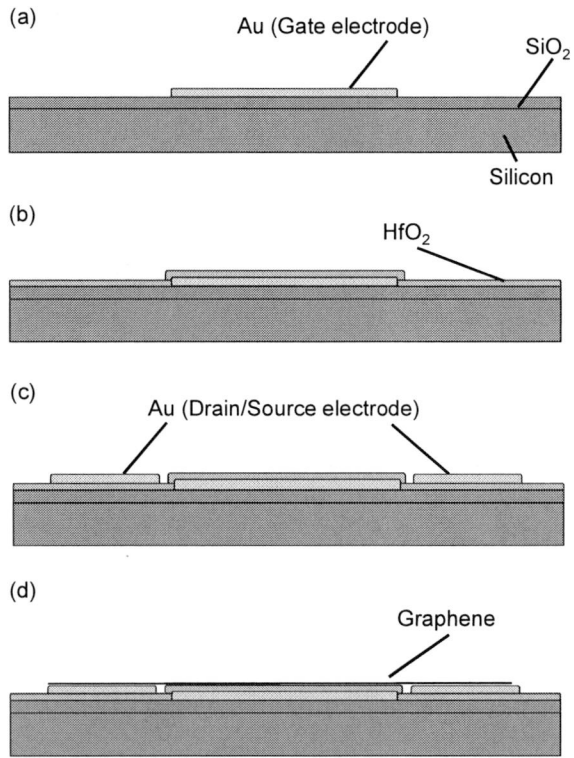

Figure 2: Fabrication process. (a) Deposition and patterning of 5/45 nm Cr/Au gate electrode, (b) deposition of 20 nm HfO$_2$ dielectric layer using ALD, (c) fabrication of drain and source electrodes using lift-off, and (d) transfer of graphene.

EXPERIMENTS
Fabrication

The nanosensor was fabricated on a SiO$_2$-coated silicon substrate. After cleaning by piranha, a layer of 5/45 nm Cr/Au was deposited using thermal evaporation (BOC 306 Thermal Evaporator, Edward). A layer of photoresist (S1811, Shipley) was then spin-coated on top of Au at 5000 rpm for 1 min, and baked at 115 °C for 1 min. Photolithography (MA6, Suss MicroTec) was then used to pattern the shape of the gate electrode on the wafer. The wafer was then developed in developer (AZ MIF 300, AZ Electronic Materials) and local wet etched in gold and chrome etchant sequentially (Fig. 2a). The wafer was cleaned with piranha solution followed by oxygen plasma. Next, a 20 nm HfO$_2$ layer was deposited on top of the gate electrode using atomic layer deposition (ALD, Savannah 200, Cambridge Nano Tech) at 3.6×10^{-1} torr and the temperature as high as 200 °C (Fig. 2b). Another layer of photoresist (S1811, Shipley) was spin-coated and patterned to define the shape of source and drain electrodes, followed by deposition of 5/45 nm Cr/Au. Lastly, the wafer was immersed in photoresist stripper (AZ MIF 400 Stripper) and acetone sequentially to dissolve the photoresist and shape the drain and source electrodes (Fig. 2d).

After the completion of the fabrication process, a single-layer graphene sheet synthesized by chemical vapor deposition (CVD) was subsequently transferred onto the sensor to cover the source, drain and gate electrodes (Fig. 3a). A Raman spectrum (Fig. 3c) was taken to confirm the monolayer graphene sheet throughout the conducting channel (30-μm × 40-μm in size). A polydimethylsiloxane (PDMS)-based microchamber was used to confine the liquid sample on top of the graphene.

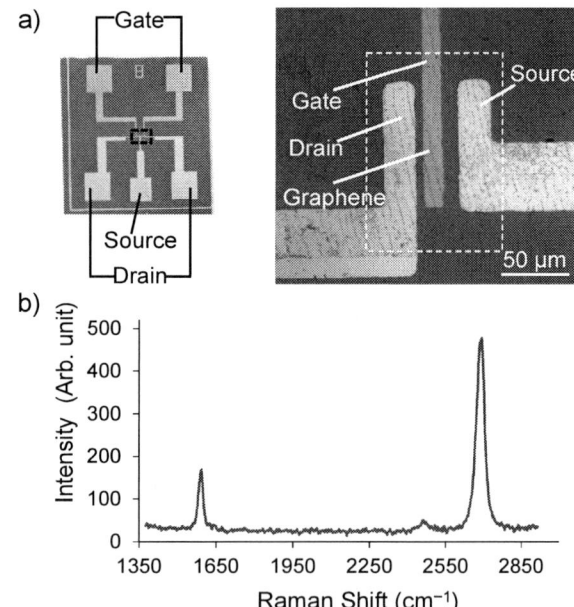

Figure 3: (a) Photo- and micrograph images of a fabricated sensor. The sensor chip is 1 cm × 1 cm in size. Dashed box indicates the graphene sensing region. (b) Raman spectrum of the single-layer CVD graphene.

Results and Discussion

We first measured the drain-source current (I_{DS}) while the solid-gate voltage (V_{BG}) was swept from -0.2 to 1.9 V to obtain transfer characteristics of bare graphene in air. An ambipolar curve was observed with the Dirac point of 0.7 V, which is defined as the gate voltage at which of the minimum drain-source current (Fig. 4), confirming that the conductivity of the graphene was being altered by the field effect via the dielectric layer.

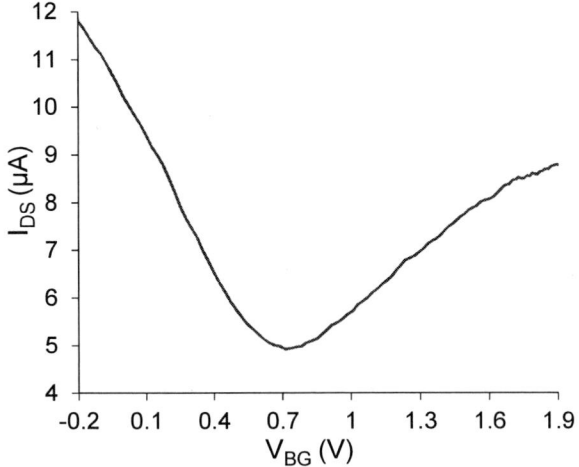

Figure 4: Transfer characteristic for graphene in air. The ambipolar curve was observed where the minimum drain-source current, defined as the Dirac point, was located at $V_{BG} = 0.7$ V.

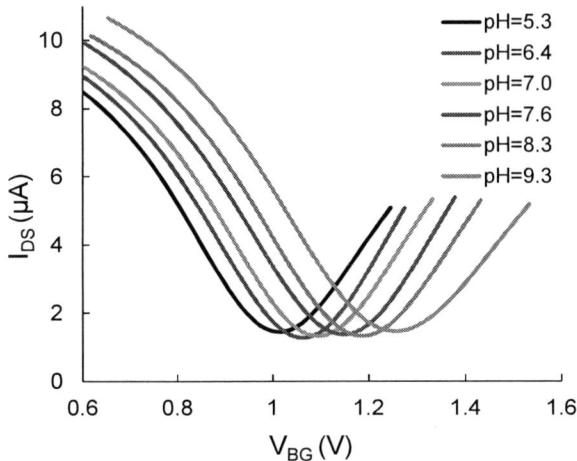

Figure 5: Transfer characteristics obtained with different pH buffer solutions. The Dirac point shifts to higher gate voltages with increasing pH.

We then tested our nanosensor for pH sensing in liquid media. Samples at various pH values (5.3 to 9.3) were prepared by mixing NaOH or HCl with phosphate buffered saline (PBS) buffer. A sample solution was incubated with our nanosensor, during which I_{ds} values were measured with the gate voltage V_{BG} swept from 0.6 V to 1.6 V. We found that the Dirac points at all pH levels were located at $V_{BG} < 1.5$ V (Fig. 5). These significantly reduced gate voltage values, compared to 40~50 V for SiO_2 based solid gated sensor, were attributed to the high gate capacitance and therefore the high transconductance provided by the high-κ HfO_2 dielectric layer. V_{BG} at the Dirac point was found to linearly increase with the pH value at a rate of

~57.5 mV/pH (Fig. 6a), suggesting that the electrostatic potential (E_{TG}) on the top of the graphene surface, varies with changes in pH following the Nernst equation [8], When pH increases, the electrostatic potential on the top of the graphene increases, therefore, the curve shifts to the right to compensate for the variation in the electrostatic potential.

Although the underlying mechanism of graphene's electrical response to pH changes is still under debate, it has been widely accepted that variations in the carrier concentration caused by ions adsorbed on graphene changes the conductivity of graphene. It has been recently reported that the charging of the EDL capacitor [9], instead of the surface transfer doping [7], is responsible for the mechanism of GFET-based pH and ion sensing. Our solid gated sensor, which avoids the influence of the externally applied gate voltage on the EDL, can be used to investigate the effect of either the EDL capacitor charging or the surface charge transfer. The constant slope in Fig. 6b implies a constant carrier mobility within graphene in spite of the pH variation. Therefore, the surface transfer doping [10] is not a dominant effect, which would otherwise the carrier mobility significantly [7]. This is in agreement with the study of Xia et al. [11].

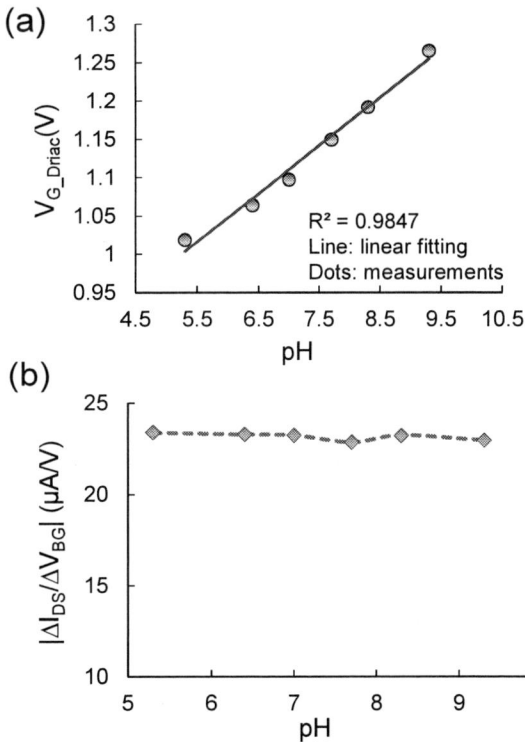

Figure 6: (a). Dependence of the Dirac point on. The Dirac point shifts linearly to higher gate voltages with the increase of pH (sensitivity: 57.5 mV/pH). (b). The ratio of ΔI_{DS} to ΔV_{BG}, i.e. the slope, of the linear section of each transfer characteristic curve at different pH.

To demonstrate the ability of the sensor to conduct real-time pH monitoring, we measured the source-drain current I_{ds} at a fixed gate voltage ($V_{BG} = 0.75$ V), while sequentially introducing samples with increasing pH values (Fig. 7). Experimental results showed that I_{ds}

increased with pH in the range between 5.3 and 9.3, implying that when the solution becomes more alkaline (i.e., pH increases), the EDL capacitance charges and the E_{TG} increases. With a constant V_{BG}, the effective $V_{GS} = |V_{BG} - E_{TG}|$ will accordingly decrease, which results in an increase in I_{ds}, in agreement with the relationship between I_{ds} and V_{BG} as shown in Fig. 5.

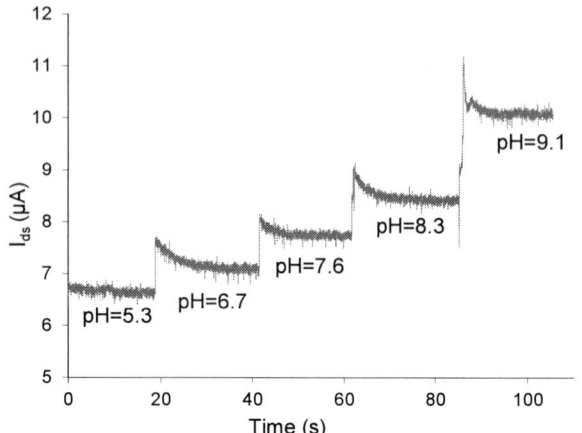

Figure 7: Real time measurement of pH changes. The responses were demonstrated by the changes of the source-drain current I_{DS} at a fixed gate voltage $V_{BG} = 0.75V$. The consistent increase of I_{ds} with pH at $V_{BG} = 0.75V$ is in agreement with the results in Fig. 5.

CONSLUSION

We have described a high-κ solid-gated GFET nanosensor in liquid media. The embedded gate allows removal of external gate electrode and thereby is amenable to *in vivo* sensing. The use of a high-κ material allows the device to operate at low gate voltages and avoids errors caused by gate capacitance variations. This sensor is capable of measuring pH in a range of 5.3 to 9.3 with a sensitivity of ~57.5 mV/pH. We also find the pH dependent electrical responses are attributable to the charging of the double layer capacitor instead of surface transfer doping. These results have demonstrated that the GFET nanosensor can be potentially enable highly integrated sensing of chemical and biological analytes.

ACKNOWLEDGMENTS

This work was partially supported by the National Institutes of Health (award number: 1DP3 DK101085-01). We also appreciate the helpful discussions with Lei Wang and Xian Zhang from Columbia University. In addition, Y. Zhu and C. Wang gratefully acknowledge the National Scholarships (award numbers 201206250034 and 201206200032) from the China Scholarship Council.

REFERENCES

[1] Z. G. Cheng, Q. Li, Z. J. Li, Q. Y. Zhou, and Y. Fang, "Suspended Graphene Sensors with Improved Signal and Reduced Noise," *Nano Letters*, vol. 10, pp. 1864-1868, May 2010.

[2] W. G. Xu, X. Ling, J. Q. Xiao, M. S. Dresselhaus, J. Kong, H. X. Xu, *et al.*, "Surface enhanced Raman spectroscopy on a flat graphene surface," *Proceedings of the National Academy of Sciences of the United States of America*, vol. 109, pp. 9281-9286, Jun 12 2012.

[3] X. H. Kang, J. Wang, H. Wu, J. Liu, I. A. Aksay, and Y. H. Lin, "A graphene-based electrochemical sensor for sensitive detection of paracetamol," *Talanta*, vol. 81, pp. 754-759, May 15 2010.

[4] Y. Ohno, K. Maehashi, Y. Yamashiro, and K. Matsumoto, "Electrolyte-Gated Graphene Field-Effect Transistors for Detecting pH Protein Adsorption," *Nano Letters*, vol. 9, pp. 3318-3322, Sep 2009.

[5] T. Y. Chen, T. K. L. Phan, C. L. Hsu, Y. H. Lee, J. T. W. Wang, K. H. Wei, *et al.*, "Label-free detection of DNA hybridization using transistors based on CVD grown graphene," *Biosensors & Bioelectronics*, vol. 41, pp. 103-109, Mar 15 2013.

[6] W. J. Yuan and G. Q. Shi, "Graphene-based gas sensors," *Journal of Materials Chemistry A*, vol. 1, pp. 10078-10091, 2013.

[7] J. H. Chen, C. Jang, S. Adam, M. S. Fuhrer, E. D. Williams, and M. Ishigami, "Charged-impurity scattering in graphene," *Nature Physics*, vol. 4, pp. 377-381, May 2008.

[8] A. S. Feiner and A. J. Mcevoy, "The Nernst Equation," *Journal of Chemical Education*, vol. 71, pp. 493-494, Jun 1994.

[9] B. Mailly-Giacchetti, A. Hsu, H. Wang, V. Vinciguerra, F. Pappalardo, L. Occhipinti, *et al.*, "pH sensing properties of graphene solution-gated field-effect transistors," *Journal of Applied Physics*, vol. 114, Aug 28 2013.

[10] H. T. Liu, Y. Q. Liu, and D. B. Zhu, "Chemical doping of graphene," *Journal of Materials Chemistry*, vol. 21, pp. 3335-3345, 2011.

[11] J. L. Xia, F. Chen, P. Wiktor, D. K. Ferry, and N. J. Tao, "Effect of Top Dielectric Medium on Gate Capacitance of Graphene Field Effect Transistors: Implications in Mobility Measurements and Sensor Applications," *Nano Letters*, vol. 10, pp. 5060-5064, Dec 2010.

CONTACT

*Yibo Zhu, yz2471@columbia.edu

MULTI-MODAL GRAPHENE POLYMER INTERFACE CHARACTERIZATION PLATFORM FOR VAPORIZABLE ELECTRONICS

V. Gund[1], A. Ruyack[1], K. Camera[2], S. Ardanuc[1], C. Ober[2], and A. Lal[1]
[1]*Sonic*MEMS Laboratory, Cornell University, Ithaca, NY, USA
[2]Ober Group, Cornell University, Ithaca, NY, USA

ABSTRACT

Characterization of physical and chemical properties of organic thin films such as polycarbonates is essential for their application-specific design. In this paper, we report a novel micromechanical resonant membrane platform with both graphene resistivity and mass-dependent frequency sensing to measure the thermal response, vaporization, and subsequent mass change of thin-film analytes deposited on top of the membrane. Graphene transferred on silicon nitride (Si_xN_y) is below the analyte thin film, and the graphene resistance changes due to dangling bond interactions between the analyte film and graphene surface. At the same time, the resonance frequency of the analyte film/graphene/Si_xN_y composite membrane changes with variations in temperature, mechanical properties, mass and elasticity of the analyte film. This device provides a bi-modal characterization of the organic thin film, both in the electrical and mechanical domain, which allows one to identify and optimize the organic film formulation. As an example, we demonstrate differentiating formulations of a polymer blend to realize low degradation temperature, which is critical for vaporizable electronics enabling low-power transience.

INTRODUCTION

The design and synthesis of novel polycarbonates using experimental and theoretical techniques find use in various applications. The polycarbonates are used in electronic components for insulation and as dielectrics, for macro and micro tubing, and in niche applications such as polycarbonate lenses and sterile medical storage. In MEMS microfabrication, these polymers are used in photoresists for lithography and as protective films [1]. However, no single application accounts for more than 10% of the commercial market volume for polycarbonates because of the wide variety of properties offered by these materials [2]. More importantly, the capability to modify and tune their physical and chemical properties, specific to the application, with only minor changes in composition makes them extremely versatile.

Small weight-percent (wt%) additives in polymer blends can drastically change their physical and material properties, and alter their response to temperature and pressure. For vaporizable polycarbonates (VPC), exposure to light of various frequencies can also change chemical properties such as surface bonds and their energies. This enables their use as passive sensors with secure physical encoding of material properties for quantitatively recording light flux intensity and temperature changes [3]. Designing polymers for such applications requires optimization of their glass-transition temperature (T_g) and decomposition temperature (T_d). The polymer effect on electrical conductivity of surrounding materials also needs monitoring. To quickly characterize these blends while exploring a large design space for the VPC, it is impractical to synthesize large quantities of these materials. Small quantities also allow efficient use of reagents. Hence, a microscale characterization platform that can enable polymer characterization with high specificity is essential.

The traditional method of measuring thermal-response of analytes is Thermo-Gravimetric Analysis (TGA). Micromechanical TGA devices use resonant vibrating cantilevers, with integrated piezoresistive sensors, for temperature cycling and measuring corresponding mass-change with shifts in resonance frequencies [4]. However, they do not provide information on dynamic surface interactions. Desorption kinetics of thin-film analytes deposited on microhotplate arrays have been studied using mass spectroscopy [5] but do not have mass-sensing capability. Our device is based on a multi-modal sensing approach that can measure mass-loss and surface-interaction changes simultaneously.

DEVICE CONCEPT

For the characterization platform, we use graphene transferred on a Si_xN_y membrane, as shown in Figure 1. The graphene film is patterned to define electrodes for sensing its resistivity and to selectively heat the suspended region for polymer characterization. A PZT plate is attached to drive the structure at its resonance modes [6].

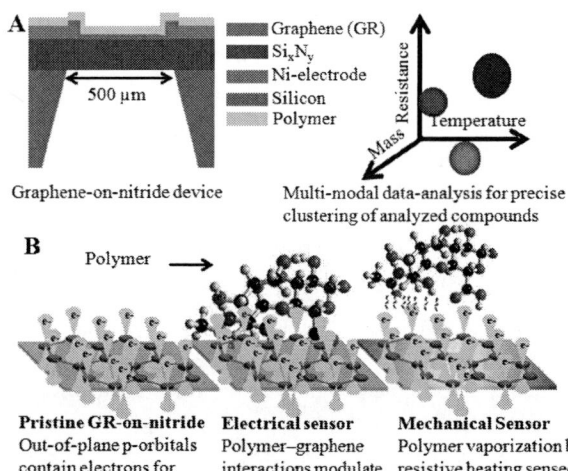

Figure 1: A). Characterization platform with spun-on analyte polymer for thermal cycling B). Polymer surface-interactions with the atomically-thin graphene are sensed electrically with graphene resistivity modulation. The mass-loss and vaporization are sensed mechanically with membrane resonance-frequency shift.

Electrical Sensing

Graphene is an atomically thin sheet of carbon atoms arranged in a honeycomb lattice and shows outstanding electrical properties as a zero-band gap semiconductor with high electrical conductivity [7]. Carbon atoms connected to each other in the lattice are sp^2 hybridized. The remaining p-orbital, with a single electron, is free for conduction. The resistivity of graphene can be, hence, reversibly modulated by molecules adsorbed and desorbed on its surface via charge-transfer. This has been used to study graphene's interactions with a variety of molecules [8] using graphene FETs. Our device measures the modulation of graphene's resistance for sensing surface-modifications during temperature ramp as graphene's atomic interactions with the spun-on polycarbonate change. The graphene also serves the purpose of a resistive heater for temperature ramping with Joule heating.

Mechanical Sensing

The suspended Si_xN_y membrane is the structural layer, actuated at resonance, for mass-sensing in our device. The fundamental resonance frequency of a square membrane, f, in the stress-dominated regime is [9]:

$$f = \frac{1}{a}\sqrt{\frac{\sigma}{2\rho}} = \frac{1}{a}\sqrt{\frac{\sigma V}{2m}} \qquad (1)$$

Here, a is the length of the square, σ is the in-built stress and ρ is the density, also expressed as the ratio of mass m to volume V. The second expression yields the relationship for mass-loss Δm to frequency-shift Δf:

$$\frac{\Delta f}{f} = -\frac{\Delta m}{2m} \qquad (2)$$

In addition, there is a temperature dependence of the membrane-frequency during heating, based on changes in σ and the membrane dimensions. Since, the membrane is clamped on all sides, and heating is confined to the center of the membrane, the length is constant to first order. However, an effective thermal expansion coefficient (TEC) for the thickness, which can change at the center of the membrane, needs to be accounted for. The temperature dependence, derived from (1), is:

$$\frac{1}{f}\frac{df}{dT} = \frac{1}{2\sigma}\frac{d\sigma}{dT} + \frac{3}{2}\alpha \qquad (3)$$

Here, T is the temperature, α is the effective TEC accounting for the Si_xN_y, metal electrodes and spun-on VPC, and t is the thickness of the layers.

The thermal time constant for membrane heating is a few milliseconds due to its small thermal mass. On the other hand, the time to raise the temperature of spun-on VPC and overcome its relatively high heat of vaporization is on the scale of minutes. Hence, during temperature ramp, we expect fast negative shifts in resonance frequency due to membrane heating, a result of effective spring-softening. At sufficiently high input power, we expect this to be followed by a slow positive shift as the VPC heats up and vaporizes, leading to mass-loss.

DEVICE FABRICATION

We have previously demonstrated graphene on Si_xN_y devices for resonator frequency trimming [10]. The fabrication process-flow described therein was used in this work with some modifications. 45 nm thick nickel-electrodes were evaporated and patterned on 360 nm thick Si_xN_y for contact with graphene, which was transferred on top of electrodes. Graphene was also patterned by oxygen plasma-etch only on top of the suspended Si_xN_y to minimize heat losses to the substrate for power-savings.

POLYCARBONATE SYNTHESIS

The polycarbonate used in this study was synthesized based on [3]. VPC was polymerized using the bis(carbonylimidazolide) of 2,5-dimethyl-2,5-hexanediol and 80% 1,4- and 20% 1,3-benzenedimethanol. Two different VPC test solutions were made. One consisted of 5 wt% VPC in dichloromethane (DCM) and the other contained 5 wt% VPC + 5wt% (with respect to polymer content) photoacid generator (PAG) (VPC + PAG) in DCM. A PAG, after exposure to ultraviolet (UV) light, provides an acid source, which catalyzes polycarbonates decomposition [11]. By adding PAG to one solution and not the other, we create two different polymer blends with different thermal properties. Each solution was spin coated onto a silicon wafer followed by a post-spin bake of 120 °C for 2 minutes to form a thin film of ~500 nm. The VPC + PAG film was exposed to 254 nm UV light with a dose of 300 mJ/cm². Each film was removed from the wafer and the T_d was verified using a commercial TGA. VPC had a T_d of 209 °C, while the VPC + PAG had a lower T_d of 97 °C. After confirmation of different thermal properties, thin films of each blend were spun on graphene devices using the same procedure as above.

TESTING & RESULTS
Unloaded Device - Characterization

Square membranes of side 500 μm were tested for all experiments. Device resonance modes were excited with backside AC PZT actuation and identified with the Polytek MSA 500 laser interferometer. The fundamental mode at 293 kHz with a Q-factor of 1050 in 3 mbar vacuum, shown in Figure 2A, was used. As a control experiment, voltage was ramped across the graphene heater on an unloaded device while simultaneously monitoring membrane resonance frequency shift and graphene resistance change. This is plotted in Figure 2B. With no spun-on polymer, device resonance frequency decreased over millisecond time scales, as expected, due to spring-softening of the membrane with increasing heat. Graphene-resistance varied non-monotonically and approached a value of 8 kΩ at high input-power.

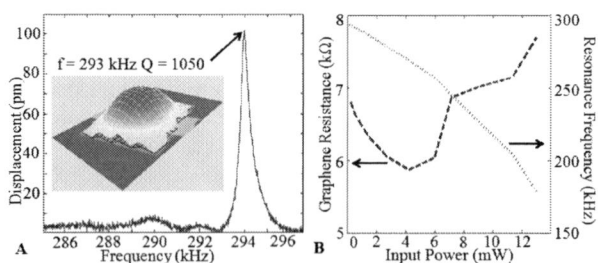

Figure 2: A) Unloaded device fundamental resonance frequency measurement B) Control-measurement of unloaded membrane resonance frequency shift and graphene resistance change vs. input power.

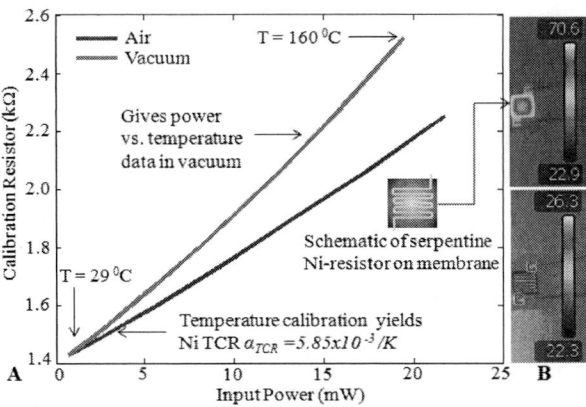

Figure 3: A). Temperature calibration of Si_xN_y membrane devices with serpentine calibration nickel resistors to obtain temperature versus power data in air and vacuum. B). IR camera images of the membrane with calibration resistor, during voltage-ramp in air (blue curve in A)

For temperature calibration, a Si_xN_y membrane with serpentine nickel resistor of baseline resistance 1.46 kΩ at room temperature, and no graphene, was heated by voltage-ramp across the resistor. The membrane surface temperature was monitored with a FLIR T300 infra-red camera while monitoring the nickel resistance, as shown in Figure 3. Using this highly linear calibration data ($r^2 = 0.99$), the thermal expansion coefficient (TEC) of deposited nickel was extracted as $\alpha_{TCR} = 5.85 \times 10^{-3}/K$ with the variation from baseline resistance R_0 given by:

$$R = R_0(1 + \alpha_{TCR}T)$$

The same device was then tested for voltage-ramp in 3 mbar vacuum to obtain a temperature versus power calibration, to account for reduced heat-losses in vacuum.

Polymer-loaded Devices - Resistive Sensing

With spun-on VPC (Device-1) and VPC+PAG (Device-2) polymer, devices were tested with voltage-ramp to heat the polymer while tracking the resonance frequency and graphene resistance. The resistances of the

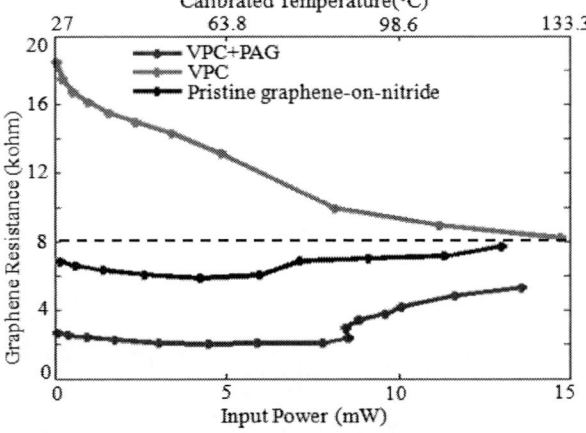

Figure 4: Graphene resistance change due to spun-on polymer, before thermal degradation, varies by 10x between VPC and VPC+PAG. After mass-loss at high input-powers, the graphene is cleaned and both resistances converge to pristine graphene-on-nitride resistance.

graphene on Device-1 and Device-2 before voltage-ramp were 18.6 kΩ and 2.76 kΩ. This indicated high sensitivity of the graphene to small differences in the polymer, in this case the PAG, as a result of different surface interactions. Graphene resistance was modulated with increasing voltage, as VPC degraded thermally. Eventually, as the VPC is "cleaned" off the surface, resistances in both devices approach that of pristine graphene on unloaded devices as shown in Figure 4.

Polymer-loaded Devices - Mass Sensing

The initial resonance frequencies for Devices 1 and 2 were 208.9 kHz and 233.2 kHz respectively. These are lower than the unloaded device frequency of 293 kHz due to different wt% of spun-on polymers. The resonance frequency measured versus time, shown in Figure 5, initially decreased on a scale of milliseconds during steps in voltage-ramp, just like the unloaded devices. However, with higher input-power, the analytes heated up sufficiently and slow-mass degradation was observed on a scale of minutes even as the power was kept constant over those intervals. This measurement is limited by device Q-factor degradation at large input power > 25 mW.

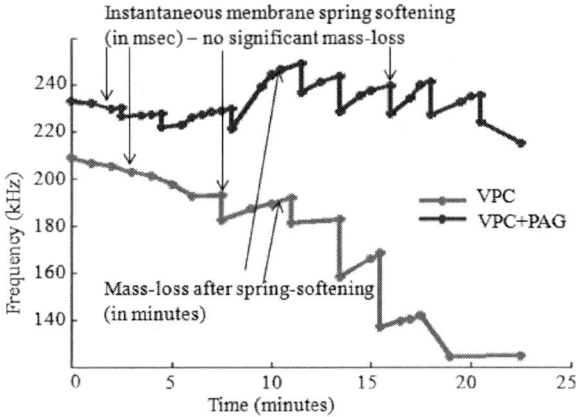

Figure 5: Resonance frequency vs. time during voltage-ramp of the graphene. Regions of only spring softening (low input power), and spring softening followed by mass-degradation (high input power) are shown

To measure mass-degradation versus temperature, the frequency shifts due to spring-softening were subtracted to calculate mass-loss % using (2). This enabled determination of the thermal response of VPCs to increasing input power, and hence, temperature using the calibration data from Figure 3. Simultaneously, the membrane surface was optically monitored with the MSA-500 microscope (20x) to observe VPC vaporization. Figure 6 shows stages in the mass-degradation process mapped to corresponding images of the membrane for Device-2. The measured mass losses for Device-1 and 2 were 33 and 70 % respectively, in reasonable agreement with values of 43 and 60 % obtained by image processing results (using ImageJ software). Residual polymer on the platform, such as in Figure 6D, is due to inefficient heating near the anchors as a result of substrate losses. This can be improved by co-optimization of the electrode design and graphene patterning.

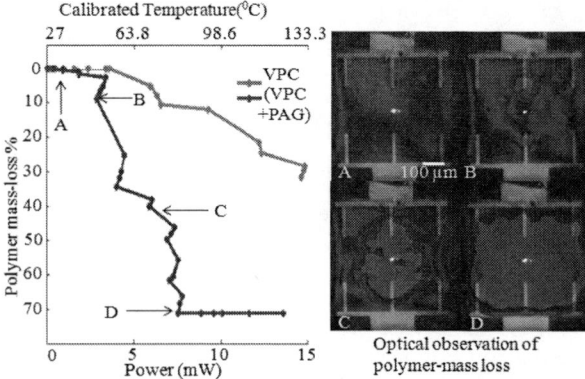

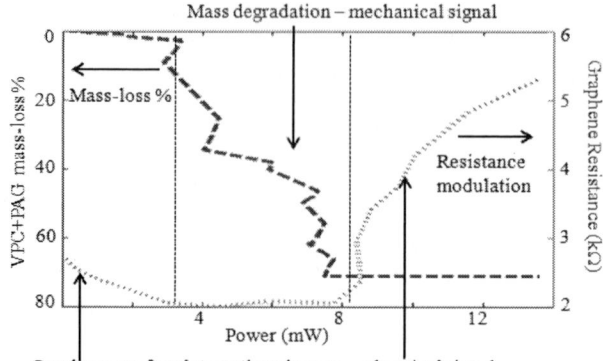

Figure 6: *Polymer mass-loss % vs. input power - A) Low input-power with no material loss B) Onset of material loss, also seen optically C) Rapid polymer degradation sensed as a large frequency-shift D) Polymer degradation completed on the membrane. Measured 70 and 33% mass-loss was verified by image processing*

Figure 7: *Multi modal polymer characterization - The platform gives complementary electrical and mechanical signals in different input-power regimes, shown by dotted vertical lines, for sensing graphene-polymer surface interactions, mass-degradation, and surface cleaning.*

Multi-modal Sensing

By combining frequency and resistance measurements, it is possible to get a 2-dimensional signal for the thermal response of the material, for mass-loss and surface-changes. The signals for Device-2 are shown in Figure 7. The electrical (green curve) and mechanical (blue curve) signals are complementary to each other in this case, highlighting the advantage of sensing with more than one modality. For low input-power, there is almost no mass-degradation producing negligible frequency shifts. However, due to graphene-polymer nanoscale surface interactions, the graphene resistance is modulated significantly. At higher powers, mass-degradation is observed causing large frequency shifts but only small changes in graphene resistance showing minimized surface interactions. At very high powers when mass-degradation is almost complete, no further frequency shifts are seen. However, as the graphene is "cleaned" by high current-drive annealing [12], its resistance approaches that of the unloaded device.

CONCLUSIONS

We have presented a multi-modal characterization platform for polymer analysis using graphene on Si_xN_y as our sensor. While our application is targeted towards VPCs for novel vaporizable electronics, the platform is broadly applicable and can be extended to the analysis of other polymers and spun-on analytes, within the device temperature operation range. As a mW range analysis tool, it is efficient at analyzing picogram quantities of polymers. For thermally degradable polymers, the device can be cleaned by ensuring that the graphene-resistance approaches that of pristine graphene while resonance frequency approaches that of the unloaded device resonance, allowing for cost effectiveness and ease of use.

ACKNOWLEDGEMENTS

The authors acknowledge funding support by the DARPA VAPR program. All device fabrication was done at the Cornell NanoScale Facility (CNF), a member of the National Nanotechnology Infrastructure Network.

REFERENCES

[1] I. Blakey, *et al.*, "Polycarbonate based nonchemically amplified photoresists for extreme ultraviolet lithography," *Proceedings of SPIE*, Vol. 7636, 2010.

[2] Legrand and Bendler, *Handbook of Polycarbonate Science and Technology*, Marcel Dekker, 1999.

[3] F. M. Houlihan, *et al.*, "Thermally Depolymerizable Polycarbonates. 2. Synthesis of Novel Linear Tertiary Copolycarbonates by Phase-Transfer Catalysis", *Macromolecules*, pp13-19, 1986.

[4] Berger, *et al.*, "Micromechanical thermogravimetry," *Chemical Physics Letters*, 1989.

[5] Semancik, *et al.*, "Microhotplate platforms for chemical sensor research," *Sensors and Actuators B: Chemical*, pp. 579-591, 2001.

[6] V. Kaajakari and A. Lal, "Ultrasonically driven surface micromachined motor," *Proc. of 13th IEEE International Conference on Micro Electro Mechanical Systems*, 2000.

[7] K. S. Novoselov, *et al.*, "Electric Field Effect in Atomically Thin Carbon Films," *Science*, Vol 306, pp. 666-669, 2004.

[8] V. Georgakilas, *et al.*, "Functionalization of Graphene: Covalent and Non-Covalent Approaches, Derivatives and Applications," *Chemical Reviews*, 2009.

[9] Graff, *Wave Motion in Elastic Solids*, Dover, 1991.

[10] H. Hosseinzadegan and A. Lal, "Tip-based graphene etching for MEMS resonator frequency trimming", in *Tech. Digest of the 17th International Conference on Solid-State Sensors, Actuators and Microsystems (Transducers '13)*, 2013.

[11] J. M. Frechet *et al.*, "Thermally Depolymerizable Polycarbonates V. Acid Catalyzed Thermolysis of Allylic and Benzylic Polycarbonates: A new Route to Resist Imaging", *Polymer Journal*, 1987.

[12] J. Moser, *et al.*, "Current-induced cleaning of graphene", *Applied Physics Letters*, 2007.

CONTACT

*V. Gund, tel: +1-650-521-1172; vvg3@cornell.edu

TWO-DIMENSIONAL MoS₂ NANOMECHANICAL RESONATORS FREELY-SUSPENDED ON MICROTRENCHES ON FLEXIBLE SUBSTRATE

Rui Yang[1][†], Zenghui Wang[1], Peng Wang[1], Rene Lujan[2], Tse Nga Ng[2], Philip X.-L. Feng[1][†]

[1]Electrical Engineering, Case Western Reserve University, Cleveland, OH 44106, USA
[2]Palo Alto Research Center (PARC), Palo Alto, CA 94304, USA
[†]Email: rui.yang@case.edu; philip.feng@case.edu

ABSTRACT

This digest paper reports on the first high-frequency nanomechanical resonators based on molybdenum disulfide (MoS₂) crystalline flakes freely-suspended on microtrenches (~13μm wide and 14μm deep) fabricated on flexible substrate, with bendability and stretchability. Through investigations of the device resonances via optical excitation and detection by ultrasensitive laser interferometry, we first observe multimode resonances up to ~50MHz with the polydimethylsiloxane (PDMS) substrate under different bending and stretching conditions. The device resonance frequencies (f_{res}) first increase and then stabilize while quality (Q) factors are enhanced with PDMS trench widening of up to 161% in the region away from the MoS₂ flake and 61% at the MoS₂ flake edge, without breaking the device. This platform could facilitate investigations of the strain limits in devices, strain-induced bandgap tuning in two-dimensional (2D) crystals, and strain-engineered performance enhancement (*e.g.*, f_{res}, Q, and mobility) in 2D resonators. Furthermore, it is well suited for exploring new 2D flexible and wearable electronic components such as strain gauges and resonant transducers.

INTRODUCTION

Semiconducting transition metal dichalcogenides (TMDCs) such as MoS₂ have emerged as a new material platform among 2D layered materials, with thickness-dependent bandgap and mobility, unique spin and valley properties, and excellent mechanical properties [1,2]. Field-effect transistors (FETs) [3,4,5] and optoelectronic devices [6] based on MoS₂ have been reported, showing promising performance for future device applications. Furthermore, the excellent electromechanical properties such as high Young's modulus (E_Y), high strain limits (ε_{int}), and strain-sensitive band structures [1,2] have also spurred intense interests in developing micro/nano-electromechanical systems (MEMS/NEMS) incorporating ultrathin 2D semiconducting materials such as MoS₂. Recently, 2D MoS₂ nanomechanical resonators have been demonstrated on conventional rigid substrates (SiO₂ on Si) [7,8]; but these substrates are not strain-tolerant and these 2D NEMS cannot achieve high strain levels.

MoS₂ transistors on flexible substrates with different bending radius have been reported [9,10], demonstrating stable FET performance under various bending conditions, while suspended MoS₂ resonators on flexible substrates have not been explored. Since straining has been found to enhance both f_{res} and Q in conventional resonators [11], it is highly intriguing to explore such effects in ultrathin layered materials. Resonators on flexible substrates, combined with transistors and other

devices [12], shall open new avenues in flexible electronics and systems.

In this work, we demonstrate the first MoS₂ resonators on flexible substrates that are tolerant to large amount of bending and stretching. The MoS₂ resonators could be suspended on the PDMS trench by either direct exfoliation or aligned dry-transfer process, with the assistance of the O₂ plasma treatment. We show that the device could sustain large bending and straining after clamping the MoS₂ with silicone adhesive. The device resonances are recorded under different bending conditions, showing no observable sliding or delamination. This creates a unique platform for exploring strain effect on the performance of 2D resonators and future applications in flexible electronics.

DEVICE FABRICATION PROCESS

We fabricate the MoS₂ resonators on PDMS substrates with microtrenches (Fig. 1a), to demonstrate highly flexible devices (Fig. 1b). Scanning electron microscope (SEM) image (Fig. 1c) shows that a MoS₂ flake is suspended across the trench, and optical microscope image confirms that we clamp a MoS₂ flake with silicone adhesive (Fig. 1d).

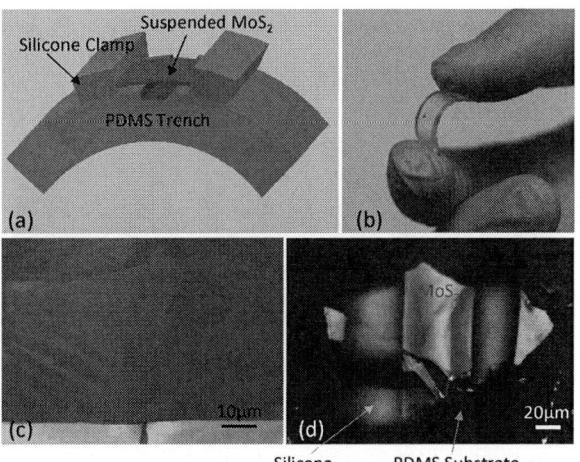

Figure 1: (a) A 3D illustration of the MoS₂ resonator freely-suspended on a microtrench on flexible PDMS substrate. (b) Photograph of a MoS₂ resonator on PDMS, showing the substrate is flexible. (c) SEM image of a device suspended on the PDMS microtrench. (d) Optical microscope image of another MoS₂ resonator clamped with silicone adhesive.

The microtrenches on the PDMS substrate (Sylgard 184, Dow Corning) are molded from a silicon mask patterned by reactive ion etching (RIE, Fig. 2a & 2b). We employ two methods (direct exfoliation and dry-transfer process) to align the MoS$_2$ flakes over the PDMS microtrenches. Before the transfer, we use O$_2$ plasma to treat both PDMS and the tape (direct exfoliation) or polymer stamp (dry-transfer) to enhance adhesion (Fig. 2c & 2d). For direct exfoliation, we apply the tape with MoS$_2$ onto the PDMS substrate. For dry-transfer process, we align the desired MoS$_2$ flake on the polymer stamp onto the PDMS trench before the PDMS substrate loses the adhesiveness (Fig. 2e). After the MoS$_2$ is suspended on the PDMS trench, we use a micropositioner to apply silicone adhesive to clamp the MoS$_2$ to prevent sliding during the bending experiments (Fig. 2f).

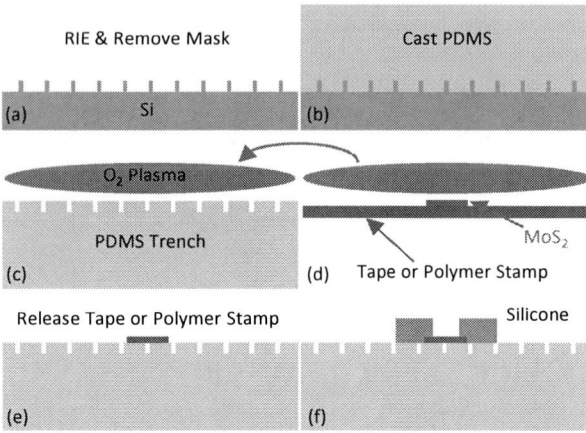

Figure 2: Fabrication process of the MoS$_2$ resonators on PDMS substrate with microtrenches. (a) RIE of Si and removal of the etch mask. (b) Casting PDMS onto the Si mold and then peeling off PDMS. (c) O$_2$ plasma treatment of the PDMS trench substrate. (d) O$_2$ plasma treatment of the tape (for direct exfoliation) or the polymer stamp (for dry-transfer) with MoS$_2$. (e) Engaging the tape or polymer stamp in contact with the PDMS substrate, and then releasing them. (f) Clamping the MoS$_2$ with silicone glue.

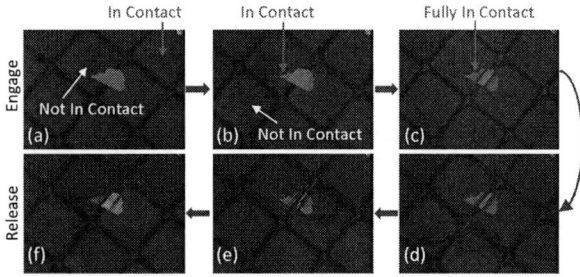

Figure 3: Dry-transfer process of positioning a multilayer MoS$_2$ onto the PDMS microtrench substrate. (a)-(c) The process of engaging, by pressing the polymer stamp onto the PDMS substrate. (d)-(f) The process of releasing.

For the dry-transfer process, we first exfoliate MoS$_2$ onto a polymer stamp mounted on the glass slide, and then identify the appropriate flake under optical microscope. After O$_2$ plasma treatment of the PDMS substrate and MoS$_2$, we mount the glass slide with the polymer stamp on a micropositioner and align the MoS$_2$ flake with the PDMS trench under optical microscope. In the transfer process, we observe the polymer stamp and PDMS substrate gradually come into contact as we engage (Fig. 3a-c) and release the polymer stamp (Fig. 3d-f). The transfer process is verified by the difference in optical contrast (Fig. 3a versus Fig. 3f).

MoS$_2$ RESONANCES ON PDMS TRENCH

We measure the resonance of the MoS$_2$ resonators on PDMS by using a specially engineered multi-laser interferometric detection system [13]. We drive the device motion with an AC modulated 405nm blue laser that periodically heats up the device and result in motion due to thermal expansion. We focus a 633nm laser with laser power of 0.35mW onto the MoS$_2$, and the reflected light intensity is modulated by device motions. The signal is captured by a high-speed photodetector, and is recorded by a network analyzer. The device is mounted inside a vacuum chamber (~50mTorr) at room temperature.

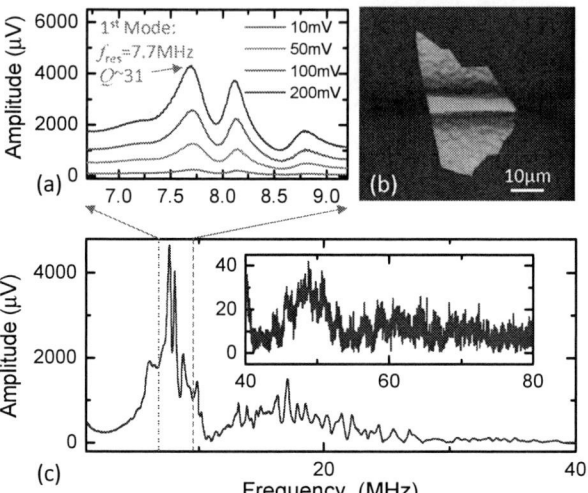

Figure 4: Driven resonance of a multilayer MoS$_2$ resonator on PDMS shown in Fig. 3. (a) The first three resonances with different driving amplitudes (different colors). (b) Optical microscope image of the device without silicone clamp. (c) Resonance spectrum (0–40MHz) showing multiple modes. Inset shows resonance up to ~50MHz.

Figure 4a shows the first three driven resonances measured from the dry-transferred device shown in Fig. 3 and Fig. 4b. Fitting to the driven damped harmonic resonator model results in f_{res}=7.7MHz and Q=31 for the first resonance mode. The different modulation amplitude of the blue laser is shown by different colors, and we observe that the device resonance response increases almost linearly with higher driving amplitude. Figure 4c shows the whole resonance spectrum of the device, where multimode resonances are measured by the precise interferometric system, up to ~50MHz (Fig. 4c inset).

BENDING WITHOUT CLAMP

Figure 5 shows the optical images of a device fabricated with direct exfoliation and without applying silicone clamp, as we gradually bend the PDMS substrate and measure the device resonance. The PDMS trench widening away from the MoS$_2$ flake and at edge of the MoS$_2$ flake can be estimated by carefully comparing the optical images under different bending conditions. The images show that the MoS$_2$ starts to slide on the PDMS substrate at PDMS trench widening of 57% off the MoS$_2$

flake and 33% at edge of the MoS$_2$ flake. At 94% of trench broadening, the MoS$_2$ clearly slides and displaces from the original position. We note that the widening of PDMS trench at edge of the MoS$_2$ flake does not necessarily represent the actual strain in the MoS$_2$ device; instead, it is likely originated from the direct exfoliation process, where the MoS$_2$ could be pressed into the trench and result in a slack belt or membrane as the exfoliation tape is released from the PDMS substrate. To further quantify the strain induced in MoS$_2$, extra measurements such as Raman spectroscopy and photoluminescence calibration in real time may be needed.

under bending for the device in Fig. 5, respectively. We observe multimode resonances from this device, and find that the resonance frequency of the first prominent resonance decreases from 13.2MHz before bending to 3.35MHz under bending, with PDMS trench widening of 30% away from MoS$_2$ and 9% at edge of MoS$_2$ flake. Figure 6c summarizes f_{res} and Q for the first resonance with different amount of broadening of the PDMS trench at edge of MoS$_2$. We find that f_{res} first decreases then stabilized and then decreases again due to sliding. The Q of the device has a trend of decreasing with bending.

BENDING WITH CLAMP

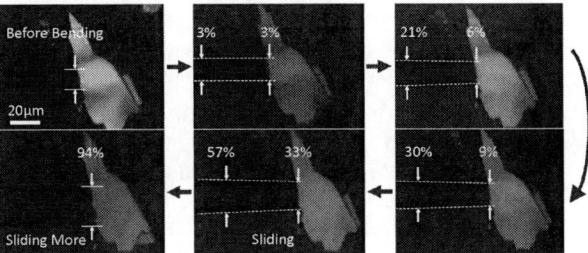

Figure 5: Optical microscope images showing the widening of the PDMS trench away from the multilayer MoS$_2$ (white) and at edge of the MoS$_2$ device (yellow) with continuously increasing bending, for a device fabricated with direct exfoliation without silicone clamp. The bending sequence is indicated by blue arrows.

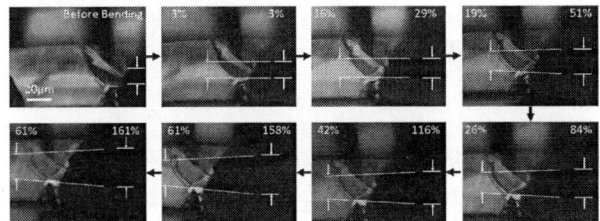

Figure 7: Optical microscope images showing the widening of the PDMS trench off the multilayer MoS$_2$ (white) and near the MoS$_2$ boundary (yellow) with continuously increasing bending, for a device fabricated by direct exfoliation and clamped with silicone adhesive sealant, as shown in Fig. 1(d).

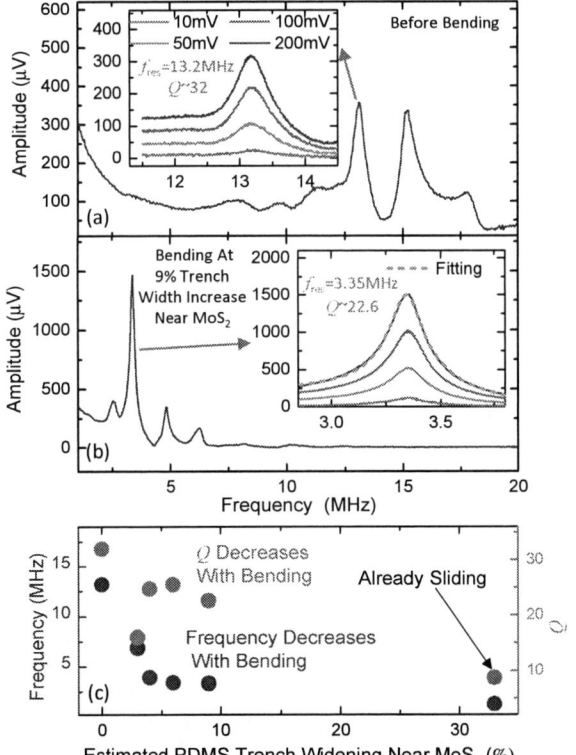

Figure 6: Resonance data with different bending for the device shown in Fig. 5. (a)-(b) Resonance spectra of the device (a) before bending, and (b) with bending. Insets: The first resonances with different driving amplitude. (c) f_{res} & Q of the first resonance at varying amount of broadening of the PDMS trench at the MoS$_2$ edge.

Figure 6a and 6b show resonance spectra before and

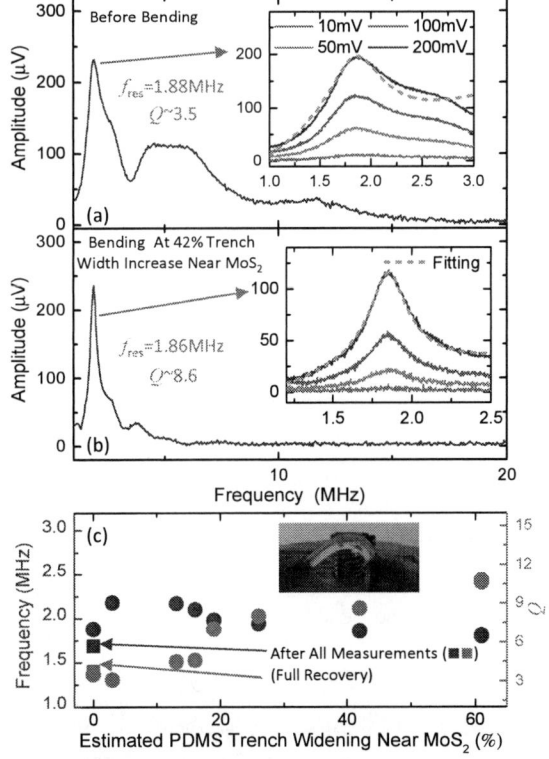

Figure 8: Resonance data with different bending for the device shown in Fig. 1d and Fig. 7. (a)-(b) Resonance spectra of the device (a) before bending, and (b) with bending. Insets in (a) and (b): The first resonances with different driving amplitude. (c) f_{res} & Q of the 1st resonance at varying broadening of PDMS trench at edge of MoS$_2$ (with recovery). Inset in (c): Photograph of the MoS$_2$ device on PDMS substrate under tensile straining.

978-1-4799-7956-1/15 $31.00 © 2015 IEEE

In order to prevent the MoS$_2$ from sliding on the PDMS under bending, we apply silicone adhesive on both sides of the trench. Optical images are recorded in the bending process (Fig. 7), showing no obvious sliding between MoS$_2$ and PDMS substrate when the widening of the PDMS trench is 161% off the MoS$_2$ flake and is 61% at the MoS$_2$ edge, which is larger trench widening than that achieved in devices without silicone clamping. Again the device is likely a slack MoS$_2$ flake suspended on the trench after the direct exfoliation process, as we can infer from the optical contrast in Fig. 1d and Fig. 7. Upon bending, the PDMS trench widens and the slack device could end up under tension, and beyond that it becomes difficult to keep increasing the trench width, as shown in the final bending images in Fig. 7. Since MoS$_2$ is mechanically stronger than PDMS, when the MoS$_2$ is under tension, the trench width under the MoS$_2$ flake may not follow the yellow dashed lines shown in Fig. 7; but at edge of the MoS$_2$ flake, the estimation of the trench width using the dashed lines is reasonable and valid.

Figure 8a and 8b show resonance spectra before and under bending for the device in Fig. 7, respectively. We find no obvious change in f_{res} before bending and after PDMS trench widens for 42% near the MoS$_2$ device and 116% away from the MoS$_2$, while Q is enhanced from 3.5 to 8.6. We summarize f_{res} and Q for the first resonance of the device with different amount of PDMS trench broadening near MoS$_2$ flake, as shown in Fig. 8c. We observe that frequency first increases, then slightly decrease and stabilizes, while Q is enhanced with increasing tensile strain. When the bending is released, the device resonance restores well, very close to the initial frequency value (Fig. 8c).

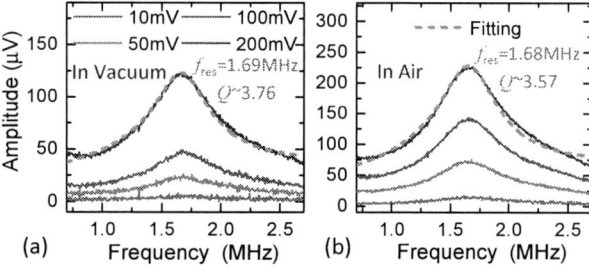

Figure 9: Resonance spectra of the first resonance of the device in Fig. 7 measured in (a) vacuum and (b) air, after all of the bending experiments.

We also measure the device resonance under different pressures after the bending experiment (Fig. 9). We find that the device f_{res} is quite similar in vacuum and in air, and Q only decreases by 5% when the device is measured in air compared to the Q in vacuum. This suggests that air damping effect is not the dominant factor in affecting the Q of the device, and other damping factors such as clamping loss are more significant.

CONCLUSIONS

In summary, we have demonstrated the first high-frequency MoS$_2$ nanomechanical resonators suspended on microtrenches patterned on a highly bendable and stretchable PDMS substrate. The devices show multimode resonances at high frequency and are tolerant to large tensile strain which enhances the Q factor in clamped resonators. The fabrication and measurement techniques can be readily applied to other 2D materials such as graphene, boron nitride and black phosphorus. The device platform is promising for extending the functionalities of 2D NEMS resonators onto flexible substrates.

ACKNOWLEDGEMENTS

We thank support from Case School of Engineering, National Academy of Engineering (NAE) Grainger Foundation Frontier of Engineering (FOE) Award (FOE 2013-1005), CWRU Provost's ACES+ Advance Opportunity Award, the T. Keith Glennan Fellowship, the CSC Fellowship (No. 2011625071), and PARC.

REFERENCES

[1] Q. H. Wang, *et al.*, "Electronics and Optoelectronics of Two-Dimensional Transition Metal Dichalcogenides", *Nature Nanotech.*, vol. 7, pp. 699-712, 2012.

[2] D. Jariwala, *et al.*, "Emerging Device Applications for Semiconducting Two-Dimensional Transition Metal Dichalcogenides", *ACS Nano*, vol. 8, pp. 1102-1120, 2014.

[3] B. Radisavljevic, A. Radenovic, J. Brivio, V. Giacometti, A. Kis, "Single-Layer MoS$_2$ Transistors", *Nature Nanotech.*, vol. 6, pp. 147-150, 2011.

[4] R. Yang, Z. Wang, P. X.-L. Feng, "Electrical Breakdown of Multilayer MoS$_2$ Field-Effect Transistors with Thickness-Dependent Mobility", *Nanoscale*, vol. 6, pp. 12383-12390, 2014.

[5] R. Yang, *et al.*, "Multilayer MoS$_2$ Transistors Enabled by a Facile Dry-Transfer Technique and Thermal Annealing", *J. Vac. Sci. Technol. B*, vol. 32, art. no. 061203, 2014.

[6] O. Lopez-Sanchez, *et al.*, "Ultrasensitive Photodetectors based on Monolayer MoS$_2$", *Nature Nanotech.*, vol. 8, pp. 497-501, 2013.

[7] J. Lee, *et al.*, "High Frequency MoS$_2$ Nanomechanical Resonators", *ACS Nano*, vol. 7, pp. 6086-6091, 2013.

[8] A. Castellanos-Gomez, *et al.*, "Single-Layer MoS$_2$ Mechanical Resonators", *Adv. Mater.*, vol. 25, pp. 6719-6723, 2013.

[9] G.-H. Lee, *et al.*, "Flexible and Transparent MoS$_2$ Field-Effect Transistors on Hexagonal Boron Nitride-Graphene Heterostructures", *ACS Nano*, vol. 7, pp. 7931-7936, 2013.

[10] H.-Y. Chang, *et al.*, "High-Performance, Highly Bendable MoS$_2$ Transistors with High-K Dielectrics for Flexible Low-Power Systems", *ACS Nano*, vol. 7, pp. 5446-5452, 2013.

[11] S. S. Verbridge, *et al.*, "Macroscopic Tuning of Nanomechanics: Substrate Bending for Reversible Control of Frequency and Quality Factor of Nanostring Resonators", *Nano Lett.*, vol. 7, pp. 1728-1735, 2007.

[12] T. N. Ng, *et al.*, "Scalable Printed Electronics: An Organic Decoder Addressing Ferroelectric Non-Volatile Memory", *Sci. Rep.*, vol. 2, art. no. 585, 2012.

[13] R. Yang, *et al.*, "Smart-Cut 6H-Silicon Carbide (SiC) Microdisk Torsional Resonators with Sensitive Photon Radiation Detection", in *Proc. 27th IEEE Int. Conf. on MEMS (MEMS 2014)*, pp. 793-796, San Francisco, January 26-30, 2014.

978-1-4799-7956-1/15 $31.00 © 2015 IEEE

A SENSOR FOR STIFFNESS CHANGE SENSING BASED ON THREE WEAKLY COUPLED RESONATORS WITH ENHANCED SENSITIVITY

*Chun Zhao[1, *], Graham S. Wood[1], Jianbing Xie[2], Honglong Chang[2],*
Suan Hui Pu[1, 3], Harold M. H. Chong[1] and Michael Kraft[4]

[1]Nano Research Group, University of Southampton, UK
[2]MOE Key Laboratory of Micro and Nano System for Aerospace, Northwestern Polytechnical University, China
[3]University of Southampton Malaysia Campus, Malaysia
[4]University of Liege, Montefiore Institute, Belgium

ABSTRACT

This paper reports on a novel MEMS resonant sensing device consisting of three weakly coupled resonators that can achieve an order of magnitude improvement in sensitivity to stiffness change, compared to current state-of-the-art resonator sensors with similar size and resonant frequency. In a 3 degree-of-freedom (DoF) system, if an external stimulus causes change in the spring stiffness of one resonator, mode localization occurs, leading to a drastic change of mode shape, which can be detected by measuring the modal amplitude ratio change. A 49 times improvement in sensitivity compared to a previously reported 2DoF resonator sensor, and 4 orders of magnitude enhancement compared to a 1DoF resonator sensor has been achieved.

INTRODUCTION

Over the last couple of decades, micro- and nano-fabricated resonant devices have been widely used to sense small changes in the properties of the resonator [1]. Among these, sensing devices that detect stiffness change have been used for many applications, such as accelerometers [2], imaging microscopy [3] and others.

For sensing a change in stiffness, an amplitude modulation sensing paradigm with two weakly coupled resonators [4] was previously proposed to enhance the sensitivity compared to conventional single resonator sensors with frequency shift as output [5]. By combining two identical resonators and a weak coupling element in between, the change in mode shapes is more pronounced than the shift in frequency for the same stiffness perturbation [6].

The device reported here employed a novel approach based on three weakly coupled resonators arranged in a chain. Unlike previous work using 2DoF resonators, for which identical resonators were used, we intentionally designed the suspension system of the middle resonator stiffer than that of the other two identical resonators; in this way, an enhancement in sensitivity could be achieved [7].

THEORY

System Model

The lumped parameter block diagram of a 3DoF resonator system is shown in Fig. 1. Each resonator is modelled as a mass and spring; damping is neglected for the analysis. The springs between the resonators are the coupling springs.

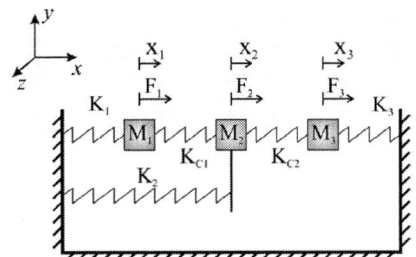

Figure 1: Mass-damper-spring lumped parameter model of a 3DoF resonator sensing device

Suppose the mass of all resonators are identical, i.e. $M_1=M_2=M_3=M$, the two coupling springs are also identical, $K_{c1}=K_{c2}=K_c$, whereas the spring stiffness of the resonators are asymmetrical with $K_1=K$, $K_3=K+\Delta K$. In addition, the stiffness of the resonator in the middle is K_2. Further, assuming all springs are linear, and no movement in the y and z-axis, the equations of motions in the x-axis after Laplace transform are given by:

$$H_1(s)X_1(s) = F_1(s) + K_c X_2(s) \quad (1)$$

$$H_2(s)X_2(s) = F_2(s) + K_c[X_1(s) + X_3(s)] \quad (2)$$

$$H_3(s)X_3(s) = F_3(s) + K_c X_2(s) \quad (3)$$

where the transfer functions are defined as:

$$H_1(s) = Ms^2 + K + K_c \quad (4)$$

$$H_2(s) = Ms^2 + K_2 + 2K_c \quad (5)$$

$$H_3(s) = Ms^2 + K + K_c + \Delta K \quad (6)$$

If the system is actuated by $F_1(s)$ only, the displacement $X_1(s)$ and $X_3(s)$ can be computed as a function of $F_1(s)$:

$$X_1(s) = \frac{F_1(s)[H_2(s)H_3(s) - K_c^2]}{H_1(s)H_2(s)H_3(s) - K_c^2[H_1(s) + H_3(s)]} \quad (7)$$

$$X_3(s) = \frac{F_1(s)K_c^2}{H_1(s)H_2(s)H_3(s) - K_c^2[H_1(s) + H_3(s)]} \quad (8)$$

In the ideal case with negligible damping and $\Delta K=0$, there are three distinctive modes: in the first mode, all three resonators vibrate in-phase; in the second mode, resonators 1 and 3 are out-of-phase, with the resonator in the middle being stationary; in the third mode, resonators 1 and 3 are in-phase, but are out-of-phase with respect to resonator 2 [8]. When a perturbation in stiffness is introduced, $\Delta K \neq 0$, the three modes are disturbed resulting in amplitude changes and mode localization occurs [9]. The modes of interest in this work are the first two modes due to higher sensitivity than the third mode, which will be referred to as in-phase and out-of-phase modes, respectively.

In this work, the amplitude ratio $|X_1(s)/X_3(s)|$ is used to gauge the mode localization caused by stiffness perturbation.

Amplitude Ratio and Sensitivity Analysis

Assuming a weak coupling stiffness of $K_c < K/10$ and the stiffness of resonator 2 being more than twice than that of resonator 1, so that the following condition is satisfied:

$$| K_c | < \frac{K}{10} < \frac{K_2 - K}{10} \quad (9)$$

Let s=jω, the frequencies of the in-phase and out-of-phase modes can be approximated as:

$$\omega_{ip} \approx \sqrt{\frac{1}{M}[K + K_c + \frac{1}{2}(\Delta K - \alpha - \sqrt{\Delta K^2 + \alpha^2})]} \quad (10)$$

$$\omega_{op} \approx \sqrt{\frac{1}{M}[K + K_c + \frac{1}{2}(\Delta K - \alpha + \sqrt{\Delta K^2 + \alpha^2})]} \quad (11)$$

where ω_{ip} and ω_{op} denote the frequencies of the in-phase and out-of-phase modes, respectively, and

$$\alpha = \frac{2K_c^2}{K_2 - K + K_c} \quad (12)$$

Substituting (10) and (11) into (7) and (8), we can estimate the amplitude ratios for the in-phase and out-of-phase modes as:

$$\left| \frac{X_1(j\omega_{ip})}{X_3(j\omega_{ip})} \right| \approx \left| \frac{\sqrt{\gamma_3^2(\Delta K / K)^2 + 4} + \gamma_3(\Delta K / K)}{2} \right| \quad (13)$$

$$\left| \frac{X_1(j\omega_{op})}{X_3(j\omega_{op})} \right| \approx \left| -\frac{\sqrt{\gamma_3^2(\Delta K / K)^2 + 4} - \gamma_3(\Delta K / K)}{2} \right| \quad (14)$$

where,

$$\gamma_3 = \frac{2K}{\alpha} = \frac{K(K_2 - K + K_c)}{K_c^2} \quad (15)$$

To verify the results, an equivalent electrical RLC model was constructed as shown in Figure 2 [8].

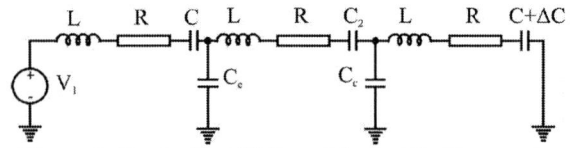

Figure 2: Equivalent RLC model of a 3DoF resonator system

The electrical model was simulated using values listed in Table 1 representing our designed device. Small resistors were added so that the PSpice simulation converges. The resulting simulated quality factor was 10^5, which is a good approximation of the undamped system. The simulated resonant frequencies are compared to theoretical values calculated using (10) and (11) in Figure 3, and the theoretical amplitude ratios computed by

(13) and (14) are verified in Figure 4.

Table 1: Values used in the simulation of the electrical equivalent model

Component	Value	Mechanical model equivalent
L	0.489MH	M
C	0.254fF	K
C_2	84.8aF	$K_2/K=3$
C_c	19.07fF	$K/K_c 75$, $\gamma_3=11324$
R	0.44MΩ	$Q=10^5$
ω_0	14.27kHz	Resonant frequency of single resonator

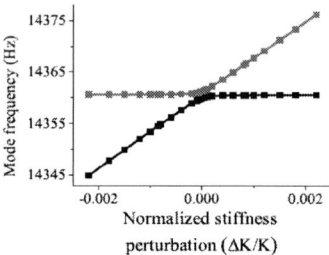

Figure 3: Simulation results showing the in-phase (black) and out-of-phase (red) mode frequencies as a function of a normalized stiffness perturbation. The theoretically calculated mode frequencies match well with the simulated values.

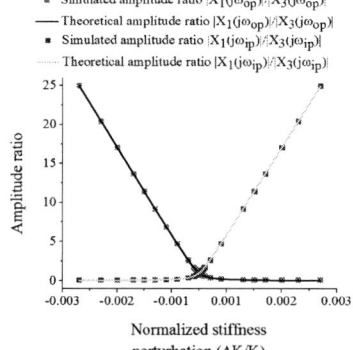

Figure 4: Simulated and calculated (using (13) and (14)) amplitude ratios of in-phase and out-of-phase modes as a function of normalized stiffness perturbation. The theoretically calculated amplitude ratios match well with simulated values.

It can be seen from Figures 3 and 4 that the theoretical estimations of mode frequencies and amplitude ratios match well with the simulated results (within 1%).

Due to the symmetry as shown in Figure 4, without loss of generality, the amplitude ratio of the out-of-phase mode for $\Delta K/K<0$ is chosen for the following sensitivity analysis.

It can be seen from Figure 4 that for negative stiffness perturbations the amplitude ratio is approximately a linear function of stiffness perturbation. Assuming $|\gamma_3\Delta K/K|>10$, the mathematical amplitude ratio

(14) can be linearized as:

$$\left| \frac{X_1(j\omega_{op})}{X_3(j\omega_{op})} \right| \approx -\frac{\gamma_3 \Delta K}{K} \qquad (16)$$

The linear sensitivity of the sensor (the ratio of the change in amplitude ratio to the normalized stiffness change) can therefore be expressed as:

$$S_{3DoF} = \partial \left(\left| \frac{X_1(j\omega_{op})}{X_3(j\omega_{op})} \right| \right) \Big/ \partial \left(\frac{\Delta K}{K} \right) \approx -\gamma_3 \qquad (17)$$

where γ_3 is defined in (15).

EXPERIMENTAL RESULTS
Device Fabrication

A 3DoF resonator device was fabricated using a single mask silicon-on-insulator (SOI) process with a structural layer of 30µm thickness. The process involves the following steps: 1) photoresist deposition and patterning; 2) DRIE etching of the device layer to define the structure; 3) overetch by DRIE, utilizing the notching effect to dry-release the majority of the proof mass, thereby avoiding stiction of the proof mass to the handle wafer during the wet release step; 4) removing the photoresist, followed by dicing; 5) HF wet release the resonator structures. A summary of the process flow is shown in Figure 5, a more detailed description is provided in [10].

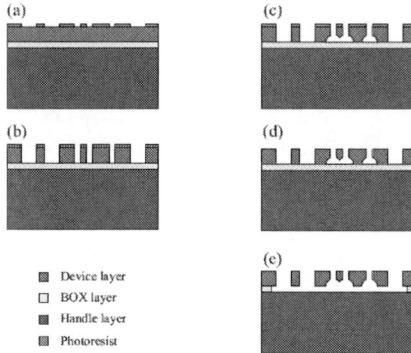

Figure 5: The process flow of the single mask SOI process: a) deposition and patterning of photoresist, b) DRIE etching, c) overetching, d) photoresist removal and dicing, e) HF solution release.

An SEM micrograph of a fabricated chip of the 3DoF MEMS resonator sensor is shown in Figure 6.

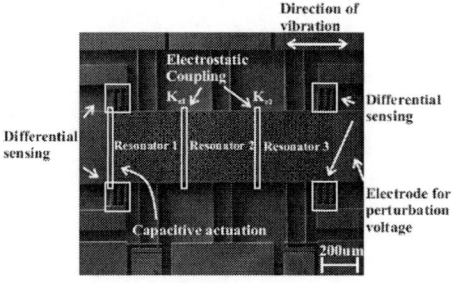

Figure 6: SEM image of a micro-fabricated prototype 3DoF resonator sensing device

Experimental Configuration

The system consists of three resonators. Electrostatic springs were used as coupling elements between the resonators [6], allowing variable coupling strength. Identical bias voltages of 30V were applied to resonators 1 and 3, whereas resonator 2 was grounded, to ensure $K_{c1}=K_{c2}$. To demonstrate the sensitivity of the 3DoF device to stiffness perturbations, another variable DC voltage was applied on the electrode on the right. Actuation of the resonators was realized by applying an AC voltage to the electrode on the left. Differential motional currents were obtained through the differential sensing comb fingers attached to resonators 1 and 3. The configuration of the device for characterization is shown in Figure 7.

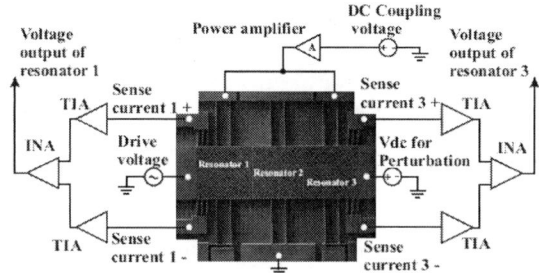

Figure 7: Test configuration of the prototype 3DoF resonator sensing device

The chip was wire bonded to the contacts of a chip carrier, and tested electrically on printed circuit board, which consisted of standard transimpedance amplifiers (TIA) and instrumentation amplifiers (INA). The chip and the circuit board were placed into a customized vacuum chamber with electrical feedthroughs. The ambient pressure for the testing was 20µTorr, so a high quality factor could be obtained. The experimental set-up is shown in Figure 8.

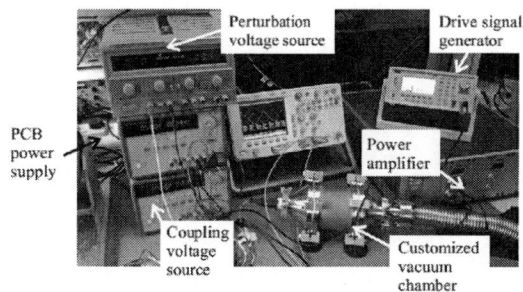

Figure 8: Experimental set up for 3DoF sensor characterization

RESULTS AND DISCUSSION
Frequency Response

The frequency response of the device was measured with various perturbation voltages applied. A typical frequency response of resonators 1 and 3 of the sensing device is shown in Figure 9. The measured 3-dB bandwidth and the quality factor of the out-of-phase mode

978-1-4799-7956-1/15 $31.00 © 2015 IEEE   883

were 2.40Hz and 6221, respectively. The frequency difference between the in-phase and out-of-phase modes was 4.99Hz, which is greater than twice of the 3-dB bandwidth of the out-of-phase mode, indicating weak damping, which thus can be neglected.

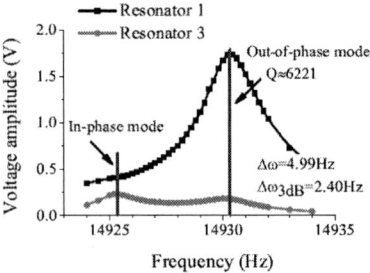

Figure 9: Typical frequency response of resonators 1 and 3, with 30V coupling voltage and 4.15V perturbation voltage.

Sensitivity

Upon finding the mode frequency of the out-of-phase mode, mode amplitudes of resonators 1 and 3 were averaged and recorded. The amplitude ratios of the out-of-phase mode were then computed. Figure 10 shows the measured amplitude ratio (quotient of modal amplitudes of resonators 1 and 3) at the out-of-phase mode for different stiffness perturbations with 30V coupling voltage. The measurement results are presented together with a linear fit. The linear sensitivity to normalized stiffness change extracted from the measured data was found to be 13558, whereas the theoretical calculated value was 17073. The discrepancy was due to fabrication variances. Table 1 lists a comparison of sensitivity between state-of-the-art resonator sensors (1Dof and 2DoF) for stiffness change sensing and our work.

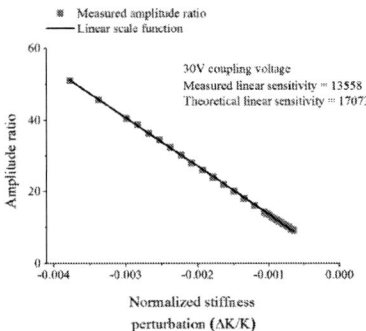

Figure 10: Measured amplitude ratio at the out-of-phase mode of the 3DoF resonator sensor, for different stiffness perturbations.

Table 2: Sensitivity comparison

Reference	Sensor output	Measured sensitivity	Sensor type
[5]	Frequency shift	0.5	Single resonator
[6]	Eigenstate shift	275	Two resonators
This work	Amplitude ratio change	13558	Three resonators

CONCLUSIONS AND OUTLOOK

In this paper, we have reported a novel 3DoF resonator device for stiffness change sensing applications. The measured sensitivity of a prototype sensor represents an improvement by over 49 times compared to the state-of-the-art stiffness change sensors consisting of two weakly coupled resonators. In the future, the effect of damping will be included in the analysis. In addition, other specifications of the sensor, such as dynamic range, linearity and resolution will also be investigated.

ACKNOWLEDGEMENT

The authors would like to thank the support from the Programme of Introducing Talents of Discipline to Universities, China (Grant No.B13044).

REFERENCES

[1] M. Schmidt and R.T. Rowe, "Silicon resonant microsensors," *14th Automotive Materials Conference: Ceramic Engineering and Science Proceedings*, Volume 8, Issue 9/10, pages 1019–1034. Wiley Online Library.
[2] A. Seshia, M. Palaniapan, T.A Roessig, R.T. Howe, R. W. Gooch, T. R. Schimert, and S. Montague, "A vacuum packaged surface micromachined resonant accelerometer," *Microelectromechanical Systems, Journal of*, 11(6):784–793, Dec 2002.
[3] U. Durig, J. K. Gimzewski and D. W. Pohl, "Experimental Observation of Forces Acting during Scanning Tunneling Microscopy", *Phys. Rev. Lett.*, 57:2403–2406, Nov 1986.
[4] M. Spletzer, A. Raman, A. Q. Wu, X. Xu and R. Reifenberger, "Ultrasensitive mass sensing using mode localization in coupled microcantilevers", *Applied Physics Letters*, 88, 254102, 2006.
[5] F. J. Giessibl, "A direct method to calculate tip–sample forces from frequency shifts in frequency-modulation atomic force microscopy", *Applied Physics Letters*, 78(1), 2001.
[6] P. Thiruvenkatanathan, J. Yan and A. A. Seshia, "Ultrasensitive mode-localized micromechanical electrometer," In *Frequency Control Symposium (FCS), 2010 IEEE International*, pages 91–96, June 2010.
[7] C. Zhao, G. S. Wood, S. H. Pu and M. Kraft, "Design of an ultra-sensitive MEMS force sensor utilizing mode localization in weakly coupled resonators", In *23rd MME Workshop, 2012*.
[8] C. T.-C. Nguyen, "Frequency-selective mems for miniaturized low-power communication devices," *Microwave Theory and Techniques, IEEE Transactions on*, 47(8):1486–1503, Aug 1999.
[9] C. Pierre, "Mode localization and eigenvalue loci veering phenomena in disordered structures," *Journal of Sound and Vibration, 126(3):485 – 502, 1988*.
[10] J. B. Xie, Y. C. Hao, H. L. Chang, and W. Z. Yuan, "Single mask selective release process for complex soi mems device," *Key Engineering Materials, 562:1116–1121, 2013*.

CONTACT

*Chun Zhao, tel: +44 77 30043696; cz1y10@soton.ac.uk

CHIP-SCALE AEROSOL IMPACTOR WITH INTEGRATED RESONANT MASS BALANCES FOR REAL TIME MONITORING OF AIRBORNE PARTICULATE CONCENTRATIONS

M. Maldonado-Garcia[1], E. Mehdizadeh[1], V. Kumar[1], J.C. Wilson[2] and S. Pourkamali[1]
[1]Electrical Engineering Department, University of Texas at Dallas, Richardson TX, USA
[2]Mechanical and Materials Engineering Department, University of Denver, Denver CO, USA

ABSTRACT

This work presents chip-scale integration of MEMS resonant mass balances along with aerosol inertial impactors (airborne micro/nanoparticle collectors). A three mask microfabrication process has been developed to produce the main components; mass balance, impactor nozzle, and impaction micro-chamber on a single SOI substrate. In addition to extreme miniaturization of a conventionally bulky setup and allowing real-time particulate mass concentration data collection, this approach addresses assembly challenges for discrete versions of such systems, e.g. misalignment between MEMS resonators and nozzles. Furthermore, small nozzle diameters achievable through microfabrication, minimizes the air flow and therefore pump capacity requirements.

INTRODUCTION

Atmospheric particulate matter (PM) also known as "atmospheric aerosol" is comprised of tiny pieces of solid or liquid matter suspended in the air. The proven health effects caused by excessive levels of PM in the air include asthma, cancers, cardiovascular issues, respiratory diseases, and birth defects. Therefore, monitoring of PM concentrations in high risk environments such as polluted urban areas, dusty work environments (e.g. mines), operating rooms, etc. is of great importance [1,2]. Aerosol inertial impactors are one of the most popular tools for PM concentration monitoring that can collect particles down to a few nanometers in diameter on a substrate by passing the air through narrow orifices (nozzles) directed towards an impaction substrate (Fig. 1). This is achieved by forming a partial vacuum on one side of the nozzles using a vacuum pump. As a result, particles smaller than the stage cut-point diameter will escape collision with the impaction plate and particles larger than the cut-point are deposited on the impaction plate [3,4].

In conventional aerosol impactors, after collection of enough particle mass, the impaction substrate is to be taken out of the system and manually weighed to determine the mass of collected particles. Such impactors weigh hundreds of grams and require relatively large vacuum pumps to maintain enough air flow and collect enough particulate mass within the desired timeframe (Fig. 2a). In an effort to miniaturize aerosol impactors and equip them with real-time PM monitoring capability, MEMS resonant mass balances were embedded within small size impactors (Fig. 2b) where the incoming PM is collected on the resonator surface shifting its resonance frequency [4,5]. The current work presents the ultimate miniaturization of aerosol impactors by integration of the mass balance and impactor components on a single chip (Fig 2c) that can potentially be integrated within a smart phone or a very small clip-on device.

Figure 2: (a) A conventional aerosol impactor without real-time measurement capability (b) First generation of real-time impactor with embedded MEMS resonant mass balance [5] (c) Chip-scale real-time impactor, this work.

MEMS BALANCE DESCRIPTION

The chip-scale impactor consisting of the resonator, the nozzles and the micro-chamber (distance between resonator backside surface and nozzle) are fabricated out of a non-conventional SOI substrate with a buried nitride layer using a three-mask fabrication process (Fig. 3). The SOI substrate comprises of a 5 μm thick silicon device layer, 0.5 μm thick silicon nitride layer (instead of buried oxide), and 500 μm thick silicon handle substrate. The fabrication process starts with thermally growing a 2 μm thick oxide layer followed by low pressure chemical vapor deposition (LPCVD) of 1 μm low-stress silicon nitride. This is to minimize the thermal stress issues as well as to begin with a thick enough hardmask that can endure the long etch process for creating the micro-chambers, as explained later on. The resonant structure

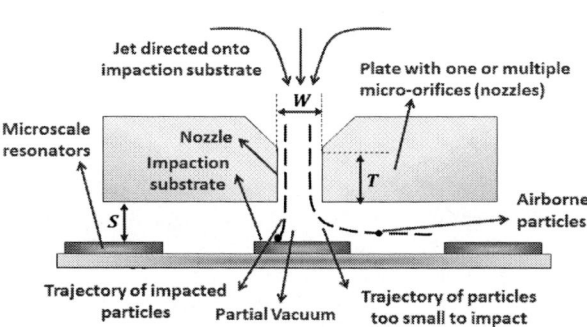

Figure 1: Schematic view of an aerosol inertial impactor integrated with microscale balances as the impaction substrates, showing particle trajectories for different particle sizes [4].

patterns were created by etching the top nitride layer half way through (~500 nm deep, as shown in Fig. 3a). The openings to create the micro-chambers were then patterned on the remaining top nitride and oxide layers followed by deep reactive ion etching (DRIE) of the silicon device layer (Fig. 3b). To protect the sidewalls of the silicon device layer during the subsequent etch steps, a 1 µm thick oxide layer was thermally grown (Fig. 3c). Backside lithography was then performed on the substrate to pattern the nozzles. Nozzles were then etched in the handle layer all the way up to the buried nitride layer via DRIE from backside (Fig. 3d). Next, the buried nitride was etched back from the frontside of the wafer through the previously formed openings in the silicon device layer. The micro-chambers were carved out of the silicon handle layer by a customized isotropic silicon dry etch recipe (Fig. 3e). The isotropic silicon etch recipe is based on sulfur hexafluoride (SF_6) and Argon (Ar) mixture gas that has a selectivity of 1:300 to LPCVD Si_3N_4 at an etch rate of 5 µm/min. The frontside nitride and oxide layers were then partially etched back leaving behind only resonant structure patterns. Finally, the devices were formed by DRIE of the silicon device layer and etching away the exposed buried nitride layer from the frontside to fully release the devices (Fig. 3f). To remove any remaining nitride and oxide hard mask layers on the resonant structures, the samples were dipped in phosphoric acid (H_3PO_4) at 150 ^0C and hydrofluoric acid (HF) respectively.

Figure 4 shows the SEM views of fabricated 5.2 MHz resonant structures. The silicon thermal-piezoresistive resonators are suspended on top of the impaction chamber with one nozzle underneath each resonator plate (Fig. 4a). The SEM view of the chip-scale impactor array containing four resonant devices (with 50×50 µm^2 plate size) is illustrated in Figure 4b.

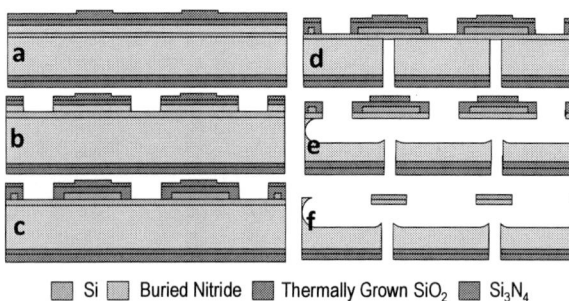

Si ☐ Buried Nitride ☐ Thermally Grown SiO₂ ☐ Si₃N₄

Figure 3: Fabrication process flow for the chip-scale impactors (a) patterning resonators by dry etching silicon nitride half way through,; (b) impactor micro-chamber patterning by dry etching oxide, nitride and silicon, (c) thermally growing a thin oxide film on silicon device layer sidewalls; (d) forming the nozzles from the wafer backside via DRIE of the silicon handle layer; (e) isotropic dry etching of silicon handle layer from top to generate the micro chambers (f) nitride and oxide removal.

The mass loading of the collected particles on the resonator surface can be theoretically calculated based on the resonance frequency and mass of the resonator as follows:

$$f = \frac{1}{2\pi}\sqrt{\frac{k}{m}} \rightarrow \frac{\partial f}{\partial m} = -\frac{f}{2m} \qquad (1)$$

where k, m and f are effective stiffness, effective mass and resonance frequency of the resonator respectively. The thermal-piezoresistive resonators are comprised of a narrow beam connecting two plates where the particles passing through the associated nozzles will be collected (Fig. 4a). The short narrow beam acts as a thermal actuator as well as piezoresistive sensor. An AC electrical current passing through the narrow beam is responsible for generation of a fluctuating temperature and in turn alternating thermal stress in the beam. The resulting thermal stress will actuate the resonator in its in-plane extensional resonant mode. Modulation in the resistance of the same actuator beam due to the mechanical vibrations via the piezoresistive effect allows monitoring of the resonator vibration amplitude. [6].

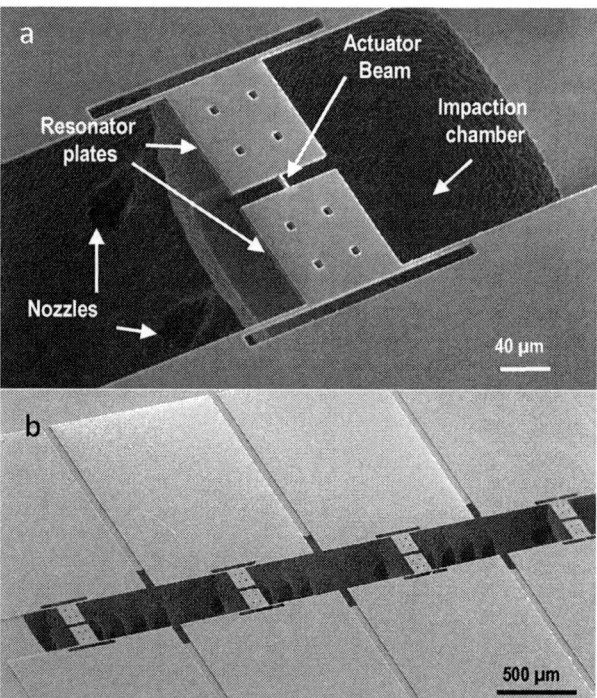

Figure 4: SEM views of the fabricated chip-scale impactors (a) thermal-piezoresistive resonator with one nozzle underneath each plate (b) impactor array chip containing four resonators.

AEROSOL IMPACTOR DESIGN

The inertial aerosol impactor design is based on the Stokes (St_{50}, with 50 % probability) and Reynolds (Re) numbers, which predict the probability of a particle colliding on an impaction substrate and the efficiency with which the particles will be collected respectively [7]. Particles smaller than the stage cut-point diameter (D_{p50}) will escape collision with the impaction plate and will continue being airborne while particles larger than the cut-point will be deposited on the impaction plate. The stage cut point (D_{p50}) is given by the following equation:

$$D_{p50} = \sqrt{\frac{9\mu W}{\rho_p C V_o}} \sqrt{St_{50}} \qquad (2)$$

where ρ_p is the particle density, V_o is the average air velocity at the nozzle exit, μ is air viscosity, W is the nozzle diameter and C the Cunningham slip correction factor [4]. The Reynolds number efficiency Re, considered for the current design was ~3000 which shows sharp collection efficiency [7]. The parameters related with Re are shown in equation (3).

$$Re = \frac{\rho VoW}{\mu} = \frac{4\rho Q}{\pi n \mu W} \qquad (3)$$

where Q is the volumetric flow rate through the nozzles, ρ is the air density and n is the number of nozzles. To achieve the desired Q values, a Parker's T2-05 pump was utilized along with a 13.5 mm wide micro diaphragm. Table 1 shows the design parameters for the current single stage aerosol impactor. After solving the slip correction factor equation [5] and equation (3), the cut-point converges to 30 nm. The calculations were based on a nozzle diameter (W) of ~ 0.034 mm, measured by SEM after microfabrication.

Table 1: Design parameters

Parameter	Value	Unit
Q	0.147	LPM
P_2	0.3	atm
V_0	1350	m/s
W	0.034	mm
n	2	-
$\sqrt{St_{50}}$	0.47	-
D_{50}	30	nm

The theoretical impactor efficiency collection curves for the Reynolds number (Re) [7] are strongly correlated with S, jet to plate distance and W, jet width or diameter (Fig, 1). In the current work a highly controlled S/W ratio for the micro scale impactor with minimal misalignment can be achieved. In addition small microlithography defined nozzle diameters allow smaller pump capacity than previous works.

RESULTS

The chip-scale impactor was placed face up on a specifically designed printed circuit board (PCB). An opening smaller than the size of the chip was drilled in the PCB underneath the chip (Fig, 5). One of the resonators was wire bonded to the metal traces on the PCB with the electrical components schematically shown in Figure 5. Air flow through the nozzle is maintained by a sealed connection to the pump from the top surface of the chip and particles are collected on the resonator backside (Fig, 5).

Particle collection was performed in four different environments: laboratory, air purifier output, cleanroom, and gowning area of cleanroom. The frequency shifts were recorded for one hour periods in each environment while collecting particles on the resonator surface. After exposure to airborne particles, the device plates were broken off the chip by sticking a piece of tape to the top resonator surface and peeling it off. This is to inspect the collected particles on the resonator backside.

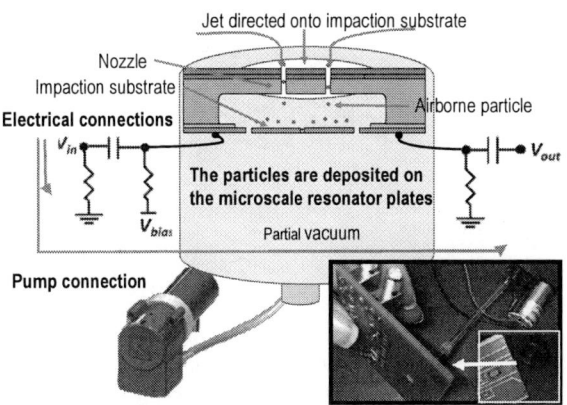

Figure 5: Schematic side view of the experimental setup including the integrated chip scale impactor, electrical components, and the vacuum pump. The electrical connections between the thermal-piezoresistive resonator and the printed circuit board are formed by aluminum wirebonds. The vacuum pump establishes the airflow through the nozzles by generating a partial vacuum within the micro-chamber of the impactor.

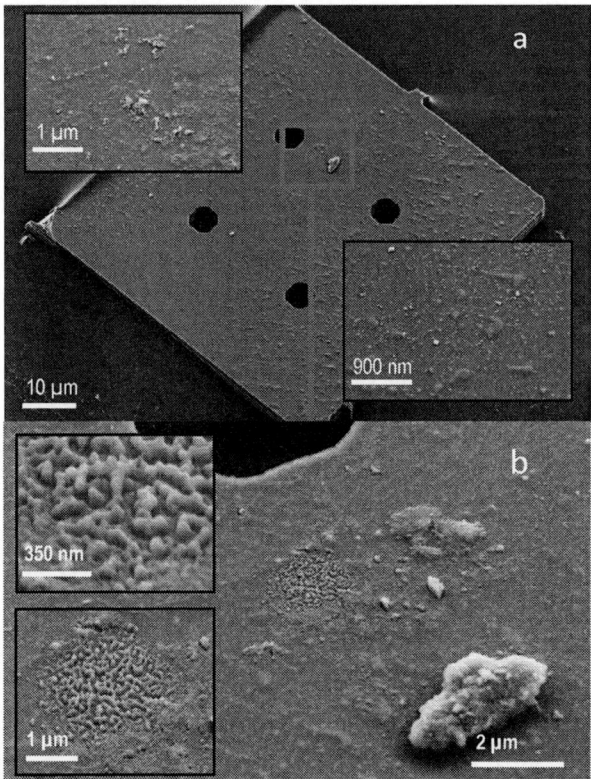

Figure 6: SEM view of (a) backside of the thermal-piezoresistive resonator of Figure 4 after airborne particle deposition. (b) Close-up view of particles in µm to nm scale size.

Figure 6 shows SEM views of the device plate backside where the particles are collected.

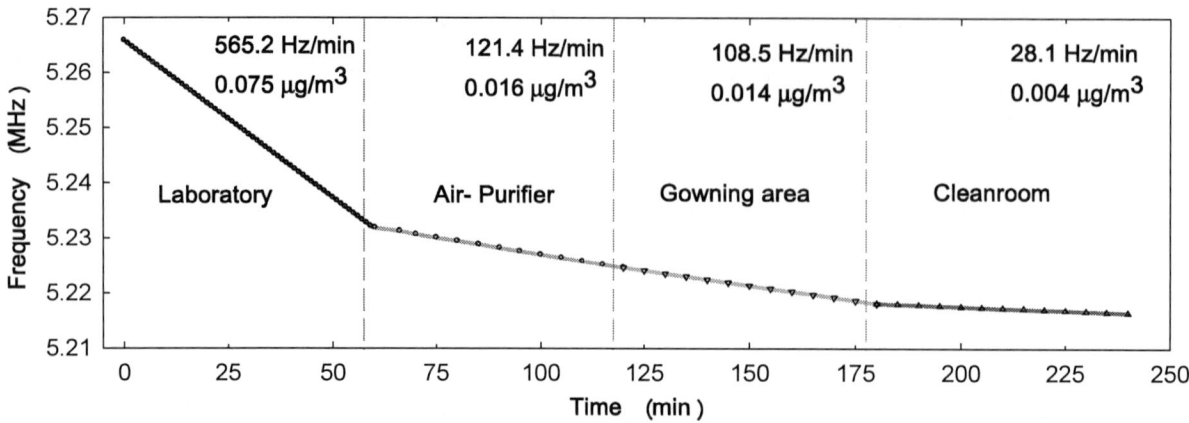

Figure 7: Measured resonant frequency of the thermally actuated MEMS resonator during exposure to airborne particles in different environments. As expected frequency shift rate is smaller for cleaner air samples

It can be seen in Figure 6a that the particles spread all over the plate and particle sizes range from micrometers to tens of nanometer which agree with cut-point calculation. More zoomed-in views of the deposited particles are shown in Figure 6b. Particles as small as a few tens of nanometers are collected using a single stage with a couple of nozzles with 30 μm diameter. Figure 7 shows the recorded resonant frequency of the resonant mass balance as the particles are collected in different environments. The slope of each graph is directly correlated to the particulate mass concentration in each environment clearly showing how the rate of frequency shift decreases as the air samples become cleaner. The particulate mass concentration in regular laboratory air, output of a consumer grade air purifier, gowning area of the cleanroom and inside the cleanroom were measured to be 75, 16, 14 and 4 ng/m³ respectively.

The results demonstrate that the gowning area and the output of the air purifier are similar in terms of particulate mass concentrations. The total frequency shift of 0.9% (50 kHz) over 250 minutes corresponds to 12 ng of collected particle mass, which is ~ 1.8% of the resonator original mass.

CONCLUSIONS

Chip scale particulate collectors with integrated MEMS resonant mass balances were fabricated on a SOI substrate. Collection and real-time mass concentration measurements for particle sizes ranging from a few microns to tens of nanometers were successfully demonstrated. The rate of resonator mass loading is clearly correlated to cleanliness of air samples. Future work includes demonstration of multi-stage chip-scale impactors with particle size segregation capability.

ACKNOWLEDGEMENTS

This work has been supported by the National Science Foundation under award 1300143. Authors would like to thank CONACYT for the 199880 scholarship.

REFERENCES

[1] U. Poschl, Atmospheric aerosols: Composition, transformation, climate and health effects *J. Atmospheric Chem. Sci.*, vol. 44, pp.7520-40, 2005.

[2] M. Kumala, H. Vehkanaki, T. Petaja, M.Dal Maso, A. Lauri, V. Kerminem, V.M. Kerminem, W. Birmili, P.H. McMurry, Formation and growth rates of ultrafine atmospheric particles: A review observations, *Journal of aerosol science*, , vol 35, pp 143-176, 2004.

[3] T.T. Mercer, M.I. Tillery, G.J. Newton, A multi-stage, low flow rate cascade impactor, *Journal of Aerosol Sci*, vol. 1, pp 9-15, 1970.

[4] E. Mehdizadeh, J. C. Wilson, A. Hajjam, A. Rahafrooz, and S. Pourkamali, "Aerosol impactor with embedded MEMS resonant mass balance for real-time particulate mass concentration monitoring", *International Conference on Solid State Sensors, Actuators, and Microsystems* (Transducers), 2013.

[5] E. Mehdizadeh, V. Kumar, J. Gonzales, R. Abdolvand, and S. Pourkamali, "A Two-Stage Aerosol Impactor with Embedded MEMS Resonant Mass Balances for Particulate Size Segregation and Mass Concentration Monitoring", *IEEE Sensors Conference*, 2013.

[6] A. Rahafrooz and S. Pourkamali, "High-frequency thermally actuated electromechanical resonators with piezoresistive readout," *IEEE Transactions on Electron Devices*, vol. 58, no. 4, pp. 1205-1214, 2011.

[7] Marple, V.A. and Willeke, K., "Impactor design." *Atmospheric Environment*, vol. 10, pp. 891-896, 1976.

CONTACT

* M. Maldonado-Garcia, tel: +1 (214) 770-9378; maribelmg@utdallas.edu

HARBOR SEAL WHISKER INSPIRED FLOW SENSORS TO REDUCE VORTEX-INDUCED VIBRATIONS

A.G.P. Kottapalli[1,2], M. Asadnia[1,2], J.M. Miao[1] and M.S. Triantafyllou[2,3]*

[1]School of Mechanical and Aerospace Engineering, Nanyang Technological University, SINGAPORE

[2]Center for Environmental Sensing and Modeling (CENSAM), Singapore-MIT Alliance for Research and Technology (SMART), SINGAPORE

[3]School of Mechanical Engineering, Massachusetts Institute of Technology, USA

ABSTRACT

Some of the biological sensors found in nature portray the best designs with incomprehensible sensing features. Recent studies reveal that Harbor seals (*Phoca vitulina*) are capable of performing underwater wake tracking using their vibrissae alone. It is believed that the unique geometry of the whiskers may play a role in suppressing vortex-induced vibrations. In this work, we developed piezoelectric MEMS flow sensors that utilize high aspect ratio micro-whiskers to sense flows underwater with high sensitivity. Through a comparative experimental study, the contribution of a whisker-like geometry towards the reduction of vortex-induced vibrations as compared to a cylindrical geometry is validated.

INTRODUCTION

In the past few decades there has been a considerable research focus on developing sensors for flow sensing on unmanned underwater vehicles (UUVs) and unmanned robotic vehicles (URVs). The next generation of UUVs and URVs, due to their smarter and innovative designs, pose stringent demands on the sensors that enable them explore their environment. Self-powered and light-weight sensors offer huge benefits in sensing, especially, on URVs due to their small size and the presence of moving actuators. Many aquatic animals have already developed the best set of flow sensors that enable them to swim with extremely high maneuverability and stability. Blind cave fish, using the lateral-line of sensors on their body, attain astounding performance in realizing hydrodynamic vision, energy-efficient maneuvering and high control and maneuverability [1,2]. The current abilities of UUVs could undergo a sea-change when arrays of such artificial sensors become available. In view of the same, in this work we develop ultra-sensitive, self-powered, light-weight and surface-mountable MEMS flow sensors for underwater sensing applications.

Researchers in the past developed MEMS flow sensors featuring micro-hair cells inspired by the hair cells of the biological neuromast sensors of the fish [3-8]. Tiny polymer hair cells that extend into the surrounding flow respond to variations in the flow and in turn elicit a resistance change in the piezoresistors embedded on a MEMS sensing membrane [6-8]. Most of these sensors feature circular cylindrical hair cells that extend into the flow and interact with the flow. Often, such hair cells

when subject to flow, experience vortex-induced vibrations other than the drag pressure induced by the steady flow. Harbor seals, due to their survival in dark and turbid waters utilize their whiskers to analyze and track hydrodynamic wake signatures left by other animals underwater [9]. The seal is capable of scanning for wake signatures in water while swimming at a speed of nearly one meter per second [10]. Biological experiments show that seals are able to detect the wake of a fish up to 35s after the fish has passed [11]. The seal extends its whiskers into the flow with the length of the whiskers perpendicular to the swimming direction. Seal whiskers feature a unique undulatory asymmetric geometry along the length of the whisker which is believed to have a contribution towards the suppression of vortex-induced vibrations (VIV) [10,12]. In this work, we present the design, fabrication and experimental characterization of flow sensors that achieve ultra-high sensitivity due to the ability of the whisker-inspired hair cell geometries to suppress VIV.

MEMS WHISKER SENSOR

The MEMS artificial seal whisker sensor is designed on similar structural principles as that of the real seal whisker sensor. The whiskers have an extremely complex 3D geometry with an elliptical cross section and undulations on major and minor axes with an offset angle between leading edge and trailing edge and a variation of the offset angle along the length of the whisker. Conventionally, hair cell sensors were developed through SU-8 photo-patterning [3]. However it is possible to develop cylindrical SU-8 hair cells of high aspect ratio, it is difficult to fabricate undulatory features along the length of the hair cell using SU-8 photo-patterning method. We employed stereolithography (SLA) technique to develop artificial whisker sensors using polycarbonate material. The Solidworks model of the whisker structure is as shown in figure 1. All the key dimensional features of the undulations on the whisker are kept the same as those described in the case of real seal whiskers in [10] with the exception for the length of the whisker which is kept 4mm. Too tall whiskers result in high bending in the piezoelectric sensing membrane causing it to break. The real seal whisker does not possess any sensing nerves along the length of the whisker [10]. The sensing occurs at the cheeks of the seal when the whiskers bend under external force. In the same way, in the current sensor design, the piezoelectric sensing element is located at the base of the polymer whisker. The whisker structure is mounted on a MEMS sensing membrane consisting of 27μm thick piezoelectric Pb(Zr$_{0.52}$Ti$_{0.48}$)O$_3$ (PZT) film bonded to a 20μm thick silicon membrane.

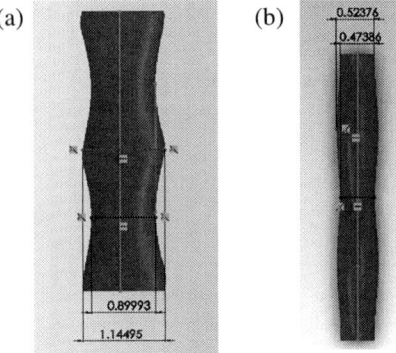

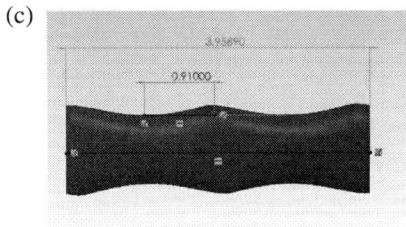

Figure 1: Artificial seal whisker design (a), (b) and (c) show the various geometrical dimensions of the artificial seal whisker mounted on the PZT MEMS sensing membrane.

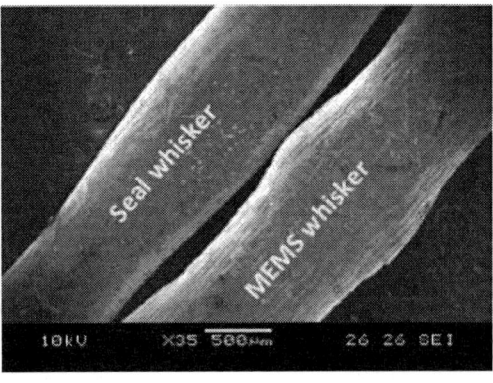

Figure 2: An SEM image of the artificial and real seal whiskers.

These tiny polymer whiskers extend into the external flow and respond by deflecting in compliance to flow variations. Displacement of the whiskers causes the underlying piezoelectric sensing membrane to bend which produces charges and thereby provides flow information. The basic sensor structure, fabrication methods and fundamental flow characterizations of the MEMS artificial whisker sensor are discussed in detailed in [13].

A scanning electron microscope (SEM) image of the artificial and real seal whisker are shown in figure 2.

WHISKER GEOMETRY AND VIV REDUCTION

Two major experiments are conducted using the MEMS artificial seal whisker sensor. The first experiment is designed to determine the sensitivity and accuracy of the artificial whisker sensor to flow variations. The results of this experiment are presented in [13]. The second major goal, which is targeted in this paper, is to evaluate the contribution of the morphology of the whisker towards suppressing vortex-induced vibrations, which would otherwise cause significant noise in sensing low flow velocities that exist within the wake. We conducted a comparative experimental study by analyzing the vortex-induced vibrations experienced by the artificial whisker structure as compared to a circular cylinder. Two MEMS sensors are fabricated -one possessing a circular cylindrical hair cell and the other processing a hair cell with whisker-like undulations. The circular cylindrical structure has a height of 4mm (equal to that of the whisker structure) and a diameter of 0.5mm (approximately equal to the average of the diameters of the widest and the narrowest undulations along the minor axis of the whisker).

Experimental results

Two MEMS sensors with identical piezoelectric sensing base but one bearing a circular cylinder and the other bearing an artificial whisker are used for the experiments.

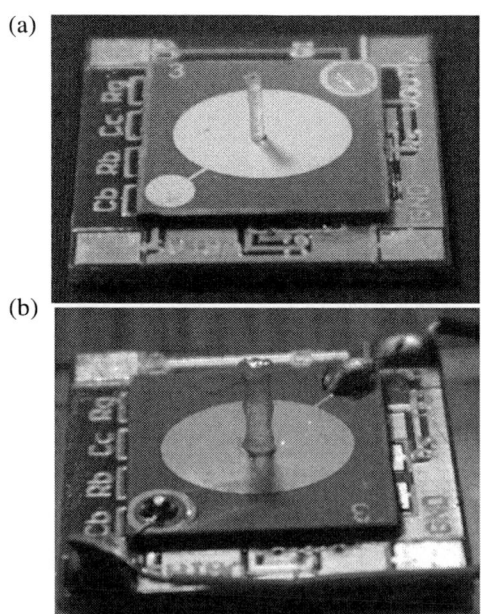

Figure 3: Photographs of MEMS piezoelectric membranes with (a) circular cylinder (b) whisker structure mounted on at the center of the membrane.

The sensors are tested in a Long Win LW-3457 model closed circuit water tunnel with a test-section of dimensions 0.3(W) ×0.4(H) × 1(L) m³. The water tunnel is fitted with turbulence reducing steel screen and two honeycomb layers. Both the sensors are mounted at the center of the test-section of the water tunnel in such a way that the direction of water flow is perpendicular to the long axis of the whisker and the cylinder. Figure 3 shows the photographs of the two MEMS sensors that are used in the experiment.

The steady-state flow velocity in the water tunnel is set to 0.425m/s. The outputs from both the sensors are simultaneously acquired through National Instruments Data acquisition card (NI-DAQ) and are recorded in LABVIEW. As the water flows past the structures vortex induced vibrations cause the structure to vibrate which in turn causes the piezoelectric membrane to bend. Analyzing the spectral content of the output of the sensors, the frequency at which vortex-induced vibrations occur can be detected. The expected VIV frequency for the circular cylinder and the whisker will not be the same due to the difference in the diameters. The experimentally determined VIV frequencies for the circular cylinder and the whisker structure match very closely with those of the theoretically estimated values. Figure 4 shows the frequency content of the data collected from both the sensors.

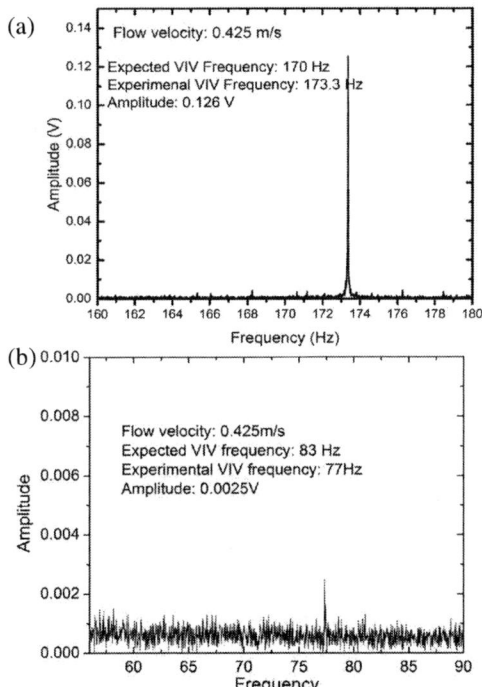

Figure 4: Experimental results that show a VIV suppression due to the whisker-like geometry. Two PZT MEMS sensors- (a) one featuring a hair cell with plain cylindrical geometry and (b) the other hair cell with whisker-like undulations are placed in steady flow generated in a water tunnel.

The VIV frequency peaks for the circular cylinder and whisker occur at 173.3 Hz and 77 Hz which closely compare to theoretically estimated values for the same which are 170 Hz and 83 Hz respectively. It can be observed from the amplitude of the VIV frequency peak obtained experimentally that the VIV frequency peak in case of the whisker structure is 50 times smaller than that of the circular cylinder structure (figure 4).

CONCLUSION

Due to the dark and turbid water environments where the Harbor seals live, their whiskers play a significant sensory role. The whiskers can sense flow velocities within the wake and track their prey through the wake signatures left by them as they swim. The unique undulatory geometry on the seal whiskers is believed to suppress the VIV-generated noises thereby rendering them to be sensitive to flow signals of importance. In this work, we report the development of self-powered piezoelectric MEMS flow sensors which feature a seal-whisker-inspired structural geometry. The results presented experimentally evaluate the VIV reduction abilities of the whisker-like geometry as compared to a circular cylindrical geometry.

ACKNOWLEDGEMENTS
This research was funded by the Singapore National Research Foundation (NRF) through the Singapore-MIT Alliance for Research and Technology (SMART) Centre, Center for Environmental Sensing and Modeling (CENSAM) IRG.

REFERENCES
[1] J. C. Montgomery, S. Coombs, and M Halstead, "Biology of the mechanosensory lateral line in fishes," *Rev. Fish Biol. Fisher.*, vol. 5, pp. 399-416, 1995.

[2] J. C. Montgomery, S. Coombs, and C. F. Baker, "The mechanosensory lateral line system of the hypogean form of *Astyanax fasciatus*," *Evol. Biol. Fish.*, vol. 62, pp. 87-96, 2001.

[3] N. Chen, C. Tucker, J. M. Enge, Y. C. Yang, S. Pandya and C. liu "Design and characterization of artificial haircel sensor for flow sensing with ultrahigh velocity and angular sensitivity,", *J. Microelectromech. Syst.* Vol. 16, pp. 999-1014, 2007.

[4] A. G. P. Kottapalli, M. Asadnia, J. M. Miao, G. Barbastathis and M. Triantafyllou, " A flexible liquid crystal polymer MEMS pressure sensor array for fish-like underwater sensing," *Smart. Mater. Struct.*, vol. 21, p. 115030, 2012.

[5] M. Asadnia, A.G.P. Kottapalli, Z. Shen, J. M. Miao, and M. Triantafyllou, "Flexible, and surface-mountable piezoelectric sensor arrays for underwater sensing in marine vehicles," *IEEE Sensors J.* vol. 13, pp. 3918-3925, 2013.

[6] A. G. P. Kottapalli, M. Asadnia, J. M. Miao, and M. Triantafyllou, "Electrospun nanofibrils encapsulated in hydrogel cupula for biomimetic MEMS flow sensor development," in *Proc. IEEE MEMS'13 Conference,* Taipei, January 20-24, 2013, pp. 25-28.

[7] A. G. P. Kottapalli, M. Asadnia, J. M. Miao, G. Barbastathis, and M. S. Triantafyllou, "Polymer

MEMS pressure sensor arrays for fish-like underwater sensing applications," Micro. Nano. Lett., vol. 7, pp. 1189-1192, 2013.

[8] A. G. P. Kottapalli, M. Asadnia, J. M. Miao, and M. Triantafyllou, "Touch at a distance sensing: Lateral-line inspired MEMS flow sensors," *Bioinspir. Biomim.* vol. 9, p. 046011, 2014.

[9] G. Dehnhardt, B. Mauck and H. Bleckmenn, "Seal whiskers detect water movements," *Nature,* vol. 394, pp. 235-236, 1998.

[10] W. Hanke et.al., "Harbor seal vibrissa morphology supresses vortex-induced vibrations," *J. Exp. Biol., vol. 213,* pp. 2194-2200, 2010.

[11] S. Wieskotten, G. Dehnhardt, B. Mauck, L. Miersch, and W. Hanke, "Hydrodynamic determination of the moving direction of an artificial fin by a Harbour seal (Phoca vitulina)," *J. Exp. Biol., vol. 213,* pp. 2665-2672, 2010.

[12] H. Beem, M. Hildner and M. Triantafyllou, "Characterization of harbor seal whisker-inspired flow sensor," in *Proc. Oceans'12 Conference,* Virginia, October 14-19, 2012, pp. 1-4.

[13] A. G. P. Kottapalli, M. Asadnia, J. M. Miao, and M. Triantafyllou, "Harbor seal inspired MEMS artificial micro whisker sensors," in *Proc. IEEE MEMS'14 Conference,* San Francisco, January 26-30, 2014, pp. 741-744

CONTACT

*A.G.P. Kottapalli, tel: +65-93722843; Center for Environmental Sensing and Modeling (CENSAM), Singapore-MIT Alliance for Research and Technology (SMART), ajay@smart.mit.edu

ISOTROPIC 3D SILICON HALL SENSOR

Christian Sander, Carsten Leube, Taimur Aftab, Patrick Ruther and Oliver Paul
Department of Microsystems Engineering (IMTEK), University of Freiburg, Germany

ABSTRACT

This paper reports the first three-dimensional (3D) Hall sensor with isotropic sensitivity for the three spatial components of the magnetic field. The silicon device has the shape of a hexagonal prism with symmetric sets of three contacts on its top and bottom surfaces. Sending currents obliquely across the device allows one to operate it as three mutually crossing, identical, and effectively orthogonal Hall sensors. The sensitivity vectors of these Hall sensors are easily converted into sensitivities in the Cartesian coordinate system xyz. We demonstrate a design achieving sensitivities of $S_x = 33.0$ mV/VT, $S_y = 33.9$ mV/VT and $S_z = 33.3$ mV/VT.

INTRODUCTION

Magnetic field sensors based on the Hall-effect in silicon (Si) are nowadays most widely used for contactless position measurements. Traditional planar Hall-effect transducers are sensitive to only one component of the magnetic flux density B. For the purpose of measuring all three spatial components (B_x, B_y, and B_z) of B, three separate, orthogonally aligned devices have been used, as illustrated in figure 1(a). Drawbacks of this approach are the costly assembly, the challenging orthogonal alignment during assembly, the spatial resolution limit due to the distance among sensor chips, and the handling of 12 contacts [1]. An alternative concept is to co-integrate a planar Hall plate with two vertical Hall sensors (VHS) [2,3]. However, the vertical Hall structures show lower sensitivities and higher offsets and thus require separate signal conditioning circuitry for compensation [4,5]. Furthermore, the number of contacts remains an issue. Integrated devices using only one active region may have fewer contacts and potentially offer increased spatial resolution. However, the lack of a fourfold symmetry, the inherently different sensitivities of the individual channels and cross-sensitivities between channels lower the overall performance [6,7]. No semiconductor magnetic field sensor is currently able to measure three orthogonal components of B with equal sensitivities and similar offsets.

In this paper we present the first 3D Hall device fulfilling these requirements. Section 2 introduces the novel sensor concept. The fabrication process is illustrated in Section 3, following with the experimental results in Section 4.

SENSOR DESIGN

The ideal planar Hall-effect device is a rotational symmetric plate with four peripheral contacts. Its output voltage can be written as

$$V_{out}(B) = V_{Hall}(B) + V_{off} = SV_{in}B + V_{off} \qquad (1)$$

where B, $S = V_{in}^{-1}\partial V_{out}/\partial B$, V_{Hall}, V_{in}, and $V_{off} = V_{out}(B=0)$ denote the out-of-plane magnetic flux density, the voltage-related sensitivity, the Hall voltage, the

drive voltage and the offset of the device in the absence of a magnetic field, respectively. Usually, V_{Hall} is obtained by the spinning-current method, where the biasing and sensing contacts are cyclically permuted four times and the output voltage is averaged, thus effectively eliminating V_{off} [8].

With properly dimensioned contacts, one achieves maximum S with a value of $0.47\mu_H$ [9,10], where μ_H denotes the Hall mobility. One such optimal sensor geometry is a square shaped device with four identical contacts centered on its four sides. This suggests that an ideal 3D magnetic field sensor with isotropic sensitivities could be a conductive cube with one contact centered on each of its six faces as shown in Fig. 2(a). The sensor can be operated as three equivalent, orthogonal Hall plates in three orthogonal planes defined by three groups of four contacts. The corresponding sensitivity vectors $\boldsymbol{S}_{x'} = (S_{x'},0,0)$, $\boldsymbol{S}_{y'} = (0,S_{y'},0)$, and $\boldsymbol{S}_{z'} = (0,0,S_{z'})$ are mutually orthogonal, have identical magnitude, and allow the measurement of the components of B, i.e., $B_{x'}$, $B_{y'}$, and $B_{z'}$. Since deep buried contacts cannot be realized in standard IC technology, we deformed the cube so that its contacts are relocated to the top and bottom surfaces of a wafer. This deformation is illustrated in figures 2(b) and (c). Contacts 1, 3, and 5 adjacent to one corner of the cube thereby move to the top surface, while contacts 2, 4, and 6 adjacent to the diagonally opposite corner are relocated to the bottom surface. The diagonal axis connecting these corners becomes an axis perpendicular to the wafer. As a result, the cube is deformed into a hexagonal prism and the sensor is symmetric under $120°$ rotation in the xy plane. Again, the device can be operated in three planes, as indicated in

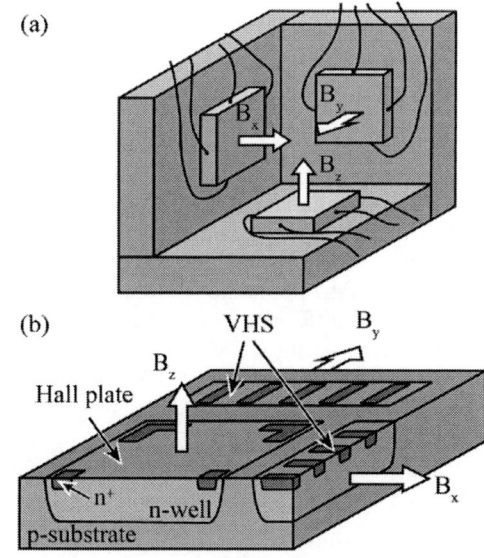

Figure 1: (a) Classic implementation of a three-axis magnetic field sensor consisting of three orthogonal Hall plates and (b) schematic of an integrated three-axis magnetic field sensor using a Hall plate und two orthogonal VHS.

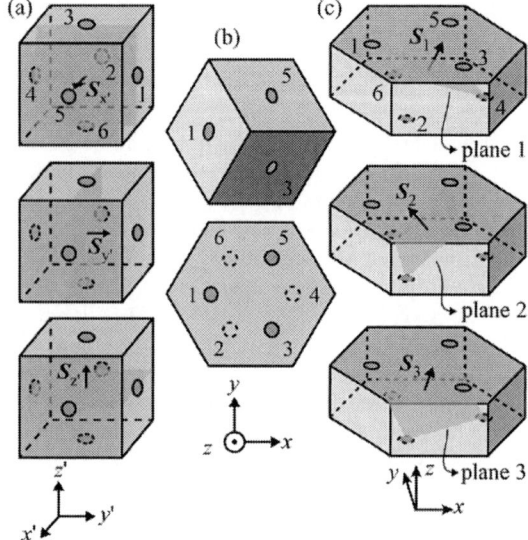

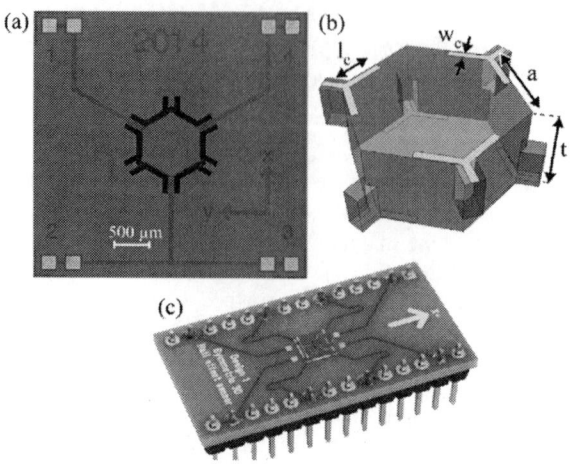

Figure 3: (a) Micrograph of the novel sensor, (b) schematic with relevant geometry parameters and (c) assembled device.

Figure 2: Schematic of (a) an ideal 3D Hall cube with all three operation planes and respective sensitivity vectors $S_{x'}$, $S_{y'}$ and $S_{z'}$, (b) a diagonal view of the cube and its projection onto two parallel xy-planes and (c) the novel hexagonal 3D Hall device with its three operation planes and respective sensitivity vectors S_1, S_2 and S_3.

Table 2: Geometry parameters of the device.

Parameter	t	a	l_c	w_c
(µm)	525	492	442	10

figure 2(c). For a properly dimensioned device, the sensitivity vectors $S_1 = (S_{1x}, S_{1y}, S_{1z})$, $S_2 = (S_{2x}, S_{2y}, S_{2z})$, $S_3 = (S_{3x}, S_{3y}, S_{3z})$ corresponding to these three planes are again orthogonal; using $S_{ij} = (\partial V_{\text{Hall},i}/\partial B_j)/V_{\text{in}}$, with planes $i = 1,2,3$ and directions $j = x,y,z$. In view of the device symmetry they have the same magnitude S_0. In the *xyz* coordinate system the normalized sensitivity vectors have the components listed in Table 1. These coefficients are the direction cosines of the sensitivity vectors with respect to the *x*, *y*, and *z* axes. They constitute the ideal orthogonal transformation matrix O mapping the normal base vectors of the *xyz* coordinate system onto those defined by the normalized sensitivities.

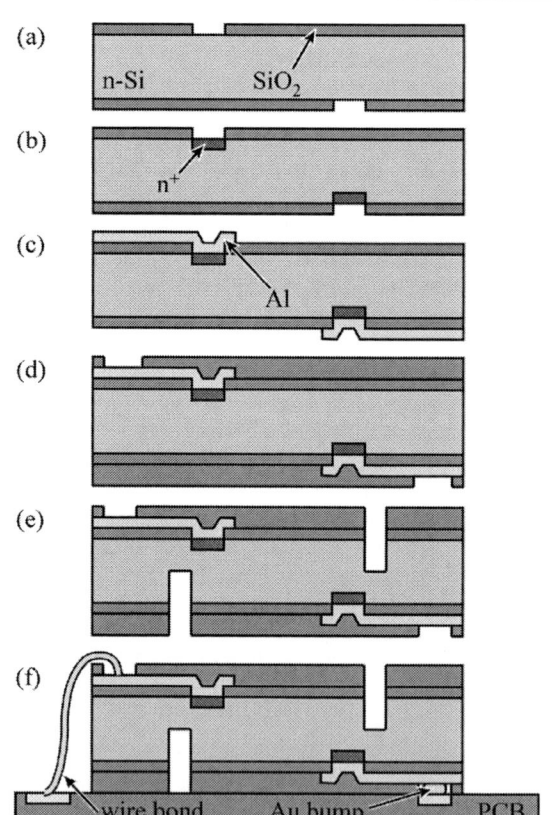

Figure 4: Fabrication process and assembly of the 3D Hall sensor including (a) insulation, (b) ion implantation, (c) metalization, (d) passivation, (e) DRIE, and (f) flip-chip and wire-bonding.

FABRICATION AND ASSEMBLY

Figure 3 shows a micrograph of a 3D Hall-effect sensor fabricated according to this novel concept, a schematic defining the relevant geometrical parameters listed in Table 2, and a photograph of a printed circuit board (PCB) -mounted device ready for characterization. The fabrication process is summarized in figure 4. It combines insulation, contact diffusion, metallization, and dual-side deep reactive ion etching (DRIE). A patterned thermal oxide layer (700 nm) on the front and rear of an n-doped Si wafer {Fig. 4(a)} serves as an implantation mask for

n^+ contacts {Fig. 4(b)}. The implantation energy is 80 keV with a dose of 2×10^{16} cm^{-2} for phosphorus ions. An Al metallization (500 nm) is then sputtered on the front and rear and patterned using wet etching {Fig. 4(c)}.

Table 1: Normalized values of the sensitivity matrix O for the ideal hexagonal 3D Hall sensor.

normalized sensitivity components	sensitivity direction j		
	x	y	z
S_{1j}/S_0	0.408	0.707	0.577
S_{2j}/S_0	−0.816	0	0.577
S_{3j}/S_0	0.408	−0.707	0.577

$\left.\right\} = O$

978-1-4799-7956-1/15 $31.00 © 2015 IEEE   894

Subsequently, a low temperature oxide (LTO) passivation (2 μm) is deposited and structured using reactive ion etching (RIE) to expose the contact pads {Fig. 4(d)}. The hexagonal shape is defined by DRIE to the depth of 270 μm from both sides {Fig. 4(e)}. As a result of the mask design, the prism is suspended by six bridges with a height and width of about 255 μm and 100 μm, respectively. The rear electrical interconnection is performed by flip-chip bonding the sensor chip onto a PCB, whereas the front contacts are wire-bonded to it {Fig. 4(f)}.

EXPERIMENTAL RESULTS

The electrical characterization of the novel device was performed using the parameter analyzer 4156C from Agilent. A 3D Helmholtz coil setup served to extract the electric field dependent sensitivities of the devices up to 3.6 mT.

As mentioned, the device can be operated in three planes defined by three groups of four contacts. The offset characteristic for one plane is shown in figure 5. For each plane the measurement is performed in four modes applying the spinning-current method [8]. The single-mode offset V_{off} increases linearly with the input voltage V_{in}. For all modes $|V_{off}|$ is about 1% of V_{in}. The residual offset V_{res} of about 6.3 μV at $V_{in} = 1$ V obtained by averaging over the four modes is smaller than $|V_{off}|$ by a factor of more than 1500. The input resistances R_{in} for Modes 1 and 3, and Modes 2 and 4 are 259.3 Ω and 262.5 Ω, respectively. The deviation is likely due to fabrication imperfections.

The Hall response $V_{Hall} = (V_{Hall,1}, V_{Hall,2}, V_{Hall,3})$ after current spinning for all three operation planes ($i = 1,2,3$) at $V_{in} = 1$ V is shown in figure 6. For each plane, three sensitivity values using again $S_{ij} = (\partial V_{Hall,i}/\partial B_j)/V_{in}$, with $j = x$, y, and z, are extracted. They constitute the sensitivity matrix T_{meas} listed in Table 3 together with the normalized values forming the measured transformation matrix O_{meas}. The elements of O_{meas} are again the direction cosines of the sensitivity vectors, now experimentally determined, with respect to the x, y, and z coordinate axes. These values are in good agreement with those of O in Table 1. This reflects the fact that the directions of the measured sensitivity vectors S_1 to S_3 are close to those of the ideal device in figure 1(c). To evaluate the sensitivities and angular error with respect to the xyz coordinate system, the measured sensitivity vectors S_1 to S_3 are transformed using the ideal

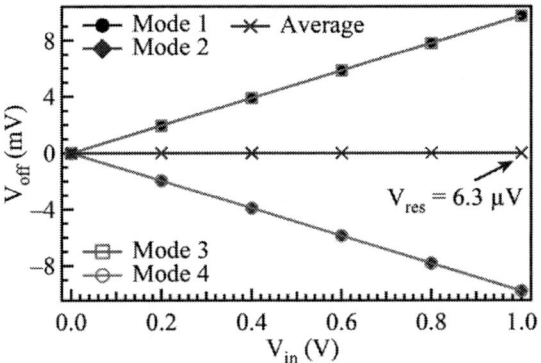

Figure 5: Offset voltage V_{off} and residual offset V_{res} after current-spinning as a function of the input voltage V_{in} for a representative plane.

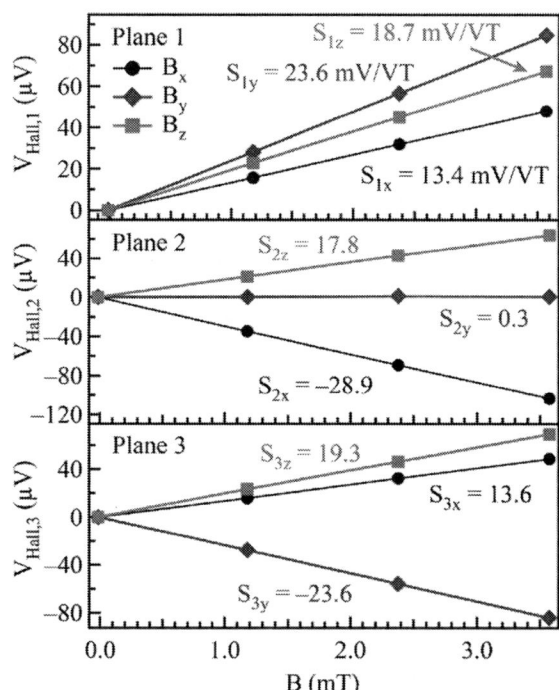

Figure 6: Hall voltage $V_{Hall,i}$ for each plane ($i = 1,2,3$) as a function of the spatial components of the magnetic field B_j ($j = x,y,z$), including the extracted sensitivities $S_{ij} = (\partial V_{Hall,i}/\partial B_j)/V_{in}$.

Table 3: Measured and normalized values of the sensitivities for the realized hexagonal 3D Hall sensor.

sensitivity components	sensitivity direction j		
	x	y	z
S_{1j} (mV/VT)	13.4	23.6	18.7
S_{2j} (mV/VT)	−28.9	0.3	17.8
S_{3j} (mV/VT)	13.6	−23.6	19.3
normalized values			
$S_{1j}/\lvert S_1\rvert$	0.406	0.716	0.568
$S_{2j}/\lvert S_2\rvert$	−0.851	0.008	0.525
$S_{3j}/\lvert S_3\rvert$	0.408	−0.706	0.579

($\} = T_{meas}$ for the first three data rows, $\} = O_{meas}$ for the normalized rows)

transformation matrix O. The transformed sensitivities are calculated as $S_x = (33.0, -0.1, -0.4)$ mV/VT, $S_y = (-1.3, 33.9, -1.7)$ mV/VT and $S_z = (0,0,33.3)$ mV/VT, respectively. The angular errors between these sensitivity vectors and the corresponding axes of the xyz coordinate system are found to be $\Delta x = 0.7°$, $\Delta y = 3.6°$ and $\Delta z = 0.1°$, showing the already high degree of isotropy and orthogonality of this initial device.

In practice, the three components of the magnetic field are accurately calculated by multiplying the inverse of the measured sensitivity matrix T_{meas} with the vector of the three Hall voltages after current spinning V_{Hall}. For illustration we applied a reference field vector B with a magnitude of 3.5 mT and rotated it in the xy, xz and yz plane from 0° to 90° in steps of 5°. At each step we extracted the measured field vector by $B_{meas} = T_{meas}^{-1} V_{Hall}$. The result is shown in figure 7. The overall deviation of the measured B-field magnitude from the nominal value of 3.5 mT is only 0.1±0.9%. The extracted B vectors show an angular

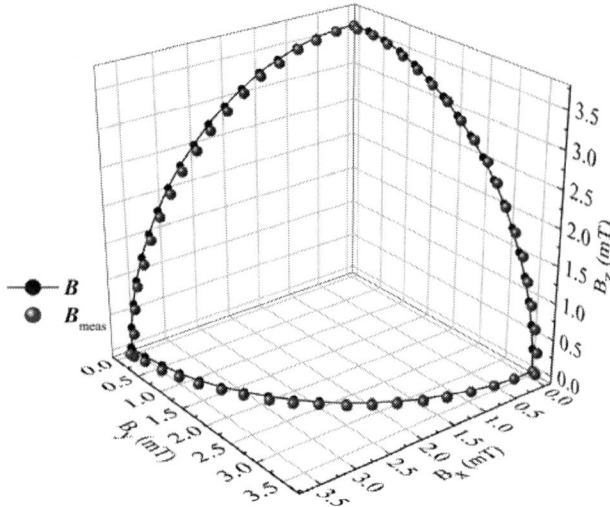

*Figure 7: Measured magnetic field vector **B**_meas for an applied reference field **B**.*

deviation from the direction of the applied **B** field of $0.8 \pm 0.4°$.

CONCLUSION

The novel 3D Hall-effect sensor shows a highly isotropic response to the magnetic field. In view of its symmetry, it can be operated with the same drive and compensation circuitry for its three sensitive planes. It has to be noted that the current device is a discrete component lending itself for hybrid integration with drive and signal conditioning circuitry. This can possibly be achieved by multi chip stacking assembly. Co-integration with on-chip circuitry will likely require additional technological developments, since the low doping of the substrate used for the sensor is not the standard doping of CMOS substrates.

REFERENCES

[1] E. Ramsden, *Hall Effect Sensors – Theory and Applications*, 2nd ed., Elsevier, Netherlands, 2006.

[2] F. Burger, P-A. Besse, and R. S. Popovic, "New fully integrated 3-D silicon Hall sensor for precise angular-position measurements", *Sensors and Actuators A: Physical*, vol. 67, pp. 72-76, 1998.

[3] P. Kejik, E. Schurig, F. Bergsma, R.S. Popovic, "First fully CMOS-integrated 3D Hall probe", in *Digest Tech. Papers Transducers'05 Conference*, Seoul, 2005, pp. 317-320.

[4] J. Pascal, L. Hébrard, V. Frick, J.-B. Kammerer, J.-P. Blondé, "Intrinsic limits of the sensitivity of CMOS integrated vertical Hall devices", *Sensors and Actuators A: Physical*, vol. 152, pp. 21-28, 2009.

[5] O. Paul, R. Raz, T. Kaufmann, "Analysis of the offset of semiconductor vertical Hall devices", *Sensors and Actuators A: Physical*, vol. 174, 2012.

[6] Ch. Roumenin, K. Dimitrov, A. Ivanov, "Integrated vector sensor and magnetic compass using a novel 3D Hall structure", *Sensors and Actuators A: Physical*, vol. 92, pp 119-122, 2001.

[7] S. Lozanova, Sv. Noykov, Ch. Roumenin, "A novel 3-D Hall magnetometer using subsequent measurement method", *Sensors and Actuators A: Physical*, vol. 153, pp. 205-211, 2009.

[8] P. Munter, "A low-offset spinning-current Hall plate", *Sensors and Actuators A: Physical*, vol 22, pp. 743-746, 1989.

[9] R.S. Popovic, *Hall Effect Devices*, 2nd ed., The Adam Hilger Series on Sensors, IOP Publishing Ltd., Bristol, 1991.

[10] O. Paul, M. Cornils. "Explicit connection between sample geometry and Hall response", *Appl. Phys. Lett.*, vol. 95, 232112, 2009.

CONTACT

C. Sander, tel: +49-761-2037193;
christian.sander@imtek.de

MASS-FABRICATION COMPATIBLE MECHANISM FOR CONVERTING IN-PLANE TO OUT-OF-PLANE MOTION

Inbar Hotzen, Orna Ternyak, Shai Shmulevich, and David Elata

Faculty of Mechanical Engineering, Technion - Israel Institute of Technology, Haifa, ISRAEL

ABSTRACT

We present a mechanism that converts in-plane to out-of-plane motion, which is fully compatible with standard mass-fabrication methods. The mechanism harnesses the well-established in-plane actuation achieved by comb-drives, and converts it to out-of-plane motion. The motion conversion ratio is constant (i.e. linear conversion), and it can be easily tuned by adding or subtracting modular elements, in an otherwise unchanged design planform. We experimentally demonstrate the linearity of the mechanism, and use dedicated test devices to show the tunability of the conversion ratio. With a different test device, we demonstrate parallel out-of-plane motion of a flat stage. The measurements of this device show good agreement with model predictions.

INTRODUCTION

Parallel out-of-plane motion is desired in a variety of fields such as: confocal microscopy [1, 2], infrared sensors [3], light modulators [4-6] and more.

Electrostatic actuation is prevalent in MEMS. Out-of-plane displacements can be easily achieved by using gap-closing electrostatic actuators. However, gap-closing actuators suffer from a limited range of stable motion (up to the pull-in point), and their response is inherently nonlinear. One other limitation, which is often overlooked, is that gap-closing actuators can be driven in one direction only: electrostatic forces cannot increase the gap beyond its nominal value [7, 8]. These limitations can be easily overcome by using comb-drive actuators. Double-sided comb-drive actuators offer a linear response over a large stable range of motion, and may be driven in both attitudes along their primary axis of motion [9, 10]. However, comb-drive actuators are optimal only for in-plane motions. In order to harness the excellent properties of in-plane comb-drive actuation, and achieve a well-behaved out-of-plane motion, motion conversion mechanisms with a linear response are coveted.

An elegant design of such a motion-conversion mechanism was introduced by Ando et al. [11, 12]. That mechanism was based on beams with slanted cross-sections. When a horizontal force is applied to a beam with a slanted cross-section, the resulting displacement has an out-of-plane component [13]. A schematic view of a motion conversion mechanism that is based on beams with a slanted cross-section is illustrated in Fig. 1.

Beams with slanted cross-sections may be easily produced in single-crystalline silicon (SCS) by using anisotropic wet etching. Beams with straight (vertical) cross-sections, which are necessary for comb-drives and suspensions, can be readily produced by DRIE micromachining. However, it is prohibitively difficult to combine the two processes, because each would damage the results of the other. Therefore, a structure design which combines beams of straight and of slanted cross-sections is incompatible with mass-fabrication.

In fact, Ando has used a post DRIE, focused ion beam (FIB) milling to produce beams with slanted cross-sections [11, 12]. But this is also incompatible with mass fabrication.

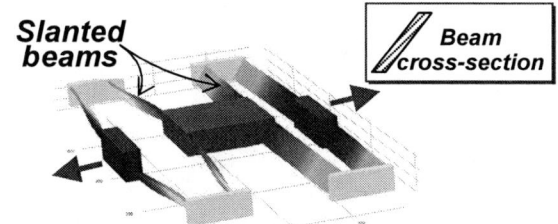

Figure 1: Motion conversion mechanism based on beams with slanted cross sections [11, 12]. In-plane forces, marked by the blue arrows, induce an out-of-plane displacement of the stage (red=elevation). , This design is incompatible with mass-fabrication techniques.

A possible solution to this problem is to use beams with an 'L'-shaped cross-section. Such beams respond mechanically as beams with a slanted cross-section, but they may be produced by DRIE with two depths of etching. However, beams with an 'L'-shaped cross-section are very stiff, and their deformation would require excessive forces.

Recently we have presented a new design of a motion conversion mechanism which utilizes ladder-shaped elements [14]. These elements are constructed from two long beams of different heights, which are fastened together by a small number of rigid connectors. All the cross-sections in this new design are straight and are therefore fully compatible with mass-fabrication. These elements emulate the response of beams with 'L'-shaped cross-sections, but they are less stiff and their deformation requires smaller forces.

In the present work, we present devices with similar design as in [14], but with improved response and a more detailed characterization. We also present new devices that were specifically designed to demonstrate the tunability of the conversion ratio. In these devices several different mechanisms are connected in series such that their relative performance can be accurately assessed.

DESIGN

A schematic view of our new conversion mechanism is illustrated in Fig. 2. The mechanism is constructed from beams with straight cross-sections only, and therefore is fully compatible with standard mass-fabrication processes. The basic building block of the mechanism is a ladder-shaped element with dual-height beams. Pairs of beams with different heights are fastened together by a small number of rigid connectors, as detailed in Fig. 2. The rigid connectors couple the in-plane and out-of-plane responses, and effectively create a slanted cross-section.

978-1-4799-7956-1/15 $31.00 © 2015 IEEE

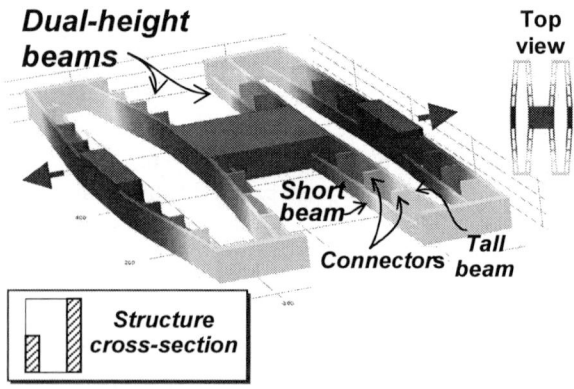

Figure 2: Conversion mechanism based on ladder-shaped elements with dual-height beams [14, 15]. All cross-sections are straight and therefore the mechanism is fully compatible with standard mass-fabrication techniques. In-plane forces marked by blue arrows induce out-of-plane displacement of the stage (red=elevation).

Comparison device

In previous work we demonstrated by simulations, that the conversion ratio is affected by the number of rigid connectors. The conversion ratio increases with the number of connectors, and saturates when a sufficient number is used [14]. In this work, in order to comparatively quantify the performance of mechanisms with different number of connectors, we designed and fabricated a special comparison device (Fig. 3). In this device all mechanisms share a common main shuttle, such that their in-plane motion is precisely the same.

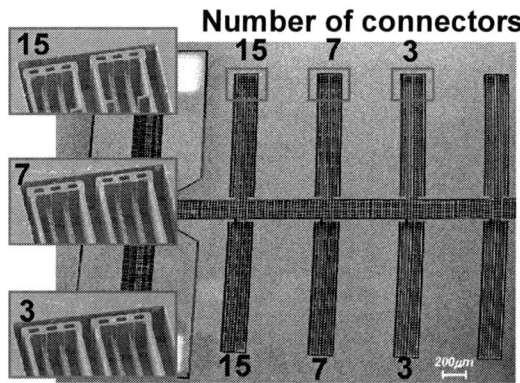

Figure 3: SEM image of a new test device designed to demonstrate the tunability of the conversion ratio. All conversion structures are connected to the same main shuttle. Each adjacent pair of conversion structures has the same number of connectors. In each pair one tip moves up and the other moves down (slight differences between the insets). The device is symmetric by design such that no tilting motion is induced on the main shuttle.

The device consists of pairs of open-ended structures, each with a different number of rigid connectors. All the structures are connected to the same main shuttle. When an in-plane motion is induced to the main shuttle, the tips of the beams curl out-of-plane. The pairs are designed

such that one tip moves up and the other down (see insets in Fig. 3). In this way it is easier to identify tip motions, and it is possible to identify common bias motions.

Parallel out-of-plane motion of a flat stage

A second type of device is presented in Fig. 4. This is the same type that is schematically shown in Fig. 2. In this device, four conversion mechanisms are connected to the corners of a flat central stage. When the comb-drives are actuated in opposite directions, they induce in-plane forces on the mechanisms, which are converted to out-of-plane motion. The stage can be either elevated or lowered (depending on the forces applied and the mechanism design). The devices were fabricated in a single-sided SOI process, as detailed in the following. The stage perforation was necessary for releasing the sacrificial BOX. A back-side opening would have enabled a solid stage.

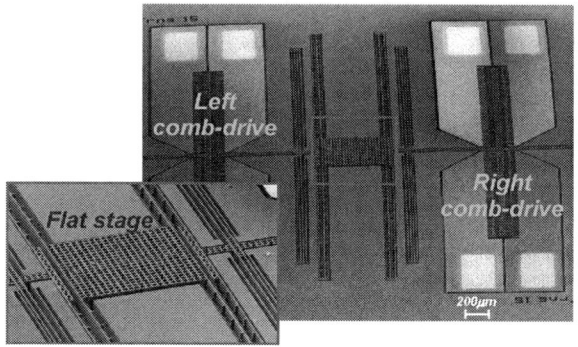

Figure 4: SEM image of a device of the type presented in Fig. 2, with 15 connectors in each ladder-shaped element. The central stage is connected to the conversion structures at the corners, and is uniformly elevated when in-plane forces are applied by the comb-drives.

FABRICATION PROCESS

A standard one-sided SOI process was used, which is fully compatible with mass-fabrication. The process flow includes an additional step for reducing the height of the short beam [16, 17], and includes a method for eliminating miss-alignment. The main steps of the fabrication process are illustrated in Fig. 5.

The starting material is an SOI wafer with a thick BOX layer of 4μm and a device layer of 20 μm (Fig. 5a). We start with SiO2 hard-mask deposition using PECVD (Fig. 5b). The hard-mask is etched in two RIE steps. In the first step (Fig. 5c), we implement our strategy for eliminating the effects of possible miss-alignment. 10μm wide openings are etched where we will later define the recessed beams, even though all beams are only 5μm wide. This allows us to accommodate a misalignment of up to 2.5μm in the next lithography step, where the device is defined by the second hard-mask etching step (Fig. 5d). Now, the Silicon device layer is etched using DRIE milling (Fig. 5e). After etching through the device layer, the photo resist is stripped (Fig. 5f) leaving the beams we wish to recess unprotected in the final short Silicon etch (Fig. 5g). The top layer of Oxide is removed and gold pads are evaporated using a shadow mask (Fig. 5h). In the final step the sacrificial BOX is etched using vapor HF (Fig. 5i) to release the structures.

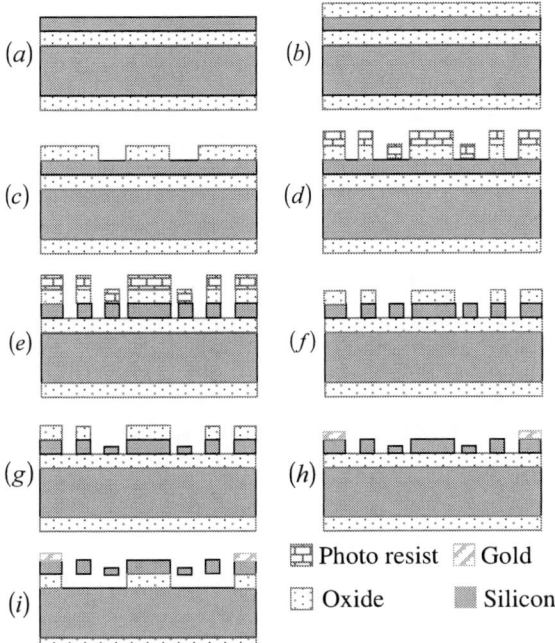

Figure 5: Main steps of the fabrication process. (a) Starting material - SOI wafer with a thick BOX layer of 4μm. (b) Oxide deposition - PECVD. (c) First lithography and SiO₂ etch - RIE. (d) Second lithography and SiO₂ etch - RIE. (e) Silicon etch - DRIE. (f) Photoresist stripping. The entire device is protected by the hard mask except for the recessed beams. (g) Short DRIE step for producing the recessed beams. (h) Gold evaporation for metal pads. (i) BOX removal – vapor HF.

Figure 6 is a SEM image of the fabricated device, the inset shows a close-up of the dual-height beam structure. In this image the tall beam height is 20μm and the short beam height is 8μm, this is an optimal ratio according to an extensive analysis which will be detailed elsewhere.

Figure 6: SEM photo of a fabricated device, the inset shows a close-up of the dual-height beam elements. The device layer thickness and tall beam height is 20μm, the short beam height is 8μm.

EXPERIMENT RESULTS
Comparison device

Figure 7 shows the relation between measured in-plane and measured out-of-plane motions of the comparison device. The relative motion of the tips of the conversion mechanism is shown, for different number of connectors. All measurements were performed using an optical profilometer. A typical measurement is shown in the inset of Fig. 7. All conversion structures have the same planform, but in each pair a different number of connectors is used (Fig. 3). It is evident that increasing the number of connectors increases the conversion ratio. These results show that the conversion ratio can be simply tuned by changing the number of connectors while maintaining the rest of the design parameters.

In addition, it is evident that the relation between in-plane and out-of-plane motion is linear. We attribute the slight deviations from linearity to inaccuracies in interpretation of the profilometer images.

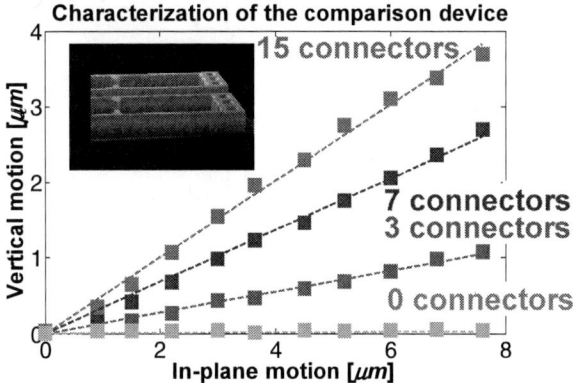

Figure 7: Measured relative out-of-plane motion between the tips of adjacent conversion mechanisms, shown in Fig. 3. It is evident that the conversion ratio can be tuned by changing the number of connectors. The relation between in-plane and out-of-plane displacement is linear. The inset shows a typical optical profilometer measurement.

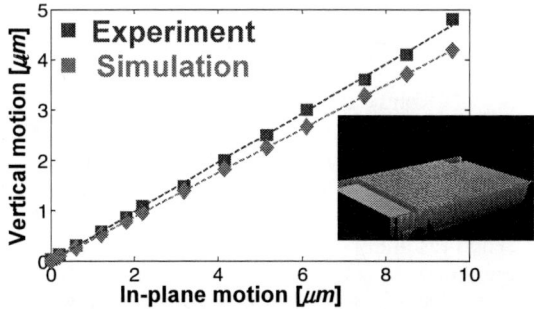

Figure 8: Experiment and simulation results of the device shown in Fig. 4. with 7 connectors. There is a linear relation between in-plane and out-of-plane displacements. The slight difference between experiment and simulation is due to fabrication inaccuracies. The inset shows a typical optical profilometer measurement.

978-1-4799-7956-1/15 $31.00 © 2015 IEEE

Parallel stage

Figure 8 shows out-of-plane displacements of the flat-stage device of the type shown in Fig. 4. The inset shows a typical profilometer image. The measured results are marked by blue squares, and model predictions simulated with COMSOL_4.3 are marked by red diamonds. In this device 7 connectors were used in each ladder-shaped element. The measured displacements are 10% larger than the model predictions. This difference is partially due to geometrical differences between fabricated devices and model, and partially due to electrostatic levitation of the comb-drives as a result of fringing fields [18]. The measured conversion ratio is 1:2 - for an in plane displacement of 9.6 μm we achieve a 4.8 μm out-of-plane displacement.

SUMMARY

We present a motion conversion mechanism based on beams with two different heights. All the cross sections in the design are straight and therefore the mechanism is fully compatible with standard mass-fabrication techniques. The device offers a linear relation between in-plane and out-of-plane motions. The conversion ratio can be easily tuned by simply changing the number of rigid connectors between the beams.

In future work we hope to demonstrate simultaneous motion of a solid stage, in both in-plane and out-of-plane directions. Also, we intend to adapt the design to achieve well-controlled tilting motions using in-plane actuation.

ACKNOWLEDGEMENTS

This study was partially supported by the Russell Berrie Nanotechnology Institute (RBNI), and by the Micro and Nano Fabrication Unit (MNFU) at the Technion, Israel Institute of Technology.

REFERENCES

[1] S. Kwon and L. P. Lee, "Stacked two dimensional micro-lens scanner for micro confocal imaging array," *IEEE-MEMS 2002*. Las-Vegas, 2002.

[2] S. Kwon, V. Milanovic, and L. P. Lee, "Large-displacement vertical microlens scanner with low driving voltage," *IEEE Photonics Technology Letters*, **14**, 1572-1574, 2002.

[3] S. H. Kong, D. D. L. Wijngaards, and R. F. Wolffenbuttel, "Infrared micro-spectrometer based on a diffraction grating," *Sensors and Actuators A: Physical*, **92**, 88-95, 2001.

[4] D. Lee, U. Krishnamoorthy, K. Yu, and O. Solgaard, "High-resolution, high-speed microscanner in single-crystalline silicon actuated by self-aligned dual-mode vertical electrostatic combdrive with capability for phased array operation," *IEEE Transducers 2003*, Boston, 2003.

[5] A. P. Lee, C. F. McConaghy, G. Sommargren, P. Krulevitch, and E. W. Campbell, "Vertical-actuated electrostatic comb drive with in situ capacitive position correction for application in phase shifting diffraction interferometry," *JMEMS*, **12**, 960-971, 2003.

[6] S.-W. Chung and Y.-K. Kim, "Design and fabrication of 10X10 micro-spatial light modulator array for phase and amplitude modulation," *Sensors and Actuators A: Physical*, **78**, 63-70, 1999.

[7] J. A. Pelesko and D. H. Bernstein, Modeling MEMS and NEMS. Boca Raton, FL: Chapman & Hall/CRC, 2003.

[8] D. Elata, "Modeling the Electromechanical Response of Electrostatic Actuators," in MEMS/NEMS : handbook techniques and applications, *Sensors and Actuators*, C. T. Leondes, Ed. New York: Springer, 2006, **4** , 93-119.

[9] W. C. Tang, T.-C. H. Nguyen, and R. T. Howe, "Laterally Driven Polysilicon Resonant Microstructures," *Sensors and Actuators*. **20**, 25-32, 1989.

[10] C. Marxer, O. Manzardo, H. P. Herzig, R. Dandliker, and N. De-Rooij, "An Electrostatic actuator with large dynamic range and linear displacement-voltage behavior for a miniature spectrometer," *IEEE Transducers 1999*, Sendai, Japan, 1999.

[11] Y. Ando, T. Ikehara, and S. Matsumoto, "Design, fabrication and testing of new comb actuators realizing three-dimensional continuous motions," *Sensors and Actuators A: Physical*, 2002.

[12] Y. Ando, "Fabrication and testing of three-dimensional stages providing displacement of up to 8um," presented at Solid-State Sensors, Actuators and Microsystems. *IEEE Transducers 2005*, Seoul, Korea, 2005.

[13] J. M. Gere and S. Timoshenko, Mechanics of materials, 3rd ed. Boston: PWS-KENT Pub. Co., 1990.

[14] I. Hotzen, O. Ternyak, S. Shmulevich and D. Elata, "Selective stiffening for producing motion conversion mechanisms," *Eurosensoes 2014*, Brecsia, Italy, 2014.

[15] D. Elata and A. Hirshberg, "Motion conversion mechanisms." Patent Pending, 2013.

[16] Y. Mita, M. Mita, A. Tixier, J.-P. Gouy, and H. Fujita, "Embedded-mask-methods for mm-scale multi-layer vertical/slanted Si structures," *IEEE MEMS 2000*, Miyazaki, Japan, 2000.

[17] V. Milanovic, "Multilevel beam SOI-MEMS fabrication and applications," *Electronics, Circuits and Systems*, 2002.

[18] W. C. Tang, M. G. Lim, and R. T. Howe, "Electrostatic Comb Drive Levitation And Control Method," *JMEMS*, **1**, 170-178, 1992.

MONOLITHIC INTEGRATION OF MICRO MAGNETIC PILLAR ARRAY WITH ANISOTROPIC MAGNETO-RESISTIVE (AMR) STRUCTURE FOR OUT-OF-PLANE MAGNETIC FIELD DETECTION

Wei-Ming Lai[1], Fu-Ming Hsu[1], Wei-Lun Sung[1], Rongshun Chen[1], and Weileun Fang[1,2]
[1]Power Mechanical Eng. dept., [2]NEMS Inst., National Tsing Hua University, HsinChu, TAIWAN

ABSTRACT

A novel integrate micro magnetic pillar array with anisotropic magneto-resistive (AMR) structure for out-of-plane magnetic field detection has been proposed and demonstrated. The merits of this study are: (1) The magnetic pillar converts magnetic field from out-of-plane to in-plane and provide magnetic gain (G) where G≈1/N (N:demagnetization factor) [1]; (2) The micro magnetic pillar array can be batch fabricated and defined at appropriate locations, and further monolithically integrated with AMR sensor; and (3) The magnetic conversion efficiency can be easily further enhanced by increasing the density of magnetic pillar array. The AMR sensor with different magnetic pillar density has been implemented and measurements indicate that the AMR sensor with larger density of magnetic pillar array has better sensitivity and resolution for detecting out-of-plane magnetic field.

INTRODUCTION

Presently, the magnetic sensors are of importance to detect magnetic field and further extract information from sensing signals. The magnetic sensors of different dynamic ranges and resolutions have been investigated for various applications [2-3] (e.g., electronic compasses, GPS navigation, position sensing, current detection, non-contact switch, etc.). Magnetic sensors with different sensing mechanisms have been reported, such as anisotropic magneto-resistive (AMR) magnetometers [4], Hall devices, [5], fluxgate magnetometer [6], and magneto-impedance (MI) magnetometer [7]. The AMR magnetometers have the advantages of simple operation principle. The external magnetic field will change the magnetization and resistance of AMR structures so that high sensitivity in low magnetic field is achieved. Thus, the AMR magnetic sensors have found many applications in electronic compasses, angular position sensors, reading heads etc. [3].

Due to the shape anisotropy effect, the AMR sensors generally are thin film structure which only sensitive to the in-plane magnetic field. There are some approaches like structure design and bias design to increase the performance of AMR sensor [8-9]. It remains challenge to detect out-of-plane magnetic field using AMR film. Some out-of-plane AMR sensors have been reported [10-12]. However, the process is not straightforward to deposit AMR film on anisotropic-etched V-shaped groves [10].The alignment and size limit are critical concerns for the approaches to assemble bulk permalloy plate aside AMR element [11], and bulk magnetic flux concentrator above AMR sensor [12]. Thus, this study monolithically integrates the AMR film with micro magnetic pillar array using micro fabrication processes to realize out-of-plane AMR sensor.

The proposed AMR sensor design exploits the photolithography technology to pattern photoresist as the mold for electroplating. The nickel pillar array is then electroplated as the magnetic concentrator. Through this method, the magnetic pillar array can be monolithically integrated with AMR element at proper places to effectively convert the magnetic field to implement the AMR sensor for out-of-plane field detection.

DESIGN CONCEPT

Fig.1a shows the design concept, including micro magnetic pillar array as magnetic concentrator and AMR sensor of meandering shape for sensing the magnetic field from the hard axis of meandering shape. Simulation shows the magnetic pillar acting as magnetic concentrator to provide magnetic conversion near the bottom edge of pillar from out-of-plane magnetic field into in-plane one. Furthermore, the magnetic pillar provide a magnetic gain (G) and the gain is reciprocal to the demagnetization factor N which can be modulated by varying the aspect ratio of magnetic pillar [13] (i.e. G≈1/N = 4(H/D)/√π+1, where H is the height and D is the diameter of pillar). Therefore the magnetic gain G can be enhanced by increasing the aspect ratio of pillar H/D. Fig.1b shows five different pillar array designs for comparison. As the density of the micro magnetic array increases, the magnetic conversion efficiency will be enhanced. Moreover, due to low H/D, the efficiency of a whole magnetic plate is reduced.

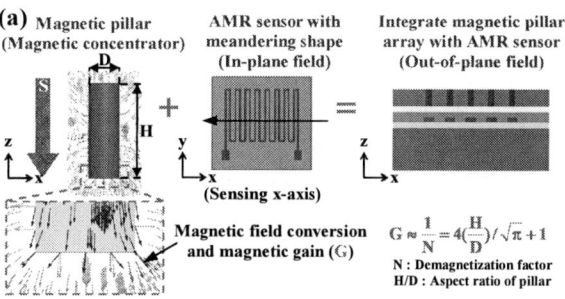

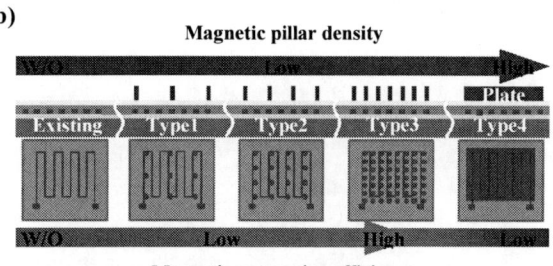

Figure 1: (a) Design concept of AMR sensor with micro magnetic pillar array, and (b) magnetic conversion efficiency with density of pillar array.

FABRICATION AND RESULTS

Fig.2 shows fabrication process steps. In Fig.2a, the AMR film of permalloy (NiFe) was evaporated and patterned as meandering shape on a silicon wafer with thermal oxide as insulation layer and annealed at 200°C for enhancing the AMR properties. Fig.2b shows the deposition of PE oxide as insulation layer and the Cu film as electroplating seed layer. In Fig.2c, the PR mold for electroplating was patterned by photolithography. The shape and location of the pillar array was precisely defined in this step. Fig.2d shows the Nickel pillar array was batch fabricated using the electroplating process. These Ni pillar array was monolithically integrated with AMR structure as magnetic concentrator.

Fig.3a shows four types of fabricated devices (as indicated in figure, existing, type1, type2, and type3) with different magnetic pillar density. In addition, the meandering AMR sensing element underneath can be observed in the top-left micrograph. Fig.3b shows the magnetic pillar number and area ration (pillar area per unit area) for these four designs. Fig.3c shows the SEM micrograph of magnetic pillar array after removing the PR mold. It shows the magnetic pillar array has been batch fabricated and properly arranged. The zoom-in micrograph in Fig.3d shows the single cylindrical pillar with diameter of 25μm and height of 28μm. The nickel pillar with vertical sidewall was successfully defined. Fig.3e shows the SEM micrograph of meandering shape AMR film, and Fig.3f indicates the zoom-in micrograph to display the microstructure of AMR thin film. The grain of the permalloy film was re-arranged after 200°C annealing. The grain growth and the decrease of grain boundary are clearly observed

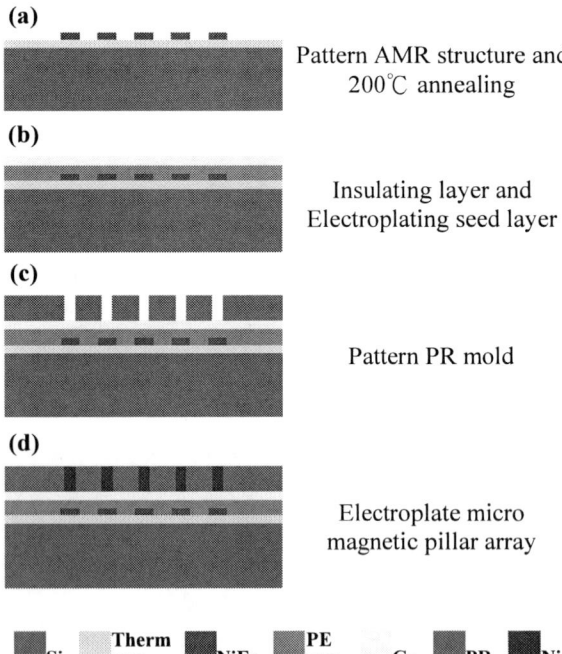

(a) Pattern AMR structure and 200℃ annealing

(b) Insulating layer and Electroplating seed layer

(c) Pattern PR mold

(d) Electroplate micro magnetic pillar array

Si　Therm SiO₂　NiFe　PE SiO₂　Cu　PR　Ni

Figure 2: Fabrication process steps, (a) Evaporate and pattern of NiFe (AMR) film, (b) deposition of insulating layer and electroplating seed layer, (c) pattern PR mold, and (d) electroplate Ni as micro magnetic pillar array.

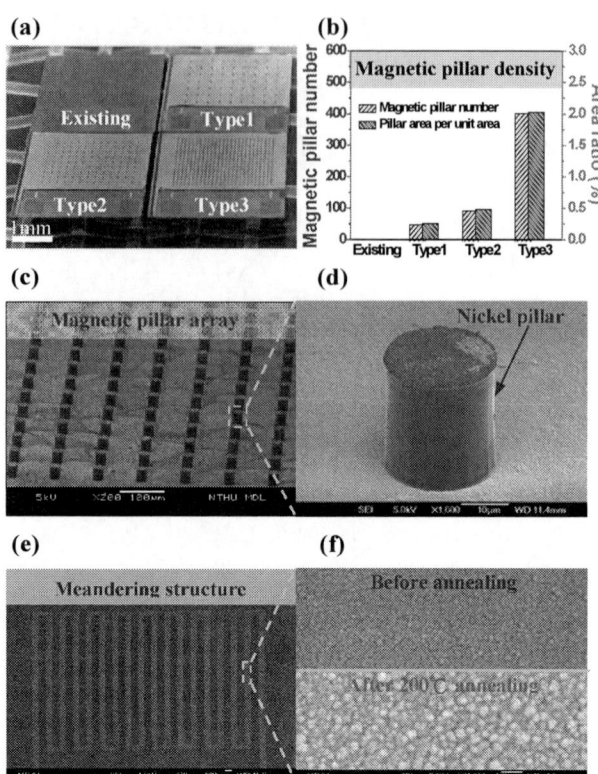

Figure 3: (a) Micrograph of fabricated AMR structure with different magnetic pillar array designs, (b) magnetic pillar number and area ratio (pillar area per unit area), (c-d) micrographs of magnetic pillar and array, (e) micrographs of AMR structure, and (f) microstructure of the AMR film.

MEASUREMENT RESULT

Firstly, this study performed the M-H curve testing for the existing device (only permalloy thin film structure on substrate) and the type1 device (permalloy thin film with nickel pillar array). Fig.4a shows the measurement setup of VSM (vibrating sample magnetometer) for magnetic property characterization. The M-H curves in Figs.4b-c indicate the magnetic saturation is increased in 3-axes for devices with magnetic pillar array and the remanence in 3-axes of type1 device are similar.

The resistance change of the AMR sensors associated with external applied magnetic fields was tested. Fig.5a shows the measurement setup in this study. The Helmholtz coil was employed to specify magnetic field on test device. The Gauss meter was used to calibrate the magnitude of magnetic field. Finally, the resistance change of the AMR sensors was measured by the source meter. Measurement results in Fig.5b show the calibration curves of in-plane magnetic field for different AMR sensors. The results indicate that the existing, type1, type2, and type3 devices have similar performance. In short, the resistance of these designs decreases as the external magnetic field increases, and a near 1% resistance change is achieved before saturation. However, the type4 device with non-patterned nickel plate has different trend of resistance change. Moreover, the measurements in Fig.5c indicate only devices with magnetic pillar can detect the out-of-plane magnetic field. The type3 sensor design with the largest number of magnetic pillars has the best sensing performance.

Test setup in Fig.6a was established to characterize the voltage change with applied magnetic fields. Again, magnetic field was specified by Helmholtz coil and monitored by Gauss meter. The input voltage was given by source meter. Finally, after magnified by the inverse OP amplifier, the output voltage was then detected by

(a)

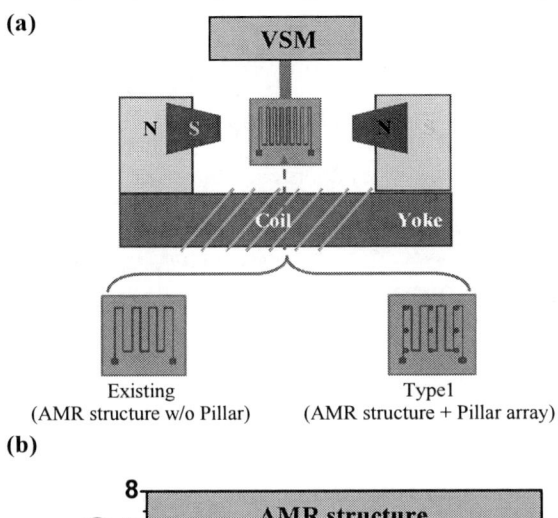

(b)

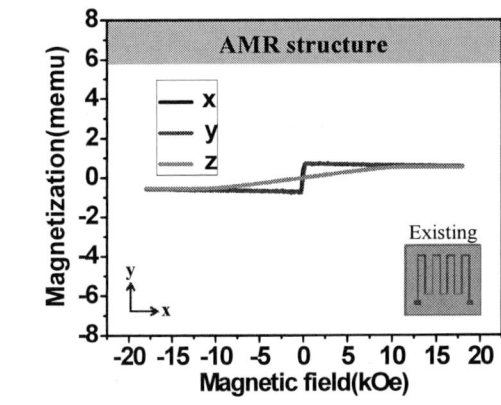

(c)

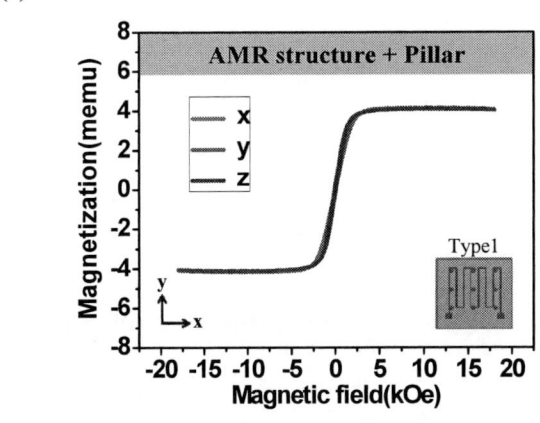

Figure 4: M-H curve measurement by VSM (a) The schema of measurement setup. (b) Existing(AMR structure). (c) Type1(AMR structure + Magnetic pillar array).

(a)

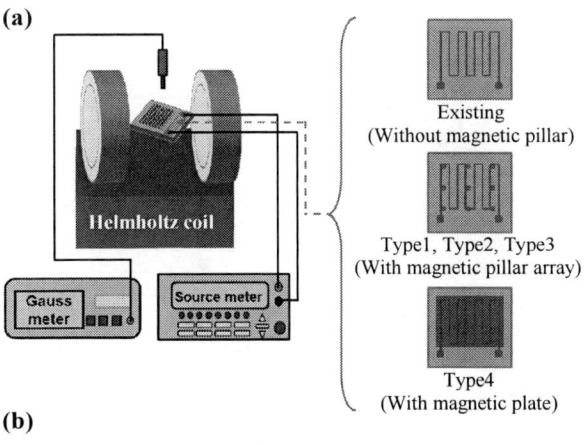

(b)

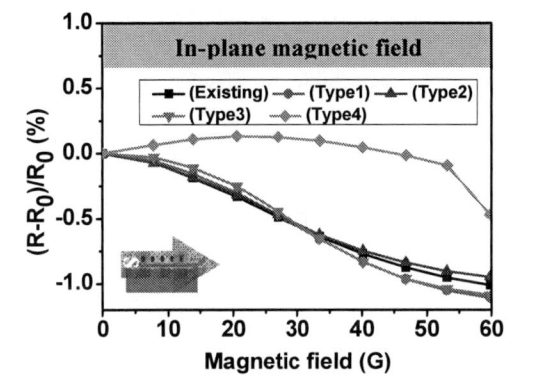

(c)

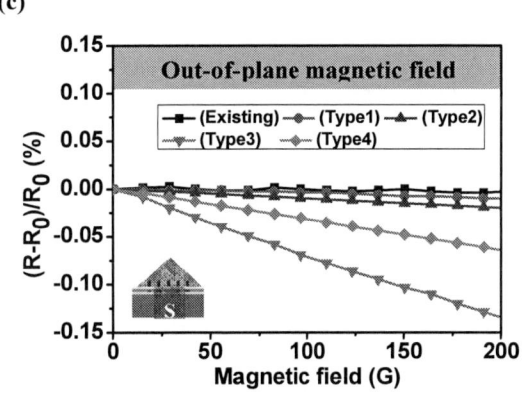

Figure 5: Resistance change measurement in applied magnetic field (a) The schema of measurement setup; (b) In-plane magnetic field measurement; (c) Out-of-plane magnetic field measurement

another source meter. Measurements in Fig.6b show the voltage change by out-of-plane magnetic field; and indicate that only the devices with magnetic pillar can detect the magnetic field. The result also indicates that the sensitivity of proposed sensor is improved by increasing the density of magnetic pillar array. Similarly, the resolution of proposed sensor has an enhancement with increase of magnetic pillar density. In all of the proposed design, type3 has the best performance, and even better than the type4 device with non-patterned magnetic plate as magnetic concentrator. The above results support the claims in Fig.1b.

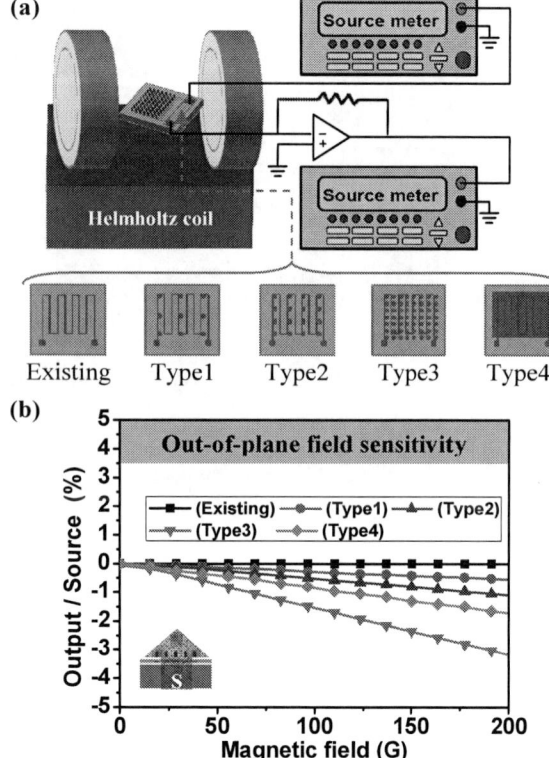

(a)

Source meter

Helmholtz coil

Source meter

Existing   Type1   Type2   Type3   Type4

(b)

Out-of-plane field sensitivity

- ■ (Existing) ● (Type1) ▲ (Type2)
- ▼ (Type3) ◆ (Type4)

Output / Source (%)

Magnetic field (G)

Figure 6: Voltage sensitivity measurement in applied magnetic field (a) The schema of measurement setup; (b) Out-of-plane magnetic field measurement

Table 1: Summary of the performance for different magnetic pillar array designs.

	Magnetic pillar	Ms (memu)	Sensitivity (mV/G)	Resolution ($G/Hz^{1/2}@1Hz$)
Exist	-	0.33	-	-
Type1	63	4.282	0.014	0.295
Type2	117	6.942	0.029	0.142
Type3	494	7.214	0.084	0.047
Type4	-	346.274	0.018	0.183

CONCLUSIONS

In summary, this study has successfully implemented the monolithic integration of micro magnetic pillar array with anisotropic magneto-resistive structure for out-of-plane magnetic field detection. Various types of the magnetic pillar density with different magnetic conversion efficiency have been demonstrated through photolithography technology. Measurements indicate that the magnetic pillar array can effectively convert the magnetic field and can successfully detect the out-of-plane magnetic field after integration with AMR structure. The sensitivity and resolution of the sensors for out-of-plane magnetic field will be enhanced with the increase of the density of the magnetic pillar array. The measurement result also shows magnetic pillar array has better magnetic conversion efficiency than a whole magnetic plate. It can further enhance the performance of the sensors by increasing the density of magnetic pillar array or the aspect ratio of magnetic pillar.

ACKNOWLEDGMENTS

This paper was partially supported by Nation Science Council, Taiwan, under contract 103N2061E1, NSC 101-2221-E-007-069-MY3, NSC 102-2221-E-007-027 -MY3 and NSC 103-2811-E-007-008-MY3. The authors would like to express his appreciation to the Nano Science and Technology Center of National Tsing Hua University in providing the fabrication facilities.

REFERENCES

[1] R. S. Popovic, P. M. Drljaca, C. Schott, R. Racz, "Integrated Hall Sensor/Flux Concentrator Microsystem", *MIDEM Conference*, 31 (2001)4, str. pp. 215-219.

[2] J. Lenz and A. S. Edelstein, "Magnetic Sensors and Their Applications", *IEEE Sensors Journal*, vol. 6, No. 3, pp. 631-646, 2006.

[3] R. S. Popovic, J. A. Flanagan, and P. A. Besse, "The future of magnetic sensors", *Sensors and Actuators A*, 56, pp. 39-55, 1996.

[4] S. Tumanski, *Thin Film Magnetoresistive Sensors*. Bristol, U.K.: Inst. Physics, 2001

[5] R. S. Popovic, *Hall Effect Devices*. Bristol, U.K.: Inst. Physics, 2004.

[6] P. Ripka, "Advances in fluxgate sensors," Sensors and Actuators A, vol. 106, pp. 8-14, 2003.

[7] M. Mohri, N. Bushida, M. Noda, H. Yoshida, L. V. Panina, and T. Uchiyama, "Magneto-Impedance Element", *IEEE Transactions on Magnetics*, vol. 31, No. 4, pp. 2455-2460, 1995.

[8] K.E. Kuijk, W.J. van Gestel and F.W. Gorter, "The Barber Pole, A Linear Magnetoresistive Head", *IEEE Transactions on Magnetics*, Vol. Mag-11, No. 5, pp. 1215-1217, 1975.

[9] D. J. Mapps, M. L. Watson, and N. Fry, "A Double Bifilar Magneto-Resistor for Earth's Field Detection", *IEEE Transactions on Magnetics*, Vol. Mag-23, No. 5, pp. 2413-2415, 1987.

[10] F. C. S. da Silva, S. T. Halloran, L. Yuan, and D. P. Pappas, "A z-component magnetoresistive sensor", *Applied Physics Letters*,92(2008)142502.

[11] M. Suzuki, T. Fukutani, T. Hirata, S. Aoyagi, and S. Shingubara, "**Triaxis Magnetoresistive (MR) Sensor Using Permalloy Plate of Distorting Magnetic Field**", *IEEE Conference MEMS*, pp. 671-674, 2010.

[12] Y. Cai, C. Byun, Y. Zhao, and L. Jiang, "Monolithic Tri-axis AMR Sensor and Manufacturing Method Thereof", *US Patent 0206137A1, 2012.*

[13] M. Sato, and Y. Ishii, "**Simple and approximate expression of demagnetizing factors of uniformly magnetized rectangular rod and cylinder**", *J. Appl. Phys.*, 66 (1989), pp. 983-985

CONTACT

*Weileun Fang, Department of PME/Institute of NEMS, National Tsing Hua University, Hsinchu, 30013, Taiwan, Tel: +886-3-574-2923; Fax: +886-3-573-9372; E-mail: fang@pme.nthu.edu.tw

MULTI-COLOR IMAGING WITH SILICON-ON-INSULATOR DIODE UNCOOLED INFRARED FOCAL PLANE ARRAY USING THROUGH-HOLE PLASMONIC METAMATERIAL ABSORBERS

Daisuke Fujisawa[1], Shinpei Ogawa[1], Hisatoshi Hata[1], Mitsuharu Uetsuki[1], Koji Misaki[1], Yousuke Takagawa[2], and Masafumi Kimata[2]

[1]Mitsubishi Electric Corporation, JAPAN
[2]Ritsumeikan University, Shiga, JAPAN

ABSTRACT

This paper reports a silicon-on-insulator diode uncooled infrared focal plane array (IRFPA) with through-hole plasmonic metamaterial absorbers (TH-PLMAs) for multi-color imaging with a 320×240 array format. Through-holes formed on the PLMA can reduce the thermal mass while maintaining both the single-mode and high absorption due to plasmonic metamaterial structures, which results in fast response and high responsivity. The detection wavelength of the PLMA with through-holes can be controlled over a wide range of the IR spectrum by varying the size of the micropatches on the top layer.

INTRODUCTION

Information such as the shape, position and average radiant intensity of objects can be obtained through a typical uncooled infrared (IR) sensor, while color (wavelength) information cannot. If an uncooled IR sensor with a wavelength selective function could be fabricated, then advanced uncooled IR sensors with color information and multicolor imaging could be developed, which would be advantageous for a wide range of applications, such as fire detection, gas analysis, medical imaging, and material recognition [1]. Typical methods for realizing wavelength selectivity include a narrow band-pass filter attached to a sensor [2], control of absorbing materials, multilayer structures [3], and optical resonant structures [4] that require gap or thickness control. These typical approaches cannot easily integrate different pixels in an array, which is a serious disadvantage for multicolor imaging.

Recently, we have realized wavelength or polarization selective uncooled IR sensors using microelectromechanical system (MEMS)-based thermopiles with two-dimensional plasmonic absorbers [5-8]. However, high-performance detectors and absorbers are highly desirable for uncooled thermal imaging.

In this study, we report multi-color imaging using a silicon-on-insulator (SOI) diode uncooled infrared focal plane array (IRFPA) with through-hole plasmonic metamaterial absorbers (TH-PLMAs) for multi-color imaging. The SOI diode is one of the most advanced uncooled IR sensors owing to its high responsivity and fast response [9]. Through-holes were formed on PLMAs to achieve reduced thermal mass and small volume. The effect of through-holes on wavelength selectivity was calculated theoretically. A fabrication procedure was developed for the SOI diode structure with TH-PLMAs using MEMS-based technology.

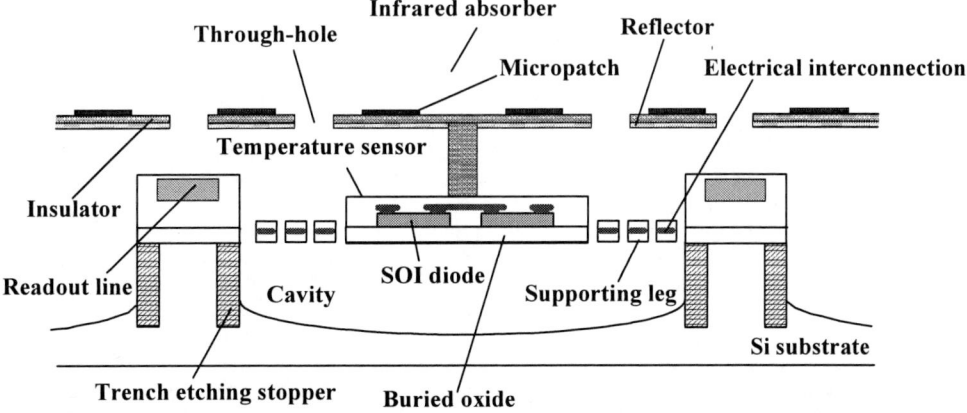

Figure 1: Cross-sectional pixel structure of the SOI diode uncooled infrared sensor using micropatch-based PLMAs with through-holes.

SENSOR DESIGN
SOI Diode Uncooled IRFPA

Figure 1 illustrates the cross-sectional pixel structure of the multi-color SOI diode uncooled IRFPA. The p-n junction diodes fabricated in the SOI layer are used as a temperature sensor. Figure 2 shows the structure of a conventional p^+n vertical diode. Conventionally, each p^+n vertical diode is formed between a p^+ diffusion layer and an n-body in each SOI active area, and ten diodes are serially interconnected in order to increase the temperature coefficient. The temperature sensor with SOI diodes is thermally isolated from the Si substrate by a cavity formed underneath and supporting legs including electrical interconnections [9]. The infrared absorber, which is constructed with a TH-PLMA, is located above the temperature sensor and the supporting legs. The detection principle of the temperature sensor is based on the absorption of incident infrared rays by the TH-PLMA, which causes a change in the SOI diode temperature. The temperature sensor is active when a constant current flows through the diodes under forward bias conditions. The temperature change in the SOI diodes is detected as a forward voltage change in the SOI diodes. The detection wavelength depends on the size of the micropatch, regardless of the array period.

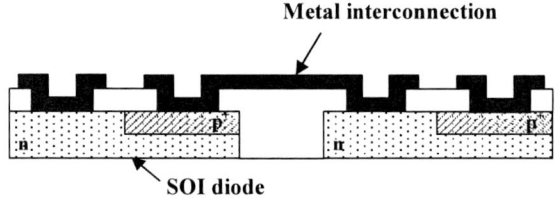

Figure 2: Schematic of SOI diode structure.

Through-hole Plasmonic Metamaterial Absorbers

Figure 3(a) illustrates one unit of TH-PLMAs. PLMAs are metal-insulator-metal (MIM) structures [10-12]. Aluminum (Al) is used as metal and SiO_2 is used as insulator in this study. Al is suitable for commercial use owing to its low cost. The diameter of the through-holes and micropatches is 1.5 μm. The periods of the through-holes and micropatches are 4.0 and 8.0 μm, respectively. Figure 3(b) shows calculated absorption spectra for PLMAs with and without through-holes. A strong unity absorption of over 90% is obtained for both structures, which means that the present configuration of through-holes should have less influence on absorption properties. Moreover, it indicates that the plasmonic resonance is strongly localized at the micropatches and that the waveguide mode is also strongly localized in the SiO_2 sandwiched between the micropatch and the metal directly beneath it. The influence of through-holes can be negligible for both resonance modes.

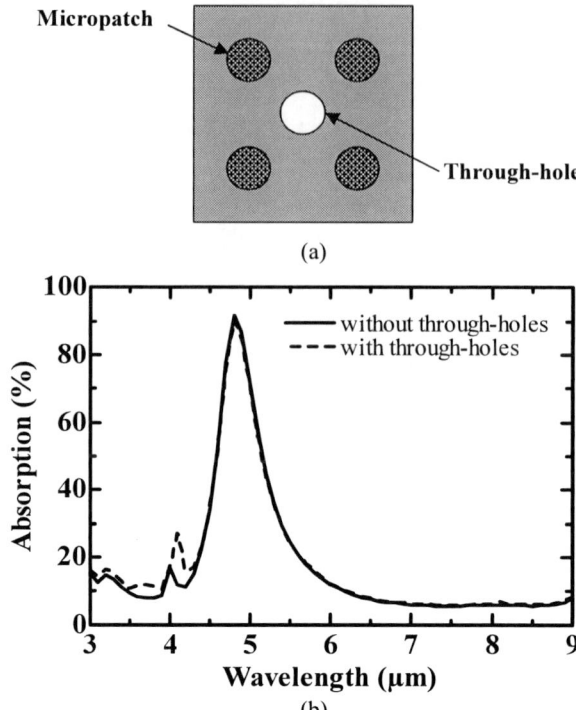

Figure 3: Calculated absorption of PLMA with and without through-holes.

FABRICATION

The pixel structure shown in Fig. 1 is fabricated with a fully dry bulk/surface combined micromachining process using organic sacrificial layers. The process flow is shown in Fig. 4. Each pixel has a trench etching stopper surrounding it, fabricated using the Si-LSI process for the readout circuits. After the fabrication of SOI diodes and local interconnections, the first etching hole, which reaches the surface of the Si substrate, is opened to form thermal isolation legs (Fig. 2(a)). Following the formation of the infrared absorber, the organic sacrificial layer is coated and annealed. Next, the micropatch-based PLMA is located over the temperature sensor and the supporting legs (Fig. 2(b)). Then, the infrared absorber with through-holes, which connects to the temperature sensor area with the pillar, is formed. Subsequently, the etching hole, which reaches the surface of the Si substrate, is opened near the center of the pixel. Next, a dry etching process using XeF_2 gas through the etching hole forms the cavity in the Si substrate. Since the etching process proceeds isotropically, the cavity takes on a concave shape, which enlarges the margin to the depth of the trench etching stopper. The buried oxide protects the SOI diodes from Si etching (Fig. 2(c)). Finally, the organic sacrificial layers are removed using O_2 plasma ashing (Fig. 2(d)).

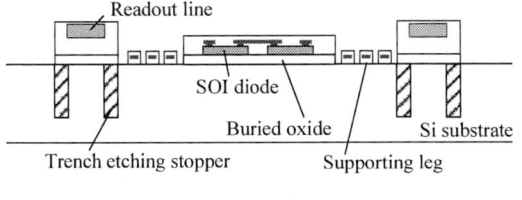

(a)

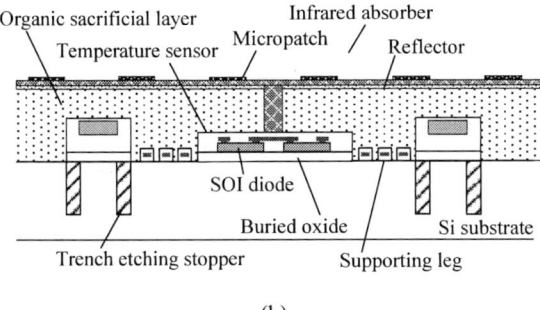

(b)

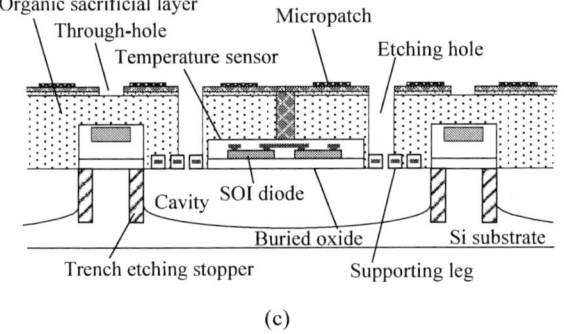

(c)

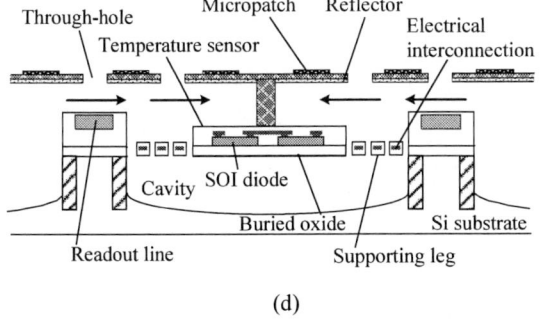

(d)

Figure 4: Process flow of the SOI diode uncooled infrared sensor using micropatch-based PLMA with through-holes.

To develop the multicolor SOI diode uncooled IRFPA using TH-PLMAs, a test chip was fabricated and assessed. Figure 5 shows a scanning electron microscopy (SEM) image of the TH-PLMA. The thermally isolated structure was successfully formed. Figure 6 shows a photograph of the fabricated IRFPA with a size of 20.0×19.0 mm.

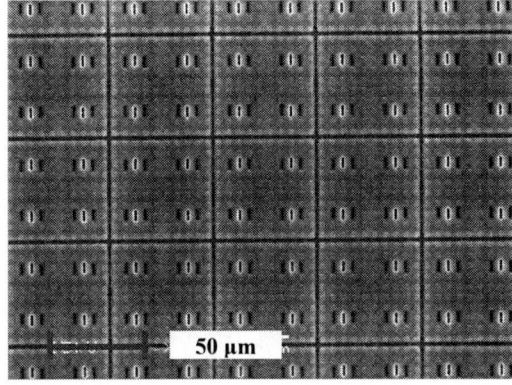

Figure 5: SEM micrograph of through-hole PLMA.

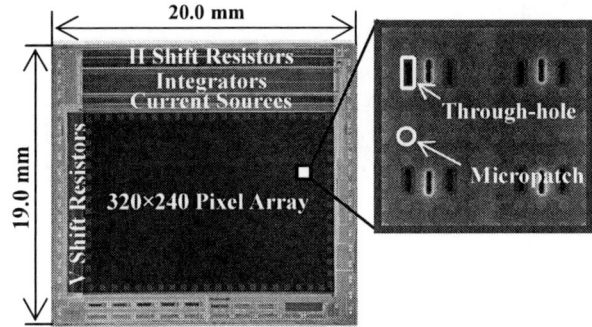

Figure 6: Photograph of 50 µm pixel pitch 320×240 SOI diode IRFPA with micropatch-based PLMA.

MEASUREMENT

To verify the proposed TH-PLMAs, we fabricated micropatch-based PLMAs. Figure 7 shows the reflection spectrum obtained for these PLMAs. A sufficient absorption of over 90% was obtained with 1.8-µm-diameter micropatches. The absorption wavelength was controlled by the size of the micropatch.

Furthermore, a 50 µm pixel pitch 320×240 SOI diode uncooled IRFPA was fabricated and favorable FPA operations were successfully verified. The diameter of the micropatch was about 1.2 µm.

978-1-4799-7956-1/15 $31.00 © 2015 IEEE

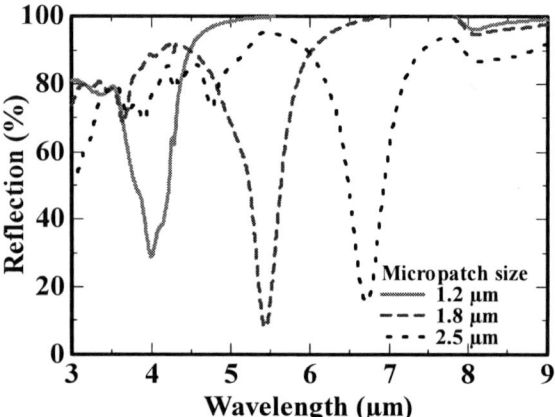

Figure 7: Measured reflection spectra for the micropatch-based PLMA.

Figure 8 shows the normalized responsivity of the developed sensor. The sensor has strong responsivity peaks in the MWIR region.

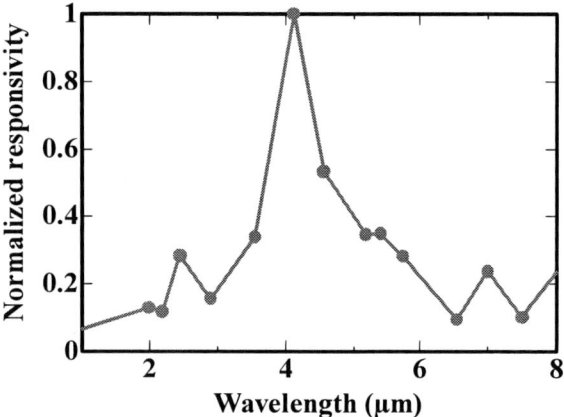

Figure 8: Spectral responsivity of the developed sensor using micropatch-based PLMAs with through-holes.

CONCLUSIONS

We have developed a SOI diode uncooled IRFPA with TH-PLMAs in order to fabricate a multicolor image sensor. Measurement of the spectral responsivity demonstrated that selective enhancement is achieved according to the TH-PLMA. The detection wavelength was governed by the micropatch size. The results obtained here will contribute to the development of novel multicolor imaging for IR sensors.

In conclusion, the proposed and developed SOI diode uncooled IRFPA with TH-PLMAs proved highly promising as a multicolor image sensor.

ACKNOWLEDGEMENTS

The authors thank Yasuhiro Kosasayama, Masashi Ueno, Takahiro Ohnakado, and Tetsuya Satake of the Advanced Technology R&D Center of Mitsubishi Electric Corporation for their helpful discussions and assistance.

REFERENCES

[1] M. Vollmer and K.-P. Mollmann, *Infrared Thermal Imaging: Fundamentals, Research and Applications.* Weinheim: Wiley-VCH, 2010.

[2] R. Haidar, G. Vincent, S. Collin, N. Bardou, N. Guérineau, J. Deschamps, *et al.*, "Free-standing subwavelength metallic gratings for snapshot multispectral imaging," *Appl. Phys. Lett.*, 96, (2010).

[3] S. W. Han, J. W. Kim, Y. S. Sohn, and D. P. Neikirk, "Design of infrared wavelength-selective microbolometers using planar multimode detectors," *Elec. Lett.*, 40, (2004), 1410.

[4] R. P. Shea, A. S. Gawarikar, and J. J. Talghader, "Midwave thermal infrared detection using semiconductor selective absorption," *Opt. Exp.*, 18, (2010), 22833.

[5] S. Ogawa, K. Okada, N. Fukushima, and M. Kimata, "Wavelength selective uncooled infrared sensor by plasmonics," *Appl. Phys. Lett.*, 100, (2012), 021111.

[6] S. Ogawa, J. Komoda, K. Masuda, and M. Kimata, "Wavelength selective wideband uncooled infrared sensor using a two-dimensional plasmonic absorber," *Opt. Eng.*, 52, (2013), 127104.

[7] S. Ogawa, K. Masuda, Y. Takagawa, and M. Kimata, "Polarization-selective uncooled infrared sensor with asymmetric two-dimensional plasmonic absorber," *Opt. Eng.*, 53, (2014), 107110.

[8] K. Masuda, S. Ogawa, Y. Takagawa, and M. Kimata, "Optimization of Two-Dimensional Plasmonic Absorbers Based on a Metamaterial and Cylindrical Cavity Model Approach for High-Responsivity Wavelength-Selective Uncooled Infrared Sensors," *Sens. Mater.*, 26, (2014), 215.

[9] D. Fujisawa, T. Maegawa, Y. Ohta, Y. Kosasayama, T. Ohnakado, H. Hata, *et al.*, "Two-million-pixel SOI diode uncooled IRFPA with 15μm pixel pitch," presented at the SPIE, Baltimore, Maryland, USA 2012.

[10] H. Tao, N. I. Landy, C. M. Bingham, X. Zhang, R. D. Averitt, and W. J. Padilla, "A metamaterial absorber for the terahertz regime: Design, fabrication and characterization," *Opt. Exp.*, 16, (2008), 7181.

[11] X. Liu, T. Starr, A. F. Starr, and W. J. Padilla, "Infrared Spatial and Frequency Selective Metamaterial with Near-Unity Absorbance," *Phys. Rev. Lett.*, 104, (2010), 207403.

[12] T. Maier and H. Brueckl, "Multispectral microbolometers for the midinfrared," *Opt. Lett.*, 35, (2010), 3766.

PASSIVE WIRELESS TEMPERATURE SENSING WITH PIEZOELECTRIC MEMS RESONATORS

Hediyeh Fatemi, Mohammad J. Modarres-Zadeh, and Reza Abdolvand
University of Central Florida, Orlando, Florida, USA

ABSTRACT

For the first time, a lateral-extensional piezoelectric MEMS resonator is utilized to enable wireless passive temperature sensing with an accuracy of less than 0.1°C at 1m distance. The high quality factor (2900) and low motional resistance (147Ω) of a 990MHz thin-film piezoelectric-on-silicon (TPoS) resonator are exploited to accurately determine the temperature of the subject by monitoring the induced change in the resonance frequency. The resonator is directly connected to an antenna and is excited wirelessly by a pulse-modulated sinusoidal signal. The decaying signal from the resonator is then recorded and the resonance frequency is determined from the frequency spectrum of the resonator's decaying response. The wireless temperature measurement was extended to a 3m distance with an interrogation signal power of 500mW using 5dBi gain antennas.

INTRODUCTION

Resonant sensors have become a common solution for sensing many physical parameters such as pressure, temperature, viscosity, mass, etc. The primary advantage of resonant sensors is that their output is a frequency or phase change and can be directly fed into digital circuits unlike analogue sensors [1].

Wireless temperature sensors are growing in popularity in industrial, commercial, healthcare, and even residential settings [2]. Furthermore, there is a high demand for passive wireless temperature sensing in a variety of applications like inaccessible locations and hazardous environments where long lifetime is essential [3] or harsh environments where battery lifetime is excessively shortened. The first remote temperature sensors introduced by Bao *et al* in 1987, employed surface acoustic wave (SAW) devices [4]. Since then a variety of such devices have been developed and are widely used [3, 5, 6].

Conventionally, piezoelectric crystals such as LiNbO$_3$ or quartz were exploited to fabricate resonant temperature sensors [7]. The high temperature sensitivity of specific crystalline cuts of quartz has enabled sensors with very high resolution (0.001°C) [8]. The drawback, however is their relatively large size and incompatibility of their fabrication process with the mature silicon micro-fabrication technology. In this sense, silicon micromachined sensors are of particular interest due to their small size and the potential for high-volume production at very low cost [9].

Temperature-dependent quartz, SAW, and MEMS resonators have all been incorporated in oscillator circuits to yield temperature sensors with very high resolution [10, 8, 11]. The drawback of an oscillator-based sensor is the relatively large power dissipated in the circuit compared to the power dissipated in the low-loss resonator [1]. On the other hand, the electronic circuitry may limit the performance of the sensor by increasing the noise floor.

In summary, the most important characteristics of a resonator that yields to an efficient passive wireless temperature sensor are as follows: (I) There is no need for bias voltages for operation of the resonator to simplify the passive operation, (II) The temperature-dependency of the resonant frequency has to be substantial to guarantee high resolution, (III) The motional resistance of the resonator has to be small in order to minimize the power loss in the resonator and consequently increase the sensing distance, (IV) The quality factor (Q) of the resonator should be high to improve the signal to noise ratio and consequently achieve a higher temperature resolution.

Piezoelectrically-transduced MEMS resonators do not require a bias voltage to operate and offer larger electromechanical coupling coefficient relative to that of electrostatic resonators, which ultimately results in a lower motional resistance [12, 13]. Thin-film piezoelectric-on-silicon (TPoS) resonators are a subcategory of piezoelectric resonators that exhibit high quality factors [14]. The resonant frequency of TPoS resonators is defined by the lateral dimensions of the resonator which enables fabrication of a wide frequency range on a single substrate. Furthermore, the temperature coefficient of frequency (TCF) can be engineered by changing the orientation of the resonator or doping the silicon layer [15]. Correspondingly, TPoS technology is a suitable candidate for implementing passive wireless temperature sensors.

While thin-film piezoelectric MEMS resonators offer smaller size and lower manufacturing/packaging cost compared to their SAW counterparts, no sensor system has been reported that utilizes MEMS resonators in a passive or a wireless configuration. This work is the first to demonstrate a passive wireless temperature sensor using a 990MHz TPoS resonator.

The paper is organized as the following. The next section will discuss the TPoS resonator and its frequency and temperature characteristics. The measurement setup and the sensor readout scheme will be explained afterwards followed by the experimental results. Discussion and conclusion will be in the succeeding sections.

TPOS RESONATOR

The TPoS resonator comprises a thin piezoelectric film sandwiched between two metal layers stacked on top of a silicon film (Figure 1a). Readers can refer to [16] for the detailed fabrication process. The suspended resonator is anchored to the substrate by tethers placed at zero-displacement nodes. The top metal is patterned so that the top electrodes mimic the stress field in the structure at a targeted resonant mode as depicted in Figure 1. The piezoelectric film is actuated by the electric field applied between the two metal layers and consequently an acoustic wave is launched in the bulk of the device.

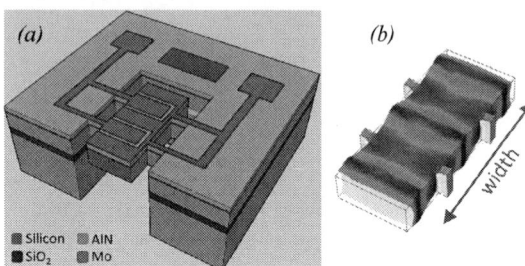

Figure 1: (a) The schematic of a 5th-order TPoS resonator and (b) the stress profile of a silicon slab with two pairs of fixed tethers at the 5th-order width-extensional mode.

TPoS resonators offer high Q and low motional resistance all in a very small footprint [12, 14]. These resonators can be designed in either one-port or two-port configuration. In a two-port design, the resonator is excited through one of the ports and the other port is used to sense the vibration. Whereas in a one-port configuration, both excitation and sensing are done through one of the ports while the second port can be exploited to tune the resonance frequency [17].

Figure 2 shows the frequency response of a 21st-order resonator used in this work which exhibits Q of 2900 in air and motional resistance of 147Ω at 990MHz. The impulse response of a resonator decays exponentially ($e^{-t/\tau}$) where the time constant τ is equal to $\frac{2Q}{\omega_0}$ by definition [18]. This implies that the envelope of the response decreases by %98 in 4τ seconds. The high Q of this resonator results in a relatively long decay time of 4μs.

The sensitivity of the resonator to temperature is a function of device's TCF:

$$f = (TCF . \Delta T + 1) . f_0 \qquad (1)$$

which is in turn a function of the stack materials and could be approximated according to equation (2):

$$TCF_{total} = \frac{TCF_{Mo}.t_{Mo} + TCF_{AlN}.t_{AlN} + TCF_{Si}.t_{Si}}{t_{Mo} + t_{AlN} + t_{Si}} \qquad (2)$$

where t denotes the corresponding film thickness.

In order to realize a temperature sensor with high resolution, large TCF values are desirable. This is achieved by using an un-doped or very lightly-doped silicon as the

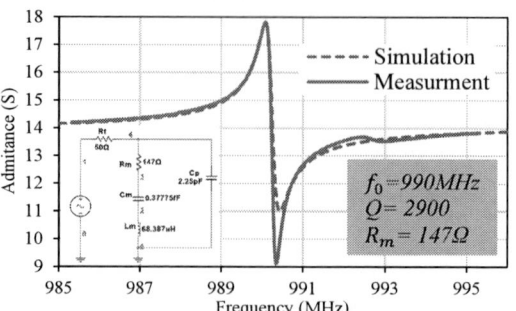

Figure 2: One-port frequency characteristics of the TPoS resonator. The inset is the equivalent circuit model.

starting substrate [15]. The resonator of this work is fabricated on a 5μm Boron-doped silicon substrate with fairly low resistivity of 1-20 Ω.cm. To characterize the resonator's TCF, it was placed in a temperature-controlled chamber and heated up from room temperature to +85°C in 5°C increments while the resonance frequency was recorded using a network analyzer (R&S-ZNB8) in a one-port configuration. A linear TCF of $-30\,^{ppm}/_{°C}$ was measured for this temperature range.

WIRELESS MEASUREMENT SETUP

To utilize a resonator as a wireless temperature sensor, the resonator is interrogated by a pulse-modulated sinusoidal signal which forces the resonator into oscillation. Next, the resonator decay response is captured and evaluated to determine the temperature.

A schematic view of the setup is shown in Figure 1. The RF signal generator transmits a pulse-modulated sine wave through a 3dBi antenna. The device under test (DUT) is interrogated during the pulse on-time and its response is analyzed during the off-time. The DUT, directly connected to a 5dBi antenna with no additional circuitry, is forced into oscillation upon receiving the interrogation signal. The interrogation period (on-time) is long enough for the resonator to reach its maximum oscillation amplitude. If the frequency of the sine wave is equal to the natural resonance frequency of the resonator, the amplitude of oscillations will be maximized. The frequency of the sine wave can be dynamically tuned during the interrogation to match the resonance frequency in order to maximize the amplitude of the resonator response and thus guarantee a larger signal to noise ratio. At the end of the pulse, the resonator's oscillation amplitude starts to decay over a period of 4τ as explained in the previous section. An oscilloscope with a high sampling rate (R&S-RTO1024) is used to collect the signal in real time (see the signal received at the oscilloscope antenna in Figure 1). A 5dBi antenna is connected to the oscilloscope. Figure 2 exhibits the waveform received by the oscilloscope showing the interrogation signal and the resonator response. Note that the much stronger interrogation signal is directly received from the RF generator during the on-time.

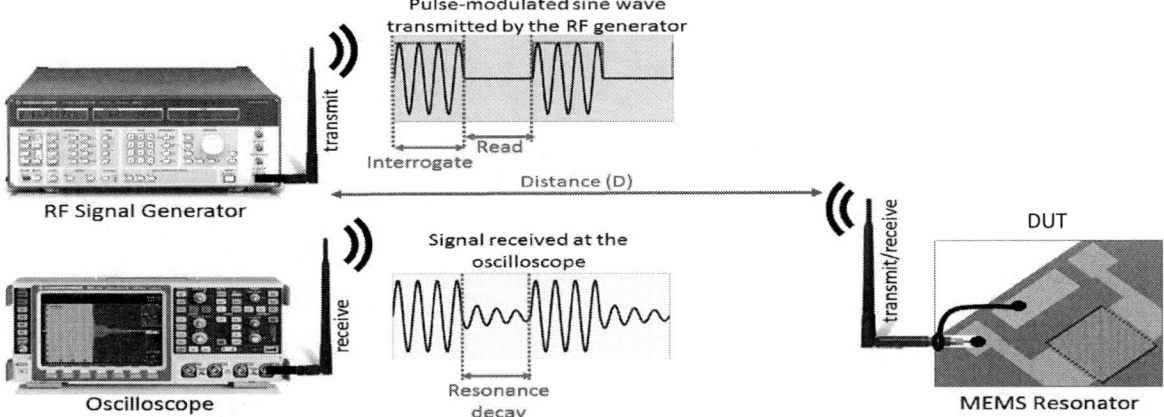

Figure 1: Measurement setup of the wireless temperature sensor. The MEMS resonator on the right is a high-order multi-tether design and is directly connected to an antenna.

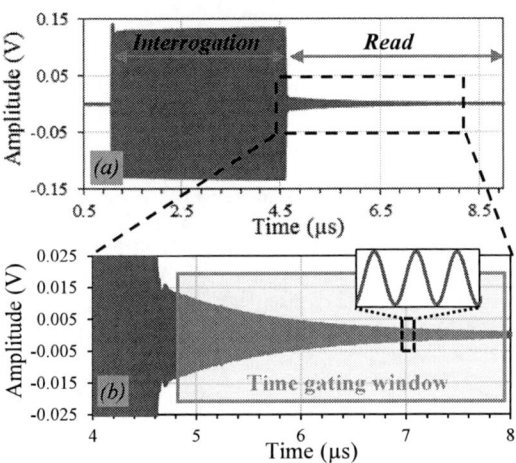

Figure 2: (a) Waveform as received by the oscilloscope showing the interrogation signal and the resonator decay response. This measurement was performed with transmit power of 0.5W at 10cm distance between the antennas. (b) waveform close-up and the time gate window used for taking FFT.

In order to eliminate the interrogation signal at the receiver, the waveform is gated in time domain (Figure 2b). An additional 200ns after the end of the interrogation pulse is ignored to account for the environmental echoes. A fast Fourier transform (FFT) is then taken on the signal using Labview software in order to extract the resonance frequency of the resonator (Figure 5). The temperature can then be calculated from the resonance frequency and TCF.

EXPERIMENTAL RESULTS

In order to characterize the passive temperature sensor, the resonator is put into the temperature-controlled chamber and connected to an antenna on the outside. The wireless configuration described above is used to measure

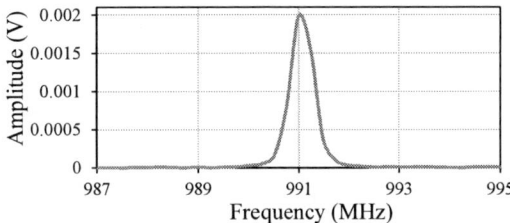

Figure 3: The spectral density of the resonator decay response after taking Fourier transform of the time-gated waveform of Figure 4b.

the frequency variations over the same temperature range as before. The interrogation signal power was set to 500mW and the distance (D) to 1m. Figure 4 compares the results of direct and wireless measurements.

The very short time required for each temperature reading (~7µs) enables collecting numerous samples in a very short period of time. Figure 5 shows the temperature histogram of 2000 samples with a mean value of 22.27°C and standard deviation (STD) of 0.07°C resulting in temperature resolution of less than 0.1°C which was also verified by experiment. In addition, a maximum D of 3m was achieved with 500mW power, though the resolution reduced to 0.6°C.

DISCUSSION

The temperature resolution is a function of signal to noise ratio at the receiving antenna (oscilloscope). Friis equation explains the dependency of the received signal power on different parameters [19]:

$$\frac{P_r}{P_t} = \frac{G_t^2 G_s^2 \lambda^4 \chi}{(4\pi D)^4} \qquad (3)$$

where P_t and P_r are the power transmitted and received by the transceiver respectively. G_t and G_r are the gain of the interrogator and sensor antennas respectively, λ is the free-space wavelength, and χ represents the mismatch losses. To further increase D without sacrificing the resolution, higher signal power could be used to interrogate the resonator. According to equation (3), the resolution achieved at 3m (0.6°C) can be maintained at twice the distance if the peak power is increased 16 times. However, the average power could easily be kept below a desired level by reducing the sampling rate.

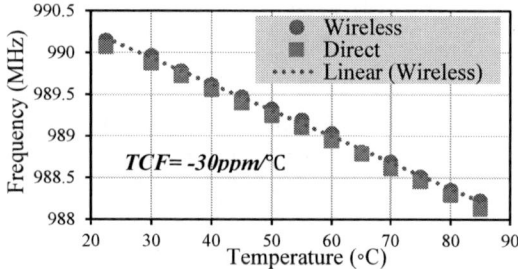

Figure 4: The resonance frequency vs. the resonator's temperature measured with direct and wireless configurations. The slight frequency shift between the two cases is due to different loading effects on the resonator.

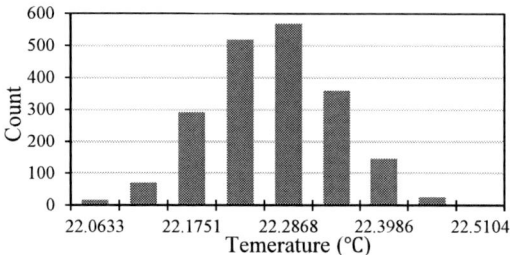

Figure 5: Histogram of 2000 temperature samples collected at room temperature.

CONCLUSION

For the first time, MEMS resonators are exploited for passive wireless temperature sensing. The temperature-sensitive component is a thin-film piezoelectric-on-silicon resonator with TCF of -30ppm/°C. The resonator is directly connected to an antenna with no additional circuitry. Temperature resolution of 0.1°C was achieved at a sensing distance of 1m and a peak power of 0.5W. Further improvement of the temperature resolution is possible by using resonators with higher TCF or higher gain antennas.

ACKNOWLEDGEMENT

Authors wish to thank Brandon Harrington for his contribution in the fabrication of devices.

REFERENCES

[1] R. M. Langdon, "Resonator sensors-a review," *J. Physics E: Scientific Instruments,* vol. 18, no.2, 1985.

[2] D. Stevens, *et al.*; "Applications of wireless temperature measurement using saw resonators," *Acoustic Wave Devices for Future Mobile Communications Systems Symp.,* 2010.

[3] L. Reindl, *et al.*, "Theory and application of passive SAW radio transponders as sensors," *Trans. Ultrasonics,*

Ferroelectrics, and Frequency Control, vol. 45, no. 5, pp. 1281-1292, 1998.

[4] X. Q. Bao, W. Burkhard, V. V. Varadan, V. K. Varadan; "SAW Temperature Sensor and Remote Reading System," *IEEE Ultrasonics Symp.,* pp.583-586, Oct. 1987.

[5] W. Buff, et al., "Remote sensor system using passive SAW sensors," *Ultrasonics Symp.,* pp.585-588, 1994.

[6] V. Kalinin, G. Bown, and A. Leigh; "Contactless Torque and Temperature Sensor Based on SAW Resonators," *Ultrasonics Symp.,* pp.1490-1493, 2006.

[7] Z. Xiangwen, W. Fei-Yue, L. Li; "Optimal selection of piezoelectric substrates and crystal cuts for SAW-based pressure and temperature sensors," *Ultrasonics, Ferroelectrics, and Frequency Control,* vol.54, no.6, pp.1207-1216, June 2007.

[8] T. Ueda, K. Fusao, T. Iino, D. Yamazaki; "Temperature Sensor Using Quartz Tuning Fork Resonator," *Frequency Control Symp.,* pp. 224-229, May 1986.

[9] C. M. Jha, et al.; "Cmos-Compatible Dual-Resonator MEMS Temperature Sensor with Milli-Degree Accuracy," *Solid-State Sensors, Actuators and Microsystems Conf.,* pp.229-232, June 2007.

[10] F. J. Azcondo, and J. Peire; "Quartz crystal oscillator used as temperature sensor," *Industrial Electronics, Control and Instrumentation,* pp.2580-2585, vol.3, 1991.

[11] T. M. Reeder and D. E. Cullen; "Surface-acoustic-wave pressure and temperature sensors," *Proceedings of the IEEE,* vol.64, no.5, pp.754-756, May 1976

[12] R. Abdolvand, H. M. Lavasani, G. K. Ho, and F. Ayazi; "Thin-film piezoelectric-on-silicon resonators for high-frequency reference oscillator applications," *Trans. Ultrasonics, Ferroelectrics and Frequency Control,* vol.55, no.12, pp.2596-2606, Dec. 2008.

[13] G. Piazza, P. J. Stephanou, A. P. Pisano; "One and Two Port Piezoelectric Contour-Mode MEMS Resonators for Frequency Synthesis," *European Solid-State Device,* pp.182-185, Sep. 2006.

[14] M. Shahmohammadi, B. P. Harrington, and R. Abdolvand; "Concurrent enhancement of Q and power handling in multi-tether high-order extensional resonators," *IEEE Microwave Symp.,* pp.1452-1455, 2010.

[15] M. Shahmohammadi, B. P. Harrington, and R. Abdolvand; "Turnover Temperature Point in Extensional-Mode Highly Doped Silicon Microresonators," *Trans. Electron Devices,* vol.60, no.3, pp.1213-1220, 2013.

[16] M. Shahmohammadi, H. Fatemi, and R. Abdolvand; "Nonlinearity reduction in silicon resonators by doping and re-orientation," *IEEE Micro Electro Mechanical Systems,* pp.793-796, Jan. 2013.

[17] M. Shahmohammadi, D. Dikbas, B. P. Harrington, and R. Abdolvand. "Passive tuning in lateral-mode thin-film piezoelectric oscillators," *Frequency Control and the European Frequency and Time Forum,* pp. 1-5, May 2011.

[18] S. D. Senturia; *Microsystem design,* vol. 3. Boston: Kluwer academic publishers, 2001.

[19] C. A. Balanis; *Antenna theory: analysis and design,* John Wiley & Sons, 2012.

978-1-4799-7956-1/15 $31.00 © 2015 IEEE

SNR IMPROVEMENT IN AMPLITUDE MODULATED RESONANT MEMS SENSORS VIA THERMAL-PIEZORESISTIVE INTERNAL AMPLIFICATION

Mohammad Mahdavi, Alireza Ramezany, Varun Kumar, and Siavash Pourkamali
Department of Electrical Engineering, University of Texas at Dallas, TX, USA

ABSTRACT

Effect of thermal-piezoresistive internal amplification on signal to noise ratio (SNR) of amplitude modulated resonant MEMS sensors (e.g. vibratory gyroscopes and Lorentz force magnetometers) has been studied in this work showing the possibility to significantly improve the detection limit. It has been shown that as the thermal-piezoresistive amplification sets in, noise rms value increases with a slower rate than the boost in vibration amplitude and output signal level, therefore the SNR increases. In addition to higher sensitivity due to internal amplification in such devices, improvement in SNR reduces the minimum detectable signal in presence of limiting Brownian and thermal noises. Preliminary measurement results show that increasing the DC bias current, which leads to a 3X increase in vibration amplitude, improves the SNR by a factor of 4.5 (6.6 dB).

INTRODUCTION

Micro/nanomechanical resonant devices can be used as highly sensitive magnetometers, accelerometers, force, or pressure sensors [1]–[5]. In such devices a mechanical deflection, a change in mechanical properties of the structure, or an applied force can be translated into a change in the resonator vibration amplitude or frequency. The thermally actuated MEMS resonator of Fig. 1, with piezoresistive readout, can be used as a highly sensitive amplitude modulated sensor in which an external force, e.g. Lorentz force or Coriolis force, can excite vibrations and modulate the vibration amplitude and consequently output current/voltage [3].

There is always a demand for sensors with higher sensitivity and better resolution. The minimum detectable signal in MEMS sensors is limited by electrical and mechanical noise regardless of their detection mechanism. In addition to the electrical-thermal noise of readout circuitry, the movement of the microscale freestanding structures is affected by different sources of mechanical noise, especially mechanical-thermal noise [6]. The mechanical-thermal noise, analogous to electrical-thermal noise, is caused by the interior random movement of molecules in structure and Brownian movement of surrounding molecules as well. Special attention to thermal noise is required when it comes to the thermal-piezoresistive MEMS resonators, which in fact turn thermal power change into mechanical movement.

It has been shown that in resonant structures with a micro to nanoscale narrow beam acting as a thermal actuator, increasing the DC bias current in the actuator beam can lead to vibration amplitude enhancement (DC to AC energy conversion, AKA energy pump) [7], [8]. This can increase the resonator vibration amplitude and its effective quality factor all the way to infinity when self-

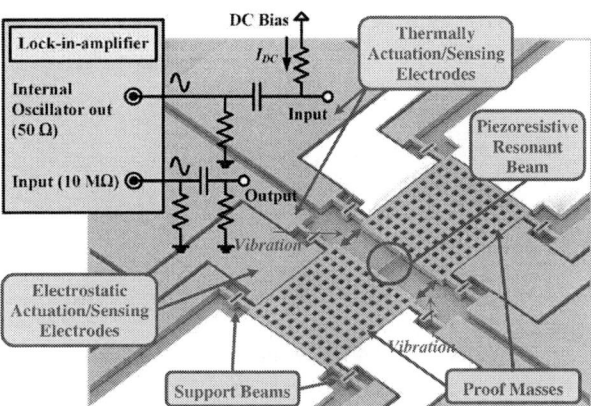

Figure 1: SEM image of a crystalline silicon thermal-piezoresistive resonator and its supporting test circuitry; the internal oscillator output of lock-in-amplifier was used as an input to one side of the actuator beam and the other side is connected to the input of the lock-in-amplifier as the sensor output signal.

sustained oscillation sets in. In practice, Qs up to tens of millions have been reported previously [7]. The internal amplification is a result of coupling between Joule's heating and piezoresistivity of the actuator beam [9]. As the actuator goes under periodic tensile and compressive stress during resonance, its electrical resistance is modulated due to the piezoresistive effect. The modulated resistance along with the DC bias current leads to an additional Joule's heating component that could help increase the vibration amplitude if it has the right phase with respect to the original actuating force (e.g. Coriolis or Lorentz force) [3]. The main unanswered question targeted in this paper is how internal amplification affects the noise behavior of such devices and whether it can lead to any improvement in the sensor minimum detectable limit.

In this study, a thermal-piezoresistive resonator was tested using a lock-in-amplifier (Fig.1). The frequency responses of the device along with noise spectra at different bias currents were measured to monitor device performance in presence of noise. The results suggest that although the noise rms value increases together with the increase in Q and voltage amplitude, SNR value improves due to the decrease in noise bandwidth (B_n), and the demonstrated SNR improvement is a unique feature of such devices, which can lead to orders of magnitude higher sensor sensitivities.

AMPLITUDE MODULATED RESONANT MEMS SENSOR

The MEMS resonant structure used in this work is

comprised of a narrow single crystalline silicon beam terminated by two suspended masses on its two sides. Applying a combination of a DC and an AC electrical signal to the beam leads to a fluctuating Joule's heating component that can actuate the beam in its longitudinal extensional resonant mode with the two plates vibrating back and forth in opposite in-plane directions [8]. The relation between displacement and the thermal force in such mass-spring resonant system can be written as below:

$$X_{th} = \frac{1}{Ms^2 + bs + K}.F_{th} \qquad (1)$$

where M is the effective moving mass, K axial stiffness of the beam, and b is the damping coefficient. Such resonator can be used as an amplitude-modulated MEMS sensor when an external source of actuation, for instance a Lorentz force, is applied to the structure instead of thermal force. The movement caused by the external force will result in an AC output voltage via the piezoresistive effect of the silicon beam. Moreover, The combination of the DC bias current and the fluctuating resistance will create an internal fluctuating stress in-phase with sensor's displacement, which can also be interpreted as an energy that is being pumped back into the system [9] (Fig. 2).

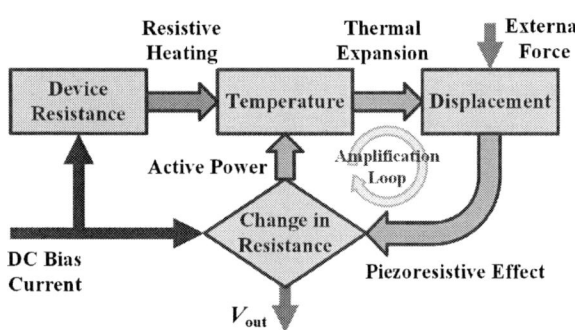

Figure 2: Block diagram of amplitude modulated MEMS sensor; illustrating thermally-piezoresistive internal amplification loop

This energy pump by amplifying the stress, which results from the increase in the bias current, enables such resonator to have a significantly higher vibration amplitude and therefore higher output signal in response to the same actuating force.

Although the thermal-piezoresistive internal amplification will result in a higher sensitivity, the question arises whether or not the minuscule vibrations caused by noise fluctuation undergo the same amplification process and diminish the minimum detectable signal of such sensors.

MECHANICAL-THERMAL NOISE

The lowest detectable signal in a sensor is limited by the noise fluctuation which is resulted from random movement of molecules within the mechanical structure and electrical circuitry. It has been shown that the presence of a damping component in the mechanical structure is the main source of mechanical-thermal noise fluctuation similar to resistors as a noise source in electrical circuit. So, solving the motion equation of the spring-mass system at equilibrium will lead to the same result as electrical white noise density in which the electrical resistance will be replaced by damping b [6]:

$$u_n = \sqrt{4k_BTb} \qquad [V/\sqrt{Hz}] \qquad (2)$$

where k_B and T are Boltzmann's constant and temperature respectively. This equation suggests that if the damping, or in other words system's loss decreases, the noise density will also decrease. At the same time increase in the temperature leads to a higher noise density. In the proposed MEMS sensor, reduction in noise density is anticipated due to the reduction in motional resistance as a result of thermally-piezoresistive internal amplification. However, increase in the bias current results in a higher temperature, and consequently higher noise density.

The measurement results of the output noise characteristics of the sensor were examined to address device's performance

RESULTS AND DISSCUSSION

The crystalline silicon thermally actuated resonator examined in this work is comprised of two 800×800 μm^2 plates coupled with a 30×1.8 μm^2 beam fabricated on a 15 μm thick n-type low resistivity device layer of a SOI substrate. To investigate the noise behavior of the sensor, 7220 signal recovery lock-in-amplifier was used. To minimize damping resulting from the surrounding environment, all of the tests were conducted under 1 mTorr vacuum. The input voltage amplitude provided by the internal lock-in-amplifier oscillator was set to 80 μV in order to avoid device saturation. To make the results comparable, the input lock-in-amplifier's sensitivity was set to 100 μV for all measurement. Both the frequency response and noise spectra of the device were collected at 4 different bias currents as illustrated in Figure 2.

As shown by the frequency responses, the structure has a resonant frequency of around 213 kHz. At higher bias currents the beam stiffness is slightly reduced due to increased temperature resulting in a decrease in resonance frequency. More importantly, changing the bias current from 3.2 mA to 4 mA changes the output voltage amplitude at resonance from 25 μV to 78 μV. At the same time, the resonator effective Q increases from 6,600 to 18,355. Improvement in signal level and effective Q is a result of thermal-piezoresistive internal amplification as stated previously. By increasing the bias current, higher quality factors up to millions are anticipated.

Although the noise spectra also show an increase in noise density from 1.3 to 2.1 $\mu V/\sqrt{Hz}$, this rate is not following amplification in signal level and Q. On the contrary, decrease in noise was anticipated by reduction in motional resistance based on the Eq. 2. There are two possible explanations for this observation; first, by increasing the bias current the structure temperature rises, which results in higher noise fluctuation thus higher dissipation. Second, by increasing the voltage gain caused by energy pump, the output transferred electrical noise generated in the biasing and beam DC resistances is

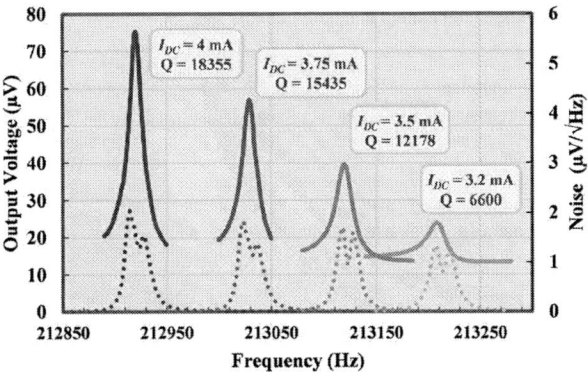

Figure 2: Measured device frequency response and noise spectrum data. The signal frequency response of the device at different DC bias currents (I_{DC}) is shown with solid lines and their corresponding noise density is shown with dotted lines.

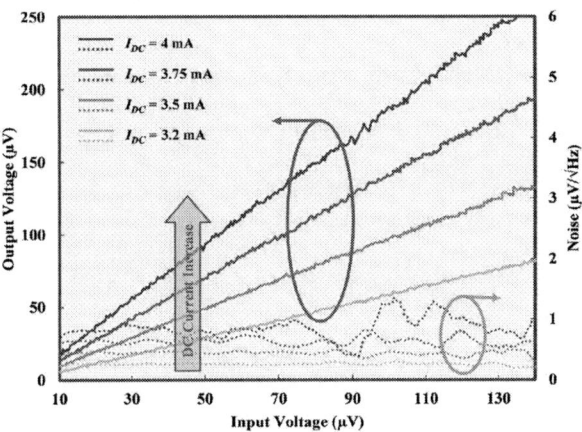

Figure 3: The measured device input-output characteristics; the oscillator frequency is set to the resonance frequency of the resonator. The left side axis shows the output voltage while the right axis illustrates the noise density. Both the output signal and noise increase by increasing the bias current.

increases. The nature of two adjacent peaks in noise density is still unclear, and needs more studies.

One of the sensor performance characterizations involves input-output voltage characteristics, which can show the device linearity (dynamic range) in response to the input signal. To perform such analysis the input voltage amplitude was set in range of 10-140 µV at the resonance frequency extracted from frequency response for the same bias currents in fig 2 (Figure 3, solid lines). These characteristics show the high dynamic range of sensor. At the same time, output noise was measured to assure the independency of noise from input voltage amplitude (Figure 3, dotted lines). Again, it is clear from the results that the increase in the bias current enhances the output voltage. Also, noise density graphs show the

increase in the output fluctuation by bias current, but the change in the input amplitude does not affect the noise density.

Signal to Noise Ratio (SNR) Calculations

In practice, the rms output noise value is obtained from noise density to determine the minimum detectable signal. As it was stated before, to calculate this rms value, the noise density is multiplied with the equivalent noise bandwidth, B_n, which can be defined by equation below [10]:

$$B_n = \frac{1}{|H(f_0)|^2} \int_0^\infty |H(f)|^2 df \qquad (3)$$

where $H(f)$ is the input/output voltage transfer function. The frequency response depicted in Figure 2 was used as a transfer function to extract the B_n at each bias current. The calculated equivalent noise bandwidths are reported in Figure 4 along with signal bandwidths (-3dB bandwidth) and noise density in order to compare their change by increase in the bias current. As one can conclude from this comparison, the noise bandwidth is always bigger than signal bandwidth, and it decreases at a higher rate than the signal bandwidth.

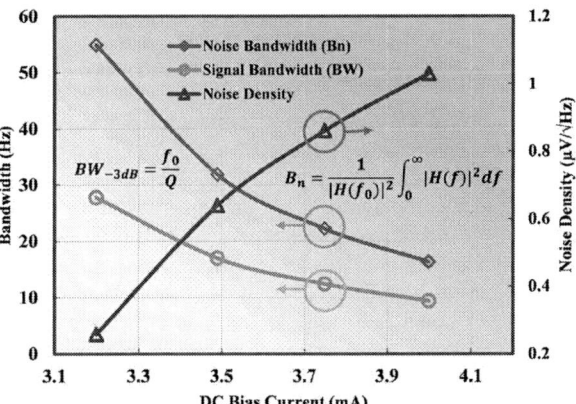

Figure 4: Both the signal and noise bandwidth are decreasing as the bias current is increasing, but noise bandwidth B_n is decreasing at a higher rate than signal bandwidth BW. As shown in previous graphs the noise density is increasing.

Having noise densities and equivalent noise bandwidths, one can simply determine the noise rms value using Equation 5 for which the results are shown in Figure 5 with circle markers:

$$v_{rms\ noise} = \sqrt{\overline{u_n^2} . B_n} \qquad (4)$$

Finally, resonance peak voltage value and the calculated rms noise were used to obtain SNR values in dB (Figure 5, square markers).

$$SNR = 20.\log\left(\frac{v_{signal}}{v_{rms\ noise}}\right) \qquad (5)$$

The calculations highlight two aspects of resonant sensors with thermal-piezoresistive internal amplification: 1) the increase in bias current makes the output rms noise

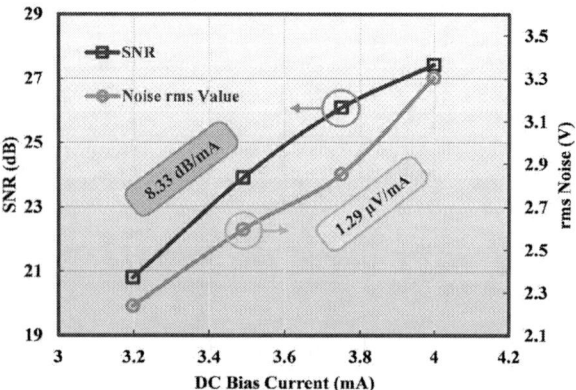

Figure 5: These graphs show the improvement in signal to noise ratio (SNR) (square markers) in spite of increase in noise rms value (circles markers) by increase in the bias current.

bigger, 2) despite the increase in output rms noise, SNR value increases. The latter makes thermal-piezoresistive amplification an effective technique to reach orders of magnitude higher sensitivities for conventional amplitude modulated MEMS sensors.

CONCLUSION

Preliminary studies were conducted on the noise behavior of thermal-piezoresisitve internally amplified MEMS resonant structures. The studies show that increasing the resonator bias current, which results in thermal-piezoresistive internal amplification, will lead to an increase in signal to noise ratio. It was shown that increase in the signal level occurs at a higher rate than the increase in noise rms value as the bias current increases. The internal thermal-piezoresistive energy pump partially compensates resonator loss by power absorbed from the DC source, thus decreasing power dissipation and consequently noise fluctuations. The high quality factor and high SNR make such devices high-sensitivity/resolution and reliable candidates for sensing applications such as accelerometer, gyroscope, magnetometer, etc.

Higher bias current is needed to achieve higher effective quality factor and SNR. The higher the DC current is, the higher the temperature and power consumption are going to be. Furthermore, the higher temperature can lead to the higher phase noise and frequency instability. Further investigation of device performance and noise behavior is required develop a deeper understanding of contributing factors to noise and SNR for such devices. Also, the nature of adjacent peaks in noise spectra is still unknown which need to be investigated as a part of future work.

ACKNOWLEDGEMENTS

This work was supported by of the US National

Science Foundation (NSF) grant # 1345161. Authors would like to thank cleanroom staff at University of Texas at Dallas.

REFERENCES

[1] M. Li, E. J. Ng, V. A. Hong, C. H. Ahn, Y. Yang, T. W. Kenny, and D. A. Horsley, "Lorentz force magnetometer using a micromechanical oscillator," *Appl. Phys. Lett.*, vol. 103, no. 17, p. 173504, Oct. 2013.

[2] C. L. Roozeboom, B. E. Hill, V. A. Hong, C. H. Ahn, E. J. Ng, Y. Yang, T. W. Kenny, M. A. Hopcroft, and B. L. Pruitt, "Multifunctional Integrated Sensors for Multiparameter Monitoring Applications," *J. Microelectromechanical Syst.*, vol. PP, no. 99, pp. 1–1, 2014.

[3] E. Mehdizadeh, V. Kumar, and S. Pourkamali, "Sensitivity Enhancement of Lorentz Force MEMS Resonant Magnetometers via Internal Thermal-Piezoresistive Amplification," *IEEE Electron Device Lett.*, vol. 35, no. 2, pp. 268–270, Feb. 2014.

[4] F. A. Levinzon, "Ultra-Low-Noise Seismic Piezoelectric Accelerometer With Integral FET Amplifier," *IEEE Sens. J.*, vol. 12, no. 6, pp. 2262–2268, Jun. 2012.

[5] E. Mehdizadeh, M. Rostami, X. Guo, and S. Pourkamali, "Atomic Resolution Disk Resonant Force and Displacement Sensors for Measurements in Liquid," *IEEE Electron Device Lett.*, vol. 35, no. 8, pp. 874–876, Aug. 2014.

[6] T. B. Gabrielson, "Mechanical-thermal noise in micromachined acoustic and vibration sensors," *IEEE Trans. Electron Devices*, vol. 40, no. 5, pp. 903–909, May 1993.

[7] A. Rahafrooz and S. Pourkamali, "Thermal-Piezoresistive Energy Pumps in Micromechanical Resonant Structures," *IEEE Trans. Electron Devices*, vol. 59, no. 12, pp. 3587–3593, Dec. 2012.

[8] A. Rahafrooz and S. Pourkamali, "High-Frequency Thermally Actuated Electromechanical Resonators With Piezoresistive Readout," *IEEE Trans. Electron Devices*, vol. 58, no. 4, pp. 1205–1214, Apr. 2011.

[9] A. Ramezay, M. Mahdavi, X. Guo, and S. Pourkamali, "Resonant MEMS Piezoresistors for narrow band electronic amplification," presented at the Solid-State Sensor, Actuator and Microsystems workshop, Hilton Head, 2014.

[10] G. Vasilescu, *Electronic Noise and Interfering Signals: Principles and Applications*. Springer Science & Business Media, 2006.

CONTACT

M. Mahdavi, tel: +1-469-360-6450; mxm@utdallas.edu

A RESONANT PIEZOELECTRIC MICROPHONE ARRAY FOR DETECTION OF ACOUSTIC SIGNATURES IN NOISY ENVIRONMENTS

Anton A. Shkel, Lukas Baumgartel, and Eun Sok Kim
University of Southern California, Los Angeles, California, USA

ABSTRACT

This paper reports a MEMS acoustic resonator array that improves Automatic Speech Recognition (ASR) and signature detection characteristics in environments with high levels of acoustic interference. The experiments with the acoustic resonator array show an increase of 62.7 percentage points in ASR transcription accuracy for signals buried under noise with -15 dB Signal-to-Noise Ratio (SNR). The results of this study support the development of resonant acoustic sensors for a variety of pattern recognition applications.

INTRODUCTION

Resonant acoustic transducers are commonly employed in applications such as acoustic ranging [1] and ultrasonic imaging [2]. For pattern recognition applications, however, emphasis has typically been placed on sensors with a flat pressure-frequency response used in conjunction with digital signal processing. These applications include speech processing [3], acoustic leak detection [4], electronic health monitoring [5], and vehicle classification [6].

While digital filtering has traditionally been seen as superior to an analog approach due to its versatility and cost effectiveness, analog filtering has been shown to have a number of advantages over a purely digital approach. Digital signal processing is power hungry, making sensors relying on digital processing difficult to implement for low-power and self-powered applications. Additionally, digital filtering suffers from limitations in sampling signals with a large dynamic range in amplitude. In very low SNR conditions, analog-to-digital converters may become saturated and the features of interest may be lost or distorted before processing.

These limitations serve as motivation for the development of high quality factor acoustic sensors for signal recognition applications in noisy environments. In this paper we will report experimental evidence for the effective use of resonant acoustic sensors as a mechanical method for improved noise filtering.

DEVICE

The acoustic resonator array used in the ASR experiments reported in this paper is the array of paddle-shaped piezoelectric cantilevers fabricated with a well-controlled fabrication method for accurate resonant frequencies and quality factors [7]. The array has 13 cantilevers with their resonant frequencies linearly-spaced between 860 and 6,260 Hz.

Fabrication

The array is fabricated on a silicon-on-insulator (SOI) wafer with 4.5 μm thick device layer (Figure 1ai). The SOI's buried oxide layer serves as an etch-stop for anisotropic potassium hydroxide (KOH) backside etching (Figure 1aii). This is followed by removing the buried oxide with wet etching in buffered hydrofluoric acid. An SOI wafer is chosen as the substrate because of the accuracy and consistency of the SOI device layer thickness. This allows for reliable control of cantilever dimension and frequency characteristics.

Fabrication of the piezoelectric sensing element consists of: (1) sputter deposition of aluminum (0.20 μm) to serve as the bottom electrode, (2) sputter deposition of piezoelectric zinc oxide (ZnO) (0.59 μm) which is the active piezoelectric sensing material, (3) plasma-enhanced chemical vapor deposition (PECVD) of silicon nitride (0.10 μm) to provide electrical isolation between the top and bottom electrode, and (4) sputter deposition of aluminum (0.22 μm) to serve as the top electrode. Deep Reactive Ion Etching (DRIE) is used for the final release of the paddles (Figure 1aiii).

A pre-amplifier circuit is designed for a voltage gain of 40 dB on each channel using a simple non-inverting op-amp configuration. The diced microphone array is wire-bonded directly to the preamplifier PCB.

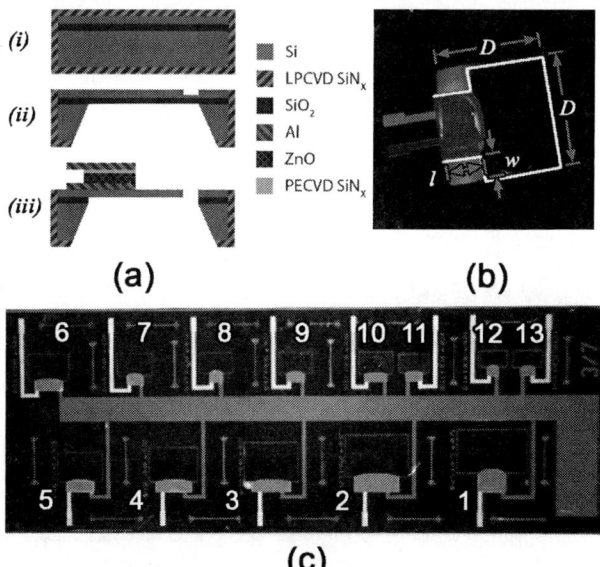

Figure 1: (a) Brief microfabrication process of the resonant piezoelectric microphone. (b) Top-view photo of the paddle structure used in the microphone. (c) Thirteen-element piezoelectric microphone array based on silicon paddles with resonant frequencies from 860 Hz to 6.26 kHz.

Measurements

Each of the 13 microphones was characterized by performing an acoustic frequency sweep in an anechoic chamber and measuring the response. The electrical output of the microphones was compared to the output of a calibrated precision microphone (G.R.A.S. type 40AO) to obtain sensitivity as a function of frequency (Figure 2). Individual microphones have an average quality factor of 43.1, with individual quality factors ranging from 21.6 to 59.3. The unamplified sensitivities at the resonant frequencies range from 202.6 mV/Pa to 10.80 mV/Pa (Table 1).

Table 1. Summary of device characteristics from 4 of 13 microphone-array elements

Mic	D [mm]	l [mm]	w [mm]	f_0 [Hz]	Sens. at f_0 [mV/Pa]
1	2.5	0.75	0.75	860	202.60
5	1.6	0.4	0.28	2591	24.62
9	1.2	0.35	0.27	4341	17.03
13	1.0	0.34	0.24	6263	10.80

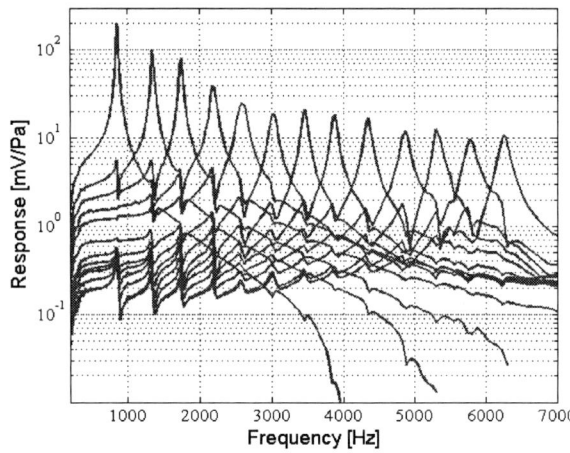

Figure 2: Plot of the sensitivity (in mV/Pa) of each of 13 microphone elements in microphone array.

EXPERIMENT

Setup

Figure 3 illustrates the experimental setup for testing of the microphone array in acoustic signature detection applications. The MEMS microphone array and pre-amplifier circuit are contained within an aluminum electro-magnetic shielding box. A commercial speaker is used as the signal source, while a calibrated precision microphone with flat response served as a reference. A pre-amplifier circuit, based on the LM386, amplifies the reference microphone output to an amplitude comparable to that of the resonant microphone array in order to ensure equivalent sampling resolution conditions. This entire assembly is placed within an anechoic chamber for acoustic isolation

A Roga 2-channel USB data acquisition system is used with a sampling resolution of 305 µV at a sampling rate of 16 kHz. A PC is used with the NI Labview development environment for data acquisition, experiment automation, and signal analysis.

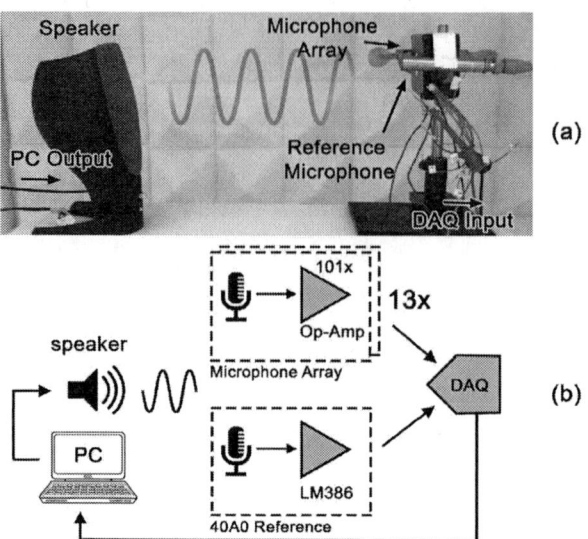

Figure 3: (a) Photograph of the measurement setup. (b) Schematic diagram of the setup.

Preliminary experiments with a pulse-width-modulated signal that is buried under noise (15 dB higher than the signal, i.e., -15 dB SNR) show that both the microphone array (without any filtering) and the reference microphone (with digital filtering) can recover the signal (Figure 4). However, it was found that the advantages of the resonant microphone array are limited in white noise conditions, as noise at the resonant frequency has significant contributions in both cases.

For this reason we investigate the effect of sinusoidal noise at a single frequency. Although not a realistic environmental condition, this noise profile mimics environments where noise is dominant in a limited bandwidth. In practice, if the spectral conditions of the noise are known, microphone resonances can be designed to capture signals with minimal acoustic interference.

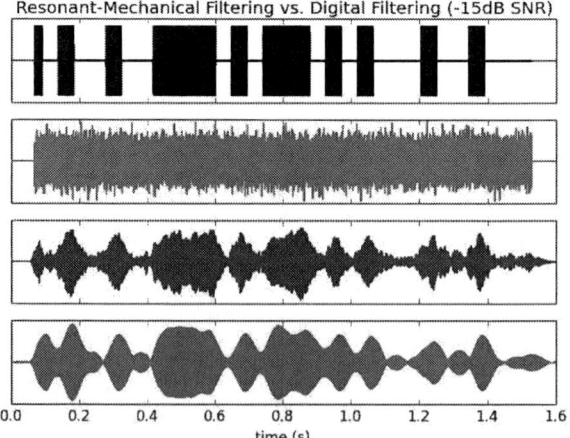

Figure 4: (a) Waveform graph of pulse-width-modulated input signal. (b) Signal with -15dB SNR additive white Gaussian noise. (c) Measured MEMS resonant microphone electrical output. (d) Measured reference microphone output with digital band-pass post-filtering.

Digital Acoustic Communication Experiment

Experiments using individual resonant microphone elements for the sensing of single-tone audio signals were performed. These experiments served as a method of assessing the pattern recognition performance of the microphone array without the additional variables present in more complex signal recognition applications.

The input signal used is a sequence of randomly-generated bits that is amplitude-modulated with a carrier frequency corresponding to the resonant frequency of the individual sensing element. Sinusoidal noise is combined additively with the signal at a SNR ranging from -72 dB to -40 dB. The microphone output is digitized and demodulated using Labview's Modulation Toolkit. For each SNR point, 800 bits are transmitted, received, and demodulated to obtain a Bit-Error Rate (BER). These results are plotted in Figure 5.

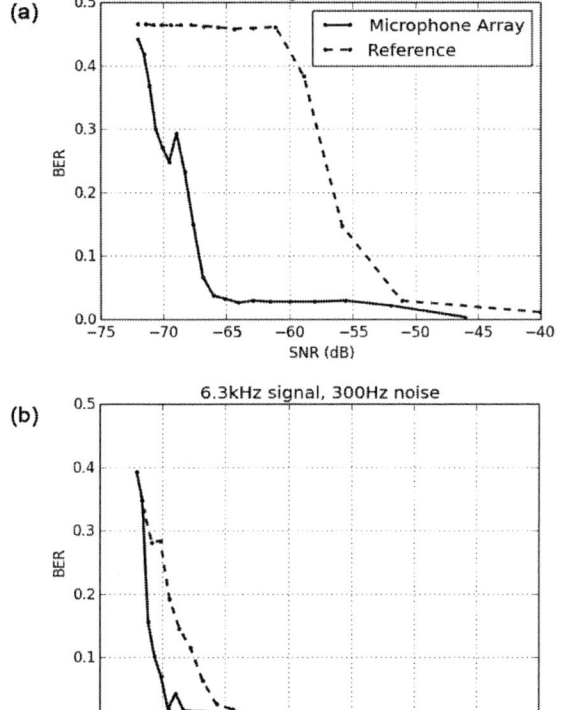

Figure 5: (a) Measured bit error rate for 860 Hz signal with 2 kHz sinusoidal noise. (b) Measured bit-error rate for 6.3 kHz signal with 300 Hz sinusoidal noise.

Two results are shown; in the first case, detection of a low-frequency signal (860 Hz) with high-frequency noise (2 kHz) is performed using the microphone #1 (of which the resonant frequency is 860 Hz). The MEMS microphone sensitivity at the noise frequency of 2 kHz is 0.66 mV/Pa, which is 0.3% of the sensitivity at resonance. In the second case, detection of high-frequency signals (6.3 kHz) with low-frequency noise (300 Hz) is performed using microphone #13 (of which the resonant frequency is 6.3 kHz). The MEMS cantilever sensitivity at 300 Hz is 0.14 mV/Pa, 1.3% of the sensitivity at resonance. These results show a significant decrease in error rate as a result of using a highly resonant acoustic

sensor rather than a wide-band microphone.

We see that in both of these cases, sensitivity is quite low for the noise frequency. BER is nearly zero down to about -70 dB SNR, where it sharply saturates to nearly 0.5, the probability of random guessing. For situations where noise frequency is much closer to the resonant frequency of the microphone, we expect to see an increase in the minimum SNR required for reliable demodulation.

In Figure 6, we see that this is indeed the case. As noise frequency approaches the resonant frequency, the sensitivity to the noise increases and thus the BER performance degrades. However we see that even with only about a 2.4% offset between noise frequency and carrier frequency, SNRs of about -15 dB can be achieved before BER reaches 25%. Thus, resonant microphones with high quality factors can effectively recover signals buried under noise that has spectral components very close to the center frequency of the microphone.

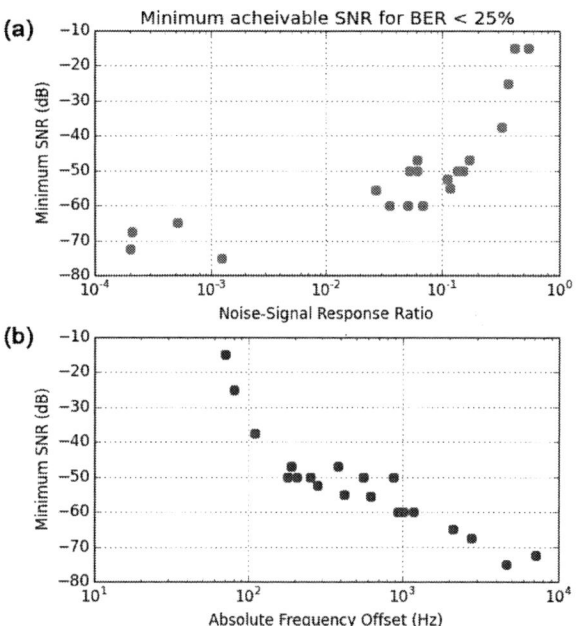

Figure 6: Experimental measurements of minimum SNR such that BER < 25% for a microphone element with 3 kHz resonant frequency. The x-axis plots (a) the ratio of the electrical response of the microphone at the noise frequency to that at the resonant frequency and (b) the absolute value of the difference between noise frequency and signal frequency.

Automatic Speech Recognition Experiment

The results of the digital acoustic communication experiment were extended to the more specific application of Automatic-Speech Recognition (ASR).

Speech processing was performed using the CMU Sphinx4 toolkit with a custom front-end based on Linear-Frequency Cepstral Coefficients (LFCC). To compute LFCC features, first the speech data is divided into 25 ms frames with 10 ms separation between data blocks. Several filters are applied for pre-emphasis and windowing. This is followed by a Fast Fourier Transform (FFT) for each data block to obtain a normalized magnitude frequency response.

The subsequent step of applying bandpass filtering is performed differently for reference microphone data and microphone array speech output. Thirteen triangular filter banks are used for the reference microphone. Each of these filters has a quality factor of 43.1 and linearly spaced center frequencies to coincide with microphone array resonances. Microphone array data is processed with thirteen rectangular filter banks to isolate resonance peaks from one another. The rectangular filter bandwidth coincides with the base dimension of the triangular filter banks used for the reference microphone. In both cases, the energy coefficients for each of the thirteen filter banks undergo a discrete cosine transform to obtain a 13-element feature vector to be directly used for ASR decoding.

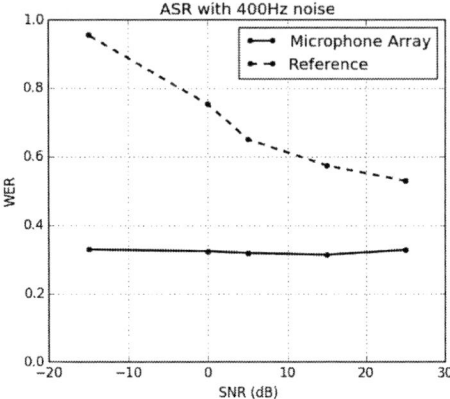

Figure 7: Measured word-error rates (WERs) of the acoustic resonator array and the reference microphone used to process automatic speech recognition with various levels of 400 Hz sinusoidal noise.

The microphone input consisted of 140 alphanumeric speech recordings from the AN4 database. This database was used due to its reduced vocabulary set, which allowed for large variations in recognition accuracy over a range of wide range of SNR conditions. A 400 Hz sinusoidal noise was additively combined with the speech data at SNRs ranging from -15 dB to 25 dB. The speech output at both microphones is processed using the Sphinx toolkit to obtain decoded transcriptions, which are then compared to the original speech text to obtain a Word-Error Rate (WER). Figure 7 shows the average WER of the resulting transcriptions as a function of the SNR. These results demonstrate that speech acquired using the resonant microphone array is relatively unaffected by large levels of out-of-band acoustic noise, while a typical flat-response microphone relying on digital band-pass filtering has degrading performance under high levels of acoustic noise.

It is hypothesized that the increased improvement in recognition accuracy of the microphone array is related to the limitations in dynamic range of the analog-to-digital conversion process. As out-of-band noise amplitude increases, the signal becomes distorted at all frequencies for a wideband microphone (Figure 8b). As a result, digital filtering is unable to accurately recover the original signal. In contrast, the signal output of the high-Q MEMS microphone array does not experience any significant distortion (Figure 8c).

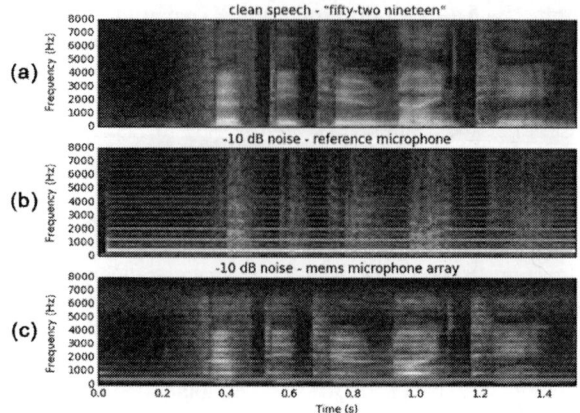

Figure 8: Spectrogram of the utterance "fifty-two nineteen" in the case of (a) a clean audio file without any filtering, (b) the reference microphone for -10 dB applied noise, and (c) a summation of the outputs from 13 microphone array elements in -10 dB noise.

CONCLUSION

These experiments make a strong case for the use of resonant acoustic transducer arrays in signature detection applications, especially noisy environments.

ACKNOWLEDGMENTS

The authors wish to thank the Chevron-USC Partnership for their financial support.

REFERENCES

[1] B.E. Boser, et al., "Ultrasonic transducers for navigation," *Proceedings of Meetings on Acoustics, Vol. 19, 030087,* 2013.

[2] B.T. Khuri-Yakub, O. Oralkan, "Capacitive micro-machined ultrasonic transducers for medical imaging and therapy," *Journal of micromechanics and microengineering, vol. 21, no. 5, 054004,* 2011.

[3] T. Virtanen, et al., *Techniques for noise robustness in automatic speech recognition,* John Wiley & Sons, 2012.

[4] J. ChunLei, W. Yuan, "The research of natural gas pipeline leak detection based on adaptive filter technology," *IEEE Int. Conf. on Measurement, Information and Control (ICMIC),* vol. 2, pp. 1229-1233, 2013.

[5] D.V.J. Pinard, M. M. Blanckenberg, "Automated pediatric cardiac auscultation," *IEEE Trans. on Biomedical Engineering, vol. 54, no. 2, pp. 244-252,* 2007.

[6] G. Padmavathi, D. Shanmugapriya, M. Kalaivani, "Digital watermarking technique in vehicle identification using wireless sensor Networks," *IEEE Int. Conf. on Advanced Computer Theory and Engineering (ICACTE),* vol. 2, pp. V2-6, 2010.

[7] Baumgartel, Lukas, et al., "Resonance-enhanced piezoelectric microphone array for broadband or prefiltered acoustic sensing," *IEEE/ASME Journal of Microelectromechanical Systems,* vol. 22, pp. 107-114, 2013.

CONTACT

*A.A. Shkel; shkel@usc.edu

978-1-4799-7956-1/15 $31.00 © 2015 IEEE

AIR-COUPLED ALUMINUM NITRIDE PIEZOELECTRIC MICROMACHINED ULTRASONIC TRANSDUCERS AT 0.3 MHZ TO 0.9 MHZ

Ofer Rozen[1]*, Scott. T. Block[1,2], Stefon E. Shelton[2], Richard J. Przybyla[2], and David A. Horsley[1]

[1] University of California Davis, USA

[2] Chirp Microsystems, Inc., USA

ABSTRACT

Air-coupled piezoelectric micromachined ultrasonic transducers (PMUTs) operating at frequencies ranging from 0.3 MHz to 0.9 MHz were designed, fabricated and characterized. We increased the fractional bandwidth by 51% and improved the piezoelectric coupling over 80% by patterning the diaphragm center into a ring or structural ribs, resulting in a reduction of the PMUT's mass. Pulse-echo testing was conducted in air using PMUTs at frequencies up to 0.9 MHz and the measured acoustic loss versus path-length was compared to theoretical models. Devices were fabricated in an industrial foundry process using wafer-level bonding of a MEMS PMUT wafer to a CMOS wafer using a conductive metal eutectic bond. This process allows for close integration of PMUT arrays and signal processing circuitry and is used here to study the effects of wafer-level packaging on acoustic performance.

INTRODUCTION

Typically, fluid-coupled micromachined ultrasonic transducers (MUTs) used in medical imaging operate at 5 MHz or more, while air-coupled ultrasonic transducers usually operate at relatively low frequencies, 40 to 200 kHz [1], in order to maximize the range by reducing the thermal losses in air [2]. Increasing the operating frequency allows the MUT diameter to be decreased and improves the angular resolution of a MUT array with a given area [3], both of which are crucial characteristics for ultrasonic sensors used in consumer electronics as well as in non-destructive evaluation (NDE) applications [4].

Here, we present piezoelectric MUTs (PMUTs) fabricated in an industrial foundry process in which the MEMS wafer is wafer-bonded to a CMOS wafer containing signal processing circuitry. While the direct bonding of CMOS to MEMS affords high signal integrity, it imposes acoustic boundary conditions on the PMUT membrane that may lead to poor performance due to squeeze film damping or other loss mechanisms. We demonstrate that proper acoustic design of the CMOS-MEMS assembly can lead to acoustic performance that is equivalent to that of previous single-chip PMUTs.

To enable small, low-cost arrays of PMUTs, we investigate the acoustic performance of PMUTs with up to 25% smaller area and 400% higher frequency than earlier 200 kHz PMUT designs. High frequency arrays are expected to have shorter range due to increased acoustic absorption in air and this was characterized experimentally

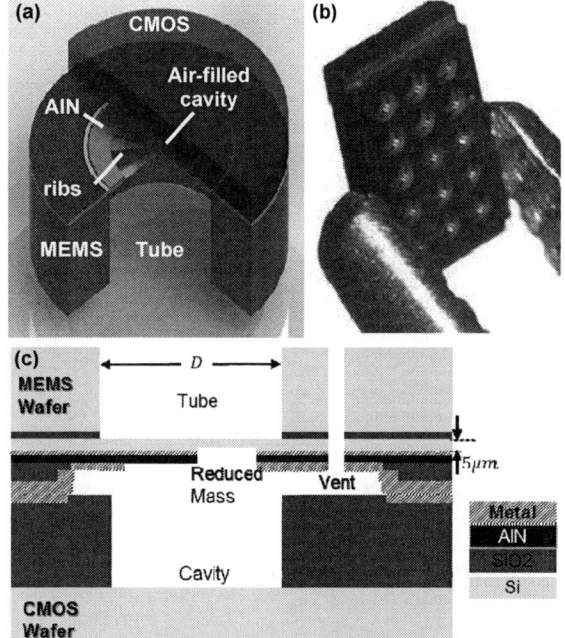

Figure 1 (a) Illustration of the bonded-wafer PMUT structure. The PMUT shown has a ring electrode and stiffening ribs. The CMOS level is shown as transparent; (b) Image of a 3x5 PMUT array measuring 2.6 by 4.5 mm. The CMOS bond-pads are visible at the top edge of the die. (c) Schematic section of the bonded MEMS and CMOS die.

using pulse-echo measurements. Because reducing the PMUT diameter leads to increased quality factor (Q) and poor pulse response, we investigate patterned membrane designs as a means to reduce Q, thereby increasing the bandwidth [5].

DESIGN

We designed PMUTs, shown in Figure 1, at four different center frequencies from 200 kHz to 900 kHz by adjusting the diameter between 360 μm to 780 μm. Three different membrane designs were investigated: (1) a flat diaphragm with a center electrode; (2) a partially-etched diaphragm with an outer ring electrode; and (3) a diaphragm with radial stiffening ribs and an outer ring electrode. The ring and ribs were designed to reduce the diaphragm's mass while enhancing its stiffness, therefore increasing the

978-1-4799-7956-1/15 $31.00 © 2015 IEEE

fractional bandwidth [5]. The ring and wedge-shaped ribs were patterned by a 3.5 μm deep partial etch of the 5 μm thick Si elastic layer.

The air-filled cavity between the MEMS and CMOS wafers results in squeeze-film damping, the real part of which is modeled by [6]

$$R_{cav} = \frac{3\pi\mu r^4}{2h^3} \qquad (1)$$

where r and h are the radius and depth of the cavity and μ is the dynamic viscosity of air. Since the damping term R_{cav} decreases with the cavity depth, the quality factor will increase. However, the backside cavity impedance also includes imaginary components corresponding to the mechanical compliance and inertial mass of the cavity, [7]

$$C_{cav} = \frac{h}{\rho_0 c^2} \frac{1}{\pi r^2} \qquad (2)$$

$$I_{cav} = \frac{\rho_0 \pi r^2 h}{3} \qquad (3)$$

where ρ_0 and c are air density and speed of sound, respectively. The effect of I_{cav} and C_{cav} is that the PMUT's resonance frequency changes with the cavity depth. The modeled and measured Q factor as a function of cavity depth are shown. Figure 2 shows that Q reaches an approximately constant value for cavity depths greater than 20 μm. Note that while low quality factor is generally a positive characteristic [3, 5] of PMUTs, as it indicates efficient transfer of mechanical energy to the air, the energy should be transmitted as the ultrasonic output, and not lost to other mechanisms such as squeeze-film damping.

FABRICATION

PMUTs were fabricated in an industrial foundry process in which the MEMS wafer is bonded to a CMOS wafer using a conductive Al-Ge eutectic bond, as shown in Figure 1. Following bonding, the MEMS wafer is DRIE

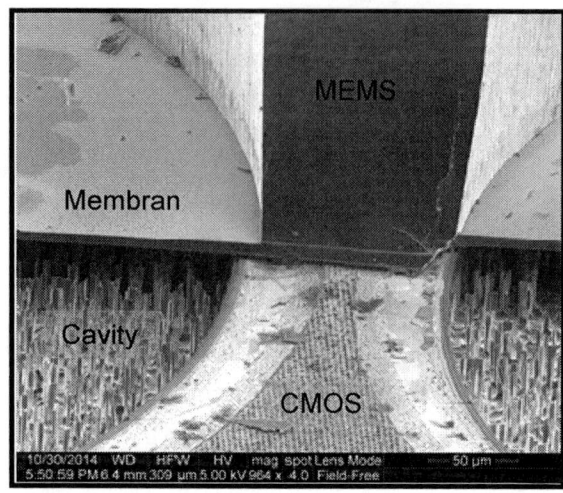

Figure 3: SEM image of a die cross section showing the cavity etched into the CMOS wafer. The pillars in the cavity area formed due to an error on the layout

etched to release the AlN-Si PMUT membrane and the bonded wafer is diced to expose bondpads on the CMOS wafer. To investigate the effect of cavity depth on squeeze-film damping, devices from two different wafers were tested. On the first wafer, the cavity between the PMUT and the CMOS wafer was approximately 2.5 μm. On the second wafer, a 50 μm cavity was etched into the CMOS wafer in the region beneath the PMUT, as shown in Figure 3.

RESULTS

The displacement frequency response was measured by laser Doppler Vibrometry (LDV). The center frequency and quality factor were found by fitting the frequency response to a second-order transfer function, and are presented in

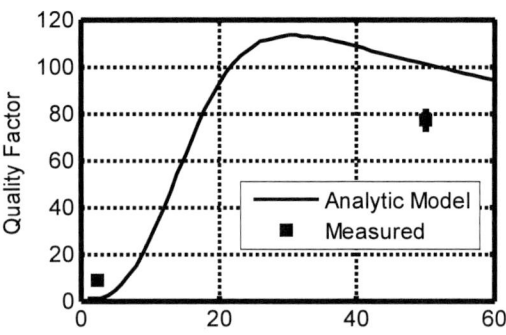

Figure 2: Simulated and measured quality factor as a function of cavity depth for a flat, 360 μm radius PMUT. Cavity depth of the measured devices is the nominal value and can vary by ±10 μm.

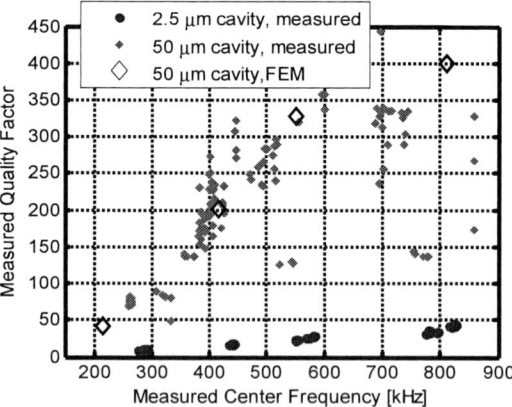

Figure 4: Simulated and measured quality factor and center frequency of for PMUTS with back side cavitys depths of 2.5μm and 50μm.

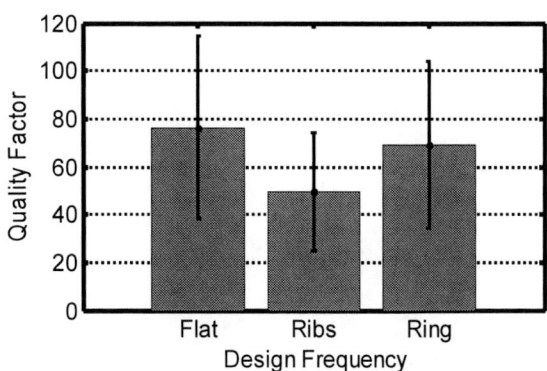

Figure 5: Measured quality factor for 300 kHz PMUTs having three different designs with cavity depth of 50μm. Error bars indicate the measured range.

Figure 4. In devices with a 2.5 μm cavity, the system bandwidth is governed by squeeze-film damping, instead of energy transfer to the acoustic output. In comparison, devices with a 50 μm cavity show high Q (up to 350 in devices operating from 600 kHz to 800 kHz), similar to MEMS devices lacking the CMOS wafer. Additionally, the devices with a cavity show more die-to-die variation in the center frequency than those without the cavity. The source of this variation is not entirely clear but may be caused by particulate in the cavity resulting from imperfect DRIE caused by a layout error.

As shown in Figure 5, the Ribs design, in which a partial etch is used to reduce the membrane mass, achieves lower quality factor than that of the Flat design. The reduction in Q achieved by the Ribs and Ring designs is summarized in Table 1. The Ribs design shows the most improvement, reducing Q by up to 51% relative to the Flat diaphragm design. Moreover, the displacement normalized by the quality factor was compared as a measure of the piezoelectric coupling. For all designs, the average normalized displacement of the ribs design was 27-80% higher than that of the flat design.

Table 1: Reduction in Q relative to flat PMUT design

	300 kHz	400 kHz	600 kHz	800 kHz
Ribs	35%,	51%	31%	13%
Ring	10%	18%	12%	14%

PMUT arrays require close matching between the center frequencies of the PMUTs in the array. Using dice selected from various locations across the 200 mm wafer, we compared the range of center frequencies of 10 identical 400 kHz Ribs designs on the same die. For six dice with 50 μm cavities and five with 2.5 cavities, the worst matching observed on a single die was 6.7% and 2.5% and the best matching was 0.8% and 1% respectively.

We also studied the frequency variation across the

Cross-Correlation of Received Signal

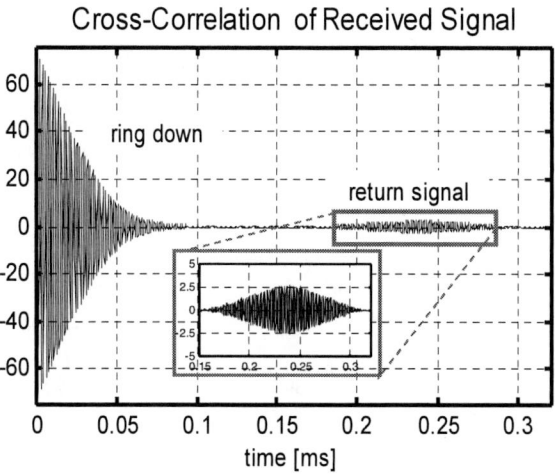

Figure 6: Typical cross-correlation of received signal measured from a PMUT. The device was excited at 7 V_{rms}. Distance is estimated using the time of flight estimated from the peak of the cross-correlation measurement. The peak of the correlation as well as the characterization of the amplifier is used to estimate the return loss (V_{RX}/V_{TX}).

entire wafer, which is an important metric for manufacturing yield. The across-wafer frequency variation was 23% for the 800 kHz devices and 29% for the 400 kHz devices. Using FEM simulations for the frequency variation induced by AlN residual stress and Si device layer thickness variations, we estimate the AlN residual stress varies by 170 MPa to 350 MPa across the wafer, and the Si device layer total thickness variation (TTV) is in the range of 0.7-1 μm, which agree well with the known fabrication tolerances.

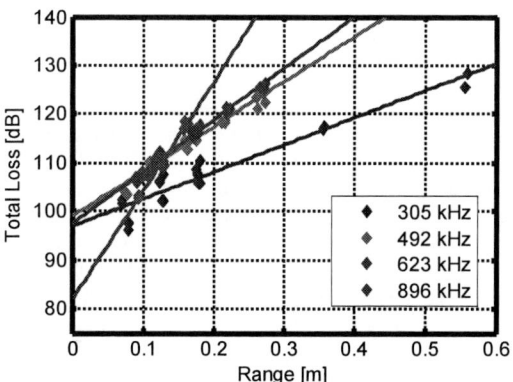

Figure 7: Acoustic loss measured for PMUTs with four different center frequencies. The slope of the best-fit line increases with frequency due to the frequency-dependent absorption of air.

978-1-4799-7956-1/15 $31.00 © 2015 IEEE   923

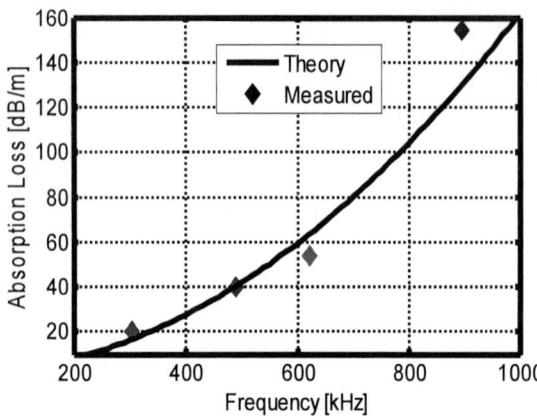

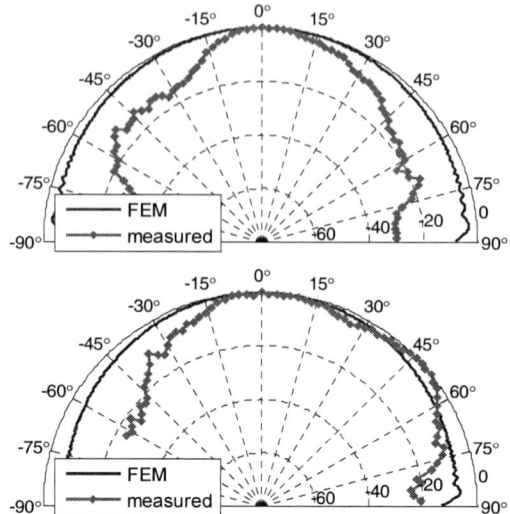

Figure 8: Measured absorption loss extracted from pulse-echo data shows good agreement with theoretical absorption

Pulse-echo measurements were conducted to evaluate the acoustic performance at each frequency. The received signal was cross-correlated with the measured transmit burst pattern and is shown in Figure 6. The propagation loss is a function of path-length and the loss increases strongly with frequency, Figure 7, e.g. a 900 kHz transducer at 15cm exhibits 35 dB greater loss than a 300 kHz transducer over the same path-length. The absorption loss at each frequency has been extracted from the difference in slope of the lines in Figure 7, and agrees well with theoretical absorption loss [7] for air at room temperature and 50% relative humidity, Figure 8.

Finally, we measured the transducers directivity, Figure 9, by rotating the PMUT relative to a flat reflecting target. The measured directivity is narrower than simulated. The origin of this effect is under investigation.

CONCLUSIONS

We have designed, fabricated, and characterized air coupled PMUTs at frequencies ranging from 0.3 MHz to 0.9 MHz. Pulse-echo measurements showed that the loss as a function of frequency agrees well with theory. We found that the Rib design decreases the quality factor by up to 51% and improves the coupling coefficient by up to 80% compared to the Flat design. A backside cavity depth of 50 µm was proven to be a sufficient in order to minimize the influence of squeeze film damping on acoustic output.

ACKNOWLEDGMENTS

The authors thank Invensense for device fabrication and Devin Truong for conducting LDV measurements.

Figure 9: Simulated and measured directivity of Flat PMUTs with 305 kHz (top) and 492 kHz (bottom) center frequency.

REFERENCES

[1]   B. A. Griffin, M. D. Williams, C. S. Coffman, and M. Sheplak, "Aluminum Nitride Ultrasonic Air-Coupled Actuator," *Microelectromechanical Systems, Journal of,* vol. 20, pp. 476-486, 2011.

[2]   R. Przybyla, A. Flynn, V. Jain, S. Shelton, A. Guedes, I. Izyumin, *et al.,* "A micromechanical ultrasonic distance sensor with> 1 meter range," in *Solid-State Sensors, Actuators and Microsystems Conference (TRANSDUCERS), 2011 16th International,* 2011, pp. 2070-2073.

[3]   R. J. Przybyla, S. E. Shelton, A. Guedes, R. Krigel, D. A. Horsley, and B. E. Boser, "In-Air Ultrasonic Rangefinding and Angle Estimation Using an Array of AlN Micromachined Transducers," in *Solid-State Sensors, Actuators & Microsystems Workshop,* 2012, p. 3.

[4]   E. Blomme, D. Bulcaen, and F. Declercq, "Air-coupled ultrasonic NDE: experiments in the frequency range 750kHz–2MHz," *NDT & E International,* vol. 35, pp. 417-426, 10// 2002.

[5]   O. Rozen, S. E. Shelton, A. Guedes, and D. A. Horsley, "Variable Thickness Diaphragm for a Stress Insensitive Wideband Piezoelectric Micromachined Ultrasonic Transducer," in *Solid-State Sensors, Actuators & Microsystems Workshop,* 2014.

[6]   M. Bao and H. Yang, "Squeeze film air damping in MEMS," *Sensors and Actuators A: Physical,* vol. 136, pp. 3-27, 5/1/ 2007.

[7]   D. T. Blackstock, *Fundamentals of Physical Acoustics*: Wiley-interscience, 2000.

CONTACT

*O. Rozen, tel: +1-530-219-8461; orozen@ucdavis.edu

SOUND FOCUSING IN LIQUID USING A VARIFOCAL ACOUSTIC MIRROR

Ryo Aoki, Nguyen Thanh-Vinh, Kentaro Noda,
Tomoyuki Takahata, Kiyoshi Matsumoto, and Isao Shimoyama
The University of Tokyo, JAPAN

ABSTRACT

This paper reports a method to focus sound in liquid to a desired point using a PDMS-based varifocal acoustic mirror. We used PDMS-air boundary as a sound reflector to concentrate sound. By adjusting the curvature radius of the acoustic mirror, we could change the position where sound was concentrated. We confirmed that our method was able to improve the output of the acoustic sensor for different sound source positions.

INTRODUCTION

Recently, MEMS-based acoustic sensors such as microphone [1] and hydrophone (underwater microphone) [2], [3] have been widely developed and used because they have advantages in batch fabrication, low manufacturing cost and miniaturization [4], [5].

When we utilize MEMS-based acoustic sensors to detect sound, it is significant to focus sound to a sensor to improve the sensor output due to the small acoustic energy contributed to the miniaturized sensors. A common way to focus sound is to use an acoustic mirror [6]. Figure 1 illustrates sound focusing using an acoustic mirror which has a spherical surface. It is known that the source-mirror distance a, the distance between the convergence point and the mirror b, and the curvature radius of the mirror R have the following relationship;

$$\frac{1}{a} + \frac{1}{b} = \frac{2}{R} \qquad (1)$$

A conventional acoustic mirror has a rigid surface with fixed curvature radius R. As long as the sound source P is far away from the mirror ($a \gg R$), the convergence point Q does not depend on the source-mirror distance a ($b = R/2$), hence it is possible to fix the location of the acoustic sensor to this point Q to improve the sensor output. However, when the source-mirror distance a is relatively small (e.g. $a \approx R$), the convergence point Q moves as the location of the source P changes; thus, it is necessary to change the acoustic sensor position to the moving convergence point to obtain the maximum output.

In this paper, instead of adjusting the location of the acoustic sensor, we propose an alternative solution using a varifocal acoustic mirror as shown in Figure 2. The varifocal acoustic mirror is a kind of acoustic mirror which can change its surface curvature radius R. By changing the curvature radius of the mirror so as not to change convergence point Q, it is possible to focus sound to the acoustic sensor without changing sensor's position.

DESIGN, PRINCIPLE AND FABRICATION

The design of the proposed varifocal acoustic mirror is shown in Figure 3. The central part of the upper PDMS

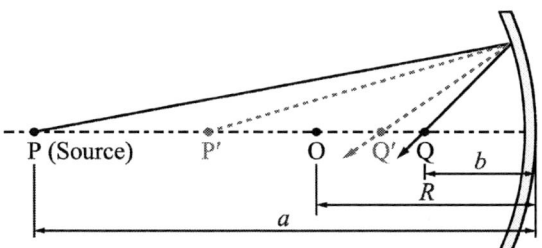

Figure 1: Sound focusing using an acoustic mirror. When a point source P moves to P', the convergence point Q also moves to Q'.

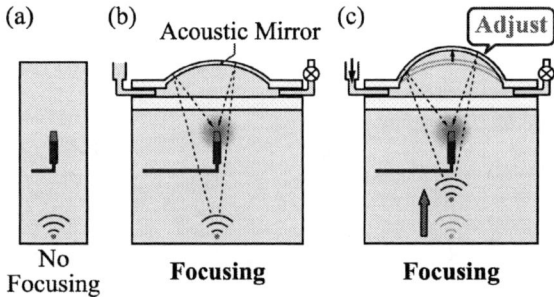

Figure 2: Sound focusing in liquid using a varifocal acoustic mirror. Compared to an acoustic sensor output without focusing (a), the sensor output improves by sound focusing using a varifocal acoustic mirror (b). When the position of the source changes, the proposed device can adjust its curvature radius and focus sound to a fixed sensor (c).

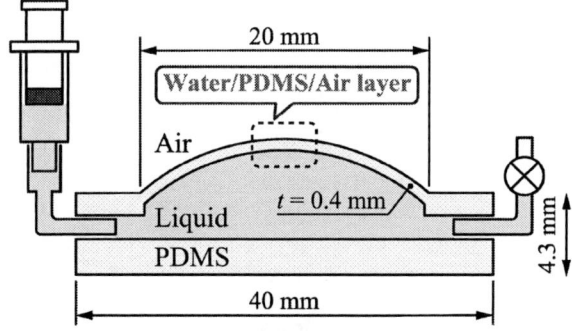

Figure 3: Design of the proposed varifocal acoustic mirror which mainly consists of two PDMS membranes, tubes, and liquid. The lower part of the PDMS membrane is thick; it does not deform largely due to the liquid (water) pressure from inside.

978-1-4799-7956-1/15 $31.00 © 2015 IEEE   925   MEMS 2015, Estoril, PORTUGAL, 18 - 22 January, 2015

Table 1: Acoustic impedance values of air, PDMS, and water, obtained as a product of density and sound velocity.

	Air	PDMS	Water
Density [kg/m³]	1.3	1.0×10^3	1.0×10^3
Sound Velocity [m/s]	3.3×10^2	1.0×10^3	1.5×10^3
Acoustic Impedance [kg/(m²s)]	4.3×10^2	1.0×10^6	1.5×10^6
Reference		[7]	

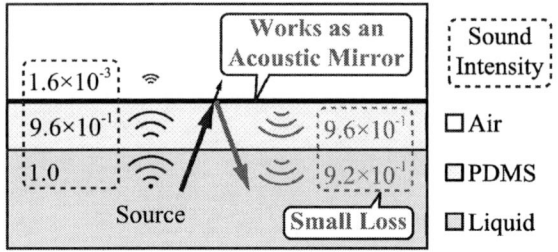

Figure 4: Propagation of sound wave emitted from the point sound source in water from the aspect of sound intensity. Sound wave is reflected at the PDMS-air boundary due to a large difference in acoustic impedance values.

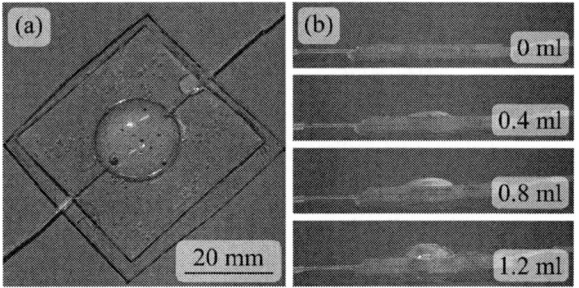

Figure 5: (a) A photograph of the fabricated varifocal acoustic mirror. (b) Photographs of the mirror deformation caused by changing the volume of containing water. The volume of water when the mirror was flat was defined as 0 ml.

membrane is thin enough to change its surface curvature radius by injecting water using the syringe.

The acoustic impedance values of air, PDMS and water are presented in Table 1. It is known that sound waves are well reflected at the boundary between two media whose acoustic values are largely different [7]. PDMS and water have close acoustic impedance values and they are more than 1000 times larger than that of air. From this, the water-PDMS boundary well transmits sound waves, but on the other hand, the PDMS-air boundary reflects almost all of the incident sound waves, hence it works as an acoustic mirror as shown in Figure 4.

Photographs of a fabricated varifocal acoustic mirror are shown in Figure 5. It was confirmed that the curvature radius of the mirror surface can be changed by injecting or withdrawing water using a syringe.

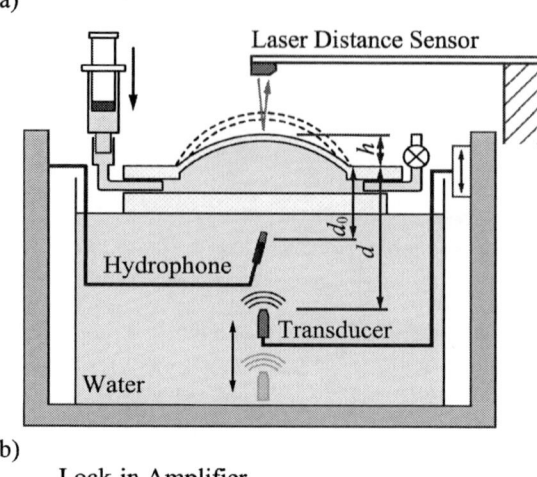

(a)

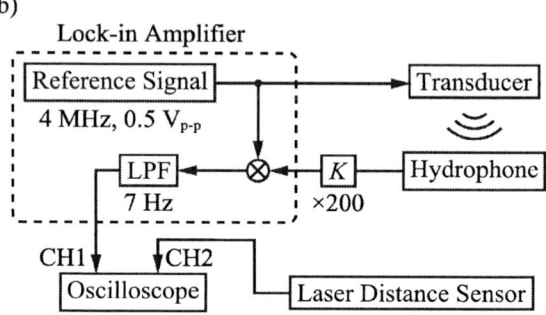

(b)

Figure 6: (a) Experimental setup to investigate the relationship between the output of the hydrophone and the curvature radius of the mirror for three different transducer positions. (b) Measuring circuit for the experiment.

EXPERIMENT AND RESULT

Figure 6(a) shows a schematic image of the experimental setup to investigate the relationship between the amplitude of an acoustic sensor output and the curvature radius of the fabricated varifocal acoustic mirror. We used a hydrophone (NH8224, Toray Engineering) as an acoustic sensor, and a transducer (V384-SU, Olympus) as a sound source used in water. A laser distance sensor (IL-065, KEYENCE) was used in order to measure inflation of the mirror. The distance between the mirror and the transducer was defined as d, the height of the mirror as h. The distance between the mirror and the hydrophone (d_0) was fixed to 20 mm. Figure 6(b) shows a measuring circuit for this experiment. We used a lock-in amplifier (HF2LI, Zurich Instruments) to measure the amplitude of the hydrophone output, which is connected to the input of the lock-in amplifier through a pre-amplifier. Reference signal was set to 4 MHz, 0.5 V_{p-p} and used both as a signal source for the transducer and a signal for input demodulator. We recorded the time series data of the amplitude of the hydrophone output and the laser distance sensor simultaneously.

Figure 7(a) is a time series data of the amplitude of the hydrophone output and the height of the mirror when the distance between the mirror and the transducer was set to 62 mm. The height of the mirror h increased as we gradually injected water into the device. The amplitude of the hydrophone output increased and reached its peak and

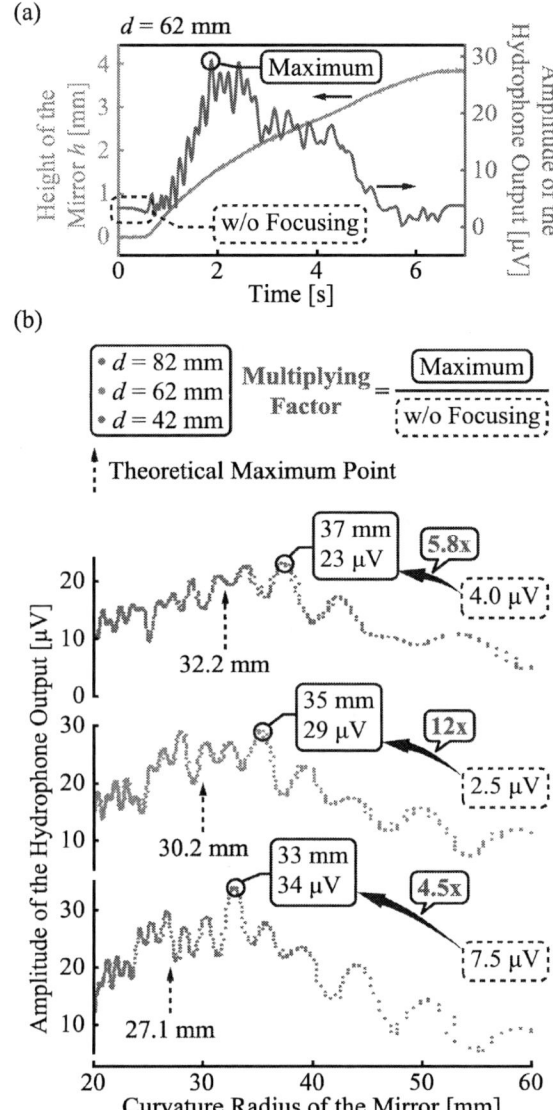

Figure 7: (a) The time series data of the experiment when d = 62 mm. (b) The relationship between the output of the hydrophone and the curvature radius of the mirror for three different positions of the transducer.

then decreased, which means that the amplitude of the hydrophone output can be maximized at a certain curvature radius of the mirror. Assuming that the surface of the mirror was spherical, the curvature radius of the mirror was calculated from the height and the base radius of the mirror geometrically.

Figure 7(b) shows the relationship between the amplitude of the hydrophone output and the curvature radius of the mirror for three different transducer positions, d = 82, 62, 42 mm. We theoretically calculated curvature radii corresponding to the maximum sound pressure level which are shown as vertical dashed arrows. These theoretical values are slightly different from the experimental values corresponding to the measured maximum amplitudes, yet the tendency that they decrease as the distance between the mirror and the transducer decreases are similar. Furthermore, in order to evaluate the

sound focusing efficiency of the mirror, we compared the two values of the maximum amplitude of the hydrophone output and the amplitude without focusing (the amplitude when the mirror was flat). The values when the mirror was flat was regarded as those without focusing. It was found that the multiplying factor of the mirror was 5.8, 12 and 4.5 times respectively and was confirmed that the mirror can focus sound from sources of different distances.

CONCLUSION

In conclusion, we proposed the varifocal acoustic mirror for improvement of the acoustic sensor output by sound focusing in water. The proposed varifocal acoustic mirror was designed to change curvature radius of its surface according to the volume of water between the two PDMS membranes. It was demonstrated that the acoustic sensor's output improved for three different sound source positions with focusing by the proposed device.

ACKNOWLEDGEMENTS

This work was supported in part by JSPS KAKENHI grant number 26286033.

REFERENCES

[1] P. R. Scheeper et al., "Fabrication of silicon condenser microphones using single wafer technology," J. Microelectromech. Syst., vol. 1, pp. 147–154, Aug. 2002.

[2] H. Wu et al., "Progress of the development and calibration of needle-type ultrasonic hydrophone," in Proc. 2013 Symp. Piezoelectricity, Acoustic Waves, and Device Applications, Changsha, 2013, pp. 1–4.

[3] C. Xue et al., "Design, fabrication, and preliminary characterization of a novel MEMS bionic vector hydrophone," Microelectronics J., vol. 38, pp. 1021–1026, Oct. 2007.

[4] J. J. Neumann, Jr., and K. J. Gabriel, "A fully-integrated CMOS-MEMS audio microphone," in Proc. 12th Int. Conf. Solid-State Sensors, Actuators and Microsystems, Boston, 2003, pp. 230–233.

[5] J. W. Weigold et al., "A MEMS condenser microphone for consumer applications," in Proc. 19th IEEE Int. Conf. Micro Electro Mechanical Systems, Istanbul, 2006, pp. 86-89.

[6] G. Zhu et al., "The non-rigid surface acoustic concave mirror model and sound field analysis," in Proc. 2011 Int. Conf. Electronics and Optoelectronics, Dalian, 2011, vol. 3, pp. 104–106.

[7] I. Leibacher et al., "Impedance matched channel walls in acoustofluidic systems," Lab Chip, vol. 14, pp. 463–470, Feb. 2014.

CONTACT

*Ryo Aoki, Department of Mechano-Informatics, Graduate School of Information Science and Technology, The University of Tokyo
Address: Eng. Bldg. 2, Rm. 81B (8th Floor), 7-3-1 Hongo, Bunnkyo-ku, Tokyo, 113-8656, Japan
E-mail: r_aoki@leopard.t.u-tokyo.ac.jp
Tel: +81-3-5841-6318, Fax: +81-3-3818-0835

978-1-4799-7956-1/15 $31.00 © 2015 IEEE

BIMORPH PMUT WITH DUAL ELECTRODES

Sina Akhbari[1], Firas Sammoura[1,2], Chen Yang[1], Maitha Mahmoud[2], Nawal Aqab[2], and Liwei Lin[1]

[1]Department of Mechanical Engineering, University of California at Berkeley, Berkeley, CA, USA
[2]Department of Electrical Engineering and Computer Science, Masdar Institute of Science and Technology, Abu Dhabi, UAE

ABSTRACT

The concept of "bimorph" piezoelectric micromachined ultrasonic transducers (pMUTs) has been demonstrated by utilizing a two active AlN layers structure constructed in a CMOS-compatible process. The prototype device has two 0.95μm-thick AlN layers sandwiched by three 0.15μm-thick Mo electrodes. In a prototype, both an inner circular and an outer annular electrode are designed on a 230 μm in radius, circular-shape diaphragm. When actuated with the inner electrode of 160μm in radius, the pMUT has a resonant frequency of 198.8 kHz and central displacement of 407.4 nm/V. Under the differential drive scheme using the dual-electrodes for large acoustic outputs at a low frequency, the measured central displacement is 13.0 nm/V, which is about 400% higher than that of a unimorph AlN-pMUT under similar actuation conditions. As such, the dual-electrode bimorph pMUT presents the improved operation as compared with the state-of-the-art flat pMUT design to achieve enhanced acoustic outputs.

INTRODUCTION

Conventional ultrasound transducers are based on PZT as the piezoelectric material and operated in the thickness mode for various types of applications, including medical ultrasonography [1]. The operation resonant frequency is determined by the thickness of the PZT layer and tight process control is required to meet the targeted frequency [2]. Considering the fabrication limitations, few products involving 2D ultrasound transducer arrays are in the market. The large transducer dimensions prohibit the applications of ultrasound transducers in devices such as cell phones. In addition, piezo-ceramics inherently have high acoustic impedance, which is difficult to match with liquid or air media [3]. Nevertheless, thickness-mode PZT sensors are the current choices for applications such as distance sensors, burglar alarms, medical imaging, and nondestructive tests.

Ultrasound transducers with small form factor, linear response, low voltage, and high acoustic pressure are desirable for next generation hand-held devices. In the past twenty years, micromachined ultrasonic transducers (MUTs) have emerged as the key candidates to replace conventional ultrasound transducers with good features in acoustic matching, large bandwidth, miniaturization, and low-cost by batch fabrication [4]. Plate flexural mode operations actuated either capacitively (cMUTs [5]) or piezoelectrically (pMUTs [6]) have been developed by many researchers. MEMS fabrication technologies can be utilized and the mechanical impedance of micromachined ultrasonic transducers can be closely matched to that of the imaging medium, resulting in improved bandwidth and system efficiency. However, cMUTs require high DC voltage and small gap with known drawbacks in dielectric charging and non-linear plate deflection with applied bias

[7]. On the other hand, pMUTs operate under low voltage bias with possible large diaphragm deformations for high acoustic pressure [8], while low electromechanical coupling has been a key drawback [9]. Therefore, there are various efforts to improve the electromechanical coupling of PMUTs by structural designs [10].

In order to improve the electromechanical coupling and output acoustic pressure per unit input voltage of pMUTs, designs such as the "dome-shape" piezoelectric actuators have been proposed to convert in-plane strain to flexural deflection [11]. In our previous work, PMUTs with multiple electrodes have shown enhanced effective electromechanical coupling factors, about 211% larger than that of the state-of-the-art, single electrode design [12]. A two-port pMUT using aluminum nitride (AlN) as the piezoelectric material and driven differentially with inner and outer electrodes has been validated to double the electromechanical coupling efficiency as compared with a conventional pMUT structure with 100% higher acoustic output per unit input voltage, and 485% larger suppression in magnitude of second harmonic mode [13].

This work extends the previous theoretical work on the "bimorph" pMUTs [14] using dual electrode designs with several accomplishments: (1) demonstration of a CMOS-compatible process with AlN as the piezoelectric layers; and (2) experimentally measured 400% higher output deformation as compared with the conventional unimorph pMUT by applying a differential drive scheme using the two active piezoelectric layers.

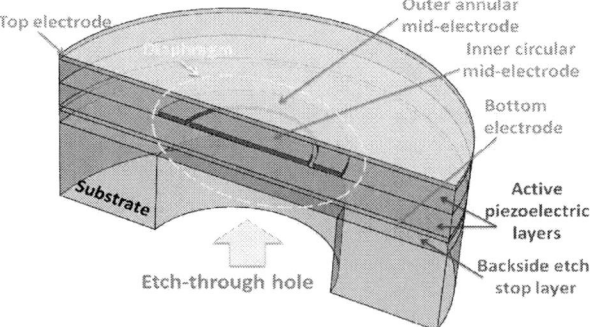

Figure 1: 3D schematic drawing showing the cross-sectional view of a bimorph pMUT with two active AlN layers and the dual electrode configuration: inner circular shape and outer annular shape electrodes.

CONCEPT

Figure 1 is a 3D schematic diagram illustrating the cross-sectional view of the bimorph pMUT design with dual-electrodes. The bimorph pMUT consists of two active piezoelectric layers, sandwiched between top, bottom, and middle electrodes.

A backside etch-through hole forms the circular diaphragm and defines its diameter. The two middle

---

978-1-4799-7956-1/15 $31.00 © 2015 IEEE   928   MEMS 2015, Estoril, PORTUGAL, 18 - 22 January, 2015

electrodes are separated via a small gap as an inner circular-shape and outer annular-shape electrodes, which are electrically isolated and can be differentially actuated with two voltage sources of the same amplitude and opposite polarity.

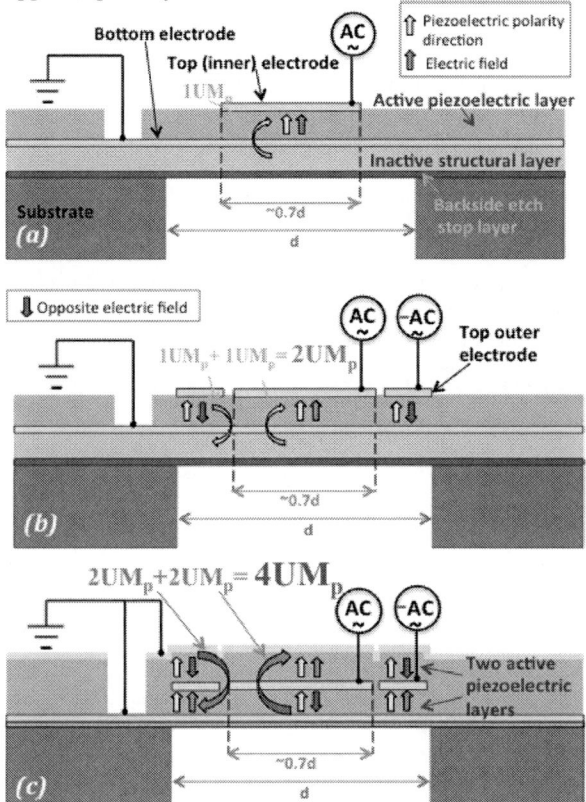

Figure 2: Cross-sectional views of (a) the conventional unimorph pMUT with one active layer; (b) the two-port unimorph pMUT with inner and outer electrodes driven differentially; and (c) the bimorph pMUT with dual electrodes to be driven differentially. It is expected to have 4X larger theoretical bending moment.

Figure 2a illustrates a conventional unimorph pMUT with a single active piezoelectric layer, a structural layer, and a top circular electrode [6]. Under an electric field between the top and bottom electrodes, an in-plane electromechanical strain is induced by the piezoelectric d_{31} effect. The strain profile generates a $1UM_p$ bending moment (UM_p is defined as an arbitrary unit of bending moment per unit input voltage) on the diaphragm and makes it move out-of-plane, i.e. in a flexural mode. A dual-electrode unimorph pMUT [13] with an inner circular and outer annular electrode is shown in Figure 2b. When driven differentially with voltage sources of the same magnitude and opposite polarity, the inner (outer) portion of the diaphragm will be in contraction while the outer (inner) portion will be in expansion to generate a $2UM_p$ bending moment. In these single active-layer devices, the electromechanical energy transformation comes from the single piezoelectric layer and some of the energy is used to mechanically deform the inactive structural layer.

Figure 3: Process flow of the dual electrode bimorph pMUT: (a) depositions of the stop AlN layer, bottom Mo electrode, bottom AlN layer, and middle Mo electrode; (b) patterning and etching of the middle electrode; (c) depositions of the top AlN layer and top Mo electrode; opening of contacts to middle and bottom electrodes; (d) backside DRIE to release the diaphragm.

In contrast to the conventional unimorph pMUT structure and inspired by well-known cantilever-type piezoelectric bimorph structures [15], the bimorph pMUT with two active layers is shown in Figure 2c with the same polarity and separated by a patterned middle/common electrode. The top and bottom electrodes are connected to ground and the middle inner and outer electrodes are driven with AC voltages with equal magnitude but opposite phase. As such, in addition to the 2x displacement produced by the differential dual electrode similar to the case in Figure 2b, the two active layers provide another 2x larger volumetric displacement as compared with the single-layer structure. In other words, each layer generates $2UM_p$ on the diaphragm while operating in the differential mode and a total of $4UM_p$ will be imposed on the diaphragm. As such, 4x higher responses are expected by replacing the bottom structural layer by another active layer and adding the dual electrode design.

FABRICATION
Process Flow

Figure 3 shows the CMOS-compatible process flow chart for the dual-electrode bimorph pMUT, starting with: (a) sputter deposition of 200 nm-thick AlN stop layer, 150 nm-thick Mo bottom electrode, first 0.95 μm-thick AlN active layer, and 150 nm-thick middle Mo electrode; (b) patterning middle electrodes by plasma etching using SF_6; (c) sputter deposition of the second 0.95 μm-thick AlN layer and 150 nm-thick top Mo electrode. The via openings to both the middle and bottom electrodes are subsequently formed using Fluorine-based plasma to etch Mo. Chlorine-based plasma is used to etch AlN, in combination with a final wet etching step with MF-319 to clear the remaining AlN and ensure minimal damage and maximum selectivity to the Mo electrodes; and finally (d) backside DRIE to release the diaphragm.

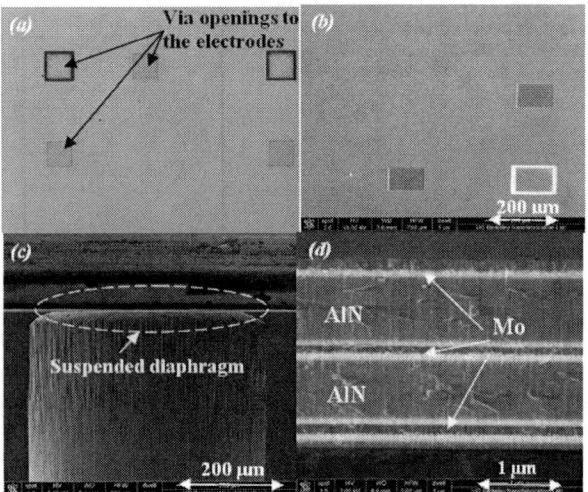

Figure 4: (a) optical top view image and (b) tilted view SEM micrograph of a dual electrode bimorph pMUT; (c) SEM image of a cleaved device showing the cross-sectional view; (d) close-up view of the diaphragm in Fig. 4c showing good crystal alignment of both AlN layers.

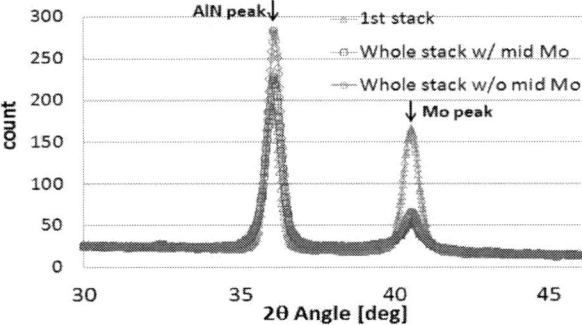

Figure 5: XRD 2θ scan of first AlN layer stack and the whole AlN stack with and without middle electrode.

Fabrication Results

Figures 4a and 4b show the respective optical and SEM top view images of a fabricated bimorph pMUT prototype with dual electrode configuration. Imprints on the top electrodes are apparent on figures, which are related to the step height variations by underlying middle electrodes. Figure 4c shows the cross-sectional SEM micrograph of a cleaved device illustrating the clamped diaphragm. Figure 4d is a close-up view of the diaphragm highlighting its layer composition, and revealing that the suspended AlN diaphragm stack has good crystal alignment in both active AlN layers.

The crystalline quality of the two active AlN layers has been further investigated using XRD (X-Ray Diffractometer), with special focus on the second AlN layer sputtered on top of the patterned middle Mo electrode as the structure has been exposed to the external environment before the deposition of the second AlN active layer. Good crystal orientation of both AlN active layers is critical for the bimorph pMUT. Figure 5 shows the XRD 2θ scans for the first sputtered single-AlN layer, and the total stack with both AlN layers. The XRD data for the total stack are collected at two different locations,

the first of which is where the middle Mo electrode is sandwiched between the AlN layers, and the second of which is where the middle Mo electrode is etched away and the two AlN layers are in direct contact. Good crystallinity data of the first AlN layer and the whole stack are observed in Figure 5 at the two different scan locations. Although the XRD measurements do not reveal explicit information about the quality of the second sputtered layer, the crystalline information of the latter can be projected by comparing the collected data of the first AlN layer to the whole stack. It is observed from Figure 5 that the whole stack has a sharper peek, higher amplitude, and slightly narrower FWHM in the (002) crystalline direction than that of the first AlN layer, which indicates that the latter must be as well oriented as the first layer. The FWHM of AlN was measured to be about 1.7° from the Rocking Curve.

RESULTS AND DISCUSSIONS

Figure 6 compares the measured, theoretically derived, and simulated low frequency center displacements of a bimorph pMUT under different driving inputs by (a) inner electrode; (b) outer electrode; (c) out-of-phase drive (differential); and (d) in-phase actuation. In all driving methods, both top and bottom electrodes are grounded while inner and/or outer middle electrodes are actuated.

The dynamic responses are experimentally measured for a pMUT prototype #1 with the aforementioned stack compositions and thicknesses with inner circular electrode radius of 160 μm, and outer annular electrode radius of 230 μm using a laser Doppler vibrometer (LDV). Under the drive of the inner electrode, the DC displacement is theoretically predicted as 7.49 nm/V and experimentally measured as 7.26 nm/V. Under the outer electrode drive, the DC displacement is theoretically predicted as 7.27 nm/V and experimentally measured as 5.75 nm/V. Theoretically, the inner and outer drive should realize similar DC displacement per unit input voltage if the electrode areas are the same (the inner circular electrode radius is 70% of the pMUT diaphragm radius). The discrepancy in the measured values is attributed to the backside over etch which results in larger diaphragm than the original design. Under the differential drive operation, the measured DC displacement is 13.01 nm/V, only 11.9% lower than the theoretically predicted value of 14.76 nm/V. The differential drive mode produced roughly 2x larger displacement per unit input voltage compared with the result from single electrode actuation. Under the in-phase actuation, the DC displacement has a negligible value of 0.11 nm/V, validating the theoretical expectations and the need of middle electrodes [8].

Figure 7 compares the simulated and measured frequency responses of two pMUT prototypes. When the radius of the pMUT is reduced to 170 μm (prototype #2), a 452nm/V center displacement is measured under resonance at 344kHz. On the other hand, the measured center displacement is 407.4 nm/V under resonance at 198.9 kHz for prototype #1. This is 51% lower than the simulated value in air medium of 841.7 nm/V but is still ~4X larger than the reported results of prior unimorph flat

AlN pMUTs with similar dimensions and frequency [13]. The reduced experimental performance is related to the high residual stress (60 MPa tensile), which increases material damping. It is noted that the residual stress effect is more prominent for pMUTs with larger diaphragms.

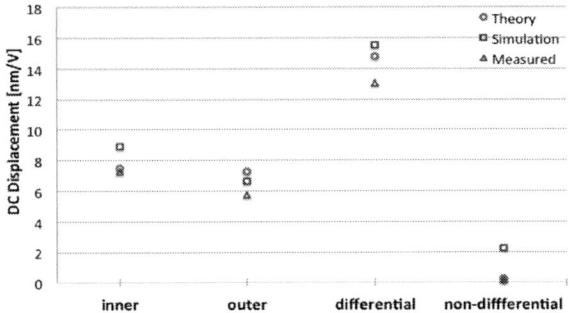

Figure 6: Measured, theoretical, and simulated low frequency displacements of a bimorph pMUT (measured resonant frequency at 206 kHz) using (1) inner electrode, (2) outer electrode, (3) differential drive, and (4) non-differential (in phase) actuation.

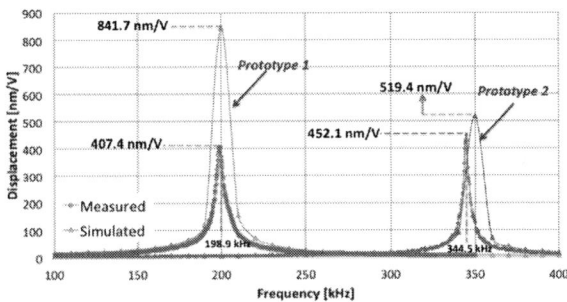

Figure 7: Measured and simulated results of two bimorph pMUTs actuated by the inner electrode with 230 and 170μm in radii and resonant frequencies of 198.9 and 344.5 kHz, respectively.

Table 1: Resonant displacement and resonant frequency of unimorph and bimorph pMUTs.

	Resonant Disp. [nm/V]	Center Freq. [kHz]
Unimorph w/annular electrode [13]	86	291
Unimorph w/ central electrode [13]	104	291
Unimorph w/ two port diff drive [13]	188	291
Bimorph (this work)	452	345

Table 1 compares the performances of previously reported unimorph pMUTs [13] with the same frequency ranges and bimorph pMUT prototypes. The table reveals that the bimorph pMUT with dual electrodes has superior performance because it has higher displacements per input voltage at higher resonant frequencies. The generated acoustic pressure is proportional to the combination of operation frequency and volumetric displacement. Thus, bimorph pMUTs are expected to generate higher pressure at higher frequency for high acoustic power transmission.

ACKNOWLEDGEMENTS
The authors would like to thank Jeffery Clarkson, Ryan Rivers, and Jay Morford from the UC Berkeley Marvell nanofabrication laboratory and Dr. Amir Heidari and Hadi Najar from the Mechanical and Aerospace Engineering Department of UC Davis for their valuable helps. This project is funded in part by GlobalFoundries Singapore Pte. Ltd. under Abu Dhabi-Singapore Twin Lab between the Institute of Microelectronics (IME) and Masdar Institute of Science and Technology, Abu Dhabi, UAE.

REFERENCES
[1] A. J. Tajik *et al.*, "Two-dimensional real-time ultrasonic imaging of the heart and great vessels. Technique, image orientation, structure identification, and validation," *Mayo Clin. Proc.*, vol. 53, no. 5, pp. 271–303, 1978

[2] G. S. Kino, Acoustic Waves: Device, Imaging, and Analog Signal Processing. Upper Saddle River, NJ: Prentice Hall, 1987.

[3] T. Ikeda, Fundamentals of Piezoelectricity, Oxford University Press, 1990.

[4] B. Khuri-Yakub, "Next-gen ultrasound," *IEEE Spectr.*, vol. 46, no. 5, pp. 44–54, May 2009.

[5] B. T. Khuri-Yakub and O. Oralkan, "Capacitive micromachined ultrasonic transducers for medical imaging and therapy," *J. Micromech. Microeng.*, vol. 21, art. no. 54004, May 2011.

[6] S. Shelton *et al.*, "CMOS-compatible AlN piezoelectric micromachined ultrasonic transducers," *in Proc. 2009 IEEE Int. Ultrason. Symp. (IUS)*, pp. 402-405, Rome, Italy, September 20-23, 2009.

[7] A. S. Ergun, G. G. Yaralioglu, and B. T. Khuri-Yakub, "Capacitive micromachined ultrasonic transducers: Theory and technology," *J. Aerosp. Eng.*, vol. 16, no. 2, pp. 76–84, 2003.

[8] K. Smyth *et al.*, "Analytic solution for N-electrode actuated piezoelectric disk with application to piezoelectric micromachined ultrasonic transducers," *IEEE Trans. Ultrason. Ferroelectr. Freq. Control*, vol. 60, no. 8, pp. 1756–1767, Aug. 2013.

[9] P. Muralt *et al.*, "Piezoelectric micromachined ultrasonic transducers based on PZT thin films," IEEE Trans. Ultrason. Ferroelectr. Freq. Control, vol. 52, no. 12, pp. 2276–2288, Dec. 2005.

[10] F. Sammoura, K. Smyth, and S. G. Kim, "Optimizing the electrode size of circular bimorph plates with different boundary conditions for maximum deflection of piezoelectric micromachined ultrasonic transducers," *Ultrasonics*, vol. 53, pp. 328–334, Feb. 2013.

[11] S. Akhbari *et al.*, "Highly Responsive Curved Aluminum Nitride PMUT," in Proc. 27th IEEE Micro Electro Mechanical Syst. Conf.*, pp. 124-127, San Francisco, CA, January 26-30, 2014.

[12] F. Sammoura, S. Akhbari, N. Aqab, M. Mahmoud, and L. Lin, "Multiple Electrode Piezoelectric Micromachined Ultrasonic Transducers," *in Proc. 2014 IEEE Int. Ultrason. Symp. (IUS)*, pp. 305-308, Chicago, IL, USA, September 3-6, 2014.

[13] F. Sammoura *et al.*, "A Two-Port Piezoelectric Micromachined Ultrasonic Transducer," *in Proc. 2014 Joint IEEE ISAF/IWATMD/PFM*, pp. 1-4, State College, PA, May 12-16, 2014.

[14] F. Sammoura *et al.*, "Theoretical Modeling and Equivalent Electric Circuit of a Bimorph Piezoelectric Micromachined Ultrasonic Transducer," *IEEE Trans. Ultrason. Ferroelectr. Freq. Control*, Vol. 59, No. 5, May 2012.

[15] A. Erturk and D. J. Inman, "An experimentally validated bimorph cantilever model for piezoelectric energy harvesting from base excitations," *Smart Mater. Struct.*, vol. 18 (2009) 025009.

CONTACT
*S. Akhbari, tel: +1-510-9267150; sina.akhbari@berkeley.edu

A LOW-NOISE SUB-500μW LORENTZ FORCE BASED INTEGRATED MAGNETIC FIELD SENSING SYSTEM

S. Brenna, P. Minotti, A. Bonfanti, G. Laghi, G. Langfelder, A. Longoni and A. L. Lacaita
Dipartimento di Elettronica, Informazione e Bioingegneria, Politecnico di Milano, Milano, Italy

ABSTRACT

This paper shows, for the first time, a complete magnetic field sensing system including a Lorentz-force sensor operating out of resonance coupled to an integrated circuit for sensing and actuating the device. Working out of resonance, the trade-off between maximum sensing bandwidth and minimum detectable magnetic field is overwhelmed, improving the resolution and enlarging the bandwidth. However, the reduction of signal amplitude makes the readout electronics a critical block. Measurements carried-out on the whole system show an achievable resolution of 180 nT·mA/$\sqrt{\text{Hz}}$ over a 150-Hz bandwidth and an overall power consumption of 460 μW. The integrated readout circuit low-noise performance does not limit the resolution, which is set by the MEMS thermomechanical noise.

INTRODUCTION

There is a growing interest in three-axis magnetic field sensors for low-cost, low-power applications such as electronic compass for smart mobile phones and portable amusements. In these systems, Lorentz-force magnetometers are advantageous over other magnetic sensors, like anisotropic magnetoresistance (AMR), magnetic tunnel junction (MTJ) or Hall-effect devices, because they are built using the same microelectromechanical system (MEMS) technology as for gyroscopes and accelerometers currently used in consumer electronics [1]. MEMS magnetometers based on the Lorentz-force principle and operating at the resonance frequency have been under scientific investigation since a decade [2-4]. However, this mode of operation shows a clear trade-off between achievable maximum sensing bandwidth and resolution [5], as both terms depend on the value of the damping coefficient b, but with opposite effects: a lower damping factor implies lower thermomechanical noise but limits the -3 dB bandwidth of the transfer function. Thus, with this approach, it is difficult to achieve the band required for future applications (50 Hz). In a recent work [4], Lorentz-force MEMS magnetometers operated slightly off-resonance were proposed, following the basic working principle of mode-split operation used for low-power large-bandwidth gyroscopes. Off-resonance operation allows to hold the same signal-to-noise ratio (SNR) per unit bandwidth, extending the maximum achievable bandwidth, but featuring the drawback of a reduced electromechanical sensitivity, defined as the capacitance variation per unit magnetic field. This implies that the readout circuit has to face a lower signal amplitude, with respect to the case of traditional resonance operation, and its contribution to the input equivalent magnetic noise has to be carefully minimized in order to benefit of the reduced minimum detectable signal with a reasonable amount of power. Typical resolutions of MEMS magnetometers from

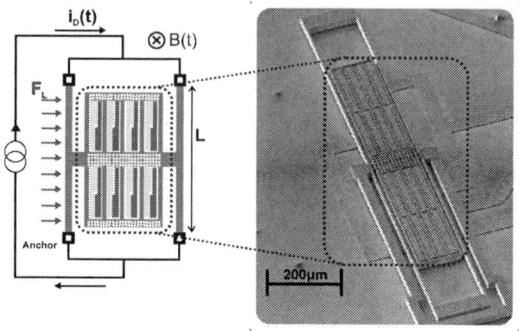

Figure 1. Schematic representation of a Lorentz-force MEMS magnetometer with a scanning electron microscope view of the implemented device.

the literature are in the order of 40 to 500 nT·mA/$\sqrt{\text{Hz}}$ [4], [6], but such values do not include the readout electronics. Today, to the best of our knowledge, no example of integrated circuits for Lorentz-force magnetometers has been reported in literature. This paper proposes, for the first time, a system prototype made of a MEMS Lorentz-force magnetometer coupled to an integrated circuit (IC) featuring a resolution comparable to the state-of-the art magnetic sensors and a selectable bandwidth up to 150 Hz.

MAGNETOMETER DESIGN

The schematic representation of a Lorentz-force magnetometer is shown in Figure 1. Suitable sustaining springs with a stiffness k, anchored to the substrate on one side, hold a central shuttle, which forms with fixed electrodes (stators) a set of two differential parallel-plate capacitors. Each capacitance has a nominal rest value C_0 and is formed by N cells of facing area A and nominal gap g between the plates. An AC drive current $i_D(t)$ flowing through the springs generates, in presence of a component B of the magnetic flux density in the z-direction, i.e. out-of-plane, a Lorentz-force distributed on the length L.

Mode of operation

In [4], the authors propose to drive the sensor out of resonance, with a slight offset frequency $\Delta f = f_s - f_D$ with respect to the resonance frequency, f_s. This method allows to break the bandwidth-resolution trade-off and thus to lower the package pressure in order to achieve a lower detectable magnetic field. In fact, denoting as Q the device quality factor, in the case of i) $Q \gg 1$, ii) $\Delta f \ll f_s$ and iii) $\Delta f \gg f_s/(2Q)$, the transfer function between the Lorentz-force and the magnetometer displacement can be considered

Table 1
Design parameters for the magnetometer of this work

Symbol	Quantity	Value
m	device mass	$1.2 \cdot 10^{-9}$ kg
k	device stiffness	19 N/m
L	spring length	1060 μm
g	gap	2.1 μm
Q	quality factor	800
p	pressure	0.25 mbar
b	damping coefficient	$2 \cdot 10^{-7}$ kg/s
f_s	resonant frequency	20 kHz
BW_{-3dB}	mechanical bandwidth	12 Hz
C_0	rest capacitance	300 fF

flat (within a suitable electronic bandwidth) and equal to [4]

$$\frac{x}{F_L} \cong \frac{1}{2k}\frac{f_s}{2\Delta f} = \frac{Q_{eff}}{2k}, \tag{1}$$

Q_{eff} being an effective quality factor that takes into account the residual increase of the gain with respect to the DC value $(1/k)$ due to the operation close to the resonance. With respect to the operation at resonance, the same Lorentz force causes a lower displacement, being $Q_{eff} \ll Q$. The minimum detectable signal can be evaluated considering that the system bandwidth is no longer set by the mechanical filtering of the sensor, but by the electronic circuit providing the signal amplification and demodulation. Assuming the electronic noise being negligible with respect to the mechanical noise, the minimum detectable signal is

$$B_{min} = \frac{4}{i_D L} \cdot \sqrt{k_B T b \cdot BW}, \tag{2}$$

where BW is the -3 dB bandwidth of the following baseband electronic filter. This expression is identical to the one holding for the device operated at resonance. In that case, the bandwidth is set by the mechanical transfer function of the device $BW = b/(4\pi m)$. On the contrary, in the off-resonance operating mode the sensing bandwidth is no longer set by the damping factor, which can be therefore minimized to reduce the thermomechanical noise by acting on the device geometry and pressure. At the same time, the bandwidth can be extended up to a relevant fraction of the frequency offset, while keeping low the intrinsic noise if b is properly reduced. However, it's worth pointing out that the minimum measurable magnetic field should also account for the electronics noise, which may reduce the minimum detectable field if the readout circuit is not accurately designed.

Mechanical design

The proposed Lorentz-force magnetometer is shown in Figure 1. This device has been fabricated through the ST-Microelectronics ThELMA (Thick Epitaxial Layer for Microactuators and Accelerometers) process. The length of the springs (anchor point to anchor point) is 1060 μm, and the overall width is 150 μm. The process height is 22 μm, the same as the one used for accelerometers and gyroscopes in this industrial process. The sensing cells are implemented by eight differential parallel plates, each formed by pairs

of capacitors having a 330-μm length and a nominal gap at rest of 2.1 μm. The device has been encapsulated at a nominal pressure of 0.25 mbar, the minimum achievable, corresponding to a damping factor of $2.1 \cdot 10^{-7}$ kg/s. This value leads to a quality factor Q of about 800 at the resonance frequency of 19800 Hz and a mechanical bandwidth of 12 Hz. The device is intentionally driven with a negative frequency offset of 200 Hz, corresponding to an effective quality factor of about 50. The gain of the sensor, normalized to the drive current i_D, is $(\Delta C / (\Delta B \cdot i_D))_{@~200~Hz} = 0.33$ zF/(nT·mA) at the 200-Hz frequency offset, while at resonance it increases to $(\Delta C / (\Delta B \cdot i_D))_{@~res} = 5.35$ zF/(nT·mA). The geometrical and mechanical parameters of the MEMS device are summarized in Table 1.

DRIVE AND SENSE ELECTRONICS

The capacitive sensing and driving circuit is integrated in a single die adopting the Austriamicrosystems (AMS) 0.35-μm process and 3-V supply voltage. The simplified schematic of the sense electronics is shown in Figure 2 together with the microphotograph of the die.

The first stage consists of a fully-differential charge amplifier, preferred to a transimpedance stage due to its lower noise, directly connected to the magnetometer stators. While the rotor is ideally at the ground potential, the stators are biased at the common-mode input voltage of the charge-amplifier that, in turn, is equal to the output common-mode voltage of the differential amplifier. This voltage can be externally tuned in order to bias the magnetometer capacitors at different voltage values changing the overall sensitivity. In fact, if the driving current flowing into the springs is $i_D \sin(2\pi f_D t)$, the differential current read by the charge amplifier is

$$i_S(t) \cong V\frac{dC}{dt} \cong 2\pi f_D V_{bias} \Delta C \cos(2\pi f_D t), \tag{3}$$

where C is the differential device capacitance and V_{bias} the stators bias voltage set by the output common-mode network of the charge amplifier. This sensing current is then amplified by the feedback capacitors, $C_F = 25$ fF. This value guarantees a sufficiently accurate gain, being the parasitic capacitance between the charge amplifier input and output of a few fF. The feedback resistors are implemented by two PMOS transistors biased in sub-threshold region [7]. These transistors lead to a resistance larger than 1 GΩ assuring that the sensing current is amplified by the feedback capacitance at frequencies close to resonance. Furthermore, this resistance is large enough to make its noise negligible with respect to the amplifier noise.

A second charge amplifier is inserted after the first stage in order to increase the signal dynamics. This second charge amplifier features a gain of 20 dB, adopting an input capacitance of 2.5 pF and a feedback capacitor of 25 fF.

The amplified signal is then demodulated by a passive mixer that is driven with a signal having the same phase/frequency of the AC current flowing into the sensor spring. This allows to downconvert the magnetic field signal that is modulated at a frequency $f_s - \Delta f$ by the drive current $i_D(t)$.

Finally, the mixer output is converted to a single-ended signal by an instrumentation amplifier before being filtered by a 2nd order g_m-C filter. Its low-pass frequency can be tuned between 10 and 150 Hz. A highly selective filter is desirable

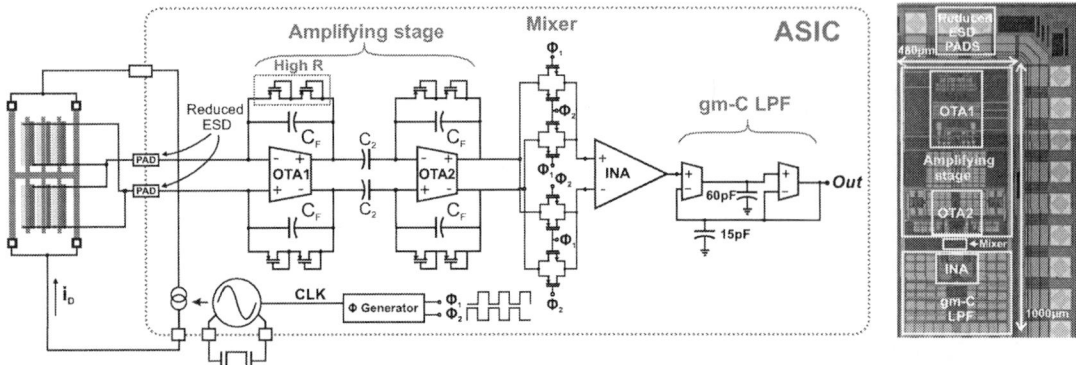

Figure 2. Schematic representation of the readout circuit and microphotograph of the electronic die.

in order to filter out the sensor noise around the resonance.

The electronic noise at the output of the overall circuit is mainly due to the input differential pair in the first charge amplifier OTA. In order to evaluate the equivalent magnetic field noise, let us refer to the electronic noise at the output of the first charge amplifier. The corresponding noise power spectral density is given by

$$\overline{E_{n,out}^2} \cong \overline{E_{n,eq}^2} \cdot \left(1 + \frac{C_T}{C_F}\right)^2 \cong \overline{E_{n,eq}^2} \cdot \left(\frac{C_0}{C_F}\right)^2 \cdot \left(1 + \frac{C_P}{C_0}\right)^2,$$
(4)

where $\overline{E_{n,eq}^2}$ is the input-referred OTA noise spectral density, C_P is the parasitic capacitance at each OTA input node and $C_T = C_P + C_0$.

The contribution of the sensing electronics to the equivalent input magnetic noise can be easily evaluated as

$$B_{min} \cong \frac{1}{i_D L} \cdot \frac{\sqrt{\overline{E_{n,eq}^2} BW} \left(1 + \frac{C_P}{C_0}\right) gk}{Q_{eff} V_{bias}}.$$
(5)

From Equation (5) it's clear that as the device is operated out of resonance, the effective quality factor is lower than at resonance and, thus, the electronics noise has a larger impact on the system resolution. Moreover, the parasitic capacitance has to be minimized and kept comparable to the MEMS capacitance in order not to amplify the electronic noise. In the present design, the parasitic capacitance has been minimized directly bonding the two dies and adopting custom-designed PADs. In this way, C_T has been kept lower than 3 pF, compared to a MEMS capacitance of 300 fF, while a standard PAD would give a C_P capacitance larger than 6 pF.

EXPERIMENTAL RESULTS

The MEMS die has been wire bonded to the electronic die and glued on a socket board, as shown in Figure 3.

The sensitivity of the overall system, evaluated as the ratio between the output voltage and the input magnetic field, has been measured adopting a Helmoltz coil generating z-axis magnetic fields from -5 to +5 mT. Measurements were performed driving the sensor spring with a 20-μA$_{rms}$ current and with a MEMS bias voltage of 2.2 V. The resulting z-axis system sensitivity is about 63 V/T, while the cross-axis rejection is larger than 48 dB, limited by manual alignment. Figure 4 shows the resulting characteristic between the output

Figure 3. Picture of the MEMS magnetometer wire bonded to the electronic die and glued on a socket board.

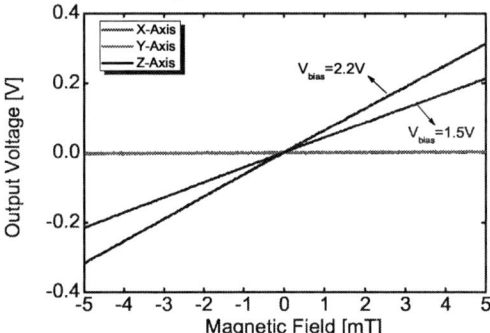

Figure 4. Magnetic field to output voltage characteristic.

voltage and the input magnetic field. Also, the input-output characteristic with a MEMS bias voltage of 1.5 V is shown. In this case, the sensitivity is lower and equal to 42 V/T. The driving current of 20 μA$_{rms}$ has been chosen considering a maximum output offset up to 300 mV (i.e. 1/10 of the available output dynamic). The origin of this offset is under investigation. Possible offset sources are simultaneous mechanical and electrical asymmetries, and/or direct feedthrough of the drive signal.

The measured system bandwidth is in the range 10-150 Hz and can be selected by simply tuning the g_m-C filter bias current. Figure 5 shows the normalized frequency response under AC z-axis magnetic field for the minimum and the maximum system bandwidth. At the largest bandwidth, the MEMS resonant peak begins to appear due to the limited cut-

978-1-4799-7956-1/15 $31.00 © 2015 IEEE

Table 2
Comparison of the presented magnetic field sensing system performance with the state-of-the-art

System	BW	Driving Current	Supply	Spring Length	FoM	IC Power
	[Hz]	[μA_{rms}]	[V_{DC}]	[μm]	[nT·mA/$\sqrt{Hz}$]	[μW]
[2]	2	5900	4	1600 μm	110	N/A
[3]	50	900	4	370 μm	450	N/A
[5]	>100	35	6	1060 μm	170	N/A
This Work	10-150	20	3	1060 μm	180	460

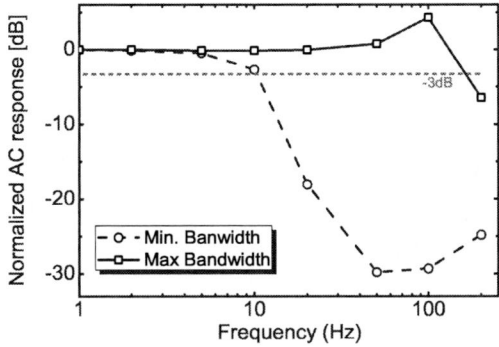

Figure 5. Normalized AC response of the overall system.

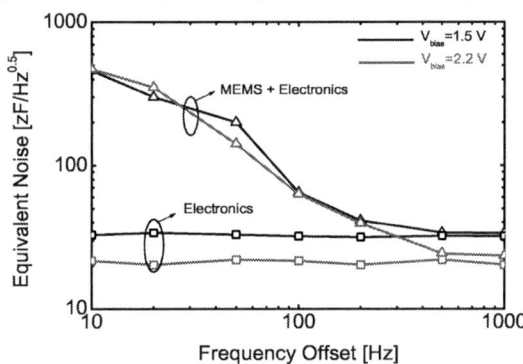

Figure 6. Measured equivalent noise of the system quoted as rms capacitance variation in a sensing bandwidth of 1 Hz.

off frequency of the low-pass filter.

Noise measurements have been performed by setting to zero both the AC drive current and the magnetic field, and by performing the demodulation at different frequency offsets. For a null bias voltage, the thermomecanical noise does not produce any current fluctuation in the capacitances, thus only the electronic noise contributes to the output noise. On the contrary, when a bias voltage is applied to the device, both noise contributions are effective. Figure 6 shows the input equivalent noise quoted as rms capacitance variation per unit sensing bandwidth zF/$\sqrt{Hz}$ at different frequency offsets and for V_{bias} = 1.5 V and 2.2 V. In the latter case, the equivalent electronics noise is negligible with respect to the thermomechanical contribution up to a frequency offset of 500 Hz. At 200-Hz offset from resonance, the equivalent noise is about 41 zF/$\sqrt{Hz}$, corresponding to a sensor resolution of

180 nT·mA/$\sqrt{Hz}$. This value well compares to the results obtained in [5], with board level electronics.

CONCLUSIONS

Here we present a low-power and low-noise magnetic field system composed of a Lorentz-force magnetometer coupled to an ASIC performing the sensor drive and readout. The magnetometer works off-resonance to break the sensing bandwidth-resolution trade-off. The performance of the whole system are compared with the state-of-the-art Lorentz-force magnetic sensors in Table 2. Despite this is the first system with integrated readout electronics for Lorentz-force MEMS magnetometers, the system resolution well compares with the other works, but featuring a lower supply voltage and a larger sensing bandwidth.

ACKNOWLEDGMENTS

The authors gratefully acknowledge ST-Microelectronics for device fabrication. This work has been performed in the context of the ENIAC JU project Lab4MEMS.

REFERENCES

[1] J. Lenz and A. S. Edelstein, "Magnetic sensors and their applications," *IEEE Sensors Journal*, pp. 631649, June 2006.

[2] M. Li and et al., "Single structure 3-axis Lorentz force magnetometer with sub-30 nT/pHz resolution," *Proc. of MEMS*, pp. 8083, 2014.

[3] H. Emmerich and M. Schofthaler, "Magnetic field measurements with a novel surface micromachined magnetic-field sensor," *IEEE Trans. Electron Devices*, vol. 47, no. 5, pp. 972977, May 2000.

[4] G. Langfelder and A. Tocchio, "Operation of Lorentz force MEMS magnetometers with a frequency offset between driving current and mechanical resonance," *IEEE Trans. on Magnetics*, vol. 50, no. 1, pp. 10431056, Jan 2014.

[5] G. Langfelder et al., "Off-resonance low-pressure operation of Lorentz force MEMS magnetometers," *Trans. on Industrial Electronics*, vol. 21, pp. 10021010, Mar 2013.

[6] J. Kynnarainen et al., "A 3D microelectromechanical compass," *Sensors and Actuators A*, vol. 142, no. 2, pp. 561568, 2008.

[7] A. Bonfanti et al., "A multi-channel low-power IC for neural spike recording with data compression and narrow-band 400-MHz MC-FSK wireless transmission," *Proc. of ESSCIRC 2010*, vol. 47, no. 4, pp. 330333, 2010.

978-1-4799-7956-1/15 $31.00 © 2015 IEEE

CHIP-SCALE ELECTRODYNAMIC SYNTHETIC JET ACTUATORS

Shashank G. Sawant[1]*, Eric A. Deem[2], Dominic J. Agentis[3], Louis N. Cattafesta[2],
and David P. Arnold[1]

[1]Interdisciplinary Microsystems Group, University of Florida, Gainesville, Florida, USA
[2]Florida Center for Advanced Aero-Propulsion, Florida State University, Tallahassee, Florida, USA
[3]Department of Mechanical Engineering, Virginia Tech, Blacksburg, Virginia, USA

ABSTRACT

We report a chip-scale electrodynamic synthetic jet actuator that integrates both a coil and a wax-bonded NdFeB permanent magnet using standard silicon micromachining processes. A copper micro-coil and a poly-dimethyl-siloxane (PDMS) diaphragm with the magnet are fabricated onto two different dies. An orifice, through which the jet is synthesized, is laser cut on the coil die. The dies are then assembled together to create a device with the dimensions 7.5 mm × 7.5 mm × 1 mm. Operating at 180 Hz and 20 mW input power, the device generates a fluidic jet with a peak velocity of 2.1 m/s.

INTRODUCTION

A synthetic jet actuator comprises an oscillating diaphragm coupled with a cavity having an orifice (Fig. 1). The oscillating diaphragm creates cyclical phases of ingestion and expulsion of fluid through the orifice, which under the right conditions, result in the formation of a jet [1]. The jet is synthesized entirely out of the ambient fluid and no net mass is added to or removed from it [2]. Consequently, synthetic jets are also referred to as zero-net-mass-flux actuators [3,4].

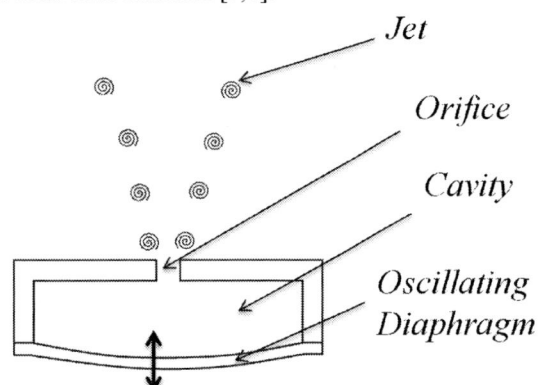

Figure 1: Fundamental components of a synthetic jet actuator.

The primary application of synthetic jet actuators is in the field of active flow control [2]. In 2000, McCormick demonstrated boundary layer separation control over a wing section, using a directed synthetic jet driven by a speaker [5]. Amitay *et al.* were able to increase the stall of an unconventional airfoil (made of a cylindrical section in the front and a NACA airfoil section in the rear) from about 5° to about 17° using synthetic jets [6]. Apart from flow control, synthetic jet actuators have also shown promise in cooling electronic devices. In 2003 Kercher *et al.* were able to reduce the thermal resistance of their heated chip setup by about 49% using a synthetic jet [7]. In 2005, Mahalingam and Glezer presented a heat sink fitted with synthetic jets and were able to achieve cooling much more efficiently than a typical heat sink with a fan [8].

For oscillating the diaphragm, various actuation mechanisms like piezoelectric, electrodynamic [2] and even electrostatic, have been explored. Of these, the electrodynamic and the piezoelectric actuation schemes are the two most popular methods of actuation [9]. Most of the piezoelectric synthetic jet actuators have a natural frequency in the range of a few thousand hertz and operate with a low stroke. Also, these actuators are known to suffer from mechanical failure of the diaphragm after prolonged use [10]. In this light, electrodynamic synthetic jet actuators offer an attractive alternative, as they can be operated with a higher stroke at an operating frequency of only a few hundred hertz [11]. Operating at a lower frequency not only makes them quieter (inefficient acoustic radiation) but also more effective, as they can leverage low frequencies of flow instabilities for enhanced flow interaction [12].

The location of the synthetic jet, relative to the point of flow separation over an airfoil, is an important factor contributing to the overall effectiveness of the actuator [6]. By fabricating a device with a small surface area, it could be possible to place the jet in confined structures or even conformably cover the outer surfaces of aircraft structures. Other benefits of micro-actuators include lower weight, lower power consumption, and individual addressability.

In 2010, Gimeno *et al.* [13] presented a compact synthetic jet device with a volume of about 1 cm^3. Operating in the frequency range of 400 Hz to 700 Hz, it was able to produce a jet with the maximum velocity of 55 m/s. The peak power consumption of the device was reported to be 500 mW. In this work, we present an electrodynamic device which is 16x smaller than the one presented by Gimeno *et al.* Operating at a natural diaphragm frequency of 180 Hz, the device consumes only 20 mW power and produces a peak output velocity of 2.1 m/s.

FABRICATION

The schematic of the actuator presented in this work is shown in Fig. 2. Two dies — the diaphragm die and the coil die — form the primary components of the actuator. The diaphragm die consists of a circular PDMS diaphragm with a concentric silicon cup. The cup holds the wax-bonded NdFeB powder magnet. The coil wafer has an electroplated copper micro-coil on its surface. It also has an orifice out of which the jet emanates.

Coil die

The fabrication steps of the coil die are shown in Fig. 3 a. First, a 400 nm thick PECVD oxide is deposited on a

550-μm-thick Si wafer to provide electrical insulation from the substrate. Then a 75-nm-thick layer of Cr is sputtered on the oxide as an adhesion layer. That is followed by an 850-nm-thick layer of Cu, which is again followed by a 75-nm-thick layer of Cr. This micron-thick metal sandwich forms the connecting layer, connecting the innermost coil turn to an externally located bond pad.

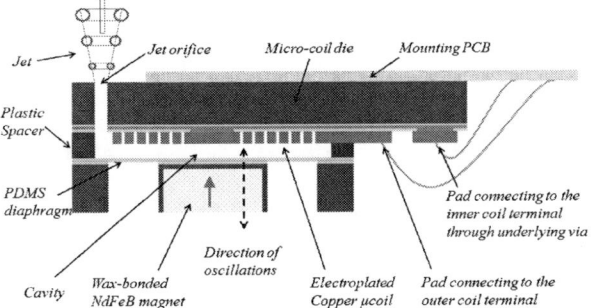

Figure 2: Schematic of the electrodynamic synthetic jet actuator.

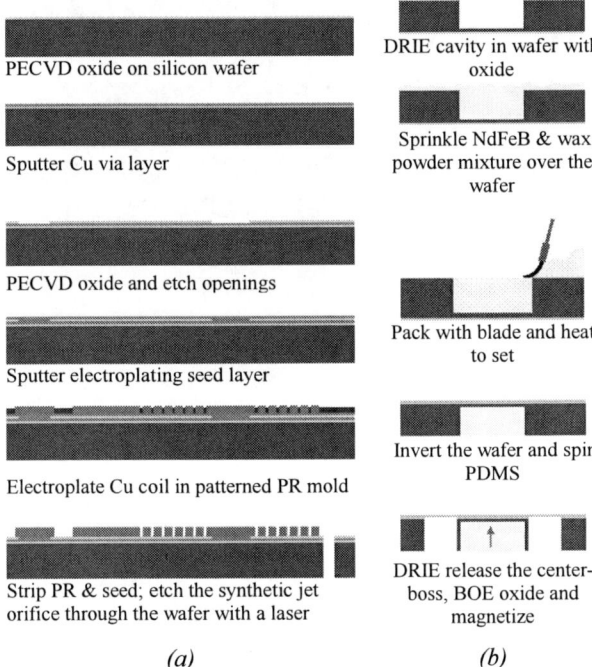

Figure 3: Fabrication steps for the actuator dies: a) coil die and b) diaphragm die.

On this connecting layer, a 400-nm-thick PECVD oxide is again deposited. This isolates the coil layer from the connecting layer. A layer of the positive photoresist (PR) AZ9260 patterned with two openings in the oxide layer. The oxide layer is then dry-etched using CF_4 and O_2 to create the two openings which help make the connection from the center of the coil to an outside bond-pad through the connecting layer. Next, the photoresist is washed away using a heated bath of N-Methyl-2-pyrrolidone (NMP) after which the wafer is washed and dried. On the etched oxide layer an electroplating seed layer is deposited by sputtering 50 nm of Cr followed by 100 nm of Cu. Then, a 35 μm thick layer of PR AZ9260 is spun and patterned on the seed layer, forming a mold for subsequent electroplating of the coil using a standard Cu

electroplating bath. After this, the PR mold and the seed layer are removed. The resulting coil is about 35 μm thick, with a trace width of 35 μm and pitch of 70 μm. It has an inner radius of 1 mm and an outer radius of 2.5 mm, with 21 turns from the center to its edge. The bond pad that connects to the inner terminal of the coil is a square with an area of 1 mm^2. Since it is electroplated along with the coil, it too has a thickness of 35 μm. The other pad, i.e. the outer terminal of the coil, is simply an extension of the coil (Fig. 4).

Figure 4: Coil and diaphragm dies compared to a penny.

Diaphragm die

Fig. 3 b shows the fabrication steps for the diaphragm die. A 400-nm-thick PECVD oxide is deposited on a 550 μm Si wafer. The other side of wafer is coated with a PR layer of AZ9260 about 12 μm thick, and patterned into circles corresponding to the magnets. The magnet cavities are then etched using deep reactive ion etching (DRIE) to a depth of about 425 μm after which the PR coat is stripped away.

Fabrication of the magnets uses a combination of the techniques presented in [14] and [15]. Two NdFeB powders, MQP-S-11-9 and MQFP-B (Magnequench, Inc.), are mixed together in equal proportions by weight. The powders have a particle size of 50 μm and 15 μm, respectively. To this, a wax powder Logitech 0CON-196 is added, to create a mixture with the proportions 47:47:6 by weight, for MQP-S-11-9 : MQFP-B : 0CON-196. The mixture is ground to a fine powder using a mortar and a pestle, and is then spread over the wafer with the DRIE cavities. The powder is packed tightly into the cavities by repeatedly wiping the wafer with a blade. Then the wax is melted and the magnet particles are bound by heating the wafer in an oven at 165 °C for 10 minutes. A test magnet fabricated using the same mixture and process is measured to have a coercivity H_{ci} = 740 kA/m, a remanent flux density B_R = 0.45 T and a maximum energy product BH_{max} = 33 kJ/m^3.

Next, a Sylgard 184 PDMS kit is used to create the PDMS membrane. The liquid mixture is spun over the oxide deposited in the first step to get a coat about 15 μm thick, followed by an oven cure at 100 °C for an hour.

The magnet side of the wafer is again coated with PR and patterned with annular rings, concentric to the circular magnets. DRIE is used to etch completely through the wafer by using the oxide as an etch-stop. The exposed oxide is removed using buffered oxide etch (BOE), and the magnets are released, now being held only through the PDMS diaphragm (Fig. 4). The diameter of the diaphragm and the center-boss with magnet are 4.4 mm and 2.8 mm, respectively.

Laser-cutting and assembly

The coil wafer and the diaphragm wafer are diced using an Oxford Lasers J-355PS laser micromachining workstation. Also using the laser, a 440 μm diameter orifice is made in the coil dies, and a double-sided tape, less than 100 μm thick, is cut to the shape of the spacer (Fig 5 a). To assemble the actuator, the coil die is secured on a PCB using two-part epoxy. The coil pads are soldered to the PCB and the spacer is placed on the coil die. The diaphragm die is visually aligned with the coil die and placed on the spacer. A thin layer of two-part epoxy is applied on the periphery to create a stronger bond between the two dies (Fig. 5 b).

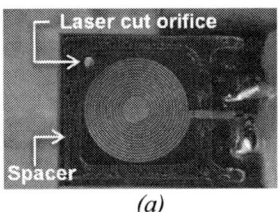 

(a)   *(b)*

Figure 5: Optical images of the actuator dies: a) coil die and b) diaphragm die.

FLUIDIC CHARACTERIZATION

2-D particle image velocimetry (PIV) is used in order to measure the fluidic output of the actuator. PIV is a technique that captures two successive images of particles suspended in the flow field (referred to as seed), and the velocity field is calculated by the particle shift determined by spatial cross-correlation and the time between the image pairs [16].

The actuator is mounted inside a large quiescent box, in which the free stream flow is negligible. Four of the six sides of the box are transparent to allow for the optical access required for PIV. A 200 mJ Evergreen Nd:YAG laser is used for particle illumination. A compound spherical lens system is used to adjust the beam waist to be as close to the actuator orifice as possible.

Additionally, a cylindrical lens is used to fan the light out into a vertical sheet. The nominal thickness of the lightsheet is approximately 0.2 mm and intersects the center of the actuator orifice. The particles are photographed by a LaVision VC Imager Pro X CCD camera with an 80 mm lens. In order to achieve uniform seeding density with sufficiently small particles, the chamber is seeded with incense smoke. In each image, the particle diameter is nominally 5 pixels. The particle density is approximately 450 particles per square millimeter. The time between PIV snapshots is set to $\Delta t = 30$ μs. A function generator is used to generate the driving

signal, which is amplified using a power amplifier. The actuator current amplitude and frequency are monitored using an oscilloscope.

The frequency response measurements are carried out at two current levels: 100 mA$_{pp}$ and 200 mA$_{pp}$. The natural frequency of the diaphragm is found to be 180 Hz by measuring its displacement response using laser displacement sensor. At this frequency, phase-locked PIV image-pairs are acquired for 12 phases, for both input levels, and 200 image pairs are recorded per phase. The trigger is generated with reference to the driving signal of the actuator by a delay generator.

In addition to the measurements at 180 Hz, measurements are also performed at three more frequencies of operation: 100 Hz, 140 Hz and 220 Hz. For these measurements, the trigger is randomized w. r. t. the signal driving the actuator, and a thousand image pairs are captured at each of these frequencies.

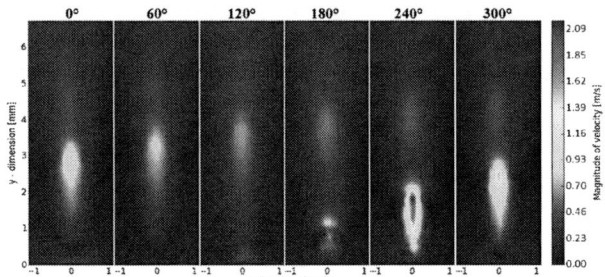

Figure 6: Evolution of the synthetic jet over a cycle with an input current of 200 mA$_{pp}$ at 180 Hz.

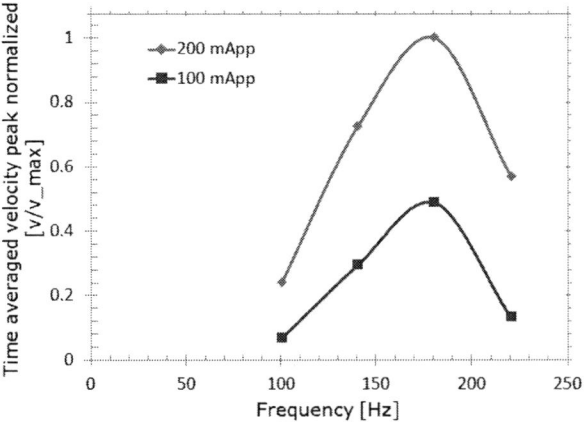

Figure 7: Output frequency response of the actuator for inputs corresponding to 100 mA$_{pp}$ and 200 mA$_{pp}$.

The average PIV flow fields obtained from the image pairs at an input of 200 mA$_{pp}$ at 180 Hz, for 6 of the 12 phases recorded are shown in Fig. 6. At this input, the peak velocities in a window 2 mm high, 440 μm wide, centered just above the orifice are averaged for all image pairs at the phase 240°, and the maximum velocity of the jet is thus measured to be 2.1±0.2 m/s. The frequency response, plotted in Fig. 7, is calculated by measuring the peak value in the flow field obtained by averaging all the image pairs (2400 phase locked images pairs at 180 Hz and 1000 image pairs every other frequency of operation) for each input case.

978-1-4799-7956-1/15 $31.00 © 2015 IEEE

Table 1: Comparison of electrodynamic synthetic jet actuators

Actuator	Peak output velocity [m/s]	Natural Frequency [Hz]	Jet orifice area [mm^2]	Device volume [cm^3]	Power consumption [W]	Peak momentum per unit power & volume [s^{2}m^{-4}]
Liang, Kuga, Taya [11]	190	220	10	801	200	0.007
Kercher et al [7]	26	99	4.45	22.5	1	3.03
Gimeno et al [13]	55	600	0.28	1	0.5	18.2
This work	2.1	180	0.15	0.062	0.02	150

CONCLUSIONS

The fabrication and fluidic characterization of a chip-scale synthetic jet actuator are presented in this work. The actuator consists of a copper micro-coil and a PDMS diaphragm connected to an integrated wax-bonded powder NdFeB magnet. Silicon micromachining techniques are used for fabricating the coil and the diaphragm, on two separate dies. Using these dies the actuator is assembled, and PIV measurements indicate a peak velocity output of 2.1 m/s for input current of 200 mA$_{pp}$ at 180 Hz with 20 mW input power.

Several other previously reported electrodynamic synthetic jet actuators are summarized in Table 1. Compared to the device presented by Gimeno *et al.* [13], perhaps the smallest electrodynamic synthetic jet previously published, the device presented in this work is 16x smaller. It is also about 8 times more efficient at producing fluidic momentum per device volume per unit power, a feature that makes the design attractive for applications like micro air vehicle control and cooling of electronics in smartphones and tablets.

ACKNOWLEDGEMENTS

This work was supported in part by the Army Research Office Grant W911NF-09-1-0511.

REFERENCES

[1] Y. Utturkar, R. Holman, R. Mittal, B. Carroll, M. Sheplak, and L. Cattafesta, "A Jet Formation Criterion for Synthetic Jet Actuators," 41st Aerospace Sciences Meeting and Exhibit, Reno, Nevada, 2003.

[2] A. Glezer and M. Amitay, "Synthetic Jets," *Annu. Rev. Fluid Mech.*, vol. 34, pp. 503–529, 2002.

[3] M. Oyarzun and L. N. Cattafesta, "Design and Optimization of Piezoelectric Zero-Net Mass-Flux Actuators," in *5th Flow Control Conference, AIAA*, 2010.

[4] S. G. Sawant, M. Oyarzun, M. Sheplak, L. N. Cattafesta, and D. P. Arnold, "Modeling of Electrodynamic Zero-Net Mass-Flux Actuators," *AIAA J.*, vol. 50, no. 6, pp. 1347–1359, Jun. 2012.

[5] D. C. McCormick, "Boundary layer separation control with directed synthetic jets," 38th Aerospace Sciences Meeting and Exhibit, Reno, Nevada, 2000.

[6] M. Amitay, D. R. Smith, V. Kibens, D. E. Parekh, and A. Glezer, "Aerodynamic Flow Control over an Unconventional Airfoil Using Synthetic Jet Actuators," *AIAA J.*, vol. 39, no. 3, pp. 361–370, Mar. 2001.

[7] D. S. Kercher, J. B. Lee, O. Brand, M. G. Allen, and A. Glezer, "Microjet cooling devices for thermal management of electronics," *Compon. Packag. Technol. IEEE Trans. On*, vol. 26, no. 2, pp. 359 – 366, Jun. 2003.

[8] R. Mahalingam and A. Glezer, "Design and thermal characteristics of a synthetic jet ejector heat sink," *J. Electron. Packag.*, vol. 127, no. 2, pp. 172–177, 2005.

[9] L. N. Cattafesta and M. Sheplak, "Actuators for Active Flow Control," *Annu. Rev. Fluid Mech.*, vol. 43, no. 1, pp. 247–272, 2011.

[10] L. Cattafesta, D. Shukla, S. Garg, and J. Ross, "Development of an adaptive weapons-bay suppression system," *AIAA Pap.*, vol. 1901, p. 1999, 1999.

[11] Y. Liang, Y. Kuga, and M. Taya, "Design of membrane actuator based on ferromagnetic shape memory alloy composite for synthetic jet applications," *Sens. Actuators Phys.*, vol. 125, no. 2, pp. 512 – 518, 2006.

[12] C. Lee, G. Hong, Q. P. Ha, and S. G. Mallinson, "A piezoelectrically actuated micro synthetic jet for active flow control," *Sens. Actuators Phys.*, vol. 108, pp. 168 – 174, 2003.

[13] L. Gimeno, A. Talbi, R. Viard, A. Merlen, P. Pernod, and V. Preobrazhensky, "Synthetic jets based on micro magneto mechanical systems for aerodynamic flow control," *J. Micromechanics Microengineering*, vol. 20, p. 075004 (7pp), May 2010.

[14] N. Wang and D. P. Arnold, "Batch-Fabricated Electrodynamic Microactuators With Integrated Micromagnets," *Magn. IEEE Trans. On*, vol. 46, no. 6, pp. 1798–1801, Jun. 2010.

[15] O. D. Oniku and D. P. Arnold, "High-energy-density permanent micromagnets formed from heterogeneous magnetic powder mixtures," in *Micro Electro Mechanical Systems (MEMS), 2012 IEEE 25th International Conference on*, 2012, pp. 436–439.

[16] E. J. Stamhuis, "Basics and principles of particle image velocimetry (PIV) for mapping biogenic and biologically relevant flows," *Aquat. Ecol.*, vol. 40, no. 4, pp. 463–479, 2006.

CONTACT

*S. G. Sawant, tel: +1-352-2261688; sgsawant@gmail.com

FULLY-POLYMERIC NEM RELAY FOR FLEXIBLE, TRANSPARENT, ULTRA-LOW POWER ELECTRONICS AND SENSORS

Yanbiao Pan, Fangzhou Yu, and Jaeseok Jeon

Rutgers, The State University of New Jersey, Piscataway, NJ 08854, USA

ABSTRACT

A fully-polymeric nano-electro-mechanical (NEM) relay based on a conductive polymer, Poly(3,4-Ethylenedi oxythiophene):Polystyrene-Sulfonate (PEDOT:PSS) and dielectric polymers is proposed for the first time to enable flexible, transparent, ultralow-power electronics and sensors, and the first functional prototype fabricated using a five-mask low-thermal-budget process is demonstrated. The prototype shows zero off-state leakage current, abrupt on/off switching, complementary switching behavior, and relatively high on/off current ratio. Exploiting the water-absorption behavior of the polymers, the potential use of the relay as a biochemical sensor is also demonstrated.

INTRODUCTION

A conductive polymer, Poly(3,4-Ethylenedioxythioph ene):Polystyrene-Sulfonate (PEDOT:PSS), has been explored for use in solar cells [1] and thin-film-transistors [2] as a candidate to replace Indium Tin Oxide (ITO) due to its optical-transparency (> 90 % in the visible spectrum [3]) and relatively high conductivity (~10^3 S·cm^{-1} [4]); and stretchable electrodes [5] and strain gauges [6] due to its plasticity and piezoresistive behavior. However, because of its hygroscopic behavior [7], which causes swelling and shrinking of the film upon absorption of moisture, and chemical incompatibility with acid-sensitive photoresists and alkaline developers and strippers used in conventional photolithography [8], fabrication of most PEDOT:PSS-based devices has had to resort to nonconventional lithographic techniques such as roll-to-roll and inkjet printing or indirect patterning methods such as lift-off [9]. Most, if not all, PEDOT:PSS-based devices reported to date are fabricated using a single mask layer with poor photolithography resolution, thus limiting their potential applications.

In this work, PEDOT:PSS is proposed as an electrode and sensing layer and polymer dielectrics as an insulator for NEM relays, to enable flexible, transparent, and ultralow-power electronics and sensors, and the functional prototype fabricated with a five-mask low-thermal-budget process is demonstrated. Since the prototype relay shows zero off-state leakage current and abrupt on/off switching and comprises polymers with a low Young's modulus (< ~5 GPa), the relay can be operated as a switch at a very low supply voltage (< 100 mV) to achieve very low dynamic power consumption without trading off zero static power consumption and thus realize a much better energy efficiency than MOS transistors. The relay can also be operated as a humidity (potentially biochemical) sensor due to the water-absorption behavior of polymers, which induces changes in on-state current and threshold voltage.

The relay structure (**Fig. 1**) comprises six terminals: a movable gate, a body, and two pairs of source/drain. The operation of the relay mimics that of MOS transistors, in that current flow between the source and drain on either

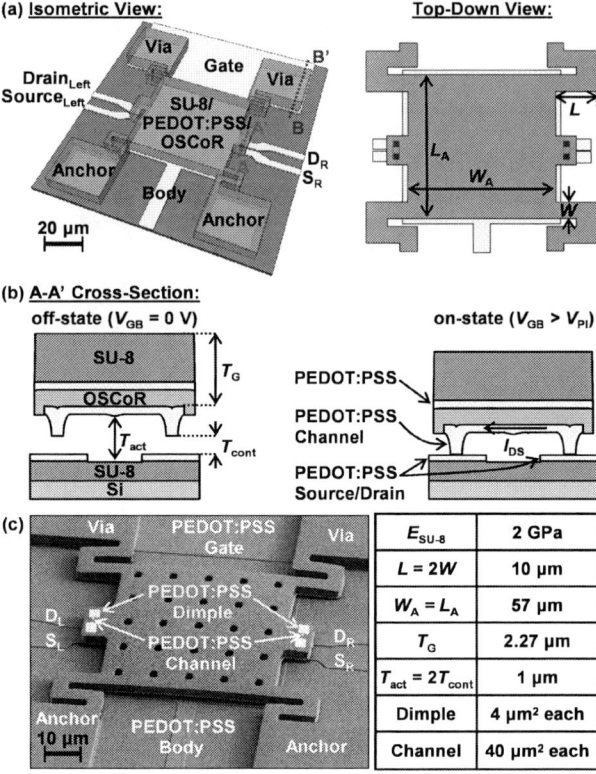

Figure 1: (a) Schematic of an all-polymer, 1-input/2-output relay. The movable stack is electrically connected to the gate through the vias. (b) Off-state: an air gap prevents current to flow between the source and drain on each side. On-state: electrostatic force between the gate and body brings both channels into contact with the pairs of source and drain. (c) SEM and design parameters.

side is controlled by the voltage between the movable gate and body electrodes (V_{GB}). As V_{GB} applied across the actuation gap (T_{act}) increases, the electrostatic attractive force (F_{elec}) between the gate and body increases parabolically, while the spring restoring force of the folded-flexures (F_{spring}) increases linearly. When F_{elec} exceeds F_{spring}, the movable gate snaps down abruptly, and the conductive polymer channel (underneath a polymer gate dielectric) is brought into contact with the pair of source and drain on either side to conduct current. The pull-in voltage (V_{PI}) at which the gate stack pulls in depends on design parameters [10]:

$$V_{PI} = (8 \cdot k_{eff} \cdot T_{act}^3 / (27 \cdot \varepsilon_0 \cdot A))^{0.5} \qquad (1)$$

where k_{eff} is the effective spring constant of the movable structure, T_{act} is the as-fabricated actuation gap thickness (for $V_{GB} = 0$ V), ε_0 is the permittivity of air, and A is the gate-to-body overlap area. When V_{GB} is lower than a release voltage (V_{RL}), F_{spring} is large enough to overcome F_{elec} and surface adhesion force at the contacts, and hence the contacts to the channel on both sides are broken.

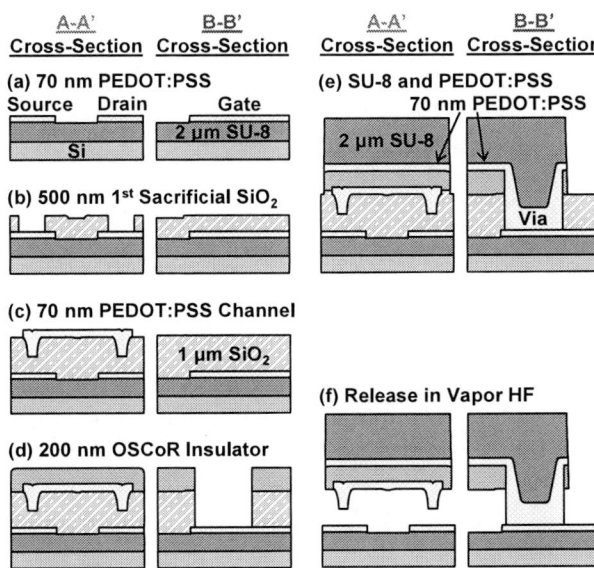

Figure 2: Low-thermal-budget (≤ 150 °C) process flow:
(a) _Mask 1:_ Formation of the gate, body, and source/drain (70 nm PEDOT:PSS) on the cross-linked SU-8.
(b) _Mask 2:_ PECVD (at 150 °C) and patterning of the 1st sacrificial SiO₂ (500 nm) to form contact dimples.
(c) _Mask 3:_ PECVD (at 150 °C) of the 2nd sacrificial SiO₂ (500 nm), followed by spinning and patterning of the PEDOT:PSS (70 nm), to form a channel on either side.
(d) _Mask 4:_ Spin-coating and patterning of the gate dielectric (200 nm OSCoR 4000 polymer) to insulate the channel on either side from the gate stack and form vias.
(e) _Mask 5:_ Spin-coating and patterning of the structural gate stack comprising SU-8, PEDOT:PSS, and OSCoR.
(f) Movable stack released in vapor HF at 50 °C.

FABRICATION PROCESS

Fig. 2 shows a five-mask low-thermal-budget process used to fabricate the polymeric relay (in **Fig. 1**). Since PEDOT:PSS (Clevios PH1000) has a finite conductivity, and is acidic and hygroscopic, a reliable multiple-layer PEDOT:PSS-based fabrication process requires proper patterning and surface treatment techniques.

Because the resistivity of pristine PEDOT:PSS is relatively high (~1 Ω·cm [11]), post-deposition treatment on the film with acidic solutions [4] or organic compounds such as Ethylene Glycol [11] and Dimethyl Sulfoxide [11] is necessary to reduce its resistivity. After the deposition of PEDOT:PSS (**Figs. 2(a) and (c)**), the substrate was dipped in methanol for a minute. The measured resistivity of the film was ~3 Ω·cm _vs._ ~$2 \cdot 10^{-3}$ Ω·cm before _vs._ after the treatment.

Because PEDOT:PSS is acidic and thus reacts with acid-sensitive photoresists and standard alkaline developers and strippers, fluorinated photoresist (OSCoR 4000) and fluoroether-based developer and stripper, which are benign to PEDOT:PSS, were used. The OSCoR 4000 was also used as a gate dielectric (**Fig. 2(d)**) to insulate the PEDOT:PSS channel on each side from the movable gate stack (SU-8/PEDOT:PSS/OSCoR).

Because PEDOT:PSS is hygroscopic, _i.e._, swells or shrinks when exposed to a water mixture during subsequent steps and thus tends to crack or delaminate, assuring strong adhesion between PEDOT:PSS and its contacting material is critical. Thus, to improve adhesion between PEDOT:PSS and the SU-8 substrate insulator (**Fig. 2(a)**), the hydrophobic SU-8 surface was first activated with an oxygen plasma, and then an adhesion promoter (Silquest A-187) was applied prior to the spin-coating of PEDOT:PSS. To promote adhesion between PEDOT:PSS and the sacrificial SiO₂ (**Fig. 2(b)**), A-187 was applied before the PECVD of SiO₂, which otherwise causes cracking of both films when the aqueous PEDOT:PSS used in the channel formation (**Fig. 2(c)**) penetrates through the pin-holes of the porous SiO₂ into the interface between SiO₂ and PEDOT:PSS and damages the underlying PEDOT:PSS.

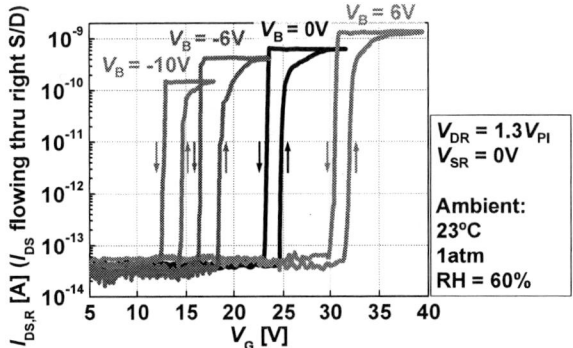

Figure 3: Measured $I_{DS,R}$ vs. V_G characteristics of the fabricated all-polymer 1-input/2-output relay (**Fig. 1**) for various body biases. Immeasurably-low off-state leakage current and abrupt switching behavior were observed.

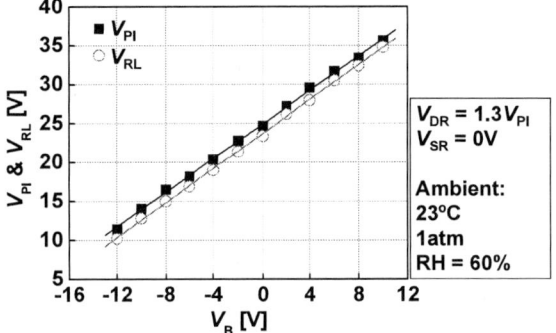

Figure 4: Measured V_{PI} and V_{RL} vs. V_B. A change in V_B results in commensurate changes to V_{PI} and V_{RL}. $V_{PI}–V_{RL}$ = 1.30 V, 1.28 V, and 1.17 V for V_B = -6 V, 0 V, and 6 V, respectively. Hysteresis behavior ($V_{PI} \neq V_{RL}$) is due to finite surface adhesion force and pull-in mode operation.

RESULTS AND DISCUSSION

Polymeric Relay as a Switching Device

The operation of the all-polymer relay as a switching device is demonstrated. **Fig. 3** shows measured I_{DS}-V_G characteristics of the fabricated relay (in **Fig. 1**). Relay switching characteristics: zero off-state leakage current due to an air gap between the channel and source/drain on either side (on the order of fA, which is the noise level of the semiconductor parameter analyzer used for testing), abrupt on/off switching behavior (< 1 mV/dec), and relatively high I_{on}/I_{off} ratio of ~10^5 were observed. The meas-

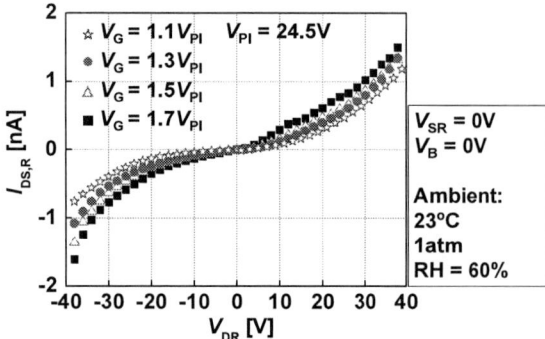

Figure 5: *Measured $I_{DS,R}$ as a function of V_{DR}. A diode behavior is seen due to a potential energy barrier between the PEDOT:PSS and the W probe tip used for testing. A stronger gate-overdrive (V_G-V_{PI}) for a given V_{DR} results in a larger on-state current, $I_{DS,R}$.*

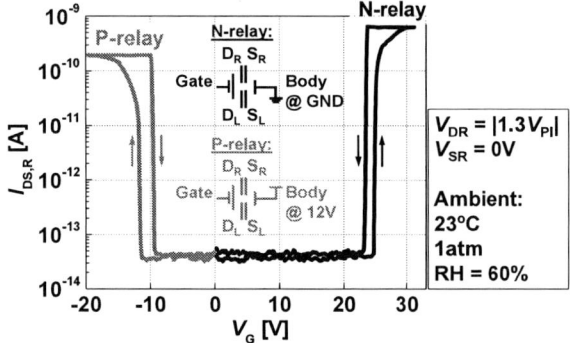

Figure 6: *Since electrostatic actuation is ambipolar, the operation of the relay mimics that of an N-channel or a P-channel MOSFET. N-relay and P-relay are achieved by biasing the body terminal at 0 V or V_{DD}, respectively. The left pair of source/drain was left floating.*

ured V_{PI} value (V_{PI} = 24.2 V at V_B = 0 V) is slightly larger than the theoretical value (V_{PI} = 22.0 V at V_B = 0 V) obtained from (1), due to a small negative strain gradient within the gate stack (SU-8/PEDOT:PSS/OSCoR), which makes the actual actuation gap thickness (T_{act}) approximately 100 nm larger than the as-fabricated T_{act} of 1 μm. Increasing the drain bias voltage (V_{DR} = 1.3·V_{PI}) decreases the on-state resistance (R_{on}): 57.7 GΩ, 50.4 GΩ, and 30.4 GΩ for V_B of -6 V, 0 V, and 6 V, respectively. These are larger than those expected for hard contact materials (< ~10 KΩ for Tungsten [12]) because of the finite conductivity of PEDOT:PSS (~10^2 S·cm^{-1} for the film treated with methanol). The conductivity of PEDOT:PSS can be further increased by mixing it with highly-conductive metallic particles such as Graphene [13] and/or treating it with formic acids [4]. The drain electrode was biased at a value 30 % larger than V_{PI} for each body bias to achieve a stable contact after the relay is pulled-in; I_{DS} still increases for V_G values above V_{PI} because the number of asperities in contact increases with increasing gate overdrive (V_G–V_{PI}). Hysteresis behavior (V_{PI}–V_{RL} > 0 V) is caused by non-zero surface adhesion force in the contact regions and pull-in mode operation, which can be remedied by using proper surface-coating materials [14] and operating the relay in non-pull-in mode with t_{dimple}/t_{act} < 1/3 [12].

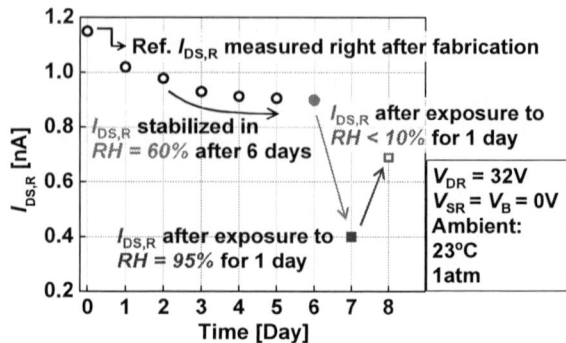

Figure 7: *Measured $I_{DS,R}$ as a function of time at room temperature and pressure. When the relay was placed in RH = 60 %, $I_{DS,R}$ were decreased gradually and stabilized eventually; in RH = 95%, $I_{DS,R}$ was dropped dramatically; and in RH < 10%, $I_{DS,R}$ was recovered partially.*

Fig. 4 shows that the body bias of the relay can be used to tune the threshold V_{PI}. A change in V_B causes commensurate changes to V_{PI} and V_{RL} because the relay is actuated electrostatically, *i.e.*, driven by the absolute voltage between V_G and V_B. The hysteresis for different V_B values is approximately 1.3 V; V_{PI}–V_{RL} = 1.30 V, 1.28 V, and 1.17 V for V_B = -6 V, 0 V, and 6 V, respectively. The right pair of source and drain was biased at 0 V and 1.3·V_{PI}, respectively, and the left pair of source and drain (V_{DL} and V_{SL}) was left floating.

Fig. 5 shows measured I_{DS}-V_{DS} characteristics of the relay. A diode behavior is observed due to a potential energy barrier to electron flow (Φ_B = ~0.8 eV) between the PEDOT:PSS (Φ_P = ~5.3 eV [15]) and the Tungsten (W) probe tip (Φ_M = ~4.5 eV) used for testing. I_{DS} increases (or R_{on} decreases) with larger gate overdrives (V_G–V_{PI}) for a given V_{DS} because the contact force between the channel and source/drain contacting surfaces increases and thus the number of contacting asperities on the surfaces increases.

The operation of the relay mimics that of MOS transistors due to the ambipolar nature of electrostatic actuation. **Fig. 6** shows that complementary switching behavior can be achieved by adjusting the body bias of the relay. By applying 0 V or V_{DD} onto the body terminal, the relay can be operated like an N-channel or a P-channel MOSFET, respectively.

Polymeric Relay as a Humidity Sensor

The potential use of the all-polymer relay as a biochemical sensor is demonstrated. The water-absorption behavior of PEDOT:PSS was exploited to use the relay as a humidity sensor. **Fig. 7** shows measured $I_{DS,R}$ of the relay over time. When the as-fabricated relay was left under ambient relative humidity (RH = 60 %), $I_{DS,R}$ were decreased gradually and stabilized after six days. This is because PEDOT:PSS absorbs moisture from the ambient air until it reaches an equilibrium moisture content, as expected for typical polymers that absorb and are permeable to moisture [11], [16]. The relay was then placed in a beaker containing water (RH = 95 %) for 24 hours, to expose the hygroscopic PEDOT:PSS (that absorbs moisture very easily) to moisture-rich environments. The $I_{DS,R}$ of the relay was decreased significantly (by ~55.6 %), due

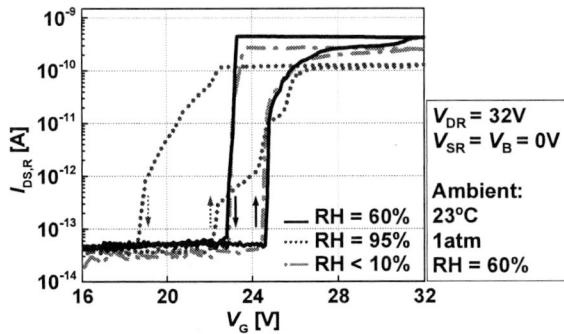

Figure 8: Measured $I_{DS,R}$ vs. V_G in various RH conditions at 23 °C and 1 atm. After the relay was exposed to RH = 95 % for a day, $I_{DS,R}$ and V_{PI} were lowered. These $I_{DS,R}$ and V_{PI} were recovered partially and fully, respectively, after the relay was stored in RH < 10 % for a day.

to phase separation between the PEDOT chains and the PSS component in the polymer mixture [16]. When the relay was placed in a desiccator (RH < 10 %) for the next 24 hours, $I_{DS,R}$ was recovered to ~76.7 % of its original value (measured in Day 6 in **Fig. 7**).

Fig. 8 shows measured $I_{DS,R}$ vs. V_G characteristics of the relay under different RH conditions. When the relay was exposed to RH = 95 % for one day (after being stored in RH = 60 % for six days), the threshold V_{PI} of the relay was dropped by ~10.6 % because the folded-flexures of the gate stack absorb moisture and become mechanically more compliant (lower effective spring constant). After the relay was stored in a desiccator (RH < 10 %) for another day, V_{PI} was recovered to ~99 % of the previous value (measured in RH = 60 %). These changes in $I_{DS,R}$ and V_{PI} shown in **Figs. 7 and 8** indicate changes in humidity levels. The relay sensor can be post-fabricated using the low-thermal-budget process (in **Fig. 3**) on top of CMOS readout circuits or integrated with relay-based readout circuits.

CONCLUSION

A fully-polymeric NEM relay comprising conductive and dielectric polymers has been proposed and demonstrated. The polymeric relay technology can be a compelling choice for flexible, transparent, and ultralow-power electronics and sensors with further design and process improvements.

ACKNOWLEDGEMENTS

Research carried out in part at the Center for Functional Nanomaterials (CFN), Brookhaven National Laboratory (BNL), which is supported by the U.S. Department of Energy, Office of Basic Energy Sciences, under Contract No. DE-AC02-98CH10886. The authors would like to thank Dr. Ming Lu for valuable discussions.

REFERENCES

[1] W. U. Huynh *et al.*, "Hybrid Nanorod-Polymer Solar Cells," *Science*, vol. 295, pp. 2425-2427, 2002.

[2] S. Allard *et al.*, "Organic Semiconductors for Solution-Processable Field-Effect Transistors (OFETs)," *Angewandte Chemie*, vol. 47, pp. 4070-4098, 2008.

[3] N. Kim *et al.*, "Highly Conductive PEDOT:PSS Nanofi brils Induced by Solution-Processed Crystallization," *Advanced Materials*, vol. 26, pp. 2268-2272, 2014.

[4] D. A. Mengistie et al., "Highly Conductive PEDOT: PSS Treated with Formic Acid for ITO-Free Polymer Solar Cells," *Applied Materials and Interfaces*, vol. 6, pp. 2292-2299, 2014.

[5] D. J. Lipomi *et al.*, "Electronic Properties of Transparent Conductive Films of PEDOT:PSS on Stretchable Substrates," *Chemistry of Materials*, vol. 24, pp. 373-382, 2012.

[6] U. Lang *et al.*, "Piezoresistive properties of PEDOT: PSS," *Microelectronic Engineering*, vol. 86, pp. 330-334, 2009.

[7] B. Friedel *et al.*, "Effects of Layer Thickness and Annealing of PEDOT:PSS Layers in Organic Photodetectors," *Macromolecules*, vol. 42, pp. 6741-6747, 2009.

[8] P. G. Taylor *et al.*, "Orthogonal Patterning of PEDOT PSS for Organic Electronics using Hydrofluoroether Solvents," *Advanced Materials*, vol. 21, pp. 2314-2317, 2009.

[9] B. Charlot *et al.*, "Micropatterning PEDOT:PSS layers," *Springer Microsystem Technologies*, vol. 19, pp. 895-903, 2013.

[10] J. Jeon *et al.*, "Multiple-Input Relay Design for More Compact Implementation of Digital Logic Circuits," *Electron Device Letters*, vol. 33, pp. 281-283, 2012.

[11] D. Alemu *et al.*, "Highly conductive PEDOT:P-SS electrode by simple film treatment with methanol for ITO-free polymer solar cells," *Energy and Environmental Science*, vol. 5, pp. 9662-9671, 2012.

[12] H. Kam *et al.*, "Design and reliability of a micro-relay technology for zero-standby-power digital logic applications," *Int'l Electron Devices Meeting*, pp. 809-811, 2009.

[13] M. Zhang *el al.*, "Solution-Processed PEDOT:PSS/ Graphene Composites as the Electrocatalyst for Oxygen Reduction Reaction," *Applied Materials and Interfaces*, vol. 6, pp. 3587-3593, 2014.

[14] R. Maboudian *et al.*, "Self-assembled monolayers as anti-stiction coatings for MEMS: characteristics and recent developments," *Sensors and Actuators A*, vol. 83, pp. 219-223, 2000.

[15] J. Huang *et al.*, "Investigation of the Effects of Doping and Post-Deposition Treatments on the Conductivity, Morphology, and Work Function of Poly(3,4-ethylenedioxythiophene)/Poly(styrenesulfonate) Films," *Advanced Functional Materials*, vol. 15, pp. 290-296, 2005.

[16] S. Taccola *et al.*, "Characterization of free-standing PEDOT:PSS/iron oxide nanoparticle composite thin films and application as conformable humidity sensors," *Applied Materials and Interfaces*, vol. 5, pp. 6324-6332, 2013.

CONTACT

Yanbiao Pan, yanbiao.pan@rutgers.edu

UHF PIEZOELECTRIC QUARTZ MEMS MAGNETOMETERS BASED ON ACOUSTIC COUPLING OF FLEXURAL AND THICKNESS SHEAR MODES

Hung D. Nguyen, Joshua A. Erbland, Logan D. Sorenson, Raviv Perahia, Lian X. Huang,
Richard J. Joyce, Yeong Yoon, Deborah J. Kirby, Tracy J. Boden, Robert B. McElwain,
and David T. Chang
HRL Laboratories, LLC, Malibu, CA, USA

ABSTRACT

This paper reports the design, fabrication, and characterization of piezoelectric quartz MEMS magnetometers based on acoustic coupling between resonance modes. The magnetic sensors described herein employ a novel transduction scheme to upconvert the desired near-DC magnetic field signal (using the fundamental flexural mode) onto frequency modulated (FM) sidebands of the primary quartz thickness shear (TS) oscillation at frequencies above 500 MHz. First-generation devices exhibit flexural and TS resonances at 2.77 kHz and at 583.31 MHz, respectively, and magnetic sensitivity of 63.6 V/T was measured with an AC loop current of 9.2 mA. This novel sensing method, intended for electronic compassing, illuminates the interactions between low and high frequency acoustic modes within resonant devices.

INTRODUCTION

Opportunities for technology insertion of micro-scale magnetometers, as a standalone sensor or as an integral component of a sensor suite, have grown with the demand for position, navigation, and timing (PNT) units and attitude and heading reference systems (AHRS). Attitude acquisition for inertial navigation requires very precise magnetometers with detection limits under 50 nT to achieve an angle resolution of less than 0.1°. Our quartz MEMS magnetometer can detect the vector component of the magnetic field, allowing its use in AHRS and PNT units as part of an integrated navigation system.

With few exceptions [1,2], most MEMS magnetometers are built on silicon to ease fabrication and CMOS integration despite silicon's high thermal instability and modest quality factor. MEMS resonant magnetometers typically rely on Lorentz force to induce either frequency modulation (FM) [3-5] or amplitude modulation (AM) [6-7] of the output signal. In contrast, our quartz MEMS magnetometer employs a hybrid approach to exploit Q-amplification of the flexural mode, greater immunity to electronic noise by FM detection, frequency and thermal stability of AT-cut quartz, and ruggedness of the high-Q ($Q_{resonator} > 9400$) low phase noise quartz UHF TS oscillator [8]. Depending on the crystal cut, quartz magnetometers can be designed to exhibit excellent long-term stability and performance. The piezoelectric sensing mechanism of quartz allows for embedded drive and sense electrodes on the resonator plate. Further, the piezoelectric quartz magnetometers have a high frequency response, operating at an output frequency > 10X higher than other Lorentz-force MEMS magnetometers.

PRINCIPLE OF OPERATION

Figure 1 illustrates the key elements of the quartz magnetometer as well as the detection mechanism and the associated Clapp oscillator circuit. The magnetometer is based on a quartz UHF micro-resonator platform. The vibrating quartz plate or active region of the magnetometer is cantilevered from a wider base plate which is anchored down to the silicon substrate. The embedded current loop used to generate the Lorentz force follows along the edges of the quartz plate. An RF signal between top and bottom sense electrodes placed near the clamped end induces thickness shear acoustic waves that propagate through the thickness of the quartz volume bounded by the electrodes.

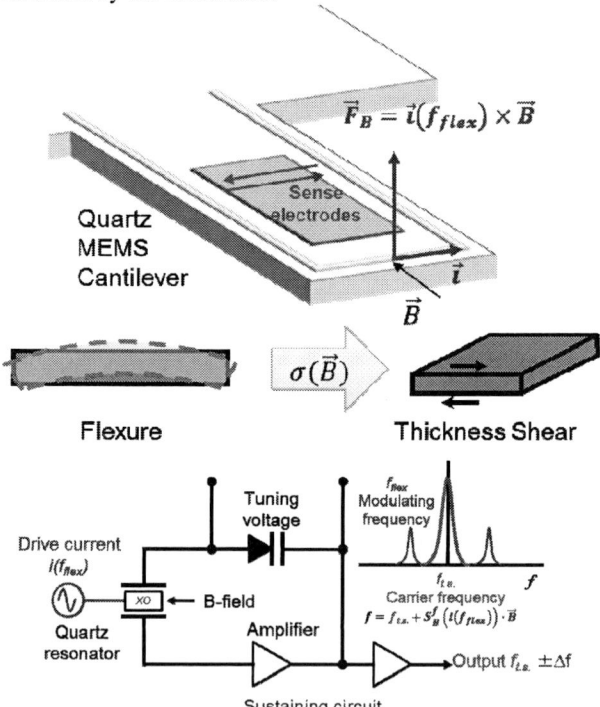

Figure 1: Schematic of quartz magnetometer. Acoustic coupling between the two resonance modes induced by Lorentz force generates FM sidebands whose amplitude is a function of the resonator frequency, sensitivity S_B^f, drive current $i(f_{flex})$, and magnetic field $\vec{B}$.

The sensing mechanism is based on the acoustic coupling between two resonance modes where one mode is driven into mechanical vibration by the Lorentz force and a second higher frequency thickness shear mode detects the low-frequency vibration and resultant bending strain in the form of a frequency shift. With the sustaining amplifier loop closed across the sense electrodes but no current excitation along the current loop, the quartz magnetometer resonates at the thickness shear mode frequency $f_{t.s.}$ and outputs a carrier signal (red peak in Figure 1) that is unperturbed by the external magnetic field. Once an AC current $\vec{\iota}$ is applied along the plate edge and interacts with the external magnetic field $\vec{B}$, the generated Lorentz force $\left(\vec{F}_B = \vec{\iota}(f_{flex})L \times \vec{B}\right)$ at the tip of the resonator drives the plate into mechanical vibration at the fundamental flexural mode frequency f_{flex}. This flexural motion modulates the stiffness of the active TS region due to strain sensitivity of the quartz elastic moduli, resulting in frequency modulation of the TS mode and sidebands signals (shown in blue in Figure 1) offset at f_{flex} from the TS frequency ($f_{t.s.}$).

DEVICE DESIGN AND OPTIMIZATION

In Figure 2, a simulated plot of the magnetic sensitivity in response to static Lorentz force highlights a unique feature of the quartz magnetometer: the anisotropy of AT-cut quartz enables bi-directional sensing represented by negative and positive frequency shifts depending on the direction of the magnetic field vector. This sensitivity is expected to increase by a factor equal to the mechanical Q assuming the magnetometer operates in flexural resonance and under vacuum. We expect the sensitivity to reach 55 kHz/T with 1 mA in the AC current loop for this prototype design.

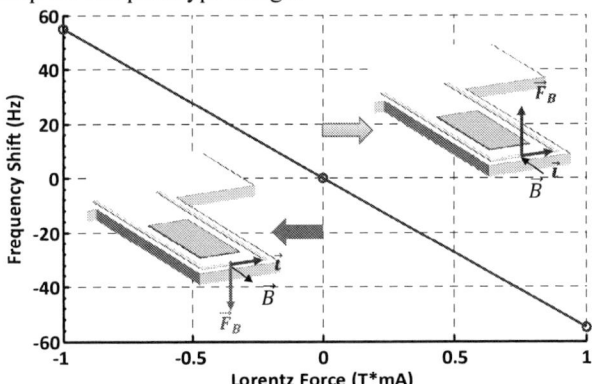

Figure 2: Bi-directional shift in frequency depending on the direction of the B-field allows quartz to be used as a vector magnetometer.

In Figure 3, the calculated modal dispersion curves of the base design show a large frequency separation of the desired first flexural driving mode from other modes, ensuring that only the flexural mode is excited by the AC current loop at its resonance frequency. This guarantees minimal cross-axis sensitivity when the current loop is driven at the fundamental flexural frequency. A thinner quartz plate is desired for maximum Lorentz force sensitivity but at lower flexural frequency. However, for the sense mode, the thickness shear frequency increases with thinner quartz plate, where a 3 μm-thick quartz magnetometer will exhibit a thickness shear frequency of 500 MHz. The plate length for the first fabrication run was chosen with these considerations to be 600 μm with a

plate thickness of approximately 2.5 μm.

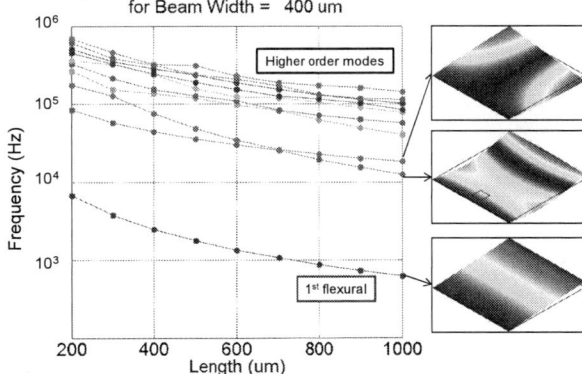

Figure 3: The fundamental flexural mode is well separated from other modes to ensure minimal cross talk.

EXPERIMENTAL RESULTS

A quartz magnetometer with a newly designed trapezoidal quartz plate is shown in Figure 4. The width of the quartz plate grows from the fixed end such that the plate is at its widest at the free end. The sensing electrodes are strategically placed close to the base where maximum bending strain is concentrated. The wedge plate design allows for a longer current line along the free end, increasing the Lorentz force as well as concentrating the resultant bending strain at the sense electrodes for enhanced magnetic field sensitivity. The fabricated magnetometers are then encapsulated and singulated for mounting onto the Clapp oscillator printed circuit board.

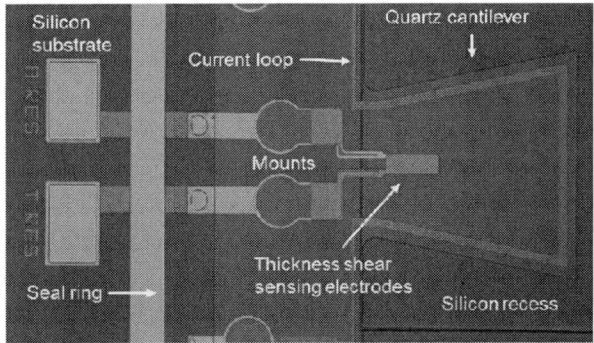

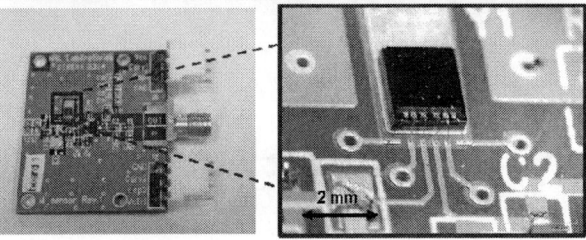

Figure 4: Optical images of the various assembly stages of the magnetometer: fabricated quartz device (top), vacuum encapsulated (bottom right), and operated as an oscillator (bottom left).

The discrete Clapp oscillator circuit designed around the magnetometer outputs a 1 mW, 583.31 MHz signal at a minimum V_{DC} = 5 V and 30 mW of supplied power.

978-1-4799-7956-1/15 $31.00 © 2015 IEEE   945

The magnetometer circuit is currently measured in an open-loop configuration where the signal output from the oscillator is fed into an HP 8563 Spectrum Analyzer. A Neodymium magnet is positioned close to the magnetometer on the oscillator board to demonstrate the detection scheme. Figure 5 shows the appearance of FM sidebands spaced at multiples of the AC current loop frequency of 2.77 kHz in the presence of the external DC magnetic field.

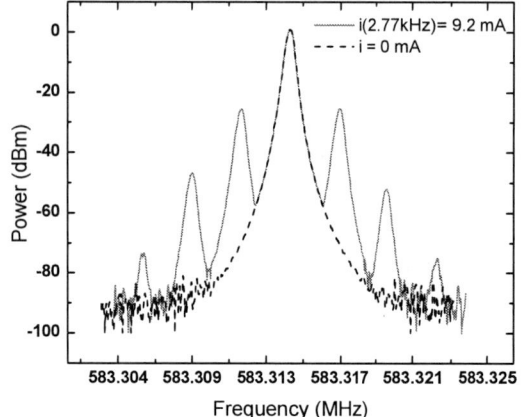

Figure 5: The output of the Clapp oscillator circuit when no AC current is applied to the current loop is 0 dBm (black). -25 dBm FM sidebands and harmonics (red) are generated when DC magnetic field couples with the AC drive current at 2.77 kHz.

A second oscillator board with a packaged quartz magnetometer is placed within a 1 cu. ft. Helmholtz coil shown in Figure 6. With a reference fluxgate magnetometer (Bartington MAG-03MSB250), the background magnetic field is nulled by adjusting the current through the three coils to generate a cancellation field. Additional current is supplied to generate a controlled field directed exclusively along any one of the three orthogonal axes to evaluate the sensor performance including scale factor and cross-axis sensitivity.

Figure 6: Magnetometer board positioned in the center of the Helmholtz coil used to generate uniform magnetic field along the three orthogonal axes.

The quartz magnetometer can detect the vector components of the field as illustrated by the transfer curves in Figure 7. The three transfer curves are obtained from the first FM sideband amplitude as the drive current is swept in frequency from DC to 25 kHz at a fixed peak amplitude of 9.2 mA. Each curve represents the sensor response to a 500 µT magnetic field along a particular axis. The plot shows three resonant peaks at 2.77, 12.3, and 19.5 kHz, with the lowest frequency peak identified as the fundamental flexural mode. As expected, sensor response at 2.77 kHz is greatest when the field is directed along z-axis or axial length of the sensor as shown in the inset diagram. The Lorentz force generated by the field along the x and y axes should not induce flexural vibrations at 2.77 kHz as indicated by the sideband amplitude residing below the noise floor. From these results, sensitivity to the cross-axis magnetic field components is suppressed at least 30 dB below the desired component of magnetic field.

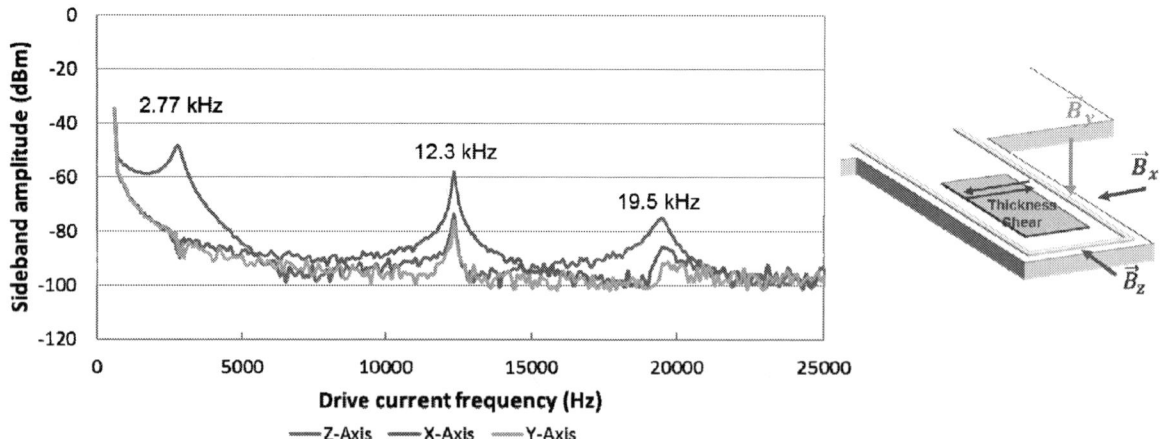

Figure 7: Transfer response obtained by sweeping AC current loop frequency and monitoring first FM sideband amplitude. The first resonant peak at 2.77 kHz corresponds to the main flexural mode, while the higher peaks correspond to higher order torsional and flexural modes.

With the magnetic field along the most sensitive axis and the sensor mechanically driven at 2.77 kHz by the AC current loop, the magnetometer's scale factor (sensitivity) is measured in Figure 7 at two different current amplitudes. From these two plots and their respective slopes, the magnetic sensitivity is 13.6 V/T and 63.6 V/T at 2 mA and 9.2 mA, respectively, and nonlinearity appears negligible up to 250 µT (limited by reference magnetometer). The same measurement was repeated for the remaining two axes to determine the cross axis sensitivity derived in Figure 9. The ratio between the z-axis voltage sensitivity the other two axes can be as high as 35:1, which is 31 dB in agreement with the measurement of Figure 7.

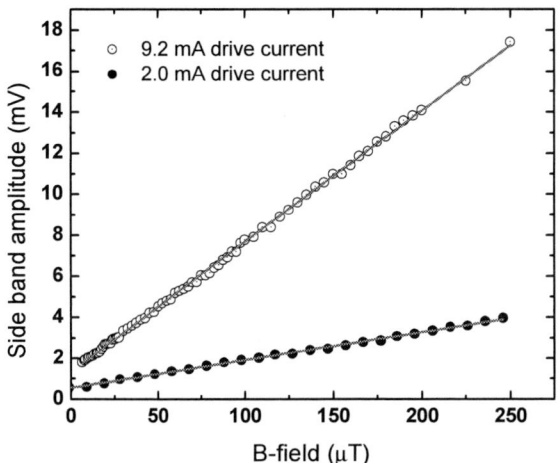

Figure 8: Magnetometer sensitivity depends on the AC current in the loop. Sensitivity increases to 63.6 V/T by raising the AC current to 9.2 mA. In both cases, the AC current loop frequency was set to 2.77 kHz to match the flexural mode resonance.

DISCUSSION AND CONCLUSIONS

We introduced a new magnetic sensing mechanism based on a crystalline quartz resonator platform. The acoustic coupling between a mechanical mode and a piezoelectric mode enables detection of the magnetic field through the applied Lorentz force. The sensor leverages the performance advantages of both frequency and amplitude modulation to achieve high sensitivity with high cross-axis immunity.

REFERENCES

[1] Y. Hui, T. Nan, N. X. Sun, and M. Rinaldi, "High Resolution Magnetometer Based on a High Frequency Magnetoelectric MEMS-CMOS Oscillator," J. MEMS, Early Access Online, 2014.

[2] D. A. Oursler et al., Johns Hopkins APL Technical Digest, vol. 20, no. 2, 1999.

[3] R. Sunier, T. Vancura, Y. Li, K.-U. Kirstein, H. Baltes, and O. Brand, "Resonant Magnetic Field Sensor With Frequency Output," J. MEMS, vol. 15, no. 5, pp. 1098–1107, Oct. 2006.

[4] B. Bahreyni and C. Shafai, "A Resonant Micromachined Magnetic Field Sensor," IEEE Sensors Journal, vol. 7, no. 9, pp. 1326–1334, Sep. 2007.

[5] M. Li, S. Sonmezoglu, and D. Horsley, "Extended Bandwidth Lorentz Force Magnetometer Based on Quadrature Frequency Modulation," J. MEMS, Early Access Online, 2014.

[6] H. Emmerich and M. Schofthaler, "Magnetic field measurements with a novel surface micromachined magnetic-field sensor," Electron Devices, IEEE Transactions on, vol. 47, no. 5, pp. 972–977, 2000.

[7] A. L. Herrera-May, P. J. García-Ramírez, L. A. Aguilera-Cortés, J. Martínez-Castillo, A. Sauceda-Carvajal, L. García-González, and E. Figueras-Costa, "A resonant magnetic field microsensor with high quality factor at atmospheric pressure," Journal of Micromechanics and Microengineering, vol. 19, no. 1, p. 015016, Jan. 2009.

[8] D. T. Chang, H. P. Moyer, R. G. Nagele, R. L. Kubena, R. J. Joyce, D. J. Kirby, P. D. Brewer, H. D. Nguyen, and F. P. Stratton, "Nonlinear UHF quartz MEMS oscillator with phase noise reduction," in 2013 IEEE 26th International Conference on Micro Electro Mechanical Systems (MEMS), 2013, pp. 781–784.

CONTACT

*H. Nguyen, tel: +1-310-317-5000; hdnguyen@hrl.com

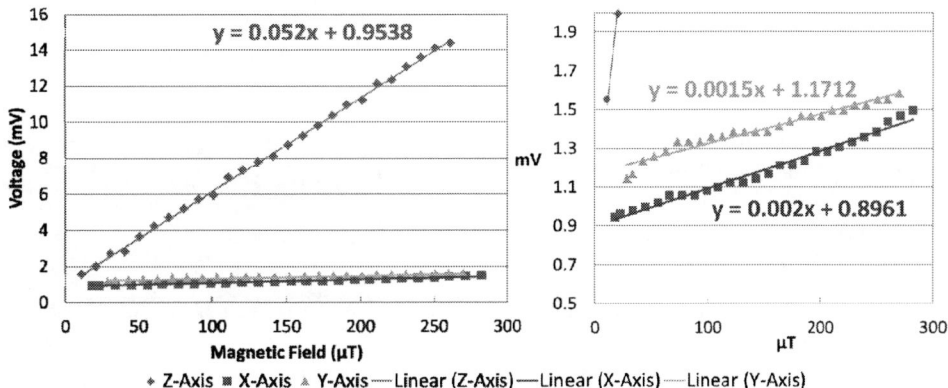

Figure 9: (left) Cross axis sensitivity shows strong response to the z-axis magnetic field. (right) Zoom in of x- and y-axis cross sensitivity showing 28 to 31 dB suppression relative to z-axis.

A 45°-TILTED 2-AXIS SCANNING MICROMIRROR INTEGRATED ON A SILICON OPTICAL BENCH FOR 3D ENDOSCOPIC OPTICAL IMAGING

Can Duan[1], Wei Wang[1], Xiaoyang Zhang[1], Jinling Ding[2], Qiao Chen[2], Antonio Pozzi[3], Huikai Xie[1]

[1]Department of Electrical and Computer Engineering, University of Florida, Gainesville, USA
[2]WiO Technology Ltd., Wuxi, CHINA
[3]Department of Small Animal Surgery, University of Florida, Gainesville, USA

ABSTRACT

This paper presents a 2-axis electrothermal single-crystal-silicon (SCS) micromirror that is tilted 45° out of plane on a silicon optical bench (SiOB). The SiOB provides mechanical support and electrical wiring to the tilted 2-axis mirror as well as an aligned trench for assembling other optical components such as optical fibers. The tilt of the mirror is achieved with the bending of a set of stressed bimorph beams and the stop is provided by the silicon sidewall. The tilt angle can be precisely controlled by properly choosing the distance from the mirror frame to the silicon sidewall and the flexure bimorph length. The mirror plate is 0.72 mm × 0.72 mm and the footprint of the entire MEMS device is 2.22 mm × 1.25 mm. The measured maximum optical scan angles of the mirror are 40.0° in both x- and y-axis.

INTRODUCTION

Microendoscopic optical imaging is an emerging field that allows high-resolution (cellular level), cross-sectional, non-invasive imaging techniques, such as optical coherence tomography (OCT), to be directly applied inside human body for in vivo early cancer detection and real-time image-guided surgery [1,2]. Electrothermally-actuated micromirrors have been extensively employed in microendoscopic OCT imaging probes, thanks to their large linear scan range, low drive voltage and high fill factor [3]. A typical side-viewing endoscopic OCT probe is illustrated in Fig. 1(a), where the micromirror is mounted on a 45° slope; the electrical connection is provided either by wire bonding [3], resulting in large probe size, or flip-chip bonding [2], suffering from serious electrical contact failure problems. A SiOB with a pre-fabricated 45° trench and a micromirror manually fixed in the trench using solder balls was demonstrated [4], but to form reliable electrical and mechanical connections is still a big challenge.

In this paper, a 45°-tilted two-axis scanning MEMS micromirror integrated with a SiOB has been proposed and fabricated to avoid the bulky probe design and overcome the assembly challenges. With this new SiOB design, the endoscopic OCT probe can be significantly reduced in size (Fig. 1(b)) and its assembly process is much simplified.

DEVICE DESIGN

Flexure bimorphs and stopper design

The proposed device is illustrated in Fig. 2(a), where a 2-axis scanning SCS mirror is integrated with a 45°-tilted SiOB with electrical wiring to the actuators and an aligned trench for easy assembly of other optical components. The 2-axis mirror is supported by a rigid silicon frame that is connected to the substrate via a curled bimorph flexure.

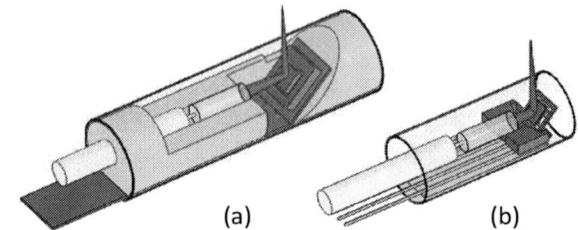

Figure 1: MEMS based side-viewing probe designs. (a) Using a regular planar mirror on a 45° slope. (b) Using the proposed 45° tilted micromirror.

The mirror and frame are made of the SOI device layer while the curled bimorph flexure consists of Al as the bottom layer and SiO_2 as the top layer. Upon release, the curled flexure bends the frame and the mirror all way back to the SOI handle layer side such that the stopper on the frame is in contact with and stopped by the sidewall of the handle layer (Fig. 2(a)). The thicknesses of the Al and SiO_2 layers are designed respectively as 0.8 μm and 1.0 μm to maximize the bimorph curvature. The designed length of the flexure bimorph is 300 μm, leading to a theoretical initial tangential angle larger than 45° without being stopped. An SEM of a fabricated device is shown in Fig. 2(b).

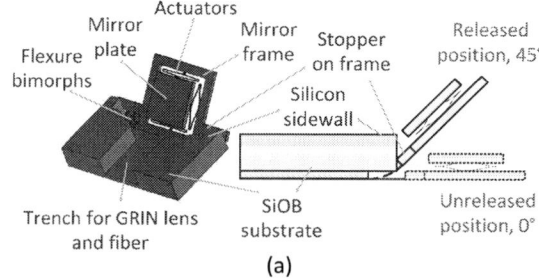

Figure 2: The proposed 45° tilted 2-axis scanning micromirror. (a) 3D view and schematic of the proposed device. (b) SEM of a fabricated device.

The frame tilt angle (e.g., 45°) is precisely controlled by the geometric design of the distance from the stopper to the silicon sidewall as well as the flexure bimorph curvature and length. As the schematic in Fig. 3 illustrates, the length of the extended anchor, or the distance between the silicon sidewall to the stopper, will determine the tilting angle of the mirror frame in case that the initial tangential angle of the flexure bimorph is larger than 45° with no stoppers. The radius of curvature of the curved flexure bimorph R_{fb} is deducted as:

$$R_{fb} = \frac{L_{fb}}{\theta_c} \qquad (1)$$

where L_{fb} is the length of the flexure bimorph, which is set as 300 μm according to theoretical calculation and simulation. θ_c represents the central angle corresponding to the curved flexure bimorph, which is the same with the tilting angle of the mirror frame, i.e., 45°.

From geometric analysis, the length of the anchor on the substrate should satisfy the following relationship with the bimorph length and rotation angle.

$$L_{out} = R_{fb} \sin \theta_c - T_d \sin \theta_c \qquad (2)$$

where T_d is the thickness of the mirror frame, i.e., the thickness of the device layer of the SOI wafers used in this work, which is 60 μm. Thus the calculated length of the anchor is 228 μm.

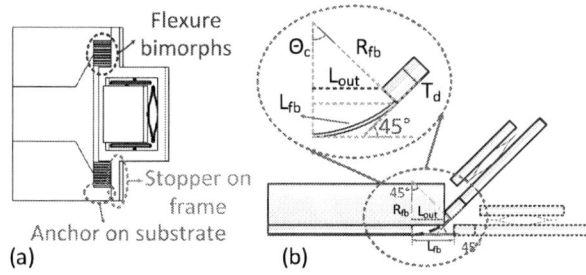

Figure 3: Geometric analysis of the tilt angle. (a) Top view and (b) Side view of the device.

Two-axis scanning mirror design

The 2-axis mirror employs a folded dual S-shaped bimorph (FDSB) actuator design which is similar to the one reported in [5]. As shown in Figs. 4(a) and (b), each FDSB actuator consists of two S-shaped, invert-series-connected curved bimorphs. Each S-shaped bimorph is made up of three parts including two active bimorphs with inversely deposited SiO2/Al layers, and an overlap section to strengthen the connection of the two active bimorphs. The FDSB actuator can convert the curling of single bimorphs to a pure vertical displacement at the tip without lateral shift, as shown in Fig. 4(c). Both piston and tip-tilt motions of the mirror plate can be obtained by controlling the phases of the drive voltages for the four pairs of FDSB actuators. The four actuator pairs are respectively wired to four pads with a common ground pad. The pads are designed to be as large as 180 μm to ensure good bonding and reliable electrical connection.

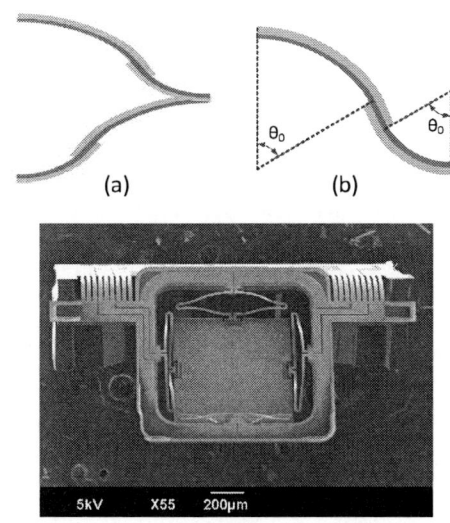

Figure 4: The two-axis scanning mirror design. Schematic view of (a) a FDSB actuator, and (b) an S-shaped bimorph. (c) SEM of a fabricated device showing SCS mirror plate supported by four FDSB actuators.

SiOB design

On the other side of the SiOB are two grooves that can hold an optical fiber tip and a GRIN lens to deliver the light to the centered mirror surface, as shown in Fig. 5(a). The light out of the GRIN lens will be reflected by the 45° tilted mirror plate to achieve a 90° output direction. The groove openings will match the sizes of the employed optical fiber and GRIN lens such that the optical axes are all alighted to the center of the scanning mirror. Note it is essential to achieve the designed tilt angle to ensure the designed height of the mirror center.

A device design containing only a groove for GRIN lens is shown in Fig. 5(b). The footprint of the designed SiOB is 2.22 mm × 1.25 mm. The height of the mirror center is approximately 520 μm at 45° tilt angle. According to the geometric relationship shown in Fig. 5(c), the calculated groove opening is 623 μm for holding a 700 μm (diameter) GRIN lens and keeping its optical axis aligned with that of the mirror plate.

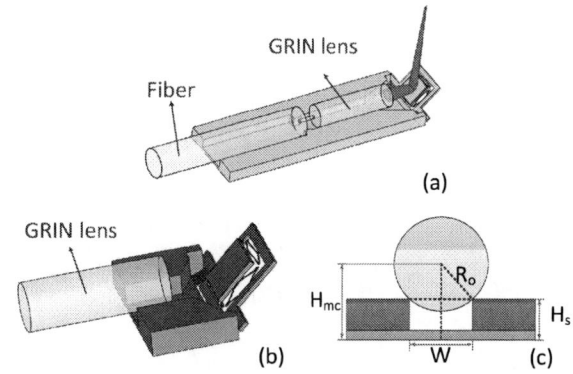

Figure 5: The SiOB design. (a) A long SiOB design containing two grooves for both optical fiber and GRIN lens. (b) A short SiOB design. (c) Cross-sectional view of the SiOB holding a GRIN lens.

DEVICE FABRICATION

The fabrication process starts with SOI wafers with a 60 μm device layer, a 2 μm buried oxide (BOX) layer and a 300 μm handling layer. First, a 1 μm PECVD SiO$_2$ layer is deposited and patterned on the front side of the SOI wafer to form the bimorph containing SiO$_2$ as the bottom layer. Buffered oxide etchant (BOE) is employed in order to form a smooth edge for providing good metal line continuation. A 0.05 μm PECVD SiO$_2$ is deposited on the front side of the SOI for insulation and adhesion enhancement. Then a Cr/Pt/Cr lift-off process is performed to form heaters along the bimorph actuators. After that, a 0.1 μm PECVD SiO$_2$ layer is deposited and patterned using dry etch in a Unaxis reactive-ion etching (RIE) system, and then a 0.8 μm Al layer is sputtered and patterned by lift-off process to serve as the other layer of the bimorph actuators as well as the embedded electrical wiring, pads and mirror surface coating. Another 1.2 μm PECVD SiO$_2$ layer is deposited and patterned by RIE dry etch to form bimorph actuators with SiO$_2$ as the top layer. Then the front side is spin coated with a 6.5 μm AZ9260 photoresist to protect the exposed Al layer. Another 8.5 μm AZ9260 is coated and patterned on the backside of the SOI to form the mask for backside DRIE silicon etch. A silicon carrier wafer is attached to the front side of the SOI wafer for DRIE etch. The DRIE etch stops at the BOX layer. Then the BOX layer is removed through RIE and a Cr/Al layer is sputtered on the backside to form a mirror surface on the backside of the device. After this step, the carrier wafer is removed by immersing it in Acetone and the SOI wafer is separated into individual die. Individual dies are glued on the carrier wafer using a thermal release tape (Nitto, 3195M) with its front side upward and now it is ready for release.

The release starts with an anisotropic DRIE etch through the device layer to open the silicon trenches (Fig. 6(a)). The final release is achieved by isotropic etching to undercut silicon underneath the bimorphs. Thin-film stresses result in initial displacement of the bimorphs after the complete removal of silicon underneath the bimorphs. These two release steps are critical. The mirror actuator bimorphs must be released (Fig. 6(b)) before the release of the flexure bimorphs (Fig. 6(c)). This is achieved by having the width of the actuator bimorphs smaller than that of the flexure bimorphs. In this way, the silicon beneath the actuator bimorphs will be etched completely while the silicon underneath the flexure bimorphs is only partially etched; so the actuator bimorphs will bend first and then the flexure bimorphs will curl over to the base and be stopped by the stopper on the base.

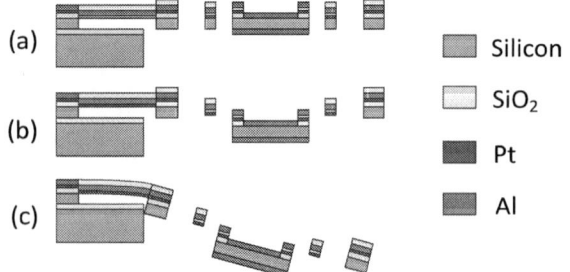

Figure 6: Final release steps of fabrication process flow. (a) Si anisotropic etching. (b) Si isotropic etching to release bimorph actuators. (c) Si isotropic etching to release flexure bimorphs.

DEVICE CHARACTERIZATION

Fig. 7(a) shows a front-view SEM of a fabricated device, where the Al-coated mirror plate is 0.72×0.72×0.06 mm^3 and the SiOB substrate is 2.22×1.25×0.36 mm^3. SEMs of the closed-up views of the curled flexure, stopper and actuator are respectively shown in Figs. 7(b), (c) and (d). The stopper is stopped at the sidewall (Fig. 7(d)). The initial radius of curvature of the flexure bimorphs is 277.2 μm, calculated from the initial displacement of the bimorph tip connected to the stopper on the mirror frame. The measured tilt angle of the mirror plate out of the SiOB substrate is 62°, which is larger than the designed 45°. This is caused by the silicon undercut of the silicon sidewall beneath the stopper during the final isotropic etch release step. As illustrated in the close-up SEM in Figs. 7(c) and (d), a 30 μm lateral undercut of silicon occurs for both the stopper on frame and the silicon on substrate. The initial elevation of the mirror plate is 155 μm and the mirror plate is parallel with the mirror frame, thanks to the symmetric design of four dual-S-shape bimorph actuators. The resistances of four actuators are around 260 Ω with difference within 3.7% due to the Pt and Al pathway difference.

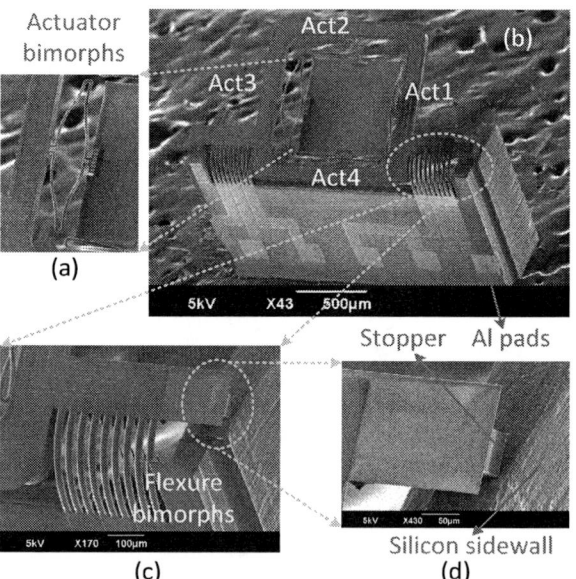

Figure 7: SEMs of a fabricated device. (a) An actuator. (b) Rear side view. (c) Curled flexure. (d) Stopper.

Static Response

The measured static tip-tilt actuations including both single actuator driving actuator and differential driving of opposing actuator pairs are shown in Fig. 8. The mirror plate has a large initial displacement of 155 μm underneath the frame, which theoretically allows a maximum mechanical scan angle of ±12.1°, corresponding to optical scan angle of ±24.2°. The measured average optical scan angle for a single actuator reaches 19.8° at only 5.5 Vdc. A wide linear range from 3° to 17° has been achieved with the drive voltage from 1.2 V to 4.6 V. The actuation characteristics of the four actuators are slightly different but the differences are within 8.5%. The maximum total optical scan range of 40.0° can be obtained by applying differential voltages with a 2.75 V bias voltage and a peak-to-peak voltage of 5.5 V to opposing actuators.

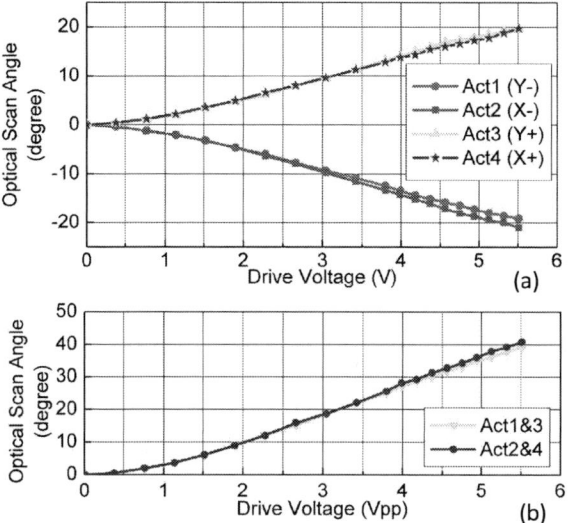

Figure 8: Quasi-static characterization of a bonded device. (a) Single actuator driving. (b) Differential driving.

Raster Scan

A raster scan pattern as well as the test set-up is shown in Fig. 9. Two sinusoidal drive signals with offset of 2.2 V and 4.4 Vpp are applied to both the fast scan actuators and the slow scan actuators. The frequencies of the driving signals for the fast-scan axis and slow-scan axis are 160 Hz and 8 Hz, respectively. A square scan area of 31° × 30° has been achieved.

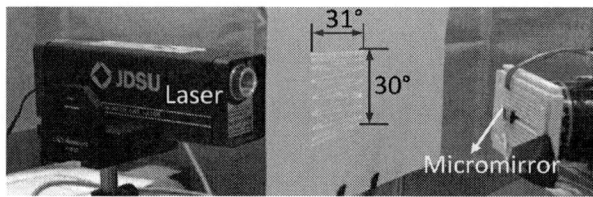

Figure 9: Test set-up and raster scan pattern (fast scan: 4.4 Vpp at 160 Hz; slow scan: 4.4 Vpp at 8 Hz).

Frequency Response

The measured response time of each actuator is 2.8 ms. Fig. 10 shows the frequency response of the scanning mirror by actuating each actuator individually. The frequency response is measured by using a similar set-up as the one shown in Fig. 9, but adding a position sensing detector (OT-301DL, ON-TRAK) and a network analyzer (Analog Kit). The first mode occurs at 191 Hz, which is identified as the frame rotation mode. The piston mode is not shown up using this rotational driving method. The second and third peaks are the rotational modes of the mirror actuators, at 693 Hz and 744 Hz, respectively. The fourth mode is another rotational mode related to the frame scanning. The two rotational modes of the mirror plate are desired for optical imaging and displays. The modes of the

mirror frame may generate some disturbing vibrations but can be eliminated by fixing the anchor and the stopper on the mirror frame with optical glue.

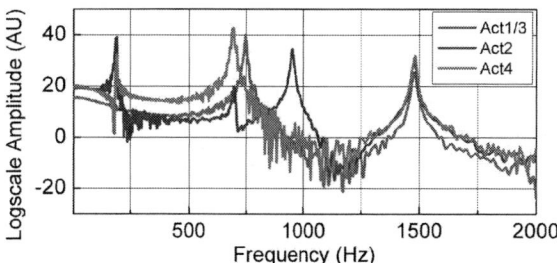

Figure 10: Frequency response of the device.

CONCLUSION

In this work, a 2-axis electrothermal SCS micromirror integrated on a SiOB with a preset tilt angle has been successfully demonstrated. Large scan angle can be achieved at low drive voltage. The tilt angle can also be adjusted with simple layout design or by controlling the silicon undercut at the last release step. The bimorph curling induced tilt can further be fixed permanently with glue. This technology has potential to enable a new class of ultra-compact microendoscopic optical imaging probes for in vivo early cancer detection.

ACKNOWLEDGEMENTS

This work was supported by the National Science Foundation under award #1002209.

REFERENCES

[1] B. J. Vakoc, D. Fukumura, R. K. Jain, and B. E. Bouma, "Cancer imaging by optical coherence tomography: preclinical progress and clinical potential", *Nat. Rev. Cancer*, vol. 12, no. 5, pp. 363-368, 2012.

[2] C. W. Sun, S. Y. Lee, and K. F. Lin, "Review: optical scanning probe for optical coherence tomography", *J. Med. Biol. Eng.*, vol. 34, no.1, pp. 95-100, 2014.

[3] J. Sun, S. Guang, L. Wu, L. Liu, S. W. Choe, B. S. Sorg, and H. Xie, "3D in vivo optical coherence tomography based on a low-voltage, large-scan-range 2D MEMS mirror", *Opt. Express*, vol. 18, no. 12, pp. 12065-12075, 2010.

[4] Y. Xu, J. Singh, C. S. Premachandran, A. Khairyanto, K. W. S. Chen, N. Chen, C. J. R. Sheppard, and M. Oliva., "Design and development of a 3D scanning MEMS OCT probe using a novel SiOB package assembly", *J. Micromech. Microeng.*, vol. 18, no. 12, pp. 125005, 2008.

[5] S. R. Samuelson, L. Wu, J. Sun, S. W. Choe, B. S. Sorg, and H. Xie, "A 2.8-mm imaging probe based on a high-fill-factor MEMS mirror and wire-bonding-free packaging for endoscopic optical coherence tomography", *J. Micromech. Microeng.*, vol. 21, no. 6, pp. 1291-1302, 2012.

CONTACT

*H. Xie, tel: +1-352-846-0441; hkxie@ufl.ece.edu

COMPACT NEAR-EYE DISPLAY SYSTEM USING A SUPERLENS-BASED MICROLENS ARRAY MAGNIFIER

Hongbae S. Park[1], Reynald Hoskinson[2], Hamid Abdollahi[2], and Boris Stoeber[1]
[1]The University of British Columbia, Vancouver, Canada
[2]Recon Instruments Inc., Vancouver, Canada

ABSTRACT

This paper reports a new approach to making a very compact near-eye display (NED) using only two layers of microlens arrays (MLA) working in conjunction as a magnifying lens (MLA magnifier). The purpose of the MLA magnifier is to generate a virtual image of a display, positioned within several centimeters from the eye, at optical infinity to minimize the optical disparity between thesurrounding scenery and the image on the display. Our MLA magnifier is about 2 mm thick with a system focal length of 5 mm and a total thickness of around 7 mm (excluding the thickness of the display) in non-folded optics configuration, which is much more compact in comparison to other popular NEDs such as Google Glass or Recon Instrument's Snow goggles having folded optics.

INTRODUCTION

In adults, the unaided human eye cannot focus on objects closer than 10 cm. Therefore all NEDs require an optical element, in one form or another [1] that generates a virtual image further away from the physical location of the display so that it can be comfortably viewed. Oftentimes the profile of the optical element can be quite thick if the optical element includes reflecting surfaces (about 13 mm in Google Glass [2]), which deteriorates the form-factor and increases weight.

This work builds on the idea of superlenses first explored by Gabor [7] and later expanded by others [3-6] for imaging applications. Previous superlenses are not suitable for NEDs as they are primarily used for collecting light onto photosensors, nor are they compact or able to form a sufficiently large "eyebox;" the eyebox is the volume in front of an NED within which the complete virtual image can be viewed. Unlike these previous approaches, we incorporate plano-concave lenslets in the superlens structure, and specifically design it for use as a magnifying lens (magnifier) in an NED system. The plano-concave MLA allows us to design a near-to-eye lens that provides a better compromise between thickness and eyebox than has previously been achieved, e.g. [8]. With the use of a plano-concave MLA, a compact NED system comprised of one plano-concave MLA and one plano-convex MLA can be formed with a sufficiently large eyebox.

SUPERLENS THEORY

A superlens, in the context of this study refers to a composite optical component with multiple layers of MLAs; each MLA may have a different lens pitch and a focal length. The proposed magnifier is a two-layer superlens. Using ray transfer matrix analysis, the superlens system can be simply represented in the matrix notation

$$\begin{bmatrix} h_{out} \\ \alpha_{out} \\ 1 \end{bmatrix} = \begin{bmatrix} M_{11} & M_{12} & \Delta h \\ M_{21} & M_{22} & \Delta \alpha \\ 0 & 0 & 1 \end{bmatrix} \begin{bmatrix} h_{in} \\ \alpha_{in} \\ 1 \end{bmatrix}, \qquad (1)$$

where h_{in} and α_{in} represent the ray height and angle of the input ray at an arbitrary distance u before entering the superlens, and h_{out} and α_{out} are the respective attributes of the rays exiting the superlens, at an arbitrary distance v after the superlens, as shown in Figure 1. Δh and $\Delta \alpha$ represent shift in ray height and angle intrinsic to the superlens, introduced by each microlens on the MLA.

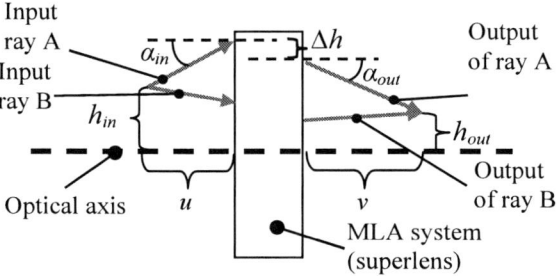

Figure 1: Definition of the ray height and ray angle for the input and output rays to the MLA system.

The entries of the 3x3 matrix in eq. (1) can be found as:

$$M_{11} = d \left(\frac{v - f_2 - vf_2}{f_1 f_2} \right) - \frac{v}{f_2} + 1 \qquad (2)$$

$$M_{12} = F \left[d \left(\frac{v}{f_1 f_2} - \frac{1}{f_1} \right) - \frac{v}{f_1} \right] - d \left(\frac{v}{f_2} - 1 \right) + v \qquad (3)$$

$$\Delta h = \frac{Np_1}{f_1} \left[d \left(\frac{v}{f_2} - 1 \right) - v \right] - \frac{Np_2 v}{f_2} \qquad (4)$$

$$M_{21} = \frac{d}{f_1 f_2} - \frac{1}{f_1} - \frac{1}{f_2} \qquad (5)$$

$$M_{22} = F \left(\frac{d}{f_1 f_2} - \frac{1}{f_1} - \frac{1}{f_2} \right) - \frac{d}{f_2} + 1 \qquad (6)$$

$$\Delta \alpha = Np_1 \left(\frac{d}{f_1 f_2} - \frac{1}{f_1} \right) - \frac{Np_2}{f_2}, \qquad (7)$$

which result from tracing the rays originating from an object at a distance F away, through the two MLA layers that make up the superlens, finally reaching the image plane at a distance v away. In eq. (2)-(7), f_1 and f_2 represent the focal lengths and p_1 and p_2 represent the microlens pitch of the first and the second MLA layer respectively. N is an integer multiplier that represents location of the Nth microlens from the optical axis. d is the gap between the two MLAs, as shown in Figure 2.

Several conditions can be imposed on the two-layer superlens system that result in the collimation of the exit rays, that is when v (image distance of the superlens) in Figure 1 is at infinity. Let us first assume v is finite. As defined in eq. (1), h_{out} is a function of h_{in}, α_{out} and Δh by default. If we assume that a Lambertian point source on the object plane radiates light rays in all directions, then there would be infinitely many input rays originating from a given h_{in} with different α_{in}. For an imaging condition, h_{out} should be independent of α_{in}, because regardless of α_{in} the light rays must focus on a common point at v; that is, all of the output rays will have the same h_{out} at v despite each having different α_{in}, as in Figure 1 where rays A and B have the same h_{out}. The ray height shift Δh intrinsic to the superlens should also not affect h_{out}. This implies that h_{out} should be a function of only h_{in}. If we now assume that v is moved toward infinity then the exiting light rays are collimated and parallel to each other, thus the α_{out} of the output light rays should be identical and be a function of only h_{in}. These conditions imply that:

$$M_{12} = 0 \tag{8}$$
$$\Delta h = 0 \tag{9}$$
$$M_{22} = 0 \tag{10}$$
$$\Delta \alpha = 0 \tag{11}$$
$$h_{out} = M_{11} h_{in} \tag{12}$$
$$\alpha_{out} = M_{21} h_{in}. \tag{13}$$

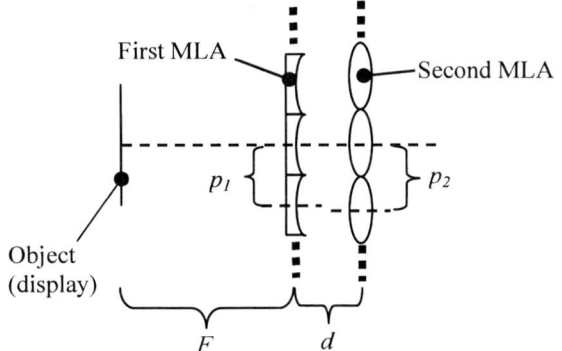

Figure 2: Definition of the variables used in ray transfer analysis of the superlens. The MLAs are only partially shown.

Rearranging the system of equations (8)-(13), we can make the following observation that

$$F = \frac{p_1(d - f_2)}{p_1 - p_2}, \tag{14}$$

demonstrating that two MLAs with lens pitches p_1 and p_2 can produce a combined focal length F of the superlens. Note that p_1 and p_2 must be different for F to be a finite number. Also, F must be positive in order for the MLA system to work as a magnifier, like any lens. One combination of p_1, p_2, f_1, and f_2 that results in F being positive is when $p_2 > p_1$, $d > f_2$, and f_1, f_2 are both positive, meaning that both MLAs have convex microlenses. In the context of this study, this shall be referred to as the "convex-convex" magnifier. This is what was used in previous systems [3-7].

One disadvantage of using a convex-convex magnifier in an NED is that the inter-MLA gap d needs to be larger than the focal length f_2 of the second MLA, so that the two MLA layers need to be separated by a minimum distance of $v_1 + f_2$, where v_1 is the image distance from the first MLA for a given F. This inherently increases the total thickness of the NED, which is against our goal of making a compact

NED. Also, the convex-convex magnifier makes the exit angle α_{out} of the collimated beam positive, which makes it harder to form a large eyebox, because it means that the bundles of the collimated light diverge from the center of the optical system as shown in Figure 3a. Thus, a positive F and a negative α_{out} would be desired.

Another combination of the lens parameters result in exactly that, when f_1 is negative and f_2 is positive, and $p_1 > p_2$. That is, the first MLA should be a concave MLA and the second MLA remains as a convex MLA as shown in Fig. 3b. Consequently, the concave first MLA now generates a virtual image of the object (the display), and the image plane is formed on the object side. Thus the second MLA can now be brought closer to the first MLA. For the rest of the paper, this combination of MLAs shall be referred to as the "concave-convex" magnifier. The differences in system compactness and the size of the eyebox between the convex-convex and concave-convex combinations are illustrated in Figure 4.

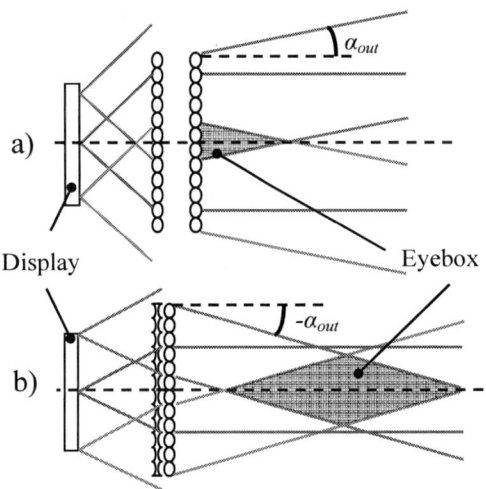

Figure 3: Effect of the exit angle on the formation of the eyebox for a) a convex-convex magnifier and b) a concave-convex magnifier.

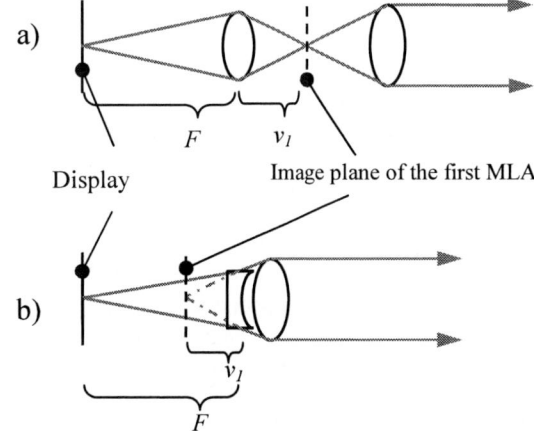

Figure 4: The difference in compactness between a) a convex-convex magnifier and b) a concave-convex magnifier.

With the equations (8)-(13) describing the optical system and the conditions for the concave-convex superlens, we can now calculate the parameters of the MLAs against performance criteria for the MLA magnifier. One criterion is the *eye-relief* which is the clearance between the pupil of the eye and the MLA magnifier. For purposes of comparison, the eye-relief was set identical to Recon Instrument's Snow goggle which has an eye-relief of 18 mm. The other criterion is to make the MLA magnifier as compact as possible; that is, to reduce F as much as we can.

Due to difficulties in fabricating suitable convex MLAs within our lab settings, a commercially available MLA, sourced from Suss MicroOptics based in Switzerland, was used as the convex MLA of our concave-convex magnifier, with a focal length of about 1.02 mm and a radius of curvature of 0.470 mm.

In order for the exit rays to be collimated, the focal plane of the second MLA must coincide with the image plane of the first MLA. In a concave-convex magnifier, the magnifier can be made most compact when the two MLA layers reside on the same plane under thin-lens approximation. Although this is not physically possible, it can be achieved at close approximation. With this assumption, the focal length of the second MLA should equal to the image distance of the first MLA. Then, the following equation can be obtained from studying the ray propagation through the second lens

$$v_1 = BFL_2 = \frac{\frac{R_2 n_2}{n_2 - n_1} - t_2}{n_2},\qquad(15)$$

where R_2, n_2, and t_2 are the radius of curvature, refractive index, and lens thickness (excluding the lens sag) of the convex MLA, respectively. In eq. 15, the convex MLA was assumed to be a thick lens and the back focal length BFL_2 of this MLA was let equal to the image distance v_1 of the first MLA. Also, from rearranging equations (8)-(11), the ratio of pitch between the second and first MLA can be found as

$$\frac{p_2}{p_1} = \frac{F}{F - f_1}.\qquad(16)$$

Since the parameters of the convex MLA are known, we can solve for the remaining parameters of the superlens from the above system of equations in terms of F and f_1. By studying the tradespace spanned by F and f_1, we can find the optimum set of parameters with the smallest F. Since the above equations assume that the MLAs are thin, a few modifications were made to the equations to take into account the actual thicknesses of the MLAs. Also, the parameters were chosen such that the concave MLA and the convex MLA can be placed right next to each other (no gap). The parameters are shown in Table 1, which are also optimized for spherical aberration using the ray tracing software Zemax (version 12).

SIMULATION

The concave-convex magnifier is simulated using Zemax to see whether the MLA magnifier with the parameters shown in Table 1 can perform as designed.

Table 1: List of MLA parameters.

Parameter	Value
f_1 (mm)	-0.41
f_2 (mm)	1.02
n_1	1.42 (PDMS)
r_1 (µm)	157
p_1 (µm)	260
p_2 (µm)	250
F (mm)	5.4
Eye-relief (mm)	16

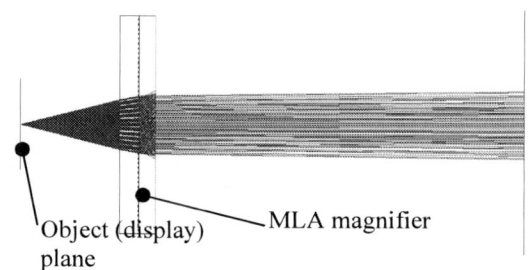

Figure 5: Simulation of the concave-convex magnifier with parameters in Table 1.

Figure 5 shows light rays launched from a point source at the center of the object plane (as in a pixel on a display) are collimated exiting the MLA magnifier. Figure 6 shows the collimation of the light launched from different heights on the object plane converging towards the optical axis of the superlens, which confirms α_{out} being negative.

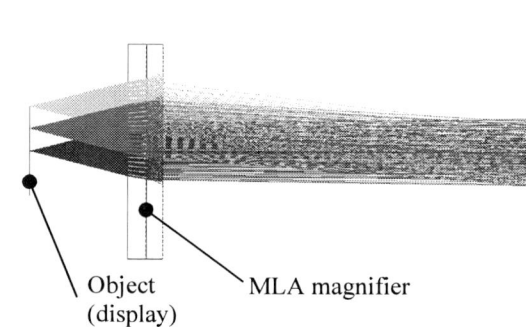

Figure 6: Simulation of the concave-convex magnifier with multiple sources of light at different object height.

FABRICATION

The concave MLA was made using photolithography and reflow processes followed by PDMS casting. The use of photolithography and reflow in making microlenses has been demonstrated in prior literatures such as [9,10]. First, the photolithography process defines periodic islands of photoresist cylinders on a silicon wafer, whose spacing is identical to the pitch of the microlenses required. The reflow process melts the photoresist cylinders and the molten photoresist beads on the wafer due to surface tension, forming spherical caps. Assuming the volume of the photoresist lens stays constant before and after the reflow, the required thickness h for the photoresist deposition during photolithography can be approximated by solving the equations

978-1-4799-7956-1/15 $31.00 © 2015 IEEE   954

Volume of cylinder = volume of spherical cap (17)

$$\pi r^2 h = \int_{R-t}^{R} \pi (R^2 - y^2)\, dy, \tag{18}$$

where R and t refer to the radius of curvature and the sag of the spherical cap. The silicon wafer with spherical caps formed after the reflow serves as the mold for forming the concave array of microlenses. PDMS resin and a curing agent are mixed in a weight ratio of 5:1, and the mixture is poured onto the wafer mold to a thickness of 1 mm and cured at 60°C. Once hardened, the PDMS is peeled from the wafer and cut into a 12 mm x 12 mm square, which is the same size as the commercial convex MLA.

TESTING

An apparatus as shown in Figure 8 was set up on an optical table to test the MLA magnifier. The apparatus consists of a small display, the MLA magnifier, and a camera to capture the test image.

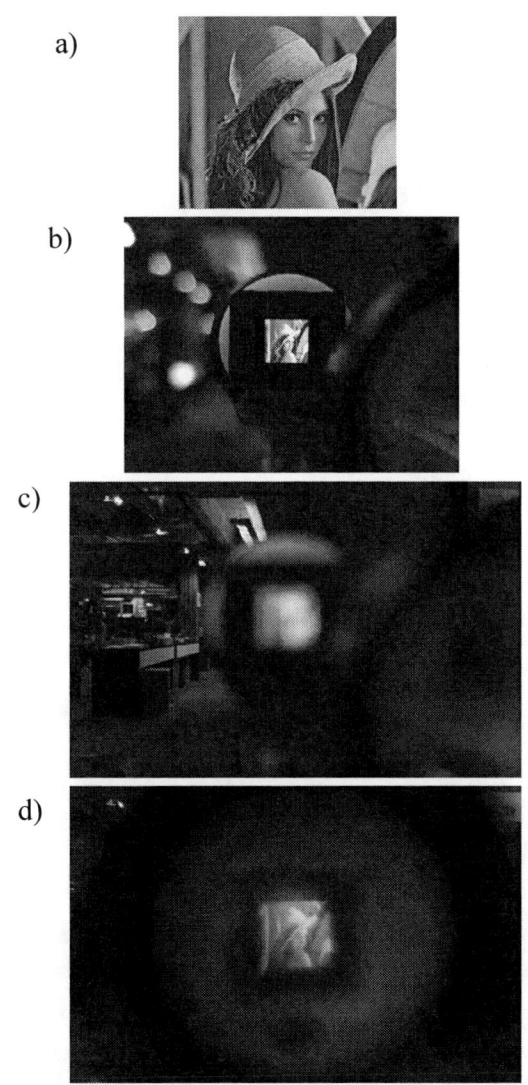

Figure 7: A test image seen through the MLA magnifier; a) image of Lena used for testing, b) display showing the Lena image, with camera focused at the display ~3 cm away from camera lens using a macro mode, c) display seen without the magnifier, d) display seen with the magnifier.

The camera used in the apparatus was focused at infinity to mimic the relaxed state of the eye's lens. A cropped image of Lena is used as the test image in Figure 7. The image is in focus from the camera's perspective which indicates that the MLA magnifier can indeed collimate light.

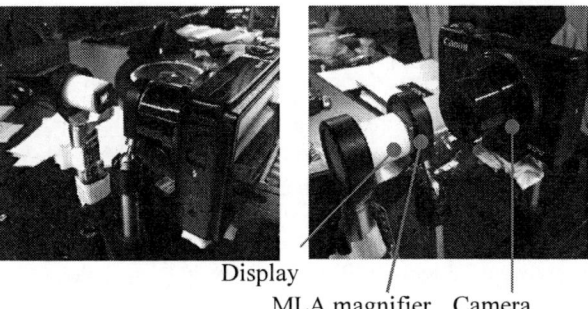

Display

MLA magnifier   Camera

Figure 8: Pictures of the test apparatus taken at different angles.

REFERENCES

[1] J. Rolland and O. Cakmakci, "Head-worn displays: the future through new eyes," Opt. Photon. News, vol. 20, no. 4, pp. 20–27, 2009.

[2] M, Olsson et al., "Wearable Display Device," U.S. Patent D659739 S1, May 15th, 2012.

[3] C. Hembd-Sölner et al., "Imaging properties of the Gabor superlens," Journal of Optics A: Pure and Applied Optics, vol. 1, no. 1, p. 94, 1999.

[4] K. Stollberg, "The Gabor superlens as an alternative waferlevel camera approach inspired by superposition compound eyes of nocturnal insects," Optics Express, Vol. 17, Issue 18, pp. 15747-15759, 2009.

[5] J. Duparré et al., "Microoptical telescope compound eye," Optics express, vol. 13, no. 3, pp. 889–903, 2005.

[6] V. Shaoulov, R. Martins, and J. P. Rolland, "Compact microlenslet-array-based magnifier," Optics letters, vol. 29, no. 7, pp. 709–711, 2004.

[7] D. Gabor, "Optical System Composed of Lenticules," U.S. Patent 2351034 A, June 13th, 1944.

[8] D. Lanman and D. Luebke, "Near-eye light field displays," ACM Transactions on Graphics, vol. 32, no. 6, pp. 1–10, Nov. 2013.

[9] H. Yang, C.-K.Chao, M.-K.Wei, and C.-P. Lin, "High fill-factor microlens array mold insert fabrication using a thermal reflow process," Journal of Micromechanics and Microengineering, vol. 14, no. 8, pp. 1197–1204, Aug. 2004.

[10] M.-H. Wu and G. M. Whitesides, "Fabrication of two-dimensional arrays of microlenses and their applications in photolithography," Journal of micromechanics and microengineering, vol. 12, no. 6, p. 747, 2002.

CONTACT

*Sam(Hongbae) Park, tel: +1-778-908-2848; sam@reconinstruments.com

OPEN-STRUCTURE ELECTROWETTING DISPLAY WITH CAPACITIVE SENSING FEEDBACK SYSTEM

Seungyul Choi and Junghoon Lee
Seoul National University, Seoul, South Korea

ABSTRACT

We report an open-structure electrowetting-based reflective display with capacitive sensing feedback that enables an effective self-dosing of ink, high contrast, and the precise control of color level. EWOD-based reflective display [1, 2] has been considered as a promising display technology due to high speed and contrast for e-paper application, but needs more improvements in effective packaging, enhanced contrast, and operation reliability. Here we introduce an EWOD display that can achieve such improvements via an open structure design and a capacitive feedback system. Our report includes the demonstration of quick and effective ink dosing process, off color area being ~ 8% of viewable area, and precision control of color area even under a large variation of interfacial tension.

INTRODUCTION

Electrowetting reflective display has been studied for a decade since EWOD (electrowetting on dielectric) was first introduced [3]. The EWOD display uses two immiscible liquids such as colored oil and conductive liquid. The oil is usually kept within a pixel of grid restricted by a wall structure [1]. This design has inherent challenges in liquid dosing process and high contrast. These issues were addressed via the introduction of a "two-story" architecture for self-dosing and off ink storage in a dimple reservoir [2]. This approach, however, requires difficult fabrication steps, and the device was thick with a limitation in flexible applications. Furthermore, this complex architecture can hardly accommodate a feedback system [4] which is required for the accurate control of color area against disturbances such as contact angle hysteresis, and temperature / interfacial tension variations.

Our device that has an open architecture with all pixel cells connected, enabling the aforementioned benefits with a simple fabrication process. Colored oil can be uniformly "self-dosed" into every pixel through the open gap. With this thin and flat architecture, the capacitance between the conducting liquid and the electrode can be readily measured and used for feedback to regulate the color area. We designed and implemented a hardware circuit that achieved an excellent feedback system.

SAMPLE PREPARATION FOR DISPLAY

Open-structure fabrication and packaging

Fabrication process is shown in Figure 6. First we pattern AZ5214 photo-resister on ITO coated glass substrate and etch ITO with etchant. Next step is to coat SU-8 (microchem) photoresistor at the thickness of 25 μm and pattern the open structure. Barriers of the open structure width is 20 μm and pixel size is 200 μm each. Final step is Parylene C coating with vaporization system (PDS 2010) at the thickness of 1 μm which is for dielectric and hydrophobic layer.

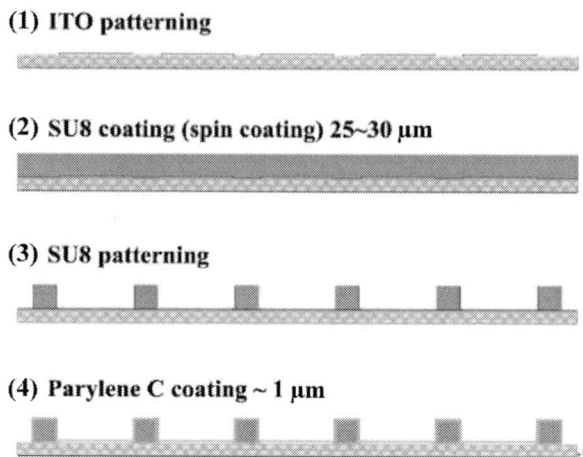

(1) ITO patterning

(2) SU8 coating (spin coating) 25~30 μm

(3) SU8 patterning

(4) Parylene C coating ~ 1 μm

Figure 1: Open-structure fabrication process.

Packaging has two steps including self-dosing. First all pixels are filled with color oil as illustrated in Figure 3. Oil (chloronaphthalene : dodecane, 9:1) is mixed with Oil Blue N dye for coloring. Next step is to introduce a certain amount of conducting liquid on the sample and squeeze the liquid droplet with top glass coated by ITO, and cover the sample simultaneously. Sample is sealed with PSA (pressure sensitive adhesion) tape.

Polyacrylic acid (PAA) is used as an electrolyte to increase conductivity of the conducting liquid. PAA shows excellent property in terms of breakdown [5]. Breakdown causes permanent damage to EWOD device. Thus it is critical to reduce this effect and PAA is one of the best materials for this goal.

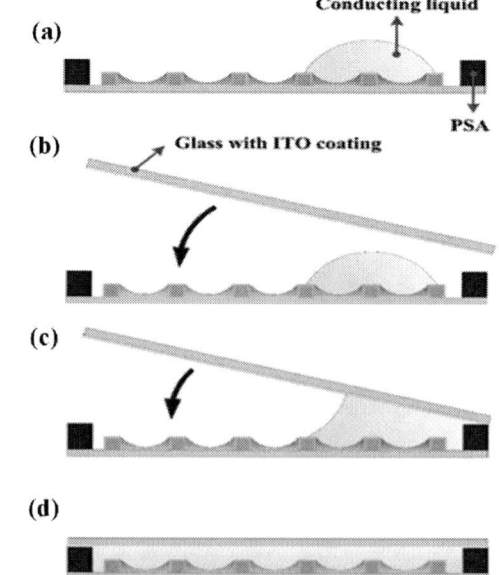

Figure 2: Packaging process

FEEDBACK SYSTEM DESIGN
The concept of feedback system

It is crucial to control color area precisely. However, such control is limited by hysteresis, temperature, or any other external effects in EWOD display. Feedback control is required to circumvent this problem. Color area is strongly related to the contact area between the conducting liquid and surface. Our target was to obtain the information for the exact contact area. The capacitance is related with the contact and it can be expressed as (1) below.

$$C = \frac{\varepsilon_0 \varepsilon_r A}{d} \qquad (1)$$

where, ε_0 is the dielectric constant of vacuum, ε_r is the equivalent relative dielectric constant of coating, d is the thickness of dielectric coating, A is the area of electrode. Area, A, in (1) is the contact area between conducting liquid and electrode at the bottom of a pixel in our system because conducting liquid works as the counter electrode.

We can measure the exact area, $A = Cd / \varepsilon_0 \varepsilon_r$, with capacitance measurement between the conducting liquid and electrode because the values of ε_0, ε_r, and d, which are material property and fabrication factor respectively, are already known.

Feedback circuit design

The type and amplitude of operation voltage and target resolution are required to be considered for capacitance measurement. AC voltage with sine wave at the frequency of 4 kHz was used as operation voltage and the range was from 0 V_{rms} to 30 V_{rms}. Target resolution can be calculated with geometric information and material properties. The size of electrode in each pixel was 200 μm × 200 μm and Parylene C dielectric layer had 1 μm thickness with the dielectric constant of 2.9. We measured capacitance value when conducting liquid was in contact with whole electrode in a pixel (200 μm × 200 μm) with LCR meter (Agilent E4980A) and the value was approximately 1.129 pF and maximum resolution was required to be 1 fF.

Figure 3 shows the circuit design. Cx is the capacitance formed by conducting liquid and electrode in a single pixel. First step was to achieve V_{rms} value of V2 AC signal. Resistor, Rc, was connected with Cx in series. V2 signal was substantially low because the impedance of resistor was much lower than that of capacitance, Cx. OpAmp (operational amplifier) was used to amplify the V2 signal. We changed AC signal to DC signal at the amplitude of V_{rms} of the V2 through modulator. The secondary anti-alising filter was also added to reduce signal noise.

The target of feedback system in EWOD display is to maintain white area constantly. This requires keeping the contact area between conducting liquid and electrode in a pixel unchanged. For instance, when our setting target white area was 50% of total pixel size, it can change due to temperature or system condition variation. If the white area deceases below 50%, operation voltage should increase to compensate for the reduction, and if the area becomes higher than 50% the operation voltage also compensates by increasing

The size of white area is the contact area of conducting liquid and electrode in a pixel which is strongly related with capacitance, Cx. Cx increases as WA becomes larger and vice versa. Here Cx directly affects V2. A large WA means higher Cx and it increases the amplitude of V2 signal (high V_{rms} of V2). WA variation can be defined by the difference between DAC voltage (V1 in Figure 3) and V_{rms} of V2. It is considered as an error in feedback circuit (DAC voltage – V_{rms} of V2). Our target V_{rms} of V2, which indicates target white area, is defined by DAC voltage that can be controlled. If there is no difference between two signal (DAC voltage and V_{rms} of V2), V_{rms} of V2 is the same as DAC voltage. If DAC voltage is higher than V_{rms} of V2 the error becomes positive. This error signal has two factors. One is sign (minus or plus) and the other is amplitude. This error signal is multiplied with unit voltage and it is summed with operational voltage. If white area is larger than target, error signal becomes negative, which decreases operation voltage to reduce white area.

RESULTS AND DISCUSSION
Self-dosing and feedback circuit test

Liquid dosing is one of the critical challenges of EWOD display. Usually it is required to introduce an exact amount of color liquid into every single pixel for proper operation. Such uniform dosing is challenging for isolated structures.

Every neighboring pixels has connection in our device and whole area is coated with hydrophobic material. Such open structure and hydrophobic property make it simple to dose liquid uniformly and easily. Color oil spreads efficiently and fast on whole pixel surface which has hydrophobic property due to capillary force shown in Figure 4. Oil rapidly filled up all cells by a capillary flow through the gaps with a single-shot pipetting, resulting in the even distribution within 1 sec.

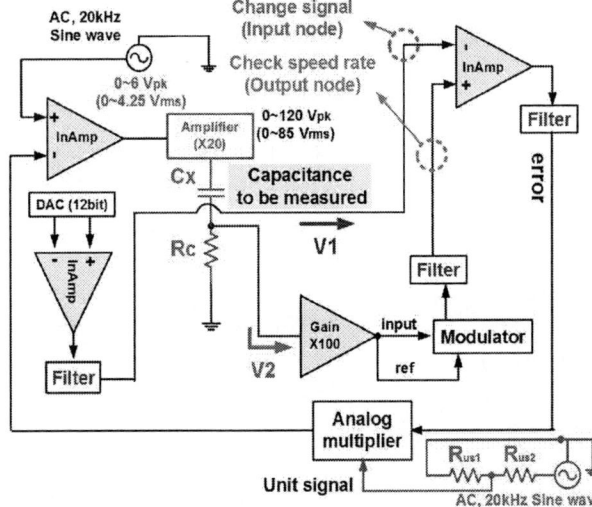

Figure 3: The schematic of feedback circuit

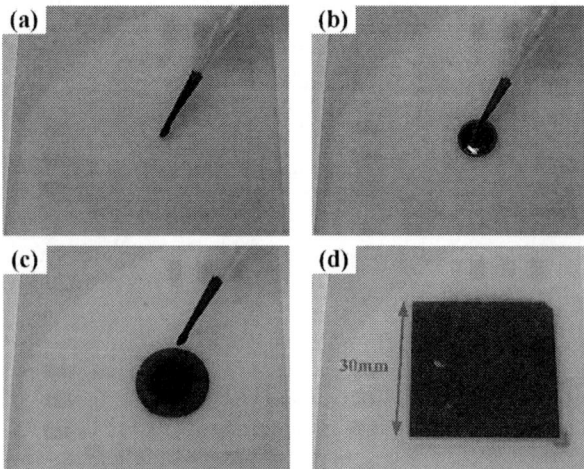

Figure 4: Self-dosing in 30 mm × 30 mm sample.

We did feedback system test with single pixel measurement sample shown in Figure 5. Fabrication process is exactly same as EWOD display device with open structure. We use gold electrode instead of ITO and electrode is placed in only one pixel which is at the center of 5 × 5 array, so we can apply operation voltage to the single pixel not others.

Two parallel experiments were done with feedback and without feedback system. First, we applied certain operation voltage, put surfactant (Tween 80), and observed WA change that is yellow part in Figure 6. The results of two cases were totally different. Experiment in case of sample without feedback system, Figure 6 (a), showed WA increased abruptly right after surfactant injection. However, sample with feedback system maintained white area almost the same even after surfactant injection, Figure 6 (b).

We quantified WA changes with image analysis. First we converted images into black and white and calculated mean gray value. We defined deviation value with (2).

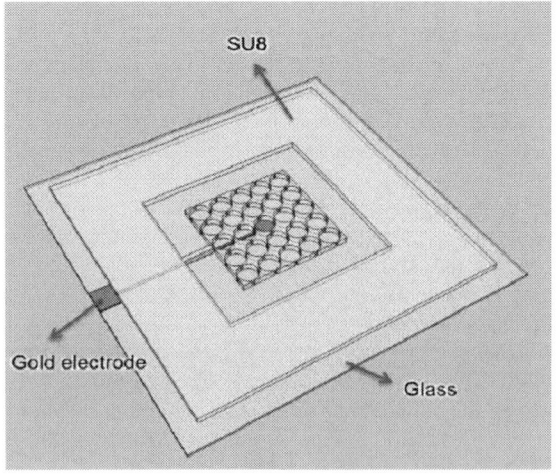

Figure 5: Single pixel measurement sample. It is designed for feedback system test with capacitance measurement.

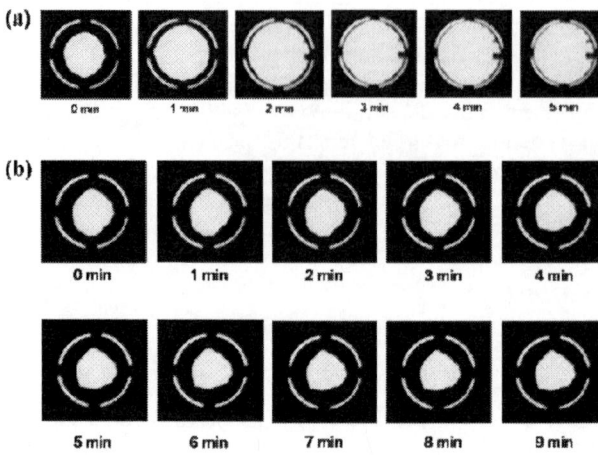

Figure 6: (a) WA change without feedback system. (b) WA change with feedback system. WA increased even within 1 min in case of non-feedback sample, but WA did change with sample integrated feedback system after 9 min.

$$Deviation = \frac{MG_t - MG_i}{MG_i} \qquad (2)$$

where, MG_i is mean gray value of initial state which is before surfactant injection and MG_t is mean gray value at certain time, t, after surfactant injection from image analysis. Deviation is based on initial MG_i value, in other words, it shows how much WA changes comparing with the initial WA.

Figure 7 is the result. Surfactant reduces the interfacial tension leading to increase the WA according to electrowetting equation and the deviation increased up to more than 95% without feedback system. However, the deviation is reduced by 85% with feedback system.

We checked the bandwidth using oscilloscope (Wavesurfer 454, Lecoy®). First we decided the input and output node in the feedback circuit and defined feedback loop, Figure 3. Bandwidth we measured was 500 Hz which is much higher than video speed (60 Hz).

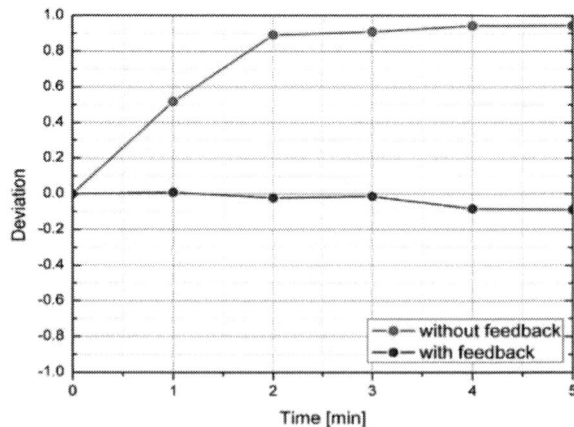

Figure 7: White area deviation of both cases with or without feedback system.

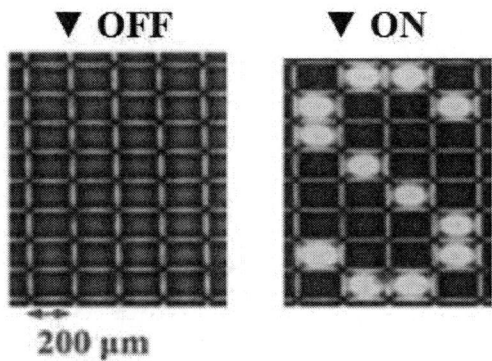

Figure 8: Operation example. Voltage at ON state is 30 V with 500 Hz square wave.

Electrowetting display with open-structure operation

We used electrode bundle that forms specific letter shape. We linked several electrodes in pixels in ITO patterning step. We can apply voltage to all electrode of pixels which are connected together at the same time. Figure 8 shows one of examples. In this case, electrodes are linked with the letter of 'S'.

We defined WAR (white area ratio) like (3) and it is important to attain clear and fine contrast display.

$$WAR[\%] = \frac{MG_V - MG_{0V}}{MG_{fw} - MG_{0V}} \times 100[\%] \qquad (3)$$

where MG_V is mean gray value (MG) at a certain voltage V, MG_{fw} is the state when a pixel is fully filled with white color, and MG_{0V} is MG at zero voltage, which is initial state.

We achieved 92.6% of WAR from the calculation at 30 V. It is relatively high WAR comparing other EWOD display. In addition, we did the demonstration of a fine gray-scale display. WA is larger in a single pixel when WAR gets higher. Figure 9 shows the gray scale expression.

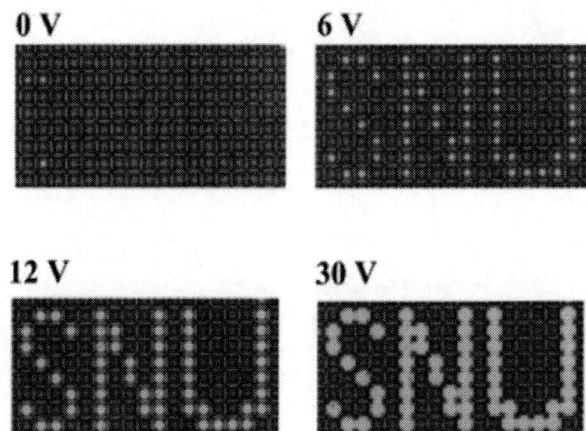

Figure 9: Gray scale expression. White area increases as the operation voltage gets higher.

We measured the operation speed of our device using high speed camera. WAR of the sample having pixel size with 200 μm × 200 μm calculated according to time. It took 90 ms WAR to change from 0 to 92% (on speed) and over 600 ms from 92 to 0% (off speed). However, we measured operation speed with a sample having smaller pixel size, 100 μm × 100 μm and the speed was much faster than 200 μm × 200 μm case. On speed from 0 to 85% was less than 8 ms and off speed is approximately 7 ms.

CONCLUSION

We report the new designed feedback system imbedded EWOD display device with open structure. Our approach in the suggested EWOD display has advantages in terms of large viewable area ratio (~92%), self-dosing, video speed (less than 10 ms), high resolution display (up to 508 ppi), and natural gray scale expression. In addition, we developed feedback system with capacitance measurement and viewing area was control precisely regardless of external environment changes. Surfactant test shows feedback system decreased the deviation by 80% in single pixel test. We verified EWOD display device with precise feedback control system at the first time and showed the high feasibility as well. It is essential to improve the reliability of EWOD reflective display in the future.

ACKNOWLEDGEMENTS

This work was supported by the ICT & Future Planning as Global Frontier Project (CISS-2012M3A6A 6054193) and the Brain Korea 21 Plus Project in 2014. And the fabrication was performed at the Interuniversity Semiconductor Research Center (ISRC) in Seoul National University.

REFERENCES

[1] R. A. Hayes and B. J. Feenstra, Video-speed electronic paper based on electrowetting. *Nature* **425**, 383-385 (2003)

[2] J. Heikenfeld, K. Zhou, E. Kreit, B. Raj, S. Yang, B. Sun, et al., Electrofluidic displays using Young–Laplace transposition of brilliant pigment dispersions. *Nature Photonics* **3**, 292-296 (2009)

[3] J. Lee, H. Moon, J. Fowler, T. Schoellhammer, and C.-J. Kim, Electrowetting and electrowetting-on-dielectric for microscale liquid handling. *Sensors and Actuators A: Physical* **95**, 259-268 (2002)

[4] Jian Gong and CJ Kim, All-electronic droplet generation on-chip with real-time feedback control for EWOD digital microfluidics. *Lab chip* **8**, 898-906 (2008)

[5] S Choi, Y Kwon, YS Choi, ES Kim, J Bae, J Lee, Improvement in the Breakdown Properties of Electrowetting Using Polyelectrolyte Ionic Solution. *Langmuir* **29(1)**, 501-509 (2013)

CONTACT

*Junghoon Lee, tel: +82-2-880-9104; jleenano@snu.ac.kr

TUNABLE METAMATERIAL LENS ARRAY VIA METADROPLETS

Q. H. Song[1,2], W. M. Zhu[2], W. Zhang[2], P. C. Wu[2], Z. X. Shen[2], Z. C. Yang[4], Y. F. Jin[4], Y. L. Hao[4],
T. Bourouina[3], Y. Leprince-Wang[1†], and A. Q. Liu[2†]

[1]Université Paris-Est, UPEM, F-77454 Marne-la-Vallée, France
[2]School of Electrical and Electronic Engineering, Nanyang Technological University
50 Nanyang Avenue, Singapore 639798
[3]Université Paris-Est, ESYCOM, ESIEE, Paris F-93162 Marne-la-Vallée, France
[4]National Key Laboratory of Science and Technology on Micro/Nano Fabrication,
Institute of Microelectronics, Peking University, Beijing 100871, China

ABSTRACT

In this paper, a tunable THz lens array based on reconfigurable metamaterials is reported. The metamaterial consists with 40 × 40 liquid metal microdroplets, in which the shape is tuned under different air pressures. The droplets are formed by injecting the liquid mercury into a pre-designed microchannel network and arranged to focus incident THz wave. The focus spot size can be tuned with different droplet geometries. As a result, a tunable THz lens is constructed and the smallest spot size of 0.54λ is achieved. The tunable metamaterial lens is realized through simple fabrication processes, which has potential applications in flat lens and imaging system.

INTRODUCTION

Metamaterials, or rationally designed artificial materials with sub-wavelength scale elements, offers a fantastic platform to control and manipulate the electromagnetic (EM) waves. The sub-wavelength elements, typically metallic patterns, can respond to both the electric field and the magnetic field. Many extraordinary physical phenomena are then induced, such as negative index [1-2], cloaking [3-4], zero epsilon [5-6], giant chirality [7-8], or exotic and useful hyperbolic dispersion anisotropy [9]. Furthermore, as the metamaterial responded frequency highly depends on the size of the sub-wavelength elements, the EM wave of different frequencies can be effectively modulated. Compared to natural materials, metamaterial also has the advantage of real-time dynamically tunable for EM wave modulation. Tunable metamaterials are widely studied using micro-electro-mechanical system (MEMS) [10-12], phase changing materials [13] and liquid crystals [14].

In the previous tunable metamaterials [15-17], the tuning flexibilities, such as the tuning range and the resonant mode switching, are highly dependent on how the sub-wavelength elements are modulated during the tuning process. Among different tuning mechanisms, changing the geometry of the metallic structure of the sub-wavelength element typically results in a dramatic change in the EM properties since the electric and the magnetic response of the metamaterial are directly dependent on the element shapes. Previous works on MEMS tunable metamaterials [18-19] target on the change of the geometry shape of the metal elements by changing the near-field coupling of the metal parts anchored on the

movable islands driven by micromachined actuators. However, it is difficult to reshape the metal structures once it is forged.

Liquid metals with sub-wavelength feature size are recently utilized to construct tunable metamaterials due to their flexibility on reshaping the geometry [20-21]. This pioneer work used a complex microfluidic system for the tuning function. Although it offers individual sub-wavelength element tuning without any metallic electrode contact that can potentially spoil the EM properties of the metamaterials and introduce extra losses, it still suffers many drawbacks due to the complexity of the system and limited tuning speed. Here, an alternative technique is applied to tune the geometry of metal liquid droplet for the THz wave modulation. Based on this technique, a tunable lens array is designed with simple fabrication processes and control systems. The swift tuning on the focus spot size is realized.

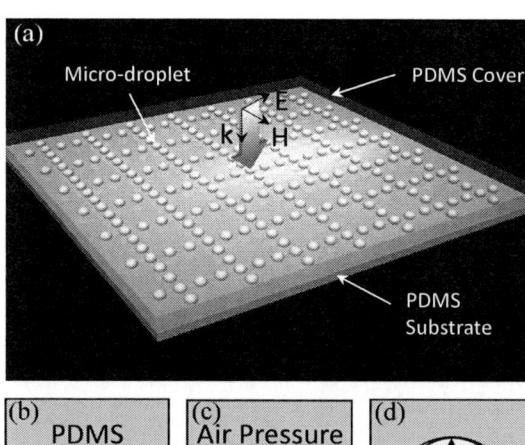

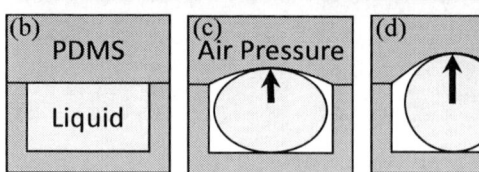

Figure 1: (a) Schematic of tunable lens array based on microdroplets. (b) The droplets are confined in PDMS channel. (c, d) The working principle of tuning method by applying air pressure.

METAMATERIAL LENS ARRAY DESIGN

Figure 1(a) shows the schematic of the lens array based on random access metamaterials formed by mercury microdroplets with a period of 300 μm. Each lens is formed by 5 mercury droplets, one of which is at the center with four others surrounded. Proper spatial phase distribution can be induced through such structure, which can effectively focus the incident light. By controlling the shape of the droplets, the focus spot can be switched between the focusing and defocusing states. Figure 1(b) shows the liquid mercury being confined in the PDMS channel. The radii of the droplets can be tuned under different air pressures. The air pressure pushes the PDMS channel up, resulting in the increase of the droplet height and decrease of the droplet radius because of surface tension. Therefore, the interaction between the incident THz wave and the droplet metamaterial is changed, which consequently tunes the resonance in the structure and the focusing state of the incident light.

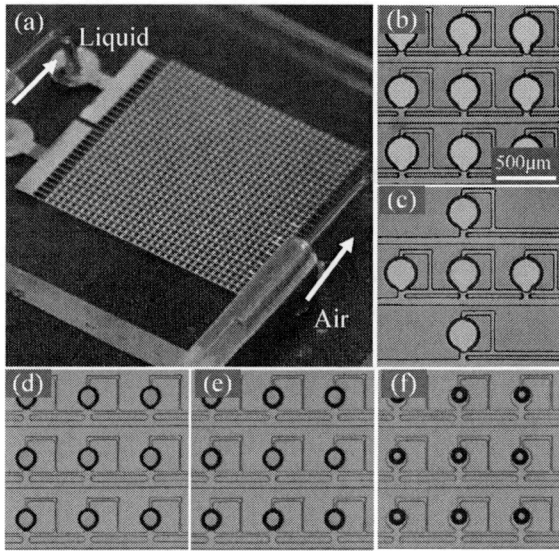

Figure 2: (a) Fabrication results of the micro-droplets with (b) uniform droplets pattern and (c) one single lens of five droplets cluster. The radii of the droplets (d, e, f) can be tuned by applying different air pressure.

The fabrication of the microfluidic system is based on the polymer soft lithography technique. A 6-inch silicon is cleaned using a piranha solution (H_2SO_4 + H_2O_2) and spin coated with a 50-μm SU-8 photoresist (MicroChem, SU-8 50) at 2000 rpm for 30 s using a spin coater (CEE, 200). The silicon substrate is soft baked using a hot plate at 65°C for 6 min and 95°C for 20 min. The substrate is then exposed to UV light for 30 s under the plastic mask using a mask aligner (OAI, J500-IR/VIS). The post expose bake is performed at 65°C for 1 min and 95°C for 5 min after the exposure. The SU-8 layer, which is used as the master of the PDMS channels, is developed using the SU-8 developer

(MicroChem) for 6 min. The PDMS channels are fabricated using the replica molding, which is the casting of PDMS prepolymer against a master and obtaining the negative replica of the master. Three masters with different patterns are fabricated for the metamolecule array layer, respectively. There are altogether three PDMS layers. Two layers with microchannels and one layer without any pattern as the substrate. The microfluidic system is fabricated by plasma bonding the three PDMS layers together.

Figure 2(a) shows the fabrication results of the tunable metamaterial chip with the uniform micro-droplets pattern shown in Fig. 2(b) and one single lens that consists of five independent micro-droplets shown in Fig. 2(c). The channels for mercury and air injection are constructed to control the shapes of the droplets. The metal liquid is injected into the microchannel by syringe pumps. When the pressure is maintained at a low level, the liquid metal fills in the wide microchannel. Then, the air is injected into the narrow channel, which breaks the liquid metal flow into separated droplets and expels the extra liquid. Figure 2(d), (e) and (f) demonstrate the continuous tuning of the droplet radius from 120 to 80 μm by applying different air pressures.

RESULTS AND DISCUSSION

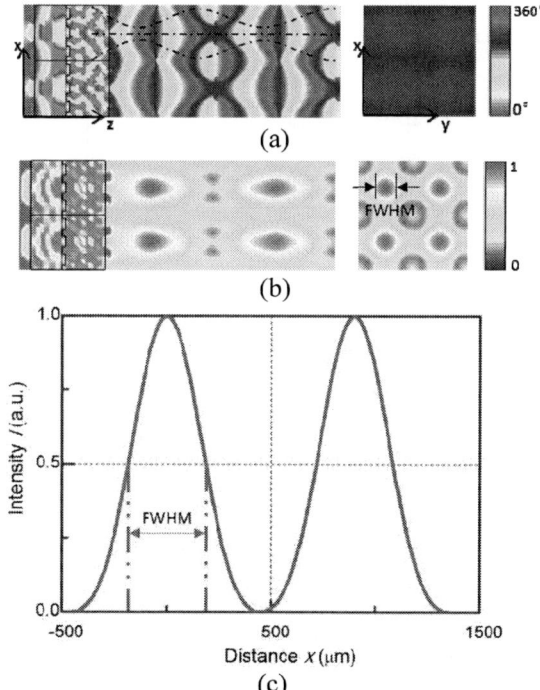

Figure 3: Simulation results of (a) phase and (b) electric field distribution at z direction and x-y cross section, respectively; (c) full width at half maximum at focus point.

Figure 3(a) shows the simulation results of the phase distribution in the propagation direction (right) and the x-y cross-section at the focus point (left). The corresponding electric field distribution is shown in Fig. 3(b). As the boundary condition of the central droplet is different with that of the four surrounded droplets, a different phase delay is formed. The phase distribution cross-section shows a larger phase delay at the center, which steers the light to the center of the cluster. The wavelength of the incident light is 670 μm (0.448 THz in frequency). When the THz wave is incident on the cluster normally with an optimized droplet radius, which is 90 μm in this design, it will be focused at the other side of the metamaterial lens because of the spatial phase modulation. Repetition of the focus is observed in z direction, which is due to the Talbot effect. Figure 3(c) shows the full width at half maximum (FWHM) at the focus point is as narrow as 362 μm (0.54λ).

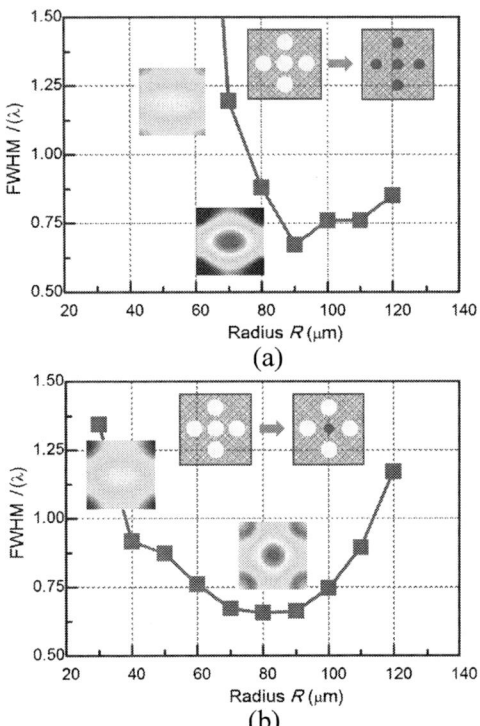

Figure 4: The full width at half maximum is changed by (a) tuning all the droplets simultaneously, and (b) tuning only the central droplets.

Figure 4 shows the simulation results of the FWHM of the focus spot in the x-y plane, which can be effectively tuned by air pressing all the droplets in Fig. 4(a) or only the central droplet of each lens in Fig. 4(b). The FWHM becomes lowest when the radius R is 90 μm for all droplets, and when R is 80 μm for the central droplet. As the droplet radius is detuned, the FWHM increases and the spot becomes defocused. Comparing the two tuning mechanisms, the all-droplet tuning approach realizes a much more abrupt FWHM change. Therefore, it switches the focusing state of the

metamaterial lens to the defocusing state more effectively.

CONCLUSIONS

In conclusions, a THz tunable lens array based on mercury droplets is designed, fabricated and numerically demonstrated. The radius of each droplet are tuned from 120 μm to 80 μm, while the FWHM can be controlled to switch the metamaterial lens between the focusing state and the defocusing state. This metamaterial tuning technique is easy and flexible for the EM wave control, which has potential applications on tunable lens array and can be used in imaging system and detectors.

ACKNOWLEDGEMENTS

The work is supported by the Environmental and Water Industry Development Council of Singapore (EWI), RPC programme (Grant No.: 1102-IRIS-05-01, 1102-IRIS-05-02, 1102-IRIS-05-04 and 1102-IRIS-05-05)

REFERENCES

[1] D. R. Smith, J. B. Pendry and M. C. K. Wiltshire, "Metamaterials and negative refractive index", *Science*, vol. 305(5685), pp.788-792, 2004.

[2] T. Xu, A. Agrawal, M. Abashin, K. J. Chau, and H. J. Lezec, "All-angle negative refraction and active flat lensing of ultraviolet light", Nature, vol. 497(7450), pp. 470-474, 2013.

[3] D. Schurig, J. J. Mock, B. J. Justice, S. A. Cummer, J. B. Pendry, A. F. Starr and D. R. Smith, "Metamaterial electromagnetic cloak at microwave frequencies", *Science*, vol. 314, pp. 977-980, 2006.

[4] W. Cai, U. K. Chettiar, A. V. Kildishev and V. M. Shalaev, "Optical cloaking with metamaterials", *Nature photonics*, vol. 1(4), pp. 224-227, 2007.

[5] M. Silveirinha and N. Engheta, "Tunneling of electromagnetic energy through subwavelength channels and bends using ε-near-zero materials", *Phys. Rev. Lett.*, vol. 97, pp. 157403, 2006.

[6] R. Mass, J. Parsons, N. Engheta and A. Polman, "Experimental realization of an epsilon-near-zero metamaterial at visible wavelengths". *Nat. Photo.*, vol. 7, pp. 907-912, 2013.

[7] A. V. Rogacheva, V. A. Fedotov, A. S. Schwanecke and N. I. Zheludev, "Giant gyrotropy due to electromagnetic-field coupling in a bilayered chiral structure". *Phys. Rev. Lett.*, vol. 97, pp. 177401, 2006.

[8] W. Zhang, W. M. Zhu, E. E. M. Chia, Z. X. Shen, H. Cai, Y. D. Gu, W. Ser, and A. Q. Liu, "A pseudo-planar metasurface for a polarization rotator". *Opt. Exp.*, vol. 22, issue 9, pp. 10446-10454, 2014.

[9] J. Elser, R. Wangberg, V. A. Podolskiy, and E. E. Narimanov, "Nanowire metamaterials with extreme optical anisotropy", *Appl. Phys. Lett.*, vol. 89(26), pp. 261102-261102, 2006.

[10] W. Zhang, A. Q. Liu, W. M. Zhu, E. P. Li, H. Tanoto, Q. Y. Wu, J. H. Teng, X. H. Zhang, M. L. J. Tsai, G. Q. Lo and D. L. Kwong, "Micromachined switchable

metamaterial with dual resonance". *Appl. Phys. Lett.*, vol. 101(15), pp. 151902-151902, 2012.

[11] W. Zhang, W. M. Zhu, H. Cai, M. L. J. Tsai, G. Q. Lo, D. P. Tsai, H. Tanoto, J. H. Teng, X. H. Zhang, D. L. Kwong and A. Q. Liu, "Resonance Switchable Metamaterials using MEMS Fabrications", *IEEE Journal of selected topics in quantum electronics*, vol. 19, pp. 4700306-4700306, 2013.

[12] W. M. Zhu, A. Q. Liu, W. Zhang, J. F. Tao, T. Bourouina, J. H. Teng, X. H. Zhang, Q. Y. Wu, H. Tanoto, H. C. Guo, G. Q. Lo and D. L, Kwong, "Polarization dependent state to polarization independent state change in THz metamaterials", *Appl. Phys. Lett.*, vol 99, 221102,2011.

[13] N. Yu, P. Genevet, M. A. Kats, F. Aieta, J. P. Tetienne, F. Capasso and Z. Gaburro, "Light propagation with phase discontinuities: generalized laws of reflection and refraction", Science, vol. 334(6054), pp. 333-337, 2011.

[14] D. H. Werner, D. H. Kwon, I. C. Khoo, A.V. Kildishev and V. M. Shalaev, "Liquid crystal clad near infrared metamaterials with tunable negative-zero-positive reflective indices". *Opt. Exp.* 15(6), 3342-3347, 2007.

[15] A. Q. Liu, W. M. Zhu, D. P. Tsai and N. I. Zheludev "Micromachined tunable metamaterials: a review", *Journal of Optics*, vol. 14(11), pp. 114009, 2012.

[16] H. T. Chen, J. P. Willie, J. M. O. Zide, A. C. Gossard, A. J. Taylor, and R. D. Averitt, "Active terahertz metamaterial devices". *Nat.*, vol. 444, pp. 597-600, 2006.

[17] H. Tao, A. C. Strikwerda, K. Fan, W. J. Padilla, X. Zhang, and R. D. Averritt. "Reconfigurable terahertz metamaterials". *Phy. Rev. Lett.* vol. 103,147401, 2009.

[18] W. M. Zhu, A. Q. Liu, T. Bourouina, D. P. Tsai, J. H. Teng, X. H. Zhang, G. Q. Lo, D. L .Kwong and N. I. Zheludev, "Microelectromechanical Maltese-cross metamaterial with tunable terahertz anisotropy", *Nat. commun.*, vol. 3, pp. 1274, 2012.

[19] W. M. Zhu, A. Q. Liu, X. M. Zhang, D. P. Tsai, T. Bourouina, J. H. Teng, X. H. Zhang, H. C. Guo, H. Tanoto, T. Mei, G. Q. Lo and D. L. Kwong. "Switchable magnetic metamaterials using micromachining processes" *Adv. Mat.*, vol. 23(15), pp. 1792-1796, 2011.

[20] T. S. Kasirga, Y. N. Ertas, M. Bayindir, "Microfluidics for reconfigurable electromagnetic metamaterials", *Appl. Phys. Lett.*, vol. 95(21), pp. 214102-214102-3, 2009.

[21] J. A. Gordon, C. L. Holloway, J. Booth, S. Kim, Y. Wang, B. J. James, R. N. David. "Fluid interactions with metafilms/metasurfaces for tuning, sensing, and microwave-assisted". *Phys. Rev. B*, vol. 83, 205130, 2011.

CONTACT

[†] Y. L. Wang, Tel:+33-623131398; yamin.leprince@u-pem.fr
[†] A. Q. Liu, Tel: +65-67904336; eaqliu@ntu.edu.sg

FABRICATION OF PATTERNED MAGNETIC MICROSTRUCTURES USING MAGNETICALLY ASSEMBLED NANOPARTICLES

Camilo Velez [1], Isaac Torres-Díaz [2], Lorena Maldonado-Camargo [3],*
Carlos Rinaldi [2,3] and David P. Arnold [1]

[1] Department of Electrical & Computer Engineering, Interdisciplinary Microsystems Group;
[2] J. Crayton Pruitt Family Dept. Biomedical Engineering; [3]Dept. Chemical Engineering,
University of Florida, USA

ABSTRACT

This work describes the modeling and experimental characterization of a fabrication method for forming magnetic microstructures using self-assembled iron oxide (Fe_3O_4) magnetic nanoparticles. This method can potentially be used in roll-to-roll production of magnetic structures patterned onto substrates or optionally lifted off to create free-floating micromagnetic actuators. This article reports: (1) the use of a selective magnetization process to create magnetic microstructures with complex, photolithographically defined shapes, (2) development of multi-physics simulations that model key fabrication steps (selective magnetization and particle assembly), and (3) experimental evaluation of the microstructure features (line width and height) as functions of process variables. The primary accomplishment is obtaining well-defined microstructures with complex shape and demonstrating their magnetic actuation when released as free-floating structures.

INTRODUCTION

Organized structures of magnetic nanoparticles are attracting growing interest in different areas such as data storage media [1-2], sensors and electronics [3], MEMS [4-5], biomedicine [6], and photonics [7-8]. A common process of creating arrays of nanoparticles is based on embedding magnetic particles in polymers [5]. However, in order to maintain photolithographic patternability and reasonable fluidic viscosities, the particle loading (volume fraction) is usually quite low, therein hindering the magnetic response. To overcome these limitations, various self-assembly techniques of magnetic nanoparticles have been envisioned [8-11]. Additionally, the particles can be cross-linked to one another [12].

A recent work of our group successfully demonstrated the combination of nanoparticle self-assembly and crosslinking for the formation of magnetic microstructures [13]. A subsequent structure release via dissolution of sacrificial layer was also demonstrated. While this work validated the basic idea, the microstructural shapes were limited to simple lines (due to the geometry of the magnetic recording head used to create the magnetic patterns). For future applications, a fabrication technology is needed to generate more complex features.

Herein, we adopt a selective magnetization technique [14] to imprint relatively arbitrary magnetic pole patterns into hard magnetic substrates using soft magnetic "magnetization masks." Magnetic structures of various geometric shapes are fabricated using Fe_3O_4 (iron oxide) nanoparticles. The structures are eventually released from the substrate where they show response to magnetic stimulus. We also present a set of multi-physics simulations that model the selective magnetization process as well as the magnetic nanoparticle self-assembly process. These models help to guide certain process parameters such as magnetization mask dimensions, magnetization fields, and nanoparticle assembly time.

FABRICATION PROCESS

Fig. 1 depicts the overall magnetic structure fabrication concept. Each sub-step is detailed as follows.

Selective Magnetization

First, using the methods in [15], a nickel mask (magnetization mask) is used to create the desired magnetic patterns (shapes) on a magnetic substrate (Hi8-MP video tape in this case). The mask is fabricated by electroplating 45-µm-thick Ni structures (squares and crosses of 175 µm and 150 µm overall size, respectively) on a silicon wafer. The tape is initially uniformly pre-magnetized out-of-plane (up) with 6 T. Then the Ni

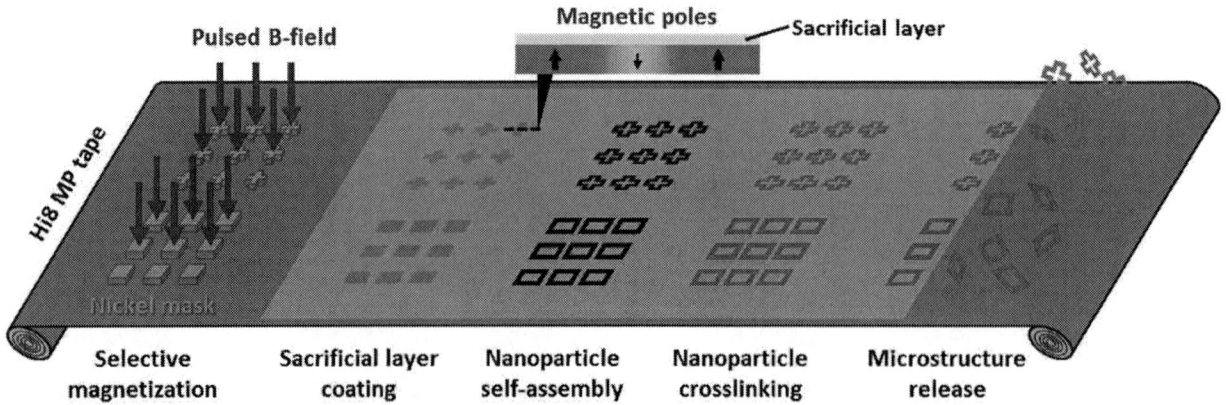

Figure 1: Roll to roll fabrication method process concept.

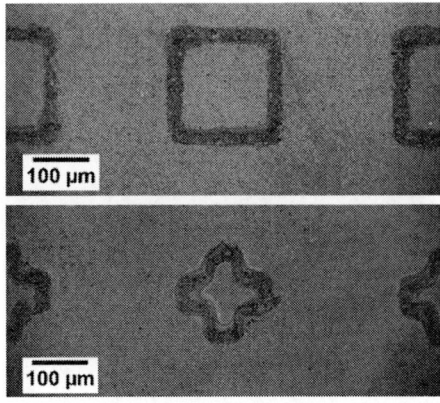

Figure 2: Examples of self-assembled and cross-linked microstructures on the substrate before release. Darker regions on the boundary of the shapes indicates nanoparticles agglomeration.

mask is placed in contact with the tape, and a reverse magnetic field pulse (down) is used to imprint the geometric shapes of the mask as magnetic pole patterns (down) in the tape.

Nanoparticles Self-assembly

After selective magnetization, the tape is coated with a 350-nm-thick sacrificial layer of LOR-3B (polydimethylglutarimide-based resist). Magnetic nanoparticles are then assembled onto the sacrificial layer, relying on their attraction to the higher magnetic gradient regions at the boundaries of the magnetic poles.

Superparamagnetic iron oxide particles (Fe_3O_4) are synthesized by thermal decomposition of an iron-oleate precursor [16]. The physical diameter is 19 ± 1 nm, measured by TEM, showing good monodispersity. Magnetic hysteresis is measured using a SQUID magnetometer and fit to the Langevin equation using the procedure suggested by Chantrell [17] to estimate the distribution of size of magnetic particles in the sample. The magnetic volume fraction of the particle suspension is estimated as 0.062 %v/v with the magnetic core diameter 9 ± 1.6 nm.

Nanoparticle assembly is made by covering the magnetized substrate with 1 mL of the particle suspension for a specific time (ranging from seconds to hours). After completing the assembly, the tape is removed from suspension, washed with deionized water by manual agitation, and left to dry at room temperature at a 45° inclination over an absorbent surface.

Nanoparticle Crosslinking

The as-synthesized particles are soluble only in organic solvents, due to the presence of the oleic acid on the particle surfaces. Crosslinking of the magnetic particles is achieved with branched polyethyleneimine (PEI), but PEI is a cationic polymer soluble in water [13]. Therefore, an additional phase-transfer procedure is necessary to suspend the synthesized particles in a water-based solution. To transfer the particles from an organic to a water-based phase, the procedure described by Wang [18] is used to cleave the double bond present in the oleic acid chain, obtaining carboxylic acid as terminal groups. These acid groups are crosslinked with primary amines in the polymer chains of a poly-cation (PEI) with molecular weight of

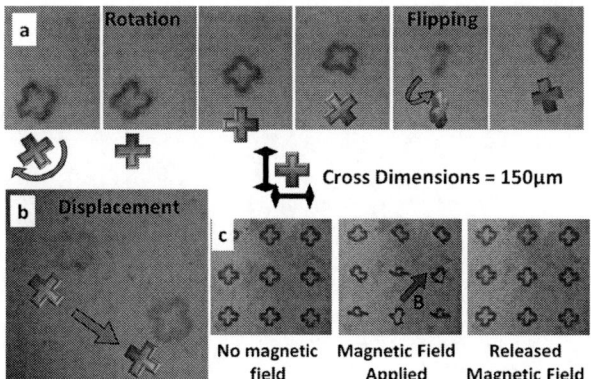

Figure 3: Rotation experiments on microstructure changing the orientation of the magnetic field. (a) Rotation and flipping, (b) superpose picture of displacement and (c) actuation with half of the structure anchored to the substrate.

2.5 kDa, promoted by 1-ethyl-3-[3-dimethylaminopropyl] carbodiimide hydrochloride (EDC) and N-hydroxysuccinimide (NHS). Specifically, the sample is immersed in the PEI, EDC/ NHS solution (2:1) and stirred for 1 hour. The sample is then washed with deionized water and left to dry at room temperature.

Fig. 2 presents two examples of structures obtained during the process before release. The particles accumulate at the perimeters of the magnetic poles, with no particles in the center regions of the shapes.

Microstructure Release

Lastly, the underlying sacrificial layer (350 nm of LOR 3B) is dissolved in AZ300 MIF developer, releasing the structures from the substrate to obtain free-floating structures into solution. Released structures proved to respond to magnetic field variations, as shown in Fig. 3. By changing the direction of an externally applied field in different axes, we generate displacement, rotation and translation of the free-floating structure.

PROCESS SIMULATIONS

COMSOL Multiphysics is used to produce a connected sequence of simulations to model the selective magnetization process and the resulting magnetic stray fields from the tape [19]. Matlab is then used to calculate the motion of the particles during self-assembly and their resultant distribution over the surface of the tape. Fig. 4 illustrates the modeling process.

Two forces acting on a nanoparticle are considered in this study—the magnetic force $\vec{F}_m$ and the Stokes' drag force $\vec{F}_d$—with a force balance equation as [9], [20]:

$$m \cdot \frac{d\vec{v}}{dt} = \vec{F}_m + \vec{F}_d = (\vec{m} \cdot \vec{\nabla})\vec{B} - 6\pi\eta\, r_{mp}\, \vec{v} \qquad (1)$$

where $\vec{m}$ is the magnetic moment of the particle (field-dependent, from measured hysteresis curve), $\vec{B}$ is the magnetic flux density produced by the magnetic pattern, η is the fluid viscosity, r_{mp} the hydrodynamic radius of the particle, and $\vec{v}$ the velocity of the nanoparticle. In this work we present a system level approximation solution, where particle-particle

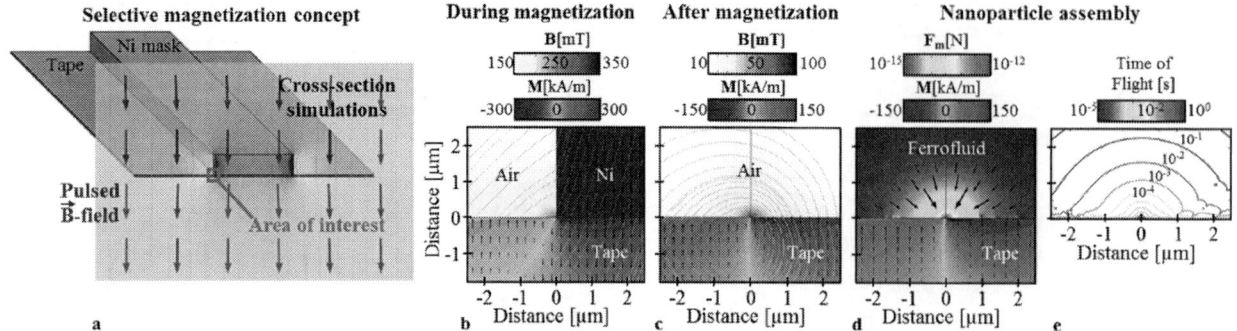

Figure 4: Simulation steps: a) Complete simulation concept showing the cross-section and the area of interest. b) Magnetization (for tape and Ni) and H-field (for air) during selective magnetization. c) Magnetization (for tape) and H-field (for air) after magnetization. d) Magnetic force produced by the selective magnetization of the tape (M for tape). e)Particles time of flight.

interactions or Brownian motion are ignored, and the inertia term is found to be negligible. Inclusion of these considerations will be presented in future publications. The velocity of the magnetic nanoparticle is given by:

$$\vec{v} = \frac{(\vec{m} \cdot \vec{\nabla})\vec{B}}{6\pi\eta\, r_{mp}} \qquad (2)$$

From the velocity relationship we can calculate the particle time of flight—the time required for a particle to reach the surface of the tape.

A cross-section general scheme of the magnetization process is illustrated in Fig. 4a with a superimposed B-field plot. Fig. 4b is a zoom-in of the magnetic fields during the selective magnetization step. The resulting stray B-field produced by the magnetized tape is shown in Fig. 4c. The magnetic force described in eq. (1) is mapped in Fig. 4d. A contour plot of the time of flight in the simulated space is plotted in Fig. 4e.

PROCESS CONTROL

While there are many different process variables associated with the fabrication process, two process variables are systematically evaluated: the magnetic flux density during selective magnetization (50–500 mT) and the particle assembly time (32–100,000 s).

An optical profilometer (Contour GT-I, Bruker) is used to measure the hill-like profile of the self-assembled microstructures, for example as shown in Fig. 5a. Post-processing analysis converts the 3D-profile map into a series of cross-sections from the structure boundary. All the cross-sections are aligned and tilt is removed to obtain

the average hill-like profile of the structures, as shown in Fig. 5b. From this, the average line widths (at 50 % of the hill height) and average line heights are calculated for different process conditions.

Fig. 6a and 6b show the measured profiles and extracted line width/height for different magnetic reversal field strengths. Our goal is to obtain strong structures that do not break during release or actuation. Making the magnetic structures as thick as possible (large line height/width aspect ratio) helps us achieve that goal. Profile measurements (Fig. 6b), in addition to optical inspection, show that a magnetic reversal field of ~200 mT maximizes this aspect ratio, with line height/width of 230 ± 80 nm / 3.9 ± 1 μm. This field value also minimizes particle deposition over undesired areas i.e. none magnetically patterned areas. This magnetic field is then used to conduct the assembly-time experiment.

Fig. 6c and 6d show the measured profiles for different assembly times. The data clearly shows an increasing trend in the line width and height with time. For 32 s the line height/width is 47 ± 19 nm / 3 ± 1.5 μm, whereas for 100,000 s the line/height is 820 ± 150 nm / 6.4 ± 1 μm. By controlling the assembly time, we can adjust the optimum structure line width and line height.

CONCLUSIONS

In this paper, a fabrication technique of magnetic microstructures by *in-situ* crosslinking of magically assembled nanoparticles first demonstrated by [13], is

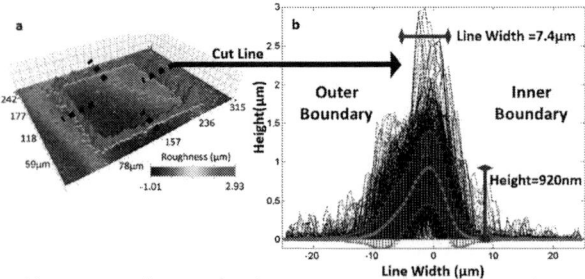

self-assembled magnetic microstructures: (a) Surface measurement, where the black lines indicates examples of the cross section analyzed during post processing. (b) Post processing mean profile curve.

enhanced by adding a selective magnetization technique to

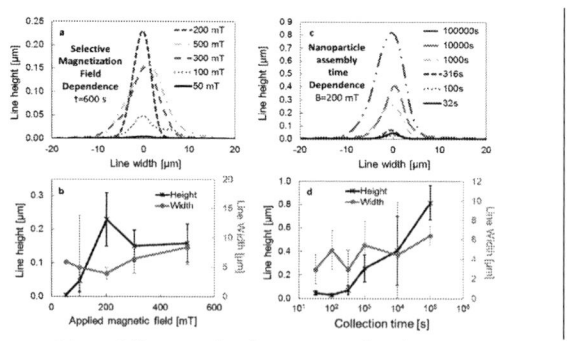

Figure 6: Two process variable influences on the resultant microstructure profiles: (a and c) Mean profile curves and (b and d) line/height and line/width. N=5 squares per sample.

define complex structural patterns. Multi-physics simulations are used to model the selective magnetization process and the nanoparticle assembly steps. Simulations results are used to guide fabrication process parameters and reconcile observed experimental results. The influence of the magnetic reversal field amplitude and particle assembly time are evaluated experimentally by measuring the microstructure dimensions. Nanoparticle assembly time is shown to provide a clear control over the microstructure line height, whereas the reversal field applied during magnetization must be optimized to avoid defects in the structure. Prototype microstructures maintain their shapes after releasing and prove to have a magnetic response, demonstrating that this method is useful in production of free-floating micromagnetic actuators.

ACKNOWLEDGEMENTS

This work is funded in part by the UF Office of Research and NIH Grant 1R21AR064402-01A1. The authors thank Mr. Ololade Oniku, Mr. Nicolas Garraud and Dr. Alexandra Garraud for their valuable discussions, as well as the staff of the UF Nanoscale Research Facility and the UF Major Analytical Instrumentation Center for their assistance in the microfabrication and material characterization.

REFERENCES

[1] Q. Dai and A. Nelson, "Magnetically-responsive self assembled composites.," *Chem. Soc. Rev.*, vol. 39, no. 11, pp. 4057–66, Nov. 2010.

[2] C. A. Ross, "Patterned Magnetic Recording Media," *Annu. Rev. Mater. Res.*, vol. 31, no. 1, pp. 203–235, 2001.

[3] A. N. Shipway, E. Katz, and I. Willner, "Nanoparticle Arrays on Surfaces for Electronic , Optical , and Sensor Applications," *ChemPhysChem*, vol. 1, pp. 18–52, 2000.

[4] N. Damean, B. a Parviz, J. N. Lee, T. Odom, and G. M. Whitesides, "Composite ferromagnetic photoresist for the fabrication of microelectromechanical systems," *J. Micromechanics Microengineering*, vol. 15, no. 1, pp. 29–34, Jan. 2005.

[5] J. Kim, S. E. Chung, S.-E. Choi, H. Lee, J. Kim, and S. Kwon, "Programming magnetic anisotropy in polymeric microactuators.," *Nat. Mater.*, vol. 10, no. 10, pp. 747–52, Oct. 2011.

[6] D. F. Emerich and C. G. Thanos, "The pinpoint promise of nanoparticle-based drug delivery and molecular diagnosis.," *Biomol. Eng.*, vol. 23, no. 4, pp. 171–84, Sep. 2006.

[7] A. Yethiraj, J. H. J. Thijssen, A. Wouterse, and A. Van Blaaderen, "Large-Area Electric-Field-Induced Colloidal Single Crystals for Photonic Applications," *Adv. Mater.*, vol. 16, no. 7, pp. 596–600, 2004.

[8] L. He, Y. Hu, H. Kim, J. Ge, S. Kwon, and Y. Yin, "Magnetic assembly of nonmagnetic particles into photonic crystal structures.," *Nano Lett.*, vol. 10, no. 11, pp. 4708–4714, Nov. 2010.

[9] X. Xue and E. P. Furlani, "Template-assisted nano-patterning of magnetic core-shell particles in gradient fields.," *Phys. Chem. Chem. Phys.*, vol. 16, no. 26, pp. 13306–17, Jul. 2014.

[10] J. Henderson, S. Shi, S. Cakmaktepe, and T. M. Crawford, "Pattern transfer nanomanufacturing using magnetic recording for programmed nanoparticle assembly.," *Nanotechnology*, vol. 23, no. 18, p. 185304, May 2012.

[11] L. F. Zanini, N. M. Dempsey, D. Givord, G. Reyne, and F. Dumas-Bouchiat, "Autonomous micro-magnet based systems for highly efficient magnetic separation," *Appl. Phys. Lett.*, vol. 99, no. 23, p. 232504, 2011.

[12] R. Barbucci, G. Giani, S. Fedi, S. Bottari, and M. Casolaro, "Biohydrogels with magnetic nanoparticles as crosslinker: characteristics and potential use for controlled antitumor drug-delivery.," *Acta Biomater.*, vol. 8, no. 12, pp. 4244–52, Dec. 2012.

[13] C. Velez, I. Torres-díaz, L. Maldonado-camargo, C. Rinaldi, and D. P. Arnold, "Fabrication of magnetic microstructures by in situ crosslinking of magnetically assembled nanoparticles," in *Solid-State Sensors, Actuators and Microsystems Workshop*, Hilton Head, SC, June 2014, pp. 331–334.

[14] A. Garraud, O. D. Oniku, W. C. Patterson, E. Shorman, D. Le Roy, N. M. Dempsey, and D. P. Arnold, "Microscale magnetic patterning of hard magnetic films using microfabricated magnetizing masks," in *MEMS 2014*, San Jose, CA. Jan 2014, pp. 520–523.

[15] O. D. Oniku, P. V. Ryiz, A. Garraud, and D. P. Arnold, "Imprinting of fine-scale magnetic patterns in electroplated hard magnetic films using magnetic foil masks," *J. Appl. Phys.*, vol. 115, no. 17, p. 17A718, Feb. 2014.

[16] J. Park, K. An, Y. Hwang, J.-G. Park, H.-J. Noh, J.-Y. Kim, J.-H. Park, N.-M. Hwang, and T. Hyeon, "Ultra-large-scale syntheses of monodisperse nanocrystals.," *Nat. Mater.*, vol. 3, no. 12, pp. 891–5, Dec. 2004.

[17] R. Chantrell, J. Popplewell, and S. Charles, "Measurements of particle size distribution parameters in ferrofluids," *IEEE Trans. Magn.*, vol. 14, no. 5, pp. 975–977, Sep. 1978.

[18] M. Wang, M.-L. Peng, W. Cheng, Y.-L. Cui, and C. Chen, "A novel approach for transfering oleic acid capped Iron Oxide Nanoparticles to water phase," *J. Nanosci. Nanotechnol.*, vol. 11, no. 4, pp. 3688–3691, 2011.

[19] O. D. Oniku, A. Garraud, E. E. Shorman, W. C. Patterson, D. P. Arnold, and I. M. Group, "Modeling of a micromagnetic imprinting process," in *Solid-State Sensors, Actuators and Microsystems Workshop*, Hilton Head,SC, June 2014, pp. 187–190.

[20] T. H. Boyer, "The force on a magnetic dipole," *Am. J. Phys.*, vol. 56, no. 8, p. 688, 1988.

CONTACT*

Camilo Velez, tel: +1-352-575-7581; camilovelez@ufl.edu

ELECTRIC CONTACT STABILITY AND READOUT RESOLUTION OF THE ANTIWEAR PROBE WITH A GROOVE AND OIL LUBRICATION SLIDING SYSTEM FOR PROBE-BASED ARCHIVE MEMORIES

Yasushi Tomizawa[1], Kiminori Toya[2], Atsuro Oonishi[1], Yongfang Li[1],
Jun Hirota[3], Moto Yabuki[3], Iwao Kunishima[3], and Hideo Shinomiya[3]

[1]Corporate Research and Development Center, Toshiba Corporation, JAPAN
[2]Corporate Manufacturing Engineering Center, Toshiba Corporation, JAPAN
[3]Semiconductor & Storage Products Company, Toshiba Corporation, JAPAN

ABSTRACT

The authors have proposed the novel concept of a sliding system involving an antiwear probe with a groove and oil lubrication for probe-based archive memories. The system has been proven to endure a meter-scale probe slide, not only in terms of electric contact stability but also in regards to the readout resolution degradation of recorded patterns. This demonstrates the possibility of bringing probe-based memories into actual products used in data archiving, which requires limited time data access.

INTRODUCTION

Probe-based memories are expected to surpass the recording density limits of existing devices, such as flash memory and hard disk drives, making them attractive as next-generation storage devices. Several studies have demonstrated high density data recording and readout using probes with nanoscale tips [1-4]. However, to bring these technologies to practical implementation, the system must have an ability to endure a very long distance probe slide while maintaining a stable electric contact of the probe tip and high readout resolution. Even data archiving-specific storage devices, which only access data at a limited incidence, required a probe that endures at least several meters of slide. Thus far, no one has succeeded in achieving this goal. To realize the meter-scale sliding endurance, the authors focused on an antiwear probe with a sidewall electrode developed for probe lithography [5, 6], and proposed the sliding system involving an antiwear probe with a groove and oil lubrication (AGO). A novel groove structure was introduced for wear debris management and oil lubrication was for wear reduction.

SYSTEM

An example of a future probe-based memory involving an AGO system is shown in Figure 1. Data were recorded in the media by applying voltage from each antiwear probe through bit-patterned top electrodes acting as media protection layers. The introduction of an oil lubricant on the media surface reduced wear. Ferroelectric substances are the most promising materials for recording media because of their potential high recording density (over 10 Tbpsi) [2]. Resistive metal oxides are also attractive recording media materials because they comprise simple constituents and possibility of multi-level recording [4]. In this case, the need for an accurate detection of media resistance demanded an enhanced level of electric contact resistance stability between the probe tip and the media.

As described previously [5, 6], the antiwear probe consisted of a nondoped silicon (Si) cantilever and a metal sidewall electrode, for which surfaces simultaneously made contact with and slid on the substrate. The electric contact area of the electrode was as small as that of a conventional sharp probe tip, but the physical contact area of the tip exceeded that of a conventional tip, reducing the wear-progression speed. In addition, a groove was formed near the contact edge of the probe to capture wear debris and prevent this debris from hindering accurate recording and readout.

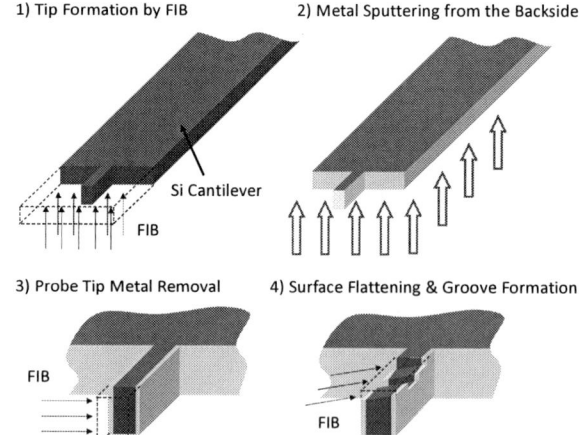

Figure 2: Fabrication process of the grooved antiwear probe.

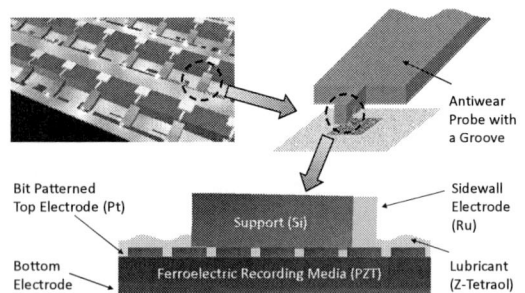

Figure 1: Possible components of a future probe-based memory using an AGO system.

978-1-4799-7956-1/15 $31.00 © 2015 IEEE

MEMS 2015, Estoril, PORTUGAL, 18 - 22 January, 2015

FABRICATION

Antiwear probes were fabricated as shown in Figure 2. To cancel the influence of process deviation caused by typical etching methods, shape formation was conducted using a focused ion beam (FIB). First, a commercial Si cantilever for scanning probe microscopy was cut by FIB at its edge to generate the probe tip. Next, the electrode metal was sputtered from the back of the cantilever, and the metal at the foremost part of the probe tip was removed by FIB to form an independent sidewall electrode. Finally, the probe surface, which touched the recording media, was flattened by FIB so that the Si support and the metal electrode surfaces belonged exactly to the same plane. At the same time, a groove was formed at the probe tip.

Scanning electron microscopy (SEM) images of the fabricated antiwear probes without and with a groove are shown in Figures 3(a) and 3(b), respectively. These probes were actually used in the following experiments.

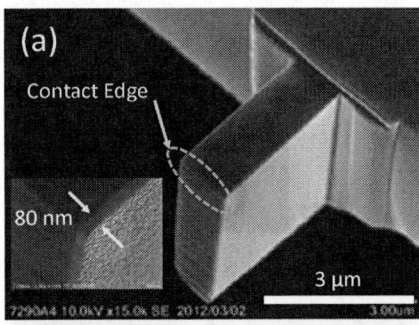

Figure 3: Fabricated antiwear probe (a) without and (b) with a groove.

EXPERIMENTS
Electric Contact Stability

Assuming that recording media consisted of resistive materials, the electric contact stability of the AGO system was evaluated by scanning spreading resistance microscopy (SSRM) [7], which measures probe contact resistance over an extremely wide range (10^3–10^{10} Ω) using a logarithmic current amplifier. An antiwear probe without a groove was slid on a platinum (Pt) film substrate, which was supposed to act as a top electrode in actual memory devices and was covered with the oil lubricant "Z-Tetraol". The probe tip was pressed with a contact force of about 500 nN and scanned the substrate within a 150×150 μm^2 observation area at a speed of 100 $\mu m/s$. SSRM images showing two-dimensional maps of the electric contact resistance distribution were obtained while scanning. Measurements were repeated at different scanning areas until the cumulative sliding distance exceeded 2 m. Ruthenium (Ru) was chosen as the material of the 80-nm-thick electrode at the probe sidewall because Ru presents similar hardness to Si cantilevers [6] and were covered with a thin native oxide film which was conductive [8]. A bias voltage of 0.01 V was applied to the probe during the scan.

Measurement results are shown in Figure 4. Deviations in electric contact resistance within the section of every 150-mm slide, which corresponds to the acquisition of one SSRM image, are plotted in box-plot style. These results demonstrated that, even without a groove on the probe tip, the sliding system endured a 2-m probe slide while maintaining the contact resistance below 100 kΩ most of the time. However, the resistance occasionally reached approximately 100 MΩ as a result of interferences between probe and substrate. There interferences may stem from the accidental adhesion of wear debris on the contacting surface of the probe tip.

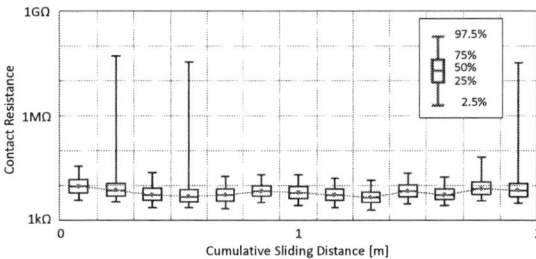

Figure 4: Changes in electric contact resistance between the probe electrode and the metal substrate during a probe slide.

To assess the effectiveness of the groove forming, additional measurements were performed under three different groove/lubricant conditions. These measurements used probes with 30-nm-thick Ru sidewall electrodes. Probe scanning speed and cumulative sliding distance were set to 40 $\mu m/s$ and 0.1 m, respectively. After each slide, probe tips were examined by SEM to evaluate wear debris adhesion.

SSRM and SEM images are shown in Figure 5. In the absence of a groove and a lubricant, the measured SSRM image was almost white. This suggests that the probe electrode could not maintain electric contact with the substrate because of debris adhesion near the sidewall electrode, as shown by SEM. However, when the probe bore a groove, the resistance during the slide remained below 10 kΩ even in the absence of a lubricant. This indicates that the groove captured the wear debris and prevented their adhesion to the probe surface. The introduction of a lubricant slightly increased the average resistance, but it helped to drastically reduce the volume of wear debris.

978-1-4799-7956-1/15 $31.00 © 2015 IEEE

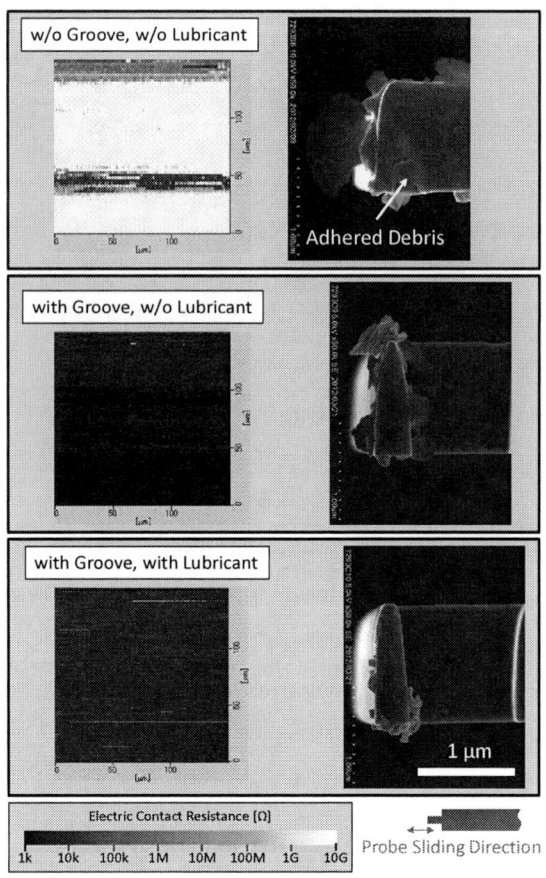

Figure 5: SSRM images obtained by probe scanning in a 150 × 150 μm² area on the metal substrate, and SEM images of probe tips after the slide.

Readout Resolution

The readout resolution of the recording system coupled with the antiwear probe was evaluated using a ferroelectric lead zirconate titanate (PZT) film substrate as the recording media and a scanning nonlinear dielectric microscope (SNDM) [2] for signal readout. An SNDM can detect the direction of ferroelectric crystal polarization at the probe tip-substrate contact position, enabling the readout of "0" and "1" stored as positive and negative polarizations of the ferroelectric recording media, respectively.

First, a 10 × 10 μm² polarization domain was formed on a PZT substrate by applying a 5-V DC bias voltage via the antiwear probe while scanning the area at 20 μm/s. Next, the same probe was scanned at 40 μm/s in a 20 × 20 μm² area including the recorded polarization domain while an SNDM image was acquired to readout the polarization direction. The probe was subsequently slid on the PZT substrate at 280 μm/s for a cumulative distance of 1 m. Finally, the domain forming and SNDM readout operations were performed again for a different scanning area, to compare between system readout

resolutions before and after the 1-m slide.

Because the PZT film was not optimized for probe data recording, its surface was far rougher (R_a = 12.5 nm) than that of conventional magnetic recording media, and its grain size was rather large (ca. 100–200 nm). To minimize the influence of this rough surface on the measurements, a probe comprising thin sidewall electrodes was not desirable. Therefore, a prototype antiwear probe evaluated in a previous study [5, 6] was chosen here because of its thick sidewall electrode (ca. 300 nm) made of tungsten. Surface flattening and groove formation were omitted. Oil lubrication was not introduced to simplify the system. The contact force of the probe tip was kept within about 1.2 μN during all operations.

SNDM images of the polarization domain before and after the probe slide are shown in Figure 6. The SNDM signals were normalized so that the readout signals from negative and positive polarization areas amounted to 100% to 0%, respectively, for direct comparison of these images. The results implied that the boundaries of the polarization domain looked rather clearer after than before the slide, but not more indistinct.

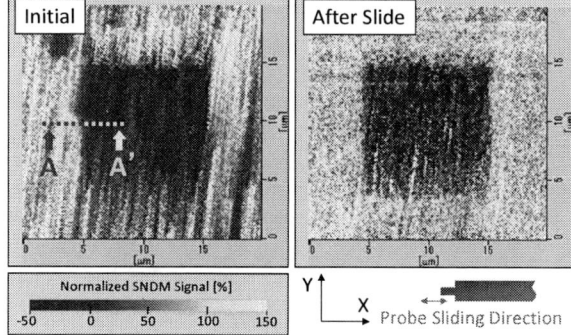

Figure 6: Normalized SNDM readout images of a polarization domain before and after the probe slide.

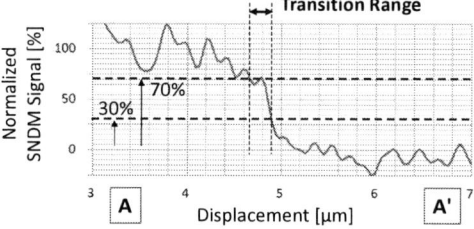

Figure 7: Transition range of the normalized SNDM readout signal at the edge of a polarization domain.

The transition range of the normalized SNDM signal was chosen as an evaluation index for a quantitative comparison of the readout resolution at the domain boundary. Figure 7 shows the cross-sectional signal profile of the SNDM image along the A–A' line in Figure 6, which intersects the left edge of the polarization domain. The transition range was defined as the distance between the points where the normalized SNDM signal equaled 30% and 70%. A shorter transition range is indicative of the fine readout resolution of the system.

Three boundaries of the polarization domain (left, right, and top edge) were addressed, and transition ranges were evaluated along five arbitrary cross-sections through each boundary, providing mean readout resolutions and their deviation.

Table 1: Averages and standard deviations (σ) of the transition range across three boundaries of the polarization domain.

[nm]	Left Edge		Right Edge		Top Edge	
	ave.	σ	ave.	σ	ave.	σ
Initial	261	186	289	309	299	125
After Slide	313	275	211	212	301	85

Transition range evaluation results are shown in Table 1. Although the probe was worn after a 1-m slide, the average transition range across the domain boundaries was maintained at ca. 300 nm, approximating the thickness of the probe sidewall electrode. This implies that the readout resolution of the evaluated system roughly depends on the electrode thickness and thus, does not change even the probe is worn.

On the other hand, the transition range at each boundary exhibited a great dispersion extending to several hundreds of nanometers. As mentioned above, the boundaries of polycrystalline PZT grains, whose size were also several hundreds of nanometers, may affect the polarization domain wall position. Further, the surface roughness of the PZT film may deteriorate the readout signal. A more precise evaluation using recording media displaying a flatter surface and a finer grain size (or monocrystalline material) is desirable to completely uncover the factors determining the readout resolution.

The transition range at the bottom edge of the polarization domain was very unstable, hindering its evaluation. Because surface flattening and groove formation steps were omitted during probe fabrication, contaminants and wear debris were likely to adhere to the probe surface, degrading the readout resolution. Further experiments using a complete AGO system are strongly needed.

CONCLUSION

A novel AGO sliding system was proposed to achieve a probe tip with a meter-scale sliding endurance for probe-based archive memories, and its endurance was evaluated.

SSRM experiments demonstrated that, even in the absence of a groove, an antiwear probe endured a 2-m slide while maintaining almost stable electric contact with the assistance of oil lubrication. The introduction of a groove for debris management was effective in enhancing the electric contact stability.

The SNDM readout experiment using a ferroelectric PZT substrate and an antiwear probe revealed that the readout resolution remained intact after a 1-m probe slide. Additional experiments using a complete AGO system and optimized recording media are needed to further investigate the factors affecting the readout resolution.

ACKNOWLEDGEMENTS

The authors would like to thank Professor H. Fujita, Professor H. Toshiyoshi, and Professor M. Sugiyama at the University of Tokyo, Professor G. Hashiguchi at Shizuoka University, and Professor Y. Ando at Tokyo University of Agriculture and Technology for their helpful suggestions and valuable discussion. We also acknowledge Mr. T. Endo, Mr. K. Katayama, and other members of Toshiba Group for their support and advice. We especially would like to express our greatest respect for Mr. A. Koga at Toshiba Corporation for his remarkable ability to promote the project. This work was partly supported by the New Energy and Industrial Technology Development Organization (NEDO).

REFERENCES

[1] E. Eleftheriou *et al.*, "Millipede — A MEMS-Based Scanning-Probe Data-Storage System," *IEEE Trans. on Magnetics*, vol. 39, no. 2, pp. 938-945, 2003.

[2] K. Tanaka *et al.*, "Scanning Nonlinear Dielectric Microscopy Nano-Science and Technology for Next Generation High Density Ferroelectric Data Storage," *Jpn. J. App. Phys.*, vol. 47, no. 5, pp. 3311-3325, 2008.

[3] C. D. Wright *et al.*, "Terabit-Per-Square-Inch Data Storage Using Phase-Change Media and Scanning Electrical Nanoprobes," *IEEE Trans. on Nanotech.*, vol. 5, no. 1, pp. 50-61, 2006.

[4] C. Yoshida *et al.*, "Direct Observation of Oxygen Movement during Resistance Switching in NiO/Pt Film," *Appl. Phys. Lett.*, vol. 93, no. 4, p. 042106, 2008.

[5] Y. Li *et al.*, "Wear-Insensitive Sidewall Microprobe with Long-Term Stable Performance for Scanning Probe Microscopy Lithography," *J. Microelectromech. Systems*, vol. 22, no. 4, pp. 901-908, 2013

[6] Y. Tomizawa *et al.*, "Electric Contact Stability of Anti-Wear Probes," *IEICE Electronics Express*, vol. 9, no. 21, pp. 1675-1682, 2012.

[7] L. Zhang *et al.*, "High-Resolution Characterization of Ultrashallow Junctions by Measuring in Vacuum with Scanning Spreading Resistance Microscopy," *Appl. Phys. Lett.*, vol. 90, no. 19, p. 192103, 2007.

[8] Y. Tomizawa *et al.*, "Influence of the Material Properties on Major Tribological Factors at a Nano-Scale Sliding Electric Contact of Probe Devices," *J. Advanced Mechanical Design, Systems, and Manufacturing*, vol. 7, no. 1, pp. 15-29, 2013.

CONTACT

*Y. Tomizawa, tel: +81-44-549-2361;
yasushi.tomizawa@toshiba.co.jp

NEMS VARIABLE OPTICAL ATTENUATOR (VOA) DRIVEN BY OPTICAL FORCE

B. Dong[1,2], H. Cai[2], Y. D. Gu[2], Z. C. Yang[3], Y. F. Jin[3], Y. L. Hao[3], D. L. Kwong[2] and A. Q. Liu[1†]

[1]School of Electrical & Electronic Engineering, Nanyang Technological University, Singapore 639798
[2]Institute of Microelectronics, A*STAR (Agency for Science, Technology and Research)
Singapore 117685
[3]National Key Laboratory of Science and Technology on Micro/Nano Fabrication
Institute of Microelectronics, Peking University, Beijing 100871, China

ABSTRACT

This paper reports a Nanoelectromechanical System (NEMS) Variable Optical Attenuator (VOA) driven by the optical gradient force. The VOA is realized via a waveguide based directional coupler. The gap in the directional coupler is controlled via an optical force driven actuator. The doubly-clamped silicon beam actuator is controlled by tuning the wavelength of control light. The signal light can be attenuated to 20% of original value in the experiment. The NEMS VOAs have the merits of small dimension, low power consumption and good capability for all optical integration, making it a good candidate for future applications in silicon photonics circuit and optical communication devices.

INTRODUCTION

An optical attenuator is a device used to reduce the power level of an optical signal, either in free space or in waveguide structures. Variable optical attenuator (VOA) plays an important role in optical fiber communication systems and MEMS integrated VOAs are developed and applied for the communication network [1].The MEMS integrated VOAs generally utilize MEMS actuator to control a shutter or a mirror, which can block or redirect light from propagating between fibres and waveguides. The advantages of these MEMS VOAs include high extinction ratio, low power consumption, mechanical stability, low cost and small size [2]. Besides the MEMS VOAs, other approaches including thermal-optical effects [3], electro-optical effects [4], and opto-optical effects [5] have been demonstrated. However, stronger demand for highly integrated and compact devices has restricted the applications of the MEMS VOAs due to their large scale actuator. Highly integrated and compact VOAs driven by NEMS actuator can be possible if one can integrate VOAs onto planar optical waveguide circuits.

This paper reports a NEMS VOA driven by optical gradient force. The VOA is realized via waveguide based optical directional coupler, whose gap can be controlled via optical force driven actuator [6-8]. The NEMS VOAs have merits such as small dimension, low power consumption and good capability for all optical integration as compared with conventional MEMS based fiber VOA [1], which makes it a good candidate for future applications in silicon photonics circuit and optical communication devices.

DESIGN AND THEORY

The schematic of the NEMS VOA is shown in Fig. 1(a), which consists of a control waveguide, a ring resonator enabled optomechanical actuator, and a directional waveguide coupler. The high power single wavelength control light, whose wavelength is close to the resonance wavelength of the ring resonator, can couple from the control waveguide to the ring resonator. Due to the high Q-factor of the ring resonator, light accumulates inside the ring resonator with optical gradient force generated between the ring resonator and the mechanical arc actuator. The power level of the signal light is modulated after passing through the directional coupler, while the attenuation is controlled by the actuation distance of the optomechanical actuator.

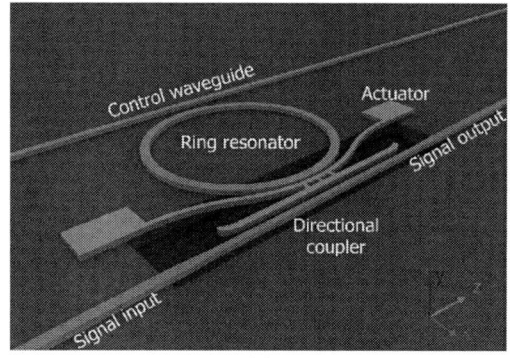

(a)

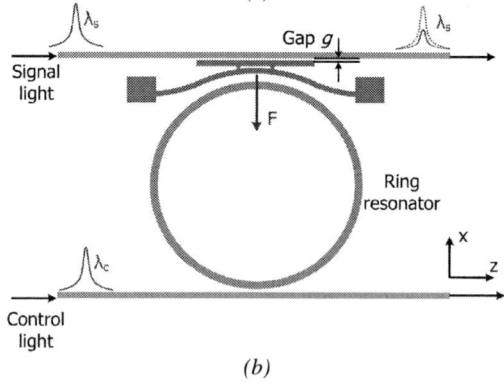

(b)

Figure 1: (a) Schematic of NEMS variable optical attenuator and (b) layout of the NEMS VOA driven by optical gradient force.

978-1-4799-7956-1/15 $31.00 © 2015 IEEE

MEMS 2015, Estoril, PORTUGAL, 18 - 22 January, 2015

The optomechanical actuator consists of a ring resonator, an arc shape doubly-clamped silicon beam, and a coupler waveguide. The actuation distance is controlled by tuning the wavelength of control light. When the wavelength of the control light is increased from blue-detuned wavelength to resonance wavelength, the optical force is increased because more energy is coupled into the ring resonator. The arc actuator therefore moves towards the ring resonator and remains stable at the new balance position. The simulated actuation of the actuator at various wavelength detuning is shown in Fig 2. When the detuning is zero, the actuator is at initial position and the actuation distance is at 26.7 nm. The actuator is more sensitive to wavelength detuning at larger gap value. The actuation distance can reach up to 149.87 nm when the wavelength detuning is 2.5 nm.

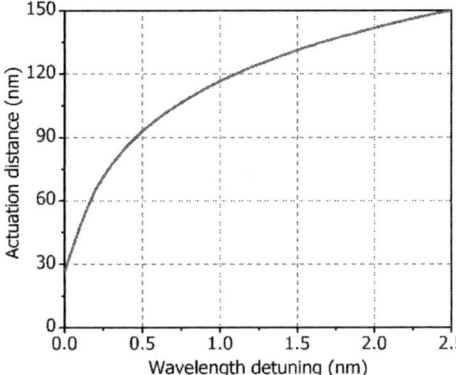

Fig. 2: Simulated actuation distance at various wavelength detuning when the optical power is 10 mW.

The optical directional coupler consists of two waveguides of equal width, one is coupler waveguide which is driven and controlled by the optomechanical actuator, and the other one is the single waveguide. Light transmitting inside the signal waveguide slowly couples to the coupler waveguide when propagating along the z direction, while the coupling efficiency depends on the gap between the coupler waveguide and single waveguide. The coupling coefficient can be expressed as

$$\kappa = C \exp(-\gamma g) \qquad (1)$$

where C is constant which depend on the refractive index of waveguide and the geometry of the waveguides, γ is the attenuation coefficient in the coupling direction and g is the gap between waveguides.

Figure 3 shows the simulated propagation of light in the directional coupler, where the waveguide width is 400 nm and the gap in between is 200 nm and 150 nm respectively. It takes 28.5 µm for 1550 nm light to be fully coupled to another waveguide when the gap is 200 nm. The light can be fully coupled to coupler waveguide at 150-nm gap.

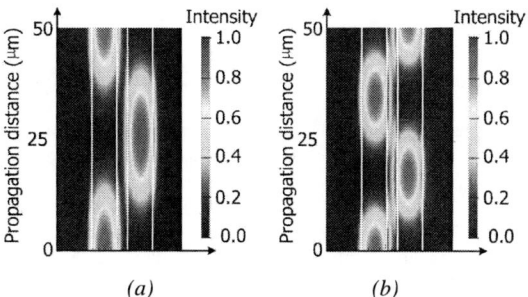

(a) *(b)*

Figure 3: Light intensity along propagation distance when the (a) gap = 200 nm and (b) = 150 nm,

The optical power attenuation A is defined as the power loss at the original signal waveguide, which is due to the coupled energy in the coupler waveguide, the attenuation deals with gap can be expressed as

$$A = \frac{P_{input} - P_{original}}{P_{input}} = 1 - \sin^2(\kappa(g) \cdot L) \qquad (2)$$

where L is the coupling length. The intensity of signal light drops when the actuation distance increases, as shown in Fig. 4. The intensity attenuation of the signal light verses the actuation distance of the optomechanical actuator is sinusoidal, but the attenuation of the signal light is treated as linear in the central. When the control light wavelength is λ_1, the actuation distance is Δg_1, and the light intensity in the signal waveguide is I_1. When the control light wavelength increases to λ_2, the actuation distance increases to Δg_2, and the light intensity in the signal waveguide is I_2.

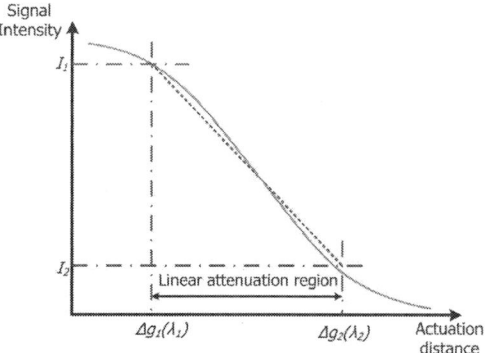

Figure 4: Intensity attenuation of signal light versus the actuation distance of the actuator.

FABRICATION

The NEMS VOA is fabricated by using nano-silicon-photonic fabrication processes. The waveguides and ring resonators are fabricated by one time RIE process using SiO$_2$ as hard mask. After a 2-μm-cladding oxide deposition, narrower trench is etched first, followed by Al$_2$O$_3$ deposition. The window is then defined by lithography followed by Al$_2$O$_3$ etch. After the second trench etch, the devices are released through HF vapor etching process.

Figure 5 shows the SEM images of the NEMS VOA. Fig. 5(a) shows the released directional coupler and nano-actuator (highlighted in yellow). Fig. 5(b) shows the VOA array, which consists of two series connected VOA, which can enhance the attenuation.

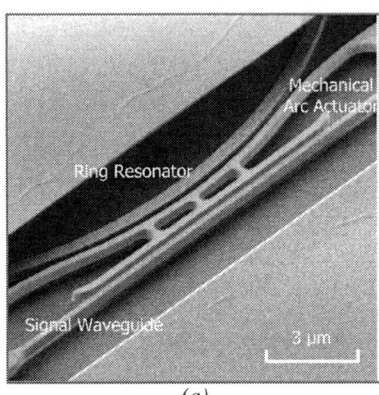

(a)

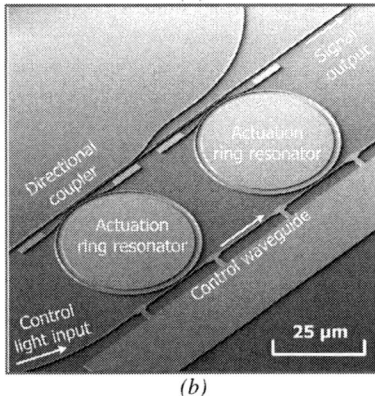

(b)

Figure 5: SEM images of the optical memory. (a) NEMS actuator and signal waveguide and (b) two series connected VOA array.

EXPERIMENTS AND DISCUSSIONS

The NEMS VOAs are measured using the setup shown in Fig. 6. Two input fibers are aligned with the control waveguide and signal waveguide, respectively, such that control light and signal light can be coupled into the device. Control light from a tunable laser source (Santec TCL510) is amplified by the EDFA (Amonics EDFA-CL-27) and passes through a 98/2 splitter. 2% of light is detected by the photo detector (PD) to monitor the power level while most of the energy is coupled into the control waveguide. The light is detected by an optical spectrum analyzer (OSA) (Yokogawa AQ6370C) for wavelength and power monitoring. The signal light from the tunable laser source is coupled into the signal waveguide through a 98/2 splitter and monitored by the PD. By tuning the wavelength of the control light, the detector at the end of the signal waveguide can measure the attenuation of the signal light

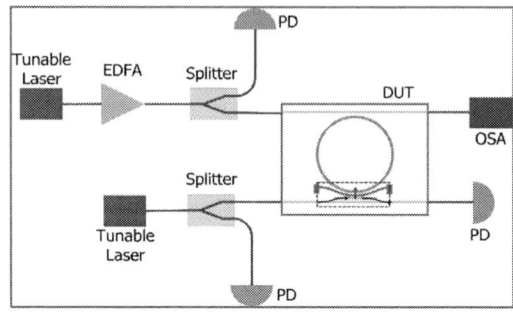

Figure 6: Experimental setup for VOA characterization.

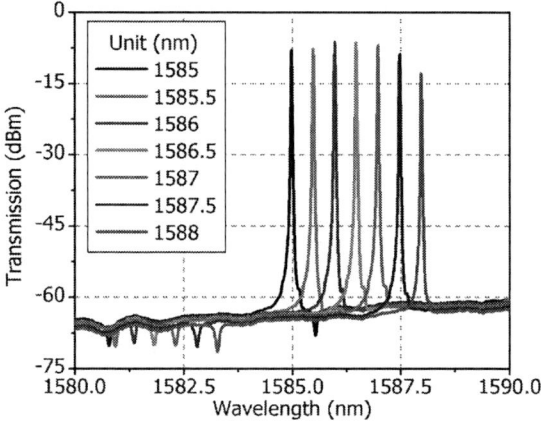

Figure 7: Transmission spectra of NEMS VOA with various input powers.

The wavelength of the control light is tuned from 1585 nm to 1588 nm while the resonance wavelengths are measured. When the wavelength of the control light is 1585 nm, the control light is not coupled into the ring resonator, the actuator is at the rest state and the resonance wavelength is 1580.78 nm. When the wavelength of the control light is increased to 1585.5 nm, light is coupled into the ring resonator, thus the actuator is pulled towards the ring resonator and the resonance wavelength is increased to 1580.94 nm. When the wavelength of the control light is increased to 1586 nm, the resonance wavelength increases to 1581.36 nm and the actuation distance is increased further. When the wavelength of control light is increased to 1588 nm, the resonance wavelength reaches 1583.28 nm.

978-1-4799-7956-1/15 $31.00 © 2015 IEEE

The optical gradient force cannot overcome the mechanical spring force when the wavelength of the control light is increased further and the actuator is pulled back to its original position.

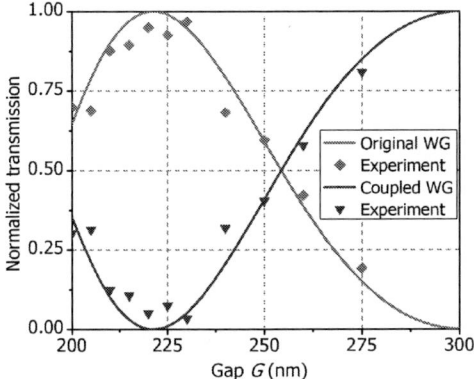

Figure 8: The normalized transmission of the NEMS VOA at various gaps.

The directional coupler of the other VOA design has a coupling length of 20 μm, while the optimized waveguide structure is 350 nm in width. The normalized transmission of the signal light is shown in Fig. 8. Both signal waveguide (original WG) and coupler waveguide (coupler WG) signal are recorded. The transmission in the signal waveguide increases till the gap between waveguide reaches 220 nm. Thereafter, the light intensity drops to 20% when the gap is close to 275 nm.

CONCLUSIONS

In conclusions, a NEMS VOA is experimentally demonstrated. The VOA is realized via a waveguide based directional coupler. The gap between the directional coupler is controlled via an optical force driven actuator. The doubly-clamped silicon beam actuator is controlled by tuning the wavelength of control light. The signal light can be attenuated to 20% of original value in experiment. A VOA array is also designed and characterized. It can improve the attenuation effects but introduce more losses. The NEMS VOAs have the merits of small dimension, low power consumption and good capability for all optical integration, making it a good candidate for future applications in silicon photonics circuit and optical communication devices.

ACKNOWLEDGMENTS

The work is supported by the Environmental and Water Industry Development Council of Singapore (EWI), RPC programme (Grant No.: 1102-IRIS-05-01, 1102-IRIS-05-02, 1102-IRIS-05-04 and 1102-IRIS-05-05).

REFERENCES

[1] X. M. Zhang, A. Q. Liu, and C. Lu, "New near-field and far-field attenuation models for free-space variable optical attenuators," Journal of Lightwave Technology, vol. 21, pp. 3417-3426, 2003.

[2] X. H. Ma and G. S. Kuo, "Optical switching technology comparison: optical MEMS vs. other technologies," Communications Magazine, IEEE, vol 41, pp. S16-S23, 2003.

[3] W. M. Zhu, T. Zhong, A. Q. Liu, X. M. Zhang and M. Yu, "Micromachined optical well structure for thermo-optic switching", Appl. Phys. Lett., vol. 91, 261106, 2007.

[4] M. Papuchon, Y. Combemale, X. Mathieu, D. B. Ostrowsky, L. Reiber, A. M. Roy, B. Sejourne, and M. Werner, "Electrically switched optical directional coupler: Cobra," Applied Physics Letters, vol. 27, pp. 289-291, 1975.

[5] X. J. Ke, M. R. Wang, and D. Li, "All-optical controlled variable optical attenuator using photochromic sol gel material," Photonics Technology Letters, IEEE, vol. 18, pp. 1025-1027, 2006.

[6] B. Dong, H. Cai, G. I. Ng, P. Kropelnicki, J. M. Tsai, A. B. Randles, M. Tang, Y. D. Gu, Z. G. Suo and A. Q. Liu, "A nanoelectromechanical systems actuator driven and controlled by Q-factor attenuation of ring resonator," Applied Physics Letters, Vol 103, 181105, 2013.

[7] M. Ren, J. Huang, H. Cai, J. M. Tsai, J. Zhou, Z. Liu, Z. Suo, and A. Q. Liu, "Nano-optomechanical actuator and pull-back instability," ACS Nano, Vol 7, pp.1676–1681, 2013.

[8] H. Cai, B. Dong, J. F. Tao, L. Ding, J. M. Tsai, G. Q. Lo, A. Q. Liu and D. L. Kwong, "A nanoelectromechanical systems optical switch driven by optical gradient force," Applied Physics Letters, Vol 102, 023103, 2013.

CONTACT

†A. Q. Liu, +65-67904336; eaqliu@ntu.edu.sg

A SUPER-REGENERATIVE OPTICAL RECEIVER BASED ON AN OPTOMECHANICAL OSCILLATOR

Turker Beyazoglu, Tristan O. Rocheleau, Alejandro J. Grine, Karen E. Grutter, Ming C. Wu and Clark T.-C. Nguyen
University of California, Berkeley, USA

ABSTRACT

A super-regenerative optical receiver detecting on-off key (OOK) modulated light inputs has been demonstrated that harnesses the radiation-pressure gain of a self-sustained electro-opto-mechanical oscillator (EOMO) to render its oscillation amplitude a function of the intensity of light coupled into the oscillator. Unlike previous electronic super-regenerative receivers, this rendition removes the need to periodically quench the oscillation signal, which then simplifies the receiver architecture and increases the attainable receive bit rate. A fully functional receiver with a compact ~90 μm EOMO comprised only of silicon-compatible materials demonstrates successful recovery of a 2 kbps bit stream from an OOK modulated 1550 nm laser input. By removing the need for the expensive III-V compound semiconductor materials often used in conventional optical receivers, this EOMO-based receiver offers a lower cost alternative for sensor network applications.

INTRODUCTION

Radiation pressure-driven optomechanical oscillators (RP-OMO) that harness light energy to produce microwave signals from on-chip micro-devices have proven useful in stand-alone oscillator [1], communications [2], and sensing applications [3]. The addition of electrodes to conventional optomechanical devices allows electrically coupled inputs [4],[5],[6] as well as optical ones that then enable new integrated electro-optomechanical systems where electrical signals modify optical properties [7]. The converse should also be true, where laser light coupled to an electro-opto-mechanical system might change the electro-mechanical properties of the device, perhaps in a way that allows electrical detection and decoding of optical signals. If possible, this might then enable an optical receiver constructed strictly in silicon compatible materials, i.e., with no need for compound semiconductor photonic devices and the associated cost and technology required to integrate them alongside silicon electronics.

Pursuant to capitalizing on this possibility, and spurred by recent demonstrations of simple low power MEMS radios based on super-regenerative reception [8],[9], this work presents for the first time a fully functional super-regenerative optical receiver based on an electro-opto-mechanical oscillator (EOMO), *cf.* Fig. 1, that detects on-off key (OOK) modulated light input and directly demodulates and recovers input bits in the electrical domain. The key enabler here is the simultaneous use of both electrical and optical input/output (I/O) ports, the former used in the positive feedback loop of a self-sustained electronic oscillator circuit; while the latter used to accept optical inputs that perturb the steady-state oscillation amplitude of the electronic oscillator. Via use of an EOMO constructed of only silicon-compatible materials, this receiver obviates the need for compound semiconductor technology while still

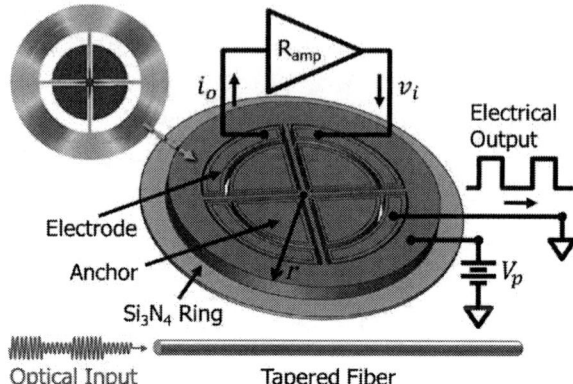

Fig. 1: Perspective-view schematic of the EOMO and basic receiver operation. Here, an electronic amplifier connects to input/output polysilicon electrodes and sustains oscillation. An amplitude modulated optical input couples to the Si_3N_4 ring of the EOMO and changes the output electrical oscillation amplitude, which indicates the received bits.

providing optical reception commensurate with the needs of massive autonomous sensor networks, for which cost is paramount [10].

DEVICE STRUCTURE AND OPERATION

The EOMO illustrated in Fig. 1 comprises a high mechanical Q_m polysilicon inner ring mechanically attached at its outer edge to a concentric high optical Q_o (but comparatively low mechanical Q_m) stoichiometric silicon nitride ring. Four radially symmetric spokes extend from a common central anchor and attach to the inner edge of the polysilicon ring to support the entire structure. The radially symmetric contour mode shape depicted in the inset of Fig. 1 imparts equal and opposite forces along the support beams that cancel on the central anchor point, and thus, generate very little displacement there. This then greatly reduces energy leakage to the substrate, which raises the mechanical Q_m.

Both the polysilicon and silicon nitride rings supply vibrational energy to the composite structure in proportion to their effective masses, defined as $m_{eff} = 2U/(\omega_m^2 \cdot \Re(r))$, where U is the total stored energy in the mechanical mode, ω_m the mechanical resonance frequency, and $\Re(r)$ the radial displacement amplitude at radius r. Because it is physically much larger, the polysilicon ring dominates the total effective mass m_{eff}, so its lateral dimensions govern to first order the EOMO's mechanical resonance frequency. This also means the polysilicon ring contributes a larger portion of the total vibrational energy U in the system while introducing very little loss (because of its high mechanical Q_m). With energy added without additional loss, the composite structure then exhibits a much higher mechanical Q_m

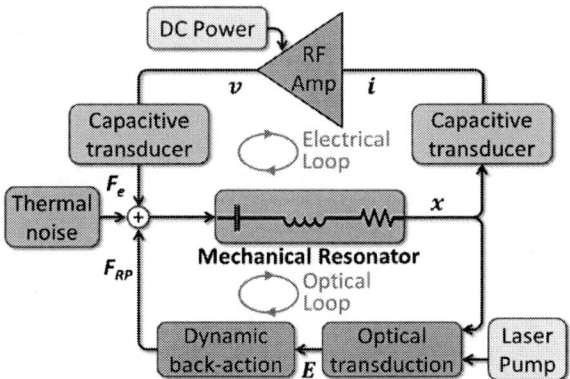

Fig. 2: *Super-regenerative optical receiver model. Light received at the proper wavelength forms an additional positive feedback loop, thereby raising the steady-state oscillation amplitude from the no light case (where only the upper branch contributes to the loop gain).*

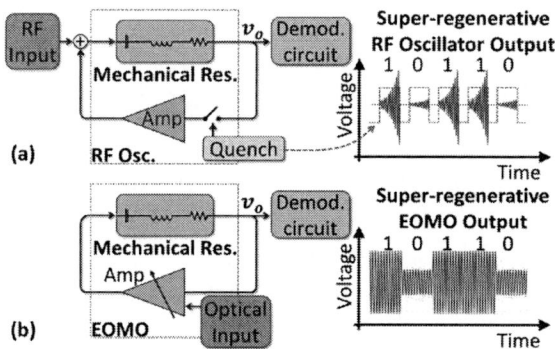

Fig. 3: *Comparison of conventional and EOMO-based super-regenerative receivers. (a) Reception of a '1' or a '0' is determined by the speed at which oscillations reach a prescribed threshold value starting from a quenched state. (b) Reception of a '1' or a '0', without quenching, is determined by the amplitude of oscillation, which can switch quickly, greatly increasing the permissible bit data rate.*

than otherwise provided by a silicon nitride-only ring, as demonstrated in [4]. This high Q_m is key to low power and low phase noise in the present wireless receiver.

Electrical I/O

To enable electrical I/O, four polysilicon electrodes flank the inner edge of the polysilicon ring with 450 nm spacings to form parallel plate capacitors that then realize capacitive-gap transducers. The electrodes anchor to underlying polysilicon interconnects that facilitate signal routing and connection to external electronic circuitry. Exciting the EOMO electrically entails applying a DC bias V_P to the conductive polysilicon ring and an AC voltage v_i to an input electrode, where $V_P \gg v_i$. The voltage difference across the capacitive gap generates a time-varying force that drives the ring into contour mode vibration when it matches the mechanical resonance frequency. The ensuing motion then generates displacement currents across each DC-biased time-varying gap, which can then serve as output signals proportional to displacement or velocity.

Optical I/O

The silicon nitride outer ring enables optical input by accepting laser light from a waveguide—in this case, tapered fiber [11]—brought close enough to evanescently couple the light into to the silicon nitride ring. This ring then serves as a high-Q_o optical cavity that supports whispering gallery mode optical resonances, where the light continuously circulates around the ring outer edges. Ideally, the light would circulate forever. In reality, of course, loss caused by scattering or coupling to nearby objects limits the time light can spend in the ring, which then limits the optical Q_o. Placement of the nitride ring at the outer edge of the structure avoids scattering from the polysilicon material or from coupling with the inner polysilicon electrodes, thereby maximizing the optical Q_o.

Once the ring structure vibrates (via either electrical or optical means to be discussed in the next section), the same fiber that delivered the input light can optically sense the ring motion as a modulation sideband spaced by the ring vibration frequency from the laser carrier. A photodiode can then demodulate the signal to isolate the mechanical resonance.

Self-Sustained Oscillation.

With two I/O modes, the EOMO device of Fig. 1 offers two methods to instigate self-sustained oscillation: electrical or optical. Fig. 2 summarizes the two methods via a simple block diagram with two feedback loops. The electrical method is the same as that used in conventional oscillators [12], where two electrodes (i.e., capacitive-gap transducers) of the EOMO connect to the input and output terminals of an electronic amplifier to create a positive feedback loop with loop gain greater than unity when $A_l = R_{amp}/R_x > 1$. Here, R_{amp} is the transresistance of the amplifier, and

$$R_x = \frac{k}{\omega_m Q_m V_p^2}\left(\frac{\partial C_i}{\partial x}\frac{\partial C_o}{\partial x}\right)^{-1}$$

is the motional resistance between the EOMO electrodes embedded in the loop, where C_i and C_o are the total static capacitances of the input and output transducers, respectively, k is the mechanical stiffness, and x is the resonator displacement. With loop gain greater than unity, regenerative amplification of ring structure's Brownian motion at its resonance frequency eventually leads to sustained oscillation with steady-state amplitude governed by nonlinearities that reduce gain as amplitude increases. The "electrical loop" in Fig. 2 summarizes the operative mechanisms in this mode of self-sustained oscillation.

The optical method, on the other hand, does not require an external amplifier, but rather just a strong enough blue-detuned laser input (or pump) to incite self-sustained optomechanical oscillation, as described in [13]. Here, the field in the high-Q_o cavity builds up to a sizable circulating optical power that generates an outward radial radiation pressure force on the silicon nitride ring. When Brownian motion (again, strongest at the ring mechanical resonance frequency) modulates the optical cavity boundary, it modulates the radiation pressure force, leading to a force at the ring mechanical resonance frequency. When the laser intensity is strong enough, the displacement-to-radiation pressure force transfer function—captured by the "optical

978-1-4799-7956-1/15 $31.00 © 2015 IEEE

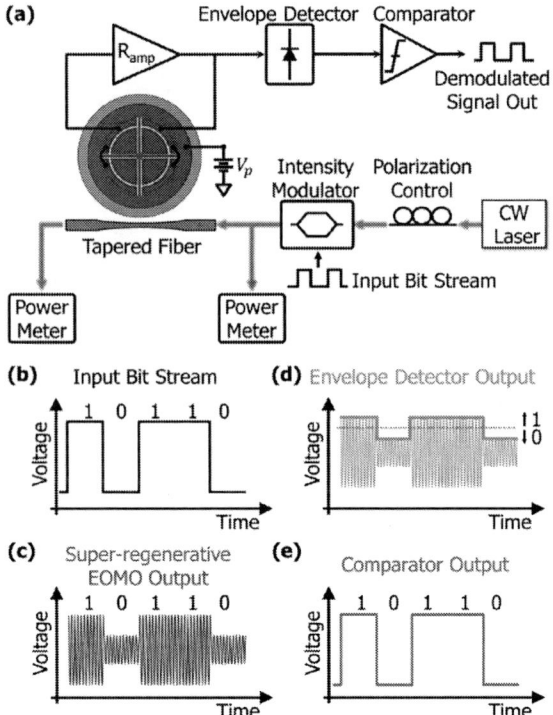

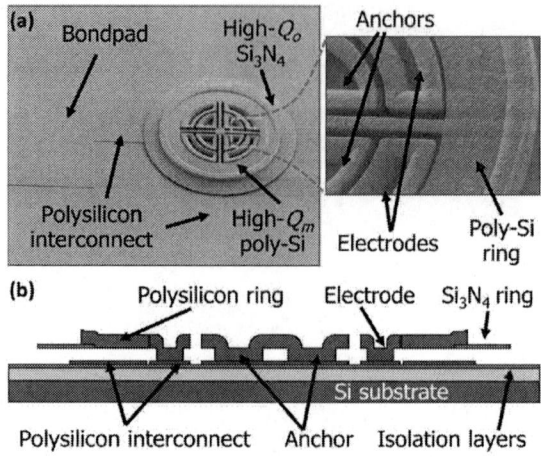

Fig. 5: (a) SEM image and (b) cross-section of the EOMO.

Fig. 4: (a) Pictorial summary of the super-regenerative receiver. An electronic amplifier placed in a positive feedback loop with the EOMO sustains oscillation while a tapered fiber couples the optical field modulated by the input bit stream (b) into the EOMO, changing the amplitude of oscillation (c). An envelope detector measuring the amplitude (d) feeds to a comparator that recovers the data (e).

transduction" and "dynamic back-action" blocks in Fig. 2—contributes sufficient gain to the "optical loop" to achieve a loop gain greater than unity. This then instigates regenerative oscillation growth in the exact same manner as the "electrical loop".

The super-regenerative optical receiver of Fig. 1 employs the gains of both Fig. 2 modes, simultaneously. It specifically uses the electrical mode to instigate and sustain a primary oscillation, and the optical mode to influence the amplitude of the oscillation. To facilitate analysis, Fig. 3(b) condenses the complexity of Fig. 2 into a simpler equivalent block diagram that lumps the electrical and optomechanical gain mechanisms into a single amplifier controlled by the optical input. Here, the stronger the optical input, the larger the amplifier gain. The larger the amplifier gain, the larger the nonlinearity required to limit oscillation growth, and the larger the displacement amplitude needed to generate that nonlinearity. Thus, the steady-state amplitude of the oscillator becomes a direct function of the laser input power, which is the crux behind the present super-regenerative optical receiver.

SUPER-REGENERATIVE OPTICAL RECEIVER

Fig. 3 compares a conventional super-regenerative receiver (a) with the EOMO-based one of this work (b). As shown, both harness the positive feedback loop gain of a closed-loop oscillator to regeneratively, i.e., cycle–by–cycle, achieve an enormous front-end gain capable of detecting tiny received signals. In the former approach, in the absence of an RF signal, the oscillation amplitude rises slowly and gets quenched before reaching a threshold value, which indicates a '0'. On the other hand, in the presence of received RF power, the oscillation amplitude rises quickly past the threshold before quenching, which indicates a '1'. In this case, reception of a '1' or a '0' is determined by the speed at which oscillations reach a prescribed threshold value after starting from a quenched state, where quenching is done once for every bit cycle. In this mode of operation, the bit rate is limited by both the speed at which oscillations grow and the speed at which they can be quenched.

The EOMO-based approach of this work differs in that it does not require quenching of the oscillation. With reference to Fig. 4(a), the EOMO's electrodes are embedded in a positive feedback loop with an electronic amplifier, providing enough gain for oscillation even in the absence of an optical input. An input light that is slightly blue-detuned from the optical resonance wavelength (corresponding to a '1' in OOK) induces radiation pressure, increasing the total force (and the loop gain) applied to the mechanical resonator, and thereby raising the steady-state oscillation amplitude from the no light case (which corresponds to a '0'). The oscillation amplitude thus indicates whether a '1' or a '0' is received. Fig. 4 illustrates this receiver operation by comparing time domain traces at the (c) EOMO amplifier, (d) envelope detector, and (e) comparator outputs, for a given input bit stream (b). Here, since the oscillator merely switches between amplitude states, the time it takes for the amplitude to grow is shorter than growing from zero, so 0-to-1 transitions can be quite fast.

EXPERIMENTAL RESULTS

Fig. 5 presents the SEM and cross-section of the EOMO fabricated via the process of [4] and used to demonstrate optical reception via the setup of Fig. 4(a). The device comprises a polysilicon ring with 30 μm-inner and 40 μm-outer radii physically attached at its outer edge to a 6 μm wide silicon-nitride ring, yielding a 36.9-MHz mechanical resonance frequency with a mechanical Q_m of 15,740.

EOMO and receiver performance measurements used

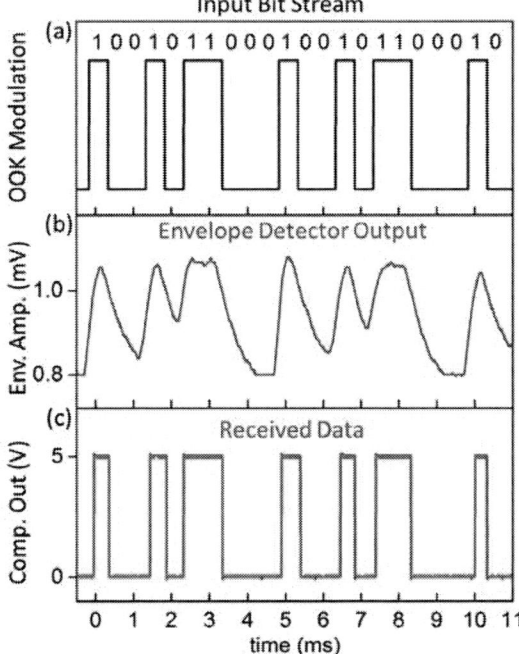

Fig. 6: Measured time-traces illustrating super-regenerative optical receiver operation. (a) Input bit stream modulating a CW laser on resonance, (b) envelope detector output showing the EOMO oscillation amplitude, and (c) output bit stream for a 1 mV threshold from comparator output. The output waveform is identical to the input, as desired, confirming successful wireless optical OOK reception with a 2 kbps data rate.

the custom-built vacuum chamber of [14] in which a sealed probe station provides easy access to device electrodes, and nano-positioning piezo stages provide precise control of optical coupling. To construct the complete optical receiver, the EOMO's electrical ports connect to a sustaining electronic amplifier realized by a Zurich Instruments' HF2LI lock-in unit. Here, the use of a lock-in amplifier provides a simple off-chip implementation with enhanced noise rejection while also conveniently serving as the next stage envelope detector.

Fig. 6 presents measured time-traces confirming receiver operation. Here, an input bit stream modulates the power of a CW 1550 nm laser between 13 μW, indicating a '0', and 750 μW, indicating a '1'. This modulated light input then couples to the EOMO, modulating its radiation pressure gain, thereby modulating the oscillation amplitude. The EOMO's electrical output then feeds an envelope detector that produces the envelope trace in Fig. 6(b). The amplitude trace is then directed to a comparator that produces the output bit stream (Fig. 6(c)) which is identical to the input stream of Fig. 6(a), confirming successful optical OOK reception with a 2 kbps data rate.

CONCLUSIONS

An integrated EOMO has realized a first super-regenerative optical receiver that operates by harnessing the radiation-pressure gain of the EOMO to render its oscillation amplitude a function of the intensity of light coupled into the oscillator. Unlike its RF analogues, this super-regenerative receiver rendition operates without the need to periodically quench the oscillation, and this simplifies the receiver architecture while increasing the attainable receive bit rate. The demonstrated recovery of a 2 kbps bit stream from an OOK modulated 1550 nm laser input by this fully functional EOMO-based optical receiver encourages expansion of this capability to versions that support faster bit rates, perhaps made possible by tweaks to the mechanical and optical Q's of the multi-material device.

By removing the need for the expensive III-V compound semiconductor materials often used in conventional optical receivers, this optical super-regenerative receiver additionally offers a lower cost alternative for sensor network applications. Indeed, the operation modes and mechanisms demonstrated by this EOMO based receiver present one plausible approach to a silicon-compatible single-chip receiver with WDM capability, where multiple devices operating at different wavelengths decode the data simultaneously, allowing channelized optical communications.

ACKNOWLEDGEMENTS

Authors would like to thank Burak Eminoglu and Kirti Mansukhani for their valuable discussion on the lock-in amplifier implementation and the DARPA ORCHID program for funding.

REFERENCES

[1] H. Rokhsari, et al., *Opt. Express*, vol. 13, no. 14, pp. 5293–5301, 2005.

[2] M. Hossein-Zadeh, et al., *IEEE J. Sel. Top. Quantum Electron.*, vol. 16, no. 1, pp. 276–287, 2010.

[3] F. Liu, et al., *IEEE Sens. J.*, vol. 13, no. 1, pp. 146–147, 2013.

[4] T. Beyazoglu, et al., *Proceedings, 2014 IEEE 27th Int. Conf. on MEMS*, pp. 1193–1196, 2014.

[5] J. Bochmann, et al., *Nat. Phys.*, vol. 9, no. 11, pp. 712–716, Sep. 2013.

[6] S. Sridaran, et al., *Opt. Express*, vol. 19, no. 10, pp. 9020–9026, 2011.

[7] R. Perahia, et al., *Appl. Phys. Lett.*, vol. 97, pp. 191112(3), 2010.

[8] B. Otis, et al., *Proceedings, ISSCC*, 2005, pp. 396-606 Vol. 1.

[9] T. O. Rocheleau, et al., *Proceedings, Solid-State Sensors, Actuators, Microsystems Work., Hilton Head*, pp. 83-87, 2014.

[10] J. M. Kahn, et al., *J. of Comm. Netw.*, vol. 2, no. 3, pp. 188–196, 2000.

[11] J. C. Knight, et al., *Opt. Lett.*, vol. 22, no. 15, pp. 1129-1131, 1997.

[12] Y.W Lin, et al., *IEEE J. Solid-State Circuits*, vol. 39, no. 12, pp. 2477–2491, 2004.

[13] M. Hossein-Zadeh, et al., *Phys. Rev. A*, vol. 74, no. 2, pp. 023813(5), 2006.

[14] T. O. Rocheleau, et al., *Proceedings, 2013 IEEE 26th Int. Conf. Micro Electro Mech. Syst.*, pp. 118–121, 2013.

CONTACT

*Turker Beyazoglu, email: turker@eecs.berkeley.edu

INTEGRATED PIEZOELECTRICALLY DRIVEN ACOUSTO-OPTIC MODULATOR

Siddhartha Ghosh[1] and Gianluca Piazza[1]
[1]Carnegie Mellon University, Pittsburgh, USA

ABSTRACT

This paper presents a new type of acousto-optic modulator based on the conjunction of a piezoelectric contour mode resonator (CMR) with a photonic whispering gallery mode resonator (WGMR). The monolithic device fabricated in aluminum nitride (AlN) exhibits the coupling of piezoelectrically-generated lateral vibrations into a traveling-wave photonic ring resonator in a fully-integrated platform with electrodes directly patterned on the CMR body. We demonstrate the optical sensing of a piezoelectrically actuated mechanical mode at 654 MHz, enabling new possibilities for MEMS-based RF-photonics applications or new degrees of control of phonon-photon interactions in the field of optomechanics.

INTRODUCTION

In recent years, there has been a great deal of interest in producing MEMS devices that interact with optical cavities for enabling the development of frequency references [1] and optical modulators [2], among other applications. These devices generally benefit from high optical readout sensitivity in order to enable the interrogation of mechanical modes which have traditionally been of interest in MEMS-based radio frequency (RF) systems. Among the variety of cavity optomechanical devices that have been produced, there are a number of examples which employ the actuation forces of light, such as radiation pressure [1],[3] and electrostriction [4]. In the area of electrically-driven actuation, a number of demonstrations have been made with electrostatic devices [2],[5]. The use of piezoelectrically-driven resonators provides a better interface to RF signals by offering the ability to scale to higher frequencies of operation without the need for complex processing to produce capacitive gaps [5]. Recently, there have been examples of piezoelectric actuation in optomechanical cavities through the use of suspended RF probes [6] and by launching acoustic waves into a photonic crystal [7]. In order to enable facile incorporation into filters, oscillators and other devices commonly produced in the RF MEMS community however, it would be advantageous to make use of a fabrication process in which electrodes are patterned directly on the resonator body. This approach would eliminate any limitations on the range of frequencies that could be excited in the mechanical resonator, and could interface with the body of work already generated for resonant piezoelectric MEMS devices [8].

In this paper, we present a piezoelectric acousto-optic modulator that consists of two AlN ring resonators connected by a mechanical coupling beam. The operational principle of the device is similar to the design presented in [2] for electrostatic disk resonators. In addition to producing an original class of acousto-optic modulators, by generating a platform that integrates optical modulation in piezoelectric thin films, we also enable the possibility to produce new types of RF-photonic mixed systems.

DESIGN PRINCIPLES

The device under consideration is shown in Figure 1. The two rings constituting the modulator are designed to allow the coexistence of mechanical and optical modes, which interact to modulate the whispering gallery mode (WGM) resonance condition of $m\lambda_o = n_{eff}2\pi R$, where m is the azimuthal mode number, λ_o is the resonant wavelength, n_{eff} is the mode effective index and R is the radius of the ring. In the first ring, which has top and bottom electrodes, piezoelectric excitation is used to generate displacements in the radial direction, which are transferred through a coupling beam to the second ring without electrodes. Simultaneously, the second ring supports the existence of a traveling-wave WGM optical resonance. Physical deformations to the WGMR induced by the coupling of mechanical vibrations from the first ring cause the optical resonance to shift. Thus when the photonic resonator is fixed at a specific wavelength, and the device is piezoelectrically actuated, changes to the transmission intensity can be detected. This process gives rise to the device's operation as an optical modulator.

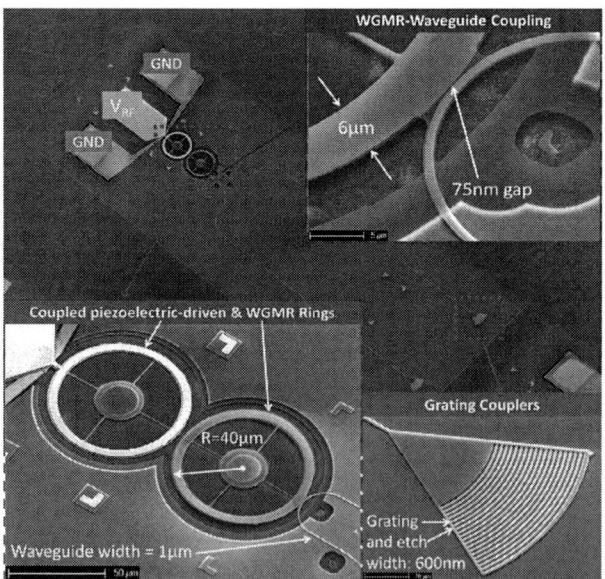

Figure 1: Scanning electron microscope images of the released device highlighting (clockwise from bottom left) the coupled microrings, waveguide to WGMR gap of 75nm and the optical input/output grating couplers.

Acoustic Resonator

In-plane displacements are generated in the first AlN ring through the application of an AC electric field across the film thickness, through the d_{31} piezoelectric coefficient. The radius and width of the rings (40μm and 6μm respectively) are selected based on the demonstration of high optical quality factors (Q_{opt}) in silicon nitride [9], a material with optical characteristics similar to AlN. A geometry matching the photonic resonator is selected for the acoustic resonator, with top Au and bottom Pt electrodes sandwiching the first ring. This allows the first ring to vibrate in a contour mode shape, with a frequency set by the width of the ring [10]. The displacement profile of the acoustic resonator's cross-section at resonance is shown in Figure 2a, which represents the peak modulation potential transferrable to the WGMR.

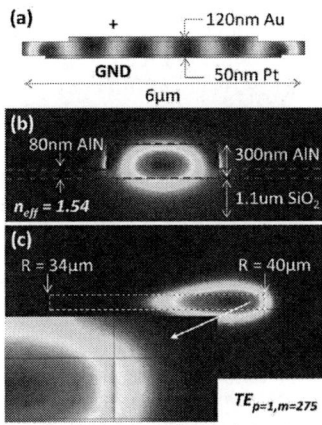

Figure 2: (a) Displacement mode shape of the CMR ring. (b) Mode confinement in rib waveguide with 80nm partially-etched AlN. (c) Axisymmetric optical mode in the WGMR, showing evanescent extension of the mode beyond ring edge, required for coupling to waveguide.

Photonic Resonator

The main considerations for the design of the photonic ring resonator are the coupling and confinement of the optical mode. In order to produce a fully-integrated design, grating couplers are used for the input and output ports for the confined light. The design of the waveguides and grating couplers is largely similar to what is described in [11] with a few modifications to facilitate the fabrication of the MEMS resonators. The AlN thickness is reduced to 300nm, which still provides strong confinement with the use of 1μm wide waveguides, while eliminating the need for a hard mask during processing. Furthermore in order to release the resonators by undercutting the silicon dioxide (SiO$_2$), without the need for a special masking layer, a rib waveguide geometry is selected for the confinement of the optical mode. This allows the remaining slab of AlN to be used as a natural masking layer during the release process. Based on the 300nm thick waveguide, 80nm is selected as an appropriate slab thickness for the rib. The resulting transverse electric (TE) optical mode is shown in Figure 2b, which is compatible with the 1.2μm pitch, 50% duty cycle, fully etched grating couplers that are used.

Coupling of the optical mode from the rib waveguide to the photonic WGMR is achieved through a gap between them. In order to evaluate an appropriate coupling gap size, we consider the leakage of the optical mode from the WGMR into the surrounding region. The optical mode in the WGMR may be simulated with an axisymmetric finite element simulation of the electromagnetic resonator [12]. As shown in Figure 2c, the mode profile in the 6μm wide ring is shown for the 1st-order radial quasi-TE mode, where m=275. Since the dielectric ring is entirely surrounded by vacuum following release, the optical mode is tightly confined, and the evanescent extension beyond the ring's border is limited to ~200nm. In order to ensure critical coupling of the optical mode from the waveguide to the photonic resonator, a gap size of 75nm is selected for the device.

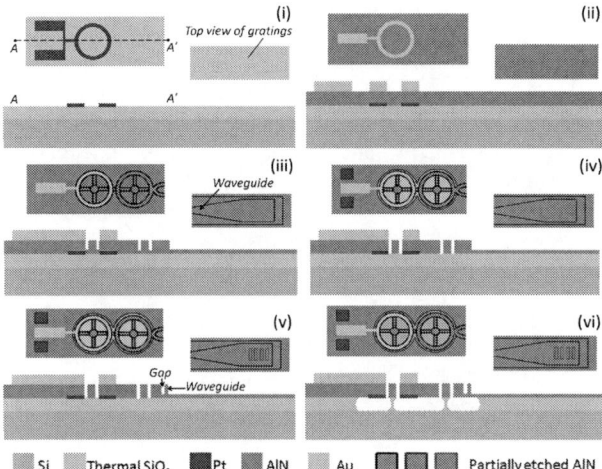

Figure 3: Process flow for concurrent production of acoustic and photonic resonators.

FABRICATION

Fabrication of the devices is completed through a 5-mask process shown pictorially in Figure 3. We begin the process with thermal oxidation of high-resistivity silicon wafers to grow a 1.1μm thick SiO$_2$ undercladding layer. This layer provides confinement of the optical mode in the AlN, and its thickness is determined through finite difference time domain (FDTD) simulation of the coupling transmission. Bottom electrodes are then patterned through stepper lithography and lift-off of 50nm Pt. The 300nm AlN layer is then deposited on the wafer, and top electrodes are patterned in a similar fashion with lift-off of 120nm Au. Once the electrodes have been patterned, stepper lithography is used once more to define the boundaries of the acoustic and photonic resonators, as well as the waveguides and grating couplers. At this point, all small features (namely gaps and gratings) are still undefined. The AlN is now partially etched down to the 80nm slab, which is used to define the rib waveguides (and subsequently serve as the release mask). Following this stage, the wafer is diced into individual dies, for processing with electron-beam lithography with the Nanometer Pattern Generation System (NPGS). First, the release windows and vias to the bottom electrodes are defined, and a short etch of AlN is performed to etch through the 80nm slab layer. This exposes the underlying SiO$_2$ (in the case of the release windows) and the bottom Pt for probing (in the case of the vias). A second e-beam lithography step is then performed to define the critical features that were previously left undefined. Namely, the

75nm WGMR-waveguide gap and the 600nm gratings are patterned in this step. This is followed by another partial etch of AlN down to the surrounding slab thickness of ~80nm. Once all the features have been defined, the resonators are released with the use of a timed vapor HF etch to undercut the rings and leave them both anchored at their central hubs.

EXPERIMENTAL RESULTS

Testing of the devices takes place with the experimental setup shown in the schematic of Figure 4. As a result of the modulator design, the acoustic and photonic resonator parts may be each tested separately, prior to characterizing the operation of the complete device. This also provides a means to compare the mechanical quality factor of the device through two (electromechanical and optomechanical) independent measurement schemes.

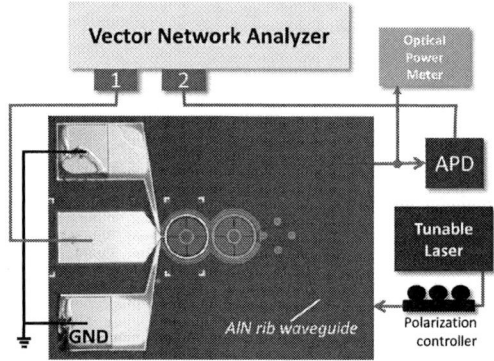

Figure 4: Experimental setup used to probe the mechanical and optical resonances of the device, via electrical (SMA connections in blue) and optical (fiber connections in red) measurements.

Acoustic Resonator Characterization

The composite acoustic resonator was tested at room temperature in ambient air, with a vector network analyzer (VNA) (Agilent N5230A) and a probe station configured to simultaneously characterize the optical response. Ground-Signal-Ground (GSG) probes were used to probe the device after performing a short-open-load calibration on a reference substrate. The S_{11} parameters were then measured over a range of frequencies and converted into equivalent admittances. A

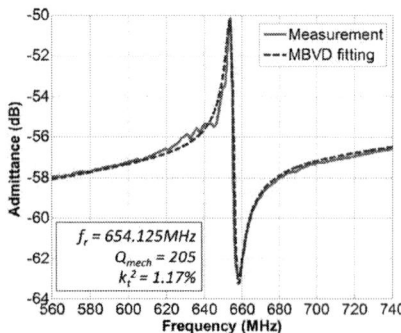

Figure 5: Measured admittance response of the composite acousto-optic modulator, actuated & sensed electrically in a one-port configuration.

single prominent resonance was measured at 654.1 MHz. This peak is shown in the admittance plot of Figure 5. The data was then fitted with a modified Butterworth van Dyke (MBVD) equivalent circuit model. The mechanical quality factor (Q_m) of the composite structure (both coupled rings) extracted from the measurement is ~205, which can later be compared with the optomechanical measurement.

Photonic Resonator Characterization

Optical responses are measured by recording the transmission characteristics of light coupled into the device through the grating couplers over a range of wavelengths. A tunable laser source (Santec TSL-510) operating in the telecommunications C-band is fed through a polarization controller to select TE mode injection into the input grating. Transmitted light collected from the fiber at the output grating is then fed to an optical power meter (Exfo PM-1103) to make insertion loss measurements for the photonic resonator. The normalized transmission (in which the losses of the gratings have been subtracted out) is shown for an optical resonance at 1552.032 nm in Figure 6. The extinction ratio approaches 10 dB, indicating that the resonator is critically coupled. The data was then least-square fit to a Lorentzian function in order to extract the full-width at half maximum (FWHM) ($\Delta\lambda$) for the resonance. This figure is used to determine the optical (loaded) quality factor $Q_{opt} = \lambda_o/\Delta\lambda$, where λ_o is the resonant wavelength. Based on the extracted FWHM, the optical quality factor for the device is ~22,500.

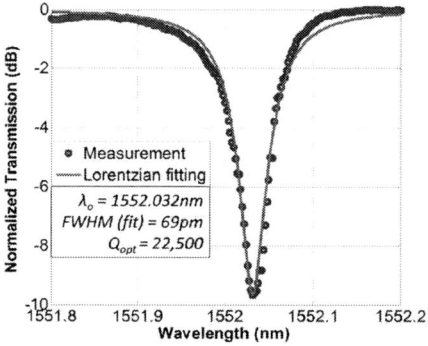

Figure 6: Normalized transmission response of the high optical quality factor (Q_{opt}) resonance of the WGMR used for interrogating the mechanical modes optically.

Acousto-optic modulation

The characterization of the acousto-optic modulator requires simultaneous utilization of the RF and photonic setups. Piezoelectric excitation is still applied through the first port. In order to detect the optical modulation however, the output fiber is fed into an avalanche photodetector (APD) (New Focus 1647) to convert the mechanically-modulated optical signal back into the electrical domain. The APD output is then placed back into the second port of the VNA to make S_{21} measurements for the device. During this process, the bias point of the wavelength in the transmission spectrum of the photonic resonator determines the maximum modulation capability. This is due to the fact that the total output modulation is directly proportional to the slope of the transmission curve. As a result, the modulation capability is also strongly affected by the Q_{opt} of the device.

The S_{21} transmission characteristic of the modulator is shown in Figure 7. Optical power from the output grating is sufficiently low to use a bias setting with a conversion gain of 14,000 V/W in the APD. The S_{21} parameter is then plotted for a frequency sweep centered around the center frequency at which the device is actuated. The laser can then be biased at a fixed wavelength, to monitor the difference in output modulation ability.

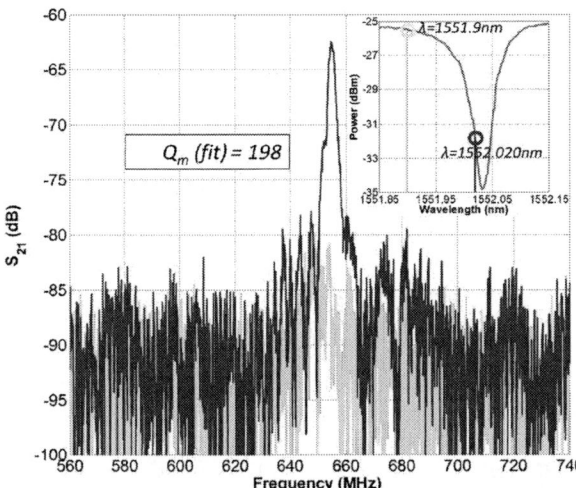

Figure 7: S_{21} response from the output of the APD, exhibiting mechanical resonance observed piezoelectrically through the biasing of the optical wavelength relative to photonic resonator transmission curve (shown in inset).

As shown in Figure 7, S_{21} output when the laser wavelength is fixed at 1551.9nm is negligible (depicted in green). This corresponds to a position on the photonic resonator transmission curve where the change in slope is relatively minimal. As the bias wavelength increases along the resonance curve however, the peak measured S_{21} value also increases until it reaches a maximum at 1552.020nm. Here, peak modulation is achieved, and the piezoelectrically-driven mechanical resonance is detected (depicted in blue) at 654.1MHz. By normalizing the peak modulation data and fitting to a Gaussian curve, we find an extracted mechanical quality factor ~198, exhibiting good agreement with the piezoelectrically generated result.

CONCLUSION

We have demonstrated a design for the concurrent production of acoustic and photonic resonators in a fully-integrated platform, which allows the interaction of the two domains to be studied in new ways. Previous piezoelectric demonstrations have lacked integration of electrodes, or produced modulation in structures with standing wave optical resonances. In addition, this design demonstrates a new type of narrow-band acousto-optic modulator, which can be interfaced with existing designs

for optical signal processing applications.

ACKNOWLEDGEMENTS

This work was completed with support from the National Science Foundation under award NSF ECCS-1201659. The devices were fabricated in the Carnegie Mellon Nanofabrication Facility.

REFERENCES

[1] T.O. Rocheleau, A.J. Grine, K.E. Grutter, R.A. Schneider, N. Quack, M.C. Wu and C.T.-C. Nguyen, "Enhancement of Mechanical Q for Low Phase Noise Optomechanical Oscillators," in *Proc. MEMS 2013*, pp. 118-121.

[2] S. Sridaran and S. A. Bhave, "Opto-acoustic oscillator using silicon MEMS optical modulator," in *Proc. Transducers 2011*, pp. 2920-2923.

[3] M. Hossein-Zadeh, H. Rokhsari, A. Hajimiri, and K.J. Vahala, "Characterization of a radiation-pressure-driven micromechanical oscillator," *Phys. Rev. A*, vol. 74, pp. 023813-1-023813-15, 2006.

[4] M. Tomes and T. Carmon, "Photonic micro-electromechanical systems vibrating at X-band (11-GHz) rates," *Phys. Rev. Lett.*, vol. 102, pp. 113601-1-113601-4, 2009.

[5] S. Tallur and S. A. Bhave, "Monolithic 2GHz electrostatically actuated MEMS oscillator with opto-mechanical frequency multiplier," in *Proc. Transducers 2013*, pp. 1472-1475.

[6] C. Xiong, L. Fan, X. Sun and H.X. Tang, "Cavity piezooptomechanics: Piezoelectrically excited, optically transduced optomechanical resonators," *Appl. Phys. Lett.*, vol. 102, 0211101-0211104, 2013.

[7] J. Bochmann, A. Vainsencher, D.D. Awschalom and A.N. Cleland, "Nanomechanical coupling between microwave and optical photons," *Nature Physics*, vol. 9, pp. 1-5, 2013.

[8] G. Piazza, V. Felmetsger, P. Muralt, R. H. Olsson III, and R. Ruby, "Piezoelectric aluminum nitride films for microelectromechanical systems," *MRS Bull.*, vol. 37, pp. 1051–1061, 2012.

[9] S. Tallur and S. A. Bhave, "A Silicon Nitride Optoemchanical Oscillator with Zero Flicker Noise," in *Proc. MEMS 2012*, pp. 19-22.

[10] G. Piazza, P.J. Stephanou, J.M. Porter, M.B.J. Wijesundara and A.P. Pisano, "Low Motional Resistance Ring-Shaped Contour-Mode Aluminum Nitride Piezoelectric Micromechanical Resonators for UHF Applications," in *Proc. MEMS 2005*, pp. 20-23.

[11] S. Ghosh, C.R. Doerr and G. Piazza, "Aluminum Nitride Grating Couplers," *Appl. Optics*, vol. 51, pp. 3763-3767, 2012.

[12] M. Oxborrow, "Traceable 2-D finite element simulation of the whispering gallery modes of axisymmetric electromagnetic resonators," *Trans. Microwave Theory and Tech.*, vol. 55, pp. 1209-1218, 2007.

CONTACT

*S. Ghosh, tel: +1-412-2686611; sidghosh@cmu.edu

SPECTRALLY SELECTIVE INFRARED DETECTOR BASED ON AN ULTRA-THIN PIEZOELECTRIC RESONANT METAMATERIAL

Yu Hui[*] and Matteo Rinaldi
Department of Electrical and Computer Engineering, Northeastern University, Boston, USA

ABSTRACT

This paper reports on the first demonstration of a spectrally selective uncooled microelectromechanical resonant infrared (IR) detector based on an ultra-thin piezoelectric resonant metamaterial. The use of an ultra-thin (*600 nm*) piezoelectric metamaterial to form the resonant body of the device eliminates the electromechanical loading effect associated with the integration of an IR absorber (guaranteeing high electromechanical performance: quality factor, $Q{\sim}1407$ and electromechanical coupling coefficient, $k_t^2{\sim}1.9\%$) and enables strong and spectrally selective absorption of long wavelength infrared (LWIR) radiation in an ultra-low volume device, resulting in a fast (thermal time constant ~650 μs) and high resolution (noise equivalent power ~7 *nW/Hz$^{1/2}$* for a *200 Hz* bandwidth) LWIR detector prototype with a ~40% absorption for an optimized spectral wavelength of *8.8 μm* with Full Width at Half Maximum (FWHM) of *1.88 μm*.

INTRODUCTION

In recent years, the interest in uncooled infrared (IR) detectors based on Micro/Nano-electromechanical system (MEMS/NEMS) resonator technologies has been steadily growing thanks to their unique combination of high sensitivity (very reduced dimensions) and low noise performance (intrinsically high quality factor) [1]. MEMS/NEMS resonant IR detectors based on quartz [2], gallium nitride [3] and aluminum nitride (AlN) [4, 5] piezoelectric resonators have been demonstrated and showed promising performance. Among these, the AlN resonant technology offers some unique advantages: ultra-thin (10s ~ 100s nm) and high quality AlN film can be directly deposited on Silicon substrates by low-temperature sputtering process [6], enabling the fabrication of ultra-low volume resonators with excellent electromechanical performance. Furthermore, the post-CMOS compatibility of the microfabrication process used for the fabrication of AlN MEMS resonant devices [7], enables the monolithic integration of resonant sensors and CMOS electronic readouts on the same substrate which is an attractive features for the implementation of ultra-miniaturized, high performance and low power sensing platforms [8]. Such unique features have been recently exploited for the first demonstration of a MEMS-CMOS uncooled resonant IR detector prototype, based on an AlN nano-plate resonator, showing the best performance reported to date among MEMS resonant IR detectors (noise equivalent power, *NEP~371 pW/Hz$^{1/2}$*). [5].

Despite the demonstrated potential of the AlN nano-plate resonator technology for the development of uncooled broadband IR detectors, high performance and spectrally selective MEMS resonant IR detectors have not been demonstrated so far, due to the lack of naturally occurring materials with strong absorption coefficients over ultra-thin thicknesses that are also compatible with conventional transduction and microfabrication techniques. In this work, a stepping stone towards the development of high performance and spectrally selective MEMS resonant IR detectors is set by demonstrating an ultra-thin (*600 nm*) Piezoelectric Resonant Metamaterial (PRM). Sensing and actuation of a high frequency (*167 MHz*) bulk acoustic mode of vibration in a free-standing ultra-thin piezoelectric metamaterial is demonstrated for the first time and exploited for the implementation of a novel MEMS resonator with a unique combination of optical and electromechanical properties. The use of an ultra-thin piezoelectric metamaterial to form the resonant body of the device eliminates the electromechanical loading effect associated with the integration of an IR absorber (guaranteeing high electromechanical performance: quality factor, $Q{\sim}1407$ and electromechanical coupling coefficient, $k_t^2{\sim}1.9\%$) and enables strong and spectrally selective absorption of LWIR radiation in an ultra-low volume device (due to the properly engineered IR absorption properties of the piezoelectric metamaterial forming the resonant body of the device); resulting in a fast (thermal time constant ~650 μs) and high resolution (noise equivalent power ~7 *nW/Hz$^{1/2}$* at 200 Hz bandwidth) LWIR detector prototype with a ~40% absorption for an optimized spectral wavelength of *8.8 μm* with Full Width at Half Maximum (FWHM) of *1.88 μm*. The demonstrated PRM detector technology marks a milestone towards the implementation of a new class of high performance, miniaturized and low power IR spectroscopy and multi-spectral imaging systems.

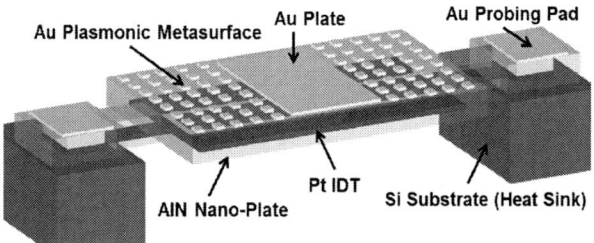

Figure 1: 3-dimentional representation (not to scale) of the proposed spectrally selective infrared detector based on an ultra-thin AlN piezoelectric resonant metamaterial.

DESIGN AND FABRICATION

The most important parameters that ought to be considered for the design of a high performance resonant IR detector are the IR absorptance, η, the device responsivity, R_s, and the noise induced frequency fluctuation, f_n, as these parameters are directly related to the noise equivalent power (NEP) of the detector, which represents the minimum incident IR power that can be detected,

$$NEP = \frac{f_n}{\eta \cdot R_s} \qquad (1)$$

$$R_s = R_{th} \cdot TCF \qquad (2)$$

where R_{th} is the thermal resistance of the detector and TCF is the temperature coefficient of frequency. It has been demonstrated that high device responsivity can be achieved by taking advantage of the unique scaling capability of the AlN piezoelectric technology [6] to fabricate high quality factor (which guarantees low noise performance) piezoelectric resonant nano-plates that are extremely well isolated from the heat sink (R_{th} as high as $\sim 10^5$ K/W) and whose resonance frequency is highly sensitive to temperature (TCF as high as -30 ppm/K) [5]. Even though these resonant structures are characterized by a large thermal resistance, the thermal time constant, τ, of these detectors is maintained below ~ms, thanks to the small thermal capacitance associated with such ultra-low volume resonant nano-plates [9].

$$\tau = R_{th} \cdot C_{th} \qquad (3)$$

Despite the excellent thermal detection capabilities of these AlN resonant nano-plates, the IR absorptance in such ultra-thin structures is almost null which drastically limits the fraction of incident IR power that can be effectively detected. In order to boost the IR absorptance of the structure, IR absorbing materials (such as silicon nitride or carbon based thin-films) are conventionally integrated on top of the resonant device. Nevertheless, the slight improvement in IR absorptance achieved using this conventional approach is typically offset by a substantial deterioration of the electromechanical and thermal performance of the device due to the electrical and mechanical loading effects of the IR absorbing material (which negatively affects Q, k_t^2, TCF and C_{th}).

To overcome this fundamental issue, in this paper we propose a transformative approach in which, instead of adding an IR absorbing layer to the material stack forming the resonator, the IR absorption properties of the piezoelectric nano-plate itself are properly tailored to achieve strong spectrally selective absorption of long-wavelength infrared (LWIR) radiation over such deeply subwavelength thickness, resulting in novel ultra-thin piezoelectric resonant metamaterial (PRM) with a unique combination of optical and electromechanical properties not found in nature.

The core of the proposed spectrally selective uncooled resonant IR detector is an AlN piezoelectric nano-plate (500 nm thick) sandwiched between a 50 nm thick platinum (Pt) bottom inter-digital transducer (IDT) and a top electrically floating layer (50 nm thick gold, Au) patterned with the goal of confining the electric field induced by the bottom IDT across the piezoelectric nano-plate, and, at the same time, enabling absorption of IR radiation in the ultra-thin piezoelectric nano-plate thanks to suitably tailored plasmonic resonances. The proper patterning of plasmonic nanostructures in the top metal layer of the device significantly enhances field concentration, enabling spectrally selective absorption of IR radiation over the ultra-thin structure.

When an alternating current (AC) signal is applied to the bottom IDT, the top electrically floating metal plate

acts to confine the electric field across the AlN piezoelectric nano-plate and a high order contour-extensional mode of vibration is excited through the equivalent d_{31} piezoelectric coefficient of AlN [10]. The resonance frequency of such PRM is defined by the pitch, W_0, of the IDT and the equivalent Young's modulus, E_{eq}, and density, ρ_{eq}, of the resonant material stack, given by (4),

$$f = \frac{1}{2W_0}\sqrt{\frac{E_{eq}}{\rho_{eq}}} \qquad (4)$$

The incident IR radiation absorbed in the plasmonically enhanced resonant nano-plate causes a large and fast increase of the device temperature, ΔT, due to the excellent thermal isolation ($R_{th} \sim 10^4$ K/W) and extremely low thermal mass of the freestanding resonant nano-plate (released from the substrate). Such IR induced temperature rise results in a shift in the mechanical resonance frequency of the resonator due to the intrinsically large temperature coefficient of frequency ($TCF \sim$ -30 ppm/K) of the device. Therefore, incident optical power can be readily detected by monitoring the resonance frequency of the device.

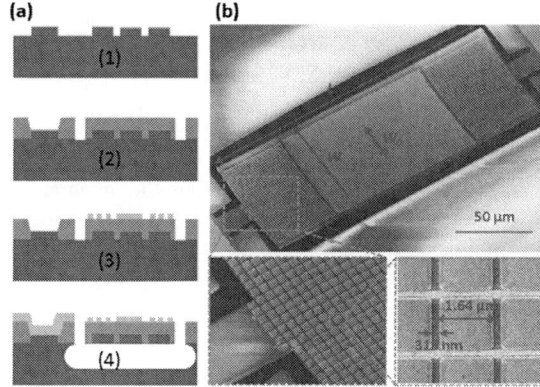

Figure 2: (a) 4-mask photo-lithography and 1-e-beam lithography fabrication process and (b) scanning electron microscopy images of the AlN PRM IR detector. The dimensions of the device are: $L = 200~\mu m$, $W = 75~\mu m$, $W_0 = 25~\mu m$.

For this first prototype, the pitch, W_0, of the IDT was set to 25 μm, targeting a resonance frequency of ~160 MHz, while the critical dimensions of the unit cell forming the plasmonic metasurface were optimized by 3D Finite Integration Technique (FIT) simulations using CST and set to $a=1640$ nm and $b=313$ nm (Fig. 2) for a maximum spectrally selective IR absorptance centered at a spectral wavelength $\lambda \sim 9~\mu m$. The top metal coverage ratio between the solid Au plate and Au plasmonic metasurface was set to 1:1, which guarantees high electromechanical coupling coefficient and sufficient IR absorption, at the same time.

The device was fabricated using a combination of 4-mask photo-lithography and 1-e-beam lithography micro/nano-fabrication process (Fig. 2-a). The fabrication started with a high resistivity silicon (Si) wafer: (1) 50 nm thick platinum was sputter-deposited and patterned by lift-off process to define the bottom IDT; (2) 500 nm thick high quality c-axis oriented AlN film was sputter-deposited and wet-etched by phosphoric acid (H_3PO_4) to open vias to access to the bottom electrodes, the shape of the AlN nano-plate was defined by dry-etching in Cl_2 based chemistry; (3) 50 nm thick gold was deposited by e-beam evaporation and patterned by e-beam lithography and lift-off process to define the plasmonic nanostructures and the solid metal plate on the top surface of the nano-plate; (4) 100 nm thick gold was deposited by e-beam evaporation and patterned by lift-off process to define the top probing pad. Finally, the Si substrate underneath the device was completely etched using xenon difluoride (XeF_2) isotropic etching to release the structure. The fabricated AlN PRM IR detector is shown in the SEM image in Fig. 2-b.

EXPERIMENTAL RESULTS

The electromechanical performance of the fabricated AlN PRM IR detector was characterized by measuring its admittance amplitude versus frequency using an Agilent E5071C network analyzer after performing an open-short-load calibration on a standard substrate. The measured resonance frequency was found to be *166.7 MHz* (Fig. 3). High electromechanical performance (mechanical quality factor, *Q,* of *1407* and electromechanical coupling coefficient, k_t^2, of *1.9%*), comparable to the one of conventional ultra-thin film AlN MEMS resonators, was achieved, indicating that the use of an ultra-thin piezoelectric metamaterial to form the resonant body of the device does not deteriorate the electromechanical performance of the resonator (unlike the integration of conventional IR absorbers). The *TCF* of the device was measured using a temperature controlled RF probe-station and found to be *~4.42 kHz/K* (*-26.5 ppm/K*), which is comparable to the typical values recorded for conventional 500 nm thick AlN contour-mode resonators [1].

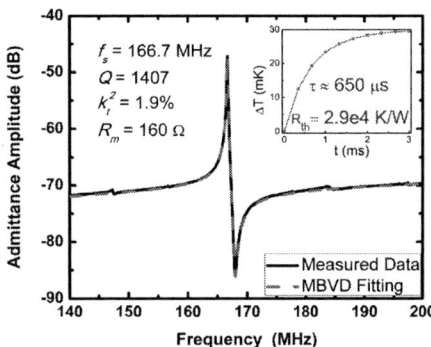

Figure 3: Measured admittance amplitude versus frequency and modified Butterworth-Van Dyke (MBVD) fitting of the fabricated AlN PRM IR detector. The inset shows the 3D FEM simulated thermal time constant, τ, and thermal resistance, R_{th}.

The thermal properties of the AlN PRM IR detector were evaluated by 3D finite element method (FEM) simulation using COMSOL multiphysics. The thermal resistance of the device was evaluated by applying different levels of thermal power to the device and monitoring its temperature rise. A R_{th}~2.9×10^4 *K/W* was extracted from the FEM simulation. The responsivity of

the detector was calculated by multiplying the measured *TCF* and the simulated R_{th}, and it was found to be *~128 Hz/μW*. The thermal time constant of the IR detector was estimated by simulating the transient temperature increase of the device in response to a thermal power of *1 μW*. A thermal time constant *τ~650 μs* was extracted from 3D FEM simulation. Such short response time is due to the extremely small volume of the nano-plate resonator (reduced thermal mass).

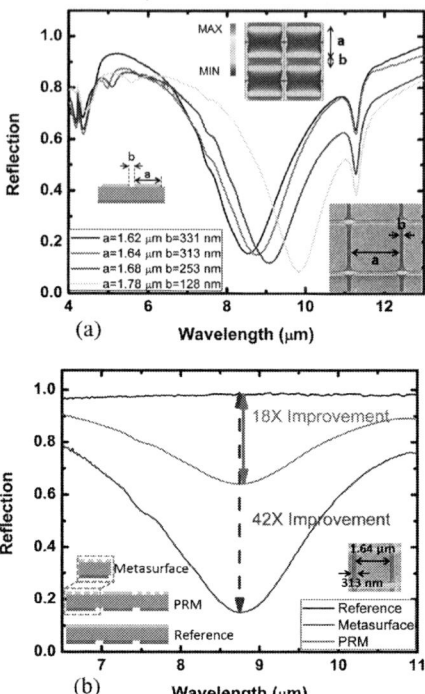

Figure 4: (a) Measured reflectance FTIR spectra of 4 fabricated plasmonic metasurfaces with different unit cell dimensions. The inset shows the 3D FIT simulated electric field distribution at the absorptance peak frequency. (b) Measured reflectance FTIR spectra of the fabricated AlN PRM IR detector and a conventional AlN MEMS resonator (reference).

The reflection spectra, *R*, of four fabricated piezoelectric plasmonic metasurfaces with different unit cell dimensions were measured using a Bruker V70 Fourier transform infrared (FTIR) spectrometer and Hyperion 1000 microscope. The spectral absorptance, *A*, was estimated as *A=1-R*, assuming no transmission through the resonant structure (which is a reasonable assumption given the large metal coverage of both the top Au plasmonic structures and bottom Pt IDT). The measured results in Fig. 4-a show that a strong and spectrally selective absorption of LWIR radiation, with lithographically determined center frequency and peak values higher than *85%* (over the device area covered by the plasmonic metasurface), was readily achieved.

Similarly, the reflection spectrum of the fabricated AlN PRM IR detector (unit cell dimension of *a=1640 nm* and *b=313 nm*) was measured and compared with the one of a reference AlN resonator with the same geometry as the AlN PRM but a solid Au plate covering the entire resonant body (Fig. 4-b). Despite the partial coverage of

978-1-4799-7956-1/15 $31.00 © 2015 IEEE

the plasmonic metasurface (inclusion of a solid metal plate needed to achieve high electromechanical coupling) an absorption peak value of ~36% was achieved, which represents a ~18X enhancement in absorptance compared to a conventional AlN device. The measured data indicate that the IR absorptance of the device could be significantly improved (to a maximum of ~85% for the current design) by optimizing the coverage of the plasmonic metasurface and considering a trade-off with the minimum tolerable value of k_t^2.

The responses of the fabricated AlN PRM IR detector and the reference device to IR radiation were characterized using a 1500 K globar (2 -16 μm emission) as an illumination source coupled to a Bruker Hyperion 1000 microscope. The measured frequency responses in Fig. 5 show that despite the relatively broadband emission of the globar (compared to the *1.88 μm* FWHM of the PRM), a *~5.5X* enhanced responsivity was recorded for the PRM detector, thanks to its properly engineered absorption properties. Based on the measured noise spectral density of *~0.9 Hz/Hz$^{1/2}$*, the NEP of the detector was calculated by dividing the noise spectral density by the responsivity, and found to be *~7 nW/Hz$^{1/2}$*.

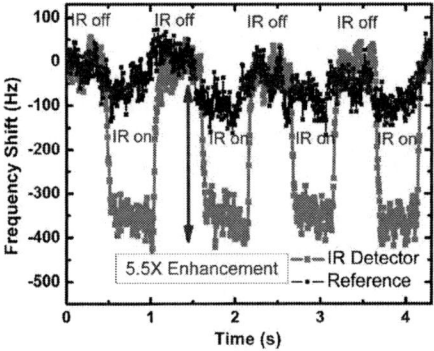

Figure 5: Measured responses of the fabricated AlN PRM IR detector and a conventional AlN MEMS resonator as a reference to a broadband IR radiation (2-16 μm).

CONCLUSIONS

This paper introduces a transformative approach to the development of high resolution, fast and spectrally selective uncooled resonant IR detectors in which an ultra-thin piezoelectric metamaterial is used to form the vibrating body of a MEMS resonator with a unique combination of optical and electromechanical properties not found in nature. High electromechanical performance (quality factor, Q~1407 and electromechanical coupling coefficient, k_t^2~1.9%) and strong and spectrally selective absorption of LWIR radiation *(~40% absorption for an optimized spectral wavelength of 8.8 μm with FWHM of 1.88 μm)* in an ultra-low volume piezoelectric resonant device are simultaneously achieved, resulting in the first prototype of a high resolution (*NEP~7 nW/Hz$^{1/2}$*) and fast (thermal time constant *τ~650 μs*) spectrally selective piezoelectric MEMS resonant IR detector. The demonstrated piezoelectric resonant metamaterial (PRM) detector technology marks a milestone towards the implementation of a new class of high performance, miniaturized and low power IR spectroscopy and multi-spectral imaging systems.

ACKNOWLEDGEMENTS

The authors wish to thank the staff of the George J. Kostas Nanotechnology and Manufacturing Facility, Northeastern University for their support in device fabrication. This project was supported by the DARPA Young Faculty Award N66001-12-1-4221 and the NSF CAREER Award ECCS-1350114.

REFERENCES

[1] Y. Hui, and M. Rinaldi, "Fast and High Resolution Thermal Detector based on an Aluminum Nitride Piezoelectric Microelectromechanical Resonator with an Integrated Suspended Heat Absorbing Element", *Appl. Phys. Lett.*, Vol. 102, pp. 093501, 2013.

[2] M. B. Pisani, K. Ren, P. Kao, and S. Tadigadapa, "Application of Micromachined Y-Cut-Quartz Bulk Acoustic Wave Resonator for Infrared Sensing," *J. Microelectromech. Syst.*, vol. 20, pp. 288-296, 2011.

[3] V. J. Gokhale, and M. Rais-Zadeh, "Uncooled Infrared Detectors Using Gallium Nitride on Silicon Micromechanical Resonators," *J. Microelectromech. Syst.*, vol. 23, pp. 803-810, 2014.

[4] Y. Hui, and M. Rinaldi, "High Performance NEMS Resonant Infrared Detector based on an Aluminum Nitride Nano-Plate Resonator," in *Digest Tech. Papers Transducers'13 Conference*, Barcelona, June 16-20, 2013, pp. 968-971.

[5] Y. Hui, Z. Qian, G. Hummel, and M. Rinaldi, "Pico-Watts Range Uncooled Infrared Detector Based on a Freestanding Piezoelectric Resonant Microplate with Nanoscale Metal Anchors", in *Digest Tech. Papers Hilton Head'14 Conference*, Hilton Head Island, June 8-12, 2014, pp. 387-390.

[6] U. Zaghloul, and G. Piazza, "Synthesis and Characterization of 10 nm Thick Piezoelectric AlN Films with High c-axis Orientation for Miniaturized Nanoelectromechanical Devices," *Appl. Phys. Lett.*, vol. 104, pp. 253101, 2014.

[7] R. H. Olsson, K. E. Wojciechowski, M. S. Baker, M. R. Tuck, and J. G. Fleming, "Post-CMOS-Compatible Aluminum Nitride Resonant MEMS Accelerometers," *J. Microelectromech. Syst.*, vol. 18, pp. 671-678, 2009.

[8] Y. Hui, T. Nan, N. X. Sun, and M. Rinaldi, "High Resolution Magnetometer Based on a High Frequency Magnetoelectric MEMS-CMOS Oscillator," *J. Microelectromech. Syst.*, DOI: 10.1109/JMEMS.2014.2322012, 2014.

[9] Y. Hui, and M. Rinaldi, "Aluminum Nitride Nano-Plate Resonant Infrared Sensor with Self-Sustained CMOS Oscillator for Nano-Watts Range Power Detection", in *Digest Tech. Papers IFCS'13 Conference*, Prague, July 21-25, 2013, pp. 559-561.

[10] M. Rinaldi, and G. Piazza, "Effects of Volume and Frequency Scaling in AlN Contour Mode NEMS Resonators on Oscillator Phase Noise," in *Digest Tech. Papers IFCS'11 Conference*, San Francisco, May 1-5, 2011, pp. 1-5.

CONTACT

*Y. Hui, tel: +1-617-758-9518; y.hui@husky.neu.edu

A CMOS-MEMS ARRAYED RGFET OSCILLATOR USING A BAND-TO-BAND TUNNELING BIAS SCHEME

Chi-Hang Chin[1], Cheng-Syun Li[1], Ming-Huang Li[1], and Sheng-Shian Li[1,2]
[1]Inst. of NanoEngineering and MicroSystems
[2]Dept. of Power Mechanical Engineering, National Tsing Hua University, Hsinchu, Taiwan

ABSTRACT

In this work, a CMOS-MEMS arrayed resonant gate field effect transistor (RGFET) oscillator is demonstrated for the first time. With the mechanically coupled array approach and deep submicron gap spacing, the proposed resonator with Q of 1,800 under purely capacitive transduction achieves the record-low motional impedance R_m of 1.1 kΩ among all CMOS-MEMS resonators. By using the FET readout, a CMOS-MEMS arrayed RGFET oscillator is realized through a closed-loop configuration, demonstrating phase noise performance of -96 dBc/Hz at 1 kHz offset and -122 dBc/Hz at far-from-carrier offset, respectively. In particular, a novel band-to-band tunneling bias scheme is employed for the proposed CMOS-MEMS RGFET without the need of manual switch charging or complicated biasing circuits. The proposed device is fabricated by a standard 0.35 µm CMOS process together with a maskless release process.

INTRODUCTION

The first MEMS resonant gate field effect transistor (RGFET) which combines a resonant transducer and a transistor into a single device was reported in 1960's, thus launching the development of capacitive MEMS resonators, filters, and oscillators [1]. As compared to the conventional quartz oscillators [2] which currently are still the mainstream technology in the timing markets, MEMS oscillators greatly benefit the miniaturization and possibly IC integration for timing reference devices [3]. Moreover, the wearable devices and the upcoming Internet of Things (IoT) demanding a significant number of sensors and wireless systems necessitate miniacturized oscillators where MEMS oscillators in an SoC configuration are the most welcome. Thanks to the inherent integration capability of CMOS-MEMS technology, the CMOS-MEMS oscillators featuring small form factor, fast turnaround time, and simple post process [4] find great opportunities in the abovementioned wearable and IoT applications. Therefore, this work proposes a CMOS-MEMS RGFET oscillator fabricated through a previously developed CMOS-MEMS platform to address the current bottleneck on miniaturization and integration of the timing reference components.

In our previous work, we presented the first CMOS-MEMS RGFET by employing a standard TSMC 0.35 µm 2-poly-4-metal (2P4M) CMOS process [5]. A clamped-clamped beam type metal/oxide composite RGFET can be realized by utilizing a simple post-CMOS process without any additional photolithography step. This device under purely capacitive transduction possesses decent performance, including low motional impedance and high quality factor. Furthermore, by the help of the FET's intrinsic transconductance gain, the RGFET detection behaves better than the purely capacitive readout.

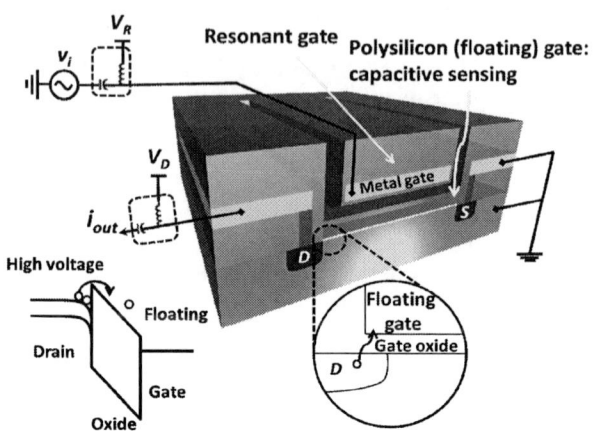

Figure 1: Perspective-view schematic of a CMOS-MEMS RGFET driven by an electrostatic force and sensed by either transistor or capacitive readout. Through a band-to-band tunneling charging technique, the FET can be operated at its saturation region without additional voltage supply on the floating (polysilicon) gate.

However, the first prototype of the CMOS-MEMS RGFET requires a manual switch to charge its gate for the saturation-mode operation. In order to satisfy the practical implementation, this manual switch should be replaced by a more realistic biasing scheme.

To resolve the biasing issue of the FET, a band-to-band tunneling bias scheme which mimics the operation of a conventional flash memory is employed to replace the previously used manual switch charging. Moreover, to achieve a better capacitive transduction and thus smaller motional impedance, a clamped-clamped beam mechanically-coupled array to serve as the resonant gate together with a multi-finger FET for the readout is also designed in this work. To investigate the detail characteristics of the proposed RGFET, both the static *I-V* analysis and the frequency characteristics for the resonant behavior under such a unique bias scheme are carried out in this work. The time-elapsed monitoring is also performed, which ensures that the charging technique is suitable for the operation of the RGFET. Finally, a closed-loop configuration of the proposed RGFET attains a self-sustained oscillation under a proper biasing condition while the oscillating signal is measured using a voltage buffer outside the oscillation loop.

DESIGN
CMOS MEMS arrayed RGFET

A perspective-view schematic of the proposed CMOS-MEMS arrayed RGFET is shown in Figure 1. The device is fabricated using the mature CMOS-MEMS platform [5] with the SEM picture shown in Figure 2

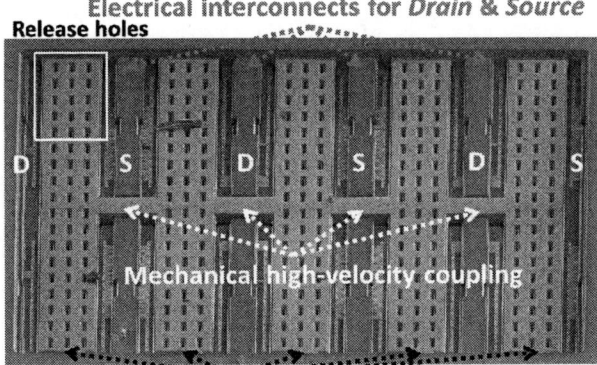

Array resonant gate (clamped-clamped beam)

Figure 2: Global SEM view of the CMOS-MEMS arrayed RGFET oscillator. The electrical interconnects are placed underneath the resonant structures.

where a clamped-clamped beam array is utilized to serve as the resonant gate of the RGFET. The mechanically-coupled array structure (i.e., resonant gate) consists of five resonant beams coupled with their adjacent one at the high-velocity locations, which greatly reduce the motional impedance R_m (ideally by 5X) and at the same time effectively suppress spurious modes of the resonant gate. To reduce the wet etching time, the release slots are placed on the resonant gate structure. After the polysilicon sacrificial layer is removed, a tiny 175-nm air gap can be achieved, which is key to enabling efficient capacitive transduction.

To operate the device of Figure 1, a combined dc bias voltage V_R and ac signal v_i through a bias-T generates an electrostatic force between the resonant gate and the underneath polysilicon gate to excite the resonant gate into vibration. A multi-finger FET is placed at the bottom with its polysilicon gate right beneath the MEMS arrayed structure. The polysilicon gate serves two important roles: (i) the sensing electrode of the capacitive transducer and (ii) the gate of the underneath FET. The capacitive output signal can be directly measured from the polysilicon gate to extract the motional impedance R_m for a conventional operation of capacitive resonators. To verify the FET readout, the motional current on the polysilicon gate would be amplified by the FET's intrinsic transconduction gain. Please note another bias-T is utilized at the drain node (i.e., output port) to provide appropriate biasing for the FET to work in its saturation region. The source and body nodes of the FET are connected together as an ac ground to avoid the body effect.

Band-to-band tunneling bias scheme

To attain a sufficient transconductance gain of the FET, the polysilicon gate should be biased at a proper level to inverse the FET's channel. Unfortunately, if the polysilicon gate is connected to a voltage source (e.g., power supply), the motional signal would be effectively shorted to an ac ground. For the proper RGFET operation, the polysilicon gate should be floating to provide a high impedance node while maintaining a sufficient bias voltage for operating in its saturation region. To solve this issue, we previously used a manual switch to charge the

polysilicon gate [5]; however this manual switch is not a practical solution for real system implementation.

In contrast, to charge the polysilicon gate without the switch, a simple and effective solution which mimics the operation of a non-volatile memory is employed. As a result, a hot-electron tunneling concept [6] is utilized to charge the floating polysilicon gate. This band-to-band tunneling (BTBT) phenomenon usually occurs at the interface of polysilicon gate/silicon dioxide/bulk silicon when the electron (or electron hole) attains enough energy to cross the barrier of the insulator. The inset band diagram in Figure 1 depicts the working principle of our device. First, we apply a large dc voltage at the drain node while keeping the polysilicon gate floating. With sufficient energy, the electron holes gain enough momentum to move across the gate oxide and reach the floating polysilicon gate. The threshold energy is determined by the thickness of the gate oxide and the doping concentration of the silicon. With the BTBT charging, the charge would be trapped in the gate for a long period of time since the polysilicon is surrounded by the insulating silicon dioxide.

DEVICE CHARACTERIZATION
Static IV characteristics of the BTBT operation

To characterize the BTBT charging scheme of the proposed RGFET, a normal I_D-V_D sweep measurement during various sweep cycles are carried out as shown in Figure 3. The BTBT charging occurs when the drain voltage is close to 9 V while the drain current abruptly increases exponentially since the polysilicon gate is charged to invert the FET's channel. During the charging operation, we clamp the current at 1 mA for protecting the device. At the second sweep, the initial current level indicates that the polysilicon possesses certain bias level since the drain current is far away from zero. We also observe that the BTBT occurs again even when the drain voltage is only half of the first sweep. With existing bias of the polysilicon gate from the first sweep, the threshold of the second BTBT charging becomes much lower since the electron holes can be transferred into the polysilicon gate not only from the drain but also from the channel. After a

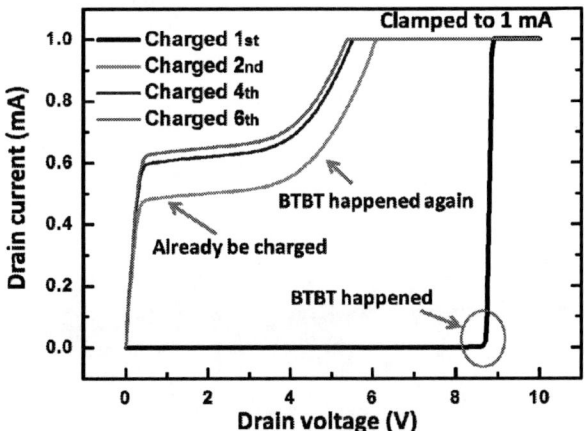

Figure 3: Behavior of the drain current under the band-to-band tunneling (BTBT) charging operation. Six charging cycles are carried out for I_D-V_D comparison. After the polysilicon gate is charged, the current level shows the FET is at its saturation state once the applied $V_D > V_{DSAT}$.

couple of sweep cycles, the charging bias level of the polysilicon gate eventually reach a saturated value which is not infinite.

Frequency characteristics of the RGFET

Figure 4 shows the frequency characteristics of the proposed device. We measure the capacitive sensing signal from the polysilicon gate as shown in Figure 4(a). The resonance frequency is around 1.1 MHz with an insertion loss of -21.4 dB and a stopband rejection of 20 dB. The extracted motional impedance R_m is 1.1 kΩ at a bias voltage V_R of 35 V. As compared to the state-of-the-art CMOS-MEMS resonators, the measured motional impedance of the resonator is the record-low value owing to its tiny air gap spacing [5] and mechanically-coupled array design (i.e., large transduction area) [7].

Figure 4(b) presents the measured responses using the FET readout through the BTBT biasing operation. As compared to the capacitive detection, the insertion loss is reduced by 10 dB when the drain current is 4.6 mA at V_D of 2 V even with a smaller V_R of 20 V. The measured frequency of the FET detection is higher than that of the capacitive sensing due to less electrical stiffness. The quality factor is around 1,800 for both capacitive sensing and FET readout. Although the bias level of the polysilicon gate is solely determined by the BTBT charging phenomenon, the transconductance gain of the FET can still be adjusted by the drain voltage V_D. The gain control capability governed by V_D is also observed in Figure 4(b) where the transmission peak decreases as the transconductance gain of the FET becomes lower. With zero drain voltage, the device can be turned off.

Time-elapsed measurement

The charge leakage of the floating polysilicon gate is critical since it affects the practical usage of the RGFET. The long-term transmission measurement in Figure 5 indicates that the charges are well preserved in the polysilicon gate greater than one day. The monitoring characteristics include the static I_D current and frequency transmission.

In the preceding measurement, every electrical source is maintained at proper bias levels while the network

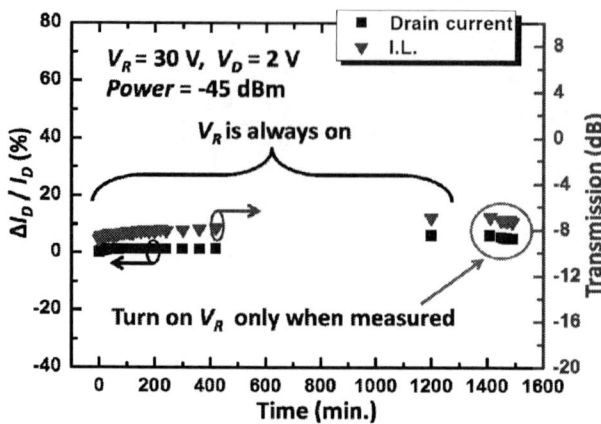

Figure 5: Long-term S21 transmission measurement for the charged RGFET. The signal amplitude starts to decrease when V_R is off, indicating that the V_R bias helps to keep the charges staying at the floating (polysilicon) gate.

analyzer is triggered only when the frequency response measurement is performed. During one day monitoring, not only I_D but also the transmission signal gently increases when the time elapses. After one day operation, the transmission peak increases 2 dB and I_D increases 5%. This variation might be due to another charging effect since the composite material structure mostly composed of silicon dioxide layer. After one day monitoring, we turn off V_R for most of the time and only turn it on whenever the frequency measurement is carried out. Obviously V_R is another key to preserving the bias state of the polysilicon gate since the performance of the RGFET starts to degrade.

RGFET SELF-SUSTAINED OSCILLATOR

RGFET oscillator setup

To achieve self-oscillation, a MEMS resonator integrated with an op amplifier in a closed-loop configuration is needed [8]. In an RGFET device, it is straightforward to convert the device into an oscillator since the RGFET combines a resonant transducer and an active circuit into a single device. To meet the Barkhausen criteria in the proposed CMOS-MEMS RGFET, a unity-gain buffer is connected at the output of the FET to alleviate the impedance mismatch issue between the RGFET and the 50Ω-based test instruments while offering 0-degree loop phase [5]. The self-sustained oscillator measurement setup is depicted in Figure 6. By utilizing the bias-T, the closed-loop configuration can be realized.

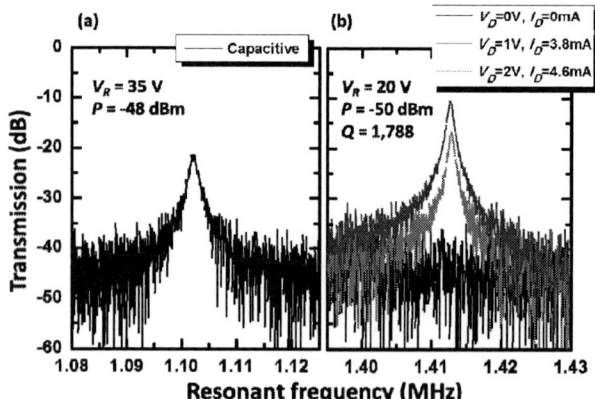

Figure 4: Frequency characteristics of the RGFET measured by (a) capacitive sensing and (b) FET detection. The FET transduction shows the motional signal is enhanced with a gain-control capability.

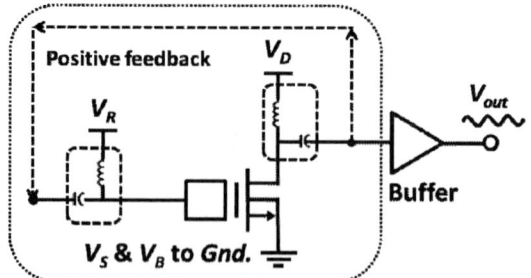

Figure 6: CMOS-MEMS RGFET oscillator setup. The unity-gain buffer is connected right after the FET.

978-1-4799-7956-1/15 $31.00 © 2015 IEEE   990

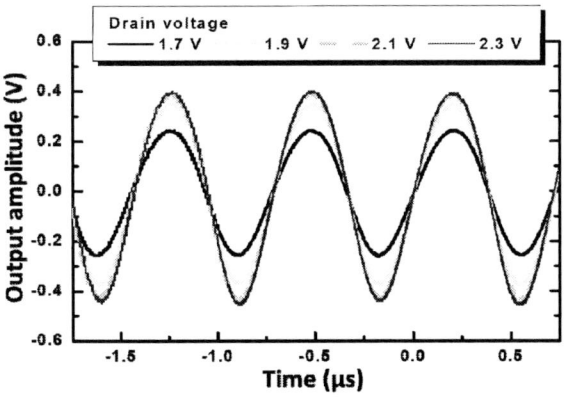

Figure 7: Measured time-domain oscillation waveform of the proposed CMOS-MEMS RGFET oscillator.

Characterization of CMOS-MEMS RGFET oscillator

The output sine wave of the CMSO-MEMS RGFET oscillator is measured by an oscilloscope as shown in Figure 7. The peak-to-peak amplitude is more than 0.8 V. The drain voltage control is carried out again to verify that the peak-to-peak output voltage varies with V_D. As considering the gain of the FET, the amplification is not linear in Figure 7. The V_D voltage varying from 1.7 V to 2.3 V indicates the FET operation starts from the triode region to the saturated region. As a result, the amplitude of the oscillator stays stable between 2.1 V and 2.3 V due to the FET saturation mode.

The measured phase noise (PN) performance is shown in Figure 8. The resonant system starts to oscillate at V_R of 20.6 V with the output carrier power of -15.2 dBm under power consumption of 1.95 mW. Referring to the phase noise model of MEMS oscillators [9], the close-in PN is quite reasonable (-96 dBc/Hz at 1 kHz) since the quality factor of the composite device is not very high. In addition, the resonant gate is driven into its non-linear vibration region, thus degrading the PN. Thanks to the array design of the proposed CMOS-MEMS RGFET oscillator, its low motional impedance leads to a decent PN at far-from-carrier offset (-122 dBc/Hz at 1 MHz).

CONCLUSION

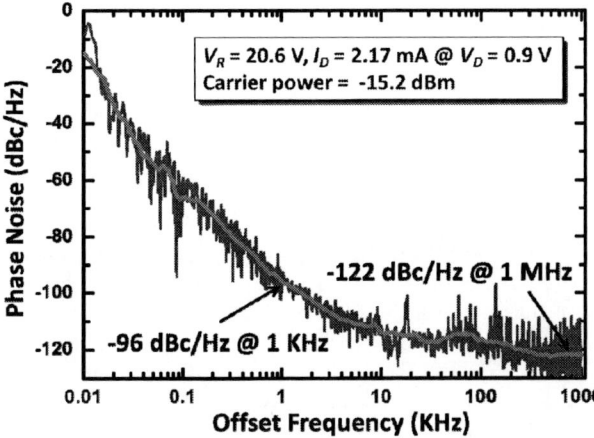

Figure 8: Measured phase noise showing decent performance at the far-from-carrier offset.

A CMOS-MEMS arrayed RGFET oscillator with a reliable BTBT charging bias scheme is presented in this work. After the BTBT charging, the resonant signal is greatly enhanced by the FET detection even under a small dc bias V_R. In addition, the proposed charging device is stable, which is verified by the time-elapsed measurement. The record-low motional impedance (1.1 kΩ) leads to a decent phase noise at the far-from-carrier offset (-122 dBc/Hz).

ACKNOWLEDGEMENTS

This research was sponsored by the MOST of Taiwan (MOST-103-2221-E-007-113-MY3). The chip fabrication was supported by CIC and TSMC, Taiwan. The authors would also like to appreciate the Center for Nanotechnology, Materials Science and Microsystems of National Tsing Hua University for the use of fabrication and measurement facilities.

REFERENCES

[1] H. C. Nathanson, W. E. Newell, R. A. Wickstrom, and J. R. Davis, Jr., "The resonant gate transistor," *IEEE Transactions on Electron Devices*, vol. 14, pp. 117-133, 1967.

[2] J. Rutman, "Characterization of phase and frequency instabilities in precision frequency sources: fifteen years of progress," *Proceedings of the IEEE*, vol. 66, pp. 1048-1075, 1978.

[3] C. T. C. Nguyen, "Frequency-selective MEMS for miniaturized low-power communication devices," *IEEE Transactions on Microwave Theory and Techniques*, vol. 47, pp. 1486-1503, 1999.

[4] J. Verd, *et al.*, "Monolithic CMOS MEMS oscillator circuit for sensing in the attogram range," *IEEE Electron Device Letters*, vol. 29, pp. 146-148, 2008.

[5] C.-H. Chin, C.-S. Li, M.-H. Li, Y.-L. Wang, and S.-S. Li, "Fabrication and characterization of a charge-biased CMOS-MEMS resonant gate field effect transistor," *Journal of Micromechanics and Microengineering*, vol. 24, p. 095005, 2014.

[6] R. Bez, E. Camerlenghi, A. Modelli, and A. Visconti, "Introduction to flash memory," *Proceedings of the IEEE*, vol. 91, pp. 489-502, 2003.

[7] S. Lee and C. T.-C. Nguyen, "Mechanically-coupled micromechanical resonator arrays for improved phase noise," in *Proceedings of the 2004 IEEE International Frequency Control Symposium*, pp. 144-150.

[8] M.-H. Li, C.-Y. Chen, C.-S. Li, C.-H. Chin, and S.-S. Li, "A monolithic CMOS-MEMS oscillator based on an ultra-low-power ovenized micromechanical resonator," *Journal of Microelectromechanical Systems*, in press.

[9] S. Lee and C. T.-C. Nguyen, "Influence of automatic level control on micromechanical resonator oscillator phase noise," in *Proceedings of the 2003 IEEE International Frequency Control Symposium*, pp. 341-349.

CONTACT

* S.-S. Li, Tel: +886-3-516-2401; ssli@mx.nthu.edu.tw

ACTIVE REFLECTORS FOR HIGH PERFORMANCE LITHIUM NIOBATE ON SILICON DIOXIDE RESONATORS

Lisha Shi[1], and Gianluca Piazza[1]
[1] Carnegie Mellon University, Pittsburgh, USA

ABSTRACT

This paper reports on the design and demonstration of active reflectors for enhancing the electromechanical coupling (k_t^2) and suppressing spurious modes in Laterally Vibrating Resonators (LVRs) based on X-cut ion-sliced Lithium Niobate (LN) thin film on silicon dioxide (SiO$_2$). By adding electroded quarter wavelength ($\lambda/4$) regions at the two ends of the resonant plate, active reflectors (since an electrical signal is applied to them) are formed to improve the device performance. Optimized active reflectors that resort to 100% metal coverage of the $\lambda/4$ extensions enable: (i) a considerable improvement of k_t^2, (ii) spurious mode suppression, and robustness to processing (iii) misalignment and (iv) over/under-etching. 2X improvement in k_t^2 and significant suppression of in-band spurious vibrations were attained with respect to the conventional design (without active reflectors) despite 0.5 µm misalignment and more than 0.5 µm overetch in the fabrication process.

INTRODUCTION

The rapidly growing demand for multi- functional wireless communication systems is driving the development of monolithic frequency-agile RF front ends. The envisioned RF transceivers require miniature filtering modules that can cover the current commercial operating bands and potentially offer adaptation to future standards. Therefore, research activities on high frequency MEMS resonators have focused on the development of devices for programmable RF front-ends with a particular attention to reconfigurable filters [1-3]. The need to meet the requirements dictated by wireless standards and simultaneously achieve reconfiguration has pushed researchers to look at devices and materials with higher electromechanical coupling, k_t^2 [4-5].

LN LVRs have exhibited extraordinary high k_t^2 (> 15%) and Qs around 1,000 [6-8]. Because of the high k_t^2, it is extremely complicated to control the various modes of vibration and ensure that a single mode is excited. Process variations also render any of the techniques previously implemented for confining energy in a single mode [7-8] ineffective. We introduce the active reflector design as a method to achieve high k_t^2 and suppress spurious modes even in the presence of misalignment and over/under-etching errors. These high k_t^2, spurs free, process-variation resistant and multi-frequency LN thin film resonators can be used to form filters covering the existing communication standards and potentially offer adaptive frequency response.

INTRODUCTION

Device Operation and Modeling

The LN LVR consists of metal interdigital fingers (IDT) on top of a mechanically suspended LN/SiO$_2$ thin film (Fig. 1). The electric field induced by the IDTs,

which are alternatively connected to ground and signal, excites lateral expansion and compression in adjacent fingers (mode known as S0 lamb wave). The mechanical resonance is a function of the lateral dimension of the excitation electrode and is given by:

$$f = \frac{1}{\lambda}\sqrt{\frac{E_p}{\rho}} = \frac{1}{2w}\sqrt{\frac{E_p}{\rho}} \tag{1}$$

where w *is* the pitch of the IDTs, E_p the equivalent Young's modulus, and ρ the equivalent density of the resonator stack. λ is the wavelength of the standing wave in the resonator, which is set to 12 µm in this work to ensure operation around 500 MHz.

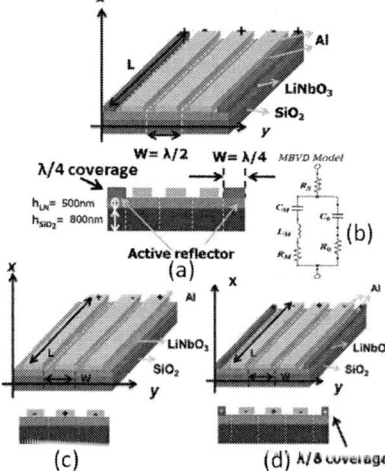

Figure 1: (a), (c) and (d), 3D schematic and cross section views of type A, D and B (see Table I) LVRs, respectively. In (a), characteristic geometrical parameters of 100% covered active reflector design are shown. The equivalent MBVD model of the resonator is displayed in (b).

The resonator equivalent electrical circuit is described by the Modified Butterworth-Van Dyke (MBVD) model (Fig.1). The values of the motional parameters of the MBVD model are given by the following expressions [9]:

$$R_m = \frac{\pi^2}{8}\frac{1}{\omega_s C_0}\frac{1}{k_t^2 Q} \tag{2}$$

$$L_m = \frac{\pi^2}{8}\frac{1}{\omega_s^2 C_0}\frac{1}{k_t^2} \tag{3}$$

$$C_m = \frac{8}{\pi^2}C_0 k_t^2 \tag{4}$$

where C_0 (static capacitance) is a function of finger pitch (w), length (L), and spacing between electrodes (w_s). k_t^2 refers to the effective electromechanical coupling related to the ratio of the motional and static capacitances. Q is the quality factor of the series resonance frequency. In Fig. 1(b) R_s and R_o represent parasitic series and parallel resistances coming from the electrodes and the LN film.

A previously demonstrated LN resonator design [6]

that consisted of a number of evenly spaced IDT metal fingers with identical width (Fig. 1(c), referred as conventional design in this work) introduced unwanted spurious modes and overtone resonances due to the piezoelectric coefficients used to excite vibrations and the material's high intrinsic k_t^2. Different techniques [7-8] have been implemented to help confining energy in a single mode, but suffer from process variations such as misalignment and over-etching. Following the concepts presented in prior demonstrations, in this work active reflectors were introduced for LN/SiO$_2$ LVRs and a systematic approach was used to understand and demonstrate the impact of electroded quarter wavelength ($\lambda/4$) regions at the two ends of the resonant plate (Fig. 1).

Modeling of the electromechanical coupling of LiNbO$_3$ LVRs

Finite element method (FEM) (Eigen frequency simulation) was used to derive the electromechanical coupling factor for specific modes of vibrations as a function of in plane orientation. The coupling coefficient of the laterally vibrating devices is derived by comparing the phase velocity, v_m, of the thin film stack having electrodes patterned on the top surface with the phase velocity, v_0, of an unmetalized surface. The LN and SiO$_2$ thicknesses were set to 500 nm and 800 nm respectively to compensate large temperature coefficient of frequency exhibited by stand-alone LN devices to be around –20 ppm/K.

The 2D FEA of the film stack is performed by imposing periodic boundary conditions at the two ends of the cross section having a width equal to the acoustic wavelength [10]. Note that this simulation assumes infinite periodicity and therefore the effect of the edges is not considered. Through this FEA we can derive the value of k_t^2 as:

$$k^2 \approx \frac{v_0^2 - v_m^2}{v_0^2} \qquad (5)$$

$$k_t^2 = \frac{k^2}{1+k^2} \qquad (6)$$

where k^2 is the intrinsic electromechanical coupling. This technique permits us to rapidly evaluate different material orientations. The FEM data suggests that, a maximum coupling for the S0 mode that exceeds 10 % is found for an in-plane rotation of 60° with respect to the +z axis (see Fig.5).

FEM (frequency response analysis) was also used to optimize the metal coverage ratio for $\lambda/2$ resonator fingers to attain maximum k_t^2. Metal coverage and polarity (signal or ground) of $\lambda/4$ active reflectors were also designed via FEA (frequency response analysis) to attain maximum k_t^2 and minimum sensitivity to misalignment and over/under-etching. Note that in these cases frequency response analysis were conducted in order to take into account the impact of the edges. The k_t^2 is extracted from the simulated admittance response of each configuration as:

$$k_t^2 = \frac{\pi^2}{8} \frac{f_p^2 - f_s^2}{f_p^2} \qquad (7)$$

where f_p and f_s are the resonator series and parallel resonances. The FEA suggested that 50% covered active reflector renders the maximum k_t^2 (13.6%) surpassing by

2X the k_t^2 (6.6%) of conventional design and 100% covered active reflector yield a slightly smaller value (11.5%). However, once subjected to misalignment (0.5 µm ($\lambda/24$) in the simulation), devices with 100% covered active reflector demonstrate the most stable k_t^2 and complete suppression of spurs. Different levels of over-etching were simulated to reveal the robustness of LVR that use 100% covered active reflectors. The analysis shows that for LVR with 100% covered active reflectors, the k_t^2 is practically unaffected by over-etching as long as it is kept below 0.6 µm ($\lambda/20$). Over-etching below 0.1 µm ($\lambda/120$) barely introduces spurious modes. When over-etching becomes larger than 0.5 µm ($\lambda/24$), spurs close to the main resonance peak occur. Additionally, spurs grow seriously once combined with misalignment. The FEA shows that 100% coverage and polarity opposite to adjacent $\lambda/2$ electrode optimize the overall device performance. The simulation results were validated experimentally by building a matrix of devices with different parameters (Table I).

Table I: Characteristic geometrical parameters of the main LN/SiO$_2$ LVRs studied in this work. Type A, B and C refer to active reflector design with 100% (Fig. 1a), 50% (Fig.1 (d)) and no metal coverage (0%), respectively, for resonator vibrating around 500MHz. Type D refers to the conventional design (Fig.1(c)).

Device type	Electrode pitch [µm]	Metal strip width on main finger [µm]	Metal strip width on active reflector[µm]	Electrode aperture [µm]	Electrode coverage of main finger [%]	Electrode coverage of active reflector[%]
A1	6	2	3	80,120	30%	100%
A2	6	3	3	80,120	50%	100%
A3	6	4	3	80,120	67%	100%
B1	6	2	1.5	80,120	30%	50%
C1	6	2	0	80,120	30%	0%
D1	6	2	NA	80,120	30%	NA
D2	6	3	NA	80,120	50%	NA
D3	6	4	NA	80,120	67%	NA

FABRICATION METHOD

As shown in Fig.2, a three mask fabrication process was developed for the making of LN thin film on SiO$_2$ LVRs. IDT fingers formed by 10 nm Cr and 100 nm AlSiCu (~95%Al, <5% Si & Cu) were deposited on the LN thin film by sputtering and patterned by lift-off. Au was deposited on the pad and frame area for reducing series resistance as well as facilitating post processing such as wire bonding. A Cl$_2$/BCl$_3$ - ICP step was conducted to define the boundaries of the resonant structure. Lastly, a XeF$_2$ dry etch is used to remove the Si layer under the composite structure of LN and SiO$_2$.

In this work, The LN ICP etch recipe based on Cl$_2$/BCl$_3$ chemistry that was developed in [11] was re-qualified for anisotropic etching of X-cut LN. SiO$_2$ served as masking layer for LN etching and displayed a good selectivity of 1:2. An Al mask (thickness of 0.3µm) was used to pattern the SiO$_2$ layer for etching of LN and oxide. Thick PR (~4 µm) was used on top of Al during SiO$_2$ etching to obtain a clean etch and straight sidewall angles. Overall, an etch rate around 200nm/min and 80° sidewall angle definition were obtained for LN etching together with a smooth surface free from re-deposition.

Devices having various orientations (90°, ±70°, ±60°,

±50°, ±30°, ±10° with respect to +Z axis) were fabricated on the same die to verify the k_t^2 dependency on orientation. The fabrication resulted in an average of 0.5 μm misalignment and 0.5 μm over-etching in both x and y directions. The SEM image of the active reflector device with 100% coverage is shown in Fig.3.

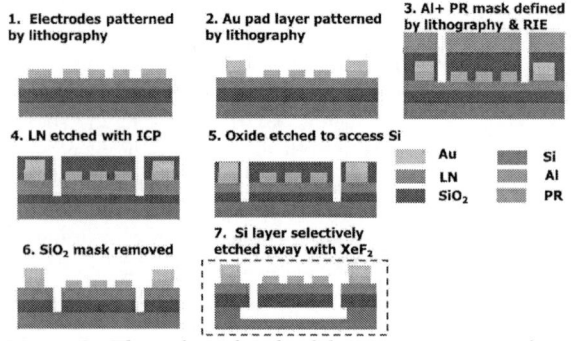

Figure 2: Flow chart for the fabrication process of X-cut ion-sliced LN thin film on SiO_2 MEMS resonators with the final released device circled in blue dashed lines.

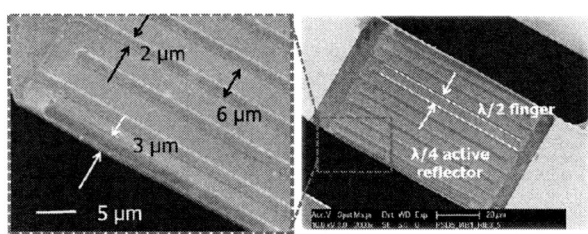

Figure 3: SEM of active reflector design (10 80 μm long electrode and 6 μm pitch with 2 μm metal strips) for device type A1. Active reflectors are highlighted in red.

EXPERIMENTAL RESULTS

The fabricated LN on SiO_2 devices were tested in an RF probe station via an Agilent (N5230A) Network Analyzer. Q was extracted from the admittance plot by measuring the 3dB bandwidth of the resonance peak. The six equivalent circuit parameters were determined by fitting the measured results to the MBVD equivalent circuit model after de-embedding of the parasitic capacitance coming from the pad layout.

Characterization of k_t^2

The designed S0 mode is present in the admittance response of devices placed at all orientations, and is dominant (highest Q in the range of 200 to 1100) when the devices are placed at all orientations except from -50°, -60° and -70° to +z axis. As shown in Fig. 4, a k_t^2 of 12.5% was measured for a resonator of type A1 with center frequency of 489.3 MHz and placed 60° to the +z axis. The average value of the measured k_t^2 for LVRs of type A1 are shown in Fig.5, which displayed the same trend of the k_t^2 estimated with Eqs. (5) and (6), but showed a value that fluctuated within 50% of the predicted one for certain orientations. We believe that the reason of this discrepancy might be that k_t^2 is affected by processing as x-y misalignment errors have a different impact on devices with different orientations and location on the chip. Experimental data also confirms that a maximum coupling for the S0 mode that exceeds 10 % is

obtained for an in-plane rotation of 60° with respect to the +z axis.

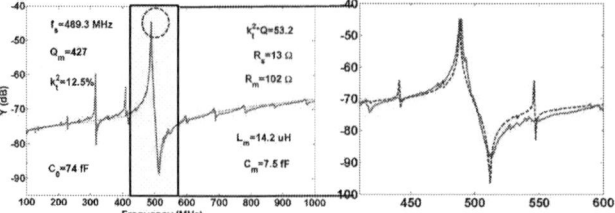

Figure 4: The measured and fitted admittance response for LN on SiO_2 LVR placed at 60° to the +z axis (type A1). The comparison of the measured and 2D COMSOL simulated responses (misalignment of 0.5 μm and over etching of 0.5 μm)for the same devices are shown in the zoomed-in plots on the right.

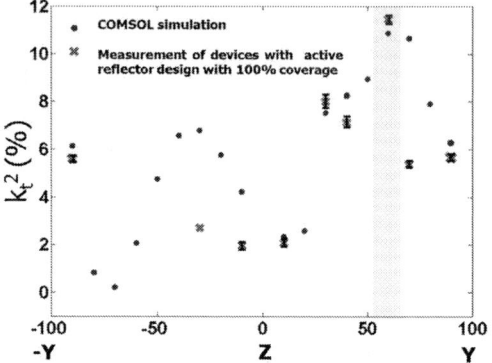

Figure 5: The comparison of the averaged measured and simulated k_t^2 (simulation assumes infinite periodicity and therefore the effect of the edges is not there) for LN on SiO_2 LVRs of type A1 for various orientations.

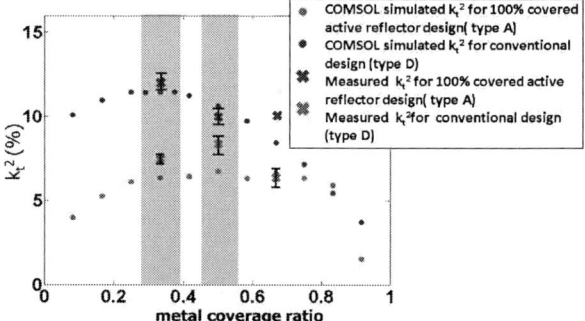

Figure 6: Average experimental measurements validated COMSOL simulation on metal coverage ratio for λ/2 resonator fingers for conventional design (type D) and active reflector (type A).

Devices oriented at -60° to the +z axis exhibited dominant resonance around 300 MHz. For -70°, -50° orientation to +z axis, no working devices are available for measurement since they either exhibited cracks after the XeF$_2$ release due to residual stress in the composite plate or were damaged by misalignment and over-etching during processing.

λ/2 Electrode Coverage Optimization

Metal coverage ratio for λ/2 resonator fingers was explored to maximize k_t^2 for both conventional design and active reflectors design. Devices of type A and D are compared to reveal the dependency of k_t^2 on metal coverage ratio for $\lambda/2$ resonator fingers. The average value of the measured k_t^2 (Fig.6) displayed the same trend of the simulated results for both classes of devices.

Active Reflector Coverage Optimization

The same designed devices with different metal coverage ratio (0%, 50% and 100%) of $\lambda/4$ active reflectors were studied to attain maximum k_t^2 and minimum sensitivity to misalignment and over/under etching. As shown in Fig.7, with the presence of a considerable misalignment (0.5 µm on average in both x and y directions) and over-etching (0.5 µm on average in both directions), 100% covered reflector design is more robust and significantly subdues spurious modes. The dependence of k_t^2 on active reflector coverage was also studied experimentally. The results of this analysis are reported in Table II.

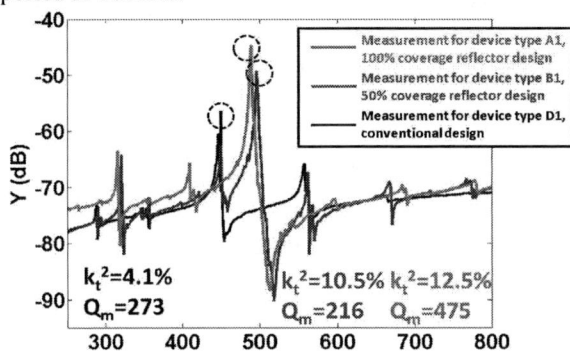

Figure7: The measured responses of LN on SiO$_2$ LVRs vibrating around 500 MHz (120 µm aperture at -60° to the +z axis) of device types A1, B1 and D1. All devices exhibited the same amount of misalignment (0.5 µm in x and y) and over etching (0.5 µm in both directions).

Table II: Comparison of the measured and simulated k_t^2 for LN on SiO$_2$ LVRs (80 µm aperture at -60° to the +z axis)

Device type (# of measured devices)	Active reflector Metal coverage	COMSOL ideal k_t^2	COMSOL with 0.5 µm misalignment and 0.5µm over-etching	Average measured k_t^2
A1(5)	100%	11.5%	10.6%	11.3%
B1(6)	50%	13.6%	11.0%	9.4%
C1(5)	0%	10.7%	10.3%	10.3%

CONCLUSION

This paper presented the design, fabrication and experimental verification of active reflectors for enhancing k_t^2 and suppressing spurious modes in LVRs based on X-cut ion-sliced LN thin films on SiO$_2$. With 800 nm SiO$_2$ added to 500 nm LN, the micromechanical resonators exhibited high k_t^2 up to 12.5%, and Q of 475. The 2X improvement in k_t^2 and complete suppression of in-band spurious vibrations clearly indicate the effect of active reflectors over conventional designs. By selecting an appropriate metal coverage ratio for both active reflector and $\lambda/2$ resonator fingers, a significant spurious mode suppression and robustness to misalignment and

over/under-etching were achieved.

Future work will mainly focus on improving the performance of the resonators. Specifically, attention will be on Q improvement by optimizing the fabrication process, reducing residual stress, and defining resonator anchors that better confine the energy in the targeted resonance.

REFERENCES

[1] R. H. Olsson III, J. Nguyen and T. Pluym, "A Programmable Bandwidth Aluminum Nitride Microresonator Filter," Govt. Microcircuit App. and Critical Tech. Conf., March 2013..

[2] R. Aigner, "Filter Technologies for converged RF-fronted Architectures: SAW, BAW and Beyond", Silicon Monolithic Integrated Circuits in RF Systems (SiRF), 2010.pp136-139

[3] X. Lu, J.Galipeau, K.Mouthaan., E.H. Briot, B.Abbott, " Reconfigurable multiband SAW filters for LTE applications" Radio and Wireless Symposium (RWS), 20-23 January 2013

[4] M. Kadota, T. Ogami, T. Kimura, and K. Daimon, "Tunable Filters Using Wide-Band Elastic Resonators", Trans. Ultrason. Ferroelec. Freq. Cont, 2013, vol.60, pp10.

[5] A. Konno, M. Sumisaka, A. Teshigahara, K. Kano, K.Hashimo, H. Hirano, M. Esashi, M. Kadota, S. Tanaka, "4ScAlN Lamb Wave Resonator in GHz Range Released by XeF$_2$ Etching" in Ultrasonic Symp., July 2013, pp. 1378 – 1381.

[6] S. Gong, L. Shi, and G. Piazza, "High electromechanical coupling resonators using ion sliced X-cut LiNbO$_3$ thin film," MTT-S, June, 2012, pp. 1-3.

[7] S. Gong, and G. Piazza, "Weighted Electrode Configuration for electromechanical coupling enhancement in a New Class of micromachined Lithium Niobate Laterally Vibrating Resonators" IEDM Dec.2012, pp. 165-168.

[8] R. Wang, S. Bhave, and K. Bhattacharjee, "Thin-film high k_t^2Q, multi-frequency lithium niobate resonators,"IEEE MEMS, January 20-24, 2013, pp. 165-168.

[9] G. Piazza, P. Stephanou, and A. Pisano, "Piezoelectric aluminum nitride vibrating Contour-Mode MEMS resonators," J. Microelectromech. Syst.,vol. 15, no. 6, pp. 1406 –1418, December 2006.

[10] J.J.Campbell, W.R.Jones "A method for estimating optimal crystal cuts and propagation directions for excitation of piezoelectric surface waves" IEEE Transactions on Sonics and Ultrasonics, vol. su-15, no. 4, October. 1968

[11] L. Shi and G. Piazza, "Ion-Sliced Lithium Niobate on Silicon Dioxide for Engineering the Temperature Coefficient of Frequency of Laterally Vibrating Resonators", IEEE Intl. Freq. Control Symp., July, 2013,pp.417-420

CONTACT

L. Shi, tel: +1-412-6384585; lishashi@andrew.cmu.edu
G. Piazza, tel: +1-412-268-7762; piazza@ece.cmu.edu

APPLICATION OF STATISTICAL ELEMENT SELECTION TO 3D INTEGRATED ALN MEMS FILTERS FOR PERFORMANCE CORRECTION AND YIELD ENHANCEMENT

Albert Patterson[1], Enes Calayir[1], Gary K. Fedder[1], Gianluca Piazza[1], Bo Woon Soon[2], and Navab Singh[2]

[1]Carnegie Mellon University, USA

[2]Institute of Microelectronics, Agency for Science, Technology and Research (A*STAR), Singapore

ABSTRACT

By 3D integration of an array of 12 nominally identical AlN MEMS sub-filters with a CMOS switching matrix and application of statistical element selection to the same system, we have built a self-healing filter offering 495 unique filter responses and a tuning range of 500 kHz for both center frequency and bandwidth. The demonstrated system enables correction of intrinsic, fabrication-induced variation in filter performance that would otherwise constitute a severe yield limitation to the manufacture of standalone filters.

INTRODUCTION

The advent of MEMS resonators and filters in recent years has brought about exciting possibilities for size and cost reduction as a replacement for more conventional filtering elements as well as a means of achieving new functionality by using MEMS devices as building blocks in future, more complex systems. The adoption of thin-film bulk acoustic resonators (FBAR) over conventional surface acoustic wave (SAW) components [1] serves as validation of RF MEMS as replacement technology, but realizing new functionality is still the subject of much research.

As an example, RF MEMS resonators offer the prospect of designing many devices of varying frequencies on a single chip, which could be applied in software-defined cognitive radio. As described in [2], building a true software-defined cognitive radio without excessive power consumption in the analog-to-digital conversion (ADC) requires front-end narrow-band filtering. Such filtering can be achieved via a programmable frequency gate synthesized from an array of MEMS filters, such as laterally vibrating MEMS resonant filters including electrostatic polysilicon [3], piezoelectric-on-diamond [4] and piezoelectric [5] devices, where the latter is utilized in this work.

However, despite the exciting prospects of RF MEMS devices, both the succession of previous generation technology and the facilitation of future technology are hindered by fabrication induced variations. At the micro and nanoscale, uniformity among devices is very difficult to achieve, resulting in small physical differences that manifest as appreciable variation in electrical properties. To realize the potential impact of RF MEMS devices, it is critical to develop devices of high reliability that are robust against manufacturing variations.

To that end, this work presents experimental results, including tuning ranges and typical response properties, from a RF piezoelectric sub-filter array 3D integrated with a CMOS switch matrix and exploiting statistical element selection (SES) for performance correction. Frequency response reconfigurability and yield improvement vs. standalone filters will be presented, and limitations of the device and design improvements will be discussed.

STATISTICAL ELEMENT SELECTION

In this work, we studied the fabrication-induced performance variation in two-port AlN contour-mode resonator filters. Table 1 shows the measured statistics for the filter center frequency (f_0), insertion loss (*IL*), bandwidth (*BW*) and out of band rejection (*OBR*). While inter-die variations may be corrected through trimming, a method for correcting intra-die variations must be developed in order to make MEMS filters reliable and robust against performance variations.

Table 1: Performance statistics of standalone MEMS filters. 36 total samples measured across 3 dies.

	f_0	IL	BW	OBR
Pooled Mean	1.15 GHz	4.44 dB	3.83 MHz	24.81 dB
Pooled STD	0.46 %	14.72 %	5.91 %	7.68 %
Intra-die STD	0.02 %	14.06 %	1.54 %	2.14 %

To address the intra-die variation, we borrow the concept of SES [6][7], where a subset k of N nominally identical sub-filter elements are selected in parallel to construct a high-yield, self-healing filter. For even a modest N, this method provides a large number of selectable configurations with slightly varying performance due to the small variations within each sub-filter, allowing the filter performance to be tuned into the device specification bounds.

As an example of the impact of variations and the utility of SES, consider that due to just the intra-die variation of the center frequency of MEMS resonator-filters, the yield of the standalone filters would be very low. A typical filter designed to have a center frequency of 1 GHz and a bandwidth of 3.8 MHz and required to be within 100 kHz of the frequency target, would have a yield below 36%. However, application of SES with N of 12 and k of 4 provides 495 unique, selectable frequency responses, resulting in a dramatic increase in yield, as illustrated in Fig. 1. Though this demonstration only considers the center frequency, the methodology readily extends to the other performance properties.

978-1-4799-7956-1/15 $31.00 © 2015 IEEE

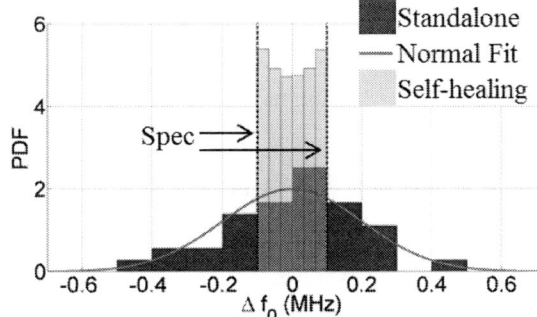

Figure 1: Probability density function (PDF) of measured center frequency offset (Δf_0) of standalone filters vs. simulated distribution for a self-healing filter applying SES to tune the absolute center frequency offset below 100 kHz.

SELF-HEALING FILTER DESIGN

In this work, the self-healing filter is achieved by 3D integration of a CMOS switch array with a bank of 12 nominally identical sub-filters, fabricated in a separate process. The conceptual circuit diagram of the device is shown in Fig. 2.

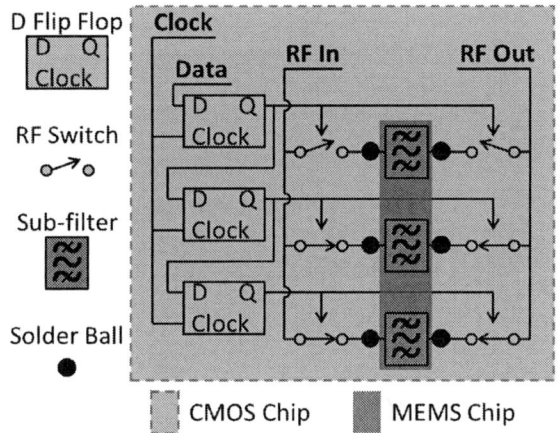

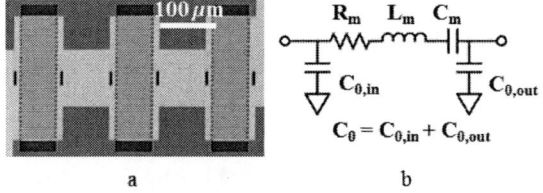

Figure 2: Conceptual circuit of 3D integrated self-healing filter.

The sub-filters in this implementation are self-coupled, two-port AlN laterally vibrating resonators with hermetically sealed thin-film encapsulation (TFE), fabricated at A*STAR Institute of Microelectronics (IME). An image of a sub-filter prior to encapsulation is shown in Fig. 3, along with the equivalent circuit model of each of the three resonators in the sub-filter.

Figure 3: a) Sub-filter element prior to encapsulation. b) Simplified Butterworth-Van Dyke resonator circuit model (feedthrough capacitance is not shown).

In this implementation, a sub-filter center frequency (f_0) of 1.25 GHz was targeted. Also, it was desired to set a static capacitance (C_0) (defined as the sum of the input and output transducer capacitances in [5]) of 1.3 pF to meet 50 Ω impedance conditions, but, due to size constraints, C_0 of 1.1 pF was targeted, which translates to 60 Ω filter impedance for $k = 4$. Geometric design parameters and estimated single resonator equivalent Butterworth-Van Dyke circuit parameters are shown in table 2, where a Q of 1000 and a k_t^2 of 1.9% were assumed based on prior experimental results. Resonator and filter performance calculations are detailed in [5].

Table 2: a) Design parameters for resonators in sub-filter array. b) Approximate resonator circuit parameters. Z_0 is input impedance, R_m, C_m and L_m are the equivalent motional resistance, capacitance and inductance.

Finger pitch, W	Finger length, L	AlN thickness, T	# of fingers on input, n_{in}	# of fingers on output, n_{out}
4 μm	65 μm	1 μm	27	28

a

f_0	C_0	Z_0	R_m	C_m	L_m
1.25 GHz	1.1 pF	240 Ω	30 Ω	4.4 fF	3.7 μH

b

The devices fabricated at IME are shown in Fig. 4, where Fig. 4a shows the cross section of the thin film encapsulated resonator, and Fig. 4b shows the 3D integrated self-healing filter with the probe pads visible at the front of the CMOS chip and solder balls visible at the interface.

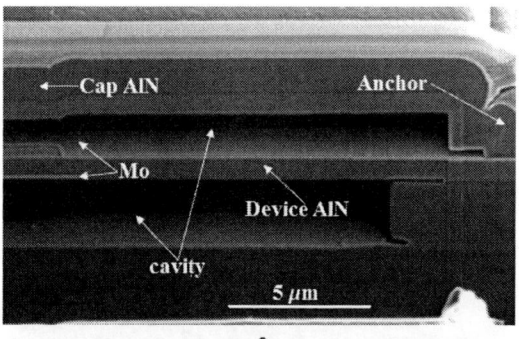

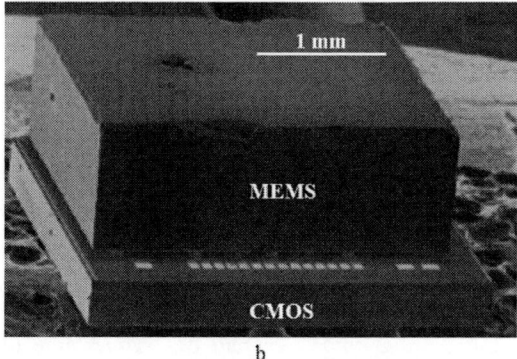

Figure 4: a) Cross section of thin film encapsulated resonator. b) 3D integrated self-healing filter.

978-1-4799-7956-1/15 $31.00 © 2015 IEEE

The number of sub-filter elements, N, was set to 12 in this implementation to achieve over 90% theoretical yield and due to the fixed size of the mated CMOS chip.

FABRICATION

MEMS fabrication was performed at IME, Singapore on 8" high resistivity SOI wafers as schematically shown in Fig. 5. The first step was the definition of the isolation trench, which confined the etching of the Si device layer. Next, the bottom metal layer of 150 nm thick molybdenum was deposited, patterned and planarized, followed by deposition of 1 μm of AlN. After formation of vias to the bottom metal through the AlN, 150 nm of molybdenum was deposited and patterned to form the top metal electrodes, which completed the fabrication of the MEMS resonator filters. To form the TFE, a 4 μm layer of sacrificial amorphous Si was deposited to eventually form the resonator cavity. This was followed by a 1 μm layer of AlN. Removal of the sacrificial Si and SiO_2 followed by deposition of 3 μm of plasma enhanced chemical vapor deposited SiO_2 to seal the encapsulation completed the TFE fabrication. Lastly, 100 nm of Ti and 0.5 μm of Au under bump metallization (UBM) was performed to complete the fabrication process. The wafer was then diced to separate the chips. For 3D heterogeneous integration, the MEMS chip was flipped and bonded onto the solder bumped CMOS chip.

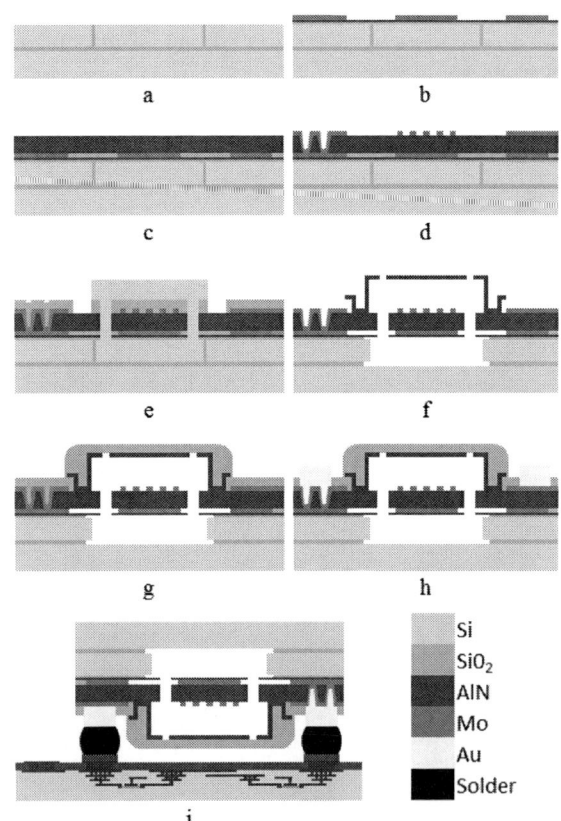

Figure 5: a) Formation of isolation. b) Formation of bottom metal. c) Deposition of AlN. d) Formation of top metal. e) Formation of sacrificial Si. f) Formation of cap and removal of sacrificial Si and SiO_2. g) Deposition of seal. h) UBM. i) Flip chip bonding to CMOS

Fabrication of the CMOS chip was performed in IBM's 65 nm process and solder ball deposition and dicing was performed by Tag and Label Manufacturers Institute (TLMI, Gloucester, MA). The CMOS chip features an array of RF switches, individually addressable through a shift register as well as electrostatic discharge protection and commercial RF probe pads.

TESTING

Due to a non-standard layout of the probe pads, probing of the self-healing filter was performed using a custom mixed signal probe from Cascade Microtech which included pins for power and logic signals for the shift register and a pair of ground-signal RF probes for measurement of the frequency response. Calibration of the mixed signal probe for two-port measurements was performed using a custom script to execute a short-open-load calibration as described in [8]. Logic signals were provided by a DAQ digital acquisition board (National Instruments) for configuring the self-healing filter through the shift registers. This configuration facilitated measurement of multiple frequency responses without lifting the probe simply by applying the appropriate signal through the DAQ.

RESULTS

Frequency response measurements were made for both the self-healing filter and for standalone filters. Fig. 6 shows measurements of three of the available 495 unique frequency responses of the self-healing filter operating at 1.14 GHz with insertion loss of 7 dB and rejection of 16 dB vs. the response of a characteristic standalone filter with insertion loss of 4.4 dB and rejection of 24 dB, where all devices were fabricated in the same fab run. In this implementation, the routing parasitics from the CMOS degraded RF performance. A 5.2 nH shunt inductor was used to partially cancel parasitics, but this can be obviated in future implementations by routing on a low loss RF substrate such as the MEMS chip. In this first implementation, we have demonstrated the availability of many unique frequency responses and the ability to readily select any of them with the self-healing filter.

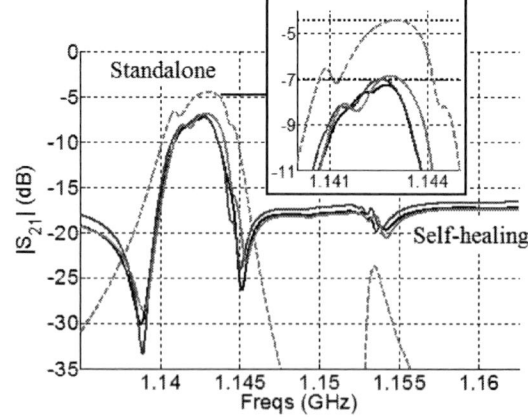

Figure 6: Three of the 495 selectable frequency responses of the self-healing filter vs. the frequency response of a characteristic standalone filter.

978-1-4799-7956-1/15 $31.00 © 2015 IEEE

The fine tuning of properties facilitated by the self-healing filter's many slightly varying frequency responses is illustrated in Fig. 7, which shows the full spread of center frequency offset and bandwidth compared against the listed specifications, demonstrating that both properties may be finely adjusted over a span of around 500 kHz to enhance yield by tuning the self-healing filter into a set of specification bounds. For this implementation, only a subset of frequency responses were measured, and then all other frequency responses were calculated from the measurements.

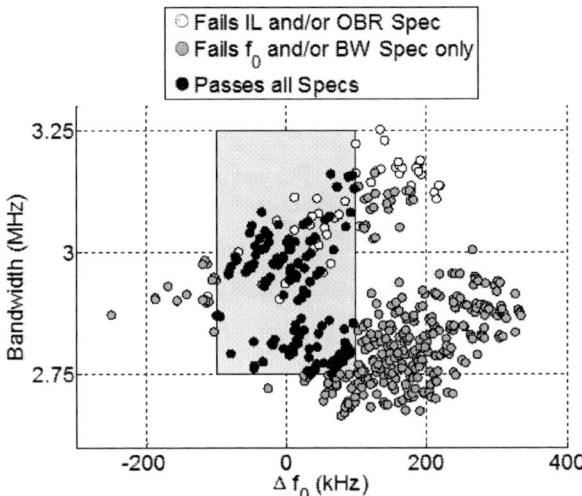

Figure 7: Center frequency offset (Δf_0) and bandwidth of each response available from the self-healing filter, matched to 40 Ω and assuming 5.2 nH shunt inductor. The many black dots indicate a variety of responses that pass the specifications: f_0 within 100 kHz of target, BW between 2.75 MHz and 3.25 MHz, IL below 8 dB and OBR above 15 dB.

CONCLUSIONS

In this work, we have characterized the fabrication-induced variation in the performance of two-port AlN contour-mode filters, demonstrating the substantial hurdle that such variations represent for the implementation of MEMS filters. To counteract these variations and realize high yield and reliability with MEMS filters, we have demonstrated a self-healing filter, capable of finely tuning both the frequency and bandwidth over a range of 500 kHz to adjust performance to within specification bounds. In this initial implementation, design of the routing on the CMOS chip is suboptimal and introduces considerable transmission loss and feedthrough capacitance that degrade filter performance. Nevertheless, we have demonstrated the application of statistical element selection to RF MEMS filters for performance correction and yield enhancement.

ACKNOWLEDGEMENTS

This work was supported by the Intelligence Advanced Research Program Agency (IARPA) and Space and Naval Warfare Systems Center Pacific under Contract No. N66001-12-C-2008. Any opinions, findings and conclusions or recommendations expressed in this material are those of the authors and do not necessarily reflect the views of IARPA and Space and Naval Warfare Systems Center Pacific.

REFERENCES

[1] Ruby, R., "A decade of FBAR success and what is needed for another successful decade," Symposium on Piezoelectricity, Acoustic Waves and Device Applications (SPAWDA), pp.365-369, 9-11 Dec. 2011

[2] Nguyen, C.T.-C., "MEMS-based RF channel selection for true software-defined cognitive radio and low-power sensor communications," *IEEE Communications Magazine*, vol.51, no.4, pp.110-119, April 2013

[3] Clark, J.R.; Wan-Thai Hsu; Abdelmoneum, M.A.; Nguyen, C.T.-C., "High-Q UHF micromechanical radial-contour mode disk resonators," *IEEE J. Microelectromechanical Systems*, vol.14, no.6, pp.1298-1310, Dec. 2005

[4] Fatemi, H.; Zeng, H.; Carlisle, J.A.; Abdolvand, R., "High-Frequency Thin-Film AlN-on-Diamond Lateral–Extensional Resonators," *IEEE J. Microelectromechanical Systems*, vol.22, no.3, pp.678-686, June 2013

[5] Rinaldi, M.; Zuniga, C.; Chengjie Zuo; Piazza, G., "Super-high-frequency two-port AlN contour-mode resonators for RF applications," *IEEE Trans. Ultrasonics, Ferroelectrics, and Frequency Control*, vol.57, no.1, pp.38-45, Jan. 2010

[6] Keskin, G.; Proesel, J.; Pileggi, L., "Statistical modeling and post manufacturing configuration for scaled analog CMOS," IEEE Custom Integrated Circuits Conference (CICC), pp.1-4, 19-22 Sept. 2010

[7] Fa Wang; Keskin, G.; Phelps, A.; Rotner, J.; Xin Li; Fedder, G.K.; Mukherjee, T.; Pileggi, L.T., "Statistical design and optimization for adaptive post-silicon tuning of MEMS filters," 49th ACM/EDAC/IEEE Design Automation Conference (DAC), pp.176-181, 3-7 June 2012

[8] ZhenYu Chen; You Lin Wang; Yu Liu; Ning Hua Zhu, "Two-port calibration of test fixtures with OSL method," Proc. 3rd Int'l Conference Microwave and Millimeter Wave Technology (ICMMT), pp.138-141, 17-19 Aug. 2002

DAMPING IN 1 GHZ LATERALLY-VIBRATING COMPOSITE PIEZOELECTRIC RESONATORS

Jeronimo Segovia-Fernandez, and Gianluca Piazza
Carnegie Mellon University, Pittsburgh, USA

ABSTRACT

This paper focuses on experimentally verifying the physics of damping in 1 GHz laterally-vibrating composite piezoelectric resonators. This work confutes a previously developed theory of interfacial dissipation, a slip phenomenon occurring at the interface between dissimilar materials, which associated damping to a stress jump (or difference in Young's moduli (ΔE)) of the materials forming the interface. This work finds that damping in 1 GHz laterally-vibrating AlN resonators could be attributed to either interfacial dissipation due to an acoustic velocity jump (Δv) or thermoelastic dissipation (TED) in the electrodes.

INTRODUCTION

RF MEMS resonators are the natural replacement for quartz crystals and surface acoustic wave (SAW) devices to build stable frequency references in modern wireless communication systems. Major advantages of the RF MEMS technology with respect to its counterparts are its small form factor and suitability to be integrated with IC technologies [1-7]. On the other hand, most RF MEMS resonators exhibit a limited quality factor (Q), which is a key parameter for improving phase noise and reducing power consumption in oscillators.

Laterally-vibrating composite piezoelectric resonators are an emerging class of RF MEMS transducers whose main advantage compared to other topologies is the capability to fabricate multiple frequency devices on a single chip. Moreover, they have already shown low motional resistance (50 Ω) and the ability to operate at very high resonance frequencies (f_r) (up to several GHz) [8]. In terms of Q, the values that have been reported at these frequencies are generally smaller than those exhibited at lower frequencies [9]. A better understanding of the dominant damping mechanisms that this class of resonators experience around and above 1 GHz is required in order to improve Q.

Previous studies performed on laterally-vibrating AlN resonators have suggested that at 1 GHz dissipation due to both phonon-phonon and electron-phonon interactions can be excluded as main sources of damping. The estimated Q provided for the former at room temperature is 25,000 [10] and the trend of Q versus temperature reported for the latter is opposite to what we have observed experimentally [11]. Additionally, air damping can be ignored considering the large separation (>10s of μm) that exists between the device and its surroundings. In a similar manner, anchor losses, which are significant in most devices operating at lower f_r, can be considered less relevant when the resonator operates at 1 GHz and the acoustic wavelength becomes smaller with respect to the size of the resonator [9]. Therefore, we suspect that the unloaded Q (Q_u) (after excluding electrical losses) of these devices is likely related to interfacial dissipation between the electrodes and the AlN film [12] or thermoelastic dissipation (TED) in the electrodes [11].

This work aims at experimentally investigating the effect of the electrodes on the acoustic damping of 1 GHz laterally-vibrating AlN resonators. For this purpose, devices having three different electrode-to-finger-width ratios (W_m/W_f= {0.75, 0.5, 0.25}) and top electrode metals (Au, Pt, and Al) have been designed, fabricated, and tested. For each electrode configuration, both resonator body dimensions (100x100 μm²) and anchor scheme (fully-anchored) were kept fixed. Special attention was paid to both fabrication and testing steps as a way to obtain reliable results that ensure robust conclusions. The implemented electrode variations allowed us to plot $1/Q_u$ as a function of mechanical, geometrical, and thermal parameters of the resonators and gather further insights in the sources of damping for 1 GHz laterally vibrating composite piezoelectric resonators.

DEVICE DESIGN AND FABRICATION

The principle of operation of laterally-vibrating composite piezoelectric resonator has been introduced in previous papers [7, 8]. All the devices employed in this study are formed by a vibrating AlN plate sandwiched between a bottom metal plate and a patterned top electrode. Interdigitated metal lines connected alternatively to signal and ground voltages form the top electrode. These electrodes generate electric field lines across the piezoelectric film that are confined in the bottom floating plate. The field lines are primarily directed along the thickness of the film producing a lateral mode of vibration (or symmetric lamb wave mode) in the piezoelectric plate (Figure 1).

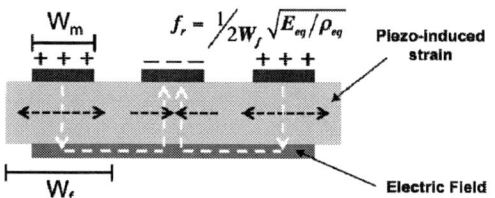

Figure 1: Cross section of a 3-finger laterally-vibrating composite piezoelectric resonator and f_r equation.

The resonance frequency (f_r) of these devices is set by the electrode pitch (or finger width (W_f)) and the equivalent acoustic velocity of the resonator stack ($E_{eq}/\rho_{eq})^{1/2}$. For this work, all the devices were designed to have the same pitch (W_f= 4 μm) which corresponds to a frequency around 1 GHz (exact value depends on metal electrode), AlN plate dimensions (100x100 μm²), and number of fingers (25). Three different W_m/W_f (0.75, 0.5, and 0.25) and electrode metals (Au, Pt, and Al) were used to study the electrodes impact on Q. The selected metals were representative of materials having different electrical, mechanical and thermal properties (Table 1).

978-1-4799-7956-1/15 $31.00 © 2015 IEEE

Table 1: Summary of electrical, mechanical, and thermal properties of the metals used in the study. The resistivity (ρ_e) corresponds to the actual thin film values measured at 10 and 300 K. Instead, the density (ρ_m), both standard (E) and relative variation of adiabatic and isothermal Young's modulus ($\Delta E^T/E$), specific heat capacity (C_m), and thermal conductivity (κ) correspond to the bulk material values found in [11].

	$\rho_e^{10\,K}$ [Ωnm]	$\rho_e^{300\,K}$ [Ωnm]	ρ_m [kg/m³]	E [GPa]	$\Delta E^T/E$ [%]	C_m [J/KgK]	κ [W/mK]
AlN			3260	357		740	70
Au	37.1	55	19300	79	3.3	130	313
Pt	81.6	202	21450	168	3.8	134	67
Al	26	55	2700	70	2	920	238

To exclude the impact of any process variation on damping, all the devices under test were built on the same wafer. The resonant structure is formed by a rectangular AlN plate (1.1 µm-thick) sandwiched between a Pt bottom plate (100 nm-thick) and a top electrode layer (100 nm-thick). The devices are fabricated according to the process sequence shown in Figure 2.

Figure 2: Schematic representation of the 3 mask fabrication process used for the making of AlN resonators with different electrode metals: (a) sputtering and patterning of the bottom metal plate, deposition of the AlN film and dry etching of AlN in Cl₂-based chemistry using an oxide hard mask; (b) sputtering and patterning of the top electrode; (c) release of the AlN resonator in XeF₂.

Figure 3 shows the lumped electrical model of the laterally-vibrating piezoelectric resonator. The equivalent stiffness, damping, and mass of the device are represented by the motional capacitance (C_M), resistance (R_M), and inductance (L_M), respectively. This circuit is known as the modified Butterworth Van-Dyke (mBVD) model as it also includes the parasitic components that account for the electrical losses in both pads and electrodes (R_S), and substrate and the dielectric polarization of AlN (R_0), as the device also behaves as a standard capacitor (C_0).

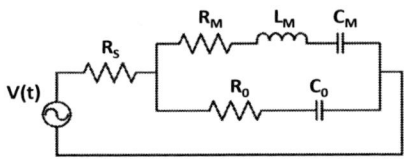

Figure 3: Equivalent mBVD model of AlN resonator including motional and purely electrical components.

EXPERIMENTAL TESTING

Experimental data were collected over a total sample of 72 resonators (24 for each W_m/W_f and electrode material). All the devices tested came from 2 different chips belonging to the same wafer. As a result, 8 identical resonators (4 per chip) were measured to ensure a minimum level of statistics. An Agilent N5230A network analyzer was used to record the electrical response (admittance) of the resonators. At f_r the loaded Q (Q_l) of the device was extracted as the ratio f_r/f_{-3dB}. To exclude

the impact of electrical loading we compare the unloaded Qs of the resonators (Q_u) that were computed as follows:

$$Q_u = \frac{Q_l}{1 - \dfrac{R_S}{R_M + R_S}} \qquad (1)$$

Table 2 shows the averages for the main characteristic parameters of each resonator that was tested at room temperature (300 K). In terms of Q_u a relative variation between the maximum and minimum values (($Q_u^{MAX} - Q_u^{MIN}$)/Q_u^{MIN}) of 310 % is extracted across the sample. This result confirms that the electrodes are one of the main sources of energy dissipation for laterally-vibrating AlN resonators operating at 1 GHz.

Table 2: Summary of the experimental data measured at 300 K for a sample of 72 1 GHz laterally-vibrating AlN resonators.

	W_m/W_f	R_s [Ω]	f_r [MHz]	C_0 [fF]	k_t^2 [%]	Q_l	R_m+R_s [Ω]	Q_u
				AVERAGE				
Au	0.75	1.02	992.76	333	0.96	778	84.32	788
	0.5	1.53	1037.61	244	1.30	1034	57.32	1062
	0.25	3.06	1056.37	180	1.07	1502	65.32	1576
Pt	0.75	3.73	983.14	332	0.86	1247	54.84	1336
	0.5	5.6	1042.69	246	1.04	1451	50.84	1631
	0.25	11.2	1073.35	179	0.76	1651	83.84	1915
Al	0.75	1.02	1072.88	340	1.06	2424	21.32	2584
	0.5	1.53	1078.5	249	1.35	2451	22.32	2632
	0.25	3.06	1082.99	191	1.12	2898	29.32	3234

The previous measurements were repeated at cryogenic (10 K) temperature to investigate the effect of temperature on Q. To lower the sample temperature, a cryogen-free micro-manipulated probe station (model CRX-VF Lakeshore) was employed. Table 3 reports the average and standard deviation of Q_u recorded at 10 and 300 K. Moreover, it shows a relative variation of Q_u that is within the range of 64 to 180 %. This variation demonstrates that the dissipation mechanism concerning the electrodes is temperature dependent and hence of thermoelastic nature. On the other hand, a different trend of Q_u vs. W_m/W_f for Pt and Al is observed at 10 K. In our opinion, this indicates that other loss mechanisms (i.e. anchor losses) start playing a more important role at low temperature. For this reason, we will analyze the data at 300 K and fit them to two proposed models of energy dissipation in the metal electrodes.

Table 3: Average, standard deviation and relative variation of Q_u recorded at 10 and 300 K. At 10 K all the devices with Au and 0.75 coverage stopped functioning and no data could be collected.

	W_m/W_f	Q_u at 10 K		Q_u at 300 K		$\dfrac{AVG^2 - AVG^1}{AVG^1}$
		AVG²	STD DEV	AVG¹	STD DEV	
Au	0.75	NA	NA	788	569	NA
	0.5	2346	817	1062	315	1.21
	0.25	2579	742	1576	239	0.64
Pt	0.75	3741	765	1336	868	1.80
	0.5	2713	1252	1631	524	0.66
	0.25	3343	1559	1915	578	0.75
Al	0.75	6115	2525	2584	564	1.37
	0.5	4685	2886	2632	928	0.78
	0.25	7851	1716	3234	399	1.43

978-1-4799-7956-1/15 $31.00 © 2015 IEEE

POSSIBLE CAUSES OF DISSIPATION

Damping in laterally-vibrating piezoelectric MEMS resonators is not well understood. Either interfacial dissipation or damping in the electrodes have been identified as the main loss mechanisms affecting the Q of this class of resonators. For the former, the stress jump (or ΔE) displayed at the interface between each metal layer and the piezoelectric plate has been considered as the source of interfacial slip [12]. Figure 4 plots the extracted $1/Q_u$ versus ΔE^2, which is proportional to the energy lost (E_{lost}), for different electrode metals at a fixed W_m/W_f. Although there are two metal-to-AlN interfaces in the resonator, the impact on $1/Q_u$ of the common interface (AlN to bottom Pt) is assumed to be equivalent across the entire sample as it should only affect the y-intercept of our fittings (the same applies to other constant loss mechanisms such as anchor losses). Even though this assumption did not prove to be consistent across all samples, we will use it to simplify our analysis. Despite the lack of rigor, we believe this assumption is justified by the correlation values (R^2) exhibited by our linear fittings. By using a simple linear regression to fit the experimental data to the stress jump theory, a weak correlation ($R^2 \sim 0$) is found. These results demonstrate that, for this study, the stress jump at the interface is not the cause of dissipation.

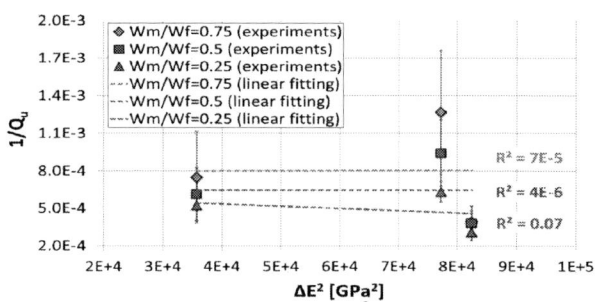

Figure 4: Average $1/Q_u$ vs. ΔE^2 for different W_m/W_f assuming stress jump.

Interfacial dissipation due to velocity jump

To explain the experimental findings and correlate them to interfacial dissipation, we need to postulate an alternative theory that can be explained from the perspective of material science [13]. This theory involves the existence of a relative displacement between the atomic layers of metal and AlN due to their different v $(E/\rho)^{1/2}$. This phenomenon creates an in-plane strain at the interface (ε_{int}), which changes the crystal structure of the interface and produces energy dissipation (Figure 5).

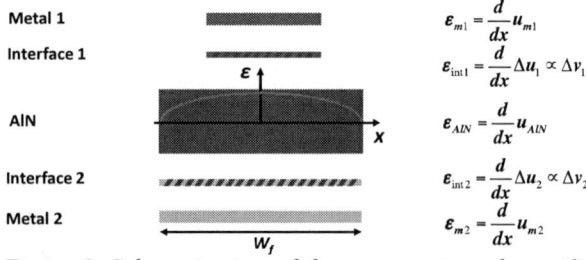

Figure 5: Schematic view of the cross section of one AlN resonator finger that consists of two metal layers (bottom plate and top electrode), one AlN layer, and two solid-solid interfaces formed between them.

As previously formulated in [14], E_{lost} due to interfacial dissipation will be proportional to the following integral:

$$E_{lost} \propto \int_0^T \int_S (\dot{\varepsilon}_{int})^2 \, dS \, dt \qquad (2)$$

where S represents the metal/AlN contact area ($W_m \cdot L$), and T is the period of vibration. Note that $\dot{\varepsilon}_{int}$ represents the time derivative of ε_{int} and makes E_{lost} to be proportional to f_r after integrating over one period of vibration, T (this was verified experimentally in [9]). Considering a 1D lateral mode of vibration to model ε_{int} (see Figure 5) we can express Q_u assuming uniquely interfacial dissipation due to a velocity jump (Δv) (difference in v between AlN and metal)) as follows:

$$\frac{1}{Q_u} \propto \Delta v^2 \left[\frac{W_m}{W_f} + \frac{1}{\pi} \sin\left(\pi \frac{W_m}{W_f} \right) \right] \qquad (3)$$

Figure 6 plots the average $1/Q_u$ vs. Δv^2 for different electrode metals at a fixed W_m/W_f and reveals that, for all the cases, a strong correlation exists between these two parameters ($0.89 < R^2 < 1$). Moreover, Figure 7 plots $1/Q_u$ as a function of W_m/W_f according to Eq. (3) for different electrode metals and W_m/W_f. Hence a strong correlation between $1/Q_u$ and the strain distribution on the top electrode-to-AlN interface is found (R^2 of 0.87).

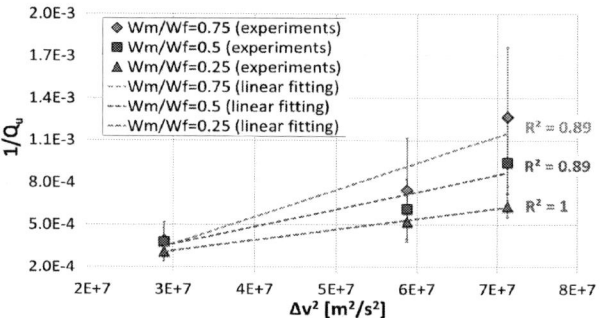

Figure 6: Average $1/Q_u$ vs. Δv^2 for different W_m/W_f assuming interfacial dissipation due to a velocity jump.

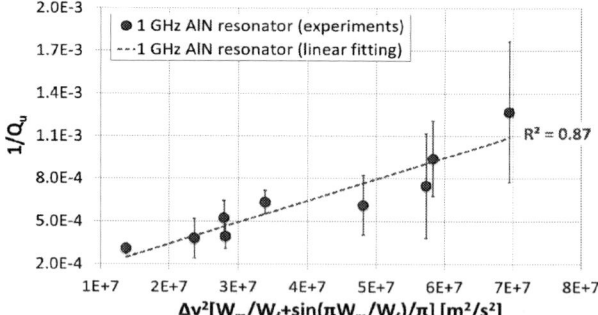

Figure 7: Average $1/Q_u$ vs. the expression in Eq. (3), which describes interfacial dissipation due to a velocity jump.

Thermoelastic dissipation (TED)

TED arises from the nonlinear interactions that occur between acoustic phonons and thermal phonons when a solid is subjected to intrinsic deformations. This damping mechanism becomes more important for propagating longitudinal modes over transversal modes (as they involve a greater change of volume) and for metals over semiconductors (as they present larger κ). The attenuation coefficient for TED (α) can be calculated assuming a standard viscoelastic medium in which there is a propagating longitudinal (1D) mode of vibration and is expressed as follows [11]:

$$\alpha = \frac{\Delta E^T}{2E}\left[\frac{\omega^2\tau^2}{1+\omega^2\tau^2}\right]\frac{1}{\rho_m v \tau} \quad (4)$$

where:

$$\tau = \frac{3\kappa}{C_m v^2} \quad (5)$$

τ is the phonon relaxation time constant and ω is the resonator frequency of excitation. For all the metals under study $\omega\tau$ is $\ll 1$. As a result, Eq. (4) can be combined with Eq. (5) and reduced to the following expression:

$$\alpha = \frac{3}{2}\frac{\Delta E^T}{E}\frac{\omega^2\kappa}{\rho_m C_m v^3} \quad (6)$$

Assuming intrinsic viscoelastic dissipation, the E_{lost} in the electrodes will be proportional to the amount of stored energy in the metal multiplied by α as shown in [15]:

$$E_{lost} \propto 2\lambda\alpha\iiint_V E\varepsilon^2 dV \quad (7)$$

where λ is the wavelength of the mode excited in the resonator (λ is equal to $2W_f$ in this study). Finally, the acoustic damping in the electrodes due to TED can be expressed as a function of thermal, mechanical and geometrical parameters such as follows:

$$\frac{1}{Q_u} \propto \frac{\Delta E^T \kappa}{\rho_m C_m v^3}\left[\frac{W_m}{W_f}+\frac{1}{\pi}\sin\left(\pi\frac{W_m}{W_f}\right)\right] \quad (8)$$

Figure 8 plots the average $1/Q_u$ vs. Eq. (8) using the theoretical parameters shown in Table 1 for all the cases under study. Considering the uncertainty associated with the parameter values used in Eq. (8) a relatively good correlation (R^2 approaching 0.8) between experimental and analytical data is found.

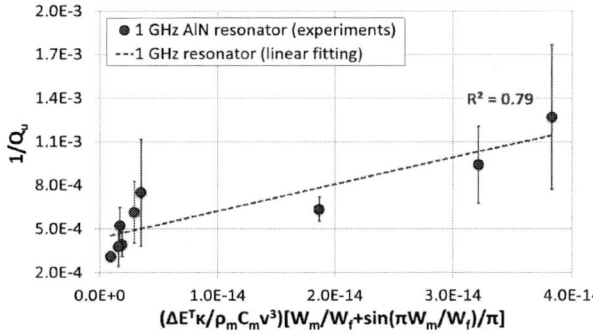

Figure 8: Average $1/Q_u$ vs. Eq. (8) assuming TED.

CONCLUSSION

The reported experimental results demonstrate that there is an impact of the metal electrodes on the Q of 1 GHz laterally-vibrating composite AlN resonators. According to the experiments this electrode dissipation mechanism is also temperature dependent. As a result, two different types of damping of thermoelastic nature are analyzed: interfacial dissipation between electrodes and AlN plate and thermoelastic dissipation in the electrodes (TED). By fitting the experimental Qs to the difference in either Young's moduli or acoustic velocities at the interface we conclude that interfacial dissipation (if accepted as the main physical phenomenon of damping) cannot be caused by a stress jump but rather a velocity jump. In parallel, we use the TED theory to analyze the electrode damping mechanism and find that a good agreement exists between theory and experimental data. Further analysis on different resonator/electrode geometries are required to unveil the source of damping and will be the subject of future work.

ACKNOWLEDGEMENTS

The authors would like to thank the DEFYS DARPA contract # FA86501217624 for funding.

REFERENCES

[1] R. Wang, S. A. Bhave, and K. Bhattacharjee, *IEEE Int. Conf. on MEMS*, pp. 165-8, Jan 2013.

[2] H. Fatemi, H. Zeng, J. A. Carlisle, and R. Abdolvand, *Journal of MEMS*, vol. 22, no. 3, pp. 678-86, 2013.

[3] J. L. Fu, R. Tabrizian, and F. Ayazi, *IEEE Trans. on Electron Devices*, vol. 61, no. 2, pp. 591-7, 2014.

[4] T.-T. Yen, A. P. Pisano, C.T.-C. Nguyen, *IEEE Int. Conf. on MEMS*, pp. 114-7, Jan. 2013.

[5] L.C. Popa, D. Weinstein, *IEEE Frequency Control Symp.* May 2014.

[6] A. Ansari and M. Rais-Zadeh, *IEEE Trans. on Electron Devices*, vol. 61, no. 4, pp. 1006–13, 2014.

[7] G. Piazza, P. J. Stephanou, A. P. Pisano, *Journal of MEMS*, vol. 15, no. 6, pp. 1406-18, 2006.

[8] M. Rinaldi, C. Zuniga, C. Zuo, and G. Piazza, *IEEE Trans. on Ultrasonics, Ferroelectrics, and Frequency Control*, vol. 57, n. 1, pp. 38-45, 2010.

[9] J. Segovia-Fernandez, M. Cremonesi, C. Cassella, A. Frangi, and G. Piazza, *Transducers*, 2013.

[10] S. A. Chandorkar, M. Agarwal, R. Melamud, R. N. Candler, K. E. Goodson and T. W. Kenny, *IEEE Int. Conf. on MEMS*, pp. 74-77, Jan 2008.

[11] M. T. Wauk, Ph.D. dissertation, Dept. of Applied Physics, Stanford University, CA, 1969.

[12] Z. Hao, and B. Liao, *Sensors and Actuators A: Physical*, vol. 163, no. 1, pp. 401-9, 2010.

[13] I.N. Mastorakos, H.M. Zbib, D.F. Bahr, *Applied Physics Letters* 94, 173114 (2009).

[14] A. Frangi, M. Cremonesi, A. Jaakkola, T. Puensala, *IEEE Int. Ultrasonics Symp.*, Oct. 2012.

[15] B. A. Auld, *Acoustic Fields and Waves in Solids, 2nd ed.*, Krieger publishing company, 1990.

CONTACT

*J. Segovia-Fernandez; jsegovia@andrew.cmu.edu

DUAL-CLOCK WITH SINGLE AND MONOLITHICAL 0-LEVEL VACUUM PACKAGED MEMS-on-CMOS RESONATOR

A. Uranga[1], G. Sobreviela[1], N. Barniol[1], E. Marigó[2], C. Tay-Wee-Song[2], M. Shunmugam[2],
A. A. Zainuddin[2], A. Kumar-Kantimahanti[2], V. Madhaven[2] and M. Soundara-Pandian[2]

[1]Universitat Autònoma de Barcelona, SPAIN
[2]Silterra Malaysia Sdn. Bhd., MALAYSIA

ABSTRACT

This paper demonstrates the feasibility of a novel fabrication approach of MEMS resonators above standard CMOS circuitry and with zero-level vacuum package. As a proof of concept a monolithical CMOS-MEMS-closed loop oscillator showing dual-clock capabilities (11.9 MHz and 24.5 MHz) is presented. These two frequencies correspond to two different resonator modes, specifically the torsional and vertical out of plane, of a paddle shaped MEMS resonator.

INTRODUCTION

The implementation of microelectromechanical devices is replacing the quartz oscillator as frequency references due to its smaller size and reduced power consumption and production cost [1-6]. Additionally, multifrequency reconfigurable oscillators are being demanded nowadays to fulfill the multiband RF front-ends specifications [7].

Different examples of MEMS oscillators can be found in the literature, but most of them lacks of an easy CMOS integration at chip level, where both the MEMS and the circuitry share the same substrate. The system is formed by two independent dies that are wire-bonded together.

Opposite to this approach, the monolithical integration offers as a result a single die with both the MEMS and the CMOS circuitry. According to when the MEMS is implemented three solutions are distinguished: MEMS-first, MEMS-last and Intra-CMOS solution [2]. The use of an Intra-CMOS implementation, through the use of the inherent layers of the CMOS process, offers as a benefit a good matching with the interfacing circuitry, an easy and fast MEMS implementation with a good yield, although a reduction of the properties of the MEMS is obtained, due to the limitations of the materials that form the MEMS. Several groups have been reported on CMOS-MEMS oscillators following this approach [2, 4, 6].

As an example of a MEMS-first, in [3] a single-crystal silicon NEMS resonator is monolithically co-integrated with a CMOS circuitry to develop an oscillator. The solution requires several challenges in terms of development of a technological platform that makes compatible the deposition of the silicon MEMS and the following standard CMOS technology (without any process modification).

As a third solution, the MEMS can be stacked over the CMOS die (MEMS-last or MEMS-on-CMOS solution), opening the possibilities for the material selection. However, thermal and process compatibility should be assured. As an example, the use of AlN films is well known to implement piezoelectric resonators as a part of the oscillator [8]. Moving to capacitively sensed resonators in [5] a nickel micromechanical resonator disk array over finished foundry 0.35 μm CMOS circuitry is reported. Recently, Silicon Labs company offers a commercial MEMS-based oscillator based on a technology (CMEMS) that allows the integration of Poly-SiGe resonators on top of a 0.13 μm CMOS substrate [9]. Our approach is aimed to develop a design process of MEMS on top of a commercial CMOS technology (Silterra 0.18um), that can be easily exported to any other technological node. Moreover, we will take advantage of the process in order to provide with a zero-level package at wafer level. As a proof of concept the MEMS will be embedded into a sustaining amplifier that will allow to obtain a dual clock.

MEMS FABRICATION PROCESS

The main focus in our integration scheme is to simplify the manufacturing technique and achieve a well-controlled nano-gap. The reduction of the gap implies an increase of the electromechanical coupling factor and, in terms of implementation of an oscillator a reduction of the motional resistance that the sustaining amplifier needs to compensate, alleviating the circuitry performance. Different MEMS structures have been implemented.

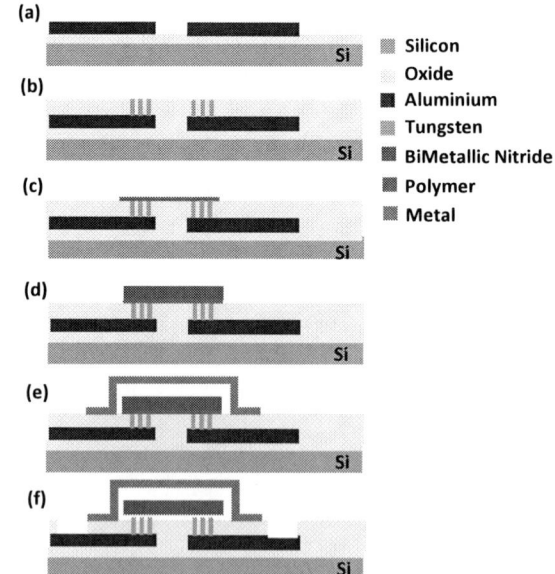

Figure 1: a) Deposit and pattern CMOS top metal layer b) Via plug process c) Sacrificial polymer coat, pattern and etch d) Deposit and pattern resonator layer e) Release and zero-level packaging f) Release and bond pad.

978-1-4799-7956-1/15 $31.00 © 2015 IEEE

The fabrication process of the resonator is shown in Fig. 1. Electrodes were fabricated using the "via" layer of the technology. A sacrificial polymer is coated, patterned and thinned down to 90nm. The structural layer is a bi-metallic thin film of 500nm. The nano-gap is formed by ashing the sacrificial layer. Vertical gaps up to 25 nm have been achieved. Finally, a metallic cover is implemented on top of the MEMS to assure a hermetic vacuum sealing of the structure. Figure 2 shows a cross-sectional SEM view of the final device where the package, on top of the resonator is observed.

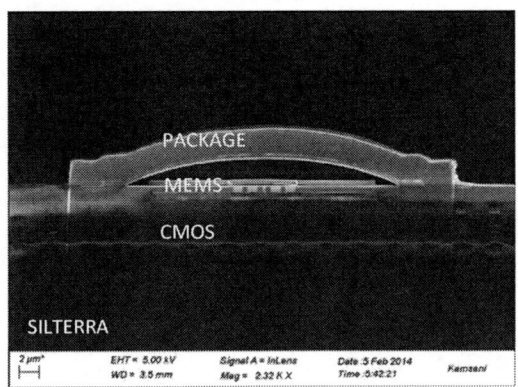

Figure 2: X-SEM view of a fabricated device showing vacuum package.

SUSTAINED AMPLIFIER DESIGN

An integrated amplifier has been implemented in order to allow self-sustained oscillations. The circuitry is based on a low noise CMOS transimpedance amplifier (TIA) in charge of integrating the motional current of the MEMS resonator followed by a 50Ω buffer. The resonator is electrostatically excited and, in the closed loop configuration, the excitation comes from the amplifier output (see figure 3). The gain of the amplifier has been designed to reach the oscillation conditions. In particular, taking into account the input capacitance of the amplifier, $135dB\Omega$ @24MHz transimpedance gain is achieved.

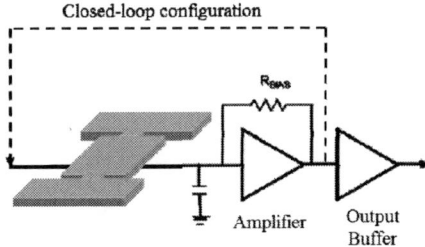

Figure 3: Schema of the implemented full system, both in open and closed-loop configuration. For the open configuration the MEMS is externally excited while the output of the amplifier is feedback to achieve self-oscillation

The circuit has been implemented using the 0.18 um CMOS commercial technology from Silterra Malaysia Sdn Bhd. Figure 4 shows an optical image of the MEMS

and the sustaining amplifier system. It can be seen how the MEMS and the circuit are monolithically integrated on the same die.

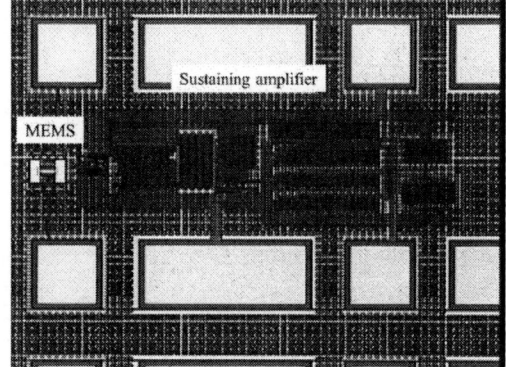

Figure 4: Optical image of the CMOS-MEMS oscillator.

EXPERIMENTAL RESULTS
Characterization of the MEMS

The frequency response of a stand-alone resonator has been characterized for different vacuum level conditions. In particular a comb shaped MEMS resonator, without any zero-level package has been chosen. Electrical characterization has been performed using a network analyzer (Agilent E5100A). Fig 5 shows the evolution of the quality factor, Q, with the vacuum level of the resonator. A Q value around 1200 is achieved when the vacuum reaches 1 mbar and below.

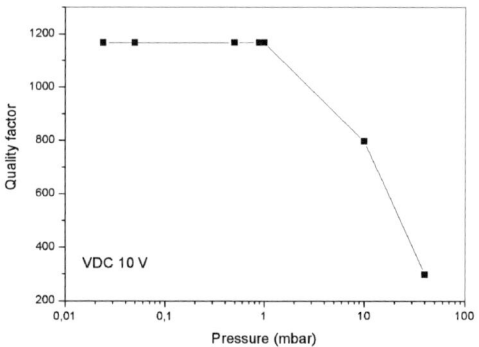

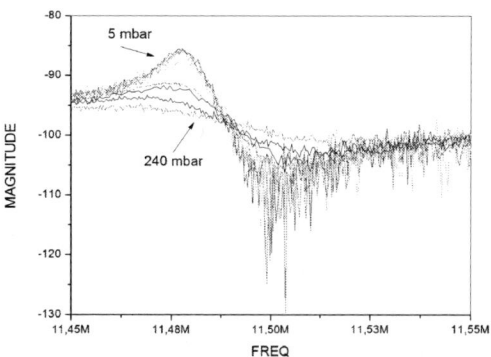

Figure 5: (a) Measured evolution of the quality factor (Q) (a) and frequency response (b) of a comb-shaped resonator with the vacuum level

978-1-4799-7956-1/15 $31.00 © 2015 IEEE

A paddle shaped out-of-plane resonator device has been chosen to implement the oscillator. This structure presents two resonant modes at 11.9 MHz and 24.5 MHz that correspond to the torsional and vertical movement. Figure 6 shows a SEM image of the resonator with main dimensions.

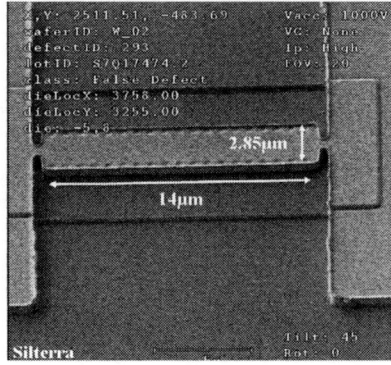

Figure 6: SEM image of the implemented MEMS paddle resonator without vacuum packaging.

The frequency response of the two resonant modes of the packaged resonator has been characterized under atmospheric pressure to compare with the quality factor of the previous results. A DC voltage of 15V (vertical resonant mode, fig 7) and 20 V (torsional mode, fig 8) has been applied to the paddle with 90nm of vertical gap. The AC excitation voltage is applied to the excitation electrode, while the electrical readout signal is acquired from the readout electrode being amplified through the implemented TIA.

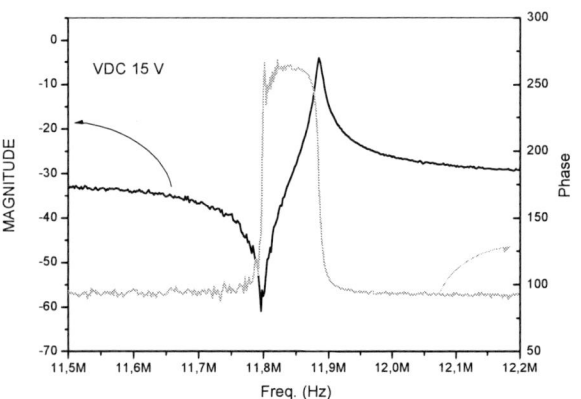

Figure 7: Measured frequency response of the MEMS paddle resonator showing the vertical resonant mode

Figures 7 and 8 show both, the magnitude and phase response of the open loop system for the vertical and torsional modes. A quality factor of 1500 is achieved in both modes, which assures that the MEMS is working under vacuum conditions thanks to the implemented zero level package. Compared with previous CMOS-MEMS oscillators [2] this new MEMS-on-CMOS resonator presents high Q*f product (3.6 *10^{10}) with small system area (180μm x 60μm) including packaging.

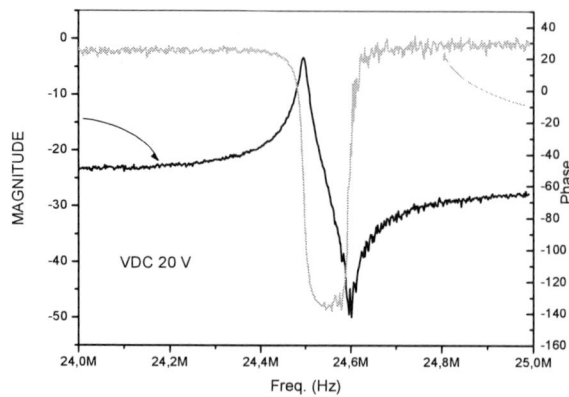

Figure 8: Measured frequency response of the MEMS paddle resonator showing the torsional resonator mode.

Taking into account the buffer losses (-6 dB) both torsional and vertical modes accomplish the Barkhausen criteria in magnitude but only the torsional mode reaches the 0° phase condition.

Characterization of the MEMS oscillator

The closed loop design has been tested to corroborate the self-sustained oscillation. Taking into account the previous open loop results, a DC voltage of 20V has been applied to the paddle while, in this case, the amplifier output is directly connected to the excitation electrode (on-chip close loop oscillator). Only the buffer output is connected to the oscilloscope or to the spectrum analyzer using a bias tee for the electrical characterization.

Figure 9 plots the measured oscillator time domain response, proving the viability of the system to achieve oscillation with a polarization voltage of 20V. It can be seen how the oscillation frequency corresponds to the torsional resonant frequency of the structure, where both the magnitude and phase response accomplish the Barkhausen criteria.

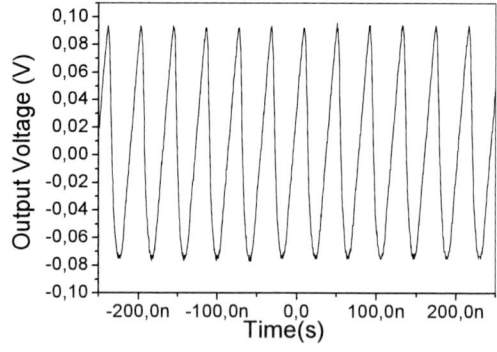

Figure 9: Measured time domain response of the oscillator acquired using an oscilloscope.

The measured output power spectrum along with oscillator phase noise is showed in figure 10. A phase noise of -70 dBc/Hz is achieved at 1 kHz, while the floor noise corresponds to -120 dBc/Hz.

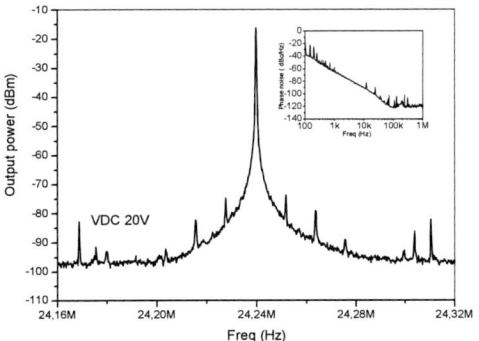

Figure 10: Measured output power spectrum of the MEMS paddle oscillator in torsional mode where the inset corresponds to the measured phase-noise.

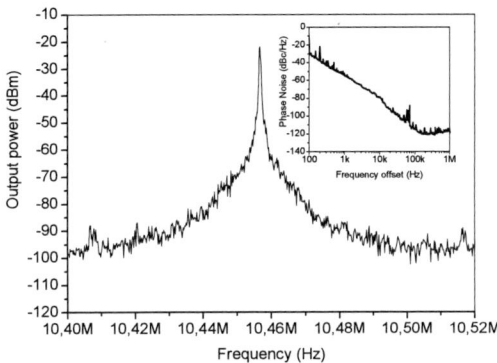

Figure 11: Measured output power spectrum response and phase noise of the vertical mode of the MEMS paddle oscillator. An external phase shifter has been used to allow to close the loop.

The vertical resonant mode has been also proved, using an external phase-shifter to close the loop. Figure 11, presents the power spectrum response and phase noise, corresponding to the vertical resonant mode (10.46 MHz.)

Thanks to the possibility to excite these two resonant modes, the system opens the possibility to easily obtain a dual clock from the same resonator (see block diagram in Figure 12). In this case, only an inverter is needed to add the 180° phase that accomplish the Barkhausen criteria.

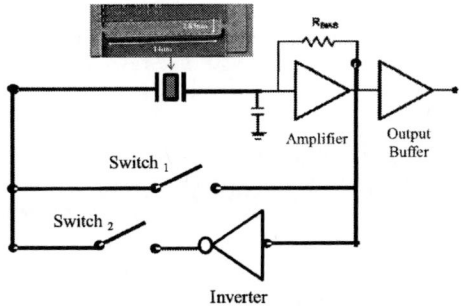

Figure 12: Block diagram of the dual clock. The inverter allows to achieve oscillation in the vertical mode (adding 180 °)

CONCLUSIONS

We have provided a simple and cost-effective MEMS manufacturing technique that allows to design a MEMS on top of a CMOS chip, achieving nano-gaps as small as 25 nm. Moreover, the process allows to implement a zero-level package at wafer level.

A paddle shaped resonator has been built on top of a sustaining amplifier that allows to get a dual clock, taking profit of the two vertical and torsional resonant modes. Therefore, an easy reconfigurable oscillator in terms of resonant frequency is proposed.

REFERENCES

[1] J. T. M. van Beek and R. Puers. A review of MEMS oscillators for frequency reference and timing applications. J. Micromech. Microeng. 22, 2012 013001 (35pp) doi:10.1088/0960-1317/22/1/013001

[2] A. Uranga, J. Verd and N. Barniol, "CMOS–MEMS resonators: From devices to applications", Microelectronic Engineering 132, pp. 58–73, 2015

[3] J. Philippe, G. Arndt, E. Colinet, M. Savoye, T. Ernst, E. Ollier and J. Arcamone, "Fully monolithic and ultra-compact NEMS-CMOS self-oscillator based-on single-crystal silicon resonators and low-cost CMOS circuitry", in *Int. Conf. on Micro Electro Mechanical Systems*, San Francisco, January 26-30, 2014, pp. 1071 - 1074

[4] M-H Li, C-Y Chen, C-S Li, C-H Chin, C-C Chen, and S-S Li, "Foundry-CMOS Integrated Oscillator Circuits Based on Ultra-low Power Ovenized CMOS-MEMS Resonators" in Int. Electron Devices Meeting, December 9-11, 2013, pp. 18.4.1 - 18.4.4.

[5] W-L Huang, R. Zeying, L. Yu-Wei, C. Hsien-Yeh, J. Lahann and C.T.C. Nguyen, "Fully Monolithic CMOS Nickel micromechanical resonator oscillator" in Int. Conf. on Micro Electro Mechanical Systems, Tucson, January 13-17, 2008, pp 10–13

[6] J. Verd, A. Uranga, J. Segura, N. Barniol. "A 3V CMOS-MEMS oscillator in a 0.35 um CMOS technology" Int. Conf. on Solid-State Sensors, Actuators and Microsystems, Transducers'2013, pp. 806-809.

[7] C. Zuo, J. Van der Spiegel and G. Piazza "Dual-Mode Resonator and Switchless Reconfigurable Oscillator Based on Piezoelectric AlN MEMS Technology", IEEE Tran. on Elect. Dev.,vol. 58, n°. 10, pp. 3599-3603, 2011

[8] K. E. Wojciechowski, R. H. Olsson III, M. R. Tuck, E. Roherty-Osmun and T. A. Hill, "Single-chip precision oscillators based on multi-frequency, high-Q aluminum nitride MEMS resonators," in TRANSDUCERS 2009 - 15th International Conference on Solid-State Sensors, Actuators and Microsystems, 2009, pp. 2126-2130.

[9] Silicon Labs white paper "CMEMS®Technology: Leveraging High-Volume CMOS Manufacturing for MEMS-Based Frequency Control" Retrieved November 2014, http:// www.silabs.com .

CONTACT

* arantxa.uranga@uab.cat

EXPERIMENTAL INVESTIGATION ON MODE COUPLING OF BULK MODE SILICON MEMS RESONATORS

*Yushi Yang[1], Eldwin Ng[1], Pavel Polunin[2], Yunhan Chen[1], Scott Strachan[2], Vu Hong[1],
Chae Hyuck Ahn[1], Ori Shoshani[2], Steven Shaw[2], Mark Dykman[2], and Thomas Kenny[1]*
[1] Stanford University, CA, USA   [2] Michigan State University, MI, USA

ABSTRACT

This paper demonstrates the effect of nonlinear elasticity on the coupling between different bulk modes of silicon MEMS resonators. From experimental data, we observe that the coupling has a strong dependence on the resonant mode order, the mode shape of the coupled modes, as well as the doping type / concentration, and crystal orientation, leading to a variety of complex and potentially useful phenomena.

INTRODUCTION

Nonlinear mode coupling arises from the strain induced within a resonator when one mode is excited to large vibration amplitude and the large strain affects the other modes that are simultaneously driven. To explore the effect of mode coupling, two resonant modes are excited simultaneously, of which one can be denoted as the 'driven mode' (DR) and the other as 'detection mode' (DET). When the DR mode is excited to large amplitude, a frequency shift of the DET mode, which is proportional to the square of the DR mode's displacement amplitude, can be observed.

Previous research has analyzed the effect of nonlinear mode coupling in NEMS clamped-clamped beams for both in-plane and out-of-plane flexural modes, and it was quantitatively explained as a geometrical stiffening effect [1-4]. Such effects can be useful for understanding resonator parameters, as well as providing tunability for resonant frequencies and quality factor (Q) [5]. The nonlinear coupling effect has also been observed in bulk mode resonators [6], where material effects play a much larger role than geometrical effects. This work will extend previous research and discuss the effect of doping on the mode coupling behavior for bulk mode silicon resonators.

This paper builds on the work presented in [7], where the nonlinear elasticity of silicon was observed to be doping dependent. In this paper, we will focus on analyzing the effect of doping, device orientation, and order of resonant mode on the nonlinear mode coupling effect in bulk mode MEMS resonators.

DESIGN AND FABRICATION

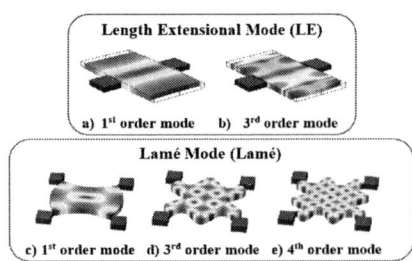

Figure 1: Simulated mode shapes for the LE mode resonators (a - b) and Lamé mode resonator (c - e).

Two types of bulk mode resonators are used in this study: length extensional resonators (LE) and Lamé mode resonators (Lamé). The simulated mode shapes are shown in Fig. 1 and the resonator parameters are listed in Table 1.

Table 1: Resonator design parameters

	LE	Lamé
Length (L)	600μm	300μm
Width (W)	300μm	
Height (h)	40μm	17μm
Gap size (d)	0.7μm	0.2μm (pull-in)
Device Orientation	<100>, <110>	<100>

To examine the resonant modes more carefully, the measured frequency responses for these devices are plotted in Fig. 2. It can be seen that the devices exhibits few spurious modes in this frequency range. For each resonator, we chose the 1st order mode as the DR mode due to the relatively higher quality factor, and their respective higher order mode as the DET mode.

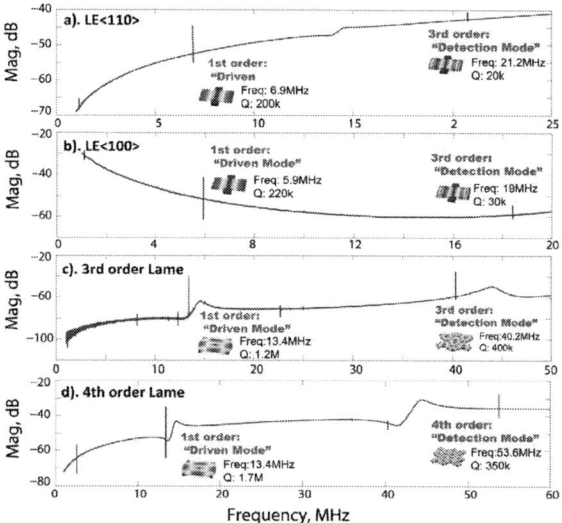

Figure 2: Frequency responses showing modes of interest for the LE (a, b) and Lamé resonators (c, d).

The LE resonators are fabricated using a wafer-level encapsulation process (*epi-seal*), where a detailed description of the fabrication process can be found in [8]. The *epi-seal* encapsulation process was proposed by researchers at the Robert Bosch Research and Technology Center in Palo Alto and then demonstrated in a close collaboration with Stanford University. This collaboration is continuing to develop improvements and extensions to this process for many applications, while the baseline process has been brought into commercial production by SiTime Inc., recently acquired by

978-1-4799-7956-1/15 $31.00 © 2015 IEEE

Megachips Inc. To study the effect of doping and device orientation on mode coupling, four different doping types / concentrations are investigated, with the measured device layer resistivity listed as follows: Phosphorus (*N-H*) 1.78mΩ-cm, Antimony (*N-L*) 17.1mΩ-cm, Boron (*P-H*) 0.7mΩ-cm and Boron (*P-L*) 15.8mΩ-cm. The Lamé resonators were fabricated in the process presented in [9], where specially designed pull-in electrodes are used to excite the 3rd and 4th order resonant modes. The device layer resistivity is measured to be 0.7mΩ-cm (Boron doped).

RESONATOR MODELING

We begin the analysis of the nonlinear coupling between oscillator modes by modeling the system of interest as two coupled Duffing resonators, one for each mode, with equations of motion

$$m_i\ddot{x}_i + 2c_i\dot{x}_i + k_{0i}x_i + k_{1i}x_i^3 + \frac{\partial U_{int}}{\partial x_i} = F_i, \quad (1)$$

where x_i is a modal amplitude, index $i = 1, 2$ denotes the DR and DET modes respectively, m_i is the effective (modal) mass, c_i is the linear modal damping coefficient, k_{ji} are the linear and nonlinear stiffness constants, F_i is the force exciting i^{th} mode and $U_{int} = U_{int}(x_1, x_2)$ is the interaction potential which is given by

$$U_{int}(x_1, x_2) = \frac{1}{2}\alpha_1 x_1^2 x_2^2 \quad (2)$$

Note that there are many possible interaction terms; here we focus on the so-called dispersive coupling term that is important in cases without internal resonance [11–13]. Both symmetric (quartic in potential) and non-symmetric (cubic in potential) terms can contribute to this coupling, but these combine to give the same effect described by the present model. When internal resonances occur, with frequency ratios 1:n for n=1,2,3,4, additional nonlinear terms of different orders must be included in the model and the analysis becomes more complicated [13].

The frequency-amplitude dependence of Eq. (1) can be determined by considering the dynamics of the undamped, unforced system [13]. Taking $c_i = 0$ and $F_i = 0$ and rescaling, the dynamics of the DR and DET modes are described by

$$\ddot{x}_1 + \omega_{01}^2 x_1 + \gamma_1 x_1^3 + \beta_1 x_1 x_2^2 = 0, \quad (3)$$

$$\ddot{x}_2 + \omega_{02}^2 x_2 + \gamma_2 x_2^3 + \beta_1 m_{12} x_1^2 x_2 = 0, \quad (4)$$

where $\beta_1 = \frac{\alpha_1 \omega_{01}^2}{k_{01}}$ and $m_{12} = \frac{m_1}{m_2} = \frac{k_{01}}{k_{02}}\frac{\omega_{02}^2}{\omega_{01}^2}$ is the effective mass ratio.

In the case of weak nonlinearities, perturbation methods can be applied [13]. Employing frequency relations $\omega_{02} \approx n\omega_{01}, n = 3,4$, we obtain expressions for frequency shifts ($\delta\omega_i \equiv \omega_i - \omega_{0i}$) in the DR and DET modes as

$$\delta\omega_1 = \frac{3}{8}\frac{\gamma_1}{\omega_{01}}A_1^2 + \frac{1}{4}\frac{\beta_1}{\omega_{01}}A_2^2 \quad (5)$$

$$\delta\omega_2 = \frac{3}{8}\frac{\gamma_2}{\omega_{02}}A_2^2 + \frac{1}{4}\frac{\beta_1 m_{12}}{\omega_{02}}A_1^2 \quad (6)$$

Eqs. (5) and (6) reveal that the frequency of each mode is indeed a function of the amplitudes of both modes and can increase or decrease depending on the signs and magnitudes of coefficients γ_j, β_j and amplitudes A_i.

In the experiments, we excite the DR mode to a large amplitude relative to that of the DET mode ($A_1 \gg A_2$). Thus, Eqs. (5)-(6) simplify as

$$\delta\omega_2 \cong \frac{1}{4}\frac{m_{12}}{\omega_{02}}A_1^2\beta_1 = KA_1^2, \quad (7)$$

$$\frac{\delta\omega_2}{\delta\omega_1} \cong \frac{2}{3}\frac{m_{12}}{\gamma_1}\frac{\omega_{01}}{\omega_{02}}\beta_1 = \lambda. \quad (8)$$

It is clear from Eqs. (7) and (8) that the frequency shift of the DET mode is proportional to the squared amplitude of the DR mode, and that the ratio between the frequency shifts of the DET and DR modes is proportional to the relative mode order, $\frac{\omega_{02}}{\omega_{01}}$. This result is very useful as it allows one to estimate the strength of the modal coupling $\propto x_1^2 x_2^2$.

EXPERIMENT SETUP

The experiment builds on the concept demonstrated in [14], in which by varying the operating phase, stable oscillation beyond the critical bifurcation point can be obtained. Using this concept, this paper adapts the experimental method presented in [7] with slight variations. The setup used for characterizing these resonators is shown in Fig. 3. Dual phase-lock loops, which are independent of each other on the Zurich HF2LI-PLL, are used to simultaneously excite and monitor the responses of both the DR and DET modes separately. A DC voltage is applied on the resonator body to bias the device. To combine the two PLL output signals, the AC output signals from the two PLLs are summed using a 0° combiner, and the combined output signal is then used to excite the resonator. The resonator output, after amplification through a transimpedance amplifier, is divided using a splitter and fed back to the two PLLs individually, and internal filters within the PLL are used to obtain the corresponding frequency signal.

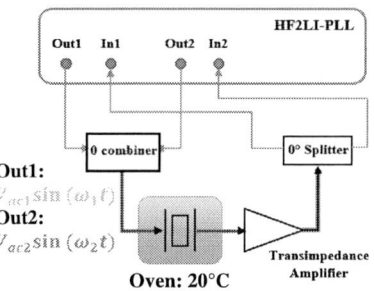

Figure 3: Experiment Setup using dual PLLs to excite and monitor the two modes simultaneously.

In the experiment, the DET mode is excited using a small AC signal to operate in the linear oscillation regime, while the excitation signal of the DR mode is gradually increased to drive the device into nonlinear regime (Fig.4).

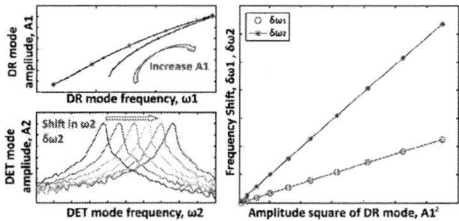

Figure 4: Example of increasing in A_1 causing changes in both $\delta\omega_1$ and $\delta\omega_2$.

RESULTS AND DISCUSSION

• Effect of resonant mode order

To study the effect of resonant mode order on coupling behavior, we first gradually increased the excitation to the Lamé resonator DR mode (1st order) while maintaining a constant excitation to the DET mode (3rd and 4th order). The frequency shifts of the DR and DET mode are simultaneously recorded. The raw data is plotted in Fig. 5a and 5c, where no distinction is made between electrostatic nonlinearity and material nonlinearity. We then switched the DR mode and DET mode and repeated the experiment (Fig. 5b and 5d).

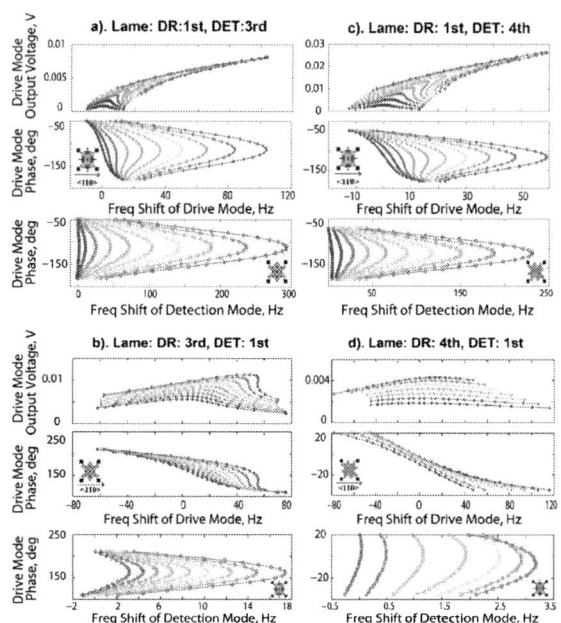

Figure 5: Closed loop frequency amplitude response of the 3rd and 4th order Lamé mode to study the effect of mode order on coupling behavior.

From these measured results, the frequency shift of the DR mode ($\delta\omega_1$) versus the frequency shift of the DET mode ($\delta\omega_2$) at maximum amplitude can be extracted and shown in Fig. 6. By applying linear least-squares fit, it can be seen that the slope of the fitted line λ, has a strong correlation with the relative mode orders between the DR and DET modes, and the comparisons are listed in Table 2. The obtained results perfectly match the behavior of the modal frequency shifts derived above in Eq. 8.

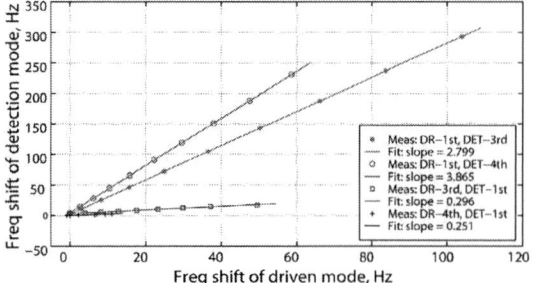

Figure 6: Fitted result using $\delta f_2 \cong \lambda \delta f_1$ for the 3rd and 4th order Lamé mode resonators.

Table 2: Fitted $\delta f_2 \cong \lambda \delta f_1$ showing relation between the frequency shifts of the DR mode and DET mode

	Slope: λ	Mode order: $\frac{\omega_{02}}{\omega_{01}}$
DR: 1st, DET: 3rd	2.799	3
DR: 1st, DET: 4th	3.865	4
DR: 3rd, DET: 1st	0.296	0.33
DR: 4th, DET: 1st	0.251	0.25

• Effect of doping and crystal orientation

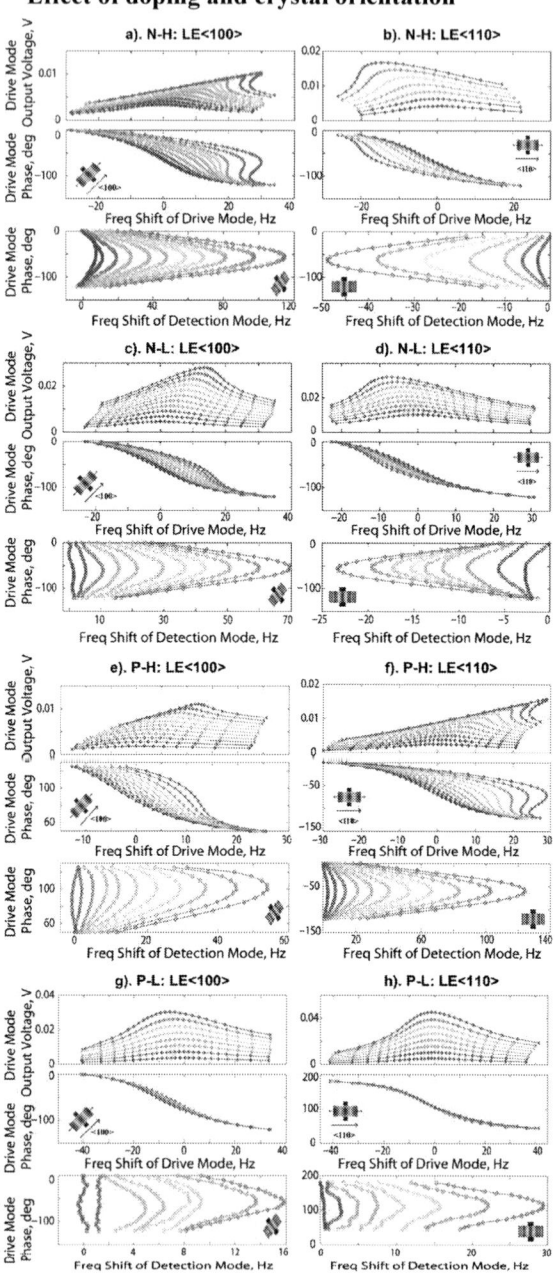

Figure 7: Closed-loop frequency amplitude response of the LE mode resonators to study the effect of doping and crystal orientation on the coupling behavior.

Similar analysis can be applied to the LE mode resonators to study the effect of doping and crystal orientation (Fig. 7). Here we use the 1st order mode as the DR mode and 3rd order mode as the DET mode. The extracted results using $\delta f_2 = \frac{KA_1^2}{2\pi}$ from Eq. (7) are shown in Fig. 8.

It is clear that for the fixed device geometry and the wafer orientation, the coupling constant α_1 depends on the doping type/concentration significantly. Additionally we have found that the coupling coefficient also strongly depends on the device alignment to the crystal structure (<110> vs. <100>) while the device geometry and doping type/level are kept fixed. In particular, we see from Fig. 8 that for the N-type doping devices α_1 changes its sign. This allows us to anticipate that other coupling constants may also be very sensitive to the doping type/concentration.

The measured result is analogous to the ones reported in [7], where the doping affects the nonlinear elasticity of silicon, leading to a difference in both A-f effect and coupling behavior. Utilizing this mode coupling behavior, separation of the material nonlinearity versus electrostatic nonlinearity may be possible. For example, for the P-L doped LE resonators (Fig.7g and 7h), even though the frequency- amplitude response for the DR mode shows a softening (decreasing frequency) behavior, the DET mode appears hardening (increasing frequency). This seems to justify the statement in [7] that electrostatic softening effects may be significant.

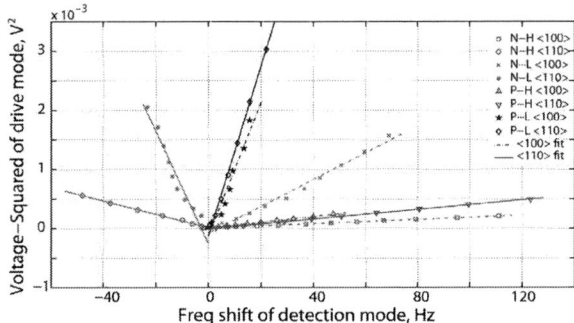

Figure 8: Fitted result using $\delta f_2 = \frac{KA_1^2}{2\pi}$ for the LE resonators under different doping and crystal orientation.

CONCLUSION

In conclusion, the nonlinear modal coupling in bulk mode silicon MEMS resonators has been analyzed and demonstrated. We have shown analytically and in the experiment that symmetric nonlinear coupling term $\propto x_1^2 x_2^2$ introduces the dependence of the frequency of one mode on the amplitude of the other one. We have also observed that the coupling strength α_1 strongly depends on doping type and concentration as well as the crystal orientation. These results show that the dynamics of the system of interest cannot be fully described by geometric and electrostatic nonlinearities – the doping-dependent material nonlinear elasticity must be also taken into account. These results are important for the system identification purposes as they provide a nice way for the estimation of the coupling constants. The future work will focus on developing a protocol for characterizing the coupling term constants and creating a thorough model accounting for higher-order effects.

ACKNOWLEDGEMENTS

This work was supported by DARPA grant FA8650-13-1-7301, "Mesodynamic Architectures (MESO)," managed by Dr. Dan Green and Dr. Jeff Rogers. The fabrication work was performed at the Stanford Nanofabrication Facility (SNF) which is supported by National Science Foundation through the NNIN under Grant ECS-9731293. The authors would like to thank the staffs at SNF for their help during the fabrication process.

REFERENCES

[1] P. A. Truitt, et al. "Linear and nonlinear mode coupling between transverse modes of a nanomechanical resonator", *APL* vol. 114, 114307 (2013).

[2] M. H. Matheny, et al. "Nonlinear mode coupling in nanomechanical systems", *Nano Lett,* vol.13, pp 1622-1626 (2013).

[3] H. J. R. Westra, et al. "Nonlinear modal interactions in clamped-clamped mechanical resonators", *PRL* vol. 105, 117205 (2010)

[4] K. J. Lulla, et al. "Nonlinear modal coupling in a high-stress doubly clamped nanomechanical resonator", *New J. of Phys,* vol.14 (2012)

[5] W. J. Venstra, et al. "Q-factor control of a microcantilever by mechanical sideband excitation", *APL,* vol. 99, 151904 (2011)

[6] T. Dunn, et al. "Anharmonic modal coupling in a bulk Micromechanical resonator", *APL,* vol 97, 123109 (2010)

[7] Y. Yang, et al. "Measurement of the nonlinear elasticity of doped bulk-mode MEMS resonators", *2014 Hilton Head Workshop,* pp. 285-288.

[8] R.N. Candler, et al. "Long-Term and Accelerated Life Testing of a Novel Single-Wafer Vacuum Encapsulation for MEMS Resonators," *JMEMS,* vol. 15, No. 6, pp. 1446-1456, (2006).

[9] E. J. Ng, et al. "An etch hole-free process for temperature-compensated, high Q, encapsulated resonators", 2014 *Hilton Head Workshop,* pp. 99-100.

[10] L. D. Landau, et al. "Resonance in non-linear oscillations", Mechanics, vol. 1, *Course of Theoretical Physics, 3 ed.,* MA USA: Butterworth-Heinemann, pp. 87-92, (1982).

[11] D. Antonio, et al. "Frequency stabilization in nonlinear micromechanical oscillators", *Nature Communications* 3, 806 (2012).

[12] D. Czaplewski, et al. "Phase noise reduction in an oscillator through coupling to an internal resonance mode", 2014 *Hilton Head Workshop,* pp. 80-82.

[13] A. H. Nayfeh and D.T. Mook, *Nonlinear Oscillations,* John Wiley & Sons, pp. 388-391 (2008).

[14] H.K. Lee, et al, "Stable operation of MEMS oscillators far above the critical vibration amplitude in the nonlinear regime", *JMEMS,* vol.20, no.6, (2010).

CONTACT

*Yushi Yang, +1-765-404-0884; ysyang88@stanford.edu

MEMS-BASED RF PROBES FOR ON-WAFER MICROWAVE CHARACTERIZATION OF MICRO/NANOELECTRONICS

Jaouad Marzouk*, Steve Arscott, Abdelhatif El Fellahi, Kamel Haddadi,
Tuami Lasri, Lionel Buchaillot and Gilles Dambrine

Institute of Electronics, Microelectronics and Nanotechnology, University of Lille 1, FRANCE

ABSTRACT

We demonstrate a radio frequency (RF) probe based on microelectromechanical systems (MEMS) design and processing technologies. The probe responds to the current needs of microelectronics requiring microwave characterization of nanoscale devices and systems having micrometer pad sizes. The use of MEMS technologies enables the probe contact pad area dimensions to be reduced by a three orders of magnitude compared to existing commercial RF probes. On-wafer RF measurements prove the feasibility of the approach to 30 GHz at very low contact resistance <<1 Ω. A contact aging study demonstrates that the probes are capable of forming this contact for 6000 contact cycles.

INTRODUCTION

In the context of modern microelectronics, the scientific community is focusing on the development of new miniaturized devices and integrated circuits (IC) using new materials and technologies [1,2]. These future nanoscale devices necessitate trade-offs in terms of scaling, power consumption, performance and cost. Therefore, the development of these miniaturized devices will partly depend on our ability to accurately measure the electrical performances of devices having ever diminishing sizes at increasingly higher working frequencies. For long time, the characterization of radio frequency (RF) integrated devices has been made by mounting a single chip in a test fixture [3]. RF characterization using direct on-chip probing - via coplanar probes - is frequently used for RF integrated circuits. Thus, calibration and measurement are made easier without the requirement of having to de-embed the device - this permits rapid and accurate RF measurements. For many years, the scientific community has been using conventional macroscopic ground-signal-ground (GSG) probes for the characterization of high frequency integrated devices [4]. Using such probes, the characterization of devices in the GHz-THz range has been successfully achieved [5,6].

Current conventional macroscopic assembled RF microwave probes, e.g. Cascade Microtech®, will no longer meet the needs of the industry's evolving standards for on-chip characterization [7]. Such commercial probes display a contact surface of the order of hundreds of square micrometers associated with a pitch (spacing between contacts) length of tens of micrometers. These dimensions require devices with pad sizes of comparable dimensions for on-wafer characterization. Large probes coupled with large pads inevitably incur measurement problems associated with parasitic elements. In addition, such commercial probes are costly, making device characterization relatively expensive. MEMS technologies offer a solution to these problems by enabling the fabrication of miniature RF probes compatible with miniature electronic devices and circuits. MEMS technology can provide two major advantages over the traditional approach used to make "macroscopic" RF probes. MEMS technology allows miniaturization of probes. This can lead to micrometer dimensions, even nanometer. Currently, the conventional RF probes have minimum dimensions of several hundreds of micrometers. The miniaturization of RF probes provides several benefits: for example, the increase in frequency by minimizing parasitic elements and a reduction of the surface of the wafer components dedicated to calibration and characterization. For some years, several research groups have been interested in the micromachined RF probes [8,12]. We note the development of ultra-thin silicon chips for sub-millimeter wave mixing applications using SOI [8]. In 2008, Liu *et al.*, demonstrated successful Ni plated CMOS-MEMS electrothermal conductive probes for memory intensive self-configuring ICs, by using MEMS technologies [9]. Novel elastic substrate RF-MEMS probes were developed based on the PDMS [10]. Recently, miniaturized probes have been developed for the range 500 GHz-750 GHz [11]. These probes show a contact area around 40×10 μm^2 with a pitch of the order of 30 μm. More recently, by using MEMS technology, a miniaturized probe with integrated sensor have been developed with a piezoresistive sensor [12].These sensors allow the control of the inclination of the probe relative to the surface - which enables an optimized contact between the probe tips and the pads of the device under test (DUT) and second optimization of the contact force in order to obtain minimum contact resistance - without damaging the probe. However, despite this size reduction of the pads and the pitch, these probes are still not compatible with the dimension brought into play in modern microelectronic devices.

In this paper we present a new miniaturized GSG probe dedicated to RF characterization of microelectronics. The proposed probe is based on MEMS technologies which make use of the advantages of SOI processes - especially the mechanical properties of Si and Cr/Au/Ni to obtain a low contact resistance between the probe tip and the DUT pads. This approach allows the development of an RF probe with micrometer and sub-micrometer contacts dimension - the pitch is also of the same size. The probes tips have a pitch of <5 μm and a contact pad area of the order of 1 μm^2. This reduction of the contact pad area of three orders of magnitude, compared to existing conventional RF probes. This scaling can allow a significant reduction of the parasitic effects and a reduction of the access pad size and therefore a significant increase of the on-wafer device integration. For the probe characterization (DC and RF), the probe has been successfully integrated onto a printed circuit board (PCB) by bonding interconnections for input/output (I/O) interface with automatic test equipment in a scanning electron microscope (SEM).

PROBE DESIGN AND FABRICATION

In order to achieve optimal electrical contact between the probe and the DUT pad, the probe must provide enough mechanical deformation to allow good physical contact but without inducing mechanical damage to the probe materials. Furthermore, the probe design must deliver enough force to generate the lowest possible contact resistance between the probe tips and the DUT pads. Thus, the tip contacts should be parallel to the DUT pads surface during measurement. When the cantilever tips press on the pads of the DUT, the cantilever deflection should be such that the contact area is maximal - see Fig. 1. It has been reported that a low contact resistance can be achieved between two metals if we have a flat contact area due to the cold welding of metals [13].

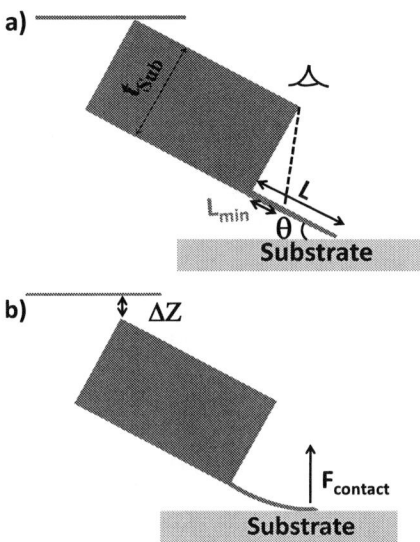

Figure 1: Side view of the probe illustrating the transition to the optimal probing position. a) Initial state – probe angle(θ) and b) the state of the cantilever (length L) after a displacement of the probe by Δz enabling minimal contact resistance between the substrate and the probe due to cantilever deflection and sufficient force ($F_{contact}$) on the probe contacts.

In order to avoid damage to the probe, the probe material should provide excellent elastic properties with high yield strength. In MEMS technologies, silicon is commonly used as a cantilever material [14] due to the high yield strength - which is in the order of 4.5 GPa [15]. This gives a large safety margin during the probe manipulation in case of possible user error. For the tip material, contact metallization should be robust to allow repeated measurements. To this end, we have used nickel for the tip contacts due to its high hardness and good electrical properties which should increase tip robustness and give a low contact resistance; as reported in the literature [16].The cantilever is designed using a 20 μm thick device layer (DL) of an SOI wafer, the cantilever's

dimensions are: length (L) =400 μm and a base width (w) = 300 μm - the probe pitch is 4.5 μm with three contact areas. Fig. 2 shows the stress distribution on the cantilever when a probing force of 2 mN is exerted at the tip obtained by numerical modeling by employing commercially available software (®COMSOL). We can note that, the maximum stress is around 88 MPa situated at the base of the cantilever - which is much lower than the failure stress of the silicon. Under this load, (2 mN) per tip, the predicted tip displacement is ~1.7 μm; this causes a flat surface between the probe tip and DUT pads when the cantilever tips press on the pads of the DUT – thus allowing minimal contact resistance. Fig. 3 show the influence of the cantilever thickness, cantilever length on the mechanical properties of the probe. One can see that the thickness of the cantilever greatly influences the stress and the displacement of the tip – see Fig. 3a. By increasing the cantilever length the displacement δ increases for a given force – see Fig. 3b.

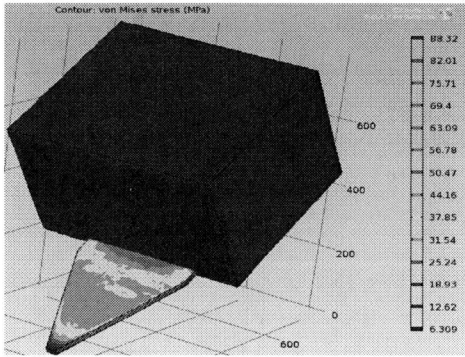

Figure 2: MEMS design of the RF probe showing stress field on probe cantilever when a probing force of 2 mN is exerted at the tip.

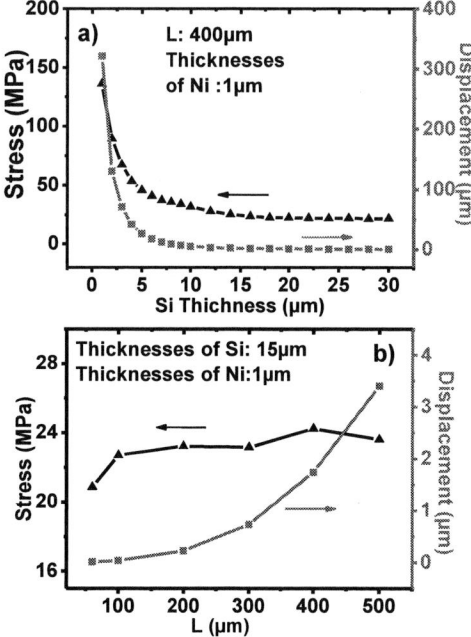

Figure 3: Numerical modeling of the tip displacement and mechanical stress as a function of cantilever thickness (a) and cantilever length (b).

The RF probe design is based on an optimized metallized coplanar line technology [17]. The probe fabrication was performed by using SOI material – see Fig. 4. We start with a 3-inch diameter (100) oriented SOI substrate (Si-Mat, Germany), the double-side polished SOI present a high-resistivity (>1000 Ω cm) with a Handle thickness was 400 μm, the BOX thickness is 2 μm and the DL is 20 μm thick - also high resistivity (>1000 Ω cm). Next, 500 nm of gold (Au) is evaporated onto wafer surface to define the coplanar line. Nickel (100 nm) was evaporated onto the wafer surfaced to define the probe tips. We have defined the probe mesa by deep reactive ion etching (DRIE) using the Advanced Silicon Etch (ASE) technique (Surface Technology Systems, UK). The 400 μm thick silicon Handle wafer was etched from the back side using DRIE etching (STS, UK). The BOX oxide layer (2 μm) of the SOI substrate was then etched away using HF based wet etching. Finally, each individual probe is successfully obtained by manually breaking the two supporting arms. Fig. 5 shows images of the results of the microfabrication.

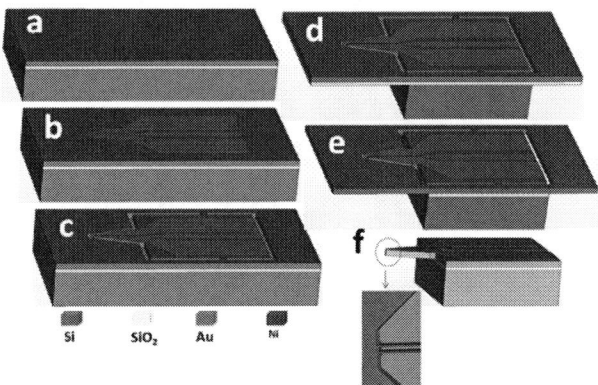

Figure 4: Probe fabrication process. (a) SOI wafer, (b) coplanar line and tip metallization, (c) mesa etching, (d) cantilever etching, (e) cantilever releasing and (f) probe dicing.

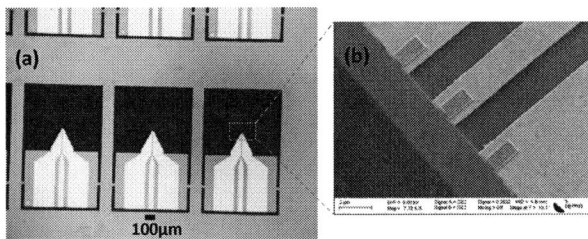

Figure 5: Images of the finished microprobes. (a) Photograph of a portion of SOI showing individual fabricated probes, (b) SEM image - shows zoom of probe tip with Nickel pads.

PROBE CHARACTERIZATION

The probe has been successfully integrated onto a PCB by bonding interconnections for I/O interface with automatic test equipment - the PCBs are designed on coplanar waveguide technology. The characterization setup involves a nano-positioner, with high resolution displacement (<10 nm), inserted into as SEM – see Fig. 6. The micromachined RF probes are mounted onto the PCB

board, the connection is obtained via Au-to-Au thermo-compression wire bonding. The assembled RF probe-PCB board is shown in Fig. 6. The bonded probe-PCB board is fixed on the nano-positioner for mounting into the SEM Fig. 6 [inset (b)].

DC measurements were able to characterize the contact resistance as a function of contact force applied by the probe and the number of contact cycles. The contact resistance of the probe was measured inside the SEM after the mounting of the probe onto a coplanar waveguide PCB board by bonding and mounting on the nano-positioner. Fig. 7 demonstrates that the contact resistance decrease with increasing displacement of the nano-positioner (applied force), approaching a resistance of 0.02 Ω - which is comparable with commercial probes. An ageing study (see inset to Fig. 7) showed that the probes are capable of forming this DC contact for over 6000 contact cycles.

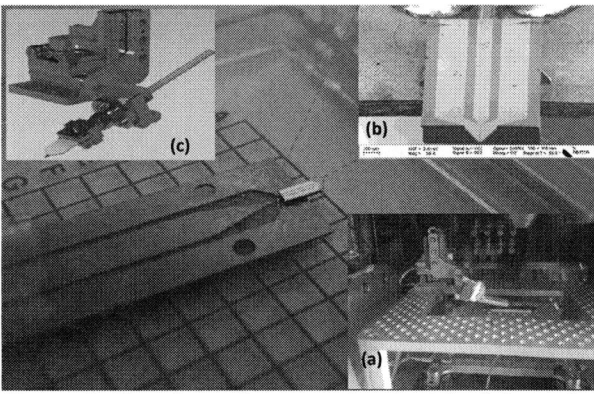

Figure 6: Experimental setup - photograph of the probe-PCB. Insets – (a) SEM setup, (b) SEM image of the bonded Probe-PCB board - the probe chip dimension is 1mm × 3.2 mm and (c) nano-positioner

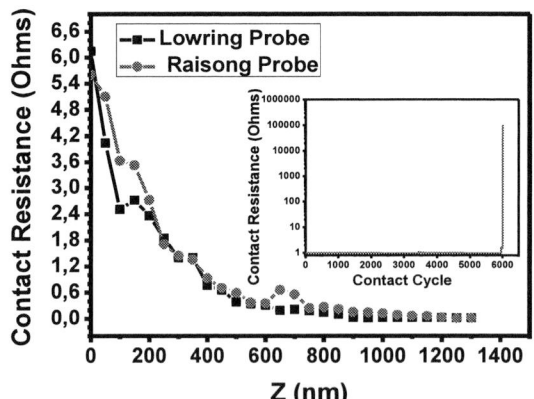

Figure 7: Plot of the contact resistance plotted versus the displacement of the nano-positioner. Inset - Contact resistance over 6000 contact cycles on gold pad.

RF measurements were made using the probes with a N5245A vector network analyzer (VNA) (Agilent Technologies, USA) and two GSG commercial RF probes (SUSS MicroTec). We have measured the reflection coefficient directly on-wafer. Fig. 8 shows the reflection coefficient as a function of the cantilever length (100 and

400 μm). These reflection coefficients corresponding to an open ended probe. We can note, a relative good behavior in both cantilever length over the frequency range considered. Thus, the propagation losses do not exceed -2 dB and -4 dB at 30 GHz respectively for a cantilever length of 100 and 400 μm.

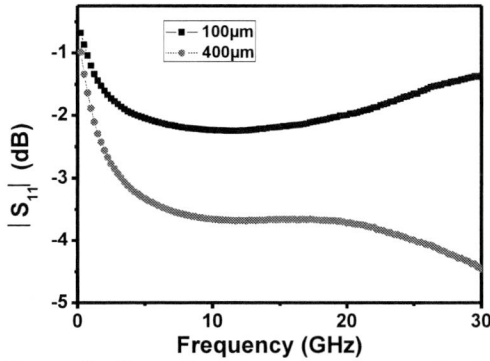

Figure 8: On-wafer microwave characterization using a VNA. Measured reflection coefficient of the micromachined probe length in opened ended probe structure as a function of cantilever length.

CONCLUSION

We describe a prototyping of a miniature, microcantilever-based RF microwave probe. The probes have been developed using MEMS technology on SOI wafers. The probes incorporate silicon microcantilevers having a coplanar signal line width of 2 μm having a GSG pitch of 4.5 μm. The contact resistance measurement shows a low contact resistance - around 0.02Ω - which is comparable with commercial probes. The preliminary RF measurement data of the miniaturized probe shows that the probe is compatible with the trends of microwave characterization of microelectronics having diminishing dimensions.

ACKNOWLEDGEMENTS
This work was supported by the EQUIPEX 'ExCELSiOR' project.

REFERENCES
[1] C. A. Mack, "Fifty Years of Moore's Law", *IEEE Trans Semicon. Manufacturing*, 24(2), pp. 202-207(2011).

[2] W. Yanqing, D. B. Farmer, X. Fengnian and P. Avouris, "Graphene Electronics: Materials, Devices, and Circuits", *Proceedings of the IEEE*, 101(7), pp. 1620-1637 (2013).

[3] S. A. Wartenberg, "Selected topics in RF coplanar probing", *IEEE Trans. Microwave Theory Tech.*, 51, pp.1413–1421 (2003).

[4] A. Rumiantsev, R.Doerner, "RF probe technology: History and Selected Topics", *Microwave Magazine IEEE*, 14, pp. 46-58 (2013).

[5] R. V. Tuyl, C. Liechti, R. Lee, and E. Gowen, "4-GHz frequency deviation with GaAs MESFET ICs," *in IEEE Int. Solid- State Circuits Conf. Dig. Tech. Papers, Philadelphia, PA*, pp. 198–199 (1977).

[6] M. F. Bauwensu, N. Alijabbari, A. W Lichtenberge N. S. Barke, R. M Weikle, and C. L. Brown , "A 1.1 THz micromachined on-wafer probe", International Microwave symposium (IMS), IEEE MTT-S, pp.1-4 (2014).

[7] M. J. H. van Dal, N. Collaerta, "Highly manufacturable FinFETs with sub-10nm fin width and high aspect ratio fabricated with immersion lithography" ,*VLSI Technology, IEEE symposium*, pp. 110-111 (2007).

[8] R. B. Bass, J. C. Schultz, A. W. Lichtenberger, R. M. Weiklel , S-K. Pan, E. Bryerton, C. K. Walker and J. Kooi, "Ultra-Thin Silicon Chips for Submillimeter-Wave Applications", *15th International Symposium on Space Terahert: Technology*, pp. 392-399 (2004).

[9] J. Liu, M. Noman, J.A. Bain, T.E. Schlesinger and G.K. Fedder, " CMOS-MEMS Probes for Reconfigurable IC'S", *IEEE 21st International Conference on MEMS*, pp. 515-518 (2008).

[10] T. Yuan, D. Chen, J. Chen, H.Fu, S. Kurth, T. Otto, T. Gessner, "Design, fabrication and characterization of MEMS probe card for fine pitch IC testing", *Sensors and Actuators A*, 204, pp. 67-73(2013).

[11] T.J. Reck, L. Chen, C. Zhang, A. Arsenovic, C. Groppi, A. W. Lichtenberger, R. M. Weikle, and N. S. Barker, "Micromachined probes for submillimeter-wave on-wafer measurements-Part I:mechanical design and characterization," *IEEE Trans. Terahertz Sci. Technol.*, 1(2), pp. 349-356, (2011).

[12] Q. Yu, M. Bauwens, C. Zhang, R. Weikle, and N. Scott Barker, "Improved Micromachined Terahertz On-Wafer Probe Using Integrated Strain Sensor", *IEEE Trans.on Microwave Theory and Techniques*, 61(12), pp. 4613 – 4620(2013).

[13] J. F. Archarj, "Contact and Rubbing of Flat Surfaces", *J. Appl. Phys.*, 24, pp. 981-988(1953).

[14] B. H. Kim, H. C. Kim, S. D. Choi, K. Chun, J. B. Kim, and J. H. Kim, "A robust MEMS probe card with vertical guide for a fine pitch test, " *J. Micromech. Microeng.* 17, pp. 1350–1359 (2007).

[15] K. E. Petersen, "Silicon as a mechanical material", *Proc. IEEE*, 70(5), pp. 420–457 (1982).

[16] K. Kataoka, S. Kawamura, T. Itoh, K. Ishikawa, H. Honma, and T. Suga, "Electroplating Ni micro-cantilevers for low contact-force IC probing," *Sens. Actuat. A*, 103, pp. 116–121 (2003).

[17] J. Marzouk, S. Arscott, K. Haddadi, T. Lasri and G. Dambrine, "Miniaturized MEMS-based GSG Probes for Microwave Characterization" *in Proc. 44th European Microwave Conf.*, pp. 1–4 (2014).

CONTACT
*J.Marzouk, tel: +33 3 20 19 79 43; jaouad.marzouk@iemn.univ-lille1.fr

MICROMECHANICAL RING RESONATORS WITH A 2D PHONONIC CRYSTAL SUPPORT FOR MECHANICAL ROBUSTNESS AND PROVIDING MASK MISALIGNMENT TOLERANCE

B. Figeys[1,2]*, B. Nauwelaers[2], H.A.C. Tilmans[1], and X. Rottenberg[1]
[1]Imec, Heverlee, BELGIUM
[2]ESAT, KU Leuven, Leuven, BELGIUM

ABSTRACT

This paper reports on the design of ring-type electrostatically transduced bulk acoustic wave resonators designed for increased shock and vibration resistance. This was achieved through a 2D Phononic Crystal (PnC) support. The PnC is designed to operate in its bandgap so that it acts as a non-propagating medium, hereby achieving simultaneously a mechanically strong and acoustically well-confined support. We manufactured SiGe-resonators at 137.8MHz with a Q-factor of around 17,000. Another feature of this design is the process tolerance of the Q-factor (within 5%) and the resonance frequency towards mask misalignment ($<7\mu m$) for the center support.

KEYWORDS

MEMS, 2D phononic crystal, SiGe ring resonator,

INTRODUCTION

Micromachined bulk acoustic wave (BAW) resonators are considered for a wide variety of applications ranging from timing devices to chemical sensors [1]. This work considers electrostatically transduced contour mode ring resonators [2], belonging to the class of surface micromachined BAW resonators [1]. In previous research the ring resonators' Q-factor was maximized by using notched spoke supports to support the ring via nodal points [3,4]. Further research explored the use of quarter wavelength spokes [5], as shown in the left of Figure 1.

This work considers an option to make a mechanically more robust design of the resonator while limiting the support losses. For increased shock and vibration resistance one would ideally see the "hollow disk" filled with a stiff medium as is shown in the right part of Figure 1, which is similar to the design of [2]. However for achieving high Q-factors research typically used tiny critically designed support spokes to limit support losses, which is in contrast to the ideal criterion for mechanical robustness. This paper proposes to fill the hollow disk with a 2D phononic crystal (PnC), which acts, if well-designed, as a mechanically robust and non-propagating support.

The success of 2D PnCs to define resonating cavities has already been demonstrated [6,7]. Similarly have phononic crystal strips (~1D PnCs) been used to reduce support losses for wineglass mode ring resonators [8].

To further improve the mechanical robustness of our design the size of the support's stem can be increased by reducing the number of unit cells of the PnC. It is shown that this has no drastic impact on the device's Q-factor. This work also assesses the tolerance to misalignment of the support stem with the center of the device, which likewise has only a small impact on the device's Q-factor.

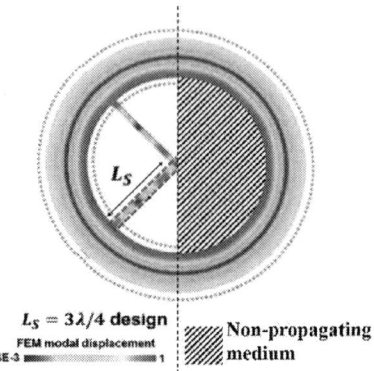

Figure 1: Sketch of conceptual idea. Switch from (left) a spoke supported ring resonator [5] to (right) a non-propagating medium support, here a 2D PnC.

DESIGN

Ring resonator design

The acoustic cavity of the ring shaped BAW resonator is defined by a free edge as outer boundary and a 2D PnC as inner boundary. This work opted for a honeycomb lattice of circular holes in the released layer as PnC because of its large bandgap and low degree of anisotropy [9].

Figure 2 shows the PhC's band structure for in-plane modes in a 4um thick SiGe layer (E =130GPa, ν = 0.22, ρ = 4400kg/m³) made of circular holes of 8um in diameter and 10um apart.

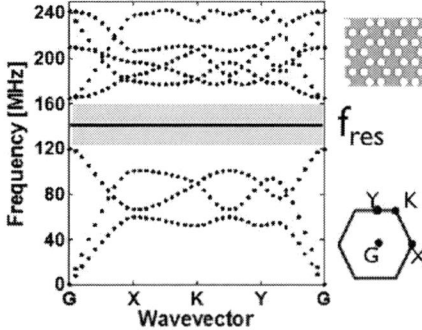

Figure 2: Band diagram of the phononic crystal, honeycomb lattice of circular holes. [6] The grey area indicates the bandgap for in-plane modes.

The width of the ring is hard to define, but the disk radius is chosen such that the ring resonance frequency f_{res} is situated in the PnC's bandgap, as shown in Figure 2. A deviation from the ring width with free-free boundaries

Table 1. Assessment of the mechanical robustness against shock and vibrations: the resonance frequencies for the different modes of acceleration (Figure 5) are proportional to the mechanical robustness.

Modes of acceleration	Resonance frequencies for corresponding modes [kHz]			
	Stem area:100µm²	100µm² / 700µm²	100µm²	700µm² / 2 500µm²
Out-of-plane rotation (a)	297	394 / 538	214	363 / 432
Out-of-plane translation (b)	450	520 / 600	295	443 / 505
In-plane rotation (c)	533	956 / 2 600	220	531 / 557
In-plane translation (d)	4 840	5 940 / 7 300	2 860	3 240 / 3 340

[3-5] is expected, because of the phase shift at the PnC boundary similar to what has been described for unreleased resonators [10]. Figure 3a shows the result of an eigenmode simulation with COMSOL multiphysics for a disk radius R_{disk}=77µm, for which the resonance frequency f_{res} is in the center of the bandgap, 140.2MHz.

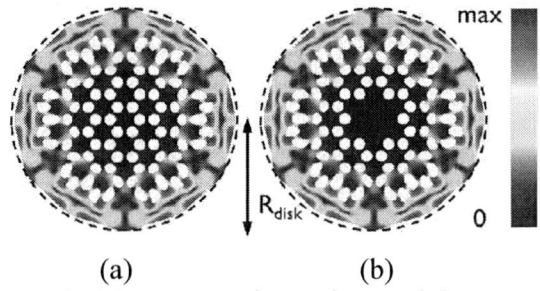

(a)   (b)

Figure 3: FEM eigenmode simulation of the ring-type resonators with a PnC support. Both have the same resonance frequency f_{res}=140.2MHz. The colors represent total displacement.

The 10x10µm² support stem will be located in the center of the disk, as shown in red in Figure 4a. Simulations show that it is possible to remove the inner holes from the structure, Figure 3b, and yet conserve the resonance; the wave decays fast enough in the PnC and the resonance frequency is not affected according to eigenmode simulations. Such a structure opens the opportunity to increase the support's stem in size (up to 700µm²) to make a mechanically stronger connection to the substrate.

A potential mask misalignment leading to a shift of the support's stem away from the center of the ring, as depicted in Figure 4, will be evaluated. In order to evaluate this effect on the device's performance the support's stem (100µm²) was shifted towards different directions and with different distances d (0µm, 6µm, 6.5µm, 7.5µm, 9µm).

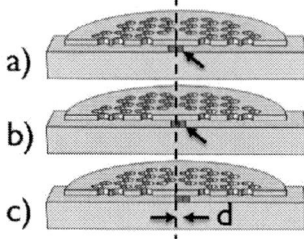

Figure 4: Out-of-plane cross-section of the resonator that depicts a mask misalignment. The support's stem (red) is shifted from the center.

Projected mechanical robustness improvement

The improved shock and vibration resistance is only demonstrated via simulations. The two devices in Figure 3 are compared with quarter wavelength spoke supported ring resonators of similar size and frequency [5].

The figure of merit for shock and vibration resistance is proportional to the resonance frequency of the translational and rotational modes of vibration relative to the support, as shown in Figure 5 [11].

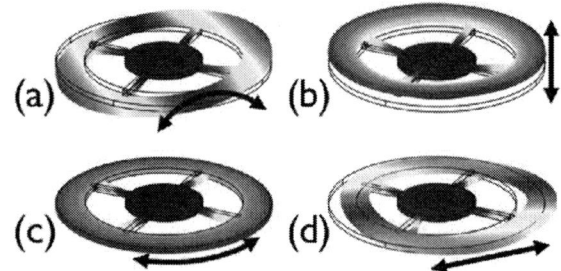

Figure 5: Eigenmodes of vibration for shock and vibrations resistance assessment in Table 1. The colors represent total displacement, with the same color scale as Figure 3.

Table 1 shows the simulated resonance frequencies for in-plane and out-of-plane, translational and rotational resonance frequencies for the following structures:
1. Device 1 is shown in Figure 3a, with a 100µm² surface area for the support's stem.
2. Device 2 is shown in Figure 3b, with a 100µm² or a stronger 700µm² large support.
3. Device 3 is a ring with outer radius 77µm, and designed for 140MHz with long "quarter" wavelength spokes that are 4µm wide and 48µm long. The stem's surface area is also 100µm² large.
4. Device 4 is identical to number 3, but with shorter quarter wavelength spokes, 29µm, and thus stiffer. The devices are compared with a stem area of 700µm² or 2500µm².

As expected, the improvement is in general the largest for in-plane vibrations. Only a slight improvement for out-of-plane vibrations is visible between device 1 and 3, and a doubling of the resonance frequency for in-plane vibrations. The wider support of the device 2 compared to device 1 gives indeed a mechanically more robust structure. Device 4 can be thought of as the strongest spoke supported device, because of its short spokes, and has a similar performance to device 1, but is bested by device 2. In conclusion, the 2D PnC support indeed improves the shock and vibration resistance of the devices. Despite this result one should note that this design has only half the transduction area, i.e. the inner side of the circle is not accessible anymore for transduction.

RESULTS

The devices measured and reported on in this paper were produced in imec's SiGeMEMS technology. Figure 6 shows pictures of the fabricated devices. Not visible in these pictures are the $1\mu m^2$-sized release holes in the structure. Also not visible in the pictures are the $20\mu m^2$ dummy structures in the holes of the PnC to achieve etch uniformity.

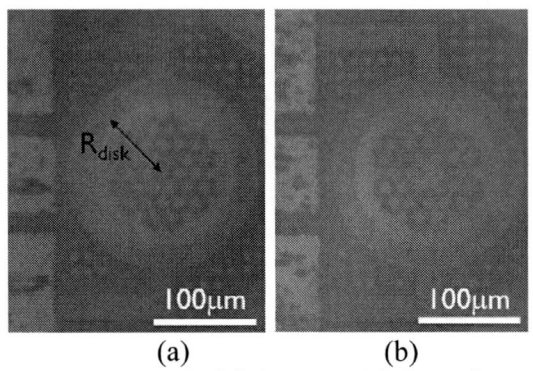

(a)   (b)

Figure 6: Pictures of a full devices with holes in the center (a), and without holes in the center (b).

The devices were designed as one-port resonators with a 500nm transduction gap, and measured with 80V DC-bias, shown in Figure 7, in 0.01mbar vacuum.

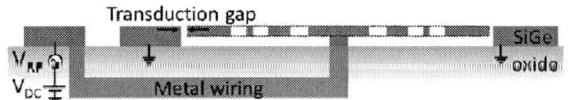

Figure 7: Electromechanical cross-section of a device.

A typical S_{11}-measurement is shown in Figure 8, resonance is found around 137.8MHz. An average Q-factor of 17,000 and an average motional resistance of 370kOhm were extracted for devices with R_{disk} =77μm as shown in Figure 3a and Figure 6a.

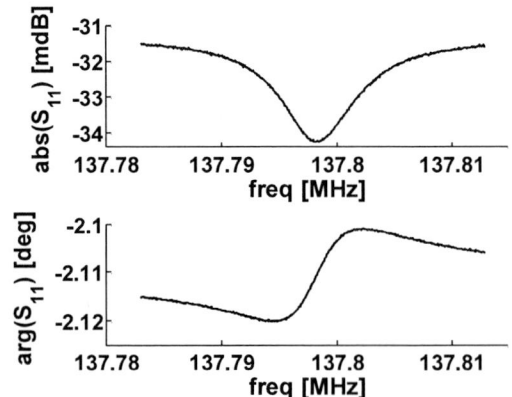

Figure 8: Typical measurement S_{11} measurement of a device as shown in Figure 3a with R_{disk} = 77μm.

To evaluate the effect of resonance frequency location in the PnC bandgap several devices with different resonance frequencies were fabricated and measured. The resonance frequency was swept over part of the expected PnC bandgap by varying the disk radius R_{disk}, equivalent to a variation in the ring width. Figure 9 (green) shows that an increase in disk radius R_{disk} indeed decreases the resonance frequency. The trend is the same for simulation and measurement. The resonance frequency variation for one device across dies and wafers is on the order of 0.2% and thus not visible on this graph.

The device with disk radius R_{disk}=76μm does not follow the trend because the release holes have been omitted for this device. It reveals that the relatively large frequency shift between measurement and simulation is mainly due to the release holes, which were not considered in simulations, in the other devices.

The Q-factor of the measured devices in blue in Figure 9 shows a larger spread than was expected. A variation in Q-factor was expected due to process variations between wafers and between different die positions on wafer. We investigate further the potential effect of badly positioned dummy structures in the PnC holes on the Q-factor of these devices.

At high frequencies the Q-factor decreases when the resonance frequency approaches the bandgap edge (160MHz). At low frequencies the dashed blue line, the average of blue dots, shows a slight decrease but not as pronounced as for high frequencies. We believe the decrease in Q-factor is due to a decrease in reflection for finite crystals near the bandgap edges, and thus caused by an increase of support losses. The bandgap might be slightly lower in frequency than simulated because of the not included release holes in simulations, which would explain the shift in maximal Q-factor.

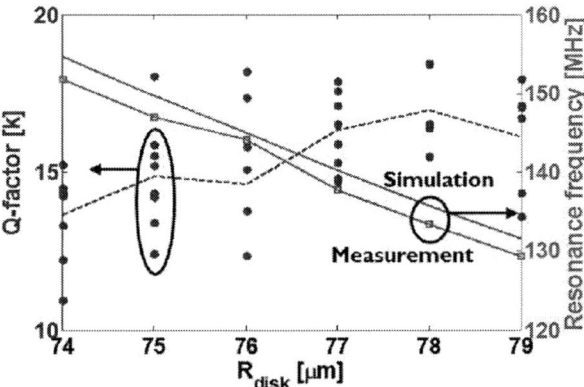

Figure 9: The resonance frequency (green, right axis) decreases with increasing disk radius, i.e. increasing ring width. The two lines represent measurement and simulations. The Q-factor (blue, left axis) of measured devices is shown with dots, we showed a dashed trend line through the average Q-factors. The Q-factor decreases for resonance frequencies close to the top PnC bandgap edge.

Figure 10 compares the Q-factor for different designs to evaluate the effect of a misalignment of the support's stem as depicted in Figure 4. From left to right could the devices be considered more "defective", because of a more pronounced loss of symmetry, the average Q-factor decreases slightly as was expected. The resonance frequency however is not affected.

The removal of holes in the center, shown in Figure 3, decreases the Q-factor by roughly 5% (family 1 vs. 2 in Figure 10). However we do not yet understand why there is an increased spread in Q-factor for family 2, the 14 devices in the boxplot might not be enough to make statistical relevant statements.

The support's stem is shifted by resp. $0\mu m$, $6\mu m$, $6.5\mu m$, $7.5\mu m$ and $9\mu m$, as shown in Figure 4, and also varied in orientation of the shift. To a first order approximation the Q-factor does not degrade significantly compared to the initial spread in the devices' Q-factors across dies and wafers. Family 6 has a larger spread, it however remains an open question whether this is a similar artefact comparable to family 2 or whether this is the result of a too large shift of the support stem.

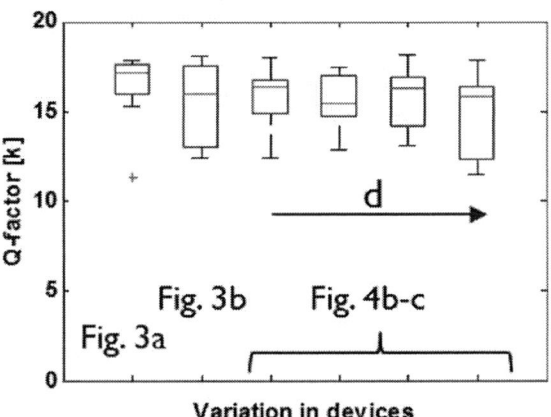

Figure 10: Q-factor of several families of devices, shown in Figure 3 and Figure 4. In total measurements of 72 data devices are used for this graph. We expect the second family (Fig. 3b) to be defective because of its unexpected large variation in Q-factor.

CONCLUSIONS

This paper reports on the design of ring-type electrostatically transduced bulk acoustic wave resonators designed for increased shock and vibration resistance. This was achieved through a 2D Phononic Crystal (PnC) support. The PnC is designed to operate in its bandgap so that it acts as a non-propagating medium, hereby achieving simultaneously a mechanically strong and acoustically well-confined support.

We manufactured SiGe-resonators at 137.8MHz with a Q-factor around 17,000 and a motional resistance of 350kOhm at 80V DC-bias. Furthermore did we demonstrate that one needs only a few rows of PnC unit cells. The removal of one row of unit cells aimed at improving the mechanical robustness lead to roughly 5% decrease in average Q-factor. We could not observe a clear effect on the resonance frequency or decrease in Q-factor for a small ($<7\mu m$) misalignment of the support's stem.

REFERENCES

[1] C. T.-C. Nguyen, "MEMS technology for timing and frequency control," *IEEE Transactions on Ultrasonics, Ferroelectrics, and Frequency Control*, vol. 54, no. 2, pp. 251–270, Feb. 2007.

[2] J. Wang, J. E. Butler, T. Feygelson, and C. T.-C. Nguyen, "1.51-GHz nanocrystalline diamond micromechanical disk resonator with material-mismatched isolating support," in *Micro Electro Mechanical Systems, 2004. 17th IEEE International Conference on. (MEMS)*, 2004, pp. 641–644.

[3] S.-S. Li, Y.-W. Lin, Y. Xie, Z. Ren, and C. T.-C. Nguyen, "Micromechanical 'hollow-disk' ring resonators," in *Micro Electro Mechanical Systems, 2004. 17th IEEE International Conference on. (MEMS)*, 2004, pp. 821–824.

[4] W.-L. Huang, S.-S. Li, Z. Ren, and C. T.-C. Nguyen, "UHF Nickelmicromechanical Spoke-Supported Ring Resonators," in *Solid-State Sensors, Actuators and Microsystems Conference, 2007. TRANSDUCERS 2007. International*, 2007, pp. 323–326.

[5] T. L. Naing, T. Beyazoglu, L. Wu, M. Akgul, Z. Ren, T. O. Rocheleau, and C. T.-C. Nguyen, "2.97-GHz CVD diamond ring resonator with Q >40,000," in *Frequency Control Symposium (FCS), 2012 IEEE International*, 2012, pp. 1–6.

[6] S. Mohammadi, A. A. Eftekhar, and A. Adibi, "Support loss-free micro/nano-mechanical resonators using phononic crystal slab waveguides," in *Frequency Control Symposium (FCS), 2010 IEEE International*, 2010, pp. 521–523.

[7] S. Mohammadi, A. A. Eftekhar, A. Khelif, and A. Adibi, "A high-quality factor piezoelectric-on-substrate phononic crystal micromechanical resonator," in *Ultrasonics Symposium (IUS), 2009 IEEE International*, 2009, pp. 1158–1160.

[8] J. Hsu, F. Hsu, T. Huang, C. Wang, and P. Chang, "Reducing anchor loss in micromechanical resonators using phononic crystal strips," in *Ultrasonics Symposium (IUS), 2011 IEEE International*, 2011, pp. 2483–2486.

[9] S. Mohammadi, A. A. Eftekhar, A. Khelif, H. Moubchir, R. Westafer, W. D. Hunt, and A. Adibi, "Complete phononic bandgaps and bandgap maps in two-dimensional silicon phononic crystal plates," *Electronics Letters*, vol. 43, no. 16, pp. 898–899, Aug. 2007.

[10] W. Wang and D. Weinstein, "Deep Trench capacitor drive of a 3.3 GHz unreleased Si MEMS resonator," in *Electron Devices Meeting (IEDM), 2012 IEEE International*, 2012, pp. 15.1.1–15.1.4.

[11] S. Sundaram, M. Tormen, B. Timotijevic, R. Lockhart, T. Overstolz, R. P. Stanley, and H. R. Shea, "Vibration and shock reliability of MEMS: modeling and experimental validation," *J. Micromech. Microeng.*, vol. 21, no. 4, p. 045022, Apr. 2011.

CONTACT

*B. Figeys, tel: +32 16/28.35.53;
Bruno.Figeys@imec.be

SILICON-MICROMACHINED SPACERS FOR UHF CAVITY RESONATORS

Dimitra Psychogiou[1], Michael D. Sinanis[1] and Dimitrios Peroulis[1]

[1] School of Electrical and Computer Engineering, Birck Nanotechnology Center, Purdue University,
West Lafayette, IN 47907, USA

ABSTRACT

This paper reports on a novel hybrid integration concept that enables the realization of high-quality (Q) factor, low-frequency coaxial cavity resonators with well-defined capacitive-loading and variable center frequency. It is based on a silicon-micromachined spacer that is mounted on top of a conventional CNC-machined metallic cavity to functionalize the resonator's capacitance. For the first time, it is demonstrated that low-frequency resonators with micrometer-scale gaps (10s of microns), relatively large Q-factor (459-505) and tunable response (18.5%) can be constructed without the need for post-fabrication tuning. To demonstrate these benefits, a resonator assembly was designed, built and experimentally tested at the UHF band and for a frequency tuning range between 1424-1711 MHz.

INTRODUCTION

Coaxial cavity resonators are fundamental components of RF-filters. Their unique advantages of: a) preserving a relatively high quality (Q)-factor in a wide tuning range (>50%), b) high RF-power handling (>1 W) and of c) spurious-free RF response, have made them indispensable components of various microwave and millimeter-wave communication systems [1]-[6]. In order to reduce their size and enable wide frequency tuning, a capacitive-gap (capacitive loading) is typically introduced between the post and the cavity upper wall [2]-[6]. For small form-factor and widely-tunable resonators, micrometer-scale gaps need to be created and controlled [4]-[6]. This is particularly critical for UHF (300-3000 MHz) applications due to the requirement for resonator geometries with cm-scale dimensions (wavelength at 600 MHz: 50 cm). These resonators are typically manufactured by CNC-machining and assembled with mechanical screws [1]-[2]. Major shortcomings of this approach include the inability to: (a) accurately create and control the capacitive-gap between the post and the cavity upper wall due to the large fabrication (250-500 μm) and assembly tolerances that create the need for post-fabrication tuning (b) realize small capacitive-gaps (10s of μm) due to the large surface roughness of the CNC-machined parts (2.5-10 μm). Consequently, precise pre-fabrication assembly of these devices is difficult—leading to narrower-than-expected tuning, degraded insertion loss and low fabrication yield.

With the aim of overcoming the drawbacks mentioned above, a silicon-micromachined spacer that can be mounted on top of a CNC-machined cavity has been developed and reported in this paper for the first time. It allows the realization of low-frequency (UHF-band), high-Q resonators with well-defined capacitive-loading and variable center frequency. A resonator assembly with tunable frequency (1424-1711 MHz) and high-Q factor (459-505) has been manufactured and measured as a proof-of-concept demonstrator.

RESONATOR CONCEPT

A schematic illustration of the cavity resonator concept with well-defined and controllable capacitive gaps is shown in Figure 1. It is based on: 1) a silicon-micromachined spacer that functionalizes the resonator's capacitance, 2) a conventional CNC-machined metallic cavity that primarily sets the resonator's inductance and 3) a commercially available piezoelectric-actuator that tunes the resonant frequency of the resonator by modifying the capacitive gap of the silicon-micromachined spacer at will. A detailed description of the silicon-micromachined spacer is depicted in Figure 1(b). It consist of: (1) the 200-μm-thick diaphragm-piece that resembles the resonator upper

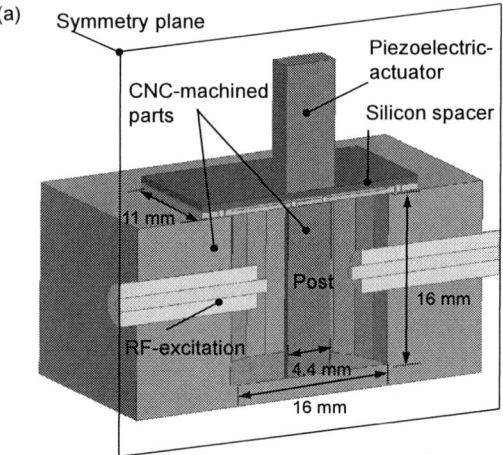

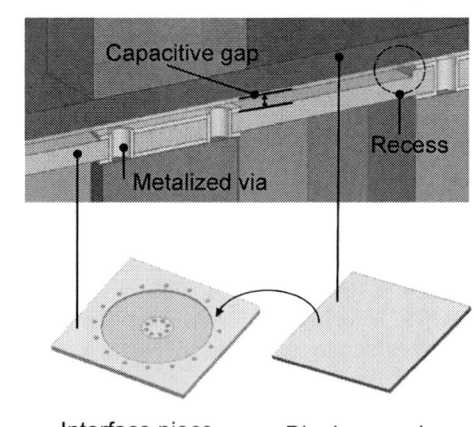

Figure 1: (a) Schematic view of the proposed cavity resonator with well-defined and controllable capacitive gaps. The gap is defined in-between the interface and the diaphragm piece by introducing a 40 μm-thick recess in the interface piece and by transferring the resonator post and outer wall with metalized via holes and thick metallization areas. It is then controlled to smaller values with a commercially-available piezoelectric actuator. (b) Detail of the silicon-micromachined spacer.

978-1-4799-7956-1/15 $31.00 © 2015 IEEE

wall and (2) the 200-μm-thick interface-piece that facilitates the capacitive-gap transfer from the cavity resonator to the silicon-spacer and the realization of a high-Q capacitor with small surface-roughness walls. Noteworthy is that the resonator capacitance is created in-between the two pieces (interface and diaphragm) by first introducing a ~40 μm-recess in the interface-piece and secondly by transferring the post area and the cavity outer-wall to the silicon-spacer with the aid of gold (Au)-metalized areas and thru-silicon vias. This gap is tuned to smaller values with the aid of the piezoelectric-actuator that is placed on top of the diaphragm. With this approach, small capacitive gaps (10s of μm) can be achieved due to the small surface roughness of the Au-metallized silicon areas (~0.5 μm for electroplated-Au). It is important to note that for a high-Q operation, thick metallization layers need to be created (>10 μm) in order to avoid conductive losses due to the skin-depth effect (note the Au skin-depth at 1500 MHz is 2 μm) Moreover, for the realization of the interface piece a relatively thin (~200 μm) and highly resistive (>3000 Ω·cm) silicon wafer needs to be utilized in order to minimize the dielectric losses form the E-field penetration within the silicon interface piece.

FABRICATION

Standard microfabrication processes and conventional CNC-machining were used for the fabrication of the proposed resonator assembly. It involves the realization of three main parts: the interface piece and the diaphragm piece that structure the silicon-micromachined spacer and the coaxial cavity—as shown in Figure 2 and described in-detail in the next sections.

Interface piece

The fabrication of the interface piece is performed on a 200-μm-thick, high-resistivity silicon (Si) wafer in order to minimize the dielectric losses of the RF-signal, as well as to enable the realization of thru-silicon vias with Au-metalized side-walls that are electrically connected to the upper and bottom metallization layer. The fabrication starts by growing 1-μm-thick SiO₂ on both sides of the 200 μm-thick Si wafer. Afterwards, a layer of AZ9260 positive photoresist layer is spin-coated on the wafer so as to act as an etching mask. It is then etched using a Buffered Oxide Etch (BOE) solution. Then, the wafer is cleaned using a solvents cleaning method. Furthermore, it is immersed in Tetramethylammonium Hydroxide (TMAH) solution to etch the recess that is needed for the definition of the desired capacitive gap (40 μm). Note that a TMAH solution with 20ppm Triton X100 surfactant was chosen in order to obtain a mirror surface finish at the bottom of the etched cavity. The remaining SiO₂ is striped-off using BOE and is later on regrown so as to serve as hard mask for the via holes opening. A thick layer of AZ9260 positive photoresist is utilized for the via-holes formation due to the non-uniform profile of the wafer (40 μm recess). Subsequently, the oxide is etched using Deep Reactive Ion Etching (DRIE) and the remaining photoresist is removed. This is then followed by the via-holes opening through DRIE process. At the end of the process, the oxide-masking layer is removed through wet etching with BOE. Next, an

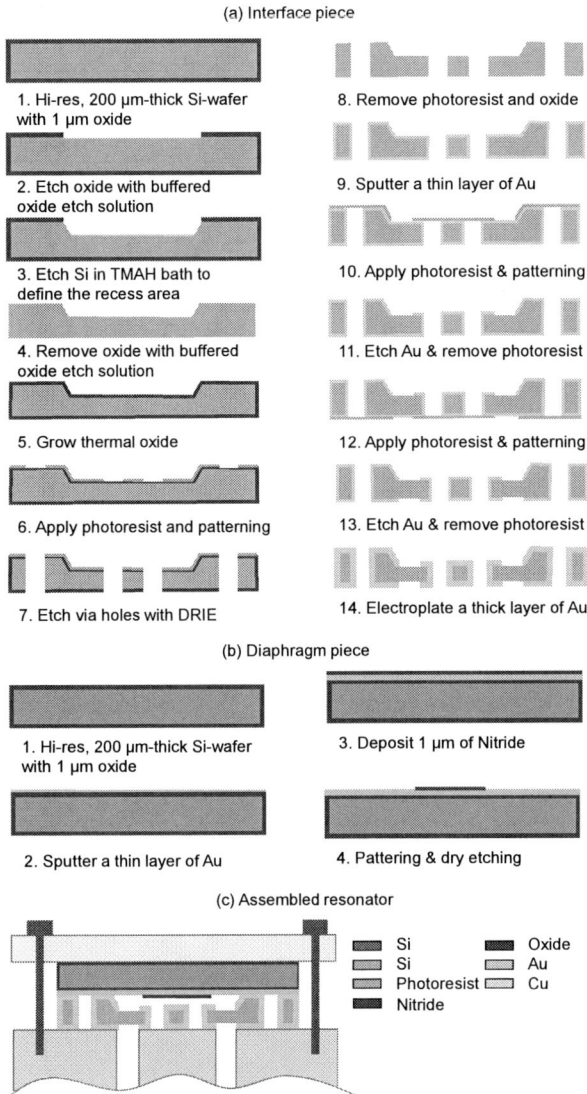

Figure 2: (a) Fabrication process flow of the silicon interface piece. (b) Fabrication process flow of the silicon diaphragm piece. (c) Resonator assembly.

oxide layer is grown on both sides of the wafer and a thin layer of titanium/ gold (Ti/Au) is sputtered so as to serve as seed layer for the subsequent electroplating process. In order to protect the walls of the metalized thru-silicon vias a negative dry-film photoresist is applied and patterned. The Au is then removed by sequential etching of the seed layer of Ti/Au. Lastly, the wafer is diced in individual chips and each die is electroplated with a thick layer of Au (>10 μm).

Diaphragm piece

For the fabrication of the diaphragm piece a 200-μm-thick, low-resistivity Si wafer is employed. The fabrication starts by growing 1 μm-thick SiO₂ on both sides of the 200 μm-thick Si wafer—to create a dielectric layer and is followed by sputtering a thin layer of Ti/Au on one side. Afterwards, a thick layer of Au is electroplated as described at the previous section. Then, 1-μm-thick layer of nitride is deposited through Plasma Enhanced Chemical Vapor

Deposition (PECVD) process in order to prevent potential RF shorts between the diaphragm and the interface piece during actuation. A positive photoresist is spin-coated and patterned and the PECVD nitride is selectively removed by dry etching using plasma technology. Lastly, the wafer is diced in individual chips that are ready for assembly.

Coaxial cavity

The coaxial cavity is structured in copper and was manufactured using conventional CNC-machining. In order to reduce the surface roughness of the upper surface of the coaxial cavity and enable a good electrical contact with the silicon-micromachined spacer chemical mechanical polishing (CMP) is performed on its upper surface. This is then followed by sputtering a thin layer of Au.

CONCEPT VALIDATION

In order to evaluate the feasibility of the proposed concept a tunable resonator geometry was designed, manufactured and measured for a frequency range within the UHF band. A three-dimensional view of the simulated model including realistic dimensions—specified by electromagnetic simulations in ANSYS HFSS—is illustrated in Figure 1. The details of the manufactured silicon-spacer (outer dimensions: 22 mm x 22 mm), the resonator assembly and the utilized experimental setup are presented in Figure 3. As can be seen, the silicon-spacer is mounted on top of the cavity with mechanical pressure. A mechanical structure was furthermore added on the sides of the cavity to facilitate the actuator mounting. For proof-of-concept demonstration purposes a commercially available piezoelectric stacked-actuator from STEMiNC was utilized. Note that in an alternative manner, piezoelectric

disk actuators that do not require external mechanical support—as the ones in [5] can be directly attached on the silicon spacer.

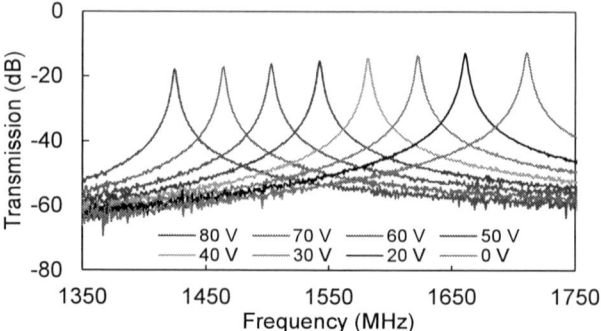

Figure 4: RF-measured response (transmitted power) of the assembled weakly-coupled cavity resonator for various DC-biasing states of the piezoelectric actuator between 0 and 80 V.

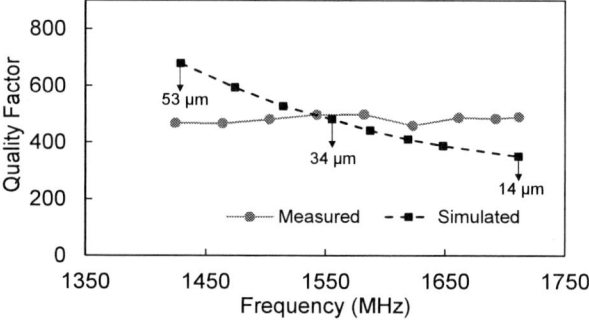

Figure 5: RF-measured and HFSS-simulated quality factor for various resonant frequencies of the resonator assembly. The figure illustrates a comparison between the measured tuning states of Figure 4 and the HFSS-simulated states for capacitive gap variations between 53 and 14 μm.

The RF-performance of the resonator was experimentally evaluated in terms of scattering parameters (transmitted/reflected power) with an Agilent-E8361A network-analyzer and DC biasing provided by a 2400 Keithley source meter. Various DC-biasing states between 0-80 V (<40 μm displacement) that result in a center frequency tuning between 1711-1424 MHz were examined, Figure 4. For this tuning range, the achieved capacitive-gaps were determined between 53-14 μm after fitting the simulated resonator response (in ANSYS HFSS) to the one obtained by RF measurements. The measured Q factor of this resonator was calculated from the measured 3-dB bandwidth Δf of each tuning state (Figure 4) using (1)—in which IL denotes insertion loss at the center frequency f_o and was found between 459 and 505. As can be seen in Figure 5, the measured Q is in a fair agreement with the one predicted in ANSYS HFSS-simulations (405-667), successfully demonstrating the feasibility of the proposed resonator concept. All-in-all, this is the first UHF resonator with such a high-Q and electronically-controllable micro-meter scale gaps. More details on this

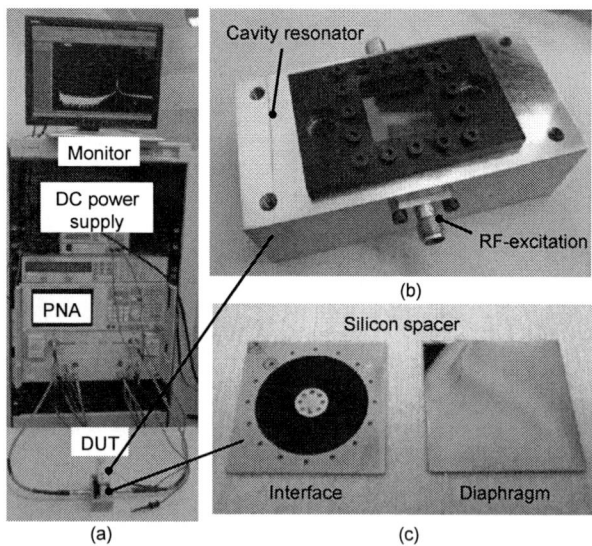

Figure 3: (a) Experimental setup and manufactured prototype. The resonant frequency of the resonator is tuned by applying a DC bias voltage between 0-80 V to a STEMiNC (SMPAK157742D50) piezoelectric-actuator that is mounted on top of the silicon-micromachined spacer. (b) Assembled cavity resonator and silicon-micromachined spacer. (c) Silicon-micromachined spacer.

978-1-4799-7956-1/15 $31.00 © 2015 IEEE

concept will be presented at the conference.

$$Q = \frac{Q_m}{1 - 10^{-IL/10}}, \qquad Q_m = \frac{f_o}{\Delta f} \qquad (1)$$

CONCLUSION

A silicon-micromachined spacer and a hybrid-integration concept that facilitate the realization of low-frequency (UHF band) and high-Q coaxial cavity resonators with well-defined, electronically-controllable micrometer-scale (10s of microns) gaps was reported in this paper. Using the proposed hybrid integration technique, a tunable resonator was experimentally validated. It exhibited a variable center frequency between 1424-1711 MHz while preserving a relatively high-Q between 459 and 505.

REFERENCES

[1] V. E. Boria and B. Gimeno, "Waveguide Filters for Satellites", *IEEE Microw. Magazine*, vol. 8, no. 5, pp.60-70, 2007.

[2] G. F. Craven and R. F. Skedd, Evanescent Mode Microwave Components, *Artech House*, 1987.

[3] S.-J. Park, I. Reines, C. Patel, and G. M. Rebeiz, "High-Q RF-MEMS 4–6-GHz tunable evanescent-mode cavity filter", *IEEE Trans. Microwave Theory Tech.*, vol. 58, no. 2, pp. 381–389, Feb. 2010.

[4] D. Psychogiou, D. Peroulis, Y. Li and C. Hafner, "V-band bandpass filter with continuously variable centre frequency", *IET Microw., Antennas Propag.*, vol. 7, no. 8, pp. 701-707, April 2013.

[5] D. Psychogiou and D. Peroulis, "Reconfigurable bandpass filter with center frequency and bandwidth control," *Microw. Opt. Technol. Lett.*, vol. 55, no. 11, pp. 2745–2750, Nov. 2013.

[6] M. S. Arif, and D. Peroulis, "A 6 to 24 GHz continuously tunable, micro-fabricated, high-Q cavity resonator with electrostatic MEMS actuation", in *IEEE MTT-S Int. Microwave Symp. Digest*, pp.1-3, 17-22 June 2012.

CONTACT

*D. Psychogiou, tel: +1-765-496-0326;
pdimitra@purdue.edu

SIMULTANEOUS MULTI-FREQUENCY SWITCHABLE OSCILLATOR AND FSK MODULATOR BASED ON A CAPACITIVE-GAP MEMS DISK ARRAY

Thura Lin Naing, Tristan O. Rocheleau, and Clark T.-C. Nguyen
University of California, Berkeley, USA

ABSTRACT

An electromechanical circuit constructed from array-composites of capacitive-gap micromechanical resonators with differing frequencies, wired in closed-loop feedback with a single ASIC amplifier, provides a first MEMS-based multi-frequency oscillator generating simultaneous oscillation outputs in the vicinity of 61 MHz. The use of only one amplifier for all frequencies (as opposed to one for each frequency) saves substantial power and is made possible by exploiting softening and damping non-linearities in the MEMS resonators, often considered a limitation, but here providing amplitude limiting that prevents amplifier desensitization to other frequencies. Furthermore, electrical stiffness-based frequency tuning enables Frequency-Shift Keyed (FSK) modulation of the output waveform, offering a space and power-efficient multichannel transmitter, as desired for mobile applications requiring long battery life, such as wireless sensor nodes. Indeed, while capable of multiple simultaneous and independent frequency outputs, this oscillator consumes only 137 µW, which is one-third that of previous multi-frequency efforts that only produce one frequency at a time [1].

INTRODUCTION

Wireless technology, which already plays a major part in our daily lives, is expected to expand to networks of billions of autonomous sensors in coming years: the so-called Internet of Things [2]. In one vision, sensors employing tiny, low-cost wireless motes collect and transmit data through a mesh network while operating only on scavenged or battery power. Here, small form-factor, spectrum efficient, low-power wireless communication links are essential. Vibrating RF MEMS technology, with already available products ranging from compact low-phase-noise MEMS-based reference oscillators [3, 4], to band-selecting RF front-end duplexers [5], offers a compelling potential route towards such a vision.

Indeed, capacitive-gap transduced MEMS resonators already offer space and power savings over conventional oscillators [3], where the high Q's >100,000 exhibited by these on-chip resonators allow for noise performance exceeding even the challenging GSM specifications with only ~100 µW of power consumption. MEMS-based radios [6] offer even more interesting possibilities, especially considering the low power and blocker resilience they provide in such small sizes. While impressive, these previous MEMS circuits have limited frequency tuning capability (~100 kHz range), and lack the ability to simultaneously communicate on separate channels.

Pursuant to solving these deficiencies, this work presents an oscillator (*cf.* Fig. 1) that combines a single amplifier with a plurality of MEMS resonators capable of not only outputting multiple independent frequencies in the vicinity of 61MHz; but also, by exercising voltage-controlled electrical stiffness tuning [7] of individual

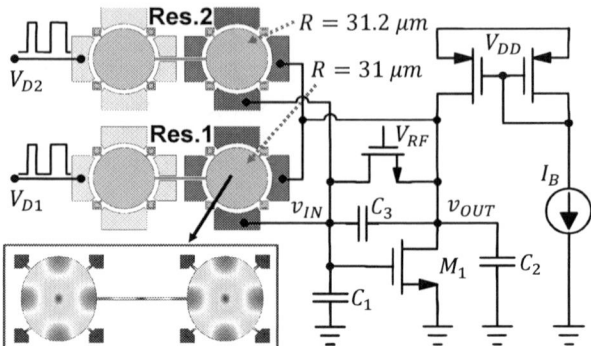

Fig. 1: Schematic of the Pierce topology multi-oscillator circuit used in this work. Independent tuning voltages and input bit streams are applied to Res. 1 and Res. 2. In each array-composite resonator, electrodes with the same color are electrically connected together. Inset shows FEM mode shape simulation of the two-disk array.

resonator array-composites, Frequency Shift Key (FSK) modulating each frequency to generate waveforms suitable for simultaneous wireless transmission in multiple channels. The chosen frequency is ideal for long-range unlicensed operation in the 52-74 MHz band white-space [8] and ISM bands at 27.12 MHz and 40.68 MHz [9].

DEVICE DESIGN AND OPERATION

The multi-oscillator system here comprises two main parts: a sustaining amplifier circuit and a multi-frequency MEMS array-composite resonator circuit. The ability to use only one amplifier with multiple resonators saves considerable power and derives from the ability of MEMS resonators to limit the oscillation amplitude [10]. In particular, unlike the vast majority of oscillators in which the sustaining amplifier rails out to limit the oscillation amplitude, thereby desensitizing it to any other frequency; the multiple oscillation amplitudes (at different frequencies) of this MEMS oscillator limit via spring softening and damping nonlinearities in the MEMS resonators. This then allows the amplifier to remain linear and provide gain at multiple frequencies.

The use of array-composites like that of Fig. 2, rather than just single resonators, allows for additional electrodes through which more control of the total array-composite resonator is obtained, e.g., for frequency pulling, strong input/output (I/O), "on/off" switching, etc. The array-composite of Fig. 2 specifically uses wine-glass-mode disk resonators to take advantage of their ability to attain the needed frequencies while allowing accurate specification via CAD layout of multiple unique frequencies on the same die. Each resonator comprises a 2 µm-thick, ~31 µm-radius polysilicon disk supported by beams at quasi nodal points and electrically coupled along their sidewalls to input-output electrodes by tiny 50 nm capacitive gaps.

In each two-disk array-composite, a half-wavelength

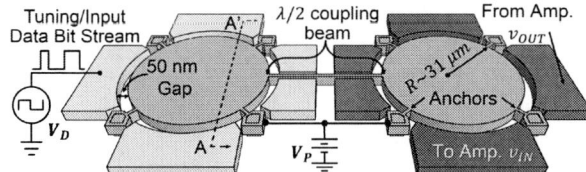

Fig. 2: Perspective view of a single MEMS two-disk array-composite comprising two suspended, mechanically-coupled disk resonators (orange) anchored at their 4 nodal points. Tuning electrodes (green) on one disk allow for electronic frequency control while input (blue) and output electrodes (purple) on the second one connect to bus bars to provide a multi-port multi-resonator device.

beam mechanically couples the disks, where sizing to half the acoustic wavelength forces the individual resonator disks to move in-phase at a single resonance frequency. Effectively, the array of disks behaves as one disk with twice the sidewall surface area with which to electrically interrogate or control it. To excite the composite resonator into motion, a bias voltage V_P on the disk structure combines with an ac drive voltage applied to the input electrodes (blue and labeled "From Amp") around the right-hand disk to produce forces across the input electrode-to-resonator gaps that, at resonance, excite the wine-glass (i.e., compound (2, 1)) mode shape, shown in the inset of Fig. 1. Here, disk radius R primarily sets the resonance frequency [11, 12]. A 0.2 µm difference in disk radii separates the Fig. 1 resonator frequencies by 300 kHz around 61 MHz.

While the disk on the right provides the I/O interface (to the sustaining amplifier), the disk on the left enables control of frequency via the voltage-controllable electrical stiffness, which influences the frequency via [7, 12]:

$$f_{ot} = f_{nom} \sqrt{1 - \frac{k_e}{k_{mre}}}$$

$$f_{nom} = \frac{1}{2\pi} \sqrt{\frac{k_{mre}}{m_{mre}}}; \quad k_e = \frac{\alpha^2 C_o}{d_0^2} V_{PG}^2 \quad (1)$$

where k_{mre} and m_{mre} are the effective dynamic mechanical stiffness and mass at the highest displacement location, k_e is electrical stiffness, the C_o is the total electrode-to-resonator overlap capacitance of a disk, d_o is the gap spacing, V_{PG} is the bias voltage across the gap, and α is a dimensionless constant based on mode and electrode shape, equal to 0.787 for the design used here [12]. This effect enables both frequency tuning via adjustment of the voltage on tuning electrodes (green in Fig. 2) producing the typical tuning response shown in Fig. 3(a), as well as a simple FSK modulation of the steady-state oscillation to be discussed.

To shut an array-composite off, just set its V_P to zero.

OSCILLATOR DESIGN AND LIMITING

Upon connection of the I/O electrodes of two or more Fig. 2-like array-composites to a suitable sustaining amplifier, oscillation ensues for those devices given sufficiently large dc-bias voltages V_P's. Whether or not the V_P is large enough depends on the relative magnitude of the resulting motional resistance R_x of the array-composite in question versus the effective transresistance gain R_{amp} of the sustaining amplifier. If $R_{amp} > R_x$, oscillations start

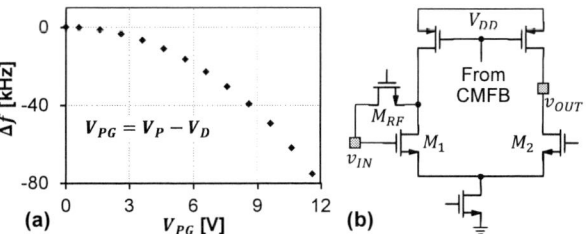

Fig. 3: (a) Oscillation frequency tuning vs. applied voltage across the capacitance gap of the tuning resonator. (b) TIA amplifier circuit schematic [6].

up and continue to grow until some form of nonlinearity reduces the loop gain to 1, at which point growth stops and steady-state oscillation ensues with a constant amplitude. In the vast majority of cases, including quartz crystal oscillators, electronic amplifier nonlinearity is responsible for limiting. However, unlike macroscopic resonators, the MEMS-based frequency setting element in the present oscillator can actually go nonlinear before the amplifier, at which point it ends up limiting the steady-state oscillation amplitude. Whether or not this happens depends upon the type and linearity of the amplifier itself.

Amplifier Limiting

Among MEMS-based oscillator types, series resonant and Pierce topologies have been most popular and successful. Series resonant oscillators often employ TransImpedance Amplifiers (TIA's) [6], such as shown in Fig. 3(b), that amplify an input current to an output voltage by a gain factor set by the value of resistance simulated by the shunt-shunt feedback transistor M_{RF}. While the ability of this topology to cancel common-mode noise is beneficial [13], it exhibits stronger nonlinearity than alternative circuits, because: 1) it uses three stacked transistors between the supply and ground, which sacrifices output voltage swing; and 2) the resistance of the feedback transistor M_{RF} can vary significantly as the output voltage changes. As a result, a TIA-based series resonant oscillator generally limits oscillation at amplitudes smaller than required to incite sufficient resonator nonlinearity, i.e., the amplifier limits the oscillation, not the resonator.

Conversely, the simpler Pierce topology [3] of Fig. 1 allows larger output swings, as it has only two stacked transistors between the supply and ground and does not employ feedback. By staying linear under larger voltage swings, a Pierce topology makes possible the resonator-limited operation needed to achieve a multi-oscillator.

Resonator Limiting

The disk resonator nonlinearity responsible for limiting oscillator amplitude generally manifests as a combination of stiffness nonlinearity that generates the well-known Duffing response [14]; and damping nonlinearity that increases resonator loss at large displacement amplitudes. In MEMS-based resonators, Duffing nonlinearity appears as either a hardening nonlinearity typically caused by stress, where the frequency response bends forwards, i.e., towards higher frequencies; or a softening nonlinearity, caused in capacitive-gap transduced resonators by higher-order components of electrical stiffness that bend the frequency response backwards, i.e., towards lower frequencies.

For the tiny-gap disks of this work, the softening

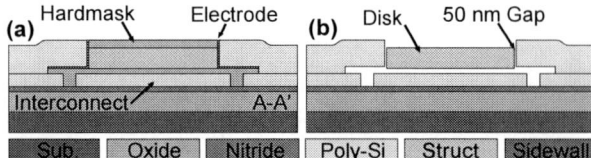

Fig. 4: Cross-sections of the fabricated disk resonators (a) before release and (b) after release in 49% HF.

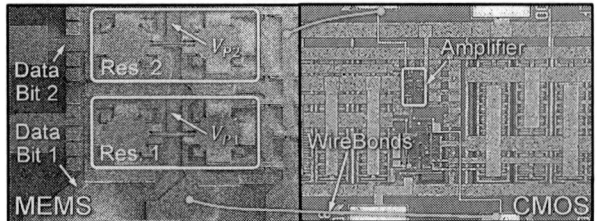

Fig. 5: SEMs of the fabricated MEMS circuit and die photo of the custom-made CMOS amplifier IC. Wirebond connections shown in orange.

nonlinearity dominates. Pursuant to modeling this nonlinearity, the differential equation governing resonator motion is

$$m_{mre}\frac{\partial^2 x}{\partial t^2} + (\gamma_0 + \eta x^2)\frac{\partial x}{\partial t} + kx + k_1 x^2 + \cdots$$
$$k_2 x^3 = \frac{\alpha C_o}{2d_0}V_{PG}v_{IN}\cos(\omega t) = F_\omega \cos(\omega t) \quad (2)$$

where F_ω is the magnitude of the drive force acting on the resonator, k_1 and k_2 model the softening nonlinearity derived from capacitive-gap interactions, η is an empirically-determined term describing non-linear damping [15], and

$$k = k_{mre} - 2k_e; \; k_1 = \frac{6k_e}{2d_0}; \; k_2 = -\frac{4k_e}{d_0^2}; \; \gamma_0 = \frac{\omega_o m_{mre}}{Q} \quad (3)$$

where Q is the resonator's unloaded quality factor, and $\omega_o = \sqrt{k/m_{mre}}$. Using the perturbation method of [15], this nonlinear system yields an approximate displacement amplitude as a function of frequency given by

$$X_0 = \frac{F_\omega/m_{mre}}{\sqrt{(\omega^2 - \omega_o'^2)^2 + (\omega\omega_o'/Q')^2}}$$
$$Q' = \frac{2\omega_o m_{mre}}{2\gamma_0 + \eta X_0^2}; \; \omega_o' = \omega_o + \kappa \cdot X_0^2; \; \kappa = \left[\frac{3k_2}{8k} - \frac{5k_1^2}{12k^2}\right]\omega_o \quad (4)$$

Equation (4) captures the drive amplitude-induced bending and peak lowering of the frequency response responsible for oscillator amplitude limiting.

EXPERIMENTAL RESULTS

Pursuant to demonstrating a multi-frequency oscillator and FSK-generator, two-disk array-composites like that of Fig. 2 were designed to operate around 61 MHz and fabricated using a process similar to that of [6] to achieve the final cross-section shown in Fig. 4. Here, POCl₃-doped polysilicon deposited via low-pressure chemical-vapor deposition (LPCVD) at 615 °C provided all resonator structure, electrode, and electrical interconnect material. A high-temperature oxide (HTO) sidewall sacrificial deposition defined the 50-nm resonator-to-electrode gaps. Structures were released in 49% HF to yield the final test devices.

Amplifier ICs were designed and fabricated using a

Table 1: Design and Extracted Parameters of the Fabricated MEMS Resonator Array-Composite.

Q	40k	α	0.787	k_{mre}[N/m]	1.61×10^6
d_0[nm]	50	m_{mre}[pg]	11.33	k_e[N/m]	9.73×10^2
V_{PG}[V]	8.4	γ_0[N.s/m]	1.07×10^{-7}	k_1[N/m²]	5.84×10^{10}
C_o[fF]	55.66	η[N.s/m³]	4×10^8	k_2[N/m³]	-1.56×10^{18}

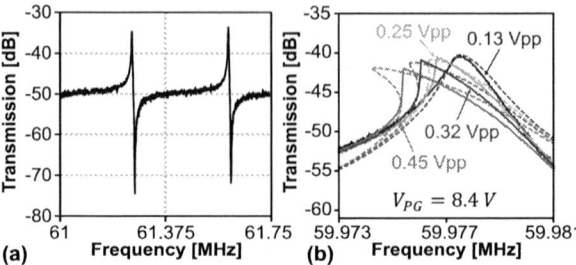

Fig. 6: (a) Frequency response as measured by an Agilent E5071C network analyzer for a two-disk array-composite with both resonators turned 'on' via applied bias voltage. (b) Measured (solid lines) and theoretical (dashed lines) spring softening and damping nonlinear response as a function of increasing drive voltage.

0.35 μm CMOS technology. To construct a complete multi-frequency oscillator, wire bonds connect released two-disk array-composites with the CMOS ASIC shown in Fig. 5, both of which are then mounted on a circuit board to provide needed bias and signal voltages.

Fig. 6(a) presents the frequency response of a two-disk array-composite under 10 V dc-bias and in vacuum. Fig. 6(b) presents the measured (solid lines) and theoretical (dotted lines, using (4)) the frequency response behavior of a single such disk with increasing drive voltage measured with forward going frequency, showing both spring softening (generating the saw-tooth shape) and damping (causing a decrease in peak height) nonlinearities. Again, the latter damping nonlinearity limits the oscillation amplitude. Here, the model can be seen to closely match measurement, verifying that capacitive-gap derived phenomena govern the stiffness non-linearity and that significant amplitude-limiting loss manifests when the drive voltage surpasses ~130 mVpp.

Fig. 7(a,b) present measured output spectra from a Pierce-based multi-oscillator like that of Fig. 1, where application or removal of V_P's ~10 V on/off switch the MEMS array-composites sequentially. Fig. 7(c) and Fig. 8 demonstrate operation with bias applied to both resonators, showing two simultaneous oscillation frequencies, each independently amplitude-limited as expected when using the linear sustaining amplifier of a Pierce oscillators. Here, the ASIC operates off a 2.8 V supply drawing ~49 μA. To verify the need for MEMS-based amplitude limiting, Fig. 9 presents the output of another such oscillator built instead using the TIA of Fig. 3(b) designed to limit at drive amplitudes above 10 mVpp. Unlike the Pierce design, when both resonators are "on", one of them grows in amplitude faster than the other and causes amplifier-induced (rather than resonator-induced) limiting, desensitizing the amplifier to other frequencies, and making it impossible to achieve the desired multi-frequency output.

Finally, to gauge the efficacy of this mechanical circuit as a multi-channel transmitter, Fig. 10 presents the output

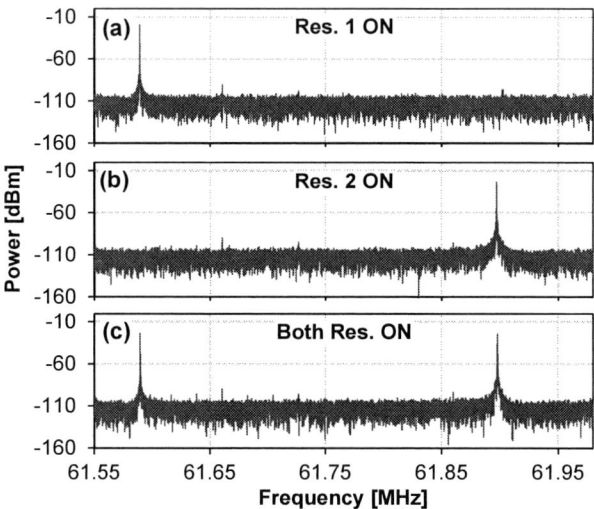

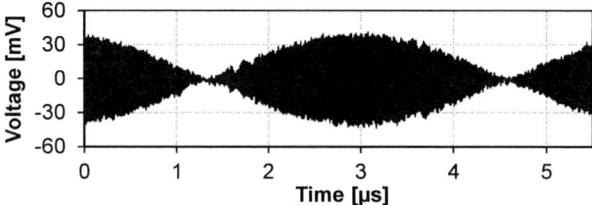

Fig. 7: Pierce oscillator output spectra measured on an Agilent N9030A spectrum analyzer. (a) shows oscillator output with resonator 1 turned on via an applied bias of 9.35 V, (b) with 11.6 V applied to resonator 2, and (c) simultaneous oscillation with bias voltages applied to both.

Fig. 8: Measured oscillator output waveform of the Pierce-based circuit with two resonators active.

with one oscillator active with a binary FSK input signal applied to the tuning electrode of its array-composite. This realizes electrical stiffness-induced switching of the oscillator frequency, thereby producing the fast-response, continuous-phase FSK modulation shown, at a minimum-shift keyed bitrate of 40 kbps.

CONCLUSIONS

The multi-frequency oscillator in this work is the first of its kind to generate simultaneous oscillation outputs around 61 MHz using capacitive-gap MEMS resonator array-composites while employing only a single amplifier. This MEMS-based circuit provides not only independently switchable and tunable oscillation outputs at multiple frequencies, but also a multi-channel FSK transmitter, all in a power- and space-saving package commensurate with the needs of long-term mobile applications, such as wireless sensor nodes. The ability to simultaneously transmit on multiple channels using only one amplifier enables a high degree of wireless multiplexing with substantially less power than competing multi-amplifier approaches—clearly desirable for tomorrow's autonomous wireless networks.

Acknowledgement: This work was supported by DARPA.

REFERENCES

[1] M. Rinaldi, *et al.*, "Reconfigurable ...," *IEEE Tran. on Electron Devices*, vol. 58, no. 5, pp. 1281-1286, May 2011.
[2] "Facts and Forecasts: Billions of Things, Trillions of

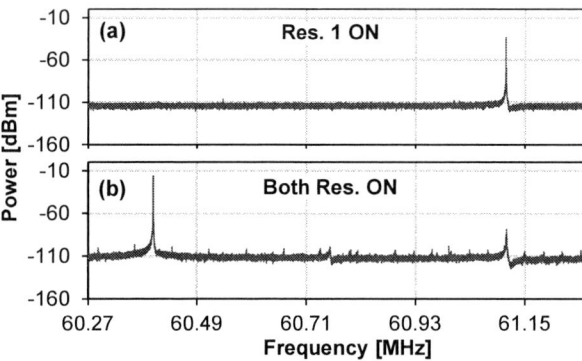

Fig. 9: TIA-based oscillator output spectra with (a) one resonator turned on and (b) both resonators on and loop gain increased by ~30% in an attempt to produce simultaneous oscillation. Oscillation at one frequency desensitizes the TIA, suppressing output at the other.

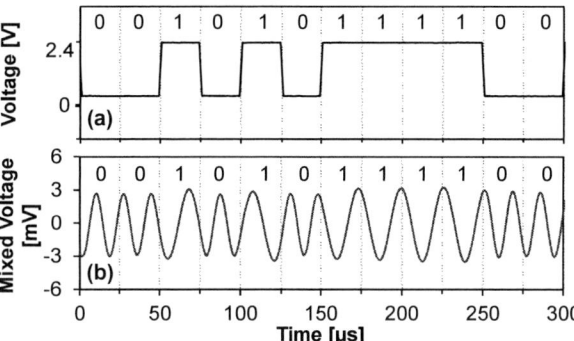

Fig. 10: Applying the 40 kbps modulation bit stream of (a) to the tuning electrodes on one disk array-composite generates the measured FSK modulated waveform of (b), shown mixed down to ~20 kHz to facilitate visualization.

Dollars," [Online]. Available: http://www.siemens.com.
[3] T. L. Naing, *et al.*, "A 78-microwatt GSM phase noise-compliant...," in *Proceedings, IFCS*, 2013, pp. 562-565.
[4] H. Lee, *et al.*, "Low Jitter and Temperature Stable MEMS Oscillators," in *Proceedings, IFCS*, 2012, pp. 271-275.
[5] Avago, LTE Band 7 Duplexer, Part No. ACMD-6007.
[6] T. O. Rocheleau, *et al.*, "A MEMS-based tunable RF channel...," in *Tech. Digest, Hilton-Head Conf.*, 2014.
[7] H. C. Nathanson, *et al.*, "The resonant gate...," *IEEE Trans. on Electron Devices*, vol. 14, no. 3, pp. 117- 133, Mar. 1967.
[8] "FCC, Et Docket No. 08-260, Second Report and Order," 2008. Available: http://hraunfoss.fcc.gov/edocs_public/attachmatch/DOC-286566A1.pdf.
[9] "FCC Online Table of Frequency Allocations," [Online]. Available: http://transition.fcc.gov.
[10] S. Lee, *et al.*, "Phase noise amplitude dependence in self-limiting ...," in in *Tech. Digest, Hilton-Head Conf.*, 2004.
[11] M. Onoe, "Contour vibrations of isotropic ...," *J. Acoust. Soc. Amer.*, vol. 28, no. 6, p. 1158–1162, Nov. 1956.
[12] M. Akgul, *et al.*, "A negative capacitance ...," *IEEE Trans. on UFFC*, vol. 61, no. 5, pp. 849-869, May 2014.
[13] Y.-W. Lin, *et al.*, "Series-resonant VHF micromechanical ...," *IEEE JSSC*, vol. 39, no. 12, pp. 2477-2491, Dec. 2004.
[14] G. Duffing, Erzwungene Schwingungen bei veranderlicher Eigenfrequenz und ihre ...: Vieweg & Sohn, 1918.
[15] R. Lifshitz, *et al.*, "Nonlinear ...," in *Reviews of Nonlinear Dynamics and Complexity*, Wiley, 2009, pp. 1-52.

CONTACT: Thura Lin Naing, tel: +1-510-384-4285; thura@eecs.berkeley.edu.

978-1-4799-7956-1/15 $31.00 © 2015 IEEE

TAPERED PHONONIC CRYSTAL SAW RESONATOR IN GAN

Siping Wang, Laura C. Popa and Dana Weinstein
HybridMEMS Group, Massachusetts Institute of Technology

ABSTRACT

This paper presents a new Phononic Crystal (PnC) resonator design in which a tapered PnC is used to confine a 970 MHz SAW resonance in a GaN-on-Si platform. Like other SAW resonator designs, the proposed resonator eliminates the release step common to most MEMS devices, leading to higher yield and simpler design and packaging.

However, the use of a tapered PnC reflector in this work (Fig. 1) reduces the footprint of SAW resonators by $> 100\times$ relative to conventional metal grating reflectors while maintaining high Q. A $3.5\times$ improvement in Q is experimentally demonstrated relative to resonators with uniform PnC reflectors of comparable dimensions. These devices can be integrated seamlessly with GaN MMIC technology.

INTRODUCTION

GaN MEMS resonators benefit from high piezoelectric coefficients, low acoustic losses, high electron mobility, radiation hardness, 2D electron gas (2DEG) sensing and transducer switching [1], and capacity for monolithic integration with HEMTs and passives in GaN MMIC technology [2]. In particular, many groups have studied GaN SAW devices for sensors and RF front-end components due to ease of fabrication on multiple substrates (e.g. SiC, Sapphire). However, one drawback of conventional SAW resonators is their large size, limited by low acoustic impedance mismatch of periodic metal gratings used to confine acoustic energy. Typically, hundreds of gratings are required, resulting in mm-scale devices. Shallow grooves (depth $h \ll \lambda_{SAW}$) etched into the piezoelectric material can increase acoustic impedance mismatch with 2-3$\times$ reduction in footprint. In principle, deep etched grooves would provide stronger reflection, but resonance degrades rapidly with groove depth for $h/\lambda_{SAW} > 0.025$ due to wave scattering into the substrate [3].

Alternately, a Phononic Crystal (PnC) composed of deep etched periodic holes can be implemented to form the SAW reflector. Such micro-scale PnC structures have been shown to provide good acoustic confinement [4]. Researchers have demonstrated PnCs in metal gratings [5] and band gaps for SAWs using etched holes [6]. However, just as in the case of deep-etched groove reflectors, a deep-etched uniform PnC provides too high an impedance contrast to the resonant cavity, resulting in substrate scattering at the resonator-PnC interface (Fig. 2(a)). In this work, we modify the PnC reflector to generate a gradual impedance change using tapered hole dimensions and reduce SAW scattering at the resonator boundaries (Fig. 2(b)). As can be seen, only 20 layers of reflectors are necessary for effective energy confinement, leading to SAW resonator design with significantly smaller footprint. This tapered PnC is analogous to Gradient Index (GRIN) photonic crystals. This method of tapering reflectors

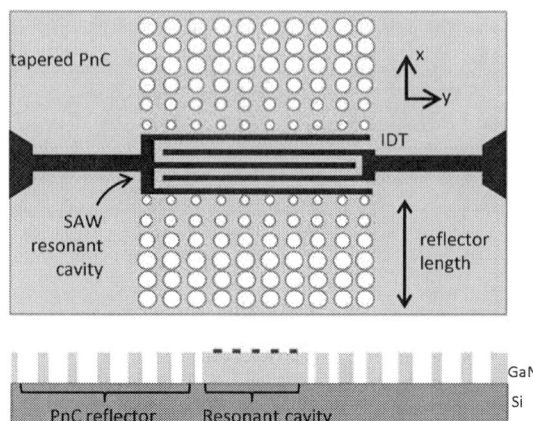

Figure 1: Top view and cross section schematics of tapered Phononic Crystal (PnC) SAW resonator.

cannot be applied to deep-etched grooves, since the very first free boundary will prevent acoustic waves from penetrating through, thus no subsequent tapering structures are of any effect.

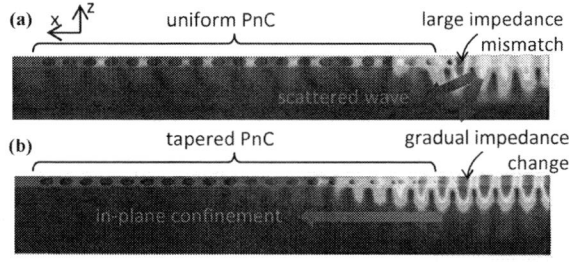

Figure 2: Simulation of SAW resonators with (a) uniform PnC reflector and (b) tapered PnC reflectors. The tapered design eliminates scattering of the SAW wave into the substrate due to abrupt acoustic impedance mismatch.

DESIGN AND SIMULATION

A square lattice PnC was chosen to form reflectors for the SAW resonator and to provide a band gap for effective confinement of acoustic energy. The PnC unit cell is a square pillar of infinite height composed of 1.56μm of GaN and infinite Si substrate underneath. A circular void is etched through the GaN layer from the top, providing scattering and reflection for the incoming SAW wave. The PnC band structure (or, dispersion relationship) of the unit cell is given in Fig. 3(a) and the irreducible Brillouin zone (IBZ) is shown in Fig. 3(b).

The band structure was derived in COMSOL FEM simulation by calculating the eigenfrequencies of the unit cell for each specific wave vector k, which is applied as a Floquet periodic boundary condition. In simulation, the unit cell is made tall enough (in the z-axis) to imitate the behavior of an infinite pillar. It has been verified that increasing the unit cell height does not affect bands below the sound cone, which contains an infinite number of modes that propagate in the Si substrate. The first 25 eigenmodes were calculated at each k to determine the PnC dispersion relations as shown in Fig. 3(a). Both modes outside the sound cone and modes populating the sound cone can be seen.

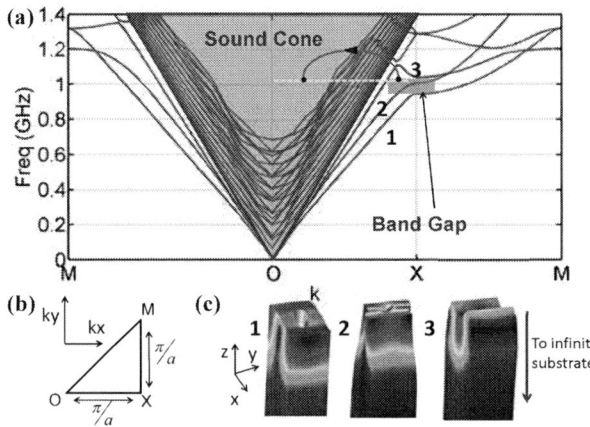

Figure 3: (a) Band structure of the PnC unit cell ,(b) its irreducible Brillouin Zone and (c) mode shapes for the 3 most relevant modes falling outside the sound cone, which do not propagate in the Si substrate.

It should be noted that the PnC does not have a complete band gap across all wave propagation directions (k). Rather, there is only partial band gap (between band 3 and band 1) from O to X. Band 2 cuts across the gap, but its displacement field is mostly along y-axis, as opposed to bands 1 and 3 whose displacement fields are along the x-axis and z-axis, as shown in Fig. 3(c). As a result the SAW wave propagating along the x-axis cannot couple with band 2. This partial band gap is sufficient for the designed resonator since the PnC needs to be reflective only in the resonant dimension and the targeted 1GHz resonance is located in the middle of the gap. If the PnC were directly applied to define the resonator boundaries, there would be an abrupt transition from solid GaN to PnC reflector resulting in broken translational symmetry along the x-axis. This would result in a coupling of modes between those in the PnC band gap and the infinite number of modes inside the sound cone, causing scattering into the substrate (illustrated in Fig. 3(a) using the red curve). The tapered PnC provides an adiabatic change in acoustic impedance such that the SAW propagates only in plane. This condition reduces scattering and isolates the surface modes from those of the sound cone.

Table 1 tabulates the performance of SAW resonators using various reflector designs for comparison with the proposed tapered PnC resonator. A figure of merit (FOM) of $f \cdot Q/L'$ is defined to capture the trade-off between reflector size and performance. Here L' is the reflector length normalized by $\lambda_{SAW}/2$. The 1GHz tapered PnC design in

SAW Reflector	f (MHz)	Q	L'	FOM	Comment
Metal grating	34.6	3850	400	333	Al on LiNbO$_3$ [8]
Shallow etched grating (grooves)	170	12000	300	6840	Y-Z LiNbO$_3$ [3]
	174	30000	600	8700	Y-Z LiNbO$_3$ [3]
	175	30000	600	8750	ST-quartz [3]
	786	11250	1000	8843	rotated Y-cut quartz [7]
Deep etched uniform PnC	980	246	20	12054	1.56um GaN on Si (simulation)
Deep etched tapered PnC	979	2600	20	1.27e5	1.56um GaN on Si (simulation)

Table 1: Comparison of SAW resonator performance. FOM is defined as $f \cdot Q/L'$, where L' is reflector length normalized by $\lambda_{SAW}/2$.

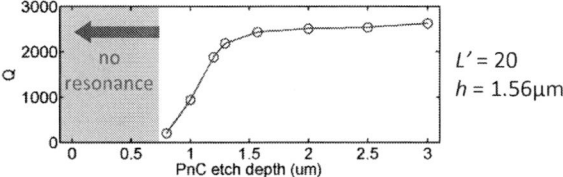

Figure 4: Simulated dependence of Q on etch depth of a 1 GHz SAW resonator with tapered PnC reflectors in GaN on (111) Si.

GaN-on-Si exhibits 14× higher FOM than state of the art Quartz [7][3] and LiNbO3 [8][3] SAW resonators. This result is based on FEM simulation assuming 1.56μm GaN on (111) Si with PnC etched through the GaN layer. The quality factor extracted from FEM modeling is due to substrate losses simulated through Perfectly Matched Layers (PMLs). In the case of the tapered PnC, a fixed PnC pitch was used while hole diameter was varied linearly to adjust the local acoustic impedance in the PnC. A maximum hole radius of $r = 0.21 \times a$ (where a is the PnC pitch) was chosen based on lithography and etching limitation for microfabrication. Unlike the case of etched grooves and uniform PnC, a deeper etch does not necessarily result in more scattering, since we can achieve an adiabatic impedance change between the resonant cavity and the reflector using varying hole dimensions. The tapering structure does lead to a slight increase in the total footprint, but this price is insignificant compared to the drastic increase in Q compared with the uniform PnC or the hundreds of reflecting grooves in conventional SAW resonators.

As seen in Fig.4, simulated Q for the tapered PnC resonator increases with etch depth, but saturates at depths $h \sim \lambda_{SAW}$. However there are practical constraints on the etch depth defined by fabrication limitations. Correspondingly, a minimum hole size of 0.3μm and linear taper over 8 unit cells were chosen for 1GHz SAW design (pitch $a = 1.92\mu$m, max hole diameter = 0.81μm). The devices in this work are based on a design with etch depth of 1.56μm.

In practice, it is expected that etch depth will be affected by the hole diameter. As the voids become smaller, the etch will be shallower and thus non-uniform along the tapered PnC. However, the tapered SAW resonator design is robust to the etch depth and hole size correlation, since all that is required is an adiabatic transition from homogeneous mate-

978-1-4799-7956-1/15 $31.00 © 2015 IEEE

rial to uniform PnC reflectors. A gradual transition that involves both growing hole diameter and deeper etch achieves this goal and can be optimized by adjustment of PnC tapering to compensate for any effect of reduced etch depth for smaller hole size.

FABRICATION

Folded PnC resonators were fabricated in MIT's Microsystems Technology Lab using Raytheon's MMIC GaN-on-Si heterostructure, comprised of AlGaN(25 nm)/GaN(1.7 μm) grown on (111)-Si using Molecular Beam Epitaxy (MBE).

A shallow AlGaN etch was used to remove the 2D electron gas (2DEG) between the Al-GaN/GaN layers and allow for efficient piezoelectric transduction in the GaN layer. A 100 nm layer of Ni (which can be used as the gate metal for GaN HEMTs) was then deposited and patterned to define piezoelectric IDTs. The choice of Ni for the electrodes is a departure from conventional Au electrodes found in GaN MMICs, as Au is mechanically lossy and is known to reduce resonator Q. However, Au-free GaN HEMT technology is starting to become mainstream as foundries develop heterogeneous processes with CMOS or simply want to run GaN processes in the same facilities as CMOS. Since these devices are processed side by side with GaN HEMTs, a PECVD Si_3N_4 layer (150 nm) was deposited to passivate the surface and protect the 2DEG channel in the HEMTs. A deep Cl_2 GaN etch then defined the PnCs and acoustic cavities.

Metal pads (50 nm Ti/300 nm Au) were then connected to the gate electrodes through vias in the passivation layer. In our typical GaN MEMS processing [1], a XeF_2 etch would then be used to release the MEMS structures from the Si substrate. This step is not necessary for SAW devices. Fig. 5 shows the device cross section.

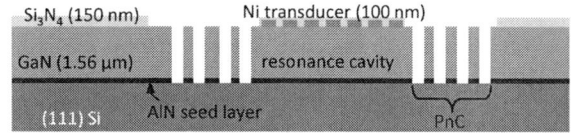

Figure 5: Cross section of the fabricated GaN SAW resonator using MBE GaN on (111) Si.

RESULTS AND ANALYSIS

Devices were characterized in vacuum using a standard 1-port measurement using an Agilent 5225A Network Analyzer, with on-chip Open de-embedding. SEMs for GaN SAW resonators with metal grating, uniform PnC, and tapered PnC reflectors, along with their measured frequency response are shown in Fig. 6. All three resonators have identical overall dimensions ($L' = 20$). The metal grating SAW resonator shows no discernable peak as the dimension of the reflector is too small to successfully confine a mode. The uniform PnC provides Q of 248 at 947 MHz. In comparison, the tapered PnC resonator demonstrates Q of 880 at 976 MHz, a 3.5$\times$ improvement in performance. Measured Q is lower than predicted by simulation due to exposure and etch variations resulting in smaller and shallower holes in the PnC than designed. However the comparison between the

different resonator designs on the same chip clearly demonstrates the benefits of tapered PnC design over uniform PnC and metal grating reflector for SAW resonators.

A major benefit of SAW resonators is the large contact region with the substrate, which enables excellent heat dissipation. As shown in Fig. 7(b), the resonant frequency barely shifts (<3ppm) when input power level is increased from -7.5dBm to +10dBm. Significant $A - f$ effects have been observed in other MEMS resonators anchored using tethers [9], which introduce large thermal resistance and significant frequency shift due to self heating at high power levels. In comparison, Fig. 7(a) shows the frequency response of a tethered device fabricated on the chip at increasing power levels and the frequency shift is as high as 128ppm. A slight increase of the floor (0.03dB) is observed for the GaN SAW resonator, likely due to the nonlinear permittivity of GaN.

CONCLUSION

This paper presents a compact GaN SAW resonator design compatible with GaN MMIC technology. Applying a gradual taper to a PnC for confining the SAW, this design reduces the scattering loss of deep etched PnC reflectors by 3.5$\times$ and reduces the necessary footprint of SAW resonators by 100$\times$ compared with conventional designs using metal gratings or shallow-etched grooves. This work offers a solution for monolithic timing and RF wireless communication applications including GHz MEMS front end band pass filters with small footprint and excellent power handling in GaN MMIC technology.

ACKNOWLEDGEMENT

The authors thank Brian Schultz and Thomas Kazior at Raytheon for GaN growth and process discussions, and Prof. Steven Johnson at MIT for helpful discussion. This work was funded by DARPA DAHI Foundry N66001-13-1-4022 and NSF Career EECS-1150493.

References

[1] L. Popa and D. Weinstein, "Switchable piezoelectric transduction in AlGaN/GaN MEMS resonators," in *Transducers*, 2013, pp. 2461–2464.

[2] A. Ansari, M. Rais-Zadeh, and et al., "Monolithic integration of GaN-based micromechanical resonators and HEMTs for timing applications," in *IEDM*, Dec 2012, pp. 15.5.1–15.5.4.

[3] R. Li, R. Williamson, and et al., "Experimental Exploration of the Limits of Achievable Q of Grooved Surface-Wave Resonators," in *IEEE Ultrasonics Symposium*, 1975, pp. 279–283.

[4] M. Su, R. Olsson, El-Kady, and et al., "Realization of a Phononic Crystal Operating at Gigahertz Frequencies," *APL*, vol. 96, no. 5, pp. 053 111–3, Feb 2010.

[5] V. Yantchev and V. Plessky, "Analysis of Two Dimensional Composite Surface Grating Structures with Applications to Low Loss Microacoustic Resonators," *JAP*, vol. 114, no. 7, p. 074902, 2013.

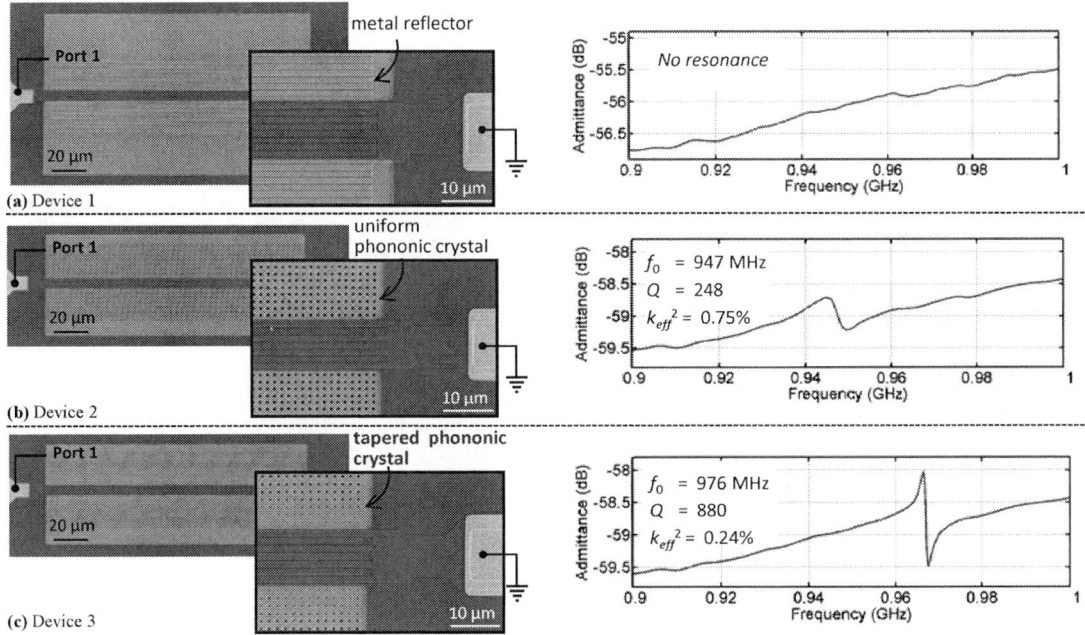

Figure 6: SEMs and measured frequency response of three types of GaN SAW resonators with identical foot print using $L' = 20$ (38.4 μm). While resonance could not be established by metal grating reflectors as shown in (a), tapered PnC resonator (c) provides $3.5\times$ improvement in Q over uniform PnC (b).

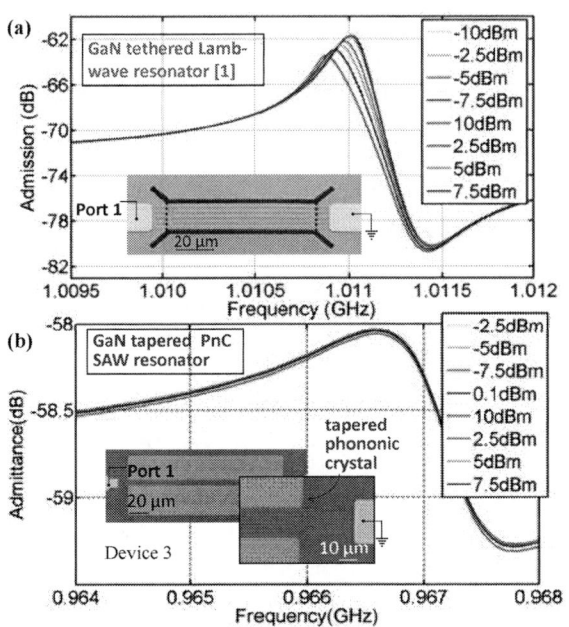

Figure 7: Admittance of the GaN tethered Lamb-wave resonator and GaN tapered PnC SAW resonator measured at various RF power levels.

[6] S. Benchabane, V. Laude, and et al., "Evidence for Complete Surface Wave Band Gap in a Piezoelectric Phononic Crystal," *Physical Review E*, vol. 73, no. 6, p. 065601, 2006.

[7] P. S. Cross, W. R. Shreve, and T. S. Tan, "Synchronous IDT SAW resonators with Q above 10,000," in *Ultrasonics Symposium.* IEEE, 1979, pp. 824–9.

[8] P. S. Cross, "Properties of Reflective Arrays for Surface Acoustic Resonators," *Ultrasonics Transactions*, vol. 23, no. 4, pp. 255–262, 1976.

[9] A. Tazzoli, M. Rinaldi, and G. Piazza, "Experimental Investigation of Thermally Induced Nonlinearities in Aluminum Nitride Contour-Mode MEMS Resonators," *Electron Device Letters, IEEE*, vol. 33, no. 5, pp. 724–6, May 2012.

ENHANCED CONTROLLABILITY IN MEMS METAMATERIAL

Prakash Pitchappa[1,2], Chong Pei Ho[1,2], You Qian[1,2], Yu-Sheng Lin[1],
Navab Singh[2], and Chengkuo Lee[1*]

[1]Department of Electrical and Computer Engineering, National University of Singapore,
SINGAPORE
[2]Institute of Microelectronics, Agency for Science, Technology and Research (A*STAR),
SINGAPORE

ABSTRACT

In this report, we demonstrate a method to enhance the controllability of MEMS tunable metamaterial by isolating the electrical routing of alternate lines in the metamaterial unit cell array. The metamaterial consists of alternate lines of split ring resonators with two released heights. This allows for two independent tuning characteristics for a single MEMS metamaterial by selecting between the two external control ports. This technology can be further improved to provide line or pixel wise control, and can even be programmed to have one of many functionalities such as tunable filter, multicolor spatial modulator, gradient metamaterial or random metamaterial.

INTRODUCTION

Tunability in electromagnetic (EM) metamaterial adds in a whole new dimension of tailoring and actively manipulating the exotic EM properties enabled by metamaterial. Popular approaches to achieve tunable metamaterial are reported by either changing the material properties of a part of the metamaterial unit cell or the medium surrounding the metamaterial. Some of the control inputs includes optical [1], electrical [2], thermal [3] or magnetic [4] means. Most of these approaches utilize exotic materials that provides limited and non-linear performance and also demands bulky setup to provide the control signal. This limits the usage of these approaches in potential applications.

Microelectromechanical system (MEMS) based tunable metamaterials, uses a part of unit cell as active element that will mechanically deform to external control signals, which will then provide the desired tunability. MEMS metamaterials are extremely small, significantly fast and can be realized using CMOS compatible materials and processes. The major research interest in MEMS metamaterial is the integration of different types of MEMS actuators to provide various functionalities such as improved tuning range, switching of responses, active reconfiguration of unit cell, and lot more. The type of MEMS actuator used will define the specific functionalities of the MEMS metamaterial. Some of the popular terahertz MEMS metamaterials reported so far includes the use of out-of- plane thermally deformable bimorph structures [5], comb drive based in-plane reconfigurable metamaterials [6], stress beam based out of plane movable cantilevers [7-10] and electrostatically tuned parallel plate actuators [11] based tunable filters and modulators.

The next big paradigm shift in tunable metamaterial will be brought by realizing improved controllability in the tunable metamaterial. The best form of tunable metamaterial will be the one in which each of the unit cell can be addressed independently. This will pave the way for programming a single metamaterial to be used as a tunable filter, electro-optical switch, spatial modulator, multicolor metamaterial, gradient metamaterial or random metamaterial. However as a first step towards the realization of pixelated metamaterial, in this report we experimentally demonstrate the possibility of independently controlling the alternate lines of MEMS metamaterial by isolating the routing lines between the adjacent cells.

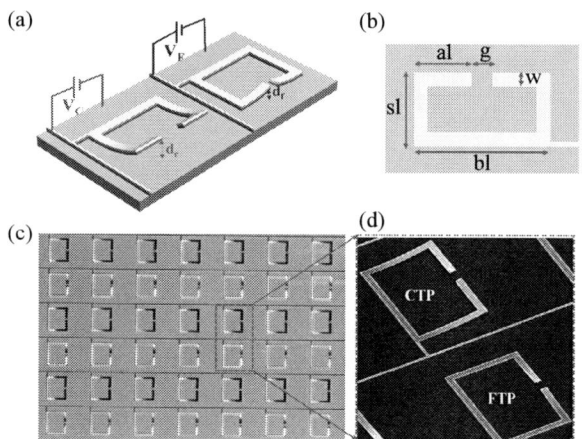

Figure 1: (a) Schematic drawing of MEMS metamaterial super cell with FTP and CTP part that can be independently controlled, (b) Top view of the SRR unit cell and its geometrical definitions, (c) and (d) shows the OM image of the fabricated metamaterial array and SEM image of the metamaterial super cell, respectively.

DESIGN AND SIMULATION

The proposed MEMS metamaterial consists of an array of 240 μm x 120 μm sized super-cell. The super cell comprises of two identical split ring resonators (SRR) as shown in Figure 1(a). Each of the SRRs in the super cell has a pitch of 120 μm x 120 μm. The base length (bl), side length (sl), gap side arm length (al), width (w) and gap (g) of the SRRs is 80 μm, 55 μm, 35 μm, 5 μm and 7 μm, respectively as shown in Figure 1(b). The difference between the two SRRs in the super cell is the portion of SRR that is released. In one of the SRRs, only gap side arms are released and is termed as Fine Tuning Part (FTP), while in the other SRR, side arms along with gap side arms are released and is termed as Coarse Tuning Part (CTP) as shown in Figure 1(d). More importantly, FTP and CTP are electrically isolated, and so can be actuated independently.

The SRRs are made of 0.5 μm aluminum with a 40 nm

aluminum oxide insulating layer, fabricated on a silicon (Si) substrate using CMOS compatible process [7]. After the release process, the cantilevers bends up due to the residual stress in the bimaterial layers and this initial state of the cantilevers is defined as state "0". Figure 1(c) and (d) shows the OM image of the metamaterial array and SEM image of the fabricated metamaterial super cell, respectively. The released cantilevers are actuated using electrostatic force, by applying voltage across the released cantilevers and Si substrate. As the applied voltage increases, the electrostatic force increases and brings the released cantilever closer to Si substrate. Beyond a critical voltage, called the pull-in voltage, the released cantilever will come in physical contact with Si substrate and this defines state "1" of the released cantilevers. As the FTP and CTP can be actuated independently, the MEMS metamaterial can have one of four possible configurations: S_{F0-C0}, S_{F1-C0}, S_{F0-C1}, and S_{F1-C1}.

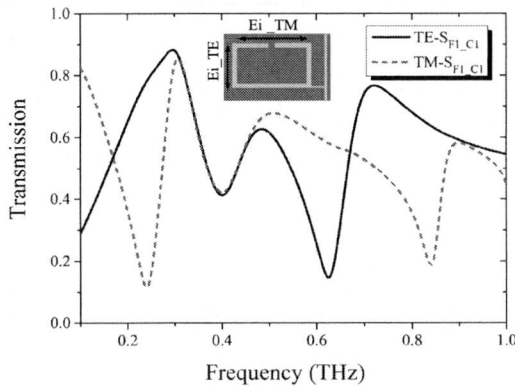

Figure 2: Simulated TE (black-solid) and TM (red-dash) mode transmission response for the MEMS metamaterial when both the FTP and CTP released cantilevers are in State "1" (S_{F1_C1}).

In order to determine the resonance mechanisms, finite-difference time-domain simulations were carried out for the metamaterial in S_{F1-C1} configuration. Aluminum was modelled as lossy metal with conductivity of 1e7 S/m, while Al_2O_3 and Si was modelled as dielectric with refractive index of 3.07 and 3.41, respectively in the THz spectral region. The incoming THz waves is incident normally and the electric field of the incoming waves that vary along gap side of SRR and side arm of SRR is considered as TM and TE mode, respectively. Figure 2 shows the simulated transmission response for TE (black-solid) and TM (red-dash) for the metamaterial in S_{F1-C1} configuration. The simulated surface current distribution in the SRR unit cell provides the clear understanding of physical mechanisms causing these resonances as shown in Figure 3.

In case of TE mode (E field along SRR side arms as shown in Figure 2 inset), there are two resonances at 0.4 THz and 0.63 THz. The circulating current at 0.4 THz suggests the excitation of LC resonance caused due to SRR like structures formed between the desired unit cell and metal lines as shown in Figure 3(a). At 0.63 THz, the dipolar resonance induces the surface currents along both the side arm length of the SRR as shown in Figure 3(b). In case of TM mode (E field along the gap bearing side of SRR as shown in Figure 2 inset), there are three resonances excited at 0.24 THz, 0.4 THz and 0.84 THz. The circulating current configuration in the SRR unit cell, shows the excitation of fundamental LC mode resonance at 0.24 THz as shown in Figure 3(c). At 0.4 THz, the resonance is caused due to the inclusion of the routing metal lines as shown in Figure 3(d). At 0.84 THz, the current configuration in the base length and gap side arm length are in the same direction as the incident E field and so suggests dipolar mode resonance.

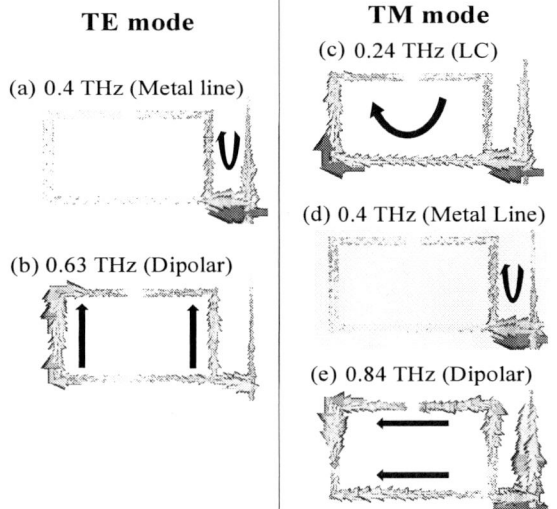

Figure 3: Simulated surface current distribution at different resonance frequency for TE and TM modes in MEMS metamaterial in S_{F1_C1} configuration.

To achieve tunability, air gap is introduced between the part of SRR unit cell and Si substrate. As the air gap increases, the resonance frequency will be red shifted owing to the increased net electrical capacitance of the metamaterial structure [7-11]. The understanding of these resonance mechanisms is of great importance, as depending upon the part of SRR actuated, only particular resonances will be influenced with changing air gap.

CHARACTERIZATION

In order to demonstrate the independent actuation of FTP and CTP part of the MEMS metamaterial and to study its electromagnetic response in four different configurations, two sets of experiments were carried out separately as discussed in the following two sub sections:

Electromechanical Characterization

The deflection of the released cantilever with respect to applied voltage was measured using Lyncee Tec Reflection Digital Holographic Microscope. Ideally the initial tip displacement for the cantilevers after the release process should be 100 nm, which corresponds to the thickness of the sacrificial layer used. However due to the residual stress in the bimaterial layers forming the cantilevers, the tip displacement is much higher. The measured initial tip displacement for FTP and CTP was 2 μm and 4.25 μm, respectively. In order to determine the pull-in voltages of FTP and CTP cantilevers, voltage was

978-1-4799-7956-1/15 $31.00 © 2015 IEEE   1033

applied independently for the FTP and CTP suspended cantilevers with respect to Si substrate in steps of 5 V. Figure 4 shows the initial deformation of the 35 μm long released FTP cantilevers (black cross circle), with the tip displacement of approximately 2 μm. When the applied voltage was increased to 35 V, pull-in occurred and the FTP cantilever comes in physical contact with the substrate as shown in the inset in Figure 4. Similarly, Figure 5 shows that the tip displacement of the 55 μm long released CTP cantilever to be 4.25 μm and the measured pull-in voltage is 25 V.

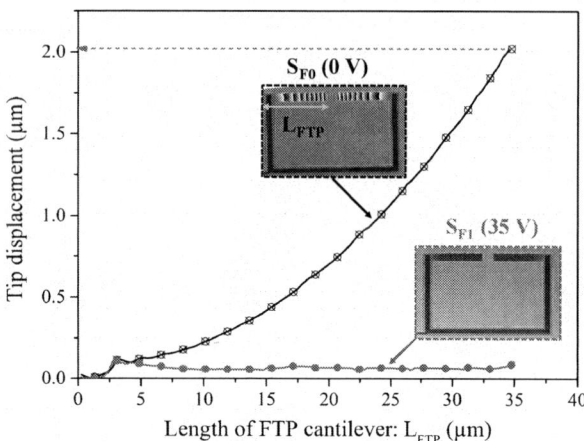

Figure 4: Measured FTP cantilever deflection after release, S_{F0} without applying voltage (black cross circle) and after applying pull-in voltage of 35 V, S_{F1} (red solid circle). The inset shows the phase variation with respect to vertical deformation for before and after applying 35 V.

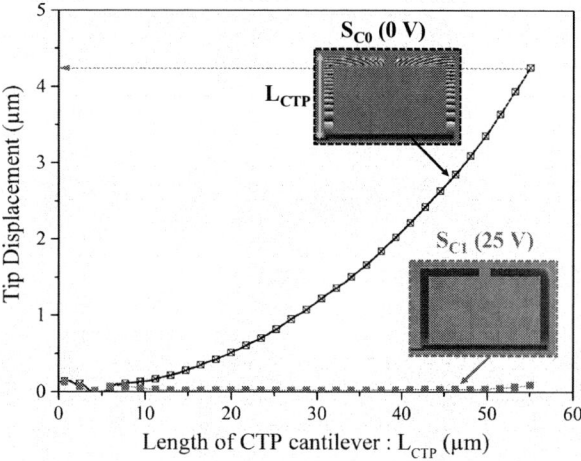

Figure 5: Measured CTP cantilever deflection after release, S_{C0} without applying voltage (black cross square) and after applying pull-in voltage of 25 V, S_{C1} (red solid circle). The inset shows the phase variation with respect to vertical deformation for before and after applying 25 V.

It is important to note that during pull in, the Al_2O_3 dielectric layer beneath the Al layer in the cantilever acts as a insulator between the cantilever and Si substrate, and thereby preventing electrical short. Thus for the EM experiments, 35 V and 25 V was used to achieve State "1"

for FTP and CTP, respectively. Interesting, the snapped down FTP and CTP cantilevers return back to the original State "0", when the voltage is removed, thereby ensuring repeatable operation for the devices.

Electromagnetic Characterization

The EM response of the proposed metamaterial at different configurations were measured using Teraview TPS 3000 system. The incident THz wave is incident normally on to the sample and transmission spectra were measured.

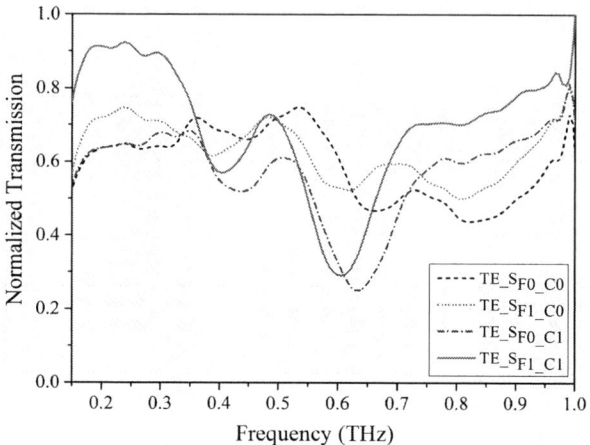

Figure 6: Measured THz transmission spectra in TE mode for different configuration of the MEMS metamaterial: S_{F0-C0} (black-dash), S_{F1-C0} (red-dot), S_{F0-C1} (blue-dash dot), and S_{F1-C1} (green-solid).

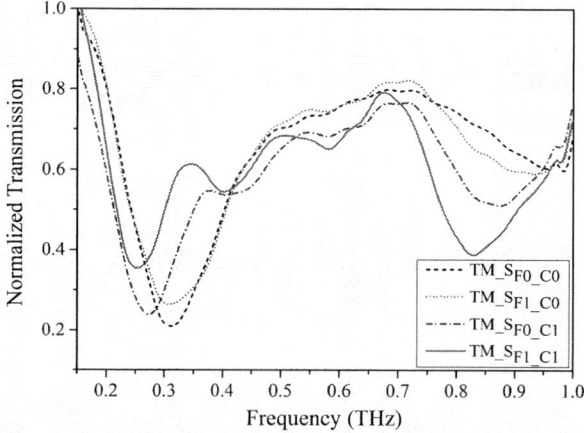

Figure 7: Measured THz transmission spectra in TM mode for different configuration of the MEMS metamaterial: S_{F0-C0} (black-dash), S_{F1-C0} (red-dot), S_{F0-C1} (blue-dash dot), and S_{F1-C1} (green-solid).

The resonance frequency of interest in TE mode is the dipolar resonance that occurs at 0.82 THz for the S_{F0-C0} configuration as show in Figure 6 (black-dash). When FTP SRR is switched to state "1", the dipolar resonance does not shift, because the released part of the FTP SRR is the gap side arms and does not influence the excitation of dipolar resonance in TE mode as shown in Figure 6 (red-dot). However on the other hand, when CTP SRR is switched to State "1", the resonance frequency shifts to

0.64 THz as shown in Figure 6 (blue - dash dot), because the CTP part includes the side arms of the SRR, which is the region where the resonance excitation occurs. Additionally when both the SRRs are actuated the resonance frequency shift to 0.61 THz which is in good agreement with the simulation results in Figure 2 (black-solid). For the TM mode measurement shown in Figure 7, the metamaterial configuration in $S_{F0\text{-}C0}$ (black-dash) has two resonant frequencies of interest at 0.31 THz and 0.95 THz. When only the FTP SRR is actuated, the shift in both these resonances are relatively negligible; around 0.005 THz and 0.02 THz, respectively as shown in $S_{F1\text{-}C0}$ (red-dot) in Figure 7. This is due to the smaller actuation range of FTP released cantilevers. However when the CTP SRR is actuated to $S_{F0\text{-}C1}$ (blue-dash dot), the resonance frequencies shift to 0.26 THz and 0.86 THz. When both the FTP and CTP SRRs are actuated, the resonance frequencies shift to 0.25 THz and 0.83 THz, respectively as shown in Figure 7 (green-solid) and is good agreement with the simulation results for $S_{F1\text{-}C1}$ configuration shown in Figure 2 (red-dot).

The proposed MEMS metamaterial provides two modes of tuning range – Fine and Coarse, which can be selected by appropriately choosing between the two SRRs actuation lines. The tuning range in TE mode for FTP and CTP cases was 0.002 THz and 0.2 THz, respectively. In the TM mode the tuning range for LC mode resonance was 0.005 THz for FTP and 0.05 THz for CTP, and for the dipolar resonance the shift was 0.02 THz for FTP and 0.13 THz for CTP cases, respectively. In the future works, we will improve the level of controllability in metamaterial by providing line wise or pixel wise control; so that a single metamaterial can be programmed to achieve numerous functionalities such as tunable filters, multi-color filters or modulators, electro-optic switches, gradient metamaterial, or random metamaterial.

CONCLUSIONS:

We experimentally demonstrate a method to enhance the controllability of MEMS metamaterial by isolating the alternate routing lines of the metamaterial. In order to demonstrate the concept; a super cell with two identical SRRs with different released part was used. By independently actuating the fine and coarse tuning split ring resonators, the tuning range was altered by approximately 10 times for the fundamental LC mode resonance. This simple method of enhancing the controllability opens up whole new exciting research possibilities and enables the realization of a single tunable metamaterial, where each unit cell can be addressed independently and further programmed to realize various functionalities.

ACKNOWLEDGEMENTS

The authors acknowledge the financial support from research grant AcRF Tier 2 - MOE2012-T2-2-154 at the National University of Singapore.

REFERENCES

[1] H. T. Chen, W. J. Padilla, M. J. Cich, A. K. Azad, R. D. Averitt, and A. J. Taylor, "A metamaterial solid-state terahertz phase modulator," Nature Photonics, vol. 3, no. 3, pp. 148–151, Mar. 2009.

[2] H. T. Chen, W. J. Padilla, J. M. O. Zide, A. C. Gossard, A. J. Taylor, and R. D. Averitt, "Active terahertz metamaterial devices," Nature, vol. 444, no. 7119, pp. 597–600, Nov. 2006.

[3] R. Singh, Z. Tian, J. Han, C. Rockstuhl, J. Gu, and W. Zhang, "Cryogenic temperatures as a path toward high-Q terahertz metamaterials," Appl. Phys. Lett., vol. 96, no. 7, pp. 071114-1–071114-3, Feb. 2010.

[4] B. Jin, C. Zhang, S. Engelbrecht, A. Pimenov, J. Wu, Q. Xu, C. Cao, J. Chen, W. Xu, L. Kang, and P. Wu, "Low loss and magnetic field-tunable superconducting terahertz metamaterial," Opt Exp., vol. 18, no. 18, pp. 17504–17509, Aug. 2010

[5] H. Tao, A. C. Strikwerda, K. Fan, W. J. Padilla, X. Zhang, and R. D. Averitt, "Reconfigurable terahertz metamaterials," Phys. Rev. Lett., vol. 103, no. 14, pp. 147401-1–147401-4, Oct. 2009.

[6] W. M. Zhu, A. Q. Liu, T. Bourouina, D. P. Tasi, J. H. Teng, X. H. Zhang, G. Q. Lo, D. L. Kwong, and N. I. Zheludev, "Microelectromechanical maltese-cross metamaterial with tunable terahertz anisotropy," Nature Commun., vol. 3, no. 1274, pp. 1–6, Dec. 2012.

[7] Y. S. Lin, Y. Qian, F. Ma, Z. Liu, P. Kropelnicki, and C. Lee, "Development of stress-induced curved actuators for a tunable THz filter based on double split-ring resonators," Appl. Phys. Lett., vol. 102, no. 11, pp. 111908-1–111908-5, Mar. 2013.

[8] P. Pitchappa, C. P. Ho, Y. S. Lin, P. Kropelnicki, C. Y. Huang, N. Singh, and C. Lee, "Micro-electro-mechanically tunable metamaterial with enhanced electro-optic performance," Appl. Phys. Lett., vol. 104, no. 15, pp. 151104-1–151104-5, Apr. 2014.

[9] C. P. Ho, P. Pitchappa, Y. S. Lin, C. Y. Huang, P. Kropelnicki, and C. Lee, "Electrothermally actuated microelectromechanical systems based omega-ring terahertz metamaterial with polarization dependent characteristics," Appl. Phys. Lett., vol. 104, no. 16, pp. 161104-1–161104-5, Apr. 2014.

[10] F. Ma, Y. S. Lin, X. Zhang, and C. Lee, "Tunable multiband terahertz metamaterials using a reconfigurable electric split-ring resonator array," Light, Sci. Appl., vol. 3, no. e171, pp. 1–8, May. 2014.

[11] Z. Han, K. Kohno, H. Fujita, K. Hirakawa, and H. Toshiyoshi, "MEMS reconfigurable metamaterial for terahertz switchable filter and modulator", Opt. Exp., vol. 22, no. 18, pp. 21326-21339, Sep. 2014..

CONTACT

*C. Lee, tel: +65 6516-5865; elelc@nus.edu.sg

BI-DIRECTIONAL EXTENDED RANGE PARALLEL PLATE ELECTROSTATIC ACTUATOR BASED ON FEEDBACK LINEARIZATION

E.E. Moreira[1], F.S. Alves[1], R.A. Dias[2], M. Costa[2], H. Fonseca[2], J. Cabral[1], J. Gaspar[2] and
L.A. Rocha[1,2],

[1]CMEMS, University of Minho, Guimarães, PORTUGAL
[2]INL, International Iberian Nanotechnology Laboratory, Braga, PORTUGAL

ABSTRACT

In this paper, we present a bi-directional extended range parallel-plate electrostatic actuator using feedback linearization control. The actuator can have stable displacements up to 90% of the full-gap (limited by mechanical stoppers) on both directions, i.e., the device can move ±2µm within a ±2.25µm gap. The system has successfully tracked references until 1kHz (limited by the dynamics of the device) and it presents a capacitor tuning range of 17, using an actuation voltage from 0 to 10V. The results presented here are a clear advance in respect to the current state-of-the-art in terms of tracking capabilities, total stable displacement and tuning range.

INTRODUCTION

Although commonly used, electrostatic parallel-plate actuation is limited to displacements up to 1/3 of the gap due to the pull-in effect [1]. Several techniques have been developed to overcome this limitation. The simplest of all consists on the design of an actuator with a gap 3 times larger than the desired one [2]. This is a simple technique, but the desired displacement is achieved at the cost of a higher voltage, which in most cases is a limitation by itself. Main techniques reported in literature to extend the travel range of electrostatic parallel-plate actuators can be classified into four main groups, namely: geometry leverage [3], series feedback capacitor [4], current drive methods [5] and closed-loop voltage control [6-8].

Extended operation of parallel-plate electrostatic actuators using feedback voltage control has already been demonstrated in [6-8]. While in [6, 7] the control algorithm is based on feedback linearization, the work presented in [8] uses an on-off control algorithm to achieve extended operation. In all these works [6-8], extended ranges up to 90% of the available gap were experimentally demonstrated. The work reported in [1] also demonstrated tracking capabilities up to 20Hz (for 90% of the gap).

The work presented here extends the state-of-the-art on feedback voltage control by including bi-directionality and an improved control algorithm enabling reference tracking up to 500Hz (for 90% of the gap). Tracking beyond 500Hz is also achieved, but the displacements are limited by the dynamics of the mechanical structure used.

ACTUATOR CONCEPT

The use of feedback voltage control in parallel-plate electrostatic actuators enables stable operation beyond 1/3 of the gap, i.e., the closed-loop system can be made stable for displacements over the full-gap by proper control of the applied voltage.

The work presented here uses feedback linearization

in the control loop to achieve extended range in parallel-plate electrostatic actuators. Feedback linearization attempts to cancel the nonlinearities of a nonlinear system so that the closed-loop dynamic behavior is in a linear form. The application of the technique results in a linear input-output relation (the nonlinearities are cancelled out). This technique can be applied to a class of nonlinear systems (in which MEMS devices are included) that are described by the so-called companion form, or controllability canonical form [9]:

$$x^{(n)} = f(x) + g(x)u \qquad (1)$$

where u is the scalar control input, x is the scalar output of interest and $X = \left[x, \dot{x}, ..., x^{(n-1)} \right]$ are the state variables. The linearized control input, when using this technique, becomes:

$$u = \frac{1}{g(x_{meas})}[c - f(x_{meas})] \qquad (2)$$

Here c is the response of the linear controller and x_{meas} is the measured output. In the case of a 1-DOF parallel-plate MEMS actuator, the equation of motion is [8]:

$$\dot{x} = y$$

$$\dot{y} = \frac{C_0 d_0 V^2}{2m(d_0 - x)^2} - \frac{k}{m}x - \frac{b(x)}{m}y \qquad (3)$$

where C_0 is the actuator rest capacitance, d_0 is the gap at rest between electrodes, m is the mass of the actuator, k is the spring constant and $b(x)$ the nonlinear damping coefficient. The resulting linearization function is:

$$V = \sqrt{\frac{2m(d_0 - x)^2}{C_0 d_0}\left[c + \frac{k}{m}x + \frac{b(x)}{m}\dot{x}\right]} \qquad (4)$$

A diagram of the closed-loop system used in this work is presented in Figure 1. The displacement is measured by sensing the changes in a capacitive transducer and the system is linearized using equation (4). The linear controller block of Figure 1 is responsible for the linear control response c and is implemented using linear control techniques such as proportional control or PID control.

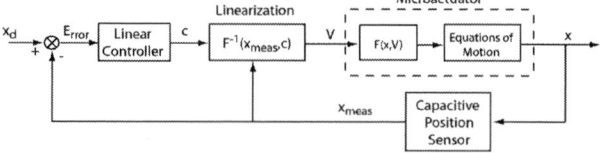

Figure 1: Diagram of the feedback linearization control loop for the 1-DOF MEMS device.

In the case the actuator enables bi-directional drive (see Figure 2) the same feedback control loop and linearization function applies for both sides. In this case, the side to be actuated (left or right) is chosen according to the reference value used (x_D in Figure 1). The inclusion of bi-directionality is particularly interesting for variable capacitors, since it enables extension of the tuning range.

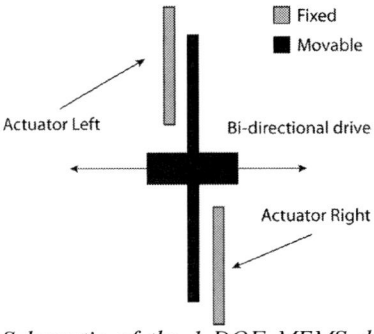

Figure 2: Schematic of the 1-DOF MEMS device used, showing bi-directional drive.

For a practical implementation of the feedback linearization control technique, the linear controller and the linearization function should be implemented in the digital domain. Also, the details of the actuator must be known for the linearization function. A close inspection to equation (4) reveals that the nonlinear damping coefficient and the displacement derivative are the more complex terms of the linearization function. One of the simplifications introduced here is disregarding the damping coefficient from the linearization expression in order to simplify the nonlinear control. This appears justified considering the fact that the damping force tends to zero as the velocity decreases and ultimately plays no role when the device remains at a constant position. The important part is that the device goes to the desired unstable position as long as the appropriate voltage level is applied.

SYSTEM IMPLEMENTATION

A block diagram of the closed-loop system implemented is shown in Figure 3. The key component of the system is the electrostatically actuated parallel-plate structure (Figure 4). The devices were fabricated in a 25 μm SOI process using two masks only (one for definition of the metal pads and other for definition of the structure). After DRIE the devices are released with HF vapor. The devices have a measured resonance frequency of 665Hz, a quality factor of 1.75 and a pull-in voltage of 5.7V. The actuators have a rest gap of 2.25μm, with physical mechanical stoppers at ±2μm limiting the maximum displacement to 90% of the full-gap (on both sides). The device has separated electrodes for actuation, enabling actuation on both directions (bi-directional drive), and differential capacitive sensing. A digital to analog converter (DAC) and an analog to digital converter (ADC), both operating at 1MHz sampling frequency, are responsible for actuating the device and reading the charge amplifier-based readout circuit, respectively. In order to guarantee control efficiency and flexibility, the controller was implemented on a field programmable gate array (FPGA). The FPGA also has modules for the communication protocol (UART) with a PC, and for the control of the actuation and acquisition systems.

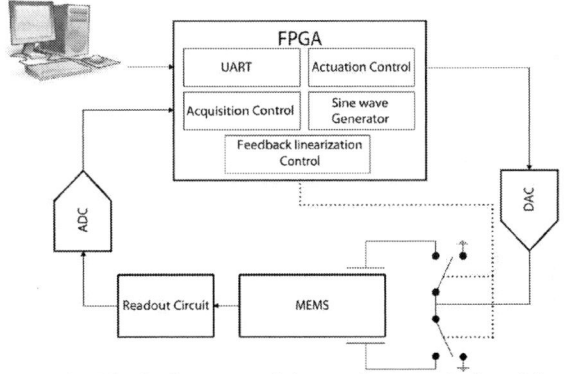

Figure 3: Block diagram of the implemented closed-loop control system.

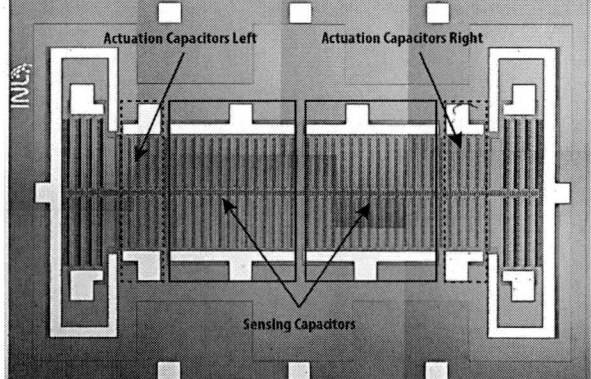

Figure 4: Microscope picture of the MEMS actuator used in this work.

Implementation of a closed-loop voltage controlled actuator using the feedback linearization technique requires that the characteristics of the device and the device displacement are known. In order to implement the linearization function, the devices were characterized using the method described in [10]. From both pull-in voltage and resonance frequency measurements, the values of interest (m, k, d_0 and C_0) were estimated. The response of the readout circuit was measured by applying a known acceleration to the actuator (using a shaker) while measuring the voltage output. Since the mass and spring of the device were estimated, the transfer curve indicating the displacement for a given output voltage can be computed (Figure 5). A lookup table describing the measured voltage output was implemented in the FPGA with 180 positions.

EXPERIMENTAL RESULTS

After system implementation, several tests were initially performed to check the suitability of the technique with focus on the details of the linear controller. The results presented during this section were acquired using a data acquisition board from National Instruments (NI USB-6363). For comparison (with the references used), the measured output voltages are presented as displacements (computed using the measured transfer curve shown in Figure 5).

978-1-4799-7956-1/15 $31.00 © 2015 IEEE

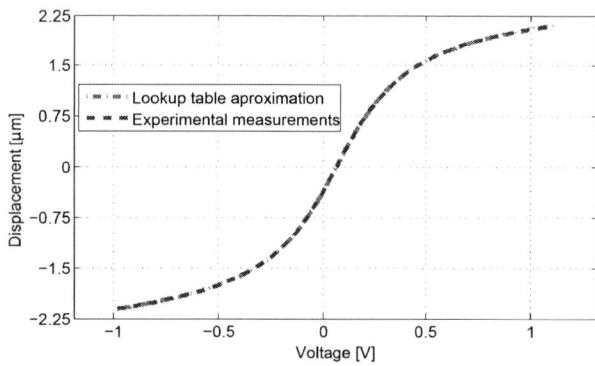

Figure 5: Measured readout output voltage and curve used to retrieve device position (x_{meas}).

Figure 6 shows operation details of the implemented electrostatic actuator for a reference change beyond the pull-in limit using a proportional controller with a gain of 10 (K_p=10). The results clearly demonstrate that the actuator can follow the reference and stabilizes at the desired position (1.75μm), beyond the pull-in limit. The measurement uncertainty, due to noise of the readout circuit is below 5nm.

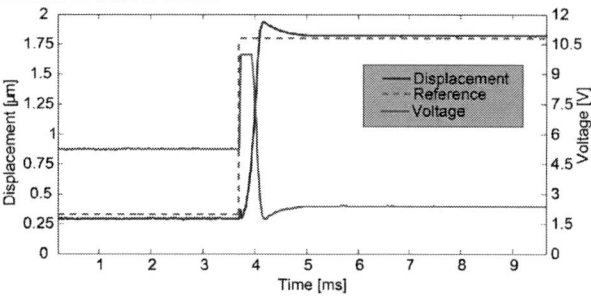

Figure 6: Measured details (voltage and displacement) of the electrostatic actuator for a reference change beyond the pull-in limit.

Figure 7 shows results for a second set of tests where the controller characteristics were verified. Both proportional (Figure 7) and proportional-integral linear controllers were experimentally verified. As expected, integral gain reduces the permanent error (to the reference) at the expense of a slower system response.

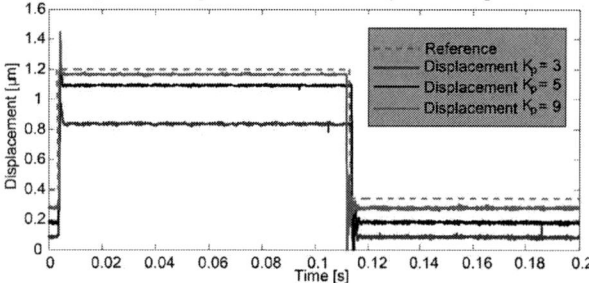

Figure 7: Measured actuator response for different linear controller proportional gains. The higher the gain, the lower is the error to the reference, but oscillations during transitions increase.

Next, the tracking of sinusoidal references was performed along the full available gap (on both directions) for several frequencies. Figures 8, 9 and 10 show sinusoidal reference tracking for 3 frequencies, 10, 500 and 1000 Hz. The results clearly demonstrate that the proposed electrostatic actuator is capable of bi-directional tracking up to 1000Hz (limited by the device dynamics).

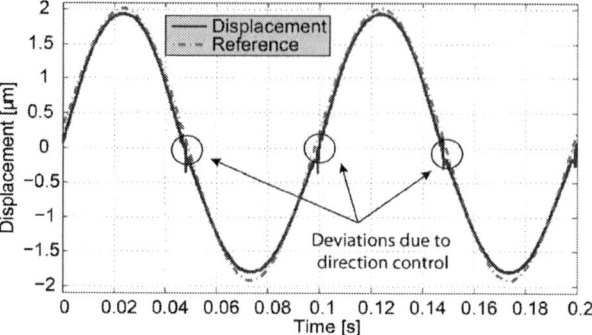

Figure 8: Experimental demonstration of bi-directional tracking at 10Hz using a Kp of 5. Some problems remain on the transition from the left to right side.

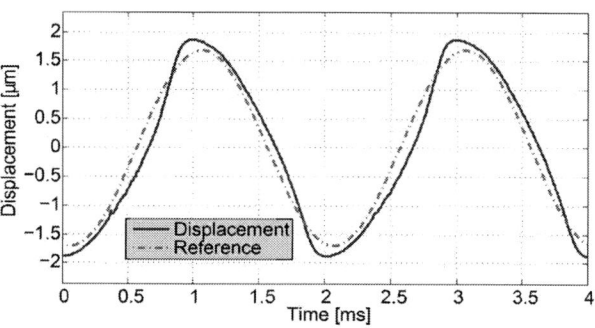

Figure 9: Experimental demonstration of bi-directional tracking at 500 Hz using a Kp of 5. Deformation of the displacement due to the MEMS dynamics (resonance of 665Hz) is already visible.

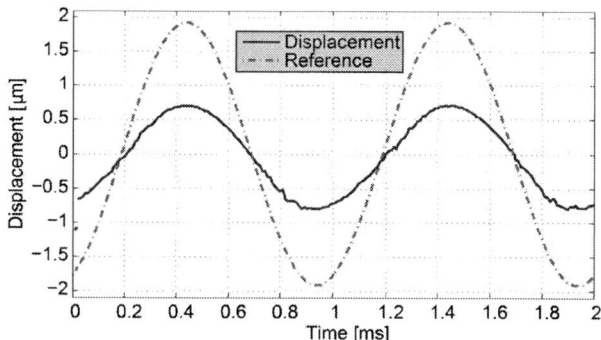

Figure 10: Experimental demonstration of bi-directional tracking at 1 kHz using a Kp of 5. The displacement does not follow the reference due to the dynamics of the MEMS.

Finally, the capability of the actuator to follow complex reference signals was tested. Figure 11 presents results showing the actuator tracking a complex reference. Some oscillations are visible during fast transitions due to the gain used in the controller.

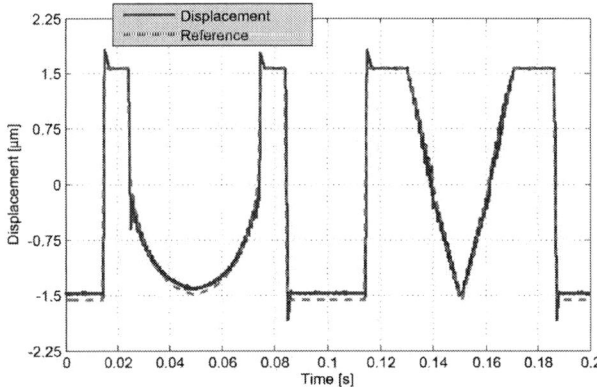

Figure 11: Tracking of complex references using a Kp of 5 at 5Hz signal repetition.

Discussion

During the tests, the feedback linearization technique used performed extremely well, and stable displacements beyond the pull-in limit were achieved even when the device parameters (m, k, d_0 and C_0) where modified (with deviations in excess of 50%). In this case, the error between the desired reference and displacement increased, but the controller was still able to maintain the actuator in the unstable region.

The tests performed also demonstrated that the linear controller used and the gains can dramatically change the response of the actuator. High proportional gains increase oscillations during transitions while reducing the permanent error. It was also noticeable that the oscillations during transitions, for the same proportional gain, were different according to the actuator position. This is likely due to the nonlinear behavior of the damping coefficient, given that the damping coefficient was neglected on the implementation of the linearization function. The use of dynamic gains might be an interesting solution to minimize these effects.

CONCLUSIONS

A bi-directional extended range parallel-plate electrostatic actuator based on feedback linearization was presented here. The parallel-plate electrostatic actuator includes bi-directionality and an improved control algorithm enabling reference tracking up to 500Hz (for 90% of the gap). Tracking beyond 500Hz was also achieved, but the displacements are limited by the dynamics of the mechanical structure used. For microstrostures with higher resonance frequencies, this limit can be overcome. The inclusion of bi-directionality is particularly interesting for variable capacitors, since it enables extension of the tuning range. Considering the sense capacitors of the device used, a tuning range of 17 (0.75 – 13pF), well above that of MEMS variable capacitors presented in literature (5.7 in [11] and 4.4 in [12]), was achieved.

REFERENCES

[1] L.A. Rocha, E. Cretu and R.F. Wolffenbuttel, "Analysis and Analytical Modeling of Static Pull-In With Application to MEMS-Based Voltage Reference and Process Monitoring", *J. Microelectromech. Syst.*, vol. 13, pp. 342-354, 2004.

[2] T.G.S.M Rijks, J.T.M van Beek, P.G. Steeneken, M.J.E Ulenaers, J. De Coster and R. Puers, "RF MEMS Tunable Capacitors with Large Tuning Ratio", in *Proc. International Conference on Micro Electro Mechanical Systems 2004*, Maastricht, The Netherlands, January 25-29, 2004, pp. 777-780.

[3] E.S. Hung and S.D. Senturia, "Extending the Travel Range of Analog-Tuned Electrostatic Actuators", *J. Microelectromech. Syst.*, vol. 8, pp. 497-505, 1999.

[4] E.K. Chan and R.W. Dutton, "Electrostatic Micromechanical Actuator with Extended Range of Travel", *J. Microelectromech. Syst.*, vol. 9, pp. 321-328, 2000.

[5] R. Nadal-Guardia, A. Dehé, R. Aigner and L.M. Castañer, "Current Drive Methods to Extend the Range of Travel of Electrostatic Microactuators Beyond the Voltage Pull-In Point", *J. Microelectromech. Syst.*, vol. 11, pp. 255-263, 2002.

[6] D. Piyabongkarn, Y. Sun, R. Rajamani, A. Sezen, and B. J. Nelson, Travel range extension of a MEMS electrostatic microactuator, *IEEE Trans. Control Syst. Technol.*, vol. 13, no. 1, pp. 138–145, 2005.

[7] S. Towfighian, A. Seleim, E. M. Abdel-Rahman and G R Heppler, "A large-stroke electrostatic micro-actuator" J. Micromech. Microeng. Vol. 21, 075023 (12pp), 2011.

[8] L. A. Rocha, E. Cretu and R.F. Wolffenbuttel, "Using dynamic voltage drive in a parallel-plate electrostatic actuator for full-gap travel range and positioning *J. Microelectromech. Syst.*, vol. 15, pp. 69–83, 2006.

[9] Jean-Jacques E. Slotine, *Applied Nonlinear Control*, Prentice-Hall, Englewood Cliffs, 1991.

[10] L.A. Rocha, R. A. Dias, E. Cretu, L. Mol and R.F. Wolffenbuttel, "Auto-calibration of Capacitive MEMS Accelerometers based on Pull-In Voltage" *Microsystem Technologies*, vol. 17, pp. 429-436, 2011.

[11] M. Bakri-Kassem and R. R. Mansour, "High Tuning Range Parallel Plate MEMS Variable Capacitors with Arrays of Supporting Beams", in Proc. International Conference on Micro Electro Mechanical Systems 2006, Istanbul, Turkey, January 22-26 Jan. 2006, pp. 666 – 669.

[12] A.M. Elshurafa and E. El-Masry, "A novel 3-in-1 MEMS variable capacitance device with a customizable tuning range*", 4th International Design and Test Workshop (IDT)*, Riyadh, Saudi Arabia, November 15-17, 2009, pp. 1-4.

CONTACT

*L.A. Rocha, tel: +351-253-510192; lrocha@dei.uminho.pt

BIOCOMPATIBLE CIRCUIT-BREAKER CHIP FOR TEMPERATURE REGULATION OF ELECTROTHERMALLY DRIVEN SMART IMPLANTS

Yi Luo, Masoud Dahmardeh, and Kenichi Takahata
University of British Columbia, Vancouver, CANADA

ABSTRACT

This paper presents a thermoresponsive micro circuit breaker for biomedical applications specifically targeted at implantable microsystems. The circuit breaker is micromachined to have a shape-memory-alloy cantilever actuator as a normally closed temperature-sensitive switch to protect the device of interest from overheating, a critical safety feature for smart implants including those that are electrothermally driven. The micro breaker is constructed using biocompatible materials in a form of 1.5×2.0×0.46 mm³ Ti-packaged chip and exhibits a cold-state resistance of 14 Ω. The chip functionality is tested in combination with a wireless resonant heater powered by radio-frequency electromagnetic radiation, demonstrating self-regulation of heater temperature. The developed breaker chip operates in a fully passive manner, contributing to the miniaturization of smart implants in which thermal management is essential.

INTRODUCTION

Rapid development and innovations in microsystems and biomedical engineering have led to the emergence of electronic medical implants and wearable microdevices [1]. "Smart" implants have been demonstrated with great potential for clinical applications, including miniaturized drug delivery devices [2] and telemetric stents [3]. With the growth of micro-electro-mechanical systems (MEMS) along with integrated-circuit technology, multifunctional circuitry mixed with sophisticated mechanical functions has been applied to many commercial products and is extending its contribution to medical and implantable devices. Thermal management is one of the critical issues in these application areas. The device circuitry, powered either internally (with batteries or other power sources) or externally (e.g., radio-frequency (RF) power transfer), generates heat during its operation based on Joule's first law. For smart implants, temperature regulation is an essential task to prevent them from overheating and ensure their safe operations. This task is even more important for the devices whose active functions are achieved by electrothermal means [4, 5]. Although thermal management could be accomplished using separate sensors and electronics, the space available for medical devices is typically limited, and this is especially true for implants. The use of passive components that offer a temperature regulation function will be a promising path to addressing these issues. MEMS-based temperature-sensitive switches were developed towards achieving higher isolation, speed, and repeatability; however, the studies have been limited to the devices with complex designs and/or relatively large sizes [6, 7]. Shape-memory-alloy (SMA) based electrothermal actuators have been actively investigated to show promising applications. The attributes of large actuation force and biocompatibility of Ni-Ti SMA, Nitinol, constitute advantages over other actuation

techniques, making them a favorable option in various engineering areas [8, 9]. Although SMA actuation is relatively slow as it relies on thermal phase transition of the alloy, this drawback is not a major concern in many cases including a majority of biomedical applications. Moreover, being a thermoresponsive material, SMA can be used to design a smart switch with a built-in sensing function that triggers its actuation in response to ambient temperature.

This paper reports a normally closed, SMA-based circuit-breaker switch microfabricated within a biocompatible chip package for its integration with implantable microdevices as a power circuit component (Figure 1). An out-of-plane Nitinol cantilever is utilized as a temperature-sensitive actuator in the chip that responds to heat generated in the device, enabling reversible breaking of the device's circuitry to automatically limit its temperature and prevent the implant from overheating. The characteristics of microfabricated prototype are studied, along with its demonstration in temperature control of a selected test device that is wirelessly powered.

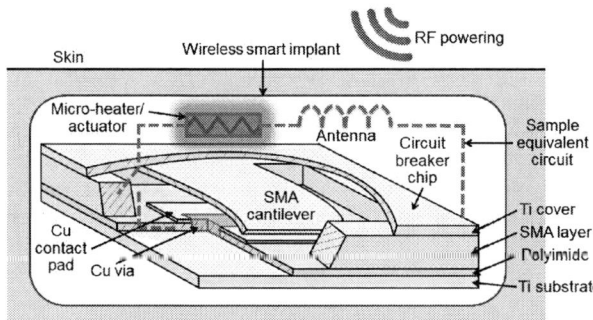

Figure 1: Conceptual illustration of circuit-breaker chip and its application for temperature regulation of wireless smart implant driven by RF-to-thermal power conversion.

DEVICE PRINCIPLE AND DESIGN

The circuit breaker investigated is inserted into a power line of the circuit of interest in close proximity to the region that behaves as the main heat source so that produced heat is directly transferred to the breaker switch. The cantilever actuator designed for the breaker is essentially a bimorph structure that is comprised of bulk-micromachined Nitinol cantilever coated with a thin-film reset layer. The SMA cantilever is designed to have the flat state as its memorized shape, which is restored when temperature of the material is elevated to exceed the threshold level at which the SMA enters its austenitic phase. At low temperatures with which the SMA is in its martensitic phase, the material becomes compliant, and a mismatch between the coefficients of thermal expansion of the SMA and the reset layer material is exploited to bend the cantilever down; this causes the cantilever's tip to make contact with the electrode pad created on the substrate, forming a normally closed switch. In this study, thin-film

978-1-4799-7956-1/15 $31.00 © 2015 IEEE   1040   MEMS 2015, Estoril, PORTUGAL, 18 - 22 January, 2015

SiO$_2$ is used as the reset layer formed on top of the Nitinol cantilever (refer to Figure 2 for their dimensions) to generate a compressive stress and induce the downward bending. Once heat flux coming from the circuit becomes high enough to trigger the actuation of Nitinol cantilever, the structure returns to its flat shape and temporarily breaks the circuit and its powering so that the circuit cools down. Lowering the circuit temperature leads to the recovery of the cantilever switch to its bent and closed condition, restoring the circuit's operation.

The switch component is sealed within a chip-based package for its protection and the ease of integration with an external circuit. As shown in Figure 1, the Nitinol cantilever is held by a rectangular-shaped frame of the same Nitinol; this frame serves as a platform for bonding with other layers to complete the chip. To form a contact switch, a Cu electrode pad is patterned on the central portion of the polyimide substrate on which the Nitinol layer is bonded. The tip of bent Nitinol cantilever touches on this pad to close the switch while the polyimide layer isolates the pad from the Nitinol layer. To physically support the polyimide layer and establish electrical access from outside to the Cu pad, a Ti plate is fixed on the backside of the polyimide layer and electrically connected to the pad through a via created in the layer, serving as an electrical terminal to be connected with the external circuit. Another Ti plate/cap is bonded on top of the Nitinol frame to seal the cantilever inside. The top Ti layer serves as the other electrical terminal of the breaker. The outer surfaces of the completed chip consist of Ti, Nitinol, and polyimide that are all considered as biocompatible materials [10, 11]. The chip may be directly mounted on a terminal of the external circuit to connect the bottom or top Ti surface of the chip to the circuit, and a wire could be bonded between the other Ti side of the chip and another terminal of the circuit to insert the circuit breaker in series.

FABRICATION

The fabrication of the Nitinol layer (Figure 2, steps 1-3) starts from patterning of a cavity (1.3-mm wide, 1.7-mm long, and 180-μm deep) at the center of a 190-μm-thick Nitinol plate (Alloy M, Memry, Germany; austenitic start temperature (A_s) = 56.5 °C) with the size of 2.0×1.5 mm^2 using a micro-electro-discharge-machining (μEDM) system (EM203, SmalTec International, IL, USA). The shape of the cantilever is further patterned with μEDM by cutting through the bottom layer of the cavity, creating spaces between all sides of the cantilever (except its fixed end) and the outer frame of the Nitinol plate. Next, a 4-μm-thick SiO$_2$ reset layer is deposited on top of the Nitinol cantilever (backside of the cavity region, while masking the outer frame) with plasma-enhanced chemical vapor deposition at 250 °C. (After this step, the cantilever is bent due to a compressive stress induced by the SiO$_2$ deposited.) Then, a 200-nm-thick Cu layer (together with a 15-nm-thick Cr adhesion layer) is selectively deposited on the backside of the cantilever with electron-beam evaporation. The fabrication of the substrate component (Figure 2, steps 4-6) uses a single-sided Cu-clad polyimide film (G2300, Shehdahl, MN, USA) with 50-μm thickness prepared to have the size of the breaker chip. The Cu-clad

layer (5-μm thickness) is lithographically patterned using dry-film photoresist (PM240, DuPont, DE, USA) to define the shape of the switch's electrode pad with an opening (500-μm square) in its center, after which the polyimide is wet etched to create a via hole through the opening. With the polyimide film aligned to a 100-μm-thick Ti plate (μEDMed to have the size of the chip), electroplating is performed to fill the via hole with Cu and electrically connect the Ti plate to the patterned Cu pad on the polyimide. The top Ti cap is shaped in a similar manner as the other (bottom) Ti plate, except for creating a 50-μm-deep cavity in the center to avoid potential contact of the cantilever with the bottom of the cap when combined. Finally, the chip is assembled by bonding the Ti cap, the Nitinol actuator layer, and the bottom Ti plate together using conductive epoxy with precise alignment while applying a force (of 10 N) to the combination (Figure 2, step 7), completing the fabrication (Figure 3(b)). A scanning electron microscope (SEM) image of the Nitinol cantilever (Figure 3(a)) indicates that the tip of the cantilever touches down on the Cu electrode pad underneath at room temperature, establishing a normally closed switch as designed. The final thickness of the completed chip is measured to be 460 μm (suggesting an average bonding layer thickness of 15 μm).

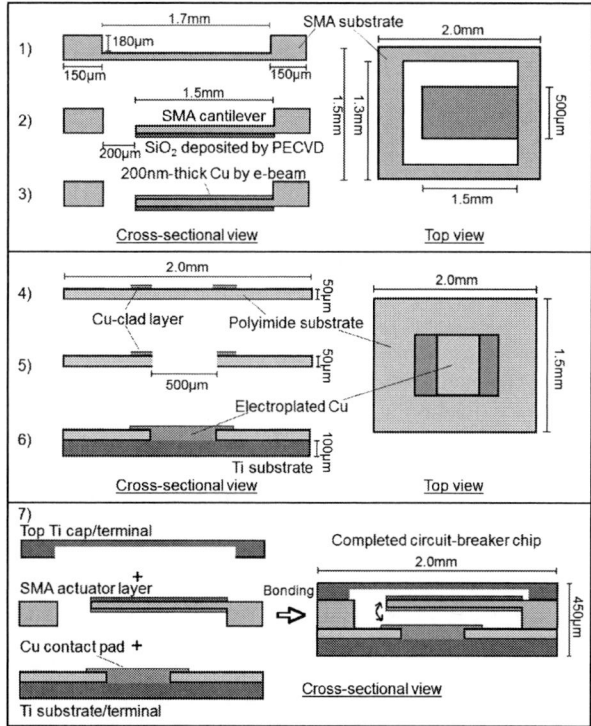

Figure 2: Fabrication process developed.

EXPERIMENTAL RESULTS

The mechanical response of the fabricated Nitinol cantilever to thermal stimulation was first evaluated, by increasing temperature of the component (before packaging, up to 80 °C on a hot plate) while measuring out-of-plane displacements of the cantilever using a laser displacement sensor (LK-G32, Keyence, ON, Canada; 10-nm sensing resolution) whose laser spot was aligned at

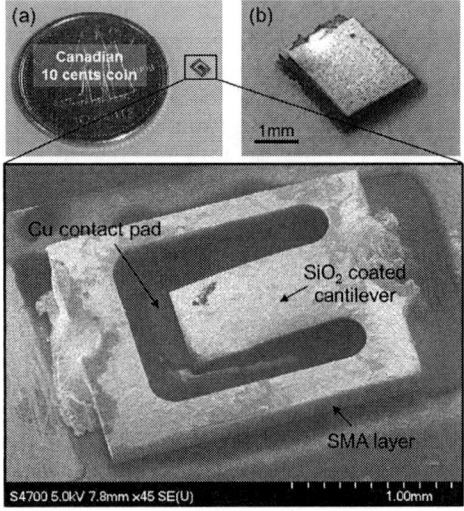

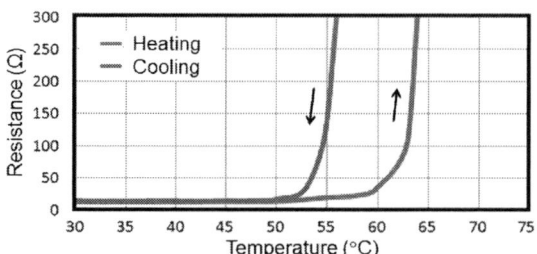

Figure 5: Electrical resistance of fabricated breaker chip measured as a function of temperature.

Figure 3: Fabricated samples of (a) Nitinol cantilever component patterned using μEDM with a close-up SEM image and (b) completed circuit-breaker chip.

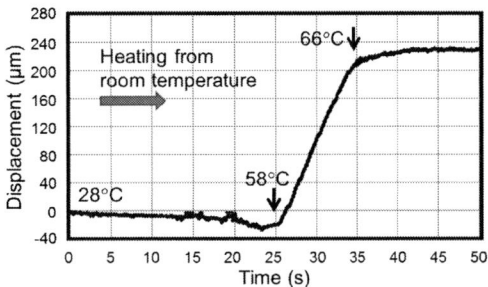

Figure 4: Measured temporal displacement of fabricated Nitinol cantilever with elevated temperature.

the free end of the cantilever. Figure 4 shows a typical measurement result of the displacement tracked with time. As shown, the cantilever remained almost stationary until its temperature reached ~58 °C (at ~25 seconds), after which the cantilever displaced upward and returned to its flat shape, leading to an approximate displacement of 220 μm when temperature reached ~66 °C (at ~35 seconds). These characteristic temperatures measured match well with the corresponding transitional temperatures (A_s of 56.5 °C and austenitic peak temperature (A_p) of 68.5 °C) of the Nitinol material used. The electrical behavior of the fabricated chip in response to applied heat was also characterized while recording the electrical resistance between the two terminals of the breaker chip as well as its temperature (elevated with the hot plate) using a thermal camera (Jenopik VarioCam HiRes 1.2M, Jena, Germany). As can be seen in the measured result shown in Figure 5, the self-resistance of the chip was measured to be approximately 14 Ω at room temperature. This base resistance stably remained until the chip temperature rose to ~55 °C, at which the resistance showed gradual increases, followed by a rapid rise to infinity representing the complete open state of the switch as the temperature passed around 63 °C. As the switch cooled down, the resistance remained at infinity until the temperature decreased to ~56 °C, after which the self-resistance dropped rapidly and back to its initial state at ~50 °C. The

observed hysteresis of the switch's behavior is consistent with that of the Nitinol's phase transition.

One of the potential applications of the proposed circuit breaker is the temperature regulation of wireless heaters that have been utilized to enable active functions in different types of smart implants [4, 5]. These wireless heaters are essentially inductor-capacitor (*LC*) resonant tanks; a radiation of RF electromagnetic field to the tank results in inductive heating when the field frequency matches the resonant frequency of the tank [4]. After implantation, these devices are operated by applying the field through the skin using an external transmission antenna placed on the skin. An apparent challenge in their operations is that depending on the location of the implant, the intensity of arrival field at the implant varies. If the field intensity becomes high under a certain condition, it could lead to a hazardous situation of overheating. The integration of the circuit breaker with the wireless heater is expected to physically limit and automatically regulate inductive heating "on site" in any kind of unexpected circumstance, ensuring that the implant device works safely and reliably in an open-loop and passive manner. Having this application focus, temperature regulation of RF resonant heating was studied using an experimental set-up shown in Figure 6. A test *LC*-tank heater coupled with a fabricated circuit-breaker chip (formed by series connecting a 420-nH solenoid coil, a discrete 10-pF capacitor, and the chip) was wirelessly excited with an RF field generated using an external coil antenna that was inductively coupled with the *LC* tank. The antenna was powered using an RF signal generator (HP8657A) through a power amplifier with an output power of 320 mW. The heater was activated by tuning the frequency of the RF signal/field to the resonant frequency of the circuit (61 MHz). The temperatures of both the wireless heater (inductor of the circuit) and the breaker chip coupled adjacent to it were simultaneously monitored and analyzed through the thermal camera.

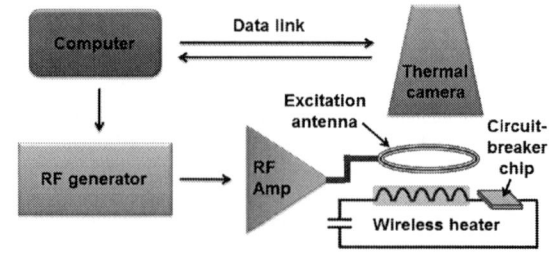

Figure 6: Experimental set-up used for wireless heating tests with an RF resonant heater.

The anticipated behavior of this wireless system is as follows: With a tuned RF excitation, an electromotive force is generated in the tank circuit as it resonates, causing an RF current that generates heat in the inductor of the circuit. This heat is transferred to the circuit-breaker chip connected with the inductor to raise chip temperature, and, when it exceeds the designed threshold level, the chip opens the circuit to temporarily terminate the current and heating in the inductor. The breaker closes the circuit again when its temperature becomes low enough, permitting the flow of RF current and resultant heating again. With continuous field radiation, this open-close cycle of the breaker is self-sustained to regulate inductor's temperature within the threshold level defined by the breaker. The measurement result shown in Figure 7 validates the designed function of the circuit breaker described above. As RF heating was initiated and the circuit breaker was heated up to ~63 °C, the Nitinol cantilever actuated to open the circuit. This terminated the RF heating of the inductor, lowering temperature of the breaker down to ~51 °C, at which the breaker closed the circuit to restore its resonance and RF heating again. Overall, temperature of the wireless heater was limited to ~52 °C while the breaker exhibited an oscillated temperature cycle between 51 °C and 63 °C, achieving on-site regulation of wireless induction heating.

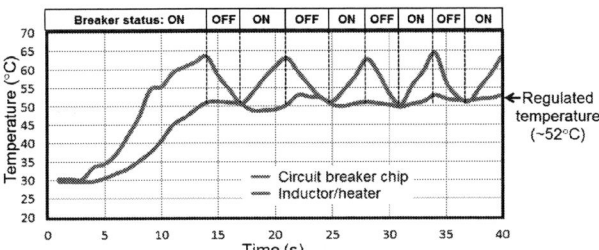

Figure 7: Measured thermal response of circuit-breaker-coupled wireless heater, showing self-regulated on-off cycle of breaker chip and resultant temperature regulation observed in the heating inductor.

CONCLUSION

A biocompatible micro circuit breaker has been designed, fabricated, and demonstrated experimentally with a focus on its application to electronic medical implants. The device design was centered on a micromachined Nitinol cantilever actuator that was used as a normally closed, thermoresponsive smart switch for opening/closing a circuit of interest depending on heating condition of the circuit in a passive manner. The circuit breaker was microfabricated with a combination of μEDM and lithography-based processing and Ti packaged in a form of chip. The thermomechanical and electrical behaviors of the fabricated breaker chip were experimentally characterized. The designed function of the developed breaker was successfully demonstrated by coupling it with a circuit platform of wireless RF resonant heater, showing that the heater temperature was well regulated at ~52 °C through self-sustained on-off switching cycles of the breaker. These results verify the intended functionality of the developed circuit breaker as a safety temperature limiter. The promising outcome from this study encourages further design optimization and improvement in the performance and miniaturization of the breaker chip towards realizing reliable thermal management in a variety of implantable microsystems.

ACKNOWLEDGMENTS

This work was partially supported by the Natural Sciences and Engineering Research Council of Canada, the Canada Foundation for Innovation, the British Columbia Knowledge Development Fund, and the Canadian Microelectronics Corporation. K. Takahata is supported by the Canada Research Chairs program. The authors acknowledge travel support provided by ICICS.

REFERENCES

[1] K. Menon, R.A. Roy, N. Sood and R.K. Mittal, "The applications of bioMEMS in diagnosis, cell biology, and therapy: A review," *BioNanoScience*, Vol. 3, pp. 356–366, 2013.

[2] N.M. Elman, H.L. Ho Duc and M.J. Cima, "An implantable MEMS drug delivery device for rapid delivery in ambulatory emergency care," *J. Biomed. Sci.*, Vol. 11, pp.625–631, 2009.

[3] X. Chen, D. Brox, B. Assadsangabi, Y. Hsiang and K. Takahata, "Intelligent telemetric stent for wireless monitoring of intravascular pressure and its *in vivo* testing," *Biomed. Microdev.*, Vol. 16, pp. 745–759, 2014.

[4] R. Rahimi, E.H. Sarraf, G.K. Wong and K. Takahata, "Implantable drug delivery device using frequency-controlled wireless hydrogel microvalves," *Biomed. Microdev.*, Vol. 13, pp. 267–277, 2011.

[5] Y. Luo, M. Dahmardeh, X. Chen and K. Takahata, "Selective RF heating of resonant stent toward wireless endohyperthermia for restenosis inhibition," in *Proc. IEEE Int. Conf. Micro Elec. Mech. Syst.*, San Francisco, Jan. 26-30, 2014, pp. 877–880.

[6] X.Q. Sun, K.R. Farmer and W.N. Carr, "A bistable microrelay based on two-segment multimorph cantilever actuators," in *Proc. IEEE Int. Conf. Micro Elec. Mech. Syst.*, Heidelberg, Jan. 25-29, 1998, pp. 154–159.

[7] M.C. Geear, E.M. Yeatman, A.S. Holmes, R.R.A. Syms and A.P. Finlay, "Microengineered electrically resettable circuit breaker," *J. Microelectromech. Syst.*, Vol. 13, pp. 887–894, 2004.

[8] C.J. Qin, P.S. Ma and Y. Qin, "A prototype micro-wheeled-robot using SMA actuator," *Sensor. Actuator. A Phys.*, Vol. 113, pp. 94–99, 2004.

[9] J. Szewczyk, E. Marchandise, P. Flaud, L. Royon and R. Blanc, "Active catheters for neuroradiology," *J. Robot. Mechatron.*, Vol. 23, pp. 105–115, 2011.

[10] J. Ryhanen, "Biocompatibility of Nitinol," *Min. Invas. Ther. & Allied Technol.*, Vol. 9, pp. 99–106, 2000.

[11] J.M. Se, S.J. Kim, H. Chung, E.T. Kim, H.G. Yu and Y.S. Yu, "Biocompatibility of polyimide microelectrode array for retinal stimulation," *Mater. Sci. Eng. C*, Vol. 24, pp. 185–189, 2004.

CONTACT

Yi Luo, tel: +1-604-3633468; luoyikey@ece.ubc.ca

CONTROLLABLE 'SOMERSAULT' MAGNETIC SOFT ROBOTICS

T. S. Zhang[1,2], A. Kim[1,2], M. Ochoa[1,2], and B. Ziaie[1,2]

[1] Birck Nanotechnology Center, Purdue University, West Lafayette, IN, USA
[2] School of Electrical and Computer Engineering, Purdue University, West Lafayette, IN, USA

ABSTRACT

This paper reports the fabrication and characterization of magnetic soft robotics with controllable somersault linear motion. The structures consist of polymeric magnet discs embedded in an elastomer and cut into various shapes. A cyclic magnetic field induces linear motion of the actuators on an inclined acrylic surface (0–90°). We investigate the effect of friction, geometry, and mass distribution on the controllability of their trajectory. The optimized actuator exhibits controllable linear movement in a flip-and-forward manner for an extended distance of 260 mm on horizontal surfaces and 80 mm on vertical surfaces.

INTRODUCTION

Soft magnetic actuators offer a convenient modality for controlled motion at the millimeter/micrometer scale while remaining sufficiently soft for interfacing with other emerging flexible systems. The reliability of magnetic actuation at short distances has been demonstrated by the numerous MEMS-based ones developed over that past two decades [1], [2]. When coupled with the flexibility and stretchability of elastomeric materials, the resulting combination of properties is ideal for creating novel biomedical and environmental microsystems with biomimetic functionality and non-contact (remote) control. Due to their vast potential applications, researchers have developed various approaches for creating different types of soft actuators as part of a growing field of soft robotics [3]-[6]. These systems boast a broad array of mechanical actuation capabilities from simple material deformations to more complex bending [3] and twisting [4]. However, their fabrication procedure is often equally complex or expensive, requiring either clean room processes [5] or magnetic-anisotropy shaping methods [3]-[6]. Additionally, their overall motion is usually limited to straight linear motion on horizontal surfaces. Reduction of their cost and enhancement of their actuation properties can be achieved simultaneously by designing a careful interplay among the material composition, geometry, and interfacial properties (e.g., friction and hydrophobicity) between the actuator and its target operation medium/surface. This approach enables the creation of more sophisticated actuators motions without making the structures too convoluted.

In this paper we present a controllable soft magnetic actuator which innovatively incorporates the elastomer's flexibility and high friction to achieve multiple sophisticated movements including somersault motion as well as climbing inclined surfaces towards or away from a cyclic magnetic field. All these movements are actuated and modulated by the rotation of a rectangular permanent magnet at a fixed position. Additionally, the fabrication process of the presented soft actuator is straightforward and inexpensive. Figure 1 illustrates the mode of actuation; here, a soft magnetic actuator shown flip-flopping along a plane in response to a rotating permanent magnet. The incline angle of this plane is adjustable between 0° and 90°.

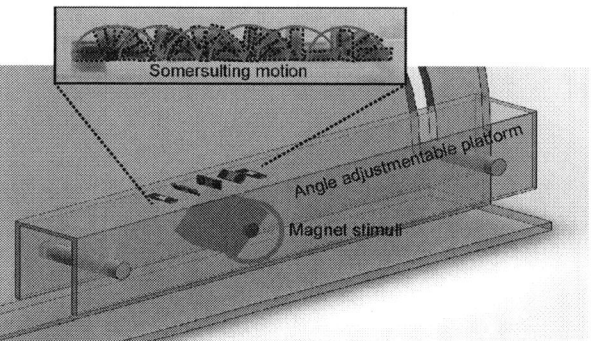

Figure 1: Schematic view of a magnetic soft-robotic moving on a Plexiglas (acrylic) platform.

OPERATION MECHANISM

A key parameter governing the motion of the actuator is the interface properties between the actuator and its operating surface. Proper friction, in particular, is critical for enabling 3D motion and climbing; hence, it is important to understand the delicate balance between friction and magnetic forces. Figure 2 shows the sequence of intermediate phases during a flip-and-forward 'somersault' linear motion. When a strong cyclic magnetic field is applied (e.g., from a rotating permanent magnet), four forces are at play: the gravitational force, the normal force, the magnetic force, and the frictional force. Their specific response to a time-varying magnetic field creates a delicate interaction which is used to create complex motion.

In Figure 2a, the actuator starts at rest and is pulled up by the magnetic force. The magnetic force is still pulling the actuator but the friction prevents it from sliding while the contact surface acts as a pivot when it flips, Figure 2b. Once the actuator is set to a vertical position (Figure 2c), the direction of the magnetic force will be reversed due to the exposure of the polymeric magnet's opposite polarity, and that repulsive force will balance the gravitational force, Figure 2d. In Figure 2e, the actuator finishes one flip cycle and the magnetic force applied on it is identical to the initial phase, but in a different location. This somersault cycle can be repeated for continued forward motion until the soft actuator reaches the region where the magnetic force dissipates. The speed of motion slows down as the actuator travels farther due to a lower magnetic force. Intriguingly, if the magnet stimulus stops rotating abruptly, the actuator will also 'freeze' at the corresponding intermediate motion phase. This reveals that the four forces (gravitational force, normal force, magnetic force, and frictional force) are in equilibrium at any instant time. It is the continuous gradient in magnetic force that breaks the stability and drives the actuator into steady movement.

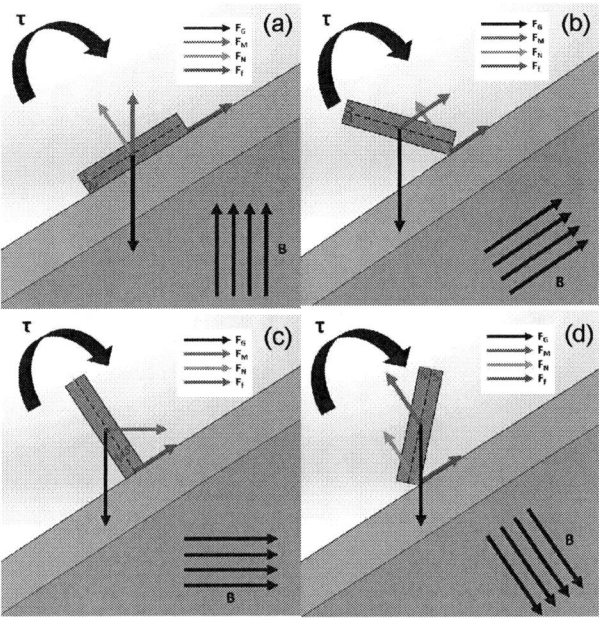

Figure 2: Decomposition of various forces during one cycle of linear movement.

For actuators with mass of about 2g (as in our case), the same physics applies to motion along both, vertical and horizontal planes, given that the four forces can still balance each other. However, a greater magnetic driving force and friction are required to overcome the greater component of the actuator's gravitational force and maintain the balance. As a consequence the maximum travel distance decreases as the surface becomes more inclined; this is discussed more in detail in the results section.

FABRICATION PROCESS

The fabrication process of the soft magnetic actuator is straightforward and adaptable to batch processing. It consists of embedding laser-micromachined polymeric magnets into silicone elastomer sheets in various configurations and laser machining the structures into specific geometries, Figure 3. First, 5 mm-diameter discs are cut from a polymeric magnet sheet ($BaFe_{12}O_{19}$ mixed with plastic binder, 0.8 mm, Master Magnet, Model# 96794) using a 10.6 μm CO_2 laser (Universal Laser

Systems, Scottsdale, AZ), Figure 3a. Circular magnets are chosen to provide omnidirectional and anisotropic magnetic field [7]. Next, the polymer magnetic discs are transferred to a 1 mm deep acrylic mold. An alignment mat with permanent magnets arranged (and fixed) in a specific configuration is placed under the acrylic mold to align the polymeric magnets in the desired shape and locations, Figure 3b. A silicone elastomer pre-polymer (e.g. Polydimethylsiloxane (PDMS) or Eco-flex® (Smooth-on Inc.)) is then poured into the mold and allowed to crosslink, Figure 3c. After complete curing, the different shapes are cut out by the same laser system, Figure 3d.

In order to further evaluate the effect of friction on the actuator's performance, two other actuators of the same dimensions but different materials were prepared: a bare polymeric magnet, and a polyethylene terephthalate (PET) film (with attached polymer magnet discs), Figure 3e. The friction coefficients of the materials are ranked as following.

$$\mu_{PET} < \mu_{bare\ magnet} < \mu_{PDMS} < \mu_{Eco\text{-}flex®} \quad (1)$$

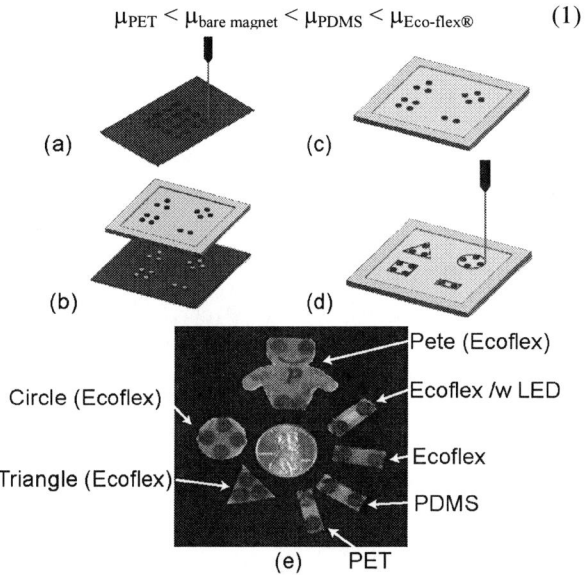

Figure 3: Fabrication procedure: (a) polymeric magnetic sheet machined with laser, (b) place and position magnets in a mold, (c) casting Eco-flex®, (d) laser machining the soft robots, and (e) optical photograph of various shape designs and material.

EXPERIMENTAL PROCEDURE

The soft magnetic actuators were placed on a horizontal acrylic surface and were actuated by the permanent magnet stimuli. The movements were video recorded and analyzed frame by frame. For material validation, rectangular Eco-flex® actuator, rectangular PDMS actuator, bare polymeric magnet, and polyethylene terephthalate (PET) with polymeric magnet attached were tested on horizontal acrylic surfaces, Figure 4. All actuators were positioned at the same initial position along the edge of the acrylic platform and approximately 50 mm away from external rotational magnetic stimulus. The actuators were also tested on an inclined plane (0°–

90°); for this, each actuator was placed at an initial position on the inclined plane and the maximum traveling distance (without slipping) was measured for each angle of inclination.

RESULTS AND DISCUSSION

Among the different materials, only the Eco-flex® based actuator showed a perfect somersault motion pattern (up to 260 mm without sliding or shifting), Figure 4a. The PDMS based actuator showed similar results; however, the surface of PDMS has a relatively low friction compared to Eco-Flex®, which caused some degree of sliding. In particular, when close to the magnetic stimulus, its motion degraded to simply cyclic motion (following the magnet below) since magnetic force dominated over friction, Figure 4b. Meanwhile, those materials with low friction (i.e. bare polymeric magnet and PET with polymeric magnet) presented uncontrollable motion. The bare polymeric magnet immediately slid to the location of the external magnetic stimuli as soon as it was positioned at initial point, and it proceeded to follow the cyclic magnetic field without any linear motion, Figure 4c. The PET with polymeric magnet attached showed similar motion as the bare polymer magnet; however, the higher rigidity of PET helped the motion, allowing it to somersault in place above the stimulus location.

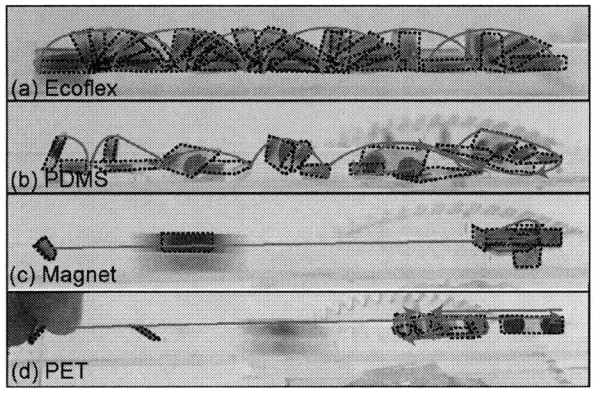

Figure 4: Snapshots showing the performance of various materials.

Figure 5 illustrates the shape validation of the proposed soft actuators. For these experiments, we used Eco-Flex® due to its superior performance compared to the other materials tested. Using Eco-Flex®, we created different shapes (e.g., circular, triangular, rectangular and square) using the same technique described in fabrication section. Circular actuators showed folding and uncontrollable motion as shown in Figure 5a. Once the actuator set to vertical position, its round edge steered to an unpredictable path as it approached the magnet stimulus. Once it reached the maximum magnetic field, its forward motion ceased, resulting in only a circular path, Figure 5a. On the other hand, the square shape showed better performance. However, it folded intermittently

during the motion, which caused its travel path to deviate slightly from the centerline path due to folding-induced changes in mass distribution, Figure 5b. The triangular actuator had a sharp edge that made it feasible to travel farther, and its dimension was small enough to prevent folding; however, its uneven weigh distribution resulted in an unpredictable trajectory, Figure 5c. We also designed a more humanoid shape, which we named Pete. Pete had two arms on sides, so when it is folded, most of the weight concentrated at the bottom. Weight concentration improved folding uniformity, enabling it to produce flip-and-forward motion, Figure 5d. Lastly, rectangular samples with sharp edges and small dimensions (to prevent folding) were able to move in controllable and predictable as a linear path, Figure 5e. In summary, shapes with straight edges and uniform weight distributions (e.g. rectangles) showed controllable and predictable motion; shapes with larger dimension tended to fold during motion due to low stiffness.

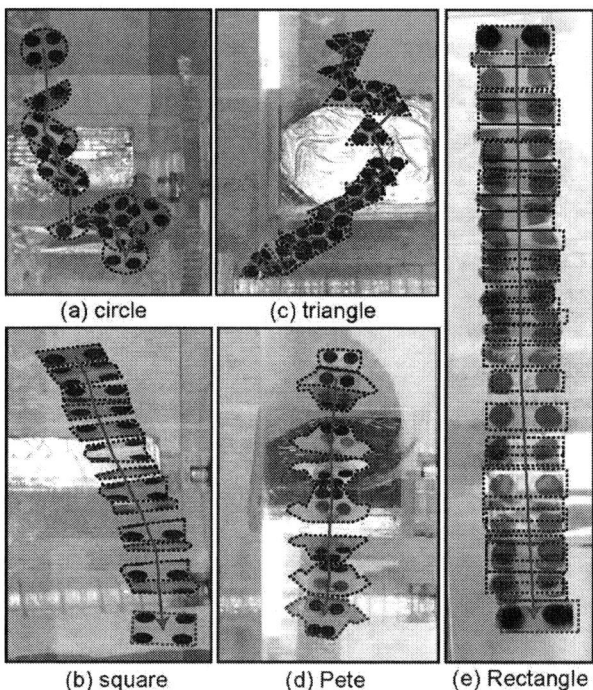

Figure 5: Motion pattern of different designs.

The performance of the various material actuators on an inclined plane is shown in Figure 6. As the incline angle increases, the interaction of gravity with the actuator motion also increases, which in turn reduces the maximum climbing distance. As shown in figure 6, the Eco-flex® actuator traveled the furthest distance at 90° (100 mm), while PET with polymeric magnet were trapped at the location of the magnetic stimulus and could not travel. Although the PDMS and the Eco-flex® actuators follow a similar trend for angles greater than 15°, the Eco-flex® actuator exhibited superior performance for all cases. However, PDMS actuator climbed slightly longer at 15° due to balancing of the magnetic and gravity forces to compensate insufficient friction. The result confirms the external magnetic force and friction overcome a greater component of the actuator's gravity.

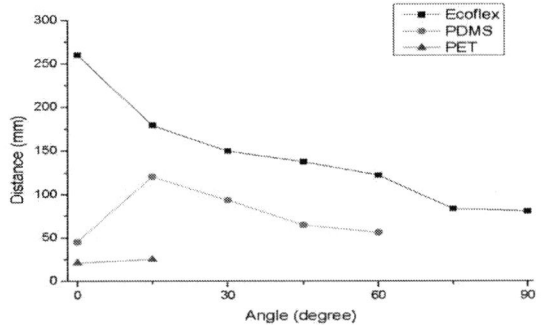

Figure 6: Angle of incline vs. travelled distance for magnetic soft robotics.

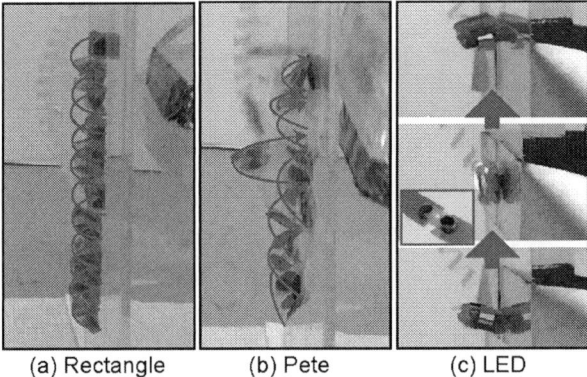

(a) Rectangle (b) Pete (c) LED

Figure 7: (a) rectangle and (b) Pete climbing a vertical surface, (c) embed LED lights on when it passes across a connector bridge.

Figure 7 shows snapshots of the rectangular and humanoid shape (Pete) actuator climbing a vertical wall (angle of 90°). The rectangular actuator moved as in the flat acrylic platform, with somersault motion, Figure 7a. However, since the gravitational force is larger in this vertical configuration, its travelling distance was not as much as on a flat surface, Figure 6. Similarly, the Pete actuator could climb with somersault motion, but the gravitational force interferes with folding, Figure 7b. Its trajectory became increasingly difficult to predict the farther it travelled.

As a more application specific demonstration, we created a climbing actuator with an embedded LED whose leads extend to the sides of the actuator, Figure 7c. As the climber moves along a connector bridge (exposed electrodes with a voltage source), the LED lights up. We expect this LED climber to shine light on future possible applications.

CONCLUSION

In this paper, we introduced a soft magnetic robotics actuator with simple structure of polymeric magnets embedded in high friction elastomer, Eco-flex®. Its fabrication is straightforward and can be applied for batch processing. The device is capable of performing controlled linear motion and steering on smooth horizontal and vertical surfaces. All motions are driven by the rotation of a fixed permanent magnet. We evaluated the impact of friction on the actuator's and concluded that the best material is one with friction and mass density close to those of Eco-Flex®. We showed that geometry plays an important role in the controllability of motion; in particular, small shapes with straight edges and uniform mass distribution showed the best performance. Finally, we demonstrated the actuator's ability to climb up an incline in a somersaulting fashion, including up to 100 mm up a vertical wall (away from the cyclic magnetic stimulus). These investigations set the groundwork for future implementations of magnetically controlled climbing soft robotics for biomimetic applications.

ACKNOWLEDGEMENTS

The authors thank the staff of Birck Nanotechnology Center at Purdue University for their support and assistance in fabrication.

REFERENCES

[1] D. Niarchos, "Magnetic MEMS: Key Issues and Some Applications", *Sensors and Actuators A: Physical,* vol. 106, pp. 255-262, 2003.

[2] O. Cugat, G. Reyne, J. Delamare, H. Rostaing, "Novel Magnetic Micro-actuators and Systems (MAGMAS) Using Permanent Magnets", *Sensors and Actuators A: Physical*, vol. 129, pp. 265–269, 2006.

[3] J. Kim, S.E. Chung, S. Choi, H. Lee, J. Kim, S.Kwon, "Programming Magnetic Anisotropy in Polymeric Microactuators", *Nature Materials*, vol. 10, pp. 747–752, 2011.

[4] S. H. Kim, K. Shin, S. Hashi, K. Ishiyama, "Magnetic Fish-robot Based on Multi-motion Control of a Flexible Magnetic Actuator", *Bioinspiration & Biomimetics*, vol. 7, pp. 036007, 2012.

[5] M. Khoo, C. Liu, "Micro Magnetic Silicone Elastomer Membrane Actuator", *Sensors and Actuators A: Physical*, vol. 89, pp. 259–266, 2001.

[6] R. Fuhrer, E.K. Athanassiou, N.A. Luechinger, W.J. Stark, "Crosslinking Metal Nanoparticles Into the Polymer Backbone of Hydrogels Enables Preparation of Soft, Magnetic Field-driven Actuators with Muscle-like Flexibility", *Small*, vol. *5*, pp. 383–388, 2009.

[7] R. Ravaud, G. Lemarquand, V. Lemarquand, C., Depollier, "Analytical Calculation of the Magnetic Field Created by Permanent-Magnet Rings". *IEEE Transactions on Magnetics*, vol. 44, pp. 1982-1989, 2008.

CONTACT

*T.S. Zhang, tel: +1-765-4181366; zhan1476@purdue.edu

CYBORG BEETLE: THRUST CONTROL OF FREE FLYING BEETLE VIA A MINIATURE WIRELESS NEUROMUSCULAR STIMULATOR

T.T. Vo Doan, Y. Li, F. Cao, and H. Sato[*]

School of Mechanical and Aerospace Engineering,
Nanyang Technological University, SINGAPORE

ABSTRACT

We demonstrate free-flight thrust control of the cyborg beetle which consisted of a living beetle platform (*Mecynorrhina torquata*), a microcomputer backpack (neuromuscular electrical stimulator) and a micro battery. A pair of thin wire electrodes coming from outputs of the backpack was implanted into the left or right subalar muscle, a major flight muscle of the beetle. The implanted muscle was stimulated in free-flight on demand when the commands sent to the backpack wirelessly by custom software running on the operator laptop. By varying the stimulation frequency, we were able to grade the induced thrust for decelerating the beetle in the air. The achievement of free-flight thrust control would open a new realm of flight control to further complex maneuver.

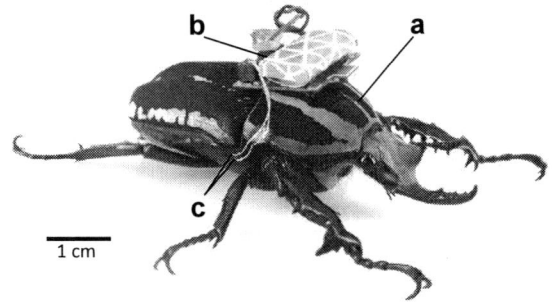

Figure 1: Photograph of a cyborg beetle; (a) live beetle platform (Mecynorrhina, ~6 cm, ~10 g), (b) wireless backpack assembly, (c) stimulation electrodes implanted into subalar muscle.

INTRODUCTION

The early stage of insect-machine hybrid systems, often referred to as Cyborg Insects, consisted of a living insect with a miniature analog circuit of surface mount electronics assembly mounted on, and researchers could remotely record muscle potentials of the living insect [1-4]. The advancement in micro-processor technology has allowed us to not only record but also stimulate multiple neuromuscular sites of living insect via a microcomputer tiny enough to mount on the insect. With such a stimulator, nowadays researchers can attempt to induce their desired motor actions and behaviors from the insects toward walking and flight control. Recent progress in the walking control by stimulating the nervous system of spider [5] and cockroach [6] showed the feasibility of employing living insects as the platform for a terrain legged robot. Furthermore, finding of graded response in the induced leg motion by altering the frequency of electrical stimulation of leg muscles allowed to develop a closed loop control system to regulate the insect leg to move along pre-determined motion paths [7]. For the flight control, many research groups have succeeded in flight initiation

and cessation, left-right turns of living insects. The flight initiation and cessation were demonstrated by stimulating the optic lobes of the beetle [8-10] while the turning control was achieved by stimulating the flight muscles in beetle and moth [8-11] or stimulating the nervous system to induce abdomen movements in moth [12]. A challenge ahead for more sophisticated flight control is thrust, and we have attempted to grade the speed of freely flying insect by altering stimulation parameters.

In this study, we have implemented the wireless stimulator (Fig. 2), set up a motion capture room for real-time flight path tracking (Fig. 4), and successfully achieved the graded thrust control in freely flying beetle by altering the frequency of the electrical stimulation to the subalar muscle (Fig. 3b), a major flight muscle directly inserted to the wing base.

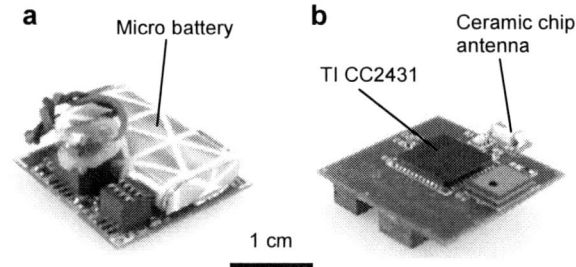

Figure 2: Wireless backpack assembly (a) top view shows two micro header for power connection and stimulus outputs with a micro battery (Fullriver, 3.7 V, 10 mAh) connected, (b) bottom view shows the Chipcon TI CC2431 microcontroller and the ceramic chip antenna (AN3216, 2.4 GHz) mounted on the PCB. Total weight is 1250 mg; microcontroller = 130 mg, battery = 345 mg, PCB + other components = 673 mg, retro reflective tape + adhesive = 102 mg.

EXPERIMENTAL PROCEDURE

Wireless Neuromuscular Stimulator Backpack

For the microcomputer backpack (neuromuscular electrical stimulator, Fig. 2) , a Chipcon TI CC2431 wireless microcontroller (130 mg, 6 x 6 mm, 2.4 GHz), an AN3216 ceramic chip antenna (2.4 GHz) and other components were assembled on a custom designed printed circuit board (PCB). The backpack was mounted on the pronotum of the beetle platform with a double sided adhesive tape (Fig. 1). A micro Li-polymer battery (3.7 V, 10 mAh, 345 mg) was attached onto the backpack for the power supply (Figs. 1b and 2a).

Electrode Implantation

Mecynorrhina torquata beetle (~ 10 g, ~ 8 cm, Fig. 1a) was employed as the living insect platform. The Teflon-coated silver wire electrodes (A-M system, 127 μm bare diameter, 178 μm coated diameter) were burned to expose the bare silver at both terminals. Two electrodes were implanted into each subalar muscle of the beetle (Fig. 3b) with a depth of ~ 3 mm and secured with beeswax. The other terminals of the electrodes were inserted into the outputs of the backpack mounted on the beetle.

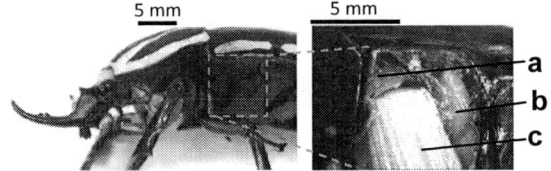

Figure 3: Anatomy of Mecynorrhina direct flight muscles; (left) external view of the beetle, (right) internal anatomy of the thorax exposing the direct flight muscles, (a) wing-folding muscle, (b) subasalar muscle, (c) basalar muscle. These direct flight muscles inserted directly to the wing hinges and work as the micro actuators to modulate the wing kinematics.

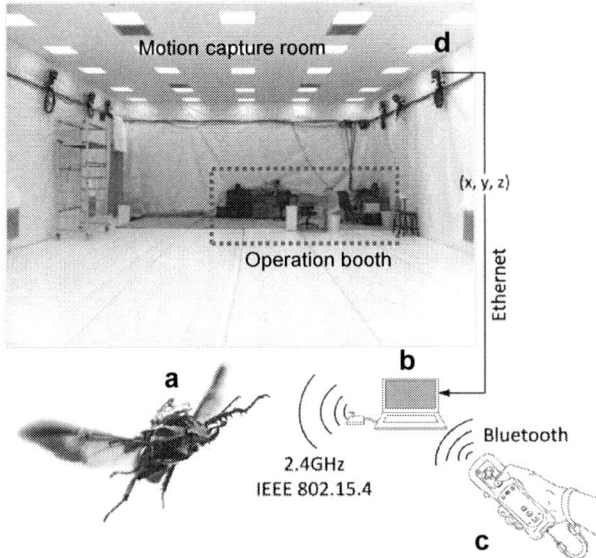

Figure 4: Free flight experiment setup for cyborg beetle in the flight capture room (9 m X 16 m X 4 m). (a) Freely flying cyborg beetle was controlled by a custom software, BeetleCommander v1.8, running on the operator laptop placed at operation booth. (b) The laptop sends control signal to the cyborg beetle wirelessly using the CC2431's built-in 2.4 GHz IEEE 802.15.4 transceiver via a USB / serial-interfaced base station. (c) A Nintendo Wii remote controller was used as user input to issue the command to the laptop via a Bluetooth transceiver. (d) 3D motion capture system of 18 T160 and T40s VICON cameras connected to the laptop via Ethernet port is used for capturing the beetle's positions. When the operator presses the Wii remote controller, it sends the command to the laptop. This command is processed by BeetleCommander v1.8 and then sent to the cyborg beetle's backpack. After receiving the command, the backpack generated pulse train to electrically stimulate the flight muscles.

Free Flight Experiment

The cyborg beetle was commanded by custom

software BeetleCommanderv1.8e running on the operator laptop that was able to manipulate the electrical stimulus parameters as multiple pulses trains (monophasic pulses, 3 V and 3 ms pulse width). The stimulating commands were sent to the backpack via the wireless communication of the backpack and the base station that connected to the operator laptop via a serial connection. For user interaction, a Nintendo Wii remote controller was implemented to send the user commands to the BeetleCommander wirelessly via a Bluetooth transceiver connected to the laptop. When the experimentalist presses a flight command button of the remote controller, BeetleCommander sends the command with predefined stimulus parameters to the backpack, which then generates an electrical stimulus to the left or right subalar muscle. For energy saving, the backpack is able to switch active/sleep mode by facing up/down the remote controller, which is automatically detected by the inertial measurement unit (IMU) of the remote controller.

After the implantation, the cyborg beetle was gently released to fly in a closed motion capture room (9 x 16 x 4, Fig. 4d) that equipped of T160 and T40s VICON cameras system. The system detected the retro-reflective taped wrapped around the battery (Fig. 2a) as a marker and reconstructed its positions in real-time. The positions of the marker were fed to the BeetleCommander for synchronizing with the stimulation commands.

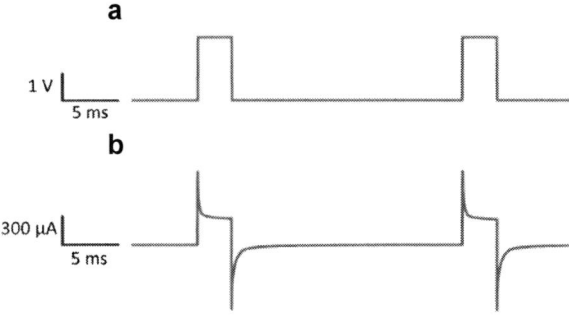

Figure 5: Stimulation waveform and electrical current flow through the subalar muscle. (a) The stimulation pulse train (3V, 3 ms pulse width and 50 Hz) from the waveform function generator was applied on the subalar muscle via the two implanted electrodes. (b) The current waveform passed through the subalar muscle.

FREE FLIGHT THRUST CONTROL

The left or right subalar muscle of a flying beetle was stimulated by 1 second multiple pulses train (3 V, 3 ms pulse width and 40 – 100 Hz, Fig. 5) for each trial (N = 14 beetles, n = 706 trials). Instead of performing left-right turnings as stimulating the basalar muscle [9, 10], the subalar stimulated beetle showed a clear decrease of flight speed during the stimulation period compared to that before the stimulation (Fig. 6). The estimated horizontal acceleration exhibited a clear deceleration. While 50 Hz – 70 Hz showed a slight decelerating effect of ~ 0.5 m/s², the most effective range 80 Hz – 100 Hz show clearer effect with a graded response as higher frequency induced more deceleration to the flying beetle (Fig. 7).

978-1-4799-7956-1/15 $31.00 © 2015 IEEE   1049

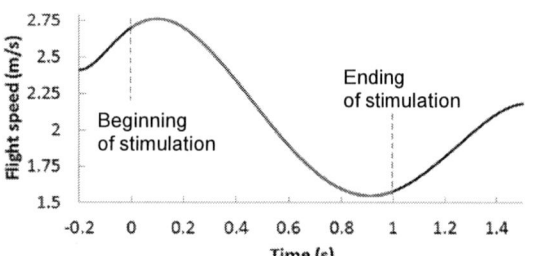

Figure 6: Representative speed profile of freely flying cyborg beetle during the stimulation of subalar muscle. The red line indicates the speed of the beetle during the stimulation while the black one shows the speed of the beetle without stimulation. The speed of the beetle reduced during the stimulation and started to recover when the stimulation ended. The beetle was stimulated by 1s stimulus train of multiple pulses (3V, 3 ms, 100 Hz).

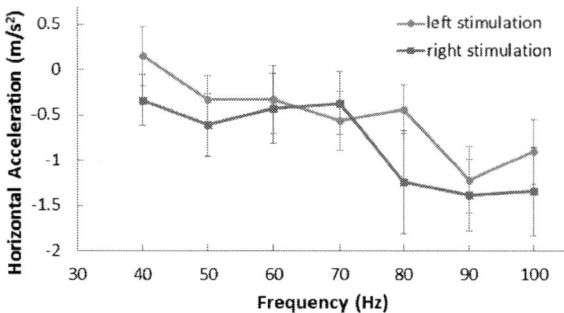

Figure 7: Horizontal acceleration of cyborg beetle induced by the stimulation of subalar muscle in free flight. The stimulation decelerated the flight speed of the beetle with the most effective range from 80 Hz to 100 Hz. The deceleration of flight speed is associated with the reduction of beetle's thrust. Error bars denoted 90% confidence interval (N=14 beetles, n = 706 trials).

The stimulation of subalar muscle also exposed an ultralow power consumption that is ~ 200 μW (N = 3 beetles, n = 150 trials). It would help the system save energy efficiently and enhance the working duration of the cyborg insect when using the micro battery.

CONCLUSION

We demonstrated the achievement of thrust control by wirelessly stimulating the subalar muscle for decelerating the untethered flying beetle. This capability would open a new realm of flight control to further complex maneuver as hovering and landing.

ACKNOWLEDGEMENTS

This material is based on the work supported by Nanyang Assistant Professorship (NAP), Agency for Science, Technology and Research (A*STAR) Public Sector Research Funding (PSF), A*STAR-JST (The Japan Science and Technology Agency) joint grant. The authors offer their appreciation to Mr. Chew Hock See, Mr. Cheo Hock Leong, Ms. Chia Hwee Lang, Mr. Roger Tan Kay Chia at School of MAE, NTU, Professor Michel M. Maharbiz and Berkeley Sensor and Actuator Center (BSAC) at UC Berkeley for their continuous support.

REFERENCES

[1] W. Kutsch, G. Schwarz, H. Fischer, and H. Kautz, "Wireless Transmission of Muscle Potentials During Free Flight of a Locus," *J. Exp. Biol.*, vol. 185, pp. 367-373, 1993.

[2] N. Ando, I. Shimoyama, and R. Kanzaki, "A Dual-Channel FM Transmitter for Acquisition of Flight Muscle Activities from the Freely Flying Hawkmoth, Agrius Convolvuli," *J. Neurosci. Methods*, vol. 115, pp. 181-187, 2002.

[3] H. Fischer, H. Kautz, and W. Kutsch, "A Radiotelemetric 2-Channel Unit for Transmission of Muscle Potentials During Free Flight of the Desert Locust, Schistocerca Gregaria," *J. Neurosci. Methods*, vol. 64, pp. 39-45, 1996.

[4] P. Mohseni, K. Nagarajan, B. Ziaie, K. Najafi, and S. B. Crary, "An Ultralight Biotelemetry Backpack for Recording EMG Signals in Moths," *IEEE Trans. Biomed. Eng.*, vol. 48, pp. 734-737, 2001.

[5] Z. Yang, K. Y. Chun, J. Xu, F. Lin, D. Li, and H. Ren, "A Preliminary Study of Motion Control Pattern for Biorobotic Spiders," in *Proceedings IEEE 11th International Conference on Control & Automation (ICCA)* 2014.

[6] T. Latif, E. Whitmire, T. Novak, and A. Bozkurt, "Towards Fenceless Boundaries for Solar Powered Insect Biobots," in *Proceedings 36th Annual International Conference of the IEEE Engineering in Medicine and Biology Society*, 2014.

[7] F. Cao, Z. Chao, T. T. Vo Doan, Y. Li, D. H. Sangi, J. S. Koh, *et al.*, "A Biological Micro Actuator: Graded and Closed-Loop Control of Insect Leg Motion by Electrical Stimulation of Muscles," *PLoS ONE*, vol. 9, 2014.

[8] M. M. Maharbiz and H. Sato, "Cyborg Beetles," *Sci. Am.*, vol. 303, pp. 94-99, 2010.

[9] H. Sato, Y. Peeri, E. Baghoomian, C. W. Berry, and M. M. Maharbiz, "Radio-Controlled Cyborg Beetles: A Radio-Frequency System for Insect Neural Flight Control," in *Proceedings IEEE 22nd International Conference on Micro Electro Mechanical Systems (MEMS)*, 2009, pp. 216-219.

[10] H. Sato, C. W. Berry, Y. Peeri, E. Baghoomian, B. E. Casey, G. Lavella, *et al.*, "Remote Radio Control of Insect Flight," *Front. Neurosci.*, vol. 3, 2009.

[11] A. Bozkurt, R. F. Gilmour, and A. Lal, "Balloon-Assisted Flight of Radio-Controlled Insect Biobots," *IEEE Trans. Biomed. Eng.*, vol. 56, pp. 2304-2307, 2009.

[12] A. J. Hinterwirth, B. Medina, J. Lockey, D. Otten, J. Voldman, J. H. Lang, *et al.*, "Wireless Stimulation of Antennal Muscles in Freely Flying Hawkmoths Leads to Flight Path Changes," *PloS ONE*, vol. 7, 2012.

CONTACT

*H. Sato, tel: +65 6790 5010; hirosato@ntu.edu.sg

CYLINDRICAL HALBACH MAGNET ARRAY FOR ELECTROMAGNETIC VIBRATION ENERGY HARVESTERS

Iman Shahosseini and Khalil Najafi
Center for Wireless Integrated MicroSensing and Systems (WIMS²)
University of Michigan, Ann Arbor, MI, USA

ABSTRACT

This paper reports the design, optimization, and test results of a new magnetic structure for kinetic energy harvesters allowing seven-fold increase in power density compared to single-magnet configuration. Electromagnetic energy harvesters with "single cylindrical Halbach array" and "double-concentric Halbach array" magnetic structures composed of NdFeB magnets with 1.4 T residual flux density were fabricated and tested: respectively 5 mW and 15 mW power was measured and generated at low-amplitude (1 mm) low-frequency (<10 Hz) vibrations. These structures yield respectively a maximum power density of 14 mW/cm³/g² and 26 mW/cm³/g².

INTRODUCTION

With recent progress in microelectronic technology, the size and the power consumption of electronic components and integrated electronic systems have been significantly reduced. These features enable applications requiring high mobility and long-term autonomy. To power such systems, a reliable power source is required. Whilst batteries are usually selected to accomplish this role, their need for regular maintenance and frequent replacement, especially in extreme operating temperatures, motivates the search for other alternatives. To procure a cost-effective solution and a reliable power source, energy harvesters turn out as an attractive replacement.

Kinetic energy harvesting has been widely studied in the past few decades to generate power from ambient vibrations found excessively in most civil infrastructures, marine/air-ground transportation, and human motions. The challenge of generating useful amounts of power becomes more acute when the harvester has to operate under low-amplitude low-frequency vibrations. Numerous design approaches have been proposed to improve the output power of kinetic energy harvesters by: using a resonant oscillating mechanism [1], deploying mechanical frequency up-conversion [2], or using rotational movement of an eccentric mass [3]. In electromagnetic energy harvesters, however, few reports concentrate on power improvements via configuring the magnetic structure instead of the abovementioned approaches [4-8].

With a focus on electromagnetic harvesters, this paper studies two new magnetic structures with the goal of achieving a higher power density, a feature which becomes attractive for industrial applications requiring compact harvester units. This paper first compares different magnetic structures in terms of the change rate of magnetic flux using finite element method (FEM). Second, with the goal of improving power density, the most attractive magnetic structure is geometrically optimized. Then, dynamic measurements are carried out on built harvester prototypes, and finally the tests results are discussed.

MAGNETIC STRUCTURE

The energy transduction of electromagnetic harvesters is based on Faraday's law of induction which relates the generated electromotive force ε to the changing rate of the magnetic flux $\delta\Phi_B/dt$ passing through an N-turn coil:

$$|\varepsilon| = N \, d\Phi_B/dt \tag{1}$$

When assuming an energy harvester with R_{Coil} and R_{load} as the source and the load resistance respectively, the harvester output power P_{out} becomes proportional to:

$$P_{out} \propto (N^2 R_{Coil}/(R_{Coil}+R_{Load})^2) \, (dz/dt)^2 \, (d\Phi_B/dz)^2 \tag{2}$$

where $d\Phi_B/dz$ is the variation of the magnetic flux with coil-magnet linear displacement and dz/dt is the vibration velocity.

As (2) shows, the output power of electromagnetic energy harvesters is governed by three main factors: 1) the coil parameters (number of turns, electrical resistance, relative positioning to magnets, etc.), 2) the vibration velocity (coil-magnet relative motion), and 3) the flux linkage created by magnet(s). Our previous works [9-10] showed that the first and second factors could be improved respectively through design optimization, and by coupling the harvester to a mechanical amplifier, respectively. Since the output power varies with square of magnetic flux, the last factor is also essential to maximize the output power.

Instead of using a single magnet (Fig. 1-a), other magnetic structures have been studied for micro/macro-scale energy harvesters: axial magnets with like-poles facing each other (Fig. 1-b), and with a ferromagnetic spacer between permanent magnets (Fig. 1-c), an array of flat magnets with poles alternated up/down [6], planar Halbach arrays [7], and circular Halbach arrays for rotary energy harvesters [8]. Arranging magnets in Halbach configuration is attractive because the flux density intensifies on one side of the array, the coil side, and attenuates to near zero on the other side where electronics may be placed to reduce device size, or eliminate the need for magnetic shielding.

To design more efficient electromagnetic harvesters, a cylindrical stack of axial and radial magnets configured in Halbach order can be used instead. Fig. 1-d illustrates a proper rotating pattern of four magnetizations (toward center, up, out-of-center, and down stacked up on top of each other) to intensify the magnetic field near the coils. Compared with structures of Figs. 1-a to 1-c, this structure improves channeling of the flux linkage through coil windings which are aligned with radial magnets.

For better channeling of flux linkage through all coil layers, a ferromagnetic tube can be installed around the outer edge of coils. However, depending on its permeability and saturation magnetic flux density, a minimum tube thickness is necessary to modify the spatial distribution of magnetic flux created by permanent magnets. When the tube is replaced with a second array of magnets properly oriented, not only better magnetic channeling is achieved, but also the magnetic flux density is increased due to the added magnets. This is the reason for which a second cylindrical Halbach array with inward intensified field and in 180° phase difference with the first array is added on the outer coil edge (Fig. 1-e). This architecture enhances the channeling of magnetic flux through coil windings while superimposing the Halbach effects of 1st and 2nd arrays to enhance the output.

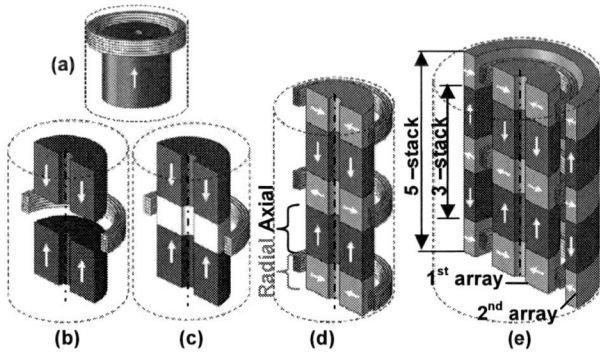

Figure 1: Schematic presentation of analyzed magnetic structures: (a) single magnet, (b) axial magnets with like-poles facing each other, (c) same structure incorporating soft magnet spacers, (d) single cylindrical Halbach array, (e) double-concentric Halbach array

To compare these structures, FEM simulations were performed using ANSYS. Dimensions of axial magnets were set at 19 mm in diameter and 10 mm in height. The spacer (or the radial magnet in case of Halbach arrays) was sized 19 mm in diameter and 6 mm in height. With respect to NdFeB permanent magnets of grade N50M, a residual magnetic flux density of 1.4 T was defined for both axial and radial magnets. Fig. 2 compares the change rate of the magnetic flux for each of Fig. 1

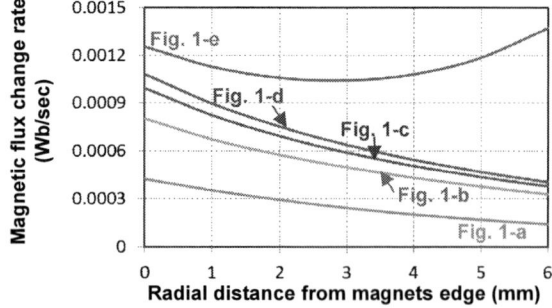

Figure 2: FEM analysis of magnetic flux rate as a function of radial distance from magnets edge for Fig. 1 structures at 5 Hz and 1 mm amplitude

structures as a function of radial distance from the magnets edge. To introduce the term of velocity, sinusoidal vibrations with 1 mm peak amplitude and 5 Hz frequency were considered. As Fig. 2 shows, the magnetic structure with like-poles facing each other is better by almost a factor of 2x than the magnetic flux rate of single-magnet structure (curves 1-b vs 1-a). By introducing a soft magnet as spacer, e.g. low carbon steel, between hard magnets (Fig. 1-c) further 20% increase is obtained in the flux change rate. The increase relative to the magnetic structure with like-poles facing each other is 30% when using single cylindrical Halbach magnet array (curves 1-d vs 1-b). However, when using the structure of double-concentric Halbach array, the flux change rate increases not only in the vicinity of inner magnets edge (first 3 mm in radial distance), but also near the outer magnets edge (3-6 mm following radial distance in Fig. 2) causing 3-10x increase relative to a single-magnet structure.

When comparing the performance of described magnetic structures at low-frequency low-vibration amplitudes (Fig. 3), it is noticed that relative to a single magnet, the single-cylindrical and double-concentric Halbach magnet arrays can improve the power density by as much as 3.5 and 7 respectively. The inset of Fig. 3 illustrates the spatial distribution of magnetic flux density created by the double-concentric Halbach array structure with the coil sandwiched between magnet arrays and initially aligned with radial magnets.

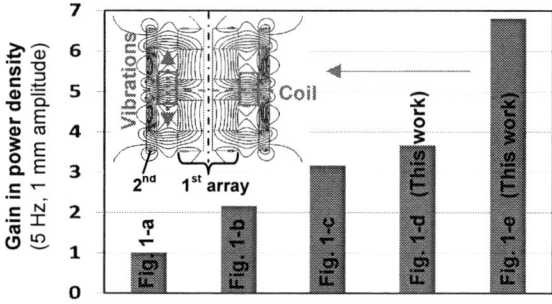

Figure 3: Calculated gain in power density for structures of Fig. 1, inset: distribution of magnetic flux density for double-concentric Halbach array

DESIGN OPTIMIZATION

In this section, the dimensions of coils and magnets composing the double-concentric Halbach array structure are calculated to maximize power density. For this purpose, first FEM simulations are performed to provide the cartography of magnetic flux density for different geometries of magnets. Then, with the help of analytical calculations, for each configuration the coil length is adjusted in a way to obtain the maximum output power. For all analyzed configurations the inner diameter of the 1st magnet array was set at 3 mm (equal to the diameter of the rod which is used later to assemble inner magnets). In the same way the outer diameter of the 2nd magnet array was set at 38 mm. The available space between the 1st and 2nd arrays is used to integrate the coil which is spaced as much as 1 mm from each magnet edge to prevent mechanical contacts with magnets.

Fig. 4 shows the variation of the power density with height of axial magnets at 5 Hz and 1 mm vibration amplitude. The curves represent different height ratios of radial to axial magnets (from 1/3 to 2/1). For all these configurations, the 1st array outer diameter is 19 mm, the 2nd array inner diameter is 32 mm, and the coil thickness remains constant at 4.5 mm. It is worth noting that on this graph there is more than one specific configuration which yields the maximum power density around 0.11 mW/cm^3 (i.e. the peaks on 1/2 and 2/1 curves). To find the maximum power density, FEM simulations and analytical calculations were also performed for magnet arrays with other diameters. Among all we selected one of the most attractive configurations to build prototypes: 1st array with 19 mm outer diameter, 2nd array with 32 mm inner diameter, axial magnets with 10 mm height, radial magnets with 5 mm height (height ratio: ½), and a coil section of 4.5 mm x 8 mm.

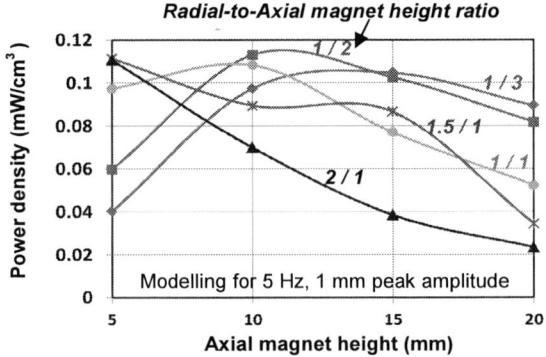

Figure 4: Optimization of double-concentric Halbach array, 1st array 19 mm in diameter, 2nd array 32 mm and 38 mm in inner and outer diameter respectively

FABRICATION AND TEST RESULTS

Based on optimized dimensions of magnets in the previous step, custom-made NdFeB magnets with grade N50M were used to fabricate energy harvester prototypes. Due to the high aspect ratio of axial magnets of the 2nd array, an assembly of four 90° arc segments was used instead. Owing to unavailability of radially magnetized NdFeB magnets, magnet arc segments were also assembled for the 1st and 2nd magnet arrays. Knowing that the radial segments are magnetized along a straight axis, but not along the radial axis, the assembly of smaller arc segments was preferred, i.e. twelve juxtaposed 30° segments, to approximate as much as possible perfect radial magnets. To mount the magnets, metallic jigs and fixtures were designed and machined. Loctite instant adhesive was used for chemical bonding. Fig. 5 shows the assembled axial and radial magnets under green magnetic field viewing film with indicated magnetization direction.

After assembling axial and radial magnets one-by-one, they were stacked up on top of each other in Halbach rotating pattern. Figs. 5-a and 5-b show 3-stack magnets for the 1st and 2nd arrays, respectively. Inserting the 1st array inside the 2nd array completes the double-concentric

Halbach array structure. For the coil, a 39 AWG copper wire was wound around a 70-μm thick Teflon tube over a length of 8 mm resulting in 0.9 kΩ resistance. A thin brass rod connects the coil to the external vibrating source (Fig. 5-c). Fig. 5-d shows the complete electromagnetic energy harvester next to a D-cell battery with inserted coil which vibrates inside and between two cylindrical magnet arrays.

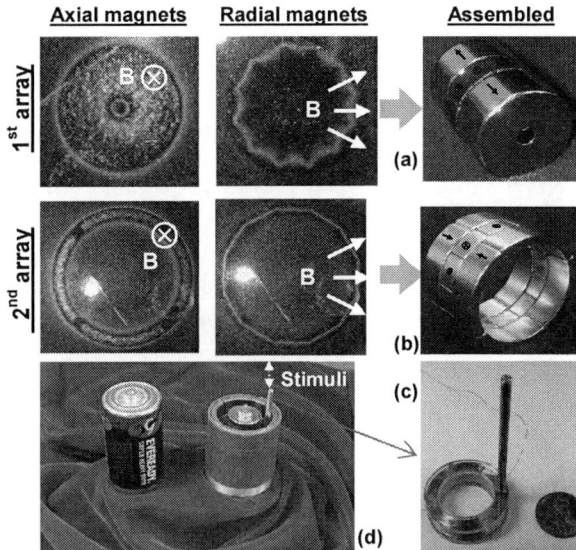

Figure 5: Harvester fabrication in double-concentric Halbach array configuration: (a) assembly of 1st magnet array, (b) assembly of 2nd magnet array, (c) copper coil, (d) harvester beside a D-cell battery

For dynamic tests, other prototypes based on Figs. 1-a, 1-b, and 1-d were also prepared. Then the magnet part was fixed to a stationary ring stand, and the coil was connected via the rod to APS 113 Electro-Seis shaker. The output power was measured at 1 mm peak vibration amplitude for each of magnetic structures across a load with a resistance identical to the coil. The results are presented in Fig. 6.

At 10 Hz, 5 mW and 15 mW were respectively generated with "single cylindrical Halbach array" and "double-concentric Halbach array". The latter in average increases the output power by as much as 15x over the

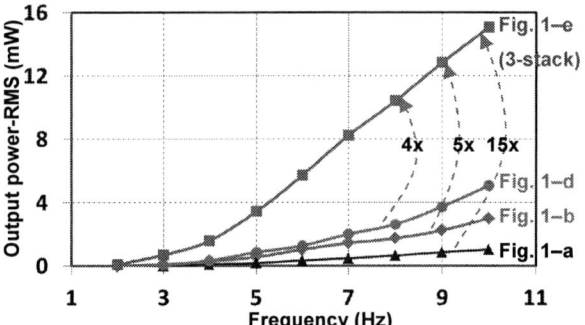

Figure 6: Measured output power for prototypes based on Fig. 1 magnetic structures at low-frequency low-vibration amplitude (1 mm peak), indicated gain factor by switching to double-concentric Halbach array

single-magnet structure. It, furthermore, improves the generated power by a factor of 5 over the magnetic structure with like-poles facing each other, and by 4 in average relative to single cylindrical Halbach array.

Fig. 7 compares the performance of double-concentric Halbach array with state of the art in terms of normalized power density. Despite being a non-resonant harvester, the developed magnetic structure provides power densities higher than most of other resonant electromagnetic generators. In good correlation with Fig. 3, by replacing single-magnet structure with double-concentric Halbach array, a gain ranging from 5 to 7 is achieved in power density. This new magnetic architecture can also be implemented in resonant harvesters to improve the power density of $26 \text{ mW/cm}^3/\text{g}^2$, or be employed in electromechanical actuators requiring high force densities.

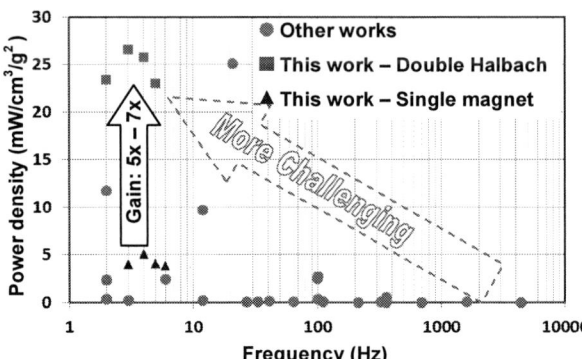

Figure 7: Comparison of normalized power density for electromagnetic energy harvesters

CONCLUSION

This paper presented two cylindrical magnetic structures based on Halbach array to be used in electromagnetic energy harvesters. With goal of higher power generation especially in low-amplitude low-frequency vibrations, the proposed magnetic structures were optimized in terms of power density and their prototypes were fabricated using custom-made NdFeB magnets.

The dynamic tests measured 5 mW and 15 mW output power, respectively for "single cylindrical Halbach array" and "double-concentric Halbach array" structures at 10 Hz and 1 mm peak vibration amplitude. These built energy harvesters present respectively maximum of $14 \text{ mW/cm}^3/\text{g}^2$ and $26 \text{ mW/cm}^3/\text{g}^2$ power densities at low frequencies.

In good correlation with simulations, the measurements showed the double-concentric Halbach magnet array yields seven-fold increase in power density compared to a single-magnet structure. The gain in generated power reaches 15 as a result of improved channeling of magnetic flux density and higher flux change rate for coil windings.

Although this work uses cylindrical Halbach arrays in direct-force energy harvesters, the proposed magnetic structure can be exploited in other types of energy harvesters like resonant-based systems, or even in electromechanical actuators to provide high driving power densities. With the goal of miniaturization, further investigations should also be conducted to study the influence of downscaling on the harvester performance.

ACKNOWLEDGEMENTS

The authors thank Pascal Sautier and Patrice Levallard at Hutchinson Research & Development Center, Montargis, for technical discussions, and acknowledge the support of Christophe Dominiak for the project. They also thank Robert Gordenker at WIMS[2] Research Center, University of Michigan, for his technical assistance with experimental setup. This work is funded and supported by Hutchinson SA, Centre de Recherche, France.

REFERENCES

[1] E. Sardini and M. Serpelloni, "An Efficient Electromagnetic Power Harvesting Device for Low-Frequency Applications", *Sensors and Actuators*, A 172, pp. 475-482, 2011.

[2] H. Kulah, K. Najafi, "Energy Scavenging from Low-Frequency Vibrations by Using Frequency Up-Conversion for Wireless Sensor Applications", *IEEE Sens. J.*, 8(3), pp. 261-268, 2008.

[3] E. Romero, M.R. Neuman, and R.O. Warrington, "Rotational Energy Harvester for Body Motion", in *Digest MEMS Conference*, pp. 1325-1328, 2011.

[4] H. Liu, Y. Qian, N. Wang, and C. Lee, "Study of the Wideband Behavior of an In-Plane Electromagnetic MEMS Energy Harvester", in *Digest MEMS Conference*, pp. 829-832, 2013.

[5] C. R. Saha, T. O'Donnell, N.Wang, and P. McCloskey, "Electromagnetic Generator for Harvesting Energy from Human Motion", *Sensors and Actuators*, A 147, pp. 248-253, 2008.

[6] T. Shirai, Y. Wakasa, T. Nakagawa, K. Nomura, and H. Yagyu, "Electromagnetic Energy Harvester with High Efficiency Using Micro-machining Si Springs", in *Digest MEMS Conference*, pp. 378-381, 2014.

[7] D. Zhu, S. Beeby, J. Tudor, and N.R. Harris, "Electromagnetic Vibration Energy Harvesting Using an Improved Halbach Array", in *Digest PowerMEMS Conference*, pp. 251-254, 2012.

[8] Y. J. Wang, C. D. Chen, C. K. Sung, and C. Li, "Natural Frequency Self-Tuning Energy Harvester Using a Circular Halbach Array Magnetic Disk", *Intel. Mat. Sys. & Struc. J.* 23(8), pp. 933-943, 2012.

[9] I. Shahosseini and K. Najafi, "Mechanical Amplifier for Translational Kinetic Energy Harvesters", in *Digest PowerMEMS Conference*, 2014.

[10] I. Shahosseini, R. L. Peterson, E. E. Aktakka, and K. Najafi, "Electromagnetic Generator Optimization for Non-resonant Energy Harvester", in *Digest IEEE Sensors Conference*, pp. 178-181, 2014.

CONTACT

* I. Shahosseini, tel: +1-734-647-3984, fax: +1-734-763 9324; imanshah@umich.edu

* K. Najafi, tel: +1-734-763-6650, fax: +1-734-763 9324; najafi@umich.edu

FLUID SEPARATED VOLUMETRIC FLOW CONVERTER (FSVFC) FOR HIGH SPEED AND PRECISE CELL POSITION CONTROL

Takumi Monzawa[1], Shinya Sakuma[2], Fumihito Arai[2] and Makoto Kaneko[1]
[1]Department of Mechanical Engineering, Osaka University, Suita, Japan
[2]Department of Micro-Nano Systems Engineering, Nagoya University, Nagoya, Japan

ABSTRACT

This paper reports the on-chip Fluid Separated Volumetric Flow Converter (FSVFC) capable of high speed cell position control with high resolution. There are various situations where we need to avoid mixing working fluid and actuation fluid for biological considerations, and such demand motivates the development of FSVFC. The proposed on-chip comb shaped FSVFC is composed of two separated groups of microfluidic channels arranged in parallel, and the main advantage is that actuation can be transmitted through the FSVFC without fluids being mixed. We succeeded in manipulating the position of a cell in microfluidic channel in the working area with the FSVFC as well as an online high speed vision sensor and a high speed PZT actuator. According to the experimental results, the average rise time of 12 miliseconds and the position control within 240 nanometers are achieved.

INTRODUCTION

It is well known that there is a close relationship between the deformability of Red Blood Cell (RBC) and decease[1], such as malaria[2], sepsis[3], diabetes[4, 5], and hypertension[6]. For example, if we are suffering from malaria, RBCs are infected with plasmodium falciparum. Finally, they become stiffer and lose deformability.

Many methods for evaluating the deformability of single RBCs have been reported. For example, RBCs are directly evaluated by Atomic force microscopy (AFM)[7] or optical tweezers.[8] But in these methods, it requires long time to measure the deformability of cells, thus they are limited from evaluating many cells in short time. RBCs deformability can also be evaluated by using microfluidic devices.[9-15] For the evaluation in microfluidic devices, blood usually has to be firstly diluted by proper solution, such as saline because of the high RBC density. From the viewpoint of knowing the natural environment of RBC, the deformability test should be done by using its own plasma for dilution. However, it's not preferable for patient to be withdrawn large quantity of blood since blood, the actuation fluid, needs to be filled inside the syringe which controls RBCs, as shown in Figure 1(a). Furthermore, using plasma as actuation fluid is not appropriate from the sanitary and economical point of view because it directly contacts with the actuator.

Figure 1(b) shows Fluid Separated Volumetric Flow Converter (FSVFC), which is proposed and developed here aiming to solve these issues and to achieve a clean, accurate RBC deformability with minimum quantity of whole blood. Before manipulating cells by FSVFC, the number of RBC should be reduced. Such dilution can be achieved by an on-chip dilution as reported in literatures.[16-20] Next, each RBC will be examined by passing through the constriction channel, and at the same time, the RBC motion is controlled by external syringe

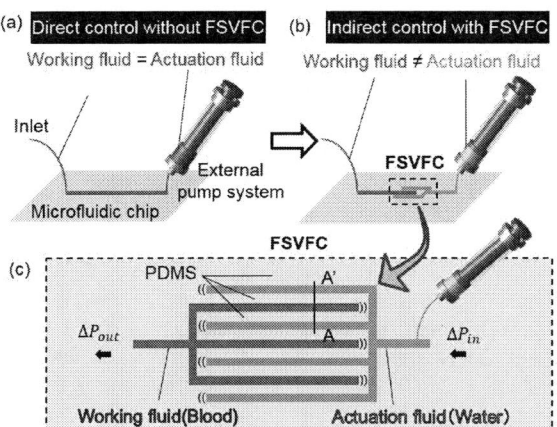

Figure 1: Schematic view of position control system. (a) Conventional method which directly control flow by external syringe pump without FSVFC. (b) Proposed method which indirectly controls fluid flow through FSVFC. (c) The enlarged view of proposed FSVFC.

pump with the proposed FSVFC. In this paper, we particularly focus on the principle and design of FSVFC as shown in Figure 1(c).

Different works on isolating the actuation and working fluid by multiple layers have been discussed[21]. But the overall response has not yet been deeply discussed, particularly the dynamic response and the resolution of cell positioning. In addition, using multiple layers requires accurate fabrication. Therefore, in this work we proposed such a FSVFC that can control flow precisely and convert flow in single layer.

Based on these backgrounds, this paper will start with the discussion on the design orientation of the converter with respect to the key parameter controlling the ratio between the input and the output volumetric flow. We then show the complete set of experimental results on two different geometries of converter with dynamic response and resolution.

THE WORKING PRINCIPLE OF FSVFC

Figure 1 shows the schematic view of flow control system. Figure 1(a) is conventional method where external syringe pump is directly connected to evaluation channel of microfluidic chip. Against it, Figures 1(b) and (c) show the proposed on-chip FSVFC and an enlarged view, respectively. External syringe pump is connected to the core channel, from which several sub-channels are branched. The end of sub-channel is closed so that we can keep the pressure change. Due to the pressure increase, for example, the pressure in the actuation side increases, which eventually brings the volumetric increase of PDMS. This

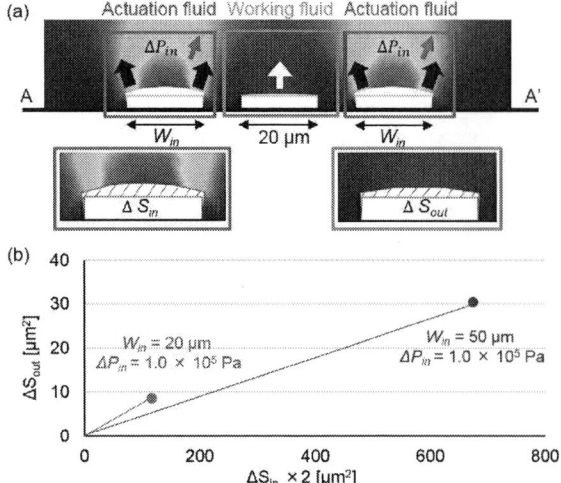

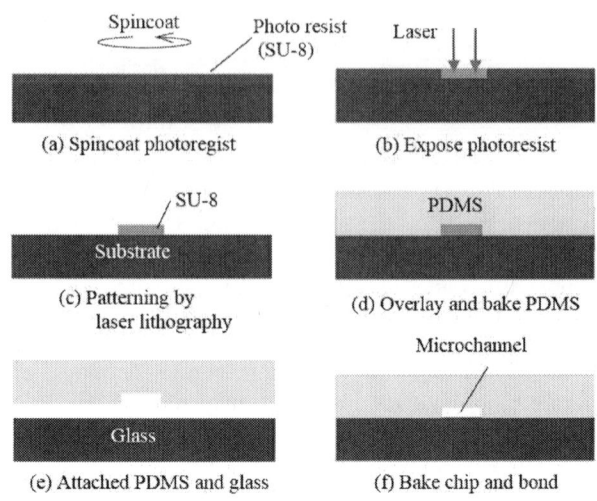

Figure 2: Simulation of deformation on A-A' cross section where Young's modulus: 7.5×10^5 [Pa], Poison ratio: 0.5, Density: 9.2×10^2 [kg/m³] [22], Simulation tool: COMSOL a)W: 20 μm, ΔP_{in}: 1.0×10^5 Pa (b) Relationship between ΔS_{in} and ΔS_{out}.

Figure 4: Steps for microfluidic chip fabrication.

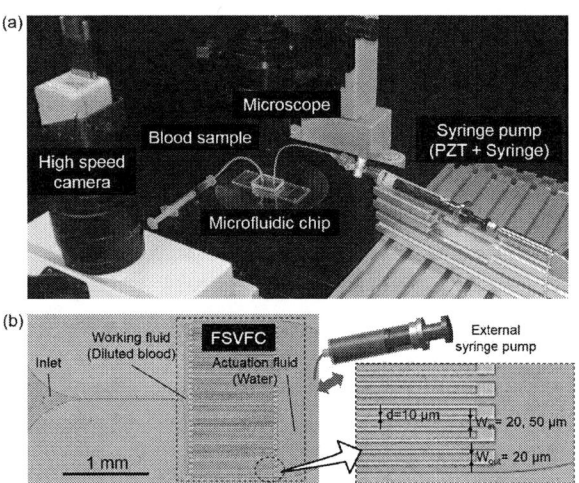

Figure 3: Experimental system. (a) An overview of the experimental system. (b) Actual images of FSVFC in microfluidic chip captured through a microscope.

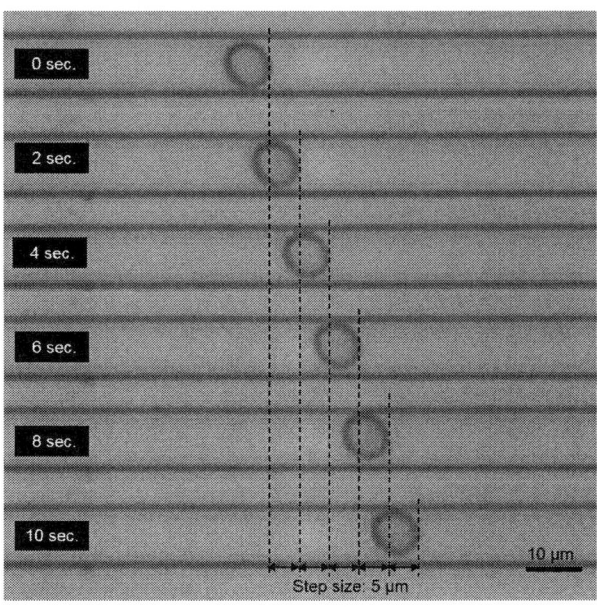

Figure 5: An example of step control of RBC position with FSVFC.

volumetric change affects the working side. As a result, an actuator motion causes the flow change at the working side.

Figure 2 shows the simulation results of the deformation under input pressure, ΔP_{in}, of A-A' cross section as shown in Figure 2(a). ΔS_{in} and ΔS_{out} denote the increase of cross-sectional area of actuation and working sides, respectively. By these results in Figure 2(b), we can see that the wider channel W_{in} makes the larger volumetric flow, where the distance between the channels of actuation and working side is fixed by 10 micron.

EXPERIMENTAL SETUP
Experimental setup

Figure 3(a) shows the experimental system composed of a microfluidic chip including the fabricated FSVFC, an online vision (Photron: FASTCAMMH4-10K) to observe cell position through the microscope, and a PZT actuator (MESTEC Co.: PSt 150/5/40 VS10) for controlling a RBC in the channel on the working side, respectively.

Figure 3(b) shows the actual images of FSVFC in microfluidic chip. The FSVFC have 31 comb shaped chambers for actuation fluid and 30 comb shaped chambers for working fluid. The microchip is made by soft lithography method which is briefly summarized in Figure 4.

Step control of RBC position

Figure 5 shows the example of manipulation of RBC

978-1-4799-7956-1/15 $31.00 © 2015 IEEE

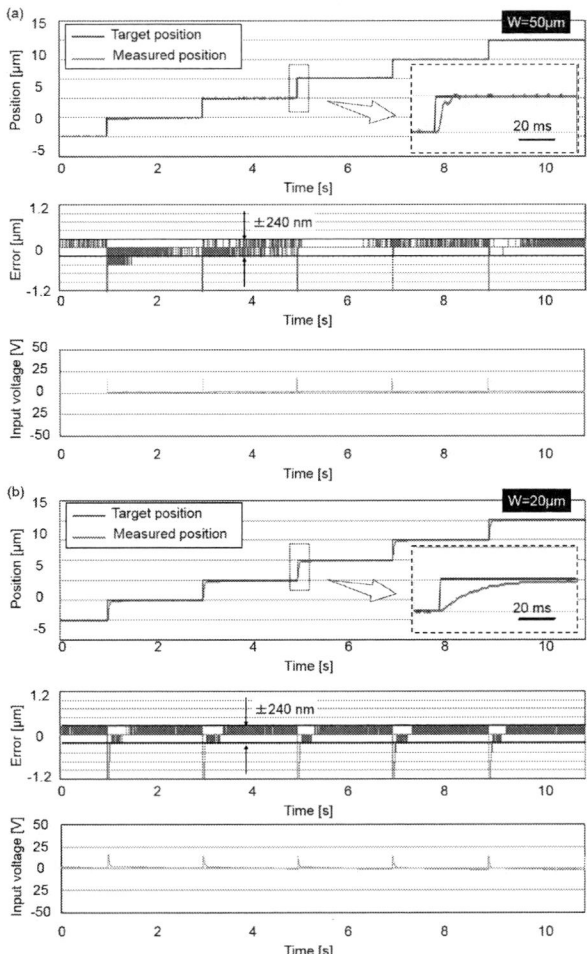

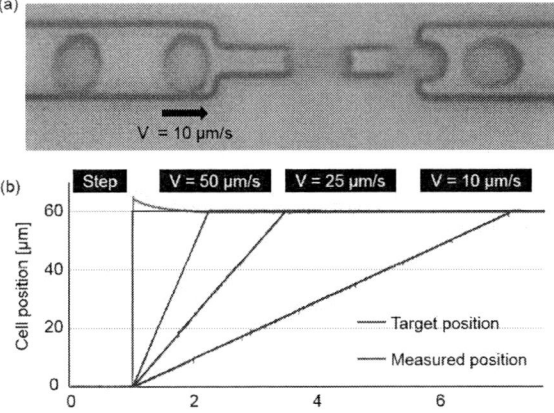

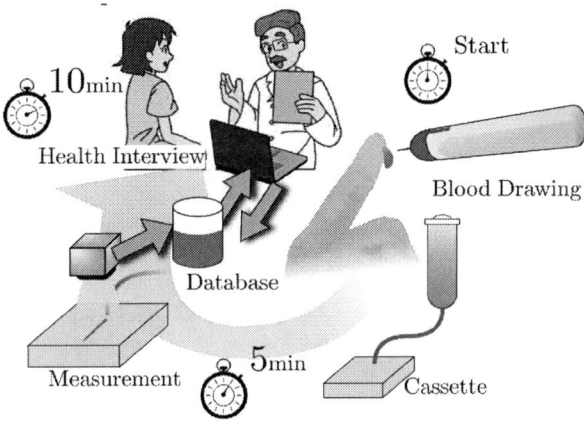

Figure 7: An example of the deformability evaluation on a RBC with FSVFC. (a)A series RBC images through the constriction channel. (b) The RBC is controlled at the various constant velocity through the constriction channel.

Figure 6: Position control of RBCs with FSVFC which have two different widths of actuation chambers, W_{in}. (a) W_{in}=50μm (b) W_{in}=20μm.

position. The RBCs are moved in steps with step size of 5μm. Figure 6 shows the experimental results where blue, red, green and cyan lines are the target position, actual position, position error and input actuator voltage, respectively. The blood diluted by saline with 2% is utilized as the working fluid in both Figures 6(a) and (b). Figures 6(a) and (b) are done under W_{in} = 20 and W_{in}= 50μm, respectively.

According to the results, the RBC position is controlled within 240nm, equivalent to one pixel, in the steady state. The rise time is 12ms for W_{in} = 50μm, while it is 53ms for W_{in} = 20μm. We find that the response under W_{in} = 50μm is about 4.5 times faster than that under W_{in} = 20μm. This result leads to the phenomenon that a larger W_{in} would result in a better sensitivity. The results match to the tendency observed in simulations.

Constant velocity control

Figure 7(a) shows images of a RBC manipulated through a constriction channel. The cell position can still be fairly controlled even with the geometric constraints and cell deformation. Figure 7(b) shows the behavior of RBC in channel, where the blue and the red lines are the target

Figure 8: Schematic view of 10 min health interview using the RBC deformability checker with FSVFC.

position and the measured position, respectively. The blue and the red lines closely overlapped with each other under constant velocity control in the constriction channel. It means that the velocity control which is the key for evaluation the deformability can be achieved.

CONCLUSIONS

The on-chip Fluid Separated Volumetric Flow Converter (FSVFC) capable of high speed cell position control with high resolution is proposed and developed. Three concluding remarks are:

1. We found that a larger W_{in}, which is the width of actuation side comb of FSVFC, would make a better sensitivity of volumetric flow convert.
2. We achieved the precise position control of cells in the stepwise motion with 5 micron step size. The rise time is only 12ms, and the steady state error of cell position is within 240 nm.
3. We achieved the precise velocity control in the demonstration of RBC control at various constant

speed for deformability evaluation.

For the future work, we would examine the deformability of RBC from whole blood with proposed constant velocity control with FSVFC. The ultimate goal is that medical doctors can do health interview in only 10 minutes from the moment patient's blood is withdrawn, as illustrated in Figure 8.

ACKNOWLEDGEMENTS

This work is supported by Grant#23106003, #26630098, #26820086 and "Creating Hybrid Organs of the future" of The Ministry of Education, Culture, Sports, Science and Technology (MEXT) of Japan.

REFERENCES

[1] G. Y. Lee and C. T. Lim, "Biomechanics approaches to studying human diseases," Trends in Biotechnology, vol. 25, no. 3, pp. 111–118, Mar. 2007.

[2] C. T. Lim, "Single Cell Mechanics Study of the Human Disease Malaria," Journal of Biomechanical Science and Engineering, vol. 1, no. 1, pp. 82–92, 2006.

[3] O. K. Baskurt, D. Gelmont, and H. J. Meiselman, "Red Blood Cell Deformability in Sepsis," Am J Respir Crit Care Med, vol. 157, no. 2, pp. 421–427, Feb. 1998.

[4] K. Tsukada, E. Sekizuka, C. Oshio, and H. Minamitani, "Direct Measurement of Erythrocyte Deformability in Diabetes Mellitus with a Transparent Microchannel Capillary Model and High-Speed Video Camera System," Microvascular Research, vol. 61, no. 3, pp. 231–239, May 2001.

[5] C. D. Brown, H. S. Ghali, Z. Zhao, L. L. Thomas, and E. A. Friedman, "Association of reduced red blood cell deformability and diabetic nephropathy," Kidney Int, vol. 67, no. 1, pp. 295–300, Jan. 2005.

[6] G. Cicco and A. Pirrelli, "Red blood cell (RBC) deformability, RBC aggregability and tissue oxygenation in hypertension," Clinical Hemorheology and Microcirculation, vol. 21, no. 3, pp. 169–177, Jan. 1999.

[7] K. E. Bremmell, A. Evans, and C. A. Prestidge, "Deformation and nano-rheology of red blood cells: An AFM investigation," Colloids and Surfaces B: Biointerfaces, vol. 50, no. 1, pp. 43–48, Jun. 2006.

[8] M. M. Brandão, A. Fontes, M. L. Barjas-Castro, L. C. Barbosa, F. F. Costa, C. L. Cesar, and S. T. O. Saad, "Optical tweezers for measuring red blood cell elasticity: application to the study of drug response in sickle cell disease," European Journal of Haematology, vol. 70, no. 4, pp. 207–211, Apr. 2003.

[9] C.-H. D. Tsai, S. Sakuma, F. Arai, and M. Kaneko, "A New Dimensionless Index for Evaluating Cell Stiffness-Based Deformability in Microchannel," IEEE Transactions on Biomedical Engineering, vol. 61, no. 4, pp. 1187–1195, Apr. 2014.

[10] S. Sakuma, K. Kuroda, C.-H. D. Tsai, W. Fukui, F. Arai, and M. Kaneko, "Red blood cell fatigue evaluation based on the close-encountering point between extensibility and recoverability," Lab Chip, vol. 14, no. 6, pp. 1135–1141, Feb. 2014.

[11] Y. Zheng, E. Shojaei-Baghini, A. Azad, C. Wang, and Y. Sun, "High-throughput biophysical measurement of human red blood cells," Lab Chip, vol. 12, no. 14, pp. 2560–2567, Jun. 2012.

[12] C.-H. D. Tsai, S. Sakuma, F. Arai, T. Taniguchi, T. Ohtani, Y. Sakata, and M. Kaneko, "Geometrical alignment for improving cell evaluation in a microchannel with application on multiple myeloma red blood cells," RSC Adv., vol. 4, no. 85, pp. 45050–45058, Sep. 2014.

[13] A. Adamo, A. Sharei, L. Adamo, B. Lee, S. Mao, and K. F. Jensen, "Microfluidics-Based Assessment of Cell Deformability," Anal. Chem., vol. 84, no. 15, pp. 6438–6443, Aug. 2012.

[14] J. P. Shelby, J. White, K. Ganesan, P. K. Rathod, and D. T. Chiu, "A microfluidic model for single-cell capillary obstruction by Plasmodium falciparum-infected erythrocytes," PNAS, vol. 100, no. 25, pp. 14618–14622, Dec. 2003.

[15] C. T. Lim, E. H. Zhou, and S. T. Quek, "Mechanical models for living cells—a review," Journal of Biomechanics, vol. 39, no. 2, pp. 195–216, 2006.

[16] M. Kersaudy-Kerhoas and E. Sollier, "Micro-scale blood plasma separation: from acoustophoresis to egg-beaters," Lab Chip, vol. 13, no. 17, pp. 3323–3346, Jul. 2013.

[17] A. Lenshof, A. Ahmad-Tajudin, K. Järås, A.-M. Swärd-Nilsson, L. Åberg, G. Marko-Varga, J. Malm, H. Lilja, and T. Laurell, "Acoustic Whole Blood Plasmapheresis Chip for Prostate Specific Antigen Microarray Diagnostics," Anal. Chem., vol. 81, no. 15, pp. 6030–6037, Aug. 2009.

[18] M. Kersaudy-Kerhoas, D. M. Kavanagh, R. S. Dhariwal, C. J. Campbell, and M. P. Y. Desmulliez, "Validation of a blood plasma separation system by biomarker detection," Lab Chip, vol. 10, no. 12, pp. 1587–1595, Jun. 2010.

[19] V. Doyeux, T. Podgorski, S. Peponas, M. Ismail, and G. Coupier, "Spheres in the vicinity of a bifurcation: elucidating the Zweifach-Fung effect," Journal of Fluid Mechanics, vol. 674, pp. 359–388, May 2011.

[20] J. S. Yoon, J. T. Germaine, and P. J. Culligan, "Visualization of particle behavior within a porous medium: Mechanisms for particle filtration and retardation during downward transport," Water Resour. Res., vol. 42, no. 6, p. W06417, Jun. 2006.

[21] W. H. Grover, M. G. von Muhlen, and S. R. Manalis, "Teflon films for chemically-inert microfluidic valves and pumps," Lab Chip, vol. 8, no. 6, pp. 913–918, May 2008.

[22] D. Armani, C. Liu, and N. Aluru, "Re-configurable fluid circuits by PDMS elastomer micromachining," in Twelfth IEEE International Conference on Micro Electro Mechanical Systems, 1999. MEMS '99, 1999, pp. 222–227.

CONTACT

*T.Monzawa, tel:+81-6-6879-7333;
monzawa@hh.mech.eng.osaka-u.ac.jp

ON-CHIP ENUCLEATION USING AN UNTETHERED MICROROBOT INCORPORATED WITH AN ACOUSTICALLY OSCILLATING BUBBLE

Il Song Park, Young Rang Lee, Sung Jin Hong, Kang Yong Lee and Sang Kug Chung
Myongji University, Yongin, Gyeonggido, South Korea

ABSTRACT

This paper presents a novel on-chip enucleation method using a magnetically driven microrobot incorporated with an acoustically excited bubble. The proposed microrobot mainly consists of a compressible bubble for the manipulation of a cell and twin neodymium magnets for the motion control of the microrobot in an aqueous medium. First, the two-dimensional (2D) motion control of the microrobot – horizontal, vertical, and rotational motions – is demonstrated by using an external magnetic controller attached beneath the bottom of a chip owing to the interaction forces induced by twin magnets installed inside both the microrobot and controller. Second, the enucleation of a cell is separately investigated using an acoustically oscillating bubble. When a bubble is acoustically excited at its natural frequency, it oscillates and simultaneously generates cavitational microstreaming and radiation forces around it. The flow pattern and strength of the microstreaming in different frequencies and voltages are studied using a microscopic PIV system. The flow strength is proportional to the voltage and strongly dependent on the frequency and maximum at its natural frequency. To investigate the effects of the microstreaming on the inside of a cell through a narrow slit, particle extraction test is conducted using a small PDMS cylinder chamber. It shows that most of the particles initially filled in the chamber are extracted by an acoustically excited bubble and the chamber becomes empty within 160 seconds. Finally, as proof of concept, the embryo extraction from a fish egg (1.7 mm diameter) dyed by methylene blue is successfully achieved using the proposed microrobot. This on-chip enucleation technique may improve the efficiency of enucleation processes with minimizing the contact damage between a cell and a micromanipulation tool.

INTRODUCTION

Various kinds of animals – sheep, goats, cattle, mice, pigs, cats and rabbits – have been cloned by somatic cell nuclear transfer (SCNT) since Dolly, the first clone of an adult animal, was born in early 1997[1]. However the success probability of the cloning technology based on SCNT is extremely low. Dolly was the only success among 277 attempts, which shows a 0.3 % success probability. And other cloning adult animals also showed the similar success probability. One of the main causes is the micromanipulation tools and procedures applied in the cloning technology[2, 3].

For cell manipulation various micromanipulation tools and devices have been developed over the past 100 years. Marshall A. Barber firstly developed the principles of microinjection applied to the early pipette system and demonstrated the manipulation of bacterial cells at the beginning of the 20th century[4]. Since then the various types of pipettes such as an automated pipette and a computer-aided pipette have been developed for improving the accuracy and efficiency of the micromanipulation processes and used as one of the most popular tools for cell manipulation in biomedical fields.

However, the use of mechanical micromanipulation techniques causes an unwanted contact problem – the solid parts in the systems have to come into physical contact with the targeted objects, which thereby damages them. In particular, the contacts induced by invasive mechanical techniques may critically damage biological objects such as cells[5, 6]. To prevent contact damages during micro-object manipulations, alternative techniques based on optic, electric, and magnetic forces have been developed[7, 8].

(a)

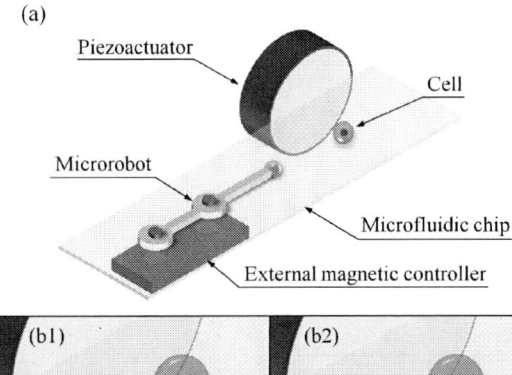

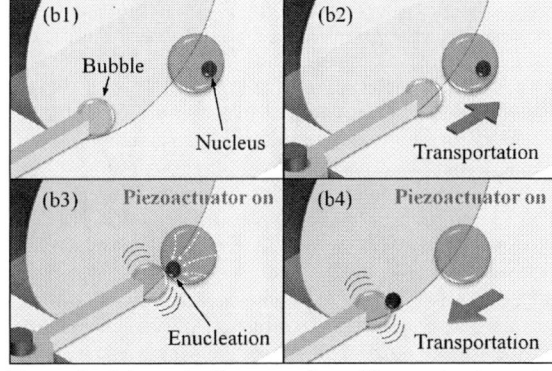

Figure 1: Schematic diagram of on-chip enucleation using an untethered microrobot: (a) The overall experimental setup; (b1) Initial state; (b2, b3) First, a microrobot is closely transported to a cell using an external magnetic controller. And then a piezoactuator is turned on to excite a bubble attached on the tip end of the microrobot. The excited bubble induced microstreaming with radiation forces pulls and eventually extracts a nucleus through the prepared slit of the cell; (b4) The extracted nucleus is carried to a target place by the microrobot.

978-1-4799-7956-1/15 $31.00 © 2015 IEEE

This paper presents a novel on-chip enucleation method using an untethered microrobot incorporated with an acoustically excited bubble, which will allow minimally invasive cell surgery for the cloning technology and biomedical applications. The proposed microrobot mainly consists of a compressible bubble for the manipulation of a cell and twin permanent magnets for the motion control of the microrobot in an aqueous medium. Arai research group previously developed magnetically-driven-microtools (MMT) actuated by permanent magnets for on-chip manipulation along with experimental verification[9]. However, the MMT can damage target objects such as biological cells with soft membranes due to the physical contact with the solid parts of the MMT, similar to micropipettes and microgrippers. On the other hand, our proposed microrobot can minimize the contact damage by applying a compressible bubble incorporated with ultrasound with maintaining high throughput and repeatability.

When a bubble attached on the tip end of a microrobot is acoustically excited by a piezoactuator attached on the side of a chip around its natural frequency, it oscillates and simultaneously generates cavitational microstreaming and radiation forces, so-called Bjerknes forces, resulting in pulling and eventually extracting a nucleus through the slit of the cell, as shown in Fig. 1[5]. The extracted nucleus can be transported to any location with the microrobot actuated by an external magnetic controller, consisting of a precisely controllable traverse and twin permanent magnets.

FABRICATIONS AND EXPERIMENT SETUPS

In order to prove the concept of a proposed on-chip microrobot, it is microfabricated by using the standard MEMS technology, as shown in Fig. 2. For a lift-off process, a chromium layer with a thickness of 400 Å as a sacrificial layer is deposited on a clean glass slide $(7.5(L) \times 2.5(W)$ cm^2) by using a sputter in Fig. 2(a). And a SU-8 layer (SU-8 2100, MicroChem Corp.) with a thickness of 200 μm as a structure layer is deposited on the top of the chromium layer by using a spin coater and soft-baked for 1 hour at 95 °C in Fig. 2(b-c). For patterning processes, the SU-8 layer is patterned by using a mask aligner (MDA-400M, Midas System Co., Ltd) with the designed mask, developed in a SU-8 developer (AZ 1500 thinner, AZ Electronic Materials Ltd.), and hard-baked for 20 minute at 150 °C in Fig. 2(d-f). And then the designed SU-8 structure is finally obtained from the lift-off process by submerging the substrate into chromium etchants (651826, Sigma-Aldrich Co., LLC) in Fig. 2(g). Lastly, two cylinder-type neodymium magnets (1 mm(D) × 1 mm(H)) are installed on the designed spots in the microrobot by using epoxy in Fig. 2(h).

For acoustic excitation, a sine wave voltage was generated by a function generator (33210A, Agilent Co.), and amplified up to a few hundred voltages by a voltage amplifier (BA4825, NF Co.). The amplified voltage signal was transmitted to a cylinder-type piezoactuator (PIC151, Physik Instruments Inc.) attached to the side of a water chamber

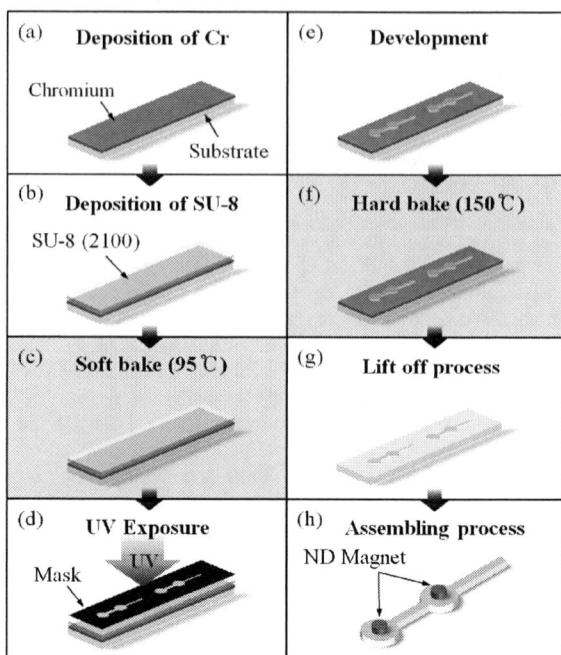

Figure 2: Microfabrication processes.

$(7.5(L) \times 2.5(W) \times 0.5(H)$ cm^3). The visual results during the experiments were captured by a charge coupled device (CCD) camera (EO-1312C, Edmund Optics) integrated with a zoom lens (VZMTM 450i eo, Edmund Optics) and saved on a PC in Fig. 3.

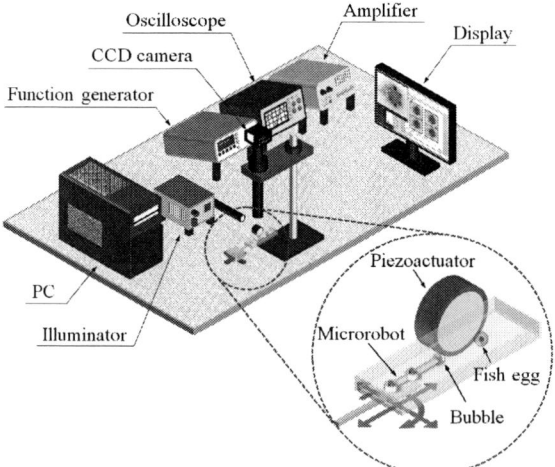

Figure 3: Schematic diagram of experimental setups.

EXPERIMENTAL RESULTS

The two-dimensional (2D) motion control of a microrobot made up a microfabricated polymer body and twin neodymium magnets is firstly demonstrated on a chip filled with an aqueous medium, as shown in Fig. 4. When an external magnetic controller attached beneath the bottom of the chip moves, the microrobot follows the motion of the controller owing to the magnetic interaction forces induced by two identical magnets installed inside both the microrobot and controller with the same distance between front and rear magnets. The rotational motion of the microrobot is also demonstrated by rotating the controller on the rear magnet in Fig. 4(f).

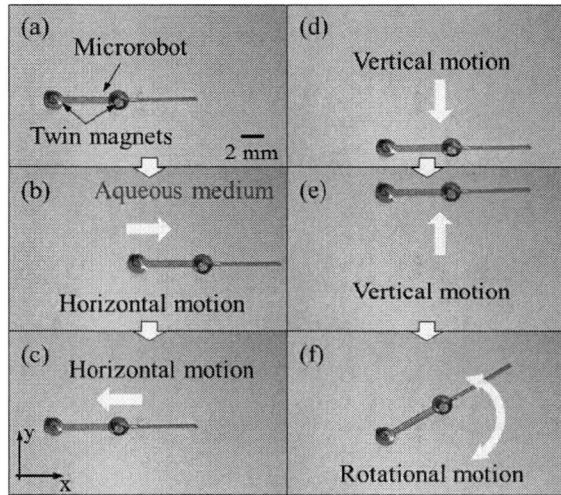

Figure 4: Two-dimensional (2D) control of a microrobot consists of a polymer body and twin permanent magnets by an external magnetic controller attached on the chip bottom.

Cavitational microstreaming induced by an acoustically oscillating bubble is experimentally investigated. When a bubble in a water chamber is acoustically excited by a piezoactuator at 14 kHz, it oscillates in a harmony with the applied frequency due to its compressibility and simultaneously generates an inward directional microstreaming around it in Fig. 5. For measuring the velocity fields in different frequencies and voltages, a microscopic PIV system mainly consisted of a microscope, a high-speed camera, and a laser is used. The flow velocity is founded to be strongly dependent on the applied frequency and proportional to the applied voltage. The velocity is maximum at the bubble's natural frequency and decreases as the applied frequency deviates. And the maximum velocity often occurs at 0.4 D (D is a bubble diameter) above the center rim of the bubble.

To investigate the effects of an oscillating bubble induced microstreaming on the inside of a cell through a narrow slit, particle extraction is tested using a small PDMS cylinder chamber in Fig. 6. When a bubble is excited by a piezoactuator at 15 kHz, it generates an inward directional

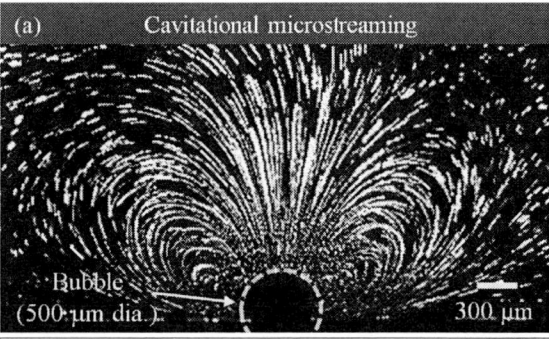

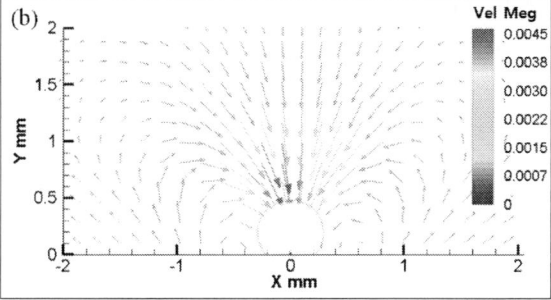

Figure 5: Cavitational microstreaming: (a) Flow visualization image. For flow visualization, fluorescent particles (15 μm diameter) are seeded and suspended in the water; (b) Particle image velocimetry (PIV) image.

microstreaming around it, extracting fluorescent particles (15 μm diameter) from the chamber through a prepared narrow channel (100 μm diameter). After 160 seconds, most of the particles are extracted, and the chamber becomes empty.

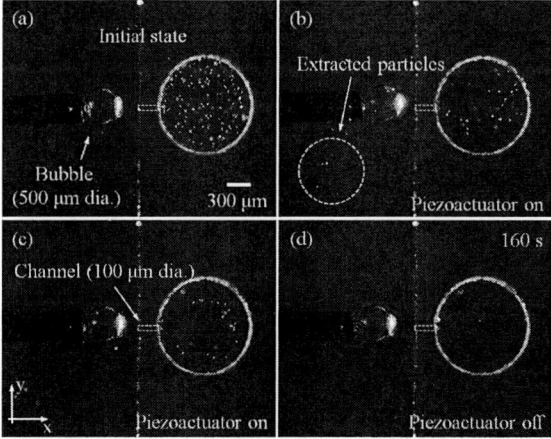

Figure 6: Particle extraction: (a) A bubble (500 μm diameter) and cylinder chamber (1 mm diameter) filled with fluorescent particles (15 μm diameter) are initially placed about 200 μm apart; (b-c) When the bubble is acoustically excited at 15 kHz, it generates microstreaming and simultaneously extracts the particles from the chamber; (d) After 160 seconds, the chamber becomes empty without the particles.

978-1-4799-7956-1/15 $31.00 © 2015 IEEE

As proof of concept, the embryo extraction from a fish egg (1.7 mm diameter) with a soft membrane wall as a cell is experimentally demonstrated using the microrobot, as shown in Fig. 7. First, a microrobot actuated by an external magnetic controller is transported and closed to the fish egg dyed by methylene blue and confined in a cage made of 6 piles to prevent unwanted movement during the experiment in Fig. 7(a-b). When a bubble attached on the tip end of the microrobot is excited by a piezoactuator at 15 kHz, the oscillating bubble generates microstreaming, and the embryo from the egg is simultaneously spilling out through the prepared slit (about 100 μm diameter) of the egg in Fig. 7(c-e). And then the extracted embryo is carried to a target place by the microrobot in Fig. 7(f).

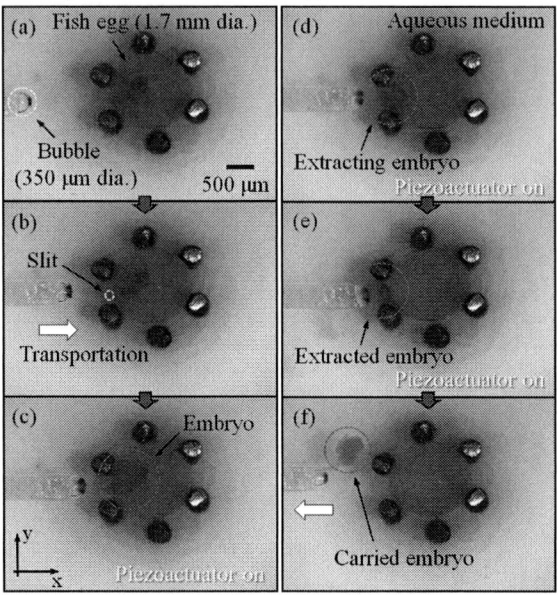

Figure 7: Embryo extraction: (a) Initial state; (b) A microrobot is closely transported to a fish egg using an external magnetic controller; (c-e) A piezoactuator is turned on to excite a bubble attached on the tip end of the microrobot. The excited bubble induced microstreaming with radiation forces pulls and eventually extracts an embryo through the prepared slit of the fish egg. (f) The extracted embryo is carried to a target place by the microrobot. Note that the red circles indicate the embryo.

CONCLUSIONS

This paper describes a novel on-chip enucleation method using a microrobot. Two actuation schemes, magnetic actuation for the propulsion of the microrobot and acoustic actuation for the enucleation of a cell, are used. First, the two-dimensional motion control of the microrobot is demonstrated by using an external magnetic controller owing to the interaction forces induced by twin magnets installed inside both the microrobot and controller. Second, the enucleation of a cell is separately investigated using an acoustically oscillating bubble. An oscillating bubble induced cavitational microstreaming in different frequencies and voltages is quantitatively studied using a microscopic PIV system. Finally, integrating the two actuation, the embryo extraction from a fish egg (1.7 mm diameter) is achieved on a chip. This non-invasive micromanipulation technique can be applied to bio-cell manipulation and micro-assembly.

ACKNOWLEDGEMENTS

This research was supported by Basic Science Research Program through the National Research Foundation of Korea (NRF) funded by the Ministry of Education, Science and Technology (2011-0025039).

REFERENCES

[1] J. L. Edwards, *et al.*, "Cloning adult farm animals: a review of the possibilities and problems associated with somatic cell nuclear transfer," *American Journal of Reproductive Immunology*, vol. 50, pp. pp. 113-123, 2003.

[2] J. P. Desai, *et al.*, "Engineering approaches to biomanipulation," *Annual Review of Biomedical Engineering*, vol. 9, pp. 35-53, 2007.

[3] E. W. H. Jager, *et al.*, "Microrobots for micrometer-size objects in aqueous media: potential tools for single-cell manipulation," *Science*, vol. 288, pp. 2335-2338, 2000.

[4] V. Korzh and U. Strähle, "Marshall Barber and the century of microinjection: from cloning of bacteria to cloning of everything," *Differentiation*, vol. 70, pp. pp. 221-226, 2002.

[5] J. O. Kwon, *et al.*, "Electromagnetically actuated micromanipulator using an acoustically oscillating bubble," *Journal of Micromechanics and Microengineering*, vol. 21, p. 115023, 2011.

[6] J. H. Lee, *et al.*, "On-Chip Micromanipulation by AC-EWOD Driven Twin Bubbles," *Sensors and Actuators A: Physical*, vol. 195, pp. PP. 167-174, 2013.

[7] J. O. Kwon, *et al.*, "Micro-object manipulation in a microfabricated channel using an electromagnetically driven microrobot with an acoustically oscillating bubble," *Sensors and Actuators A: Physical*, vol. 215, pp. pp. 77-82, 2014.

[8] J. Voldman, "Electrical forces for microscale cell manipulation," *Annual Review of Biomedical Engineering*, vol. 8, pp. 425-454, 2006.

[9] T. K. M. Hagiwara, *et al.*, "On-chip magnetically actuated robot with ultrasonic vibration for single cell manipulations," *Lab on a chip*, vol. 11, pp. pp. 2049 - 2054, 2011.

CONTACT

S. K. Chung, Tel: +82-31-330-6346; skchung@mju.ac.kr

OUT-OF-PLANE MICRO-FORCE FUNCTION GENERATOR WITH INHERENT SELF-FEEDBACK FOR MICRO-DEFORMATION MODIFYING

Xinghua Wang, Dingbang Xiao, Xuezhong Wu, Zhanqiang Hou, Zhihua Chen, and Hanhui He

[1]College of Mechatronics and Automation, National University of Defense Technology, CHINA

ABSTRACT

Many micro-electro-mechanical structures are always subject to residual stress and can easily cause mechanical deformation. The warpage of device substrate could directly affect the performance and should be effectively controlled. This paper mainly reports a novel concept of out-of-plane micro-force function generator for micro-deformation modifying. The proposed generator is based on batch fabricated polymer thermal actuators array and could actively modify micro-substrate warpage. Experimental results showed that the out-of-plane micro-force function generator was able to achieve accurate rectifying of substrate micro-deformation. This strategy constructively utilizes the inherent self-feedback for in-situ deformation control and has the potential for solving stress-induced problems of micro-fabricated devices.

INTRODUCTION

Microelectromechanical systems (MEMS) devices are always subjected to residual or thermal mechanical stresses [1,2]. Consequently, structure geometrical parameters may be deviated from desired design values and cause stress-induced instability, resulting in frequency variation, bias and scale factor drift or even breakage of the structure [3]. Numerous studies have elucidated the basic thermo-mechanical response of MEMS devices. The out-of-plane deformation due to misfit strains is a critical issue, and it should be controlled and maintained at a level low enough to avoid impact to the performance [4-6]. Much impressive work has been done to solve this problem and there are vast literatures on topics relating to the deformation and stress control [7]. Unfortunately due to some inherent reasons, traditional strategies such as circuit temperature compensation cannot be basically effective [8]. Some comparatively effective techniques, take electrostatic calibration for example, are limited to insufficient rectifying capability despite of resolution and flexibility [9]. This prevents these strategies from being used in high performance levels.

Thermal actuation has advantages over other types of micro-actuation, which features compactness, high force, and low driving voltage [10]. One obvious advantage of the electro-thermal actuators is that it can provide large force and displacements at CMOS comparative voltages [11]. Thus, we proposed a technique by employing an out-of-plane micro-force function generator (MFFG) to rectify the micro-deformation due to mechanical stress [12]. Schematic view of the device is depicted in Figure 1. Arrays of micro-fabricated polymer thermal actuators work in parallel to actively modify micro-substrate warpage. Furthermore, the approach constructively utilizes the inherent self-feedback of electrical-thermal resistances for in-situ deformation observing and control.

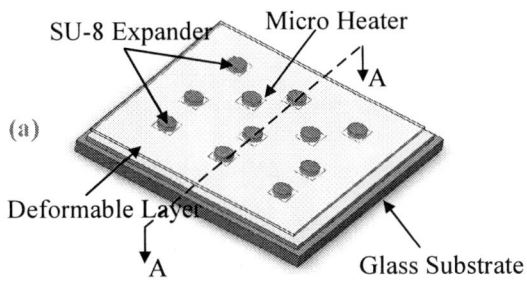

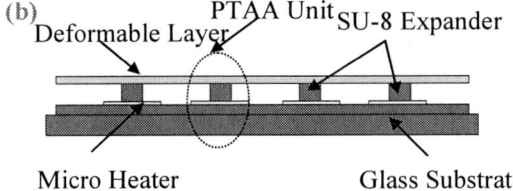

Figure 1: Schematic view of the out-of-plane micro-force function generator. (a) System components of the device; (b) A-A cross section view.

Figure 2 shows working principle of the proposed out-of-plane MFFG. When DC voltage is applied to the micro-heater, the generated joule heat is efficiently transferred to the surrounding structures. Thus, bottom of SU-8 expander would experience higher temperature (Figure 2(a)). Since there are no or lower currents pass through, the thermal actuators array would push different deformation, see Figure 2(b). Consequently, the out-of-plane mechanical deformation of the deformable silicon plate above the actuators array would be modified.

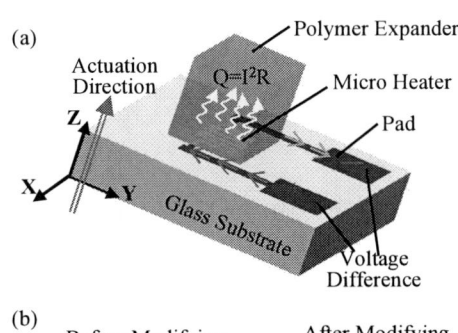

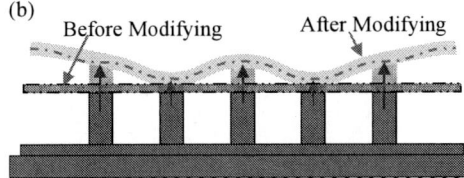

Figure 2: Schematic view of the out-of-plane micro-force function generator. (a) Single actuator unit; (b) Operation principle of the MFFG.

DEVICE FABRICATION

This out-of-plane micro-force function generator is fabricated using three-masks process, which mainly consists of three basic steps (Figure 3): fabrication of the micro-heater, defining of SU-8 expander and integrating with the aimed devices.

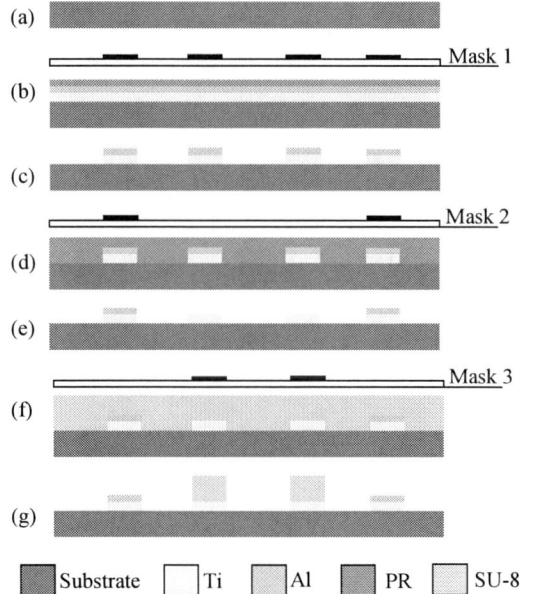

Figure 3: Fabrication process of the micro-fabricated out-of-plane micro-force function generator. (a) preparation of device substrate; (b~c) deposition of Ti/Al metal layer and photoresist patterned using photolithography through Mask-1; (d~e) Photoresist patterned using photolithography by Mask-2; (f~g) SU-8 polymer spin coated and exposed to UV light using Mask-3.

As can be seen in Figure 3, the fabrication process started with a Pyrex glass substrate. Pyrex glass is selected as the bulk substrate because of its lower thermal conductivity and higher electrical resistivity compared to silicon, and thus achieving good heat confinement and low power consumption. A metal layer of Ti was deposited on the substrate and then a layer of Al was deposited onto the Ti layer. Positive photo-resist was spin coated onto the metal layers and patterned using photolithography through a designed Mask-1. After solvent and DI-water clean, the Pyrex glass substrate with Ti/Al metals was spin-coated and patterned using photolithography by another designed Mask-2. The Al in the designed area was removed and zigzag Ti was exposed as the resistor.

Defining of the SU-8 expanders array started with spin-coating of SU-8-2100 (intended thickness of 200μm) on Pyrex glass substrate with fabricated micro heaters. Then the SU-8 layer was soft baked and exposed to UV light using a designed mask to define the thermal expander (Figure 8). In preparation for the final step, a silicon deformable layer with the thickness of 150μm was first fabricated, and then it was mounted to the surface of the SU-8 thermal actuators array using epoxy. The fabricated out-of-plane micro-force function generator can be seen in Figure 9.

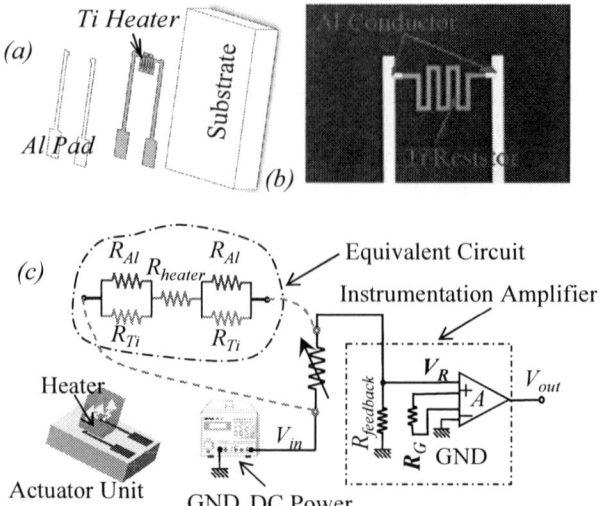

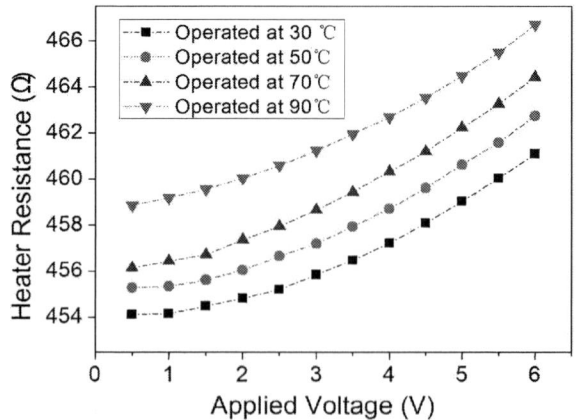

Figure 4: The micro-fabricated Al/Ti heater and resistance measuring circuit. (a~b) the fabricated Al/Ti micro heater; (c) equivalent circuit of the micro-heater and the schematic diagram of the circuit prototype.

The fabricated micro-heater consists of 200nm Ti layer and 70nm Al layer, see figure 4(a). The Ti layer acts as both the resistor and the adhesion layer, while the Al layer acts as both the conductor and contact pad. In order to observe the transient Al/Ti micro-heater resistance, a custom circuitry was designed. Figure 4(c) shows the equivalent circuit model of the micro-heater and the schematic diagram of the circuit prototype.

EXPERIMENTAL PERFORMANCE

To validate the micro-force function generator model, experiments were carried out. Electro-thermal-mechanical behaviors of the out-of-plane MFFG sample under resistive heating were performed. It is obvious that the resistance of the micro-fabricated heater is an important parameter for the MFFG sample. When the MFFG is operated at different temperatures, the initial resistance would show different values. Figure 5 shows the variation of the micro-heater resistance with external applied voltage from 0V to 6V at different ambient environment temperatures.

Figure 5. Variation of the micro-heater resistance with applied voltage at different ambient temperature.

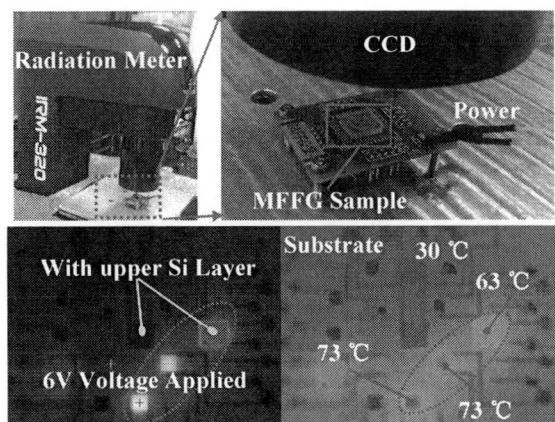

Figure 6: Operating temperature of the out-of-plane micro-force function generator unit.

The operating temperature of thermal actuators is another key parameter of the micro-force function generator system. In order to characterize the sample, a 6V DC voltage is applied. The average in-situ temperature was monitored through an active infrared radiation temperature meter, IRM-320. The operating temperatures of SU-8 expander with and without a silicon die were measured, and experimental results are shown in Figure 6. We can find that the working temperature mainly fluctuated from 50℃ to 80℃. As polymer SU-8 is thermally insulated, the operating temperature of the top surface is lower, which is about 73℃ at 6V.

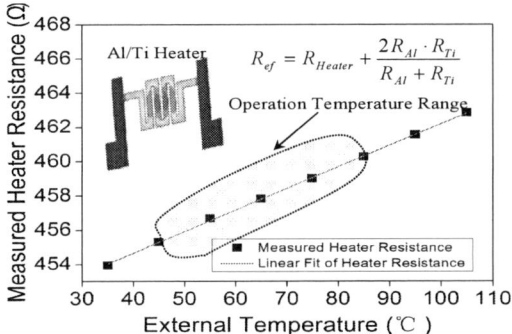

Figure 7: Variation of the Al/Ti micro-heater resistance with external ambient temperature.

Variation of the micro-heater resistance with different external temperatures can be seen in Figure 7. We can find that during the main operation temperature range (50℃ ~80 ℃), good linear relationship between the heater resistance and external temperature can be easily found.

The electrical induced displacement of a typical actuator unit was measured using a non-contact displacement sensor, LK-G10. As mentioned above, the resistance of the micro-heater can be used as an inherent parameter for temperature observing and displacement feedback. In order to validate it, the in-situ actuator tip displacement and real-time thermal-resistance of the micro heater were observed at the same time. Results were shown in Figure 8, and we can see that the tip displacement can be determined indirectly by observing the resistance of the micro-fabricated heater. The ratio coefficient between the displacement and resistance is 0.43μm/Ω.

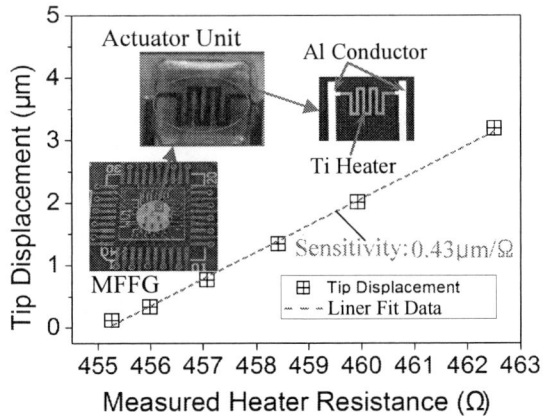

Figure 8: Variation of the tip displacement of the thermal actuator unit with the micro-heater thermal resistance.

RESULTS AND DISCUSSION

The out-of-plane micro-force function generator was driven using a multi-channels voltage control strategy. When the thermal actuators array was driven by different distributed voltages, the silicon layer above the actuators array would present different shape. Thus, certain warpage of the micro substrate can be modified by applying certain distributed voltages. Based on this setup and strategy, experiment was carried out to actively modify the 150μm thickness silicon layer. Based on our previous experiment, we select 6V as a typical operation voltage value. In order to validate the proposed prototype, two different distributed modes were performed. The out-of-plane deformation of the MFFG can be measured by a non-contact optical interferometer. Figure 9(a) is the center actuation mode and the maximum displacement in the center point is 0.29μm at 6V. Figure 9(b) is the H-type actuation mode and the maximum displacement is about 0.70μm at the edge of the deformable layer.

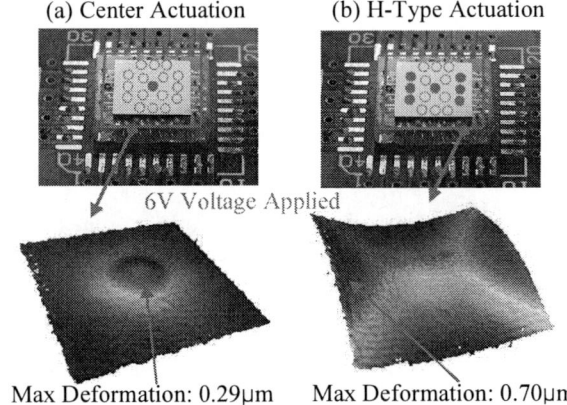

Figure 9: Two different types of 3D optical image of the out-of-plane micro-force function generator with two different actuation modes (6V applied voltage). (a) center actuation mode and (b) H-type actuation mode.

Finally, we find that the micro-deformation could be actively and effectively modified by employing this out-of-plane micro-force function generator, and it has potential for solving stress-induced problems of micro-fabricated MEMS devices.

CONCLUSION AND FUTURE WORK

We reported a novel out-of-plane micro-force function generator which was based on batch fabricated thermal actuators array. The thermal actuator was fabricated utilizing SU-8 polymer as the functional expander and Ti/Al electrode as the micro-heater. Experiments have been performed to characterize the MFFG. And results showed that the warpage of the micro-substrate can be rectified by employing this out-of-plane MFFG. This research provides a more flexible technology for MEMS devices based on multilayered structures and several different materials.

Improvements in fabrication process for better characterization and optimization of the out-of-plane micro-force function generator are now under way. Additional characterization of the fully assembled MFFG will also be carried out.

ACKNOWLEDGEMENTS

This work was supported by National Natural Science Foundation of China (NSFC, Grant No. 51175506), and carried out at the Microsystem Laboratory, College of Mechatronics and Automation, National University of Defense Technology (NUDT). The authors are grateful to Yanglin Peng and Fujing Tian for their help in device fabrication and characterization.

REFERENCES

[1] Chia-Fang Chiang, Andrew B. Graham, Brain J. Lee, Chae Hyuck Ahn, Eldwin J. Ng, Gary J. O'Brien and Thomas W. Kenny, "Resonant pressure sensor with on-chip temperature and strain sensors for error correction," in MEMS'13, Taipei, 2013, pp. 45-48.

[2] Jiachou Wang, Lijian Yang, and Xinxin Li, "On-chip integrated PS3 (packaging-stress suppressed suspension) for thermal-stress fress package of pressure sensor." In MEMS'13,Taipei, 2013, pp. 49-52.

[3] Jin-Won Joo, and Sung-Hoon Choa, "Deformation behavior of MEMS gyroscope sensor package subjected to temperature change," IEEE Transaction on Components and Packaging Technologies, 2007, 30, (2), pp. 346-354.

[4] Martin L. Dunn, Yanhang Zhang, Victor M. Bright, "Deformation and structural stability of layered plate microstructures subjected to thermal loading," Journal of Microelectromechanical Systems, 2002, 11, (4), pp. 372-384.

[5] Hou, Z.Q., Xiao, D.B., Wu, X.Z., Dong, P.T., Chen, Z.H., Niu, Z.Y., and Zhang, X.: 'Effect of Axial Force on the Performance of Micromachined Vibratory Rate Gyroscopes', Sensors-Basel, 2011, 11, (1), pp. 296-309.

[6] Hou, Z.Q., Xiao, D.B., Wu, X.Z., Jian, B.S., Chen, Z.H., and Zhang, X.: 'Effect of die attachment on key dynamical parameters of micromachined gyroscopes', Microsyst Technol, 2012, 18, (4), pp. 507-513.

[7] Gang Dai, Mei Li, Xiaoping He, Lianming Du, Beibei Shao, Wei Su, "Thermal drift analysis using a multiphysics model of bulk silicon MEMS capacitive accelerometer," Sensors and Actuators A, 2011, 172, pp. 369-378.

[8] Dunzhu Xia, Shuling Chen, Shourong Wang, and Hongsheng Li, "Microgyroscope temperature effects and compensation-control methods," Sensors-Basel, 2009, 9, (10), pp. 8349-8376.

[9] S. Sonmezoglu, S.E. Alper, and T. Akin, "A high performance automatic mode-matched MEMS gyroscope with an improved thermal stability of the scale factor," Transducers 2013, Barcelona, SPAIN, pp. 2519-2522.

[10] Gih-Keong Lau, Johannes F.L. Goosen, Fred van Keulen, Trinh Chu Duc, and Pasqulina M. Sarro, "Polymeric thermal microactuator with embedded silicon skeleton: Part I-design and analysis," Journal of Microelectromechanical Systems, 2008, 17, (4), pp. 809-821.

[11] K.L. Zhang, S.K. Chou, S.S. Ang, "Fabrication, modeling and testing of a thin Au/Ti microheater," International Journal of Thermal Sciences, 2007, 46, pp. 580-588.

[12] Xinghua Wang, Dingbang Xiao, Xuezhong Wu, Zhanqiang Hou, Zhihua Chen, "Out-of-plane micro-force function generator based on polymeric thermal actuators for modifying micro-deformation," Applied Physics Letters, 2013,103, pp. 151902.

CONTACT

*Xinghua Wang, tel: +86-731-84574958;
wangxinghua87@nudt.edu.cn

A PAPER-LIKE MICRO-SUPERCAPACITOR WITH PATTERNED BUCKYPAPER ELECTRODES USING A NOVEL VACUUM FILTRATION TECHNIQUE

Cheng-Wen Ma, Po-Cheng Huang, and Yao-Joe Yang
National Taiwan University, Taipei, TAIWAN

ABSTRACT

This study reports a paper-like micro-supercapacitor (SC) with in-plane interdigital buckypaper electrodes on a filter membrane substrate. A vacuum filtration method assisted by lithography techniques is proposed for patterning buckypaper. The proposed micro-SC features advantages including a flexible structure, simple fabrication, easy chip integration, and high specific capacitance. Increasing the aspect ratio of the patterned buckypaper electrodes effectively enhances the specific capacitance of the SC. The specific capacitance measured by cyclic voltammetry was 107.27 mF/cm^2 at a scan rate of 20 mV/sec. The measured charge-discharge behaviors at various discharge rates show the electrochemical stability of the device.

INTRODUCTION

Rapid advances in microelectronics have increased the demand for miniaturized power sources for effective on-chip integration with electronic components. Supercapacitors (SCs) have significant potential in this regard due to their high power density and robust cycling stability [1, 2], and excellent performance has been observed through the use of a material matrix formed with carbon nanotubes (CNTs), in which the pore structures provide excellent ion accessibility and transport [3, 4]. Therefore buckypaper, which is a thin sheet formed with an aggregate of CNTs, has been reported as a promising electrode material for SCs [5, 6].

Najafabadi et al. demonstrated a composite electrode consisting of carbon nanohorns (CNHs) and thin films of CNTs (i.e., buckypaper) for a SC. The pore structure of the composite electrode was tailored for electrolyte retention to deliver higher power [7]. A composite CNT film fabricated by depositing vanadium oxide onto CNT buckypaper using supercritical fluid deposition was proposed as the electrodes for SC applications [8]. Che et al. proposed a method to realize cellular CNT buckypaper materials with excellent mechanical, electrical and electrochemical properties. The fabricated material can be employed as the electrodes of light-weight and high-temperature SCs [9].

SC structures are classified as sandwich type [10], roll type [11], or interdigital type [12]. The interdigital type features several advantages over the others. For example, its charge storage capacity can be extended by increasing its area or its electrode thickness. Also, electrode patterns can be designed to easily adjust the device's internal impedance [13]. Using micromachining techniques, various SCs with interdigital structures have been proposed.

Shen et al. proposed a three-dimensional micro-SC with high-aspect-ratio interdigital structure realized by using deep etching techniques and self-supporting active materials. The fabrication method of the proposed device is applicable for chip integration of SCs [14]. An ultrahigh power micrometre-sized SCs based on onion-like carbon was proposed [15]. The device was realized by the electrophoretic deposition of onion-like carbon particles onto interdigital gold film on a silicon wafer. Beidaghi et al. proposed a method for fabricating interdigitated electrodes for micro-SC application based on reduced graphene oxide (rGO) and CNT composites. The electrodes were realized by electrostatic spray deposition and photolithography lift-off techniques [16]. A flexible micro-SC with a patterned polyaniline (PANI) nanowire-array microelectrode was demonstrated [17]. The interdigital microelectrode was fabricated using photolithography techniques on a flexible PET chip.

This work presents the development of a paper-like micro-SC with interdigital buckypaper electrodes. A novel method based on vacuum filtration and lithography techniques is proposed to pattern buckypaper into high-aspect-ratio interdigital shapes. The specific capacitance of the proposed micro-SC is enhanced by the relatively large thickness of the interdigital buckypaper electrodes. In addition, the proposed micro-SC features advantages including flexibility, light weight, and a simple fabrication process.

DEVICE DESIGN

Figure 1 shows the schematic of the proposed micro-SC (without electrolyte). The proposed device consists of a nylon membrane filter (MS® Nylon membrane filter, pore size: 0.8 μm) structure and a pair of interdigital buckypaper electrodes, which are patterned using a typical photolithography process with vacuum filtration. A gold film is deposited on the top surface of the electrodes.

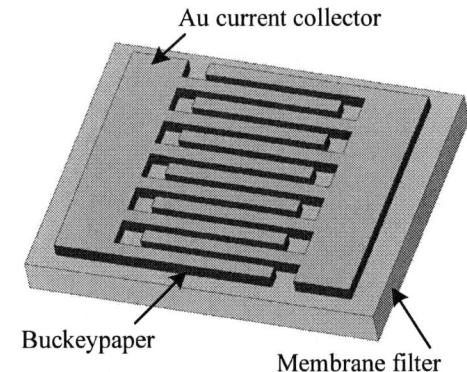

Figure 1: Schematic of the proposed micro supercapacitor.

Figure 2 provides two schematic illustrations of SC devices, one with a typical sandwich structure, and the other with an in-plane interdigital structure. For the sandwich structure (Fig. 2(a)), increasing the active material thickness (h_1) will improve the SC charge capacity per unit of substrate area. However, increasing h_1 will directly increase the ion migration distance, which will significantly reduce the charge/discharge rates of SCs.

For the proposed in-plane interdigital structure, as shown in Fig. 2(b), the SC charge capacity can be increased by increasing the electrode thickness (h_2). However, increasing h_2 will only slightly increase the ion migration distance. Therefore, SCs with the in-plane interdigital electrode structure possibly feature better charge capacity without performance deterioration. In addition, the proposed in-plane device, using filtration paper as the substrate, is flexible and can be easily integrated into circuit chips.

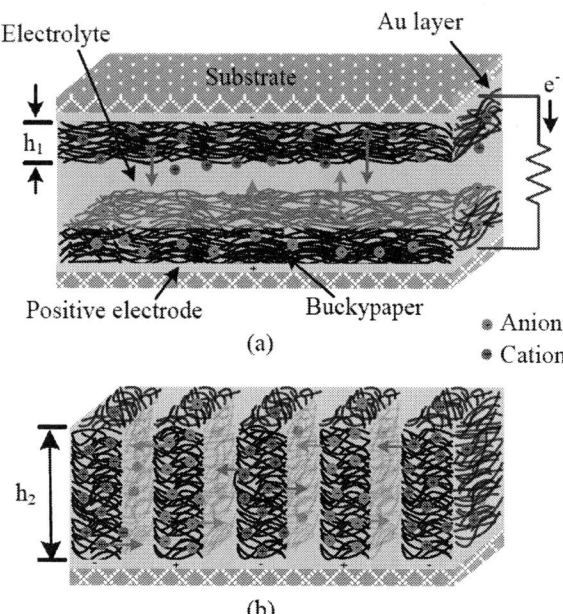

Figure 2: Two schematics of supercapacitor devices. (a) Sandwich structure. (b) Interdigital structure.

FABRICATION

Figure 3 describes the fabrication process of the proposed device. The SU-8 photoresist (SU-8 2050, MicroChem Corporation) was spin-coated onto a supporting plane, as shown in Fig. 3(a). The starting material is a nylon membrane filter placed on top of the SU-8 film, as shown in Fig. 3(b). SU-8 photoresist is then spin-coated on the membrane filter (Fig. 3(c)). After soft baked, a standard lithography process was performed and the SU-8 photoresist was patterned, as shown in Fig. 3(d) and (e). Then, vacuum filtration using an MWCNT-dispersed solution was then performed, and MWCNT molecules filled the SU-8 trenches, as shown in Fig. 3(f). Note that the concentration of the MWCNT solution was 0.01 wt%. The solution was subjected to ultrasound for 2 h to reduce the CNT bundling tendency prior to the filtration process. After vacuum filtration, the 200 nm Au layer is then deposited on the top of the MWCNT film as the current collector (Fig. 3(g)). Finally, SU-8 was removed using remover PG (Microchem, USA)

[16], and the patterned buckypaper electrodes were fabricated (Fig. 3(h)). PVA-KOH gel electrolyte was synthesized by mixing 5g PVA and 2.8 g KOH with 50 ml DI water in a beaker at 85°C until clear.

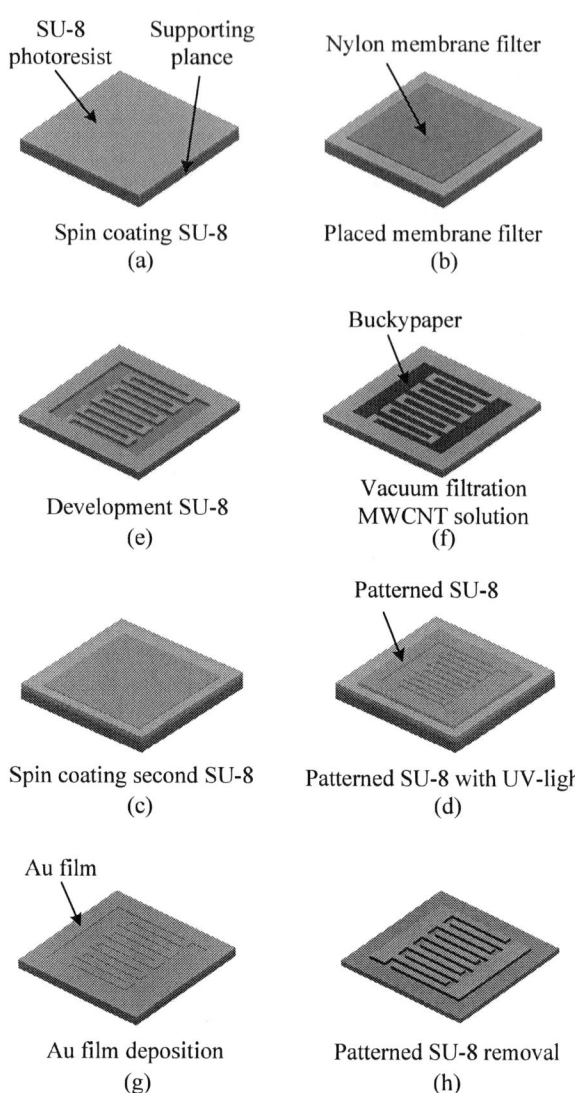

Figure 3 - The fabricated process of the supercapacitor with an interdigital structure by patterned buckypaper electrode.

Figure 4 shows the buckypaper thickness vs. the volume of MWCNT solution consumed during the filtration process. Figure 5 shows the fabrication results. Figure 5(a) shows the fabricated device (no packaging and electrolyte). The total size is about 4×4 mm^2. Each interdigital electrode has 7 fingers, each measuring 2.73 mm in length, with a width of 80 μm. The gap between each finger is 100 μm. The thicknesses of the buckypapers with or without the patterned SU-8 structures on the filter membrane are almost identical. Figure 5(b) shows the SEM picture of the interdigital electrodes after removing SU-8. Figure 5(c) shows the SEM picture of the interdigital electrodes after removing SU-8. The thickness was about 100 μm. Figure 5(d) shows the SEM image of the top of the patterned buckypaper. These figures also show that the

978-1-4799-7956-1/15 $31.00 © 2015 IEEE   1068

buckypaper was a tailored pore structure for excellent ion accessibility.

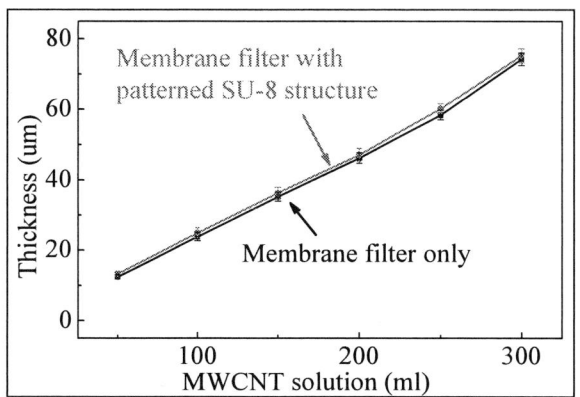

Figure 4: The buckypaper thickness vs. the volume of MWCNT solution.

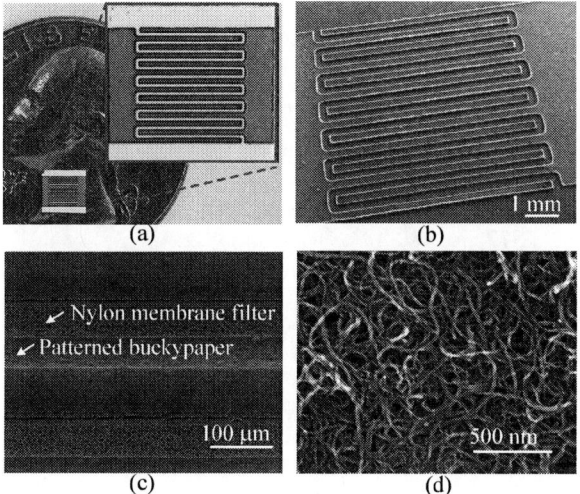

Figure 5: (a) The fabricated micro-SC. (b) The SEM picture of micro-SC. (c) The SEM picture of the patterned buckypaper. (d) The SEM image on the top of patterned buckypaper.

MEASUREMENT AND DISCUSSION

Figure 6 shows the CV curves at various scanning rates using the CHI 627D electrochemical station (CH Instruments, Taiwan). These curves assume nearly rectangular shapes, which implies excellent and reversible capacitive behavior. From the CV curves, the specific capacitances can be calculated according to the following equation [1].

$$C_s = \frac{AREA_{CV}}{s \cdot A \cdot \Delta V} \tag{1}$$

where $AREA_{CV}$ is the area integral of voltammogram (i.e., the total voltammetric charge obtained by the integration of the positive and negative sweeps in a cyclic voltammogram), s is the scan rate, A is the total geometric surface area of a single electrode, and ΔV is the scanned potential window. Table I shows the calculated specific capacitance on various scan rates. The maximum specific capacitance was calculated as 107.27 mF/cm^2 at a scan rate

of 20 mV/sec.

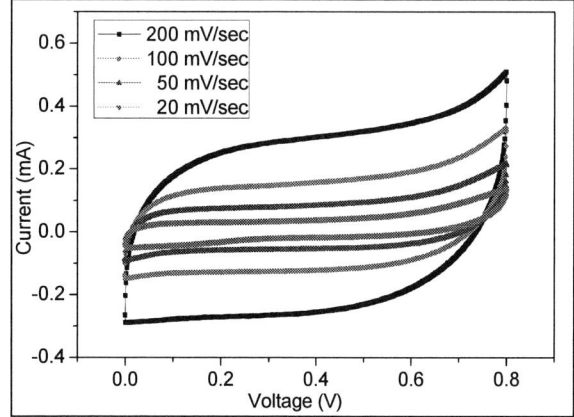

Figure 6: Cyclic voltammetry curve of the prototype at various scanning rates from 20 to 200 mV/sec.

TABLE I. Specific capacitance on various scan rates.

Scan rate (mV/sec)	200	100	50	20
C_s (mF/cm^2)	85.03	89.83	94.19	107.3

Figure 7 shows the galvanostatic charge/discharge curves for current densities of 1, 2, 5, and 10 mA/cm^2. The curves in the figure are quite linear and their triangular shapes indicate good electrochemical capacitive characteristics. Each charge/discharge process of the curves starts with a sharp voltage drop (iR drop) because of the SC internal resistance. The capacitance of the SC can be obtained by calculating the slope of the discharge curve [3].

$$c = \frac{i \cdot \Delta t}{\Delta V} \tag{2}$$

where i is the applied current that is a constant for each charging process, and ΔV is the voltage range. For the device, the maximum specific capacitance was 119.8 mF/cm^2 at a charging current of 1 mA/cm^2. Table II lists the calculated specific capacitance ($C_d = c/A$) for different charging currents.

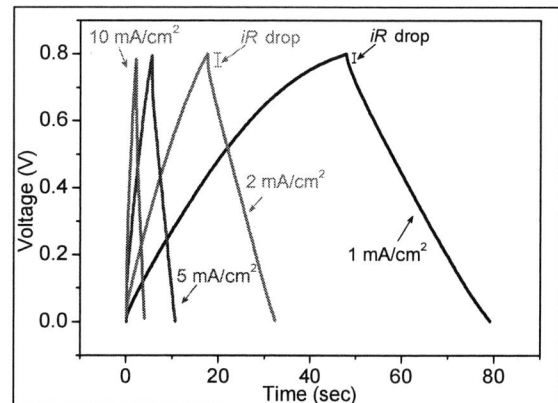

Figure 7: Galvanostatic charge/discharge curves at various charge/discharge rates.

TABLE II. Specific capacitance at various discharge current density.

Current density (mA/cm^2)	10	5	2	1
Δt (sec)	2.2	5.6	17.6	47.9
C_d (mF/cm^2)	55.0	70.0	88.0	119.8

CONCLUSION

In conclusion, this work presents the development of a micro-supercapacitor with in-plane interdigital buckypaper electrodes. By using a novel vacuum filtration method with lithography techniques, SU-8 photoresist was patterned on the flexible membrane filter and served as the filtration mask. Cyclic voltammetry and charge-discharge experiments show the device's good electrochemical stability. The specific capacitance was measured by cyclic voltammetry as 107.27 mF/cm^2 at a scan rate of 20 mV/sec. The advantages of the proposed device include a flexible structure, simple fabrication, high specific capacitance, and easy integration with micro chips. The proposed micro-supercapacitor potentially can be used in the development of energy storage devices for microsystems.

ACKNOWLEDGEMENTS

This work was supported in part by the National Science Council, Taiwan, R.O.C. (Contract No: NSC 100-2221-E-002-075-MY3).

REFERENCES

[1] A. Ramadoss and S. J. Kim, "Improved activity of a graphene–tio2 hybrid electrode in an electrochemical supercapacitor", *Carbon,* vol. 63, no. 0, pp. 434-445, 2013.

[2] L.-F. Chen, Z.-H. Huang, H.-W. Liang, W.-T. Yao, Z.-Y. Yu, and S.-H. Yu, "Flexible all-solid-state high-power supercapacitor fabricated with nitrogen-doped carbon nanofiber electrode material derived from bacterial cellulose", *Energy & Environmental Science,* vol. 6, no. 11, pp. 3331-3338, 2013.

[3] M. Kaempgen, J. Ma, G. Gruner, G. Wee, and S. G. Mhaisalkar, "Bifunctional carbon nanotube networks for supercapacitors", *Applied Physics Letters,* vol. 90, no. 26, 264104, 2007.

[4] Z. Niu, W. Zhou, J. Chen, G. Feng, H. Li, Y. Hu, W. Ma, H. Dong, J. Li, and S. Xie, "A repeated halving approach to fabricate ultrathin single-walled carbon nanotube films for transparent supercapacitors", *Small,* vol. 9, no. 4, pp. 518-524, 2013.

[5] H. Chen, J. Di, Y. Jin, M. Chen, J. Tian, and Q. Li, "Active carbon wrapped carbon nanotube buckypaper for the electrode of electrochemical supercapacitors", *Journal of Power Sources,* vol. 237, no. 0, pp. 325-331, 2013.

[6] R. K. Das, B. Liu, J. R. Reynolds, and A. G. Rinzler, "Engineered macroporosity in single-wall carbon nanotube films", *Nano Letters,* vol. 9, no. 2, pp. 677-683, 2009.

[7] A. Izadi-Najafabadi, T. Yamada, D. N. Futaba, M. Yudasaka, H. Takagi, H. Hatori, S. Iijima, and K. Hata, "High-power supercapacitor electrodes from single-walled carbon nanohorn/nanotube composite", *ACS Nano,* vol. 5, no. 2, pp. 811-819, 2011.

[8] D. Quyet Huu, Z. Changchun, Z. Chuck, W. Ben, and Z. Jim, "Supercritical fluid deposition of vanadium oxide on multi-walled carbon nanotube buckypaper for supercapacitor electrode application", *Nanotechnology,* vol. 22, no. 36, p. 365402, 2011.

[9] G. Che, B. B. Lakshmi, C. R. Martin, E. R. Fisher, and R. S. Ruoff, "Chemical vapor deposition based synthesis of carbon nanotubes and nanofibers using a template method," *Chemistry of Materials,* vol. 10, no. 1, pp. 260-267, 1998.

[10] C. Meng, C. Liu, L. Chen, C. Hu, and S. Fan, "Highly flexible and all-solid-state paperlike polymer supercapacitors," *Nano Letters,* vol. 10, no. 10, pp. 4025-4031, 2010.

[11] H. Ji, Y. Mei, and O. G. Schmidt, "Swiss roll nanomembranes with controlled proton diffusion as redox micro-supercapacitors," *Chemical Communications,* vol. 46, no. 22, pp. 3881-3883, 2010.

[12] J. Chmiola, C. Largeot, P.-L. Taberna, P. Simon, and Y. Gogotsi, "Monolithic carbide-derived carbon films for micro-supercapacitors," *Science,* vol. 328, no. 5977, pp. 480-483, 2010.

[13] W. Gao, N. Singh, L. Song, Z. Liu, A. L. M. Reddy, L. Ci, R. Vajtai, Q. Zhang, B. Wei, and P. M. Ajayan, "Direct laser writing of micro-supercapacitors on hydrated graphite oxide films," *Nat Nano,* vol. 6, no. 8, pp. 496-500, 2011.

[14] C. Shen, X. Wang, W. Zhang, and F. Kang, "A high-performance three-dimensional micro supercapacitor based on self-supporting composite materials," *Journal of Power Sources,* vol. 196, no. 23, pp. 10465-10471, 2011.

[15] D. Pech, M. Brunet, H. Durou, P. Huang, V. Mochalin, Y. Gogotsi, P.-L. Taberna, and P. Simon, "Ultrahigh-power micrometre-sized supercapacitors based on onion-like carbon," *Nat Nano,* vol. 5, no. 9, pp. 651-654, 2010.

[16] M. Beidaghi and C. Wang, "Micro-supercapacitors based on interdigital electrodes of reduced graphene oxide and carbon nanotube composites with ultrahigh power handling performance," *Advanced Functional Materials,* vol. 22, no. 21, pp. 4501-4510, 2012.

[17] W. Liu, X. Yan, J. Chen, Y. Feng, and Q. Xue, "Novel and high-performance asymmetric micro-supercapacitors based on graphene quantum dots and polyaniline nanofibers," *Nanoscale,* vol. 5, no. 13, pp. 6053-6062, 2013.

CONTACT

Cheng-Wen Ma, tel: +886-2-33664941#802,
E-mail: jeson@mems.me.ntu.edu.tw

A POTASSIUM ELECTRET ENERGY HARVESTER FOR 3D-STACK ASSEMBLY

Kensuke Misawa[1], Sugiyama Tatsuhiko[2], Gen Hashiguchi[2], Hiroyuki Fujita[3]
and Hiroshi Toshiyoshi[1,3]

[1]Research Center for Advanced Science and Technology (RCAST), the University of Tokyo
[2]Research Institute of Electronics, Shizuoka University
[3]Institute of Industrial Science, the University of Tokyo

ABSTRACT

We report an electrostatic energy harvester based on the potassium ion (K+) electret that could be stacked up into a 3D structure to multiply the output power. Vertical comb electrodes are implemented in a silicon-on-insulator (SOI) wafer with a relatively heavy mass in the handle layer to lower the resonance. A single substrate formation exhibited a 0.34 µW output at 310 Hz for a load resistance of 1 MΩ.

INTRODUCTION

Downsizing of microelectronics predicts a new era of distributed wireless sensor network called trillion-sensors [1] as a potential application of the More-than-Moore type integrated MEMS. Besides the importance of the sensors for specific applications, autonomous energy source is an indispensable element. Recent R&D on MEMS energy harvesters (EHs) demonstrated mainly three different schemes including electromagnetic [2], piezoelectric [3], triboelectric [4] and electrostatic [5]. Due to the potential compatibility with the microelectronics, we have chosen the electrostatic type based on the permanent electrical charge called "electret"; nonetheless, the electret density remains the technical challenge to improve the output capacity.

Potassium electret [6] used in this work is included in a thermally grown silicon oxide and electrically activated through the biased field at temperature of 650 °C or higher. Unlike other activation methods such as soft-X ray radiation [7] and surface corona discharge [8], the presented method can form electrets on the surface of a deep trench, thereby increasing the output power in a given device footprint.

ENERGY HARVESTER WITH POTTASIUM ELECTRETS

Figure 1(a) shows the schematic device of a single unit cell EH. The chip is intentionally made in a disk shape such that it could be stacked up to make a battery-like cylinder. The device is composed of three parts: a movable mass, suspension springs, and comb electrodes. The movable vertical electrodes are connected into a piece with a central mass made in the handle layer to lower the first-mode resonance to a few hundreds Hz. The supporting springs and the electrodes are designed to be point symmetry such that the vertical vibration is preferably chosen to make mutual displacement of electrodes in the out-of-plane directions as illustrated in Figure 1(b). The surface of the electrodes are conformally covered with thermally grown silicon oxide after the sacrificial releasing step; high density potassium ions are included within the silicon oxide, and electrically activated by the bias voltage applied at a high temperature [6]. All the comb-electrodes are electrically connected into one on the round chip frame to collect generated electrical currents. The fixed electrodes also take a form of suspended structure such that it would pick up different frequency at its own resonance, thereby increasing the bandwidth to receptacle wide range of external vibrations.

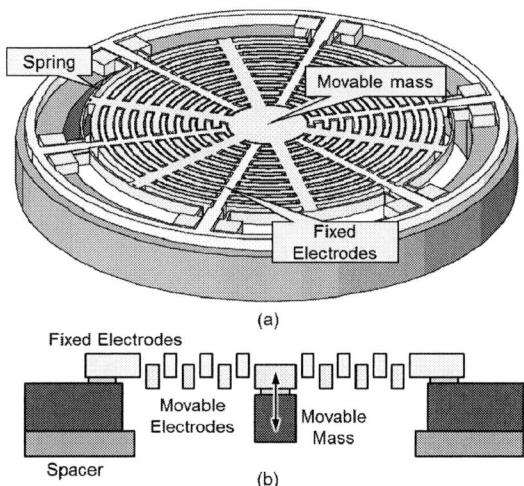

Figure 1: Schematic of (a) a single unit cell for the electret type energy harvester, and (b) its cross sectional view. An SOI wafer of 25-µm-thick active layer is used to make the electrodes and suspensions.

Figure 2 shows the scanning electron microscope (SEM) images of the developed EH device. The disk was 1 cm in diameter, and a 40 µm gap was made between the fixed and the movable electrodes. A silicon-on-insulator (SOI) wafer with a 25 µm thick active layer was processed by the high-aspect ratio deep reactive ion etching (DRIE). The electrode patterns on the front surface were combined into a pair of electrical interconnecting lines that lead to the grip on the chip, where wire-bonding would be performed to complete the electrical connection off the chip. The backside of the chip was processed by the double-step DRIE to make a space accommodating the electrodes when stacked up in the vertical direction. One would have to assemble the chips by placing the chips in every 60 degrees due to the point symmetry of the electrode designs. The additional mass made behind the central disk was designed to attach onto the bottom chip such that they would vibrate in synchronization.

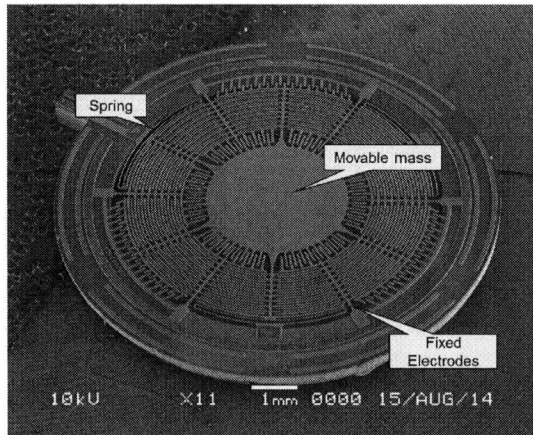

(a)

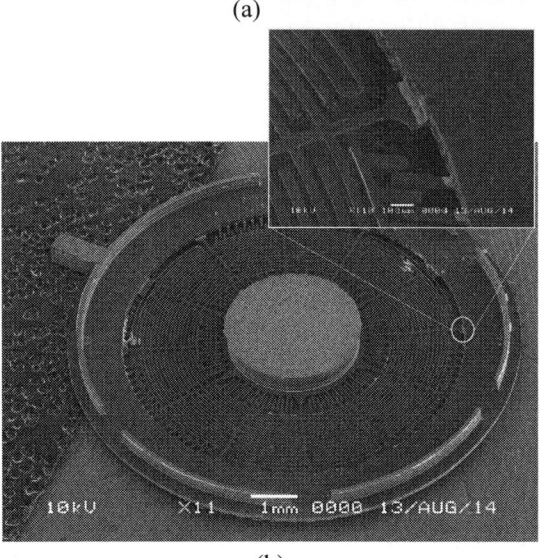

(b)

Figure 2: Scanning electron microscope image of (a) the front side of the EH cell composed of center mass, comb electrodes and spring, and (b) its backside view. Double-step DRIE was used to form the recess on the backside to make a space that accommodated the structure on the chip placed underneath.

FABRICATION PROCESS

Fabrication process steps are schematically shown in Figure 3. In (a), an SOI wafer was first protected with an LPCVD (low-pressure chemical vapor deposition) silicon nitride that was later used as a LOCOS (local oxidation) mask for the silicon. The front side SOI was 25 μm, the buried oxide (BOX) 2 μm, and the handle wafer 500 μm typically. (b) An aluminum layer was deposited on the backside of the wafer and patterned into the shape of the bottom mask, while the front side is processed by the DRIE with a photoresist mask (Shipley S-1818). (c) The metal mask and an additional photoresist mask on the backside were used as an etching mask for the double step DRIE. (d) With the progress of the backside DRIE, the more susceptible photoresist mask worn out, and the double-stepped recesses were formed with the aluminum mask. (e) The DRIE process was stopped at the interface to the BOX, (f) which was selectively removed in an HF acid to release the structure.

The wafer was then put in an oxidation furnace to form a thermally grown oxide of 0.5 μm in thickness by bubbling a 40wt% KOH solution with a nitrogen gas for 18 hours at 950 °C. The chip was then taken out of the wafer and individually polarized with a dc voltage of 150 V at 650 °C for 5 minutes. The chip was then cooled down to 100 °C with the voltage on. A self-assembled monolayer of HMDS (hexamethyldisilazane) was put as a passivation film to delay the degradation of the built-in electret due to the moisture in air. The chip was then wire-bonded onto a step and sealed in a vacuum.

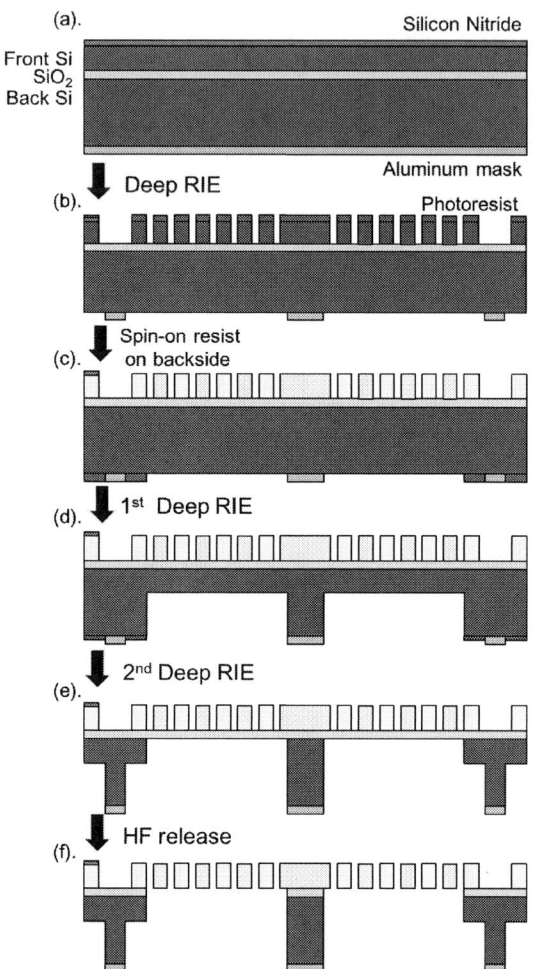

Figure 3: Process flow of the surface micromachining. Thermal oxidation and polarization processes will follow to make the electret film.

EXPERIMENTAL RESULTS

The developed EH chip was tested on a vibration bench (C-300 DSP Controller) with a laser Doppler vibrometer (LDV), as schematically shown in Figure 4. A function generator was used to give a command voltage to the vibration bench, and the out-of-plane motion of the EH's central mass in the Z-direction was monitored with the LDV. Total four different types of signals were simultaneously monitored to correlate the mechanical and electrical behavior: the FG output voltage, the calibrated

978-1-4799-7956-1/15 $31.00 © 2015 IEEE

acceleration signal from the vibrator, the Z-direction motion of the device, and the generated voltage. A 1 MΩ resistance was typically used as a load to the EH.

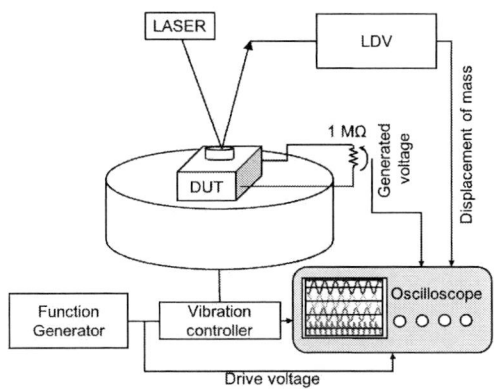

Figure 4: Schematically illustrated set-up for EH characterization. Voltage from the function generator was used to shake the vibration bench. Acceleration was monitored with a sensor built-in the system. The mechanical out-of-plane motion was monitored with the laser Doppler vibrometer to correlate with the generated voltage. Experiment was performed at room temperature.

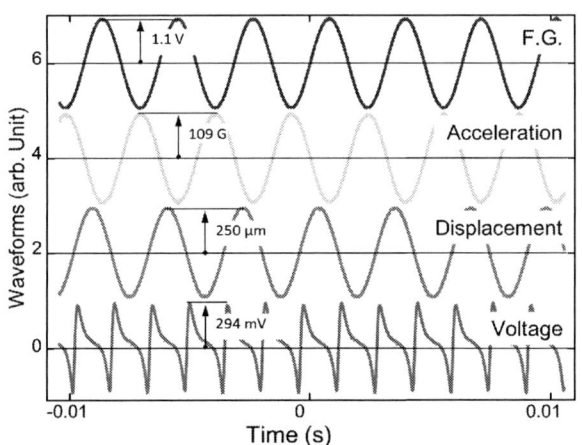

Figure 5: Waveforms of Function generator (F.G.), Acceleration of shaker, Displacement of center of mass and Generated voltage by vibrating. F.G. generates voltage at 310 Hz.

The signals were monitored as shown in Figure 5. Due to the nature of the electrostatic energy harvesting, where voltage was generated when the movable electrodes were on the way to the rest position, the frequency component in the output voltage was double the input mechanical vibration. Maximum output voltage of 0.3 V was obtained at a resonant frequency of 310 Hz and at a peak amplitude of 250 μm for a load resistance of 1 MΩ. Typical output power at this condition was 0.34 μW.

Figure 6 shows the frequency response of the EH after calibrating the shaker's response. The response amplitudes were plot with respect to the excitation frequency of the function generator, and hence the resonant peaks shown at mechanical 108, 150, and 310 Hz were in fact generated

electrical 216, 300, and 620 Hz, respectively. From this result, the most principal mode for energy scavenge was at a mechanical 310 Hz, which could be reduced further for the environmental vibration spectrum that typically distributed in 100 Hz or lower.

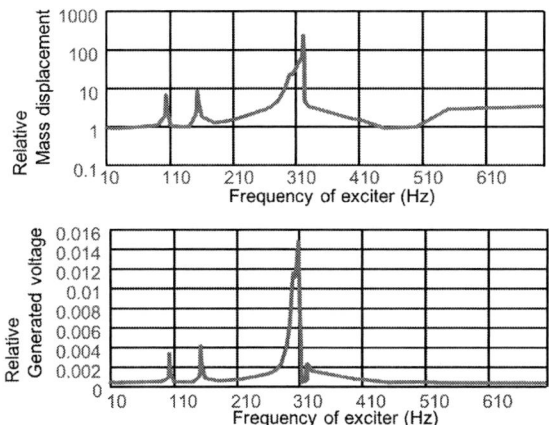

Figure 6: Frequency response of the EH calibrated by the shaker's reference output. Electrical peak frequencies were 108, 150, and 310 Hz.

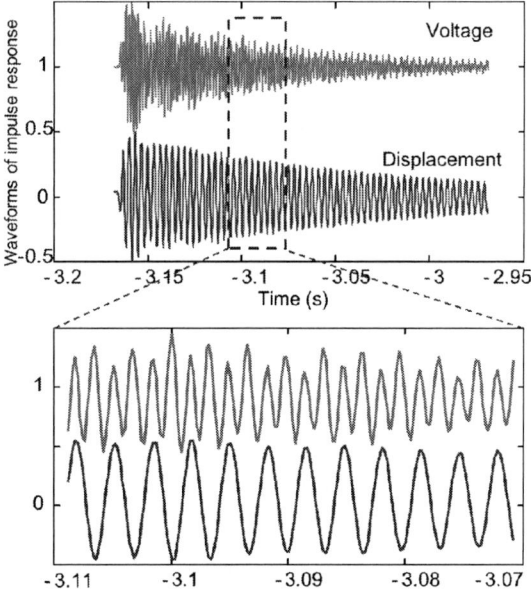

Figure 7: Impulse response of the EH when hit with a metallic bar on the ceramic package. Lower box indicates the close-up view of the vibration, where the output voltage has double the frequency of mechanical vibration.

Figure 7 is the impulse response of the open-circuit EH when hit with a metallic bar on the ceramic package. The waveform clearly showed that the output electrical voltage oscillated twice in one mechanical cycle. The figure also shows the detailed view of the waveforms. Due to the parasitic modes of the suspended structures, the response seemed to contain other frequency components to cause a beat envelope.

Figure 8 shows the fast Fourier transform (FFT) analysis results of the generated voltage and the LDV-measured mechanical vibration. The most significant mode at the mechanical 302 Hz was the out-of-plane motion of the suspended mass, and it corresponded to the electrical outputs around 615 Hz. Besides, the identical frequency also generate another electrical output component in the same frequency plausibly through the electrodes' lateral vibration, where no frequency doubling effect would be seen.

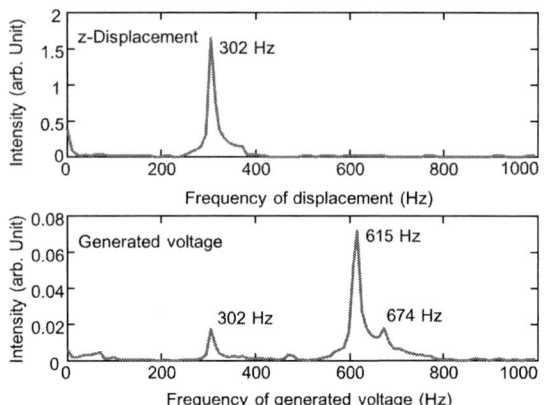

Figure 8: Impulse response in the frequency domain, comparing the mechanical vibration of the center mass and the output voltage.

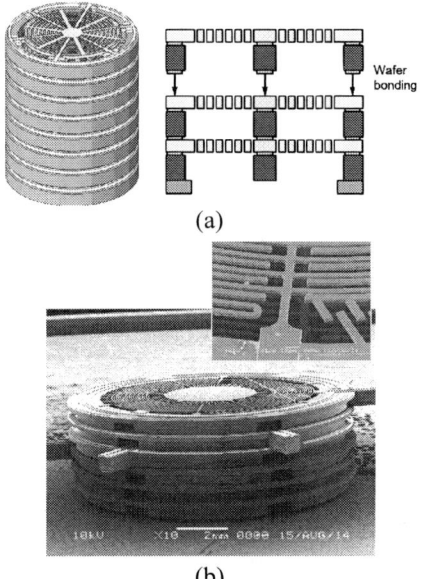

(a)

(b)

Figure 9: Demonstration of stacked version of electret EH cells. Chips were bonded with dispensed photoresist.

For demonstration of output multiplication, we have developed a 3D-stack of five unit modules as schematically shown in Figure 9. The center masses on the chips were all mechanically connected into one such that they would respond as one to the external vibration. The figure also shows the SEM image of manually assembled chips; glue drops were dispensed on the chip surface. Resonant frequency should be maintained for the amount of the glue was negligible compared with the mass of the suspended structures. Output current could be multiplied by the parallel connection of the output ports, while output voltage could be enhanced by employing narrower air gaps.

CONCLUSION

We demonstrated an electrostatic type energy harvester based on the potassium ion electrets embedded in the thermally grown silicon oxide. The MEMS unit cell was made to be compatible with vertical stack for higher output current or voltage. Vertical comb electrodes were implemented in an SOI wafer by using the high-aspect ratio DRIE processes. A single substrate formation showed an output voltage of 0.3 V at a resonant frequency of 310 Hz for a load resistance of 1 MΩ. Typical output power at this condition was 0.34 µW.

ACKNOWLEDGEMENTS

This work is supported by the JSPS Core-to-Core Program.

REFERENCES

[1] "TSsensor Summits for Trillion Sensor Roadmap", URL available at http://www.tsensorssummit.org/

[2] Peihong Wang, Katsuhiko Tanaka, Susumu Sugiyama, Xuhan Dai, Xiaolin Zhao, and Jingquan Liu, "A micro electromagnetic low level vibration energy harvester based on MEMS technology", *Microsyst Technolt.*, vol. 15, pp. 941-951, 2009.

[3] R Elfrink, T M Kamel, M Goedbloed,S Matova, D Hohlfeld, Y van Andel, and R van Schaijk, "Vibration energy harvesting with aluminum nitride-based piezoelectric devices", *Microsyst Technolt.*, vol. 15, pp. 941-951, 2009.

[4] Te-ChienHou, YaYang, Hulin Zhang, Jun Chen, Lih-Juann Chen, Zhong and Lin Wang, "Triboelectric nanogenerator built inside shoe insole for harvesting walking energy", *Nano Energy*, vol. 2, pp. 856-862, 2013.

[5] Yuji Suzuki, Daigo Miki, Masato Edamoto, and Makoto Honzumi, "A MEMS electret generator with electrostatic levitation for vibration-driven energy-harvesting applications", *J. Micromech. Microeng.*, vol. 20, 104002 (8pp), 2010.

[6] Tatsuhiko Sugiyama, Mitsuru Aoyama, Yasushi Shibata, Masato Suzuki, Takashi Konno, Manabu Ataka, Hiroyuki Fujita, and Gen Hashiguchi, "SiO2 Electret Generated by Potassium Ions on a Comb-Drive Actuator", *Applied Physics Express*, vol. 4, 114103, 2011.

[7] Makoto Honzumi, Kei Hagiwara, Yoshinori Iguchi, and Yuji Suzuki, "High-speed electret charging using vacuum UV photoionization", *Applied Physics Letters*, vol. 98, 052901, 2011.

[8] Jen-Shih Chang, Phil A. Lawless, and Toshiaki Yamamoto, "Corona Discharge Processes", *IEEE TRANSACTIONS ON PLASMA SCIENCE*, vol. 19, No. 6, Dec. 1991.

CONTACT

Hiroshi Toshiyoshi, mail: hiro@iis.u-tokyo.ac.jp

FABRICATION OF A THREE DIMENSIONAL CANTILEVERED VIBRATIONAL ENERGY HARVESTER USING SILVER INK

Shih-Jui Chen, Yao-Yun Feng, and Shu-Yu Liu
Department of Mechanical Engineering, National Central University, TAIWAN

ABSTRACT

A novel 3D micro electromagnetic energy harvester capturing mechanical vibrational energy from environmental activities is proposed. The harvester is built on a cantilever microstructure, on which 3D coils locate. The energy harvester is aimed to generate power in the relative low frequency range, which is below 500 Hz. By injecting multi-layer of spiral coils on a cantilever structure, the resonant frequency can be reduced and the output voltage and power can be increased.

INTRODUCTION

Nowadays, with the advances in the portable electronic devices, substitutive power sources from the environment, such as light, thermal, and kinetic energies, have become important to extend the battery life. Among all, kinetic energy source is found to be the best energy scavenging solution [1]. Vibrational energy harvester fabricated using MEMS technology draws wide interests in the academia and industry due to the small size and portability. When an energy harvester experiences vibrational motion, the electric power can be generated through different transduction mechanisms. For low frequency operation purposes, energy harvester using electromagnetic transduction is selected, especially for a low impedance load.

Williams [2] proposed the analysis of a micro-electric generator for microsystems. Maximum amount of power generation can be predicted for a generator operating at resonance. Environmental vibrations typically occur at low frequencies, so it is important to design the energy harvester with low resonant frequency. If the vibrational frequency doesn't match the resonant frequency, the output power will be greatly attenuated. Although vibrational energy can be an infinite power source, it still suffers from limited bandwidth and output power for practical operations. Many techniques, including resonance frequency tuning, multimodal energy harvesting, and frequency up-conversion, are used in broadband vibration-based energy harvesting [3]. Sari [4] used an array of parylene-based micro generators to increase the usable frequency range by combining 35 different lengths of cantilevers. This is an effective method to increase the bandwidth.

This paper reports a vibrational electromagnetic energy harvester with 3D conductive coils stacked on a cantilever microstructure. If there is relative movement between the coils and external permanent magnets, electromotive force will be induced due to the Faraday's law of induction. Compared to state of the art, multiple layers of conductive coils are dispensed on the micro-machined cantilever diaphragm by an injecting machine. Schematic of the energy harvester with two layers of coils is shown in Fig.1.

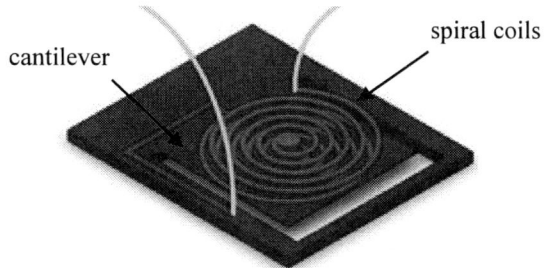

Figure 1: Schematic of the energy harvester with two layers of coils.

FABRICATION

For fabricating 3D coils, a 3-axis injecting machine is used to stack coils with conductive silver ink and insulation material. The parameters of the dispenser including needle size, pressure, temperature, speed and material property are well controlled for appropriate shape and size of the coil and insulation layers. Detailed device structures are shown in Fig. 2.

The fabrication process flow of the 3D energy harvester is schematically shown in Fig. 3. First, a 20-μm-thick silicon diagram is formed by KOH etching on the backside of the wafer. The cantilever diaphragm is released by RIE of the silicon diaphragm from the front side. Then, conductive coils are formed by using an injecting machine. To form the first layer of spiral coil, silver ink of 20 μm is delineated on the cantilever. On top of the coil, thick silicone oil of 15 μm is dispensed to form an insulation layer for the second layer of 20-μm-thick spiral coil. Finally, a second insulation layer is dispensed on the second coil layer. The insulation layer can be further removed by silicone oil remover to form the released structure of the energy harvester.

Figure 2: Photos of the electromagnetic energy harvester with two layers of coils: (upper) 5 loops for two coil layers and single insulation layer; (lower) 5 and 3 loops for the first and second coil layers without insulation layers.

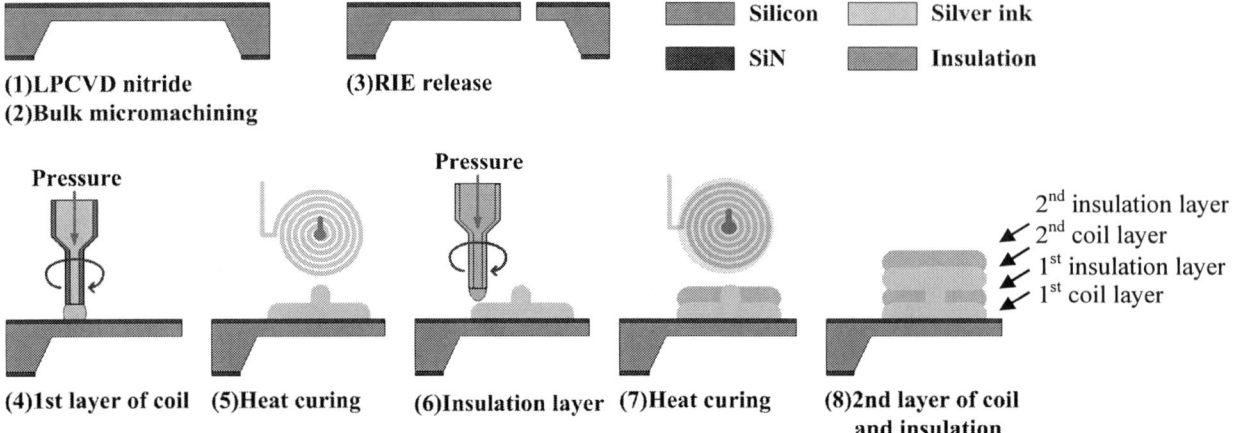

(1)LPCVD nitride
(2)Bulk micromachining

(3)RIE release

Silicon   Silver ink

SiN   Insulation

2nd insulation layer
2nd coil layer
1st insulation layer
1st coil layer

(4)1st layer of coil **(5)Heat curing** **(6)Insulation layer** **(7)Heat curing** **(8)2nd layer of coil and insulation**

Figure 3: Fabrication process flow of the energy harvester.

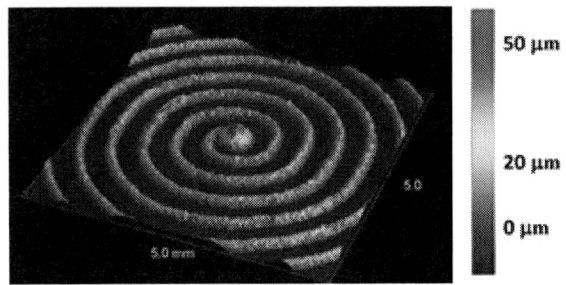

Figure 4: Profile of the first layer of spiral coil.

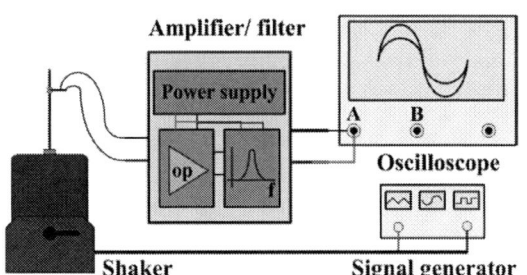

Figure 5: Testing setup for emf measurement.

THEORY

A typical energy harvester can be modeled as a second order system. For a given sinusoidal displacement input, the average power generation at resonance can be expressed by [2, 5]:

$$P = \frac{m\zeta_{em}\omega_n^3 Y^2}{4\zeta^2} \qquad (1)$$

where ζ_{em} and ζ are the electromagnetic transducer damping ratio and overall damping ratio, respectively. For the energy harvester under a given natural frequency (ω_n), input amplitude (Y), and mass (m), the average power generation will be decided by the damping ratios. Increasing the number of coil loops can effectively increase the electromagnetic transducer damping and the output power.

EXPERIMENTAL RESULTS

The fabricated energy harvester with a magnet was characterized with a testing system including a shaker, a signal generator, an oscilloscope, and a variable resistor. The vibrational excitation is applied to the energy harvester by the shaker, and the excitation frequency is controlled by the signal generator. The electromotive force (emf) measurement setup is shown in Fig. 5. Both measurements of the emf and the output power, generated by the energy harvester, are performed with respect to different vibrational frequencies.

Two measurements of the electromotive force, or open circuit voltage, across output terminals of the energy harvester are performed. In case of the energy harvester with single coil layer and single insulation layer, the peak output voltage is 9.3 mV at 259 Hz in response to a vibration acceleration of 6 g (Fig.6). In case of the energy harvester with double coil layers and double insulation layers, the peak output voltage is 19.2 mV at 160 Hz in response to a vibration acceleration of 6.9 g (Fig.7). Comparing double-layer case with single-layer case, the resonant frequency drops about 40% and the output voltage doubles.

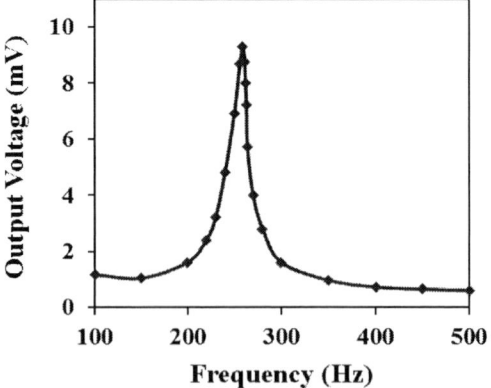

Figure 6: Measured open circuit output voltages for the energy harvester with single coil layer and single insulation layer.

978-1-4799-7956-1/15 $31.00 © 2015 IEEE

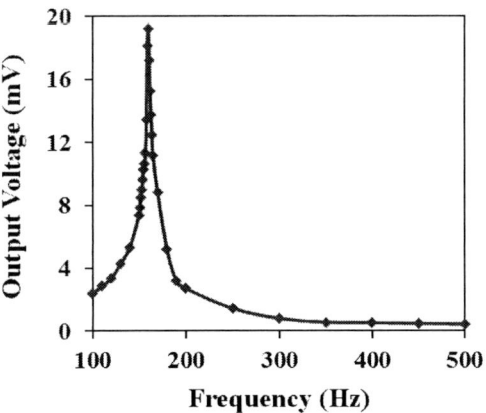

Figure 7: Measured open circuit output voltages for energy harvesters with double coil layers and double insulation layers.

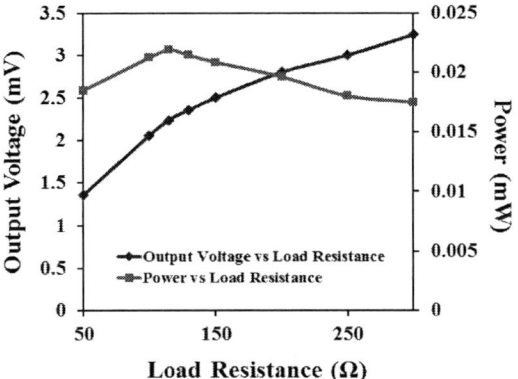

Figure 8: Measured output voltage and power for energy harvesters with single coil layer and single insulation layer.

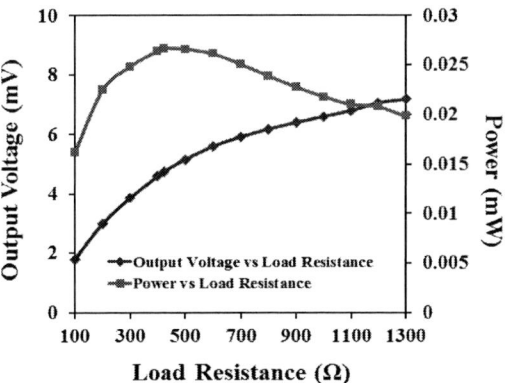

Figure 9: Measured output voltage and power for energy harvesters with double coil layers and double insulation layers.

The power outputs at the fundamental frequency of the energy harvester are measured as a function of the load resistance. When the load resistance is equal to the coil resistance, maximum power output would occur. In case of the energy harvester with single coil layer and single insulation layer, a maximum power of 22 µW is delivered

to a 115 Ω resistor (Fig. 8). In case of the energy harvester with double coil layers and double insulation layers, a maximum power of 27 µW is delivered to a 424 Ω resistor (Fig. 9).

SUMMARY

A 3D micro electromagnetic energy harvester fabricated by MEMS fabrication process and injecting technique is presented in this paper. Compared to state of the art, multiple layers of conductive coils are dispensed on the micro-machined cantilever diaphragm by an injecting machine. In case of the energy harvester with single coil layer and single insulation layer, an emf of 9.3 mV at 259 Hz is generated and a maximum power of 22 µW is delivered to a 115 Ω resistor. In case of the energy harvester with double coil layers and double insulation layers, an emf of 19.2 mV at 160 Hz is generated and a maximum power of 22 µW is delivered to a 424 Ω resistor. Comparing double-layer case with single-layer case, the resonant frequency drops about 40%, the output voltage doubles, and the power increases. By adding the proof mass, the performance of the energy harvester can be further improved due to the increasing of the vibration amplitude. These results show improvement in the usable output power for applications of vibrational energy harvesting in wireless sensor networks.

ACKNOWLEDGEMENTS

This material is based upon work supported by Ministry of Science and Technology, Taiwan, R.O.C. under contract no. NSC 101-2221-E-008-029 and MOST 103-2221-E-008-013. We are grateful to the National Center for High-performance Computing, Taiwan, R.O.C. for providing software and facilities.

REFERENCES

[1] S. Roundy, P.K. Wright, and J. Rabaey, "A study of low level vibrations as a power source for wireless sensor nodes," *Computer Communications*, 26, pp. 1131–1144, 2003.

[2] C.B. Williams and R.B. Yates, "Analysis of a micro-electric generator for microsystems," *Sensors and Actuators A*, vol. 52, pp. 8-11, 1996.

[3] L. Tang, Y. Yang and C.K. Soh, "Toward broadband vibration-based energy harvesting," *Journal of Intelligent Material Systems and Structures*, vol. 92, pp. 1867–1897, 2010.

[4] I. Sari, T. Balkan and H. Kulah, "An electromagnetic micro power generator for wideband environmental vibrations," *Sensors and Actuators A: Physical*, vol. 145–146(0), pp. 405-413, 2008.

[5] S.P. Beeby, M.J. Tudor, E. Koukharenko, N.M. White, T. O'Donnell, C. Saha, S. Kulkarni, S. Roy, "Design and performance of a microelectromagnetic vibration powered generator", *IEEE International Conference on Solid-State Sensors and Actuators*, Seoul, Korea, June 5-9, pp.780-783, 2005.

CONTACT

*S.J. Chen, tel: +886-3-4267374; raychen@ncu.edu.tw

978-1-4799-7956-1/15 $31.00 © 2015 IEEE

WAFER-LEVEL FABRICATION OF A TRIBOELECTRIC ENERGY HARVESTER

Mengdi Han[1], Bocheng Yu[1], Zongming Su[1], Bo Meng[1], Xiaoliang Cheng[1], Xiao-Sheng Zhang[1] and Haixia Zhang[1]*

[1] Science and Technology on Micro/Nano Fabrication Lab, Institute of Microelectronics, Peking University, Beijing, CHINA

ABSTRACT

We present a wafer-level fabrication method for triboelectric energy harvester (TEH), which fabricates the TEH completely in batch fabrication process, without any manually assembly step. Finite element method (FEM) simulation was conducted to investigate the open-circuit potential distribution and short-circuit charge distribution. Experimental measurements show that this device can produce 235 mV peak voltage at the frequency of 30 Hz, under the 100 MΩ external resistance. Compared with previous TEHs, the proposed device can be batch fabricated in CMOS-compatible process and the reduced size allows it to be easily integrated with other electronic devices (*e.g.*, keyboards).

INTRODUCTION

Vibration energy harvesters scavenge energy from the ambient environment [1], which is a promising substitute for traditional batteries to power portable electronic devices. The energy harvesters can be designed based on electromagnetic [2], piezoelectric [3], and electrostatic [4] transduction mechanisms. Compared with the previous two methods (*i.e.*, electromagnetic and piezoelectric), electrostatic energy harvesters do not rely on functional materials such as permanent magnets (*e.g.*, NdFeB [2], CoNiMnP [5]) and piezoelectric materials (*e.g.*, PZT [3], PVDF [6]), which offers more flexibility in the fabrication process and material selection.

However, the requirement of external power supply greatly limits the application field of electrostatic energy harvesters [7]. To solve this problem, the TEH was first proposed in 2012 [8], and develops rapidly due to its unique mechanism for obtaining surface charge. Due to the contact electrification effect, when two different materials contact with each other, charges with the same quantity but opposite polarity will be accumulated on the surface of two materials according to the triboelectric series [9]. Based on this principle, TEH can generate electricity without electret or external power supply compared to traditional electrostatic energy harvesters. Moreover, TEHs have been utilized not only as a power source to drive electronic devices [10,11], but also as an active sensor to detect the mechanical/chemical change [12,13]. Notwithstanding, recent-developed TEHs are only partially based on MEMS process and the requirement of folding, pasting, or clamping [14-16] greatly limits the fabrication efficiency and further development.

Therefore, in this work, we designed a TEH which can be batch fabricated at wafer scale using traditional MEMS process. In the fabrication process, Au and parylene were utilized as two materials for contact electrification. Photoresist with 8 μm thickness was used as sacrificial layer to form the gap. In addition, FEM simulations and experimental measurements were carried out to investigate the output performance of the TEH.

DESIGN AND FABRICATION

3D schematic structure and exploded view of the TEH is shown in Figure 1(a)(b), which includes the top/bottom Au electrodes, flexible parylene diaphragm, Si proof mass, and SiO$_2$ insulation layer. An 8 μm gap exists between the electrode and parylene diaphragm, as shown in Figure 1(c). In this structure, the parylene diaphragm, which has low Young's modulus, is utilized as the vibration part for low frequency vibration energy harvesting. The Si mass further reduces the resonant frequency of the TEH and can also promote the output performance.

Based on this structure, when applying an external vibration/force, the Au electrode will contact with the parylene layer and get charged due to the contact electrification effect. According to the triboelectric series, the positive charges will be accumulated on the Au surface and negative charges will be obtained on the parylene layer. Then, as the gap distance changes, electrons will flow in the external circuit owing to the electrostatic induction, which is similar to traditional electrostatic energy harvesters.

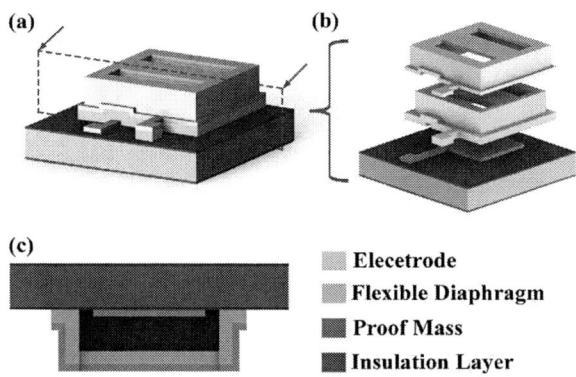

■ Elecetrode
■ Flexible Diaphragm
■ Proof Mass
■ Insulation Layer

Figure 1: Structure of the TEH. (a) 3D view, (b) exploded view, and (c) cross-section of the structure.

Fabrication of the TEH was conducted using MEMS process. The process begins with a 4-inch Si wafer which has 3000 Å SiO$_2$ layer on both sides. Firstly, 300 Å Cr layer was sputtered as the adhesion layer followed by 3000 Å Au layer (Figure 2(a)). Then, photoresist with the thickness of 2 μm was spin coated and patterned as a mask to etch the Cr/Au layer (Figure 2(b)). Next, as shown in Figure 2(c), photoresist with the thickness of 8 μm was patterned as the sacrificial layer followed by 5 μm parylene coating. Afterwards, another Cr/Au electrode (300 Å/3000 Å) was sputtered onto the parylene and wet etched using a

12 μm photoresist layer as the mask. The parylene was also patterned using O_2 plasma with the same mask (Figure 2(d)). After removing the photoresist (both the 8 μm layer and the 12 μm layer), a movable structure based on parylene diaphragm was obtained as shown in Figure 2(e). Finally, the wafer is diced and every single device was turned upside down utilizing the silicon as proof mass (Figure 2(f)).

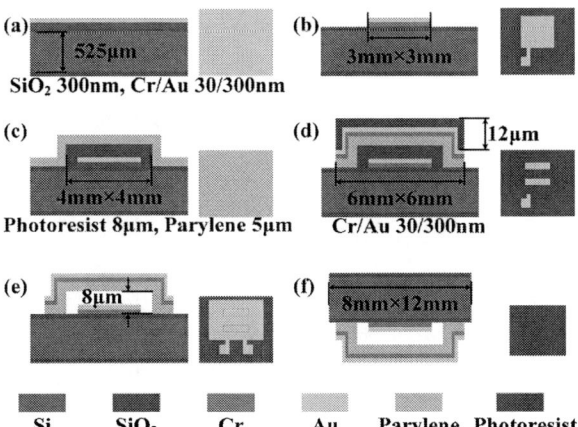

Figure 2: Fabrication process with cross-section view on the left of each subfigure and top view on the right of each sub figure. (a) Cr/Au sputtering, (b) pattern of the first electrode, (c) coating parylene using photoresist as the sacrificial layer, (d) pattern the second electrode and parylene layer, (e) removing photoresist, (f) dicing and placing upside down.

It has to be mentioned that the layout for each layer should be carefully designed in order to complete the whole fabrication process. For example, to pattern the 8 μm photoresist layer, two rectangular holes were designed for the following photoresist removing process. In addition, the thickness of the sacrificial photoresist layer and the last photoresist layer was set as 8 μm and 12 μm, respectively, in order to provide enough gap distance for the TEH and guarantee good step coverage.

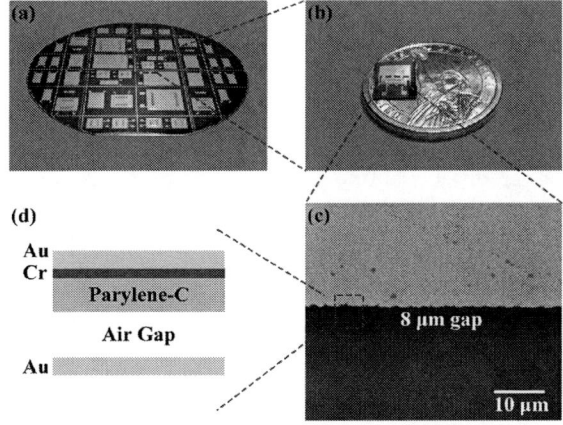

Figure 3: (a) Photo of the fabricated TEH wafer. (b) Enlarged photo of a single TEH die. (c)SEM image of the gap area with 30° tilt view. (d) Schematic illustration of the gap area.

Using the proposed process, the TEH has been fabricated in wafer scale, as shown in Figure 3(a). The enlarged photo of a single device is shown in Figure 3(b). SEM image with 30° tilt view and illustration of the 8 μm gap area are shown in Figure 3(c)(d), respectively.

SIMULATION

Output performance of the TEH was first analyzed through FEM simulation using a 2D electrostatic model. The dimension of the model was set according to the actual size of the TEH as indicated in Figure 2, except that the 3000 Å SiO_2 layer is omitted and the electrode of Cr/Au is simplified by a 1 μm Au layer. An air box with the size of 10 mm × 2.5 mm was established as the simulation area. The electric potential at infinite was set as zero.

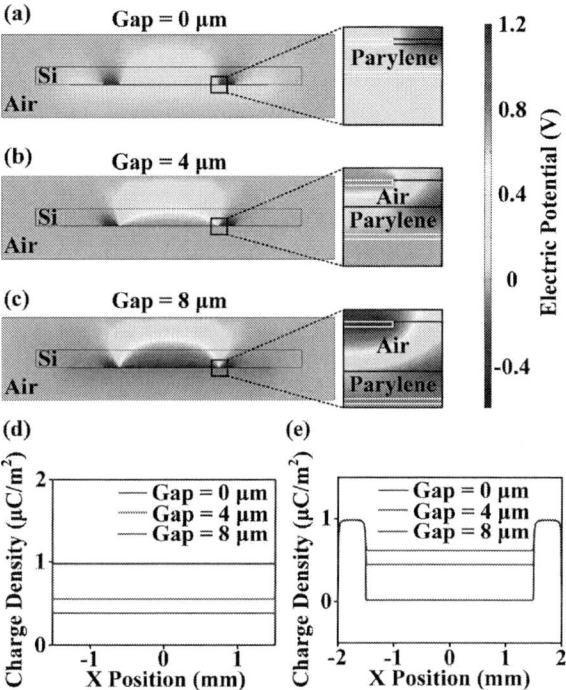

Figure 4: Potential distribution with the gap distance of (a) 0 μm, (b) 4 μm, and (c) 8 μm; the white rectangles in the enlarged view indicate the electrode. Charge distributions at the inner surface of the (d) top electrode, and (e) bottom electrode with different gap.

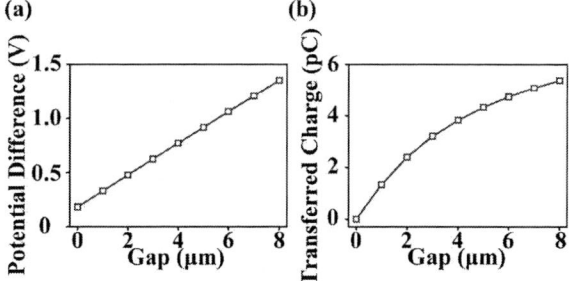

Figure 5: Simulation results of the (a) open-circuit potential difference, and (b) short-circuit charge transfer against different gap distances.

To investigate the open-circuit condition, the surface charge density on the parylene layer was set as 1 μC/m², the total charge on the top electrode was set as 12 pC, which is the product of the surface charge density and area of the parylene layer. Meanwhile, the total charge on the bottom electrode was set as zero. As shown in Figure 4(a)-(c), when the gap distance between the top electrode and parylene layer changes from 0 to 8 µm, electric potential of the top electrode keeps increasing, while the potential at the bottom electrode decreases monotonously. For the short-circuit condition, the only difference is that the total charges on the top and bottom electrode together were set as 12 pC. In this case, with a larger gap, the surface charge density of top electrode decreases while the bottom electrode shows an opposite tendency, as shown in Figure 4(d)(e).

Relationship between the output performance and gap distance is summarized in Figure 5(a)(b). Although the open-circuit potential difference increases almost linearly with the gap distance (Figure 5(a)), the increasing rate of transferred charge at short-circuit condition slows down at larger gap distances. In the real application, the transferred charge determines the ability of energy harvesters to charge a capacitor or battery. Therefore, the small gap distance (*i.e.*, 8 µm) of this TEH will not greatly decrease the output performance compared to those manually assembled TEHs with larger gap distance.

MEASURMENT AND APPLICATION

Experimental measurement was conducted using a vibration system, which includes an oscilloscope with wave generation module (Agilent DSO-X 2014A), a modal shaker (JZK-10), and a power amplifier (SINOCERA YE5871A).

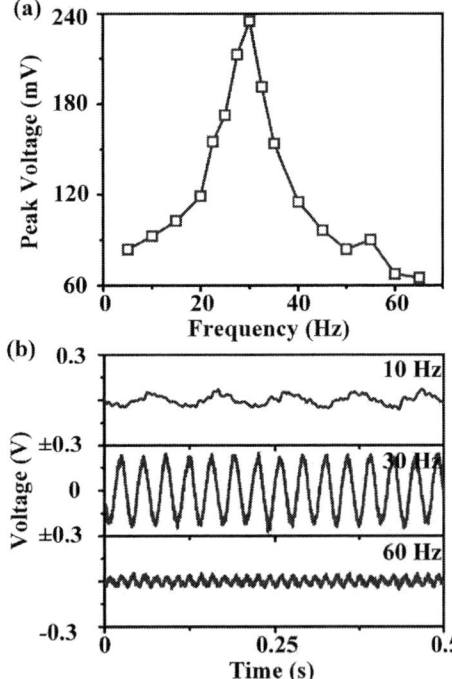

Figure 6: (a) Measured voltage output of the TEH under different vibration frequency. (b) Detailed time-domain output at 10 Hz, 30 Hz, and 60 Hz.

Frequency response of the TEH was tested at 0.1 g acceleration with a 100 MΩ probe. Thanks to the flexible parylene diaphragm, the TEH can produce large output at low frequencies. As shown in Figure 6(a), the peak output voltage of the TEH reaches 235 mV at 30 Hz. Detailed real time output voltage curves at 10 Hz, 30 Hz, and 60 Hz are shown in Figure 6(b). Due to the complete fabrication process, the total volume of the TEH can be reduced to 51.6 mm³, which is much smaller than previous TEHs [11,13]. Thus, the output power at resonance reaches 10.7 nW/cm³.

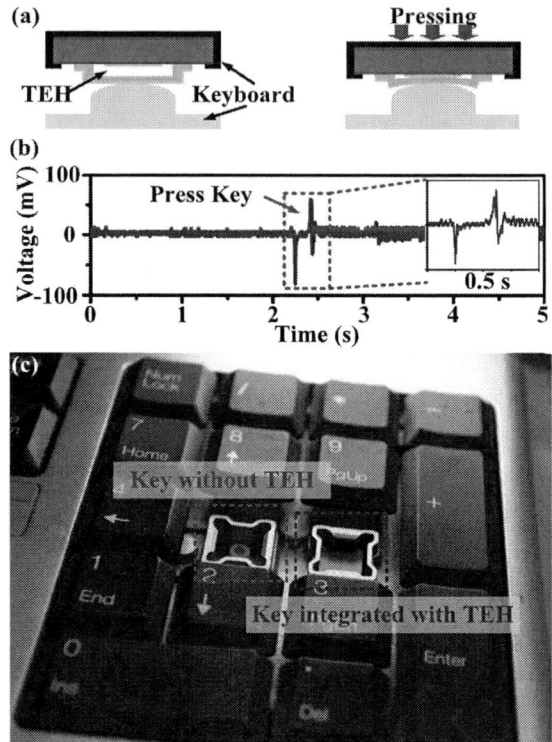

Figure 7: (a) Schematic illustration of the integrated TEH. (b) Output voltage under key pressing. (c) photo of the keyboard integrated with the TEH.

Compared with other TEHs, it's easier to integrate this device with other COMS circuits and electronic devices, owing to the CMOS-compatible process and miniature size. In this work, an example is demonstrated by integrating the TEH into a normal computer keyboard. The schematic illustration is shown in Figure 7(a). When pressing the key, the gap distance of the TEH will decrease thus inducting electron flow in the external circuit through electrostatic induction. When the key is released, the gap distance increases, which will cause the electron to flow in the opposite direction. The testing result is shown in Figure 7(b), indicating that a 168.8 mV peak-to-peak voltage can be obtained through daily typing. Photo of the integrated TEH is shown in Figure 7(c). In the number keypad, the TEH was placed under number "6". The internal structure for unplaced key (*i.e.*, number "5") is also shown for comparison. Through this integration, the mechanical energy from daily typing can be effectively harvested and converted into electricity.

CONCLUSION

In summary, we have designed a TEH which can be fabricated completely in MEMS process. In the fabrication process, three masks were designed in order to pattern the first electrode, prepare the sacrificial layer, and pattern the second electrode. The thickness for each photoresist layer is 2 μm, 8 μm, and 12 μm, respectively. Parylene was utilized as one friction material as well as the flexible diaphragm. Meanwhile, Au was used as another friction material and the electrodes as well. The silicon wafer was utilized as the proof mass after dicing and placing upside down. Based on this structure, devices have been fabricated and investigated through both FEM simulation and experimental measurement. Through simulation, the relationship between open-circuit potential distribution, short-circuit charge distribution, and the gap distance was obtained. Besides, testing results show that this TEH can produce peak output voltage of 235 mV at the frequency of 30 Hz at 100 MΩ external resistance, with a corresponding power density of 10.7 nW/cm^3. Benefit from the miniature size, the TEH was also integrated to a computer keyboard to harvest the mechanical energy from daily typing.

ACKNOWLEDGEMENTS

This work is supported by the National Natural Science Foundation of China (Grant No. 61176103, 91023045 and 91323304), the National Hi-Tech Research and Development Program of China ("863" Project) (Grant No. 2013AA041102), and the National Ph. D. Foundation Project (Grant No. 20110001110103) and the Beijing Natural Science Foundation of China (Grant No. 4141002).

REFERENCES

[1] S. P. Beeby, M. J. Tudor, N. M. White, "Energy Harvesting Vibration Sources for Microsystems Applications", *Meas. Sci. Technol.*, vol. 17, pp. R175–R195, 2006.

[2] M. Han, W. Liu, B. Meng, X. S. Zhang, X. Sun, H. Zhang, "Springless Cubic Harvester for Converting Three Dimensional Vibration Energy", in *Digest Tech. MEMS'14 Conference*, San Francisco, January 26-30, 2014, pp. 425–428.

[3] A. Hajati, S. G. Kim, "Ultra-Wide Bandwidth Piezoelectric Energy Harvesting", Appl. Phys. Lett., vol. 99, pp. 083105-1–083105-4, 2011.

[4] L. G. W. Tvedt, D. S. Nguyen, E. Halvorsen, "Nonlinear Behavior of an Electrostatic Energy Harvester Under Wide- and Narrowband Excitation", *J. Microelectromech. Syst.*, vol. 19, pp. 305–316, 2010.

[5] M. Han, Q. Yuan, X. Sun, H. Zhang, "Design and Fabrication of Integrated Magnetic MEMS Energy Harvester for Low Frequency Applications", *J. Microelectromech. Syst.*, vol. 23, pp. 204–212, 2014.

[6] M. Han, X. S. Zhang, W. Liu, X. Sun, X. Peng, H. Zhang, "Low-Frequency Wide-Band Hybrid Energy Harvester based on Piezoelectric and Triboelectric Mechanism", *Sci. China Tech. Sci.*, vol. 56, pp. 1835–1841, 2013.

[7] T. Krupenkin, J. A. Taylor, "Reverse Electrowetting as a New Approach to High-Power Energy Harvesting", *Nat. Commun.*, vol. 2, pp. 448-1–448-7, 2011.

[8] F. R. Fan, Z. Q. Tian, Z. L. Wang, "Flexible Triboelectric Generator", *Nano Energy*, vol. 1, pp. 328–334, 2012.

[9] A. F. Diaz, R. M. Felix-Navarro, "A Semi-Quantitative Tribo-Electric Series for Polymeric Materials: the Influence of Chemical Structure and Properties", *J. Electrostat.*, vol. 62, pp. 277–290, 2004.

[10] G. Zhu, J. Chen, T. Zhang, Q. Jing, Z. L. Wang, "Radial-Arrayed Rotary Electrification for High Performance Triboelectric Generator", *Nat. Commun.*, vol. 5, pp. 3426-1–3426-9, 2014.

[11] M. Han, X. S. Zhang, B. Meng, W. Liu, W. Tang, X. Sun, W. Wang, H. Zhang, "r-Shaped Hybrid Nanogenerator with Enhanced Piezoelectricity", *ACS Nano*, vol. 7, pp. 8554–8560, 2013.

[12] Z. H. Lin, G. Zhu, Y. S. Zhou, Y. Yang, P. Bai, Z. L. Wang, "A Self-Powered Triboelectric Nanosensor for Mercury Ion Detection", *Angew. Chem. Int. Edit.*, vol. 52, pp. 5065-5069, 2013.

[13] M. Han, X. S. Zhang, X. Sun, B. Meng, W. Liu, H. Zhang, "Magnetic-Assisted Triboelectric Nanogenerators as Self-Powered Visualized Omnidirectional Tilt Sensing System", *Sci. Rep.*, vol. 4, pp. 4811-1–4811-7, 2014.

[14] B. Meng, W. Tang, X. S. Zhang, M. Han, X. Sun, W. Liu, H. Zhang, "A High Performance Triboelectric Generator for Harvesting Low Frequency Ambient Vibration Energy", in *Digest Tech. MEMS'14 Conference*, San Francisco, January 26-30, 2014, pp. 346–349.

[15] Y. Lu, X. Wang, X. Wu, J. Qin, R. Lu, "A non-resonant, gravity-induced micro triboelectric harvester to collect kinetic energy from low-frequency jiggling movements of human limbs", *J. Micromech. Microeng.*, vol. 24, pp. 065010-1–065010-8, 2014.

[16] L. Dhakar, F. E. H. Tay, C. Lee, "Development of a Broadband Triboelectric Energy Harvester With SU-8 Micropillars", *J. Microelectromech. Syst.*, doi: 10.1109/JMEMS.2014.2317718, 2014.

CONTACT

*H.X. Zhang, Tel: +86-10-62766570; zhang-alice@pku.edu.cn

DISPLACEMENT MAGNIFICATION OF GEL ACTUATOR USING pNIPAAm-SU8 HYBRID STRUCTURE

Takanobu Kogure[1], Ryoya Okada[1], Shingo Maeda[1] and Sumito Nagasawa[1,]*
[1] Dept. of Eng. Sci. and Mech., Shibaura Institute of Technology, Tokyo, JAPAN

ABSTRACT

In this paper, a fabrication method for a hybrid structure of the poly-N-Isopropylacrylamide (pNIPAAm) gel as a soft material and the SU-8 as a solid material is proposed. This pNIPAAm-SU8 hybrid structure can be applied to various applications such as gel actuators. These hybrid structures have a serious problem that it is easily broken by the difference of swelling rate. Our proposed hybrid structure was not broken though repeating swelling-shrinking states for 10 or more times. A small displacement of the gel is occurred by the phase transition, then it is magnified by combining with the SU-8 solid structure. We demonstrated that the expanded tip displacement of the SU-8 solid structure reached 145μm.

INTRODUCTION

Hydrogel is focused in various fields ranging from medicine, biology, chemistry, material science and engineering as adaptive material and devices. Hydrogel has three dimensional network structures. And also, hydrogel changes its properties and functions in response to external stimuli such as pH, temperature, light irradiation and so on [1-4]. A pNIPAAm (p-N-Isopropylacrylamide) gel, which has thermal sensitivity, is a prominent example of a stimuli responsive hydrogel. A phase transition is occurred at about 32 °C. In 32 °C or less, the pNIPAAm gel becomes a hydrophilic and swells. In 32 °C or more, by the phase transition the pNIPAAm gel becomes a hydrophobic and shrinks. This phase transition temperature is called the LCST (lower critical solution temperature). At the LCST, the gel volume expands rapidly about 1000% from the initial volume [3, 4]. Micro fabrication techniques that apply these soft materials to the microactuator have been studied [4-8]. Three different methods for shaping the hydrogel have been investigated. In the first method, the hydrogel is polymerized in a silicon micro casting mold which is fabricated by the wet or dry etching [4, 5]. In the second method, the hydrogel is etched using an oxygen plasma [6]. In the third method, the hydrogel is patterned using the photolithography techniques [7, 8].

However, since most of the hydrogels are easily broken, a high aspect ratio hydrogel structure is difficult to fabricate using a micro casting mold. The etching or photolithography techniques are applied to fabricate only two dimensional micro hydrogel structures. By assembling a silicon high aspect ratio structure and a hydrogel, a high aspect ratio ciliary structure was successfully demonstrated [9].

We propose a fabrication method for a novel hybrid structure, in which a solid microstructure is embedded in the gel, and that hybrid structure can magnify a small displacement of the gel actuator. A schematic diagram of our study is shown in figure 1. The SU-8 is a thick photo resist, and it is patterned using a photolithography. In order to anchor the SU-8 solid structure in the hydrogel, the root part of the SU-8 was shaped into a comb-like structure. Furthermore, the SU-8 solid structure get various functions by devising its shape. By combining the solid structure and the hydrogel, it is possible to make a hydrogel actuator with a movable frame. Using this movable frame fixed in the hydrogel, we can use advantages of the soft property of the hydrogel effectively.

pNIPAAm-SU8 HYBRID STRUCTURE
Methodology

The stimuli responsive hydrogel is a soft material and easily changed its volume by external stimulus. Simply inserting the solid structure into the hydrogel, the solid structure falls off from the gel because of the volume change difference between the solid structure and the gel. Therefore, that structure does not work as a hybrid structure.

In order to fabricate the hybrid structure of the hydrogel and the solid structure, there are two possible approaches. In the first method, the solid structure is fixed in the hydrogel by the chemical bonding as shown in figure 2a. Using polyglycerol methacrylate （PGMA） as an anchoring layer, the solid structure was chemically grafted onto the hydrogel by the cross-linking [9]. In the other method, like a proposed method in this study, the root part

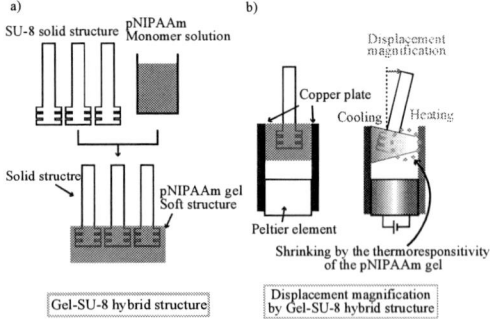

Figure1: (a) Schematic diagram of gel-SU-8 hybrid structure. (b) Displacement magnification mechanism of hybrid structure.

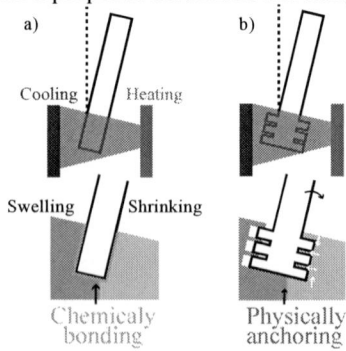

Figure 2: (a) Chemical bonding method for the hybrid structure. (b) Physically anchoring method for our pNIPAAm-SU8 hybrid structure.

of the solid structure is patterned into a comb-like structure. The comb-like structure works as a physically anchor to prevent detachment of the solid structure by the swelling-shrinking of the hydrogel.

In this study, the pNIPAAm-SU8 hybrid structure was fabricated. The hybrid structure consists of the pNIPAAm gel which has the thermal sensitivity, and the SU-8 solid structure which has the comb-like anchor as shown in figure 2b. The fabrication process of the hybrid structure is shown in following 3 steps. In the 1st step the SU-8 solid structure is fabricated, in the 2nd step the pNIPAAm monomer solution is prepared and finally in the 3rd step the SU-8 solid structure and the pNIPAAm gel are integrated.

SU-8 solid structure

Figure 3a shows the fabrication process of the SU-8 solid structure. LOR-30B was spin coated on a SiO_2 substrate and baked in the 180 °C oven for 10 minutes. This LOR-30B layer was used as a sacrificial layer. Then the SU-8 3050 was spin coated on the LOR sacrificial layer, and pre-backed on the 95 °C hotplate for 45 minutes. After that, the SU-8 was exposed with the h-line and developed using the SU-8 developer. After the development, dipping the sample into the NMD-3 (2.38% TMAH) for 24 hours to remove the LOR sacrificial layer. As a result, the SU-8 solid structure was released from the SiO_2 substrate. The released SU-8 structure was dried on the 140 °C hotplate for 10 minutes.

Figure 3b and 3c show the SEM photos. The SU-8 solid structure consists of a main part and a handling part as shown in figure 3b. The root part of the main part embedded in the pNIPAAm gel has a comb-like structure. The root part of the SU-8 solid structure was tightly meshed with the pNIPAAm gel because the pNIPAAm gel filled into gaps of the comb structure. The handling part was cut and removed after finishing the integration of the pNIPAAm gel and the SU-8 solid structure.

pNIPAAm monomer solution

The pNIPAAm gel was used as an actuator of the pNIPAAm-SU8 hybrid structure. The pNIPAAm gel is a hydrogel that shows the thermal responsive. The phase transition is occurred at the LCST of about 32 °C and it caused the large volume change. The embedded SU-8 solid structure magnifies the small displacement caused in the

pNIPAAm gel. The pNIPAAm gel was manufactured by polymerizing a monomer solution of the pNIPAAm. The monomer solution of the pNIPAAm were prepared as follows: 1.433g of N-Isopropylacrylamide (NIPAAm), 0.06g of N,N'-Methylenebisacrylamide (MBAA) as a cross-linker and 0.064g of 2,2'-azobis (isobutyronitrile) (AIBN) as a initiator were dissolved in 9ml of O_2-free methanol. Oxygen inhibits the polymerization of the gel and it makes the polymerization process failure. Dissolved oxygen was replaced with nitrogen by N_2 bubbling.

The SU-8 solid structure was placed at designed position in the monomer solution of the pNIPAAm, then the monomer was polymerized, the pNIPAAm-SU8 hybrid structure was realized.

pNIPAAm-SU8 hybrid structure

The SU-8 solid structure and the pNIPAAm monomer solution were prepared by the above process to assemble the pNIPAAm-SU8 hybrid structure.

Figure 4 shows the fabrication process of the pNIPAAm-SU8 hybrid structure. The pNIPAAm-SU8 hybrid structure was produced by polymerizing the pNIPAAm gel in a mold. The SU-8 solid structure was also placed at designed position in the mold. The mold was designed using a 3D-CAD software and a desktop numerically controlled milling machine (Roland MDX-20) cut a modeling wax. This mold has an insertion point for placing the SU-8 solid structure, and a chamber for polymerizing the pNIPAAm gel. The SU-8 solid structure was placed in the mold, then the mold was sealed using a silicone rubber sheet with a silicone grease. The monomer solution of the pNIPAAm was injected through monomer inlet using a syringe. Filled the mold with the monomer solution, monomer inlet was covered using silicone rubber sheet with silicone grease.

The mold was sealed completely using glass plate and clamps. After that, the monomer solution was polymerized at 60 °C for 12 hours. After polymerization, the hybrid structure was released from the mold and the handling part of the SU-8 solid structure was cut and removed from main part. To remove unreacted monomer, the hybrid gel was soaked in pure methanol for a week. The hybrid gel was carefully hydrated through dipping it in a graded series of methanol-water mixtures, for 1 day each in 75, 50, 25 and 0%. Finally the pNIPAAm shows the thermo sensitivity. Our pNIPAAm-SU8 hybrid structure is shown in figure 5a and 5b. Figure 5a shows that the SU-8 solid structure was embedded in the pNIPAAm gel. It was confirmed that the pNIPAAm-Su8 hybrid structure is established. Figure 5b is magnified photo of the root part of the hybrid structure.

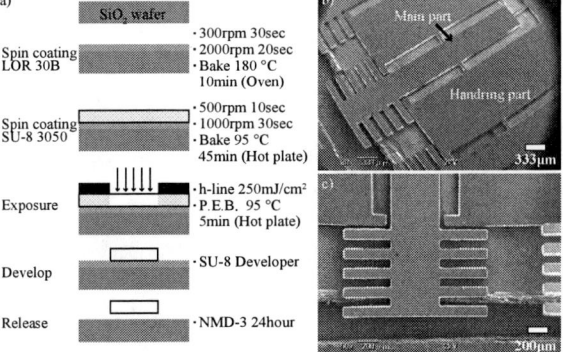

Figure 3: (a) Fabrication process of SU-8 solid structure. (b) Our SU-8 solid structure. (c) Comb-like root part of SU-8 solid structure.

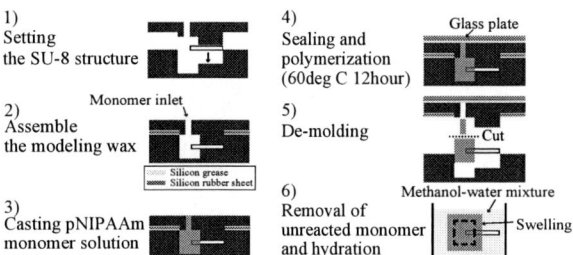

Figure 4: Fabrication process of the pNIPAAm-SU8 hybrid structure.

978-1-4799-7956-1/15 $31.00 © 2015 IEEE

Although the pNIPAAm gel was swelled because meshed with pNIPAAm gel by the comb like structure of SU-8.

EXPERIMENTAL RESULTS

pNIPAAm gel characteristic measurement

Since the pNIPAAm gel was utilized as a thermal actuator, we evaluated the thermal responsive characteristics of the pNIPAAm gel. A measurement sample of the pNIPAAm gel was shaped in rectangular solid which is 1mm in width, 1mm in height and 5mm in length. The sample was dipped in a thermo static bath and its temperature was changed from 10 °C to 50 °C. Until the gel deformation reaches in a steady-state, the temperature is kept for 20 minutes, and then the deformation displacement was measured.

Figure 6a shows that shrinkage ratio of the pNIPAAm gel relative to the initial size at 10 °C. There is a tendency to shrink with increasing temperature. This pNIPAAm gel occured the phase transition at from 33 °C to 35°C rapidly. By the phase transition the transparent pNIPAAm gel was turned to cloudy white. Therefore the LCST of the pNIPAAm gel our used is around 34 °C.

Durability of pNIPAAm-SU8 hybrid structure

Since the pNIPAAm and the SU-8 solid structure have quite difference of the thermal volume change caused by the phase transition, the durability confirmation is very important in this hybrid structure. This durability had been confirmed by repeating the temperature change from 33°C and 35 °C around the LCST. The micrographs of the hydrogel before and after the volume change are shown in figure 6b. Although changing the gel temperature

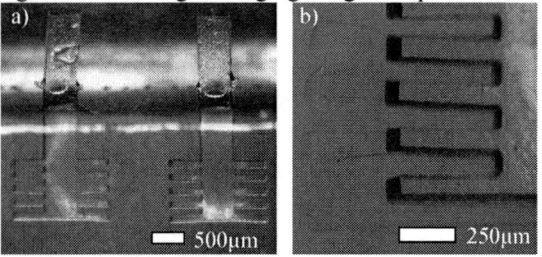

Figure 5: (a) Photo of our hybrid structure. (b) Although the gel is swelled by thermal sensitivity, the SU-8 structure is tightly meshed.

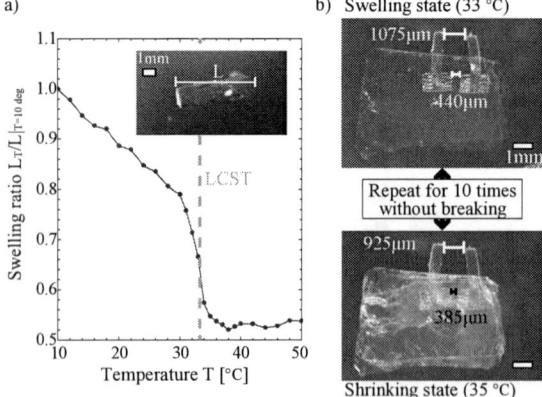

Figure 6: (a) Temperature-Swelling ratio graph of the pNIPAAm gel. (b) Swelling state and shrinking state of our hybrid structure.

repeatedly for 10 or more times, the pNIPAAm-SU8 hybrid structure was maintained its whole structure without break down.

Displacement magnification

As shown in figure 1a, in our displacement magnification mechanism the pNIPAAm gel with the SU-8 solid structure is placed between the two plates of copper. Since these two plates are connected to the both sides of a Peltier element, one side of the gel is heated and the other side is cooled by these plates. A difference of the small volume change in both sides of the solid structure root part is occurred, and then this difference is magnified by the SU-8 solid structure applying the principle of leverage. Figure 7 shows an experimental setup. The copper plates for the heat transfer were bonded with a thermally conductive grease on the both sides of the Peltier element. As a heat sink was attached to the cooling side of the copper plate, the cooling plate temperature was fixed to the water bath temperature (25 °C). Two thermocouple thermometers were attached on the copper plates to measure the plate temperatures. When a voltage of 1.5V was supplied to the Peltier element, the gel was heated from 25 °C to 36 °C. The displacement of the tip of the SU-8 solid structure was recorded using a microscope and measured by the video analysis. Figure 8 shows a photo of the pNIPAAm-SU8 hybrid structure before heating, and a photo of the SU-8 solid structure tip reached to the maximum displacement. At 0min, whole pNIPAAm gel was transparent. After10min from the start of heating, the pNIPAAm gel on the heating plate side turned clouded. Shrinkage area reached to approximately one-half of the gel. In this time, 145µm of the maximum tip displacement

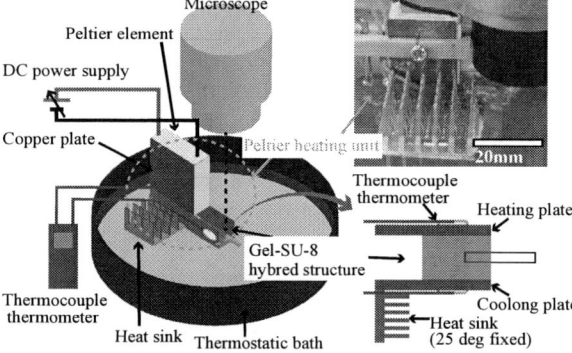

Figure 7: Illust. of experimental setup for actuating the gel using a Peltier element.

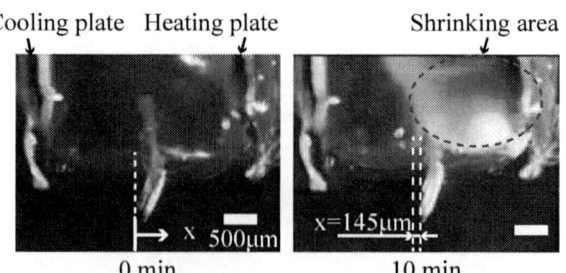

Figure 8: Photos of the hybrid structure during local heating. 145µm of displacement was archived at the tip of the SU-8 structure.

was obtained. Figure 9 shows a graph of the tip displacement of the SU-8 solid structure versus the temperature of the heating plate. As the heating plate temperature is raised, the tip displacement increases. At 10min after starting, since the temperature of the heating plate was beyond the LCST, the tip displacement was reached 145μm.

However the temperature of the heating plate increased slowly, the tip displacement decreased. Since the shrinkage area of the pNIPAAm gel had spread through the whole area of the gel with the continuous heating, there is no difference in volumes on the both sides of the SU-8 solid structure, and the displacement is no longer expanded.

CONCLUTION

In our study, the fabrication process of the pNIPAAm-SU8 hybrid structure which consists of the pNIPAAm gel as a soft material and SU-8 micro actuator as a solid material is proposed. Furthermore, pNIPAAm-SU8 hybrid structure was heated using the Peltier element. The small volume change of pNIPAAm gel caused by thermal sensitivity was magnified by the pNIPAAmSU-8 hybrid structure.

The SU-8 solid structure was fabricated using surface micromachining. LOR-30B was spin coated on the SiO_2 substrate as a sacrificial layer. Then SU-8 3050 was spin coated on the sacrifice layer and patterned. Finally the sacrifice layer was removed using NMD-3, the SU-8 solid structure was released from the substrate.

The pNIPAAm-SU8 hybrid structure was fabricated assembling the pNIPAAm gel and the SU-8 solid structure. The SU-8 solid structure was settled in the casting mold and the gel was polymerized. The root of the SU-8 solid structure was shaped in comb like structure. Even though the hybrid structure was given temperature change and occurred the volume change repeatedly, the hybrid structure was maintained without broken.

Our pNIPAAm-SU8 hybrid structure was heated using the Peltier element. The temperature of the heating plate was reached to the LCST and the shrinkage area was spread approximately one-half of the gel, the tip displacement of the SU-8 solid stricture was reached to 145μm in maximum. After that, despite of continuing to heat, the tip displacement decreased. This decrease of the tip displacement was occurred due to excess expansion of the gel shrink area.

By applying the method of fabricating the hybrid structure with a combination of soft materials and solid materials, more complex hybrid structure can be established. Furthermore the hybrid structure to be driven by external stimulus except temperature can be realized by using another type of hydrogel. Production technique of the hybrid structure of the soft materials and the solid materials contribute to the realization of the soft material MEMS actuator.

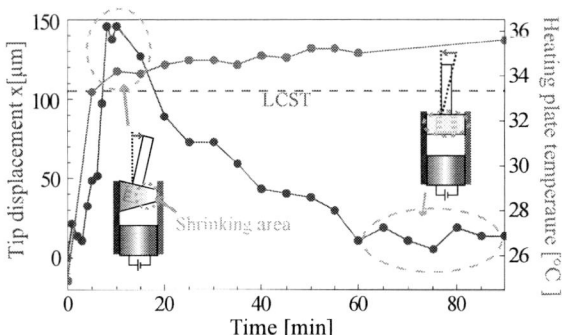

Figure 9: Graph of tip displacement and heating plate temperature. Displacement increases with increasing temperature of the heating plate.

REFERENCES

[1] T. Tanaka "Gels", *Sci. Am.*, vol. 244, No. 1, pp. 124-138, 1981.

[2] Y. Takashima, S. Hatanaka, M. Otsubo, M. Nakahata, T. Kakuta, A. Hashizume, H. Yamaguchi, A. Harada, "Expansion–contraction of photoresponsive artificial muscle regulated by host–guest interactions.", *Nature communications*, 3, 1270, 2012.

[3] S. Hirotsu, Phase transition of a polymer gel in pure and mixed solvent media.", *Journal of the Physical Society of Japan*, 56.1, pp. 233-242, 1987.

[4] J. Kim, B. Kim, J, Ryu, Y. Jeong, J. Park, H. C. Kim, K, Chun, "Potential of thermo-sensitive hydrogel as an actuator.", *Japanese journal of applied physics*, 44(7S), 5764, 2005.

[5] F. Chiellini, R. Bizzarri, C. K., Ober, D. Schmaljohann, T. Yu, R. Solaro, E. Chiellini, "Patterning of polymeric hydrogels for biomedical applications.", *Macromolecular Rapid Communications*, 22(15), pp. 1284-1287, 2001.

[6] M Lei, Y Gu, A Baldi, RA Siegel, B Ziaie, "Soft mold-dry etch: a novel hydrogel patterning technique for biomedical applications.", in *Engineering in Medicine and Biology Society, 2004. IEMBS'04. 26th Annual International Conference of the IEEE*, Vol. 1, pp. 1983-1986, 2004.

[7] K Ito, S Sakuma, Y Yokoyama, F Arai, "Photoprocessible thermo-sensitive gel actuator for functional microfluidic devices.", In *Solid-State Sensors, Actuators and Microsystems (TRANSDUCERS & EUROSENSORS XXVII), 2013 Transducers & Eurosensors XXVII: The 17th International Conference on*, pp. 1811-1814, 2013.

[8] N Bassik, BT Abebe, KE Laflin, DH Gracias, "Photolithographically patterned smart hydrogel based bilayer actuators.", *Polymer*, 51(26), 6093-6098, 2010.

[9] A Sidorenko, T Krupenkin, A. Taylor, P Fratzl, J. Aizenberg, "Reversible switching of hydrogel-actuated nanostructures into complex micropatterns.", *Science*, 315(5811), 487-490, 2007.

CONTACT

*S. Nagasawa, tel: +81-3-5859-8063;
nagasawa@shibaura-it.ac.jp

LIQUID-TOLERANT ELECTRET USING SUPER-LYOPHOBIC PILLAR SURFACE

Yu-Chung Chen[1], Ki-Young Song[1], Kenichi Morimoto[1] and Yuji Suzuki[1]
[1] Department of Mechanical Engineering, The University of Tokyo, JAPAN

ABSTRACT

For the first time, we have realized electret that can be used in the liquid environment toward better performance of electret-based generators and actuators. By using super-lyophobic overhanging pillar surface, stable Cassie-Baxter (C-B) state is sustained even with low-surface-tension hexadecane liquid and with high surface potential. Pillar surface with a SiO_2 electret layer has been successfully charged with soft X-ray photoionization we previously developed. The pillar surface significantly suppressed the charge decay, and surface potential as high as 160 V has been maintained after 5-day contact with hexadecane droplet.

INTRODUCTION

Recently, electret is used in various MEMS devices including droplet manipulation using liquid-dielectro-phoresis on electret (L-DEPOE) [1] and vibration energy harvesters [2]. In L-DEPOE, non-uniform electrostatic field with electret and electrodes can be driven with low operation voltage compatible with CMOS devices. Energy density of these electret devices is proportional to $\varepsilon\varepsilon_0 E^2/2$ (*E*: electric field strength), so that their performance can be much improved with increasing the permittivity of the medium ε between the electret and the counter electrode by inserting high-permittivity dielectric liquid.

However, charges in electret layers easily disappear at the contact with liquid. We previously propose protection layers with fluorinated polymer and parylene-C for the electret layer [1], but the charge decay is still significant. Up to now, there is no proven method for stabilizing electret in the liquid environment [3].

In the present study, super-lyophobic pillar surface is proposed for novel liquid-tolerant electret with minimal direct contact between liquid and the charged surface. As shown in Fig. 1, overhanging SiO_2 pillars are formed on the surface and the whole surface is charged to form electret. With the aid of overhanging structures, even low-surface-tension liquid [4] can be suspended on air, so that direct contact between the liquid and the electret surface is avoided. We employ soft X-ray charging [5] we previously developed, which enables uniform charging even on the sidewalls of pillars. For sustaining the Cassie-Baxter (C-B) state under a high electric field, we design the pillar dimensions based on our stability model for the air-liquid interface [6].

Here we present the results from our prototyped super-lyophobic pillar surface, showing that the lifetime of electret can be significantly improved in the liquid environment. The interface behavior of C-B droplets of both polar and non-polar liquids is systematically examined by changing the surface potential on the pillar.

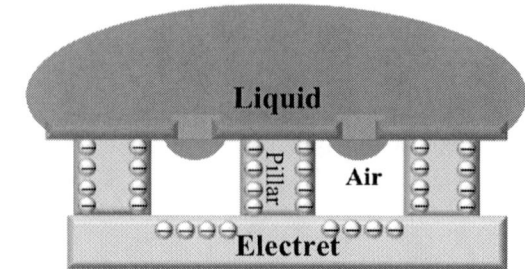

Figure 1: Super-lyophobic electret surface.

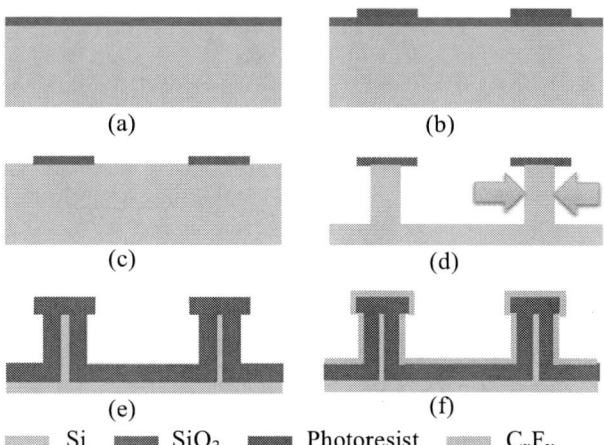

Si SiO_2 Photoresist C_xF_y

Figure 2: Fabrication process.

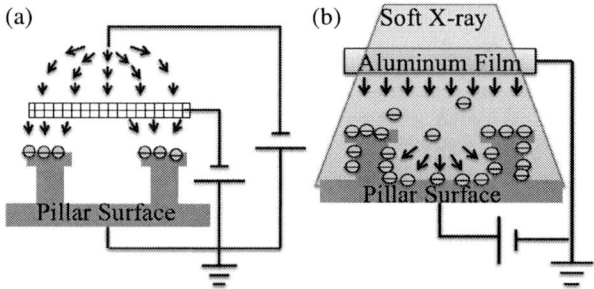

Figure 3: (a) Corona charging (b) Soft X-ray Charging.

DESIGN OF THE PILLAR

To achieve our goal, the C-B state should be sustained even under a high electric field. We designed the pillar dimensions to achieve high critical voltage above -500 V based on our stability model, in which a force balance among surface tension force, electrostatic force, and pinning force is considered [6]. The surface potential under the balance of these forces can be expressed as:

$$V = 8th_0 \frac{F_p + F_s}{\pi\varepsilon\varepsilon_0\left[(p\sqrt{2}-D)^2 + 4(t-h_0)^2\right]}, \quad (1)$$

where F_p and F_s represent the pinning force and the surface tension force, and h_0, p, t, and D denote the

978-1-4799-7956-1/15 $31.00 © 2015 IEEE   1086   MEMS 2015, Estoril, PORTUGAL, 18 - 22 January, 2015

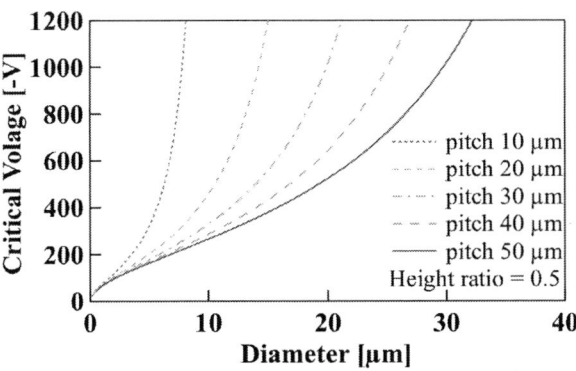

Figure 4: Calculated critical voltage versus pillar diameter for constant height ratio.

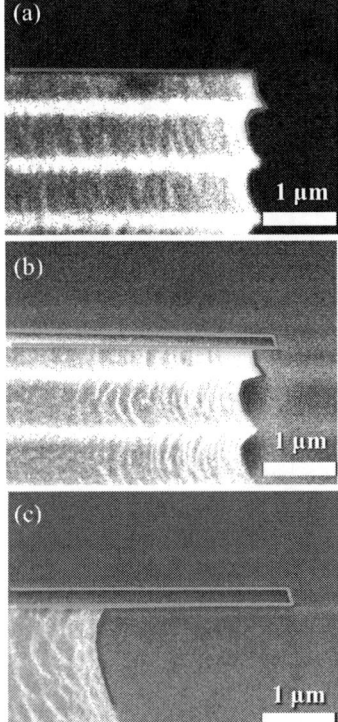

Figure 5: Silicon pillar edge: (a) without cap, (b) with SiO₂ cap, (c) with large overhanging cap.

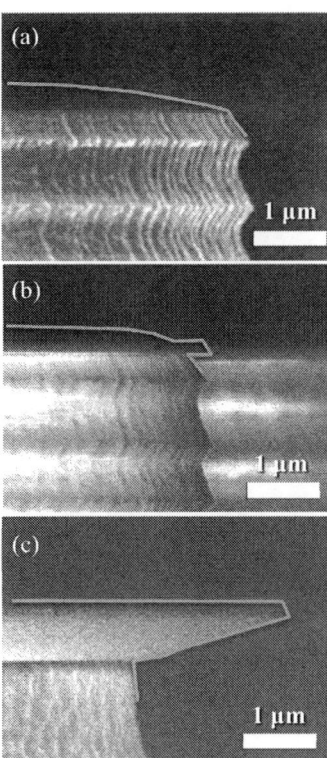

Figure 6: SiO₂ pillar edge: (a) round edge, (b) sharper edge, (c) large overhanging edge.

Figure 7: Large overhanging pillar surface, Pitch=40μm, Diameter=17μm, Height=21μm. Solid fraction=0.15.

distance between liquid surface and the bottom solid surface, the pitch, the height, and the diameter of the pillars, respectively. F_p and F_s are given by

$$F_S = \pi D \gamma_{lv} \cos\theta_0 = \pi D \gamma_{lv} \frac{4(p\sqrt{2}-D)(t-h_0)}{(p\sqrt{2}-D)^2 + 4(t-h_0)^2} \ , \quad (2)$$

$$F_p = \pi D \gamma_{lv} \left(\cos\theta_{adv} - \cos\theta_{smooth} \right) \ , \quad (3)$$

where γ_{lv} represents the interfacial tension between a liquid and vapor, and θ_{adv} and θ_{smooth} denote the maximum advancing contact angle and the contact angle on a flat surface, respectively.

The critical voltage V_c of a given pillar geometry (p, t, D), with which the Cassie-Baxter to Wenzel transition occurs due to the increased electrostatic force, can be obtained as:

$$V_C(p,t,D) = \max_{0 \le h_0 \le t} V(h_0,p,t,D) \ . \quad (4)$$

From Eq. (4), the critical voltage is obtained as a function of geometrical parameters. Here, the height ratio, t/p, is set constant at 0.5. When the desired critical voltage is specified, a set of the diameter and the pitch can be uniquely obtained, while considering the condition for the contact angle. In the present design, the contact angle is estimated from the C-B model:

$$\cos\theta_{CB} = r_S f_S \cos\theta_{smooth} + f_S - 1 \ , \quad (5)$$

where θ_{CB} and θ_{smooth} denote the contact angle on the C-B and smooth surface, respectively, and the solid fraction is defined as $f_S = \pi D^2 / 4p^2$. Here, the roughness factor on the pillar top surface is assumed as $r_S = 1$.

EXPERIMENTAL METHOD

Figure 2 shows the fabrication process. A 200 nm-thick SiO_2 layer is grown on the Si wafer using thermal oxidation (Fig. 2a). Then, using EB lithography, EB resist ZEP520A is patterned on the surface (Fig. 2b). The SiO_2 layer is etched by ICP-RIE with CHF_3 plasma (Fig. 2c). This is followed by DRIE with Bosch process to form Si pillars with SiO_2 caps (Fig. 2d). Different edges are formed by controlling the side-etching under the SiO_2 cap. With the second thermal oxidation, 0.67 μm-thick SiO_2 layer is obtained on the whole surface (Fig. 2e). The SiO_2 on the backside was removed by BHF. The pillars are finally coated with a hydrophobic C_xF_y layer by CVD (Fig. 2f). Before charging, the sample is annealed in 350°C with N_2 environment to increase the stability.

Corona charging (Fig 3a) is often used for electrets, but it cannot be used for the present 3-D structures due to charge accumulation at the pillar tips that impedes ion penetration into the pillar bottom. Instead, soft X-ray charging [5] (Fig. 3b) enables uniform charging even on the sidewalls of pillars. With this technique, electrons generated through photoionization among the pillars are dragged onto the pillar surface with the bias voltage until the magnitude of the surface potential reaches the bias voltage.

RESULTS AND DISCUSSION

Figure 4 shows the critical voltage for design parameters calculated from Eq. (4). In the present study, the design values of the critical voltage and the contact angle are set as V_C = -500 V and $\theta_{CB} >$ 150°, and the pitch and the diameter is determined as p = 40 μm and D = 17 μm, respectively.

SEM images of the Si pillar edge and the SiO_2 pillar edge are respectively shown in Figs. 5 and 6. Green lines represent the edge of the SiO_2, and orange lines show the edge of the Si pillar. Three kinds of Si pillars with different edges were obtained by changing the DRIE recipe. The deformation of the pillars after thermal oxidation might be caused by the thermal stress at the interface between Si and the SiO_2 cap. The edge of the Si pillar without cap resulted in a round shape. The SiO_2 had no significant deformation during oxidation, so that a sharp edge can be formed with the cap. Figure 7 shows an array of pillars with large overhanging SiO_2 cap with sharp edge.

The present pillar surface shows the Cassie-Baxter (CB) state for glycerol droplet, a polar liquid (Fig. 8a). The measured contact angle on the pillar surface with the large overhanging cap is as large as 171°, corresponding to the CB state. As previous described shown in Fig. 8b, the CB state is collapsed when the applied voltage is over its critical value.

Table 1 shows the critical voltage measured for the pillars with different edges but with the same pitch, the diameter, and the height. On the pillar surface with the round edge, the critical surface voltage is -81.1±3.0 V, which is in good agreement with the critical surface potential of -81.9±2.7 V after the soft X-ray charging. It is increased to -125.3±13.8 and -256.3±29.6 V for the sharp edge cap and the large overhanging cap, respectively. The

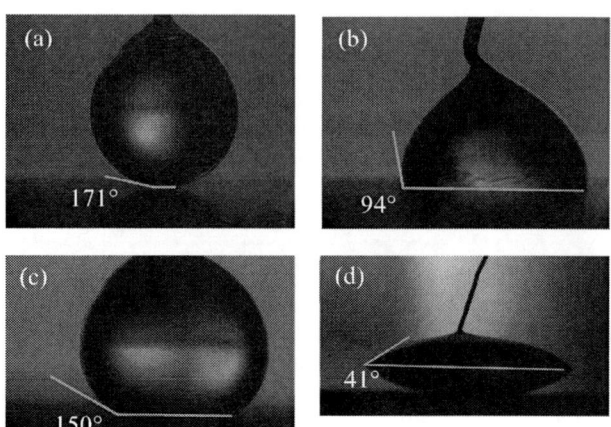

Figure 8: Wetting behavior on the pillar surface: (a) Glycerol droplet in the CB state, (b) Glycerol droplet collapsed to the Wenzel state, (c) Hexadecane droplet in the CB-state, (d) Hexadecane on a plane surface.

Table 1: Critical voltage measured with applied voltage

Pillar Edge Type	Critical Voltage
Round edge	-81.1±3.0 V
Sharp edge cap	-125.3±13.8 V
Large overhanging cap	-256.3±29.6 V

critical voltage measured for Si pillars is in good agreement with the present design value of -500 V, while the critical voltage for the SiO_2 pillars is much smaller. This is because our previous model does not consider the effect of the pillar deformation.

The pillar with large overhanging cap realizes super-lyophobicity with a large contact angle of 150° for hexadecane (Fig. 8c), which has a small contact angle of 41° on a flat surface (Fig. 8d).

In order to avoid fast decay of the charges in SiO_2 electret [7], we applied the post-annealing process [8]. Figure 9 shows the decay of the surface potential of SiO_2 electret with different thickness and annealing time. The potential decay was measured in dry air for flat samples with 0.67 μm- and 2.28 μm-thick SiO_2. Without annealing, 0.67 μm-thick SiO_2 exhibits faster decay than 2.28 μm-thick SiO_2. With the annealing, the decay rate is significantly reduced, and the effect of annealing is larger for 0.67 μm-thick SiO_2. The surface potential remains over -490 V after 4-day elapsed time, which is equivalent to the surface charge density as high as 25.5 mC/m^2.

Figure 10 shows the decay of the surface potential at the contact with glycerol. The surface potential decreased from -110 V to zero within 1 minute for the plane surface. For the pillar surface, on the other hand, the surface potential decay becomes slower; after 2 and 4 hours, the surface potential is -64 V and -57 V, respectively. Glycerol is a polar liquid with electrical conductivity. If the droplet removes the charge only from the top surface of the pillar, 90 % of the charges should remain in the present pillar design, and 53% if the charges on sidewalls are also removed. However, as shown in Fig. 10, the surface potential is gradually decreased with time and

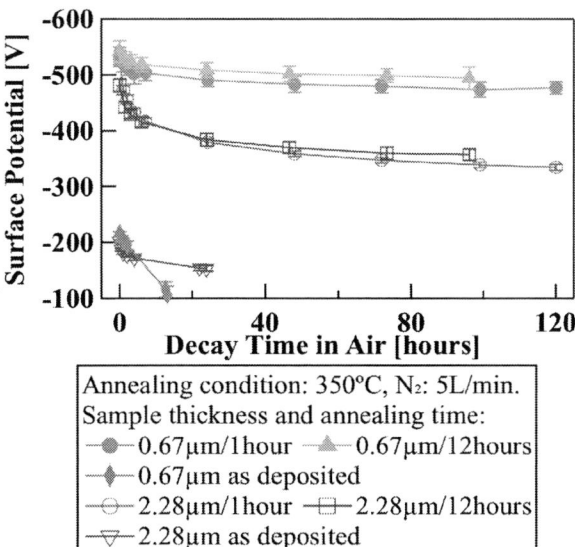

Figure 9: Effect of annealing on the stability of SiO₂ electret.

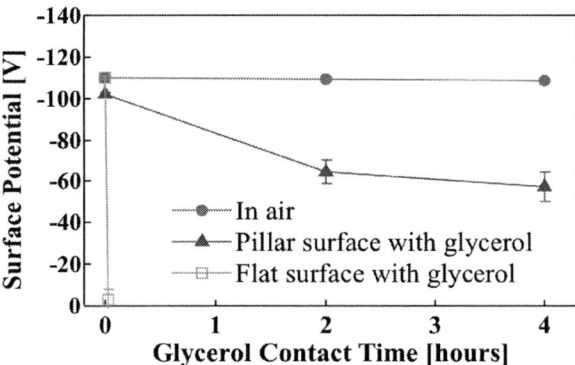

Figure 10: Decay of the surface potential in contact with glycerol.

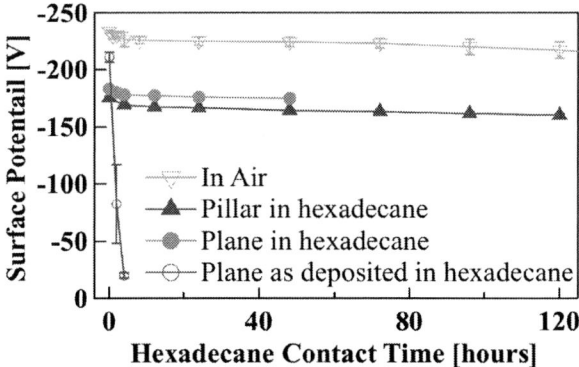

Figure 11: Decay of the surface potential in contact with hexadecane.

eventually approaches to zero, indicating that charges on the bottom surface is also affected by the presence of droplet.

Figure 11 shows the decay curve for hexadecane. Started from -180V, the surface potential is only slightly decreased in 5-day contact with droplet. However, the similar trend of decay is also observed on a plane surface at least up to 40 hours after the contact. It is noted that without the annealing step, the surface potential of the plane surface is rapidly decreased in contact with hexadecane. More investigation is necessary to clarify the discharge phenomena with nonpolar dielectric liquids such as hexadecane.

CONCLUSION

SiO₂ pillar surface with large overhanging cap is developed and charged with soft X-ray. With the present super-lyophobic pillar surface, direct contact between the liquid and the electret surface is avoided, and the lifetime of electret is significantly improved in liquid environment. The present work should be an important step towards electret devices with higher energy density and applications in droplet operation.

This work is partially supported through JSPS NEXT Program. Photo-mask is made using the University of Tokyo VLSI Design and Education Center (VDEC)'s 8-inch EB writer F5112+VD01 donated by ADVANTEST Corporation. KS was supported through Post-doctoral Fellowships for Foreign Researchers by JSPS.

REFERENCES

[1] T. Wu, Y. Suzuki and N. Kasagi, "Low-voltage Droplet Manipulation Using Liquid Dielectrophoresis on Electret," *J. Micromech. Microeng.*, Vol. 20, 085043, 2010.

[2] Y. Suzuki, "Recent Progress in MEMS Electret Generator for Energy Harvesting," *IEEJ Trans. Electr. Electr. Eng.*, Vol. 6, pp. 101-111, 2011.

[3] G. M. Sessler, *Electrets, 3rd ed.*, Laplacian Press, California, 1998.

[4] A. Tuteja, W. Choi, J. M. Mabry, G. H. McKinley, and R. E. Cohen, "Robust omniphobic surfaces," *Proc. Natl. Acad. Sci.* , vol. 105 , pp. 18200–18205, 2008.

[5] K. Hagiwara, M. Goto, Y. Iguchi, T. Tajima, Y. Yasuno, H. Kodama, K. Kidokoro, and Y. Suzuki, "Electret Charging Method Based on Soft X-ray Photoionization for MEMS Transducers," *IEEE Tran. Dielectr. Electr. Insul.*, Vol. 19, pp. 1291–1298, 2012.

[6] K.-Y. Song, K. Morimoto, and Y. Suzuki, "New Mathematical Model for Electrostatic Stability of the Cassie State on MEMS-Based Pillared Surface," *17th Int. Conf. Minituarised Syst. Chem. Life Sci. (MicroTAS 2013), Freiburg*, 2013.

[7] U. Mescheder, B. Müller, S. Baborie and P. Urbanovic, "Properties of SiO₂ electret films charged by ion implantation for MEMS-based energy harvesting systems," *J. Micromech. Microeng.*, vol. 19, 94003, 2009.

[8] T. Minami, T. Utsubo, T. Yamatani, T. Miyata, and Y. Ohbayashi, "SiO₂ electret thin films prepared by various deposition methods," *Thin Solid Films*, vol. 426, pp. 47–52, 2003.

CONTACT

*Yu-Chung Chen, tel: +81-3-5841-6419;
ychen@mesl.t.u-tokyo.ac.jp

GRAPHENE ONE-SHOT MICRO-VALVE: TOWARDS VAPORIZABLE ELECTRONICS

V. Gund, A. Ruyack, S. Ardanuc, and A. Lal

*Sonic*MEMS Laboratory, Cornell University, Ithaca, NY, USA

ABSTRACT

One-shot micro-valves are key to applications that require leak-proof sealing of micro-cavities before they are irreversibly triggered to expose the cavity content to the outside environment. Here, we report a novel micro-scale one-shot valve made of graphene transferred on to silicon-nitride (Si_xN_y) membranes. The valve triggers thermo-mechanically in 15.4±3.9 msec consuming 142.1±13.5 mW electrical power, corresponding to 2.2 mJ input energy. The valve membrane can sustain a differential pressure of 2 bars. The graphene serves multiple purposes in the device, acting as an atomically thin barrier to gas-diffusion, a resistive heater, and a source of stress in triggering the valve owing to its negative thermal expansion coefficient (TEC). The valve is designed to trigger and expose arrays of nanogram micro-packets of reactive alkali metal atoms such as rubidium, exposing it to atmospheric oxygen causing it generate heat. This heat can be used for massively parallel vaporization of vaporizable polymer electronics. A 1 mm^3 Li-battery can trigger 1580 valves on a single chip for enabling transient electronics.

INTRODUCTION

Low-cost wireless microsystems are becoming increasingly pervasive for environmental monitoring. As parts of wireless sensor networks, such microsystems are being deployed on the battlefield for secure data acquisition. A challenge facing the use of such sensor-nodes is tracking and recovering each and every device after its intended use. When left alone, these electronics can be collected by adversaries and potentially reverse engineered. When used for environmental monitoring, the sensors left behind are a source of contamination as electronic waste. A potential solution to eliminate unwanted acquisition and to the environmental pollution problem is to create transient micro-sensors nodes which can be triggered remotely to vanish while leaving behind minimal residue. Such transience has been demonstrated via liquid chemical dissolution [1] but depends on large fuel reservoirs for the transience which can be defeated by adversaries and can leave behind contaminants.

Our proposed sensor and electronics transience is achieved by massively parallel vaporization of a new vaporizable polycarbonate (VPC) which vaporizes at <150 °C using heat released by controlled reaction of rubidium stored in an array of polymer micro-packets. We have demonstrated and done thermal characterization for such polymers [2]. Figure 1 shows our micro-packet concept and time-evolution of the chip vaporization, where, initially, rubidium micro-droplets are embedded in the VPC and sealed with the graphene-on-nitride valve. When the valves are triggered, rubidium reacts with ambient oxygen exothermically to vaporize the VPC substrate. The CMOS components can react with xenon

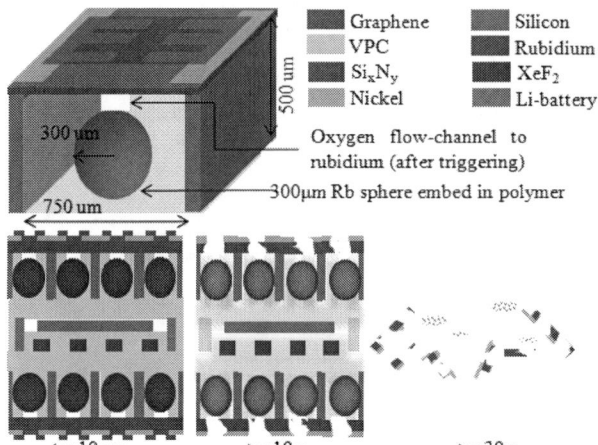

Figure 1: Proposed vaporizable electronics building block (top): Alkali metal micro packet embedded in VPC with one-shot graphene-on-nitride valve. Micro packet array vaporization (bottom): t=10ms – Valves are triggered and rubidium is exposed to air; t=10s – Metal reacts with O_2 heating substrate to ~200 °C vaporizing the substrate with the battery exposed to air; t=30s – Package vaporizes, Si_xN_y and metal residue left behind

difluoride (XeF_2), that is also stored in the substrate, and the lithium battery is consumed in a self-sustaining reaction at temperatures >120 °C generated by the rubidium. Consequently, only very little metal residue and unreacted Si_xN_y is left behind. For this vaporizable electronics architecture, the micro-packet needs to be such that oxygen does not leak/diffuse to react with rubidium before the trigger is applied, preventing unexpected device failure. When transience is triggered, rubidium must be quickly exposed to oxygen with little battery-energy available at the end of the device-life. Hence, a one-shot valve that is impermeable to gas-diffusion, and one which has low trigger-energy, is critical for sealing and eventually opening each micro-packet.

ONE-SHOT VALVE

Previous one-shot valves have used combustible explosive, ignited by integrated heaters [3]. Such valves need relatively low trigger-energy but have little controllability of opening due to fuel-volatility, and fuel-degradation over time is a potential failure mode for triggering. Another type of valve, made of a few tens of microns of polyethylene (PE) and polyethylene terepthalate (PET), has proved to be a reliable seal and operates by melting thermosetting adhesives for valve-delamination [4]. However, triggering in this case relies on over-pressure of stored fluids inside the sealed cavity. In addition, they need high energy to overcome heat of fusion of the valve-adhesive, typically an alloy. One-shot valves, which rely on melting solder or alloys, have also

978-1-4799-7956-1/15 $31.00 © 2015 IEEE

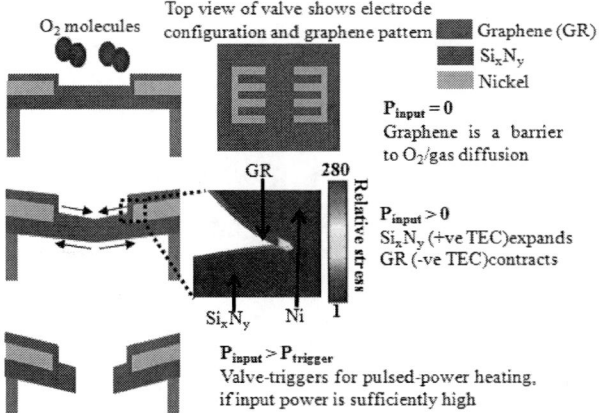

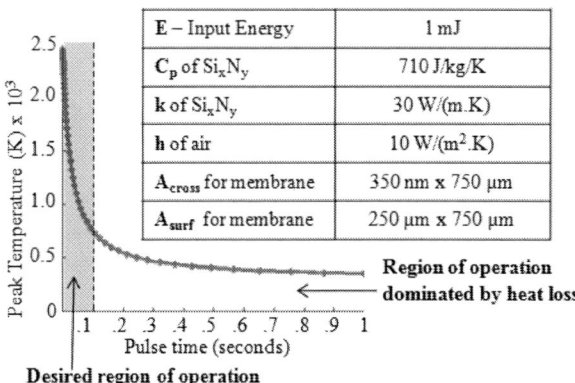

Figure 2: With no current through the graphene P_{input}=0 during regular device operation, the valve is a nanoscale diffusion-barrier for standard gases. With Joule-heating P_{input}>0, the valve surface deforms due to TEC mismatch of the graphene, Si_xN_y and nickel. Thermo-mechanical stresses result in valve-fracture when P_{input} >$P_{trigger}$.

Figure 3: Analytical calculation results showing maximum surface temperature (T) and trigger pulse-time for fixed input energy (E). Pulsed-power operation is critical for achieving low-energy triggering. The table in the inset shows parameters used for this model

been shown to sustain high-pressure differentials but they require long time-duration heating to temperatures >500°C [5]. Our novel graphene-based valve achieves leak-proof seal owing to graphene atomic impermeability, while requiring significantly lower energy for triggering without reaching very high temperatures, owing to the negative TEC of graphene in combination with the positive TECs of Si_xN_y and nickel.

Valve Design

The valve structure is shown in Figure 2. Graphene has demonstrated outstanding properties as an atomically thin sheet to diffusion of standard gases [6]. Graphene film on top of silicon-nitride provides a barrier to gases that might otherwise diffuse through to the rubidium. It also serves as the planar resistive heater for raising the temperature of the graphene-on-nitride valve. Remarkably, graphene has negative TEC of -8 ppm/°C[7], in contrast to the positive TECs of Si_xN_y and nickel. During heating, the opposing stresses in the graphene and the layers beneath graphene create a strong bimorph bending, responsible for cracking the membrane and valve opening. The thermal masses of the graphene and Si_xN_y layers are also low enough to heat up to high temperatures with very small input energy. Thus, the valve achieves thermo-mechanically actuated fracturing and triggering at relatively low temperatures.

Time-Dependent Valve Trigger

Valve-operation relies on thermo-mechanical stresses generated in the layer stack. Hence, it is important to optimize the heating methodology to minimize power consumption while still generating the required stresses to trigger the valve. When the valve surface is heated slowly, it experiences conductive heat-loss through the silicon substrate as well as convective loss to ambient air. The energy balance equation for valve-heating is [8]:

$$E = mC_p\Delta T + hA_{surface}\Delta Tdt + \frac{kA_{cross}}{L}\Delta Tdt \qquad (1)$$

Here, E is the total energy supplied via Joule heating. Terms on the right hand-side of the equation, in order, represent: 1) Heat absorbed by the valve of total mass m and heat capacity C_p for raising the valve temperature by ΔT, 2) convective loss to air through the exposed valve surface-area $A_{surface}$ in time dt, and 3) conductive loss to the substrate through the valve cross-section A_{cross} in time dt respectively. The constants h and k are convective heat transfer and thermal conductivity coefficients. To concentrate heat in raising the valve-temperature for fixed input energy E, the first term must be maximized. This is possible if the time of heating is very small i.e. via pulsed heating. Figure 3 shows an analytical plot of valve surface-temperature achieved versus time taken to deliver a fixed amount of energy, 1mJ in this case, for this model. The desired region of operation is shaded in grey, where the dt term is minimized.

DEVICE FABRICATION

We have previously demonstrated devices with CVD graphene transferred on Si_xN_y membranes for resonator frequency-trimming applications [9]. Device-fabrication follows the process-flow described therein with some modifications. 45-nm thick nickel electrode-fingers were evaporated and patterned on a 360 nm thick low-stress Si_xN_y. The suspended Si_xN_y membranes tested for this report are 750 x 750-μm^2, enough to seal the embedded rubidium in VPC. The electrode fingers were patterned on the valve surface to provide uniform Joule-heating, as shown in Figure 2. The graphene was patterned only on top of the suspended membrane by oxygen plasma-etch. The patterned area was chosen to minimize heat loss through the graphene film to the silicon substrate.

VALVE-CHARACTERIZATION

Valves were first characterized for heating with slow current-ramp through the graphene with a Keithley 2400 SourceMeter. A FLIR T300 infra-red (IR) camera was used to simultaneously monitor the valve temperature and estimate the power required to trigger the valve. Typical

978-1-4799-7956-1/15 $31.00 © 2015 IEEE

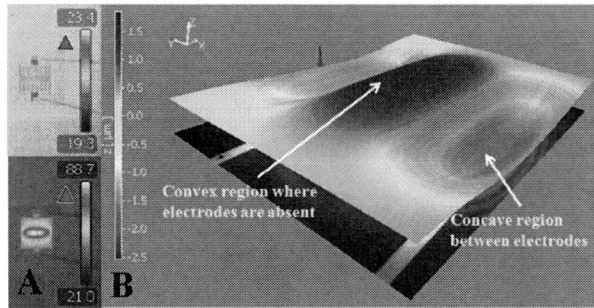

Figure 4: A) IR images of the valve surface with 0 mA (top) and 2.5 mA (bottom) input current through the graphene. B) Surface deformation verifies targeted TEC mismatch effects for producing high thermal stresses

sheet resistance of CVD-grown graphene, which is transferred on our devices, is 2.1 kΩ/□ and the values of the resistors formed were 3 kΩ. Device-to-device resistance variation is about 15% due to graphene grain-size variation and depends on presence of grain-boundaries. Figure 4A shows IR-camera images during current ramp. Graphene resistance reduces at higher current drive, perhaps due to current-induced annealing [10] as surface impurities are burnt off. In addition, static white-light interferometry was conducted with the Polytek MSA 500 operating in the optical profilometry mode to quantify the valve-surface deformation due to thermal stresses. Figure 4B presents a 3D view of the valve surface deformation at 3 mA input current and 30 mW power applied to a typical valve showing local curvatures, in qualitative agreement with FEM simulations. The peak-to-peak deformation along the Z-axis is >4 μm corresponding to a displacement to valve-thickness ratio >10:1. This is critical for producing high-stress, and ultimately, cleaving the valve surface.

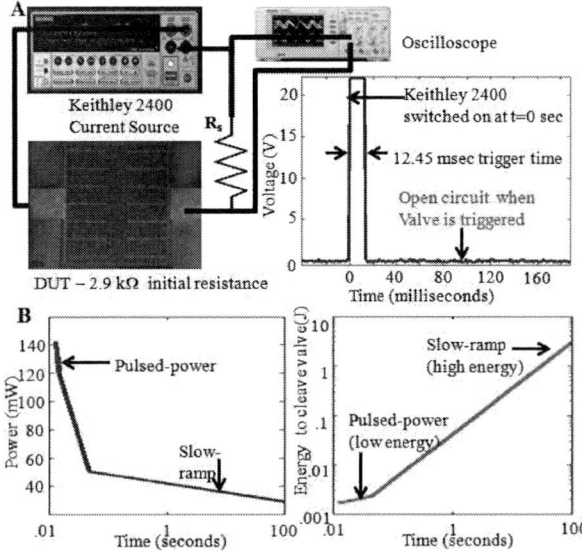

Figure 5: A) Experimental setup for measuring valve trigger time B) Input-power plotted versus measured valve-opening time show that pulsed-power triggering is 1000x energy-efficient than slow power-ramp when the pulse duration is reduced from ~100 sec to 15.4 msec

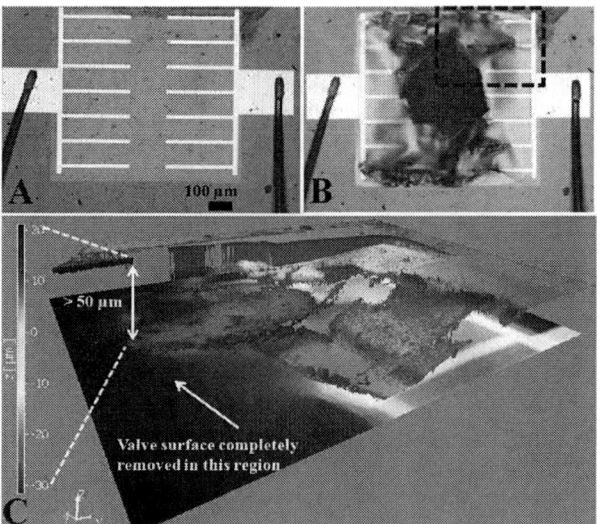

Figure 6: A) Optical microscope images of a valve before and B) after triggering C) Surface profilometry of triggered valve plotted in 3D for the region indicated with dashed-line in B. The dark region in the center has valve-surface completely removed, lost to ambient air.

VALVE TRIGGER-TESTING

Pulsed-power triggering was used to measure valve opening-time for different input powers. The setup for measuring valve trigger-time is shown in Figure 5A - a ceramic resistor R_s is connected in series with the valve, which is then triggered. The series-resistor conducts for the time that the valve is intact after current is pulsed, and then stops conducting once the valve breaks. Figure 5B shows time for actuation as a function of power and total energy consumed, indicating that large input powers significantly reduce total energy consumption, which was predicted by the thermal model in (1).

VALVE SEALING TESTS

Optical Verification

The valves were imaged with an optical microscope before and after triggering, as shown in Figure 6A and 6B. Figure 6C shows a 3D image of surface profilometry for the corner of the valve shown with the dotted line in 6B. The peak-to-peak deflection of the valve surface fragments is >50 μm as seen on the scale. The large-deflection opening can be attributed to built-in stress (about 220 MPa) in the Si_xN_y film, before it is triggered. These are relieved when the valve is triggered.

Valve Maximum Pressure Limit

To test valve durability across pressure differences, a syringe with 3 ml of air at atmospheric pressure was attached to the back of the valve with epoxy glue to make a sealed cavity. The air in the syringe was then compressed slowly, temperature being kept constant, so that the pressure was inversely proportional to volume of air sealed. The valve cracked when the air was compressed to approximately 1 ml indicating that the pressure inside the sealed cavity was 3 bar, so the pressure-difference sustained across the valve just before breaking was 2 bar.

Table 1: Comparison of our device performance with previously reported one-shot valves in literature

Work	Time of Opening (sec)	Power (W)	Trigger Temp (^{0}C)	Sustained Pressure Diff. (bar)
[3]	25×10^{-3}	72	300	-
[4]	1.2	0.2-0.4	120	2
[5]	~10	13	~500	300
This Work	15×10^{-3}	0.14	~280	2

Table 1 compares the performance of our device with other one-shot-valves reported in literature.

Valve Sealing Test

To further verify valve-sealing, a syringe with 3 ml air at atmospheric pressure, corresponding to 0.12 millimoles (mmol), was adhesively bonded to the backside of a valve with epoxy glue. This assembly was placed in a vacuum-chamber with a total volume of 1650 ml. The chamber was then pumped to 4 mbar (0.25 mmol air excluding the syringe-valve assembly) and the valve triggered, releasing the sealed gas. The schematic of the setup is shown in Figure 7 and analysis for triggering is given in Table 2.

Table 2: Sealing analysis – before and after triggering

Before Triggering			
Region	**Pressure (bar)**	**Volume (ml)**	**Millimoles of air(mmol)**
Chamber	4×10^{-3}	1650	0.25
Valve-Syringe	1	3	0.12
After Triggering			
Final total volume (ml)			1650
Total air in chamber (mmol)			0.37
Expected pressure (bar)			5.9×10^{-3}
Observed instantaneous pressure (bar)			5.4×10^{-3}

The expected and observed increase in pressure matches closely, and the deviation is due to potential outgassing of materials and finite time taken for air-stored in the syringe to be detected by the gauge. The final chamber in the pressure was 6 mbar.

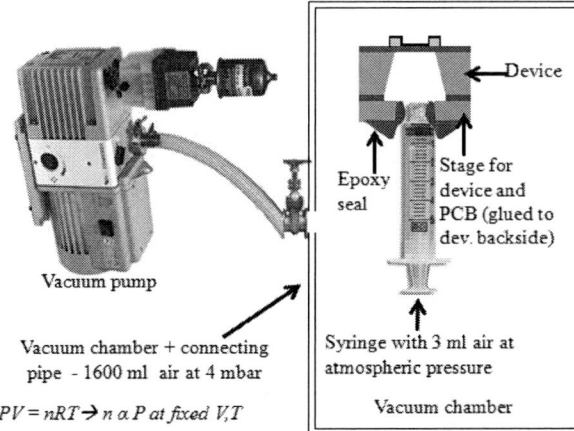

Figure 7: Schematic of setup for measuring the sealing test in vacuum with the valve-syringe assembly

CONCLUSIONS

A novel one-shot graphene-based membrane valve, which is compatible with triggering transience in vaporizable electronics, has been demonstrated. We have demonstrated the use of graphene to simultaneously provide a molecular barrier to gas diffusion and achieve thermomechanical cleavage via temperature gradient-induced deformation across the membrane exceeding $4\mu m_{pp}$. Pulsed operation of the valve has been shown to reduce the energy required for triggering, and the results provide pathways to a design space for higher efficiency valves. Apart from the need in vaporizable electronics, such valves are central in leak-proof sealing of micro-cavities and applications that require durability under pressure, and ease of triggering.

ACKNOWLEDGEMENTS

The authors would like to acknowledge funding support provided by the DARPA VAPR program. All device fabrication was completed at the Cornell Nanofabrication Facility (CNF), a member of the National Nanotechnology Infrastructure Network.

REFERENCES

[1] N. Banerjee, *et al.*, "Microfluidic Device For Triggered Chip Transience", *IEEE Sensors*, pp1-4, 2013.

[2] V. Gund, *et al.,* "Multi-Modal Graphene Polymer Interface Characterization Platform For vaporizable Electronics", *Proc. of 28th IEEE International Conference on Micro Electro Mechanical Systems (MEMS '15)*, 2015.

[3] C. Rossi, *et al.*, "Pyrotechnic actuator: a new generation of Si integrated actuator", *Sensors and Actuators*, pp 211-215, 1999.

[4] Ph. Renaud, *et al.,* "Miniature One-Shot Valve", *Proc. of 11th IEEE International Conference on Micro Electro Mechanical Systems*, 1998.

[5] J. Bejhed, *et al.*, "Demonstration of a single use microsystem valve for high gas pressure applications", *Journal of Micromechanics and Microengineering*, pp 472-481, 2007.

[6] J.S. Bunch, *et al.*, "Impermeable Atomic Membranes from Graphene Sheets", *Nano Letters*, 2008.

[7] D. Yoon, *et al.*, "Negative Thermal Expansion Coefficient of Graphene Measured by Raman Spectroscopy", *Nano Letters*, 2011.

[8] W. J. Parker, *et al.*, "Flash Method of Determining Thermal Diffusivity, Heat Capacity, and Thermal Conductivity", *Journal of Applied Physics*,1961.

[9] H. Hosseinzadegan and A. Lal, "Tip-based graphene etching for MEMS resonator frequency trimming", in *Tech. Digest of the 17th International Conference on Solid-State Sensors, Actuators and Microsystems (Transducers '13)*, 2013.

[10] J. Moser, A Barreiro, and A. Bachtold, "Current-induced cleaning of graphene", *Applied Physics Letters*, 2007.

CONTACT

*V. Gund, tel: +1-650-521-1172; vvg3@cornell.edu

PIEZOELECTRIC MICRO ENERGY HARVESTERS EMPLOYING ADVANCED (MG,ZR)-CODOPED ALN THIN FILM

L. V. Minh[1], M. Hara[1], H. Kuwano[1], T. Yokoyama[2], T. Nishihara[2], and M. Ueda[3]

[1]Graduate School of Engineering, Tohoku University, Sendai, JP
[2]Taiyo Yuden Co. Ltd., Akashi, JP
[3]Taiyo Yuden Mobile Technology Co. Ltd., Yokohama, JP

ABSTRACT

We report the new doped-AlN thin film, (Mg,Zr)AlN, based micro energy harvester. By co-doping Mg and Zr into AlN crystal, (Mg,Zr)AlN shows giant piezoelectricity and preserves low permittivity. (Mg,Zr)AlN has higher figure of merit (FOM = $e_{31}^2/(\varepsilon_0\varepsilon)$) than conventional PZT. The 13 at.%-(Mg,Zr)AlN had the experimental FOM of up to 16.7 GPa. The micromachining harvester provided the high normalized power density of 3.72 mW.g^{-2}.cm^{-3}. This achievement was 1.5-fold increase compared to state of the art.

INTRODUCTION

Autonomous sensor nodes are essential components to build the networks of things. Those sensors are rapidly developed more functionality and higher performance. They continually decrease in their size and power consumption. However, batteries powering the sensors prevent the scale-down progress and limit the lifetime. A vibration-driven energy harvester using a piezoelectric material is a feasible solution to the powering issue.

There are many works focused on developing both macro and micro energy harvesters [1-4]. Although their performance is promising, the higher performance is still a strong desire for practical applications. A piezoelectric thin film serving as an electromechanical material plays a crucial role to enhance power capacity of a harvester. A material with high power of figure of merit (FOM = $e_{31}^2/(\varepsilon_0\varepsilon)$) is more preferable for energy harvesting [2, 3].

Conventionally, PZT and AlN materials are well-known for fabricating energy harvesters. PZT has high piezoelectric coefficient and permittivity [4]. In contrast, AlN possesses low piezoelectric coefficient and permittivity [3]. In term of the FOM factor, those materials are applicable for energy harvesting. However, AlN is more preferable than PZT due to its non-toxic material and complementary metal-oxide-semiconductor (CMOS) process compatibility. For practical application, AlN needs to increase further FOM. Hence, increasing the piezoelectric coefficient of the low permittivity material might tackle the current requirement.

By doping a metal element into the Al site of the wurtzite AlN crystal, the piezoelectric coefficient might be adjustable [5-8]. Both theoretical and experimental studies prove that doping Sc into AlN has giant piezoelectric coefficient [6,7]. The 6-fold increase in piezoelectric coefficient might be achieved at the high Sc-concentration dopant of up to 40% [6,7]. However, scandium is a rare, expensive transition metal. Investigating alternative dopants that are abundant in nature are crucial for future application development. Recent work shows that co-doping Mg and Zr into AlN crystal also improves piezoelectric coefficient in the similar trend [9]. Co-doping

Mg and Zr might be an attractive alternative for high performance energy harvesters.

In this paper, we develop the MgZr-codoped AlN-based micro energy harvester. The 13 at.%-(Mg,Zr)AlN thin film was applied to the silicon-micromachined energy harvester with the cantilever structure. The microstructure, piezoelectric properties, and the device fabrication and evaluation are presented in this work.

THIN FILM CHARACTERIZATIONS

(Mg,Zr)AlN thin film was synthesized on silicon-on-insulator (SOI) substrate with the buffered layer ruthenium. A cleaned SOI was prepared (002)-oriented Ru of 200 nm in thick for both growing (Mg,Zr)AlN thin film and functioning bottom electrode of the harvester. A 2 μm-(Mg,Zr)AlN thin film was deposited on the substrate by reactively sputtering with Al, Zr, and Al-Mg targets in Ar/N$_2$ ambient. The dopant concentration was controlled by the target composition.

Microstructure of the as-deposited thin film was analyzed using X-ray diffraction (XRD) and field-emission electron scanning microscopy (FE-SEM). Figure 1(a) shows the XRD patterns and the rocking curve of the 13 at.%-(Mg,Zr)AlN thin film. The film had the highly orientation of (002) at 35.8°. That is left shifted compared to the peak (002) of the pure AlN. The phenomenon might be explained to the stretching of the AlN crystal when Al site is substituted by Mg or Zr atom [8,9]. The 13 at.%-(Mg,Zr)AlN thin film grew highly c-axis orientation with the small FWHM of 2.343°. Moreover, the FE-SEM images of the surface and cross-sectional view confirms the highly dense growth and free-void thin film (Fig. 1(b)).

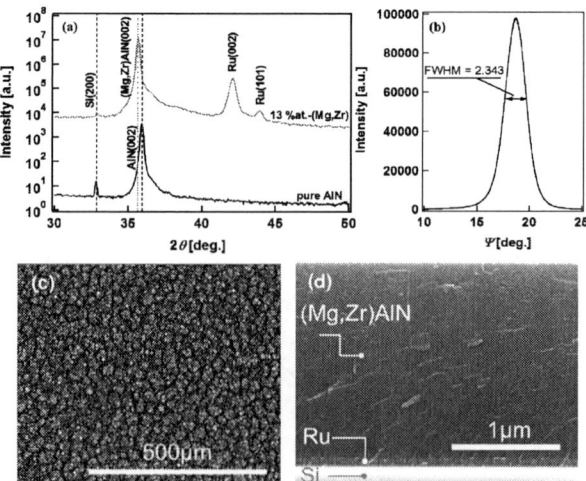

Figure 1: The XRD patterns: 13 at.-%(Mg,Zr)AlN and pure-AlN XRD patterns (a), rocking curve at (0002)-peak of 13 at.%-(Mg,Zr)AlN (b).

Table 1. Piezoelectric properties of the (Mg,Zr)AlN thin film and comparison to the published AlN materials

Parameter	This work (Mg,Zr)AlN	Bulk AlN [12]	Thin film AlN [3]
$-e_{31}$ [C/m^2]	1.27	0.60	1.00
ε_{31}	10.89	9.18	10.22
FOM [GPa]	16.70	4.43	10.95

The piezoelectric coefficient of the doped AlN was evaluated using d_{31}-unimorph micro cantilevers (Fig. 2 (a)). The dimension of the cantilever is 1000 μm×300 μm×12 μm. The thickness of the cantilever includes 2 μm-(Mg,Zr)AlN and 10 μm silicon.

The piezoelectric coefficient is deduced from the tip displacement under the given applied voltage [11]. Figure 2(b) shows the tip displacement and piezoelectric coefficient as a function of the applied voltage. The piezoelectric coefficient $-d_{31}$ of the 13 at.% (Mg,Zr)AlN was 3.1 pm/V equivalent to the piezoelectric coefficient $-e_{31}$ of 1.27 C/m^2. The value doubles the value of bulk AlN thin film and higher than that of the poly-crystal pure AlN thin film (Table 1). With regard to permittivity, the value determined by a HP 4194A impedance analyzer shows small change compared to that of pure AlN. Hence, by co-doping Mg and Zr into AlN crystal, the FOM of the 13 at.%-(Mg,Zr)AlN was up to 1.67-fold that of the poly-crystal pure AlN thin film (Table 1).

DEVICE FABRICATION

Micro energy harvester is developed from the successful 13 at.% (Mg,Zr)AlN thin film on SOI substrate. The micro energy harvester is used to verify the applicability of the thin film for energy harvesting. The harvester has a cantilever structure suspended at one end by a silicon proof mass. The 13 at.% (Mg,Zr)AlN thin film covered on the cantilever serves as the electromechanical material. The geometric configuration is listed in Table 2.

Figure 3 shows a fabrication flow in detail. The 13 at.%-(Mg,Zr)AlN thin film of 2 μm thick was deposited on 200 nm-Ru/SOI substrates. The (Mg,Zr)AlN was patterned by fast atomic beam (FAB) with SF$_6$ gas and Cr mask with the etching rate of 14 nm/min. Using the neutral atoms and radicals produced from the neutralization of SF$_6$ plasma to etch material, SF$_6$ FAB allows to pattern small and high-aspect-ratio pattern without destroying material properties [2,12]. However, there is low etching selectivity between (Mg,Zr)AlN and Ru bottom electrode. Therefore,

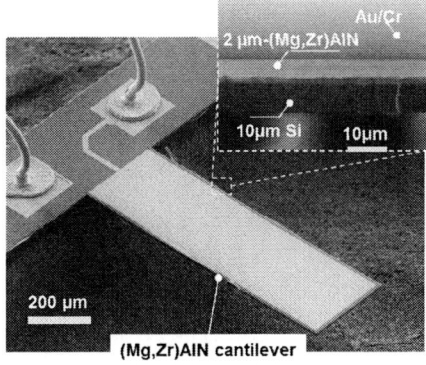

(a)

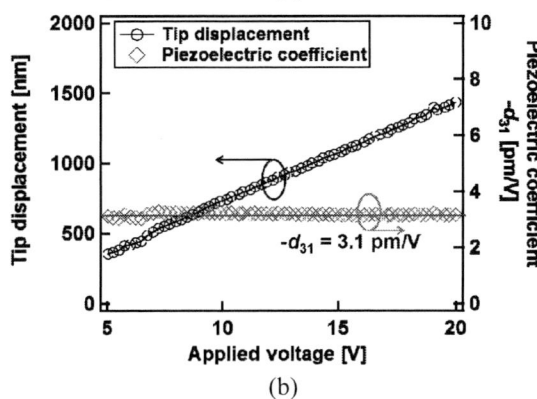

(b)

Figure 2: The d_{31}-unimorph (Mg,Zr)AlN micro cantilever (a), tip displacement of the cantilever and piezoelectric coefficient as a function of the applied voltage (b).

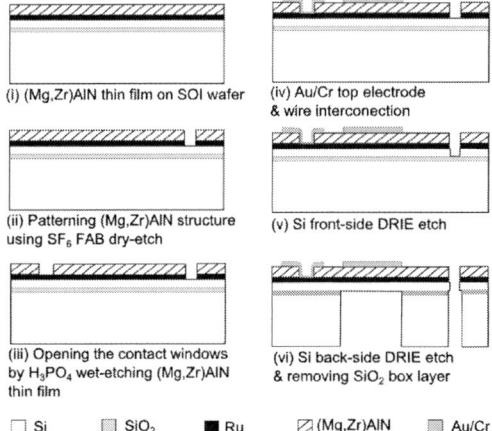

Figure 3: Fabrication flow of the micro energy harvester implementing the (Mg,Zr)AlN thin film.

Table 2. The geometric parameters of the harvester

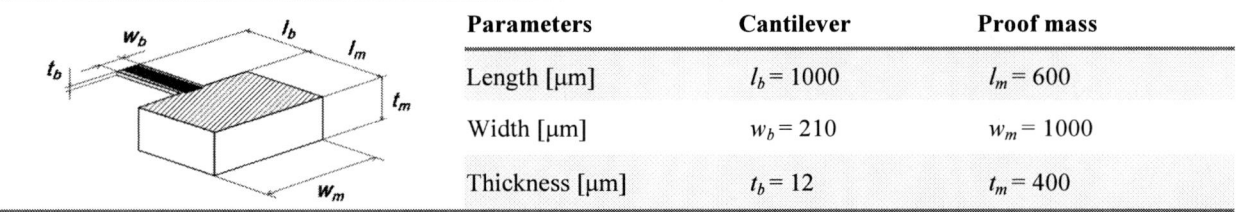

Parameters	Cantilever	Proof mass
Length [μm]	$l_b = 1000$	$l_m = 600$
Width [μm]	$w_b = 210$	$w_m = 1000$
Thickness [μm]	$t_b = 12$	$t_m = 400$

wet-etching with hot HP_3O_4 at 95 °C was used to selectively etch the (Mg,Zr)AlN over Ru for opening a contact window. The wet etching rate of the doped AlN was up to 240 nm/min. The Au/Cr top electrode of 200 nm was fabricated by a lift-off process. The device layer of the substrate was etched by deep reactive ion etching (DRIE) to shape the cantilever and the proof mass. Finally, the handle layer was removed to release the cantilever using DRIE. Figure 4 shows the SEM image of the fabricated device.

RESULTS AND DISCUSSIONS

The fabricated (Mg,Zr)AlN-based micro energy harvester was evaluated power characteristics using the resistive connection and the shaker [2]. Figure 5 shows the power spectra of the (Mg,Zr)AlN harvester under various accelerations at the optimal resistance of 3.35 MΩ. Small right shifting of the resonant frequency when increasing the acceleration is due to the stiffening effect under the large displacement of the cantilever. At the acceleration of 10 m/s², the harvester had the resonant frequency of 573 Hz. The root-mean-squared output power was up to 951 nW. We define that the normalize power density (NPD) of the harvester is the ratio between the output power over the squared input acceleration and the effective volume. The effective volume is the sum of both the cantilever and proof mass volumes. Hence, the 13 at.%-(Mg,Zr)AlN harvester achieved the NPD of 3.72 mW.g^{-2}.cm^{-3}.

Table 3 summarizes the performance of energy harvesters developed by using various piezoelectric materials. (Mg,Zr)AlN harvester shows the highest performance. The NPD of the (Mg,Zr)AlN harvester is 4.5-fold larger than that of the PZT-based micro energy harvester [4]. Our achievement also increases 1.5-fold compared to the highest NPD of the published AlN-based micro energy harvester [3].

Because the harvesters shown in Table 3 operate at different resonant frequencies, the normalized energy density that is a product between NPD and the frequency might be a norm for performance comparison. We plot the NPD as a function of the frequency as shown in Fig. 6. Therefore, each dashed line on this plot expresses the same value of the normalized energy density. It is also confirmed that our (Mg,Zr)AlN micro energy harvester has the best performance.

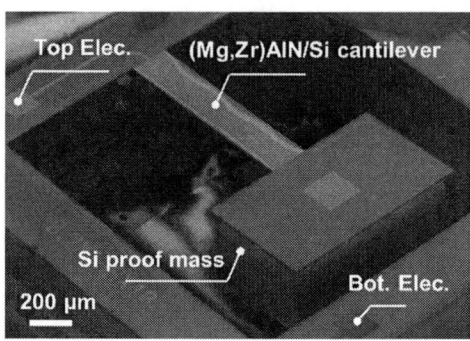

Figure 4: SEM image of the 13 at.%-(Mg,Zr)AlN-based micro energy harvester.

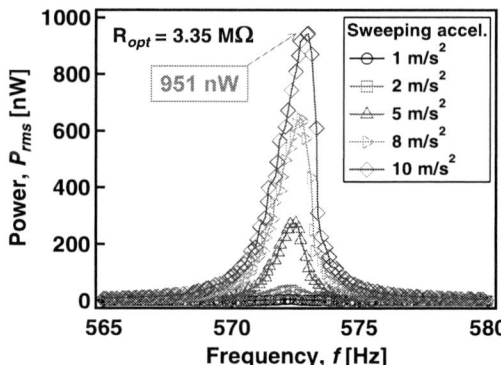

Figure 5: The power spectra depend on the acceleration at the optimal resistance.

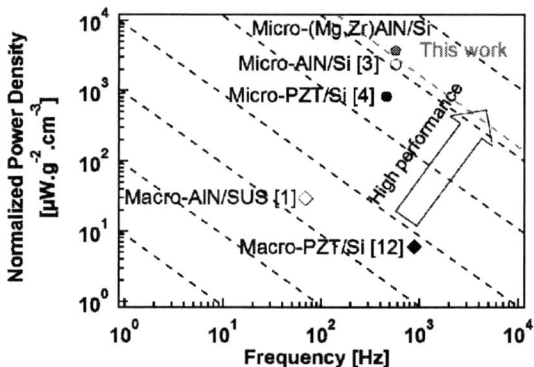

Figure 6: Performance of (Mg,Zr)AlN-based micro energy harvester compared to state of the art.

Table 3. Performance table for the published cantilever type AlN/PZT energy harvesters

	Structure	V_{eff} [mm³]	a_0 [g]	f [Hz]	load R [kΩ]	P [μW]	NPD [μW/g²/cm³]	FOM [GPa]
This work	(MgZr)AlN/Si	0.25		573	3400	0.93	3720	16.70
R. Elfrink et. al. [3]	AlN/Si	12.73		572	500	30.0	2360	10.95
Z. Cao et. al. [1]	AlN/SUS (Macro)	114.25	1	74.1	752	5.13	44	N/A
I. Kanno et. al. [11]	PZT/Si (Macro)	137.36		1036	1.51	1.00	7.3	11.20 to 15.95
D. Shen et. al. [4]	PZT/Si	0.65	2	463	6.00	2.16	820	N/A

CONCLUSION

This work presents the new doped-AlN, (Mg,Zr)AlN, energy harvester. By co-doping Mg and Zr into the wurtzite AlN crystal, the (Mg,Zr)AlN enhances piezoelectric coefficient and preserves the permittivity. We succeeded in depositing (Mg,Zr)AlN on the SOI wafer with the 13 at.% dopants. The deposited (Mg,Zr)AlN had the piezoelectric coefficient of 1.27 (C/m^2) and dielectric constant of 10.89, leading to the high FOM of 16.7. The micromachining process was established successfully for applying the (Mg,Zr)AlN thin film for micro energy harvester. The micromachining harvester was confirmed with the high normalized power density of 3.72 mW.g^{-2}.cm^{-3}. This achievement was 1.5-fold increase compared to state of the art. The achievement opens a door for high performance micro energy harvester with the cheap-dopant AlN thin film.

ACKNOLWEDGEMENT

A part of this work was performed at micro/nano-machining research and education center, Tohoku University, Japan. This work was partly conducted under the project "R&D Center of Excellence for Integrated Microsystems, Tohoku University", funded by the Ministry of Education, Culture, Sports, Science and Technology (MEXT).

REFERENCES

[1] Z. Cao, J. Zhang, H. Kuwano, "Design and Characterization of Miniature Piezoelectric Generators with Low Resonant Frequency", *Sensors and Actuators A.*, vol. 179, pp. 178-184, 2012.

[2] L. V. Minh, M. Hara, F. Horikiri, K. Shibata, T. Mishima, H. Kuwano, "Bulk Micromachined Energy Harvesters Employing (K,Na)NbO$_3$ Thin Film", *J. Micromech. Microeng.*, vol. 23, 035029-6, 2013.

[3] R. Elfrink, T. M. Kamel, M. Goedbloed, S. Matova, D. Hohfeld, Y. Andel, R. Schajk, "Vibration Energy Harvesting with Aluminum Nitride-based Piezoelectric Devices", *J. Micromech. Microeng.*, vol. 19, 094005-8, 2009.

[4] D. Shen, J. H. Park, J. Ajitsaria, S. Y. Choe, H. C. Wikle, D. J. Kim, "The Design, Fabrication and Evaluation of a MEMS PZT Cantilever with an Integrated Si Proof Mass for Vibration Energy Harvesting", *J. Micromech. Microeng.*, vol. 18, 055017-7, 2008.

[5] M. Akiyama, K. Umeda, A. Honda, T. Magase, "Influence of Scandium Concentration on Power Generation Figure of Merit of Scandium Aluminum Nitride Thin Films", *Appl. Phys. Lett.*, vol. 102, 021915-4, 2014.

[6] F. Tasnadi, I. A. Abrikosov, I. Katardjiev, "Significant Configurational Dependence of the Electromechanical Coupling Constant of B$_{0.125}$Al$_{0.875}$N", *Appl. Phys. Lett.*, vol. 94, 151911-3, 2009.

[7] F. Tasnadi, B. Alling, C. Hoglund, G. Wingqvist, J. Birch, L. Hultman, I. A. Abrikosov, "Origin of the Anomalous Piezoelectric Response in Wurtzite Sc$_x$Al$_{1-x}$N Alloys", *Physical Review Letters*, vol. 104, 137601-4, 2010.

[8] M. Akiyama, T. Kamohara, K. Kano, A. Teshigahara, Y. Takeuchi, N. Kawahara, "Enhancement of Piezoelectric Response in Scandium Aluminum Nitride Alloy Thin Films Prepared by Dual Reactive Cosputtering", *Adv. Mater.*, vol. 21, 593-596, 2009.

[9] T. Yokoyama, Y. Iwazaki, Y. Onda, T. Nishihara, M. Ueda, "Highly Piezoelectric Co-doped AlN Thin Films for Bulk Acoustic Wave Resonators", *UFFC, EFTF, and PFM Symposium*, pp.1382-1385, 2013.

[10] K. Yamanouchi, "Acoustic Wave Device Technology", *Ohmasa*, pp. 136, 2004.

[11] I. Kanno, T. Ichida, K. Adachi, H. Kotera, K. Shibata, T. Mishima, "Power-generation Performance of Lead-free (K,Na)NbO$_3$ Piezoelectric Thin-film Energy Harvesters", *Sensors and Actuators A.*, vol. 179, 132-136, 2012.

[12] F. Shimokawa, H. Tanaka, Y. Uenishi, R. Sawada, "Reactive-fast-atom Beam Etching of GaAs using Cl$_2$ gas ", *J. Appl. Phys.*, vol. 66, pp. 2613-2618, 1989.

CONTACT

Le Van Minh,
Tel (Fax): +81-22-795-4771
E-mail: minhlv@nanosys.mech.tohoku.ac.jp

PULSE POLING WITHIN 1 SECOND ENHANCE THE PIEZOELECTRIC PROPERTY OF PZT THIN FILMS

T. Kobayashi[1], Y. Suzuki[1,2], N. Makimoto[1], H. Funakubo[3], and R. Maeda[1]

[1]National Institute of Advanced Industrial Science and Technology (AIST), Tsukuba, Japan,
[2]Ibaraki University, Hitachi, Japan, [3]Tokyo Institute of Technology, Yokohama, Japan

ABSTRACT

We have succeeded in enhancing the piezoelectric property of lead zirconate titanate (PZT) thin films by employing "pulse poling" technique. We have fabricated MEMS-based microcantilevers using $Pb(Zr_{0.52},Ti_{0.48})O_3$ (MPB-PZT) thin films. By applying 1 kHz and 100 V of unipolar triangle pulse voltage, the transverse piezoelectric constant $-d_{31}$ of the PZT thin film has been enhanced as high as 105 pm/V, which is larger than the dc-poled PZT thin films (78 pm/V). The $-d_{31}$ of pulse-poled the PZT thin films measured under the unipolar actuation at 0-12 V reaches 158 pm/V. The results indicate that only 1 second of pulse poling gives better piezoelectric property of the PZT thin films compared to several minutes of dc poling.

INTRODUCTION

Morphotropic phase boundary composition lead zirconate titanate $(Pb(Zr_{0.52},Ti_{0.48})O_3$, MPB-PZT) thin films are used as actuators for inkjet heads [1], optical microscanners [2], RF-switches [3], electric field sensors [4] and so on. PZT thin films usually need several minutes of dc voltage (10-20 V/μm) application for activation, which can be called as dc poling. However, such a several minutes of dc poling reduce the throughput of device fabrication. One of the promising solution for reducing the poling time is "pulse poling", where pulse voltage with higher electric field than dc voltage is applied. Recently, we have demonstrate that the pulse poling enhance the piezoelectric property of tetragonal composition $Pb(Zr_{0.3},Ti_{0.7})O_3$ thin films [5,6]. Then, in the present study, we have employed the pulse poling to MPB-PZT thin films.

MICROCANTILEVER FABRICATION

We have prepared 1.9-μm-thick MPB-PZT thin films with (001)/(100) orientation by sol-gel deposition [7]. Commercially available PZT solution (PZT-20,Kojundo Chemical Co., Saitama, Japan) with a Pb/Ti/Zr molar ratio of 120/52/48 was deposited by spin coating. Then, the as-deposited films were heated at 250°C for 5 min. on a hot plate and crystallized by rapid thermal annealing (RTA) at 650°C for 2min. The ramping rate for RTA is 100°C /sec.

We have fabricated piezoelectric microcantilevers using MPB-PZT thin films through MEMS microfabrication process, which is just the same as that reported in our previous study [5]. Figure 1 shows the MEMS microfabrication process. The process began from Pt/Ti/PZT/Pt/Ti/SiO₂ multilayer deposition on silicon-on-insulator (SOI) wafers with a 10-μm-thick structural Si layer. The SOI wafers, at first, were oxidized at 1100°C. After the oxidation, Pt (0.1 μm)/Ti (0.05 μm) thin films as a bottom electrode were deposited by DC magnetron sputtering. Then, 1.9-μm-thick MPB-PZT thin films were formed by sol-gel deposition as described above. Finally, Pt (0.15 μm)/Ti (0.05 μm) thin films were

deposited by DC magnetron sputtering.

The obtained multilayers were etched by dry and wet etching to form microcantilevers. Through mask 1, Pt/Ti thin films as a top electrode were etched using Ar ions to determine the top electrodes. Through mask 2, PZT thin films were wet-etched using a mixture of HF, HNO₃ and HCl. Through mask 3, Pt/Ti thin films as a bottom electrode were etched using Ar ions and thermal SiO₂ thin films were etched by reactive ion etching (RIE) with CHF₃ gas to pattern bottom electrodes. Through mask 4, structural Si and buried oxide (BOX) layers were etched by RIE with SF₆ (for Si) and CHF₃ (for BOX) gases. Finally, Si substrates and BOX were etched from their back surface to form the piezoelectric microcantilevers. The fabricated microcantilevers are 1.1 mm long, 300 μm wide and 10 μm thick, respectively. Figure 2 shows the 4 inch wafer after whole microfabrication process.

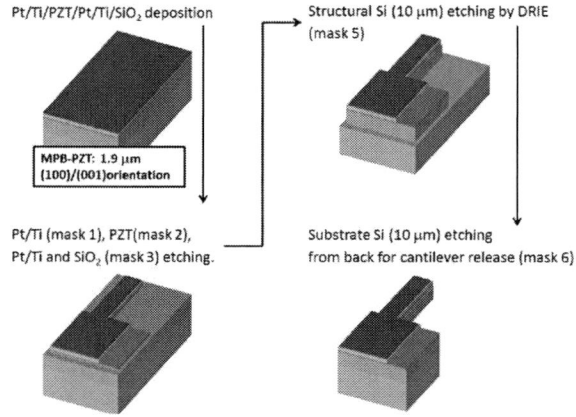

Figure 1: MEMS microfabrication process for piezoelectric microcantilevers.

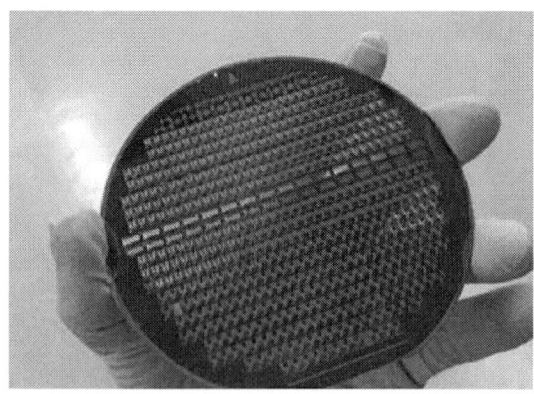

Figure 2: Photograph of 4 inch wafer after whole MEMS microfabrication process.

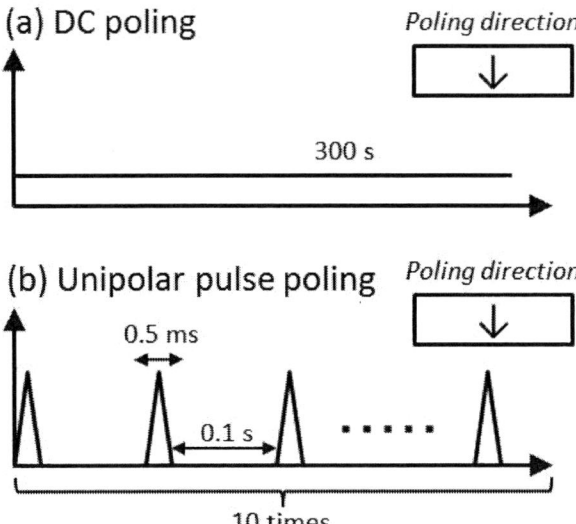

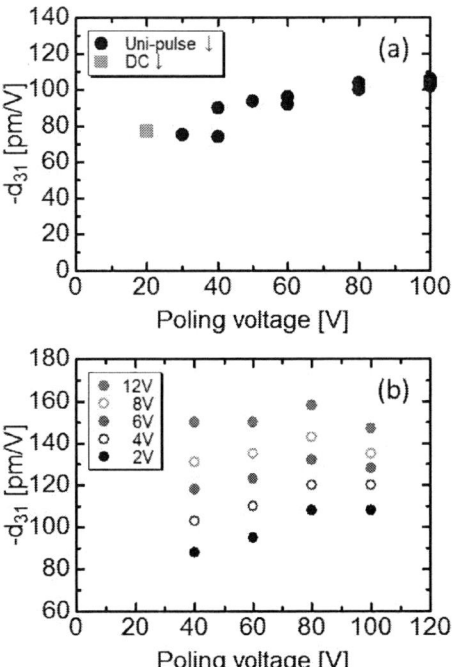

Figure 3: Waveforms of voltage for (a) dc poling and (b) unipolar pulse poling. The poling direction was downward for both cases.

The MPB-PZT thin films on the fabricated microcantilevers were poled by applying dc and unipolar triangle pulse voltage as shown in figures 3(a,b). The top and bottom electrodes were connected to signal and ground lines, respectively. The dc voltage was 20 V, which means that the poling direction is downward. For pulse poling, positive intermittent triangle waves of 40-100 V and 1 kHz were applied 10 times. The poling direction is also downward.

RESULTS

The MPB-PZT thin films on the fabricated microcantilevers were poled by applying dc and unipolar triangle pulse voltage as shown in figures 2(a,b). The top and bottom electrodes were connected to signal and ground lines, respectively. The dc voltage was 20 V, which means that the poling direction is downward. For pulse poling, positive intermittent triangle waves of 40-100 V and 1 kHz were applied 10 times. The poling direction is also downward. In order to estimate the transverse piezoelectric constant d_{31}, the tip displacement of the microcantilevers was measured. We have estimated the piezoelectric constant d_{31} using the following equations [8],

$$\delta = \frac{3AB}{K} d_{31} L^2 V \qquad (1),$$

$$A = s_{11}^{Si} s_{11}^{PZT} \left(s_{11}^{Si} t_{PZT} + s_{11}^{PZT} t_{Si} \right) \qquad (2),$$

$$B = t_{Si}(t_{Si} + t_{PZT}) / \left(s_{11}^{Si} t_{PZT} + s_{11}^{PZT} t_{Si} \right) \qquad (3),$$

$$K = \left(s_{11}^{Si}\right)^2 t_{PZT}^4 + 4 s_{11}^{Si} s_{11}^{PZT} t_{Si} t_{PZT}^3 + 6 s_{11}^{Si} s_{11}^{PZT} t_{Si}^2 t_{PZT}^2 + 4 s_{11}^{Si} s_{11}^{PZT} t_{Si}^3 t_{PZT} + (s_{11}^{PZT})^2 t_{Si}^4 \qquad (4),$$

where s_{11}, t, and L are compliance of Si (5.9×10^{-12} GPa^{-1}) and PZT (1.4×10^{-11} GPa^{-1}), thickness of Si (10 μm) and PZT (1.9 μm), length of the microcantilevers (1.1 mm), respectively. V is applied voltage to actuate the microcantilevers. In the present study, V was 1-3 V_{pp} for small displacement measurement under bipolar actuation and 0 to 2-12 V for large displacement one under unipolar

Figure 4: Piezoelectric constant $-d_{31}$ as a function of poling voltage measured under (a) bipolar actuation at 1-3 V_{pp} and (b) unipolar actuation at 2-12 V.

actuation, where frequency was 800 Hz. The tip displacement was measured by laser displacement meter. We have estimated the d_{31} of PZT thin films dc and pulse-poled at 20 V and 40-100 V.

Figure 4(a) shows the estimated $-d_{31}$ values as a function of dc and pulse poling voltage measured under bipolar actuation at 1-3 V_{pp}. The $-d_{31}$ of dc-poled PZT thin films is 78 pm/V, while that of pulse-poled PZT thin films exceeds 100 pm/V. The results indicate that only 1 second of pulse poling gives larger piezoelectric constant than several minutes of dc poling. Figure 4(b) shows $-d_{31}$ values as a function of pulse poling voltage measured under unipolar actuation at 2-12 V. The larger actuation voltage resulted in the larger piezoelectric constant $-d_{31}$. The highest $-d_{31}$ is as high as 158 pm/V. Note that the dependence on pulse poling voltage is different for actuation voltage: the $-d_{31}$ reaches its maximum at the pulse poling voltage of 100 V for 2 and 4 V actuation, while at the pulse poling voltage of 80 V for 6, 8 and 12 V actuation.

DISCUSSION

In the following, we discuss the difference in the dependence on pulse poling voltage. It is well known that the piezoelectric response of PZT thin films is a summation of intrinsic and extrinsic contribution. The former originate from the piezoelectric response of single domains, and the latter from the domain wall motion. Since the coercive voltage of the present PZT thin films is 7 V, the d_{31} measured at 2 V originate from intrinsic contribution, while that measured at 12 V from both intrinsic and extrinsic contributions. Therefore, we assume that the piezoelectric constant measured at 12 V $d_{31,12V}$ is expressed as follows,

$$d_{31,12V} = d_{31,2V} + (d_{31,12V} - d_{31,2V}) \qquad (5),$$

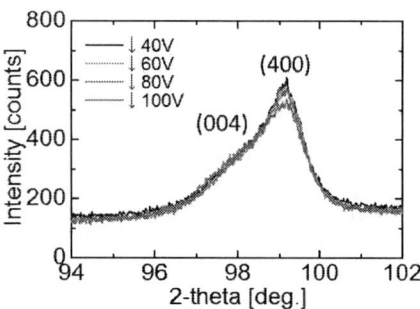

Figure 5: XRD patterns of the PZT thin films pulse-poled at 40-100 V around PZT(004)/(400) peak.

Table 1: $d_{31,12V}$, $d_{31,2V}$, $d_{31,12V}$-$d_{31,2V}$, $V_{(001)}$ and $V_{(100)}$ of PZT thin films pulse-poled at 40-100 V

Poling [V]	$d_{31,12V}$ [pm/V]	$d_{31,2V}$ Intrinsic piezo.	$d_{31,12V}$-$d_{31,2V}$ Extrinsic piezo.	$V_{(001)}$ [%]	$V_{(100)}$ [%]
40	150	58.7	41.3	37.4	62.6
60	150	63.3	36.7	38.8	61.2
80	158	68.5	31.5	39.0	61.0
100	147	73.5	26.5	41.6	58.4

where $d_{31,2V}$ is that measured at 2 V. In equation (5), the first term corresponds to the intrinsic piezoelectric effect, and the second term to the extrinsic piezoelectric effect.

As described above, the present PZT thin films consist of (001) and (100) oriented domain. The intrinsic piezoelectric effect is related to lattice distortion of (001) domain, while the extrinsic one to switching of (100) domain to (001) domain. Thus, we have investigated the degree of (001) and (100) orientation of the PZT thin films by using X-ray diffractometry (XRD). Figure 5 shows XRD patterns of PZT thin films around PZT(004)/(400) peak. From the XRD pattern, we have estimated the degree of (001) and (100) orientation $V_{(001)}$ and $V_{(100)}$ by the pseudo-Voigt function fitting.

Table 1 summarize $d_{31,12V}$, $d_{31,2V}$, $d_{31,12V}$-$d_{31,2V}$, $V_{(001)}$ and $V_{(100)}$ of PZT thin films pulse-poled at 40-100 V. The degree of (001) orientation $V_{(001)}$ increased with pulse poling voltage, which leads to the increase in the intrinsic contribution to piezoelectric constant $d_{31,2V}$. On the other hand, the extrinsic contribution to piezoelectric constant $d_{31,12V}$-$d_{31,2V}$ decrease with increasing pulse poling voltage. With increase in unipolar actuation voltage, the extrinsic contribution to the total piezoelectric constant becomes more dominant. Thus, the large decrease in $V_{(100)}$ occurred by 100 V of pulse poling have resulted in the lower $d_{31,12V}$ than that of the PZT thin films pulse-poled at 40-80 V.

The other possibility of the lower $d_{31,12V}$ of the PZT thin films pulse-poled at 100 V is crack formation observed in the case of tetragonal composition PZT thin films [5,6]. Figures 6(a-d) show cross-sectional SEM images of the PZT thin films pulse-poled at 40-100 V. Only the PZT thin films pulse-poled at 100 V have crack which go along the thickness direction (shown by white arrow in the picture). In the present case, 14 similar cracks were observed in 300 μm. The cracks may also lower the $d_{31,12V}$ of the PZT thin films pulse-poled at 100 V. As shown in the SEM images, the cracks go through along the small pores generated during PZT deposition. If the deposition conditions are

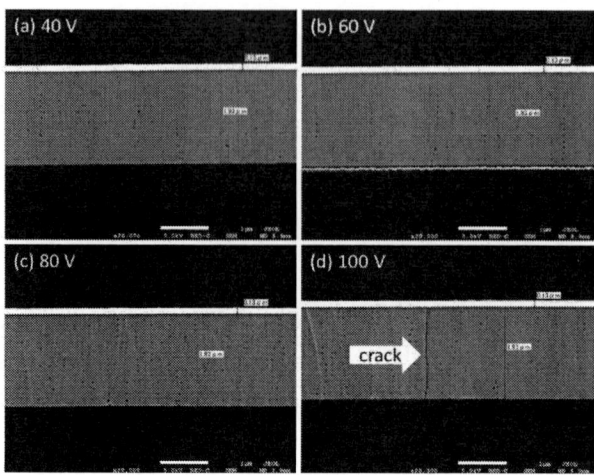

Figure 6: Cross-sectional SEM images of the PZT thin films pulse-poled at 40-100 V.

improved to reduce the generation of pores, the crack formation can be avoided. The improvement of the deposition condition is now underway.

CONCLUSION

In the present study, we have employed pulse poling to activate the MPB-PZT thin films on the MEMS-based microcantilevers. The transverse piezoelectric constant -d_{31} of the PZT thin film, measured under the actuation condition of 1-3V_{pp}, has increased with pulse poling voltage and reached 105 pm/V at the pulse poling voltage of 100 V. While the -d_{31} measured under the unipolar actuation at 12V has reached its maximum (158 pm/V) at the pulse poling voltage of 80 V. Through XRD analysis, it has been turned out the that large decrease in $V_{(100)}$ occurred by 100 V of pulse poling has led to lower $d_{31,12V}$ than that of the PZT thin films pulse-poled at 40-80 V. Although the mechanism of the difference in the dependence on pulse poling voltage is not fully understood, the results indicate that only 1 second of pulse poling gives better piezoelectric property of the PZT thin films compared to several minutes of dc poling.

ACKNOWLDGEMENTS

This research is granted by the Japan Society for the Promotion of Science (JSPS) through the "Funding Program for World-Leading Innovative R&D on Science and Technology (FIRST Program)," initiated by the Council for Science and Technology Policy (CSTP).

CONTACT

takeshi Kobayashi: takeshi-kobayashi@aist.go.jp

REFERENCES

[1] M. Murata, T. Kondoh, T. Yagi, N. Funatsu, K. Tanaka, H. Tsukuni, K. Ohno, H. Usami, R. Nayve, N. Inoue, S. Seto, and N. Morita, "High-resolution piezo inkjet printehead fabricated by three dimensional electrical connection method using through glass via", *Proc. MEMS 2009*, pp. 507-510.

[2] M. Tani, M. Akamatsu, Y. Yasuda, and H. Toshiyoshi, "A two-axis piezoelectric tilting micromirror with a

newly developed PZT-meandering actuator", *Proc. MEMS 2007*, pp. 699-702.

[3] K. Matsuo, M. Moriyama, M. Esashi, and S. Tanaka, "Low-voltage PZT-actuated MEMS switch monolithically integrated with CMOS circuit", *Proc. MEMS 2012*, pp. 1153-1156.

[4] T. Kobayashi, S. Oyama, H. Okada, N. Makimoto, K. Tanaka, T. Itoh, and R. Maeda, "An electrostatic field sensor driven by self-excited vibration", *Proc. MEMS 2012*, pp. 527-530.

[5] T. Kobayashi, N. Makimoto, T. Oikawa, A. Wada, H. Funakubo, and R. Maeda, "Linear actuation piezoelectric microcantilever using tetragonal composition PZT thin films", *Proc. MEMS 2013*, pp. 413-416.

[6] T. Kobayashi, N. Makimoto, H. Funakubo, T. Oikawa, A. Wada, T. Itoh, and R. Maeda, "Enhanced performance of sensor/actuator integrated piezoelectric microcantilever by using tetragonal composition PZT thin films", *Proc. Transducers 2013*, pp. 1879-1812.

[7] T. Kobayashi, M. Ichiki, J. Tsaur, and R. Maeda, "Effect of multi-coating process on the orientation and microstructure of lead zirconate titanate (PZT) thin films derived by chemical solution deposition", *Thin Solid Films*, vol. 489, pp. 74-78, 2005

[8] J. G. Smits and W. S. Choi, "The constituent equations of piezoelectric heterogeneous bimorphs", IEEE Trans. Ultrason. Ferroelectr. Freq. Control, vol. 38, pp. 256-270, 1991

A BETAVOLTAIC MICROCELL BASED ON SEMICONDUCTING SINGLE-WALLED CARBON NANOTUBE ARRAYS/SI HETEROJUNCTIONS

M.G. Li and J. Zhang[*]

National Key Lab of Micro/ Nanometer Fabrication Technology, Institute of Microelectronics, Peking University, Beijing, China

ABSTRACT

In this paper, a novel betavoltaic (BV) microcell based on semiconducting single-walled carbon nanotube (s-SWCNT) arrays/Si heterojunctions are demonstrated for the first time. The aligned arrays of p-type s-SWCNTs were prepared on n-type silicon to form the p-n heterojunctions as energy conversion by the traditional micro-fabrication process and dielectrophoretic (DEP) technology. The s-SWCNT arrays/Si p-n heterojunction displays better rectification characteristics than the SWCNT-based Schottky junctions such as Au/s-SWCNT/Ti and SWCNTs thin film/Si in our previous works. Under 7.8mCi/cm^2 ^{63}Ni irradiation, the open circuit voltage (V_{OC}) of 62mV, short current density (J_{SC}) of 3.8μA/cm^2, fill factor (FF) of 33.4% and energy conversion efficiency (η) of 9.8% were achieved. The results indicated that this s-SWCNT arrays/Si microcell has a huge potential for application in BV microcells.

INTRODUCTION

BV cell has drawn significant attention for decades due to their high energy density, long lifetime and durability in extreme environments with high radioactivity. It might become an alternative of widely-used solar cell in many applications, such as wireless sensor nodes, astronautics and various kinds of Micro Electro Mechanical Systems (MEMS). As the critical part of a BV cell, the energy conversion part has been made of diverse semiconductive materials, such as GaN, GaAs, and SiC [1-3]. However, most of these microcells are limited to the low current density and energy conversion efficiency.

Because of the high aspect ratios and large surface area, which can enhance the dissociation of the electron-hole pairs (EHPs) and the charge carrier transportation, nanotechnologies and nanomaterials offer innovative opportunities to make radioisotope power generators more efficient. Compared with traditional materials, single-walled carbon nanotube (SWCNT) is a perfect material due to the unique one-dimensional structure, defect-free property (which greatly lowers carrier recombination), high carrier mobility as well as low carrier scattering[4, 5]. Our group has applied the SWCNTs material to the BV microcell as the energy conversion devices for the first time, such as the Schottky junction formed between metallic SWCNT (m-SWCNT) and semiconductor like SWCNT film/Si [6] and the asymmetric metal electrodes and semiconducting SWCNT (s-SWCNT) like Au/s-SWCNT/Ti [7]. The energy conversion efficiency (η) in [6] was 0.15% and that in [7] increased to 4.29%, but it's not a huge rise compared to the other conventional BV microcells. The reason may be that the SWCNTs-based Schottky junctions do not have a good performance in the separation and collection of the beta-excited carriers.

Here, we report a novel BV microcell based on s-SWCNT/Si p-n junction structure, considering that the p-n junction exhibits better rectification property than the Schottky one [8]. It is the first time that the s-SWCNTs/Si p-n heterojunctions is implemented as the energy conversion in BV microcell. Detailed design and fabrication process are presented, and the test results are discussed and compared with those of the Schottky junctions in the following section. It showed that the s-SWCNTs arrays/Si BV microcell can achieve better performance than our previous ones.

DESIGN AND THOERY

Our device structure consists of an energy conversion part and a radiation source part, as shown in Figure 1. In the energy conversion part, we utilize the aligned arrays of s-SWCNTs which are innately p-type when exposed in oxygen contacting with the n-type silicon substrate (2-4Ωcm) to form p-n junctions. The front and back electrode made of gold (Au) forming ohmic contact with s-SWCNTs and Si respectively are employed to transport carriers to external circuit. The front Au electrode and the Ti electrode were prepared for DEP assembly of SWCNT, and the front Ti electrode functions as the sacrificial layer, which is described in the fabrication section. To improve the assembly productivity of SWCNTs, the front electrodes are designed into 30 opposite finger pairs which are 30μm in length, 3μm in width, 5μm in the lateral interval and 1μm in the gap of the opposite finger pairs. SiO$_2$ layer is used to insulate the front electrode from the Si substrate. In the radiation source part, we choose ^{63}Ni as the radiation source because of its high purity of emitted beta particles, low-energy radiation (the average energy of 17.6 keV with a maximum value of 66.7keV), long half-time (100 years) and high safety for human handling.

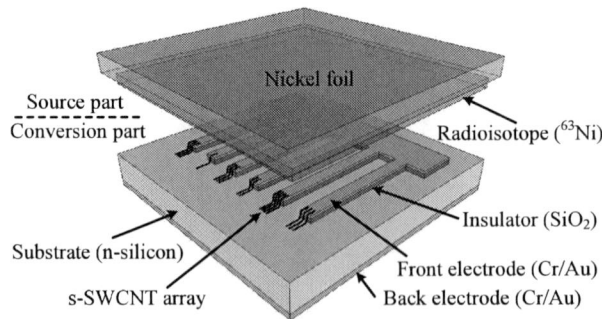

Figure 1: A three-dimension schematic view of the s-SWCNT arrays/Si heterojunctions betavoltaic (BV) microcell.

In the BV microcell, the difference of work function between the p-type s-SWCNT and n-type silicon is of vital importance. Energy band diagram of the s-SWCNT/Si p-n heterojunction under irradiation (Figure 2) illustrates the working mechanism of the energy conversion. The Fermi level of s-SWCNT is higher than that of Au and lower than that of Si, resulting in the energy band bending of s-SWCNT. Therefore, Si and s-SWCNT forms p-n junction. At the same time, Au and s-SWCNT forms ohmic contact according to the explanation by François Léonard and J. Tersoff [9], which is that the Schottky barrier height of nanotube-metal junction is mainly influenced by the work function of metals instead of the "Fermi level pinning effect".

Therefore, when the β radiation particles penetrate into the energy conversion part, they will collide with those extranuclear electrons of silicon and carbon atoms, which are then ionized, and EHPs are created. The EHPs generated in the depletion region will be completely separated by the built-in electric field, and those in the diffusion region will be transported into the depletion region and then partly separated, while those in bulk cannot be separated. Eventually, holes were collected by s-SWCNT and electrons were collected by silicon, giving rise to an electrical potential difference between s-SWCNT and silicon.

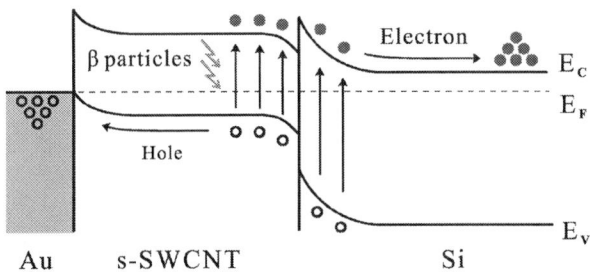

Figure 2: The energy band diagram of s-SWCNT/Si structure illustrating the energy band bending, and EHPs generation and separation under the irradiation.

FABRICATION

The fabrication process started with an N<100> silicon wafer as shown in Figure 3. After 300nm thick silicon dioxide (SiO_2) layers were thermally grown on both sides of Si wafer, the back oxide layer were wet-etched by buffered hydrofluoric acid (BHF) firstly. Then the back of silicon substrate were heavily implanted by the phosphorus ions ($\sim 10^{19} cm^{-3}$) and annealed at 1000°C. And Cr/Au (20nm/150nm) as back electrode were sputtered on the back side of Si wafer and sintered at 320°C to form a good ohmic contact (Figure 3(a)). Next, Cr/Au (20nm/150nm) and Ti (170nm) were prepared and patterned by photo lithography, sputtered and lifted off into finger pairs in sequence (Figure 3(b)). Finally, the whole wafer was diced to 6980μm×6980μm dies as samples.

In the following process (Figure 3(c-d)) s-SWCNTs are self-assembled by DEP technique between Au and Ti electrode tips. The semiconducting-enriched (99%) SWCNT purchased from NanoIntegris was used in this paper, which have a diameter range of 1.2~1.7nm with an

average of 1.4nm and a length distribution of 0.1~4μm with an average of 1μm. The concentration of s-SWCNTs solution was diluted with deionized water down to 0.1μg/ml. To minimize the SWCNTs aggregation, the s-SWCNT solution was ultrasonically dispersed for 30 minutes before DEP processes. An AC voltage source (VPP=5V and f=5MHz) was loaded between Au and Ti electrodes so that the s-SWCNTs would automatically align in the direction of the electric field and assembled between the gap. Then, a drop (~3μl) of solution was dripped onto the sample surface by the capillary tube Figure 3(c). After 60 seconds it was blow-dried gently with nitrogen gas. Subsequently the AC source was turned off and the s-SWCNTs were successfully bridged across the Au-Ti asymmetric electrodes (Figure 3(d)). Finally, a staircase-increased negative drain bias voltage (from 0V to -20V with step of 0.05V per 50ms) was applied to selectively burn off metal-SWCNTs with positive voltage of 15V on the back electrode.

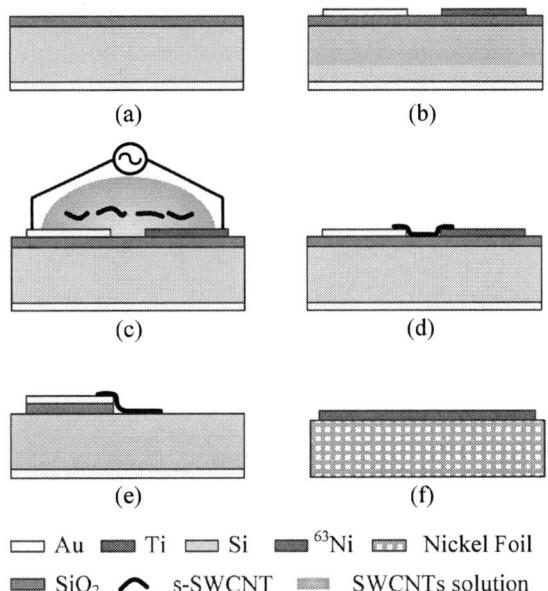

Figure 3: The fabrication process flow of the s-SWCNT arrays/Si BV microcell. (a) SiO_2 is thermally grown on the front side of Si wafer. Cr/Au back electrode is sputtered on the back side. (b) Front electrodes Cr/Au and Ti are deposited and patterned respectively. (c) SWCNTs are assembled by DEP technique. (d) S-SWCNTs were successfully bridged across the Au and Ti finger electrodes (see Figure 4). (e) The Ti and uncovered SiO_2 layer were etched by BHF to form s-SWCNTs/Si p-n heterojunction. (f) Thin-film β radiation source ^{63}Ni electroplated on the nickel foil.

Next, the Ti and uncovered SiO_2 layer were etched by BHF so that the parallel and aligned s-SWCNTs adhered tightly to Si substrate by the van der Waals force (Figure 3(e)). Figure 4 shows a scanning electron microscopy (SEM) image of one part sample. It is clearly seen that many isolated s-SWCNT bundles have formed aligned arrays contacting with the silicon substrate. The 7.8mCi/cm² ^{63}Ni thin film was electroplated on a nickel foil to serve as β radiation source as shown in Figure 3(f).

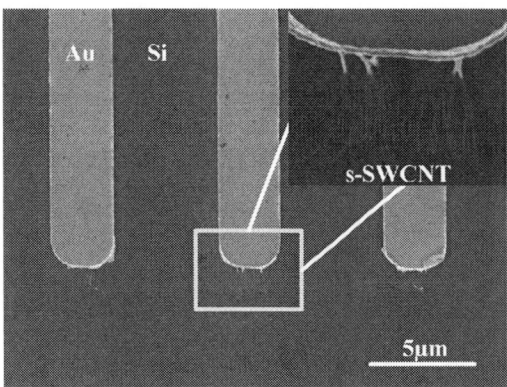

Figure 4: SEM images of the s-SWCNT arrays.

RESULTS AND DISCUSSION

For the convenience of testing, we packaged our BV microcell with the conversion part and the source part fixed to the PCB face to face with the interval of 1mm in our testing system (Figure 5). The front electrode of the conversion part was wire bonded to a pad on the PCB board and the back electrode was electrically connected to another pad by conductive tape. The whole device was tested inside an Agilent Test Fixture for shielding the light and electromagnetic interferences. I-V characteristics both in dark and under β irradiation was measured by HP 4156B semiconductor parameter analyzer.

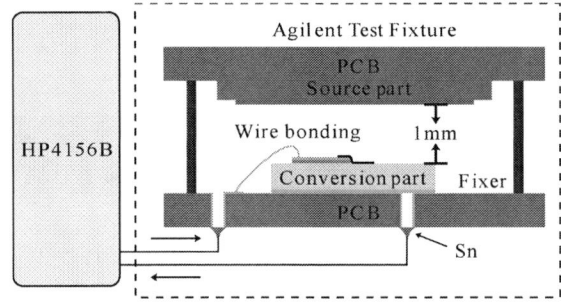

Figure 5: The testing system.

Figure 6 shows the I-V curve of the BV microcell in dark when the voltage ranging from -1V to 1V was loaded on the front Au electrode and the back Au electrode was grounded. It reveals good rectifying characteristics (inset of Figure 6). The ratio of forward current (at 0.4V) to reverse current (at -0.4V) is about 4 orders of magnitude. By fitting the experimental I-V curve with the ideal diode equation

$$I = I_0[\exp(qV / nkT) - 1] \qquad (1)$$

where I_0 is the leakage current, q is electron charge, n is ideality factor of a diode, k is Boltzmann constant and T is temperature, we deduced that I_0 is 1.5pA and n is 1.83, displaying that the s-SWCNT/Si p-n junction has much better rectification performance than our foregoing two kinds of SWCNTs-based Schottky junctions [6, 7].

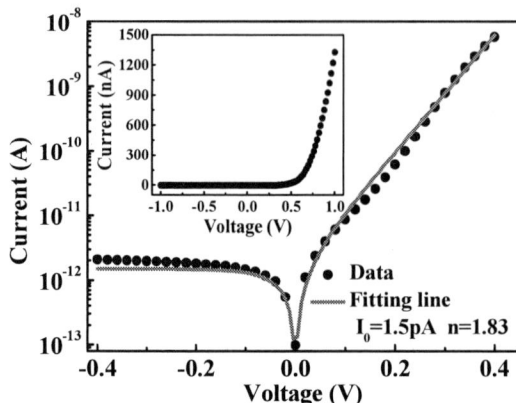

Figure 6: The Logarithmic plot of dark I-V curves of the s-SWCNT arrays/Si BV microcell in low bias. Inset is the linear plot of the I-V curve.

Under the irradiation of ^{63}Ni, the I-V curve apparently moves downwards as shown in Figure 7. The current density-voltage (J-V) curve and power density-voltage (P-V) curve were extracted as shown in Figure 8. The intercepts of y-axis and x-axis of the J-V curve show the open circuit voltage V_{OC} =62mV and the short current density J_{SC} =3.8μA/cm^2 respectively, which are both higher than those reported in our previous works [6, 7]. From the P-V curve, the fill factor (FF) can be calculated according to equation

$$FF = \frac{V_m J_m}{V_{OC} J_{SC}} \qquad (2)$$

where J_m and V_m are the current density and the voltage of the peak point of P-V curve. Thus the FF of the s-SWCNTs/Si-based BV microcell is 33.4%.

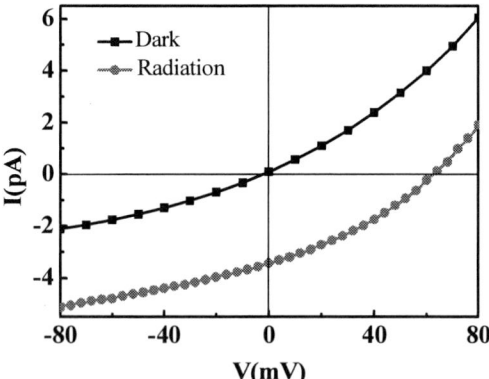

Figure 7: I-V curves measured in dark and under ^{63}Ni radiation.

The power conversion efficiency (η) of the BV microcell can be calculated by

$$\eta = \frac{P_{out}}{P_{in}} = \frac{V_m J_m}{E_{avrg} R} \qquad (3)$$

where E_{avrg} is the average electron energy of radioisotope, R is the radioactivity densities. Using the extracted V_m and

J_m, η reaches 9.8% which is about 2 times higher than that of our previous work [6, 7]. It is indicated that the s-SWCNTs/Si p-n heterojunction has a better performance in the separation and collection of the beta-excited carriers than the Schottky junction.

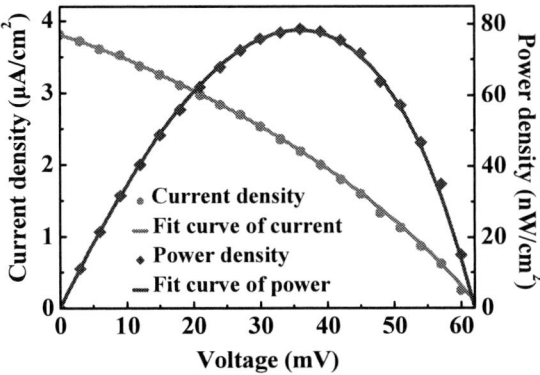

Figure 8: The Current density-voltage (J-V) and power density-voltage (P-V) curves under ^{63}Ni radiation.

CONCLUSION

In this paper, we present a novel BV microcell based on s-SWCNT arrays/Si p-n heterojunctions. The microcell is Si-based and compatible with traditional MEMS fabrication process. The SWCNT bundles were assembled between Au and Ti finger electrodes by DEP technology, and the m-SWCNTs were electrically eliminated by high current density with s-SWCNTs protected by carrier depletion. Finally the s-SWCNTs/Si p-n heterojunctions were accomplished by wet-etching Ti electrode and uncovered SiO_2 layer. The I-V curves in the dark and irradiation were measured and the performance of the BV microcell was extracted. It is indicated that the p-n heterojunction displays better rectification characteristics than our SWCNT-based Schottky junctions with I_0=1.5pA and n=1.83. Under the irradiation of 7.8mCi/cm^2 ^{63}Ni source, the microcell achieves higher BV performance with the V_{OC} of 62mV, J_{SC} of 3.8μA/cm^2, FF of 33.4% and η of 9.8%. These results suggest that the s-SWCNTs/Si p-n heterojunctions could be a promising candidate for energy conversion device of the betavoltaic microcell.

ACKNOWLEDGMENTS

This work is supported by a grant from the National Basic Research Program of China (973 Program) (No. 2015CB352100).

Thank China Institute of Atomic Energy for providing the source and testing environment.

REFERENCES

[1] M. Lu, et al., "Gallium Nitride Schottky betavoltaic nuclear batteries", *Energy Conversion and Management*, vol. 52, pp. 1955-1958, 2011.

[2] H. Chen, et al., "Design optimization of GaAs betavoltaic batteries", *Journal of Physics D-Applied Physics*, vol. 44, 2011.

[3] C. J. Eiting, et al., "Demonstration of a radiation resistant, high efficiency SiC betavoltaic", *Applied Physics Letters*, vol. 88, 2006.

[4] M. Freitag, et al., "Hot carrier electroluminescence from a single carbon nanotube", *Nano Letters*, vol. 4, pp. 1063-1066, 2004.

[5] J. D. Guo, et al., "Efficient visible photoluminescence from carbon nanotubes in zeolite templates", *Physical Review Letters*, vol. 93, 2004.

[6] C. C. Chen, et al., "A novel betavoltaic microbattery based on SWNTs thin film-silicon heterojunction", in *the 25th IEEE International Conference on Micro Electro Mechanical Systems (MEMS '12) Conference*, Paris, Jan 29 - Feb 2, 2012, pp. 1197-2000.

[7] Y. Y. Chang, et al., "A single-walled carbon nanotubes betavoltaic microcell", in *the 26th IEEE International Conference on Micro Electro Mechanical Systems (MEMS '13) Conference*, Taipei, Jan 20 - 24, 2013, pp. 825-828.

[8] Y. B. Liu, et al., "Investigation on a radiation tolerant betavoltaic battery based on Schottky barrier diode", *Appl. Radiat. Isotopes.*, vol. 70, pp. 438-441. 2012

[9] François Léonard, et al., "Role of Fermi-level pinning in nanotube Schottky diodes", *Physical Review Letters*, vol. 84, pp. 4693-4696, 2000.

CONTACT

*Jinwen Zhang, tel: +86-10-6276-6597; zhangjinwen@pku.edu.cn

A MICROGRIPPER WITH A RATCHET SELF-LOCKING MECHANISM

Y.C. Hao, W.Z. Yuan, H.M. Zhang and H.L. Chang

Northwestern Polytechnical University - MEMS Lab, Xi'an, China

ABSTRACT

This paper reports a new design for an electrostatic actuated microgripper with a ratchet self-locking mechanism. The self-locking mechanism enables long-time gripping without continuously applying the external driving signal, such as an electrical, thermal or magnetic field, which significantly reduces the effect and damage on the gripped micro-scale objects that are induced by the external driving signals. The microgripper is fabricated using a silicon-on-insulator (SOI) wafer with a 30μm device layer. The jaw gap is 100 μm, and the ratchet locking interval is 10 μm. A metal wire is successfully gripped to demonstrate the feasibility of the ratchet self-locking mechanism.

INTRODUCTION

Microgrippers with inherent advantages in operating micro-scale objects have found important applications in micro-assembly and biological research. Numerous microgrippers with different actuating mechanisms, such as electrostatic[1], thermal[2], piezoelectric[3], magnetic[4], pneumatic[5], and direct mechanical contact approaches[6], have been reported in recent decades. However, the existing microgrippers may encounter problems in manipulating micro-devices and biological tissue in the experiments that last for hours or longer. The electrostatic, thermal, piezoelectric and magnetic actuating mechanism may affect or even damage the micro-devices and biological tissues with its electrical, thermal and magnetic power. Accurately controlling the small movement on a micron scale is difficult using the pneumatic and direct mechanical contact method.

In this paper, we introduced a ratchet self-locking mechanism into our existing microgripper[7] to reduce the damage caused by the additional electrical, thermal or magnetic fields. The ratchet mechanism is a one-way intermittent motion mechanism and has been successfully used in micro electro mechanical systems (MEMS) actuators[8] as a locking mechanism. However, the ratchet pawl in [8] is a passive mechanism and cannot separate from the ratchet tooth by control. In our design, the ratchet pawl can separate from the ratchet tooth when an electrostatic actuator is used to release the objects. The designed ratchet self-locking mechanism that enables the microgripper only requires a driving signal in the moment of gripping and releasing the objects, which significantly reduces the duration of exposure of the objects to electrical, thermal or magnetic fields, and it is particularly remarkable in long-time experiments.

DESIGN

The operating principle of the ratchet self-locking mechanism is schematically shown in Figure 1: (a) original state of the ratchet self-locking mechanism; (b) rotation or movement of the ratchet tooth forces the ratchet pawl to move upward; (c) when it crosses the tip of the ratchet tooth, the ratchet pawl engages with the ratchet tooth and inhibits the inverse of the ratchet mechanism; and (d) when the ratchet pawl and ratchet tooth are separated by an external force, the ratchet mechanism is free and can return to the initial position. Thus, the ratchet self-locking mechanism can lock the microgripper when objects are manipulated without a driving signal.

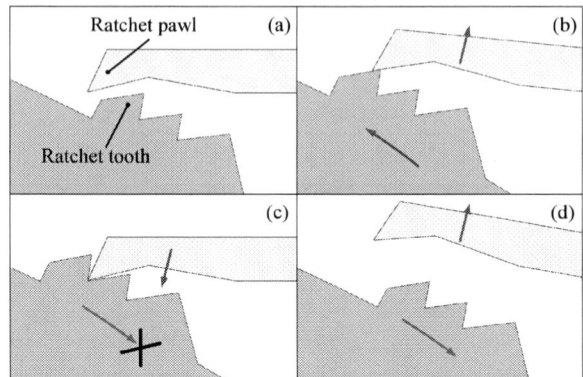

Figure 1: Operating principle of the ratchet self-locking mechanism.

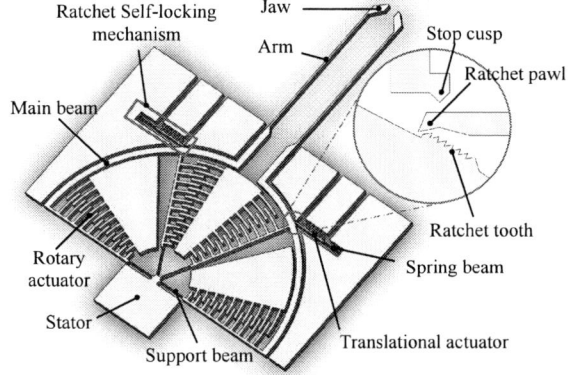

Figure 2: Schematic view of the microgripper.

As shown in Figure 2, the designed microgripper consists of an anchor that fixes the entire microgripper on the substrate; support beams that suspend the remainder of the microgripper and supply the elastic deformation, which is caused by the rotary actuator that drives the main beam; the main beams and the arms, which transfer and amplify the displacement; jaws that directly contact the targets; and the ratchet self-locking mechanism, which restricts the opening of the jaws and dominates the release of the target. In detail, the ratchet self-locking mechanism consists of multiple parts: spring beam, translational actuator, ratchet pawl, ratchet tooth and stop cusp. The engagement of the ratchet pawl and ratchet tooth prevents the jaws from inverse movement and locks the targets, and the translational actuator can set the targets free by separating the ratchet pawl and ratchet tooth. The stop cusp is designed to prevent the pulling-in of the translational actuator.

The intermittent displacement of jaws is determined based on the size of the ratchet pawl and ratchet tooth, which is confined using the MEMS fabrication process. Considering the mentioned confine, the geometric structure is optimized. As shown in Figure 3, the ratchet teeth are designed to be 3 µm tall and 4 µm wide. The angle of the pawl is 5 degree smaller than the tooth root angle of 60° to easily engage and separate. The tip of the ratchet pawl is on the root circle of the ratchet teeth, and the root circle coincides with the rotation center of the microgripper. The angle between the tip of the ratchet pawl and the root of the first ratchet tooth is 0.4°, and the angle for each tooth is 0.2°. For the designed microgripper, the jaws are closed with a 50 µm movement when the main beam rotates 1°. Thus, the first engagement occurs when the jaw moves 20 µm, but subsequent engagement occurs when the jaw moves 10 µm. The engagement and separation occur when the jaw reaches a displacement of 20 µm, 30 µm, 40 µm and 50 µm. The space between the jaws ranges from 0 to 100 µm, so the microgripper can grip a target with a size of 20 µm, 40 µm and 60 µm. Considering the elastic deformation of the slender arms, the designed microgripper can also grip objects that are slightly larger than 20 µm, 40 µm and 60 µm.

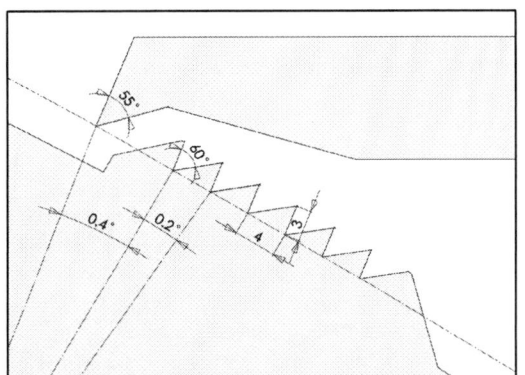

Figure 3: Detailed geometrical specification of the ratchet pawl and ratchet tooth.

FABRICATION

The microgripper is fabricated on an SOI wafer. The SOI wafer has a 400 µm thick substrate, 4 µm thick SiO_2 and 30 µm thick device layer. The fabrication process flow is shown in Figure 4.

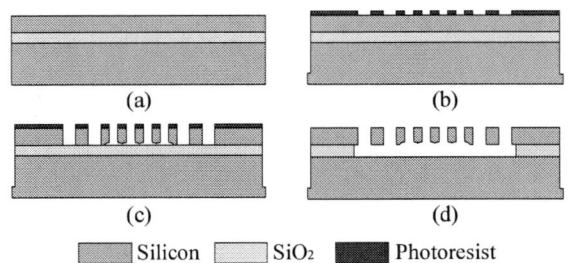

Figure 4: Microgripper fabrication process based on an SOI wafer.

In step (a), the SOI wafer is cleaned with a hydrofluoric acid solution to enhance the adhesiveness of the photoresist; step (b) is patterning by photo resist and dicing but reserving 150 µm; step (c) is a high-aspect-ratio deep reactive ion etching (DRIE) process with several minutes over etching; and step (d) removes the photoresist and is the wet-release step. The fabricated microgripper can freely move after the releasing by the hydrochloric acid solution. Then, the chip is split, and the fabrication is completed.

A scanning electron microscope (SEM) image of the fabricated microgripper is shown in Figure 5: (a) shows the global view of the fabricated microgripper, and the general size is 3,000 x 4,000 µm; (b) is the local view of the ratchet mechanism and shows the morphology of the ratchet tooth and ratchet pawl; and (c) shows the jaws, and the space between the jaws is 100 µm.

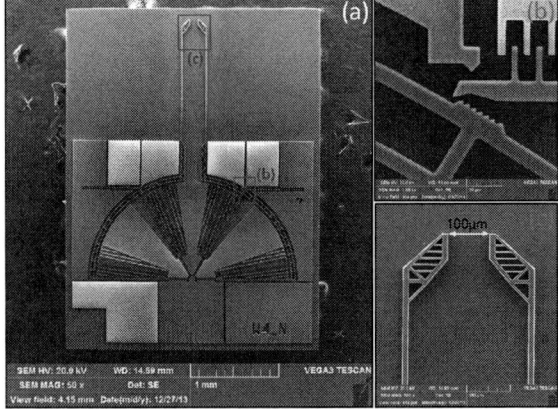

Figure 5: SEM of the fabricated microgripper.

EXPERIMENTAL SETUP AND RESULTS

A peripheral control and observing system is necessary to operate the microgripper. As shown in Figure 6, the control and observing system consists of a printed circuit board (PCB) as the microgripper carrier, a position platform with three degrees of freedom (3-DOF) to control the microgripper position, a transport probe to inject samples, a voltage source to supply the driving voltage, a high voltage amplifier to increase the driving signal, and a microscope for observing.

Figure 6: Apparatus setup and arrangement.

Without gripped targets, the voltage is recorded every time the ratchet pawl engages or separates with a ratchet tooth. As presented in Figure 7, the red dots are the measured value of the driving voltage when engagement occurs, and the black line is the analytical relationship between the voltage and the displacement. The dotted line is the fitting curve of the measured value of the driving voltage. Table 1 shows the measured voltage that separates the ratchet tooth and ratchet pawl for different position of the jaws. The deviation between the analytical value and the measured results mainly comes from the approximation in the formula derivation, machining error in the MEMS fabrication and, particularly, the effect of adhesion and friction, which is caused by the contact of the ratchet tooth and ratchet pawl.

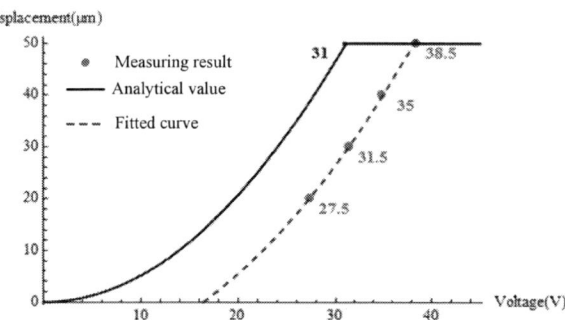

Figure 7: Relationship between the driving voltage and displacement of the jaw.

Table 1: Voltage to separate the ratchet tooth and ratchet pawl.

Displacement of jaw (μm)	20	30	40	50	Analytical value
Separation voltage (V)	58.5	65	70.5	76	22

The performance of the designed microgripper is assessed by gripping a metal wire with a diameter of 1.6 mil (40.64 μm). The experiment is shown in Figure 8: (a) positioning the microgripper and preparing to handle the metal wire; (b) the rotary actuator is applied with a driving voltage of 35.5 V to grip the metal wire; and (c) the metal wire is released with a voltage of 77 V on the translational actuator.

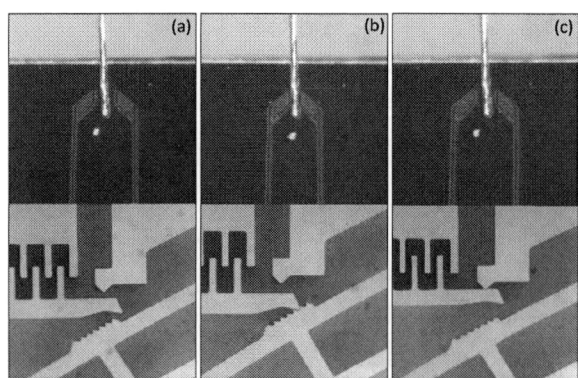

Figure 8: Experimental demonstration of gripping.

The diameter of the metal wire is 0.64 μm larger than 40 μm, and the actuator must overcome the elastic force produced by the arm deformation to lock and release the metal wire. Thus, the voltage applied on the rotary actuator and the translational actuator is larger than the corresponding value in Figure 7 and Table 1. Hence, the deformation of the slender arm can tolerate the size deviation to some extent and make the microgripper more practicable and flexible.

CONCLUSION

An electrostatic microgripper with a ratchet self-locking mechanism is presented. The feasibility of the microgripper has been successfully demonstrated by gripping a metal wire. Currently, the microgripper can handle targets with a size of approximately 20 μm, 40 μm and 60 μm, but the range can be expanded to several microns or hundreds of microns according to the requirement by adjusting the position of the ratchet tooth and the space of the jaws.

ACKNOWLEDGMENTS

This work was supported by the Chinese National Science Foundation (Grant no 61273052); in part by the Fundamental Research Funds for the Central Universities (Grant no 3102014JC02010505) and 111 project (Grant no B13044).

REFERENCES

[1] C.-J. Kim, A. P. Pisano, R. S. Muller, and M. G. Lim, "Polysilicon microgripper," IEEE 4th Tech. Dig. Solid-State Sens. Actuator Work., 1990.

[2] K. Kim, X. Liu, Y. Zhang, and Y. Sun, "Nanonewton force-controlled manipulation of biological cells using a monolithic MEMS microgripper with two-axis force feedback," J. Micromechanics Microengineering, vol. 18, no. 5, p. 055013, May 2008.

[3] D. H. Wang, Q. Yang, and H. M. Dong, "A Monolithic Compliant Piezoelectric-Driven Microgripper: Design, Modeling, and Testing," IEEE/ASME Trans. Mechatronics, vol. 18, no. 1, pp. 138–147, Feb. 2013.

[4] T. Ger, H. Huang, W. Chen, and M. Lai, "Magnetically-controllable zigzag structures as cell microgripper.," Lab Chip, vol. 13, no. 12, pp. 2364–9, Jun. 2013.

[5] A. Alogla, P. Scanlan, W. M. Shu, and R. L. Reuben, "A scalable syringe-actuated microgripper for biological manipulation," Sensors Actuators A Phys., vol. 202, pp. 135–139, Nov. 2013.

[6] B. A. Wester, S. Rajaraman, J. D. Ross, M. C. Laplaca, and M. G. Allen, "Development and characterization of a packaged mechanically actuated microtweezer system," Sensors Actuators A Phys., vol. 167, no. 2, pp. 502–511, Jun. 2011.

[7] H. Chang, H. Zhao, J. Xie, Y. Hao, F. Zhang, and W. Yuan, "Design and fabrication of a rotary comb-actuated microgripper with high driving efficiency," in 2012 IEEE 25th International Conference on Micro Electro Mechanical Systems (MEMS), 2012, no. February, pp. 1145–1148.

[8] S. Barnes, S. Miller, M. Rodgers, and F. Bitsie,

"Torsional ratcheting actuating system," in 2000 International Conference on Modeling and Simulation of Microsystems - MSM 2000, 2000, no. 505, pp. 273–276.

CONTATCT

*Honglong Chang, tel: +86-029-8849-2841; changhl@nwpu.edu.cn

A HIGH-EFFICIENT BROADBAND ENERGY HARVESTER BASED ON NON-CONTACT COUPLING TECHNIQUE FOR AMBIENT VIBRATIONS

X. Wu and D. W. Lee

MEMS and Nanotechnology Laboratory, School of Mechanical Engineering,
Chonnam National University, Gwangju, SOUTH KOREA

ABSTRACT

In this paper, a high-efficient piezoelectric energy harvester based on non-contact coupling technique is proposed and characterized, which allows it, for the first time, to take advantage of multi-cantilevers and frequency up-conversion technique to enhance the power generation efficiency for ambient excitation. The unique energy harvester can effectively scavenge environmental vibration energy with a wide bandwidth. Aiming for high space efficiency and less installation constrains, folded cantilevers are designed for this energy harvester. With a load resistance of 50 kΩ, a maximum output power of 18.45 μW is achieved with the fabricated energy harvester, which is a suitable power supply for the sensor nodes in WSN applications.

INTRODUCTION

With the increasing demand of real-time information communication in the modern world, wireless sensor networks (WSNs) technology plays a key role in the information era [1]. To avoid the extremely large maintenance cost and inconvenience caused by the limited life of batteries employed in commercial WSNs, various energy harvesters have been developed as substitutes, which can scavenge unlimited ambient energy for powering the sensor nodes in WSNs [2, 3].

Among many options, considering its sufficiency as a power source and its potential for miniaturization, piezoelectric energy harvesting has been regarded as an alternative way to replace the battery used for WSN applications [4]. However, since the ambient vibration often lies around a broad low frequency range, i.e., in the range of 0-40 Hz, the high resonant frequency (~hundreds or thousands of Hertz) of many vibration energy harvesters leads to inefficient power generation owing to the frequency mismatching [5]. To overcome the frequency gap between the environment and energy harvesters, frequency up-conversion technique is proposed and applied on the energy harvesters [6, 7]. Unfortunately, the performance of this type of energy harvester relies too heavily on the driving element, which is designed for harvesting ambient low frequency vibration. Being subjected to the narrow bandwidth of the driving part, the power generation efficiency will be confined under the broad vibration range in our environment. On the other hand, the physical impact during operation in this technique reduces the reliability and service life of the device. Aiming for a wide bandwidth to match the environmental vibration, energy harvesters combining numbers of piezoelectric cantilevers with various resonant frequencies are developed [8, 9]. However, for such a multi-cantilever (or cantilever array) based energy harvester, not only will the narrow response bandwidth of each cantilever restrict the total frequency range of the device, but the increasing number of cantilevers also make a drawback in the space efficiency. Moreover, the frequency selecting mechanism (only one cantilever can achieve the resonance for each excitation frequency) leads to low power generation efficiency when under a wide frequency range of excitation.

In this work, a high-efficient energy harvester based on non-contact coupling technique is proposed, which can realize a broadband under real environmental vibrations. The energy harvester takes advantage of the frequency up-conversion technique and multi-cantilevers to enhance the power generation efficiency for ambient low frequency excitation. In addition, a folded cantilever structure is utilized to improve the space and the power efficiency.

DESIGN AND MODELING

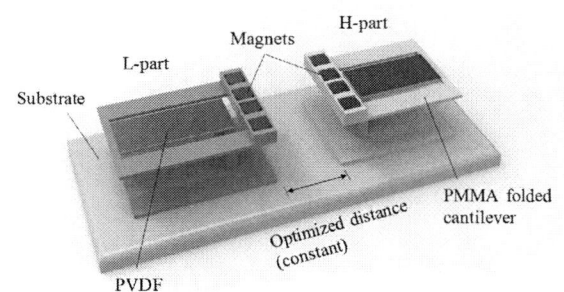

Figure 1: The designed high-efficient broadband energy harvester based on non-contact coupling technique.

As shown in Fig. 1, one piezoelectric energy harvester with low resonant frequency (L-part) is coupled with another piezoelectric energy harvester with high resonant frequency (H-part) through a non-contact magnetic force. The non-contact magnetic force is provided by 8 identical permanent NdFeB magnets, which are mounted on the end of cantilevers with the same magnetic pole arrangements. Polyvinylidene fluoride (PVDF) film is attached uniformly on the cantilevers for generating power during vibrations. D_{31}-mode is utilized in order to obtain higher power output [10]. By adjusting the dimension parameters (i.e., thickness and length of the cantilever), resonant frequencies of both the L-part and H-part are designed within a range of 0-40 Hz (L-part: 18 Hz, H-part: 32 Hz), which satisfy the ordinary range of ambient vibration.

In this case, both parts have their own high-efficient range for environmental vibration in normal conditions. Moreover, due to the non-contact magnetic force coupling, the two parts can enhance each others' output performance. This is described in detail as the following two conditions. a) Under an environmental excitation with low vibration frequency, the L-part can generate power with high efficiency due to the matching between ambient excitation

978-1-4799-7956-1/15 $31.00 © 2015 IEEE 1110 MEMS 2015, Estoril, PORTUGAL, 18 - 22 January, 2015

and its own resonant frequency, meanwhile driving the H-part to generate power with high efficiency because of the frequency up-conversion mechanism. b) On the other hand, when the ambient vibration frequency is close to the high-efficient zone of the H-part (near its natural frequency), the H-part can also promote the deformation/stress of the L-part owing to the magnetic force, which contributes to the output of the L-part. An expanded high-efficient bandwidth can thus be achieved for ambient vibration. Apart from the aforementioned advantages, due to the non-contact driving method, the lifespan of this device can be largely enhanced. In this design, the separation between the L-part and H-part is maintained with an optimized distance to achieve a maximum output. The two parts can be mounted on a substrate with this optimized separation during operating. In addition, a folded cantilever structure is designed and employed to enhance the space efficiency of this energy harvester.

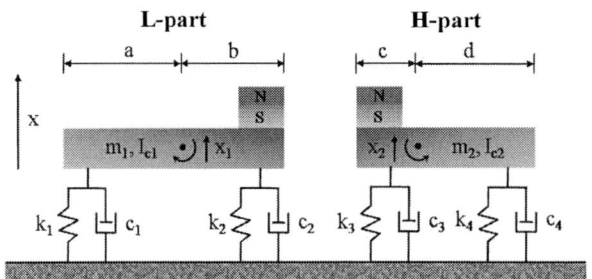

Figure 2: The simplified dynamic model of the designed energy harvester.

In order to investigate the characteristics of the designed energy harvester, a simplified dynamic model is established. As shown in Fig. 2, each part of the designed energy harvester can be modeled as a damped two degree-of-freedom (DOF) mass-spring system. Supported by damped springs with spring constant k_i and damping coefficient c_i (where i=1, 2, 3, 4), the equation of motion for the complete model can be expressed as:

$$
\begin{bmatrix} m_1 & 0 & 0 & 0 \\ 0 & I_{c1} & 0 & 0 \\ 0 & 0 & m_2 & 0 \\ 0 & 0 & 0 & I_{c2} \end{bmatrix}\begin{bmatrix} \ddot{x}_1 \\ \ddot{\theta}_1 \\ \ddot{x}_2 \\ \ddot{\theta}_2 \end{bmatrix} + \begin{bmatrix} c_1 & 0 & 0 & 0 \\ 0 & c_2 & 0 & 0 \\ 0 & 0 & c_3 & 0 \\ 0 & 0 & 0 & c_4 \end{bmatrix}\begin{bmatrix} \dot{x}_1 \\ \dot{\theta}_1 \\ \dot{x}_2 \\ \dot{\theta}_2 \end{bmatrix} +
$$
$$
\begin{bmatrix} k_1+k_2 & -(k_1a-k_2b) & 0 & 0 \\ -(k_1a-k_2b) & k_1a^2+k_2b^2 & 0 & 0 \\ 0 & 0 & k_3+k_4 & -(k_3d-k_4c) \\ 0 & 0 & -(k_3d-k_4c) & k_3c^2+k_4d^2 \end{bmatrix}\begin{bmatrix} x_1 \\ \theta_1 \\ x_2 \\ \theta_2 \end{bmatrix} = \begin{bmatrix} F_{G1}+F_{mag} \\ -F_{mag}b \\ F_{G2}-F_{mag} \\ F_{mag}c \end{bmatrix} \quad (1)
$$

where m_j and I_{cj} (where j=1, 2), are the mass and moment of inertia around the centroid of each cantilever, respectively. F_{Gj} is the inertial forces acting on the mass m_j. Considering that the magnetic force contributes to the coupling effect and output performance, the magnetic force between the L-part and H-part could be derived as:

$$
F_{mag} = -8\pi K_d R^2 \sum_{i=1}^{4}\sum_{j=5}^{8}\int_{0}^{+\infty} J_0(\frac{r_{ij}q}{R})\frac{J_1^2(q)}{q}\sinh(q\tau_i)\sinh(q\tau_j)e^{-q\varsigma}dq \quad (2)
$$

where K_d and R are the magnetostatic energy constant and

radius of a single permanent magnet, respectively. r_{ij} is the lateral separation between magnet i and j. τ, ζ, and e represent the aspect ratio of each magnet, the reduced distance between the centers of the two magnet groups, and the total magnetic energy, respectively. J_0 and J_1 are the modified Bessel function of the first type for integer orders α=0, 1 [11].

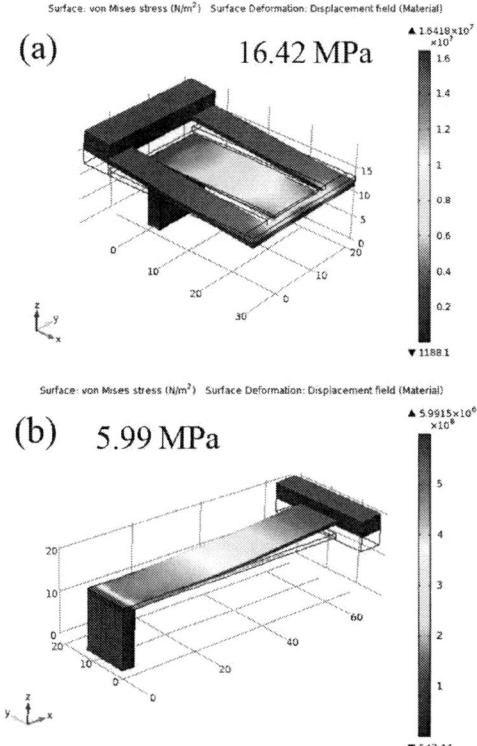

Figure 3: The comparison between folded cantilever and conventional straight cantilever (before folding) based on simulation results. Stress distribution: (a) and (b). Higher stress distribution can be achieved by the folded cantilever, which benefits the power generation.

For the piezoelectric energy harvester, a high stress distribution on the piezoelectric material can benefit the output. Therefore, high stress distribution is desired for the cantilever that attached with PVDF film in this device. For this reason, a folded cantilever is designed and adopted to this energy harvester. To verify the advantages of utilizing the designed folded structure as the cantilever in the designed energy harvester, Finite Element Method (FEM) simulations for both folded cantilever and conventional straight cantilever are carried out using COMSOL Multiphysics software with the same input conditions. For convenient comparison, the length of the conventional straight cantilever is equal to that of the folded cantilever before folding. Furthermore, the material and parameters of the folded cantilever model established in the simulation remain identical with those of the L-part. The magnets on L-part are simplified as a proof-mass but with the same physical properties. As can be seen from Fig. 3(a) and (b), a much higher stress distribution can be achieved by using this folded cantilever than by using a straight cantilever, which will effectively enhance the power generation capability and the efficiency for low frequency vibration energy harvesting.

EXPERIMENT AND RESULTS

In order to verify the output performance of the design, a prototype of this energy harvester is manufactured using 3D printing technology, which can conveniently produce the integrated total structure. A ProJet HD 3500 Plus Professional 3D printer (3D Systems, Corp, Rock Hill, USA) is employed to manufacture the energy harvester as well as its folded cantilevers. The geometric dimensions and material details of the fabricated energy harvester are listed in Table 1. During the performance test, the energy harvester is mounted on a shaker TIRA Vib BAA 120 (TIRA Gmbh, Inc, Germany) and is vertically vibrated with sinusoidal excitation at various frequencies, which is provided by the Agilent 33120A function generator (Agilent Technologies, Inc, USA). The output voltage and power generation frequency of the energy harvester is measured using a Tektronix TDS 2014B oscilloscope (Tektronix, Inc, USA).

Table 1: Parameters of the fabricated energy harvester.

Parameters	L-part	H-part
Total size	$39 \times 28 \times 15$ mm^3	$30 \times 28 \times 15$ mm^3
Material	PMMA	PMMA
Natural frequency	Before coupling: 18 Hz After coupling: 17 Hz	Before coupling: 32 Hz After coupling: 30 Hz
End-magnets	Diam. 5 mm, thickness 2.5 mm	NdFeB, 0.14 T

Considering that the distance between the L-part and H-part significantly affects the strength of magnetic coupling, a suitable separation of the two parts should be selected for realizing effective coupling mechanism. Since the two magnet groups will be attracted to each other owing to the strong magnetic force when the distance is lower than 6 mm and thereby leading to invalid working performance, the range of this distance is selected as 6-15 mm. When the distance is small in this range, i.e. 6mm, due to the constraints of magnetic repulsive force, the output of the two parts is even lower than that for the condition without magnetic coupling. In this condition, the L-part and H-part cannot completely vibrate under excitation. On the contrary, when the distance is increased to 14 mm or larger, the magnetic force is too weak to maintain the coupling mechanism, thereby resulting in an output performance near that without coupling. With various gap distances between the two parts, a favorable output performance can be achieved by the energy harvester with 11 mm distance under different frequencies of vibration. Therefore, aiming for high output capability in the same conditions, 11 mm is utilized as the optimized separation between L-part and H-part when mounted on substrate in further experiments.

To verify the influence of a non-contact magnetic coupling for the energy harvester, a comparison experiment on output waveform characterization is carried out. The permanent magnets are replaced by 8 steel mass cylinders, which eliminate the influence of the magnetic coupling effect. The weight of mass cylinders used is exactly same as the total mass of the permanent magnets. In the experiment, for convenient observation, the excitation frequencies are set to the resonant frequencies of the two folded cantilever beams (before coupling: 18 Hz and 32 Hz, after coupling: 17 Hz and 30 Hz), respectively. As shown in Fig. 4(a) and (b), it can be found that the waveform phase (or frequency) of the two parts maintain exactly same when the energy harvester vibrate under the condition of the non-magnetic coupling. On the other hand, once the two parts are magnetically coupled with each other, not only the frequency up-conversion mechanism can be realized at a low frequency range, but the output voltage of the two parts can also be improved, which is shown in Fig. 4(c) and (d), respectively. It should be noted that the original resonant frequencies of the two parts are slightly lowered after coupling. This may be attributed to the influence of magnetic force.

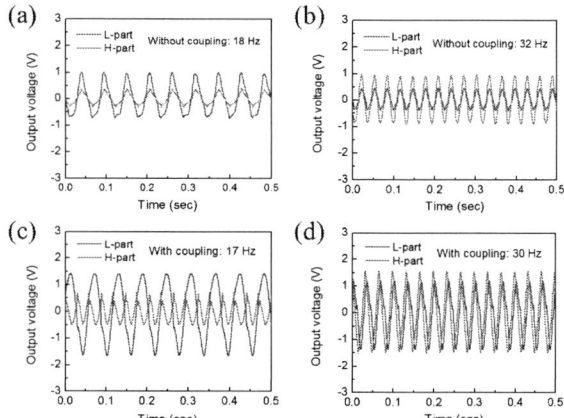

Figure 4: The output waveform comparison between the energy harvesters without and with non-contact magnetic coupling technique. (a) and (b) exhibit the waveform without coupling, (c) and (d) show the waveform with coupling.

The main advantage of the proposed design is aiming for a wide working bandwidth under ambient vibrations. To determine the coupling influence on the bandwidth of the energy harvester, the bandwidth characterization before and after magnetic coupling should be carried out. Considering the stress distribution and output will be increased with the increase in displacement for a piezoelectric cantilever, the displacement characterization can be utilized to investigate the bandwidth of this energy harvester. As shown in Fig. 5, the displacement characterization for the two folded cantilevers is conducted in a frequency range of 0-40 Hz. Assuming 4 mm as the high-efficiency threshold, we can see that a wide high-efficient zone is achieved after non-contact magnetic force coupling, which is much broader than that before coupling. This wide band will allow the energy harvester to maintain a high-efficient output performance under random vibration conditions in real applications. In addition, because of the magnetic repulsive force, the maximum deformation of each piezoelectric cantilever become larger than that before coupling; thereby further enhancing the peak output voltage. A maximum voltage of 1.5 V can be achieved for L-part and H-part at their own resonant frequency, respectively. Therefore, owing to the non-contact magnetic coupling technique, not only enormous improvement on working bandwidth can be produced for the energy harvester, but the output performance can also be enhanced. Furthermore, a durable lifespan will be obtained. In this way, the high-efficient energy harvester proposed in this work can effectively convert ambient vibration energy into electricity in a broad bandwidth. Additionally, according to the displacement characterization, it can be seen that a minimum suspended height of 10 mm is desired for avoiding any collision between the cantilever end and the substrate, which has been carefully ensured during design process of this energy harvester.

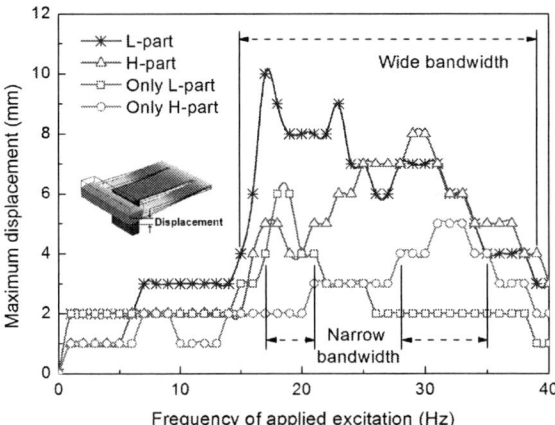

Figure 5: The maximum displacement of the cantilever end from its original position under various frequencies of excitations.

In order to investigate the power output of this energy harvester in practical applications, a load resistance of 50 kΩ is connected to the device. With a sinusoidal excitation of 30 Hz, a peak output power of 11.25 µW and 7.2 µW are achieved by the H-part and L-part (18.45 µW in total), which is suitable for the minimum power requirement of the WSN application (<10µW) [12].

CONCLUSIONS

In this study, a high-efficient piezoelectric energy harvester based on a non-contact magnetic coupling technique is designed and experimentally characterized. The two parts can effectively improve each others' output performance through the coupling mechanism in the ordinary vibration environment. Folded cantilevers are designed for this energy harvester to benefit space and power efficiency, which is verified in the FEM simulation results. An appropriate separation between the two parts is selected to ensure a favorable coupling mechanism and output performance. In the performance testing, a wide bandwidth for environmental low frequency vibration is achieved due to the non-contact coupling technique. With a load resistance of 50 kΩ, a peak output power of 18.45 µW

is achieved by the energy harvester, which satisfies the power requirement of a sensor node in WSN applications.

ACKNOWLEDGEMENTS

This research was supported by National Research Foundation of Korea (NRF) Grant funded by the MEST of the Korean government (No. 2012K1A3A1A20031500) and International Collaborative R&D Program through KIAT grant funded by the MOTIE (N0000894).

REFERENCES

[1] I. F. Akyildiz, "Wireless sensor networks: A survey", *Comput. Netw.*, vol. 38, pp. 393-422, 2002.

[2] Q. C. Tang, Y. L. Yang and X. X. Li, "Repulsively driven frequency-increased-generators for durable energy harvesting from ultra-low frequency vibration", *Rev. Sci. Instrum.*, vol. 85, pp. (045004) 1-5, 2014.

[3] X. Wu, M. Parmar and D. W. Lee, "A seesaw-structured energy harvester with superwide bandwidth for TPMS application", *IEEE/ASME Trans. Mechatronics*, vol. 19, no. 5, pp. 1514-1522, 2014.

[4] S. Roundy and P. K. Wright, "A piezoelectric vibration based generator for wireless electronics", *Smart Mater. Struct.*, vol. 13, no. 5, pp. 1131-1142, 2004.

[5] T. Galchev and K. Najafi, "Micro power generator for harvesting low-frequency and nonperiodic vibrations", *J. Microelectromech. Syst.*, vol. 20, no. 4, pp. 852-865, 2011.

[6] H. C. Liu, C. Lee, T. Kobayashi, C. J. Tay and C. Quana, "Piezoelectric MEMS-based wideband energy harvesting systems using a frequency-up-conversion cantilever stopper", *Sensor. Actuat. A: Phys.*, vol. 186, pp. 242-248, 2012.

[7] M. A. Halim, S. Khym and J. Y. Park, "Frequency up-converted wide bandwidth piezoelectric energy harvester using mechanical impact", *J. Appl. Phys.*, vol. 114, pp. (044902) 1-5, 2013.

[8] S. M. Shahruz, "Design of mechanical band-pass filters for energy scavenging", *J. Sound. Vib.*, vol. 292, pp. 987-998, 2006.

[9] J. Liu, H. Fang, Z. Xu, X. Mao, X. Shen, D. Chen, H. Liao and B. Cai, "A MEMS-based piezoelectric power generator array for vibration energy harvesting", *Microelectron. J.*, vol. 39, no. 5, pp. 802-806, 2008.

[10] B. S. Lee, S. C. Lin, W. J. Wu, X. Y. Wang, P. Z. Chang and C. K. Lee, "Piezoelectric MEMS generators fabricated with an aerosol deposition PZT thin film", *J. Micromech. Microeng.*, vol. 19, no. 6, pp. 1-8, 2009.

[11] D. Vokoun, M. Beleggia, L. Heller and P. Sittner, "Magnetostatic interactions and forces between cylindrical permanent magnets", *J. Magn. Magn. Mater.*, vol. 321, no. 22, pp. 3758-3763, 2009.

[12] R. Elfrink, "First autonomous wireless sensor node powered by a vacuum-packaged piezoelectric MEMS energy harvester," in *Proc. IEEE Electron. Dev. Meeting*, Dec. 2009, vol. 22, no. 5, pp. 1-4.

CONTACT

*D. W. Lee, tel: +82-062-530-1684; mems@jnu.ac.kr

BIDIRECTIONAL THERMOELECTRIC ENERGY GENERATOR BASED ON A PHASE-CHANGE LENS FOR CONCENTRATING SOLAR POWER

M.S. Kim, M.K. Kim, H.R. Ahn, and Y.J. Kim[*]

School of Mechanical Engineering, Yonsei University, Republic of Korea

ABSTRACT

This paper reports a bidirectional thermoelectric energy generator (TEG) with double type lenses for concentrating solar power. When solar power was applied to the TEG, solar energy is concentrated by PMMA lens firstly. The concentrated energy is absorbed as heat energy through phase-change of phase change material (PCM). And then, the liquid PCM lens focuses energy on the TEG. After removing energy source, the latent heat in PCM is released. Therefore, the proposed TEG generates energy steadily.

INTRODUCTION

Generally, there are three ways of converting solar energy to electrical energy through the use of energy materials: photovoltaic, solar thermal, and thermoelectric technology. The photovoltaic systems provide the best present solution for highly distributed, domestic power generation, which is also the most fitting market for thermoelectrics [1]. Thermoelectrics are materials which generate a voltage in the presence of temperature gradient [2]. The thermoelectric technology has been of great interest in energy application because of its well-known advantages such as good stability, high reliability, and simple fabrication process [3]. However, the low conversion efficiency is the need to improve. The efficiency of a thermoelectric material is governed by its figure of merit, $zT = S^2\sigma T/\kappa$, where S is the Seebeck coefficient, σ the electrical conductivity, T the absolute temperature, and κ the thermal conductivity. Many researches are actively being conducted to improve the zT values. However, it has low conversion efficiencies and limitation to various applications yet [4-6]. For improving the conversion efficiency, various studies on the energy generator design are in progress actively [7-9]. The simplest design of a thermoelectric energy generator (TEG) for solar power consists of unicouple TEG, a solar absorber that absorbs the incident solar radiation, and optical components that concentrate the solar power [9]. But, realizing solar TEG structures requires a large and complex optical concentrator system that contains vacuum enclosure.

In this paper, we developed a bidirectional thermoelectric energy generator with double type lenses for concentrating solar power. Lenses are composed of a general dome-shaped PMMA lens and a cylindrical-shaped lens based on phase-change materials (PCM). This design helps to improve the energy generation efficiency without a large and complex optical concentrator system. After removing energy source, the latent heat in PCM is released. Therefore, the proposed TEG generates energy steadily even though the energy source was removed.

DESIGN AND METHOD

Design

Figure 1 shows the conceptual view of the proposed a phase-change lens based thermoelectric energy generator. The proposed solar TEG consisted of TEG part and lens part. The TEG part was composed of column-type thermoelectric semiconductors and commercial solar absorber. Thermoelectric semiconductors were inserted in PDMS mold with each diameter and height of 1.5mm and 3mm. Each inserted thermoelectric semiconductor had cylindrical shape. The proposed solar TEG had two pairs of p-type and n-type thermoelectric semiconductor. P-type and n-types thermoelectric materials were arranged in sequential order which were electrically connected by copper wire. And the commercial solar absorber was located on the top of PDMS mold. It also does top electrode role because its substrate was metal. The lens part was composed of a PMMA lens and PCM lens. An interspace between solar absorber and PDMS mold is filled PCM. From among various PCMs, paraffin, a widely used PCM, is used in this study. And then, a dome-shaped PMMA film covered the PCM part. The dome-shaped PMMA lens is fabricated by commercial flat PMMA film, and the PMMA lens thickness is 110μm.

A thermoelectric energy generator is an electronic device that converts thermal gradient between hot junction and cold junction into electric voltage. When heat energy is applied to the hot junction of thermoelectric material, electric voltage is generated through the Seebeck effect. The generated electrical signal is proportional to the number of thermocouple and the thermal gradient of the each junction.

$$V = m \cdot \alpha \cdot \Delta T \qquad (1)$$

When V is the generated electrical voltage, m is the number of thermocouple, α is appropriate piezoelectric coefficient for the axis of applied stress, ΔT is the thermal gradient of the each junction, respectively.

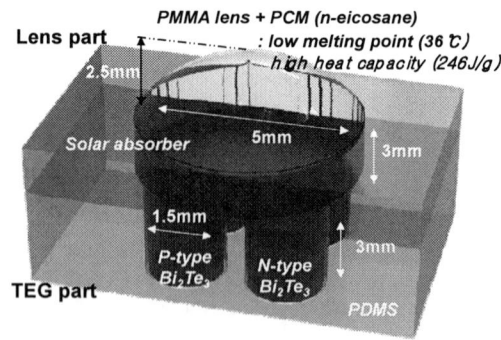

Figure 1: The conceptual view of the proposed a phase -change lens based thermoelectric energy generator

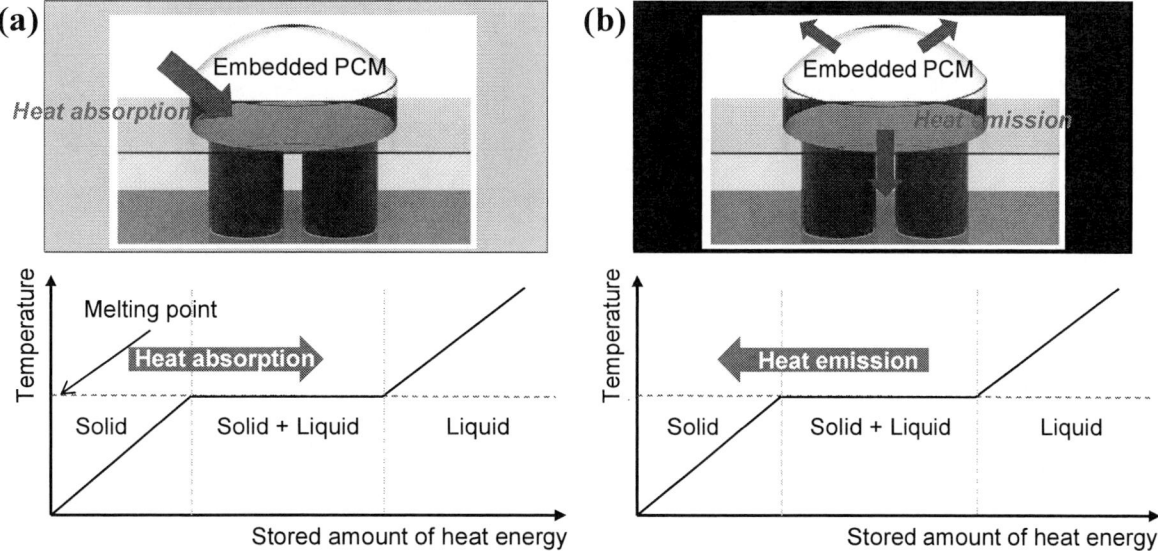

Figure 2: The latent heat effect of the embedded PCM on proposed solar TEG when (a) the solar power is applied and (b) the solar power is removed.

Effect of phase change material

Among the various thermal energy storage methods, latent thermal energy storage employing PCMs are the most effective way of the thermal energy storage. PCMs have advantages such as high energy storage density and isothermal operating characteristics. PCMs undergo solid-solid, liquid-gas, and solid-liquid phase change. Of these, solid-liquid PCMs are useful because they store a relatively large quantity of heat over a narrow temperature range, without a corresponding large volume change [10].

Paraffins have been widely used due to their high latent heat storage capacity and appropriate thermal properties such as little or no supercooling low vapor pressure, good thermal and chemical stability, and self-nucleating behavior [11]. In addition, paraffin in the liquid state is clean and transparent. The liquid PCM has good refractive index at 1.47 that can be used as a lens. We use n-eicosane (Sigma-Aldrich Co. LLC.) which is one of the types of n-paraffin wax. It has a low melting point of less than 36 °C and a large heat capacity about 246 J g^{-1}.

Figure 2 shows the latent heat effect of the embedded PCM on proposed solar TEG. When the PCM temperature reaches the melting point of the PCM, the phase of the PCM begins to change from solid to liquid by absorbing heat energy. When the heat source is removed, the stored heat energy in the PCM is released. Therefore, the TEG can continue to generate electrical energy through the stored heat energy in the PCM.

Energy harvesting mechanism

When solar energy was applied to the TEG, solar energy is concentrated by PMMA lens firstly. And then, hot junction temperature of the TEG is increased according to the amount of concentrated energy. The concentrated energy is absorbed as heat energy through phase-change of the PCM. At the end of the phase-change, a melted PCM became transparent. Since then, the liquid PCM plays a role as a second lens. The concentrated energy by the PMMA lens is re-focused through the liquid PCM lens. This mechanism helps to improve the energy generation efficiency. After removing energy source, the latent heat in the PCM is released. The latent heat energy plays a role as another energy source at the absence of sunlight. Therefore, the proposed solar TEG generates energy steadily even though the energy source was removed.

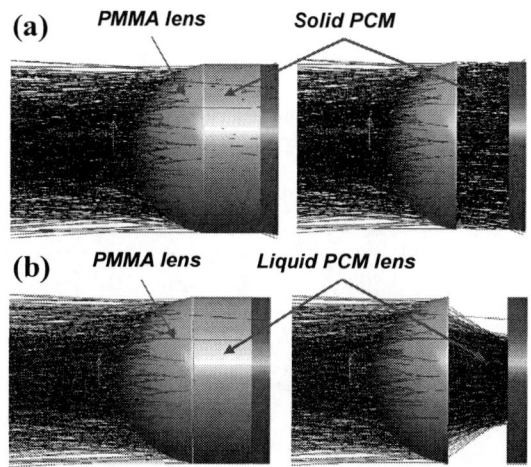

Figure 4: The results of a computational simulation of the light trace through each lens. (a) PMMA lens: solar energy 1st concentration, (b) Generation of PCM (Liquid) lens: solar energy 2nd concentration

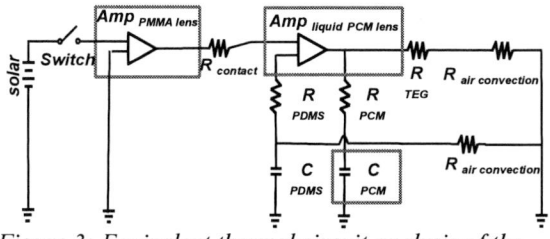

Figure 3: Equivalent thermal circuit analysis of the proposed TEG

The effect of the lenses can be explained more clearly using an equivalent thermal circuit as shown in figure 3 boxes. In thermal circuit, the solar energy is expressed as a DC voltage source, and the switch represents the presence or absence of solar energy. PMMA lens and liquid PCM lens work as an amplifier in the thermal circuit. The heat energy stored in PCM can be expressed in the thermal capacitance. The equivalent thermal capacitance of the PCM , C_{PCM}, is very large because the PCM has a very high latent heat.

Figure 4 shows the simulation results of the light trace through each lens using LightTools 7.2. In this paper, we want to improve the performance of the TEG when applying the each lens, so the TEG and solar absorber parameters were selected arbitrarily. The TEG and solar absorber diameter fixed to 1.5mm and 5mm, respectively. And then, the simulation is conducted according to the curvature of PMMA lens and the thickness of the liquid PCM lens. It was calculated that the amount of light reaching the TEG through lens from a light source. When the lens and TEG radius are equal, the amount of light was the most collected. Figure 4 (a) shows the first concentration of solar energy through the PMMA lens. Figure 4 (b) shows the second concentration of solar energy through liquid PCM lens. By adopting the double type lenses, it show that the amount of energy arriving per unit area was significantly increased.

EXPERIMENTS
Fabrication

The TEG part was fabricated based on PDMS mold. Firstly, thermoelectric materials were injected into PDMS mold using dispenser printing technology. After printing, the commercial solar absorber (mirotherm, Alanod-Solar GmbH & Co. KG) was attached on the top of PDMS mold. And then, PDMS mold for PCM was attached on the solar absorber. The melted n-paraffin (n-eicosane, Sigma -Aldrich Co. LLC.) was poured into the space of PDMS body. And then, a dome-shaped PMMA film covered the PCM part. The dome-shaped PMMA lens is fabricated by commercial flat PMMA film (C200, Sejin T.S. Co., Ltd.).

Experimental setup

The output characteristic of the fabricated solar TEG was evaluated by an experimental setup shown in figure 6. It consisted of a 50-W halogen lamp and a digital oscilloscope (U1460B, Agilent). The fabricated solar TEG was heated in controlled radiation using a 50-W halogen lamp in the dark room. To determine the power output of the fabricated solar TEG, the output voltages were

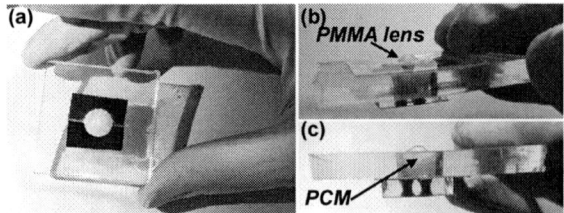

Figure 5: The optical photographs of the fabricated device. (a) a top view, (b) a diagonal view, (c) a side view

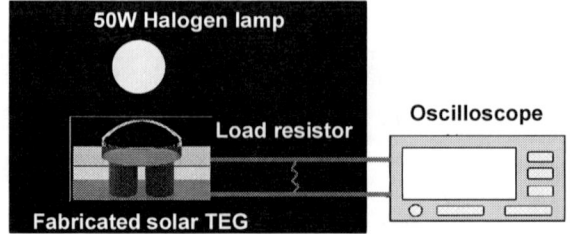

Figure 6: The experimental setup for the evaluation of the fabricated solar TEG

measured under 30Ω load resistances.

RESULTS AND DISCUSSION
First power generation

Figure 7 shows the output voltage of the TEG with and without PMMA lens and PCM. Because the PMMA lens induces first energy concentration, the output voltages of proposed device are slightly larger than that of the other until the end of the phase-change (0 ~ 225 sec). When the phase-change of the PCM has been completed, the output voltage was raised rapidly (225 sec ~). The output power of the conventional TEG is approximately 0.36 μW at 3.4 mV. The output power of the proposed solar TEG is approximately 0.6 μW at 4.3 mV. The maximum output power of the proposed TEG was improved 56 % greater than that of the other.

Second power generation

Figure 8 shows the output voltages of each TEG in the dark room. After removing the halogen lamp, the power generation of all TEG was gradually decreased as shown in Figure 8 (a). There was no significant difference in the power generation reduction zone. The phase change of PCM occurs from a liquid to a solid while heat is released in response to the passage of time. When the phase of PCM occurred, the latent heat in the PCM is released. The latent heat energy plays a role as another energy source. The proposed solar TEG generated the power for half an hour without light source. Experimental results show that the solar TEG can produce electrical power without light energy source.

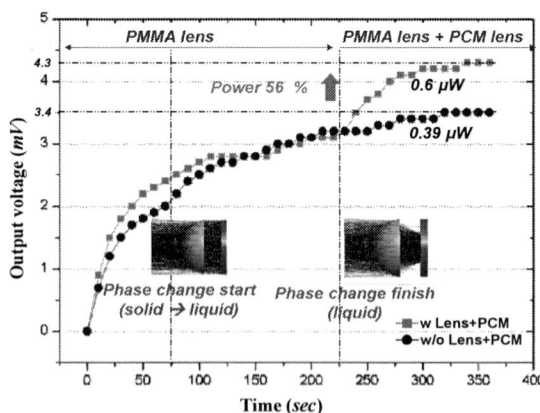

Figure 7: The output voltage of the TEG with and without lenses in the light (The light source: 50W halogen lamp)

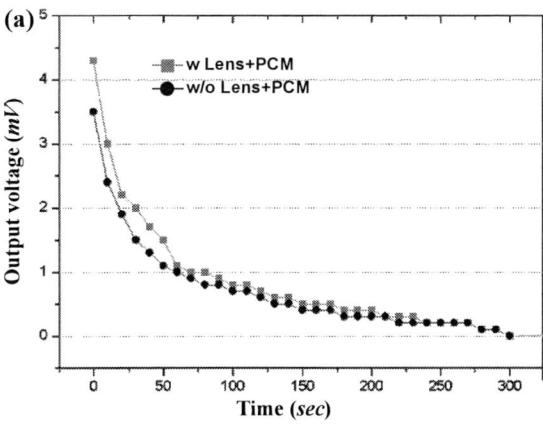

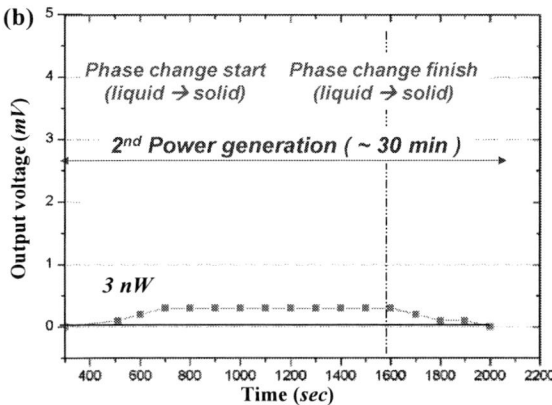

Figure 8: The output voltage of the TEG with and without lenses in the dark. (a) Heat dissipation section, (b) PCM phase-change section (liquid → solid): 2ⁿᵈ power generation

CONCLUSION

In this paper, we have proposed and successfully demonstrated a bidirectional thermoelectric energy generator based on phase-change lens. From the experimental results it was confirmed that the solar TEG shows high power generation by applying double type lenses. It was also shown that the solar TEG continued to produce electrical power when the light source was removed by using the stored heat in the PCM.

However, the solar TEG has an insufficient energy generation capacity to use to various applications. To further enhance their performance, optimization of the double lenses and adopting PCMs with large heat capacity may be necessary.

ACKNOWLEDGEMENTS

This research was supported by the Pioneer Research Center Program through the National Research Foundation of Korea funded by the Ministry of Science, ICT & Future Planning (2010-0019313).

REFERENCES

[1] D. Kraemer et al., "High-performance flat-panel solar thermoelectric generators with high thermal concentration", *Nature Mater.*, vol. 10, pp. 532-538, 2011.

[2] Goldsmid, Hiroshi Julian, *Applications of thermoelectricity*, London: Methuen, 1960.

[3] P. Li et al., "Design of a concentration solar thermoelectric generator", *J. Electron. Mater.*, vol. 39, pp. 1522-1530, 2010.

[4] G. J. Snyder and E, S, Toberer, "Complex thermoelectric materials", *Nature Mater.*, vol. 7, pp. 105-114, 2008.

[5] A. J. Minnich et al., "Bulk nanostructured thermoelectric materials: current research and future prospects", *Energy Environ, Sci.*, vol. 2, pp. 466-479, 2009.

[6] J. R. Sootsman et al., "New and old concepts in thermoelectric materials", *Angew. Chem., Int. Ed.*, vol. 48, pp. 8616-8639, 2009.

[7] Kenneth McEnaney et al., "Modeling of concentrating solar thermoelectric generators", *J. Appl. Phys.*, vol. 110, pp. 074502, 2011.

[8] Lauryn L. Baranowski et al., "Concentrated solar thermoelectric generators", *Energy Environ, Sci.*, vol. 5, pp. 9055-9067, 2012.

[9] Daniel Kraemer et al., "Modeling and optimization of solar thermoelectric generators for terrestrial applications", *Solar Energy*, vol. 86, pp. 1338-1350, 2012.

[10] S. M. Hasnain, "Review on sustainable thermal energy storage technologies, part 1: heat storage materials and techniques", *Energy Convers. Mgmt.*, vol. 39, pp. 1127-1138, 1998.

[11] Hui Li, Xu Liu, Guiyin Fang, "Preparation and characteristics of n-nonadecane/cement composites as thermal energy storage materials in buildings", *Energy and Buildings*, vol. 42, pp. 1661-1665, 2010.

CONTACT

*Y. J. Kim, tel: +82-2-2123-2844; yjk@yonsei.ac.kr

BIOTEMPLATED HIERARCHICAL NICKEL OXIDE SUPERCAPACITOR ELECTRODES

Sangwook Chu[1,2], Konstantinos Gerasopoulos[1], and Reza Ghodssi[1,2]
[1]MEMS Sensors and Actuators Laboratory (MSAL),
Institute for Systems Research, [2]Department Electrical and Computer Engineering,
University of Maryland, College Park, Maryland, USA

ABSTRACT

This paper presents fabrication and characterization of *Tobacco mosaic virus* (TMV)-templated hierarchical nickel oxide (NiO) supercapacitor electrodes. The hierarchical NiO electrode was created by integrating silicon (Si) micropillars with thermally oxidized nickel (Ni)-coated TMVs. SEM images taken after the high temperature oxidation process verified the robustness of the bio-nanotemplates, while the formation of Ni(core)/NiO(shell) structure around the rod-like virus was confirmed by scanning transmission electron microscopy (STEM). Electrochemical characterization showed that the capacitance for TMV structured Ni/NiO electrodes continuously increases for the first 500 cycles, with significantly higher, non-linear increase for the hierarchical versus TMV-only (nanostructured) electrodes. Further electrochemical and SEM analysis support the hypothesis that the capacity increase is due to the increase in electroactive sites of the nanostructured NiO throughout the initial charge/discharge cycles.

INTRODUCTION

Supercapacitors, including electric double-layer capacitors (EDLCs) and pseudocapacitors, have attracted significant attention in energy storage devices due to their critical role of bridging conventional dielectric capacitors and batteries in terms of power and energy density. Especially, the high power capability, long cycle life, and minimal safety concern expand their possible applications for electric vehicles, portable electronics, and industrial high power-required systems. Various transition metal oxides (RuO_2, NiO, MnO_2, Co_3O_4, etc.) have been proved as electrode materials for supercapacitors, and the major focus of the research has been on cost reduction, achieving higher energy densities (F/g or F/cm^2) with environmental friendly materials [1].

Previously, our group and collaborators have demonstrated the use of genetically modified TMVs as nanostructures for various types of energy storage devices [2-5]. The unique bio-nanotemplates have been uniformly coated with active materials using thin film deposition techniques resulting Ni(core)/active material(shell) electrode structures. These core/shell nanostructures were further integrated with gold micropillars to form hierarchical architectures with significant increase in energy density [4]. The current work focuses on the fabrication and characterization of NiO supercapacitor electrodes based on this hierarchical approach. Here, we leverage the Ni coating to create Ni/NiO active supercapacitor materials using a simple thermal oxidation process that does not require any additional deposition. The nanostructures are combined with three-dimensional Si micropillars to further enhance the areal capacitance.

MATERIALS AND CHEMISTRY

Ni-coated Tobacco mosaic virus

The *Tobacco mosaic virus* is a plant virus with high aspect ratio cylindrical coat-protein structure (300nm long, 18nm diameter). Genetic modifications that introduce cysteine groups in the coat proteins facilitate self-assembly and electroless metal deposition. The self-assembly process and the virus metallization have been well characterized in previous research [6].

Here, we functionalize the surface of the Ni-coated virus via a high temperature annealing process (Figure 1a). As prepared Ni on TMV is metallic with residual oxide (NiO) and hydroxide ($Ni(OH)_2$) on its surface [6]. Thermal oxidation at temperatures above 300°C result in removal of $Ni(OH)_2$ and increase in the NiO content [7-8].

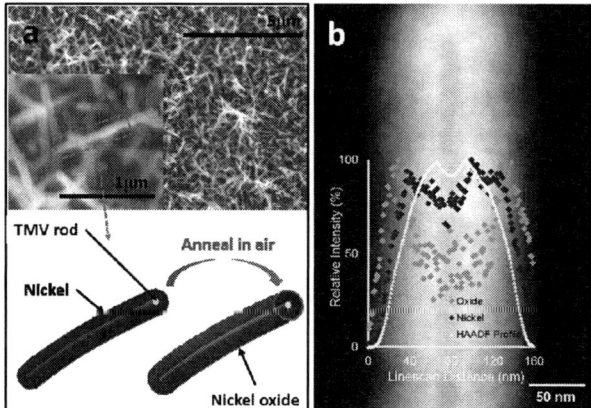

Figure 1: (a) (top) SEM images of Ni coated TMVs self-assembled on gold substrate, and (bottom) Schematic description of NiO layer formation on Ni-coated TMV via high temperature annealing in air. (c) STEM image and EELS analysis of the TMV-centered Ni(core)/NiO(shell) nanorod.

Nickel oxide supercapacitor electrode

NiO is one of the transition metal oxides being studied as a supercapacitor active material. It is a low cost and environmentally benign material with very high theoretical gravimetric capacitance (2328F/g) [1]. The generally accepted rechargeable energy storage mechanism of nickel oxide is shown in the equation (1).

$$NiO + OH^- \Leftrightarrow NiOOH + e^- \qquad (1)$$

In an alkaline electrolyte solution, NiO is charged to nickel oxyhydroxide (NiOOH) during oxidation, and discharged back to NiO during the reduction.

The formation of NiO on Ni-coated TMV was achieved by one step annealing process. The Ni-coated TMVs were annealed in a box furnace at 300°C for 2 hours, and the nanoscale active layer was formed without eliminating the conductive Ni layer. The STEM image in Figure 1b shows the oxidized nickel coated TMV nanorod, with electron energy loss spectroscopy (EELS) analysis confirming the formation of Ni(core)/NiO(shell) architecture around the TMV (TMV/Ni/NiO).

ELECTRODE FABRICATION

The hierarchical electrode was fabricated by integrating Si micropillars with the self-assembled viruses. The fabrication process is described in Figure 2a. The micropillars were fabricated on a p-type Si wafer using photolithography and deep reactive ion etching (DRIE). AZ5214 negative photoresist was used as the DRIE mask and the pillar arrays (14 μm diameter and spacing) were etched down to 70 μm (2.5 μm/min) resulting in micro structures with 5:1 aspect ratio. The microstructure enhances the surface area by factor of 4.4 compared to a flat surface based on its dimensions as described at the bottom of Figure 2a. After stripping the remaining photoresist, the surface of the Si pillar arrays was passivated with a PECVD silicon nitride film for substrate isolation, and chrome (30nm)/gold (200nm) were deposited using sputtering to function both as current collectors and substrate for TMV assembly.

After the micro pillar fabrication, the wafer was diced into die (16 x 8.5 mm²). The virus self-assembly and metallization process were performed as described in [6]. Briefly, the electrodes were immersed in a 0.2mg/ml virus solution and incubated overnight at room temperature. The surface of the self-assembled virus was activated with

a palladium catalyst for 5 hours followed by electroless Ni deposition for 5 minutes. Finally, the fabricated hierarchical TMV/Ni electrode was annealed in an aired box furnace at 300°C for two hours with 5°C/min ramp. The hierarchical electrodes after the annealing are shown in the SEM images of Figure 2b.

RESULTS AND ANALYSIS
Electrochemical experiment setup

Electrochemical experiments were performed using two molar (2M) potassium hydroxide (KOH) solution in a three-electrode configuration (reference electrode: silver/silver chloride (Ag/AgCl), counter electrode: platinum (Pt) foil). NiO electrodes were prepared in three different geometries (planar, nanostructured, and hierarchical) to evaluate the respective performance. The electrodes were examined with galvanostatic charge/discharge cycling at 2mA/cm² between 0 and 0.5V.

Galvanostatic charge/discharge tests

The areal discharge capacitance (mF/cm²) of the three different electrodes was compared over the first 500 charge/discharge cycles. As shown in Figure 3, a significant non-linear increase of areal capacitance was observed over the 500 cycles for the nanostructured and hierarchical Ni/NiO electrodes (26% and 193 %, respectively). Comparing the nanostructured and planar Ni/NiO electrodes, an approximately 9-fold increase in charge capacitance was observed which is in good agreement with expected increase in surface area and reported results [2,4]. For the hierarchical electrode, the initial capacitance was lower than expected. However, the capacitance of the electrode reached 585.9mF/cm² after the non-linear increase by the end of the 500 charge/discharge cycle test. This exceeded that of the nanostructured (174.8mF/cm²) and planar-Ni/NiO (19.7mF/cm²) by 3.4 and 29.7 times, respectively.

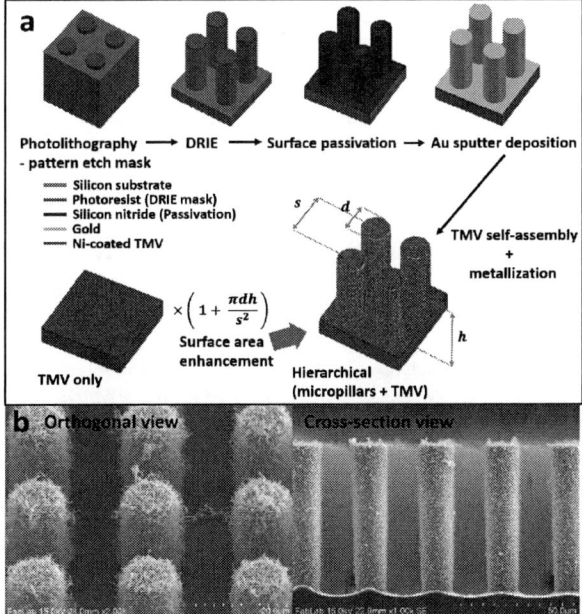

Figure 2: (a) Hierarchical electrode fabrication process with (bottom) schematic explanation of surface area enhancement factor from micropillar geometries. (b) SEM images showing (left) orthogonal and (right) cross-sectional view of the hierarchical NiO electrode.

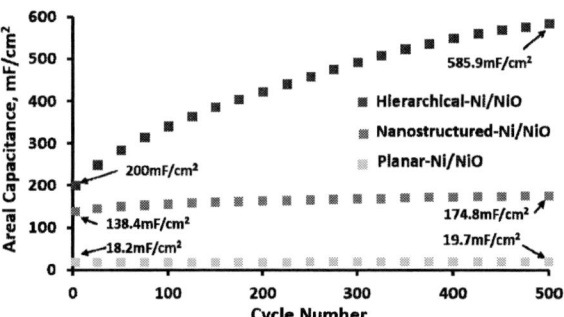

Figure 3: Areal discharge capacitance of the three different electrodes (hierarchical, nanostructured, and planar Ni/NiO) over initial 500 charge/discharge cycles at 2mA/cm².

Charge capacity increase analysis

In pseudocapacitors, there are two major charge storage mechanisms (double layer charge adsorptions and faradaic bulk reactions), and the mechanisms can be understood from the discharge plots shown in Figure 4a [9]. The plots for the TMV structured electrodes were analyzed to elucidate the capacitance increase phenomena.

978-1-4799-7956-1/15 $31.00 © 2015 IEEE   1119

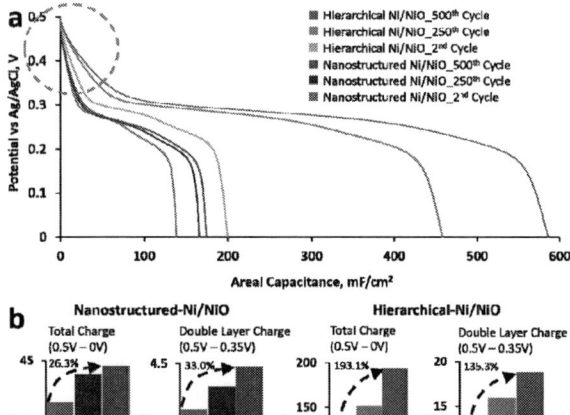

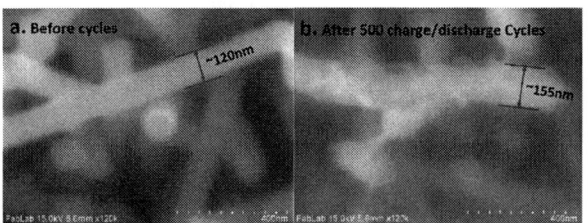

Figure 4: (a) Discharge plot of hierarchical and nanostructured-Ni/NiO electrodes at 2nd, 250th, and 500th cycle showing non-linear capacitance increase. (b) Comparison between double layer and total charge capacity of the two electrodes showing similar-trend-wise increase over the cycles.

First, the amount of the double layer charge was estimated from the initial linear discharge region indicated by the red dotted circle in Figure 4a (0.5V – 0.35V). The amount of total charge for the entire discharge potential window (0.5V – 0V) was calculated (the current was integrated over the discharge duration throughout the potential regions). Figure 4b compares both calculated values at three different cycles (2nd, 250th, and 500th) for nanostructured and hierarchical-Ni/NiO electrodes. The plot shows a similar increasing trend between the double layer charge and total charge for both electrodes. Since the amount of double layer charge adsorption depends on the surface area, the result indicates that there is an increase in faradaic reaction active area/sites during the charge/discharge cycle tests. By comparing the increase in charge between the cycles, it is expected that the activation process will saturate during subsequent cycles.

Figure 5 shows SEM images of the TMV/Ni/NiO nanorods taken before (Figure 5a) and after (Figure 5b) the charge/discharge experiment. The TMV/Ni/NiO nanorod after 500 charge/discharge cycles showed a change in

surface morphology supporting the discharge plot analysis results in Figure 4. The image shown in Figure 5b implies that the increase in mesoscale porosity throughout the electrochemical cycles may play an important role in charge capacity increase.

Charge capacity saturation

Hierarchical electrodes with Si micropillars possessing a 3.3:1 aspect ratio geometry were fabricated to understand the kinetics of the capacitance increase and saturation. Two electrode samples were tested at different current densities ($2mA/cm^2$ and $0.2mA/cm^2$), and Figure 6a shows the areal capacitance increase/saturation in the initial 500 charge/discharge cycles. The electrodes tested at lower current density ($0.2mA/cm^2$) reached saturation after ~200 cycles, while the electrode tested at higher current density ($2mA/cm^2$) required ~ 1500 cycles to reach saturation. Figure 6b shows the areal capacitances for subsequent cycles, showing a stable capacitance following saturation.

The result indicates that a lower current density is more efficient in activating the electrode. The areal capacitance of the electrode can be tested at $2mA/cm^2$ can be further compared with the values from Figure 3. The saturated areal capacitance resulted ~$470mF/cm^2$, which scales up according to the expected level (~$540mF/cm^2$) for micropillars with a 3.3:1 aspect ratio. Higher saturated areal capacitance was measured (~$700mF/cm^2$) with the $0.2mA/cm^2$ current density, which is attributed to the rate capability dependence of the capacitance for energy storage active materials.

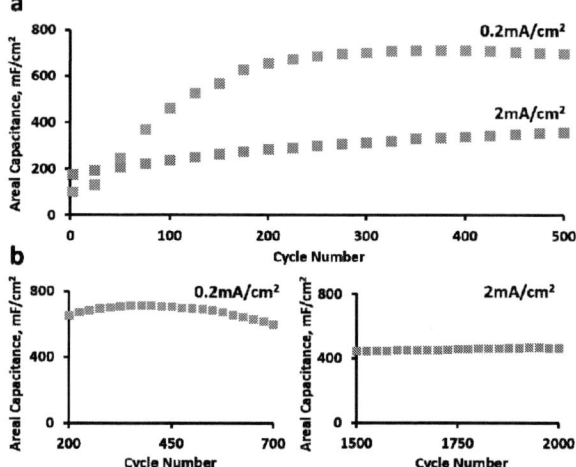

Figure 6: Areal capacitance measurement for hierarchical electrodes (3.3:1 Si micropillar aspect ratio) over many cycles for saturation of the areal capacitance increase. (a) Areal capacitance of the two electrodes tested at two different current densities (0.2 and 2 mA/cm^2) for initial 500 cycles. (b) Areal capacitances at two different current densities, (left) $0.2mA/cm^2$, and (right) $2mA/cm^2$, after reaching the saturation.

Figure 5: SEM images comparing morphology of TMV/Ni/NiO nanorods before and after galvanostatic charge/discharge cycles. (a) The nanorods after immersion in electrolyte (without cycling). (b) Morphology change of the nanorods after 500 charge/discharge cycles.

DISCUSSION

The increase in charge capacity for the nanostructured Ni/NiO electrode requires further study to understand the mechanism. It is hypothesized that an increase in mesoscale porosity results in an increase in reactive surface area. This is explained from the results of this work, particularly through the increase in double layer charge and the surface morphology change during the electrochemical cycling. Another possible reason for the increase can be progressive conversion of the Ni conductive layer to NiO throughout the cycling which was suggested in our previous work [2]. Further analysis using electrochemical impedance spectroscopy and scanning transmission electron microscopy can be used to study changes in charge transfer kinetics for the TMV/Ni/NiO during cycling.

CONCLUSION

This work reported hierarchical-Ni/NiO supercapacitor electrodes based on TMV nanostructured templates. The energy storage active layer was formed on the Ni-coated TMVs through the thermal oxidation process without using additional active layer deposition steps. The successive formation of the Ni(core)/NiO(shell) nanorod was confirmed by STEM and EELS analysis. The active nanostructures were integrated with Si micropillars (5:1 aspect ratio) resulting in hierarchical electrode architectures with enhanced energy storage performances. The capacitance for TMV structured electrodes continuously increased for the first 500 cycles, with a significantly higher, non-linear increase for hierarchical versus nanostructured electrodes. After 500 cycles, an ultra-high areal capacitance of 585.9mF/cm^2 was achieved with hierarchical-Ni/NiO electrodes, exceeding the capacitance of nanostructured only and planar Ni/NiO by a factor of 3.4 and 29.7, respectively.

Current density dependence of the capacitance increase and saturation was studied with hierarchical electrodes. The results indicated that the cycle activation of nanostructured NiO is more efficient at lower current density (0.2mA/cm^2) reaching the saturation region at lower cycle numbers (~200) with higher areal capacitance (~700mF/cm^2) compared to 2mA/cm^2 current density.

ACKNOWLEDGEMENT

This research was funded by Nanostructures for Electrical Energy Storage, an Energy Frontiers Reseach Center funded by the U.S. Department of Energy, Office of Science, Office of Basic Energy Sciences. The authors would like to thank Professor James N. Culver for providing access to their lab for TMV purification, and the staff of the FabLab and NISPLab at the Maryland Nanocenter for providing access to their clean-room and spectroscopy/imaging facilities.

REFERENCES

[1] G. Wang et al., "A review of electrode materials for electrochemical supercapacitors," *Chem. Soc. Rev.,* vol. 41, pp. 797-828, 2012.

[2] K. Gerasopoulos et al., "Nanostructured nickel electrodes using the *Tobacco mosaic virus* for microbattery application," *J. Micromech. Microeng.,* vol. 18, pp. 104003-104010, 2008.

[3] M. Gnerlich et al., "Solid flexible electrochemical supercapacitor using *Tobacco mosaic* virus nanostructures and ALD ruthenium oxide," *J. Micromech. Microeng.,* vol. 23, pp. 114014-114020, 2013.

[4] K. Gerasopoulos et al., "Hierarchical three-dimensional microbattery electrodes combining bottom-up self-assembly and top-down micromachining," *ACS Nano,* vol. 6, pp. 6422-6432, 2012.

[5] X. Chen et al., "Virus-enabled silicon anode for lithium-ion batteries," *ACS Nano,* vol. 4, pp. 5366-5372, 2010.

[6] E. Royston et al., "Self-assembly of virus-structured high surface area nanomaterials and their application as battery electrodes," *Langmuir,* vol. 24, pp. 906-912, 2008.

[7] S. V. Kumari et al., "Surface oxidation of nickel thin films," *J. Mater. Sci. Lett.,* vol. 11, pp. 761-762, 1992.

[8] F. Zhang et al., "Nanocrystalline NiO as an electrode material for electrochemical capacitor," *J. Mater. Chem. Phys.,* vol. 83, pp. 260-264, 2004.

[9] Z. Lu et al., "Stable ultrahigh specific capacitance of NiO nanorod arrays," *Nano Res.,* vol. 4, pp. 658-665, 2011.

CONTACT

*R. Ghodssi, tel: +1-301-4051897; ghodssi@umd.edu

DOUBLY RE-ENTRANT CAVITIES
TO SUSTAIN BOILING NUCLEATION IN FC-72

Tingyi "Leo" Liu[*] and Chang-Jin "CJ" Kim
Department of Mechanical and Aerospace Engineering
University of California, Los Angeles (UCLA), USA

ABSTRACT

We report a micro- and nano-machined surface cavity on which boiling nucleation resumes after ceasing in perfluorohexane (i.e., FC-72, a Fluorinert™ from 3M™) for a short time. To the best of our knowledge, this is the first such success with FC-72, which has the lowest surface tension of all available liquids and completely wets any existing material including Teflon® so that all surface cavities get flooded once nucleation stops. Despite the wide usage in electronics cooling, FC-72 was never considered for boiling heat transfer because once cooled boiling would not restart without an excessive heating. Our result experimentally supports the half-century-old idea of doubly re-entrant cavities envisioned to facilitate more stable nucleation, encouraging further development.

INTRODUCTION

Heat transfer with phase changes is the most efficient mode of heat transfer and anticipated to satisfy some of the most challenging needs of cooling in modern engineering, such as the power dissipation of integrated circuits [1]. Unlike evaporation, which is also a phase change process, boiling involves creation of new liquid-vapor interfaces at discrete sites on a heated surface [2]. Among the entire boiling regimes, nucleate boiling, where vapor bubbles are generated and released from a heated surface, is the most efficient [2]. Many researchers have been developing surfaces that expand the range of nucleate boiling to promote and sustain the nucleation [3–5]. Recent enhancement involved combination of surface micro and nano structuring and a hydrophobic coating to promote the boiling nucleation of water [6].

While found effective for water and some solvents, the abovementioned enhancement, by using surface roughness in combination with surface hydrophobicity, has not been valid for dielectric fluorocarbon liquids (e.g., FC liquids from 3M™), which are the typical coolants for electronics. With distinctively low surface tensions, FCs wet all known materials whether hydrophilic or hydrophobic, flooding all the microcavities on a surface as soon as boiling stops and preventing vapor trapping needed for future nucleation.

Combining the long-proposed idea that microscale cavities with a re-entrant shape would be more resistant against flooding [3–5] and the recent fabrication process to micromachine doubly re-entrant post structures that can suspend all liquids including FC-72 [7], here we develop doubly re-entrant *cavities* and test them for FC boiling for the first time. We compare the doubly re-entrant cavity (Fig. 1C) with other artificial cavities (Fig. 1A&B) to directly verify its effectiveness in sustaining nucleation. In order to focus on the cavity stability and not affected by heat transfer, the tests are performed in a uniformly heated pool of FC-72 rather than heating the surfaces.

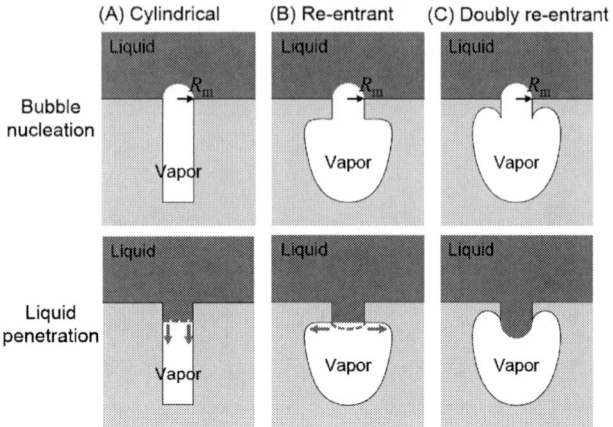

Figure 1: Three types of cavities in the literature to start boiling nucleation. If the cavity mouth radius R_m is equal, all three are to nucleate at the same superheat. However, completely wetting liquids like FC-72 may penetrate into the cavity (meniscus indicated by the red dashed line) during a bubble departure even if the meniscus is stable under static conditions. Our goal is to study if the doubly re-entrant cavity provides more stable nucleation.

THEORY AND EXPERIMENT DESIGN
Superheat Required for Boiling Nucleation

For a boiling bubble embryo with radius R_b, the incipient boiling temperature – i.e., the temperature at which nucleation starts – is derived as follows. We assume that (i) vapor inside the bubble behaves as an ideal gas and (ii) the temperature is continuous across the meniscus, i.e., the liquid temperature right outside the bubble equals the vapor temperature right inside the bubble. When boiling occurs, the liquid right outside the bubble should at least be at the saturation pressure p_{sat} so that it can be turned into vapor that feeds the bubble to grow. The pressure of the vapor bubble over the liquid is given by the Laplace pressure [8],

$$\Delta p = p_v - p_l = \frac{2\gamma}{R_b} \qquad (1)$$

where p_v is the vapor pressure inside the bubble (assumed uniform), p_l is the liquid pressure right outside the bubble ($p_l \geq p_{sat}$), and γ is the surface tension of the liquid-vapor interface at saturation point. According to the ideal gas law, since the pressure of the vapor is higher than the saturation pressure, the temperature of the vapor should be higher than the saturation temperature. Since we assume there is no temperature jump across the liquid-vapor interface, the temperature of the liquid right outside the bubble equals the temperature of the vapor inside the bubble. Thus, the liquid right outside the bubble is at a temperature higher than its saturation temperature, i.e., the liquid is superheated, as shown in Fig. 2.

978-1-4799-7956-1/15 $31.00 © 2015 IEEE   1122   MEMS 2015, Estoril, PORTUGAL, 18 - 22 January, 2015

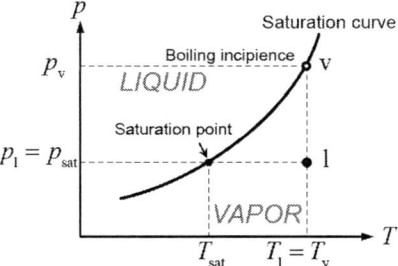

Figure 2: Pressure-temperature (p-T) relation of a boiling bubble. Since the liquid outside the bubble has the same temperature but a lower pressure compared with the vapor inside (l vs. v in the graph), the liquid should be superheated by at least $T - T_{sat}$ for boiling nucleation to occur.

By assuming an isothermal process and the liquid density much greater than the vapor density so that $1/\rho_v - 1/\rho_l \approx 1/\rho_v$, the Clausius-Clapeyron equation [4] gives

$$\Delta T = T_l - T_{sat} = \frac{2\gamma}{R_b} \frac{T_{sat}}{\rho_v h_{lv}} \qquad (2)$$

where T_l is the temperature of the liquid, T_{sat} is the saturation temperature, ρ_v is the density of the vapor, h_{lv} is the latent heat of vaporization. For a circular cavity with mouth radius R_m (Fig. 1), the maximum superheat is given by the smallest radius of the bubble, which equals the cavity mouth radius R_m. Therefore, the superheat required for nucleation on these cavities is

$$\Delta T = \frac{2\gamma}{R_m} \frac{T_{sat}}{\rho_v h_{lv}} \qquad (3)$$

As shown in the first row of Fig. 1, the superheat required for nucleation is the same for the cylindrical, re-entrant, and doubly re-entrant cavities because they have the same mouth radius R_m. However, this static analysis does not guarantee the vapor embryo can be sustained during boiling since a film of liquid may be pushed into the cavity after a bubble leaves [2]. Therefore, the level of resistance that a cavity can provide to prevent liquid penetration determines the stability of nucleate boiling. As shown in the second row of Fig. 1, cylindrical (or conical) cavities provide no resistance to prevent liquid flooding, and re-entrant cavities provide little resistance ($\Delta p = 0$). In contrast, doubly re-entrant cavities provide the best resistance, as they can prevent liquid penetration even when $\Delta p < 0$, i.e., by allowing the meniscus to form convex into the cavity.

The Test Chip with Artificial Cavities

To help observing the active nucleation, we designed each test chip to have a collection of cavities (~50 nucleation sites) only at the center of the chip. To investigate the effect on different cavity sizes, i.e., different mouth radii, we included $\phi 5$ μm and $\phi 10$ μm openings (i.e., R_m ~2.5 μm and ~5 μm, respectively) for each of the three cavity types shown in Fig. 1. Following the fabrication similar (but different) to the one reported in our recent work [7], Fig. 3 shows the fabricated microscale cavities ($\phi 10$ μm opening) with nanoscale details of a nearly perfect doubly re-entrant shape [5].

Figure 3: SEM images of doubly re-entrant cavities fabricated ($\phi 10$ μm or R_m = 5.17 μm): (A) angled view, (B) cross-sectional view, (C) zoomed-in view at doubly re-entrant section.

Boiling Experiment Setup

Figure 4 shows the apparatus for boiling experiment. By controlling the temperature of the FC-72 pool with an outer water bath, the FC-72 and the chip remain at the same temperature at all times. This simple immersion experiment allows a direction comparison with Eq. 3 free of any temperature gradient effect because the boiling nucleation is solely controlled by liquid superheat. For each test, the FC-72 pool was superheated (~5 K above boiling point) and verified no nucleation at random spots (e.g., caused by dirt) before placing the chip in the FC-72 pool so that nucleation would only start on the designated microcavities once the test chip was loaded. In addition, the chip was preheated on a hot plate before the loading to avoid capillary condensation inside the cavities. When the FC-72 was being cooled in a slow and controlled manner (0.08 K/min) to ensure uniform temperature in the pool, the nucleation was monitored by a CCD camera (Dino-Lite Pro AM411) with the superheat measured in sync (OMEGA thermistor 44031, ±0.1°C interchangeability). Once nucleation ceased, we stopped the cooling and maintained the temperature. After several minutes, we restarted heating to see if nucleation could be restarted. For each chip, three experiments were performed to ensure repeatability.

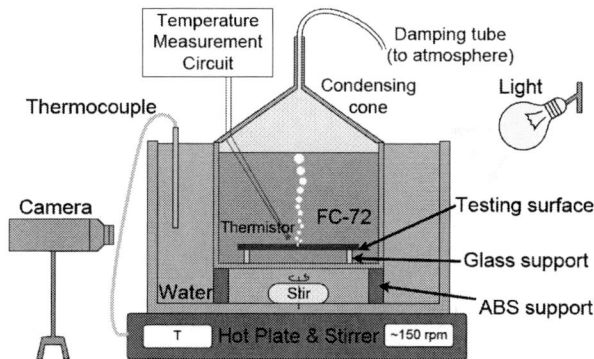

Figure 4: Experimental setup of FC-72 pool boiling test. A water bath is feedback-heated and served as a uniform heat source for FC-72 pool. FC-72 is degassed before experiment. A thermistor (±0.1°C accuracy) is placed near the test chip to measures the liquid/chip temperature and connected to outside via a feedthrough. A condensing cone with a damping tube is used to recollect FC-72 condensate while providing atmospheric pressure. A custom program is used to sync the temperature measurement with the CCD camera.

RESULTS AND DISCUSSION

We found that nucleation on the doubly re-entrant cavities persisted down to a very small superheat (e.g., 1.88 K for $\phi 5\ \mu m$ opening), perfectly matching Eq. 3, as shown in Fig. 5. In comparison, nucleation on cylindrical and re-entrant cavities ceased early at a larger superheat, suggesting that FC-72 invaded the cavities during a bubble departure [2], as explained in Fig. 1. When heated after leaving the chips for ~5 minutes without nucleation, the doubly re-entrant cavities resumed nucleation at the temperature they ceased to nucleate during cooling, as shown in Fig. 6. In stark comparison, the cylindrical and re-entrant cavities could not reactivate nucleation even when random spots inside the FC-72 pool started to nucleate given a much larger superheat (~7 K). Separate study showed that the cylindrical and re-entrant cavities required a superheat similar to that required for a smooth surface (~30 K [9]), confirming they were flooded after the nucleation ceased.

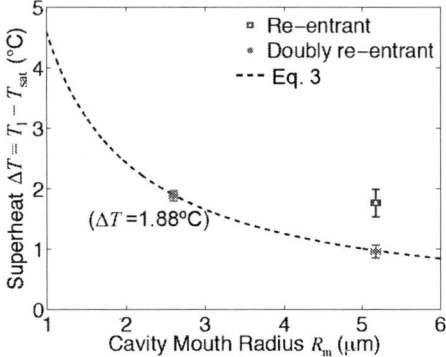

Figure 5: Comparison of liquid superheats required to maintain nucleation on different cavities. On the doubly re-entrant cavities the experimental data matched Eq. 3 perfectly for both R_m =2.60 μm and 5.17 μm. On the re-entrant cavities, a slightly higher superheat was needed than Eq. 3 for R_m = 5.17 μm. On the re-entrant cavities of R_m = 2.60 μm and the cylindrical cavities of both R_m = 2.60 μm and 5.17 μm, nucleation ceased as soon as the heating stopped and therefore no data were obtained.

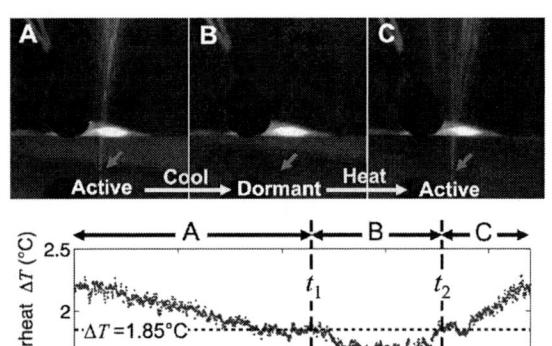

Figure 6: Nucleation ceasing and resuming on doubly re-entrant cavities. Snapshots show the nucleation stopped upon cooling and restarted upon heating on $\phi 5\ \mu m$ doubly re-entrant cavities (red arrow). Note that at both t_1 (stop) and t_2 (restart), the liquid superheat is ~1.85 K, matching Eq.3 and the measurement shown in Fig. 5.

CONCLUSIONS

We have presented the first investigation of nucleate boiling of FC-72 on doubly re-entrant cavities. Rather than studying heat transfer, we have focused on comparing the stabilities of boiling nucleation on different types of cavities in a pool of FC-72 under uniform temperature. Compared with cylindrical and re-entrant cavities, doubly re-entrant cavities have been found to provide a more stable nucleation and match the theoretical prediction perfectly. The doubly re-entrant cavities were able to prevent cavity flooding for a short period (~5 min) without an active nucleation and restart nucleation upon reheating. In contrast, other cavities were flooded as soon as the nucleation ceased and could not restart nucleation unless a very large superheat was applied. Our success of restarting FC-72 nucleation has confirmed the old hypothesis that doubly re-entrant cavities would enhance nucleate boiling of completely wetting liquids. Although the current result is limited to scientific value, it encourages further engineering investigation to find if the success can be extended to significant subcooling.

ACKNOWLEDGEMENTS

We thank Prof. Daniel Attinger and Dr. Christophe Frankiewicz for their helpful discussion on the FC-72 heat transfer experiments.

REFERENCES

[1] E. Pop, "Energy dissipation and transport in nanoscale devices," *Nano Res.*, vol. 3, pp. 147–169, 2010.

[2] V. K. Dhir, "Boiling heat transfer," *Annu. Rev. Fluid Mech.*, vol. 30, pp. 365–401, 1998.

[3] R. L. Webb, "The evolution of enhanced surface geometries for nucleate boiling," *Heat Transf. Eng.*, vol. 2, pp. 46–69, 1981.

[4] V. P. Carey, *Liquid-vapor phase-change phenomena: an introduction to the thermophysics of vaporization and condensation processes in heat transfer equipment*. Washington, DC: Hemisphere, 1992.

[5] C.-J. Kim, "Structured Surfaces for Enhanced Nucleate Boiling," M.S. Thesis, Iowa State University, 1985.

[6] A. R. Betz, J. Jenkins, C.-J. Kim, and D. Attinger, "Boiling heat transfer on superhydrophilic, superhydrophobic, and superbiphilic surfaces," *Int. J. Heat Mass Transf.*, vol. 57, pp. 733–741, 2013.

[7] T. Liu and C.-J. Kim, "Turning a surface superrepellent even to completely wetting liquids," *Science*, vol. 346, 2014 (In Press).

[8] A. W. Adamson and A. P. Gast, *Physical Chemistry of Surfaces*, 6th ed. New York: Wiley, 1997.

[9] S. M. You, A. Bar-Cohen, and T. W. Simon, "Boiling incipience and nucleate boiling heat transfer of highly wetting dielectric fluids from electronic materials," *IEEE Trans. Compon. Hybrids Manuf. Technol.* vol. 13, pp. 1032–1039, 1990.

CONTACT

*T. Liu, tel: +1-310-825-3977; leolty@ucla.edu

EXPERIMENTALLY VERIFIED MODEL OF ELECTROSTATIC ENERGY HARVESTER WITH INTERNAL IMPACTS

Binh Duc Truong, Cuong Phu Le, and Einar Halvorsen

Department of Micro- and Nano Systems Technology, Buskerud and Vestfold University College, Campus Vestfold, Raveien 215, 3184 Borre, NORWAY

ABSTRACT

This paper presents experimentally verified progress on modeling of MEMS electrostatic energy harvesters with internal impacts on transducing end-stops. The two-mechanical-degrees-of-freedom device dynamics are described by a set of ordinary differential equations which can be represented by an equivalent circuit and solved numerically in the time domain using a circuit simulator. The model accounts for the electromechanical nonlinearities, nonlinear damping upon impact at strong accelerations and the nonlinear squeezed-film damping force of the in-plane gap-closing transducer functioning as end-stop. The comparison between simulation and experimental results shows that these effects are crucial and gives good agreement for phenomenological damping parameters. This is a significant step towards accurate modeling of this complex system and is an important prerequisite to improve performance under displacement-limited operation.

INTRODUCTION

Vibration energy harvesting to power wireless sensor systems and microelectronic applications can eliminate use of batteries which have negative impacts on device size, achievable operational lifetimes and the environment. By scavenging surrounding vibration energy and converting into electrical energy, the system is enabled to operate autonomously [1-2]. Commonly, a harvester's architecture is a spring-mass systems with a transducer based on either three mechanisms: piezoelectric, electrostatic and electromagnetic conversion. Proof mass motion under ambient vibrations induces electromechanical transduction that generates power. For such resonant device structures with high mechanical quality factor Q, use of rigid end-stops to confine maximum amplitude of the proof mass displacement under strong vibration is applicable. Many works have exploited impact on the end-stops to enhance system bandwidth. However, an undesirable consequence remains power saturation when the displacement limit is reached [3-6].

In our previous works, we have experimentally demonstrated a device concept that utilizes sub-transducers functioning as soft end-stops to collect extra power adding to that of the main transducers [7-8]. The technique can be considered as an alternative to the load optimization used for the canonical generators [9]. The measurement result shows that achieved power continues to increase even when large vibration drives the proof mass displacement to the limit. Hence, this approach

overcomes a major limitation of conventional harvester designs that experience saturated power with use of rigid end-stops. The prototype performance exhibits strong nonlinearities because of the collisions between the main proof mass and the end-stop transducer in the impact regime. These severe and complex nonlinearities make the impact device hard to model. A simple lumped-model was studied to predict the nonlinearities for this type of device in [10]. Although the model is able to capture the main aspects of device behavior, there are still significant differences between the measured and simulated results, which can be attributed to inadequately modelled damping.

In this work, we investigate effects of the nonlinear damping caused by the impact force and the squeeze-film air damping in the gap-closing capacitance structures of the end-stop transducer.

ANALYSIS AND MODELING

Impact-based device concept

The device concept with transducing end-stop is shown in Figure 1a. The motion of the main proof mass is limited by maximum amplitude of X_{max}, while its relative distance to the transducing end-stop at equilibrium position is x_1. The maximum possible displacement amplitude of the end-stop transducer is $x_2 = X_{max} - x_1$. When the main proof mass reaches x_1, the end-stop transducer is activated and generates additional electrical power beyond that generated by the main transducers.

Figure 1b shows key features of the impact device design using electrostatic conversion for all transducers. The main transducer consists of two anti-phase transducers with overlap-varying capacitances. The end-stop transducer is desired to effectively harvest energy from the impact. Thus, the end-stop transducer uses gap-closing capacitance to have high coupling. Two moving structures are suspended by linear folded-springs and have the same direction of in-plane movement. Figure 1c displays a close-up view of the fabricated device using SOIMUMPs process with a layer thickness of $t=25$ μm and an active area of 4x5 mm^2. All design parameters of the prototype can be found in [7].

Lumped-model

All transducers operate in the continuous mode, separately biased by a voltage source V_p for the main transducers and a voltage source V_s for the end-stop transducer. All fixed electrodes are connected to the external loads.

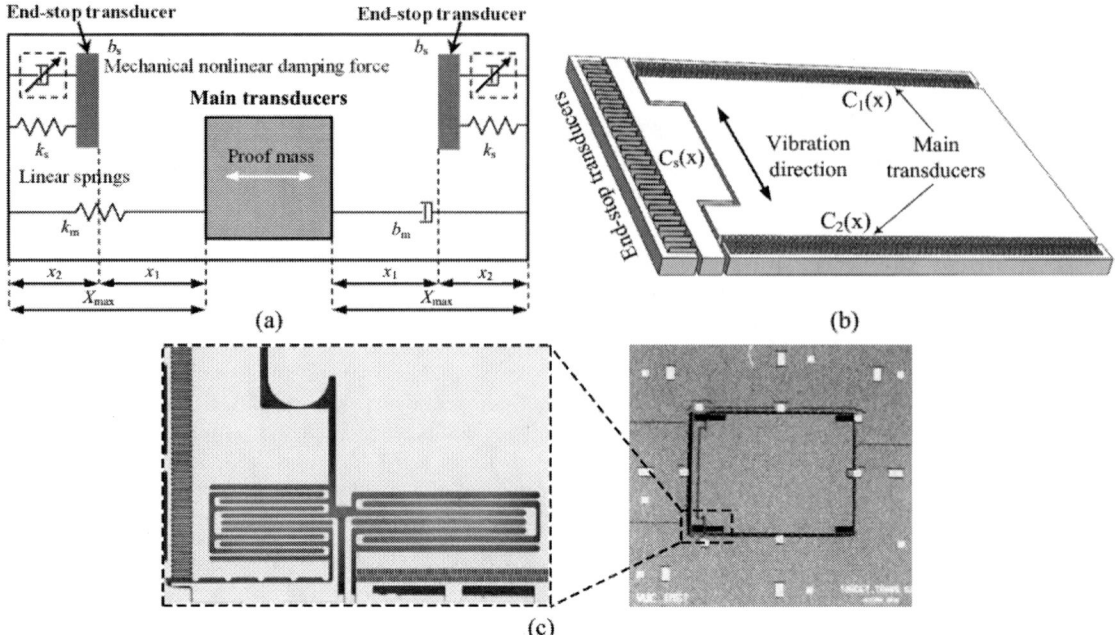

Figure 1: (a) A drawing illustration of device concept with use of end-stop transducer as an additional harvester, (b) key features of the electrostatic device and (c) a close-up view of the MEMS prototype fabricated in the SOIMUMPs process

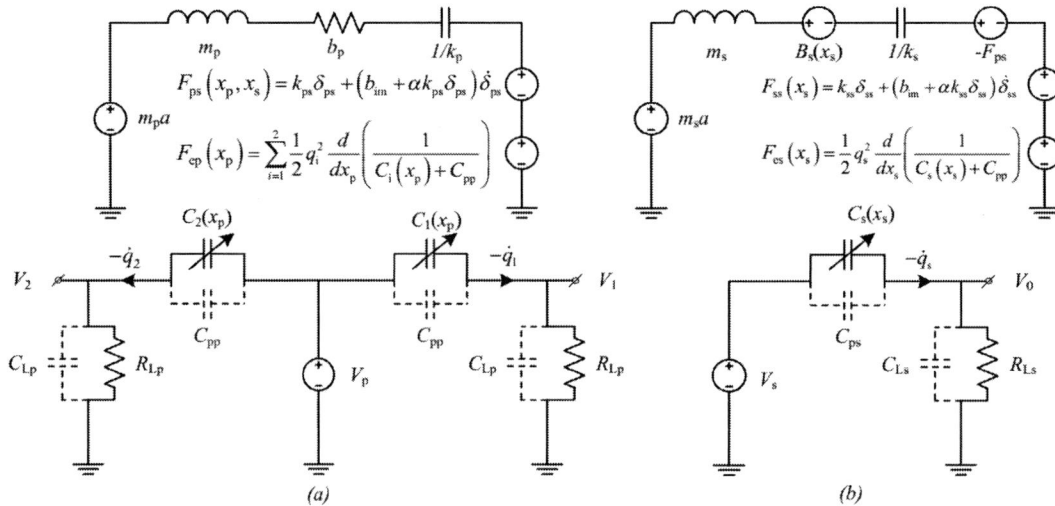

Figure 2: Equivalent-circuit model of the impact device: (a) main transducers and (b) end-stop transducer

The lumped-model of the impact device has the equivalent circuits shown in Figure 2. The model has two degrees of freedom: displacement x_p of the main proof mass and displacement x_s of the end-stop mass. The mechanical and electrical domains are captured by

$$m_p\ddot{x}_p + b_p\dot{x}_p + k_p x_p + F_{ps}(x_p, x_s) + F_{ep}(x_p) = m_p a \quad (1)$$

$$V_p = -\frac{q_{1/2}}{C_{1/2}(x_p) + C_{pp}} + V_{1/2} \quad (2)$$

for the main transducers and by

$$m_s\ddot{x}_s + B_s(x_s) + k_s x_s - F_{ps}(x_p, x_s) + F_{ss}(x_s) + F_{es}(x_s) = m_s a \quad (3)$$

$$V_s = -\frac{q_s}{C_{1/2}(x_p) + C_{ps}} + V_0 \quad (4)$$

for the end-stop transducer. $q_{1/2}$ and q_s are the charges on the main transducers and the end-stop transducer

respectively. The transducer electrostatic forces are $F_{ep}(x_p)$ and $F_{es}(x_s)$ while the mutual force that arises from the internal impacts is denoted $F_{ps}(x_p, x_s)$. $F_{ss}(x_s)$ is the impact force between the movable and the rigid end-stops at the ultimate displacement. δ_{ps} and δ_{ss} are the deformation displacements between the main proof mass and the end-stop proof mass, and between the end-stop proof mass and the rigid end-stops during impact respectively. Stray capacitances are also included in the model. The mechanical damping force in the end-stop transducer is $B_s(x_s)$ while that of the main transducers is dominated by slide film damping $b_p\dot{x}_p$.

All nonlinear forces are implemented as behavioral sources in the circuit simulator. There are three types of sources of nonlinearities considered in the lumped-model: i) the electrostatic forces $F_{ep}(x_p)$ and $F_{es}(x_s)$, ii) the damping force $B_s(x_s)$ due to squeeze-film damping in the gap-closing end-stop transducer and iii) the impact forces $F_{ps}(x_p, x_s)$, $F_{ss}(x_s)$.

Squeeze-film damping force

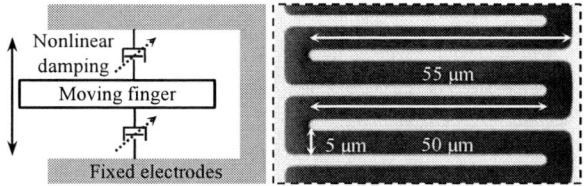

Figure 3: Squeeze film air damping between moving and fixed electrodes in end-stop transducer

Figure 3 shows the electrode structure of the end-stop transducer. In the layout, the nominal gap between fingers is g_0=5.0 µm and the minimum gap is 1.0 µm. When the impact force drives the gap size sufficiently small, the squeeze-film damping force is a major factor opposing motion of the end-stop transducer. In modelling, we consider the thin air film incompressible. The damping force follows the linearized Reynolds equation in the operational frequency range. Including the squeeze-film damping of the gap-closing capacitor structure [11] and an additional phenomenological linear damping, the total damping force of the end-stop transducer is

$$B_s(x_s) = b_g(x_s)\dot{x}_s \qquad (5)$$

where

$$b_g(x_s) = b_l + \frac{b_n}{\left(1 - \dfrac{x_s}{g_0}\right)^3} + \frac{b_n}{\left(1 + \dfrac{x_s}{g_0}\right)^3} \qquad (6)$$

$$b_n = 2 N_g \mu \frac{L t^3}{g_0^3} \qquad (7)$$

N_g is the number of fingers, L is the nominal finger overlap and b_l is the linear damping coefficient.

Impact forces

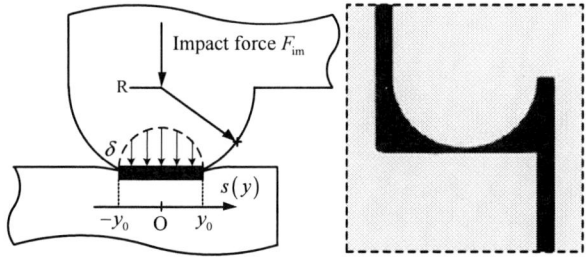

Figure 4: Impact region between the main proof mass and the end-stop proof mass

Figure 4 shows the impact region between the two masses. The impact shape is a line contact between semi-cylindrical bumps and flat surfaces. Thus, the impact force can be found based on Hertz's contact problem and nonlinear viscoelastic model that simultaneously accounts for the coefficient of restitution and the initial impact velocity. However, the latter component is very hard to define due to extremely complex interplay between two moving structures. We assume operation of the impact device in regimes of low frequency and small impact velocity. Hence, the nonlinear impact force can be

simplified as [12]

$$F_{im} = k_{im}\delta + \left(b_{im} + \alpha k_{im}\delta\right)\dot{\delta} \qquad (8)$$

where k_{im} is the impact stiffness, b_{im} and α are damping coefficients accounting for the impact losses.

While the damping coefficients of the impact force are found by fitting to measurement, the impact stiffness is estimated by static analysis of the Hertz's contact, giving by the following equation [13]

$$\delta = \frac{2\lambda F_{im}}{t}\left[1 + \ln\frac{t^3}{2\lambda F_{im}R}\right] \qquad (9)$$

where $\lambda = \dfrac{1 - \upsilon^2}{\pi E}$, E is the Young's modulus, υ is the Poison's ratio and R the radius of cylinder. For a small range of the deformation, Eq. (9) can be linearized by an approximated stiffness $k_{im} \approx 0.058\dfrac{\pi E t}{1 - \upsilon^2}$ [14]. For comparison, the linear damping written in a spring-damper model of the impact force $F_{im} = k_{im}\delta + b_{im}\dot{\delta}$ is studied in simulation.

RESULTS AND DISCUSSION

The model parameters are found based on the design layout and fitting to the measurement results in both the linear regime and the impact regime. All parameters of the prototype model are given in Table 1.

Table 1: Model parameters of the impact device

Parameters	Main transducer	End-stop transducer
Proof mass	m_p=1.15 mg	m_s=0.05 mg
Spring stiffness	k_p=18.2 N/m	k_p=16.3 N/m
Thin-film air damping	b_p=1.50e-5 Ns/m	b_l=1.10e-5 Ns/m, b_n=6.18e-5 Ns/m
Impact damping	b_{im}=5.38 Ns/m, α = 4.5	
Nominal capacitance	C_{0p}=1.08 pF	C_{0s}=1.30 pF
Parasitic capacitance	C_{pp}=7.50 pF	C_{ps}=4.00 pF
Load resistance	R_{Lp}=18.5 MΩ	R_{Ls}=18.5 MΩ
Load capacitance	C_{Lp}=4.20 pF	C_{Ls}=4.20 pF
Impact stiffness	k_{ps}=3.26e6 N/m, k_{pp}=8.08e6 N/m	

Figure 5 compares measurement and simulation results for various excitation levels. For A=0.021 g and 0.057 g, the main proof mass displacement is far below the limit, giving no impact. The linear response of the impact device provides a subset of the model parameters, while the remaining parameters of the nonlinear damping force and the impact force are obtained from fitting to experiment results in the impact regime. Higher accelerations A=0.318 g and A=0.707 g leads to increased impact intensities. The strong impacts make the displacement waveforms irregular. This irregularity is manifest also in the average power in the figure. The simulation shows that the dynamics are still well-captured by the model with the nonlinear damping impact force, but there is a slight deviation observed for the model of the linear damping impact force. The difference indicates that the nonlinear model of the impact force has made improvement over the linear one in reproducing the complex behavior of the impact device.

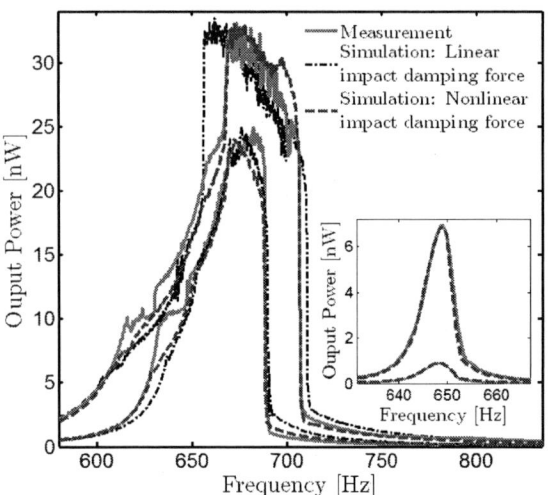

Figure 5: Total power in impact regime for A=0.318 g and 0.707 g at bias voltage of $V_p=V_s=9.2$ V. Subfigure shows frequency responses in linear regime for A=0.021 g and 0.057 g

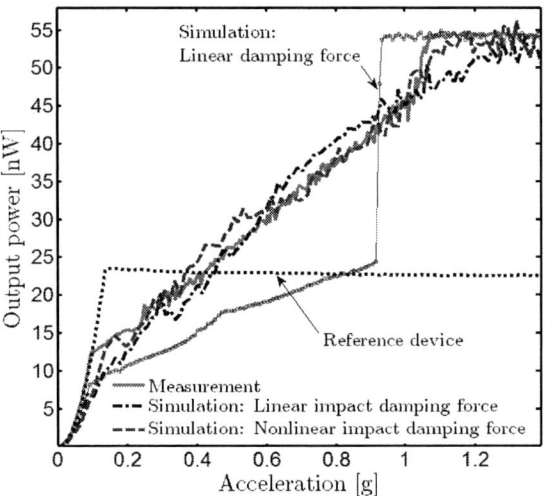

Figure 6: Total power of the impact and reference devices at their resonant frequencies under acceleration sweep for $V_p=9.2$ V and $V_s=14.5$ V

The obtained power of the impact device is enhanced in the impact regime when the transducer end-stop is more compliant. This can be done by electric control of the bias voltage V_s as demonstrated in our previous work [9]. Figure 6 shows validation of the model under the acceleration sweep when the effective stiffness of the end-stop transducer is significantly reduced when $V_s=14.5$ V. Both experiment and simulation results show power about 2.4 times higher than that of a standard reference device of the same size and with identical operating conditions. At small excitation (linear regime), the main proof mass displacement is too small to engage end-stops, giving total power mostly from the main transducers. Sufficient acceleration amplitude leads to the impact and extra power from the end-stop transducer contributing to the total power which grows like linearly before the

maximum value is reached. The simulation agrees with the measurement when the effects of the impact nonlinearity and the nonlinear squeeze-film damping force are included, while a considerably different response exhibiting a clear jump phenomenon results from use of a mere linear damping-force model for the end-stop transducer.

CONCLUSION

The impact device with the end-stop transducer has proven advantages over conventional designs with rigid end-stops. The improved power is achieved from the internal impact mechanism to overcome the well-known phenomenon of power saturation. The prototype performance was verified by a lumped-model built that is able to capture the complex dynamics in the impact regime. The nonlinearities from the electromechanical coupling, the squeeze-film damping force and the impact force are all accounted for the model. The good agreement between the simulation and experiment results indicates that the model is useful to predict the intricate behavior of the impact device and to further optimize the system in future work.

ACKNOWLEDGEMENTS

This work was supported by the Research Council of Norway through Grants no. 191282 and 229716/E20.

REFERENCES

[1] P. D. Mitcheson et al., *IEEE Proceedings*, **96** (9), pp. 1457-1486, 2008.

[2] S. Roundy, P. K. Wright, J. Rabaey, *Computer Communications*, **26**(11), pp.1131-44, 2003.

[3] M. S. M. Soliman et al., *J. Micromech. Microeng.*, **18**, 115021, 2008.

[4] D. Homann, B. Folkmer and Y. Manoli, *J. Micromech. Microeng.* **19**, 094001, 2009.

[5] H. Liu et al., *J. Microelectromech. Syst.*, **20**(5), 1131–42, 2011.

[6] M. Borowiec, G. Litak, S. Lenci, *Nonlinear Systems and Complexity*, **6**, pp. 315-321, 2014.

[7] C. P. Le, E. Halvorsen, O. Søråsen and E. M. Yeatman, *Proc. PowerMEMS* 2011, pp. 122-125, 2011.

[8] B. D. Truong, C. P. Le, E. Halvorsen, *Proc. PowerMEMS* 2014, (to be published) 2014.

[9] P. D. Mitcheson et al., *J. Microelectromech. Syst.*, **13**, pp. 429-440, 2004.

[10] C. P. Le and E. Halvorsen, *Small-Scale Energy Harvesting*, *INTECH*, pp. 265-282, 2012.

[11] M. Bao and H. Yang, *Sensors and Actuators A*, **136**, pp. 3–27, 2007.

[12] M. Machado et al., *J. Mechanism and Machine Theory*, **53**, pp. 99-121, 2012.

[13] B. N. Norden, *NBSIR* **73-243**, Institute for Basic. Standards, National Bureau of Standards, 1973.

[14] C. P. Le and E. Halvorsen, *J. Micromech. Microeng.*, **22**, 074013, 2012.

CONTACT

*Einar Halvorsen, tel: +47 33 03 77 25; Einar.Halvorsen@hbv.no

MONOLITHIC 2-AXIS IN-PLANE PZT LATERAL BIMORPH ENERGY HARVESTER WITH DIFFERENTIAL OUTPUT

Sachin Nadig, Serhan Ardanuç, and Amit Lal

*Sonic*MEMS Laboratory, School of Electrical and Computer Engineering
Cornell University, Ithaca, NY, USA

ABSTRACT

We report a 2-axis (X-Y) piezoelectric energy harvester, whose sensitive axis in-plane is rotationally invariant, a result achieved by spiral in-place bimorph, which can be modeled as a cascade of lateral bimorphs. This is different than conventional piezoelectric energy harvesters that are sensitive only along one axis or can realize multi-axis sensitivity through package level assembly of multiple devices at different orientations. Rotational invariance of the sensitive axis is useful when deployed in applications where the device orientation with respect to the vibration is stochastic in nature. At 1-g applied acceleration, our device generates maximum differential voltages of ~ ±6.8Vpp across a 1MΩ load, at a resonance frequency of ~163.5Hz. The harvester has a Q-factor of 129.5 and electromechanical coupling coefficient of k_t^2 ~3% for the fundamental lateral mode and a peak normalized output power density of ~0.89µWatt/mm³g⁻².

INTRODUCTION

One of the key challenges in energy harvesting is the ability to generate electrical energy from vibration regardless of its direction. This can be achieved in part by manual packaging of single harvesters oriented along different directions or wafer-level integration of energy harvesters which are sensitive along different axis [1]. While both approaches need additional circuitry to combine the generated powers, manual assembly approach also suffers from increased system volume and cost.
For piezoelectric energy harvesters,

$$Power(resonance) \propto \frac{k_{31}^2 Q^2 m_{eff} a^2}{f_r} , \qquad (1)$$

where, '*a*' is the applied acceleration and '*f_r*' the resonance frequency [2]. Recent work on thin film PZT harvesters such as sol-gel, sputtered AlN and also thinned PZT on silicon [2], which have reported good Q (32-1800) and k_{31}^2 (0.05-0.35), generally aim to reduce resonance frequencies to improve Normalized Power Densities (NPD) in µWatt/mm³g⁻² [2]. Although high values for NPD (6.309 µWatt/mm³g⁻² [2]) are reported in the literature, most devices rely on resonant structures sensitive along a single sensitive axis (generally out of plane) and are not monolithic. High NPD and low resonance frequency is achieved by steps requiring die-level alignment and bonding of high density materials, such as tungsten in [2], which leads to a multi-step process flow. This work uses a single-step, laser-micro machined, spiral shaped beam structure to get around single-axis limitation while still achieving high normalized power densities at low resonance frequencies. The energy harvester is fabricated using a laser micromachining process, which is a subtractive process requiring no additional deposition. The entire device is fabricated using direct-write micro patterning technique involving precision removal of PZT and/or electrode material [3].

The constraints considered for our energy harvester design are as summarized in Figure 1. Our design aims to maximize power output for a given volume for lower resonance frequency while being responsive to any in-plane (X-Y) vibration. The operation of the key transducer element of our 2-axis bulk-micro machined PZT harvester is sketched in Figure 2. The design exploits lateral bimorphs [3], generating differential voltage from in-plane motion at high efficiencies. The monolithic nature of the device eliminates the need for additional circuitry to combine the outputs of multiple harvesters for the purpose of multi-directional operation.

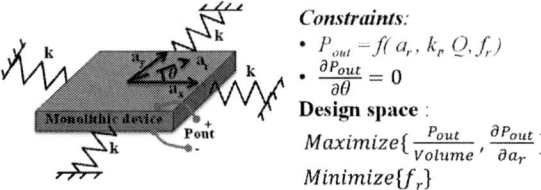

Figure 1: Metrics considered in the design of a 2-axis energy harvester under arbitrary planar acceleration a_r. k_t: electromechanical coupling coefficient, f_r: resonance frequency, k: spring constant.

DESIGN

The vibrational energy harvester has the lateral PZT bimorph as its fundamental building block [3]. The lateral bimorph generates differential voltages for in-plane bending due to equal and opposite tensile and compressive strains as shown in Figure 2 and owing to voltage coefficient g_{31}. The equal and opposite polarity voltages generated on top electrodes due to compressive and tensile stress can be useful, post rectification, in analog electronics that require symmetric ±V supply for operation, which is otherwise realized by having additional capacitors and inductors.

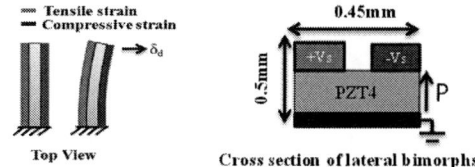

Figure 2: Basic building block of the harvester.

Here, we test two designs. First, we characterize the performance of a single bimorph element as energy harvester. For this purpose, we choose a common design that includes lateral bimorph with mass on the tip, as shown in Figure 3, all cut out from a single PZT substrate. The mass at the tip was chosen so as to lower the resonance frequency while still maintaining the fundamental resonance to be in-plane. Next, a second device structure with a spiral design as shown in Figure 4 is explored. This design allows improvements in performance and metrics described in Figure 1.

PZT4 used in both designs was chosen for its higher coupling coefficients and Curie temperature (324°C), the latter of which is important to minimize any potential heat induced local depolarization of bulk PZT during laser micromachining. Furthermore, bulk PZT is also a good choice as a tip mass to lower resonance frequency due to its high density.

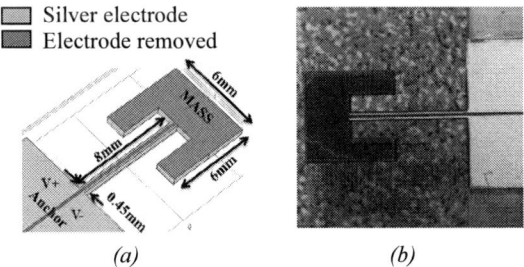

Figure 3: A generic energy harvester design with lateral PZT bimorph with tip mass. (a) Schematic of the device (b) Fabricated device on PCB

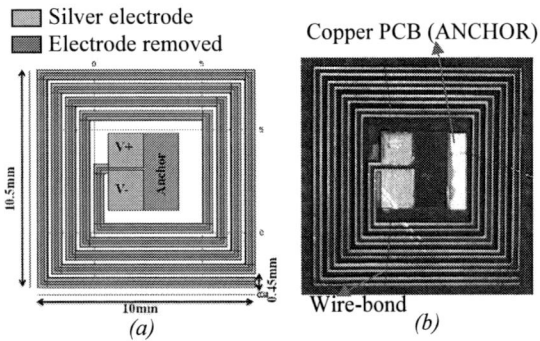

Figure 4: Two axis spiral PZT energy harvester. (a) Schematic (b) Fabricated device anchored to a PCB

FABRICATION

The energy harvester is fabricated using laser micromachining, which involves precision removal of PZT and/or electrode material [3], by repeated and selective scanning of a 355nm UV laser (LPKF ProtoLaser U) on 0.5 mm thick PZT-4 plates. The scan speeds are optimized for through cut of PZT and/or metal without depoling piezoelectric regions that are part of the device. The process can achieve an aspect ratio of 1:5 (0.1mm wide, 0.5 mm

thick) bulk PZT structures. For the energy harvester, PZT-4 plates with silver as top and bottom electrodes are laser-machined [3, 4] to form 0.45mm wide and 0.5mm thick lateral bimorphs, with electrode patterns as shown in Figure 2. Typical cut rates for PZT are ~16.6 µm/min in thickness, and 250 mm/s laterally leading to beams with minimum width of 100µm. After laser-micromachining, the PZT structures were cleaned with acetone followed by IPA (Iso-Propyl Alcohol) to clear any debris around the processed regions. Described process enables both the active energy harvesting element and the mass to be cut out from single PZT substrate.

MODELLING AND EXPERIMENTAL RESULTS
LATERAL BIMORPH WITH TIP MASS
Impedance Analysis

For characterization, the device shown in Figure 3b, which is adhered to PCB with silver acrylic paint, is used. This also ensured electrical contact to bottom electrode (ground). The two electrodes were then wire-bonded to pads on the PCB. The impedance response of the device is shown in Figure 5. The device has a resonance frequency of ~626.5 Hz and the mode is simulated to be in-plane using COMSOL with a measured Q-factor of 45 with $k_t^2 \sim 3.55\%$.

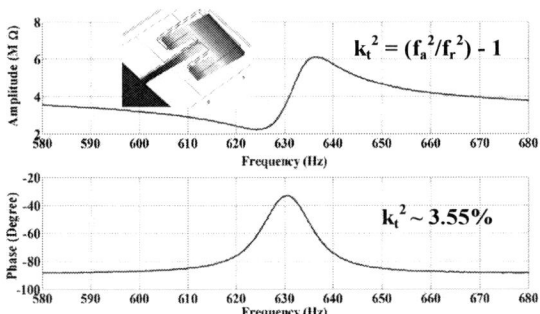

Figure 5: Impedance response of the lateral bimorph with a tip mass. Plot shows resonance at ~626.5Hz measured using impedance analyzer HP4194A. The Q factor is measured to be ~45. The mode is simulated to be in-plane using COMSOL with the displacement contour shown in the inset.

Power Output

The device was clamped on a shaker table to measure the peak power for 1g acceleration at its resonance. The device generates 4.5µWatts when connected across a load resistor of 198kΩ, which was found to be the optimal load for maximum power. The corresponding normalized power density is ~ 0.305µW/mm^3/g^2.

SPIRAL ENERGY HARVESTER

For improvement in NPD, Q-factor and to achieve rotational invariance, we pursued the spiral shaped energy harvester as shown in Figure 4. The lateral bimorphs were

978-1-4799-7956-1/15 $31.00 © 2015 IEEE   1130

micromachined in the form of a rectangular spiral coil of volume 26mm^3. Cascading of several such lateral bimorphs in a spiral fashion enables the device to be sensitive to any in-plane vibrations, making the energy harvester's sensitive axis rotationally invariant.

Design optimization

Near rotational symmetry of the spiral shaped beam structures allows the sensitive axis of the harvester to be rotationally invariant in-plane. Figure 6 shows the simulated curves for optimal spiral design as a function of number of coils, N. Also for N<2, it is seen from simulations that the fundamental mode is no longer in-plane. N=4 is chosen due to the trade-off between resonance frequency and NPD.

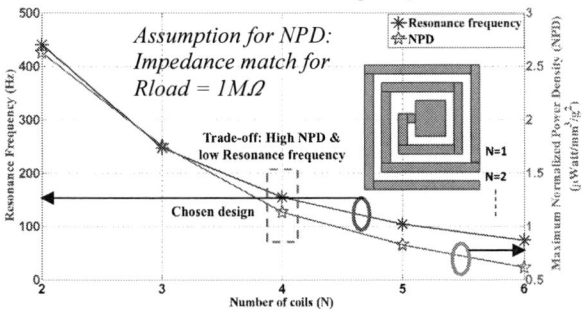

Figure 6: Simulated relations between NPD, resonance frequency versus number of coils N. Coils (N) is as shown in inset sketch

Impedance Analysis

It is shown in [5] that the energy harvester's load, peak power and its frequency is strongly dependent on K^2Q and is referred to as the coupling efficiency figure of merit. This is evident from the expression for impedance of the device, which is essential for calculating resonance frequency and the optimal load to be connected to the energy harvester for maximum power transfer. The impedance Z of the device can be expressed as, [5]

$$Z = \frac{-j}{\omega_0 C} \frac{j\frac{\omega}{\omega_0} + Q\left(1 - \frac{\omega^2}{\omega_0^2}\right)}{\frac{\omega}{\omega_0}\left[j\frac{\omega}{\omega_0} + Q\left(1 - \frac{\omega^2}{\omega_0^2}\right) + QK^2\right]}, \quad (2)$$

where, ω_0 is the resonance frequency, K is the generalized electromechanical coupling coefficient (K=k_{31},is experimentally determined to be 2.71%), and C is the capacitance of the device ~3nF. The measurements were done using impedance analyzer and further verified with lock-in amplifier frequency sweeps which yielded an in-plane Q-factor of 129.5. Analytical expression for Z is in close agreement to experimentally determined impedance curve shown in Figure 7(a). Figure 7(b) shows the equivalent circuit which closely models the device [5]. For maximum power transfer from the piezoelectric vibrational energy harvester, load impedance must be the complex conjugate of the device impedance (Z= $Z_{load}{}^*$), which is

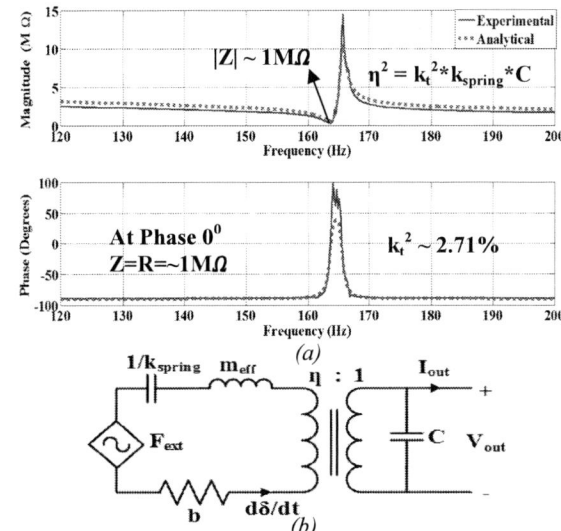

Figure 7: Impedance response of the device. (a) Plot showing resonance frequency of ~163.5Hz measured using Impedance analyzer (HP4194A) & the analytical curve for Z with extracted parameters. The measured Q factor for the mode is 129.5 (b) Shows the equivalent circuit model of the device.

purely real at resonance (163.5Hz) that corresponds to zero phase. The impedance at this frequency is 1MΩ as seen in Figure 7a. Hence the load resistance is chosen to be the same. The fabricated spiral energy harvester here falls under the strongly coupled case [5] of $K^2Q > 2$ ($K^2Q = 3.88$), with two operation frequencies for maximum power, corresponding to 163.5Hz and 165.5Hz, where impedance has zero phase for the tested device.

In-plane nature of the mode shape is verified by FEM in COMSOL as seen in Figure 8, which shows the surface potential distribution for 1g acceleration along the device's X axis. Similar results were obtained for acceleration applied along any axis in X-Y plane. The differential strain

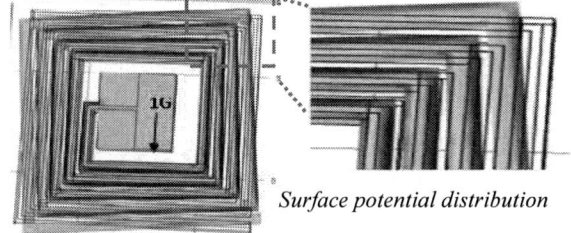

Surface potential distribution

▬ Peak tensile strain (surface potential ≈+7.7V)
▬ Peak compressive strain (surface potential≈ -7.7V)
▬ Minimum strain

Figure 8: COMSOL simulation of mode shape and the associated surface potential profile. Simulation shows peak open circuit voltage is 7.7V for 1g acceleration along X-axis of the device at the simulated device resonance frequency of 155.3Hz

due to equal and opposite compression and tension during in-plane bending of the cascaded bimorphs generates differential voltages. The strain is maximum at the L-shaped corners, which act as apparent anchors. This is shown in the zoomed picture at the inset of Figure 8. It is seen in simulations that the contributions to strain/voltage from subsequent outer rings reduce, which is in agreement with the results of Figure 6.

Power Output

The experimental setup to characterize the energy harvester is shown in Figure 9. The device was clamped on a commercial shaker table from Vibrations Research at varying angular orientations to measure output versus shaker frequency, amplitude, and harvester mounting angle. Experiments are performed for both 1MΩ load and at near infinite load by measuring open-circuit voltages using a JFET buffer to cancel the effect of oscilloscope finite input impedance. Experimental data show an average voltage of ~ 6.4Vpp across 1MΩ load. Table 1 lists the relatively constant output (within ± 7.4%) and power as a function of the planar orientation with respect to vibration. The NPD of the device is measured to be 0.89μWatt/mm^{3}g^{-2}. Compared to the simple bimorph energy harvester with tip mass, the spiral energy harvester has ≈2.9 times higher NPD because of higher Q and lower resonance frequency. The maximum power generated for 1MΩ load is ~23.14 μWatt for applied acceleration of 1g at 163.5 Hz.

Figure 9: Experimental setup showing the device mounted vertically on a Z-axis shaker table to excite the device's in-plane resonance. The device is mounted at different angles about its XY plane to test its dependence on vibration orientation.

Table 1: Measured voltages at resonance frequency for 1g applied acceleration for different mount angles about the XY plane of the device on the shaker table

Load	JFET Buffer*	1MΩ load	
Metric	**V pp(V)**	**Vpp(V)**	**Power(μWatts)**
X	±8V	±6.8V	23.12
20°	±7.8V	±6.4V	20.48
45°	±7.6V	±6.3V	19.85
70°	±7.6V	±6.3V	19.85
Y	±7.6V	±6.8V	23.12

*Measurements were done with JFET buffer to measure open circuit voltages generated for applied 1g accelerations to compare with COMSOL model (Figure 8)

CONCLUSIONS

A piezoelectric vibration energy harvester whose sensitive axis is shown to be rotationally symmetric with power output variations to within 7.4% is presented. Fabricated using only a single step laser micromachining of bulk PZT, this harvester has a resonance frequency of 163.5Hz and an NPD of ~0.89 μWatt/mm^{3}g^{-2}. COMSOL simulation results for the in-plane mode frequency of 155.3Hz matches experimental data to within ~5%. Such a power harvester could be useful in applications where in-plane motion directionality changes with time such as motion of a hovering aircraft or on top of a head mounted helmet display.

ACKNOWLEDGEMENTS

We would like to thank DARPA PASCAL program for funding this work. This work was performed in part at the Cornell NanoScale Facility, a member of the National Nanotechnology Infrastructure Network, which is supported by the National Science Foundation (Grant ECCS-0335765). Also, the authors would like to thank Abhijit Lavania for his help in COMSOL simulations.

REFERENCES

[1] Fu, J. L., et al. "Multi-axis AlN-on-Silicon vibration energy harvester with integrated frequency-upconverting transducers." *Micro Electro Mechanical Systems (MEMS), 2012 IEEE 25th International Conference on.* IEEE, 2012.
[2] Aktakka, Ethem Erkan, Rebecca L. Peterson, and Khalil Najafi. "A CMOS-compatible piezoelectric vibration energy scavenger based on the integration of bulk PZT films on silicon." *Electron Devices Meeting (IEDM), 2010 IEEE International.* IEEE, 2010.
[3] Nadig, Sachin, Serhan Ardanuc, and Amit Lal. "Planar laser-micro machined bulk PZT bimorph For in-plane actuation." *Applications of Ferroelectric and Workshop on the Piezoresponse Force Microscopy (ISAF/PFM), 2013 IEEE International Symposium on the.* IEEE, 2013.
[4] Nadig, Sachin, Serhan Ardanuc, and Amit Lal. "Monolithic piezoelectric in-plane motion stage with low cross-axis-coupling." *Micro Electro Mechanical Systems (MEMS), 2014 IEEE 27th International Conference on.* IEEE, 2014.
[5] Lei, Anders, et al. "Impedance Based Characterization of a High-Coupled Screen Printed PZT Thick Film Unimorph Energy Harvester." *Journal of Microelectromechanical Systems"* 23.4 (2014): 842-854.

SOLID-STATE FLEXIBLE MICRO SUPERCAPACITORS BY DIRECT-WRITE POROUS NANOFIBERS

Caiwei Shen[1]*, Guoxi Luo[1], Alina Kozinda[1], Mohan Sanghadasa[2] and Liwei Lin[1]

[1]University of California at Berkeley, Berkeley, CA, USA

[2]Aviation and Missile Research, Development, and Engineering Center, US Army, Redstone Arsenal, AL, USA

ABSTRACT

Solid-state flexible micro supercapacitors based on porous and conducting polymer nanofibers via the direct-write, near-field electrospinning process have been constructed. Testing results have shown a capacitance of $0.3mF/cm^2$, 30X larger as compared with those of flat electrodes. Key innovations of this work include: (1) densely-packed, porous 3D nanostructures with conductive nanofibers via the near-field electrospinning process; (2) flexible solid-state micro electrodes with high energy density using the pseudocapacitive effect; and (3) simple yet versatile manufacturing process compatible with various substrates and surfaces. As such, this technology is readily available to make practical MEMS energy storage devices.

INTRODUCTION

Emerging electronic systems such as wearable electronics, implantable medical devices, and active radio frequency identification (RFID) systems all have strong interests in the possibility of harvesting and storing energy from the environment. In the considerations of power devices for micro systems, the follow characteristics are desirable: (1) small size compatible with other components in the micro system; (2) good flexibility to fit in with different form-factors; (3) high energy conversion and storage capacity; and (4) low manufacturing cost with the possibility to directly integrate with other micro devices. Specifically, micro energy storage units are indispensable in storing electricity from energy harvesters, while batteries and supercapacitors are the most common solutions. Among these two systems, supercapacitors are safer in operation with longer cycle lives and higher charge/discharge rates but generally with lower energy density [1-4].

Supercapacitors are categorized by the two charge storage mechanisms [1, 2]. One is based on the electrical double layer (EDL) effect which stores static charges on the interface between a high-surface-area electrode and an ion containing electrolyte. The other kind mainly relies on fast and reversible redox reactions that happen at the surface of active materials, and is known as pseudo capacitor. Transient metal oxides or conducting polymers are generally used in pseudo capacitors to achieve high specific capacitances as compared with those of EDL capacitors. However, pseudo capacitors with flexible electrodes are difficult to make as the bulk materials used in pseudo capacitors are usually stiff or brittle. On the other hand, nanostructured carbon materials with large surface areas, including carbon nanotubes (CNTs) and graphene have been used to make flexible, EDL-type supercapacitor electrodes [5-7] with lower capacitances when compared with pseudo capacitors. Therefore, a key goal of this work is to make flexible pseudo capacitors with large capacitance on flexible substrates.

Different technologies have been demonstrated to make integrated on-chip micro supercapacitors. For example, chemical vapor deposition (CVD) has been used for the direct growth of CNTs [8], carbide derived carbon structures [9], and silicon carbide nanowires [10] as the supercapacitor electrodes on silicon wafers for possible integration with microelectronics. However, the typical high temperature process requirements in making these nanostructures prevent their direct integration, especially, with polymer substrates for flexible electronics. The carbonization of patterned photoresist has been used for the synthesis of micro-sized carbon electrodes but high temperature process again limits the possibility for direct integration [11-13]. Electroplating, on the other hand, has been used to grow conducting polymers on the surfaces of 3D electrodes [14] as well as to assemble onion-like carbon particles to form interdigital electrodes [15]. Both techniques involve wet processes in which the shot substrates are immersed in complex electrolytes which can limit the direct integration processes. Printing [16] and laser writing [17] have also been used to make micro supercapacitors. Although they are applicable for various substrates, only certain carbon materials have been demonstrated with limited performances.

Here we propose and demonstrate the solid-state flexible micro supercapacitors to address some of the drawbacks in aforementioned micro energy storage systems: (1) the use of nanofibers on electrodes for large surface areas and improved capacitance; (2) direct-write conducting polymer deposition on arbitrary, flexible substrates via near-field electrospinning; and (3) 30X vastly improved capacitance on the interdigital micro electrodes as compared with that of flat electrodes.

DESIGN AND FABRICATION

Electrospinning is a simple and versatile technique that utilizes high electrostatic force to generate continuous fibers with diameters from submicron down to nanometer [18]. Different kinds of functional fibers, most of which are made of chain polymers, can be produced by electrospinning as long as a uniform solution with proper viscosity and surface tension can be prepared. Conducting polymers, such as polypyrrole, are a group of polymers that can conduct electrons when are doped [19]. They can work as pseudo capacitive materials to store energy by the doping/de-doping (charging/discharging) effect in ion-containing electrolyte [19, 20]. Electrospun conducting polymer nanofibers can facilitate ion transport in electrodes with enlarged surface areas. Moreover, since nanofibers are mechanically flexible, brittle bulk pseudo capacitor materials based on transient metal oxides or electrically conducting polymers could be made flexible for supercapacitor applications.

978-1-4799-7956-1/15 $31.00 © 2015 IEEE   1133   MEMS 2015, Estoril, PORTUGAL, 18 - 22 January, 2015

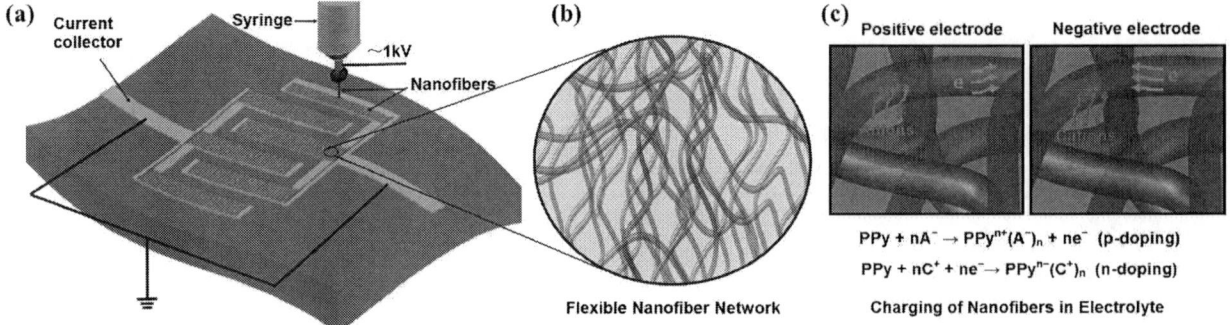

Figure 1: Schematics of (a) direct-write functional nanofibers onto patterned interditigal current collector on flexible substrate using near-field electrospinning; (b) magnified view of nanofiber network on the current collector; (c) nanofibers containing conducting polymer (e.g. PPy) in electrolyte with anions (A⁻) and cations (C⁺): when they are charged, p-doping and n-doping occur at positive and negative electrodes, respectively.

DESIGN

We applied near-field electrospinning (NFES) [21], in which the electrode-to-collector distance is only 0.5-5mm, to directly write nanofibers onto metal electrodes on flexible substrates, as shown in figure 1(a). NFES can orderly deposit nanofibers onto micro-sized patterns. As such, the resulting micro energy storage devices can be easily integrated with other flexible miniaturized systems. The nanofiber network (figure 1(b)) is highly porous, which provides pathway for ion transport and high surface area for the ion doping and de-doping processes. The p- and n-doping processes occur at the positive and negative electrode, respectively, when the device is being charged (figure 1(c)).

MATERIAL PREPARATION

Polypyrrole (PPy) dispersed in water (5wt%), polyethylene oxide (PEO, Mw~300,000), Triton X-100, polyvinyl alcohol (PVA, Mw~90,000), and H_3PO_4 are all supplied from Sigma-Aldrich.

The preparation of polymer solution is important to a successful NFES process. In a typical process, 0.5g of PEO, serving as a polymer carrier that makes the whole solution easier for electrospinning, is dissolved in 2g of de-ionized water and stirred for 3hr. Then 5g of PPy suspension and 0.25g of Triton, which is a surfactant used to improve the quality of the electrospun fibers, are added into the solution and stirred overnight. The mixture is prepared 30min before the NFES.

The solid electrolyte is prepared by first dissolving 1g of PVA in 8g of de-ionized water at 80°C under stirring until it becomes clear and 1g of H_3PO_4 is added into the solution. The viscous solution becomes solid electrolyte by evaporating the water content.

FABRICATION DETAILS

The flexible substrate used in this work is the Kapton® polyimide film. First, photoresist is spin-coated onto a Kapton film and photolithography is conducted to define the interdigital electrodes. A layer of 100nm thick gold is deposited and the photoresist is removed by the lift-off process to construct gold patterns on the Kapton substrate. Each finger is 200μm wide, 1300μm long, with 200μm wide gap in between as shown in Fig. 2(a). The total comb-shape interdigital electrode area is 2.2mm X 1.9mm.

During the NFES process, the Kapton substrate is placed on top of an X-Y motion stage controlled by a computer. The polymer solution is filled into a syringe equipped with a 22 gauge flat-end needle. The syringe is put right above the gold pattern, and the distance between the needle and the substrate is around 3-5mm. The needle is connected to a high voltage source (~1kV) and the gold pattern on the substrate is grounded. The solution in the syringe is dragged out of the tip of the needle by the electrostatic force and nanofibers are deposited onto the gold electrodes. The motion stage is preprogrammed during the process to control the nanofibers right on the "fingers", as depicted in figure 1(a). Figures 2(b) and 2(c) show optical photos of a gold electrodes without and with nanofibers on top, and a magnified view is in figure 2(d). Nanofibers on polymer substrate don't have clear images under SEM, so sample with electrospun nanofibers on a

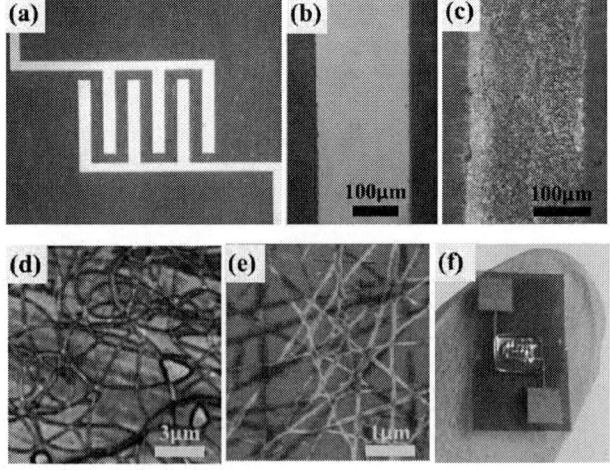

Figure 2: Optical microscopic photos of (a) patterned gold electrodes; (b) a blank electrode surface; (c) an electrode that is covered by electrospun nanofibers; (d) a magnified image of the nanofiber network; (e) SEM image of nanofibers that are electrospun on a silicon wafer; and (f) Optical photo of a micro supercapacitor cell with the solidified polymer electrolyte covering the active area.

978-1-4799-7956-1/15 $31.00 © 2015 IEEE   1134

silicon substrate is prepared and its SEM image is in figure 2(e). The average diameter of the nanofibers is around 200nm and the thickness of the nanofiber network is about 0.5~2μm for the prototype. The thickness can be adjusted by the number of written layers on the substrate.

The electrolyte solution mentioned previously is dried under 50°C for 1hr to form a solid gel. It is then cut and pressed onto the active electrode area in Figure 2(f). The cell is finally heated to 50°C for 10min to allow the penetrations of electrolyte into the nanofiber network, and is ready for electrochemical characterizations. A control cell is also fabricated following the same protocols, except that no fibers are written on the gold electrode.

RESULTS AND DISCUSSION

The supercapacitor performance is characterized and recorded by a Gamry Reference 600 potentiostat. Figure 3(a) compares the cyclic voltammetric curves of a cell with nanofibers and a control cell without nanofibers. The control cell shows a small current caused by the electrical double layer effect of the gold layer. The prototype with nanofibers shows 30 times larger current density than the control cell under the same scan rate of 20mV/s, which proves that the supercapacitor performance is mainly

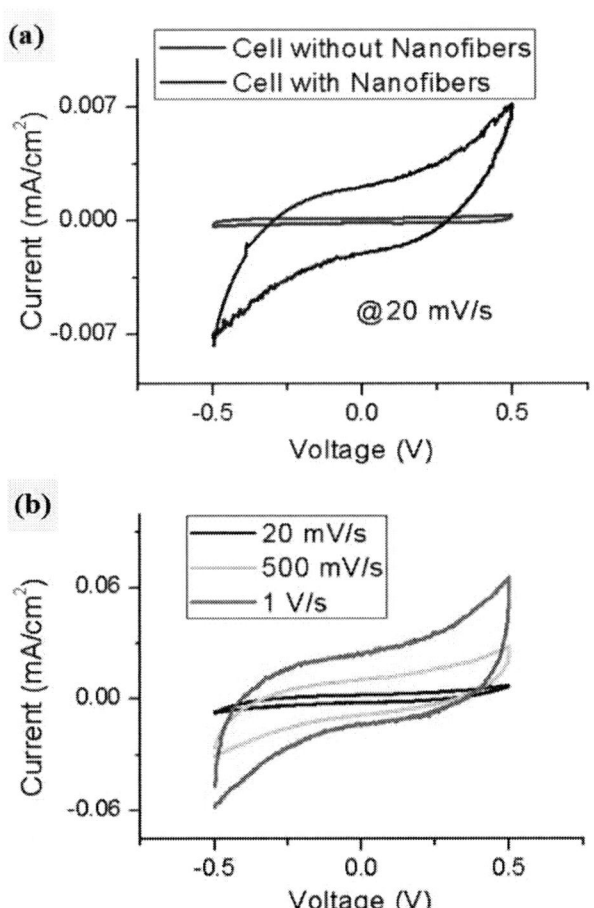

Figure 3: Cyclic voltammetric curves of a micro supercapacitor cell (a) with and without nanofibers; (b) under different scan rates.

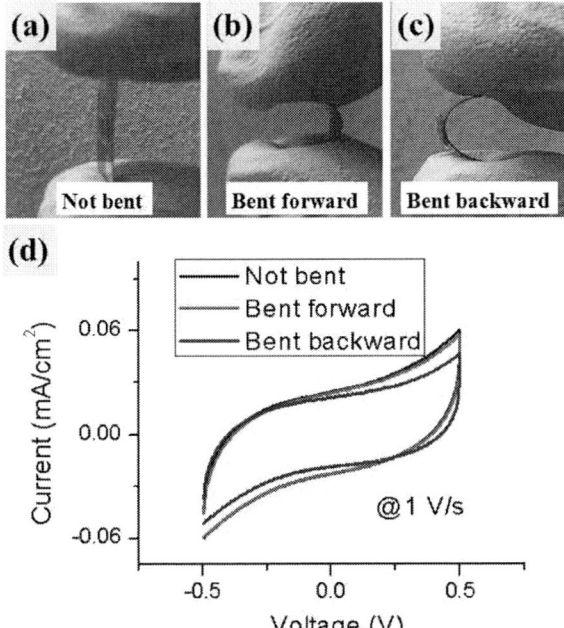

Figure 4: Photos of the micro supercapacitor with electrodes facing left when it's (a) not bent, (b) bent forward, and (c) bent backward. (d) Cyclic voltammetric curves of the cell when it's bent at different directions.

contributed by electrospun nanofibers instead of the gold electrodes. The symmetric parallelogram shape of the curve indicates the capacitive energy storage of the cell.

The capacitance of the device can be estimated by dividing the current with scan rate, and the result is about $0.3mF/cm^2$ for the prototype. A number of papers have demonstrated that performance including the capacitance and power per unit area can be improved, by better designing of the electrode sizes and by increasing the thickness of the functional materials, [13, 15]. The capacity of the prototype characterized in this paper can also be directly increased by repeating the NFES process, i.e. by depositing thicker layer of nanofibers. In addition, the solid electrolyte used in this work may not cover the whole surface area of the nanofibers, and as a result, part of the potentially functional material may not contribute to the energy storage. It is also possible to increase the capacity of the device by simply designing a better process to infiltrate the solid electrolyte into the electrodes or instead using a liquid electrolyte.

A theoretical capacitance of the micro prototype can be expected by the following calculation: assuming that the volumetric capacitance of bulk conducting polymer is around $200F/cm^3$ [19]. In a porous electrode constructed by electrospun nanofibers, about 25% of the volume is occupied by the conducting polymer. Then, if two electrodes are separated in a plane, each has an area of $0.5cm^2$ and an average thickness of 1μm, it would have a capacitance of $200F/cm^3 \times 25\% \times 0.5cm^2 \times 10^{-4}$ cm = 2.5 mF. As two electrodes are in series, the total capacitance of the device would be 2.5/2=1.25mF. Therefore, a capacitance of over $1mF/cm^2$ per 1μm thickness could be expected for such a device by proper design, which is a

978-1-4799-7956-1/15 $31.00 © 2015 IEEE

high value compared with state-of-art on-chip micro supercapacitors.

Figure 3(b) shows the cyclic voltammetric curves of the same prototype under different scan rates. The curves show that the current density is almost proportional to the scan rate, which is a typical behavior of a well performed supercapacitor. The device is charging and discharging at higher rate when the scan rate is higher. An excellent capacitive performance is preserved at 1V/s indicates a good power performance of the device.

We have also conducted a few bending tests of the prototype to evaluate its flexibility. Photos in figure 4(a) to (c) are the bending conditions we have applied, in which the prototype is kept straight, bent forward, and bent backward, respectively. Cyclic voltammetric curves are recorded at each bending state, as plotted in figure 4(d). The curves show that the performance of the prototype doesn't change much when it's bent. Since the prototype has not been sealed during the test to secure the polymer electrolyte on the electrode structure, a slight detachment of the polymer electrolyte has occurred at the edge when the device was bent backward and this has resulted a minor drop in the current.

CONCLUSION

We have successfully demonstrated a solid-state flexible micro supercapacitors based on polymer nanofibers. The nanofibers can be directly written onto micro-sized patterns on various substrates including flexible polymer films. The porous network formed by conducting polymer nanofibers stores energy by the pseudo capacitive effect, and it shows 30X larger capacitance as compared with that from a flat electrode. Future improvements can be made by further improving the polymer fiber networks for better supercapacitor performances, designing electrode patterns of appropriate sizes, thickening the nanoporous network with more layers of nanofibers, using better electrolyte, and so on. In summary, our approach is simple, versatile, and compatible with different substrates for practical MEMS energy storage devices.

ACKNOWLEDGEMENTS

This work is supported in part by BSAC, Berkeley Sensor and Actuator Center, and National Science Foundation Industry/University Cooperative Research Center.

REFERENCES

[1] B. E. Conway, *Electrochemical Supercapacitors: Scientific Fundamentals and Technological Applications,* Kluwer Academic/Plenum Publishers, 1999.

[2] P. Simon, and Y. Gogotsi, "Materials for Electrochemical Capacitors", *Nat. Mater.*, vol. 7, pp. 845-854, 2008.

[3] M. Armand, and J. M. Tarascon, "Building Better Batteries", *Nature*, vol. 451, pp. 652-657, 2008.

[4] J. M. Miller, *Ultracapacitor Applications*, The Institution of Engineering and Technology, London, 2011.

[5] M. Kaempgen, C.K. Chan, J. Ma, Y. Cui, and G. Gruner, "Printable Thin Film Supercapacitors Using Single-Walled Carbon Nanotubes", *Nano Lett.*, vol. 9, pp. 1872–1876, 2009.

[6] M. F. El-Kady, *et al.,* "Laser Scribing of High-Performance and Flexible Graphene-Based Electrochemical Capacitors", *Science*, vol. 335, pp.1326-1330, 2012.

[7] Y. Jiang, A. Kozinda and L. Lin, "Flexible Energy Storage Devices Based on Carbon Nanotube Forests with Built-in Metal Electrodes," *Sensors and Actuators - A Physical*, vol. 195, pp. 224-230, 2013.

[8] Y. Jiang, Q. Zhou, and L. Lin, "Planar MEMS Supercapacitor Using Carbon Nanotube Forests", *Proc. MEMS 2009*, pp. 587-590.

[9] J. Chmiola, *et al.,* "Monolithic Carbide-Derived Carbon Films for Micro-Supercapacitors", *Science*, vol. 328, pp. 480-483, 2010.

[10] J. Alper, et al., "Silicon Carbide Nanowires as Highly Robust Electrodes for Microsupercapacitors", *J. Power Sources*, vol. 230, pp. 298-302, 2013.

[11] M. Beidaghi, W. Chen, and C. Wang, "Electrochemically Activated Carbon Micro-electrode Arrays for Electrochemical Micro-capacitors", *J. Power Sources*, vol. 196, pp. 2403-2409, 2011.

[12] C. Shen, *et al.,* "Direct Prototyping of Patterned Nanoporous Carbon: A Route from Materials to On-chip Devices", *Sci. Rep.*, 3:2294, 2013.

[13] B. Hsia, *et al.,* "Photoresist-derived Porous Carbon for On-chip Micro-supercapacitors", *Carbon*, vol. 57, pp. 395-400, 2013.

[14] W. Sun, and X. Chen, "Fabrication and Tests of a Novel Three Dimensional Micro-supercapacitor", *Microelectron. Eng.*, vol. 86, pp. 1307–1310, 2009.

[15] D. Pech, *et al.,* "Ultrahigh-Power Micrometre-sized Supercapacitors based on Onion-like Carbon", *Nat. Nanotech.*, vol. 5, pp. 651-654, 2010.

[16] D. Pech, *et al.,* "Elaboration of a Microstructured Inkjet-Printed Carbon Electrochemical Capacitor", *J. Power Sources*, vol. 195, pp. 1266-1269, 2010.

[17] W. Gao, *et al.,* "Direct Laser Writing of Micro-supercapacitors on Hydrated Graphite Oxide Films", *Nat. Nanotech.*, vol. 6, pp. 496-500, 2011.

[18] C. Zhang and S. Yu, "Nanoparticles meet electrospinning: recent advances and future prospects", *Chem. Soc. Rev.*, vol. 43, pp. 4423-4448, 2014.

[19] G. Snook, P. Kao, and A. Best, "Conducting-Polymer-based Supercapacitor Devices and Electrodes", *J. Power Sources*, vol. 196, pp. 1-12, 2011.

[20] I. Chronakis, S. Grapenson, and A. Jakob, "Conductive Polypyrrole Nanofibers via Electrospinning: Electrical and Morphological Properties", *Polymer*, vol. 47, pp. 1597-1603, 2006.

[21] D. Sun, C. Chang, S. Li, and L. Lin, "Near-Field Electrospinning", *Nano Lett.*, vol. 6, pp. 839-842, 2006.

CONTACT

*Caiwei Shen, tel: +1-510-642-8983; shencw10@berkeley.edu

STRETCHABLE WIRELESS POWER TRANSFER WITH A LIQUID ALLOY COIL

Seung Hee Jeong[1] and Zhigang Wu[1,2]*

[1]Microsystems Technology, Uppsala University, SWEDEN
[2]State Key Laboratory of Digital Equipment and Manufacturing, Huazong University of Science and Technology, CHINA

ABSTRACT

An integrated stretchable wireless power transfer device was demonstrated by packaging rigid electronic chips onto a liquid alloy coil patterned on a half-cured polydimethylsiloxane (PDMS) surface. To obtain low enough resistance, the long liquid alloy coil with a large cross section was made with a tape transfer masking followed by spray deposition of the liquid alloy. The measured results indicated the wireless power transfer efficiency reached 10% at 140 kHz and good performance under 25% overall strain. Different sizes of liquid alloy coils and a soft magnetic composite core were tested to improve the efficiency of the system.

INTRODUCTION

Stretchable electronics provides advanced integrated circuits by combining thin film technology and transfer processes, which already demonstrated great potential in conformably wearable and implantable devices [1]. To maximize the user experience, a monolithic integrated energy source is necessary for such a stretchable system. However, energy sources require an area or a volume for energy storage or energy harvesting. A wireless solution with antennas might be a feasible candidate as such a smart power source.

Several small scale antennas for a wireless power transfer (WPT) system for emerging electronic applications have been reported with a metal wire coil [2-4] as well as a stretchable thin metal film coil [5]. A gallium-based liquid alloy [6] can provide a highly stretchable functionality as a high performance conducting material. This liquid alloy has low resistivity and allows for a large cross section when stretched, which provides high electrical conductivity, compliance and reliability at high strain, and the gliding contact between a rigid device and the liquid alloy conductor does not fracture when strained [7,8]. In our previous work [7], metal stencil printing requires that all parts of the mask are connected and robust, otherwise they will easily deform. A new approach for liquid alloy patterning was proposed with a tape transfer mask [8], which allows the use of isolated parts with fragile and fine structures, and liquid alloy spraying for wireless power transfer coil fabrication. A stretchable wireless transfer coil was integrated with rigid circuits and successfully demonstrated with strain. A magnetic composite [9] as a core of the liquid alloy coil was tested, and different sizes of the liquid alloy coils were tested for a higher efficiency system. A stretchable WPT system with a liquid alloy coil may enable implantable or wearable electronic devices to be free from a bulky battery or wiring to an external power source.

DESIGN OF SOFT LIQUID ALLOY COIL

A liquid alloy, Galinstan (Geratherm Medical AG), coil was designed to be used as an electromagnetically resonant coil by inductive coupling for a wireless power transfer system. The coil has a non-equilateral octagon shape [3,4] and sized with outer dimensions of 30×30 mm and inner dimensions of 13.5×13.5 mm, which was based on the transmitting coil size. A smaller size of the liquid alloy coil, which was half sized in the all dimensions that was the line width of 250 µm and the distance of 175 µm between coil turns, was used for comparison of the efficiency with the large one.

The transmitting coil from a commercial product (Würth elektronik) was prepared with a simple transformer circuit. Impedance matching of the coils and circuits with resistance and capacitance at the given inductance of the liquid alloy coil (4µH) and the resonant frequency of 140 kHz was achieved by a parametric study of resistance, capacitance and frequency in the designed circuits. The bottom layer of PDMS (Wacker Chemie) packaging of the liquid alloy coil was made with thin layer (100 µm) to minimize electromagnetic coupling loss in the packaging layer.

DEVICE FABRICATION

The stretchable WPT device, Figure 1, was fabricated with a liquid alloy coil and a circuit with five chips including a rectifier, capacitors, a resistor and an LED integrated in a PDMS package. The planar liquid alloy coil was 82 cm long, 550 µm wide and 120 µm thick.

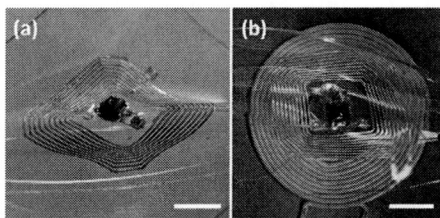

Figure 1: Photographs of a stretchable wireless power transfer coil fabricated with a liquid alloy patterning and chip integration in PDMS packaging (a) a stretched, free standing, integrated liquid alloy coil device and (b) a stretchable wireless power transfer device working with LED lightening. (Scale bars indicate 10 mm).

The liquid alloy coil was fabricated by spraying a liquid alloy through a tape transfer masking technique [8] as described in Figure 2. A tape mask was used for patterning of a liquid alloy, which was prepared with a cutting plotter, and a transfer tape was employed to transfer the cut mask onto a semi-cured PDMS substrate. After peeling off the tape mask, chips were placed on liquid alloy pads of the circuit and multilayer fabrication was done to connect the coil with the integrated circuit. Finally, the uncured PDMS was poured over to encapsulate and create an elastic package. An LED was integrated in the circuit to visualize the device working.

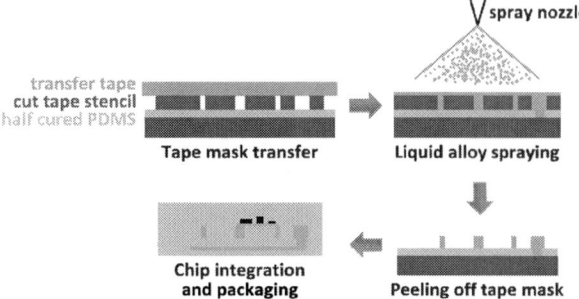

Figure 2: Fabrication process of the stretchable liquid alloy coil device with tape transfer mask, spraying, chip integration and packaging.

A magnetic powder, cobalt iron oxide (Sigma-Aldrich), was mechanically mixed with PDMS as 50 wt% and cured at 70°C for one day to make a core material. The cured magnetic composite was soft and elastic. The magnetic composite core was structured from a metal mold with the diameter of 13 mm and the height of 2 mm. A reference PDMS core with the same dimension as the magnetic composite core was prepared, which was used to make the same gap between the transmitting and receiving coil in case of no magnetic core under the liquid alloy coil for the magnetic core effect comparison.

RESULTS AND DISCUSSION
Stretchability of the Integrated Wireless Power Transfer Device

The fabricated liquid alloy WPT coil was tested as a receiving (R_x) part coupled with a transmitting (T_x) part by changing strains and distances between the T_x and R_x coil. The stretchable WPT device which had a liquid alloy coil and integrated chips showed an efficiency of 10% and worked with 25% strain, Figure 3.

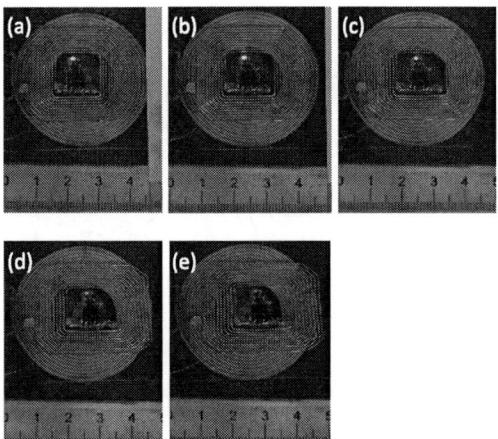

Figure 3: Stretchability test of the fabricated liquid alloy coil integrated with chips in PDMS packaging during wireless power transfer operation with various strains from (a) 0% to (e) 25%.

Strain and Positioning Effect

Without chip integration in the package but with copper wires inserted through a PDMS package for measurements, stretching of the liquid alloy coil up to 50% caused changes of the resistance and the power efficiency of the device as shown in Figure 4. The transferred power efficiency was calculated from the measured voltages and currents as

$$\eta(\%) = \frac{Power(R_x)}{Power(T_x)} \times 100 \qquad (1)$$

and the inductively induced voltage and current to the liquid alloy coil in the receiving part was measured after rectification.

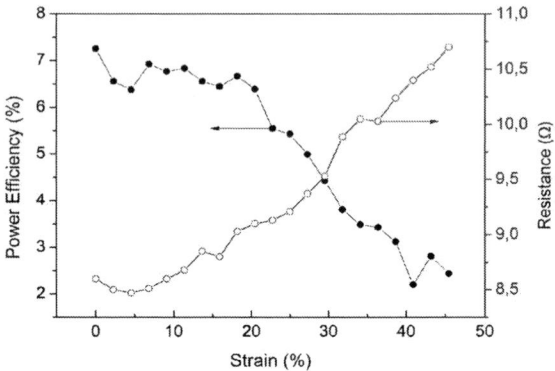

Figure 4: Strain effect of the stretchable liquid alloy coil on power transfer efficiency and resistance.

By increasing the strain up to 50% of the initial length of the soft liquid alloy coil, the resistance of the coil was increased to 26.5% of its initial resistance, and the power efficiency was reduced by 71.4% of the initial efficiency. The reduction of the power efficiency was caused by the resistance increase of the liquid alloy coil by strain, by the change of the coil area which is connected to the Q factor and the coupling factor, and by the misalignment of the coaxial position between the T_x and R_x coil. Whereas, when the coaxial alignment of two coils was maintained, the power efficiency was sustained with small changes, which can be originated from a manually controlled setup for alignment and stretching, as shown in Figure 5.

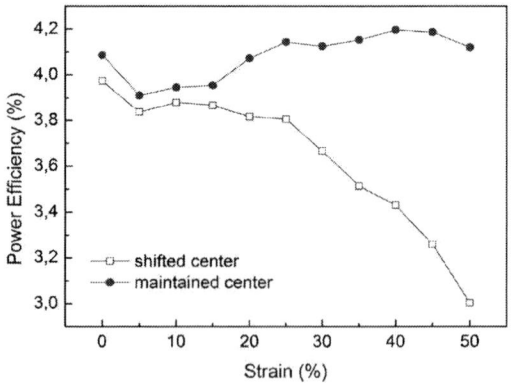

Figure 5: Alignment effect of coaxial positioning of the T_x coil and the R_x coil on the efficiency when the liquid alloy coil is stretched.

Figure 6 shows the efficiency change of the soft liquid alloy device in the wireless power transfer system with the distance changes of the T_x and R_x coil in the different directions, which are the distance of the two coils' axis and the distance of the two coils' planar surfaces. Both of the cases shows power efficiency reduction by the distance increase with the similar behaviors.

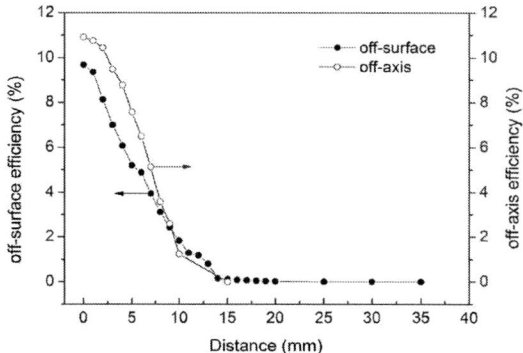

Figure 6: Wireless power transfer efficiency changes according to the positions of the liquid alloy coil relative to the transmitting coil.

Coil Size Effect

Different sizes of the liquid alloy coils were fabricated to compare their efficiencies. A smaller size of a liquid alloy coil compared to the original one was fabricated with half dimensions of the original one. The resistance of the large coil and the small coil was 11 and 9.8 , respectively. During the tests of the two coils, the air gaps between T_x coil and R_x coil were the same. The efficiencies of the two liquid alloy coils were tested with the same transmitting coil and the same circuits at 140 kHz as shown in Table 1.

Table 1: Transferred power efficiency of two different size coils of a liquid alloy.

Coil size	Tx Power (W)	Rx Power (mW)	Efficiency (%)
Large size	4.96	489	9.86
Small size	2.90	269	2.97

Figure 7 shows the efficiency changes of the two different size coils by strain changes. The smaller coil showed lower efficiency when the strain was increased. The power efficiency loss was calculated with the equation (1) as

$$\eta_{loss}(\%) = \frac{\eta_i - \eta_s}{\eta_i} \times 100 \qquad (2)$$

where η_i is the power efficiency of the system without strain in the liquid alloy coil and η_s is the power efficiency with strain in the liquid alloy coil.

An optimal design of a receiving coil size corresponding to a transmitting coil size will give higher efficiency with a higher Q factor in its RLC circuit and a higher coupling factor from its electromagnetic field interaction. A further study of the optimal design of the coil for a specific application in the future will be required.

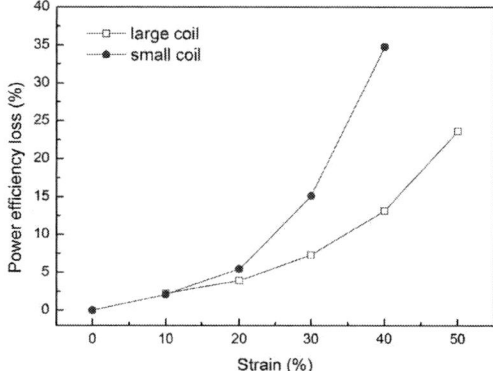

Figure 7: Coil size effect under strain conditions on wireless power transfer efficiency with the liquid alloy coils of two different sizes.

Magnetic Core Effect

To study a magnetic core effect, a magnetic composite core was applied under the large size coil of the liquid alloy to enhance the electromagnetic coupling between the T_x coil and the liquid alloy R_x coil. Figure 8 shows the composite core of cobalt iron oxide increased the wireless power transfer efficiency less than 1.5%, compared to the liquid alloy coil with a reference PDMS core, when the coaxial alignment of the T_x coil and the liquid alloy R_x coil was shifted with different distances between the axis of the two coils. The magnetic core effect on the efficiency was relatively higher at the larger shifted distance between the axis of the T_x and R_x coil.

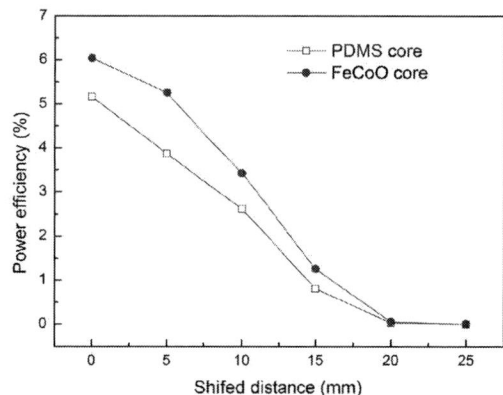

Figure 8: Magnetic composite core effect on power transfer efficiency with shifted coaxial distances of the Tx and Rx coil.

For a further application of the magnetic composite core to the resonant inductive coupling system, the characterization of magnetic properties, such as magnetic field strength, permeability and hysteresis, and the size of a magnetic composite core, will be needed in the future. Magnetic composite core integration with a stronger magnetic material in the elastomer package may help electromagnetic coupling in the WPT system.

CONCLUSIONS

A stretchable wireless power transfer coil with a liquid alloy and integration of rigid chips in PDMS packaging was demonstrated. The liquid alloy based wireless transfer coil showed 10% efficiency in the WPT system and strained up to 25% in stretch with chip integration. Different sizes of the liquid alloy coils and a magnetic composite core affected to the wireless power transfer efficiency under different test conditions. Optimization of the liquid alloy coil design and circuit impedance matching should be studied for a higher efficiency of the stretchable wireless power transfer system. The liquid alloy coil can be applied to soft electronic systems as a wireless power source for a smart package and this system can provide user convenience by a wireless energy transfer function without any wired connection through a package to the wearable electronics or biomedical systems.

ACKNOWLEDGEMENTS

The authors thank to Johan Sundqvist at Nanospace AB for his help on the circuit design and discussion on the wireless power transfer system. We acknowledge the Swedish Research Council (No. 2010-5443) for financial supporting. Wu thanks the support of Chinese central government through its 1000 youth talent program.

REFERENCES

[1] J. A. Rogers, T. Someya, Y. Huang, "Materials and Mechanics for Stretchable Electronics", *Science*, vol. 327, no. 5973, pp. 1603-1607, 2010.

[2] S. Kim, R. Harrison, F. Solzbacher, "Influence of system integration and packaging for a wireless neural interface on its wireless powering performance", *30th Annual International IEEE EMBS Conference,* Vancouver, August 20-24, 2008, pp. 3182-3185.

[3] R. Bosshard, J. Mühlethaler, J. W. Kolar, I.

Stevanovic, "Optimized Magnetic Design for Inductive Power Transfer Coils", in *Proceedings of the 28th Applied Power Electronics Conference and Exposition (APEC 2013)*, Long Beach, California, March 17-21, 2013, pp. 1812-1819.

[4] D. Hendrickx, J. Pannier, T. Nobels, F. Petré, "Wireless power transfer for industrial applications through resonant magnetic induction", *International Workshop on Wireless Energy Transport and Harvesting*, Eindhoven, June 26-28, 2011, pp. 1-6.

[5] S. Xu, Y. Zhang, J. Cho, J. Lee, X. Huang, L. Jia, J. A. Fan, Y. Su, J. Su, H. Zhang, H. Cheng, B. Lu, C. Yu, C. Chuang, T. -I. Kim, T. Song, K. Shigeta, S. Kang, C. Dagdeviren, I. Petrov, "Stretchable batteries with self-similar serpentine interconnects and integrated wireless recharging systems", *Nature Communications*, vol. 4, no. 1543, 2013.

[6] R. C. Chiechi, E. A. Weiss, M. D. Dickey, G. M. Whitesides, "Eutectic Gallium–Indium (EGaIn): A Moldable Liquid Metal for Electrical Characterization of Self-Assembled Monolayers", *Angew. Chem. Int. Ed.*, vol. 47, pp. 142 –144, 2008.

[7] S. H. Jeong, A. Hagman, K. Hjort, M. Jobs, J. Sundqvist, Z. G. Wu, "Liquid alloy printing of microfluidic stretchable electronics", *Lab Chip,* vol. 12, pp. 4657-4664, 2012.

[8] S. H. Jeong, K. Hjort, Z. G. Wu, "Tape transfer printing of a liquid metal alloy for stretchable RF electronics", *Sensors*, vol. 14, no. 9, pp. 16311-16321, 2014.

[9] W. Wang, Z. Yao, J. C. Chen, J. Fang, "Composite elastic magnet films with hard magnetic feature", *J. Micromech. Microeng.*, vol. 14, pp. 1321-1327, 2004.

CONTACT

*Z.G. Wu, tel: +46-18-471-1086;
Zhigang.Wu@angstrom.uu.se

THREE-AXIS PIEZOELECTRIC VIBRATION ENERGY HARVESTER

Ethem Erkan Aktakka, and Khalil Najafi
Center for Wireless Integrated MicroSensing & Systems (WIMS²)
University of Michigan, Ann Arbor, MI, USA

ABSTRACT

This paper reports for the first time a piezoelectric inertial harvester for scavenging vibrational energy in all three dimensions. The harvester is formed of optimized bulk-PZT/Si crab legs such that the first three resonance vibrational modes are in-plane and out-of-plane translational modes with closely-spaced frequencies (387-398 Hz). Partitioned electrodes on the PZT arms collect mechanical energy in the transverse (d_{31}) piezoelectric mode, and have phase difference in their voltage outputs according to the axis of the applied vibration. The harvester is realized using a planar micro-fabrication method, with internal and packaged device volumes of 2.5 and 6.5 cm³, respectively. The device has generated 5-53 μW from 50 mg acceleration applied in in-plane and out-of-plane directions at the respective resonance frequencies, and has a half-power bandwidth of 8-16 Hz. The normalized power density (0.3-3.2 mW/cm³/g²) compares favorably with respect to previously reported multi-axis inertial energy harvesters.

INTRODUCTION

Inertial energy harvesting from ambient mechanical vibrations is a promising technology to enable next-generation wireless sensor nodes, which are self-powered, maintenance free, and thus truly autonomous [1]. There have been a large number of micro and meso-scale inertial harvesters reported up to date. Most can only operate at a single vibrational axis. Harvesting electrical energy from vibrations applied along any spatial direction can both improve their power output and extend the practical applications. One way to achieve this goal is by using three individual harvesters aligned along the three different axes assembled in a single package [2]. This will, however, decrease the total power density and increase the overall cost due to the enlarged device size. Previously, single-transducer three–axis energy harvesters were reported only for electrostatic and electromagnetic resonators [3-4], although with very limited performance in terms of power output (16-25 nW) and power density (<125 nW/cm³/g²), resonance frequency (1.5-25 kHz), and frequency split (100-1000 Hz). Until now, only 2-axis piezoelectric inertial harvesters have been reported, based on configurations such as asymmetric inertial mass [5], multiple mass-spring combinations [6], three-dimensional (L or U shaped) beam design [7-8], permanent-magnet and ball-bearing combination [9], and non-linear motion of a circular cantilever rod due to surrounding permanent magnet architecture [10]. In addition to the limited number of operational axes, the architectures used in these devices require mostly three-dimensional structures with manual assembly, which prevent further device miniaturization. This paper reports for the first time a piezoelectric transducer that can harvest energy from all three axes (Figure 1). The

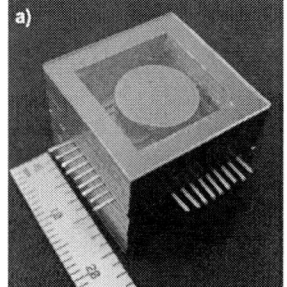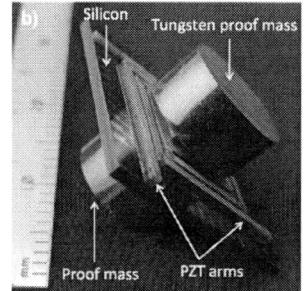
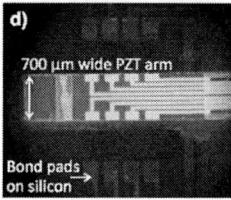

Figure 1: a) Three-axis piezoelectric inertial energy harvester packaged in 6.5 cm³, b) Device structure with balanced proof mass on both sides, c) PZT-Si unimorph harvester platform, d) Wire-bond pads on PZT & Si.

reported device architecture can be further miniaturized through existing planar micro-fabrication methods.

DESIGN OF THREE-AXIS HARVESTER

The harvester is formed of four 700-μm wide and 1.5 mm thick PZT/Si unimorph, crab-leg suspensions that symmetrically hold an 8×8 mm² sized center platform and enable its in-plane and out-of-plane motions. There is a common ground electrode underneath the piezoelectric layer, while each crab-leg suspension has 8 partitioned top electrodes to harvest energy in the transverse (31) piezoelectric mode (Figure 2). Depending on the applied acceleration direction, and thus the created mechanical strain on the beams, the outputs from these top electrodes have different voltage polarity. This is unlike a conventional piezoelectric cantilever beam, where a single top and bottom electrode is used to harvest only from out-of-plane vibrations. The challenges introduced by these partitioned electrodes on the required power management circuitry are the decreased capacitance of (and collected charge from) each electrode, need for a full-wave rectifier and load matching circuitry for each electrode, and the cost of increased IC chip area to accommodate the number of input pads. The partitioned electrodes enable the presented harvester to be utilized for harvesting energy from not only translational but also rotational periodic motions (tilting around X, Y, or Z axes), although these are not studied in this paper due to the required test setup.

For maximum efficiency, the presented multi-axis harvester is designed for resonant operation within a

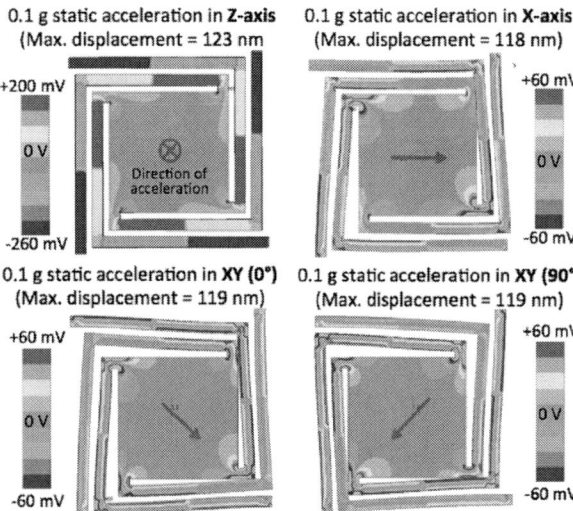

Figure 2: Finite element simulation results for open circuit output when a static acceleration of 100 mg amplitude is applied in different axes.

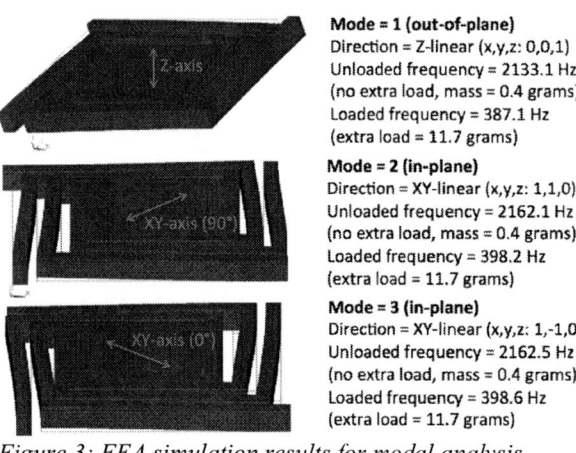

Mode = 1 (out-of-plane)
Direction = Z-linear (x,y,z: 0,0,1)
Unloaded frequency = 2133.1 Hz
(no extra load, mass = 0.4 grams)
Loaded frequency = 387.1 Hz
(extra load = 11.7 grams)

Mode = 2 (in-plane)
Direction = XY-linear (x,y,z: 1,1,0)
Unloaded frequency = 2162.1 Hz
(no extra load, mass = 0.4 grams)
Loaded frequency = 398.2 Hz
(extra load = 11.7 grams)

Mode = 3 (in-plane)
Direction = XY-linear (x,y,z: 1,-1,0)
Unloaded frequency = 2162.5 Hz
(no extra load, mass = 0.4 grams)
Loaded frequency = 398.6 Hz
(extra load = 11.7 grams)

Figure 3: FEA simulation results for modal analysis.

limited frequency range. Therefore, the beam dimensions are optimized via finite element modeling (ANSYS) such that the resonance frequencies for translational motion in both in-plane (XY-axes) and out-of-plane (Z-axis) directions are closely matched (Figure 3). Without any additional mass, the center stage has a weight of 0.4 grams, and the frequencies associated with the considered modes of the platform are simulated to be in the range of 2.1 kHz. These values are calculated to decrease down to 385-400 Hz when an external proof mass of 11.7 grams is attached, which is in a more practical frequency range for industrial applications.

DEVICE FABRICATION

The fabrication process is summarized in Fig. 4. After e-beam evaporation and lift-off patterning of Cr/Au metal electrodes on a 500 μm thick polished bulk-PZT substrate, a ~4 μm thick parylene film is evaporated and RIE-patterned as an electrical isolation layer. Next, the electrical interconnects are deposited and patterned on this

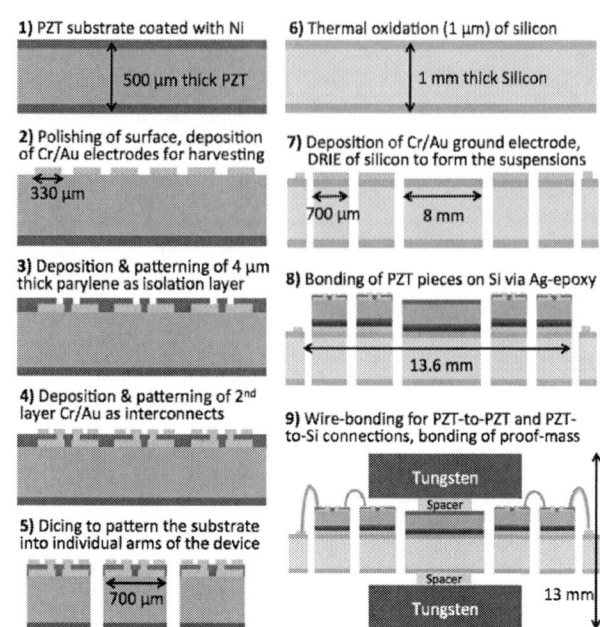

Figure 4: Fabrication process of the piezoelectric three-axis vibration energy harvester.

parylene layer allowing individual connections to the partitioned energy-harvesting electrodes underneath. Finally, the piezoelectric substrate is diced into individual pieces to be utilized as the suspension arms, which are manually aligned and assembled via conductive epoxy on a 1 mm thick, oxidized, metallized and DRIE-patterned silicon die (Figure 1c).

Electrical connections from one PZT piece to the adjacent piece on the same crab-leg suspension are achieved via ball bonding between pads on these arms. Similarly, the electrical connections from a crab-leg suspension to the silicon die are achieved via wire bonding (Figure 1d). The wire bonds are covered by epoxy to provide mechanical protection during vibration tests. Two similarly sized tungsten proof-masses with 10-mm diameter (11.7 grams in total) are attached to the top and bottom sides of the center platform with 0.7 mm thick spacers in-between (Figure 1b). Compared to the case of using a single proof mass at either side of the platform, this configuration prevents unbalanced motions due to unequal weight distribution. For instance, without this measure, a translational vibration input along X-axis would also create tilting motion around Y-axis since the centroid of the proof mass is not located at the center of the moving platform.

For handling and testing purposes, the harvester is packaged in a non-hermetic silicon frame with glass covers on its top and bottom sides (Figure 1a). The packaged device volume is 6.5 cm^3, while the internal device consumes 2.5 cm^3 (13.6×13.6×13.6 mm^3). Since the presented piezoelectric device architecture is planar, it can be further miniaturized via a previously reported microfabrication method, which involves solid-state bonding, lapping, and etching of a bulk-PZT substrate on silicon [11].

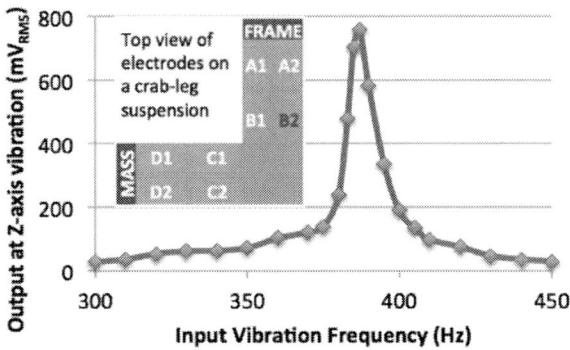

Figure 5: Measured frequency response from a top electrode (B2) when harvester is excited out-of-plane (Z).

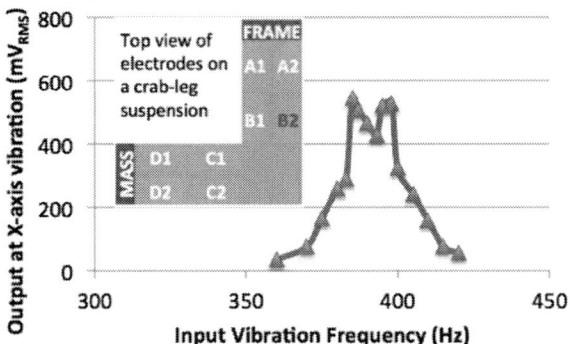

Figure 6: Measured frequency response from a top electrode (B2) when harvester is excited in-plane (X).

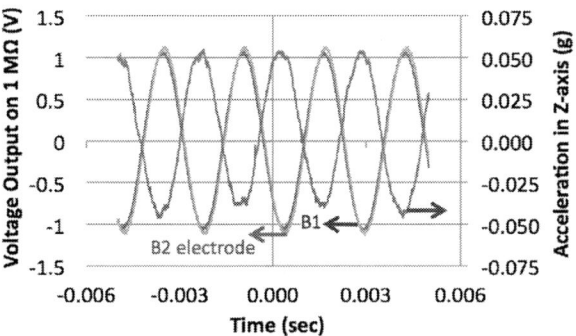

Figure 7: Measured voltage outputs from two electrodes on 1 MΩ loads for 50 mg acceleration amplitude at 387 Hz in the Z-axis direction.

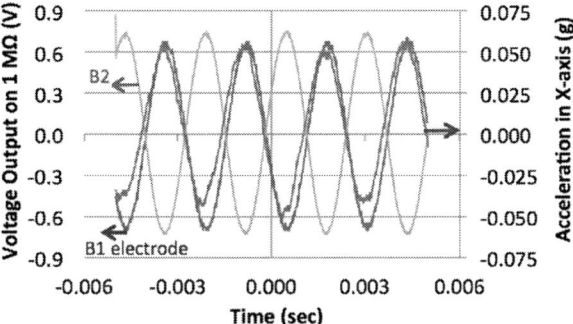

Figure 8: Measured voltage outputs from two electrodes on 1 MΩ resistive loads for 50 mg acceleration amplitude at 398 Hz in the X-axis direction.

EXPERIMENTAL RESULTS

The harvester is fixed on and excited by an electromagnetic shaker table at 50 mg acceleration amplitude in the X-, Y-, and Z-axes individually. The measured resonance frequency for the out-of-plane mode (Z) is ~387 Hz with a half-power bandwidth of 8 Hz (Figure 5). Due to the rectangular shape of the device package and the utilized mechanical clamping method, the harvester had to be excited across X or Y axes to detect its in-plane modes, although the 2nd and 3rd modes are actually 45° misaligned from these directions. In this test, the resonance frequencies are measured as ~385 Hz and ~398 Hz for the in-plane modes (XY-axis 0° and 90°) (Figure 6). Since the in-plane mode resonance frequencies are closely spaced and the tested input vibration has vector components in both directions, this results in an extended bandwidth of 16 Hz.

Depending on the input vibration axis, it is observed that there is amplitude and phase difference between the voltage outputs of individual electrodes as expected (Figures 7-8). The optimum external resistive load to obtain maximum power output from each electrode is measured to be 1 MΩ. The sum of measured power output from 32 individual electrodes is 52.9 µW for Z-axis, and 4.9 µW for X-axis, and 5.1 µW for Y-axis vibration inputs. The calculated normalized power densities (0.3-3.2 mW/cm^3/g^2) compare favorably with previously reported multi-axis harvesters, although the values are slightly lower than single-axis, high-performance, microfabricated bulk-PZT cantilever beam energy harvesters (3.5-10 mW/cm^3/g^2) [12-13].

The energy harvested from in-plane vibrations is measured as considerably lower compared to the energy obtained from out-of-plane vibration. This is mostly due to the undesired confinement of most of the mechanical stress on the silicon material instead of the piezoelectric layer during lateral bending. This problem can be solved in a future device architecture, where a PZT/PZT bimorph beam cross-section with partitioned electrodes on both top and bottom surfaces is used instead of the presented PZT/Si unimorph beam design. Another cause is that lateral bending creates varying amount of surface charge (and thus electrical potential) from the center of the piezoelectric beam to its lateral edge, while a single electrode is used to cover this region and average out the potential. This results in un-optimized external-load matching and electrical damping on the beam motion. Finally, the 45° misalignment between the actuation direction and the tested in-plane resonance mode causes only one component in the vector basis of the vibration input to be effective in the power output.

CONCLUSION

A three-axis resonant inertial energy harvester, based on transverse-mode piezoelectric crab-leg suspensions

978-1-4799-7956-1/15 $31.00 © 2015 IEEE   1143

and partitioned top electrodes, is presented. For fast prototyping purpose, the introduced planar device architecture is fabricated via individually assembling processed and diced bulk-PZT legs on a patterned silicon die, and forming electrical connections between individual pieces through wire bonding. The multi-axis harvester can scavenge 5-50 µW power from vibrations with 50 mg acceleration amplitude in the X, Y or Z spatial directions at the first three modal frequencies of the device (387-398 Hz). Future work will focus on improving power output in in-plane vibration modes.

ACKNOWLEDGEMENTS

This research is partially supported by DARPA PASCAL award #W31P4Q-12-1-0002. The fabrication process was performed at the University of Michigan's Lurie Nanofabrication Facility (LNF), a member of the National Nanotechnology Infrastructure Network (NNIN), which is funded in part by the National Science Foundation (NSF).

REFERENCES

[1] S. Chamanian, S. Baghaee, H. Ulusan, O. Zorlu, H. Kulah, E. Uysal-Biyikoglu, "Powering-up wireless sensor nodes utilizing rechargeable batteries and an electromagnetic vibration energy harvesting system," *Energies*, vol. 7, pp. 6323-6339, 2014.

[2] J. L. Fu, Y. Nakano, L. D. Sorenson, F. Ayazi, "Multi-axis AlN-on-Silicon vibration energy harvester with integrated frequency up-converting transducers," *Proc. of IEEE MEMS*, pp. 1269-1272, 2012.

[3] A. G. Fowler, S. O. R. Moheimani, S. Behrens, "A 3-DOF SOI MEMS ultrasonic energy harvester for implanted devices," *Journal of Physics: Conference Series, PowerMEMS 2013*, vol. 476, 012002, pp. 1-5, 2013.

[4] H. Liu, B. W. Soon, N. Wang, C. J. Tay, C. Quan, C. Lee, "Feasibility study of a 3D vibration-driven electromagnetic MEMS energy harvester with multiple vibration modes," *J. Micromech. Microeng.*, vol. 22, 125020, pp. 1-11, 2012.

[5] J. C. Park, J. Y. Park, "A two-dimensional vibration energy harvester using a piezoelectric bimorph cantilever with an asymmetric inertial mass," *Proc. PowerMEMS*, pp. 281-284, 2011.

[6] S. Hashimoto, et al., "Multi-mode and multi-axis vibration power generation effective for vehicles," *IEEE Int. Symp. Industrial Electronics (ISIE)*, pp. 1-6, 2013.

[7] H. Wu, L. Tang, Y. Yang, C. K. Soh, "Feasibility study of multi-directional vibration energy harvesting with a frame harvester," *Proc. of SPIE, Active and Passive Smart Structures and Integrated Systems*, vol. 9057, 905703, pp. 1-12, 2014.

[8] M. D. Dawson, D. A. W. Barton, "Vibrational energy harvesting using a multi-degree-of-freedom device," *Proc. Int. Conf. Noise and Vibration (ISMA)*, pp. 3691-3704, 2010.

[9] S. Moss, J. McLeod, I. Powlesl and, S. Galea, "Bi-axial vibration energy harvesting," Unclassified technical report by Air Vehicles Division, Defense Science and Technology Organization, Commonwealth of Australia, DSTO-TR-2649, pp. 1-41, July 2012.

[10] J. Yang, Y. Wen, P. Li, X. Yue, Q. Yu, X. Bai, "A two-dimensional broadband vibration energy harvester using magnetoelectric transducer," Applied Physics Letters, vol. 103, 243903, 2013.

[11] E. E. Aktakka, R. L. Peterson, K. Najafi, "Wafer-level integration of high-quality bulk piezoelectric ceramics on silicon," *IEEE Transactions on Electron Devices*, vol. 60, pp. 2022-2030, 2013.

[12] E. E. Aktakka, K. Najafi, "A micro inertial energy harvesting platform with self-supplied power management circuit for autonomous wireless sensor nodes," *IEEE Journal of Solid State Circuits*, vol. 49, pp. 1-13, 2014.

[13] E. E. Aktakka, R. L. Peterson, K. Najafi, ""Multi-layer PZT stacking process for piezoelectric bimorph energy harvesters" *11th International Conference on Micro and Nanotechnology for Power Generation and Energy Conversion Applications (PowerMEMS'11)*, Seoul, Republic of Korea, pp. 139-142, Nov. 2011.

CONTACT

*Ethem Erkan Aktakka, Tel: +1-734-272-3170; aktakka@umich.edu. Khalil Najafi, Tel: +1-734-763-6650; najafi@umich.edu.

VERTICAL CAPACITIVE ENERGY HARVESTER POSITIVELY USING CONTACT BETWEEN PROOF MASS AND ELECTRET PLATE - STIFFNESS MATCHING BY SPRING SUPPORT OF PLATE AND STICTION PREVENTION BY STOPPER MECHANISM -

Tomokazu Takahashi[1], Masato Suzuki[1], Toshio Nishida[2], Yasuhiro Yoshikawa[2], and Seiji Aoyagi[1]
[1]Kansai University, Osaka, JAPAN
[2]ROHM Co. Ltd., Kyoto, JAPAN

ABSTRACT

We proposed the vertical capacitive energy harvester using dielectric plate and polymer electret [1-3]. In this study, two methods to effectively use the contact between proof mass and electret; one is to match the stiffness between plate and mass, which is effective to increase their contact duration. For this purpose, instead of a gel for shock-absorber, a soft spring for supporting the plate is employed. Another is to study detach mass from their contact, opposing electrostatic attraction. We contrived a stopper mechanism. The output power was improved 5 times up to 50 μW when acceleration and frequency was 0.2 *g* and 18 Hz.

INTRODUCTION

A conventional capacitive harvester has the comb shaped electret and electrodes, and a laterally movable mass, which is supported by spring [4-10]. The electrode faces charge-implanted electret with small gap distance [4, 5]. When the mass is motion-driven, the overlapping area between electrode and electret is changed. Thus, the attracted holes in electrode are changes and the electric current is produced. The advantage of comb shape is frequency up-conversion.

However, the laterally type has some limitations. The gap is limited, because the discharge possibility was increased. Thus the gap should be rather large, limiting the change in capacitance. The high charge density in electret is difficult due to electrostatic repulsion effect [5, 10]. The comb shaped electret and electrodes should be aligned with high accuracy [4, 5]. A slight misalignment causes a rapid decrease of output power. The reported highest power of the lateral type still remains below milliwatt order [4]. We previously proposed a lateral device based on the fringe electric field change inside ferroelectric. It achieved a small gap and one side wiring [11]. Its output power was, however, limited to microwatt order.

There have been some reports on the vertical type harvester [12-15]. An upper electrode faces an electret on a lower electrode. The capacitance changes when the upper electrode is moved. Thus the current are produced. However, the amplitude of electrode was limited to a small value, because the discharge occurs at small gap between the upper electrode and the electret. The small amplitude causes the small change in capacitance and output power.

In the previous articles, we proposed the vertical type harvester using dielectric, as shown in Fig. 1 [1-3]. The discharge is avoided due to using the dielectric beneath the counter electrode. Thus, the small gap or even contact state

is feasible. The device achieves the large change in capacitance, i.e., generates the high output power. The electret needs no comb shaped patterning, which is effective for realizing stable and high surface potential. The high accuracy is not necessary to alignment between mass and electret, due to no patterning.

PRINCIPLE

Figure 1(a) shows a principle of vertical type harvester. The implanted charge in electret is balanced with the sum of plus charge in top and bottom electrodes. When the gap between electret and dielectric is large, the pulse charge in bottom electrode is equal to implanted charge, since the pulse charge in top electrode is approximately zero. When the gap is small, the implanted charge attracts the pulse charge, which flow from bottom to top electrode. When the gap is zero, the harvester is equivalent of RC circuit, which has the constant resistance and capacitance. In contact, the current decay exponentially. The longer contact time contributes to keep high current for a longer time.

Our previous harvester used a gel to support the electret plate. It could absorb the collision shock and reduce the radial vibration. However, its stiffness did not match with that of mass vibration system, making the contact instantaneous (Fig. 1(b)). In the presence harvester, the electret plate is supported by a spring (Fig. 1(c)). The stiffness of spring is adjusted so that the resonant frequency of plate is matched with that of mass, which mitigates the repelling force and ensures the stable contact for a long duration.

A stopper mechanism is added, which limits the upward movement of electret plate (Fig. 2). It effectively works especially under high surface potential and low acceleration.

STRUCTURE

Figure 3 shows the structure of proposed energy harvester. The brass mass was supported by a double phosphor bronze spring for suppressing the unwanted second vibration mode. A compounded TiO_2-BaO plated was bonded on the mass, since the dielectric material prevents electret from a discharge. The CYTOP (Asahi Glass Co., Ltd.) electret film were formed on cupper plate. The surface potential of electret was −480 V. The electret plate was supported by cupper spring. The spring constant was 1 N/mm, which was expected by a numerical analysis. In the initial position, this spring and acrylic stopper were in contact. The cases were made of acrylic. Figure 4 shows the fabricated miniature harvester.

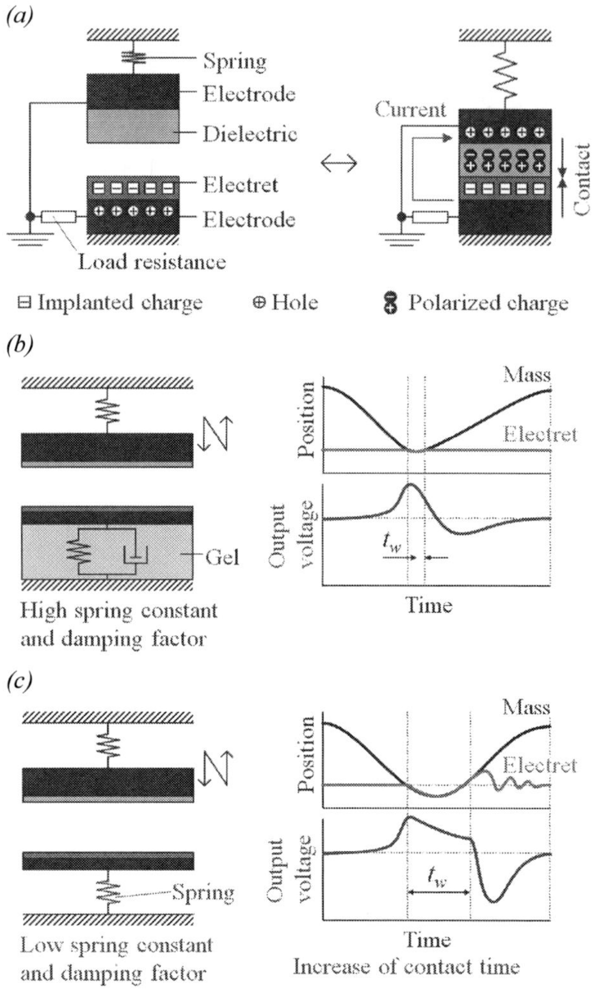

(a)

(b)

(c)

Figure 1: Principle of using contact between mass and electret using support instead of gel for stiffness matching.

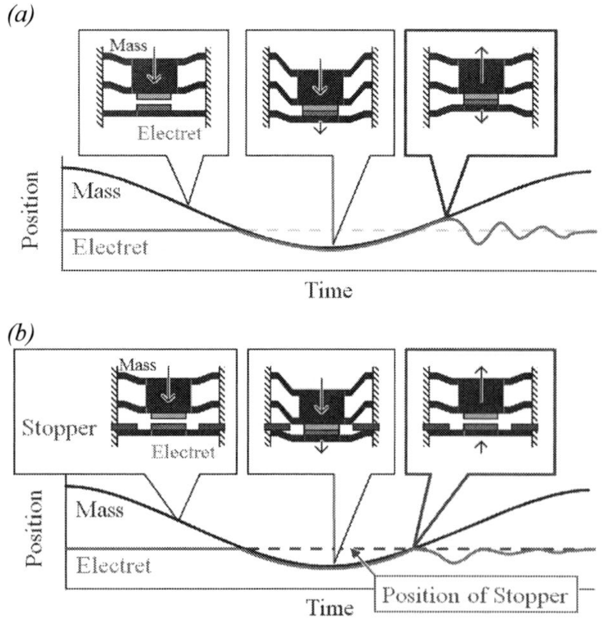

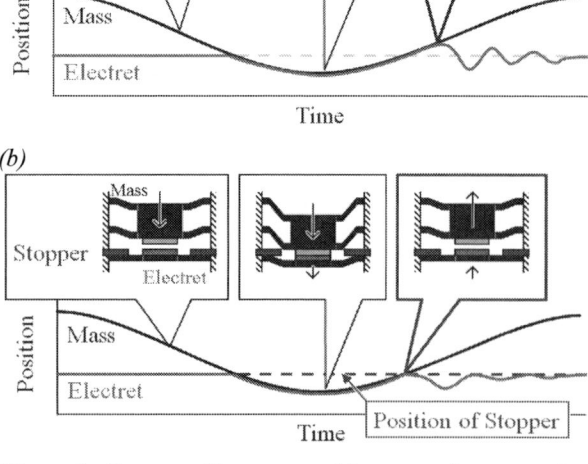

Figure 2: Concept of harvester with stopper for preventing absorption between electret and dielectric.

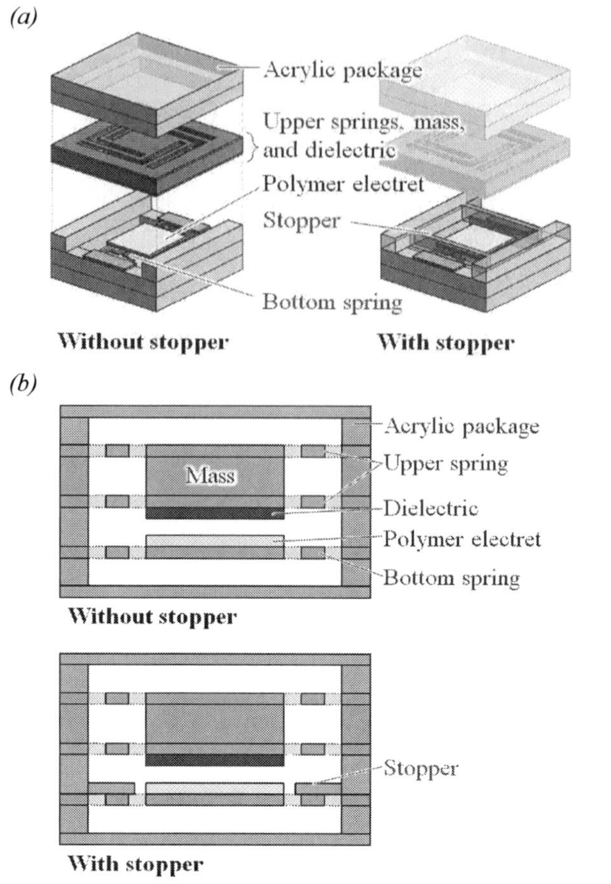

(a)

(b)

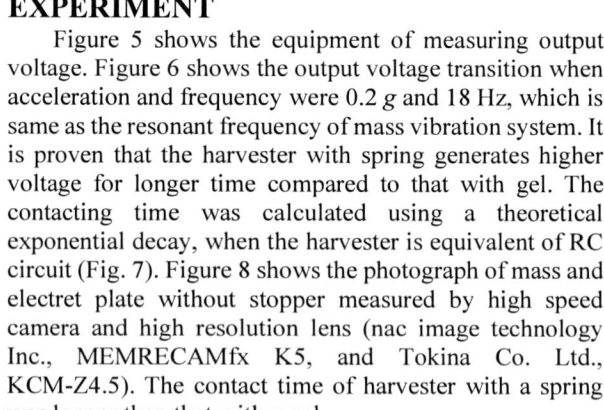

Figure 3: Structure of proposed harvester using spring support for electret: (a) bird eye's view, and (b) cross-sectional view.

EXPERIMENT

Figure 5 shows the equipment of measuring output voltage. Figure 6 shows the output voltage transition when acceleration and frequency were 0.2 g and 18 Hz, which is same as the resonant frequency of mass vibration system. It is proven that the harvester with spring generates higher voltage for longer time compared to that with gel. The contacting time was calculated using a theoretical exponential decay, when the harvester is equivalent of RC circuit (Fig. 7). Figure 8 shows the photograph of mass and electret plate without stopper measured by high speed camera and high resolution lens (nac image technology Inc., MEMRECAMfx K5, and Tokina Co. Ltd., KCM-Z4.5). The contact time of harvester with a spring was longer than that with a gel.

Figure 9 shows the relationship between output power and input acceleration. The output power was calculated by the output voltage and resistance. The remarkable effect of matching stiffness was confirmed. It is noted that the harvester can generate the power even at small acceleration of 0.05 g, when the mass and electret were not in contact. The achieved output of 50 μW is not less than reported ones, even taking account of its size and weight (Table 1) [7, 8].

978-1-4799-7956-1/15 $31.00 © 2015 IEEE

(a)

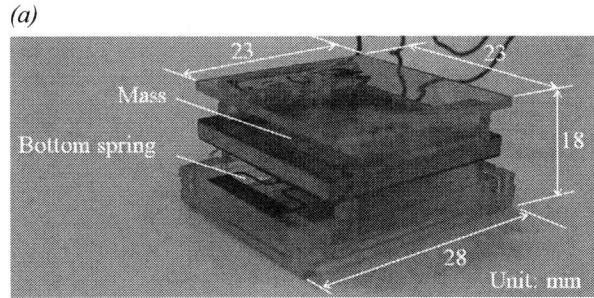

(b)

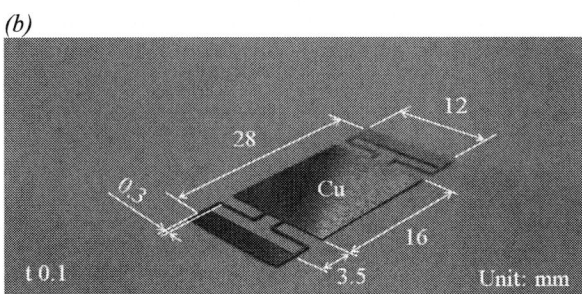

(c)

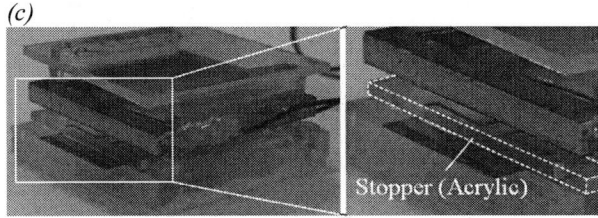

Figure 4: Photograph of (a) fabricated harvester, (b) bottom spring, (c) stopper.

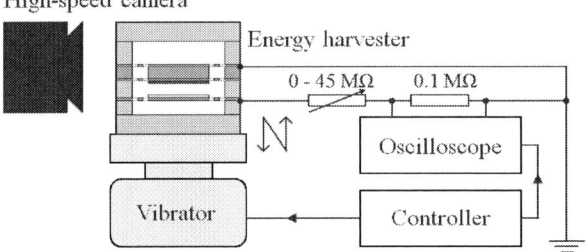

Figure 5: Equipment of measuring output voltage.

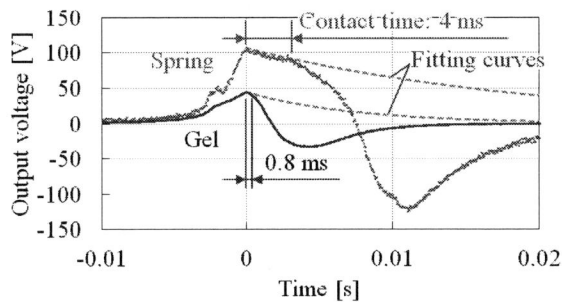

Figure 6: Experimental results and fitting curves of relationship between time and voltage.

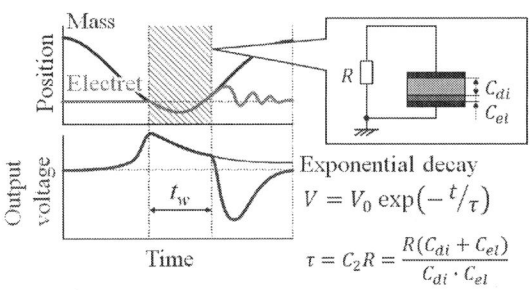

Figure 7: Calculation of fitting curve.

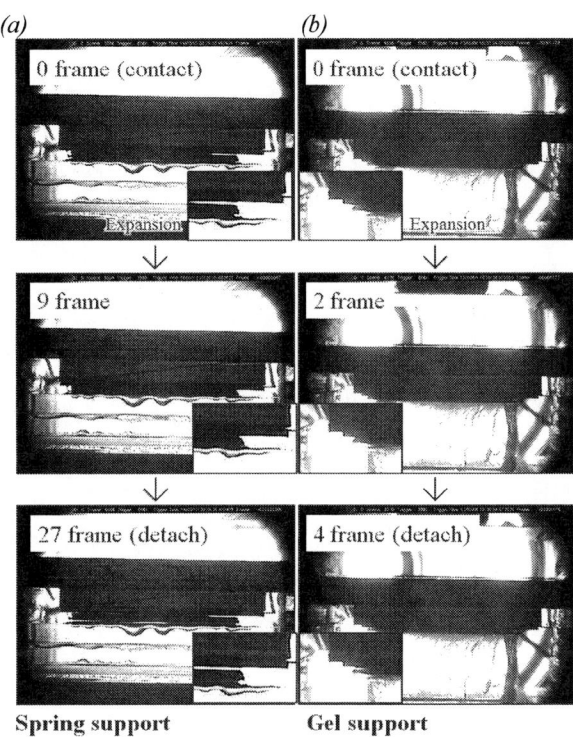

Spring support　　**Gel support**

Figure 8: High speed camera photograph of mass and electret plate without stopper.

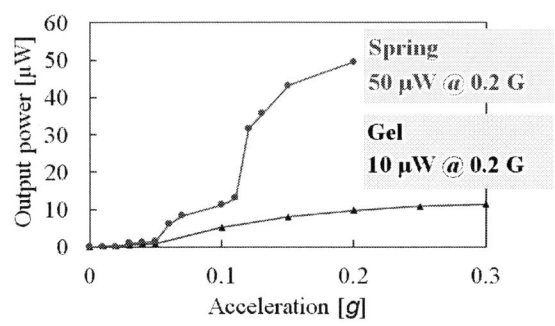

Figure 9: Experimental results of relationship between acceleration and output power.

We preliminarily characterized the device with proposed stopper. Output voltage transition is shown in Fig. 10 (normalized so that the peak is 1). It is proven that the mass is detached from the plate quickly when the stopper is applied. It is noted the minus peak voltage is increased, which contributes to increasing the output power. The output power of harvester with stopper was twice as high as that without stopper.

Table 1: Performance comparison (capacitive vibratory).

	Power	Device (mass size and weight)	Acceleration
Our device	50 μW	$28 \times 23 \times 15$ mm^3 ($15 \times 15 \times 2$ mm^3, 4 gf)	0.2 g (18 Hz /150 μm$_{p-p}$)
Y. Suzuki et al. [7]	6 μW	18.5×16.5 mm^2 (11.6×10.2 mm^2, 0.4 gf)	1.4 g (40 Hz / -)
T. Fujita et al. [8]	0.23 μW	12×13 mm^2 ($10 \times 10 \times 5$ mm^3, 0.12 gf)	0.1 g (10 Hz /200 μm$_{p-p}$)

Experimental results (0.2 g, 18 Hz)

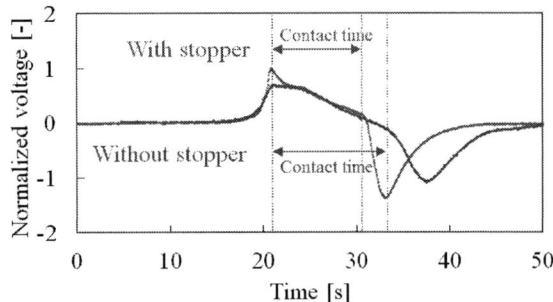

Figure 10: Effect of stopper for Experimental results of relationship between acceleration and output power.

CONCLUSIOIN

We improved output power by two methods. One is to match the stiffness between plate and mass, which is effective to increase their contact duration. For this purpose, the electret plate was supported by metal spring. The harvester with spring generated higher voltage for longer time compared to that with a gel of previous device. The output power was improved by 5 times up to 50 μW. Another is a stopper mechanism for avoiding a sticking between mass and electret. The mass was detached from electret plate quickly when the stopper is applied.

REFERENCES

[1] T. Takahashi *et al.*, "Milliwatt order vibratory energy harvesting using electret and ferroelectric –discharge does not occur with small gap and only one wiring is required," in Digest Tech, MEMS 2012 (2012) p. 1265-1269.

[2] T. Takahashi *et al.*, "Application of paraelectric to a miniature capacitive energy harvester realizing several tens micro watt –relationship between polarization hysteresis and output power–", *in Digest Tech. MEMS 2013* (2013) pp.877-880.

[3] T. Takahashi *et al.*, "A miniature harvester of vertical vibratory capacitive type achieving several tens microwatt for broad frequency of 20-40 Hz," *in Digest Tech. Transducers 2013*, (2013) pp.1340-1343.

[4] Y. Sakane *et al.*, "The development of a high-performance perfluorinated polymer electret and its application to micro power generation", *J. Micromech. Microeng.*, vol. 18, 2008, p. 104011 (6 pages).

[5] Y. Naruse *et al.*, "Electrostatic micro power generation from low-frequency vibration such as human motion", *J. Micromech. Microeng.*, vol. 19, 2009, p. 094002 (5 pages).

[6] Y. Suzuki, "Development of a MEMS Energy Harvester with High-perfomance Polymer electrets", *in Digest Tech. PowerMEMS 2010*, 2010, pp. 47-52.

[7] K. Matsumoto *et al.*, "Vibration-powered Battery-less Sensor Node Using MEMS Electret Generator," *in Digest Tech. PowerMEMS 2011*, 2011, pp. 134-137.K. Fujii, T.

[8] K. Fujii, *et al.*, "Electret Based Energy Harvester using a Shared Si Electrode", in Digest Tech. Transducers 2011, 2011, pp. 2634-2637.

[9] T. Fujita *et al.*, "Evaluation of the Electret Based Energy Harvester by Using Multipurpose Data Logging Device", *in Digest Tech. PowerMEMS 2011*, 2011, pp. 130-133.

[10] T. Genda *et al.*, "Charging method of micropatterned electrets by contact", *J. J. Appl. Phys.*, vol. 44, no. 7A, 2005, pp. 5062-5067.

[11] T. Takahashi *et al.*, "Electret Energy Harvesting Based on Fringe Electrical Filed Change inside Trenched Ferroelectric", *in Digest Tech. MEMS 2011*, 2011, pp.1305-1308.

[12] T. Suzuki *et al.*, "Novel Vibration-Driven Micro-Electrostatic Induction Energy Harvester with Asymmetric Multi-Resonant Spring", *in Digest Tech. IEEE SENSORS 2010 Conference*, 2010, pp. 1161-1164.

[13] Y. C. Lee and Y. Chiu, "Low-cost Out-of-plane Vibrational Electret Energy Harvester", *in Digest Tech. PowerMEMS 2011*, 2011, pp. 302-305.

[14] Y. Chiu *et al.*, "Flat and robust out-of-plane vibrational electret energy harvester", *J. Micromech. and Microeng.*, vol. 23, no. 1, 2013, 015012.

[15] H. Asanuma *et al.*, "Ferroelectric dipole electrets prepared from soft and hard PZT ceramics in electrostatic vibration energy harvester", *J. Phys.: Conference Series*, vol. 47, 2013, 012041 (5 pages).

CONTACT

*T. Takahashi, tel: +6-6368-1835;
t.taka@kansai-u.ac.jp

MEMS 2015 Keyword Index

2D Crystal	877
2D Phononic Crystal	1016
3D	284
3D Cell Culture	443, 643
3D Fabrication	1071
3D Hall Sensor	893
3D Integration	304, 996
3D Microfluidics	196
3D Morphology Reconstruction	358
3D Photolithography	841
3D Printing	261
3D Structure	405, 539
4H-SiC	276
6-Axis Sensor	730

A

Accelerometer (s)	849, 865
Acoustics	921
Acousto-Optic Modulators	980
Actuator	114
Adherent Cells	439
Adhesion	393, 849
Adhesive Bonding	280
Adipocyte	643
Affinity Binding	488
AFM (Atomic Force Microscope)	150, 732, 752
Air Flow	710
Air-Molding	348
ALD (Atomic Layer Deposition)	771, 805
Alginate	26
Alignment	465
Allan Deviation	180
Allan Variance	37
Aluminum Nitride - AlN	73, 140, 373, 837, 921, 928, 984, 1094
Amplitude-Modulated Sensor	913
AMR Sensor	901
Amyloid	604
Anchor Loss	797
Anode	130
Anodic Silicon Dioxide	385
Anodized Metal	265
Aptamer	488, 569, 581
Aqueous Two Phase Extraction	547
Array (s)	917, 952

Assembling	14
Autofluorescence	358
Autorotation	308

B

Baking Powder	504
Band-to-Band Tunneling	988
BAW Gyroscope	789
Bead (s)	431, 604
Betavoltaic Microcell	1102
Bimorph	928
Bio-Dissolvable	636
Bio-Inspired	500, 706, 889
Bio-MEMS	336, 354
Biochemical Sensor	612
Biodegradable	340
Biomarker	455
Biomimetics	500, 889
Biosensor	447, 573
Biosignal Detection	833
Biotemplate	1118
Black Silicon	365
Blood Cell Measurement	435
Body-on-a-Chip	535
Boiling	1122
Bovine Serum Albumin (BSA)	470
Breast Cancer Cell	698
Breath	126
Broadband	146
Bubble	655
Buckypaper	1067

C

C-Reactive Protein	581
Cantilever	144, 718
Capacitive Feedback	956
Capacitive Humidity Sensor	767
Capacitive Transduction	150, 219
Capillary Filling	10
Capillary Force	413
Capillary Number	459
Capillary Vessel	480
Carbon Nanotubes & Devices	249, 312, 369, 401, 417, 744, 1102
Carbonization	817
Castellated Electrode	508
Cavitational Microstreaming	1059
Cavity	1122
Cell Fiber	643
Cell Filter	340
Cell Lysing	188
Cell Manipulation	200, 1059
Cell Migration	624, 628
Cell Origami	589
Centrifugal	192, 504, 523
Chemical Analysis	608

Chemomechanical Transducer	219
Circuit-Breaker Chip	1040
Circulating Tumor Cell	192, 340, 459
CMOS-MEMS	732, 736, 767, 853, 988
Cold Switching	41
Collagen	465
Comb-Drive	14, 215
Commercialization	1
Communication	53
Complete Blood Count	435
Consumer Electronics	61
Contact Angle Hysteresis	484
Contact Electrification	102
Continuous Particle Sorting	508
Core-Shell	472
Core-Shell Microfiber	463
Cryogenic Measurement	14
Crystal	242
Crystal Silicon	389
Current Source	551
Curved PMUT	837
Curved Surface	351
Cyborg Insect	1048
Cylindrical	801

D

Deep Reactive Ion Etching	288, 365
Deep Silicon Etching	885
Deposition	467
Di-Electrowetting	496
Diabetic Retinopathy	154
Diagnosis	447
Diamond	801
Dielectrophoresis	196, 354, 508
Digital Microfluidics	443, 492, 519
Dilatant Fluid	421
Dilution	543
Direct Laser Writing	470
Displacement Magnification	1082
Dissolvable Film	451
Doppler Effect	640
Double Re-Entrant	6, 1122
Droplet Detection	785
Droplet Vibration	176
Drug Cocktail Optimization	658
Drug Delivery	158
Drug Screening	566
Dry Etching	77
Dual Resonance Modes	146
Durotaxis	628

E

Elastomer Packaging	1137
Electret	527, 1071, 1086, 1145
Electric Contact	968
Electric Field	238
Electric Field Trapping	81
Electrical Double Layer	118
Electrical Impedance Measurement	435
Electrical Stimulation	649
Electro Tactile Display	649
Electrochemical Etching	649
Electrochemical Impedance	620, 662
Electrochemical Reaction	555
Electrodynamics	936
Electrokinetic Force	512
Electroless Plating	373
Electromagnetic	122
Electromagnetic	1051, 1075
Electromechanical Coupling	992
Electromicrofluidic	516
Electron Beam	512
Electronic Skin	760
Electrophoresis	488
Electroplating	272, 401
Electrospray	134
Electrostatic	45, 1125
Electrostatic Actuator (s)	22, 276, 897, 1036
Electrostatic Lens	93
Electrowetting	496, 956
Electrowetting-on-Dielectric (EWOD)	265, 519, 547
Embeded Passives	427
Encapsulated	393
End-Stop Transducer	1125
Endoscopic Optical Imaging	948
Endothelial Cell	480
Energy Harvesting	122, 126, 118, 527, 1051, 1071, 1075, 1078, 1094, 1110, 1129, 1145
Energy Pump	913
Energy Storage	1133
Enhanced Controllability	1032
Entropic Effect	300
Enzyme-Linked Immunosorbent Assay	523
Epitaxial	393
Epi-Seal	1
Etching	284
Eutectic Bonding	409
Exfoliation	280
Extracellular Matrix	463

F

Fabrication	1078
Ferrofluid	122
Ferromagnetic Resonance	208
Fiber Array	162
Fiber Sensor	775
Field Effect Transistor	417, 581

Fixation 405
Flapping Wing 22
Flexible 110, 636, 726, 845
Flexible Device (s) 81, 106
Flexible Electronics 1133
Flexible Sensors/Actuators 377
Flexible Substrate 877
Flexion Sensing 760
Flexural Rigidity 238
Flow 620
Flow Control 2
Flow Focusing 539
Flow Lithography 666
Flow Sensor 500, 889
Folded-Beam Suspensions 215
Food Texture 646
Force Measurement 702
Force/Moment Transducer 736
Force/Torque Sensor 257
Free Molecular Flow, Knudsen Diffusion 397
Free-Standing Structures 304
Frequency Characteristic 212
Frequency Modulation 33
Frequency Tuning 789
Frictional Force 245, 253
FSK Modulator 1024
Functional Paper 861
Functional Surface 829
Fused Silica 821

G

Gallium Nitride 1028
Gas Chromatography 771
Gas Sensor (s) 312, 779
Gauge Factor 361
Glass 332
Glass Blowing 805
Glass Substrate 427
Glass Transition 405
Gold Nanoparticles 81
Graphene 324, 569, 600, 865, 869, 873, 1090
Graphene Oxide 2, 381
Graphite on Paper 825
Gyroscope (s) 33, 37, 801, 813, 821

H

H_2S 779
Halbach Array 1051
Hall-Effect 893
Hanging Drop 226, 535
Hard Mask Free 77
Heat Spreader 616
Hemolysin 235
Hierarchical Electrode 1118
High Aspect Ratio 77, 358, 427, 805

High Frequency	150
High Power	118
High Speed Flow	196
High Topography	328
Human Unbilical Vein Endothelial Cell	476
Humidity Sensing	783, 857
Hybrid Parylene	397
Hybrid Silicon	682
Hybrid Structure	1082
Hydrocephalus	620, 662
Hydrogel (s)	26, 585
Hysteresis	531

I

Impactor	885
Impedance	604
Impedance Spectroscopy	226
Implants	620
In Vitro Fertilization (IVF)	519
Induced Pluripotent Stem Cell	463
Inductor	208
Inductors	261
Inertial Harvester	1141
Infrared Sensing	73, 905, 984
Injection	655
Injection Molding	373
Inkjet Printing	632
Instrumentation	849
Integrated Light Source	162
Integration	417
Interface	496
Interface Dynamics	547
Interfacial Dissipation	1000
Interferometric	69
Internal Impact	1125
Intracellular Calcium Level	651
Ion Sensor	785
Ion Source	93
Ion Trap	292
Ionic Liquid	93, 97, 118, 783
IPL Photoresist	470

K

| Keyboard | 1078 |
| Kinesin | 694 |

L

Lab on a Disc	504, 523
Lab-on-a-Chip	10, 451, 563
Large Displacements	14
Laser	817
Laser Doppler Velocimeter	748
Laser Machined	332
Lateral Flow Device	10
Leading-Edge Vortex	308

Length Extensional Mode 785
Lens Array 960
Li-Ion Battery 130
Linear Springs 215
Lipophilic Membranes 523
Liquid 144
Liquid Alloy Patterning 1137
Liquid Metal 261
Liquid Phase Detection 612
Liquid Sensing 718
Liquid Spring 122
Lithium Niobate 992
Live Cancer 324
Load Sensor 833
Lorentz Force 204, 944
Low-Melting-Point-Alloy 18
Low-Noise Low-Power Electronics 932
Low-Temperature Bonding 413
Lung Cancer 600
Lung Fibroblast 476

M

Magnesium 340
Magnetic 604
Magnetic Actuator 1044
Magnetic Devices 936
Magnetic Field Sensor 893
Magnetic Flux Concentrator 901
Magnetic Forces 964
Magnetic Micropillars 624
Magnetic Particles 964
Magnetic Resonance Force Microscopy 344
Magnetic Resonance Imaging 674
Magnetic Structure 1051
Magnetic Thin Films 208
Magnetometer 204, 932, 944
Magnetophoresis 192
Magnifier 952
Mass Sensing 180
Mass Spectrometer 134
Material Reduction 775
Material Transfer 280
MATRIX 467
Mechanotransduction 624
Membrane Sensor 219
Membrane Transport 439
MEMS on CMOS 1004
Metamaterial 960, 984
Metastability 531
Micro Airflow Sensor 714
Micro Electro Mechanical System 37, 180, 292, 296, 381, 714, 718, 932,
.... 992, 1012, 1016, 1032, 1036, 1098
Microactuators 936
Microbial Fuel Cell 573, 577
Microcantilever 752

Microchamber 344
Microchannel 666, 702, 841
Microcoil 674
Microcomponent (s) 300, 516
Microdisk 698
Microdispenser 658
Microdroplet 960
Microelectro Mechanical System Micromirrors 948
Microelectro Mechanical System Technology 110
Microelectrode 686
Microfabrication 268, 292
Microfiltration 459
Microfluidic (s) 2, 180, 188, 332, 348, 431, 472, 480, 504, 531, 543, 551, 581
Microfluidic Chip 200
Microgenerator 1094
Microgripper 1106
Microlens 952
Micromachining 284
Micromanipulator 1059
Micromirror 401
Micromixer 658
Micropatterning 336, 829
Microphones 917
Micropillar 628
Microplasma 268
Micropore 235
Micropump 535
Microrobotics 296
Microstructure 964
Microsupercapacitor 1067, 1133
Microtissue Spheroid 226
Microtube 184
Microtubule 238, 694
Microvalve 563
Microwave Characterization 1012
Microwell Device 439
Mode Coupling 1008
Mode Localization 881
Mode Matching 789
Molecular Detection 230
Molecular Sorter 238
Molybdenum Disulfide (MoS2) 877
Monolithic Integration 771
MOSFET 686
Motion Conversion 897
Motion Sensor 106
Multi-Axis 1141
Multi-Frequency 1024
Multi-Roof Tile-Shaped Vibration Modes 718
Multi-Step-Assay 451
Multilayer Interconnection 265
Multilayer Sidewall Transfer Lithography 288
Myoelectric 678

N

n-p-n	276
Nanoactuator	694
Nanocomposite	130, 744
Nanofiber Forests	857
Nanofluidic (s)	512
Nanofluidic Crystal	593
Nanomaterial	447
Nanoparticle	8, 608
Nanopatterns	555
Nanopillars	320
Nanopore	512
Nanoporous Anodic Aluminum Oxide	69
Nanoscale Gap	85
Nanosensor	869
Nanotribology	968
Nanowire	361
Negative Dielectrophoresis	492
NEMS	57, 288, 686, 865, 940, 972
Neural Probe	158, 616, 636, 682
Neurodegeneration	455
Neuromodulation	651
Neuromuscular Stimulation	1048
Neurons	328
Nickel Oxide Supercapacitor	1118
Non-Immuno Protein Assay	455
Nonlinear Damping	1125

O

One-Shot Valve	1090
One-Step Polymerization	472
Open Microfluidics	535
Optical Cross Connect	57
Optical Force	57
Optical Microsensor	748
Optical Receivers	976
Optical Resonators	976
Optical Stimulation	162
Optical Waveguide	166
Optofluidic Lithography	551
Optogenetic (s)	158, 162, 616
Optomechanical	45, 49
Optrode	682
Organ Hardness	253
Organ on a Chip	566
Oscillating Bubble	1059
Oscillator (s)	41, 793, 976, 988, 1004, 1024
Out-of-Plane	1063
Out-of-Plane Motion	897
Ovenization	793, 809
Overhanging Pillar	1086
Oxygen Transporter	154

P

P-N Heterojunction	1102
Package	421
Packaging	1
Paper-Based Devices	825
Paper-Based Microfluidics	10, 447
Paper-Based Sensor	577
Parametric Drive	29
Parametric Resonators	172
Particle Focusing	492
Particle Processing Chip	559
Particulate Matter	885
Parylene	268, 328, 358, 397, 405, 467, 612, 662, 682
Parylene Photonics	682
Passive	909
Patency Sensor	662
PDMS	316, 397, 431, 467, 476, 539, 563, 624, 817
PDMS Fiber	423
Perfluorohexane	1122
Perfusable Channel	351
Phase Change Lens	1114
Phononic Crystal	797, 1028
Photolithography	97
Photonics	53, 980
Photoresist	97
Piezo Electric Micromachined Ultrasound Transducer (pMUT)	140, 921, 928
Piezoelectric	944, 1129, 1141
Piezoelectric Actuators	65
Piezoelectric Coefficient	861
Piezoelectric Gyroscope	789
Piezoelectric Material	377
Piezoelectric Paper	861
Piezoelectric Resonators	1000
Piezoelectric Thin Film	1094
Piezoelectric Transducer	651
Piezoresistive Cantilever	176, 670, 706
Piezoresistor	144, 245, 361, 730, 825
Pirani	89
Pixel	69
Planer Bilayer	235
Plant-on-a-Chip	702
PLL	793
Pneumatic Balloon Actuator	166
Pneumatic Pump	666
Point-of-Care-Test	219
Poling	1098
Poly(3,4-Ethylenedioxythiophene): Polystyrene-Sulfonate	940
Poly(vinylidenefluoride-co-trifluoroethylene)	110
Polycrystalline Diamond	616
Polyethylene Glycol	636
Polymer Characterization	873
Polymer Distinguishing	102
Polymerase Chain Reaction	488
Polysilicon	361
Position Control	1055

Potassium Ion 1071
Pressure Sensing 722
Pressure Sensor 89
Pressure Stability 484
Printed Sensor 756
Probe-Based Memory 968
Process Monitors 312
Processing 435
Protein 242
Protein Cross-Linking 470
Protein Pattern 328
Pseudo Inverse 736
Pulse Wave 670
PVDF Nanofiber 678
PZT 146, 1098
PZT Thin Film 377

Q

Q Enhancement 204
Quadrature Error 37
Quadruple Filter 134
Quality Factor 797, 821, 853, 1000
Quantitative Assay 543
Quantum Information Processing 292
Quartz 944
Quartz Crystal Resonator 833

R

Ratchet 1106
Rate Integrating Gyroscope 29
Reactive Ion Etching 93
Red Blood Cell 431, 1055
Reference Plane 253
Reflective Display 956
Reflective Micro Objective 632
Reflow Process 427
Relay 114, 940
Reliability 114
Resonator (s) 1, 41, 73, 172, 184, 204, 215, 612, 698, 785, 797, 801, 809,
.... 821, 853, 877, 881, 885, 909, 913, 984, 992, 1020, 1008
Resorbable Electronics 168
Resorcinol-Formaldehyde Aerogel 767
Respiration 126
Retinal Ischemia 154
RF Ground-Signal-Ground (GSG) Probes 1012
RF MEMS 976, 1020
RF Passives 261
RF Resonator 1020
RGFET 988
ROADM 53
Root Growth Characterization 702
Running 257

S

SAW	144, 1028
Scale Factor Stability	29, 33
Scanning Confocal Fluorescence Microscopy	632
Scanning Electron Microscopy	344
Scanning Probe Microscopy	732, 968
Screen Printing	272
Selectivity	756
Self-Assembly	8, 300, 600, 608, 964
Self-Excited	41
Self-Excited Vibration	22
Self-Healing	81, 996
Self-Locking Mechanism	1106
Self-Mixing	547
Self-Powered Sensor	106
Self-Propagating Exothermic Reaction	373
Sensitivity	881
Sensor (s)	324, 593, 600, 752, 917
Sensor Fusion	212
SERS	223, 608
Serum	6
Shape-Memory-Alloy Actuator	1040
Shear Stress Sensor	710
Sheathless Focusing	196
Sheet	465
Shrink Polymer	324
Side-Wall Doping	710
SiGe Ring Resonator	1016
Signal Processing	61
Silicon	385
Silicon Carbide (SiC)	284, 698
Silicon Microelectrode Array	636
Silicon Nanoparticles	130
Silicon on Nothing	184
Silicon Optical Bench	948
Silk	168
Silver Nanoparticles	775
Single Cell Analysis	589
Single Crystalline	276
Single Molecule Sensing	596
Skin-Equivalent	351
Smart Glove	760
Smart Implant	1040
SNR	913
Soft Actuator	18, 26
Soft Material	1082
Soft Robotics	1044
Soft X-Ray Charging	1086
SOI MEMS Devices	409
SOI Technology	1012
Solid-Gate	869
Sound Focusing	925
Spherical Resonator	805
Spike Pin	257
Spring	26

SRAM 49
Statistical Element Selection 996
Stencil 272
Stencil Lithography 632
Stiction 85, 393
Stiffness Gradient 628
Strain Sensor (s) 678, 744, 756
Stress Effects on MEMS 813
Stress Engineering 296, 837
Stress Mapping 736
Stretchable 744, 817
Stretchable Electronics 1137
Stretchable Spring 423
SU-8 308
Superhydrophobic 6
Superlens 952
Superlyophobic 484, 1086
Surface Enhanced Raman Scattering 320
Surface Roughness 245
Surface Tension 496
Surface Texturing 385
Switch 114
Switchable Sensitivity 674
Synthetic Jet Actuators 936

T

Tactile Sensor (s) 245, 249, 253, 421, 726, 740, 825
Tantalum Pentoxide 265
Tau Protein 455
Teeth Vibration 646
TEM Liquid Cell 8
Temperature 1063
Temperature Compensation 809
Temperature Effect 389
Temperature Regulation 1040
Temperature Sensor 756, 909
Temperature Stable 793
Tensile Test 381, 369, 389
Terahertz 960
TFT Susbtrate 354
Thermal Sensor 845
Thermal-Piezoresistive 732, 885
Thermocouple 841
Thermoelastic Dissipation 1000
Thermoelectric Energy Generator 1114
Thin Film Edge Electrode Lithography 555
Thin Film Encapsulation 996
Thin Si Film 344
Tissue Engineering 348, 351, 480
Torque 730
Touch Sensor 861
Transducers 917
Transient Electronics 873, 1090
Transmitter 1024
Trench Sidewall 841

Triboelectric 1078
Triboelectric Energy Harvester 106
Triboelectric Generator 102
Tunable Metamaterial 1032
Tunable Oscillator 45
Tunable Resonator (s) 172, 1020
Tunable Ring Resonator 53

U

U-Shaped 775
Ultrafine Particle Counter 559
Ultrasonic Receiver 640
Ultrasonic Transducer 140, 146
Ultrasound 837, 928
Ultrasound Imaging 140
Ultrasound Stimulation 651
Underwater 925
User Experience 61

V

Vacuum Cavity 845
Vacuum Interface 134
Valve 666
Vapor-Liquid-Solid Growth 686
Variable Capacitors 1036
Varifocal Acoustic Mirror 925
Varifocal Liquid Lens 65
Velcro-Like Surfaces 413
Velocity Control 694
Vertical-Parallel-Plate Array 767
Vibration Energy Harvesting 1141
Vibration-Induced Flow 200
Viscosity 176

W

Wafer Level Packaging 409, 833
Wafer-Scale Synthesis 312
Water-Electrode 126
Wet Anisotropic Etching 385
Wet Etching Process 377
Whole Angle Mechanization 29
Wicking 332
Wireless 909
Wireless Power Transfer 1137
Wrinkle 829

X

XYZ Microstage 14

Z

Zebrafish ECG 690
ZnO Nanowire 764, 779

MEMS 2015 Author Index

A

Abdollahi, H.	952
Abdolvand, R.	909
Aftab, T.	893
Agache, V.	180
Agah, M.	771
Agarkova, I.	226
Agarwal, M.	1
Agentis, D.J.	936
Ahmad, M.M.	288
Ahn, C.H.	1, 29, 33, 393, 809, 1008
Ahn, H.R.	1114
Ahn, J.	292
Ahrens, G.	97
Aimé, J.P.	150
Akbar, M.	771
Akhbari, S.	837, 928
Akin, T.	409
Akita, I.	686
Akita, S.	756
Aktakka, E.E.	1141
Alam, R.	523
Alivisatos, A.P.	8
Allain, M.	65
Alper, S.E.	409
Alves, F.S.	1036
Amjadi, M.	744
An, B.	168
An, Q.L.	857
An, Z.	401
Anderson, K.	523
Aoki, R.	925
Aoyagi, S.	1145
Aqab, N.	928
Arai, F.	200, 431, 833, 1055
Ardanuç, S.	873, 1090, 1129
Ardito, R.	849
Arie, T.	756
Arnold, D.P.	936, 964
Arscott, S.	1012
Asadnia, M.	500, 678, 889
Asai, K.	686
Ashok, A.	385
Ayazi, F.	789

B

Bahl, G.	1
Bakhtina, N.A.	97
Baléras, F.	180
Barniol, N.	1004
Barz, F.	636
Baumgartel, L.	917
Becker, F.	736
Becker, M.F.	616
Beebe, T.	690
Bergbreiter, S.	726
Bergman, L.A.	752
Berthelot, A.	37
Berthet, C.	180
Beyazoglu, T.	976
Bharadwaja, V.	527
Bittner, A.	718
Blanche, T.J.	682
Block, S.T.	921
Boden, T.J.	944
Bolis, S.	65
Bonfanti, A.	932
Boser, B.E.	33, 140, 146
Bourouina, T.	960
Boyle, D.	2
Brenckle, M.A.	168
Brenna, S.	932
Bridoux, C.	65
Bright, V.M.	805
Brooke, H.	825
Brugger, J.	632
Buchaillot, L.	1012
Buckley, A.	551
Buisson, L.	150
Bulović, V.	85
Bürgel, S.C.	226
Burger, R.	504

C

Cabral, J.	1036
Cai, H.	45, 49, 57, 972
Calayir, E.	996
Camera, K.	873
Candler, R.N.	1
Cao, C.	381
Cao, F.	1048
Cattafesta, L.N.	936
Cha, W.	865
Chamanzar, M.	682
Chang, D.T.	944
Chang, H.L.	881, 1106

Chang, J.H.-C.	154, 340
Chang, K.-T.	154
Chang, K.-W.	581
Chang, W.-H.	581
Chapin, C.A.	284
Charalambides, A.	726
Chen, B.	381
Chen, C.-C.	519
Chen, D.P.	320, 857
Chen, H.T.	102
Chen, K.	555
Chen, K.L.	1
Chen, P.-H.	272
Chen, Q.	948
Chen, R.	740
Chen, R.S.	901
Chen, S.	265
Chen, S.-J.	1075
Chen, S.-Y.	516
Chen, T.N.	45
Chen, T.-Y.	740
Chen, X.	110, 268
Chen, Y.	612
Chen, Y.	1, 809, 1008
Chen, Y.-C.	89
Chen, Y.-C.	1086
Chen, Z.	1063
Cheng, C.-L.	423, 767
Cheng, J.	726
Cheng, X.L.	102, 1078
Cheon, H.	292
Chiang, C.F.	1
Chiang, M.-Y.	516
Chiao, M.	624
Chikasawa, T.	480
Chikkadi, K.	312
Chin, C.-H.	988
Chiou, P.-Y.	196
Cho, D.D.	292
Cho, H.	752
Cho, I.-J.	158, 651
Cho, J.Y.	821
Cho, S.K.	496
Choi, D.-H.	126
Choi, G.	577
Choi, J.-K.	219
Choi, S.	573, 577
Choi, S.	956
Choi, Y.S.	261
Chong, H.M.H.	881
Chou, B.C.S.	89
Chu, A.	551

Chu, C.-H.	581
Chu, S.	1118
Chuang, S.-T.	740
Chun, Y.E.	651
Chung, L.Y.	519
Chung, M.	551
Chung, S.K.	1059
Chung, V.P.J.	767
Chung, Y.-C.	740
Chung, Y.-H.	519
Clemens, D.L.	196
Cochet, M.	180
Colinge, C.	280
Collins, S.D.	188
Connell, L.	188
Costa, M.	1036
Cui, T.	324, 600

D

Dahmardeh, M.	1040
Dambrine, G.	1012
Dao, D.V.	825
De Masi, B.	849
Deem, E.A.	936
Dellea, S.	37, 849
Denman, D.J.	682
Deotare, P.B.	85
Dhakar, L.	106
Dias, R.A.	1036
Ding, J.	948
Dinh, T.	825
Dohi, T.	674
Dong, B.	45, 49, 57, 972
Dowling, K.M.	284
Du, Y.	304
Duan, C.	948
Ducatteau, D.	150
Ducrée, J.	2, 192, 504, 523
Duraffourg, L.	361
Dykman, M.	1008

E

El Fellahi, A.	1012
Elata, D.	41, 172, 215, 897
Elezgaray, J.	150
Elmlinger, P.	162
Eminoglu, B.	33
Erbland, J.A.	944
Ernst, T.	361
Errando-Herranz, C.	53
Everhart, C.L.M.	1

F

Fan, B.	616
Fan, C.-C.	89
Fan, J.	543
Fan, S.-K.	516
Fang, W.	69, 423, 767, 901
Fanget, S.	65
Fatemi, H.	909
Faucher, M.	150
Fedder, G.K.	813, 996
Feng, P.X.-L.	698, 877
Feng, Y.-Y.	1075
Figeys, B.	1016
Figueroa, C.	73
Filleter, T.	381
Finne-Wistrand, A.	472
Flader, I.	1, 809
Fluri, D.A.	226
Fonseca, H.	1036
Foroutan, V.	296
Forsberg, F.	280
Fraiwan, A.	573, 577
Frey, O.	226, 535
Fujisawa, D.	905
Fujita, H.	118, 354, 455, 555, 829, 1071
Fujita, K.	373
Fujito, T.	373
Funakubo, H.	1098
Fung, S.	140, 146

G

Gardner, D.S.	208
Gaspar, J.	1036
Gaughran, J.	2
George, S.M.	443
Gerasopoulos, K.	1118
Gertsch, J.	805
Ghodssi, R.	1118
Ghosh, S.	980
Giacci, F.	37
Giner, J.	805
Glick, C.C.	261, 551
Glynn, M.	192
Glynn, M.T.	504
Gokhale, V.J.	73
Goldkorn, A.	340
Goryu, A.	555
Goswami, A.	527
Goto, T.	480
Gowda, S.	527
Graham, A.B.	1

Gray, J.M.	805
Grine, A.J.	976
Gruetzner, G.	97
Grutter, K.E.	976
Gu, Y.D.	45, 49, 57, 972
Gund, V.	873, 1090
Guo, H.	110, 268
Guo, X.	204
Gylfason, K.B.	53

H

Habasaki, S.	666
Haddadi, K.	1012
Hadji, C.	180
Haluska, M.	312
Halvorsen, E.	1125
Hamano, H.	336
Han, C.-H.	114, 126
Han, J.S.	559
Han, M.D.	102, 1078
Hansson, J.	10, 563
Hao, Y.C.	1106
Hao, Y.L.	45, 49, 57, 960, 972
Haq, M.	616
Hara, M.	93, 1094
Harada, S.	756
Haraldsson, T.	10, 472, 563
Hashida, T.	401
Hashiguchi, G.	1071
Hassett, D.J.	577
Hata, H.	905
Hayakawa, T.	200
Hayashi, H.	344
Hayashi, T.	476
Hayashida, Y.	748
He, H.	1063
He, Q.	110
Heidari, A.	801, 837
Heinz, D.B.	1, 393
Hentz, S.	361
Hida, H.	377, 702
Hierlemann, A.	226, 535
Hierold, C.	312
Higashiyama, T.	702
Higurashi, E.	748
Hillmering, J.	563
Hirai, Y.	389, 417
Hirota, J.	968
Ho, C.P.	1032
Hodjat-Shamami, M.	789
Hokazono, K.	417
Honda, W.	756

Hone, J.	865, 869
Hong, S.	292
Hong, S.J.	1059
Hong, V.A.	1, 29, 33, 393, 1008
Hoorfar, M.	492
Hopcroft, M.A.	1
Horsley, D.A.	140, 146, 801, 837, 921
Hoshino, T.	512
Hoskinson, R.	952
Hotzen, I.	41, 172, 215, 897
Hou, Z.	1063
Houmadi, S.	150
Hsiai, T.	690
Hsiao, A.Y.	643
Hsu, C.-H.	519
Hsu, F.-M.	901
Hsu, W.	261
Huang, H.	779
Huang, H.-Y.	519
Huang, J.G.	45
Huang, L.X.	944
Huang, P.-C.	1067
Huang, W.-Y.	658
Hui, Y.	984
Humayun, M.S.	154
Hung, CP	234
Hurst, A.M.	865
Hwang, J.	559
Hwang, S.W.	168

I

Icard, B.	180
Ikeda, K.	463
Inoue, K.	373
Ishida, M.	686
Ishido, H.	257
Isono, Y.	369
Isozaki, A.	405, 706
Isozaki, N.	238
Ito, J.	480, 539
Ito, T.	748
Iwai, K.	551
Iwase, E.	81
Izyumin, I.I.	33

J

Jacot-Descombes, L.	632
Jacquet, F.	65
Jalabert, L.	328
Jeon, J.	940
Jeong, B.	752
Jeong, S.H.	1137

Jia, H. 698
Jiang, H. 332
Jin, W. 662
Jin, Y.F. 45, 49, 57, 960, 972
Joss, D. 632
Joyce, R.J. 944

K

Kan, T. 405
Kanao, K. 756
Kaneko, M. 431, 722, 1055
Kaneko, T. 670
Kang, D. 154, 340, 397
Kang, J.Y. 604
Kanno, I. 377, 702
Kao, W.-J. 581
Kaplan, D. 168
Karsten, S.L. 455
Kawano, T. 686
Kazama, R. 710
Kelm, J.M. 226
Kenny, T.W. 1, 29, 33, 393, 809, 1008
Keshavarzi, S. 413
Keum, H. 752
Khademolhosseini, F. 624
Khomenko, A. 616
Kilcawley, N.A. 504
Kim, A. 585, 1044
Kim, B. 1
Kim, B.J. 620, 662
Kim, C.-J. 6, 265, 1122
Kim, E.S. 122, 917
Kim, H.-L. 559
Kim, J. 158
Kim, J. 184
Kim, J. 488
Kim, J. 752
Kim, M. 292
Kim, M.J. 604
Kim, M.K. 1114
Kim, M.S. 1114
Kim, P. 845
Kim, S. 168, 752
Kim, S. 230
Kim, T. 292
Kim, T.G. 158
Kim, T.S. 651
Kim, Y.J. 1114
Kim, Y.-J. 354
Kim, Y.-J. 559
Kimata, M. 905
Kimbrell, T.S. 393

Kinahan, D.J. 192, 504, 523
Kirby, D.J. 944
Kitamura, N. 649
Kline, M.H. 33
Ko, K. 651
Ko, S.-D. 114
Kobayashi, T. 166, 1098
Kogure, T. 1082
Koh, K. 208
Konishi, S. 166
Korner, K. 551
Korvink, J.G. 97
Kosanic, D. 632
Koshi, T. 81
Kotera, H. 238, 455, 476, 694
Kottapalli, A.G.P. 500, 678, 889
Kozai, R. 245
Kozinda, A. 1133
Kraft, M. 881
Kucera, M. 718
Kumagai, S. 841
Kumar, V. 204, 885, 913
Kumar-Kantimahanti, A. 1004
Kung, Y.-C. 196
Kunishima, I. 968
Kuno, T. 369
Kuroda, Y. 373
Kuwano, H. 93, 1094
Kwon, H.-B. 559
Kwon, K.-Y. 616
Kwong, D.L. 45, 49, 57, 972

L

Lacaita, A.L. 932
Lacour, S.P. 760
Ladner, C. 361
Laghi, G. 932
Lai, W.-M. 901
Lal, A. 873, 1090, 1129
Lammel, G. 61
Lang, J.H. 85
Langfelder, G. 37, 849, 932
Lapatki, B. 736
Larramendy, F. 328
Lasri, T. 1012
Lawlor, D. 504
Le, C.P. 1125
Lee, B.-Y. 196
Lee, C. 106, 1032
Lee, C.J. 158, 651
Lee, C.-W. 223
Lee, D.W. 1110

Lee, G. 732
Lee, G.-B. 581, 658
Lee, H. 573
Lee, H.J. 158, 651
Lee, H.K. 1
Lee, J. 184
Lee, J. 219, 230, 628, 956
Lee, J. 752
Lee, J. 764
Lee, J.-H. 651
Lee, J.E.-Y. 797
Lee, K.Y. 1059
Lee, M. 292
Lee, S. 628
Lee, S. 865
Lee, S.-M. 559
Lee, S.H. 604
Lee, W.C. 8, 465
Lee, Y. 316, 764
Lee, Y.-K. 459
Lee, Y.R. 1059
Legrand, B. 150
Lei, C. 320, 857
Lell, J.A. 531
Lenk, G. 451
Leprince-Wang, Y. 960
Lesecq, S. 65
Leube, C. 893
Li, B. 543
Li, C.-S. 853, 988
Li, D. 320, 857
Li, D. 775
Li, J. 265
Li, M.-H. 853, 988
Li, M.G. 1102
Li, S. 130
Li, S.-S. 853, 988
Li, T. 726
Li, W. 616
Li, X. 130
Li, X. 447
Li, X. 612, 779, 785
Li, Y. 555, 968
Li, Y. 1048
Li, Z. 358
Liang, C. 459
Liang, T. 857
Lim, A.V. 276
Lim, C.J. 624
Lim, Y. 316, 764
Lin, C.-H. 272
Lin, C.-L. 581

Lin, J.X. 57
Lin, K.C.-H. 89
Lin, L. 22, 208, 261, 551, 801, 837, 928, 1133
Lin, Q. 488, 569, 869
Lin, T.-H. 249
Lin, W.-C. 89
Lin, Y.-S. 1032
Liu, A.Q. 45, 49, 57, 960, 972
Liu, C. 551
Liu, C.C. 624
Liu, D. 288
Liu, D. 459
Liu, G. 569
Liu, J. 110, 268
Liu, M. 304
Liu, M.C.-M. 89
Liu, P. 77
Liu, S. 714
Liu, S.-Y. 1075
Liu, T. 6, 1122
Liu, W. 312
Liu, X. 447
Liu, Y. 154, 340, 690
Liu, Y. 358, 467
Lo, C.-Y. 740
Lo, P.-H. 69
Longoni, A. 37, 932
Lu, Y. 140, 146
Lu, Z.-R. 698
Lujan, R. 877
Luo, B. 427
Luo, G. 1133
Luo, G.-L. 69
Luo, K. 77
Luo, Y. 1040
Lutz, M. 1

M

Ma, C.-W. 249, 1067
Ma, M.Y 427
Ma, S. 459
Mabuchi, K. 512
MacKinnon, N. 97
Madhaven, V. 1004
Maeda, R. 1098
Maeda, S. 1082
Maeda, Y. 253
Mahadeva, S.K. 861
Maharbiz, M.M. 682
Mahdavi, M. 204, 913
Mahmoud, M. 928
Mairiaux, E. 150

Majumdar, R.	296
Makimoto, N.	1098
Makino, H.	686
Maldonado-Camargo, L.	964
Maldonado-Garcia, M.	885
Mansour, R.R.	732
Mao, H.Y.	320, 857
Marelli, B.	168
Marigó, E.	1004
Marzouk, J.	1012
Matsui, D.	608
Matsui, R.	640
Matsuki, S.	397
Matsumoto, H.	589
Matsumoto, K.	144, 176, 212, 257, 405, 421, 640, 646, 670, 706, 710, 730, 783, 925
Matsunaga, Y.	354
McElwain, R.B.	944
McFarland, D.M.	752
Mehdizadeh, E.	204, 885
Melamud, R.	1
Meng, B.	102, 1078
Meng, E.	620, 662
Merzeau, P.	150
Mescheder, U.	413
Messana, M.W.	1
Meyhöfer, E.	238
Miao, J.M.	500, 678, 714, 889
Michaud, H.O.	760
Miki, N.	649
Millis, J.	188
Ming, A.J.	320, 857
Minh, L.V.	1094
Minh-Dung, N.	144, 212, 640, 670
Minotti, P.	932
Misaki, K.	905
Misawa, K.	1071
Mishra, R.	523
Mitsuya, H.	118
Miura, T.	476
Miwa, K.	118
Modarres-Zadeh, M.J.	909
Monzawa, T.	1055
Moon, H.	443, 547
Moore, J.	551
Moreau, J.E.	168
Moreira, E.E.	1036
Mori, N.	351, 589
Mori, S.	566
Morimoto, K.	1086
Morimoto, Y.	18, 351, 566, 589, 596, 666
Morishita, Y.	730
Morita, N.	748

Mukherjee, T. ... 813
Mukouyama, Y. ... 666
Murakami, R. ... 431
Murali, K. ... 154
Muramatsu, Y. ... 166
Murashige, K. ... 674
Murozaki, Y. ... 833
Murphy, J. ... 2

N

Nadig, S. ... 1129
Nagasawa, S. ... 308, 1082
Naing, T.L. ... 1024
Najafi, K. ... 821, 1051, 1141
Najar, H. ... 801
Nakahara, T. ... 694
Nakai, A. ... 257, 730
Nakamura, Y. ... 480
Nam, E.R. ... 465
Namazu, T. ... 373
Nansai, H. ... 480
Nauwelaers, B. ... 1016
Ng, E.J. ... 1, 29, 33, 393, 809, 1008
Ng, T.N. ... 877
Nguyen, C.T.-C. ... 976, 1024
Nguyen, H.D. ... 944
Nguyen, J. ... 435
Nguyen, N.-T. ... 825
Nguyen, T.-Q. ... 348
Nicolas, S. ... 65
Niklaus, F. ... 53, 280
Niroui, F. ... 85
Nishida, T. ... 1145
Nishihara, T. ... 1094
Nishiyama, K. ... 476
Nobukawa, A. ... 596
Noda, K. ... 925
Nogami, H. ... 748
Norford, L.K. ... 714
Norouzpour-Shirazi, A. ... 789
Notaguchi, M. ... 702
Nuckolls, C. ... 869
Nwankire, C. ... 192

O

O'Brien, G.J. ... 1, 393
O'Brien, K.P. ... 208
Ober, C. ... 873
Ochoa, M. ... 332, 817, 1044
Ogawa, S. ... 905
Oguchi, H. ... 93
Oh, S.-J. ... 651

Okabe, U. 300
Okada, R. 1082
Okada, Y. 439
Okano, T. 300
Okitsu, T. 463, 643, 666
Oku, H. 480
Olfat, M. 732
Olsen, T. 488
Omenetto, F.G. 168
Omi, S. 655
Ono, S. 118
Ono, T. 14, 344, 401
Onoe, H. 18, 26, 463
Oonishi, A. 968
Ortiz, S. 531
Osaki, T. 235, 336, 596
Ou, W. 320, 857
Ou, Y. 320, 857
Ouerghi, I. 361
Ozoe, K. 702

P

Pal, P. 385
Pan, S. 714
Pan, T. 543
Pan, Y. 940
Paprotny, I. 296
Park, H.S. 952
Park, I. 744
Park, I.S. 1059
Park, J. 8, 154
Park, J. 340, 690
Park, J.H. 585
Park, W.-T. 348
Partridge, A. 1
Patterson, A. 996
Paul, O. 162, 328, 636, 736, 893
Pei, R. 488
Peng, S. 551
Perahia, R. 944
Peroulis, D. 1020
Petrone, N. 869
Pettit, C. 752
Pfusterschmied, G. 718
Phan, H.-P. 825
Philippe, J. 361
Piazza, G. 980, 992, 996, 1000
Pitchappa, P. 1032
Polcawich, R.G. 146
Polunin, P. 1008
Popa, L.C. 1028
Pourkamali, S. 204, 885, 913

Pouydebasque, A. 65
Pozzi, A. 948
Przybyla, R.J. 921
Psychogiou, D. 1020
Pu, S.H. 881

Q

Qamar, A. 825
Qi, M. 22
Qian, Y. 1032
Qu, V. 1
Quang-Khang, P. 144, 670

R

Rachet, B. 632
Rahimi, R. 332, 817
Rais-Zadeh, M. 73, 793
Ramezany, A. 913
Rechenberg, R. 616
Reinecke, H. 413
Rey, P. 37
Rezaei Nejad, H. 492
Rinaldi, C. 964
Rinaldi, M. 984
Rismani Yazdi, S. 535
Rizzini, F. 849
Rocha, L.A. 1036
Rocheleau, T.O. 976, 1024
Rogers, J.A. 168
Roman, C. 312
Rottenberg, X. 1016
Roxhed, N. 280, 451
Roy, D. 527
Rozen, O. 146, 921
Ruiz-Díez, V. 718
Ruther, P. 162, 636, 893
Ruyack, A. 873, 1090

S

Saha, B. 628
Saito, D. 801
Sakuma, S. 200, 431, 833, 1055
Sakurai, Y. 377
Salvetat, J.P. 150
Salvia, J.C. 1
Samiei, E. 492
Sammoura, F. 837, 928
Sánchez-Rojas, J.L. 718
Sander, C. 736, 893
Sando, S. 324
Sanghadasa, M. 1133
Sarkar, N. 732

Sasaki, M.	841
Sato, H.	1048
Sawada, R.	748
Sawahata, H.	686
Sawant, S.G.	936
Scheiblin, P.	361
Schmid, U.	718
Schmid, Y.	226
Schmidt, F.	736
Schwaerzle, M.	162
Scianmarello, N.	154
Ségard, B.-D.	354
Segovia-Fernandez, J.	1000
Sekiguchi, T.	480, 539
Sekine, R.	480
Sen, P.	527
Senesky, D.G.	284
Senkal, D.	29
Seo, M.-H.	114, 126
Serien, D.	328, 470, 589
Shadmani, A.	535
Shaffer, E.	632
Shahosseini, I.	1051
Shakeel, H.	771
Shang, J.	427
Shankar, A.	284
Shaw, S.	1008
Shelton, S.E.	921
Shen, C.	1133
Shen, H.-H.	519
Shen, Z.X.	960
Shi, L.	992
Shibata, M.	841
Shibata, S.	845
Shikida, M.	845
Shimokawa, F.	245, 253
Shimoyama, I.	144, 176, 212, 257, 405, 421, 640, 646, 670, 706, 710, 730, 783, 925
Shin, H.	316, 764
Shin, K.-S.	604
Shinomiya, H.	968
Shintaku, H.	238, 455, 476, 694
Shkel, A.A.	917
Shkel, A.M.	29, 805
Shmulevich, S.	41, 172, 215, 897
Shoji, S.	480, 539
Shoshani, O.	1008
Shunmugam, M.	1004
Sinani, M.D.	1020
Singh, N.	996, 1032
Sletten, E.M.	85
Smith, G.L.	146
Smith, R.L.	188

Sobreviela, G. 1004
Sochol, R.D. 551
Son, Y. 158
Song, K. 459
Song, K.-Y. 1086
Song, Q.H. 960
Song, Y.-H. 114
Song, Z. 304
Soon, B.W. 996
Sorenson, L.D. 944
Soundara-Pandian, M. 1004
Stehle, J. 393
Stemme, G. 53, 280, 451
Stoeber, B. 861, 952
Stojanovic, M.N. 488
Strachan, S. 1008
Strathearn, D. 732
Su, Z. 1078
Subramaniyan, S.P. 455
Sugano, K. 369, 608
Sugiyama, T. 1071
Sun, C. 775
Sun, Y. 381, 435
Sun, Y. 472
Sung, W.-L. 423, 901
Suria, A.J. 284
Suzuki, C. 646
Suzuki, H. 300, 439
Suzuki, M. 1145
Suzuki, T. 245, 253
Suzuki, Y. 1086, 1098
Swager, T.M. 85
Syms, R.R.A. 134, 288
Syu, T. 242

T

Tabata, O. 389, 417, 608
Tabrizian, R. 789
Tai, Y.-C. 154, 340, 397, 690
Takagawa, Y. 905
Takahashi, H. 212, 257, 405, 706, 710
Takahashi, K. 655
Takahashi, T. 1145
Takahata, K. 1040
Takahata, T. 212, 257, 421, 640, 646, 670, 710, 925
Takane, K. 373
Takao, H. 245, 253
Takasawa, S. 242
Takeda, N. 480, 539
Takei, A. 829
Takei, K. 756
Takei, Y. 640, 646, 670, 783

Takeuchi, S. 8, 18, 235, 328, 336, 351, 463, 465, 470, 566, 589, 596, 636, 643, 666
Takigawa-Imamura, H. 476
Tan, Q.L. 857
Tanaka, M. 686
Tang, C. 459
Tang, H. 698
Tang, H.-Y. 140, 146
Tang, L.C. 320, 857
Tao, H. 168
Tarhan, M.C. 455
Tatar, E. 813
Tay, F.E.H. 106
Tay-Wee-Song, C. 1004
Tayebi, N. 208
Teixidor, J. 760
Terao, K. 245, 253
Ternyak, O. 897
Thanh-Vinh, N. 144, 176, 925
Théron, D. 150
Tilmans, H.A.C. 1016
Tixier-Mita, A. 354
Tocchio, A. 849
Toda, M. 14, 344, 401
Tomizawa, Y. 968
Tomoike, F. 235
Tonooka, T. 235, 596
Torres-Díaz, I. 964
Torunbalci, M.M. 409
Toshiyoshi, H. 354, 555, 1071
Toya, K. 968
Triantafyllou, M.S. 500, 678, 889
Tripathy, A. 527
Truong, B.D. 1125
Tsai, C.-H.D. 722
Tsai, J.M.L. 73
Tseng, F.-G. 223
Tsuchiya, T. 389, 417, 608
Tsugane, M. 439

U

Ueda, M. 1094
Uesugi, A. 389
Uetsuki, M. 905
Uranga, A. 1004

V

Vakakis, A.F. 752
Veale, M. 551
Velez, C. 964
Vembadi, A. 504
Vo Doan, T.T. 1048
Voigt, A. 97

W

Walter, B.	150
Walus, K.	861
Wang, A.I.	85
Wang, B.	593
Wang, C.	365, 569, 869
Wang, C.	435
Wang, H.-S.	89
Wang, J.	785
Wang, K.	658
Wang, P.	877
Wang, S.	1
Wang, S.	1028
Wang, W.	77
Wang, W.	77, 358, 467, 593
Wang, W.	948
Wang, W.B.	857
Wang, X.	110, 268, 1063
Wang, X.	130
Wang, Y.	77
Wang, Y.	122
Wang, Y.	320, 857
Wang, Y.-L.	581
Wang, Z.	268
Wang, Z.	304
Wang, Z.	484
Wang, Z.	877
Watanabe, R.	212
Weber, A.J.	616
Wei, X.	573
Wei, Y.	435
Weinstein, D.	1028
Weitz, D.A.	8
Wijethunga, P.A.L.	547
Wijngaart, van der, W.	10, 472, 563
Wilson, J.C.	885
Wood, G.S.	881
Wu, D.	304
Wu, J.H.	45, 57
Wu, J.-K.	223
Wu, M.C.	976
Wu, P.C.	960
Wu, S.-Y.	261
Wu, T.	484
Wu, W.	467
Wu, X.	698
Wu, X.	1063
Wu, X.	1110
Wu, Z.	793
Wu, Z.	1137

X

Xiao, D.	1063
Xie, H.	948
Xie, J.	881
Xie, S.	632
Xing, X.	508
Xiong, J.J.	320, 857
Xu, K.	775
Xu, P.	612, 779, 785
Xu, R.	459
Xu, T.	340
Xu, Y.	340
Xue, F.	714
Xue, G.	14

Y

Yabuki, M.	968
Yagami, S.	377
Yama, G.	1, 393
Yamada, S.	118
Yamagiwa, S.	686
Yamaguchi, T.	841
Yamamoto, G.	401
Yamamoto, Y.	756
Yamane, H.	308
Yamanishi, Y.	242, 655
Yamauchi, K.	369
Yan, X.	22
Yang, B.	110, 268
Yang, C.	208, 261, 801, 837, 928
Yang, C.S.	110, 268
Yang, F.	77, 358
Yang, J.	569
Yang, M.	168
Yang, R.	877
Yang, S.	304
Yang, W.	573
Yang, Y.	1, 29, 33, 393, 809, 1008
Yang, Y.-J.	249, 1067
Yang, Z.	714
Yang, Z.C.	45, 49, 57, 960, 972
Yao, D.-J.	519
Yasuga, H.	10
Yeh, Y.-C.	33
Yip, M.-C.	767
Yobas, L.	508
Yokokawa, R.	238, 455, 476, 694
Yokoyama, T.	1094
Yoneoka, S.	1
Yoo, J.-Y.	126
Yoon, D.H.	480, 539

Yoon, E.-S.	158, 651
Yoon, J.-B.	114, 126
Yoon, Y.	944
Yoon, Y.-H.	114
Yoshida, K.	26
Yoshida, R.	93
Yoshida, S.	18, 328, 589
Yoshikawa, Y.	1145
Yu, B.	1078
Yu, F.	612, 785
Yu, F.	940
Yu, H.	612, 775
Yu, J.	869
Yu, L.	620, 662
Yu, S.	775
Yu, W.	817
Yuan, J.	496
Yuan, W.Z.	1106
Yun, G.-S.	114
Yun, K.-S.	651

Z

Zainuddin, A.A.	1004
Zarudniev, M.	65
Zhang, B.	324, 600
Zhang, D.	77, 358
Zhang, H.	365, 1078
Zhang, H.M.	1106
Zhang, H.X.	102
Zhang, J.	1102
Zhang, L.	358, 467
Zhang, Q.	122
Zhang, R.	593
Zhang, T.S.	1044
Zhang, W.	960
Zhang, X.	340, 690
Zhang, X.	948
Zhang, X.-S.	365, 1078
Zhang, Y.	714
Zhao, C.	447
Zhao, C.	459
Zhao, C.	881
Zhao, F.	276
Zhao, L.	122
Zhao, W.	593
Zhao, X.	365
Zheng, D.	779
Zheng, L.	18
Zheng, Y.	435
Zhou, X.	472
Zhu, F.Y.	102, 365
Zhu, H.	797

Zhu, J. 488
Zhu, W.M. 960
Zhu, Y. 569, 869
Zhu, Y. 825
Ziaie, B. 332, 585, 817, 1044
Zohar, Y. 459

9781479979561

2017 40th International Convention on Information and Communication Technology, Electronics and Microelectronics (MIPRO 2017)

Opatija, Croatia
22-26 May 2017

Pages 1-782

IEEE Catalog Number: CFP1739K-POD
ISBN: 978-1-5090-4969-1